ISOTOPES

ISOTOPES

Principles and Applications

Third Edition

GUNTER FAURE AND TERESA M. MENSING

JOHN WILEY & SONS, INC.

Published by John Wiley & Sons, Inc., Hoboken, New Jersey
Published simultaneously in Canada

Library of Congress Cataloging-in-Publication Data

Faure, Gunter.
Isotopes : principles and applications.—3rd ed. / Gunter Faure and Teresa M. Mensing.
p. cm.
Rev. ed. of: Principles of isotope geology. 2nd ed. c1986.
Includes bibliographical references and indexes.
ISBN 0-471-38437-2 (acid-free paper)
1. Isotope geology. I. Mensing, Teresa M. II. Faure, Gunter. Principles of isotope geology. III. Title.
QE501.4.N9 F38 2005
551.9—dc22

2003022089

11

For my grandchildren,
Chloë and Sarah, Ryan and Rachel, and William

Gunter Faure

In loving memory of my mother,
W. Gene Mensing

Teresa M. Mensing

Contents

Part II Radiogenic Isotope Geochronometers 73

The author index and the end-of-chapter problems are available on the worldwide web at <www.wiley.com/college/faure>.

Preface

This book contains a comprehensive presentation of all aspects of isotope geoscience and is intended to be used in universities and colleges as a textbook in lecture courses for senior undergraduate and first-year graduate students in the earth sciences. In addition, this book will serve as a source of information for professionals in the earth sciences and related disciplines. The encyclopedic function of this book is enhanced by a detailed table of contents and by author and subject indexes. The book also contains a large number of original diagrams with extended captions that facilitate browsing and convey information without requiring close reading of the text. In addition, each chapter contains cross-references to other parts of the book where related information can be found.

The third edition of this book has been completely rewritten, reorganized, and expanded in order to bring it up to date and to make it useful to students and professionals in the earth sciences and related fields. As presently configured, the third edition can be used in support of several lecture courses, including isotope geochronometry, environmental isotope geochemistry, environmental radioactivity, stable isotope geochemistry, and graduate-level seminars in all of these subjects. In addition, this book contains information that is relevant to igneous petrology, ore deposits, stratigraphy, oceanography, hydrology, atmospheric science, soil science, archeology/anthropology, medical geochemistry, and meteoritics.

The presentation of the subject matter in each chapter starts with the relevant basic principles and proceeds from there to the derivation of the equations and to statements of the assumptions on which the equations are based. All of the substantive chapters contain applications of isotope systematics to the solution of interesting problems in geology and related disciplines. Each chapter ends with a summary of important insights and with lists of references.

The references at the ends of most chapters have been subdivided in order to help students to pursue a topic of their choosing beyond the material presented in the text. In addition, the references to subjects that are not fully developed in the text or that support data tables are presented in compressed form in appropriate places in the text and are not repeated in the end-of-chapter reference lists. Most of the papers used in this book were published between about 1990 and 2002. However, some references date back to the 1960s and 1970s, when many seminal papers were published that are the foundation of current research.

Additional indexes, a solutions manual, and a set of numerical problems that illustrate the principles presented in each chapter are available at www.wiley.com/college/faure. The solution of numerical problems by students is an important learning experience that will give them the necessary confidence to test the calculations and conclusions of research reports in the scientific literature. Future earth and environmental scientists must be able to make such calculations in order to understand in quantitative terms whatever phenomenon they are investigating. The emphasis in this book is on figuring out how something works and to use that insight to solve problems by means of calculations. However, this book does not teach mathematical modeling or how to program computers to interpret large data sets.

In addition, this book does not contain detailed information on analytical procedures and on the instrumentation that is required to make accurate and precise measurements of isotope abundances and decay rates. Students who are planning a research career in isotope geoscience should serve an apprenticeship with a master scientist in order to learn the skills that are necessary in their chosen field and to acquire the work habits that are required to achieve success.

Although most universities and colleges do not have operational isotope-research laboratories, isotope geoscience should be a standard item of the curriculum in all Earth Science Departments because isotopic data have become a major source of information in all branches of the earth sciences and related fields. In fact, the wide range of applications of isotopic data provides an overview of many subjects of earth science that are traditionally taught as separate entities.

The material in this book is presented in five parts. The first part (Chapters 1–4) is introductory and presents important background information. Students who are well prepared should be able to bypass these chapters without difficulty. Nevertheless, we have included them to assure that all students can build on a common base in the more substantive chapters that follow. The second part (Chapters 5–15) presents the dating methods based on the accumulation of radiogenic isotopes of Sr, Ar, Ca, Nd, Pb, Hf, Os, Ce, and Ba. The third part (Chapters 16–19) is devoted to isotope geochemistry, including mixing theory and applications of radiogenic isotopes to the origin of igneous rocks, to the transport of these isotopes from the continents to the oceans, and to the isotope compositions of Sr, Nd, Pb, Os, and Hf in the oceans. The fourth sequence of chapters (20–25) deals with the natural radioactivity resulting either from the presence of short-lived daughters of long-lived radioactive parents or from the presence of radionuclides produced by interactions with cosmic rays, by induced fission of ^{235}U in nuclear reactors and nuclear weapons, and by capture of neutrons released by uranium fission. The last section of the book (Chapters 26–30) contains summaries of the isotopic compositions of H, B, Li, C, N, O, Si, Cl, and S and other elements (Cu, Fe, Se, and Br) whose atoms are fractionated in nature. These elements are among the most abundant in the crust of the Earth and provide useful information on a wide range of geological materials, primarily in terms of isotopic equilibration temperatures and as monitors of certain biological, chemical, and physical processes in nature.

The scope of this book reflects the applicability of isotopic data in a wide range of terrestrial and extraterrestrial settings. It also reflects the ingenuity of the scientists who work in this field. Isotope geoscience is still at the forefront of the earth sciences and therefore continues to attract the brightest young scientists in the world today.

We are grateful to our colleagues who have shared their insights with us by sending us reprints of their papers and by speaking to us about their work. Others sent us comments about the second edition of this book and made suggestions for its improvement. We thank all of them by listing their names in alphabetical order and apologize in advance to anyone whose name we failed to mention: G. Åberg, F. Albarède, L. T. Aldrich, C. J. Allègre, H. L. Allsopp, E. Anders, J. Andersson, R. L. Armstrong, H. Baadsgaard, H.-M. Bao, E. Bard, E. Barrera, M. Barton, K. Bell, K. Benkala, S. M. Bergström, R. A. Berner, M. Bielsky-Zyskind, J. M. Bigham, B. A. B. Blackwell, P. D. Boger, M. D. Bottino, J. R. Bowman, O. van Breemen, D. J. Brookins, D. Buchanan, H. S. Cao, D. H. Carr, L. M. Centeno, S. Chaudhuri, W. A. Cassidy, J.-F. Chen, S. E. Church, N. Clauer, R. N. Clayton, W. Compston, U. G. Cordani, H. Craig, J. H. Crocket, G. Crozaz, G. K. Czamanske, R. D. Dallmeyer, E. J. Dasch, J. L. DeLaeter, C. F. Davidson, G. M. Denton, P. Deines, D. J. DePaolo, S. Deutsch, T. Ding, R. S. Dietz, B. R. Doe, R. Doig, A. Dreimanis, R. Eastin, H. Elderfield, D. H. Elliot, W. C. Elliott, C. Emiliani, S. Epstein, W. G. Ernst, H. W. Fairbairn, H. Faul, R. Felder, R. J. Fleck, R. V. Fodor, K. A. Foland, E. Fortner III, B. M. French, F. A. Frey, G. M. Friedman, S. J. Fritz, C. D. Frost, R. Fuge, P. D. Fullagar, H. Furness, W. S. Fyfe, E. M. Galimov, M. Gascoyne, P. W. Gast, T. W. Gevers, B. J. Giletti, S. S. Goldich, A. M. Goodwin, I. M. Gorokhov, N. K. Grant, P. Grootes, R. Grün, B. L. Gulson, J. D. Gunner, U. Haack, A. N. Halliday, M. Halpern, G. N. Hanson, R. S. Harmon, S. R. Hart, R. Harvey, C. J. Hawkesworth, K. S. Heier, A. Heimann, J. M. Hergt, L. F. Herzog, H. H. Hess, M. Hochella, J. Hoefs, A. W. Hofmann, A. Holmes, J. Hower, P. M. Hurley, E. Iberall-Robbins, M. Ikramuddin, E. Jäger, B. M. Jahn, L. M. Jones, R. I. Kalamarides, F. Kalsbeek, A. Karam, E. Keppens, S. E. Kesler, R. W. Kistler, L. Koch, C. Koeberl, L. N. Kogarko, G. T. Kovatch, T. E. Krogh, G. Kurat, M. D. Kurz, P. R. Kyle, D. Lal, M. A. Lanphere, G. Lee, W. P. Leeman, W. E. LeMasurier, P. C. Lightfoot, J. F. Lindsay, H. J. Lippolt, D. Long, L. E. Long, K. R. Ludwig, G. W. Lugmair, W. B. Lyons, J. D. Macdougall, R. Mauche, J. McArthur, M. T. McCulloch, R. H. McNutt, M. A. Menzies, J. H. Mercer, M. Mohlzahn, R. Montigny, S. Moorbath, S. M. Morse, L. A. Munk, S. B. Mukasa, V. R. Murthy, E.-R. Neumann, L. O. Nicolaysen, K. Nishiizumi, K. Notsu, J. D. Obradovich, R. K. O'Nions, V. M. Oversby, L. B. Owen, J. M. Palais, R. J. Pankhurst, J. J. Papike, W. J. Pegram, Z. E. Peterman, N. Piatak, R. T. Pidgeon, W. H. Pinson, Jr., F. A. Podosek, J. L. Powell, D. E. Pride, H. N. A. Priem, P. Pushkar, D. C. Rex, H. Sakai, V. J. M. Salters, S. M. Savin, M. Schidlowski, J.-G. Schilling, W. R. Van Schmus, C. C. Schnetzler, J. W. Schopf, J. Schutt, H. P. Schwarcz, N. R. Shaffer, D. M. Shaw, S. B. Shirey, E. T. C. Spooner, A. Starinsky, R. H. Steiger, G. Steinitz, P. Stille, A. M. Stueber, E. Stump, K. V. Subbarao, J. F. Sutter, M. Tatsumoto, H. P. Taylor, Jr., S. R. Taylor, J. Teller, M. Thiemens, R. N. Thompson, G. R. Tilton, K. K. Turekian, R. J. Uffen, H. C. Urey, J. Veizer, J. W. Valley, P. Vidal, T. A. Vogel, F. A. Wade, M. Wadhwa, H. Wänke, R. M. Walker, R. K. Wanless, G. J. Wasserburg, P. N. Webb, K. H. Wedepohl, K. S. Wehn, I. Wendt, G. W. Wetherill, J. T. Wilson, P. J. Wyllie, H. S. Yoder, D. York, R. E. Zartman, and A. Zindler. We are especially grateful to Betty Heath for her tireless efforts and skill in wordprocessing the text. We also thank Karen Tyler and Pei Jian Zhou who worked with us on this book. Last but not least, we thank James Harper of John Wiley and Sons for making the necessary decisions that enabled us to publish this book.

Gunter Faure and Teresa M. Mensing

PART I

Principles of Atomic Physics

The science of geochronometry is based on the decay of long-lived radioactive nuclides to stable daughters. Chapters 1–4 present the basic principles of atomic physics and chemistry that govern this phenomenon and demonstrate the applications of these principles to the earth sciences.

CHAPTER 1

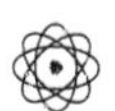

Nuclear Systematics

ISOTOPE geology is concerned with the measurement and interpretation of the variations of the isotope composition of certain elements in natural materials. These variations are the result of two quite different processes:

1. the spontaneous decay of the nuclei of certain atoms to form stable nuclei of other elements and the accumulation of these radiogenic daughter atoms in the minerals in which they formed and
2. the enrichment (or depletion) of certain stable atoms of elements of low atomic number in the products of chemical reactions as a result of changes in state such as evaporation and condensation of water and during physical processes such as diffusion.

The interpretation of the changes in the isotope compositions of the affected elements has become a powerful source of information in all branches of the earth sciences. The use of this investigative tool requires a thorough understanding of geological, hydrological, biospheric, and atmospheric processes occurring on the Earth and elsewhere in the solar system. In addition, the interpretation of isotopic data requires knowledge of the relevant principles of atomic physics, physical chemistry, and biochemistry. The decay of unstable atoms is accompanied by the emission of nuclear particles and radiant energy, which together constitute the phenomenon of radioactivity. The discovery of this process near the end of the nineteenth century was a milestone in the history physics and greatly increased our understanding of the Earth.

1.1 DISCOVERY OF RADIOACTIVITY

The rise of geology as a science is commonly associated with the work of James Hutton in Scotland. He emphasized the importance of very slow but continuously acting processes that shape the surface of the Earth. This idea conflicted with Catastrophism and foreshadowed the concept of Uniformitarianism, developed by Hutton in his book *Theory of the Earth*, published in 1785. His principal point was that the same geological processes occurring at the present time have shaped the history of the Earth in the past and will continue to do so in the future. He stated that he could find "no vestige of a beginning—no prospect of an end" for the Earth. Hutton's conclusion regarding the history of the Earth was not well received by his contemporaries. However, as time passed, geologists accepted the principle of Uniformitarianism, including the conviction that very long periods of time are required for the deposition of sedimentary rocks whose accumulated thickness amounts to many miles. In 1830, Charles Lyell published the first volumes of his *Principles of Geology*. By the middle of the nineteenth century, geologists seemed to be secure in their conviction that the Earth was indeed very old and that long periods of time are required for the deposition of the great thickness of sedimentary rocks that had been mapped in the field.

The apparent antiquity of the Earth and the principle of Uniformitarianism were unexpectedly attacked by William Thomson, better known as Lord Kelvin (Burchfield, 1975). Thomson was Britain's most prominent physicist during the second half of the nineteenth century. His invasion into geology profoundly influenced geological opinion regarding the age of the Earth for about 50 years. Between 1862 and 1899 Thomson published a number of papers in which he set a series of limits on the possible age of the Earth. His calculations were based on considerations of the luminosity of the Sun (Thomson, 1862), the cooling history of the Earth, and the effect of lunar tides on the rate of rotation of the Earth. He initially concluded that the Earth could not be much more than 100 million years old. In subsequent papers, he further reduced the age of the Earth. In 1897, Lord Kelvin (he was raised to peerage in 1892) delivered his famous lecture, "The Age of the Earth as an Abode Fitted for Life" (Thomson, 1899), in which he narrowed the possible age of the Earth to between 20 and 40 million years.

These and earlier estimates of the age of the Earth by Lord Kelvin and others were a serious embarrassment to geologists. Kelvin's arguments seemed to be irrefutable, and yet they were inconsistent with the evidence as interpreted by geologists on the basis of Uniformitarianism. Ironically, one year before Lord Kelvin presented his famous lecture, the French physicist Henri Becquerel (1896) had announced the discovery of radioactivity. Only a few years later it was recognized that the disintegration of radioactive elements is an exothermic process. Therefore, the natural radioactivity of rocks produces heat, and the Earth is not merely a cooling body, as Lord Kelvin had assumed in his calculation.

Becquerel's discoveries attracted the attention of several young scientists, among them Marie (Manya) Sklodowska who came to Paris in 1891 from her native Poland to study at the Sorbonne. On July 25, 1895, she married Pierre Curie, a physics professor at the Sorbonne. After Becquerel reported his discoveries regarding salts of uranium, Marie Curie decided to devote her doctoral dissertation to a systematic search to determine whether other elements and their compounds emit similar radiation (Curie, 1898). Her work was rewarded when she discovered that thorium is also an active emitter of penetrating radiation. Turning to natural uranium and thorium minerals, she noticed that these materials are far more active than the pure salts of these elements. This important observation suggested to her that natural uranium ore, such as pitchblende, should contain more powerful emitters of radiation than uranium. For this reason, Marie and Pierre Curie requested a quantity of uranium ore from the mines of Joachimsthal in Czechoslovakia and, in 1898, began a systematic effort to find the powerful emitter whose presence she had postulated. The search eventually led to the discovery of two new active elements, which they named polonium and radium. Marie Curie coined the word "radioactivity" on the basis of the emissions of radium. In 1903, the Curies shared the Nobel Prize for physics with Henri Becquerel for the discovery of radioactivity.

1.2 INTERNAL STRUCTURE OF ATOMS

Every atom contains a small, positively charged nucleus in which most of its mass is concentrated. The nucleus is surrounded by a cloud of electrons that are in motion around it. In a neutral atom, the negative charges of the electrons exactly balance the total positive charge of the nucleus. The diameters of atoms are of the order of 10^{-8} centimeters (cm) and are conveniently expressed in angstrom units (1 Å $= 10^{-8}$ cm). The nuclei of atoms are about 10,000 times smaller than that and have diameters of 10^{-12} cm, or 10^{-4} Å. The density of nuclear matter is about 100 million tons per cubic centimeter. The nucleus contains a large number of different elementary particles that interact with each other and are organized into complex patterns within the nucleus. It will suffice for the time being to introduce only two of these, the proton (p) and the neutron (n), which are collectively referred to as nucleons. Protons and neutrons can be regarded as the main building blocks of the nucleus because they account for its mass and electrical charge. Briefly stated, a proton is a particle having a positive charge that is equal in magnitude but opposite in polarity to the charge of an electron. Neutrons have a slightly larger mass than protons and carry no electrical charge. Extranuclear neutrons are unstable and decay spontaneously to form protons and electrons with a "halflife" of 10.6 min. The other principal components of atoms are the electrons, which swarm around the nucleus. Electrons at rest have a small mass (1/1836.1 that of hydrogen atoms) and a negative electrical charge. The number of extranuclear electrons in a neutral atom is equal to the number of protons. The protons in the nucleus of an atom therefore determine how many electrons that atom can have when it is electrically neutral. The number of electrons and their distribution about the nucleus in turn determine the chemical properties of that atom.

1.2a Nuclear Systematics

The composition of atoms is described by specifying the number of protons and neutrons that are present in the nucleus. The number of protons (Z) is called the *atomic number* and the number of neutrons (N) is the *neutron number*. The atomic number Z also indicates the number of extranuclear electrons in a neutral atom. The sum of protons and neutrons in the nucleus of an atom is the *mass number* (A). The composition of the nucleus of an atom is represented by the simple relationship

$$A = Z + N \tag{1.1}$$

Another word for atom that is widely used is *nuclide*. The composition of any nuclide can be represented by means of a shorthand notation consisting of the chemical symbol of the element, the mass number written as a superscript, and the atomic number written as a subscript. For example, ${}^{14}_{6}C$ identifies the nuclide as an atom of carbon having 6 protons (therefore 6 electrons in a neutral atom) and a total of 14 nucleons. Equation 1.1 indicates that the nucleus of this nuclide contains $14 - 6 = 8$ neutrons. Similarly, ${}^{23}_{11}Na$ is a sodium atom having 11 protons and $23 - 11 = 12$ neutrons. Actually, it is redundant to specify Z when the chemical symbol is used. For this reason, the subscript (Z) is sometimes omitted in informal usage.

A great deal of information about nuclides can be shown on a diagram in which each nuclide is represented by a square in coordinates Z and N. Figure 1.1 is a part of such a chart of the nuclides. Each element on this chart is represented by several nuclides having different neutron numbers arranged in a horizontal row. Atoms which have the same Z but different values of N are called *isotopes*. The isotopes of an element have identical chemical properties and differ only in their masses. Nuclides that occupy vertical columns on the chart of the nuclides have the same value of N but different values of Z and are called *isotones*. Isotones are therefore atoms of different elements. The chart also contains nuclides that occupy diagonal rows. These have the same value of A and are called *isobars*. Isobars have different values of Z and N and are therefore atoms of different elements. However, because they contain the same number of nucleons, they have similar but not identical masses.

1.2b Atomic Weights of Elements

The masses of atoms are too small to be conveniently expressed in grams. For this reason, the *atomic mass unit* (amu) is defined as one-twelfth of the mass of ${}^{12}_{6}C$. In other words, the mass of ${}^{12}_{6}C$ is arbitrarily fixed at 12.00... amu, and the masses of all other nuclides and subatomic particles are expressed by comparison to that of ${}^{12}_{6}C$. The masses of the isotopes of the elements have been measured by mass spectrometry and are known with great precision and accuracy.

The total number of different nuclides is close to 2500, but only 270 of these are stable, including long-lived radioactive isotopes that still occur naturally because of their slow rate of decay. The stable nuclides, along with a small number of naturally occurring long-lived unstable nuclides, make up the elements in the periodic table. Many elements have two or more naturally occurring isotopes, some have only one, and two elements (technetium and promethium) have none. These

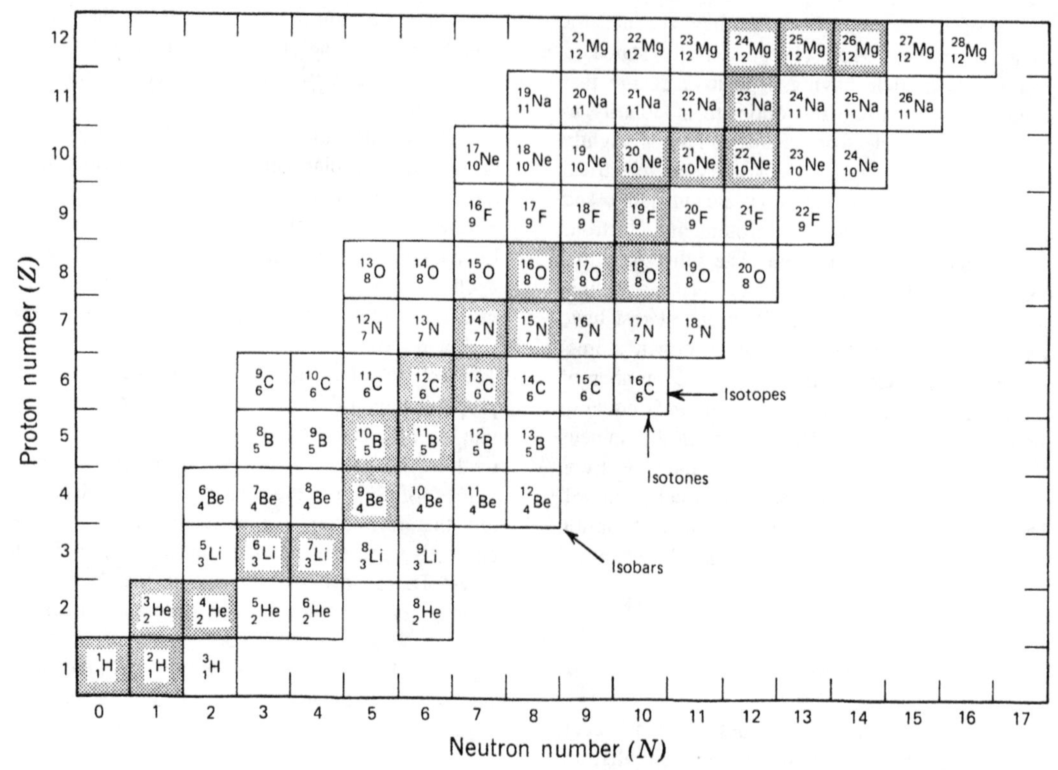

FIGURE 1.1 Partial chart of the nuclides. Each square represents a particular nuclide defined in terms of number of protons (Z) and neutrons (N) that make up its nucleus. The shaded squares represent stable atoms, whereas the white squares are the unstable or radioactive nuclides. Isotopes are atoms having the same Z but different values of N. Isotones have the same N but different values of Z. Isobars have the same A but different values of Z and N. Isotopes are atoms of the same element and therefore have identical chemical properties.

two elements therefore do not occur naturally on the Earth. However, they have been identified in the optical spectra of certain stars where they are synthesized by nuclear reactions.

The relative proportions of the naturally occurring isotopes of an element are expressed in terms of percent by number. For example, the statement that the isotopic abundance of $^{85}_{37}$Rb is 72.15 percent means that in a sample of 10,000 Rb atoms 7215 are the isotope $^{85}_{37}$Rb. When the masses of the naturally occurring isotopes of an element and their abundances are known, the atomic weight of that element can be calculated. The atomic weight of an element is the sum of the masses of its naturally occurring isotopes weighted in accordance with the abundance of each isotope expressed as a decimal fraction. For example, the atomic weight of chlorine (Cl) is calculated from the masses and abundances of its two naturally occurring isotopes:

Isotope	Mass × Abundance
$^{35}_{17}$Cl	34.96885 × 0.7577 = 26.4958
$^{37}_{17}$Cl	36.96590 × 0.2423 = 8.9568
	Atomic weight = 35.4526 amu

The abundances of the naturally occurring isotopes of the elements and their measured masses are listed in tables such as those of the *Handbook of Chemistry and Physics* (Lide and Frederikse, 1995).

Although the atomic weights of the elements are expressed in atomic mass units, it is convenient to define the *gram atomic weight*, or *mole*, which is the atomic weight of an element in grams. One mole of an atom or a compound contains a fixed number of atoms or molecules, respectively. The number of atoms or molecules in one mole is given by Avogadro's number, which is equal to 6.022045×10^{23} atoms or molecules per mole.

1.2c Binding Energy of Nucleus

The definition of the atomic mass unit provides an opportunity to calculate the mass of a particular nuclide by adding the masses of protons + electrons ($M_H = 1.00782503$ amu) and of the neutrons ($M_n = 1.00866491$ amu) of which it is composed. These calculated masses are consistently greater than the measured masses. It appears, therefore, that the mass of an atom is less than the sum of its parts. This phenomenon is an important clue to an understanding of the nature of the atomic nucleus. The explanation of the observed *mass defect* is that some of the mass of the nuclear particles is converted into *binding energy* that holds the nucleus together. The binding energy (E_B) is calculated by means of Einstein's equation:

$$E_B = \Delta m\ c^2 \qquad (1.2)$$

where Δm is the mass defect and c is the speed of light in a vacuum ($2.99792458 \times 10^{10}$ cm/s).

The calculation of the binding energy requires a review of the relationship between units of mass and energy. The basic unit of energy in the cgs system (centimeter, gram, second) is the erg. However, the amount of energy released by a nuclear reaction involving a single atom is only a small fraction of one erg. For this reason, the *electron volt* (eV) is defined as the energy acquired by any charged particle carrying a unit electronic charge when it is acted upon by a potential difference of one volt. One electron volt is equivalent to 1.60210×10^{-12} erg. It is convenient to define two additional units, the kiloelectron volt (keV) and the million electron volt (MeV), where 1 keV = 10^3 eV and 1 MeV = 10^6 eV.

The conversion of the amu into grams follows from the definition of the atomic mass unit:

$$1 \text{ amu} = \frac{1}{12} \times \frac{12.000}{A} = \frac{1}{A} \text{gram}$$

where A is Avogadro's number. The amount of energy equivalent to 1 amu is obtained from Equation 1.2 by substituting $1/A$ for the mass of 1 amu in grams and by converting ergs to million electron volts by means of the appropriate conversion factor given above:

$$E = \frac{(2.99792458 \times 10^{10})^2}{6.022045 \times 10^{23} \times 1.60210 \times 10^{-12} \times 10^6}$$
$$= 931.5 \text{ MeV/amu}$$

The result of this calculation indicates that 1 amu of mass is equivalent to 931.5 MeV of energy. The binding energy of a nucleus (E_B) in million electron volts can now be calculated from the mass defect (Δm) in amu by means of the equation

$$E_B = 931.5\ \Delta m \qquad (1.3)$$

For example, the theoretical mass of $^{27}_{13}$Al is $13 \times 1.00782503 + 14 \times 1.00866491 = 27.22303413$ amu. Its measured mass is only 26.981538 amu. Therefore, the mass defect and binding energy of $^{27}_{13}$Al are

$$\Delta m = 0.24149613 \text{ amu}$$
$$E_B = \frac{0.24149613 \times 931.5}{27} = 8.332 \text{ MeV/nucleon}$$

The binding energies per nucleon of the atoms of most elements have values ranging from about 7.5 to 8.8 MeV. The binding energy per nucleon rises slightly with increasing mass number and reaches a maximum value for $^{56}_{26}$Fe. Thereafter, the binding energies decline slowly with increasing mass number. The binding energies of the atoms of H, He, Li, and Be are lower than the binding energies of the other elements.

1.2d Nuclear Stability and Abundance

It is reasonable to expect that a relationship exists between the stability of the nucleus of an atom and the abundance of that atom in nature. Conversely, one can use the observed abundances of the elements in the solar system and the abundances of their naturally occurring isotopes to derive information about the apparent stabilities of different kinds of atoms.

Given that only about 270 of nearly 2500 known nuclides are stable, nuclear stability is the exception rather than the rule. The point is illustrated in Figure 1.2, which is a schematic plot of the nuclides in coordinates of N and Z. In this diagram, the stable nuclides (shown in black) form a band flanked on both sides by unstable nuclides (shown in white). The existence of such a region of stability indicates that only those nuclei are stable in which Z and N are nearly equal. Actually, the ratio of N to Z increases from 1 to about 3 with increasing values of A. Only ${}^{1}_{1}H$ and ${}^{3}_{2}He$ have fewer neutrons than protons.

Another interesting observation is that most of the stable nuclides have even numbers of protons and neutrons. Stable nuclides with even Z and odd N or vice versa are much less common, whereas nuclides having odd Z and odd N are rare. These facts are shown in Table 1.1, where long-lived radioactive nuclides having halflives greater than 500×10^6 years are considered to be stable (Walker et al., 1989a). The relations in Table 1.1 suggest that the nucleons within the nucleus are arranged into regular patterns consisting of even numbers of protons and neutrons.

A careful study of this phenomenon combined with theoretical considerations has led to the concept of magic numbers for Z and N. Nuclides having magic proton numbers or magic neutron numbers or both are unusually stable, as indicated by their greater abundance or, in the case of unstable nuclides, by their slower decay rates. The magic numbers for Z and N are 2, 8, 10, 20, 28, 50, 82, and 126.

Calcium in Figure 1.3 is a good example of the effect of the magic number. It has a magic proton number (20) and six stable isotopes whose mass numbers are 40, 42, 43, 44, 46, and 48. All but one of these stable isotopes have even neutron numbers. The nucleus of ${}^{40}_{20}Ca$ is doubly magic because it also has a magic number of neutrons ($N = 20$). Its isotopic abundance is 96.94 percent. The nucleus of ${}^{48}_{20}Ca$ is also doubly magic ($N =$ 28), but its abundance is only 0.19 percent, which is nevertheless remarkably high considering that it is some distance from the region of stability. Figure 1.3 also demonstrates that the halflives of the unstable (radioactive) isotopes of Ca increase from $A = 35$ to $A = 41$ and then decline from $A = 45$ to $A = 53$. The halflife of an unstable isotope is the time required for one-half of the atoms of that isotope to decay. In the context of Figure 1.3, the halflife is a measure of the relative stability of the unstable Ca isotopes: The longer the halflife, the more "stable" the nucleus. Therefore, Figure 1.3 illustrates the phenomenon that the stable isotopes of a given element are confined to a limited range of their mass numbers and that the radioactive isotopes at greater and smaller values of A become increasingly unstable (i.e., have progressively shorter halflives).

The chart of the nuclides (Figures 1.1 and 1.2) as well as the isotopes of Ca in Figure 1.3 reveal that most known nuclides are not stable but decompose spontaneously until they achieve a stable nuclear configuration. These are the so-called radioactive nuclides, or radionuclides. The spontaneous transformations that occur in their nuclei give rise to the phenomenon of *radioactivity*. Most of the known radionuclides do not occur naturally because their decay rates are rapid compared with the age of the solar system. However, they can be produced artificially by means of nuclear reactions in the laboratory. Radionuclides occur in nature for several reasons:

1. They have not yet completely decayed because their decay rates are very slow (${}^{238}_{92}U$, ${}^{235}_{92}U$, ${}^{232}_{90}Th$, ${}^{87}_{37}Rb$, ${}^{40}_{19}K$, and others).
2. They are produced by the decay of long-lived, naturally occurring, radioactive parents (${}^{234}_{92}U$, ${}^{230}_{90}Th$, ${}^{226}_{88}Ra$, and others).
3. They are produced by nuclear reactions occurring in nature (${}^{14}_{6}C$, ${}^{10}_{4}Be$, ${}^{32}_{14}Si$, and others).

There is also a fourth group of radionuclides that can now be found in nature because they have been produced artificially, mainly as a result of the operation of nuclear fission reactors and by the testing of explosive fission and fusion devices. The dispersal of these radionuclides into the atmosphere

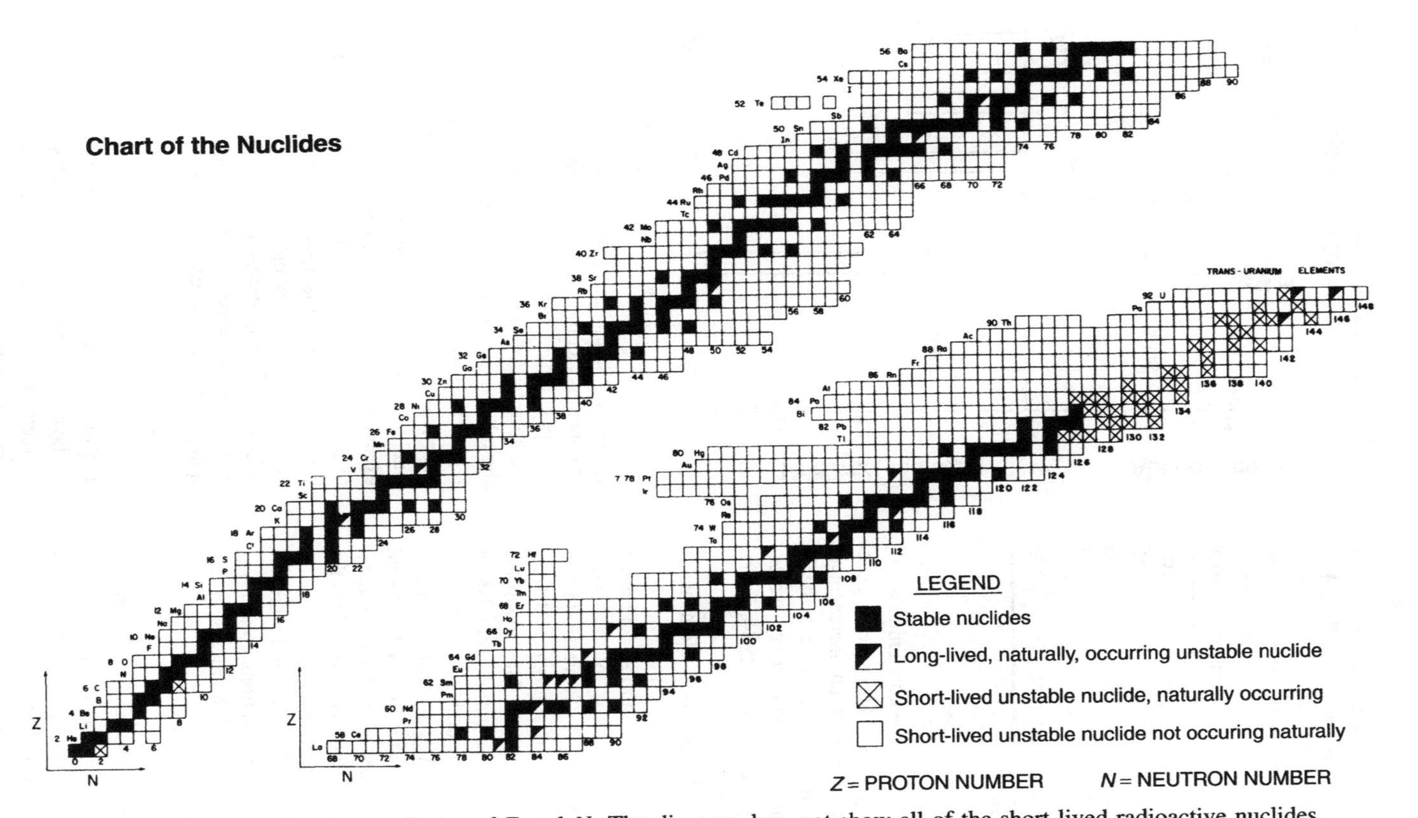

FIGURE 1.2 Chart of the nuclides in coordinates of Z and N. The diagram does not show all of the short-lived radioactive nuclides. After Walker et al. (1989b).

Table 1.1. Distribution of Stable Nuclides Depending on "Evenness" or "Oddness" of *A*, *Z*, and *N*

A	*Z*	*N*	Number of Stable Nuclides
Even	Even	Even	161
Odd	Even	Odd	55
Odd	Odd	Even	50
Even	Odd	Odd	4
Total number of stable nuclides			270

Source: Walker et al., 1989a.

results in "fallout" and may cause contamination of food crops and drinking water. The safe disposal of radioactive waste products is an increasingly serious problem that will become more acute in the future as the quantity of radioactive "waste" increases (Winograd, 1981; Brookins, 1984; Faure, 1998).

The abundances of the chemical elements in the solar system given in Table 1.2 and Figure 1.4 are expressed in terms of the number of atoms of each element per 10^6 atoms of Si. They were compiled by Anders and Grevesse (1989) from spectroscopic analyses of sunlight and from chemical analyses of meteorites. The logarithm to the base 10 of the elemental abundances in Figure 1.4 forms a "sawtooth" pattern with increasing atomic number because the elements with even Z are more abundant than the adjacent elements that have odd values of Z. This statement is known as the Oddo–Harkins rule after the codiscoverers of this phenomenon.

The pattern of variation displayed in Figure 1.4 permits several additional observations concerning the abundances of the chemical elements in the solar system:

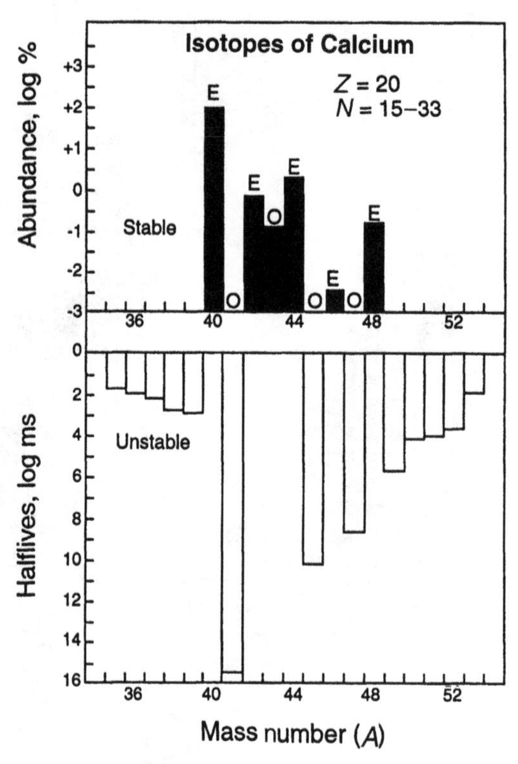

FIGURE 1.3 Stable and unstable (radioactive) isotopes of calcium. The abundances of the stable isotopes (black) are expressed as the logarithm to the base ten of the percent by number, whereas the relative stability of the unstable isotopes is represented by the logarithm to the base ten of their halflives expressed in milliseconds. The stable isotopes have even (E) mass numbers, except for $^{43}_{20}Ca$, whose abundance is less than that of its neighbors at $A = 42$ and $A = 44$. The gaps between the stable isotopes are occupied by unstable isotopes having odd (O) values of A. The halflives of the unstable isotopes decline with increasing "distance" from the central region of stability. Data from Walker et al. (1989b).

1. The elements H ($Z = 1$) and He ($Z = 2$) are by far the most abundant in the solar system.
2 The abundances of Li ($Z = 3$), Be ($Z = 4$), and B ($Z = 5$) are anomalously low.
3. The abundances of the elements whose atomic numbers are greater than 6 (C) decrease exponentially with increasing values of Z.
4. Iron ($Z = 26$) is exceptionally abundant, whereas the abundance of F ($Z = 9$) is less than expected.
5. Technetium (Tc) and promethium (Pm) do not occur naturally in the solar system because all of their isotopes are unstable and have short halflives.

Table 1.2. Abundance of Elements in the Solar System (atoms per 10^6 Si atoms)

Z	Element	Abundance	Z	Element	Abundance
1	H	2.79×10^{10}	44	Ru	1.86
2	He	2.72×10^{9}	45	Rh	3.44×10^{-1}
3	Li	5.71×10^{1}	46	Pd	1.39
4	Be	7.3×10^{-1}	47	Ag	4.86×10^{-1}
5	B	2.12×10^{1}	48	Cd	1.61
6	C	1.01×10^{7}	49	In	1.84×10^{-1}
7	N	3.13×10^{6}	50	Sn	3.82
8	O	2.38×10^{7}	51	Sb	3.09×10^{-1}
9	F	8.43×10^{2}	52	Te	4.81
10	Ne	3.44×10^{6}	53	I	9.0×10^{-1}
11	Na	5.74×10^{4}	54	Xe	4.7
12	Mg	1.074×10^{6}	55	Cs	3.72×10^{-1}
13	Al	8.49×10^{4}	56	Ba	4.49
14	Si	1.00×10^{6}	57	La	4.460×10^{-1}
15	P	1.04×10^{4}	58	Ce	1.136
16	S	5.15×10^{5}	59	Pr	1.669×10^{-1}
17	Cl	5.24×10^{3}	60	Nd	8.279×10^{-1}
18	Ar	1.01×10^{5}	61	Pm	—
19	K	3.770×10^{3}	62	Sm	2.582×10^{-1}
20	Ca	6.11×10^{4}	63	Eu	9.73×10^{-2}
21	Sc	3.42×10^{1}	64	Gd	3.30×10^{-1}
22	Ti	2.400×10^{3}	65	Tb	6.03×10^{-2}
23	V	2.93×10^{2}	66	Dy	3.42×10^{-1}
24	Cr	1.35×10^{4}	67	Ho	8.89×10^{-2}
25	Mn	9.50×10^{3}	68	Er	2.508×10^{-1}
26	Fe	9.00×10^{5}	69	Tm	3.78×10^{-2}
27	Co	2.250×10^{3}	70	Yb	2.479×10^{-1}
28	Ni	4.93×10^{4}	71	Lu	3.67×10^{-2}
29	Cu	5.22×10^{2}	72	Hf	1.54×10^{-1}
30	Zn	1.260×10^{3}	73	Ta	2.07×10^{-2}
31	Ga	3.78×10^{1}	74	W	1.33×10^{-1}
32	Ge	1.19×10^{2}	75	Re	5.17×10^{-2}
33	As	6.56	76	Os	6.75×10^{-1}
34	Se	6.21×10^{1}	77	Ir	6.1×10^{-1}
35	Br	1.18×10^{1}	78	Pt	1.34
36	Kr	4.5×10^{1}	79	Au	1.87×10^{-1}
37	Rb	7.09	80	Hg	3.4×10^{-1}
38	Sr	2.35×10^{1}	81	Tl	1.84×10^{-1}
39	Y	4.64	82	Pb	3.15
40	Zr	1.14×10^{1}	83	Bi	1.44×10^{-1}
41	Nb	0.698×10^{-1}	90	Th	3.35×10^{-2}
42	Mo	2.55	92	U	9.00×10^{-3}
43	Tc	—			

Source: Anders and Grevesse, 1989.

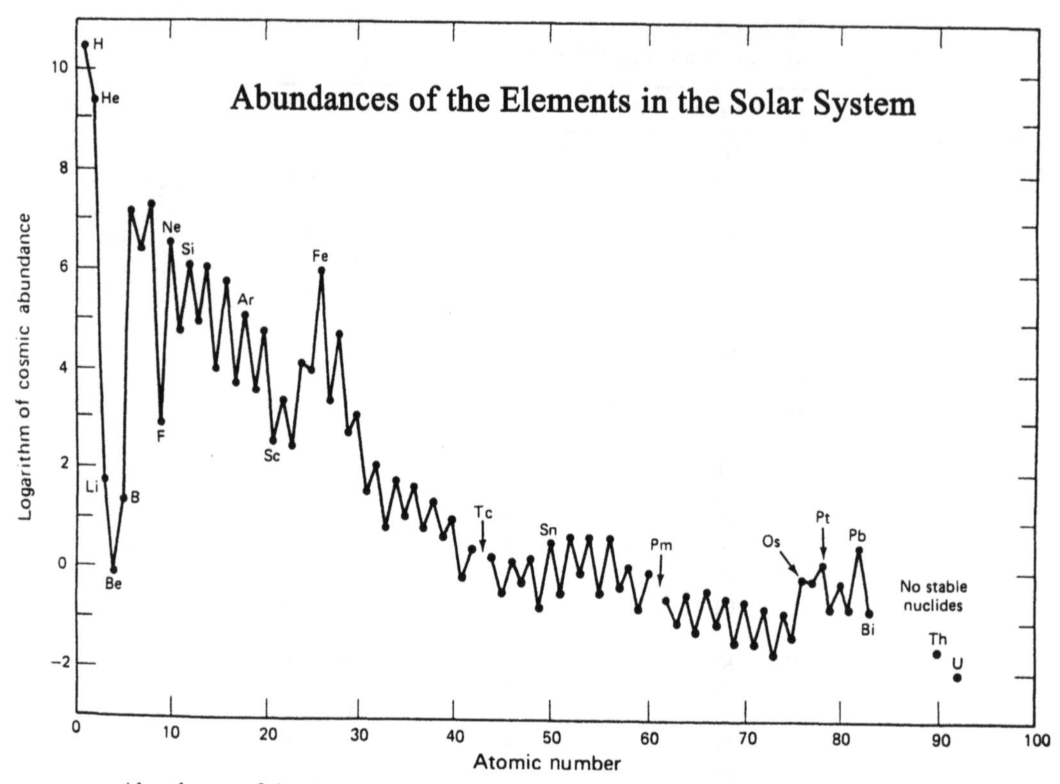

FIGURE 1.4 Abundances of the elements in the solar system versus their atomic number. The abundances are expressed as the logarithm to the base 10 of the number of atoms of each element relative to 10^6 atoms of silicon. Data are listed in Table 1.2. After Anders and Grevesse (1989).

6. The abundance of Pb ($Z = 82$) is somewhat greater than expected because its isotopes are the stable end products of the decay of the isotopes of U and Th.
7. Uranium has the lowest abundance of any element in the solar system, except as noted in 5 above.
8. The abundances of elements that lie between Bi ($Z = 83$) and Th ($Z = 90$) (e.g., Ra) and at $Z = 91$ are exceedingly low because these elements are the unstable daughters of the decay of the naturally occurring isotopes of U and Th.

The observed pattern of abundances of the chemical elements constrains the process of nucleosynthesis in the ancestral stars, which contributed matter to the solar nebula from which the Sun and the planets of the solar system formed. The nuclear reactions that occur in stars during their evolution and during their final explosions as supernovas produced all of the known nuclides (except ${}^1_1H$), including both the stable and the unstable isotopes of the elements. Consequently, the Sun, the Earth, and all of the other planets and their satellites in the solar system are composed of the comparatively small number of stable and long-lived radioactive nuclides that have survived to the present time.

1.3 ORIGIN OF THE ELEMENTS

The nuclear reactions in the interiors of stars start at temperatures in excess of 10×10^6K with fusion of H nuclei (protons and deuterons) to form nuclei of He. The Sun has been generating the energy it radiates into space by the H fusion reaction, which means that it is producing only helium and deuterium. Therefore, all of the other elements in the Sun originated from ancestral stars whose terminal explosions formed the cloud of gas and dust in interstellar space from which the Sun and the planets of the solar system formed 4.6×10^9 years ago.

The fusion of H nuclei to form He takes place by the proton–proton chain, which includes ^2_1H and ^3_2He as intermediate products:

$$^1_1H + ^1_1H \rightarrow ^2_1H + \beta^+ + \nu + 0.422 \text{ MeV} \quad (1.4)$$

$$\beta^+ + e^- \rightarrow 1.02 \text{ MeV} \quad \text{(annihilation)} \quad (1.5)$$

$$^2_1H + ^1_1H \rightarrow ^3_2He + \gamma + 5.493 \text{ MeV} \quad (1.6)$$

$$^3_2He + ^3_2He \rightarrow ^4_2He + 2\ ^1_1H + 12.859 \text{ MeV} \quad (1.7)$$

where β^+ is a positron (positively charged electron) and ν is a neutrino. The stable intermediate products ^2_1H and ^3_2He are consumed in reaction 1.6 and 1.7 to form the stable nucleus of ^4_2He, which is identical to the alpha-particle (α-particle) named by Ernest Rutherford in 1899. The total amount of energy released by the synthesis of one helium nucleus is 19.794 MeV. Therefore, the formation of 1 mol of ^4_2He by the proton–proton chain releases an amount of heat (Q) equal to

$$Q = 19.794 \times 6.022045 \times 10^{23} \times 1.60210 \times 10^{-6} \times 10^{-7} \text{ J} = 190.97 \times 10^{10} \text{ J/mol}$$

where 1 MeV $= 1.60210 \times 10^{-6}$ erg, 1 erg $= 10^7$ joule (J), and Avogadro's number $= 6.022045 \times 10^{23}$ atoms per mole.

Stars like the Sun that inherited $^{12}_6C$ and other nuclides from a previous cycle of stellar evolution and nucleosynthesis also synthesize ^4_2He by the so-called CNO cycle, in which $^{12}_6C$ acts as a catalyst (Bethe, 1939, 1968):

$$^{12}_6C + ^1_1H \rightarrow ^{13}_7N + \gamma \quad (1.8)$$

$$^{13}_7N \rightarrow ^{13}_6C + \beta^+ + \nu \quad \text{(positron decay)} \quad (1.9)$$

$$^{13}_6C + ^1_1H \rightarrow ^{14}_7N + \gamma \quad (1.10)$$

$$^{14}_7N + ^1_1H \rightarrow ^{15}_8O + \gamma \quad (1.11)$$

$$^{15}_8O \rightarrow ^{15}_7N + \beta^+ + \nu \quad \text{(positron decay)} \quad (1.12)$$

$$^{15}_7N + ^1_1H \rightarrow ^{12}_6C + ^4_2He \quad (1.13)$$

The nucleus of $^{12}_6C$ is released and is available for another iteration of the CNO cycle. In this way, the CNO cycle converts four protons into one ^4_2He nucleus, thereby releasing the same amount of heat energy as the proton–proton chain.

The entire sequence of nuclear reactions by means of which stars generate energy at different stages in their evolution was revealed in a famous paper by Burbidge et al. (1957). A critical step in the nucleosynthesis process is the formation of Li, Be, and B, whose abundances in the solar system (Figure 1.4) are anomalously low by more than five orders of magnitude compared to C ($Z = 6$). The "gap" in the pattern of abundances of the elements is bridged by the so-called triple-α process:

$$^4_2He + ^4_2He \rightarrow ^8_4Be \quad \text{(very unstable)} \quad (1.14)$$

$$^8_4Be + ^4_2He \rightarrow ^{12}_6C \quad (1.15)$$

The nucleus of ^8_4Be has a halflife of only 7×10^{-17} s, which requires that it must capture an α-particle almost immediately after its formation (hence the name triple-α process). The low probability of this reaction is overcome by the high density and high temperature of He in the cores of stars in the red giant stage of their evolution. The triple-α process has a profound effect on the evolution of stars and makes possible the synthesis of all chemical elements having atomic numbers of 6 and higher.

The nucleosynthesis of the chemical elements in stars progresses from $^{12}_6C$ by means of fusion of nuclei having Z values of 6, 8, and 10 as well as by capture of α-particles, protons, and neutrons. Many nuclides having atomic numbers greater than 26 (Fe) form by neutron capture on two different timescales. In the s-process (slow timescale) the radioactive products of neutron capture decay to a stable isotope of another element before the next neutron capture occurs. In the r-process (rapid timescale), neutron capture occurs rapidly before the product nuclides have time to decay. The nuclear reactions by means of which the naturally occurring isotopes of the elements were synthesized in stars were identified by Anders and Grevesse (1989).

1.4 SUMMARY

The chemical elements were synthesized by nuclear reactions in stars starting with the fusion of four hydrogen nuclei to form helium. This is the process by means of which the Sun has been producing the energy it radiates into space since its formation from the solar nebula more than 4.6×10^9 years ago. The solar nebula originally contained about 2500 different nuclides, most of which were unstable and have since decayed to form stable products. Only 270 nuclides have survived to the present either because they are stable or because they decay very slowly.

The stable and unstable nuclides are conveniently inventoried in the chart of the nuclides in coordinates of atomic number (Z) and neutron number (N). The chart reveals that the stable isotopes of the elements occur in a band that extends diagonally across it with the unstable nuclides arranged on either side of it. The stable nuclides predominantly have even values of Z and N, whereas only a small number of stable nuclides have odd values of both Z and N. The even–odd criterion also applies to the abundances of the stable isotopes of a given element and is recognized by the Oddo–Harkins rule, which applies to the abundances of the stable isotopes as well as to the abundances of the chemical elements in the solar system.

The abundances and masses of the naturally occurring isotopes of the elements determine their atomic weights and hence their gram-atomic weights or moles. The transformation of moles to numbers of atoms or molecules is based on Avogadro's law, which plays an important role not only in physical chemistry but also in the derivation of the law of radioactivity.

REFERENCES

Anders, E., and N. Grevesse, 1989. Abundances of the elements: Meteoritic and solar. *Geochim. Cosmochim. Acta*, 53:197–214.

Becquerel, H., 1896. Sur les radiations invisibles emisses par phosphorescence; Sur les radiations invisibles emisses par les corps phosphorescents; Sur les radiations invisibles emisses par les sels d'uranium. *Compt. Rend.*, 122:420, 501, 689.

Bethe, H. A., 1939. Energy production in stars. *Phys. Rev.*, 55:434–456.

Bethe, H. A., 1968. Energy production in stars. *Science*, 161:541–547.

Brookins, D. G., 1984. *Geochemical Aspects of Radioactive Waste Disposal.* Springer-Verlag, New York.

Burbidge, E. M., G. R. Burbidge, W. A. Fowler, and F. Hoyle, 1957. Synthesis of the elements in stars. *Rev. Modern Phys.*, 29:547–650.

Burchfield, J. D., 1975. *Lord Kelvin and the Age of the Earth.* Science History Publications, New York.

Curie, M. S., 1898. Rayons emis par les composes de l'uranium et du thorium. *Compt. Rend.*, 126:1101.

Faure, G., 1998. *Principles and Applications of Geochemistry*, 2nd ed. Prentice-Hall, Upper Saddle River, New Jersey.

Lide, D. R., and H. P. R. Frederikse, 1995. *CRC Handbook of Chemistry and Physics*, 76th ed. CRC Press, Boca Raton, Florida.

Thomson, J. W. (Lord Kelvin), 1862. On the age of the sun's heat. *Popular Lectures and Addresses*, 1:349.

Thomson, W. (Lord Kelvin), 1899. The age of the earth as an abode fitted for life. *Philos. Mag.*, 47(5): 66–90.

Walker, F. W., J. R. Parrington, and F. Feiner, 1989a. *Nuclides and Isotopes; Information Booklet Accompanying the Chart of the Nuclides*, 14th ed. General Electric Co., Nuclear Energy Operations, San Jose, California.

Walker, F. W., J. R. Parrington, and F. Feiner, 1989b. *Chart of the Nuclides*, 14th ed. Knolls Atomic Power Laboratory, General Electric Co., San Jose, California.

Winograd, I. J., 1981. Radioactive waste disposal in thick unsaturated zones. *Science*, 212:1457–1464.

CHAPTER 2

Decay Modes of Radionuclides

The nuclei of unstable atoms undergo spontaneous transformations that involve the emission of particles and radiant energy. These processes give rise to the phenomenon of radioactivity. There are several different ways in which unstable atoms can decay. Some atoms decay in two or three different ways, but most do so in only one particular way. In either case, radioactive decay results in changes of Z and N of the parent and thus leads to the transformation of an atom of one element into that of another element. The radiogenic daughter may itself be radioactive and, in turn, decays to form an isotope of yet another element. This process continues until at last a stable nucleus is produced.

Soon after radioactivity was discovered at the end of the nineteenth century, Rutherford and others demonstrated that it involved the emission of three different types of rays, which he named alpha (α), beta (β), and gamma (γ). The β-rays were subsequently shown to be streams of particles identical to electrons. It is customary to restrict the term β-particle to electrons emitted by the nucleus of an atom. The β-particle is represented by β^- or β^+, depending on its electrical charge. The existence of positively charged β-particles (positrons) was first predicted by P. A. M. Dirac and was subsequently confirmed by Anderson in 1932 on the basis of cosmic-ray tracks in a Wilson cloud chamber. In addition, electrons are involved in a third mode of decay called electron capture. Alpha-particles are the nuclei of ^4_2He and the γ-rays turned out to be electromagnetic radiation having shorter wavelengths than x-rays.

2.1 BETA-DECAY

Radionuclides located to the right of the band of stable nuclei in the chart of the nuclides (Figure 1.2) have an excess of neutrons and decay by emitting β^--particles because the number of neutrons decreases during this process. Radionuclides located on the left side of the band of stable nuclides are deficient in neutrons and decay by positron emission and/or electron capture, both of which increase the number of neutrons. Decay by emission of α-particles is available to radionuclides having $Z > 58$ and to a few radionuclides of low atomic number (e.g., some or all of the radionuclides of He, Li, Be, and B).

2.1a Beta- (Negatron) Decay

A large group of unstable atoms decays by emitting a negatively charged β-particle (negatron) and neutrinos from the nucleus, often accompanied by the emission of radiant energy in the form of γ-rays. According to a theory by E. Fermi first published in 1934, β-decay can be regarded as a transformation of a neutron in the nucleus into a proton and an electron. The electron is then expelled from the nucleus as a negative β-particle.

As a result of such a β-decay, the atomic number of the atom is increased by 1 while its neutron number is reduced by 1:

	Atomic Number	Neutron Number	Mass Number
Parent	Z	N	$Z + N = A$
Daughter	$Z + 1$	$N - 1$	$Z + 1 + N - 1 = A$

Consequently, the daughters of radioactive parents that decay by negatron emission are isobars and can be located on a chart of the nuclides, as shown in Figure 2.1.

In 1914 Chadwick observed that the β-particles emitted by a particular radioactive nuclide have a continuous energy distribution. This means that the β-particles have kinetic energies that range continuously from nearly zero to some maximum value, as illustrated in Figure 2.2 by the energy spectrum of β^--particles emitted by $^{40}_{19}K$. Careful studies of this phenomenon have shown that most of the β-particles emitted by a particular nuclide have kinetic energies equal to about one-third of the maximum energy and only a relatively small number of the β-particles are emitted with the maximum kinetic energy. The question is, when a β-particle is emitted with less than the maximum energy, what happens to the rest of the kinetic energy? Beta-decay seemed to violate the principle of conservation of mass and energy, which is one of the cornerstones of science. It also appeared to be violating other kinds of conservation laws, all of which resulted in a serious embarrassment for nuclear physicists.

This crisis was resolved in 1931 by W. Pauli, who suggested the existence of a hypothetical particle that is produced during β-decay and has a series of postulated properties such that the conservation laws are satisfied (unpublished address at the University of Tübingen). This mysterious particle was named *neutrino* by Fermi, (1934), who incorporated it into his theory of β-decay. The neutrino has no charge and a very small rest mass, but it can have varying amounts of kinetic energy. It interacts "sparingly" with matter and consequently carries its energy into intergalactic space. According to Fermi's theory, each β-decay includes the emission of a β-particle and neutrino having complementary kinetic

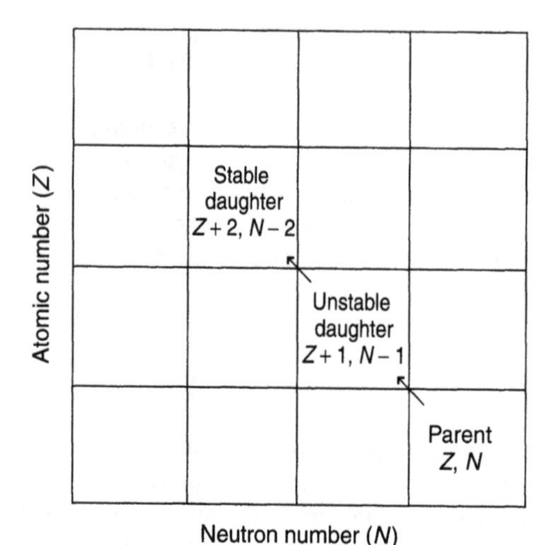

FIGURE 2.1 Schematic representation of sequential β^--decay in coordinates of Z and N of the chart of the nuclides (Figure 1.2). The diagram demonstrates that the daughters of β^--decay are isobars of their parent and isotopes of different elements.

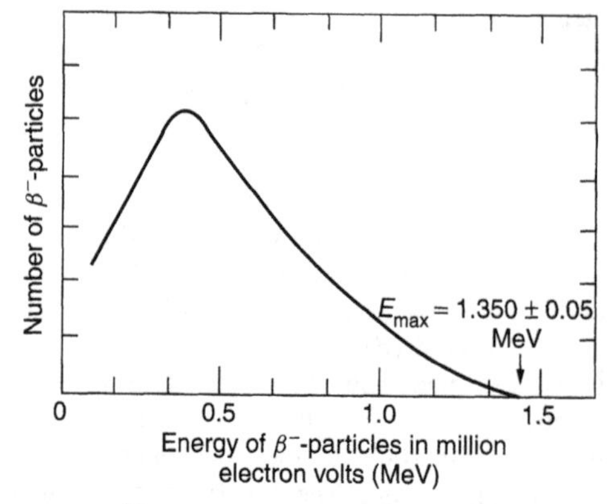

FIGURE 2.2 Energy spectrum of β^--particles emitted by $^{40}_{19}K$. Most of the β^--particles are emitted with energies that are about one-third of the maximum endpoint energy. During each β^--decay, an antineutrino is emitted with a kinetic energy equal to the difference between the maximum energy and the kinetic energy of the associated β^--particle. Data from Dzelepow et al. (1946).

energies such that their sum is equal to the maximum kinetic energy of the suite of β^--particles emitted by a particular nuclide.

The existence of the elusive neutrinos was confirmed in 1954 by F. Reines and C. L. Cowan (Reines, 1979). Current theories indicate that the Earth is exposed to a continuous flux of neutrinos of the order of $10^{10}-10^{11}$ neutrinos/cm^2/s, most of which originate in the Sun. However, initial efforts (Bahcall, 1969) to measure the flux density of neutrinos by means of an elaborate detection system inside the Homestake Gold Mine in Lead, South Dakota, suggested a somewhat lower value. The apparent deficiency of solar neutrinos posed a problem to astrophysicists because the neutrino flux is a direct monitor of the nuclear reactions responsible for energy production in the Sun. The shortage of solar neutrinos indicated that either our understanding of the nuclear reactions in the interior of the Sun was faulty, that the Sun is currently producing less energy in its interior than is compatible with the amount of energy being radiated into space, or that the physical properties of neutrinos are not understood.

The deficiency in the solar neutrino flux was eventually explained by the discovery that there are three types (or flavors) of neutrinos: the electron neutrino, the muon neutrino, and the tau-neutrino (τ-neutrino) and their corresponding antineutrinos. Only one of these is detectable by the formation of $^{37}_{18}$Ar from $^{37}_{19}$Cl in chlorine-rich liquids used in conventional neutrino detectors. In addition, the three types of neutrinos "oscillate," which means that they can change their identity from one flavor to another. Such transformations are only possible if neutrinos have a small amount of mass when they are at rest. Consequently, neutrinos not only carry energy but also contain some of the mass of the universe.

Neutrino detectors must be shielded from cosmic rays and therefore have been built in underground mines. For example, the first successful neutrino detector was a 387,500-L tank of perchloroethylene set up about 1500 m below the surface in the Homestake Gold Mine in Lead, South Dakota. This detector became operational in 1967 and is still in use. Another neutrino detector containing 50,000 tons of pure water is located in an active underground zinc mine at Kamioka about 250 km north of Tokyo, Japan. Additional neutrino observatories exist in a salt mine near Cleveland, Ohio, and at Sudbury, Ontario. Even the East Antarctic ice sheet is being used to detect neutrinos. The newer detectors count neutrinos by recording the flash of Cerenkov radiation that is emitted when a neutrino collides with a proton in a water molecule. This type of radiation is emitted when charged particles travel faster than the speed of light in a medium other than a perfect vacuum.

The nuclear transformations during negatron decay are exemplified by the decay of short-lived $^{24}_{11}$Na to stable $^{24}_{12}$Mg with a halflife of 14.97 h. The equation representing this transformation is

$$^{24}_{11}\text{Na} \rightarrow {}^{24}_{12}\text{Mg} + \beta^- + \bar{\nu} + 2\gamma + E \qquad (2.1)$$

where β^- is the β-particle, $\bar{\nu}$ is the complementary antineutrino, γ represents the γ-rays emitted by the product nucleus, and E is the total decay energy in units of million electron volts. The energies of the β^--particles and γ-rays emitted by $^{24}_{11}$Na are (Lide and Frederikse, 1995)

Maximum energy of $\beta^-(E_{max})$:	1.389 MeV
Energy of γ-rays(E_γ) :	1.3686 MeV
	2.7541 MeV
Total decay energy (E) :	5.514 MeV

The total decay energy E of $^{24}_{11}$Na is the sum of the maximum kinetic energy of the β^--particles (E_{max}) and the energies of the γ-rays (E_γ):

$$E = E_{max} + 2E_\gamma \qquad (2.2)$$

The sequence of events associated with the decay of $^{24}_{11}$Na to $^{24}_{12}$Mg is indicated in coordinates of Z and E by the decay scheme diagram in Figure 2.3. The emission of β-particles and the complementary neutrinos by $^{24}_{11}$Na leaves the nucleus of $^{24}_{12}$Mg in an excited state at an energy of 4.125 MeV above the ground state ($E - E_{max}$). The nucleus of $^{24}_{12}$Mg subsequently de-excites by emitting two γ-rays whose combined energy is 4.1227 MeV. The energy of the neutrino emitted during the β-decay of $^{24}_{11}$Na is the difference

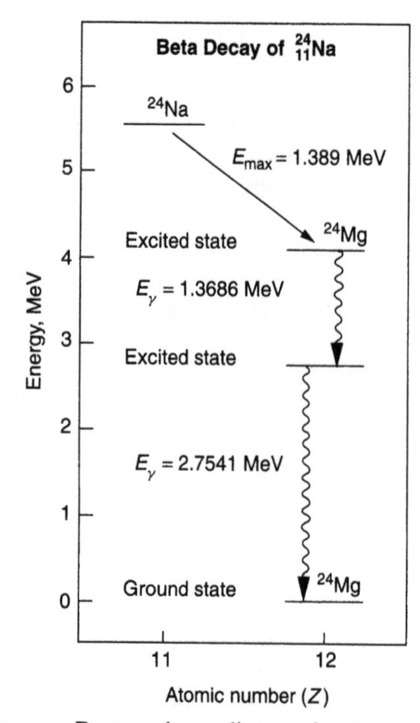

FIGURE 2.3 Decay scheme diagram for the β^--decay of short-lived $^{24}_{11}$Na to stable $^{24}_{12}$Mg. The decay takes place by the emission of one suite of β-particles ($E_{max} = 1.389$ MeV) followed by two γ-rays that de-excite the nucleus of $^{24}_{12}$Mg to the ground state. The halflife of $^{24}_{11}$Na is 14.97 h. Data from Lide and Frederikse (1995).

between E_{max} and the kinetic energy of a particular β^--particle.

In some cases, the nucleus of the product of β-decay de-excites by transferring the excess energy to an extranuclear electron that is then ejected from the atom with a kinetic energy equal to the difference between the excitation energy of the nucleus and the binding energy of the electron. These electrons are superimposed on the continuous spectrum of the β-particles and give rise to a line spectrum. This process is called internal conversion. The loss of an extranuclear electron causes the atom to emit x-rays that form when the vacancy in its electronic structure is filled by other electrons.

Some nuclei in excited states may remain in a metastable condition for measurable lengths of time. Such excited nuclei are called isomers. They decay to their ground state by emission of γ-rays or by internal conversion with halflives ranging from 10^{-11} s up to 241 years, as in the case of ^{192m}Ir. The superscript "m" identifies the nuclide as an isomer of ^{192}Ir, which is itself radioactive and decays by β^--emission to stable ^{192}Pt. The identification of isomers has become somewhat confused by technological improvements in the measurement of short intervals of time, and the number of known short-lived isomers has been increasing correspondingly.

Metastable nuclei produced by β-decay may also emit neutrons, protons, or even α-particles (Walker et al., 1989). For example, $^{17}_{7}$N decays by β^--emission to several short-lived excited states of $^{17}_{8}$O that subsequently emit delayed neutrons having several different energies. Similarly, $^{25}_{14}$Si decays by β^+-emission to excited states of $^{25}_{13}$Al that de-excite by emitting delayed protons.

Many β^--emitters have complex β-energy spectra, meaning that they emit two or more suites of β-particles such that each suite has a different endpoint or E_{max} energy. For example, Lide and Frederikse (1995) recorded the following data for $^{27}_{12}$Mg, which decays to $^{27}_{13}$Al by β^--decay with a halflife of 9.45 m, as shown in Figure 2.4:

Particle	Energy MeV	Frequency, %
β_1	1.59	41
β_2	1.75	58
β_3	2.65	0.3
γ_1	0.17068	1[a]
γ_2	0.84376	70[a]
γ_3	1.01443	29[a]
Total decay energy	2.610	

[a]Recalculated to 100%

The three pathways of the β-decay of $^{27}_{12}$Mg must release identical amounts of energy equal to the total decay energy (2.610 MeV) within the precision of measurement:

1. 1.59 MeV, β_1
   0.17068 MeV, γ_1
   0.84376 MeV, γ_2
   2.604 MeV

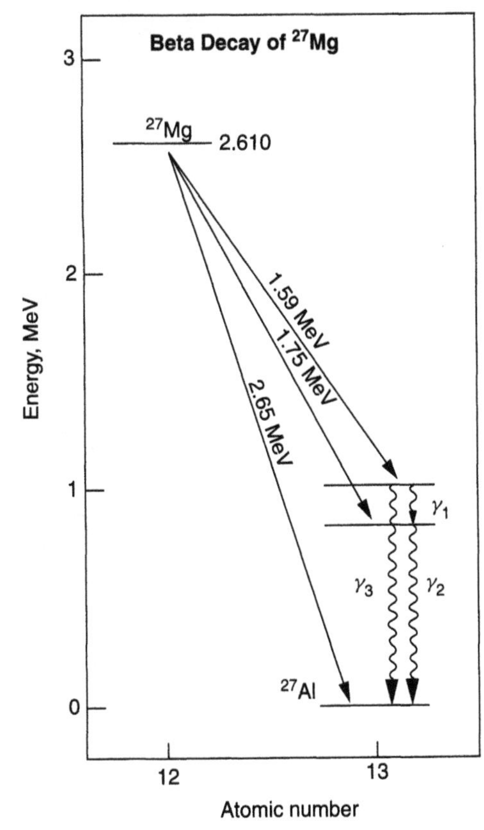

FIGURE 2.4 The β^--decay of $^{27}_{12}$Mg to stable $^{27}_{13}$Al. The parent emits three suites of β-particles, only one of which reaches the ground state of $^{27}_{13}$Al. The two excited states of the progeny nucleus de-excite by emitting three γ-rays. Data from Lide and Frederikse (1995).

2. 1.59 MeV, β_1
   1.01443 MeV, γ_3
   2.604 MeV
3. 1.75 MeV, β_2
   0.84376 MeV, γ_2
   2.594 MeV
4. 2.65 MeV, β_3
   0.00 MeV, no gamma
   2.65 MeV

The energy released by decay path 4 (2.65 MeV) differs from the decay energy (2.610 MeV) by only 1.5 percent, which is attributable to analytical errors in the data. Since β^--decay liberates energy, the parent radionuclide must have either more mass or more energy than the product atom. In the case under consideration, the mass of $^{27}_{13}$Al is 26.981538 amu and the mass (Δm) equivalent to the decay energy of $^{27}_{12}$Mg is

$$\Delta m = \frac{2.610}{931.5} = 0.002801 \text{ amu}$$

Therefore, the mass of unstable $^{27}_{12}$Mg is 26.981538 + 0.002801 = 26.984339 amu.

A final point about negatron decay is that extranuclear neutrons are unstable and decay to protons by emitting a β^--particle. The decay energy is 0.78235 MeV, which is equal to E_{max} of the β-particles that are emitted by the neutron. The halflife is 10.4 min. Protons appear to be stable and therefore live forever.

2.1b Positron Decay

Radionuclides located to the left of the band of stable isotopes of the elements in the chart of the nuclides (Figure 1.2) have a deficiency of neutrons (or an excess of protons). Therefore, in Fermi's theory of β-decay positron emission is regarded as the transformation of a proton in the nucleus into a neutron, a positron, and a neutrino. The configuration of the product nucleus relative to that of its parent can be determined from the following statements:

	Atomic Number	Neutron Number	Mass Number
Parent	Z	N	$Z + N = A$
Daughter	$Z - 1$	$N + 1$	$Z - 1 + N - 1 = A$

Consequently, the product nuclides of positron decay are isobars of their parents and hence are isotopes of different chemical elements. The location of the product nuclides relative to their parent in the chart of the nuclides is indicated in Figure 2.5. Positron decay (like negatron decay) can occur sequentially until a stable product nucleus is formed.

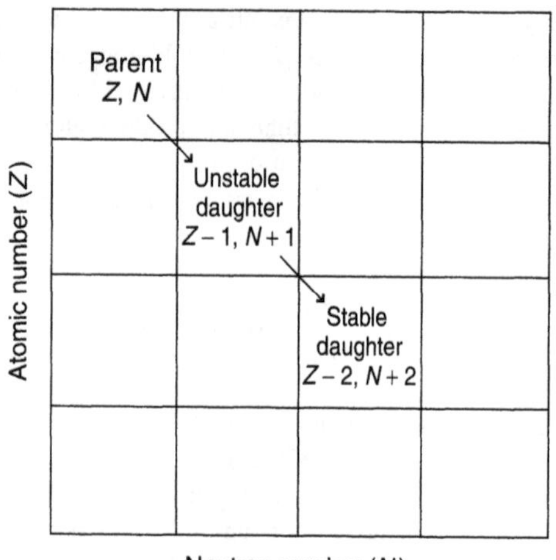

FIGURE 2.5 Schematic representation of the position of daughter nuclides produced by positron emission of the parent. Note that the atomic number (Z) decreases by 1 while the neutron number (N) increases by 1.

Positrons are equal in mass but have a charge of opposite polarity compared to ordinary electrons. The kinetic energies of the positrons emitted by the nucleus of a particular atom range up to a maximum value referred to as E_{max}. Many radionuclides that are subject to positron decay emit two or more suites of positrons having different values of E_{max}.

Positron decay can be represented by an equation that identifies the parent nuclide, the progeny of the decay, and the associated nuclear particles and γ-rays. For example,

$$^{18}_{9}\text{F} \rightarrow {}^{18}_{8}\text{O} + \beta^+ + \nu + \gamma + E \qquad (2.3)$$

where β^+ is the positron, ν is the complementary neutrino emitted with each positron, γ represents the γ-rays that are emitted when the nucleus of the product is in an excited state, and E is the total decay energy. In the positron decay of $^{18}_{9}$F, $E =$ 1.655 MeV, $E_{max} = 0.635$ MeV, and no γ-rays are emitted by $^{18}_{8}$O (Lide and Frederikse, 1995). Note that the total decay energy (E) is greater than the maximum energy of the positrons (E_{max}), which requires that additional energy is released by the annihilation of the positron. After its emission from the nucleus, the positron loses energy as a result of inelastic collisions with the nuclei of nearby atoms and ultimately comes to rest next to an ordinary electron. The rest masses of the positron and the electron are each converted into two annihilation γ-rays of 0.511 MeV that are emitted in opposite directions. Therefore, the total decay energy of positron decay of $^{18}_{9}$F is

$$E = E_{max} + E_\gamma + 1.02 \text{ MeV} \quad \text{(annihilation)}$$

This explains why the total decay energy (E) of the positron decay of $^{18}_{9}$F is

$$E = 0.635 + 1.02 = 1.655 \text{ MeV}$$

The emission of two suites of positrons by $^{14}_{8}$O followed by one γ-ray is illustrated in Figure 2.6 based on data reported by Lide and Frederikse (1995):

Particles	E_{max}, MeV	Frequency, %
β_1^+	1.811	99
β_2^+	4.12	0.6
γ_1	2.312	99
Total decay energy	5.1430	

The two decay pathways available to $^{14}_{8}$O yield the same energy change equal to the total decay energy within analytical errors:

1. 1.811 MeV β_1^+
   2.312 MeV γ_1
   1.02 MeV 2γ, annihilation
   5.143 MeV
2. 4.12 MeV β_2^+
   1.02 MeV 2γ, annihilation
   5.14 MeV

The mass of $^{14}_{8}$O must be greater than that of $^{14}_{7}$N by at least two electron masses. Since the total decay energy of $^{14}_{8}$O is equivalent to $5.143/931.5 = 0.005521$ amu and the mass of $^{14}_{7}$N is 15.003074 amu, the mass of $^{14}_{8}$O is $14.003074 + 0.005521 = 14.008595$ amu (Lide and Frederikse, 1995).

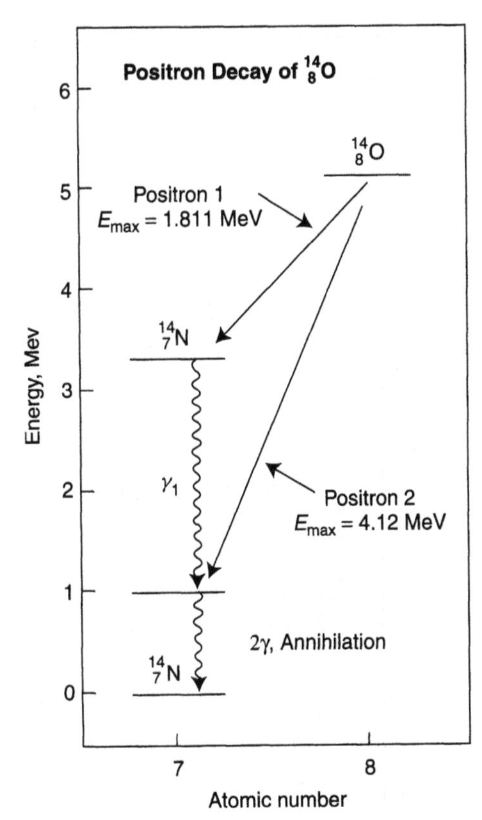

FIGURE 2.6 Positron decay of $^{14}_{8}O$ to $^{14}_{7}N$ with a halflife of 70.6 s. Two suites of positrons are emitted, one of which leaves the product nucleus in an excited state and is followed by the emission of an energetic γ-ray. The positrons of both suites are accompanied by complementary neutrinos and all of the positrons are annihilated by interacting with an ordinary electron, resulting in the formation of two γ-rays having a combined energy of 1.02 MeV. Data from Lide and Frederikse (1995).

2.1c Electron Capture Decay

An alternative mechanism whereby a nucleus can decrease its proton number and increase its neutron number is by capturing one of its extranuclear electrons. The probability of capture is greatest for the electrons in the K shell (quantum number 1) because they are closest to the nucleus. However, electrons from other shells (L, M, etc.) can also be captured. Electron capture also requires the emission of a neutrino from the nucleus. However, the neutrinos are monoenergetic because the electrons are captured from a definite energy state.

Electron capture can be visualized as the reaction between an extranuclear electron and a proton in the nucleus to form a neutron and a neutrino. The product nucleus has the following configuration compared with its parent:

	Proton Number	Neutron Number	Mass Number
Parent	Z	N	$Z + N = A$
Daughter	$Z - 1$	$N + 1$	$Z - 1 + N + 1 = A$

The daughter is isobaric with its parent and occupies the same position on the chart of the nuclides relative to its parent as the daughter of a positron decay, as illustrated in Figure 2.5.

Electron capture may leave the product nucleus in an excited state and is then followed by the emission of a γ-ray. Removal of an extranuclear electron from the K shell or from higher energy shells leaves a vacancy that is subsequently filled by other electrons that fall into the vacant position. In the process, these electrons emit a series of x-rays. If a vacancy in the K shell is filled by an electron from the next higher shell (L), the resulting x-ray may interact with another electron that is then ejected with a kinetic energy equal to the difference between the energy of the K x-ray and its own binding energy. Such electrons are called *Auger electrons*.

The emission of characteristic x-rays following electron capture decay has been used to construct portable sources of x-rays. For example, $^{125}_{53}I$ decays by electron capture to an excited state of $^{125}_{52}Te$ (Te = tellurium) that decays to the ground state by internal conversion and electron emission. During this process, a series of x-rays are emitted with energies ranging from 27.2 to 31.5 keV. Meyers (1962) used this decay pair to construct a point source of x-rays in a thin plastic tube weighing about 1 g with which he made medical radiographs of good quality. Such portable isotopic x-ray sources may also be useful in geology for radiography of rocks and minerals.

2.1d Branched Beta-Decay

According to a rule formulated by the Austrian physicist J. Mattauch in 1934, two adjacent isobars cannot both be stable. The reason is that adjacent isobars have different masses and binding energies, which makes possible a spontaneous reaction whereby one isobar is converted into the other by a suitable β-decay that liberates energy. Mattauch's isobar rule implies that two stable isobars must be separated by a radioactive isobar that can undergo branched decay and thus form two stable isobaric daughters.

An important example of branched decay is provided in Figure 2.7 by naturally occurring radioactive $^{40}_{19}K$ whose isobaric neighbors ($A = 40$) are both stable, namely $^{40}_{18}Ar$ and $^{40}_{20}Ca$. Therefore, $^{40}_{19}K$ decays to $^{40}_{18}Ar$ by means of positron emission and electron capture and to $^{40}_{20}Ca$ by negatron decay. The relevant data provided by Lippolt (1987) are as follows:

Particle	Energy, MeV	Frequency, %
1. Positron Emission to $^{40}_{18}Ar$		
Positron	0.483	0.001
γ	1.02	Annihilation
2. Electron Capture to $^{40}_{18}Ar$		
Capture 1	0.0442	10.32
γ_1	1.4608	
Capture 2	1.505	0.16
Total decay	1.505	
3. Negatron Emission to $^{40}_{20}Ca$		
Negatron	1.312	89.52
Total decay	1.312	

The principal decay mode of $^{40}_{19}K$ to $^{40}_{18}Ar$ is electron capture decay ($10.32 + 0.16 = 10.48\%$), whereas the frequency of positron decay is only 0.001 percent. Consequently, 10.48 percent of $^{40}_{19}K$ atoms decay by electron capture and positron emission to stable $^{40}_{18}Ar$ and 89.52 percent decay by β^--emission to stable $^{40}_{20}Ca$. The decay scheme diagram in Figure 2.7 makes clear that one of the two electron capture modes (capture 1) leaves the $^{40}_{18}Ar$ nucleus in an excited state, which then de-excites by means of an energetic γ-ray (1.4608 MeV).

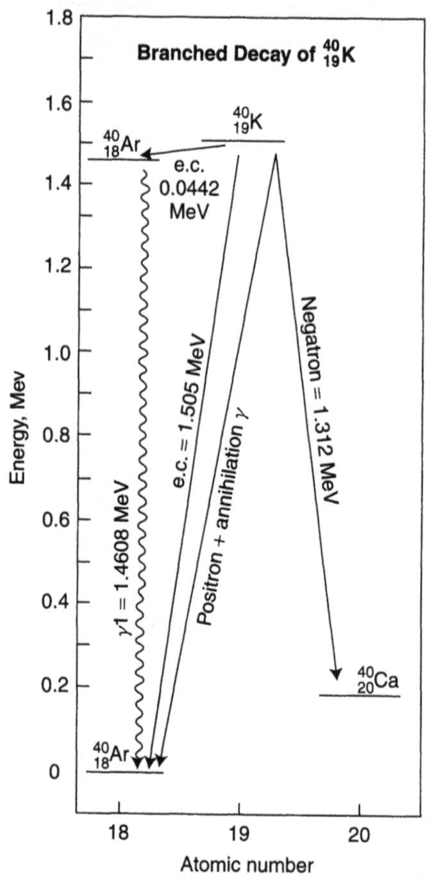

FIGURE 2.7 Branched β-decay by $^{40}_{19}K$ to $^{40}_{18}Ar$ and $^{40}_{20}Ca$. The decay to $^{40}_{18}Ar$ proceeds by means of electron capture (e.c.) and positron decay. The maximum energy of the positrons (0.483 MeV) has been combined in this diagram with the energy of the annihilation γs (1.02 MeV). The decay by negatron emission leads to the ground state of $^{40}_{20}Ca$. The energetic γ-ray (1.4608 MeV) that characterizes the radiochemical properties of $^{40}_{19}K$ is associated with the dominant electron capture mode. Data from Lippolt (1987), Lide and Frederikse (1995), and Dalrymple and Lanphere (1969).

The second electron capture mode (capture 2) and the positron decay both reach the ground state of $^{40}_{18}Ar$ directly. The energy of the annihilation γ-rays (1.02 MeV) has been combined in Figure 2.7

with the maximum energy of the suite of positrons (0.483 MeV) for a total of 1.505 MeV (Dalrymple and Lanphere, 1969).

In conclusion, the reader can verify that each of the three pathways leading to $^{40}_{18}$Ar in Figure 2.7 is accompanied by the same change in energy. In addition, the diagram illustrates the basis for Mattauch's rule because $^{40}_{20}$Ca cannot decay to $^{40}_{18}$Ar due to the existence of unstable $^{40}_{19}$K, which undergoes branched decay. Additional examples of branched decay of naturally occurring long-lived radionuclides exist among isobars having $A = 50$ ($^{50}_{23}$V), $A = 138$ ($^{138}_{57}$La), $A = 176$ ($^{176}_{71}$Lu), and $A = 180$ ($^{180}_{73}$Ta).

2.1e Energy Profiles of Isobaric Sections

The existence of the band of stable nuclides that extends across the chart of the nuclides implies that these nuclides reside along the bottom of an "energy valley." The radionuclides that occupy the walls in any isobaric section of this valley lose energy as they decay by the appropriate mode of β-decay until a stable nuclide at the bottom of the valley is reached. Radionuclides that have an excess of neutrons emit negatrons and antineutrinos, whereas radionuclides with a deficiency of neutrons emit positrons and neutrinos or decay by electron capture followed by the emission of monoenergetic neutrinos.

The energy profile of the isobars having $A = 38$ in Figure 2.8 was constructed by using the total decay energies of the radionuclides and their atomic numbers. The only stable nuclide in this isobaric section is $^{38}_{18}$Ar, which is therefore located at the bottom of the energy profile and given an energy of zero. The decay energies of the negatron emitters are $^{38}_{17}$Cl ($E = 4.917$ MeV) and $^{38}_{16}$S ($E = 2.94$ MeV). The other β^--emitters in this section ($^{38}_{15}$P, $E = 12.7$ MeV, and $^{38}_{14}$Si, $E = 37.994$ MeV) are not shown to avoid distorting the scale of the diagram. Accordingly, $^{38}_{17}$Cl is located at $Z = 17$ and $E = 4.917$ MeV, whereas $^{38}_{16}$S is placed at $Z = 16$ and $E = 4.917 + 2.94 = 7.857$ MeV above $^{38}_{18}$Ar. The positron emitters are $^{38}_{19}$K ($E = 5.913$ MeV) and $^{38}_{20}$Ca ($E = 6.74$ MeV), where the energies of the annihilation γs are included in the total decay energies. In addition, neither of these radionuclides is subject to electron capture decay (Walker et al., 1989).

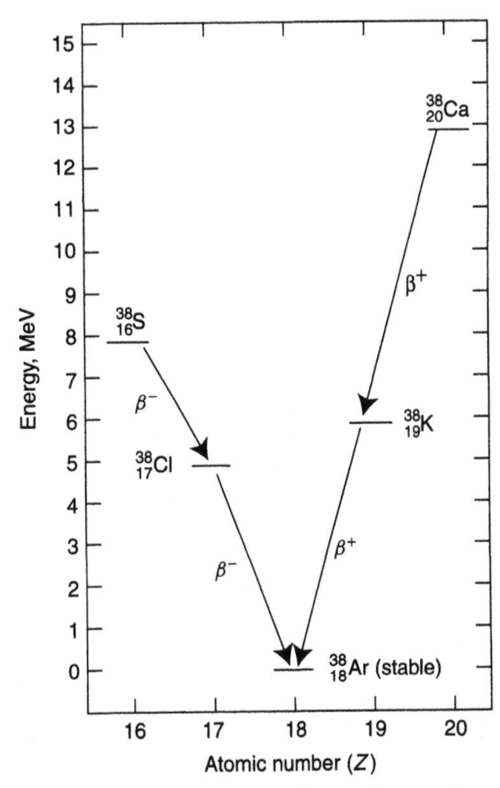

FIGURE 2.8 Schematic decay-scheme diagram for isobars having $A = 38$. The radionuclides $^{38}_{20}$Ca and $^{38}_{19}$K decay by positron emission to stable $^{38}_{18}$Ar, while $^{38}_{16}$S and $^{38}_{17}$Cl decay by negatron emission. The nuclide $^{38}_{18}$Ar is the only stable nuclide in this set of isobars, which is consistent with the fact that it lies at the bottom of this energy profile. Data from Walker et al. (1989).

The only exceptions to the internal consistency of β-decay in isobaric energy profiles occur in cases where one of the radionuclides decays by emitting α-particles ($Z = 2$, $A = 4$) and forms a product nucleus that is located in a different isobaric section. In addition, a few of the heaviest radionuclides (e.g., $^{234}_{92}$U, $^{235}_{92}$U, and $^{238}_{92}$U) decay by spontaneous fission to form products having a wide range of mass numbers, from $A = 72$ to $A = 167$.

2.2 ALPHA-DECAY

A large group of radionuclides decays by the spontaneous emission of α-particles from their nuclei. This mode of decay is available to nuclides having atomic number of 58 (cerium) or greater and to a few nuclides of low atomic number, including ${}^{5}_{2}\mathrm{He}$, ${}^{5}_{3}\mathrm{Li}$, ${}^{6}_{4}\mathrm{Be}$, and ${}^{8}_{4}\mathrm{Be}$, all of which have short halflives and therefore do not occur naturally. Long-lived α-emitters include rare-earth elements—${}^{147}_{62}\mathrm{Sm}$ [$T_{1/2} = 1.06 \times 10^{11}$ years (y)], ${}^{148}_{62}\mathrm{Sm}$ ($T_{1/2} = 7 \times 10^{15}$ y), and ${}^{152}_{64}\mathrm{Gd}$ ($T_{1/2} = 1.1 \times 10^{14}$ y)—as well as the platinum-group metals—${}^{186}_{76}\mathrm{Os}$ ($T_{1/2} = 2 \times 10^{15}$ y) and ${}^{190}_{78}\mathrm{Pt}$ ($T_{1/2} = 6 \times 10^{11}$ y). However, the most prominent α-emitters in nature are the isotopes of U (${}^{238}_{92}\mathrm{U}$, ${}^{235}_{92}\mathrm{U}$, and ${}^{234}_{92}\mathrm{U}$) and ${}^{232}_{90}\mathrm{Th}$ as well as their respective unstable daughters. In addition, some of the isotopes of the transuranium elements (neptunium, plutonium, americium, and curium) decay by α-emission. These elements do not occur naturally (except in trace amounts in uranium ore) but are synthesized in the fuel rods of nuclear reactors primarily from ${}^{238}_{92}\mathrm{U}$ by neutron capture reactions (Walker et al., 1989).

2.2a Parent–Daughter Relations

Alpha-particles are identical to the nuclei of ${}^{4}_{2}\mathrm{He}$ because they are composed of two protons and two neutrons and therefore have a charge of +2 electron units. The configuration of the product nucleus can be derived from its parent by the following simple relation:

	Atomic Number	Neutron Number	Mass Number
Parent	Z	N	$Z + N = A$
Daughter	$Z - 2$	$N - 2$	$Z + N - 4 = A - 4$

The emission of an α-particle reduces both the atomic number and the neutron number by 2 and the mass number by 4. The daughter is an isotope of a different element and is *not* an isobar of its parent, as in the case of β-decay or electron capture. The position on the chart of the nuclides of the daughter resulting from the α-decay of its parent is illustrated in Figure 2.9.

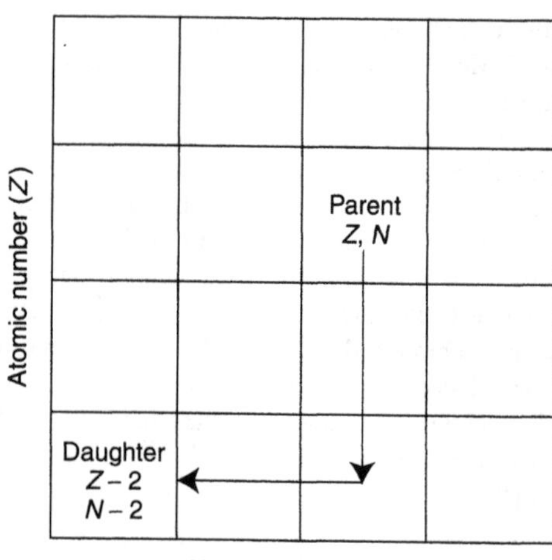

FIGURE 2.9 Location of the daughter nuclide produced by α-decay of its parent in coordinates of Z and N. Both the atomic number and the neutron number of the daughter are reduced by 2, thus reducing its mass number by 4. The daughter may itself be subject to further decay by α-emission or negatron emission or both.

Alpha-decay can be represented by an equation that identifies the product nucleus and includes the α-particle and associated γ-rays:

$$ {}^{238}_{92}\mathrm{U} \rightarrow {}^{234}_{90}\mathrm{Th} + {}^{4}_{2}\mathrm{He} + \gamma + E \qquad (2.4) $$

Note that A and Z of the parent must be equal to the respective sums of A and Z of the product nucleus and the α-particle. The α-particles are monoenergetic, but many radionuclides emit several suites of α-particles having specific kinetic energies. In cases where the product nucleus is left in an excited state, each α-particle is followed by a γ-ray that reduces the energy of the product nucleus to a lower state or to the ground state. However, no neutrinos are produced because all α-particles belonging to a particular suite have the same energy.

2.2b Alpha-Recoil Energy

When an α-particle is ejected from the nucleus, it imparts an appreciable amount of recoil energy to the product nucleus, in accordance with Newton's third law of motion: Action equals reaction. Recoil can be neglected in β-decay because of the small mass of the electron. The amount of recoil energy depends on the kinetic energy of the α-particle and on the masses of the product nucleus and of the α-particle.

The magnitude of the recoil energy can be derived from the conservation of momentum:

$$M_a v_a = M_p v_p \tag{2.5}$$

where M_a and v_a are the mass and velocity of the α-particle, respectively, and M_p and v_p are the mass and velocity of the product nucleus, respectively.

It follows that

$$v_p = \frac{M_a v_a}{M_p}$$

The total energy of the α-particle (E_a) is the sum of the measured kinetic energy (KE) of the α-particle and the recoil energy of the product nucleus:

$$E_a = \tfrac{1}{2} M_a v_a^2 + \tfrac{1}{2} M_p v_p^2 \tag{2.6}$$

Substituting for v_p from equation 2.5 yields

$$E_a = \tfrac{1}{2} M_a v_a^2 + \tfrac{1}{2} M_p \left(\frac{M_a v_a}{M_p}\right)^2$$

$$= \tfrac{1}{2} M_a v_a^2 + \tfrac{1}{2} M_a v_a^2 \left(\frac{M_a}{M_p}\right)$$

Therefore, $E_a = \mathrm{KE}_a + \mathrm{KE}_a(M_a/M_a)$. In other words, the recoil energy E_r imparted to the product nucleus by the α-particle is

$$E_r = \mathrm{KE}_a \left(\frac{M_a}{M_p}\right) \tag{2.7}$$

In most cases, the recoil energy can be estimated with sufficient accuracy by substituting the mass numbers for the actual masses. For example, $^{238}_{92}\mathrm{U}$ emits three suites of α-particles having kinetic energies and frequencies of 4.196 MeV (77 percent), 4.147 MeV (23 percent), and 4.039 MeV (0.23 percent) (Lide and Frederikse, 1995). The recoil energy imparted to the product nucleus of $^{234}_{90}\mathrm{Th}$ (equation 2.7) by the most energetic α-particle emitted by $^{238}_{92}\mathrm{U}$ is

$$E_r = \frac{4.196 \times 4}{234} = 0.0717 \text{ MeV}$$

The recoil energy must be added to the measured kinetic energy of the α-particle in order to obtain its actual kinetic energy:

$$E_a = 4.196 + 0.0717 = 4.268 \text{ MeV}$$

2.2c Decay Scheme Diagrams

The decay scheme diagram for the α-decay of $^{238}_{92}\mathrm{U}$ in Figure 2.10 is based on the data provided by Lide and Frederikse (1995) augmented by the recoil energies calculated from equation 2.7:

Particle	Energy, MeV (Including Recoil)	Frequency, %
α_1	4.268	77
α_2	4.217	23
α_3	4.087	0.23
γ	0.04955	
Total decay	4.268	

The diagram demonstrates that the emission of the second suite of α-particles (α_2) by $^{238}_{92}\mathrm{U}$ is followed by a weak γ-ray (0.04955 MeV) by means of which the product nucleus ($^{234}_{90}\mathrm{Th}$) reaches its ground state. The infrequently emitted third α-suite (α_3, 0.23 percent) is followed by the release of L x-rays as the nucleus of $^{234}_{90}\mathrm{Th}$ de-excites by internal conversion.

A more complicated case is the α-decay of $^{228}_{90}\mathrm{Th}$ (a daughter of naturally occurring long-lived $^{232}_{90}\mathrm{Th}$), which emits four suites of α-particles in its decay to $^{224}_{88}\mathrm{Ra}$. The total kinetic energies of the α-particles (including the recoil energy)

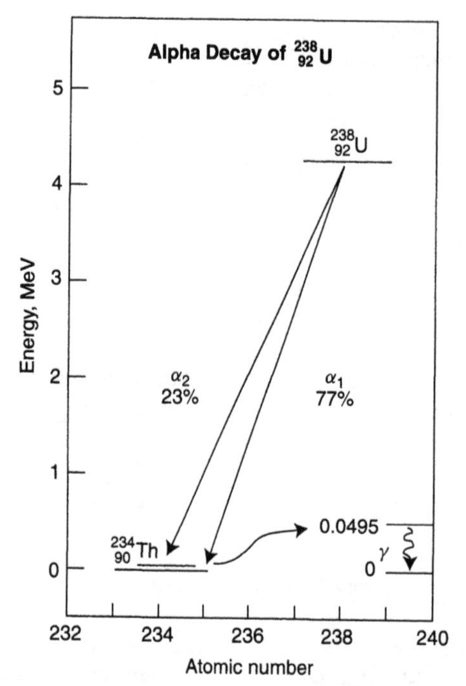

FIGURE 2.10 Alpha-decay of $^{238}_{92}U$ to $^{234}_{90}Th$ with a halflife of 4.468×10^9 y. The emission of α-particles of the α_2 suite leaves the product nucleus in an excited state which reaches the ground state by emitting a weak γ-ray (0.04955 MeV). A third suite of α-particles (α_3), which is emitted infrequently (0.23%), is not shown here but is discussed in the text. Data from Lide and Frederikse (1995).

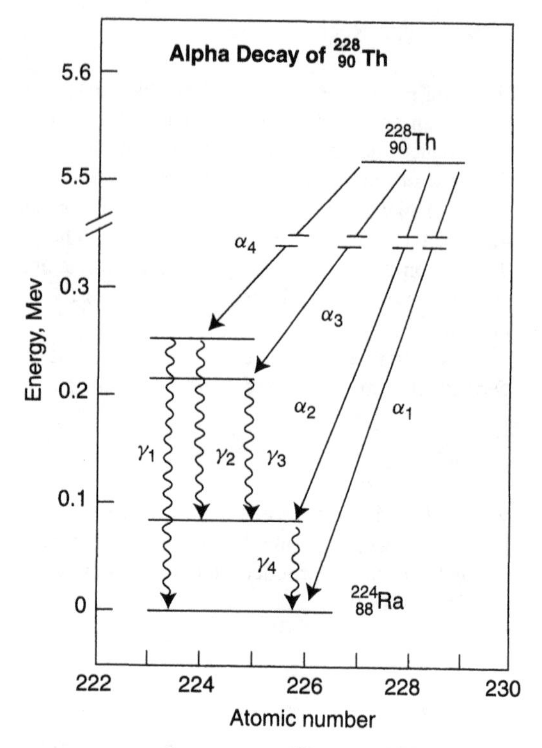

FIGURE 2.11 Alpha-decay of $^{228}_{90}Th$ to $^{224}_{88}Ra$ with a halflife of 1.913 y. In this case, four suites of α-particles are emitted followed by four γ-rays. The five different pathways release the same amount of energy. Data from Friedlander et al. (1981).

and the energies of associated γ-rays reported by Friedlander et al. (1981) are as follows:

Particles	Energies, MeV (Including Recoil)	Frequency, %
α_1	5.517	71
α_2	5.433	28
α_3	5.301	0.4
α_4	5.265	0.2
γ_1	0.217	
γ_2	0.169	
γ_3	0.133	
γ_4	0.084	
Total decay	5.517	

All of the suites of α-particles and γ-rays were used to construct the decay scheme diagram of $^{228}_{90}Th$ in Figure 2.11. The five pathways available to $^{228}_{90}Th$ release identical amounts of energy equal to the total α-decay energy of 5.517 MeV.

The naturally occurring long-lived radioactive isotopes of U($^{235}_{92}U$, $^{238}_{92}U$) and Th (^{232}Th) give rise to long chains of unstable daughters that decay by α- or negatron emission or both. All of the daughters of U and Th have short halflives but, nevertheless, occur in nature because they are continually produced by the decay of their long-lived parents. For example, $^{234}_{92}U$ occurs naturally because it is a member of the $^{238}_{92}U$ decay series. It has a small but measurable abundance of 0.0055

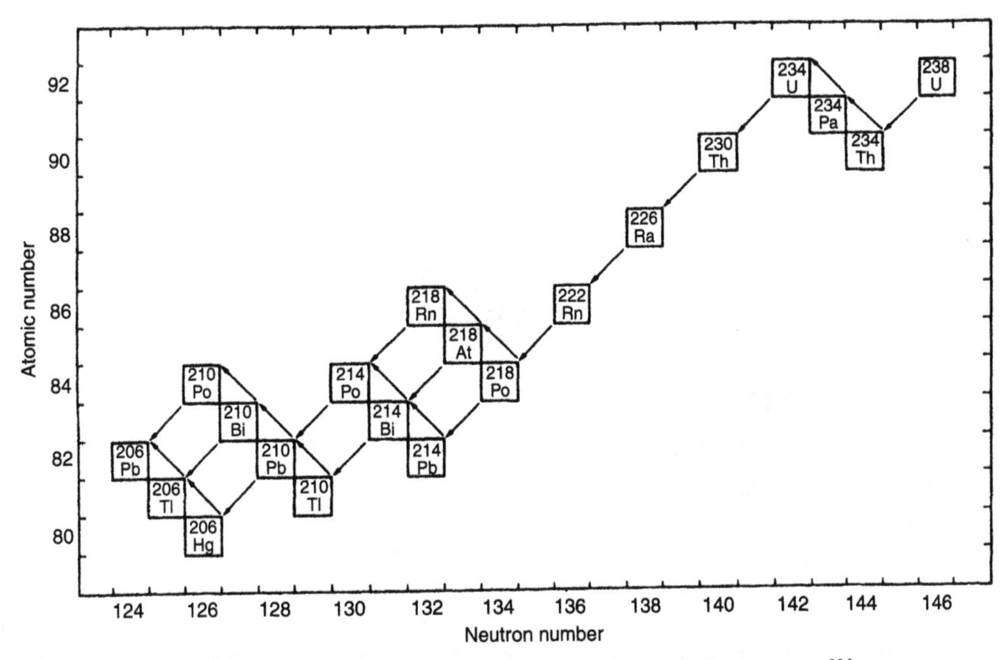

FIGURE 2.12 Chain of short-lived intermediate daughters in the decay of $^{238}_{92}U$ to stable $^{206}_{82}Pb$.

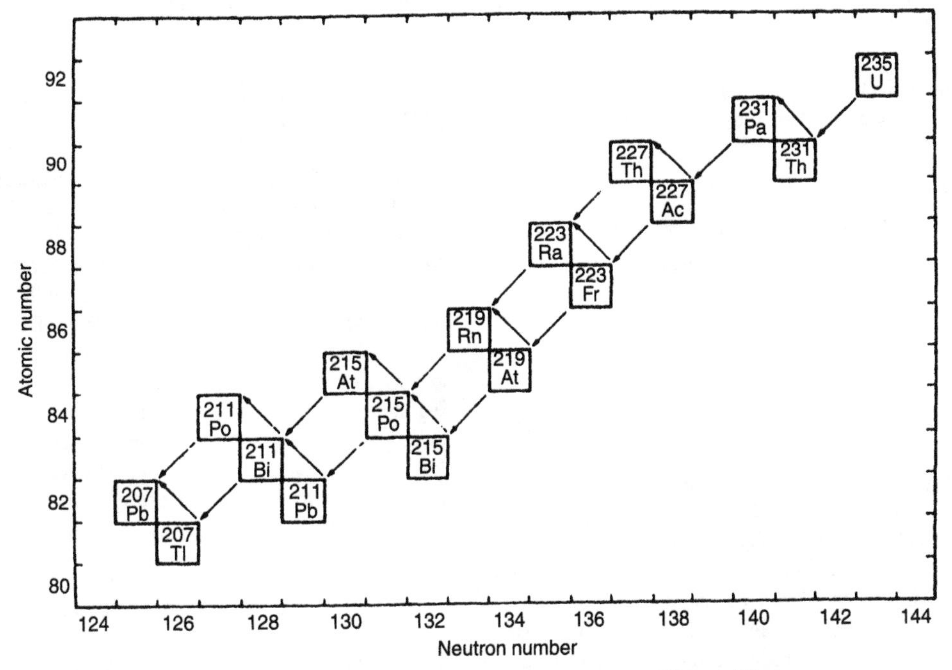

FIGURE 2.13 Chain of short-lived intermediate daughters in the decay of $^{235}_{92}U$ to $^{207}_{82}Pb$.

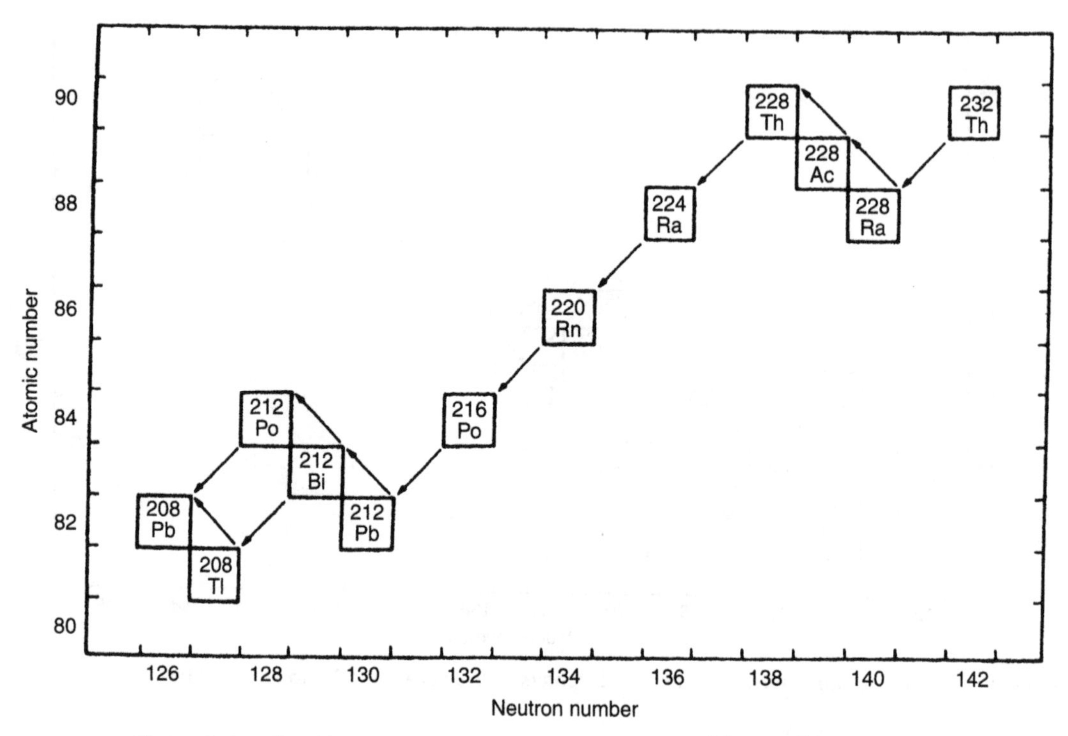

FIGURE 2.14 Chain of short-lived intermediate daughters in the decay of $^{232}_{90}Th$ to $^{208}_{82}Pb$.

percent and a halflife of 2.46×10^5 years. All of the decay series end when they reach stable isotopes of Pb:

Parent		Stable Product
$^{238}_{92}U$	$\rightarrow$	$^{206}_{82}Pb$
$^{235}_{92}U$	$\rightarrow$	$^{207}_{82}Pb$
$^{232}_{90}Th$	$\rightarrow$	$^{208}_{82}Pb$

All three decay series in Figures 2.12–2.14 split repeatedly because some radionuclides in each chain can decay by both α-emission and negatron decay. However, all of the decay paths in a particular series always lead to the formation of the same stable end product. In addition, although the three decay chains repeatedly split and reunite, the intermediate daughters of one chain are not produced by the other decay chains.

2.3 SPONTANEOUS AND INDUCED FISSION

The nuclei of atoms can be compared to drops of liquid that assume a spherical shape in response to surface tension. Large drops are easily disrupted by small perturbations and may even break apart spontaneously. Bohr and Wheeler (1939) used this analogy to explain why the nuclei of some of the isotopes of heavy elements disintegrate spontaneously and why others can be induced to split by bombarding them with neutrons, protons, α-particles, and even γ-rays or x-rays. The Bohr–Wheeler model for nuclear fission indicates that the energy barrier (ΔE) against fission is related to Z^2/A (Friedlander et al., 1981):

$$\Delta E = 19.0 - 0.36\frac{Z^2}{A} \qquad (2.8)$$

where ΔE is the energy in million electron volts required to energize a nucleus sufficiently to cause it to fission. The graph of equation 2.8 in Figure 2.15 indicates that the activation energy required for nuclear fission (ΔE), or the barrier against fission, decreases with increasing atomic number (Z) from Pb ($Z = 82$) to Cm ($Z = 96$). In addition, the activation energy for fission of the isotopes of a particular element decreases with decreasing mass number A. Since Z is constant for the isotopes of an element, the decrease in the activation energy for fission is caused by the decrease of the neutron number (N).

2.3a Spontaneous Fission

The nuclei of certain isotopes of elements having high atomic numbers disintegrate spontaneously and form isotopes of two fission product elements. Spontaneous nuclear fission of uranium isotopes was first reported by Flerov and Petrzhak (1940). It is accompanied by the emission of neutrons, γ-rays, and other nuclear particles and releases a large amount of energy, about 200 MeV per fission event. Spontaneous fission also occurs in some of the isotopes of the transuranium elements, including plutonium ($_{94}$Pu), americium ($_{95}$Am), curium ($_{96}$Cm), berkelium ($_{97}$Bk), californium ($_{98}$Cf), einsteinium ($_{99}$Es), fermium ($_{100}$Fm), mendelevium ($_{101}$Md), nobelium ($_{102}$No), lawrencium ($_{103}$Lr), rutherfordium ($_{104}$Rf), and hahnium ($_{105}$Ha), as well as elements having still higher atomic numbers (Walker et al., 1989).

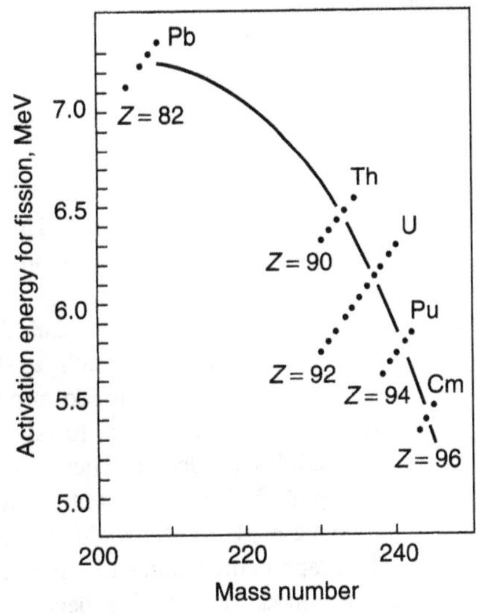

FIGURE 2.15 Activation energy required for nuclear fission according to equation 2.8. The decrease in ΔE with increasing atomic number implies an increase in the susceptibility of these nuclides to nuclear fission (Friedlander et al., 1981).

Most of the radionuclides that decay by spontaneous fission are also subject to α-decay. In such cases, the probabilities of spontaneous fission and α-decay are expressed by separate decay constants. The halflives for spontaneous fission decrease with increasing atomic number from about 10^{18} years (Th) to 10^{6} years (Cf). For example, the halflife for spontaneous fission of $^{238}_{92}$U is 8.19×10^{15} years, although a value of 10.11×10^{15} years has also been reported.

The availability of spontaneous fission as a decay mode is affected by whether the atomic number of the element is even or odd. The data of Walker et al. (1989) in Figure 2.16 indicate clearly that a larger number of the isotopes of the transuranium elements having even Z can decay by spontaneous fission than is possible for isotopes having odd Z.

2.3b Induced Fission

Induced fission was discovered by Hahn and Strassmann (1939) as a result of their attempt to produce transuranium elements by bombarding U with neutrons. The radioactive isotope of U produced by this procedure was then expected to decay by negatron emission to a nucleus of Np having atomic number 93:

$$^{238}_{92}\mathrm{U} + ^{1}_{0}\mathrm{n} \rightarrow ^{239}_{92}\mathrm{U} + \gamma \qquad (2.9)$$

$$^{239}_{92}\mathrm{U} \rightarrow ^{239}_{93}\mathrm{Np} + \beta^{-} + \overline{\nu} + \gamma \qquad (2.10)$$

After some very careful chemical detective work, Hahn and Strassmann discovered that the U target they had irradiated with neutrons contained $^{139}_{56}$Ba and $^{140}_{57}$La. They concluded from this evidence that the nuclei of U atoms had been split. In that case, lighter fission products should also have been produced. A further search by Hahn and Strassmann

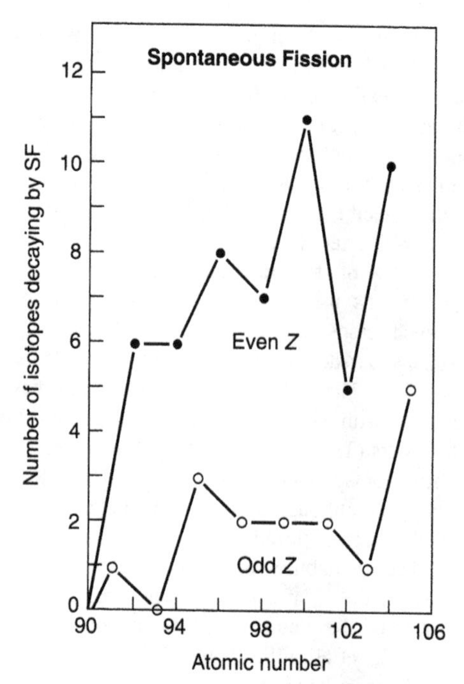

FIGURE 2.16 Prevalence of spontaneous fission (SF) among the isotopes of Th (90), Pa (91), U (92), Np (93), Pu (94), Am (95), Cm (96), Bk (97), Cf (98), Es (99), Fm (100), Md (101), No (102), Lw (103), Rf (104), and Ha (105). The graph demonstrates that SF is more common among the isotopes of elements having even Z than among elements having odd Z. Data from Walker et al. (1989).

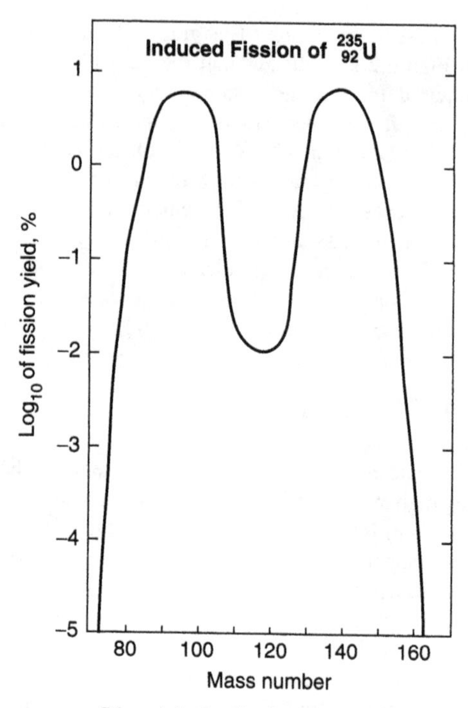

FIGURE 2.17 Bimodal distribution of the products of induced fission of $^{235}_{92}U$. The fission yield is expressed as the logarithm of the percent abundance of each product. Adapted from Friedlander et al. (1981).

did reveal the presence of isotopes of Sr ($Z = 38$), Y ($Z = 39$), Kr ($Z = 36$), and Xe ($Z = 54$) in the irradiated U target. They also demonstrated that the amounts of energy released during induced nuclear fission of U are much larger than occur during α-decay (Frisch, 1939; Kaplan, 1955).

Present knowledge indicates that slow neutrons can induce fission of $^{235}_{92}U$ but not of $^{238}_{92}U$, whereas fast neutrons having kinetic energies greater than 1 MeV induce fission in both nuclides. In addition, the nuclei of many other isotopes of Th, Pa, and U and of the transuranium elements fission when they are bombarded by high-energy α-particles, protons, deuterons (nuclei of $^{2}_{1}H$), and γ-rays.

The products of induced fission of $^{235}_{92}U$ have mass numbers between 72 and 167 and plot to the right of the band of stability in the chart of the nuclides. Accordingly, these nuclides have an *excess* of neutrons and decay by negatron emission until they reach a stable isobar. The abundances of the fission products of $^{235}_{92}U$ range widely from 10^{-5} to about 6 percent of all nuclides produced. Therefore, the abundances in Figure 2.17 are represented by the logarithm to the base 10 of the yield expressed in percent. The graph demonstrates that the fission products have a bimodal distribution because two complementary nuclides are produced in each fission event. Consequently, fission products having intermediate mass numbers between 110 and 120 have low abundances of the order of only 0.01 percent. During each fission event, several neutrons are released having a wide range of kinetic energies from about 0.05 to 17 MeV (Kaplan, 1955). These neutrons can cause additional fissions of $^{235}_{92}U$ after their energies have been reduced by collisions with nuclei of other atoms.

2.3c Nuclear Power Reactors

The release of about 200 MeV of energy as a result of the induced fission of a single atom of $^{235}_{92}U$ soon led to efforts to harness this nuclear energy. Since each fission event releases 2.5 neutrons on average, Enrico Fermi suggested that these neutrons could be used to cause additional fissions of $^{235}_{92}U$ leading to a self-sustaining chain reaction. He determined that the kinetic energies of the neutrons could be reduced by collisions with atoms of carbon and therefore set up an experimental nuclear-fission reactor using graphite as the moderator. This reactor, built under the stands of the football stadium (Stagg Field) of the University of Chicago, achieved a chain reaction on December 2, 1942. The first research reactor, producing 3.8 megawatts of power, went into operation only one year later at the Oak Ridge National Laboratory in Tennessee.

The reactor fuel of modern nuclear reactors, such as the pressurized-water reactor in Figure 2.18, consists of U that has been enriched in $^{235}_{92}U$ from about 10 percent to as much as 90 percent in fuel used in the reactors of submarines. The chain reaction is controlled by means of neutron absorbers (e.g., B or Cd) in the form of rods that can be moved in and out of the reactor core. In many reactors, water is used to transport the heat generated in the core to a heat exchanger, where it is transferred to a steam generator.

Nuclear reactors can be classified on the basis of different sets of criteria:

1. Type of fuel used: natural U, U enriched in $^{235}_{92}U$, $^{239}_{94}Pu$, or $^{233}_{92}U$.
2. Energy of neutrons used to cause fission: fast, partially moderated, or thermal.
3. Moderator: water, heavy water, graphite, or beryllium.
4. Coolant: water, air, helium, or liquid metal (Na).
5. Purpose: thermal energy production, neutron source, formation of fissionable isotopes (e.g., $^{239}_{94}Pu$ or $^{233}_{92}U$), or research.

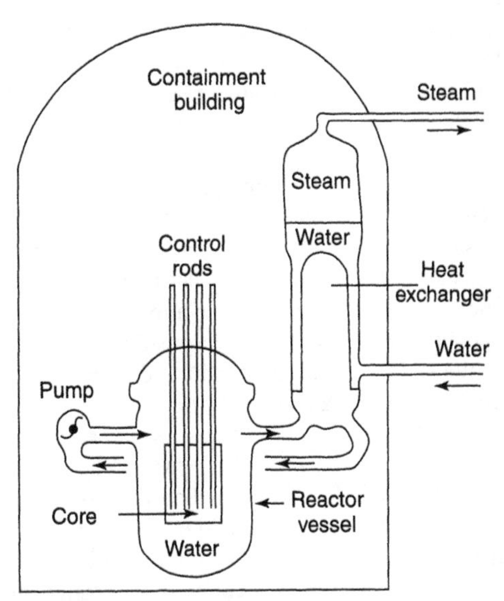

FIGURE 2.18 Schematic design of a pressurized water reactor. Hot water under pressure generates steam in the heat exchanger. The steam is used to drive a turbine and returns to the heat exchanger in the liquid state. In this way, the water, which moderates the neutrons and cools the core of the reactor, does not come in contact with the steam used to generate electricity. Adapted from Nisbet (1991).

The production of electricity by nuclear reactors in the U.S. started in 1962 and rose steadily until the early 1990s. It has since stabilized at about 600 billion kilowatt-hours per year or about 20 percent of the total annual production. The amount of nuclear electric power produced in the United States will decline in the future because no new power reactors have been built since 1978 (Holland and Petersen, 1995). Certain other countries have likewise reduced their dependence on nuclear energy (e.g., Sweden, Germany, and Switzerland). However, other countries (e.g., France, South Korea, North Korea, and Japan) continue to expand the production of electricity by nuclear reactors. In France almost 75 percent of the electricity is generated by nuclear reactors and additional reactors are under construction (Haefele, 1990).

Nuclear power reactors provide an alternative to generating electricity by burning fossil fuel, which contaminates the atmosphere and contributes to global warming. However, even though nuclear reactors do not emit carbon dioxide and other contaminants into the atmosphere, they do generate radioactive waste products that must be sequestered in subsurface repositories for at least 10,000 years (Section 25.3). In addition, accidents involving nuclear reactors caused by equipment failure or operator error, can cause widespread contamination of the surface of the Earth with radioactive fission products. Such an accident occurred on March 28, 1979, when a pressurized water reactor on Three Mile Island near Harrisburg, Pennsylvania, malfunctioned. As a result, radioactive noble gases (^{133}Xe and ^{95}Kr) were released into the atmosphere. Although there were no fatalities attributable to this accident, the subsequent clean-up cost about 2 billion dollars. In addition, this accident contributed to the decline of public support for nuclear power in the United States.

A more serious reactor accident occurred on April 26, 1986, at Chernobyl in the Ukraine (Section 25.4). As a result of operator error, the power output of the reactor increased rapidly causing a steam explosion which destroyed the reactor core and breached the containment building. Fires broke out and burned for ten days while a large amount of radionuclides escaped into the atmosphere reaching a height of 7.5 km and contaminating most of the northern hemisphere. Thirty-one persons died and 116,000 persons had to be permanently evacuated from the vicinity of Chernobyl. As a result of this accident, the soil and the groundwater within a radius of about 30 km of Chernobyl are contaminated with radionuclides of strontium, cesium, and plutonium.

2.3d Nuclear Waste

The fuel rods of nuclear reactors become highly radioactive because of the accumulation of fission-product radionuclides (Figure 2.19) and their decay products. In addition, neutron capture by $^{238}_{92}$U in the fuel followed by negatron decay of the products causes the formation of radioactive isotopes of the transuranium elements Np, Pu, Am, and Cm (Section 25.2). The principal pathways in

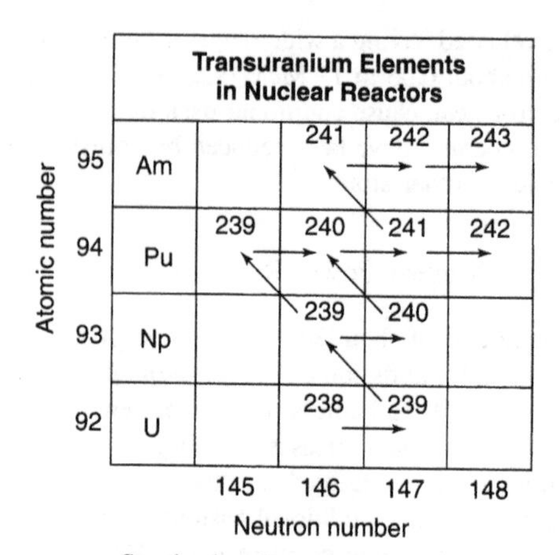

FIGURE 2.19 Synthesis of transuranium elements in the fuel rods of nuclear reactors by successive capture of thermal neutrons followed by negatron emission. The process is affected by the inability of some of the isotopes of Pu and Am to decay by negatron emission. Data from Walker et al., (1989).

Figure 2.19 lead to $^{239}_{93}$Np and $^{240}_{93}$Np, which decay by negatron emission to $^{239}_{94}$Pu and $^{240}_{94}$Pu, respectively. Additional neutron captures form $^{241}_{94}$Pu, which undergoes negatron decay to $^{241}_{95}$Am from which $^{242}_{95}$Am, $^{243}_{95}$Am, and $^{245}_{95}$Am are produced by successive neutron captures. Americium-244 can decay by negatron emission to ^{244}Cm. The synthesis of transuranium elements in Figure 2.19 is affected by the inability of some of the product nuclei to decay by negatron emission, including the Pu isotopes 239, 240, and 242 and the Am isotopes 241, 242, and 244. These radionuclides decay only by α-emission and by spontaneous fission.

The accumulation of radionuclides in the fuel rods of nuclear reactors constitutes a serious environmental problem primarily because of the intense flux of energetic gamma rays they emit. After the spent fuel is removed from the core of the reactor, it is stored temporarily in steel-lined concrete pools at the reactor sites in order to allow the γ-ray flux to decline. The Nuclear Waste Policy Act of 1982 passed by the U.S. Congress and later amended in 1987 requires that high-level nuclear waste must be disposed of in an underground geological repository. Extensive testing is in progress to determine whether Yucca Mountain in western Nevada is an acceptable site for the safe disposal of the rapidly accumulating reactor waste. The weight of spent U fuel in the United States in the year 2000 exceeded 40,000 metric tons, but most of it still resides in temporary storage at more than 100 reactor sites in 35 states.

2.4 SUMMARY

Most of the unstable nuclides that formed by nucleosynthesis in the ancestral stars prior to the origin of the solar system have long since decayed. Only a few long-lived radionuclides still exist naturally. These radionuclides and the stable isotopes of the elements compose the matter of which all living and inanimate objects are composed.

The nuclei of the unstable nuclides decay spontaneously by emitting α- and/or β-particles as well as γ-rays. The process stops only when the protons and neutrons that remain form a stable nucleus in the ground state. The stable atoms are located along a band that stretches across the chart of the nuclides. The energies released by β-decay demonstrate that the stable atoms occur at the bottom of isobaric energy sections as required by thermodynamics.

Beta-decay is accompanied by the emission of neutrinos that interact sparingly with matter but nevertheless carry varying amounts of energy from the Earth into the universe at large. In addition, one form of β-decay involves the emission of positively charged electrons that are annihilated when they interact with normal extranuclear electrons.

Nuclei that contain 90 or more protons are so large that many of them can break up spontaneously into two fragments. These heavy nuclei can also be induced to split by bombarding them with energetic neutrons, protons, α-particles, and γ-rays. The discovery by Otto Hahn and Fritz Strassmann that uranium atoms can be split by irradiating them with neutrons was a milestone in the evolution of science during the twentieth century. The energy liberated by self-sustaining chain reactions based on induced fission of uranium and plutonium has improved the standard of living in many countries. However, nuclear fission has also been used to construct weapons of unprecedented destructive power. In addition, the waste products of nuclear fission are a burden humans will have to bear for the next hundred centuries and beyond.

REFERENCES

Anderson, C. D., 1932. The positive electron. *Science*, 76:238.

Bahcall, J. B., 1969. Neutrinos from the sun. *Sci. Am.*, 221:29–37.

Bohr, N., and J. A. Wheeler, 1939. The mechanism of nuclear fission. *Phys. Rev.*, 5:426–450.

Dalrymple, G. B., and M. A. Lanphere, 1969. *Potassium-argon dating*. W. H. Freeman, San Francisco.

Dzelepow, B., M. Kopjova, and E. Vorobjov, 1946. β-Spectrum of K^{40}. *Phys. Rev.*, 69:538–539.

Fermi, E, 1934. Versuch einer Theorie der β- Strahlen. *Z. Phys.*, 88:161–177.

Flerov, G. N., and K. A. Petrzhak, 1940. Spontaneous fission of uranium. *J. Phys. (U.S.S.R.)*, 3:275–380.

Friedlander, G. J., W. Kennedy, and J. M. Miller, 1981. *Nuclear and Radiochemistry*. Wiley, New York.

Frisch, O. R., 1939. Physical evidence for the division of heavy nuclei under neutron bombardment. *Nature*, 143:276.

Häfele, W., 1990. Energy from nuclear power. *Scientific American*, 263:136–144.

Hahn, O., and F. Strassmann, 1939. Über den Nachweis und das Verhalten der bei der Bestrahlung des Urans mittels Neutronen entstehenden Erdalkalimetalle. *Naturwissenschaften*, 27:11–15.

Holland, H. D., and U. Petersen, 1995. *Living Dangerously: The Earth, Its Resources, and the Environment*. Princeton University Press, Princeton, New Jersey.

Kaplan, I., 1955. *Nuclear Physics*. Addison-Wesley, Reading, Massachusetts.

Lide, D. R., and H. P. R. Frederikse, 1995. *CRC Handbook of Chemistry and Physics*, 76th ed. CRC Press, Boca Raton, Florida.

Lippolt, H. J., 1987. Personal communication with Gunter Faure.

Mattauch, J., 1934. Zur Systematik der Isotopen. *Z. Phys.*, 91:361–371.

Meyers, W. G., 1962. On a new source of x-rays. *Ohio State Medical J.*, 58:772–773.

Nisbet, E. G., 1991. *Leaving Eden: To Protect and Manage the Earth*. Cambridge University Press, New York, 358.

Reines, F., 1979. The early days of experimental neutrino physics. *Science*, 203:11–16.

Walker, F. W., J. R. Parrington, and F. Feiner, 1989. *Chart of the Nuclides*, 14th ed. Knolls Atomic Power Laboratory, General Electric Co., San Jose, California.

CHAPTER 3

Radioactive Decay

RUTHERFORD and Soddy (1902a, b, c, d), working at McGill University in Montreal, dissolved thorium nitrate in water and then reprecipitated the thorium as the hydroxide. They found that the radioactivity of the thorium was greatly diminished but the remaining solution contained a highly radioactive substance, which they called ThX, later identified as $^{224}_{88}Ra$. Rutherford and Soddy made careful measurements of the activities of thorium and ThX over a period of about one month and observed that the activity of ThX decreased exponentially to zero in that period of time while the activity of thorium recovered its previous intensity. From the results of these experiments, Rutherford and Soddy concluded that radioactivity involved the spontaneous decomposition of the atoms of one element to form atoms of another element. They suggested that radioactivity is a property of certain atoms and that the rate of disintegration is proportional to the number of atoms remaining. This explanation of radioactive decay has never been seriously challenged or modified and is known as the law of radioactivity.

3.1 LAW OF RADIOACTIVITY

According to the theory of Rutherford and Soddy, the rate of decay of an unstable nuclide is proportional to the number of atoms (N) remaining at any time (t). Translating this statement into mathematical language results in

$$-\frac{dN}{dt} \propto N \tag{3.1}$$

where dN/dt is the rate of change of the number of parent atoms and the minus sign is required because the rate decreases as a function of time. The proportionality expressed above is transformed into an equality by the introduction of the proportionality constant (λ), called the decay constant. The numerical value of λ is characteristic of the particular radionuclide under consideration and is expressed in units of reciprocal time. The decay constant represents the probability that an atom will decay within a stated unit of time. The equation describing the rate of decay of radionuclide therefore is

$$-\frac{dN}{dt} = \lambda N \tag{3.2}$$

It will be helpful later to remember that λN means "the rate of decay."

The next step is to rearrange the terms of equation 3.2 and to integrate:

$$-\int \frac{dN}{N} = \lambda \int dt \tag{3.3}$$

which yields

$$-\ln N = \lambda t + C \tag{3.4}$$

Where $\ln N$ is the logarithm to the base e of N ($e = 2.718\ldots.$) and C is the constant of integration. This constant can be evaluated from the condition that $N = N_0$ when $t = 0$. Therefore,

$$C = -\ln N_0 \tag{3.5}$$

Substituting into equation 3.4 yields

$$-\ln N = \lambda t - \ln N_0 \tag{3.6}$$

$$\ln N - \ln N_0 = -\lambda t$$

$$\ln \frac{N}{N_0} = -\lambda t$$

$$\frac{N}{N_0} = e^{-\lambda t}$$

$$N = N_0 e^{-\lambda t} \tag{3.7}$$

Equation 3.7 gives the number of radioactive parent atoms (N) that remain at any time (t) of an original number of atoms (N_0) that were present when $t = 0$. It is the basic equation describing all radioactive decay processes.

The decay constant (λ) of a radionuclide is related to its halflife ($T_{1/2}$) by a relation derivable from equation 3.7. The halflife is the time required for one-half of a given number of atoms of a radionuclide to decay. It follows therefore that, when $t = T_{1/2}$, $N = \frac{1}{2}N_0$. Substituting these values into equation 3.7 yields

$$\tfrac{1}{2}N_0 = N_0^{-\lambda T_{1/2}}$$

$$\ln\left(\tfrac{1}{2}\right) = -\lambda T_{1/2}$$

$$\ln 2 = \lambda T_{1/2}$$

$$T_{1/2} = \frac{\ln 2}{\lambda} = \frac{0.693}{\lambda} \tag{3.8}$$

Equation 3.8 provides a convenient relationship between the halflife of a radionuclide and its decay constant.

Another parameter that is sometimes used to describe the decay of a radioactive species is the mean life, which is defined as the average life expectancy of a radioactive atom. The mean life (τ) is defined as

$$\tau = -\frac{1}{N_0}\int_{t=0}^{t=\infty} (t)\, dN \tag{3.9}$$

From equation 3.2

$$-dN = \lambda N\, dt$$

Therefore,

$$\tau = \frac{1}{N_0}\int_0^{\infty} \lambda N t\, dt$$

Substituting

$$N = N_0 e^{-\lambda t},$$

$$\tau = \lambda \int_0^{\infty} t e^{-\lambda t}\, dt = -\left[\frac{\lambda t + 1}{\lambda} e^{-\lambda t}\right]_0^{\infty}$$

$$= \frac{1}{\lambda} \tag{3.10}$$

Thus the mean life (τ) is equal to the reciprocal of the decay constant. It is longer than the halflife by the factor 1/0.693. The activity of a radioactive nuclide is reduced by a factor equal to $1/e$ of its initial value during each mean life. Therefore, the decay of a radioactive nuclide can be described either in terms of its halflife or in terms of its mean life. In isotope geochemistry the halflife has been used traditionally to describe the decay of radionuclides and the growth of their daughters.

For this reason, time was expressed in multiples of the halflife in the construction of Figure 3.1, which is an illustration of the decay curve of a radionuclide. In accordance with the definition of the halflife, the number of radionuclides remaining was reduced to 1/2 (or 2^{-1}) after one halflife, to 1/4 (or 2^{-2}) after two halflives, and so on, until after n halflives only 2^{-n} of the original number of atoms of the radionuclide remain.

Since the rate of decay is equal to λN, equation 3.7 can be modified to yield the activity (A), which is defined as the rate of decay of a radionuclide:

$$\lambda N = \lambda N_0 e^{-\lambda t}.$$

Since $A = \lambda N$,

$$A = A_0 e^{-\lambda t} \tag{3.11}$$

The activity of a radionuclide as defined above differs from the counting rate (CR), which

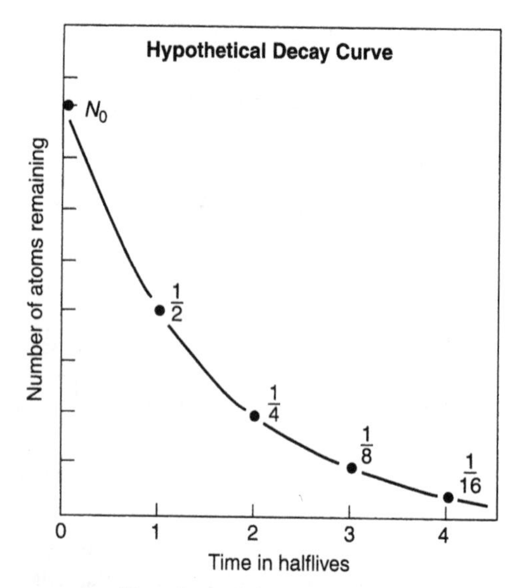

FIGURE 3.1 Hypothetical decay curve of a radionuclide as a function of time measured in multiples of its halflife. The initial number of atoms (N_0) is reduced to 1/2 after one halflife, 1/4 after two halflives, and so on.

is measured with a suitable radiation detector. The relation between the activity and the counting rate is

$$CR = cA \tag{3.12}$$

where (c) is the efficiency of the detector. The efficiency depends on the design of the detector, the type and energy of the radiation, the geometry of the counter and sample, and other experimental factors. The numerical value of the efficiency must be determined by an appropriate calibration of the detector using standard sources of the kind of radiation to be detected.

Equation 3.11 can be linearized by taking logarithms to the base e of both sides:

$$\ln A = \ln A_0 - \lambda t \tag{3.13}$$

When the decay rate A of a particular radionuclide is measured at known intervals of time, a plot of $\ln A$ (y-coordinate) versus t (x-coordinate) defines a straight line such that $\ln A_0$ is the intercept on the y-axis and λ is the slope of the line. Therefore, equation 3.13 has been used to measure the decay constant of most of the radionuclides included in the chart of nuclides (Figure 1.2).

The decay of a radionuclide described by equations 3.11 and 3.13 is illustrated in Figure 3.2 by the decay of $^{24}_{11}Na$, which has a halflife of 14.96 h (Walker et al., 1989). This radionuclide is produced when stable $^{24}_{11}Na$ is irradiated by thermal neutrons in a nuclear reactor:

$$^{23}_{11}Na + ^{1}_{0}n \rightarrow ^{24}_{11}Na + \gamma \tag{3.14}$$

This reaction is conveniently expressed in the form

$$^{23}_{11}Na\left(^{1}_{0}n, \gamma\right)^{24}_{11}Na \tag{3.15}$$

where $^{23}_{11}Na$ is the target atom, $^{24}_{11}Na$ is the product, $^{1}_{0}n$ is the neutron going into the nucleus of $^{23}_{11}Na$, and γ is the γ-ray emitted by $^{24}_{11}Na$. The product nucleus ($^{24}_{11}Na$) subsequently decays by β- (negatron) decay to stable $^{24}_{12}Mg$.

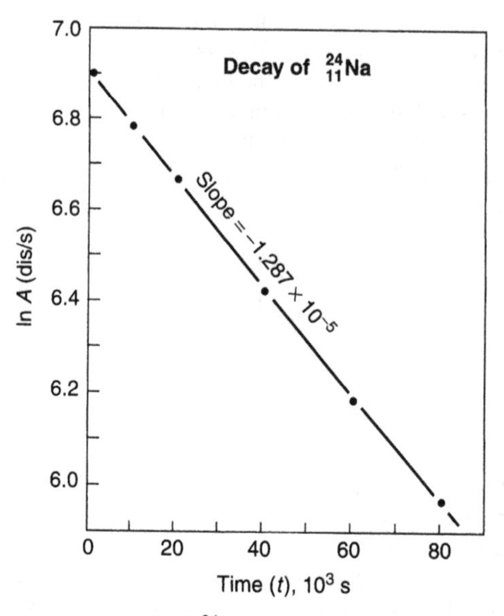

FIGURE 3.2 Decay of $^{24}_{11}Na$ in coordinates of $\ln A$ and t, where A is the rate of decay in disintegrations per second and t is the elapsed time measured in seconds. In accordance with equation 3.13, the slope of the straight line is the decay constant λ of $^{24}_{11}Na$.

Table 3.1. Decay of $^{24}_{11}$Na with Halflife $T_{1/2} = 14.96$ h ($\lambda = 1.287 \times 10^{-5}$ s^{-1}) for $A_0 = 1000$ dis/s ($\ln A_0 = 6.907$) Based on Equation 3.13

A, dis/s	$\ln A$	t, s
1000	6.907	0
878	6.777	10×10^3
772	6.649	20×10^3
616	6.423	40×10^3
461	6.134	60×10^3
356	5.877	80×10^3

The decay rate of $^{24}_{11}$Na in Figure 3.2 is expressed in disintegrations per second (dis/s) derived by solution of equation 3.13 for specified values of time measured in seconds. The initial activity A_0 was set equal to 10^3 disintegrations per second and the decay constant of $^{24}_{11}$Na from equation 3.8 is

$$\lambda = \frac{\ln 2}{14.96 \times 60 \times 60} = 1.287 \times 10^{-5} \text{ s}^{-1}$$

The activity A and its natural logarithm at different values of t are listed in Table 3.1. A least-squares regression of these data points to a straight line yields a slope $m = 1.287 \times 10^{-5}$ s^{-1}, which is the decay constant of $^{24}_{11}$Na. Although this example is contrived, Figure 3.2 does demonstrate the method by which decay constants of radionuclides are determined.

3.2 RADIATION DETECTORS

When energetic α- and β-particles as well as γ-rays and x-rays interact with matter, ions are produced in the absorber. Therefore, radiation detectors are designed to measure the rate of formation of ions in an absorber either by means of an ionization chamber or by means of scintillation counters (Faul, 1954; Smales and Wager, 1960).

3.2a Geiger–Müller Counters

The popular radiation detector in Figure 3.3 designed by Geiger and Müller (1929) includes an ionization chamber that consists of a thin-walled tube composed of glass or aluminum containing a central metal electrode. The tube contains a mixture of argon and small amounts of another volatile compound such as methane, alcohol, or halogens. The central electrode is the anode (+), whereas the walls of the chamber are the cathode (−). When a negatively charged β-particle or a γ-ray penetrates the wall of the ionization chamber, ions form in the gas mixture. Negatively charged ions or electrons are attracted to the central electrode and thereby cause a small electrical current to flow across the ionization chamber. These electrical signals are counted electronically and are displayed as a counting rate either in analog or digital form.

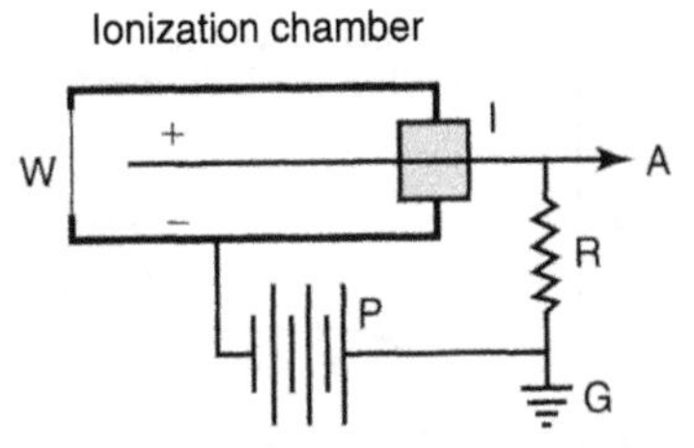

FIGURE 3.3 Design of the Geiger–Müller counter, including an ionization chamber equipped with a thin window (W) and a high-voltage adjustable DC power supply (P). In addition, I = insulator, R = resistor, G = ground, + = anode, − = cathode. Adapted from Kaplan (1955).

The sensitivity of Geiger–Müller counters depends on the voltage difference between the electrodes in the ionization chamber. The response of a Geiger–Müller counter in Figure 3.4 to a *constant* flux of energetic particles or electromagnetic radiation increases with the magnitude of the voltage applied to the electrodes of the ionization chamber. The voltage in Figure 3.4 increases from an initial threshold value (1), then rises rapidly in the proportional region (2), until it reaches the stable operating plateau (3). When the voltage is increased to still higher values, the ionization chamber goes into continuous discharge (4) and quickly overheats.

To improve the efficiency of Geiger–Müller counters, the ionization chamber can be equipped with a window that is transparent to x-rays and β-particles made of beryllium, a plastic material called Mylar, or muscovite. Since α-particles are

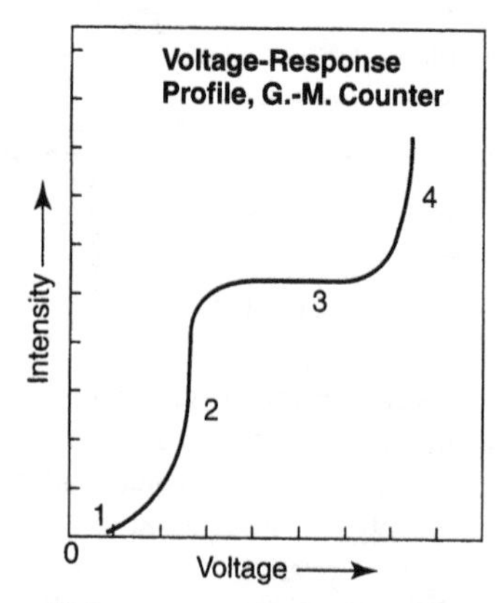

FIGURE 3.4 Voltage–response profile of a Geiger–Müller counter: 1 = threshold voltage; 2 = proportional region; 3 = operating plateau (also called the Geiger region); 4 = continuous discharge. The ionization chamber is assumed to be exposed to a constant flux of negatrons, γ-rays, or x-rays as the voltage is increased. Adapted from Moorbath (1960).

easily absorbed, α-emitters are placed inside the ionization chamber because even thin windows of Geiger tubes significantly retard α-particles.

The Geiger–Müller counter also detects cosmic rays and radiation emitted by radionuclides in the environment. This background counting rate must be subtracted from the total observed counting rate, especially in cases where the total counting rate is low. The background radiation can be reduced by shielding the ionization chamber and sample holder by means of suitable absorbers (e.g., lead or steel plates).

The background counting rate attributable to cosmic rays (high-energy protons, neutrons, muons, and γ-rays) increases with the elevation above sea level, whereas the environmental background depends on the composition of the walls of the building and of the local bedrock. Buildings constructed of U- and Th-bearing rocks and located at high elevations are characterized by high levels of background radiation.

The accurate determination of decay rates requires careful consideration of several phenomena associated with the experimental conditions. These include the dead time of the counter, which is the time interval between two successive ionization events that can be detected by the Geiger tube. The dead time causes a decrease of the observed counting rate when the counting rate reaches high values. In addition, α- and β- (negatron) particles may be absorbed by the sample being counted (self-absorption) as well as by the molecules of the air in the space between the sample and the detector. The effect of self-absorption on the observed counting rate can be reduced by depositing the sample as a thin film on a planchet. The absorption of β-particles and x-rays is reduced by evacuating the space between the source and the detector or by replacing the air with helium.

A further consideration in analytical radiochemistry is the geometry of the sample in relation to the detector. In cases where the sample is a three-dimensional object, it emits radiation in all directions, whereas the ionization chamber placed above or beneath the specimen is exposed to only one-half of the radiation or less (2π geometry). The geometrical aspects of radiochemical analysis can be improved by placing the sample inside the ionization chamber (4π geometry). In some cases (e.g., detection of $^{14}_{6}C$ in carbon dioxide), the gas to be analyzed is placed inside the ionization chamber together with a mixture of argon and methane.

In practice, Geiger–Müller counters must be calibrated at regular intervals by use of radiation standards whose decay rate can be calculated (e.g., a known weight of Ra). Survey meters used for monitoring radioactivity in nature or in the laboratory display the observed counting rate either in counts per second or in terms of radiation dose in millirems per hour or both.

3.2b Scintillation Counters

When energetic α-particles strike a surface coated with zinc sulfide, flashes of light (scintillations) are observed. Rutherford (1911) used this phenomenon to detect α-particles scattered by gold foil in his famous experiment that resulted in the discovery of the atomic nucleus. The scintillation counter is an elaboration of the fluorescent screen used by Rutherford in which the light flashes are detected by a photomultiplier tube and the scintillations are counted electronically. In addition, different kinds of materials are used to improve the sensitivity of scintillation detectors for counting β-particles, γ-rays, neutrons, and cosmic rays. Scintillation counters equipped with thallium-activated NaI or germanium crystals are especially sensitive to γ-rays whose detection efficiency by Geiger–Müller counters is only about 1 percent.

Scintillation detectors have several other desirable properties:

1. The intensity of the flash of light is proportional to the energy of the incident γ-ray.
2. The efficiency for detecting γ-rays approaches 100 percent.
3. The dead time of NaI crystals is about 2×10^{-7} s, compared to 10^{-4} s for ionization chambers.
4. The efficiency of detecting γ-radiation is improved by placing the sample into a well cut into the scintillation crystal, thereby approaching 4π geometry.
5. The signal-to-noise ratio of scintillation detectors is enhanced by cooling the crystal with liquid nitrogen.

The relation between the energy of γ-photons and the magnitude of the scintillation that results from their absorption is used in the γ-ray spectrometer to detect and register γ-quanta having different energies. As a result, γ-rays emitted by samples containing several radionuclides can be distinguished from each other and counted separately. Modern γ-ray spectrometers have up to 8000 "channels" that register γs having narrowly defined energies. The radionuclides present in a sample can be identified by the energies of the γ-rays they emit. For example, the presence of $^{40}_{19}K$ is easily detected by its energetic γ-ray having an energy of 1.4608 MeV (Figure 2.7).

The rate of decay of a known radionuclide, derived from the observed counting rate, can be converted to the concentration of the radionuclide in conventional units such as parts per million (ppm) by means of the relation

$$N = \frac{A}{\lambda} \qquad (3.16)$$

provided that the rate of decay A and the decay constant λ are expressed in the same units of time. The conversion of the counting rate to the corresponding weight of the radionuclide makes use of Avogadro's number (6.022×10^{23} atoms/mol) and the atomic mass of the nuclide in grams.

3.3 GROWTH OF RADIOACTIVE DAUGHTERS

Many radionuclides decay to form daughters that are themselves radioactive and thereby initiate a decay chain that continues until it reaches a stable nuclide. This phenomenon is exemplified by the daughters of uranium and thorium (Figures 2.12–2.14) as well as by the decay of fission product nuclides, all of which give rise to decay series of varying length and complexity.

3.3a Decay to an Unstable Daughter

When a radionuclide (N_1) decays to a radioactive daughter (N_2) that subsequently decays to a third daughter (N_3), the rate of decay of N_2 is the difference between the rate at which N_2 is produced by decay of its parent and its own rate of decay. The rate of decay of the parent (N_1) is given by equation 3.2:

$$-\frac{dN_1}{dt} = \lambda_1 N_1$$

Similarly, the rate of decay of the first daughter (N_2) is

$$-\frac{dN_2}{dt} = \lambda_2 N_2$$

Therefore, the net rate of change of the number of daughters N_2 is

$$\frac{dN_2}{dt} = \lambda_1 N_1 - \lambda_2 N_2 \qquad (3.17)$$

where λ_1 and λ_2 are the decay constants of radionuclides 1 and 2, respectively, and N_1 and N_2 are the numbers of atoms remaining at any time t. The number of atoms of N_1 remaining at any time t is given by equation 3.7:

$$N_1 = N_1^0 e^{-\lambda_1 t}$$

where N_1^0 indicates the number of atoms of N_1 at $t = 0$. This equation is substituted into equation 3.17 and, by rearrangement of terms, yields

$$\frac{dN_2}{dt} + \lambda_2 N_2 - \lambda_1 N_1^0 e^{-\lambda_1 t} = 0 \tag{3.18}$$

This is a linear differential equation of the first order solved by Bateman (1910).

The solution to equation 3.18 for the case under consideration is

$$N_2 = \frac{\lambda_1}{\lambda_2 - \lambda_1} N_1^0 (e^{-\lambda_1 t} - e^{-\lambda_2 t}) + N_2^0 e^{-\lambda_2 t} \tag{3.19}$$

The first term of this equation gives the number of atoms of N_2 that have formed by decay of N_1 but have not yet decayed. The second term represents the number of atoms of N_2 that remain from an initial number N_2^0. If $N_2^0 = 0$, equation 3.19 reduces to

$$N_2 = \frac{\lambda_1}{\lambda_2 - \lambda_1} N_1^0 \left(e^{-\lambda_1 t} - e^{-\lambda_2 t}\right) \tag{3.20}$$

The general solution of a radioactive decay series $N_1 \rightarrow N_2 \rightarrow N_3 \cdots N_n$ for the case that $N_2^0 = N_3^0 = \cdots = N_n^0 = 0$ takes the form

$$N_n = C_1 e^{-\lambda_1 t} + C_2 e^{-\lambda_2 t} + \cdots + C_n e^{-\lambda_n t} \tag{3.21}$$

where

$$C_1 = \frac{\lambda_1 \lambda_2 \cdots \lambda_{n-1} N_1^0}{(\lambda_2 - \lambda_1)(\lambda_3 - \lambda_1) \cdots (\lambda_n - \lambda_1)}$$

$$C_2 = \frac{\lambda_1 \lambda_2 \cdots \lambda_{n-1} N_1^0}{(\lambda_1 - \lambda_2)(\lambda_3 - \lambda_2) \cdots (\lambda_n - \lambda_2)}$$

$$C_3 = \frac{\lambda_1 \lambda_2 \cdots \lambda_{n-1} N_1^0}{(\lambda_1 - \lambda_3)(\lambda_2 - \lambda_3) \cdots (\lambda_n - \lambda_3)}$$

The general solution expressed in equation 3.21 can be adapted to provide a specific solution for any nuclide in the decay chain. For example, if $n = 3$,

$$N_3 = C_1 e^{-\lambda_1 t} + C_2 e^{-\lambda_2 t} + C_3 e^{-\lambda_3 t} \tag{3.22}$$

In this case

$$C_1 = \frac{\lambda_1 \lambda_2 N_1^0}{(\lambda_2 - \lambda_1)(\lambda_3 - \lambda_1)}$$

$$C_2 = \frac{\lambda_1 \lambda_2 N_1^0}{(\lambda_1 - \lambda_2)(\lambda_3 - \lambda_2)}$$

$$C_3 = \frac{\lambda_1 \lambda_2 N_1^0}{(\lambda_1 - \lambda_3)(\lambda_2 - \lambda_3)}$$

The complete equation is obtained by substituting C_1, C_2, C_3 into equation 3.22:

$$N_3 = \frac{\lambda_1 \lambda_2 N_1^0 e^{-\lambda_1 t}}{(\lambda_2 - \lambda_1)(\lambda_3 - \lambda_1)} + \frac{\lambda_1 \lambda_2 N_1^0 e^{-\lambda_2 t}}{(\lambda_1 - \lambda_2)(\lambda_3 - \lambda_2)} + \frac{\lambda_1 \lambda_2 N_1^0 e^{-\lambda_3 t}}{(\lambda_1 - \lambda_3)(\lambda_2 - \lambda_3)} \tag{3.23}$$

A numerical example illustrates the behavior of a three-component decay series. Assume that the halflife of the parent (N_1) is 1 h ($\lambda_1 = 0.693\ \text{h}^{-1}$) and that of its first daughter (N_2) is 5 h ($\lambda_2 = 0.1386\ \text{h}^{-1}$). The third daughter ($N_3$) is stable (i.e., $\lambda_3 = 0$). In addition, assume that $N_1^0 = 100$, $N_2^0 = 0$, and $N_3^0 = 0$. Figure 3.5, calculated by means of equations 3.7, 3.20, and 3.23, demonstrates that N_1 decreases exponentially while N_2 increases and reaches a maximum value in about 3 h. Thereafter, N_2 declines while N_3 continues to increase until it approaches the value of N_1^0 asymptotically as t goes to infinity.

3.3b Secular Equilibrium

If the parent (N_1) of a decay chain has a longer halflife than its unstable daughter (N_2), then

$$\lambda_1 \ll \lambda_2 \quad \text{and} \quad \lambda_2 - \lambda_1 \simeq \lambda_2$$

Consequently, equation 3.20 becomes

$$N_2 = \frac{\lambda_1 N_1^0}{\lambda_2} \left(e^{-\lambda_1 t} - e^{-\lambda_2 t}\right)$$

In addition, $e^{-\lambda_2 t}$ approaches zero faster than $e^{-\lambda_1 t}$ as t increases. Therefore, equation 3.20 can be

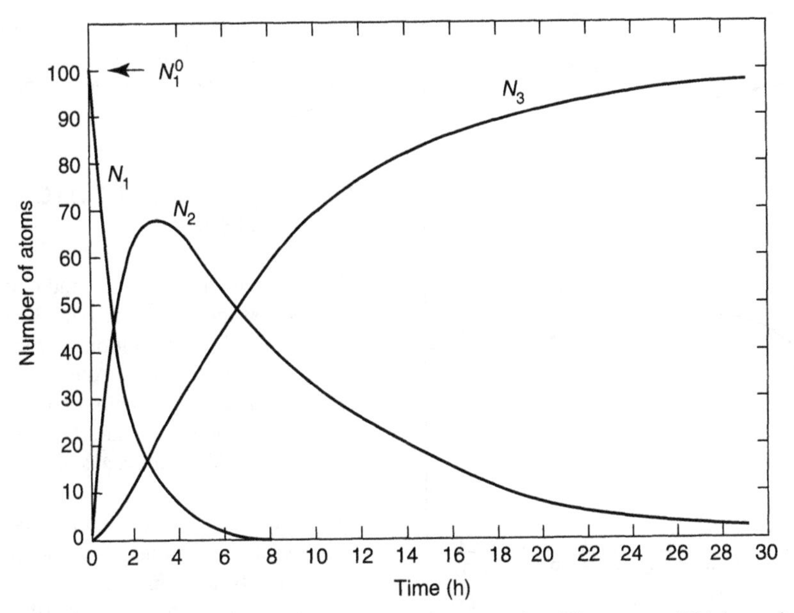

FIGURE 3.5 Radioactive decay and growth in a three-component series. The parent (N_1) has a halflife of 1 h; the first daughter (N_2) has a halflife of 5 h. The third daughter (N_3) is stable. Initially only atoms of the parent are present, that is, $N_1^0 = 100$, $N_2^0 = 0$, $N_3^0 = 0$. The number of atoms of N_2 initially increases, reaches a maximum about 3 h after decay began, and then decreases. The number of stable daughters (N_3) increases and approaches N_1^0 as t goes to infinity. Adapted from Kaplan (1955, p. 200).

further modified to a good approximation by

$$N_2 = \frac{\lambda_1 N_1^0 e^{-\lambda_1 t}}{\lambda_2} \tag{3.24}$$

Since $N_1^0 e^{-\lambda_1 t} = N_1$,

$$N_2 = \frac{\lambda_1 N_1}{\lambda_2}$$

$$N_1\lambda_1 = N_2\lambda_2 \tag{3.25}$$

Equation 3.25 indicates that, with increasing time, the rate of decay of the unstable daughter ($N_2\lambda_2$) becomes equal to that of its long-lived parent ($N_1\lambda_1$). This condition is known as secular equilibrium.

The change with time of the number of parent atoms (N_1, $T_{1/2} = 10$ h) and of their unstable daughter atoms (N_2, $T_{1/2} = 1.0$ h) is illustrated in Figure 3.6 in log-linear coordinates. The number of long-lived parent atoms (N_1) decreases with time, whereas the number of unstable daughters (N_2) initially increases and then decreases at the same rate as the parent.

In the decay series arising from $^{238}_{92}U$, $^{235}_{92}U$, and $^{232}_{90}Th$ (Figures 2.12–2.14), the condition of secular equilibrium is propagated along each chain such that

$$N_1\lambda_1 = N_2\lambda_2 = N_3\lambda_3 = \cdots = N_n\lambda_n \tag{3.26}$$

When secular equilibrium has been established, the number of atoms present of any of the unstable daughters is

$$N_n = \frac{\lambda_1 N_1}{\lambda_n} \tag{3.27}$$

and the ratios of the disintegration rates of the daughters are

$$\frac{\lambda_1 N_1}{\lambda_2 N_2} = \frac{\lambda_2 N_2}{\lambda_3 N_3} = \cdots = 1.000 \tag{3.28}$$

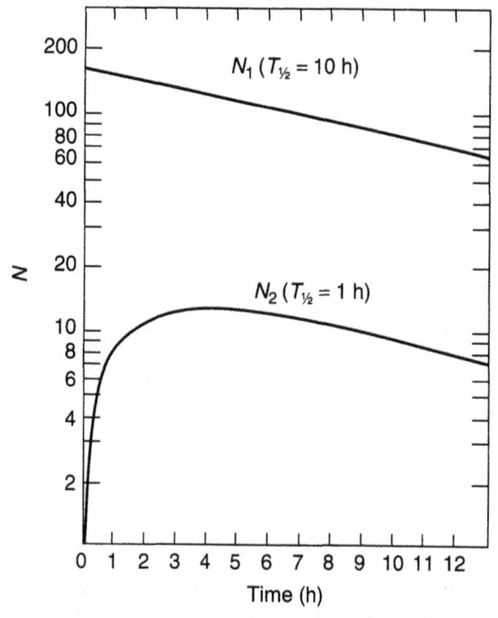

FIGURE 3.6 Decay of long-lived parent (N_1) to a short-lived daughter (N_2). The halflife of the parent is 10 h, while that of the daughter is 1.0 h. The number of daughter atoms increases as a function of time from an initial value of zero.

Equation 3.28 provides a test for the existence of secular equilibrium because, in that case, the ratios of the decay rates (activities) of the parent and any of its intermediate daughters are equal to unity.

In addition, the rate of decay of the long-lived parent at the head of the chain is equal to the rate of production of the stable daughter that ends the chain. Therefore, in the decay of $^{238}_{92}U$ to $^{206}_{82}Pb$, $^{235}_{92}U$ to $^{207}_{82}Pb$, and $^{232}_{90}Th$ to $^{208}_{82}Pb$, the intermediate unstable daughters can be eliminated from consideration provided secular equilibrium was established and none of the intermediate daughters escaped from the sample or were added from external sources (Section 10.2). Under these conditions, the number of stable radiogenic daughters atoms (D^*) in a unit weight of a rock or mineral is equal to the number of parent atoms that have decayed since the rock or mineral came into existence:

$$D^* = N_1^0 - N_1 \tag{3.29}$$

This relation is the key to the measurement of the ages of rocks and minerals based on the decay of long-lived parents to stable radiogenic daughters (Section 4.1).

3.4 UNITS OF RADIOACTIVITY AND DOSAGE

The rate of decay of a radionuclide can be expressed as the specific activity defined in terms of the number of disintegrations per second per gram of the chemical element. Alternatively, the rate of decay is measured in units of becquerels (Bq) or curies (Ci). These units are defined as the amount of a radionuclide that decays at a specified rate:

One becquerel is the amount of a radionuclide that decays at a rate of one disintegration per second.
One curie is the amount of a radionuclide that decays at the rate of 3.7×10^{10} disintegrations per second, which is the rate of decay of one gram of radium.

Decay rates expressed in becquerels tend to be large numbers requiring the use of scientific notation, such as 5×10^{10} Bq. For this reason, the becquerel is modified by prefixes scaled in powers of 10^3:

kilo (k) = 10^3
mega (M) = 10^6
giga (G) = 10^9
tera (T) = 10^{12}
pita (P) = 10^{15}
exa (E) = 10^{18}

For example, a decay rate of 5×10^{10} Bq can be expressed as 50 GBq.

Similarly, the curie is a rather large unit whose magnitude can be reduced by a second set of prefixes that scale its value downward in powers of 10^{-3}:

milli (m) = 10^{-3}
micro (μ) = 10^{-6}
nano (n) = 10^{-9}
pico (p) = 10^{-12}
femto (f) = 10^{-15}
atto (a) = 10^{-18}

Therefore, a decay rate of 5×10^{-10} Ci can also be expressed as 0.5 nCi.

The effect of ionizing radiation on an absorber depends on the dose, which is defined as the amount of energy transmitted by the radiation to one gram of the absorber. Therefore, the dose depends on the energy of the radiation and the duration of the exposure. The absorption of ionizing radiation by an inanimate absorber is measured in rads and grays defined as follows:

one rad: Dose corresponding to 100 ergs/g of absorber.
one gray (Gy) = 100 rads.

When the radiation is absorbed by biological tissue, the effect is measured similarly in units of rems (roentgen equivalent man) and sieverts:

one rem: Dose corresponding to 100 ergs/g of tissue.
one sievert (Sv) = 100 rems.

The effect of ionizing radiation on biological tissue depends not only on the dose but also on the type of radiation. Therefore, rads are converted into rems by means of an equivalence factor whose magnitude depends on the type of radiation or particles that are absorbed:

$$\text{rem} = \text{rad} \times \text{equivalence factor} \tag{3.30}$$

The values of the equivalence factors associated with different kinds of energetic particles or electromagnetic radiation are as follows:

Radiation	Equivalence Factor
Beta-particles, γ-rays, and x-rays	1
Neutrons (slow)	3
Neutrons (fast)	10
Alpha-particles (internal exposure)	20

Accordingly, a dose of 5 rads resulting from the exposure of an inanimate absorber to slow neutrons corresponds to a dose of 15 rems by biological tissue. The numerical value of the equivalence factor for slow neutrons takes into consideration that the damage that thermal neutrons do to tissue is increased by the formation of radionuclides by the process of neutron activation (Section 3.8). Similarly, the dose imparted to tissue by α-particles is magnified by a factor of 20 because internal exposure to α-particles causes much more biological damage than the effect the same α-particles have on an inorganic absorber.

The evaluation of the radiation dose in rads is based on the rate of decay of the source, the energy of the radiation, the duration of the exposure, and the weight of the absorber.

$$\text{Dose} = \frac{A \times t \times E \times 1.6021 \times 10^{-6}}{100 \times W} \quad \text{rads} \tag{3.31}$$

where A = rate of decay of radionuclide in the source, Bq
t = duration of the exposure,
E = energy of absorbed radiation, MeV ($1 \text{ MeV} = 1.6021 \times 10^{-6}$ ergs)
W = weight of absorber, g

Alternatively, the dose can be expressed as a rate in terms of units such as millirems per hour (mrems/h), which means that an exposure of 1 h provides a dose of 10^{-3} rem.

3.5 MEDICAL EFFECTS OF IONIZING RADIATION

The medical effects of exposure to ionizing radiation by humans depend on many factors:

1. type of radiation (α-, β-, γ-, x-ray, etc.),
2. dose of radiation absorbed,
3. type of exposure (external or internal), and
4. body part exposed to the radiation.

The α- and β-particles are absorbed by air and can barely penetrate human skin. Therefore, external sources of α- and β-particles are relatively harmless. However, internal exposure caused by inhalation, ingestion, or injection of α- and β-emitters causes extensive radiation damage because these particles can destroy molecules by breaking chemical bonds. As a result, the affected cells die unless the damage is repaired.

The medical damage also depends on the body parts that are exposed. In general, hands are less sensitive than eyes and reproductive organs. Whole-body irradiation may lead to radiation sickness and death, especially when a large amount of radiation is absorbed in a short interval of time. Symptoms of radiation damage may be suppressed in cases where small doses are absorbed over long periods of time, allowing damaged cells to be repaired.

The medical consequences of exposure to ionizing radiation include

1. radiation sickness (5–300 rems)—skin burns, ulcers, nausea, vomiting, diarrhea, hair loss, and changes in blood chemistry;
2. cancer (e.g., leukemia);
3. cataracts;
4. mental retardation;
5. chromosome aberrations and genetic disorders; and
6. weakening of the immune system allowing other diseases to develop.

Children and fetuses are especially vulnerable and should not be exposed to ionizing radiation. The best policy is to avoid exposure unless it is required for medical diagnosis or for the treatment of a life-threatening medical condition. The permissible dose for occupational workers is 5 rems per year and for the general population the limit is 0.5 rem per year.

The effect of ionizing radiation on human health is indicated in Figure 3.7. Curve *A* represents the proportion of individuals who suffer radiation sickness as a result of exposure to doses from 0 to 300 rems. Curve *B* expresses the death rate among individuals after exposure to between 300 and 1000 rems. The data demonstrate that an exposure to 300 rems causes radiation sickness in all cases and leads to the death of 20 percent of the victims within one month. Exposure to between 500 and 700 rems causes death in nearly all cases depending on the promptness and adequacy of medical treatment. Exposure to 1000 rems or more results in a death rate of 100 percent.

Lethal doses of ionizing radiation occurred in 1945 in Hiroshima and Nagasaki, Japan, as a result of the explosion of two nuclear bombs during World War II. Since that time, accidental exposures to lethal doses of radiation have been rare. A summary of such accidents was published by Eisenbud and Gesell (1997). The worst case of accidental exposure occurred in 1986 at Chernobyl, Ukraine, where 203 persons received more than 100 rems, causing 29 deaths. A much larger group of 24,200 persons received lower doses of radiation, which are expected to cause a significant increase in the cancer rate in the affected population.

The annual exposure of humans caused by radiation in the environment varies widely depending on lifestyles and place of residence, as indicated in Table 3.2. When cigarettes, eye glasses, and enameled jewelry are excluded, the average annual dose is about 360 mrems per person. The largest contributors to this dose are radon (56 percent), internal self-irradiation (11 percent), medical x-rays (11 percent), rocks and soil (8 percent), cosmic rays (8 percent), nuclear medicine (4 percent), consumer products (3 percent), and other sources (2 percent) (Anonymous, 1992).

Ionizing radiation in the environment on the surface of Earth has contributed to the evolution of life throughout geological time by causing mutations and by stimulating the development of cellular repair processes. Karam and Leslie (1999) estimated that the radiation dose from rocks and soil has decreased from about 1.6 milligray per year (mGy/y) at 4 gigayears (Ga) to 0.66 mGy/y at the present time. Similarly, the self-irradiation dose of organisms containing 250 mmol/L of K has decreased from 5.5 to 0.70 mGy/y. The decrease in the environmental radiation dose was caused by the decrease in the abundances of the naturally occurring radionuclides of U, Th, and K and by the evolution of the chemical composition of the continental crust. As a consequence of this heritage, all organisms can cope with low levels of environmental radiation and have the ability to repair cellular damage caused by exposure to ionizing radiation in the environment.

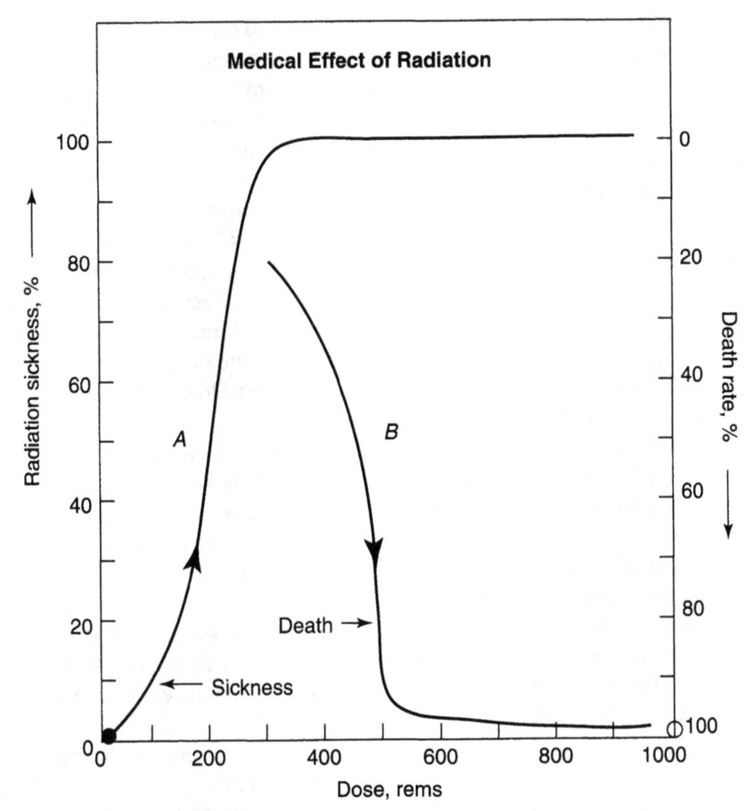

FIGURE 3.7 Effect of ionizing radiation on humans expressed in rems. Curve *A* indicates the incidence of radiation sickness between 0 and 300 rems. Curve *B* presents the fraction of the population that will die as a consequence of exposures of between 300 and 1000 rems. Note that the death rate is scaled on the right side of the diagram. Data from Anonymous (1992).

Table 3.2. Sources of Ionizing Radiation and Resulting Dose Rate to Humans

Activity and Source	Dose, mrem/y
1. Smoking one pack of cigarettes per day (^{210}Po and ^{210}Pb)	2000–5000 (to lungs)
2. Wearing eye glasses (U and Th in the glass)	500
3. Inhalation of Rn (national average)	100–200 (to lungs)
4. Exposure to ^{40}K and U in fertilizer and food	100–200
5. Reduced atmospheric shielding of flight crews	170
6. Exposure to U in glaze of enameled jewelry depending on duration and presence or absence of metal backing	25–4000
7. Excess exposure to cosmic radiation when living at high elevations compared to living at sea level	70
8. Living in house constructed of bricks rather than wood (U and Th in the bricks)	10

Source: Anonymous, 1992.

3.6 SOURCES OF ENVIRONMENTAL RADIOACTIVITY

The background radioactivity that is registered by radiation detectors originates from both natural and anthropogenic sources. The *natural sources* of ionizing radiation include the following:

1. decay of long-lived radionuclides such as $^{40}_{19}K$, $^{87}_{37}Rb$, $^{147}_{62}Sm$, $^{176}_{71}Lu$, $^{232}_{90}Th$, $^{235}_{92}U$, and $^{238}_{92}U$;
2. decay of the unstable daughters of long-lived naturally occurring isotopes of U and Th;
3. cosmogenic radionuclides, including ^{3}H, ^{10}Be, ^{14}C, ^{26}Al, and many other short-lived nuclides that form in the atmosphere and in rocks and soil exposed at the surface of the Earth;
4. cosmic rays composed of nuclear particles and of the nuclei of atoms emitted by the Sun and by sources in the galaxy; and
5. ultraviolet (uv) radiation having a range of wavelength from 400 to <290 nm: UV-A 400–320 nm; UV-B 320–290 nm; UV-C <290 nm.

These radiation sources are part of the natural environment in which life on Earth has evolved (e.g., Karam and Leslie, 1999).

The *anthropogenic sources* of environmental radioactivity constitute a significant hazard to human health. Therefore, appropriate precautions must be taken to limit the exposure of humans to the radiation emitted by these sources. Although the hazards associated with anthropogenic sources of ionizing radiation are widely recognized, accidental exposures can result from ignorance, sabotage, or carelessness or because of the pervasive character of the problem.

An important anthropogenic source of radiation arises from the use of certain kinds of appliances in the manufacturing industry, in research institutes, and in the home. These appliances include different kinds of particle accelerators, γ-ray sources, x-ray sources, and household appliances such as television screens and computer monitors. In addition, luminous dials on alarm clocks and uranium-bearing glazes on ornamental and household ceramics can be sources of ionizing radiation, although most have probably been eliminated in response to concerns for public health.

The most serious threat to public health undoubtedly originates from the operation of nuclear fission reactors and from the high-level nuclear waste they generate in the form of spent fuel rods (Sections 2.3c, 2.3d). In addition, the explosion of fission and fusion devices in the atmosphere has released large quantities of fission products in the form of fallout. To avoid global contamination of the surface of Earth, nuclear weapon tests are supposed to be carried out in the subsurface—if it all.

The problems associated with the safe storage of nuclear waste involve technical, political, legal, and sociological considerations. A large number of geochemists are participating in the testing of proposed nuclear repositories and of the various options for treating and packaging the waste in order to assure public safety for more than 10,000 years in the future. This subject has been adequately discussed in all of its many ramifications by Eisenbud and Gesell (1997), Brookins (1984), and many others, including a brief summary by Faure (1998).

The injection of radionuclides into the atmosphere as a result of the explosion of nuclear devices has provided an opportunity to study the global distribution pattern of the resulting fallout. The observed distribution of $^{90}_{38}Sr$ and $^{137}_{55}Cs$ (Sections 25.5 and 25.6), both of which are fission product nuclides with halflives of about 30 years, indicates that fallout radionuclides are transported by the prevailing winds in the troposphere (0–10 km) and are deposited in a belt that circles the globe. For example, most of the fallout from weapons tests with yields in the kiloton range in Nevada in 1953 remained within 20° north and south of the latitude of the test site (Machta et al., 1956). The deposition of fallout within this band varied locally depending on the timing of meteoric precipitation in relation to the cloud of radioactive dust passing overhead. Nevertheless, regions of the Earth more than 40° latitude north or south of the Nevada test site received only small amounts of fallout. More powerful explosions releasing the equivalent of several megatons of TNT injected radioactive dust into the stratosphere (10–50 km), allowing it to spread globally from pole to pole. However, the amount of fallout that reached the polar regions after the explosions of thermonuclear devices in the Marshall Islands in 1952 was significantly less than 10 mCi/100 km^2, compared to more than 10,000 mCi/100 km^2 in the central Pacific Ocean near the test site (Machta et al., 1956).

3.7 NUCLEAR REACTIONS

Nuclear reactions can be induced by bombarding the nucleus of any atom with a variety of nuclear particles, such as protons, deuterons (nuclei of deuterium), α-particles, neutrons, and so forth. In the case of bombardment with positively charged particles, nuclear reactions can occur only when the particles have sufficient energy to overcome the electrostatic repulsion by the protons in the target nucleus. Therefore, the projectiles must have a certain minimum energy that depends on their electric charge and the atomic number of the target nucleus. These limitations do not apply to neutrons, which have no charge and can interact with the nucleus of any atom.

When a nuclear particle enters the nucleus, a reaction occurs that manifests itself by the emission of particles or of radiant energy from the product nucleus. Depending on the nature of the reaction, the product may be an isotope of the same element as the target or it may be an isotope of a different element. The kind of reaction that takes place depends on the nature of the projectile, its energy, and the nature of the target nucleus. Such nuclear reactions make possible elemental transformations of the atoms of one element into atoms of another element.

Nuclear reactions can be described by equations similar to those used in Chapter 2 to represent radioactive decay. For example, the reaction

$$^{A}_{Z}\text{X} + {}^{1}_{1}\text{H} \rightarrow {}_{Z+1}{}^{A}\text{Y} + {}^{1}_{0}\text{n} \tag{3.32}$$

indicates that a target atom ($^{A}_{Z}$X) reacts with a proton ($^{1}_{1}$H) and forms a product atom (${}_{Z+1}{}^{A}$Y) and a neutron ($^{1}_{0}$n), which is emitted. In this case, the product atom is an isobar of the target atom and therefore is an isotope of another element. The same reaction can also be represented by the shorthand notation (equation 3.15)

$$^{A}_{Z}\text{X(p, n)}\ {}_{Z+1}{}^{A}\text{Y} \tag{3.33}$$

The nuclide in front of the brackets is the target, while the nuclide following the brackets is the product. Inside the brackets are shown the proton (p), which enters the target nucleus, and the neutron (n), which comes out of the product nucleus. The two particles are separated by a comma to indicate which particle goes in and which comes out. The notation

$$^{A}_{Z}\text{X}\ (\alpha, 2\text{n})\ {}^{A+2}_{Z+2}\text{Y} \tag{3.34}$$

means that one α-particle goes in and two neutrons come out. As a result, the target atom ($^{A}_{Z}$X) is transformed into the product atom ($^{A+2}_{Z+2}$X), which in this case is an atom of another element but is not an isobar of the target. The variety of artificial nuclear reactions that are possible and the relative positions of target and product nuclides are summarized in Figure 3.8.

3.8 NEUTRON ACTIVATION ANALYSIS

Reactions caused by the irradiation of atoms with neutrons are of great interest in isotope geology and geochemistry because such reactions are used to measure the concentrations of trace elements in geological materials. Neutrons are produced in large numbers by the controlled fission of ^{235}U in nuclear reactors and are therefore readily available for irradiation purposes. The neutrons produced during fission of a uranium atom are initially emitted with high velocities and are called "fast" neutrons. To sustain the chain reaction, the fast neutrons must be slowed down because the fission reaction is strongly favored when it is induced by "slow" rather than fast neutrons. The slowing down of the neutrons is achieved by a "moderator" with which the fast neutrons can collide without being absorbed. The first experimental reactors used

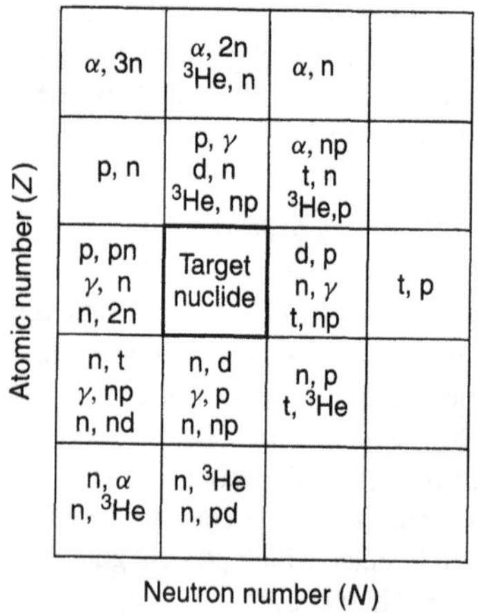

Atomic number (Z)

α, 3n	α, 2n ^{3}He, n	α, n	
p, n	p, γ d, n ^{3}He, np	α, np t, n ^{3}He,p	
p, pn γ, n n, 2n	Target nuclide	d, p n, γ t, np	t, p
n, t γ, np n, nd	n, d γ, p n, np	n, p t, ^{3}He	
n, α n, ^{3}He	n, ^{3}He n, pd		

Neutron number (N)

FIGURE 3.8 Displacement of product nuclides relative to the target nuclide caused by nuclear bombardment reactions. The symbol in front of the comma indicates the particle or radiation that goes in; the symbol after the comma indicates the particle or radiation that comes out of the product nucleus: p = proton ($^1_1$H), n = neutron, γ = γ-ray, α = α-particle ($^4_2$He), d = deuteron ($^2_1$H), and t = tritium ($^3_1$H). After Walker et al. (1989).

graphite (C) or heavy water (D_2O) as moderators, although ordinary water (H_2O) also serves this purpose in the so-called swimming pool reactors (Section 2.3c).

Slow neutrons, having velocities corresponding to the ambient temperature, are readily absorbed by the nuclei of most of the stable isotopes of the elements. The mass number of the product nucleus is increased by 1 compared with the target nucleus, but Z remains unchanged so that the product is an isotope of the same element as the target. The product nucleus is left in an excited state and de-excites by emission of γ-rays. The absorption of a slow neutron by the nucleus of an atom can be represented by the reaction

$$^A_Z\mathrm{X} + ^1_0\mathrm{n} \rightarrow {}^{A+1}_{\;\;Z}\mathrm{Y} + \gamma \tag{3.35}$$

or

$$^A_Z\mathrm{X}(\mathrm{n}, \gamma)\ {}^{A+1}_{\;\;Z}\mathrm{Y}$$

The product of an "(n, γ)" reaction may be either stable or unstable. When the product nuclides produced by neutron irradiation are unstable, the sample becomes radioactive as the product nuclides decay with their characteristic halflives. Thus a slow-neutron irradiation of a sample composed of the stable atoms of a variety of elements leads to the formation of radioactive isotopes of these elements, and the irradiated sample then becomes radioactive—hence the term *neutron activation*. The induced activity of a specific radioactive isotope of an element in the irradiated sample depends on several factors, including the weight of the target element in the sample. This fact is the basis for using neutron activation as an analytical tool (Mapper, 1960; Brunfelt and Steinnes, 1971; DeSoete et al., 1972).

The production of a radionuclide (P) by a nuclear reaction occurring at a constant rate (R) is analogous to the decay of a long-lived radionuclide (N_1) to an unstable daughter (N_2). In this case, the decay of the parent is replaced by the production of radioactive atoms by a nuclear reaction caused by irradiating the target with particles produced by a nuclear reactor, a cyclotron, or a linear particle accelerator. Since the reactor or particle accelerator can be operated at a constant rate, the production rate R does not decrease with time, meaning that the decay constant λ_1 of the hypothetical parent is equal to zero. Therefore, equation 3.20 can be adapted to the production and decay of a radionuclide produced by a nuclear reaction:

$$N_2 = \frac{\lambda_1}{\lambda_2 - \lambda_1} N_1^0 (e^{-\lambda_1 t} - e^{-\lambda_2 t})$$

Because $\lambda_1 = 0$, $\lambda_2 - \lambda_1 = \lambda_2$ and $e^{-\lambda_1 t} = 1$. In addition, $\lambda_1 N_1^0$ (the rate of decay of the parent nuclide) is replaced by the rate of production of the radionuclide (R). With these modifications, equation 3.20 becomes

$$P = \frac{R}{\lambda}(1 - e^{-\lambda t}) \tag{3.36}$$

where P = number of product radionuclides
R = rate at which P is produced by the nuclear reaction caused by the irradiation of the target
λ = decay constant of product nuclide P
t = duration of irradiation

As the irradiation time approaches infinity,

$$\lim_{t\to\infty}(1 - e^{-\lambda t}) = 1$$

and hence from equation 3.36

$$\lambda P = R \qquad (3.37)$$

Evidently, the rate of decay of the product nuclide eventually approaches its production rate, which is the maximum or saturation disintegration rate that can be achieved. If one must pay for an irradiation based on the irradiation time, it is obviously important to limit the duration of the irradiation to something less than the time required to achieve saturation. The most advantageous irradiation time can be determined by rewriting equation 3.36 in the form

$$\frac{\lambda P}{R} = (1 - e^{-\lambda t}) \qquad (3.38)$$

where $\lambda P/R$ is the disintegration rate of the product expressed as a fraction of the maximum attainable rate. The ratio $\lambda P/R$ in Figure 3.9 rises with increasing irradiation time (in multiples of the halflife) and approaches unity. The diagram demonstrates that irradiations should be limited to three halflives of the desired product or less before the decay rate of the product approaches the production rate.

When the irradiation is terminated, the product nuclide continues to decay and the number of atoms remaining at any time is

$$P = \frac{R}{\lambda}(1 - e^{-\lambda t_i})e^{-\lambda t_d} \qquad (3.39)$$

where t_i is the length of time the sample was irradiated and t_d is the decay time measured from the time the irradiation was terminated.

The decay rate of the product radionuclide is determined by means of a suitable radiation detector, such as a γ-ray spectrometer, which resolves the energy spectrum of the γ-rays emanating from the activated target. Therefore, the rates of decay of different product radionuclides in the target can be determined separately. The decay rate of each product radionuclide (A) obeys the equation

$$A = R(1 - e^{-\lambda t_i})e^{-\lambda t_d} \qquad (3.40)$$

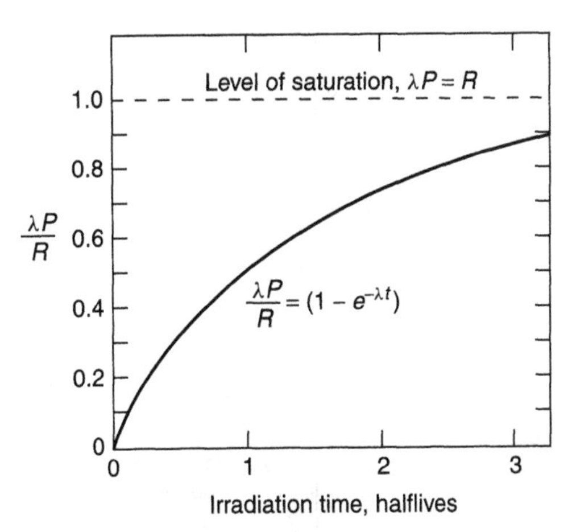

FIGURE 3.9 The function $\lambda P/R$ is the ratio of the disintegration rate of a radionuclide (λP) to the production rate (R). Since the maximum disintegration rate is equal to the production rate, the ratio $\lambda P/R$ approaches unity with increasing irradiation times, here expressed in terms of halflives of the product nuclide. The graph shows that after an irradiation time equal to one halflife the activity is one-half of the maximum. After two halflives, the activity is three-quarters of the maximum activity. Therefore, an irradiation time between two and three halflives of the desired product nuclide is sufficient to produce an appreciable fraction of the maximum activity.

where λ is the decay constant of the desired radionuclide. The rate of decay of a product radionuclide in Figure 3.10 initially increases during the irradiation of the target. When the irradiation is terminated, the decay rate of the radionuclide decreases exponentially in accordance with the law of radioactivity expressed by equation (3.11);

$$A = A_0 e^{-\lambda t}$$

The production rate R for an irradiation with slow neutrons is defined as

$$R = Na\sigma F \qquad (3.41)$$

where N is the number of target atoms of a particular element in the irradiated sample, a is

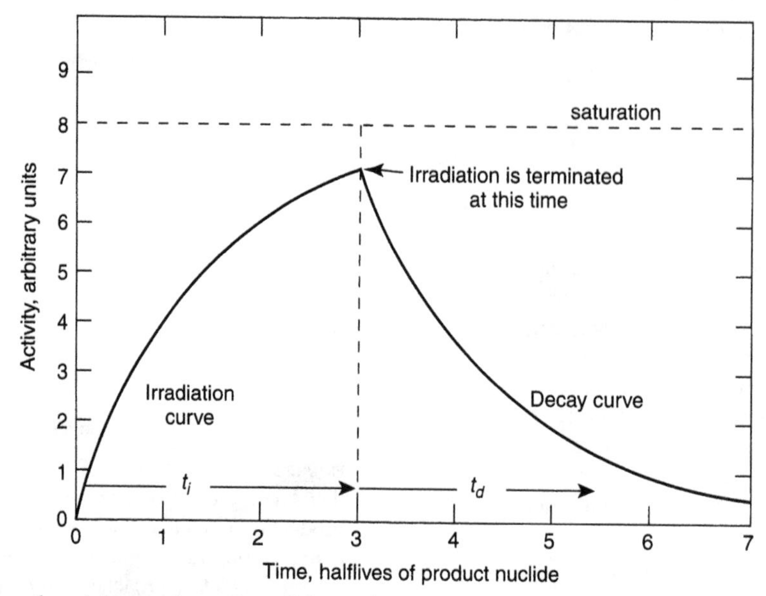

FIGURE 3.10 Growth and decay of a radionuclide produced by irradiating a target (e.g., a sample of rock, minerals, plant material) with nuclear particles such as slow neutrons in a nuclear reactor. The duration of the irradiation and of subsequent decay is conveniently expressed as multiples of the halflife of a specific radionuclide produced in the target.

the isotopic abundance of the target isotope, σ is the neutron capture cross section in units of barns (where 1 barn $= 10^{-24}$ cm^2) of the target nuclide, and F is the neutron flux in units of neutrons per square centimeter per second. The cross section of a nuclear reaction (σ) is a measure of the probability that a particular reaction will occur. This probability is related to the cross-sectional area of the nucleus. Since the radius of an atomic nucleus is of the order of 10^{-12} cm, its cross-sectional area is of the order of 10^{-24} cm^2. Hence, the reaction probability is measured in multiples of 10^{-24} cm^2. The flux F is expressed as the number of neutrons that cross an area of 1 cm^2 in 1 s. The slow-neutron flux available for irradiations at many nuclear reactors is of the order of 10^{12}–10^{13} neutrons/cm^2 s. Even higher flux densities are available in some of the largest reactors.

To determine the concentration of an element in a sample of matter, a known weight of the sample is irradiated with slow neutrons along with a standard containing a known amount of the same element. When the irradiation is completed, the activity due to decay of the desired product nuclide is determined at intervals of time in both the sample and the standard. The measurement of the decay rate of a particular radionuclide produced by the irradiation is often complicated by the presence of other radionuclides. If the desired radionuclide is a γ-emitter, it is usually possible to screen out β- or α-particles by means of suitable absorbers or to measure the decay rate of the desired radionuclide by means of γ-ray spectrometry.

If interfering radionuclides cannot be avoided by instrumental methods, chemical separations may have to be performed on the irradiated sample. Such separations usually require dissolution of the target followed by wet-chemical separations of the desired elements based on the use of "carriers," which are small amounts of the unactivated element that facilitate precipitation reactions but do not interfere directly with measurement of the induced radioactivity. The chemical processing of irradiated samples requires careful consideration of the safety of the analyst and proper disposal of radioactive chemical waste.

After the rates of decay of the desired radionuclide in the irradiated target and in the standard have been determined at known times, the resulting data are plotted in coordinates of ln A and t_d to define straight lines based on equation 3.13:

$$\ln A = \ln A_0 - \lambda t$$

These straight lines are used in Figure 3.11 to determine the initial activities of the product radionuclide in the sample (A_0^{spl}) and in the standard (A_0^{st}). The initial activities are related to the weights of the target element (X_w) by

$$\frac{A_0^{\text{spl}}}{A_0^{\text{st}}} = \frac{X_W^{\text{spl}}}{X_W^{\text{st}}} \quad (3.42)$$

It follows that

$$X_W^{\text{spl}} = \left(\frac{A_0^{\text{spl}}}{A_0^{\text{st}}}\right) X_W^{\text{st}}$$

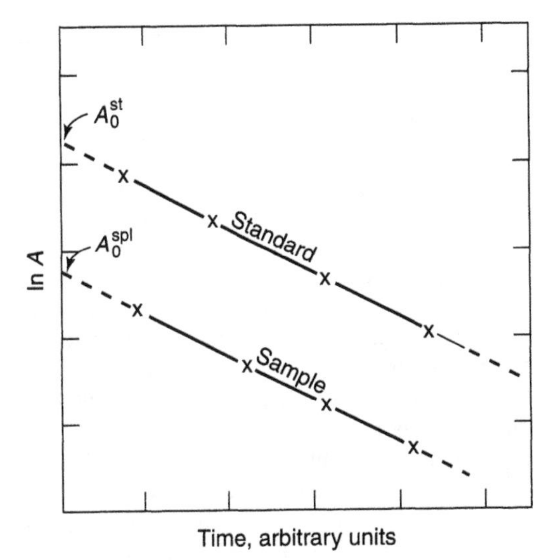

FIGURE 3.11 Plot of the rate of decay of a single radionuclide after neutron irradiation of a sample and a standard in coordinates of ln A and t. The resulting straight lines are extrapolated back to A_0, which is the activity of the radionuclide in the sample and the standard at the end of the irradiation. The slopes of the two lines are identical and equal to $-\lambda$ of the radionuclide.

from which the concentration of element X in the irradiated target can be calculated based on the known weight of the sample.

The standard that is irradiated with the sample may be a known weight of a rock or mineral in powdered form whose chemical composition was determined by other analytical methods. Alternatively, the standard may be a known weight of a compound of the element of interest (the analyte) or a mixture of compounds of several elements. Regardless of the type of standard used, several elements can be determined by a single irradiation provided the rates of decay of their product radionuclides can be measured without interference. When several elements are to be determined, the initial activities of their product radionuclides in the standard and in the sample must be determined as illustrated in Figure 3.11 and their amounts are calculated from equation 3.42.

The calibration can be improved by irradiating three or more standards that contain different amounts of each analyte. In this case, the initial decay rates of each analyte in each standard must be determined. The resulting data define points in coordinates of A_0^{st} and the amount of each analyte in the standards. In most cases, these data points define straight lines whose equation in the slope–intercept form is

$$y = mx + b$$

where y is the A_0 value and x is the corresponding weight of the analyte. The slope (m) of the straight line in Figure 3.12 is the calibration factor, whereas the intercept (b) should be equal to zero. The weight of the analyte (X) in a sample that was irradiated with the standards is calculated by substituting its initial activity (A_0^{spl}) into the equation of the calibration line in Figure 3.12:

$$X_w^{\text{spl}} = \frac{A_0^{\text{spl}}}{m} - b \quad (3.43)$$

Before samples and standards are irradiated in a nuclear reactor, the analyst is required to estimate the rate of decay of the samples when they are withdrawn from the reactor. Such estimates are based on the chemical composition and on the expected production rates R of the product

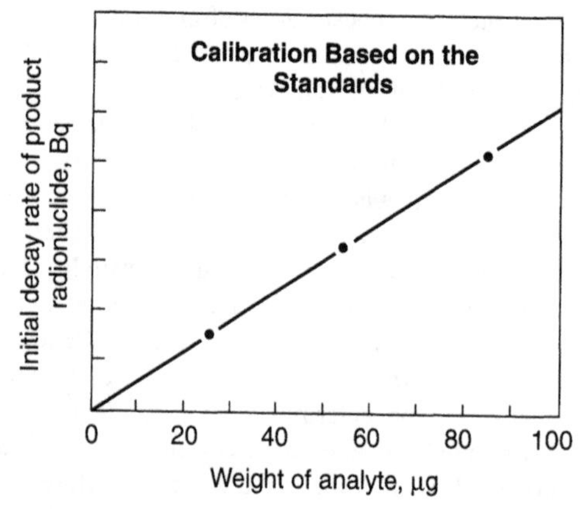

FIGURE 3.12 Calibration based on the initial decay rate of the radionuclide produced by the slow-neutron irradiation of three standards containing different weights of the element to be determined (the analyte).

radionuclides and on the irradiation time t_i in accordance with equations 3.40 and 3.41.

For example, the irradiation of a sample containing sodium results in the formation of unstable $^{24}_{11}Na$ from the stable $^{23}_{11}Na$ by the reaction

$$^{23}_{11}\text{Na}\ (n, \gamma)\ ^{24}_{11}\text{Na}$$

If the concentration of sodium in a sample weighing 0.5 g is 1.2 percent and the neutron flux $F = 1 \times 10^{12}$ neutrons/cm^2 s, the production rate R of $^{24}_{11}Na$ (equation 3.40) is

$$R = Na\sigma F$$

In the case under consideration, the number of sodium atoms in the sample being irradiated is

$$N = \frac{1.2 \times 0.5 \times 6.022 \times 10^{23}}{100 \times 22.98977}$$
$$= 1.57 \times 10^{20} \text{ atoms}$$

Since $a = 100\%$, $\sigma = 0.53 \times 10^{-24}$ cm^2, and $F = 1 \times 10^{12}$ neutrons/ cm^2/ s,

$$R = 1.57 \times 10^{20} \times 1.00 \times 0.53 \times 10^{-24} \times 1 \times 10^{12}$$
$$= 0.832 \times 10^{8} \text{ atoms/s}$$

If the duration of the irradiation is 30 min (0.5 h), the rate of disintegration ($A = \lambda P$) of $^{24}_{11}Na$ calculated from equation 3.39 is

$$A = R(1 - e^{-\lambda t_i})$$

Since the halflife of $^{24}_{11}Na$ is 14.96 h, its decay constant (equation 3.8) is

$$\lambda = \frac{0.6931}{14.96} = 0.04633 \text{ h}^{-1}$$

Therefore, the decay rate of $^{24}_{11}Na$ after a half-hour irradiation is

$$A = 0.832 \times 10^{8}\ (1 - e^{-0.04633 \times 0.5})$$
$$= 0.832 \times 10^{8}\ (1 - 0.9771)$$

Note that $e^{-\lambda t}$ is dimensionless provided λ and t are expressed in the same units. Therefore,

$$A = 1.905 \times 10^{6} \text{ dis/s (or Bq)}$$

A decay rate 1.905×10^6 Bq is convertible into curies:

$$A = \frac{1.905 \times 10^{6}}{3.7 \times 10^{10}} = 0.5148 \times 10^{-4} \text{ Ci}$$
$$= 51.48\ \mu\text{Ci}$$

To avoid potentially harmful exposure of the analyst to such levels of radiation, irradiated samples are allowed to "cool" in a lead-lined box before being transferred to the radiation detector. In the case under consideration, the decay rate of $^{24}_{11}Na$ is reduced by a factor of 2^{-1} for every 14.96 h of elapsed decay time.

The thermal neutron capture cross sections of the stable isotopes of the elements vary widely from small fractions to several thousand barns. Elements with isotopes that have high neutron capture cross sections (e.g., the rare earths) can be determined with exceptional sensitivity by neutron activation. Other desirable attributes of this analytical method include the following:

1. A high degree of accuracy exists, especially when the standards are mixtures of purified reagents having stoichiometric compositions.

2. The high sensitivity for elements having large neutron capture cross sections permits analysis of small samples or determining elements having low concentrations.
3. The samples are not destroyed physically and can be used for other purposes after the decay rate has subsided.
4. The method is insensitive to contamination following irradiation because only the radioactive product nuclides are included in the analysis.

3.9 SUMMARY

The rate of decay of radionuclides is governed by the law of radioactivity, which was enunciated by Rutherford and Soddy in 1902. This law describes the rate of decay of a radionuclide by the decay constant λ, which has a specific value characteristic of each radionuclide. The decay curve of a radionuclide can be plotted by expressing time in multiples of the halflife, which is derived from the decay constant.

The rate of decay of an unstable nuclide is measured by means of ionization chambers (Geiger–Müller counter) and by scintillation counters using crystals of NaI or Ge. The operation of such radiation detectors reveals the presence of background radiation caused by natural and anthropogenic environmental radioactivity. The decay of the unstable daughters of the long-lived unstable isotopes of U and Th constitutes an important component of this background radioactivity. The mathematical analysis of the decay and growth of the series of unstable daughters of U and Th leads to the concept of secular equilibrium when the rate of decay of the short-lived daughters is equal to the rate of decay of the long-lived parents in each of the three decay chains initiated by the isotopes of U and Th.

The rate of decay is measured in units of becquerels and curies, both of which are modified by appropriate prefixes. The absorption of high-energy γ-rays or α- and β-particles causes ionization of atoms and molecules in the absorber by the disruption of chemical bonds. The exposure of biological tissue to ionizing radiation causes injury resulting from damage to cells. In case of exposure by humans, the severity of the injury depends on the dose of radiation absorbed and on the type of exposure. Internal exposure to α- and β-particles is especially harmful and must be avoided. External exposure to γ-rays may cause radiation sickness and death. The most serious cases of radiation exposure by humans have resulted from the explosion of nuclear weapons and the accidental release of radionuclides from nuclear power reactors.

Nuclear reactors generate heat by the controlled induced fission of ^{235}U. The heat is used to make steam, which drives electric generators (dynamos). The high flux of neutrons available in nuclear reactors permits the manufacture of radioactive isotopes for medical use and for neutron activation of samples for analytical purposes.

REFERENCES

Anonymous, 1992. *Ionizing radiation*, DOE/RW-0362 SR. Unit 2, Science, Society, and America's Nuclear Waste. Department of Energy, Washington, D.C.

Bateman, H., 1910. Solution of a system of differential equations occurring in the theory of radioactive transformations. *Proc. Cambridge Philos. Soc.*, 15: 423.

Brunfelt, A. O., and E. Steinnes, 1971. Activation analysis in geochemistry and cosmochemistry. *Proc. Nato Adv. Study Inst.*, Oslo Universitetsforlaget.

Brookins, D. G., 1984. *Geochemical Aspects of Radioactive Waste Disposal.* Springer-Verlag, New York.

DeSoete, D., R. Gijbels, and J. Hoste, 1972. *Neutron Activation Analysis.* Wiley, New York.

Eisenbud, M., and R. Gesell, 1997. *Environmental Radioactivity from Natural, Industrial, and Military Sources.* Academic, San Diego, California.

Faul, H. (Ed.), 1954. *Nuclear Geology.* Wiley, New York.

Faure, G., 1998. *Principles and Applications of Geochemistry*, 2nd ed. Prentice-Hall, Upper Saddle River, New Jersey.

Friedlander, G., J. W. Kennedy, and J. M. Miller, 1981. *Nuclear and Radiochemistry*, 3rd ed. Wiley, New York.

Geiger, H., and W. Müller, 1929. Das Electronenzählrohr. *Phys. Z.*, 30:489–493.

Kaplan, I., 1955. *Nuclear Physics.* Addison-Wesley, Reading, Massachusetts.

Karam, P. A., and S. A. Leslie, 1999. Calculations of background beta-gamma radiation dose through geologic time. *Health Phys.*, 77:662–667.

Machta, L., R. J. List, and L. F. Hubert, 1956. Worldwide travel of atomic debris. *Science*, 124:474–477.

Mapper, D., 1960. Radioactivation analysis. In A. A. Smales and L. R. Wager (Eds.), *Methods of Geochemistry*, pp. 297–357. Interscience, New York.

Moorbath, S., 1960. Radiochemical methods. In A. A. Smales and L. R. Wager (Eds.), *Methods in Geochemistry*, pp. 247–296. Interscience, New York.

Rutherford, E., 1911. The scattering of α and β particles by matter and the structure of the atom. *Philos. Mag.*, 21:669.

Rutherford, E., and F. Soddy, 1902a. The cause and nature of radioactivity, Pt. I. *Philos. Mag.*, 4 (Ser. 6): 370–396.

Rutherford, E., and F. Soddy, 1902b. The cause and nature of radioactivity, Pt. II. *Philos. Mag.*, 4 (Ser. 6): 569–585.

Rutherford, E., and F. Soddy, 1902c. The radioactivity of thorium compounds. I. An investigation of the radioactive emanation. *J. Chem. Soc. Lond.*, 81:321–350.

Rutherford, E., and F. Soddy, 1902d. The radioactivity of thorium compounds. II. The cause and nature of radioactivity. *J. Chem. Soc. Lond.*, 81:837–650.

Smales, A. A., and L. R. Wager (Eds.), 1960. *Methods in Geochemistry*. Interscience, New York.

Walker, F. W., J. R. Parrington, and F. Feiner, 1989. *Chart of the Nuclides*, General Electric Co., San Jose, California.

CHAPTER 4

Geochronometry

The long-lived radioactive isotopes of certain elements decay to stable daughters that have accumulated in rocks and minerals. These parent–daughter combinations provide a means of measuring the ages of terrestrial and extraterrestrial rocks based on the law of radioactivity. In addition, the accumulation of radiogenic daughters in the various geological reservoirs of the Earth (e.g., continental crust, lithospheric mantle) provides information concerning the origin of igneous rocks and the consequent growth of the continental crust (Faure, 2001). The parent–daughter isotopes that are most useful for geochronometry and for the study of the petrogenesis of igneous rocks are listed in Table 4.1.

Table 4.1. Long-Lived and Naturally Occurring Radionuclides and Their Stable Radiogenic Daughters That Are Useful for Measuring Ages of Terrestrial and Extraterrestrial Rocks

Parents	Decay Modes[a]	Halflives (10^9 years)	Daughters
$^{40}_{19}K$	e,β^+	11.9	$^{40}_{18}Ar$
$^{40}_{19}K$	β^-	1.39	$^{40}_{20}Ca$
$^{87}_{37}Rb$	β^-	48.8	$^{87}_{38}Sr$
$^{147}_{62}Sm$	α	106	$^{143}_{60}Nd$
$^{176}_{71}Lu$	β^-	36	$^{176}_{72}Hf$
$^{187}_{75}Re$	β^-	41	$^{187}_{76}Os$
$^{232}_{90}Th$	α,β^-	14	$^{208}_{82}Pb$
$^{235}_{92}U$	α,β^-	0.704	$^{207}_{82}Pb$
$^{238}_{92}U$	α,β^-	4.47	$^{206}_{82}Pb$

Source: Walker et al., 1989.

[a] e = electron capture decay; β^+ = positron decay; β^- = negatron decay; α = α-decay.

4.1 GROWTH OF RADIOGENIC DAUGHTERS

When a long-lived radionuclide (N) decays in a closed system, the number of stable radiogenic daughters (D^*) that accumulate in a unit weight of the rock or mineral is equal to the number of parent atoms that have decayed:

$$D^* = N_0 - N \tag{4.1}$$

where N_0 is the number of parent atoms present initially. Substituting the law of radioactivity expressed by equation 3.7

$$N = N_0 e^{-\lambda t}$$

into equation 4.1 yields

$$D^* = N_0 - N_0 e^{-\lambda t}$$

from which it follows that

$$D^* = N_0(1 - e^{-\lambda t}) \tag{4.2}$$

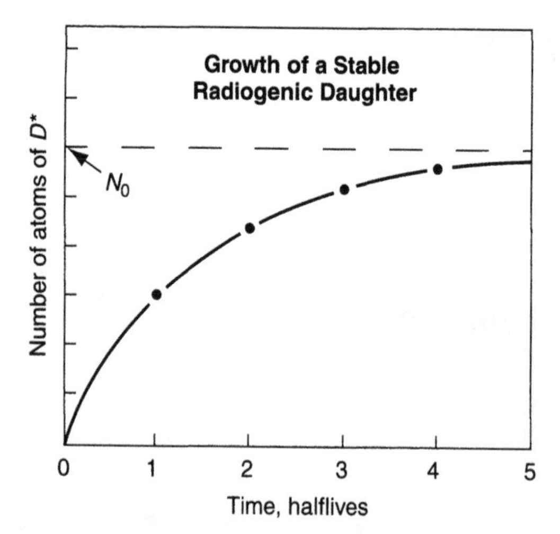

FIGURE 4.1 Growth of a stable daughter in a closed system by decay of a radioactive parent. In the limit as $t \rightarrow \infty$, the number of daughter atoms approaches the number of parent atoms (N_0) present initially.

This is the equation of the growth curve of a stable daughter in Figure 4.1. As t approaches infinity,

$$\lim_{t \to \infty} (1 - e^{-\lambda t}) = 1 \tag{4.3}$$

and, therefore, in the limit ($t \rightarrow \infty$)

$$D^* = N_0 \tag{4.4}$$

In other words, as decay of the parent nuclide (N) in a closed system continues, the number of stable daughters approaches the number of parent atoms that were present initially.

The problem with equation 4.2 is that the number of parent atoms present initially (N_0) in a unit weight of rock or mineral is not measurable and can be determined only if the number of parent atoms remaining (N) and the age of the sample are known. Therefore, equation 4.2 is not suitable for geochronometry.

The impasse is avoided by replacing N_0 in equation 4.1 by

$$N_0 = Ne^{\lambda t} \tag{4.5}$$

obtained by a simple algebraic manipulation of equation 3.7:

$$\begin{aligned} D^* &= Ne^{\lambda t} - N \\ &= N(e^{\lambda t} - 1) \end{aligned} \tag{4.6}$$

This equation contains two measurable parameters (D^* and N) and therefore can be solved for t if the value of the decay constant of the parent (λ) is known.

A unit weight of rock or mineral may contain a certain number of daughter atoms that were included in it at the time of formation. Therefore, the total number of daughter atoms in a unit weight of rock or mineral at the present time is

$$D = D_0 + D^* \tag{4.7}$$

where D_0 is the initial number of daughter atoms. Consequently, a complete description of the number of radiogenic daughter atoms is

$$D = D_0 + N(e^{\lambda t} - 1) \tag{4.8}$$

This equation is the basis for age determinations of terrestrial and extraterrestrial rocks and minerals by the decay of long-lived and naturally occurring radionuclides to stable daughters. Both D and N in equation 4.8 can be measured by analysis of the sample to be dated, and the equation can then be solved for time t provided that the decay constant λ of the parent is known and that an appropriate value is used for D_0. From equation 4.8

$$t = \frac{1}{\lambda} \ln \left(\frac{D - D_0}{N} + 1 \right) \tag{4.9}$$

When several rock or mineral samples from the same body of igneous rock are analyzed, the resulting values of D and N satisfy equation 4.8, provided that all of the samples have the same value of D_0 and the same age (t). In that case, the measured values of D and N define a straight line called an *isochron* because all points on that line represent rock or mineral systems that have the same age. The isochron method of dating is illustrated in Figure 4.2 and is discussed in Section 4.2d.

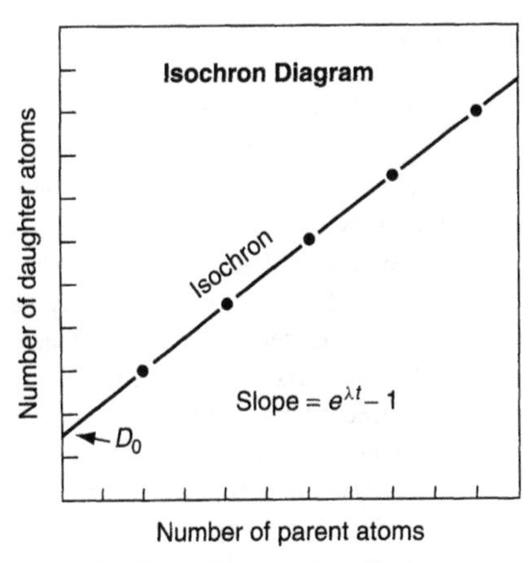

FIGURE 4.2 Isochron diagram based on equation 4.8. The coordinates of the solid circles are the measured values of D and N for a suite of cogenetic igneous rocks having a range of chemical compositions. The age of the rocks that define the isochron is calculated from its slope. The intercept of the isochron is equal to D_0.

4.2 ASSUMPTIONS FOR DATING

The numerical value of t obtained from equation 4.9 is a date in the past. The interpretation of this date depends on certain assumptions about the geological history of the rock or mineral being dated and about the numerical values of D_0 and λ being used in the calculation. The assumptions are as follows:

1. The rock or mineral sample being dated has not gained or lost parent or daughter atoms except by decay of the parent to the stable daughter.
2. The decay constant of the parent nuclide is independent of time and is not affected by the physical conditions to which the nuclide may have been subjected and its value is known accurately.
3. An appropriate value of D_0 is used in the calculation based on either knowledge of the chemical properties of the daughter element or its isotope composition in the terrestrial reservoir from which the rock or mineral originated.
4. The measured values of D and N are accurate and representative of the rock or minerals being dated.

These assumptions are discussed in general terms in the sections that follow. More specific information is provided in the presentations of each of the several dating methods.

4.2a Closed System

The extent to which rocks and minerals have actually remained closed to parent and daughter elements depends on many factors, including

1. the retentivity of the minerals for the parent and daughter elements,
2. the physical and chemical properties of the parent and daughter elements,
3. the history of metamorphic alteration of the rock or minerals in response to episodic changes of pressure and temperature,
4. the rate of cooling following the last thermal episode, and
5. interactions with aqueous solutions.

Each of the parent–daughter couples listed in Table 4.1 responds differently to these kinds of disturbances depending on their severity and duration. Therefore, the dates derived for a particular rock or mineral by different applicable geochronometers are discordant in many cases. The reason is that each parent–daughter couple responds in its own way to a given set of perturbations.

The all-important issue to be addressed in such cases is to identify the date that actually records the age of the rock or mineral being dated based on knowledge of the origin of different kinds of rocks. For example, the age of igneous rocks is the time elapsed since the minerals of which they are composed crystallized from magma. The age of sedimentary rocks is the time since they were deposited by settling of sediment or by precipitation of minerals from water. The definition of the age of metamorphic rocks is more complex because they form by recrystallization of pre-existing igneous or sedimentary protoliths. In addition, metamorphic rocks may have been recrystallized repeatedly, causing changes not only in the texture of the rocks but also in the minerals they contain and in their chemical composition.

Each of the principal rock types presents a different challenge for geochronometry. For example, plutonic igneous rocks cool slowly and allow daughter atoms of certain elements to escape until the temperature has decreased sufficiently to prevent the loss of the mobile radiogenic daughter (e.g., ^{40}Ar). Consequently, K–Ar dates of micas in igneous or metamorphic rocks are, in many cases, lower than Rb–Sr or Sm–Nd dates of the same minerals. In addition, the chemical and mineralogical compositions of plutonic igneous rocks near the contact with the country rock may be altered by interactions with heated groundwater. As a result, the initial abundances of the daughter isotope (e.g., ^{87}Sr and of other radiogenic isotopes) may be changed, thereby introducing systematic errors into dates calculated by the Rb–Sr and other methods.

Sedimentary rocks are difficult to date by any of the available geochronometers because

1. clastic sedimentary rocks (sandstones and shales) consist of pre-existing mineral particles whose ages depend on their provenance and
2. chemically precipitated sedimentary rocks (carbonate rocks and evaporites) either have unfavorable chemical compositions (e.g., limestone) or do not remain closed to parent and daughter elements (evaporites).

Successful age determinations of sedimentary rocks require the presence of authigenic clay minerals (e.g., glauconite or fine-grained illite) or of volcanic rocks either in discrete layers (e.g., lavas flows or tuff) or as dispersed crystals of minerals that resist weathering (e.g., sanidine, muscovite, zircon).

Dating of metamorphic rocks is complicated by their geological histories, which include episodes of recrystallization followed by slow cooling. In addition, certain uranium-bearing minerals (e.g., zircon) resist alteration during thermal metamorphism and, in some cases, yield dates that exceed the age of the protoliths and of the subsequent episode of metamorphism. Therefore, the geological histories of high-grade metamorphic rocks (e.g., granitic gneisses) are revealed only when several different isotopic geochronometers are used to date them.

4.2b Decay Constants

The decay of radionuclides by emission of negatrons, positrons, and α-particles and by spontaneous fission is a nuclear process that is not affected by the electron density in the vicinity of the nucleus or by external pressure and temperature of the environment to which these radionuclides may be subjected. Therefore, the decay constants of negatron, positron, and α-emitters are invariant with respect to time and unaffected by environmental conditions. The same statement applies to the decay constants of spontaneous fission, although in this case fission may be induced in some cases by exposure to neutrons, energetic α-particles, and γ-rays.

The only decay modes that may be affected by the extranuclear environment are electron capture decay and internal conversion of isomers. In both types of decay, the electron density in the vicinity of the nucleus may affect the probability of decay (Hahn et al., 1976). However, variations of the decay constants of isomeric transitions are of no consequence in geochronometry because the "parent" is merely an excited nuclear state of the same isotope as its "daughter."

Several studies have shown that the decay constants of ^{7}Be and ^{131}Ba, which decay by electron capture, increase very slightly when these atoms are subjected to high pressures of the order of 100 kbars or more. A case in point is the work of Hensley et al. (1973) with ^{7}Be, which decays by electron capture of ^{7}Li. These authors observed an increase of 0.59 percent in the decay constant of ^{7}Be when this nuclide (in the form of BeO) was subjected to a pressure of 270 ± 10 kbars in a diamond-anvil press. The only naturally occurring radioactive nuclide used for dating that decays by electron capture is ^{40}K, which forms ^{40}Ar. However, there is no evidence that potassium now residing in the crust of the Earth has been subjected to pressures of the order of several hundred kilobars for a sufficient length of time to affect the amount of radiogenic ^{40}Ar produced. Consequently, there is no reason to doubt that the decay constants of the naturally occurring long-lived radioactive isotopes used for dating are invariant and independent of the physical and chemical conditions to which they may have been subjected since nucleosynthesis. Obviously, this statement does not apply to atoms in the Sun or in other stars where electron capture and internal conversion decay modes may be inhibited by the removal of all or most electrons from the atoms.

4.2c Initial Abundance of Radiogenic Daughters

The dates calculated from equation 4.9 require information concerning the number of radiogenic daughters (D_0) per unit weight that were present at the time of crystallization of the rock or mineral being dated. This parameter can be estimated in some cases, whereas in others it is actually determined from the measurements from which the date is calculated. For example, when the decay of ^{40}K to ^{40}Ar is used for dating K-rich minerals (e.g., biotite, muscovite, lepidolite), virtually all of the ^{40}Ar present in these minerals is radiogenic, and therefore $D_0 = 0$. In other words, no ^{40}Ar was incorporated into the minerals at the time of their crystallization because Ar is a noble gas whose solubility in silicate melts is very low. Similarly, zircon crystals exclude Pb^{2+} ions but admit U^{4+} and Th^{4+} ions, in accordance with Goldschmidt's rules of substitution (Faure, 1998). Therefore, virtually all of the Pb in zircon crystals is a decay product of U and Th, making U–Pb and Th–Pb dates insensitive to the value of D_0 (i.e., $^{206}Pb_0$, $^{207}Pb_0$, and $^{208}Pb_0$). The initial abundances of radiogenic daughters can also be determined by the isochron method of dating illustrated in Figure 4.2.

4.2d Isochrons

Age determinations by the Rb–Sr, Sm–Nd, Lu–Hf, and Re–Os methods are calculated in most cases from analytical data derived from *suites* of whole-rock samples of igneous rocks assumed to be cogenetic. Samples that satisfy this criterion have the same age (t) and are assumed to have the same value of D_0. Therefore, any two samples of cogenetic igneous rock satisfy equation 4.8:

$$D = D_0 + N(e^{\lambda t} - 1)$$

which can be used to calculate both t and D_0. In actual practice, suites of five or more cogenetic rock samples are analyzed, and the resulting values of D and N are used to define isochrons from which D_0 and t are determined by a statistical procedure explained below.

Equation 4.8 represents a family of straight lines in the slope–intercept form

$$y = mx + b \qquad (4.10)$$

where b is the intercept on the y-axis and $m = e^{\lambda t} - 1$ is the slope. Therefore, sets of measured values of N and D, derived from a suite of cogenetic igneous rocks, define a straight line whose slope ($e^{\lambda t} - 1$) and y-intercept (D_0) can be determined by least-squares regression (Figure 4.2). Since the slope of a straight line is constant, all rock samples that lie on the straight line have the same age t. Therefore, this line is called the isochron. In addition, all rock samples that define an isochron also have the same value of D_0. The time elapsed since all specimens in that suite of rocks had the same value of D_0 is calculated from the slope of the isochron, $m = e^{\lambda t} - 1$:

$$t = \frac{1}{\lambda} \ln(m + 1) \qquad (4.11)$$

The isochron method of dating also permits an evaluation of the assumption that all of the specimens in the suite of rocks being dated actually have the same values of D_0 and t. Any specimen that deviates from the isochron by more than the analytical errors of D and N is violating one or more of the assumptions of dating. In addition, data points may scatter above or below the isochron because the rocks do not all have the same value of D_0 even though they are cogenetic. Experience indicates that, in some cases, lava flows erupted by the same volcano in a short interval of time have different values of D_0. This phenomenon reduces the accuracy of dates derivable by the isochron method but is an important clue in the petrogenesis of volcanic and plutonic igneous rocks. Although clastic sedimentary and high-grade metamorphic rocks in some cases also form straight lines on the isochron diagram, the significance of dates derivable from them is open to question. Therefore, the interpretation of such dates is deferred to the chapters concerned with the various isotopic dating methods.

4.2e Terminology

The dates derived from equation 4.9 or 4.11 are expressed in units of kiloyears (ka), megayears (Ma), and gigayears (Ga):

$$1 \text{ ka} = 10^3 \text{ y} \qquad 1 \text{ Ma} = 10^6 \text{ y} \qquad 1 \text{ Ga} = 10^9 \text{ y}$$

These units specify dates in the past and should not be followed by the word *ago*. In addition, they do not represent intervals of time. For example, the preferred expression is "The age of this rock is 100 Ma, which means that it has existed for 100 million years (My)."

In addition, the word *date* is not synonymous with the word *age* because a suite of rocks can have a spectrum of dates that record events in their geological history as revealed by different isotopic chronometers. In some cases, none of the isotopic dates corresponds to the age of the rocks, depending on how that term is defined by the investigator. Instead, different isotopic chronometers can provide information concerning magma sources or protoliths, the date of crystallization of the minerals, dates of subsequent isotopic re-equilibration during episodes of metamorphism, cooling of the rocks after initial crystallization or after their last episode of metamorphism, as well as evidence for hydrothermal alteration and/or chemical weathering. Therefore, isotopic methods of dating can provide much more information about a body of cogenetic rocks than merely their age.

4.3 FITTING OF ISOCHRONS

The dating of rocks or minerals begins with the selection of a suite of samples that, on the basis of prior geological evaluation, are likely to satisfy the assumption that all specimens chosen for analysis formed at the same time, had the same initial abundance of the radiogenic daughters, and are likely to have remained closed systems. Samples that show evidence of chemical weathering or of other forms of postdepositional alteration must be excluded at the outset. After the values of D and N of the samples have been determined and have been plotted on the isochron diagram, the problem arises of fitting the "best" straight line to the data points. The fit of the data points to a straight line is never "perfect" because of errors arising from the analyses of the samples. These analytical errors give rise to corresponding uncertainties in the date and initial value of D that can be derived from the analytical data.

The term *analytical error* generally means the deviation of a measured value from its true value. Such errors may be random or systematic. Random errors have a "normal," or Gaussian, distribution about the arithmetic mean of the measurements that approaches the true value as the number of measurements increases. Systematic errors, on the other hand, are consistent differences between the true value and a set of measurements such that their arithmetic mean is displaced from the true value. Therefore, random errors determine the precision of a set of measurements while systematic errors limit their accuracy. Systematic errors in the measurements of D and N lead to similar systematic errors in the value of D_0 and in the date derived from an isochron. Such errors must obviously be eliminated as nearly as possible to assure the accuracy of the results. The important considerations in fitting isochrons are

1. how to use the analytical errors of the coordinates in determining the best slope and intercept, and
2. how to decide whether a particular point fits the isochron within the analytical errors.

If it can be shown that a point does not fit the isochron within random analytical errors, then either the coordinates of that data point have systematic analytical error or the sample does not satisfy the prerequisite assumptions of dating by the isochron method. It may have a different age or a different value of D_0 compared with other samples in the suite or it may not have remained a closed system.

The simplest method of fitting an isochron to data points is to draw the best straight line by eye on a piece of graph paper. If the scale is sufficiently large, the slope and intercept can be determined from the graph with enough accuracy to serve at least as a good first approximation. This method is useful for obtaining preliminary estimates of the slope and intercept but should be replaced by a more objective approach in any serious evaluation of isochrons.

4.3a Unweighted Regression

A somewhat better method is the *least-squares* regression procedure, which consists of minimizing the deviations of either the x- or the y-coordinate from the best line in the slope–intercept form. The equations for calculating the best slope m and intercept b of a straight line are

$$m = \frac{\sum XY - (\sum X)(\sum Y)/n}{\sum Y^2 - (\sum X)^2/n} \tag{4.12}$$

$$b = \frac{(\sum X)(\sum XY) - \sum Y(\sum X^2)}{(\sum X)^2 - n\ (\sum X^2)} \tag{4.13}$$

where y represents the N-coordinate in equation 4.8, x is the D-coordinate in the same equation, and n is the number of data points. Some electronic calculators have built-in linear regression functions that can be executed with a few key strokes. The simple least-squares regression is based on the assumption that the deviations of the data points from the best straight line are due only to errors in the y-coordinates and that the x-coordinates are free of error. This is clearly not a good assumption for fitting isochrons because in geochronometry both the x- and y-coordinates have analytical errors.

The preferred method of calculating the slope and intercept of the isochron must take into account the known analytical errors of the coordinates of the data points on the isochron diagram. This is accomplished by the use of weighting factors that are calculated from the reciprocals of the analytical errors of each coordinate. The problem of fitting isochrons and of determining the best slope and intercept from analytical data has been treated elsewhere:

> York, 1966, *Can. J. Phys.*, 46:1845–1847. McIntyre et al., 1966, *J. Geophys. Res.*, 71:5459–5468. York, 1967, *Earth Planet. Sci. Lett.*, 2:479–482. York, 1969, *Earth Planet. Sci. Lett.*, 5:320–324. Williamson, 1969, *Can. J. Phys.*, 46:1845–1847. Brooks et al., 1972, *Rev. Geophys. Space Phys.*, 10:551–577. Cameron et al., 1981, *Geochim. Cosmochim. Acta*, 45:1087–1097, and Powell et al., 2002, *Chem. Geol.*, 185:191–204 and additional references cited therein. Serious students of geochronometry may wish to acquire a copy of the *Geochronological Toolkit* prepared by Ludwig (2000).

4.3b Weighted Regression

The formulation of the weighted regression equation by York (1966, 1969) is based on minimizing a parameter S that is the weighted residual sum of squares of data points relative to a straight line $y = mx + b$ (Harmer and Eglington, 1990). The value of this parameter is

$$S = \sum[(Y_i - mX_i - b)^2\ Z_i] \tag{4.14}$$

where Y_i, X_i = measured values of the X and Y parameters of each data point
m = slope of the best-fit straight line
b = intercept on the Y-axis of the best-fit straight line
Z_i = weighting term for each sample in the regression

The parameter Z_i is defined by the equation

$$Z_i = \frac{W_{X_i} W_{Y_i}}{m^2 W_{Y_i} + W_{X_i} - 2m\ r_i\ (W_{X_i} W_{Y_i})^{1/2}} \tag{4.15}$$

where W_{X_i}, W_{Y_i} = weighting factors for X- and Y-coordinates of any data point i
r_i = correlation between analytical errors of X and Y for any sample i

The weighting factors for the X- and Y-coordinates are expressed as

$$W_{x \text{ or } y} = \frac{1}{\sigma^2_{x \text{ or } y}} \tag{4.16}$$

where σ^2 is the variance of the analytical errors of X and Y. Therefore Z_i can be expressed in terms of the analytical errors:

$$Z_i = \frac{1}{m^2\sigma^2_{X_i} + \sigma^2_{Y_i} - 2mr_i\sigma_{X_i}\sigma_{Y_i}} \tag{4.17}$$

The regression line must pass through the weighted center of gravity of the measurements whose coordinates X_m and Y_m are

$$X_m = \frac{\sum(Z_i X_i)}{\sum Z_i} \tag{4.18}$$

$$Y_m = \frac{\sum(Z_i Y_i)}{\sum(Z_i)} \tag{4.19}$$

The slope of the weighted regression line obtained by minimizing the parameter S takes the form

$$m = \frac{\sum[Z_i^2 V_i(U_i/W_{Y_i} + mV_i/W_{X_i} - r_i V_i/\alpha_i)]}{\sum[Z_i^2 U_i(U_i/W_{Y_i} + mV_i/W_{X_i} - mr_i V_i/\alpha_i)]} \tag{4.20}$$

where

$$\alpha_i = (W_{X_i} W_{Y_i})^{1/2} \tag{4.21}$$

$$U_i = X_i - X_m \tag{4.22}$$

$$V_i = Y_i - Y_m \tag{4.23}$$

and all summations in equations 4.18–4.20 include all of the samples used in the regression (n).

Equation 4.20 must be solved by iteration, which means that the equation is first solved by substituting an estimated value of m into the right side of equation 4.20 and the results are used to refine m until the differences between successively calculated values are reduced to the desired level of precision.

The intercept b of the regression line is calculated from the equation

$$Y_m = mX_m + b \tag{4.24}$$

by substituting the best value of m obtained above and using the coordinates of the weighted center of gravity. Therefore,

$$b = Y_m - mX_m \tag{4.25}$$

The variances σ^2 of the slope and intercept of the weighted regression line are calculated from

$$(\sigma_m)^2 = \frac{1}{\sum(Z_i U_i^2)} \tag{4.26}$$

$$(\sigma_b)^2 = (\sigma_m)^2 \frac{\sum(Z_i X_i^2)}{\sum Z_i} \tag{4.27}$$

Note that Harmer and Eglington (1990) pointed out that equation 4.27 was incorrectly printed in the original formulation of York (1969, p. 323).

This presentation of the equations required to carry out a weighted regression of analytical data illustrates the fact that these calculations are laborious and therefore require the use of appropriate computer software. All of the currently active research groups engaged in geochronometry by the abundances of stable radiogenic isotopes use computer programs that, in many cases, are based on the equations of York (1966, 1969). Some of the computer codes being used were originally provided by D. York (University of Toronto), G. A. McIntyre (CSIRO, Canberra), I. Wendt (Bundesanstalt für Bodenforschung, Hannover), and C. Brooks (Carnegie Institution of Washington). The dates obtained from a given data set by these weighted regression codes may differ slightly in some cases because of differences in computational strategies and undocumented magnitudes of the analytical errors. Harmer and Eglington (1990) recommended that the calculation of dates by weighted-regression procedures should be standardized to permit accurate comparisons of dates reported by different laboratories. Computer codes that could be adopted for this purpose have been published by Eglington and Harmer (1989) and Ludwig (2000).

4.3c Goodness of Fit

The goodness of fit of a data set to a straight line is indicated by the *mean sum of weighted deviations* (MSWD), defined as

$$\text{MSWD} = \frac{S}{n-2} \tag{4.28}$$

where S is defined by equation 4.14, $n-2$ is the number of degrees of freedom of S, and n is the number of samples being regressed (Harmer and Eglington, 1990). The numerical value of MSWD calculated from equation 4.28 depends on both the number of samples being regressed (n) and the magnitudes of the analytical errors that determine the weighting factors used to calculate Z_i.

When the regression is based on a very large number of samples ($n = \infty$) and the analytical errors are based on a large number of replicates, the value of the MSWD should be equal to or less than 1. In that case, the scatter of the data points above and below the best-fit line is consistent with the analytical errors of X and Y and the line is an isochron.

In cases where the number of samples and the number of replicates are both small, the limiting value of the MSWD required for the line to qualify as an isochron increases. Brooks et al. (1972) constructed a table that gives the expected values of MSWD at the 95 percent confidence limit for different values of the number of samples (n_s) and the number of replicates (n_r). For example, the table predicts that a well-defined isochron based on five samples and five replicates has an MSWD equal to 5.41 at the 95 percent confidence limit. In cases where the MSWD calculated from such a data set ($n_s = 5$, $n_r = 5$) is greater than 5.41, the scatter of the data points is greater than expected. In such cases, the deviations of the data points from the best-fit line are caused not only by analytical errors but also by geological errors. In other words, the samples that were analyzed do not satisfy the assumptions of dating. In this case, the best-fit line is not an isochron and the date derived from it is not a reliable age determination. Instead, such lines are called "errorchrons" and dates derived from them are suspect.

The graph in Figure 4.3 demonstrates the relation between the limiting value of the MSWD and the number of samples n_s for different numbers of duplicates n_r. If the analytical errors are well known from a large number of duplicates (e.g., $n_r = 120$), then the limiting value of the MSWD for 10 samples is 2.02 at the 95 percent confidence limit. However, if only 10 duplicates were analyzed, the limiting MSWD is 3.07. The dependence of the limiting value of the MSWD on the number of replicates n_r requires that the magnitudes of analytical errors should be determined for each data set by replicating the

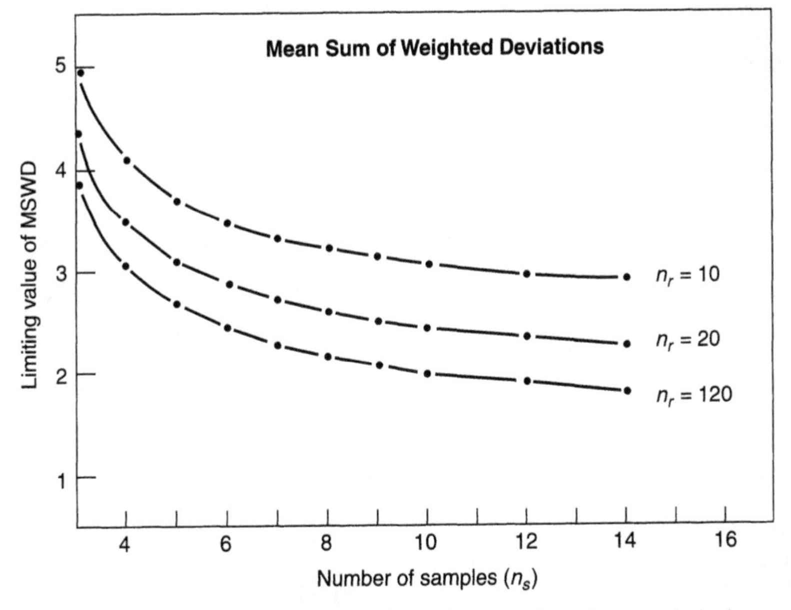

FIGURE 4.3 Dependence of the MSWD of isochrons defined by samples whose analytical errors account for the scatter of data points about the best-fit straight line at the 95 percent confidence limit. In cases where the MSWD calculated from equation 4.28 exceeds the limiting value of the MSWD for the same number of samples (n_s) and duplicates (n_r), the line is not an isochron but is called an errorchron and the date derived from it is questionable. Based on calculations by Brooks et al. (1972) and Harmer and Eglington (1990).

analyses many times. This requirement is difficult to satisfy because of the excessive amount of analytical labor. Therefore, geochronometry laboratories assign analytical errors based on past experience and determine limiting values of MSWD for arbitrarily selected values of n_r (e.g., 20, 40, 60).

In summary, the evaluation of the goodness of fit of a given data set to a best-fit line relies on the comparison between the MSWD calculated from equation 4.28 and a statistically predicted value of the MSWD whose magnitude depends on the number of samples and the number of replicates. This procedure becomes flawed when the magnitudes of the analytical errors associated with each sample are not determined by replication but are assigned on the basis of past experience.

4.4 MASS SPECTROMETRY AND ISOTOPE DILUTION

The development of isotopic methods of dating rocks and minerals was made possible by the invention of mass spectrometers in the early part of the twentieth century. The first instrument was built by J. J. Thomson at the Cavendish Laboratory of the University of Cambridge and was used to demonstrate that neon has two isotopes of differing masses ($^{20}_{10}Ne$ and $^{22}_{10}Ne$). Thomson's "positive-ray apparatus" used photographic emulsions to detect the neon isotopes and therefore was a mass spectrograph. Modern versions of this instrument use electronic methods of measuring the intensities of the ion beams and are called mass spectrometers.

Thomson's work was followed up by F. W. Aston in England and by A. J. Dempster at the University of Chicago. Aston (1919) and Dempster (1918) designed mass spectrographs they used in subsequent years to discover the naturally occurring isotopes of most of the elements in the periodic table and to measure their masses and their abundances. The design of mass spectrographs was further improved in the 1930s by K. T. Bainbridge, J. Mattauch, and R. Herzog. At the end of that decade, the work of discovering the naturally occurring isotopes of the elements and of measuring their masses and abundances was virtually complete. Since then, mass spectrometers have been employed in a wide range of research problems in physics, chemistry, and biology. In addition, mass spectrometers based on a design by Nier (1940) made possible the measurement and interpretation of variations in the isotopic compositions of certain elements in natural materials and thus permitted the spectacular growth of isotope geology. The design of modern mass spectrometers has been treated by Duckworth (1958), McDowell (1963), Milne (1971), and others.

4.4a Principles of Mass Spectrometry

A mass spectrometer is an instrument designed to separate charged atoms and molecules on the basis of their masses based on their motions in magnetic fields. Most of the mass spectrometers currently in use in isotope geology have evolved from the work of Dempster and Bainbridge and follow the design of Nier (1940), whose mass spectrometers achieved a high level of accuracy and reliability of operation.

The modern Nier-type mass spectrometer consists of three essential parts: (1) a source of a monoenergetic beam of ions, (2) a magnetic analyzer, and (3) an ion collector, identified in Figure 4.4. All three parts of the mass spectrometer are evacuated to pressures of the order of $10^{-6}-10^{-9}$ mm Hg. Both gaseous and solid samples can be analyzed, depending on the design of the ion source. For a mass analysis of a gaseous sample, such as Ar or CO_2, the sample gas is allowed to leak into the source through a small orifice while the system is being pumped (dynamic analysis) or a small amount of gas is admitted into the mass spectrometer with the pump valves closed (static analysis). The molecules are then ionized by bombardment with electrons. The resulting positively charged ions are accelerated by a high-voltage electric field and are collimated into a beam by means of suitably spaced slit plates. For the mass analysis of solid samples, a salt of the element is deposited on a filament that is then mounted in the source. The filament (composed of Ta, Re, or W) is heated electrically to a temperature sufficient to volatilize the element to be analyzed. The high temperature of the evaporating filament or of an adjacent incandescent filament causes ionization of the atoms in the vapor. The resulting ions are then accelerated and collimated into a beam as before.

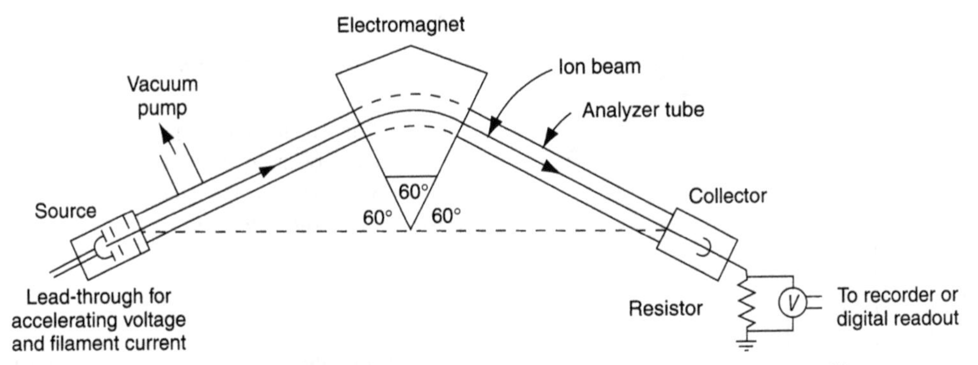

FIGURE 4.4 Schematic diagram of a 60°-sector mass spectrometer showing arrangement of ion source, electromagnet, and collector.

The ion beam enters a magnetic field generated by an electromagnet whose pole pieces are shaped and positioned in such a way that the magnetic field lines are perpendicular to the direction of travel of the ions. The magnetic field deflects the ions into circular paths whose radii are proportional to the masses of the isotopes; that is, the heavier ions are deflected less than the lighter ones. The separated ion beams continue through the analyzer tube to the collector, where they generate a positive electrical charge.

The ion collector consists of a metal cup positioned behind a slit plate. The accelerating voltage in the source and the magnetic field are adjusted in such a way that one of the ion beams is focused through the collector slit and enters the detector cup while the other ion beams collide with a grounded slit plate or with the metallic walls of the flight tube and are neutralized. The beam that enters the collector cup is neutralized by electrons that flow from ground to the collector through a large resistor ($10^{10}-10^{12}$ Ω). The voltage difference generated across the terminals of this resistor is amplified and measured with a digital or analog voltmeter or can be displayed on a stripchart recorder.

A mass analysis of an element (or compound) consisting of several isotopes (or isotopic masses) is obtained by varying either the magnetic field or the accelerating voltage in such a way that the separated ion beams of the element are focused into the collector in succession. The resulting signal consists of a series of peaks and valleys that form the mass spectrum of the element. Each peak represents a discrete mass-to-charge ratio that identifies each isotope in the mass spectrum of the element being analyzed. The height of these peaks is proportional to the relative abundances of the isotopes. Figure 4.5 is such a mass spectrum of Sr consisting of the stable isotopes ^{88}Sr, ^{87}Sr, ^{86}Sr, and ^{84}Sr.

In recent years, the design of mass spectrometers has been improved by the refinement of ion optics and by incorporation of automated data collection systems using computers:

Hagan and deLaeter, 1966, *J. Sci. Instrum.*, 43:662–664. Weichert et al., 1967, *Can. J. Phys.*, 45:2609–2619. Arriens and Compston, 1968, *Int. J. Mass Spectrom. Ion Phys.*, 1:471–481. Wasserburg et al., 1969, *Rev. Sci. Instrum.*, 40:288–295. Stacey et al., 1971, *Can. J. Earth Sci.*, 8:371–377. Stacey et al., 1972, *Can. J. Earth Sci.*, 9:824–834. Schoeller and Hayes, 1975, *Anal. Chem.*, 47:408–415. Sherrill and Dalrymple, 1979, U.S. Geol. Surv. Prof. Paper 1129-A.

In addition, the newest commercially built mass spectrometers are equipped with multiple collectors in which several isotopic ion beams are detected simultaneously. As a result, the precision with which isotope ratios can be measured is improved because fluctuations in beam intensity are averaged out much more effectively than in instruments in which the beam is "scanned" either continuously or stepwise.

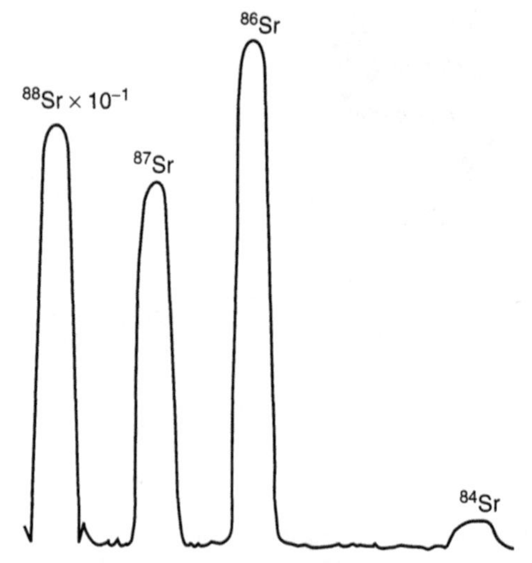

FIGURE 4.5 Mass spectrum of strontium obtained on a 60°-sector, 15.24-cm-radius, Nier-type mass spectrometer equipped with a single-filament source for analysis of solid samples (Nuclide Corp., Model 6-60-S). The spectrum was scanned by continuous variation of the magnetic field at a constant rate and was traced on a linear strip chart recorder. Note that the height of the ^{88}Sr peak has been reduced by a factor of $\frac{1}{10}$. The heights of the peaks are proportional to the abundances of the respective isotopes.

4.4b Equations of Motion of Ions

When an ion of mass m and charge e is acted upon by a potential difference of V volts, it acquires an energy E equal to

$$E = eV = \tfrac{1}{2}mv^2 \tag{4.29}$$

where v is its velocity. All ions having the same charge emerge from the exit slit of the source with the same kinetic energy because they were accelerated by the same potential difference V. However, ions of different masses have differing velocities:

$$v = \left(\frac{2eV}{m}\right)^{1/2} \tag{4.30}$$

When the ions enter the magnetic field, they are deflected into circular paths subject to the condition

$$Bev = m\frac{v^2}{r} \tag{4.31}$$

where B is the strength of the magnetic field and r is the radius of the path of an ion.

The square of the velocity v of the ions, according to equation 4.30, is

$$v^2 = \frac{2eV}{m}$$

In addition, the velocity of the ions derived from equation 4.31 is

$$v = \frac{Ber}{m}$$

$$v^2 = \frac{B^2e^2r^2}{m^2}$$

Therefore, the radius r of the circular path of the ions traveling in the magnetic field is derived from the equation

$$\frac{2eV}{m} = \frac{B^2e^2r^2}{m^2} \tag{4.32}$$

$$r = \frac{1}{B}\left(\frac{2Vm}{e}\right)^{1/2} \tag{4.33}$$

When B is expressed in gauss, r in centimeters, m in atomic mass units, V in volts, and e in units of electronic charge, equation 4.33 takes the form

$$r = \frac{143.95}{B}\left(\frac{mV}{e}\right)^{1/2} \tag{4.34}$$

Equation 4.34 can be used to calculate the magnetic field strength required to focus ions of $^{40}Ca^+$ into the collector of a mass spectrometer whose analyzer tube has a radius of 30.48 cm (12 in.), assuming $V = 1000$ V. Since the charge of the ^{40}Ca ion is $e = 1$ electronic charge unit and since its mass $m =$ 39.9626 amu, the strength of the magnetic field is $B = 944$ G. Equation 4.34 indicates that either the accelerating voltage or the magnetic field can be adjusted to force any ion of mass m and charge e into a path of radius r. The equation also shows that if B and V are constants, the radius of the path of ions having unit charge is proportional to the square root of the mass:

$$r \propto \sqrt[2]{m} \qquad (4.35)$$

Consequently, ions of large mass are deflected into paths having larger radii than lighter ions. In other words, heavy isotopes are deflected from a straight-line path less than light isotopes.

An important feature of Nier-type mass spectrometers is the wedge-shaped magnetic field that is achieved by the design of the pole pieces of the electromagnet. To focus the ion beam into the collector, the exit slit, the sector axis of the pole pieces, and the collector slit must lie on a straight line, as shown in Figure 4.4.

Nier used his mass spectrometers to determine the isotopic compositions of many elements, including Ca, Ti, S, Ar, Sr, Ba, Bi, Tl, Hg, Ne, Kr, Rb, Xe, C, N, O, and K (Nier, 1938a, b, 1950a, b). In addition, Nier and Gulbransen (1939) first reported the variation of the isotopic composition of C in nature. Subsequently, Nier designed mass spectrometers for the simultaneous detection of two ion beams, which permit isotope ratios to be measured instantaneously and precisely (Nier 1947; Nier et al., 1947). Nier's isotope ratio mass spectrometer was further improved by McKinney et al. (1950) and has made possible the precise determination of isotope ratios of D/H, $^{13}C/^{12}C$, $^{18}O/^{16}O$, and $^{34}S/^{32}S$ in geological and biological materials (see Chapters 26, 27, 28, 29, and 30).

4.4c Ion Microprobes

Conventional mass spectrometers generally require that the chemical element to be analyzed must be separated from the other elements in the sample in which it resides. This requirement makes it very difficult to analyze individual mineral grains or to study the variation of the isotopic composition of an element within such a grain. Moreover, it is increasingly difficult to control contamination during chemical processing as the number of atoms in the sample decreases. The ion microprobe was designed to avoid these problems and to permit the study of small mineral grains in terrestrial and extraterrestrial samples.

In the ion microprobe, a beam of negatively charged oxygen atoms is focused onto the gold- or carbon-coated surface of the sample. The size of the illuminated spot ranges from 3 to 10 μm, but scattering of the primary beam may enlarge it. The ions that are sputtered from the surface of the sample are accelerated by an electric field, as in conventional mass spectrometers, and are separated magnetically. The instrument can provide information about both the chemical composition and the isotopic composition of selected elements in the sample (Andersen and Hinthorne, 1972; Meyer et al., 1974; Shimizu et al., 1978).

Early models of ion probe mass spectrometers encountered difficulties that limited their usefulness in isotopic studies: (1) isobaric interference from ionized molecular species and multiply charged ions, (2) inadequate resolution, (3) instability of ion beams and low measurement precision, and (4) scattering of the primary beam causing emission of ions from neighboring mineral grains or from inclusions outside of the illuminated spot. Most of these problems have been reduced or eliminated and the ion microprobe has become an important tool for the study of very small samples.

For example, Hutcheon (1982) and Huneke et al. (1983) used an ion microprobe to detect very large enrichments of stable ^{26}Mg in certain Ca–Al-rich inclusions of the carbonaceous chondrite Allende (Section 24.2). The anomalous abundance of ^{26}Mg is attributable to the decay of its parent ^{26}Al in the inclusions, which therefore must have formed in the solar nebula before the short-lived ^{26}Al ($T_{1/2} = 0.716 \times 10^6$ y) had decayed. Hence, these inclusions are among the oldest objects in the solar system and have preserved information about its earliest history.

Another interesting application of the ion microprobe mass spectrometer is a study by Hart et al. (1981) of a galena crystal from the Buick mine of southeastern Missouri (Section 11.5a). Their data revealed that the crystal (12 × 13 mm in size) is concentrically zoned in terms of the isotope ratios of lead and that the range of variation is nearly as large as that previously reported for the entire mining district. This crystal, therefore, contains a record of the chronological changes of lead isotope compositions in the ore-forming fluid and constrains the hydrological models that have been proposed to explain the formation of these important lead–zinc deposits.

4.4d Tandem-Accelerator Mass Spectrometers

The problems with isobaric interference from ionized molecules having the same mass number as the isotope of a particular element have been completely eliminated by ultrasensitive tandem mass spectrometers using high-voltage accelerators. Such instruments have been designed and built by groups of scientists at the University of Rochester, the University of Toronto, Oxford University, and the University of Arizona (Purser et al., 1982). In this instrument, negatively charged ions are generated by bombarding a solid sample with a beam of cesium ions accelerated to energies of 30 keV. Ions having the desired mass are separated magnetically in the first of two mass spectrometers and are then accelerated by an electric field of several million volts. At that point, all particles having the same mass as the desired isotope are stripped of several electrons in a differentially pumped gas cell or by collision with a metal foil. This causes molecules to disintegrate into their constituent monoatomic ions. The resulting mixture of positively charged ions is further accelerated and then sorted in the second mass spectrometer, which allows only ions with the desired mass-to-charge ratio to pass into a solid-state nuclear detector that is capable of distinguishing among isobaric ions on the basis of their energy, rate of energy loss, and mass number. This equipment is capable of identifying and counting individual ions as they arrive in the nuclear detector.

An important application of this instrument has been to determine the abundances of short-lived radioactive atoms produced by cosmic rays in the atmosphere. The best-known of these cosmogenic radionuclides is ^{14}C (Section 23.1), which is used for dating of carbon-bearing materials that formed on the surface of the Earth in the past 35,000 years or so. In a conventional mass spectrometer, ^{14}C would be difficult to detect because of its exceedingly low abundance and because ^{14}C would be masked by the presence of ^{14}N. The tandem-accelerator mass spectrometer avoids the interference from ^{14}N by using negatively charged ions in its first stage because negatively charged ^{14}N is not stable. Isobaric molecules such as $^{12}CH_2$ ($A = 14$), which would be difficult to separate from ^{14}C, are destroyed by stripping several electrons from them. As a result, ^{14}C is enhanced relative to the background by a factor of up to 10^{17} during its transit through the instrument. Consequently, tandem-accelerator mass spectrometers are used to date carbon samples weighing less than 1 mg, whereas the conventional radiochemical method requires much larger amounts depending on the ^{14}C content of the sample.

The power of this new instrumentation was demonstrated in spectacular fashion when Purser and his colleagues (1982) were able to date 0.9 mg of carbon from 4 mg of muscle tissue of a baby woolly mammoth named Dima, whose frozen body was found in June of 1977 during an excavation in the Magadan region of northeastern Siberia. Using their newly built ultrasensitive mass spectrometer, they determined that Dima had died 27,000 ± 1000 years ago. Additional applications of accelerator mass spectrometry have been discussed by Brown (1984) and by Litherland (1979).

4.4e Isotope Dilution Analysis

The concentrations of the parent and daughter elements in rocks and minerals to be dated can be determined by one of several methods: instrumental neutron activation (INA; Section 3.8), x-ray fluorescence (XRF), inductively coupled plasma mass spectrometry (ICP-MS), inductively coupled plasma optical spectrometry (ICP-OES), atomic absorption spectrometry (AAS), and flame photometry (FP). The choice depends on the concentrations of the elements to be determined, the amount of sample available for analysis, the lower limit of detection (sensitivity) for the elements to be determined, and the availability of the required instrumentation. In spite of the desirable qualities of the analytical tools listed above, isotope dilution is still the preferred method of analysis for use in geochronometry due to the following features:

1. Very low concentrations of many elements can be measured with great accuracy.

2. Small amounts of sample weighing only a few milligrams can be analyzed provided that background contamination is minimized.
3. The accuracy of the method depends on the measured weights of primary standards.
4. The method is free of interferences by other elements because the analyte is recovered in pure form by ion exchange chromatography prior to being analyzed on the mass spectrometer.
5. Several elements can be determined from a single dissolution.
6. The isotope composition of the daughter element can be calculated from the results of an isotope dilution analysis.

Isotope dilution is based on the determination of the isotopic composition of an element in a mixture of a known quantity of a "spike" with an unknown quantity of the normal element. The spike is a solution (liquid or gaseous) containing a known concentration of a particular element whose isotopic composition has been changed by the enrichment of one of its isotopes. The sample to be analyzed contains an unknown concentration of the element whose isotopic composition is known. Therefore, when a known weight of a sample is dissolved and the solution is mixed with a known amount of spike, the isotopic composition of the mixture can be used to calculate the amount of the element in the sample. Isotope dilution analysis is possible for elements that have two or more isotopes provided that a spike enriched in one of the isotopes of that element is available (Moore et al., 1973). The isotope enrichment is usually achieved by means of a large electromagnetic separator called a calutron that is capable of separating the isotopes of an element quantitatively.

When the element to be analyzed is nonvolatile, the spike is prepared in the form of a solution whose concentration and isotopic composition are verified separately. A known weight (or volume) of this spike is then added to a known weight (or volume) of the sample in solution, and the mixture is stirred to achieve complete isotopic homogenization. The element to be determined (analyte) is separated from the spiked solution by cation exchange chromatography, and the mixture of the normal element and the spike is then analyzed on a mass spectrometer to determine its isotopic composition. The result, expressed in terms of the ratio of the abundance of two isotopes, is used to calculate the concentration of the analyte in the sample.

Let R be the ratio of the abundances of two isotopes A and B of an element and let N and S be the numbers of atoms of the normal element and of the spike, respectively, in the mixture. Then

$$R_m = \frac{\mathrm{Ab}_N^{\mathrm{A}} N + \mathrm{Ab}_S^{\mathrm{A}} S}{\mathrm{Ab}_N^{\mathrm{B}} N + \mathrm{Ab}_S^{\mathrm{B}} S} \tag{4.36}$$

where $\mathrm{Ab}_N^{\mathrm{A}}$ is the abundance of isotope A in the normal element and so forth and R_m is the ratio of isotopes A and B in the mixture. The two isotopes (A and B) of the analyte are chosen such that A is the isotope which is enriched in the spike and B is a nonradiogenic isotope of the same element. After R_m has been measured on a mass spectrometer, the number of atoms of the analyte in the sample (N) can be calculated by solving equation 4.36:

$$N = S\left[\frac{\mathrm{Ab}_S^{\mathrm{A}} - R_m \mathrm{Ab}_S^{\mathrm{B}}}{R_m \mathrm{Ab}_N^{\mathrm{B}} - \mathrm{Ab}_N^{\mathrm{A}}}\right] \tag{4.37}$$

where N and S are in terms of numbers of atoms. Therefore, the corresponding weights of N and S in grams are obtained by dividing each by Avogadro's number (A) and by multiplying the result by the atomic weights of the analyte in the sample (W_N) and in the spike (W_S). The atomic weights of N and S are not identical because of the difference in their isotopic composition:

$$N_W = \frac{N \times W_N}{A}$$

$$N = \frac{N_W A}{W_N} \quad \text{and} \quad S = \frac{S_W A}{W_S}$$

When these expressions are substituted into equation 4.37, Avogadro's number cancels:

$$\frac{N_W}{W_N} = \frac{S_W}{W_S}\left[\frac{\mathrm{Ab}_S^{\mathrm{A}} - R_m \mathrm{Ab}_S^{\mathrm{B}}}{R_m \mathrm{Ab}_N^{\mathrm{B}} - \mathrm{Ab}_N^{\mathrm{A}}}\right]$$

from which it follows that

$$N_W = S_W\left(\frac{W_N}{W_S}\right)\left[\frac{\mathrm{Ab}_S^{\mathrm{A}} - R_m \mathrm{Ab}_S^{\mathrm{B}}}{R_m \mathrm{Ab}_N^{\mathrm{B}} - \mathrm{Ab}_N^{\mathrm{A}}}\right] \tag{4.38}$$

where N_W and S_W are the weights of the normal element and the spike in the mixture in units of grams, milligrams, or micrograms, depending on the most convenient way of expressing the concentration of the spike solution. The concentration of the analyte in the sample is obtained by dividing N_W in micrograms (10^{-6} g) by the weight of the sample in grams. The resulting concentration units are microgram per gram or parts per million (ppm). Alternatively, the weight of the element (N_W) can also be expressed in nanograms (10^{-9} g) or picograms (10^{-12} g) and the corresponding units of concentration are parts per billion (ppb) or parts per trillion (ppt), respectively.

For example, the Rb concentration of a rock or mineral sample can be determined by use of a spike enriched in one of its two naturally occurring isotopes, whose abundances are ${}^{87}_{37}\mathrm{Rb}_N = 27.83\%$ and ${}^{85}_{37}\mathrm{Rb}_N = 72.17\%$. The spike is enriched in ${}^{87}_{37}\mathrm{Rb}$ rather than in ${}^{85}\mathrm{Rb}$ in order to enhance the sensitivity of the analysis and has the following isotope composition: ${}^{87}_{37}\mathrm{Rb}_S = 99.40\%$ and ${}^{85}_{37}\mathrm{Rb}_S = 0.60\%$. The spike solution contains 7.50 μg of Rb_S per gram of solution. This value was determined by analyzing a mixture of the spike solution and a standard solution of normal Rb prepared by dissolving a known weight of a stoichiometric Rb salt (e.g., RbCl) in a known a volume of deionized water.

In the example discussed here, 3.50 g of the spike solution was added to 0.25 g of the powdered rock sample, which was then digested in a mixture of hydrofluoric acid (HF) and perchloric acid ($HClO_4$). The residue was dissolved in 2 N HCl and Rb was separated from the solution by cation exchange chromatography. After the normal and spike components of Rb had mixed during the digestion of the rock sample, complete recovery of Rb from the ion exchange column was not required. Therefore, an aliquot of the isotopically mixed Rb was analyzed on a "solid source" mass spectrometer (Faure and Powell, 1972). The results indicated that the mixture had a ${}^{87}\mathrm{Rb}/{}^{85}\mathrm{Rb}$ ratio (R_m) of 1.55

According to the statement of the problem, $S_W = 26.26$ μg, $\mathrm{Ab}_N^{\mathrm{A}} = 0.2783$ (^{87}Rb), $\mathrm{Ab}_N^{\mathrm{B}} = 0.7217$ (^{85}Rb), $\mathrm{Ab}_S^{\mathrm{A}} = 0.9940$, $\mathrm{Ab}_S^{\mathrm{B}} = 0.0060$, $W_N = 85.4677$, and $W_S = 86.8971$. The atomic weight of the spike rubidium must be calculated from its isotopic composition and the atomic masses of its naturally occurring isotopes. Substituting into equation 4.33 yields

$$N_W = S_W\left(\frac{W_N}{W_S}\right)\left(\frac{\mathrm{Ab}_S^{\mathrm{A}} - R_m\mathrm{Ab}_S^{\mathrm{B}}}{R_m\mathrm{Ab}_N^{\mathrm{B}} - \mathrm{Ab}_N^{\mathrm{A}}}\right)$$

$$= 26.25\left(\frac{85.4677}{86.8971}\right)\left(\frac{0.9940 - 1.55 \times 0.0060}{1.55 \times 0.7217 - 0.2783}\right)$$

$$= 30.25\ \mu\mathrm{g}$$

Therefore, the concentration of Rb in the rock sample is 30.25/0.25 = 121.0 ppm.

The advantages of isotope dilution over other analytical methods have already been identified in this section. The method also has certain difficulties:

1. The calibration of spike solutions is subject to errors that may arise from the possible nonstoichiometric composition of the compound containing the spike, from isotope fractionation during the isotope analysis of the spike on a mass spectrometer, and from errors in weighing and diluting the spike.
2. The measured value of the isotope ratio of the mixture may be affected by fractionation in the mass spectrometer. If the element has only two naturally occurring isotopes, this error cannot be corrected.
3. The normal element and the spike must mix completely. Complete mixing may be difficult to achieve in some geological samples in which the element to be determined resides in refractory minerals or when the element is incorporated into an insoluble precipitate during solution of the sample.
4. The concentration of the spike solution may change as a function of time because of evaporation or by adsorption of the element onto the walls of the container used to store the spike.
5. The isotopic composition of the spike or of the mixture of the normal element and the spike may be changed during the processing of the sample as a result of isotope exchange reactions with the isotopes of that element in the walls of the container.

6. The ratio N/S must be optimized so as to avoid magnification of errors or loss of sensitivity of the functional relationship between R_m and N/S in equation 4.33.
7. The method is time consuming, which discourages replication of analyses and thereby results in inadequate documentation of analytical errors.

Nevertheless, isotope dilution is indispensable in age determinations of rocks and minerals based on radioactive decay.

Isotope dilution has an important advantage over other analytical methods. When elements are determined that have more than two naturally occurring isotopes, two or more isotopic ratios can be measured from which not only the concentration but also the isotopic composition of the normal element can be calculated. Such treatment is especially useful in studies of the isotopic compositions of Sr and Pb, both of which have four naturally occurring isotopes. The procedures for making such calculations have been discussed elsewhere:

Long, 1966, *Earth Planet. Sci. Lett.*, 1:289–292. Boelrijk, 1968, *Chem. Geol.*, 3:323–325. Krogh and Hurley, 1968, *J. Geophys. Res.*, 73:7107–7125. Dodson, 1970, *Geochim. Cosmochim. Acta*, 34:1241–1244. Gale, 1970, *Chem. Geol.*, 6:305–310. Russell, 1971, *J. Geophys. Res.*, 76:4949–4955. Russell, 1977, *Chem. Geol.*, 20:307–314. Cumming, 1973, *Chem. Geol.*, 11:157–165. Hamelin et al., 1985, *Geochim. Cosmochim. Acta*, 49:173–182.

4.5 SUMMARY

The theory and practice of geochronometry presented in this chapter apply to all of the parent–daughter pairs listed in Table 4.1. These geochronometers continue to be used to date terrestrial and extraterrestrial rocks and minerals. The resulting dates have provided information about the age and geological history of the Earth and about the origin of the solar system. Whereas the basic principles are not difficult to learn, considerable skill and experience are required to analyze the samples. In addition, the mass spectrometer and related analytical facilities require a substantial investment of funds that is beyond the means of most universities and research institutes. Therefore, geochronometry has become a team effort combining the skills of field geologists who collect and document the samples with the expertise of analytical chemists and physicists who make the necessary measurements. The final interpretation of the data should be done jointly because only the field geologist can assess the geological significance of the dates whose accuracy and precision depend on the skill of the technical experts.

REFERENCES

Andersen, C. A., and J. R. Hinthorne, 1972. Ion microprobe mass analyzer. *Science*, 175:853–860.

Aston, F. W., 1919. A positive ray spectrograph. *Philos. Mag.*, 38(Ser. 6): 707–714.

Brooks, C., S. R. Hart, and I. Wendt, 1972. Realistic use of two-error regression treatments as applied to rubidium-strontium data. *Rev. Geophys. Space Phys.*, 10:551–577.

Brown, L., 1984. Applications of accelerator mass spectrometry. *Ann. Rev. Earth Planet. Sci.*, 12:39–59.

Dempster, A. J., 1918. A new method of positive ray analysis. *Phys. Rev.*, 11:316–325.

Duckworth, H. E., 1958. *Mass Spectroscopy*. Cambridge University Press, Cambridge, United Kingdom.

Eglington, B. M., and R. E. Harmer, 1989. *GEODATE: A Program for the Processing and Regression of Isotope Data Using IBM-Compatible Microcomputers*, CSIR Manual EMA-H 8901. EMAtek, CSIR, Pretoria, South Africa.

Faure, G., 1998. *Principles and Applications of Geochemistry*, 2nd ed. Prentice-Hall, Upper Saddle River, New Jersey.

Faure, G., 2001. *Origin of Igneous Rocks; the Isotopic Evidence*. Springer-Verlag, Heidelberg.

Faure, G., and J. L. Powell, 1972. *Strontium Isotope Geology*. Springer-Verlag, Heidelberg.

Hahn, H. P., H. J. Born, and J. I. Kim, 1976. Survey of the rate perturbation of nuclear decay. *Radiochim. Acta*, 23:23–37.

Harmer, R. E., and B. M. Eglington, 1990. A review of the statistical principles of geochronometry: Towards a more consistent approach for reporting geochronological data. *S. Afr. J. Geol.*, 93:845–856.

Hart, S. R., N. Shimizu, and D. A. Sverjensky, 1981. Lead isotope zoning in galena: An ion microprobe study of a galena crystal from the Buick Mine, southeast Missouri. *Econ. Geol.*, 76:1873–1878.

Hensley, W. K., W. A. Bassett, and J. R. Huizenga, 1973. Pressure dependence of the radioactive decay constant of beryllium-7. *Science*, 181:1164–1165.

Huneke, J. C., J. T. Armstrong, and G. J. Wasserburg, 1983. FUN with PANURGE: High mass resolution ion microprobe measurements of Mg in Allende inclusions. *Geochim. Cosmochim. Acta*, 47:1635–1650.

Hutcheon, I. D., 1982. Ion probe magnesium isotopic measurements of Allende inclusions. In L. A. Currie (Ed.), *Nuclear and Chemical Dating Techniques: Interpreting the Environmental Record*, American Chemical Society Symposium Series, No. 176, pp. 96–128. American Chemical Society, Washington, D.C.

Litherland, A. E., 1979. Dating methods of Pleistocene deposits and their problems: The promise of atom counting. *Geosci. Can.*, 6:80–82.

Ludwig, K. R., 2000. *Isoplot/Ex Version 2.4. A Geochronological Tool Kit for Microsoft Excel*, Spec. Pub. 56. Berkeley Geochronological Centre, Berkeley, California.

McDowell, C. A., 1963. *Mass Spectrometry*. McGraw-Hill, New York.

McKinney, C. R., J. M. McCrea, S. Epstein, H. A. Allen, and H. C. Urey, 1950. Improvements in mass spectrometers for the measurements of small differences in isotope abundance ratios. *Rev. Sci. Instrum.*, 21:724–730.

Meyer, C. Jr., D. H. Anderson, and J. G. Bradley, 1974. Ion microprobe mass analyses of plagioclase from "non-mare" lunar samples. *Geochim. Cosmochim. Acta*, 1(Suppl. 5): 685–706.

Milne, G. W. (Ed.), 1971. *Mass Spectrometry: Techniques and Applications*. Wiley-Interscience, New York.

Moore, L. J., J. R. Moody, I. L. Barnes, J. W. Gramlich, T. J. Murphy, P. J. Paulsen, and W. R. Shields, 1973. Trace determination of rubidium and strontium in silicate glass standard reference materials. *Anal. Chem.*, 45:2384–2387.

Nier, A. O. , 1938a. The isotope constitution of calcium, titanium, sulphur, and argon. *Phys. Rev.*, 53:282–286.

Nier, A. O., 1938b. Isotopic constitution of strontium, barium, bismuth, thallium, and mercury. *Phys. Rev.*, 54:275–278.

Nier, A. O., 1940. A mass spectrometer for routine isotope abundance measurements. *Rev. Sci. Instrum.*, 11:212–216.

Nier, A. O., 1947. A mass spectrometer for isotope and gas analysis. *Rev. Sci. Instrum.*, 18:398–411.

Nier, A. O, 1950a. A redetermination of the relative abundances of the isotopes of carbon, nitrogen, oxygen, argon, and potassium. *Phys. Rev.*, 77:789–793.

Nier, A. O., 1950b. A redetermination of the relative abundances of the isotopes of neon, krypton, rubidium, xenon, and mercury. *Phys. Rev.*, 79:450–454.

Nier, A. O., and E. A. Gulbransen, 1939. Variations in the relative abundance of the carbon isotopes. *J. Am. Chem. Soc.*, 61:697–698.

Nier, A. O., E. P. Ney, and M. G. Inghram, 1947. A null method for the comparison of two ion currents in a mass spectrometer. *Rev. Sci. Instrum.*, 18:294–297.

Purser, K. H., C. J. Russo, R. B. Liebert, H. E. Gove, D. Elmore, R. Ferraro, A. E. Litherland, R. P. Beukens, K. H. Chang, L. R. Kilius, and H. W. Lee, 1982. The application of electrostatic tandems to ultrasensitive mass spectrometry and nuclear dating. In L. A. Currie (Ed.), *Nuclear and Chemical Dating Techniques: Interpreting the Environmental Record*, American Chemical Society Symposium Series, No. 176, pp. 45–74. American Chemical Society, Washington, D.C.

Shimizu, N., M. P. Semet, and C. J. Allègre, 1978. Geochemical applications of quantitative ion-microprobe analyses. *Geochim. Cosmochim. Acta*, 42:1321–1334.

Walker, F. W., J. R. Parrington, and F. Feiner, 1989. *Chart of the Nuclides*. General Electric Co., San Jose, California.

York, D., 1966. Least-squares fitting of a straight line. *Can. J. Phys.*, 44:1079–1086.

York, D., 1969. Least-squares fitting of a straight line with correlated errors. *Earth Planet. Sci. Lett.*, 5:320–324.

PART II

Radiogenic Isotope Geochronometers

Under favorable conditions, the long-lived radionuclides and their stable radiogenic daughters listed in Table 4.1 satisfy the assumptions for dating discussed in Chapter 4. Therefore, these parent–daughter pairs are useful for dating cogenetic igneous and high-grade metamorphic rocks based on the analysis of whole-rock or mineral samples. However, each method has certain unique features that arise from the chemical properties of the parent and daughter elements. These properties govern their abundances in different kinds of rocks and minerals and thereby predispose each geochronometer for dating of certain kinds of rocks.

CHAPTER 5

The Rb–Sr Method

ALTHOUGH the natural radioactivity of rubidium (Rb) was demonstrated in 1906 by Campbell and Wood, more than 30 years elapsed before ^{87}Rb was identified as the naturally occurring radioactive isotope (Hahn et al., 1937; Mattauch, 1937). The feasibility of dating Rb-bearing minerals by the decay of ^{87}Rb to ^{87}Sr was subsequently discussed by Hahn and Walling (1938), and the first age determination by this method followed a few years later (Hahn et al., 1943). However, the Rb–Sr method of dating did not come into wide use until the 1950s, when mass spectrometers based on Nier's design became available for the isotopic analysis of solids, and the concentrations of Rb and Sr could be measured by isotope dilution combined with the separation of these elements by cation exchange chromatography. The entire subject of dating by the Rb–Sr method, including its history, theoretical basis, and applicability, was presented in detail by Faure and Powell (1972). The isotope geochemistry of Sr in the oceans is presented in Sections 19.1 and 19.2.

5.1 GEOCHEMISTRY OF Rb AND Sr

Rubidium is an alkali metal belonging to group IA, which consists of Li, Na, K, Rb, Cs, and Fr. The ionic radius of Rb^+ (1.48 Å) is sufficiently similar to that of K^+ (1.33 Å) to allow it to substitute for K^+ in all K-bearing minerals. Consequently, Rb is a dispersed element that does not form minerals of its own, but it occurs in easily detectable amounts in common K-bearing minerals such as the micas (muscovite, biotite, phlogopite, and lepidolite), K-feldspar (orthoclase and microcline), certain clay minerals, and evaporite minerals such as sylvite and carnallite.

Rubidium has two naturally occurring isotopes $^{85}_{37}Rb$ and $^{87}_{37}Rb$, whose isotopic abundances are 72.1654 and 27.8346 percent, respectively. Its atomic weight is 85.46776 amu (Catanzaro et al., 1969). Rubidium-87 is radioactive and decays to stable ^{87}Sr by emission of a negative β-particle:

$$^{87}_{37}\text{Rb} \rightarrow {}^{87}_{38}\text{Sr} + \beta^- + \overline{\nu} + Q \qquad (5.1)$$

where β^- is the β-particle, $\overline{\nu}$ is an antineutrino, and Q is the decay energy (Section 2.1a). The decay energy is only 0.275 MeV, which has caused problems in the determination of the specific decay rate of this isotope.

Strontium is a member of the alkaline earths of group IIA, which consists of Be, Mg, Ca, Sr, Ba, and Ra. The ionic radius of Sr^{2+} (1.13 Å) is slightly larger than that of Ca^{2+} (0.99 Å), which it can replace in many minerals. Strontium is also a dispersed element and occurs in Ca-bearing minerals such a plagioclase, apatite, and calcium carbonate, especially aragonite. The ability of Sr^{2+} to replace Ca^{2+} is somewhat restricted by the fact that Sr ions favor eightfold coordinated sites, whereas Ca ions can be accommodated in both six- and eightfold coordinated lattice sites because of their smaller size. Moreover, Sr^{2+} ions can be captured in place of K^{+} ions by K-feldspar, but the replacement of K^{+} by Sr^{2+} must be coupled by the replacement of Si^{4+} by Al^{3+} to preserve electrical neutrality. Strontium is the major cation in strontianite ($SrCO_3$) and celestite ($SrSO_4$), both of which occur in certain hydrothermal deposits and in carbonate rocks.

Strontium has four naturally occurring isotopes ($^{88}_{38}Sr$, $^{87}_{38}Sr$, $^{86}_{38}Sr$, and $^{84}_{38}Sr$), all of which are stable. Their isotopic abundances are approximately 82.53, 7.04, 9.87, and 0.56 percent, respectively. The isotopic abundances of Sr isotopes are variable because of the formation of radiogenic ^{87}Sr by the decay of naturally occurring ^{87}Rb. For this reason, the isotopic composition of Sr in a rock or mineral that contains Rb depends on the age and Rb/Sr ratio of that rock or mineral.

The average concentrations or Rb, K, Sr, and Ca in different kinds of igneous and sedimentary rocks in Table 5.1 illustrate the general geochemical coherence of Rb with K and of Sr with Ca. The Rb concentrations of common igneous and sedimentary rocks range from less than 1 ppm in ultramafic rocks and carbonates to more than 170 ppm in low-Ca granitic rocks. The concentrations of Sr range from a few parts per million in ultramafic rocks to about 465 ppm in basaltic rocks and reach very high values in carbonate rocks (up to 2000 ppm or more). Evidently, most common rocks contain appreciable concentrations of Rb and Sr of the order of tens to several hundred parts per million. The Rb-rich rocks generally have low Sr concentrations and vice versa. Therefore, the Rb/Sr ratios of common igneous rocks range between wide limits from 0.06 (basaltic rocks) to 1.7 or more in highly differentiated granitic rocks having low Ca concentrations.

During the fractional crystallization of magma, Sr tends to be concentrated in plagioclase, whereas Rb remains in the liquid phase. Consequently, the Rb/Sr ratio of the residual magmatic liquid increases gradually in the course of progressive crystallization of magma. Suites of differentiated igneous rocks therefore tend to have increasing Rb/Sr ratios with increasing degree of differentiation. The highest Rb/Sr ratios, amounting to 10 or higher, occur in late-stage differentiates, including pegmatites.

Table 5.1. Average Concentrations (ppm) of Rb, K, Sr, and Ca in Igneous and Sedimentary Rocks

	Concentration			
Rock Type	Rb	K	Sr	Ca
Ultramafic	0.2	40	1	25,000
Basaltic	30	8,300	465	76,000
High-Ca granitic	110	25,200	440	25,300
Low-Ca granitic	170	42,000	100	5,100
Syenite	110	48,000	200	18,000
Shale	140	26,600	300	22,100
Sandstone	60	10,700	20	39,100
Carbonate	3	2,700	610	302,300
Deep-sea carbonate	10	2,900	2000	312,400
Deep-sea clay	110	25,000	180	29,000

Source: Turekian and Wedepohl, 1961.

5.2 PRINCIPLES OF DATING

The growth of radiogenic ^{87}Sr in a Rb-rich mineral can be described by an equation derivable from the law of radioactivity (Chapter 4). The total number of atoms of ^{87}Sr in a mineral whose age is t years is obtained from equation 4.8:

$$D = D_0 + N\left(e^{\lambda t} - 1\right)$$

Substitution of the appropriate parent and daughter isotopes of Rb and Sr yields

$$^{87}Sr = {}^{87}Sr_0 + {}^{87}Rb\left(e^{\lambda t} - 1\right) \qquad (5.2)$$

where ^{87}Sr = total number of atoms of this isotope in a unit weight of mineral at the present time

$^{87}Sr_0$ = number of atoms of this isotope that were incorporated into the same unit weight of this mineral at the time of its formation

^{87}Rb = number of atoms of this isotope in a unit weight of the mineral at the present time

λ = decay constant of ^{87}Rb in reciprocal years (y^{-1})

t = time elapsed in years since the time of formation of the mineral

It is helpful to divide each term of equation 5.2 by the number of ^{86}Sr atoms in a unit weight of the sample. This is permissible because the number of ^{86}Sr atoms in a unit weight of rock or mineral remains constant even though the abundance of this isotope decreases with time as the number of radiogenic ^{87}Sr atoms increases by decay of ^{87}Rb:

$$\frac{^{87}Sr}{^{86}Sr} = \left(\frac{^{87}Sr}{^{86}Sr}\right)_0 + \frac{^{87}Rb}{^{86}Sr}(e^{\lambda t} - 1) \tag{5.3}$$

Equation 5.3 is the basis for age determinations of rocks and minerals by the Rb–Sr method. The validity of dates calculated from this equation is subject to the assumptions discussed in Section 4.2.

The halflife of ^{87}Rb was difficult to determine because of the low maximum energy of the β^--particles it emits (0.275 MeV) and because of the slow rate of decay. The presently accepted value of the halflife is based on the work of Neumann and Huster (1974), who obtained a value of $(4.88 + 0.06 - 0.10) \times 10^{10}$ years by direct counting of the β-particles emitted by ^{87}Rb. Davis et al. (1977) reported a similar value of $(4.89 \pm 0.04) \times 10^{10}$ from measurements of the number of radiogenic ^{87}Sr atoms that accumulated in 20 g of purified Rb perchlorate in 19 years.

These and other measurements discussed by Steiger and Jäger (1977) caused the Subcommission on Geochronology of the International Union of Geological Sciences (IUGS) to accept a value of 1.42×10^{-11} y^{-11} for the decay constant of ^{87}Rb. This value corresponds to a halflife of 48.8×10^9 years for ^{87}Rb. The subcommission also recommended the adoption of the following isotope ratios: $^{86}Sr/^{88}Sr = 0.1194$, $^{84}Sr/^{86}Sr = 0.056584$, and $^{85}Rb/^{87}Rb = 2.59265$.

Prior to the adoption of the present value of the decay constant of ^{87}Rb in 1977, two different values were in use: $\lambda = 1.39 \times 10^{-11}$ y^{-1} and $\lambda = 1.47 \times 10^{-11}$ y^{-1}. Careful attention to the original sources is required so that dates calculated with these old constants can be recognized and corrected. Since isotopic dates are calculated from equation 4.9,

$$t = \frac{1}{\lambda} \ln\left(\frac{D - D_0}{N} + 1\right)$$

dates calculated relative to $\lambda = 1.39 \times 10^{-11}$ y^{-1} are recalculated to $\lambda = 1.42 \times 10^{-11}$ y^{-1} by the equation

$$t_{1.42} = \frac{\text{date} \times 1.39 \times 10^{-11}}{1.42 \times 10^{-11}} \tag{5.4}$$

where $t_{1.42}$ is the date relative to 1.42×10^{-11} y^{-1}, whereas the original "date" was calculated relative to 1.39×10^{-11} y^{-1}. It follows that $t_{1.42} < t_{1.39}$ and $t_{1.42} > t_{1.47}$.

The increase of the $^{87}Sr/^{86}Sr$ ratio as a function of time expressed by equation 5.3 can be simplified to the form of a straight line by expanding $e^{\lambda t}$ as a power series:

$$e^{\lambda t} = 1 + \lambda t + \frac{(\lambda t)^2}{2!} + \cdots \tag{5.5}$$

Since the decay constant of ^{87}Rb is a very small number ($\lambda = 1.42 \times 10^{-11}$ y^{-1}),

$$\frac{(\lambda t)^2}{2!} \ll \lambda t \tag{5.6}$$

even for large values of t. For example, if $t = 1 \times 10^9$ years,

$$\frac{(\lambda t)^2}{2!} = \frac{(1.42 \times 10^{-11} \times 10^9)^2}{1 \times 2} = 1.0 \times 10^{-4}$$

whereas $\lambda t = 1.42 \times 10^{-2}$. Therefore, to a good approximation,

$$e^{\lambda t} - 1 = 1 + \lambda t - 1 = \lambda t$$

and equation 5.3 can be rewritten as

$$\frac{^{87}Sr}{^{86}Sr} = \left(\frac{^{87}Sr}{^{86}Sr}\right)_0 + \frac{^{87}Rb}{^{86}Sr}\lambda t \tag{5.7}$$

The relation between the measured Rb/Sr ratio and the corresponding $^{87}Rb/^{86}Sr$ ratio is expressed by the equation

$$\left(\frac{^{87}\mathrm{Rb}}{^{86}\mathrm{Sr}}\right)_{\mathrm{atom}} = \left(\frac{\mathrm{Rb}}{\mathrm{Sr}}\right)_{\mathrm{conc}} \times \frac{\mathrm{Ab}\ ^{87}\mathrm{Rb} \times \mathrm{WSr}}{\mathrm{Ab}\ ^{86}\mathrm{Sr} \times \mathrm{WRb}} \tag{5.8}$$

If the $^{87}Sr/^{86}Sr$ ratio is greater than 0.700 but less than 0.800, the value of the conversion factor is approximately equal to 2.89. Therefore, equation 5.3 can be restated in the convenient form

$$\frac{^{87}\mathrm{Sr}}{^{86}\mathrm{Sr}} = \left(\frac{^{87}\mathrm{Sr}}{^{86}\mathrm{Sr}}\right)_0 + 2.89\left(\frac{\mathrm{Rb}}{\mathrm{Sr}}\right)\lambda t \tag{5.9}$$

5.2a Fractionation Correction

The measured isotope ratios of Sr and other elements differ from their true values because of fractionation in the mass spectrometer. The resulting distortion of the isotope ratios can be corrected if the element being analyzed has three or more naturally occurring isotopes. In the case of Sr, which has four stable isotopes, the $^{86}Sr/^{88}Sr$ ratio is assumed to have a value of 0.11940, which was originally reported by Nier (1938) and was later adopted by the Subcommission on Geochronology of the IUGS (Steiger and Jäger, 1977). In addition, the assumption is made that the fractionation of Sr isotopes is a linear function of the difference in the masses of the isotopes (Herzog et al., 1958). Accordingly, Faure and Hurley (1963) multiplied the measured $^{87}Sr/^{86}Sr$ ratios by a factor f, which they defined by the equation

$$f = \frac{M}{(M + 0.1194)/2} = \frac{2M}{M + 0.1194} \tag{5.10}$$

where M is the measured $^{86}Sr/^{88}Sr$ ratio and $(M + 0.1194)/2$ is the midpoint between the measured $^{86}Sr/^{88}Sr$ ratio and 0.1194. The rationale for this formulation is that the mass difference between ^{87}Sr and ^{86}Sr is only one-half that between ^{86}Sr and ^{88}Sr. Therefore, the correction factor changes the $^{87}Sr/^{86}Sr$ ratio by only one-half the amount that is required to change the measured $^{86}Sr/^{88}Sr$ ratio from M to 0.1194. Following the publication of the paper by Faure and Hurley (1963), this procedure for correcting measured $^{87}Sr/^{86}Sr$ ratios was adopted by all investigators worldwide and has been used ever since. An exponential fractionation law was later proposed by Hart and Zindler (1989) but has not been widely used.

The correction also eliminates the effects of fractionation of Sr isotopes that may occur in the course of geochemical processes in nature, such as crystallization of minerals from magma or from aqueous solutions. The measured isotope ratios of other elements (e.g., Nd, Pb, Hf, and Os) are also corrected for fractionation that takes place during the analysis in the mass spectrometer. These procedures are discussed in the appropriate chapters.

5.2b Interlaboratory Isotope Standards

In spite of efforts to eliminate the effect of mass-dependent fractionation from isotope ratios of Sr, small interlaboratory differences in the measured $^{87}Sr/^{86}Sr$ ratios persist. The causes for these differences probably include details of instrumental design, problems of resolution of mass peaks particularly between ^{88}Sr and ^{87}Sr, uncontrollable variations in the purity of the Sr salt that is prepared from the samples, differences in measurement protocols, and other factors.

The problem with interlaboratory discrepancies of fractionation-corrected $^{87}Sr/^{86}Sr$ ratios is controlled by means of two Sr isotope standards that are analyzed regularly by most investigators:

1. Eimer and Amend (E&A) Sr carbonate, lot number 292327, which was informally distributed by Professor W. H. Pinson at MIT, and
2. NBS 987 Sr carbonate, distributed by the U.S. National Bureau of Standards.

The E&A standard has an $^{87}Sr/^{86}Sr$ ratio of 0.7080 after correction for fractionation in the mass spectrometer to $^{86}Sr/^{88}Sr = 0.11940$. The corrected $^{87}Sr/^{86}Sr$ ratio of NBS 987 is 0.71025. A lower value of 0.71014 ± 0.00020, given on the Certificate of Analysis (November 8, 1971), was later found to be incorrect. Most investigators either report the average $^{87}Sr/^{86}Sr$ ratio of one or both standards in their published reports or state that the measured $^{87}Sr/^{86}Sr$ ratios of the samples are compatible with 0.7080 for E&A or 0.71025 for NBS 987. Additional information about the history of these standards is available in Faure (2001).

5.2c Rb–Sr Dates of Minerals

The Rb–Sr method is not only used to date suites of cogenetic whole-rock specimens (Section 4.2d) but is also capable of dating separated Rb-rich and Sr-poor minerals such as the micas and K-feldspar. For this purpose, the desired minerals are separated from sieved size fractions of crushed rock samples by appropriate techniques, including the use of heavy liquids, as well as magnetic and electrostatic separators, followed by removal of impurities by hand under a binocular microscope.

The Rb and Sr concentrations of the separated mineral fractions are determined by isotope dilution (Section 4.3e) or by other methods such as x-ray fluorescence (XRF) and inductively-coupled plasma spectrometry (ICP). In addition, the $^{87}Sr/^{86}Sr$ ratio is measured on a suitable mass spectrometer using a pure Sr salt obtained from the mineral by dissolving it in acid followed by the separation of Sr by cation exchange chromatography. The ratio of the measured concentrations of Rb to Sr (Rb/Sr) is converted into the $^{87}Rb/^{86}Sr$ ratio by use of equation 5.8:

$$\left(\frac{^{87}Rb}{^{86}Sr}\right)_{atom} = \left(\frac{Rb}{Sr}\right)_{conc.} \times \frac{Ab\ ^{87}Rb \times WSr}{Ab\ ^{86}Sr \times WRb}$$

Where $^{87}Rb/^{86}Sr$ is the ratio of these isotopes in terms of numbers of atoms in a unit weight of the mineral at the present time, Rb/Sr is the ratio of the concentrations of these elements, Ab ^{87}Rb and Ab ^{86}Sr are the isotopic abundances of ^{87}Rb and ^{86}Sr, respectively, and WRb and WSr are the respective atomic weights. The abundance of ^{86}Sr and the atomic weight of Sr depend on the abundance of ^{87}Sr, and therefore appropriate values must be calculated for each sample.

The abundances of the Sr isotopes are calculated from the isotope ratios expressed relative to ^{88}Sr. If the measured $^{87}Sr/^{86}Sr$ ratio is 0.7090 and $^{86}Sr/^{88}Sr = 0.1194$, the $^{87}Sr/^{88}Sr$ ratio is calculated from the equation

$$\frac{^{87}Sr}{^{88}Sr} = \frac{^{87}Sr}{^{86}Sr} \times \frac{^{86}Sr}{^{88}Sr}$$

$$\frac{^{87}Sr}{^{88}Sr} = 0.7090 \times 0.1194 = 0.08465$$

The isotope ratios are then used to calculate the abundances of the isotopes as summarized in Table 5.2. Since the sum of the isotope ratios relative to ^{88}Sr is 1.21080, the abundance of each of the isotopes is expressed relative to that sum (e.g., Ab $^{87}Sr = 0.08465/1.21080 = 0.06991$). The sum of the abundances of all of the Sr isotopes must add up to 1.0000. The results indicate that the abundance of ^{86}Sr (Ab ^{86}Sr in equation 5.8) in this example is 9.861 percent by number or "atom percent." The atomic weight of this sample of Sr is calculated from the abundances of its naturally occurring isotopes and from their isotopic masses expressed in atomic mass units (amu), as demonstrated in Table 5.3. Therefore, the atomic weight of the Sr having an $^{87}Sr/^{86}Sr$ ratio of 0.7090 is 87.616 after rounding to five significant figures. In general, the atomic weights of Sr, Nd, Pb, Hf, Ce, and Os depend on the abundances of their radiogenic isotopes and therefore are not constant for all rock or mineral samples in which these elements occur.

Although the abundance of ^{87}Rb and the atomic weight of Rb have also changed continuously since nucleosynthesis, all samples of terrestrial, meteoritic, and lunar Rb have the same isotopic composition and atomic weight at the present time. This fact implies that Rb in the solar nebula from which the Sun and the planets of the

Table 5.2. Calculation of Abundances of Sr Isotopes from Measured Isotopic Ratios Relative to ^{88}Sr

Isotopic Ratios	Mass	Abundance
87/88 = 0.08465	87	0.06991
86/88 = 0.11940	86	0.09861
84/88 = 0.00675	84	0.00557
88/88 = 1.00000	88	0.82590
Sum = 1.21080		Sum = 0.99999

Table 5.3. Calculation of the Atomic Weight of Sr from Abundances of Its Naturally Occurring Isotopes

Mass Number	Abundance	Mass, amu	Mass × Abundance
88	0.82590	87.905618	72.60124991
87	0.06991	86.908883	6.07580001
86	0.09861	85.090266	8.47151272
84	0.00557	83.913429	0.46739780
			Sum = 87.61596044 amu

Note: See Table 5.2. The atomic masses are expressed in amu, where 1 amu = $\frac{1}{12}$ the mass of an atom of $^{12}_{6}C$ (Walker et al., 1989).

solar system originated was isotopically homogeneous.

Similarly, Sr is assumed to have had a uniform isotopic composition in the solar nebula even though large differences have been discovered in the isotopic compositions of other elements in refractory inclusions of meteorites (Shima, 1986; Section 24.5).

When the $^{87}Sr/^{86}Sr$ ratio of Sr in a Rb-bearing mineral has been measured and its $^{87}Rb/^{86}Sr$ ratio has been determined from equation 5.8, a date can be calculated from equation 5.3 provided that an appropriate value is substituted for the initial $^{87}Sr/^{86}Sr$ ratio:

$$t = \frac{1}{\lambda} \ln \left[\frac{^{87}Sr/^{86}Sr - (^{87}Sr/^{86}Sr)_0}{^{87}Rb/^{86}Sr} + 1 \right] \quad (5.11)$$

When the mineral to be dated is strongly enriched in radiogenic ^{87}Sr, the date calculated from equation 5.11 is insensitive to the value of the initial $^{87}Sr/^{86}Sr$ ratio. Therefore, one may select an initial $^{87}Sr/^{86}Sr$ ratio of 0.703, which is representative of Sr derived from the upper mantle of the Earth. However, in cases where the minerals being dated are not strongly enriched in radiogenic ^{87}Sr, the uncertainty of the assumed initial $^{87}Sr/^{86}Sr$ ratio reduces the accuracy of calculated mineral dates. For this reason, Rb–Sr mineral dates are not as reliable as age determinations based on suites of cogenetic igneous and high-grade metamorphic rocks.

Igneous rocks of granitic composition may contain both mica minerals and K-feldspar, all of which can be dated by the Rb–Sr method. Ideally, all minerals of an igneous rock should indicate the same date, which can then be regarded as the age of the rock. When mineral dates obtained from one rock specimen or from a suite of cogenetic igneous rocks are in agreement, they are said to be "concordant." Unfortunately, "discordance" of mineral dates is more common than concordance. The reason is that the constituent minerals of a rock may gain or lose radiogenic ^{87}Sr as a result of reheating during regional or contact metamorphism after crystallization from a magma. In such cases, the mineral dates generally are not reliable indicators of the age of the rock but instead provide information about the last episode of metamorphism and about the subsequent rate of cooling.

5.3 Rb–Sr ISOCHRONS

Suites of whole-rock samples of cogenetic igneous and high-grade metamorphic rocks form isochrons in coordinates of $^{87}Sr/^{86}Sr$ and $^{87}Rb/^{86}Sr$, in accordance with equation 5.3 and Section 4.2d, provided they satisfy the assumptions that all specimens had the same initial $^{87}Sr/^{86}Sr$ ratio at the same time in the past and that they did not gain or lose Rb or Sr since that time. Isochrons that are defined by rock samples that meet these criteria yield both the initial $^{87}Sr/^{86}Sr$ ratio and the date in the past when all specimens had the same initial $^{87}Sr/^{86}Sr$ ratio (Sections 4.2 and 4.3). The initial $^{87}Sr/^{86}Sr$ ratio is the intercept of the isochron with the y-axis, and the date is calculated from the slope m of the isochron, in accordance with equation 4.11:

$$m = e^{\lambda t} - 1 \qquad t = \frac{1}{\lambda} \ln(m + 1)$$

In cases where the cogenetic igneous rocks that define a Rb–Sr isochron formed by fractional crystallization of an isotopically homogeneous magma, the date calculated from the slope of the isochron is the age of these rocks. However, rock samples that formed from locally contaminated magma may deviate from the isochron and thereby degrade the reliability of the date calculated from its slope. In addition, suites of cogenetic igneous rocks that were isotopically homogenized during an episode of regional metamorphism may yield whole-rock Rb–Sr isochron dates that are younger than their crystallization age.

The isotopic evolution of Sr in a suite of three hypothetical igneous rocks that formed from a common magma with different Rb/Sr ratios is illustrated in Figure 5.1. At the time of crystallization, all three rocks plot as points on a straight line whose slope is zero because they all have the same $^{87}Sr/^{86}Sr$ ratio. After cooling to a temperature at which they become closed systems, their $^{87}Sr/^{86}Sr$ ratios begin to increase as a result of decay of ^{87}Rb to ^{87}Sr. Each decay of ^{87}Rb reduces the $^{87}Rb/^{86}Sr$ ratio and increases the $^{87}Sr/^{86}Sr$ ratio by the same amount. Consequently, these ratios change along straight lines with a slope of -1 in such a way that the rock samples remain on the isochron as its slope increases as a function of time. The value of the y-intercept, however, remains constant and indicates the initial $^{87}Sr/^{86}Sr$ ratio of the suite of rocks.

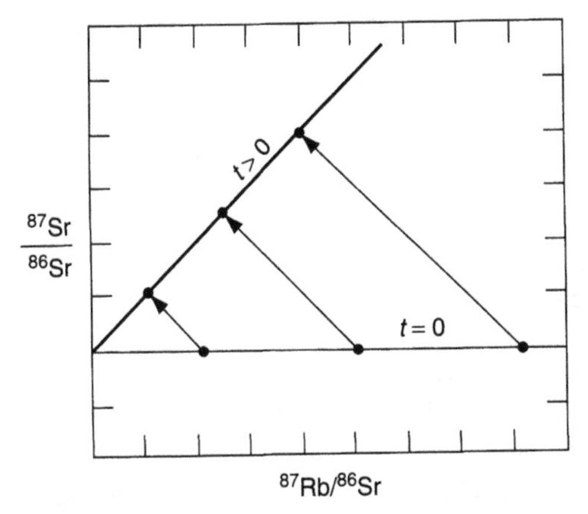

FIGURE 5.1 Rb–Sr isochron diagram showing time-dependent isotopic evolution of rock systems after their crystallization from a homogeneous magma.

5.3a Mesozoic Granite Plutons of Nigeria

Alkali-rich anorogenic granite plutons in central Nigeria of West Africa have yielded Rb–Sr dates (based on $\lambda = 1.42 \times 10^{-11}$ y^{-1}) ranging from 172 ± 4 Ma in the north to 151 ± 4 Ma in the south (van Breemen et al., 1975). These plutons were intruded into crystalline basement rocks of Precambrian to late Paleozoic age and have an average diameter of about 16 km. The Liruei pluton consists of biotite granites, albite–riebeckite granite, as well as smaller amounts of fayalite–granite porphyry and riebeckite–arfvedsonite granite.

Whole-rock samples of these rock types were analyzed for Rb and Sr by isotope dilution, and the isotope composition was measured on a single-filament (Ta) solid-source mass spectrometer at the Scottish Universities Research and Reactor Centre in East Kilbride, Scotland. The $^{87}Sr/^{86}Sr$ ratio of the E&A standard obtained by van Breemen et al. (1975) on this instrument in the course of this study was 0.70812 ± 0.00009 (2σ). The data were regressed by the least-squares method of McIntyre et al. (1966). The analytical errors were ± 0.7 percent for the $^{87}Rb/^{86}Sr$ ratio and ± 0.00037 for $^{87}Sr/^{86}Sr$ based on replicate analyses. The decay constant for ^{87}Rb was taken to be 1.39×10^{-11} y^{-1}.

The alkali-rich granites of the Liruei pluton have characteristically high Rb concentrations, ranging from 126 to 1570 ppm, but low Sr concentrations, between 0.948 and 26.3 ppm. The resulting Rb/Sr ratios vary widely from 9.62 to 663. The low Sr concentrations are a consequence of magmatic differentiation by the crystallization of feldspar in crustal magma chambers. Rocks that have low Sr concentrations (e.g., <10 ppm) are sensitive to contamination by Sr derived from groundwater or even rain, which not only increases the concentration of Sr but also affects its isotopic composition. However, the anorogenic alkali granites of Nigeria dated by van Breemen et al. (1975) appear not to have been contaminated in this way.

The unusually high Rb/Sr ratios of the alkali granites of the Liruei pluton of Nigeria have caused their $^{87}Sr/^{86}Sr$ ratios to reach very high values in excess of 8.6. Such high values have also been reported in mica (biotite and muscovite) of Precambrian age but otherwise are not typical of ordinary crustal rocks whose $^{87}Sr/^{86}Sr$ ratios in most cases range only from 0.700 to about 0.800.

Seven whole-rock samples of the biotite granite from the Liruei pluton of Nigeria form an isochron in Figure 5.2, which yields a date of 167 ± 2 Ma (recalculated to $\lambda = 1.42 \times 10^{-11}\ y^{-1}$) and an initial $^{87}Sr/^{86}Sr$ ratio of 0.729 ± 0.009, which has not been corrected to 0.7080 for E&A because only three significant figures are available. The calculated MSWD of the isochron is 1.68, compared to 3.33 predicted by Brooks et al. (1972) for 7 samples and 10 duplicates. Therefore, the regression line is an isochron and the date derived from its slope is the age of these rocks.

The biotite granite of the Liruei pluton of Nigeria has a high initial $^{87}Sr/^{86}Sr$ (0.729 ± 0.009) ratio, which is typical of granitic rocks of Precambrian age in the continental crust. However, it is much higher than the $^{87}Sr/^{86}Sr$ of magmas of basaltic composition, which have low $^{87}Sr/^{86}Sr$ ratios of about 0.703 ± 0.001 (Faure, 2001). The difference between the $^{87}Sr/^{86}Sr$ ratios of granitic rocks in the crust and of basaltic or gabbroic rocks has arisen because crustal rocks have higher Rb/Sr ratios than the ultramafic rocks of the mantle where basaltic magmas form. Consequently, the Sr in crustal rocks has been enriched in radiogenic ^{87}Sr by decay ^{87}Rb compared to the Sr in the ultramafic rocks of the lithospheric and asthenospheric mantle.

These insights were used by Faure and Hurley (1963) to propose that the initial $^{87}Sr/^{86}Sr$ ratios of igneous rocks provide information about their origin. For example, igneous rocks having elevated

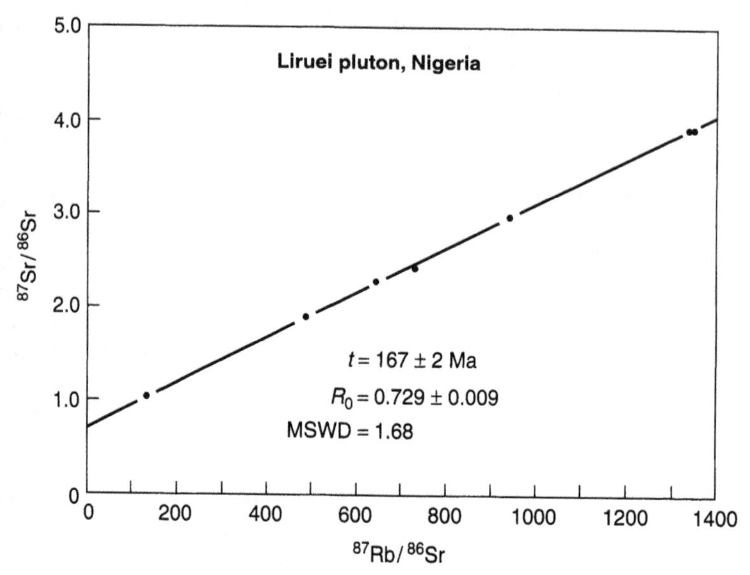

FIGURE 5.2 Whole-rock Rb–Sr isochron of biotite granite, Liruei pluton, central Nigeria. The slope and intercept of the isochron were calculated by means of the weighted-regression procedure of McIntyre et al. (1966). The date was recalculated to $\lambda = 1.42 \times 10^{-11}\ y^{-1}$ and the average $^{87}Sr/^{86}Sr$ ratio of E&A measured in the course of this study was 0.70812 ± 0.00009. Data and interpretation from van Breemen et al. (1975).

initial $^{87}Sr/^{86}Sr$ ratios formed by melting of protoliths in the continental crust, whereas igneous rocks having low initial $^{87}Sr/^{86}Sr$ ratios similar to those of recently erupted basalt originated from sources in the mantle. This criterion, stated here in its simplest form, has become an important source of information about the origin of igneous rocks in different tectonic settings. It was the subject of a book by Faure (2001) and will be used in Chapter 17 of this book to discuss the petrogenesis of igneous rocks.

Thousands of age determinations of suites of igneous rocks have been published in scientific journals, such as *Journal of Petrology*, *Contributions to Mineralogy and Petrology*, *Geochimica et Cosmochimica Acta*, *Canadian Journal of Earth Science*, *Precambrian Research*, *Lithos*, *Bulletin of the Geological Society of America*, *Journal of the Geological Society of London*, and many others. Consequently, the interested reader can easily find examples of age determinations of igneous rocks by the Rb–Sr and other methods based on radiogenic isotopes. The examples cited in Sections 5.3b, 5.3c, and 5.3d concern Rb–Sr dating of meteorites and martian as well as lunar rocks, respectively. Although these rocks are not typical of terrestrial igneous rocks, their study has yielded important information about the origin of the solar system and about the geological history of Mars and the Moon.

5.3b Stony and Iron Meteorites

Meteorites are composed of silicate and oxide minerals similar to those that form mafic igneous rocks on the Earth (e.g., pyroxene, olivine, plagioclase, magnetite, ilmenite). In addition, they contain varying amounts of metallic iron and nickel in the form of dispersed grains or in nearly pure form. Consequently, meteorites are subdivided on the basis of their composition into stony, stony iron, and metallic iron objects. The stony meteorites constitute about 95 percent of all meteorites that were collected shortly after their fall was witnessed, whereas iron meteorites make up only about 4 percent and stony irons about 1 percent of the observed falls. Stony meteorites are subdivided on the basis of textural, mineralogical, and chemical criteria into chondrites and achondrites. Chondrites are stony meteorites that contain small spherical pellets about 1 mm in diameter known as chondrules. Many chondrites are polymict breccias that display the effects of a wide range of thermal and shock metamorphism. Stony meteorites that lack chondrules are called achondrites. Achondrites whose chemical compositions and textures are similar to those of mafic or ultramafic igneous rocks on the Earth are referred to as the *basaltic achondrites.*

Most meteorites originated from the asteroids that orbit the Sun in the space between Mars and Jupiter. The asteroids are themselves fragments of larger parent bodies that originated either as large planetesimals or as small solid planets that formed when the Sun and the planets of the solar system came into existence by the condensation of the solar nebula and the capture of the resulting planetesimals. In addition, a small number of stony meteorites originated from the Moon and from Mars after being ejected by impacts of meteorites from the asteroidal belt (Gladman et al., 1996).

All meteorites contain small concentrations of long-lived radioactive isotopes, including ^{87}Rb, and therefore can be dated by the Rb–Sr and other isotopic methods. A compilation of data by Faure and Powell (1972) indicated that chondrites contain between 9 and 12 ppm Sr and between 1 and 4 ppm Rb, giving them Rb/Sr ratios ranging from 0.083 to 0.44. The achondrites have elevated Sr concentrations between 70 and 85 ppm, but their Rb concentrations range from only 0.001 ppm (Ca-rich achondrites) to 0.1 ppm (Ca-poor achondrites). In spite of the differences in the Rb and Sr concentrations of chondrites and achondrites, both are datable by the Rb–Sr method provided they satisfy the necessary assumptions outlined in Section 4.2.

In the absence of *a priori* information about the origin of chondrites and achondrites, three strategies for dating them by the Rb–Sr method have been used:

1. Whole-rock specimens of chondrites and achondrites were used to construct Rb–Sr isochrons (e.g., Pinson et al., 1965).
2. Whole-rock specimens of chondrites and achondrites belonging to selected classes were dated separately (e.g., Gopalan and Wetherill, 1968, 1969, 1970).

3. Individual specimens of chondrites and achondrites were dated by means of internal mineral isochrons (e.g., Kempe and Müller, 1969).

The intensive research devoted to the dating of stony meteorites by the Rb–Sr and other isotopic methods has demonstrated that, regardless of the strategy used, most stony meteorites yield Rb–Sr isochron dates of about 4.55×10^9 years with narrow error limits. For example, Kempe and Müller (1969) reported that the age of the amphoterite chondrite Krähenberg is $(4.600 \pm 0.014) \times 10^9$ years ($\lambda = 1.42 \times 10^{11}$ y^{-1}) based on an internal mineral isochron. The Ca-rich achondrites analyzed by Papanastassiou and Wasserburg (1969) have a Rb–Sr whole-rock isochron date of $(4.30 \pm 0.26) \times 10^9$ years ($\lambda = 1.42 \times 10^{-11}$ y^{-1}), which is in satisfactory agreement with dates derived from other kinds of meteorites. Even the silicate inclusions of seven different iron meteorites analyzed by Burnett and Wasserburg (1967) yielded internal mineral isochron dates whose average is $(4.55 \pm 0.077) \times 10^9$ years ($\lambda = 1.42 \times 10^{-11}$ y^{-1}). The reliability of this date is enhanced because of the shielding of the silicate inclusions by the iron matrix in which they were embedded after they crystallized from a silicate liquid. A few years later, Sanz et al. (1970) reported an identical date of $(4.51 \pm 0.04) \times 10^9$ years for minerals in the silicate inclusions of the iron meteorite Colomera.

All other isotopic dating methods have repeatedly confirmed that the silicate minerals of stony and iron meteorites crystallized at 4.55 Ga within a relatively short interval of time lasting only a few tens of millions of years. Although most stony meteorites subsequently experienced short episodes of shock metamorphism caused by collisions among the asteroids, the Rb–Sr systematics of most meteorite specimens were not detectably affected. Therefore, most meteorites have remained undisturbed while they resided in space for 4.55×10^9 years until they impacted on the Earth.

The whole-rock Rb–Sr isochron of 10 LL-type chondrites in Figure 5.3 corresponds to a precisely defined date of 4.493 ± 0.018 Ga ($\lambda = 1.42 \times 10^{-11}$ y^{-1}) and yields an initial $^{87}Sr/^{86}Sr$ ratio of 0.69893 ± 0.00008 relative to a value of 0.71025 for the $^{87}Sr/^{86}Sr$ ratio of the NBS 987 Sr isotope standard (Minster and Allègre, 1981). These results demonstrate that the chondrites in Figure 5.3 had the same $^{87}Sr/^{86}Sr$ ratio at 4.493 ± 0.018 Ga. Since most stony meteorites have yielded similar Rb–Sr dates and initial $^{87}Sr/^{86}Sr$ ratios, the isotope composition of Sr at the time of formation of the solar system appears to have been homogeneous. In that case, the Sr that became part of the Earth at about 4.55 Ga probably also had the same isotopic composition as the Sr in asteroids and meteoroids at the time of their formation.

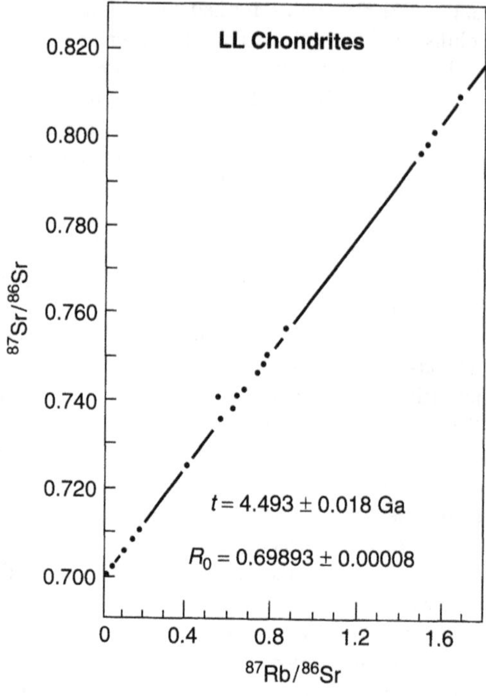

FIGURE 5.3 Whole-rock Rb–Sr isochron of 10 chondrites of the LL class. The date is relative to $\lambda = 1.42 \times 10^{-11}$ y^{-1} for the decay constant of ^{87}Rb and the initial ratio was adjusted to an $^{87}Sr/^{86}Sr$ ratio of 0.71025 for the NBS 987 Sr isotope standard. Data from Minster and Allègre (1981).

The isotopic composition of the primeval Sr that existed in the solar system at the time the terrestrial planets formed was determined by studies of achondrite meteorites because they have lower Rb/Sr ratios than the chondrites. As a result, the Sr in achondrites contains less radiogenic ^{87}Sr than the Sr in chondrites, which means that their initial $^{87}Sr/^{86}Sr$ ratio can be precisely determined on a Rb–Sr isochron diagram. For this reason, Papanastassiou and Wasserburg (1969) analyzed seven whole-rock achondrite meteorites and determined a value of 0.698990 ± 0.000047 for the initial $^{87}Sr/^{86}Sr$ ratio, which is known as BABI (basaltic achondrite best initial). The numerical value of BABI was later adjusted to 0.69897 ± 0.00003 relative to 0.71014 for NBS 987. After the $^{87}Sr/^{86}Sr$ ratio of NBS 987 was increased to 0.71025, the value of BABI rose correspondingly to 0.69908.

The BABI value is not actually the lowest initial $^{87}Sr/^{86}Sr$ ratio derived from achondrite meteorites. For example, the achondrite Angra dos Reis has an initial $^{87}Sr/^{86}Sr$ ratio of only 0.69884 ± 0.00003 known as ADOR (Wasserburg et al., 1977; Lugmair and Galer, 1992). In addition, a basaltic clast in the polymict breccia meteorite Kapoeta yielded an initial $^{87}Sr/^{86}Sr$ ratio of 0.69885 ± 0.00004 (Dymek et al., 1976). Wetherill et al. (1973) reported that Ca–Al-rich inclusions in the carbonaceous chondrite Pueblito de Allende have an initial $^{87}Sr/^{86}Sr$ ratio of 0.69880, and Gray et al. (1973) obtained an even lower value of 0.69877 ± 0.00002 for Ca–Al-rich chondrules in the carbonaceous chondrite Allende. Although these values have not been adjusted to 0.71025 for NBS 987, the small range of initial $^{87}Sr/^{86}Sr$ ratios of different meteorites indicates that their parent bodies did not all crystallize at the same time. Instead, the preferred interpretation is that the differences in the initial $^{87}Sr/^{86}Sr$ ratios of different kinds of meteorites were caused by small differences in the rates of cooling of their parent bodies whose $^{87}Sr/^{86}Sr$ were increasing with time because of decay of ^{87}Rb. Therefore, BABI is the $^{87}Sr/^{86}Sr$ ratio of the parent body of the achondrites at a certain time in the early history of the solar system. According to the theory of the origin of the terrestrial planets, BABI is also the $^{87}Sr/^{86}Sr$ ratio of Sr that was incorporated into the Earth at the time of its formation at about 4.55 Ga. Therefore, the evolution of the isotopic composition of Sr in the Earth, caused by the decay of ^{87}Rb, started with BABI at $^{87}Sr/^{86}Sr = 0.69908$ relative to 0.71025 for NBS 987.

5.3c Martian Meteorites

A small group of achondrite meteorites originally attracted attention because they yielded surprisingly low dates by the Rb–Sr and other methods (Podosek and Huneke, 1973). These meteorites are now known to have originated from Mars because the noble gases they contain have the same abundances as the noble gases in the atmosphere of Mars measured by instruments on the Viking landers in 1975. The group of martian meteorites includes 14 specimens most of which resemble Shergotty, Nakhla, or Chassigny. Therefore, the martian meteorites are referred to as the SNC group of achondrites. Shergotty, which fell on August 25, 1865, in India, was the subject of a study by a large group of experts who published their results in an issue of *Geochimica et Cosmochimica Acta* (vol. 50, No. 6, 1986; see Laul, 1986). The fall of Nakhla was witnessed on June 29, 1911, in Egypt, and Chassigny fell on October 3, 1815, in France. At least six of the known martian meteorites were collected on the East Antarctic ice sheet (Whillans and Cassidy, 1983).

The meteorite Nakhla was dated by Papanastassiou and Wasserburg (1974), who reported a Rb–Sr mineral isochron date of only 1.34 ± 0.02 Ga (recalculated to $\lambda = 1.42 \times 10^{-11}$ y^{-1} from 1.39×10^{-11} y^{-1}) with an initial $^{87}Sr/^{86}Sr$ ratio of 0.70275 ± 0.00007. Subsequently, Gale et al. (1975) obtained a similar Rb–Sr date of 1.21 ± 0.01 Ga for Nakhla and an initial $^{87}Sr/^{86}Sr$ ratio 0.70246 ± 0.00003 (relative to 0.7080 for E&A), whereas Nakamura et al. (1982) reported a date of 1.26 ± 0.07 Ga based on the Sm–Nd method. Gale and Nakamura and their collaborators interpreted the Rb–Sr date of Nakhla to be the time of crystallization of the minerals from a silicate melt. Consequently, Nakhla must have originated from a body in the solar system that was large enough to have remained geologically active long after the parent bodies of most meteorites had cooled and had become inactive (Borg et al., 1997). The other "nakhlite" meteorites (Lafayette and Gov. Valadares) as well as Chassigny also have crystallization ages of about 1.3 Ga.

The isotopic evolution of Sr in Nakhla depicted in Figure 5.4 is based on equation 5.9 and started at 4.55 Ga with an $^{87}Sr/^{86}Sr$ ratio equal to

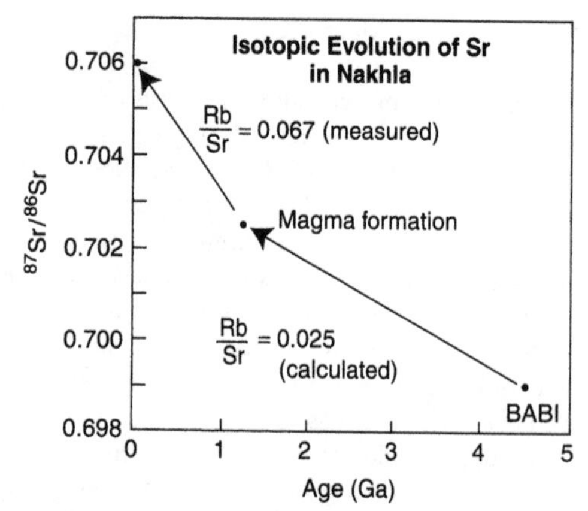

FIGURE 5.4 Evolution of $^{87}Sr/^{86}Sr$ ratio of Nakhla and its magma source on Mars. In this two-stage model, the $^{87}Sr/^{86}Sr$ ratio of the magma source increased from BABI at 4.55 Ga to 0.70246 at 1.21 Ga when partial melting occurred. The resulting magma crystallized below the surface of Mars to form the Nakhla rock, part of which was ejected from the surface of Mars by the impact of a meteorite at about 12 Ma. The Rb/Sr ratio of the magma source (0.025) increased to 0.067 in the Nakhla rock as a result of the formation and crystallization of magma. Data from Gale et al. (1975).

BABI (0.6990). Subsequently, the $^{87}Sr/^{86}Sr$ ratio of the magma source increased to 0.70246 at 1.21 Ga when magma formed by partial melting at depth below the surface of Mars. The magma may have intruded the overlying rocks and crystallized in a short interval of time to form igneous rocks with an average Rb/Sr ratio of 0.067. The $^{87}Sr/^{86}Sr$ ratio of the Nakhla rock increased because of decay of ^{87}Rb and reached its present average value of 0.70597 relative to 0.7080 for E&A (Gale et al., 1975).

The Rb/Sr ratio of the magma source of Nakhla at 1.21 Ga can be estimated by means of equation 5.9:

$$0.70246 = 0.6990 + 2.89\left(\frac{\text{Rb}}{\text{Sr}}\right) \times 1.42$$

$$\times 10^{-11}\, 3.34 \times 10^{9}$$

where the decay time $t = (4.55 - 1.21) \times 10^9 = 3.34 \times 10^9$ y.

The result is

$$\frac{\text{Rb}}{\text{Sr}} \simeq 0.025$$

Evidently, the Nakhla rock had a significantly higher Rb/Sr ratio (0.067) than the magma source from which it originated (0.025). The apparent enrichment of the magma in Rb relative to Sr as a result of partial melting is a common phenomenon in the petrogenesis of igneous rocks (Faure and Hurley, 1963; Faure, 2001).

Although there is no longer any doubt that Nakhla and the other SNC achondrites originated from Mars, the events that caused them to be transported to the Earth are not easy to reconstruct. The proposal that has been discussed is that the rocks were dislodged by the impacts of meteorites and achieved sufficient velocity to escape from the gravitational field of Mars. After spending several million years in orbit around the Sun, the martian rocks were eventually captured by the gravitational field of the Earth, much like ordinary meteorites from the asteroidal belt. The feasibility of this hypothesis hinges on the question whether rocks can be ejected from the surface of Mars by the impact of a meteorite without being vaporized or melted (Wasson and Wetherill, 1979; McSween, 1984, 1985, 1994; Gladman et al., 1996).

The possibility that rocks can be dislodged from the surface of Mars and the Moon (Wänke, 1966) was confirmed in 1981 when the first lunar meteorite (ALHA 81005) was found near the Allan Hills of Antarctica. This specimen (weight 31.4 g) was identified as a lunar highland breccia by Palme et al. (1983) at a special symposium during the Fourteenth Lunar and Planetary Science Conference in Houston, Texas. In fact, Palme and his colleagues cited a remarkable statement by Georg Christoph Lichtenberg in 1797: "The Moon is a rude neighbor because he throws stones at us." The lunar origin of ALHA 81005 helped to convince the scientific community that the SNC meteorites actually did originate from Mars, in spite of the apparent difficulty of ejecting rocks from the surface of this planet.

The low age of Nakhla and other nakhlites was surpassed by the achondrite Shergotty for which Nyquist et al. (1979) reported a Rb–Sr mineral isochron date of only 161 ± 11 Ma with an initial $^{87}Sr/^{86}Sr$ ratio of 0.72260 ± 0.00012. Similar dates were later reported by Shih et al. (1982) for the SNC meteorites Zagami ($t = 176 \pm 4$ Ma, $R_0 = 0.72145 \pm 0.00005$) and ALHA 77005 ($t = 183 \pm 12$ Ma, $R_0 = 0.71037 \pm 0.00005$), where R_0 is the initial $^{87}Sr/^{86}Sr$ ratio. Additional age determinations of Shergotty were published by Jagoutz and Wänke (1986), whereas Jagoutz (1989) dated ALHA 77005, which also belongs to the shergottite group.

The Shergotty meteorite contains evidence for severe shock metamorphism, suggesting that the isotopic composition of Sr in the minerals was homogenized during the impact of a meteorite on the surface of Mars. However, Nyquist also demonstrated experimentally that the passage of a shock wave (350 kbars) does not homogenize the isotope ratios of Sr in the minerals of a lunar basalt from the Sea of Tranquillity. Therefore, Nyquist et al. (1979) concluded that the body of rock from which Shergotty was ultimately derived had been heated to a temperature of less than 400°C (Duke, 1968) by an impact of a meteorite on Mars and that the Shergotty rocks had then cooled slowly for about 10,000 years under a cover of overburden several hundred meters thick. A study of cosmogenic radionuclides by Nishiizumi et al. (1986) later confirmed that the Shergotty meteorite was not exposed to cosmic rays while it resided on Mars and that it was ejected at about 2.2 Ma. The other "shergottites" (Zagami, EETA 9001, QUE 94201, ALHA 77005, LEW 88516, and Y793605) all have formation ages of about 170 Ma and were also ejected at about 3 Ma except EETA 9001, which has a lower exposure age of 0.7 Ma.

The Rb–Sr dates of the SNC meteorites cited above are illustrated in Figure 5.5 by means of isochron lines, including one whose slope corresponds to an age of 4.55 Ga shared by most stony and iron meteorites. The diagram also emphasizes the fact that all SNC meteorites included in this diagram have elevated initial $^{87}Sr/^{86}Sr$ ratios relative to BABI. Therefore, these SNC meteorites formed from pre-existing rocks on Mars having a range of Rb/Sr ratios rather than from primordial matter having an $^{87}Sr/^{86}Sr$ ratio equal to BABI. The evidence concerning the ages, mineralogy, and petrogenesis of martian rocks has been summarized and interpreted by Jones (1986), Jagoutz (1991), and McSween (1994).

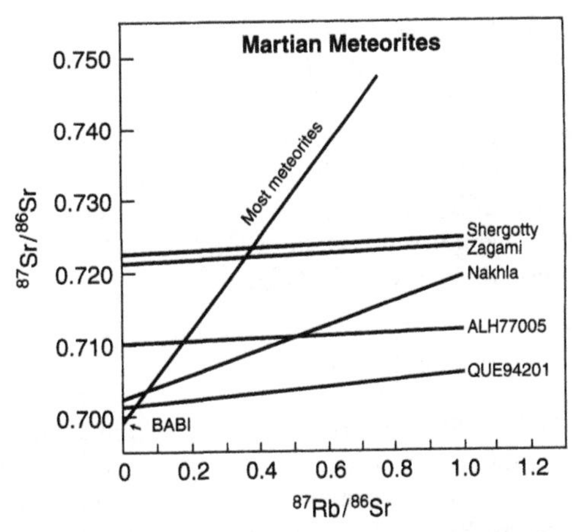

FIGURE 5.5 Rb–Sr mineral isochrons of the SNC meteorites Shergotty, Zagami, Nakhla, ALH 77005, and QUE 94201 compared to the isochron defined by meteorites from the asteroidal belt. Note that the Rb–Sr dates of all of the martian meteorites are lower than those of most meteorites and their initial $^{87}Sr/^{86}Sr$ ratios are higher than BABI. Data from Papanastassiou and Wasserburg, (1974), Gale et al. (1975), Nyquist et al. (1979), Shih et al. (1982), Jagoutz and Wänke (1986), and Borg et al. (1997).

Actually, the most important martian meteorite (ALH 84001) was not recognized for more than 10 years after it was collected in 1984 in the Allan Hills because it is an orthopyroxenite that does not resemble any of the SNC groups of meteorites (Mittlefehldt, 1994). This ultramafic igneous rock crystallized from magma below the surface of Mars at 4.5 Ga, was altered by shock metamorphism at 4.0 Ga, and was ejected from the surface of Mars at about 16 Ma. This specimen contains fractures that are partially filled with carbonates of Ca, Mg, and Fe (Gleason et al., 1997). In addition, some of the fractures contain worm-like objects 20–100 nm long that resemble terrestrial nanobacteria (McKay et al., 1996). Although this evidence does not prove the former existence of bacterial life-forms on Mars, the age of ALH 84001 (4.5 Ga) indicates that this rock formed at a time when liquid water occurred on the surface of

Mars. Since the presence of water is a prerequisite to life (*ubi aqua, ibi vita*), fossilized bacteria may be present in the oldest rocks of Mars. In any case, the presence of fracture-filling carbonates in ALH 84001 and of hydroxyl-bearing minerals in other SNC meteorites (e.g., nakhlites) demonstrates that hydrothermal activity occurred in the martian crust (Romanek et al., 1994).

5.3d Lunar Rocks

On July 20, 1969, Neil A. Armstrong and Edwin E. Aldrin landed on the surface of the Moon in the Sea of Tranquillity (Mare Tranquillitatis). The rock samples they collected were subsequently analyzed for dating by the Rb–Sr and other applicable methods. All of the rock specimens collected by the American astronauts in the course of the Apollo program were boulders and cobbles of mafic igneous rocks and related breccias. Since none of the lunar samples were collected from outcrops, there is no *a priori* evidence that they are comagmatic in origin. Therefore, these rocks had to be dated by means of Rb–Sr isochrons defined by the various Rb-bearing mineral fractions (Papanastassiou et al., 1970). The first age determinations of rocks from the Sea of Tranquillity were published in *Science* (vol. 167, No. 3918, January 1970). A summary and evaluation of Rb–Sr dates and initial $^{87}Sr/^{86}Sr$ ratios of lunar rocks was subsequently prepared by Nyquist (1977). An example of an "internal" Rb–Sr isochron in Figure 5.6 of sample 12002,147 (olivine-rich basalt) from the Ocean of Storms (Apollo 12) yielded a date of 3.36 ± 0.10 Ga and an initial $^{87}Sr/^{86}Sr$ ratio of 0.69949 ± 0.00005 (Papanastassiou and Wasserburg, 1970). The most Rb-rich and Sr-poor mineral phases in this rock are ilmenite and olivine, whereas pyroxene and plagioclase are Rb poor and Sr rich and therefore have lower $^{87}Rb/^{86}Sr$ ratios than the whole rock. The authors interpreted this date as the time of crystallization of the basalt. This age determination therefore indicates that the basin of the Ocean of Storms was flooded by basalt lava at about this time.

The initial $^{87}Sr/^{86}Sr$ ratio derived from the isochron in Figure 5.6 (0.69949 ± 0.00005) is significantly greater than BABI (0.69908 ± 0.00003).

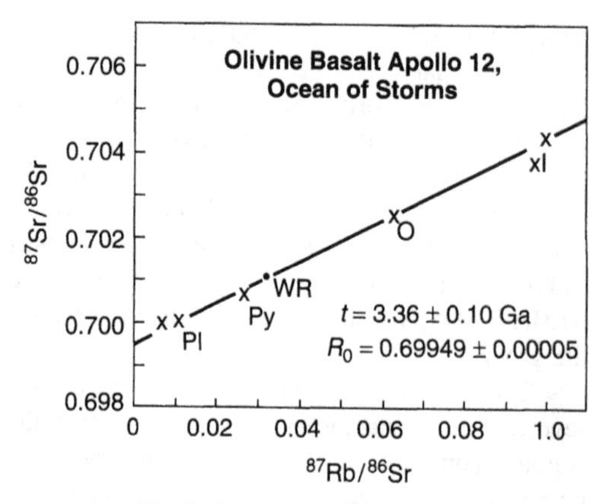

FIGURE 5.6 Rb–Sr isochron of the constituent minerals of olivine basalt 12002 (Apollo 12), Ocean of Storms, Moon. The minerals are identified by letter: I = ilmenite, O = olivine; Py = pyroxene, and Pl = plagioclase. The whole-rock sample is designated WR. The date is the time of crystallization of this volcanic rock from lava that had been extruded into the mare basin. Data from Papanastassiou and Wasserburg (1970).

This fact demonstrates that the magma that formed sample 12002,147 originated from sources containing Sr whose $^{87}Sr/^{86}Sr$ ratio had increased by decay of ^{87}Rb prior to the formation of the magma. Eight whole-rock samples of basalt from the Ocean of Storms analyzed by Papanastassiou and Wasserburg (1970, 1971) define a straight line in Figure 5.7 whose slope yields a date of 4.34 ± 0.15 Ga ($2\bar{\sigma}$, $N = 8$). The initial $^{87}Sr/^{86}Sr$ ratio of 0.6990 is indistinguishable from BABI. The goodness of fit of the data points to the straight line in Figure 5.7 requires that the Rb/Sr ratios of the basalt samples are virtually identical to those of their respective magma sources, which implies a high degree of partial melting. In addition, these results indicate that the Rb/Sr ratios of the magmas were not changed by fractional crystallization or by assimilation of rocks having higher or lower Rb/Sr ratios. O'Nions and Pankhurst (1972) confirmed these conclusions when they obtained a whole-rock Rb–Sr isochron date of 4.31 ± 0.04 Ga for basalt samples from the Apollo 14 site on the Moon with an initial $^{87}Sr/^{86}Sr$ ratio of 0.69905 ± 0.00008.

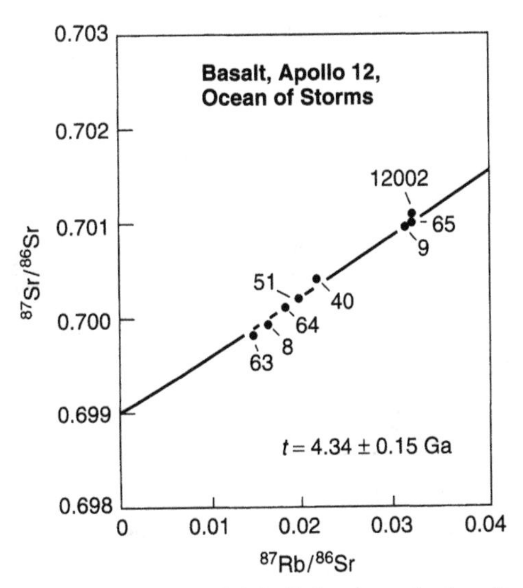

FIGURE 5.7 Whole-rock Rb–Sr isochron for basalt samples on the Ocean of Storms on the Moon. The samples are identified by number (i.e., 12002 is the olivine-rich basalt in Figure 5.6). Sample 12070 was omitted because it has abnormally high $^{87}Sr/^{86}Sr$ and $^{87}Rb/^{86}Sr$ ratios of 0.70803 ± 0.00006 and 0.1444, respectively. Data from Papanastassiou and Wasserburg (1970, 1971).

These pioneering studies have been succeeded by ongoing investigations of the petrogenesis of lunar rocks based on the isotopic compositions of Sr, Nd, Pb, Hf, and Os as well as on trace-element distributions and on the physical chemistry of silicate systems. The ferroan anorthosites that occur as clasts in lunar breccias have received special attention because they originated from the lunar highlands and are among the oldest rocks on the Moon:

Dixon and Papike, 1975, *Proc. 16th Lunar Sci. Conf.*, pp. 263–291. Ryder, 1982, *Geochim. Cosmochim. Acta*, 46:1591–1601. Hanan and Tilton, 1987, *Earth Planet. Sci. Lett.*, 84:15–21. Carlson and Lugmair, 1988, *Earth Planet. Sci. Lett.*, 90:119–130. James et al., 1989, *Proc. 19th Lunar Sci. Conf.*, pp. 219–243. Floss et al., 1998, *Geochim. Cosmochim. Acta*, 62:1255–1284. James et al., 1991, *Proc. 21st Lunar Sci. Conf.*, pp. 63–87. Borg et al., 1999, *Geochim. Cosmochim. Acta*, 63:2679–2691.

5.4 DATING METAMORPHIC ROCKS

Rocks may be subjected to physical and chemical processes as a result of which their mineralogical and chemical compositions and even their textures are changed. These processes constitute metamorphism in the broadest sense of the word. Metamorphism almost invariably involves an increase in the ambient temperature, which promotes recrystallization of existing minerals or may cause formation of new minerals at the expense of existing ones. These mineralogical changes imply considerable mobility of the chemical constituents of rocks either by virtue of the presence of an aqueous phase or by diffusions of ions or both. Metamorphism may also be accompanied by metasomatism as a result of which both the bulk chemical composition as well as the trace-element composition of the rocks are changed.

5.4a Isotopic Homogenization

Thermal metamorphism exerts a profound effect on the parent–daughter relationships of all naturally occurring radiogenic and radioactive isotopes that may be present in a rock. In fact, even a modest increase in temperature of 100–200°C may have drastic effects on the parent–daughter relationships of certain natural decay schemes (e.g., Rb–Sr and K–Ar) without necessarily being reflected in the usual mineralogical or textural criteria for metamorphism. The apparent sensitivity of isotopic systems in rocks to increases in temperature is probably related to the fact that rates of diffusion of ions through a crystal lattice and across grain boundaries are sensitive functions of the temperature. Moreover, the daughter atoms produced by decay in a mineral are isotopes of different elements and have different ionic charges and radii compared with their parents. The energy released during the decay may also produce dislocations or even destroy the crystal lattice locally, thus making it all the more easy for the radiogenic daughters to escape.

For these reasons, it is not surprising that Rb–Sr decay schemes in minerals are profoundly affected by even modest increases in temperature during metamorphism. The observed behavior of minerals can generally be treated as though it had been caused solely by the migration of radiogenic ^{87}Sr among the constituent minerals of a rock. However, this is undoubtedly an oversimplification, and it is likely that the concentrations of Rb and Sr in the minerals are also affected. Nevertheless, the presentation that follows is based on the assumption that radiogenic ^{87}Sr is the mobile component and the concentrations of Rb and Sr of the minerals remain essentially constant during regional contact metamorphism.

The changes that occur in the ^{87}Sr/^{86}Sr ratios of minerals of a rock during metamorphism can be visualized by means of equation 5.7:

$$\frac{^{87}\mathrm{Sr}}{^{86}\mathrm{Sr}} = \left(\frac{^{87}\mathrm{Sr}}{^{86}\mathrm{Sr}}\right)_0 + \frac{^{87}\mathrm{Rb}}{^{86}\mathrm{Sr}}\lambda t$$

which is the equation of a straight line in the familiar slope–intercept form in coordinates of ^{87}Sr/^{86}Sr and t. Equation 5.7 indicates that the ^{87}Sr/^{86}Sr ratio of a Rb-bearing rock or mineral increases as a linear function of time at a rate controlled by the slope m, which is related to the ^{87}Rb/^{86}Sr ratio:

$$m = \frac{^{87}\mathrm{Rb}}{^{86}\mathrm{Sr}}\lambda \qquad (5.12)$$

In accordance with equation 5.7, the ^{87}Sr/^{86}Sr ratios of four rock or mineral samples in Figure 5.8 increase with time at rates proportional to their respective ^{87}Rb/^{86}Sr ratios.

The effect of an episode of thermal metamorphism on the ^{87}Sr/^{86}Sr ratios of the minerals of an igneous rock is described by a model proposed by Fairbairn et al. (1961) and is illustrated in Figure 5.9. The igneous rock in question is composed of two Rb-rich minerals (e.g., biotite and microcline or orthoclase) and one Rb-poor but Sr-rich mineral such as apatite. This rock formed at t_i by crystallization of a magma containing Sr of uniform isotopic composition. Consequently, all

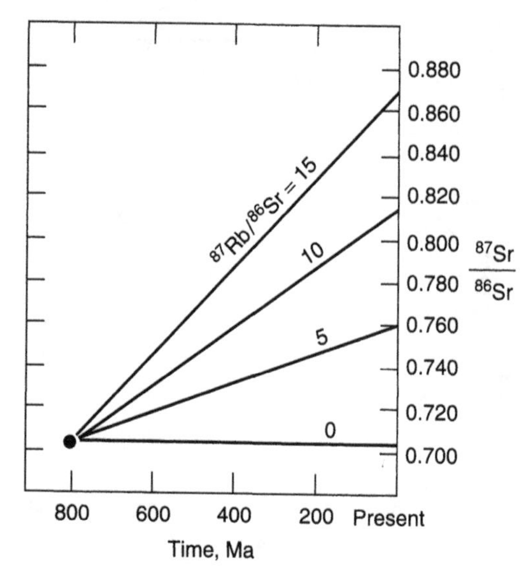

FIGURE 5.8 Evolution of the isotope composition of strontium in a suite of four rocks or minerals having different ^{87}Rb/^{86}Sr ratios. The four rocks or minerals came into existence at 800 Ma when each had the same ^{87}Sr/^{86}Sr ratio of 0.704. Thereafter, the ^{87}Sr/^{86}Sr ratios evolved along diverging straight lines whose slopes depend on the ^{87}Rb/^{86}Sr ratio of each system. The rock or mineral whose ^{87}Rb/^{86}Sr ratio is zero has a constant ^{87}Sr/^{86}Sr ratio equal to the initial value of this ratio.

of the minerals in this rock had the same initial ^{87}Sr/^{86}Sr ratio.

Some time after crystallization, the rock was subjected to an increase in temperature for a short interval of time Δt. Subsequently, the rock cooled to the ambient temperature t_m years ago and remained undisturbed to the present time. Starting t_i years ago when the rock had cooled sufficiently for its minerals to retain radiogenic ^{87}Sr (Dodson, 1973), their ^{87}Sr/^{86}Sr ratios evolved along straight lines at rates controlled by their ^{87}Rb/^{86}Sr ratios. When the temperature was increased, the Rb-rich phases (biotite and K-feldspar) lost radiogenic ^{87}Sr, and their ^{87}Sr/^{86}Sr ratios decreased until they became identical with that of the rock as a whole. The radiogenic ^{87}Sr lost by the Rb-rich phases entered the apatite, causing its ^{87}Sr/^{86}Sr

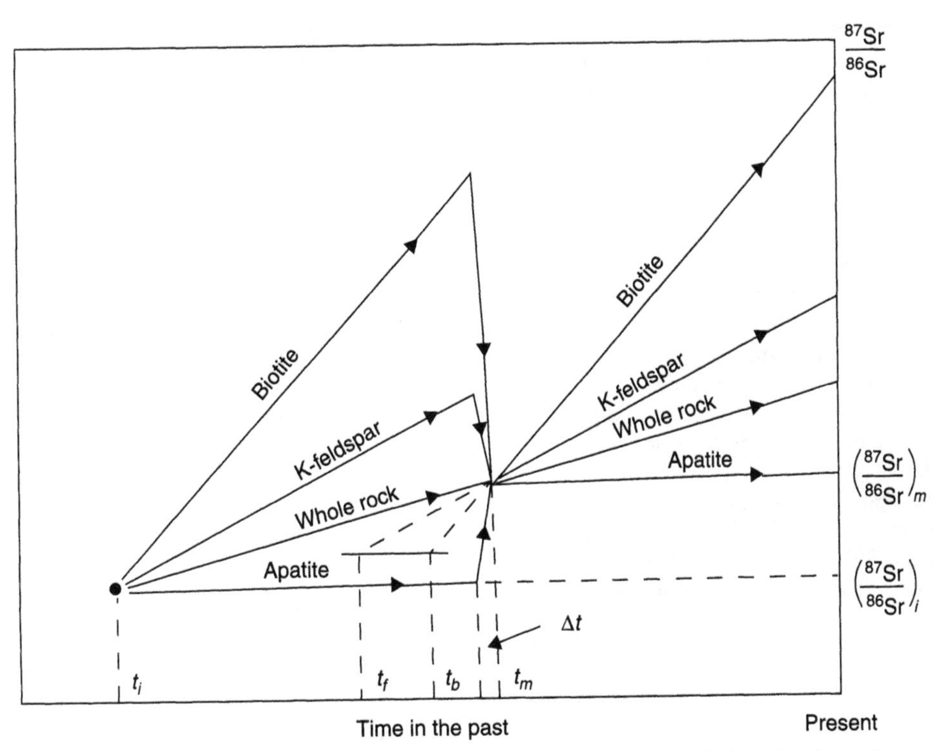

FIGURE 5.9 Strontium isotope evolution caused by isotopic homogenization of minerals of a rock as a result of thermal metamorphism: t_i = time elapsed since initial crystallization and cooling when the isotopic composition of strontium in all minerals was $(^{87}Sr/^{86}Sr)_i$; t_m = time elapsed since closure of the minerals following isotopic re-equilibration of strontium to $(^{87}Sr/^{86}Sr)_m$ by thermal metamorphism; t_b and t_f are fictitious model dates of biotite and feldspar, respectively, calculated relative to an arbitrary and inappropriate choice of the initial $^{87}Sr/^{86}Sr$ ratio.

ratio to increase until it too was equal to that of the whole rock. As a result of these changes brought about by the increase in temperature, Sr was isotopically homogenized so that all minerals once again had the same $^{87}Sr/^{86}Sr$ ratio equal to $(^{87}Sr/^{86}Sr)_m$. Note especially that $(^{87}Sr/^{86}Sr)_m$ is greater than $(^{87}Sr/^{86}Sr)_i$ and that the whole rock remained closed while radiogenic ^{87}Sr was redistributed among the minerals until the $^{87}Sr/^{86}Sr$ ratios of all the minerals were equalized. It is entirely possible that the concentrations of Rb and Sr of the minerals were also affected by this process, although Figure 5.9 was constructed with the assumption that they were not. Following isotopic homogenization, the temperature eventually decreased sufficiently for the minerals to become closed systems at t_m. Their $^{87}Sr/^{86}Sr$ ratios then increased to the present time at rates consistent with their $^{87}Rb/^{86}Sr$ ratios.

There are two meaningful dates in the history outlined above: t_i, the time elapsed since crystallization, and t_m, the time elapsed since metamorphism of the igneous rock. The Rb-rich minerals of the rock record only one of these, namely t_m. It can be obtained by solving two simultaneous equations of the form of equation 5.7 for the biotite and the K-feldspar. Alternatively, the $^{87}Sr/^{86}Sr$ ratio of a Sr-rich mineral such as apatite can be used as the initial ratio to calculate model dates for Rb-rich phases provided that the Sr in the apatite actually had the same $^{87}Sr/^{86}Sr$ ratios as the Rb-bearing minerals. If isotopic homogenization was complete and is accurately reflected by the $^{87}Sr/^{86}Sr$ ratio of the apatite, the model dates of the biotite and the K-feldspar are concordant. On the other hand, if model dates are calculated for these two minerals by using an inappropriate initial $^{87}Sr/^{86}Sr$ ratio, the resulting dates indicated by t_b and t_f in Figure 5.9 are discordant and meaningless. This illustrates the general unreliability of model dates derived by analysis of separated minerals. Such dates may even exceed the age of the rock, depending on the initial $^{87}Sr/^{86}Sr$ ratio that is assumed in the calculation. At best, the minerals of a metamorphosed igneous rock can be used to determine the time elapsed since closure of the minerals following thermal metamorphism.

The isotopic evolution of Sr of a metamorphic rock described above and illustrated in Figure 5.9 suggests that the $^{87}Sr/^{86}Sr$ ratio of the rock as a whole was not affected by the redistribution of radiogenic ^{87}Sr among the minerals. Therefore, whole-rock samples collected from the same body of metamorphic rocks are expected to define an isochron whose slope can yield the time of initial crystallization when all such whole-rock samples had the same initial $^{87}Sr/^{86}Sr$ ratio. The effects of thermal metamorphism of igneous rocks and their constituent minerals are depicted in Figure 5.10 in the coordinates of a Rb–Sr isochron diagram.

The crosses marked R1, R2, and R3 are whole rocks, while the open circles marked M2 are minerals of R2. The rocks and minerals initially lay on an isochron of slope equal to zero and thus $t = 0$. Their subsequent history is similar to that described in Figure 5.9. All systems initially moved along straight-line paths, as shown by the arrows. After an interval of time equal to $t_i - t_m$, the rocks were heated for a short period of time. The whole-rock samples remained closed, but radiogenic ^{87}Sr was redistributed among the minerals M2 until they all had the same $^{87}Sr/^{86}Sr$ ratio as R2. Therefore, at a time $t = t_m$ years ago, the

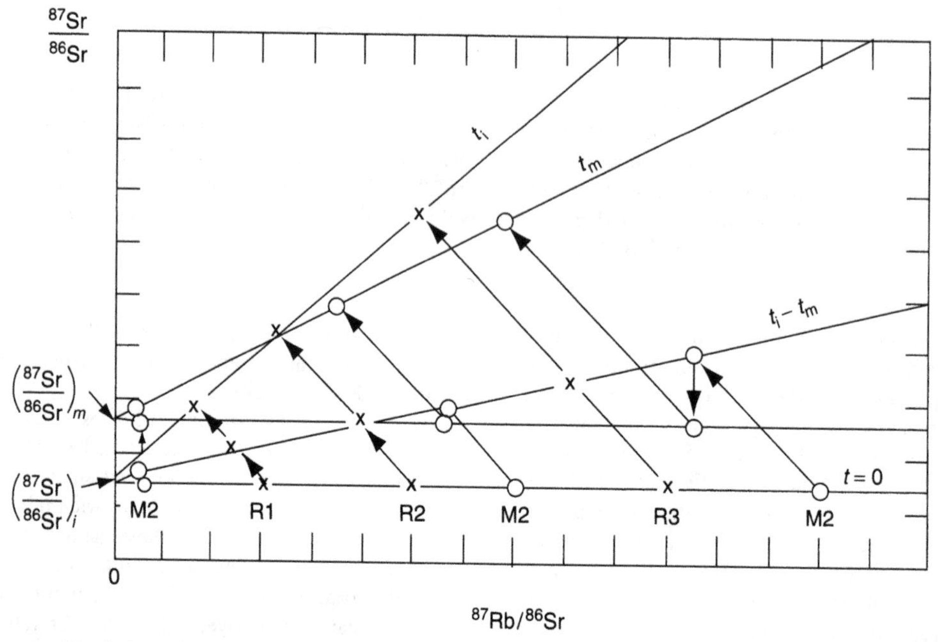

FIGURE 5.10 Evolution of strontium in three whole rocks (R1, R2, and R3) and in the minerals of R2 (M2). The strontium in the minerals was isotopically homogenized by an episode of thermal metamorphism of short duration. The slope of the whole-rock isochron corresponds to t_i, the time elapsed since crystallization of these rocks. The slope of the mineral isochron indicates t_m, the time elapsed since the end of the thermal metamorphism.

minerals of R2 had realigned themselves on a new isochron having a slope equal to zero. The whole-rock systems continued their evolution without interruption and formed an isochron whose slope corresponds to t_i, which is the time elapsed since initial crystallization. In the meantime the minerals evolved on their own isochron, which still includes R2, but whose slope represents t_m, the time elapsed since the minerals became closed systems after being re-equilibrated by thermal metamorphism. The y-intercepts of the whole rock and the mineral isochrons correspond to $(^{87}Sr/^{86}Sr)_i$ and $(^{87}Sr/^{86}Sr)_m$, respectively. Although Figure 5.10 includes only the minerals of whole-rock sample R2, the minerals of the other rocks in this suite form similar isochrons having identical slopes but different initial $^{87}Sr/^{86}Sr$ ratios whose numerical values depend on the $^{87}Sr/^{86}Sr$ ratios of their respective whole-rock samples at the time of the episode of thermal metamorphism.

5.4b Carn Chuinneag Granite, Scotland

A granite complex at Carn Chuinneag (pronounced Carn Coon-e-ag) in the northern Scottish Highlands provides a good example of an igneous rock that was later regionally metamorphosed (Long, 1964). Figure 5.11 is an isochron formed by four whole-rock specimens from this and a related intrusive. The age of these rocks is 548 ± 10 Ma (recalculated to $\lambda^{87}Rb = 1.42 \times 10^{-11}\ y^{-1}$) and their initial ratio is 0.710 ± 0.002. The dashed line in Figure 5.11 is the isochron defined by the minerals (muscovite, biotite, and K-feldspar) of whole-rock sample 20782. This mineral isochron passes through the whole-rock point, which confirms that the $^{87}Sr/^{86}Sr$ ratios of the minerals were homogenized to the $^{87}Sr/^{86}Sr$ ratio of rock sample 20782. The slope of the mineral isochron yields a date of only 403 ± 5 Ma, which is 145 million years (m.y.) less than the crystallization age of the Carn Chuinneag intrusion. The initial $^{87}Sr/^{86}Sr$ ratio of the minerals is 0.782 ± 0.30 and represents the isotope composition of Sr in whole-rock sample 20782 at the time of metamorphism at 403 Ma. The original $^{87}Sr/^{86}Sr$ ratio of this rock sample at the time of crystallization at 548 ± 10 Ma was 0.710 ± 0.02, as indicated by the whole-rock isochron.

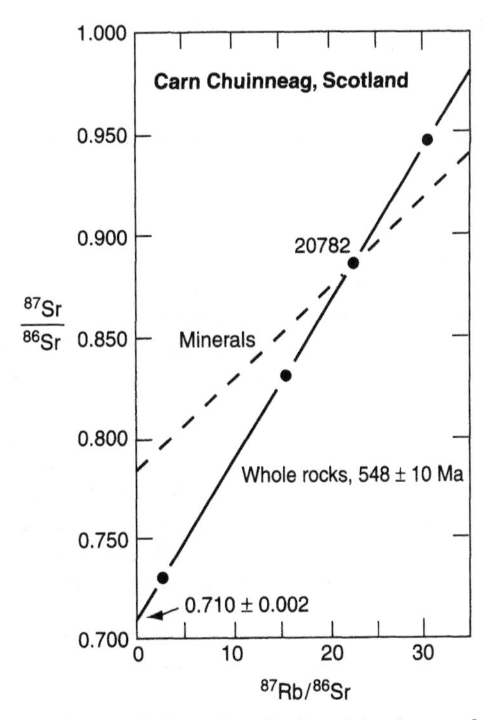

FIGURE 5.11 Whole-rock and mineral isochrons of granitic rocks from the Carn Chuinneag complex in the Highlands of northern Scotland. The date derived from the whole-rock isochron is the crystallization age of the rocks. The minerals of whole-rock sample 20782 define a separate isochron that dates an episode of thermal metamorphism. Additional age determinations of this complex were reported by Pidgeon and Johnson (1974). Data from Long (1964).

All of the foregoing discussion is predicated on the assumption that the minerals were isotopically homogenized during thermal metamorphism and that whole-rock samples of the size of conventional hand specimens remained closed. If homogenization of the minerals was incomplete, the time of metamorphism cannot be determined from them. In this case, the whole rocks may still tell the time of initial crystallization. However, if the rocks were chemically altered so that Rb and/or Sr were either added or lost at any time after their formation, they cannot be dated by the Rb–Sr method. Pidgeon and Johnson (1974) compared U–Pb dates of zircon in the Carn Chuinneag granites to whole-rock Rb–Sr dates of these rocks.

5.4c Amitsoq Gneiss, Southwest Greenland

An example of dating granitic gneisses was presented by Moorbath et al. (1972) at Oxford University for samples of the Amitsoq gneiss from the Godthaab district in southwestern Greenland. The Amitsoq gneiss was formed by deformation, metamorphism, and migmatization of a complex of igneous rocks of granitic composition. Several suites of these rocks from the Godthaab district were dated by the whole-rock Rb–Sr method. Figure 5.12 shows the isochron of Amitsoq gneisses from the Qilangarssuit area. The slope of this isochron yields a date of 3660 ± 99 Ma and an initial $^{87}Sr/^{86}Sr$ ratio of 0.7009 ± 0.0011. The Amitsoq gneiss is one of the oldest terrestrial rocks known. The date indicated by the isochron may be the time of crystallization of the igneous rocks or it may reflect the metamorphic event that produced the Amitsoq gneiss. The latter is preferred in this case.

The antiquity of the Amitsoq gneiss was subsequently confirmed by studies based on a variety of isotopic dating techniques (Baadsgaard, 1973; Moorbath et al., 1975, 1977). In addition, Baadsgaard et al. (1976) dated biotite from the Amitsoq gneiss by the K–Ar and Rb–Sr methods. The results indicate that the micas record the end of a major metamorphic event at about 1525 Ma. After publication of these results, other examples of very old dates were reported for rocks in South Africa, Zimbabwe, Labrador, and Minnesota. These results have contributed significantly to the establishment of the time scale for the Precambrian part of Earth history.

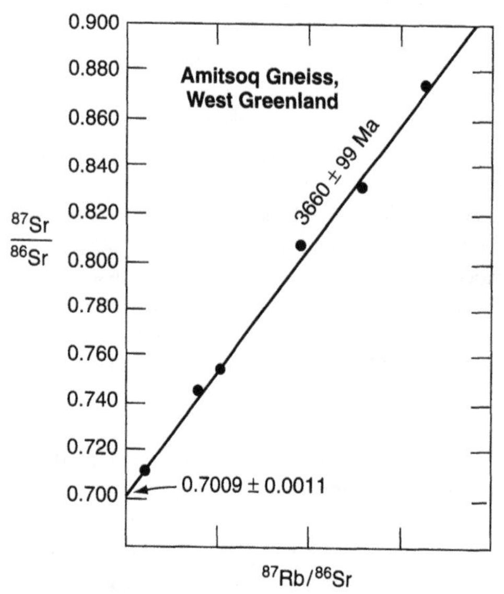

FIGURE 5.12 Whole-rock isochron for the Amitsoq gneiss from the Qilangarssuit area of the Godthaab district of southwestern Greenland. These are among the oldest terrestrial rocks known. The date probably represents the time elapsed since metamorphism of igneous rocks of granitic composition to form the Amitsoq gneiss. Data from Moorbath et al. (1972).

5.4d La Gorce Formation, Wisconsin Range, Antarctica

Metamorphic rocks also form from sequences of sedimentary rocks containing minerals that are unstable at elevated temperatures and therefore recrystallize to form new mineral assemblages when the temperature is elevated. The Rb-bearing minerals of such metamorphic rocks (primarily micas and K-feldspars) can be dated either by the isochron method or by calculating model dates. Such dates generally reflect the time elapsed since the end of the last metamorphic episode to which the rocks were subjected. Whole-rock isochrons may likewise indicate the age of the metamorphic event during which the sediment was recrystallized. This point is illustrated in Figure 5.13 by an isochron for samples of phyllite and slate from the La Gorce Formation of the Wisconsin Range in the Transantarctic Mountains (Montigny and Faure, 1969). The La Gorce Formation consists of fine-grained detrital sedimentary rocks that were isoclinally folded and metamorphosed to the greenschist facies during the Ross Orogeny in Late Cambrian to Early Ordovician time. However, several lines of field evidence leave little room for doubt that these rocks were deposited in late Precambrian time. Nevertheless, the isochron defined by whole-rock

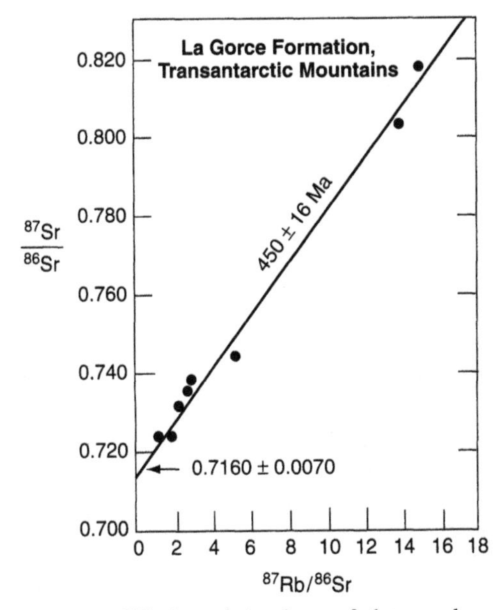

FIGURE 5.13 Whole-rock isochron of slates and phyllites from the La Gorce Formation of probable late Precambrian age from the Wisconsin Range of the Transantarctic Mountains. The date indicated by the slope of the isochron represents the time of metamorphism and extensive homogenization of the isotopic composition of Sr in these rocks. The scatter of points about the isochron is greater than expected from analytical errors alone and suggests that homogenization was not complete or approached different values in different parts of the formation. Data from Montigny and Faure (1969).

samples of the La Gorce Formation indicates a date of only 450 ± 16 Ma (Middle Ordovician), which is the time of metamorphism of these rocks rather than their depositional age. Evidently, the Sr in these samples was isotopically re-equilibrated on a large scale during metamorphism, thereby resetting the Rb–Sr clocks so that they no longer indicate the time elapsed since deposition.

5.5 DATING SEDIMENTARY ROCKS

The ages of sedimentary rocks are especially important because the fossil assemblages they contain have been used to subdivide geological time of the Phanerozoic Eon. Unfortunately, sedimentary rocks are difficult to date by the Rb–Sr method and by the other decay schemes discussed in this book. The reasons vary but include the fact that mud, silt, and sand are composed of mineral particles whose ages predate the time of deposition of the sediment and that clay minerals are altered by interacting with pore water and by recrystallization during deep burial, tectonic deformation of sedimentary rocks, and low grades of regional metamorphism. In addition, sedimentary rocks composed of biogenic or chemical precipitates either have low concentrations of the long-lived parent radionuclides (e.g., limestone and chert) or do not quantitatively retain the parent and daughter isotopes (evaporite rocks). These circumstances have made it difficult to establish a geological timescale for the Phanerozoic Eon.

5.5a Geological Timescale

The voluminous literature concerned with dating sedimentary rocks continues to grow and has been used repeatedly to refine the geological timescales starting with the original proposal by Arthur Holmes in 1913:

Holmes, 1947, *Trans. Geol. Soc. Glasgow*, 21:117–152. Holmes, 1960, *Trans. Edinburgh Geol. Soc.*, 17:183–216. Kulp, 1961, *Science*, 133:1105–1114. Wager, 1964, *J. Geol. Soc. Lond.*, Suppl. 120s. Harland and Francis, 1971, *Geol. Soc. Lond.*, Spec. Pub. No. 5. Cohee et al., 1978, *Am. Assoc. Petrol. Geol.*, Studies in Geology, No. 6. Faul, 1978, *Am. Sci.*, 66:159–165. Palmer, 1983, *Geology*, 11:503–504. Carr et al., 1984, *Geology*, 12:274–277.

Many critically important dates of sedimentary rocks are based on age determinations of interbedded volcanic rocks (lava flows or tuffs); on cross-cutting relationships of intrusive dikes, sills, and plutons; and on the ages of crystalline basement rocks that directly underlie sedimentary rocks. Comprehensive evaluations of available age determination of sedimentary rocks were published in book form by Odin (1982) and Harland et al.

(1989). In addition, the International Commission on Stratigraphy of the IUGS continues to update the geological timescale based on new information. The Appendix at the end of this book contains the dates that define the timescale for the Phanerozoic as of 2003 based on the recommendation of the IUGS.

5.5b Glauconite

Glauconite is a three-layer clay that forms in marine environments by alteration of a variety of pre-existing materials (Faure, 1998). It occurs in the form of dark green pellets (1–2 mm in diameter) composed of aggregates of small platy crystallites whose diameters range from 1 to 10 μm. Glauconite contains varying amounts of K^+ and Rb^+ that neutralize negative charges in the octahedral layer where Al^{3+} is replaced by Fe^{2+} and Mg^{2+}. The presence of K and Rb and evidence that the glauconite pellets formed in marine environments close to the time of deposition of their sedimentary host rocks has motivated attempts to use them to date sedimentary rocks by the Rb–Sr (and K–Ar) methods. However, the first results indicated that Rb–Sr and K–Ar dates of glauconites underestimate the depositional ages of their sedimentary host rocks by 10–20 percent (Hurley et al., 1960).

The origin and chemical evolution of glauconite was further investigated by Odin and Matter (1981), Clauer et al. (1992), Stille and Clauer (1994), and Clauer and Chaudhuri (1995). The results indicate that the glauconite pellets gain K and exclude detrital clays as they evolve and only glauconites that contain more than 6 percent K_2O yield Rb–Sr and K–Ar dates that agree with their stratigraphic ages. However, Rb–Sr dates of glauconites that were altered by deep burial, tectonic deformation of the host rocks, and interaction with circulating brines in the subsurface in many cases postdate the time of deposition. Therefore, the geological history of the host rock and the state of evolution of the glauconite pellets must be evaluated before an age determination is attempted.

Odin and Matter (1981) used x-ray diffraction patterns to define four evolutionary stages of glauconite pellets ranging from low-K glauconitic smectite to K-rich glauconitic mica:

State	K_2O, %
Highly evolved	8.5 and higher
Evolved	6.5–8.5
Little evolved	5.0–6.5
Nascent	>3.0 to <5.0

Experience has shown that nascent and little evolved glauconites, in some cases, yield dates that exceed the depositional age because they still contain detrital clays that are later assimilated or eliminated in the more evolved glauconitic pellets. Therefore, Odin and Matter (1981) proposed the term "glaucony" for the green pellets, which range in composition from glauconitic smectite to glauconitic mica.

The use of glauconitic mica for dating marine sedimentary rocks has yielded many disappointments and a few successes. As a result, some geochronologists have disregarded glauconite dates in the construction of the Phanerozoic timescale (Obradovich and Cobban, 1975; Obradovich, 1988), whereas others emphasize that Rb–Sr isochron dates of preselected glauconitic mica are reliable (Odin, 1982a).

Samples of glauconitic pellets to be used for dating are prepared by repeatedly passing the crushed host rock through a magnetic separator (Odin et al., 1974; Odin, 1982b). The glauconite concentrates may be further purified by heavy-liquid separation in mixtures of bromoform and acetone, by leaching them with ammonium acetate solution or with cold dilute hydrochloric acid, and by manually removing impurities under a binocular microscope (Odin, 1982a; Harris and Fullagar, 1991). The concentrates obtained by these procedures should be analyzed by x-ray diffraction and their K concentrations should be determined before samples composed of glauconitic micas are selected for dating by the Rb–Sr (or K–Ar) method. Five or more selected samples of glauconitic mica from a single stratigraphic unit are used to plot a Rb–Sr isochron that yields not only a date but also the value of the initial $^{87}Sr/^{86}Sr$ ratio.

The interpretation of the resulting dates is exemplified by the work of Laskowski et al. (1980) and Grant et al. (1984), who reported a Rb–Sr isochron date for well-ordered glauconites in marine carbonate rocks of the Belfast Member of

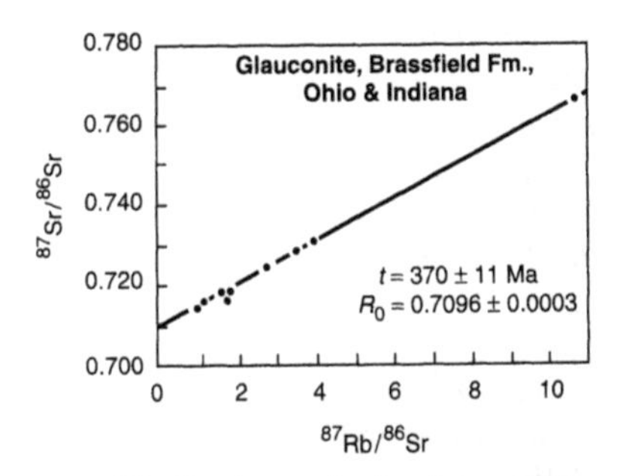

FIGURE 5.14 Rb–Sr glauconite isochron of the Lower Silurian Brassfield Formation of Ohio and Indiana. The calculated date (370 ± 11 Ma) is less than the stratigraphic age of the Brassfield Formation and probably refers to isotopic re-equilibration of Sr in the glauconite by a Sr-bearing aqueous fluid. Data from Grant et al. (1984).

the Brassfield Formation (Silurian; Llandoverian) in Ohio and Indiana. Their results in Figure 5.14 yielded a Rb–Sr isochron date of 370 ± 11 Ma and an initial $^{87}Sr/^{86}Sr$ ratio of 0.7096 ± 0.003 that is similar to the average $^{87}Sr/^{86}Sr$ ratio (0.7097 ± 0.0002) of calcite in vugs and veins of the Brassfield Formation. The good agreement between the initial $^{87}Sr/^{86}Sr$ ratio of the glauconites and the $^{87}Sr/^{86}Sr$ ratio of vein calcite in the Brassfield Formation supports the conclusion of Grant et al. (1984) that the Sr in the glauconites was re-equilibrated isotopically by circulating groundwater or brines. Since the stratigraphic age of the Brassfield Formation is about 430 Ma, the isotopic re-equilibration of Sr in the glauconites occurred about 60 million years after deposition.

Glauconites in the Davis Formation (Late Cambrian) near the town of Bunker, Viburnum Trend, Missouri, also yielded a Rb–Sr isochron date (387 ± 21 Ma) that is younger than the stratigraphic age of the host (Grant et al., 1984). The initial $^{87}Sr/^{86}Sr$ ratio of the glauconite isochron (0.7136 ± 0.0080) exceeds the $^{87}Sr/^{86}Sr$ ratio of the carbonate matrix but is similar to that of calcite in the Pb–Zn ore (0.7137 ± 0.0004) of this region. Therefore, in this case the glauconite isochron also dates the time of deposition of the ore minerals by circulating brines as proposed by Kish and Stein (1979), who reported a Rb–Sr isochron date of 358 ± 6 Ma for glauconite from the Magmont Mine, Viburnum Trend, Missouri.

Well-crystallizated glauconites in the Lion Mountain Member (sandstone) of the Riley Formation (Late Cambrian) exposed in the Llano Uplift of central Texas also record the time of diagenetic alteration rather than the time of deposition. Morton and Long (1980) analyzed nine glauconite fractions of varying purity, some of which had been leached with ammonium acetate or with dilute hydrochloric acid (0.1 and 2.5 N) for different lengths of time (4 and 100 h). The Rb–Sr model dates, calculated by assuming that the initial $^{87}Sr/^{86}Sr$ ratio of the glauconites was equal to that of marine Sr during the Late Cambrian Epoch (see Burke et al., 1982), were all less than the stratigraphic age by about 17 percent. Leaching with 2.5 N HCl caused a decrease of the Sr concentrations of the glauconites from about 14.9 to 1.35 ppm, but lowered their Rb concentrations only slightly, from 275 to 260 ppm. Leaching with 0.1 N HCl and ammonium acetate resulted in smaller changes. Consequently, leaching with 2.5 N HCl caused the $^{87}Rb/^{86}Sr$ ratios of the glauconites in Figure 5.15 to increase from 53.3 to 556, whereas their $^{87}Sr/^{86}Sr$ ratios increased from 1.0303 to 4.0914. However, their model Rb–Sr dates changed only from 422 ± 3 to 427 ± 5 Ma after 100 h of leaching with 2.5 N HCl. Therefore, the loosely bound Sr in these glauconites presumably had the same $^{87}Sr/^{86}Sr$ ratio as the Sr that existed in the glauconites at the time of their final recrystallization.

The age of recrystallization of the glauconites in the Lion Mountain Member is indicated by the slope of the Rb–Sr isochron in Figure 5.16 defined by all nine samples analyzed by Morton and Long (1980). The date is 429 ± 17 Ma and the initial $^{87}Sr/^{86}Sr$ ratio is 0.7070 ± 0.0033. The date records the time of diagenetic alteration caused by burial after deposition. Therefore, the results of Grant et al. (1984) and Morton and Long (1980) confirm the conclusion of Hurley et al. (1960) that Rb–Sr and K–Ar dates of Paleozoic glauconites underestimate their depositional ages by 10–20 percent.

The apparent resistance of well-crystallized glauconites to alteration documented by Morton and Long (1980) bodes well for Rb–Sr dates of glauconites that occur in unmetamorphosed clastic

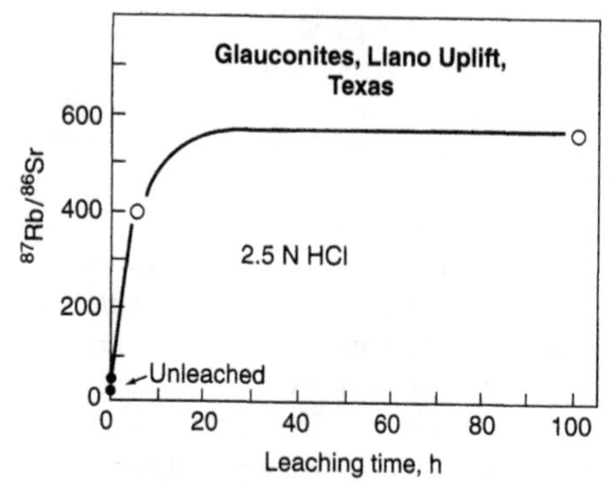

FIGURE 5.15 Effect of leaching glauconite of the Lion Mountain Member (Late Cambrian), Llano Uplift, central Texas, with 2.5 N HCl for up to 100 h. As a result, the Sr concentration decreased significantly, whereas Rb was virtually unaffected. In spite of the increase of the $^{87}Rb/^{86}Sr$ ratio, the calculated model dates remained about 17 percent lower than the stratigraphic age. Data from Morton and Long (1980).

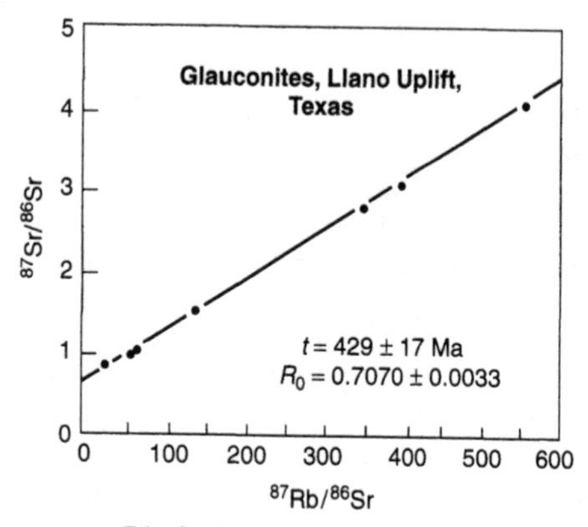

FIGURE 5.16 Rb–Sr isochron for Late Cambrian glauconites of the Lion Mountain Member (sandstone) of the Riley Formation, Llano Uplift, central Texas. The date of 429 ± 17 Ma is the time elapsed since the glauconites recrystallized during deep burial after deposition at about 515 Ma. Data from Morton and Long (1980).

sedimentary rocks of Late Proterozoic age that are underlain by stable granitic basement rocks of Precambrian shield areas of northern and central Australia (McDougall et al., 1965; Cooper et al., 1971) and western North America (Gulbrandsen et al., 1963; Obradovich and Peterman, 1968). Most of the Precambrian glauconites in Australia and North America form linear arrays on the Rb–Sr isochron diagram and yield dates that are compatible with the stratigraphic ages of their Proterozoic sedimentary host rocks. The glauconites in the Late Proterozoic Belt Series in the Sun River area of western Montana define a Rb–Sr isochron in Figure 5.17 from which Obradovich and Peterman (1968) derived a date of 1072 ± 22 Ma ($\lambda = 1.42 \times 10^{-11}$ y^{-1}) and an initial $^{87}Sr/^{86}Sr$ ratio of 0.7089 ± 0.0135. Since the initial $^{87}Sr/^{86}Sr$ ratio of the glauconites is consistent with the isotope composition of marine Sr of Late Proterozoic age, Obradovich and Peterman (1968) concluded that the glauconites had formed in isotopic equilibrium with seawater and that the date therefore represents the time of deposition of the Belt Series in the Sun River area of western Montana.

5.5c Authigenic Feldspar

Adularia, an authigenic feldspar (Cerny and Chapman, 1986), occurs in small concentrations

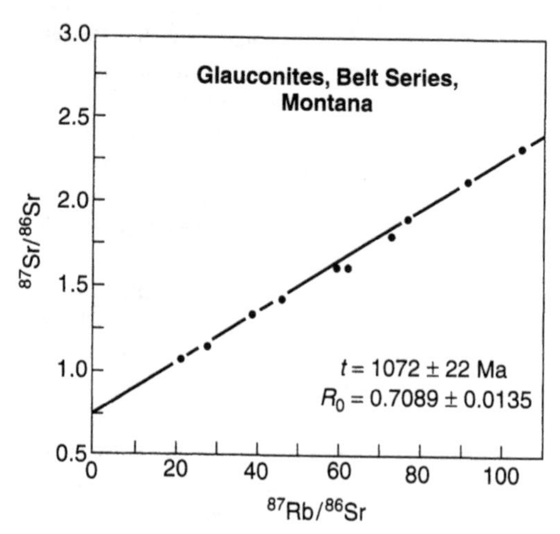

FIGURE 5.17 Rb–Sr isochron of glauconite in the McNamara, Shepard, and Empire formations of the Belt Series, Sun River area, western Montana, United States. Data from Obradovich and Peterman (1968: $\lambda = 1.42 \times 10^{-11}$ y^{-1}).

in sedimentary rocks of marine origin, such as sandstone, siltstone, shale, limestone, dolomite, and evaporites (Kastner, 1971; Buyce and Friedman, 1975; Ali and Turner, 1982). The authigenic feldspar crystals form during diagenesis of sediment by chemical reactions between kaolinite [$Al_2Si_2O_5(OH)_4$] and bottom seawater or pore water (Helgeson, 1972; Kastner and Siever, 1979, Bowers et al., 1984; Faure, 1998). In addition, authigenic feldspar overgrowths can form on detrital feldspar grains (Waugh, 1978) and can be separated from the detrital substrate, thereby allowing both to be dated by the Rb–Sr or K–Ar method and by the isotope composition of Pb (Girard et al., 1988, 1989; Girard and Onstott, 1991; Hearn et al., 1987; Faure, 1992). In addition, overgrowths of authigenic quartz on detrital quartz grains were successfully separated by Lee and Savin (1985) for isotope analysis of oxygen.

Authigenic feldspar also occurs in the granitic gneisses of the Precambrian basement that underlie the Paleozoic sedimentary rocks of the midwestern United States (Bass and Ferrara, 1969; Mensing and Faure, 1983; Faure and Barbis, 1983) as well as in hydrothermal quartz veins in California, New Zealand, and Europe and in felsic spilites of Spain:

> Silberman et al., 1972, *Econ. Geol.*, 67:597–604. Steiner, 1970, *Mineral. Mag.*, 7:916–922. Blattner, 1975, *Am. J. Sci.*, 275:785–800. Pivec, 1973, *Acta Univ. Carolinae Geol.*, 3:171–177. Halliday and Mitchell, 1976, *Earth Planet. Sci.* Lett., 29:227–237. Lippolt et al., 1985, *Neues Jahrbuch Mineral. Monatshefte*, 2:49–57. Munhá et al., 1980, *Contrib. Mineral. Petrol.*, 75:15–19.

The Middle Proterozoic basement rocks of Ohio consist in part of granitic gneisses that contain a fresh salmon-colored feldspar. Mensing and Faure (1983) demonstrated by cathodoluminescence that this feldspar consists of adularia containing remnants of microcline. The adularia is most abundant within 6 m of the former Precambrian erosion surface but grades into microcline at greater depth. Feldspar concentrates recovered from a core of the Precambrian basement in Scioto County, Ohio, forms two straight lines in the Rb–Sr isochron diagram in Figure 5.18. The adularia–microcline

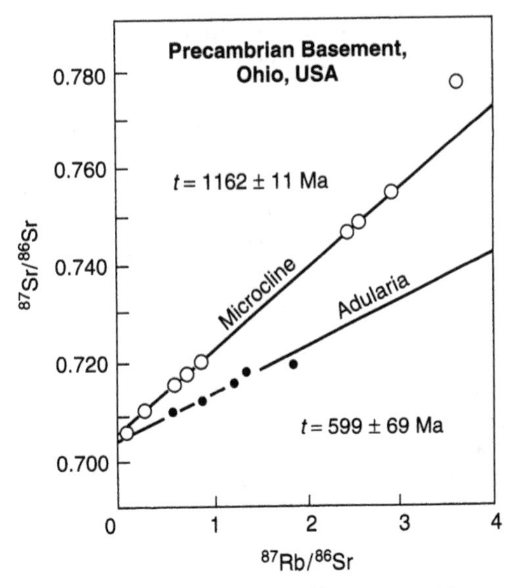

FIGURE 5.18 Rb–Sr dates of feldspar in granitic gneisses of the Precambrian basement, Scioto County, Ohio. The adularia occurs within 6 m of the former erosion surface, whereas the microcline dominates at greater depth. Data from Mensing and Faure (1983).

mixtures in the upper part of the basement yielded a date of 599 ± 69 Ma, whereas the microclines extracted from rocks more than 6 m below the nonconformity crystallized at 1162 ± 11 Ma. Mensing and Faure (1983) concluded that the adularia developed from kaolinite that had formed by chemical weathering of microcline in Precambrian time when the granitic gneiss was exposed at the surface:

$$\underset{\text{kaolinite}}{Al_2Si_2O_5(OH)_4} + 2K^+ + \underset{\text{silicic acid}}{4H_4SiO_4} \rightarrow \underset{\text{K-feldspar}}{2KAlSi_3O_8} + 9H_2O + 2H^+ \quad (5.13)$$

The necessary K^+ and silicic acid were provided by seawater or related brines during the early Paleozoic Era. Therefore, the adularia formed by alteration of Precambrian weathering products that were buried by the marine transgression during the Cambrian Period. Since the adularia contains varying amounts of residual microcline, its age is probably less than the date derived from the slope of the line in Figure 5.18.

Adularia also occurs in the Precambrian basement rocks in other counties of Ohio, for which a whole-rock Rb–Sr date of 563 ± 10 Ma was reported by Faure and Barbis (1983). However, a gneissic quartz monzonite in the basement of Sandusky County yielded a Precambrian date of 1197 ± 101 Ma. These results indicate that the distribution of adularia is discontinuous, presumably because the kaolinite from which it formed accumulated only in topographic depressions on the Precambrian weathering surface. Additional age determinations of biotite and a few feldspars in the Precambrian basement of the midwestern United States (including Ohio) were published by Lidiak et al. (1966).

5.5d Detrital Minerals

All clastic sedimentary rocks contain mineral or rock particles that were derived by weathering and erosion of rocks on the continents. The minerals that survive this process because they resist chemical and mechanical weathering include muscovite, K-feldspar, illite, zircon, and others, some of which are suitable for dating by isotopic methods. These minerals can be separated from the sedimentary rocks in which they occur and may be dated by appropriate isotopic methods to determine the age of the rocks from which they were derived. Such "provenance dates" can be used to identify specific sources of sediment for paleogeographic reconstructions of depositional basins. Heller et al. (1985) used this method to determine the provenance of the Eocene Tyee Sandstone of Oregon based on isotopic dates of detrital muscovite and K-feldspar. Another example is the work of Gaudette et al. (1981), who dated individual zircon grains from the Potsdam Sandstone of New York by the U–Pb method.

In cases where a particular detrital mineral was derived from two or more source rocks having significantly different ages, the resulting mixture of mineral grains yields a mixed date that has no real time significance but, instead, reflects the proportions of grains derived from the contributing sources. Stratigraphic variations of mixed dates may be caused by changes in the relative proportions of mineral grains derived from multiple sources or they may signal the introduction of minerals from a new source. Systematic regional variations of mixed dates in a specific stratigraphic unit may be contoured to reveal the locations of source rocks that contributed mineral grains of a particular age. Most of these possibilities have not yet been tested.

Age determinations of detrital minerals by the Rb–Sr method rely on the validity of several assumptions:

1. The mineral particles remained closed to Rb and Sr during weathering and transport.
2. The minerals were not altered during diagenesis and do not have overgrowths formed during diagenesis.
3. The initial $^{87}Sr/^{86}Sr$ ratios of minerals derived from sources of different ages are similar.

These assumptions are probably not strictly satisfied by any of the common detrital minerals datable by the Rb–Sr method. Nevertheless, Taylor and Faure (1981), Faure and Taylor (1981), and Faure et al. (1983) obtained useful geological information by dating feldspar in glacial sediment in both North America and the Transantarctic Mountains.

The dating of detrital feldspar in till is facilitated by systematic variations of the K-feldspar/plagioclase ratios of different grain size fractions. In many cases, the Rb/Sr ratios of feldspar concentrates increase with grain size because the K-feldspar/plagioclase ratios of coarse-sand fractions of till are higher than those of the fine-sand or silt fractions. The presence of quartz as an impurity in the feldspar concentrates does not interfere with dating by the Rb–Sr method because quartz has very low concentrations of Rb and Sr and therefore does not alter the Rb/Sr ratio of the feldspar. Two linear arrays formed by grain size fractions of feldspar concentrates derived from late Cenozoic till in the Transantarctic Mountains appear in Figure 5.19.

In some cases, sandstones contain flakes of detrital muscovite that can be dated by the Rb–Sr, K–Ar, or $^{40}Ar/^{39}Ar$ methods. A case in point is the Meridian Sand (Eocene) of Mississippi and Alabama. Wermund et al. (1966) reported that

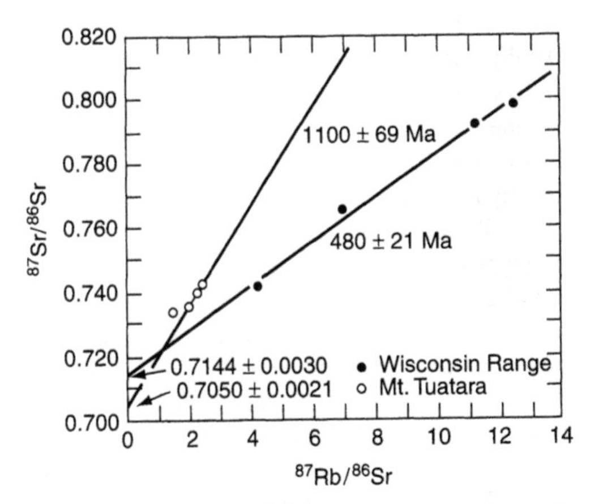

FIGURE 5.19 Rb–Sr dates of grain-size fractions of feldspar–quartz concentrates derived from late Cenozoic till in the Transantarctic Mountains. The date of 480 ± 21 Ma, derived from detrital feldspar in till of the Wisconsin Range (85° 45′ S, 125° 00′ W), is indistinguishable from the age of the local granitic basement rocks. The date of 1100 ± 69 Ma of feldspar in till on Mt. Tuatara (80° 30′ S, 156° 00′ W), adjacent to the Byrd Glacier, indicates that the feldspar is a mixture of a local component (~500 Ma) and feldspar that originated from the Precambrian Shield of East Antarctica. Data from Faure et al. (1983) and Faure and Taylor (1981).

the 60 samples they examined contained from 0.25 to 18.4 percent muscovite with diameters up to 7 mm. Two samples of muscovite in this Eocene sandstone yielded dates of 308 ± 6 and 290 ± 6 Ma (Pennsylvanian), thus demonstrating that the muscovite is detrital in origin and probably originated by erosion of source rocks in the southern Appalachian Mountains located east of the outcrop belt of the Meridian Sand.

Another example of provenance dating was reported by Dallmeyer (1987), who obtained a $^{40}Ar/^{39}Ar$ date of 504.1 ± 1.2 Ma for muscovite extracted from an Early Ordovician (505–478 Ma) sandstone recovered in a core at a depth of 1282 m in Marion County, Florida. The early Paleozoic sedimentary rocks in the subsurface of this area are part of the Suwanee Terrane. These sedimentary rocks in Florida have been correlated with sedimentary rocks of similar age in West Africa that are underlain by metamorphic basement rocks containing muscovite that cooled between 510 and 500 Ma. Therefore, Dallmeyer (1987) concluded that the muscovite in the Early Ordovician sandstone of Florida originated from metamorphic rocks of Senegal and Guinea in West Africa prior to the opening of the Iapetus Ocean.

5.5e Bentonite and Tuff

Layers of volcanic ash in some cases contain crystals of minerals that can be dated by the Rb–Sr and K–Ar methods (e.g., biotite and sanidine) or by the U–Pb method (e.g., zircon). For example, sanidine crystals recovered from pyroclastic deposits of Late Carboniferous (Pennsylvanian) age in Europe were dated by Lippolt et al. (1984), who used both the K–Ar and $^{40}Ar/^{39}Ar$ methods. Similarly, bentonite beds consisting of smectite clay derived by alteration of volcanic ash may also contain resistant minerals that can be dated by the appropriate isotopic methods. The principles of dating bentonite beds by analysis of highly purified fractions of resistant minerals were summarized by Baadsgaard and Lerbekmo (1982). The same authors dated biotite, sanidine, and zircon crystals from bentonite beds associated with Paleocene coal seams of Montana, Saskatchewan, and Alberta (Baadsgaard and Lerbekmo, 1980, 1983; Baadsgaard et al., 1988). These bentonite beds are located about 1 m above the K-T boundary layer in which Lerbekmo and St. Louis (1986) recorded significant Ir concentrations up to about 3 ng/g (e.g., Nevis Coal, Scollard Canyon, Red Deer Valley, Alberta; Baadsgaard et al., 1988).

The Rb–Sr isochron for biotite and sanidine from the bentonite in the Z coal at Hell Creek, Montana (47°32′N, 106°56′W), in Figure 5.20 yielded a date of 63.7 ± 0.6 Ma (MSWD = 1.07), which is in good agreement with a K–Ar date of 64.6 ± 1.0 Ma for sanidine and a U–Pb date of 63.9 + 0.6, −0.8 Ma of associated zircon crystals. The authors leached the biotite crystals with cold 2 N HCl in order to remove coatings of limonite and carbonate and noted that biotite samples that had lost more than 30 percent of the Rb

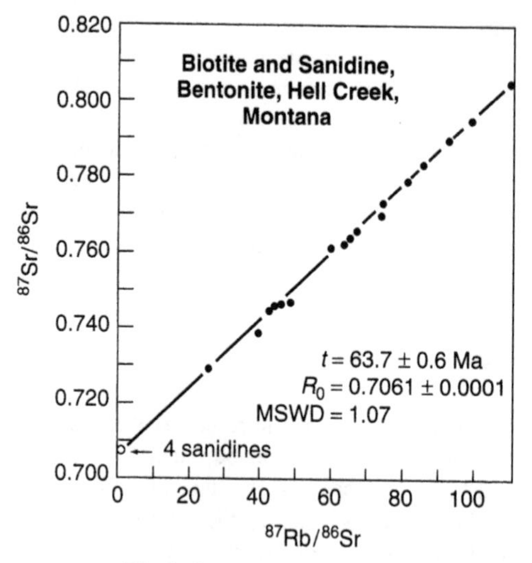

FIGURE 5.20 Rb–Sr isochron of biotite and sanidine extracted from bentonite in the Z coal at Hell Creek, Montana. The bentonite lies about 1 m above the K–T boundary at this location. Data from Baadsgaard and Lerbekmo (1983) and Baadsgaard et al. (1988).

they originally contained plotted below the Rb–Sr isochron formed by the Rb-rich biotites.

Other examples of successful age determinations of bentonite beds include the following:

Yanagi et al., 1988, *Can. J. Earth Sci.*, 25:1123–1127. Harris and Fullagar, 1989, *Geol. Soc. Am. Bull.*, 101:573–577. Lanphere and Tailleur, 1983, *Cretaceous Res.*, 4:361–370. Elliott and Aronson, 1987, *Geology*, 15:735–739. Fullagar and Bottino, 1969, *Southeast Geol.*, 10:247–256.

5.5f Shale

Shale and other fine-grained clastic sedimentary rocks contain the clay mineral illite, which can have sufficient concentrations of Rb (and K) to make them attractive for dating by the Rb–Sr and K–Ar methods (Spears, 1974). Illite consists of various polytypes that are recognizable by their x-ray diffraction patterns: 1Md, 1M, 2M, and 3T (Bailey, 1988 a, b). According to Clauer and Chaudhuri (1995), the 1Md and 1M polytypes of illite form at low temperature during diagenesis of marine sediment, whereas the 2M polytype forms above 250°C during low-grade regional metamorphism and hydrothermal alteration of sedimentary rocks (e.g., Zwingmann et al., 1999). The 3T illite polytype occurs in hydrothermally altered rocks but is not common.

Most marine shales contain mixtures of diagenetic (1M and 1Md) and detrital illite (2M) in varying proportions. Consequently, whole-rock Rb–Sr (and K–Ar) dates of illitic shale are difficult to interpret because the presence of varying proportions of the 2M polytype causes the dates to exceed the depositional age of the rock. On the other hand, the 1M and 1Md polytypes may form during late-stage diagenesis or burial metamorphism and, in that case, yield Rb–Sr (and K–Ar) dates that are younger than the stratigraphic age of the rocks.

For example, Gebauer and Grünenfelder (1974) demonstrated that Rb–Sr isochron dates of clay-rich sedimentary rocks of early Paleozoic age in the Montagne Noir of France indicate the time of folding and low-grade metamorphism rather than the time of deposition. However, rock samples from the Montagne Noir that contained significant amounts of detrital minerals (e.g., muscovite) yielded dates that approached the age of the source rocks and exceeded the age of deposition. The authors concluded that whole-rock Rb–Sr dates of clay-rich rocks record the time of late-stage diagenesis and low-grade metamorphism rather than the time of deposition.

All clay minerals in shale contain varying amounts of sorbed Sr, whose presence alters the isotope composition of Sr in the clay minerals (e.g., illite) and in the rocks as a whole. In addition, clay minerals in marine shale sorb varying amounts of Rb from the environment in which they are deposited. The presence of sorbed Sr and Rb in marine shales in many cases lowers their Rb/Sr ratios and violates one of the key assumptions of dating (closed system). Therefore, the reliability of Rb–Sr dates is improved by leaching powdered samples of shale with various solvents that can remove the exchangeable Rb and Sr without loss of the strongly held Rb and the associated Sr.

The solvents used for this purpose include dilute (1 N) solutions of HCl, NH_4Cl, humic acid, NH_4–ethylenediaminetetraacetic acid (EDTA), ammonium acetate, and acetone (Clauer et al., 1992, 1993; Clauer and Chaudhuri, 1995). These solvents can also remove certain mineral impurities (e.g., calcite and ferric oxide) from the powdered shale samples and thereby increase their Rb/Sr ratios. Therefore, the residues of such pretreatment procedures are expected to yield more meaningful isochron dates than unleached samples of shale.

The effectiveness of dating different grain size fractions of illite that were leached prior to analysis was demonstrated by Morton (1985) for the case of the Woodford Shale (Late Devonian) of west Texas. The clay minerals in the 1–2-μm fraction of this shale contain the 2M polytype. After treatment with ammonium acetate to remove exchangeable Rb and Sr, the coarse clay fractions formed a straight line corresponding to a date of 539 ± 18 Ma (Cambrian), which is significantly older than the depositional age of the Woodford Shale, thought to be between 360 and 374 Ma (Late Devonian). Morton (1985) interpreted the Rb–Sr date of the 2M illites as the age of the source regions from which the coarse illite originated. The finer clay fractions (<0.2 μm) of the Woodford Shale contain the 1Md polymorph and formed a second line that yielded a date of only 301 ± 4 Ma (Pennsylvanian). This date may represent an episode of diagenesis related to the occurrence of the Ouachita orogeny in mid-Pennsylvanian time during which Sr in the fine-grained authigenic clay was isotopically re-equilibrated with Sr in pore water.

This interpretation is compatible with the initial $^{87}Sr/^{86}Sr$ ratios of the two isochrons. The authigenic clay fractions have an initial $^{87}Sr/^{86}Sr$ ratio of 0.7086 ± 0.0004 that is similar to the $^{87}Sr/^{86}Sr$ ratio of marine limestones of Paleozoic age. Therefore, the pore water in the Woodford Shale at the time of diagenesis contained Sr derived from underlying marine carbonates of Paleozoic age. The detrital clays have a much lower initial ratio of 0.7037 ± 0.0008 found only in young volcanic rocks derived from magma sources in the mantle of the Earth.

The effect of leaching clay samples with various kinds of solvents was reviewed by Clauer et al. (1993) based on the information that had accumulated in the literature. Their own experimental results in Figure 5.21 were derived from the study of a plastic claystone of the Early Cambrian Lontova Formation collected in a quarry near Kunda, Estonia. Clauer et al. (1993) demonstrated that both coarse (2–0.8 μm) and fine illite fractions (<0.4 μm) of this claystone lost varying proportions of Sr depending on the solvent, ranging up to nearly 50 percent by leaching with NH_4–EDTA, whereas the losses of Rb did not exceed 10 percent. Consequently, leaching with NH_4–EDTA increased the Rb/Sr ratio of the coarse clay fraction from 1.47 to 2.60 and of the fine fraction from 2.06 to 3.59.

The $^{87}Sr/^{86}Sr$ ratios of the clay residues increased in all cases as a result of leaching,

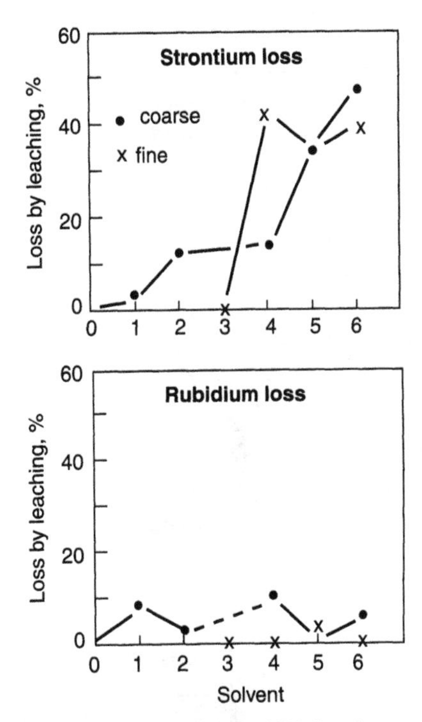

FIGURE 5.21 Fractions of Sr and Rb lost by leaching coarse (2–0.8-μm) and fine (<0.4-μm) fractions of a plastic claystone of Cambrian age from Estonia by solvents identified by number: 0, unleached; 1, humic acid; 2, NH_4Cl; 3, acetone; 4, 1 N HCl; 5, ion exchange resin; 6, NH_4–EDTA. Data from Clauer et al. (1993).

whereas the leachates had uniformly low $^{87}Sr/^{86}Sr$ ratios of 0.7120 (coarse fraction) and 0.7132 (fine fraction). In addition, the leachates had low Rb/Sr ratios that averaged 0.081. As a result, all of the leachates formed a cluster of data points close to the y-axis on a Rb–Sr isochron diagram. The leached coarse and fine clay fractions of the Estonian claystones define two straight lines in Figure 5.22 that converge to this cluster of leachates and yield dates of 638 ± 9 Ma (coarse) and 508 ± 10 Ma (fine), respectively.

Clauer et al. (1993) favored the explanation that the different solvents they used selectively removed certain materials present in the claystone. This conclusion is supported by apparent losses of CaO, Ba, and Zr from the fine-grained clay fraction as a result of leaching with all of the solvents

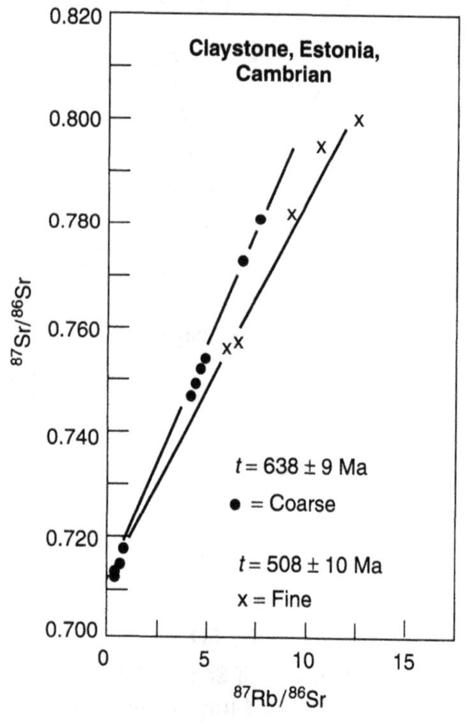

FIGURE 5.22 Rb–Sr dates of leached coarse and fine fractions of a claystone of Cambrian age in Estonia. The leachates form a cluster of data points close to the point of intersection of the two straight lines. Data from Clauer et al. (1993).

except acetone. In addition, some losses may have occurred from the fine fraction by particles that remained in suspension in the leachate. The $^{87}Sr/^{86}Sr$ ratio of the Sr in the material removed by leaching was similar to that which existed in the clay minerals at the time they became closed systems, indicating that, in this case, leaching did not preferentially remove radiogenic ^{87}Sr from the clay minerals.

The date derivable from the coarse clay fractions (638 ± 9 Ma) exceeds the stratigraphic age of the Estonian claystone, whereas the date calculated from the fine fractions (508 ± 10 Ma) is less than the stratigraphic age. Clauer et al. (1993) considered the straight line defined by the coarse fractions to be a mixing line between detrital 2M illite and diagenetic illite. Therefore, the date derivable from the slope of this line is older than the depositional age of the claystone. Since the claystone was never buried more than a few hundred meters and since the clay particles had the characteristic shapes of authigenic clay minerals, the fine fractions of the leached clay define a proper Rb–Sr isochron (Clauer et al., 1993). However, the date is the time elapsed since closure of the clay sediment, which apparently occurred at least 20 million years after deposition. Therefore, the Rb–Sr isochron of the fine fractions underestimated the depositional age of the Estonian claystone. Clauer et al. (1993) also reported that the coarse and fine leached fractions of clay yielded K–Ar dates of 555 ± 35 and 460 ± 41 Ma, respectively. Both values are lower than the Rb–Sr dates of the same fractions and are not reliable determinations of the depositional age of Estonian claystone. The extent to which Sr in the detrital mineral grains of sediment is isotopically homogenized during diagenesis and burial metamorphism remains controversial, probably because differences in local conditions affect the outcome of the process (Allsopp and Kolbe, 1965; Bofinger and Compston, 1967; Perry and Turekian, 1974; Cordani et al., 1978; Awwiller, 1994).

Whole-rock Rb–Sr dates of unmetamorphosed clay-rich sedimentary rocks of Precambrian age have yielded dates that are consistent with the stratigraphic position of such rocks and with the geological history of the region. For example, Moorbath (1969) obtained a date of 915 ± 24 Ma for the Lower Torridonian sedimentary rocks of northwest

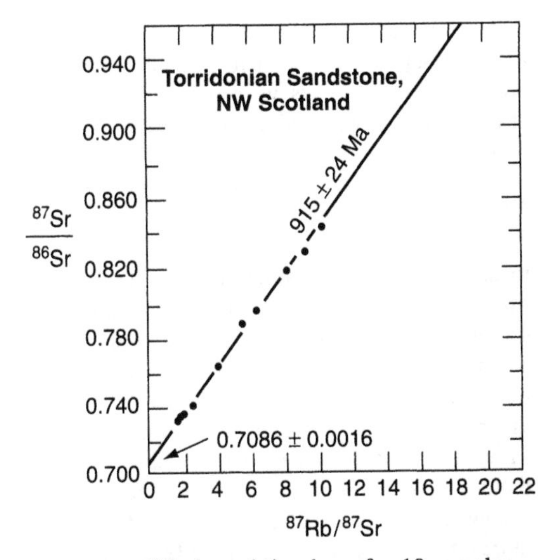

FIGURE 5.23 Whole-rock isochron for 10 samples of unmetamorphosed siltstone from the Lower Torridonian of Stoer in northwest Scotland. The good fit of these samples on the isochron indicates that they had the same initial $^{87}Sr/^{86}Sr$ ratio at 915 ± 24 Ma, which can be interpreted as the time of diagenesis soon after deposition and compaction of these sedimentary rocks. Data from Moorbath (1969).

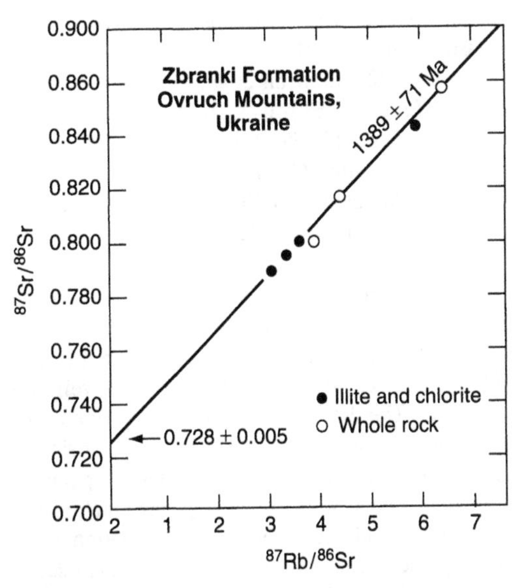

FIGURE 5.24 Rb–Sr isochron of the <2-μm clay fractions and whole-rock samples of the Zbranki Formation from the Ovruch Mountains, Ukraine. Data and interpretation from Gorokhov et al. (1981).

Scotland in Figure 5.23 and a date of 744 ± 17 Ma for the Upper Torridonian Applecross Formation, which lies uncomfortably on the Lower Torridonian formations. He interpreted the dates as being representative of the time of diagenesis, which followed closely after deposition and compaction of these rocks.

Another example of dating clay fractions of unmetamorphosed Precambrian sedimentary rocks in Figure 5.24 is based on the work of Gorokhov et al. (1981). Size fractions (<2 μm) composed of illite and chlorite with minor smectite and kaolinite as well as whole-rock samples of unmetamorphosed sedimentary rocks collected from the Zbranki Formation in the Ovruch Mountains of the Ukraine yielded a Rb–Sr isochron date of 1389 ± 71 Ma. The crystallinity index of the illite indicates that the illite is either authigenic or transformed during diagenesis. Therefore, the date refers to the time of diagenesis shortly after deposition of the sediment.

Other examples of Rb–Sr dating of clay-rich sedimentary rocks of early Paleozoic or Precambrian age include the following:

Compston and Pidgeon, 1962, *J. Geophys. Res.*, 67:3493–3502. Whitney and Hurley, 1964, *Geochim. Cosmochim. Acta*, 28:425–436. Chaudhuri and Faure, 1967, *Econ. Geol.*, 62:1011–1033. Bofinger et al., 1968, *Geochim. Cosmochim. Acta*, 32:823–833. Allsopp et al., 1968, *Can. J. Earth Sci.*, 5:605–619. Chaudhuri and Brookins, 1969, *Geol. Soc. Am. Bull.*, 80:2605–2610. Chaudhuri and Brookins, 1969b, *J. Sed. Pet.*, 39:364–366. Fairbairn et al., 1969, *Can. J. Earth Sci.*, 6:489–497. Hurley et al., 1972, *Earth Planet. Sci. Lett.*, 14:360–366. Clauer, 1973, *Geochim. Cosmochim. Acta*, 37:2243–2256. Clauer, 1974, *Earth Planet. Sci. Lett.*, 22:404–412. Clauer and Kröner, 1979, *Earth Planet. Sci. Lett.*, 42:117–131. Clauer, 1981, *Precamb. Res.*, 15:331–352. Clauer et al., 1982, *Precamb. Res.*, 18:53–71.

Bonhomme, 1982, *Precamb. Res.*, 18:5–25. Bonhomme et al., 1982, *Precamb. Res.*, 18:87–102. Hofmann et al., 1974, *Geol. Soc. Am. Bull.*, 85:639–644. Kralik, 1982, *Precamb. Res.*, 18:157–170. Stille and Clauer, 1986, *Geochim. Cosmochim. Acta*, 50:1141–1146. Cingolani and Bonhomme, 1982, *Precamb. Res.*, 18:119–132.

In general, whole-rock Rb–Sr dates of unmetamorphosed Precambrian shales provide useful information in cases where the stratigraphic ages of these rocks are unknown or are not well constrained. The Rb–Sr dating of fine-grained illite fractions extracted from fossiliferous shales of Phanerozoic age requires a great deal of painstaking preparatory work, including size fractionation of the clay minerals, their characterization by x-ray diffraction and scanning electron microscopy, and leaching with a variety of solvents. In spite of the best efforts of the most experienced clay geochronologists, authigenic illite (and glauconite) in virtually all cases records the time of final closure of the Rb–Sr systems that occurred after the original time of deposition. Although such dates cannot be used to refine the geological timescale, they do contribute to the reconstruction of the depositional history and structural evolution of sedimentary basins. The recent study of Zwingmann et al. (1999) on the formation of illite in the Rotliegend Sandstone (Permian, northern Germany) is a good example of this approach.

5.6 SUMMARY

The Rb–Sr method of dating encompasses most of the principles that apply also to other methods of dating by means of radiogenic isotopes. However, the interpretation of the resulting dates requires that the minerals and rocks satisfy certain assumptions that were discussed in Chapter 4.

Given the wide distribution of Rb and Sr in many rock-forming minerals, the Rb–Sr method can be used to date a wide variety of igneous, metamorphic, and sedimentary rocks. The dates refer in all cases to the time elapsed since final homogenization of the isotope composition of Sr. Therefore, whole-rock Rb–Sr isochron dates of unmetamorphosed volcanic and plutonic igneous rocks closely approach their crystallization ages provided that the rocks were not metamorphosed or hydrothermally altered and that plutonic igneous rocks cooled rapidly. In addition, such rocks must have formed from magma having the same $^{87}Sr/^{86}Sr$ ratio throughout.

Whole-rock Rb-Sr dates of metamorphic rocks record the time of metamorphism rather than the time of initial crystallization of igneous protoliths or of deposition in the case of sedimentary precursors. The minerals of polymetamorphic rocks may yield discordant Rb–Sr dates depending on the retentivity of different Rb-bearing minerals. In some cases, the least retentive minerals (e.g., biotite) reflect the time elapsed since cooling following the last metamorphic event, whereas more retentive minerals (e.g., hornblende) may only have been partly reset and yield dates that are intermediate between the time of crystallization and metamorphism.

The dating of sedimentary rocks by the Rb–Sr method is especially important for the refinement of the geological timescale. Unfortunately, Rb–Sr dates of clastic sedimentary rocks (e.g., shale and mudstone) refer, in most cases, to the time of diagenesis or to episodes of isotopic re-equilibration by interaction with hot brines. Similarly, authigenic Rb-bearing minerals (e.g., glauconite and illite) are sensitive to postdepositional alteration. The most reliable method of dating sedimentary rocks is by means of well-crystallized minerals (e.g., biotite and sanidine) in layers of volcanic ash or bentonite interbedded with terrigeneous sedimentary rocks.

Age determinations of stony meteorites (e.g., chondrites and achondrites) by the Rb–Sr method have provided a reliable estimate of the age of the solar system (4.55×10^9 years) and of the $^{87}Sr/^{86}Sr$ ratio of the solid particles from which the planets and other solid objects of the solar system formed. In addition, the Rb–Sr dates of lunar rocks have established the geological history of the Moon from about 4.5 to 3.0 Ga. The study of the geological history of the Moon was made possible by the analysis of rock samples recovered by the American and Soviet space programs and of lunar meteorites collected on the surface of the East Antarctic ice sheet. Similarly, Rb–Sr dating of martian meteorites has provided evidence for magmatic activity on

Mars extending from 4.5 to 1.2 Ga as well as for the occurrence of geological activity as recently as about 170 Ma.

REFERENCES FOR Rb–Sr DATING (SECTIONS 5.1–5.3a)

Brooks, C., S. R. Hart, and I. Wendt, 1972. Realistic use of two-error regression treatments as applied to rubidium-strontium data. *Rev. Geophys. Space Phys.*, 10(2): 551–577.

Campbell, N. R., and A. Wood, 1906. The radioactivity of the alkali metals. *Proc. Cambridge Philos. Soc.*, 14:15–21.

Catanzaro, E. J., T. J. Murphy, E. L. Garner, and W. R. Shields, 1969. Absolute isotopic abundance ratio and atomic weight of terrestrial rubidium. *J. Res. Natl. Bur. Std. Phys. and Chem.*, 73A: 511–516.

Davis, D. W., J. Gray, G. L. Cumming, and H. Baadsgaard, 1977. Determination of the ^{87}Rb decay constant. *Geochim. Cosmochim. Acta*, 41:1745–1749.

Faure, G., 2001. *Origin of Igneous Rocks; the Isotopic Evidence*. Springer-Verlag, Heidelberg.

Faure, G., and P. M. Hurley, 1963. The isotopic composition of strontium in oceanic and continental basalt: Application to the origin of igneous rocks. *J. Petrol.*, 4:31–50.

Faure, G., and J. L. Powell, 1972. *Strontium Isotope Geology*. Springer-Verlag, Heidelberg.

Hahn, O., F. Strassman, J. Mattauch, and H. Ewald, 1943. Geologische Altersbestimmungen mit der Strontiummethode. *Chem. Zeitung*, 67:55–56.

Hahn, O., F. Strassman, and E. Walling, 1937. Herstellung wägbarer Mengen des Strontiumisotops 87 als Umwandlungsprodukt des Rubidiums aus einem kanadischen Glimmer. *Naturwissenschaften*, 25:189.

Hahn, O., and E. Walling, 1938. Über die Möglichkeit geologischer Altersbestimmungen rubidiumhaltiger Mineralen and Gesteine. *Z. Anorg. Allgem. Chem.*, 236:78–82.

Hart, S. R., and A. Zindler, 1989. Isotope fractionation laws: A test using calcium. *Int. J. Mass Spectrom. Ion Process.*, 89:287–301.

Herzog, L. F., W. H. Pinson, and R. F. Cormier, 1958. Sediment age determination by Rb/Sr analysis of glauconite. *Am. Assoc. Petrol. Geol. Bull.*, 42:717–733.

Mattauch, J., 1937. Das Paar Rb^{87}-Sr^{87} und die Isobarenregel. *Naturwissenschaften*, 25:189–191.

McIntyre, G. A., C. Brooks, W. Compston, and A. Turek, 1966. The statistical assessment of Rb-Sr isochrons. *J. Geophys. Res.*, 71(22): 5459–5468.

Neumann, W., and H. Huster, 1974. The half-life of ^{87}Rb measured as a difference between the isotopes ^{87}Rb and ^{85}Rb. *Z. Phy.*, 270:121–127.

Nier, A. O., 1938. Isotopic constitution of strontium, barium, bismuth, thallium, and mercury. *Phys. Rev.*, 54:275–278.

Shima, M., 1986. A summary of extremes of isotopic variations in extra-terrestrial materials. *Geochim. Cosmochim. Acta*, 50:577–584.

Steiger, R. H., and E. Jäger, 1977. Subcommission on Geochronology: Convention on the use of decay constants in geo- and cosmochronology. *Earth Planet. Sci. Lett.*, 36:359–362.

Turekian, K. K., and K. H. Wedepohl, 1961. Distribution of the elements in some major units of the Earth's crust. *Geol. Soc. Am. Bull.*, 72:175–182.

van Breemen, O., J. Hutchinson, and P. Bowden, 1975. Age and origin of the Nigerian Mesozoic granites: A Rb-Sr isotopic study. *Contrib. Mineral. Petrol.*, 50:157–172.

Walker, F. W., J. R. Parrington, and F. Feiner, 1989. *Chart of the Nuclides*, 14th ed. General Electric Co., San Jose, California.

REFERENCES FOR STONY AND IRON METEORITES (SECTION 5.3b)

Burnett, D. S., and G. J. Wasserburg, 1967. ^{87}Rb-^{87}Sr ages of silicate inclusions in iron meteorites. *Earth Planet. Sci. Lett.*, 2:397–408.

Dymek, R. G., A. L. Albee, A. A. Chodos, and G. J. Wasserburg, 1976. Petrography of isotopically-dated clasts in the Kapoeta howardite and petrologic constraints on the evolution of its parent body. *Geochim. Cosmochim. Acta*, 40:1115–1130.

Faure, G., and J. L. Powell, 1972. *Strontium Isotope Geology*. Springer-Verlag, Heidelberg.

Gladman, B. J., J. A. Burns, M. Duncan, P. Lee, and H. F. Levinson, 1996. The exchange of impact ejecta between terrestrial planets. *Science*, 271:1387–1392.

Gopalan, K., and G. W. Wetherill, 1968. Rubidium-strontium age of hypersthene (L) chondrites. *J. Geophys. Res.*, 73:7133–7136.

Gopalan, K., and G. W. Wetherill, 1969. Rubidium-strontium age of amphoterite (LL) chondrites. *J. Geophys. Res.*, 74:4349–4358.

Gopalan, K., and G. W. Wetherill, 1970. Rubidium-strontium studies on enstatite chondrites: Whole meteorites and mineral isochron. *J. Geophys. Res.*, 75:3457–3467.

Gray, C. M., D. A. Papanastassiou, and G. J. Wasserburg, 1973. The identification of early condensates from the solar nebula. *Icarus*, 20:213–239.

Kempe, W., and O. Müller, 1969. The stony meteorite Krähenberg: Its chemical composition and the Rb-Sr age of the light and dark portions. In P. M. Millman (ed.), *Meteorite Research*, pp. 418–428. Reidel, Dortrecht, The Netherlands.

Lugmair, G. W., and S. J. G. Galer, 1992. Age and isotopic relationships among angrites Lewis Cliff 86010 and Angra dos Reis. *Geochim. Cosmochim. Acta*, 56:1673–1694.

Minster, J. F., and C. J. Allègre, 1981. ^{87}Rb-^{87}Sr dating of LL chondrites. *Earth Planet. Sci. Lett.*, 56:89–106.

Papanastassiou, D. A., and G. J. Wasserburg, 1969. Initial strontium isotopic abundances and the resolution of small time differences in the formation of planetary objects. *Earth Planet. Sci. Lett.*, 5:361–376.

Pinson, W. H., C. C. Schnetzler, E. Beiser, H. W. Fairbairn, and P. M. Hurley, 1965. Rb-Sr age of stony meteorites. *Geochim. Cosmochim. Acta*, 29:455–466.

Sanz, H. G., D. S. Burnett, and G. J. Wasserburg, 1970. A precise ^{87}Rb/^{87}Sr age and initial ^{87}Sr/^{86}Sr for the Colomera iron meteorite. *Geochim. Cosmochim. Acta*, 34:1227–1239.

Wasserburg, G. J., F. Tera, D. A. Papanastassiou, and J. C. Huneke, 1977. Isotopic and chemical investigations on Angra dos Reis. *Earth Planet. Sci. Lett.*, 35:294–316.

Wetherill, G. W., R. K. Mark, and C. Lee-Hu, 1973. Chondrites: Initial strontium^{87}/strontium^{86} ratios and the early history of the solar system. *Science*, 182:281–283.

REFERENCES FOR MARTIAN METEORITES (SECTION 5.3c)

Borg, L. E., L. E. Nyquist, L. A. Taylor, H. Wiesmann, and C. -Y. Shih, 1997. Constraints on martian differentiation processes from Rb-Sr and Sm-Nd isotopic analyses of the basaltic shergottite QUE 94201. *Geochim. Cosmochim. Acta*, 61:4915–4931.

Duke, M. B., 1968. The Shergotty meteorite: Magmatic and shock metamorphic features. In B. M. French and N. M. Short (Eds.), *Shock Metamorphism of Natural Materials*, pp. 613–621. The Mono Book Corp., Baltimore, Maryland.

Faure, G., 2001. *Origin of Igneous Rock; the Isotopic Evidence*. Springer-Verlag, Heidelberg.

Faure, G., and P. M. Hurley, 1963. The isotopic composition of strontium in oceanic and continental basalt: Application to the origin of igneous rocks. *J. Petrol.*, 4:31–50.

Gale, N. H., J. W. Arden, and R. Hutchison, 1975. The chronology of the Nakhla achondrite meteorite. *Earth Planet. Sci. Lett.*, 26:195–206.

Gladman, B. J., J. A. Burns, M. Duncan, P. Lee, and H. F. Levinson, 1996. The exchange of impact ejecta between terrestrial planets. *Science*, 271:1387–1392.

Gleason, J. D., D. A. Kring, D. H. Hill, and W. V. Boynton, 1997. Petrography and bulk chemistry of martian orthopyroxenite ALH 84001: Implications for the origin of secondary carbonates. *Geochim. Cosmochim. Acta*, 61:3503–3512.

Jagoutz, E., 1989. Strontium and neodymium isotopic systematics is ALHA 77005: Age of shock metamorphism in shergottites and magmatic differentiation on Mars. *Geochim. Cosmochim. Acta*, 53:2429–2441.

Jagoutz, E., 1991. Chronology of SNC meteorites. *Space Sci. Rev.*, 56:13–22.

Jagoutz, E., and H. Wänke, 1986. Strontium and neodymium isotopic systematics of Shergotty meteorite. *Geochim. Cosmochim. Acta*, 50:939–953.

Jones, J. H., 1986. A discussion of isotopic systematics and mineral zoning in the shergottites: Evidence for a 180 m.y. igneous crystallization age. *Geochim. Cosmochim. Acta*, 50:969–977.

Laul, J. C., 1986. The Shergotty consortium and SNC meteorites: An overview. *Geochim. Cosmochim. Acta*, 50:875–887.

McKay, D. S., E. K. Gibson, Jr., K. L. Thomas-Keprta, H. Vali, C. S. Romanek, S. J. Clemett, X. D. F. Chillier, C. R. Maechling, and R. Zare, 1996. Search for past life on Mars: Possible relic biogenic activity in Martian meteorite. *Science*, 273:924–930.

McSween, H. Y., Jr., 1984. SNC meteorites: Are they martian rocks? *Geology*, 12:3–6.

McSween, H. Y., Jr., 1985. SNC meteorites: Clues to martian petrologic evolution? *Rev. Geophys.*, 23:391–416.

McSween, H. Y. Jr., 1994. What we have learned about Mars from SNC meteorites. *Meteoritics*, 29:757–779.

Mittlefehldt, D. W., 1994. ALH 84001, a cumulate orthopyroxenite member of the martian meteorite group. *Meteoritics*, 29:214–221.

Nakamura, N., D. M. Unruh, M. Tatsumoto, and R. Hutchison, 1982. Origin and evolution of the Nakhla meteorite inferred from the Sm-Nd and U-Pb systematics and REE, Ba, Sr, and K abundances. *Geochim. Cosmochim. Acta*, 46:1555–1573.

Nishiizumi, K., J. Klein, R. Middlemost, D. Elmore, P. W. Kubik, and J. R. Arnold, 1986. Exposure history of shergottites. *Geochim. Cosmochim. Acta*, 40:1017–1021.

Nyquist, L. E., J. L. Wooden, B. M. Bansal, H. Wiesmann, G. McKay, and D. D. Bogard, 1979. Rb-Sr age of the Shergotty achondrite and implications for metamorphic resetting of isochron ages. *Geochim. Cosmochim. Acta*, 43:1057–1074.

Palme, H., B. Spettel, G. Weckwerth, and H. Wänke, 1983. Antarctic meteorite ALHA 81005, a piece of the ancient lunar highland crust. Abstract. In K. Keil and J. J. Papike (Cochairs), *Meteorites from the Earth's Moon*, pp. 25–26. Fourteenth Lunar and Planetary Science Conference, Lunar and Planetary Science Institute, Houston, Texas.

Papanastassiou, D. A., and G. J. Wasserburg, 1974. Evidence for late formation and young metamorphism in the achondrite Nakhla. *Geophys. Res. Lett.*, 1:23–26.

Podosek, F. A., and J. C. Huneke, 1973. ^{40}Ar-^{39}Ar chronology of four calcium-rich achondrites. *Geochim. Cosmochim. Acta*, 37:667–684.

Romanek, C. S., M. M. Grady, I. P. Wright, D. W. Mittlefehldt, R. A. Socki, C. T. Pillinger, and E. K. Gibson, Jr., 1994. Record of fluid-rock interaction on Mars from the meteorite ALH 84001. *Nature*, 372:655–657.

Shih, C. -Y., L. E. Nyquist, D. D. Bogard, G. A. McKay, J. L. Wooden, B. M. Bansal, and H. Wiesmann, 1982. Chronology and petrogenesis of young achondrites, Shergotty, Zagami, and ALHA 77005: Late magmatism on a geologically active planet. *Geochim. Cosmochim. Acta*, 46:2323–2344.

Wänke, H., 1966. Der Mond als Mutterkörper der Bronzit Chondrite. *Z. Naturforschung*, 21a: 93–110.

Wasson, J. T., and G. W. Wetherill, 1979. Dynamical, chemical, and isotopic evidence regarding the formation locations of asteroids and meteorites. In T. Gehrels (Ed.), *Asteroids*, pp. 926–974. University of Arizona Press, Tucson.

Whillans, I. M., and W. A. Cassidy, 1983. Catch a falling star: Meteorites and old ice. *Science*, 222:55–57.

REFERENCES FOR LUNAR ROCKS (SECTION 5.3d)

Nyquist, L. E., 1977. Lunar Rb-Sr chronology. *Phys. Chem. Earth*, 10:103–142.

O'Nions, R. K., and R. J. Pankhurst, 1972. A 4.3 aeon pre-Imbrium event. *Nature*, 237:446–447.

Papanastassiou, D. A., and G. J. Wasserburg, 1970. Rb-Sr ages from the Ocean of Storms. *Earth Planet. Sci. Lett.*, 8:269–278.

Papanastassiou, D. A., and G. J. Wasserburg, 1971. Lunar chronology and evolution from Rb-Sr studies of Apollo 11 and 12 samples. *Earth Planet. Sci. Lett.*, 11:37–62.

Papanastassiou, D. A., G. J. Wasserburg, and D. S. Burnett, 1970. Rb-Sr ages of lunar rocks from the Sea of Tranquillity. *Earth Planet. Sci. Lett.*, 8:1–9.

REFERENCES FOR DATING METAMORPHIC ROCKS (SECTION 5.4)

Baadsgaard, H., 1973. U-Th-Pb dates on zircons from the early Precambrian Amitsoq gneisses, Godthaab District, West Greenland. *Earth Planet. Sci. Lett.*, 19:22–28.

Baadsgaard, H., R. St. J. Lambert, and J. Krupicka, 1976. Mineral isotopic age relationships in the polymetamorphic Amitsoq gneisses, Godthaab District, West Greenland. *Geochim. Cosmochim. Acta*, 40:513–527.

Dodson, M. H., 1973. Closure temperature in cooling geochronological and petrological systems. *Contrib. Mineral. Petrol.*, 40:259–274.

Fairbairn, H. W., P. M. Hurley, and W. H. Pinson, Jr., 1961. The relation of discordant Rb-Sr mineral and rock ages in an igneous rock to its time of subsequent $^{87}Sr/Sr^{86}$ metamorphism. *Geochim. Cosmochim. Acta*, 23:135–144.

Long, L. E., 1964. Rb-Sr chronology of the Carn Chuinneag Intrusion, Rosshire, Scotland. *J. Geophys. Res.*, 69:1589–1597.

Montigny, R., and G. Faure, 1969. Contribution au probleme de l'homogeneisation isotopique du strontium des roches totales au course de metamorphisme: Cas du Wisconsin Range, Antarctique. *C.R. Acad. Sci. Paris*, 268:1012–1015.

Moorbath, S., J. H. Allaart, D. Bridgwater, and V. R. McGregor, 1977. Rb-Sr ages of Early Archaean supracrustal rocks and Amitsoq gneisses at Isua. *Nature*, 270:43–44.

Moorbath, S., R. K. O'Nion, and R. J. Pankhurst, 1975. The evolution of early Precambrian crustal rocks at Isua, West Greenland—geochemical and isotopic evidence. *Earth Planet. Sci. Lett.*, 27:229–239.

Moorbath, S., R. K. O'Nion, R. J. Pankhurst, N. H. Gale, and V. R. McGregor, 1972. Further rubidium-strontium age determinations on the very early Precambrian rocks of the Godthaab district, West Greenland. *Nature Phys. Sci.*, 240:78–82.

Pidgeon, R. T., and M. R. W. Johnson, 1974. A comparison of zircon U-Pb and whole-rock Rb-Sr systems in three phases of the Carn Chuinneag Granite, Northern Scotland. *Earth Planet. Sci. Lett.*, 24:105–112.

REFERENCES FOR GEOLOGICAL TIMESCALE (SECTION 5.5a)

Harland, W. B., R. L. Armstrong, L. E. Craig, A. V. Cox, A. G. Smith, and D. G. Smith, 1989. *A Geologic Time*

Scale 1989. Cambridge University Press, Cambridge, United Kingdom.

Holmes, A., 1913. *The Age of the Earth*. Harper, New York.

Odin, G. S. (Ed.), 1982. *Numerical Dating in Stratigraphy, Parts I and II*. Wiley, Chichester.

REFERENCES FOR GLAUCONITE (SECTION 5.5b)

Burke, W. H., R. E. Denison, E. A. Heatherington, R. B. Koepnick, N. F. Nelson, and J. B. Otto, 1982. Variation of seawater $^{87}Sr/^{86}Sr$ throughout Phanerozoic time. *Geology*, 10:516–519.

Clauer, N., and S. Chaudhuri, 1995. *Clays in Crustal Environments; Isotopic Dating and Tracing*. Springer-Verlag, Heidelberg.

Clauer, N., E. Keppens, and P. Stille, 1992. Sr isotopic constraints on the process of glauconitization. *Geology*, 20:133–136.

Cooper, J. A., A. T. Wells, and T. Nicholas, 1971. Dating of glauconite from the Ngalia Basin, Northern Territory, Australia. *J. Geol. Soc. Austral.*, 18(2): 97–106.

Faure, G., 1998. *Principles and Applications of Geochemistry*, 2nd ed. Prentice-Hall, Upper Saddle River, New Jersey.

Grant, N. K., T. E. Laskowski, and K. A. Foland, 1984. Rb-Sr and K-Ar ages of Paleozoic glauconites from Ohio-Indiana and Missouri, USA. *Isotope Geosci.*, 2:217–239.

Gulbrandsen, R. A., S. S. Goldich, and H. H. Thomas, 1963. Glauconite from the Precambrian Belt Series, Montana. *Science*, 140:390–391.

Harris, W. B., and P. D. Fullagar, 1991. Middle Eocene and late Oligocene isotopic dates of glauconitic mica from the Santee River area, South Carolina. *Southeast. Geol.*, 32(1): 1–19.

Hurley, P. M., R. F. Cormier, J. Hower, H. W. Fairbairn, and W. H. Pinson, Jr., 1960. Reliability of glauconite for age measurements by K-Ar and Rb-Sr methods. *Am. Assoc. Petrol. Geol. Bull.*, 44:1793–1808.

Kish, S. A., and H. J. Stein, 1979. The timing of ore mineralization. Viburnum Trend, southeast Missouri lead-zinc district: Rb-Sr glauconite dating. Abstract with Program. *Geol. Soc. Am.*, 11:458.

Laskowski, T. E., R. H. Fluegeman, and N. K. Grant, 1980. Rb-Sr glauconite systematics and the uplift of the Cincinnati arch. *Geology*, 8:368–370.

McDougall, I., P. R. Dunn, W. Compston, A. W. Webb, J. R. Richards, and V. M. Bofinger, 1965. Isotopic age determinations on Precambrian rocks of the Carpentaria region, Northern Territory, Australia. *J. Geol. Soc. Austral.*, 12(1): 67–90.

Morton, J. P., and L. E. Long, 1980. Rb-Sr dating of Paleozoic glauconite from the Llano region, central Texas. *Geochim. Cosmochim. Acta*, 44:663–672.

Obradovich, J. D., 1988. A different perspective on glauconite as a chronometer for geologic time scale studies. *Paleoceanography*, 3:757–770.

Obradovich, J. D., and W. C. Cobban, 1975. A time-scale for the Late Cretaceous of the western interior of the United States. *Geol. Assoc. Can., Spec. Paper*, 13:31–54.

Obradovich, J. D., and Z. E. Peterman, 1968. Geochronology of the Belt Series, Montana. *Can. J. Earth Sci.*, 5:737–747.

Odin, G. S. (Ed.), 1982a. *Numerical Dating in Stratigraphy*. Wiley, Chichester.

Odin, G. S., 1982b. How to measure glaucony ages. In G. S. Odin (Ed.), *Numerical Dating in Stratigraphy*, pp. 387–403. Wiley, Chichester.

Odin, G. S., J. C. Hunziker, E. Keppens, P. G. Laga, and E. Pasteels, 1974. Analyse radiometrique de glauconies par les méthods au strontium et a l'argon; l'Oligo-Miocene de Belgique. *Bull. Soc. Belge Géol.*, 83(1): 35–48.

Odin, G. S., and A. Matter, 1981. De glauconiarum origine. *Sedimentology*, 28:611–641.

Stille, P., and N. Clauer, 1994. The process of glauconitization: Chemical and isotopic evidence. *Contrib. Mineral. Petrol.*, 117:253–262.

REFERENCES FOR AUTHIGENIC FELDSPAR (SECTION 5.5c)

Ali, A. D., and P. Turner, 1982. Authigenic K-feldspar in the Bromsgrove Sandstone Formation (Triassic) of central England. *J. Sediment. Petrol.*, 52:187–197.

Bass, M. N., and G. Ferrara, 1969. Age of adularia and metamorphism, Ouachita Mountains, Arkansas. *Am. J. Sci.*, 267:491–498.

Bowers, T. S., K. J. Jackson, and H. C. Helgeson, 1984. *Equilibrium Activity Diagrams*. Springer-Verlag, Berlin.

Buyce, M. R., and G. M. Friedman, 1975. Significance of authigenic K-feldspar in Cambrian-Ordovician carbonate rocks of the proto-Atlantic shelf of North America. *J. Sediment. Petrol.*, 45:808–821.

Cerny, P., and R. Chapman, 1986. Adularia from hydrothermal vein deposits: Extremes in structural state. *Can. Mineral.*, 24:717–728.

Faure, G., 1992. Isotopic records in detrital and authigenic feldspars in sedimentary rocks. In N. Clauer and S. Chaudhuri (Eds.), 1992. *Isotopic Signatures and Sedimentary Records*, pp. 215–238. Springer-Verlag, Heidelberg.

Faure, G., 1998. *Principles and Applications of Geochemistry*, 2nd ed. Prentice-Hall, Upper Saddle River, New Jersey.

Faure, G., and F. C. Barbis, 1983. Detection of neoformed adularia by Rb-Sr age determinations of granitic rocks in Ohio. In S. S. Augustithis (Ed.), *Leaching and Diffusion in Rocks and Their Weathering Products*, pp. 307–320. Theophrastus Publications, Athens, Greece.

Girard, J. P., J. L. Aronson, and S. M. Savin, 1988. Separation, K/Ar dating, and $^{18}O/^{16}O$ ratio measurements of diagenetic K-feldspar overgrowths: An example from the Lower Cretaceous arkoses of the Angola margin. *Geochim. Cosmochim. Acta*, 51:2207–2214.

Girard, J. P., and T. C. Onstott, 1991. Application of $^{40}Ar/^{39}Ar$ laser probe and step-heating techniques to the dating of diagenetic K-feldspar overgrowths. *Geochim. Cosmochim. Acta*, 52:2207–2214.

Girard, J. P., S. M. Savin, and J. L. Aronson, 1989. Diagenesis of the Lower Cretaceous arkoses of the Angola margin: Petrologic, K/Ar dating, and $^{18}O/^{16}O$ evidence. *J. Sediment. Petrol.*, 59:519–538.

Hearn, P. P., J. F. Sutter, and H. E. Belkin, 1987. Evidence for late Paleozoic brine migration in Cambrian carbonate rocks of the central and southern Appalachians: Implications for Mississippi-Valley type sulfide mineralization. *Geochim. Cosmochim. Acta*, 51:1323–1334.

Helgeson, H. C., 1974. Chemical interaction of feldspars and aqueous solutions. In W. S. McKenzie and J. Zussman (Eds.), The Feldspars, *Proc. NATO Adv. Study Inst.*, pp. 184–215. Manchester University Press, Manchester, United Kingdom.

Kastner, M., 1971. Authigenic feldspars in carbonate rocks. *Am. Mineral.*, 56:1403–1442.

Kastner, M., and R. Siever, 1979. Low temperature feldspars in sedimentary rocks. *Am. J. Sci.*, 279:435–479.

Lee, M., and S. M. Savin, 1985. Isolation of diagenetic overgrowths on quartz sand grains for oxygen isotope analysis. *Geochim. Cosmochim. Acta*, 49:497–501.

Lidiak, E. G., R. F. Marvin, H. H. Thomas, and M. N. Bass, 1966. Geochronology of the midcontinent region, United States. 4. Eastern area. *J. Geophys. Res.*, 71(22): 5427–5438.

Mensing, T. M., and G. Faure, 1983. Identification and age of neoformed Paleozoic feldspar (adularia) in a Precambrian basement core from Scioto County, Ohio, USA. *Contrib. Mineral. Petrol.*, 82:327–333.

Waugh, B., 1978. Authigenic K-feldspar in British Permo-Triassic sandstones. *J. Geol. Soc. Lond.*, 135:51–56.

REFERENCES FOR DETRITAL MINERALS (SECTION 5.5d)

Dallmeyer, R. D., 1987. $^{40}Ar/^{39}Ar$ age of detrital muscovite within Lower Ordovician sandstone in the coastal plain basement of Florida: Implications for West African terrane linkages. *Geology*, 15:998–1001.

Faure, G., and K. S. Taylor, 1981. Provenance of some glacial deposits in the Transantarctic Mountains based on Rb-Sr dating of feldspars. *Chem. Geol.*, 32:271–290.

Faure, G., K. S. Taylor, and J. H. Mercer, 1983. Rb-Sr provenance dates of feldspar in glacial deposits of the Wisconsin Range, Transantarctic Mountains. *Geol. Soc. Am. Bull.*, 94:1275–1280.

Gaudette, H. E., A. Vitrac-Michard, and C. J. Allègre, 1981. North American Precambrian history recorded in a single sample: High resolution U-Pb systematics of the Potsdam Sandstone detrital zircon, New York State. *Earth Planet. Sci. Lett.*, 54:248–260.

Heller, P. L., Z. E. Peterman, J. R. O'Neil, and M. Shafiqullah, 1985. Isotopic provenance of sandstones from the Eocene Tyee Formation, Oregon Coast Range. *Geol. Soc. Am. Bull.*, 96:770–780.

Taylor, K. S., and G. Faure, 1981. Rb-Sr dating of detrital feldspar: A new method to study till. *J. Geol.*, 89:97–107.

Wermund, E. G., W. H. Burke, Jr., and G. S. Kenny, 1966. K-Ar ages of detrital muscovite in the Meridian Sand of Alabama and Mississippi. *Geol. Soc. Am.*, 77:319–322.

REFERENCES FOR BENTONITE AND TUFF (SECTION 5.5e)

Baadsgaard, H., and J. F. Lerbekmo, 1980. A Rb-Sr age for the Cretaceous-Tertiary boundary (Z coal), Hell Creek, Montana. *Can. J. Earth Sci.*, 17:671–673.

Baadsgaard, H., and J. F. Lerbekmo, 1982. The dating of bentonite beds. In G. S. Odin (Ed.), *Numerical Dating in Stratigraphy*, pp. 423–440. Wiley, Chichester.

Baadsgaard, H., and J. F. Lerbekmo, 1983. Rb-Sr and U-Pb dating of bentonites. *Can. J. Earth Sci.*, 20:1282–1290.

Baadsgaard, H., J. F. Lerbekmo, and I. McDougall, 1988. A radiometric age for the Cretaceous-Tertiary boundary based upon K-Ar, Rb-Sr, and U-Pb ages of bentonites from Alberta, Saskatchewan, and Montana. *Can. J. Earth Sci.*, 25:1088–1097.

Lerbekmo, J. F., and R. M. St Louis, 1986. The terminal Cretaceous iridium anomaly in the Red Deer Valley, Alberta, Canada. *Can. J. Earth Sci.*, 23:120–124.

Lippolt, H. J., J. C. Hess, and K. Burger, 1984. Isotopische Alter von pyroklastischen Sanidinen aus Kaolin-Kohlentonsteinen als Korrelationsmarken für das mitteleuropäische Oberkarbon. *Fortschr. Geol. Rheinland Westf.*, 32:119–150.

REFERENCES FOR SHALE (SECTION 5.5f)

Allsopp, H. L., and P. Kolbe, 1965. Isotopic age determination of the Cape Granite and intruded Malmesburg sediments, Cape Peninsula, South Africa. *Geochim. Cosmochim. Acta*, 29:1115–1130.

Awwiller, D. N., 1994. Geochronology and mass transfer in Gulf Coast mudrocks (south-central Texas, U.S.A.): Rb-Sr, Sm-Nd, and REE systematics. *Chem. Geol.*, 116:61–84.

Bailey, S. W., 1988a. Polytypism of 1 : 1 layer silicates. In S. W. Bailey (Ed.), Hydrous Phyllosilicates (exclusive of micas). Rev. Mineral. 19: 9–27. Mineralogical Society of America, Washington, D.C.

Bailey, S. W., 1988b. X-ray diffraction identification of the polytypes of mica, serpentine, and chlorite. *Clays and Clay Minerals*, 36:193–213.

Bofinger, V. M., and W. Compston, 1967. A reassessment of the age of the Hamilton Group, New York and Pennsylvania, and the role of inherited radiogenic Sr^{87}. *Geochim. Cosmochim. Acta*, 31:2353–2359.

Clauer, N., and S. Chaudhuri, 1995. *Clays in Crustal Environments; Isotopic Dating and Tracing*. Springer-Verlag, Heidelberg.

Clauer, N., S. Chaudhuri, M. Kralik, and C. Bonnot-Courtois, 1993. Effects of experimental leaching on Rb-Sr and K-Ar isotopic systems and REE contents in diagenetic illite. *Chem. Geol.*, 103:1–16.

Clauer, N., S. M. Savin, and S. Chaudhuri, 1992. Isotopic compositions of clay minerals as indicators of the timing and conditions of sedimentation and burial diagenesis. In N. Clauer and S. Chaudhuri (Eds.), *Isotopic Signatures and Sedimentary Records*, pp. 238–286. Springer-Verlag, Heidelberg.

Cordani, U. G., K. Kawashita, and A. Thomaz-Filho, 1978. Applicability of the Rb-Sr method to shales and related rocks. *Am. Assoc. Petrol. Geol. Spec. Publ.*, 6:93–117.

Gebauer, D., and M. Grünenfelder, 1974. Rb-Sr wholerock dating of late diagenetic to anchimetamorphic Paleozoic sediments in southern France (Montagne Noire). *Contrib. Mineral. Petrol.*, 47:113–130.

Gorokhov, I. M., N. Clauer, E. S. Varshavskaya, E. P. Kutyavin, and A. S. Drannik, 1981. Rb-Sr ages of Precambrian sediments from the Ovruch Mountain Range, northwestern Ukraine (USSR). *Precambrian Res.*, 16:55–65.

Moorbath, S., 1969. Evidence for the age of deposition of the Torridonian sediments of northwest Scotland. *Scott. J. Geol.*, 5 (Pt. 2): 154–170.

Morton, J. P., 1985. Rb-Sr dating of diagenesis and source age of clays in Upper Devonian black shale of Texas. *Geol. Soc. Am. Bull.*, 96:1043–1049.

Perry, E. A., and K. K. Turekian, 1974. The effects of diagenesis on the redistribution of strontium isotopes in shales. *Geochim. Cosmochim. Acta*, 38:929–935.

Spears, D. A., 1974. The Rb-Sr age dating of some Carboniferous shales. *Geochim. Cosmochim. Acta*, 38:235–244.

Zwingmann, H., N. Clauer, and R. Gaupp, 1999. Structure-related geochemical (REE) and isotopic (K-Ar, Rb-Sr, $\delta^{18}O$) characteristics of clay minerals from Rotliegend Sandstone reservoirs (Permian, northern Germany). *Geochim. Cosmochim. Acta*, 63:2805–2823.

CHAPTER 6

The K–Ar Method

THE isotopic composition of K was first studied by Aston (1921), who discovered ^{39}K and ^{41}K. The radioactivity of K salts was suggested by J. J. Thomson in 1905 and was subsequently demonstrated by Campbell and Wood (1906) and Campbell (1908). However, the naturally occurring radioactive isotope of K was not identified until 1935, when A. O. Nier presented conclusive evidence for the existence of ^{40}K using a much more sensitive mass spectrometer than had been available to Aston. The possible modes of decay open to ^{40}K were discussed by the German physicist Von Weizsäcker (1937), who concluded that ^{40}K undergoes branched decay to ^{40}Ca and to ^{40}Ar, based partly on the fact that the abundance of Ar in the atmosphere of the Earth is about 1000 times greater than expected when compared to the "cosmic" abundances of the other noble gases. Von Weizsäcker also postulated that radiogenic ^{40}Ar should be present in old K-bearing minerals. Ten years later, Aldrich and Nier (1948) confirmed Von Weizsäcker's prediction by demonstrating that four geologically old minerals (orthoclase, microcline, sylvite, and langbeinite) did, in fact, contain radiogenic ^{40}Ar. The theoretical basis for the K–Ar method of dating was therefore established by about 1950 and, since then, it has become an important and widely used method of measuring the ages of K-bearing rocks and minerals. Excellent accounts of the principles, techniques, and applications of the K–Ar method of dating have been provided by Dalrymple and Lanphere (1969), Schaeffer and Zähringer (1966), McDougall (1966), Damon (1970), Hunziker (1979), and Dalrymple (1991).

6.1 PRINCIPLES AND METHODOLOGY

Potassium ($Z = 19$) is an alkali metal (group IA), along with Li, Na, Rb, and Cs. It is one of the eight most abundant chemical elements in the crust of the Earth and is a major constituent of many important rock-forming minerals, such as the micas, the feldspars, the feldspathoids, clay minerals, and certain evaporite minerals (Heier and Adams, 1964).

Potassium has three naturally occurring isotopes whose abundances are $^{39}_{19}K = 93.2581 \pm 0.0029\%$, $^{40}_{19}K = 0.01167 \pm 0.00004\%$, and $^{41}_{19}K = 6.7302 \pm 0.0029\%$. The atomic weight of K calculated from these data is 39.098304 ± 0.000058 (Garner et al., 1975). The isotopic composition of Ar in the terrestrial atmosphere was measured by Nier (1950): $^{40}_{18}Ar = 99.60\%$, $^{38}_{18}Ar = 0.063\%$, and $^{36}_{18}Ar = 0.337\%$. The atomic weight of Ar is 39.9476 and its $^{40}Ar/^{36}Ar$ ratio is $99.60/0.337 = 295.5$.

The decay of naturally occurring ^{40}K to stable ^{40}Ar by electron capture and by positron emission was described in Section 2.1d and is illustrated in Figure 2.7. About 10.48 percent of the ^{40}K atoms that decay form ^{40}Ar and the rest (89.52

percent) form ${}^{40}_{20}Ca$. Therefore, the decay of ^{40}K is used in the K–Ar and K–Ca methods of dating. In addition, the decay of ^{40}K to ^{40}Ar is the basis for the $^{40}Ar/^{39}Ar$ method of dating, which uses an unconventional approach to the problem of measuring the K concentrations of rocks and minerals (Chapter 7).

The growth of radiogenic ^{40}Ar and ^{40}Ca in a K-bearing system closed to K, Ar, and Ca during its lifetime is expressed by an equation analogous to equation 4.6:

$$^{40}Ar^* + {}^{40}Ca^* = {}^{40}K\,(e^{\lambda t} - 1) \qquad (6.1)$$

where λ is the total decay constant of ^{40}K. Each branch of the decay scheme gives rise to separate decay constants λ_e and λ_β such that the total decay constant is

$$\lambda = \lambda_e + \lambda_\beta$$

where λ_e refers to the decay of ^{40}K to ^{40}Ar and λ_β represents the decay to ^{40}Ca. The decay constants recommended by the IUGS Subcommission on Geochronology (Steiger and Jäger, 1977) are

$$\lambda_e = 0.581 \times 10^{-10}\ y^{-1} \qquad \lambda_\beta = 4.962 \times 10^{-10}\ y^{-1}$$

The total decay constant of ^{40}K is

$$\lambda = (0.581 + 4.962) \times 10^{-10} = 5.543 \times 10^{-10}\ y^{-1}$$

which corresponds to a halflife of

$$T_{1/2} = \frac{0.693}{5.543 \times 10^{-10}} = 1.250 \times 10^9\ y$$

The branching ratio R is defined as λ_e/λ_β and has a value of 0.117. The fraction of ^{40}K atoms that decay to ^{40}Ar is given by $(\lambda_e/\lambda)\ {}^{40}K$, which is used to express the growth of radiogenic ^{40}Ar atoms in a K-bearing rock or mineral:

$$^{40}Ar^* = \frac{\lambda_e}{\lambda}\ {}^{40}K(e^{\lambda t} - 1) \qquad (6.2)$$

The total number of ^{40}Ar atoms is

$$^{40}Ar = {}^{40}Ar_0 + {}^{40}Ar^* \qquad (6.3)$$

where $^{40}Ar_0$ is the number of atoms of this isotope per unit weight of sample that were incorporated into the rock or mineral at the time of its formation. Since Ar is a noble gas and since its solubility in silicate melts is low, A_0 is assumed to be equal to zero. Therefore, equation 6.2 is used to calculate dates by the conventional K–Ar method.

To date a K-bearing rock or mineral by the K–Ar method, the number of ^{40}Ar and ^{40}K atoms in a unit weight of sample must be measured. The K concentration is determined by dissolving a known weight of powdered sample in high-purity hydrofluoric acid. The resulting solution is analyzed by FP, AAS, ICP spectrometry, or isotope dilution using a spike enriched in one of the isotopes of K. Alternatively, K concentrations can be measured nondestructively by XRF or neutron activation. (Cooper, 1963; Müller, 1966; Dalrymple and Lanphere, 1969). The number of ^{40}K atoms per gram can be calculated from the measured K concentration of the sample, as demonstrated for Rb and Sr by equation 5.8.

The amount of ^{40}Ar per gram of sample is determined by isotope dilution using a spike enriched in ^{38}Ar. The procedure involves fusing a known weight of the purified mineral concentrate (or powdered rock) in a Mo crucible sealed in a vacuum system. The Ar released by the sample is mixed with a known quantity of the ^{38}Ar spike, and the mixture of gases is purified by removal of all reactive gases, including H_2, CO_2, H_2O, and N_2. The remaining mixture of noble gases is then introduced into the source of a mass spectrometer where the $^{40}Ar/^{38}Ar$ and $^{38}Ar/^{36}Ar$ ratios are measured.

The Ar isotopes in the mixture originated from the sample being analyzed, from the Ar spike that was added, and from atmospheric Ar that was sorbed to the mineral grains of the sample and to the walls of the vacuum system. The ^{40}Ar contributed by the atmospheric contaminant can be subtracted from the total ^{40}Ar in the mixture by assuming that all of the ^{36}Ar is of atmospheric origin. Since the $^{40}Ar/^{36}Ar$ ratio of atmospheric Ar is 295.5,

$$^{40}Ar_a = 295.5\ {}^{36}Ar_a$$

where $^{36}Ar_a$ is the height of the Ar peak at mass 36 on the same scale as the Ar peak at mass 40.

The equation relating the amount of radiogenic ^{40}Ar to the measured $^{40}Ar/^{38}Ar$ and $^{38}Ar/^{36}Ar$ ratios and to the isotope ratios of the Ar spike and to atmospheric Ar was derived by Dalrymple and Lanphere (1969).

The number of radiogenic ^{40}Ar and ^{40}K atoms in a unit weight of the sample are used to solve equation 6.2 for t:

$$t = \frac{1}{\lambda} \ln \left[\frac{^{40}\text{Ar}^*}{^{40}\text{K}} \left(\frac{\lambda}{\lambda_e} \right) + 1 \right] \qquad (6.4)$$

The value of t so calculated is the *age* of the mineral *only* when the following assumptions are satisfied:

1. No radiogenic ^{40}Ar produced by decay of ^{40}K in the mineral during its lifetime has escaped.
2. The mineral became closed to ^{40}Ar soon after its formation, which means that it must have cooled rapidly after crystallization, unless it formed at a low temperature.
3. No ^{40}Ar was incorporated into the mineral either at the time of its formation or during a later metamorphic event.
4. An appropriate correction is made for the presence of atmospheric ^{40}Ar.
5. The mineral was closed to K throughout its lifetime.
6. The isotopic composition of K in the mineral is normal and was not changed by fractionation or other processes except by decay of ^{40}K.
7. The decay constants of ^{40}K are known accurately and have not been affected by the physical or chemical conditions of the environment in which the K existed since it was incorporated into the Earth (Section 4.2b).
8. The concentrations of ^{40}Ar and K were determined accurately.

These assumptions require careful evaluation in each case and place certain restrictions on the geological interpretation of K–Ar dates.

The last three assumptions are quite general in scope and express certain fundamental conditions of dating by any method based on radioactivity. The isotopic composition of K in terrestrial samples is constant, even though fractionation of K isotopes has been observed on a small scale across contacts of igneous intrusions (Verbeek and Schreiner, 1967; Morozova and Alferovsky, 1974). The date calculated from equation 6.4 is expressed in terms of ka (10^3 years ago), Ma (10^6 years ago), and Ga (10^9 years ago), as discussed in Section 4.2e.

The calculation of K–Ar dates from equation 6.4 is illustrated by a set of measurements obtained for a sample of muscovite from a pegmatite in the Wisconsin Range of the Transantarctic Mountains:

$$\text{K} = 8.378\% \qquad ^{40}\text{Ar}^* = 0.3305 \text{ ppm}$$

The atomic weight of K is 39.098304, the abundance of ^{40}K is 0.0001167 (expressed as a decimal fraction), the atomic weight of ^{40}Ar is 39.9623, and A is Avogadro's number (6.022×10^{23} atoms/mol). Therefore, in terms of numbers of atoms,

$$\frac{^{40}\text{Ar}^*}{^{40}\text{K}} = \frac{0.3305 \times 39.098304 \times A}{39.9623 \times 8.378 \times 10^4 \times 0.0001167\ \text{A}}$$

Note that the factor of 10^4 is required to convert the concentration of K from percent to parts per million (i.e., 1.0% = 10^4 ppm) and that Avogadro's number cancels. Therefore, in the case under consideration, the atomic $^{40}Ar^*/^{40}K$ ratio is

$$\frac{^{40}\text{Ar}^*}{^{40}\text{K}} = 0.03307$$

Substituting into equation 6.4,

$$t = \frac{1}{5.543 \times 10^{-10}} \ln \left[0.03307 \left(\frac{5.543}{0.581} \right) + 1 \right]$$

$$= 494.7 \times 10^6 \text{ y or } 494.7 \text{ Ma.}$$

6.2 RETENTION OF ^{40}Ar BY MINERALS

Minerals to be dated by the K–Ar method must have retained all of the radiogenic ^{40}Ar produced within them by decay of ^{40}K (assumption 1) and they must not contain any excess ^{40}Ar (assumption 3). Argon loss from minerals may occur because Ar is a noble gas and therefore does not form bonds with other atoms in a crystal lattice. In general, Ar loss can be attributed to the following causes:

1. inability of a mineral lattice to retain Ar even at low temperature and atmospheric pressure;
2. either partial or complete melting of rocks followed by crystallization of new minerals from the resulting melt;
3. metamorphism at elevated temperatures and pressures, resulting in complete or partial Ar loss by diffusion depending on the temperature and duration of the event;
4. increase in temperature due to deep burial or contact metamorphism, causing Ar loss from most minerals without producing any other physical or chemical changes in the rock;
5. chemical weathering and alteration by aqueous fluids, leading not only to Ar loss but also to changes in the K concentration of minerals;
6. solution and redeposition of water-soluble minerals such as sylvite; and
7. mechanical breakdown of minerals, radiation damage, and shock metamorphism. Even excessive grinding during preparation of samples for dating by the K–Ar method may cause partial Ar loss.

6.2a Idaho Springs Gneiss, Colorado

A good demonstration of Ar loss due to an increase in temperature in a contact metamorphic zone was presented by Hart (1964) and Hart et al. (1968), who dated hornblende, biotite, and K-feldspar of rocks from the Idaho Springs Formation (Precambrian) collected at increasing distances from the intrusive contact of the Eldora quartz monzonite stock (Tertiary) in the Front Range of Colorado. The Idaho Springs Formation consists of a quartz–feldspar–biotite gneiss with some biotite–quartz–feldspar schist and amphibolite. Age determinations reviewed by Hart (1964) suggest that these rocks were regionally metamorphosed at about 1350–1400 Ma. The Eldora stock was intruded into these rocks at about 55 Ma and produced very minor mineralogical contact metamorphic effects in the Idaho Springs Formation. However, Figure 6.1 shows that the K–Ar dates

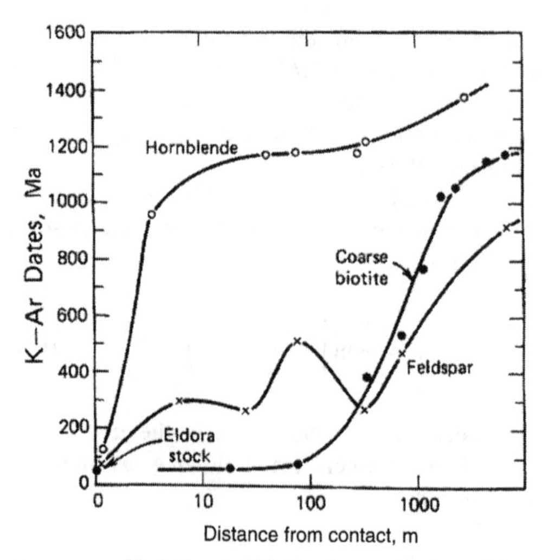

FIGURE 6.1 Variation of K–Ar dates of minerals from the Idaho Springs Formation of the Front Range of Colorado in a contact metamorphic zone produced by intrusion of the Eldora stock. The Idaho Springs Formation was regionally metamorphosed at 1350–1400 Ma. The age of the Eldora stock is 55 Ma. Data from Hart (1964).

of hornblende, biotite, and K-feldspar were profoundly affected as a result of losses of varying amounts of radiogenic ^{40}Ar. The fraction of Ar lost from each of the minerals decreases as a function of distance from the contact and reflects the differing retentivities of these minerals for radiogenic Ar.

The coarse biotite apparently lost all of its radiogenic Ar out to a distance of about 30 m from the contact, and the effects of Ar loss can be traced for more than 2135 m beyond that distance. The K–Ar dates of the biotite reported by Hart (1964) stabilize at 1200 Ma at a distance of about 4270 m from the contact, whereas those of the K-feldspar increase somewhat erratically away from the contact and show the effects of Ar loss even at a distance of 6860 m. The hornblende, on the other hand, displays a superior retentivity for Ar. Loss of Ar is confined primarily to a distance of about 3 m from the contact and then diminishes rapidly. Approximately 30 m from the contact the K–Ar dates of the hornblende reach a plateau value of 1200 Ma and ultimately rise to 1375 Ma about 4270 m from the contact. It is not clear whether the 1375-million-year date reflects the response of hornblende to the contact metamorphism or whether it is due to the earlier regional metamorphism of the Idaho Springs Formation. Nevertheless, these results clearly illustrate the fact that contact metamorphism causes systematic loss of radiogenic Ar from minerals and that under identical physical and chemical conditions hornblende retains its radiogenic Ar better than biotite.

Potassium feldspar is not suitable for dating by the K–Ar method because it loses Ar readily, even at relatively low temperatures. However, McDowell (1983) has shown that it is difficult to extract radiogenic ^{40}Ar from alkali feldspar even at a temperature of 1670°C partly because of the absence of structural volatiles, such as H_2O, that help to flush the Ar from the melt. The apparent poor retentivity of K-feldspar may therefore be caused by experimental problems rather than by the "leakiness" of its crystal lattice. Hart (1964) also dated biotite from the Idaho Springs Formation by the Rb–Sr method and observed a pattern of variation similar to that of the K–Ar dates. However, the Rb–Sr dates are consistently higher than the K–Ar dates, which suggests that, in this case, biotite retained radiogenic ^{87}Sr somewhat better than ^{40}Ar. It should be clear, of course, that K–Ar dates in such a contact metamorphic aureole primarily reflect variable losses of radiogenic Ar and such dates cannot be used to determine the "age" of the rocks. In the case illustrated in Figure 6.1, only the dates in the immediate vicinity of the intrusive and those sufficiently removed from it have geological significance.

6.2b Snowbank Stock, Minnesota

In a similar study, Hanson and Gast (1967) measured K–Ar and Rb–Sr dates of biotite, muscovite, and K-feldspar at increasing distances from the intrusive contact of the Duluth Gabbro (1.05 Ga) with the granitic rocks of the Snowbank stock (2.60 Ga) of Minnesota. The samples were collected primarily from outcrops along the shores of Snowbank Lake and its principal islands. The results in Figure 6.2 demonstrate that the K–Ar dates of the minerals increase with distance from 1.05 Ga (the age of the Duluth Gabbro) to 2.6 Ga

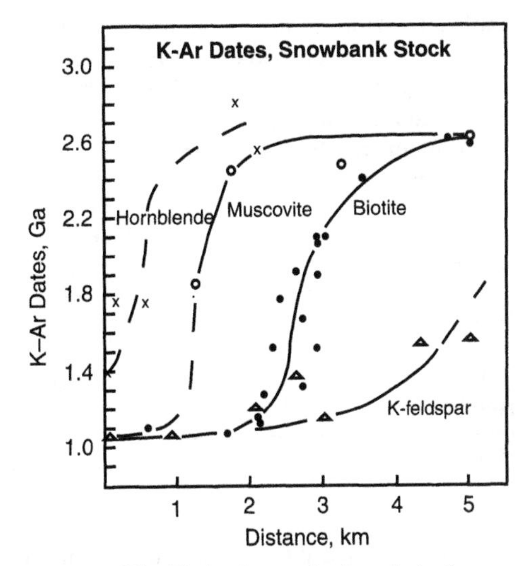

FIGURE 6.2 The K–Ar dates of minerals in the Snowbank stock (2.6 Ga) as a function of distance from the intrusive contact of the Duluth Gabbro (1.05 Ga) in Minnesota. Data from Hanson and Gast (1967).

(the age of the Snowbank stock). The increase in temperature following the intrusion of the Duluth Gabbro at 1.05 Ga caused virtually complete loss of radiogenic ^{40}Ar from K-feldspar, biotite, and muscovite near the contact and smaller fractional losses farther away. The width of the metamorphic aureole defined by partial ^{40}Ar loss from the minerals of the Snowbank stock is indicated in Table 6.1

Hanson and Gast (1967) also estimated the maximum temperature to which the rocks of the Snowbank stock were heated by the intrusion of the Duluth Gabbro based on heat flow models of Jaeger (1957) and assuming a temperature of 1150°C at the contact. The results in Table 6.1 are plausible approximations of the "closing or blocking temperatures" of the K-bearing minerals. Accordingly, hornblende retains radiogenic ^{40}Ar below about 550°C, muscovite becomes retentive at about 450°C, biotite does so at 350°C, whereas K-feldspar loses ^{40}Ar even at temperatures less than about 300°C. The concept of the closure temperature of K-bearing minerals and of the resulting "cooling ages" was treated quantitatively by Dodson (1973, 1979).

Table 6.1. Width of the Metamorphic Aureole within the Snowbank Stock of Minnesota as Defined by Partial Loss of Radiogenic ^{40}Ar from Different Minerals Caused by the Intrusion of the Duluth Gabbro

Mineral	Width of Aureole, km	Maximum Temperature,[a] °C
K-feldspar	>5.0	~300
Biotite	4.5–5.0	~350
Muscovite	2.0–3.0	~450
Hornblende	1.5–2.0	~550

[a]Estimates from Figure 12b of Hanson and Gast (1967).

The estimated distances and maximum temperatures for complete retention of ^{40}Ar by the minerals in Table 6.1 can be interpreted as exemplified by biotite. All biotites in the Snowbank pluton located less than 4.5–5.0 km from the contact with the Duluth Gabbro were heated to maximum temperatures in excess of about 350°C and therefore lost varying fractions of the ^{40}Ar that had accumulated previously. For the same reason, biotite that crystallizes from magma at high temperature does not begin to retain radiogenic ^{40}Ar until the ambient temperature decreases to less than about 350°C. In this way, the data of Hart (1964) and Hanson and Gast (1967) help to define the meaning of K–Ar dates of biotite and other K-bearing silicate minerals in both contact metamorphic aureoles and igneous rocks. Hanson et al. (1971) subsequently investigated the loss of radiogenic Pb from sphenes in the Giants Range Granite (2.7 Ga), Minnesota, caused by the intrusion of the Duluth Gabbro (1.05 Ga).

To be useful for dating by the K–Ar method, minerals and rocks must retain Ar quantitatively (although all minerals begin to lose Ar with increasing temperature), they must be resistant to chemical alteration, and they must contain K (although not necessarily as a major constituent). In addition, they should be sufficiently common to be useful for geological investigations. Table 6.2 is a compilation of the minerals (and rocks) that meet these criteria and are most useful for dating by the K–Ar method.

6.2c Excess ^{40}Ar

The total number of ^{40}Ar atoms in a unit weight of a K-bearing mineral or rock that has retained all of the radiogenic ^{40}Ar produced by decay of ^{40}K in accordance with equations 6.2 and 6.3 is

$$^{40}\mathrm{Ar} = {}^{40}\mathrm{Ar}_0 + \left(\frac{\lambda_e}{\lambda}\right) {}^{40}\mathrm{K}(e^{\lambda t} - 1) \qquad (6.5)$$

In this equation, $^{40}Ar_0$ represents the number of ^{40}Ar atoms per unit weight of sample that were not formed by decay of ^{40}K in the rock or mineral. This nonradiogenic component may consist of Ar contributed by one or several possible sources, including (1) Ar that was dissolved in the magma and that may have originated from the mantle of the Earth or by outgassing of old K-bearing minerals of the crust, (2) Ar that was evolved during later thermal metamorphism of the rocks and that diffused into the minerals during that event, and (3) atmospheric Ar adsorbed on grain boundaries and microfractures while the rock was exposed to the atmosphere in the field and in the laboratory or firmly held within the crystal lattice. In conventional K–Ar dating the assumption is made that all of the initial ^{40}Ar is of atmospheric origin and can be subtracted from the total ^{40}Ar in the sample on the basis of its $^{40}Ar/^{36}Ar$ ratio of 295.5. If minerals contain "magmatic" and "metamorphic" Ar (sources 1 and 2 above), the calculated K–Ar date is older than the age of the minerals, which are then said to contain "excess" ^{40}Ar.

One of the samples analyzed by Hanson and Gast (1967) was a pyroxene concentrate from the Snowbank pluton (2.6 Ga) collected at the contact with the Duluth Gabbro (1.05 Ga). The K–Ar date of this pyroxene (2.97 Ga) actually exceeds the age of the Snowbank pluton, indicating that the pyroxene contained excess ^{40}Ar even though all other minerals collected within 1 km of the contact *lost* varying amounts of ^{40}Ar (Figure 6.2). The excess ^{40}Ar was presumably incorporated into the pyroxene either at the time of crystallization of the Snowbank pluton at 2.6 Ga or during the later intrusion of the Duluth Gabbro at 1.05 Ga. The evidence for loss of ^{40}Ar by biotite, muscovite, and

Table 6.2. Common Rock-Forming Minerals Suitable for Dating by the K–Ar Method

	Rock Type			
	Volcanic	Plutonic	Metamorphic	Sedimentary
Feldspars				
Sanidine	⊗			
Anorthoclase	⊗			
Plagioclase	⊗			
Adularia			×	×
Feldspathoids				
Leucite	×			
Nepheline	×	×		
Mica				
Biotite	⊗	⊗	⊗	
Phlogopite		⊗	⊗	
Muscovite		⊗	⊗	⊗[a]
Lepidolite		⊗		
Glaucony				×
Illite				×
Amphibole				
Hornblende	⊗	⊗	⊗	
Pyroxene	×	×		
Whole rock	⊗		×	×

Note: ⊗, often useful; ×, sometimes useful.
[a] Detrital.
Source: Adapted from Dalrymple and Lanphere (1969).

K-feldspar at 1.05 Ga implies that ^{40}Ar had moved into the grain boundaries of the Snowbank pluton and therefore could have diffused into pyroxene and into other minerals whose crystal structures contain sites that can be occupied by ^{40}Ar. Other examples of the presence of excess ^{40}Ar in minerals were published by Giletti (1974a, b) and by the following:

> Hart and Dodd, 1962, *J. Geophys. Res.*, 67:2998–2999. McDougall and Green, 1964, *Norsk Geol. Tidsskr.*, 44:183–196. York and MacIntrye, 1964, *Trans. Am. Geophys. Union*, 46:177. Livingston et al., 1967, *J. Geophys. Res.*, 72:1361–1375. Laughlin, 1969, *J. Geophys. Res.*, 74:6684–6690. Schwartzman and Giletti, 1977, *Contrib. Mineral. Petrol*, 60:143–159.

The presence of excess ^{40}Ar has been detected not only in pyroxene but also in certain other minerals that have low K concentrations, including diamonds (Ozima et al., 1983), beryl, cordierite, and tourmaline (Damon and Kulp, 1958a, b), as well as in fluid inclusions of quartz and fluorite (Rama et al., 1965). High-K minerals such as the micas, feldspars, and feldspathoids rarely contain excess ^{40}Ar, whereas hornblende does so in a few cases. Even volcanic rocks and the inclusions they contain may harbor excess ^{40}Ar (Damon et al., 1967).

The occurrence of excess ^{40}Ar in mafic rocks of the Kola Peninsula of Russia was discovered by E. K. Gerling and his colleagues in the Laboratory of Precambrian Geology of St. Petersburg (Gerling et al., 1962). These investigators reported K–Ar dates ranging from 3.8 to 10.0 Ga for samples of pyroxenite, peridotite, and dunite xenoliths in the gabbro intrusions of the Monche-Tundra Hill area. Since the age of the Earth was already known to be 4.5 Ga, Gerling and his associates concluded that ultramafic xenoliths yielding K–Ar dates of 10.0 Ga contain excess ^{40}Ar. Nevertheless, the excessively old dates were, at the time, a serious embarrassment for the new science of geochronometry and therefore stimulated efforts to date these rocks by other methods (Faure, 2001). Professor Gerling described this exciting period in his life in an autobiographical paper published in an international journal (Gerling, 1984). The most recent work of Tolstikhin et al. (1992) yielded a Sm–Nd date of 2.492 ± 0.048 Ga for the Sopcha pyroxenite on the Kola Peninsula, which means that all ultramafic rocks in this area with K–Ar dates in excess of 2.5 Ga contain excess ^{40}Ar.

In retrospect, the presence of excess ^{40}Ar in certain minerals and even in volcanic rocks is fairly common and was thoroughly documented by Damon et al. (1967). These authors applied the term "extraneous" to any ^{40}Ar in a rock or mineral that had not formed by in situ decay of ^{40}K and was not of atmospheric origin. Dalrymple and Lanphere (1969) devoted an entire chapter of their book to the occurrence of extraneous (or excess) ^{40}Ar in different kinds of silicate minerals. They defined "inherited" ^{40}Ar as a contaminant that originated either from the presence of xenoliths or xenocrysts in volcanic and plutonic igneous rocks or from unrelated mineral grains picked up during processing of the sample in the laboratory. The presence of inherited ^{40}Ar is especially troublesome for dating young volcaniclastic rocks like those at Olduvai Gorge in Tanzania and at other archeological or anthropological sites (Curtis, 1966).

6.3 K–Ar ISOCHRONS

The isochron method of dating can avoid the problem caused by the presence of excess ^{40}Ar under certain favorable circumstances. Equation 6.5 can be modified by dividing each term by the number of ^{36}Ar atoms per unit weight of sample:

$$\frac{^{40}\mathrm{Ar}}{^{36}\mathrm{Ar}} = \left(\frac{^{40}\mathrm{Ar}}{^{36}\mathrm{Ar}}\right)_0 + \left(\frac{\lambda_e}{\lambda}\right)\frac{^{40}\mathrm{K}}{^{36}\mathrm{Ar}}(e^{\lambda t} - 1) \quad (6.6)$$

Equation 6.6 expresses the measured ^{40}Ar/^{36}Ar ratio as a sum of two terms. The first term represents the Ar contributed by extraneous sources, whereas the second term represents the radiogenic component that has accumulated since closure of the mineral or rock. The magnitude of the $(^{40}\mathrm{Ar}/^{36}\mathrm{Ar})_0$ ratio depends on the isotopic composition of the different Ar components present and may thus differ from that of atmospheric Ar.

Suites of comagmatic volcanic rocks or samples of a mineral in a suite of plutonic igneous or metamorphic rocks (e.g., biotite, muscovite) should, under favorable conditions, have the same initial ^{40}Ar/^{36}Ar ratio and the same closure age t. Therefore, the measured ^{40}Ar/^{36}Ar and the ^{40}K/^{36}Ar ratios of such suites of samples can define an isochron whose slope is

$$m = \left(\frac{\lambda_e}{\lambda}\right)(e^{\lambda t} - 1) \quad (6.7)$$

from which the closure age of the samples can be calculated:

$$t = \frac{1}{\lambda}\ln\left[m\left(\frac{\lambda}{\lambda_e}\right) + 1\right] \quad (6.8)$$

The intercept of the straight line with the y-axis is the initial ^{40}Ar/^{36}Ar ratio, whose numerical value provides information about the sources from which the excess Ar originated. In addition, the discussion in Section 6.2 makes clear that the K–Ar dates of minerals in igneous and metamorphic rocks depend on their closure temperatures and on the rate of cooling. Therefore, K–Ar isochrons can be constructed only by combining data derived from suites of samples of the same mineral all of which must have experienced the same thermal history. The slope and intercept of K–Ar isochrons can be calculated by the same weighted-regression procedure outlined in Section 4.3. In cases where the retentivity of the same mineral (e.g., biotite) and the thermal history experienced by a suite of that mineral are not sufficiently uniform to satisfy the assumptions for dating by the isochron method, dates may have to be calculated separately for each sample.

The K–Ar isochrons have been used by McDougall et al. (1969) and by Hayatsu and

Carmichael (1977), among others, and were evaluated in detail by Shafiqullah and Damon (1974), Roddick (1978), and Armstrong (1978). The initial ratios reported by Hayatsu and Carmichael (1977) and later by Hayatsu and Palmer (1975) are significantly higher than the atmospheric value. A possible explanation for this result is that the Ar that was initially incorporated into the minerals was not entirely of atmospheric origin but included components derived by outgassing of old rocks in the crust and mantle. The Ar contributed from these sources is highly enriched in ^{40}Ar relative to ^{36}Ar and causes the initial ^{40}Ar/^{36}Ar ratios obtained from K–Ar isochrons to be higher than the value of this ratio in atmospheric Ar. The presence of such extraneous radiogenic ^{40}Ar causes conventional K–Ar dates of minerals to be older than the age of the rocks in which they occur and gives rise to the phenomenon of "excess" Ar discussed in the previous section.

Fitch et al. (1976) used the K–Ar isochron technique to reinterpret the K–Ar data for several suites of volcanic rocks and minerals, including those of Curtis and Hay (1972) for an ignimbrite-ash fall at Olduvai Gorge in Tanzania in Figure 6.3. Nine samples yielded an average date of 1.976 ± 0.034 Ma. Fitch et al. (1976) subsequently used the measurements to construct a K–Ar isochron for these rocks. The slope of the isochron was calculated by the regression procedure of York

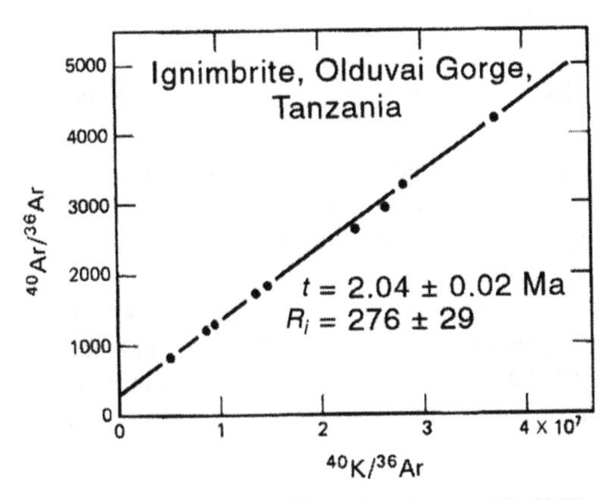

FIGURE 6.3 Whole rock K–Ar isochron of Tuff IB from Olduvai Gorge, Tanzania. Data from Fitch et al. (1976) and Curtis and Hay (1972).

(1969) and corresponds to a date of 2.04 ± 0.02 Ma. The initial ^{40}Ar/^{36}Ar ratio is 276 ± 29. Both dates were recalculated using the decay constants of Steiger and Jäger (1977). The precise age of the volcanic rocks at Olduvai Gorge is of paramount importance because the remains of early humans have been found there.

6.4 VOLCANIC ROCKS OF TERTIARY AGE

The possibility of dating volcanic rocks by analysis of whole-rock specimens (Table 6.2) is of great interest. Several studies summarized by Dalrymple and Lanphere (1969) have demonstrated the reliability of K–Ar dates of whole-rock samples of basalt and other volcanic rock types. Such rocks consist primarily of plagioclase and pyroxene, both of which retain Ar satisfactorily. However, samples that have been altered or that contain devitrified glass, secondary minerals (zeolites, calcite, or clay minerals), and xenoliths or xenocrysts should be avoided. The Ar retentivity of anhydrous volcanic glass is fairly good. However, the Ar retentivity of devitrified or hydrated glass is questionable. The presence of devitrified glass and secondary alteration products results in a lowering of K–Ar dates. On the other hand, xenoliths and xenocrysts are likely to contain excess radiogenic ^{40}Ar whose presence in a sample increases the measured K–Ar date. Basalts that were extruded on the ocean floor under high hydrostatic pressure also contain excess ^{40}Ar that is concentrated in the glassy crusts of individual "pillows" apparently because Ar in solution in the lava is trapped there by high pressure and rapid quenching. The amount of excess argon *decreases* inward from the glassy rind of the pillows and *increases* as a function of water depth, which suggests that it is controlled primarily by the hydrostatic pressure. For these reasons, only holocrystalline samples taken from the centers of pillows should be used for dating submarine pillow basalts. The presence of excess ^{40}Ar in pillow basalt and examples of geological interpretations of K–Ar dates of volcanic rocks include the following studies:

Dating Volcanic Rocks: Armstrong et al., 1975, *Am. J. Sci.*, 275:225–251. DeLong and

McDowell, 1975, *Geology*, 3:691–694. Dalrymple et al., 1975, *Geol. Soc. Am. Bull.*, 86:1463–1467. Wellman and McDougall, 1974a, *J. Geol. Soc. Austral.*, 32:247–272. Wellman and McDougall, 1974b, *Tectonophysics*, 23:49–65.

Excess ^{40}Ar in Pillow Basalt: Dalrymple and Moore, 1968, *Science*, 161:1132–1135. Fisher, 1971, *Earth Planet. Sci. Lett.*, 12:321–324. Fisher, 1981, *Nature*, 290:42–43. Mellor and Mussett, 1975, *Earth Planet. Sci. Lett.*, 26:312–318. Ozima et al., 1977, *Roy. Astron. Soc. Geophys. J.*, 51:479–485. Seidemann, 1977, *Geol. Soc. Am. Bull.*, 88:1666.

6.4a Rate of Motion of the Hawaiian Islands

The Hawaiian Islands in the Pacific Ocean (Figure 6.4) are the erosional remnants of former volcanoes that formed above a mantle plume (also called a hot spot). These volcanoes became extinct when the northwesterly motion of the Pacific plate moved them away from their magma sources associated with the plume. The volcano Kilauea on the island of Hawaii (the "Big Island") has been active in historical time because it is presently located above the plume, which has remained stationary. In addition, the small submarine volcano Loihi, located off the southeast coast of Hawaii, appears destined to grow in size as the island of Hawaii moves off the hot spot (Staudigel et al., 1984; Kent et al., 1999).

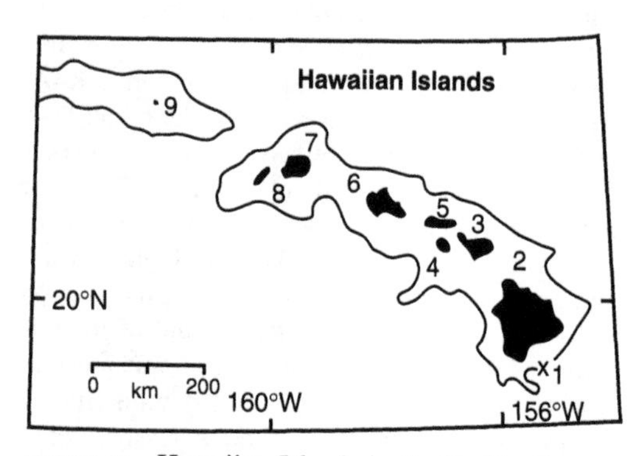

FIGURE 6.4 Hawaiian Islands in the Pacific Ocean: 1. Loihi; 2. Hawaii; 3. Maui; 4. Kahoolawe; 5. Molokai; 6. Oahu; 7. Kauai; 8. Niihau; 9. Nihoa. The contour represents a depth of 4000 m. Adapted from J. B. Garver, Jr. J. F. Shupe, J. F. Dorr, O. G. A. M. Payne, and M. B. Hunsiker (Eds.), *Atlas of the World*, 6th ed., National Geographic Society, Washington, D.C., 1990.

This insight into the "inner workings" of the Earth (first proposed by Wilson, 1963) is based, at least in part, on the K–Ar dates of the basalt lavas from the Hawaiian Islands as well as from the submerged seamounts of the Hawaiian Ridge and the Emperor Chain, which extend all the way to the Aleutian Trench (Jackson et al., 1972). The petrogenesis of the basalts on the Hawaiian Islands and other oceanic island was reviewed by Faure (2001) from the perspective of the isotope compositions of Sr, Nd, Pb, Hf, Re, and O.

McDougall and Duncan (1980) plotted the K–Ar dates of volcanic rocks on the Hawaiian Islands versus the distance from the volcano Kilauea on the Big Island and regressed these data to a straight line whose equation is

$$t = 0.01035d - 0.9444 \qquad (6.9)$$

where t is the age in millions of years and d is the distance from Kilauea in kilometers. The rate of movement of the Hawaiian volcanoes implied by these data is the reciprocal of the slope of the line in Figure 6.5:

$$\text{Rate of motion} = \frac{1000 \times 100}{0.01035 \times 10^6} = 9.66 \text{ cm/y}$$

When the standard deviation of the slope is included, the rate of motion of the Hawaiian Islands is 9.66 ± 0.27 cm/y. Dalrymple et al. (1981) later added the available K–Ar dates of basalts of the Emperor Seamounts to the average K–Ar dates of the Hawaiian Islands and derived a combined rate of 8.2 ± 0.2 cm/y. Other island chains in Polynesia have yielded similar rates (McDougall and Duncan, 1980):

Marquesas Islands	10.4 ± 1.8 cm/y
Society Islands	10.9 ± 1.0 cm/y
Austral Islands	10.7 ± 1.6 cm/y

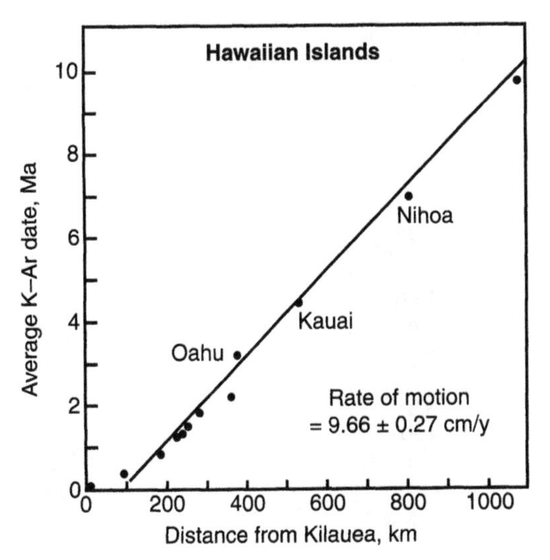

FIGURE 6.5 Average K–Ar date of the shield-building stage of volcanic activity of the Hawaiian Islands and distance from the volcano Kilauea on Hawaii (Figure 6.4). Data from McDougall and Duncan (1980).

Additional K–Ar dates of volcanic rocks on the Hawaiian Islands and Emperor Seamounts have been published:

McDougall, 1964, *Geol. Soc. Am. Bull.*, 75:107–128. Clague and Dalrymple, 1973, *Earth Planet. Sci. Lett.*, 17:411–415. Dalrymple et al., 1974, *Geol. Soc. Am. Bull.*, 85: 727–738. Clague and Dalrymple, 1975, *Geophys. Res. Lett.*, 2:305–308. Clague et al., 1975, *Geol. Soc. Am. Bull.*, 86:991–998. Dalrymple and Clague, 1976, *Earth Planet. Sci. Lett.*, 31:313–329. Bonhomme et al., 1977, *Geol. Soc. Am. Bull.*, 88:1282–1286. Dalrymple et al., 1977, *Earth Planet. Sci. Lett.*, 37:107–116. McDougall, 1979, *Earth Planet. Sci. Lett.*, 46:31–42.

The seamounts of the Emperor Chain were the subject of volume 55 of the *Initial Reports of the Deep Sea Drilling Project* (DSDP Leg 55) edited by E. D. Jackson and I. Koisumi.

6.4b Magnetic Reversal Chronology

The first indication that the Earth's magnetic field may have reversed its polarity was presented in 1906 by the French physicist Bernard Brunhes. Twenty years later, M. Matuyama concluded that the Earth's magnetic field had a reversed polarity during the early Quaternary Period and that it has been normal since then. However, no systematic study of the history of polarity reversals was possible until the K–Ar method became available for dating of very young volcanic rocks. The year 1963 was a turning point in the continuing efforts to study the history of reversals of the Earth's magnetic field as recorded by sequences of volcanic rocks whose ages can be measured by the K–Ar method. In that year, Cox et al. (1963a, b) and McDougall and Tarling (1963) proposed timescales for reversals of the Earth's magnetic field, and Vine and Mathews (1963) suggested that the linear magnetic anomalies along midocean ridges could be explained by seafloor spreading and periodic polarity reversals of the magnetic field. Since then, the K–Ar method of dating has played a key role in the study of the history of the magnetic field and the establishment of a timescale for polarity reversals.

The reversal chronology consists of extended intervals of time called *epochs* during which the field had predominantly one polarity. Such epochs are interrupted by shorter *events* of opposite polarity. The distinction between epochs and events is based primarily on the length of time involved. Epochs are named after deceased scientists who made significant contributions to studies of the Earth's magnetic field. Events are named after the location at which a reversal was first recognized. The reversal timescale for the past 5.0 million years contains four epochs, starting with the present: Brunhes (normal), Matuyama (reversed), Gauss (normal), and Gilbert (reversed). These are shown in Table 6.3 along with several events of shorter duration.

The polarity timescale has been extended into the Mesozoic Era by the analysis of deep-sea sediment cores and by magnetic anomaly patterns in the oceans. The results reveal the occurrence of frequent polarity reversals lasting from 10^4 to 10^7 years. The frequency of geomagnetic

Table 6.3. Geomagnetic Polarity Timescale from the Present to 5 Ma[a]

Epoch	Event	Polarity	Time Interval, Ma	Anomaly Number
Brunhes	—	Normal	0–0.72	1
	Laschamp?	Reversed	0.045–0.025[b]	
	Blake?	Reversed		
Matuyama	—	Reversed	0.72–2.47	
	Jaramillo	Normal	0.91–0.97	1′
	Olduvai	Normal	1.66–1.87	2
	Réunion	Normal	2.01–1.04	
		Normal	2.12–2.14	
Gauss	—	Normal	2.47–3.40	2.1′ 2.2′ 2.3′
	Kaena	Reversed	2.91–2.98	
	Mammoth	Reversed	3.07–3.17	
Gilbert	—	Reversed	3.17–5.41	
	Cochiti	Normal	3.86–3.98	3.1
	Nunivak	Normal	4.12–4.26	3.2
	Sidufjall	Normal	4.41–4.49	3.3
	Thvera	Normal	4.59–4.79	3.4

[a]The timescale is based on K–Ar dates of volcanic rocks recalculated to the decay constants and atomic abundances of K recommended by the Subcommission on Geochronology of the IUGS and presented in Section 1 of this chapter (Mankinen and Dalrymple, 1979; Ness et al., 1980).
[b]Hall and York (1978).

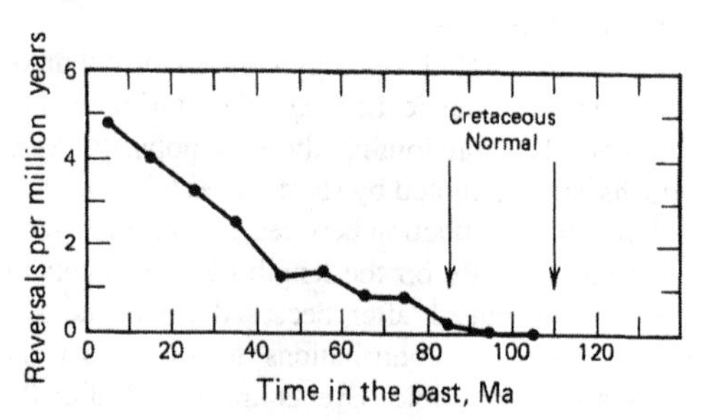

FIGURE 6.6 Frequency of geomagnetic polarity reversals per million years calculated for 10-million-year intervals from the data of Ness et al. (1980). Note that the frequency of reversals has increased since late Mesozoic time. A similar compilation to 150 Ma was published by McElhinny (1978).

polarity reversals has varied systematically in Figure 6.6, which demonstrates that a "quiet" interval occurred during the Cretaceous Period followed by a "disturbed" interval with many reversals that extends into the present. The high frequency of reversals during the Cenozoic Era complicates the use of geomagnetic polarity reversals for the correlation of rock units. However, the existence of quiet intervals lasting 10 million years or longer may be useful for worldwide correlation of all kinds of rock units. The quiet intervals were summarized by McElhinny (1978) as follows:

Cretaceous Normal (KN), 85–110 Ma
Jurassic Normal (JN), 145–164 Ma
Triassic Normal (TRN), 205–220 Ma
Permo-Carboniferous Reversed (PCR), 227–313 Ma
Ordovician Reversed, Early to Middle Ordovician
Cambrian Reversed, Middle to Late Cambrian
Latest Precambrian Reversed, Precambrian–Cambrian boundary

The history of polarity reversals throughout Phanerozoic and Precambrian time is being studied by the analyses of sedimentary rocks. These studies have been facilitated by the development of highly sensitive magnetometers called SQUIDs (superconducting quantum interference devices) by means of which the remnant magnetism of virtually all kinds of sedimentary rocks can be analyzed.

However, a word of caution is in order. Sedimentary rocks are imperfect recorders of the Earth's magnetic field because of distortions that are introduced during deposition of sediment particles or that are later superimposed on the rocks by chemical alteration. McElhinny (1978) cautioned that magnetostratigraphic sequences should be based on multiple sampling of rock layers and that unstable components of the remanent magnetism must be removed by appropriate laboratory methods.

The combination of magnetic polarity reversals and seafloor spreading at midocean ridges has given rise to linear magnetic anomalies in the ocean basins. These anomalies result from the polarity changes of the magnetism implanted in the basalt flows of the oceanic crust at the time of their extrusion at spreading ridges. The K–Ar dates of basalts recovered by drilling in the oceans initially provided a basis for relating the magnetic anomalies on the floor of the oceans to the reversal chronology in Table 6.3, which had been worked out on land. Later, these patterns were extended to progressively older rocks back in time to the Cretaceous Period.

The magnetic polarity and K–Ar dates of basalt of the oceanic crust have been used to contour the bottom of the ocean basins in terms of isochrons (i.e., lines of equal K–Ar dates). Instead of plotting the boundaries of polarity reversals, isochron lines are drawn with a contour interval of 10 million years such that the present crest of the adjacent spreading ridge is the zero isochron and the isochrons become virtual growth lines of the seafloor. Isochron maps like Figure 6.7 can now be used to reconstruct the tectonic history of ocean basins (Heirtzler et al., 1968) and to constrain the ages of volcanic islands that have formed in the ocean basins (Dupuy et al., 1993; Calmant, 1987; McNutt et al., 1989).

6.4c Argon from the Mantle

The presence of excess ^{40}Ar in basalt erupted on the seafloor is an important phenomenon with

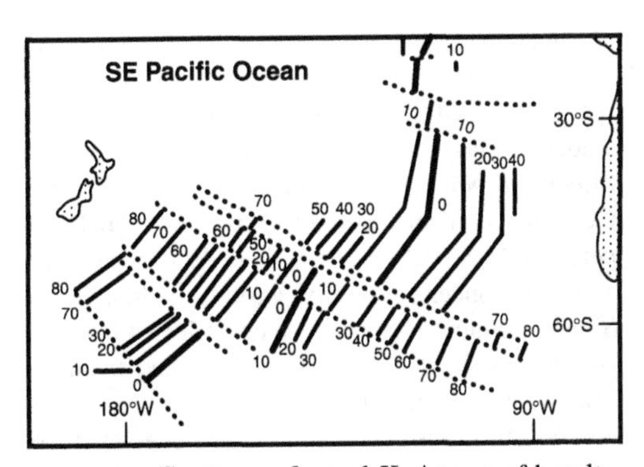

FIGURE 6.7 Contours of equal K–Ar age of basalt in the Southeast Pacific Ocean. The heavy line is the zero isochron drawn along the crest of the spreading ridge. The contour interval is 10 Ma and the dotted lines identify fracture zones. Adapted from Heirtzler et al. (1968).

implications for theories about outgassing of the mantle of the Earth and the evolution of its atmosphere (Ozima and Podosek, 1983). The Ar in vesicular glassy basalt extruded on the seafloor is a mixture of two components derived from the mantle and from the atmosphere of the Earth (Fisher, 1981). The challenge is to determine the $^{40}Ar/^{36}Ar$ ratio of the mantle-derived Ar in the presence of varying proportions of atmospheric Ar. A study by Allègre et al. (1983) demonstrated that the $^{40}Ar/^{36}Ar$ ratio of Ar in oceanic basalt reaches values of about 25,000 and that the magnitude of this ratio varies depending on the geochemical history of the magma sources.

Fisher (1981) analyzed basalt glasses from single sites in the Pacific and Atlantic Oceans, respectively, and obtained $^{40}Ar/^{36}Ar$ ratios of about 10,000. In addition, Marty et al. (1983) analyzed gases trapped in the vesicles of the so-called popping rocks dredged from the Mid-Atlantic Ridge at 36° 49.3′ N latitude and 33° 15′ W longitude. The rocks recovered at this site are so gas rich that they exploded with a loud "popping" sound when they were first brought to the surface from a depth of 2490 m (Hekinian et al., 1973). The explosions continued for three days and caused the fragments to jump up to 150 cm into the air. Petrographic examination indicated that the popping rocks are composed of tholeiite glass containing a large number of vesicles whose diameters are less than 1 mm. One specimen was found to contain an unusually large amount of gas (0.881 mL/g) composed primarily of CO_2 (38.8 percent by volume), H_2 (26.7 percent), CO (16.8 percent), SO_2 (10.3 percent), HCl (6.6 percent), H_2S (0.2 percent), $N_2 + Ar$ (0.1 percent), and low concentrations of various hydrocarbons. Similar popping rocks have been collected from the continental slope of Baja California and from the Mid-Atlantic Ridge at 45° N.

Small blocks of glass from one of these popping rocks were heated stepwise to release Ar and other gases for isotope analysis. The data indicate that the Ar released from vesicles at 750°C has a measured $^{40}Ar/^{36}Ar$ ratio of $10{,}524 \pm 1047$. However, the gas released by melting the glass at 1500°C has a much lower $^{40}Ar/^{36}Ar$ ratio of only 930 ± 160, and a powdered sample yielded a still lower value of 385 ± 14. Marty et al. (1983) concluded from these results that the Ar dissolved in the glass was contaminated by atmospheric Ar but the gas in the vesicles is likely to be derived from the mantle. Their data therefore strengthen the conclusion that Ar in the mantle of the Earth is strongly enriched in radiogenic ^{40}Ar compared with Ar in the atmosphere.

The growth of radiogenic ^{40}Ar by decay of ^{40}K in the mantle of the Earth and its subsequent transfer to the atmosphere were modeled by Ozima (1975). He concluded that the atmosphere could not have formed by degassing of the Earth at a constant rate because such models require that the $^{40}Ar/^{36}Ar$ ratio of the mantle today is less than 2000 and its K concentration is less than 50 ppm. Both results are unacceptable. However, he found that models based on an episode of degassing early in the history of the Earth followed by later continuous degassing yield satisfactory results. For example, a present-day $^{40}Ar/^{36}Ar$ ratio in the mantle of about 10,000 is predicted by a model in which about 95 percent of the initial Ar was lost from the mantle by degassing at about 4.45 Ga. Ozima's calculations support earlier discussions of this problem by Fanale (1971), Schwartzman (1973), and others who favored the conclusion that most of the Ar and other noble gases were lost from the mantle by degassing very soon after formation of the Earth. The high $^{40}Ar/^{36}Ar$ ratio of the mantle today therefore results from the early removal of ^{36}Ar followed by the growth of radiogenic ^{40}Ar by decay of ^{40}K in the interior of the Earth.

6.5 DATING SEDIMENTARY ROCKS

The problems of dating sedimentary rocks by means of radiogenic isotopes have already been discussed in Section 5.5 in connection with applications of the Rb–Sr method of dating. Attempts to date sedimentary rocks by the K–Ar method have yielded disappointing results because sedimentary rocks and the minerals they contain are difficult to date for two reasons:

1. The most abundant kinds of sedimentary rocks (sandstone, siltstone, and shale) are either entirely composed of or contain mineral particles derived by erosion of older rocks. Dates obtainable from them may be used to study their provenance, but they generally do not indicate the time of deposition.
2. Sedimentary rocks composed primarily of authigenic minerals either do not contain sufficient K for dating or do not quantitatively retain the radiogenic ^{40}Ar that forms by decay of ^{40}K after deposition.

These difficulties eliminate carbonate rocks, evaporites, iron formations, and phosphorites from consideration. The opportunity to date sedimentary rocks is therefore limited to K-bearing minerals in volcanic rocks that may be interbedded with sedimentary rocks and to certain minerals, such as sanidine and glauconite, that form by chemical reactions in the oceans. However, clay minerals may be contaminated by inclusions of older detrital minerals or may be altered after deposition in the course of diagenesis, burial metamorphism, tectonic deformation, and weathering (e.g., glauconite, illite, and others). The feasibility of dating sedimentary rocks and the K-bearing minerals they contain by the K–Ar method was originally discussed by Hurley (1966), Dalrymple and Lanphere (1969), Dalrymple (1991), and most recently Clauer and Chaudhuri (1995).

6.5a Shale

The effect of diagenesis on the K–Ar dates derivable from shale was studied by Perry (1974) and Aronson and Hower (1976), among others. Both studies demonstrated that whole-rock K–Ar dates of shale exceed the time of deposition but decrease with increasing depth of burial. This phenomenon is caused by the loss of radiogenic ^{40}Ar from detrital K-feldspar and mica that become unstable. The decomposition of these minerals coincides with an increase with depth in the abundance of illite in illite–smectite mixed-layer clay. These reactions constitute "burial metamorphism," which results in the redistribution of K from detrital K-feldspar and mica to the clay minerals. Potassium is essentially conserved in this process, whereas the radiogenic ^{40}Ar is lost, thus lowering the K–Ar dates of whole-rock shale samples and of those size fractions that contain detrital K-feldspar and mica.

The decrease in the K–Ar dates with depth of shale in the Gulf Coast region of the United States cannot be attributed to loss of ^{40}Ar by geothermal heating. Aronson and Hower (1976) pointed out that the decrease of the K–Ar dates in a core from Harris County, Texas, ceased at a depth of about 3700 m where the temperature was 95°C. Moreover, the smallest grain size fraction (<0.1 μm), which is most susceptible to Ar loss, actually contained more radiogenic ^{40}Ar below 3700 m than above that depth. They demonstrated that the increase in the ^{40}Ar concentration of this fraction coincided with the increase in its K concentration and therefore was caused by decay of ^{40}K since burial metamorphism. In fact, Aronson and Hower (1976) used the increases in the concentrations of K and ^{40}Ar of the <0.1 μm fraction to calculate a date of about 18 Ma (Burdigalian Stage of the Miocene Epoch) for the age of burial metamorphism in the core. This date is compatible with the late Oligocene to early Miocene age of the rocks.

The effect of burial metamorphism on K–Ar dates of Miocene shale from the Gulf Coast of Louisiana is illustrated in Figures 6.8*a*, *b* by K–Ar dates of different-size fractions of clay that decrease with depth of burial and tend to converge (Perry, 1974). The K–Ar dates of the <0.5 μm fraction, composed primarily of illite–smectite mixed-layer clay, also decrease with depth because of the substantial increase in the K concentration of that fraction during burial metamorphism. However, the dates of different size fractions do not converge to the depositional age. Therefore, in this case neither the shale as a whole nor the clay minerals yield K–Ar dates that correspond to the time of deposition.

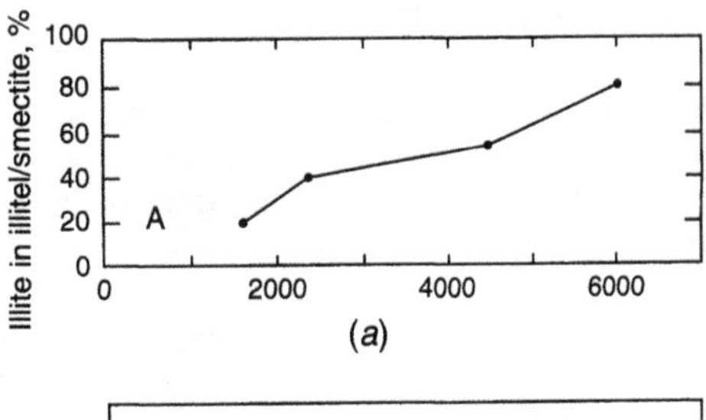

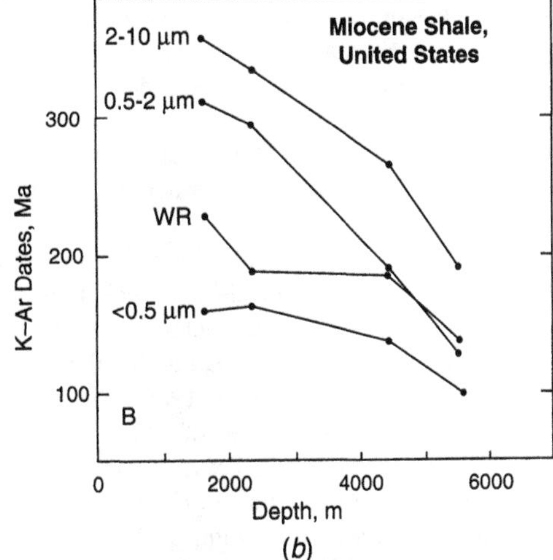

FIGURE 6.8 (*a*) Increase of the abundance of illite layers in illite–smectite mixed-layer clay in shale of Miocene age from the Gulf coast of Louisiana. (*b*) Decrease of K–Ar dates of whole-rock samples and grain size fractions from the shale identified in (*a*). The lowering of the K–Ar dates is caused by the loss radiogenic ^{40}Ar from detrital K-feldspar and mica as a result of burial metamorphism. The process involves redistribution of K from feldspar and mica to clay minerals and may be caused by the conversion of smectite to illite. The grain sizes are expressed in micrometers. Data from Perry (1974).

6.5b Potassium-Rich Bentonites

In some cases, sedimentary rocks are interbedded with layers of bentonite that formed by the alteration of volcanic ash deposited by eruptions that occurred hundreds of kilometers away. The volcanic ash was originally composed primarily of shards and fragments of glass that subsequently altered to smectite and other clay minerals by chemical reactions controlled by the environment of deposition (Person, 1982). Bentonites are useful for the correlation of stratigraphic sections because they were deposited rapidly over large areas downwind from their source (Huff et al., 1992; Carey and Sparks, 1986; Rose and Chesner, 1990; McKerrow et al., 1991).

Whole-rock samples of bentonite and size fractions of the clay minerals of which they are composed, in some cases, contain sufficient K to make them amenable to dating by the K–Ar method (Aronson and Douthitt, 1986). Since bentonites are typically composed of smectite (three-layer clay minerals) having low K concentrations, the K–Ar dates record the time of K enrichment and the resulting increase in the ratio of illite to smectite (Elliott et al., 1991). The illitization has been attributed to alteration by seawater, burial metamorphism, contact metamorphism, and hydrothermal activity (Altaner et al., 1984; Elliott and Aronson, 1987).

An example of illitization of a bentonite bed (2.5 m thick) as a consequence of burial metamorphism was described by Altaner et al. (1984). The bentonite bed in question is interbedded with marine shale of the Late Cretaceous Marias River Formation in the Sun River Canyon of the Sawtooth Range of Montana. The authors observed that the smectite of the bentonite bed had been partially converted to illite (up to 70 percent) at the upper and lower contacts presumably by diffusion of K from the shale into the bentonite at temperatures between 100 and 200°C as a result of burial under thrust sheets during the Laramide orogeny. The K was liberated by the decomposition of K-feldspar and mica in the shale and caused the concentration of K_2O in the bentonite to increase from about 2.25 percent in the center of the bed up to 5.0 percent at the lower contact. The K–Ar dates of illite–smectite (<1.0–0.2 μm) clay fractions of the bentonite bed reported by Altaner et al. (1984) ranged from about 56 to 50 Ma and are, in all cases, younger than the stratigraphic age of the bentonite by about 30 million years (Hoffman et al., 1976). In addition, the K–Ar dates of low-illite clay fractions from the center of the bentonite bed are lower (50 Ma) than dates from the illite-rich upper and lower contacts (56 and 54 Ma, respectively). Altaner et al. (1984) concluded that the K–Ar date of the lower contact (54 Ma) represents the time elapsed since the temperature reached its maximum value and the difference in the K–Ar dates at the contact and in the center of the bed (3–4 million years) measures the slow rate of diffusion of K from the interbedded shale into the bentonite bed.

However, the enrichment of bentonite beds in K is not, in all cases, controlled by the depth of burial and the resulting increase of the temperature. Elliott and Aronson (1987) reported that the 1–2 μm and <1.0 μm clay fractions of Middle Ordovician K-rich bentonites in the southern Appalachian Mountains yielded uniform K–Ar dates between 272 and 303 Ma regardless of the depth of burial. Since these dates coincide with the Alleghenian orogeny (Late Pennsylvanian to Early Permian), the illitization of the bentonites was apparently caused by hot brines that were squeezed from the sediment in the Appalachian Basin during the Alleghenian orogeny into the rocks along the western edge of the basin. Such brines may also have caused the deposition of ore and associated gangue minerals of the Mississippi Valley Pb–Zn deposits mentioned in Section 5.5c. A genetic relation between the formation of illite and the deposition of ore minerals in the Kupferschiefer of the Zechstein basin of Poland was also inferred by Bechtel et al. (1999) from K–Ar dates of illites.

However, as in the case of Rb–Sr dates of clay minerals, the K–Ar dates of illite–smectite mixed-layer clays underestimate the stratigraphic ages of sedimentary rocks that have been deeply buried, hydrothermally altered, metamorphosed, or exposed to hot brines during orogenies even though the rocks were not themselves deformed.

6.5c Volcanogenic Minerals in Sedimentary Rocks

Bentonites and other kinds of pyroclastic rocks also contain residual volcanic minerals such as sanidine, plagioclase, anorthoclase, quartz, pyroxene, olivine, amphibole, biotite, sphene, apatite, allanite, and zircon, several of which can be dated by isotopic methods. Sanidine yields the most reliable K–Ar dates because of its high K concentration, resistance to weathering, and low initial ^{40}Ar content. (Damon and Teichmüller, 1971; Hess and Lippolt, 1980; Hellmann and Lippolt, 1981; Hess et al., 1983). In addition, sanidines are suitable for dating by the ^{40}Ar/^{39}Ar method to be presented in Chapter 7. Unaltered biotite, containing more than 7 percent K_2O, is commonly found in bentonites and also yields reliable K–Ar dates (Baadsgaard and Lerbekmo, 1982). The reliability of K–Ar dates of volcanogenic crystals in bentonites and other kinds of sediments depends on the assumption that the crystals formed a short time prior to their deposition and that detrital grains derived from continental sources are absent (Dymond, 1966).

A good example of K–Ar dating of biotite crystals in bentonite beds was reported by Lanphere and Tailleur (1983). These authors dated five biotite samples extracted from drillcores of bentonites of the Late Cretaceous (Turonian) Shale Wall Member of the Seabee Formation in northern Alaska. The K_2O concentrations of the biotites ranged from 6.71 to 8.785 percent, and the resulting K–Ar dates of the five samples varied only from 91.5 to 93.6 Ma with a mean of 92.8 ± 0.7 Ma. In this case, the conventional K–Ar dates of the biotites were concordant with a ^{40}Ar/^{39}Ar date of 93.4 ± 1.3 Ma for a biotite concentrate from one of the three drillcores. These K–Ar dates constrain the range of the pelecypod *Inoceramus labiatus*, which occurs in the Shale Wall Member of the Seabee Formation.

Layers of volcanic pyroclastics also occur in marine sediment deposited in the vicinity of submarine volcanoes or along midocean ridges. The pyroclastic layers may be extensively altered to zeolite but still contain well-crystallized minerals (e.g., anorthoclase, sanidine, plagioclase, amphibole, and biotite) as well as unaltered shards of volcanic glass. Dymond (1966) dated mineral concentrates

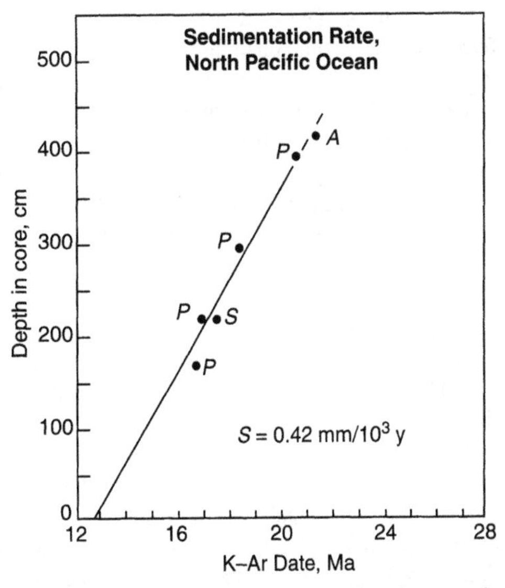

FIGURE 6.9 Sedimentation rate at 28° 19′ N, 133° 12′ W in the North Pacific Ocean in 4420 m of water between 169 and 416 cm in piston core Ris 120p based on minerals in volcanic ash layers: P = plagioclase, S = sanidine, A = anorthoclase. Data from Dymond (1966).

derived from volcanic ash layers in two piston cores taken in the northern Pacific Ocean at 28° 19′ N and 133° 12′ W about 1850 km west of Baja California at a water depth of 4420 m. The results in Figure 6.9 for one of the cores (Ris 120P) define a straight line with a slope of 42.46 cm/10^6 y that corresponds to a sedimentation rate S given by

$$S = \frac{42.46 \times 10 \times 10^3}{10^6} = 0.42 \text{ mm}/10^3 \text{ y}$$

This core was composed of red clay, which characterizes the deep-sea sediment deposited on the abyssal plain of the northern Pacific Ocean. The sediment did not contain calcium carbonate or opal. The absence of calcium carbonate in core Ris 120P indicates that the depth of water at this site exceeds the carbonate compensation depth (i.e., carbonate shells dissolve as they sink through the water column). The low sedimentation rate indicated by the K–Ar dates of the volcanogenic minerals is typical of large areas of the ocean basins.

In addition, Dymond (1966) demonstrated the feasibility of dating unaltered shards of volcanic glass in marine sediment collected in the Pacific Ocean off the west coast of Central America in water depths ranging from 2450 to 3700 m. The glass was of rhyolitic composition and had K concentrations ranging from 3.32 to 3.95 percent. The resulting dates were all less than 1.0 Ma, with a range from 0.98 ± 0.10 to 0.49 ± 0.04 Ma. Similar glass at the Guadelupe site close to the west coast of Baja California but in deep water (3566 m) indicated a sedimentation rate of 1.1 cm/10^3 y.

These results demonstrate that well-crystallized K-bearing minerals in volcanic ash layers and bentonites as well as unaltered K-rich volcanic glass shards are suitable for dating by the K–Ar method and are capable of yielding precise and accurate dates. The use of such samples for dating by the ^{40}Ar/^{39}Ar method will come up for discussion in Chapter 7. The use of glauconite for dating of sedimentary rocks by the K–Ar method was adequately covered in Section 5.5b, devoted to dating by the Rb–Sr method.

Similarly, the feasibility of dating detrital minerals (e.g., muscovite and microcline) in clastic sedimentary rocks (e.g., sandstones, siltstones, till) was discussed in Section 5.5d and therefore requires no further elaboration. In contrast to glauconite and volcanogenic minerals in bentonites and pyroclastic rocks, detrital minerals predate the deposition of the sedimentary rocks in which they occur and are used to determine the provenance of the sediment rather than the depositional age of their sedimentary host rocks.

6.5d Metasedimentary Rocks

The evidence that virtually all K-bearing minerals can lose radiogenic ^{40}Ar at elevated temperatures was used by Harper (1964) to propose that K–Ar dates of slates record the occurrence of regional metamorphic events. In the ideal case, the metamorphic temperatures rise sufficiently to allow all of the accumulated ^{40}Ar to escape from the shale during its conversion to slate or phyllite. Subsequently, as the slate cools through the blocking temperature, radiogenic ^{40}Ar formed by in situ decay of ^{40}K begins to accumulate and is quantitatively retained. The resulting K–Ar dates of slates depend not only on the maximum temperature to which they were subjected (which controls the loss of accumulated ^{40}Ar) but also on the rate of cooling, which determines the time when newly formed ^{40}Ar begins to accumulate.

The maximum temperatures and subsequent cooling rates of rocks in a particular orogenic belt vary regionally as a function of distance from the center of most intense metamorphism. In addition, the thermal histories of sedimentary rocks in orogenic belts depend on the depth of burial and hence on their stratigraphic position. Therefore, Harper (1964) concluded that the K–Ar dates of slates, formed by regional metamorphism during an orogeny, reflect the time elapsed since cooling through their blocking temperatures rather than the age of the orogeny or the time of deposition of the shale protolith. Since the cooling rate of slates is related to their stratigraphic position, only the K–Ar dates of upper level slates, which were completely outgassed during the orogeny and subsequently cooled rapidly, approach the age of the orogeny.

Harper (1964, 1967a) applied these principles to the K–Ar dates of the Dalradian Series of metasedimentary rocks in the Highlands of Scotland that were metamorphosed during the Caledonian orogeny. The stratigraphic age of the Dalradian Series ranges from Late Proterozoic to Early Cambrian. Harper's conclusion was that the peak of the Caledonian orogeny occurred at 470 Ma and was recorded by the most retentive minerals (muscovite and hornblende) in the metamorphic rocks of the Dalradian Series. Cooling continued irregularly to 440 Ma as recorded by biotite and whole-rock samples.

The effect of progressive regional metamorphism on clay-rich sedimentary rocks is recorded by the crystallinity index (CI) of illite, which is defined as the width of the [001] reflection (9.98–10.02 Å) at half height on an x-ray diffraction scan of the untreated mineral (Clauer and Chaudhuri, 1995). Experience indicates that the width of this illite peak at half height decreases with increasing metamorphic temperature as the mineral recrystallizes. Therefore, Bonhomme et al. (1980) used the CI of illite in Late Jurassic (Oxfordian) and Late Triassic (Rhaetian) sedimentary rocks to demonstrate the decrease of the K–Ar dates of illite as a function of increasing metamorphic grade in the southwestern Alps of Europe.

The clay fractions (<2 μm) of the unmetamorphosed Rhaetian sediment yielded a K–Ar isochron date of 155 ± 9 Ma and an initial $^{40}Ar/^{36}Ar$ ratio of 305 ± 91, which is consistent with atmospheric Ar. However, the K–Ar date is lower than the stratigraphic age of 209.5–208 Ma assigned by Harland et al. (1989) to rocks of Rhaetian age. Bonhomme et al. (1980) attributed the low K–Ar age of these rocks to the effects of late-stage diagenesis.

The clay fractions (<2 μm) of the unmetamorphosed Oxfordian sediment define a K–Ar isochron in Figure 6.10 that corresponds to a date of 163 ± 6 Ma and an initial $^{40}Ar/^{36}Ar$ ratio of 655 ± 22. Although the date is compatible with its stratigraphic age of 157.1–154.7 Ma (Harland et al., 1989), the high initial $^{40}Ar/^{36}Ar$ ratio reveals the presence of excess ^{40}Ar in detrital minerals of the clay fractions. This finding does not invalidate the date provided that the excess ^{40}Ar occurred in all of the clay fractions at the time of deposition.

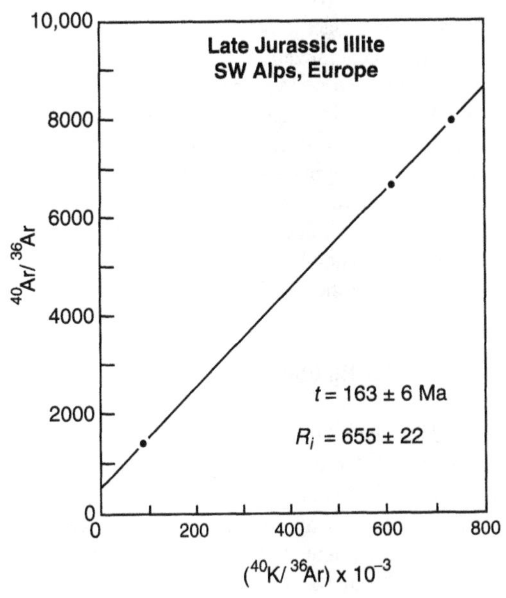

FIGURE 6.10 The K–Ar isochron defined by unmetamorphosed illite (<2 μm) in sedimentary rocks of Late Jurassic age in the southwestern Alps of Europe. Data from Bonhomme et al. (1980).

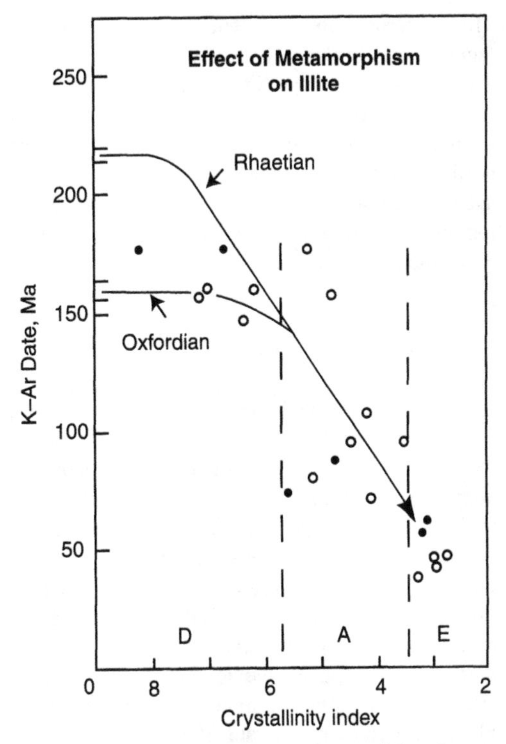

FIGURE 6.11 Decrease of K–Ar dates of the clay fraction (<2 μm) of Late Triassic (Rhaetian) and Late Jurassic (Oxfordian) sedimentary rocks with increasing metamorphic grade in the southwestern Alps of Europe: D = diagenesis or burial metamorphism; A = anchizone; E = epizone of regional metamorphism. The crystallinity index of illite decreases with increasing metamorphic grade as defined in the text. Data from Bonhomme et al. (1980).

The K–Ar dates of the clay fractions of both the Rhaetian and Oxfordian sediment in Figure 6.11 decrease with increasing metamorphic grade and converge to low values at the highest metamorphic grade (CI < 3.5):

Rhaetian metasediments: 43.8 ± 4.5 Ma (Eocene).
Oxfordian metasediments: 60.5 ± 5.0 Ma (Paleocene).

These dates record an episode of greenschist facies metamorphism that affected the rocks of the southwestern Alps (Frey et al., 1974). Similar results were reported by Liewig et al. (1981) for phengites (sericite) in the schistes lustrés of the northern Cottic Alps. In addition, Frey et al. (1973) used glauconite to reveal the effect of low-grade regional metamorphism in the Alps of eastern Switzerland.

6.6 METAMORPHIC VEIL

In volcanic rocks and shallow igneous intrusives, which cool rapidly, the K-bearing minerals begin to accumulate radiogenic Ar almost immediately after crystallization. However, in deep-seated plutonic and metamorphic rocks, which cool slowly, Ar retention may be delayed until the temperature drops to some critical value below which diffusion of Ar becomes ineffective and the accumulation of radiogenic ^{40}Ar begins. The blocking temperature at which the loss of ^{40}Ar by diffusion out of a particular mineral becomes negligible compared with its rate of accumulation was first mentioned in Section 6.2. As a consequence of the ease with which radiogenic ^{40}Ar is lost when K-bearing minerals are heated, K–Ar dates represent the time elapsed since the dated mineral cooled through its blocking temperature, assuming that it remained closed to Ar and K after that event. Studies of Ar diffusion in contact metamorphic zones and in the laboratory indicate that the minerals used for dating by the K–Ar method have different blocking temperatures. The importance of diffusion of radiogenic daughters (including ^{40}Ar) to geochronometry has been reviewed by Giletti (1974). A mathematical analysis of this phenomenon was made by Dodson (1973, 1979).

Since hornblende has a higher blocking temperature than biotite (Section 6.2a and 6.2b), the K–Ar dates of hornblende in plutonic igneous and high-grade metamorphic rocks that cooled slowly are older than the K–Ar dates of biotite in the same rocks. The difference between the K–Ar dates of coexisting hornblende and biotite in such rocks can be used to determine the cooling rate provided the respective blocking temperatures are known. In addition, the K–Ar dates of a K-bearing mineral (e.g., hornblende, muscovite, biotite) of a suite of igneous or high-grade metamorphic rocks collected regionally from a batholith or an orogenic belt, respectively, can be contoured to form an imaginary surface. Such a contoured surface acts like a veil because it obscures the ages of the rocks (Armstrong, 1966).

In the case of plutonic igneous rocks, the rate of cooling of batholiths is rarely uniform, causing the contoured surface of K–Ar dates to have "topography": Hills occur where the rocks cooled rapidly and depressions form where the rocks cooled slowly. In the case of high-grade metamorphic rocks (schists and gneisses) in orogenic belts, the cooling rates depend on the rate of uplift, which may also vary regionally. In fact, the veil of K–Ar dates covering a block of regionally metamorphosed rocks can have slope because of tilting of the block during the uplift. In addition, K–Ar dates derived from the core of a mountain range can be younger than those on the flanks because the rocks in the core were heated to a higher temperature at greater depth and therefore cooled more slowly than equivalent rocks along the flanks of the orogenic belt (Harper, 1967b).

These generalizations require that all previously accumulated ^{40}Ar was purged from the minerals during metamorphic recrystallization and that no excess ^{40}Ar was incorporated into the minerals from the magma in case of igneous rocks or from the "environment" in case of orogenic belts. In addition, the mineral samples used for dating must have remained closed to K and Ar after they cooled through the blocking temperature.

6.6a Idaho Batholith

The Idaho batholith in northwestern Montana and Idaho is a large composite body of igneous rocks of granitic composition ($\sim 4 \times 10^4$ km^2) that was intruded primarily during the Cretaceous Period (Larsen and Schmidt, 1958). The K–Ar dates of 33 samples of biotite from this batholith measured by Criss et al. (1982) range from 93 to 37 Ma, in good agreement with dates reported previously by Armstrong (1975), McDowell and Kulp (1969), and Percious et al. (1967). The Idaho batholith (Figure 6.12) includes several

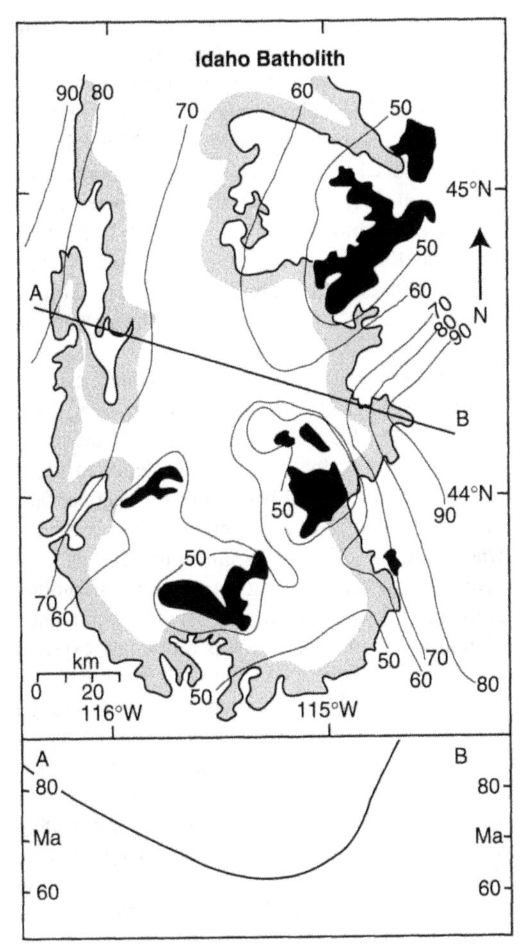

FIGURE 6.12 Contour map of K–Ar dates of biotites in the Idaho batholith on northwestern Montana and Idaho based primarily on the data of Criss et al. (1982). The areas in black are Eocene plutons that intruded the Cretaceous rocks of the batholith and caused partial losses of ^{40}Ar from biotite in the contact zone. The profile of K–Ar dates was drawn along the line A–B. Adapted from Criss et al. (1982).

The contours in Figure 6.12 are based primarily on the K–Ar dates reported by Criss et al. (1982). In addition, the diagram contains a profile of the K–Ar dates along the line A–B drawn through the center of the Idaho batholith. This profile demonstrates that the K–Ar dates of biotite decrease from about 90 Ma along the eastern and western contacts to about 60 Ma in the center of the batholith. This distribution of dates can be attributed to differences in the cooling rates because the rocks in the center of the batholith presumably cooled more slowly than those along the eastern and western contacts. In addition, the intrusion of the Eocene plutons at about 45 Ma caused variable and localized losses of ^{40}Ar from the biotite in the Idaho batholith and may have contributed to the lowering of K–Ar dates in the center of the batholith.

A further complication arises because the K–Ar dates of biotites in the Idaho batholith depend on the elevation of the collecting site. For example, in Figure 6.13 the oldest dates (63–64 Ma) between 115° 00′ and 115° 20′ W longitude occur from 2000 to 2300 m above sea level (a.s.l.), whereas the biotites with the lowest dates (53–54 Ma) are located at only 1500 m above sea level. Criss et al. (1982) concluded from this evidence that the upper part of the Idaho batholith originally was a dome that caused rocks close to the comparatively small plutons of Eocene age that caused ^{40}Ar loss from biotite in the contact zone surrounding them. The best estimate of the age of the Idaho batholith is >97 Ma based on a K–Ar date of biotite in granodiorite along the eastern border (Armstrong, 1975).

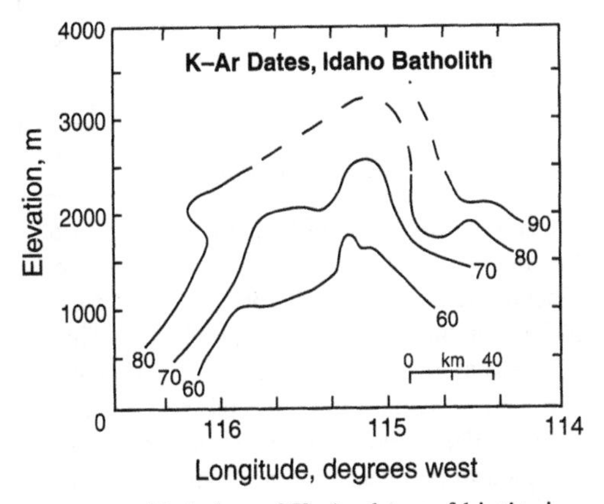

FIGURE 6.13 Variation of K–Ar dates of biotite in the Idaho batholith with elevation of the collecting sites. The contours represent lines of equal dates expressed in million years before present. Adapted from Criss et al. (1982).

top to cool more rapidly than rocks at deeper levels. Subsequently, the top of the batholith was dissected by erosion, thereby exposing rocks that had cooled at different rates depending on their elevation in the resulting topography.

6.6b Continental Crust

An interesting consequence of the temperature dependence of Ar retention is that K-bearing minerals cannot retain radiogenic ^{40}Ar below a certain depth in the crust, depending on the geothermal gradient. The depths in the continental crust where K-feldspar, biotite, and hornblende are at their blocking (or closure) temperatures are indicated in Figure 6.14. The blocking temperatures were calculated by Berger and York (1981) from diffusion rates of Ar for a cooling rate of 5°C/10^6 y. If the geothermal gradient of the continental crust is 23°C/km, the blocking temperature of K-feldspar (230°C) is reached at a depth of 9.3 km, that of biotite (373°C) at 15.5 km, and that of hornblende (685°C) at 29.1 km. Consequently, these results indicated that biotite below a depth of 15.5 km in this model of the continental crust cannot retain ^{40}Ar and therefore has a K–Ar date of zero. Similar statements apply to K-feldspar and hornblende in Figure 6.14.

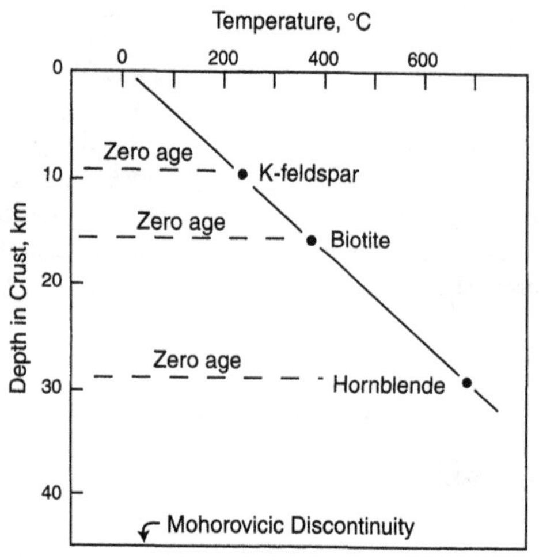

FIGURE 6.14 Retention of ^{40}Ar by minerals as a function of depth in the continental crust. The blocking temperatures are 230°C (K-feldspar), 373°C (biotite), and 685°C (hornblende). The geothermal gradient is 23°C/km, the average surface temperature is 15°C, and the relation between depth and the geothermal temperature is assumed to be linear. The resulting depth for zero K–Ar dates are 9.3 km (K-feldspar), 15.5 km (biotite), and 29.1 km (hornblende). The Mohorovicic Discontinuity was placed at a depth of 45 km. Blocking temperatures from Berger and York (1981).

In cases where a block of continental crust is uplifted as a result of tectonic activity (e.g., by a rising plume of asthenospheric mantle), increased erosion at the surface reduces the depth to rocks that previously resided below the zero-age surface of a particular K-bearing mineral. For example, when biotite-bearing rocks are uplifted, they may cool through the blocking temperature of biotite, allowing in situ produced ^{40}Ar to accumulate. Therefore, when such a rock is ultimately exposed at the surface, the K–Ar dates of biotite record the time elapsed since the rocks cooled through the blocking temperature as a result of uplift and erosion of the overlying rocks. Therefore, the K–Ar dates of biotites in Precambrian gneisses are a measure of the stability of the continental crust because they indicate the time elapsed since the rocks were uplifted through the isotherm corresponding to the blocking temperature of biotite (Harper, 1967b).

6.7 PRECAMBRIAN TIMESCALES

The K–Ar dates of biotite and muscovite from igneous and metamorphic rocks have played an important role in the development of the Precambrian timescale used by the Geological Survey of Canada. Starting in 1959, the Geochronology Laboratories of the Geological Survey of Canada began a program of dating rocks from the Canadian Precambrian shield to identify major orogenic events and to assist in the construction of a Precambrian timescale. The initial phase of this program consisted primarily of measurements of K–Ar dates of biotite and muscovite with some work on hornblende and whole-rock samples of mafic dikes. By the end of 1967, R. K. Wanless and his colleagues had dated more than 750 samples from all over the Canadian shield. These dates were used by Stockwell (1968, 1972) to construct a timescale for the Precambrian rocks exposed in Canada.

Stockwell showed that the K–Ar dates tend to cluster around certain values, which he identified as being the ages of orogenies. He defined an "orogeny" as a process of folding affecting large segments of the crust commonly associated with virtually contemporaneous regional metamorphism and emplacement of granitic bodies (Stockwell, 1972). He estimated the ages of these orogenies by the arithmetic means of the K–Ar dates in each cluster and applied to them the following names: Grenvillian (945 Ma), Elsonian (1370 Ma), Hudsonian (1735 Ma), and Kenoran (2490 Ma). Stockwell used these orogenies to define intervals of time, shown in Figure 6.15, each of which starts with the deposition of sediment on the erosion surface cut into the rocks subjected to the preceding orogeny and terminates with folding, metamorphism, uplift, and erosion during the next orogeny. For example, rocks of Aphebian age were deposited on a basement that was metamorphosed during the Kenoran orogeny. The Aphebian Era ended with the folding and metamorphism of the sedimentary rocks during the Hudsonian orogeny. Sedimentary and volcanic rocks of Helikian age were then deposited on a basement of rocks deformed and metamorphosed in the Hudsonian orogeny, and so on.

The timescale that was originally based on K–Ar dates of micas was later modified by redefining the ages of the orogenies on the basis of whole-rock Rb–Sr isochron dates and U–Pb dates of zircons (Stockwell, 1972, 1982). These dates reflect more accurately the time of crystallization of rocks during an orogeny than do K–Ar dates of micas. Use of Rb–Sr and U–Pb dates to determine the ages of orogenies is preferable because K–Ar dates depend on the rates of uplift and cooling and thus usually postdate metamorphism. Stockwell's most recent estimates of ages of the orogenies are indicated in Figure 6.15 and were slightly modified by Douglas (1980). However, the basic outline of his timescale remains the same and is in general use by the Geological Survey of Canada.

Eon	Era	Sub-Era		Orogeny	Age, Ma
	Paleozoic				~570
Proterozoic	Hadrynian	Late		Avalonian	~620
		Early			
	Helikian	Neo	Late	**Grenvillian**	1000
			Early	Elsevirian	1200
		Paleo	Late	Elsonian	1400
			Early	Killarnian	~1500
				Hudsonian	~1750
	Aphebian	Late		Moranian	1870
		Middle		Blezardian	2140?
		Early		**Kenoran**	~2510
Archean		Late		Laurentian	2670
		Late Middle		Wanipigowan	~2900
		Early Middle			
				Uivakian	~3400
		Early			

FIGURE 6.15 Precambrian timescale used by the Geological Survey of Canada. The ages of the orogenies are based primarily on K–Ar dates of micas but have been supplemented by Rb–Sr and U–Pb dates. Modified after Table 2, p. 14, of Stockwell (1982).

An important aspect of this classification is the recognition of eight structural provinces on the Canadian Shield. Each province was stabilized at the end of a specific orogeny as indicated by the clustering of K–Ar mica dates in each province around a "magic number." For example, the Superior Province, identified in Figure 6.16, is characterized by K–Ar dates that cluster around a mean of 2500 Ma corresponding to the Kenoran orogeny. Evidently, the rocks of the Superior Province were metamorphosed during the Kenoran orogeny and have not been reheated since then. The rocks of the Churchill Province were metamorphosed for the last time during the Hudsonian orogeny and

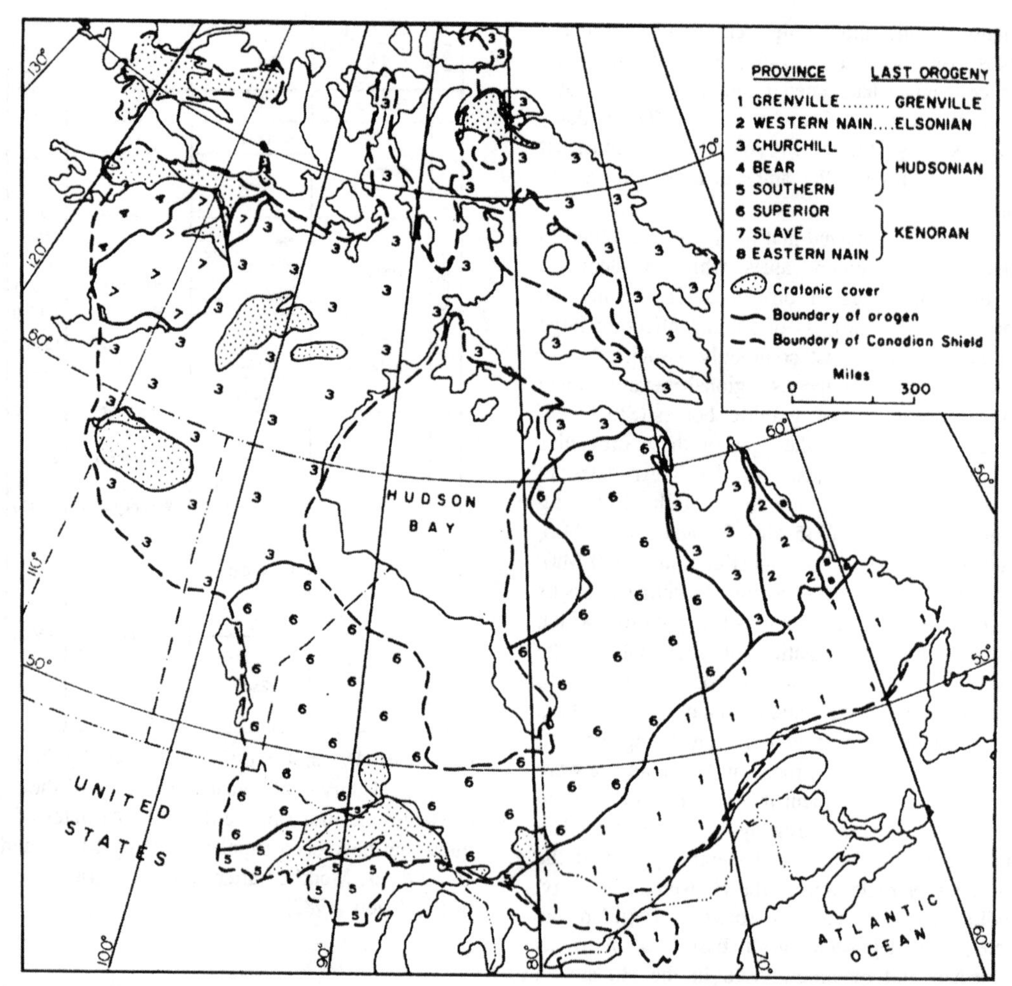

FIGURE 6.16 Orogenies and structural provinces of the Canadian Shield as shown originally by Stockwell (1968). A colored map published by Stockwell (1982) contains more detail but does not differ significantly from the earlier version. Reproduced by permission of the National Research Council of Canada from the *Canadian Journal of Earth Sciences*, Vol. 5, pp. 693–698, 1968.

those of the Grenville Province by the Grenvillian orogeny. In this way, the entire shield has been subdivided into structural provinces that are separated from each other by major unconformities or orogenic fronts.

The K–Ar dates of micas in each province form a metamorphic veil that hides the true ages of the rocks. It is known, for example, that the Grenville Province contains rocks that formed long before the Grenville orogeny (Doig, 1975; Krogh and Hurley, 1968; Krogh and Davis, 1969; Grant, 1964). However, all of the rocks in this province bear the strong metamorphic and structural imprint of the Grenvillian orogeny and, therefore, are properly considered to form a coherent structural province. Similar statements can be made about the other provinces, although their preorogenic histories have not yet been revealed in as much detail in all cases.

The timescale used in Minnesota (Goldich et al., 1961; Goldich, 1968) and adjacent areas around Lake Superior in the United States is based on a threefold division of Precambrian time (Early, Middle, and Late). Early Precambrian time ended with the Algoman orogeny (2400–2750 Ma), whereas the Middle Precambrian was terminated by the Penokean orogeny (1600–1900 Ma.). The Algoman orogeny of Minnesota corresponds to the Kenoran orogeny of Canada, and the Penokean orogeny corresponds to the Hudsonian orogeny. The Elsonian and Grenvillian orogenies are not recognized in Minnesota because these orogenies did not affect the rocks of this region. However, the Keweenawan igneous event, which occurred from 1000 to 1200 Ma, is probably synchronous with the Grenvillian orogeny of Canada. Different timescales based on age determinations have been proposed or adopted for other areas and correlations between them are therefore made difficult (Harland et al., 1989).

Table 6.4. Timescale for the Precambrian[a]

Eon	Era	Age of Boundary, Ma
		570
Proterozoic	Late (Neo)	
		1000
	Middle (Meso)	
		1600
	Early (Paleo)	
		2500
Archean	Late	
		3000
	Middle	
		3400
	Early	
		3800?

[a] This timescale is to be used in The Geology of North America, sponsored by the Geological Society of America in connection with the celebration of the society's centennial (Harrison and Peterman, 1982; Palmer, 1983).

The most recent timescale for the Precambrian of North America in Table 6.4 was compiled by Palmer (1983) based on recommendations by Harrison and Peterman (1982). This timescale was established to assist in the preparation of a comprehensive synthesis of the geology of North America sponsored by the Geological Society of America. It divides the Precambrian into two eons, each of which is subdivided into three eras, as shown in Table 6.4.

Another timescale for the Precambrian history of Earth was assembled by Harland et al. (1989) in Table 6.5. These authors documented the history of the attempts to set up a timescale that would be applicable globally and would be accepted by geologists everywhere. For the purpose of this presentation, the timescale in Table 6.4 will be used because it is generally understood by North American geologists, whereas the timescale in Table 6.5 is in use in Europe.

Table 6.5. Timescale for the Precambrian History of the Earth

Era	Period	Date, Ga
Paleozoic	Cambrian	
		0.570
Sinian	Vendian	
		0.610
Sinian	Sturtian	
		0.80
Riphean		
		1.65
Animikean		
		2.2
Huronian		
		2.45
Randian		
		2.8
Swazian		
		3.5
Isuan		
		3.8
Hadean		
		4.56

Source: Harland et al., 1989.

6.8 SUMMARY

The K–Ar method of dating is applicable to certain K-bearing minerals and rocks that retain radiogenic Ar quantitatively after cooling through their respective blocking temperatures. The minerals that are most suitable for dating include biotite, muscovite, and hornblende. In addition, volcanic and shallow intrusive rocks may be dated as whole-rock samples provided they are free of foreign inclusions and do not contain hydrated or devitrified glass. The K–Ar dates of well-crystallized glauconite in unmetamorphosed sedimentary rocks, in most cases, postdate the time of deposition but may record late-stage diagenesis, deep burial, structural deformation, and alteration by hot subsurface brines. In contrast to glauconite and other clay minerals, sanidine and biotite in bentonite and volcanic ash beds have yielded K–Ar dates that are in good agreement with the stratigraphic ages of the sedimentary rocks with which the bentonite and tuff beds are interbedded. The K-bearing sheet-silicate minerals in shale undergo transformations during burial metamorphism that release radiogenic ^{40}Ar but tend to conserve K. Consequently, whole-rock samples and grain-size fractions of shale give neither the age of their source regions nor the time of deposition.

In spite of these limitations, the K–Ar method of dating has played an important role in the development of the earth sciences. For example, K–Ar dates of volcanic rocks of Tertiary age have been used to determine the rate of motion of lithospheric plates above the heads of stationary asthenospheric plumes. In addition, the timescale for the reversal of the Earth's magnetic feld was established by K–Ar dates of volcanic rocks in which these reversals were recorded

Another important insight was derived from the present ^{40}Ar/^{36}Ar ratio of rocks in the mantle of the Earth. The high value of this ratio ($\sim$10,000) implies that a large fraction of ^{36}Ar was lost from the interior of the Earth at an early stage in its evolution in pre-Archean time, thereby allowing the ^{40}Ar/^{36}Ar ratio of mantle rocks to rise by decay of ^{40}K to their present high value. The Ar that was lost from the mantle in pre-Archean time now resides in the atmosphere of the Earth. Consequently, the ^{40}Ar/^{36}Ar ratio of the present atmosphere is comparatively low ($\sim$295) because the radiogenic ^{40}Ar that has escaped from the mantle of the Earth since Early Archean time has mixed with the Ar of the pre-Archean atmosphere, which had a low ^{40}Ar/^{36}Ar ratio.

The K–Ar method has also played a significant role in the reconstruction of the history of the Earth because it has recorded the occurrence of orogenic events. The ages of orogenies are the basis for the Precambrian geological timescale used in Canada to delineate structural provinces of the Precambrian Shield of North America.

The temperature dependence of ^{40}Ar retention by K-bearing minerals has given rise to the metamorphic veil, which is defined on the basis of the K–Ar dates of a selected mineral in large regions of Precambrian shields. For example, the metamorphic veil formed by the K–Ar dates of biotite in the granitic gneisses of the Grenville structural province of Canada records the time when the rocks, which are now exposed at the surface, cooled through the blocking temperature while they were many kilometers below the surface. This interpretation of K–Ar dates provides information about rates of uplift and erosion of the continental crust and thus about the stability of the crust.

In spite of the fact that the K–Ar method of dating has been widely used to derive important insights about the geological history of the Earth, its reliability as a geochronometer is limited. These limitations of the conventional K–Ar method are overcome by the so-called ^{40}Ar/^{39}Ar (or Ar/Ar) method, which is the subject of the next chapter.

REFERENCES FOR PRINCIPLES AND METHODOLOGY (INTRODUCTION, SECTION 6.1)

Aldrich, L. T., and A. O. Nier, 1948. Argon 40 in potassium minerals. *Phys. Rev.*, 74:876–877.

Aston, F. W., 1921. The mass spectra of the alkali metals. *Philos. Mag.*, 42(Ser. 6): 436–441.

Campbell, N. R., 1908. The radioactivity of potassium, with special reference to solutions of its salts. *Proc. Cambridge Philos. Soc.*, 14:557–567.

Campbell, N. R., and A. Wood, 1906. The radioactivity of the alkali metals. *Proc. Cambridge Philos. Soc.*, 14:15–21.

Cooper, J. A., 1963. The flame photometric determination of potassium in geological materials used for potassium-argon dating. *Geochim. Cosmochim Acta*, 27:525–546.

Dalrymple, G. B., 1991. *The Age of the Earth*. Stanford University Press, Stanford, California.

Dalrymple, G. B., and M. A. Lanphere, 1969. *Potassium-Argon Dating*. W.H. Freeman, San Francisco.

Damon, P. E., 1970. A theory of "real" K–Ar clocks. *Eclogae Geol. Helv.*, 63:69–76.

Garner, E. L., T. J. Murphy, J. W. Gramlich, P. J. Paulsen, and I. L. Barnes, 1975. Absolute isotopic abundance ratios and the atomic weight of a reference sample of potassium. *J. Res. Natl. Bur. Stand.–A. Phys. Chem.*, 79A: 713–725.

Heier, K. S., and J. A. S. Adams, 1964. The geochemistry of the alkali metals. In *Physics and Chemistry of the Earth*, Vol. 5, pp. 253–381. Pergamon, Oxford.

Hunziker, J. C., 1979. Potassium-argon dating. In E. Jäger and J. C. Hunziker (Eds.), *Lectures in Isotope Geology*, pp. 52–76. Springer-Verlag, Berlin.

McDougall, I., 1966. Precision methods of potassium-argon isotopic age determination on young rocks. In *Methods and Techniques in Geophysics*, Vol. 2, pp. 279–304. Wiley-Interscience, New York.

Morozova, I. M., and A. A. Alferovsky, 1974. Fractionation of lithium and potassium isotopes in geological processes. *Geokhimiya*, 1:30–39.

Müller, O., 1966. Potassium analysis. In O. A. Schaeffer and J. Zähringer (Eds.), *Potassium Argon Dating*, pp. 40–66. Springer-Verlag, New York.

Nier, A. O., 1935. Evidence for the existence of an isotope of potassium of mass 40. *Phys. Rev.*, 48:283–294.

Nier, A. O., 1950. A redetermination of the relative abundances of the isotopes of carbon, nitrogen, oxygen, argon, and potassium. *Phys. Rev.*, 77:789–793.

Schaeffer, O. A., and J. Zähringer (Eds.), 1966. *Potassium Argon Dating*. Springer-Verlag, New York.

Steiger, R. H., and E. Jäger, 1977. Subcommission on Geochronology: Convention on the use of decay constants in geo- and cosmochronology. *Earth Planet. Sci. Lett.*, 36:359–362.

Thomson, J. J., 1905. On the emission of negative corpuscles by the alkali metals. *Philos. Mag.*, 10(Ser. 6): 584–590.

Verbeek, A. A., and G. D. L. Schreiner, 1967. Variations in ^{39}K:^{41}K ratio and movement of potassium in a granite-amphibolite contact region. *Geochim. Cosmochim. Acta*, 31:2125–2163.

Von Weizsäcker, C. F., 1937. Über die Möglichkeit eines dualen-Zerfalls von Kalium. *Phys. Zeitschrift*, 38:623–624.

REFERENCES FOR RETENTION OF ^{40}Ar BY MINERALS (SECTIONS 6.2a–6.2b)

Dodson, M. H., 1973. Closure temperature in cooling geochronological and petrological systems. *Contrib. Mineral. Petrol.*, 40:259–274.

Dodson, M. H., 1979. Theory of cooling ages. In E. Jäger and J. C. Hunziker (Eds.), *Lectures in Isotope Geology*, pp. 194–202. Springer-Verlag, Berlin.

Hanson, G. N., E. J. Catanzaro, and D. H. Anderson, 1971. U-Pb ages for sphene in a contact metamorphic zone. *Earth Planet. Sci. Lett.*, 12:231–237.

Hanson, G. N., and P. W. Gast, 1967. Kinetic studies in contact metamorphic zones. *Geochim. Cosmochim. Acta*, 31:1119–1153.

Hart, S. R., 1964. The petrology and isotopic mineral-age relations of a contact zone in the Front Range, Colorado. *J. Geol.*, 72:493–525.

Hart, S. R., G. L. Davis, R. H. Steiger, and G. R. Tilton, 1968. A comparison of the isotopic mineral-age variations and petrologic changes induced by contact metamorphism. In E. I. Hamilton and R. M. Farquhar (Eds.), *Radiometric Dating for Geologists*, pp. 73–110. Wiley-Interscience, New York.

Jaeger, J. C., 1957. The temperature in the neighborhood of a cooling intrusive sheet. *Am. J. Sci.*, 255:306–318.

McDowell, F. W., 1983. K-Ar dating: Incomplete extraction of radiogenic argon from alkali feldspar. *Chem. Geol. Isotope Geosci.*, 1:118–126.

REFERENCES FOR EXCESS ^{40}Ar (SECTION 6.2c)

Curtis, G. H., 1966. The problem of contamination in obtaining accurate dates of young geologic rocks. In O. A. Schaeffer and J. Zähringer (Eds.), *Potassium-Argon Dating*, pp. 151–162. Springer-Verlag, New York.

Dalrymple, G. B., and M. A. Lanphere, 1969. *Potassium-Argon Dating*. W.H. Freeman, San Francisco.

Damon, P. E., and J. L. Kulp, 1958a. Excess helium and argon in beryl and other minerals. *Am. Mineral.*, 42:443–459.

Damon, P. E., and J. L. Kulp, 1958b. Inert gases and the evolution of the atmosphere. *Geochim. Cosmochim. Acta*, 13:280–292.

Damon, P. E., A. W. Laughlin, and J. K. Percious, 1967. The problem of excess argon-40 in volcanic rocks. In *Symposium on Radioactive Dating and Methods of Low-Level Counting*, SM-87/45, International Atomic Energy Agency, Vienna, Austria.

Faure, G., 2001. *Origin of Igneous Rocks; the Isotopic Evidence*. Springer-Verlag, Heidelberg.

Gerling, E. K., 1984. Reminiscences about some works connected with the study of noble gases, their isotopic composition and geochronology. *Chem. Geol.*, 46:271–289; *Isotope Geosci.*, 2(4): 271–289.

Gerling, E. K., Yu. A. Shukolyukov, T. V. Koltsova, I. I. Matseeva, and S. Z. Yakovleva, 1962. The determination of the age of basic rocks by the K-Ar method. *Geokhimiya*, 11:931–938 (in Russian).

Giletti, B. J., 1974a. Diffusion related to geochronology. In A. W. Hofmann, B. J. Giletti, H. S. Yoder, Jr., and R. A. Yund (eds.), *Geochemical Transport and Kinetics*, Pub. 634, pp. 61–76. Carnegie Institution of Washington, Washington, D.C.

Giletti, B. J., 1974b. Studies in diffusion I: Argon in phlogopite mica. In A. W. Hofmann, B. J. Giletti, H. S. Yoder, Jr., and R. A. Yund (eds.), *Geochemical Transport and Kinetics*, Pub. 634, pp. 107–116. Carnegie Institution of Washington, Washington, D.C.

Hanson, G. N., and P. W. Gast, 1967. Kinetic studies in contact metamorphic zones. *Geochim. Cosmochim. Acta*, 31:1119–1153.

Ozima, M., S. Zashu, and O. Nitosh, 1983. $^3He/^4He$ ratio, noble gas abundance, and K-Ar dating of diamonds. An attempt to search for the records of early terrestrial history. *Geochim. Cosmochim. Acta*, 47:2217–2224.

Rama, S. N. I., S. R. Hart, and E. Roedder, 1965. Excess radiogenic argon in fluid inclusions. *J. Geophys. Res.*, 70:508–511.

Tolstikhin, I. N., V. S. Dokuchaeva, I. L. Kamensky, and Yu. V. Amelin, 1992. Juvenile helium in ancient rocks: II, U-He, K-Ar, Sm-Nd, and Rb-Sr systematics in the Monche pluton. $^3He/^4He$ ratios frozen in uranium-free ultramafic rocks. *Geochim. Cosmochim. Acta*, 56:987–999.

REFERENCES FOR K–Ar ISOCHRONS (SECTION 6.3)

Armstrong, R. L., 1978. Removal of atmospheric argon contamination and the use and misuse of the K-Ar isochron method: Discussion. *Can. J. Earth Sci.*, 15:325–326.

Curtis, G. H., and R. L. Hay, 1972. Further geological studies and potassium-argon dating at Olduvai Gorge and Ngorongoro Crater. In W. W. Bishop and J. A. Miller (Eds.), *Calibration of Hominoid Evolution*, pp. 289–301. Scottish Academic Press, Edinburgh.

Fitch, F. J., J. A. Miller, and P. J. Hooker, 1976. Single whole-rock K-Ar isochrons. *Geol. Mag.*, 113:1–10.

Hayatsu, A., and C. M. Carmichael, 1977. Removal of atmospheric argon contamination and the use and misuse of the K-Ar isochron method. *Can. J. Sci.*, 14:337–345.

Hayatsu, A., and H. C. Palmer, 1975. K-Ar isochron study of the Tudor Gabbro, Grenville Province, Ontario. *Earth Planet. Sci. Lett.*, 25:208–212.

McDougall, I., H. A. Polach, and J. J. Stipp, 1969. Excess radiogenic argon in young subaerial basalts from the Auckland volcanic fields, New Zealand. *Geochim. Cosmochim. Acta*, 33:1485–1520.

Roddick, J. C., 1978. The application of isochron diagrams in ^{40}Ar–^{36}Ar dating: A discussion. *Earth Planet. Sci. Lett.*, 41:223–244.

Shafiqullah, M., and P. E. Damon, 1974. Evaluation of K-Ar isochron methods. *Geochim. Cosmochim. Acta*, 38:1341–1358.

Steiger, R. H., and E. Jäger, 1977. Subcommission on Geochronology: Convention on the use of decay constants in geo- and cosmochronology. *Earth Planet. Sci. Lett.*, 36:359–362.

York, D., 1969. Least-squares fitting of a straight line with correlated errors. *Earth Planet. Sci. Lett.*, 5:320–324.

REFERENCES FOR RATE OF MOTION OF HAWAIIAN ISLANDS (SECTION 6.4a)

Dalrymple, B. G., and D. D. Lanphere (1969). *Potassium-Argon Dating*. W. H. Freeman, San Francisco.

Dalrymple, B. G., D. A. Clague, M. O. Garcia, and S. W. Bright, 1981. Petrology and K-Ar ages of dredged samples from Laysan Island and Northampton Bank volcanoes, Hawaiian Ridge, and evolution of the Hawaiian-Emperor chain: Summary. *Geol. Soc. Am. Bull.*, 92:315–318.

Faure, G., 2001. *Origin of Igneous Rocks; the Isotopic Evidence*. Springer-Verlag, Heidelberg.

Jackson, E. D., E. A. Silver, and G. B. Dalrymple, 1972. Hawaiian-Emperor Chain and its relation to Cenozoic circumpacific tectonics. *Geol. Soc. Am. Bull.*, 83:601–618.

Kent, A. J. R., D. A. Clague, M. Honda, E. M. Stolper, I. D. Hutcheon, and M. D. Norman, 1999. Widespread assimilation of a seawater-derived component at Loihi seamount, Hawaii. *Geochim. Cosmochim. Acta*, 63:2749–2761.

McDougall, I., and R. A. Duncan, 1980. Linear volcanic chains—recording plate motions? *Tectonophysics*, 63: 275–295.

Staudigel, H., A. Zindler, S. R. Hart, T. Leslie, C. -Y. Chen, and D. Clague, 1984. The isotope systematics of a juvenile intraplate volcano: Pb, Nd, and Sr isotope ratios of basalts from Loihi Seamount, Hawaii. *Earth Planet. Sci. Lett.*, 69:13–29.

Wilson, J. T., 1963. A possible origin of the Hawaiian Islands. *Can. J. Phys.*, 41:863–870.

REFERENCES FOR MAGNETIC REVERSAL CHRONOLOGY (SECTION 6.4b)

Calmant, S., 1987. The elastic thickness of the lithosphere in the Pacific Ocean. *Earth Planet. Sci. Lett.*, 85:277–288.

Cox, A., R. R. Doell, and G. B. Dalrymple, 1963a. Geomagnetic polarity epochs and Pleistocene geochronometry. *Nature*, 198:1049–1051.

Cox, A., R. R. Doell, and G. B. Dalrymple, 1963b. Geomagnetic polarity epochs: Sierra Nevada II. *Science*, 142:382–385.

Dupuy, C., P. Vidal, R. C. Maury, and G. Guille, 1993. Basalts from Mururoa, Fangataufa, and Gambier islands (French Polynesia): Geochemical dependence on the age of the lithosphere. *Earth Planet. Sci. Lett.*, 117:89–100.

Hall, C. M., and D. York, 1978. K-Ar and ^{40}Ar/^{39}Ar age of the Laschamp geomagnetic polarity reversal. *Nature*, 274:462–464.

Heirtzler, J. R., G. O. Dickson, E. J. Herron, W. C. Pittman, and X. LePichon, 1968. Marine magnetic anomalies, geomagnetic field reversals, and motions of the ocean floor and continents. *J. Geophys. Res.*, 73:2119–2136.

Mankinen, E. A., and G. B. Dalrymple, 1979. Revised geomagnetic polarity time scale for the interval 0–5 m.y. B.P., *J. Geophys. Res.*, 84(B2): 615–626.

McDougall, I., and D. H. Tarling, 1963. Dating of polarity zones in the Hawaiian Islands. *Nature*, 200:54–56.

McElhinny, M. W, 1978. The magnetic polarity time scale: Prospects and possibilities in magnetostratigraphy. In G. V. Cohee, H. D. Hedberg, and M. F. Glaesssner (Eds.), *Contributions to the Geologic Time Scale*, pp. 57–65, Studies in Geology No. 6. American Association of Petroleum Geologists, Tulsa, Oklahoma.

McNutt, M., K. Fischer, S. Kruse, and J. Natland, 1989. The origin of the Marquesas fracture zone ridge and its implications for the origin of hotspots. *Earth Planet. Sci. Lett.*, 91:381–393.

Ness, G. S., S. Levi, and R. Couch, 1980. Marine magnetic anomaly timescales for the Cenozoic and Late Cretaceous: A précis, critique, and synthesis. *Rev. Geophys. Space Phys.*, 18:753–770.

Vine, F. J., and D. H. Matthews, 1963. Magnetic anomalies over oceanic ridges. *Nature*, 199:947–949.

REFERENCES FOR ARGON FROM THE MANTLE (SECTION 6.4c)

Allègre, C. J., T. Staudacher, Ph. Sarda, and M. Kurz, 1983. Constraints on evolution of Earth's mantle from rare gas systematics. *Nature*, 303:762–766.

Fanale, F. P., 1971. A case for catastrophic early degassing of the Earth. *Chem. Geol.*, 8:79–105.

Fisher, D. E., 1981. Quantitative retention of argon in glassy basalt. *Nature*, 290:42–43.

Hekinian, R., M. Chaigneau, and J. L. Cheminee, 1973. Popping rocks and lava tubes from the Mid-Atlantic Rift Valley at 36° N. *Nature*, 245:371–373.

Marty, B., S. Zashu, and M. Ozima, 1983. Two noble gas components in a Mid-Atlantic Ridge basalt. *Nature*, 302:238–240.

Ozima, M., 1975. Ar isotopes and Earth atmosphere evolution models. *Geochim. Cosmochim. Acta*, 38:1127–1134.

Ozima, M., and F. A. Podosek, 1983. *Noble Gas Geochemistry*. Cambridge University Press, New York.

Schwartzman, D. W., 1973. On argon degassing models of the Earth. *Nature Phys. Sci.*, 245:20–21.

REFERENCES FOR DATING SEDIMENTARY ROCKS (SHALE AND BENTONITE) (SECTIONS 6.5–6.5b)

Altaner, S. P., J. Hower, G. Whitney, and J. L. Aronson, 1984. Model for K-bentonite formation: Evidence from zoned K-bentonites in the disturbed belt, Montana. *Geology*, 12:412–415.

Aronson, J. L., and C. B. Douthitt, 1986. K-Ar systematics of an acid-treated illite/smectite: Implications for evaluating age and crystal structure. *Clays and Clay Minerals*, 34:473–482.

Aronson, J. L., and J. Hower, 1976. Mechanism of burial metamorphism of argillaceous sediment 2. Radiogenic argon evidence. *Geol. Soc. Am. Bull.*, 87:738–744.

Bechtel, A., W. C. Elliott, J. M. Wampler, and S. Oszczepalski, 1999. Clay mineralogy, crystallinity, and K-Ar ages of illites with the Polish Zechstein Basin: Implications for the age of Kupferschiefer mineralization. *Econ. Geol.*, 94:261–272.

Carey, S., and R. S. J. Sparks, 1986. Quantitative models of the fallout and dispersal of tephra from volcanic eruption columns. *Bull. Volcanol.*, 48:109–125.

Clauer, N., and S. Chaudhuri, 1995. *Clays in Crustal Environments; Isotopic Dating and Tracing*. Springer-Verlag, Heidelberg.

Dalrymple, G. B., 1991. *The Age of the Earth*. Stanford University Press, Stanford, California.

Dalrymple, G. B., and M. A. Lanphere, 1969. *Potassium-Argon Dating*. W. H. Freeman, San Francisco.

Elliott, W. C., and J. L. Aronson, 1987. Alleghenian episode of K-bentonite illitization in the southern Appalachian Basin. *Geology*, 15:735–739.

Elliott, W. C., J. L. Aronson, G. Matisoff, and D. L. Gautier, 1991. Kinetics of the smectite to illite transformation in the Denver Basin: Clay mineral, K-Ar data, and mathematical model results. *Am. Assoc. Petrol. Geol. Bull.*, 75:436–462.

Hoffman, J., J. Hower, and J. L. Aronson, 1976. Radiometric dating of time of thrusting in the disturbed belt of Montana. *Geology*, 4:16–20.

Huff, W. D., S. M. Bergström, and D. R. Kokata, 1992. Gigantic Ordovician ash fall in North America and Europe: Biological, tectonomagmatic, and event-stratigraphic significance. *Geology*, 20:875–878.

Hurley, P. M., 1966. K-Ar dating of sediments. In O. A. Schaeffer and J. Zähringer (Eds.), *Potassium-Argon Dating*, pp. 134–150. Springer-Verlag, Heidelberg.

McKerrow, W. S., J. F. Dewey, and C. R. Scotese, 1991. The Ordovician and Silurian development of the Iapetus Ocean. *Spec. Pap. Palaeontol.*, 44:165–178.

Perry, E. A., Jr., 1974. Diagenesis and the K-Ar dating of shales and clay minerals. *Geol. Soc. Am. Bull.*, 85:827–830.

Person, A., 1982. The genesis of bentonites. In G. S. Odin (Ed.), *Numerical Dating in Stratigraphy*, Vol. 1, pp. 407–421. Wiley, Chichester.

Rose, W. I., and C. A. Chesner, 1990. Worldwide dispersal of ash and gases from earth's largest known eruption: Toba, Sumatra, 75 ka. *Paleo. Paleo. Paleo.*, 89:269–275.

REFERENCES FOR VOLCANOGENIC MINERALS IN SEDIMENTARY ROCKS (SECTION 6.5c)

Baadsgaard, H., and J. F. Lerbekmo, 1982. The dating of bentonite beds. In G. S. Odin (Ed.), *Numerical Dating in Stratigraphy*, Part 1, pp. 423–440. Wiley, New York.

Damon, P. E., and R. Teichmüller, 1971. Das absolute Alter des sanidinführenden kaolinischen Tonsteins im Flöz Hagen 2 des Westfal C im Ruhrrevier. *Fortschritte Geol. Rheinland Westfalen*, 18:53–56.

Dymond, J. R., 1966. Potassium–argon geochronology of deep-sea sediment. *Science*, 152:1239–1241.

Hellmann, K. N., and H. J. Lippolt, 1981. Calibration of the Middle Triassic timescale by conventional K-Ar and $^{40}Ar/^{39}Ar$ dating of alkali feldspars. *J. Geophys.*, 50:73–88.

Hess, J. C., S. Backfisch, and H. J. Lippolt, 1983. Konkordantes Sanidin-und discordante Biotitalter eines Karbontuffs der Baden-Badener Senke, Nordschwarzwald. *Neues Jahrbuch Geol. Paläont.*, 1983(5): 277–292.

Hess, J. C., and H. J. Lippolt, 1980. Das Vorkommen von Sanidin in Kaolin–Kohlentonsteinen des Saarkarbons. *Oberrhein. Geol. Abhandl.*, 29:71–80.

Lanphere, M. A., and I. L. Tailleur, 1983. K-Ar ages of bentonites in the Seabee Formation, northern Alaska: A Late Cretaceous (Turonian) timescale point. *Cretaceous Res.*, 4:361–370.

REFERENCES FOR METASEDIMENTARY ROCKS AND METAMORPHIC VEILS (SECTIONS 6.5d–6.6)

Armstrong, R. L., 1966. K-Ar dating of plutonic and volcanic rocks in orogenic belts. In O. A. Schaeffer and J. Zähringer (Eds.), *Potassium Argon Dating*, pp. 117–131. Springer-Verlag, New York.

Armstrong, R. L., 1975. The geochronometry of Idaho. *Isochron/West*, No. 14.

Berger, G. W., and D. York, 1981. Geothermometry from $^{40}Ar/^{39}Ar$ dating experiments. *Geochim. Cosmochim. Acta*, 45:795–811.

Bonhomme, M. G., P. Saliot, and Y. Pinault, 1980. Interpretation of potassium-argon isotopic data related to metamorphic events in the southwestern Alps. *Schweiz Mineral. Petrogr. Mitteilungen*, 60:81–98.

Clauer, N., and S. Chaudhuri, 1995. *Clays in Crustal Environments: Isotope Dating and Tracing*. Springer-Verlag, Heidelberg.

Criss, R. E., M. A. Lanphere, and H. P. Taylor, Jr., 1982. Effects of regional uplift, deformation, and meteoric-hydrothermal metamorphism on K-Ar ages of biotite in the southern half of the Idaho Batholith. *J. Geophys. Res.*, 87(B8): 7029–7046.

Dodson, M. H., 1973. Closure temperatures in cooling geochronological and petrological systems. *Contrib. Mineral Petrol.*, 40:259–274.

Dodson, M. H., 1979. Theory of cooling ages. In E. Jäger and J. C. Hunziker (Eds.), *Lectures in Isotope Geology*, pp. 194–202. Springer-Verlag, Heidelberg.

Frey, M., J. C. Hunziker, W. Frank, J. Boquet, G. V. Dal Piaz, E. Jäger, and E. Niggli, 1974. Alpine metamorphism of the Alps: A review. *Schweiz. Mineral. Petrograph. Mitteil.*, 54:247–290.

Frey, M., J. C. Hunziker, P. Roggwiller, and C. Schindler, 1973. Pregressive niedriggradige Metamorphose glaukonitführender Horizonte in den helvetischen Alpen der Ostschweiz. *Contrib. Mineral. Petrol.*, 39:185–218.

Giletti, B. J., 1974. Diffusion related to geochronology. In A. W. Hofmann, B. J. Giletti, H. S. Yoder, and R. A. Yund (Eds.), *Geochemical Transport and Kinetics*, Pub. 643, Carnegie Institute of Washington, Washington, D.C.

Harland, W. B., R. L. Armstrong, A. V. Cox, L. E. Craig, A. G. Smith, and D. G. Smith, 1989. *A Geologic Time Scale 1989*. Cambridge University Press, Cambridge.

Harper, C. T., 1964. Potassium-argon ages of slates and their geologic significance. *Nature*, 203:468–470.

Harper, C. T., 1967a. The geological interpretation of potassium-argon ages of metamorphic rocks from the Scottish Caledonides. *Scott. J. Geol.*, 3:46–66.

Harper, C. T., 1967b. On the interpretation of potassium-argon ages from Precambrian shields and Phanerozoic orogens. *Earth Planet. Sci. Lett.*, 3:128–132.

Larsen, E. S. Jr., and R. G. Schmidt, 1958. A reconnaissance of the Idaho batholith and comparison with the southern California batholith. *U.S. Geol. Surv. Bull.*, 1070A: 1–33.

Liewig, N., J. -M. Caron, and N. Clauer, 1981. Geochemical and K-Ar isotopic behaviour of alpine sheet silicates during polyphased deformation. *Tectonophysics*, 78:273–290.

McDowell, F. W., and J. L. Kulp, 1969. Potassium-argon dating of the Idaho batholith. *Geol. Soc. Am. Bull.*, 80:2379–2382.

Percious, J. K., P. E. Damon, and H. J. Olson, 1967. *Radiometric Dating of Idaho Batholith Porphyries*, Annual Prog. Rep. 000-689-76, Appendix A–X. U.S. Atomic Energy Commission, Washington, D.C.

REFERENCES FOR PRECAMBRIAN TIMESCALES (SECTION 6.7)

Doig, R., 1975. A syenite pluton of Archean age within the Grenville Province of Quebec. *Can. J. Earth Sci.*, 12:890–893.

Douglas, R. J. W., 1980. *Proposals for Time Classification and Correlation of Precambrian Rocks and Events in Canada and Adjacent Areas of the Canadian Shield*, Paper 80-24, Geological Survey of Canada. Ottawa, Ontario, Canada.

Goldich, S. S., 1968. Geochronology in the Lake Superior region. *Can. J. Earth Sci.*, 5:715–724.

Goldich, S. S., A. O. Nier, H. Baadsgaard, J. H. Hoffman, and H. W. Krueger, 1961. The Precambrian geology and geochronology of Minnesota. *Minn. Geol. Surv. Bull.* 41.

Grant, J. A., 1964. Rubidium-strontium isochron study of the Grenville front near Lake Timagami, Ontario. *Science*, 146:1049–1053.

Harland, W. B., R. L. Armstrong, A. V. Cox, L. E. Craig, A. G. Smith, and D. G. Smith, 1989. *A Geologic Time Scale 1989*. Cambridge University Press, Cambridge.

Harrison, J. E., and Z. E. Peterman, 1982. North American Commission on Stratigraphic Nomenclature. Report 9, Adoption of geochronometric units for divisions of Precambrian time. *Am. Assoc. Petrol. Geol. Bull.*, 66:801–802.

Krogh, T. E., and G. L. Davis, 1969. Old isotopic ages in the northwestern Grenville Province, Ontario. *Geol. Assoc. Can.*, Spec. Paper 5:189–192.

Krogh, T. E., and P. M. Hurley, 1968. Strontium isotope variation and whole-rock isochron studies, Grenville Province of Ontario, *J. Geophys. Res.*, 73:7107–7125.

Palmer, A. R., 1983. The decade of North American geology. 1983 geologic time scale. *Geology*, 11:503–504.

Stockwell, C. H., 1968. Geochronology of stratified rocks on the Canadian Shield. *Can. J. Earth Sci.*, 5:693–698.

Stockwell, C. H., 1972. *Revised Precambrian Time Scale for the Canadian Shield*, Paper 72-52. Geological Survey of Canada. Ottawa, Ontario, Canada.

Stockwell, C. H., 1982. *Proposals for Time Classification and Correlation of Precambrian Rocks and Events in Canada and Adjacent Areas of the Canadian Shield*, Part 1, Paper 80-19. Geological Survey of Canada. Ottawa, Ontario, Canada.

CHAPTER 7

The $^{40}Ar^*/^{39}Ar$ Method

THE conventional K–Ar method of dating depends on the assumption that the sample contained no Ar at the time of its formation and all radiogenic Ar produced within it was quantitatively retained. Because Ar may diffuse out of minerals even at temperatures well below their melting point, K–Ar dates represent the time elapsed since cooling to temperatures at which diffusion loss of Ar is insignificant. However, under certain circumstances excess radiogenic ^{40}Ar may also be present, which causes K–Ar dates to be too old. Another problem is that in the conventional K–Ar method the concentrations of ^{40}Ar and K are measured separately on different aliquots of the sample. Therefore, the sample being dated must be homogeneous with respect to both elements. This requirement may not be satisfied in all cases, especially by fine-grained or glassy volcanic rocks.

The $^{40}Ar^*/^{39}A$ method of dating, first described in detail by Merrihue and Turner (1966), can overcome some of the limitations of the conventional K–Ar method because K and Ar are determined on the same sample and only measurements of the isotope ratios of Ar are required. The problem of inhomogeneity of samples and the need to measure the absolute concentrations of K and Ar are thus eliminated. This method is therefore well suited to the dating of small or valuable samples such as meteorites or lunar rocks and minerals, especially when the samples are heated stepwise with a continuous laser.

7.1 PRINCIPLES AND METHODOLOGY

The $^{40}Ar^*/^{39}Ar$ method of dating is based on the formation of ^{39}Ar by the irradiation of K-bearing samples with thermal and fast neutrons in a nuclear reactor. The desired reaction is

$$^{39}_{19}K\ (n,p)\ ^{39}_{18}Ar \qquad (7.1)$$

Argon-39 is unstable and decays to ^{39}K by β-emission with a halflife of 269 years. Because of its slow rate of decay, ^{39}Ar can be treated as though it were stable during the short period of time involved in the analyses. Whereas Wänke and König (1959) actually used a counting technique to determine the amount of ^{39}Ar produced by this reaction, Merrihue (1965) proposed that the $^{40}Ar^*/^{39}Ar$ ratio could be measured by mass spectrometry. Subsequently, Merrihue and Turner (1966) described such a procedure and reported dates for several stony meteorites that appeared to be in good agreement with conventional K–Ar dates of the same meteorites. The principles of the $^{40}Ar/^{39}Ar$ method of dating have been presented by Dalrymple and Lanphere (1971), McDougall (1974), Dallmeyer (1979), McDougall and Harrison (1988), and Dalrymple (1991).

When a K-bearing sample is irradiated with neutrons in a nuclear reactor, isotopes of Ar are formed by several reactions involving K, Ca, and Cl in the target. In the ideal case, ^{39}Ar is produced only by the n, p reaction with ^{39}K. In the formulation of Mitchell (1968), the number of ^{39}Ar atoms formed in the sample by the neutron irradiation is

$$^{39}\text{Ar} = {}^{39}\text{K}\ \Delta T \int \phi(\varepsilon)\sigma(\varepsilon)\,d\varepsilon \qquad (7.2)$$

where ^{39}K is the number of atoms of this isotope in the irradiated sample, ΔT is the duration of the irradiation, $\phi(\varepsilon)$ is the neutron flux density at energy ε, $\sigma(\varepsilon)$ is the capture cross section of ^{39}K for neutrons having energy ε, and the integration is carried out over the entire energy spectrum of the neutrons. The number of radiogenic ^{40}Ar atoms in the irradiated sample due to decay of ^{40}K during its lifetime is given by equation 6.2:

$$^{40}\text{Ar}^* = \frac{\lambda_e}{\lambda}\,{}^{40}\text{K}\,(e^{\lambda t} - 1) \qquad (7.3)$$

where the asterisk identifies radiogenic ^{40}Ar, λ_e is the decay constant of ^{40}K for electron capture, and λ is the total decay constant of ^{40}K. After neutron irradiation of a sample, its $^{40}Ar^*/^{39}Ar$ ratio is obtained by dividing equation 7.3 by equation 7.2:

$$\frac{^{40}\text{Ar}^*}{^{39}\text{Ar}} = \frac{\lambda_e}{\lambda}\,\frac{^{40}\text{K}(e^{\lambda t} - 1)}{^{39}\text{K}\ \Delta T \int \phi(\varepsilon)\sigma(\varepsilon)\,d\varepsilon} \qquad (7.4)$$

The neutron flux density and the capture cross sections are difficult to evaluate from first principles because the energy spectrum of the incident neutrons and the cross sections of ^{39}K for capture of neutrons of varying energies are not well known. However, equation 7.4 can be simplified by introducing the parameter J, defined as

$$J = \frac{\lambda}{\lambda_e}\,\frac{^{39}\text{K}\ \Delta T}{^{40}\text{K}} \int \phi(\varepsilon)\sigma(\varepsilon)\,d\varepsilon \qquad (7.5)$$

which leads to

$$\frac{^{40}\text{Ar}^*}{^{39}\text{Ar}} = \frac{e^{\lambda t} - 1}{J} \qquad (7.6)$$

Equation 7.6 suggests that J can be determined by irradiating a sample of *known* age (the flux monitor) together with samples whose ages are unknown. After the $^{40}Ar^*/^{39}Ar$ ratio of the monitor has been measured, J can be calculated from equation 7.6:

$$J = \frac{e^{\lambda t_m} - 1}{(^{40}\text{Ar}^*/^{39}\text{Ar})_m} \qquad (7.7)$$

where t_m is the known age of the flux monitor and $(^{40}Ar^*/^{39}Ar)_m$ is the measured value of this ratio in the monitor.

The energy spectrum of the neutron flux to which a particular sample is exposed during the irradiation depends on its position in the sample holder. For this reason, several samples of the flux monitor are inserted into the sample holder at known positions between unknown samples. The entire package is then irradiated for several days in a nuclear reactor to allow ^{39}Ar to be produced. After the irradiation, the Ar in the flux monitors is released by fusion in a vacuum system and their $^{40}Ar^*/^{39}Ar$ ratios are measured by mass spectrometry. The J values are then calculated from equation 7.7 and are plotted as a function of their position in the sample holder. The respective J values of the unknown samples are obtained by interpolation of the resulting graph from their known positions in the holder.

The $^{40}Ar^*/^{39}Ar$ ratios of the irradiated unknown samples are determined similarly by melting them individually in a vacuum chamber and by measuring the $^{40}Ar^*/^{39}Ar$ ratio of the released Ar in a gas source mass spectrometer. The resulting $^{40}Ar^*/^{39}Ar$ ratios of the unknown samples are then used to calculate dates from equation 7.6:

$$t = \frac{1}{\lambda}\ln\left(\frac{^{40}\text{Ar}^*}{^{39}\text{Ar}}J + 1\right) \qquad (7.8)$$

Several different mineral concentrates have been used as flux monitors. Their ages must be accurately known because they are used for calculating the value of J in equation 7.7. An error in the age of the monitor is therefore propagated from equations 7.7 and 7.8 and results in a corresponding systematic error in the calculated $^{40}Ar^*/^{39}Ar$ dates of samples that were irradiated with that monitor. Roddick (1983) published precise K–Ar dates of four widely used flux monitors (Hb3gr, MMHb-1, LP-6, and FY12a).

The estimated analytical error in the calculated date, according to Dalrymple and Lanphere (1971), is

$$\sigma \simeq \left[\frac{J^2F^2(\sigma_F^2 + \sigma_J^2)}{t^2\lambda^2(1+FJ)^2}\right]^{1/2} \quad (7.9)$$

where $F = {}^{40}Ar^*/{}^{39}Ar$; σ_F^2 and σ_J^2 are the variances of F and J, respectively, expressed in percent; t is the age of the sample; and λ is the total decay constant of ^{40}K. The dates obtained in this manner are referred to as *total argon release dates*. They are subject to the same limitations as conventional K–Ar dates because they depend on the assumption that no radiogenic ^{40}Ar has escaped from the sample and no excess ^{40}Ar is present. However, such dates avoid the problems arising from the inhomogeneous distribution of K and Ar in a sample and require only the measurement of isotope ratios of Ar.

In the ideal case outlined above, it is assumed that all of the ^{40}Ar in the irradiated sample is either radiogenic or atmospheric, all of the ^{36}Ar is atmospheric, and ^{39}Ar is produced only by $^{39}K(n, p)\ ^{39}Ar$. In this case, the measured values of the $^{40}Ar/^{39}Ar$ and $^{36}Ar/^{39}Ar$ ratios can be used to calculate the desired ratio of radiogenic ^{40}Ar to ^{39}Ar:

$$\frac{^{40}Ar^*}{^{39}Ar} = \left(\frac{^{40}Ar}{^{39}Ar}\right) - 295.5\left(\frac{^{36}Ar}{^{39}Ar}\right) \quad (7.10)$$

where 295.5 is the $^{40}Ar/^{36}Ar$ ratio of atmospheric Ar.

Actually, Ar isotopes are also produced by several interfering reactions caused by interactions of neutrons with the isotopes of Ca, K, and Cl in the sample. Therefore, a series of corrections must be made that are especially important for young samples ($\sim 10^6$ years) and those having Ca/K >10. The interfering reactions are listed in Table 7.1 and can be studied by reference to Figure 7.1. Detailed discussions of the corrections have been given by Mitchell (1968), Brereton (1970), Turner (1971), Dalrymple and Lanphere (1971), and Tetley et al. (1980).

The most important interfering reactions are those involving the isotopes of Ca. Starting at the top of Table 7.1, ^{36}Ar is produced by $^{40}Ca(n, n\alpha)\ ^{36}Ar$ but is removed by $^{36}Ar(n, \gamma)\ ^{37}Ar$. These reactions interfere with the atmospheric Ar correction, which is based on ^{36}Ar. Calcium is also primarily responsible for the production of ^{37}Ar by means of $^{40}Ca(n, \alpha)\ ^{37}Ar$. The abundance of ^{37}Ar in an irradiated sample is an indication of the extent of Ca interference because the ^{37}Ar yields from $^{39}K(n, nd)\ ^{37}Ar$ and $^{36}Ar(n, \gamma)\ ^{37}Ar$ are small. Argon-37 is radioactive and decays by electron capture to stable ^{37}Cl with a short halflife of 35.1 days. For this reason, a correction for decay of ^{37}Ar after irradiation must be made because the ^{40}Ca abundance derived from ^{37}Ar is used to estimate the contributions to ^{40}Ar by the reactions $^{43}Ca(n, \alpha)\ ^{40}Ar$ and $^{44}Ca(n, n\alpha)\ ^{40}Ar$,

Table 7.1. Interfering Nuclear Reactions Caused by Neutron Irradiation of Mineral Samples

Argon Produced	Calcium	Potassium	Argon	Chlorine
^{36}Ar	$^{40}Ca(n, n\alpha)$			
^{37}Ar	$^{40}Ca(n,\alpha)$	$^{39}K(n, nd)$	$^{36}Ar(n, \gamma)$	
^{38}Ar	$^{42}Ca(n, n\alpha)$	$^{39}K(n, d)$	$^{40}Ar(n, nd, \beta^-)$	$^{37}Cl(n, \gamma, \beta^-)$
		$^{41}K(n, \alpha, \beta^-)$		
^{39}Ar	$^{42}Ca(n, \alpha)$	$^{39}K(n, p)$[a]	$^{38}Ar(n, \gamma)$	
	$^{43}Ca(n, n\alpha)$	$^{40}K(n, d)$	$^{40}Ar(n, d, \beta^-)$	
^{40}Ar	$^{43}Ca(n, \alpha)$	$^{40}K(n, p)$		
	$^{44}Ca(n, n\alpha)$	$^{41}K(n,d)$		

Source: Brereton, 1970.

[a] This is the principal reaction on which the $^{40}Ar^*/^{39}Ar$ method is based.

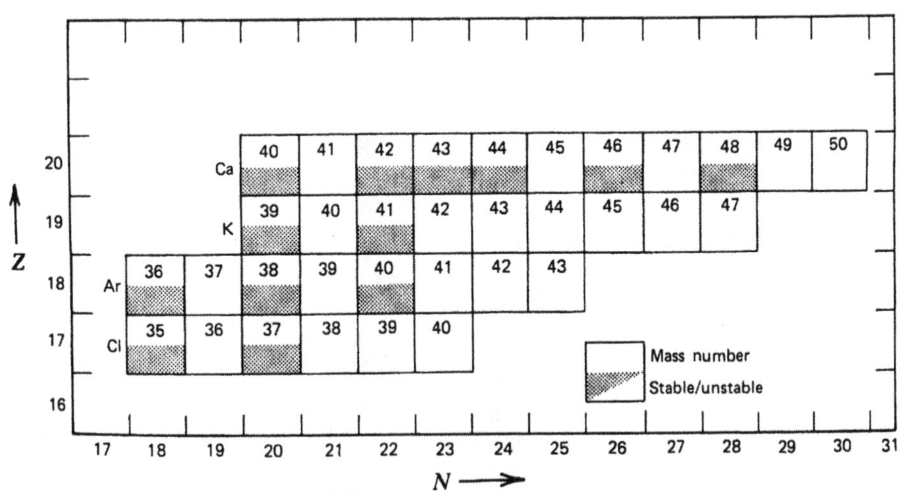

FIGURE 7.1 Segment of the chart of the nuclides showing most of the stable and unstable isotopes of Ca, K, Ar, and Cl that participate in nuclear reactions with neutrons.

whereas ^{39}Ar is produced by ^{42}Ca (n,α) ^{39}Ar and by ^{43}Ca (n, nα) ^{39}Ar. The production of ^{38}Ar is of little consequence unless an ^{38}Ar spike is used for measuring absolute quantities of Ar. According to measurements by Dalrymple and Lanphere (1971), the correction factors for Ca- and K-derived Ar in the TRIGA reactor of the U.S. Geological Survey are $(^{36}\text{Ar}/^{37}\text{Ar})_{\text{Ca}} = (2.72 \pm 0.014) \times 10^{-4}$, $(^{39}\text{Ar}/^{37}\text{Ar})_{\text{Ca}} = (6.33 \pm 0.043) \times 10^{-4}$, and $(^{40}\text{Ar}/^{39}\text{Ar})_{\text{K}} = (5.9 \pm 0.42) \times 10^{-3}$. The production ratio of ^{40}Ar/^{37}Ar due to Ca is 4.6×10^{-3}. By far the most important corrections for Ca-rich samples result from the production of ^{39}Ar by ^{42}Ca (n, α) ^{39}Ar and of ^{36}Ar by ^{40}Ca(n, nα) ^{36}Ar. For young samples having low Ca/K ratios, corrections are also necessary to remove ^{40}Ar produced by ^{40}K(n,p) ^{40}Ar and ^{41}K(n,d) ^{40}Ar. Brereton (1970) derived an equation that relates the age of an irradiated sample to its ^{40}Ar*/^{39}Ar ratio corrected for all interfering reactions. Dalrymple and Lanphere (1971) developed a more general expression for their parameter $F = {}^{40}\text{Ar}^*/^{39}\text{Ar}$:

$$F = \frac{A - C_1B + C_1C_2D - C_3}{1 - C_4D} \tag{7.11}$$

where A = measured value of the ^{40}Ar/^{39}Ar ratio
B = measured value of the ^{36}Ar/^{39}Ar ratio
C_1 = ^{40}Ar/^{39}Ar ratio in the atmosphere, =295.5
C_2 = ^{36}Ar/^{37}Ar ratio produced by interfering neutron reactions with Ca, $2.72 \pm 0.014 \times 10^{-4}$
C_3 = ^{40}Ar/^{39}Ar ratio produced by interfering neutron reactions with K, $(5.9 \pm 0.42) \times 10^{-3}$
C_4 = ^{39}Ar/^{37}Ar ratio produced by interfering neutron reactions with Ca, $6.33 \pm 0.043 \times 10^{-4}$
D = ^{37}Ar/^{39}Ar ratio in sample after correcting for decay of ^{37}Ar

7.2 INCREMENTAL HEATING TECHNIQUE

The principles of the ^{40}Ar*/^{39}Ar method permit dates to be calculated from the measured ^{40}Ar*/^{39}Ar ratio of gas released by melting the sample in a vacuum. Consequently, a series of dates can be obtained for a single sample by releasing Ar from it in steps at increasing temperatures. If the sample has been closed to Ar and K since the time of initial cooling, the ^{40}Ar*/^{39}Ar ratios and thus the dates calculated at each step should be constant. However, if radiogenic Ar was lost from some crystallographic sites but not from others some time after initial cooling, the $^{40}Ar^{*}/^{39}Ar$ ratios of the gas released at different temperatures may vary, and a spectrum of dates will then result from which the time elapsed since initial cooling may be inferred.

Turner (1968, 1969) used a model, illustrated in Figure 7.2, to predict the $^{40}Ar^{*}/^{39}Ar$ spectrum of a K-bearing mineral grain that experienced partial loss of radiogenic ^{40}Ar due to heating by a metamorphic event. Figure 7.2*a* shows the uniform distribution of the $^{40}Ar^{*}/^{40}K$ ratio expected in a closed system. Figure 7.2*b* illustrates the effect of partial loss of ^{40}Ar due to heating as a result of which the $^{40}Ar^{*}/^{40}K$ ratio at the edge of the grain is reduced to zero but remains unaffected at its center. Figure 7.2*c* shows the distribution of the $^{40}Ar^{*}/^{40}K$ ratios in the grain at a later time assuming that it became a closed system again after the episode of metamorphism. When grains of uniform size of a specific K-bearing mineral that has experienced such a history are analyzed by incremental heating, the $^{40}Ar^{*}/^{39}Ar$ ratios of gas fractions released at different temperatures vary in a systematic fashion. The first gas to be released at the lowest temperature originates from the surfaces of the grains and from sites that lose Ar readily. This gas has a low $^{40}Ar^{*}/^{39}Ar$ ratio, corresponding to the time elapsed after accumulation of radiogenic ^{40}Ar resumed after metamorphism. Gas fractions released at higher temperatures have higher $^{40}Ar^{*}/^{39}Ar$ ratios because this Ar is removed from more retentive sites that lost smaller fractions of $^{40}Ar^{*}$ during metamorphism. Ultimately, the $^{40}Ar^{*}/^{39}Ar$ ratios may reach a plateau corresponding to a date that approaches the time elapsed since original cooling of the mineral without subsequent loss of ^{40}Ar.

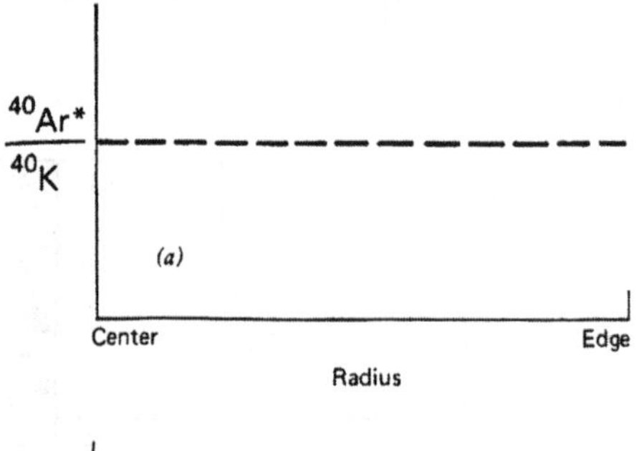

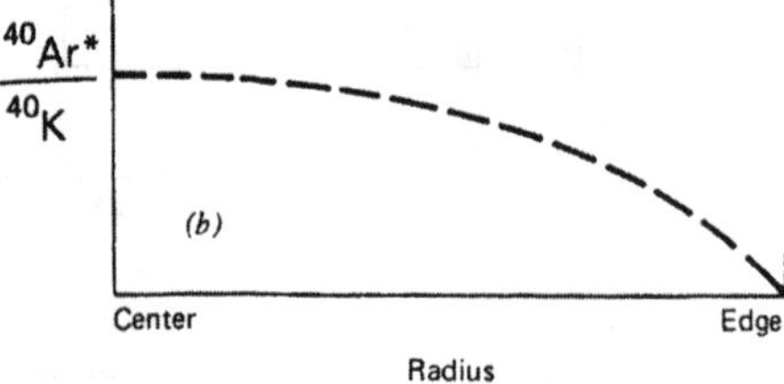

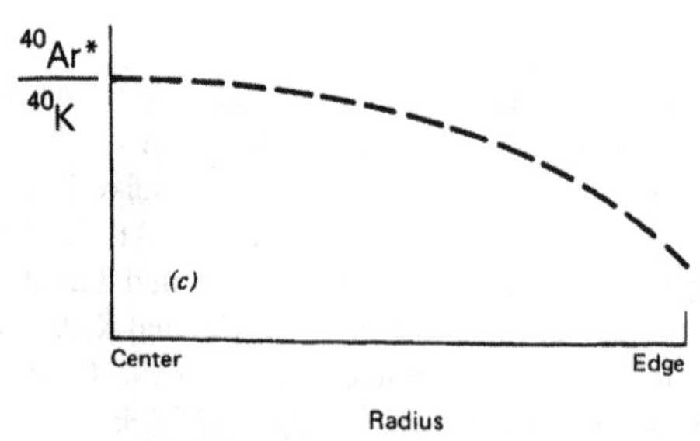

FIGURE 7.2 (*a*) Schematic diagram showing the uniform distribution of the $^{40}Ar^{*}/^{39}Ar$ ratio in a spherical mineral grain that has been a closed system since the time of its formation. (*b*) Partial loss of $^{40}Ar^{*}$ due to diffusion as a result of heating a spherical mineral grain. Note that the $^{40}Ar^{*}/^{40}K$ ratio is reduced to zero on the grain surface but remains unchanged at its center. (*c*) Distribution of $^{40}Ar^{*}/^{40}K$ ratios some time after partial Ar loss after the system has again become closed to $^{40}Ar^{*}$. Note that the value of $^{40}Ar^{*}/^{40}K$ on the grain surface reflects the time elapsed since partial outgassing while the value in the center of the grain corresponds to the age of the mineral. Adapted from Turner (1968).

Stepwise heating of a mineral sample that has experienced partial loss of radiogenic Ar thus permits the calculation of a series of dates that, in the ideal case, include the time of metamorphism (low-temperature release) and the time of initial cooling (high-temperature plateau). Such theoretical release patterns are shown in Figure 7.3 for a sample that is 4.5 billion years old (cooling age) and that experienced partial Ar loss at 500 Ma. The curves are based on calculations of Turner (1968) for uniform spheres (Figure 7.3*a*) and for spheres having log-normal size distributions (Figure 7.3*b*). Note that the date calculated by total release of all gas has a meaningless intermediate value. The principal advantage of the $^{40}Ar^{*}/^{39}Ar$ method of dating lies in this stepwise degassing technique because it permits the determination of a date that approaches

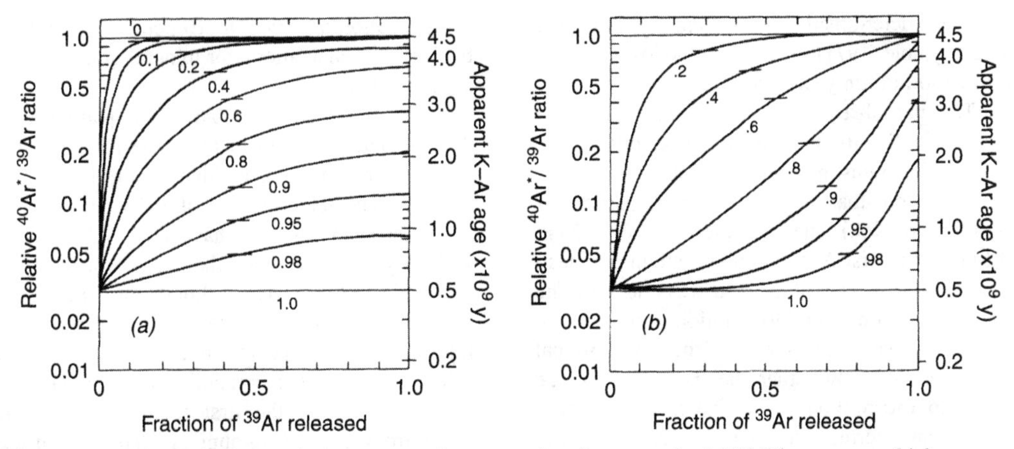

FIGURE 7.3 (*a*) Theoretical Ar release patterns for a sample whose age is 4.55 billion years, which was partially outgassed at 0.5 Ga. The numbers on the curves are the fraction of ^{40}Ar lost and the horizontal bars indicate the average $^{40}Ar^*/^{39}Ar$ ratio and the corresponding total-release date of the sample. Note that the plateau dates for ^{40}Ar loss greater than about 20 percent are less than the age of the sample. The mineral grains were assumed to be spheres of uniform size. (*b*) Theoretical Ar release patterns for the same samples assuming a lognormal distribution of grain radii ($\sigma = 0.33$). Note that the release curves become concave upward in cases where the partial Ar loss is about 80 percent or more. The $^{40}Ar^*/^{39}Ar$ ratio of gas released at low temperatures from samples with large Ar loss indicates the date of outgassing, while the high-temperature fraction approaches the age of the sample where Ar loss has been small. From G. Turner, in L. H. Ahrens (Ed.), *Origin and Distribution of the Elements*, Pergamon, Oxford, 1968, pp. 387–398. Adapted by permission of the publisher.

but may underestimate the original cooling age of K-bearing rocks and minerals that experienced partial loss of radiogenic ^{40}Ar during a metamorphic episode.

The incremental heating technique has been widely used to date terrestrial rocks and minerals from metamorphic terranes:

Bryhni et al., 1971, *Norsk Geol. Tidsskrift*, 51:391–406. Brereton, 1972, *Geophys. J. Roy. Astron. Soc.*, 27:449–478. Lanphere and Albee, 1974, *Am. J. Sci.*, 274:545–555. Dallmeyer et al., 1975, *Geol. Soc. Am. Bull.*, 86:1435–1443. Dallmeyer, 1975, *Geochim. Cosmochim. Acta*, 39:1655–1669. Berger, 1975, *Earth Planet. Sci. Lett.*, 26:387–408. Hanson et al., 1975, *Geochim. Cosmochim. Acta*, 39:1269–1277. Albarède, 1978, *Earth Planet. Sci. Lett.*, 38:387–397. Claesson and Roddick, 1983, *Lithos*, 16:61–73. Lopez-Martinez et al., 1984, *Nature*, 307:352–354. Baksi et al., 1991, *Earth Planet. Sci. Lett.*, 104:292–297. Harlan et al., 1996, *Can. J. Earth Sci.*, 33:1648–1654.

7.2a Marble Mountains, California

The crystalline complex of the Marble Mountains in southeastern California formed between 1400 and 1450 Ma based on a Rb–Sr date of biotite (1410 ± 30 Ma; Lanphere, 1964) and a U–Pb zircon date of 1450 Ma (Silver and McKinney, 1962). However, the conventional K–Ar date of biotite is 1152 ± 30 Ma and that of microcline is only 992 ± 26 Ma (Lanphere and Dalrymple, 1971). The K–Ar dates of both minerals are less than the crystallization age of the complex, which indicates that both lost radiogenic ^{40}Ar during a later thermal event. The Ar loss of the biotite is about 20 percent. The microcline lost even more ^{40}Ar, consistent with its lower retentivity noted before (Chapter 5).

Other age determinations reviewed by Lanphere and Dalrymple (1971) suggest that the loss occurred 160 to 180 million years ago.

The partial-release spectrum of $^{40}Ar^*/^{39}Ar$ dates of *biotite* from a sample of granite in the Marble Mountains in Figure 7.4 fits the pattern for a mineral that has experienced partial Ar loss (Turner, 1968). The dates increase from 223 ± 30 Ma for the first Ar fraction and attain a plateau after about 30 percent of the ^{39}Ar was released. The average plateau date of 1300 million years is about 150 million years greater than the conventional K–Ar date but is lower than the crystallization age by 100 to 150 million years. The total-Ar release date has an intermediate value of 1251 ± 61 Ma. The plateau date of this biotite is less than its known crystallization age by 5 to 10 percent as predicted by Turner's calculations for mineral grains that lost 20 percent Ar.

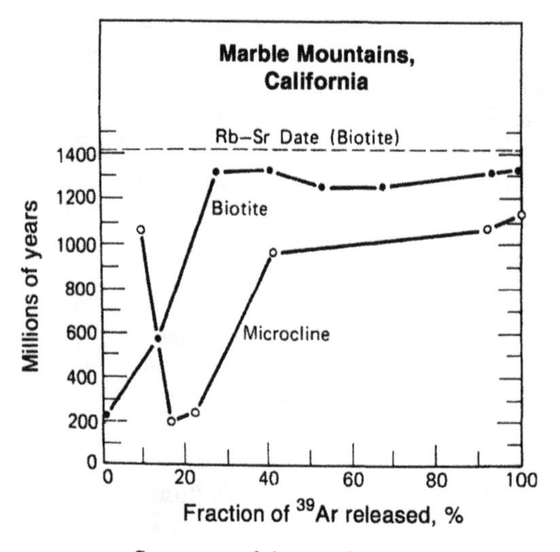

FIGURE 7.4 Spectrum of dates calculated from $^{40}Ar^*/^{39}Ar$ ratios of gas fractions released by incremental heating of neutron-irradiated biotite and microcline from the Precambrian basement complex of the Marble Mountains in southeastern California. The age of these rocks lies between 1400 and 1450 Ma. The high-temperature gas fractions of the biotite indicate a date of 1300 Ma, while its conventional K–Ar date is only 1152 ± 30 Ma. The plateau of the microcline dates is not well established, which suggests significantly greater losses of ^{40}Ar from this mineral than from the biotite. Data from Lanphere and Dalrymple (1971).

The spectrum of $^{40}Ar^*/^{39}Ar$ partial-release dates of *microcline* from the same rock specimen differs significantly from that of the biotite. The first fraction yielded a date of 1063 million years, while the next fractions have dates of about 200 million years. The dates then rise to a maximum value of 1133 ± 10 Ma without achieving a stable plateau. In such cases, the maximum date at the highest release temperature may be regarded as a lower limit to the crystallization age. The high date obtained for the first fraction of gas may result from a small amount of ^{40}Ar that diffused from crystal lattice to nonlattice sites from which it is readily removed at low temperature. The ^{40}Ar released at low temperature may also have been picked up from ambient fluids and stored in sites from which it is released at low temperature. The lowest dates of both the microcline and the biotite are overestimates of the age of the thermal event that caused Ar loss from these minerals. Nevertheless, the $^{40}Ar^*/^{39}Ar$ plateau date of the biotite approaches the presumed age of this mineral and hence that of the granite in the Marble Mountains.

7.2b Diabase Dikes in Liberia, West Africa

The presence of excess ^{40}Ar in some minerals and whole-rock samples causes serious errors in conventional K–Ar dates that cannot be corrected. Lanphere and Dalrymple (1971) therefore analyzed two whole-rock samples of diabase dikes in Liberia that are known to contain excess ^{40}Ar in order to determine whether the stepwise heating technique can overcome this difficulty.

The diabase samples were taken from dikes that intrude crystalline rocks of Precambrian age in western Liberia. Dating by the Rb–Sr method indicated that the basement rocks are from 2700 to 3400 million years old. Conventional K–Ar dates of whole-rock samples and plagioclase from the diabase dikes are highly inconsistent and range from 1200 to 186 Ma. The K–Ar dates on similar dikes, intruding Paleozoic sedimentary rocks elsewhere in Liberia, give more concordant dates ranging from

173 to 193 Ma, and paleomagnetic pole positions of all of the dikes suggest that they are post-Triassic and pre-Tertiary in age (Hurley et al., 1971; Dalrymple et al., 1975; Onstott and Dorbor, 1987; Dupuy et al., 1988; Mauche et al., 1989).

The conventional K–Ar date of a sample of these dikes determined by Lanphere and Dalrymple (1971) is 853 ± 26 Ma and its total-release $^{40}Ar^*/^{39}Ar$ date is 838 ± 42 Ma. Both dates are well in excess of the age of this rock. The spectrum of $^{40}Ar^*/^{39}Ar$ partial-release dates in Figure 7.5 has a saddle-shaped pattern with an excessively old date of 3827 ± 136 Ma for gas released at 400°C, which represents only 0.4 percent of the total ^{39}Ar in this sample. Gas released stepwise at increasing temperatures yields progressively lower $^{40}Ar^*/^{39}Ar$ dates that decrease to 268 ± 4 Ma and then rise again to 1995 ± 16 Ma for ^{39}Ar released at 1100°C. Although none of these dates equals the known age of the Liberian diabase dikes (175–190 Ma; Lanphere and Dalrymple, 1971), the presence of excess ^{40}Ar in these rocks is clearly indicated by the old date of the Ar released at the lowest temperature. The spectrum of partial $^{40}Ar^*/^{39}Ar$ release dates of a second sample of diabase from Liberia analyzed by Lanphere and Dalrymple (1971) is similar to that in Figure 7.5. Therefore, the authors concluded that the excess ^{40}Ar originated from the Precambrian basement rocks into which these dikes were intruded. This conclusion was later confirmed by Dalrymple et al. (1975). Brereton (1972) reported a similar result for a diabase dike from the Frederickshab area of southern West Greenland that also contains excess ^{40}Ar. In addition, Lee et al. (1990) and Radhakrishna et al. (1999) studied the effects of contact metamorphism on $^{40}Ar^*/^{39}Ar$ dates of diabase dikes intruding Precambrian basement rocks in the Superior structural province of Ontario and of the granulite terrain in southern India, respectively.

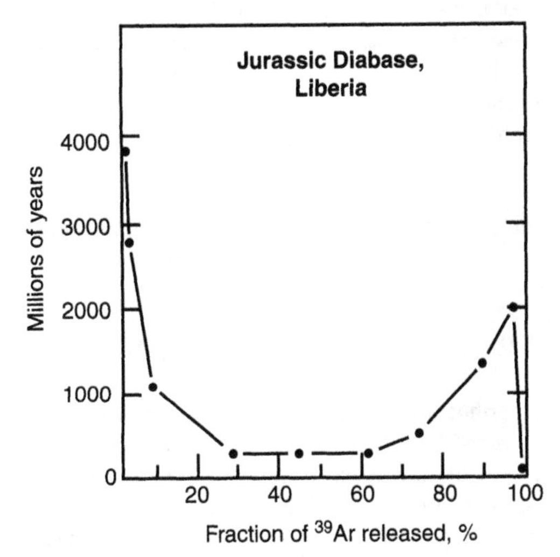

FIGURE 7.5 Spectrum of dates obtained by incremental heating of a neutron-irradiated whole-rock sample of diabase from Liberia. This dike is post-Triassic to pre-Tertiary in age. It was contaminated with ^{40}Ar during intrusion into the granitic basement rocks, which range in age from 2700 to 3400 Ma. The conventional K–Ar date of this rock is 853 ± 26 Ma, which is clearly in excess of its geological age. The spectrum of dates reveals the presence of excess ^{40}Ar in the low-temperature fractions but does not permit a geologically meaningful interpretation. Data from Lanphere and Dalrymple (1971).

7.3 EXCESS ^{40}Ar

The diabase dikes of Liberia exemplify rocks and minerals that contain varying amounts of ^{40}Ar that did not form by in situ decay of ^{40}K within the rocks or minerals under consideration. In some cases, the excess ^{40}Ar enters the minerals of igneous rocks at the time of their crystallization from a magma. In other cases, the ^{40}Ar collects in grain boundaries and fractures in minerals of igneous and metamorphic rocks or diffuses into the crystal lattice (Maluski et al., 1990).

Conventional K–Ar dates of minerals containing excess ^{40}Ar are greater than their cooling ages and are not valid age determinations. The incremental heating technique of the $^{40}Ar^*/^{39}Ar$ method can sometimes detect the presence of excess ^{40}Ar by the "saddle-shaped" spectrum of dates like that shown in Figure 7.5. Such patterns have also been reported for biotite, pyroxene, hornblende, and plagioclase from rocks known to contain excess ^{40}Ar (Lanphere and Dalrymple, 1976; Harrison and McDougall, 1981). Although these patterns reveal the presence of excess ^{40}Ar in the material being dated, its age cannot be inferred from the spectrum or from a plot of data points in coordinates of $^{40}Ar^{*}/^{36}Ar$ and $^{39}Ar/^{36}Ar$. Even more disturbing are reports that biotites from metamorphic rocks known to contain excess ^{40}Ar have partial-release spectra with flat plateaus that exceed the known geological ages of the rocks (Pankhurst et al. 1973; Roddick et al., 1980; Foland, 1983.)

In these circumstances, the incremental heating technique does not distinguish between the radiogenic ^{40}Ar that formed by decay of ^{40}K in the rock and the excess ^{40}Ar that was added after or during crystallization. The reason is that the excess ^{40}Ar is distributed uniformly with respect to the K in the sample. Evidently, just because the spectrum of partial-release dates of a sample has a well-developed plateau does not necessarily indicate that its $^{40}Ar^{*}/^{39}Ar$ plateau date is geologically valid (Foland, 1983).

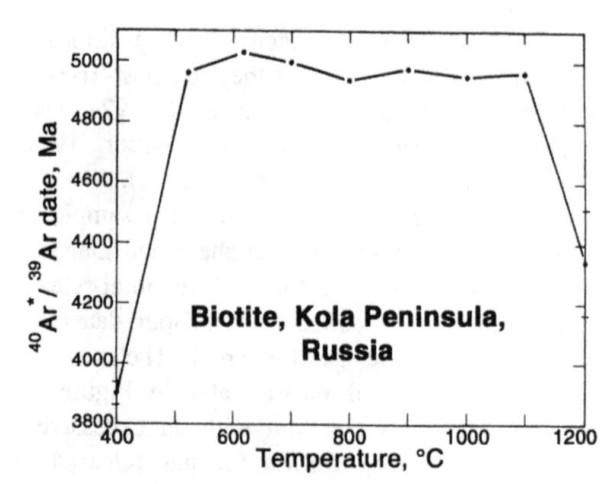

FIGURE 7.6 Stepwise Ar release pattern of biotite from granitic rocks intruding gneisses of the Kola Peninsula of Russia. The spectrum of $^{40}Ar^{*}/^{39}Ar$ dates of this mineral has a well-developed plateau with an average date of 4982 Ma. The anomalously old date is attributable to the presence of excess ^{40}Ar that was not recognized by the $^{40}Ar^{*}/^{39}Ar$ method. The plateau date in this example exceeds the age of the Earth. Data from Ashkinadze et al. (1978).

7.3a Kola Peninsula, Russia

A spectacular example of excess Ar is illustrated in Figure 7.6 for a biotite from granitic rocks that intrude the gneisses along the Voronaya River in the Kola Peninsula of Russia. The age of these granitic rocks is 2736 ± 65 Ma based on the decay of U and Th to Pb. However, the conventional K–Ar date of the biotite shown in Figure 7.6 is 5200 Ma, which exceeds the age of the Earth. A stepwise release of Ar from this biotite yielded a normal-looking pattern of dates with a good plateau and an average $^{40}Ar^{*}/^{39}Ar$ date of 4982 Ma. Ashkinadze et al. (1978), who reported these results, concluded that this and other biotites from the Kola Peninsula contain excess ^{40}Ar, that the $^{40}Ar^{*}/^{39}Ar$ method in this case does not indicate the presence of the excess ^{40}Ar, and that the plateau date is not the age of the biotites.

The presence of excess ^{40}Ar in mafic rocks of the Kola Peninsula was originally discovered by Gerling et al. (1962) and confirmed by Kaneoka (1974) and Kirsten and Gentner (1966). This phenomenon was later discussed by Gerling (1984) and was mentioned in Section 6.2c. The work of Ashkinadze et al. (1978) demonstrated that excess ^{40}Ar can occur not only in the K-poor minerals of mafic rocks on the Kola Peninsula but also in K-rich minerals such as biotite.

7.3b Anorthoclase, Mt. Erebus, Antarctica

Anorthoclase-bearing phonolite lavas have been extruded by Mt. Erebus, an active volcano on Ross Island off the coast of southern Victoria Land in Antarctica (Kyle, 1994; Faure, 2001). Nevertheless, the anorthoclase phenocrysts of lava flows that were erupted in 1984 have yielded K–Ar dates up to 400,000 years because of the presence of excess ^{40}Ar (Armstrong, 1978).

Esser et al. (1997) demonstrated that the integrated $^{40}Ar^{*}/^{39}Ar$ dates of anorthoclase increase from about 49,000 years (1 percent melt) to 179,000 years (10 percent melt) and to 640,000 years (30 percent melt). Therefore, the excess ^{40}Ar is apparently concentrated in melt inclusions within the anorthoclase crystals. Esser et al. (1997) attempted to remove the melt inclusions by crushing the anorthoclase and by leaching the fragments with hydrofluoric acid. However, small melt inclusions within the anorthoclase grains persisted. The Ar trapped in these melt inclusions was released primarily at temperatures above 1200°C when the incongruent melting of the anorthoclase host allowed the Ar in the melt inclusions to escape. Consequently, anorthoclase samples containing less than 1 percent melt yielded low-temperature (<1200°C) plateau dates as low as 8000 ± 2000 years, whereas the high-temperature Ar released at >1200°C yielded dates in excess of 100,000 years.

In retrospect, the presence of small amounts of excess ^{40}Ar in present-day volcanic rocks is consistent with the derivation of magma from the mantle of the Earth where the Ar has much higher $^{40}Ar/^{36}Ar$ ratios than the present atmosphere (Section 6.4c). The presence of mantle-derived ^{40}Ar increases the K–Ar and $^{40}Ar^*/^{39}Ar$ dates, especially in samples that are young or have low K concentrations or both.

7.4 ARGON ISOTOPE CORRELATION DIAGRAM

The conventional K–Ar and $^{40}Ar^*/^{39}Ar$ methods of dating require a correction for the presence of atmospheric ^{40}Ar. This correction is based on the assumption that ^{36}Ar is of atmospheric origin and that the $^{40}Ar/^{36}Ar$ ratio of the atmospheric component is 295.5. This assumption may not be strictly valid because the $^{40}Ar/^{36}Ar$ ratio of Ar that was incorporated into minerals at the time of their crystallization may differ significantly from that of Ar in the modern atmosphere (Musset and Dalrymple, 1968). Therefore, the observed amount of ^{36}Ar in a rock or mineral does not necessarily permit an accurate correction to be made for the presence of nonradiogenic ^{40}Ar. If the $^{40}Ar/^{36}Ar$ ratio of the inherited Ar is greater than 295.5, an apparent excess of radiogenic ^{40}Ar will result. If the $^{40}Ar/^{36}Ar$ ratio of the inherited Ar is less than 295.5, there will be an apparent deficiency of radiogenic ^{40}Ar, which is then erroneously attributed to partial loss of ^{40}Ar.

This difficulty can be avoided by use of the Ar isotope correlation diagram, which was originally suggested by Merrihue and Turner (1966). The measured $^{40}Ar/^{36}Ar$ ratio of gas released during the stepwise heating of an irradiated sample (after correction for interfering reactions) is

$$\left(\frac{^{40}\mathrm{Ar}}{^{36}\mathrm{Ar}}\right)_m = \frac{^{40}\mathrm{Ar}_c + {}^{40}\mathrm{Ar}^*}{^{36}\mathrm{Ar}_c} \tag{7.12}$$

where the subscript c identifies contaminant Ar, which includes both the atmospheric component and Ar that was either occluded during crystallization or entered the mineral at a later time. Equation 7.12 can be rewritten as

$$\left(\frac{^{40}\mathrm{Ar}}{^{36}\mathrm{Ar}}\right)_m = \left(\frac{^{40}\mathrm{Ar}}{^{36}\mathrm{Ar}}\right)_c + \left(\frac{^{40}\mathrm{Ar}^*}{^{39}\mathrm{Ar}}\right)_k \left(\frac{^{39}\mathrm{Ar}}{^{36}\mathrm{Ar}}\right)_m \tag{7.13}$$

where the subscript m identifies measured ratios and k refers to Ar produced by K in the sample. Equation 7.13 represents a family of straight lines in coordinates of $(^{40}Ar/^{36}Ar)_m$ and $(^{39}Ar/^{36}Ar)_m$ whose slopes are equal to the $^{40}Ar^*/^{39}Ar$ ratio. It follows that the measured $^{40}Ar/^{36}Ar$ and $^{39}Ar/^{36}Ar$ ratios of incremental gas fractions released from undisturbed mineral or rock samples define a series of points that fit a straight line. This line is an isochron whose slope is the $^{40}Ar^*/^{39}Ar$ ratio, which is related to the age of the sample by equation 7.8. The intercept of the isochron is the $^{40}Ar/^{36}Ar$ ratio of the contaminant, that is, the ratio of these isotopes in the nonradiogenic fraction of this gas associated with a given sample. The use of the Ar isotope correlation diagram was discussed by Dalrymple et al. (1988) and Heizler and Harrison (1988). Dalrymple and Lanphere (1974) demonstrated that the isochron dates of 11 undisturbed samples are identical, within experimental errors, to their respective plateau dates and that, in most cases, the initial $^{40}Ar/^{36}Ar$ ratios do not differ significantly from the atmospheric value of this ratio (295.5). An alternative interpretation of Ar isochrons discussed by Roddick (1978) is that they are mixing lines whose slopes yield incorrect dates.

7.4a Portage Lake Volcanics, Michigan

The reliability of Ar isochron dates was tested by Dalrymple and Lanphere (1974) by analyzing minerals (muscovite, biotite, hornblende, sanidine, and plagioclase) and whole-rock samples of igneous rocks (dacite, diabase, and basalt) that had not been disturbed after initial crystallization. One of the samples they analyzed was a plagioclase concentrate obtained from a drill core of the Portage Lake Volcanics on the Keweenaw Peninsula of Michigan.

The Portage Lake Volcanics are part of a large basalt plateau that formed as a result of rifting of the North American continent at about 1.1 Ga (Hutchinson et al., 1990; Nicholson and Shirey, 1990; Nicholson et al., 1992). The plateau basalts and associated sedimentary rocks on the Keweenaw Peninsula contain large deposits of native Cu and strata-bound Cu sulfide deposits that made this region one of the principal sources of Cu in the United States during the late-nineteenth and early-twentieth centuries (Broderick, 1956; White, 1960a, b). The basalt flows of the Portage Lake Volcanics crystallized between 1096.2 ± 1.8 and 1094.0 ± 1.5 Ma, as indicated by U–Pb zircon dates reported by Davis and Paces (1990).

The partial-release $^{40}Ar^*/^{39}Ar$ spectrum of dates of the plagioclase in Figure 7.7*a* is flat, as expected for presumably unaltered minerals, and yields a plateau date of 1048 ± 11 Ma for 68.8 percent of the total ^{39}Ar that was released. This date is based on values of the decay constants of ^{40}K that are no longer used: $\lambda_e = 0.585 \times 10^{-10}\ y^{-1}$ and $\lambda_\beta = 4.72 \times 10^{-10}\ y^{-1}$ with a total decay constant $\lambda = 5.305 \times 10^{-10}\ y^{-1}$. The Ar isochron in Figure 7.7*b* is based on the same data that define the plateau and indicates a date of 1057 ± 12 Ma with an intercept at $^{40}Ar/^{36}Ar = 188 \pm 116$. The date derived from the plateau and from the isochron are indistinguishable from each

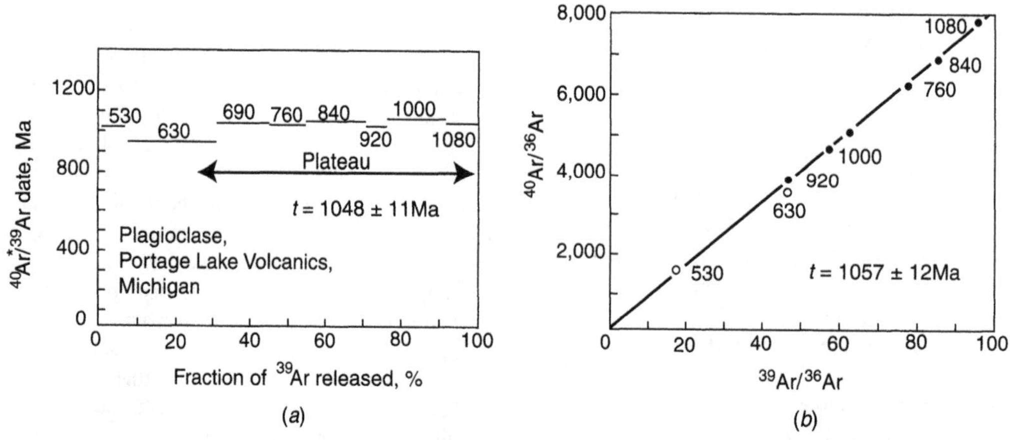

FIGURE 7.7 (*a*) Spectrum of $^{40}Ar^*/^{39}Ar$ partial-release dates of plagioclase in tholeiite basalt of the Portage Lake Volcanics on the Keweenaw Peninsula of Michigan. The numbers above the gas increments are release temperatures in degrees Celsius. The plateau date of 1048 ± 11 Ma is the weighted average of $^{40}Ar^*/^{39}Ar$ dates and represents 68.8 percent of the ^{39}Ar that was released. The decay constants of ^{40}K used in this study, $\lambda_e = 0.585 \times 10^{-10}\ y^{-1}$, $\lambda_\beta = 4.72 \times 10^{-10}\ y^{-1}$, and $\lambda = 5.305 \times 10^{-10}\ y^{-1}$, have been replaced by the values recommended by the IUGS Subcommission on Geochronology (Steiger and Jäger, 1977). (*b*) Argon isotope correlation diagram based on measured values of $^{40}Ar/^{36}Ar$ and $^{39}Ar/^{36}Ar$ ratios after correcting for interfering neutron capture reactions with Ca and K. The slope of the straight line was calculated from the coordinates of the same gas fractions that define the plateau in part (*a*). The regression was based on the method of York (1969) with correlated errors. The slope of the isochron line yields a date of 1057 ± 12 Ma based on the decay constants of ^{40}K stated in part (*a*). Although the plateau date and the isochron date are identical within the stated errors, both are younger than the crystallization age of the basalt flows of the Portage Lake Volcanics. Data from Dalrymple and Lanphere (1974).

other and are consistent with the Rb–Sr dates of related rocks on the Keweenaw Peninsula reported by Chaudhuri and Faure (1967) and Bornhorst et al. (1988), but it is lower than the U–Pb dates of Davis and Paces (1990) and others. The explanation suggested by Dalrymple and Lanphere (1974) is that the plagioclase had lost radiogenic ^{40}Ar during a low-grade metamorphic event recognized by Jolly and Smith (1972) that may have been associated with the deposition of the Cu ores by hydrothermal fluids. These fluids also deposited chlorite, K-feldspar, epidote, and calcite in the vesicles of the Portage Lake Volcanics between 1060 ± 20 and 1047 ± 33 Ma based on Rb–Sr isochron dates reported by Bornhorst et al. (1988). Consequently, the $^{40}Ar^*/^{39}Ar$ plateau date of plagioclase in Figure 7.7*a* underestimates the age of the Portage Lake Volcanics and does not reveal the loss of ^{40}Ar that occurred during the alteration of the flows.

The $^{40}Ar^*/^{39}Ar$ dates calculated by Dalrymple and Lanphere (1974) for the plagioclase in the Portage Lake Volcanics can be converted from $\lambda = 5.305 \times 10^{-10}$ y^{-1} to $\lambda = 5.543 \times 10^{-10}$ y^{-1} (Steiger and Jäger, 1977; Begemann et al., 2001) by first recalculating the J value (0.00937) determined from the flux monitor, whose age was 160 Ma. Since J is defined by equation 7.7,

$$J = \frac{e^{\lambda t} - 1}{^{40}\text{Ar}^*/^{39}\text{Ar}}$$

where t is the age of the flux monitor and λ is the decay constant of ^{40}K, the J value for the plagioclase ($\lambda = 5.543 \times 10^{-10}$ y^{-1}) is

$$J = \frac{0.00937 \times 0.09273}{0.08858} = 0.009808$$

where $0.08858 = e^{\lambda t} - 1$ for $\lambda = 5.305 \times 10^{-10}$ y^{-1} and $0.09273 = e^{\lambda t} - 1$ for $\lambda = 5.543 \times 10^{-10}$ y^{-1}. The date calculated from the slope of the isochron by means of equation 7.8 ($\lambda = 5.543 \times 10^{-10}$ y^{-1}) is

$$t = \frac{1}{5.543 \times 10^{-10}} \ln(80.25 \times 0.009808 + 1)$$
$$= 1047.4 \text{ Ma}$$

where 80.25 is the $^{40}Ar^*/^{39}Ar$ ratio (slope of the isochron) as determined by Dalrymple and Lanphere (1974) by the regression method of York (1969). The recalculated date (relative to $\lambda = 5.543 \times 10^{-10}$ y^{-1}) is only about 1 percent lower than the value reported by Dalrymple and Lanphere (1974) because the decrease in the date is partly offset by the increase in the J value. As a result, the conclusions concerning the $^{40}Ar^*/^{39}Ar$ dates of plagioclase in the basalt of the Portage Lake Volcanics are not significantly altered by the recalculation.

7.4b Lunar Basalt and Orange Glass

The partial-release technique and the Ar isotope correlation diagram have been extensively used to date both lunar rocks and meteorites by the $^{40}Ar^*/^{39}Ar$ method. The partial-release spectrum of $^{40}Ar^*/^{39}Ar$ dates of a lunar basalt from the Taurus-Littrow valley in Figure 7.8 has a well-defined plateau date of 3.77 ± 0.05 Ga based on three gas fractions that contained 46.2 percent

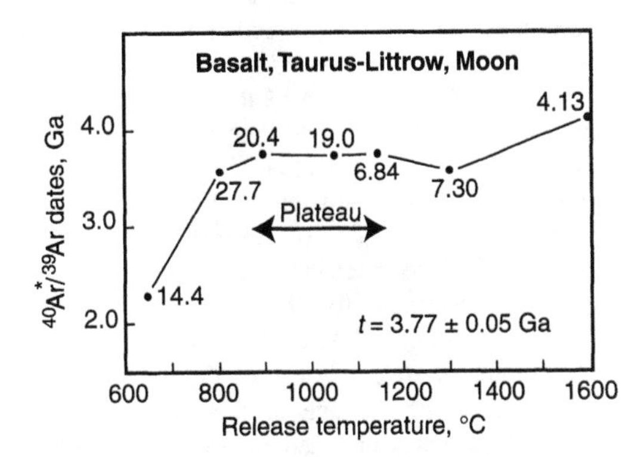

FIGURE 7.8 Partial-release $^{40}Ar^*/^{39}Ar$ spectrum of dates for a basalt sample (75083, 2.5) collected by the astronauts of Apollo 17 in the Taurus-Littrow valley of the Moon. The plateau date of 3.77 ± 0.05 Ga is based on three gas fractions that yielded similar $^{40}Ar^*/^{39}Ar$ dates. The fraction of ^{39}Ar (in percent) released at each incremental temperature is indicated by the number beside the data point. Data from Husain and Schaeffer (1973).

of the ^{39}Ar (Husain and Schaeffer, 1973). The dates in Figure 7.8 are plotted versus the release temperature, whereas the fraction of ^{39}Ar associated with each step is indicated in the diagram. The dates of the low-temperature gas fractions at 650 and 800°C are lower than the plateau date because of loss of ^{40}Ar*. The last gas to be released (4.13 percent at 1600°C) yields a date of 4.14 ± 0.29 Ga, which is not significantly higher than the plateau date. The decay constant of ^{40}K was taken to be $\lambda = 5.305 \times 10^{-10}$ y^{-1} rather than the currently used value of $\lambda = 5.543 \times 10^{-10}$ y^{-1}.

Husain and Schaeffer (1973) also dated a sample of orange glass that was collected by the astronauts of Apollo 17 in the vicinity of Shorty Crater in the Taurus-Littrow valley. At the time it was collected, this material was thought to have resulted from recent volcanic activity on the Moon and therefore attracted much attention. Husain and Schaeffer (1973) analyzed Ar from an 85-mg sample of this glass by the stepwise outgassing procedure. One of the problems of dating material from the Moon by the ^{40}Ar/^{39}Ar method is that both ^{36}Ar and ^{40}Ar are implanted in rocks exposed to the solar wind (Davis, 1977). To calculate the ^{40}Ar*/^{39}Ar ratios of Ar released from lunar samples, a correction must therefore be made for the excess ^{40}Ar due to the solar wind. The ^{40}Ar/^{39}Ar ratio of this component was determined by Husain and Schaeffer (1973) by means of the Ar correlation diagram shown in Figure 7.9*a*. A least-squares fit of the data points gives an intercept value of 3.3 ± 0.2 and a slope of 154 ± 3. The intercept was used to correct each gas increment for the presence of excess ^{40}Ar. The resulting ^{40}Ar*/^{39}Ar ratios give rise to a spectrum of dates shown in Figure 7.9*b*. The best estimate of the age of this sample is the mean of the dates obtained for fractions released between 1150 and 1350°C. The plateau date of the orange glass in Figure 7.9*b* is 3.71 ± 0.06 Ga relative to $\lambda(^{40}\text{K}) = 5.305 \times 10^{-10}$ y^{-1}. This date is identical to the age of basalt fragments in the soil that covers the floor of the Taurus-Littrow valley. Therefore, the orange glass is not the product of recent volcanic activity.

The slope of the Ar isotope correlation diagram for the orange glass in Figure 7.9*a* (^{40}Ar*/^{39}Ar = 154 ± 3) provides an alternative estimate of the age of this material based on equation 7.8:

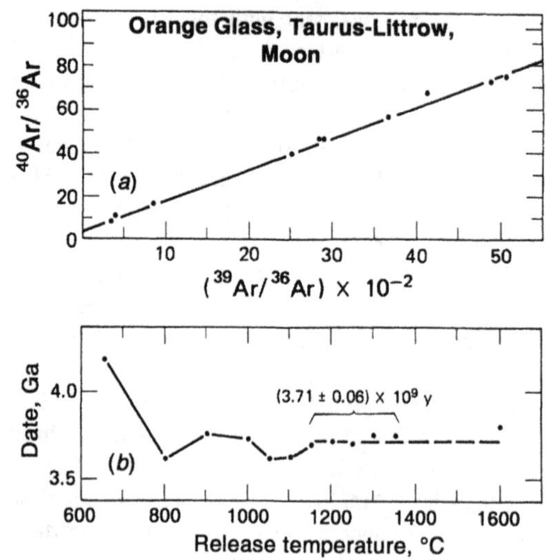

FIGURE 7.9 (*a*) Argon isotope correlation diagram for orange glass (74220, 39) from the Taurus-Littrow valley of the Moon (Apollo 17). The coordinates of the data points are isotope ratios of Ar released by stepwise degassing of a neutron-irradiated sample weighing only 85 mg. The slope of the correlation line is the ^{40}Ar*/^{39}Ar ratio and has a value of 154 ± 3 that corresponds to an age of 3.7 Ga. (*b*) Spectrum of partial-release ^{40}Ar*/^{39}Ar dates for gas fractions released at different temperatures from the same orange glass. The ^{40}Ar/^{39}Ar ratios were corrected for excess ^{40}Ar due to ion implantation by the solar wind using the intercept value of the correlation line [part (*a*)], which is ^{40}Ar/^{36}Ar = 3.3 ± 0.2. The average plateau date is 3.71 ± 0.06 Ga, calculated from the gas fractions released between 1150 and 1350°C. The total decay constant of ^{40}K is $\lambda = 5.305 \times 10^{-10}$ y^{-1}. Data from Husain and Schaeffer (1973).

$$t = \frac{1}{\lambda} \ln \left(\frac{^{40}\text{Ar}^*}{^{39}\text{Ar}} J + 1 \right)$$

The numerical value of J can be recovered from the dates and ^{40}Ar*/^{39}Ar ratios reported by Husain and Schaeffer (1973) for any of the gas fractions released by the orange glass. For example, the gas released at 650°C has a corrected $^{40}Ar^*/^{39}Ar$ ratio of 205 ± 13, from which the authors calculated a date of 4.18 ± 0.2 Ga. Accordingly, the value of J is given by equation 7.7:

$$J = \frac{e^{\lambda t} - 1}{^{40}Ar^*/^{39}Ar} = \frac{9.1842 - 1}{205} = 0.03992$$

The $^{40}Ar^*/^{39}Ar$ date indicated by the slope of the isochron in Figure 7.9*a* is

$$t = \frac{1}{5.305 \times 10^{-10}} \ln(154 \times 0.03992 + 1)$$
$$= 3.70 \pm 0.03\text{Ga}$$

This date is identical to the plateau date of the orange glass in Figure 7.9*b*.

Additional examples of $^{40}Ar/^{39}Ar$ dating of lunar basalt and meteorites include the following:

Podosek, 1971, *Geochim. Cosmochim. Acta*, 35:157–173. Bogard and Hirsch, 1980, *Geochim. Cosmochim. Acta*, 44:1667–1682. Turner, 1970, *Earth Planet. Sci. Lett.*, 9:177–180. Podosek, 1972, *Geochim. Cosmochim. Acta*, 36:755–772. Sutter et al., 1971, *Earth Planet. Sci. Lett.*, 11:249–253. Turner et al., 1971, *Earth Planet. Sci. Lett.*, 12:19–35. Turner and Cadogan, 1975, *Proc. 6th Lunar Sci. Conf.*, pp. 1509–1538. Schaeffer and Schaeffer, 1977, *Proc. 8th Lunar Sci. Conf.*, pp. 2253–2300. Guggisberg et al., 1979, *Proc. 10th Lunar Sci. Conf.*, pp. 1–39. Eichhorn et al., 1979, *Proc. 10th Lunar Sci. Conf.*, pp. 763–788. Herzog et al., 1980, *Proc. 11th Lunar Sci. Conf.*, pp. 959–976. Husain et al., 1971, *Science*, 173:1235–1236. Husain et al., 1972, *Science*, 175:428–430. Husain, 1974, *J. Geophys. Res.*, 79:2588–2608.

The incremental heating technique has also been used to date meteorites that originated from Mars (e.g., Shergotty by Bogard et al., 1979; ALH 84001 by Turner et al., 1997) and meteorite impact craters on the Earth (Layer, 2000). The results have permitted significant advances in the study of the solar system. For example, Podosek and Huneke (1973) first reported anomalously low $^{40}Ar^*/^{39}Ar$ partial-release dates of about 1.4 ± 0.1 Ga for six gas fractions released from 630 to 1195°C by the Ca-rich achondrite Lafayette. This result was the first indication that the age of Lafayette differs from those of most other achondrite meteorites and ultimately led to the conclusions that Lafayette and the other SNC meteorites originated from Mars (Section 5.3c).

7.5 LASER ABLATION

The power of the $^{40}Ar^*/^{39}Ar$ method of dating was increased greatly by the use of lasers to release Ar from single grains of neutron-irradiated minerals and rocks. A laser was first used by Megrue (1971) to date individual clasts of a lunar breccia by melting small volumes of rock in discrete pits. This method was subsequently refined by Schaeffer (1982), who used a pulsed ruby laser to date mineral grains embedded in polished surfaces of terrestrial and extraterrestrial rocks. Schaeffer and his associates melted minerals in pits having diameters between 10 and 100 μm and analyzed the gases in a conventional gas source mass spectrometer that was connected directly to the vacuum chamber containing the sample. The release of Ar from small volumes of minerals (~0.2 μg) has many advantages over the conventional method of melting larger amounts (~100 mg) of irradiated rocks or minerals:

1. Single grains of minerals can be dated without interference from inclusions of other minerals.
2. The small volume of melt greatly reduces the volume of the Ar blank and thereby reduces analytical errors.
3. Different minerals or several grains of the same mineral in a polished specimen of rock can be dated.
4. Mineral separations, which are time consuming and not always effective, are unnecessary.

However, the dates obtained by melting small volumes of minerals in laser pits of larger grains are equivalent to total-release $^{40}Ar^*/^{39}Ar$ dates that are sensitive to loss of radiogenic ^{40}Ar.

This limitation of the laser ablation method was eliminated by York et al. (1981), who used a continuous 15-W Ar ion laser to heat selected grains of minerals and fine-grained rocks. The mineral grains were irradiated individually rather than as part of whole-rock specimens. York et al. (1981) demonstrated that individual mineral grains of biotite, muscovite, and hornblende can be heated with a laser until they melt completely, thereby yielding total-release $^{40}Ar^{*}/^{39}Ar$ dates that are identical to dates obtained by the conventional method of fusion. In addition, York et al. (1981) released Ar stepwise from a grain of slate ($\sim$3 mm in diameter) by illuminating a spot ($\sim$600 μm) with laser light stepwise at increasing power until the entire grain melted. The plateau date of the slate (2.56 ± 0.01 Ga) obtained by the stepwise laser ablation method is identical to the plateau date obtained by the conventional stepwise release method of a larger sample of the same slate (2.55 ± 0.01 Ga).

The stepwise laser ablation method has become the standard procedure of $^{40}Ar^{*}/^{39}Ar$ dating because it permits selected mineral or rock grains to be dated with excellent precision and sensitivity:

Glass et al., 1986, *Chem. Geol. (Isotope Geosci. Sect.)*, 59:181–186. LoBello et al., 1987, *Chem. Geol. (Isotope Geosci. Sect.)*, 66:61–71. Layer et al., 1987, *Geophys. Res. Lett.*, 14:757–760. Dalrymple and Duffield, 1988, *Geophys. Res. Lett.*, 15:436–466.

7.5a Dating Meteorite Impact Craters

The Earth has been bombarded by large meteoroids and comets throughout its history. A partial record of these impacts has been preserved by more than about 160 craters that formed by the explosions that followed the impact of meteoroids or comets on the continents and continental shelves of the Earth. These meteorite impact craters are characterized by the presence of shocked minerals (e.g., quartz and zircon) and of melt rocks composed of glass.

The El'gygytgyn impact crater at 67° 27′ N and 172° 05′ E in the Okhotsk-Chukotka region of northeastern Russia has a diameter of about 18 km and is presently occupied by a lake. The bedrock surrounding the crater consists of volcanic rocks (rhyodacite tuff and ignimbrite with minor amounts of rhyolite, andesite, and basalt) of Cretaceous age. Layer (2000) dated 16 samples of meltrock, collected primarily from the rim of the crater, by the laser ablation partial-release $^{40}Ar/^{39}Ar$ method described by Layer et al. (1987). Ten of the meltrock samples yielded plateau dates ranging from 3.38 ± 0.08 to 3.74 ± 0.12 Ma with a weighted average of 3.58 ± 0.04 Ma. Two samples had somewhat older plateau dates of 4.10 ± 0.20 and 4.38 ± 0.14 Ma, three samples failed to yield $^{40}Ar^{*}/^{39}Ar$ plateaus, and a Cretaceous volcanic rock suffered extensive ^{40}Ar loss as a result of partial resetting of the K–Ar systematics at 3.58 Ma. The narrowly defined age of the meltrock (3.5 ± 0.04 Ma), together with the presence of shocked quartz, identifies Lake El'gygytgyn as a relatively recent impact crater. Samples of unaltered volcanic rocks from the vicinity of the crater yielded $^{40}Ar^{*}/^{39}Ar$ plateau dates ranging from 89.3 to 83.2 Ma and are clearly of Cretaceous (Turonian to Santonian) age.

Age determinations by the $^{40}Ar^{*}/^{39}Ar$ method of shocked rocks and impact melts from many other known impact craters have been reported in the literature, including the Brent Crater in Ontario (Davis, 1977), Manicouagen Lake in Quebec (Wolfe, 1971), Mien and Siljan craters in Sweden (Bottomley et al., 1978), Nördlinger Ries crater in Germany (Jessberger et al., 1978), and many others.

7.5b Sanidine Crystals, Yellowstone Park, Wyoming

Another important refinement of the $^{40}Ar^{*}/^{39}Ar$ method is the dating of suites of individual sanidine crystals extracted from rhyolite tuffs of Pleistocene age. The crystals are irradiated with fast neutrons and are then fused by means of a 6-W Ar ion laser (Gansecki et al., 1996). This methodology is being used increasingly to date volcano-sedimentary rocks of Quaternary age (e.g., Lippolt et al., 1984, 1986; van den Bogard et al., 1987, 1989; van den Bogaard, 1995; Lanphere et al., 1999).

Gansecki et al. (1998) used this technique to date sanidine crystals taken from three rhyolite tuff (ignimbrite) layers in Yellowstone Park of Wyoming. One suite of 28 sanidine crystals from Member B of the Lava Creek Tuff yielded total $^{40}Ar^{*}/^{39}Ar$ fusion dates ranging from 0.554 ± 0.003 to 0.607 ± 0.006 Ma. The weighted average of these dates is 0.602 ± 0.004 Ma. A few older sanidine crystals that originated from the same stratigraphic unit are probably xenocrysts that were entrained into the rhyolite magma during its explosive eruption.

The $^{40}Ar^{*}/^{39}Ar$ dates of sanidine crystals reported by Gansecki et al. (1998) confirm the dates published previously by Obradovich (1992) and Obradovich and Izett (1991), who obtained dates of 0.617 ± 0.008 and of 0.61 ± 0.02 Ma, respectively, for sanidines of the Lava Creek Tuff. These results exemplify the current state of the art, which has made it possible to date very young rocks whose ages approach historical time (e.g., Chen et al., 1996; Spell and Harrison, 1993).

7.5c Intercalibrations

The high precision and versatility of the $^{40}Ar^{*}/^{39}Ar$ method of dating have inevitably led to comparisons of $^{40}Ar^{*}/^{39}Ar$ dates with age determinations of cogenetic minerals and rocks by the U–Th–Pb methods, which are equally precise and similarly versatile. The principal sources of uncertainty in the numerical values of isotopic dates are the measured values of the relevant decay constants and, in the case of the $^{40}Ar^{*}/^{39}Ar$ method, the age (or $^{40}Ar^{*}/^{39}K$ ratio) of the flux monitors used in the neutron irradiations (Roddick, 1983). For this reason, Villeneuve et al. (2000) proposed that the age of biotite phenocrysts (MAC-83) from an Oligocene Tuff (Cerro Huancahuancane Formation, Cordillera de Carabaya, southeastern Peru) is fixed by cogenetic monazite crystals whose $^{235}U–^{207}Pb$ date is 24.21 ± 0.10 Ma. Use of this date establishes a direct link between $^{40}Ar^{*}/^{39}Ar$ dates and U–Pb dates. An age determination of sanidine in the Fish Canyon Tuff, which has been widely used as a flux monitor in $^{40}Ar^{*}/^{39}Ar$ dating, yielded a $^{40}Ar^{*}/^{39}Ar$ date of 27.98 ± 0.15 Ma relative to 24.21 ± 0.10 Ma for MAC-83. A previous K–Ar age determination of sanidine of the Fish Canyon Tuff by Renne et al. (1997) yielded a date of 28.03 ± 0.09 Ma, which is identical to the date of 27.98 ± 0.15 Ma obtained by Villeneuve et al. (2000). However, K–Ar dates of biotites from the Fish Canyon Tuff range from 27.42 ± 0.31 Ma (preparation FCT-3) to 27.55 ± 0.24 Ma (FCT-2) (Lanphere and Baadsgaard, 1997, 1998). The effect of various flux monitors on the accuracy of $^{40}Ar^{*}/^{39}Ar$ dates was also considered by Renne et al. (1994), Baksi et al. (1996), and Renne et al. (1998a, b), whereas Renne et al. (1997) extended dating by the $^{40}Ar^{*}/^{39}Ar$ method into the historical time of Pliny the Younger.

7.6 SEDIMENTARY ROCKS

The importance of dating sedimentary rocks has already come up for discussion in Sections 5.5 (Rb–Sr) and 6.5 (K–Ar). Consequently, all of the strategies of the $^{40}Ar^{*}/^{39}Ar$ method have been used to date sedimentary rocks and the authigenic K-bearing minerals they contain (e.g., glauconite and illite).

7.6a Loss of ^{39}Ar by Recoil

The attempt to date clay minerals by the $^{40}Ar^{*}/^{39}Ar$ method initially yielded unexpectedly old dates. For example, Brereton et al. (1976) reported excessively old total-release $^{40}Ar^{*}/^{39}Ar$ dates for Early Ordovician glauconite from Västergotland in Sweden, whereas their conventional K–Ar dates are about 8 percent lower than their known geological age. Brereton and his colleagues attributed the anomalous $^{40}Ar^{*}/^{39}Ar$ dates of these glauconites to the loss of ^{39}Ar by recoil during the irradiation and concluded that $^{40}Ar^{*}/^{39}Ar$ dates of glauconites are not geologically meaningful. These conclusions were subsequently confirmed by Foland et al. (1984) and other investigators.

The assumptions of the incremental heating technique do not take into account the possible loss or relocation of ^{39}Ar as a result of the recoil of the nucleus of ^{39}K caused by the emission of a proton during the (n, p) reaction. If a significant fraction of the ^{39}Ar is lost from the irradiated sample, its

$^{40}Ar^*/^{39}Ar$ ratio is increased correspondingly, and its total-release date would be older than its geological age. The loss of ^{39}Ar from a K-rich mineral grain occurs from a surface layer whose thickness is about 0.08 μm, according to calculations by Turner and Cadogan (1974). The effect of ^{39}Ar loss by recoil therefore depends essentially on the size of the K-bearing minerals. Fine-grained samples, or those in which the K-bearing minerals are highly fractured, lose a larger proportion of ^{39}Ar by recoil during the irradiation than do coarse-grained rocks in which the K-bearing minerals are well crystallized and not fractured. These considerations suggest that the $^{40}Ar^*/^{39}Ar$ method may not be reliable for dating K-rich clay minerals such as glauconite, which commonly occurs in the form of green pellets that consist of aggregates of platy crystals about 1 μm in thickness.

The effect of ^{39}Ar loss by recoil on the shapes of partial-release spectra is not well understood. Turner and Cadogan (1974) considered the possibility that the actual increase of the $^{40}Ar^*/^{39}Ar$ ratio of the K-rich phases depends on their grain size and therefore on their surface-to-volume ratio. A sample containing small grains of K minerals having a large surface-to-volume ratio may exhibit a greater increase of the $^{40}Ar^*/^{39}Ar$ ratio because of ^{39}Ar loss by recoil than a similar sample containing larger grains of K-bearing minerals. Therefore, the partial Ar-release pattern of a sample affected by the recoil phenomenon should show an elevated $^{40}Ar^*/^{39}Ar$ ratio in the low-temperature gas fraction. The magnitude of this effect should be inversely proportional to the size of the K-bearing minerals. However, the effect may be masked by loss of radiogenic ^{40}Ar from grain boundaries as a result of diffusion after crystallization.

The ^{39}Ar expelled from the surface layer of the K-bearing minerals may be embedded in neighboring K-poor minerals, such as pyroxenes, which are generally more retentive than K-rich minerals, such as K-feldspar or mica. Therefore, the argon released at high temperature may have low $^{40}Ar^*/^{39}Ar$ ratios because it originates from pyroxenes that contain the ^{39}Ar expelled from the K minerals during the neutron irradiation of the sample. This effect may explain why the stepwise release spectra of some rocks from the Moon show a decrease of the $^{40}Ar^*/^{39}Ar$ ratio at the highest gas release temperatures.

7.6b Glauconite and Illite

When the $^{40}Ar^*/^{39}Ar$ method is used to date glauconite and illite, the problems that must be overcome concern not only the time of closure of these minerals (Sections 5.5b and 5.5f) but also the loss of ^{39}Ar as a result of recoil. One way to minimize the loss of ^{39}Ar is to encapsulate the sample and to measure the amount of ^{39}Ar released during the irradiation. When this gas is added to the Ar released during step heating, the resulting "integrated $^{40}Ar^*/^{39}Ar$ dates" are equivalent to total-release $^{40}Ar^*/^{39}Ar$ and to conventional K–Ar dates. However, the use of lasers allows single glauconite pellets and single illite flakes to be dated, thereby improving the measurement precision and assuring the homogeneity of the material.

Smith et al. (1993) used the laser ablation technique to date single glauconite pellets ranging in age from Ordovician to Tertiary. The samples were irradiated in evacuated ampules ($\sim 10^{-7}$ Torr) composed of quartz glass (~0.2 mL). After the irradiation, the ampules were opened within the gas inlet system of the mass spectrometer and both the amount and isotope composition of the recoiled gas were measured. Subsequently, the glauconite pellets were heated by irradiating them stepwise with a continuous 20-W Ar ion laser. In addition, Smith et al. (1993) irradiated a second set of the same glauconites in air for comparison of the resulting $^{40}Ar^*/^{39}Ar$ dates. The authors reported that the recoiled Ar in the evacuated ampules was composed predominantly of ^{39}Ar and did not contain significant amounts of ^{40}Ar. The fractions of recoiled ^{39}Ar ranged from 27 to 64 percent of the amounts expected from the K concentrations. Consequently, the $^{40}Ar^*/^{39}Ar$ dates calculated from the recoiled Ar were close to zero.

The partial-release $^{40}Ar^*/^{39}Ar$ dates of a Cretaceous glauconite pellet in Figure 7.10 increase from near zero (recoiled ^{39}Ar) to about 110 Ma. The integrated date of this in vacuo irradiated glauconite pellet is 83.8 ± 1.1 Ma, compared to its conventional K–Ar date of 89.8 ± 3.6 Ma reported by Odin (1982). The fact that the $^{40}Ar^*/^{39}Ar$ date

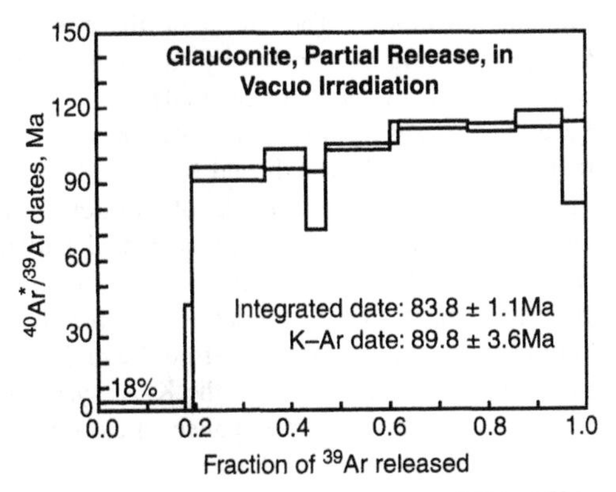

FIGURE 7.10 Partial-release spectrum of $^{40}Ar^{*}/^{39}Ar$ dates of a Cretaceous glauconite pellet irradiated in an evacuated ampule and heated stepwise by a continuous 20-W Ar ion laser. The loss of ^{39}Ar by recoil increased the measured $^{40}Ar^{*}/^{39}Ar$ ratios, causing the calculated dates to exceed the age of this mineral. However, when the recoiled ^{39}Ar is added to the rest of the Ar released by laser heating, the "integrated $^{40}Ar^{*}/^{39}Ar$ date" closely approaches its conventional K–Ar date. Adapted from Smith et al. (1993).

was measured on a single glauconite pellet, rather than on an assemblage of pellets as in the conventional K–Ar method, may improve its reliability as an estimate of the age of this glauconite.

In general, the results reported by Smith et al. (1993) indicate that the glauconite pellets that were irradiated in evacuated ampules yielded lower integrated $^{40}Ar^{*}/^{39}Ar$ dates in all cases than the same samples irradiated in air. Recent studies by Foland et al. (1992), Dong et al. (1997), Onstott et al. (1997), and Lin et al. (2000) likewise addressed the problems caused by the loss of ^{39}Ar from lamellar mineral grains with thicknesses of less than about 1.0 μm.

The reliability of $^{40}Ar^{*}/^{39}Ar$ dates determined on single glauconite pellets after irradiation in evacuated ampules was evaluated by Smith et al. (1998). Their results indicate that the integrated $^{40}Ar^{*}/^{39}Ar$ dates of individual glauconite pellets taken from the same layer of rock range over an interval of from 6 to 7 million years during which some of the glauconite pellets continued to mature. Consequently, Smith et al. (1998) concluded that the oldest dates approach the stratigraphic age of the host rock of the glauconite pellets, excluding only relict grains that yielded anomalously old dates. The remaining dates have a polymodal distribution, which implies that the process of glauconitization was modulated, perhaps because of time-dependent variation in sea level. Therefore, Smith et al. (1998) proposed that gaps in the spectrum of glauconite $^{40}Ar^{*}/^{39}Ar$ dates represent intervals of low-sea-level stands on the continental margins.

A statistical treatment of the $^{40}Ar^{*}/^{39}Ar$ dates of glauconites analyzed by Smith et al. (1998) in Table 7.2 yielded age estimates for three stratigraphic boundaries that agree with the values given by Harland et al. (1989) within the stated errors.

The existence of a formation interval for glauconite pellets revealed by their $^{40}Ar^{*}/^{39}Ar$ dates explains why $^{40}Ar^{*}/^{39}Ar$ and K–Ar dates derived from bulk samples containing hundreds of individual pellets are, in most cases, less than the stratigraphic age of the host rock and why they are not as reproducible as expected from the errors of measurement. The multimodal spectrum of $^{40}Ar^{*}/^{39}Ar$ dates of suites of individual glauconite pellets contrasts sharply with the unimodal distribution of dates that Smith et al. (1998) recorded for 49 individual crystals of sanidine for

Table 7.2. $^{40}Ar^{*}/^{39}Ar$ Dates of Single Glauconite Pellets Irradiated in Evacuated Ampules

Boundary	Best Estimate (Ma)	
	Smith et al., 1998	Harland et al., 1989[a]
Burdigalian to Aquitanian (Miocene)	20.5 + 2.6, −2.5	21.5
Bartonian to Lutetian (Eocene)	43.4 + 2.9, −1.7	42.1
Cenomanian to Albian (Late to Early Cretaceous)	96.7 + 3.1, −2.7	97.0

Source: Smith et al., 1998.

[a]The dates refer to the ages of the boundaries between these stages.

the Taylor Creek Rhyolite, which yielded a mean of 27.92 ± 0.05 Ma (not shown). Consequently, $^{40}Ar^{*}/^{39}Ar$ dates or Rb–Sr and U–Pb dates derived from individual crystals of high-temperature minerals in tuffaceous sediment (e.g., sanidine, muscovite, phlogopite, biotite, and zircon) are the most reliable indicators of the ages of sedimentary rocks.

7.7 METASEDIMENTARY ROCKS

The $^{40}Ar^{*}/^{39}Ar$ and K–Ar methods have been widely used to study whole-rock samples of metasedimentary rocks (Dodson and Rex, 1971; Adams et al., 1975) and separated minerals (e.g., phengite, biotite, hornblende, garnet) because the stepwise release procedure can provide information about the history of metamorphism of such rocks and minerals. For example, Scaillet et al. (1990) demonstrated that partial-release $^{40}Ar^{*}/^{39}Ar$ dates of phengites (sericite) in the Dora Maira nappe of the western Alps of Italy record several tectonometamorphic episodes both in the form of partial-release plateau dates and by the range of spot-fusion dates on single flakes of phengite. Similar results were reported by Wijbrans et al. (1990) for phengites from the Cyclades of Greece. Additional references to $^{40}Ar^{*}/^{39}Ar$ dating of low-grade metamorphic rocks include the following:

Sission and Onstott, 1986, *Geochim. Cosmochim. Acta*, 50:2111–2118. Reuter and Dallmeyer, 1987, *Contrib. Mineral. Petrol.*, 97:352–360. Ross and Sharp, 1988, *Contrib. Mineral. Petrol.*, 100:213–221. Dunlap et al., 1991, *Geology*, 19(12):1213–1216.

7.7a Meguma Group, Nova Scotia

The slates of the Meguma Group in Nova Scotia have attracted attention because they contain gold-bearing quartz veins (Kontak et al., 1998). The sedimentary rocks of the Meguma Group were deposited in early Paleozoic time and were subsequently subjected to low-grade regional metamorphism during the Acadian orogeny followed by intrusion of the granitic rocks of the South Mountain batholith. The slates of the Meguma Group were originally dated by the conventional whole-rock K–Ar method (Reynolds et al., 1973). Spectra of partial-release $^{40}Ar^{*}/^{39}Ar$ dates of these rocks reported by Reynolds and Muecke (1978) in Figure 7.11*a* include a sample of hornfels from the contact zone with a well-constrained partial-release $^{40}Ar^{*}/^{39}Ar$ date of 385 ± 1 Ma, which is in good agreement with K–Ar dates of biotite in the South Mountain batholith. Therefore, Reynolds and Muecke (1978) concluded that the K–Ar systematics of this and other samples collected close to the

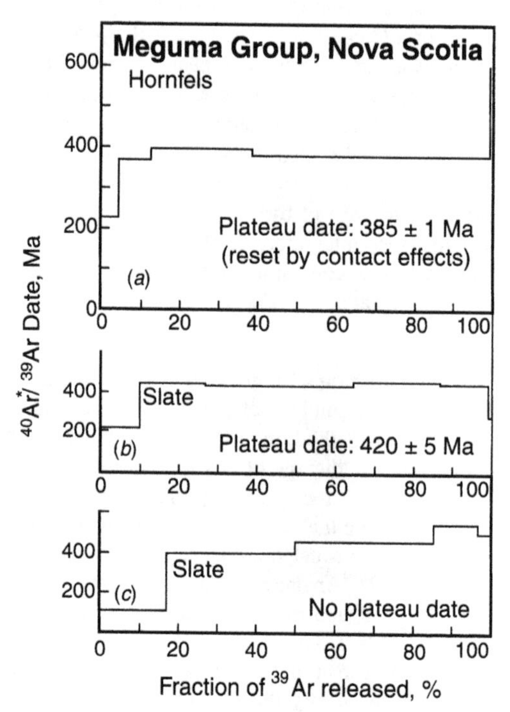

FIGURE 7.11 (*a*) Partial-release $^{40}Ar^{*}/^{39}Ar$ dates of hornfels from the contact between metasedimentary rocks of the early Paleozoic Meguma Group and the granitic intrusives of the South Mountain batholith in Nova Scotia, Canada. (*b*) Spectrum of $^{40}Ar^{*}/^{39}Ar$ dates of slate of the Meguma Group unaffected by the intrusion of granites. (*c*) Spectrum of $^{40}Ar^{*}/^{39}Ar$ dates of slate containing detrital microcline. No plateau date can be calculated. Data from Reynolds and Muecke (1978).

contact were completely reset by the intrusion of the granitic rocks.

A second group of slates collected farther from the intrusive contact also yielded well-defined $^{40}Ar^*/^{39}Ar$ partial-release plateaus, including an older date of about 420 ± 5 Ma in Figure 7.11*b*. The plateau dates of these slates exceed the age of the batholith but underestimate the age of the Acadian orogeny in this region. Two additional samples of slate, one of which is shown in Figure 7.11*c*, contained detrital K-feldspar and formed steplike profiles that did not permit a reliable age determination.

The results of the study of Reynolds and Muecke (1978) can be summarized as follows:

1. Slates unaffected by later contact metamorphism yield age spectra with well-defined plateaus that reflect the time of regional metamorphism.
2. Slates that were reheated by nearby magmatic activity have plateau dates that reflect the time of reheating even though this event is not recorded by mineralogical changes.
3. The $^{40}Ar^*/^{39}Ar$ dates of slates containing K-feldspar increase with increasing gas release temperature and fail to form a stable plateau.
4. The plateau dates of slates that do not contain K-feldspar are slightly higher than their total-release and conventional K-Ar dates because the low-temperature gas fractions appear to be depleted in radiogenic ^{40}Ar.

An important point to be made in conclusions is that the $^{40}Ar^*/^{39}Ar$ plateaus of slates can be interpreted to date the time of regional metamorphism only when it can be shown from field evidence that the rocks were not reheated in the course of magmatic activity subsequent to their metamorphic crystallization.

7.7b Barberton Greenstone Belt, Swaziland

Metasedimentary rocks of low grade contain minerals with different blocking temperatures for radiogenic ^{40}Ar. Therefore, spectra of $^{40}Ar^*/^{39}Ar$ dates obtained by stepwise release of Ar from whole-rock samples can record the occurrence of more than one metamorphic episode. In addition, the $^{40}Ar^*/^{39}Ar$ method can be applied not only to very young minerals (e.g., sanidine in Pleistocene tuff) but also to metasedimentary and metavolcanic rocks of Early Archean age, such as the Isua supracrustals in southern West Greenland and the low-grade metamorphic rocks of the Barberton greenstone belt in Swaziland, South Africa.

The metavolcanic rocks (komatiites) of the Onverwacht Group in the Barberton greenstone belt yielded $^{40}Ar^*/^{39}Ar$ plateau dates between 3.11 and 3.49 Ga (Lopez-Martinez et al., 1984). The oldest $^{40}Ar^*/^{39}Ar$ dates between 3.45 and 3.49 Ga are not distinguishable from the whole-rock Sm–Nd isochron date of 3.540 ± 0.015 Ga. Since the $^{40}Ar^*/^{39}Ar$ dates reflect the time of last alteration of the K–Ar systematics of these rocks, Lopez-Martinez et al. (1984) concluded that the komatiite flows of the Onverwacht Group were subjected to low-grade regional metamorphism less than 100 million years after their eruption.

The metavolcanic rocks of the Onverwacht Group are overlain by the metasedimentary rocks of the Fig Tree and Moodies Groups. The latter contains the oldest record of tides on the Earth (Eriksson and Simpson, 2000). The Fig Tree Group is composed of greywackes and shales but includes also conglomerates, sandstones, banded iron formation, chert, and barite. Alexander (1975) obtained a saddle-shaped profile of $^{40}Ar^*/^{39}Ar$ dates for a sample of chert with a total-release date of 2.976 ± 0.026 Ga ($\lambda = 5.5305 \times 10^{-10}$ y^{-1}). Other age determinations reviewed by de Ronde et al. (1991) indicate that the age of the Fig Tree Group is less than 3.453 ± 0.018 Ga but older than about 3.23 Ga, whereas the age of the Moodies Group is 3.164 Ga.

De Ronde et al. (1991) used a continuous Ar-ion laser to heat selected grains of irradiated barite, sandstone, and shale at stepwise-increasing temperatures. The isotope composition of the Ar released during each step was measured by mass spectrometry. All corrections for the presence of atmospheric Ar and for interfering neutron capture reactions were made by a preprogrammed dedicated computer. The Ca/K ratios of the samples were calculated from the $^{37}Ar/^{39}Ar$ ratios because ^{37}Ar is produced by the ^{40}Ca (n, α) ^{37}Ar reaction. Similarly, the $^{38}Ar/^{39}Ar$ ratio yielded the Cl/K ratio of the sample because ^{38}Ar is formed during the irradiation by the $^{37}Cl(n, \gamma, \beta^{-})\ ^{38}Ar$ reaction. The Ca/K and Cl/K ratios calculated by this method provide information about the minerals from which the Ar in the sample was derived at different temperatures.

The results, illustrated in Figure 7.12 for a sample of barite in the Fig Tree Group, indicate that the high-temperature gas containing about 20 percent of the released ^{39}Ar originated from a mineral that had a higher Ca/K ratio (0.3 to 1.0) than the lower temperature gas (0–80 percent of ^{39}Ar), which originated from a mineral having a low Ca/K ratio (<0.1). The plateau date of the low-temperature gas is 2.673 ± 0.0018 Ga. The $^{40}Ar^{*}/^{39}Ar$ ratios of the high-temperature gas yielded an older date of about 2.820 Ga. A sandstone sample that was closely associated with the barite recorded a plateau date of only 2.408 ± 0.0014 Ga calculated for the low-temperature gas derived from illite or sericite (not shown).

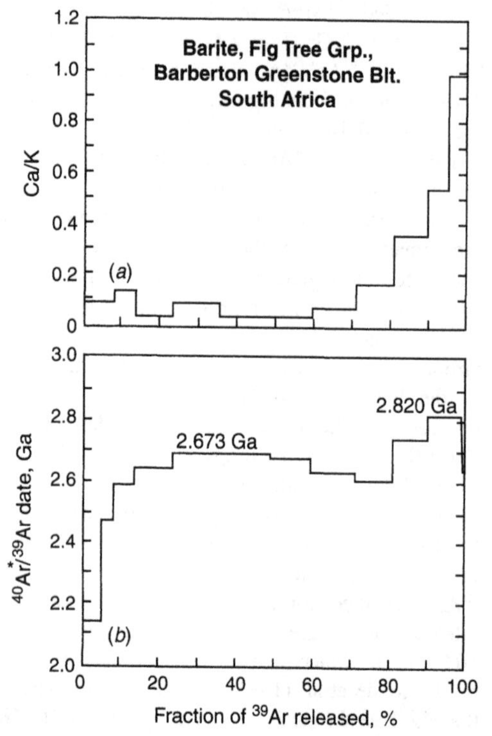

FIGURE 7.12 (*a*) Spectrum of Ca/K ratios of barite in the Fig Tree Group of the Barberton greenstone belt in Swaziland, South Africa. The Ca/K ratios were calculated from the isotope composition of Ar released during stepwise heating, as explained in the text. The increase of the Ca/K ratio of the high-temperature fractions of Ar was caused by the presence of a second mineral that has a higher blocking temperature (i.e., is more retentive) than barite. (*b*) Spectrum of $^{40}Ar^{*}/^{39}Ar$ dates of the same barite showing two plateaus with dates of 2.673 ± 0.0018 and 2.820 Ga. The age of the metasedimentary rocks of the Fig Tree Group lies between 3.45 and 3.23 Ga. Adapted from de Ronde et al. (1991).

Other mineral phases in metasedimentary rocks of the Fig Tree Group (e.g., quartz, tourmaline, barite, and illite/sericite) yielded dates of about 2.350–2.400 and 2.025–2.090 Ga. De Ronde et al. (1991) concluded that these $^{40}Ar^{*}/^{39}Ar$ dates record the effects of thermal overprinting associated with the intrusion of granitic plutons (2.673 ± 0.018 Ga) and with the occurrence of large-scale tectonothermal activity that accompanied the intrusion of the Bushveld Complex (2.025–2.090 Ga) composed of differentiated gabbros and ultramafic rocks (Faure, 2001).

The wide range of $^{40}Ar^{*}/^{39}Ar$ plateau dates reported by Lopez-Martinez et al. (1984) and de Ronde et al. (1991) for rocks of the Barberton greenstone belt (3.49 and 2.025 Ga) is attributable both to the tectonothermal conditions of rock samples in different parts of this belt and to differences in the blocking temperatures of the principal K-bearing minerals they contain (e.g., amphiboles, barite, illite, quartz, and tourmaline). Metasedimentary rocks containing two or more K-bearing minerals with different blocking temperatures can yield complex $^{40}Ar^{*}/^{39}Ar$ partial-release spectra. In cases where the intensity of metamorphic episodes decreased with time, the most retentive minerals record the earliest event, whereas the least retentive minerals record the last episode of metamorphism. Consequently, the spectra of $^{40}Ar^{*}/^{39}Ar$ dates of rocks containing minerals having different blocking temperatures may have two valid plateaus.

7.7c Dating of Low-K Minerals

A final point to be made here is that the $^{40}Ar^{*}/^{39}Ar$ dates reported by Lopez-Martinez et al. (1984)

and de Ronde et al. (1991) were obtained from rocks whose minerals had remarkably low K concentrations (e.g., 80 ppm for the komatiites of the Onverwacht Group). Both studies were carried out by members of the research group of Derek York in the Physics Department of the University of Toronto. York and his associates have demonstrated that even minerals having low K concentrations may be amenable to $^{40}Ar^{*}/^{39}Ar$ dating by the laser step-heating procedure, such as pyrite (York et al., 1982), pyroxene (Hanes et al., 1985), hornblende (Layer et al., 1987), and magnetite (Özdemir and York, 1990).

The possibility of dating common sulfide minerals by the $^{40}Ar^{*}/^{39}Ar$ method is especially noteworthy because minerals such as pyrite (FeS_2) occur in certain kinds of ore deposits as well as in shales, limestones, and many kinds of igneous rocks and meteorites. For example, Niemeyer (1979) reported that he had dated troilite (FeS) from the iron meteorites Pitts and Mundrabilla by the $^{40}Ar^{*}/^{39}Ar$ partial-release method. The spectrum of dates he obtained was complex and required careful interpretation. Nevertheless, these results suggested that even K-poor sulfide minerals may be datable by the $^{40}Ar^{*}/^{39}Ar$ method.

York et al. (1982) subsequently demonstrated that pyrite from the Geco Mine at Manitouwadge in northern Ontario is datable by the $^{40}Ar^{*}/^{39}Ar$ method. They irradiated six pyrite samples and plotted the measured $^{40}Ar/^{36}Ar$ and $^{39}Ar/^{36}Ar$ ratios on an isotope correlation diagram. The results in Figure 7.13 are linearly correlated and correspond to a date of 2500 ± 120 Ma. This date is similar to plateau dates of two biotites from the Geco ore body.

In a related study, Czamanske et al. (1978) dated the K-bearing sulfide minerals rasvumite (KFe_2S_3) and bartonite ($K_3Fe_{10}S_{14}$) by the total-release $^{40}Ar^{*}/^{39}Ar$ method. These minerals occur in an alkali-rich mafic diatreme of Tertiary age that intruded a melange of Franciscan rocks at Coyote Peak about 42 km north of the town of Arcata in northern California. Rasvumite was originally discovered in the alkali-rich Khibina intrusion on the Kola Peninsula of Russia. The specimen of rasvumite at Coyote Peak contained 16.3 percent K, 44.8 percent Fe, 37.5 percent S, and 0.06 percent Na. The K concentration of the

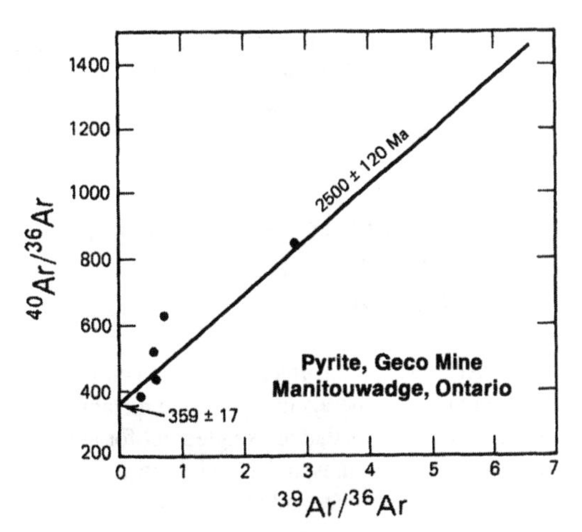

FIGURE 7.13 Isotope correlation diagram for Ar released by fusion of pyrite samples from the sulfide ore body at the Geco Mine, Manitouwadge, Ontario. The $^{40}Ar^{*}/^{39}Ar$ ratio given by the slope of the line corresponds to a date of 2500 ± 120 Ma and the initial $^{40}Ar/^{36}Ar$ ratio is 359 ± 17. The date derived from the pyrite is indistinguishable from $^{40}Ar^{*}/^{39}Ar$ plateau dates of biotites from the Geco Mine. Data from York et al. (1982).

bartonite at Coyote Peak was 9.91 percent (Cl poor) and 10.7 percent (Cl rich). It also contained Fe (50.5 and 49.0 percent), S (38.2 and 37.3 percent), and small amounts of Cu, Ni, Cl, and Na. The $^{40}Ar^{*}/^{39}Ar$ total-release dates of the rasvumite and bartonite in Table 7.3 are in good agreement with a $^{40}Ar^{*}/^{39}Ar$ total-release date of phlogopite from Coyote Peak (Czamanske et al., 1978). Although

Table 7.3. $^{40}Ar^{*}/^{39}Ar$ Dates of Fe–K Sulfide Minerals Rasvumite and Bartonite at Coyote Peak, California

Mineral	$^{40}Ar^{*}/^{39}Ar$ (Ma)	K–Ar (Ma)
Phlogopite	28.3 ± 0.4	30.2 ± 1.0
Rasvumite	26.5 ± 0.5	
Bartonite	29.4 ± 0.5	

Source: Czamanske et al., 1978.

rasvumite and bartonite are rare minerals and are K rich rather than K poor, their existence provided an opportunity to date sulfide minerals by the $^{40}Ar^{*}/^{39}Ar$ method.

7.8 METAMORPHIC ROCKS: BROKEN HILL, N.S.W., AUSTRALIA

The $^{40}Ar^{*}/^{39}Ar$ partial-release method continues to be used to date minerals in high-grade metamorphic rocks. The minerals that are singled out for analysis include not only muscovite and biotite but also hornblende, pyroxene, and plagioclase:

> Dallmeyer, 1975a, *Geochim. Cosmochim. Acta*, 39:1655–1669. Dallmeyer et al., 1975, *Geol. Soc. Am. Bull.*, 86:1435–1443. Hanson et al., 1975, *Geochim. Cosmochim. Acta*, 39:1269–1278. Berger, 1975, *Earth Planet. Sci. Lett.*, 26:387–408. Dallmeyer and Rivers, 1983, *Geochim. Cosmochim. Acta*, 47:413–428. Claesson and Roddick, 1983, *Lithos*, 16:61–73. Berry and McDougall, 1986, *Chem. Geol. (Isotope Geosci. Sect.)*, 59:43–58. Lippolt and Kirsch, 1994, *Chemie der Erde*, 54:179–198. Pickles et al., 1997, *Geochim. Cosmochim. Acta*, 61:3809–3834. Hames and Cheney, 1997, *Geochim. Cosmochim. Acta*, 61:3863–3872. Kelley et al., 1997, *Geochim. Cosmochim. Acta*, 61:3873–3878. Siebel et al., 1998, *Geology*, 26:31–34.

In addition, the $^{40}Ar^{*}/^{39}Ar$ method has been used to date ore deposits:

> Dallmeyer, 1975b, *Econ. Geol.*, 70:341–345. Harrison and McDougall, 1981, *Earth Planet. Sci. Lett.*, 55:123–149. Wu, 1992, *Mineral. Deposits*, 10(4):358–369. Kontak et al., 1998, *Can. J. Earth Sci.*, 35:746–761. Bierstein et al., 1999, *Austral. J. Earth Sci.*, 46:301.

Additional reports on $^{40}Ar^{*}/^{39}Ar$ dating of ore deposits have been published in *Economic Geology, Mineralium Deposita*, and other earth science journals.

The Pb–Zn ore at Broken Hill occurs in high-grade metamorphic rocks of the Willyama Complex, which consists of metasedimentary and metavolcanic rocks with minor granitic, mafic, and ultramafic intrusives (Vernon, 1969). The rocks exposed at the surface range in grade from andalusite–muscovite gneiss in the north to two-pyroxene granulites in the south. These rocks were subjected to three episodes of folding followed by the intrusion of granitic rocks of the Mundi Mundi type and by ultramafic rocks.

The whole-rock Rb–Sr method used by Pidgeon (1967) and Shaw (1969) yielded a date of 1660 ± 10 Ma ($\lambda = 1.42 \times 10^{-11}$ y^{-1}) for the age of the high-grade metamorphic rocks of the Mine Sequence which hosts the ore at Broken Hill. Pidgeon (1967) also obtained a Rb–Sr date of 1490 ± 20 Ma for whole-rock samples and muscovite of the Mundi Mundi granitoids. However, K–Ar as well as Rb–Sr dates of biotites in the Mundi Mundi granitoids indicate the occurrence of a thermal event at about 520 Ma and the neoformation of muscovite at about 484 Ma in shear zones that cut the Mine Sequence (Richards and Pidgeon, 1963; Binns and Miller, 1963).

The conventional K–Ar dates reported by Harrison and McDougall (1981) for minerals in the metamorphic rocks at Broken Hill in Figure 7.14 range from 529.3 ± 9.0 Ma (biotite) to 6150.0 ± 62 Ma (plagioclase). The occurrence of K–Ar dates in excess of 3.5 Ga in several samples of plagioclase and pyroxene indicates the presence of excess ^{40}Ar, whereas the uniformly low biotite dates (529.3–540.0 Ma) reflect the time of final cooling of the Willyama Complex to the blocking temperature of this mineral. The K–Ar dates of hornblendes that are clustered between 1.5 and 2.0 Ga in Figure 7.14 actually have a bimodal distribution with means of 1660 ± 27 and 1850 ± 35 Ma. Although the average K–Ar hornblende date of 1660 ± 27 Ma is identical to the whole-rock Rb–Sr isochron date (1660 ± 10 Ma) of Pidgeon (1967) and therefore presumably dates the time of high-grade metamorphism, the wide range of K–Ar dates and the evidence for excess ^{40}Ar in several minerals do not permit a definitive interpretation of these dates in terms of the geological history of the ore deposit at Broken Hill.

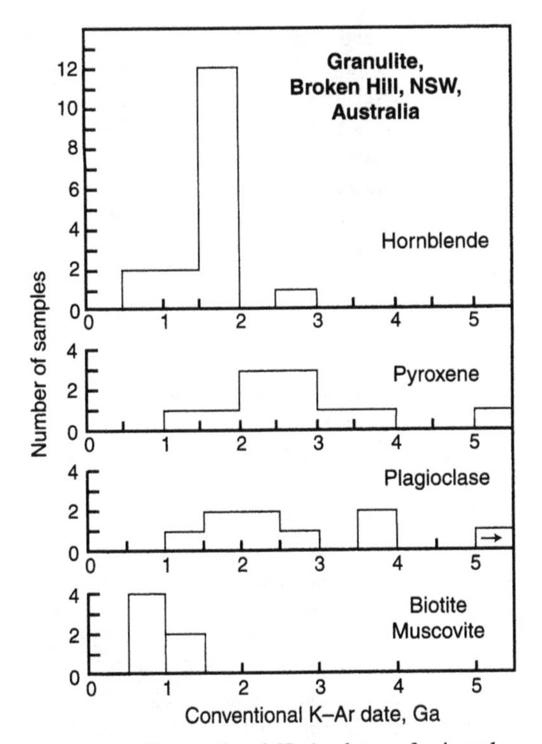

FIGURE 7.14 Conventional K–Ar dates of minerals (hornblende, pyroxene, plagioclase) in granulites and in granitic intrusives (biotite and muscovite) at Broken Hill, New South Wales, Australia. The anomalously old dates (>2.0 Ga) are caused by the presence of excess ^{40}Ar in all of the minerals in the granulite gneisses. Data from Harrison and McDougall (1981).

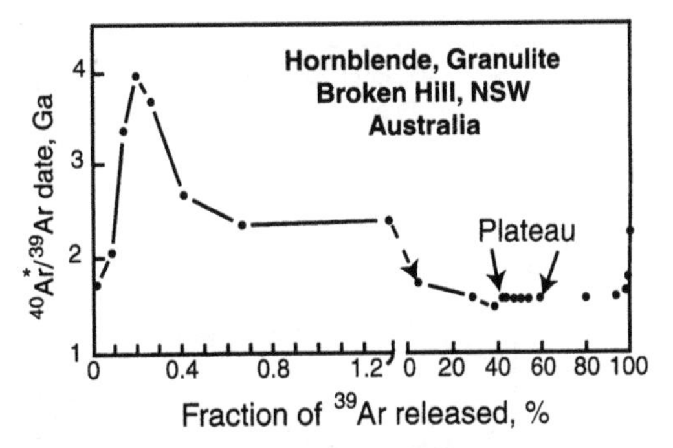

FIGURE 7.15 Spectrum of $^{40}Ar^*/^{39}Ar$ dates of hornblende in the granulite at Broken Hill, N.S.W., Australia. The dates calculated from the low-temperature gas (1.3 percent) are inflated because of the presence of excess ^{40}Ar. The plateau date is 1577 ± 3 Ma. Data from Harrison and McDougall (1981).

The $^{40}Ar^*/^{39}Ar$ spectrum of dates reported by Harrison and McDougall (1981) for one of the hornblendes (K–Ar date: 1628 ± 46 Ma) in Figure 7.15 clearly demonstrates the presence of excess ^{40}Ar, especially in the first 1.30 percent of ^{39}Ar released at $T < 860$°C, which yielded dates approaching 4.0 Ga. However, about 20 percent of the gas released between 940 and 1020°C formed a plateau with an average date of 1577 ± 3.0 Ma. Therefore, the conventional K–Ar date of this hornblende (1628 Ma) is in error because of the presence of excess ^{40}Ar. The dates calculated from the high-temperature gas (1050–1120°C) increased stepwise to 2307 Ma. The six gas fractions that form the plateau in Figure 7.15 also define a K–Ar isochron (not shown) corresponding to a date of 1573 ± 5 Ma and an initial $^{40}Ar/^{39}Ar$ ratio of 348 ± 58, in agreement with atmospheric Ar. Therefore, Harrison and McDougall (1981) concluded that this hornblende cooled through its blocking temperature of about 500°C at the time indicated by the plateau and the isochron at about 1575 Ma. The anomalously old dates of the low-temperature gas indicate that the excess ^{40}Ar diffused from grain boundaries into the surface layer of this and other hornblendes and that the $^{40}Ar/^{36}Ar$ ratio of this Ar was greater than 1500. The plagioclase in the same sample of granulite yielded a conventional K–Ar date of 1971.0 ± 18.0 Ma, which exceeds the $^{40}Ar^*/^{39}Ar$ plateau date of the hornblende in this rock. Since plagioclase does not retain ^{40}Ar as well as hornblende, the plagioclase also contains excess ^{40}Ar.

A plagioclase in a rock from a nearby location in the same layer of granulite yielded the saddle-shaped pattern in Figure 7.16 in which the $^{40}Ar^*/^{39}Ar$ dates of the low-temperature gas (<550°C, 7.38 percent of ^{39}Ar) reach a value of 5748 ± 23 Ma before declining to 599.8 ± 17 Ma for gas released at 840°C (7.4 percent of total ^{39}Ar). The dates then increase to 5598 ± 18 Ma ($T =$ 1250°C, 2.7 percent of total ^{39}Ar). The minimum

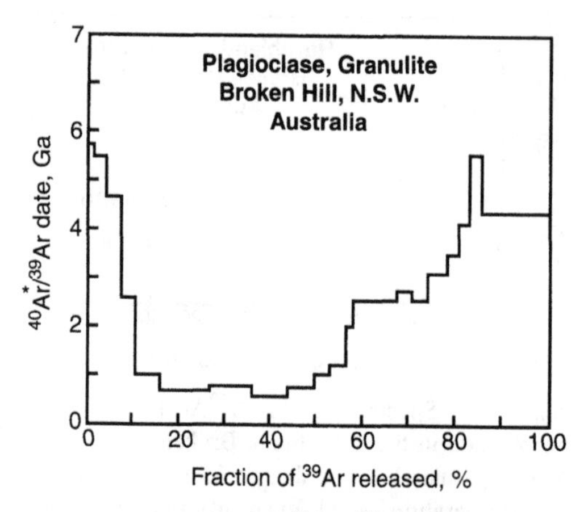

FIGURE 7.16 Spectrum of $^{40}Ar^{*}/^{39}Ar$ dates of plagioclase in granulite at Broken Hill, N.S.W., Australia. The saddle shape of the spectrum is typical of all plagioclase spectra at this locality. The presence of excess ^{40}Ar is indicated by the anomalous dates of the low-temperature gas (11 percent). The dates in the saddle are older than the time of complete outgassing and subsequent introduction of excess ^{40}Ar. The excessive old dates of the high-temperature gas (>50 percent) result from the presence of lamellae formed by Huttenlocher exsolution. Data from Harrison and McDougall (1981).

date of this feldspar is an overestimate of the time when this mineral lost all previously accumulated radiogenic ^{40}Ar and also predates the time when the excess ^{40}Ar entered the plagioclase by diffusion from the grain boundaries.

An important question considered by Harrison and McDougall (1981) is why the dates of high-temperature gas in this and all other feldspars at Broken Hill rise to values exceeding the age of the Earth. The presence of highly retentive accessory minerals, which may explain the saddle-shaped spectrum of the whole-rock sample of diabase in Figure 7.5, does not apply here because the saddle-shaped pattern in Figure 7.16 is for a highly purified sample of plagioclase. Harrison and McDougall (1981) attributed the anomalously old $^{40}Ar^{*}/^{39}Ar$ dates of the Ar released at high temperature to the effects of Huttenlocher exsolution, which causes the formation of alternating anorthite-rich and albite–anorthite lamellae. Each pair of anorthite–albite lamellae is about 50 Å wide and is separated from the neighboring pair by a surface along which rapid diffusion can occur. The existence of lamellae causes loss of ^{39}Ar by recoil, which results in an increase in the $^{40}Ar^{*}/^{39}Ar$ ratio of the gas released at high temperature primarily from the anorthite-rich lamellae. Similar complex spectra of $^{40}Ar^{*}/^{39}Ar$ dates may occur in pyroxene, which also contains exsolution lamellae. The problems of dating feldspars by the $^{40}Ar^{*}/^{39}Ar$ method have been discussed by Harrison (1990), Harrison et al. (1991), and Reddy et al. (1999).

The thermal history of this region revealed by the age determinations includes the following events (Harrison and McDougall, 1981):

1. High-grade regional metamorphism at 1660 Ma and $T \sim 800°C$ (whole-rock Rb–Sr isochron).
2. Cooling to $T \sim 500°C$ at 1575 Ma, implying a cooling rate of about of 3.5°C per 10^6 years. ($^{40}Ar^{*}/^{39}Ar$ plateau date of hornblende).
3. Continued cooling to $T \sim 350°C$ at about 1450 Ma at a rate of 1.2°C per 10^6 years (K–Ar dates, muscovite, Mundi Mundi granitoids).
4. After continued cooling, the temperature rose to 300 or 350°C at 520 Ma (K–Ar and Rb–Sr dates of biotite).
5. Continued cooling to 100°C at 280 Ma (fission-track date of apatite). (See Section 22.2).

7.9 THERMOCHRONOMETRY: HALIBURTON HIGHLANDS, ONTARIO, CANADA

Age determinations of minerals by the conventional K–Ar and $^{40}Ar^{*}/^{39}Ar$ methods may be used to determine the age as well as certain physical parameters relating to the diffusion of Ar. This extension of the K–Ar methods of dating was discussed by Albaréde (1978) and was later developed by Berger and York (1981) and Harrison (1981). Berger and York gave the name "argon isotope thermochronometry" to such studies. The basic objective of thermochronometry is to determine not only dates for different K-bearing minerals in a rock but also the temperatures that existed in the rock at the times recorded by the minerals. In the ideal case, this information may be combined to construct a cooling curve for the rocks and, hence, its thermal history (McDougall and Harrison, 1988).

To accomplish this objective, one must assume that Ar is, in fact, transported by diffusion and is not released during loss of water, phase transformations, or melting of the mineral. To assure that Ar was released by diffusion, Harrison (1981) heated hornblendes under hydrothermal conditions at different temperatures and calculated the diffusion coefficients for Ar from the fraction of the radiogenic ^{40}Ar released as a function of temperature using a formula derived by Fechtig and Kalbitzer (1966). The relationship between the diffusion coefficient D (in cm^2/s) and the temperature T (in kelvin) is given by the Arrhenius equation

$$D = D_0 e^{-E/RT} \tag{7.14}$$

where D_0 is the frequency factor (in cm^2/s), E is the activation energy (in kcal/mol or kJ/mol), and R is the gas constant (1.987 cal/deg/mol or 8.31441 J/mol/K). This equation is transformed into a straight line by taking natural logarithms of both sides:

$$\ln D = \ln D_0 - \frac{E}{RT} \tag{7.15}$$

To make the activation energy E (expressed in kcal/mol) compatible with R (cal/deg/mol), E is multiplied by 10^3:

$$\ln D = \ln D_0 - \frac{10^3 E}{RT} \tag{7.16}$$

Therefore, the numerical values of diffusion coefficients, calculated from the partial-Ar-release data at stepwise-increasing temperatures, generate a series of data points that define a straight line in coordinates of ln D and $10^3/T$. The activation energy E is determined from the slope m of this line by the relationship $E = mR$, whereas the frequency factor $\ln D_0$ is given by its intercept with the ordinate. The Arrhenius equation can also be written in terms of D/a^2, where a is the radius of

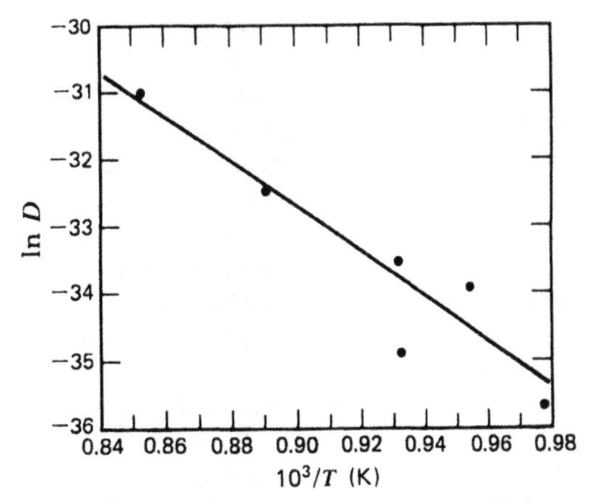

FIGURE 7.17 Arrhenius plot of diffusion coefficients for Ar in hornblende (sample 77-600) at different temperatures between 750 and 900°C. The slope of the straight line yields an activation energy $E = 66.1 \pm 3.9$ kcal/mol and a frequency factor $D_0 = 0.061 + 0.36, -0.10$ cm^2/s. The hornblende was heated under hydrothermal conditions to assure that the Ar was released only by diffusion and not by loss of water or melting of the mineral. Data from Harrison (1981).

the grains from which Ar is diffusing. This procedure for determining D_0 and E is illustrated in Figure 7.17 by diffusion data for hornblende from Harrison (1981).

By combining diffusion coefficients for Ar in hornblende obtained experimentally and geologically, Harrison (1981) obtained an activation energy $E = 64.1 \pm 1.7$ kcal/mol and $D_0 = 0.024 + 0.053, -0.011$ cm^2/s. He used these results to calculate closure temperatures (T_C) between 578 and 490°C for hornblende grains having effective diameters of 80 μm and cooling rates ranging from 500 to 5°C per 10^6 years using Dodson's formula:

$$T_C = \frac{E/R}{\ln[(ART_C^2 D_0/a^2)/(E\, dT/dt)]} \tag{7.17}$$

where A is a geometry factor equal to 55 for spherical grains and 27 for cylindrical grains and dT/dt is the cooling rate at the closure temperature T_C (Dodson, 1973, 1976, 1979). Equation 7.17 can be solved iteratively by substituting an arbitrarily selected value of T_C into the logarithmic term, which is then used to calculate T_C for appropriate values of $A, R, D_0/a^2, E$, and dT/dt. The value of T_C obtained in this way is then back-substituted into the logarithmic term until the calculated value of T_C becomes constant with successive iterations. Harrison's work demonstrated that the closure temperatures of hornblendes increase with increasing grain size and cooling rate but appear to be independent of the chemical composition of the hornblende.

Berger and York (1981) used the results of step heating for dating by the $^{40}Ar^*/^{39}Ar$ method to determine frequency factors and activation energies for the diffusion of radiogenic ^{40}Ar in hornblende, biotite, K-feldspar, and plagioclase of rocks from the Grenville structural province of Canada. They calculated diffusion coefficients from the relative volumes of radiogenic ^{40}Ar released at different temperatures and applied two criteria for selecting reliable data: (1) at least five points must form a "reliable" plateau in the spectrum of $^{40}Ar^*/^{39}Ar$ dates and (2) the same points must form a straight line in the Arrhenius plot in coordinates of $\ln D_0$ and $10^3/T$. The results, summarized in Table 7.4, confirm that the closure temperatures for radiogenic ^{40}Ar of these minerals decrease sequentially from hornblende, biotite, K-feldspar, and plagioclase. However, the numerical values of the closure temperatures depend on the assumption that Ar was, in fact, lost by diffusion and on other factors, including (1) the geometry of the grains, (2) the effective radius of the grains, and (3) the rate of cooling at the closure temperature.

Among the 18 mineral concentrates Berger and York (1981) analyzed in their study, 11 met both of their criteria for data selection. They used these reliable data points to construct a cooling curve for the Haliburton Highlands of the Grenville Province of Ontario, shown in Figure 7.18. The cooling rates implied by this curve decrease with time from about 20°C/10^6 y at 975 Ma to about 2°C/10^6 y at 925 Ma and about 0.3°C/10^6 y at 700 Ma. By using the metamorphic grade of the rocks of the Haliburton Highlands to determine lithostatic pressure, Berger and York (1981) estimated that the depth of burial was about 21 km at about 1000 Ma when the temperature was about 700°C. These results imply a high geothermal gradient for the region of about 33°C/km compared to the present gradient between 15°C/km and 20°C/km. The cooling curve preferred by Berger and York (1981) indicates that the rocks had reached a temperature of about 140°C at 550 Ma. The geothermal gradient given above indicates that the rocks had been uplifted from a depth of 21 km at 1000 Ma to about 4.3 km below the surface at 550 Ma, which implies at rate of uplift of 37 m/10^6 y, or 0.037 mm/y. Subsequently, the uplift rate decreased to a time-integrated value of about 8 m/10^6 y in the period from 550 Ma to the present. These deductions are compatible with the geology of the Grenville Province and may be taken as confirmation of the validity of Ar isotope thermochronometry.

Table 7.4. Summary of Closure Temperatures Calculated from Rate Parameters for Diffusion of Ar in Different Minerals at Cooling Rates of 5°C/10^6 y

Mineral	Closure Temperature, °C	Grain Size, μm	Reference
Hornblende	490	80	Harrison, 1981
Hornblende	685 ± 53^a	210–840	Berger and York, 1981
Biotite	373 ± 21^a	500–1410	Berger and York, 1981
K-feldspar	230 ± 18^b	125–840	Berger and York, 1981
Plagioclase	$176 \pm 54^{a,b}$	125–210	Berger and York, 1981
Microcline	132 ± 13	125–250	Harrison and McDougall, 1982

[a] Average of several determinations judged to be "reliable" by the authors.
[b] Cooling rate assumed to be 0.5°C/10^6 y.

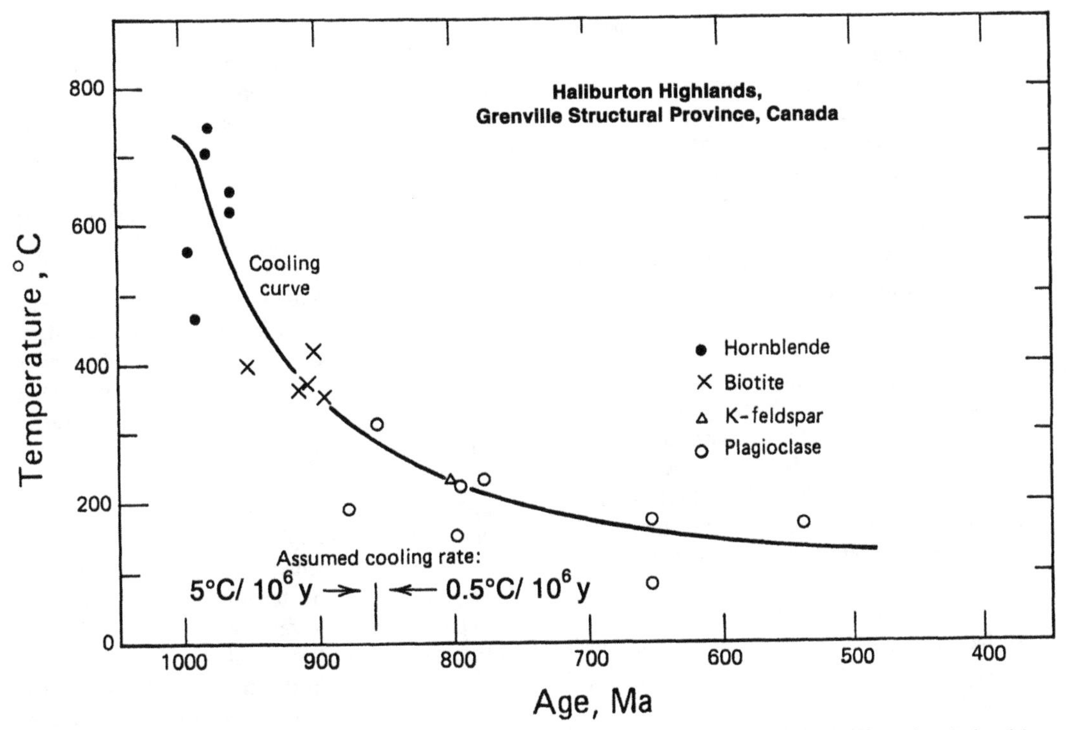

FIGURE 7.18 Cooling curve for the Haliburton Highlands, Grenville structural province, Ontario, derived by determining both a date and a closure temperature during step heating of minerals for dating by the $^{40}Ar^{*}/^{39}Ar$ method. The error bars have been omitted to avoid obscuring the cooling curve. The results imply a high geothermal gradient of about 33°C/km at about 1000 Ma and cooling rates that decreased from about 20°C/10^6 y at 975 Ma to about 2°C/10^6 y at 925 Ma and to about 0.3°C/10^6 y at 700 Ma. By combining the thermal history with estimates of lithostatic pressure at 1000 Ma, time-integrated uplift rates of about 37 m/10^6 y between 1000 and 550 Ma and about 8 m/10^6 y from 550 Ma to the present are indicated. Adapted from Berger and York (1981).

Additional application of this technique have been presented elsewhere:

Harrison and McDougall, 1982, *Geochim. Cosmochim. Acta*, 46:1811–1820. Harrison and Bé, 1983, *Earth Planet. Sci. Lett.*, 64:244–256. Harrison et al., 1986, *J. Geophys. Res.*, 91(B2):1899–1908. Scaillet et al., 1996, *Geochim. Cosmochim. Acta*, 60:4673–4688. Holm and Lux, 1998, *Can. J. Earth Sci.*, 35:1143–1151. Foster et al., 1989, *J. Geol.*, 97(2):232–244.

A summary of studies on cooling histories was published by York (1984).

7.10 SUMMARY

The $^{40}Ar^{*}/^{39}Ar$ method is based on analytical procedures that permit its application not only to extraterrestrial rocks from the asteroids, the Moon, and Mars but also to dating single grains of minerals in terrestrial rocks whose ages approach historical time:

1. Potassium is determined by ^{39}Ar formed by an n, p reaction on ^{39}K in a nuclear reactor.
2. The neutron flux to which the samples are exposed during the irradiation is determined by a monitor consisting of a K-bearing mineral

whose age has been accurately determined by another method.

3. The minerals or whole-rock samples are heated stepwise at increasing temperatures and the $^{40}Ar^*/^{39}Ar$ ratios of the gas released at each step are used to calculate a spectrum of dates.
4. The spectrum of $^{40}Ar^*/^{39}Ar$ dates can be interpreted to detect
   a. loss of $^{40}Ar^*$ from lattice sites that release Ar at low temperature because they have low activation energies;
   b. presence of excess ^{40}Ar in the form of diffusion profiles that extend from the surfaces into the interiors of mineral grains;
   c. presence of a "plateau" consisting of a significant fraction of the released Ar, yielding constant dates within analytical errors; and
   d. presence of two or more mineral components having different Ca/K ratios that release Ar yielding different $^{40}Ar^*/^{39}Ar$ dates.
5. Individual mineral grains in a polished rock surface can be dated by releasing gas from small pits by means of a pulsed laser to obtain multiple total-release dates.
6. Separated grains of minerals and of fine-grained rocks can be heated stepwise by means of a continuous laser whose power can be adjusted.
7. The $^{40}Ar/^{36}Ar$ and $^{39}Ar/^{36}Ar$ ratios of gas that was released at increasing temperatures are linearly correlated and define a straight line whose slope is the $^{40}Ar^*/^{39}Ar$ ratio of the specimen, whereas the intercept is the $^{40}Ar/^{36}Ar$ ratio of atmospheric and/or contaminant Ar.
8. Argon-39 formed during neutron irradiation of glauconite pellets and other minerals containing platy crystallites with a thickness of 1 μm or less may be lost because of recoil. The recoiled ^{39}Ar can be recovered by encapsulating the mineral grains during the irradiation.

The procedures listed above can be used to date not only K minerals (e.g. biotite, muscovite, sanidine) but also minerals having low K concentrations (e.g., hornblende, pyroxene, plagioclase, pyrite, magnetite). If the minerals remained closed to K and Ar after the time of crystallization and after they cooled through the blocking temperature, the plateau dates approach but may underestimate the crystallization age of the minerals. Minerals or rocks containing excess ^{40}Ar, in many cases, have saddle-shaped spectra of $^{40}Ar^*/^{39}Ar$ dates, but the lowest dates in the middle of the saddle, in most cases, overestimate the age of the minerals or rocks.

The $^{40}Ar^*/^{39}Ar$ method has been widely used to date high-grade metamorphic and igneous rocks that contain several K-bearing minerals with different blocking temperatures (e.g., hornblende, muscovite, biotite, K-feldspar). The blocking temperatures of these minerals and their $^{40}Ar^*/^{39}Ar$ dates can be used to reconstruct the cooling histories of the rocks, which reveal not only episodic heating by the intrusion of magma but also uplift and erosion of blocks of continental crust.

The miniaturization of sample size and of the gas-handling equipment has greatly reduced the contamination of samples by atmospheric Ar, which has improved the sensitivity of the equipment and the precision of the measurements. As a result, samples with ages on the order of 10^4 years can now be dated successfully. In addition, the use of computers in the control of the equipment and in the reduction of the data has facilitated the analysis of samples to such an extent that several laboratories are now equipped to operate continuously in automatic mode.

REFERENCES FOR PRINCIPLES AND METHODOLOGY (INTRODUCTION, SECTIONS 7.1–7.2)

Brereton, N. R., 1970. Corrections for interfering isotopes in the $^{40}Ar/^{39}Ar$ dating method. *Earth Planet. Sci. Lett.*, 8:427–433.

Dallmeyer, R. D., 1979. $^{40}Ar/^{39}Ar$ dating: Principles, techniques, and applications in orogenic terranes. In E. Jäger and J. E. Hunziker (Eds.), *Lectures in Isotope Geology*, pp. 77–105. Springer-Verlag, Berlin.

Dalrymple, G. B., 1991. *The Age of the Earth.* Stanford University Press, Stanford, California.

Dalrymple, G. B., and M. A. Lanphere, 1971. $^{40}Ar/^{39}Ar$ technique of K/Ar dating: A comparison with the conventional technique. *Earth Planet. Sci. Lett.*, 12:300–308.

McDougall, I, 1974. The $^{40}Ar/^{39}Ar$ method of K–Ar age determination of rocks using HIFAR reactor. *Atom.*

Energy Austral., Austral. Atom. Energy Comm., 17(3): 3–12.

McDougall, I., and T. M. Harrison, 1988. *Geochronology and Thermochronology by the $^{40}Ar/^{39}Ar$ Method.* Oxford University Press, New York.

Merrihue, C. M., 1965. Trace-element determinations and potassium-argon dating by mass spectroscopy of neutron-irradiated samples. *Trans. Am. Geophys. Union*, 46:125.

Merrihue, C. M., and G. Turner, 1966. Potassium-argon dating by activation with fast neutrons. *J. Geophys. Res.*, 71:2852–2857.

Mitchell, J. G., 1968. The argon-40/argon-39 method for potassium-argon age determination. *Geochim. Cosmochim. Acta*, 32:781–790.

Roddick, J. C., 1983. High precision intercalibration of ^{40}Ar-^{39}Ar standards. *Geochim. Cosmochim. Acta*, 47:887–898.

Tetley, N., I. McDougall, and H. R. Heydegger, 1980. Thermal neutron interferences in the $^{40}Ar/^{39}Ar$ dating technique. *J. Geophys. Res.*, 85(B12): 7201–7205.

Turner, G., 1968. The distribution of potassium and argon in chondrites. In L. H. Ahrens (Ed.), *Origin and Distribution of the Elements*, pp. 387–398. Pergamon, London.

Turner, G., 1969. Thermal histories of meteorites by the ^{39}Ar-^{40}Ar method. In P. M. Millman (Ed.), *Meteorite Research*, pp. 407–417. Reidel Dordrecht.

Turner, G., 1971. Argon-40-argon-39 dating: The optimization of irradiation parameters. *Earth Planet. Sci. Lett.*, 10:227–234.

Wänke, H., and H. König 1959. Eine neue Methode zur Kalium-Argon-Altersbestimmung und ihre Anwendung auf Steinmeteorite. *Z. Naturforsch.*, 14a: 860–866.

REFERENCES FOR MARBLE MOUNTAINS, CALIFORNIA (SECTION 7.2a)

Lanphere, M. A., 1964. Geochronologic studies in the eastern Mojave Desert, California. *J. Geol.*, 72:381.

Lanphere, M. A., and G. B. Dalrymple, 1971. A test of the $^{40}Ar/^{39}Ar$ age spectrum technique on some terrestrial materials. *Earth Planet. Sci. Lett.*, 12:359–372.

Silver, L. T., and C. R. McKinney, 1962. U-Pb isotopic age studies of a Precambrian granite, Marble Mountains, San Bernardino County, California (abstract). *Geol. Soc. Am. Spec. Pap.*, 73:65.

Turner, G., 1968. The distribution of potassium and argon in chondrites. In L. H. Ahrens (Ed.), *Origin and Distribution of the Elements*, pp. 387–398. Pergamon, London.

REFERENCES FOR DIABASE DIKES, LIBERIA, WEST AFRICA (SECTION 7.2b)

Brereton, N. R., 1972. A reappraisal of the $^{40}Ar/^{39}Ar$ stepwise degassing technique. *Geophys. J. Roy. Astronom. Soc.*, 27:449–478.

Dalrymple, G. S., C. S. Gromé, and R. W. White, 1975. Potassium-argon age and paleo-magnetism of diabase dikes in Liberia: Initiation of central Atlantic rifting. *Geol. Soc. Am. Bull.*, 86:399–411.

Dupuy, C., J. Marsh, J. Dostal, A. Michard, and S. Testa, 1988. Asthenospheric and lithospheric sources for Mesozoic dolerites from Liberia (Africa): Trace element and isotopic evidence. *Earth Planet. Sci. Lett.*, 87:100–110.

Hurley, P. M., G. W. Leo, R. W. White, and H. W. Fairbairn, 1971. Liberian age province (about 2700 m.y.) and adjacent age provinces in Liberia and Sierra Leone. *Geol. Soc. Am. Bull.*, 82:3483–3490.

Lanphere, M. A., and G. B. Dalrymple, 1971. A test of the $^{40}Ar/^{39}Ar$ age spectrum technique on some terrestrial materials. *Earth Planet. Sci. Lett.*, 12:359–372.

Lee, J. K. W., T. C. Onstott, and A. J. Crawford, 1990. An $^{40}Ar/^{39}Ar$ investigation of the contact effects of a dyke intrusion, Kapuskasing Structural Zone, Ontario. A comparison of laser microprobe and furnace extraction technique. *Contrib. Mineral. Petrol.*, 105(1):87–105.

Mauche, R., G. Faure, L. M. Jones, and J. Hoefs, 1989. Anomalous isotopic compositions of Sr, Ar, and O in the Mesozoic diabase dikes of Liberia, West Africa. *Contrib. Mineral. Petrol.*, 101:12–18.

Onstott, T. C., and J. Dorbor, 1987. $^{40}Ar/^{39}Ar$ and paleomagnetic results from Liberia and the Precambrian APW data base for the West African shield. *J. Afr. Earth Sci.*, 6:537–552.

Radhakrishna, T., H. Maluski, J. G. Mitchell, and M. Joseph, 1999. $^{40}Ar/^{39}Ar$ and K/Ar geochronology of the dykes from the south Indian granulite terrain. *Tectonophysics*, 304:109–130.

REFERENCES FOR EXCESS ^{40}Ar (SECTION 7.3 INTRODUCTION)

Foland, K. A., 1983. $^{40}Ar/^{39}Ar$ incremental heating plateaus for biotites with excess argon. *Isotope Geosci.*, 1:3–21.

Harrison, T. M., and I. McDougall, 1981. Excess ^{40}Ar in metamorphic rocks from Broken Hill, New South

Wales: Implications for $^{40}Ar/^{39}Ar$ age spectra and the thermal history of the region. *Earth Planet. Sci. Lett.*, 55:123–149.

Lanphere, M. A., and G. B. Dalrymple, 1976. Identification of excess ^{40}Ar by the $^{40}Ar/^{39}Ar$ age spectrum technique. *Earth Planet. Sci. Lett.*, 32:141–148.

Maluski, H., P. Monie, J. R. Kienast, and A. Rahmani, 1990. Location of extraneous argon in granulitic-facies minerals: A paired microprobe-laser probe $^{40}Ar/^{39}Ar$ analysis. *Chem. Geol. (Isotope Geosci.)*, 11(3): 193–218.

Pankhurst, R. J., S. Moorbath, D. G. Rex, and G. Turner, 1973. Mineral age patterns in ca. 3700 m.y. old rocks from West Greenland. *Earth Planet. Sci. Lett.*, 20:157–170.

Roddick, J. C., R. A. Cliff, and D. C. Rex, 1980. The evolution of excess argon in alpine biotites—a $^{40}Ar/^{39}Ar$ analysis. *Earth Planet. Sci. Lett.*, 48:185–208.

REFERENCES FOR ULTRAMAFIC ROCKS, KOLA PENINSULA, RUSSIA (SECTION 7.3a)

Ashkinadze, G. Sh. , B. M. Gorokhovskiy, and Yu. A. Shukolyukov, 1978. $^{40}Ar/^{39}Ar$ dating of biotite containing excess ^{40}Ar. *Geochem. Int.*, 1977:172–176.

Gerling, E. K., 1984. Reminiscences about some works connected with the study of noble gases, their isotopic composition, and geochronology. *Chem. Geol. (Isotope Geosci. Sect.)*, 2:271–289.

Gerling, E. K., Yu. A. Shukolyukov, T. V. Koltsova, I. I. Matveeva, and S. Z. Yakovleva, 1962. The determination of the age of basic rocks by the K-Ar method. *Geokhimiya*, No. 11: 931–938 (in Russian).

Kaneoka, I., 1974. Investigation of excess argon in ultramafic rocks from the Kola Peninsula by the $^{40}Ar/^{39}Ar$ method. *Earth Planet. Sci. Lett.*, 22:145–156.

Kirsten, T., and W. Gentner, 1966. K-Ar Altersbestimmungen an Ultrabasiten des Baltischen Schildes. *Z. Naturforschung*, 21:119.

REFERENCES FOR ANORTHOCLASE, MT. EREBUS, ANTARCTICA (SECTION 7.3b)

Armstrong, R. L., 1978. K-Ar dating: Late Cenozoic McMurdo Volcanic Group and dry valley glacial history, Victoria Land, Antarctica. *New Zealand J. Geol. Geophys.*, 21:685–698.

Esser, R. P., W. C. McIntosh, M. T. Heizler, and P. R. Kyle, 1997. Excess argon in melt inclusions in zero-age anorthoclase feldspar from Mt. Erebus, Antarctica, as revealed by the $^{40}Ar/^{39}Ar$ method. *Geochim. Cosmochim. Acta*, 61:3789–3802.

Faure, G., 2001. *Origin of Igneous Rocks; the Isotopic Evidence*. Springer-Verlag, Heidelberg.

Kyle, P. R. (Ed.), 1994. *Volcanological and Environmental Studies of Mount Erebus, Antarctica*, Antarctic Research Series, Vol. 66, American Geophysical Union, Washington, D.C.

REFERENCES FOR ARGON ISOTOPE CORRELATION DIAGRAM (SECTION 7.4 INTRODUCTION)

Dalrymple, G. B., and M. A. Lanphere, 1974. $^{40}Ar/^{39}Ar$ age spectra of some undisturbed terrestrial samples. *Geochim. Cosmochim. Acta*, 38:715–738.

Dalrymple, G. B., M. A. Lanphere, and M. S. Pringle, 1988. Correlation diagrams in $^{40}Ar/^{39}Ar$ dating: Is there a correct choice? *Geophys. Res. Lett.*, 15:589–596.

Heizler, M. T., and T. M. Harrison, 1988. Multiple trapped argon isotope components revealed by $^{40}Ar/^{39}Ar$ isochron analysis. *Geochim. Cosmochim. Acta*, 52(5): 1295–1318.

Merrihue, C. M., and G. Turner, 1966. Potassium-argon dating by activation with fast neutrons. *J. Geophys. Res.*, 71:2852–2857.

Musset, A. E., and G. B. Dalrymple, 1968. An investigation of the source of air Ar contamination in K-Ar dating. *Earth Planet. Sci. Lett.*, 4:422–426.

Roddick, J. C., 1978. The application of isochron diagrams in ^{40}Ar-^{39}Ar dating: A discussion. *Earth Planet. Sci. Lett.*, 41:233–244.

REFERENCES FOR PORTAGE LAKE VOLCANICS, MICHIGAN (SECTION 7.4a)

Begemann, F., K. R. Ludwig, G. W. Lugmair, K. Min, L. E. Nyquist, P. J. Patchett, P. R. Renne, C.-Y. Shih, I. M. Villa, and R. J. Walker, 2001. Call for an improved set of decay constants for geochronological use. *Geochim. Cosmochim. Acta*, 65:111–121.

Bornhorst, T. J., J. B. Paces, N. K.Grant, J. D.Obradovich, and N. K. Huber, 1988. Age of native copper mineralization, Keweenaw Peninsula, Michigan. *Econ. Geol.*, 83:619–625.

Broderick, T. M., 1956. Copper deposits of the Lake Superior region. *Econ. Geol.*, 51:285–287.

Chaudhuri, S., and G. Faure, 1967. Geochronology of the Keweenawan rocks, White Pine, Michigan. *Econ. Geol.*, 62:1011–1033.

Dalrymple, G. B., and M. A. Lanphere, 1974. $^{40}Ar/^{39}Ar$ age spectra of some undisturbed terrestrial samples. *Geochim. Cosmochim. Acta*, 38:715–738.

Davis, D. W., and J. B. Paces, 1990. Time resolution of geologic events on the Keweenaw Peninsula and implications for development of the Midcontinent Rift System. *Earth Planet. Sci. Lett.*, 97:54–64.

Hutchinson, D. R., R. S. White, W. F. Cannon, and K. J. Schulz, 1990. Keweenaw hot spot: Geophysical evidence for a 1.1 Ga mantle plume beneath the midcontinent Rift System. *J. Geophys. Res.*, 95(B7): 10869–10884.

Jolly, W. T., and R. E. Smith, 1972. Degradation and metamorphic differentiation of the Keweenawan tholeiite lavas of northern Michigan, U.S.A. *J. Petrol.*, 13:273–309.

Nicholson, S. W., W. F. Cannon, and K. J. Schulz, 1992. Metallogeny of the Midcontinent Rift System of North of America. *Precamb. Res.*, 58:355–386.

Nicholson, S. W., and S. B. Shirey, 1990. Midcontinent Rift volcanism in the Lake Superior region: Sr, Nd, and Pb isotopic evidence for a mantle plume origin. *J. Geophys. Res.*, 95(B7): 10851–10868.

Steiger, R. H., and E. Jäger, 1977. Subcommission on Geochronology: Convention on the use of decay constants in geo- and cosmochronology. *Earth Planet. Sci. Lett.*, 36:359–362.

White, W. S., 1960a. The White Pine copper deposit. *Econ. Geol.*, 55:402–409.

White, W. S., 1960b. The Keweenawan lavas of Lake Superior, an example of flood basalts. *Am. J. Sci.*, 258A: 367–374.

York, D., 1969. Least-squares fitting of a straight line with correlated errors. *Earth Planet. Sci. Lett.*, 5:320–324.

REFERENCES FOR LUNAR BASALT, ORANGE GLASS, AND METEORITES (SECTION 7.4b)

Bogard, D. D., L. Husain, and L. E. Nyquist, 1979. ^{40}Ar-^{39}Ar age of the Shergotty achondrite and implications for its post-shock thermal history. *Geochim. Cosmochim. Acta*, 43:1047–1055.

Davis, P. K., 1977. Effects of shock pressure on ^{40}Ar-^{39}Ar radiometric age determinations. *Geochim. Cosmochim. Acta*, 41:195–205.

Husain, L., and O. A. Schaeffer, 1973. Lunar volcanism: Age of the glass in the Apollo 17 orange soil. *Science*, 180:1358–1360.

Layer, P. W., 2000. Argon-40/argon-39 age of the El'gygytgyn impact event, Chukotka, Russia. *Meteor. Planet. Sci.*, 35:591–599.

Podosek, F. A., and J. C. Huneke, 1973. Argon 40–argon 39 chronology of four calcium-rich achondrites. *Geochim. Cosmochim. Acta*, 37:667–684.

Turner, G., S. F. Knott, R. D. Ash, and J. D. Gilmour, 1997. Ar-Ar chronology of the Martian meteorite ALH84001: Evidence for the timing of the early bombardment of Mars. *Geochim. Cosmochim. Acta*, 61:3835–3850.

REFERENCES FOR LASER ABLATION (SECTION 7.5 INTRODUCTION)

Megrue, G. H., 1971. Distribution and origin of helium, neon, and argon isotopes in Apollo 12 samples by in situ analysis with a laser-probe mass spectrometer. *J. Geophys. Res.*, 76:4956–4968.

Schaeffer, O. A., 1982. Laser microprobe argon 39-argon 40 dating of individual grains. In L. A. Currie (Ed.), *Nuclear and Chemical Dating Techniques*, American Chemical Society, Symposium Series 176, pp. 13–148. American Chemical Society, Washington, D.C.

York, D., C. M. Hall, Y. Yanase, J. A. Hanes, and W. J. Kenyon, 1981. $^{40}Ar/^{39}Ar$ dating of terrestrial minerals with a continuous laser. *Geophys. Res. Lett.*, 8:1136–1138.

REFERENCES FOR DATING METEORITE IMPACT CRATERS (SECTION 7.5a)

Bottomley, R. J., D. York, and R. A. F. Grieve, 1978. ^{40}Ar-^{39}Ar ages of Scandinavian impact structures: 1 Mien and Siljan. *Contrib. Mineral. Petrol.*, 68:79–84.

Davis, P. K., 1977. Effects of shock pressure on ^{40}Ar-^{39}Ar radiometric age determinations. *Geochim. Cosmochim. Acta*, 41:195–206.

Jessberger, E. K., Th. Staudacher, B. Domink, T. Kirsten, and O. A. Schaeffer, 1978. Limited response of the K–Ar system to the Nördlinger Ries giant meteorite impact. *Nature*, 271:338–339.

Layer, P. W., 2000. Argon-40/argon-39 age of the El'gygytgyn impact crater, Chukotka, Russia. *Meteor. Planet. Sci.*, 35:591–599.

Layer, P. W., C. M. Hall, and D. York, 1987. The derivation of $^{40}Ar/^{39}Ar$ age spectra of single grains of hornblende and biotite by laser step heating. *Geophys. Res. Lett.*, 14:757–760.

Wolfe, S. H., 1971. Potassium-argon ages of the Manicouagan-Mushalagan lakes structure. *J. Geophys. Res.*, 76:5424–5436.

REFERENCES FOR SANIDINE CRYSTALS (SECTION 7.5b)

Chen, Y., P. E. Smith, N. M. Evensen, D. York, and K. R. Lajoie, 1996. The edge of time: Dating young volcanic ash layers with the ^{40}Ar-^{39}Ar laser probe. *Science*, 274:1176–1178.

Gansecki, C. A., G. A. Mahood, and M. O. McWilliams, 1996. $^{40}Ar/^{39}Ar$ geochronology of rhyolites erupted following collapse of the Yellowstone caldera: Implications for crustal contamination. *Earth Planet. Sci. Lett.*, 142:91–107.

Gansecki, C. A., G. A. Mahood, and M. O. McWilliams, 1998. New ages for the climactic eruptions at Yellowstone: Single crystal $^{40}Ar/^{39}Ar$ dating identifies contamination. *Geology*, 26:343–346.

Lanphere, M. A., D. E. Champion, M. A. Clynne, and L. J. P. Muffler, 1999. Revised age of the Rockland tephra, northern California. Implications for climate and stratigraphic reconstructions in the western United States. *Geology*, 27:135–138.

Lippolt, H. J., U. Fuhrmann, and H. Hradetzky, 1986. $^{40}Ar/^{39}Ar$ age determinations on sanidines of the Eifel volcanic field (Federal Republic of Germany): Constraints on age and duration of a middle Pleistocene cold period. *Chem. Geol.*, 59:187–194.

Lippolt, H. J., J. C. Hess, and K. Burger, 1984. Isotopische Alter von pyroklastischen Sanidinen aus Kaolin-Kohlentonsteinen als Korrelationsmarken für das mitteleuropäische Oberkarbon. *Fortschr. Geol. Rheinld. Westf.*, 32:119–150.

Obradovich, J. D., 1992. *Geochronology of the late Cenozoic Volcanism of Yellowstone National Park and Adjoining Areas, Wyoming and Idaho*, U.S. Geol. Surv. Open File Rept. 92-408. U.S. Geological Survey, Washington, D.C.

Obradovich, J. D., and G. A. Izett, 1991. $^{40}Ar/^{39}Ar$ ages of upper Cenozoic Yellowstone Group tuffs. *Geol. Soc. Am. Abstr.*, 23(2): 84.

Spell, T. L., and T. M. Harrison, 1993. $^{40}Ar/^{39}Ar$ geochronology of post-Valles caldera rhyolites, Jemez volcanic field, New Mexico. *J. Geophys. Res.*, 98:8031–8051.

van den Bogaard, P., 1995. $^{40}Ar/^{39}Ar$ ages of sanidine phenocrysts from Laacher See tephra (12,900 yr BP): Chronostratigraphic and petrologic significance. *Earth Planet. Sci. Lett.*, 133:163–174.

van den Bogaard, P., C. M. Hall, H.-U. Schmincke, and D. York, 1987. $^{40}Ar/^{39}Ar$ laser dating of single grains: Ages of Quaternary tephra from the East Eifel volcanic field. *Geophys. Res. Lett.*, 14:1211–1214.

van den Bogaard, P., C. M. Hall, H.-U. Schmincke, and D. York, 1989. Precise single-grain $^{40}Ar/^{39}Ar$ dating of a cold to warm climate transition in central Europe. *Nature*, 342:523–535.

REFERENCES FOR INTERCALIBRATIONS (SECTION 7.5c)

Baksi, A. K., D. A. Archibald, and E. Farrar, 1996. Intercalibration of $^{40}Ar/^{39}Ar$ dating standards. *Chem. Geol. (Isotope Geosci. Sect.)*, 129:307–324.

Lanphere, M. A. and H. Baadsgaard, 1997. The Fish Canyon Tuff; a standard for geochronology. Abstract, *EOS*, AGU 1997 Spring Meeting, 78: 326.

Lanphere, M. A., and H. Baadsgaard, 1998. The Fish Canyon Tuff—a standard for geochronology. *Program with Abstracts, GAC/MAC*, 23: A102.

Renne, P. R., A. L. Deino, R. C. Walter, B. D. Turrin, C. C. Swisher III, T. A. Becker, G. H. Curtis, W. D. Sharp, and A. R. Jasouni, 1994. Intercalibration of astronomical and radioisotope time. *Geology*, 22:783–786.

Renne, P. R., W. D. Sharp, A. L. Deino, G. Orsi, and L. Civetta, 1997. $^{40}Ar/^{39}Ar$ dating into the historical realm: Calibration against Pliny the Younger. *Science*, 277:1279–1280.

Renne, P. R., D. B. Karner, and K. R. Ludwig, 1998a. Absolute ages aren't exactly. *Science*, 282:1840–1841.

Renne, P. R., C. C. Swisher, A. L. Deino, D. B. Karner, T. L. Owen, and D. J. DePaolo, 1998b. Intercalibration of standards, absolute ages and uncertainties in $^{40}Ar/^{39}Ar$ dating. *Chem. Geol.*, 145:117–152.

Roddick, J. C., 1983. High precision intercalibration of ^{40}Ar-^{39}Ar standards. *Geochim. Cosmochim. Acta*, 47:887–898.

Villeneuve, M., H. A. Sandeman, and W. J. Davis, 2000. A method for intercalibration of U-Th-Pb and ^{40}Ar-^{39}Ar ages in the Phanerozoic. *Geochim. Cosmochim. Acta*, 64:4017–4030.

REFERENCES FOR SEDIMENTARY ROCKS (SECTION 7.6)

Brereton, N. R., R. J. Hooker, and J. A. Miller, 1976. Some conventional potassium-argon and $^{40}Ar/^{39}Ar$ age studies of glauconite. *Geol. Mag.*, 113:329–340.

Dong, H., C. M. Hall, A. N. Halliday, and D. R. Peacor, 1997. Laser ^{40}Ar-^{39}Ar dating of microgram-size illite samples and implications for thin-section dating. *Geochim. Cosmochim. Acta*, 61:3803–3808.

Foland, K. A., F. A. Hubacher, and G. B. Arehart, 1992. ^{40}Ar/^{39}Ar dating of very fine grained samples; an encapsulated-vial procedure to overcome the problem of ^{39}Ar recoil loss. *Chem. Geol.*, 102:269–276.

Foland, K. A., J. S. Linder, T. E. Laskowski, and N. K. Grant, 1984. ^{40}Ar/^{39}Ar dating of glauconites: Measured ^{39}Ar recoil loss from well crystallized specimens. *Isotope Geosci.*, 2:241–264.

Harland, W. B., R. L. Armstrong, A. V. Cox, L. E.Craig, A. G. Smith, and D. G. Smith, 1989. *A geologic Time Scale 1989*. Cambridge University Press, Cambridge, United Kingdom.

Lin, H.-H., T. C. Onstott, and H. Dong, 2000. Backscattered ^{39}Ar loss in fine-grained minerals; implications for ^{40}Ar/^{39}Ar geochronology of clay. *Geochim. Cosmochim. Acta*, 64:3965–3974.

Odin, G. S. (Ed.), 1982. *Numerical Dating in Stratigraphy*, vols. 1 and 2. Wiley, Chichester.

Onstott, T. C., C. Mueller, P. J. Vroluk, and D. R. Pevear, 1997. Laser ^{40}Ar/^{39}Ar microprobe analyses of fine grained illite. *Geochim. Cosmochim. Acta*, 61:3851–3862.

Smith, P. E., N. M. Evensen, and D. York, 1993. First successful ^{40}Ar-^{39}Ar dating of glauconies: Argon recoil in single grains of cryptocrystalline material. *Geology*, 21:41–44.

Smith, P. E., N. M. Evensen, D. York, and S. G. Odin, 1998. Single-grain ^{40}Ar-^{39}Ar ages of glauconies: Implications for the geologic time scale and global sea level variations. *Science*, 279:1517–1518.

Turner, G., and P. H. Cadogan, 1974. Possible effects of ^{39}Ar recoil in ^{40}Ar-^{39}Ar dating. *Geochim. Cosmochim. Acta*, Suppl. 5(2): 1601–1615.

REFERENCES FOR METASEDIMENTARY ROCKS (SECTION 7.7 INTRODUCTION)

Adams, C. J. D., C. T. Harper, and M. G. Laird, 1975. K-Ar ages of low grade metasediments of the Greenland and Waiuta Groups in Westland and Buller, New Zealand. *New Zealand J. Geol. Geophys.*, 18:39–48.

Dodson, M. H., and D. C. Rex, 1971. Potassium-argon ages of slates and phyllites from south-west England. *Q. J. Geol. Soc. Lond.*, 126:465–499.

Scaillet, S., G. Féraud, Y. Lagabrielle, M. Ballevre, and G. Ruffet, 1990. ^{40}Ar/^{39}Ar laser-probe dating by step heating and spot fusion of phengites from the Dora Maira nappe of the western Alps, Italy. *Geology*, 18(8): 741–744.

Wijbrans, J. R., M. Schliestedt, and D. York, 1990. Single grain argon laser probe dating of phengites from the blueschist to greenschist transition of Sifnos (Clyclades, Greece). *Contrib. Mineral. Petrol.*, 104(5): 582–593.

REFERENCES FOR MEGUMA GROUP, NOVA SCOTIA (SECTION 7.7a)

Kontak, D. J., R. J. Horne, H. Sandeman, D. Archibald, and J. K. W. Lee, 1998. ^{40}Ar/^{39}Ar dating of ribbon textured veins and wall-rock material from Meguma lode gold deposits, Nova Scotia: Implications for timing and duration of vein formation in slate-belt hosted vein gold deposits. *Can. J. Earth, Sci.*, 35:746–761.

Reynolds, P. H., E. E. Kublick, and G. K. Muecke, 1973. Potassium-argon dating of slates from the Meguma Group, Nova Scotia. *Can. J. Earth Sci.*, 10:1059–1067.

Reynolds, P. H., and G. K. Muecke, 1978. Age studies on slates: Applicability of the ^{40}Ar/^{39}Ar stepwise outgassing method. *Earth Planet. Sci. Lett.*, 40:111–118.

REFERENCES FOR BARBERTON GREENSTONE BELT, SWAZILAND (SECTION 7.7b)

Alexander, E. C., Jr., 1975. ^{40}Ar-^{39}Ar studies of Precambrian cherts; an unsuccessful attempt to measure the time evolution of the atmospheric ^{40}Ar/^{39}Ar ratio. *Precamb. Res.*, 2:329–344.

de Ronde,C. E. J., C. M. Hall, D. York, and E. T. C. Spooner, 1991. Laser step-heating ^{40}Ar/^{39}Ar age spectra from early Archean (~3.5 Ga) Barberton greenstone belt sediments: A technique for detecting cryptic tectono-thermal events. *Geochim. Cosmochim. Acta*, 55:1933–1953.

Eriksson, K. A., and E. L. Simpson, 2000. Quantifying the oldest tidal record: The 3.2 Ga Moodies Group, Barberton Greenstone belt, South Africa. *Geology*, 28:831–834.

Faure, G., 2001. *Origin of Igneous Rocks; the Isotopic Evidence*. Springer-Verlag, Heidelberg.

Lopez-Martinez, M., D. York, C. M.Hall, and J. A.Hanes, 1984. Oldest reliable ^{40}Ar/^{39}Ar ages for terrestrial rocks: Barberton Mountain komatiites. *Nature*, 307:352–353.

REFERENCES FOR DATING OF LOW-K MINERALS (SECTION 7.7c)

Czamanske, G. K., M. A. Lanphere, R. C. Erd, and M. C. Blake, Jr., 1978. Age measurements of potassium-bearing sulfide minerals by the $^{40}Ar/^{39}Ar$ technique. *Earth Planet. Sci. Lett.*, 40:107–110.

de Ronde, C. E. J., C. M. Hall, D. York, and E. T. C. Spooner, 1991. Laser step-heating $^{40}Ar/^{39}Ar$ age spectra from Early Archean (~3.5 Ga) Barberton greenstone belt sediments: A technique for detecting cryptic tectono-thermal events. *Geochim. Cosmochim. Acta*, 55:1933–1952.

Hanes, J. A., D. York, and C. M. Hall, 1985. An $^{40}Ar/^{39}Ar$ geochronological and electron microprobe investigation of an Archean pyroxenite and its bearing on ancient atmospheric compositions. *Can. J. Earth Sci.*, 22:947–958.

Layer, P. W., C. M. Hall, and D. York, 1987. The derivation of $^{40}Ar/^{39}Ar$ age spectra of single grains of hornblende and biotite by laser step-heating. *Geophys. Res. Lett.*, 14:757–760.

Lopez-Martinez, M., D. York, C. M. Hall, and J. A. Hanes, 1984. Oldest reliable $^{40}Ar/^{39}Ar$ ages for terrestrial rocks: Barberton Mountain komatiites. *Nature*, 307:352–353.

Niemeyer, S., 1979. ^{40}Ar-^{39}Ar dating of inclusions from IAB iron meteorites. *Geochim. Cosmochim. Acta*, 43:1829–1840.

Özdemir, Ö., and D. York, 1990. $^{40}Ar/^{39}Ar$ laser dating of a single grain of magnetite. *Tectonophysics*, 184(1): 21–34.

York, D., A. Masliwec, P.Kuylida, J. A. Hanes, C. M. Hall, W. J. Kenyon, E. T. C. Spooner, and S. D. Scott, 1982. $^{40}Ar/^{39}Ar$ dating of pyrite. *Nature*, 300:52–53.

REFERENCES FOR METAMORPHIC ROCKS: BROKEN HILL, N.S.W., AUSTRALIA (SECTION 7.8)

Binns, R. A., and J. A. Miller, 1963. Potassium-argon age determinations on some rocks from the Broken Hill region of New South Wales. *Nature*, 199:274–275.

Harrison, T. M., 1990. Some observations on the interpretation of feldspar $^{40}Ar/^{39}Ar$ results. *Chem. Geol. (Isotope Geosci.)*, 11(3):219–230.

Harrison, T. M., O. M. Lovera, and M. T. Heizler, 1991. $^{40}Ar/^{39}Ar$ results for alkali feldspars containing diffusion domains with differing activation energy. *Geochim. Cosmochim. Acta.*, 55:1435–1448.

Harrison, T. M., and I. McDougall, 1981. Excess ^{40}Ar in metamorphic rocks from Broken Hill, New South Wales: Implications for $^{40}Ar/^{39}Ar$ age spectra and the thermal history of the region. *Earth Planet. Sci. Lett.*, 55:123–149.

Pidgeon, R. T., 1967. A rubidium-strontium geochronological study of the Willyama Complex, Broken Hill, Australia. *J. Petrol.*, 8:283–324.

Reddy, S. M., G. J. Potts, S. P. Kelley, and N. O. Arnaud, 1999. The effects of deformation-induced microstructures on intragrain $^{40}Ar/^{39}Ar$ ages in potassium feldspar. *Geology*, 27(4):363–366.

Richards, J. R., and R. T. Pidgeon, 1963. Some age measurements on micas from Broken Hill Australia. *J. Geol. Soc. Austral.*, 10:243–260.

Shaw, S. E., 1969. Rb-Sr isotopic studies of the mine sequence rocks at Broken Hill. Monograph. *Australasian Inst. Mining Metallurgy*, 1969:185–198.

Vernon, R. H., 1969. The Willyama Complex. *J. Geol. Soc. Austral.*, 16:20–55.

REFERENCES FOR THERMOCHRONOMETRY: HALIBURTON HIGHLANDS, ONTARIO, CANADA (SECTION 7.9)

Albarède, F., 1978. The recovery of spatial isotope distributions from stepwise degassing data. *Earth Planet. Sci. Lett.*, 39:387–397.

Berger, G. W., and D. York, 1981. Geothermometry from $^{40}Ar/^{39}Ar$ dating experiments. *Geochim. Cosmochim. Acta*, 45:795–811.

Dodson, M. H., 1973. Closure temperature in cooling geochronological and petrological systems. *Contrib. Mineral. Petrol.*, 40:259–274.

Dodson, M. H., 1976. Kinetic processes and thermal history of slowly cooling solids. *Nature*, 259:551–553.

Dodson, M. H., 1979. Theory of cooling ages. In E. Jäger and J. C. Hunziker (Eds.), *Lectures in Isotope Geology*, pp. 194–202. Springer-Verlag, Berlin.

Fechtig, H., and S. Kalbitzer, 1966. The diffusion of argon in potassium-bearing solids. In O. A. Schaeffer and J. Zähringer (Eds.), *Potassium-Argon Dating*, pp. 68–107. Springer-Verlag, New York.

Harrison, T. M., 1981. Diffusion of ^{40}Ar in hornblende. *Contrib. Mineral. Petrol.*, 78:324–331.

Harrison, T. M., and I. McDougall, 1982. The thermal significance of potassium feldspar K-Ar ages inferred from ^{40}Ar/^{39}Ar age spectrum results. *Geochim. Cosmochim. Acta*, 46:1811–1820.

McDougall, I., and T. M. Harrison, 1988. *Geochronology and Thermochronology by the ^{40}Ar/^{39}Ar Method*. Oxford University Press, New York.

York, D. 1984. Cooling histories from ^{40}Ar/^{39}Ar age spectra: Implications for Precambrian plate tectonics. *Annu. Rev. Earth Planet. Sci.*, 12:383–409.

CHAPTER 8

The K–Ca Method

THE decay of ^{40}K to ^{40}Ar by electron capture and positron emission depicted in Figure 2.7 has become the basis for the conventional K–Ar and ^{40}Ar*/^{39}Ar methods of dating (Chapters 6 and 7). The utility of the geochronometer based on the complementary decay of ^{40}K to ^{40}Ca by negatron emission is limited by the high abundance of ^{40}Ca (it is doubly magic) and by the fractionation of Ca isotopes during analysis in the mass spectrometer.

Calcium (atomic number 20) has six stable isotopes whose abundances are ^{40}Ca = 96.9821%, ^{42}Ca = 0.6421%, ^{43}Ca = 0.1334%, ^{44}Ca = 2.0567%, ^{46}Ca = 0.0031%, and ^{48}Ca = 0.1824% (Nier, 1938). Its atomic weight is 40.076 (Russell et al., 1978). The possibility of fractionation of Ca isotopes is enhanced by the considerable difference in their masses, which amounts to about 20 percent for ^{48}Ca relative to ^{40}Ca.

The need for improvement in the decay constants of ^{40}K, discussed in Section 7.5c, prompted Nägler and Villa (2000) to attempt a redetermination of the branching ratio of ^{40}K. The method they chose was to date gem-quality muscovite and sanidine by the K–Ca and ^{40}Ar/^{39}Ar methods. The results indicated that 10.67 percent of ^{40}K atoms (by number) decay to ^{40}Ar and 89.33 percent decay to ^{40}Ca, yielding a branching ratio (BR) of

$$\text{BR} = \frac{10.67}{89.33} = 0.1194$$

This value is about 2.0 percent higher than the branching ratio (0.1170) recommended by Steiger and Jäger (1977). However, the value of the branching ratio determined by Nägler and Villa (2000) is similar to 0.1187 reported by Endt and Van der Leun (1973) and endorsed by Min et al. (2000) based on their comparison of ^{40}Ar*/^{39}Ar dates of alkali feldspars and U–Pb dates of zircon in the Palisade rhyolite of Minnesota.

8.1 PRINCIPLES AND METHODOLOGY

The total decay constant of ^{40}K (λ) is the sum of λ_e and λ_β:

$$\lambda = \lambda_e + \lambda_\beta \tag{8.1}$$

where $\lambda_e = 0.581 \times 10^{-10}\ \text{y}^{-1}$ (decay to ^{40}Ar)
$\lambda_\beta = 4.962 \times 10^{-10}\ \text{y}^{-1}$ (decay to ^{40}Ca)

Hence, $\lambda = 5.543 \times 10^{-10}\ \text{y}^{-1}$ (Steiger and Jäger, 1977). Consequently, the ^{40}Ca/^{42}Ca ratio of a K-bearing mineral or rock is expressed by the equation

$$\frac{^{40}\text{Ca}}{^{42}\text{Ca}} = \left(\frac{^{40}\text{Ca}}{^{42}\text{Ca}}\right)_i + \left(\frac{\lambda_\beta}{\lambda}\right)\frac{^{40}\text{K}}{^{42}\text{Ca}}(e^{\lambda t} - 1) \tag{8.2}$$

where $\left(\frac{^{40}\text{Ca}}{^{42}\text{Ca}}\right)$ = ratio of these isotopes at the present time

$\left(\frac{^{40}Ca}{^{42}Ca}\right)_i$ = initial ratio of these isotopes at the time of formation or after the last isotopic homogenization of the system

$\frac{^{40}K}{^{42}Ca}$ = ratio of these isotopes in the sample at the present time

t = time elapsed since formation or last isotopic homogenization

The decay constants recommended by Steiger and Jäger (1977) have been widely used, which has helped to standardize the calculation of isotopic dates based on the decay of ^{40}K. However, the universal acceptance of these decay constants does not guarantee their accuracy. The improvement in the precision of K–Ar, $^{40}Ar^*/^{39}Ar$, and K–Ca dates now requires a re-evaluation of the accuracy of the decay constants of ^{40}K (Min et al., 2000; Nägler and Villa, 2000; Begemann et al., 2001).

The K–Ca geochronometer can be used very much like the Rb–Sr geochronometer (Chapter 5). Cogenetic rocks or minerals having the same initial $^{40}Ca/^{42}Ca$ ratio and the same age lie on an isochron in coordinates of $^{40}Ca/^{42}Ca$ and $^{40}K/^{42}Ca$ whose slope is

$$m = \left(\frac{\lambda_\beta}{\lambda}\right)(e^{\lambda t} - 1) \quad (8.3)$$

and whose intercept is equal to the initial $^{40}Ca/^{42}Ca$ ratio.

Analytical problems associated with the measurement of the isotopic composition of Ca have inhibited the use of the K–Ca geochronometer for many years (Ahrens, 1950). Polevaya et al. (1958) attempted to date K-bearing evaporite minerals of Paleozoic age by this method and were able to demonstrate that the K–Ca dates agreed better with their depositional ages than dates based on the decay of ^{40}K to ^{40}Ar. Later, Coleman (1971) used the K–Ca method to date a suite of micas from pegmatites in the Scottish Highlands. However, his results are only in moderately good agreement with their Rb–Sr dates. Subsequent research by Wilhelm and Ackermann (1972) and Heumann et al. (1979) concentrated on dating K-bearing minerals in sedimentary rocks such as sylvite, langbeinite, and K-feldspar. More recently, the K–Ca method of dating K-rich minerals and rocks has experienced a modest revival justified by the use of a double spike that can eliminate the effects of Ca isotope fractionation (Eugster et al., 1969), by the reduction of laboratory contamination, by improvements in the design of mass spectrometers (e.g., use of multiple collectors), and by the availability of high-speed data acquisition and processing. For example, Marshall et al. (1986) used the K–Ca method to date authigenic sanidine at the base of the St. Peter Sandstone (Middle Ordovician) of Wisconsin, and Baadsgaard (1987) analyzed K-bearing marine evaporite minerals (carnallite and sylvite) of the Prairie Formation (~380 Ma, Middle Devonian) in Saskatchewan, Canada. His results indicated that the minerals had recrystallized sporadically between 86 and 2.5 Ma (Late Cretaceous to Pliocene).

Another noteworthy application of the K–Ca decay scheme to dating muscovite and biotite of Archean age in the Yilgarn block of Western Australia was reported by Fletcher et al. (1997). Their results indicated that the K–Ca dates of muscovite rosettes are more precise but consistently lower than Rb–Sr dates of the same material. The authors attributed the difference in the dates to the greater mobility of $^{40}Ca^*$ compared to $^{87}Sr^*$ and concluded that K–Ca dates are not reliable indicators of the time of crystallization of Archean muscovites.

8.1a Pikes Peak Granite, Colorado

The improvements made in the mass spectrometric analysis of Ca by Russell et al. (1978) were used by Marshall and DePaolo (1982) to date granitic rocks and their minerals by the K–Ca method. The Pikes Peak batholith of Colorado (Barker et al., 1975) was intruded into Proterozoic gneisses (1.8 Ga) as well as into grandodiorites (1.70–1.65 Ga) and granites (1.4 Ga) (Peterman and Hedge, 1968). The analytical data derived by Marshall and DePaolo (1982) for a whole-rock sample of granite and its constituent minerals (plagioclase, K-feldspar, and biotite) of the Pikes Peak batholith define an isochron in Figure 8.1 in coordinates of $^{40}Ca/^{42}Ca$ and $^{40}K/^{42}Ca$. The slope m of this isochron (0.6990), determined

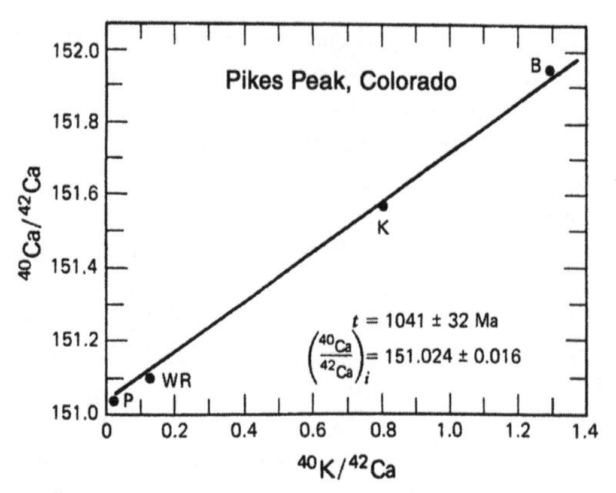

FIGURE 8.1 The K–Ca isochron for minerals and one whole-rock sample of granite from the Pikes Peak batholith of Colorado, United States. The $^{40}Ca/^{42}Ca$ ratios were corrected for isotope fractionation to 0.31221 for the $^{42}Ca/^{44}Ca$ ratio. Data from Marshall and DePaolo (1982).

by Marshall and DePaolo (1982) by the method of York (1969), can be used to calculate the K–Ca date from equation 8.3:

$$m = \frac{\lambda_\beta}{\lambda}(e^{\lambda t} - 1) \qquad t = \frac{1}{\lambda}\ln\left[\left(\frac{\lambda}{\lambda_\beta}m\right) + 1\right]$$

Since $\lambda_\beta = 4.962 \times 10^{-10}\ y^{-1}$ and $\lambda = 5.543 \times 10^{-10}\ y^{-1}$ (Steiger and Jäger, 1977):

$$t = \frac{1}{5.543 \times 10^{-10}}\ln\left[\left(\frac{5.543}{4.962} \times 0.6990\right) + 1\right]$$

$$= 1041 \times 10^6 \text{ y or } 1041 \text{ Ma}$$

The authors reported a date of 1041 ± 32 Ma and an initial $^{40}Ca/^{42}Ca$ ratio of 151.024 ± 0.016. The date is in good agreement with the whole-rock Rb–Sr isochron date of 1008 ± 13 Ma ($\lambda = 1.42 \times 10^{-11}\ y^{-1}$) reported by Barker et al. (1976) and Hedge (1970).

The initial $^{40}Ca/^{42}Ca$ ratio of the Pikes Peak batholith (151.024 ± 0.016) is indistinguishable from the initial $^{40}Ca/^{42}Ca$ ratios of the stony meteorites Murchison (carbonaceous chondrite) and Norton County (achondrite) and of three mantle-derived rocks and minerals analyzed by Marshall and DePaolo (1982). The initial $^{40}Ca/^{42}Ca$ of all these samples overlap within analytical errors and have a mean value of 151.016 ± 0.11. Evidently, the K/Ca ratio of the mantle of the Earth is so low that it has preserved the primordial isotopic composition of Ca.

The time-dependent increase of the $^{40}Ca/^{42}Ca$ ratios of the minerals and whole-rock samples of the Pikes Peak batholith in Figure 8.2 is visibly nonlinear because the comparatively short halflife of ^{40}K decaying to ^{40}Ca (1.396×10^9 years) invalidates the approximation that $e^{\lambda t} - 1 = \lambda t$, which works well for the Rb–Sr decay scheme ($T_{1/2} = 48.8 \times 10^9$ y).

8.1b Lunar Granite

The K–Ca method was also used by Shih et al. (1994) to date a small fragment of a lunar granite (#12033, 576) by a procedure described by Shih et al. (1993). Granitic rocks on the Moon formed only by extreme fractional crystallization of basalt magma and therefore are very rare. Only a few small fragments of granite have been recovered (Ryder, 1976). The specimen analyzed by Shih et al. (1994) in Figure 8.3 yielded a well-defined K–Ca mineral isochron date of 3.62 ± 0.11 Ga and an initial $^{40}Ca/^{44}Ca$ ratio of 47.160 ± 0.006.

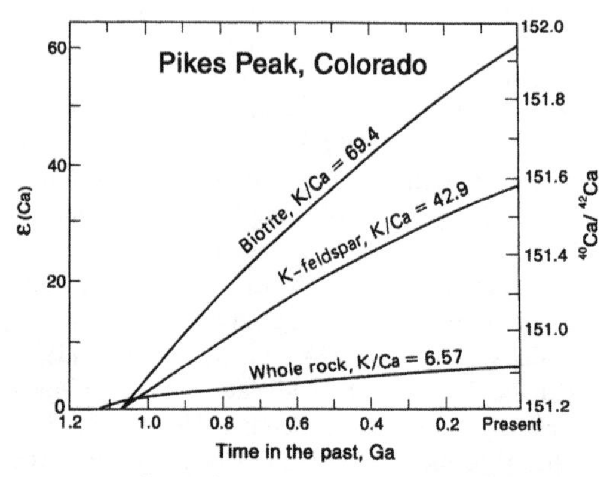

FIGURE 8.2 Isotopic evolution of Ca as a function of time in minerals and a whole-rock sample of granite from the Pikes Peak batholith of Colorado. Note the curvature of the evolution lines. Data from Marshall and DePaolo (1982).

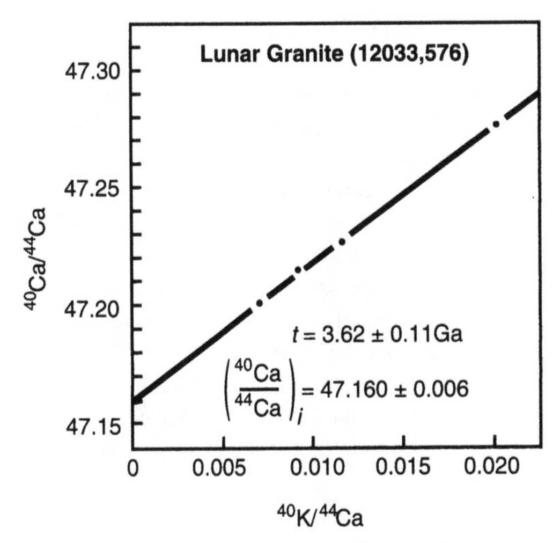

FIGURE 8.3 The K–Ca mineral–whole rock isochron of a small specimen of lunar granite (12033, 576) collected from "soil" at the north rim of Head Crater, Ocean of Storms, Apollo 12. Data from Shih et al. (1994).

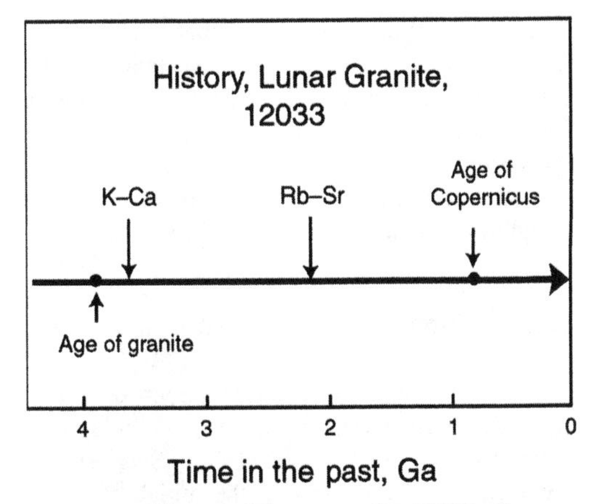

FIGURE 8.4 History of lunar granite 12033, Ocean of Storms, Apollo 12. The K–Ca and Rb–Sr dates of this granite were altered by diffusion of ^{40}Ca and ^{87}Sr following the formation of the crater Copernicus by impact of a meteorite. Data from Bogard et al. (1994) and Shih et al. (1994).

Although the isotope composition of Ca was expressed as the $^{40}Ca/^{44}Ca$ ratio rather than as the $^{40}Ca/^{42}Ca$ ratio as in equation 8.2, the fractionation correction was based on $^{40}Ca/^{42}Ca = 0.31221$, in agreement with Marshall and DePaolo (1982). In addition, Shih et al. (1994) reported $^{40}Ca/^{44}Ca = 47.134 \pm 0.004$ for the JSC (Johnson Space Center) Ca standard and $^{40}K/^{41}K = 0.001735 \pm 0.000003$ for the NBS 985 K standard.

The Rb–Sr system in this sample of lunar granite shows evidence of alteration and yielded a poorly defined isochron date of 2.15 ± 0.65 Ga (recalculated to $\lambda = 1.42 \times 10^{-11}$ y^{-1}) with an initial ratio of 0.775 ± 0.022, which is unusually high for lunar rocks. Uranium–Pb dates of microscopic zircon crystals in granite 12033 reported by Meyer et al. (1989) indicated that they crystallized at 3.898 ± 0.010 Ga and were subsequently reheated at 760 ± 200 Ma. This date was confirmed by a $^{40}Ar^*/^{39}Ar$ date of 796 ± 15 Ma by Bogard et al. (1994), who interpreted it as the age of the lunar impact crater Copernicus. Therefore, the evidence compiled in Figure 8.4 indicates that specimen 12033 crystallized at about 3.40 Ga and was reheated at about 800 Ma following the formation of the crater Copernicus.

A second clast of a lunar granite (14303, 204; Fra Mauro, Apollo 14) analyzed by Shih et al (1994) has a better defined Rb–Sr isochron date of 3.87 ± 0.38 Ga ($\lambda = 1.42 \times 10^{-11}$ y^{-1}) and an initial ratio of 0.7046 ± 0.0051. In addition, the Rb–Sr isochron date of this specimen is not distinguishable from its K–Ca isochron date of 4.04 ± 0.64 Ga.

8.2 ISOTOPE GEOCHEMISTRY OF CALCIUM

The isotope composition of Ca in K-bearing rocks and minerals changes as a function of time because of the accumulation of radiogenic ^{40}Ca, which causes the abundances of the other Ca isotopes to decrease because of the closed-table effect (i.e., the abundances must sum to 100). However, the addition of radiogenic ^{40}Ca does not change the abundance ratios among the other Ca isotopes. In addition, mass-dependent fractionation may alter the isotope composition of Ca in certain kinds of terrestrial and extraterrestrial processes (e.g., precipitation of biogenic Ca carbonate and evaporation/condensation of Ca vapor in a vacuum). A third cause for variations of the isotopic composition of Ca in meteorites and interplanetary dust particles is the nuclear reactions that occurred in the ancestral stars and during the subsequent formation of solid particles in the solar nebula.

8.2a Radiogenic ^{40}Ca in Terrestrial Rocks

The use of the isotope compositions of Ca to study the petrogenesis of igneous rocks was originally proposed by Holmes (1932). Such studies are facilitated by comparing the initial $^{40}Ca/^{42}Ca$ ratio of igneous rocks to the $^{40}Ca/^{42}Ca$ ratios of the *chondritic uniform reservoir* (CHUR) by means of the so-called epsilon-parameter (ε-parameter Section 9.2), which represents the isotope composition of Ca and other elements in the magma sources of the mantle of the Earth. The initial $^{40}Ca/^{42}Ca$ ratios of stony meteorites and rocks that originated from the mantle of the Earth range only between narrow limits (Backus et al., 1964; Barnes et al., 1972) and have an average value of 151.016 ± 0.011 (Marshall and DePaolo, 1982). Consequently, it is convenient to express $^{40}Ca/^{42}Ca$ ratios of terrestrial rocks in terms of the ε-Ca parameter:

$$\varepsilon(\text{Ca}) = \left[\frac{(^{40}\text{Ca}/^{42}\text{Ca})_{\text{spl}}}{(^{40}\text{Ca}/^{42}\text{Ca})_{\text{mantle}}} - 1\right] \times 10^4 \quad (8.4)$$

In this way, the ε(Ca) parameter indicates the abundance of radiogenic ^{40}Ca in any sample of terrestrial (or meteoritic) Ca. For example, the whole-rock sample of the Pikes Peak granite has a present $^{40}Ca/^{42}Ca$ ratio of 151.109, which means that its ε(Ca) value is

$$\varepsilon(\text{Ca}) = \left(\frac{151.109}{151.016} - 1\right) \times 10^4 = 6.2$$

Additional advantages of the ε-notation are that the values are positive except under very unusual circumstances and interlaboratory discrepancies are diminished provided each research group uses its own average $^{40}Ca/^{42}Ca$ ratio for Ca in meteorites and mantle-derived rocks to calculate ε-values.

For example, Nelson and McCulloch (1989) reported a value of 151.078 for the $^{40}Ca/^{42}Ca$ ratio of a midocean ridge basalt (MORB) in the FAMOUS (French American Undersea Study) area of the Mid-Atlantic Ridge and used it to calculate ε-values for a variety of terrestrial rocks. Therefore, the ε(Ca) values of Marshall and DePaolo (1982, 1989) and Nelson and McCulloch (1989) in Figure 8.5 express the enrichment in radiogenic ^{40}Ca of the terrestrial samples they analyzed.

The average ε(Ca) values of volcanic rocks of Tertiary to Recent age analyzed by Marshall and DePaolo (1989) and Nelson and McCulloch (1989) reflect the differences in their magma sources and petrogenesis:

Midocean ridge basalt (3) -0.1 ± 0.5
Ocean island basalt (1) $+1.6 \pm 0.8$
Island-arc volcanic (8) $+1.5 \pm 0.6$
Shale, Timor Trough (1) $+2.3 \pm 0.9$

The oceanic island basalt (Hawaii) and the island-arc volcanics (Guam, New Britain, Aleutian, and Banda Islands) contain more radiogenic ^{40}Ca than MORB, in agreement with $^{87}Sr/^{86}Sr$ ratios of volcanic rocks in these tectonic settings (Faure, 2001). The shale in the Timor Trough is enriched in radiogenic ^{40}Ca because it contains mineral grains eroded from the Precambrian rocks of northwestern Australia.

The ε(Ca) values of 11 carbonatites from seven countries appear to be independent of their ages (2040–17 Ma) and have a mean of $+1.3 \pm 0.8$ (Marshall and DePaolo, 1982; Nelson and McCulloch, 1989). Similarly, the ε(Ca) values of 9 ultrapotassic rocks in Australia, Antarctica, and southeastern Spain (0–20 Ma) have a mean of $+0.5 \pm 0.5$ (Nelson and McCulloch, 1989). Both rock types originated from sources in the mantle and contain Ca whose isotope composition is similar to that of oceanic island basalt and volcanic rocks erupted in island arcs. However, kimberlites (587 Ma) and lamproites (1227 Ma) of Precambrian age in central West Greenland have ε(Ca) values of $+2.7 \pm 0.1$ and $+4.8 \pm 0.5$, respectively, which indicates enrichment in radiogenic ^{40}Ca (Nelson and McCulloch, 1989).

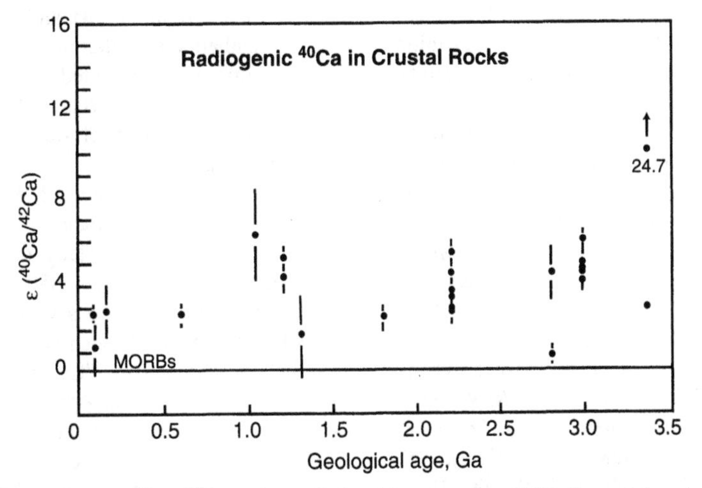

FIGURE 8.5 Epsilon $^{40}Ca/^{42}Ca$ values (defined by equation 8.4) of crustal rocks at the present time. The ε(Ca) values of granites from Marshall and DePaolo (1989) were plotted versus the age of the Precambrian basement rocks. Volcanic rocks of Tertiary to Recent age and carbonatites were excluded. The data demonstrate the presence of radiogenic ^{40}Ca in the rocks of the continental crust. The ε-values of Marshall and DePaolo (1982, 1989) are relative to $^{40}Ca/^{42}Ca = 151.016$, whereas Nelson and McCulloch (1989) used 151.078. The ε(Ca) values of MORBs reported by both research groups average -0.1 ± 0.5 and are indistinguishable from zero. Data from Marshall and DePaolo (1982, 1989) and Nelson and McCulloch (1989).

The same authors also demonstrated that gypsum in lacustrine evaporites on the Precambrian Shield of Western Australia is enriched in radiogenic ^{40}Ca derived from the underlying basement rocks:

Gypsum, Proterozoic basement (1) 2.9 ± 0.6
Gypsum, Archean basement (4) 4.9 ± 0.8

In fact, the ε(Ca) values of Archean metamorphic rocks in the Mt. Narryer region of Western Australia range from 2.9 ± 0.1 to 24.7 ± 1.6 depending on their K/Ca ratios. The enrichment of these rocks in radiogenic ^{40}Ca is exceeded only by the K-rich minerals of the Pikes Peak granite for which Marshall and DePaolo (1982) reported ε(Ca) values of 37.1 ± 1.5 (K-feldspar) and 61.3 ± 2.9 (biotite).

Epsilon-parameters can be defined for any desired isotope ratio of Ca. For example, Fletcher et al. (1997) used the $^{40}Ca/^{44}Ca$ ratio and calculated ε(Ca) values relative to the $^{40}Ca/^{44}Ca$ ratio of calcium carbonate in the shell of the giant clam *Tridacna gigas*. The average $^{40}Ca/^{42}Ca$ ratio of this material reported by Nelson and McCulloch (1989) (151.079 ± 0.028) is indistinguishable from the $^{40}Ca/^{42}Ca$ ratio of Ca in MORBs (151.078 ± 0.008). Since Nelson and McCulloch (1989) normalized the measured isotope ratios of Ca to $^{42}Ca/^{44}Ca$ 0.31221, the average $^{40}Ca/^{44}Ca$ ratio of the *Tridacna* standard measured by Nelson and McCulloch (1987) is

$$\left(\frac{^{40}\mathrm{Ca}}{^{44}\mathrm{Ca}}\right) = \left(\frac{^{42}\mathrm{Ca}}{^{44}\mathrm{Ca}}\right) \times \left(\frac{^{40}\mathrm{Ca}}{^{42}\mathrm{Ca}}\right)$$
$$= 0.31221 \times 151.079 = 47.168$$

8.2b Mass-Dependent Isotope Fractionation

Fractionation of Ca isotopes in nature was predicted because of the large mass differences among its isotopes. However, attempts to detect differences in the isotope ratios of Ca in natural samples initially produced contradictory and inconclusive results:

> Pinson et al., 1957, *Geol. Soc. Am. Bull.*, 68:1781–1782. Corless and Winchester, 1964, *Pure Appl. Chem.*, 8:317–323. Hirt and Epstein, 1964a, *Helv. Phys. Acta*, 37:179. Hirt and Epstein, 1964b, *Trans. Am. Geophys. Union*, 45:113. Artemov et al., 1966, *Geochem. (USSR)*, 3:1082–1086. Corless, 1966, *Anal. Chem.*, 38:810–813. Miller et al., 1966, *Geochem. Int.*, 3:929–933. Meshcheryakov and Stolbov, 1967, *Geochem. (USSR)*, 4:1001–1003. Letolle, 1968, *Earth Planet. Sci. Lett.*, 5:207–208. Stahl, 1968, *Earth Planet. Sci. Lett.*, 5:171–174. Stahl and Wendt, 1968, *Earth Planet. Sci. Lett.*, 5:184–186. Möller and Papendorf, 1971, *Earth Planet. Sci. Lett.*, 11:192–194. Heumann and Lieser, 1972, *Z. Naturforsch.*, 276:126–133. Heumann and Lieser, 1973, *Geochim. Cosmochim. Acta*, 37:1463–1471. Heumann and Luecke, 1973, *Earth Planet. Sci. Lett.*, 20:341–346.

Natural fractionation of Ca isotopes is potentially important because the variation of isotope ratios of Ca may convey information about the processes that have acted on a particular sample. Interpretations of this kind are the basis of the isotope geology of certain elements with low atomic numbers, such as H, C, O, N, and S (Chapters 26, 27, 28, 29). The analytical techniques that have been used to detect isotope fractionation of Ca have included neutron activation (Corless, 1968), but mass spectrometry is clearly the best method for this purpose because it is direct and highly sensitive. However, mass spectrometry itself causes isotope fractionation in the element being analyzed and thereby obscures any natural isotope fractionation that may exist in the sample. This problem can be controlled by standardizing operating conditions (Heumann and Lieser, 1973) but must be *eliminated* if isotope fractionation of Ca is to be detected and used to solve problems in the earth sciences. The only way to eliminate this problem is the double-spike isotope dilution technique that was used first by Hirt and Epstein (1964a, b) and later by Russell et al. (1978), Lee et al. (1978), and Schmitt et al. (2001). The mathematical relationships of the double-spike technique were presented by Eugster et al. (1969) in the context of a search for variations in the isotope composition of Ba in meteorites and terrestrial samples.

The definitive study of W. A. Russell uncovered some startling facts about isotope fractionation of Ca in industrially purified reagents. For example, Russell et al. (1980) reported that during evaporation of CaF_2 in a vacuum at 1200°C, ^{40}Ca is lost preferentially such that the first Ca to be distilled is enriched in ^{40}Ca, whereas later fractions are depleted in ^{40}Ca. Another example of isotope fractionation turned up in two different lots of spectrographically pure $CaCO_3$ studied by Russell et al. (1978). One lot of $CaCO_3$ had been prepared from Ca metal that had been purified by a vacuum evaporation–distillation process. The second lot was obtained from Derbyshire chalk in England and was not subjected to purification by vacuum distillation. The sample of Ca metal purified by vacuum distillation was found to be enriched in ^{40}Ca, thus confirming that vacuum distillation causes fractionation of Ca isotopes.

Another important discovery reported by Russell and Papanastassiou (1978) is that Ca isotopes are fractionated by ion exchange used to separate Ca from other elements before analysis on the mass spectrometer. The effect is caused by the preferential retention of the lighter isotopes on the column. As a result, the first Ca eluted from the column is deficient in ^{40}Ca compared with ^{44}Ca. The problem can be avoided either by recovering all of the Ca placed on the ion exchange column or by adding the double spike before the Ca-bearing solutions are passed through the column.

To make use of the full precision with which isotope ratios can be measured on modern mass spectrometers, Russell et al. (1978) formulated an "exponential" fractionation law in which the correction per mass unit decreases with increasing mass difference. This empirical relationship was shown to fit experimental data better than a linear mass fractionation equation or one based on Rayleigh distillation. This conclusion was later confirmed by Hart and Zindler (1989).

These technical improvements enabled Russell et al. (1978) to compare isotopic compositions of

Ca in a wide range of terrestrial and extraterrestrial samples. Their results in Figure 8.5 indicate that the $^{40}Ca/^{42}Ca$ ratios vary only within quite narrow limits. These results established that mass-dependent fractionation of Ca isotopes in nature is less than expected.

More recently, Richter et al. (1999) studied the diffusion of Ca isotopes in molten $CaO–SiO_2–Al_2O_3$ at 1500°C and 1.0 gigapascals (GPa) and concluded that the mass-dependent diffusion coefficients (D) of two isotopes (1 and 2) are related by the equation

$$\frac{D_1}{D_2} = \left(\frac{m_2}{m_1}\right)^{\beta} \qquad \text{where } \beta = 0.05, \ldots, 0.10 \tag{8.5}$$

The authors predicted that diffusion of Ca isotopes between silicate melts whose Ca concentrations differ by factors of 5–10 can cause variations of the $^{44}Ca/^{40}Ca$ ratio of 2–5‰. A difference of 5‰ in the $^{44}Ca/^{40}Ca$ ratios has not yet been detected in naturally occurring volcanic rocks, although it is certainly resolvable with current analytical techniques. Richter et al. (1999) suggested that the predicted fractionation of Ca isotopes may be restricted to the boundary layers separating melts having large differences in their Ca concentrations, such as magmas of basaltic and rhyolitic composition.

A study of mass-dependent isotope fractionation of Ca by Skulan et al. (1997) demonstrated that the $\delta^{44}Ca$ values of marine and nonmarine samples range from +0.49 to −2.88‰, where $\delta^{44}Ca$ is defined as

$$\delta^{44}\text{Ca} = \left[\frac{(^{44}\text{Ca}/^{40}\text{Ca})_{\text{spl}}}{(^{44}\text{Ca}/^{40}\text{Ca})_{\text{std}}} - 1\right] \times 10^3‰ \tag{8.6}$$

The standard is an ultrapure Ca carbonate whose $^{40}Ca/^{44}Ca$ ratio is 47.144 ($^{44}Ca/^{40}Ca = 0.021211$). The definition of the $\delta^{44}Ca$ parameter is consistent with the way the isotope compositions of H, C, N, O, and S are conventionally expressed (section 26.2). The isotope compositions of Ca reported by Skulan et al. (1997) were measured by means of the double-spike technique, were normalized to $^{42}Ca/^{44}Ca = 0.31221$, and have a precision of ±0.15‰.

The conversion of $^{40}Ca/^{44}Ca$ ratios to $\delta^{44}Ca$ values defined above is illustrated below for data from Russell et al. (1978). These authors reported $^{40}Ca/^{44}Ca = 47.109 \pm 0.007$ for Ca in seawater. Since the δ-parameter is traditionally expressed in terms of the ratios of the heavy isotope divided by a lighter one, the $^{40}Ca/^{44}Ca$ ratios are inverted:

$$\frac{^{44}\text{Ca}}{^{40}\text{Ca}} = \begin{cases} 0.0212273 \pm 0.0000777 & \text{(seawater)} \\ 0.0212116 & \text{(standard)} \end{cases}$$

The $\delta^{44}Ca$ value of seawater reported by Russell et al. (1978) is

$$\delta^{44}\text{Ca} = \left[\frac{0.0212273}{0.0212116} - 1\right] \times 10^3$$
$$= +0.74 \pm 0.15‰$$

which is in good agreement with the $\delta^{44}Ca$ value (+0.92 ± 0.18‰) reported by Skulan et al. (1997) about 20 years later. Additional determinations by Zhu and Macdougall (1998) indicate that the isotopic composition of Ca in seawater in the Atlantic, Pacific, and Indian Oceans is constant with an average $^{44}Ca/^{40}Ca$ ratio of 0.0217470 ± 0.0000013. However, the $\delta^{44}Ca$ value of average seawater reported by Zhu and Macdougall (1998) differs from that obtained by Russell et al. (1978) and Skulan et al. (1997) by about 2.4% even though all three groups used the double-spike technique to correct for isotope fractionation in the mass spectrometer.

The discrepancy arises partly because Russell et al. (1978) and Skulan et al. (1997) corrected the Ca isotope ratios to $^{42}Ca/^{44}Ca = 0.31221$, whereas Zhu and Macdougall (1998) normalized to $^{44}Ca/^{40}Ca = 0.0217$, which yields a $^{42}Ca/^{44}Ca$ ratio of only 0.308799. The discrepancy highlights the need for the adoption of uniform data reduction procedures and for the use of interlaboratory standards such as those used to standardize measured isotope ratios of Sr (Section 5.2b). In fact, Zhu and Macdougall (1998) recommended seawater as an interlaboratory Ca standard because it is readily available, has a high Ca concentration, and has a constant isotopic composition. A study by Schmitt et al. (2001) of the isotope composition of Ca in modern seawater yielded a $^{40}Ca/^{44}Ca$ ratio of 45.143 ± 0.003.

Skulan et al. (1997) reported that the δ^{44}Ca values of volcanic rocks (basalt, dacite, rhyolite) and of marine carbonate (carbonate ooze and chalk) range only within narrow limits from +0.34‰ (alkali basalt, Mauna Kea, Hawaii) to −0.37‰ (Miocene carbonate ooze) with an average of 0 ± 0.2‰.

The δ^{44}Ca values of skeletal carbonate and bone in Figure 8.6 range widely from $+0.49 \pm 0.01$‰ (warm-water foraminifer) to -2.88 ± 0.06‰ (deer bone). The data imply that the Ca that is incorporated into shells and bones is depleted in ^{44}Ca relative to ^{40}Ca or is enriched in ^{40}Ca relative to the heavier isotopes of Ca. The progressive decrease of δ^{44}Ca values with increasing body weight of adult specimens suggests that the effect is propagated within the food web. The enrichment in ^{40}Ca of skeletal carbonate of marine organisms causes seawater to become depleted in ^{40}Ca (i.e., δ^{44}Ca is positive) because Ca is primarily removed from seawater by the deposition of biogenic carbonate. Since unfractionated Ca on land has δ^{44}Ca values close to 0.0‰ compared to about +1.0‰ for Ca in seawater, land animals should have lower (or more negative) δ^{44}Ca values on average than marine animals. In addition, both Skulan et al. (1997) and Zhu and Macdougall (1998) found evidence in their data that the enrichment of marine biogenic carbonate in ^{40}Ca increases with decreasing temperature.

The Ca in some of the major rivers of the world is variably enriched in ^{40}Ca (or depleted in ^{44}Ca), as expected from the diversity of the rocks weathering in their drainage basins (Zhu and Macdougall, 1998). In addition, calcareous ooze and other forms of biogenic carbonates are even more enriched in ^{40}Ca than river water entering the oceans. Therefore, Zhu and Macdougall (1998) concluded from the data available to them that the isotopic composition of Ca in river water and of marine carbonate are inconsistent with the isotopic

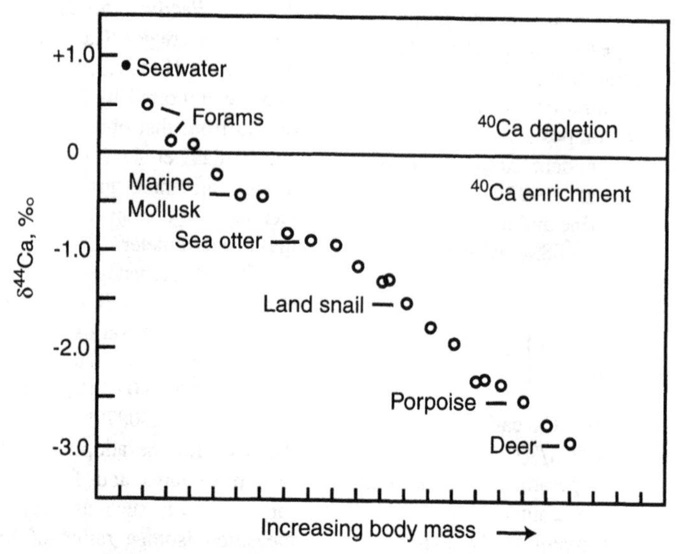

FIGURE 8.6 Mass-dependent fractionation of Ca in skeletal Ca carbonate of marine and continental organisms. The progressive decrease of the δ^{44}Ca values with increasing body weight of the adult organisms implies progressive enrichment in ^{40}Ca relative to ultrapure $CaCO_3$, whose ^{40}Ca/^{44}Ca ratio is 47.144. The δ^{44}Ca values of volcanic rocks and marine carbonate ooze and chalk cluster closely around 0 ± 0.2‰. The δ^{44}Ca value of seawater is +0.92‰. Data from Skulan et al. (1997).

composition of Ca in seawater. Much work remains to be done before the isotope geochemistry of Ca is completely understood.

Since Ca is an essential element in the formation of bones, its absorption in the intestines and its distribution in bones are being studied by medical researchers. In some cases, the Ca concentrations are measured by isotope dilution based on the ingestion of Ca spikes (Fitzgerald, 1975; Price et al., 1990).

8.2c Isotope Anomalies in the Solar Nebula

The isotope compositions of Ca and other elements in meteorites, interplanetary dust particles, lunar rocks, and other types of extraterrestrial material may differ from those of terrestrial samples because the elements may have been affected by several kinds of processes (Section 24.5):

1. nucleosynthesis reactions in the ancestral stars of the solar nebula,
2. nuclear reactions in the solar nebula before and after formation of planetary objects by irradiation with solar and galactic cosmic rays as well as by low-energy neutrons,
3. mass-dependent isotope fractionation during repeated episodes of condensation and volatilization in the solar nebula,
4. decay of short-lived radionuclides to stable daughters in the solar nebula prior to and during formation of solid objects, and
5. implantation of the nuclei of atoms into solid objects exposed to the solar wind in interplanetary space and on the surface of the Moon.

A review of the literature by Shima (1986) indicated that at least 32 elements in meteorites exhibit significant variations in their isotope compositions. The list of affected elements includes not only Ca but also Ar, K, Sr, Nd, Sm, Hf, Os, and Pb, all of which have either stable radiogenic or long-lived radioactive isotopes.

The anomalous isotopic compositions of Ca in bulk samples of stony meteorites are not evident in Figure 8.7 (Russell et al., 1978) and in the results reported by other investigators (e.g., Backus et al., 1964; Barnes et al., 1971; Marshall and DePaolo, 1982) because they occur primarily in Ca–Al-rich and other types of inclusions in carbonaceous chondrites (Jungk et al., 1984). These

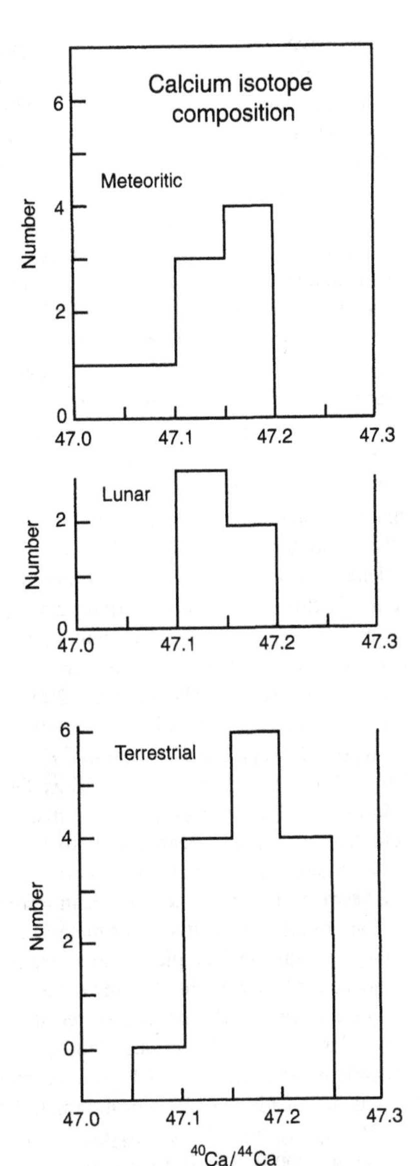

FIGURE 8.7 Range of $^{40}Ca/^{44}Ca$ ratios of Ca in stony meteorites, lunar basalt and minerals, and terrestrial samples, including inorganic and biogenic materials. Data from Russell et al. (1978).

variations have been attributed to fractionation during volatilization and condensation (high Al inclusion, Allende; Lee et al., 1978, 1979), to nucleosynthesis reactions (Ca–Al-rich aggregates and chondrules, Allende; Niederer and Papanastassiou, 1979, 1984), and to nuclear spallation reactions caused by cosmic rays (iron meteorites; Shima et al., 1969; Hintenberger et al., 1965). In addition, Young et al. (1981) measured the isotope composition of Ca in cosmic rays.

A recent study by Humayun and Clayton (1995) of the isotope composition of K in chondrite and achondrite meteorites, separated chondrules, Ca–Al-rich inclusions (CAIs), and lunar rocks found no evidence for isotope fractionation. The authors therefore concluded that the absence of isotope fractionation of K does not support the hypothesis that the observed depletion of meteorites and planets in alkali metals and other volatile elements occurred by vaporization during the formation of the solar system but may have occurred instead during an earlier period, when the solar nebula was still hot (Wang et al., 2001).

In a related investigation, Podosek et al. (1999) reported the presence of excess ^{40}K in some leachates of the carbonaceous chondrites Orgueil and Murchison. The authors proposed that a certain amount of ^{40}K had been added to the solar nebula by a late supernova. Consequently, the difference between the abundance of ^{40}K in terrestrial K and that of certain constituents of meteorites was caused by incomplete isotopic homogenization of the newly added ^{40}K with the K in the solar nebula.

Some primitive carbonaceous meteorites contain presolar grains of various kinds, including diamonds and graphite (C), silicon carbide (SiC), silicon nitride (Si_3N_4), corundum (Al_2O_3), spinel ($MgAl_2O_4$), hibonite ($CaAl_{12}O_{19}$), and TiO_2 (Hoppe et al., 2000). The isotope compositions of the elements in these grains differ greatly from those found on Earth, which indicates that the grains formed in stars rather than in the solar nebula (Hoppe and Zinner, 2000; Ebel and Grossman, 2001). For example, C, N, Mg, Si, and Ca in silicon carbide grains in the Murchison carbonaceous chondrite have widely varying isotope composition (e.g., $^{12}C/^{13}C$ ratios range from 18 to 6800, $^{14}N/^{15}N$ ratios vary from 13 to 200). The Ca in these silicon carbide inclusions is enriched in ^{44}Ca, which formed by β-decay of radioactive ^{44}Ti with a halflife of 60 years (Hoppe et al., 2000). In addition, the presence of short-lived ^{41}Ca in hibonite grains of Murchison and Allende was reported by Sahijpal et al. (2000).

8.3 SUMMARY

The K–Ca geochronometer is not widely used even though about 89 percent of ^{40}K atoms decay to ^{40}Ca while only 11 percent decay to ^{40}Ar. The principal reasons are that ^{40}Ca is by far the most abundant isotope of Ca (96.9821 percent) and that Ca isotopes are fractionated by ion exchange and during analysis on solid-source mass spectrometers. These analytical difficulties have been largely overcome and several successful age determinations of igneous and metamorphic rocks and of their K-bearing minerals have been published.

The isotope composition of Ca can be used to study the origin of igneous rocks based on the abundance of radiogenic ^{40}Ca. The available data are less precise than the isotope ratios of Sr, Nd, and Pb. Nevertheless, some progress has been made in outlining the isotope geochemistry of Ca in igneous and metamorphic rocks in the continental crust of the Earth.

An important aspect of the geochemistry of Ca is the fractionation of its isotopes during uptake and fixation by organism. After nearly half a century of uncertainty, the available evidence now indicates that organisms preferentially incorporate ^{40}Ca, that this effect increases up the food chain, and that it has depleted seawater in ^{40}Ca.

The isotope composition of Ca in bulk samples of stony meteorites does not differ detectably from that of terrestrial samples. However, small Ca–Al-rich grains and other kinds of exotic particles in carbonaceous chondrites contain Ca and K whose isotope compositions were affected by nuclear reactions in the ancestral supernovas that contributed to the formation of the solar nebula. The anomalous isotope compositions of Ca and many other chemical elements in these inclusions are an important clue to the synthesis of the elements in stars and the conditions in the solar nebula prior to the formation of the solar system.

REFERENCES FOR PRINCIPLES AND METHODOLOGY (INTRODUCTION, SECTION 8.1 INTRODUCTION)

Ahrens, L. H., 1950. The feasibility of a calcium method in the determination of geologic age. *Geochim. Cosmochim. Acta*, 1:312–316.

Baadsgaard, H., 1987. Rb-Sr and K-Ca isotope systematics in minerals from potassium horizons in the Prairie Evaporite Formation, Saskatchewan, Canada. *Chem. Geol. (Isotope Geosci. Sect.)*, 66:1–15.

Begemann, F., K. R. Ludwig, G. W. Lugmair, K. Min, L. E. Nyquist, P. J. Patchett, P. R. Renne, C. -Y. Shih, I. M. Villa, and R. J. Walker, 2001. Call for an improved set of decay constants for geochronological use. *Geochim. Cosmochim. Acta*, 65:111–121.

Coleman, M. L., 1971. Potassium-calcium dates from pegmatitic micas. *Earth Planet. Sci. Lett.*, 12:399–405.

Endt, P. M., and C. Van der Leun, 1973. Energy levels of $A = 21-44$ nuclei (V). *Nucl. Phys. A*, 214:1–625.

Eugster, O., F. Tera, and G. J. Wasserburg, 1969. Isotopic analysis of barium in meteorites and in terrestrial samples. *J. Geophys. Res.*, 74:3897–3908.

Fletcher, I. R., N. J. McNaughton, R. T. Pidgeon, and K. J. R. Rosman, 1997. Sequential closure of K-Ca and Rb-Sr isotopic systems in Archaean micas. *Chem. Geol.*, 138:289–301.

Heumann, K. G., E. Kubassek, W. Schwabenbauer, and I. Stadler, 1979. Analytical method for the K/Ca age determination of geological samples. *Fresenius Z. Anal. Chem.*, 297:35–43.

Marshall, B. D., H. H. Woodard, and D. J. DePaolo, 1986. K-Ca systematics of authigenic sanidine from Waukau, WI, and the diffusivity of argon. *Geology*, 14:936–938.

Min, K., R. Mundil, P. R. Renne, and K. R. Renne, 2000. A test for systematic errors in $^{40}Ar/^{39}Ar$ geochronology through comparison with U-Pb analysis of a 1.1 Ga rhyolite. *Geochim. Cosmochim. Acta*, 64:73–98.

Nägler, Th. F., and I. M. Villa, 2000. In pursuit of the ^{40}K branching ratio: K-Ca and $^{39}Ar/^{40}Ar$ dating of gem silicates. *Chem. Geol.*, 169:5–16.

Nier, A. O., 1938. The isotopic constitution of calcium, titanium, sulphur, and argon. *Phys. Rev.*, 53:282–286.

Polevaya, N. I., N. E. Titov, V. S. Belyaev, and V. D Sprintsson, 1958. Application of the Ca method in the absolute age determination of sylvites. *Geochemistry*, 8:897–906.

Russell, W. A., D. A. Papanastassiou, and T. A.Tombrello, 1978. Ca isotope fractionation on the Earth and solar system materials. *Geochim. Cosmochim. Acta*, 42:1075–1090.

Steiger, R. H., and E. Jäger, 1977. Subcommission on Geochronology: Convention on the use of decay constants in geo- and cosmo-chronology. *Earth Planet. Sci. Lett.*, 36:359–362.

Wilhelm, H. G., and W. Ackermann, 1972. Age determination of sylvite from the upper Zechstein deposits of the Werra District by the potassium-calcium method. *Z. Naturforsch. P. A*, 27A:1256–1259.

REFERENCES FOR PIKES PEAK GRANITE, COLORADO (SECTION 8.1a)

Barker, F., H. T. Millard, Jr., C. E. Hedge, and J. R. O'Neil, 1976. Pikes Peak batholith: Geochemistry of some minor elements and isotopes, and implications for magma genesis. In R. C. Epis and R. J. Weimer (Eds.), *Studies in Colorado Field Geology*, Golden, Colorado. Prof. Contrib. No. 8, pp. 44–56. Colorado School of Mines.

Barker, F., D. R. Wones, W. N. Sharp, and G. A. Desborough, 1975. The Pikes Peak batholith, Colorado Front Range, and a model for the origin of the gabbro-anorthosite-syenite-potassic granite suite. *Precamb. Res.*, 2:97–160.

Hedge, C. E., 1970. Whole-rock Rb-Sr age of the Pikes Peak batholith, Colorado. *U.S. Geol. Surv. Prof. Paper*, 700-B:86–89.

Marshall, B. D., and D. J. DePaolo, 1982. Precise age determinations and petrogenetic studies using the K-Ca method. *Geochim. Cosmochim. Acta*, 46:2537–2545.

Peterman, Z. E., and C. E. Hedge, 1968. Chronology of Precambrian events in the Front Range, Colorado. *Can. J. Earth Sci.*, 5:749–756.

Russell, W. A., D. A. Papanastassiou, and T. A. Tombrello, 1978. Ca isotope fractionation on the Earth and other solar system materials. *Geochim. Cosmochim. Acta*, 42:1075–1090.

Steiger, R. H., and E. Jäger, 1977. Subcommission on Geochronology: Convention on the use of decay constants in geo- and cosmo-chronology. *Earth Planet. Sci. Lett.*, 36:359–362.

York, D., 1969. Least squares fitting of a straight line with correlated errors. *Earth Planet. Sci. Lett.*, 5:320–324.

REFERENCES FOR LUNAR GRANITE (SECTION 8.1b)

Bogard, D. D., D. H. Garrison, C. Y. Shih, and L. E. Nyquist, 1994. $^{39}Ar/^{40}Ar$ dating of two lunar granites:

The age of Copernicus. *Geochim. Cosmochim. Acta*, 58:3093–3100.

Marshall, B. D., and D. J. DePaolo, 1982. Precise age determinations and petrogenetic studies using the K-Ca method. *Geochim. Cosmochim. Acta*, 46:2537–2545.

Meyer, C., Jr., I. S. Williams, and W. Compston, 1989. $^{207}Pb/^{206}Pb$ ages of zircon-containing rock fragments indicate continuous magmatism in the lunar crust from 4350 to 3900 million years (Abstract). *Lunar Planet. Sci. Conf.*, 20:691–692.

Ryder, G., 1976. Lunar sample 15405: Remnant of a KREEP basalt–granite differentiated pluton. *Earth Planet. Sci. Lett.*, 29:255–268.

Shih, C.-Y., L. E. Nyquist, and H. Wiesmann, 1993. K-Ca chronology of lunar granites. *Geochim. Cosmochim. Acta*, 57:4827–4841.

Shih, C.-Y., L. E. Nyquist, and H. Wiesmann, 1994. K-Ca and Rb-Sr dating of two lunar granites: Relative chronometer resetting. *Geochim. Cosmochim. Acta*, 58:3101–3116.

REFERENCES FOR RADIOGENIC ^{40}Ca IN TERRESTRIAL ROCKS (SECTION 8.2a)

Backus, M. M., W. H. Pinson, and L. F. Herzog, 1964. Calcium isotope ratios in the Homestead and Pasamonte meteorites and a Devonian limestone. *Geochim. Cosmochim. Acta*, 28:735–742.

Barnes, I. L., et al., 1972. Isotopic abundance ratios and concentrations of selected elements in Apollo 14 samples. *Proc. Third Lunar Sci. Conf., Geochim. Cosmochim. Acta*, Suppl. 3:1465–1472

Faure, G., 2001. *The Origin of Igneous Rocks; the Isotopic Evidence*. Springer-Verlag, Heidelberg.

Fletcher, I. R., N. J. Naughton, R. T. Pidgeon, and K. J. R. Rosman, 1997. Sequential closure of K-Ca and Rb-Sr isotopic systems in Archaean micas. *Chem. Geol.*, 138:289–301.

Holmes, A., 1932. The origin of igneous rocks. *Geol. Mag.*, 69:543–558.

Marshall, B. D., and D. J. DePaolo, 1982. Precise age determinations and petrogenetic studies using the K-Ca method. *Geochim. Cosmochim. Acta*, 46:2537–2545.

Marshall, B. D., and D. J. DePaolo, 1989. Calcium isotopes in igneous rocks and the origin of granite. *Geochim. Cosmochim. Acta*, 53:917–922.

Nelson, D. R., and M. T. McCulloch, 1989. Petrogenetic applications of the $^{40}K/^{40}Ca$ radiogenic decay scheme—A reconnaissance study. *Chem. Geol.*, 79: 257–293.

REFERENCES FOR MASS-DEPENDENT ISOTOPE FRACTIONATION (SECTION 8.2b)

Corless, J. T. 1968. Observations on the isotopic geochemistry of calcium. *Earth Planet. Sci. Lett.*, 4: 475–478.

Eugster, O., F. Tera, and G. J. Wasserburg, 1969. Isotopic analysis of barium in meteorites and in terrestrial samples. *J. Geophys. Res.*, 74:3897–3908.

Fitzgerald, E. R., 1975. Calcium-isotope effects in mechanical spectra of cancellous bone. *Med. Biol. Eng.*, 3:717–719.

Hart, S. R., and A. Zindler, 1989. Isotope fractionation laws: A test using calcium. *Int. J. Mass Spectrom. Ion Process.*, 89:287–301.

Heumann, K. G., and K. H. Lieser, 1973. Untersuchung von Isotopenfeinvariationen des Calciums in der Natur an rezenten Karbonaten and Sulfaten. *Geochim. Cosmochim. Acta*, 37:1463–1471.

Hirt, B., and S. Epstein, 1964a. Die Isotopenzusammensetzung von natürlichem Calcium. *Helv. Phys. Acta*, 37:179.

Hirt, B., and S. Epstein, 1964b. A search for isotopic variations in some terrestrial and meteoritic calcium. *Trans. Am. Geophys. Union*, 45:113.

Lee, T., D. A. Papanastassiou, and G. J. Wasserburg, 1978. Calcium isotopic anomalies in the Allende meteorite. *Astrophys. J. Lett.*, 220:L21–L25.

Price, R. I., G. N. Kent, K. J. R. Rosman, D. H. Gutteridge, J. Reeve, J. P. Allen, B. G. A. Stuckey, M. Smith, G. Guelfi, J. Hickling, and S. L. Blakeman, 1990. Kinetics of intestinal calcium absorption in humans using stable isotopes and high-precision thermal ionization mass spectrometry. *Biomed. Environ. Mass Spectrom.*, 19:353–359.

Richter, F. M., Y. Liang, and A. M. Davis, 1999. Isotope fractionation by diffusion in molten oxides. *Geochim. Cosmochim. Acta*, 63:2853–2861.

Russell, W. A., and D. A. Papanastassiou, 1978. Calcium isotope fractionation in ion exchange chromatography. *Anal. Chem.*, 50:1151–1154.

Russell, W. A., D. A. Papanastassiou, and T. A. Tombrello, 1978. Ca isotope fractionation on the Earth and other solar system materials. *Geochim. Cosmochim. Acta*, 42:1075–1090.

Russell, W. A., D. A. Papanastassiou, and T. A. Tombrello, 1980. The fractionation of Ca isotopes by sputtering. *Radiation Effects*, 52:41–52.

Schmitt, A.-D., G. Bracke, P. Stille, and B. Kiefel, 2001. The calcium isotope composition of modern seawater determined by thermal ionisation mass spectrometry. *Geostandards Newsletter*, 25:267–275.

Skulan, J., D. J. DePaolo, and T. L. Owens, 1997. Biological control of calcium isotopic abundance in the global calcium cycle. *Geochim. Cosmochim. Acta*, 61:2505–2510.

Zhu, P., and J. D. Macdougall, 1998. Calcium isotopes in the marine environment and the oceanic calcium cycle. *Geochim. Cosmochim. Acta*, 62:1691–1698.

REFERENCES FOR ISOTOPE ANOMALIES IN THE SOLAR NEBULA (SECTION 8.2c)

Backus, M. M., W. H. Pinson, L. F. Herzog, and P. M. Hurley, 1964. Calcium isotope ratios in the Homestead and Pasamonte meteorites and a Devonian limestone. *Geochim. Cosmochim. Acta*, 28:735–742.

Barnes, I. L., B. S. Carpenter, E. L. Garner, J. W. Gramlich, E. C. Kuehner, L. A. Machlan, E. J. Maienthal, J. R. Moody, L. F. Moore, T. J. Murphy, P. J. Paulson, K. M. Sappenfield, and W. R. Shields, 1971. Isotopic abundance ratios and concentrations of selected elements in Apollo 14 samples. *Proc. Third Lunar Sci. Conf., Geochim. Cosmochim. Acta*, Suppl. 3:1465–1472.

Ebel, D. S., and L. Grossman, 2001. Condensation from supernova gas made of free atoms. *Geochim. Cosmochim. Acta*, 65:469–477.

Hintenberger, H., H. Voshage, and H. Sarker, 1965. Durch die kosmische Strahlung produziertes Lithium und Calcium in Eisenmeteoriten. *Z. Naturforsch.*, 20a: 965–967.

Hoppe, P., R. Strebel, P. Eberhardt, S. Amari, and R. S. Lewis, 2000. Isotopic properties of silicon carbide X grains from the Murchison meteorite in the size range 0.5 to 1.5 μm. *Meteoritics Planet. Sci.*, 35:1157–1176.

Hoppe, P., and E. Zinner, 2000. Presolar dust grains from meteorites and their stellar sources. *J. Geophys. Res., Space Phys.*, 105:10371–10385.

Humayun, M., and R. N. Clayton, 1995. Potassium isotope cosmochemistry: Genetic implications of volatile element depletion. *Geochim. Cosmochim. Acta*, 59: 2131–2148.

Jungk, M. H. A., T. Shimamura, and G. W. Lugmair, 1984. Ca isotope variations in Allende. *Geochim. Cosmochim. Acta*, 48:2651–2658.

Lee, T., D. A. Papanastassiou, and G. J. Wasserburg, 1978. Calcium isotopic anomalies in the Allende meteorite. *Astrophys. J. Lett.*, 220:L21–L25.

Lee, T., W. A. Russell, and G. J. Wasserburg, 1979. Ca isotopic anomalies and the lack of ^{26}Al in an unusual Allende inclusion. *Astrophys. J. Lett.*, 228:L93–L98.

Marshall, B. D., and D. J. DePaolo, 1982. Precise age determinations and petrogenetic studies using the K-Ca method. *Geochim. Cosmochim. Acta*, 46:2537–2545.

Niederer, F. R., and D. A. Papanastassiou, 1979. Ca isotopes in Allende and Leoville inclusions. *Lunar Planet Sci.*, 10:913–915.

Niederer, F. R., and D. A. Papanastassiou, 1984. Ca isotopes in refractory inclusions. *Geochim. Cosmochim. Acta*, 48:1279–1293.

Podosek, F. A., R. H. Nichol, Jr., J. C. Brannon, B. S. Meyer, U. Ott, C. L. Jennings, and N. Luo, 1999. Potassium, stardust, and the last supernova. *Geochim. Cosmochim. Acta*, 63:2351–2362.

Russell, W. A., D. A. Papanastassiou, and T. A. Tombrello, 1978. Ca isotope fractionation on the Earth and other solar system materials. *Geochim. Cosmochim. Acta*, 42:1075–1090.

Sahijpal, S., J. N. Goswami, and A. M. Davis, 2000. K, Mg, Ti, and Ca isotopic compositions and refractory trace element abundances in hibonites from CM and CV meteorites: Implications for early solar system processes. *Geochim. Cosmochim. Acta*, 64:1989–2005.

Shima, M., 1986. A summary of extremes of isotopic variations in extra-terrestrial materials. *Geochim. Cosmochim. Acta*, 50:577–584.

Shima, M., M. Imamura, H. Matsuda, and M. Honda, 1969. Some stable and long-lived nuclides produced by spallation in meteoritic iron. In P. M. Millman (Ed.), *Meteorite Research*, pp. 335–347. Reidel, Dordrecht, The Netherlands.

Wang, J., A. M. Davis, R. N. Clayton, T. K. Mayeda, and A. Hashimoto, 2001. Chemical and isotopic fractionation during evaporation of the FeO-MgO-SiO_2-CaO-Al_2O_3-TiO_2 rare earth element melt system. *Geochim. Cosmochim. Acta*, 65:479–494.

Young, J. S., P. S. Freier, C. J. Waddington, N. R. Brewster, and R. K. Fickel, 1981. The elemental and isotopic composition of cosmic rays: Silicon to nickel. *Astrophys. J.*, 246:1014–1030.

CHAPTER 9

The Sm–Nd Method

SAMARIUM (Sm, $Z = 62$) and neodymium (Nd, $Z = 60$) are rare-earth elements (REEs) that occur in many rock-forming silicate, phosphate, and carbonate minerals. One of the Sm isotopes ($^{147}_{62}Sm$) is radioactive and decays by α-emission to a stable isotope of Nd ($^{143}_{60}Nd$). Although the halflife of ^{147}Sm is very long ($T_{1/2} = 1.06 \times 10^{11}$ y, $\lambda = 6.54 \times 10^{-12}$ y^{-1}), this decay scheme is useful for dating terrestrial rocks, stony meteorites, and lunar rocks. Moreover, the growth of radiogenic ^{143}Nd and ^{87}Sr together provides new insights into the geochemical evolution of planetary objects and the genesis of igneous rocks. The isotope geochemistry of Nd in the oceans is presented in Section 19.3.

9.1 GEOCHEMISTRY OF Sm AND Nd

The REEs generally form ions with a charge of +3 whose radii decrease with increasing atomic number from 1.15 Å in lanthanum (La, $Z = 57$) to 0.93 Å in lutetium (Lu, $Z = 71$). The REEs occur in high concentrations in several economically important minerals such as bastnaesite ($CeFCO_3$), monazite ($CePO_4$), and cerite [(Ca, $Mg)_2(Ce)_8(SiO_4)_7 \cdot 3H_2O$]. In addition, they occur as trace elements in common rock-forming minerals in which they replace major ions. They may also reside in inclusions of certain accessory minerals in the common rock-forming silicates. Minerals exercise a considerable degree of selectivity in admitting REEs into their crystal structures. Feldspar, biotite, and apatite tend to concentrate the light REEs (Ce group) whereas pyroxenes, amphiboles, and garnet concentrate the heavy REEs (Gd group). The selectivity of the rock-forming minerals for the light or heavy rare earths affects the REE concentrations of the rocks in which these minerals occur.

Neodymium and Sm belong to the light REEs. Their abundances in the solar system are 8.279×10^{-1} atoms of Nd per 10^6 atoms of Si, whereas that of Sm is $2.582 \times 10^{-1}/10^6$ Si (Anders and Grevesse, 1989). The lower abundance of Sm compared to Nd is consistent with the general decrease in the cosmic abundances of the elements with increasing atomic number. The atomic Sm/Nd ratio in the solar system is 0.31.

The concentrations of Sm and Nd in calc-alkaline plutonic and volcanic igneous rocks in Table 9.1 and Figure 9.1 range from <1.0 ppm in ultramafic rocks to about 8 ppm Sm and 45 ppm Nd in granite. Alkali-rich igneous rocks have consistently higher Sm and Nd concentrations than the calc-alkaline suite, ranging up to about 15 ppm Sm and 85 ppm Nd. The Sm/Nd ratios of igneous rocks decrease from about 0.32 in MORBs and komatiites to <0.20 in granites and to even lower values in alkali-rich rocks (e.g., Sm/Nd = 0.11 in syenite). In fact, all of the rock types that characterize the continental crust (granite, granulite, shale, graywacke, sandstone, limestone, etc.) have

Table 9.1. Average Concentrations of Sm and Nd in Common Terrestrial and Extraterrestrial Rocks

Rock Type	Sm, ppm	Nd, ppm	Sm/Nd
	Volcanic Rocks		
Komatiites	1.14(15)	3.59(15)	0.317
Tholeiites (Archean)	1.96(23)	6.67(23)	0.293
MORB (tholeiite)	3.30(73)	10.3(78)	0.320
Continental tholeiites	5.32(657)	24.2(657)	0.220
Calc-alkaline basalt	6.07(24)	32.6(25)	0.186
Alkali basalt	8.07(65)	41.5(62)	0.194
Trachyte	14.1(8)	73.2(8)	0.192
Leucite basalt, phonolite	15.1(29)	81.4(30)	0.185
Andesites	3.90(28)	20.6(40)	0.185
Dacite, rhyodacite	5.05(32)	24.9(33)	0.202
Rhyolite	4.65(10)	21.6(19)	0.215
	Plutonic Rocks		
Ultramafic	0.582(19)	2.28(19)	0.255
Gabbro	1.78(52)	7.53(56)	0.236
Intermediate rocks	5.65(14)	26.2(41)	0.215
Granite	8.22(587)	43.5(588)	0.188
Granulite (metamorphic)	4.96(16)	31.8(30)	0.156
Syenite	9.5(8)	86.0(8)	0.11
Kimberlite	8.08(37)	66.1(37)	0.122
Carbonatite	38.7(47)	178.8(47)	0.216
	Sedimentary Rocks		
Shale	10.4(949)	49.8(949)	0.209
Graywacke	5.03(6)	25.5(6)	0.197
Sandstone	8.93(183)	39.4(319)	0.227
Limestone	2.03(125)	8.75(125)	0.232
Phosphate rock	341(102)	1228(102)	0.266
Seawater	0.545×10^{-6} (12)	2.58×10^{-6} (12)	0.211

Note: Data from Herrmann (1970) and other sources in the literature. The number of samples included in each average is indicated in parentheses.

lower Sm/Nd ratios than mafic igneous rocks (tholeiite basalt, gabbro, ultramafic rocks, etc.).

The unusual geochemical behavior of Sm and Nd arises from the effects of the lanthanide contraction, which results from the filling of the f electronic orbitals. Consequently, the ionic radius of Sm ($Z = 62$) is smaller than that of Nd ($Z = 60$). Even though the difference in the radii is small ($Nd^{3+} = 1.08$ Å; $Sm^{3+} = 1.04$ Å), Nd is preferentially concentrated in the liquid phase during partial melting of silicate minerals, whereas Sm remains in the residual solids. For this reason, basalt magmas have *lower* Sm/Nd ratios than the source rocks from which they formed. The preferential partitioning of Nd into the melt phase has caused the rocks of the continental crust to be enriched in Nd relative to Sm compared to the residual rocks in the lithospheric mantle.

The concentrations of Sm and Nd reach high values in certain accessory phosphate minerals

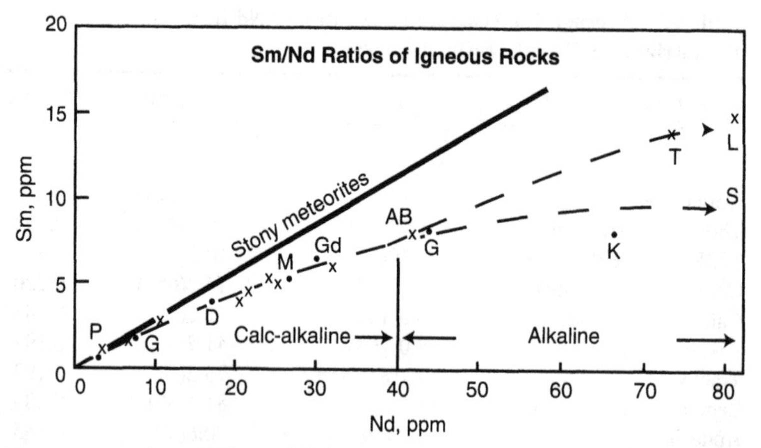

FIGURE 9.1 Average concentrations of Sm and Nd in igneous rocks: (●) plutonic rocks; (×) volcanic rocks. The major rock types are identified by letter: P = peridotite; G = gabbro; D = diorite; M = monzonite; Gd = granodiorite; G = granite; AB = alkali basalt; K = kimberlite; T = trachyte; L = leucitite; S = syenite. The alkaline plutonic and volcanic rocks have higher Sm and Nd concentrations than calc-alkaline rocks. Most magmatic differentiates have lower Sm/Nd ratios than stony meteorites. Data from Herrmann (1970) and other sources.

(e.g., apatite and monazite) and in carbonatites. However, even these minerals and carbonatites are more enriched in Nd than in Sm and hence their Sm/Nd ratios are less than 0.32. Among the rock-forming silicate minerals, garnet is the only one with a high Sm/Nd ratio (0.54) even though its concentrations of Sm (1.17 ppm) and Nd (2.17 ppm) are both low. Several other rock-forming silicate minerals (e.g., K-feldspar, biotite, amphibole, and clinopyroxene) have higher Sm and Nd concentrations than garnet, but their Sm/Nd ratios are less than 0.32 in most cases.

The lanthanide contraction causes the distribution of Sm and Nd to be opposite to that of Rb and Sr. Consequently, differences have developed in the isotope compositions of Nd and Sr in the rocks of the continental crust and the lithospheric mantle. The preferential partitioning of Sm into the residual solids during partial melting in the mantle has caused the residual rocks of the mantle to become enriched in the radiogenic ^{143}Nd compared to the rocks of the continental crust. Since Rb is concentrated into the melt phase, the residual solids in the mantle contain less radiogenic ^{87}Sr than the rocks of the continental crust. The complementary geochemical properties of the Sm–Nd and Rb–Sr decay schemes enhance the effectiveness of the isotope compositions of Nd and Sr in the study of the petrogenesis of igneous rocks.

Samarium and Nd each have seven naturally occurring isotopes whose mass numbers (A) and abundances are listed in Table 9.2. The corresponding atomic weights are Nd = 144.24, Sm = 150.36.

Table 9.2. Abundances and Mass Numbers (*A*) of Naturally Occurring Isotopes of Sm and Nd

Neodymium (Z = 60)		Samarium (Z = 62)	
A	Abundance, %	*A*	Abundance, %
142	27.1	144	3.1
143	12.2	147	15.0
144	23.9	148	11.2
145	8.3	149	13.8
146	17.2	150	7.4
148	5.7	152	26.7
150	5.6	154	22.8
Sum	100.0	Sum	100.0

Source: Lide and Frederikse, 1995.

Samarium-147 is radioactive and decays by α-emission to ^{143}Nd:

$$^{147}_{62}\text{Sm} \rightarrow {}^{143}_{60}\text{Nd} + {}^{4}_{2}\text{He} + E \qquad (9.1)$$

where $^{4}_{2}$He is the α-particle and E is the total decay energy. The halflife and decay constant of $^{147}_{62}$Sm are

$$T_{1/2} = 1.06 \times 10^{11} \text{ y}$$

$$\lambda = \frac{\ln 2}{1.06 \times 10^{11}} = 6.54 \times 10^{-12} \text{ y}^{-1}$$

based on reviews of the literature by Lugmair and Marti (1978) and Begemann et al. (2001).

9.2 PRINCIPLES AND METHODOLOGY

The isotope composition of Nd is expressed by the ^{143}Nd/^{144}Nd ratio, which increases as a function of time because of the decay of ^{147}Sm in accordance with equation 4.8:

$$\frac{^{143}\text{Nd}}{^{144}\text{Nd}} = \left(\frac{^{143}\text{Nd}}{^{144}\text{Nd}}\right)_i + \frac{^{147}\text{Sm}}{^{144}\text{Nd}}(e^{\lambda t} - 1) \qquad (9.2)$$

The ^{147}Sm/^{144}Nd ratios of rocks or minerals are calculated from the measured concentrations of Sm and Nd:

$$\frac{^{147}\text{Sm}}{^{144}\text{Nd}} = \left(\frac{\text{Sm}}{\text{Nd}}\right)_c \times \frac{\text{at. wt. Nd} \times \text{Ab}^{147}\text{Sm}}{\text{at. wt. Sm} \times \text{Ab}^{144}\text{Nd}} \qquad (9.3)$$

The present abundance of ^{147}Sm and the atomic weight of Sm in terrestrial rocks are constant, whereas the present abundance of ^{144}Nd and the atomic weight of Nd in equation 9.3 depend, strictly speaking, on the abundance of radiogenic ^{143}Nd and hence on the age and Sm/Nd ratio of the sample being analyzed. In reality, the slow rate of decay of ^{147}Sm and the low Sm/Nd ratios of most rocks and minerals cause only small changes in the isotope composition of Nd. Therefore, equation 9.3 can be rewritten to a good approximation as

$$\frac{^{147}\text{Sm}}{^{144}\text{Nd}} = \left(\frac{\text{Sm}}{\text{Nd}}\right)_c \times \frac{144.24 \times 15.0}{150.36 \times 23.9}$$

$$= \left(\frac{\text{Sm}}{\text{Nd}}\right)_c \times 0.602 \qquad (9.4)$$

The concentrations of Sm and Nd are measured by isotope dilution (Section 4.4e) but can also be determined by ICP-MS.

In view of the long halflife of ^{147}Sm, equation 9.2 can be linearized by expanding $e^{\lambda t}$ as a power series, as shown in Section 5.2. In addition, the ^{147}Sm/^{144}Nd ratio can be expressed in terms of the Sm/Nd concentration ratio (equation 9.4) to yield the approximate relation

$$\frac{^{143}\text{Nd}}{^{144}\text{Nd}} = \left(\frac{^{143}\text{Nd}}{^{144}\text{Nd}}\right)_i + 0.602\left(\frac{\text{Sm}}{\text{Nd}}\right)_c \lambda t \qquad (9.5)$$

According to this equation, the ^{143}Nd/^{144}Nd ratio of a closed volume of Sm-bearing rocks increases linearly as a function of time at a rate determined by its Sm/Nd ratio.

Equation 9.5 was used in Figure 9.2 to illustrate the evolution of the isotope composition of Nd in the mantle of the Earth as a result of the withdrawal of basalt magma. Partial melting of undepleted rocks in the mantle enriches the residue in Sm and causes the melt-depleted rocks of the mantle to acquire elevated ^{143}Nd/^{144}Nd ratios with increasing time. The basalt magma that is produced by partial melting in the mantle is erupted at the surface or crystallizes as gabbroic plutons in the continental crust. In either case, these rocks are depleted in Sm and their ^{143}Nd/^{144}Nd ratios increase slowly because of their comparatively low Sm/Nd ratios. The significance of CHUR is explained in Section 9.2a below.

9.2a Isotope Fractionation and CHUR

The numerical values of the ^{143}Nd/^{144}Nd ratios in Figure 9.2 depend on the way in which the measured isotope ratios are corrected for fractionation and other procedural issues, such as

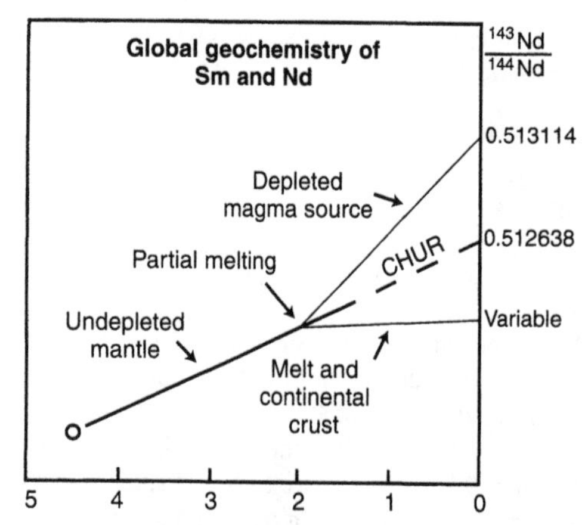

FIGURE 9.2 Effect of partial melting in the mantle of the Earth on the isotopic evolution of Nd in the rocks of the resulting continental crust and in the residual (melt-depleted) mantle. The undepleted mantle is assumed to have the same Sm/Nd ratio as average chondritic meteorites used to define CHUR (chondritic uniform reservoir). The numerical values of the $^{143}Nd/^{144}Nd$ ratios that characterize CHUR and depleted magma sources in the mantle are discussed in the text.

1. the proper calibration of spike solutions used to determine concentrations of Nd and Sm by isotope dilution,
2. corrections for interfering isobaric ions in the mass spectra of Sm and Nd,
3. the form of the isotope fractionation law used to correct for mass-dependent fractionation in the mass spectrometer, and
4. the numerical value of the isotope ratio chosen for this purpose.

These problems were discussed in detail by Wasserburg et al. (1981) and are a concern primarily for isotope geologists who use the Sm–Nd methods of dating. However, it is necessary for all users of these results to be aware that the absolute values of measured $^{143}Nd/^{144}Nd$ ratios depend on such technical details.

Scientists at the California Institute of Technology initially normalized measured $^{143}Nd/^{144}Nd$ ratios to $^{150}Nd/^{142}Nd = 0.2096$ (Jacobsen and Wasserburg, 1980) but later decided to use $^{146}Nd/^{142}Nd = 0.636151$ (Wasserburg et al., 1981). Other laboratories have used $^{148}Nd/^{144}Nd = 0.241578$ or $^{146}Nd/^{144}Nd = 0.7219$. Not all of these isotope ratios are compatible. Table 9.3 contains two sets of isotope ratios that were normalized to $^{146}Nd/^{142}Nd = 0.636151$ (row 1) and $^{146}Nd/^{144}Nd = 0.7219$ (row 2), respectively. Both sets of ratios assume $^{18}O/^{16}O = 0.00211$ and $^{17}O/^{16}O = 0.000387$ to correct NdO^+ mass spectra.

The isotope evolution of Nd in the mantle of the Earth in Figure 9.2 is described in terms of a reservoir known as CHUR which stands for "chondritic uniform reservoir" (DePaolo and Wasserburg, 1976; Jacobsen and Wasserburg, 1980). The model assumes that the Nd in the mantle has evolved in a uniform reservoir whose Sm/Nd ratio is equal to that of chondritic meteorites. The present value of the $^{143}Nd/^{144}Nd$ ratio of this reservoir in Table 9.3 is 0.512638 when the measured value of this ratio is corrected for fractionation to $^{146}Nd/^{144}Nd = 0.7219$ (row 2). Alternatively, the present value of the $^{143}Nd/^{144}Nd$ ratio of CHUR is 0.511847 in cases where the measured value of this ratio is corrected to $^{146}Nd/^{142}Nd = 0.636151$ (row 1). This value of the $^{146}Nd/^{142}Nd$ ratio is obtained from the data in

Table 9.3. Compatible Sets of Isotope Ratios of Nd in a Reservoir of Chrondritic Composition at the Present Time

Row	$^{142}Nd/^{144}Nd$	$^{143}Nd/^{144}Nd$	$^{145}Nd/^{144}Nd$	$^{146}Nd/^{144}Nd$	$^{148}Nd/^{144}Nd$	$^{150}Nd/^{144}Nd$
1[a]	1.138305	0.511847	0.348956	0.724134	0.243075	0.238619
2[b]	1.141827	0.512638	0.348417	0.7219	0.241578	0.236418

Source: Wasserburg et al., 1981.

[a]Row 1 contains isotope ratios of Nd in the chondritic uniform reservoir today after correcting NdO^+ mass scans for oxygen having $^{18}O/^{16}O = 0.00211$ and $^{17}O/^{16}O = 0.000387$ and corrected for isotope fractionation to $^{146}Nd/^{142}Nd = 0.636151$.
[b]Row 2 contains the same isotope ratios as row 1 renormalized to $^{146}Nd/^{144}Nd = 0.7219$.

row 1 by dividing the $^{146}Nd/^{142}Nd$ ratio (0.724134) by the $^{142}Nd/^{144}Nd$ ratio (1.138305):

$$\frac{^{146}\mathrm{Nd}/^{144}\mathrm{Nd}}{^{142}\mathrm{Nd}/^{144}\mathrm{Nd}} = \frac{^{146}\mathrm{Nd}}{^{142}\mathrm{Nd}} = \frac{0.724134}{1.138305} = 0.636151$$

To convert a $^{143}Nd/^{144}Nd$ ratio that was normalized to any ratio in row 1 of Table 9.3 to its corresponding value consistent with row 2, it must be multiplied by the factor 0.512638/0.511847 = 1.00154.

For the sake of consistency, all $^{143}Nd/^{144}Nd$ ratios cited in this book have been corrected to $^{146}Nd/^{144}Nd = 0.7219$ (row 2 in Table 9.3), and the present value of the $^{143}Nd/^{144}Nd$ ratio of CHUR is 0.512638. In addition, the $^{147}Sm/^{144}Nd$ ratio of CHUR is 0.1967 based on the average Sm/Nd ratio of stony meteorites.

9.2b Model Dates Based on CHUR

The $^{143}Nd/^{144}Nd$ ratio of CHUR at any time t in the past can be calculated by means of equation 9.2:

$$\left(\frac{^{143}\mathrm{Nd}}{^{144}\mathrm{Nd}}\right)^0_{\mathrm{CHUR}} = \left(\frac{^{143}\mathrm{Nd}}{^{144}\mathrm{Nd}}\right)^t_{\mathrm{CHUR}} + \left(\frac{^{147}\mathrm{Sm}}{^{144}\mathrm{Nd}}\right)^0_{\mathrm{CHUR}} (e^{\lambda t} - 1)$$

where the superscript zero signifies the present (i.e., $t = 0$) and the superscript t refers to a time in the past (i.e., t years ago). Therefore,

$$\left(\frac{^{143}\mathrm{Nd}}{^{144}\mathrm{Nd}}\right)^t_{\mathrm{CHUR}} = \left(\frac{^{143}\mathrm{Nd}}{^{144}\mathrm{Nd}}\right)^0_{\mathrm{CHUR}} - \left(\frac{^{147}\mathrm{Sm}}{^{144}\mathrm{Nd}}\right)^0_{\mathrm{CHUR}} (e^{\lambda t} - 1) \tag{9.6}$$

Substituting the appropriate values and setting $t = 4.5 \times 10^9$ years,

$$\left(\frac{^{143}\mathrm{Nd}}{^{144}\mathrm{Nd}}\right)^t_{\mathrm{CHUR}} = 0.512638 - 0.1967(e^{0.02943} - 1) = 0.506764$$

This value is indistinguishable from the average initial $^{143}Nd/^{144}Nd$ ratio in Table 9.4 of six achondrite meteorites (0.506754 ± 0.000063) whose average Sm–Nd isochron age is 4.58 ± 0.02 Ga. The agreement is expected because CHUR was defined on the basis of analytical data derived from stony meteorites.

The CHUR model can be used to date rocks of the continental crust by determining the time at which the Nd they contain separated from the "chondritic reservoir." For this purpose, equation 9.2 is used to express the $^{143}Nd/^{144}Nd$ ratio of the rock (R) at some time t in the past analogous to equation 9.6 for CHUR:

$$\left(\frac{^{143}\mathrm{Nd}}{^{144}\mathrm{Nd}}\right)^t_R = \left(\frac{^{143}\mathrm{Nd}}{^{144}\mathrm{Nd}}\right)^0_R - \left(\frac{^{147}\mathrm{Sm}}{^{144}\mathrm{Nd}}\right)^0_R (e^{\lambda t} - 1)$$

Similarly, for CHUR,

$$\left(\frac{^{143}\mathrm{Nd}}{^{144}\mathrm{Nd}}\right)^t_{\mathrm{CHUR}} = \left(\frac{^{143}\mathrm{Nd}}{^{144}\mathrm{Nd}}\right)^0_{\mathrm{CHUR}} - \left(\frac{^{147}\mathrm{Sm}}{^{144}\mathrm{Nd}}\right)^0_{\mathrm{CHUR}} (e^{\lambda t} - 1)$$

Since

$$\left(\frac{^{143}\mathrm{Nd}}{^{144}\mathrm{Nd}}\right)^t_R = \left(\frac{^{143}\mathrm{Nd}}{^{144}\mathrm{Nd}}\right)^t_{\mathrm{CHUR}}$$

the right sides of both equations must also be equal, which yields

$$\left(\frac{^{143}\mathrm{Nd}}{^{144}\mathrm{Nd}}\right)^0_{\mathrm{CHUR}} - \left(\frac{^{147}\mathrm{Sm}}{^{144}\mathrm{Nd}}\right)^0_{\mathrm{CHUR}} (e^{\lambda t} - 1) = \left(\frac{^{143}\mathrm{Nd}}{^{144}\mathrm{Nd}}\right)^0_R - \left(\frac{^{147}\mathrm{Sm}}{^{144}\mathrm{Nd}}\right)^0_R (e^{\lambda t} - 1)$$

Solving for $e^{\lambda t} - 1$ yields

$$(e^{\lambda t} - 1)\left[\left(\frac{^{147}\mathrm{Sm}}{^{144}\mathrm{Nd}}\right)^0_R - \left(\frac{^{147}\mathrm{Sm}}{^{144}\mathrm{Nd}}\right)^0_{\mathrm{CHUR}}\right] = \left(\frac{^{143}\mathrm{Nd}}{^{144}\mathrm{Nd}}\right)^0_R - \left(\frac{^{143}\mathrm{Nd}}{^{144}\mathrm{Nd}}\right)^0_{\mathrm{CHUR}}$$

$$e^{\lambda t} - 1 = \frac{(^{143}\mathrm{Nd}/^{144}\mathrm{Nd})^0_R - (^{143}\mathrm{Nd}/^{144}\mathrm{Nd})^0_{\mathrm{CHUR}}}{(^{147}\mathrm{Sm}/^{144}\mathrm{Nd})^0_R - (^{147}\mathrm{Sm}/^{144}\mathrm{Nd})^0_{\mathrm{CHUR}}}$$

Solving for t and substituting the values for the isotope ratios of CHUR yields

$$t = \frac{1}{\lambda} \ln \left[\frac{(^{143}\text{Nd}/^{144}\text{Nd})^0_R - 0.512638}{(^{147}\text{Sm}/^{144}\text{Nd})^0_R - 0.1967} + 1 \right] \quad (9.7)$$

The model dates derived from equation 9.7 are based on the following assumptions:

1. Mantle-derived magmas contained Nd whose isotope composition was accurately represented by CHUR throughout geological time.
2. The Sm/Nd ratios of the rocks being dated did not change during their residence in the continental crust.
3. The measured ^{143}Nd/^{144}Nd ratios of the rocks and of CHUR were corrected for isotope fractionation to the same reference ratio.
4. The ^{143}Nd/^{144}Nd ratios of the rocks (after correction for fractionation) are free of instrumental bias and the concentrations of Sm and Nd are accurate.

Although such model dates can be calculated for single samples of rocks and minerals, their interpretation is clouded by the uncertainties concerning the relevance of the CHUR model and by the possible alteration of the Sm/Nd ratios of the samples. Nevertheless, Nd model dates are used to estimate the crustal residence ages of rocks that experienced several episodes of metamorphism during which the Rb–Sr and K–Ar geochronometers were reset.

The calculation of Nd model dates is illustrated in Figure 9.3 by means of a hypothetical example involving a whole-rock sample of syenite whose present Sm/Nd ratio is 0.110 and whose ^{143}Nd/^{144}Nd ratio is 0.510485 (corrected to ^{146}Nd/^{144}Nd = 0.7219). The Nd model date of this specimen is determined from equation 9.7:

$$t = \frac{1}{6.54 \times 10^{-12}} \ln \left[\frac{0.510485 - 0.512638}{0.110 \times 0.602 - 0.1967} + 1 \right]$$

$$= 2.50 \text{ Ga}$$

Figure 9.3 demonstrates that the Nd that now resides in the specimen of syenite could have originated from CHUR at 2.50 Ga provided that the Sm/Nd ratio of the syenite did not change later.

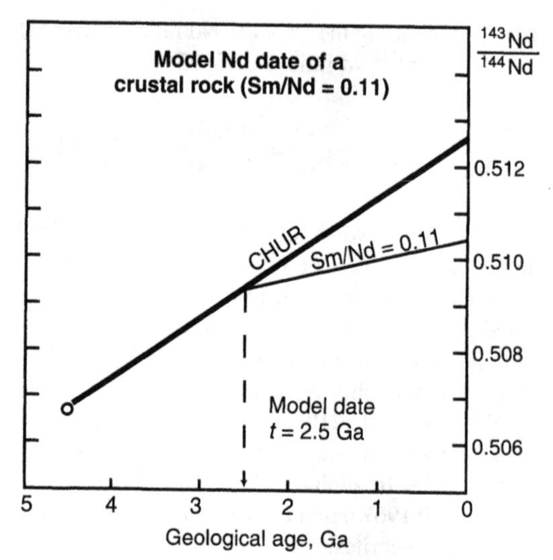

FIGURE 9.3 Hypothetical model date relative to CHUR for a sample of syenite (Sm/Nd = 0.11) whose present ^{143}Nd/^{144}Nd ratio is 0.510485 corrected to ^{146}Nd/^{144}Nd = 0.7219 (row 2, Table 9.3). The diagram demonstrates that the Nd in this rock could have separated from CHUR by magma formation at 2.5 Ga.

However, the model on which this date is based does not preclude the possibility that the age of the syenite is less than 2.50 Ga and that it formed from a crustal magma that originated by partial melting of mantle-derived volcanic rocks or volcanogenic sediment. In that case, the mantle separation date of 2.50 Ga is valid only if the protolith had the same Sm/Nd ratio as the syenite.

9.2c Isotope Standards

Most researchers who measure isotope ratios of Nd in geological materials report their results for certain isotope standards which facilitates interlaboratory comparisons of ^{143}Nd/^{144}Nd ratios provided that fractionation corrections are made the same way. In addition, replicate analyses of standards serve as a test of the performance of mass spectrometers and indicate the reproducibility of the corrected ^{143}Nd/^{144}Nd ratios.

The materials that are in use as Nd isotope standards include BCR-1, a powdered sample of

Columbia River Basalt prepared and distributed by the U.S. Geological Survey, a high-purity Nd_2O_3 reagent sold by Johnson and Mathey (J&M catalog No. 321), and the Nd_2O_3 known as the LaJolla Nd standard prepared by Lugmair and Carlson (1978). The latter was replaced by Tanaka et al. (2000) with a new standard called JNdi-1, which consists of Nd_2O_3 purified by the Shin-Etsu Chemical Co. of Japan (lot No. ND04-010). The $^{143}Nd/^{144}Nd$ ratio of this standard was measured 133 times on 11 mass spectrometers alternating with the LaJolla Nd standard, which was analyzed 145 times. The results indicate that the average $^{143}Nd/^{144}Nd$ ratio of JNdi-1 is 0.512115 ± 0.000007. This value exceeds the $^{143}Nd/^{144}Nd$ ratio of the LaJolla Nd standard (0.511858 ± 0.000007; Lugmair and Carlson, 1978) by a factor of 1.000502 (Tanaka et al., 2000). The $^{143}Nd/^{144}Nd$ ratios of JNdi-1 and LaJolla Nd cited above were corrected for isotope fractionation to $^{146}Nd/^{144}Nd = 0.7219$ (JNdi-1) and to $^{146}Nd/^{144}Nd = 0.721906$ (LaJolla).

The Nd_2O_3 marketed by J&M 321 was analyzed 12 times by Jahn et al. (1980), who reported an average $^{143}Nd/^{144}Nd$ ratio of 0.511137 ± 0.000008, compared to 0.512650 ± 0.000040 for BCR-1 after both ratios were corrected to $^{146}Nd/^{144}Nd = 0.72190$.

The LaJolla Nd standard and BCR-1 have both been analyzed hundreds of times. A random sampling of the literature yielded average $^{143}Nd/^{144}Nd$ ratios of 0.512642 ± 0.000006 ($2\overline{\sigma}$, $N = 11$) for BCR-1 and 0.511848 ± 0.00005 ($2\overline{\sigma}$, $N = 11$) for the LaJolla Nd standard relative to 0.72190 for $^{146}Nd/^{144}Nd$ (row 2, Table 9.3).

The use of BCR-1 as a Nd isotope standard is especially appropriate because the Nd must be separated by the same procedure that is used to analyze unknown rocks and minerals. Therefore, replicate analyses of BCR-1 reflect not only the characteristics and condition of the mass spectrometer but also the effectiveness of the chemical processing of silicate rocks and minerals.

9.2d Epsilon Notation

The $^{143}Nd/^{144}Nd$ ratios of terrestrial rocks are evaluated by comparison to the present isotope ratio of CHUR. This comparison is facilitated by the ε-Nd notation defined by DePaolo and Wasserburg (1976):

$$\varepsilon^0(\mathrm{Nd}) = \left[\frac{(^{143}\mathrm{Nd}/^{144}\mathrm{Nd})^0_R - (^{143}\mathrm{Nd}/^{144}\mathrm{Nd})^0_{\mathrm{CHUR}}}{(^{143}\mathrm{Nd}/^{144}\mathrm{Nd})^0_{\mathrm{CHUR}}}\right] \times 10^4 \tag{9.8}$$

where the superscript zero refers to the present time ($t = 0$) and

$$\left(\frac{^{143}\mathrm{Nd}}{^{144}\mathrm{Nd}}\right)^0_{\mathrm{CHUR}} = 0.512638 \quad \text{relative to} \quad \frac{^{146}\mathrm{Nd}}{^{144}\mathrm{Nd}} = 0.7219$$

The numerical value of $\varepsilon^0(\mathrm{Nd})$ of a terrestrial rock or mineral may be positive, negative, or zero. If $\varepsilon^0(\mathrm{Nd})$ is positive, the Nd in the samples is enriched in radiogenic ^{143}Nd relative to CHUR and therefore originated from a magma-depleted source in the mantle, as illustrated in Figure 9.2. If $\varepsilon^0(\mathrm{Nd})$ is negative, the Nd originated from a magma whose Sm/Nd ratio was less than that of CHUR as a result of partial melting in the mantle (Figure 9.2). The case that $\varepsilon^0(\mathrm{Nd}) = 0$ arises only when the present $^{143}Nd/^{144}Nd$ ratio of a terrestrial rock is equal to 0.512638. In summary, igneous rocks derived from depleted mantle sources have positive $\varepsilon^0(\mathrm{Nd})$ values, whereas rocks of the continental crust have negative $\varepsilon^0(\mathrm{Nd})$ values.

The initial $^{143}Nd/^{144}Nd$ ratios determined from whole-rock Sm–Nd isochrons of igneous or metamorphic rocks can also be compared to the $^{143}Nd/^{144}Nd$ ratio of CHUR at the time t specified by the slope of the isochron. In this case, the $^{143}Nd/^{144}Nd$ ratio of CHUR at time t is calculated from equation 9.9:

$$\left(\frac{^{143}\mathrm{Nd}}{^{144}\mathrm{Nd}}\right)^t_{\mathrm{CHUR}} = 0.512638 - 0.1967\,(e^{\lambda t} - 1)$$

where t is the date indicated by the whole-rock Sm–Nd isochron of the rocks being dated. The comparison of the initial $^{143}Nd/^{144}Nd$ ratio derived from an isochron and the $^{143}Nd/^{144}Nd$ ratio of CHUR at time t is made by means of $\varepsilon^t(\mathrm{Nd})$ defined as before:

$$\varepsilon^t(\text{Nd}) = \left[\frac{(^{143}\text{Nd}/^{144}\text{N})^t_R - (^{143}\text{Nd}/^{144}\text{Nd})^t_{\text{CHUR}}}{(^{144}\text{Nd}/^{144}\text{Nd})^t_{\text{CHUR}}}\right] \times 10^4 \qquad (9.9)$$

where $\left(\frac{^{143}\text{Nd}}{^{144}\text{Nd}}\right)^t_R$ = initial ^{143}Nd/^{144}Nd ratio derived from a whole-rock Sm–Nd isochron

$\left(\frac{^{143}\text{Nd}}{^{144}\text{Nd}}\right)^t_{\text{CHUR}}$ = ^{143}Nd/^{144}Nd ratio of CHUR at the time t and calculated from equation 9.6

In cases where ε^t(Nd) is positive, the *protoliths* of the rocks of age t originated from magma-depleted sources in the mantle. If ε^t(Nd) is negative, the rocks of age t formed from protoliths in the continental crust. These insights provide information about the growth of the continental crust by addition of volcanic or plutonic rocks derived from the mantle [positive ε^t(Nd)] and about recycling of crustal rocks during orogenies by regional metamorphism and partial melting of terrigenous sediment [negative ε^t(Nd)]. However, care must be taken to assure that the fractionation correction applied to the measured ^{143}Nd/^{144}Nd ratios is compatible with the ^{143}Nd/^{144}Nd ratio of CHUR used to calculate ε-values.

9.3 DATING BY THE Sm–Nd METHOD

The Sm–Nd method is most frequently used to date cogenetic mafic igneous rocks by the conventional isochron method (Section 5.3) by equation 9.2:

$$\frac{^{143}\text{Nd}}{^{144}\text{Nd}} = \left(\frac{^{143}\text{Nd}}{^{144}\text{Nd}}\right)_i + \frac{^{147}\text{Sm}}{^{144}\text{Nd}}(e^{\lambda t} - 1)$$

The equation forms a straight line in coordinates of ^{147}Sm/^{144}Nd and ^{143}Nd/^{144}Nd in the slope–intercept form

$$y = b + mx$$

such that the intercept $b = (^{143}\text{Nd}/^{144}\text{Nd})_i$ and the slope $m = e^{\lambda t} - 1$. Rocks or minerals that formed at the same time and had the same initial ^{143}Nd/^{144}Nd ratio are represented by points that define the straight line, which is called an isochron. The equation of the isochron is derived from the analytical data by least-squares cubic regression weighted by the reciprocals of the analytical errors (Section 4.3). The date is calculated from the slope m of the isochron:

$$m = e^{\lambda t} - 1 \qquad t = \frac{1}{\lambda}\ln(m + 1)$$

The interpretation of Sm–Nd dates depends on the assumptions discussed in Section 4.2, which also apply to Rb–Sr and K–Ar dates in Chapters 5 and 6, respectively. The Sm–Nd method is especially well suited to dating mafic volcanic and plutonic rocks of Precambrian age because these kinds of rocks tend to have higher Sm/Nd ratios than igneous rocks of sialic composition (e.g., rhyolites and granites). In addition, the Sm–Nd system is more resistant to alteration than the Rb–Sr and K–Ar systems because the trivalent REEs (Sm^{3+} and Nd^{3+}) form stronger bonds with anions in ionic crystals than Sr^{2+}, Rb^{+}, and K^{+}. Consequently, whole-rock Sm–Nd isochron dates are generally older than Rb–Sr isochron dates derived from the same rocks and are a better measure of the crystallization age of such rocks.

9.3a Onverwacht Group, South Africa

The power of the Sm–Nd method of dating is illustrated by the volcanic rocks of the Onverwacht Group in southern Africa mentioned in Section 7.7b in connection with dating by the ^{40}Ar*/^{39}Ar method. The Onverwacht Group forms the basal portion of the early Precambrian Barberton greenstone belt in Swaziland and Transvaal of South Africa. It consists of a lower ultramafic unit composed of peridotitic and basaltic komatiites with lesser amounts of basic and acid tuffs and small felsic porphyry intrusions that are concordant and penecontemporaneous with the flows. The upper part of the Onverwacht Group is known as the "mafic-to-felsic unit." It is less mafic in character and contains more chemical sedimentary rocks than the lower ultramafic unit.

The age of the volcanic rocks of the Onverwacht Group was not known accurately for many years because low-grade regional metamorphism has disturbed the Rb–Sr and U–Pb systems in these rocks. For example, Allsopp et al. (1973) obtained a Rb–Sr whole-rock isochron date for felsic volcanic rocks of only 2570 ± 40 Ma, whereas Hurley et al. (1972) reported a date of 3280 ± 70 Ma for sediment from the "Middle Marker Horizon" that separates the lower ultramafic unit from the upper mafic-to-felsic unit. Zircons and sulfide minerals from the felsic rocks of the upper unit yielded an even older date of about 3310 Ma by the U–Pb method (Van Niekerk and Burger, 1969). The oldest Rb–Sr whole-rock isochron date of 3430 ± 200 Ma was obtained by Jahn and Shih (1974) for mineral fractions of different density recovered from a basaltic komatiite taken from the Komati Formation at the top of the lower ultramafic unit.

The Sm–Nd data of Hamilton et al. (1979a) for ultramafic-to-felsic volcanic rocks from the lower ultramafic unit define a straight line in Figure 9.4 from which they calculated a date of 3540 ± 30 Ma and an initial $^{143}Nd/^{144}Nd$ ratio of 0.50809 ± 0.00004. These results were subsequently confirmed by Jahn et al. (1982), who reported a Sm–Nd isochron date of 3560 ± 240 Ma and an initial $^{143}Nd/^{144}Nd$ ratio of 0.50818 ± 0.00023 relative to $^{146}Nd/^{144}Nd = 0.72190$.

The Sm–Nd isochron defined by the komatiites and felsic differentiates of the Onverwacht Group in Figure 9.4 has a slope of 0.02342 ± 0.00019 and therefore yields a date t of

$$t = \frac{1}{6.54 \times 10^{-12}} \ln(0.02342 + 1)$$
$$= 3540 \pm 30 \text{ Ma}$$

The $\varepsilon^t(\text{Nd})$ value that corresponds to the initial $^{143}Nd/^{144}Nd$ ratio (0.50809 ± 0.00004) is calculated from equation 9.9:

$$\varepsilon^t(\text{Nd}) = \left[\frac{(^{143}\text{Nd}/^{143}\text{Nd})^t_R}{(^{143}\text{Nd}/^{144}\text{Nd})^t_{\text{CHUR}}} - 1\right] \times 10^4$$

where $(^{143}\text{Nd}/^{144}\text{Nd})^t_R = 0.50809 \pm 0.00004$ and $(^{143}\text{Nd}/^{144}\text{Nd})^t_{\text{CHUR}}$ is calculated from

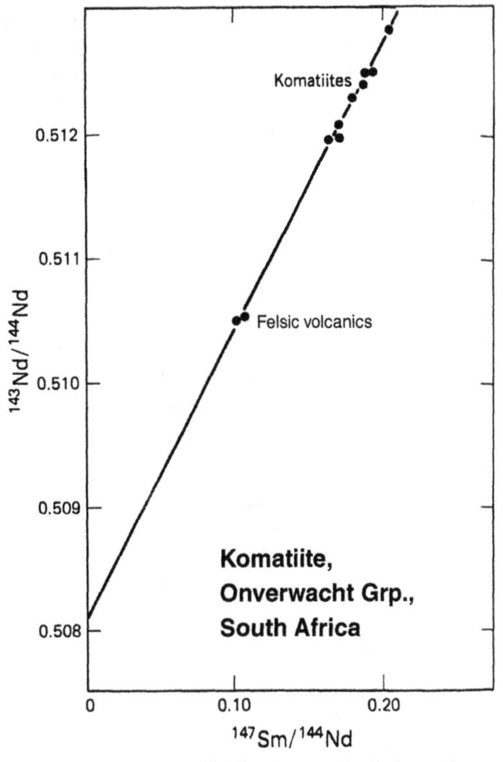

FIGURE 9.4 The Sm–Nd isochron of whole-rock samples from the lower ultramafic unit of the Onverwacht Group in South Africa. The rocks range in composition from peridotitic komatiites to sodic porphyry. The former have higher Sm/Nd ratios than the latter and consequently have higher $^{143}Nd/^{144}Nd$ ratios. The age of these rocks is 3540 ± 30 Ma and their initial $^{143}Nd/^{144}Nd$ ratio is 0.50809 ± 0.00004. Data from Hamilton et al. (1979a).

$$\left(\frac{^{143}\text{Nd}}{^{144}\text{Nd}}\right)^t_{\text{CHUR}} = 0.512638 - 0.1967\,(e^{\lambda t} - 1)$$

Since

$$e^{\lambda t} - 1 = 0.02342 \qquad \left(\frac{^{143}\text{Nd}}{^{144}\text{Nd}}\right)^t_{\text{CHUR}} = 0.50803$$

then

$$\varepsilon^t(\text{Nd}) = \left(\frac{0.50809}{0.50803} - 1\right) \times 10^4 = +1.2 \pm 0.8$$

The range of ε^t(Nd) values results from the uncertainty of ±0.00004 in the initial $^{143}Nd/^{144}Nd$ ratio. Since the ε^t(Nd) value is virtually indistinguishable from zero, the volcanic rocks of the Onverwacht Group originated from undepleted magma sources in the underlying mantle at 3.54 ± 0.03 Ga.

The Rb–Sr data in Figure 9.5 of komatiite basalts from the Komati Formation of the Onverwacht Group scatter widely above and below a 3.54-Ga reference isochron (Jahn et al., 1982). The failure of these rocks to define an isochron was caused by the alteration of the Rb–Sr systematics during low-grade regional metamorphism. These data therefore illustrate the point that the Sm–Nd system in metavolcanic rocks is more robust than the Rb–Sr system.

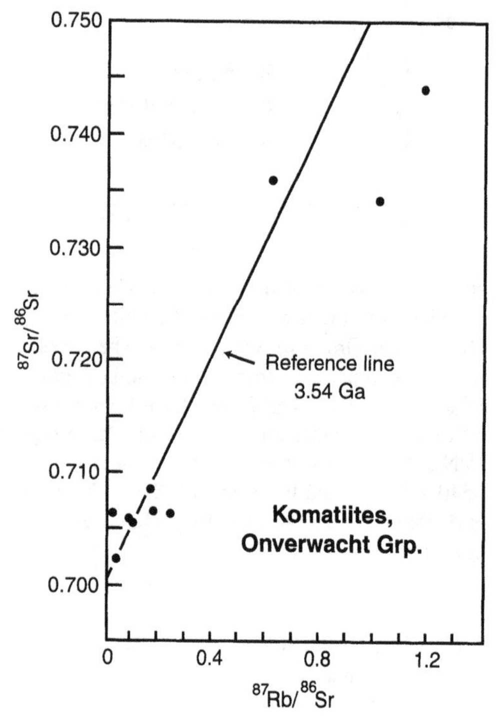

FIGURE 9.5 The Rb–Sr systematics of komatiites in the Onverwacht Group of Swaziland, South Africa. The reference line is based on the whole-rock Sm–Nd date of metavolcanic rocks in the Onverwacht Group reported by Hamilton et al. (1979a). The Rb–Sr data scatter widely, indicating alteration during low-grade metamorphism. Data from Jahn et al. (1982).

Several other suites of basaltic volcanic rocks of early Precambrian age that had resisted dating by the Rb–Sr or U–Pb method have been successfully dated by the Sm–Nd method, such as the komatiites in the greenstone belts of Zimbabwe (Hamilton et al., 1977), of Munro Township in Ontario, Canada (Zindler et al., 1978), and of the Yilgarn block of Western Australia (McCulloch and Compston, 1981). In addition, the Sm–Nd method has been used successfully to confirm age determinations of supracrustal rocks at Isua in Greenland (Hamilton et al., 1978) and the Lewisian gneisses in Scotland (Hamilton et al., 1979b).

9.3b Growth of the Continental Crust

Conventional methods of dating crustal rocks by isotopic methods, in most cases, reveal only the age of the last recrystallization event and do not indicate the time when the material first separated from magma reservoirs in the mantle. This limitation of conventional methods of dating is overcome by Sm–Nd dates derived by comparison with CHUR as outlined in Section 9.2b. These so-called Nd model dates are calculated from equation 9.7 and represent the time since the Nd in a terrestrial rock separated from a magma source in the mantle represented by CHUR. The dates are based on the assumption that the Sm/Nd ratio of the material was not changed by any of the geological processes to which it was subjected (e.g., fractional crystallization of the magma, weathering of the resulting igneous rocks, transport and deposition of sediment, diagenesis, metamorphic recrystallization, and water–rock interaction). Although this condition may not be met in all cases, the general resistance of the Sm–Nd systematics of crustal rocks to alteration has been widely noted in the geochemical literature.

McCulloch and Wasserburg (1978) used Nd model dates of composite samples of rocks, representing different structural provinces of the Precambrian Shield of North America, in order to ascertain whether the rocks in these provinces had actually been added to the crust at the time indicated by their Rb–Sr and K–Ar dates. Their data were recalculated to make the $(^{143}Nd/^{144}Nd)^0_R$ ratios compatible with $^{146}Nd/^{144}Nd = 0.72190$ and to convert the $f(\text{Sm/Nd})$ values into the $^{147}\text{Sm}/^{144}\text{Nd}$ ratios, given that

$$f\left(\frac{\text{Sm}}{\text{Nd}}\right) = \left[\frac{(\text{Sm/Nd})^0_R}{(\text{Sm/Nd})_{\text{CHUR}}} - 1\right] \qquad (9.10)$$

and that McCulloch and Wasserburg (1978) stated that $(^{147}\text{Sm}/^{144}\text{Nd})_{\text{CHUR}} = 0.1936$ rather than 0.1967.

Therefore, the data for the composite sample from New Quebec yield a Nd model date based on equation 9.7:

$$t = \frac{1}{\lambda}\ln\left[\frac{(^{143}\text{Nd}/^{144}\text{Nd})^0_R - (^{143}\text{Nd}/^{144}\text{Nd})^0_{\text{CHUR}}}{(^{147}\text{Sm}/^{144}\text{Nd})^0_R\ (^{147}\text{Sm}/^{144}\text{Nd})^0_{\text{CHUR}}} + 1\right]$$

$$= \frac{1}{6.54 \times 10^{-12}}\ln\left(\frac{0.511130 - 0.512638}{0.1076 - 0.1936} + 1\right)$$

$$= 2.66 \text{ Ga}$$

This result indicates that the Nd in the rocks of New Quebec in Figure 9.6 of the Superior tectonic province was withdrawn from magma reservoirs (CHUR) in the underlying mantle at about 2.66 Ga, which is indistinguishable from the time indicated by their crystallization ages of 2.5–2.7 Ga. In other words, the Superior tectonic province in New Quebec consists of "new crust" rather than of recycled crust. In addition, the intrusion and/or eruption of the igneous rocks and their stabilization in the continental crust took about 100 million years or less and is not resolved by the available dates. These conclusions are confirmed by the Nd model date of 2.50 ± 0.05 Ga for the rocks of Northern Quebec in Figure 9.6. Similarly, the model dates for the gneisses and granites of the Fort Enterprise areas in the Slave tectonic province (2.62 ± 0.05 and 2.49 ± 0.04 Ga) are indistinguishable from the geological age of the rocks (2.5–2.7 Ga), indicating that this province is also composed of new crust.

The rocks in the Churchill tectonic province (1.8–1.9 Ga) are represented in Figure 9.6 by rocks from Baffin Island and Saskatchewan. The Nd

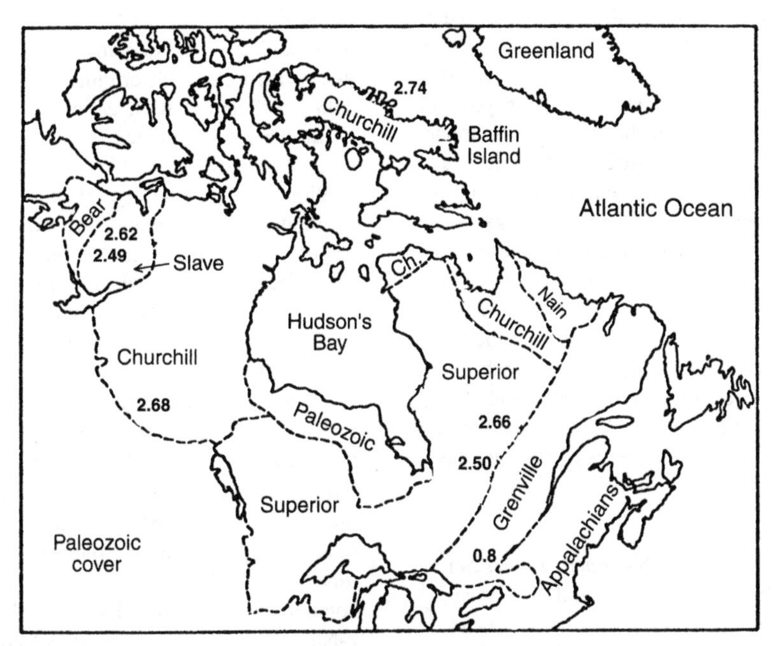

FIGURE 9.6 Neodymium model dates relative to CHUR of composite rock samples from the principal tectonic provinces of the Precambrian Shield of Canada. The dates are in units of billions of years. The model dates of rocks in the Churchill Province (Baffin Island and Saskatchewan) reveal that the Nd in these rocks has resided in the crust since the Late Archean Era. Adapted from McCulloch and Wasserburg (1978).

model dates of these composites (2.74 ± 0.05 and 2.68 ± 0.06 Ga, respectively) are *older* than the crystallization ages of the rocks in the Churchill province. Consequently, the Nd in these rocks left the magma sources in the mantle between 2.74 and 2.68 Ga and resided in the crust for about 900 million years before the present rocks formed. These results therefore reveal that the Churchill tectonic province consists, at least in part, of recycled or rejuvenated rocks of Late Archean age. Additional insight was provided by Chauvel et al. (1987), who demonstrated that the Nd model dates of rocks in the Trans-Hudson orogen (Churchill tectonic province) decrease systematically with increasing distance from the Superior tectonic province. This gradual transition implies that the rocks of the Trans-Hudson orogen contain varying amounts of Nd derived from the Archean rocks of the Superior Province. The geological history and tectonic evolution of the Churchill Province and its relation to the Archean crust of the Superior Province was summarized by St-Onge et al. (2000) and Bickford et al. (1986).

The rocks of the Grenville tectonic province (Figure 9.6) yield a model Nd date of 0.80 ± 0.04 Ga compared to a geological age of 0.9–1.2 Ga. Although the Nd model date is younger than the geological age in this case, the Sr model date (1.01 ± 0.01 Ga) reported by McCulloch and Wasserburg (1978) is consistent with the age of this tectonic province. Therefore, the rocks of the Grenville Province are younger, in most cases, than the rocks of the Superior and Churchill tectonic provinces. However, rejuvenated rocks of Archean age do occur locally along the suture between the Grenville and Superior provinces, which forms the so-called Grenville Front (Frith and Doig, 1975; Gower and Tucker, 1994). Neodymium model dates of sedimentary rocks have also been reported by:

Miller and O'Nions, 1984, *Earth Planet Sci. Lett.*, 68:459–470. Allègre and Rousseau, 1984, *Earth Planet. Sci. Lett.*, 67:19–34. Michard et al., 1985, *Geochim. Cosmochim. Acta*, 49:601–610. Frost and Winston, 1986, *J. Geol.*, 95:309–327. Frost and Coombs, 1989, *Am. J. Sci.*, 289:744–770.

The interpretation of model Nd dates based on CHUR is controversial for a number of reasons (Arndt and Goldstein, 1987):

1. The mantle is not "uniform" in composition but consists of several components having characteristic isotope compositions of Sr, Nd, Pb, Hf, Os, and other elements (Hart, 1988).
2. Magmas may originate not only in the mantle but also from a mixture of sources in the continental crust and mantle or entirely by partial melting in the continental crust.
3. Consequently, Nd model dates may be fictitious and should not be used to determine the rate of growth of the continental crust.

The problem is illustrated in Figure 9.7, which depicts the isotopic evolution of Nd in a hypothetical crustal rock whose present $^{143}Nd/^{144}Nd$ ratio is 0.5100 and whose initial $^{143}Nd/^{144}Nd$ ratio (0.5090 at 2.0 Ga) was determined by means of a Sm–Nd isochron. An extrapolation of the evolution line of this rock yields a model Nd date of 3.25 Ga. Figure 9.7 demonstrates that if the magma that formed this rock originated from a mixture of "old crust" and CHUR, the model date of 3.25 Ga is, at best, an average of the mantle separation date of the old crust (3.65 Ga) and the crystallization date of the hypothetical rock (2.0 Ga). In addition, the mantle-derived magma may have originated not from CHUR but from a depleted lithospheric mantle whose present $^{143}Nd/^{144}Nd$ ratio of about 0.51315 is indicated by MORBs, which are derived from a depleted-MORB mantle (DMM) (Hart, 1988).

The best way to date suites of cogenetic crustal rocks is by means of whole-rock Sm–Nd isochrons that yield a date as well as the initial $^{143}Nd/^{144}Nd$ ratio of the rocks. The numerical value of the initial ratio can then be compared to the $^{143}Nd/^{144}Nd$ ratio of an appropriate magma source in the mantle at the time of crystallization of the rock samples. This comparison is embodied in the calculation of the ε-Nd value at time t by means of equation 9.9:

$$\varepsilon^t(\mathrm{Nd}) = \left[\frac{(^{143}\mathrm{Nd}/^{144}\mathrm{Nd})^t_R}{(^{143}\mathrm{Nd}/^{144}\mathrm{Nd})^t_{\mathrm{CHUR}}} - 1\right] \times 10^4$$

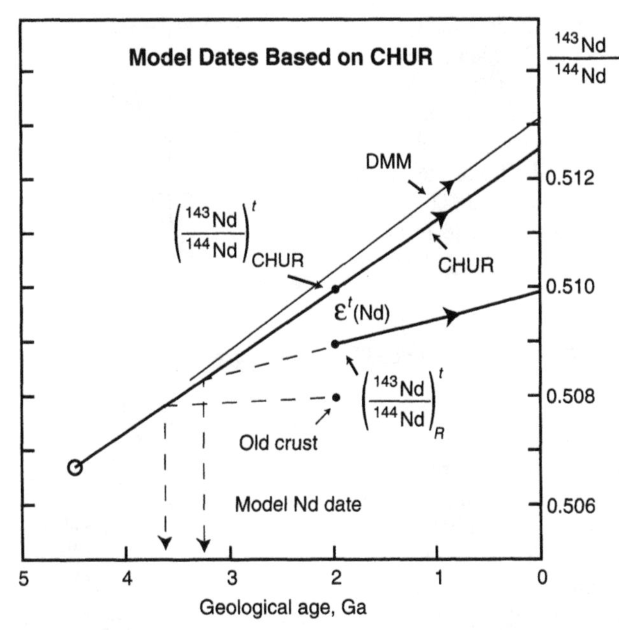

FIGURE 9.7 The ε(Nd) values for crustal rocks at the time of their crystallization at time t relative to CHUR. In this case, $(^{143}Nd/^{144}Nd)^t_R < (^{143}Nd/^{144}Nd)^t_{CHUR}$, causing ε^t(Nd) to be negative, in accordance with equation 9.9. The protoliths of the crustal rocks separated from CHUR at 3.25 Ga and resided in the crust for 1250 million years before assuming their present form at 2.0 Ga.

In this expression, CHUR serves as a reference that can be replaced by DMM or another mantle component depending on the circumstances. The interpretation of ε^t(Nd) arises primarily from its algebraic sign rather than from its numerical value because rocks that contain Nd derived from pre-existing crustal rocks yield negative ε^t(Nd) values. For example, in Figure 9.7 the initial $^{143}Nd/^{144}Nd$ ratio of the hypothetical rock sample is convertible into an ε-value:

$$\varepsilon^t(\mathrm{Nd}) = \left[\frac{0.5090}{0.5100} - 1\right] \times 10^4 = -19.6$$

The limitations of Nd model dates derived from CHUR were taken into consideration elsewhere:

Cordani and Sato, 1999, *Episodes*, 22:167–173. Patchett and Bridgwater, 1984, *Contrib. Mineral. Petrol.*, 87:311–318. Patchett and Kouvo, 1986, *Contrib. Mineral. Petrol.*, 92:1–12. Patchett and Arndt, 1986, *Earth Planet. Sci. Lett.*, 78:329–338. Wilson et al., 1985, *Earth Planet. Sci. Lett.*, 72:376–388. Huhma, 1986, *Geol. Surv. Finland Bull.*, 337:1–48. Kalsbeek and Taylor, 1985, *Earth Planet Sci. Lett.*, 73:65–80.

9.4 METEORITES AND MARTIAN ROCKS

Many stony meteorites have been dated by the Sm–Nd method by analyzing mineral fractions separated from individual specimens. The results are interpreted on Sm–Nd isochron diagrams to determine both the age and the primordial $^{143}Nd/^{144}Nd$ ratio of these meteorites. A mineral isochron in Figure 9.8 for the achondrite (eucrite) Moama published by Hamet et al. (1978) indicates an age of 4.58 ± 0.05 Ga for this meteorite and a primordial $^{143}Nd/^{144}Nd$ ratio of 0.50684 ± 0.00008. The age and initial $^{143}Nd/^{144}Nd$ ratio of Moama agree well with those of other achondrites listed in Table 9.4 and with Rb–Sr dates (Section 5.3b). The average initial $^{143}Nd/^{144}Nd$ ratio of the achondrites in Table 9.4 is 0.506754 ± 0.000063 ($2\bar{\sigma}$, $N = 7$). This value is indistinguishable from the $^{143}Nd/^{144}Nd$ ratio of CHUR, as pointed out in Section 9.2b.

The Sm–Nd method has also been used to date martian rocks (e.g., Shergotty, Nakhla, Zagami, Allan Hills 77005) discussed in Section 5.3c. For example, Figure 9.9 contains a Sm–Nd mineral/whole-rock isochron for Nakhla from which Nakamura et al. (1982) obtained a date of 1.26 ± 0.07 Ga. Shih et al. (1982) determined a similar date of 1.34 ± 0.06 Ga from a whole-rock isochron of Shergotty, Zagami, and Allan Hills 77005. These results confirm the Rb–Sr dates of 1.21 ± 0.01 Ga (Gale et al., 1975) and 1.34 ± 0.02 Ga (Papanastassiou and Wasserburg, 1974) for Nakhla.

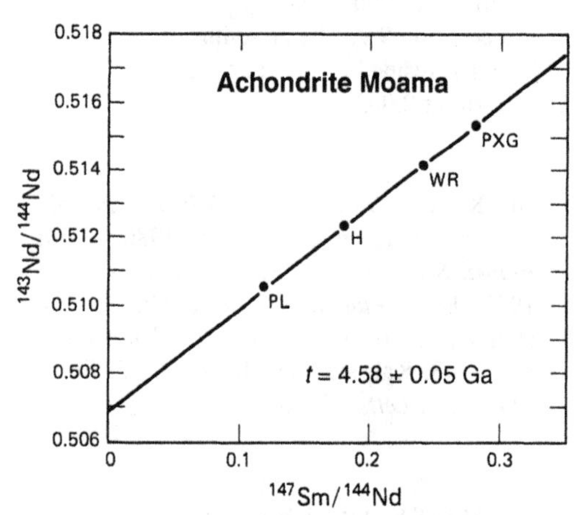

FIGURE 9.8 The Sm–Nd isochron based on mineral fractions of the achondrite (eucrite) Moama. The age of this meteorite is 4.58 ± 0.05 Ga and the initial (primordial) $^{143}Nd/^{144}Nd$ ratio is 0.50684 ± 0.00008. The measured $^{143}Nd/^{144}Nd$ ratios were corrected for isotope fractionation to $^{150}Nd/^{144}Nd = 0.236433$ (row 2, Table 9.3). Abbreviations: PXG = greenish pyroxene; WR = whole rock, H = heavy fraction >3.2 g/cm^3, PG = plagioclase. Data from Hamet et al. (1978).

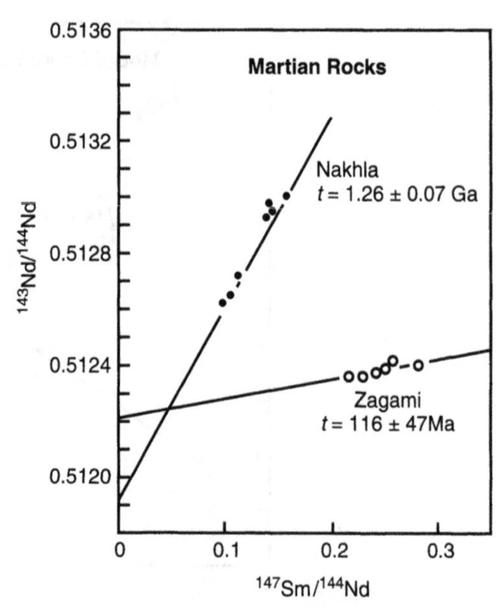

FIGURE 9.9 The Sm–Nd mineral isochrons for the martian rocks Nakhla and Zagami. The Sm–Nd date and initial $^{143}Nd/^{144}Nd$ ratio of Nakhla are 1.26 ± 0.07 Ga and 0.51181 ± 0.00007, respectively. The Zagami isochron yields a date of 116 ± 47 Ma and an initial $^{143}Nd/^{144}Nd$ of 0.512208 ± 0.000074 (recalculated to row 2 of Table 9.3). Data from Nakamura et al. (1982) for Nakhla and from Shih et al. (1982) for Zagami.

Figure 9.9 also contains a Sm–Nd isochron of the minerals of Zagami that yields a date of only 116 ± 47 Ma. This age determination refers to an episode of internal redistribution of radiogenic ^{143}Nd as a result of shock metamorphism of the Zagami rock. The Sm–Nd date is somewhat lower than the Rb–Sr internal isochron date of Zagami (180 ± 4 Ma) and is also lower than the Rb–Sr dates of Shergotty (165 ± 11 Ma) and ALH 77005 (187 ± 12 Ma). Shih et al. (1982) concluded that Zagami, Shergotty, and ALH 77005 experienced shock metamorphism at 180 ± 4 Ma while all three rocks resided on their parent body now known to be the planet Mars. Cosmic-ray exposure dates indicate that ALH 77005 was ejected from the surface of Mars at about 3.0 Ma followed by

Table 9.4. Ages and Primordial $^{143}Nd/^{144}Nd$ Ratios of Stony Meteorites and Martian Rocks Determined by the Sm–Nd Method

Meteorite	Age (Ga)	$\left(\frac{^{143}Nd}{^{144}Nd}\right)_i^a$	Reference[b]
	Achondrites		
Juvinas	4.56 ± 0.08	0.50677 ± 10	1
Juvinas	4.60	0.506616 ± 22	6
Angra dos Reis	4.55 ± 0.04	0.50682 ± 5	2
Angra dos Reis	4.562 ± 0.031	0.506664 ± 37	7
Pasamonte	4.58 ± 0.12	0.50681 ± 14	3
Moore County	4.60 ± 0.03	0.50676 ± 7	4
Moama	4.58 ± 0.05	0.50684 ± 8	5
	Chrondrites		
Murchison, Allende, 4.60 Guareña, Peace River, St. Severin, Juvinas (achondrite)		0.506609 ± 8	6
	Martian Rocks		
Nakhla	1.26 ± 0.07	0.51181 ± 7	8
Shergotty, Zagami, Allan Hills 7705	1.34 ± 0.06	0.51020 ± 10	9

[a] All initial $^{143}Nd/^{144}Nd$ have been normalized to an isotope ratio in row 2 of Table 9.3 and are compatible with $^{146}Nd/^{144}Nd = 0.7219$.

[b] 1. Lugmair et al., 1976, Proc. 7th Lunar Planet. Sci. Conf., *Geochim. Cosmochim. Acta*, Suppl: 2009–2033. 2. Lugmair and Marti, 1977, *Earth Planet. Sci. Lett.*, 35:273–284. 3. Unruh et al., 1977, *Earth Planet. Sci. Lett.*, 37:1–12. 4. Nakamura et al., 1977, *Proc. 8th Lunar Sci. Conf.*, Lunar Planet. Sci. Inst., Houston, TX, pp. 712–713. 5. Hamet et al., 1978, Proc. 9th Lunar Sci. Conf., *Geochim. Cosmochim. Acta*, Suppl.: 1115–1136. 6. Jacobsen and Wasserburg, 1980, *Earth Planet. Sci. Lett.*, 50:139–155. 7. Jacobsen and Wasserburg, 1981, *Proc. 12th Lunar Sci. Conf.*, Lunar Planet. Inst., Houston, TX, pp. 500–502. 8. Nakamura et al., 1982, *Geochim. Cosmochim. Acta*, 46:1555–1573. 9. Shih et al., 1982, *Geochim. Cosmochim. Acta*, 46:2323–2344.

Zagami and Shergotty at 2.5 Ma. The petrogenesis of the martian meteorites and hence the evolution of the mantle of Mars have been discussed elsewhere:

> Jones, 1986, *Geochim. Cosmochim. Acta*, 50:969–977. Jones, 1989, *Proc. 19th Lunar Sci. Conf.*, Lunar Planet. Sci. Inst., Houston, TX, pp. 465–475. Longhi, 1991, *Proc. 21st Lunar Sci. Conf.*, Lunar Planet. Sci. Inst., Houston, TX, pp. 465–475. Jagoutz, 1991, *Space Sci. Rev.*, 56:13–22. McSween et al., 1996, *Geochim. Cosmochim. Acta*, 22:4563–4569. Borg et al., 1997, *Geochim. Cosmochim. Acta*, 61:4915–4931.

9.5 LUNAR ROCKS

The lunar rocks collected by the Apollo astronauts must be dated by analyzing the minerals they contain because all of the specimens were loose rocks and are not products of the same magma. Therefore, investigators devised a variety of procedures for separating mineral fractions based on their magnetic properties, density, and physical appearance. The resulting internal Sm–Nd isochron dates are generally in good agreement with Rb–Sr, $^{40}Ar^*/^{39}Ar$, and U-Pb dates of the same specimens (Nyquist and Shih, 1992). However, in some cases the Rb–Sr and $^{40}Ar^*/^{39}Ar$ dates of impact melts were reset by thermal metamorphism (Sections 5.3d and 7.4c).

A medium-grained basalt (specimen 75075) from the top of a large boulder at the edge of Camelot crater in the Taurus-Littrow valley (Apollo 17) was analyzed by Lugmair et al. (1975) for dating by means of an internal Sm–Nd mineral isochron. The results in Figure 9.10 are based on analyses of plagioclase (Pl), ilmenite (I), whole rock (WR), and pyroxene (Py). The slope m of the isochron is 0.02449, which corresponds to a date of

$$t = \frac{1}{6.54 \times 10^{-12}} \ln(1.02449) = 3.70 \text{ Ga}$$

The initial $^{143}Nd/^{144}Nd$ ratio is 0.50825 ± 0.00012 relative to $^{146}Nd/^{144}Nd = 0.721869$.

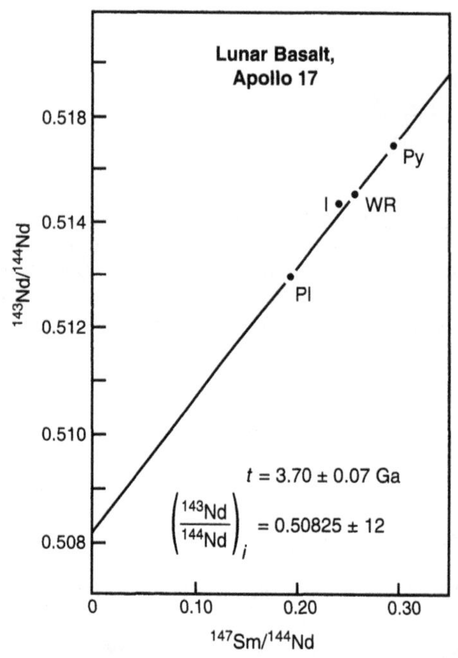

FIGURE 9.10 Internal Sm–Nd isochron of basalt sample 75075 collected in the Taurus-Littrow valley (Apollo 17) of the Moon. Data from Lugmair et al. (1975).

The $^{147}Sm/^{144}Nd$ ratio (0.2566 ± 0.0003) of the basalt dated by Lugmair et al. (1975) is significantly higher than that of CHUR (0.1967). The apparent enrichment of this rock in Sm relative to Nd implies that its magma source was either a cumulate of pyroxene crystals or the residue of a previous episode of magma formation. In that case, the initial $^{143}Nd/^{144}Nd$ ratio of the basalt at 3.70 Ga should be greater than the $^{143}Nd/^{144}Nd$ ratio of CHUR at that time and its $\varepsilon^t(\text{Nd})$ value should be positive. Since

$$\varepsilon^t(\text{Nd}) = \left[\frac{(^{143}\text{Nd}/^{144}\text{Nd})^t_R}{(^{143}\text{Nd}/^{144}\text{Nd})^t_{\text{CHUR}}} - 1\right] \times 10^4$$

the $\varepsilon^t(\text{Nd})$ value of lunar basalt 75075 is

$$\varepsilon^t(\text{Nd}) = \left[\frac{0.50825}{0.50782} - 1\right] \times 10^4 = +8.5$$

The positive sign of $\varepsilon^t(\text{Nd})$ confirms the conjecture that the magma, from which this and other lunar basalt flows formed, originated from sources that had been enriched in Sm and therefore had a higher Sm/Nd ratio than chondrite meteorites.

The validity of this conclusion depends on the assumption that the mantle of the Moon is adequately represented by stony meteorites on the basis of which CHUR was defined. Lugmair et al. (1975) discussed the isotopic evolution of basalt 75075 with reference to the achondrite Juvinas whose initial $^{143}Nd/^{144}Nd$ ratio at 4.56 Ga was 0.50687 and whose $^{147}Sm/^{144}Nd$ ratio is 0.1849 (Sm/Nd = 0.3072). The authors demonstrated that the magma source of basalt 75075 formed at 4.35 Ga by an increase of its Sm/Nd ratio from 0.3072 (Juvinas) to 0.4262. Subsequently, basalt magma formed in this Sm-enriched reservoir at 3.70 ± 0.07 Ga and crystallized at the surface without additional differentiation or alteration until 143 ± 5 Ma when specimen 75075 was ejected by a meteorite impact. Most recently, Nyquist and Shih (1992) reconstructed the history of magmatic activity on the Moon based on K–Ar, $^{40}Ar^*/^{39}Ar$, Rb–Sr, Sm–Nd, U–Pb, and Pb–Pb dates of plutonic rocks. Additional discussions of the mantle of the Moon have been published by:

Papanastassiou et al., 1977, *Proc. 8th Lunar Sci. Conf.*, Lunar Planet Sci. Inst., Houston, TX, pp. 1639–1672. Lugmair and Marti, 1978, *Earth Planet. Sci. Lett.*, 39:349–357; Lugmair and Carlson, 1978, *Proc. 9th Lunar Sci. Conf.*, Lunar Planet. Sci. Inst., Houston, TX, pp. 689–704. Nyquist et al., 1979, *Proc. 10th Lunar Sci. Conf.*, Lunar Planet. Sci. Inst., Houston, TX, pp. 77–114.

9.6 SUMMARY

The Sm–Nd method is widely used to date mafic igneous rocks of Precambrian age by means of whole-rock isochrons. Such dates are more reliable estimates of the ages of igneous and metamorphic rocks than Rb–Sr, K–Ar, and $^{40}Ar^{*}/^{39}Ar$ dates because the concentrations of Sm and Nd of rocks are more resistant to alteration during metamorphism and hydrothermal alteration than the concentrations of Rb, Sr, K, and Ar.

The isotopic evolution of Nd in the mantle of the Earth is assumed to be represented by a chondritic uniform reservoir CHUR whose present $^{143}Nd/^{144}Nd$ ratio is 0.512638 after correction for isotope fractionation to $^{146}Nd/^{144}Nd = 0.72190$. The present $^{147}Nd/^{144}Nd$ ratio is 0.1967. CHUR can be used to calculate model dates for samples of crustal rocks based on the time in the past when CHUR and the rock samples had identical $^{143}Nd/^{144}Nd$ ratios. However, the resulting model dates do not necessarily indicate the time when the Nd in a crustal rock was removed from the mantle and therefore are not reliable indicators of crustal growth.

CHUR serves a more useful purpose in the calculation of ε-values based on the initial $^{143}Nd/^{144}Nd$ ratios determined from Sm–Nd isochrons of igneous and metamorphic rocks. If $\varepsilon^{t}(Nd)$ is negative, the rocks contain crustal material that was recycled at the time the present rocks formed. If $\varepsilon^{t}(Nd)$ is positive, the rocks originated from magma sources in the mantle that were enriched in Sm during a previous episode of partial melting.

The Sm–Nd method has been effective in dating stony meteorites and rocks from the Moon and Mars. In addition, the Sm–Nd systematics of extraterrestrial rocks provide insights into the petrogenesis of such rocks and into the evolution of their magma sources during the early history of the solar system.

REFERENCES FOR PRINCIPLES AND METHODOLOGY (SECTIONS 9.1–9.2)

Anders, E., and N. Grevesse, 1989. Abundances of the elements: Meteoritic and solar. *Geochim. Cosmochim. Acta*, 53:197–214.

Begemann, F., K. R. Ludwig, G. W. Lugmair, K. Min, L. E. Nyquist, P. J. Patchett, P. R. Renne, C.-Y. Shih, I. M. Villa, and R. J. Walker, 2001. Call for an improved set of decay constants for geochronological use. *Geochim. Cosmochim. Acta*, 65:111–121.

DePaolo, D. J., and G. J. Wasserburg, 1976. Nd isotopic variations and petrogenetic models. *Geophys. Res. Lett.*, 3:249–252.

Herrmann, A. G., 1970. Yttrium and the lanthanides. In K. H. Wedepohl (Ed.), *Handbook of Geochemistry*, Vol. 2(5), pp. 57-71B–M. Springer-Verlag, Berlin.

Jacobsen, S. B., and G. J. Wasserburg, 1980. Sm-Nd isotopic evolution of chondrites. *Earth Planet. Sci. Lett.*, 50:139–155.

Jahn, B.-M., J. Bernard-Griffiths, R. Charlot, J. Cornichet, and F. Vidal, 1980. Nd and Sr isotopic compositions and REE abundances of Cretaceous MORB (Holes 417D and 418A Legs 51, 52, and 53). *Earth Planet. Sci. Lett.*, 48:171–184.

Lide, D. R., and H. P. R. Frederikse, 1995. *Handbook of Chemistry and Physics*, 76th ed. CRC Press, Boca Raton, Florida.

Lugmair, G. W. and K. Marti, 1978. Lunar initial $^{143}Nd/^{144}Nd$: Differential evolution of the lunar crust and mantle. *Earth Planet. Sci. Lett.*, 39:349–357.

Lugmair, G. W., and R. W. Carlson, 1978. The Sm-Nd history of KREEP, Proc. 9th Lunar Planet. Sci. Conf. *Geochim. Cosmochim. Acta*, 684–704.

Tanaka, T., et al., 2000. JNdi-1: A neodymium isotopic reference in consistency with LaJolla neodymium. *Chem. Geol.*, 168:279–281.

Wasserburg, G. J., S. B. Jacobsen, D. J. DePaolo, M. T. McCulloch, and T. Wen, 1981. Precise determination of Sm/Nd ratios, Sm and Nd isotopic abundances in standard solutions. *Geochim. Cosmochim. Acta*, 45:2311–2323.

REFERENCES FOR METAVOLCANIC ROCKS OF THE ONVERWACHT GROUP, SOUTH AFRICA AND OTHERS. (SECTION 9.3a)

Allsopp, H. L., M. J. Viljoen, and R. P. Viljoen, 1973. Strontium isotopic studies of the mafic and felsic rocks of the Onverwacht Group of the Swaziland sequence. *Geologische Rundschau*, 62:902–917.

Hamilton, P. J., N. M. Evensen, R. K. O'Nions, H. S. Smith, and A. J. Erlank, 1979a. Sm-Nd dating of Onverwacht Group volcanics, Southern Africa. *Nature*, 279:298–300.

Hamilton, P. J., N. M. Evensen, R. K. O'Nions, and J. Tarney, 1979b. Sm-Nd systematics of Lewisian gneisses: Implications for the origin of granulites. *Nature*, 277:25–28.

Hamilton, P. J., R. K. O'Nions, and N. M. Evensen, 1977. Sm-Nd dating of Archaean basic and ultrabasic volcanics. *Earth Planet. Sci. Lett.*, 36:263–268.

Hamilton, P. J., R. K. O'Nions, N. M. Evensen, D. Bridgwater, and J. H. Allaart, 1978. Sm-Nd isotopic investigations of the Isua Supracrustals and implications for mantle evolution. *Nature*, 272:41–43.

Hurley, P. M., W. H. Pinson, Jr., B. Nagy, and T. M. Teska, 1972. Ancient age of the Middle Marker Horizon: Onverwacht Group, Swaziland sequence, South Africa. *Earth Planet. Sci. Lett.*, 14:360–366.

Jahn, B.-M., G. Gruau, and A. Y. Glickson, 1982. Komatiites of the Onverwacht Group, S. Africa: REE geochemistry, Sm/Nd age, and mantle evolution. *Contrib. Mineral. Petrol.*, 80:25–40.

Jahn, B.-M., and C.-Y. Shih, 1974. On the age of the Onverwacht Group, Swaziland Sequence, South Africa. *Geochim. Cosmochim. Acta*, 38:873–885.

McCulloch, M. T., and W. Compston, 1981. Sm-Nd age of Kambalda and Kanowna greenstones and heterogeneity in the Archean mantle. *Nature*, 294:322–327.

Van Niekerk, C. B., and A. J. Burger, 1969. The U-Pb isotopic dating of South African acid lavas. *Bull. Volcanol.*, 32:481.

Zindler, A., C. Brooks, N. T. Arndt, and S. R. Hart, 1978. Nd and Sr isotope data from komatiitic and tholeiitic rocks of Munro Township, Ontario. In R. E. Zartman (Ed.), *International Conference on Geochronology, Cosmochronology, and Isotope Geology*, U. S. Geol. Survey Open-File Rept. 78–701, pp. 469–471. U.S. Geological Survey, Washington, D.C.

REFERENCES FOR GROWTH OF CONTINENTAL CRUST (SECTION 9.3b)

Arndt, N. T., and S. L. Goldstein, 1987. Use and abuse of crust-formation ages. *Geology*, 15:893–895.

Bickford, M. E., W. R. Van Schmus, and I. Zietz, 1986. Proterozoic history of the midcontinent region of North America. *Geology*, 14:492–496.

Chauvel, C., N. T. Arndt, S. Kielinczuk, and A. Thom, 1987. Formation of Canadian 1.9 Ga old continental crust: (1) Nd isotopic data. *Can. J. Earth Sci.*, 24:396–406.

Frith, R. A., and R. Doig, 1975. Pre-Kenoran tonalitic gneisses in the Grenville province. *Can. J. Earth Sci.*, 12:844–849.

Gower, C. F., and R. D. Tucker, 1994. Distribution of pre-1400 Ma crust in the Grenville province; implications for rifting in Laurentia-Baltica during geon 14. *Geology*, 22:827–830.

Hart, S. R., 1988. Heterogeneous mantle domains: Signatures, genesis, and mixing chronologies. *Earth Planet. Sci. Lett.*, 90:273–296.

McCulloch, M. T., and G. J. Wasserburg, 1978. Sm-Nd and Rb-Sr chronology of continental crust formation. *Science*, 200:1003–1011.

St.-Onge, M. R., D. J. Scott, and S. R. Lucas, 2000. Early partitioning of Quebec: Microcontinent formation in the Paleoproterozoic. *Geology*, 28:323–326.

REFERENCES FOR METEORITES, MARTIAN ROCKS, AND LUNAR ROCKS (SECTIONS 9.4–9.5)

Gale, N. H., J. W. Arden, and R. Hutchison, 1975. The chronology of the Nakhla achondrite meteorite. *Earth Planet. Sci. Lett.*, 26:195–206.

Hamet, J., N. Nakamura, D. M. Unruh, and M. Tatsumoto, 1978. Origin and history of the adcumulate eucrite, Moama as inferred from REE abundances, Sm-Nd, and U-Pb systematics. Proc. 9th Lunar Science Conf. *Geochim. Cosmochim. Acta*, 1115–1136.

Lugmair, G. W., N. B. Scheinin, and K. Marti, 1975. Sm-Nd age and history of Apollo 17 basalt 75075: Evidence for early differentiation of the lunar exterior. In *Proc. 6th Lunar Sci. Conf.*, pp. 1419–1429. Lunar and Planet, Science Institute, Houston, Texas.

Nakamura, N., D. M. Unruh, and M. Tatsumoto, 1982. Origin and evolution of the Nakhla meteorite inferred from the Sm-Nd and U-Pb systematics and REE, Ba, Sr, Rb and K abundances. *Geochim. Cosmochim. Acta*, 46:1555–1573.

Nyquist, L. E., and C.-Y. Shih, 1992. The isotopic record of lunar volcanism. *Geochim. Cosmochim. Acta*, 56:2213–2234.

Papanastassiou, D. A., and G. J. Wasserburg, 1974. Evidence for late formation and young metamorphism in the achondrite Nakhla. *Geophys. Res. Lett.*, 1: 23–26.

Shih, C.-Y., L. E. Nyquist, D. D. Bogard, G. A. McKay, J. L. Wooden, B. M. Bansall, and H. Wiesmann, 1982. Chronology and petrogenesis of young achondrites, Shergotty, Zagami, and ALHA 77005: Late magmatism on a geologically active planet. *Geochim. Cosmochim. Acta*, 46:2323–2344.

CHAPTER 10

The U–Pb, Th–Pb, and Pb–Pb Methods

THE decay of U and Th to stable isotopes of Pb is the basis for several important methods of dating. These arise not only from the transformation of U and Th to Pb but also derive from the time-dependent evolution of common Pb, from the decay of the intermediate daughters of U (Chapter 20), and from the resulting isotopic composition of He (Chapter 21) as well as from the accumulation of radiation damage in crystals (Chapter 22). Age determinations of rocks based on the decay of U and the resulting accumulation of Pb and He were first attempted in the early years of the twentieth century by Ernest Rutherford and B. B. Boltwood. Subsequently, A. Holmes used chemical U–Pb and U–He dates to propose the first geological timescale in his book on the age of the Earth published in 1913.

After the invention of mass spectrometers by J. J. Thomson and F. W. Aston prior to 1920 (Section 4.4a), age determinations based on the isotopic composition of Pb became possible based not only on the decay of U and Th to Pb but also on the isotope ratios of common Pb. As a result of the continuing refinement of analytical procedures and instrumentation, the U,Th–Pb methods of dating have become the most precise and most accurate geochronometers for determining the ages of terrestrial and extraterrestrial rocks.

10.1 GEOCHEMISTRY OF U AND Th

Uranium and Th are members of the actinide series of elements in which the $5f$ orbitals are progressively filled with electrons. Because of similar electron configurations, Th ($Z = 90$) and U ($Z = 92$) have similar chemical properties. Both elements occur in nature in the tetravalent oxidation state and their ions have similar radii (U^{4+} = 1.05 Å, Th^{4+} = 1.10 Å). Consequently, the two elements can substitute extensively for each other, which explains their geochemical coherence. However, under oxidizing conditions, U forms the uranyl ion (UO_2^{2+}) in which U has a valence of +6. The uranyl ion forms compounds that are soluble in water. Therefore, U is a mobile element under oxidizing conditions and is separated from Th, which exists only in the tetravalent state and whose compounds are generally insoluble in water.

The average concentrations of U and Th in chondritic meteorites in Table 10.1 are 1×10^{-2} and 4×10^{-2} ppm, respectively. These values may be taken as an indication of the very low abundance of these elements in the solar system. In the course of partial melting of rocks in the mantle of the Earth, U and Th are concentrated in the liquid phase and thus become incorporated into the more silica-rich products. For that reason, progressive geochemical

Table 10.1. Average Concentration of U, Th, and Pb in Igneous, Sedimentary, and Metamorphic Rocks

	Concentration (ppm)			Th/U
Rock Type	U	Th	Pb	Ratio
Chondrites	0.01	0.04	1.0	4.0
Achondrites	0.07	0.36	0.4	5.1
Iron meteorites	0.008	0.01	0.1	1.2
Ultramafic rocks	0.014	0.05	0.3	3.6
Gabbro	0.84	3.8	2.7	4.5
Basalt	0.43	1.6	3.7	3.7
Andesite	~2.4	~8	5.8	3.3
Nepheline syenite	8.2	17.0	14.4	2.1
Granitic rocks	4.8	21.5	23.0	4.5
Shale	3.2	11.7	22.8	3.7
Sandstone	1.4	3.9	13.7	2.8
Carbonate rocks	1.9	1.2	5.6	0.63
Granitic gneiss	3.5	12.9	19.6	3.7
Granulite	1.6	7.2	18.7	4.5

Note: Data for U and Th from compilations by Rogers and Adams (1969a, b) and for Pb from Wedepohl (1974). The concentrations given in this table are in most cases unweighted arithmetic means. The concentrations of these elements in the various rock types vary widely depending primarily on the mineral compositions of the rocks.

differentiation of the upper mantle of the Earth has enriched the rocks of the continental crust in U and Th compared to those of the upper mantle.

The concentrations of U, Th, and Pb in Table 10.1 increase from basaltic rocks to granites, yet their Th/U ratios remain virtually constant. Granitic rocks are enriched in Th relative to U, perhaps because some of the U is removed in aqueous solutions as uranyl ion during the final stages of crystallization of granitic magmas. The Th/U ratios of sedimentary rocks are similar to those of igneous rocks, with the exception that sedimentary carbonates are enriched in U and have a low average Th/U ratio of about 0.63. The U enrichment of carbonate rocks results from the fact that U occurs in the oceans as the uranyl ion, which coprecipitates with calcium carbonate, while Th is associated primarily with water-insoluble sediment.

The concentrations of U and Th in the common rock-forming silicate minerals are uniformly low, on the order of a few parts per million or less. Instead, these two elements occur primarily in certain accessory minerals in which they either are major constituents or replace other elements. The list of these minerals includes uraninite and thorianite (oxides); zircon, thorite, and allanite (silicates); monazite, apatite, and xenotime (phosphates); and sphene (titanosilicate). Many other U and Th minerals exist (Heinrich, 1958). Most of these occur under oxidizing conditions and contain the uranyl ion. Additional information on the geochemistry of U, Th, and Pb can be found in the *Handbook of Geochemistry* (Wedepohl, 1978).

10.2 DECAY OF U AND Th ISOTOPES

Uranium has three naturally occurring isotopes, ^{238}U, ^{235}U, and ^{234}U, all of which are radioactive. Thorium exists primarily as ^{232}Th. Five additional radioactive isotopes of Th occur in nature because they are short-lived intermediate daughters of ^{238}U, ^{235}U, and ^{232}Th. The isotope abundances, halflives, and decay constants ^{238}U, ^{235}U, ^{234}U, and ^{232}Th are listed in Table 10.2.

The principal isotopes of U and Th are each the parent of a chain of radioactive daughters ending with stable isotopes of Pb. The decay of ^{238}U gives rise to the uranium series, which includes ^{234}U as an intermediate daughter and ends in stable ^{206}Pb. The decay of ^{238}U to ^{206}Pb can be summarized by the equation

$$^{238}_{92}U \rightarrow {}^{206}_{82}Pb + 8^{4}_{2}He + 6\beta^{-} + Q \qquad (10.1)$$

where $Q = 47.4$ MeV/atom or 0.71 cal/g y (Wetherill, 1966). Each atom of ^{238}U that decays ultimately produces one atom of ^{206}Pb by emission of eight α-particles and six β-particles. The parameter Q represents the sum of the decay energies of the entire

Table 10.2. Abundances, Halflives, and Decay Constants of Principal Naturally Occurring Isotopes of U and Th

Isotope	Abundance (%)	Halflife (years)	Decay Constant (y^{-1})	Reference[a]
^{238}U	99.2743	4.468×10^9	1.55125×10^{-10}	1
^{235}U	0.7200	0.7038×10^9	9.8485×10^{-10}	1
^{234}U	0.0055	2.45×10^5	2.829×10^{-6}	2
^{232}Th	100.00	14.010×10^9	4.9475×10^{-11}	1

[a] 1— Steiger and Jäger, 1977; 2— Lide and Frederikse, 1995.

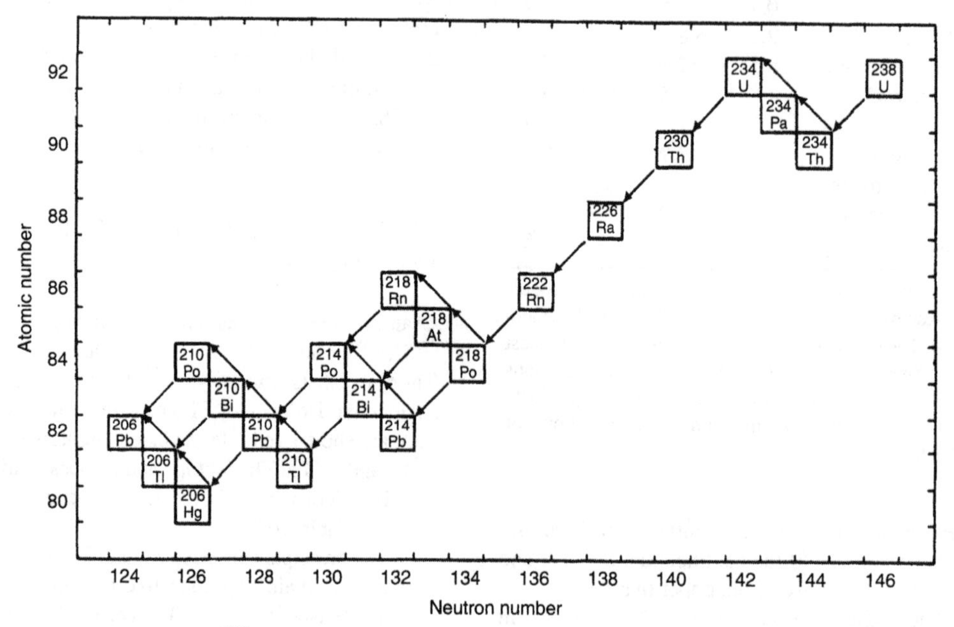

FIGURE 10.1 Decay chain of ^{238}U resulting from the successive emission of α- and β-particles. The final decay product is stable ^{206}Pb.

series in units of million electron volts. Several intermediate daughters in this series in Figure 10.1 undergo branched decay involving the emission of either an α-particle or a β-particle. The chain therefore splits into separate branches, but ^{206}Pb is the stable end product of all possible decay paths.

The decay of ^{235}U gives rise to the actinium series, which ends with stable ^{207}Pb after emission of seven α-particles and four β-particles:

$$^{235}_{92}U \rightarrow ^{207}_{82}Pb + 7\,^4_2He + 4\beta^- + Q \qquad (10.2)$$

where $Q = 45.2$ MeV/atom or 4.3 cal/g y (Wetherill, 1966). This series also branches as shown in Figure 10.2.

The decay of ^{232}Th results in the emission of six α-particles and four β-particles, leading to the formation of stable ^{208}Pb (Figure 10.3). This decay series therefore can be written as

$$^{232}_{90}Th \rightarrow ^{208}_{82}Pb + 6\,^4_2He + 4\beta^- + Q \qquad (10.3)$$

where $Q = 39.8$ MeV/atom or 0.20 cal/g y (Wetherill, 1966).

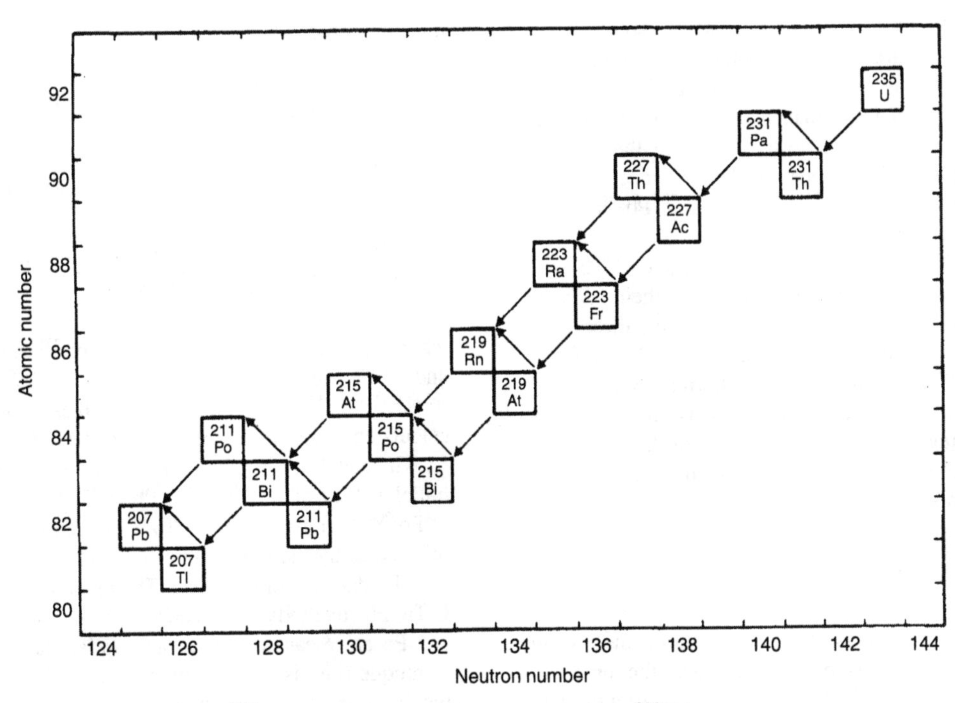

FIGURE 10.2 Decay chain of ^{235}U ending with the formation of stable ^{207}Pb.

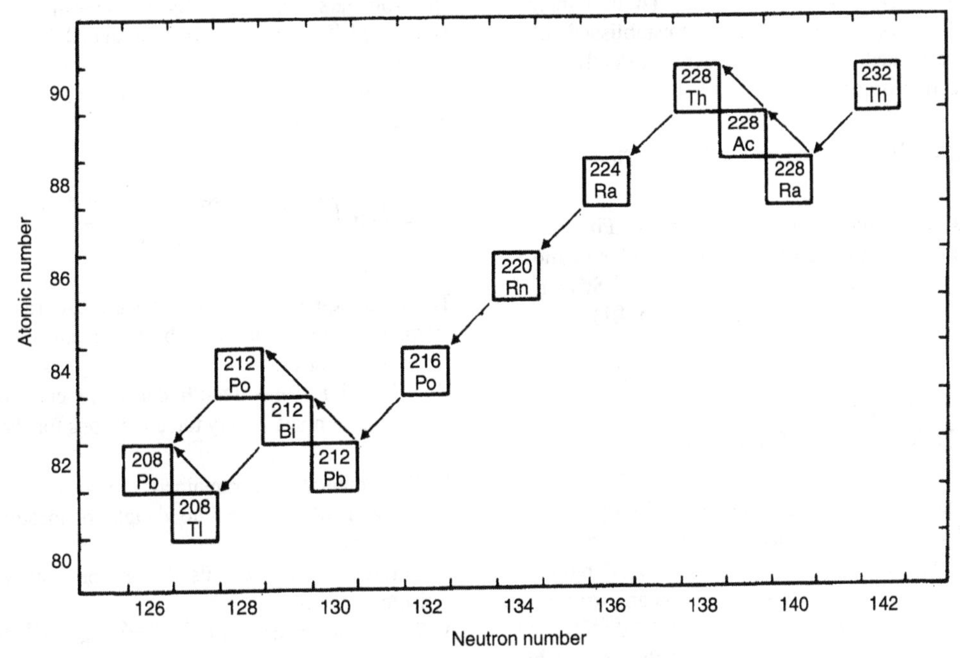

FIGURE 10.3 Decay chain of ^{232}Th resulting in the formation of stable ^{208}Pb.

In spite of the fact that 43 isotopes of 12 elements are formed as intermediate daughters in these decay series (not counting ^{4}He), none is a member of more than one series. In other words, each decay chain always leads to the formation of a specific isotope of Pb. The decay of ^{238}U produces ^{206}Pb, ^{235}U produces ^{207}Pb, and ^{232}Th produces ^{208}Pb.

The halflives of ^{238}U, ^{235}U, and ^{232}Th are all very much longer than those of their respective daughters (Table 10.2). Therefore, these decay series satisfy the prerequisite condition for the establishment of secular equilibrium (Section 3.3b). When secular equilibrium exists in a U- or Th-bearing mineral, the decay rates of the intermediate daughters are equal to those of their respective parents:

$$N_1\lambda_1 = N_2\lambda_2 = N_3\lambda_3 = \cdots \quad (10.4)$$

If a mineral is a closed system and secular equilibrium has been established, the production rate of the stable daughter at the end of a particular decay chain is equal to the rate of decay of its parent at the head of the chain. Therefore, the decay of the isotopes of U and Th in minerals in which secular equilibrium has established itself can be treated as though it occurred directly to the respective isotopes of Pb:

$$^{238}U \rightarrow {}^{206}Pb \qquad ^{235}U \rightarrow {}^{207}Pb \qquad ^{232}Th \rightarrow {}^{208}Pb$$

As a result, the growth of radiogenic Pb can be described by means of equations similar to those used to represent the decay of ^{87}Rb to ^{87}Sr and of ^{147}Sm to ^{143}Nd (equation 4.8, Section 4.1).

10.3 PRINCIPLES AND METHODOLOGY

The accumulation of the radiogenic isotopes of Pb by decay of their respective parents is governed by equations derivable from the law of radioactivity (Section 4.1). These equations are written in terms of the atomic $^{206}Pb/^{204}Pb$, $^{207}Pb/^{204}Pb$, and $^{208}Pb/^{204}Pb$ ratios because ^{204}Pb is the only stable nonradiogenic isotope of Pb:

$$\frac{^{206}Pb}{^{204}Pb} = \left(\frac{^{206}Pb}{^{204}Pb}\right)_i + \frac{^{238}U}{^{204}Pb}(e^{\lambda_1 t} - 1) \quad (10.5)$$

$$\frac{^{207}Pb}{^{204}Pb} = \left(\frac{^{207}Pb}{^{204}Pb}\right)_i + \frac{^{235}U}{^{204}Pb}(e^{\lambda_2 t} - 1) \quad (10.6)$$

$$\frac{^{208}Pb}{^{204}Pb} = \left(\frac{^{208}Pb}{^{204}Pb}\right)_i + \frac{^{232}Th}{^{204}Pb}(e^{\lambda_3 t} - 1) \quad (10.7)$$

where λ_1, λ_2, λ_3 are decay constants of ^{238}U, ^{235}U, and ^{232}Th, respectively (Table 10.2); $^{238}U/^{204}Pb$, $^{235}U/^{204}Pb$, $^{232}Th/^{204}Pb$ are ratios of these isotopes calculated from the measured concentrations of U, Th, and Pb; and the subscript i refers to the initial values of the $^{206}Pb/^{204}Pb$, $^{207}Pb/^{204}Pb$, and $^{208}Pb/^{204}Pb$ ratios. The $^{238}U/^{204}Pb$ ratio is also referred to by the Greek letter mu (μ).

To date U minerals (or Th minerals) by the U,Th–Pb methods, the concentrations of U,Th, and Pb are measured by an appropriate analytical technique (i.e., isotope dilution, Section 4.4e), and the isotope composition of Pb is determined on a solid-source mass spectrometer, although ion probe mass spectrometers and ICP mass spectrometers are being used increasingly. The U–Pb and Th–Pb dates are calculated by means of equations 10.5–10.7 using assumed values of the initial isotope ratios of Pb (Ludwig, 1993). For example, equation 10.5 yields

$$t_6 = \frac{1}{\lambda_1} \ln \left(\frac{(^{206}Pb/^{204}Pb) - (^{206}Pb/^{204}Pb)_i}{^{238}U/^{204}Pb} + 1 \right) \quad (10.8)$$

The other equations are solved similarly, resulting in three independent dates based on each of the three decay series.

The U–Pb and Th–Pb dates are concordant only if the samples satisfy the conditions for dating:

1. The mineral has remained closed to U, Th, Pb, and all intermediate daughters throughout its history.
2. Correct values are used for the initial Pb isotope ratios.
3. The decay constants of ^{238}U, ^{235}U, and ^{232}Th are known accurately.

4. The isotopic composition of U is normal and has not been modified by isotope fractionation or by the occurrence of a natural chain reaction based on induced fission of ^{235}U.
5. All analytical results are accurate and free of systematic errors.

The assumption that the samples being dated remained closed to U, Th, Pb, and all intermediate daughters is satisfied *only in rare cases* because U is a mobile element in oxidizing environments and therefore tends to be lost during chemical weathering. In addition, the emission of α-particles causes radiation damage in crystals, which facilitates the loss of Pb and the other intermediate daughters in each of the three decay chains. Consequently, U–Pb and Th–Pb dates of rocks and minerals are rarely concordant. Fortunately, procedures have been devised to overcome this problem.

The choice of the initial isotope ratios of Pb is a problem only for dating rocks and minerals that have low U/Pb (or Th/Pb) ratios and, in addition, are young. The numerical values of the initial isotope ratios of Pb do not significantly affect the calculated U–Pb (and Th–Pb) dates of Precambrian rocks and minerals having high U/Pb (and Th/Pb) ratios because their present isotope ratios of Pb, in most cases, reach large values. For example, Krogh (1973) reported a $^{206}Pb/^{204}Pb$ ratio of 126,000 for crystals of Precambrian zircon. Therefore, the isotope abundances of common Pb can be used in such cases to calculate initial isotope ratios of Pb:

$$\frac{^{206}Pb}{^{204}Pb} = 17.21 \qquad \frac{^{207}Pb}{^{204}Pb} = 15.78$$

$$\frac{^{208}Pb}{^{204}Pb} = 37.43$$

The decay constants of the naturally occurring long-lived isotopes of U and Th in Table 10.2 were fixed by the IUGS Subcommission on Geochronology during the Twenty-Fifth International Geological Congress in Sydney, Australia (Steiger and Jäger, 1977). In addition, the subcommission adopted a value of 137.88 for the atomic $^{238}U/^{235}U$ ratio. These values are now used by all isotope geologists to avoid the confusion caused by the use of different constants in the calculation of isotopic dates. The numerical values of the decay constants of the isotopes of U and Th are probably more accurate than those of other long-lived radionuclides because of their importance in the nuclear industry. Therefore, certain refractory U-bearing minerals that yield concordant U–Pb and Th–Pb dates can be used to refine the decay constants of other radionuclides used in geochronometry (Begemann et al., 2001).

The question about the abundances of the U isotopes merits attention because real differences in the isotopic composition of terrestrial and extraterrestrial U have been reported. For example, U deposits of Precambrian age at Oklo in Gabon, Africa, are significantly depleted in ^{235}U because this isotope was consumed by neutron-induced fission when the deposits became natural fission reactors at about 1.8 Ga (Lancelot et al., 1975; Cowan, 1976; Kuroda, 1982). The abundance of ^{235}U in the ore mined at Oklo is as low as 0.3 percent, compared to 0.72 percent in normal U. Fortunately for isotope geochronometry, natural fission reactors appear to be rare. Therefore, in the absence of compelling evidence to the contrary, age determinations by the U–Pb method of terrestrial and lunar rocks and meteorites are based on a value of 137.88 for the present-day $^{238}U/^{235}U$ ratio.

The effect of Pb loss on U–Pb dates can be minimized by calculating a date based on the $^{207}Pb/^{206}Pb$ ratio which is insensitive to recent Pb loss provided that the Pb that was lost from the mineral had the same isotope composition as the Pb that remained (i.e., no isotope fractionation). The relation between the $^{207}Pb/^{206}Pb$ ratio and time results from the difference in the decay constants (or halflives) of ^{238}U and ^{235}U. The desired equation is obtained by combining equations 10.5 and 10.6:

$$\frac{^{207}Pb/^{204}Pb - (^{207}Pb/^{204}Pb)_i}{^{206}Pb/^{204}Pb - (^{206}Pb/^{204}Pb)_i} = \frac{^{235}U}{^{238}U}\left(\frac{e^{\lambda_2 t} - 1}{e^{\lambda_1 t} - 1}\right) \tag{10.9}$$

This equation has several interesting properties:

1. It involves the $^{235}U/^{238}U$ ratio which is a constant (= 1/137.88) for all U of normal isotope composition in the Earth, the Moon, Mars, and meteorites at the present time.

2. The equation does not require knowledge of the concentrations of U and Pb and involves only isotope ratios of Pb.
3. The left side of equation 10.9 is equal to the $^{207}Pb/^{206}Pb$ ratio of radiogenic Pb:

$$\frac{^{207}\text{Pb}/^{204}\text{Pb} - (^{207}\text{Pb}/^{204}\text{Pb})_i}{^{206}\text{Pb}/^{204}\text{Pb} - (^{206}\text{Pb}/^{204}\text{Pb})_i} = \left(\frac{^{207}\text{Pb}}{^{206}\text{Pb}}\right)^* \quad (10.10)$$

where the asterisk identifies the radiogenic isotopes.
4. Equation 10.9 cannot be solved for t by algebraic methods because it is transcendental, but it can be solved by iteration or by interpolation in a table.

A difficulty arises in the solution of equation 10.9 when $t = 0$, which yields the indeterminate result 0/0. This difficulty is overcome by means of l'Hôpital's rule, which states that

$$\lim_{t \to 0} \frac{f(t)}{g(t)} = \lim_{t \to 0} \frac{f'(t)}{g'(t)}$$

where $f'(t)$ and $g'(t)$ are the first derivatives of the functions f and g with respect to t. Therefore, according to l'Hôpital's rule,

$$\lim_{t \to 0} \frac{e^{\lambda_2 t} - 1}{e^{\lambda_1 t} - 1} = \lim_{t \to 0} \frac{\lambda_2 e^{\lambda_2 t}}{\lambda_1 e^{\lambda_1 t}} = \frac{\lambda_2}{\lambda_1}$$

Therefore, the value of $(^{207}Pb/^{206}Pb)^*$ at the present time ($t = 0$) is

$$\left(\frac{^{207}\text{Pb}}{^{206}\text{Pb}}\right)^* = \frac{^{235}\text{U}\,\lambda_2}{^{238}\text{U}\,\lambda_1} \quad (10.11)$$

Equation 10.11 indicates that the $(^{207}Pb/^{206}Pb)^*$ ratio which forms by decay of ^{238}U and ^{235}U at the present time is equal to the rates of decay of the two U isotopes at the present time. This result is consistent with the statement of the law of radioactivity (Section 3.1). Therefore,

$$\left(\frac{^{207}\text{Pb}}{^{206}\text{Pb}}\right)^*_{t=0} = \frac{1}{137.88} \times \frac{9.8485 \times 10^{-10}}{1.55125 \times 10^{-10}} = 0.04604$$

The numerical values of $e^{\lambda_1 t} - 1$ and $e^{\lambda_2 t} - 1$ in Table 10.3 yield the $(^{207}Pb/^{206}Pb)^*$ ratios for increasing values of t ranging from $t = 0$ to $t = 4.6 \times 10^9$ y. This table can be used to solve equation 10.9 for t by linear interpolation based on the $(^{207}Pb/^{206}Pb)^*$ ratio calculated from equation 10.10.

Although U and Th occur in a large number of minerals, only a few are suitable for dating by the U, Th–Pb methods. To be useful for dating,

Table 10.3. Numerical Values of $e^{\lambda_1 t} - 1$ and $e^{\lambda_2 t} - 1$ and of the Radiogenic $^{207}Pb/^{206}Pb$ Ratio as a Function of Age t in Wetherill's Concordia

t, $\times 10^9$ y	$e^{\lambda_1 t} - 1$	$e^{\lambda_2 t} - 1$	$^{207}Pb^*/^{206}Pb$
0	0.0000	0.0000	0.04604
0.2	0.0315	0.2177	0.05012
0.4	0.0640	0.4828	0.05471
0.6	0.0975	0.8056	0.05992
0.8	0.1321	1.1987	0.06581
1.0	0.1678	1.6774	0.07250
1.2	0.2046	2.2603	0.08012
1.4	0.2426	2.9701	0.08879
1.6	0.2817	3.8344	0.09872
1.8	0.3221	4.8869	0.11004
2.0	0.3638	6.1685	0.12298
2.2	0.4067	7.7292	0.13783
2.4	0.4511	9.6296	0.15482
2.6	0.4968	11.9437	0.17436
2.8	0.5440	14.7617	0.19680
3.0	0.5926	18.1931	0.22266
3.2	0.6428	22.3716	0.25241
3.4	0.6946	27.4597	0.28672
3.6	0.7480	33.6556	0.32634
3.8	0.8030	41.2004	0.37212
4.0	0.8599	50.3878	0.42498
4.2	0.9185	61.5752	0.48623
4.4	0.9789	75.1984	0.55714
4.6	1.0413	91.7873	0.63930

Note:

λ_1 (^{238}U) $= 1.55125 \times 10^{-10}$ y^{-1}

λ_2 (^{235}U) $= 9.8485 \times 10^{-10}$ y^{-1}

$$\left(\frac{^{207}\text{Pb}}{^{206}\text{Pb}}\right)^* = \frac{1}{137.88}\,\frac{e^{\lambda_2 t} - 1}{e^{\lambda_1 t} - 1}$$

a mineral must be retentive with respect to U, Th, Pb, and the intermediate daughters, and it should be widely distributed in a variety of rocks. The minerals that satisfy these conditions are considered elsewhere:

Zircon: Beakhouse et al., 1988, *Chem. Geol. (Isotope Geosci. Sect.)*, 72:337–351. Anderson et al., 1999, *Precamb. Res.*, 98:151–171.

Baddeleyite: Krogh et al., 1987, *Geol. Assoc. Can. Spec. Paper* 34:147–152. Heaman and LeCheminant, 1993, *Chem. Geol.*, 110:95–126. Reischmann, 1995, *S. African J. Geol.*, 98:98.

Monazite: Crowley and Ghent, 1999, *Chem. Geol.*, 157:285–302. Mezger et al., 1991, *J. Geol.*, 99:415–428. Parrish, 1990, *Can. J. Earth Sci.*, 27:1431–1450. Copeland et al., 1988, *Nature*, 333:760–762.

Apatite: Berger and Braun, 1997, *Chem. Geol.*, 142:23–40. Baadsgaard, 1983, *Grönlands Geol. Unders.*, 112:35–42. Oosthuyzen and Burger, 1973, *Earth Planet. Sci. Lett.*, 18: 29–36.

Sphene (Titanite): Essex and Gromet, 2000, *Geology*, 28:419–422. Pidgeon et al., 1996, *Earth Planet. Sci. Lett.*, 141:187–198. Mezger et al., 1991, *J. Geol.*, 99:415–428. Barrie, 1990, *Can. J. Earth Sci.*, 27:1451–1456. Catanzaro and Hanson, 1971, *Can. J. Earth Sci.*, 8:1319–1324. Hanson et al., 1971, *Earth Planet. Sci. Lett.*, 12:231–237. Tilton and Grünenfelder, 1968, *Science*, 159:1458–1461.

Garnet: Mezger et al., 1989a, *Contrib. Mineral. Petrol.*, 101:136–148. Mezger et al., 1991, *J. Geol.*, 99:415–428.

Rutile: Mezger et al., 1989b, *Earth Planet. Sci. Lett.*, 96:106–1118.

Perovskite: Smith et al., 1989, *Chem. Geol. Isotope Geosci. Sect.*, 10:137–146.

Ilmenite: Burton and O'Nions, 1990, *Geochim. Cosmochim. Acta*, 54:2593–2602. Burton and O'Nions, 1993, 57:4533–4535.

Cassiterite: Gulson and Jones, 1992, *Geology*, 20:355–358.

Calcite: Richards et al., 1998, *Geochim. Cosmochim. Acta*, 62:3683–3688. Smith et al., 1991, *Earth Planet. Sci. Lett.*, 105:474–491.

All of these minerals contain trace amounts of U and Th but have low concentrations of Pb, giving them high U/Pb and Th/Pb ratios favorable for dating.

The concentrations of U and Th in zircon range from a few hundred to a few thousand parts per million and average 1350 and 550 ppm, respectively. Zircons in pegmatites contain more U and Th than those in ordinary igneous rocks. The presence of these elements in zircon can be attributed both to isomorphous replacement of Zr^{4+} (ionic radius 0.87 Å) by U^{4+} (1.05 Å) and Th^{4+} (1.10 Å) and to the presence of inclusions of thorite ($ThSiO_4$). The substitution of Zr^{4+} by U^{4+} and Th^{4+} is limited by differences in their ionic radii. Whereas U^{4+} and Th^{4+} are admitted into zircon crystals, Pb^{2+} is excluded because of its large radius (1.32 Å) and its low charge (+2). Therefore, zircon contains very little Pb at the time of formation and has high U/Pb and Th/Pb ratios, which enhances its sensitivity as a geochronometer. For this reason, zircon is used most frequently for dating by the U, Th–Pb isotopic methods (Doe, 1970).

10.4 U,Th–Pb DATES, BOULDER CREEK BATHOLITH, COLORADO

The dating of zircon by the U,Th–Pb methods is illustrated by data reported by Stern et al. (1971) for samples from the Boulder Creek batholith of Colorado collected near Gross Dam (GP509): U = 792.1 ppm; Th = 318.6 ppm; Pb = 208.2 ppm; $^{204}Pb = 0.048\%$ (atom); $^{206}Pb = 80.33\%$; $^{207}Pb = 9.00\%$; $^{208}Pb = 10.63\%$. The authors assumed that the ordinary Pb incorporated into this zircon at the time of crystallization had an isotopic composition given by 204 : 206 : 207 : 208 = 1.00 : 16.25 : 15.51 : 35 : 73. This composition corresponds approximately to that of Pb in galena (PbS) at 1500 Ma. The atomic weight of Pb in this zircon is 205.94.

These data can be used to calculate dates based on the decay of ^{238}U to ^{206}Pb (equation 10.5), ^{235}U to ^{207}Pb (equation 10.6), ^{232}Th to ^{208}Pb (equation 10.7), and by the Pb–Pb method (equation 10.9). The time-dependent increase of the $^{206}Pb/^{204}Pb$ ratio by decay of ^{238}U is governed by equation 10.5:

$$\frac{^{206}Pb}{^{204}Pb} = \left(\frac{^{206}Pb}{^{204}Pb}\right)_i + \frac{^{238}U}{^{204}Pb}(e^{\lambda_1 t} - 1)$$

The $^{238}U/^{204}Pb$ ratio is calculated from the U/Pb ratio of the zircon:

$$\frac{^{238}U}{^{204}Pb} = \frac{792.1}{208.2} \times \frac{205.94}{238.03} \times \frac{99.27}{0.048} = 6807.4$$

where 238.03 is the atomic weight of U and 99.27 is the abundance of ^{238}U in percent by number. The $^{206}Pb/^{204}Pb$ ratio of the Pb is

$$\frac{^{206}Pb}{^{204}Pb} = \frac{80.33}{0.048} = 1673.54$$

and the initial $^{206}Pb/^{204}Pb$ ratio is

$$\frac{^{206}Pb}{^{204}Pb} = 16.25$$

Substitution of these values into equation 10.5 yields

$$1673.54 = 16.25 + 6807.4(e^{\lambda_1 t} - 1)$$

Since $\lambda_1 = 1.55125 \times 10^{-10}\ y^{-1}$,

$$t_6 = \frac{1}{1.55125 \times 10^{-10}} \ln\left(\frac{1673.54 - 16.25}{6807.4} + 1\right)$$
$$= 0.14046 \times 10^{10} = 1.405\ \text{Ga}$$

The $^{235}U/^{204}Pb$ ratio can be calculated from the data,

$$\frac{^{235}U}{^{204}Pb} = \frac{792.1}{208.2} \times \frac{205.94}{238.03} \times \frac{0.72}{0.048} = 49.37$$

or it can be derived from the fact that $^{238}U/^{235}U = 137.88$. Therefore,

$$\frac{^{235}U}{^{204}Pb} = \frac{^{238}U}{^{204}Pb} \times \frac{1}{137.88} = \frac{6807.4}{137.88} = 49.37$$

The ^{235}U–^{207}Pb date is calculated from equation 10.6:

$$\frac{^{207}Pb}{^{204}Pb} = \left(\frac{^{207}Pb}{^{204}Pb}\right)_i + \frac{^{235}U}{^{204}Pb}(e^{\lambda_2 t} - 1)$$

Since $t_2 = 9.8485 \times 10^{-10}\ y^{-1}$,

$$t_7 = \frac{1}{9.8485 \times 10^{-10}} \ln\left(\frac{187.5 - 15.51}{49.37} + 1\right)$$
$$= 0.15235 \times 10^{10} = 1.524\ \text{Ga}$$

Similarly, the ^{232}Th–^{208}Pb date is calculated from equation 10.7:

$$\frac{^{208}Pb}{^{204}Pb} = \left(\frac{^{208}Pb}{^{204}Pb}\right)_i + \frac{^{232}Th}{^{204}Pb}(e^{\lambda_3 t} - 1)$$

$$t_8 = \frac{1}{4.9475 \times 10^{-11}} \ln\left(\frac{221.45 - 35.73}{2829.4} + 1\right)$$
$$= 1.285\ \text{Ga}$$

The ratio of radiogenic ^{207}Pb to ^{206}Pb of this zircon is

$$\left(\frac{^{207}Pb}{^{206}Pb}\right)^* = \frac{187.5 - 15.51}{1673.54 - 16.25} = 0.10377$$

The Pb–Pb date is derived by interpolation in Table 10.3:

$$t_{7/6} = 1.6 + \frac{0.2(0.10377 - 0.09872)}{0.11004 - 0.09872}$$
$$= 1.689\ \text{Ga}$$

The dates calculated above are discordant:

$$t_6 = 1.405 \text{ Ga} \qquad t_7 = 1.524 \text{ Ga}$$

$$t_8 = 1.285 \text{ Ga} \qquad t_{7/6} = 1.689 \text{ Ga}$$

The most likely reason for the discordance is that the zircons in the Boulder Creek batholith lost radiogenic Pb either during an episode of thermal metamorphism or by continuous diffusion at elevated temperatures or both. If the loss of Pb occurred in the recent past without fractionation of Pb isotopes, the 207/206 date probably comes closest to the age of the zircon. Gain of Pb is an unlikely possibility because zircon cannot accommodate Pb^{2+} as noted above.

In addition, discordance of U,Th–Pb dates of minerals may also be caused by either gain or loss of U and Th and by radiochemical disequilibrium such as an initial excess or deficiency of ^{230}Th that ultimately causes an excess or deficiency of ^{206}Pb (Figure 10.1). In addition, the isotopes of Rn (a noble gas) that occur in each of the three decay series may be lost by diffusion, thus preventing secular equilibrium from establishing itself. The loss of ^{222}Rn is especially likely because it has a longer halflife (3.8235 days) than the other Rn isotopes. Therefore, its decay constant is

$$\lambda = \frac{0.693}{3.8235} = 0.1812 \text{ d}^{-1}$$

and its mean life (τ) is (equation 3.10)

$$\tau = \frac{1}{\lambda} = \frac{1}{0.1812} = 5.52 \text{ d}$$

The effect of radioactive disequilibrium on U,Th–Pb dates of minerals was investigated by Ludwig (1977), Schärer (1984), Romer (2001), and Amelin and Zaitsev (2002).

10.5 WETHERILL'S CONCORDIA

The effect of the loss of Pb or U and the gain of U on U,Th–Pb dates of minerals can be compensated by a graphical procedure developed by Ahrens (1955) and Wetherill (1956, 1963) as outlined below.

Equations 10.5 and 10.6, which govern the time-dependent increase of the $^{206}Pb/^{204}Pb$ and $^{207}Pb/^{204}Pb$ ratios of U-bearing minerals or rocks, can be rearranged to yield the ratios of radiogenic ^{206}Pb to ^{238}U and of radiogenic ^{207}Pb to ^{235}U. For equation 10.5,

$$\frac{^{206}\text{Pb}/^{204}\text{Pb} - (^{206}\text{Pb}/^{204}\text{Pb})_i}{^{238}\text{U}/^{204}\text{Pb}} = \frac{^{206}\text{Pb}^*}{^{238}\text{U}} = e^{\lambda_1 t} - 1 \qquad (10.12)$$

and, by a similar manipulation, equation 10.6 is rephrased in the form

$$\frac{^{207}\text{Pb}^*}{^{235}\text{U}} = e^{\lambda_2 t} - 1 \qquad (10.13)$$

where the asterisk is used to identify the radiogenic origin of the Pb. Values of $e^{\lambda_1 t} - 1$ and $e^{\lambda_2 t} - 1$ for different values of t are listed in Table 10.3 and were used to plot the curve in Figure 10.4. The coordinates of all points on this curve are the $^{206}Pb^*/^{238}U$ and $^{207}Pb^*/^{235}U$ ratios that yield concordant U–Pb dates. Therefore, the curve in Figure 10.4 is known as *concordia* and is associated with the name of its inventor Dr. G. W. Wetherill (Wetherill, 1956, 1963) in order to distinguish it from a different concordia developed later by Tera and Wasserburg (1972) presented in Section 10.10. Uranium-bearing minerals that contain no radiogenic ^{206}Pb and ^{207}Pb yield $t = 0$ because if $^{206}Pb^* = 0$ in equation 10.12, $^{206}Pb^*/^{238}U = 0$ and $e^{\lambda_1 t} = 1$. Taking natural logarithms yields $\lambda_1 t = 0$, which requires that $t = 0$. Uranium-bearing minerals or rocks having ages of 1.0 Ga, 1.5 Ga, and so on, are located sequentially on Wetherill's concordia.

Figure 10.4 illustrates a hypothetical history of zircon grains that originally crystallized from magma and were later subjected to an episode of thermal metamorphism. At the original time of crystallization, the zircons plotted at the origin of the concordia diagram because they did not contain radiogenic Pb. During the subsequent 2.5 billion years, the $^{206}Pb^*/^{238}U$ and $^{207}Pb^*/^{235}U$ ratios of the zircons increased by decay of ^{238}U and ^{235}U, causing them to move along concordia. After 2.5 billion years, the zircon grains lost varying

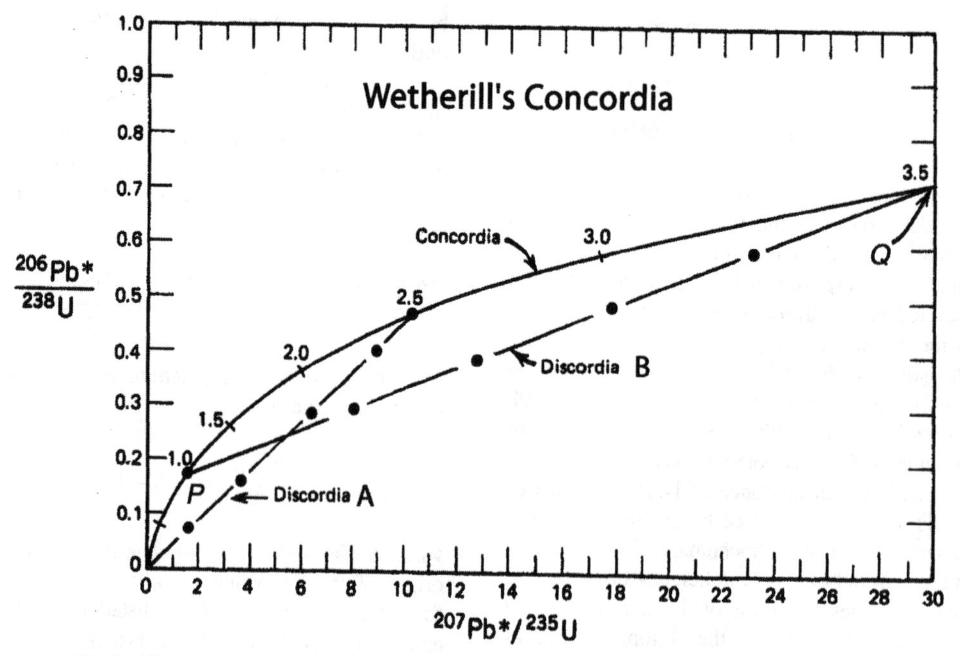

FIGURE 10.4 Wetherill's concordia diagram for the interpretation of U-bearing minerals that lost radiogenic Pb and therefore yield discordant dates.

amounts of radiogenic Pb during an episode of thermal metamorphism. Some of the crystals lost all of the accumulated radiogenic Pb and therefore returned to the origin. Others lost only varying fractions of radiogenic Pb and defined a straight-line chord, labeled *discordia A* in Figure 10.4, that extended from the point on concordia equivalent to 2.5 Ga to the origin. All of the zircons on this chord would have yielded discordant dates, which explains why it is called "discordia." At the end of the short episode of metamorphism, all of the zircon grains resumed their evolution by the decay of U isotopes and the accumulation of radiogenic Pb. The zircons that had lost all of the accumulated radiogenic Pb moved along concordia, and all of the zircon crystals that had lost only varying fractions of radiogenic Pb maintained the linear relationship to each other. At the present time, 1 billion years after the end of the episode of metamorphism, *discordia B* extends from a point on concordia equivalent to 3.5 Ga (i.e., 2.5 + 1.0 Ga) to a point representative of 1.0 Ga (i.e., the time elapsed since metamorphism).

Therefore, 1 billion years after the end of the episode of thermal metamorphism, the data points that previously defined discordia A form discordia B, which intersects concordia in two points, labeled *P* and *Q* in Figure 10.4. The coordinates of point *Q* yield concordant dates that represent the time elapsed since the original crystallization of the zircon crystals that now define discordia B.

The coordinates of point *P* also yield concordant dates, but the interpretation of this date depends on the circumstances. If the loss of radiogenic Pb actually occurred during an episode of thermal metamorphism, the date calculated from the coordinates of point *P* is the time elapsed since the end of that episode. In this case, the episodic loss of radiogenic Pb from zircons should also have caused loss of radiogenic ^{40}Ar from biotite or even muscovite. Therefore, the U–Pb date derived from point P should be confirmed by K–Ar dates of micas from the same rock specimen. Even the Rb–Sr date of micas in this rock sample may have been affected, whereas K–Ar dates of hornblende and whole-rock Rb–Sr and Sm–Nd isochron dates may not have responded to the episodic increase of the temperature.

Alternatively, Pb loss may have occurred by continuous diffusion at elevated temperature. In this case, the trajectory of U–Pb minerals follows a straight line that becomes nonlinear near the origin. As a result, a linear extrapolation of discordias yields a lower intercept with concordia that may correspond to a fictitious date. Therefore, the date calculated from the coordinates of the lower intercept (P) of discordia B in Figure 10.4 must be confirmed by K–Ar dates of micas before it can be interpreted as the age of an episode of thermal metamorphism. Ludwig (1988) and Davis (1982) devised a statistical method for fitting discordias to analytical data and for calculating the coordinates of the points of intersection with concordia.

10.5a Gain or Loss of U and Pb

The concordia diagram indicates in what way U-bearing minerals on a discordia line were altered. For example, the discordia line in Figure 10.5 intersects concordia at points Q and P as before but also contains a point M located at an arbitrarily selected position on the discordia. The position of point M depends on the value of a parameter R defined as the ratio ℓ/L, where ℓ is the distance PM along the discordia in Figure 10.5 and L is the length of the discordia between points P and Q. The parameter R measures the extent to which the daughter-to-parent ratios (i.e., $^{206}Pb^*/^{238}U$ and $^{207}Pb^*/^{235}U$) of the system represented by point M have been affected as a result of either loss of radiogenic Pb or gain or loss of U. The effect of the addition of Pb to the system is not predictable unless the isotopic composition of the new Pb is specified. The model includes a further constraint that the Pb loss must occur without discrimination against the isotopes of Pb on the basis of their masses. With these provisions, the parameter R is the ratio of daughter to parent immediately after the change divided by the ratio before the change:

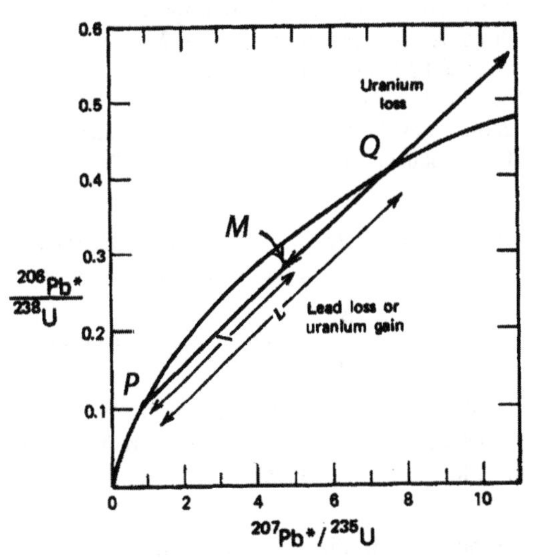

FIGURE 10.5 Concordia diagram illustrating the effects of Pb loss and U gain or loss on U–Pb systems.

$$R = \frac{(\mathrm{D/P})_a}{(\mathrm{D/P})_b} = \frac{\mathrm{D}_a \times \mathrm{P}_b}{\mathrm{P}_a \times \mathrm{D}_b} \qquad (10.14)$$

where D_a and D_b are the amounts of radiogenic Pb in the system after loss (a) and before loss (b), respectively, whereas P_a and P_b are the amounts of U after and before the alteration, respectively.

If a U–Pb system lost one-half of its radiogenic Pb but experienced no gain or loss of U,

$$D_a = \tfrac{1}{2} D_b \qquad \text{and} \qquad P_a = P_b$$

Therefore, equation 10.14 yields $R = \frac{1}{2}$ and $\ell = L/2$. In other words, point M is located on discordia halfway between P and Q in Figure 10.5. If the U concentration doubled but Pb remained unchanged, $D_a = D_b$ and P_a and $2P_b$. Hence, $R = \frac{1}{2}$ and $\ell =$ L/2 as before. Evidently, gain of U has the same effect as loss of Pb. In case one-third of the U is lost without loss of Pb, $D_a = D_b$, $P_a = 2P_b/3$, and $R = \frac{3}{2}$. Therefore,

$$\frac{\ell}{L} = \frac{3}{2} \qquad \text{and} \qquad \ell = \frac{3L}{2}$$

This result indicates that loss of U causes point M to move beyond point Q on an extension of discordia. Such behavior is displayed in some cases by whole-rock samples and by certain minerals (e.g., monazite) that yield coordinates that place them on

Table 10.4. Discordant U–Pb Dates, $^{206}Pb^*/^{238}U$, and $^{207}Pb^*/^{235}U$ Ratios of Zircons from the Morton and Montevideo Gneisses of Southwestern Minnesota

	Dates (10^9 y)				
Sample	t_{206}	t_{207}	$t_{7/6}$	$^{206}Pb^*/^{238}U$	$^{207}Pb^*/^{235}U$
1	2.614	2.872	3.055	0.5002	15.93
2	2.797	3.030	3.184	0.5434	18.77
3	2.885	3.091	3.226	0.5645	20.00

Note: Data from Catanzaro (1963) recalculated using constants in Tables 10.2 and 10.3.

an extension of discordia defined by zircons in the same rock (Rosholt et al., 1973). However, all minerals that are represented by points on discordia yield discordant U–Pb dates regardless of whether they are located below concordia or above it.

10.5b Morton Gneiss, Minnesota

The power of concordia diagrams to unscramble discordant U–Pb dates is illustrated in Figure 10.6 by data published by Catanzaro (1963) for zircons in the granitic gneiss at Morton and Granite Falls in Minnesota. All three zircons in Table 10.4 yield discordant dates ranging from 2.614 to 3.226 Ga. The pattern of variation ($t_{7/6} > t_7 > t_6$) is typical of discordant U–Pb dates (e.g., Section 10.4). The $^{206}Pb^*/^{238}U$ and $^{207}Pb^*/^{235}U$ ratios of these zircons define a discordia in Figure 10.6 that intersects concordia at two points whose coordinates yield dates of 3.550 and 1.850 Ga. The date corresponding to the upper intercept is the crystallization age of the zircons, whereas the lower intercept may date the time elapsed since episodic Pb loss during thermal metamorphism. Age determinations by the K–Ar and Rb–Sr method of biotite and feldspar from Precambrian rocks of this area indicate events at 1800 ± 100 and 2500 ± 100 million years. Therefore, there is independent evidence in this case for an event at about 1800 Ma that caused Pb loss from the zircons.

The fraction of Pb lost from the zircons in Figure 10.6 can be derived by measuring the lengths of ℓ and L defined in Figure 10.5. For

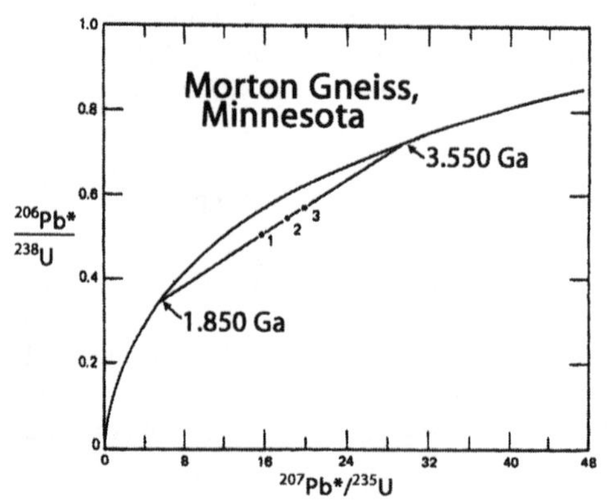

FIGURE 10.6 Concordia diagram for three discordant zircons from gneiss at Morton and Granite Falls in southern Minnesota. The discordia line intersects concordia in two points from whose coordinates the two dates can be calculated. Data from Catanzaro (1963) shown also in Table 10.4.

zircon 1, $R = \ell/L = 0.436$. If $P_a = P_b$ (i.e., no change in U concentration), equation 10.14 yields

$$R = \frac{D_a \times P_b}{P_a \times D_b} = 0.436 \qquad D_a = 0.436 D_b$$

In other words, only 43.6 percent of the Pb remains, which means that 56.4 percent of the radiogenic Pb was lost from this zircon.

10.5c U–Th–Pb Concordia Diagrams

The concept of U–Pb concordias can be extended to include the decay of ^{232}Th combined with the decay of either ^{238}U or ^{235}U. The systematics of such U–Th–Pb concordias have been discussed by Steiger and Wasserburg (1966) and Allègre (1967). It turns out that the concordia curve based on $^{208}Pb/^{232}Th$ and $^{206}Pb/^{238}U$ ratios is quite straight because of the similarity of the halflives of ^{232}Th and ^{238}U. Consequently, the intersections of discordias with this concordia have large errors that make this kind of diagram unfavorable for use in geochronometry. However, the concordia based on the decay of ^{232}Th to ^{208}Pb and ^{235}U to ^{207}Pb is shaped very much like the conventional U–Pb concordia when $^{208}Pb^*/^{232}Th$ is plotted along the ordinate and $^{207}Pb^*/^{235}U$ along the abscissa.

When a U–Th–Pb system experiences loss of parents or daughters or gain of parents, its response in the U–Th–Pb concordias is analogous to that described previously for U–Pb concordias provided that no fractionation of parents or daughters occurs. For example, if the system gains U and Th in such a way that its Th/U ratio remains constant, then the point representing that system moves onto a trajectory directed toward the origin. In case of episodic alteration, the lower intersection of discordia with the U–Th–Pb concordia indicates the time of alteration. Similarly, if U and Th are lost without fractionation, the point moves into the area above concordia and lies on a trajectory passing through the origin. Loss of Pb, without isotope fractionation, also places the system onto a discordia directed at the origin, as discussed previously. In all such cases, the significance of the date derived from the lower point of intersection is subject to the same ambiguity discussed before regarding the nature of the process (i.e., either episodic or continuous alteration).

When the alteration of a U–Th–Pb system involves preferential gain or loss of parents or disproportionate losses of ^{208}Pb or ^{207}Pb, the response is quite different. Because of the differences in the geochemical properties of U and Th, changes in the Th/U ratios are quite likely both during episodic alteration and in the case of continuous gain or loss of parents. Similarly, it is possible that ^{206}Pb and ^{207}Pb may be lost at different rates than ^{208}Pb. As a result of such fractionation, discordant U–Th–Pb systems may lie on discordias that do not pass through the origin. The lower intersections of the resulting discordias with U–Th–Pb concordias have no geological time significance because they reflect the changes that have occurred in the Th/U ratios or in the Pb isotope ratios as a result of preferential gain or loss of parents or because of preferential loss of certain specific Pb isotopes (Davis and Krogh, 2000). Both Steiger and Wasserburg (1966) and Allègre (1967) reinterpreted published analyses of zircons by means of U–Th–Pb concordias in coordinates of $^{208}Pb^*/^{232}Th$ and $^{207}Pb^*/^{235}U$. Examples of the use of this kind of interpretation include studies by Sinha (1972) and Baadsgaard (1973).

10.6 ALTERNATIVE Pb LOSS MODELS

The concordia diagram developed by Wetherill (1956) was quickly adopted by geochronologists and has been widely used ever since. The interpretation of discordant U–Pb dates by means of Wetherill's concordia combined with refinements of the analytical techniques to be described in Section 10.8 has greatly improved the accuracy and precision of U–Pb dates of refractory U-bearing minerals. The ability to date such minerals precisely and reliably has permitted significant advances in the study of the history of the Earth.

10.6a Continuous Diffusion

A few years after the publication of Wetherill's concordia model, Tilton (1960) pointed out that U-bearing minerals from five continents that have 207–206 dates greater than 2300 Ma plot on a single discordia line that indicates a crystallization age of 2800 Ma and episodic Pb loss around 600 Ma. This result, illustrated in Figure 10.7, is remarkable because it implies the occurrence of a worldwide metamorphic event between 500 and 600 Ma for which there is little or no evidence. Moreover, the episodic model suggests that negligible Pb loss occurred in these minerals from 2800 to 600 Ma in spite of evidence provided by other dating methods that magmatic activity occurred during this time interval.

Tilton (1960) offered an alternative explanation for the loss of radiogenic Pb based on continuous diffusion of Pb from crystals at a rate governed

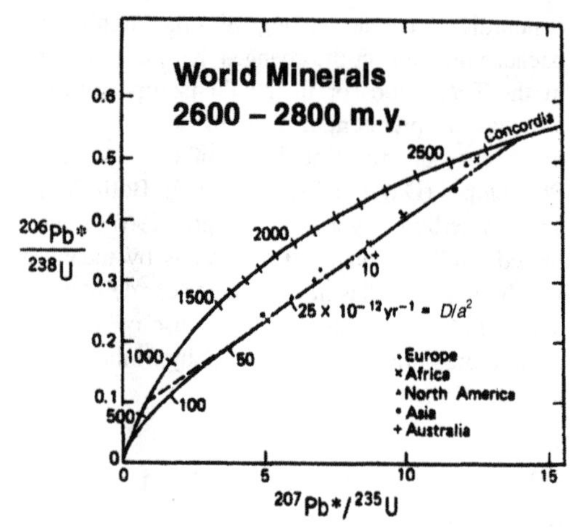

FIGURE 10.7 Concordia plot for minerals from different continents that fit the continuous-diffusion model. The position of the U–Pb system on the diffusion trajectory depends on the value of the diffusion parameter D/a^2. A linear extrapolation of the trajectory results in a fictitious date equal to 600 million years. Reproduced by permission from G. R. Tilton, *Journal of Geophysical Research*, Vol. 65, No. 9, pp. 2933–2945, 1960. Copyrighted by the American Geophysical Union.

by a diffusion coefficient D, the effective radius a, and the concentration gradient of Pb. He assumed that the crystals are spheres of effective radius a, that U is uniformly distributed throughout these spheres, that diffusion of U and intermediate daughters is negligible compared to that of Pb, that the diffusion coefficient D is a constant independent of time, and that diffusion is governed by Fick's law. With these assumptions, Tilton (1960) derived an equation relating the radiogenic-daughter-to-parent ratio of a mineral to D/a^2 and t, where t is the age of the mineral. The solutions of the diffusion equation generate curves on the concordia diagram that are the loci of points that represent U–Pb systems of specific ages that have suffered continuous Pb loss governed by the parameter D/a^2. Figure 10.7 contains such a curve for U–Pb systems that are 2800 million years old. It is apparent that the curve is very nearly a straight line for systems having $D/a^2 < 50 \times 10^{-12}$ y^{-1}. For values greater than this, the curve deviates from linearity and approaches the origin.

Tilton (1960) therefore demonstrated that a linear array of data points representing discordant U–Pb systems on a concordia diagram can be interpreted in two fundamentally different ways. If Pb loss was due to an episode of metamorphism, the date calculated from the coordinates of the lower point of intersection on concordia indicates the time since closure after Pb loss. On the other hand, if the Pb was lost continuously by diffusion, the date calculated from the coordinates of the lower intersection is fictitious. The problem is that both episodic and continuous loss of Pb results in daughter-to-parent ratios that appear to fit a straight-line discordia. Only U–Pb systems having high D/a^2 values, which therefore lost most of their Pb, deviate appreciably from linearity. This is the reason why the date derived from the lower intercept of discordia with concordia must be confirmed by K–Ar or Rb–Sr dates of micas before it can be accepted as the time elapsed since cooling after an episode of thermal metamorphism.

Tilton's treatment of Pb loss by continuous diffusion contained the implicit assumption that the temperature remained constant throughout the history of the zircons and that D/a^2 is therefore invariant with time. The general case in which the diffusion coefficient is a function of time and both Pb and U are allowed to diffuse was considered later by Wasserburg (1963) and Wetherill (1963).

10.6b Dilatancy Model

An alternative interpretation of the discordance of U–Pb dates of U-bearing minerals was proposed by Goldich and Mudrey (1972). They pointed out that such minerals suffer radiation damage as a result of the α-decay of U, Th, and their daughters. The extent of radiation damage increases with age and with U and Th concentrations of the minerals. The apparent relationship between the radioactivity of zircons and the discordance of U–Pb dates was first demonstrated by Silver and Deutsch (1961) and was used by Wasserburg (1963), who related the diffusion parameter D/a^2 to the radiation damage of zircon crystals. Goldich and Mudrey (1972) postulated that radiation damage leads to the formation of microcapillary channels that permit water to enter the crystal. This water is very tightly held until uplift and erosion cause the pressure on the minerals to be released. The resulting dilatance of the zircons allows the water to escape together with dissolved radiogenic Pb. Consequently, the loss of radiogenic Pb may be related to uplift and erosion of the crystalline basement complexes of Precambrian shields. Such relatively recent Pb loss is consistent with the observation that the 207–206 dates commonly approach the true age of U-bearing minerals.

The dilatancy model therefore provides a rational explanation for the observation emphasized by Tilton (1960) that U minerals from different continents all seem to have lost radiogenic Pb 500–600 million years ago, even though no worldwide metamorphic event is recognized in that period of the Earth's history. According to the dilatancy model, the date corresponding to the lower intercept of discordia based on cogenetic suites of U-bearing minerals from a particular region indicates the time of uplift and erosion of the rocks of that region.

10.6c Chemical Weathering

The rocks and minerals used for dating are usually collected from surface outcrops where they have been exposed to chemical weathering. Therefore, the discordance of model dates may also be the result of disturbances of the daughter-to-parent ratios caused by chemical weathering. This problem was investigated by Stern et al. (1966) by dating zircons removed from residual clay formed by chemical weathering of the Morton Gneiss near Redwood Falls, Minnesota. Zircons from this rock had been dated previously by Catanzaro (1963) and were discussed in Section 10.5b.

The model dates of three weathered zircons from the Morton Gneiss in Figure 10.8 were found to be grossly discordant, whereas the 207–206 dates are the oldest in all cases and approach the known age of these zircons. The weathered zircons lie on straight-line chords that start from discordia and lead to the origin. The displacement of the weathered zircons from their original position on discordia reveals that they lost up to 85 percent

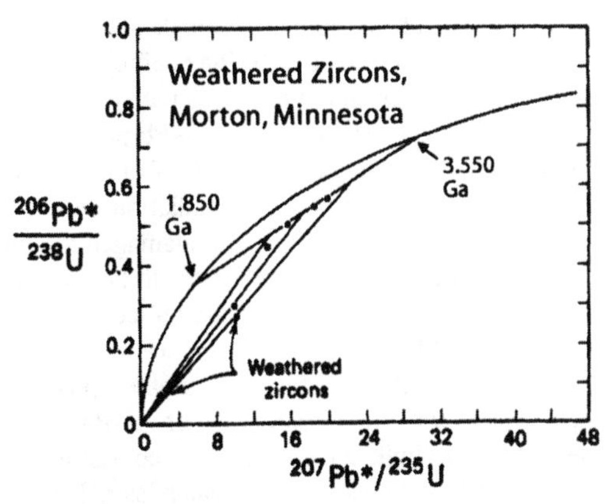

FIGURE 10.8 Concordia diagram showing the effect of Pb loss due to chemical weathering of three zircon samples recovered from residual clay derived from the Morton Gneiss near Redwood Falls, Minnesota. Data from Catanzaro (1963) and Stern et al. (1966).

of their radiogenic Pb as a result of ongoing chemical weathering.

10.6d Cores and Overgrowths

Zircon crystals in high-grade metamorphic rocks and in igneous rocks of granitic composition, in some cases, contain xenocrystic cores that predate the overgrowths which formed later during regional metamorphism or during crystallization of magma (Silver and Deutsch, 1963; Bickford et al., 1981). *Sensitive high-resolution ion microprobe* (SHRIMP) mass spectrometers permit dating of the cores and overgrowths separately, aided by back-scattered electron images of the crystals. For example, Nasdala et al. (1999) reported that overgrowths of zircons in monzonites of the Meissen Massif in Germany yielded concordant dates that agree with $^{40}Ar^*/^{39}Ar$ dates on amphiboles of the host rocks. In contrast to the overgrowths, the U–Pb dates of the cores ranged from slightly discordant (low U and Th) to highly discordant (high U and Th) and indicated a Proterozoic age. The relation between the discordance of the U–Pb systematics of the cores and their U and Th concentrations is caused by radiation damage of the crystal lattice, which increases the diffusion rate of radiogenic Pb in zircon. Cherniak and Watson (2000) confirmed that the diffusion coefficient of Pb in zircon is very small but increases with increasing radiation damage. Therefore, the U–Pb parameters measured on zircon cores deviate from concordia depending on the extent of radiation damage. The low diffusion rate of Pb in undamaged zircon overgrowths causes their U–Pb dates to be concordant and gives such zircons closure temperatures in excess of 900°C.

These insights suggest yet another possible mechanism for the discordance of U–Pb dates derived from bulk zircon samples consisting of several hundred crystals each. Such samples are mixtures of cores and overgrowths in different proportions that define straight lines between points on Wetherill's concordia representing cores (upper intercept) and overgrowths (lower intercept). In such cases, the position of a particular zircon sample on discordia is controlled by the proportion of cores and overgrowths in that sample.

10.7 REFINEMENTS IN ANALYTICAL METHODS

The concentration and purification of zircon crystals from rocks is a time-consuming and painstaking process. After crushing and sieving, the zircon fraction is recovered by means of a wet shaking table and heavy liquids (tetrabromoethane, bromoform, methylene iodide, or Clerici's solution) that separate minerals by their specific gravities. The zircon concentrate is then purified on the basis of magnetic susceptibility using a magnetic separator and by handpicking and sorting on the basis of color, size, and shape of individual grains. Krogh (1982a) demonstrated that magnetic sorting of zircons to select the least paramagnetic grains tends to move data points along discordia toward the upper point of intersection with concordia.

10.7a Purification of Zircon Grains

The precision of U–Pb dates of zircon and other U-bearing accessory minerals is significantly improved by abrading them in a stream of air using pyrite and other kinds of solids as abrasives (Krogh, 1982b; Goldich and Fischer, 1986). This procedure removes highly altered (metamict) grains because they tend to be softer than clear, crack-free grains, which are most desirable for dating. In addition, the abrasion removes the surface layers of zircon grains where Pb loss is most prevalent and thereby reduces the discordance of U–Pb dates. The dates derived from precleaned zircon grains are also more precise because error magnification caused by the extrapolations of discordia lines to concordia is avoided.

Zircon and other U-bearing refractory minerals are difficult to dissolve in mixtures of hydrofluoric (HF) and other inorganic acids. This is a serious problem because incomplete recovery of the parent and daughter elements causes systematic errors in the calculated dates. Zircon is particularly troublesome in this regard because it resists acid attack that readily decomposes other silicate minerals. One way to overcome the resistance of zircons is to fuse them with a carbonate or borate flux. However, this method is no longer used because the flux is a potential source of contamination. The preferred method of dissolving zircon samples is to decompose them at 220°C in 48 percent HF in Teflon-lined stainless steel pressure vessels (Krogh, 1973). This procedure reduces contamination because only small quantities of reagents of high purity are used. The low blank combined with the miniaturization of sample-handling techniques and improvements in the data acquisition systems of computer-operated mass spectrometers permits samples weighing less than 1 mg to be analyzed successfully.

The zircons in Figure 10.9 were recovered from an Archean granite (sample 504 of Krogh, 1982b) and were subsequently purified by abrasion and magnetic sorting. The initial concentrate, represented by the solid point on discordia, is 18.6 percent discordant. Sample 1 was purified by removal of about 75 percent of its weight by abrasion. Samples 2 and 3 were abraded for 5 h and then split on the basis of their magnetic susceptibility into a "magnetic" fraction (sample 2) and a "nonmagnetic" fraction (sample 3). The latter is only 3.8 percent discordant. The best estimate of the age of these zircons is 2668.2 ± 2.1 Ma based on the intersection of the discordia with concordia using the method of Ludwig (1980). The uncertainty of this date is only 0.08 percent of its value. These results demonstrate that the magnetic sorting and abrasion

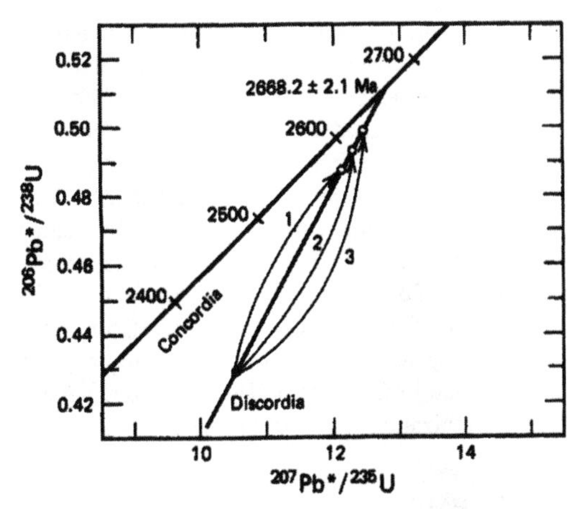

FIGURE 10.9 Improvement in U–Pb dates of zircon crystals caused by magnetic sorting and abrasion. Data from Krogh (1982b).

techniques concentrate the zircons that are most nearly concordant and therefore yield optimum results in terms of precision and accuracy.

The U,Th–Pb methods of dating refractory U-bearing minerals have been refined to such an extent that dates as low as 1.0 Ma can be measured with uncertainties of less than 20,000 years (e.g., Getty and DePaolo, 1995). In addition, U–Pb concordia dates of shocked zircons collected thousands of kilometers from the Chicxulub crater on the Yucatan Peninsula of Mexico have yielded reliable dates that clearly identify them as products of the impact at the K/T boundary (e.g., Krogh et al., 1993a, b; Kamo and Krogh, 1995). In addition, U–Pb dates of zircons in volcaniclastic rocks interbedded with well-dated marine sediment in the southern Alps have permitted a refinement of the Middle Triassic timescale (Mundil et al., 1996). These and many other examples in the literature demonstrate that the U,Th–Pb methods are capable of yielding dates with high precision and accuracy for all of geological time from the Early Archean to the Quaternary.

10.7b SHRIMP

The opportunity to date very small samples of lunar rocks by the U–Pb method originally led to the use of ion probe mass spectrometers (e.g., Andersen and Hinthorne, 1972; Hinthorne et al., 1979; Compston et al., 1984). This method of dating worked especially well with baddeleyite (ZrO_2), which occurs in mafic lunar and terrestrial rocks (Keil and Fricker, 1974). This mineral has a high closure temperature and concentrates U but excludes Pb (Heaman and LeCheminant, 1993). The use of secondary-ion mass spectrometry (SIMS) subsequently expanded following the design of a SHRIMP mass spectrometer by Dr. W. Compston at the Australian National University in Canberra (Compston, 1999). This instrument (mentioned in Section 10.6d) is capable of measuring $^{206}Pb/^{238}U$ and $^{207}Pb/^{235}U$ ratios by focusing a collimated O_2^- ion beam on a small spot of the polished cross section of a zircon grain (Smith et al., 1998; Williams, 1998). The accuracy of the results is based on the analysis of standards of known age (e.g., CZ-3, 564 Ma; Pidgeon et al., 1995). The precision of the dates arises from the ability to analyze a large number of spots on a single zircon grain.

Back-scattered electron images of polished zircon cross sections reveal that zircon crystals in high-grade metamorphic rocks and in granites have complex internal structures consisting of rounded cores surrounded by homogeneous or zoned overgrowths. The use of SHRIMPs permits the cores and overgrowths of complex zircon grains to be dated separately. In this way, the provenance of the zircon cores is determined by their U–Pb dates, whereas the time of magmatism or regional metamorphism is obtained by dating the overgrowths. This information is difficult to obtain by the conventional solution method even when applied to single grains or to fragments of grains because it requires the physical separation of cores and overgrowths by the selective abrasion technique developed by Krogh (1982). Examples of concordant U–Pb dates of inherited zircon cores have been reported elsewhere:

Friedl et al., 2000, *Geology*, 28:1035–1038. Qiu et al., 2000, *Geology*, 28:11–14. Hartmann et al., 1999, *Geology*, 27:947–950. Goodge and Fanning, 1999, *Geology*, 27:1007–1010. Brown and Fletcher, 1999, *Geology*, 27:1035–1038. Andersson et al., 1999, *Precamb. Res.*, 98:151–171.

Zircon is by no means the only mineral suitable for dating by the U–Pb method using SHRIMP: Monazite ($CePO_4$), sphene ($CaTiSiO_5$), rutile (TiO_2), perovskite ($CaTiO_2$), and uraninite (UO_2) are all potential targets. However, baddeleyite, which was the first mineral to be dated by means of an ion probe, has a flaw. Wingate and Compston (2000) reported that the measured values of $^{206}Pb/^{238}U$ ratios of this mineral vary by up to ±10 percent depending on the orientation of the crystals relative to the ion beam. The authors also tested monazite ($CePO_4$) and zircon (SL 13) (Claoué-Long et al., 1995) but found no indication that the orientation of crystals affected the results. Similarly, Stern and Berman (2000) successfully dated monazite inclusions in a Late Archean metapelite from the Western Churchill structural province of Canada.

10.7c LA-ICP-MS

Inductively coupled plasma mass spectrometry (ICP-MS) has become an important analytical tool for the chemical analysis of aqueous solutions because a large number of elements can be determined with excellent sensitivity. This equipment is also used to analyze solid samples either by dissolving them in acids or by vaporizing them with a laser. The use of laser ablation (LA) permits single mineral grains to be analyzed in situ, which eliminates the need to separate and dissolve them prior to analysis. Therefore, LA-ICP-MS is also applicable to dating U-bearing accessory minerals.

The adaptation of this instrument for use in geochronometry has been delayed because of problems with elemental fractionation and isotopic mass discrimination observed by Fryer et al. (1995), Longerich et al. (1996), and Eggins et al. (1998). These problems require the use of standards such as zircon crystals or glass standards (Fryer et al., 1993; Hirata and Nesbitt, 1995).

In addition, the diameter of the spot irradiated with the laser must be controlled during the analysis of standards and unknowns because of elemental fractionation reported by Horn et al. (2000), who demonstrated that the measured $^{206}Pb/^{238}U$ and $^{208}Pb/^{232}Th$ ratios increase with time during the laser irradiation of zircon while the $^{207}Pb/^{206}Pb$ ratios remain constant. In addition, the extent of elemental fractionation increases as the diameter of the spot decreases. These relations can be used to correct the measured $^{206}Pb/^{238}U$ and $^{208}Pb/^{232}Th$ ratios by using the diameter and the depth of laser craters, where the depth of the craters is a function of the number of pulses and hence of the irradiation time. These and other procedural improvements yielded a U–Pb concordia date of 2057 ± 8 Ma for baddeleyite of the Phalaborwa carbonatite in South Africa compared to 2059.8 ± 0.8 Ma reported by Heaman and LeCheminant (1993) by conventional thermal ionization mass spectrometry.

10.7d EMP

The electron microprobe (EMP) is capable of mapping the concentrations of a selected element in a polished surface of a mineral grain. This capability has led to a revival of the chemical U, Th–Pb methods of dating minerals such as monazite (Suzuki and Adachi, 1991; Williams et al., 1999; Poitrasson et al., 2000) and zircon (Geisler and Schleicher, 2000). This method of dating is based on measurements of the concentrations of U, Th, and Pb and relies on the following assumptions (Montel et al., 1996; Cocherie et al., 1998):

1. Lead is predominantly of radiogenic origin.
2. Uranium has a normal isotope composition.
3. The concentrations of U, Th, and Pb in the mineral have changed only by decay of U and Th isotopes to Pb.

The suitability of monazite for dating by this method arises from its high U and Th concentrations and its comparatively low Pb concentrations (Parrish, 1990).

The concentration of Pb in monazite increases with time in accordance with an equation formulated by Montel et al. (1996):

$$\mathrm{Pb} = \frac{\mathrm{Th}}{232}(e^{\lambda_3 t} - 1) \times 208 + \frac{\mathrm{U} \times 0.9928}{238.04}(e^{\lambda_1 t} - 1) \times 206 + \frac{\mathrm{U} \times 0.0072}{238.04}(e^{\lambda_2 t} - 1) \times 207 \quad (10.15)$$

where the concentrations of Pb, Th, and U are expressed in ppm (μg/g) and λ_1, λ_2, and λ_3 are the decay constants of ^{238}U, ^{235}U, and ^{232}Th, respectively.

The EMP method of dating based on U, Th, and Pb maps of monazite grains in metamorphic rocks has revealed the presence of inherited cores with younger overgrowths and has provided dates for these components (Williams et al., 1999). The application of this method to zircons of Proterozoic age indicated that the U–Th–Pb system was severely disturbed in parts of the crystals that had been depleted in Pb and enriched in Ca as a result of aqueous alteration depending on the extent of radiation damage (Geisler and Schleicher, 2000).

10.8 DATING DETRITAL ZIRCON GRAINS

The refinement of analytical procedures and of data acquisition by thermal ionization mass spectrometry (TIMS) has made it possible to analyze very small samples of zircon ranging down to single grains. In fact, even fragments of an individual zircon crystal that weighed only 0.5 mg before being broken into 11 pieces have been dated by the U–Pb method using a concordia diagram (Schärer and Allègre, 1982). Other investigators have studied zircon crystals by mounting them on the filament in the source of a mass spectrometer (Gentry et al., 1982) or by using SHRIMP mass spectrometry and LA-ICP-MS (Compston and Williams, 1984). For example, Froude et al. (1983) dated grains of zircon from the Mt. Narryer quartzite in West Australia by means of SHRIMP mass spectrometry. They discovered four grains with nearly concordant U–Pb dates between 4100 and 4200 Ma. The existence of these very old mineral grains indicates that zircon-bearing granitic rocks occurred at the surface of the Earth even before the formation of the Amitsoq gneisses in Greenland and of granitic rocks of similar age elsewhere.

The dating of single zircon crystals by TIMS is an art developed in the laboratory of C. J. Allègre at the University of Paris. The effort to date single zircon crystals was motivated by the discovery that a large sample of rock may contain zircon grains that differ in color, shape, trace-element content, and origin (Allègre et al., 1974). Therefore, a more complete record of the geological history of a rock can be assembled by dating individual grains, or fragments of grains, representing each of the different kinds of zircon crystals that may be present.

Zircon grains in sedimentary rocks derived from source regions underlain by igneous–metamorphic complexes are most likely to be heterogeneous and should, individually, record the full range of orogenic and magmatic histories of their source areas. This approach was taken by Gaudette et al. (1981), who dated different types of zircons extracted from the Upper Cambrian Potsdam Sandstone in eastern New York state.

10.8a Potsdam Sandstone, New York

The Potsdam Sandstone was deposited directly on the igneous and metamorphic rocks of the Adirondack Mountains, which are an extension of the Grenville structural province of the Canadian Precambrian Shield (Figure 6.16). There is good reason to assume, therefore, that the zircon grains in the Potsdam Sandstone originated primarily from sources within the Grenville Province located a short distance west of the collecting site. Age determinations of rocks from the Adirondack Mountains, reviewed by Gaudette et al. (1981), have yielded dates between 1000 and 1200 Ma, although some detrital zircons have suggested dates in excess of 2700 Ma.

Zircons extracted from a 35-kg sample of Potsdam Sandstone, collected near Whitehall, New York, were divided into clear and brown grains, each of which was further subdivided into rounded and elongate grains. Three "bulk" samples of the zircon concentrate defined a discordia with intersections at 1300 and 530 Ma. However, the individual grains and small samples representing the four different varieties of zircon gave quite different results summarized in Table 10.5. The 1300-Ma date indicated by the "bulk" zircon samples could be misinterpreted to suggest that the zircons in the Potsdam Sandstone originated from source rocks having that age. However, the dates derived from the different varieties of zircon demonstrate that the 1300-Ma date is an artifact produced by mixing zircons of differing ages in varying proportions.

Table 10.5. Summary of U–Pb Dates Derived from Selected Varieties of Zircon Recovered from the Upper Cambrian Potsdam Formation East of the Adirondack Mountains, New York

	Concordia Intercept Dates	
Variety of Zircon	Lower, Ma	Upper, Ma
Bulk sample	530	1300
Brown, elongate	325	1170 ± 100
Brown, rounded	380	1320 ± 80
Clear, rounded	985 ± 40	2160 ± 500
Clear, elongate	985 ± 15	2700 ± 250

Source: Gaudette et al., 1981.

The brown, elongate zircons, which appear to be least abraded, yield a date of 1170 ± 100 Ma that agrees well with the age of 1130 Ma of the Ticonderoga Gneiss in the Adirondack Mountains. The brown, rounded grains are distinctly older at 1320 ± 80 Ma and may have originated from other sources within the Grenville Province. The clear zircons yield still older dates (2160 ± 500 and 2700 ± 500 Ma) representative of sources in the Superior structural province of Canada. The lower intercept dates of the clear zircons (985 Ma) coincide with the time of high-temperature metamorphism of the Grenville rocks. Therefore, the clear zircons may have been incorporated into pre-Grenville sediment and later experienced episodic Pb loss during the Grenville orogeny. The lower intercept dates of the brown zircons (325 and 380 Ma) may be the result of Pb loss during slow cooling following the Grenville orogeny. However, the 530-Ma date indicated by the lower concordia intercept of the "bulk" samples is probably an artifact of mixing and cannot be regarded as evidence of Pb loss by chemical weathering of the zircons during deposition in Late Cambrian time.

The study of the Potsdam Sandstone illustrates the important conclusion that mixtures of two or more varieties of zircons of different origin, derived from a single sample of rock, may yield misleading results. This difficulty is avoided by dating single grains or small samples of selected grains each of which may reveal its own history. As a result, much more information about the geological history of the rock can be recovered from single zircon grains than can be learned from large samples composed of thousands of grains.

10.8b Pontiac Sandstone, Abitibi Belt, Ontario/Quebec

The Abitibi greenstone belt in northern Ontario and Quebec formed during the Archean Eon as a result of volcanic activity along an island arc and associated back-arc basin (Dimroth et al., 1982; Goodwin, 1982; Ludden et al., 1986; Corfu, 1993). The region has great economic importance because of the presence of large ore deposits of Cu, Zn, and Au in Timmins, Kirkland Lake, Rouyn-Noranda, Val d'Or, and Chibougamau. The geological history of this region has been reconstructed by careful mapping of the rocks, insightful tectonic modeling, interpretation of trace-element suites, and precise U–Pb dating of U-bearing minerals, including zircon and baddeleyite (Cattell et al., 1984; Corfu et al., 1989).

The Pontiac Group of the southern Abitibi belt consists of a thick sequence of deformed and metamorphosed sandstones and interbedded pelites that were deposited on a basement of basaltic lava flows. The sandstones are at least 1000 m thick and are composed primarily of quartz grains. The grain size, bed thickness, and sand/shale ratios decrease systematically from north to south, suggesting that the source of the sandstone was located north of the present Abitibi belt (Gariepy et al., 1984).

The existence of quartz sandstone in the southern Abitibi belt implies that a granitic basement complex or pre-existing sandstones occurred nearby and may have interacted tectonically with the island arc along which komatiites and tholeiite basalts were erupted between 2740 and 2700 Ma. Therefore, Gariepy et al. (1984) separated zircon grains from two large samples of the Pontiac Sandstone hoping to detect old zircons derived from granitic basement rocks or sedimentary rocks that predated the start of volcanic activity at about 2740 Ma.

The zircons they recovered defined several discordia lines (Figure 10.10) that yielded a range of upper and lower intercept dates (Table 10.6) depending on the shapes of the grains. The

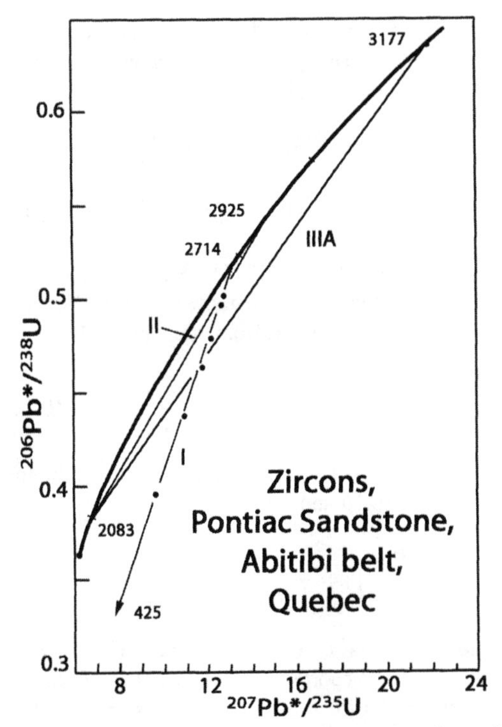

FIGURE 10.10 Concordia diagram for three suites of zircon grains extracted from two samples of the Pontiac Sandstone in the southern Abitibi greenstone belt of Ontario and Quebec. Only the data points for suite I in Table 10.6 are shown. The dates corresponding to the intersections of discordias I, II, and IIIA are in units of Ma. Data from Gariepy et al. (1984).

translucent needles, which exhibit no evidence of abrasion, yielded an upper intercept date on concordia of 2714 + 33, −23 Ma. This date and the lack of abrasion together indicate that these zircons crystallized in acid igneous rocks that may have been cogenetic with the island-arc lavas. The original host rock of the translucent zircons was exposed to weathering, the zircon crystals were transported a short distance, and were deposited in the Pontiac sedimentary basin.

The thick prisms are significantly abraded and yielded an upper intercept date of 2925 + 99, −85 Ma that is significantly older than the age of the mafic lava flows of the Abitibi belt. Consequently, these zircons originated from pre-existing igneous or sedimentary rocks on a continent adjacent to the volcanic island arc. The discordia line defined by the thick zircon prisms also yielded a lower intercept with concordia at 2083 + 143, −166 Ma, which represents an episode of Pb loss from these zircons.

The rounded zircon grains are severely abraded and metamict (damaged by radiation). These features imply that these zircons also predate the deposition of the Pontiac Sandstone and that they originated from source rocks located outside of the sedimentary basin. Five of these metamict grains yielded an upper intercept date of 3177 + 132, −102 Ma and a lower intercept date of 2095 + 97, −115 Ma. A second suite of the rounded and metamict zircons defined a discordia with an intercept at 2690 + 14, −11 Ma. Two additional grains coincide with the 3177-Ma discordia, and one grain appeared to be older than 3177 Ma.

To summarize, the U–Pb concordia dates of two suites of individual detrital zircon grains in the Pontiac Sandstone in Figure 10.10 predate the eruption of volcanic rocks of the Abitibi belt. These grains originated from sources on a nearby landmass and were transported into the depositional

Table 10.6. U–Pb Concordia Dates of Detrital Zircon Grains from Metasandstone of the Pontiac Group in the Southern Part of the Abitibi Belt in Ontario and Quebec

Group	Shape of Zircon Grains	Upper Intercept, Ma	Lower Intercept, Ma
I	Translucent needles	2714 +33, −23	425 +376, −391
II	Thick prisms	2924 +99, −85	2083 +143, −166
III	Rounded grains A	3177 +132, −102	2095 +97, −115
	Rounded grains B	2690 +14, −11	280 +285, −280

Source: Gariepy et al., 1984.

basin of the Pontiac Group. The host rocks from which the two older suites of detrital zircons originated may have been granitic basement rocks and/or pre-existing sedimentary rocks of Middle to Late Archean age. The presence of these kinds of rocks strongly implies the existence of continental crust in the area adjacent to the volcanic island-arc and back-arc basin of the Abitibi greenstone belt. However, the presence of older continental crust does not preclude the deposition of the mafic lavas on oceanic crust. In other words, the Abitibi greenstone belt itself is probably not underlain by older granitic basement rocks.

The U–Pb concordia method has been used successfully by several other research teams to date morphologically selected grains of detrital zircons:

Lancelot et al., 1973, *Comptes rendu des seances de l' Academie des Sciences de Paris, 227* (Ser. D):2116–2120. Lancelot et al., 1976, *Earth Planet. Sci. Lett.*, 29:357–366. Gebauer and Grünenfelder, 1977, *Contrib. Mineral. Petrol.*, 65:29–37. Schärer and Allègre, 1982, *Can. J. Earth Sci.*, 19:1910–1918. Goldstein et al., 1997, *Chem. Geol.*, 139:271–286. Mueller et al., 1998, *Precamb. Res.*, 91:295–307. Sircombe and Freeman, 1999, *Geology*, 27:879–882.

10.9 TERA–WASSERBURG CONCORDIA

The U, Th–Pb dates of lunar rocks reported by Tatsumoto and Rosholt (1970) and Tera and Wasserburg (1972, 1973) are significantly older than Rb–Sr and K–Ar dates of the same rocks. For example, a lunar basalt (14310) analyzed by Tera and Wasserburg (1972) yielded the following whole-rock U–Pb and Th–Pb model dates: 4.24 Ga (^{238}U–^{206}Pb), 4.27 Ga (^{235}U–^{207}Pb), and 4.13 Ga (^{232}Th–^{208}Pb), compared to Rb–Sr and K–Ar dates of only 3.88 Ga. The reason for the discrepancy is that this and other lunar rocks contain excess radiogenic ^{206}Pb, ^{207}Pb, and ^{208}Pb that was incorporated into the lunar basalt at the time of crystallization. Tera and Wasserburg (1972) therefore devised a new concordia that does not require prior knowledge of the initial ^{206}Pb/^{204}Pb and ^{207}Pb/^{204}Pb ratios.

The number of ^{206}Pb and ^{207}Pb atoms in a unit weight of U-bearing rocks or minerals is expressed by the equations

$$^{206}\text{Pb} = {}^{206}\text{Pb}_i + {}^{238}\text{U}\,(e^{\lambda_1 t} - 1) \quad (10.16)$$

$$^{207}\text{Pb} = {}^{207}\text{Pb}_i + \frac{^{238}\text{U}}{137.88}(e^{\lambda_2 t} - 1) \quad (10.17)$$

Tera and Wasserburg (1972) used these equations to define a concordia in parametric form where the x-coordinate is derived from equation 10.16,

$$\frac{^{238}\text{U}}{^{206}\text{Pb}^*} = \frac{1}{e^{\lambda_1 t} - 1} \quad (10.18)$$

and the y-coordinate is obtained by combining equations 10.16 and 10.17,

$$\left(\frac{^{207}\text{Pb}}{^{206}\text{Pb}}\right)^* = \frac{1}{137.88}\left(\frac{e^{\lambda_2 t} - 1}{e^{\lambda_1 t} - 1}\right) \quad (10.19)$$

The concordia is constructed by solving equations 10.18 (x-coordinate) and 10.19 (y-coordinate) for selected values of t. The results are listed in Table 10.7. The resulting graph in Figure 10.11 is the locus of all points representing U–Pb systems that yield concordant dates.

The discordia line in Figure 10.11 intersects the Tera–Wasserburg concordia at two points corresponding to dates t_1 (3.2 Ga) and t_2 (0.2 Ga). Extrapolation of this line beyond t_1 yields an intersection (I_0) on the y-axis (^{238}U/^{206}Pb* = 0). The numerical value of I_0 is the radiogenic ^{207}Pb/^{206}Pb ratio that formed in the interval of time between t_1 and t_2 (Tera and Wasserburg, 1974):

$$I_0 = \frac{1}{137.88}\frac{e^{\lambda_2 t_1} - e^{\lambda_2 t_2}}{e^{\lambda_1 t_1} - e^{\lambda_1 t_2}} \quad (10.20)$$

where λ_1 and λ_2 are the decay constants of ^{238}U and ^{235}U, respectively, and t_1 and t_2 are identified in Figure 10.11.

The U–Pb system whose geological history is depicted in Figure 10.11 originally contained no radiogenic Pb (i.e., ^{238}U/^{206}Pb$^* = \infty$) when it formed at $t_1 = 3.2$ Ga. Subsequently, radiogenic ^{207}Pb and ^{206}Pb accumulated by decay of ^{235}U and ^{238}U, respectively, until the system recrystallized or differentiated at $t_2 = 0.2$ Ga. The radiogenic

Table 10.7. Coordinates of Points That Define Tera–Wasserburg Concordia (Equations 10.18 and 10.19) Where x-Coordinate Is $1/(e^{\lambda_1 t} - 1)$ and y-Coordinate Is $\dfrac{1}{137.88}\left(\dfrac{e^{\lambda_2 t} - 1}{e^{\lambda_1 t} - 1}\right)$

t, Ga	x	y	t, Ga	x	y
0.2	31.746	0.05012	2.6	2.012	0.1743
0.4	15.625	0.05575	2.8	1.838	0.1968
0.6	10.256	0.05992	3.0	1.687	0.2226
0.8	7.570	0.06581	3.2	1.555	0.2524
1.0	5.959	0.0725	3.4	1.439	0.2867
1.2	4.887	0.0801	3.6	1.336	0.3263
1.4	4.122	0.0887	3.8	1.245	0.3721
1.6	3.549	0.0987	4.0	1.162	0.4249
1.8	3.104	0.1100	4.2	1.088	0.4862
2.0	2.748	0.1229	4.4	1.021	0.5571
2.2	2.458	0.1378	4.6	0.9603	0.6393
2.4	2.216	0.1548			

Note: The x- and y-values were calculated from data in Table 10.3.

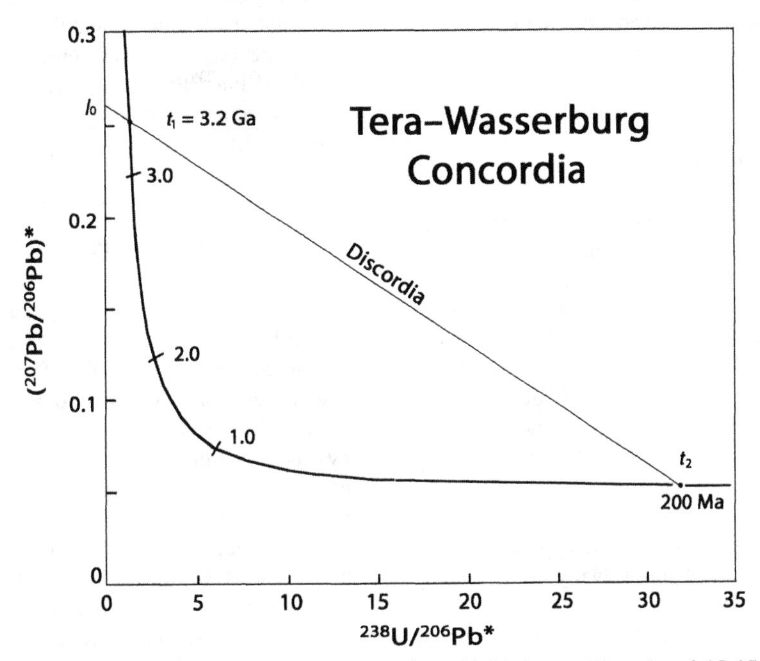

FIGURE 10.11 Tera–Wasserburg concordia based on equations 10.14 (x-coordinate) and 10.15 (y-coordinate) plotted from values in Table 10.7 assuming that λ_1 (^{238}U) = 1.55125×10^{-10} y^{-1}, λ_2 (^{235}U) = 9.8485×10^{-10} y^{-1}, and ^{238}U/^{235}U = 137.88.

^{207}Pb/^{206}Pb ratio of the Pb that had accumulated from $t_1 = 3.2$ Ga to $t_2 = 0.2$ Ga is equal to I_0 in Figure 10.11 and is expressed by equation 10.20. If a U-free mineral (e.g., plagioclase) formed during the recrystallization event at t_2, it would contain Pb whose radiogenic ^{207}Pb/^{206}Pb ratio is equal to I_0.

The U-bearing system represented by t_2 on the Tera–Wasserburg concordia was Pb free at the end of the metamorphic event (i.e., ^{238}U/^{206}Pb* $= \infty$). All of the Pb it contains at the present time formed by decay of U isotopes after the end of the recrystallization event at $t_2 = 0.2$ Ga. This interpretation implies that both t_1 and t_2 are valid dates in the geological history of a volume of U-bearing rocks.

Alternatively, the radiogenic ^{207}Pb/^{206}Pb ratio represented by I_0 may be the result of a complex process unrelated to the U–Pb system that recrystallized at t_2. In that case, the discordia is the locus of U–Pb systems that formed by mixing of two components. One of these components is I_0 and the other component is the U–Pb system represented by the point of intersection at t_2 in Figure 10.11. In that case, the date derived from the coordinates of point t_1 has no geological significance (Tera and Wasserburg, 1974).

10.9a Lunar Basalt 14053

The rock samples returned by the Apollo 14 astronauts from the Fra Mauro region of the Moon included a specimen of basalt numbered 14053. This specimen (whole rock #1) yielded highly discordant and improbable U,Th–Pb dates: ^{238}U–^{206}Pb = 5.60 Ga, ^{235}U–^{207}Pb = 5.18 Ga, ^{232}Th–^{208}Pb = 5.48 Ga, and ^{207}Pb/^{206}Pb = 5.01 Ga (Tera and Wasserburg, 1972). The measured isotope ratios of Pb were corrected for the laboratory blank and for the presence of primeval Pb which the Moon inherited from the solar nebula. This treatment of the isotope ratios is appropriate even though the Moon did not actually form directly from the solar nebula.

The same specimen of basalt had previously yielded an internal Rb–Sr isochron date of 3.88 ± 0.04 Ga ($\lambda = 1.42 \times 10^{-11}$ y^{-1}) based on analyses of plagioclase, "cristobalite," "quintessence," and the whole-rock sample (Papanastassiou and Wasserburg, 1971). The 'quintessence' fraction, obtained by handpicking, had a high ^{87}Sr/^{86}Sr ratio of 0.74470 ± 0.00027 relative to 0.70908 ± 0.00005 for seawater. The initial ^{87}Sr/^{86}Sr ratio of this specimen of lunar basalt is 0.69948 ± 0.00006.

Lunar basalt 14053 was also dated by Turner et al. (1971) by the ^{40}Ar*/^{39}Ar method applied both to the whole-rock sample and to a plagioclase concentrate. The partial-release spectra indicated well-defined plateau dates of 3.95 Ga (plagioclase) and 3.93 Ga (whole rocks) for $\lambda_\beta = 4.72 \times 10^{-10}$ y^{-1}, $\lambda_e = 0.584 \times 10^{-10}$ y^{-1}, and ^{40}K/K $= 1.19 \times 10^{-4}$. These values differ from those adopted later by Steiger and Jäger (1977). The ^{40}Ar*/^{39}Ar dates of lunar basalt 14053 can be recalculated from the data reported by Turner et al. (1971) as demonstrated in Section 7.4.

The U,Th–Pb data reported by Tera and Wasserburg (1972) for lunar basalt 14053 define a discordia line that intersects the y-axis (^{238}U/^{206}Pb* = 0) at 1.46 in Figure 10.12. The slope of the discordia line is -0.88366 based on an unweighted linear regression of three data points representing two whole-rock samples and the magnetic fraction. The slope (m) of the discordia is related to the initial (^{207}Pb/^{206}Pb)* ratio and to the age of the U–Pb system by the equation (Tera and Wasserburg, 1972):

$$m = \frac{e^{\lambda_2 t} - 1}{137.88} - \left(\frac{^{207}\text{Pb}}{^{206}\text{Pb}}\right)_i (e^{\lambda_1 t} - 1) \qquad (10.21)$$

This equation was solved graphically in Figure 10.13 for $(^{207}\text{Pb}/^{206}\text{Pb})_i = 1.46$ and values of t between 3.87 and 4.00 Ga. The graph indicates that a slope of -0.88366 corresponds to a date of 3.91 Ga. A more accurate date is obtainable by interpolating in a table containing values of the slope for selected values of t or by numerical iteration.

The interpretation of the U–Pb data for lunar basalt 14053 by means of the Tera–Wasserburg concordia yielded a date that is in good agreement with the Rb–Sr and ^{40}Ar*/^{39}Ar dates of this rock. The essential feature of this concordia is that it permits an explicit determination of the radiogenic ^{207}Pb/^{206}Pb ratio at ^{238}U/^{206}Pb* = 0 without requiring an estimate of the initial isotope

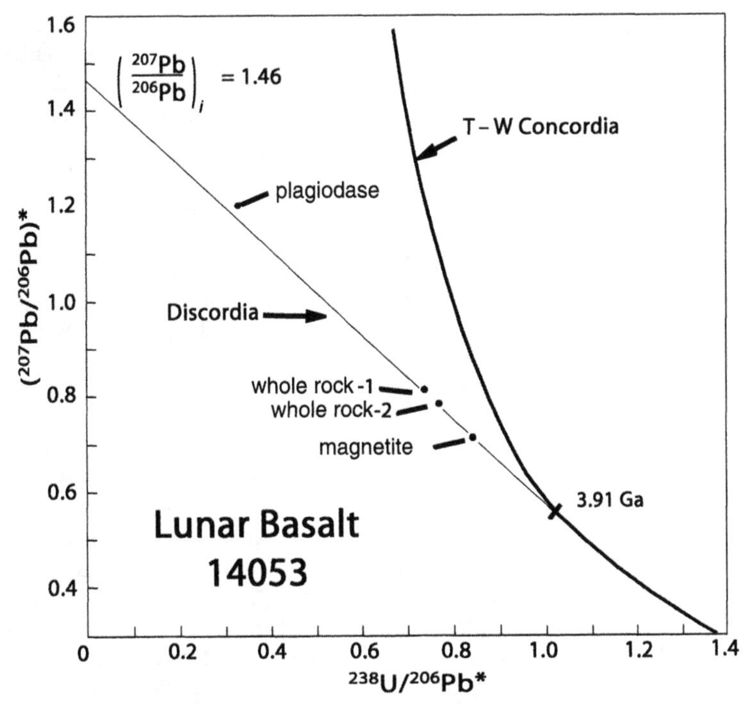

FIGURE 10.12 Tera–Wasserburg concordia diagram for U–Pb data of lunar basalt 14053. The measured abundances of the Pb isotopes were corrected for the laboratory blank and for the presence of primeval Pb: $^{206}Pb/^{204}Pb = 9.346$, $^{207}Pb/^{204}Pb = 10.218$, and $^{208}Pb/^{204}Pb = 28.960$. The slope of the discordia line is $m = -0.88366$ and the intercept on the y-axis for $^{238}U/^{206}Pb = 0$ is 1.46. The date that corresponds to the point of intersection on the Tera–Wasserburg concordia was determined graphically from equation 10.21 in Figure 10.14. Data from Tera and Wasserburg (1972).

ratios of Pb at the time of crystallization of the basalt.

10.9b Other Applications of Tera–Wasserburg Concordia

The Tera–Wasserburg concordia was extended by Wendt (1984, 1989) to a three-dimensional representation of discordant U–Pb systems. The three-dimensional U–Pb discordia method of dating was further elaborated by Zheng (1989, 1992) and was reviewed by Jahn and Cuvellier (1994).

Wendt and Carl (1985) used the extended Tera–Wasserburg concordia to date a suite of torbernites $[Cu(UO)_2P_2O_8 \cdot 12H_2O]$ in a U deposit at Höhenstein, Germany. The results indicated dates of 66 ± 27 Ma (upper intercept) and 137 ± 11 ka (lower intercept), in agreement with a date of 132 ± 4 ka determined by the decay of ^{234}U to ^{230}Th. In addition, Carl et al. (1989) dated the Falkenberg Granite in northeast Bavaria, which yielded a date of 307 ± 22 Ma by the Tera–Wasserburg concordia method. Other applications of the Tera–Wasserburg concordia include the work of Smith et al. (1991) and Richards et al. (1998), who used it to date calcite in Paleozoic limestone and Quaternary speleothems, respectively. The Tera–Wasserburg concordia was also used by Bacon et al. (2000) to date a late Pleistocene granodiorite that underlies Crater Lake in Oregon. These authors used an ion microprobe (SHRIMP) to analyze zircon crystals that yielded a date of $101 + 78$, -80 ka (uncorrected for ^{230}Th deficit).

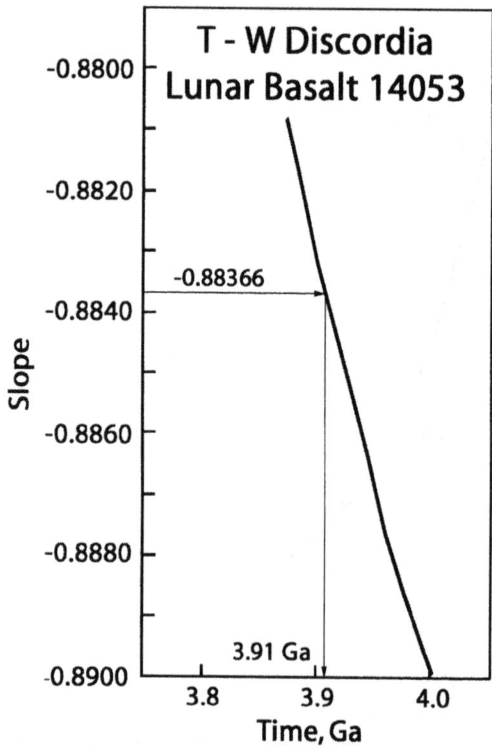

FIGURE 10.13 Graphical solution of equation 10.21 for $(^{207}Pb/^{206}Pb)_i^* = 1.46$ and based on λ_1 (^{238}U) $= 1.55125 \times 10^{-10}$ y^{-1} and λ_2 (^{235}U) $= 9.8485 \times 10^{-10}$ y^{-1}. The slope of the discordia line defined by lunar basalt 14053 in Figure 10.12 corresponds to a date of about 3.91 Ga. Data from Tera and Wasserburg (1972).

10.10 U–Pb, Th–Pb, AND Pb–Pb ISOCHRONS (GRANITE MOUNTAINS, WYOMING)

Equations 10.5–10.7, which govern the accumulation of the radiogenic isotopes of Pb, can be used to plot three independent isochrons analogous to the Rb–Sr and Sm–Nd isochrons (Chapters 5 and 9, respectively). The slopes of the ^{238}U–^{206}Pb, ^{235}U–^{207}Pb, and ^{232}Th–^{208}Pb isochrons yield dates that are concordant *only* when the samples remained closed and had identical initial isotope ratios of Pb. In most cases, U–Pb isochrons based on whole-rock samples have not been successful, primarily because rocks exposed to chemical weathering lose a significant fraction of U. However, Th–Pb isochrons can be useful because Th is retained quantitatively, in contrast to U.

10.10a U,Th–Pb Isochrons

The $^{208}Pb/^{204}Pb$ and $^{232}Th/^{204}Pb$ ratios of a suite of whole-rock samples (plus one feldspar) from the Granite Mountains of Wyoming scatter only moderately above and below a straight line in Figure 10.14*a* and yield a date of 2.79 ± 0.08 Ga. In marked contrast, the same samples in Figure 10.14*b* demonstrate significant losses of U of up to 88 percent. However, in spite of the loss of U, the $^{207}Pb/^{204}Pb$ and $^{206}Pb/^{204}Pb$ ratios define a satisfactory Pb–Pb isochron in Figure 10.14*c* (Rosholt and Bartel, 1969; Rosholt et al., 1973).

Therefore, the U–Pb isochron method of dating igneous and metamorphic rocks composed of silicate minerals does not work in this and most other cases because of variable losses of U by chemical weathering, which occurs not only at the surface of the Earth but also in the subsurface, where rocks are in contact with oxygenated groundwater. Additional examples of U–Pb and Th–Pb isochrons of granitic rocks have been published by Ulrych and Reynolds (1966) and Farquharson and Richards (1970).

10.10b Pb–Pb Isochrons

Igneous and metamorphic rocks that have lost U by recent chemical weathering may also have lost Pb. However, the isotope ratios of the remaining Pb may not have changed if the isotopes of Pb were not fractionated (Section 10.3). In other words, the isotope ratios of Pb in weathered rocks are not changed if the Pb that was lost had the same isotope composition as the Pb that was present before the loss occurred. Consequently, a date may be calculated based on the slope of the Pb–Pb isochron illustrated in Figure 10.14*c*.

The equation for Pb–Pb isochrons is derived by combining equations 10.5 and 10.6 to yield

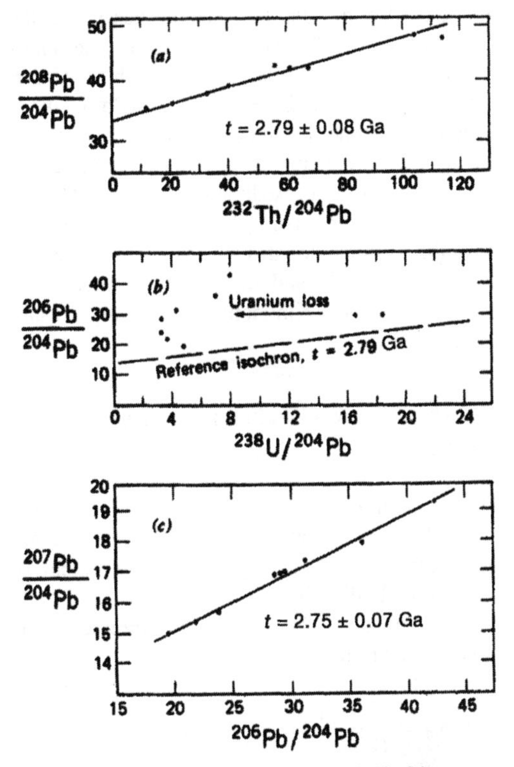

FIGURE 10.14 The Th–Pb, U–Pb, and Pb–Pb whole-rock isochrons for granitic rocks from the Granite Mountains of Wyoming. (*a*) These rocks have experienced variable gain or loss of Th, but on average the loss of Th or Pb has been small. (*b*) The data points are strongly displaced from the reference isochron drawn for $t = 2.79 \times 10^9$ y, indicating significant losses of U ranging from about 37 to 88 percent. (*c*) The isotope ratios of Pb define a Pb–Pb isochron with a slope of 0.1911 ± 0.0087 that corresponds to a date of 2.75 ± 0.07 Ga. Data by Rosholt et al. (1973).

equation 10.9, which expresses the ratio of radiogenic ^{207}Pb to ^{206}Pb:

$$\left(\frac{^{207}\text{Pb}}{^{206}\text{Pb}}\right)^* = \frac{^{207}\text{Pb}/^{204}\text{Pb} - (^{207}\text{Pb}/^{204}\text{Pb})_i}{^{206}\text{Pb}/^{204}\text{Pb} - (^{206}\text{Pb}/^{204}\text{Pb})_i} = \frac{^{235}\text{U}}{^{238}\text{U}}\left(\frac{e^{\lambda_2 t} - 1}{e^{\lambda_1 t} - 1}\right)$$

This is the equation of a straight line in coordinates of $^{206}Pb/^{204}Pb$ (x) and $^{207}Pb/^{204}Pb$ (y) whose slope m is

$$m = \frac{^{235}\text{U}}{^{238}\text{U}}\left(\frac{e^{\lambda_2 t} - 1}{e^{\lambda_1 t} - 1}\right) = \left(\frac{^{207}\text{Pb}}{^{206}\text{Pb}}\right)^* \qquad (10.22)$$

Age determinations by the Pb–Pb isochron method rely on the assumptions that all samples that define the isochron had the same initial isotope ratios of Pb, formed at the same time, and remained closed to U, Th, and Pb until the recent past, when they were exposed to chemical weathering. In addition, the $^{238}U/^{204}Pb$, $^{235}U/^{204}Pb$, and $^{232}Th/^{204}Pb$ ratios of the samples must have sufficient range to allow Pb having different isotope ratios to form within them. The slope of Pb–Pb isochrons can be used for dating by solving equation 10.22 for t by interpolating within Table 10.3. Alternatively, equation 10.22 can be solved by iteration on a computer to any desired level of precision. The use of Table 10.3 for the calculation of Pb–Pb dates is illustrated for the Pb–Pb isochron for rocks of the Granite Mountains in Figure 10.14*c*, which has a slope of 0.1911 ± 0.0087 (Rosholt et al., 1973). This $(^{207}Pb/^{206}Pb)^*$ value lies between 0.17436 and 0.19680 in Table 10.3 and therefore corresponds to a date t of

$$t = 2.6 + \frac{0.2(0.1911 - 0.17436)}{0.19680 - 0.17436} = 2.6 + 0.149 = 2.75 \text{ Ga}$$

The uncertainty of the slope (±0.0087) translates into an uncertainty of ±0.07 Ga of the Pb–Pb date of the rocks in the Granite Mountains. Therefore, the age of these rocks (2.75 ± 0.07 Ga) is in good agreement with the Th–Pb date.

The Pb–Pb method has been widely used for dating igneous and metamorphic rocks of Precambrian age, exemplified by the work of Moorbath et al. (1973) on the iron formation at Isua in southern West Greenland, Sinha (1972) on the greenstones of the Onverwacht Group in South Africa, Oversby (1975) on the acid intrusives in Western Australia, and Reynolds (1971), who used Pb–Pb isochrons to date the ore and metamorphic country rocks at Broken Hill, Australia. In these and many other cases, the Pb–Pb method has yielded the time elapsed since the isotopic homogenization of Pb and subsequent closure of the rocks to U, Th, and their intermediate daughters.

The simultaneous decay of ^{238}U to ^{206}Pb and ^{232}Th to ^{208}Pb in a suite of whole-rock samples (or minerals) that formed at the same time t and have remained closed to U, Th, and Pb is represented by an equation that is analogous to equation 10.9:

$$\frac{^{208}Pb/^{204}Pb - (^{208}Pb/^{204}Pb)_i}{^{206}Pb/^{204}Pb - (^{206}Pb/^{204}Pb)_i} = \frac{^{232}Th}{^{238}U}\frac{e^{\lambda_3 t} - 1}{e^{\lambda_1 t} - 1} \quad (10.23)$$

Equation 10.23 differs from equation 10.9 because it contains the $^{232}Th/^{238}U$ ratio, which is not a known constant like the $^{235}U/^{238}U$ ratio in equation 10.9. Since cogenetic rocks (and minerals) can have a range of Th/U ratios, their measured $^{208}Pb/^{204}Pb$ and $^{206}Pb/^{204}Pb$ ratios scatter and, in most cases, do not define a Pb–Pb isochron. Instead, equation 10.23 can be used to calculate an average Th/U ratio for rocks (or minerals) that have yielded a date from the slope of a conventional Pb–Pb isochron provided that the same rocks permit the slope of a straight line to be calculated in coordinates of $^{208}Pb/^{204}Pb$ and $^{206}Pb/^{204}Pb$. An example of such a calculation is presented in Section 10.11c based on data provided by Jahn et al. (1990).

10.11 Pb–Pb DATING OF CARBONATE ROCKS

Carbonate rocks contain U, Th, and Pb in widely varying concentrations (Table 10.1). In addition, evidence first presented by Moorbath et al. (1987) suggested that carbonate rocks, at least in some cases, retain these elements for long periods of geological time and therefore may be datable by the U–Pb, Th–Pb, and Pb–Pb methods. The potential of the Pb–Pb isochron method for dating limestones and dolomites of Precambrian age and thereby to refine the geological timescale has stimulated a great deal of research that was reviewed by Jahn and Cuvellier (1994).

In principle, carbonate rocks can be dated by means of whole-rock isochrons arising from the decay of ^{238}U to ^{206}Pb, ^{235}U to ^{207}Pb, and ^{232}Th to ^{208}Pb. In addition, carbonate rocks may be dated by the Pb–Pb method based on the decay of U to Pb (Section 10.10b). The kinds of carbonate rocks that are amenable to dating by the U–Pb, Th–Pb, and Pb–Pb methods include limestone and dolomite, marble, and secondary calcite precipitated from brines after deposition and burial of sedimentary rocks. The U, Th–Pb methods of dating have also been used to date the carbonatite at Fen, Norway, (Andersen and Taylor, 1988) and uraniferous hydrocarbons in Wales, United Kingdom (Parnell and Swainbank, 1990).

10.11a Marine Geochemistry of U, Th, and Pb

The concentrations of U, Th, and Pb in seawater in Table 10.8 are affected in differing degrees by sorption to colloidal particles. The extent of sorption of U dissolved in seawater (3.2 ppb) and sorbed to deep-sea clay (2 ppm) is expressed by its effective distribution coefficient (K_d):

$$K_d = \frac{(U)_{solid}}{(U)_{water}} = \frac{2 \times 10^3}{3.2} = 625 \quad (10.24)$$

where (U) is the concentration of this element sorbed to mineral grains and in solution in seawater expressed in identical units. The comparatively low value of K_d indicates that U is not scavenged efficiently by clay minerals in the oceans and is consistent with its long oceanic residence time of 10^6 years.

Table 10.8. Concentrations of U, Th, and Pb in Seawater and Deep-Sea Clay

Element	Seawater, ppb	Deep-Sea Clay, ppm	K_d(solid /liquid)
Uranium	3.2	2	6.2×10^2
Thorium	1×10^{-2}	10	1×10^6
Lead	3×10^{-3}	2×10^2	67×10^6
U/Pb	$\sim 10^3$	10^{-2}	—
Th/Pb	~ 3	$\sim 5 \times 10^{-2}$	—
Th/U	3.2×10^{-4}	5.0	—

Source: Data from Martin and Whitfield (1983) as reported by Jahn and Cuvellier (1994).

Thorium in solution in seawater (0.01 ppb) is strongly sorbed to deep-sea clay (10 ppm) and has an effective K_d value of 10^6 and a short oceanic residence time of less than 100 years. Evidently, Th is much more strongly sorbed by clay minerals than U, as indicated by the Th/U ratios of 5.0 in deep-sea clay and 3.1×10^{-4} in seawater.

The concentration of Pb in solution in seawater (0.003 ppb) is very low, partly because it is sorbed to clay minerals and other sediment particles ($K_d \sim 67 \times 10^6$) and partly because Pb coprecipitates with carbonates and phosphates of Ca. Consequently, the residence time of Pb in the oceans is less than about 50 years. These geochemical properties combine to cause seawater to have a high U/Pb ratio of about 1000, whereas its Th/Pb ratio is only about 3.

The concentration of U in corals appears to decrease with age from about 2.75 ppm for Holocene, Pleistocene, and Pliocene specimens to 1.75 ppm in samples of Miocene age (Broecker and Peng, 1982). Calcium carbonate of other marine organisms has even lower U concentrations than corals (e.g., green algae, ~2 ppm; mollusks, foraminifera, etc., <1 ppm). These values are in good agreement with the average U concentration of 1.9 ppm in carbonate rocks in Table 10.1. In contrast to U, the average concentration of Pb in carbonate rocks (5.6 ppm, Table 10.1) is much *higher* than that of modern marine calcite, which contains only a few nanograms of Pb per gram. Therefore, the U/Pb ratio of modern marine carbonates is of the order of 10^3, whereas typical limestones in Table 10.1 have U/Pb ratios of only 0.34. A similar discrepancy applies to the Th/Pb ratios of modern marine carbonate (Th/Pb ~ 40) and limestones (Th/Pb ~ 0.21).

The data reviewed above imply that the U/Pb ratios of marine carbonates are drastically reduced after deposition by loss of U, gain of Pb, or both. Given that Paleozoic and Mesozoic limestones and modern marine carbonates have similar U concentrations, the low U/Pb ratios of limestones must be caused by the addition of Pb to carbonates during diagenesis or by water–rock interactions with brines having high concentrations of Pb such as those in the Salton Sea of California (White, 1968).

10.11b Mushandike Limestone, Zimbabwe

The Archean craton of Zimbabwe includes several stratigraphic units listed in Table 10.9 in order of increasing age. In addition, remnants of Precambrian sedimentary rocks, including small and highly deformed lenses of the Mushandike stromatolitic limestone of uncertain age, occur within the Mushandike National Park of Zimbabwe. The Mushandike Limestone was deposited on a granitic basement complex that includes the Mushandike Granite. Moorbath et al. (1987) reported a whole-rock Rb–Sr errorchron date of 2917 ± 171 Ma for the Mushandike *Granite*. The same samples of granite also define the Pb–Pb errorchron in Figure 10.15, which yields a date of 2946 ± 130 Ma (MSWD = 5.2 for 12 data points). Although the Rb–Sr and Pb–Pb whole-rock errorchron dates are similar, they are not sufficiently precise to resolve closely spaced geological events in the geological history of the Zimbabwe craton.

Moorbath et al. (1987) also measured the isotope ratios of Pb of *carbonate samples* taken from a short drill core (25 cm in length) of the Mushandike Limestone. The Pb was extracted from the acid-soluble calcite, the silicate residue, and from whole-rock samples, including both calcite and the silicate residue fractions. All of these samples form a set of colinear points in coordinates of $^{206}Pb/^{204}Pb$ and $^{207}Pb/^{204}Pb$ in Figure 10.16 from which Moorbath et al. (1987) calculated an errorchron date of 2839 ± 33 Ma.

The remarkable aspect of this data set is that the isotope ratios of Pb vary widely (e.g.,

Table 10.9. Stratigraphic Units of the Archean Craton of Zimbabwe

Stratigraphic Unit (Group)	Approximate Age, Ga
Upper Bulawayan (upper greenstone)	2.7
Shamvaian	
Lower Bulawayan (lower greenstone)	2.9
Sebakwian	3.5

Source: Moorbath et al., 1987.

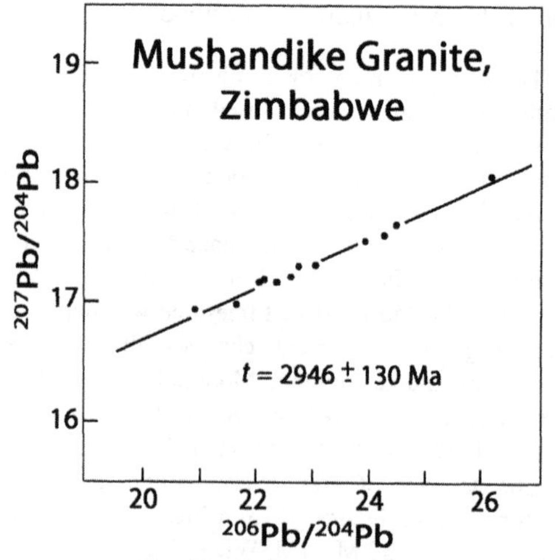

FIGURE 10.15 The Pb–Pb isochron of whole-rock samples of the Mushandike granite in the Archean craton of Zimbabwe. Data from Moorbath et al. (1987).

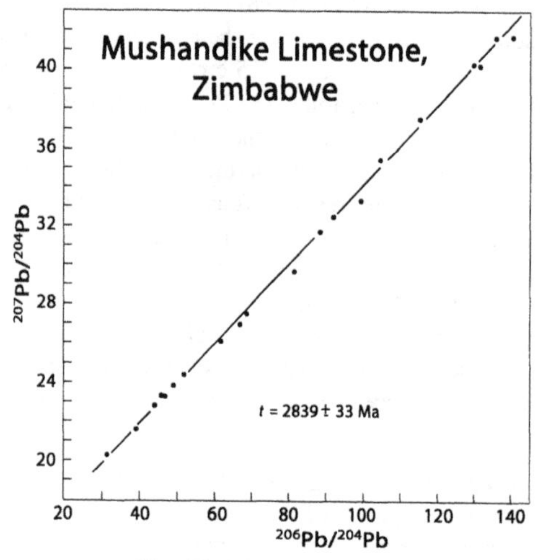

FIGURE 10.16 The Pb–Pb errorchron of the Mushandike Limestone in Mushandike National Park of Zimbabwe. Data from Moorbath et al. (1987).

$^{206}Pb/^{204}Pb$ from 31.468 to 140.499) and that centimeter-sized samples of the limestone have remained closed to U and Pb for nearly 2.9 billion years. In addition, the wide range of Pb isotope ratios requires a similarly wide range of U/Pb ratios in this limestone that far exceeds the range of U/Pb ratios in igneous and metamorphic rocks. Moorbath et al. (1987) interpreted the Pb–Pb errorchron date of the Mushandike Limestone as the age of deposition but did not rule out that it records a subsequent low-grade metamorphic event during which the isotope ratios of Pb were re-equilibrated. In any case, the Pb–Pb date clearly demonstrates that the Mushandike Limestone is of the same age as the Lower Bulawayan Group in Table 10.9 and that it is about 600 million years younger than the rocks of the Sebakwian Group.

10.11c Transvaal Dolomite, South Africa

The Kaapvaal craton of South Africa includes the metasedimentary rocks of the Transvaal Supergroup, which contains the extensive Transvaal Dolomite and the overlying Pretoria Group. Jahn et al. (1990) analyzed a suite of stromatolitic limestones from the Schmidtsdrif Formation of Griqualand that is correlated with the Transvaal Dolomite. The concentrations of U in these samples range only from 0.042 to 0.065 ppm, whereas the concentrations of Pb extend from 0.47 to 1.78 ppm and yield U/Pb ratios that vary from 0.029 to 0.12. The average U and Pb concentrations of these rocks are $U = 0.051 \pm 0.005$ ppm (2σ, $N = 9$) and $Pb = 0.82 \pm 0.27$ ppm (2σ, $N = 10$). Both values are lower than the U and Pb concentrations for typical carbonate rocks in Table 10.1.

Nevertheless, the $^{206}Pb/^{204}Pb$ and $^{207}Pb/^{204}Pb$ ratios of these carbonate rocks define a straight line in Figure 10.17 from which Jahn et al. (1990) calculated a date of 2557 ± 49 Ma. This date demonstrates a Late Archean age for the Transvaal Dolomite and implies that a lengthy erosional interval occurred between the time of deposition of the Transvaal Dolomite and the overlying Pretoria Group, which was deposited at 2224 ± 21 Ma or about 300 million years later than the Transvaal Dolomite (Walraven et al., 1990).

The $^{208}Pb/^{204}Pb$ and $^{206}Pb/^{204}Pb$ ratios of the Transvaal Dolomite scatter widely (not shown) but yield a slope of 1.1 ± 0.4 by a regression

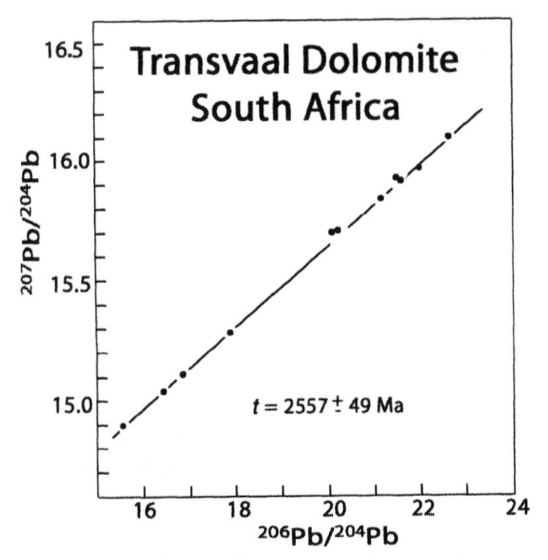

FIGURE 10.17 The Pb–Pb isochron diagram for carbonate rocks of the Transvaal Dolomite of South Africa. Data from Jahn et al. (1990).

of the data points to a straight line (Jahn et al., 1990). Accordingly, the average Th/U ratio of these carbonate rocks derivable from equation 10.20 is

$$1.1 = \frac{^{232}\text{Th}}{^{238}\text{U}} \frac{e^{\lambda_3 t} - 1}{e^{\lambda_1 t} - 1}$$

Since $\lambda_1 = 1.55125 \times 10^{-10}$ y^{-1}, $\lambda_3 = 4.9475 \times 10^{-11}$ y^{-1}, and $t = 2557 \times 10^6$ y, the ^{232}Th/^{238}U ratio is

$$\frac{^{232}\text{Th}}{^{238}\text{U}} = 1.1 \frac{e^{\lambda_1 t} - 1}{e^{\lambda_3 t} - 1} = 3.9724$$

$$\frac{\text{Th}}{\text{U}} = 3.9724 \times 0.9677 = 3.84 \pm 1.40$$

The Th/U ratio indicated by this calculation is similar to that of granitic gneisses and shale in Table 10.1 (i.e., 3.69 and 3.65, respectively).

The ^{208}Pb/^{204}Pb and ^{206}Pb/^{204}Pb ratios of several other lightly metamorphosed Precambrian carbonate deposits scatter widely or do not form straight lines. These include the dolomites of the Wittenoom and Carawine Formations of the Hamersley Group in Western Australia as well as the carbonates of the Tuanshanzi Formation in the Hebei Province of China, all of which were analyzed by Jahn and Cuvellier (1994). However, the ^{208}Pb/^{204}Pb and ^{206}Pb/^{204}Pb ratios of the acid-soluble carbonate fractions of the Mushandike Limestone of Zimbabwe dated by Moorbath et al. (1987) do form a straight line that yields a Th/U ratio of 2.2 ± 0.1.

Other examples of Pb–Pb dating of marine carbonates of Precambrian age include the Sukhaya Tunguska Formation of Russia (Ovchinnikova et al., 1995) and the Late Proterozoic carbonate rocks of the Pickelhaube Formation in the Gariep and Saldania belts of southern Namibia and South Africa (Fölling et al., 2000). The latter achieved significant improvements in the precision of Pb–Pb dates by using a double-spike technique to measure isotope ratios of Pb and by dating the carbonate and residue fractions separately. The Pb–Pb isochron of the carbonates of the Pickelhaube Formation (728 ± 32 Ma, MSWD $= 1.5$) dates the time of early diagenesis, whereas the Pb–Pb isochron of the silicate residues (543 ± 13 Ma, MSWD $= 7.4$) indicates the time of metamorphism of these rocks. However, in several other cases the carbonate and residue fractions formed a single Pb–Pb isochron (Jahn, 1988; Taylor and Kalsbeek, 1990).

10.12 U–Pb AND Th–Pb ISOCHRONS OF CARBONATE ROCKS

Although U is lost from igneous and metamorphic rocks by chemical weathering and other forms of aqueous alteration, carbonate rocks appear to be more retentive and have been dated successfully by means of U–Pb and Th–Pb isochrons. Such dates are especially informative when they are based on secondary calcite that was deposited by brines that may migrate through these rocks in response to hydraulic gradients during late-stage diagenesis accompanied by the deposition of sulfide minerals of Pb, Zn, and Cu.

10.12a Lucas Formation (Middle Devonian), Ontario

An example of U–Pb isochron dating of carbonate rocks was presented by Smith and Farquhar (1989)

Table 10.10. U and Pb Concentrations of Carbonate Materials in the Middle Devonian Lucas Formation of Ontario, Canada

Carbonate Material[a]	U, ppm	Pb, ppm	U/Pb
Dolomitic limestone (2)	1.552	0.464	3.34
Limestone (5)	0.801	0.260	3.08
Rugose coral (1)	0.256	0.226	1.13
Sparry calcite (uncategorized (12)	0.355	0.018	19.7
Sparry calcite (CL purple) (6)	0.496	0.022	22.5
Sparry calcite (CL dull orange) (4)	0.698	0.038	18.4
Sparry calcite (CL bright orange) (2)	17.6	0.227	77.5
Carbonate rocks (from Table 10.1)	1.9	5.6	0.33

Source: Smith et al., 1991.

[a]CL = cathodoluminescence. The number of samples included in each average is indicated in parentheses.

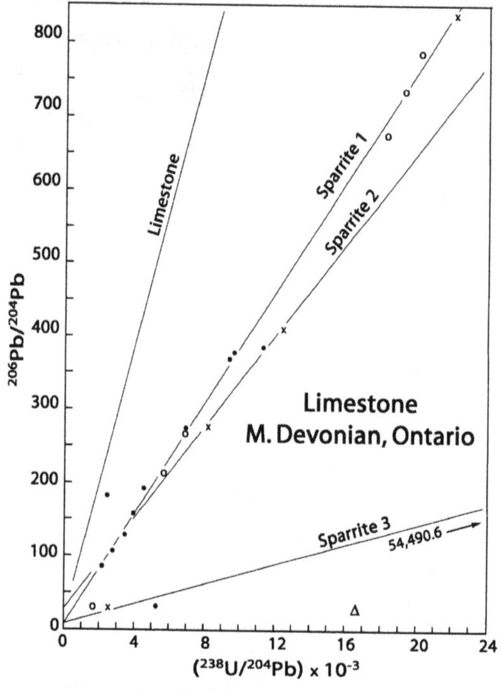

FIGURE 10.18 The U–Pb errorchrons of limestone and sparry calcite in the Lucas Formation (Middle Devonian) of southwestern Ontario, Canada. The dates derived from the slopes of the errorchrons are: limestone = 359 ± 27 Ma, sparrite 1 = 236 Ma, sparrite 2 = 194 Ma, and sparrite 3 = 44 Ma. (●) Uncategorized sparrite; (○) purple luminescene (CL); (×) dull-orange CL; (△) bright-orange CL. The date of sparrite 3 was calculated from a single sample (not shown) having $^{238}U/^{204}Pb = 54{,}490.6$, $^{206}Pb/^{204}Pb = 379.8$, and $^{206}Pb/^{204}Pb = 17.21$ (assumed). The data points that define the limestone errorchron were omitted to avoid crowding near the origin of the diagram. Data from Smith et al. (1991) and Jahn and Cuvellier (1994).

and Smith et al. (1991) for the case of the Middle Devonian Lucas Formation (Detroit River Group) in southwestern Ontario. The limestone of this formation contains pockets and veinlets of sparry calcite that luminesce in purple and two shades of orange. Although the concentrations of U and Pb in the limestone and sparry calcite of the Lucas Formation in Table 10.10 are lower in most cases than those of average carbonate rocks in Table 10.1, their U/Pb ratios are much higher and range from 1.13 (rugose coral) to 77.5 (sparry calcite, bright orange cathodoluminescence).

The U–Pb data of Smith et al. (1991) define a ^{238}U–^{206}Pb errorchron in Figure 10.18 from which Jahn and Cuvellier (1994) derived a date of 359 ± 37 Ma (MSWD = 571). The numerical value of the date is less than the stratigraphic age of these rocks (Middle Devonian: 387–374 Ma; Palmer, 1983). However, the large statistical uncertainty permits a wide range of age assignments from Early Devonian (396 Ma) to Late Mississippian (322 Ma). The most remarkable aspect of the Lucas Formation is the presence of pockets and veinlets of clear sparry calcite that apparently formed during several episodes when brines migrated through these rocks. The sparry calcite data define dates in Figure 10.18 of 236 Ma (sparrite 1), 194 Ma (sparrite 2), and 44 Ma (sparrite 3). The oldest date (236 Ma) was derived from 14 samples of sparry calcite that either were uncategorized or had purple luminescence. Sparrites with orange luminescence yielded the younger dates. Sparrite 3 in Figure 10.18 is a single sample with bright orange luminescence and an extremely high $^{238}U/^{204}Pb$ ratio of 54,490.6. Another sample with dull orange luminescence had a high $^{206}Pb/^{204}Pb$ ratio of 824.4. These results suggest that U–Pb dates of secondary calcite may be useful for the study of the evolution of sedimentary basins.

Additional examples of dating sedimentary carbonate rocks and fossils by the U–Pb method have been reported elsewhere:

DeWolf and Halliday, 1991, *Geophys. Res. Lett.*, 18:1445–1448. Smith et al., 1994, *Geochim. Cosmochim. Acta*, 58:313–322. Jones et al., 1995, *Earth Planet. Sci. Lett.*, 134:409–423. Hoff et al., 1995, *J. Sed. Pet.*, 65A:225–233. Winter and Johnson, 1995, *Earth Planet. Sci. Lett.*, 131:177–187. Israelson et al., 1996, *Earth Planet. Sci. Lett.*, 141:153–159. Rasbury et al., 1997, *Geochim. Cosmochim. Acta*, 61:1525–1529.

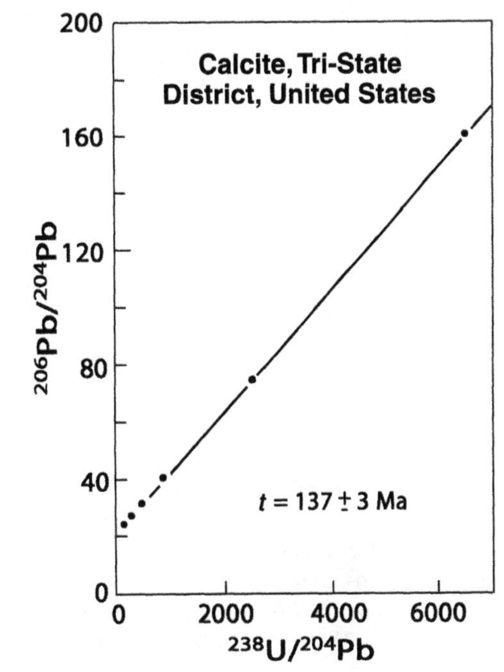

FIGURE 10.19 The U–Pb isochron for calcite in the Zn–Pb deposits at Oronogo, Missouri, in the Tri-State District, United States. The initial $^{206}Pb/^{204}Pb$ ratio is 21.98 ± 0.14 and MSWD is 2.9. Data from Coveney et al. (2000).

10.12b Zn–Pb Deposits, Tri-State District, United States

The Zn–Pb deposits in Missouri, Kansas, and Oklahoma occur in carbonate rocks of Paleozoic age. They contain not only sphalerite and galena but also secondary calcite crystals containing fluid inclusions. The filling temperatures of these fluid inclusions range from 122 to 52°C and indicate that the calcites were deposited by hydrothermal brines of varying salinities (0–24 percent NaCl) and temperatures. The age of calcite in the Zn–Pb ore deposits of the Tri-State District was determined by Brannon et al. (1996a, b) using the Th–Pb and U–Pb methods.

In a subsequent study, Coveney et al. (2000) dated calcite from Oronogo, Missouri, from Picher, Oklahoma, and from the Bendelari and Admiralty Mines in Kansas and Oklahoma by the ^{238}U–^{206}Pb and ^{232}Th–^{208}Pb isochron methods. The results indicate that the calcite at Oronogo, Missouri, has a wide range of $^{238}U/^{204}Pb$ ratios (104.2–6500) and $^{206}Pb/^{204}Pb$ ratios (24.22–161.7) that define the isochron in Figure 10.19, yielding a date of 137 ± 3 Ma (Early Cretaceous). Brannon et al. (1996b) had previously obtained a Th–Pb isochron date of 251 ± 11 Ma (Late Permian) for calcite from the Jumbo Mine in Missouri.

In addition, calcites at Picher, Oklahoma, and from the Bendelari and Admiralty Mines located along the border between Kansas and Oklahoma have Th–Pb isochron dates of 67 ± 3 Ma (Picher) and 39 ± 2 (Bendelari and Admiralty). Consequently, Coveney et al. (2000) concluded that the calcites and the associated ore minerals in Figure 10.20 formed episodically during a period of more than 200 million years by the influx of hydrothermal solutions whose temperatures decreased with time from 122 to 65°C at 251 Ma, from 80 to 53°C at 137 Ma, from 82 to 73°C at 66 Ma, from 73 to 52°C at 39°C, and finally to <52°C at the present time.

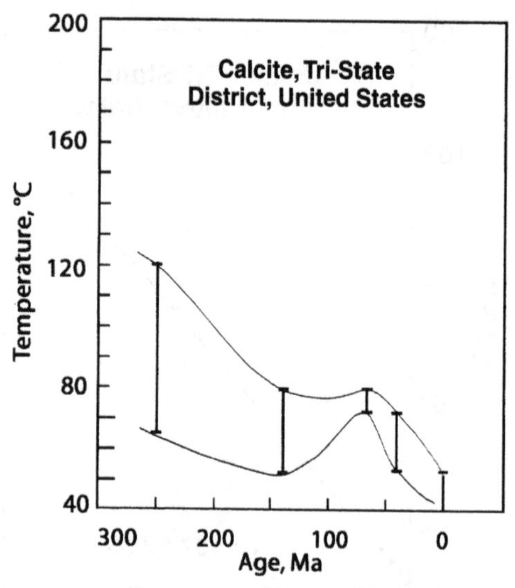

FIGURE 10.20 Temperature history of hydrothermal solutions that deposited calcite and other minerals in the Zn–Pb ore deposits of Missouri, Kansas, and Oklahoma, United States. Data from Coveney et al. (2000).

Table 10.11. Average Concentrations of U and Pb in Calcite of a Speleothem Compared to Sparry Calcite and Carbonate Rocks

Material	U, ppm	Pb, ppm
Speleothem, Winnats Head cave, Peak District, United Kingdom	25.0	0.018
Sparry calcite Lucas Formation, Table 10.6	1.9	0.040
Carbonate rocks, Table 10.1	1.9	5.6

Source: Smith et al., 1991; Richards et al., 1998.

10.12c Speleothems of Quaternary Age

Calcite deposited in stalactites and stalagmites in caves in some cases has high U/Pb ratios that permit dating by the ^{238}U–^{206}Pb method. For example, Richards et al. (1998) reported that samples of speleothems from Winnats Head cave in Carboniferous limestone of the Peak District in the United Kingdom have high average U concentrations (25.0 $\pm$ 5.8 ppm, $2\bar{\sigma}$, $N = 6$) but low average concentrations of Pb (0.0178 $\pm$ 0.0245 ppm, $2\bar{\sigma}$, $N = 6$), giving them an average U/Pb ratio of about 1400. Even more important is the fact that the U/Pb ratios of individual calcite samples range widely from about 250 to about 12700, whereas the noncarbonate residues have low U/Pb ratios between 11.8 and 25.8.

The calcites in the speleothems of Winnats Head cave in Table 10.11 have much higher concentrations of U than the sparry calcite of the Middle Devonian Lucas Formation in Table 10.10 (Smith et al., 1991). However, the average concentrations of Pb in the sparry calcite in the Lucas Formation and the speleothem at Winnats Head cave are similar, and both are lower than the average concentration of Pb in carbonate rocks (Table 10.1).

A weighted regression of the $^{238}U/^{204}Pb$ and $^{206}Pb/^{204}Pb$ ratios of calcite and residues in a stalactite from Winnats Head cave carried out by Richards et al. (1998) yielded a slope of $(2.864 \pm 0.14) \times 10^{-5}$ (MSWD = 15.4) and an initial $^{206}Pb/^{204}Pb$ ratio of 17.28 $\pm$ 0.33. The slope of the errorchron in Figure 10.21 corresponds to a date of 185 $\pm$ 9 ka, in agreement with ^{234}U–^{230}Th dates published by Ford et al. (1983) that range from 176 $+8, -7$ to 191 $+15, -13$ ka.

The validity of the U–Pb errorchron date of the speleothem depends on the assumption that ^{238}U was in secular equilibrium with its daughters when it was incorporated into the calcite of the stalactite. This assumption is questionable because the $^{234}U/^{238}U$ activity ratio in natural water is greater than unity in most cases because of the preferential dissolution of ^{234}U from U-bearing minerals (Section 20.5c).

In addition, the isotopes of Th and Pa (^{234}Th, ^{230}Th, ^{234}Pa in Figure 10.1) are removed from aqueous solution by sorption on the surfaces of suspended mineral particles and the isotopes of Rn (^{222}Rn and ^{218}Rn) escape into the atmosphere because Rn is a noble gas. Therefore, the ^{238}U and ^{235}U decay chains were probably fragmented at the time the U was deposited in the speleothems. The effect of initial disequilibrium of the daughters of U with their respective parents on calculated

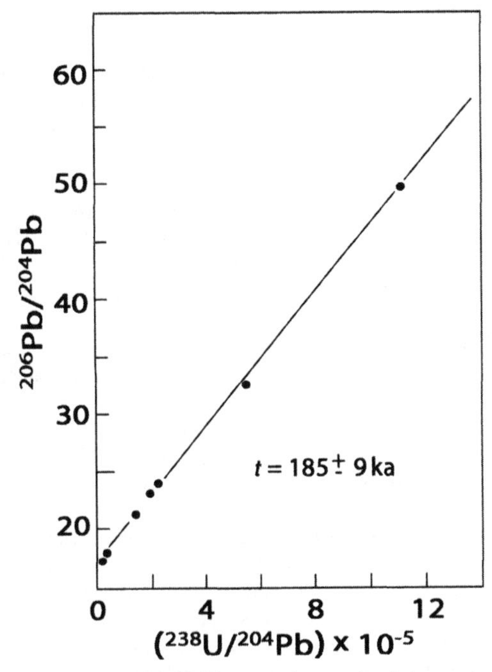

FIGURE 10.21 The U–Pb errorchron of calcite and residues of a stalactite in the Winnats Head cave, Peak District, United Kingdom. These U-rich calcite samples yield a date of 185 ± 9 ka and an initial $^{206}Pb/^{204}Pb$ ratio of 17.28 ± 0.33. A correction for the presence of excess ^{234}U increased the date to 248 ± 10 ka. Data from Richards et al. (1998).

U–Pb dates was evaluated by Ludwig (1977). In most cases, the loss of intermediate daughters at the time of deposition does not significantly affect the U–Pb dates because most of the daughters have short halflives of only days or a few years and therefore are present only in small amounts. The principal exceptions are ^{226}Ra ($T_{1/2} = 1600$ y) and ^{234}U ($T_{1/2} = 247{,}000$ y).

Since ^{226}Ra is the isotope of one of the alkaline-earth elements that form insoluble carbonates, ^{226}Ra may be coprecipitated with U in calcite. The excess ^{226}Ra ultimately forms excess $^{206}Pb^*$ and thereby increases the measured $^{206}Pb/^{204}Pb$ ratio. However, the initial presence of excess ^{226}Ra does not affect ^{238}U–^{206}Pb dates significantly in cases where the samples are older than about 2×10^5 years. Richards et al. (1998) determined that the initial $^{234}U/^{238}U$ activity ratio of the speleothems in Winnats Head cave was 1.32 + 0.08, −0.07 and therefore increased the ^{238}U–^{206}Pb dates of the calcite to 248 ± 10 ka. This date agrees with a ^{234}U–^{230}Th date of 256 + 53, −40 ka the authors determined by α-spectrometry of the same material.

In general, the results of the study of Richards et al. (1998) demonstrate that the U–Pb isochron method can be used to date U-rich calcites whose ages are also measurable by the U-series disequilibrium method to be discussed in Chapter 20. The validity of U–Pb dates depends on the requirement that the samples remained closed to U, Pb, and to all of the intermediate daughters. In addition, the initial lack of secular equilibrium of the U-decay chains must be considered, including the presence of excess ^{234}U.

10.13 SUMMARY

The U–Pb and Th–Pb methods are useful for dating minerals as well as terrestrial and extraterrestrial rocks that contain U and Th but exclude Pb. The U, Th-bearing minerals amenable to dating include primarily zircon, baddeleyite, sphene, rutile, and monazite because these minerals are widely distributed in igneous and metamorphic rocks. These minerals also resist alteration during chemical weathering, erosion, transport, and deposition and therefore occur in clastic sedimentary rocks.

In spite of their resistance to alteration, the U–Pb and Th–Pb dates of U,Th-bearing minerals are discordant in most cases because of loss of Pb and gain or loss of U during episodes of thermal metamorphism. This problem has been overcome by the use of Wetherill's concordia, which permits the determination of the crystallization age of minerals that yield discordant U–Pb and Th–Pb dates.

The precision and accuracy of U,Th–Pb dates have been improved by pretreatment of mineral grains and the refinement of analytical procedures to such an extent that single mineral grains can be analyzed for dating by TIMS. In addition, instrumental methods permit the analysis of single grains by SHRIMP mass spectrometry and LA-ICP-MS.

The refinements of the analytical procedures have been accompanied by the development of the Tera–Wasserburg concordia and its elaboration into a method of interpreting isotopic U–Pb data in three-dimensional space. The Tera–Wasserburg concordia is especially effective for dating lunar rocks, which inherited high radiogenic $^{207}Pb/^{206}Pb$ ratios at the time of their formation.

Although the conventional U–Pb isochron method of dating minerals and cogenetic suites of igneous and metamorphic rocks has failed in many cases because of recent losses of U and Pb, the Pb–Pb isochron method has been successful in dating both igneous and metamorphic rocks. The Pb–Pb isochron method is also useful for dating sedimentary carbonate rocks of Archean to Phanerozoic age. In addition, calcite crystals in Phanerozoic limestone and speleothems of Quaternary age can be dated by the U–Pb method in cases where their U/Pb ratios have sufficient range. The precision and accuracy of U–Pb dates can be further improved by eliminating the fractionation of Pb isotopes during analysis in the mass spectrometer.

REFERENCES FOR GEOCHEMISTRY AND PRINCIPLES (SECTIONS 10.1– 10.3)

Amelin, Y., and A. N. Zaitsev, 2002. Precise geochronology of phoscorites and carbonatites: The critical role of U-series disequilibrium in age interpretations. *Geochim. Cosmochim. Acta*, 66:2399–2419.

Begemann, F., K. R. Ludwig, G. W. Lugmair, K. Min, L. E. Nyquist, P. J. Patchett, P. R. Renne, C.-Y. Shih, I. M. Villa, and R. J. Walker, 2001. Call for an improved set of decay constants for geochronological use. *Geochim. Cosmochim. Acta*, 65:111–121.

Boltwood, B. B., 1907. On the ultimate disintegration products of the radioactive elements. *Amer. J. Sci.*, 23(4):77–88.

Cowan, G. A., 1976. A natural fission reactor. *Sci. Am.*, No. 7, pp. 36–47.

Doe, B. R., 1970. *Lead Isotopes*. Springer-Verlag. New York.

Heinrich, E. W., 1958. *Mineralogy and Geology of Radioactive Raw Materials*. McGraw-Hill, New York.

Holmes, A. 1913. *The Age of the Earth*. Harper and Brothers, London.

Krogh, T. E., 1973. A low-contamination method for hydrothermal decomposition of zircon and extraction of U and Pb for isotopic age determinations. *Geochim. Cosmochim. Acta*, 37:485–494.

Kuroda, P. K, 1982. *The Origin of the Chemical Elements and the Oklo Phenomenon*. Springer-Verlag, New York.

Lancelot, J. R., A. Vitrac, and C. J. Allègre, 1975. The Oklo natural reactor: Age and evolution studies of U–Pb and Rb–Sr systematics. *Earth Planet. Sci. Lett.*, 25:189–196.

Lide, D. R., and H. P. R. Frederikse, 1995. *Handbook of Chemistry and Physics*. CRC Press, Boca Raton, Florida.

Ludwig, K. R., 1993. *Pb Dat 1.24: A Computer Program for Processing Pb–U–Th Isotope Data*, USGS Open-File Rep. 88–542, p. 32. U.S. Geological Survey, Washington, D.C.

Rogers, J. J. W., and J. A. S. Adams, 1969a. Thorium. In K. H. Wedepohl (Ed.), *Handbook of Geochemistry*, Vol. II-5, Chapter 90. Springer-Verlag, Heidelberg.

Rogers, J. J. W., and J. A. S. Adams, 1969b. Uranium. In K. H. Wedepohl (Ed.), *Handbook of Geochemistry*, Vol. II-5, Chapter 90. Springer-Verlag, Heidelberg.

Rutherford, E., 1906. *Radioactive Transformations*. Charles Scribner's Sons, New York.

Steiger, R. H., and E. Jäger, 1977. Subcommission on Geochronology: Convention on the use of decay constants in geo- and cosmochronology. *Earth Planet. Sci. Lett.*, 36:359–362.

Wedepohl, K. H., 1974. Lead. In K. H. Wedepohl (Ed.), *Handbook of Geochemistry*, Sections C–D, Vol. II-5, Chapter 82. Springer-Verlag, Heidelberg.

Wedepohl, K. H., 1978. *Handbook of Geochemistry*, Part II, Vol. 5. Springer-Verlag, Berlin.

Wetherill, G. W., 1966. Radioactive decay constants and energies. In S. E. Clarke (Ed.), *Handbook of Physical Constants*, Geol. Soc. Am. Mem. 97, pp. 514–519. Geological Society of America, Boulder, Colorado. D.C.

REFERENCES FOR U–Pb AND Th–Pb DATES, BOULDER CREEK BATHOLITH, COLORADO (SECTION 10.4)

Ludwig, K. R., 1977. Effect of initial radioactive-daughter disequilibrium on U–Pb isotope apparent ages of young minerals. *J. Res., U.S.G.S.*, 5:663–667.

Romer, R. L., 2001. Isotopically heterogeneous initial Pb and continuous ^{222}Rn loss in fossils: The U–Pb systematics of Brachiosaurus brancai. *Geochim. Cosmochim. Acta*, 65:4201–4213.

Schärer, U., 1984. The effect of initial ^{230}Th disequilibrium on young U–Pb ages: The Makalu case, Himalaya. *Earth Planet. Sci. Lett.*, 67:191–124.

Stern, T. W., G. Phair, and M. R. Newell, 1971. Boulder Creek Batholith, Colorado, Part II. Isotopic ages of emplacement and morphology of zircon. *Geol. Soc. Am. Bull.*, 82:1615–1634.

REFERENCES FOR WETHERILL'S CONCORDIA AND Pb LOSS MODELS (SECTIONS 10.5–10.6)

Ahrens, L. H., 1955. Implications of the Rhodesia age pattern. *Geochim. Cosmochim. Acta*, 8:1–15.

Allègre, C. J., 1967. Methode de discussion geochronologique concordia generalisee. *Earth Planet. Sci. Lett.*, 2:57–66.

Baadsgaard, H., 1973. U–Th–Pb dates on zircons from the early Precambrian Amitsoq gneisses, Godthaab district, West Greenland. *Earth Planet. Sci. Lett.*, 19:22–28.

Bickford, M. E., R. B. Chase, B. K. Nelson, R. D. Shuster, and E. C. Arruda, 1981. U–Pb studies of zircon cores and overgrowths, and monazite: Implications for age and petrogenesis of the northeastern Idaho batholith. *J. Geol.*, 89:433–457.

Catanzaro, E. J., 1963. Zircon ages in southwestern Minnesota. *J. Geophys. Res.*, 68:2045–2048.

Cherniak, D. J., and E. B. Watson, 2000. Pb diffusion in zircon. *Chem. Geol.*, 172:5–24.

Davis, D. W., 1982. Optimum linear regression and error estimation applied to U–Pb data. *Can. J. Earth Sci.*, 19:2141–2149.

Davis, D. W., and T. E. Krogh, 2000. Preferential dissolution of ^{234}U and radiogenic Pb from α-recoil damaged lattice sites in zircon: Implications for thermal histories and Pb isotopic fractionation in the near surface environment. *Chem. Geol.*, 172:41–58.

Goldich, S. S., and M. G. Mudrey, Jr., 1972. Dilatancy model for discordant U–Pb zircon ages. In A. I. Tugarinov (Ed.), *Contributions to Recent Geochemistry and Analytical Chemistry*, A. P. Vinogradov vol., pp. 415–418. Nauka, Moscow.

Ludwig, K. R., 1988. ISOPLOT—*A Plotting and Regression Program for Radiogenic-Isotope Data for IBM-Compatible Computers*, Version 2.10, USGS Open-File Rept. 88–557. U.S. Geological Survey, Washington, D.C.

Nasdala, L., Th. Wenzel, R. T. Pidgeon, and A. Kronz, 1999. Internal structures and dating of complex zircons from Meissen Massif monzonites. *Saxony. Chem. Geol.*, 156:331–341.

Rosholt, J. N., R. E. Zartman, and I. T. Nkomo, 1973. Lead isotope systematics and uranium depletion in the Granite Mountains, Wyoming. *Geol. Soc. Am. Bull.*, 84:989–1002.

Silver, L. T., and S. Deutsch, 1963. Uranium-lead isotopic variations in zircons: A case study. *J. Geol.*, 71:721–758.

Sinha, A. K., 1972. U–Th–Pb systematics and the age of the Onverwacht Series, South Africa. *Earth Planet. Sci. Lett.*, 16:219–227.

Steiger, R. H., and G. J. Wasserburg, 1966. Systematics in the Pb^{208}–Th^{232}, Pb^{207}–U^{235}, and Pb^{206}–U^{238} systems. *J. Geophys. Res.*, 71:6065–6090.

Stern, T. W., S. S. Goldich, and M. F. Newell, 1966. Effects of weathering on the U–Pb ages of zircon from the Morton Gneiss, Minnesota. *Earth Planet. Sci. Lett.*, 1:369–371.

Tera, F., and G. J. Wasserburg, 1972. U–Th–Pb systematics in three Apollo 14 basalts and the problem of initial Pb in lunar rocks. *Earth Planet. Sci. Lett.*, 14:281–304.

Tilton, G. R., 1960. Volume diffusion as a mechanism for discordant lead ages. *J. Geophys. Res.*, 65:2933–2945.

Wasserburg, G. J., 1963. Diffusion processes in lead-uranium systems. *J. Geophys. Res.*, 68:4823–4846.

Wetherill, G. W., 1956. Discordant uranium-lead ages. *Trans. Am. Geophys. Union*, 37:320–326.

Wetherill, G. W., 1963. Discordant uranium-lead ages—Pt 2; discordant ages resulting from diffusion of lead and uranium. *J. Geophys. Res.*, 68:2957–2965.

REFERENCES FOR PURIFICATION OF ZIRCON GRAINS (SECTIONS 10.7–10.7a)

Getty, S. R., and D. J. DePaolo, 1995. Quaternary geochronology using the U–Th–Pb method. *Geochim. Cosmochim. Acta*, 59:3267–3272.

Goldich, S. S., and L. B. Fischer, 1986. Air-abrasion experiments in U–Pb dating of zircon. *Chem. Geol.*, 58:195–215.

Kamo, S. L., and T. E. Krogh, 1995. Chicxulub crater source for shocked zircon crystals from the Cretaceous–Tertiary boundary layer, Saskatchewan: Evidence from new U–Pb data. *Geology*, 23:281–284.

Krogh, T. E., 1973. A low-contamination method for hydrothermal decomposition of zircon and extraction of U and Pb for isotopic age determinations. *Geochim. Cosmochim. Acta*, 37:485–494.

Krogh, T. E., 1982a. Improved accuracy of U–Pb zircon dating by selection of more concordant fractions using a high gradient magnetic separation technique. *Geochim. Cosmochim. Acta*, 46:631–635.

Krogh, T. E., 1982b. Improved accuracy of U–Pb zircon ages by the creation of more concordant systems using an air abrasion technique. *Geochim. Cosmochim. Acta*, 46:637–649.

Krogh, T. E., S. L. Kamo, and B. F. Bohor, 1993a. Fingerprinting the K/T impact site and determining the time of impact by U–Pb dating of single shocked zircons from distal ejecta. *Earth Planet. Sci. Lett.*, 119:425–429.

Krogh, T. E., S. L. Kamo, and B. F. Bohor, 1993b. U–Pb ages of single shocked zircons linking distal K/T ejecta to the Chicxulub crater. *Nature*, 366:731–734.

Ludwig, K. R., 1980. Calculation of uncertainties of U–Pb isotopic data. *Earth Planet. Sci. Lett.*, 46:212–220.

Mundil, R., P. Brack, M. Meier, H. Rieber, and F. Oberli, 1996. High resolution U–Pb dating of Middle Triassic volcaniclastics: Time-scale calibration and verification of tuning parameters for carbonate sedimentation. *Earth Planet. Sci. Lett.*, 141:137–151.

REFERENCES FOR SHRIMP (SECTION 10.7b)

Andersen, C. A., and J. R. Hinthorne, 1972. U, Th, Pb and REE abundances and $^{207}Pb/^{206}Pb$ ages of individual minerals in returned lunar material by ion microprobe mass analysis. *Earth Planet. Sci. Lett.*, 14:195–200.

Claoué-Long, J. C., W. Compston, J. Roberts, and C. M. Fanning, 1995. Two Carboniferous ages: A comparison of SHRIMP zircon ages with conventional zircon ages and $^{40}Ar/^{39}Ar$ analysis. In *Geochronology, Time Scales, and Global Stratigraphic Correlation*, SEPM Spec. Pub. 54, pp. 3–21. Society of Economic Paleontologists and Mineralogists, Tulsa, Oklahoma.

Compston, W., 1999. Geological age determination by instrumental analysis: The 29th Halmond Lecture. *Mineral. Mag.*, 63:297–311.

Compston, W., I. S. Williams, and C. Meyer, 1984. U–Pb geochronology of zircons from lunar breccia 72171 using a sensitive high mass-resolution ion microprobe. *J. Geophys. Res.*, 89B:525–534.

Heaman, L. M., and A. N. LeCheminant, 1993. Paragenesis and U–Pb systematics of baddeleyite (ZrO_2). *Chem. Geol.*, 110:95–126.

Hinthorne, J. R., C. A. Andersen, R. L. Conrad, and J. F. Lovering, 1979. Single-grain $^{207}Pb/^{206}Pb$ and U/Pb age determinations with a 10 μm spatial resolution using the ion-microprobe mass analyzer (IMMA). *Chem. Geol.*, 25:271–303.

Keil, K., and P. E. Fricker, 1974. Baddeleyite (ZrO_2) in gabbroic rocks from Axel Heiberg Island, Canadian arctic archipelago. *Am. Mineral.*, 59:249–253.

Krogh, T. E., 1982. Improved accuracy of U–Pb zircon ages by the creation of more concordant systems using an air abrasion technique. *Geochim. Cosmochim. Acta*, 46:637–649.

Pidgeon, R. T., D. Furfaro, A. K. Kennedy, A. A. Nemchin, and W. Van Bronswijk, 1995. *Calibration of Zircon Standards for the Curtin SHRIMP II*, USGS Circular 1107, p. 251. U.S. Geological Survey, Washington, D.C.

Smith, J. B., M. E. Barley, D. L. Groves, B. Krapez, N. J. McNaughton, M. J. Bickle, and H. J. Chapman, 1998. The Sholl shear zone, West Pilbara: Evidence for a domain boundary structure from integrated tectonostratigraphic analyses, SHRIMP U–Pb dating and isotopic and geochemical data of granitoids. *Precamb. Res.*, 88:143–171.

Stern, R. A., and R. G. Berman, 2000. Monazite U–Pb and Th–Pb geochronology by ion microprobe, with an application to in-situ dating of an Archean metasedimentary rock. *Chem. Geol.*, 172:113–130.

Williams, I. S., 1998. U–Th–Pb geochronology by ion microprobe. In M. A. McKibben et al. (Eds.), *Applications of Microanalytical Techniques to Understanding Mineralizing Processes. Rev. Econ. Geol.*, 7:1–35. Economic Geology Pub. Co., SanDiego, California.

Wingate, M. T. D., and W. Compston, 2000. Crystal orientation effects during ion microprobe U–Pb analysis of baddeleyite. *Chem. Geol.*, 168:75–97.

REFERENCES FOR LA-ICP-MS (SECTION 10.7c)

Eggins, S. M., L. P. J. Kinsley, and J. M. M. Shelley, 1998. Deposition and element fractionation processes during atmospheric pressure laser sampling for analysis by ICP-MS. *Appl. Surf. Sci.*, 127/129:278–286.

Fryer, B. J., S. E. Jackson, and H. P. Longerich, 1993. The application of laser ablation microprobe inductively-coupled plasma mass-spectrometry (LAM-ICP-MS) to in situ (U)–Pb geochronology. *Chem. Geol.*, 109:1–8.

Fryer, B. J., S. E. Jackson, and H. P. Longerich, 1995. The design, operation and role of the laser-ablation microprobe coupled with an inductively coupled plasma mass-spectrometer (LAM-ICP-MS) in earth sciences. *Can. Mineral.*, 33:303–312.

Heaman, L. M., and A. N. LeCheminant, 1993. Paragenesis and U–Pb systematics of baddeleyite (ZrO_2). *Chem. Geol.*, 110:95–126.

Hirata, T., and R. W. Nesbitt, 1995. U–Pb isotope geochronology of zircon: Evaluation of laser-probe inductively coupled plasma mass spectrometry technique. *Geochim. Cosmochim. Acta*, 59:2491–2500.

Horn, I., R. L. Rudnick, and W. F. McDonough, 2000. Precise elemental and isotope ratio determination by simultaneous solution nebulization and laser-ablation ICP-MS: Application to U–Pb geochronology. *Chem. Geol.*, 167:405–425.

Longerich, H. P., D. Günther, and S. E. Jackson, 1996. Elemental fractionation in laser-ablation inductively coupled plasma mass-spectrometry. *Fresenius J. Anal. Chem.*, 355:538–542.

REFERENCES FOR EMP (SECTION 10.7d)

Cocherie, A., O. Legendre, J. J. Peucat, and A. Kouamelan, 1998. Geochronology of polygenic monazite constrained by in-situ electron microprobe Th-U-total Pb determination: Implications for Pb behavior in monazite. *Geochim. Cosmochim. Acta*, 62:2475–2497.

Geisler, Th., and H. Schleicher, 2000. Improved U-Th-total Pb dating of zircons by electron microprobe using a simple new background modeling procedure and Ca as a chemical criterion of fluid-induced U–Th–Pb discordance in zircon. *Chem. Geol.*, 163:269–285.

Montel, J., S. Foret, M. Veschambre, C. Nicollet, and A. Provost, 1996. Electron microprobe dating of monazite. *Chem. Geol.*, 131:37–53.

Parrish, R. R., 1990. U–Pb dating of monazite and its application to geological problems. *Can. J. Earth Sci.*, 27:1435–1450.

Poitrasson, F., S. Chenery, and R. J. Shepherd, 2000. Electron microprobe and LA-ICP-MS study of monazite hydrothermal alteration: Implications for U–Th–Pb geochronology and nuclear ceramics. *Geochim. Cosmochim. Acta*, 64:3283–3297.

Suzuki, K., and M. Adachi, 1991. Precambrian provenance and Silurian metamorphism of the Tsunosawa paragneiss in the South Kitakami terrane, northeast Japan, revealed by the chemical Th-U-total Pb isochron ages of monazite, zircon, and xenotime. *J. Geochem.*, 25:357–376.

Williams, M. L., M. J. Jercinovic, and M. P. Terry, 1999. Age mapping and dating of monazite on the electron microprobe: Deconvoluting multistage tectonic histories. *Geology*, 27:1023–1026.

REFERENCES FOR DATING DETRITAL ZIRCON GRAINS (SECTION 10.8)

Allègre, C. J., F. Albarède, M. Grünenfelder, and V. Köppel, 1974. $^{238}U/^{206}Pb$–$^{235}U/^{207}Pb$–$^{232}Th/^{208}Pb$ zircon geochronology in alpine and non-alpine environments. *Contrib. Mineral. Petrol.*, 43:163–194.

Cattell, A., T. E. Krogh, and N. T. Arndt, 1984. Conflicting Sm–Nd whole-rock and U–Pb zircon ages for Archean lavas from Newton Township, Abitibi belt, Ontario. *Earth Planet. Sci. Lett.*, 70:280–290.

Compston, W., and I. S. Williams, 1984. U–Pb geochronology of zircons from lunar breccia 73217 using a sensitive high mass-resolution ion microprobe. *J. Geophys. Res.*, 89(Suppl.):B525–B534.

Corfu, F., 1993. The evolution of the southern Abitibi greenstone belt in light of precise U–Pb geochronology. *Econ Geol.*, 88:1323–1340.

Corfu, F., T. E. Krogh, Y. Y. Kwok, and L. S. Jensen, 1989. U–Pb zircon geochronology in the southwestern Abitibi greenstone belt, Superior Province. *Can. J. Earth Sci.*, 26:1747–1763.

Dimroth, E., L. Imreh, M. Rocheleau, and N. Goulet, 1982. Evolution of the south-central part of the Archean Abitibi belt, Quebec. Part I: Stratigraphy and paleogeographic model. *Can. J. Earth Sci.*, 19:1729–1758.

Froude, D. O., T. R. Ireland, P. D. Kinny, I. S. Williams, and W. Compston, 1983. Ion microprobe identification of 4,100–4,200 Myr-old terrestrial zircons. *Nature*, 304:616–618.

Gariepy, C., C. J. Allègre, and J. Lajoie, 1984. U–Pb systematics in single zircons from the Pontiac sediments, Abitibi greenstone belt. *Can. J. Earth Sci.*, 21:1296–1304.

Gaudette, H. E., A. Vitrac-Michard, and C. J. Allègre, 1981. North American Precambrian history recorded in a single sample: High-resolution U–Pb systematics of the Potsdam sandstone detrital zircons, New York State. *Earth Planet. Sci. Lett.*, 54:248–260.

Gentry, R. V., T. J. Sworski, H. S. McKown, D. H. Sith, R. E. Eby, and W. H. Christie, 1982. Differential lead retention of zircons: Implications for nuclear waste containment. *Science*, 216:296–298.

Goodwin, A. M., 1982. Archean volcanoes in southwestern Abitibi Belt, Ontario and Quebec: Form, composition, and development. *Can. J. Earth Sci.*, 19:1140–1155.

Ludden, J., C. Hubert, and C. Gariepy, 1986. The tectonic evolution of the Abitibi greenstone belt of Canada. *Geol. Mag.*, 123:153–166.

Schärer, R., and C. J. Allègre, 1982. Uranium-lead system in fragments of a single zircon grain. *Nature*, 295:585–587.

REFERENCES FOR TERA–WASSERBURG CONCORDIA (SECTION 10.9)

Bacon, C. R., H. M. Persing, J. L. Wooden, and T. R. Ireland, 2000. Late Pleistocene granodiorite beneath Crater Lake caldera, Oregon, dated by ion microprobe. *Geology*, 28:467–470.

Carl, C., I. Wendt, and J. I. Wendt, 1989. U/Pb whole-rock and mineral dating of the Falkenberg granite in northeast Bavaria. *Earth Planet. Sci. Lett.*, 94:236–244.

Jahn, B.-M., and H. Cuvellier, 1994. Pb–Pb and U–Pb geochronology of carbonate rocks: An assessment. *Chem. Geol. (Isotope Geosci. Sect.)*, 115:125–151.

Papanastassiou, D. A., and G. J. Wasserburg, 1971. Rb–Sr ages of igneous rocks from the Apollo 14 mission and the age of the Fra Mauro Formation. *Earth Planet. Sci. Lett.*, 12:36.

Richards, D. A., S. H. Bottrell, R. A. Cliff, K. Ströhle, and P. J. Rowe, 1998. U–Pb dating of a speleothem of Quaternary age. *Geochim. Cosmochim. Acta*, 62:3683–3688.

Smith, P. E., R. M. Farquhar, and R. G. Hancock, 1991. Direct radiometric age determination of carbonate diagenesis using U–Pb in secondary calcite. *Earth Planet. Sci. Lett.*, 105:474–491.

Steiger, R. H., and E. Jäger, 1977. Subcommission on Geochronology: Convention on the use of decay constants in geo- and cosmochronology. *Earth Planet. Sci. Lett.*, 36:359–362.

Tatsumoto, M., and J. N. Rosholt, 1970. Age of the Moon: An isotopic study of U–Th–Pb systematics of lunar samples. *Science*, 167:461.

Tera, F., and G. J. Wasserburg, 1972. U–Th–Pb systematics in three Apollo 14 basalts and the problem of initial Pb in lunar rocks. *Earth Planet. Sci. Lett.*, 14:281–304.

Tera, F., and G. J. Wasserburg, 1973. A response to a comment on U–Pb systematics in lunar basalts. *Earth Planet. Sci. Lett.*, 19:213–217.

Tera, F., and G. J. Wasserburg, 1974. U–Th–Pb systematics on lunar rocks and inferences about lunar evolution and the age of the Moon. *Proc. 5th Lunar Conf., Geochim. Cosmochim. Acta*, 2(Suppl. 5):1571–1599.

Turner, G., J. C. Huneke, F. A. Podosek, and G. J. Wasserburg, 1971. ^{40}Ar–^{39}Ar ages and cosmic-ray exposure ages of Apollo 14 samples. *Earth Planet. Sci. Lett.*, 12:19.

Wendt, I., 1984. A three-dimensional U–Pb discordia plane to evaluate samples with common lead of unknown isotopic composition. *Chem. Geol., Isotope Geosci. Sect.*, 2:1–12.

Wendt, I., 1989. Geometric considerations of the three-dimensional U/Pb data presentation. *Earth Planet. Sci. Lett.*, 94:231–235.

Wendt, I., and C. Carl, 1985. U/Pb dating of discordant 0.1 Ma old secondary U minerals. *Earth Planet. Sci. Lett.*, 73:278–284.

Zheng, Y. F., 1989. On the use of a three-dimensional method in solving the U–Pb two-stage model. *Geochim. J.*, 23:37–43.

Zheng, Y. F., 1992. The three-dimensional U–Pb method: Generalized models and implications for U–Pb two-stage systematics. *Chem. Geol.*, 100:3–18.

REFERENCES FOR U–Pb, Th–Pb, AND Pb–Pb ISOCHRONS(GRANITE MOUNTAINS, WYOMING) (SECTION 10.10)

Farquharson, R. B., and J. R. Richards, 1970. Whole-rock U–Th–Pb and Rb–Sr ages of the Sybella micro-granite and pegmatite, Mount Isa, Queensland. *J. Geol. Soc. Austral.*, 17:53–58.

Jahn, B.-M., J. Bertrand-Sarfati, N. Morin, and J. Macé, 1990. Direct dating of stromatolitic carbonates from the Schmidtsdrif Formation (Transvaal Dolomite), South Africa, with implications on the age of the Ventersdorp Supergroup. *Geology*, 18:1211–1214.

Moorbath, S., R. K. O'Nions, and R. J. Pankhurst, 1973. Early Archean age for the Isua Iron Formation, West Greenland. *Nature*, 245:138–139.

Oversby, V. M., 1975. Lead isotopic systematics and ages of Archean acid intrusives in the Kalgoorlie-Norseman area, Western Australia. *Geochim. Cosmochim. Acta*, 39:1107–1125.

Reynolds, P. H., 1971. A U–Th–Pb lead isotope study of rocks and ores from Broken Hill, Australia. *Earth Planet. Sci. Lett.*, 12:215–223.

Rosholt, J. N., and A. J. Bartel, 1969. Uranium, thorium and lead systematics in Granite Mountains, Wyoming. *Earth Planet. Sci. Lett.*, 7:141–147.

Rosholt, J. N., R. E. Zartman, and I. T. Nkomo, 1973. Lead isotope systematics and uranium depletion in the Granite Mountains, Wyoming. *Geol. Soc. Am. Bull.*, 84:989–1002.

Sinha, A. K., 1972. U–Th–Pb systematics and the age of the Onverwacht Series, South Africa. *Earth Planet. Sci. Lett.*, 16:219–227.

Ulrych, T. J., and P. H. Reynolds, 1966. Whole-rock and mineral leads from the Llano Uplift, Texas. *J. Geophys. Res.*, 71:3089–3094.

REFERENCES FOR Pb–Pb AND U–Pb ISOCHRONS (CARBONATE ROCKS) (SECTIONS 10.11–10.12)

Andersen, T., and P. N. Taylor, 1988. Pb isotope geochemistry of the Fen carbonatite complex, southeast Norway: Age and petrogenetic implications. *Geochim. Cosmochim. Acta*, 52:209–215.

Brannon, J. C., S. C. Cole, and F. A. Podosek, 1996a. Radiometric dating of Mississippi Valley-type Pb–Zn deposits. In D. F. Sangster (Ed.), *Carbonate-Hosted Lead-Zinc Deposits*, Spec. Pub. 4, pp. 536–545. Society Economic Geologists.

Brannon, J. C., S. C. Cole, F. A. Podosek, V. M. Ragan, R. M. Coveney, Jr., M. W. Wallace, and A. J. Bradley, 1996b. Th–Pb and U–Pb dating of ore-stage calcite and Paleozoic fluid flow. *Science*, 271:491–493.

Broecker, W. S., and T.-H. Peng, 1982. *Tracers in the Sea.* Lamont-Doherty Geological Observatory, Columbia University, Palisades, New York.

Coveney, R. M., Jr., V. M. Ragan, and J. C. Brannon, 2000. Temporal benchmarks for modeling Phanerozoic flow of basinal brines and hydrocarbons in the southern midcontinent based on radiometrically dated calcite. *Geology*, 28:795–798.

Fölling, P. G., R. E. Zartman, and H. E. Frimmel, 2000. A novel approach to double-spike Pb–Pb dating of carbonate rocks: Examples from Neoproterozoic sequences in southern Africa. *Chem. Geol.*, 171:97–122.

Ford, T. D., M. Gascoyne, and J. S. Beck, 1983. Speleothem dates and Pleistocene chronology in the Peak District of Derbyshire. *Cave Sci.*, 10:103–115.

Jahn, B.-M., 1988. Pb–Pb dating of young marbles from Taiwan. *Nature*, 332:429–432.

Jahn, B.-M., J. Bertrand-Sarfati, N. Morin, and J. Macé, 1990. Direct dating of stromatolitic carbonates from the Schmidtsdrif Formation (Transvaal Dolomite), South Africa, with implication on the age of the Ventersdorp Supergroup. *Geology*, 18:1211–1214.

Jahn, B.-M., and H. Cuvellier, 1994. Pb–Pb and U–Pb geochronology of carbonate rocks: An assessment. *Chem. Geol. (Isotope Geosci. Sect.)*, 115:125–151.

Ludwig, K., 1977. Effect of initial radioactive-daughter disequilibrium on U–Pb isotope apparent ages of young minerals. *J. Res. U.S. Geol. Surv.*, 5:663–667.

Martin, J. M., and M. Whitfield, 1983. The significance of the river input of chemical elements to the oceans. In C. S. Wong, E. Boyle, K. W. Bruland, J. D. Burton, and E. D. Goldberg (Eds.), *Trace Metals in Seawater*, pp. 265–296. Plenum, New York.

Moorbath, S., P. N. Taylor, J. L. Orpen, P. Treloar, and J. F. Wilson, 1987. First direct radiometric dating of Archaean stromatolitic limestone. *Nature*, 326:865–867.

Ovchinnikova, G. V., M. A. Semikhatov, I. M. Gorokhov, B. V. Belyatskii, I. M. Vasilieva, and L. K. Levskii, 1995. U–Pb systematics of Precambrian carbonates: The Riphean Sukhaya Tunguska Formation in the Turukhansk uplift, Siberia. *Lith. Mineral Resources*, 30:477–487.

Palmer, A. R., 1983. The decade of North American geology. 1983 geologic time scale. *Geology*, 11:503–504.

Parnell, J., and I. Swainbank, 1990. Pb/Pb dating of hydrocarbon migration into a bitumen-bearing ore deposit, North Wales. *Geology*, 18:1028–1030.

Richards, D. A., S. H. Bottrell, R. A. Cliff, K. Ströhle, and P. Rowe, 1998. U–Pb dating of a speleothem of Quaternary age. *Geochim. Cosmochim. Acta*, 62:3683–3688.

Smith, P. E., and R. M. Farquhar, 1989. Direct dating of Phanerozoic sediments by the ^{238}U–^{206}Pb method. *Nature*, 341:518–521.

Smith, P. E., R. M. Farquhar, and R. G. Hancock, 1991. Direct radiometric age determination of carbonate diagenesis using U–Pb in secondary calcite. *Earth Planet. Sci. Lett.*, 105:474–491.

Taylor, P. N., and F. Kalsbeek, 1990. Dating of metamorphism of Precambrian marbles: Examples from Proterozoic mobile belts of Greenland. *Chem. Geol. (Isotope Geosci. Sect.)*, 86:21–28.

Walraven, F., R. A. Armstrong, and F. J. Kruger, 1990. A chronological framework for the middle to late Precambrian stratigraphy of the Transvaal, South Arica. *Tectonophysics*, 171:23–48.

White, D. E., 1968. Environments of generation of some base-metal ore deposits. *Econ. Geol.*, 63:301–335.

CHAPTER 11

The Common-Lead Method

LEAD is widely distributed throughout the Earth and occurs not only as the radiogenic daughter of U and Th but also forms its own minerals from which U and Th are excluded. Therefore, the isotopic composition of Pb in terrestrial rocks and minerals varies between wide limits from the highly radiogenic Pb in old U, Th-bearing minerals (Chapter 10) to the common Pb in galena (PbS) and other minerals that have low U/Pb and Th/Pb ratios.

The isotope composition of common Pb in galena is assumed to have evolved by decay of U and Th in reservoirs in the crust and lithospheric mantle of the Earth. The radiogenic isotopes of Pb subsequently mixed with the primeval Pb that was incorporated into these reservoirs at the time the Earth formed. Therefore, the Pb that was later withdrawn from the reservoirs is a mixture of the primeval Pb with varying additions of radiogenic Pb depending on the U/Pb and Th/Pb ratios of a particular reservoir. This mixed Pb subsequently crystallized as galena during ore-forming events in the history of the Earth. Therefore, the common Pb in galena and other Pb-rich but U,Th-poor minerals records the isotopic composition of its source within which it evolved for an interval of time ending with its removal from the reservoir and its crystallization as galena in the rocks of the continental crust. The common-Pb method of dating, which was proposed by Holmes (1946) and Houtermans (1946), was later presented in books by Rankama (1954, 1963), Russell and Farquhar (1960), Hamilton and Farquhar (1968), and Doe (1970). The isotope geochemistry of Pb in the oceans is discussed in Section 19.4.

11.1 THE HOLMES–HOUTERMANS MODEL

The isotopic composition of common Pb in galena and other U,Th-poor minerals (e.g., sulfide minerals of other metals and K-feldspar in which Pb^{2+} replaces K^{+}) was originally determined by Aston (1927). At that time, the atomic weight of common Pb was thought to be constant, which implied that its isotopic composition did not vary. However, Nier (1938) and Nier et al. (1941) reported that, contrary to expectations, common Pb from different sources has a range of isotopic compositions.

The suggestion of Nier et al. (1941) that common Pb is a mixture of primeval and radiogenic Pb became the basis of a model for the isotopic evolution of common Pb proposed independently by Holmes (1946) and Houtermans (1946). This model is based on certain assumptions:

1. Originally the Earth was fluid and homogeneous.
2. At that time U, Th, and Pb were uniformly distributed.
3. The isotopic composition of primeval Pb was everywhere the same.
4. Subsequently, the Earth became rigid and small regional differences arose in the U/Pb ratio.

5. The U/Pb ratio in any given region changed only as a result of decay of U to Pb.
6. At a later time, the Pb was separated from U and Th and its isotopic composition remained constant thereafter.

11.1a Decay of U to Pb

The single-stage Pb evolution model described by these assumptions can be expressed by means of equations arising from the law of radioactivity (Chapter 4). The $(^{206}Pb/^{204}Pb)_g$ ratio of Pb that was removed from a U-bearing reservoir t years ago and was crystallized as galena is equal to

$$\left(\frac{^{206}Pb}{^{204}Pb}\right)_g = \left(\frac{^{206}Pb}{^{204}Pb}\right)_p + \left(\frac{^{206}Pb}{^{204}Pb}\right)_T - \left(\frac{^{206}Pb}{^{204}Pb}\right)_t \tag{11.1}$$

where $(^{206}Pb/^{204}Pb)_p$ = value of this ratio of primeval Pb that was incorporated into the reservoir when it formed T years ago

$(^{206}Pb/^{204}Pb)_T$ = value of this ratio that formed in the same reservoir by decay of ^{238}U starting T years ago and extending to the present time

$(^{206}Pb/^{204}Pb)_t$ = value of this ratio that formed in the same reservoir by decay of ^{238}U starting t years ago and extending to the present $(t < T)$.

The value of the $(^{206}Pb/^{204}Pb)_T$ ratio of the reservoir at the present time can be calculated from the equation

$$\left(\frac{^{206}Pb}{^{204}Pb}\right)_T = \frac{^{238}U}{^{204}Pb}(e^{\lambda_1 T} - 1) \tag{11.2}$$

where $^{238}U/^{204}Pb$ is the present value of this ratio in the reservoir and λ_1 is the decay constant of ^{238}U. The value of $(^{206}Pb/^{204}Pb)_t$ is expressed similarly:

$$\left(\frac{^{206}Pb}{^{204}Pb}\right)_t = \frac{^{238}U}{^{204}Pb}(e^{\lambda_1 t} - 1) \tag{11.3}$$

Therefore, the $^{206}Pb/^{204}Pb$ ratio of the Pb that was withdrawn from this reservoir at time t is obtained by substituting equations 11.2 and 11.3 into equation 11.1:

$$\left(\frac{^{206}Pb}{^{204}Pb}\right)_g = \left(\frac{^{206}Pb}{^{204}Pb}\right)_p + \frac{^{238}U}{^{204}Pb}(e^{\lambda_1 T} - 1) - \frac{^{238}U}{^{204}Pb}(e^{\lambda_1 t} - 1)$$

which reduces to

$$\left(\frac{^{206}Pb}{^{204}Pb}\right)_g = \left(\frac{^{206}Pb}{^{204}Pb}\right)_p + \frac{^{238}U}{^{204}Pb}(e^{\lambda_1 T} - e^{\lambda_1 t}) \tag{11.4}$$

Similar equations can be written to represent the decay of ^{235}U to ^{207}Pb and ^{232}Th to ^{208}Pb:

$$\left(\frac{^{207}Pb}{^{204}Pb}\right)_g = \left(\frac{^{207}Pb}{^{204}Pb}\right)_p + \frac{^{235}U}{^{204}Pb}(e^{\lambda_2 T} - e^{\lambda_2 t}) \tag{11.5}$$

$$\left(\frac{^{208}Pb}{^{204}Pb}\right)_g = \left(\frac{^{208}Pb}{^{204}Pb}\right)_p + \frac{^{232}Th}{^{204}Pb}(e^{\lambda_3 T} - e^{\lambda_3 t}) \tag{11.6}$$

where λ_2 is the decay constant of ^{235}U and λ_3 is the decay constant of ^{232}Th (Table 10.2).

The present $^{238}U/^{204}Pb$ ratio of the reservoir from which the galena Pb was withdrawn is symbolized by the Greek letter μ (mu). Since the present $^{238}U/^{235}U$ ratio is a constant equal to 137.88 (Steiger and Jäger, 1977),

$$\frac{^{235}U/^{204}Pb}{^{238}U/^{204}Pb} = \frac{^{235}U/^{204}Pb}{\mu} = \frac{1}{137.88}$$

Hence,

$$\frac{^{235}U}{^{204}Pb} = \frac{\mu}{137.88} \tag{11.7}$$

In addition, $^{232}Th/^{204}Pb = \omega$ and $^{232}Th/^{238}U = \kappa$.

Equations 11.4 and 11.5 can be combined:

$$\frac{(^{207}Pb/^{204}Pb)_g - (^{207}Pb/^{204}Pb)_p}{(^{206}Pb/^{204}Pb)_g - (^{206}Pb/^{204}Pb)_p} = \frac{1}{137.88}\left(\frac{e^{\lambda_2 T} - e^{\lambda_2 t}}{e^{\lambda_1 T} - e^{\lambda_1 t}}\right) \tag{11.8}$$

This equation can be used to calculate the age T of the Earth using the isotope composition of Pb in a specimen of galena of known age t and the isotope composition of primeval Pb (Gerling, 1942; Moorbath, 1962).

11.1b Decay of Th to Pb

The $(^{208}Pb/^{204}Pb)_g$ in equation 11.6 can be combined with the $(^{206}Pb/^{204}Pb)_g$ ratio of equation 11.4 to link the decay of ^{232}Th in the reservoir to the decay of ^{238}U. The resulting equation is analogous to equation 11.8:

$$\frac{(^{208}Pb/^{204}Pb)_g - (^{208}Pb/^{204}Pb)_p}{(^{206}Pb/^{204}Pb)_g - (^{206}Pb/^{204}Pb)_p} = \frac{^{232}Th}{^{238}U}\left(\frac{e^{\lambda_3 T} - e^{\lambda_3 t}}{e^{\lambda_1 T} - e^{\lambda_1 t}}\right) \qquad (11.9)$$

where λ_3 is the decay constant of ^{232}Th (Table 10.2).

The $^{232}Tb/^{238}U$ ratio is expressed in terms of μ and ω as

$$\frac{^{232}Th}{^{238}U} = \frac{^{232}Th/^{204}Pb}{^{238}U/^{204}Pb} = \frac{\omega}{\mu} = \kappa \qquad (11.10)$$

The $^{232}Th/^{238}U$ ratio in equation 11.9 refers to the present Th/U ratio of the reservoir in which Pb evolved that was later deposited as galena during the formation of an ore deposit at a known time t in the past.

11.1c Analytical Methods

The interpretation of the isotope ratios of common Pb by the Holmes–Houtermans model (or by other models) may be confounded by systematic measurement errors (Catanzaro, 1967). For this reason, the analytical methods have been extensively improved in order to eliminate the effects of instrumental isotope fractionation and interlaboratory discrepancies. The effects of isotope fractionation are significantly reduced by the use of double or even triple spikes, which allow the isotope ratios of Pb to be corrected for isotope fractionation (Compston and Oversby, 1969; Dallwitz, 1970; Dodson, 1970; Gale, 1970; Ozard and Russell, 1970; Hofmann, 1971; Russell, 1971, 1975). The propagation of experimental errors in the double-spike methods was explained by Cumming (1973), whereas Richards (1981) warned against systematic errors of measurement caused by isotope fractionation in the mass spectrometer that may generate linear arrays of Pb isotope ratios of replicate analyses of the same sample. Such "pseudoisochrons" yield erroneous dates that confuse the geological interpretation of the isotopic data.

The subsequent application of Pb isotope geochemistry to the study of volcanic rocks (Chapter 17) has necessitated re-examination of the theory and practice of double spiking:

Hamelin et al., 1985, *Geochim. Cosmochim. Acta*, 49:173–182. Woodhead et al., 1995, *Analyst*, 120:35–39. Todt et al., 1996, *Am. Geophys. Union, Monograph* 95:429–437. Woodhead and Hergt, 1997, *Chem. Geol.*, 138:311–321. Galer and Abouchami, 1998, *Min. Mag.*, 62A:491–492. Powell et al., 1998, *Chem. Geol.*, 148:95–104. Galer, 1999, *Chem. Geol.*, 157:255–274. Thirlwall, 2000, *Chem. Geol.*, 163:299–322.

Another strategy to standardize isotope ratios of Pb is to define isotope standards that can be analyzed by all laboratories where such measurements are made:

Kollar et al., 1960, *Nature*, 187:754–756. Richards, 1962, *J. Geophys. Res.*, 67:869–884. Delevaux, 1963, *U.S. Geol. Surv. Prof. Paper*, 475:B160–B161. Catanzaro, 1968, *Earth Planet. Sci. Lett.*, 3:343–346. Catanzaro et al., 1968, *J. Res. Natl. Bur. Stand. (Phys. Chem.)*, 72A:261–267. Cooper et al., 1969, *Earth Planet. Sci. Lett.*, 6:467–478. Horn et al., 1997, *Geostand. Newsletter*, 12:191–203.

Although isotope dilution has been the preferred method to determine the concentration of Pb in rocks and minerals, ICP-MS is coming into wide use. This technique, combined with laser ablation, can be used to measure the isotope ratios of Pb with good precision and accuracy:

Longerich et al., 1987, *Spectrochim. Acta*, 42B:39–48. Walder and Furata, 1993, *Anal. Sci.*, 9:675–680. Fryer et al., 1995, *Can. Mineral.*, 33:303–312. Hirata, 1996, *Analyst*, 121:1407–1411. Promiès et al., 1998, *Chem. Geol.*, 144:137–149. Rehkämper and Halliday, 1998, *Int. J. Mass Spectrom Ion Proc.*, 58:123–133. Belshaw et al., 1998, *Int. J. Mass Spectrom Ion Proc.*, 181:51–58. White et al., 2000, *Chem Geol.*, 167:257–270. Poitrasson et al., 2000, *Geochim. Cosmochim. Acta*, 64:3283–3297.

In addition, ion microprobe mass spectrometers have been used to measure isotope ratios of Pb in galena (Cannon et al., 1963; Hart et al., 1981).

11.1d Primeval Pb in Meteorites

Meteorites are fragments of larger parent bodies that formed during the early history of the solar system, evolved rapidly by partial melting and chemical differentiation, and then solidified. During this process, an Fe sulfide phase formed, known as troilite (FeS), that contains appreciable concentrations of Pb but is virtually free of U and Th. Therefore, the isotopic composition of Pb in troilite has remained very nearly constant since the time of crystallization of the parent bodies of the meteorites and of the terrestrial planets that formed at the same time. For these reasons, the isotope ratios of Pb in troilite approach those of primeval Pb that was incorporated into the Earth at the time of its formation.

The isotope composition of Pb in stony and iron meteorites varies widely depending on their concentrations of U, Th, and Pb summarized in Table 11.1 (Rogers and Adams, 1969a, b; Wedepohl, 1974). The average U/Pb ratio of *chondritic meteorites* is

$$\frac{U}{Pb} = \frac{1.33 \times 10^{-2}}{1.10} = 0.012$$

A compilation of U and Pb concentrations of chondrite meteorites by Tilton (1973) yielded a similar average U/Pb ratio of 0.013.

Table 11.1. Average Concentrations of U, Th, and Pb in Stony and Iron Meteorites

Type	U, $\times 10^2$ ppm	Th, $\times 10^2$ ppm	Pb, ppm
Stony meteorites			
Chondrites	1.33 (21) ± 0.14	4.55 (12) ± 0.15	1.10 (36) ± 0.45
Achondrites	5.94 (12) ± 3.13	35.8 (16) ± 13.6	0.45 (6) ± 0.13
Iron meteorites			
Metal	<0.033 (6) ± 0.022	0.0010 (3) ± 0.0005	0.136 (19) ± 0.045
Troilite (FeS)	<0.767 (6) ± 0.300	—	5.91 (13) ± 3.08

Source: Rogers and Adams, 1969a, b; Wedepohl, 1974.
Note: The number of determinations is indicated in parentheses.

The average U/Pb ratio of *troilites* in Table 11.1 is less than 0.0013 and, in some cases, reaches values as low as 0.0005. For example, troilite in the iron meteorite Canyon Diablo, whose impact is recorded by Meteor Crater near Winslow in northern Arizona, has $U = 0.37 \times 10^{-2}$ ppm (Goles and Anders, 1962) and Pb = 6.5 ppm (Wedepohl, 1974) giving it a low U/Pb ratio of 0.00057. Therefore, troilite of Canyon Diablo was chosen to determine the isotope composition of primeval Pb. The results reported by Tatsumoto et al. (1973) in Table 11.2 were later confirmed by Chen and Wasserburg (1983). These results are also in good agreement with the isotope composition of Pb in the Pb-rich chondrite Mezö-Madaras (Pb = 5.27 ppm, U = 0.0113 ppm, U/Pb = 0.002) analyzed by Tilton (1973).

11.1e The Age of Meteorites and the Earth

The $^{206}Pb/^{204}Pb$ and $^{207}Pb/^{204}Pb$ ratios of stony meteorites have increased with time from their primeval values at rates that depend on the U/Pb ratios of the analyzed specimens in accordance with equation 11.8. When this equation is applied to meteorites, $t = 0$ because the decay of U has continued to the present time. Therefore,

Table 11.2. Isotope Ratios of Primeval Pb in Troilite of the Iron Meteorite Canyon Diablo and in the Pb-Rich Chondrite Mezö-Madaras

Sample	$^{206}Pb/^{204}Pb$	$^{207}Pb/^{204}Pb$	$^{208}Pb/^{204}Pb$	References
		Canyon Diablo		
Troilite	9.307	10.294	29.476	Tatsumoto et al., 1973
	9.3066	10.293	29.475	Chen and Wasserburg, 1983
		Mezö-Madaras		
Chondrite	9.310	10.296	(29.57)*	Tilton, 1973

*Estimated by assuming that $^{232}Th/^{238}U = 3.8$.

equation 11.8 reduces to

$$\frac{(^{207}Pb/^{204}Pb)_m - (^{207}Pb/^{204}Pb)_p}{(^{206}Pb/^{204}Pb)_m - (^{206}Pb/^{204}Pb)_p} = \frac{1}{137.88}\left(\frac{e^{\lambda_2 T} - 1}{e^{\lambda_1 T} - 1}\right) \quad (11.11)$$

where the subscripts m and p identify the isotope ratios of meteoritic and primeval Pb, respectively, and T is the age of meteorites and hence the age of the Earth. The isotope ratios of Pb of all meteorites that have remained closed to U and Pb since their formation at time T and that originally contained only primeval Pb satisfy equation 11.11 and therefore plot on a straight line in coordinates of $^{206}Pb/^{204}Pb$ (x) and $^{207}Pb/^{204}Pb$ (y). This line is a Pb–Pb isochron whose slope m is

$$m = \frac{1}{137.88}\left(\frac{e^{\lambda_2 T} - 1}{e^{\lambda_1 T} - 1}\right) \quad (11.12)$$

This equation can be solved for T by interpolating in Table 10.3 in Chapter 10.

Patterson (1956) used the isotope compositions of Pb in three stony and two iron meteorites to define the Pb–Pb isochron in Figure 11.1 whose slope yielded a date of 4.55 ± 0.07 Ga. The data array includes Pb in a sample of terrestrial marine sediment that fits the Pb–Pb meteorite isochron and thereby supports the theory that the Earth formed at the same time as the meteorites and that it originally contained primeval Pb.

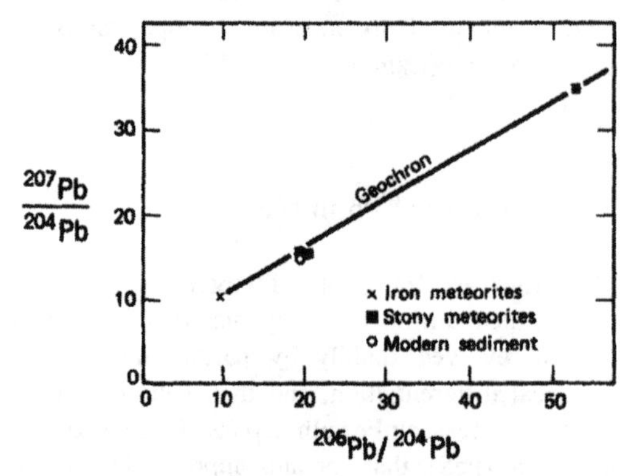

FIGURE 11.1 Lead isochron for meteorites and modern ocean sediment. The slope of this isochron (Equation 11.12) indicates an age of $T = 4.55 \pm 0.07 \times 10^9$ y for meteorites, based on the decay constants used by Patterson (1956).

Many other meteorites and their constituent minerals have been analyzed since Patterson's pioneering work:

Huey and Kohman, 1973, *J. Geophys. Res.*, 78:3227–3244. Tatsumoto et al., 1976, *Geochim. Cosmochim. Acta*, 40:617–634. Chen and Tilton, 1976, *Geochim. Cosmochim. Acta*, 40:635–643. Manhès and Allègre, 1978, *Meteoritics*, 13:543–548. Gale, 1979, *Geochim. J.*, 13:191. Gale et al., 1980, *Earth Planet. Sci. Lett.*, 48:311. Unruh et al., 1977, *Earth Planet. Sci. Lett.*, 37:1–12. Unruh et al., 1979, Proc. 10th Lunar

Planet. Conf., *Geochim. Cosmochim. Acta*, Suppl.:1011–1030.

A reconsideration by Allègre et al. (1995) of the age of meteorites based on Pb–Pb systematics yielded the following results:

Refractory inclusions, Allende	4.568–4.565 Ga
Phosphate, equilibrated chondrites	4.563–4.504 Ga
Basaltic achondrites	4.558–4.53 Ga

These results indicate that the parent bodies of stony meteorites crystallized and cooled during a period of about 64 million years between 4.50 and 4.57 Ga. Therefore, the re-examination of the data by Allègre et al. (1995) confirmed the date of 4.55 ± 0.07 Ga originally reported by Patterson (1956) based on different decay constants than those recommended later by Steiger and Jäger (1977).

The isotope analyses of Pb in meteorites also indicate that the U–Pb systematics of some meteorites were disturbed after their initial crystallization. For example, Hanan and Tilton (1985) reported a Pb–Pb date of 4.480 ± 0.011 Ga for separated phases and a whole-rock sample of Mezö-Madaras and a similar date of 4.472 ± 0.005 Ga for the meteorite Sharps. The U–Pb data of both meteorites form discordia chords on the concordia diagram indicating that their U/Pb ratios were recently disturbed. Hanan and Tilton (1985) and Minster and Allègre (1979) reported that the Rb–Sr systematics of Sharps and Mezö-Madaras were also disturbed.

11.2 DATING COMMON LEAD

In the Holmes–Houtermans model, the Pb in galena and other Pb-rich but U,Th-poor minerals is assumed to be a mixture of primeval Pb with varying amounts of radiogenic Pb that formed by decay of ^{238}U, ^{235}U, and ^{232}Th in the continental crust and lithospheric mantle. The Pb was separated from U and Th at some time (t) in the past and was deposited as galena (or some other Pb-rich and U-poor mineral), thereby preserving its isotopic composition. The isotope ratios of Pb generated by this single-stage process are expressed by equations 11.4–11.6 and depend on the time of separation (t), the age of the Earth (T), and the U/Pb and Th/Pb ratios of the reservoirs from which the Pb was extracted.

11.2a The Geochron

The time-dependent evolution of the $^{206}Pb/^{204}Pb$ and $^{207}Pb/^{204}Pb$ ratios in reservoirs having different $^{238}U/^{204}Pb$ ratios (μ) is illustrated in Figure 11.2 based on equations 11.4 and 11.5. In addition, the isotope composition of primeval Pb was assumed to be $^{206}Pb/^{204}Pb = 9.307$, $^{207}Pb/^{204}Pb = 10.294$, and $^{208}Pb/^{204}Pb = 29.476$ (Tatsumoto et al., 1973), and the age of the Earth (T) was taken to be 4.55 Ga (Patterson, 1956).

The isotopic evolution lines of galena Pb of the Holmes–Houtermans model in Figure 11.2 form a set of curved trajectories that fan out from the point representing primeval Pb depending on the μ-values of the reservoirs. Experience has shown that most samples of common Pb evolved in U–Pb systems having μ-values between 8 and 10. For example, point P in Figure 11.2 represents a sample of common Pb that evolved in a reservoir whose present μ-value is 10 and whose single-stage model date is 3.0 Ga. Therefore, the age of the mineral in which this sample of common Pb resides at the present time is 3.0 Ga or less.

In addition, Figure 11.2 contains a set of straight-line isochrons that are defined by the isotope ratios of single-stage leads that were separated at the specified dates from U–Pb reservoirs having different values of μ. One of these isochrons is called the *geochron* because it is the locus of Pb samples that have reservoir-separation dates of $t = 0$ years. The Pb samples on the geochron have resided in their respective reservoirs for 4.55×10^9 years or were removed from these reservoirs only recently. The stony meteorites in Figure 11.1 lie on the geochron because the Pb they contain is assumed to have evolved by the uninterrupted decay of U and Th in closed systems for 4.55×10^9 years. The slope of the geochron is calculated from

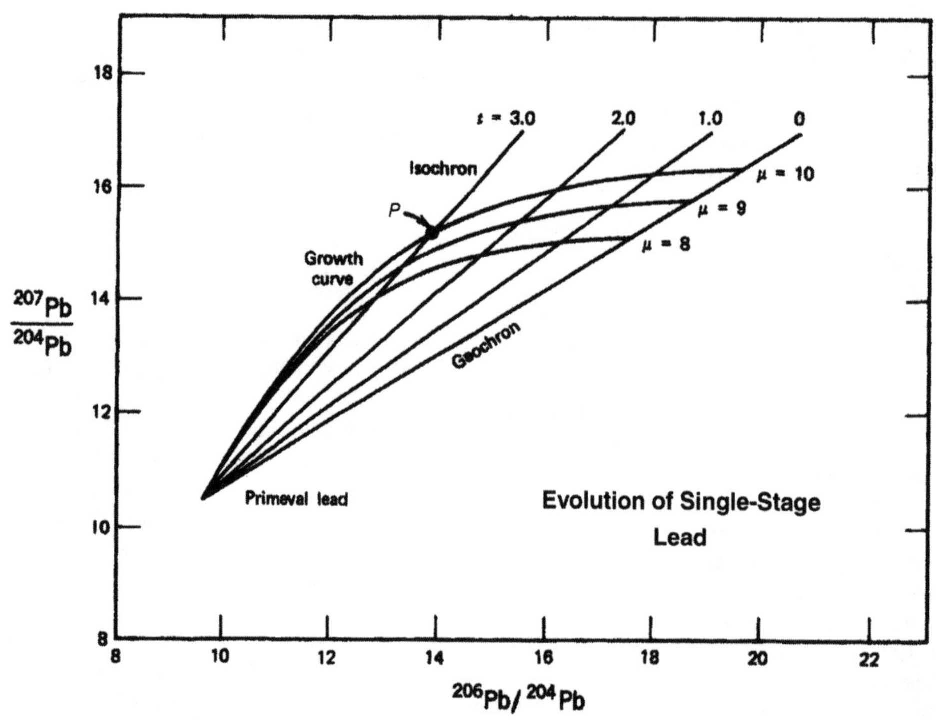

FIGURE 11.2 Graphical representation of the Holmes–Houtermans model. The curved lines are Pb growth curves for U–Pb systems having present-day μ-values of 8, 9, and 10. The straight lines are isochrons for selected values of t. For example, the coordinates of point P are the $^{207}Pb/^{204}Pb$ and $^{206}Pb/^{204}Pb$ ratios of a galena Pb that was withdrawn 3.0×10^9 years ago from a source region whose present $^{238}U/^{204}Pb$ ratio (μ) is 10.0. This diagram was constructed by solving equations 11.4 and 11.5 assuming that the age of the Earth is $T = 4.55 \times 10^9$ y.

equation 11.11:

$$m = \frac{1}{137.88}\left(\frac{e^{\lambda_2 T} - 1}{e^{\lambda_1 T} - 1}\right) = 0.61761$$

where $\lambda_1 = 1.55125 \times 10^{-10}$ y^{-1}, $\lambda_2 = 9.8485 \times 10^{-10}$ y^{-1}, and $T = 4.55 \times 10^9$ y.

11.2b Dating Single-Stage Leads

The *reservoir-separation date* of a sample of common Pb that satisfies the Holmes–Houtermans model can be calculated from its measured isotope ratios by means of equation 11.8. The procedure is to calculate the slope of the Pb–Pb isochron by means of the $^{206}Pb/^{204}Pb$ and $^{207}Pb/^{204}Pb$ ratios of the specimen of common Pb and of the primeval Pb (Table 11.2):

$$m = \frac{(^{207}Pb/^{204}Pb)_g - (^{207}Pb/^{204}Pb)_p}{(^{206}Pb/^{204}Pb)_g - (^{206}Pb/^{204}Pb)_p} \quad (11.13)$$

The result is then used to calculate the reservoir separation date (t) from the equation

$$m = \frac{1}{137.88}\left(\frac{e^{\lambda_2 T} - e^{\lambda_2 t}}{e^{\lambda_1 T} - e^{\lambda_1 t}}\right) \quad (11.14)$$

where $T = 4.55 \times 10^9$ y. Equation 11.14 can be solved by interpolation in Table 11.3 or by numerical iteration to any desired level of precision. After t has been determined in this way, the present

Table 11.3. Slopes of Isochrons (m) and Corresponding Model Dates (t) of Single-Stage Pb Based on Equation 11.14

Age (t), Ga	Slope (m)	Age (t), Ga	Slope (m)
0	0.61761	2.4	0.98101
0.2	0.63705	2.6	1.03410
0.4	0.65509	2.8	1.09295
0.6	0.67624	3.0	1.15830
0.8	0.69923	3.2	1.23100
1.0	0.72426	3.4	1.31205
1.2	0.75158	3.6	1.40257
1.4	0.78144	3.8	1.50387
1.6	0.81415	4.0	1.61743
1.8	0.85005	4.2	1.74498
2.0	0.88952	4.4	1.88849
2.2	0.93300	4.6	2.05025

Note: $\lambda_1 = 1.55125 \times 10^{-10}$ y^{-1}, $\lambda_2 = 9.8485 \times 10^{-10}$ y^{-1}, $T = 4.55 \times 10^9$ y.

value of μ of the U–Pb reservoir is calculated from equation 11.4:

$$\left(\frac{^{206}\text{Pb}}{^{204}\text{Pb}}\right)_g = \left(\frac{^{206}\text{Pb}}{^{204}\text{Pb}}\right)_p + \mu\,(e^{\lambda_1 T} - e^{\lambda_1 t})$$

$$\mu = \frac{(^{206}\text{Pb}/^{204}\text{Pb})_g - (^{206}\text{Pb}/^{204}\text{Pb})_p}{e^{\lambda_1 T} - e^{\lambda_1 t}} \quad (11.15)$$

The geological significance of the resulting values of t and μ depends on the extent to which the single-stage model reproduces the actual geological history of the samples of common Pb that was analyzed. Experience has shown that only a small number of ore deposits contain Pb that satisfies the assumptions of the single-stage model and are therefore referred to as "ordinary" leads.

Ore deposits that contain single-stage Pb can be identified by the following criteria:

1. All samples of common Pb in a particular ore deposit yield the same model date within analytical errors unless there is evidence for episodic mineralization spanning an interval of time.
2. The model dates have positive numerical values.
3. The single-stage Pb dates are in general agreement with isotopic dates of other minerals in the ore and in the surrounding country rock.

However, exact agreement with dates determined by other methods is not expected because the model Pb dates of galena refer to a different event than the isotopic dates of associated silicate minerals.

11.2c Lead from Cobalt, Ontario

The isotope ratios of Pb in the ore deposits of the Cobalt–Noranda area of Ontario and Quebec (Kanasewich and Farquhar, 1965) were included by Doe and Stacey (1974) in a list of 13 ore deposits that appear to fit the Holmes–Houtermans model. The ore deposits at Cobalt are hydrothermal veins that occur in the Early Proterozoic rocks of the Cobalt Group, which consists of conglomerate, tillite, graywacke, arkose, and sandstones (Thorpe, 1974). These metasedimentary rocks are unconformably underlain by intensely folded metavolcanic rocks of Archean age and by granitic batholiths that intrude the metavolcanics. The rocks of the Cobalt Group were themselves intruded by sills of the Nipissing Diabase to which the mineralization of the veins in the Cobalt area is genetically related (Hriskevich, 1968).

The Nipissing Diabase was dated by the whole-rock Rb–Sr isochron method, which yielded a date of 2109 ± 80 Ma (Van Schmus, 1965). In addition, Fairbairn et al. (1969) obtained a date of 2240 ± 87 Ma for graywacke and argillite of the Cobalt Group by the same method.

The average isotope ratios of Pb in galena in the ore at Cobalt are (Doe and Stacey, 1974; Stacey and Kramers, 1975):

$$\frac{^{206}\text{Pb}}{^{204}\text{Pb}} = 14.87 \qquad \frac{^{207}\text{Pb}}{^{204}\text{Pb}} = 15.16$$

$$\frac{^{208}\text{Pb}}{^{204}\text{Pb}} = 34.44$$

The slope of the Pb–Pb isochron in coordinates of ^{206}Pb/^{204}Pb and ^{207}Pb/^{204}Pb ratios (equation 11.13) is

$$m = \frac{15.16 - 10.294}{14.87 - 9.307} = 0.87470$$

The reservoir separation date t obtained from the slope of the Pb–Pb isochron by interpolation in Table 11.3 is

$$t = 1.8 + \frac{0.2 \times 0.02465}{0.03947} = 1.92 \text{ Ga}$$

The value of μ is then calculated from equation 11.15:

$$\mu = \frac{14.87 - 9.307}{2.0255 - 1.3479} = 8.2$$

The single-stage Pb–Pb date of the ore at Cobalt, Ontario, is lower than the geological age of the Cobalt Series, which ranges from 2240 ± 87 to 2109 ± 80 Ma. The discrepancy is a signal that the Pb in the ore at Cobalt, Ontario, does not quite satisfy the assumptions of the single-stage model. However, the μ value of the source region from which the Pb was derived ($\mu = 8.2$) is well within the range of values of this ratio in other mineral deposits that contain common Pb. A date based on the two-stage model of Pb evolution is calculated in Section 11.2f.

11.2d Limitations of the Single-Stage Model

Only a small number of ore deposits contain Pb that actually satisfies the single-stage model. Doe and Stacey (1974) listed 13 such deposits whose ages range from 3.2 Ga (Barberton, South Africa) to 0 Ga (White Island, New Zealand). However, the single-stage model dates of these selected deposits in Figure 11.3 are up to 400 million years younger than dates determined by other methods, which means that ore deposits of Phanerozoic age, in some cases, yield negative Pb–Pb model dates.

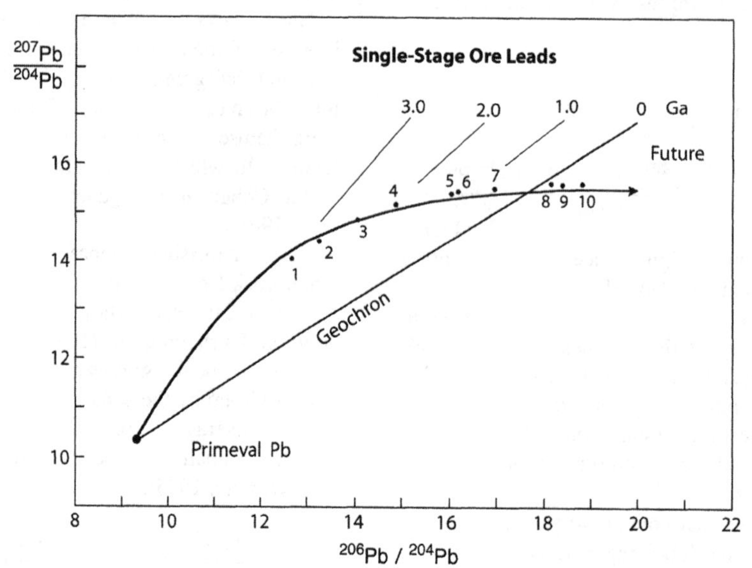

FIGURE 11.3 Isotopic evolution of common leads that meet the assumptions of the single-stage Holmes–Houtermans model for $\mu = 8.1$, $T = 4.55 \times 10^9$ y, based on equations 11.4 and 11.5. The ore deposits are identified by number: 1. Barberton, South Africa; 2. Manitouwadge, Ontario; 3. Geneva Lake, Ontario; 4. Cobalt, Ontario; 5. Broken Hill, N.S.W., Australia; 6. Mount Isa, Queensland, Australia; 7. Balmat, New York; 8. Captain's Flat, N.S.W., Australia; 9. Hall's Peak, N.S.W., Australia; 10. White Island, New Zealand. Data from Doe and Stacey (1974) and Tatsumoto et al. (1973).

To avoid this problem, Doe and Stacey (1974) reduced the value of T in equation 11.14 from 4.55 to 4.43 Ga and tabulated the slopes of single-stage Pb–Pb isochrons based on this modification. However, the necessity for making this ad hoc alteration diminishes the validity of the Holmes–Houtermans model.

Another noteworthy feature of the ore deposits selected by Doe and Stacey (1974) is that their μ-values vary only within narrow limits and have a mean of 8.10 ± 0.24 for 10 deposits. This value was used to construct the Pb growth curve in Figure 11.3. The measured $^{206}Pb/^{204}Pb$ and $^{207}Pb/^{204}Pb$ ratios of all 13 deposits follow the growth curve closely, but only 10 are shown to avoid crowding the diagram. The points representing Pb from Captain's Flat (8), Hall's Peak (9), and White Island (10) lie to the right of the geochron in a part of the diagram representing the future. These and other deposits (not shown) yield negative single-stage model Pb dates that invalidate the Holmes–Houtermans model.

The evidence that all of the deposits in Figure 11.3 have similar μ-values indicates that the reservoirs from which they originated had similar U/Pb ratios even though the ore deposits formed on different continents during a time interval of about 3 billion years. This evidence therefore supports the view that the reservoirs of common Pb on the Earth are homogeneous with respect to their U/Pb ratios. Alternatively, the evidence implies extensive isotopic homogenization of Pb during its removal from reservoirs that may have had a wide range of U/Pb ratios.

Stanton and Russell (1959) originally recognized that ore deposits containing apparent single-stage Pb occur in stratigraphic sequences of volcanic and marine sedimentary rocks that were deposited in volcanic island arcs. The ore deposits in these kinds of rocks are conformable with the structure of their host rocks and were emplaced as a result of volcanic activity without contamination by radiogenic Pb from the crust. Armstrong (1968) proposed that these single-stage leads originated from marine sediment of crustal origin, that the Pb was removed from its source during subduction of the sediment and was deposited syngenetically by precipitation of galena and other sulfides from hot aqueous solutions on the sea floor along spreading ridges, in subduction zones, and/or back-arc basins. According to this scenario, single-stage leads were homogenized isotopically during removal from heterogeneous sources.

The identification of single-stage leads and the fitting of their isotopic evolution line has been the subject of several papers:

Ostic et al., 1967, *Can. J. Earth Sci.*, 4:245–269. Kanasewich, 1968, in Hamilton and Farquhar (Eds.), *Radiometric dating for Geologists*, pp. 147–223. Wiley, New York. Sinha and Tilton, 1973, *Geochim. Cosmochim. Acta*, 37:1823–1849. Cumming and Richards, 1975, *Earth Planet. Sci. Lett.*, 28:155–171. Albarède and Juteau, 1984, *Geochim. Cosmochim. Acta*, 48:207–212.

The consensus is that single-stage leads could not have evolved continuously since the Earth formed at 4.55 Ga as postulated in the Holmes-Houtermans model. In addition, experience has shown that most samples of common Pb do not have single-stage histories and are therefore classified as being anomalous.

11.2e The Stacey–Kramers Model

The discrepancies between single-stage Pb dates of many ore deposits and their geological ages require a refinement of the Holmes–Houtermans model for the isotopic evolution of common Pb (Richards, 1971; Oversby, 1974). The fact that the single-stage Pb dates of many ore deposits of late Precambrian and Phanerozoic age are younger than their geological ages suggests that the U/Pb ratios of the Pb reservoirs in the Earth have increased either continuously or episodically with time. The reduction of the parameter T from 4.55 to 4.43 billion years proposed by Doe and Stacey (1974) is not a good solution to the problem because it violates the assumption that meteorites and the Earth have the same age (i.e., 4.55×10^9 years).

The refinement of the Holmes–Houtermans model proposed by Stacey and Kramers (1975) is that Pb in ore deposits formed by a two-stage process. The resulting Pb isotope ratios in Table 11.4 were used to plot Figure 11.4. The first

Table 11.4. Isotopic Evolution of Pb in the Two-Stage Model of Stacey and Kramers (1975) with the Decay Constants Later Adopted by Steiger and Jäger (1977)

t, Ga	$^{206}Pb/^{204}Pb$	$^{207}Pb/^{204}Pb$	$^{208}Pb/^{204}Pb$
	First Stage		
4.57	9.307	10.294	29.487
4.30	9.906	11.391	30.036
4.00	10.544	12.313	30.637
3.70	11.152	12.998	31.230
	Second Stage		
3.70	11.152	12.998	31.230
3.50	11.680	13.481	31.666
3.25	12.317	13.965	32.204
3.00	12.931	14.343	32.736
2.75	13.520	14.639	33.261
2.50	14.088	14.870	33.780
2.25	14.634	15.051	34.293
2.00	15.159	15.192	34.799
1.75	15.664	15.303	35.299
1.50	16.149	15.389	35.793
1.25	16.617	15.457	36.280
1.00	17.066	15.509	36.762
0.75	17.499	15.551	37.238
0.50	17.915	15.583	37.708
0.25	18.315	15.608	38.172
0.00	18.700	15.628	38.630

stage started at $T = 4.57 \times 10^9$ y with primeval Pb in a reservoir having $\mu = 7.192$ and $\omega = 32.208$. This stage ended at $T' = 3.70 \times 10^9$ y when μ increased from 7.192 to 9.735 and ω rose from 32.208 to 36.837. The isotope ratios of common Pb at $T' = 3.70 \times 10^9$ y were

$$\left(\frac{^{206}\text{Pb}}{^{204}\text{Pb}}\right)_{T'} = 11.152 \qquad \left(\frac{^{207}\text{Pb}}{^{204}\text{Pb}}\right)_{T'} = 12.998$$

$$\left(\frac{^{208}\text{Pb}}{^{204}\text{Pb}}\right)_{T'} = 31.230$$

During the second stage, the isotope ratios of Pb in the reservoir shown in Figure 11.4 increased by decay of U and Th from $T' = 3.70 \times 10^9$ y to the present time. The present isotope ratios of Pb in the Stacey–Kramers model are

$$\left(\frac{^{206}\text{Pb}}{^{204}\text{Pb}}\right)_{t=0} = 18.700 \qquad \left(\frac{^{207}\text{Pb}}{^{204}\text{Pb}}\right)_{t=0} = 15.628$$

$$\left(\frac{^{208}\text{Pb}}{^{204}\text{Pb}}\right)_{t=0} = 38.630$$

All common leads that evolved in this reservoir and were removed from it at some time t in the past have isotope ratios that define a point on the growth curve in Figure 11.4.

The reservoir-separation date of a sample of common Pb that evolved by the two-stage process of the Stacey–Kramers model is derivable from the equation that describes the second stage:

$$\left(\frac{^{206}\text{Pb}}{^{204}\text{Pb}}\right)_g = 11.152 + 9.735\,(e^{\lambda_1 T'} - e^{\lambda_1 t}) \tag{11.16}$$

$$\left(\frac{^{207}\text{Pb}}{^{204}\text{Pb}}\right)_g = 12.998 + \frac{9.735}{137.88}(e^{\lambda_2 T'} - e^{\lambda_2 t}) \tag{11.17}$$

and hence for the U–Pb decay

$$\frac{(^{207}\text{Pb}/^{204}\text{Pb})_g - 12.998}{(^{206}\text{Pb}/^{204}\text{Pb})_g - 11.152} = \frac{1}{137.88}\left(\frac{e^{\lambda_2 T'} - e^{\lambda_2 t}}{e^{\lambda_1 T'} - e^{\lambda_1 t}}\right) \tag{11.18}$$

where the subscript g identifies the isotope ratios of Pb in galena. The Th–Pb system is defined by a similar equation:

$$\frac{(^{208}\text{Pb}/^{204}\text{Pb})_g - 31.230}{(^{206}\text{Pb}/^{204}\text{Pb})_g - 11.152} = \left(\frac{\omega}{\mu}\right)\left(\frac{e^{\lambda_3 T'} - e^{\lambda_3 t}}{e^{\lambda_1 T'} - e^{\lambda_1 t}}\right) \tag{11.19}$$

where λ_3 is the decay constant of ^{232}Th ($\lambda_3 = 4.9475 \times 10^{-11}$ y^{-1}).

The two-stage model of Stacey and Kramers (1975) is used to date ordinary Pb by means of equation 11.18. This equation is more flexible than equations 11.16 and 11.17 because it does not require that the second-stage reservoir of the Pb samples being dated had the μ-value (9.735) specified by the model. The slopes of the secondary Pb–Pb isochrons are used to solve equation 11.18 for t by interpolating in Table 11.5. After t has been determined, the ω/μ ratio of the

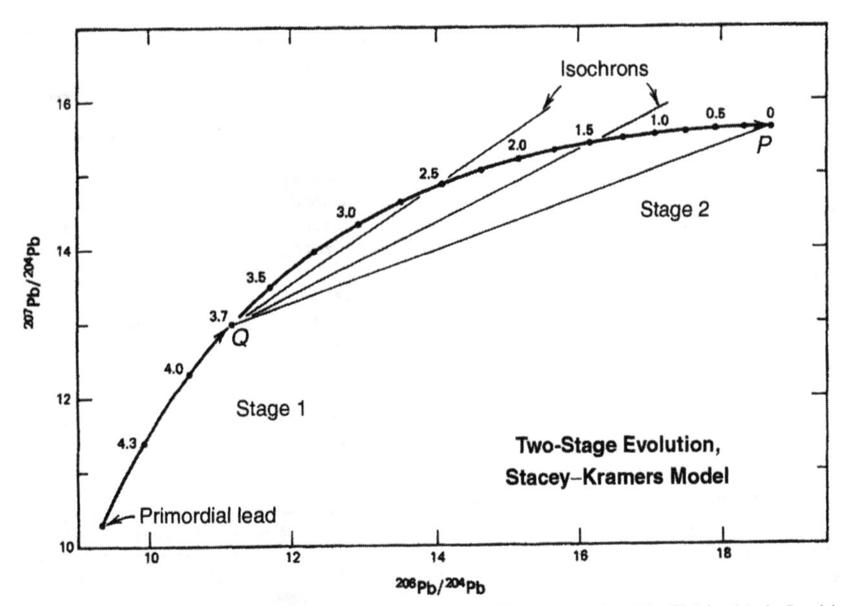

FIGURE 11.4 Two-stage Pb model of Stacey and Kramers (1975) summarized in Table 11.4. In this model, Pb evolves from primeval isotope ratios between 4.57 and 3.70 Ga in a reservoir with $^{238}U/^{204}Pb = 7.192$. At point Q ($t = 3.70$ Ga) on the evolution line, the $^{238}U/^{204}Pb$ ratio of the reservoir changed by geochemical differentiation to 9.735. Lead evolution then continued undisturbed to point P, representing average crustal Pb. Straight lines connecting any point on the evolution line between 3.70 Ga and the present to Q are isochrons. The slopes of such isochrons are related by equation 11.18 to the time elapsed since a Pb sample was isolated from the reservoir. Plotted from data in Table 9 of Stacey and Kramers (1975).

second reservoir is obtained from equation 11.19. The value of this ratio should be close to 3.78, as required by the Stacey–Kramers model (i.e., $\omega/\mu = 36.837/9.735 = 3.78$).

Table 11.5. Slopes of Isochrons (m) and Corresponding Model Dates (t) of the Two-Stage Model of Stacey and Kramers (1975)

Age (t), Ga	Slope (m)	Age (t), Ga	Slope (m)
0	0.34840	1.8	0.51781
0.2	0.36103	2.0	0.54764
0.4	0.37482	2.2	0.58079
0.6	0.38991	2.4	0.61769
0.8	0.40645	2.6	0.65884
1.0	0.42460	2.8	0.70482
1.2	0.44457	3.0	0.75630
1.4	0.46659	3.2	0.81404
1.6	0.49090	3.4	0.87892

11.2f Balmat, St Lawrence County, New York

The Pb–Zn ore deposit at Balmat, New York, is composed of sulfide minerals hosted by Late Proterozoic (Grenville) marble whose age is about 1060 Ma (Doe, 1962a, b). The isotope ratios of Pb in this deposit approach the single-stage growth curve in Figure 11.3 (Doe and Stacey, 1974). The single-stage Pb–Pb isochron has a slope $m = 0.68314$, which yields a date of 0.66 Ga by interpolation in Table 11.2. This date significantly underestimates the age of the sulfide ore at Balmat.

According to Doe and Stacey (1974), the isotope composition of Pb in the ore of Balmat is

$$\frac{^{206}Pb}{^{204}Pb} = 16.935 \qquad \frac{^{207}Pb}{^{204}Pb} = 15.505$$

$$\frac{^{208}Pb}{^{204}Pb} = 36.423$$

The slope of the isochron relative to point Q in Figure 11.4 is

$$m = \frac{15.505 - 12.998}{16.935 - 11.152} = 0.43351$$

Interpolating in Table 11.5 yields

$$t = 1.0 + \frac{0.2 \times 0.00891}{0.019857} = 1.098 \text{ Ga}$$

The date derived from the Stacey–Kramers model differs only by about 30 million years from the age of the Balmat ore as determined by other isotopic methods.

The Stacey–Kramers model also improves the concordance of the date derivable from the galena at Cobalt, Ontario (Section 11.2c). The slope of its Pb–Pb isochron relative to point Q in Figure 11.4 is

$$m = \frac{15.16 - 12.998}{14.87 - 11.152} = 0.58149$$

Interpolation in Table 11.5 yields

$$t = 2.2 + \frac{0.2 \times 0.0007}{0.0369} = 2.204 \text{ Ga}$$

This date exceeds the single-stage date (t = 1.92 Ga) by about 280 million years.

Although the two-stage model of Stacey and Kramers (1975) yields more accurate dates than the single-stage model, it is only applicable to a limited number of ore deposits. It does not work for anomalous leads that formed by mixing of two-stage Pb with radiogenic Pb derived from granitic basement rocks prior to the deposition of galena.

11.3 DATING K-FELDSPAR

Microcline or orthoclase crystals in high-grade metamorphic and igneous rocks capture Pb^{2+} in place of K^+ ions because of the similarity of their radii (Pb^{2+} = 1.20 Å; K^+ = 1.35 Å), which gives Pb^{2+} a charge-to-radius ratio of 1.67 compared to only 0.74 for K^+. Consequently, K-feldspars in Table 11.6 have elevated concentrations of Pb but low concentrations of U and Th. As a result, the isotopic composition of Pb in feldspar is virtually invariant with time. However, since feldspars are a sink for ambient Pb during metamorphism, the isotope composition of Pb in K-feldspar may be altered in some cases by the addition of varying amounts of radiogenic Pb.

Table 11.6. Concentrations of Pb, U, and Th in K-Feldspar

	Concentration (ppm)			
Source	Pb	U	Th	References[a]
Granites and gneisses	43.1 (8) ± 13.4	0.053 (5) ± 0.025	0.224 ± 0.163	1
Pegmatites	206 (12) ± 115	<0.1	—	1, 4
Marine sediment, detrital	52.0 (6) ± 3.1	0.121 (6) ± 0.034	0.31 (2) ± 0.06	2
Volcanic rocks	32.9 (3) ± 4.0	—	—	3
Plutonic rocks	51.2 (9) ± 19.8	0.29 (1)	0.31 (1)	3
Metamorphic rocks	34.9 (1)	—	—	3

[a] 1. Doe et al., 1965, *J. Geophys. Res.*, 70:1947–1968. 2. Patterson and Tatsumoto, 1964, *Geochem. Cosmochim. Acta*, 28:1–22. 3. Doe and Tilling, 1967, *Am. Mineral.*, 52:805–816. 4. Catanzaro and Gast, 1960, *Geochim. Cosmochim. Acta*, 19:113–126.

The isotopic composition of Pb in microcline of Precambrian pegmatites was determined by Catanzaro and Gast (1960), who also referenced the work of their predecessors. They reported that single-stage model dates calculated from the isotope ratios of Pb in feldspar are in good agreement with dates determined by other methods. However, a white microcline from a pegmatite (2600 Ma) on the southern slope of the Bridger Mountains and about 20 km north of Bonneville, Wyoming, yielded an anomalously low date of 1350 Ma. After leaching the microcline with 50 percent HNO_3 for 15 min, the single-stage model date increased to 2550 ± 100 Ma, which is indistinguishable from the "accepted" age of this pegmatite. The authors suggested that the feldspar of the Bonneville pegmatite had been contaminated after crystallization by the addition of ambient crustal Pb. Contamination of feldspar with crustal Pb was also detected by Doe et al. (1965), who reported that the isotope composition of Pb in the feldspar of Precambrian gneisses near Baltimore, Maryland, was altered during Paleozoic regional metamorphism.

Sinha and Tilton (1973) developed an open-system model for Pb evolution in which μ increased linearly with time. However, to achieve agreement with dates based on other isotopic methods, the authors were forced to increase the value of T from 4.55×10^9 to 4.66×10^9 years. Nevertheless, Sinha and Tilton (1973) compiled an interesting set of isotopic ratios of Pb in galena and feldspar from several regions after normalizing the isotope ratios of Pb to the same interlaboratory standard.

The *galena* samples from the Superior structural province of Canada in Table 11.7 yielded an average model date of 2951 ± 30 Ma, whereas five samples of *K-feldspar* from the same region averaged 2877 ± 75 Ma. However, age determinations by the K–Ar, Rb–Sr, and U–Pb methods reviewed by Sinha and Tilton (1973) indicate that magmatic activity and regional metamorphism in the Superior Province occurred at about 2750 ± 50 Ma. Therefore, the model Pb dates of the galenas and K-feldspar are between 200 and 125 million years older, respectively, than the conventional isotopic dates.

The isotopic compositions of Pb in the Superior structural province of Canada compiled by Sinha and Tilton (1973) were also evaluated by the two-stage Stacey–Kramers model in Table 11.7. The resulting average dates of the galenas (2796 ± 39 Ma) and feldspars (2702 ± 134 Ma) are consistently lower than those calculated by Sinha and Tilton (1973) and are in better agreement with the age of igneous activity and regional metamorphism at 2750 ± 50 Ma.

The feldspars in the pegmatites of the Baltimore area analyzed by Doe et al. (1965) and included in the compilation of data by Sinha and Tilton (1973) yield a single-stage date of −220 Ma. The two-stage model date of this feldspar based on the Stacey–Kramers model (Table 11.5) is 414 Ma, which agrees with the date of 425 Ma reported by Wetherill et al. (1966) for the post-Glenarm pegmatites of the Maryland Piedmont.

Table 11.7. Isotope Ratios of Pb and Corresponding Two-Stage Model Dates (Stacey and Kramers, 1975) of Galena and Feldspars from the Superior Structural Province of Canada (Sinha and Tilton, 1973)

Locality or Number	$^{206}Pb/^{204}Pb$	$^{207}Pb/^{204}Pb$	Slope[a]	Model I Date, Ma	Model II Date, Ma
		Galena			
Willroy Mine	13.30	14.54	0.71787	2850	2980
Geco Mine	13.30	14.51	0.70391	2796	2975
	13.407	14.576	0.69977	2778	2925
Lun Echo	13.404	14.565	0.69582	2760	2925
Average date of galenas				2796 ± 39	2951 ± 30
		Feldspar			
MG-19	13.69	14.65	0.65090	2561	2770
MG-41	13.45	14.51	0.65796	2595	2850
MG-45	13.41	14.63	0.72276	2869	2975
RN-IT	13.28	14.53	0.71992	2858	2950
KA-82	13.47	14.54	0.66522	2627	2840
Average date of feldspars:				2702 ± 134	2877 ± 75

Note: Model I: Stacey and Kramers (1975). Model II: Sinha and Tilton (1973).

[a] Relative to point Q in Figure 11.4 whose coordinates are: $^{206}Pb/^{204}Pb = 11.152$; $^{207}Pb/^{204}Pb = 12.998$; $^{208}Pb/^{204}Pb = 31.230$.

11.4 ANOMALOUS LEADS IN GALENA

Many ore deposits contain Pb whose isotopic composition is anomalous because it is not consistent with the single- and two-stage models of Pb evolution (Farquhar and Russell, 1957; Stanton and Russell, 1959). These deposits may have negative model Pb dates that represent the future rather than the past because they contain excess amounts of radiogenic Pb.

Russell and Farquhar (1960) and Kanasewich (1968) devised mathematical procedures by means of which additional information can be extracted from suites of anomalous leads (Doe, 1970). These procedures take advantage of the fact that the isotope ratios of anomalous leads in a particular ore body or mining district, in many cases, form straight lines in coordinates of the $^{206}Pb/^{204}Pb$ (x) and $^{207}Pb/^{204}Pb$ (y) ratios. These so-called anomalous Pb lines are attributable to mixing of two-stage Pb with varying amounts of radiogenic Pb. This interpretation yields useful information about the history and origin of suites of anomalous Pb derived from

1. the coordinates of the point of intersection of the anomalous Pb line and the growth curve of the Stacey–Kramers model and
2. the slope of the anomalous Pb line based either on the "instantaneous" or the "continuous" decay of the U isotopes.

11.4a Two-Stage Model Dates

A hypothetical example of anomalous leads in Figure 11.5 illustrates the effect of mixing of

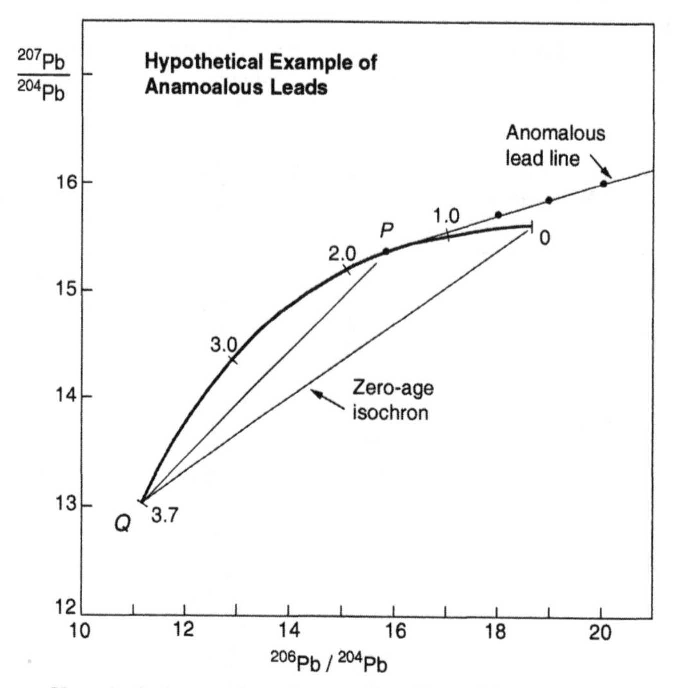

FIGURE 11.5 Hypothetical examples of anomalous Pb (solid circles) that formed by the addition of varying amounts of radiogenic Pb to common Pb at point P on the second stage of the Stacey–Kramers growth curve. The slope of the anomalous Pb line is $R = 0.160$. The growth curve was plotted from a tabulation by Stacey and Kramers (1975).

two-stage Pb with a component of radiogenic Pb. The resulting anomalous leads define a straight line of slope R that intersects the Pb growth curve of the Stacey–Kramers model at point P. The two-stage model date of the Pb represented by P is calculated from the slope of the Pb–Pb isochron relative to point Q on the Stacey–Kramers growth curve in Figure 11.4. In this example, the coordinates of point P are $^{206}Pb/^{204}Pb = 15.879$ and $^{207}Pb/^{204}Pb = 15.335$. Therefore, the slope m of the line PQ and the corresponding two-stage model date t from Table 11.5 are

$$m = \frac{15.335 - 12.998}{15.879 - 11.152} = 0.49439$$

$$t = 1.6 + \frac{0.2(0.49439 - 0.49090)}{0.51781 - 0.49090} = 1.63 \text{ Ga}$$

This date is the time in the past when the Pb was withdrawn from the second Stacey–Kramers reservoir and was thereby separated from U and Th. Subsequently, this Pb was mixed with radiogenic Pb and the resulting mixtures were then deposited in the form of galena.

11.4b Instantaneous Growth of Radiogenic Pb

The slope of the anomalous Pb line is the $^{207}Pb/^{206}Pb$ ratio of the radiogenic Pb that was added to the Pb, represented by point P in Figure 11.5. The time in the past when radiogenic Pb was formed having the $^{207}Pb/^{206}Pb$ ratio equal to the slope R of the anomalous Pb line is

$$R = \frac{^{235}U_t \lambda_2}{^{238}U_t \lambda_1} \tag{11.20}$$

where $^{235}U_t$ and $^{238}U_t$ are the abundances of the U isotopes at any time t in the past. The past abundances of the U isotopes are related to their present abundances ($^{238}U_p$ and $^{235}U_p$) by the law of radioactivity:

$$^{238}U_p = {}^{238}U_t e^{-\lambda_1 t}$$

and hence for both ^{238}U and ^{235}U

$$^{238}U_t = {}^{238}U_p e^{\lambda_1 t} \qquad {}^{235}U_t = {}^{235}U_p e^{\lambda_2 t}$$

Substitution into equation 11.20 yields

$$R = \left(\frac{\lambda_2}{\lambda_1}\right)\frac{^{235}U_p e^{\lambda_2 t}}{^{238}U_p e^{\lambda_1 t}} = \left(\frac{\lambda_2}{\lambda_1}\right)\frac{e^{t(\lambda_2 - \lambda_1)}}{137.88} \tag{11.21}$$

Solving for t yields

$$t = \frac{1}{\lambda_2 - \lambda_1}\ln\left(\frac{137.88 R \lambda_1}{\lambda_2}\right) \tag{11.22}$$

In the case under consideration, the slope of the anomalous Pb line is

$$R = \left(\frac{^{207}Pb}{^{206}Pb}\right)^* = 0.160$$

Accordingly, the instant in time in the past when the decay of ^{238}U and ^{235}U produced radiogenic ^{206}Pb and ^{207}Pb having the $^{207}Pb/^{206}Pb$ ratio of 0.160 is calculated from equation 11.22:

$$t = \frac{\ln 3.47482}{8.29725 \times 10^{-10}} = 1.50 \text{ Ga}$$

Evidently, the radiogenic Pb originated from rocks or minerals of Precambrian age that predate the origin of the ore deposit.

11.4c Continuous Growth of Radiogenic Pb

The radiogenic Pb could also have accumulated during an extended interval of time. In this case, the isotopic evolution of the radiogenic Pb is expressed by an equation of the form

$$\left(\frac{^{207}Pb}{^{206}Pb}\right)^* = \left(\frac{^{235}U}{^{238}U}\right)_p \left(\frac{e^{\lambda_2 t_r} - e^{\lambda_2 t}}{e^{\lambda_1 t_r} - e^{\lambda_1 t}}\right) \tag{11.23}$$

where $\left(\frac{^{235}U}{^{238}U}\right)_p = \frac{1}{137.88}$

t_r = present age of the basement rocks in which the radiogenic Pb evolved

t = time in the past when the radiogenic Pb was removed from the basement rocks (i.e., $t_r > t$)

Equation 11.23 can be solved for t based on the measured slope R of the anomalous Pb line provided that the age of the basement rocks (t_r) is known. If $t_r = 2.0 \times 10^9$ y, interpolation in a table (not shown) of selected values of t indicates that $R = 0.160$ corresponds to $t = 0.883$ Ga. Therefore, the radiogenic Pb component could have accumulated in Precambrian basement rocks between $t_r = 2.0$ and $t = 0.88$ Ga during an interval of about 1.12 billion years.

Alternatively, if the age t of the ore deposit is known from dating by other methods, equation 11.23 can be solved for t_r, which is an overestimate of the present age of the basement rocks in which the radiogenic component of Pb accumulated.

These interpretations have provided the following information about the hypothetical ore deposits in which the anomalous Pb occurs:

1. The two-stage Pb originated from the second stage of the Stacey–Kramers model ($\mu = 9.735$) at 1.63 Ga.
2. The Pb was subsequently contaminated with radiogenic Pb that was produced by decay of U isotopes at 1.50 Ga (instantaneous model).
3. Alternatively, the radiogenic contaminant could have accumulated between $t_r = 2.0$ Ga and $t = 0.88$ Ga (continuous model) in the basement rocks underlying the ore deposit.
4. The age of the ore deposit in question is probably less than 0.88 Ga, as indicated by the instantaneous model.

However, all of the dates derivable from the suite of anomalous leads are strongly model dependent.

11.4d Pb–Pb Isochrons

A special case arises in the calculation of the $(^{207}Pb/^{206}Pb)^*$ ratio from equation 11.23 when t is set equal to zero. The resulting equation,

$$\left(\frac{^{207}Pb}{^{206}Pb}\right)^* = \frac{1}{137.88}\left(\frac{e^{\lambda_2 t_r} - 1}{e^{\lambda_1 t_r} - 1}\right) \qquad (11.24)$$

is identical to equation 10.18 in Section 10.11a. In other words, the straight line defined by a suite of anomalous galena leads is the Pb–Pb isochron of the rocks in which the radiogenic component formed by decay of ^{238}U and ^{235}U. The value of t_r obtained from equation 11.24 is the present age of these basement rocks.

In the hypothetical example under consideration, the slope of the anomalous Pb line ($R = 0.160$) is used to solve equation 11.24 by interpolation in Table 10.3:

$$t_r = 2.4 + \frac{0.2 \times 0.00518}{0.01954} = 2.45 \text{ Ga}$$

This date is older than the age of the basement rocks assumed in Section 11.4c ($t_r = 2.0$ Ga) because t_r is a maximum for any specified value of the $(^{207}Pb/^{206}Pb)^*$ ratio when $t = 0$ in equation 11.24.

The basic reason for this relationship is that the radiogenic $^{207}Pb/^{206}Pb$ ratio, which is generated by decay of the U isotopes in Table 10.3, has decreased with time and has reached its lowest value of 0.04604 at the present time ($t = 0$). Consequently, t_r must rise to a maximum when $t = 0$ in order to generate a particular value of the $(^{207}Pb/^{206}Pb)^*$ ratio. For the same reason, t_r decreases when t increase until ultimately t_r becomes equal to t. In that case, equation 11.23 converts to equation 11.21 for the instantaneous formation of the $(^{207}Pb/^{206}Pb)^*$ ratio.

11.4e Thorogenic Lead

Anomalous leads in ore deposits also generate straight lines in coordinates of $^{206}Pb/^{204}Pb$ (x) and $^{208}Pb/^{204}Pb$ (y). Such anomalous Pb lines may intersect the Pb growth curve of the Stacey–Kramers model at a point P' in Figure 11.6. The equation of the line $P'-Q$ is

$$\frac{(^{208}Pb/^{204}Pb)_{P'} - (^{208}Pb/^{204}Pb)_Q}{(^{206}Pb/^{204}Pb)_{P'} - (^{208}Pb/^{204}Pb)_Q} = \frac{^{232}Th}{^{238}U}\left(\frac{e^{\lambda_3 t_1} - e^{\lambda_3 t_2}}{e^{\lambda_1 t_1} - e^{\lambda_1 t_2}}\right) \qquad (11.25)$$

where P', Q = isotope ratios of Pb at these points in Figure 11.6

$t_1 = 3.70 \times 10^9$ years

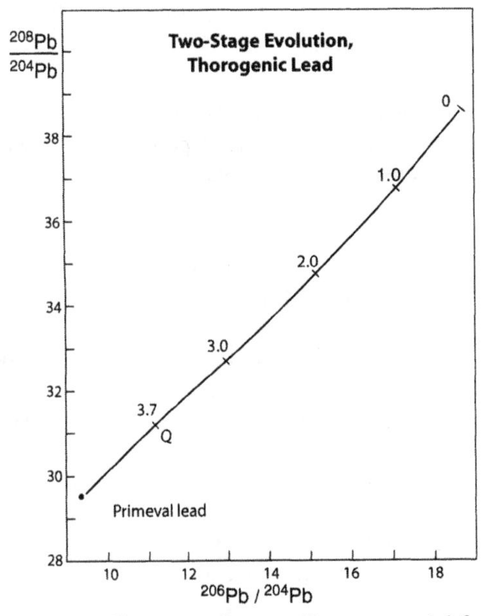

FIGURE 11.6 Two-stage Stacey–Kramers model for the evolution of thorogenic Pb. The point Q at 3.7 Ga marks the transition from the first stage to the second stage when the $^{232}Th/^{204}Pb$ ratio (ω) changed from 32.208 to 36.837 and the $^{238}U/^{204}Pb$ ratio (μ) changed from 7.192 to 9.735. Time in the past is expressed in billion years. Point P' is not shown.

t_2 = date that corresponds to any point P' on the Pb growth curve

$^{232}Th/^{238}U$ = present value of this ratio in the second stage of Stacey–Kramers model

λ_3 = decay constant of ^{232}Th (4.9475×10^{-11} y^{-1})

λ_1 = decay constant of ^{238}U (1.55125×10^{-10} y^{-1})

The Th/U ratio of the second stage of the Stacey–Kramers model has a value of 3.78, from which it follows that

$$\frac{^{232}\text{Th}}{^{238}\text{U}} = \frac{3.78 \times 238.0289 \times 100}{232.0381 \times 99.2743} = 3.90$$

The coordinates of a point P' in Figure 11.6 can be used to calculate a date t_2 from equation 11.25 based on the $^{232}Th/^{238}U$ ratio of the Pb growth curve. The result should be concordant with the date calculated from the $^{207}Pb/^{204}Pb$ and $^{206}Pb/^{204}Pb$ coordinates of point P in Figure 11.5.

The slope R' of the anomalous Pb line is the $^{208}Pb/^{206}Pb$ ratio of the radiogenic Pb that contaminated the two-stage Pb at a point P' in Figure 11.6. The equation for the instantaneously generated $(^{208}Pb/^{206}Pb)^*$ ratio is analogous to equation 11.21:

$$R' = \left(\frac{^{208}\text{Pb}}{^{206}\text{Pb}}\right)^* = \frac{\lambda_3}{\lambda_1}\left(\frac{^{232}\text{Th}}{^{238}\text{U}}\right)_p e^{t(\lambda_3-\lambda_1)} \tag{11.26}$$

where $(^{232}Th/^{238}U)_p$ is the present value of this ratio that can be determined by substituting the value of t derived by means of equation 11.21.

Similarly, the $(^{208}Pb/^{206}Pb)^*$ ratio can be generated by continuous decay during an interval of time as in equation 11.23:

$$\left(\frac{^{208}\text{Pb}}{^{206}\text{Pb}}\right)^* = \left(\frac{^{232}\text{Th}}{^{238}\text{U}}\right)_p \left(\frac{e^{\lambda_3 t_r} - e^{\lambda_3 t}}{e^{\lambda_1 t_r} - e^{\lambda_1 t}}\right) \tag{11.27}$$

This equation can be used to calculate the $(^{232}Th/^{238}U)_p$ ratio based on values of t_r and t determined from equation 11.23. Alternatively, the equation can be used to set limits on t_r and t by substituting the value of the $(^{232}Th/^{238}U)_p$ ratio obtained from equation 11.26. However, the uncertainties of the dates derived from anomalous Pb lines are compounded when such dates are used to calculate the Th/U ratios of the rocks that produced the component of radiogenic Pb.

11.4f Unresolved Issues

The two-stage model dates of common Pb whose isotope ratios fit the Stacey–Kramers Pb growth curve are in good agreement with dates determined by other methods. However, only a relatively small number of Pb deposits fit this model. In addition, the reason for the increase of the value of μ at 3.70 Ga remains unexplained. In reality, the evolution of the isotope composition of Pb in the crust of the Earth is controlled by geological processes that homogenize Pb having a wide range of isotopic compositions and thereby create the illusion that the Pb evolved in a closed system.

The straight lines defined by anomalous Pb may be the result of mixing of common Pb with varying amounts of radiogenic Pb. In cases where the anomalous Pb line intersects the Stacey–Kramers growth curve, the coordinates of the point of intersection define the isotope composition of the component of two-stage Pb and yield its model date. This date exceeds the age of the Pb deposit but may not, in all cases, correspond to a recognizable event in the geological history of the region. Likewise, the dates derivable from the slopes of anomalous Pb lines by either the instantaneous or the continuous model do not record specific geological events or processes and may be fictitious.

The decay of ^{232}Th to ^{208}Pb plays a significant role in the isotopic evolution of Pb in ore deposits and generates growth curves in coordinates of $^{206}Pb/^{204}Pb$ (x) and $^{208}Pb/^{204}Pb$ (y). The equation relating the decay products of ^{238}U and ^{232}Th can be used to calculate either the Th/U ratio of the Pb growth reservoir or the time of separation of the Pb from that reservoir, but not both. If the Pb fits the Stacey–Kramers model, the Th/U ratio of the reservoir is fixed by that model. If the Pb does not fit the U–Pb or Th–Pb growth curves, the reliability of the model date is questionable and so is the Th/U ratio calculated by use of that date.

Ore deposits containing Pb whose isotope composition is anomalous have been reported by many investigators:

Tilton and Steiger, 1969, *J. Geophys. Res.*, 74:2118–2132. Zartman and Stacey, 1971, *Econ. Geol.*, 66:849–860. Rye et al., 1974, *Econ. Geol.*, 69:814–822. Kuo and Folinsbee, 1974, *Econ. Geol.*, 69:806–813. Tugarinov et al., 1975, *Geokhimiya*, 8:1156–1163. Sato et al., 1981, *Geochem. J.*, 15:135–140. Godwin et al., 1972, *Econ. Geol.*, 77:82–94. Fletcher and Farquhar, 1982, *Econ. Geol.*, 77:464–473. Kish and Feiss, 1982, *Econ. Geol.*, 77:352–363. LeHuray, 1982, *Econ. Geol.*, 77:335–351. LeHuray, 1984, *Econ. Geol.*, 79:1561–1573. Sundblad and Stephens, 1983, *Econ. Geol.*, 78:1090–1107. Large et al., 1983, *Mineral. Deposita*, 18:235–244. Gulson et al., 1988, *Appl. Geochem.*, 3;243–254.

In addition, Zartman (1984) discussed the isotope composition of Pb in ore deposits of the Rocky Mountains of the United States.

11.5 LEAD–ZINC DEPOSITS, SOUTHEASTERN MISSOURI

The most celebrated example of anomalous Pb occurs in the Pb–Zn deposits of Missouri, Kansas, and Oklahoma, known as the Tri-State District (Section 10.12b). The literature on these and other Mississippi Valley Pb–Zn deposits cited by Goldhaber et al. (1995) includes contributions concerning the isotope compositions of Pb, S, and Sr of ore and gangue minerals:

Brown, 1967, *Econ. Geol. Monogr.*, 3:410–425. Richards et al., 1972, *Mineral. Deposita*, 7:285–291. Hart et al., 1981, *Econ Geol.*, 76:1873–1878. Kessen et al., 1981, *Econ. Geol.*, 76:913–920. Deloule et al., 1986, *Econ. Geol.*, 81:1307–1321. Crocetti et al., 1988, *Econ. Geol.*, 83:355–376. Crocetti and Holland, 1989, *Econ. Geol.*, 84:2196–2216. Goldhaber and Mosier, 1989, *U.S. Geol. Surv. Circular* 1043:8–9.

Other aspects concerning these deposits include:

Age of the Ore: Bickford et al., 1981, *Geol. Soc. Am. Bull.*, 92:323–341. Hearn and Sutter, 1985, *Science*, 228:1529–1531. Brannon et al., 1992, *Nature*, 356:509–511. Aleinikoff et al., 1993, *Geology*, 21:73–76.

Transport of Pb by Saline Groundwater: Sverjensky, 1984, *Econ. Geol.*, 79:23–37. Sverjensky, 1987, *Econ. Geol.*, 82:1130–1141. Sverjensky, 1989, *Geol. Assoc. Can. Spec. Paper* 36:127–134. Clendenin and Duane, 1990, *Geology*, 18:116–119. Bethke and Marshak, 1990, *Annu. Rev. Earth Sci.*, 18:287–315. Brannon et al., 1991, *Geochim. Cosmochim. Acta*,

55:1407–1419. Garven et al., 1993, *Am. J. Sci.*, 293:497–568.

Origin of the Ore Deposits: Sverjensky, 1981, *Econ. Geol.*, 76:1848–1872. Diehl et al., 1992, *U.S. Geol. Surv. Bull.*, 2012:A1–A13; Diehl and Goldhaber, 1993, *U.S. Geol. Surv. Bull.*, 1989:F1–F17.

The sulfide minerals of the Tri-State District consist primarily of sphalerite, galena, and chalcopyrite. These ore minerals and the associated pyrite and calcite were deposited by replacement and cavity filling in the marine carbonate rocks of Late Cambrian to Early Ordovician age. Information presented in Section 10.13b indicates that the ore and gangue minerals were deposited episodically between about 250 and 40 Ma from hot brines, remnants of which have been preserved in fluid inclusion in the calcite.

The ore deposits of southeastern Missouri occur in the dolomites of the Bonneterre Formation and in the underlying sandstones of the Lamotte Formation, both of which are Early Cambrian in age and were deposited on Precambrian basement rocks that crop out in the St. Francois Mountains in the center of the mining district. The ore deposits of southeastern Missouri have been subdivided into the Old Lead Belt, the Viburnum Trend, the Indian Creek, and several smaller districts (Goldhaber et al., 1995).

11.5a Lead in the Ore Minerals

The isotope compositions of Pb in the mines of southeastern Missouri were first determined by Nier et al. (1941) and subsequently by Doe and Delevaux (1972), Richards et al. (1972), Heyl et al. (1974), and Sverjensky et al. (1979). A comprehensive study by Goldhaber et al. (1995) provided nearly 150 new isotope analyses of Pb in the ore minerals and pyrite from the principal mines of this region as well as of Pb in the host rocks.

The isotope ratios of galena and chalcopyrite in the mines of the Viburnum Trend define an elongated area in Figure 11.7 that is displaced from the zero-age isochron and the growth curve of Pb in the Stacey–Kramers model. Therefore, the Pb in the galena and chalcopyrite of the Viburnum Trend contains excess radiogenic Pb. In fact, the anomalous Pb here and elsewhere in the world is referred to as J-type lead after the town of Joplin, Missouri.

The 36 isotope ratios of Pb in galena and chalcopyrite of the Viburnum Trend in Figure 11.7 define a poorly constrained anomalous-Pb line with a correlation coefficient of 0.586:

$$\frac{^{207}\text{Pb}}{^{204}\text{Pb}} = 0.17505\left(\frac{^{206}\text{Pb}}{^{204}\text{Pb}}\right) + 12.178$$

This line intersects the growth curve of the Stacey–Kramers model in Figure 11.8 at point P, whose coordinates are $^{206}\text{Pb}/^{204}\text{Pb} = 13.10$ and $^{207}\text{Pb}/^{204}\text{Pb} = 14.45$. The slope m of the Pb–Pb isochron $(P-Q)$ is

$$m = \frac{14.45 - 12.998}{13.10 - 11.152} = 0.74537$$

Interpolation in Table 11.5 yields

$$t = 2.8 + \frac{0.2 \times 0.04055}{0.05148} = 2.96 \text{ Ga}$$

This is the date when the two-stage Pb was withdrawn from the second-stage reservoir of the Stacey–Kramers model and was subsequently mixed with varying amounts of radiogenic Pb.

The age of the component of radiogenic Pb is derivable from the slope of the anomalous Pb line $(R = 0.17505)$ by the "instantaneous" and "continuous" models. The instantaneous model represented by equation 11.22,

$$t = \frac{1}{\lambda_2 - \lambda_1} \ln\left(\frac{137.88R\lambda_1}{\lambda_2}\right)$$

yields a date of

$$t = \frac{\ln 3.80167}{8.29725 \times 10^{-10}} = 1.61 \text{ Ga}$$

This date is the time in the past when the $^{207}\text{Pb}/^{206}\text{Pb}$ ratio of the radiogenic Pb was equal to 0.17505.

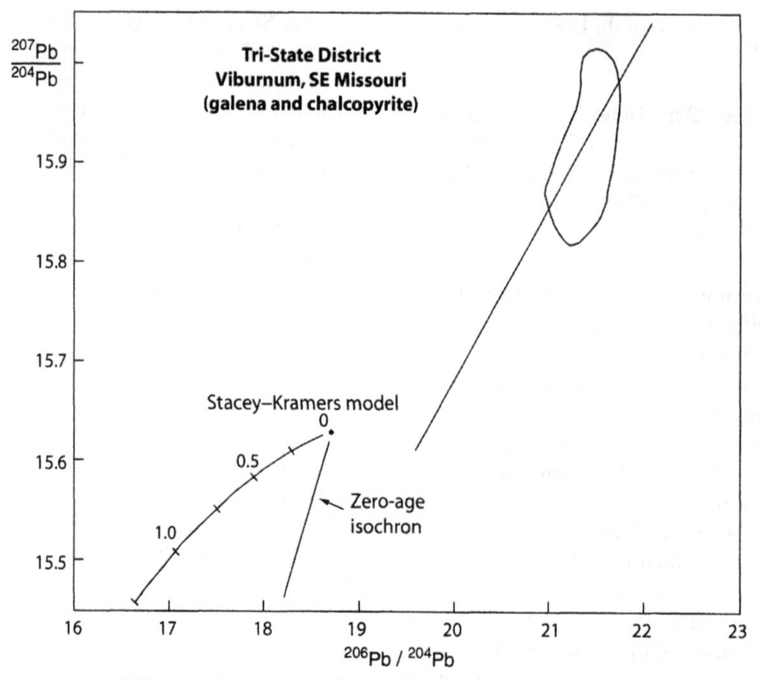

FIGURE 11.7 Isotope ratios of Pb in galena and chalcopyrite in the ore bodies of the Viburnum Trend in southeastern Missouri. These anomalous leads define a straight line whose slope (0.17505) was determined by least-squares regression of 36 data sets. The Pb growth curve represents the two-stage model of Stacey and Kramers (1975). Data from Goldhaber et al. (1995).

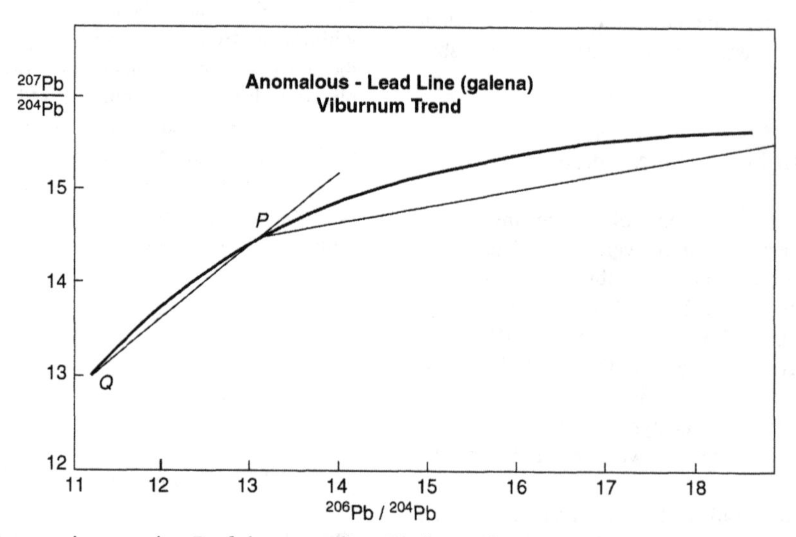

FIGURE 11.8 Intersection at point P of the anomalous Pb line defined by galena and chalcopyrite of the Viburnum Trend in Figure 11.7 with the growth curve of the Stacey–Kramers model. Data from Stacey and Kramers (1975) and Goldhaber et al. (1995).

The slope of the anomalous Pb line in Figure 11.7 can also be interpreted in accordance with the continuous model expressed by equation 11.23:

$$R = \frac{1}{137.88}\left(\frac{e^{\lambda_2 t_r} - e^{\lambda_2 t}}{e^{\lambda_1 t_r} - e^{\lambda_1 t}}\right)$$

where t_r is the age of the source rocks in which the radiogenic Pb accumulated and t is the date when the radiogenic Pb was separated from U and Th in these source rocks. According to independent evidence presented in Section 10.13b, the deposition of calcite gangue in the Bonneterre Formation started at 0.25 Ga, which therefore provides a lower limit for the value of t.

The date t_r, when the component of radiogenic Pb began to accumulate, was determined from equation 11.23 by interpolation assuming that $t = 0.25 \times 10^9$ y and $R = 0.17505$:

$$t_r = 2.4 + \frac{0.2 \times 0.01026}{0.02022} = 2.50 \text{ Ga}$$

This date is an overestimate of the age of the reservoir in which the radiogenic Pb evolved because t in equation 11.23 is probably greater than 0.25×10^9 years.

A literal interpretation of these dates implies that the two-stage Pb in the mines of the Viburnum Trend stopped evolving isotopically in Late Archean time (2.96 Ga) and that it was mixed more than 1350 million years later with radiogenic Pb that formed by decay of U isotopes at 1.61 Ga (instantaneous model). Alternatively, the radiogenic component evolved continuously from 2.50 to 0.25 Ga and then mixed with the two-stage Pb just prior to or during the deposition of galena in the Bonneterre Formation.

The dates derived above gain validity in case the two-stage Pb resided in K-feldspar in the Precambrian basement rocks of the midcontinent region. These rocks also contain zircon and other minerals within which the radiogenic Pb evolved by decay of the isotopes of U and Th. The two components of Pb were ultimately leached from the basement rocks by warm brines that transported the Pb into the overlying Paleozoic carbonate rocks and sandstones and mixed them isotopically.

11.5b Lead in Pyrite

Trace amounts of pyrite occur widely in the Bonneterre and Lamotte Formations of southeastern Missouri (Goldhaber et al., 1995). The isotope ratios of Pb in the pyrite of the Bonneterre Formation also define an anomalous Pb line whose equation is

$$\frac{^{207}\text{Pb}}{^{204}\text{Pb}} = 0.09343\left(\frac{^{206}\text{Pb}}{^{204}\text{Pb}}\right) + 13.903$$

with a correlation coefficient of 0.906 for 25 sets of data. Pyrite in the Lamotte Formation defines a similar anomalous Pb line:

$$\frac{^{207}\text{Pb}}{^{204}\text{Pb}} = 0.071100\left(\frac{^{206}\text{Pb}}{^{204}\text{Pb}}\right) + 14.313$$

which has a correlation coefficient of 0.973 for 21 data pairs.

The anomalous Pb line of disseminated pyrite in the Bonneterre Formation intersects the Stacey–Kramers growth line in Figure 11.9 at point P, whose approximate coordinates are $^{206}\text{Pb}/^{204}\text{Pb} = 18.150$ and $^{207}\text{Pb}/^{204}\text{Pb} = 15.598$. The slope m of the Pb–Pb isochron ($P-Q$) and the corresponding model date t are

$$m = \frac{15.598 - 12.998}{18.150 - 11.152} = 0.37153$$

$$t = 0.2 + \frac{0.2 \times 0.0105}{0.01379} = 0.35 \text{ Ga}$$

This result indicates that the two-stage Pb of pyrite in the Bonneterre Formation was removed from the Stacey–Kramers reservoir much more recently than the two-stage Pb in the galena of the Viburnum Trend in Figure 11.7.

The age of the radiogenic component is derivable from the slope R of the anomalous Pb line by means of equation 11.22 (instantaneous model):

$$t = \frac{1}{\lambda_2 - \lambda_1}\ln\left(\frac{137.88 R \lambda_1}{\lambda_2}\right)$$

$$= \frac{\ln(2.02908)}{8.29725 \times 10^{-10}} = 0.85 \text{ Ga}$$

Evidently, the Pb in the pyrite of the Bonneterre Formation is a mixture of two-stage Pb (which

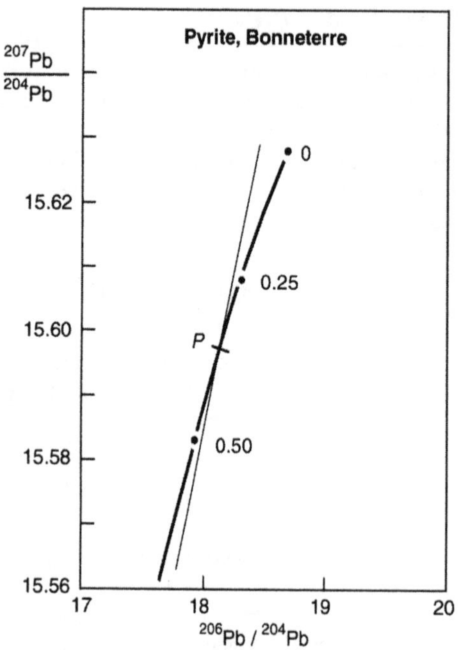

FIGURE 11.9 Anomalous Pb line of disseminated pyrite in the Bonneterre Formation (Early Cambrian) in southeastern Missouri. The Pb growth curve is based on the Stacey–Kramers model and the dates marked along it are in units of billion years. The coordinates of point *P* specify the isotope ratios of the two-stage Pb component. Data from Stacey and Kramers (1975) and Goldhaber et al.(1995).

evolved until its separation from the Stacey–Kramers reservoir during the Paleozoic Era) with radiogenic Pb that was generated in the Late Proterozoic Era.

The anomalous Pb line of pyrites in the Lamotte Formation does not intersect the Stacey–Kramers growth curve but plots above it. The two-stage Pb of pyrite in this formation therefore appears to have evolved in rocks having a μ-value greater than that of the Stacey–Kramers model.

11.5c Synthesis

The ore deposits in southeastern Missouri and elsewhere in the Tri-State District of North America were deposited by warm brines that flowed in aquifers within the Paleozoic carbonate rocks and sandstones of the Mississippi Valley region between about 250 and 40 Ma. The Pb in these ore deposits originated from the Precambrian basement rocks and from the Lamotte Sandstone, which not only was a source of Pb but also acted as a conduit that channeled the flow of the brines (Goldhaber et al., 1995). The two-stage Pb resided primarily in K-feldspar of the local Precambrian basement rocks (Section 11.3), whereas the radiogenic component originated in U-rich accessory minerals and occurred in grain boundaries and in microfractures in minerals of these basement rocks. Davis and Krogh (2000) recently demonstrated experimentally that damage to the lattice of zircon crystals by α-particle recoil promotes dissolution of ^{234}U (a daughter of ^{238}U) and of radiogenic ^{206}Pb. Therefore, mixing of the two-stage Pb and the radiogenic Pb could have occurred during leaching from the source rocks and during subsequent transport of the Pb by the brines.

The deposition of the metals Pb, Zn, and Cu as sulfide minerals required the presence of hydrogen sulfide and the availability of open cavities in the host rocks. The hydrogen sulfide (H_2S) could have been generated by the reduction of sulfate ions by bacteria (*Desulfovibrio*) feeding on petroleum in the host rocks. Bacterial reduction of sulfate enriches the resulting H_2S in ^{32}S and depletes it in ^{34}S relative to the isotope composition of S in the sulfate (Chapter 29). Goldhaber et al. (1995) reported that the $\delta^{34}S$ values of galena and chalcopyrite in the ore of the Viburnum Trend range from +1.40 to +3.2‰. These values imply strong enrichment in ^{32}S compared to $\delta^{34}S \sim +30‰$ in marine sulfate of Cambrian age. Therefore, the isotope composition of S in the ore is consistent with a bacteriogenic origin of the sulfide.

The process outlined above was activated by the movement of saline groundwater on a regional scale in response to hydraulic gradients and geothermal heat. Such conditions have occurred in many places in the geological past, which accounts for the economic importance and wide distribution of the so-called Mississippi Valley Pb–Zn deposits.

11.6 MULTISTAGE LEADS

The enrichment of Pb in radiogenic isotopes may occur not only by mixing but also by multistage isotope evolution caused by the Pb passing through two or more stages characterized by different values of μ. For example, the present isotope composition of a three-stage Pb is governed by equations of the form

$$\left(\frac{^{206}\text{Pb}}{^{204}\text{Pb}}\right)_{t=0} = \left(\frac{^{206}\text{Pb}}{^{204}\text{Pb}}\right)_p + \mu_1(e^{\lambda_1 T} - e^{\lambda_1 t_1}) + \mu_2(e^{\lambda_1 t_1} - e^{\lambda_1 t_2}) + \mu_3(e^{\lambda_1 t_2} - 1) \quad (11.28)$$

where T = age of the Earth

t_1 and t_2 = age of geological events that cause μ_1 to change to μ_2 and μ_2 to μ_3, respectively

$(^{206}\text{Pb}/^{204}\text{Pb})_p$ = primeval value of this ratio

$(^{206}\text{Pb}/^{206}\text{Pb})_{t=0}$ = present value of this ratio

Similar equations can be formulated for the ^{207}Pb/^{204}Pb and ^{208}Pb/^{204}Pb ratios. The number of stages and their μ-values may be varied without limit. However, the resulting equations cannot be solved for $t_1, t_2, \ldots$ or $\mu_1, \mu_2, \ldots$. The only known parameters are the age of the Earth (T), the isotope ratios of primeval Pb (Section 11.1c), and the measured isotope ratios of Pb and the μ-value of the rocks in which the Pb resides at the present time. Equation 11.28 can be simplified by setting

$$\mu_1(e^{\lambda_1 T} - e^{\lambda_1 t_1}) = \left(\frac{^{206}\text{Pb}}{^{204}\text{Pb}}\right)_1$$

$$\mu_2(e^{\lambda_1 t_1} - e^{\lambda_1 t_2}) = \left(\frac{^{206}\text{Pb}}{^{204}\text{Pb}}\right)_2$$

Hence,

$$\left(\frac{^{206}\text{Pb}}{^{204}\text{Pb}}\right)_{t=0} = \left(\frac{^{206}\text{Pb}}{^{204}\text{Pb}}\right)_p + \left(\frac{^{206}\text{Pb}}{^{204}\text{Pb}}\right)_1 + \left(\frac{^{206}\text{Pb}}{^{204}\text{Pb}}\right)_2 + \mu_3(e^{\lambda_1 t_2} - 1)$$

The sum of the ^{206}Pb/^{204}Pb ratios is the initial ratio of the Pb that was incorporated into the rocks in which it presently resides. Therefore,

$$\left(\frac{^{206}\text{Pb}}{^{204}\text{Pb}}\right)_i = \left(\frac{^{206}\text{Pb}}{^{204}\text{Pb}}\right)_p + \left(\frac{^{206}\text{Pb}}{^{204}\text{Pb}}\right)_1 + \left(\frac{^{206}\text{Pb}}{^{204}\text{Pb}}\right)_2 + \cdots$$

and equation 11.28 becomes

$$\left(\frac{^{206}\text{Pb}}{^{204}\text{Pb}}\right)_{t=0} = \left(\frac{^{206}\text{Pb}}{^{204}\text{Pb}}\right)_i + \mu(e^{\lambda_1 t} - 1) \quad (11.29)$$

where the subscripts of μ and t have been omitted because they are no longer relevant. In cases where μ has a range of values but $(^{206}\text{Pb}/^{204}\text{Pb})_i$ and t have the same values for all specimens in a suite of igneous or metamorphic rocks, the measured ^{206}Pb/^{204}Pb ratios have a range of values determined by equation 11.29.

A similar development applies to the measured ^{207}Pb/^{204}Pb ratios of a suite of rocks:

$$\left(\frac{^{207}\text{Pb}}{^{204}\text{Pb}}\right)_{t=0} = \left(\frac{^{207}\text{Pb}}{^{204}\text{Pb}}\right)_i + \frac{\mu}{137.88}(e^{\lambda_2 t} - 1) \quad (11.30)$$

Equations 11.29 and 11.30 can be combined to yield the ^{207}Pb/^{206}Pb ratio of the radiogenic Pb that has accumulated in the rocks since the time of their formation t years ago:

$$\left(\frac{^{207}\text{Pb}}{^{206}\text{Pb}}\right)^* = \frac{(^{207}\text{Pb}/^{204}\text{Pb})_{t=0} - (^{207}\text{Pb}/^{204}\text{Pb})_i}{(^{206}\text{Pb}/^{204}\text{Pb})_{t=0} - (^{206}\text{Pb}/^{204}\text{Pb})_i} = \frac{1}{137.88}\left(\frac{e^{\lambda_2 t} - 1}{e^{\lambda_1 t} - 1}\right) \quad (11.31)$$

Equation 11.31 represents a family of straight lines all of which pass through a common point whose coordinates are $(^{206}\text{Pb}/^{204}\text{Pb})_i$ and $(^{207}\text{Pb}/^{204}\text{Pb})_i$ and whose slopes are the radiogenic ^{207}Pb/^{206}Pb ratios.

Equation 11.31 is identical to equation 10.17 for Pb–Pb isochrons of U-bearing igneous and metamorphic rocks and sedimentary carbonate rocks. The beauty of these equations arises from the fact that Pb–Pb isochron dates are based entirely on the measured isotope composition of Pb and do not require knowledge of the concentrations of U and Pb and its prior history extending back in time to primeval Pb.

The point to be made here is that the isotope compositions of multistage leads can vary widely and therefore deviate from the narrowly defined single- or two-stage Pb growth models. In addition, multistage leads may contain excess radiogenic Pb relative to the Stacey–Kramers model and thus could be classified as being anomalous. In other words, not all anomalous leads formed by mixing of two components because some may have acquired their isotope compositions as a result of multistage histories.

11.7 SUMMARY

The isotopic composition of common Pb in minerals that contain no U or Th can be accounted for by evolutionary models of varying complexity. In the simplest case, represented by the single-stage Holmes–Houtermans model, Pb evolved in a homogeneous reservoir having characteristic U/Pb and Th/Pb ratios until it was withdrawn from that reservoir and was sequestered in a Pb mineral such as galena. The equations describing Pb evolution in this model require knowledge of the isotope ratios of primeval Pb and of the age of the Earth, both of which were determined from analyses of Pb in stony and iron meteorites.

The single-stage model was found to apply only to a small number of conformable ore deposits associated with sequences of volcanic and sedimentary rocks deposited near island arcs and subduction zones. However, even for conformable ore deposits the dates calculated from the single-stage model do not agree well with the geological ages of the associated rocks. Moreover, the Pb in most other metallic ore deposits yields single-stage model dates that are highly discrepant and, in some cases, even negative (i.e., associated with the future rather than with the past).

The two-stage model of Stacey and Kramers is based on a selected data set and proposes that ore leads evolved in two reservoirs having different U/Pb and Th/Pb ratios. The change in these ratios took place at 3.7 Ga and was caused by the geochemical differentiation of the Earth. The mantle of the Earth may have been the first reservoir but, after 3.7 Ga, Pb evolution occurred in frequently mixed continental sediment from which Pb was extracted during subduction into the mantle. The two-stage model yields dates for Pb in galena and in K-feldspar that agree much better with their geological ages than dates derived from the single-stage model.

The isotope ratios of so-called anomalous leads that do not fit the single- or two-stage model form linear arrays in the Pb evolution diagram. Such anomalous-Pb lines may be caused by the addition of varying amounts of radiogenic Pb to ordinary single- or two-stage Pb. If that is true, the slope of the anomalous Pb line can be used to derive information about the age and Th/U ratio of the source rocks that provided the radiogenic lead. In addition, the time of withdrawal of the single- or two-stage Pb from its reservoir can be estimated. However, anomalous-Pb lines may also result from mixing of two common leads of differing isotope compositions. In this case, the slope of an anomalous-Pb line does not convey information about the age of such Pb.

All of the dates that are derivable from the isotope composition of common Pb are model dependent and therefore are not as accurate or as precise as U,Th–Pb dates derivable from zircon and other U,Th-bearing minerals.

The Pb in galena and other sulfide minerals in Mississippi Valley–type Pb–Zn deposits originated from the granitic gneisses of the Precambrian basement rocks that underlie the carbonate host rocks. The so-called two-stage Pb resided in K-feldspar of the basement rocks, which crystallized at the date indicated by the point of intersection of the anomalous Pb line with the two-stage growth curve. That date also records the start of the accumulation of radiogenic Pb in U-bearing (and Th-bearing) minerals of the same basement rocks (e.g., zircon, monazite).

Eventually, both types of Pb were leached from the basement rocks (or their clastic derivatives) by geothermally heated groundwater (warm brines) whose movement was energized by hydraulic gradients and density differences. The two isotopic components of Pb were mixed in varying proportions during leaching and transport until the Pb and the associated metals (Zn and Cu) were deposited in the overlying sedimentary host rocks by cavity filling and replacement.

REFERENCES FOR THE HOLMES–HOUTERMANS MODEL (INTRODUCTION, SECTIONS 11.1– 11.1a)

Aston, F. W., 1927. The constitution of ordinary lead. *Nature*, 120:224.

Doe, B. R., 1970. *Lead Isotopes*. Springer-Verlag, Heidelberg.

Gerling, E. K., 1942. Age of the Earth according to radioactivity data. *Compt. Rend. (Doklady) Acad. Sci. USSR. 34*, 9:259–261.

Hamilton, E. I., and R. M. Farquhar, 1968. *Radiometric Dating for Geologists*. Interscience, London.

Holmes, A., 1946. An estimate of the age of the earth. *Nature*, 157:680–684.

Houtermans, F. G., 1946. Die Isotopenhäufigkeiten im natürlichen Blei und das Alter des Urans. *Naturwissenschaften*, 33:185–186, 291.

Moorbath, S., 1962. Lead isotope abundance studies on mineral occurrences in the British Isles and their geologic significance. *Philos. Trans. Roy. Soc.*, 254:295–360.

Nier, A. O., 1938. Variations in the relative abundances of the isotopes of common lead from various sources. *J. Am. Chem. Soc.*, 60:1571–1576.

Nier, A. O., R. W. Thompson, and B. F. Murphey, 1941. The isotopic constitution of lead and the measurement of geologic time. *III Phys. Rev.*, 60:112–116.

Rankama, K., 1954. *Isotope Geology*. Pergamon, London.

Rankama, K., 1963. *Progress in Isotope Geology*. Interscience, London.

Russell, R. D., and R. M. Farquhar, 1960. *Lead Isotopes in Geology*. Interscience, New York.

Steiger, R. H., and E. Jäger, 1977. Subcommission on geochronology: Convention on the use of decay constants in geo- and cosmochronology. *Earth Planet. Sci. Lett.*, 36:359–362.

REFERENCES FOR ANALYTICAL METHODS (SECTION 11.1c)

Cannon, R. S., Jr., A. P. Pierce, and M. H. Delevaux, 1963. Lead isotope variation with growth zoning in a galena crystal. *Science*, 142:574–576.

Catanzaro, E. J., 1967. Triple-filament method for solid-sample lead isotope analysis. *J. Geophys. Res.*, 72(4):1325–1327.

Compston, W., and V. M. Oversby, 1969. Lead isotope analyses using a double spike. *J. Geophys. Res.*, 74:4338–4348.

Cumming, G. L., 1973. Propagation of experimental errors in lead isotope-ratio measurements using a double spike method. *Chem. Geol.*, 11:157–165.

Dallwitz, M. J., 1970. Fractionation correction in lead isotopic analysis. *Chem. Geol.*, 6:311–314.

Dodson, M. H., 1970. Simplified equations for double-spike isotopic analyses. *Geochim. Cosmochim. Acta*, 34:1241–1244.

Gale, N. H., 1970. A solution in closed form for lead isotopic analysis using a double spike. *Chem. Geol.*, 6:305–310.

Hart, S. R., N. Shimizu, and D. A. Sverjensky, 1981. Lead isotope zoning in galena: An ion microprobe study of a galena crystal from the Buick Mine, southeast Missouri. *Econ. Geol.*, 76:1873–1878.

Hofmann, A. W., 1971. Fractionation corrections for mixed-isotope spikes of Sr, K, and P. *Earth Planet. Sci. Lett.*, 10:397–402.

Ozard, J. M., and R. D. Russell, 1970. Discrimination in solid-source lead isotope abundance measurement. *Earth Planet. Sci. Lett.*, 8:331–336.

Richards, J. R., 1981. Some thoughts on the time-dependence of lead isotope ratios. *Geochem. Int.*, 1981:17–36.

Russell, R. D., 1971. The systematics of double spiking. *J. Geophys. Res.*, 76:4949–4955.

Russell, R. D., 1975. Mass discrimination in the measurement of lead isotope reference samples. *Geochem. J.* 9:47–50.

REFERENCES FOR PRIMEVAL Pb IN METEORITES AND THE EARTH (SECTIONS 11.1D– 11.1e)

Allègre, C. J., G. Manhès, and C. Göpel, 1995. The age of the Earth. *Geochim. Cosmochim. Acta*, 59:1445–1456.

Chen, J. H., and G. J. Wasserburg, 1983. The least radiogenic Pb in iron meteorites. Paper presented at the Fourteenth Lunar and Planetary Science Conference, Abstracts, Part 1, Lunar and Planetary Institute, Houston, Texas, pp. 103–104.

Goles, G. G., and E. Anders, 1962. Abundance of iodine, tellurium, and uranium in meteorites. *Geochim. Cosmochim. Acta*, 26:723.

Hanan, B. B., and G. R. Tilton, 1985. Early planetary metamorphism in chondritic meteorites. *Earth Planet. Sci. Lett.*, 74:209–219.

Minster, J. F., and C. J. Allègre, 1979. ^{87}Rb–^{87}Sr dating of L chondrites: Effects of shock and brecciation. *Meteoritics*, 14:235.

Patterson, C. C., 1956. Age of meteorites and the earth. *Geochim. Cosmochim. Acta*, 10:230–237.

Rogers, J. J. W., and J. A. S. Adams, 1969a. Abundance of uranium in meteorites and tektites. In K. H. Wedepohl (Ed.), *Handbook of Geochemistry*, pp. 92C-1–92C-4. Springer-Verlag, Heidelberg.

Rogers, J. J. W., and J. A. S. Adams, 1969b. Abundance of thorium in meteorites, tektites, and lunar materials. In K. H. Wedepohl (Ed.), *Handbook of Geochemistry*, pp. 90C-1–90C-6. Springer-Verlag, Heidelberg.

Steiger, R. H., and E. Jäger, 1977. Subcommission on Geochronology: Convention on the use of decay constants in geo- and cosmochronology. *Earth Planet. Sci. Lett.*, 36:359–362.

Tatsumoto, M., R. J. Knight, and C. J. Allègre, 1973. Time differences in the formation of meteorites as determined from the ratio of lead-207 to lead-206. *Science*, 180:1279–1283.

Tilton, G. R., 1973. Isotopic lead ages of chondritic meteorites. *Earth Planet. Sci. Lett.*, 19:321–329.

Wedepohl, K. H., 1974. Abundance (of lead) in the cosmos, meteorites, tektites, and lunar samples. In K. H. Wedepohl, ed. *Handbook of Geochemistry*, II-5: p. 82C-O. Springer-Verlag, Heidelberg.

REFERENCES FOR DATING COMMON LEAD (SECTION 11.2)

Armstrong R. L., 1968. A model for the evolution of strontium and lead in a dynamic earth. *Rev. Geophys.*, 6:175–199.

Doe, B. R., 1962a. Distribution and composition of sulfide minerals at Balmat, New York. *Geol. Soc. Am. Bull.*, 73:833–854.

Doe, B. R., 1962b. Relationship of lead isotopes among granites, pegmatites, and sulfide ores near Balmat, New York, *J. Geophys. Res.*, 67:2895–2906.

Doe, B. R., and J. S. Stacey, 1974. The application of lead isotopes to the problems of ore genesis and ore prospect evaluation: A review. *Econ. Geol.*, 69:757–776.

Fairbairn, H. W., P. M. Hurley, K. D. Card, and C. J. Knight, 1969. Correlation of radiometric ages of Nipissing diabase and Huronian metasediments with Proterozoic orogenic events in Ontario. *Can. J. Earth Sci.*, 6:489–497.

Hriskevich, M. E., 1968. Petrology of the Nipissing diabase sill of the Cobalt area, Ontario, Canada. *Geol. Soc. Am. Bull.*, 79:1387–1404.

Kanasewich, E. R., and R. M. Farquhar, 1965. Lead isotope ratios from the Cobalt–Noranda area, Canada. *Can. J. Earth Sci.*, 2:361–384.

Oversby, V. M., 1974. A new look at the lead isotope growth curve. *Nature*, 248:132.

Patterson, C. C., 1956. Age of meteorites and the earth. *Geochim. Cosmochim. Acta*, 10:230–237.

Richards, J. R., 1971. Major lead orebodies—Mantle origin? *Econ. Geol.*, 66:425–434.

Stacey, J. S., and J. D. Kramers, 1975. Approximation of terrestrial lead isotope evolution by a two-stage model. *Earth Planet. Sci. Lett.*, 26:207–221.

Stanton, R. L., and R. D. Russell, 1959. Anomalous leads and the emplacement of lead sulfide ores. *Econ. Geol.*, 54:588–607.

Steiger, R. H. and E. Jäger, 1977. Subcommission on geochronology: Convention on the use of decay constants in geo- and cosmochronology. *Earth Planet. Sci. Lett.*, 36:359–362.

Tatsumoto, M., R. J. Knight, and C. J. Allègre, 1973. Time differences in the formation of meteorites as determined from the ratio of lead-207 to lead-206. *Science*, 180:1279–1283.

Thorpe, R., 1974. Lead isotope evidence on the genesis of silver-arsenide vein deposits of the Cobalt and Great Bear Lake areas, Canada. *Econ. Geol.*, 69:777–791.

Van Schmus, R., 1969. The geochronology of the Blind River-Bruce Mines area, Ontario, Canada. *J. Geol.*, 73:755–780.

REFERENCES FOR DATING K-FELDSPAR (SECTION 11.3)

Catanzaro, E. J., and R. W. Gast, 1960. Isotopic composition of lead in pegmatitic feldspar. *Geochim. Cosmochim. Acta*, 19:113–126.

Doe, B. R., G. R. Tilton, and C. A. Hopson, 1965. Lead isotopes in feldspars from selected granitic rocks associated with regional metamorphism. *J. Geophys. Res.*, 70:1947–1968.

Sinha, A. K., and G. R. Tilton, 1973. Isotopic evolution of common lead. *Geochim. Cosmochim. Acta*, 37:1823–1849.

Stacey, J. S., and J. D. Kramers, 1975. Approximation of terrestrial lead isotope evolution by a two-stage model. *Earth Planet. Sci. Lett.*, 26:207–221.

Wetherill, G. W., G. R. Tilton, G. L. Davis, S. R. Hart, and C. A. Hopson, 1966. Age measurements in the Maryland Piedmont. *J. Geophys. Res.*, 71:2139–2155.

REFERENCES FOR ANOMALOUS PB IN ORE DEPOSITS (SECTION 11.4)

Doe, B. R., 1970. *Lead Isotopes*. Springer-Verlag, Berlin.

Farquhar, R. M., and R. D. Russell, 1957. Anomalous leads from the upper Great Lakes region of Ontario. *Trans. Am. Geophys. Union*, 38:552–556.

Kanasewich, E. R., 1968. The interpretation of lead isotopes and their geological significance. In E. I. Hamilton and R. M. Farquhar (Eds.), *Radiometric Dating for Geologists*, pp. 147–223. Wiley-Interscience, New York.

Russell, R. D., and R. M. Farquhar, 1960. *Lead Isotopes in Geology*. Wiley-Interscience, New York.

Stanton, R. L., and R. D. Russell, 1959. Anomalous leads and the emplacement of lead sulfide ores. *Econ. Geol.*, 54:588–607.

Zartman, R. E., 1974. Lead isotopic provinces in the Cordillera of the western United States and their geologic significance. *Econ. Geol.*, 69:792–805.

REFERENCES FOR LEAD–ZINC DEPOSITS, SOUTHEASTERN MISSOURI (SECTION 11.5)

Davis, D. W., and T. E. Krogh, 2000. Preferential dissolution of ^{234}U and radiogenic Pb from α-recoil-damaged lattice sites in zircon: Implications for thermal histories and Pb isotopic fractionation in the near surface environment. *Chem. Geol.*, 172:41–58.

Doe, B. R., and M. H. Delevaux, 1972. Source of lead in southeast Missouri galena ores. *Econ. Geol.*, 67:409–435.

Goldhaber, M. B., S. E. Church, B. R. Doe, J. N. Aleinikoff, J. C. Brannon, F. A. Podosek, E. L. Mosier, C. D. Taylor, and C. A. Gent, 1995. Lead and sulfur isotope investigation of Paleozoic sedimentary rocks from the southern midcontinent of the United States: Implications for paleohydrology and ore genesis of the southeast Missouri lead belt. *Econ. Geol.*, 90:1875–1910.

Heyl, A. V., G. P. Landis, and R. D. Zartman, 1974. Isotopic evidence of the origin of Mississippi-Valley type mineral deposits: A review. *Econ. Geol.*, 69:992–1006.

Nier, A. O., R. W. Thompson, and B. F. Murphy, 1941. The isotopic constitution of lead and the measurement of geologic time. III. *Phys. Rev.*, 60:112–116.

Richards, J. R., A. K. Yonk, and C. W. Keighn, 1972. Upper Mississippi Valley lead isotopes reexamined. *Mineral. Deposita*, 7:285–291.

Stacey, J. S. and J. D. Kramers, 1975. Approximation of terrestrial lead isotope evolution by a two-stage model. *Earth Planet. Sci. Lett.*, 26:207–221.

Sverjensky, D. A., D. M. Rye, and B. R. Doe, 1979. The lead and sulfur isotopic composition of galena from a Mississippi Valley-type deposit in the New Lead Belt, southeastern Missouri. *Econ. Geol.*, 74:149–153.

CHAPTER 12

The Lu–Hf Method

LUTETIUM (Lu) is the last of the rare-earth elements (REEs). Like all REEs, Lu is widely distributed and occurs primarily in phosphate, silicate, and oxide minerals. Lutetium has become an important element in geochronometry because one of its naturally occurring isotopes (^{176}Lu) is unstable and decays by β-emission to stable ^{176}Hf. This phenomenon is the basis for the Lu–Hf method of dating and has also made Hf a useful isotopic tracer for the study of the origin of igneous rocks. The isotope geochemistry of Hf in the oceans and on the continents is presented in Section 19.7.

12.1 GEOCHEMISTRY OF Lu AND Hf

Lutetium ($Z = 71$) and hafnium ($Z = 72$) are members of the sixth period in the periodic table of the elements and have similar electron configurations specified in Table 12.1. Their valences (+3 for Lu and +4 for Hf) also affect their ionic radii (0.93 Å for Lu and 0.81 Å for Hf). The ionic radius of Lu^{3+} is similar to that of Ca^{2+} (0.99 Å), which causes Lu^{3+} to be captured by crystals in place of Ca^{2+}. Although Lu has the highest atomic number among the REEs ($Z = 71$), its ionic radius (0.93 Å) is smaller than that of the other REEs. For example, the ionic radius of Lu^{3+} is nearly 20% smaller than that of La^{3+} ($Z = 57$, $r = 1.15$ Å).

The ionic radius of Hf^{4+} (0.81 Å) is virtually identical to that of Zr^{4+} (0.80 Å), meaning that Hf is camouflaged in Zr-bearing minerals (e.g., baddeleyite, zircon, and zirkelite). Therefore, both Lu and Hf are dispersed elements that do not form their own minerals in most geological environments.

The geochemical properties of Lu and Hf are similar to those of Sm and Nd because Hf is concentrated relative to Lu in silicate liquids formed by partial melting in the mantle. Therefore, basaltic magmas derived from the mantle generally have *lower* Lu/Hf ratios than their source rocks. The residual solids that remain behind after extraction of the magma are correspondingly depleted in Hf and acquire higher Lu/Hf ratios than the rocks in the mantle before melting. Garnet, which has a strong affinity for Lu, plays an important role in this process because it inhibits Lu from entering the melt, thereby increasing the Lu/Hf ratios of the residual solids. Patchett et al. (1981a) concluded that during partial melting (5–50%) of an ultramafic rock (60% olivine, 30% orthopyroxene, and 10% clinopyroxene) Hf is enriched in the melt about 2.3 times as much as Lu.

The isotope composition and concentration of Lu in meteorites and certain rock standards were investigated by McCulloch et al. (1976), who used this information to discuss the nucleosynthesis of ^{176}Lu and ^{176}Hf by neutron capture on a slow timescale (s-process). The average concentrations of Lu and Hf in ordinary rock-forming silicate minerals in Table 12.2 are generally low except in alkali-rich minerals such as arfvedsonite (amphibole) and aegirine (pyroxene). In addition,

Table 12.1. Summary of the Chemical and Physical Properties of Lu and Hf

Properties	Lu	Hf
Atomic number	71	72
Atomic weight	174.967	178.49
Electron formula	Lu: [Xe] $4f^{14}5d^{1}6s^{2}$ Hf: [Xe] $4f^{14}5d^{2}6s^{2}$	
Valence	+3	+4
Ionic radius, Å	0.93	0.81
Electronegativity	1.2	1.3
Abundance, 10^6 Si atoms (solar system)	3.67×10^{-2}	1.54×10^{-1}

Note: The abundances of Lu and Hf in the solar system are from Anders and Grevesse (1989).

Table 12.2. Average Lu and Hf Concentration of Rock-Forming and Accessory Minerals

	Concentration (ppm)		
Minerals	Lu	Hf	Lu/Hf
Rock-Forming Minerals			
Plagioclase	0.062	0.31	0.20
Pyroxene	0.46	3.6	0.13
Diopside	0.60	2.9	0.21
Hornblende	1.07	0.61	1.75
Garnet	2.2	2.34	0.94
Biotite	2.7	1.0	2.7
Arfvedsonite	6.8	33	0.21
Accessory Minerals			
Zircon	23.65	15,177	0.0016
Eudialyte[a]	60.0	1,736	0.034
Baddeleyite[b]	70.0	13,340	0.005
Zirkelite[c]	—	4,700	—
Terrestrial Rocks			
Peridotite	0.039	1.14	0.034
Tholeiite basalt	0.50	2.45	0.20
Alkali basalt	0.65	5.2	0.12
Rhyolite	1.66	12.84	0.13
Granitic rocks	1.43	5.08	0.28
Carbonatites	2.4	10.0	0.24
Meteorites			
Chondrites	0.032	0.198	0.16
Achondrites			
Ca rich	0.35	0.73	0.48
Ca poor	0.026	0.072	0.36
Lunar Rocks			
Basalt	1.82	12.1	0.15
Soil	1.6	4.78	0.33

Source: Herrmann, 1970; Erlank et al., 1978; Patchett and Tatsumoto, 1980a, b, c, 1981a, b; Patchett et al., 1981, 1983a, b; Pettingill et al., 1981.

[a] $(Na,Ca,Fe)_6\ Zr\ (OH,Cl)\ (SiO_3)_6$.

[b] ZrO_2.

[c] $(Ca,Fe)\ (Zr,Ti,Th)_2\ O_5$.

sphene, Cr-spinel, and ilmenite are known to have elevated Hf concentrations ranging from 10 to 25 ppm (Erlank et al., 1978).

The highest concentrations of Hf occur in Zr-bearing minerals listed in Table 12.2, including zircon (15,200 ppm), zirkelite (4700 ppm), baddeleyite (13,340 ppm), and eudialyte (1740 ppm). These minerals also have elevated Lu concentrations ranging from about 25 ppm in zircon to 70 ppm in baddeleyite and eudialyte. However, the Lu/Hf ratios of these Zr-bearing minerals are very low (0.0016–0.034).

The highest Lu concentrations occur in certain rare-earth oxides (e.g., euxenite), carbonates (e.g., bastnaesite), phosphates (e.g., xenotime and monazite), and silicates (e.g., gadolinite and allanite). All of these minerals are relatively rare and occur primarily in complex pegmatites, alkali-rich intrusives, and carbonatites (e.g., Nilssen, 1973). Therefore, these rare-earth minerals are not important for dating by the Lu–Hf method, although their presence as accessory minerals significantly increases the Lu/Hf ratios of their host rocks. However, garnet has shown promise for Lu–Hf dating because of its elevated Lu/Hf ratio and its resistance to alteration (Duchene et al., 1997; Blichert-Toft et al., 1999; De Sigoyer et al., 2000; Scherer et al., 2000).

The terrestrial rocks in Table 12.2 and Figure 12.1 include peridotite (Lu = 0.039 ppm, Hf = 1.14 ppm, Lu/Hf = 0.034), tholeiite basalt

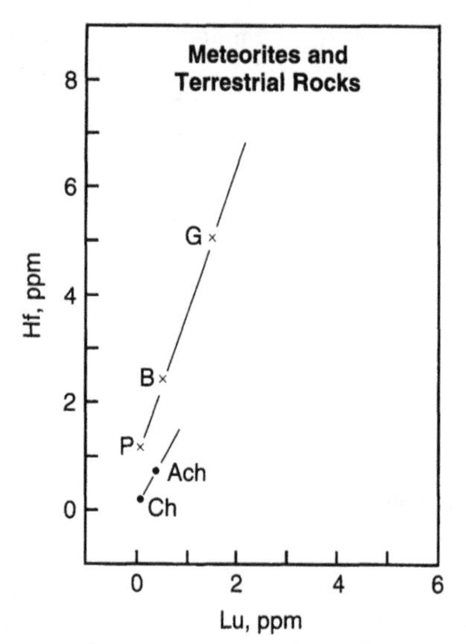

FIGURE 12.1 Average concentrations of Lu and Hf in terrestrial igneous rocks and meteorites: P = peridotite, B = tholeiite basalt, G = granitic rocks, Ch = chondrites, Ach = Ca-rich achondrites. Data for Lu from Herrmann (1970) and for Hf from Erlank et al. (1978).

(Lu = 0.50 ppm, Hf = 2.45 ppm, Lu/Hf = 0.20), and granitic rocks (Lu = 1.43 ppm, Hf = 5.08 ppm, Lu/Hf = 0.28). The Lu/Hf ratios of these and other terrestrial rocks remain low because geochemical differentiation tends to increase the concentrations of both Lu and Hf, although not at the same rate (Herrmann, 1970; Erlank et al., 1978).

The concentrations of Lu and Hf of chondritic meteorites in Table 12.2 are low (Lu = 0.032 ppm, Hf = 0.198 ppm, Lu/Hf = 0.16) compared to those of Ca-rich achondrites (Lu = 0.35 ppm, Hf = 0.73 ppm, Lu/Hf = 0.48). Lunar basalt has comparatively high concentrations of both elements (Lu = 1.82 ppm, Hf = 12.1 ppm, Lu/Hf = 0.15).

12.2 PRINCIPLES AND METHODOLOGY

Lutetium has two naturally occurring isotopes whose abundances are $^{175}_{71}\text{Lu} = 97.40$ and $^{176}_{71}\text{Lu} = 2.59$ at.%. Lutetium-176 in Figure 12.2 is

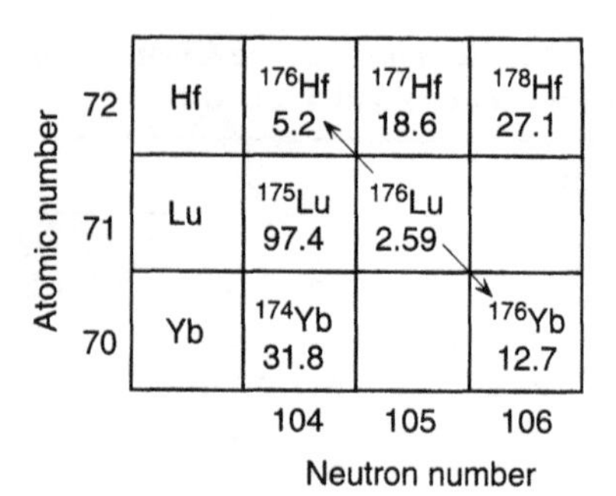

FIGURE 12.2 Branched decay of $^{176}_{71}\text{Lu}$ to $^{176}_{72}\text{Hf}$ by emission of β^--particles and to $^{176}_{70}\text{Yb}$ by electron capture.

radioactive and decays to stable $^{176}_{72}\text{Hf}$ by β-emission and to stable $^{176}_{70}\text{Yb}$ by electron capture. The branched decay of ^{176}Lu is required by Mattauch's rule (Section 2.1d) but can be disregarded in this case because fewer than $3 \pm 1\%$ of $^{176}_{71}\text{Lu}$ atoms decay by electron capture (Dixon et al., 1954). The β-decay of $^{176}_{71}\text{Lu}$ is followed by the emission of γ-rays from the nucleus of $^{176}_{72}\text{Hf}$.

The halflife of ^{176}Lu was determined by direct counting experiments (Herr et al., 1958; Prodi et al., 1969; Sguigna et al., 1982), and indirectly by analysis of Lu-bearing minerals and rocks of known age (e.g., Boudin and Deutsch, 1970). The best value of 35.7×10^9 years was derived from the measured slope of a Lu–Hf isochron of 13 Ca-rich achondrite meteorites whose age was assumed to be 4.55×10^9 years (Patchett et al., 1981a; Tatsumoto et al., 1981). Accordingly, the decay constant of ^{176}Lu is

$$\lambda = (1.94 \pm 0.07) \times 10^{-11}\ \text{y}^{-1}$$

The abundance of ^{176}Hf, expressed as the $^{176}\text{Hf}/^{177}\text{Hf}$ ratio, increases with time at a rate that depends on the atomic $^{176}\text{Lu}/^{177}\text{Hf}$ ratio of the rocks of minerals:

$$\frac{^{176}\text{Hf}}{^{177}\text{Hf}} = \left(\frac{^{176}\text{Hf}}{^{177}\text{Hf}}\right)_i + \frac{^{176}\text{Lu}}{^{177}\text{Hf}}(e^{\lambda t} - 1) \qquad (12.1)$$

where $^{176}\text{Hf}/^{177}\text{Hf}$ = ratio of these isotopes in a sample of rock or minerals at the present time

$(^{176}Hf/^{177}Hf)_i$ = ratio of these isotopes at the time of formation of the Lu-bearing rocks or minerals

$^{176}Lu/^{177}Hf$ = ratio of these isotopes in the Lu-bearing rocks or minerals at the present time

The $^{176}Lu/^{177}Hf$ ratio is calculated from the measured concentrations of Lu and Hf based on data in Table 12.3:

$$\frac{^{176}Lu}{^{177}Hf} = \left(\frac{Lu}{Hf}\right)_{conc.} \times \frac{\text{at. wt. Hf} \times \text{Ab } ^{176}Lu}{\text{at. wt. Lu} \times \text{Ab } ^{177}Hf}$$

$$= \left(\frac{Lu}{Hf}\right)_{conc.} \times 0.142$$

Although the atomic weight of Lu and the abundance (Ab) of ^{176}Lu in terrestrial and extraterrestrial rocks and minerals are constant, the atomic weight of Hf and the abundance of ^{177}Hf are not constant, strictly speaking, because they depend on the abundance of radiogenic ^{176}Hf.

In principle, equation 12.1 can be used to date samples of Lu-bearing rocks or minerals based on the measured concentrations of Lu and Hf and on the $^{176}Hf/^{177}Hf$ ratio determined by mass spectrometry. In addition, the initial $^{176}Hf/^{177}Hf$ ratio must be assumed, such as that of ordinary Hf:

$$\frac{^{176}Hf}{^{177}Hf} = \frac{5.2}{18.6} = 0.279$$

This method of dating is only applicable to minerals (or rocks) having high Lu/Hf ratios whose measured $^{176}Hf/^{177}Hf$ ratios are so large that the uncertainty of the assumed initial $^{176}Hf/^{177}Hf$ ratio does not affect the numerical values of the calculated dates.

The preferred method of dating is by the construction of Lu–Hf isochrons based on the measured $^{176}Lu/^{177}Hf$ and $^{176}Hf/^{177}Hf$ ratios of suites of cogenetic whole-rock samples of igneous or metamorphic rocks and their constituent minerals. The slope m of the isochron,

$$m = e^{\lambda t} - 1$$

yields the date t, whereas the intercept on the y-axis is the initial $^{176}Hf/^{177}Hf$ ratio. In addition, the goodness of fit of the data points to the straight line confirms but does not prove the assumption that all of the samples have the same age and the same initial $^{176}Hf/^{177}Hf$ ratio.

The analytical procedure of Patchett and Tatsumoto (1980a) calls for dissolution of rock powders by a mixture of HF and HNO_3 at 160°C for four days in sealed Teflon-coated bombs. This treatment is necessary to assure that Hf-rich minerals, primarily zircon, are decomposed. Hafnium and Lu are separated and purified by ion exchange techniques using HF and other acids as eluants. The concentrations of Lu and Hf are determined by isotope dilution (Owen and Faure, 1974). The measured $^{176}Hf/^{177}Hf$ ratios are corrected for isotope fractionation to $^{179}Hf/^{177}Hf = 0.7325$. A high-purity HfO_2 reagent (Johnson Matthey Co. 475) has been recommended as an interlaboratory Hf standard and is available from the U.S. Geological Survey in Denver, Colorado, U.S.A. The $^{176}Hf/^{177}Hf$ ratio of this standard is 0.282195 ± 0.000015 according to Patchett and Tatsumoto (1980a, b). Alternatively, Hf concentrations can also be measured by neutron activation (Boudin and Hanappe, 1967; Brooks, 1969; Rebagay and Ehmann, 1970). The analytical chemistry of Zr and Hf has been presented by Elinson and Petrov

Table 12.3. Abundances and Isotopic Masses of the Naturally Occurring Isotopes of Lu and Hf

Isotope	Abundance, at. %	Mass, amu
	Lutetium (at. wt. = 174.967)	
$^{175}_{71}Lu$	97.40	174.940770
$^{176}_{71}Lu$	2.59	175.942679
	Hafnium (at. wt. = 178.49)	
$^{174}_{72}Hf$	0.16	173.940044
$^{176}_{72}Hf$	5.2	175.941406
$^{177}_{72}Hf$	18.6	176.943217
$^{178}_{72}Hf$	27.1	177.943696
$^{179}_{72}Hf$	13.74	178.945812
$^{180}_{72}Hf$	35.2	179.946545

Source: Lide and Frederikse, 1995.

(1969) and Mukherji (1970). The concentration of Lu is determined by isotope dilution (Patchett and Tatsumoto, 1980a), which is more sensitive than neutron activation (Boudin and Dehon, 1969) and atomic absorption spectrometry (Boudin, 1967).

The classical analytical techniques based on thermal ionization mass spectrometry (TIMS) are being displaced by sensitive high-resolution ion microprobe (SHRIMP) mass spectrometry and inductively coupled plasma mass spectrometry (ICP-MS, Section 12.5). These techniques circumvent the analytical problems caused by the high first ionization potential of Hf, by the difficulty in separating Hf from Zr and Ti in solution, and by the fact that the isotope ratios of Lu cannot be corrected for fractionation during isotope dilution analysis because only two isotopes are available (Blichert-Toft et al., 1997).

12.3 CHUR AND EPSILON

The parameters that define the chondritic uniform reservoir (CHUR) for the isotopic evolution of Hf were determined by Patchett and Tatsumoto (1980, 1981) and Tatsumoto et al. (1981) with a summary by Patchett (1983). The primeval $^{176}Hf/^{177}Hf$ ratio was obtained as the initial $^{176}Hf/^{177}Hf$ ratio of a Lu–Hf isochron formed by 13 achondrite meteorites. In addition, the Lu/Hf ratio of CHUR was determined by analysis of the carbonaceous chondrite Murchison. The Lu/Hf ratio of CHUR and the primeval $^{176}Hf/^{177}Hf$ ratio were used to calculate the present $^{176}Hf/^{177}Hf$ ratio of CHUR. These three parameters were used in Figure 12.3 to represent the isotopic evolution of terrestrial Hf by means of the Patchett–Tatsumoto version

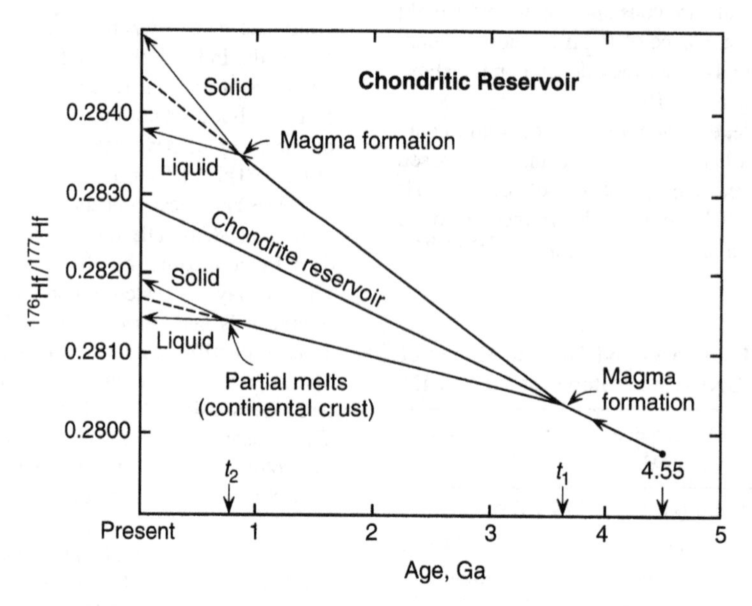

FIGURE 12.3 Isotopic evolution of Hf in a chondritic reservoir. The primordial $^{176}Hf/^{177}Hf$ ratio is 0.27978, its present $^{176}Lu/^{177}Hf$ ratio is 0.0334, the present $^{176}Hf/^{177}Hf$ ratios is 0.28286, and the age of the reservoir is 4.55×10^9 years (Patchett et al. 1981). Partial melting in the reservoir at t_1 produces liquids having lower Lu/Hf ratios, whereas the residual solids have higher Lu/Hf ratios than the chondritic reservoirs. Subsequent melting of the residual solids (depleted mantle) at t_2 produces magmas whose $^{176}Hf/^{177}Hf$ ratios are greater than those of the chondrite reservoir and their ε^t (Hf) values are therefore positive. Renewed melting of crustal rocks at t_2 results in magmas that are depleted in ^{176}Hf and have negative ε^t (Hf) values. The blending of melts derived from the depleted mantle and the continental crust results in intermediate ε-values for igneous rocks that form from such contaminated magmas.

of CHUR:

$$\left(\frac{^{176}\text{Hf}}{^{177}\text{Hf}}\right)^0_{\text{CHUR}} = 0.28286 \quad \text{at } t = 0 \text{ (present time)}$$

$$\left(\frac{^{176}\text{Lu}}{^{177}\text{Hf}}\right)^0_{\text{CHUR}} = 0.0334 \quad \text{at } t = 0 \text{ (present time)}$$

$$\left(\frac{^{176}\text{Hf}}{^{177}\text{Hf}}\right)^T_{\text{CHUR}} = 0.27978 \pm 0.00009 \quad \text{at } T = 4.55 \text{ Ga}$$

A slightly different set of parameters was reported by Blichert-Toft and Albarède (1997), which they derived from the ^{176}Hf/^{177}Hf and Lu/Hf ratios of 23 chondrite meteorites:

$$\left(\frac{^{176}\text{Hf}}{^{177}\text{Hf}}\right)^0_{\text{CHUR}} = 0.282772 \pm 0.000029 \quad \text{at } t = 0 \text{ (present time)}$$

$$\left(\frac{^{176}\text{Lu}}{^{177}\text{Hf}}\right)^0_{\text{CHUR}} = 0.0332 \pm 0.0002 \quad \text{at } t = 0 \text{ (present time)}$$

$$\left(\frac{^{176}\text{Hf}}{^{177}\text{Hf}}\right)^T_{\text{CHUR}} = 0.279742 \pm 0.000029 \quad \text{at } T = 4.56 \text{ Ga}$$

In the discussion that follows, the parameters of Patchett and Tatsumoto will continue to be used to define CHUR.

The equation for the isotopic evolution of Hf in CHUR takes the form of equation 12.1:

$$\left(\frac{^{176}\text{Hf}}{^{177}\text{Hf}}\right)^0_{\text{CHUR}} = 0.27978 + 0.0334(e^{\lambda T} - 1) \tag{12.2}$$

where T is the age of the Earth. The present ^{176}Hf/^{177}Hf ratio ($t = 0$) of a rock or mineral may be higher or lower than that of CHUR depending on the Lu/Hf ratio of the geological reservoirs through which the Hf passed before it was incorporated into a rock or mineral. The comparison of the ^{176}Hf/^{177}Hf ratio of a rock or mineral with the ^{176}Hf/^{177}Hf ratio of CHUR is expressed by the ε-value, defined as

$$\varepsilon^0(\text{Hf}) = \left[\frac{(^{176}\text{Hf}/^{177}\text{Hf})^0_{\text{spl}}}{(^{176}\text{Hf}/^{177}\text{Hf})^0_{\text{CHUR}}} - 1\right] \times 10^4 \tag{12.3}$$

where

$$\left(\frac{^{176}\text{Hf}}{^{177}\text{Hf}}\right)^0_{\text{CHUR}} = 0.28286 \quad \text{at } t = 0$$

Alternatively, the comparison can be made at some time t in the past. In that case

$$\varepsilon^t(\text{Hf}) = \left[\frac{(^{176}\text{Hf}/^{177}\text{Hf})^t_{\text{spl}}}{(^{176}\text{Hf}/^{177}\text{Hf})^t_{\text{CHUR}}} - 1\right] \times 10^4 \tag{12.4}$$

where $(^{176}\text{Hf}/^{177}\text{Hf})^t_{\text{CHUR}}$ is calculated from equation 12.2 and $(^{176}\text{Hf}/^{177}\text{Hf})^t_{\text{spl}}$ is the intercept of the Lu–Hf isochron of a suite of cogenetic rocks or minerals. Positive ε-values indicate that the sample is enriched in radiogenic ^{176}Hf compared to the chondritic reservoir and therefore originated from a source that had a higher Lu/Hf ratio than chondrites. Similarly, negative ε-values are caused by a deficiency of ^{176}Hf and imply derivation from a source with a lower Lu/Hf ratio than the chondritic reservoir, as illustrated in Figure 12.3.

12.4 MODEL Hf DATES DERIVED FROM CHUR

The data in Table 12.2 demonstrate that zircon, baddeleyite, and other Zr-bearing minerals have high concentrations of Hf in excess of 1000 ppm. Although these minerals also contain Lu (20–70 ppm), the Lu/Hf ratios of Zr-bearing minerals are low (0.0016–0.034). Consequently, the isotope composition of Hf in these minerals is virtually independent of time.

Modern instrumental methods (SHRIMP and ICP-MS) are capable of measuring the $^{176}Hf/^{177}Hf$ ratios of zircon with good precision and accuracy and thereby provide an opportunity to calculate model dates related to CHUR, as demonstrated for Nd in Section 9.2b. Therefore, in accordance with equation 9.6, the $^{176}Hf/^{177}Hf$ ratio of CHUR at any time t in the past is

$$\left(\frac{^{176}Hf}{^{177}Hf}\right)^t_{CHUR} = \left(\frac{^{176}Hf}{^{177}Hf}\right)^0_{CHUR} - \left(\frac{^{176}Lu}{^{177}Hf}\right)^0_{CHUR}(e^{\lambda t} - 1) \tag{12.5}$$

Substituting the appropriate values for $(^{176}Hf/^{177}Hf)^0_{CHUR}$ and $(^{176}Lu/^{177}Hf)^0_{CHUR}$ yields

$$\left(\frac{^{176}Hf}{^{177}Hf}\right)^t_{CHUR} = 0.28286 - 0.0334(e^{\lambda t} - 1) \tag{12.6}$$

Similarly, from equation 12.1 for a zircon crystal (subscript Z) the $(^{176}Hf/^{177}Hf)^t_Z$ ratio is

$$\left(\frac{^{176}Hf}{^{177}Hf}\right)^t_Z = \left(\frac{^{176}Hf}{^{177}Hf}\right)^0_Z - \left(\frac{^{176}Lu}{^{177}Hf}\right)^0_Z(e^{\lambda t} - 1) \tag{12.7}$$

The model date is calculated by setting

$$\left(\frac{^{176}Hf}{^{177}Hf}\right)^t_{CHUR} = \left(\frac{^{176}Hf}{^{176}Hf}\right)^t_Z$$

expressed by equations 12.6 and 12.7:

$$0.28286 - 0.0334(e^{\lambda t} - 1) = \left(\frac{^{176}Hf}{^{177}Hf}\right)^0_Z - \left(\frac{^{176}Lu}{^{177}Hf}\right)^0_Z(e^{\lambda t} - 1)$$

Solving for t yields

$$(e^{\lambda t} - 1)\left[\left(\frac{^{176}Lu}{^{177}Hf}\right)^0_Z - 0.0334\right] = \left(\frac{^{176}Hf}{^{177}Hf}\right)^0_Z - 0.28286$$

and thus

$$t = \frac{1}{\lambda}\ln\left[\frac{(^{176}Hf/^{177}Hf)^0_Z - 0.28286}{(^{176}Lu/^{177}Hf)^0_Z - 0.0334} + 1\right] \tag{12.8}$$

The date derived from equation 12.8 is the time when the Hf was separated from CHUR and entered the zircon crystal where it resided until the present time ($t = 0$).

A special case arises when the $^{176}Lu/^{177}Hf$ ratio of the sample is very much less than that of CHUR. In that case, the present $^{176}Hf/^{177}Hf$ ratio is virtually identical to the initial value at the time of crystallization. If $(^{176}Lu/^{177}Hf)^0_Z \simeq 0$ in equation 12.7, then

$$\left(\frac{^{176}Hf}{^{177}Hf}\right)^t_Z = \left(\frac{^{176}Hf}{^{177}Hf}\right)^0_Z$$

and the model date derived from

$$0.28286 - 0.0334(e^{\lambda t} - 1) = \left(\frac{^{176}Hf}{^{177}Hf}\right)^t_Z$$

is

$$t = \frac{1}{\lambda}\ln\left[\frac{0.28286 - (^{176}Hf/^{177}Hf)^t_Z}{0.0334} + 1\right]$$

$$\text{where } \lambda = 1.94 \times 10^{-11}\ y^{-1} \tag{12.9}$$

Equation 12.9 also applies in cases where the initial $^{176}Hf/^{177}Hf$ ratio has been derived from a Lu–Hf isochron.

Model dates calculated relative to CHUR based on the isotope composition of Hf are subject to the same limiting assumptions as Nd model dates presented in Section 9.2b.

12.5 APPLICATIONS OF Lu–Hf DATING

The principles of dating by the Lu–Hf method are identical to those that apply to other radiogenic isotope methods. The only differences arise from the geochemical properties of Lu and Hf and from analytical problems that affect the precision and accuracy of the resulting dates. In addition, the accumulation of radiogenic isotopes of Sr, Nd, Pb, and Hf in various terrestrial reservoirs plays an important role in the interpretation of the isotope compositions of these elements in igneous rocks (Faure, 2001).

The increasing use of ICP-MS instruments combined with laser ablation, equipped with multiple collectors, and using large electromagnets is facilitating the use of the Lu–Hf method not only for dating igneous and metamorphic rocks but also for the study of the origin of igneous rocks:

Instrumental Techniques: Walder et al., 1993, *J. Anal. Atom. Spectrom.*, 8:19–23. Salters, 1994, *Anal. Chem.*, 66:4186–4189. Thirlwall and Walder, 1995, *Chem. Geol.*, 122:241–247. Blichert-Toft et al., 1997, *Contrib. Mineral. Petrol.*, 127:248–260. David et al., 1999, *Chem. Geol.*, 157:1–12.

Chemical Separation of Hf: Barovich et al., 1995, *Chem. Geol.*, 121:303–308. Gruau et al., 1988, *Chem. Geol.*, 72:353–356. Parrish, 1987, *Chem. Geol. (Isotope Geosci. Sect.)*, 66:99–102.

Evolution of Mantle and Crust: Beard and Johnson, 1993, *Earth Planet. Sci. Lett.*, 119: 495–509. Johnson and Beard, 1993, *Nature*, 362:441–444. Johnson et al., 1996, *Geol. Soc. Am. Spec. Paper*, 315:339–352; Salters and Hart, 1991, *Earth Planet. Sci. Lett.*, 104:364–380. Salters and Zindler, 1995, *Earth Planet. Sci. Lett.*, 129:13–30. Stevenson and Patchett, 1990, *Geochim. Cosmochim. Acta*, 43:1683–1697. Vervoort and Patchett, 1996, *Geochim. Cosmochim. Acta*, 60:3717–3733. Vervoort et al., 1996, *Nature*, 379:624–627.

Examples of Lu–Hf dating using the conventional chemical separation technique and TIMS include the work of Scherer et al. (1997) on garnet-bearing xenoliths at Kilbourne Hole, New Mexico; Corfu and Nobel (1992) on the Abitibi greenstone belt of Ontario/Quebec; Corfu and Stott (1993) on igneous rocks in the Superior structural providence of Canada; and Patchett et al. (1984) on crustal recycling of Hf.

12.5a Amitsoq Gneiss, Godthåb Area, West Greenland

The Amitsoq Gneiss is one of several remnants of crustal rocks of Early Archean age according to all applicable isotopic geochronometers, including the Rb–Sr method (Section 5.4c). Pettingill et al. (1981) used the analytical procedures developed by Patchett and Tatsumoto (1980) to test the Lu–Hf method by dating a suite of whole-rock samples and separated zircons from the Amitsoq Gneiss of West Greenland. The Lu–Hf data form an isochron in Figure 12.4 from which Pettingill et al. (1981) calculated a date of 3.59 ± 0.22 Ga and an initial $^{176}Hf/^{177}Hf$ ratio of 0.280482 ± 0.000033. The date agrees with previous age determinations by other methods summarized by Baadsgaard et al. (1976) and mentioned in Section 5.4c. The initial $^{176}Hf/^{177}Hf$ ratio is higher than the primeval value (0.27978 ± 0.00009; Tatsumoto et al., 1981), as expected by the model for the isotopic evolution of Hf in CHUR in Figure 12.3.

The reproducibility of the measurements is not as good as expected from analytical errors alone because of the inhomogeneous distribution of Lu and Hf in granitic rocks. Lutetium resides primarily in sphene, allanite, and apatite, whereas Hf occurs mostly in zircon. These accessory minerals are unevenly distributed in the rocks and, in addition, are sensitive to alteration during metamorphism. In addition, zircon crystals may contain detrital cores whose age predates the crystallization of igneous rocks in which they reside or they may acquire overgrowths during metamorphism of their host. The Hf in such zircons may have a range of isotopic compositions, which contributes to the scatter of data points about the isochron and may introduce systematic errors into the date derived from it. On the other hand, zircons have low Lu/Hf ratios and, therefore, extend the range of Lu/Hf ratios of data points on isochrons and can be used to calculate initial $^{176}Hf/^{177}Hf$ ratios of igneous rocks using dates determined by other methods.

The initial $^{176}Hf/^{177}Hf$ ratio of the Lu–Hf isochron in Figure 12.4 provides an opportunity to calculate a model Hf date for the Amitsoq Gneiss by means of equation 12.9. Since $(^{176}Hf/^{177}Hf)_i =$

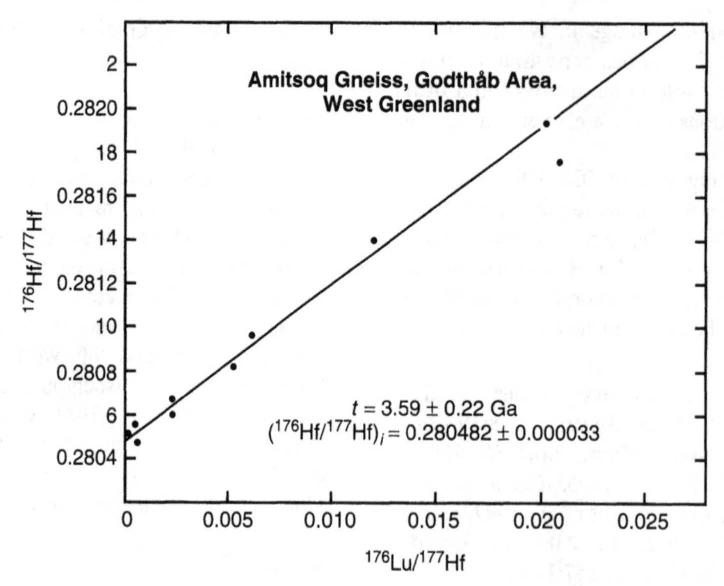

FIGURE 12.4 The Lu–Hf isochron for whole-rock samples and separated zircons from the Amitsoq Gneiss of Western Greenland. Some data points have been omitted to avoid crowding. Data and interpretation by Pettingill et al. (1981).

0.280482 ± 0.000033,

$$t = \frac{1}{1.94 \times 10^{-11}} \ln \left(\frac{0.28286 - 0.280482}{0.0334} + 1 \right)$$

$$= \frac{\ln (1.071197)}{1.94 \times 10^{-11}} = 3.55 \pm 0.5 \text{ Ga}$$

The model date is in good agreement with the Lu–Hf isochron date ($t = 3.59 \pm 0.22$ Ga), which means that the Hf originated from the chondritic reservoir in the mantle of the Earth less than about 100 million years prior to the formation of the Amitsoq Gneiss during one of the earliest orogenies on record.

12.5b Detrital Zircons, Mt. Narryer, Western Australia

The elevated Hf concentrations and low Lu/Hf ratios (Table 12.2) of zircon crystals make them and other refractory Hf-bearing minerals suitable for dating by comparison to CHUR using data obtained by SHRIMP. Kinny et al. (1991) demonstrated the feasibility of such measurements and documented sample SL-7 (Shri Lanka) as a zircon standard for the isotope composition of Hf. Replicate analyses by TIMS of SL-7 yielded

$$^{176}\text{Hf}/^{177}\text{Hf} = 0.28162 \pm 0.00003, \ 2\sigma$$

normalized to $^{178}\text{Hf}/^{177}\text{Hf} = 1.46715$. The average of nine determinations by SHRIMP was

$$^{176}\text{Hf}/^{177}\text{Hf} = 0.28183 \pm 0.00008, \ 1\sigma$$

Kinny et al. (1991) used SHRIMP to measure the initial $^{176}\text{Hf}/^{177}\text{Hf}$ ratios of zircons in the Mount Narryer quartzite of Western Australia whose U–Pb age is greater than 4.1 Ga (Froude et al. 1983; Compston et al., 1985). The zircons in the quartzite at Mt. Narryer and in a conglomerate in the nearby Jack Hills (Compston and Pidgeon, 1986) are the oldest known terrestrial minerals. Detrital zircons of Hadean age are extremely rare and have so far been found only in the northwest corner of Western Australia.

The average initial $^{176}Hf/^{177}Hf$ ratio (corrected for in situ decay of ^{176}Lu) of zircons at Mt. Narryer (Kinny et al., 1991) is

$$\left(\frac{^{176}\text{Hf}}{^{177}\text{Hf}}\right)_i = 0.28036 \pm 0.00009$$

This value yields a model Hf date relative to CHUR (equation 12.9) of

$$t = \frac{1}{1.94 \times 10^{-11}} \ln\left(\frac{0.28286 - 0.28036}{0.0334} + 1\right)$$
$$= 3.72 \pm 0.13 \text{ Ga}$$

The oldest Mt. Narryer zircons analyzed by Kinny et al. (1991) have an average initial $^{176}Hf/^{177}Hf$ ratio of 0.28005 ± 15. The Hf in these grains originated from CHUR at 4.16 ± 0.21 Ga.

These results confirm the antiquity of the zircons at Mt. Narryer and reveal that the Hf in some of the grains originated from source regions whose Lu/Hf ratios were higher than those of CHUR. A hypothetical example in Figure 12.5 illustrates a possible evolutionary path for Hf that had a $^{176}Hf/^{177}Hf$ ratio of 0.28036 at 4.16 Ga and yields a fictitious model date of 3.72 Ga. The diagram demonstrates that the Hf could have evolved in CHUR from 4.55 to 4.40 Ga when the Lu/Hf ratio increased from 0.235 (CHUR) to 0.680 (secondary reservoir). Subsequently, the $^{176}Hf/^{177}Hf$ ratio increased from 0.27989 at 4.4 Ga to 0.28036 at 4.16 Ga when the Hf was incorporated into the zircon crystals, which were ultimately preserved as detrital grains in the quartzite at Mt. Narryer.

The increase of the Lu/Hf ratio could have resulted from the removal of a melt phase from CHUR at some time between 4.16 and 4.55 Ga. The available data do not constrain the Lu/Hf ratio of the reservoir and the age of this event. Therefore, the isotopic evolution of Hf in Figure 12.5 is merely one of an infinite number of possible scenarios that illustrates why some of the zircon grains of the quartzite at Mt. Narryer have anomalously low model Hf dates. The Lu/Hf ratio of the secondary Hf reservoir could be realized by a mixture of garnet (Lu/Hf = 0.94) and diopside

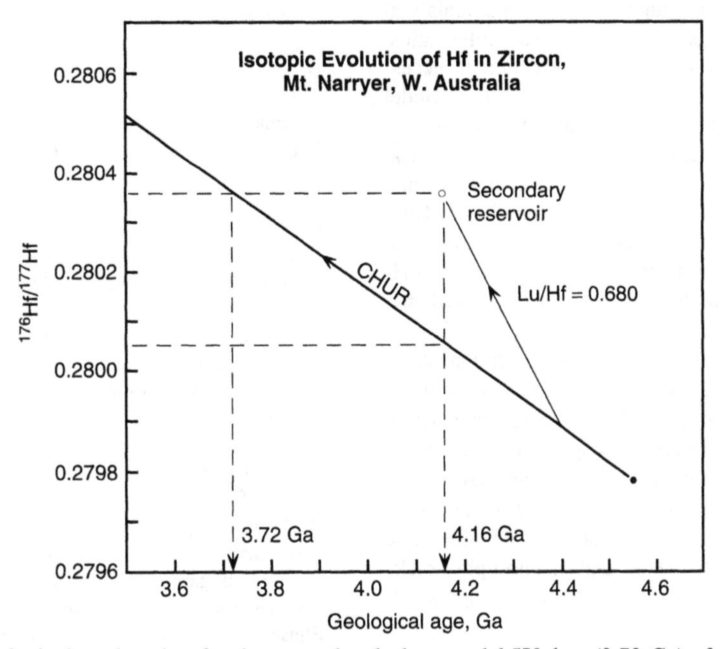

FIGURE 12.5 Hypothetical explanation for the anomalously low model Hf date (3.72 Ga) of some zircons in the quartzite at Mt. Narryer, Western Australia. Data from Kinny et al. (1991).

(Lu/Hf = 0.21) or Al-bearing pyroxene (Lu/Hf = 0.13) in Table 12.2.

12.6 SUMMARY

The development of the Lu–Hf method of dating was retarded initially by uncertainties about the value of the decay constant of ^{176}Lu and by difficulties with the analytical chemistry and mass spectrometry of Hf. After these problems had been overcome in the early 1980s, the Lu–Hf method of dating has been used to date igneous and metamorphic igneous rocks of Precambrian age as well as garnets of Cenozoic age by means of isochrons.

The isotopic evolution of Hf in the Earth can be modeled by reference to a chondritic uniform reservoir (CHUR) defined on the basis of the primeval $^{176}Hf/^{177}Hf$ ratio derived from the isochron of achondritic meteorites and by the Lu/Hf ratio of the carbonaceous chondrite Murchison. CHUR is used to define model Hf dates based on the initial $^{176}Hf/^{177}Hf$ ratios derived from Lu–Hf isochrons of terrestrial rocks.

In addition, model Hf dates can be calculated for zircon and other Zr-bearing minerals that have high Hf concentrations and low Lu/Hf ratios. In many cases, only a small correction is necessary for the in situ decay of ^{176}Lu. These model dates represent the time when the Hf was removed from CHUR and was incorporated into zircon. The model dates may be systematically in error in some cases when the Hf was not transferred directly from CHUR into zircon crystals but resided for some time in a reservoir having a higher or lower Lu/Hf ratio than CHUR prior to being incorporated into zircon crystals.

The isotope ratios of Hf in zircon crystals can be measured by instrumental methods using SHRIMP and ICP-MS instrumentation. These methods have unprecedented spatial resolution and permit cores and overgrowth to be analyzed separately. The resulting model dates of zircon crystals complement U–Pb dates obtained from the same grains not only in igneous and metamorphic rocks but also in detrital zircon grains in quartzite, sandstone, and beach sand.

The isotopic composition of Hf in different domains in the suboceanic and subcontinental lithospheric mantle is an important aspect of the study of the chemical differentiation of the Earth and the origin of igneous rocks.

REFERENCES FOR GEOCHEMISTRY AND METHODOLOGY (SECTIONS 12.1–12.2)

Anders, E., and N. Grevesse, 1989. Abundances of the elements: Meteoritic and solar. *Geochim. Cosmochim. Acta*, 53:197–214.

Blichert-Toft, J., F. Albarède, and J. Kornprobst, 1999. Lu–Hf isotope systematics of garnet pyroxenite from Beni Bousera, Morocco. *Science*, 283:1303–1305.

Blichert-Toft, J., C. Chauvel, and F. Albarède, 1997. Separation of Hf and Lu for high precision isotope analysis of rock samples by magnetic sector-multiple collector ICP-MS. *Contrib. Mineral. Petrol.*, 127:248–260.

Boudin, A., 1967. Development de la methode de mesure d'age des roches fondee sur le rapport lutetium-176/hafnium-176. In *Radioactive Dating and Methods of Low-Level Counting*, pp. 515–522. International Atomic Energy Agency, Vienna.

Boudin, A., and M. Dehon, 1969. Methodes d'analyse quantitative du lutetium dans les mineraux. *Geochim. Cosmochim. Acta*, 33:142–147.

Boudin, A., and S. Deutsch, 1970. Geochronology: Recent development in the lutetium-176/hafnium-176 dating method. *Science*, 168:1219–1220.

Boudin, A., and F. Hanappe, 1967. Methodes nouvelles d'analyse du hafnium par activation. *Radiochim. Acta*, 8:188.

Brooks, C. K., 1969. On the distribution of zirconium and hafnium in the Skaergaard intrusion, East Greenland. *Geochim. Cosmochim. Acta*, 33:357–374.

DeSigoyer, J., V. Chavagnac, J. Blichert-Toft, I. M. Villa, B. Luais, S. Guillot, M. Cosca, and G. Mascle, 2000. Dating the Indian continental subduction and collisional thickening in the northwest Himalaya: Multichronology of the Tso Morari eclogite. *Geology*, 28:487–490.

Dixon, D., A. McNair, and S. C. Curran, 1954. The natural radioactivity of lutetium. *Philos. Mag.*, 45(7): 683–694.

Duchene, S., J. Blichert-Toft, B. Luais, P. Telouk, J.-M. Lardeaux, and F. Albarède, 1997. The Lu–Hf dating of garnets and the ages of the alpine high-pressure metamorphism. *Nature*, 387:586–588.

Elinson, S. V., and K. I. Petrov, 1969. *Analytical Chemistry of Zirconium and Hafnium*. Humphrey Science, Ann Arbor, Michigan.

Erlank, A. J., H. S. Smith, J. W. Marchant, M. P. Cardoso, and L. H. Ahrens, 1978. Hafnium. In K. H. Wedepohl (Ed.), *Handbook of Geochemistry*, Section 72. Springer-Verlag, Heidelberg.

Herr, W., E. Merz, P. Eberhardt, and P. Singer, 1958. Zur Bestimmung der Halbwertszeit des Lu-176 durch den Nachweis von radiogenem Hf-176. *Z. Naturforsch.*, 13a:268–273.

Herrmann, A. G., 1970. Yttrium and the lanthanides. In K. H. Wedepohl (Ed.), *Handbook of Geochemistry*, Vol. II-5, Section 57–71. Springer-Verlag, Heidelberg.

Lide, D. R., and H. P. R. Frederikse, 1995. *CRC Handbook of Chemistry and Physics*, 76th ed. CRC Press, Boca Raton, Florida.

McCulloch, M. T., J. R. de Laeter, and K. J. R. Rosman, 1976. The isotopic composition and elemental abundance of lutetium in meteorites and terrestrial samples and the ^{176}Lu cosmochronometer. *Earth Planet. Sci. Lett.*, 28:308–322.

Mukherji, A. K., 1970. *Analytical Chemistry of Zr and Hf.* Pergamon, Elmsford, New York.

Nilssen, B., 1973. Gadolinite from Hundholmen, Tysfjord, north Norway. *Norsk Geol. Tidsskrift*, 53:343–348.

Owen, L. B., and G. Faure, 1974. Simultaneous determination of hafnium and zirconium in silicate rocks by isotope dilution. *Anal. Chem.*, 46:1323–1326.

Patchett, J. P., 1983a. Importance of the Lu–Hf isotopic system in studies of planetary chronology and chemical evolution. *Geochim. Cosmochim. Acta*, 47:81–91.

Patchett, J. P., 1983b. Hafnium isotope results from mid-ocean ridges and Kerguelen. *Lithos*, 16:47–51.

Patchett, P. J., O. Kouvo, C. E. Hedge, and M. Tatsumoto, 1981. Evolution of continental crust and mantle heterogeneity: Evidence from Hf isotopes. *Contrib. Mineral. Petrol.*, 78:279–297.

Patchett, P. J., and M. Tatsumoto, 1980a. A routine high-precision method for Lu–Hf isotope geochemistry and chronology. *Contrib. Mineral. Petrol.*, 75:263–268.

Patchett, P. J., and M. Tatsumoto, 1980b. Lu–Hf total-rock isochron for the eucrite meteorites. *Nature*, 288:571–574.

Patchett, P. J., and M. Tatsumoto, 1980c. Hafnium isotope variations in oceanic basalts. *Geophys. Res. Lett.*, 7:1077–1080.

Patchett, P. J., and M. Tatsumoto, 1981a. Lu/Hf in chondrites and definition of a chondritic hafnium growth curve. *Lunar Planet. Sci.*, 12(Pt. 2):822–824.

Patchett, P. J., and M. Tatsumoto, 1981b. The hafnium isotopic evolution of lunar basalts. *Lunar Planet. Sci.*, 13(Pt. 2):819–821.

Pettingill, H. S., P. J. Patchett, M. Tatsumoto, and S. Moorbath, 1981. Lu–Hf total-rock age for the Amitsoq Gneisses, West Greenland. *Earth Planet. Sci. Lett.*, 55:150–156.

Prodi, V., K. F. Flynn, and L. E. Glendenin, 1969. Half-life and beta spectrum of Lu176. *Phys. Rev.*, 188(4): 1930–1933.

Rebagay, T. V., and W. D. Ehmann, 1970. Simultaneous determination of zirconium and hafnium in standard rocks by neutron activation analysis. *J. Radioanal. Chem.*, 5:51–60.

Scherer, E. E., K. L. Cameron, and J. Blichert-Toft, 2000. Lu–Hf garnet geochronology: Closure temperature relative to the Sm-Nd system and the effects of trace mineral inclusion. *Geochim. Cosmochim. Acta*, 64:3413–3432.

Sguigna, A. P., A. J. Larabee, and J. C. Waddington, 1982. The half-life of ^{176}Lu by γ-γ coincidence measurement. *Can. J. Phys.*, 60:361–364.

Tatsumoto, M., D. M. Unruh, and P. J. Patchett, 1981. U–Pb and Lu–Hf systematics of Antarctic meteorites. In *Proc. 6th Symp. Antarctic Meteorites*, pp.237–249. National Institute of Polar Research, Tokyo, Japan.

REFERENCES FOR CHUR AND EPSILON (SECTION 12.3)

Blichert-Toft, J., and F. Albarède, 1997. The Lu–Hf geochemistry of chondrites and the evolution of the mantle-crust system. *Earth Planet. Sci. Lett.*, 148:243–258.

Patchett, P. J., 1983. Importance of the Lu–Hf isotopic system in studies of planetary chronology and chemical evolution. *Geochim. Cosmochim. Acta*, 47:81–91.

Patchett, P. J., and M. Tatsumoto, 1980. Lu–Hf total-rock isochron for the eucrite meteorites. *Nature*, 288:571–574.

Patchett, P. J., and M. Tatsumoto, 1981. Lu/Hf in chondrites and definition of a chondritic hafnium growth curve. *Lunar Planet. Sci.*, 12:822–824.

Tatsumoto, M., D. M. Unruh, and P. J. Patchett, 1981. U–Pb and Lu–Hf systematics of Antarctic meteorites. In *Proc. 6th Symp. Antarctic Meteorites*, pp.237–249. National Institute of Polar Research, Tokyo.

REFERENCES FOR APPLICATIONS OF Lu–Hf DATING (SECTION 12.5)

Baadsgaard, H., R. St. J. Lambert, and J. Krupicka, 1976. Mineral isotopic age relationships in the polymetamorphic Amitsoq gneisses, Godthaab District, West Greenland. *Geochim. Cosmochim. Acta*, 40:513–527.

Compston, W., D. O. Froude, T. R. Ireland, P. D. Kinny, I. S. Williams, I. R. Williams, and J. S. Myers, 1985.

The age of (a tiny part of) the Australian continent. *Nature*, 317:559–560.

Compston, W., and R. T. Pidgeon, 1986. Jack Hills, evidence of more very old detrital zircons in Western Australia. *Nature*, 321:766–769.

Corfu, F., and S. R. Noble, 1992. Genesis of the southern Abitibi greenstone belt, Superior Province, Canada: Evidence from zircon Hf isotope analyses using a single-filament technique. *Geochim. Cosmochim. Acta*, 56:2081–2097.

Corfu, F., and G. M. Stott, 1993. Age and petrogenesis of two late Archean magmatic suites, northwestern Superior Province, Canada: Zircon U–Pb and Lu–Hf isotopic relations. *J. Petrol.*, 34:817–838.

Faure, G., 2001. *Origin of Igneous Rocks; the Isotopic Evidence*. Springer-Verlag, Heidelberg.

Froude, D. O., T. R. Ireland, P. D. Kinny, I. S. Williams, W. Compston, I. R. Williams, and J. S. Myers, 1983. Ion microprobe identification of 4,100–4,200 Myr-old terrestrial zircons. *Nature*, 304:616–618.

Kinny, P. D., W. Compston, and I. S. Williams, 1991. A reconnaissance ion-probe study of hafnium isotopes in zircon. *Geochim. Cosmochim. Acta*, 55:849–859.

Patchett, P. J., and M. Tatsumoto, 1980. A routine high-precision method for Lu–Hf isotope geochemistry and chronology. *Contrib. Mineral. Petrol.*, 76:263–268.

Patchett, P. J., W. M. White, H. Feldmann, S. Kielinczuk, and A. W. Hofmann, 1984. Hafnium/rare earth element fractionation in the sedimentary system and crustal recycling into the Earth's mantle. *Earth Planet. Sci. Lett.*, 69:365–378.

Pettingill, H. S., P. J. Patchett, M. Tatsumoto, and S. Moorbath, 1981. Lu–Hf total-rock age for the Amitsoq Gneisses, West Greenland. *Earth Planet. Sci. Lett.*, 55:150–156.

Scherer, E. E., K. L. Cameron, C. M. Johnson, B. L. Beard, K. M. Barovich, and K. D. Collerson, 1997. Lu–Hf geochronology applied to dating Cenozoic events affecting lower crustal xenoliths from Kilbourne Hole, New Mexico. *Chem. Geol.*, 142:63–78.

Tatsumoto, M., D. M. Unruh, and P. J. Patchett, 1981. U–Pb and Lu–Hf systematics of Antarctic meteorites. In *Proc. 6th Symp. Antarctic Meteorites*, pp.237–249. National Institute of Polar Research, Tokyo, Japan.

CHAPTER 13

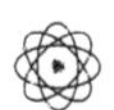

The Re–Os Method

THE Re–Os method of dating is based on the β-decay of ^{187}Re to stable ^{187}Os in certain Re-bearing minerals and rocks in the mantle and continental crust of the Earth. The development of the Re–Os geochronometer was retarded by uncertainty about the value of the decay constant of ^{187}Re and by the low abundance of Os in most rocks and minerals. These problems have been overcome, and as a result, the Re–Os method has become important for dating iron meteorites and Re-bearing terrestrial minerals and rocks. In addition, systematic variations of the isotope composition of Os are used in Chapter 17 to constrain the origin of igneous rocks and the geochemical and isotopic evolution of the crust and mantle of the Earth (Shirey and Walker, 1998). The isotope geochemistry of Os in rivers and in the oceans is presented in Sections 19.5 and 19.6, respectively.

13.1 RHENIUM AND OSMIUM IN TERRESTRIAL AND EXTRATERRESTRIAL ROCKS

Rhenium (Re, $Z = 75$) and osmium (Os, $Z = 76$), like Lu and Hf, are members of the sixth period in the periodic table of the elements. Although Re is a congener of Mn and Tc in group VIIB, its chemical properties are more similar to those of Mo ($Z = 42$), which it replaces in Mo-bearing minerals. Osmium is a member of the Pt group

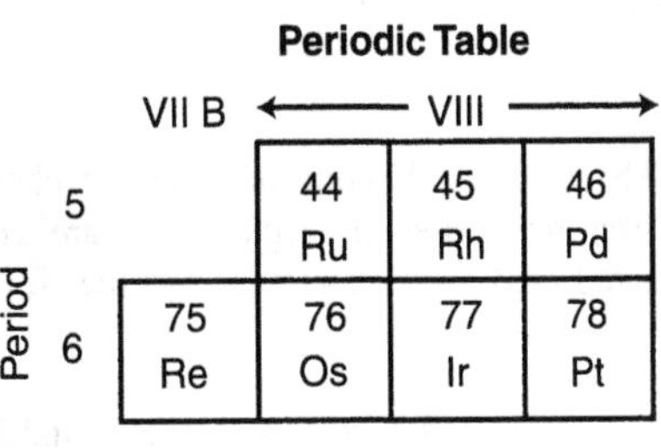

FIGURE 13.1 Location of Re and Os and the platinum group metals (PGEs) in the periodic table of the elements.

elements (PGEs) in Figure 13.1, which consists of ruthenium (Ru), rhodium (Rh), palladium (Pd), osmium (Os), iridium (Ir), and platinum (Pt). Both Re and Os have several valences consistent with their electron formulas, listed in Table 13.1. In addition, their ionic radii in the tetravalent state are identical at 0.71 Å, and both elements are strongly siderophile. The ionic radius of Re^{4+} (0.71 Å) is also similar to that of Mo^{4+} (0.68 Å), and their electronegativities are virtually identical (1.9 for Re and 1.8 for Mo), which explains why Re^{4+} can replace Mo^{4+} in molybdenite (MoS_2) and other Mo-bearing sulfide minerals. In addition, Re^{4+} tends to occur in Cu sulfide minerals presumably because Re^{4+} is captured in place of Cu^{2+} (0.69 Å, electronegativity 1.9).

Table 13.1. Physical and Chemical Properties of Re and Os

Parameter	Re	Os
Atomic number	75	76
Atomic weight	186.207	190.2286[a]
Electron formula	Re: [Xe] $4f^{14}5d^{5}\ 6s^{2}$	
	Os: [Xe] $4f^{14}5d^{6}\ 6s^{2}$	
Valences (+)	7, 6, 4, 2, −1	8, 6, 4, 3, 2
Ionic radius (Å)	0.71 (+4)	0.71 (+4)
	0.60 (+6)	—
Electronegativity	1.9	2.2
Abundance[b] (solar system, per 10^6 atoms of Si)	5.17×10^{-2}	6.75×10^{-1}

[a] Based on the isotope composition of Os in Table 13.3 reported by Shirey and Walker (1998).
[b] From Anders and Grevesse (1989).

The Re and Os concentrations of chondrites and iron meteorites in Figure 13.2 are strongly correlated and are expressed by the following equations (in ppb):

For carbonaceous chondrites (Morgan and Lovering, 1967),

$$\text{Re} = 0.0646\ \text{Os} + 8.10 \quad r^2 = 0.973 \quad N = 4$$

For ordinary chondrites (Morgan and Lovering, 1967; Chen et al., 1998),

$$\text{Re} = 0.0823\ \text{Os} + 3.83 \quad r^2 = 0.953 \quad N = 39$$

For iron meteorites (Horan et al., 1994; Shen et al., 1996, 1998),

$$\text{Re} = 0.0725\ \text{Os} + 89.57 \quad r^2 = 0.995 \quad N = 49$$

Additional determinations of the Re and Os concentrations of meteorites have been reported elsewhere:

Crocket et al., 1967, *Geochim. Cosmochim. Acta*, 31:1615–1623. Luck and Allègre, 1983, *Nature*, 302:130–132. Smoliar et al., 1996, *Science*, 271:1099–1102. Smoliar et al., 1997, *Meteoritics Planet. Sci.*, 32:A122–A123. Ebihara and Ozaki, 1995, *Geophys. Res. Lett.*, 22:2167–2170. Hirata and Nesbitt, 1997, *Earth Planet. Sci. Lett.*, 147:11–24. Horan et al., 1998, *Geochim. Cosmochim. Acta*, 62: 545–554. Morgan et al., 1992, *Earth Planet. Sci. Lett.*, 108:191–202. Morgan et al., 1995, *Geochim. Cosmochim. Acta*, 59:2331–2344. Walker and Morgan, 1989, *Science*, 243: 519–522.

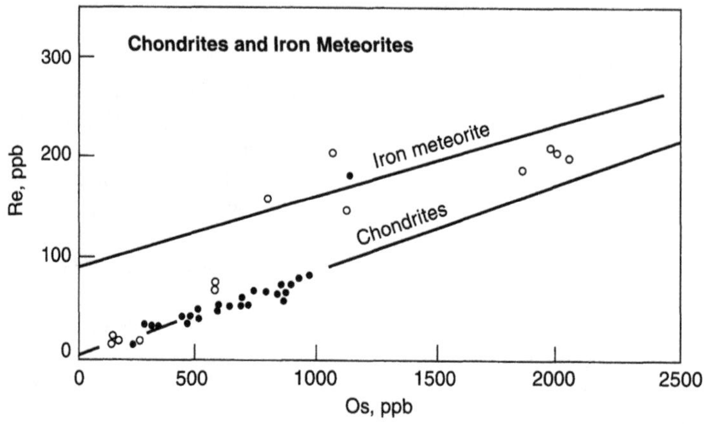

FIGURE 13.2 Concentrations of Re and Os in chondrites (solid circles) and iron meteorites (open circles). The Os concentrations of chondrites are less than about 1000 ppb, whereas those of iron meteorites exceed the scale of this diagram and range up to 67,270 ppb. The equations of the straight lines are given in the text. Data from Morgan and Lovering (1967), Horan et al. (1992), Shen et al. (1996, 1998), and Chen et al. (1998).

Metallic meteorites contain up to 67,270 ppb Os and up to 4946 ppb Re (Negrillos; Shen et al., 1996). The average concentrations of these elements in iron meteorites, listed in Table 13.2, are Os = 15,260 ppb and Re = 987 ppb, giving them a low average Re/Os ratio of 0.065. Ordinary chondrites have low average concentration of Os (690 ppb) and Re (62.6 ppb) with an average Re/Os ratio of 0.091. Rambaldi et al. (1978) demonstrated that Re and Os reside in metal grains of ordinary chondrites and that the silicate and sulfide minerals contain only about 3.90 ppb Re and 77.7 ppb Os (Re/Os = 0.050). Troilite (FeS) and schreibersite [(Fe, Ni)$_3$P] likewise have low concentrations of Re and Os, but their average Re/Os ratio is comparatively high (0.40). Papanastassiou et al. (1995) therefore investigated the feasibility of using these minerals together with metallic FeNi to obtain internal Re–Os isochrons of iron meteorites. Subsequently, Shen et al. (1996) published Re–Os dates between 3.45 ± 0.07 and 4.23 Ga for schreibersite in three iron meteorites (recalculated to $\lambda = 1.666 \times 10^{-11}\ y^{-1}$.) These dates are up to about 1 billion years younger than the Re–Os isochron dates of iron meteorites (see Table 13.4 later). A similar result was reported by Chen et al. (1998) for the Re–Os model date of troilite in the St. Severn chondrite. Lunar rocks have low concentrations of Re (0.002–2 ppb) and Os (0.02–0.9 ppb), whereas lunar soil and impact glasses have elevated Re and Os concentrations because they contain meteoritic material. The abundances of these elements in the solar system (Table 13.1) correspond to an average Re/Os ratio 0.072 (Anders and Grevesse, 1989).

The average concentrations of Re and Os in terrestrial igneous rocks in Table 13.2 are less than 1.0 ppb for Re and less than 3.0 ppb for Os. Nevertheless, Re/Os ratios of igneous rocks vary widely from less than 0.1 in ultramafic rocks to greater than 10 in sialic rocks of the continental crust. The increase of the Re/Os ratio of different kinds of igneous rocks in Figure 13.3 is caused primarily by a decrease of the Os concentration from about 2.8 ppb in ultramafic rocks to less than 0.1 ppb in granites, whereas the Re concentrations range narrowly from about 0.20 to 1.0 ppb.

Granitic gneisses and some shales have elevated Re concentrations ranging up to 200 ppb while their Os concentrations remain <10 ppb.

Table 13.2. Concentrations of Re and Os in Terrestrial and Extraterrestrial Rocks and Minerals

Material	Concentration, ppb		Re/Os	References[a]
	Re	Os		
Extraterrestrial Rocks and Minerals				
Carbonaceous chondrites	47.7 ±6.1 (5)	605 ±117 (4)	0.078	1
Ordinary chondrites				
Bulk samples	62.6 ±8.6 (40)	690 ±73 (40)	0.091	1, 2
Silicate plus sulfide	3.90 ±1.19 (22)	77.7 ±25.9 (16)	0.050	6
Metal grains	1075 ±570 (22)	12,330 ±4112 (16)	0.087	6
Iron meteorites[b]	987 ±425 (49)	15,260 ±7070 (40)	0.065	2–5
Troilite	0.716 ±0.35 (3)	1.83 ±0.70 (3)	0.39	2, 4
Schreibersite (Fe,Ni)$_3$P	16.77 ±16.3 (3)	42.6 ±41.2 (3)	0.39	4
Plagioclase	0.65 (1)	47.6 (1)	0.014	6
Terrestrial Igneous Rocks				
Peridotites	0.83 ±0.55 (8)	2.19 ±0.82 (8)	0.38	18
Ultramafic rocks	0.17 ±0.06 (16)	2.76 ±0.56 (18)	0.062 ±0.04	11, 12
Kimberlite	0.069 (1)	1.34 (1)	0.051	11
Nephelinite	0.27 (1)	0.27 (1)	1.0	11
Komatiite	0.946 ±0.145 (12)	1.568 ±0.905 (12)	0.60	17
Tholeiite	0.84 (1)	<0.03 (1)	>28	11
Olivine gabbro	0.24 ±0.04 (5)	0.073 ±0.122 (5)	3.3	14
Leucotroctolite	0.062 ±0.031 (2)	0.0053 ±0.0018 (2)	12	14
Diabase	0.42 (1)	0.26 (1)	1.6	11
Andesite	0.34 (1)	<0.02 (1)	>17	11
Granite	0.56 (1)	<0.06 (1)	>9	11
Metamorphic Rocks				
Eclogite	0.87 ±1.03 (4)	0.29 ±0.28 (4)	3	11
Granulite	0.37(1)	0.16(1)	2.3	11
Granite gneiss	100 ±109 (4)	5.90 ±6.80 (4)	17	14
Sedimentary Rocks				
Shale	232.3 ±279 (3)	1.717 ±0.932 (3)	132	9

continued overleaf

Table 13.2. ***(continued)***

Material	Concentration, ppb Re	Os	Re/Os	References[a]
		Minerals		
Molybdenite	123,110 ±90, 000 (43)	—	—	7
Molybdenite	63,860 ±42, 580 (9)	912(^{187}Os) ±1300(9)	Very large	8
Native gold	15.8 ±9.1 (8)	2890 ±2390 (9)	0.0055	10
Pyrite	2.32 ±0.49 (4)	0.385 ±0.198 (4)	6.0	10
Pentlandite	146 (1)	9.10 (1)	16.0	14
Chalcopyrite	137 (1)	6.21 (1)	22.0	14
Pyrrhotite	168 (1)	11.1 (1)	15.1	14
Gersdorffite (NiAsS)	53.4 ±0.2 (2)	10,468 ±3482 (2)	0.0051	16
Massive sulfide Cu–Ni–PGE	112 ±39 (11)	554 ±273 (13)	0.20	13
Olivine–sulfide cumulate	1.93 ±1.01 (9)	70.6 ±74 (10)	0.027	13
Magnetite	10.8 (1)	0.510 (1)	21.2	14
Ilmenite	3.68 (1)	0.182 (1)	20.2	14
Chromite	0.478 ±0.035 (21)	50.7 ±0.0 (21)	0.0094	15, 16
Garnet	12.6 (1)	—	—	16

Note: The number of samples included in each average is indicated in parentheses.

[a] 1. Morgan and Lovering, 1967a, *Geochim. Cosmochim. Acta*, 31:1893–1909. 2. Chen et al., 1998, *Geochim. Cosmochim. Acta*, 62:3379–3392. 3. Horan et al., 1992, *Science*, 255:1118–1121. 4. Shen et al., 1996, *Geochim. Cosmochim. Acta*, 60:2887–2990. 5. Shen et al., 1998, *Geochim. Cosmochim. Acta*, 62:2715–2724. 6. Rambaldi et al., 1978, *Earth Planet. Sci. Lett.*, 40:175–186. 7. Riley, 1967, *Geochim. Cosmochim. Acta*, 31:1489–1498. 8. Hirt et al., 1963, in Geiss and Goldberg (Eds.), *Earth Science and Meteoritics*, pp. 273–280, North-Holland, Amsterdam. 9. Horan et al., 1994, *Geochim. Cosmochim. Acta*, 58:257–265. 10. Kirk et al., 2001, *Geochim. Cosmochim. Acta*, 65:2149–2159. 11. Morgan and Lovering, 1967b, *Earth Planet. Sci. Lett.*, 3:219–224. 12. Meisel et al., 1996, *Nature*, 383:517–520. 13. Foster et al., 1996, *Nature*, 382:703–705. 14. Lambert et al., 2000, *Econ. Geol.*, 95:867–888. 15. Nägler et al., 1997, *Geology*, 25:983–986. 16. Walker et al., 1996, *Earth Planet. Sci. Lett.*, 141:161–173. 17. Walker et al., 1988, *Earth Planet. Sci. Lett.*, 87:1–12. 18. Roy-Barman et al., 1996, *Chem. Geol.*, 130:55–64.

[b] The Re and Os concentrations of many additional iron meteorites were determined by Pernicka and Wasson (1987).

Consequently, the granitic gneisses of the continental crust have high Re/Os ratios that average about 50 (Esser and Turekian, 1993) but may exceed 100 in some cases. The high Re/Os ratios of crustal rocks has caused them to become enriched

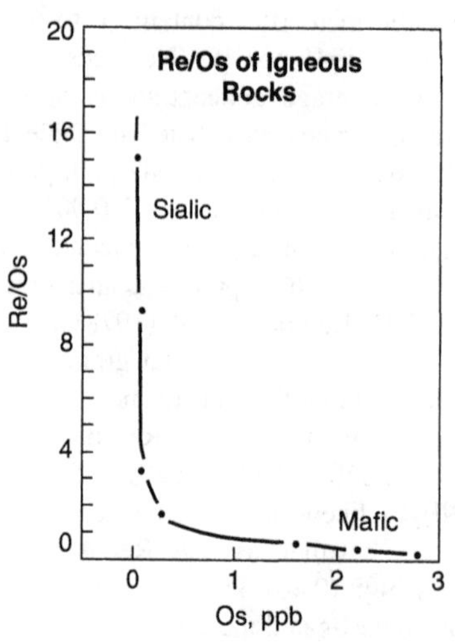

FIGURE 13.3 Variation of the Re/Os ratio of igneous rocks as a function of their Os concentrations. Data and references in Table 13.2.

in radiogenic ^{187}Os by decay of ^{187}Re over long periods of geological time. This simple fact distinguishes the rocks of the continental crust from the ultramafic rocks of the subcontinental mantle and thus contributes to the study of the origin of igneous rocks (Faure, 2001).

In addition, the low concentrations of Re and Os in rocks composed of silicate minerals compared to iron meteorites indicate that a significant fraction of both elements dissolved in molten iron during the internal differentiation of the Earth and was thereby incorporated into its core (Crocket, 1969; Morris and Short, 1969).

The minerals in Table 13.2 can be classified into three categories:

1. low Re and low Os concentrations (e.g., most rock-forming silicate minerals),
2. high Re and low Os concentrations (e.g., molybdenite and Cu-Ni-PGE sulfides), and
3. low Re and high Os concentrations (e.g., osmiridium, native gold, chromite, gersdorffite).

Several minerals have potential for dating by the Re–Os method but have not yet been investigated (e.g., magnetite, ilmenite, and garnet).

Table 13.3. Naturally Occurring Isotopes of Re and Os

Isotope	Abundance, %	Mass, amu
	Rhenium ($Z = 75$)	
^{185}Re	37.40	184.952951
^{187}Re	62.6	186.955744
	Osmium[a] ($Z = 76$)	
^{184}Os	0.0177	183.952488
^{186}Os	1.593	185.953830
^{187}Os	1.513	186.955741
^{188}Os	13.29	187.955860
^{189}Os	16.22	188.958137
^{190}Os	26.38	189.958436
^{192}Os	40.98	191.961467

[a] Johnson-Matthey Os standard corrected for isotope fractionation to $^{188}Os/^{192}Os = 0.32440$. From Shirey and Walker (1998).

13.2 PRINCIPLES AND METHODOLOGY

Rhenium ($Z = 75$) has two naturally occurring isotopes (^{185}Re and ^{187}Re) whose abundances are listed in Table 13.3. Rhenium-187 emits β^--particles as it decays to ^{187}Os, which is one of the seven naturally occurring isotopes of Os ($Z = 76$). The halflife of ^{187}Re has been difficult to determine because the endpoint energy (E_{max}) of the β-spectrum is only about 2.5 keV. Therefore, Hirt et al. (1963) determined the halflife of ^{187}Re by analyzing 10 samples of molybdenite (MoS_2) of known age and obtained a value of $(4.3 \pm 0.5) \times 10^{10}$ years. Subsequently, a direct determination by Lindner et al. (1989) yielded a value of $(4.23 \pm 0.13) \times 10^{10}$ years based on the amount of radiogenic ^{187}Os that formed by decay of ^{187}Re in known weights of a purified Re compound in one to five years. Accordingly, the halflife of ^{187}Re is

$$\lambda = \frac{\ln 2}{(4.23 \pm 0.13) \times 10^{10}} = (1.64 \pm 0.05) \times 10^{-11}\ y^{-1}$$

Shen et al. (1998) later refined the decay constant of ^{187}Re to

$$\lambda = 1.66 \times 10^{-11}\ y^{-1}$$

based on the slope of a Re–Os isochron of group IIAB iron meteorites and the knowledge that the age of the solar system is 4.56 Ga. Similarly, Shirey and Walker (1998) and Smoliar et al. (1997) used the slope of a Re–Os isochron for type IIIA iron meteorites published by Smoliar et al. (1996) and the Pb–Pb age of angrite achondrite meteorites dated by Lugmair and Galer (1992) to make a further refinement in the value of the decay constant to

$$\lambda = (1.666 \pm 0.006) \times 10^{-11}\ y^{-1}$$

The uncertainty of this value (±0.36%) does not include systematic errors arising from the calibration of the spike solution used to determine the Os concentrations by isotope dilution. Therefore, Shirey and Walker (1998) raised the uncertainty of the decay constant to ±1.0%. Accordingly, the best value of the decay constant of ^{187}Re is

$$\lambda = (1.666 \pm 0.017) \times 10^{-11}\ y^{-1}$$

The isotope composition of Os was originally determined by Nier (1937) and subsequently by many other investigators, including Masuda et al. (1987), Völkening et al. (1991), and Shirey and Walker (1998), whose results are listed in Table 13.3.

The decay of ^{187}Re to ^{187}Os is expressed by

$$^{187}_{75}Re \rightarrow ^{187}_{76}Os + \beta^- + \overline{\nu} + Q \qquad (13.1)$$

where β^- = negatively charged β-particle
$\overline{\nu}$ = antineutrino
Q = total decay energy (0.0025 MeV)

The formation of radiogenic ^{187}Os as a function of time follows from the law of radioactivity:

$$\frac{^{187}Os}{^{188}Os} = \left(\frac{^{187}Os}{^{188}Os}\right)_i + \frac{^{187}Re}{^{188}Os}(e^{\lambda t} - 1) \qquad (13.2)$$

The abundance of radiogenic ^{187}Os has also been expressed relative to ^{186}Os:

$$\frac{^{187}Os}{^{186}Os} = \left(\frac{^{187}Os}{^{186}Os}\right)_i + \frac{^{187}Re}{^{186}Os}(e^{\lambda t} - 1) \qquad (13.3)$$

The two ratios are convertible into each other by the relation

$$\frac{^{187}Os}{^{188}Os} = \frac{^{187}Os}{^{186}Os} \times \frac{^{186}Os}{^{188}Os} \qquad (13.4)$$

The numerical value of the $^{186}Os/^{188}Os$ ratio is a constant regardless of the abundances of these two isotopes, which decline as the abundance of radiogenic ^{187}Os increases. Using the data of Shirey and Walker (1998) in Table 13.3 yields

$$\frac{^{186}Os}{^{188}Os} = \frac{1.593}{13.29} = 0.11986$$

The $^{187}Os/^{188}Os$ ratio (equation 13.2) is preferred because ^{186}Os has a low abundance (1.593%), which increases the error of measurement of the $^{187}Os/^{186}Os$ ratio. In addition, the abundance of ^{186}Os in Pt-rich samples varies because ^{186}Os is a product of α-decay of naturally occurring long-lived $^{190}_{78}Pt(T_{1/2} = 4.50 \times 10^{11}$ y). In fact, $^{186}_{76}Os$ is itself a long-lived radionuclide that decays by α-emission to stable $^{182}_{74}W(T_{1/2} = 1.98 \times 10^{15}$ y) (Shirey and Walker, 1998).

The concentrations of Re and Os are determined by isotope dilution using thermal ionization mass spectrometry (Riley, 1967) or by neutron activation (e.g., Morgan, 1965; Crocket et al., 1967; Hoffman et al., 1978). Methods for separating Re and Os in aqueous solution from Mo and the PGEs have been presented elsewhere:

Morgan et al., 1991, *Talanta*, 38:259–265. Ravizza and Pyle, 1997, *Chem. Geol.*, 141: 251–268. Rehkämper and Halliday, 1997, *Talanta*, 44:663–672. Rehkämper et al., 1998, *Fresenius J. Anal. Chem.*, 361:217–219. Cohen and Water, 1996, *Anal. Chim. Acta*, 332:269–275. Brauns, 2001, *Chem. Geol.*, 176:379–384.

The dissolution of iron meteorites in Carius tubes was described in detail by Shen et al. (1996) and Shirey and Walker (1994).

The isotope composition of Os and Re is measured by means of negatively charged ions (OsO_3^-, ReO_4^-) following the work of Walczyk et al. (1991), Völkening et al. (1991), and Creaser et al. (1991). The Os isotope ratios are corrected for instrumental fractionation to $^{188}Os/^{192}Os$ = 0.32440 (Nier, 1937; Shen et al., 1996). Additional information about the analytical procedures and mass spectrometry has been provided elsewhere:

Walker et al., 1994, *Geochim. Cosmochim. Acta*, 58:4179–4197. Horan et al., 1994, *Geochim. Cosmochim. Acta*, 58:257–265. Cohen and Water, 1996, *Anal. Chim. Acta*, 332:269–275. Colodner et al., 1993, *Anal. Chem.*, 65:1419–1425.

The isotopic composition of Os has also been measured by means of ion probe mass spectrometry (Allègre and Luck, 1980), accelerator mass spectrometry (Fehn et al., 1986), and resonance ionization mass spectrometry (Walker and Fassett, 1986; Walker, 1988). Increasingly, the instrument of choice is the ICP mass spectrometer using aqueous solutions or laser ablation:

Hulbert and Gregoire, 1993, *Can. Mineral.*, 31:861–876. Pearson and Woodland, 2000, *Chem. Geol.*, 165:87–107. Hassler et al., 2000, *Chem. Geol.*, 166:1–14. Hirata et al., 1995, *Chem. Geol.*, 144:269–280. Dickin et al., 1988, *J. Anal. Atom. Spectrom.*, 3:337–342. Grégoire, 1990, *Anal. Chem.*, 62:141–146. Hirata and Masuda, 1990, *J. Anal. Atom. Spectrom.*, 5:617–630. Richardson et al., 1989, *Anal. Atom. Spectrom.*, 4:465–471. Russ et al., 1987, *Anal. Chem.*, 59:984–989. Jackson et al., 1990, *Chem. Geol.*, 83:119–132.

13.3 MOLYBDENITE AND ^{187}Re–^{187}Os ISOCHRONS

Molybdenite (MoS_2) and certain other sulfide minerals have relatively high Re concentrations ranging from about 0.245 to 1690 ppm (Hirt et al., 1963a, b; Riley, 1967; Luck and Allègre, 1982). The reason for the wide range of concentrations is not known (Riley, 1967), but it may be related to the availability of Re at the localities where this mineral crystallized.

13.3a Molybdenite

In many cases, the Os in molybdenites consists almost entirely of radiogenic ^{187}Os such that their $^{187}Re/^{188}Os$ and $^{187}Os/^{188}Os$ ratios have large values approaching infinity and therefore are difficult to measure accurately. For this reason, the growth of radiogenic ^{187}Os is expressed directly by the equation

$$^{187}\text{Os} = {}^{187}\text{Re}\,(e^{\lambda t} - 1) \qquad (13.5)$$

where the initial number of ^{187}Os atoms is assumed to be equal to zero. In addition, ^{187}Os and ^{187}Re in equation 13.5 are expressed as concentrations in micrograms per gram (Markey et al., 1998). Consequently, Re–Os dates of molybdenites are calculated from the $^{187}Os/^{187}Re$ concentration ratio by solving equation 13.5:

$$t = \frac{1}{\lambda} \ln \left(\frac{^{187}\text{Os}}{^{187}\text{Re}} + 1 \right) \qquad (13.6)$$

For example, Luck and Allègre (1982) reported the following data for molybdenite from Godthaab, Greenland:

$$\text{Re} = 55.2 \text{ ppm} \qquad {}^{187}\text{Os} = 0.98 \text{ ppm}$$

The concentration of ^{187}Re is (Tables 13.1, and 13.3)

$$^{187}\text{Re} = \frac{55.2 \times 0.626 \times 186.955}{186.207} = 34.69 \ \mu\text{g/g}$$

Hence,

$$t = \frac{1}{1.666 \times 10^{-11}} \ln \left(\frac{0.98}{34.69} + 1 \right) = 1.67 \text{ Ga}$$

This sample of molybdenite is significantly *younger* than the Archean gneisses of West Greenland, either because the mineral formed during an episode of hydrothermal activity in late Early Proterozoic time or because ^{187}Os was lost during regional metamorphism from the molybdenite that crystallized prior to 1.67 Ga.

This method of dating molybdenites was recently used by the following:

Watanabe and Stein, 2000, *Econ. Geol.*, 95:1437–1542. Torrealday et al., 2000, *Econ. Geol.*, 95:1165–1170. Stein et al., 1998a, *Mineral. Deposita*, 33:329–345. Stein et al., 1998b. *SEG Newsletter*, 32:8–15.

13.3b ^{187}Re–^{187}Os Isochrons

Equation 13.5 represents a family of straight lines in coordinates of the concentrations of $^{187}Re(x)$ and $^{187}Os(y)$. The slopes of these lines are equal to $m = e^{\lambda t} - 1$, and their intercepts on the y-axis are equal to zero. These lines are isochrons because all data points on a line having a constant slope m have the same age.

Stein et al. (2000) reported that pyrite and chalcopyrite in gold-bearing quartz veins at Harnäs in southwest Sweden have low Re concentrations ($<$1–5 ppb) as well as low radiogenic ^{187}Os concentrations (0.005–0.074 ppb). The concentrations of common Os in these minerals are so low (0.002–0.017 ppb) that the $^{187}Os/^{188}Os$ ratios are difficult to measure. The analytical data for the sulfide minerals at Harnäs in southwestern Sweden define a ^{187}Re–^{187}Os isochron in Figure 13.4 whose slope yields a date of 973 ± 34 Ma ($\lambda = 1.666 \times 10^{-11}$ y^{-1}). This date indicates that the sulfide minerals were emplaced close to the end of the Sveconorwegian (Grenvillian) orogeny, which lasted from 1.15 to 0.90 Ga (Stein et al., 2000).

One sample of pyrite in the quartz vein at Harnäs deviates significantly from the isochron and yields a date of about 616 Ma. This sample, which was analyzed in quadruplicate, dates an episode of renewed tectonic activity during which the veins were reopened allowing additional sulfide minerals to be deposited.

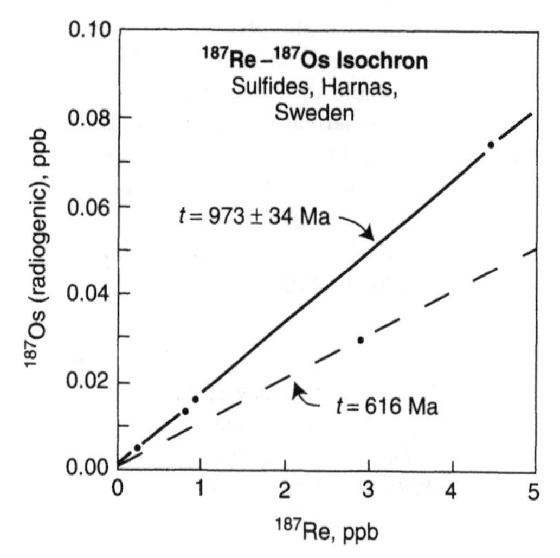

FIGURE 13.4 ^{187}Re–^{187}Os isochron defined by pyrite and chalcopyrite in a gold-bearing quartz vein at Harnäs in southwestern Sweden. The main episode of sulfide deposition occurred at 973 ± 34 Ma during the waning stage of the Sveconorwegian (Grenvillian) orogeny. A subsequent event at about 616 Ma reopened the vein system and permitted additional deposition of pyrite. Data from Stein et al. (2000) for $\lambda = 1.666 \times 10^{-11}$ y^{1}.

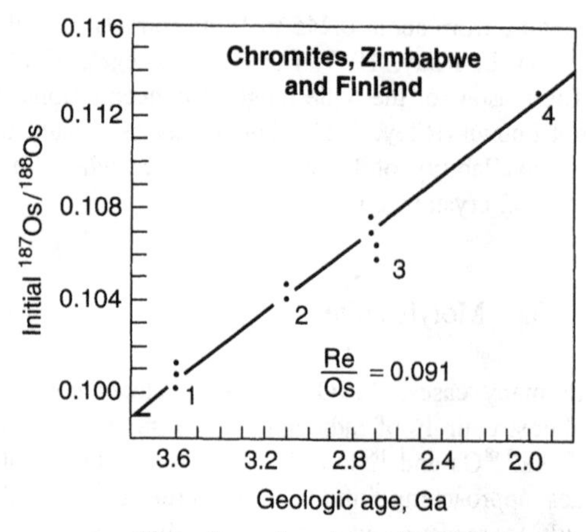

FIGURE 13.5 Time-dependent isotopic evolution of Os in the subcontinental mantle of the Earth indicated by chromites of known age from Zimbabwe and Finland. The localities are identified by number: 1-Hornet and Valley, Zimbabwe; 2-Shurugwi, Zimbabwe; 3-Inyala and Prince, Zimbabwe; 4-Outokumpu, Finland. Data from Nägler et al. (1997) (Zimbabwe) and Walker et al. (1996) (Finland).

13.3c Chromite

Samples of chromite from Zimbabwe and Finland analyzed by Nägler et al. (1997) and Walker et al. (1996) have a low average Re/Os ratio of 0.0094 ± 0.0073 and are therefore difficult to date by the conventional Re–Os isochron method. Instead, Nägler et al. (1997) calculated the initial ^{187}Os/^{188}Os ratios for chromites of known age from Zimbabwe. The results in Figure 13.5, including the data of Walker et al. (1996) for chromites at Outokumpu in Finland, define a straight line having a slope $m = 0.007463$ and an ^{187}Os/^{188}Os ratio of 0.12738 at $t = 0.0$ Ga.

The straight line in Figure 13.5 depicts the time-dependent increase of the ^{187}Os/^{188}Os ratio of the subcontinental mantle (i.e., the initial ^{187}Os/^{188}Os ratio of the chromites), as a result of the decay of ^{187}Re to ^{187}Os. The slope of the line is related to the Re/Os ratio of the source of the chromites by the equation (Section 5.2, equation 5.7)

$$\frac{^{187}\text{Os}}{^{188}\text{Os}} \cong \left(\frac{^{187}\text{Os}}{^{188}\text{Os}}\right)_i + \left(\frac{^{187}\text{Re}}{^{188}\text{Os}}\right)_{\text{atom}} \lambda t \qquad (13.7)$$

from which it follows that the slope of the line in Figure 13.5 is

$$m = \left(\frac{^{187}\text{Re}}{^{188}\text{Os}}\right)_{\text{atom}} \lambda \qquad (13.8)$$

and therefore

$$\left(\frac{^{187}\text{Re}}{^{188}\text{Os}}\right)_{\text{atom}} = \frac{m}{\lambda}$$

Since $m = 0.007463$ per 10^9 years,

$$\left(\frac{^{187}\text{Re}}{^{188}\text{Os}}\right)_{\text{atom}} = \frac{0.007463 \times 10^{-9}}{1.666 \times 10^{-11}} = 0.4479$$

The atomic $^{187}Re/^{188}Os$ ratio is converted into the Re/Os concentration ratio by

$$\left(\frac{^{187}\mathrm{Re}}{^{188}\mathrm{Os}}\right)_{\mathrm{atom}} = \left(\frac{\mathrm{Re}}{\mathrm{Os}}\right)_{\mathrm{conc}} \times \frac{\text{at. wt. Os} \times \text{Ab}\ ^{187}\mathrm{Re}}{\text{at. wt. Re} \times \text{Ab}\ ^{188}\mathrm{Os}}$$
$$= \left(\frac{\mathrm{Re}}{\mathrm{Os}}\right)_{\mathrm{conc}} \times \frac{190.2286 \times 62.6}{186.207 \times 13.29} \quad (13.9)$$

which yields

$$\left(\frac{\mathrm{Re}}{\mathrm{Os}}\right)_{\mathrm{conc}} = \left(\frac{^{187}\mathrm{Re}}{^{188}\mathrm{Os}}\right)_{\mathrm{atom}} \times 0.2078$$

Therefore, the Re/Os ratio of the source of the chromites in the subcontinental mantle based on the chromites in Figure 13.5 is

$$\left(\frac{\mathrm{Re}}{\mathrm{Os}}\right)_{\mathrm{conc}} = 0.4479 \times 0.2078 = 0.093$$

This value is indistinguishable from the average Re/Os ratio of ordinary chondrites (0.091) in Table 13.2 and thereby confirms the hypothesis that the chemical composition of the source of the chromites is similar to that of chondrite meteorites. However, Nägler et al. (1997) demonstrated that the initial $^{187}Os/^{188}Os$ ratios of the chromites from Zimbabwe are systematically lower than the $^{187}Os/^{188}Os$ ratios of CHUR between 3.5 and 2.7 Ga. The authors considered that the chromites had originated from sources in the subcontinental lithospheric mantle, which had formed during a preceding episode during which the $^{187}Os/^{188}Os$ ratio remained constant because of the absence of Re.

13.4 METEORITES AND CHUR-Os

The decay of ^{187}Re to ^{187}Os was originally investigated at the University of Cologne by W. Herr and his associates, whose pioneering studies led to an attempt by Herr et al. (1961) to date iron meteorites by this method. The subsequent report by Hirt et al. (1963) contained a Re–Os isochron for 14 iron meteorites whose slope yielded a date of 4.0 Ga based on $T_{1/2} = (4.3 \pm 0.5) \times 10^{10}$ y equivalent to $\lambda = 1.61 \times 10^{-11}\ \mathrm{y}^{-1}$. (Use of $\lambda = 1.666 \times 10^{-11}\ \mathrm{y}^{-1}$ decreases the date from 4.0 to 3.87 Ga.) Hirt et al. (1963) also reported Re and Os concentrations for 30 iron meteorites and for metallic grains of three chondrites (Table 13.2). In addition, they measured $^{187}Os/^{186}Os$ ratios of osmiridium and platinum nuggets from South Africa, Russia, and Australia and calculated Re–Os dates of nine molybdenite samples.

The modern era of Re–Os studies started with the development of an ion probe mass spectrometer described by Shimizu et al. (1978). Luck et al. (1980) used this instrument to date five iron meteorites and the metal grains of one chondrite. The slope of the resulting Re–Os isochron ($m = 0.0765 \pm 0.0035$) is equivalent to $t = 4.42 \pm 0.20$ Ga, recalculated to $\lambda = 1.666 \times 10^{-11}\ \mathrm{y}^{-1}$. The initial $^{187}Os/^{186}Os$ ratio reported by Luck et al. (1980) is 0.805 ± 0.011. Subsequently, Luck and Allègre (1983) refined these results for iron meteorites and for the metallic grains of chondrites. The insights gained by Luck and Allègre (1983) revealed that iron meteorites and metallic grains of chondrites fit the same isochron and therefore formed within a short interval of time (about 90×10^6 years) from the solar nebula, which contained Os that was isotopically homogeneous.

13.4a Iron Meteorites

Iron meteorites are subdivided into 13 groups on the basis of their concentrations of germanium (Ge), gallium (Ga), and nickel (Ni) as proposed by Wasson (1974, 1985) and Malvin et al. (1984). The systematic variations of the chemical compositions of iron meteorites are the result of fractional crystallization of iron–nickel melts that segregated from silicate and sulfide liquids of chondritic composition (Pernicka and Wasson, 1987). The Re–Os method is the only way to determine the crystallization ages of the different groups of iron meteorites because all of the other methods (Rb–Sr, K–Ar, Sm–Nd, U,Th–Pb, and Lu–Hf) are applicable only to the silicate minerals that occur in some iron meteorites (e.g., pallasites).

The Re–Os dates in Table 13.4 reveal that the different groups of iron meteorites crystallized sequentially, presumably because each group

Table 13.4. Re–Os Dates and Initial $^{187}Os/^{188}Os$ Ratios of Iron and Stony Meteorites

Material	Slope of Isochron	$(^{187}Os/^{188}Os)_i$	Date, Ga	Time Since 4.60 Ga ($\times 10^6$ years)	References[a]
		Iron Meteorites			
Group IA	0.07837 ± 0.00041	0.09570 ± 0.00016	4.529 ± 0.023	71 ± 23	1
Group IIA	0.07851 ± 0.00014	0.09544 ± 0.00007	4.537 ± 0.008	63 ± 8	2
Group IIAB	0.07824 ± 0.00053	0.09597 ± 0.00035	4.522 ± 0.029	78 ± 29	3
Group IIAB	0.07848 ± 0.00018	0.09563 ± 0.00011	4.535 ± 0.010	65 ± 10	4
Group IIIA	0.07887 ± 0.00022	0.09524 ± 0.00011	4.558 ± 0.012	42 ± 12	2
Group IIIAB	0.0775 ± 0.0018	0.09732 ± 0.00090	4.480 ± 0.100	120 ± 100	3
Group IVA	0.07721 ± 0.00046	0.09584 ± 0.00023	4.464 ± 0.026	136 ± 26	2
Group IVB	0.07834 ± 0.00052	0.09559 ± 0.00019	4.527 ± 0.029	73 ± 29	2
Group IVA	0.07951 ± 0.00078	0.09519 ± 0.00045	4.592 ± 0.044	8 ± 44	4
Group IVAB	0.07920 ± 0.00019	0.09534 ± 0.00011	4.575 ± 0.011	25 ± 11	4
Pallasites	0.07835 ± 0.00009	0.09560 ± 0.00004	4.528 ± 0.050	72 ± 50	5
		Chondrites			
St. Severin (internal isochron)	0.0797 ± 0.0026	0.0953 ± 0.0013	4.603 ± 0.150	0 ± 150	6
H group whole rock	0.07970	0.09487	4.603	0	6
Carbonaceous[b] chondrites	0.06641	0.1003	3.86	740	7

Note: All dates have been recalculated to $\lambda = 1.666 \times 10^{-11}\ y^{-1}$ and initial $^{187}Os/^{186}Os$ ratios were converted to $^{187}Os/^{188}Os$ ratios by multiplying them by 0.11986 in accordance with equation 13.4 and the abundances of Os isotopes in Table 13.3.

[a] 1. Smoliar et al., 1997, *Meteoritics Planet. Sci.*, 32:A122–A123. 2. Smoliar et al., 1996, *Science*, 271:1099–1102. 3. Horan et al., 1992, *Science*, 255:1118–1121. 4. Shen et al., 1996, *Geochim. Cosmochim. Acta*, 60:2887–2900. 5. Shen et al., 1998, *Geochim. Cosmochim. Acta*, 62:2715–2723. 6. Chen et al., 1998, *Geochim. Cosmochim. Acta*, 62:3379–3392. 7. Walker and Morgan, 1989, *Science*, 243:519–522.

[b] Excluding three analyses of Murray and one of Semarkona.

originated from a different parent body that formed by accretion of planetesimals derived from the solar nebula. Each of these ancestral parent bodies melted and differentiated by segregation of immiscible liquids in its own gravitational field (Ertel et al., 2001). Blobs of liquid iron sank to the center of the body to form a core that crystallized slowly as it cooled. Globules of sulfide liquid were trapped within the FeNi core and in the silicate liquid that formed a mantle around the core. According to this theory, the Re–Os isochrons of the groups of iron meteorites record the times when the cores of their parent bodies cooled sufficiently to allow radiogenic ^{187}Os to accumulate and for their Re/Os ratios to stabilize.

The numerical values of the Re–Os dates of iron meteorites depend on the fact that the decay constant of ^{187}Re was determined by assuming an age of 4.558 ± 0.004 Ga for achondrite (angrite) meteorites (Lugmair and Galer, 1992). However, differences in the Re–Os isochron dates of different kinds of meteorites based on this decay constant are accurate and record events that occurred primarily during the first 100 million years in the early solar system.

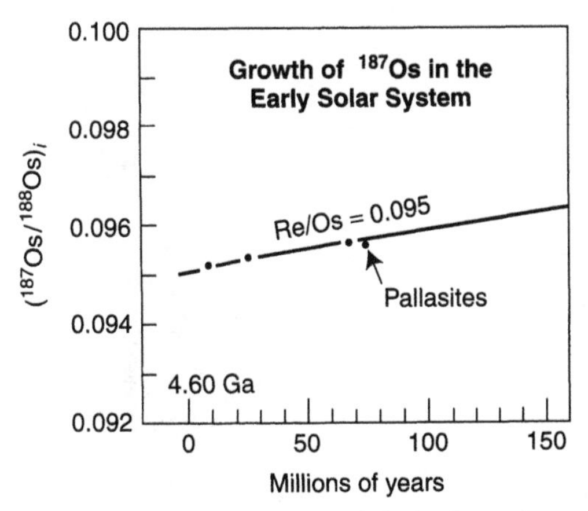

FIGURE 13.6 Isotope evolution of Os in the early solar system revealed by the initial $^{187}Os/^{188}Os$ ratios and Re–Os isochron dates of iron meteorites in Table 13.5 dated by Shen et al. (1996.) The pallasite data of Shen et al. (1998) were not included. The slope of the straight line corresponds to a Re/Os ratio of 0.095. Time was measured forward from a date of 4.60 Ga.

The initial $^{187}Os/^{188}Os$ ratios and Re–Os isochron dates of iron meteorites reported by Shen et al. (1996) in Figure 13.6 define a straight line, although the error bars were omitted for the sake of clarity of this illustration. The zero point on the time axis was arbitrarily set at 4.60 Ga, and time was measured forward from that date. A least-squares regression of the data points in Figure 13.6 to a straight line yields a slope $m = 7.635 \times 10^{-6}$ and an intercept on the y-axis (at 4.60 Ga) of 0.095137. This line expresses the time-dependent increase of the abundance of radiogenic ^{187}Os caused by decay of ^{187}Re in the solar nebula. The Re/Os ratio of the solar nebula is obtained from equation 13.2 by first calculating the $^{187}Os/^{188}Os$ ratio of iron meteorites at an arbitrarily selected value of t based on the slope and initial $^{187}Os/^{188}Os$ ratio calculated above. For example, after a decay time of 100 million years,

$$\left(\frac{^{187}\mathrm{Os}}{^{188}\mathrm{Os}}\right)_t = 0.095137 + 7.635 \times 10^{-6} \times 100$$

$$= 0.095900$$

The $^{187}Re/^{188}Os$ ratio of the reservoir follows from equation 13.2:

$$\left(\frac{^{187}\mathrm{Os}}{^{188}\mathrm{Os}}\right)_t = \left(\frac{^{187}\mathrm{Os}}{^{188}\mathrm{Os}}\right)_i + \left(\frac{^{187}\mathrm{Re}}{^{188}\mathrm{Os}}\right)_t (e^{\lambda t} - 1)$$

$$\left(\frac{^{187}\mathrm{Re}}{^{188}\mathrm{Os}}\right)_t = \frac{(^{187}\mathrm{Os}/^{188}\mathrm{Os})_t - (^{187}\mathrm{Os}/^{188}\mathrm{Os})_i}{e^{\lambda t} - 1}$$

$$= \frac{0.095900 - 0.095137}{0.0016673} = 0.4576$$

The $^{187}Re/^{186}Os$ ratio is related to the Re/Os concentration ratio by equation 13.9, and hence

$$\left(\frac{\mathrm{Re}}{\mathrm{Os}}\right)_{\mathrm{conc}} = \left(\frac{^{187}\mathrm{Re}}{^{188}\mathrm{Os}}\right)_{\mathrm{atom}} \times 0.2078$$

Therefore, the reservoir from which the iron meteorites originated was characterized by

$$\left(\frac{\mathrm{Re}}{\mathrm{Os}}\right)_{\mathrm{conc}} = 0.095$$

This value is indistinguishable from the average Re/Os ratio of ordinary chondrites (0.091) in Table 13.2 reported by Morgan and Lovering (1967) and Chen et al. (1998). Smoliar et al. (1996) obtained a similar Re/Os ratio for the reservoir of iron meteorites of 0.087 ± 0.013 based on their own data for different groups of iron meteorites. Additional noteworthy studies of the Re–Os systematics of iron meteorites include papers by Morgan et al. (1992, 1995).

In summary, these results support the conclusion that the iron meteorites crystallized from iron liquids that segregated from silicate melts of chondritic composition. In addition, the iron crystallized at different times during an interval of less than 100 million years between 4.60 and 4.50 Ga. The Re/Os concentration ratios of the solar nebula derived from analyses of iron meteorites are indistinguishable from the Re/Os ratio (0.091) of the subcontinental mantle of the Earth indicated by chromites from Zimbabwe and Finland in Figure 13.5.

13.4b Chondrites

The low concentrations of Re and Os in silicate and iron sulfide minerals are a serious obstacle

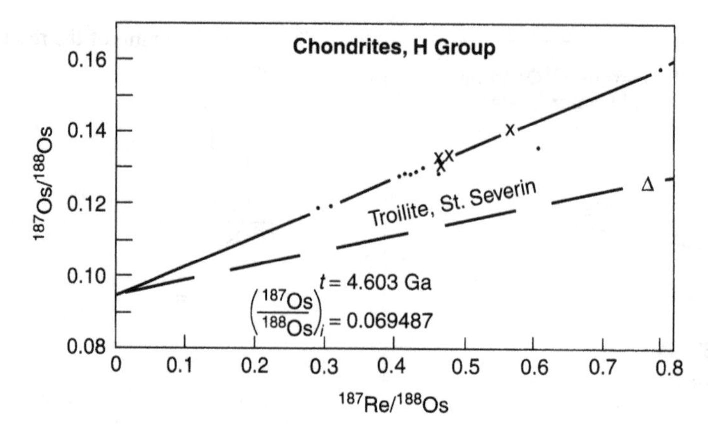

FIGURE 13.7 Re–Os isochron of H-group chondrites: solid circle = whole rock samples, cross = metal, open triangle = troilite, St. Severin. The decay constant of ^{187}Re is $\lambda = 1.666 \times 10^{-11}$ y^{-1}. Data from Chen et al. (1998).

to dating stony meteorites by the Re–Os method. Initial attempts by Luck et al. (1980) and Luck and Allègre (1983) as well as more recent work by Walker and Morgan (1989) and Walker et al. (1993) indicated that chondrites scatter above and below Re–Os isochrons of iron meteorites because of variations in their Re/Os ratios caused by late-stage alteration of stony meteorites. Alternatively, the apparent failure of stony meteorites to form Re–Os isochrons may be caused by the use of a variety of different procedures in the determination of Re and Os concentrations.

Subsequently, Chen et al. (1998) adopted a procedure for processing chondrites that yielded reproducible results. Their data for six whole-rock chondrites as well as for metal grains in Figure 13.7 define a satisfactory Re–Os isochron that is in good agreement with a Re–Os isochron for group IIAB iron meteorites analyzed by Shen et al. (1996). The slope of the chondrite isochron (0.07970) yields a date of 4.60 Ga for $\lambda = 1.666 \times 10^{-11}$ y^{1} and an initial ^{187}Os/^{188}Os ratio of 0.09487. Several whole-rock aliquots and metal grains of the St. Severn chondrite (Table 13.4) yielded an identical date and an initial ^{187}Os/^{188}Os ratio of 0.0953 ± 0.0013. The St. Severn chondrite was also analyzed by Luck and Allègre (1983) and Birck and Allègre (1994).

A sample of troilite (FeS) in the St. Severn meteorite plots far below the chondrite isochron in Figure 13.7. The concentrations of Re and Os in this mineral are Re = 0.57 ppb, Os = 2.46 ppb, Re/Os = 0.23. The date derived from the troilite (2.34 Ga) is probably fictitious because it may reflect gain of Re or loss of Os from this mineral. Similarly, Papanastassiou et al. (1995) reported that schreibersite and troilite in iron meteorites did not remain closed to Re and Os.

13.4c CHUR-Os and ε(Os)

The uniform chondritic reservoir of Os (CHUR-Os) was defined by Chen et al. (1998) based on data for H chondrites:

$$\frac{\text{Re}}{\text{Os}} = 0.088 \pm 0.002$$

$$\left(\frac{^{187}\text{Re}}{^{188}\text{Os}}\right)_{t=0} = 0.423 \pm 0.007$$

$$\left(\frac{^{187}\text{Os}}{^{188}\text{Os}}\right)_{t=0} = 0.12863 \pm 0.00046$$

The numerical values of these parameters are consistent with the average present-day ^{187}Os/^{188}Os ratio (0.1298 ± 0.0011) of the Earth reported by Meisel et al. (1996) and with the Re/Os ratio (0.0875) of chondrites obtained by Ebihara and Ozaki (1995).

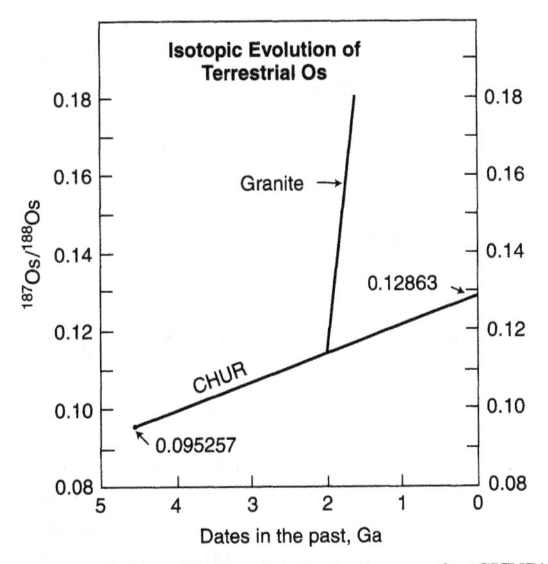

FIGURE 13.8 Isotopic evolution of terrestrial Os in the mantle (CHUR) and continental crust (granite). The $^{187}Re/^{188}Os$ ratio of granite (10) is significantly higher than that of CHUR (0.423), which has caused Os in the continental crust to become enriched in radiogenic ^{187}Os. Data from Chen et al. (1998).

The representation of CHUR-Os in Figure 13.8 is expressed by the equation

$$\left(\frac{^{187}\mathrm{Os}}{^{188}\mathrm{Os}}\right)^0_{\mathrm{CHUR}} = \left(\frac{^{187}\mathrm{Os}}{^{188}\mathrm{Os}}\right)^{\mathrm{P}}_{\mathrm{CHUR}} + \left(\frac{^{187}\mathrm{Re}}{^{188}\mathrm{Os}}\right)^0_{\mathrm{CHUR}} \times (e^{\lambda t} - 1) \qquad (13.10)$$

where $(^{187}\mathrm{Os}/^{188}\mathrm{Os})^0_{\mathrm{CHUR}}$ = present value of this ratio (i.e., $t = 0$)

$(^{187}\mathrm{Os}/^{188}\mathrm{Os})^{\mathrm{P}}_{\mathrm{CHUR}}$ = primeval value of this ratio at $t = 4.558 \times 10^9$ y

$(^{187}\mathrm{Re}/^{188}\mathrm{Os})^0_{\mathrm{CHUR}}$ = present value of this ratio

$\lambda = 1.666 \times 10^{-11}\ \mathrm{y}^{-1}$

$t = 4.558 \times 10^9$ y

Therefore, the primeval $^{187}Os/^{188}Os$ ratio is

$$\left(\frac{^{187}\mathrm{Os}}{^{188}\mathrm{Os}}\right)^{\mathrm{P}}_{\mathrm{CHUR}} = \left(\frac{^{187}\mathrm{Os}}{^{188}\mathrm{Os}}\right)^0_{\mathrm{CHUR}} - \left(\frac{^{187}\mathrm{Re}}{^{188}\mathrm{Os}}\right)^0_{\mathrm{CHUR}} (e^{\lambda t} - 1)$$

$$= 0.12863 - 0.423\ (1.07889 - 1)$$

$$= 0.095257$$

The present $^{187}Os/^{188}Os$ ratios of terrestrial rocks and minerals may be higher or lower than the present $^{187}Os/^{188}Os$ ratio of CHUR depending on the Re/Os ratios of the terrestrial reservoirs through which the Os passed after it was removed from the chondritic reservoir. The difference between the $^{187}Os/^{188}Os$ ratio of a sample of terrestrial Os and CHUR is expressed by the ε-parameter ε(Os) defined in a similar way as the ε-parameter for Sr, Nd, and Lu:

$$\varepsilon^0 = \left[\frac{(^{187}\mathrm{Os}/^{188}\mathrm{Os})^0_{\mathrm{spl}} - 0.12863}{0.12863}\right] \times 100\,\% \qquad (13.11)$$

Note that some authors have used the letter γ instead of ε and that ε(Os) is expressed in percent. In cases where ε^0 is positive, the Os in the terrestrial sample is enriched in radiogenic ^{187}Os compared to the isotope composition of Os in CHUR. If ε^0(Os) is negative, the Os has a lower $^{187}Os/^{188}Os$ ratio than CHUR.

The ε-values can also be calculated for Os at some time t in the past based on the $^{187}Os/^{188}Os$ ratios of CHUR and of the sample at that time. In most cases, the initial $^{187}Os/^{188}Os$ ratio of a suite of rocks or minerals is determined by means of a Re–Os isochron that also provides the date for the comparison with CHUR. The $^{187}Os/^{188}Os$ ratio of CHUR at t is calculated from equation 13.10:

$$\left(\frac{^{187}\text{Os}}{^{188}\text{Os}}\right)^t_{\text{CHUR}} = 0.12863 - 0.423\ (e^{\lambda t} - 1) \tag{13.12}$$

A positive value of ε^t for igneous rocks implies that the source of the magma had a higher Re/Os ratio than CHUR. In the case of high-grade metamorphic rocks, a positive ε^t-value means that the protolith had an elevated Re/Os ratio (e.g., shale, Re/Os = 135, Horan et al., 1994).

13.4d Model Dates

The Re–Os model dates of terrestrial samples are calculated as the time in the past when the initial $^{187}Os/^{188}Os$ ratio of the sample was identical to the $^{187}Os/^{188}Os$ ratio of CHUR. At this time

$$\left(\frac{^{187}\text{Os}}{^{188}\text{Os}}\right)^t_{\text{CHUR}} = \left(\frac{^{187}\text{Os}}{^{188}\text{Os}}\right)^i_{\text{spl}} \tag{13.13}$$

Combining equations 13.12 and 13.13 yields

$$\left(\frac{^{187}\text{Os}}{^{188}\text{Os}}\right)^i_{\text{spl}} = 0.12863 - 0.423\ (e^{\lambda t} - 1) \tag{13.14}$$

from which it follows that

$$t = \frac{1}{\lambda} \ln \left(\frac{0.12863 - (^{187}\text{Os}/^{188}\text{Os})^i_{\text{spl}}}{0.423} + 1 \right) \tag{13.15}$$

The initial $^{187}Os/^{188}Os$ ratio of a terrestrial rock or mineral is derived either from the Re–Os isochron or by analysis of a sample having a very low Re/Os ratio (e.g., nuggets of osmiridium.). The validity of the model dates calculated from equation 13.15 is limited by the assumptions that the Os in a sample originated from a source that is adequately represented by the chondritic reservoir and was transferred into the sample without delay and without being mixed with Os from a different source.

13.5 THE Cu–Ni SULFIDE ORES, NORIL'SK, SIBERIA

The ore deposits in the Noril'sk area of Siberia occur in plutons of gabbro that intruded genetically related plateau basalt and associated pyroclastic deposits of Late Permian to Early Triassic age (Zolotukhin and Al'Mukhamedov, 1988; Faure, 2001). The origin of the plateau basalt and the genetically related ore deposits was investigated by means of the isotopic compositions of Sr, Nd, and Pb as well as by means of trace-element concentrations:

Basalt: DePaolo and Wasserburg, 1979, *Proc. Natl. Acad. Sci. USA*, 76(7):3056–3060. Lightfoot et al., 1990, *Contrib. Mineral. Petrol.*, 104:631–644. Lightfoot et al., 1993, *Contrib. Mineral. Petrol.*, 114:171–188. Lightfoot et al., 1996, in Lightfoot and Naldrett (Eds.), *Proc. Sudbury-Noril'sk Symposium*, Spec. Paper 5, pp. 283–312, Ontario Geological Survey, Toronto. Naldrett et al., 1992, *Econ. Geol.* 87:975–1004. Sharma et al., 1991, *Geochim. Cosmochim. Acta*, 55:1183–1192. Sharma et al., 1992, *Earth Planet. Sci. Lett.*, 113:365–381; Wooden et al., 1992, *Econ. Geol.*, 86:1153–1165. Wooden et al., 1993, *Geochim. Cosmochim. Acta*, 57:3677–3704. Hawkesworth et al., 1995, *Lithos*, 34: 61–88.

Ore Deposits: Horan et al., 1993, *EOS*, 74: 555. Horan et al., 1995, *Geochim. Cosmochim. Acta*, 58:5159–5168. Brügmann et al., 1993, *Geochim. Cosmochim. Acta*, 57:2001–2018. Walker et al., 1994, *Geochim. Cosmochim. Acta*, 58:4179–4198.

The papers presented during a conference on the similarities and differences between the ore at Noril'sk in Siberia and at Sudbury, Ontario, were edited by Lightfoot and Naldrett (1994).

A specimen of ore composed of chalcopyrite and pentlandite from the Medvezhy Creek mine in the Noril'sk I intrusion has the following composition (Walker et al., 1994):

$$\text{Re} = 5.581 \text{ ppb} \qquad \text{Os} = 0.7458 \text{ ppb}$$

$$\frac{^{187}\text{Re}}{^{188}\text{Os}} = 36.88 \qquad \frac{^{187}\text{Os}}{^{188}\text{Os}} = 0.2825$$

The value of ε^0 of this specimen, derived from equation 13.11, is

$$\varepsilon^0 = \left(\frac{0.2825 - 0.12863}{0.12863}\right) \times 100 = +119.6\%$$

The large positive value of ε^0 indicates that this sample of the ore has been strongly enriched in radiogenic ^{187}Os, presumably because of in situ decay of ^{187}Re after the minerals crystallized at 246 Ma.

To calculate ε^t for this ore specimen, the value of $(^{187}\text{Os}/^{188}\text{Os})^t_{\text{CHUR}}$ is obtained from equation 13.12:

$$\left(\frac{^{187}\text{Os}}{^{188}\text{Os}}\right)^t_{\text{CHUR}} = 0.12863 - 0.423 \times 0.004106$$

$$= 0.12689$$

The $^{187}\text{Os}/^{188}\text{Os}$ ratio of the ore sample at 246 Ma is

$$\left(\frac{^{187}\text{Os}}{^{188}\text{Os}}\right)^t_{\text{spl}} = 0.2825 - 36.88 \times 0.004106$$

$$= 0.13107$$

Therefore,

$$\varepsilon^t = \left(\frac{0.13107 - 0.12689}{0.12689}\right) \times 100 = +3.29\%$$

Since ε^t is positive, the Os that was incorporated into sulfide minerals at the time of their crystallization was enriched in radiogenic ^{187}Os relative to CHUR by about 3%. In fact, the ε^t-value of all ore and rock samples from Noril'sk analyzed by Walker et al. (1994) are strongly positive.

A more representative value of ε^t for the ore at Noril'sk is obtained from the initial $^{187}\text{Os}/^{188}\text{Os}$ ratio of the Re–Os isochron obtained by Walker et al. (1994) from analyses of ore samples in the Noril'sk I and Talnakh intrusions. The isochron in Figure 13.9 has a slope of 0.0040376 ± 0.0000099, which yields a date of 241.9 ± 0.6 Ma for $\lambda = 1.666 \times 10^{-11}$ y^{-1}. The initial $^{187}\text{Os}/^{188}\text{Os}$ ratio is 0.1326 ± 0.0025. The Re–Os isochron date of the ore at Noril'sk is lower than the $^{40}\text{Ar}^*/^{39}\text{Ar}$ dates of silicate minerals for which Dalrymple et al. (1991) obtained values of 249 ± 1.6 Ma (biotite, ore vein) and 244.9 ± 0.9 *Ma* (plagioclase, basalt). The older date was subsequently confirmed by a U–Pb SHRIMP date on zircon of 248 ± 4 Ma published by Campbell et al. (1992). Nevertheless, Walker et al. (1994) preferred a date of 246 Ma because it is compatible with the Re–Os systematics.

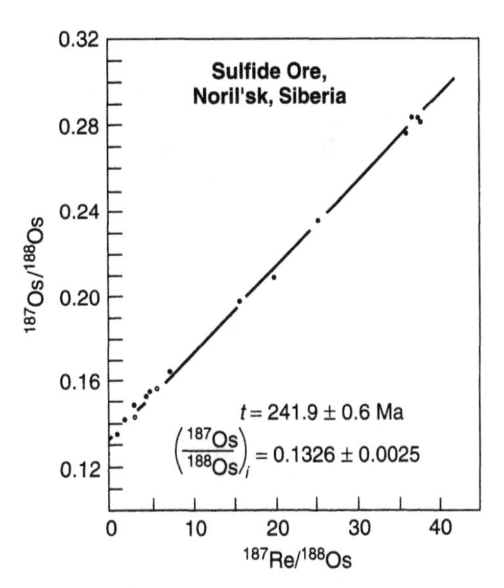

FIGURE 13.9 Re–Os whole-rock isochron defined by sulfide-bearing ore samples from the Medvezhy Creek mine and the Komsomolsky mine in the Noril'sk area of Siberia, Russia. The date was recalculated to $\lambda = 1.666 \times 10^{-11}$ y^{-1}. Data from Walker et al. (1994).

The initial $^{187}Os/^{188}Os$ ratio of the ore samples (0.1326 ± 0.0025) at $t = 241.9$ Ma is equivalent to

$$\varepsilon^t = \left(\frac{0.1326 - 0.12692}{0.12692}\right) \times 100$$
$$= +4.47 \pm 1.98\%$$

where 0.12692 is the $^{187}Os/^{188}Os$ ratio of CHUR at 241.9 Ma and the uncertainty of ε^t reflects only the uncertainty of the initial $^{187}Os/^{188}Os$ ratio of the ore.

The positive values of ε^t demonstrated above and reported by Walker et al. (1994) indicate that the Siberian flood basalts originated from sources in the mantle that were enriched in Re, most likely as a result of recycling of oceanic crust by way of an asthenospheric plume that caused large-scale decompression melting when it intruded the subcontinental lithospheric mantle.

13.6 ORIGIN OF OTHER SULFIDE ORE DEPOSITS

The isotope composition of Os in sulfide ores associated with igneous rocks is a sensitive indicator of crustal contamination (Lambert et al., 1998). The Re–Os isochron dates and initial $^{187}Os/^{188}Os$ ratios in Table 13.5 and Figure 13.10 of selected ore deposits demonstrate that the extent of crustal contamination varies widely. For example, the initial $^{187}Os/^{188}Os$ ratio of peridotite and associated Ni–Cu sulfide ores at Rankin Inlet on the northwest coast of the Hudson Bay in the Northwest Territories of Canada (1A, B in Figure 13.10) are virtually identical but differ moderately from the $^{187}Os/^{188}Os$ ratio of CHUR at 2.70 Ga (Hulbert and Grégoire, 1993). The ore of the North Rankin Ni mine has an initial $^{187}Os/^{188}Os$ ratio of 0.1262, which yields

$$\varepsilon^t = \frac{0.1262 - 0.10916}{0.10916} \times 100 = +15.6\%$$

where the $^{187}Os/^{188}Os$ of CHUR at 2.70 Ga is 0.10916 (equations 13.11 and 13.12). The peridotite at Rankin Inlet has a ε^t value of +13.2%.

Table 13.5. Initial $^{187}Os/^{188}Os$ Ratios and Dates of Sulfide Ores based on Re–Os Isochrons

Number in Figure 13.10	Deposit	$^{187}Os/^{188}Os_i$	Date, Ga ($\lambda = 1.666 \times 10^{-11}$ y^{-1})	References[a]
1A	North Rankin Ni mine, NW Territories, Canada (peridotite)	0.1236 ±0.092	2.705	1
1B	North Rankin Ni mine, NW Territories, Canada (sulfides)	0.1262 ±0.088	2.734	1
2A	Voisey's Bay intrusion, Labrador (sulfides)	1.334 ±0.194	1.302 ±0.135	2
2B	Voisey's Bay, (ultramafic igneous rocks)	0.12951	1.312	2
3	Kambalda, Western Australia	0.10889 ±0.00035	2.664 ±0.036	3
4	Witwatersrand, S. Africa (pyrite)	0.124 ±0.037	2.99 ±0.11	4
5A	Porphyry copper, Chile, El Salvador	0.78 ±0.29	0.039 ±0.002	5
5B	Porphyry copper, Chile, Chuquicamata	1.01 ±0.014	0.031 ±0.002	5
5C	Porphyry copper, Chile, Quebrada Blanca	1.28 ±0.06	0.045 ±0.003	5
6	Noril'sk, Russia (sulfides)	0.1326 ±0.0025	0.242 ±0.0006	6

[a] 1. Hulbert and Gregoire, 1993, *Can. Mineral.*, 31:861–876. 2. Lambert et al., 2000, *Econ. Geol.*, 93:121–137. 3. Foster et al., 1996, *Nature* 382:703–705. 5. Kirk et al., 2001, *Geochim. Cosmochim. Acta*, 65:2149–2159. Mathur et al., 2000, *Geology*, 28:555–558. 6. Walker et al., 1994, *Geochim. Cosmochim. Acta*, 58:4179–4198.

Evidently, the Os in the sulfide ore and in the peridotite is enriched in radiogenic ^{187}Os, which is presumably of crustal origin.

An even more compelling case for crustal contamination was reported by Lambert et al. (2000) for massive Ni–Cu–Co sulfides at Voisey's Bay on the east coast of Labrador at 56°25′ N (2A

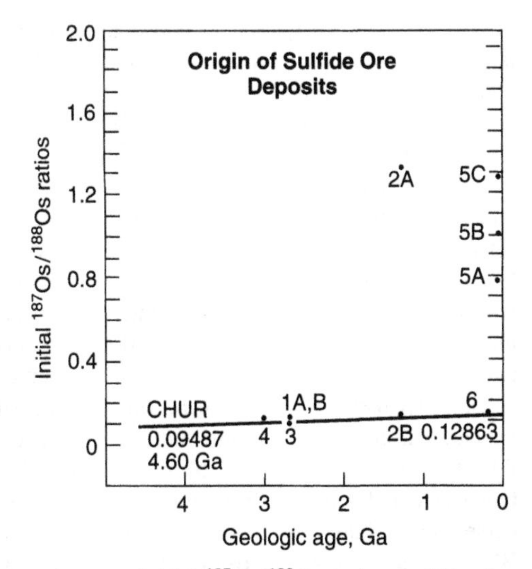

FIGURE 13.10 Initial $^{187}Os/^{188}Os$ ratios and Re–Os isochron dates of sulfide ore associated with igneous rocks: 1. North Rankin Ni mine, NW Territories, Canada; A = peridotite, B = massive sulfides. 2. Voisey's Bay, Labrador, Canada; A = massive sulfides, B = ultramafic igneous rocks. 3. Kambalda, Western Australia (sulfides, olivine–sulfide, komatiite). 4. Witwatersrand, South Africa (pyrite). 5. Porphyry copper deposits, Chile; A = El Salvador, B = Chuquicamata, C = Quebrada. 6. Noril'sk, Siberia, Russia. All dates were recalculated to $\lambda = 1.666 \times 10^{-11}\ y^{-1}$ and $^{187}Os/^{186}Os$ ratios were converted to $^{187}Os/^{188}Os$ ratios. Data sources are referenced in Table 13.5.

and 2BB in Figure 13.10). The Re–Os errorchron date of the sulfide ore (2A) is 1.302 ± 0.135 Ga, and its initial $^{187}Os/^{188}Os$ ratio is 1.334 ± 0.194. The associated ultramafic rocks (2B) have a similar Re–Os errorchron date of 1.312 Ga ($\lambda = 1.666 \times 10^{-11}\ y^{-1}$), but their initial $^{187}Os/^{188}Os$ ratio (0.12951) is much lower than that of the ore. The ε^t-value of the ore (2A) is +1018% relative to CHUR at 1.30 Ga, whereas the ultramafic rocks (2B) have $\varepsilon^t = +8.5\%$. Therefore, the Os in the sulfide ore is strongly enriched in radiogenic ^{187}Os even though it may be genetically related to the associated ultramafic rocks that originated from the mantle without extensive contamination by ^{187}Os of crustal origin.

Additional examples of Re–Os dates and initial $^{187}Os/^{188}Os$ ratios in Table 13.5 and Figure 13.10 include sulfides at Kambalda (3) in Western Australia; pyrite (4) in the gold mines of the Witwatersrand of South Africa; porphyry Cu deposits (5) of Chile; and the sulfide ore (6) in the Noril'sk district of Siberia. Close examination of the data by means of ε^t-values reveals the wide range of crustal contamination of these ore deposits.

13.7 METALLIC PGE MINERALS

The PGEs occur in the native state (e.g., platinum), as metallic alloys (e.g., osmiridium), as sulfides (e.g., laurite, RuS_2, and ehrlichmanite, OsS_2), and as sulfarsenides (e.g., ruarsite, RuAsS), all of which are associated with mafic and ultramafic igneous rocks. The metallic minerals are also concentrated as nuggets in placer deposits, alluvial sands, and clastic sedimentary rocks.

The naturally occurring metallic PGE minerals commonly contain high concentrations of Os but are virtually free of Re, which causes their Re/Os ratios to be exceedingly low. Consequently, the isotopic composition of Os in PGE nuggets is practically invariant with time and has preserved the $^{187}Os/^{188}Os$ ratio of their magma sources in the mantle. Hirt et al. (1963) originally demonstrated that samples of osmiridium from several deposits of different ages have different $^{187}Os/^{186}Os$ ratios that are significantly greater than the primeval value of this ratio determined from meteorites. Allègre and Luck (1980) subsequently measured the $^{187}Os/^{186}Os$ ratios of osmiridium nuggets of known ages in order to investigate this phenomenon more closely. Their data points (not shown) fit a straight line in coordinates of $^{187}Os/^{188}Os$ ratios and time in billions of years with a slope $m = 0.006186 \times 10^{-9}\ y^{-1}$ and a present $^{187}Os/^{188}Os$ ratio of 0.12451 (see also Sections 13.3c and 13.4a). The colinearity of the data points means that the Os in the nuggets originated from a homogeneous source in the mantle whose present Re/Os ratio is indicated by the slope of the line in accordance with equation 13.7:

$$\frac{^{187}Os}{^{188}Os} \cong \left(\frac{^{187}Os}{^{188}Os}\right)_i + \frac{^{187}Re}{^{188}Os}\lambda t$$

where the slope is

$$m = \left(\frac{^{187}\text{Re}}{^{188}\text{Os}}\right)\lambda$$

Hence, the Re/Os ratio of the source of osmiridium nuggets based on the data of Allègre and Luck (1980) is

$$\frac{\text{Re}}{\text{Os}} = \frac{0.006186 \times 10^{-9}}{1.666 \times 10^{-11}} \times 0.278 = 0.077$$

The resulting estimate of the Re/Os ratio of the mantle of the Earth is similar to the average Re/Os ratios of chondritic meteorites in Table 13.2, which range from 0.078 (carbonaceous chondrites) to 0.091 (ordinary chondrites).

13.8 GOLD DEPOSITS OF THE WITWATERSRAND, SOUTH AFRICA

The Late Archean sedimentary rocks of the Witwatersrand in South Africa contain more than 40,000 tons of native gold, but the origin of these deposits has been in doubt since these deposits were discovered in 1874 (Kirk et al., 2001). The gold, together with osmiridium, pyrite, and uraninite (UO_2), occurs primarily in the basal conglomerates of the Upper Witwatersrand Supergroup (2.89 to 2.78 Ga). In addition, gold occurs in the Dominion Group (3.09 to 3.02 Ga), which underlies the Witwatersrand Supergroup, as well as in the sedimentary rocks of the Ventersdorp and Transvaal Supergroups (2.71 to 2.02 Ga), which overlie the rocks of the Witwatersrand. The gold in the last-mentioned supergroups may have originated by erosion of the auriferous conglomerates of the Witwatersrand. Age determinations of this triad of supergroups were reviewed by Robb and Meyer (1995) and Zartman and Frimmel (1999).

The conglomerates are composed of water-worn pebbles of quartz, chert, quartzite, and slate in a sandy matrix cemented by silica. In addition, the conglomerates contain detrital grains of tourmaline, zircon, rutile, chromite, osmiridium, and graphite as well as sulfides, arsenides, and tellurides of Fe, Cu, Zn, Pb, Ni, Sb, and Co. The association of native gold with quartz–pebble conglomerates is consistent with the proposal by Ramdohr (1955, 1958) that the so-called gold reefs are fossil placers. This view was strongly contested by Davidson (1964, 1964/1965a, b), who argued that the concentrations of gold and uraninite are closely correlated, which requires that they had a similar origin. Therefore, the gold cannot be alluvial in origin because neither uraninite nor pyrite are sufficiently resistant to weathering in the presence of molecular oxygen (O_2) to survive transport by streams and deposition in placers in an oxygenated atmosphere. Davidson (1964/1965a, b, 1965) proposed instead that the gold, uraninite, and associated sulfide minerals were deposited by warm brines that had leached the metals from mineral grains dispersed throughout the sedimentary rocks of the Witwatersrand basin and from the mafic lavas of the overlying Ventersdorp Supergroup.

The detrital origin of uraninite in the gold-bearing conglomerates of the Witwatersrand has been supported by many investigators, including Liebenberg (1960), Hiemstra (1968), and Utter (1980), and was confirmed by a whole-rock U–Pb date of 3050 ± 40 Ma of the ore, which predates the time of deposition of the conglomerates at 2740 ± 19 Ma or less (Rundle and Snelling, 1977).

The evidence in favor of a detrital origin of uraninite in the conglomerates of the Witwatersrand and elsewhere (e.g., Blind River, Ontario) inevitably led to the conclusion that the concentration of O_2 in the atmosphere prior to about 2.5 Ga was considerably lower than it is at present and that the lack of O_2 prevented the oxidation of U^{4+} to U^{6+} and hence retarded the dissolution of uraninite (Holland, 1984, p. 332).

Davidson (1964, 1965) objected to the idea that the early Precambrian atmosphere was deficient in O_2 because he considered it to be a violation of the principle of Uniformitarianism. He was, after all, Professor of Geology at St. Andrews University in Scotland, where James Hutton made this principle the foundation of the science of geology.

The origin of the gold in the Witwatersrand deposits continues to be debated because it may have recrystallized during hydrothermal alteration after initial deposition in placers (Minter, 1976, 1999; Frimmel and Gartz, 1997). Some recent authors have also advocated a purely hydrothermal origin for the gold (e.g., Phillips and Law, 1994; Barnicoat et al., 1997).

The question concerning the origin of the gold and osmiridium in the conglomerates of the Witwatersrand can be settled by dating samples of both by the Re–Os method. If individual nuggets of osmiridium and gold predate the time of deposition of the rocks of the Witwatersrand, the placer hypothesis is confirmed even if the gold was later affected by low-grade metamorphism and/or hydrothermal alteration.

13.8a Osmiridium

The isotope composition of Os in small nuggets of osmiridium (50–100 μm) derived from three gold mines in the Witwatersrand district were measured by Hart and Kinloch (1989) by ion probe mass spectrometry. The mass spectra were corrected for the presence of ^{187}Re and ^{186}W and for isotope fractionation to $^{186}\text{Os}/^{188}\text{Os} = 0.12035$. The resulting $^{187}\text{Os}/^{186}\text{Os}$ ratios ranged widely from 0.872 to 0.905, with two outliers at 0.996 and 1.371. After converting to $^{187}\text{Os}/^{188}\text{Os}$ ratios by means of equation 13.4, the model Os dates were recalculated relative to CHUR-Os defined by Chen et al. (1998) using equation 13.15. The resulting dates in Figure 13.11*a* range from 3.33 to 2.79 Ga with 60% of the nuggets between 3.2 and 3.0 Ga. Therefore, most of the model dates based on CHUR-Os in Figure 13.11*a* predate the start of the deposition of the Witwatersrand Supergroup.

13.8b Gold

Osmium model dates were also used to date native gold in the rocks of the Witwatersrand because of its low average Re/Os ratio of 0.0054 in Table 13.2. The $^{187}\text{Os}/^{188}\text{Os}$ ratios of gold samples measured by Kirk et al. (2001) by negative-ion thermal ionization mass spectrometry range between 0.10533 and 0.10993. The corresponding model dates based on CHUR (Chen et al., 1998) were recalculated from equation 13.15 without making a correction for the in situ decay of ^{187}Re because of the evidence cited by Kirk et al. (2001) that Re may have been added to the gold during hydrothermal activity after deposition. In any case, the correction for

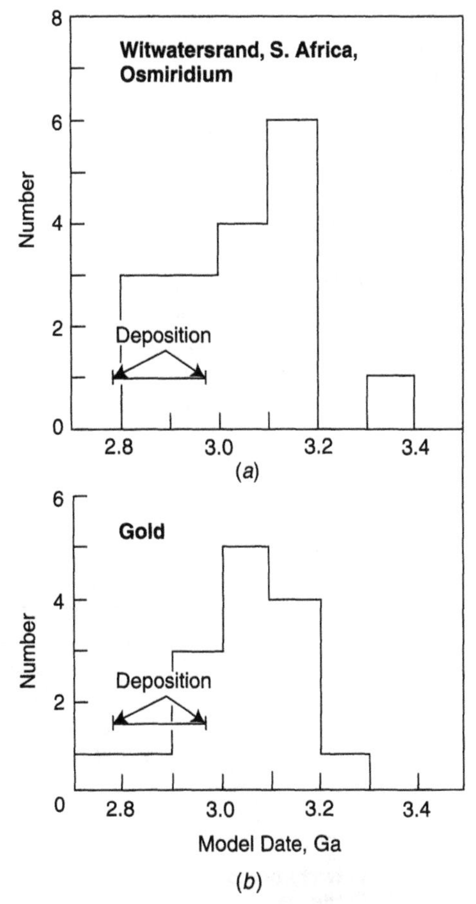

FIGURE 13.11 (*a*) Histogram of model Os dates of individual nuggets of osmiridium in three gold mines of the Witwatersrand district, South Africa. The dates were calculated from $^{187}\text{Os}/^{188}\text{Os}$ ratios by means of equation 13.4 for CHUR-Os defined by Chen et al. (1998). The depositional interval of the Witwatersrand is from Kirk et al. (2001). Data from Hart and Kinloch (1989). (*b*) Model dates of native gold in the Witwatersrand calculated relative to CHUR Os (Chen et al., 1998). Data from Kirk et al. (2001).

in situ–produced ^{187}Os increases the model dates only by about 100×10^6 years.

The results in Figure 13.11*b* demonstrate that the model dates (relative to CHUR-Os; Chen et al., 1998) of 67% of the gold samples are older than

3.0 Ga and therefore predate the deposition of the Witwatersrand Supergroup. In this regard, the model dates of gold based on the data of Kirk et al. (2001) are identical to the model dates of osmiridium nuggets derived from the data of Hart and Kinloch (1989). Both teams of investigators concluded that the osmiridium and gold in the Witwatersrand are predominantly of detrital origin. However, both also found evidence for postdepositional alteration of the nuggets by hydrothermal fluids. For example, Hart and Kinloch (1989) noted that the detrital osmiridium nuggets are, in many cases, rimmed with OsIrSAs, indicative of hydrothermal alteration and/or low-grade metamorphism.

13.8c Pyrite

Four samples of pyrite associated with native gold in the Vaal reef in the western rim of the Witwatersrand basin were analyzed for dating by the Re–Os method (Kirk et al., 2001). The results form an isochron in Figure 13.12 yielding a date of 2.99 ± 0.11 Ga ($\lambda = 1.666 \times 10^{-11}$ y^{-1}, MSWD = 0.77) and an initial $^{187}Os/^{188}Os$ ratio of 0.124 ± 0.037. The initial $^{187}Os/^{188}Os$ ratio of the pyrite places it above the line representing CHUR in Figure 13.10 (point 4). The ε^t-value of the pyrite at 3.0 Ga is $+15.9 \pm 34.6\%$. The large uncertainty arises from the statistical error of the initial $^{187}Os/^{188}Os$ ratio. In the absence of evidence to the contrary, the age of the pyrite in the gold ore of the Witwatersrand Supergroup indicates that grains of this mineral were transported into the basin in detrital form and were deposited with other high-density minerals (e.g., gold, osmiridium, uraninite, zircon) in quartz–pebble conglomerates.

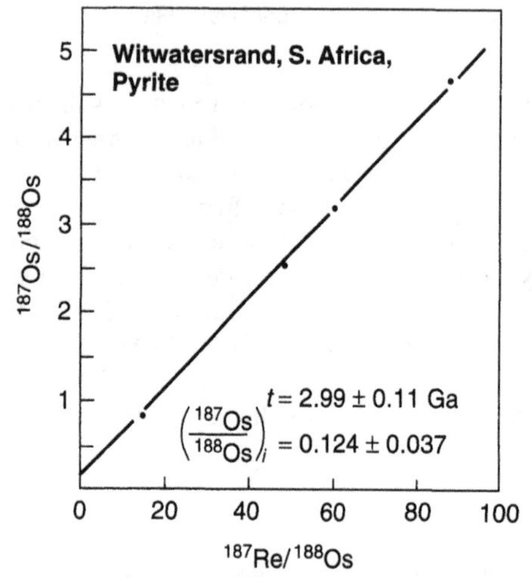

FIGURE 13.12 Re–Os isochron for pyrite in the Vaal reef of the western Witwatersrand, South Africa ($\lambda = 1.666 \times 10^{-11}$ y^{-1}). Data and interpretation by Kirk et al. (2001).

13.8d The Solution to the Problem

The results of Re–Os dating have confirmed the U–Pb dates of zircon, monazite, and whole-rock samples of uraniferous ore in the mines of Witwatersrand Supergroup of South Africa. The demonstration that the gold, osmiridium, and pyrite crystallized before they were deposited as detrital grains in the Witwatersrand basin strongly supports the fossil placer hypothesis of Ramdohr, Minter, and others. However, the mineralogical and textural evidence for a hydrothermal origin of these minerals is consistent with the evidence for loss of Pb and U and for open-system conditions of Re and Os in the ore. Although the ore probably was altered by hydrothermal fluids, Davidson's idea that the uraninite and gold were deposited by warm brines that flowed downward and spread laterally into the more permeable conglomerates in the Witwatersrand Supergroup is no longer tenable. After more than a century of effort to explain the origin of the world's greatest deposit of Au, U, and PGE, use of the Re–Os method has finally solved the problem and has thereby strengthened the evidence for the chemical evolution of the atmosphere of the Earth during the Archean Eon.

13.9 THE Pt–Os METHOD

The decay of ^{190}Pt to ^{186}Os was used by Walker et al. (1997) to study the origin of the PGE-bearing sulfide ore at Noril'sk, Russia (Section 13.5) and to date iron meteorites of groups IIA and IIB as well as of enstatite chondrites. In addition, they measured the $^{186}Os/^{188}Os$ ratios of samples of osmiridium and of the carbonaceous chondrite Allende. These data confirmed that the Pt/Os ratio of magma sources in the mantle of the Earth is similar to that of chondrite meteorites. In addition, Walker et al. (1997) determined that the decay constant of ^{190}Pt is $1.542 \pm 0.015 \times 10^{-12}$ y^{-1} and defined CHUR for the Pt–Os decay scheme as

$$\left(\frac{^{186}\text{Os}}{^{188}\text{Os}}\right)^0_{\text{CHUR}} = \left(\frac{^{186}\text{Os}}{^{188}\text{Os}}\right)^{\text{P}}_{\text{CHUR}} + \left(\frac{^{190}\text{Pt}}{^{188}\text{Os}}\right)^0 \times (e^{\lambda t} - 1) \quad (13.16)$$

where

$$\left(\frac{^{186}\text{Os}}{^{188}\text{Os}}\right)^0_{\text{CHUR}} = 0.119834 \pm 0.000002 \quad (t = 0)$$

$$\left(\frac{^{186}\text{Os}}{^{188}\text{Os}}\right)^{\text{P}}_{\text{CHUR}} = 0.119820 \quad \text{(primeval)}$$

$$\left(\frac{^{190}\text{Pt}}{^{188}\text{Os}}\right)^0_{\text{CHUR}} = 0.001659 \pm 0.000075 \quad (t = 0)$$

$$\lambda = 1.542 \pm 0.015 \times 10^{-12}\text{y}^{-1}$$

Walker et al. (1997) concluded that the Os in the mantle plume that caused the magmatic activity at Noril'sk was enriched in radiogenic ^{186}Os relative to CHUR by 0.012%. They attributed this phenomenon to either the recycling of Pt-rich oceanic crust by subduction into the mantle or by derivation of Os in a plume derived from the outer core of the Earth.

13.10 SUMMARY

Rhenium and osmium are both siderophile elements that are strongly partitioned into liquid iron rather than into sulfide or silicate melts. In addition, Re^{4+} can substitute for Mo^{4+} and Cu^{2+} in sulfide minerals. These geochemical properties allow the Re–Os method to be used for dating of iron meteorites, nuggets of osmiridium and gold, and Cu–Ni–PGE sulfide ore deposits.

The results of dating iron meteorites by the Re–Os method have provided a chronology for the crystallization of the liquid iron cores of large planetary objects during the early history of the solar system. In addition, the Re–Os systematics of chondrites define the chondritic uniform reservoir (CHUR), which is used to represent the isotopic evolution of Os in the mantle of the Earth. In addition, CHUR is the baseline for the study of the isotopic evolution of Os in the rocks of the continental crust and for the calculation of model dates based on the isotope composition of common Os in Re-free nuggets of osmiridium and gold.

Age determinations of massive sulfide ore deposits associated with mafic igneous rocks and porphyry Cu deposits not only yield dates but also reveal the extent of contamination of sulfide minerals by radiogenic ^{187}Os of crustal origin. Similarly, the Re–Os systematics of igneous rocks provide information about their origin that is complementary to the results derived from the Rb–Sr, Sm–Nd, U,Th–Pb, and Lu–Hf isotopic dating methods.

REFERENCES FOR PRINCIPLES AND METHODOLOGY (INTRODUCTION, SECTIONS 13.1–13.2)

Allègre, J. C., and J. M. Luck, 1980. Osmium isotopes as petrogenetic and geologic tracers. *Earth Planet. Sci. Lett.*, 49:148–154.

Anders, E., and N. Grevesse, 1989. Abundances of the elements: Meteoritic and solar. *Geochim. Cosmochim. Acta*, 53:197–214.

Chen, J. H., D. A. Papanastassiou, and G. J. Wasserburg, 1998. Re–Os systematics in chondrites and the fractionation of the platinum group elements in the early solar system. *Geochim. Cosmochim. Acta*, 62:3379–3392.

Creaser, R. A., D. A. Papanastassiou, and G. J. Wasserburg, 1991. Negative thermal ion mass spectrometry of Os, Re, and Ir. *Geochim. Cosmochim. Acta*, 55:397–401.

Crocket, J. H., 1969. Platinum metals. In K. H. Wedepohl (Ed.), *Handbook of Geochemistry*. II-5: 78-A-1 to 78-O Springer-Verlag, Heidelberg.

Crocket, J. H., R. R. Keays, and S. Hsieh, 1967. Precious metal abundances in some carbonaceous and enstatite chondrites. *Geochim. Cosmochim. Acta*, 31:1615–1623.

Esser, B. K., and K. K. Turekian, 1993. The osmium isotopic composition of the continental crust. *Geochim. Cosmochim. Acta*, 57:3093–3104.

Faure, G., 2001. *Origin of Igneous Rocks; the Isotopic Evidence*. Springer-Verlag, Heidelberg.

Fehn, U., R. Teng, D. Elmore, and P. W. Kubik, 1986. Isotopic composition of osmium in terrestrial samples determined by accelerator mass spectrometry. *Nature*, 323:707–709.

Hirt, B., G. R. Tilton, W. Herr, and W. Hoffmeister, 1963. The half-life of ^{187}Re. In J. Geiss and E. D. Goldberg (Eds.), *Earth Science and Meteoritics*, pp. 273–280. North-Holland, Amsterdam.

Hoffman, E. L., A. J. Naldrett, J. C. Van Loon, R. G. V. Hancock, and A. Mason, 1978. The determination of all platinum group elements and gold in rocks and ore by neutron activation analysis after preconcentration by a nickel fire-assay technique of large samples. *Anal. Chim. Acta*, 102:157–166.

Horan, M. F., J. W. Morgan, R. I. Grauch, R. M. Coveney, Jr., J. B. Murowchick, and L. J. Hulbert, 1994. Rhenium and osmium isotopes in black shales and Ni-Mo-PGE sulfide layers, Yukon Territory, Canada, and Hunan and Guizhou provinces, China. *Geochim. Cosmochim. Acta*, 58:257–265.

Horan, M. F., J. M. Morgan, R. J. Walker, and J. N. Grossman, 1992. Rhenium-osmium isotope constraints on the age of iron meteorites. *Science*, 255:1118–1121.

Lindner, M., D. A. Leich, G. P. Russ, J. M. Bazan, and R. J. Borg, 1989. Direct determination of the half-life of ^{187}Re. *Geochim. Cosmochim. Acta*, 53:1597–1606.

Lugmair, G. W., and S. J. G. Galer, 1992. Age and isotopic relationships among angrites Lewis Cliff 86010 and Angra dos Reis. *Geochim. Cosmochim. Acta*, 56:1673–1694.

Masuda, A., T. Hirata, and H. Shimizu, 1987. Determinations of the osmium isotope ratios in iron meteorites and iridosmines by ICP-MS. *Geochem. J.*, 20:233–240.

Morgan, J. W., 1965. The simultaneous determination of rhenium and osmium in rocks by neutron activation analysis. *Anal. Chim. Acta*, 32:8–16.

Morgan, J. W., and J. F. Lovering, 1967. Rhenium and osmium abundances in chondritic meteorites. *Geochim. Cosmochim. Acta*, 31:1893–1909.

Morris, D. F. C., and E. L. Short, 1969. Rhenium. In K. H. Wedepohl (Ed.), *Handbook of Geochemistry*, Vol. II-1. Springer-Verlag, Heidelberg.

Nier, A. O., 1937. The isotopic constitution of osmium. *Phys. Rev.*, 52:885.

Papanastassiou, D. A., J. J. Shen, and G. J. Wasserburg, 1995. Re–Os in FeNi sulfide and phosphide: The possible determination of internal isochrons for iron meteorites. *Meteoritics*, 30:560–561.

Pernicka, E., and J. T. Wasson, 1987. Ru, Re, Os, Pt, and Au in iron meteorites. *Geochim. Cosmochim. Acta*, 51:1717–1726.

Rambaldi, E. R., M. Cendales, and R. Thacker, 1978. Trace element distribution between magnetic and non-magnetic portions of ordinary chondrites. *Earth Planet. Sci. Lett.*, 40:175–186.

Riley, G. J., 1967. Isotopic analysis of rhenium from a thermal ionization source. *J. Sci. Instrum.*, 44:769–774.

Shen, J. J., D. A. Papanastassiou, and G. J. Wasserburg, 1996. Precise Re–Os determinations and systematics of iron meteorites. *Geochim. Cosmochim. Acta*, 60:2887–2900.

Shen, J. J., D. A. Papanastassiou, and G. J. Wasserburg, 1998. Re–Os systematics in pallasite and mesosiderite metal. *Geochim. Cosmochim. Acta*, 62:2715–2724.

Shirey, S. B., and R. W. Walker, 1994. Carius tube digestion for Re–Os chemistry: An old technique applied to new problems. *EOS, Trans. Geophys. Union*, 75:355–356.

Shirey, S. B., and R. J. Walker, 1998. The Re–Os isotope system in cosmochemistry and high-temperature geochemistry. *Annu. Rev. Earth Planet. Sci.*, 26:423–500.

Smoliar, M. I., R. J. Walker, and J. W. Morgan, 1996. Re–Os ages of group IIA, IIIA, IVA, and IVB iron meteorites. *Science*, 271:1099–1102.

Smoliar, M. I., R. J. Walker, J. W. Morgan, and S. B. Shirey, 1997. Re–Os isochron for IA meteorites—further refinement of the ^{187}Re decay constant. *Meteoritics Planet. Sci.*, 32:A122–A123.

Völkening, J., Th. Walczyk, and K. G. Heumann, 1991. Osmium isotope ratio determinations by negative thermal ionization mass spectrometry. *Intl. J. Mass Spectrom. Proc.*, 105:147–159.

Walczyk, Th., E. H. Hebeda, and K. G. Heumann, 1991. Osmium isotope ratio measurements by negative thermal ionization mass spectrometry (NTI-MS). *Fresenius J. Anal. Chem.*, 34:537–541.

Walker, R. J., 1988. Low-blank chemical separation of rhenium and osmium from gram quantities of silicate rock for measurement by resonance ionization mass spectrometry. *Anal. Chem.*, 60:1231–1234.

Walker, R. J., and J. D. Fassett, 1986. Isotopic measurement of subnanogram quantities of rhenium and osmium by resonance ionization mass spectrometry. *Anal. Chem.*, 58(14):2923–2927.

REFERENCES FOR MOLYBDENITE, CHROMITE, AND ^{187}Re–^{187}Os ISOCHRONS (SECTION 13.3)

Hirt, B., W. Herr, and W. Hoffmeister, 1963a. Age determinations by the rhenium-osmium method. In *Radioactive Dating*, pp. 35–43. International Atomic Energy Agency, Vienna.

Hirt, B., G. R. Tilton, W. Herr, and W. Hoffmeister, 1963b. The half-life of ^{87}Re. In G. Geiss and E. D. Goldberg (Eds.), *Earth Science and Meteoritics*, pp. 273–280. North-Holland, Amsterdam.

Luck, J. M., and C. J. Allègre, 1982. The study of molybdenites through the ^{187}Re-^{187}Os chronometer. *Earth Planet. Sci. Lett.*, 61:291–296.

Markey, R. J., H. J. Stein, and J. W. Morgan, 1998. Highly precise Re–Os dating of molybdenite using alkaline fusion and NTIMS. *Talanta*, 45:935–946.

Nägler, Th. F., J. D. Kramers, B. S. Kamber, R. Frei, and M. D. A. Prendergast, 1997. Growth of subcontinental lithospheric mantle beneath Zimbabwe started at or before 3.8 Ga: Re–Os study on chromites. *Geology*, 25:983–986.

Riley, G. H., 1967. Rhenium concentration in Australian molybdenites by stable isotope dilution. *Geochim. Cosmochim. Acta*, 31:1489–1498.

Stein, H. J., J. W. Morgan, and A. Scherstén, 2000. Re–Os dating of low-level highly radiogenic (LLHR) sulfides: The Harnäs gold deposit, southwest Sweden, records continental-scale tectonic events. *Econ. Geol.*, 95:1657–1671.

Walker, R. J., E. Hanski, J. Vuollo, and J. Liipo, 1996. The Os isotopic composition of Proterozoic upper mantle: Evidence for chondritic upper mantle from the Outokumpu ophiolite, Finland. *Earth Planet. Sci. Lett.*, 141:161–173.

REFERENCES FOR METEORITES AND CHUR-Os (SECTION 13.4)

Birck, J. L., and C. J. Allègre, 1994. Contrasting Re/Os magmatic fractionation in planetary basalts. *Earth Planet. Sci. Lett.*, 124:139–148.

Chen, J. H., D. A. Papanastassiou, and G. J. Wasserburg, 1998. Re–Os systematics in chondrites and the fractionation of the platinum group elements in the early solar system. *Geochim. Cosmochim. Acta*, 62:3379–3392.

Ebihara, M., and H. Ozaki, 1995. Rhenium, osmium, and iridium in Antarctic unequilibrated ordinary chondrites and implications for the solar-system abundance of Re. *Geophys. Res. Lett.*, 22:2167–2170.

Ertel, W., H. St. C. O'Neill, P. J. Sylvester, D. B. Dingwell, and B. Spettel, 2001. The solubility of rhenium in silicate melts; implications for the geochemical properties of rhenium at high temperatures. *Geochim. Cosmochim. Acta*, 65:2161–2170.

Herr, W., W. Hoffmeister, B. Hirt, J. Geiss, and F. G. Houtermans, 1961. Versuch zur Datierung von Eisenmeteoriten nach der Rhenium-Osmium Methode. *Z. Naturforsch.*, 16a:1053.

Hirt, B., W. Herr, and W. Hoffmeister, 1963. Age determinations by the rhenium-osmium method. In *Radioactive Dating*, pp. 35–44, International Atomic Energy Agency, Vienna.

Horan, M. F., J. W. Morgan, R. I. Grauch, R. M. Coveney, Jr., J. B. Murowchick, and L. J. Hulbert, 1994. Rhenium and osmium isotopes in black shales and Ni-Mo-PGE-rich sulfide layers, Yukon Territory, Canada, and Hunan and Guizhou province, China. *Geochim. Cosmochim. Acta*, 58:257–265.

Luck, J. -M., and C. J. Allègre, 1983. ^{187}Re-^{187}Os systematics in meteorites and cosmochemical consequences. *Nature*, 302:130–132.

Luck, J. -M., J. L. Birck, and C. J. Allègre, 1980. ^{187}Re-^{187}Os systematics in meteorites: Early chronology of the solar system and age of the galaxy. *Nature*, 283:256–259.

Lugmair, G. W., and S. J. G. Galer, 1992. Age and isotopic relationships among angrites Lewis Cliff 86010 and Angra dos Reis. *Geochim. Cosmochim. Acta*, 56:1673–1694.

Malvin, D. J., D. Wang, and J. T. Wasson, 1984. Chemical classification of iron meteorites-X. Multi-element studies of 43 irons, resolution of group IIIE from IIIAB, and evaluation of Cu as a taxanomic parameter. *Geochim. Cosmochim. Acta*, 48:785–804.

Meisel, T., R. J. Walker, and J. W. Morgan, 1996. The osmium isotopic composition of the Earth's primitive upper mantle. *Nature*, 383:517–520.

Morgan, J. W., and J. F. Lovering, 1967. Rhenium and osmium abundances in chondritic meteorites. *Geochim. Cosmochim. Acta*, 31:1893–1909.

Morgan, J. W., R. J. Walker, and J. N. Grossman, 1992. Rhenium-osmium isotope systematics in meteorites I: Magmatic iron meteorite groups IIAB and IIIAB. *Earth Planet. Sci. Lett.*, 108:191–202.

Morgan, J. W., M. F. Horan, R. J. Walker, and J. N. Grossman, 1995. Rhenium-osmium concentration and isotope systematics in group IIAB iron meteorites. *Geochim. Cosmochim. Acta*, 59:2331–2344.

Papanastassiou, D. A., J. J. Shen, and G. J. Wasserburg, 1995. Re–Os in FeNi, sulfide, and phosphide: The possible determination of internal isochrons for iron meteorites. *Meteoritics*, 30:560–561.

Pernicka, E., and J. R. Wasson, 1987. Ru, Re, Os, Pt, and Au in iron meteorites. *Geochim. Cosmochim. Acta*, 51:1717–1726.

Shen, J. J., D. A. Papanastassiou, and G. J. Wasserburg, 1996. Precise Re–Os determinations and systematics of iron meteorites. *Geochim. Cosmochim. Acta*, 60:2887–2900.

Shimizu, N., M. P. Semet, and C. J. Allègre, 1978. Geochemical applications of quantitative ion-microprobe analysis. *Geochim. Cosmochim. Acta*, 42:1321–1334.

Smoliar, M. I., R. J. Walker, and J. W. Morgan, 1996. Re–Os ages of Group IIA, IIIA, IVA, and IVB iron meteorites. *Science*, 271:1099–1102.

Walker, R. J., and J. W. Morgan, 1989. Rhenium-osmium isotope systematics of carbonaceous chondrites. *Science*, 243:519–522.

Walker, R. J., J. W. Morgan, M. F. Horan, and J. N. Grossman, 1993. Re–Os isotope systematics of ordinary chondrites and iron meteorites. *Lunar Planet. Sci.*, 24:1477–1478.

Wasson, J. T., 1974. *Meteorites*. Springer-Verlag, Heidelberg.

Wasson, J. T., 1985. *Meteorites: Their Record of Early Solar-System History*. Freeman, New York.

REFERENCES FOR NORIL'SK AND OTHER SULFIDE DEPOSITS (SECTIONS13.5–13.6)

Campbell, I. H., G. K. Czamanske, V. A. Fedorenko, R. I. Hill, V. Stepanov, and V. E. Kunilov, 1992. Synchronism of the Siberian traps and the Permian-Triassic boundary. *Science*, 258:1760–1763.

Dalrymple, G. B., G. K. Czamanske, and M. A. Lanphere, 1991. $^{40}Ar/^{39}Ar$ ages of samples from the Noril'sk-Talnakh ore-bearing intrusion and the Siberian flood basalts. *EOS*, 72:570.

Faure, G., 2001. *Origin of Igneous Rocks: The Isotopic Evidence*. Springer-Verlag, Heidelberg.

Hulbert, L. J., and D. C. Grégoire, 1993. Re–Os isotope systematics of the Rankin Inlet Ni ores; an example of the application of ICP-MS to investigate Ni-Cu-PGE mineralisation and the potential use of Os isotopes in mineral exploration. *Can. Mineral.*, 31:861–876.

Lambert, D. D., J. G. Foster, L. R. Frick, E. M. Ripley, and M. L. Zientek, 1998. Geodynamics of magmatic Cu-Ni-PGE sulfide deposits: New insights from the Re–Os isotope system. *Econ. Geol.*, 93:121–137.

Lambert, D. D., L. R. Frick, J. G. Foster, C. Li, and A. J. Naldrett, 2000. Re–Os isotope systematic of the Voisey's Bay Ni–Cu–Co magmatic sulfide system, Labrador, Canada: II. Implications for parental magma chemistry, ore genesis, and metal redistribution. *Econ. Geol.*, 95:867–888.

Lightfoot, P. C., and A. J. Naldrett, 1994. *Proceedings of the Sudbury-Noril'sk Symposium*, Ontario Geol. Surv., Spec. Vol. 5, Toronto.

Walker, R. J., J. W. Morgan, M. F. Horan, G. K. Czamanske, E. J. Krogstad, V. A. Fedorenko, and V. E. Kunilov, 1994. Re–Os isotopic evidence for an enriched-mantle source for the Noril'sk-type ore-bearing intrusions, Siberia. *Geochim. Cosmochim. Acta*, 58:4179–4198.

Zolotukhin, V. V., and A. I. Al'Mukhamedov, 1988. Traps of the Siberian platform. In J. D. Macdougall (Ed.), *Continental Flood Basalts*, pp. 273–310. Kluwer Academic, Dordrecht.

REFERENCES FOR GOLD AND OSMIRIDIUM, WITWATERSRAND, SOUTH AFRICA (SECTIONS 13.7–13.8)

Allègre, C. J., and J. -M. Luck, 1980. Osmium isotopes as petrogenetic and geologic tracers. *Earth Planet. Sci. Lett.*, 48:148–154.

Barnicoat, A. C., I. H. C. Henderson, R. J. Knipe, B. W. D. Yardley, R. W. Napier, N. P. C. Fox, A. K. Kenyen, D. J. Muntingh, D. Styrdom, K. S. Winkler, S. R. Lawrence, and C. Corndorf, 1997. Hydrothermal gold mineralization in the Witwatersrand basin. *Nature*, 386:820–824.

Chen, J. H., D. A. Papanastassiou, and G. J. Wasserburg, 1998. Re–Os systematics in chondrites and the fractionation of the platinum group elements in the early solar system. *Geochim. Cosmochim. Acta*, 62:3379–3392.

Davidson, C. F., 1964. Uniformitarianism and ore genesis. *Mining Mag. Lond.*, 110:176–185, 244–255.

Davidson, C. F., 1964/1965a. The mode of origin of banket orebodies. *Trans. Inst. Mining Metallurgy*, 74(6):319–338.

Davidson, C. F., 1964/1965b. The mode of origin of banket orebodies. Author's reply. *Trans. Inst. Mining Metallurgy*, 74(12):844–857.

Davidson, C. F., 1965. Geochemical aspects of atmospheric evolution. *Proc. Natl. Acad. Sci. (USA)*: 53(6):1194–1205.

Frimmel, H. E., and V. H. Gartz, 1997. Witwatersrand gold particle chemistry matches model of metamorphosed, hydrothermally altered placer deposit. *Mineralium Deposita*, 32:523–530.

Hart, S. R., and R. D. Kinloch, 1989. Osmium isotope systematics in Witwatersrand and Bushveld ore deposits. *Econ. Geol.*, 84(6):1651–1655.

Hiemstra, S. A., 1968. The geochemistry of the uraniferous conglomerate of the Dominion Reefs Mine, Klerksdorp area. *Trans. Geol. Soc. South Africa*, 71: 67–100.

Hirt, B., W. Herr, and W. Hoffmeister, 1963. Age determinations by the rhenium-osmium method. In *Radioactive Dating*, pp. 35–43. International Atomic Energy Agency, Vienna.

Holland, H. D., 1984. *The Chemical Evolution of the Atmosphere and Oceans*. Princeton University Press, Princeton, New Jersey.

Kirk, J., J. Ruiz, J. Chesley, S. Titley, and J. Walshe, 2001. A detrital model for the origin of gold and sulfides in the Witwatersrand basin based on Re–Os isotopes. *Geochim. Cosmochim. Acta*, 65:2149–2159.

Liebenberg, W. R., 1960. On the origin of uranium, gold, and osmiridium in the conglomerates of the Witwatersrand goldfields. *Neues. Jahrbuch Mineral.*, 94:831–867.

Minter, W. E. L., 1976. Detrital gold, uranium, and pyrite concentrations related to sedimentology in the Precambrian Vaal reef placer, Witwatersrand, South Africa. *Econ. Geol.*, 71:157–176.

Minter, W. E. L., 1999. Irrefutable detrital origin of Witwatersrand gold and evidence of eolian signatures. *Econ. Geol.*, 94:665–670.

Phillips, G. N., and D. M. Law, 1994. Metamorphism of the Witwatersrand gold fields: A review. *Ore Geol. Rev.*, 9:1–31.

Ramdohr, P., 1955. Neue Beobachtungen an Erzen des Witwatersrandes in Südafrica und ihre genetische Bedeutung. *Abhandlung der Akademie der Wissenschaft, Klasse Chemische Geologie*, 5:1–43.

Ramdohr, P., 1958. New observations on the ores of the Witwatersrand in South Africa and their genetic significance. *Trans. Geol. Soc. South Africa*, 61:1–50.

Robb, L. J., and F. M. Meyer, 1995. The Witwatersrand basin, South Africa: Geologic framework and mineralization processes. *Ore Geol. Rev.*, 10:67–94.

Rundle, C. C., and N. J. Snelling, 1977. The geochronology of uraniferous minerals in the Witwatersrand Triad: An interpretation of new and existing uranium-lead age data on rocks and minerals from the Dominion Reef, Witwatersrand, and Ventersdorp Supergroups. *Philos. Trans. Roy. Soc. Lond.*, A286:567–583.

Utter, T., 1980. Rounding of ore particles from the Witwatersrand gold and uranium deposits (South Africa) as an indicator of their detrital origin. *J. Sediment. Petrol.*, 50:71–76.

Zartman, R. E., and H. E. Frimmel, 1999. Rn-generated ^{206}Pb in hydrothermal sulphide minerals and bitumen from the Ventersdorp Contact Reef, South Africa. *Mineral. Petrol.*, 66:171–191.

REFERENCE FOR THE Pt–Os METHOD (SECTION13.9)

Walker, R. J., J. W. Morgan, M. I. Smoliar, E. Beary, and G. K. Czamanske, 1997. Application of the ^{190}Pt-^{186}Os isotope system to geochemistry and cosmochemistry. *Geochim. Cosmochim. Acta*, 61:4799–4808.

CHAPTER 14

The La–Ce Method

THE chart of the nuclides (Figure 1.2) contains several very long-lived radionuclides whose halflives exceed 10^{11} years. The resulting parent–daughter decay schemes are potentially useful for dating certain minerals having high parent-to-daughter ratios and for dating the terrestrial and extraterrestrial rocks in which such minerals occur. For example, the α-decay of long-lived ^{190}Pt to ^{186}Os ($T_{1/2} = 4.50 \times 10^{11}$) in Pt-bearing sulfide ore at Noril'sk, Russia, was investigated by Walker et al. (1997) and was described briefly in Section 13.9.

The very long-lived radionuclides and their daughters in Table 14.1 include ^{138}La, which undergoes branched decay to stable ^{138}Ce and ^{138}Ba. In addition, ^{50}V and ^{180}Ta also undergo branched decay as required by Mattauch's rule. The branched decay of ^{138}La gives rise to two geochronometers that arise from the β-decay of $^{138}_{57}$La to $^{138}_{58}$Ce and from the electron capture decay of $^{138}_{57}$La to $^{138}_{56}$Ba. The feasibility of dating silicate rocks by the La–Ce method was first demonstrated by Tanaka and Masuda (1982), whereas Masuda and Nakai (1983) attempted to date a sample of monazite by the La–Ba method (see Chapter 15). The development of these geochronometers has been limited by the requirement for exceptional measurement precision caused by the long halflives on the order of hundreds of billions of years and by the difficulties of measuring the decay constants for the β- and electron capture decays of ^{138}La. The abundances of both radiogenic daughters are also applicable to the study of igneous rocks in combination with Sr, Nd, Pb, Hf, and Os.

Table 14.1. Very Long-Lived Naturally Occurring Radionuclides and Their Stable Daughters

Parent	Decay	Daughter	Halflife, years
^{50}V	e	^{50}Ti	1.4×10^{17} (total)
	β^-	^{50}Cr	
^{113}Cd	β^-	^{113}In	9×10^{15}
^{115}In	β^-	^{115}Sn	4.4×10^{14}
^{123}Te	e	^{123}Sb	1.3×10^{13}
^{138}La[a]	e	^{138}Ba	$1.57 \pm 0.05 \times 10^{11}$
	β^-	^{138}Ce	$2.97 \pm 0.28 \times 10^{11}$
^{144}Nd	α	^{140}Ce	2.1×10^{15}
^{148}Sm	α	^{144}Nd	7×10^{15}
^{152}Gd	α	^{148}Sm	1.1×10^{14}
^{174}Hf	α	^{170}Yb	2.0×10^{15}
^{180}Ta	e	^{180}Hf	$>1.2 \times 10^{15}$ (total)
	β^-	^{180}W	
^{186}Os[b]	α	^{182}W	1.98×10^{15}
^{190}Pt[2]	α	^{186}Os	4.50×10^{11}

Source: Walker et al., 1989.

[a]From Makishima et al., 1993.

[b]From Walker et al., 1997, Shirey and Walker, 1998.

14.1 GEOCHEMISTRY OF La AND Ce

Lanthanum ($Z = 57$) and cerium ($Z = 58$) are REEs. Their physical and chemical properties (Table 14.2) are similar to those of samarium (Sm), neodymium (Nd), and lutetium (Lu), which were discussed in Chapters 9 (Sm–Nd) and 12 (Lu–Hf). The distribution of the electrons of La differs from that of Ce because the three valence electrons of La are located in the $5d$ and $6s$ orbitals (i.e., $5d^1 6s^2$), whereas the four valence electrons of Ce enter the $4f$ and $6s$ orbitals (i.e., $4f^2 5d^0 6s^2$). Consequently, La is trivalent and its ionic radius of 1.15 Å is the largest among the REEs. Cerium also forms trivalent ions (1.11 Å), but it can be oxidized to the tetravalent state (1.01 Å). Both elements have low electronegativities (1.1) and form predominantly ionic bonds with O^{2-} in crystals of silicate, phosphate, carbonate, and oxide minerals. The abundances of La and Ce in the solar system are 0.446 and 1.136 atoms/10^6 Si atoms, respectively (Anders and Grevesse, 1989). The greater abundance of Ce is consistent with the Oddo–Harkins rule (Faure, 1998).

The average concentrations of La and Ce in *meteorites* in Table 14.3 range from <1.0 ppm in

Table 14.2. Physical and Chemical Properties of La and Ce

Property	La	Ce
Atomic number	57	58
Atomic mass, amu	138.9055	140.12
Electron formulas	La: [Xe] $5d^1 6s^2$	Ce: [Xe] $4f^2 5d^0 6s^2$
Valence	+3	+3, +4
Ionic radius, Å	(+3) 1.15	1.11
	(+4) —	1.01
Electronegativity	1.1	1.1
Abundance[a], solar system, atoms/10^6 Si	0.446	1.136
Naturally occurring isotopes	2	4

[a] Anders and Grevesse, 1989.

Table 14.3. Average Concentrations of La and Ce in Rocks and Minerals

Rock or Mineral	Concentration, ppm La	Ce	La/Ce	References[a]
	Meteorites			
Chondrites (ordinary)	0.351 (23)	0.943 (22)	0.37	1, 2
Chondrites (carbonaceous)	0.279 (5)	0.882 (5)	0.32	1, 2
Achondrites (Ca rich)	2.48 (10)	6.60 (11)	0.38	1, 2
	Lunar Rocks			
Basalt	21.5 (2)	67.5 (2)	0.32	1
Gabbro	10.6 (2)	35.8 (2)	0.30	1
Soil(fines)	15.5 (2)	56.0 (2)	0.28	1
	Terrestrial Minerals			
Pyroxene	0.825 (2)	3.46 (2)	0.24	5
Plagioclase	2.37 (2)	3.73 (2)	0.63	5
Garnet	4.85 (1)	2.06 (1)	2.35	7
Sphene	179 (2)	918 (2)	0.19	7
Apatite	189 (2)	658 (2)	0.29	1,7
Epidote	599 (1)	1,199 (1)	0.50	7
Gadolinite	1,184 (1)	3,758 (1)	0.31	7
Allanite	32,880 (3)	49,577 (3)	0.66	7
Monazite	133,000 (1)	230,000 (1)	0.58	1
	Igneous and Metamorphic Rocks			
Peridotite	2.61 (6)	7.24 (6)	0.36	1
Midocean ridge basalt	3.60 (29)	11.4 (27)	0.32	1, 10, 11
Oceanic island and continental basalt	14.2 (19)	29.3 (21)	0.48	1,11
Island arc basalt and andesite	11.2 (13)	24.2 (13)	0.46	12
Gabbro, continental	5.10 (6)	7.49 (6)	0.68	1, 11

continued overleaf

Table 14.3. (*continued*)

Rock or Mineral	Concentration, ppm La	Ce	La/Ce	References[a]
Intermediate rocks	14.2 (4)	21.2 (4)	0.67	1
Granitic rocks	61.2 (23)	115 (22)	0.53	1
Precambrian granitic gneisses	33.2 (15)	64.2 (15)	0.52	6–8, 11
Carbonatite	411 (4)	636 (4)	0.65	1
	Sedimentary Rocks			
Shale	50.9 (14)	93.8 (15)	0.54	1
Sandstone	39.3 (9)	80.8 (8)	0.49	1
Limestone	7.2 (3)	11.3 (3)	0.64	1
Phosphorite	104 (3)	147 (3)	0.71	1
Fossil fish	1,290 (79)	2,480 (79)	0.52	1
Mn nodules	180 (22)	990 (22)	0.18	3
Chert	9.7 (7)	10.0 (7)	0.89	4

Note: The number of samples included in each average is indicated in parentheses.

[a] 1. Herrmann, 1970, in Wedepohl (Ed.) *Handbook of Geochemistry*, Sections 57 and 58, Springer-Verlag, Heidelberg. 2. Makishima and Masuda, 1993, *Chem. Geol.*, 106:197–205. 3. Amakawa et al., 1991, *Earth Planet. Sci. Lett.*, 105:554–565. 4. Shimizu et al., 1991, *Geology*, 19(4):369–371. 5. Tanaka and Masuda, 1982, *Nature*, 300:515–517. 6. Dickin, 1987a, *Nature*, 325:337–338. 7. Masuda et al., 1988, *Earth Planet. Sci. Lett.*, 89:316–322. 8. Shimizu et al., 1988, *Earth Planet. Sci. Lett.*, 91:159–169. 9. Makishima et al., 1993, *Chem. Geol. (Isotope Geosci. Sect.)*, 104:293–300. 10. Makishima and Masuda, 1994, *Chem. Geol.*, 118:1–8. 11. Tanaka et al., 1987, *Nature*, 327:113–117. 12. Shimizu et al., 1992, *Contrib. Mineral. Petrol.*, 110:242–252.

Ca-poor achondrites to 3.0 ppm La and 8.0 ppm Ce in Ca-rich achondrites. Ordinary chondrites also have low average concentrations (La = 0.355 ± 0.060 ppm, Ce = 0.956 ± 0.184 ppm) with a La/Ce ratio of 0.37. In addition, the La/Ce ratio of carbonaceous chondrites is 0.30 based on La = 0.265 ± 0.096 ppm and Ce = 0.886 ± 0.309 ppm.

Lunar basalts of the Mare Tranquillitatis have significantly higher La and Ce concentrations than chondrites (La = 21.5 ± 1.5 ppm; Ce = 67.5 ± 11.5 ppm) with La/Ce = 0.32. Lunar gabbros and the fine-grained fraction of lunar soils have similar concentrations and La/Ce ratios as lunar basalts.

The average concentrations of La and Ce in calk-alkaline igneous rocks in Table 14.3 rise progressively from peridotite to rhyolites and granitic rocks. However, the average La/Ce ratios of these rocks in Figure 14.1 rise only from about 0.30 in MORBs to 0.70 in gabbro and intermediate rocks but then decline with increasing La concentrations to 0.62 in rhyolites and 0.53 in granitic rocks. Alkali-rich igneous rocks in Table 14.3 and Figure 14.1 have higher concentrations of La and Ce than calk-alkaline rocks, but their La/Ce ratios stay close to 0.50.

The highest concentrations of La and Ce occur in phosphate minerals (apatite and monazite) and in fossilized fish bones. In addition, sedimentary phosphorite deposits have elevated concentrations of La (104 ± 60 ppm) and Ce (147 ± 45 ppm) with a La/Ce ratio of 0.71.

The point of this presentation is that the concentrations of La and Ce in terrestrial rocks and minerals range over five orders of magnitude from <10 ppm in peridotite to $>10^5$ ppm in some monazites. Nevertheless, the La/Ce ratios vary by only a factor of 2, from about 0.30 in peridotite to about 0.60 in monazite. The decline of the La/Ce ratios of rhyolites and granitic rocks in Figure 14.1 is caused by their high average Ce concentrations, which reach 115 ± 37 ppm in granitic rocks. The narrow range of variation of La/Ce ratios in terrestrial and extraterrestrial rocks limits the precision of La–Ce isochron dates.

14.2 PRINCIPLES AND METHODOLOGY

Lanthanum has two naturally occurring isotopes: ^{138}La (0.09%) and ^{139}La (99.91%). The branched decay of ^{138}La in Figure 14.2 proceeds by β^- emission to stable ^{138}Ce and by electron capture to stable ^{138}Ba:

$$^{138}_{57}\mathrm{La} \rightarrow {}^{138}_{58}\mathrm{Ce} + \beta^- + \overline{\nu} + \gamma + Q \quad (14.1)$$

$$^{138}_{57}\mathrm{La} + e^- \rightarrow {}^{138}_{56}\mathrm{Ba} + \nu + \gamma + Q \quad (14.2)$$

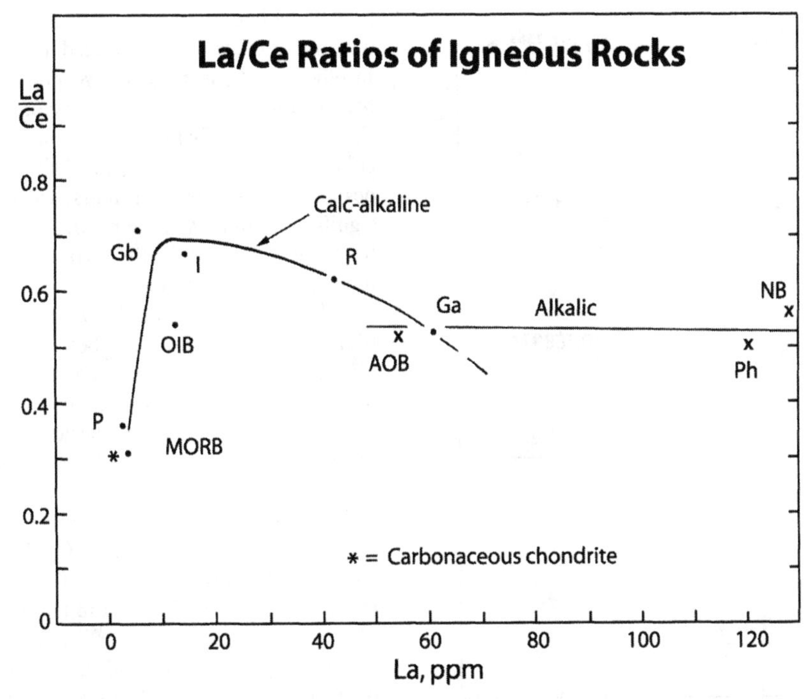

FIGURE 14.1 Variation of the La/Ce ratio with increasing La concentration in calc-alkaline (solid circles) and alkali-rich (crosses) igneous rocks identified by letters. Calc-alkaline suite: MORB = midocean ridge basalt; P = peridotite; OIB = oceanic island and continental basalt; Gb = gabbro; I = intermediate rocks; R = rhyolite; Ga = granitic rocks. Alkalic suite: AOB = alkali olivine basalt; Ph = phonolite; NB = nep-heline basalt. Trachyte is offscale at La = 168 ± 86.0 ppm, La/Ce = 0.42. All data are from a compilation by Herrmann (1970).

Cerium has four naturally occurring isotopes all of which are stable: ^{136}Ce (0.19%), ^{138}Ce (0.25%), ^{140}Ce (88.48%), and ^{142}Ce (11.08%) (Lide and Frederikse, 1995).

The decay of ^{138}La to radiogenic ^{138}Ce follows the law of radioactivity:

$$\frac{^{138}\text{Ce}}{^{142}\text{Ce}} = \left(\frac{^{138}\text{Ce}}{^{142}\text{Ce}}\right)_i + \left(\frac{\lambda_\beta}{\lambda}\right)\frac{^{138}\text{La}}{^{142}\text{Ce}}(e^{\lambda t} - 1) \tag{14.3}$$

where λ_β = the partial decay constant for β-decay
λ = the total decay constant governing the decay of ^{138}La both by β- and electron capture decay.

The isotope composition of Ce has also been expressed in terms of the ^{138}Ce/^{136}Ce ratio (e.g., Dickin, 1987a, b). The conversion of ^{138}Ce/^{136}Ce to ^{138}Ce/^{142}Ce ratios is made by

$$\begin{aligned}\frac{^{138}\text{Ce}}{^{142}\text{Ce}} &= \frac{^{138}\text{Ce}}{^{136}\text{Ce}} \times \frac{^{136}\text{Ce}}{^{142}\text{Ce}} \\ &= \frac{^{138}\text{Ce}}{^{136}\text{Ce}} \times \frac{0.19}{11.08} \\ &= \frac{^{138}\text{Ce}}{^{136}\text{Ce}} \times 0.017148\end{aligned} \tag{14.4}$$

The ^{138}La/^{142}Ce ratio is calculated from the concentrations of La and Ce:

$$\frac{^{138}\text{La}}{^{142}\text{Ce}} = \left(\frac{\text{La}}{\text{Ce}}\right)_{\text{conc}} \times \frac{\text{Ab }^{138}\text{La} \times \text{at. wt. Ce}}{\text{at. wt. La} \times \text{Ab }^{142}\text{Ce}} \tag{14.5}$$

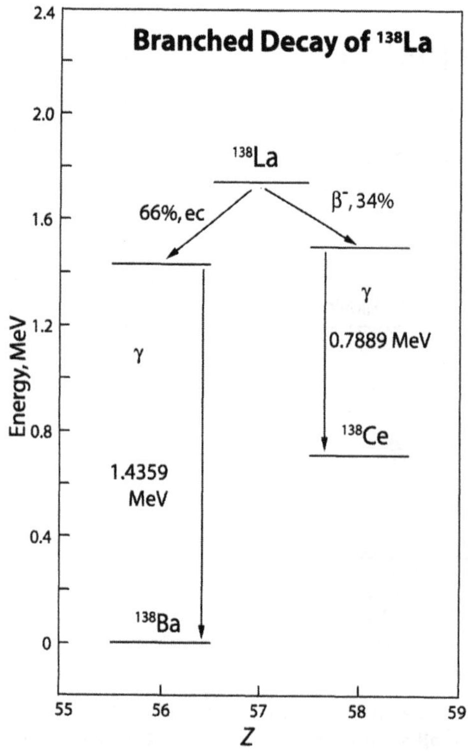

FIGURE 14.2 Decay scheme diagram for the branched decay of $^{138}_{57}La$ by β^--emission to stable $^{138}_{58}Ce$ and by electron capture to stable $^{138}_{56}Ba$. Each product nucleus emits a γ-ray (Lide and Frederikse, 1995).

For Ce that is not enriched in radiogenic ^{138}Ce the conversion factor k is given as

$$k = \frac{0.09 \times 140.12}{138.9055 \times 11.08} = 0.00819$$

The measurements of the isotope ratios of Ce used to require a large number of scans of the mass spectrum in order to achieve the necessary precision. For example, Tanaka and Masuda (1982) analyzed every sample two to four times and recorded about 100 scans each time lasting 3–4 h. Dickin (1987a) recorded up to 3800 scans for each sample requiring well over 100 h.

The time required for an isotope analysis of Ce (as well as of other elements) has been reduced by the availability of mass spectrometers equipped with multiple collectors operated in the static mode. In other words, instead of sweeping the ion beam back and forth past a single collector cup, the beam shines continuously into a set of cups each of which collects one of the isotopes. This change in the design and operation of mass spectrometers has significantly improved the precision of the data and has thereby reduced the time required to achieve the required level of precision.

The decay constants for β-emission and electron capture of ^{138}La have been determined both by direct counting of the rate of decay of ^{138}La in La_2O_3 and by analysis of rocks and minerals of known age. Tanaka and Masuda (1982) used the results of 10 direct determinations of the halflives of ^{138}La published between 1951 and 1982 to calculate weighted averages and corrected them for the presence of H_2O and CO_2 in the La_2O_3. Their result and those of other investigators in Table 14.4 indicate that large discrepancies exist in the reported values of the β-decay halflife of ^{138}La. The most recent redetermination of the β-decay halflife of ^{138}La by Makishima et al. (1993) was based on mineral-rock La–Ce isochrons of two Late Archean granites from the Yilgarn Block of Western Australia. The ages of these rocks were determined by the U–Pb, Rb–Sr, and Sm–Nd methods, which yielded concordant results, thereby demonstrating that these two rocks have remained closed to U, Pb, Rb, Sr, Sm, and Nd for $(2664 \pm 5) \times 10^6$ and $(2686 \pm 6) \times 10^6$ years, respectively. Consequently, the best value of the β-decay halflife calculated by Makishima et al. (1993) is

$$T_{1/2}(\beta) = 2.97 \pm 0.28 \times 10^{11} \text{ y}$$
$$\lambda_\beta = 2.33 \pm 0.24 \times 10^{-12} \text{ y}^{-1}$$

The halflife determination by Makishima et al. (1993) agrees with the determination of the β-decay halflife by Sato and Hirose (1981) and by Dickin (1987a). Makishima et al. (1993) also averaged the direct determinations of the electron capture halflife of Sato and Hirose (1981) and of Norman and Nelson (1983a) in Table 14.4:

$$T_{1/2}(\text{e}) = 1.57 \pm 0.05 \times 10^{11} \text{ y}$$
$$\lambda_e = 4.42 \pm 0.14 \times 10^{-12} \text{ y}^{-1}$$

Table 14.4. Halflives for β- and Electron Capture Decay of ^{138}La

β-Decay $\times 10^{11}$ years	Electron Capture $\times 10^{11}$ years	References	Explanation
3.02 ± 0.04	1.56 ± 0.05	Sato and Hirose, 1981	Gamma counting
2.69 ± 0.24	1.51 ± 0.10	Tanaka and Masuda, 1982	Weighted average of 10 determinations with $H_2O + CO_2$ correction
2.69	—	Tanaka and Masuda, 1982	La–Ce mineral isochron of gabbro, Bushveld complex, South Africa
3.19 ± 0.22	1.58 ± 0.02	Norman and Nelson, 1983a	Gamma counting
3.10 ± 0.15	—	Norman and Nelson, 1983b	Average of Norman and Nelson (1983a) and Sato and Hirose (1981)
3.02	—	Dickin, 1987a	La–Ce isochron, Lewisian gneiss, Scotland
2.58	—	Masuda et al., 1988	Recalculated from Dickin (1987a)
2.50 ± 0.18	—	Masuda et al., 1988	La–Ce and Sm–Nd isochrons of minerals in two Finnish pegmatites
2.97 ± 0.28	—	Makishima et al., 1993	La–Ce isochron, two granites, Yilgarn Block, Western Australia

The total decay constant and halflife of ^{138}La are

$$\lambda = \lambda_\beta + \lambda_e$$
$$= 2.33 \times 10^{-12} + 4.42 \times 10^{-12}\ y^{-1}$$
$$= 6.75 \pm 0.73 \times 10^{-12}\ y^{-1}$$
$$T_{1/2} = 1.03 \pm 0.10 \times 10^{11}\ y$$

All errors in this calculation were combined by the method of least squares.

Procedures for the determination of La and Ce concentrations in rock samples and for the isotope analysis of Ce have been described elsewhere:

Tanaka and Masuda, 1982, *Nature*, 300:515–517. Makishima et al., 1987, *Mass Spectrosc.*, 35:64–72. Makishima and Nakamura, 1991a, *Chem. Geol. (Isotope Geosci. Sect.)*, 94:1–11. Makishima and Nakamura, 1991b, *Chem. Geol. (Isotope Geosci. Sect.)* 94:105–110. Makishima et al., 1991, *Study Earth's Interior*, Misasa Tech. Rept., Ser. B, No. 10, Okayama University. Misasa, Japan. Makishima et al., 1993, *Chem. Geol. (Isotope Geosci. Sect.)*, 104:293–300.

14.3 La–Ce ISOCHRONS

The objective of the first age determinations by the La–Ce method by Tanaka and Masuda (1982) and Dickin (1987a) was to determine the halflife for the β-decay of ^{138}La by dating rocks that had been dated previously by other methods (e.g., the Bushveld Complex of South Africa and the Lewisian gneiss of Scotland). Dickin (1987a) emphasized that geological determinations of the halflives of long-lived radionuclides should not be based on internal mineral isochrons because the isotope systematics of minerals may have been altered by episodes of thermal metamorphism and hydrothermal alteration.

14.3a Bushveld Complex, South Africa

Tanaka and Masuda (1982) first attempted to construct a La–Ce isochron for the minerals of a specimen of gabbro from the Upper Zone of the Bushveld Complex in South Africa (Faure, 2001). The data of Tanaka and Masuda (1982) were recalculated for the presentation in Figure 14.3 by using a factor $k = 0.00819$ to convert La/Ce to $^{138}La/^{142}Ce$ ratios, and the $^{138}Ce/^{142}Ce$ ratios

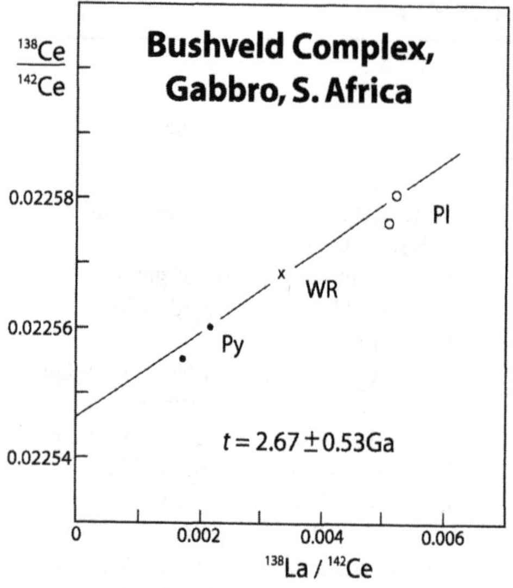

FIGURE 14.3 Application of the La–Ce method to the dating of a specimen of gabbro from the Upper Zone of the Bushveld Complex at Magnet Height, South Africa, based on $\lambda_\beta = 2.33 \times 10^{-12}\ y^{-1}$ and $\lambda = 6.75 \times 10^{-12}\ y^{1}$. Measurements by Tanaka and Masuda (1982).

were reduced by 0.0002874 to make them compatible with 0.0225685 for JMC-304 reported by Makishima and Nakamura (1991a). An unweighted least-squares regression yields a slope $m = 0.0062779$ and an intercept of 0.0225462. The La–Ce date of the gabbro from the Bushveld Complex was calculated from the slope m by the relation

$$m = \frac{\lambda_\beta}{\lambda}(e^{\lambda t} - 1) \tag{14.6}$$

thus

$$t = \frac{1}{\lambda} \ln\left(\frac{m \times \lambda}{\lambda_\beta} + 1\right)$$

Substituting appropriate values for the decay constants (Makishima et al., 1993),

$$\lambda_\beta = 2.33 \times 10^{-12}\ y^{-1} \quad \lambda = 6.75 \times 10^{-12}\ y^{-1}$$

yields

$$t = \frac{1}{6.75 \times 10^{-12}} \ln\left(\frac{0.0062779 \times 6.75}{2.33} + 1\right)$$

$$= \frac{\ln(1.0181870)}{6.75 \times 10^{-12}} = 2.67 \pm 0.53\ \text{Ga}$$

The uncertainty of the date was assigned by Tanaka and Masuda (1982). These authors also analyzed the same samples for dating by the Sm–Nd method and obtained a date of 2.05 ± 0.09 Ga. The errors are just large enough to cover the difference between the two results. Hamilton (1997) also reported a Rb–Sr date of 2.050 ± 0.024 Ga for the Bushveld Complex, which agrees very well with the Sm–Nd date obtained by Tanaka and Masuda (1982).

The initial $^{138}Ce/^{142}Ce$ ratio of the Bushveld Complex (Figure 14.3, 0.0225462) is greater than the initial $^{138}Ce/^{142}Ce$ ratio of stony meteorites (0.0225321) but less than that of the present value of the chondritic mantle (0.0225652) reported by Makishima and Masuda (1993) and discussed in Section 14.4. Therefore, the pioneering work of Tanaka and Masuda (1982) opened the door not only for dating by the La–Ce method but also for the application of the isotope composition of Ce to the study of the origin of igneous rocks.

14.3b Lewisian Gneiss, Scotland

The La–Ce method was used by Dickin (1987a) to date a suite of six whole-rock samples of Lewisian granulite gneisses that have a Sm–Nd isochron age of 2.910 ± 0.05 Ga (Hamilton et al., 1979). The La–Ce data of Dickin (1987a) were converted to $^{138}Ce/^{142}Ce$ ratios which were reduced by 0.000271 to be compatible with $^{138}Ce/^{142}Ce = 0.0225652$ for BCR-1 reported by Makishima and Nakamura (1991a). This adjustment is based on the fact that Dickin (1987b) derived a value of 0.0227823 for "bulk Earth" that is the present $^{138}Ce/^{142}Ce$ ratio of CHUR-Ce, which in turn is assumed to be identical to the $^{138}Ce/^{142}Ce$ ratio of BCR-1 (Makishima and Masuda, 1993). In addition, the La/Ce ratios reported by Dickin (1987a) were converted to the $^{138}La/^{142}Ce$ ratios by multiplying them by $k = 0.00819$ (equation 14.5). The data were used to constrain a straight line in Figure 14.4 by an unweighted least-squares regression (after omitting sample 16Y). The resulting slope $m = 0.00716611$ yields a date of 3.04 ±

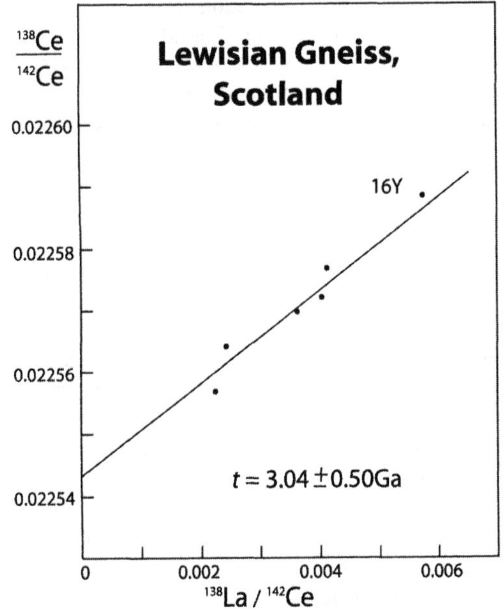

FIGURE 14.4 The La–Ce date of whole-rock samples of Lewisian gneiss, Scotland. The original data were recalculated to ^{138}Ce/^{142}Ce ratios, which were subsequently reduced by 0.0002171 as explained in the text. The La/Ce ratios were converted to ^{138}La/^{142}Ce ratios by the factor $k = 0.00819$. The date was calculated from $\lambda_\beta = 2.33 \times 10^{-12}$ y^{-1} and $\lambda = 6.75 \times 10^{-12}$ y^{-1}. Data from Dickin (1987a).

0.50 Ga from equation 14.6. The date is indistinguishable from the Sm–Nd date of 2.91 ± 0.05 Ga reported by Hamilton et al. (1979).

The initial ^{138}Ce/^{142}Ce ratio of the Lewisian gneisses (0.0225437) indicates that the Ce could have been derived from the chondritic uniform reservoir at 2.99 Ga, in good agreement with the date derived from the errorchron in Figure 14.4. (Model dates based on CHUR are discussed in Sections 14.4 and 14.6c.) These calculations confirm the conclusion by Dickin (1987a) that the Ce that now resides in the Lewisian gneiss originated from a chondritic source at 3.0 Ga. In addition, the concordance of the Sm–Nd and La–Ce dates of the whole-rock samples of Lewisian gneiss confirms the accuracy of the β-decay halflife of ^{138}La reported by Sato and Hirose (1981).

14.4 METEORITES AND CHUR-Ce

The primeval isotope composition of Ce was determined from the La–Ce systematics of stony meteorites. The first measurements of the isotope composition of Ce in stony meteorites by Shimizu et al. (1984) were not sufficiently precise to permit a reliable determination of the initial ^{138}Ce/^{142}Ce ratio. Subsequently, Makishima and Masuda (1993) used the improved methodology of Makishima and Nakamura (1991a, b) to calculate the initial ^{138}Ce/^{142}Ce ratios of the carbonaceous chondrite Murchison and of seven other stony meteorites (three chondrites and four achondrites).

The weighted average of the initial ^{138}Ce/^{142}Ce ratios of stony meteorites in Table 14.5 is

$$\left(\frac{^{138}\text{Ce}}{^{142}\text{Ce}}\right) = 0.0225321 \pm 0.000007$$

This is the primeval value of this ratio assuming an age of 4.56×10^9 years for the age of the solar system and a value of 2.33×10^{-12} y^{-1} for the β-decay constant of ^{138}La (Makishima and Masuda, 1993).

These authors also reported isotope ratios of Ce and Nd for the U.S. Geological Survey (USGS)

Table 14.5. Average Initial ^{138}Ce/^{142}Ce Ratios of Chondrite and Achondrite Meteorites Assuming an Age of 4.56×10^9 Years for the Solar System and $\lambda_\beta = 2.33 \times 10^{-12}$ y^{-1}

Class[a]	Primeval ^{138}Ce/^{142}Ce
Ordinary chondrites (3)	0.0225316 ± 0.0000011
Achondrites (4)	0.0225322 ± 0.0000010
Murchison (1)	0.0225329 ± 0.0000008
Stony meteorites (4)	0.02253254 ± 0.0000039[b]
Weighted average	0.0225321 ± 0.0000007 (2σ)

Source: Makishima and Masuda, 1993.

[a]The number of samples of each meteorite type is indicated in parentheses.

[b]From Shimizu et al. (1984), recalculated by Makishima and Masuda (1993).

standard rock BCR-1:

$$\left(\frac{^{138}\text{Ce}}{^{142}\text{Cd}}\right)_{\text{BCR-1}} = 0.0225652 \pm 0.0000018$$

$$\left(\frac{^{143}\text{Nd}}{^{144}\text{Nd}}\right)_{\text{BCR-1}} = 0.512640 \pm 0.000017$$

DePaolo and Wasserburg (1976a, b) determined that the ε^0 (Nd) value of BCR-1 is close to zero, which implies that its $^{143}Nd/^{144}Nd$ ratio does not differ significantly from the present $^{143}Nd/^{144}Nd$ ratio of CHUR-Nd. Therefore, Makishima and Masuda (1993) assumed that the $^{138}Ce/^{142}Ce$ ratio of BCR-1 is also close to the present value of the $^{138}Ce/^{142}Ce$ ratio of CHUR-Ce. In other words, the $^{138}Ce/^{142}Ce$ ratio of CHUR-Ce increased only by 0.15% from 0.0225321 to 0.0225652 in 4.56×10^9 years.

The equation for CHUR-Ce arising from the work of Makishima and Masuda (1993) is

$$\left(\frac{^{138}\text{Ce}}{^{142}\text{Ce}}\right)^0_{\text{CHUR}} = \left(\frac{^{138}\text{Ce}}{^{142}\text{Ce}}\right)^{\text{P}}_{\text{CHUR}} + \left(\frac{\lambda_\beta}{\lambda}\right) \times \left(\frac{^{138}\text{La}}{^{142}\text{Ce}}\right)^0_{\text{CHUR}} (e^{\lambda T} - 1) \quad (14.7)$$

Since

$$\left(\frac{^{138}\text{Ce}}{^{142}\text{Ce}}\right)^0_{\text{CHUR}} = 0.0225652,$$

$$\left(\frac{^{138}\text{Ce}}{^{142}\text{Ce}}\right)^{\text{P}}_{\text{CHUR}} = 0.0225321 \quad \text{(primeval value)},$$

$$\frac{\lambda_\beta}{\lambda} = \frac{2.33 \times 10^{-12}}{6.75 \times 10^{-12}} = 0.345,$$

and

$$T = 4.56 \times 10^9 \text{ y},$$

$$\left(\frac{^{138}\text{La}}{^{142}\text{Ce}}\right)^0_{\text{CHUR}} = \frac{0.0225652 - 0.0225321}{0.345 \times 0.03125} = 0.003069$$

$$\left(\frac{\text{La}}{\text{Ce}}\right)^0_{\text{CHUR}} = \frac{0.003069}{0.00819} = 0.375$$

The results of this calculation differ slightly from those of Makishima and Masuda (1993), who obtained $^{138}La/^{142}Ce = 0.00306 \pm 0.00006$ and $La/Ce = 0.375 \pm 0.007$. Nevertheless, the $^{138}Ce/^{142}Ce$ ratio of CHUR-Ce at any time t in the past is

$$\left(\frac{^{138}\text{Ce}}{^{142}\text{Ce}}\right)^t_{\text{CHUR}} = 0.0225652 - 0.345 \times 0.00306\,(e^{\lambda t} - 1) \quad (14.8)$$

The isotope composition of Ce can be used in combination with the isotope ratios of Sr, Nd, Pb, Hf, and Os to characterize the magma sources of igneous rocks and to detect contamination of the magma as a result of (1) partial melting of mechanically mixed source rocks, (2) mixing of magmas derived from different sources, or (3) assimilation of rocks in the continental crust (Faure, 2001). Isotope ratios of Ce in volcanic rocks have been reported elsewhere:

Dickin, 1987, *Nature*, 326:283–284. Tanaka et al., 1987, *Nature*, 327:113–117. Shimizu et al., 1992, *Contrib. Mineral. Petrol.*, 110: 242–252. Makishima and Masuda, 1994, *Chem. Geol.*, 118:1–8.

The interpretation of the isotope ratios of Ce is complicated because of interlaboratory discrepancies that have resulted from the use of different measurement protocols. In addition, different values of the present $^{138}Ce/^{142}Ce$ ratio of CHUR have been used to calculate ε-values, and even the β-decay constant of ^{138}La has been in doubt. Makishima and Nakamura (1991a) published a detailed description of their analytical procedures and mass spectrometry. In addition, they compiled isotope ratios of Ce in two interlaboratory standards reported by different investigators. This compilation in Table 14.6 includes data for JMC-304 (Johnson and Matthey CeO_2) and BCR-1 (basalt, Columbia River) and reveals the magnitude of the differences in the $^{138}Ce/^{142}Ce$ ratios that have been reported. The results published by Makishima and Nakamura (1991a) were obtained by use of a multicollector mass spectrometer operated statically which reduced the data acquisition time to

Table 14.6. Values of $^{138}Ce/^{142}Ce$ Ratios Reported by Different Investigators for Interlaboratory Standards JMC-304 and BCR-1

$^{138}Ce/^{142}Ce$	References
JMC 304[a]	
0.0228559	Tanaka and Masuda, 1982
0.0228559 ± 0.00000011	Shimizu et al., 1984
0.0225762 ± 0.00000014	Makishima et al., 1987
0.0225762 ± 0.00000013	Masuda et al., 1988
0.0225685 ± 0.0000004[b]	Makishima and Nakamura, 1991a
0.0225777 ± 0.00000020	Shimizu et al., 1992
BCR-1[c]	
0.0228520 ± 0.00000010	Tanaka et al., 1987
0.0225652 ± 0.0000004[b]	Makishima and Nakamura, 1991a
0.0225652 ± 0.0000011	Makishima and Masuda, 1994
0.0225722 (CHUR)	Shimizu et al., 1992

[a] Johnson and Matthey CeO_2 (Tanaka and Masuda, 1982).
[b] Preferred values.
[c] Basalt, Columbia River (USGSurv).

about 2 h and permitted the use of high beam intensities between 2×10^{-11} and 7×10^{-11} A for the $^{142}Ce^{16}O$ ions. These improvements in the mass spectrometry give confidence in their values of the $^{138}Ce/^{142}Ce$ ratios of JMC-304 and BCR-1. Makishima and Nakamura (1991a) recommended that BCR-1 be used to calculate ε^0 (Ce) values:

$$\varepsilon^0(\text{Ce}) = \left[\frac{(^{138}\text{Ce}/^{142}\text{Ce})_{\text{spl}}}{(^{138}\text{Ce}/^{142}\text{Ce})_{\text{BCR}-1}} - 1\right] \times 10^4 \tag{14.9}$$

The resulting ε^0(Ce) values are especially useful for the interpretation of the isotope composition of Ce because the $^{138}Ce/^{142}Ce$ ratios are small numbers with limited range of variation.

14.5 VOLCANIC ROCKS

The interpretation of the isotope ratios of Ce in volcanic rocks of Tertiary age was originally discussed by Dickin (1987b) and Tanaka et al. (1987). A sample of normal midocean ridge basalt (N-MORB) from the Pacific Ocean (09°00.10′ N; 106°06.77′ W) analyzed by Makishima and Masuda (1994) has a $^{138}Ce/^{142}Ce$ ratio of 0.0225616 compared to 0.0225652 for BCR-1. Therefore, the ε^0(Ce) value of this specimen is

$$\varepsilon^0(\text{Ce}) = \left(\frac{0.0225616 - 0.0225652}{0.0225652}\right) \times 10^4$$
$$= -1.6$$

The ε^0(Ce) values of 11 samples of N-MORBs from the Pacific Ocean analyzed by Makishima and Masuda (1994) range from -1.2 to -1.6 and have a mean value to -1.39 ± 0.08. The ε^0(Ce) value of five N-MORBs from the Atlantic Ocean is -1.24 ± 0.17. All the ε^0(Ce) values of N-MORBs relative to BCR-1 are negative, which means that the MORBs originated from magma sources in the mantle having a lower La/Ce ratio than the magma source of the Columbia River basalt.

The $^{138}Ce/^{142}Ce$ ratios of volcanic rocks from the Solomon and Bonin island arcs in the Pacific Ocean reported by Shimizu et al. (1992) were corrected by them to a value of 0.0225762 for JMC 304 reported by Makishima et al. (1987). Since Makishima and Nakamura (1991a) later reported 0.0225685 for the $^{138}Ce/^{142}Ce$ ratio of this standard, the isotope ratios of Ce of Shimizu et al. (1992) were reduced by 0.0000077 to be compatible with the data for MORBs reported by Makishima and Nakamura (1991a). Therefore, the ε^0(Ce) values of island arc volcanics of Shimizu et al. (1992) calculated relative to 0.0225652 for BCR-1 are equivalent to the ε^0(Ce) values of MORBs reported by Makishima and Masuda (1994).

The ε^0(Ce) values of island arc volcanics (IAV) from the Solomon and Bonin Islands in Figure 14.5 are primarily positive and range from $+0.22$ to $+2.30$. Negative ε^0(Ce) values occur in three volcanic rocks from the Solomon Islands ("Big" feldspar basalt, New Georgia, $\varepsilon^0 = -2.52$; two samples of hornblende andesite, Kolombangara

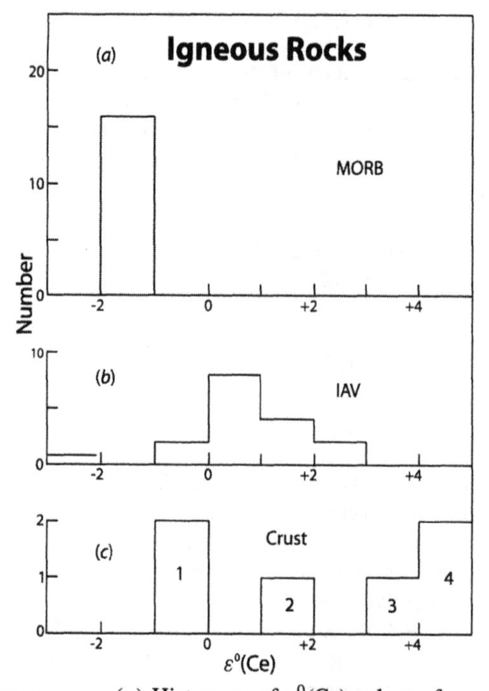

FIGURE 14.5 (*a*) Histogram of ε^0(Ce) values of MORB whose $^{138}Ce/^{142}Ce$ ratios are compatible with 0.0225652 for BCR-1 and 0.0225685 for JMC 304 reported by Makishima and Masuda (1994). (*b*) The ε^0(Ce) values of IAV analyzed by Shimizu et al. (1992) after they were reduced by 0.0000077 to be compatible with 0.0225685 for JMC 304 reported by Makishima and Nakamura (1991a). (*c*) The ε^0(Ce) values of crustal rocks after reducing the $^{138}Ce/^{142}Ce$ ratios reported by Tanaka et al. (1987) by 0.0002868 to be compatible with 0.0225652 for BCR-1: 1 = rhyolite and granodiorite, Japan; 2 = gabbro, 2050 Ma, Bushveld Complex, South Africa; 3 = granodiorite gneiss, 2100 Ma, Kenya; 4. Morton Gneiss and tonalitic gneiss, 3500 Ma, Minnesota.

Island, ε^0-values of −0.44 and −0.33). The wide range of ε^0(Ce) values of island-arc volcanics and their enrichment in radiogenic ^{138}Ce compared to MORBs indicates that some of the Ce originated from continental sediment that was subducted with the oceanic crust. The continental sediment has a higher $^{138}Ce/^{142}Ce$ ratio than MORBs because granitic rocks have higher La/Ce ratios than the depleted magma sources of MORBs.

The ε^0(Ce) values of crustal rocks in Figure 14.5*c* are based on $^{138}Ce/^{142}Ce$ ratios reported by Tanaka et al. (1987) that were reduced by 0.0002868 to be compatible with 0.0225652 for BCR-1. The resulting ε^0(Ce) values range from +1.77 (gabbro, Bushveld Complex, 2050 Ma) to +5.00 (grey tonalite gneiss, Minnesota, 3500 Ma). However, a rhyolite and a granodiorite from Japan have negative ε^0(Ce) values (−0.66 and −0.53), presumably because they formed by remelting of Tertiary basalt. Additional isotope analyses of Precambrian granitic gneisses were published by Makishima et al. (1993), Masuda et al. (1988), and Dickin (1987b).

14.6 CERIUM IN THE OCEANS

The isotopic composition of Ce in the oceans is controlled by inputs from the granitic basement rocks (and their sedimentary derivatives), from young volcanic rocks, and from hydrothermal fluids that are discharged within the ocean basins and on the surfaces of the continents. The Ce released by chemical weathering of basement rocks and their derivatives differs from that of young volcanic rocks by being enriched in radiogenic ^{138}Ce depending on the ages and La/Ce ratios of the crustal rocks. Therefore, the isotope composition of Ce in the oceans may vary regionally depending on the proportions of mixing of its two (or more) isotopic components.

14.6a Ferromanganese Nodules

Ferromanganese nodules form on the bottom of the ocean by precipitation of oxides of Fe and Mn from seawater and from water discharged by hot springs along spreading ridges. The oxide particles sorb the cations of many elements in solution in the oceans and transport them to the seafloor where they accumulate slowly in the form of ferromanganese nodules and crusts. The elements that are concentrated in these nodules include not only Fe, Mn, Cu, Ni, and Co but also the REEs, Sr, and Pb.

Amakawa et al. (1991) reported isotope ratios of Ce, Nd, and Sr of ferromanganese nodules from sites in the Atlantic and Pacific Oceans and from the Baltic and Barents Seas. They also analyzed JMC-304 and obtained a value of 0.0225781 ± 0.0000006 for the average $^{138}Ce/^{142}Ce$ ratio based on 157 determinations, whereas Makishima and Nakamura (1991a) reported a lower value of 0.0225685 for this isotope standard. Therefore, in this presentation, the $^{138}Ce/^{142}Ce$ ratios of Amakawa et al. (1991) were reduced by 0.0000096 and were then expressed as $\varepsilon^0(Ce)$ values relative to 0.0225652 for BCR-1.

The resulting $\varepsilon^0(Ce)$ values of the manganese nodules in Figure 14.6 are compatible with the $\varepsilon^0(Ce)$ values of igneous rocks in Figure 14.5. The $\varepsilon^0(Ce)$ values of ferromanganese nodules in the Atlantic Ocean range from +3.6 to −0.35 and have an average value of $+1.5 \pm 0.6$ for 12 samples. The positive sign of the $\varepsilon^0(Ce)$ values indicates that a significant fraction of Ce in the Atlantic Ocean originates from Precambrian basement rocks of the adjacent continents. The highest positive $\varepsilon^0(Ce)$ value (+3.6) occurs in a nodule collected on the abyssal plain at 23°24.0′ N, 50°45.0′ W at a depth of 5420 m. The highest negative $\varepsilon^0(Ce)$ value (−0.35) originated from 37°30.6′ N, 59°52.2′ W in 1523 m of water off the east coast of North America.

The ferromanganese nodules in the Pacific Ocean analyzed by Amakawa et al. (1991) all have negative $\varepsilon^0(Ce)$ values with a mean of -2.0 ± 0.3 for four samples. Although the available evidence is limited, it suggests that the Ce in the Pacific Ocean originates primarily from mantle-derived volcanic rocks erupted by submarine volcanoes and along midocean ridges.

In the Baltic Sea (including the Gulf of Bothnia) as well as in the Barents Sea, ferromanganese nodules have positive $\varepsilon^0(Ce)$ values ranging from +1.2 to +3.8 with means of $+2.6 \pm 0.8$ (Baltic Sea) and $+2.1 \pm 0.9$ (Barents Sea). These values are entirely consistent with the derivation of Ce from the Precambrian Shield of Scandinavia.

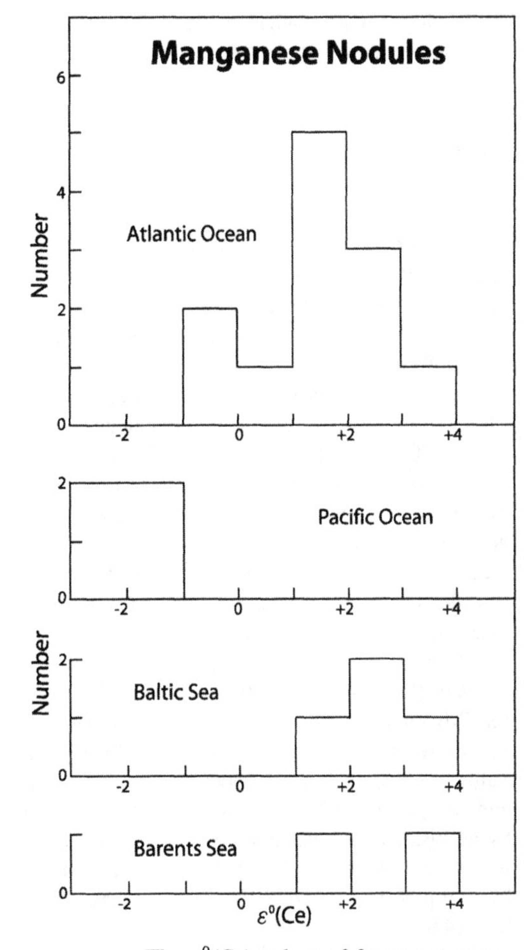

FIGURE 14.6 The $\varepsilon^0(Ce)$ values of ferromanganese nodules in the Atlantic and Pacific Oceans and in the Baltic and Barents Seas. The $^{138}Ce/^{142}Ce$ ratios measured by Amakawa et al. (1991) were reduced by 0.0000096 to be compatible with 0.0225685 for JMC-304 reported by Makishima and Nakamura (1991a). The ε-values were then expressed relative to $^{138}Ce/^{142}Ce = 0.0225652$ for BCR-1. Data from Amakawa et al. (1991) and Makishima and Nakamura (1991a).

The evidence for the existence of regional variations of $\varepsilon^0(Ce)$ values of marine ferromanganese nodules supports the hypothesis that the isotope composition of Ce in the oceans is controlled by mixing of two (or more) components derived from the Precambrian rocks of the continental crust and from volcanic rocks (and hydrothermal fluids) erupted within the ocean basins and along the volcanically active margins of the continents. Elderfield (1992) used the data of Amakawa et al. (1991)

to develop a mathematical model for this mixing process (Amakawa et al., 1992).

14.6b Chert

Marine deposits of silica containing appreciable concentrations of La (1.06–30.8 ppm) and Ce (3.52–23.9 ppm) occur in sequences of sedimentary rocks of all ages ranging from Early Archean to Cenozoic. The isotope composition of Ce in cherts has evolved by decay of ^{138}La to ^{138}Ce depending on their La/Ce ratios and ages. Nevertheless, chert samples of known age can be used to calculate their initial ^{138}Ce/^{142}Ce ratios that presumably record the isotope composition of Ce at the site of deposition in the oceans. Shimizu et al. (1991) measured the isotope compositions of Ce and Nd in seven samples of chert ranging in age from 3000 to 45 Ma in order to explore the possible use of chert as a paleo-oceanographic indicator.

Shimizu et al. (1991) corrected ^{138}Ce/^{142}Ce ratios of the chert samples to a value of 0.0225762 for JMC-304 reported by Makishima et al (1987). Therefore, in this presentation, the ^{138}Ce/^{142}Ce ratios were reduced by 0.0000077 to make them compatible with 0.0225685 for JMC-304 obtained by Makishima and Nakamura (1991a). The resulting values were recalculated as ε^t(Ce) values by comparison to CHUR-Ce expressed by equation 14.8

$$\left(\frac{^{138}\text{Ce}}{^{142}\text{Ce}}\right)^t_{\text{CHUR}} = 0.0225652 - 0.345 \times 0.00306(e^{\lambda t} - 1)$$

where $\lambda_\beta/\lambda = 0.345$ and ^{138}La/^{142}Ce of CHUR-Ce is 0.00306.

For example, Shimizu et al. (1991) reported a ^{138}Ce/^{142}Ce ratio of 0.0225782 ± 0.0000037 for a sample of chert from the Gorge Creek Group in the Pilbara Block of Western Australia and cited evidence that the age of this samples is 3000 Ma. In addition, the La/Ce ratio of Gorge Creek chert is

$$\frac{\text{La}}{\text{Ce}} = \frac{2.00}{4.33} = 0.4618$$

and the decay constants of ^{138}La from Section 14.2 are

$$\lambda_\beta = 2.33 \times 10^{-12}\text{y}^{-1} \qquad \lambda_e = 4.42 \times 10^{-12}\text{y}^{-1}$$

$$\lambda = 6.75 \times 10^{-12}\text{y}^{-1}$$

The ^{138}Ce/^{142}Ce ratio of the Gorge Creek chert reported by Shimizu et al. (1991) was reduced by 0.0000077 to obtain

$$\left(\frac{^{138}\text{Ce}}{^{142}\text{Ce}}\right)^0_{\text{chert}} = 0.0225705$$

This value differs only slightly from the ^{138}Ce/^{142}Ce ratio of common Ce, whose ^{138}Ce/^{142}Ce ratio is 0.02256. Therefore, the value of the conversion factor k in equation 14.5,

$$\frac{^{138}\text{La}}{^{142}\text{Ce}} = \left(\frac{\text{La}}{\text{Ce}}\right)_{\text{conc}} \times k$$

is, to a good approximation, $k = 0.00819$ (Section 14.2). Consequently, for the Gorge Creek chert

$$\frac{^{138}\text{La}}{^{142}\text{Ce}} = 0.4618 \times 0.00819 = 0.00378$$

and $e^{\lambda t} - 1 = 0.020456$ for $\lambda = 6.75 \times 10^{-12}$ y^{-1} and $t = 3000 \times 10^6$ y.
Hence,

$$\left(\frac{^{138}\text{Ce}}{^{142}\text{Ce}}\right)^t_{\text{chert}} = 0.0225705 - 0.345 \times 0.00378 \times 0.020456 = 0.0225438$$

Similarly, the ^{138}Ce/^{142}Ce ratio of CHUR-Ce at 3000 Ma is

$$\left(\frac{^{138}\text{Ce}}{^{142}\text{Ce}}\right)^t_{\text{CHUR}} = 0.0225436 - 0.345 \times 0.00306 \times 0.020456 = 0.0225436$$

Therefore, ε^t(Ce) of the Gorge Creek chert at 3000 Ma is

$$\varepsilon^t(\text{Ce}) = \left(\frac{0.0225438 - 0.0225436}{0.0225436}\right) \times 10^4 = +0.09$$

This value is indistinguishable from zero, which means that the Ce in the Gorge Creek chert originated from young volcanic rocks whose magma sources had similar La/Ce ratio as the magma sources of BCR-1, which was used to define the present $^{138}Ce/^{142}Ce$ ratio of CHUR.

The initial $^{138}Ce/^{142}Ce$ ratios of the cherts at their stratigraphic ages can be evaluated in Figure 14.7 by comparison to CHUR-Ce as defined by Makishima and Masuda (1993). Samples 1, 2, and 3 from the Pacific Ocean are Cenozoic in age and have a range of initial $^{138}Ce/^{142}Ce$ ratios that straddle CHUR-Ce. Samples 2 and 3 below the line have negative ε^t(Ce) values, similar to the Mn nodules in the Pacific Ocean in Figure 14.6, whereas sample 1 above the line has a positive

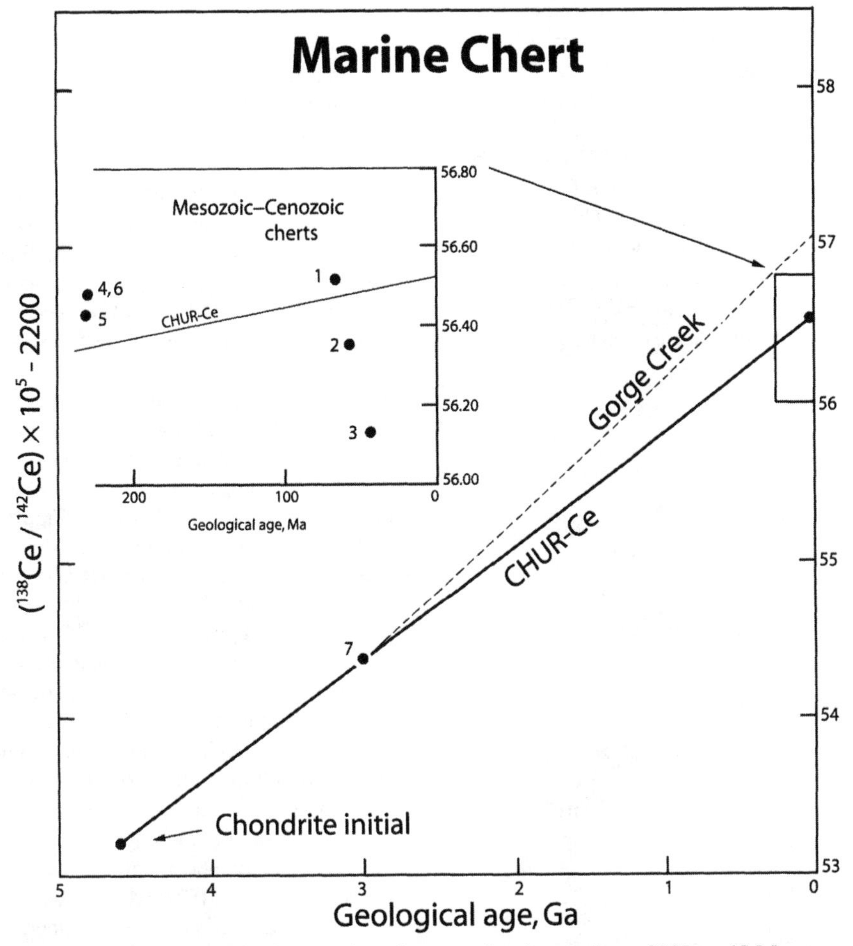

FIGURE 14.7 Isotope composition of Ce in marine chert ranging in age from 3000 to 45 Ma compared to CHUR-Ce as defined by Makishima and Masuda (1993). The y-coordinate is abbreviated in such a way that $^{137}Ce/^{142}Ce = 0.0225652$ becomes $2256.52 - 2200 = 56.52$. The samples are identified by number: 1 = central Pacific at 2°23′ N, 166°7′ W, 65 ± 30 Ma; 2 = central Pacific at 11°15′ N, 150°18′ W, 55 ± 15 Ma; 3 = Caribbean Sea at 14°47′ N, 69°19′ W, 45 ± 5 Ma; 4 = Kamiaso, Gifu, Japan, 230 ± 10 Ma; 5 = Kamiaso, Gifu, Japan, 230 ± 10 Ma; 6 = Kiryu, Gunma, Japan, 230 ± 10 Ma; 7 = Pilbara Block, Western Australia, 3000 ± 100 Ma. Data from Shimizu et al. (1991).

ε^t(Ce) value. The Triassic cherts 4, 5, and 6 also plot above the line, suggesting that the Ce they contain originated in part from continental rocks. The Gorge Creek chert (7 in Figure 14.7) plots on the CHUR-Ce evolution line consistent with $\varepsilon^t(\text{Ce}) = +0.09$ calculated above.

14.6c Model Dates for Chert

The time-dependent increase of the ^{138}Ce/^{142}Ce ratio of the Gorge Creek chert, indicated by the dashed line in Figure 14.7, reflects the fact that the La/Ce ratio of this sample (0.4618) is greater than that of CHUR (0.3736) and that the present ^{138}Ce/^{142}Ce ratio of this chert (0.0225705) is higher than the present ^{138}Ce/^{142}Ce of CHUR (0.0225652). Under these circumstances, the La–Ce systematics of Precambrian cherts can yield model dates by calculating the time when a sample of chert had the same ^{138}Ce/^{142}Ce ratio as CHUR-Ce.

The calculation is made by equating the initial ^{138}Ce/^{142}Ce ratios:

$$\left(\frac{^{138}\text{Ce}}{^{142}\text{Ce}}\right)^t_{\text{chert}} = \left(\frac{^{138}\text{Ce}}{^{142}\text{Ce}}\right)^t_{\text{CHUR}}$$

$$\left(\frac{^{138}\text{Ce}}{^{142}\text{Ce}}\right)^0_{\text{chert}} - \left(\frac{\lambda_\beta}{\lambda}\right)\left(\frac{^{138}\text{La}}{^{142}\text{Ce}}\right)^0_{\text{chert}}(e^{\lambda t} - 1)$$

$$= \left(\frac{^{138}\text{Ce}}{^{142}\text{Ce}}\right)^0_{\text{CHUR}} - \left(\frac{\lambda_\beta}{\lambda}\right)\left(\frac{^{138}\text{La}}{^{142}\text{Ce}}\right)^0_{\text{CHUR}}(e^{\lambda t} - 1) \tag{14.10}$$

Substituting the appropriate values

$$0.0225705 - 0.345 \times 0.00378(e^{\lambda t} - 1)$$
$$= 0.0225652 - 0.345 \times 0.00306(e^{\lambda t} - 1)$$

and solving for t yields

$$t = \frac{\ln 1.0213365}{6.75 \times 10^{-12}} = 3.12 \text{ Ga}$$

In other words, the Ce in the Gorge Creek chert could have originated from a chondritic magma source in the mantle at 3.12 Ga, thereby confirming its stratigraphic age of 3.0 ± 0.1 Ga assumed by Shimizu et al. (1991).

This method of dating yields inaccurate results in case the Ce had a complex history after it was removed from CHUR and before it was incorporated into the sample being dated. In addition, the La/Ce ratio of the sample must differ significantly from the La/Ce ratio of CHUR to avoid error magnification caused by acute angles between intersecting isotope evolution lines.

14.6d Seawater

The concentration of Ce in seawater is between 0.1 and 1.0 pg/g (1 pg $= 10^{-12}$ g). Nevertheless, the variation of the isotope composition of Ce in ferromanganese nodules and chert suggests that the ^{138}Ce/^{142}Ce ratio of seawater may also vary and may convey information about the sources of Ce in different parts of the world's ocean. These expectations were supported by the first measurements of ^{138}Ce/^{142}Ce ratios in seawater reported by Tanaka et al. (1990). More recently, Shimizu et al. (1994) measured ^{138}Ce/^{142}Ce ratios in six samples of seawater from three profiles in the North Pacific Ocean and corrected them to a value of 0.0225762 for JMC-304 in Table 14.6 obtained by Makishima et al. (1987). To make these data compatible with the ε^0(Ce) values of ferromanganese nodules in Figure 14.6 and of igneous rocks in Figure 14.5, the ^{138}Ce/^{142}Ce ratios reported by Shimizu et al. (1994) were reduced by 0.0000077 and the ε^0(Ce) values were then calculated relative to $(^{138}\text{Ce}/^{142}\text{Ce})^0_{\text{CHUR}} =$ 0.0225652(Makishima and Nakamura, 1991a).

The resulting ε^0(Ce) values of seawater in Figure 14.8 range from -1.5 to $+1.5$ and appear to vary with the depth of water. For example, water from 0 to 500 m depth at site DE-4 has $\varepsilon^0(\text{Ce}) = -1.5$, whereas water from 1000 to 5800 m at the same site has $\varepsilon^0(\text{Ce}) = +0.58$. Similarly, water from a depth of about 1500 m at a second site yielded $\varepsilon^0(\text{Ce}) = +0.22$, whereas water from 3000 to 5800 m at the same site yielded $+1.5$. The positive ε^0(Ce) values of seawater contrast with the negative ε^0(Ce) values of ferromanganese nodules in the Pacific Ocean in Figure 14.6 (Amakawa et al., 1991).

A possible explanation for this discrepancy is that the oxide particles of Fe and Mn form at the surface of the ocean rather than at depth

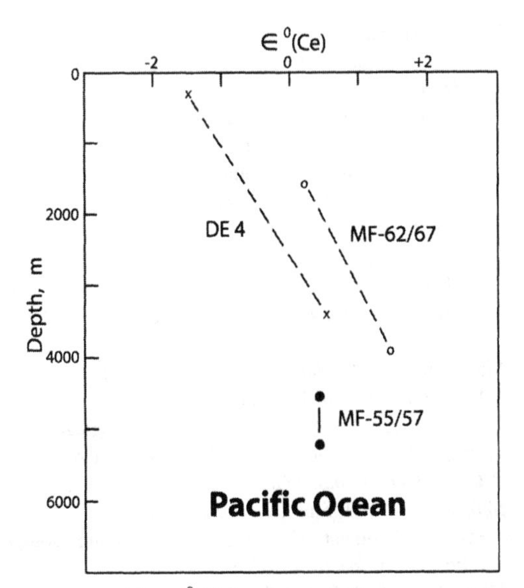

FIGURE 14.8 The ε^0(Ce) values of Ce in seawater from various depths in the Pacific Ocean. The ε^0(Ce) values are consistent with the ε^0(Ce) values in Figures 14.4 and 14.5. Data from Shimizu et al. (1994).

and then sink through the water column. Shimizu et al. (1994) also suggested that a significant fraction of Ce in the water of the Pacific Ocean is contributed by deposition of eolian dust (loess) derived from China to which they attributed a positive ε^0(Ce) value based on data from Liu et al. (1990). Therefore, the release of Ce from atmospheric dust particles in the Pacific Ocean seems to be delayed until they reach depths in excess of 1500–2500 m. As a result, only the isotope composition of Ce in the surface water (0–2500 m) maintains a record of its derivation from volcanic rocks.

These conjectures require confirmation by additional measurements of ^{138}Ce/^{142}Ce ratios of surface water collected from large areas of the ocean and from depth profiles. Such measurements are difficult because they require the collection of $10^3 - 10^4$ L of seawater per sample (Shimizu et al., 1994).

14.7 SUMMARY

The β-decay of ^{138}La to stable ^{138}Ce has not been widely used for dating of La-bearing rocks and minerals because ^{138}La has a low abundance (0.09%), its halflife is long (2.97×10^{11} years), and the La/Ce ratios of most minerals and rocks vary only within narrow limits (0.30–0.90). The highest La/Ce ratios occur in garnet and chert, whereas monazite, allanite, gadolinite, and epidote have high concentrations of both elements but low La/Ce ratios.

The halflife for β-decay of ^{138}La was determined by direct counting of La_2O_3 as well as by comparing the La–Ce systematics of rocks and minerals to dates determined by other decay schemes. Age determinations by the La–Ce method require a high precision of measurement because the ^{138}Ce/^{142}Ce ratios of typical crustal rocks (e.g., Lewisian granulite gneiss of Scotland) have increased only about 0.22% in 3.0×10^9 years.

The interpretation of ^{138}Ce/^{142}Ce ratios of terrestrial rocks and minerals is facilitated by comparing them to CHUR-Ce by means of the ε-parameter. The available data demonstrate that midocean ridge basalts (MORBs) have negative ε-values (-2 to -1), implying that they originated from depleted magma sources in the mantle. Island arc volcanics have variable ε-values that range from -2 to $+3$, presumably because they contain subducted crustal Ce, which has generally positive ε-values that rise up to $+4$. The definition of CHUR-Ce also permits the calculation of model dates (e.g., Archean chert, Pilbara, Australia).

The ε-values of manganese nodules in the Atlantic Ocean are positive, whereas those in the Pacific Ocean are negative. The difference is attributable to the greater abundance of radiogenic ^{138}Ce in the Atlantic Ocean, which receives more runoff from continents than the Pacific Ocean. The ε-values of manganese nodules in the Baltic and Barents Seas are also positive for the same reasons. These results demonstrate that the isotope composition of Ce in the oceans is the result of mixing of two or more isotopic varieties of Ce derived by weathering of crustal rocks on the continents and by volcanic activity and submarine weathering of volcanic rocks in the ocean basins. More work is needed to document the variation of the ^{138}Ce/^{142}Ce ratios of seawater sampled globally both from the surface layer and in depth profiles.

REFERENCES

Amakawa, H., J. Ingri, A. Masuda, and H. Shimizu, 1991. Isotopic compositions of Ce, Nd, and Sr in ferromanganese nodules from the Pacific and Atlantic Oceans, the Baltic and Barents Seas, and the Gulf of Bothnia. *Earth Planet. Sci. Lett.*, 105:554–565.

Amakawa, H., H. Shimizu, and A. Masuda, 1992. Reply to comment by H. Elderfield. *Earth Planet. Sci. Lett.*, 111:563–565.

Anders, E., and N. Grevesse, 1989. Abundances of the elements: Meteoritic and solar. *Geochim. Cosmochim. Acta*, 53:197–214.

DePaolo, D. J., and G. J. Wasserburg, 1976a. Nd isotopic variations and petrogenetic models. *Geophys. Res. Lett.*, 3:249–252.

DePaolo, D. J., and G. J. Wasserburg, 1976b. Inferences about magma sources and mantle structure from variations of $^{143}Nd/^{144}Nd$. *Geophys. Res. Lett.*, 3:743–746.

Dickin, A. P., 1987a. La–Ce dating of Lewisian granulites to constrain the ^{138}La β-decay half-life. *Nature*, 325:337–338.

Dickin, A. P., 1987b. Cerium isotope geochemistry of ocean island basalt. *Nature*, 326:283–284.

Elderfield, H., 1992. The Ce–Nd–Sr isotope systematics of seawater: Comment on "Isotopic compositions of Ce, Nd, and Sr in ferromanganese nodules from the Pacific and Atlantic Oceans, the Baltic and Barents Seas, and Gulf of Bothnia" by H. Amakawa, J. Ingri, A. Masuda, and H. Shimizu. *Earth Planet. Sci. Lett.*, 111:557–561.

Faure, G., 1998. *Principles and Applications of Geochemistry*. Prentice-Hall, Upper Saddle River, New Jersey.

Faure, G., 2001. *Origin of Igneous Rocks; the Isotopic Evidence*. Springer Verlag, Heidelberg.

Hamilton, P. J., 1997. Sr isotope and trace element studies of the Great Dyke and Bushveld mafic phase and their relation to Early Proterozoic magma genesis in southern Africa. *J. Petrol.*, 18:24–52.

Hamilton, P. J., N. M. Evensen, and R. K. O'Nions, 1979. Sm–Nd systematics of Lewisian gneisses; implications for the origin of granulites. *Nature*, 277:25–28.

Herrmann, A. G., 1970. Yttrium and the lanthanides. In K. H. Wedepohl (Ed.), *Handbook of Geochemistry*. Springer-Verlag, Heidelberg.

Lide, D. R., and H. P. R. Frederikse, 1995. *Handbook of Chemistry and Physics*, 76th ed. CRC Press, Boca Raton, Florida.

Liu, C.-Q., H. Shimizu, S. Nakai, G. H. Xie, and A. Masuda, 1990. Isotopic and trace element studies of Cenozoic volcanic rocks from western China: Implications for a crustal-like enriched component in the mantle. *Geochem. J.*, 24:327–342.

Makishima, A., and A. Masuda, 1993. Primordial Ce isotopic composition of the solar system. *Chem. Geol.*, 106:197–205.

Makishima, A., and A. Masuda, 1994. Ce isotope ratios of N-type MORB. *Chem. Geol.*, 118:1–8.

Makishima, A., and E. Nakamura, 1991a. Precise measurements of Ce isotope composition in rock samples. *Chem. Geol. (Isotope Geosci. Sect.)*, 94:1–11.

Makishima, A., and E. Nakamura, 1991b. Calibration of Faraday cup efficiency in a multicollector mass spectrometer. *Chem. Geol. (Isotope Geosci. Sect.)*, 94:105–110.

Makishima, A., E. Nakamura, S. Akimoto, I. H. Campbell, and R. I. Hill, 1993. New constraints on the ^{138}La β-decay constant based on a geochronological study of granites from the Yilgarn Block, Western Australia. *Chem. Geol. (Isotope Geosci. Sect.)*, 104:293–300.

Makishima, A., H. Shimizu, and A. Masuda, 1987. Precise measurement of cerium and lanthanum isotope ratios. *Mass Spectrosc.*, 35:64–72.

Masuda, A., and S. Nakai, 1983. An examination of geochronological utility of electron-capture decay of La-138. *Geochem. J.*, 17:313–314.

Masuda, A., H. Shimizu, S. Nakai, A. Makishima, and S. Lahti, 1988. ^{138}La β-decay constant estimated from geochronological studies. *Earth Planet. Sci. Lett.*, 89:316–322.

Norman, E. B., and M. A. Nelson, 1983a. Half-life and decay scheme of ^{138}La. *Phys. Rev. C.*, 27:1321–1324.

Norman, E. B., and M. A. Nelson, 1983b. On the half-life of ^{138}La. *Nature*, 306:503–504.

Sato, J., and T. Hirose, 1981. Half-life of ^{138}La. *Radiochem. Radioanalyt. Lett.*, 46:145–152.

Shimizu, H., M. Amano, and A. Masuda, 1991. La–Ce and Sm–Nd systematics of siliceous sedimentary rocks: A clue to marine environment in their deposition. *Geology*, 19(4):369–371.

Shimizu, H., H. Sawatari, Y. Kawata, P. N. Dunkley, and A. Masuda, 1992. Ce and Nd isotope geochemistry on island arc volcanic rocks with negative Ce anomaly: Existence of sources with concave REE patterns in the mantle beneath the Solomon and Bonin islands arcs. *Contrib. Mineral. Petrol.*, 110:242–252.

Shimizu, H., K. Tachikawa, A. Masuda, and Y. Nozaki, 1994. Cerium and neodymium isotope ratios of REE patterns in seawater from the North Pacific Ocean. *Geochim. Cosmochim. Acta*, 58:323–333.

Shimizu, H., T. Tanaka, and A. Masuda, 1984. Meteoritic $^{138}Ce/^{142}Ce$ ratio and its evolution. *Nature*, 307:251–252.

Shirey, S. B., and R. J. Walker, 1998. The Re–Os isotope system in cosmochemistry and high-temperature geochemistry. *Annu. Rev. Earth Phys.*, 26:423–500.

Tanaka, T., and A. Masuda, 1982. The La–Ce geochronometer. A new dating method. *Nature*, 300:515–517.

Tanaka, T., H. Shimizu, Y. Kawata, and A. Masuda, 1987. Combined La–Ce and Sm–Nd isotope systematics in petrogenetic studies. *Nature*, 327:113–117.

Tanaka, M., H. Shimizu, Y. Nozaki, Y. Ikeuchi, and A. Masuda, 1990. $^{138}Ce/^{142}Ce$ and $^{143}Nd/^{144}Nd$ ratios in seawater samples. *Geochem. J.*, 24:309–314.

Walker, F. W., J. R. Parrington, and F. Feiner, 1989. *Chart of the Nuclides*, 14th ed. General Electric Co., San Jose, California.

Walker, R. J., J. W. Morgan, E. S. Beary, M. I. Smoliar, G. K. Czamanske, and M. F. Horan, 1997. Applications of the ^{190}Pt–^{186}Os isotope system to geochemistry and cosmochemistry. *Geochim. Cosmochim. Acta*, 61:4799–4807.

CHAPTER 15

The La–Ba Method

THE isotope composition of Ba is affected by the electron capture decay of ^{138}La to stable ^{138}Ba as illustrated in Figure 14.2:

$$^{138}_{57}La + e^- \rightarrow {}^{138}_{56}Ba + \nu + 1.75 \text{ MeV} \qquad (15.1)$$

In addition, ^{138}Ba and ^{137}Ba are products of spontaneous fission of $^{238}_{92}U$ ($\lambda_{SF} = 8.46 \times 10^{-17}$ y^{-1}) with yields of 7.2 and 6.6%, respectively. Therefore, U-rich minerals having U/Ba ratios greater than about 400 become enriched in these Ba isotopes, which can cause the $^{138}Ba/^{137}Ba$ ratio to increase slightly. In addition, ^{138}Ba is the product of α-decay of long-lived ^{142}Ce whose halflife lies between 10^{15} and 10^{16} years (Nakai et al., 1986). These additional sources of ^{138}Ba are not likely to affect the $^{138}Ba/^{137}Ba$ or $^{138}Ba/^{136}Ba$ ratios of minerals that are suitable for dating by the La–Ba method.

The electron capture decay of ^{138}La has not been widely used for dating partly because most rock-forming minerals and hence most igneous and metamorphic rocks have low La/Ba ratios. Nevertheless, Masuda and Nakai (1983) were able to date two aliquots of a sample of alluvial monazite from the Eneabba district of Western Australia because of their high La/Ba ratios of 5800 and 9330.

15.1 GEOCHEMISTRY OF La AND Ba

Barium is a member of the alkaline-earth elements in group IIA of the periodic table. It has a valence of +2 and its ionic radius is 1.35 Å. Therefore, Ba^{2+} has a larger radius than La^{3+} (1.15 Å) and its charge-to-radius ratio (1.48) is lower than that of La^{3+} (2.60), which means that Ba^{2+} is excluded from rare-earth minerals and explains why monazite and other rare-earth minerals have high La/Ba ratios.

The average concentrations of La and Ba in different kinds of terrestrial rocks in Table 15.1 demonstrate that terrestrial rocks have low La/Ba ratios ranging from 0.02 in basalt to 1.8 in ultramafic rocks and to 3.0 in sandstones. The average La/Ba ratio of chondrite meteorites is 0.079.

The abundances of La and Ba in the solar system are 0.4460 and 4.49 per 10^6 atoms of Si, respectively (Anders and Grevesse, 1989). Therefore, the La/Ba ratio of matter in the solar system is

$$\left(\frac{\text{La}}{\text{Ba}}\right)_{\text{solar}} = \frac{0.4460 \times 138.91}{4.49 \times 137.34} = 0.100$$

The low La/Ba ratios of different igneous and metamorphic rocks included in the summaries of Herrmann (1970) and Puchelt (1971) imply that most of the common rock-forming silicate minerals also have low La/Ba ratios, and therefore neither the common rocks nor most of the rock-forming minerals are suitable for dating by the La–Ba method.

The concentrations of La and Ba reported by Nakai et al. (1986) for a small group of minerals in Table 15.2 demonstrate the expected reciprocal

Table 15.1. Average Concentrations of La and Ba in Terrestrial and Extraterrestrial Rocks

Rock Type	Concentration, (ppm) La	Ba	La/Ba
Terrestrial Rocks			
Ultramafic	1.3	0.7	1.8
Basalt	6.1	315	0.019
Granite			
High Ca	45	420	0.107
(Low Ca)	55	840	0.065
Shale	92	580	0.158
Sandstone	30	10	3.0
Limestone	1	10	0.10
Deep-sea clay	115	2300	0.05
Continental crust	16	250	0.064
Extraterrestrial Rocks			
Lunar rocks	—	—	0.065
Chondrites	—	—	0.079
Solar system	—	—	0.100

Source: Faure, 1998.

Table 15.2. Concentrations of La and Ba in the Rare-Earth-Bearing Minerals of the Amitsoq Gneiss, West Greenland, and a Pegmatite in Finland

Rock or Mineral	Concentration, (ppm) La	Ba	La/Ba
Amitsoq Gneiss	11.57	163.1	0.07093
Monazite[a]	—	—	7564
Allanite	2.449×10^4	49.44	495.3
Epidote	599.0	11.98	50.00
Sphene	187.1	110.2	1.697

Source: Nakai et al., 1986.

[a]From Masuda and Nakai, 1983.

relation of the concentration of La and Ba in La-rich minerals. Monazite has the highest La/Ba ratio followed by allanite, epidote, and sphene. A whole-rock sample of granitic gneiss has a low La/Ba ratio (0.071) consistent with the La/Ba ratio of low-Ca granites in Table 15.1.

15.2 PRINCIPLES AND METHODOLOGY

Barium (Z = 56) has seven stable isotopes whose abundances are (Lide and Frederikse, 1995)

$$^{130}Ba = 0.106\% \quad ^{136}Ba = 7.854\%$$
$$^{132}Ba = 0.101\% \quad ^{137}Ba = 11.23\%$$
$$^{134}Ba = 2.417\% \quad ^{138}Ba = 71.70\%$$
$$^{135}Ba = 6.592\%$$

giving it an atomic weight of 137.93

The equation governing the decay of ^{138}La to ^{138}Ba is

$$\frac{^{138}Ba}{^{137}Ba} = \left(\frac{^{138}Ba}{^{137}Ba}\right)_0 + \left(\frac{\lambda_e}{\lambda}\right)\frac{^{138}La}{^{137}Ba}(e^{\lambda t} - 1) \tag{15.2}$$

where $\lambda_e = 4.42 \times 10^{-12}\ y^{-1}$
$\lambda = 6.75 \times 10^{-12}\ y^{-1}$

The La/Ba ratio is converted into the $^{138}La/^{137}Ba$ ratio by the relation

$$\left(\frac{^{138}La}{^{137}Ba}\right)_{atom} = \left(\frac{La}{Ba}\right)_{conc} \times 0.007923 \tag{15.3}$$

based on

At. wt. La = 138.9055 At. wt. Ba = 137.93

Ab $^{138}La = 0.09\%$ Ab $^{137}Ba = 11.23\%$

Alternatively, the isotope composition of Ba can be expressed as the $^{138}Ba/^{136}Ba$ ratio, in which case the La/Ba ratio is converted into the $^{138}La/^{136}Ba$ ratio by the relation

$$\left(\frac{^{138}La}{^{136}Ba}\right)_{atom.} = \left(\frac{La}{Ba}\right)_{conc} \times 0.011329 \tag{15.4}$$

where the abundance of ^{136}Ba is 7.854%. The numerical values of the $^{138}Ba/^{136}Ba$ and $^{138}Ba/^{137}Ba$

ratios in common Ba are

$$\frac{^{138}\text{Ba}}{^{136}\text{Ba}} = \frac{71.70}{7.854} = 9.129$$

$$\frac{^{138}\text{Ba}}{^{137}\text{Ba}} = \frac{71.70}{11.23} = 6.384$$

In spite of the slow rate of electron capture decay of ^{138}La, its low abundance (0.09%), and the high abundance of ^{138}Ba, the ^{138}Ba/^{137}Ba ratios of such La-rich and Ba-poor minerals as monazite increase with time at an appreciable rate. For example, the ^{138}Ba/^{136}Ba ratio of a monazite crystal having a La/Ba ratio of 10,000 increases by 0.5024 in 1.0×10^9 years from 9.129 to 9.631, whereas the ^{138}Ba/^{137}Ba ratio increases by 0.3514 from 6.384 to 6.735, or 5.5% in both cases. Even if $t = 1.0 \times 10^8$ years, the ^{138}Ba/^{137}Ba ratio of a mineral having a La/Ba ratio of 10,000 increases by 0.0350, which amounts to about 0.55% of the ^{138}Ba/^{137}Ba ratio (6.384) of common Ba. Increases of this magnitude are measurable, given a precision of better than 0.013% reported by Nakai et al. (1986). Therefore, this example demonstrates that the La–Ba method is usable for dating La-rich and Ba-poor minerals (La/Ba = 10,000) older than about 1×10^8 years.

In view of the low average La/Ba ratio (0.100) of the bulk Earth, the ^{138}Ba/^{137}Ba ratio of common Ba has increased only by 0.000016 (0.00025%) in 4.56×10^9 years, which cannot be detected by presently available instrumentation. Therefore, the ^{138}Ba/^{137}Ba ratio of common Ba in the bulk Earth has remained virtually constant throughout geological time. For this reason, Masuda and Nakai (1983) proposed that the initial ^{138}Ba/^{137}Ba ratio for dating mineral grains or single crystals can be assumed to be a constant. In addition, samples being processed for dating by the La–Ba method are immune from contamination by common Ba because the isotope composition of the contaminant Ba is virtually identical to the initial ^{138}Ba/^{137}Ba ratio of the minerals or rocks. Under these circumstances, the addition of common Ba moves data points along the isochron and does not affect the date derived from its slope.

15.3 AMITSOQ GNEISS, WEST GREENLAND

A whole-rock sample and concentrates of allanite and epidote of the Amitsoq Gneiss (Sections 5.4c, 12.5a) define a straight line in Figure 15.1 whose slope $m = 0.107786 \pm 0.000108$ (Nakai et al., 1986). The corresponding date is 2419 ± 24 Ma, recalculated to $\lambda_e = 4.44 \times 10^{-12}$ y^{-1} and $\lambda_{total} = 6.75 \times 10^{-12}$ y^{-1}. This La–Ba date is indistinguishable from the Sm–Nd date of 2410 ± 54 Ma reported by Nakai et al. (1986) for the same samples (plus sphene). The good agreement between the La–Ba and Sm–Nd dates confirms the validity of the two decay constants of ^{138}La and of the La–Ba method of dating.

The initial ^{138}Ba/^{137}Ba ratio derived from the data in Figure 15.1 (6.38968 ± 28) is identical to the values of this ratio in Table 15.3 for common Ba analyzed by Nakai et al. (1986). These results therefore confirm that the change in isotope composition of Ba in the bulk Earth during

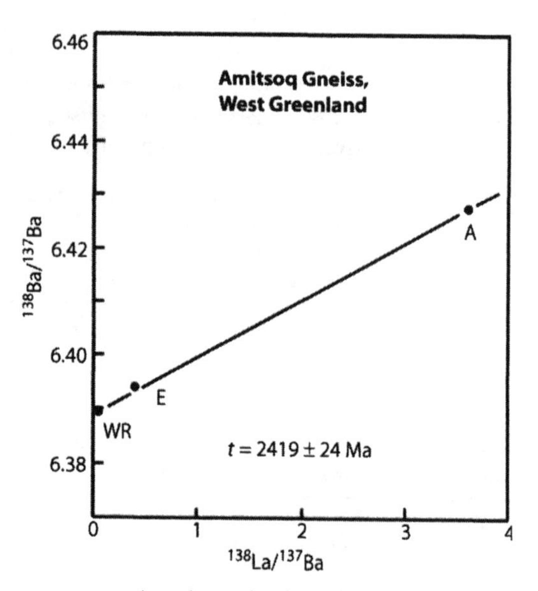

FIGURE 15.1 Age determination of the Amitsoq Gneiss, West Greenland, by the La–Ba method: A = allanite, E = epidote, WR = whole rock, $\lambda_e = 4.44 \times 10^{-12}$ y^{-1}, $\lambda_{total} = 6.75 \times 10^{-12}$ y^{-1}. Data from Nakai et al. (1986).

Table 15.3. Isotope Composition of Ba in Reagents, Standard Rocks, and the Amitsoq Gneiss Normalized to $^{136}Ba/^{137}Ba = 0.6996$

Sample	$^{138}Ba/^{137}Ba$	Age, Ma
$BaCl_2$, Merck (Art 1716)	6.38997 ± 0.00012	0
JB-1, basalt	6.38997 ± 0.00022	Tertiary
BCR-1, basalt	6.38972 ± 0.00027	6–16.5[a]
Amitsoq Gneiss	6.38968 ± 0.00028	2419

Source: Nakai et al., 1986.

[a]From Faure, 2001.

all of geological time cannot be resolved by present methods of analysis. In addition, the $^{138}Ba/^{137}Ba$ ratios are consistent with those reported for extraterrestrial Ba by Eugster et al. (1969) and Lewis et al. (1983).

15.4 MUSTIKKAMÄKI PEGMATITE, FINLAND

Allanite from the Mustikkamäki (Blueberry Hill) pegmatite has a high La/Ba ratio of 516.2 and is enriched in radiogenic ^{138}Ba with a $^{138}Ba/^{137}Ba$ ratio of 6.42033 ± 0.00030 (Nakai et al., 1986). Sphene from this pegmatite has a low La/Ba ratio (1.697), and its $^{138}Ba/^{137}Ba$ ratio is only 6.38972 ± 0.00030. Nakai et al. (1986) derived a date of 1694 ± 42 Ma from the slope of an allanite–sphene isochron line using $\lambda_e = 4.44 \times 10^{-12}$ y^{-1} and $\lambda_{total} = 6.73 \times 10^{-12}$ y^{-1}. This date is lower than the Sm–Nd isochron date of 1820 ± 29 Ma based on the allanite, sphene, and apatite of the same pegmatite. The authors attributed the discrepancy of the dates to the complex metamorphic history of the Mustikkamäki pegmatite.

The enrichment of the allanite in radiogenic ^{138}Ba provides an opportunity to calculate a La–Ba date for this mineral using equation 15.1 and the initial $^{138}Ba/^{137}Ba$ ratio of the Amitsoq Gneiss (6.38968):

$$6.42033 = 6.38968 + \left(\frac{4.42}{6.75}\right) 4.058(e^{\lambda t} - 1)$$

$$t = \frac{\ln 0.0115345}{6.75 \times 10^{-12}} = 1699.0 \text{ Ma}$$

The resulting date is indistinguishable from 1694 ± 42 Ma reported by Nakai et al. (1986).

15.5 SUMMARY

The La–Ba method is well suited to date La-rich and Ba-poor minerals such as monazite, whose La/Ba ratios approach 10,000. The resulting La–Ba mineral dates appear to be sensitive to partial loss of radiogenic ^{138}Ba during thermal metamorphism followed by slow cooling of the host rocks.

The slow rate of electron capture decay and the low average La/Ba ratio of the bulk Earth actually facilitate dating of La-rich minerals because all rocks and minerals have virtually the same initial $^{138}Ba/^{137}Ba$ ratio regardless of their age. In addition, contamination of samples with common Ba during processing in the laboratory has no effect on the calculated dates.

Individual crystals of La-rich minerals may be dated by use of laser ablation inductively coupled plasma mass spectrometry (LA-ICP-MS). Dates can be calculated based only on measured values of the La/Ba concentration ratio and on the $^{138}Ba/^{137}Ba$ isotope ratio because the initial $^{138}Ba/^{137}Ba$ ratio is a known constant. Isobaric interference by ^{138}Ce can be corrected by monitoring ^{136}Ce or ^{140}Ce, and interference by ^{138}La can be overcome by measuring ^{139}La.

REFERENCES

Anders, E., and N. Grevesse, 1989. Abundances of the elements: Meteoritic and solar. *Geochim. Cosmochim. Acta*, 53:197–214.

Eugster, O., F. Tera, and G. J. Wasserburg, 1969. Isotopic analysis of barium in meteorites and terrestrial samples. *J. Geophys. Res.*, 74:3897–3908.

Faure, G., 1998. *Principles and Applications of Geochemistry*, 2nd ed. Prentice-Hall, Upper Saddle River, New Jersey.

Faure, G., 2001. *Origin of Igneous Rocks: The Isotopic Evidence*. Springer-Verlag, Heidelberg.

Herrmann, A. G., 1970. Yttrium and the lanthanides. In K. H. Wedepohl (Ed.), *Handbook of Geochemistry*. II-5: 39 and 57–71. Springer-Verlag, Heidelberg.

Lewis, R. S., E. Anders, T. Shimamura, and G. W. Lugmair, 1983. *Science*, 222:1013–1015.

Lide, D. R., and H. P. R. Frederikse, 1995. *Handbook of Chemistry and Physics*, 76th ed. CRC Press, Boca Raton, Florida.

Masuda, A., and S. Nakai, 1983. An examination of geochronological utility of electron-capture decay of La-138. *Geochem. J.*, 17:313–314.

Nakai, S., H. Shimizu, and A. Masuda, 1986. A new geochronometer using lanthanum-138. *Nature*, 320: 433–435.

Puchelt, H., 1971. Barium. In K. H. Wedepohl (Ed.), *Handbook of Geochemistry*. II-5 : 56A–56O. Springer-Verlag, Heidelberg.

PART III

Geochemistry of Radiogenic Isotopes

The internal differentiation of the Earth has enriched the rocks of the continental crust in Rb, K, U, Th, and Re, all of which have long-lived radioactive isotopes. The resulting enrichment of the continental crust in the respective radiogenic daughters distinguishes it from the mantle, where these radiogenic isotopes have low abundances.

The opposite is true for the Sm–Nd and Lu–Hf decay schemes because the parent elements Sm and Lu are retained in the mantle, whereas their radiogenic daughter elements Nd and Hf migrate into the continental crust.

Consequently, geological processes involving the mantle and the crust of the Earth cause mixing of materials having different isotopic compositions of Sr, Nd, Pb, Hf, and Os. As a result, the isotope compositions of the daughter elements shed light on the origin of igneous rocks and on the sources of these elements in the oceans.

CHAPTER 16

Mixing Theory

SEVERAL common geological processes result in the mixing of materials having different chemical and isotopic compositions of elements such as Sr or Pb. Examples of such mixing processes are the mingling of the water of a tributary stream with that of its master stream, the discharge of river water into a lake or into the oceans, the mixing of two types of sediment in a depositional basin, and the contamination of a mantle-derived magma as a result of interactions with rocks of the Earth's crust. In all such cases, the chemical and isotopic compositions of the resulting mixtures can be related by means of simple mixing models.

The derivations of the applicable mixing equations are based on two components combined in varying proportions. Two-component mixtures may subsequently mix with a third or even with a fourth component. Such multicomponent mixtures can be treated in terms of a series of two-component mixtures.

16.1 CHEMICAL COMPOSITIONS OF MIXTURES

The chemical compositions of all components in a mixture of geological materials are assumed to be conservative. This means that the components have constant chemical compositions and that the concentrations of elements in the mixture are assumed to be expressible by the abundances of the components. Loss or gain of elements in the mixture by chemical reactions such as the formation of insoluble precipitates from aqueous solutions, sorption of ions on mineral surfaces, dissolution of minerals and removal of the resulting ions, and crystallization of minerals from silicate melts is excluded or assumed to be negligible.

16.1a Two-Component Mixtures

The abundances of two components A and B in a mixture are expressed by the mixing parameter f, defined as the weight ratio (or volume ratio) of A and B:

$$f_A = \frac{A}{A+B} \tag{16.1}$$

where A and B are the weights of these components and f_A expresses the abundances of component A in a given mixture. The abundance of component B in a two-component mixture is

$$f_B = \frac{B}{A+B} \tag{16.2}$$

It follows that

$$f_A + f_B = 1.0 \tag{16.3}$$

and

$$f_B = 1 - f_A$$

Therefore, the concentration of a conservative element X in a two-component mixture is

$$X_M = X_A f_A + X_B(1 - f_A) \tag{16.4}$$

where X_M = concentration of element X in the mixture M

X_A, X_B = concentrations of element X in components A and B

The concentration of another conservative element (Y) in a two-component mixture is expressed similarly by

$$Y_M = Y_A f_A + Y_B\ (1 - f_A) \tag{16.5}$$

Equations 16.4 and 16.5 can be combined by solving both equations for f_A and by combining the resulting expressions:

$$f_A = \frac{X_M - X_B}{X_A - X_B} = \frac{Y_M - Y_B}{Y_A - Y_B} \tag{16.6}$$

Solving for Y_M yields

$$Y_M = X_M \frac{Y_A - Y_B}{X_A - X_B} + \frac{Y_B X_A - Y_A X_B}{X_A - X_B} \tag{16.7}$$

This is the equation of a straight line of the form $y = mx + b$, which is the locus of all points representing mixtures of components A and B in differing proportions.

The mixing line defined by two components A and B in Figure 16.1 has been subdivided in 10% increments of the mixing parameter f_A. In addition, the abundance of component A in a mixture M can be determined by the lever rule:

$$f_A = \frac{MB}{AB} \tag{16.8}$$

where MB and AB are line segments measured along the mixing line.

The value to geology of these mixing relations lies in the fact that they produce linear correlations between the concentrations of pairs of conservative elements. Straight lines can therefore be fitted to the measured concentrations of two conservative elements in a suite of samples that are two-component mixtures in varying proportions. In

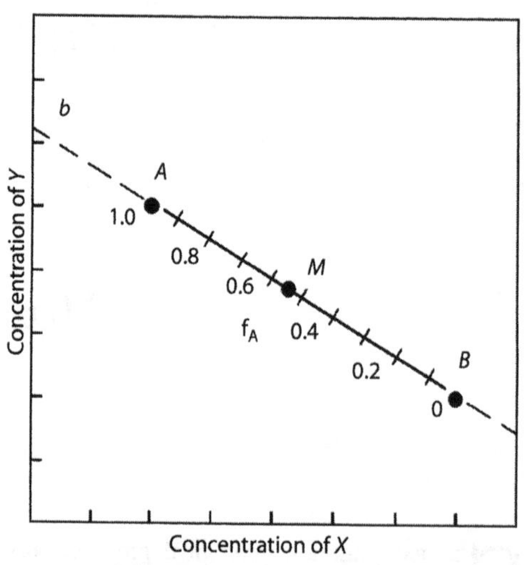

FIGURE 16.1 Mixtures of two components (A and B) in coordinates of the concentrations of two conservative chemical elements (X and Y). The mixture M is represented by a point on the straight line joining A and B. The abundance of component A (f_A) in the mixture M is equal to the ratio of the line segments: $f_A = MB/AB$.

cases where the concentrations of X or Y in one of the components are known, the concentrations of these elements in the other component can be calculated from the slope and intercept of the mixing equation. When the concentrations of X and Y in both components are known, the abundance of component A (or B) or of individual mixtures can be determined either graphically from the lever rule (equation 16.8) or by solution of equation 16.6:

$$f_A = \frac{X_M - X_B}{X_A - X_B} \quad \text{or} \quad f_A = \frac{Y_M - Y_B}{Y_A - Y_B}$$

In cases where the samples were collected at known locations, the values of f_A can be plotted on a map and contoured to reveal the flow pattern (e.g., water in subsurface or in lakes).

16.1b Sequential Two-Component Mixtures

When a two-component mixture subsequently mixes with a third component, the point representing the resulting mixture moves away from the

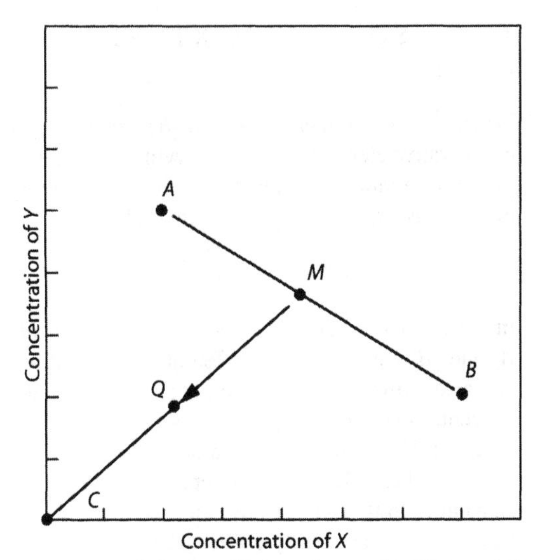

FIGURE 16.2 Dilution of the mixture M of A and B by a third component C ($X = 0$, $Y = 0$) to form sample Q. The abundance of C in Q is given by $f_C = QM/CM$.

mixing line AB and approaches point C representing the third component. If the third component does not contain the elements X and Y, it is located at the origin of the mixing diagram.

The mixture Q in Figure 16.2 formed by the addition of a certain amount of a third component C (located at the origin) to the two-component mixture M. The location of M on the mixing line AB is obtained by projecting point Q from the origin to the line AB. This procedure reduces the three-component mixture (Q) to a two-component mixture (M) of only A and B. The abundance of component A (or B) of mixture M is then obtained by the lever rule or algebraically from equation 16.6 as before. After the location of M has been determined, the abundance of component C (f_C) in mixture Q is obtained from the lever rule:

$$f_C = \frac{QM}{CM}$$

Examples of such mixtures occur in hydrogeology where deep groundwater may be diluted by recharge from the surface (Lowry et al., 1988). If the groundwater is a mixture of two components (e.g., derived from different aquifer lithologies), the addition of recharge disperses the data points from a straight line into a mixing triangle whose apeces are the X- and Y-coordinates of components A, B, and C.

The procedure described above can be applied separately to each water sample in a set. The resulting values of f_C may vary regionally or with depth below the surface. In addition, the coordinates of the projected two-component mixtures (M) yield the chemical composition of the groundwater free of the effects of dilution by surface recharge, provided that only conservative cations and anions are considered. In addition, the abundances of component A (or B) of the two-component mixtures may vary regionally and thereby reveal the source of the water component A (or B) and the direction in which it is moving.

The abundances of components A and B in mixture M and the abundance of component C in mixture Q do not sum to unity, or 100%, because they apply to two different two-component mixtures each of which sums to 100%: Q is a mixture of C and M, whereas M is a mixture of A and B. The composition of a three-component mixture can be resolved into its components only by considering it in the context of a mixing triangle.

16.1c Three-Component Mixtures

The sample M in Figure 16.3 is a mixture of three components (A, B, and C) that form a triangle. The abundances of the components of which sample M is composed are obtained by a graphical procedure because computational methods are time consuming (Petz and Faure, 1997).

The abundance of component A (g_A) of a three-component mixture M is defined as the weight (or volume) fraction:

$$g_A = \frac{A}{A + B + C} \tag{16.9}$$

The numerical value of g_A is the ratio of the length DM to DE:

$$g_A = \frac{DM}{DE}$$

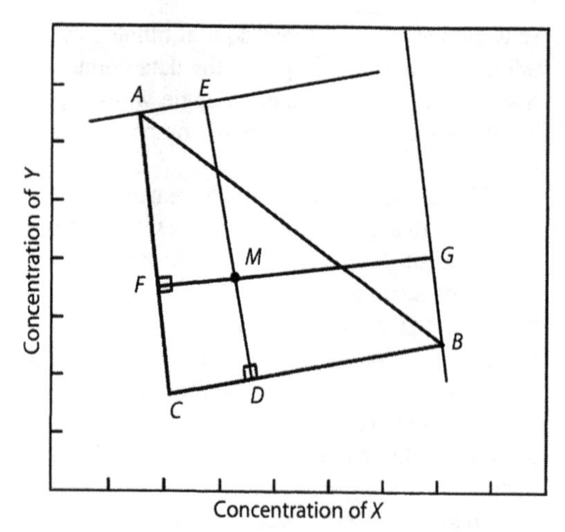

FIGURE 16.3 Resolution of a three-component mixture M into its components A, B, and C by a graphical procedure described in the text.

where DE is the height of the triangle drawn at right angles to CB from point D through M to point E located on the line AE drawn parallel to CB and at right angles to DE. Similarly, the abundance of component B (g_B) is

$$g_B = \frac{FM}{FG}$$

Since

$$g_A + g_B + g_C = 1.0 \qquad (16.10)$$

then

$$g_C = 1.0 - g_A - g_B$$

In all cases considered here, the abundances of two- and three-component mixtures are directly related to linear distances in mixing diagrams regardless of the chosen units of length. In this regard, mixing diagrams based on the concentrations of conservative elements differ from mixtures of elements having different isotopic compositions.

Examples of two- or three-component mixtures occur not only in hydrogeology but also in the formation of clastic sedimentary rocks composed of mixtures of grains of different minerals or rock types (e.g., Faure, 1998, Chapter 18).

16.2 ISOTOPIC MIXTURES OF Sr

When two components (A and B) containing a conservative element X mix in varying proportions, the concentration of element X in the resulting mixtures is expressed by equation 16.4:

$$X_M = X_A f_A + X_B(1 - f_A)$$

In cases where the element X in components A and B has not only different concentrations but also different isotope compositions, both the concentration and isotope composition of element X in a mixture M need to be accounted for. The equation that relates the isotope composition of element X to its concentration in a two-component mixture was derived by Faure (1977, Chapter 7). Element X may be Sr, Nd, Pb or any other element that contains a stable radiogenic isotope or whose isotope composition varies for other reasons (e.g., fractionation of isotopes by physical, chemical, or biological processes).

In cases where the components contain Sr having different concentrations and isotope ratios, the $^{87}Sr/^{86}Sr$ ratio of a mixture is

$$\left(\frac{^{87}Sr}{^{86}Sr}\right)_M = \left(\frac{^{87}Sr}{^{86}Sr}\right)_A f_A \left(\frac{Sr_A}{Sr_M}\right) + \left(\frac{^{87}Sr}{^{86}Sr}\right)_B \times (1 - f_A)\left(\frac{Sr_B}{Sr_M}\right) \qquad (16.11)$$

This equation contains two kinds of weighting factors: f_A and $1 - f_A$ express the abundances of components A and B, whereas Sr_A/Sr_M and Sr_B/Sr_M are the fractions of Sr in the mixture contributed by components A and B, respectively.

For example, Figure 16.4 is a plot of the $^{87}Sr/^{86}Sr$ ratios and Sr concentrations of a series of mixtures of two components that have specified concentrations and isotope ratios of Sr:

Component A :

$$Sr_A = 10 \text{ ppm} \qquad \left(\frac{^{87}Sr}{^{86}Sr}\right)_A = 0.800$$

Component B :

$$Sr_B = 100 \text{ ppm} \qquad \left(\frac{^{87}Sr}{^{86}Sr}\right)_B = 0.710$$

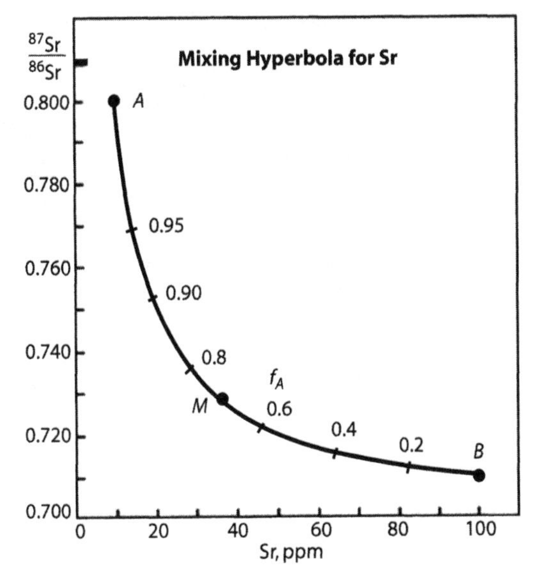

FIGURE 16.4 Hyperbola formed by mixing two components A and B having different Sr concentrations and different $^{87}Sr/^{86}Sr$ ratios.

The Sr concentrations of mixtures (Sr_M) were calculated from equation 16.4 for selected values of f_A. The results were then used in equation 16.11 to calculate the corresponding $^{87}Sr/^{86}Sr$ ratios of the mixtures:
If $f_A = 0.2$,

$$\text{Sr}_M = 10 \times 0.2 + 100 \times (1 - 0.2) = 82 \text{ ppm}$$

$$\left(\frac{^{87}\text{Sr}}{^{86}\text{Sr}}\right)_M = 0.800 \times 0.2 \times \frac{10}{82} + 0.710 \times 0.8 \times \frac{100}{82}$$

$$= 0.0195 + 0.6926 = 0.7121$$

Additional data points were calculated for increasing values of f_A, and the results were plotted in coordinates of Sr_M (x-coordinate) and $(^{87}Sr/^{86}Sr)_M$ (y-coordinate). The resulting curve in Figure 16.4 is one branch of a hyperbola rather than a straight line. If the Sr concentration of mixture M on the mixing hyperbola in Figure 16.4 is 37 ppm, the value of f_A for that mixture is calculated from equation 16.4:

$$37 = 10 f_A + 100(1 - f_A)$$

$$f_A = 0.70$$

Alternatively, f_A can be calculated from equation 16.11 given that its $^{87}Sr/^{86}Sr$ ratio $(^{87}Sr/^{86}Sr)_M$ is 0.7275 and its Sr concentration Sr_M is 37 ppm:

$$0.7275 = 0.800 f_A \frac{10}{37} + 0.710(1 - f_A)\frac{100}{37}$$

$$f_A = 0.70$$

The shape of the mixing hyperbola depends on the ratio of Sr_A/Sr_B. If $Sr_A/Sr_B < 1.0$, as in Figures 16.4 and 16.5*a*, the hyperbola is convex down. If $Sr_A/Sr_B = 1.0$, as in Figure 16.5*b*, the hyperbola turns into a straight line. If $Sr_A/Sr_B > 1.0$, as in Figure 16.5*c*, the hyperbola is convex up. If $Sr_A = Sr_B$ (Figure 16.5*b*), equation 16.11 reduces to

$$\left(\frac{^{87}\text{Sr}}{^{86}\text{Sr}}\right)_M = \left(\frac{^{87}\text{Sr}}{^{86}\text{Sr}}\right)_A f_A + \left(\frac{^{87}\text{Sr}}{^{86}\text{Sr}}\right)_B (1 - f_A) \quad (16.12)$$

In this case, equal increments of f_A change the $^{87}Sr/^{86}Sr$ ratios of the mixtures by a constant amount because equation 16.12 is a straight line in coordinates of $(^{87}Sr/^{86}Sr)_M$ and f_A:

$$\left(\frac{^{87}\text{Sr}}{^{86}\text{Sr}}\right)_M = f_A\left[\left(\frac{^{87}\text{Sr}}{^{86}\text{Sr}}\right)_A - \left(\frac{^{87}\text{Sr}}{^{86}\text{Sr}}\right)_B\right] + \left(\frac{^{87}\text{Sr}}{^{86}\text{Sr}}\right)_B \quad (16.13)$$

If $f_A = 0$, $(^{87}Sr/^{86}Sr)_M = (^{87}Sr/^{86}Sr)_B$. If $f_A = 1.0$, $(^{87}Sr/^{86}Sr)_M = (^{87}Sr/^{86}Sr)_A$.

The hyperbola in Figure 16.4 was plotted by solving equations 16.4 and 16.11 for the same values of f_A. These equations can be combined by solving both of them for f_A and equating the resulting expressions. From equations 16.4,

$$f_A = \frac{\text{Sr}_M - \text{Sr}_B}{\text{Sr}_A - \text{Sr}_B}$$

and from equation 16.11,

$$f_A = \frac{\left(\frac{^{87}\text{Sr}}{^{86}\text{Sr}}\right)_M - \left(\frac{^{87}\text{Sr}}{^{86}\text{Sr}}\right)_B \frac{\text{Sr}_B}{\text{Sr}_M}}{\left(\frac{^{87}\text{Sr}}{^{86}\text{Sr}}\right)_A \frac{\text{Sr}_A}{\text{Sr}_M} - \left(\frac{^{87}\text{Sr}}{^{86}\text{Sr}}\right)_B \frac{\text{Sr}_B}{\text{Sr}_M}}$$

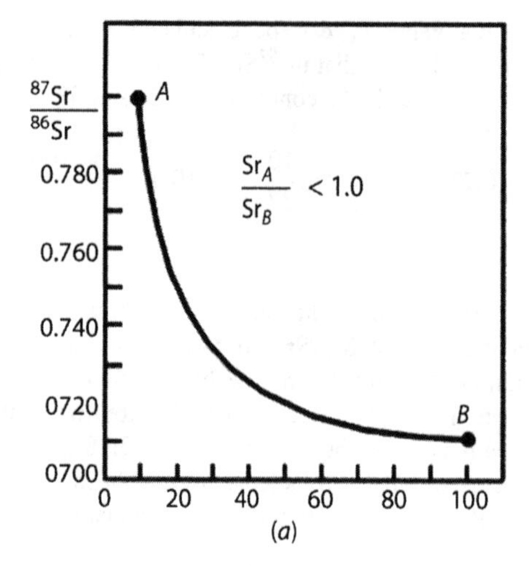

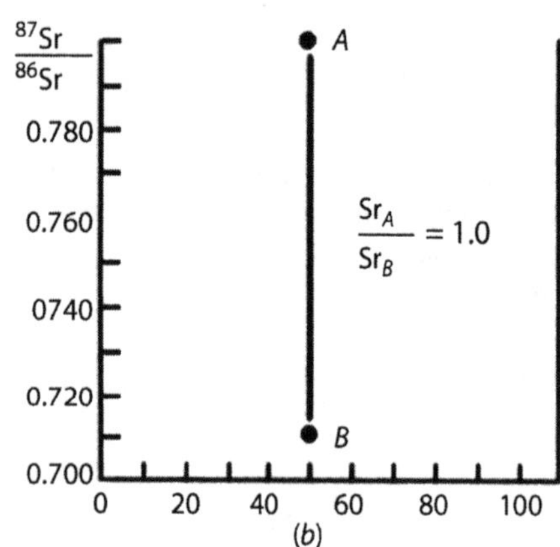

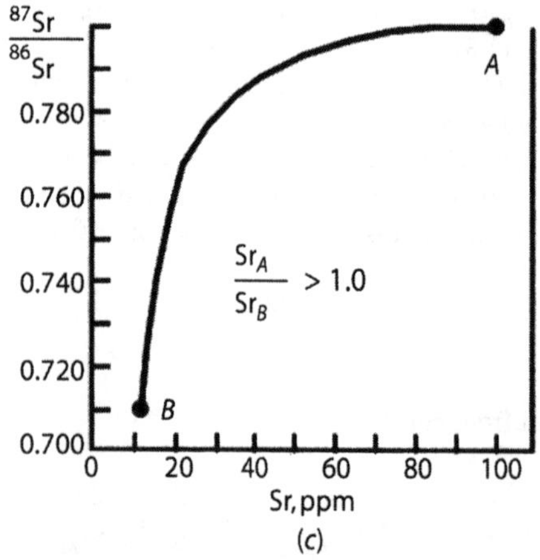

FIGURE 16.5 Effect of the Sr_A/Sr_B ratio on the shape and orientation of mixing hyperbolas.

from which it follows that

$$\left(\frac{^{87}\mathrm{Sr}}{^{86}\mathrm{Sr}}\right)_M = \frac{\mathrm{Sr}_A\,\mathrm{Sr}_B\left[(^{87}\mathrm{Sr}/^{86}\mathrm{Sr})_B - (^{87}\mathrm{Sr}/^{86}\mathrm{Sr})_A\right]}{\mathrm{Sr}_M(\mathrm{Sr}_A - \mathrm{Sr}_B)} + \frac{\mathrm{Sr}_A(^{87}\mathrm{Sr}/^{86}\mathrm{Sr})_A - \mathrm{Sr}_B(^{87}\mathrm{Sr}/^{86}\mathrm{Sr})_B}{\mathrm{Sr}_A - \mathrm{Sr}_B} \tag{16.14}$$

This is the equation of the mixing hyperbola in coordinates of the ^{87}Sr/^{86}Sr ratio and the Sr concentration in the form

$$\left(\frac{^{87}\mathrm{Sr}}{^{86}\mathrm{Sr}}\right)_M = \frac{a}{\mathrm{Sr}_M} + b \tag{16.15}$$

where

$$a = \frac{\mathrm{Sr}_A\,\mathrm{Sr}_B\left[(^{87}\mathrm{Sr}/^{86}\mathrm{Sr})_B - (^{87}\mathrm{Sr}/^{86}\mathrm{Sr})_A\right]}{\mathrm{Sr}_A - \mathrm{Sr}_B}$$

$$b = \frac{\mathrm{Sr}_A(^{87}\mathrm{Sr}/^{86}\mathrm{Sr})_A - \mathrm{Sr}_B(^{87}\mathrm{Sr}/^{86}\mathrm{Sr})_B}{\mathrm{Sr}_A - \mathrm{Sr}_B}$$

The numerical values of a and b depend entirely on the ^{87}Sr/^{86}Sr ratios and Sr concentrations of the components A and B.

Equation 16.15 is transformable into a straight line by defining the x-coordinate as $1/\mathrm{Sr}_M$. Therefore, the measured ^{87}Sr/^{86}Sr ratios and Sr concentrations of a suite of two-component mixtures define a *straight line* when the *reciprocals of the Sr concentrations* are plotted. The equation of the straight line is determined by least-squares regression, which yields the slope and intercept. The slope of the mixing line is equal to a in equation 16.15, whereas the intercept is equal to b. The conversion of a mixing hyperbola into a straight line by inverting the Sr concentrations is demonstrated in Figure 16.6. The systematic relationships of mixtures of two components that contain an element that has different isotope compositions and concentrations, demonstrated here for Sr, hold for all other elements.

16.3 ISOTOPIC MIXTURES OF Sr AND Nd

Mixing hyperbolas based on equations 16.4 and 16.11 can also be plotted for other elements,

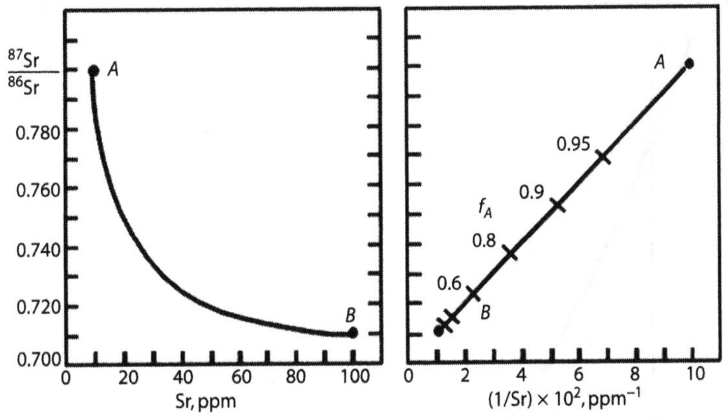

FIGURE 16.6 Conversion of a mixing hyperbola into a straight line by inverting the Sr concentrations and multiplying them by 100. The nonlinear variation of f_A along the straight line is clearly apparent.

including Nd, Pb, Hf, and Os. The equations for Nd are

$$\text{Nd}_M = \text{Nd}_A f_A + \text{Nd}_B(1 - f_A) \quad (16.16)$$

$$\left(\frac{^{143}\text{Nd}}{^{144}\text{Nd}}\right)_M = \left(\frac{^{143}\text{Nd}}{^{144}\text{Nd}}\right)_A f_A \frac{\text{Nd}_A}{\text{Nd}_M} + \left(\frac{^{143}\text{Nd}}{^{144}\text{Nd}}\right)_B \times (1 - f_A)\frac{\text{Nd}_B}{\text{Nd}_M} \quad (16.17)$$

In addition, the isotope ratios of two elements in components A and B calculated from these equations can be plotted to form a two-element isotopic mixing hyperbola.

The isotope ratios of Sr and Nd are an effective combination because the geochemical properties of the Rb–Sr system (Chapter 5) differ markedly from those of the Sm–Nd system (Chapter 9). For example, during partial melting of ultramafic rocks in the mantle of the Earth, Rb enters the melt phase in preference to Sr. Consequently, volcanic rocks generally have *higher* Rb/Sr ratios than the ultramafic rocks in the mantle from which they were derived (Faure and Hurley, 1963).

The Sm–Nd system behaves differently during partial melting because Nd is concentrated into the melt phase, whereas Sm remains with the residual solids. Consequently, volcanic rocks have *lower* Sm/Nd ratios than their magma sources in the mantle. It follows that crustal rocks have higher Rb/Sr ratios and lower Sm/Nd ratios than magma sources in the mantle. These differences have caused rocks of the continental crust to develop high $^{87}Sr/^{86}Sr$ but low $^{143}Nd/^{144}Nd$ ratios, whereas the mantle is characterized by having low $^{87}Sr/^{86}Sr$ but high $^{143}Nd/^{144}Nd$ ratios. These differences in the geochemical properties of the Rb–Sr and Sm–Nd couples and the resulting differences in the isotope compositions of Sr and Nd in the crust and mantle of the Earth are used to determine the origin of igneous rocks (Faure, 2001).

The Sr–Nd isotopic mixing hyperbola in Figure 16.7 was plotted for components that represent the mantle (component A) and continental crust (component B). The relevant data in Table 16.1 include the concentrations and isotope ratios of Sr and Nd of the components and the isotope ratios of Sr and Nd in the mixtures calculated for the selected values of f_A using a modified version of equation 16.11 in which equation 16.4 was substituted for Sr_M:

$$\left(\frac{^{87}\text{Sr}}{^{86}\text{Sr}}\right)_M = \frac{\left(\frac{^{87}\text{Sr}}{^{86}\text{Sr}}\right)_A f_A\ \text{Sr}_A + \left(\frac{^{87}\text{Sr}}{^{86}\text{Sr}}\right)_B (1 - f_A)\text{Sr}_B}{\text{Sr}_A f_A + \text{Sr}_B(1 - f_A)} \quad (16.18)$$

This equation yields the isotope ratios of Sr and Nd of mixtures directly without requiring a separate calculation of the concentrations of these elements (i.e., Sr_M and Nd_M). The resulting hyperbola in Figure 16.7 is the locus of all points

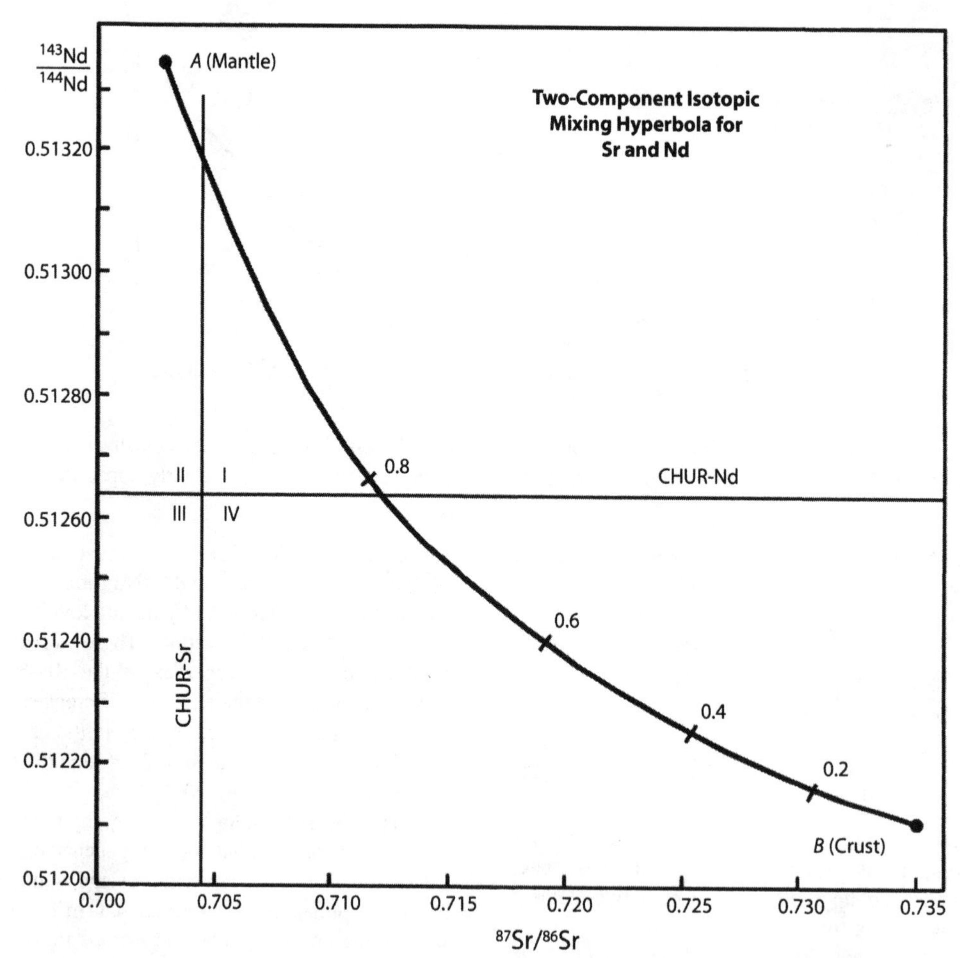

FIGURE 16.7 Two-component isotopic mixing hyperbola of Sr and Nd constructed from the data in Table 16.1.

representing mixtures of components A and B. In most cases, the concentrations of Sr and Nd of mafic volcanic rocks (component A) are higher than those of granites and granitic gneisses (component B), and the hyperbola in Figure 16.7 was constructed accordingly.

The curvature of the hyperbola in Figure 16.7 is controlled by the concentrations of Sr and Nd in accordance with a parameter r defined by Langmuir et al. (1978) and Vollmer (1976):

$$r = \frac{(^{144}\text{Nd}/^{86}\text{Sr})_A}{(^{144}\text{Nd}/^{86}\text{Sr})_B} \tag{16.19}$$

where ^{144}Nd and ^{86}Sr are the concentrations of these isotopes in components A and B or in any two data points that are well separated from each other on the mixing hyperbola. The concentrations of ^{144}Nd and ^{86}Sr are directly related to the concentrations of Nd and Sr, respectively, through constants that cancel in the ratio. Therefore, the curvature of isotopic mixing hyperbolas is determined by the concentration ratio of the elements in the components:

$$r = \frac{(\text{Nd}/\text{Sr})_A}{(\text{Nd}/\text{Sr})_B}$$

Table 16.1. Two-Component Isotopic Mixing Hyperbola for Sr and Nd in the Mantle (component *A*) and Continental Crust (component *B*) Calculated from Equation 16.18

f_A	$(^{87}Sr/^{86}Sr)_M$	$(^{143}Nd/^{144}Nd)_M$
Component B^a		
0	0.73500	0.51210
0.2	0.7306	0.51216
0.4	0.7254	0.51225
0.6	0.7192	0.512400
0.8	0.7117	0.51266
Component A^b		
1	0.7025	0.51330

[a]Concentrations component B (ppm): Sr = 160, Nd = 45.
[b]Concentrations component A (ppm): Sr = 100, Nd = 10.

The Nd and Sr concentrations of components *A* and *B* in Table 16.1 and Figure 16.7 yield a value of 0.35 for this parameter:

$$r = \frac{10/100}{45/160} = 0.35$$

The curvature of the hyperbolas in Figure 16.8 is related to the numerical value of r:

$$r < 1 \quad \text{(convex down)}$$

$$r = 1 \quad \text{(straight line)}$$

$$r > 1 \quad \text{(convex up)}$$

These relations apply to the case where $(^{143}Nd/^{144}Nd)_A > (^{143}Nd/^{144}Nd)_B$ and $(^{87}Sr/^{86}Sr)_A < (^{87}Sr/^{86}Sr)_B$ as in Table 16.1 and Figure 16.7 (equation 16.19):

$$r = \frac{(Nd/Sr)_A}{(Nd/Sr)_B}$$

Examples of two-component isotopic mixtures occur in igneous petrology when basalt magma forms by partial melting in the mantle and subsequently assimilates rocks of the continental crust. Alternatively, mixing may occur at the source when a mechanical mixture of different kinds of rocks undergoes partial melting.

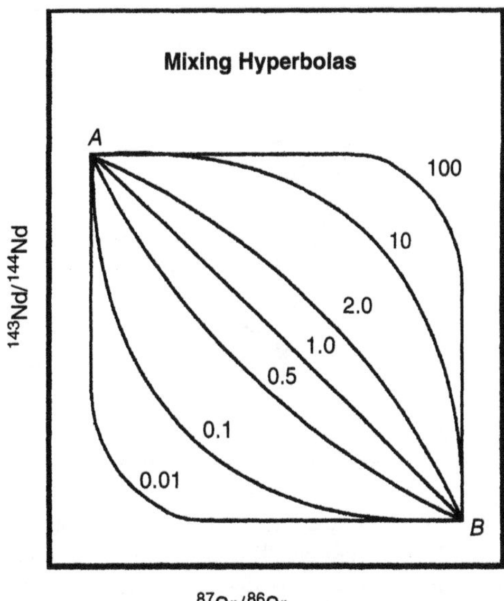

FIGURE 16.8 Curvature and orientation of isotopic mixing hyperbolas of Sr and Nd or of any two elements having different isotopic compositions. The curvature is controlled by the parameter r defined by equation 16.19. Adapted from Langmuir et al. (1978).

16.4 THREE-COMPONENT ISOTOPIC MIXTURES

Some geological materials are mixtures of three or more components each of which contains two elements (X and Y) having different isotope ratios and concentrations. Such mixtures are located within a triangle (or polygon) of mixing in coordinates of the isotope ratios of elements X and Y. For example, the mixture M in Figure 16.9 consists of three components (A, B, and C), each of which has a different isotope composition of Sr and Nd. The mixing lines between these components have curvature depending on the concentrations of Sr and Nd of the components. However, the mixing lines in Figure 16.9 have been drawn as straight lines for the sake of simplicity and because, in some cases, all three components have similar concentrations of Sr and Nd (i.e., $r = 1.0$, equation 16.19).

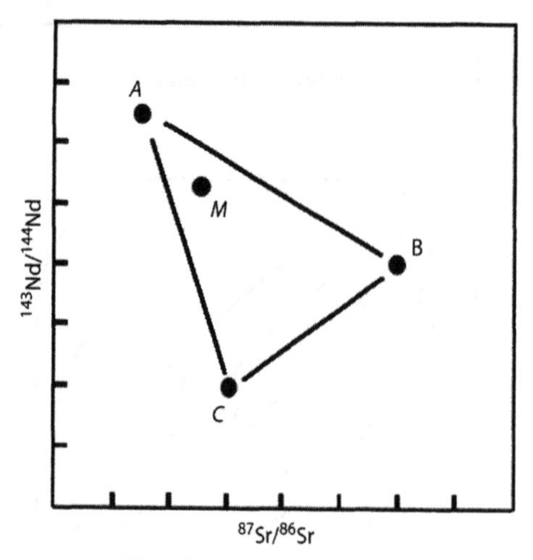

FIGURE 16.9 Three-component isotopic mixing triangle of Sr and Nd. The mixing lines forming the triangle have been drawn as straight lines even though in reality they may have curvature.

Although the location of mixture M in Figure 16.9 suggests that component A is most abundant, three-component isotopic mixtures cannot be resolved into their components as easily as chemical mixtures (e.g., Figure 16.3) because the mixing parameters f_A, f_B, and f_C vary *nonlinearly* with distance along the isotopic mixing hyperbolas. Nevertheless, the distribution of data points representing a suite of samples of three-component isotopic mixtures can be modeled to recover the proportions of mixing.

16.5 APPLICATIONS

Mixing of geological materials containing elements having different isotope compositions is a common process occurring on the surface of the Earth and in its interior. All of the topics to be discussed in Part III of this book involve mixing of water, sediment, and silicate melts. The examples that conclude this chapter are a preview of things to come.

16.5a North Channel, Lake Huron, Canada

The North Channel of Lake Huron is a body of water restricted along its southern border by Manitoulin, Cockburn, and Drummond Islands. Its northern shore is the Precambrian Shield of Canada. The chemical composition and the $^{87}Sr/^{86}Sr$ ratio of the water in the North Channel are the result of mixing of water derived from the crystalline rocks of the Precambrian Shield with the water of Lake Huron. Most of the water derived from the Precambrian Shield enters the North Channel at its western end by the discharge of St. Mary's River that drains Lake Superior. Water draining the Precambrian Shield as represented by Lake Superior has $^{87}Sr/^{86}Sr = 0.718$ (Hart and Tilton, 1966), $Sr = 21.8$ μg/L, and $Ca = 14.4$ μg/mL, whereas water in the main body of Lake Huron has $^{87}Sr/^{86}Sr = 0.7107$, $Sr = 98.2$ μg/L, and $Ca = 26.9$ μg/mL (Faure et al., 1967). The water within the North Channel is intermediate in composition, presumably because it consists of mixtures of water from Lake Superior and Lake Huron in differing proportions.

The concentrations of Ca and Sr in water from the North Channel in Figure 16.10 have the predicted linear correlation characteristic of two-component mixtures of conservative elements. The abundance of Lake Superior water in the North Channel area can be calculated by means of equation 16.4 from measured concentrations of Sr, Ca, or any other conservative element, given the concentrations of these elements in Lake Superior and Lake Huron. If Lake Superior water is taken to be component A ($Sr_A = 21.8$ μg/L) and Lake Huron water is component B ($Sr_B = 98.2$ μg/L), a sample of water in the North Channel having $Sr_M = 50.0$ μg/L contains 63% of Lake Superior water:

$$Sr_M = Sr_A f_A + Sr_B(1 - f_A)$$

$$50.0 = 21.8 f_A + 98.2(1 - f_A)$$

$$f_A = 0.63$$

The $^{87}Sr/^{86}Sr$ ratios and reciprocal Sr concentrations of water in the North Channel form a straight line in Figure 16.11*a*. The mixing equation determined by least-squares regression is

$$\left(\frac{^{87}Sr}{^{86}Sr}\right)_M = \frac{0.1996}{Sr_M} + 0.7086 \qquad (16.20)$$

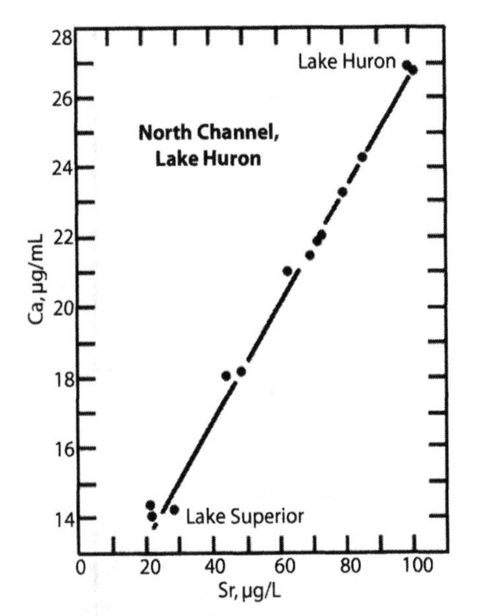

FIGURE 16.10 Mixing of Sr and Ca in water of Lake Superior (component A) and Lake Huron (component B) in the North Channel area of Lake Huron. Data from Faure et al. (1967).

This equation was replotted in Figure 16.11*b* in coordinates of $^{87}Sr/^{86}Sr$ and the Sr concentration to yield the mixing hyperbola in accordance with equations 16.14 and 16.15.

16.5b Detrital Silicate Sediment, Red Sea

Sediment accumulating in the median valley of the Red Sea is a mixture of biogenic calcium carbonate ooze and detrital grains of silicate minerals. Boger and Faure (1974) removed the carbonate fraction by leaching sediment samples with dilute HCl and then measured the $^{87}Sr/^{86}Sr$ ratios and Sr concentrations of the silicate residues from a piston core taken at 18°09′ N and 39°53′E in the median valley of the Red Sea. The results in Figure 16.12*a* define a straight line in coordinates of the $^{87}Sr/^{86}Sr$ ratio and the reciprocal Sr concentrations. The equation of this line is

$$\frac{^{87}Sr}{^{86}Sr} = \frac{0.8158}{Sr} + 0.70294 \qquad (16.21)$$

This equation was replotted in Figure 16.12*b* as a hyperbola.

The geology of the region bordering the Red Sea supports the hypothesis that the detrital silicate component of Red Sea sediment is a mixture of volcanic ash (low $^{87}Sr/^{86}Sr$, high Sr concentration) and sialic detritus derived from the Arabo-Nubian Shield (high $^{87}Sr/^{86}Sr$, low Sr concentration). Therefore, Boger and Faure (1974) used equation 16.21 to calculate the Sr concentrations of these components based on assumed

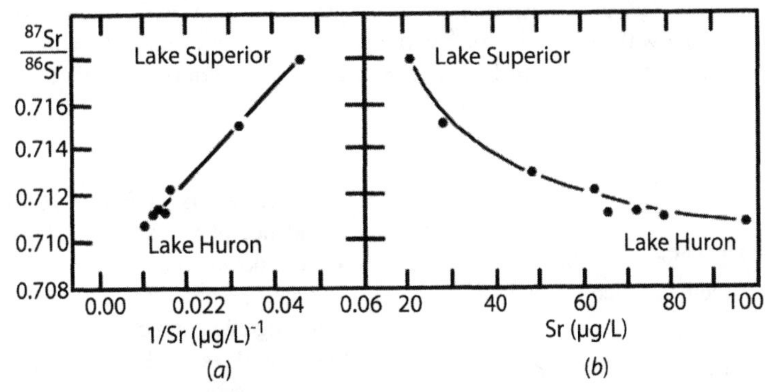

FIGURE 16.11 (*a*) Mixing of water in the North Channel which originates from Lake Superior and Lake Huron in coordinates of the $^{87}Sr/^{86}Sr$ ratio and the reciprocal Sr concentration. The equation of this line (equation 16.20) was plotted as a hyperbola in part (*b*) in coordinates of the $^{87}Sr/^{86}Sr$ and the Sr concentrations. The conversion of a mixing hyperbola into a straight line and back into a hyperbola is based on equations 16.4 and 16.15. Data from Faure et al. (1967).

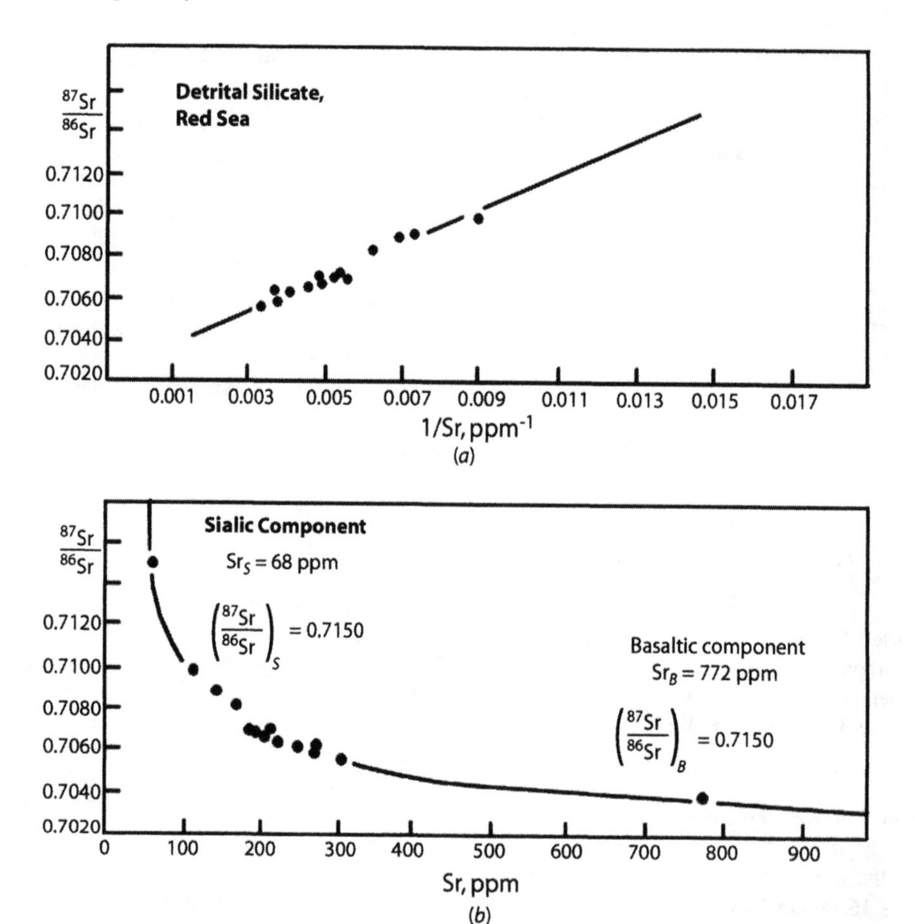

FIGURE 16.12 (a) Detrital sediment fraction from a piston core recovered at 18°09′ N and 39°53′E in the median valley of the Red Sea. (b) Transformation of the straight line in part (a) (equation 16.21) into an isotopic mixing hyperbola with assumed $^{87}Sr/^{86}Sr$ ratios of the volcanic and sialic components. From P. D. Boger and G. Faure, *Geology*, Vol. 2, pp. 181–183. Reprinted with permission from the Geological Society of America.

$^{87}Sr/^{86}Sr$ ratios:

Component	$^{87}Sr/^{86}Sr$ Assumed	Sr, ppm (equation 16.21)
Volcanic	0.7040	772
Sialic	0.7150	68

The resulting Sr concentrations of the volcanic and sialic components were then used in equation 16.4 to calculate the abundances of the volcanic component in the silicate fraction of this core.

The results in Figure 16.13 suggest that the abundance of volcanogenic detritus varies systematically down core and hence with time in the past. The stratigraphic variation of the abundance of the volcanogenic component in Figure 16.13 was used to define three kinds of layers:

1. mixed sialic–volcanic layers defined by the straight line fitted to the sediment profile,
2. sialic sediment containing less than 20% volcanogenic detritus, and

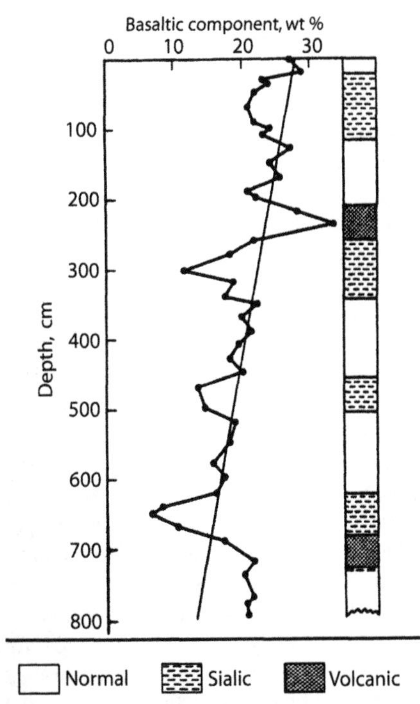

FIGURE 16.13 Stratigraphic variation of the abundance of volcanogenic sediment in the silicate fraction of a piston core collected at 18°09′ N and 39°53′ E in the median valley of the Red Sea. Adapted from Boger and Faure (1974).

3. volcanogenic sediment enriched in volcanic ash compared to the mixed layers.

The resulting lithostratigraphy records the history of sedimentation at this site. Boger and Faure (1976) and Boger et al. (1980) concluded that the layers of sialic sediment correlate approximately with interstadials of the Würm glaciation in northern Europe.

Additional examples of mixing of marine sediment have been reported elsewhere:

Dasch, 1969, *Geochim. Cosmochim. Acta*, 33:1521–1552. Biscaye and Dasch, 1971, *J. Geophys. Res.*, 76:5087–5096. Shaffer and Faure, 1976, *Geol. Soc. Am. Bull.*, 87:1491–1500. Kovach and Faure, 1977, *New Zealand J. Geol. Geophys.*, 20:1017–1026. Nardone and Faure, 1978, in Ross and Neprochnov (Eds.), *Initial Repts. Deep Sea Drilling Project*, 42(2):607–615. Ben Othman et al., 1989, *Earth Planet Sci. Lett.*, 94:1–21. Walter et al., 2000, *Geochim. Cosmochim. Acta*, 64:3813–3827.

16.5c Fictitious Rb–Sr Isochrons

Mixtures of two components having different $^{87}Sr/^{86}Sr$ and different $^{87}Rb/^{86}Sr$ ratios form points on Rb–Sr isochron diagrams that lie along a straight line joining the two components (Haack, 1990). Such mixtures may be sedimentary or igneous in origin, but the straight lines they define are mixing lines rather than isochrons.

The equation of the straight line defined by two-component mixtures has the form (Faure, 1977)

$$\left(\frac{^{87}\mathrm{Sr}}{^{86}\mathrm{Sr}}\right)_M = g\left(\frac{^{87}\mathrm{Rb}}{^{86}\mathrm{Sr}}\right) + e \qquad (16.22)$$

where

$$e = \frac{\mathrm{Rb}_B\mathrm{Sr}_A\left(^{87}\mathrm{Sr}/^{86}\mathrm{Sr}\right)_A - \mathrm{Rb}_A\mathrm{Sr}_B\left(^{87}\mathrm{Sr}/^{86}\mathrm{Sr}\right)_B}{\mathrm{Rb}_B\mathrm{Sr}_A - \mathrm{Rb}_A\mathrm{Sr}_B}$$

$$g = \frac{\mathrm{Sr}_A\mathrm{Sr}_B\left[\left(^{87}\mathrm{Sr}/^{86}\mathrm{Sr}\right)_B - \left(^{87}\mathrm{Sr}/^{86}\mathrm{Sr}\right)_A\right]}{k(\mathrm{Rb}_B\mathrm{Sr}_A - \mathrm{Rb}_A\mathrm{Sr}_B)}$$

and k is a constant whose value depends on the abundance of ^{86}Sr and the atomic weight of Sr:

$$k = \frac{(\mathrm{Ab}^{87}\mathrm{Rb})\ \text{at. wt. Sr}}{(\mathrm{Ab}^{86}\mathrm{Sr})\ \text{at. wt. Rb}} = 2.90 \pm 0.007 \quad (16.23)$$

provided that the $^{87}Sr/^{86}Sr$ ratios of Sr in components A and B are between 0.700 and 0.750.

The slope g and intercept e of equations 16.22 are calculated from the concentrations of Rb and Sr and the $^{87}Sr/^{86}Sr$ ratios of components A and B:

Component A	Component B
$Sr_A = 200$ ppm	$Sr_B = 450$ ppm
$Rb_A = 400$ ppm	$Rb_B = 40$ ppm
$(^{87}Sr/^{86}Sr)_A = 0.725$	$(^{87}Sr/^{86}Sr)_B = 0.704$

$$g = \frac{200 \times 450(0.704 - 0.725)}{2.90(40 \times 200 - 400 \times 450)}$$

$$= \frac{-1890}{-498{,}800} = 0.003789$$

Hence,

$$t = \frac{\ln 1.003789}{1.42 \times 10^{-11}} = 266 \text{ Ma}$$

The date derived from the slope of the mixing line in Figure 16.14 is fictitious because components *A* and *B* are not genetically related to each other.

Sediment composed of detrital grains derived from different sources (e.g., young volcanic rocks and old granitic rocks of the continental crust) defines mixing lines that may be mistaken for isochrons. In addition, basaltic lava flows erupted on some volcanos have a range of $^{87}Sr/^{86}Sr$ and $^{87}Rb/^{86}Sr$ ratios because basaltic magmas can assimilate varying amounts of old crustal rocks from the walls of their magma chambers.

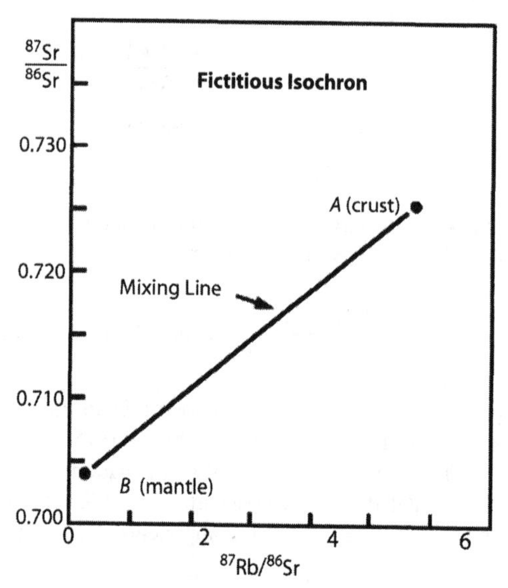

FIGURE 16.14 Fictitious isochron formed by mixtures of components *A* (crust) and *B* (mantle) in varying proportion. Such mixtures form during deposition of sediment and by assimilation of crustal rocks by mantle-derived basaltic magma.

16.5d Potassic Lavas, Toro-Ankole, East Africa

An example of correlated $^{87}Sr/^{86}Sr$ and $^{87}Rb/^{86}Sr$ ratios was reported by Bell and Powell (1969) for potassic lava flows from the Birunga and Toro-Ankole fields along the borders of Uganda, Zaire, and Rwanda of East Africa (Faure, 2001). These two volcanic centers are about 160 kilometers from each other and are located along the western rift, north of Lake Tanganyika and west of Lake Victoria. The lava flows are known to be Pliocene to Recent in age and there is evidence of volcanic activity in historical times. These volcanic rocks are therefore very young and should form an isochron having a slope approaching zero. However, Bell and Powell (1969) found a significant positive correlation between the average $^{87}Sr/^{86}Sr$ and Rb/Sr ratios of different rock types.

The $^{87}Sr/^{86}Sr$ and $^{87}Rb/^{86}Sr$ ratios of the volcanic rocks of the Toro-Ankole area form a linear array in Figure 16.15 to which a straight line has been fitted by eye. This line is not an isochron,

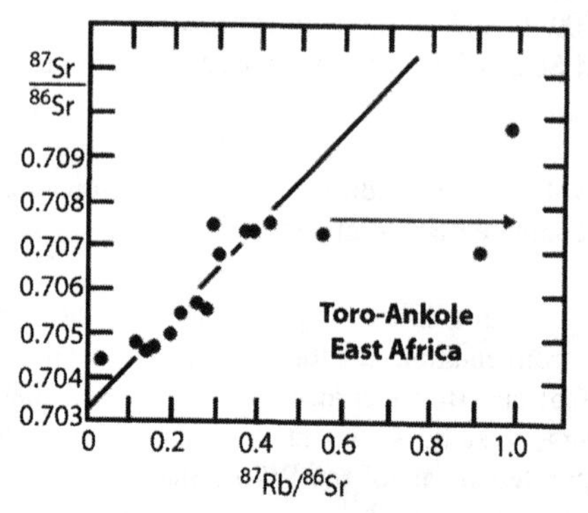

FIGURE 16.15 Potassium-rich volcanic rocks of Late Tertiary age in the Toro-Ankole area of East Africa. The correlation of $^{87}Sr/^{86}Sr$ and $^{87}Rb/^{86}Sr$ ratios is the result of assimilation of granitic basement rocks by mantle-derived magma. Hence the line is the result of mixing and is not an isochron. The Rb/Sr ratios of some samples were increased by crystallization of Sr-bearing minerals from the contaminated magma. Data from Bell and Powell (1969).

but rather it is a mixing line because the mantle-derived magma assimilated varying amounts of granitic basement rocks of the Precambrian Shield of East Africa. The contaminated magma was not homogenized by mechanical stirring and diffusion, thereby causing lava flows to have a range of $^{87}Sr/^{86}Sr$ and $^{87}Rb/^{86}Sr$ ratios. The slope of the straight line in Figure 16.15 yields a fictitious Rb–Sr date of 773 Ma. However, the date does convey the information that the crustal contaminant had elevated $^{87}Sr/^{86}Sr$ and $^{87}Rb/^{86}Sr$ ratios and may have been of Precambrian age.

The $^{87}Sr/^{86}Sr$ ratios of the contaminated magma were *not* altered by the crystallization of Sr-bearing minerals because Sr isotopes are not fractionated by this process. In any case, the effects of natural and instrumental fractionation of Sr isotopes are eliminated by the fractionation correction that is applied routinely to all measured isotope ratios of Sr (Section 5.2a). However, the crystallization of Sr-rich minerals (e.g., plagioclase) does deplete the residual magma in Sr and therefore causes its Rb/Sr ratio to rise. This explains why three samples in Figure 16.15 plot to the right of the mixing line. However, the evident colinearity of most of the data points implies that the Rb/Sr ratio of the contaminated magma, in most cases, was not affected by crystallization of plagioclase or of other Sr-bearing minerals.

Other examples of correlated $^{87}Sr/^{86}Sr$ and $^{87}Rb/^{86}Sr$ ratios of Tertiary lava flows have been reported by Dickinson et al. (1969) from southern Arabia, by Leeman and Manton (1971) from the Snake River Plain in Idaho, by Duncan and Compston (1976) from the South Pacific, and by other authors included in the monograph by Faure (2001).

16.6 SUMMARY

Mixing affects not only the chemical composition of water, sediment, and volcanic rocks but also the isotope compositions of several elements they contain. The concentrations of conservative elements in mixtures of two components define straight lines in $x-y$ diagrams. Two-component mixtures that contain varying amounts of a third component move off the two-component mixing line into a triangle of mixing. Such three-component mixtures can be resolved into their components by a simple geometric construction. The abundances of a selected component in two- and three-component mixtures can be contoured to reveal the location of the source of that component and the direction of movement of the local transport system.

Two- component mixtures containing one element having different isotope compositions form hyperbolas in coordinates of the concentration of the element (x) and the isotope ratio (y). Such single-element two-component isotopic mixing hyperbolas are convertible into straight lines by the inversion of the concentration ($1/x$). However, the abundances of the components vary nonlinearly along the length of the mixing hyperbola and the corresponding straight line.

The mixing hyperbolas of two-element, two-component isotopic mixtures cannot be converted into straight lines because the equation contains an additional term that is not compatible with the equation of a straight line. Three-component mixtures of two elements having different isotopic compositions (e.g., Sr–Nd, Sr–Pb, Nd–Pb, Nd–Os) form triangular data arrays that play an important role in the study of igneous petrogenesis.

Two-component mixtures based on isotopic and concentration ratios [e.g., $^{87}Rb/^{86}Sr$ (x) and $^{87}Sr/^{86}Sr$ (y)] form straight lines that may be mistaken for isochrons. Such pseudoisochrons are encountered in the study of detrital sedimentary rocks and in certain volcanic centers. All of the chapters in Part III contain examples of mixing of water, sediment, and silicate melts.

REFERENCES

Bell, K., and J. L. Powell, 1969. Strontium isotopic studies of alkalic rocks: The potassium-rich lavas of the Birunga and Toro-Ankole regions, east and central equatorial Africa. *J. Petrol.*, 10:536–572.

Boger, P. D., J. L. Boger, and G. Faure, 1980. Systematic variations of $^{87}Sr/^{86}Sr$ ratios, Sr composition, selected major-oxide concentrations, and mineral abundances in piston cores from the Red Sea. *Chem. Geol.*, 29:13–38.

Boger, P. D., and G. Faure, 1974. Strontium-isotope stratigraphy of a Red Sea core. *Geology*, 2:81–83.

Boger, P. D., and G. Faure, 1976. Systematic variations of sialic and volcanic detritus in piston cores from the Red Sea. *Geochim. Cosmochim. Acta*, 40:731–742.

Dickinson, D. R., M. H. Dodson, I. G. Gass, and D. C. Rex, 1969. Correlation of initial $^{87}Sr/^{86}Sr$ with Rb/Sr in some Late Tertiary volcanic rocks of South Arabia. *Earth Planet. Sci. Lett.*, 6:84–90.

Duncan, R. A., and W. Compston, 1976. Sr-isotopic evidence for an old mantle source region for French Polynesian volcanism. *Geology*, 4:728–732.

Faure, G., 1977. *Principles of Isotope Geology*, 1st ed. Wiley, New York.

Faure, G., 1998. *Principles and applications of geochemistry*. Second ed., Prentice Hall, Upper Saddle River, N.J., 599 p.

Faure, G., 2001. *Origin of Igneous Rocks; the Isotopic Evidence*. Springer-Verlag, Heidelberg.

Faure, G., and P. M. Hurley, 1963. The isotopic composition of strontium in oceanic and continental basalts: Application to the origin of igneous rocks. *J. Petrol.*, 4(1):31–50.

Faure, G., L. M. Jones, R. Eastin, and M. Christner, 1967. *Strontium Isotope Composition and Trace Element Concentrations in Lake Huron and Its Principal Tributaries*, Rept. No. 2. Laboratory for Isotope Geology and Geochemistry, The Ohio State University, Columbus, Ohio.

Haack, U., 1990. Datierung mit Rb/Sr-Mischungslinien? *Eur. J. Mineral.*, 2:86.

Hart, S. R., and G. R. Tilton, 1966. The isotope geochemistry of strontium and lead in Lake Superior sediments and water. In J. S. Steinhart and T. L. Smith (Eds.), *The Earth beneath the Continents*, Geophys. Monogr. No. 10, pp. 127–137. American Geophysical Union, Washington, D.C.

Langmuir, C. H., R. D. Vocke, Jr., G. N. Hanson, and S. R. Hart, 1978. A general mixing equation with applications to Icelandic basalts. *Earth Planet. Sci. Lett.*, 37:380–392.

Leeman, W. P., and W. I. Manton, 1971. Strontium isotopic composition of basaltic lavas from the Snake River Plain, southern Idaho. *Earth Planet. Sci. Lett.*, 11:420–434.

Lowry, R. M., G. Faure, D. I. Mullet, and L. M. Jones, 1988. Interpretation of chemical and isotopic compositions of brines based on mixing and dilution, "Clinton" sandstones, eastern Ohio, USA. *Appl. Geochem.*, 3:177–184.

Petz, T. R., and G. Faure, 1997. Mixing of water in streams: Big Walnut Creek and its tributaries. *Ohio J. Sci.*, 97(5):113–115.

Vollmer, R., 1976. Rb-Sr and U-Th-Pb systematics of alkaline rocks: The alkaline rocks from Italy. *Geochim. Cosmochim. Acta*, 40:283–286.

CHAPTER 17

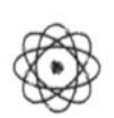

Origin of Igneous Rocks

THE formation of igneous rocks is closely related to the tectonic activity of the mantle of the Earth. Far from being homogeneous in composition and inactive, the mantle is heterogeneous and in constant convective motion. The heterogeneity of the mantle not only concerns its chemical and mineralogical composition but also includes the isotopic composition of certain elements and regional variations of its temperature.

17.1 THE PLUME THEORY

The temperature variations in the mantle result from the radioactive decay of U, Th, and K, which are recycled when oceanic crust and marine sediment are subducted along compressive plate margins (Hofmann and White, 1982). The subducted surface rocks are dehydrated and recrystallized and remain in buoyant equilibrium within the mantle for one billion years or longer. During this period of incubation, long-lived radionuclides decay, thereby generating heat and increasing the abundances of the radiogenic isotopes.

Ultimately, the increase of the temperature of the subducted rocks decreases their density relative to the rocks of the surrounding mantle, which causes them to move upward in the form of diapirs. When a diapir or plume reaches the underside of the rigid lithospheric mantle, the head spreads out by streaming laterally and may reach diameters of 1000 km or more (Schilling et al., 1992). The stress applied to the underside of the lithospheric plate causes fractures to develop that propagate upward and may result in rifting of the continental or oceanic crust. The resulting decompression causes large-scale partial melting of the rocks in the plume head and in the overlying lithospheric mantle. In addition, entrained blocks of asthenospheric mantle in the plume head may also contribute magma. The resulting basalt magma intrudes the overlying lithospheric mantle and continental crust and may break through to the surface to form volcanoes. The details of this process vary depending on local conditions but can be reconstructed by judicious interpretations of the chemical and isotopic compositions of the volcanic rocks that form at different stages in the history of a particular volcanic center.

According to this theory, plumes continue to rise from depth in the mantle at the present time. Their interaction with the lithosphere causes magma formation, uplift and rifting of the continental and oceanic crust, as well as movement of the lithospheric plates, continental drift, opening and closing of ocean basins, and formation of oceanic crust along spreading ridges and its ultimate destruction in subduction zones. This process has been operating since before Early Archean time, it is still active at present, and it will continue for hundreds of millions of years to come until the rate of heat production by radioactive decay of long-lived radionuclides decreases sufficiently to prevent plumes from forming in the mantle.

When viewed in this perspective, many geological phenomena on the surface of the Earth are recognized as manifestations of the movement of plumes in the mantle. Plumes energize the Earth and allow the process to continue by causing the heat-producing elements to be recycled. The plume theory of the Earth gives meaning to geology because it unifies a wide variety of geophysical, geochemical, tectonic, and historical observations. In that sense, it is a good theory; but, like other scientific theories, it is still being tested and will be elaborated or revised as new observations about the Earth become available.

The important role of plumes in the dynamic processes of the mantle was originally recognized by Wilson (1963a, b). Subsequently, Morgan (1971, 1972a, b, 1981) used chains of submarine volcanoes to track the motions of lithospheric plates across the tops of stationary plume heads, and Hofmann and White (1982) postulated that plumes develop from oceanic crust that was subducted as much as one billion years ago. The isotopic data that are relevant to the study of the origin of igneous rocks have been summarized by Basu and Hart (1996) and Faure (2001).

17.2 MAGMA SOURCES IN THE MANTLE

Experience has shown that the isotope ratios of Sr, Nd, Pb, Hf, and Os of rocks along midocean ridges and on volcanic islands in the oceans are correlated with each other. For example, the $^{87}Sr/^{86}Sr$ and $^{143}Nd/^{144}Nd$ ratios of MORBs form the so-called mantle array. Such correlations are evidence that basalt magmas that form in the suboceanic mantle either are mixtures of melts derived from different components or form by melting of mechanical mixtures of such components.

The available isotopic data can be explained by postulating the existence of four principal components (DMM, EM1, EM2, and HIMU) whose isotope compositions in Table 17.1 are based on the work of Zindler and Hart (1986), Hart (1988), Salters and Hart (1991), and Shirey and Walker (1998). The numerical values of isotope ratios of the magma source components can be specified because the isotopes of Sr, Nd, Pb, Hf, and Os

Table 17.1. Isotope Ratios of Sr, Nd, Pb, Hf, and Os That Characterize the Four Principal Magma Sources of Midocean Ridge and Oceanic Island Basalts

Ratios	DMM	EM1	EM2	HIMU
$^{87}Sr/^{86}Sr$	0.7022	0.7055	0.7075	0.7028
$^{143}Nd/^{144}Nd^{a}$	0.5133	0.51235	0.51264	0.51285
$^{206}Pb/^{204}Pb$	18.2	17.5	19.2	21.7
$^{208}Pb/^{204}Pb$	37.7	38.1	39.3	40.7
$^{176}Hf/^{177}Hf$	~0.28340	0.28265	0.28280	0.28290
$^{187}Os/^{188}Os$	0.125	0.152	0.136	0.150

Note: Strontium, Nd, and Pb: Hart, 1988; Hf: Salters and Hart, 1991; Os: Shirey and Walker, 1998.

[a] Relative to $^{146}Nd/^{144}Nd = 0.7219$.

are not fractionated when magma forms by partial melting of the ultramafic rocks in the mantle. Consequently, the isotope ratios of volcanic rocks in the ocean basins convey information about the magma sources in the suboceanic mantle, whereas the concentrations of trace elements of volcanic rocks depend on many factors, including the composition of the source rocks, the extent of partial melting, and the crystallization of minerals from the resulting magmas.

The evidence used by Hart (1988) to specify the isotope ratios of the principal components of magmas of oceanic island basalts (OIBs) and MORBs consisted of a series of two-dimensional diagrams of the isotope ratios of Sr, Nd, and Pb in OIBs and MORBs taken from the literature. The distribution of data points on the Nd–Sr and Sr–Pb planes in Figure 17.1 demonstrates the evidence for the existence of these components, which are identified as follows:

DMM: depleted MORB mantle
EM1: enriched mantle 1
EM2: enriched mantle 2
HIMU: high μ (i.e., $^{238}U/^{204}Pb$)

The DMM is characterized in Table 17.1 by low $^{87}Sr/^{86}Sr$ (0.7022), high $^{143}Nd/^{144}Nd$ (0.5133), and intermediate $^{206}Pb/^{204}Pb$ (18.2) ratios. The low $^{87}Sr/^{86}Sr$ and high $^{143}Nd/^{144}Nd$ ratios of DMM imply that this component was depleted in Rb and enriched in Sm in the geological past as a result of the prior removal of a partial melt from this

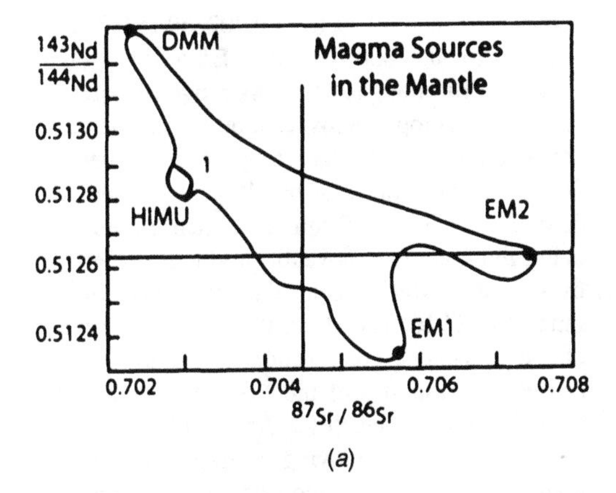

(*a*)

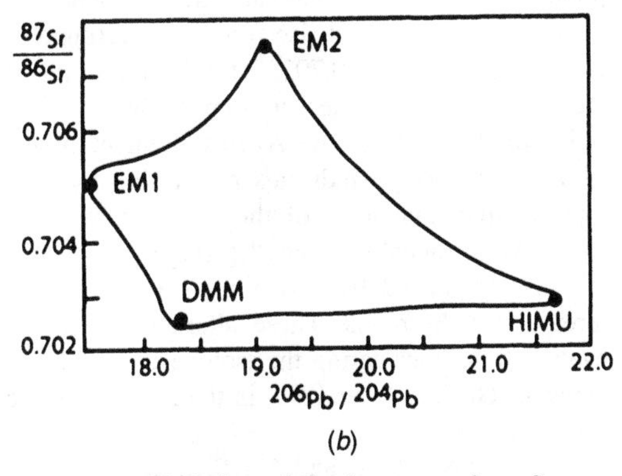

(*b*)

FIGURE 17.1 Variation of the isotope ratios of (*a*) Nd and Sr and (*b*) Sr and Pb of oceanic island basalts OIBs. The data fields identify the four principal components that participate in the formation of magma in varying proportions. Adapted from Hart (1988).

magma source. Therefore, the DMM component is presumed to be the upper (lithospheric) mantle, which participated in the formation of continental crust starting more than four billion years ago.

The EM1 and EM2 components consist primarily of basalt and sediment of the oceanic crust, which were altered by dehydration and recrystallization during subduction. From a consideration of trace element ratios and isotopic data, Weaver (1991) concluded that EM1 contains 5–10% of marine pelagic sediment, whereas EM2 contains a similar amount of terrigenous sediment. Weaver (1991) also concluded that HIMU consists of recycled basalt of the oceanic crust without a significant presence of subducted sediment. The HIMU component has high U/Pb and Th/Pb ratios that caused the $^{206}Pb/^{204}Pb$ and $^{208}Pb/^{204}Pb$ ratios of this component to rise to high values during lengthy periods of incubation in the mantle (see also Chauvel et al., 1992). Plumes may also contain entrained blocks of primitive mantle (PRIMA) whose isotope compositions are not as extreme as those of the components described above.

Stratigraphic variations of the isotope ratios of Sr, Nd, Pb, Hf, and Os in sequences of lava flows imply that time-dependent changes occurred in the proportions of the magma-source components that contributed to the formation of magma. A comprehensive review of the theory on the dynamic processes of the mantle was published by Hofmann (1997).

17.3 MIDOCEAN RIDGE BASALT

The major oceans of the Earth contain a system of midocean ridges along which tholeiite basalt flows are erupting. In places where the rate of lava production is high, large submarine basalt plateaus accumulate and the summits of the largest volcanoes form clusters of islands. The lava flows and associated intrusives along midocean ridges form the oceanic crust, which is transported away from the ridge crests by the movement of the lithospheric plates.

The volcanic rocks along the spreading ridges in the oceans are silica-rich tholeiite basalts that form as a result of large-scale (20–30%) decompression melting of the lithospheric mantle underlying the midocean ridges (Wilkinson, 1982). In general, these MORBs have low $^{87}Sr/^{86}Sr$ and high $^{143}Nd/^{144}Nd$ ratios on the basis of which the DMM component in Table 17.1 was defined. However, the isotope ratios of Sr and Nd of MORBs are not constant but vary systematically with distance *along* the midocean ridges. In addition, lava plateaus and associated volcanic islands tend to have higher $^{87}Sr/^{86}Sr$ and lower $^{143}Nd/^{144}Nd$ ratios than topographically low parts of the associated ridge.

17.3a Plumes of the Azores

The spreading ridge of the North Atlantic Ocean in Figure 17.2 contains the lava plateau of the Azore Islands and the island of Iceland, both of which are the products of vigorous volcanic activity. The $^{87}Sr/^{86}Sr$ ratios of tholeiites on the Azores platform in Figure 17.3 vary systematically with latitude between 50° and 30° N from 0.702369 to 0.703584 relative to 0.70800 for the E&A Sr isotope standard (Yu et al., 1997). This range of variation of the Sr isotope ratios cannot be attributed to contamination of the basalt by seawater because the powdered samples of basalt were leached with cold 2 N HCl for 10 min and then rinsed with deionized water to remove the products of alteration by seawater, which has a relatively high $^{87}Sr/^{86}Sr$ ratio of 0.70916. Therefore, the observed variation of the Sr isotope composition of the basalts was caused by variations of the $^{87}Sr/^{86}Sr$ ratios of the magma sources beneath the Azores plateau.

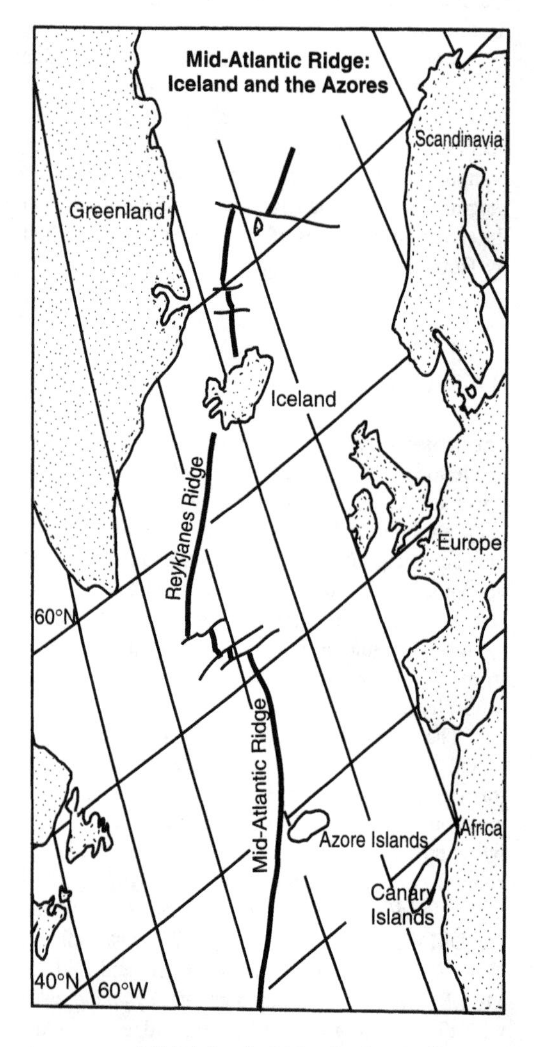

FIGURE 17.2 Mid-Atlantic Ridge in the North Atlantic Ocean. Adapted from *The Physical World*, National Geographic Society, Washington, D.C., 1975.

A three-point moving average of the data points in Figure 17.3 indicates the presence of three isotope peaks that rise from a background of low values of about 0.70255 north and south of the Azores plateau. The variation of the $^{87}Sr/^{86}Sr$ ratios of tholeiites in the Azores segment of the Mid-Atlantic Ridge indicates either that the isotope composition of Sr of the DMM component varies systematically along the ridge or that the magmas originated from two sources having different $^{87}Sr/^{86}Sr$ ratios. These alternatives can be evaluated by considering the isotope ratios of two elements (such as Sr and Nd) in the MORBs of the Azores plateau.

The $^{143}Nd/^{144}Nd$ and $^{87}Sr/^{86}Sr$ ratios of the basalt in the Azores plateau in Figure 17.4 form an example of the mantle array based on the negative correlation of these isotope ratios. The data points were regressed to a straight line represented by the equation

$$\frac{^{143}\text{Nd}}{^{144}\text{Nd}} = +0.73362 - 0.31373\left(\frac{^{87}\text{Sr}}{^{86}\text{Sr}}\right) \qquad (17.1)$$

The inverse correlation of the isotope ratios of Sr and Nd is consistent with the hypothesis that basalt magmas that erupted along the Azores plateau are mixtures of two components in the underlying mantle. One of these has high $^{143}Nd/^{144}Nd$ and low $^{87}Sr/^{86}Sr$ ratios, whereas the second component has lower $^{143}Nd/^{144}Nd$ and higher $^{87}Sr/^{86}Sr$ ratios than the depleted source.

In summary, the mantle array, defined by the isotope ratios of Sr and Nd of MORBs, is the result of mixing of magma derived from depleted and undepleted mantle. The depleted

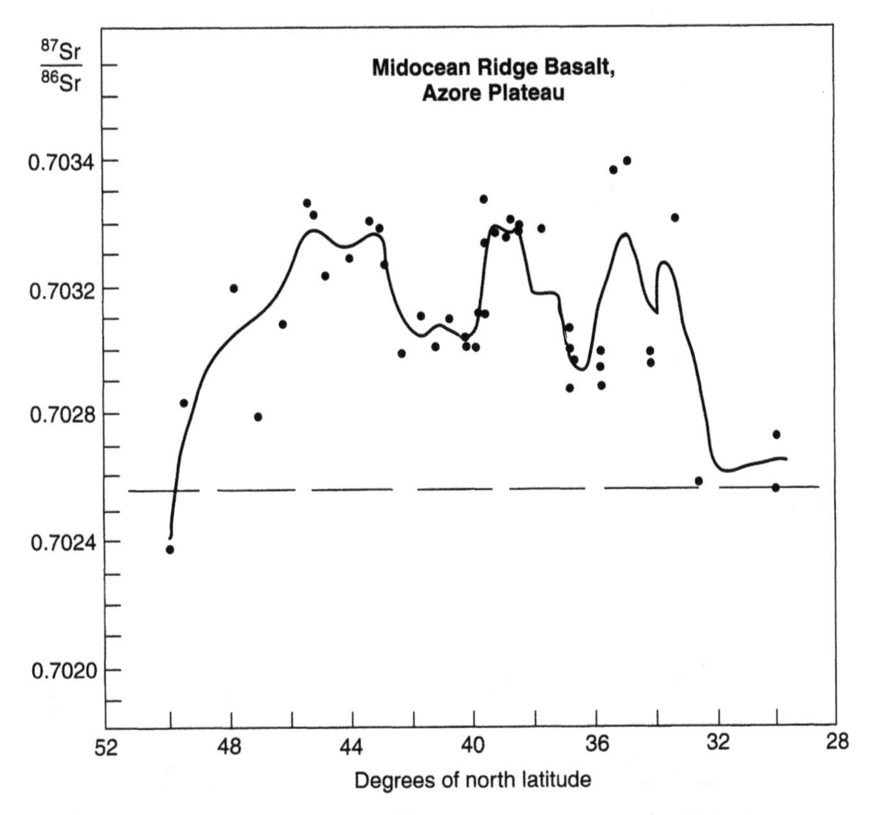

FIGURE 17.3 Three-point moving average of $^{87}Sr/^{86}Sr$ ratios of midocean ridge basalts MORBs on the Azores plateau of the Mid-Atlantic Ridge. The line fitted to the data points reveals three peaks that are attributable to the presence of plumes. The dashed line at $^{87}Sr/^{86}Sr = 0.70255$ represents Sr in the depleted lithospheric mantle at this site. Data from Yu et al. (1997).

magma source is identified as the DMM component. The second component occurs in the form of plumes composed of mixtures of subducted oceanic crust and sediment with entrained blocks of undifferentiated mantle (EM1 or EM2). The three broad peaks in Figure 17.3 represent magma derived primarily from plumes that have invaded the lithospheric mantle, whereas the lava flows that occur between these peaks originated from mixed plume–lithosphere sources. Basalt derived from depleted lithospheric mantle occurs primarily north and south of the Azores plateau and on other segments of the Mid-Atlantic Ridge.

The two types of MORBs represented in Figure 17.3 are referred to as N-MORB (normal) and P-MORB (plume). N-MORBs originate from depleted magma sources whereas P-MORBs are derived from plumes. These insights have provided the key that is needed to decipher the isotopic code encrypted in all kinds of igneous rocks (Basu and Hart, 1996).

17.3b Undifferentiated Mantle Reservoir of Sr

The straight line fitted to the data points in Figure 17.4 establishes a relationship between the isotope compositions of Sr and Nd in the magma sources underlying the Mid-Atlantic Ridge. Therefore, it can be used to calculate the $^{87}Sr/^{86}Sr$ ratio

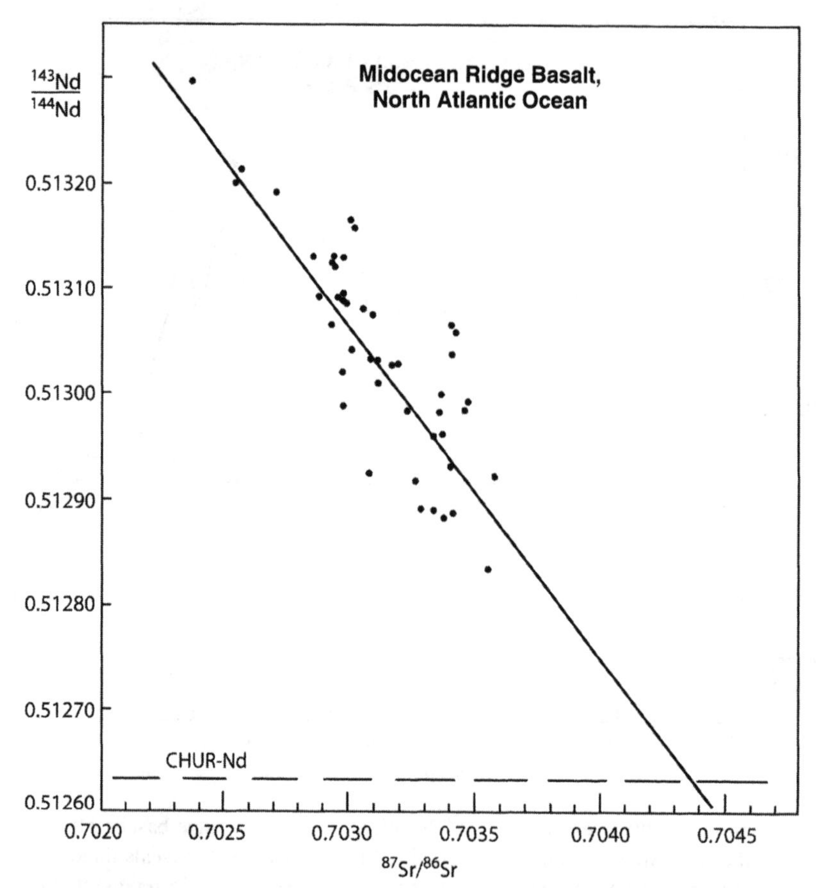

FIGURE 17.4 Isotope ratios of Sr and Nd in recently erupted MORBs on the crest of the Mid-Atlantic Ridge between latitudes of 50° and 30° N. The $^{87}Sr/^{86}Sr$ ratios were corrected for fractionation to $^{86}Sr/^{88}Sr = 0.119400$ and adjusted to 0.70800 for the E&A Sr isotope standard (Sections 5.2a and 5.2b). The $^{143}Nd/^{144}Nd$ ratios were corrected to $^{146}Nd/^{144}Nd = 0.721903$ and were adjusted to 0.511835 for the LaJolla Nd isotope standard (Sections 9.2a and 9.2c). Data from Yu et al. (1997).

of the chondritic undifferentiated reservoir (CHUR) defined for the isotopic evolution of Nd in the mantle of the Earth (Section 9.2b). Since

$$\left(\frac{^{143}Nd}{^{144}Nd}\right)^0_{CHUR} = 0.512638$$

equation 17.1 yields

$$\left(\frac{^{87}Sr}{^{86}Sr}\right)^0_{CHUR} = 0.70437$$

Other examples of the mantle array considered by DePaolo (1988) yielded a slightly higher value:

$$\left(\frac{^{87}Sr}{^{86}Sr}\right)^0_{CHUR} = 0.7045$$

Therefore, the isotopic evolution of Sr in the undifferentiated mantle is constrained by the equation

$$\left(\frac{^{87}Sr}{^{86}Sr}\right)^0_{CHUR} = \left(\frac{^{87}Sr}{^{86}Sr}\right)^T_{CHUR} + \left(\frac{^{87}Rb}{^{86}Sr}\right)^0_{CHUR} \times (e^{\lambda T} - 1) \qquad (17.2)$$

where

$$\left(\frac{^{87}\text{Sr}}{^{86}\text{Sr}}\right)^0_{\text{CHUR}} = 0.7045$$

$$\left(\frac{^{87}\text{Sr}}{^{86}\text{Sr}}\right)^T_{\text{CHUR}} = 0.69899 \quad \text{(basaltic achondrites)}$$

$$\left(\frac{^{87}\text{Rb}}{^{86}\text{Sr}}\right)^0_{\text{CHUR}} = 0.0827$$

$$T = 4.55 \times 10^9 \text{y}$$

Equation 17.2 can be used to calculate ε-Sr values and model dates as outlined in Section 9.2 for the isotope geochemistry of Nd.

17.4 BASALT AND RHYOLITE OF ICELAND

Iceland is located astride the Mid-Atlantic Ridge in Figure 17.2 between about 66° and 63° N latitude. The volcanic rocks on Iceland consist primarily of tholeiite basalt but also include rhyolite and alkali-rich basalts. The oldest exposed rocks at Breidadalsheidi in northwestern Iceland crystallized at 16.0 ± 0.3 Ma (middle Miocene) according to whole-rock K–Ar dates reported by Moorbath et al. (1968). The origin of the lava flows of Iceland has been studied intensively for many years based on their chemical compositions, trace-element distributions, and the isotopic compositions of Sr, Nd, Pb, and O (Schilling et al., 1982; Faure, 2001).

17.4a Iceland and the Reykjanes Ridge

The Reykjanes Ridge in Figure 17.2 is a segment of the Mid-Atlantic Ridge that enters Iceland from the south along the Reykjanes Peninsula. The ^{87}Sr/^{86}Sr ratios in Figure 17.5 reported by Hart et al. (1973) clearly show the transition from the MORBs of the Reykjanes Ridge (average $^{87}\text{Sr}/^{86}\text{Sr} = 0.7027 \pm$

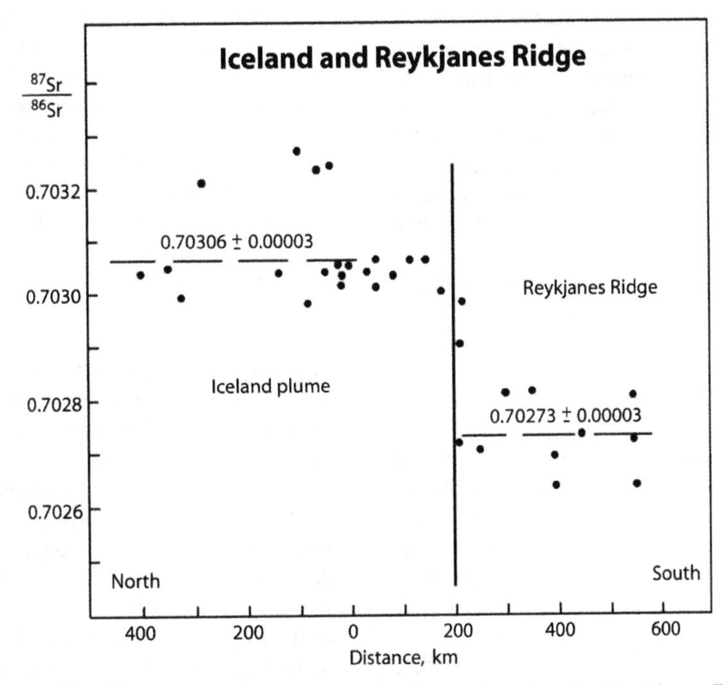

FIGURE 17.5 Profile of ^{87}Sr/^{86}Sr ratios of volcanic rocks on Iceland and on the Reykjanes Ridge south of the island. All ^{87}Sr/^{86}Sr ratios are relative to 0.70800 for E&A. The data show that the Iceland plume extends about 200 km south from the tip of the Reykjanes Peninsula. Data from Hart et al. (1973).

0.00003) to the volcanic rocks derived from the Iceland plume (average $^{87}Sr/^{86}Sr = 0.70306 \pm 0.00003$). This transition is quite abrupt and occurs about 200 km south of the tip of the Reykjanes Peninsula.

The Sr and Nd isotope ratios reported by Hémond et al. (1993) for a large number of volcanic rocks on Iceland define a mixing hyperbola in Figure 17.6. The diagram has been divided into four quadrants by reference to the present $^{87}Sr/^{86}Sr$ and $^{143}Nd/^{144}Nd$ ratios of CHUR. The rocks in quadrant II (low $^{87}Sr/^{86}Sr$ and high $^{143}Nd/^{144}Nd$ ratios) originate from mantle-derived magmas, whereas rocks of the continental crust occur in quadrant IV (high $^{87}Sr/^{86}Sr$ and low $^{143}Nd/^{144}Nd$ ratios). Therefore, mixtures of mantle-derived and crustal magmas form Sr–Nd isotope hyperbolas that extend from quadrant II to quadrant IV. Quadrants I and III involve combinations of $^{87}Sr/^{86}Sr$ and $^{143}Nd/^{144}Nd$ ratios that are only rarely realized because they are inconsistent with the geochemical properties of the Rb–Sr and Sm–Nd couples.

The data points in Figure 17.6 represent a variety of rock types of Iceland and form a segment of an isotopic mixing hyperbola that is confined to quadrant II and does not extend to quadrant IV. Therefore, the isotopic compositions of Sr and Nd imply that Iceland is not underlain by Precambrian basement rocks. Instead, the inverse correlation of the $^{87}Sr/^{86}Sr$ and $^{143}Nd/^{144}Nd$ ratios is evidence that the volcanic rocks on Iceland are mixtures of two components both of which occur within the plume above which the island has formed.

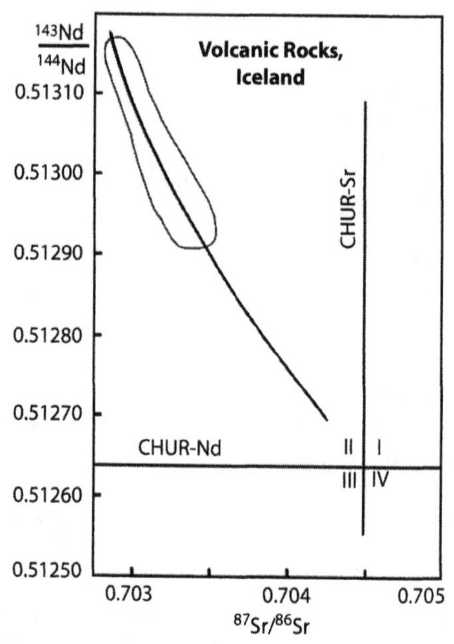

FIGURE 17.6 Strontium–neodymium isotopic mixing hyperbola formed by Late Tertiary volcanic rocks on Iceland. The diagram has been subdivided into four quadrants by the $^{87}Sr/^{86}Sr$ and $^{143}Nd/^{144}Nd$ ratios of CHUR. The existence of the data array is evidence that the Iceland plume contains two components having different $^{87}Sr/^{86}Sr$ and $^{143}Nd/^{144}Nd$ ratios. All isotope ratios were corrected for fractionation and were adjusted to $^{87}Sr/^{86}Sr = 0.71025$ for NBS 987 and to $^{143}Nd/^{144}Nd = 0.511837$ for the LaJolla Nd standard. Data from Hémond et al. (1993).

17.4b Lead in Iceland Basalt

The difference between the Iceland plume and the Mid-Atlantic Ridge is also apparent in the isotope ratios of Pb and of other elements that have radiogenic isotopes. The isotope ratios of these elements can be combined with the isotope ratios of Sr and Nd to achieve a multidimensional view of the petrogenesis of igneous rocks.

The $^{206}Pb/^{204}Pb$, $^{207}Pb/^{204}Pb$, and $^{208}Pb/^{204}Pb$ ratios of basalts along the crest of the Reykjanes Ridge in Figure 17.7 vary systematically with distance from the southern tip of the Reykjanes Peninsula. The $^{206}Pb/^{204}Pb$ and $^{208}Pb/^{204}Pb$ ratios decrease stepwise from high values in the plume-derived basalts of Iceland to low values in the MORBs of the Reykjanes Ridge, whereas the $^{207}Pb/^{204}Pb$ ratios vary only between narrow limits. All three profiles in Figure 17.7 confirm that the boundary between the Iceland plume and the depleted mantle underlying the Reykjanes Ridge is located about 200 km south of the southern tip of the Reykjanes Peninsula as deduced previously by Hart et al. (1973) from the variation of $^{87}Sr/^{86}Sr$ ratios in Figure 17.5.

The transition from plume-derived rocks to MORBs along the Reykjanes Ridge appears to

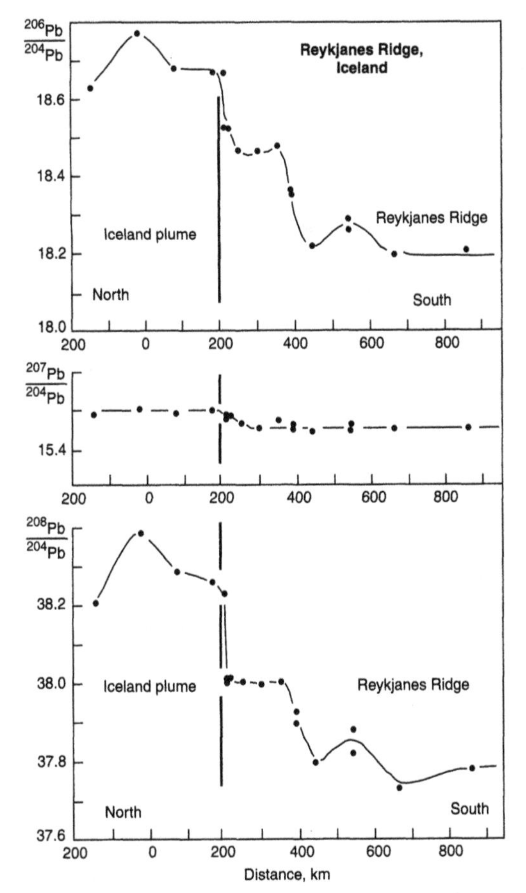

FIGURE 17.7 Variation of the isotope ratios of Pb in basalt as a function of distance along the Reykjanes Ridge measured from the southern tip of the Reykjanes Peninsula. The boundary between the Iceland plume and the depleted mantle of the Reykjanes Ridge occurs at about 200 km south, as in Figure 17.5. Data from Sun et al. (1975).

be accompanied by mixing of different isotopic components of Pb. In that case, the isotope ratios of Pb should be linearly related to the reciprocals of the concentrations of Pb in accordance with equation 16.15 in Section 16.2.

The data of Sun et al. (1975) in Figure 17.8 confirm the mixing hypothesis by forming straight lines in coordinates of the ^{206}Pb/^{204}Pb (as well as ^{208}Pb/^{204}Pb) ratios and the reciprocal Pb concentrations. The equations of these mixing lines are

$$\frac{^{206}\text{Pb}}{^{204}\text{Pb}} = 18.969 - 0.1339\left(\frac{1}{\text{Pb}}\right) \quad (17.3)$$

$$\frac{^{208}\text{Pb}}{^{204}\text{Pb}} = 38.580 - 0.1442\left(\frac{1}{\text{Pb}}\right) \quad (17.4)$$

These equations were replotted in Figure 17.8 as hyperbolas in coordinates of isotope ratios of Pb versus the concentrations of Pb. The orientation of the hyperbolas (convex upward) indicates that the magma derived from the Iceland plume has higher concentrations and higher isotope ratios of Pb than the magmas that originated from the depleted mantle beneath the Reykjanes Ridge.

The strong evidence for mixing of two kinds of Pb in the basalts of the Reykjanes Ridge presented by Sun et al. (1975) strengthened the plume hypothesis and thus contributed significantly to its acceptance by the scientific community. The mixing of the two isotopic varieties of Pb probably occurred during partial melting of mechanical mixtures of blocks of plume rocks embedded within the depleted rocks of the lithospheric mantle. However, mixing of two magmas containing the different isotopic varieties of Pb is not excluded by the data.

The ^{207}Pb/^{204}Pb ratios of basalts of Iceland were not included in Figure 17.8 because of the narrow range of variation of this ratio (15.471–15.558) reported by Sun and Jahn (1975). Nevertheless, the data in Figure 17.9 reveal that the ^{206}Pb/^{204}Pb and ^{207}Pb/^{204}Pb ratios of basalts on Iceland and the Reykjanes Ridge are positively correlated and define a single mixing line whose equation is

$$\left(\frac{^{207}\text{Pb}}{^{204}\text{Pb}}\right) = 13.7973 + 0.09075\left(\frac{^{206}\text{Pb}}{^{204}\text{Pb}}\right) \quad (17.5)$$

with a linear correlation coefficient $r^2 = 0.9309$ for 36 data sets.

This line is not a Pb–Pb isochron but a mixing line, which means that its slope is determined by the difference in the isotope ratios of Pb of the magma sources. Nevertheless, the slope of the mixing line yields a date of 1.44 Ga by interpolation in Table 10.3. This date, although fictitious, conveys

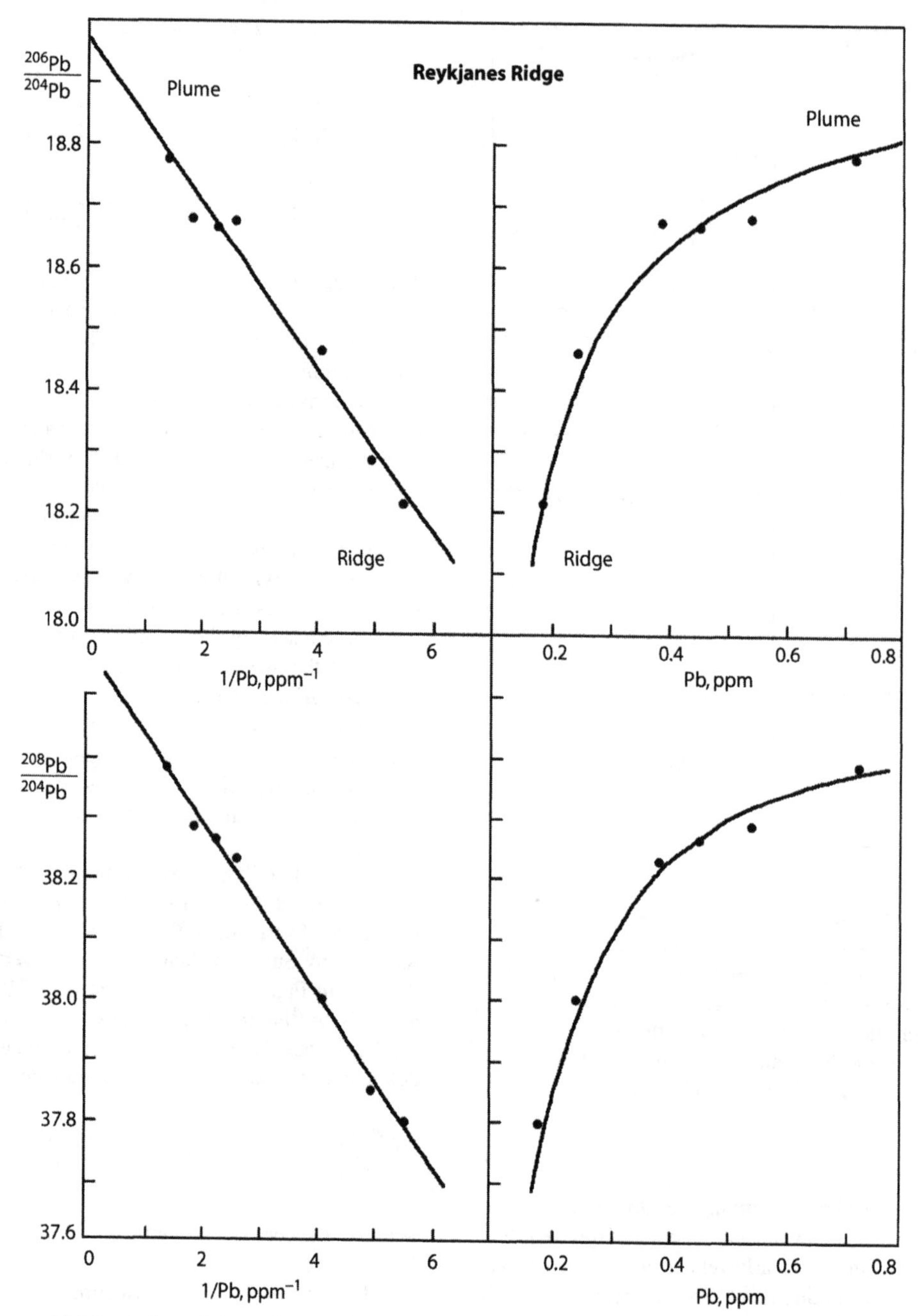

FIGURE 17.8 Evidence for mixing of two components having different isotope compositions and concentrations of Pb in basalt of the Reykjanes Ridge (Figure 17.2). The basalts derived from the Iceland plume have higher $^{206}Pb/^{204}Pb$ and $^{208}Pb/^{204}Pb$ ratios and higher concentrations of Pb than the basalts that originated from the depleted magma sources that underlie the Reykjanes Ridge. Data from Sun et al. (1975).

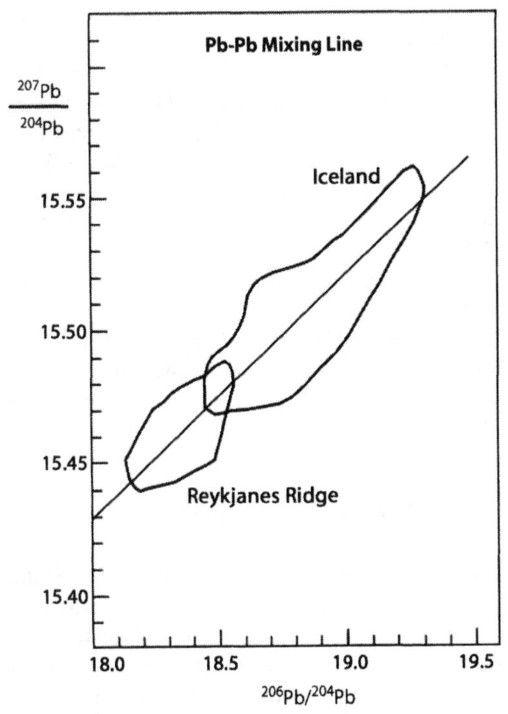

FIGURE 17.9 Isotopic mixing line defined by the $^{206}Pb/^{204}Pb$ and $^{207}Pb/^{204}Pb$ ratios of MORBs along Reykjanes Ridge and of basalts on Iceland. This mixing line is not a Pb–Pb isochron and yields a fictitious date of 1.44 Ga. Data from Sun et al. (1975) and Sun and Jahn (1975).

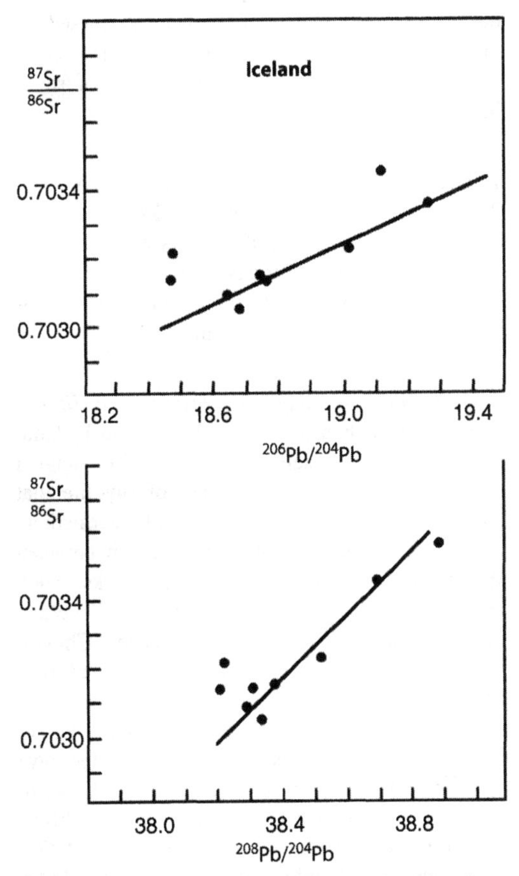

FIGURE 17.10 Isotopic mixing lines defined by the isotope ratios of Pb and Sr of post-Pleistocene basalts on Iceland. Data from Sun and Jahn (1975).

a sense of the duration of the incubation time of the U–Pb system in the rocks of the Iceland plume.

The isotope ratios of Sr and Pb in postglacial basalt on Iceland are linearly correlated in Figure 17.10 (Sun and Jahn, 1975). These data therefore demonstrate that the magma sources of the Iceland plume are heterogeneous with respect to the isotope compositions of Sr and Pb. In addition, the quasilinear correlations of the isotope ratios of Sr and Nd in Figure 17.6 and Sr and Pb in Figure 17.10 imply that the Sr/Nd and Sr/Pb ratios of the two principal components are similar. In that case, the Sr–Nd and Sr–Pb isotopic mixing lines are straight, in accordance with equation 16.19 (Langmuir et al., 1978).

The isotope ratios of Sr, Nd, and Pb of basalts dredged from the Kolbeinsey Ridge north of Iceland (Mertz et al., 1991) vary only within narrow limits, which indicates that the Iceland plume did not contribute to the formation of the MORBs on Kolbeinsey Ridge, in contrast to the Reykjanes Ridge south of Iceland (Figure 17.5).

17.4c Origin of Rhyolites

Seven of the 84 volcanic rocks analyzed by Hémond et al. (1993) in Figure 17.6 are rhyolites, whose $^{87}Sr/^{86}Sr$ and $^{143}Nd/^{144}Nd$ ratios are indistinguishable from those of the tholeiite basalts. This fact raises an interesting question concerning the origin of rhyolites that has been discussed by many petrologists (e.g., Yoder, 1973). The competing hypotheses concerning the origin of rhyolite include

1. fractional crystallization of basalt magma,
2. melting of granitic basement rocks,
3. partial melting of basaltic rocks at the base of the pile of volcanic rocks, and
4. partial melting of felsic differentiates (plagiogranites) of basaltic magma.

These hypotheses apply to the *origin of rhyolites in general* rather than just to the rhyolites of Iceland.

Fractional crystallization of basaltic magma can produce only small volumes of rhyolite that should be associated with a suite of intermediate rocks that bridge the compositional gap between the rhyolite and the parental mafic rocks. Such transitional rocks are generally absent on Iceland and in most other rhyolite occurrences (Chayes, 1963; Sigurdsson, 1977). The formation of rhyolite on Iceland by melting of granitic basement rocks was discredited by Moorbath and Walker (1965), who demonstrated that the rhyolites have low $^{87}Sr/^{86}Sr$ ratios similar to those of the associated basalt and that these ratios are very different from those of typical granitic gneisses of Precambrian age. Therefore, these authors concluded that Iceland is not underlain by granitic basement rocks. More recently, Sigurdsson (1977) suggested that the rhyolites formed by partial melting of plagiogranite that had previously formed by fractional crystallization of basalt magma at depth in the volcanic edifice.

An attractive alternative suggested by Moorbath and Walker (1965) and elaborated by O'Nions and Grönvold (1973) is that the rhyolites of Iceland formed by partial melting of mafic igneous rocks of Tertiary age. Accordingly, the rhyolite magma on Iceland originated by partial melting of the underlying oceanic crust and/or of basalt that had crystallized from mantle-derived magmas a short time before being remelted. In either case, the isotope compositions of Sr and Nd of the rhyolites are similar to those of the associated basalt because the mafic rocks have low Rb/Sr and Sm/Nd ratios and because the time between crystallization of the basalt magma and remelting to form rhyolite was short.

A detailed study by Nicholson et al. (1991) of the lava flows on the volcano Krafla (olivine tholeiite to rhyolite) in northern Iceland indicates that they are depleted in ^{18}O ($\delta^{18}O = +4.5$ to $+1.0‰$), which provides strong evidence that the magmas interacted with, or formed from, hydrothermally altered mafic rocks of the oceanic crust underlying Krafla. For example, a rhyolite of Pleistocene age on Krafla (MgO = 0.04%) has $^{87}Sr/^{86}Sr = 0.70324$ and $^{143}Nd/^{144}Nd = 0.51305$, both of which are identical to the isotope ratios (0.70326 and 0.51303) of a basalt (MgO = 9.49%) of the same age as the rhyolite.

17.4d History of the Iceland Plume

The Iceland plume arose from depth in the mantle in Late Cretaceous time and caused rifting of the overlying lithosphere which ultimately led to the opening of the North Atlantic Ocean (Figure 17.2). The interaction of the Iceland plume with the lithosphere also caused magma formation by decompression melting, which resulted in the eruption of the basalt flows and the intrusion of the mafic plutons of the North Atlantic igneous province. Different parts of this province were subsequently displaced from their original positions above the Iceland plume to their present locations, ranging from Cape Dyer on the east coast of Baffin Island (O'Nions and Clarke, 1972) to the west coast of Norway (Prestvik et al., 1999). The Atlantic igneous province also includes the Tertiary basalts of the Inner Hebrides of Scotland, the Faeroe Islands, the Antrim Plateau of Northern Ireland, the Skaergaard intrusion and related dikes and lava flows of East Greenland, as well as the basalts of Disko Island in West Greenland. The Sr–Nd isotopic mixing arrays of the igneous rocks of the North Atlantic igneous province extend into quadrant IV, indicating that the mantle-derived basalt magmas assimilated Sr and Nd from the rocks of the continental crust (Faure, 2001).

The effect of the Iceland plume on the opening of the North Atlantic Ocean and on the petrogenesis and age of the Tertiary igneous rocks of this region was the topic of the Second Arthur Holmes

European Research Meeting in Reykjavik, Iceland, in 1994. The proceedings of this conference, edited by White and Morton (1995), contain a paper by Thirlwall (1995) on the isotope composition of Pb in the Iceland plume. In addition, Taylor et al. (1997) discussed the widespread influence of the Iceland plume as recorded by the isotope compositions of Sr, Nd, Pb, and He in Tertiary basalts of the North Atlantic region.

The Iceland plume is one of several powerful plumes that continue to intrude the lithosphere from below, causing uplift, rifting, and volcanic activity on the surface of the Earth. These so-called superplumes include the Tristan de Cunha plume in the South Atlantic Ocean, the Hawaiian plume in the Pacific Ocean, the Kerguelen plume in the Indian Ocean, and the Afar plume in the Red Sea. The isotopic ratios of Sr, Nd, and Pb of the igneous rocks associated with these plumes were summarized by Faure (2001).

17.5 THE HAWAIIAN ISLANDS

The Hawaiian Islands (Figure 17.11) consist of two parallel chains of volcanoes that formed above the head of a large plume and were subsequently displaced from it by the northwestward movement of the Pacific plate. The chain of Hawaiian Islands is continued by the submerged volcanoes of the Hawaiian Ridge and by the Emperor Seamounts for a total distance of more than 6000 km. The rocks of Suiko seamount at the far end of the Emperor chain have a K–Ar date 64.7 Ma, which implies an average rate of movement of the Pacific plate of 9.3 cm per year (Faure, 2001).

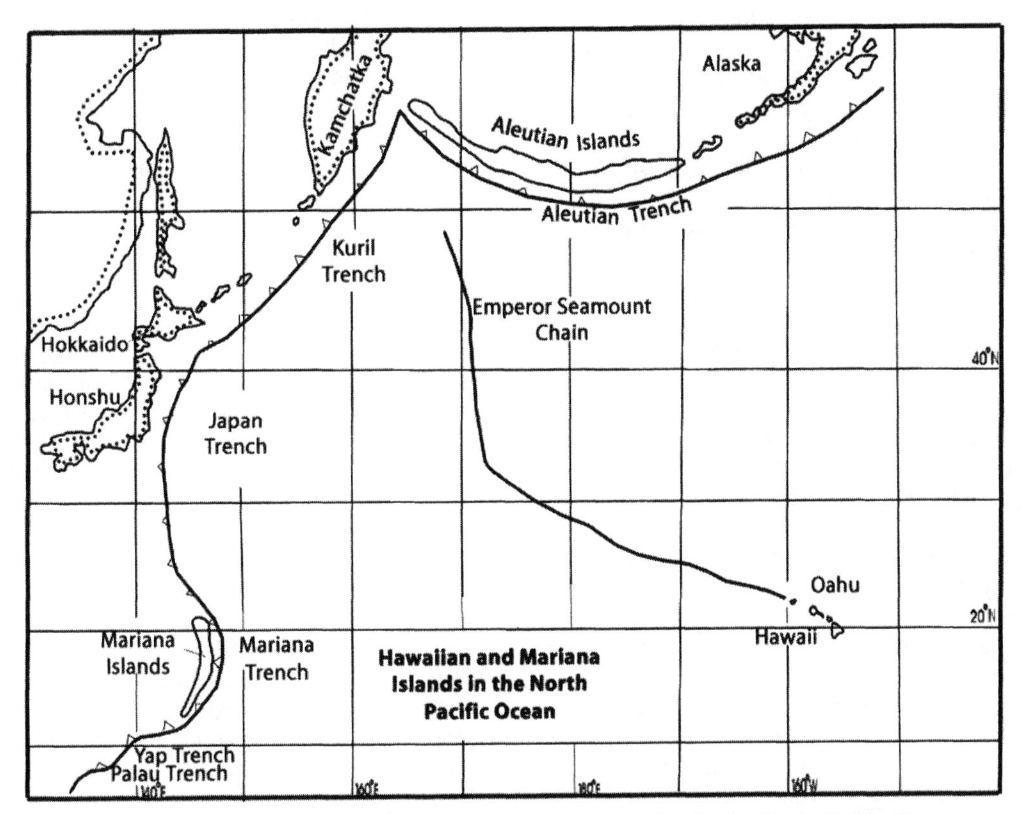

FIGURE 17.11 The Hawaiian Islands and the system of deep-sea trenches in the North Pacific Ocean. Adapted from *Pacific Ocean*, National Geographic Society, Washington, D.C., 1962.

Similar chains of volcanic islands and atolls occur in Polynesia in the South Pacific Ocean. These island chains are also associated with plumes that continue to cause volcanic activity on the seafloor at the southeastern ends of the archipelagos, including the Marquesas, Tuamotu, Society, Austral, and Cook Islands. The volcanic rocks of the islands of Tubuai, Mangaia, and Rurutu in the Austral chain have high $^{206}Pb/^{204}Pb$ ratios but low $^{87}Sr/^{86}Sr$ ratios characteristic of the HIMU component (Chauvel et al., 1992).

The volcanic activity on the Hawaiian Islands and the island chains of the South Pacific in many cases progressed through three stages:

- posterosional nephelinites (youngest),
- caldera-filling alkali volcanics, and
- shield-building tholeiites (oldest).

The volcanic rocks that formed during these sequential stages differ in their isotope compositions of Sr, Nd, and Pb as well as in major-element and trace-element compositions. The process that caused the time-dependent compositional changes of the volcanic rocks on the Hawaiian Islands was reconstructed by Chen and Frey (1985) and has been investigated by many other scientists identified by Faure (2001).

17.5a Isotopic Mixtures of Sr, Nd, and Pb

In general, the shield-building tholeiites of the Hawaiian Islands have high $^{87}Sr/^{86}Sr$ ratios ranging from 0.70310 to 0.70461 with an average of 0.70379. The caldera-filling alkali basalts have lower $^{87}Sr/^{86}Sr$ ratios between 0.70305 and 0.70440, yielding an average of 0.70342. The youngest nepheline and melilite basalts have the lowest $^{87}Sr/^{86}Sr$ ratios, extending from 0.70270 to 0.70357 with a mean of 0.70324. The progressive stratigraphic *decrease* of the average $^{87}Sr/^{86}Sr$ ratios was accompanied by an *increase* in the average Rb concentrations from 5.2 ppm in the tholeiites to 22.4 ppm in the alkali basalts and to 23.5 ppm in the nepheline-bearing lavas. The Sr concentrations also increased from 378 to 765 to 844 ppm in the same sequence (Faure, 2001).

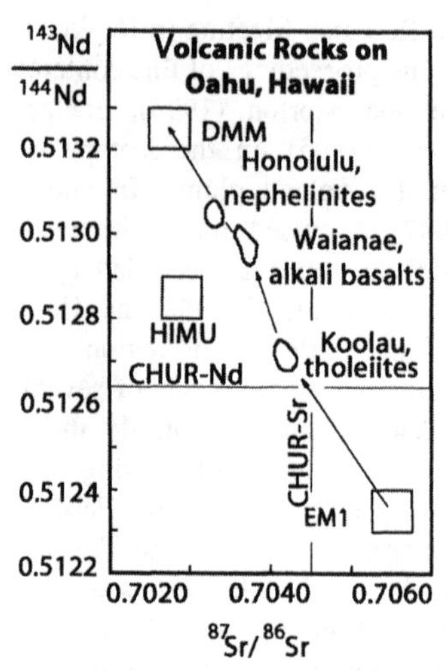

FIGURE 17.12 Isotope ratios of Sr and Nd of the principal volcanic series on the island of Oahu, Hawaii, in relation to the magma sources DMM, EM1, and HIMU identified in Table 17.1. The $^{87}Sr/^{86}Sr$ and $^{143}Nd/^{144}Nd$ ratios were corrected to 0.71025 for NBS-987 (Sr) and to 0.511860 for the LaJolla Nd standard. The measured value of the $^{143}Nd/^{144}Nd$ ratio of BCR-1 was 0.51262 ± 0.00001. Data from Stille et al. (1983).

The isotope ratios of Sr and Nd in Figure 17.12 illustrate the stratigraphic variation of these parameters in the volcanic rocks of the island of Oahu as reported by Stille et al. (1983). The data points form three clusters that are aligned between the EM1 and DMM magma sources defined in Table 17.1. The data clusters bypass HIMU, indicating that this component did not participate in the formation of the lava flows on Oahu. The distribution of data clusters in Figure 17.12 implies that the shield-building tholeiites of the Koolau Series originated primarily from the EM1 component, which is presumably contained within the Hawaiian plume. The overlying caldera-filling alkali basalts of the Waianae Series originated from mixed sources consisting of EM1 (plume rocks) and DMM (depleted lithospheric mantle). Surprisingly, the most alkali-rich rocks of the Pleistocene Honolulu Nephelinite Series originated primarily from the depleted lithospheric mantle. The elevated alkali-metal concentrations of the nephelinites resulted from a low degree of partial melting rather than by previous enrichment of the magma source in alkali elements or by fractional crystallization of basalt magma. This conclusion is justified by the fact that Rb enrichment of the magma source would have caused its $^{87}Sr/^{86}Sr$ ratio to increase with time and would have moved the nephelinites of the Honolulu Series off the EM1–DMM mixing line.

The process implied by the isotopic data in Figure 17.12 was related by Chen and Frey (1985) to the movement of the Pacific plate over the stationary Hawaiian plume. According to this theory, the shield-building tholeiites on each of the Hawaiian Islands formed when it was situated directly above the plume. At this stage, magma formed by decompression melting of plume rocks (14% melt fraction) and adjacent rocks of the depleted lithospheric mantle (about 1% melt fraction). As the movement of the plate began to displace the shield volcano away from its magma source, less plume heat was available for melting, volcanic eruptions became less frequent, and more of the magma that was produced originated from the lithospheric mantle. Hence, the $^{87}Sr/^{86}Sr$ ratios of the caldera-filling alkali basalts are lower than those of the tholeiites and their $^{143}Nd/^{144}Nd$ ratios are higher. Ultimately, small volumes of nepheline basalts formed primarily by low degrees of partial melting of the depleted lithosphere with only minor contributions from the plume.

The petrogenesis of basalts on Oahu, outlined above on the basis of their $^{87}Sr/^{86}Sr$ and $^{143}Nd/^{144}Nd$ ratios, is strongly corroborated by their $^{87}Sr/^{86}Sr$ and $^{206}Pb/^{204}Pb$ ratios in the mixing polygon in Figure 17.13. The alignment of the data fields confirms that the magmas originated from EM1 (plume) and DMM (lithospheric mantle) in changing proportions and that HIMU and EM2 did not contribute significantly to the formation of magma.

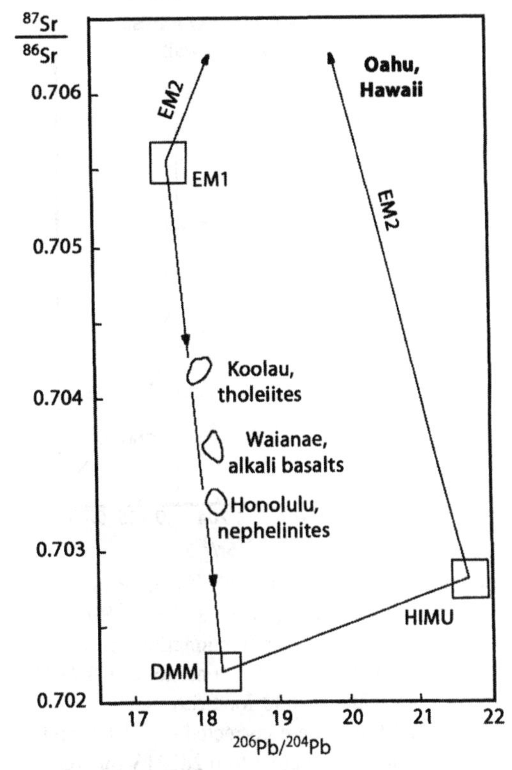

FIGURE 17.13 Isotope ratios of Sr and Pb of volcanic rocks on Oahu of the Hawaiian Islands. The placement of the data fields within the Sr–Pb mixing polygon confirms that the magmas originated from mixtures of EM1 and DMM and not from HIMU and EM2. Data from Stille et al. (1983).

17.5b Hafnium in Basalt of Oahu

Stille et al. (1983) also measured the $^{177}Hf/^{176}Hf$ ratios of the volcanic rocks of Oahu, and Stille et al. (1986) subsequently analyzed basalts on several of the other Hawaiian Islands. The $^{176}Hf/^{177}Hf$ and $^{87}Sr/^{86}Sr$ ratios of the basalts on Oahu define three data fields in Figure 17.14 to which a straight line was fitted by least-squares regression:

$$\frac{^{176}Hf}{^{177}Hf} = 0.445898 - 0.231263\frac{^{87}Sr}{^{86}Sr} \qquad (17.6)$$

This equation was used to estimate the $^{176}Hf/^{177}Hf$ ratios of DMM and EM1 based on their $^{87}Sr/^{86}Sr$ ratios listed in Table 17.1. The resulting $^{176}Hf/^{177}Hf$ ratios (0.28351 for DMM and 0.28274 for EM1) are in good agreement with the $^{176}Hf/^{177}Hf$ ratios of these components obtained by Salters and Hart (1991).

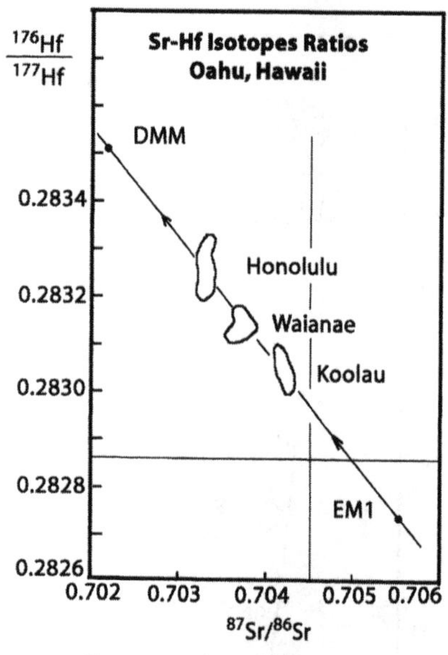

FIGURE 17.14 Isotope ratios of Hf and Sr of volcanic rocks on Oahu, Hawaii. The $^{87}Sr/^{86}Sr$ ratios were corrected for fractionation to $^{86}Sr/^{88}Sr = 0.1194$ and are consistent of 0.71025 for NBS 987 and 0.7080 for E&A. The $^{176}Hf/^{177}Hf$ ratios were corrected to $^{179}Hf/^{177}Hf = 0.7325$ and are relative to 0.282213 for the JMC-475 Hf isotope standard. The $^{176}Hf/^{177}Hf$ ratios of the magma source components were calculated from equation 17.6 for $^{87}Sr/^{86}Sr$ ratios of 0.7022 (DMM) and 0.7055 (EM1). Data from Stille et al. (1983, 1986).

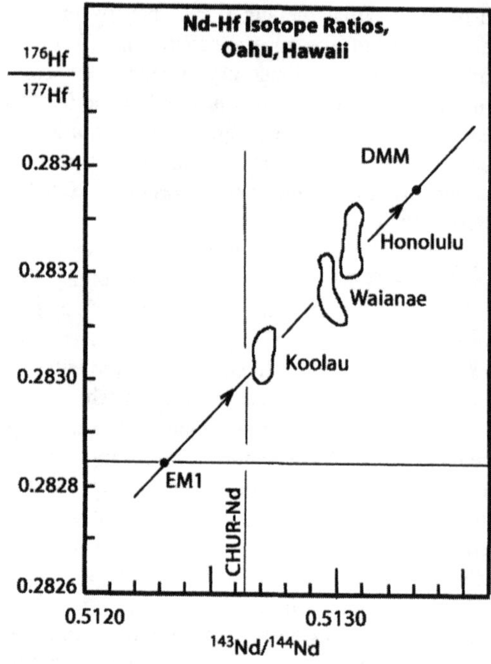

FIGURE 17.15 Isotope ratios of Hf and Nd in volcanic rocks on Oahu, Hawaii. The Nd isotope ratios were corrected for fractionation to $^{146}Nd/^{144}Nd = 0.7219$ and the $^{143}Nd/^{144}Nd$ ratios of the basalts were adjusted to $^{143}Nd/^{144}Nd = 0.511860$ for the LaJolla Nd isotope standard and yielded 0.51262 ± 0.00001 for BCR-1. The standardization of the Hf data is indicated in the caption to Figure 17.14. Data from Stille et al. (1983, 1986).

The isotope ratios of Nd and Hf in the volcanic rocks of Oahu in Figure 17.15 are positively correlated as expressed by the equation

$$\frac{^{176}Hf}{^{177}Hf} = 0.075086 \pm 0.405632\frac{^{143}Nd}{^{144}Nd} \qquad (17.7)$$

This equation yields a second set of estimates for the $^{176}Hf/^{177}Hf$ ratios of DMM and EM1, listed in Table 17.2. The $^{176}Hf/^{177}Hf$ ratios of these components derived from the Sr–Hf and Nd–Hf isotope correlations are in good agreement. The combined values are as follows:

$$\text{DMM}: \quad \frac{^{176}Hf}{^{177}Hf} = 0.28343 \pm 0.00008$$

$$\text{EM1}: \quad \frac{^{176}Hf}{^{177}Hf} = 0.28279 \pm 0.00006$$

The isotope ratios of Hf in EM2 and HIMU cannot be determined from Hawaiian basalts because these components did not contribute to their petrogenesis.

The positive correlation of the isotope ratios of Nd and Hf in the Hawaiian basalts in Figure 17.15 implies that the Sm/Nd and Lu/Hf ratios of DMM are both greater than those of the undifferentiated mantle (i.e., CHUR) because Lu as well as Sm are retained in the restite fraction during partial melting in the mantle. Consequently, basalt flows derived from depleted magma sources in the mantle

Table 17.2. Isotope Composition of Hf in the Magma-Source Components of the Mantle Based on Volcanic Rocks on the Hawaiian Islands

Component	Sr–Hf[a]	Nd–Hf[a]	Combined[a]
DMM	0.28351	0.28335	0.28343 ± 0.00008
EM1	0.28274	0.28285	0.28279 ± 0.00006

Source: Stille et al., 1983, 1986.

[a]The $^{176}Hf/^{177}Hf$ ratios of the magma components were estimated from their $^{87}Sr/^{86}Sr$ and $^{143}Nd/^{144}Nd$ ratios based on the linear correlations of the isotope ratios of Sr and Nd with those of Hf. The resulting values of the $^{176}Hf/^{177}Hf$ ratios were combined by averaging.

have higher $^{143}Nd/^{144}Nd$ and $^{176}Hf/^{177}Hf$ ratios than plume-derived volcanic rocks.

In summary, the isotope ratios of Sr, Nd, Pb, and Hf of basalts on Oahu are consistent with the stratigraphy of the lava flows and thereby support the theory of Chen and Frey (1985). The concurrent evolution of the isotope compositions of Hf and Nd in the mantle and crust of the Earth has been treated elsewhere:

Blichert-Toft and Albarède, 1999, *Geophys. Res. Lett.*, 26(7):935–938. Stevenson and Patchett, 1990, *Geochim. Cosmochim. Acta*, 54:1683–1697. Patchett, 1983, *Geochim. Cosmochim. Acta*, 47:81–91. Patchett et al., 1981, *Contrib. Mineral. Petrol.*, 78:279–297.

17.5c Osmium in Hawaiian Basalt

The $^{187}Os/^{188}Os$ ratios of tholeiite basalts that were extruded at various volcanic centers on the Hawaiian Islands during the past two million years are related to the isotope ratios of Sr, Nd, and Pb of these rocks (Martin, 1991; Bennett et al., 1996; Hauri et al., 1996). The $^{87}Sr/^{86}Sr$ and $^{187}Os/^{188}Os$ ratios form a quasilinear array located within the polygon of mixing in Figure 17.16 defined by the magma-source components in Table 17.1.

All of these tholeiites formed during the shield-building phase of volcanic activity when magma originated primarily from the plume (EM1) with lesser amounts from the depleted lithospheric mantle (DMM). The data points in Figure 17.16 define a straight line:

$$\frac{^{187}Os}{^{188}Os} = -11.0271 + 15.8624\frac{^{87}Sr}{^{86}Sr} \qquad (17.8)$$

The evident quasilinear correlation of the isotope ratios of Sr and Os confirms that the tholeiite magmas originated primarily from two components in the mantle (EM1 and DMM), as previously recognized on the basis of isotope ratios of Sr, Nd, Pb, and Hf (Figures 17.12–17.15). However, the Sr–Os data do not fit the model as well as the isotope ratios of the other elements. The misfit may be caused by

1. nonlinearity of the EM1–DMM mixing line in Sr–Os coordinates,
2. inappropriate choice of the $^{187}Os/^{188}Os$ ratios of EM1 and DMM,
3. presence of additional components in the magma sources in the mantle beneath the Hawaiian Islands, and
4. time-dependent changes in the isotope composition of Os in magmas generated during the past two million years.

Another explanation of this anomaly was suggested by Basu and Faggart (1998), who pointed out that Hawaiian volcanic rocks deviate from the Sr–Nd isotopic mantle array in a way that suggests that the DMM component underlying Hawaii has an $^{87}Sr/^{86}Sr$ ratio of 0.7030 as a result of seawater alteration. Such alteration would not have changed the isotope compositions of Nd and Os because of the low concentrations of these elements in seawater.

The $^{206}Pb/^{204}Pb$ and $^{187}Os/^{188}Os$ ratios of the same Hawaiian tholeiites in Figure 17.17 also form a straight line whose equation is

$$\frac{^{187}Os}{^{188}Os} = 0.516947 - 0.020918\frac{^{206}Pb}{^{204}Pb} \qquad (17.9)$$

The misalignment of the data in these coordinates is confined to the DMM component whose $^{206}Pb/^{204}Pb$ ratio may have been increased by seawater alteration from 18.2 to 18.75, in accordance with the hypothesis of Basu and Faggart (1998).

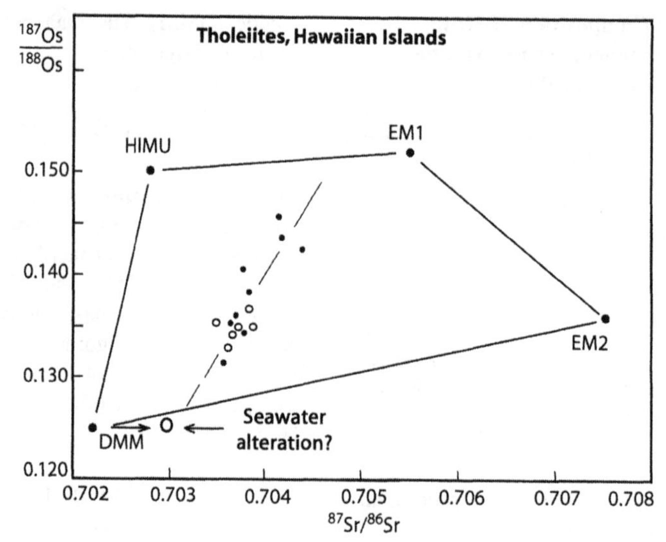

FIGURE 17.16 Correlation of $^{87}Sr/^{86}Sr$ and $^{187}Os/^{188}Os$ ratios of tholeiites erupted by various volcanoes on the Hawaiian Islands. The $^{187}Os/^{188}Os$ ratios were corrected for decay of ^{187}Re when required by the age of rock samples. Open circles represent core samples from Mauna Loa near Hilo (Hauri et al., 1996). Solid circles are samples from Koolau, Kahoolawe, Kohala, Hualalai, Mauna Loa, Loihi, and Kilauea (Bennett et al., 1996). The $^{87}Sr/^{86}Sr$ ratios reported by Hauri et al. (1996) are relative to 0.71025 for the NBS-987 Sr isotope standard. The compositions of the magma source components are from Hart (1988) and Shirey and Walker (1998).

In general, the isotope ratios of Sr, Pb, and Os in Figures 17.16 and 17.17 confirm that the Hawaiian magmas originated from the EM1 and DMM components in varying proportions. These components probably encompass material having a range of isotopic compositions of Sr, Nd, Pb, Hf, and Os, which caused the isotope ratios of the elements in the resulting volcanic rocks to scatter above and below isotopic-mixing hyperbolas.

The petrogenesis of the Hawaiian basalts continues to be investigated by means of their Os isotope ratios as exemplified by the following contributions:

Martin et al., 1994, *Earth Planet. Sci. Lett.*, 128:287–301. Hart and Ravizza, 1998, in Basu and Hart (Eds.), *Geophys. Monogr.*, 95:123–134. Okano and Tatsumoto, 1998, in Basu and Hart (Eds.), *Geophys. Monogr.*, 95:135–147. Chen et al., 1998, in Basu and Hart (Eds.), *Geophys. Monogr.*, 95:161–181.

In addition, the Hawaii Scientific Drilling Project recovered a 1060-m core near Hilo, Hawaii, and the results of the subsequent investigations were introduced by Stolper et al. (1996) in issue B5 of volume 101 of the *Journal of Geophysical Research*.

17.6 HIMU MAGMA SOURCES OF POLYNESIA

The volcanic rocks on some of the Austral-Cook islands in the South Pacific have low $^{87}Sr/^{86}Sr$ and high $^{206}Pb/^{204}Pb$ ratios, which characterizes the HIMU magma source in the mantle (Hart, 1988). The islands of this chain increase in age in a northwesterly direction and are assumed to have formed by volcanic activity caused by the

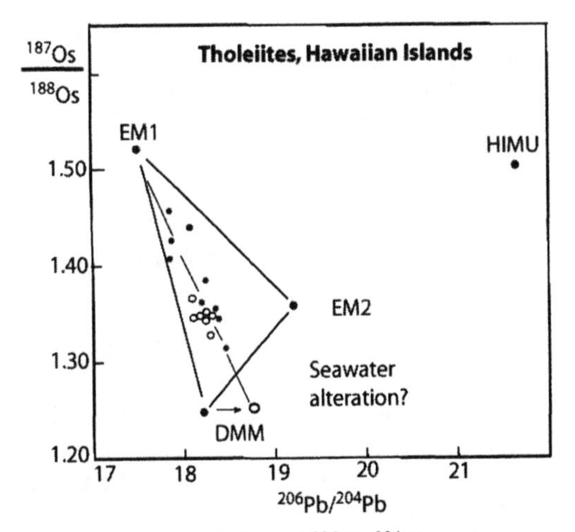

FIGURE 17.17 Correlation of $^{206}Pb/^{204}Pb$ and $^{187}Os/^{188}Os$ ratios of tholeiites erupted on Hawaiian volcanoes identified in the caption to Figure 17.16. Solid circles: Bennett et al. (1996); open circles: Hauri et al. (1996). Mantle components from Hart (1988) and Shirey and Walker (1998).

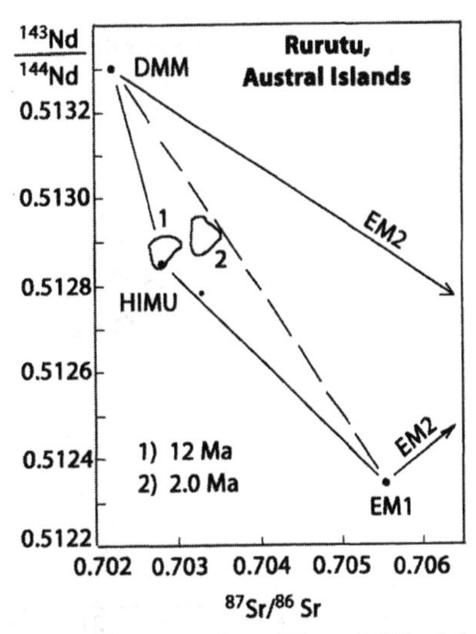

FIGURE 17.18 Isotope ratios of Sr and Nd of the two suites of volcanic rocks on the island of Rurutu, Austral Islands, Polynesia. The compositions of the mantle components are from Table 17.1 (Hart, 1988). The $^{87}Sr/^{86}Sr$ ratios are relative to 0.71025 for NBS 987. The $^{143}Nd/^{144}Nd$ ratio of the LaJolla Nd isotope standard was 0.511850. Data from Chauvel et al. (1997).

presence of a plume located at the southeastern end of the Austral chain where the submarine volcano Macdonald continues to erupt (Chauvel et al., 1997). The principal islands of the Austral chain include Marotiri, Rapa, Tubuai, and Rurutu. The colinear chain of the Cook Islands contains Mangaia, Rarotonga, and Aitutaki, to name only a few of the islands in this chain.

The volcanic rocks of Rurutu are composed of a sequence of basalt flows (about 12 Ma) overlain by alkali-rich basalts consisting of hawaiites, basanite, and nephelinites (about 2.0 Ma). The isotope ratios of Sr and Nd of the old suite of basalt in Figure 17.18 cluster close to the HIMU component, whereas the younger alkali-rich rocks can be viewed as three-component mixtures of magmas derived from the EM1–DMM–HIMU components (Chauvel et al., 1997).

The $^{87}Sr/^{86}Sr$ and $^{206}Pb/^{204}Pb$ ratios of the volcanic rocks of Rurutu form two clusters of data points in Figure 17.19 such that the old suite (12 Ma) lies close to HIMU and the young volcanics (2.0 Ma) are displaced from HIMU toward EM1. In other words, the isotope ratios of

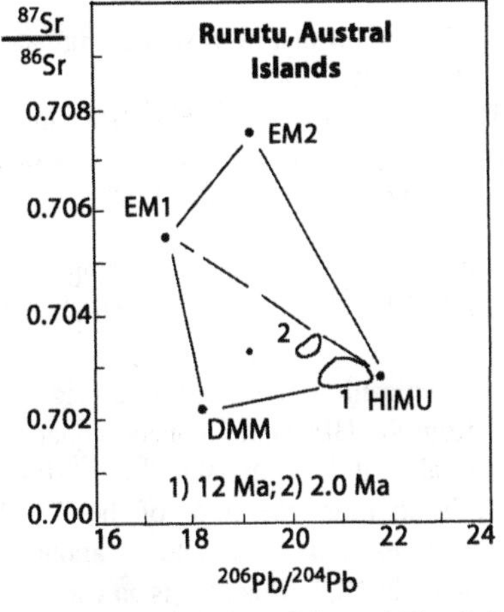

FIGURE 17.19 Isotope ratios of Sr and Pb of the volcanic rocks on Rurutu in the Austral Islands. NBS 987: $^{87}Sr/^{86}Sr = 0.71025$; NBS 981: $^{206}Pb/^{204}Pb = 16.9373$. The isotope ratios of the magma sources are from Table 17.1, after Hart (1988). Data from Chauvel et al. (1997).

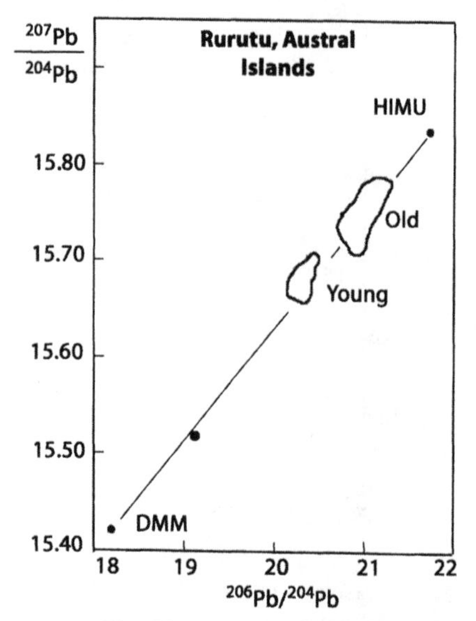

FIGURE 17.20 The Pb–Pb mixing line defined by the volcanic rocks of Rurutu, Austral Islands, Polynesia. The equation of this line provides estimates of the value of the $^{207}Pb/^{204}Pb$ ratio of HIMU and of the incubation time during which the radiogenic ^{206}Pb and ^{207}Pb accumulated. Data by Chauvel et al. (1997).

Sr and Pb confirm that both suites originated from HIMU, EM1, and DMM.

The $^{206}Pb/^{204}Pb$ and $^{207}Pb/^{204}Pb$ ratios of the old and young suites of volcanic rocks of Rurutu define a straight line in Figure 17.20:

$$\frac{^{207}Pb}{^{204}Pb} = 13.3185 + 0.115887 \frac{^{206}Pb}{^{204}Pb} \qquad (17.10)$$

Since the volcanic rocks on Rurutu originated primarily from the HIMU component, equation 17.10 yields a value of 15.83 for the $^{207}Pb/^{204}Pb$ ratio of HIMU. In addition, the slope of the Pb–Pb mixing line corresponds to a date of about 1.89 Ga based on Table 10.3. This date is an estimate of the approximate incubation time of the HIMU component in the South Pacific.

The isotope ratios of Sr, Pb, and Os of basalt on the islands of Tubuai, Rurutu, Mangaia, and Rarotonga of the Austral-Cook chain of islands in Figures 17.21*a* and *b* demonstrate that the magmas originated primarily from the HIMU and DMM components (Hauri and Hart, 1993). The same authors also demonstrated that basalt on the island of Savaii of Western Samoa were derived in part from EM2. The petrogenesis of the volcanic rocks on many other islands in the Atlantic, Pacific, and Indian Oceans was discussed by Faure (2001) based on the isotope ratios of Sr, Nd, and Pb.

17.7 SUBDUCTION ZONES

Subduction zones in the oceans cause the development of arcuate chains of volcanic islands exemplified by the Mariana Islands in the Pacific Ocean. The Mariana Island arc in Figure 17.11 is part of an extensive system of subduction zones in the western Pacific Ocean along which the Pacific plate is being subducted into the mantle. The Mariana Trench, located east of the Mariana Islands, extends south into the Yap Trench and merges northward into the Izu Trench and hence into the Japan Trench, which joins with the Kuril Trench off the coast of the Japanese island of Hokkaido. The Kuril Trench turns northeast and connects with the Aleutian Trench off the coast of Kamchatka, thereby extending this system of subduction zones to a total length of about 15,850 km. Evidently, this and the other systems of deep-sea trenches in the Pacific and Indian Oceans are major topographic features of the Earth comparable in magnitude to midocean spreading ridges and to the mountain ranges on the continents.

The volcanic rocks that are erupted on island arcs are characterized by andesite but also include a wide range of lithologies from tholeiite basalt to rhyolite. Rhyolites and ignimbrites as well as plutonic granitic rocks occur where oceanic crust is being subducted under continental crust, as for example in Japan and New Zealand and along the west coasts of North and South America. In addition, alkali-rich rocks occur where magmas form by small degrees of melting in the greatest depth in the mantle behind subduction zones (e.g., in the Sunda arc of Indonesia and in the Aleutian arc).

The volcanic rocks on island arcs form by partial melting of the lithospheric mantle located above subduction zones. Melting is promoted by the release of water from the marine sediment and from hydrothermally altered basalt of the subducted

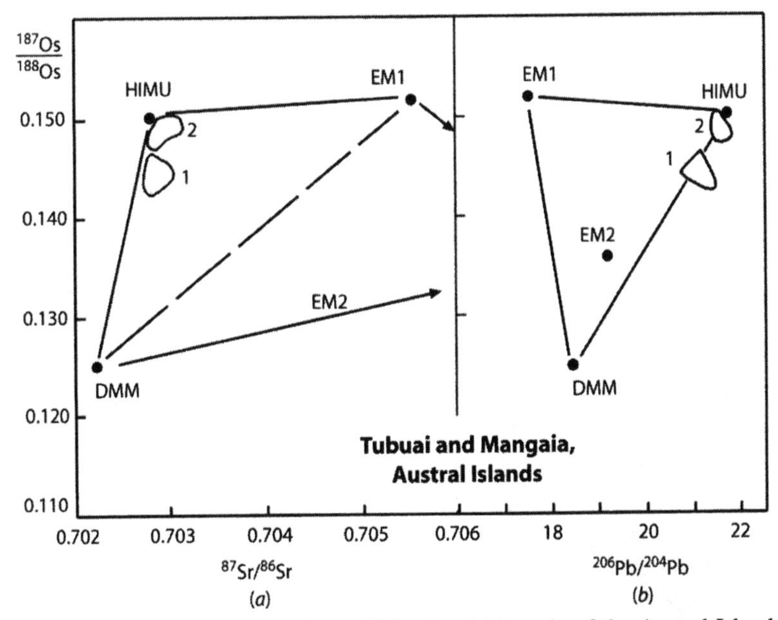

FIGURE 17.21 Petrogenesis of the volcanic rocks on Tubuai and Mangaia of the Austral Islands in Polynesia. The $^{87}Sr/^{86}Sr$ ratios were adjusted to 0.71025 for NBS 987 and the $^{143}Nd/^{144}Nd$ ratios were corrected for fractionation to $^{146}Nd/^{144}Nd = 0.7219$ and were subsequently adjusted to 0.511860 for the LaJolla Nd isotope standard. Additional information about analytical methods was provided by Hauri and Hart (1993), from whom these data were taken.

oceanic crust. The aqueous fluid that is expelled by dehydration of the down-going slab transports alkali metals, alkaline earths, rare-earth elements, and other elements into the rocks of the mantle wedge. Consequently, the isotope compositions of Sr, Nd, Pb, Hf, and Os of volcanic rocks on island arcs differ from those in the lithospheric mantle and vary regionally and stratigraphically depending on local conditions within the subduction zones. In some cases, the isotope ratios of the above-mentioned elements in volcanic rocks of island arcs resemble those of lavas on oceanic islands. Such coincidences are fortuitous and do not imply that volcanic rocks on island arcs form by the same petrogenetic processes as volcanic rocks on oceanic islands.

17.7a Mariana Island Arc

The Mariana Islands in Figure 17.11 extend north–south for a distance of about 1140 km between 12° and 20° N latitude and between 140° and 150 °E longitude in the western Pacific Ocean. The islands in the southern part of the arc, which include Guam, Rota, Saipan, and Tinian, are no longer volcanically active. The northern islands form the active part of the arc and include Anatahan, Sarigan, Guguan, Alamagan, Pagan, Agrigan, and many other islands (Ito and Stern, 1985/1986).

The volcanic activity along the Mariana arc occurred primarily between 35 and 24 Ma, 18 and 11 Ma, and from 6 Ma to Recent (Lee et al., 1995). The volcanic rocks on the Mariana Islands consist predominantly of basalt with lesser amounts of andesite, dacite, and shoshonite (Meijer, 1976; Stern, 1978, 1979; Stern et al., 1988). The submarine lavas that erupted in the back-arc basin west of the arc consist primarily of MORB-like tholeiite basalt (Gribble et al., 1998).

The $^{87}Sr/^{86}Sr$ ratios of the volcanic rocks on the northern Mariana Islands in Figure 17.22 range only within narrow limits from 0.70332 to

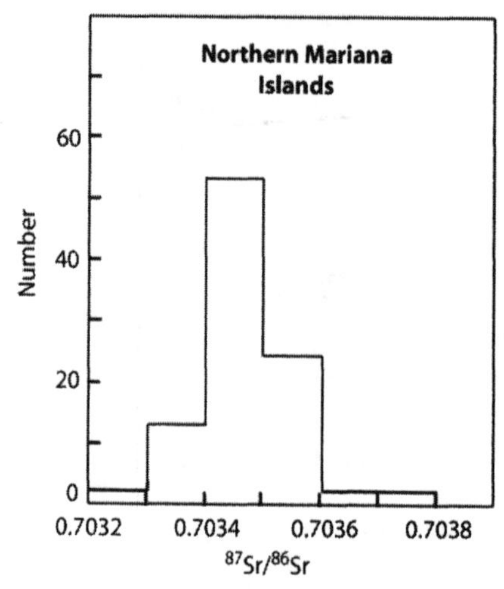

FIGURE 17.22 Isotope ratios of Sr of lava flows on the northern islands of the Mariana Island arc in the western Pacific Ocean. All $^{87}Sr/^{86}Sr$ ratios were corrected for isotope fractionation to $^{86}Sr/^{88}Sr = 0.11940$ and were adjusted to $^{87}Sr/^{86}Sr = 0.70800$ for E&A. Fifty-six percent of the 95 specimens included in the diagram have $^{87}Sr/^{86}Sr$ ratios between 0.70340 and 0.70349. Data from Meijer (1976), Woodhead and Fraser (1985), and Ito and Stern (1985/1986).

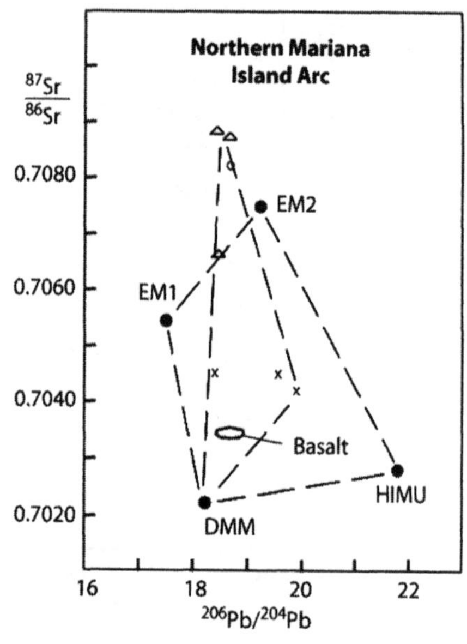

FIGURE 17.23 Isotope ratios of Pb and Sr of Late Tertiary basalt and andesite on the islands of Anatahan, Sarigan, Guguan, Alamagan, Pagan, Arigan, Asuncion, Maug, and Uracas in the northern sector of the Mariana Island arc in the western Pacific Ocean. The $^{87}Sr/^{86}Sr$ ratios were adjusted to 0.70800 for the E&A Sr isotope standard. The $^{206}Pb/^{204}Pb$ ratios were standardized by analyses of the NBS 981 Pb isotope standard. Open circles: sediment, Mariana basin; crosses: volcaniclastic sediment; open triangles: sediment, Narau basin. Data from Woodhead and Fraser (1985).

0.70394 with a mean of 0.70347 ± 0.00002 ($2\overline{\sigma}$) for 95 specimens. This value is greater than the $^{87}Sr/^{86}Sr$ ratio of normal MORBs along midocean ridges and therefore greater than the $^{87}Sr/^{86}Sr$ ratio of the depleted mantle of the lithosphere. However, the $^{87}Sr/^{86}Sr$ ratios of volcanic rocks on the northern Mariana Islands are similar to those of basalt on Oahu, Hawaii (Figure 17.12), and similar to the older suite of basalt on the island of Rurutu (Figure 17.18). This similarity was caused by the transfer of Sr from the subducted oceanic crust into the overlying wedge of the depleted lithospheric mantle.

The petrogenesis of the volcanic rocks on the northern Mariana Islands is illustrated in Figure 17.23 by the isotope ratios of Sr and Pb (Woodhead and Fraser, 1985). The data points of basalt on these islands form a small data field located within a polygon defined by the isotope ratios of DMM and of volcanogenic and pelagic marine sediment collected near the Mariana Islands. In other words, the volcanic rocks on the northern islands originated from depleted lithospheric mantle to which had been added Sr and Pb derived from the sediment component of the oceanic crust of the Pacific plate that is being subducted into the Mariana Trench. These and other elements are transferred by the aqueous fluid that forms during dehydration of the down-going slab. The water that is added to the mantle wedge by this process acts as a flux and causes magma to form by partial melting.

Figure 17.23 also contains the components that contribute to the formation of magma under oceanic islands (e.g., Iceland, the Azores, the

Hawaiian Islands, and the island chains of Polynesia). Although the isotopic data field of the basalts on the Mariana Islands is located within the polygon formed by these components, only DMM is relevant to the petrogenesis of the volcanic rocks in this and other oceanic subduction zones. The other components, in this case, are merely reference points. Nevertheless, the placement of data points representing the sediment component of the oceanic crust near the Mariana Islands is compatible with the isotope ratios of Sr and Pb in EM1, EM2, and HIMU. Consequently, the data in Figure 17.23 confirm the identification of these components as derivatives of subducted oceanic crust with varying amounts of volcanogenic and terrigenous sediment.

17.7b Andes of South America

The Andes Mountains (Figure 17.24) have formed as a result of volcanic activity and uplift caused by the eastward subduction of the Nazca plate under the continental crust of South America. The thickness of the continental crust underlying the Andes ranges from less than 30 km at the southern tip of South America to about 70 km in the Central Volcanic Zone encompassing southern Peru, southwestern Bolivia, northern Chile, and northwestern Argentina. The ages of the volcanic and plutonic rocks in this segment of the Andes decrease from about 200 Ma at the coast to less than 20 Ma about 550 km inland. The initial $^{87}Sr/^{86}Sr$ ratios of the Mesozoic igneous rocks near the coast range from about 0.7040 to 0.7060 but rise to 0.7080 in rocks of Cenozoic to Recent age farther inland (McNutt et al., 1975).

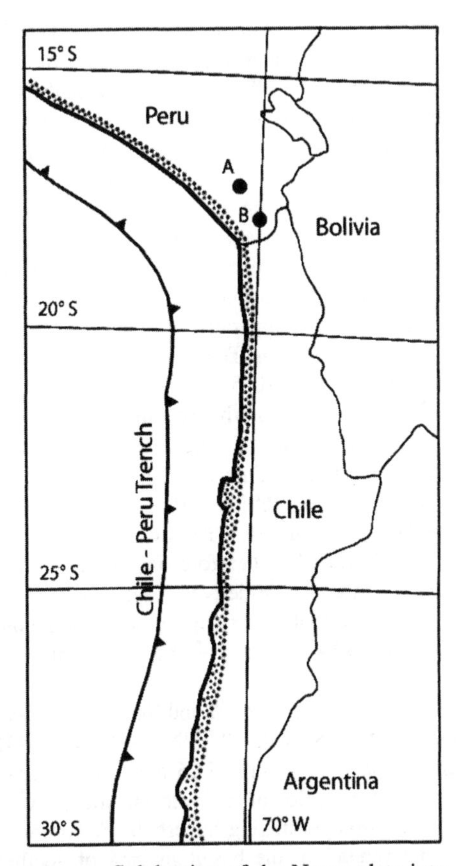

FIGURE 17.24 Subduction of the Nazca plate into the Chile–Peru Trench along the west coast of South America. Adapted from Garver, J. B., Jr., J. F. Shupe, J. F. Dorr, D. G. A. M. Payne, and M. B. Hunsiker (Eds.), *Atlas of the Word*, 6th ed., map 45, National Geographic Society, Washington, D.C., 1990.

The eastward migration of the volcanic activity in the Central Volcanic Zone of the Andes was caused both by the westward motion of the South American plate and by the easterly spreading direction of the seafloor east of the East Pacific Rise. The increase of the initial $^{87}Sr/^{86}Sr$ ratios of the igneous rocks correlates with the increasing thickness of the continental crust through which mantle-derived magmas must travel to reach the surface. Consequently, the chemical and isotope compositions of the igneous rocks are affected both by the metasomatic alteration of the magma sources in the subcontinental mantle and by the interaction of the basaltic magma with the granitic rocks of the continental crust. Moreover, heat transported by the mantle-derived magmas caused dacitic and rhyolitic magmas to form by partial melting of the granitic rocks in the continental crust. These magmas erupted explosively from shallow crustal magma chambers and deposited thick and widespread layers of ash-flow tuffs (ignimbrites) on the surface.

The petrogenesis of volcanic rocks in the Andes was the subject of a special symposium by sponsored by the American Geophysical Union in 1983 (Harmon and Barreiro, 1984). In addition, Pitcher et al. (1985) edited a book on the magmatic activity of the Andes in Peru. A summary of the origin of igneous rocks in the Andes of South America was published by Faure (2001).

The isotope ratios of Sr and Nd of volcanic rocks of Late Tertiary age at the Arequipa and Barroso volcanic centers of southern Peru (identified in Figure 17.24) plot in quadrant IV in Figure 17.25 because of their comparatively high $^{87}Sr/^{86}Sr$ and low $^{143}Nd/^{144}Nd$ ratios. The $^{87}Sr/^{86}Sr$ ratios of these rocks range from 0.7055 to 0.7080 and far exceed the $^{87}Sr/^{86}Sr$ ratios of the volcanic rocks on the Mariana Islands, most of which vary narrowly from 0.7033 to 0.7036 in Figure 17.22. In addition, the $^{143}Nd/^{144}Nd$ ratios (0.51217–0.51242) of the Peruvian lavas are all less than the $^{143}Nd/^{144}Nd$ of CHUR (0.512638). Therefore, James (1982) concluded that the mantle-derived magmas of this district had assimilated varying amounts of granitic basement rocks exemplified by the local Charcani Gneiss.

The isotope ratios of Sr and Nd of the Charcani Gneiss ($^{87}Sr/^{86}Sr = 0.740$, $^{143}Nd/^{144}Nd = 0.5115$) define a point in Figure 17.25 that can be related to the isotope ratios of the volcanic rocks (A and B) by an isotopic mixing hyperbola. According to this interpretation, the basalt magma of southern Peru formed by partial melting of the subcontinental lithospheric mantle to which Sr and Nd had been added by the aqueous fluids expelled from the subducted oceanic crust. This magma subsequently assimilated granitic basement rocks (e.g., Charcani Gneiss) while it ponded in crustal magma chambers. The amount of crustal rocks that was assimilated by the magma depends on the initial temperature of the crustal rocks, on the temperature and volume of basalt magma, and on the amount of heat produced by the crystallization of minerals from the magma (e.g., plagioclase).

The isotope ratios of Sr and Pb of the lavas in southern Peru (Arequipa and Barroso) in Figure 17.26 diverge from the isotope ratios of Late Tertiary volcanic rocks in the Central Volcanic Zone of the Andes between latitudes 26°00′ S (Cerro Galan, northwestern Argentina) and 21°52′ S (San Pedro-San Pablo, southwestern Bolivia) reported by Harmon et al. (1984). The $^{206}Pb/^{204}Pb$

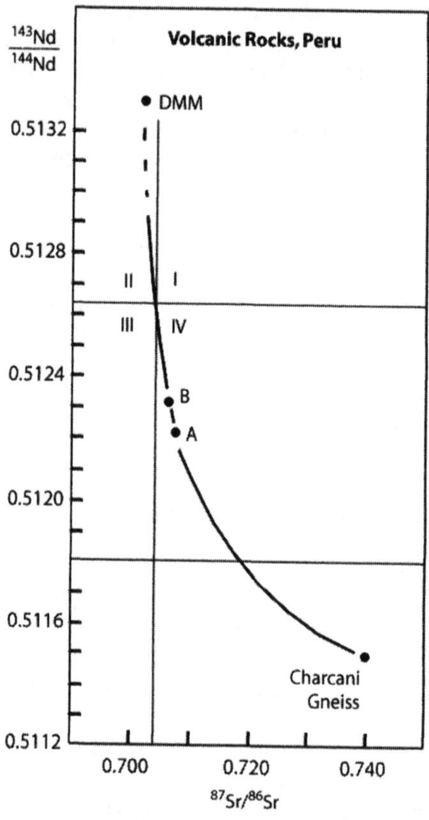

FIGURE 17.25 Average isotope ratios of Sr and Nd of Late Tertiary volcanic rocks at the Arequipa (A) and Barroso (B) volcanic centers of southern Peru in the Central Volcanic Zone of the Andes Mountains of South America. The $^{87}Sr/^{86}Sr$ ratios were adjusted to 0.7080 for the E&A isotope standard. The $^{143}Nd/^{144}Nd$ ratios were corrected to $^{146}Nd/^{144}Nd = 0.7219$ and were standardized to 0.51180 for the LaJolla Nd standard. The Sr and Nd isotope ratios of the volcanic rocks are attributable to assimilation of the local Charcani Gneiss by basalt magma derived from lithospheric mantle that may have been enriched in Sr and Nd by aqueous fluids emanating from the subducted Nazca plate. Data from James (1982).

ratios of the lavas at Arequipa and Barroso (Peru) decrease with increasing $^{87}Sr/^{86}Sr$ ratios, in agreement with magma contamination by Charcani Gneiss, whose $^{206}Pb/^{204}Pb$ ratio is 16.9 (James,

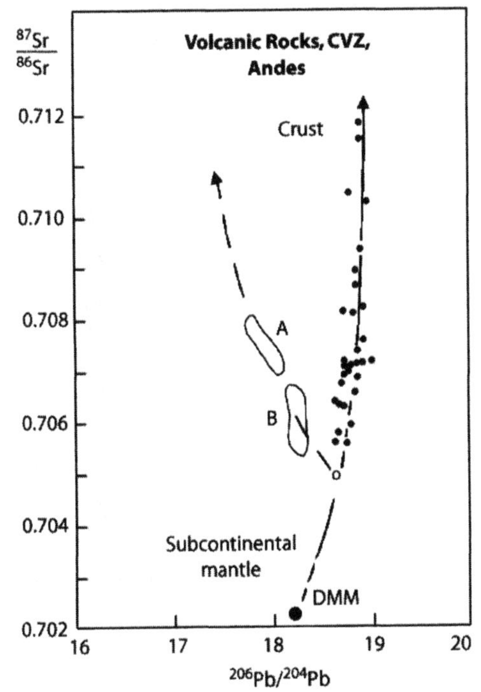

FIGURE 17.26 Isotope ratios of Pb and Sr of Late Tertiary volcanic rocks in the central Andes and at Arequipa (A) and Barroso (B) of southern Peru. The two data sets define diverging isotopic mixing hyperbolas presumably caused by the isotopic heterogeneity of Pb in the basement rocks underlying the central Andes. The two trends converge at $^{206}Pb/^{204}Pb = 18.6$ and $^{87}Sr/^{86}Sr = 0.7050$. Data from James (1982) and Harmon et al. 1984.

1982). The $^{206}Pb/^{204}Pb$ ratios of the volcanic rocks from other volcanic centers in the central Andes analyzed by Harmon et al. (1984) increase slightly from about 18.7 to 18.9. The difference in the isotope composition of the contaminant Pb is attributable to the isotopic heterogeneity of Pb in the continental crust underlying southern Peru and this segment of the Andes. The $^{87}Sr/^{86}Sr$ ratios of the lavas in the central Andes range widely from about 0.7056 to 0.7118 relative to 0.7080 for E&A.

These isotopic data are consistent with the petrogenetic model of James (1982) based on crustal contamination of basalt magmas that formed by partial melting of hydrothermally altered rocks of the mantle wedge above the subduction zone. Moreover, the least-contaminated magmas in southern Peru and elsewhere in the Central Volcanic Zone have similar $^{87}Sr/^{86}Sr$ ratios of about 0.7055. If the hypothetical Pb–Sr isotopic mixing line of the Peruvian lavas is extrapolated, as shown in Figure 17.26, it intersects the mixing line of the lavas in the central Andes at about $^{206}Pb/^{204}Pb = 18.6$ and $^{87}Sr/^{86}Sr = 0.7050$. These values may characterize the mantle-derived basalt magmas before they interacted with rocks of the continental crust of the central Andes.

17.7c Ignimbrites

Thick layers of ignimbrites of dacitic to rhyolitic composition occur throughout the Central Volcanic Zone of the Andes. The origin of these ignimbrites has been attributed to

1. crystallization of plagioclase from mantle-derived andesite magmas that had assimilated rocks of the lower continental crust (Thorpe et al., 1979; Francis et al., 1989; Siebel et al., 2001) and
2. partial melting of rocks in the lower continental crust by injection of mantle-derived mafic magmas followed by fractional crystallization (Francis et al., 1989; Hawkesworth et al., 1982).

Both explanations arise from the observation that the ignimbrites contain varying amounts of Sr, Nd, and Pb derived from the continental crust and that the contaminated magmas differentiated by crystallization of plagioclase and other minerals.

The isotope ratios of Sr and Nd in Figure 17.27 of ignimbrites at Salar de Antofalla (Argentina) and Salar de la Isla (Chile) reported by Siebel et al. (2001) are collinear with the Sr and Nd ratios of ignimbrites at Cerro Galán (Argentina) published by Francis et al. (1989). Both data sets define an isotopic mixing hyperbola extending toward the DMM (depleted mantle) component and the Charcani Gneiss of southern Peru (James, 1982), which here represents the rocks of the continental crust. All of the ignimbrite data points, like those of basalt lavas in Figure 17.26, plot in quadrant IV,

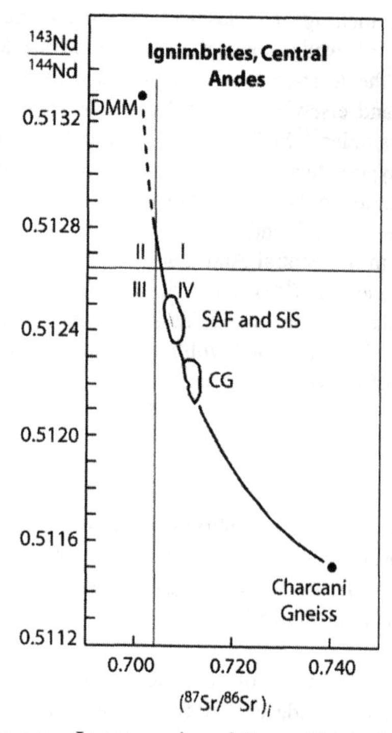

FIGURE 17.27 Isotope ratios of Sr and Nd in Late Tertiary dacitic ignimbrites in the Central Volcanic Zone of the Andes: SAF = Salar de Antofalla; CG = Cerro Gálan northwestern Argentina; SIS = Salar de la Isla, northern Chile. The data for SAF and SIS are relative to $^{87}Sr/^{86}Sr = 0.71025$ for NBS-987 and $^{143}Nd/^{144}Nd = 0.511910$ for the LaJolla Nd standard corrected for fractionation to $^{146}Nd/^{144}Nd = 0.7219$. The isotope ratios of the Charcani Gneiss are from James (1982). Data for SAF and SIS from Siebel et al. (2001) and for GC from Francis et al. (1989).

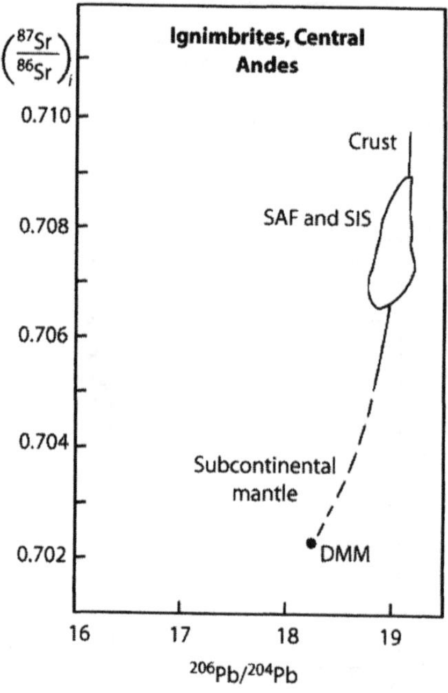

FIGURE 17.28 Isotope ratios of Sr and Pb of dacitic ignimbrites in the central Andes from Salar de Antofalla (SAF, northwestern Argentina) and Salar de la Isla (SIS, northern Chile). Data from Siebel et al. (2001).

demonstrating that the ignimbrite magmas contain significant amounts of Sr and Nd from the rocks of the continental crust. The isotopic data are therefore compatible with the hypothesis that the ignimbrites formed from contaminated magmas that differentiated by crystallization of plagioclase, which caused the magmas to become depleted in Sr (from 619 to 29.3 ppm) and enriched in Rb (from 44.1 to 224 ppm), as reported by Thorpe et al. (1979). In some cases, assimilation of crustal rocks and crystallization of plagioclase may have occurred simultaneously.

The $^{206}Pb/^{204}Pb$ ratios of dacitic ignimbrites at Salar de Antofalla (SAF, Argentina) and of Salar de la Isla (SIS, Chile) in Figure 17.28 range narrowly from 18.72 to 19.13 and correlate with increasing initial $^{87}Sr/^{86}Sr$ ratios (0.7066–0.7089), much like the andesites and basalts in Figure 17.26. The similarity of the isotope ratios of Sr and Pb of the basalts and andesites (Harmon et al., 1984) and of the ignimbrites (Siebel et al., 2001) in the central Andes supports the hypothesis that the ignimbrites originated from the same magmas that also formed andesite and basalt. The ignimbrites of intermediate (dacitic) composition in the central Andes were erupted explosively from shallow crustal magma chambers where mafic magmas could evolve chemically by crystallization of plagioclase and isotopically by assimilation of crustal rocks from the walls of the magma chamber. The rhyolite ignimbrites in the central Andes and elsewhere in Central and North America originated by differentiation of felsic magma that formed either by partial melting of mantle-derived basalt or andesite or by melting of granitic basement rocks (Faure, 2001).

17.8 CONTINENTAL FLOOD BASALT

Very large volumes of basalt magma can form by decompression melting of the ultramafic rocks in the lithospheric mantle. This phenomenon is exemplified by the basalt flows that are extruded along spreading ridges in the ocean (Section 17.3). Decompression melting also occurs in the lithospheric mantle beneath continents where rifts can form that may widen to become ocean basins. In such cases, the rifts result from the presence of large plumes that rise from depth in the mantle until they are stopped by the rigid lithospheric plate composed of continental crust and the underlying lithospheric mantle. As the initial rift widens, the basalt plateaus on opposite sides of the original rift are separated from each other and thus become located on the margins of the continents facing the ocean that has formed between them.

This application of the theory of plate tectonics explains the formation of basalt plateaus that occurred during the opening of the North Atlantic Ocean in late Mesozoic–early Cenozoic time. The basalt plateaus of Scotland, Ireland, and the Faeroe Islands and the basalts of Greenland on opposite sides of the Atlantic Ocean are all related to the Iceland plume, which caused the breakup of the supercontinent Laurasia and contributed to the subsequent opening of the North Atlantic Ocean. Other examples of this phenomenon are the basalt plateaus of Paraná, Brazil, and Namibia in southwest Africa as well as the Karoo basalt in South Africa and the Jurassic basalts of Queen Maud Land in Antarctica. The origin of these and other examples of continental basalt plateaus has been discussed by Faure (2001).

The subduction of oceanic crust and lithospheric mantle under continental crust (e.g., the Andes of South America) can cause extensional forces to act on the continental crust adjacent to the subduction zone. This phenomenon gives rise to continental basins that are the equivalent of back-arc basins in the oceans (e.g., west of the Mariana Island arc). The extension behind subduction zones causes thinning of the continental crust and decompression melting in the underlying lithospheric mantle. This tectono-magmatic process has allowed large basalt plateaus to form in the Transantarctic Mountains (Elliot, 1974) and in the states of Oregon and Washington of the USA.

17.8a Columbia River Basalt, United States

The basalt plateau of Oregon and Washington consists of about 0.3×10^6 km^3 of tholeiite basalt flows that were extruded through fissures and covered an area of 2×10^5 km^2 primarily between about 17 and 6.0 Ma. Some of the largest flows formed in a matter of only a few days. The plateau is about three kilometers thick and consists of flat-lying layers of tholeiite basalt that have been traced for up to 600 km in an east–west direction (Hooper, 1982). Most of the area of the Columbia River plateau is located east (i.e., behind) of the Cascade Range, composed of active volcanoes, including Mt. Rainier, Mt. St. Helens, Mt. Adams, Mt. Hood, and Mt. Jefferson, which derive magma from the subduction zone that was overridden by the northwesterly movement of the North American plate.

The Columbia River Basalt Group has been subdivided into five formations identified in Table 17.3. The basalt flows of the Grande Ronde Formation make up about 75% of the total volume and were erupted in about two million years

Table 17.3. Stratigraphy and K–Ar Dates of the Columbia River Basalt, Washington and Oregon, United States

Formation	K–Ar Date, Ma	Volume, % of Total
Saddle Mountain	13.5–6.0	1
Wanapum	14.5–13.5	5
Grande Ronde	16.5–14.5	75
Picture Gorge	15.9–15.3	9
Imnaha	17.0–16.5	10

Source: Hooper, 1982; Swanson et al., 1979.

at a high rate that reached 60,000 cubic kilometers per million years. Such high rates of eruption during relatively short intervals of time are typical of the development of continental basalt plateaus elsewhere in the world. The Saddle Mountain Formation contains the youngest flows (6.0–13.5 Ma) whose volume is only about 1% of the total. Nevertheless, the rocks of this formation have the most extreme isotope ratios of Sr and Nd.

The $^{87}Sr/^{86}Sr$ ratios of the basalt of the main body of the Columbia River Group (Imnaha to Wanapum, Table 17.3) range from 0.70351 to 0.70558, whereas those of the Saddle Mountain Formation (6.0–13.5 Ma) vary widely from 0.70354 to 0.71459 relative to 0.71025 for NBS-987 (Carlson et al., 1981). The $^{143}Nd/^{144}Nd$ ratios of the same samples also range widely and define a Sr–Nd isotopic mixing array in Figure 17.29. One crustal xenolith has a high $^{87}Sr/^{86}Sr$ ratio (0.76829) and a low $^{143}Nd/^{144}Nd$ ratio (0.511968) and yields model dates relative to CHUR of 1.42 Ga by Rb–Sr and 1.32 Ga by Sm–Nd (Carlson et al., 1981). These dates confirm the presence of Precambrian basement rocks beneath certain parts of the Tertiary basalt plateau of Washington and Oregon.

The isotope ratios of Sr and Nd of the basalt flows in Figure 17.29 define a Sr–Nd isotopic mixing hyperbola that extends from quadrant II into quadrant IV. The limited range of the isotope compositions of rocks representing the largest part of the volume of the basalt plateau indicates that the subcontinental mantle was either contaminated by hydrothermal fluids emanating from subducted oceanic crust or it was intrinsically heterogeneous with respect to the isotope compositions of Sr and Nd.

The enrichment of the basalt flows of the Saddle Mountain Formation in radiogenic ^{87}Sr and their low $^{143}Nd/^{144}Nd$ ratios are clear evidence that these magmas assimilated crustal rocks, presumably from the walls of crustal magma chambers. This conclusion concerning the petrogenesis of the basalt flows of the Saddle Mountain Formation raises the question whether the much more voluminous flows of the main body of the Columbia River basalt were also affected by this process. This topic was the subject of a discussion between DePaolo (1983) and Carlson et al. (1983). The resolution of this problem is complicated by the existence of regional

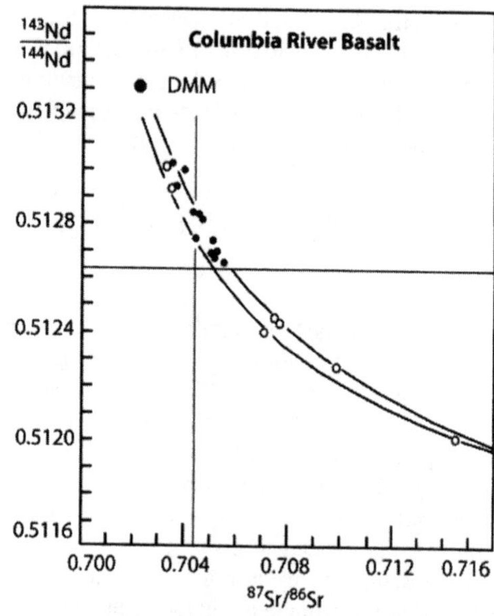

FIGURE 17.29 Isotope ratios of Sr and Nd of the Columbia River Basalt Group in Washington and Oregon, United States. The $^{87}Sr/^{86}Sr$ ratios were increased by 0.00011 to be compatible with 0.71025 for NBS 987. The $^{143}Nd/^{144}Nd$ ratios are relative to 0.511929 for the La Jolla standard. The open circles represent basalt of the Saddle Mountain Formation (13.5 to 6.0 Ma) which caps the section but makes up only about 1% of the total volume of the lavas flows (Hooper, 1982; Swanson et al., 1979). Data from Carlson et al. (1981).

variations of the isotope compositions of Sr, Nd, Pb, and O of the basalt flows. Consequently, the magmas appear to have been affected by several different processes whose influence varied regionally. For example, the presence of Precambrian basement rocks may explain why the basalt flows in the northern and eastern part of the plateau have higher $^{87}Sr/^{86}Sr$ ratios than those in the southern and western parts. In addition, the chemical and isotope compositions of magmas that pooled in crustal magma chambers may have been altered by simultaneous assimilation and fractional crystallization (AFC).

The effects of magma mixing or assimilation of crustal rocks and fractional crystallization are

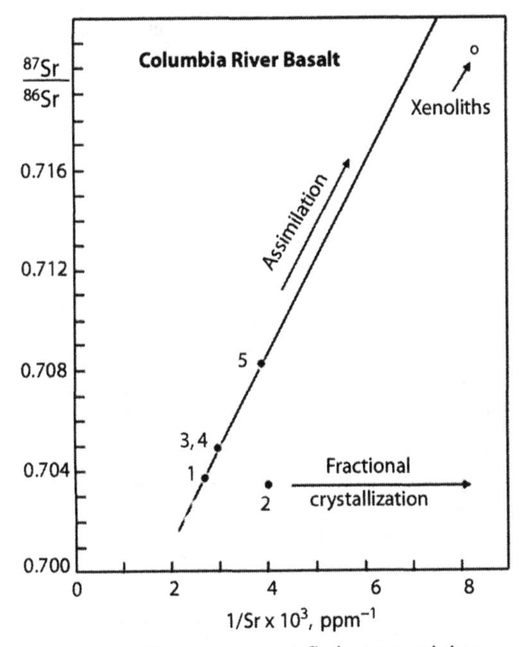

FIGURE 17.30 Two-component Sr-isotope mixing diagram based on equation 16.15 for the different formations of the Columbia River Basalt: 1 = Imnaha Formation; 2 = Picture Gorge Formation; 3 = Grande Ronde Formation; 4 = Wanapum Formation; 5 = Saddle Mountain Formation. The $^{87}Sr/^{86}Sr$ ratio of the crustal xenoliths (open circle) is the average of two samples weighted by their Sr concentrations. The arrows indicate the effects of fractional crystallization and assimilation. Data from Carlson et al. (1981) and Carlson (1984).

indicated in Figure 17.30 in coordinates of the average $^{87}Sr/^{86}Sr$ ratios and the reciprocals of the average Sr concentrations based on equation 16.15. The diagram demonstrates that the $^{87}Sr/^{86}Sr$ ratios and the reciprocals of the Sr concentrations define a straight line that passes close to a point representing two crustal xenoliths analyzed by Carlson et al. (1981) and Carlson (1984). The collinearity of these data points is evidence that the magmas of the Imnaha, Grande Ronde, Wanapum, and Saddle Mountain Formations assimilated small but increasing amounts of crustal rocks represented by the xenoliths. The distribution of the data points reveals little evidence of magma differentiation by fractional crystallization. Only the two samples of the Picture Gorge Formation analyzed by Carlson et al. (1981) deviate from the straight line in Figure 17.30 in a way that is compatible with the effects of crystallization of plagioclase, which decreases the Sr concentration of the residual magma without altering its $^{87}Sr/^{86}Sr$ ratio.

In summary, the magmas of the Columbia River basalt originated from the subcontinental mantle, which was heterogeneous with respect to the isotope ratios of Sr and Nd, because of the addition of these and other elements by aqueous fluids released by dehydration of subducted oceanic crust or because of prior alteration or both. The mantle-derived magmas subsequently pooled in crustal magma chambers where they assimilated varying amounts of Precambrian rocks and differentiated by fractional crystallization. *Assimilation* is especially prevalent in the rocks of the Saddle Mountain Formation, whereas rocks of the Picture Gorge Formation appear to show the effects of *fractional crystallization*.

The evidence in favor of crustal contamination of the mantle-derived basalt magma was strengthened by the isotope ratios of Os in Mg-rich tholeiite basalts from the Columbia River plateau, the adjacent Snake River plain, the Oregon plateau, and the Cascade Range (Hart et al., 1997; Hart, 1985). Rocks of the continental crust are strongly enriched in radiogenic ^{187}Os because they have high Re/Os ratios caused primarily by their low Os concentrations (Table 13.2). Additional data by Ripley et al. (2001) for whole-rock samples of carbonaceous shale of the Virginia Formation (1.85 Ga) in Minnesota indicate that their average Re/Os ratio is 23.6 ± 6.7 ($2\bar{\sigma}$, $N = 13$) and their average weighted $^{187}Os/^{188}Os$ ratio is 7.069 compared to $^{187}Os/^{188}Os = 0.1138$ for common Os (Table 13.3). Therefore, Os in the Virginia Formation of Minnesota is enriched in radiogenic ^{187}Os by a factor of about 62 relative to common Os.

The isotope ratios of Sr and Nd of the Late Tertiary tholeiites from the northwestern USA in Figure 17.31*a* define a hyperbolic two-element mixing curve that extends from quadrant II into quadrant IV, much like the Columbia River basalt in Figure 17.29. The $^{187}Os/^{188}Os$ ratios of these rocks in Figure 17.31*b* increase nonlinearly from

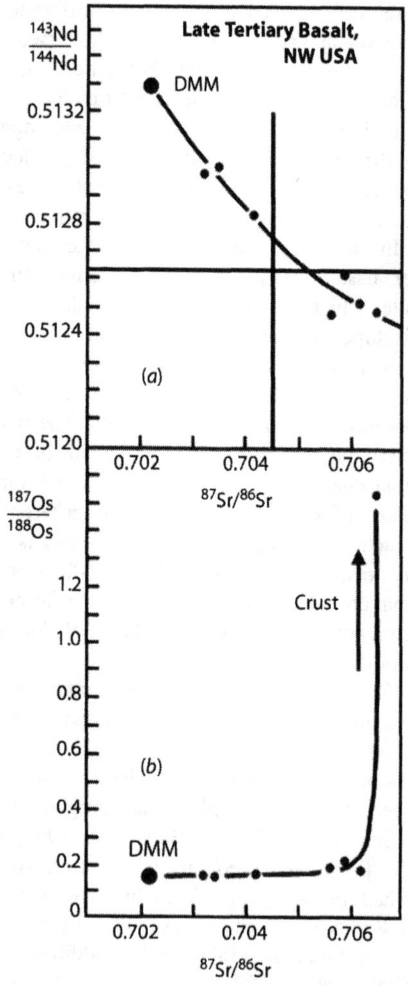

FIGURE 17.31 (*a*) Isotope ratios of Sr and Nd of Mg-rich Late Tertiary tholeiite basalt of the Columbia River plateau, the Snake River plain, the Oregon plateau, and the Cascade Range in the northwestern United States. (*b*) Isotope ratios of Sr and Os in the same rocks present clear evidence of crustal contamination of mantle-derived mafic magmas. Data from Hart et al. (1997).

0.152 to 1.53 with increasing $^{87}Sr/^{86}Sr$ ratios. The shape of the Sr–Os mixing hyperbola is convex down, which implies that the Os/Sr ratio of the crustal component was less than that of the mantle-derived basalt magma. Calculations by Hart et al. (1997) indicate that the Late Tertiary basalt of Washington and Oregon could have formed by the addition of 2–25% of crustal rocks to basalt magma derived from depleted lithospheric mantle similar to the source of mid-ocean ridge basalt.

17.8b Paraná Basalt, Brazil

The basalt flows and associated rhyolites of the Early Cretaceous Serra Geral Formation (140–110 Ma) in Figure 17.32 form an extensive plateau in the states of Paraná, Santa Catarina, and Rio Grande do Sul of southern Brazil. The volcanic rocks of the Paraná plateau were deposited on continental sandstones of Jurassic age and are overlain by similar sandstones of Cretaceous age in northern Paraná (Murata et al., 1987).

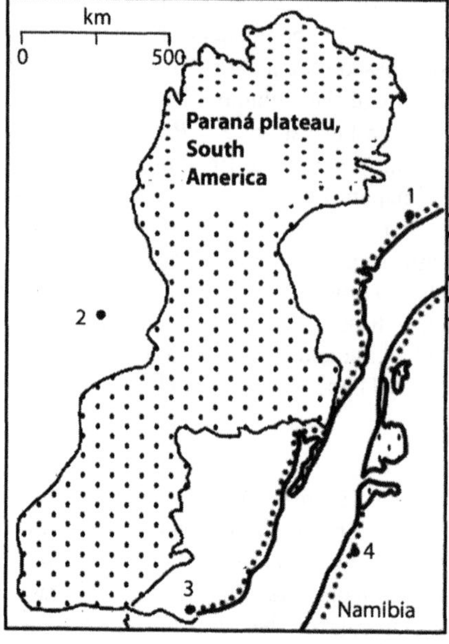

FIGURE 17.32 Outcrop pattern of Cretaceous basalt of the Paraná plateau in South America and of the Etendeka Group of Namibia in southwest Africa prior to the opening of the South Atlantic Ocean. Major towns or cities are identified by number: 1 = Sao Paulo, Brazil; 2 = Asuncion, Paraguay; 3 = Montevideo, Uruguay; 4 = Walvis Bay, Namibia. Adapted from Peate et al. (1999).

The volcanic rocks of the Paraná plateau originated from magma that formed by decompression melting caused by rifting of the continental crust of Gondwana by a large plume that is presently located in the vicinity of the island of Tristan da Cunha in the South Atlantic Ocean. When the rifts widened to form the South Atlantic Ocean, the basalt plateau that had formed above the Tristan plume was split to form the Paraná plateau of southern Brazil and the basalt plateau of the Etendeka Group of western Namibia on the opposite side of the ocean (Faure, 2001).

The tholeiite basalts of the Paraná plateau include both high-TiO_2 (>3%) and low-TiO_2 (<3%) varieties that also differ in the isotope compositions of Sr. The initial $^{87}Sr/^{86}Sr$ ratios of the low-Ti basalt range widely from about 0.7040 to 0.7160, whereas the $^{87}Sr/^{86}Sr$ ratios of the high-Ti basalts are confined between 0.7040 and 0.7060. In addition, the low-Ti basalts occur primarily in the southern part of the plateau, whereas the high-Ti basalts are more abundant in the northern part.

The initial $^{87}Sr/^{86}Sr$ and $^{143}Nd/^{144}Nd$ ratios of high-Ti basalt of the Urubici magma type (Peate et al., 1992) define a small data field in quadrant IV of Figure 17.33. The petrogenesis of the low-Ti Gramado and Esmeralda magma types was discussed by Peate and Hawkesworth (1996). In addition, two samples of the Palmas-type rhyolites in Figure 17.33 have high initial $^{87}Sr/^{86}Sr$ and low $^{143}Nd/^{144}Nd$ ratios characteristic of Precambrian basement rocks (Peate et al., 1999).

The origin of the basalt and rhyolite flows of both petrologic provinces can be explained by a theory proposed by Fodor (1987) involving the formation of Mg-rich magmas by decompression melting and their subsequent contamination as they differentiated by fractional crystallization in crustal magma chambers. According to Fodor's theory, the degree of melting depended on the proximity of the magma sources to the Tristan plume. Magmas that formed by high degrees of partial melting (e.g., 25%) close to the plume were subsequently depleted in Ti as a result of crystallization of olivine, clinopyoxene, and plagioclase. At the same time, these magmas assimilated rocks from the continental crust, which had been preheated by its proximity to the head of the Tristan plume.

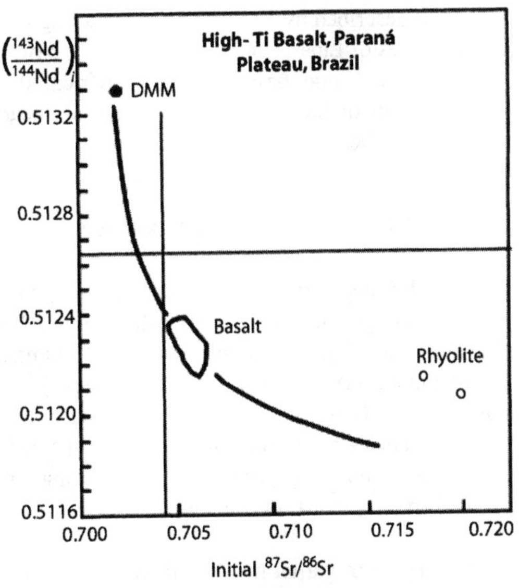

FIGURE 17.33 Initial $^{87}Sr/^{86}Sr$ and $^{143}Nd/^{144}Nd$ isotope ratios of high-Ti/Y basalt (Urubici magma type) and low-Ti/Y rhyolite (Palmas magma type) of the Paraná plateau, southern Brazil. The $^{87}Sr/^{86}Sr$ ratios are relative to 0.71025 for NBS 987 and the $^{143}Nd/^{144}Nd$ ratios are consistent with 0.51265 for BCR-1 and 0.51185 for the Johnson and Mathey Nd standard. Data from Peate et al. (1999).

Magma sources that were more distant from the Tristan plume experienced only about 10% melting and therefore formed high-Ti magmas. These magmas were not contaminated as much by assimilation of crustal rocks during fractional crystallization as the low-Ti magmas because the continental crust was not as hot.

The rhyolites in the Paraná plateau of Brazil (Figure 17.33) as well as the latites along the coast of Namibia have high initial $^{87}Sr/^{86}Sr$ and low $^{143}Nd/^{144}Nd$ ratios, presumably because they originated from magmas formed by partial melting of rocks in the continental crust (Paraná: Mantovani et al., 1985; Peate et al., 1999; Namibia: Milner et al., 1995; Ewart et al., 1998).

In general, the volcanic rocks of the Paraná plateau exemplify the extensive involvement of crustal rocks in the petrogenesis of flood basalts and associated rhyolites at many other sites in the world described by Faure (2001), including the diabase dikes of Liberia and the east coast of North America, the Karoo basalt of South Africa, the Deccan basalt of India, and the flood basalts near Noril'sk in Siberia.

17.9 ALKALI-RICH LAVAS

Alkali-rich igneous rocks occur in a wide range of tectonic settings, including oceanic islands (e.g., the Hawaiian Islands), subduction zones (e.g., Sunda arc of Indonesia), and continental rift zones (e.g., East Africa). Therefore, the tectonic setting is not the dominant factor in their petrogenesis. Instead, the processes that may participate in the formation of alkali-rich igneous rocks include

1. low degree of partial melting of source rocks in the continental crust and mantle,
2. assimilation of alkali-rich crustal rocks by mantle-derived mafic magmas,
3. fractional crystallization of magmas formed in either the continental crust or mantle, and
4. prior metasomatic enrichment of magma sources in alkali metals and volatile compounds.

Although each of the processes listed above may contribute to the formation of alkali-rich magmas, the enrichment of magma sources in alkali metals appears to be a necessary prerequisite in many cases.

In fact, the metasomatic alteration of the lithospheric mantle is directly indicated by the presence of kaersutite and phlogopite in inclusions of ultramafic rocks in mantle-derived basalt flows (e.g., Lashaine volcano, Tanzania; Eifel Mountains, Germany; Kilbourne Hole, New Mexico). Both kaersutite (amphibole) and phlogopite (mica) contain alkali metals (Na, K, and Rb) as well as water that may have originated from subducted oceanic crust in the heads of plumes rising from depth in the mantle. In addition, Chauvel et al. (1997) deduced from trace-element concentrations and isotopic data that the lithospheric mantle beneath Rurutu in the Austral Islands of Polynesia contains carbonatite that contaminated the magma of the younger suite of flows on that island (Section 17.6a). Carbonatites also occur on several oceanic islands (e.g., Fuerteventura, Canary Islands, and São Vicente, Cape Verde Islands). The presence of water and carbon dioxide in the lithospheric mantle facilitates the formation of mafic magma by decompression melting and therefore also affects the petrogenesis of flood basalt plateaus (Hawkesworth et al., 1984).

17.9a Central Italy

Alkali-rich volcanic rocks of Late Tertiary to Recent age occur in several small areas identified in Figure 17.34 along the west coast of central Italy and on the offshore islands, including Ischia, Elba, and Sicily. The rocks at these volcanic centers are noteworthy because of their high concentrations of K_2O (e.g., 10.9% at Roccamonfina) and their wide-ranging isotope ratios of Sr and Nd. The tectonic setting of this petrologic province presumably involves the African plate, which is being subducted under the Eurasian plate in the Mediterranean region. An alternative explanation proposed by Vollmer (1989, 1990) is that the Tertiary volcanic centers of Italy are located along a track formed by the counterclockwise rotation of Italy across the top of an asthenospheric plume that is presently located in Tuscany, north of Rome (Figure 17.34).

The isotope ratios of Sr and Nd of the Tertiary volcanic rocks of Italy reported by Hawkesworth and Vollmer (1979) define a series of data fields in Figure 17.35, most of which lie in quadrant IV. These data fields are aligned along hypothetical two-component isotopic mixing hyperbolas that may be the result of progressive assimilation of crustal rocks by alkali-rich magmas. However, the rhyolites of northern Tuscany ($^{87}Sr/^{86}Sr = 0.72036$; not shown in Figure 17.35) apparently formed by melting of rocks in the underlying continental crust.

The hypothesis of progressive assimilation of crustal rocks faces an obstacle because of the high Sr concentrations of the alkali-rich magmas. For example, the high-K leucitites and phonolites of the Roccamonfina volcano south of Rome have an average Sr concentration of 1782 ppm (1342–2348 ppm) compared to only 111 ppm (49.3–223 ppm) in the rhyolites of northern Tuscany. Therefore, the Sr of the alkali-rich magmas is insensitive to alteration of its isotopic composition

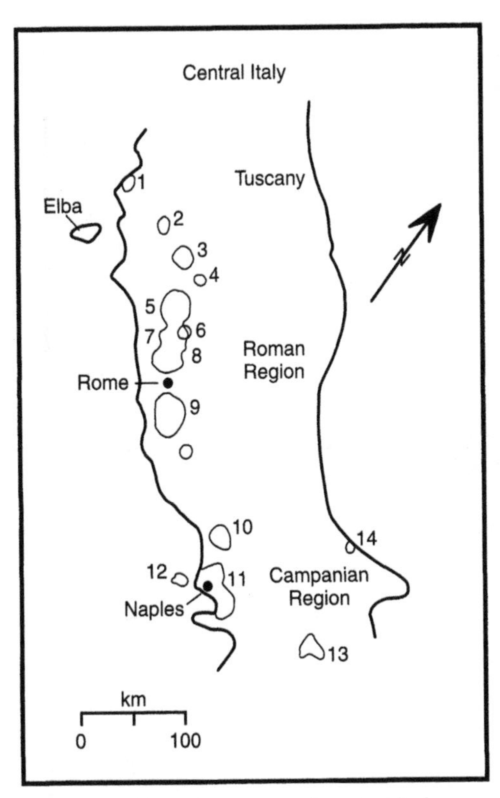

FIGURE 17.34 Volcanic centers of central Italy. Northern Tuscany: 1 = S. Vincenzo; 2 = Roccastrada. Southern Tuscany: 3 = Mt. Amiata; 4 = Radicofani; 6 = Cimino. Roman region: 5 = Vulsini; 7 = Vico; 8 = Sabatini; 9 = Alban Hills; 10 = Roccamonfina. Campanian region: 11 = Somma-Vesuvius; 12 = Ischia; 13 = Vulture; 14 = Pietre Nere. Adapted from Hawkesworth and Vollmer (1979).

by assimilation of crustal rocks like those of northern Tuscany.

This point is illustrated by the fact that in Figure 17.36 the average $^{87}Sr/^{86}Sr$ and 1/Sr ratios of the lavas at the volcanic centers of central Italy are not well correlated, as expected if they were mixtures of two components in different proportions. Although the alkali-rich rocks of southern Tuscany (Mt. Amiata, Radicofani, and Cimino) can be modeled as mixtures of magma of the Roman region with basement rocks of northern Tuscany, the lavas of Roccamonfina (low K), Somma-Vesuvius, Vulture, and the syenites of Pietre Nere deviate widely from this hypothetical mixing line. This observation does not support the hypothesis that the alkali-rich lavas of the Roman regions and at Roccamonfina originated from the same magma source as the syenites at Pietre Nere.

The Rb and Sr concentrations of the lavas in the Campanian region (including Somma-Vesuvius, the Phlegrean field, Campanian ignimbrites, and the island of Ischia) are strongly correlated and their $^{87}Sr/^{86}Sr$ ratios increase with decreasing Sr concentrations (literature data compiled by Faure, 2001). Both features are attributable to the effects of simultaneous assimilation of crustal rocks and fractional crystallization of the magmas. However, the lavas of the Roman region and of Roccamonfina are not significantly affected by fractional crystallization and crustal assimilation.

Therefore, isotopic and trace-element data reveal that different combinations of processes acted to produce the alkali-rich lavas of Italy even though the average $^{87}Sr/^{86}Sr$ and $^{143}Nd/^{144}Nd$ ratios of the lavas in the volcanic centers of this petrologic province vary systematically from south to north, as implied by the positions of the data fields in Figure 17.35. This regional pattern of the isotopic ratios is probably related to the northward movement of the subducted African plate beneath southern Italy.

Even this brief summary of some of the relevant isotopic and chemical compositions reveals the complexity of the petrogenesis of the alkali-rich lavas in the individual volcanic centers of Italy in spite of the fact that all of these centers exist within the same tectonic regime. The relevant data were summarized and reinterpreted by Vollmer (1989) but provoked a response by Peccerillo (1990), who favored a multistage subduction-related petrogenesis for the alkali-rich lavas of the Roman Province, whereas Vollmer (1990) reaffirmed his plume-track hypothesis related to the anticlockwise rotation of the Adriatic plate during the Cenozoic Era.

17.9b Leucite Hills, Wyoming, United States

The K-rich lavas of the Leucite Hills in Figure 17.37 occur in an area of about 2500 km^2 northeast of

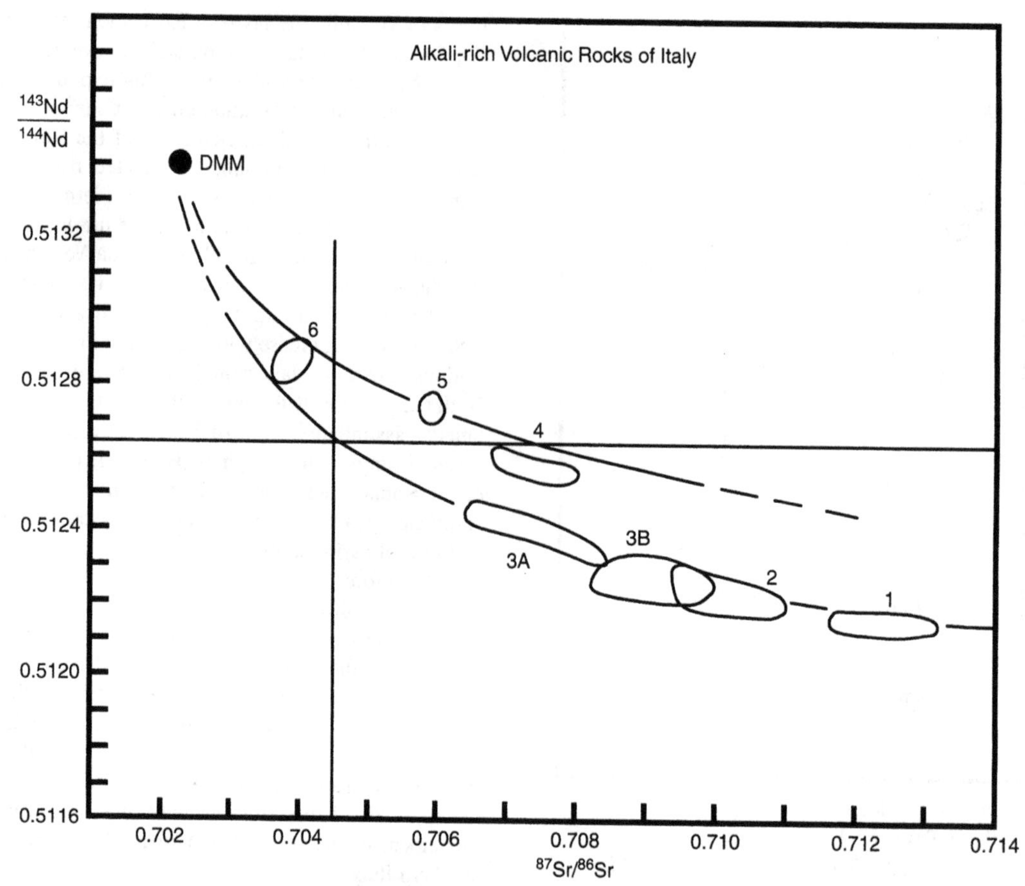

FIGURE 17.35 Isotope ratios of Sr and Nd of alkali-rich volcanic rocks starting in southern Tuscany and moving south to Mt. Vesuvius and Vulture in southern Italy: 1 = southern Tuscany, 2 = Roman region; 3A = Roccamonfina volcano, high K; 3B = Roccamonfina volcano, low K; 4 = Somma and Vesuvius; 5 = Vulture volcano; 6 = Pietre Nere on the east coast of southern Italy. The isotope ratios are relative to 0.7080 for E&A and 0.51266 for BCR-1. Data from Hawkesworth and Vollmer (1979).

Rock Springs in southwestern Wyoming. The lavas were erupted at about 1.0 Ma during the Pleistocene Epoch through sedimentary rocks of Paleozoic to Eocene age that are underlain by the Late Archean granitic gneisses of the Wyoming craton. The volcanic rocks are classified as orendite/wyomingites and madupites, whose average concentrations of K_2O range from 6.9 ± 1.3% (madupites) to 10.9 ± 1.0% (orendites), caused primarily by the presence of leucite and phlogopite (Vollmer et al., 1984).

The volcanic activity of the Leucite Hills occurred in an intracratonic setting on the northeast flank of an asymmetric dome known as the Rock Springs uplift. In addition, an east–west trending crustal suture between the Wyoming craton and metamorphic rocks of Proterozoic age is present about 200 km south of the Leucite Hills. However, neither the Rock Springs uplift nor the ancient crustal suture explains the formation of alkali-rich magmas and their eruption through the continental crust of the Wyoming craton.

The Rb and Sr concentrations of the madupites and orendite/wyomingites in the Leucite Hills in Figure 17.38 range from 165 to 350 ppm for Rb

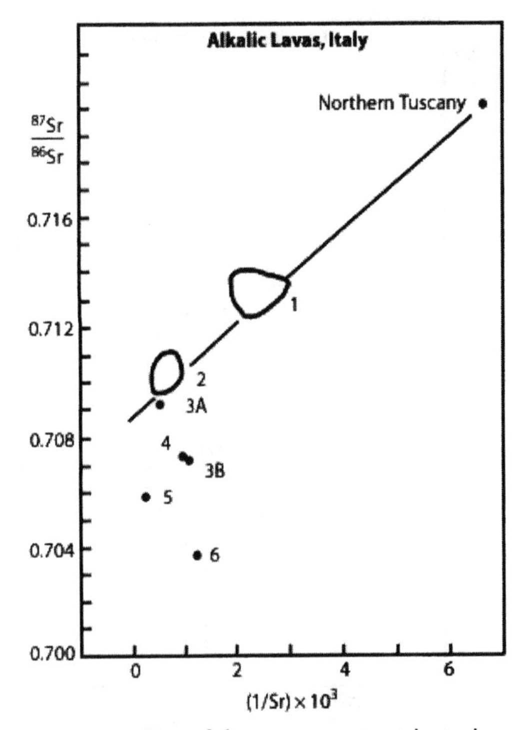

FIGURE 17.36 Test of the two-component isotopic mixing hypothesis for Sr in the alkali-rich rocks of southern Italy: 1 = southern Tuscany; 2 = Roman region; 3A = Roccamonfina, high K; 3B = Roccamonfina, low K; 4 = Somma-Vesuvius; 5 = Vulture; 6 = Pietre Nere. Data from Hawkesworth and Vollmer (1979).

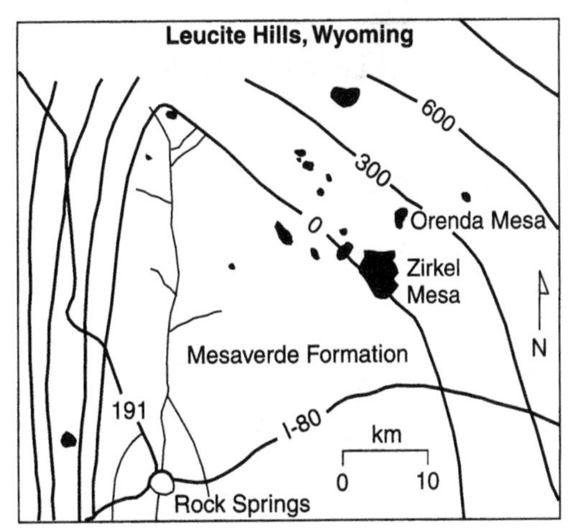

FIGURE 17.37 Leucite Hills in southwestern Wyoming, United States. The contour lines, which indicate the thickness in meters of the Late Cretaceous to Early Paleocene sedimentary rocks that overlie the Late Cretaceous Mesaverde Formation, outline the Rock Springs uplift. Zirkel Mesa is the largest of the 22 exposures of K-rich lavas in the Leucite Hills. The orendite lavas were named after Orenda Mesa. The stream is a tributary of Bitter Creek, which is an eastern tributary of the Green River. Adapted from Vollmer et al. (1984).

and from 1800 to 6300 ppm for Sr and are negatively correlated (Vollmer et al., 1984). The negative correlation of Rb and Sr concentration could have been caused by crystallization of plagioclase, which increases Rb but decreases Sr concentrations of the residual magma. This hypothesis leads to the conclusion that the parent magma had an initial Sr concentration of about 6000 ppm. However, the scatter of data points in Figure 17.38, especially of the orendites/wyomingites, is not consistent with the expected results of fractional crystallization of the parental magma. In addition, the isotope composition of Sr in such a Sr-rich magma is not sensitive to alteration by assimilation of granitic basement rocks.

The isotope ratios of Sr and Nd of the madupites and orendite/wyomingites in Figure 17.39 define two separate data fields located in quadrant IV, similar to the K-rich lavas of the Roman petrologic province in Italy in Figure 17.35. The average $^{143}Nd/^{144}Nd$ ratio of the orendites has a remarkably low value of 0.51185 ± 0.00003 ($2\bar{\sigma}$, $N = 12$), which is typical of Precambrian rocks in the continental crust (e.g., Charcani Gneiss, Figure 17.25) but is lower than the $^{143}Nd/^{144}Nd$ ratios of magma sources in the subcontinental lithospheric mantle. In addition, the separation of the two data clusters in Figure 17.39 implies that the madupite and orendite/wyomingite magmas originated from different sources.

The absence of evidence for differentiation of the magmas by fractional crystallization and for assimilation of crustal rocks suggests that the magmas did not reside in upper crustal magma

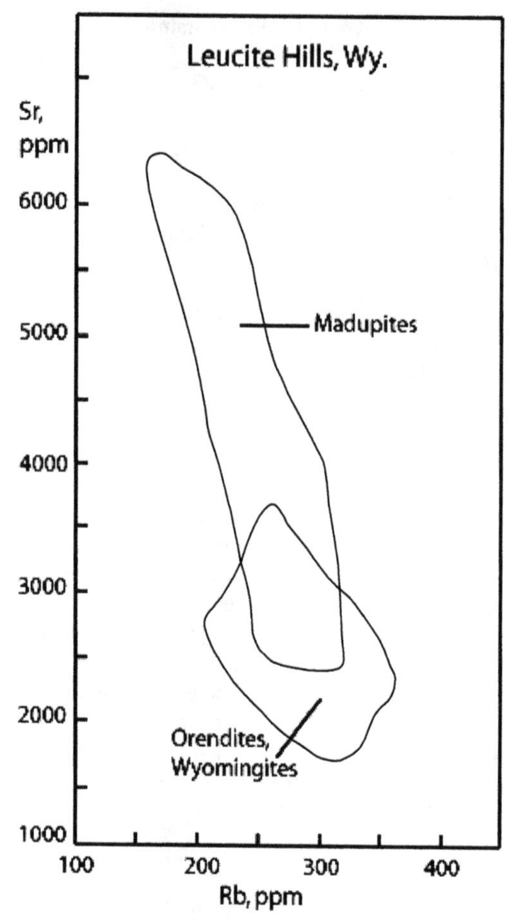

FIGURE 17.38 Concentrations of Rb and Sr of the K-rich volcanic rocks of the Leucite Hills. Data from Vollmer et al. (1984).

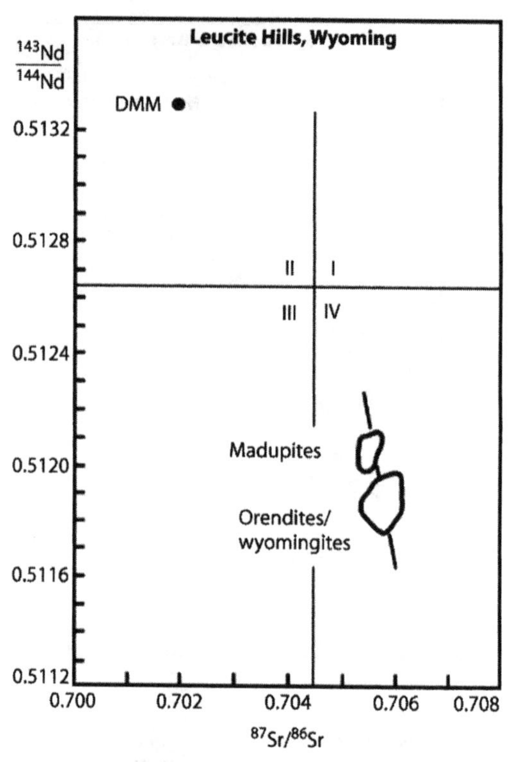

FIGURE 17.39 Isotope ratios of Sr and Nd of the K-rich Pleistocene lavas of the Leucite Hills, Wyoming. The $^{87}Sr/^{86}Sr$ ratios are relative to 0.70800 for E&A. The $^{143}Nd/^{144}Nd$ ratios were corrected for isotope fractionation to $^{146}Nd/^{144}Nd = 0.7219$ and were adjusted to 0.51265 for BCR-1. Data from Vollmer et al. (1984).

chambers but were erupted forcefully through the continental crust of the Wyoming craton. Therefore, the Sr and Nd isotope ratios of the lavas in the Leucite Hills are probably similar to those of their magma sources at the time the K-rich magmas were erupted at about 1.0 Ma.

These observations can be used to constrain the origin of the K-rich magmas of the Leucite Hills based on the assumptions that they formed by small degrees of partial melting of source rocks at depth and that these source rocks had themselves formed by crystallization of a melt that was derived from the chondritic undifferentiated reservoir CHUR during the large-scale orogeny that resulted in the formation of the Wyoming craton between 3.2 and 2.5 Ga (Vollmer et al., 1984). The partitioning of Sm and Nd between the melt phase and the residual solids caused the Sm/Nd ratio of the melt to be lower than that of the residual solids (Section 9.1). Therefore, the measured average Sm/Nd ratio of the orendite lavas is less than that of their magma source at depth.

The eruption of the orendite lava flows could have started with the formation of a magma in CHUR at about 2.85 Ga. The $^{143}Nd/^{144}Nd$ ratio of this magma (in accordance with equation 9.2) was

$$\left(\frac{^{143}\text{Nd}}{^{144}\text{Nd}}\right)^t_{\text{CHUR}} = 0.512638 - 0.1967(e^{\lambda t} - 1)$$
$$= 0.50893 \qquad (17.11)$$

where $t = 2.85 \times 10^9$ y and $\lambda = 6.54 \times 10^{-12}$ y^{-1}. The Sm/Nd ratio of the rocks that formed from this melt was sufficient to raise their average ^{143}Nd/^{144}Nd ratio from 0.50893 to 0.51185, which is the ^{143}Nd/^{144}Nd ratio of the orendite magma that formed by partial melting of these rocks at about 1.0 Ma. The Sm/Nd ratio of the source of the orendite magma is obtained from the equation

$$0.51185 = 0.50893 + 0.602\frac{\text{Sm}}{\text{Nd}}(e^{\lambda t} - 1) \quad (17.12)$$

where 0.602 is the conversion factor that relates the Sm/Nd ratio to the atomic ^{147}Sm/^{144}Nd ratio. Since $e^{\lambda t} - 1 = 0.01881$ for $t = 2.85 \times 10^9$ y and $\lambda = 6.54 \times 10^{-12}$ y^{-1}, the present value of the Sm/Nd ratio of the magma source is

$$\frac{\text{Sm}}{\text{Nd}} = \frac{0.51185 - 0.50893}{0.602 \times 0.01881} = 0.2578$$

This value is lower than the Sm/Nd ratio of CHUR by a factor of

$$\frac{(\text{Sm/Nd})^{\text{melt}}}{(\text{Sm/Nd})_{\text{CHUR}}} = \frac{0.2578}{0.3267} = 0.79$$

Similarly, the average Sm/Nd ratio of the orendites (0.1317) is lower than that of their magma source by a factor of

$$\frac{0.1317}{0.2578} = 0.51$$

The progressive decreases of the Sm/Nd ratios during each episode of partial melting is consistent with the geochemical properties of Sm and Nd and can account for the low ^{143}Nd/^{144}Nd ratio of the orendite lavas.

However, the scenario developed above based on the isotope ratios of Nd does not explain the high K_2O concentration of the orendite magmas and does not offer a plausible reason for the formation of these magmas. Therefore, Vollmer et al. (1984) favored the hypothesis that the magma source was enriched in K and other large-ion lithophile elements by fluids that emanated from subducted oceanic crust that underplated the Wyoming craton. The metasomatism did not alter the Sm–Nd systematics of the rocks in the magma source appreciably but probably did increase their Rb/Sr ratios and may have raised their ^{87}Sr/^{86}Sr ratios by the addition of radiogenic ^{87}Sr. Under these conditions, the rocks that had crystallized in the lower crust from magma that originated in the mantle at 2.85 Ga became the source of K-rich magma because of the introduction of a flux (water and carbon dioxide) and because of the addition of heat generated by radioactive decay of U, Th, and K. The resulting volatile-rich magmas forced their way to the surface in explosive eruptions similar to the intrusion of kimberlite pipes.

17.10 ORIGIN OF GRANITE

Granitic gneisses and intrusive bodies of granite are widely exposed in Precambrian shields on the continents of the world and in orogenic belts that surround them. The widespread occurrence of these kinds of rocks has attracted the attention of geologists who have attempted to explain their origin. As a result, two schools of thought emerged that advocated two different theories concerning the origin of granite:

1. Granites are igneous rocks that crystallized from magma of granitic composition.
2. Granites formed by recrystallization of preexisting sedimentary and volcanic rocks during orogenic episodes.

The so-called granites of the continental crust do not necessarily conform to the strict definition of the chemical, mineralogical, and textural criteria of granite. Therefore, these rocks are referred to as being "granitic" or "granitoid" in composition. The word *granitoid* is also used as a noun to describe coarse-grained rocks of granitic composition.

The application of isotopic and geochemical data to the study of the origin of granitic rocks resolved the two competing alternatives into a spectrum of possibilities because it became possible to distinguish "igneous" granites from "metamorphic" granites (e.g., Faure and Hurley, 1963; Faure and Powell, 1972; Faure, 2001). In addition, advances in the physical chemistry of silicate systems revealed how granitic magmas can form and how their chemical compositions evolve under conditions defined by temperature, pressure, and the content of volatiles such as water and carbon dioxide (e.g., Wyllie, 1977).

The origin of granite continues to attract attention, as exemplified by the Hutton Symposia, which were started in 1987 by the Royal Society of Edinburgh and the Royal Society of London as a celebration of the work of James Hutton. These conferences are continuing at regular intervals (e.g., Brown and Piccoli, 1995; Barbarin and Stephens, 2001).

In addition, age determinations by the Rb–Sr, K–Ar, Sm–Nd, and U–Pb methods have revealed that high-grade metamorphic rocks of granitic composition began to form only a few hundred million years after the origin of the Earth at about 4.56 Ga. The first evidence for the antiquity of certain parts of the continental crust was provided by Moorbath et al. (1972, 1973, 1975, 1977), who reported that granitic gneisses and associated metasedimentary and metavolcanic rocks of the Isua area near Godthåb in southern West Greenland formed at or prior to about 3.70 Ga. The so-called supracrustal rocks include banded iron formations and conglomerates, whose presence implies that water existed on the surface of the Earth at the time these rocks were deposited.

Granitic basement rocks and associated volcano-sedimentary complexes of Early Archean age are now known to also exist in Labrador (Hebron Gneiss, Barton, 1975), Minnesota (Morton Gneiss, Goldich and Hedge, 1974; Farhat and Wetherill, 1975), and the Slave Province (Acasta Gneiss, Wopmay orogen, Bowring et al., 1989). In addition, rocks of Early Archean age have been reported from Zimbabwe (Sebakwian greenstone belt, Moorbath et al., 1976), Western Australia, and many other sites too numerous to mention. In addition, detrital zircon grains in quartzites at Mt. Narryer and in the Jack Hills of Western Australia (Section 12.5b) have yielded U–Pb dates between 3.9 and 4.2 Ga (Froude et al., 1983; Compston and Pidgeon, 1986; Maas et al., 1992).

The Rb enrichment of the Precambrian granitic gneisses has caused their $^{87}Sr/^{86}Sr$ ratios to increase to much higher values than exist in the ultramafic rocks of the subcontinental mantle. The resulting difference in the $^{87}Sr/^{86}Sr$ ratios identifies Sr that has resided in the continental crust and distinguishes it from Sr that originated from the mantle. Therefore, Faure and Hurley (1963) proposed that the initial $^{87}Sr/^{86}Sr$ ratio of granitic rocks can be used to distinguish between rocks that contain Sr derived from the continental crust and those that originated from the underlying mantle. The application of this criterion assumes that granitic rocks that have high initial $^{87}Sr/^{86}Sr$ ratios formed by granitization of crustal rocks or are the products of mantle-derived magmas that assimilated significant amounts of continental crust containing Sr having high $^{87}Sr/^{86}Sr$ ratios.

The range of petrogenetic possibilities is exemplified by a granite pluton of the Martin Dome in the Miller Range of the Transantarctic Mountains and by the Salisbury pluton in Rowan County of North Carolina. The biotite-bearing quartz monzonite of the Martin Dome in the Miller Range at 83°15′ S and 157°00′ E intrudes quartzo-feldspathic gneisses of the Nimrod Group of Precambrian age and has a whole-rock Rb–Sr isochron date of 478 ± 17 Ma with a high initial $^{87}Sr/^{86}Sr$ ratio of 0.734 ± 0.001 (Gunner and Faure, 1972; Gunner, 1974). The initial $^{87}Sr/^{86}Sr$ ratio indicates that the Martin Dome pluton crystallized from magma that was extensively contaminated by assimilation of rocks of the continental crust. Alternatively, the Martin Dome magma could have formed by partial remelting of the rocks of the Precambrian Nimrod Group into which the magma was intruded.

A quite different set of data was reported by Fullagar et al. (1971) for the Salisbury pluton of North Carolina. This pluton is composed of white and pink adamellite (plagiogranite?) that intruded metasedimentary rocks of the so-called Charlotte Slate Belt. A whole-rock Rb–Sr isochron of the adamellite yielded a date of 402 ± 6 Ma and a low initial $^{87}Sr/^{86}Sr$ ratio of 0.7038 ± 0.0010 relative to 0.7080 for E&A. The low initial $^{87}Sr/^{86}Sr$ ratio indicates that the granitic magma of the Salisbury pluton formed either by differentiation of mantle-derived basalt magma or by partial remelting of mantle-derived volcanic or plutonic rocks.

These examples support the conclusion that igneous rocks of granitic composition can be produced by several different processes:

1. partial melting of rocks that have resided in the continental crust long enough to have significantly higher $^{87}Sr/^{86}Sr$ ratios than the subcontinental mantle (e.g., Martin Dome, Antarctica),
2. partial melting of mantle-derived volcanic or plutonic igneous rocks whose $^{87}Sr/^{86}Sr$ ratios were not increased by decay of ^{87}Rb because insufficient time had elapsed after their initial crystallization or because their Rb/Sr ratio was low or both, and
3. extreme differentiation by fractional crystallization of mantle-derived basaltic magmas.

The origin of granitic rocks from heterogeneous sources in the continental crust and underlying mantle is clearly indicated in Figure 17.40 by the wide range of their present-day Sr and Nd isotope ratios. The data points representing Mesozoic granitic rocks in the Sierra Nevada Mountains

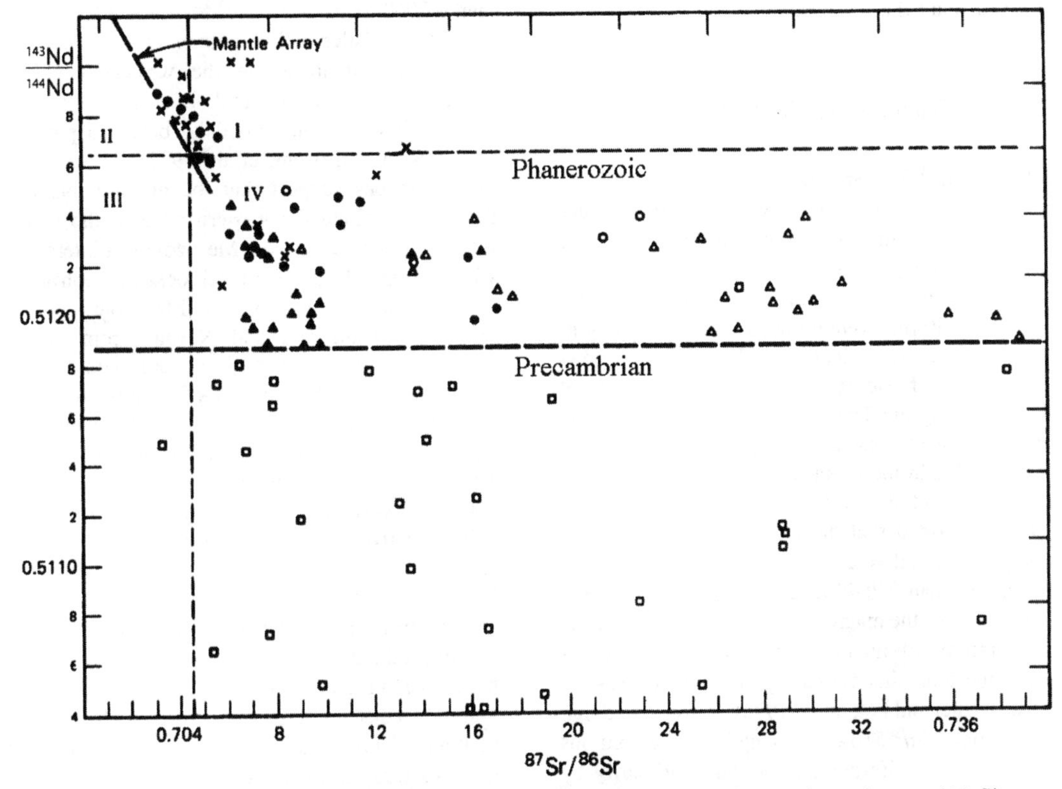

FIGURE 17.40 Present-day isotopic ratios of Nd and Sr in granitic rocks of the continental crust: (●) Sierra Nevada, California; (×) Peninsular Ranges, California and Baja California; (△) South Australia; (○) Hercynian granites, France; (▲) Caledonian granites of Scotland; (□) Precambrian granitic gneisses and metasediments. Data from DePaolo (1981, renormalized), Allègre and Othman (1980), Hamilton et al. (1980), McCulloch and Wasserburg (1978, renormalized), DePaolo et al. (1982, renormalized), and Zindler et al. (1981). All data have been corrected for isotope fractionation to $^{146}Nd/^{144}Nd = 0.7219$ or equivalent and to $^{86}Sr/^{88}Sr = 0.1194$.

and the Peninsular Range of California extend from quadrant II to quadrant IV, implying that their magmas contain mixtures of Sr and Nd derived from sources in the mantle and the continental crust. The Phanerozoic granites of South Australia, the Hercynian granites of France, and the Caledonian granites of Scotland are located in quadrant IV, where rocks of the continental crust normally occur. All granitic rocks of Precambrian age have low $^{143}Nd/^{144}Nd$ ratios less than 0.51183, whereas their $^{87}Sr/^{86}Sr$ ratios range to high values greater than 0.736. The $^{143}Nd/^{144}Nd$ ratios of granitic rocks of Precambrian age are low because their Sm/Nd ratios are lower than those of the subcontinental mantle, in accordance with the geochemical properties of Sm and Nd (Section 9.1).

17.10a Batholiths of California

The initial $^{87}Sr/^{86}Sr$ ratios of Mesozoic granitic plutons and rhyolite lava flows in California vary regionally, presumably because the magmas were affected by the presence of different kinds of crustal rocks at depth below the surface. The Mesozoic plutons of California were intruded sequentially beginning with mafic varieties and becoming progressively more felsic in composition. Emplacement began during the Triassic Period at about 210 Ma and continued episodically for 130 million years until 80 Ma in the Cretaceous Period.

Hurley et al. (1965) and Kistler and Peterman (1973) reported that the initial $^{87}Sr/^{86}Sr$ ratios of the plutonic and volcanic rocks of California range from less than 0.7040 to greater than 0.7080, perhaps because the magmas formed by partial melting of mixtures of mantle-derived and crustal rocks in proportions that varied regionally from west of east. Kistler and Peterman (1973) demonstrated that the initial $^{87}Sr/^{86}Sr$ ratios of the Mesozoic plutons and lava of California are controlled primarily by geography. Therefore, they concluded that contours of the initial $^{87}Sr/^{86}Sr$ ratios in Figure 17.41 reflect the geographic distribution of basement rocks of differing ages and chemical compositions.

For example, the trace-element concentrations of the Mesozoic granitic plutons located west of the 0.7040 contour are similar to those of oceanic tholeiites. The 0.7040 contour is also the eastern limit of exposures of ultramafic igneous rocks in California. The 0.7060 contour is the boundary between shallow-water facies of sedimentary rocks (carbonates, shales, sandstones) and deep-water facies (volcanic rocks, chert, greywacke) located west and north of the shallow-water platform sediments. In addition, Kistler and Peterman (1973) observed that quartz diorite and trondhjemite predominate where the initial $^{87}Sr/^{86}Sr$ ratios are less than 0.7040, that quartz diorite and granodiorite occur where the initial $^{87}Sr/^{86}Sr$ ratios are between 0.7040 and 0.7060, and that granodiorite and quartz monzonite appear in areas where the initial $^{87}Sr/^{86}Sr$ ratios are greater than 0.7060.

Later, Kistler and Peterman (1978) used the 0.7060 contour to locate the western edge of the Precambrian continental crust in California, Oregon, Washington, and Idaho based in part on the work of Armstrong et al. (1977). These kinds of correlations suggest that the granitic magmas that formed along the western edge of the North American plate during the Mesozoic Era interacted with or formed from the crustal rocks they intruded.

The petrogenesis of the Mesozoic granitic plutons of California and Nevada from mixed crustal and mantle sources is confirmed by the isotope ratios of Sr and Nd of granitic rocks in the Sierra Nevada Mountains and in the Peninsular Ranges of California reported by DePaolo (1980, 1981). The isotope ratios were recalculated to $t = 100 \times 10^6$ years and were expressed in terms of ε_t values relative to CHUR for Sr and Nd in order to remove the effect of in situ growth of radiogenic ^{87}Sr and ^{143}Nd. The resulting array of data points extends from quadrant II (mantle) to quadrant IV (continental crust) and fits a two-component mixing hyperbola in Figure 17.42. Therefore, the Mesozoic granitic rocks of this area formed by partial melting of mixtures of Precambrian schists and gneisses (crustal component) with mafic volcanic and plutonic rocks (mantle component). The origin of the mafic rocks is linked to subduction of the oceanic lithospheric mantle under the North American plate, which overrode a subduction zone as it moved in a northwesterly direction.

Subsequent studies have strengthened the evidence that the Mesozoic granitic plutons and lava

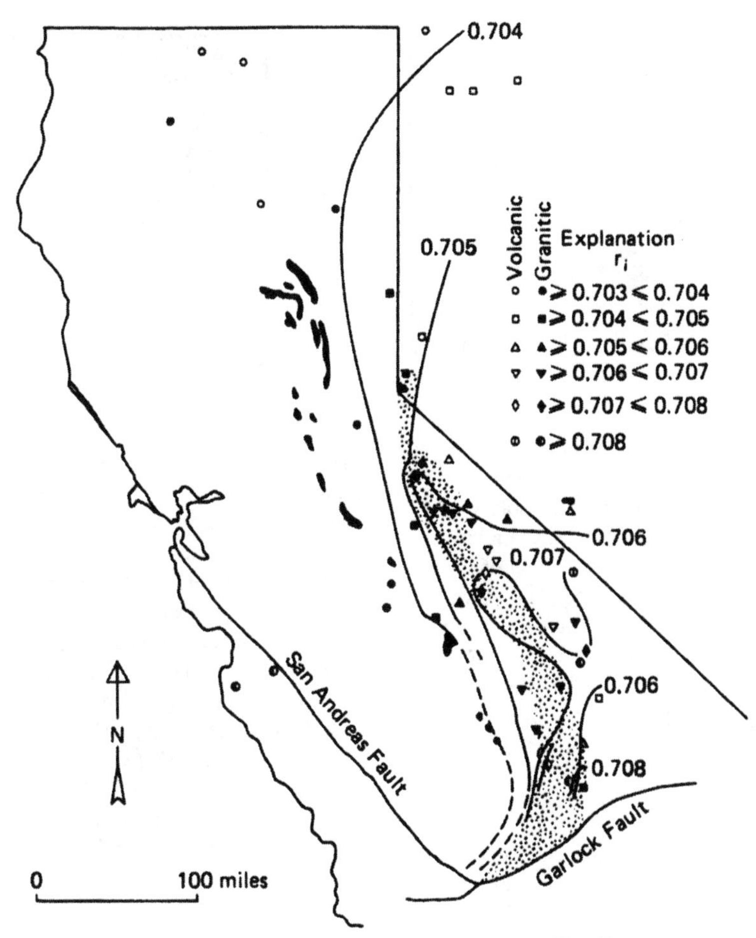

FIGURE 17.41 Contour diagram showing the regional variation of initial $^{87}Sr/^{86}Sr$ ratios of volcanic and plutonic rocks of Mesozoic age in California. From R. W. Kistler and Z. E. Peterman, *Geological Society of America Bulletin*, Vol. 84, No. 11, pp. 3489–3512. Reprinted with permission of the Geological Society of America.

flows of California originated from mixed crustal and mantle sources:

Farmer and DePaolo, 1983, *J. Geophys. Res.*, 88(B4):3379–3401. Domenick et al., 1983, *Geol. Soc. Am. Bull.*, 94:713–719. Hill et al., 1986, *Contrib. Mineral. Petrol.*, 92:351–361. Tommasini and Davies, 1997, *Earth Planet. Sci. Lett.*, 148:273–285. Mayo et al., 1998, *Int. Geol. Rev.*, 40(3):257–278. DeWitt et al., 1984, *Geol. Soc. Am. Bull.*, 95:723–739.

17.10b Genetic Classification of Granites

Regional variations of isotopic and chemical compositions, thoroughly documented in the batholiths of California, have also been reported for the batholiths of southeastern Australia. Chappell and White (1974) used mineralogical and chemical criteria to distinguish so-called I-type (igneous) from S-type (sedimentary) granites. The former are attributed to fusion of mantle-derived rocks, whereas the latter are derived from pelitic

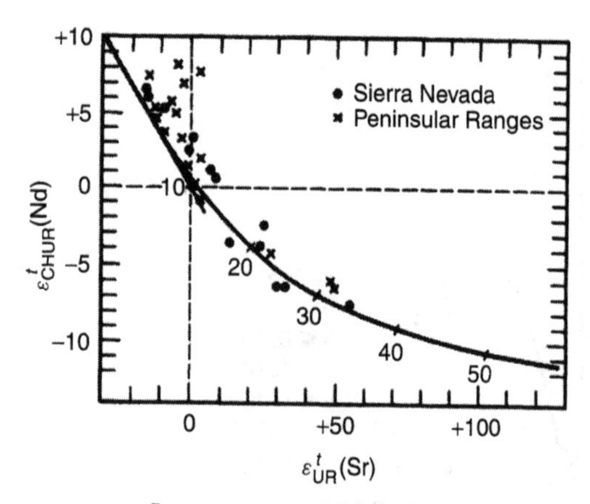

FIGURE 17.42 Isotope compositions of Nd and Sr of granitic rocks from the Sierra Nevada and Peninsular Ranges of California. The isotope ratios are expressed in terms of ε values after correction for decay since crystallization of these rocks during the Cretaceous Period. The curve represents mixtures of a Precambrian schist, into which the Sierra Nevada batholith was intruded, with a quartz diorite presumably derived from subcrustal sources as suggested by a high initial $^{143}Nd/^{144}Nd$ and low $^{87}Sr/^{86}Sr$ ratio. Data from DePaolo (1981).

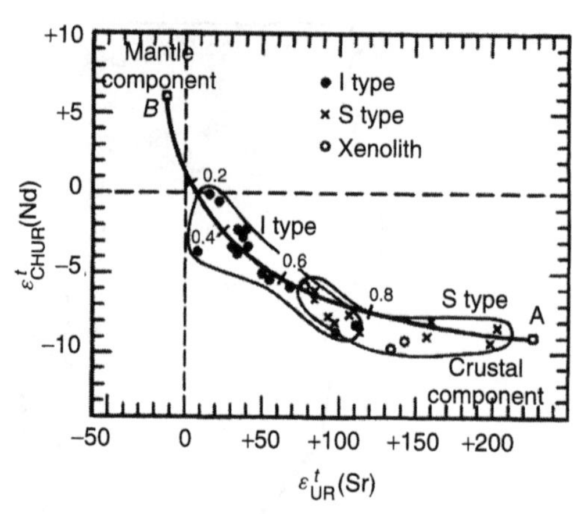

FIGURE 17.43 Epsilon-values of Nd and Sr, corrected for decay, of granitic rocks and xenoliths from the Berridale and Kosciusko batholiths of southeastern Australia. Both I-type (igneous) and S-type (sedimentary) granitic rocks fit the same mixing line, indicating that both are mixtures of two components derived from "depleted" mantle and the continental crust. The curve was fitted by using the following end-member compositions: crustal component (A): ε(Nd) $= -9.0$, Nd $= 28.0$ ppm, ε(Sr) $= 227.2$, Sr $= 140$ ppm; mantle component (B): ε(Nd) $= +6.0$, Nd $= 14.0$ ppm, ε(Sr) $= -14.20$, Sr $= 470$ ppm. Data from McCulloch and Chappell (1982).

sedimentary rocks representing recycled continental crust. Although I-type and S-type granites are the extremes of possible petrogenetic provenance, most large-scale granitic batholiths originated from mixtures of crustal and mantle-derived rocks.

In addition to S- and I-type granitoids both of which form during orogenies, granitic rocks also occur in the form of plutons in anorogenic settings. These so-called A-type granitoids are intruded during the extension of the continental crust, which causes small degrees of decompression melting of rocks in the lower crust or upper mantle. Consequently, A-type granitoids are alkali rich in many cases, and their intrusion into the upper crust is guided by the tectonic process that caused the decompression melting. As a result, plutons of A-type granitic rocks form sequentially in clusters as the tectonic activity proceeds (Section 5.3a).

Examples of A-type granitic plutons occur in the large flood-basalt provinces that are associated with rifting of the continental crust and with the subsequent opening of the Atlantic Ocean: Nigeria and Cameroon in West Africa (van Breemen et al., 1975; Jacquemin et al., 1982), northeastern Brazil (Long et al., 1986), White Mountains of New Hampshire (Eby, 1987, 1990; Foland and Allen, 1991), and many others described by Faure (2001).

The petrogenesis of S- and I-type granites is exemplified by the Berridale and Kosciusko batholiths of New South Wales in Australia. The ε_t values calculated from the initial $^{87}Sr/^{86}Sr$ and $^{143}Nd/^{144}Nd$ ratios of these rock bodies reported by McCulloch and Chappell (1982) define an isotopic mixing hyperbola in quadrant IV of Figure 17.43. Samples of I- and S-type granites derived from both batholiths form two overlapping clusters that fit the same mixing curve. Therefore, both batholiths originated from mixed sources containing mantle-derived and crustal components. The distribution of data points indicates that even specimens classified as I type on the basis of chemical, mineralogical, and textural criteria contain between 20 and 80% of the crustal component. Therefore, the evidence in Figure 17.43 confirms that the Berridale and Kosciusko batholiths are mixtures in varying proportions of source rocks derived from the mantle (I type) and from the continental crust (S type).

Additional studies by Flood and Shaw (1977) of the Carboniferous New England batholith in New South Wales revealed that S-type granites in that batholith have low initial $^{87}Sr/^{86}Sr$ ratios of about 0.7060. Low values of this ratio are supposed to characterize I-type granites, whereas S-type granites are supposed to have high initial $^{87}Sr/^{86}Sr$ ratios. The anomaly may have been caused by the derivation of the S-type granites from young volcaniclastic sediment. The isotope composition of Sr of rock samples from the New England batholith once again emphasizes the point that rocks of granitic composition form by a variety of processes that are not reducible to only two alternatives.

17.11 SUMMARY

The elements that have stable radiogenic isotopes (Sr, Nd, Pb, Hf, and Os) are useful for studies of the origin of igneous rocks because their isotope compositions, unlike the concentrations of major and trace elements, are not fractionated by partial melting and by the subsequent crystallization of minerals from cooling magmas. The isotope compositions of these elements are changed only when the magmas assimilate rocks in which these elements have different isotope compositions and in case two or more magmas having different isotope compositions mix in varying proportions. Assimilation and mixing can be recognized by the resulting correlation of isotope compositions of Sr, Nd, Pb, Hf, and Os of igneous rocks with the major-element concentrations of the rocks that form from such hybrid magmas. The reasons why the isotope compositions of Ca, Ce, and Ba are not yet widely used in the study of the petrogenesis of igneous are explained in Chapters 8, 14, and 15, respectively.

Magma forms by partial melting during decompression of hot rocks in the mantle and lower crust, by the addition of volatile compounds (e.g., H_2O and CO_2), and, more rarely, by heating of rocks to their melting temperatures. The driving force for magma formation is provided by plumes that rise from depth in the mantle until they impinge onto the underside of the lithospheric mantle. The plumes originate by subduction of oceanic crust and sediment into the mantle where they are incubated for long periods of geological time until their temperatures increase sufficiently by radioactive heating to cause them to become buoyant. The interactions of the heads of plumes with the underside of the lithosphere cause uplift and rifting of the overlying lithospheric plate, which triggers magma formation by decompression melting of hot rocks in the heads of the plumes and locally in the lithospheric mantle.

The details of this process depend on many factors:

1. the size of the plume,
2. the movement of the lithospheric plate across the head of a stationary plume,
3. the structure of the lithospheric plate (i.e., mantle plus oceanic crust or mantle plus continental crust),
4. prior depletion of the lithospheric mantle by episodes of magma formation,
5. prior metasomatic alteration of the lithosphere by fluids (H_2O and/or CO_2) that originated from the heads of rising plumes or from subducted oceanic crust that may have locally underplated the lithosphere, and
6. the opportunity for mantle-derived magmas to differentiate by assimilation of rocks of the continental crust and by simultaneous fractional crystallization in magma chambers in the continental crust (AFC).

The challenge for the earth sciences is to reconstruct the petrogenesis of volcanic and plutonic igneous rocks from their chemical and isotopic compositions and from other criteria such as the tectonic setting in which they formed, their textures, mineral compositions, flow stratigraphy, geological ages, remanent magnetization, microscopic examination of inclusions of glass and volatiles, and evidence of alteration after initial crystallization. The results of such studies have provided sufficient insights to permit the formulation of a theory concerning the dynamics of the mantle of the Earth.

The continental crust is largely composed of igneous and metamorphic rocks of granitic composition, although most of these rocks are not granites strictly speaking. The isotope ratios of Sr and Nd indicate that granitic batholiths originate from mixed sources consisting of crustal and mantle-derived components. In addition, comparatively small plutons of anorogenic granites crystallize from magmas that form by remelting of a variety of rocks in the lower crust and upper mantle in settings of extensional tectonics.

REFERENCES FOR PLUMES AND MAGMA SOURCES (SECTIONS 17.1–17.2)

Basu, A., and S. R. Hart (Eds.), 1998. *Earth Processes; Reading the Isotopic Code*, Geophys. Monograph 95. American Geophysical Union, Washington, D.C.

Chauvel, C., A. W. Hofmann, and P. Vidal, 1992. HIMU-EM: The French Polynesian connection. *Earth Planet. Sci. Lett.*, 110:99–119.

Faure, G., 2001. *The Origin of Igneous Rocks; the Isotopic Evidence*. Springer-Verlag, Heidelberg.

Hart, S. R., 1988. Heterogeneous mantle domains: Signatures, genesis, and mixing chronologies. *Earth Planet. Sci. Lett.*, 90:273–296.

Hofmann, A. W., 1997. Mantle geochemistry: The message from oceanic volcanism. *Nature*, 385:219–229.

Hofmann, A. W., and W. M. White, 1982. Mantle plumes from ancient oceanic crust. *Earth Planet. Sci. Lett.*, 57:421–436.

Morgan, W. J., 1971. Convection plumes in the lower mantle. *Nature*, 230:42–43.

Morgan, W. J., 1972a. Deep mantle convection plumes and plate motions. *Am. Assoc. Petrol. Geol. Bull.*, 56:203–213.

Morgan, W. J., 1972b. Plate motions and deep mantle convection. *Geol. Soc. Am. Mem.*, 132:203–213.

Morgan, W. J., 1981. Hotspot tracks and the opening of the Atlantic and Indian Oceans. In C. Emiliani (Ed.), *The Sea*, Vol. VII, pp. 443–487. Wiley, New York.

Salters, V. J. M., and S. R. Hart, 1991. The mantle sources of oceanic ridges, islands, and arcs; the Hf-isotope connection. *Earth Planet. Sci. Lett.*, 104:364–380.

Schilling, J. -G., R. H. Kingsley, B. B. Hanan, and B. L. McCully, 1992. Nd-Sr-Pb isotopic variations along the Gulf of Aden: Evidence for mantle plume-continental lithosphere interaction. *J. Geophys. Res.*, 97:10927–10966.

Shirey, S. B., and R. J. Walker, 1998. The Re-Os isotope system in cosmochemistry and high-temperature geochemistry. *Annu. Rev. Earth Planet. Sci.*, 26:423–500.

Weaver, B. L., 1991. The origin of oceanic island basalt end-member compositions: Trace element and isotopic constraints. *Earth Planet. Sci. Lett.*, 104:381–397.

Wilson, J. T., 1963a. A possible origin of the Hawaiian Islands. *Can. J. Phys.*, 41:863–870.

Wilson, J. T., 1963b. Evidence from islands on the spreading of the ocean floor. *Nature*, 197:536–538.

Zindler, A., and S. R. Hart, 1986. Chemical geodynamics. *Annu. Rev. Earth Planet. Sci.*, 14:493–571.

REFERENCES FOR MIDOCEAN RIDGE BASALT (SECTION 17.3)

Basu, A., and S. R. Hart (Eds.), 1998. *Earth Processes: Reading the Isotopic Code*, Geophys. Monograph 95. American Geophysical Union, Washington, D.C.

DePaolo, D. J., 1988. *Neodymium Isotope Geochemistry*. Springer-Verlag, Berlin.

Wilkinson, J. F. G., 1982. The genesis of mid-ocean ridge basalt. *Earth Sci. Rev.*, 18:1–57.

Yu, D. -M., D. Fontignie, and J. -G. Schilling, 1997. Mantle plume-ridge interaction in the central North Atlantic: A Nd isotope study of Mid-Atlantic Ridge basalts from 30° N to 50° N. *Earth Planet. Sci. Lett.*, 146:259–272.

REFERENCES FOR BASALT AND RHYOLITE OF ICELAND (SECTION 17.4)

Chayes, F., 1963. Relative abundance of intermediate members of the oceanic basalt-trachyte association. *J. Geophys. Res.*, 68:1519–1534.

Faure, G., 2001. *Origin of Igneous Rocks: The Isotopic Evidence*. Springer-Verlag, Heidelberg.

Hart, S. R., J. -G. Schilling, and J. L. Powell, 1973. Basalts from Iceland and along the Reykjanes Ridge; Sr isotope geochemistry. *Nature*, 246:104–107.

Hémond, C., N. T. Arndt, U. Lichtenstein, and A. W. Hofmann, 1993. The heterogeneous Iceland plume:

Nd-Sr-O isotopes and trace-element constraints. *J. Geophys. Res.*, 98(B9):15833–15850.

Langmuir, C. H., R. D. Vocke, Jr., G. N. Hanson, and S. R. Hart, 1978. A general mixing equation with applications to Icelandic basalts. *Earth Planet. Sci. Lett.*, 37:380–392.

Mertz, D. F., C. W. Devey, W. Todt, P. Stoffers, and A. W. Hofmann, 1991. Sr-Nd-Pb isotope evidence against plume-asthenosphere mixing north of Iceland. *Earth Planet. Sci. Lett.*, 107:243–255.

Moorbath, S., H. Sigurdsson, and R. Goodwin, 1968. K-Ar ages of the oldest exposed rock in Iceland. *Earth Planet. Sci. Lett.*, 4:197–205.

Moorbath, S., and G. P. L. Walker, 1965. Strontium isotope investigation of igneous rocks from Iceland. *Nature*, 207:837–840.

Nicholson, H., M. Condomines, J. G. Fitton, A. E. Fallick, K. Grönvold, and G. Rogers, 1991. Geochemical and isotopic evidence for crustal assimilation beneath Krafla, Iceland. *J. Petrol.*, 32:1005–1020.

O'Nions, R. K., and D. B. Clarke, 1972. Comparative trace element geochemistry of the Tertiary basalts from Baffin Bay. *Earth Planet. Sci. Lett.*, 15:436–446.

O'Nions, R. K., and K. Grönvold, 1973. Petrogenetic relationships of acid and basic rocks in Iceland: Sr-isotopes and rare-earth elements in late and post-glacial volcanics. *Earth Planet. Sci. Lett.*, 19:397–409.

Prestvik, T., T. Torske, B. Sundvoll, and H. Karlsson, 1999. Petrology of early Tertiary nephelinites off mid-Norway; additional evidence for an enriched endmember of the ancestral Iceland plume. *Lithos*, 46:317–330.

Schilling, J. -G., P. S. Meyers, and R. H. Kingsley, 1982. Evolution of the Iceland hotspot. *Nature*, 296:313–320.

Sigurdsson, H., 1977. Generation of Icelandic rhyolites by melting of plagiogranites in the oceanic layer. *Nature*, 269:25–28.

Sun, S. -S., and B. -M. Jahn, 1975. Lead and strontium isotopes in post-glacial basalts from Iceland. *Nature*, 255:527–530.

Sun, S. -S., M. Tatsumoto, and J. -G. Schilling, 1975. Mantle plume mixing along the Reykjanes Ridge axis; lead isotopic evidence. *Science*, 190:143–147.

Taylor, R. N., M. F. Thirlwall, B. J. Murton, D. R. Hilton, and M. A. M. Gee, 1997. Isotopic constraints on the influence of the Icelandic plume. *Earth Planet. Sci. Lett.*, 148:E1–E8.

Thirlwall, M. F., 1995. Generation of the Pb isotopic characteristics of the Iceland plume. *J. Geol. Soc. Lond.*, 152:991–996.

White, R. S., and A. C. Morton, 1995. The Iceland plume and its influence of the evolution of the NE Atlantic. *J. Geol. Soc. Lond.*, 152:935.

Yoder, H. S., 1973. Contemporaneous basaltic and rhyolitic magmas. *Am. Mineral.*, 58:153–171.

REFERENCES FOR HAWAIIAN ISLANDS (SECTION 17.5)

Basu, A. R., and B. E. Faggart, Jr., 1998. Temporal isotopic variations in the Hawaiian mantle plume: The Lanai anomaly, the Molokai fracture zone, and a seawater-altered lithospheric component in Hawaiian volcanism. In A. R. Basu and S. R. Hart (Eds.), *Earth Processes: Reading the Isotopic Code*, Geophys. Monograph 95, pp. 149–160. American Geophysical Union, Washington, D.C.

Bennett, V. C., T. M. Esat, and M. D. Norman, 1996. Two mantle-plume components in Hawaiian picrites inferred from correlated Os-Pb isotopes. *Nature*, 381:221–224.

Chauvel, C., A. W. Hofmann, and P. Vidal, 1992. HIMU-EM: The French Polynesian connection. *Earth Planet. Sci. Lett.*, 110:99–119.

Chen, C. -Y., and F. A. Frey, 1985. Trace element and isotopic geochemistry of lavas from the Haleakala volcano, East Maui, Hawaii: Implications for the origin of Hawaiian basalts. *J. Geophys. Res.*, 90:8743–8768.

Faure, G., 2001. *Origin of Igneous Rocks: The Isotopic Evidence*. Springer-Verlag, Heidelberg.

Hart, S. R., 1988. Heterogeneous mantle domains: Signatures, genesis, and mixing chronologies. *Earth Planet. Sci. Lett.*, 90:273–296.

Hauri, E. H., J. C. Lassiter, and D. J. DePaolo, 1996. Osmium isotope systematics of drilled lavas from Mauna Loa, Hawaii. *J. Geophys. Res.*, 101:11795–11806.

Martin, C. E., 1991. Os isotopic characteristics of mantle derived rocks. *Geochim. Cosmochim. Acta*, 55:1421–1434.

Salters, V. J. M., and S. R. Hart, 1991. The mantle sources of oceanic ridges, islands, and arcs; the Hf-isotope connection. *Earth Planet. Sci. Lett.*, 104:364–380.

Shirey, S. B. and R. J. Walker, 1998. The Re–Os isotope system in cosmochemistry and high-temperature geochemistry. *Annu. Rev. Earth Planet. Sci.*, 26:423–500.

Stille, P., D. M. Unruh, and M. Tatsumoto, 1983. Pb, Sr, Nd, and Hf isotopic evidence of multiple sources for Oahu, Hawaii, basalts. *Nature*, 304:25–29.

Stille, P., D. M. Unruh, and M. Tatsumoto, 1986. Pb, Sr, Nd, and Hf isotopic constraints on the origin of Hawaiian basalts and evidence for a unique mantle source. *Geochim. Cosmochim. Acta*, 50:2303–2319.

Stolper, E. M., D. J. DePaolo, and D. M. Thomas, 1996. Introduction to special section: Hawaii Scientific

Drilling Project. *J. Geophys. Res.*, 101(B5):11593–11598.

REFERENCES FOR HIMU SOURCES OF POLYNESIA (SECTION 17.6)

Chauvel, C., W. McDonough, G. Guille, R. Maury, and R. A. Duncan, 1997. Contrasting old and young volcanism on Rurutu Islands, Austral chain. *Chem. Geol.*, 139:125–143.

Faure, G., 2001. *Origin of Igneous Rocks: The Isotopic Evidence*. Springer-Verlag, Heidelberg.

Hart, S. R., 1988. Heterogeneous mantle domains: Signatures, genesis, and mixing chronologies. *Earth Planet. Sci. Lett.*, 90:273–296.

Hauri, E. H., and S. R. Hart, 1993. Re-Os isotope systematics of HIMU and EMII oceanic island basalts from the South Pacific Ocean. *Earth Planet. Sci. Lett.*, 114:353–371.

REFERENCES FOR MARIANA ISLAND ARC (SECTION 17.7a)

Gribble, R. F., R. J. Stern, S. Newman, S. H. Bloomer, and T. O'Hearn, 1998. Chemical and isotopic composition of lavas from the northern Mariana Trough: Implications for magma genesis in back-arc basins. *J. Petrol.*, 39:125–154.

Ito, E., and R. J. Stern, 1985/1986. Oxygen and strontium isotopic investigations of subduction zone volcanism: The case of the Volcano Arc and the Marianas Island Arc. *Earth Planet. Sci. Lett.*, 76:312–320.

Lee, J., R. J. Stern, and S. H. Bloomer, 1995. Forty million years of magmatic evolution in the Mariana arc: The tephra glass record. *J. Geophys. Res.*, 100:17671–17687.

Meijer, A., 1976. Pb and Sr isotopic data bearing on the origin of volcanic rocks from the Mariana island-arc system. *Geol. Soc. Am. Bull.*, 87:1358–1369.

Stern, R. J., 1978. Agrigan: An introduction to the geology of an active volcano in the northern Mariana Island Arc. *Bull. Volcanol.*, 41:1–13.

Stern, R. J., 1979. On the origin of andesite in the northern Mariana Island Arc: Implications from Agrigan. *Contrib. Mineral. Petrol.*, 68:207–219.

Stern, R. J., S. H. Bloomer, P. -N. Lin, E. Ito, and J. D. Morris, 1988. Shoshonitic magmas in nascent arcs: New evidence from submarine volcanoes in the northern Marianas. *Geology*, 16:426–430.

Woodhead, J. D., and D. G. Fraser, 1985. Pb, Sr, and ^{10}Be isotopic studies of volcanic rocks from the northern Mariana Islands: Implications for magma genesis and crustal recycling in the Western Pacific. *Geochim. Cosmochim. Acta*, 49:1925–1930.

REFERENCES FOR ANDES OF SOUTH AMERICA (SECTIONS 17.7b–17.7c)

Faure, G., 2001. *Origin of Igneous Rocks; the Isotopic Evidence*: Springer-Verlag, Heidelberg.

Francis, P. W., R. S. J. Sparks, C. H. Hawkesworth, R. S. Thorpe, D. M. Pyle, S. R. Tait, M. S. Mantovani, and F. McDermott, 1989. Petrology and geochemistry of volcanic rocks of the Cerro Galan caldera, northwest Argentina. *Geol. Mag.*, 126:515–547.

Harmon, R. S., and B. A. Barreiro (Eds.), 1984. *Andean Magmatism: Chemical and Isotopic Constraints*. Shiva Pub., Nantwich, United Kingdom.

Harmon, R. S., B. A. Barreiro, S. Moorbath, J. Hoefs, P. W. Francis, R. S. Thorpe, B. Déruelle, J. McHugh, and J. A. Viglino, 1984. Regional O-, Sr-, and Pb-isotope relations in Late Cenozoic calc-alkaline lavas of the Andean Cordillera. *J. Geol. Soc. Lond.*, 141:803–822.

Hawkesworth, C. J., M. Hammill, A. R. Gledhill, P. van Calsteren, and G. Rogers, 1982. Isotope and trace element evidence for late-stage intra-crustal melting in the High Andes. *Earth Planet. Sci. Lett.*, 58:240–254.

James, D. E., 1982. A combined O, Sr, Nd, and Pb isotopic and trace element study of crustal contamination in central Andean lavas, I. Local geochemical variations. *Earth Planet. Sci. Lett.*, 57:47–62.

McNutt, R. H., J. H. Crocket, and M. Zentilli, 1975. Initial $^{87}Sr/^{86}Sr$ ratio of plutonic and volcanic rocks of the Central Andes between latitude 26° and 29° south. *Earth Planet. Sci. Lett.*, 27:305–313.

Pitcher, W. S., M. P. Atherton, E. J. Cobbing, and R. D. Beckinsale (Eds.), 1985. *Magmatism at a plate edge: The Peruvian Andes*. Blackie Halsted, Glasgow, Scotland.

Siebel, W., B. W. Schnurr, K. Hahne, B. Kraemer, R. B. Trumbull, P. van den Bogaard, and R. Emmermann, 2001. Geochemistry and isotopic systematics of small- to medium-volume Neogene-Quaternary ignimbrites in the southern central Andes: Evidence for derivation from andesitic magma sources. *Chem. Geol.*, 171:213–237.

Thorpe, R. S., P. W. Francis, and S. Moorbath, 1979. Rare earth and strontium isotopes—evidence concerning the petrogenesis of north Chilean ignimbrites. *Earth Planet. Sci. Lett.*, 42:359–367.

REFERENCES FOR COLUMBIA RIVER BASALT, UNITED STATES (SECTION 17.8 INTRODUCTION, SECTION 17.8a)

Carlson, R. W., 1984. Isotopic constraints on Columbia River flood basalt genesis and the nature of the subcontinental mantle. *Geochim. Cosmochim. Acta*, 48:2357–2372.

Carlson, R. W., G. W. Lugmair, and J. D. Macdougall, 1981. Columbia River volcanism: The question of mantle heterogeneity or crustal contamination. *Geochim. Cosmochim. Acta*, 45:2483–2499.

Carlson, R. W., G. W. Lugmair, and J. D. Macdougall, 1983. "Columbia River volcanism: The question of mantle heterogeneity or crustal contamination" (reply to a comment by DePaolo). *Geochim. Cosmochim. Acta*, 47:845–846.

DePaolo, D. J., 1983. Comment on "Columbia River volcanism: The question of mantle heterogeneity or crustal contamination" by R. W. Carlson, G. W. Lugmair and J. D. Macdougall. *Geochim. Cosmochim. Acta*, 47:841–844.

Elliot, D. H., 1974. The tectonic setting of the Jurassic Ferrar Group, Antarctica. In O. Gonzalez-Ferran (Ed.), *Proceedings of the Symposium on Andean and Antarctic Volcanology Problems*, Special Series, pp. 357–372, Santiago, Chile, Int. Assoc. Volcan. and Chem. of the Earth's Interior.

Faure, G., 2001. *Origin of Igneous Rocks; the Isotopic Evidence*. Springer-Verlag, Heidelberg.

Hart, W. K., 1985. Chemical and isotopic evidence for mixing between depleted and enriched mantle, northwestern USA. *Geochim. Cosmochim. Acta*, 49:131–144.

Hart, W. K., R. W. Carlson, and S. B. Shirey, 1997. Radiogenic Os in primitive basalts from the northwestern USA: Implications for petrogenesis. *Earth Planet. Sci. Lett.*, 150:103–116.

Hooper, P. R., 1982. The Columbia River basalts. *Science*, 215:1463–1468.

Ripley, E. M., Y. -R. Park, D. D. Lambert, and L. R. Frick, 2001. Re-Os isotopic variations in carbonaceous pelites hosting the Duluth Complex: Implications for metamorphic and metasomatic processes associated with mafic magma chambers. *Geochim. Cosmochim. Acta*, 65:2965–2978.

Swanson, D. A., T. L. Wright, P. R. Hooper, and R. D. Bentley, 1979. Revisions in stratigraphic nomenclature of the Columbia River Basalt Group. *US Geol. Surv. Bull.*, 1457G:1–59.

REFERENCES FOR PARANÁ BASALT, BRAZIL (SECTION 17.8b)

Ewart, A., S. C. Milner, R. A. Armstrong, and A. R. Duncan, 1998. Etendeka volcanism of the Goboboseb Mountains and Messum igneous complex, Namibia. Part II: Voluminous quartz latite volcanism of the Awahab magma system. *J. Petrol.*, 39:227–253.

Faure, G., 2001. *Origin of Igneous Rocks; the Isotopic Evidence*. Springer-Verlag, Heidelberg.

Fodor, R. V., 1987. Low- and high-TiO_2 flood basalts of southern Brazil: Origin from picritic parentage and a common mantle source. *Earth Planet. Sci. Lett.*, 84:423–430.

Mantovani, M. S. M., L. S. Marques, M. A. de Sousa, L. Civetta, L. Atalla, and F. Innocenti, 1985. Trace element and strontium isotope constraints on the origin and evolution of Paraná continental flood basalts of Santa Catarina State (southern Brazil). *J. Petrol.*, 26:187–209.

Milner, S. C., A. R. Duncan, A. M. Whittingham, and A. Ewart, 1995. Trans-Atlantic correlation of eruptive sequences and individual silicic volcanic units within the Paraná-Etendeka igneous province. *J. Volcanol. Geotherm. Res.*, 69:137–157.

Murata, K. J., M. L. L. Formoso, and A. Roisenberg, 1987. Distribution of zeolites in lavas of southeastern Paraná basin, State of Rio Grande do Sul, Brazil. *J. Geol.*, 95:455–467.

Peate, D. W., and C. J. Hawkesworth, 1996. Lithospheric to asthenospheric transition in low-Ti flood basalts from southern Paraná, Brazil. *Chem. Geol.*, 127:1–24.

Peate, D. W., C. J. Hawkesworth, and M. S. M. Mantovani, 1992. Chemical stratigraphy of the Paraná lavas (South America): Classification of magma types and their spatial distribution. *Bull. Volcanol.*, 55:119–139.

Peate, D. W., C. J. Hawkesworth, M. S. M. Mantovani, N. W. Rogers, and S. P. Turner, 1999. Petrogenesis and stratigraphy of the high Ti/Y Urubici magma type in the Paraná basalt province and implications for the nature of "Dupal"-type mantle in the South Atlantic region. *J. Petrol.*, 40:451–473.

REFERENCES FOR ALKALI-RICH LAVAS: ITALY AND WYOMING (SECTION 17.9)

Chauvel, C., W. McDonough, G. Guille, R. Maury, and R. A. Duncan, 1997. Contrasting old and young volcanism on Rurutu Island, Austral chain. *Chem. Geol.*, 139:125–143.

Faure, G., 2001. *Origin of Igneous Rocks; the Isotopic Evidence*. Springer-Verlag, Heidelberg.

Hawkesworth, C. J., N. W. Rogers, P. W. C. van Calsteren, and M. A. Menzies, 1984. Mantle enrichment processes. *Nature*, 311:331–335.

Hawkesworth, C. J., and R. Vollmer, 1979. Crustal contamination versus enriched mantle: $^{143}Nd/^{144}Nd$ and $^{87}Sr/^{86}Sr$ evidence from the Italian volcanics. *Contrib. Mineral. Petrol.*, 69:151–165.

Peccerillo, A., 1990. On the origin of the Italian potassic magmas—comments. *Chem. Geol.*, 85:183–196.

Vollmer, R., 1989. On the origin of Italian potassic magmas: A discussion contribution. *Chem. Geol.*, 74:229–239.

Vollmer, R., 1990. On the origin of the Italian magmas—reply. *Chem. Geol.*, 85:191–196.

Vollmer, R., P. Odgen, J. -G. Schilling, R. H. Kingsley, and D. G. Waggoner, 1984. Nd and Sr isotopes in ultrapotassic volcanic rocks from the Leucite Hills, Wyoming. *Contrib. Mineral. Petrol.*, 87:359–368.

REFERENCES FOR ORIGIN OF GRANITE (SECTION 17.10)

Allègre, C. J., and D. B. Othman, 1980. Nd-Sr isotopic relationship in granitoid rocks and continental crust development: A chemical approach to orogenesis. *Nature*, 286:335–342.

Armstrong, R. L., W. H. Taubeneck, and P. O. Hales, 1977. Rb-Sr and K-Ar geochronometry of Mesozoic granitic rocks and their Sr isotopic composition, Oregon, Washington, and Idaho. *Geol. Soc. Am. Bull.*, 88:397–411.

Barbarin, B., and W. E. Stephens (Eds.), 2001. *The Fourth Hutton Symposium on the Origin of Granites and Related Rocks*, Special Papers, p. 350. Geological Society of America, Boulder, Colorado.

Barton, J. M., Jr., 1975. Rb-Sr isotopic characteristics and chemistry of the 3.6 b.y. Hebron gneiss, Labrador. *Earth Planet. Sci. Lett.*, 27:427–435.

Bowring, S. A., I. S. Williams, and W. Compston, 1989. 3.96 Ga gneisses from the Slave Province, Northwest Territories, Canada. *Geology*, 17:971–975.

Brown, M., and P. M. Piccoli (Eds.), 1995. *The Origin of Granites and Related rocks*, Circular 1129. U.S. Geological Survey, Washington, D.C.

Chappell, B. W., and A. J. R. White, 1974. Two contrasting granite types. *Pacific Geol.*, 8:173–174.

Compston, W., and R. T. Pidgeon, 1986. Jack Hills, evidence for more very old detrital zircons in western Australia. *Nature*, 321:766–769.

DePaolo, D. J., 1980. Sources of continental crust: Neodymium isotope evidence from the Sierra Nevada and Peninsular Ranges. *Science*, 209:684–687.

DePaolo, D. J., 1981. A neodymium and strontium isotopic study of the Mesozoic calc-alkaline granitic batholiths of the Sierra Nevada and Peninsular Ranges, California. *J. Geophys. Res.*, 86(B11):10470–10488.

DePaolo, D. J., W. I. Manton, E. S. Grew, and M. Halpern, 1982. Sm-Nd, Rb-Sr, and U-Th-Pb systematics of granulite-facies rocks from Fyfe Hills, Enderby Land, Antarctica. *Nature*, 298:614–618.

Eby, G. N., 1987. The Monteregian Hills and White Mountain alkaline igneous provinces, eastern North America. In J. G. Fitton and B. G. Upton (Eds.), *Alkaline Igneous Rocks*, Spec. Pub. 30, pp. 433–447. Geological Society of London, London, England

Eby, G. N., 1990. The A-type granitoids: A review of their occurrence and chemical characteristics and speculations on their petrogenesis. *Lithos*, 26:115–134.

Farhat, J. S., and G. W. Wetherill, 1975. Interpretation of apparent ages in Minnesota. *Nature*, 257:721–722.

Faure, G., 2001. *Origin of Igneous Rocks; the Isotopic Evidence*. Springer-Verlag, Heidelberg.

Faure, G., and P. M. Hurley, 1963. The isotopic composition of strontium in oceanic and continental basalt: Application to the origin of igneous rocks. *J. Petrol.*, 4:31–50.

Faure, G., and J. L. Powell, 1972. *Strontium Isotope Geology*. Springer-Verlag, Heidelberg.

Flood, R. H., and S. E. Shaw, 1977. Two "S-type" granite suites with low initial $^{87}Sr/^{86}Sr$ ratios from the New England batholith, Australia. *Contrib. Mineral. Petrol.*, 61:163–173.

Foland, K. A., and J. C. Allen, 1991. Magma sources of Mesozoic anorogenic granites of the White Mountain magma series, New England, USA. *Contrib. Mineral. Petrol.*, 109:195–211.

Froude, D. O., T. R. Ireland, P. D. Kinny, I. S. Williams, W. Compston, and J. S. Myers, 1983. Ion microprobe identification of 4100–4200 Myr old terrestrial zircons. *Nature*, 304:616–618.

Fullagar, P. D., R. E. Lemmon, and P. C. Ragland, 1971. Petrochemical and geochronological studies of plutonic rocks in the southern Appalachians. Part 1. The Salisbury pluton. *Geol. Soc. Am. Bull.*, 82:409–416.

Goldich, S. S., and C. E. Hedge, 1974. 3,800-Myr granitic gneiss in south-western Minnesota. *Nature*, 252:467–468.

Gunner, J. D., 1974. Investigations of lower Paleozoic granites in the Beardmore Glacier region. *Ant. J. US*, 9:76–81.

Gunner, J. D., and G. Faure, 1972. Rb-Sr geochronology of the Nimrod Group, central Transantarctic Mountains. In R. J. Adie (Ed.), *Antarctic Geology and Geophysics*, pp. 305–311. Universitets forlaget Oslo, Oslo.

Hamilton, P. J., R. K. O'Nions, and R. J. Pankhurst, 1980. Isotopic evidence for the provenance of some Caledonian granites. *Nature*, 287:279–284.

Hurley, P. M., P. C. Bateman, H. W. Fairbairn, and W. H. Pinson, Jr., 1965. Investigation of initial $^{87}Sr/^{86}Sr$ ratios in the Sierra Nevada plutonic province. *Geol. Soc. Am. Bull.*, 76:165–174.

Jacquemin, H., S. M. F. Sheppard, and P. Vidal, 1982. Isotope geochemistry (O, Sr, Pb) of the Golda Zuelva and Mboutou anorogenic complexes, North Cameroon: Mantle origin with evidence for crustal contamination. *Earth Planet. Sci. Lett.*, 61:97–111.

Kistler, R. W., and Z. E. Peterman, 1973. Variations in Sr, Rb, K, Na and initial $^{87}Sr/^{86}Sr$ in Mesozoic granitic rocks and intruded wall rocks in central California. *Geol. Soc. Am. Bull.*, 84:3489–3512.

Kistler, R. W., and Z. E. Peterman, 1978. *Reconstruction of Crustal Blocks of California on the Basis of Initial Strontium Isotopic Compositions of Mesozoic Granitic Rocks*, U.S. Geol. Surv. Prof. paper 1071. U.S. Geological Survey, Washington, D.C.

Long, L. E., A. N. Sial, H. Nekvasil, and G. S. Borba, 1986. Origin of the granite at Cabo de Santo Agostinho, northeast Brazil. *Contrib. Mineral. Petrol.*, 92:341–350.

Maas, R., P. D. Kinny, I. S. Williams, D. O. Froude, and W. Compston, 1992. The earth's oldest known crust: A geochronological and geochemical study of 3900–4200 Ma old detrital zircons from Mt. Narryer and Jack Hills, Western Australia. *Geochim. Cosmochim. Acta*, 56:1281–1300.

McCulloch, M. T., and B. W. Chappell, 1982. Nd isotopic characteristics of S and I-type granites. *Earth Planet. Sci. Lett.*, 58:51–64.

McCulloch, M. T., and G. J. Wasserburg, 1978. Sm-Nd and Rb-Sr chronology of continental crust formation. *Science*, 200:1003–1011.

Moorbath, S., J. H. Allaart, D. Bridgwater, and V. R. McGregor, 1977. Rb-Sr ages of Early Archaean supracrustal rocks and Amitsoq gneisses at Isua. *Nature*, 270:43–44.

Moorbath, S., R. K. O'Nions, and R. J. Pankhurst, 1973. Early Archean age for the Isua iron formation, West Greenland. *Nature*, 245:138–139.

Moorbath, S., R. K. O'Nions, and R. J. Pankhurst, 1975. The evolution of early Precambrian crustal rocks of Isua, West Greenland—geochemical and isotopic evidence. *Earth Planet. Sci. Lett.*, 27:229–239.

Moorbath, S., R. K. O'Nions, R. J. Pankhurst, N. H. Gale, and V. R. McGregor, 1972. Further rubidium-strontium age determinations on the very early Precambrian rocks of the Godthaab district, West Greenland. *Nature*, 240:78–82.

Moorbath, S., J. F. Wilson, and P. Cotterill, 1976. Early Archean age for the Sebakwian Group at Selukwe, Rhodesia. *Nature*, 264:536–538.

van Breemen, O., J. Hutchison, and P. Bowden, 1975. Age and origin of the Nigerian Mesozoic granites: A Rb-Sr isotopic study. *Contrib. Mineral. Petrol.*, 40:157–172.

Wyllie, P. J., 1977. Crustal anatexis: An experimental review. *Tectonophysics*, 43:41–71.

Zindler, A., S. R. Hart, and C. Brooks, 1981. The Shabogamo intrusive suite, Labrador: Sr and Nd isotopic evidence for contaminated mafic magmas in the Proterozoic. *Earth Planet. Sci. Lett.*, 54:217–235.

CHAPTER 18

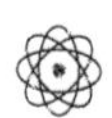

Water and Sediment

WATER and sediment play an important role in the global geological transport system that connects the continents to the oceans. The water draining the continents contains chemical elements in solution as well as particles in suspension. The suspended particles are part of the sediment load that is moved continuously by running water, wind, and continental ice sheets from the continents into the ocean basins. These fundamental geological processes not only cause the erosion of the continents and the deposition of sediment in the oceans but also affect the chemical composition of seawater.

The chemical composition of water on the continents is determined largely by the congruent and incongruent dissolution of rock-forming minerals, by the sorption of ions to the surfaces of suspended particles, and by the uptake of nutrients by plants and animals. The processes that govern the chemical composition of water on the continents are the domain of aqueous geochemistry and have been presented in textbooks by Drever (1997), Berner and Berner (1987, 1996), Langmuir (1997), Appelo and Postma (1993), Stumm (1990), Stumm and Morgan (1996), Faure (1998), and others.

The water, as well as the sediment entrained in it, contains elements that have radiogenic isotopes that formed by decay of their long-lived radioactive parent nuclides in the rocks of the continental crust. Therefore, the isotopic compositions of Sr, Nd, Pb, Hf, and Os in water and sediment vary depending on the ages and parent–daughter ratios of the bedrock exposed to weathering in the drainage basins of the continents. In cases where the geology of a drainage basin includes rocks of different ages and chemical compositions, the chemical and isotopic compositions of the water and sediment are the result of mixing of two or more components in varying proportions. The isotopic composition and concentration of the elements dissolved in the water of streams and lakes can be used to obtain information about the geology of drainage basins, about the rates of weathering of different rocks and minerals, about anthropogenic contamination of the surface and subsurface environment, and about the input of these elements to the oceans. The amount of available information for some of the elements is limited because their concentrations in water and sediment are low (e.g., Hf and Os) and because much work is still in progress.

18.1 STRONTIUM IN STREAMS

The concentrations of Sr in streams and lakes vary widely depending primarily on the mineral composition of the bedrock and on climatic factors (Palmer and Osmond, 1992). A review of the literature indicates that surface water on the Precambrian Shield of Canada and in the Appalachian Mountains contains less than 100 μg/L (ppb) of Sr in most cases, whereas water in the midwestern and southwestern states of the United States has high Sr concentrations in excess of 500 μg/L. The surface water in the midwestern states is Sr rich because this region is underlain by marine carbonate rocks of Paleozoic age that contain celestite ($SrSO_4$) and strontianite ($SrCO_3$) (Feulner and Hubble, 1960). The elevated Sr concentrations of water in the southwestern states are attributable to the effects of evaporative concentration caused by the semiarid to arid climatic conditions of this region.

The isotope composition of Sr in surface water in a drainage basin underlain by a variety of rock types depends on

1. the $^{87}Sr/^{86}Sr$ ratios and Sr concentrations of each rock type,
2. the area of the surface exposure of different rock types,
3. the susceptibility to chemical weathering of the minerals these rocks contain, and
4. mixing of water derived from different rock types in the streams draining the basin.

In most cases, the isotope composition of Sr in surface water is dominated by the rocks that are most susceptible to dissolution, have the highest Sr concentration, and have the largest surface exposure. In addition, the isotope composition of Sr in streams changes continually downstream as a result of mixing with water of tributaries, groundwater, and surface runoff. These considerations give meaning to the statement that the isotope compositions of Sr and other elements (e.g., Nd, Pb, Hf, and Os) in surface water are controlled in complex ways by the geology of the drainage basin.

The relation between the geology of a drainage basin and the isotope composition of Sr in streams and lakes leads to the insight that the $^{87}Sr/^{86}Sr$ ratios of lacustrine carbonate rocks that were deposited in the geological past can vary stratigraphically *only* as a result of changes in the geology of the basin. The geological events that can cause the $^{87}Sr/^{86}Sr$ ratios of lacustrine carbonates to change include

1. deposition of volcanic ash or extrusion of lava flows within the drainage basin,
2. exposure of granitic basement rocks or other rock types as a result of the erosion of the basin by ice sheets or streams,
3. complete removal of a distinctive Sr source by continental glaciation or stream erosion, and
4. deposition of sediment derived from distant sources by continental glaciation or wind.

Stratigraphic variations in the lacustrine carbonate rocks of the Tertiary Flagstaff Formation in Utah were first observed and interpreted by Neat et al. (1979). Similar studies of $^{87}Sr/^{86}Sr$ ratios of the lacustrine carbonates of the Eocene Green River Formation of Wyoming are by Rhodes et al. (2002) and of $^{18}O/^{16}O$ ratios by Morrill and Kock (2002). The stratigraphic variations of the isotopic composition of Sr, Nd, Pb, Hf, and Os in recently deposited lacustrine sediment also record the onset of industrial and agricultural activities, mining, and human habitation in historical time.

18.1a Rivers, Precambrian Shield, Canada

The radiogenic ^{87}Sr in Rb-bearing rocks and minerals of Precambrian cratons is released into surface water as a result of chemical weathering and is ultimately transported into the oceans (Faure et al., 1967). This phenomenon was illustrated in Figure 16.11 by mixing of the water of Lake Superior ($^{87}Sr/^{86}Sr = 0.720$) with water of Lake Huron ($^{87}Sr/^{86}Sr = 0.710$). The presence of radiogenic ^{87}Sr in surface water on the Precambrian Shield of Canada was first demonstrated by Faure et al. (1963), who analyzed the Sr in mollusk shells from lakes and rivers in northern Ontario. The results confirmed that the Sr in the surface water on the Precambrian Shield of Canada has elevated $^{87}Sr/^{86}Sr$ ratios with an arithmetic mean of 0.7163 ± 0.0019 for 15 samples relative to 0.7080 for the E&A Sr isotope standard.

In a subsequent study of the 39 largest Canadian rivers, Wadleigh et al. (1985) reported $^{87}Sr/^{86}Sr$ ratios ranging from 0.70460 (Skeena River, British Columbia) to 0.73844 (Grande Riviere de la Baleine, Nouveau Quebec) relative to 0.7080 for E&A. The three largest rivers in Canada have comparatively low $^{87}Sr/^{86}Sr$ ratios but high Sr concentrations:

St. Lawrence	0.70946, 137 μg/L
Mackenzie	0.71102, 175 μg/L
Fraser	0.71195, 80 μg/L

A comprehensive study of the isotope compositions of Sr, S, O, C, and H of water in the St. Lawrence River and its tributaries extending upstream from Quebec City to the southern end of Lake Huron was later published by Yang et al. (1996).

The combined discharge of the 39 rivers included in the study by Wadleigh et al. (1985) represents 48% of the total amount of water that drains the surface area of Canada. The average Sr concentration of this water can be calculated from the equation

$$\mathrm{Sr}_M = \mathrm{Sr}_1\, f_1 + \mathrm{Sr}_2\, f_2 + \cdots + \mathrm{Sr}_n\, f_n \quad (18.1)$$

where $\mathrm{Sr}_1, \mathrm{Sr}_2, \ldots, \mathrm{Sr}_n$ are the concentrations of Sr in water of rivers $1, 2, \ldots, n$,

$$f_1, f_2, \ldots, f_n = \frac{D_1}{\sum D_n}, \frac{D_2}{\sum D_n}, \ldots, \frac{D_n}{\sum D_n}$$

and $D_1, D_2, \ldots, D_n$ are the average annual discharges of the individual rivers included in the summation. The resulting average Sr concentration weighted by the discharge of each of the 39 rivers is 83.6 μg/L.

The average $^{87}\mathrm{Sr}/^{86}\mathrm{Sr}$ ratio of the water weighted by both the Sr concentrations and the discharges of each of the 39 rivers is obtained from the equation

$$\left(\frac{^{87}\mathrm{Sr}}{^{86}\mathrm{Sr}}\right)_M = \left(\frac{^{87}\mathrm{Sr}}{^{86}\mathrm{Sr}}\right)_1 \frac{\mathrm{Sr}_1}{\mathrm{Sr}_M} f_1 + \left(\frac{^{87}\mathrm{Sr}}{^{86}\mathrm{Sr}}\right)_2 \frac{\mathrm{Sr}_2}{\mathrm{Sr}_M} f_2 + \cdots + \left(\frac{^{87}\mathrm{Sr}}{^{86}\mathrm{Sr}}\right)_n \frac{\mathrm{Sr}_n}{\mathrm{Sr}_M} f_n \quad (18.2)$$

where $\mathrm{Sr}_M = 83.6$ μg/L and the f factors are derived from the discharge as in equation 18.1. The weighted-average $^{87}\mathrm{Sr}/^{86}\mathrm{Sr}$ ratio of the 39 Canadian rivers included in the study by Wadleigh et al. (1985) is 0.71184, which is a more appropriate estimate of the isotope composition of Sr in surface water than the unweighted arithmetic mean of 0.71703 calculated from the same data. The wide range of $^{87}\mathrm{Sr}/^{86}\mathrm{Sr}$ ratios and Sr concentrations of the Canadian rivers included in the study of Wadleigh et al. (1985) reflects the diversity of the rocks in their drainage basins. The average concentrations and isotope ratios of Sr in the rivers that drain the major tectonic provinces of Canada are listed in Table 18.1.

Table 18.1. Average Sr Concentrations and $^{87}\mathrm{Sr}/^{86}\mathrm{Sr}$ Ratios of Rivers on the Precambrian Shield of Canada in Relation to the Ages of the Rocks Underlying Their Drainage Basins

Tectonic Province	Age, Ga	Sr, μg/L	$^{87}\mathrm{Sr}/^{86}\mathrm{Sr}$
Superior[a]	>2.5	9.6	0.7295
		6–14	0.71855–0.73844
Churchill[b]	~1.8	15.6	0.7248
		8–23	0.71882–0.73013
Grenville[c]	~1.0	12.6	0.7151
		7–23	0.71307–0.71864
Paleozoic[d]	<0.6	87.5	0.71170
		59–137	0.70946–0.71563

Source: Wadleigh et al., 1985.

[a] Nottaway, Rupert, La Grande, Eastmain, Aux Feuilles, A la Baleine, Grande Rivere de la Baleine, and Arnaud.

[b] Koksoak, Churchill (Manitoba), Back, Thelon, and Kazan.

[c] Churchill (Labrador, Newfoundland), Saguenay, Manicouagan, Petit Mecatina, Moisie, Aux Outardes, and Natashquan.

[d] St. Lawrence, Fraser, Columbia, and St. John.

The results demonstrate a consistent relationship between the $^{87}\mathrm{Sr}/^{86}\mathrm{Sr}$ ratios of rivers and the crustal stabilization age of the tectonic province in which these rivers occur. For example, the highest $^{87}\mathrm{Sr}/^{86}\mathrm{Sr}$ ratios (0.71855–0.73844) exist in rivers draining the Superior tectonic province, which was subjected to the Kenoran orogeny at about 2.5 Ga (Section 6.7). Low $^{87}\mathrm{Sr}/^{86}\mathrm{Sr}$ ratios (0.70946–0.71563) occur in sedimentary rocks of Paleozoic age in the area south and west of the Hudson Bay, in the St. Lawrence lowlands, in southwestern Ontario, and in southern Alberta. Still lower $^{87}\mathrm{Sr}/^{86}\mathrm{Sr}$ ratios (0.7046–0.70542) were reported from the Skeena, Nass, and Stikine Rivers in British Columbia where volcanic rocks of Cenozoic ages are exposed. In addition, Table 18.1 demonstrates that the rivers draining igneous and metamorphic rocks have uniformly low Sr concentrations ranging from 6 to 23 μg/L, whereas the rivers draining sedimentary rocks of Paleozoic age

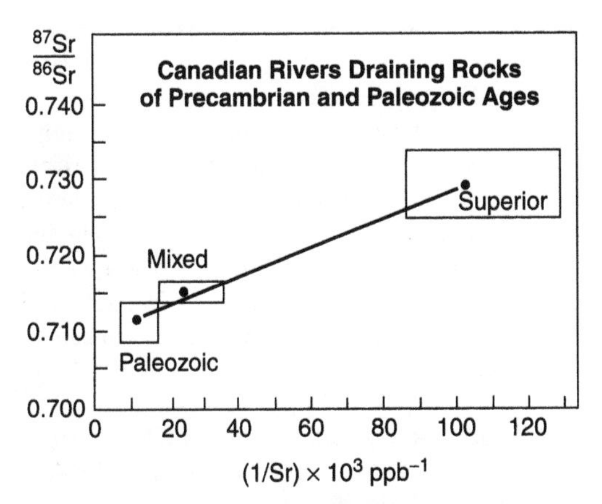

FIGURE 18.1 Mixing of water draining igneous and metamorphic rocks of the Superior tectonic province and sedimentary rocks of Paleozoic age on the Precambrian Shield of Canada. The rivers draining both types of rocks are the Nelson, Albany, Severn, Moose, Winisk, Porcupine, Hayes, Attawapiskat, and Harricana. The rivers draining only the Superior tectonic province and Paleozoic sedimentary rocks are identified in Table 18.1. Data from Wadleigh et al. (1995).

have much higher Sr concentrations between 59 and 137 μg/L.

The drainage basins of several Canadian rivers are underlain by the schists and gneisses of the Superior tectonic province as well as by sedimentary rocks (including carbonates) of Paleozoic age. The Sr isotope ratios of the water in these rivers are consistently lower than those of rivers that drain only rocks of the Superior tectonic province. The rivers that drain both types of rocks have an average $^{87}Sr/^{86}Sr$ ratio of 0.7155 ± 0.0015 ($2\overline{\sigma}$, $N = 9$) and an average Sr concentration of 39.9 ± 11.8 μg/L ($2\overline{\sigma}$, $N = 9$). These values define a point close to the Sr isotope mixing line in Figure 18.1 and thereby confirm that the Sr in these rivers is a mixture of two components originating from rocks of the Superior Province and from Paleozoic sedimentary rocks. The abundance of Sr derived from the Paleozoic rocks (f) is calculated from equation 16.4:

$$Sr_M = Sr_{Pal}\, f + Sr_{Sup}(1 - f)$$

$$39.9 = 87.5\, f + 9.6(1 - f) \qquad (18.3)$$

$$f = 0.40$$

This result demonstrates that the Sr-rich water draining Paleozoic carbonates effectively depresses the $^{87}Sr/^{86}Sr$ ratios of Sr-poor water draining igneous and metamorphic rocks of the Superior Province.

Another noteworthy result that arises from the work of Wadleigh et al. (1985) is that the data in Table 18.1 define a straight line in Figure 18.2. The equation of this line is

$$\frac{^{87}Sr}{^{86}Sr} = \left(\frac{^{87}Sr}{^{86}Sr}\right)_i + 2.89\left(\frac{Rb}{Sr}\right)_{conc} \lambda t \qquad (18.4)$$

from which the Rb/Sr ratio of the surface rocks of the Canadian Shield can be estimated:

$$\left(\frac{Rb}{Sr}\right)_{conc} = \frac{^{87}Sr/^{86}Sr - (^{87}Sr/^{86}Sr)_i}{2.89\lambda t}$$

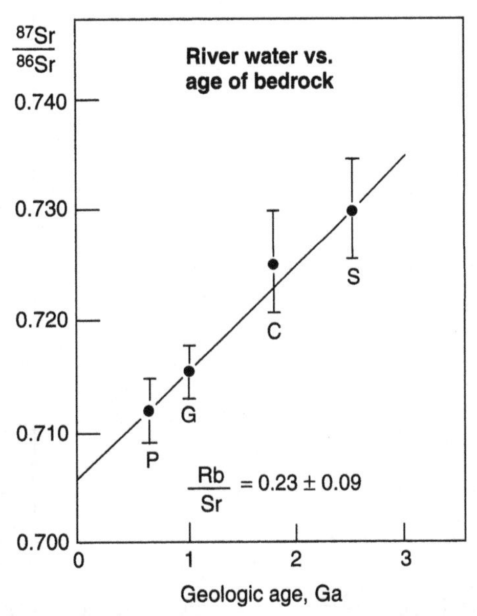

FIGURE 18.2 Average $^{87}Sr/^{86}Sr$ ratios of Canadian rivers draining tectonic provinces of different ages: S = Superior; C = Churchill; G = Grenville; P = Paleozoic. The error bars are two standard deviations of the mean, as indicated in Table 18.1. Data from Wadleigh et al. (1985).

Since the rivers draining rocks of the Superior tectonic province have $^{87}Sr/^{86}Sr = 0.7295$ and since $(^{87}Sr/^{86}Sr)_i = 0.7055$ (Figure 18.2),

$$\left(\frac{Rb}{Sr}\right)_{conc} = \frac{0.7295 - 0.7055}{2.89 \times 1.42 \times 10^{-11} \times 2.5 \times 10^9}$$

$$= 0.23 \pm 0.09$$

This value is similar to direct determinations of the Rb/Sr ratio of igneous rocks compiled in Figure 18.3.

The results of Faure et al. (1963) and Wadleigh et al. (1985) demonstrate that the rocks of the Precambrian Shield of Canada, like other cratons on the continents of the world, are releasing radiogenic ^{87}Sr into the oceans. However, the Sr-rich water draining marine carbonate rocks of Phanerozoic age masks the elevated $^{87}Sr/^{86}Sr$ ratios of the Sr-poor water draining Precambrian cratons.

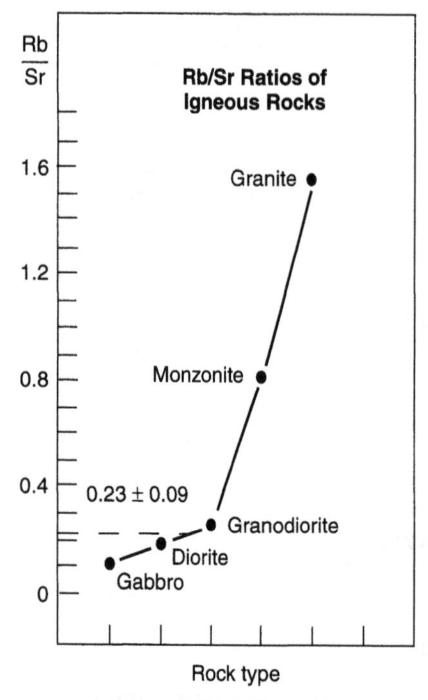

FIGURE 18.3 Average Rb/Sr ratios of igneous rocks ranging in composition from gabbro to granite. The Rb/Sr ratio of the Canadian Precambrian Shield derived in Figure 18.2 from the $^{87}Sr/^{86}Sr$ ratios of river water (0.23 ± 0.09) implies that the rocks range in composition from diorite to granodiorite. Data from Faure (1978).

The chemical compositions and the isotope ratios of Sr and other elements in the major rivers of the world have been studied in order to derive information about the rates of chemical weathering under various climatic conditions and to quantify the fluxes of dissolved elements and suspended sediment being transported from the continents to the oceans:

Weathering: Gaillardet et al., 1999, *Chem. Geol.*, 159:3–30. Benedetti et al., 1994, *Chem. Geol.*, 118:203–220. Viers et al., 2000, *Chem. Geol.*, 169:211–241. Probst et al., 2000, *Chem. Geol.*, 170:203–219.

Baltic Shield: Wickman and Åberg, 1987, *Nordic Hydrol.*, 18:21–32. Åberg and Wickman, 1987, *Nordic Hydrol.*, 18:33–42. Löfvendahl et al., 1990, *Aquat. Sci.*, 52:315–329.

France: Semhi et al., 2000, *Chem. Geol.*, 168:173–193.

Himalayas: Pande et al., 1994, *Chem. Geol.*, 116:245–259. English et al., 2000, *Geochim. Cosmochim. Acta*, 64:2549–2566. Derry and France-Lanord, 1996, *Earth Planet. Sci. Lett.*, 142:59–74.

South America: Palmer and Edmond, 1992, *Geochim. Cosmochim. Acta.*, 56:2099–2111.

18.1b Groundwater, Precambrian Shield, Canada

The chemical-weathering reactions on the Precambrian Shield of Canada may achieve a state of equilibrium when the $^{87}Sr/^{86}Sr$ ratios of the water approach those of the rocks. For this reason, McNutt et al. (1987) measured the $^{87}Sr/^{86}Sr$ ratios of a large number of whole-rock samples of the East Bull Lake intrusive in the Superior Province of Ontario and compared them to the $^{87}Sr/^{86}Sr$ ratios of groundwater collected from deep diamond-drill holes in the same intrusive.

The East Bull Lake pluton is located northwest of the town of Massey, Ontario, on the north shore of Lake Huron, close to the southern margin of the Superior Province. It is composed of layered mafic rocks ranging in composition from anorthosite to granophyre and yielded a U–Pb concordia date of 2480 ± 10 Ma based on analyses of zircon and baddeleyite (Krogh et al., 1984). The whole-rock samples of the East Bull Lake intrusive analyzed by McNutt et al. (1987) scatter widely on the Rb–Sr isochron diagram in Figure 18.4 and do not satisfy the assumptions for dating by this method. A selected set of unaltered samples yields a Rb–Sr errorchron date of 1.8 Ga, whereas the altered samples taken from four fracture zones in deep cores lead to a Rb–Sr date of 0.65 Ga. Neither one of these dates can be attributed to a recognizable geological event.

The water samples have average concentrations of Sr = 61.7 (5–148) μg/L and Rb = 26.7 (20–40) μg/L with an average Rb/Sr ratio of 0.43. Two water samples with anomalously high Sr concentrations (2310 and 2900 μg/L) were excluded from the average. The histograms in Figure 18.5 indicate that the groundwater has an average $^{87}Sr/^{86}Sr$ ratio of 0.71306 (0.71074–0.72023) for 28 samples relative to 0.7080 for E&A. The isotope composition of Sr in secondary minerals (laumontite, calcite, and gypsum) recovered from fractures in the cores is similar to that of the groundwater with an average $^{87}Sr/^{86}Sr$ ratio of 0.71289 (0.71102–0.71963) for 14 samples. The whole-rock samples of the East Bull Lake complex have a wide range of $^{87}Sr/^{86}Sr$ ratios between 0.70478 and 0.7151 with a mean of 0.70943 for 57 samples.

The $^{87}Sr/^{86}Sr$ ratios of the bedrock of the East Bull Lake complex overlap those of the groundwater and the fracture minerals. Therefore, the Sr in the groundwater could have originated from the bedrock by selective dissolution of rocks and/or minerals whose $^{87}Sr/^{86}Sr$ ratios were in the

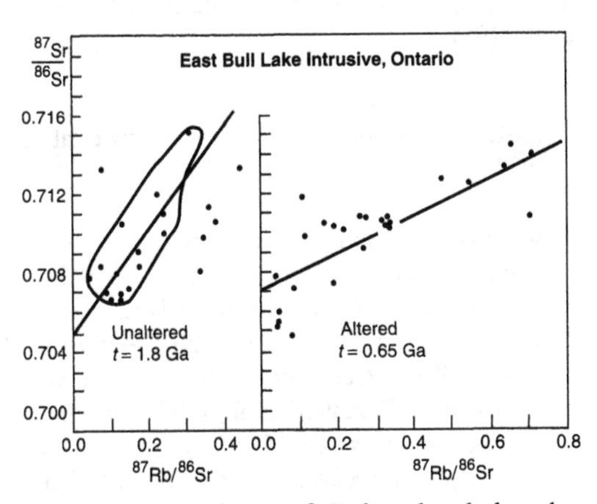

FIGURE 18.4 Errorchrons of unaltered and altered whole-rock samples recovered in drill cores from the East Bull Lake intrusive located in Ontario near the southern edge of the Superior structural province. The U–Pb concordia date of the intrusive is 2480 ± 10 Ma (Krogh et al., 1984). Data from McNutt et al. (1987).

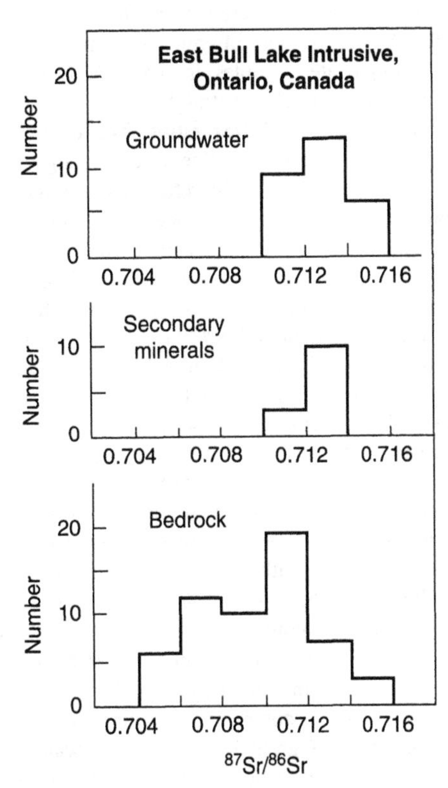

FIGURE 18.5 Histograms of $^{87}Sr/^{86}Sr$ ratios of groundwater, fracture-filling minerals, and bedrock samples of the East Bull Lake intrusion, Superior tectonic province, Ontario, Canada. The $^{87}Sr/^{86}Sr$ ratio of E&A was 0.7080. Data from McNutt et al. (1987).

upper part of the range of this ratio in the bedrock. The required selectivity of this process implies that the groundwater was not in equilibrium with all of the rock types represented in this complex. On the other hand, the good agreement between the isotope composition of Sr in the groundwater and the fracture minerals means that these minerals could have precipitated in the relatively recent past from the groundwater that occurs in the East Bull Lake intrusive at the present time.

The discrepancy between the $^{87}Sr/^{86}Sr$ ratios of the groundwater and the bedrock of the East Bull Lake complex may also be the result of mixing of two components of water that originated from the Archean gneisses of the Superior Province and from the East Bull Lake complex. If these hypothetical components are characterized by having different $^{87}Sr/^{86}Sr$ ratios and different Sr concentrations, the groundwater samples in the East Bull Lake complex should define a straight line in coordinates of the $^{87}Sr/^{86}Sr$ and 1/Sr ratios. The data points in Figure 18.6 actually define several straight lines instead of a single line. Therefore, the groundwater consists of several two-component mixtures that do not appear to be in communication with each other.

The Sr-rich waters have low $^{87}Sr/^{86}Sr$ ratios between about 0.710 and 0.712, whereas the low-Sr waters have high $^{87}Sr/^{86}Sr$ ratios ranging up to 0.720 (not shown in Figure 18.6). The Sr-rich waters probably originated by weathering of the plagioclase-bearing rocks of the East Bull Lake intrusive. The low-Sr water with high $^{87}Sr/^{86}Sr$ ratios may have originated from the Archean granitic gneisses into which the East Bull Lake complex was intruded. These groundwaters interacted primarily with microcline, muscovite, and biotite, which have lower Sr concentrations than plagioclase, are more resistant to chemical weathering than plagioclase, and are enriched in radiogenic ^{87}Sr because of in situ decay of ^{87}Rb.

The isotopic compositions of Sr and chemical compositions of groundwater, mineral water, and oilfield brines elsewhere in the world have been used to study a wide range of phenomena:

Chemical and Isotopic Evolution: Bullen et al., 1996, *Geochim. Cosmochim. Acta*, 60: 1807–1821. Bullen et al., 1997, *Geochim. Cosmochim. Acta*, 61:291–306. Négrel et al., 2001, *Chem. Geol.*, 177:287–308. Naftz et al., 1997, *Chem. Geol.*, 141:195–209.

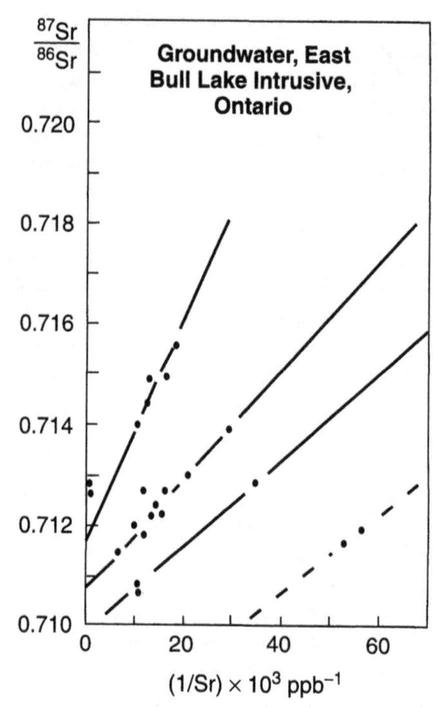

FIGURE 18.6 Strontium isotope mixing lines defined by groundwater in the East Bull Lake intrusive, Superior tectonic province, Ontario, Canada. Three samples plot off-scale. Data from McNutt et al. (1987).

Subsurface Brines: Sass and Starinsky, 1979, *Geochim. Cosmochim. Acta*, 43:885–895. Starinsky et al., 1983a, *Chem. Geol. (Isotope Geosci. Sect.)*, 1:257–267. Starinsky et al., 1983b, *Geochim. Cosmochim. Acta*, 47:687–695. Chaudhuri et al., 1987, *Geochim. Cosmochim. Acta*, 51:45–53. Lowry et al., 1988, *Appl. Geochem.*, 3:177–184. Stueber and Walter, 1991, *Geochim. Cosmochim. Acta*, 55:309–325.

Geothermal Waters: Négrel et al., 1997, *Chem. Geol.*, 135:89–101. Négrel et al., 2000, *Appl. Geochem.*, 15:1345–1367. Grimes et al., 2000, *Chem. Geol.*, 163:247–265.

Repositories of Nuclear Waste: Peterman and Wallin, 1999, *Appl. Geochem.*, 14:939–951. Stuckless et al., 1991, *Science*, 254:551–554.

Saline Groundwater in Precambrian Cratons: Gascoyne et al., 1987, in Fritz and Frape (Eds.), *Geol. Assoc. Can. Spec. Paper*, 33:53–68. Fritz and Frape, 1987, *Geol. Assoc. Canada, Spec. Paper*, 33:210pp. McNutt et al., 1990, *Geochim. Cosmochim. Acta*, 54:205–215. Nurmi et al., 1998, *Appl. Geochem.*, 3:185–203.

Fracture Minerals: Bottomley, 1987, *Appl. Geochem.*, 2:81–92. Bottomley and Veizer, 1992, *Geochim. Cosmochim. Acta*, 56:369–388. Larsson and Tullborg, 1984, *Lithos*, 17:117–125. Wallin and Peterman, 1999, *Appl. Geochem.*, 14:953–962.

Speleothems: Verheyden et al., 2000, *Chem. Geol.*, 169:131–144. Goede et al., 1998, *Chem. Geol.*, 149:37–50.

Additional references on the regional flow of groundwater in North America are listed in Sections 5.5a, 5.5b, 10.12a, and 10.12b.

18.2 SEDIMENT IN STREAMS

The sediment being transported by streams includes grains of resistant minerals (e.g., quartz, K-feldspar, muscovite, zircon, garnet) as well as weathering products (e.g., oxyhydroxides of Fe, Al, and Mn, kaolinite) and particles of organic matter. In addition, cations of chemical elements in solution are sorbed by films or organic molecules that cover the surfaces of the sediment particles. The organic matter that coats the mineral grains has negatively charged surface sites at near-neutral pH and therefore sorbs Sr^{2+} and cations of other elements from solution.

The isotopic compositions of certain elements (e.g., Sr, Nd, and Pb) that are sorbed to colloidal particles (diameters less than 1 μm) are similar to the isotopic compositions of the elements in solution. Particles of resistant minerals with diameters larger than about 1 μm in diameter contain Sr, Nd, and Pb, whose isotope compositions differ from those of the sorbed fraction because they depend on the ages of the mineral grains and on the respective parent–daughter ratios.

The elements in solution originated by the congruent or incongruent dissolution of minerals that are susceptible to weathering (e.g., calcite, sulfide minerals, plagioclase) that generally have low parent–daughter ratios (Blum et al., 1993). Therefore, the elements that are sorbed from solution to the surfaces of colloidal mineral grains are less enriched in radiogenic isotopes than the same elements contained within the coarser resistant grains. Although sediment particles of all sizes sorb ions from solution, the sorbed fraction is dominant among the colloidal particles because, collectively, they have larger surface areas than the coarse sediment grains with diameters >1.0 μm.

The sorbed fractions of Sr, Nd, and Pb in sediment can be recovered separately by leaching sediment samples with dilute acid, which causes the sorbed fraction to be released into solution without affecting the fraction of the elements that is contained within the resistant mineral grains. The isotopic compositions of sorbed Sr, Nd, and Pb provide information about the minerals that are weathering in the drainage basin and about anthropogenic contaminants that are released into streams in the form of fertilizer, municipal effluent, and industrial wastewater. The isotopic compositions of Sr, Nd, and Pb in resistant mineral grains can be used to identify their provenance by dating samples of the bulk sediment or individual minerals grains. In cases where the resistant mineral grains are derived from two or more sources having different ages or chemical compositions or both, samples of the bulk sediment can be treated as two- or three-component mixtures. An example of two-component mixtures of sediment from the Red Sea was presented in Section 16.6b.

18.2a Murray River, N.S.W., Australia

The relation between the diameters of sediment grains and the concentration and isotope ratios of Sr was demonstrated by Douglas et al. (1995) in a study of the Murray–Darling river system in New South Wales, Australia. The drainage basin of these rivers has an area of more than 1.0×10^6 km^2 and is underlain partly by granitic plutons that intruded detrital sedimentary and metamorphic rocks of early Paleozoic age. These rocks form the Great Dividing Range, which extends parallel to the southeastern coast of Australia between Melbourne in the south and Brisbane in the north. In addition, deposits of Cenozoic sand, silt, and clay are exposed in the central and western part of the basin. The climate of the Murray–Darling drainage system is subtropical in the north, temperate in the south, semiarid in the west, and alpine in the Great Dividing Range in the east.

The chemical compositions of the coarse particulate sediment fraction in Table 18.2 can be used to calculate its mineral composition (norm) by standard methods presented by Faure (1998, Section 19.2). The results indicate that the >25-μm fraction contains 53% quartz, 19.5% kaolinite, 11.3% K-feldspar, 8.0% plagioclase, 4.8% magnetite, and 2.8% miscellaneous other constituents.

Table 18.2. Chemical Analyses of Selected Size Fractions of Sediment in the Murray River at Merbein, N.S.W., Australia, Recalculated to 100%

Sediment	Concentration, %		
	>25 μm	1–0.2 μm	0.006–0.003 μm
SiO_2	78.5	54.5	28.6
TiO_2	0.58	0.60	0.06
Al_2O_3	11.7	25.4	3.64
Fe_2O_3 (t)	4.90	13.00	3.41
MnO	0.09	0.09	0.03
MgO	1.18	2.30	15.3
CaO	0.27	1.09	16.7
Na_2O	0.77	0.96	29.9
K_2O	1.93	1.99	2.23
Quartz	53.0	11.6	
Kaolinite	19.5	47.1	
K-feldspar	11.3	10.8	
Plagioclase	8.0	12.8	
Magnetite	4.8	13.7[a]	
Miscellaneous	2.8	3.9	

Source: Douglas et al., 1995.
[a] Goethite

The normative mineral compositions of the >25- and 0.2–1.0-μm sediment fractions in Table 18.2 are compatible with the geology of the drainage basin of the Murray River. However, the graphical representation of the chemical compositions in Figure 18.7 demonstrates that the concentrations of Na_2O, CaO, MgO, and K_2O rise with decreasing grain size in the colloidal sediment fraction, whereas the concentrations of SiO_2, Fe_2O_3, and Al_2O_3 decrease. Consequently, the chemical compositions of these fractions do not provide enough Al_2O_3 and SiO_2 to accommodate the rising concentrations of Na_2O, CaO, MgO, and K_2O in aluminosilicate minerals. Evidently, the chemical compositions of these colloidal sediment fractions are controlled by the sorption of ions rather than by the minerals of which the particles are composed.

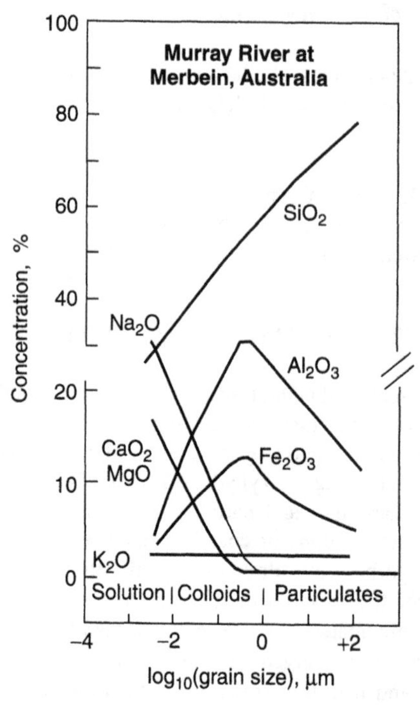

FIGURE 18.7 Variation of the concentrations of major-element oxides in size fractions of sediment in the Murray River at Merbein, N.S.W., Australia, after excluding organic matter by recalculating the analyses to 100%. Data from Douglas et al. (1995).

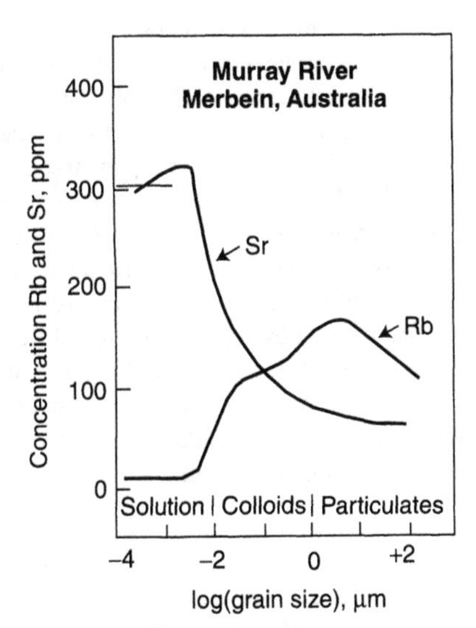

FIGURE 18.8 Variation of the concentrations of Rb and Sr with the diameters of the sediment grains in the Murray River at Merbein, Australia: particles in suspension, >25 to 1 μm; colloidal, <1 to >0.003 μm; solution, <0.003 μm. Data from Douglas et al. (1995).

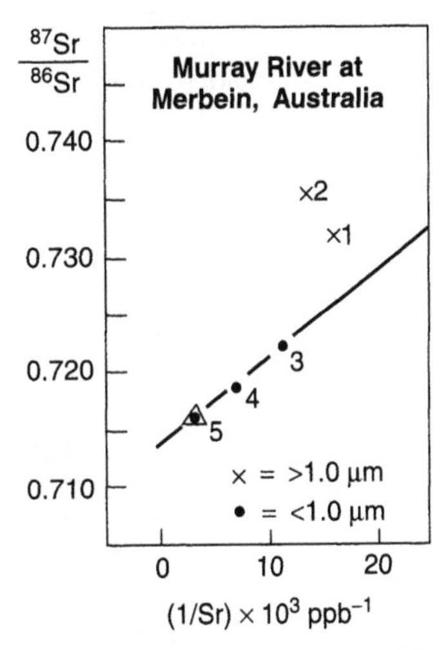

FIGURE 18.9 Isotopic mixing line defined by the $^{87}Sr/^{86}Sr$ and 1/Sr ratios of colloidal and particulate sediment in the Murray River at Merbein, Australia. The size fractions are identified by number: 1, >25 μm; 2, 1.0–25 μm; 3, 0.2–1.0 μm; 4, 0.006–0.2 μm; 5, 0.003–0.006 μm. The open triangle represents the Sr in solution in the water. Data from Douglas et al. (1995).

The concentrations of Sr in five different size fractions of sediments in the Murray River at Merbein in Figure 18.8 increase with decreasing grain size, whereas the Rb concentrations rise to a maximum in particles with diameters between 25 and 1 μm and then decline in smaller particles. The $^{87}Sr/^{86}Sr$ ratios of the various size fractions decrease from 0.73540 (1.0–25 μm) to 0.71601 in the colloidal fraction (0.003–0.006 μm).

The data in Figures 18.7 and 18.8 reveal systematic differences in the enrichment of alkali metals by sorption in the colloidal fractions relative to the particulate fractions: Na > K > Rb. In addition, the alkaline earths are enriched by sorption such that Ca > Mg > Sr. The cause for the preferential sorption of Na^+ and Ca^{2+} relative to their respective congeners is not apparent from the available data.

The $^{87}Sr/^{86}Sr$ ratios of the colloidal particles (0.003–1.0 μm) and the reciprocals of their Sr concentrations define a straight line in Figure 18.9 indicating that the Sr is a mixture of two components. The Sr concentrations and $^{87}Sr/^{86}Sr$ ratios of the smallest colloidal particles (0.003–0.006 μm) are similar to those of the dissolved Sr (<0.003 μm).

The interpretation of these data proposed by Douglas et al. (1995) is that the colloidal particles contain two isotopic varieties of Sr. One of these is sorbed from solution to organic films that coat the grains while the other is the Sr that resides within the grains and whose $^{87}Sr/^{86}Sr$ ratios were increased by decay of ^{87}Rb prior to weathering. The fraction of sorbed Sr in the colloidal particles increases with decreasing grain size because the amount of organic matter coating the grains and their surface area both increase.

The large suspended mineral particles with diameters greater than 1 μm deviate from the mixing line of the colloidal particles in Figure 18.9 because most of the Sr they contain resides within these mineral grains and only a small fraction of it is sorbed. As a result, these grains are likely to be isotopically heterogeneous because the $^{87}Sr/^{86}Sr$ ratio of each grain depends on its Rb/Sr ratio and

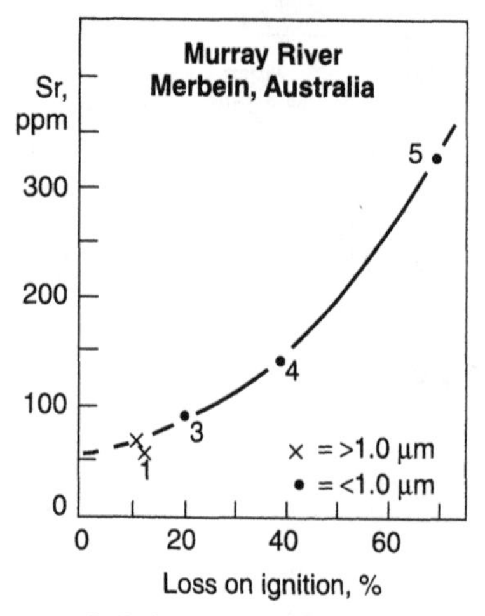

FIGURE 18.10 Relation between organic matter measured by loss on ignition and the Sr concentrations of colloidal and particulate sediment in the Murray River at Merbein, Australia. The size fractions are identified by number: 1, >25 μm; 2, 1.0–25 μm; 3, 0.2–1.0 μm; 4, 0.006–0.2 μm; 5, 0.003–0.006 μm. The Sr concentration at LOI = 0 is about 50 ppm. Data from Douglas et al. (1995).

on the geological age of the mineral of which it is composed.

The importance of the film of organic matter to sorption of Sr^{2+} and other ions on mineral grains in the Murray River at Merbein is indicated in Figure 18.10 by the relation between loss on ignition (LOI) and the Sr concentration of colloidal and particulate sediment. The data demonstrate that the high Sr concentration (326.9 ppm) of the smallest grain size (0.003–0.006 μm) is attributable to the fact that the organic film coating these grains makes up nearly 70% of their weight. The comparatively low Sr concentration (59.1 ppm) of the largest grains (>25 μm) is related to the fact that the organic film coating these grains constitutes only 12.4% of their weight even though the coating has about the same thickness as that which covers the smaller colloidal particles. An extrapolation to LOI = 0% in Figure 18.10 indicates that the minerals of the particulate sediment have an average Sr concentration of about 50 ppm.

The evidence in Figure 18.8 indicates that the colloidal sediment sorbs Sr^{2+} but not Rb^+. Therefore, the Rb/Sr ratios of the colloidal sediment decreases from 1.52 in the 0.2–1.0-μm fraction to 0.0458 in the 0.003–0.006-μm fraction. In addition, the $^{87}Sr/^{86}Sr$ ratios of the colloidal sediment decrease from 0.72208 (0.2–1.0 μm) to 0.71601 (0.003–0.006 μm) and correlate positively with the $^{87}Rb/^{86}Sr$ ratios in Figure 18.11. The straight line defined by the three colloidal fractions has a slope $m = 0.001417$ that yields a Rb–Sr date of 100 ± 2 Ma (Douglas et al., 1995). This date is fictitious because the colloidal sediment fractions define a mixing line rather than an isochron.

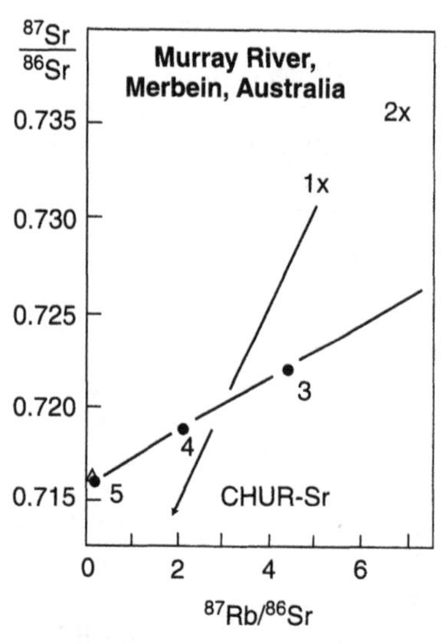

FIGURE 18.11 Mixing line defined by the colloidal fractions of sediment (solid circles) in the Murray River at Merbein, Australia, in coordinates of the $^{87}Sr/^{86}Sr$ and $^{87}Rb/^{86}Sr$ ratios. The fictitious date derived from the slope of the mixing line is 100 ± 2 Ma. The particulate fractions (crosses) of the sediment are composed primarily of minerals whose approximate Rb–Sr model date relative to CHUR is 370 Ma (Late Devonian to Early Pennsylvanian). The size fractions are identified by number: 1, >25 μm; 2, 1.0–25 μm; 3, 0.2–1.0 μm; 4, 0.006–0.2 μm; 5, 0.003–0.006. The open triangle is the soluble Rb and Sr. Data from Douglas et al. (1995).

The coarse fractions of the sediment are composed of mineral grains whose $^{87}Sr/^{86}Sr$ and $^{87}Rb/^{86}Sr$ ratios are not significantly affected by sorption. Therefore, they retain a record of the dates of their crystallization. For example, the Rb–Sr systematics of the >25-μm fraction (1 in Figure 18.11) can be used to calculate a Rb–Sr date relative to CHUR-Sr based on the equation

$$\left(\frac{^{87}\text{Sr}}{^{86}\text{Sr}}\right)^0_{\text{CHUR}} - \left(\frac{^{87}\text{Rb}}{^{86}\text{Sr}}\right)^0_{\text{CHUR}} (e^{\lambda t} - 1) = \left(\frac{^{87}\text{Sr}}{^{86}\text{Sr}}\right)^0_{\text{Sed}} - \left(\frac{^{87}\text{Rb}}{^{86}\text{Sr}}\right)^0_{\text{Sed}} (e^{\lambda t} - 1) \quad (18.5)$$

Given that

$$\left(\frac{^{87}\text{Sr}}{^{86}\text{Sr}}\right)^0_{\text{CHUR}} = 0.7045 \qquad \left(\frac{^{87}\text{Sr}}{^{86}\text{Sr}}\right)^0_{\text{Sed}} = 0.73163$$

$$\left(\frac{^{87}\text{Rb}}{^{86}\text{Sr}}\right)^0_{\text{CHUR}} = 0.0816 \qquad \left(\frac{^{87}\text{Rb}}{^{86}\text{Sr}}\right)^0_{\text{Sed}} = 5.222$$

Then

$$07045 - 0.0816(e^{\lambda t} - 1) = 0.73163 - 5.222(e^{\lambda t} - 1)$$

and

$$t = \frac{\ln 1.005279}{1.42 \times 10^{-11}} = 370 \text{ Ma}$$

This date is only an approximation of the age of the mineral grains because the sediment may be a mixture of grains having different crystallization ages, because Rb/Sr ratios of some of the grains may have been altered by weathering, and because their initial $^{87}Sr/^{86}Sr$ ratios may differ.

Nevertheless, Rb–Sr dates of individual unaltered mineral grains or of collections of grains that originated from a single source provide information concerning the provenance of the sediment. The geological value of this information is enhanced when the sediment being dated was deposited in the geological past (e.g., Muffler and Doe, 1968).

The relation between river sediment in the Namoi River, a tributary of the Darling River, and soils of its drainage basin was investigated by Martin and McCulloch (1999) using chemical analyses and isotope ratios of Sr and Nd.

18.2b Fraser River, British Columbia, Canada

The Fraser River of British Columbia drains an area of 2.38×10^5 km^2 that includes a wide variety of rocks along the active western margin of the North American continent (Cameron and Hattori, 1997). The water and sediment it discharges into the Pacific Ocean originate from four tectonic units identified in Figure 18.12, starting with the Foreland belt at the source of the Fraser River, followed by the Omineca belt, the Intermontane belt, and the Coastal belt from east to west. The rocks of the Foreland and Omineca belts consist of Late Proterozoic and Paleozoic sandstones, shales, and carbonates, but the Omineca belt also contains mafic igneous rocks derived from the mantle. The Intermontane belt is composed of several accreted terranes and consists of interbedded sedimentary

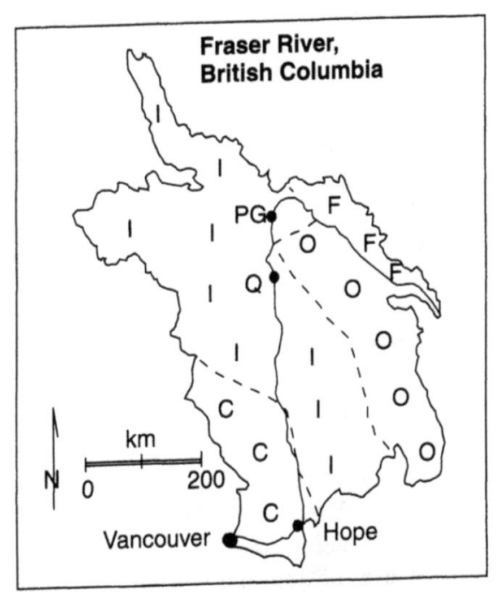

FIGURE 18.12 Drainage basin of the Fraser River, British Columbia, Canada: F = Foreland belt; O = Omineca belt; I = Intermontane belt; C = Coastal belt; PG = Prince George; and Q = Quesnel. Adapted from Cameron and Hattori (1997).

and volcanic rocks (Late Triassic to Early Tertiary) intruded by granitic plutons. The Coastal belt is underlain by metamorphic rocks of Permian to Cretaceous age intruded by quartz–diorite plutons and overlain by Andean-type volcanic rocks of Cretaceous to Tertiary age.

The Fraser River in Figure 18.12 arises near Mt. Edith Cavel in Jasper National Park and flows northwest within the Rocky Mountain Trench, which is the boundary between the Foreland belt and the Omineca belt, to the town of Prince George, where it turns southwest and crosses the Intermontane belt to the Coastal belt. It turns west at the town of Hope and enters the Pacific Ocean at Vancouver. There are no dams on the Fraser River, which means that the sediment discharged into the Pacific Ocean represents all of the rocks exposed to weathering in its drainage basin.

Cameron and Hattori (1997) collected water samples from the Fraser River and its tributaries and recovered the suspended sediment by passing the water through 0.45-μm filters. Therefore, their sediment data pertain primarily to the silt and clay size fractions. Wadleigh et al. (1985) previously reported an $^{87}Sr/^{86}Sr$ ratio of 0.71195 relative to 0.7080 for E&A and a Sr concentration of 80 μg/L in one sample of filtered water (0.45 μm) collected near the mouth of the Fraser River.

Cameron and Hattori (1997) demonstrated that the $^{87}Sr/^{86}Sr$ ratios of water and sediment differ at all of the sites along the course of the Fraser River, although the isotope ratios of both water and sediment in Figure 18.13 decline with distance from the source of the river. In the upper reaches of the Fraser River, the $^{87}Sr/^{86}Sr$ ratios of the sediment samples are higher than those of the water because of the resistance to weathering of Rb-rich mineral grains derived from source rocks of Proterozoic to Paleozoic age. In the vicinity of Quesnel, about 550 km downstream of the source of the Fraser River, the $^{87}Sr/^{86}Sr$ ratios of the sediment drop below the $^{87}Sr/^{86}Sr$ ratios of the water because of the introduction of sediment derived from mafic igneous rocks by tributaries that drain the Intermontane and Coastal belts (e.g., Nechako, Blackwater, Chilcotin, and Bridge), all of which enter the Fraser River from the west. The $^{87}Sr/^{86}Sr$ ratios of the eastern tributaries (e.g., McGregor, Quesnel, and Thompson) are similar to

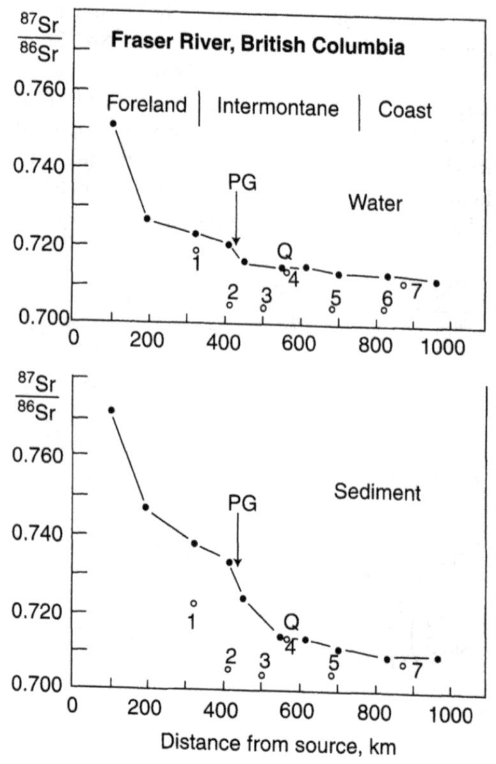

FIGURE 18.13 Longitudinal profiles of the $^{87}Sr/^{86}Sr$ ratios of water and suspended sediment (>0.45 μm) of the Fraser River, British Columbia: PG = Prince George, Q = Quesnel. The tributaries (open circles) are identified by number: 1 = McGregor (Foreland); 2 = Nechako (Intermontane); 3 = Blackwater (Intermontane); 4 = Quesnel (Intermontane and Omineca); 5 = Chilcotin (Intermontane); 6 = Bridge (coastal); 7 = Thompson (Intermontane and Omineca). Data from Cameron and Hattori (1997).

those of sediment in the Fraser River at the sites where these rivers enter their master stream.

The isotope ratios and concentrations of Sr in *water* of the Fraser River near its mouth at Alexandra Bridge, a few kilometers upstream of the town of Hope, are the result of mixing of the waters of its tributaries. The $^{87}Sr/^{86}Sr$ ratio and Sr concentration of the Fraser River at this site (0.7127, 85 μg/g) define a point in Figure 18.14 that lies on the mixing line defined by its eastern and western tributaries:

Eastern tributaries $\frac{^{87}Sr}{^{86}Sr} = 0.7149$, Sr = 85.6 μg/L

Western tributaries $\frac{^{87}Sr}{^{86}Sr} = 0.7047$, Sr = 75 μg/L

where the average $^{87}Sr/^{86}Sr$ ratios were weighted by the Sr concentrations of the water in each group of tributaries. Use of the lever rule indicates that approximately 80% of the water in the Fraser River at Alexandra Bridge near Hope originated from the eastern tributaries (i.e., the McGregor, Quesnel, and Thompson rivers).

The *sediment* in the Fraser River at Alexandra Bridge in Figure 18.14 ($^{87}Sr/^{86}Sr = 0.7098$, Sr = 249 ppm) is also a mixture of sediment contributed by the eastern and western tributaries:

Eastern tributaries $\frac{^{87}Sr}{^{86}Sr} = 0.7128$, Sr = 239 ppm

Western tributaries $\frac{^{87}Sr}{^{86}Sr} = 0.7047$, Sr = 279.6 ppm

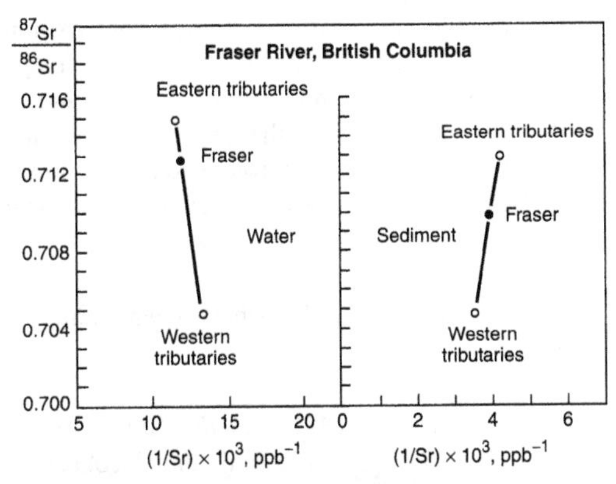

FIGURE 18.14 Two-component isotopic mixing diagrams of Sr in water and sediment of the eastern and western tributaries of the Fraser River, British Columbia. The points representing water and sediment of the Fraser River near its mouth lie on the respective mixing lines and indicate that 78% of the water and 67% of the sediment originated from the eastern tributaries. Data from Cameron and Hattori (1997).

The point representing the suspended sediment in the Fraser River at Alexandra Bridge, like the Sr in the water, is located on the mixing line in Figure 18.14 defined by the sediment of its tributaries. The estimated abundance of sediment contributed by the eastern rivers (70%) is less than that of the water derived from these rivers, either because the eastern rivers transport less sediment than the western rivers or because some of the sediment of the eastern rivers was deposited.

The isotope ratios of Sr and Nd of sediment in the Fraser River in Figure 18.15 define an isotopic mixing hyperbola that extends from quadrant II (mantle-derived rocks) to quadrant IV (crustal rocks). The sediment contributed by the western tributaries (Nechako, Blackwater, Chilcotin, and Bridge River) has high $^{143}Nd/^{144}Nd$ but low

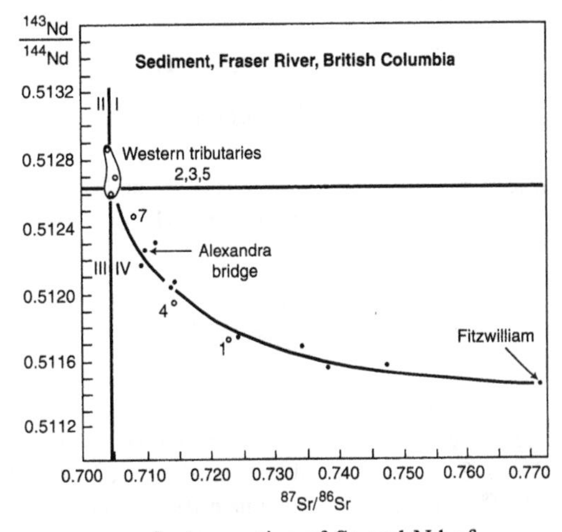

FIGURE 18.15 Isotope ratios of Sr and Nd of sediment (>0.45 μm) transported by the Fraser River, British Columbia, Canada. The tributary rivers (open circles) are identified by number. Eastern tributaries: 1 = McGregor; 4 = Quesnel; 7 = Thompson. Western tributaries: 2 = Nechako; 3 = Blackwater; 5 = Chilcotin; 6 = Bridge. E&A Sr standard: $^{87}Sr/^{86}Sr = 0.708048 \pm 0.000017$, NBS 987: $^{87}Sr/^{86}Sr = 0.710270 \pm 0.000019$. The Nd isotope analyses were corrected for isotope fractionation to $^{146}Nd/^{144}Nd = 0.7219$ and yielded $^{143}Nd/^{144}Nd = 0.511868 \pm 0.000017$ for the LaJolla Nd standard. Data from Cameron and Hattori (1997).

$^{87}Sr/^{86}Sr$ ratios, which characterizes mafic igneous rocks derived from sources in the subcontinental lithospheric mantle. The sediment near the source of the Fraser River (Fitzwilliam) has the opposite isotopic characteristics indicating a source in the Precambrian rocks of the Foreland belt.

The crustal residence age of the Precambrian basement rocks at the head of the Fraser River can be estimated by reference to CHUR-Nd based on the equation

$$\left(\frac{^{143}Nd}{^{144}Nd}\right)^0_{CHUR} - \left(\frac{^{147}Sm}{^{144}Nd}\right)^0_{CHUR}(e^{\lambda t} - 1)$$
$$= \left(\frac{^{143}Nd}{^{143}Nd}\right)^0_{Sed} - \left(\frac{^{147}Sm}{^{144}Nd}\right)^0_{Sed}(e^{\lambda t} - 1) \quad (18.6)$$

Substituting the appropriate values for the isotope ratios,

$$0.512638 - 0.1967(e^{\lambda t} - 1)$$
$$= 0.511435 - 0.113(e^{\lambda t} - 1)$$

where $(^{143}Nd/^{143}Nd)^0_{Sed} = 0.511435$ and $(^{147}Sm/^{144}Nd)^0_{Sed} = 0.113$ (Cameron and Hattori, 1997). Solving for t,

$$t = \frac{\ln 1.014372}{6.54 \times 10^{-12}} = 2.18 \times 10^9 \text{ y}$$

A second estimate of the crustal residence age of the basement rocks at the head of the Fraser River is obtainable from the $^{87}Sr/^{86}Sr$ and $^{87}Rb/^{86}Sr$ ratios of the sediment (0.7724 and 4.19, respectively):

$$0.7045 - 0.0816(e^{\lambda t} - 1) = 0.7724 - 4.19(e^{\lambda t} - 1)$$
$$t = \frac{\ln 1.016527}{1.42 \times 10^{-11}}$$
$$= 1.15 \times 10^9 \text{ y}$$

The difference between the Sm–Nd and Rb–Sr model dates is attributable to the effects of weathering on the Rb–Sr systematics of the sediment. The preferential dissolution of Sr-rich minerals (e.g., plagioclase and calcite) causes the Rb/Sr and $^{87}Sr/^{86}Sr$ ratios of the sediment to rise. The loss of plagioclase does not affect the Sm/Nd ratio of the remaining sediment because the average Sm/Nd ratio of plagioclase (0.292) is similar to the average Sm/Nd ratio (0.285) of other silicate minerals, including olivine (0.19) clinopyroxene (0.37), amphibole (0.37), biotite (0.22), K-feldspar (0.14), garnet (0.54), apatite (0.31), and monazite (0.17) (Faure, 1986). Therefore, the model Sm–Nd date of 2.18 Ga is to be preferred as an estimate of the age of the basement rocks at the head of the Fraser River.

18.3 ZAIRE AND AMAZON RIVERS

The work of Douglas et al. (1995) on the Murray River and by Cameron and Hattori (1997) on the Fraser River suggests that the water and suspended sediment in rivers may be used to derive information about several important geochemical processes affecting the continents and the oceans:

1. the effect of chemical weathering of rock-forming minerals on the isotope compositions of Sr, Nd, and Pb in solution in rivers;
2. the average chemical and isotope composition of these elements in the rocks of the drainage basin and hence in the continental crust; and
3. the rate of erosion of the drainage basin and its dependence on environmental conditions, including elevation above sea level, climate, and lithologic composition.

These kinds of problems have been addressed elsewhere:

Goldstein et al., 1984, *Earth Planet Sci. Lett.*, 70:221–236. Goldstein and Jacobsen, 1987, *Chem. Geol., Isotope Geosci. Sect.*, 66:245–272. Goldstein and Jacobsen, 1988, *Earth Planet Sci. Lett.*, 87:249–265. Åberg et al., 1989, *J. Hydrol.*, 109:65–78. Krishnaswami et al., 1992, *Earth Planet. Sci. Lett.*, 109:243–253. Asmerom and Jacobsen, 1993, *Earth Planet. Sci. Lett.*, 115:245–256. Blum et al., 1993, *Geochim. Cosmochim. Acta*, 57:5019–5025. Négrel et al., 1993, *Earth*

Planet. Sci. Lett., 120:59–76. Andersson et al., 1994, *Earth Planet. Sci. Lett.*, 124:195–210. Dupré et al., 1996, *Geochim. Cosmochim. Acta*, 60:1301–1321. Gaillardet et al., 1995, *Geochim. Cosmochim. Acta*, 59:3469–3485. Allégre et al., 1996, *Chem. Geol.*, 131:93–112. Borg and Banner, 1996, *Geochim. Cosmochim. Acta*, 60:4193–4206. Négrel and Deschamps, 1996, *Aquatic Geochim.*, 2:1–27. Henry et al., 1997, *Earth Planet. Sci. Lett.*, 146:627–644. Négrel and Grosbois, 1999, *Chem. Geol.*, 156:231–249. Négrel et al., 2000, *Chem. Geol.*, 166:271–285.

18.3a Strontium and Neodymium in Water and Sediment

The $^{87}Sr/^{86}Sr$ ratios of the water in the Zaire and Amazon basins are consistently lower in all cases than the $^{87}Sr/^{86}Sr$ ratios of the sediment suspended in the water (Allègre et al., 1996). The difference between the $^{87}Sr/^{86}Sr$ ratios of the sediment and water is expressed by the parameter

$$\Delta\left(\frac{^{87}Sr}{^{86}Sr}\right) = \left[\left(\frac{^{87}Sr}{^{86}Sr}\right)_{Sed} - \left(\frac{^{87}Sr}{^{86}Sr}\right)_{Sol}\right] \times 100 \quad (18.7)$$

The $\Delta(^{87}Sr/^{86}Sr)$ values of the Zaire and Amazon Rivers and their respective tributaries in Figure 18.16 define smooth curves with increasing values of the $^{87}Sr/^{86}Sr$ ratios of the sediment. In addition, the $\Delta(^{87}Sr/^{86}Sr)$ values of the Zaire drainage are higher than those of the Amazon, but both curves extrapolate to about 0.710 at $\Delta(^{87}Sr/^{86}Sr) = 0$. Three samples from the Amazon basin (Solimões, Madeira, and the Amazon at Santarem) deviate from this relationship.

The isotopic disequilibrium of Sr in the Zaire and Amazon Rivers is caused by the slow kinetics of dissolution of Rb-rich minerals (e.g., muscovite, microcline, illite) in these and most other drainage basins in the world. Consequently, the Sr in solution is derived primarily from low-Rb minerals that have comparatively low $^{87}Sr/^{86}Sr$ ratios but dissolve preferentially during chemical weathering (Blum et al., 1993). The Rb-rich minerals, which dominate

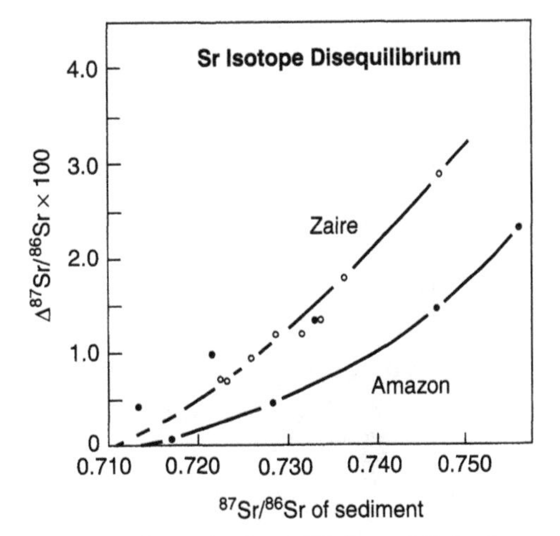

FIGURE 18.16 Isotopic disequilibrium of Sr in the suspended sediment and in solution in the water of the Zaire and Amazon Rivers as expressed by $\Delta(^{87}Sr/^{86}Sr) = [(^{87}Sr/^{86}Sr)_{sed} - (^{87}Sr/^{86}Sr)_{sol}] \times 100$. The $^{87}Sr/^{86}Sr$ ratios are relative to 0.710254 ± 0.000006 for NBS-987. Data from Allègre et al. (1996).

in the suspended sediment, resist weathering and have elevated $^{87}Sr/^{86}Sr$ ratios depending on their crystallization ages.

The isotope ratios of Sr and Nd of the suspended sediment in the Zaire and Amazon river systems define two overlapping data fields in quadrant IV of Figure 18.17. The scatter of data points reflects the heterogeneity of the parent–daughter ratios and of the crustal residence ages of the rocks in the respective drainage basins.

The suspended sediment in rivers is an imperfect sample of the bedrock exposed to weathering in their drainage basin, not only because of the preferential dissolution of certain minerals (e.g., plagioclase, calcite) but also because the topography of the basin affects the delivery of sediment to the streams and because of the preferential deposition and entrainment of sediment by streams based on the density, volume, and shape of the mineral grains.

As a result, the average isotope ratios of bulk sediment (>0.2 μm) in the drainage of the Zaire and Amazon Rivers in Table 18.3 are probably skewed in favor of minerals that resist weathering and that have physical properties that

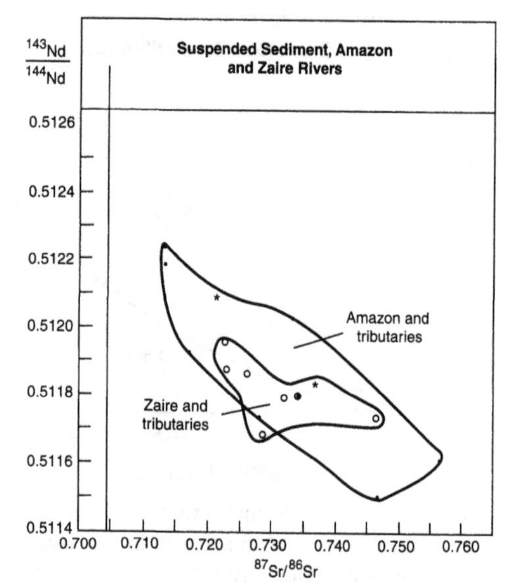

FIGURE 18.17 Isotope ratios of Sr and Nd in sediment suspended in the Zaire and Amazon Rivers and their respective tributaries. The data points are located in quadrant IV defined by CHUR. The stars are the average isotope ratios of each river system. Data from Allègre et al. (1996).

Table 18.3. Average Concentrations of Sr and Rb and Isotope Ratios of Sr and Nd of Sediment in the Zaire and Amazon Rivers and Their Tributaries

Parameter	Zaire	Amazon
Rb, ppm	58.9	60.2
Sr, ppm	61.0 (40–78)	96.7 (40–176)
$^{87}Rb/^{86}Sr$	2.79	1.79
$^{87}Sr/^{86}Sr^{a}$	0.73155	0.72341
$^{147}Sm/^{144}Nd$	0.1057	0.1025
$^{143}Nd/^{144}Nd$	0.511803	0.511996

Source: Allègre et al., 1996.

[a] Weighted by Sr concentrations. NBS 987: $^{87}Sr/^{86}Sr = 0.710254 \pm 0.000006$; LaJolla Nd standard: $^{143}Nd/^{144}Nd = 0.511850 \pm 0.000010$.

facilitate their transport in streams. Nevertheless, the data in Table 18.3 do characterize the isotopic compositions of Sr and Nd of the rocks in the large drainage basins of the Zaire and Amazon basins. In addition, these data constrain the isotope composition of crustal Sr and Nd that may enter mantle-derived magmas that intrude the continental crust.

18.3b Confluence at Manaus, Brazil

The city of Manaus is located at the confluence of the Solimões and Negro Rivers, which together form the Amazon River in Figure 18.18. (The Solimões River may also be considered to be an upstream extension of the Amazon River). Therefore, the chemical and isotope composition of water in the Amazon River downstream of Manaus is controlled by mixing of these rivers provided that the elements under consideration are conservative. The concentrations of Sr, U, and Pb in Figure 18.19 indicate that the U concentration of water in the Amazon River (0.033 ppb) downstream of the confluence is only about 7% less than predicted by the mixing model, whereas the concentration of Pb (0.07 ppb) is more than 50% lower than its predicted concentration (0.15 ppb). The apparent loss of Pb from the mixed water is caused by preferential sorption of Pb^{2+} to charged sites on the surfaces of suspended particles of minerals, ferric hydroxide, and organic matter. These results demonstrate that Pb^{2+} ions in river water are strongly nonconservative compared to the uranyl ion(UO_2^{2+}) and Sr^{2+}.

The conservative behavior of Sr^{2+} in solution in the Amazon River is confirmed in Figure 18.20 by the fact that the $^{87}Sr/^{86}Sr$ and 1/Sr ratios of the Amazon River downstream of Manaus satisfy the linear mixing equation defined by the waters of the Negro and Solimões Rivers. The point representing the Amazon River lies close to that of the Solimões River, meaning that about 90% of the water in the Amazon originates from the Solimões River.

The sediment (>0.2 μm) in suspension in the Solimões and Negro Rivers likewise mixes at the confluence at Manaus. However, the chemical composition of the sediment is altered by the preferential deposition of grains of "heavy" minerals and by the pickup of sediment from the bottom of the streams depending on variations in the velocity of the water. In spite of the resulting heterogeneity, the concentrations of

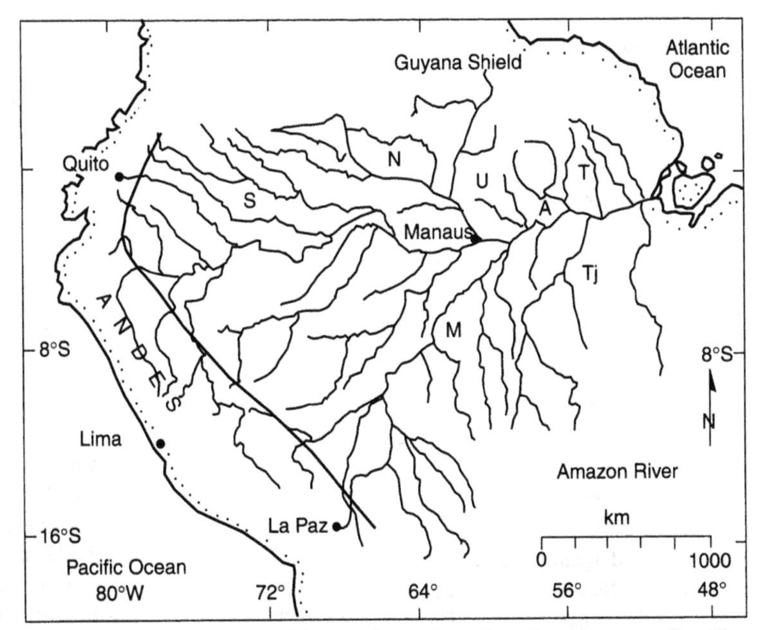

FIGURE 18.18 Drainage basin of the Amazon River and its tributaries: A = Amazon; T = Trombetas; U = Urucara; N = Negro; S = Solimões; M = Madeira; Tj = Tapajos. Adapted from Allègre et al. (1996).

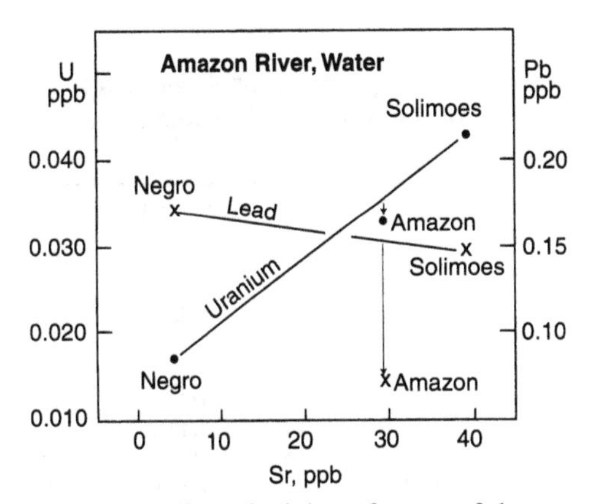

FIGURE 18.19 Effect of mixing of water of the Solimões and Negro Rivers at Manaus, Brazil, on the concentrations of Sr, U, and Pb in solution in the Amazon River downstream of the confluence. Data from Allègre et al. (1996).

Sr, U, and Pb of the sediment in the Amazon River are consistent with mixing of sediment contributed by the Solimões and Negro Rivers. If the Sr concentrations of the sediment are assumed to remain constant, the Sr–U and Sr–Pb mixing lines in Figure 18.21 both indicate that about 65% of the sediment in the Amazon River downstream of Manaus originates from the Solimões River.

18.3c Model Dates of Sediment, Amazon River

The isotope systematics of Nd and Sr in the sediment in the Amazon River and its tributaries in Table 18.4 can be used to calculate model dates with reference to CHUR. Such model dates provide a clue to the age of the rocks underlying the drainage basin and thereby identify the provenance of the sediment.

For example, the Nd model date of sediment in the Trombetas River (Table 18.4) is calculated from equation 18.6:

$$0.512638 - 0.1967(e^{\lambda t} - 1)$$
$$= 0.511500 - 0.092(e^{\lambda t} - 1)$$
$$t = \frac{\ln 1.010869}{6.54 \times 10^{-12}} = 1.65 \text{ Ga}$$

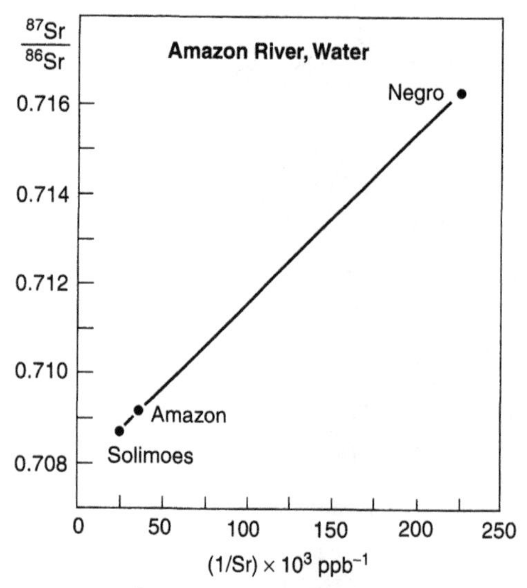

FIGURE 18.20 Isotopic mixing diagram of Sr in solution in the Solimões and Negro Rivers, which merge at Manaus to form the Amazon River. The position of the point representing the Amazon River downstream of Manaus indicates that about 90% of its water originates from the Solimões River. Data from Allègre et al. (1996).

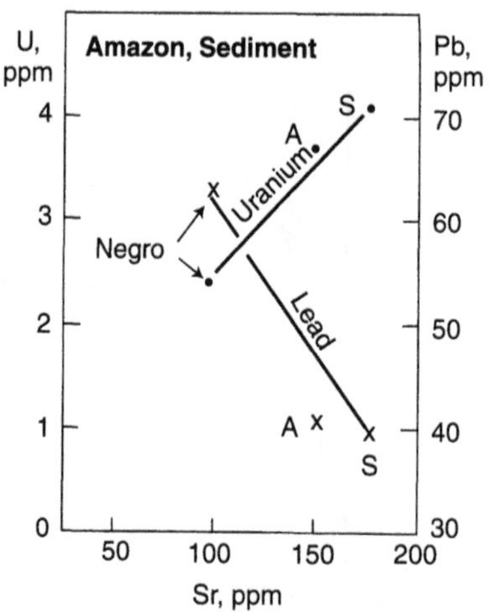

FIGURE 18.21 Mixing of suspended sediment in the Solimões (S) and Negro (N) Rivers at Manaus, Brazil. The deviations of the U and Pb concentrations of sediment in the Amazon River (A) from the mixing line are attributable to selective deposition of the "heavy" mineral fraction and resuspension of sediment from the stream beds. Data from Allègre et al. (1996).

The model date of Sr in the sediment of the Trombetas River is derived as

$$0.7045 - 0.0816(e^{\lambda t} - 1)$$
$$= 0.74683 - 1.65(e^{\lambda t} - 1)$$
$$t = \frac{\ln 1.02698}{1.42 \times 10^{-11}} = 1.87 \text{ Ga}$$

The Sm–Nd and Rb–Sr model dates of sediment in the Trombetas River are in agreement and correctly indicate that the sediment originates partly from the Precambrian craton of Guyana located north of the Amazon basin. The model dates of sediment in the rivers listed in Table 18.5 demonstrate the differences in the geology of the Solimões and Negro Rivers. The model dates of the sediment in the Solimões River (Nd: 0.87 Ga, Sr: 0.42 Ga) reflect the presence of mineral particles derived from the Mesozoic volcanic rocks of the Andes, whereas the model dates of sediment in the Negro (Nd: 1.15 Ga; Sr: 2.71 Ga) are indicative of the Precambrian rocks of the Guyana craton. The Nd model dates of sediment in the Amazon and Solimões Rivers in Table 18.5 are older than the Sr model dates. However, the reverse occurs in the Negro and Trombetas Rivers.

Table 18.4. Isotope Ratios of Sr and Nd in Sediment of the Amazon River and Selected Tributaries

River	$^{87}Sr/^{86}Sr$	$^{87}Rb/^{86}Sr$	$^{143}Nd/^{144}Nd$	$^{147}Sm/^{144}Nd$
Negro	0.71698	0.40	0.511925	0.1021
Solimões	0.71319	1.53	0.512185	0.1175
Amazon (below Manaus)	0.71327	1.84	0.512235	0.1133
Madeira	0.73352	2.13	0.512027	0.1131
Urucara	0.72835	1.87	0.511732	0.1029
Trombetas	0.74683	1.65	0.511500	0.0920
Tapajós	0.75640	3.19	0.511606	0.0605
Amazon (Santarem)	0.72146	1.75	0.512094	0.1188

Source: Allègre et al., 1996.
Note: $^{87}Sr/^{86}Sr$ of NBS 987 = 0.710254 ± 0.000006; $^{143}Nd/^{144}Nd$ of the LaJolla Nd standard = 0.511850 ± 0.000010.

Table 18.5. Model Sm–Nd and Rb–Sr Dates Relative to CHUR of Suspended Sediment in the Amazon River below Manaus and in the Solimões and Negro Rivers

River	Nd Date, Ga	Sr Date, Ga
Amazon (below Manaus)	0.74	0.35
Solimões	0.87	0.42
Negro	1.15	2.71
Trombetas	1.65	1.87

Note: Based on data by Allègre et al. (1996) listed in Table 18.4.

18.3d Lead Isotopes, Zaire and Amazon

The Pb contained in the sediment transported by rivers resides primarily in grains of K-feldspar and in certain accessory minerals such as zircon, apatite, monazite, and so on. In addition, Pb^{2+} ions are strongly sorbed by colloidal particles of minerals, by amorphous oxyhydroxides of Fe and Al, and by organic matter. The isotope composition of sorbed Pb is likely to differ from that contained in resistant mineral particles having high U/Pb ratios (e.g., zircon).

The average weighted isotope ratios of Pb in sediment of the Zaire and Amazon river systems were used to calculate the abundance of ^{204}Pb and the atomic weights of Pb. These values were used to obtain the average $^{238}U/^{204}Pb$ and $^{232}Th/^{204}Pb$ ratios in Table 18.6. The $^{206}Pb/^{204}Pb$ ratios of bulk sediment (>0.2 μm) in the Amazon River and its tributaries reported by Allègre et al. (1996) range from 18.67 (Madeira) to 19.95 (Trombetas), and their Pb concentrations vary between 23.5 and 66.3 ppm. However, the $^{206}Pb/^{204}Pb$ and 1/Pb ratios of the sediment in these rivers do not define a mixing line (not shown), presumably because of the nonconservative behavior of Pb^{2+}. The same statement applies to the Zaire River and its tributaries.

The $^{206}Pb/^{204}Pb$ and $^{207}Pb/^{204}Pb$ ratios of the sediment of the Zaire River and its tributaries loosely define a straight line in Figure 18.22. This line cannot be an isochron (or an errorchron) because the sediment samples are composed of

Table 18.6. Average Concentrations of U, Th, and Pb and Weighted Average Isotope Ratios of Pb in the Sediment (>0.2 μm) of the Zaire and Amazon Rivers and Their Tributaries

Parameter	Zaire	Amazon
U, ppm	2.5 (0.9–4.3)	3.5 (2.4–4.3)
Th, ppm	14.2 (4.7–19.2)	16.0 (12.5–19.1)
Pb, ppm	35.6 (24.6–68.1)	42.2 (23.5–66.3)
$^{206}Pb/^{204}Pb$	18.746	19.022
$^{207}Pb/^{204}Pb$	15.742	15.714
$^{208}Pb/^{204}Pb$	39.043	38.997
Th/U	5.68	4.57
$^{238}U/^{204}Pb$	4.52	5.35
$^{232}Th/^{204}Pb$	26.5	25.30
Ab ^{204}Pb	0.013417	0.013380
At. wt. Pb	207.2077	207.2023

Source: Allègre et al., 1996.

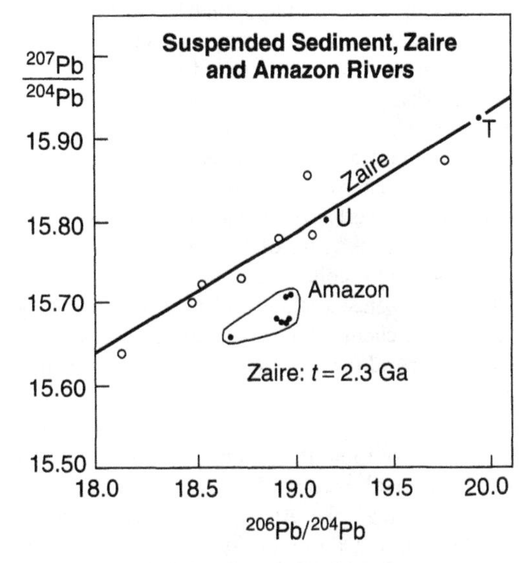

FIGURE 18.22 Mixing line defined by the $^{206}Pb/^{204}Pb$ and $^{207}Pb/^{204}Pb$ ratios of sediment (>0.2 μm) in the Zaire River and its tributaries. Two tributaries of the Amazon River (U = Urucara; T = Trombetas) also fit this array. The isotope ratios of Pb in sediment of the Amazon basin cluster in a small area of the diagram. Data from Allègre et al. (1996).

mineral grains having different ages. Instead, the linearity of this data array is the result of mixing of two principal sediment components containing differing amounts of radiogenic Pb. For example, the sediment may be a mixture of particles of Precambrian and Phanerozoic ages. The slope of this Pb–Pb array ($m = 0.14525$) yields a fictitious Pb–Pb date of about 2.3 Ga (Section 10.10a). The magnitude of this date implies a Precambrian age for one of the sediment components.

The isotope ratios of sediment of the Trombetas and Urucara Rivers in the Amazon basin (U and T in Figure 18.22) fit the Pb–Pb array of the rivers in the Zaire basin. The Precambrian provenance of the sediment in the Trombetas River is well established by their model dates in Table 18.5. Most of the other tributaries of the Amazon River cluster in a small area of Figure 18.22.

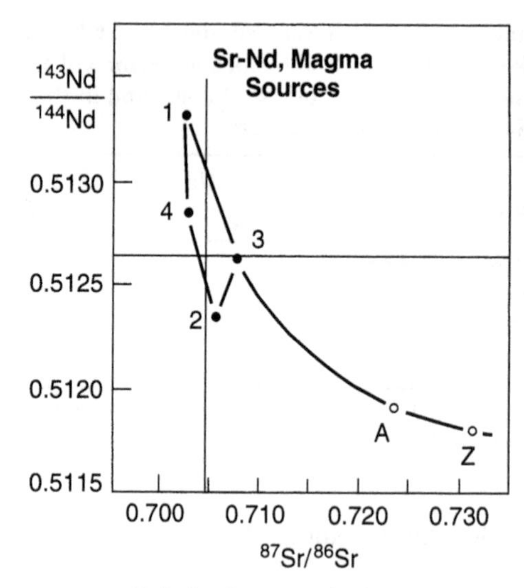

FIGURE 18.23 Relation between the isotope ratios of Sr and Nd of sediment in the Zaire (Z) and Amazon (A) Rivers and their tributaries to the isotope ratios of magma-source components in the mantle: 1 = DMM; 2 = EM1; 3 = EM2; 4 = HIMU (Table 17.1). Data from Allègre et al. (1996).

18.3e Implications for Petrogenesis

The sediment being transported by streams from the continents to the oceans accumulates in coastal basins and ultimately forms deposits of marine shale, siltstone, and sandstone. This sediment is representative (with certain limitations) of the continental crust, including both Precambrian basement rocks (granitic gneisses, metavolcanic rocks, etc.) and sedimentary, volcanic, and plutonic igneous rocks of Phanerozoic age. These kinds of rocks collectively contain Sr, Nd, and Pb, whose concentrations and isotope ratios may modify the chemical and isotopic compositions of mantle-derived magmas that intrude the continental crust.

In addition, a fraction of terrigenous sediment is incorporated into the sediment that accumulates in the abyssal basins of the oceans until it is subducted (e.g., the Aleutian Trench). Some of the Sr, Nd, and Pb of the subducted sediment is transferred to the mantle wedge and hence into magmas that form in this environment. In addition, these elements may ultimately reappear in the lavas of oceanic islands that form above plumes in the underlying mantle.

The sediment of the Zaire and Amazon Rivers is connected to the EM2 magma-source component (Table 17.1) in Figure 18.23 by a hyperbolic mixing curve. This diagram therefore illustrates the input of continental Sr and Nd into the petrogenetic processes that operate primarily within the ocean basins. The EM2 component (3 in Figure 18.23) consists of subducted oceanic crust, including terrigenous sediment. The crustal imprint is even more strongly developed when basalt magma from within the mantle-source polygon is contaminated directly by assimilating crustal rocks, represented in Figure 18.23 by sediment of the Zaire (Z) and Amazon (A) Rivers and their tributaries. This kind of petrogenesis was illustrated in Figure 17.29 by the basalt of the Saddle Mountain Formation of the Columbia River Basalt of Washington and Oregon. Additional examples of assimilation of crustal rocks are the basalts of the Paraná plateau of Brazil, the Etendeka Group of Namibia, the Karoo Group of South Africa, the Noril'sk area of Siberia, as well as others described by Faure (2001).

The isotope ratios of Sr and Pb in the sediment of the Zaire and Amazon Rivers are likewise related to EM2 by a mixing line that appears to be straight (not shown). Therefore, the isotope ratios of Pb of basalt magmas may also be altered by assimilation of crustal rocks. However, the effect is less evident because the $^{206}Pb/^{204}Pb$ ratios of crustal rocks (Zaire = 18.746, Amazon = 19.022) are similar to that of EM2.

18.4 SUMMARY

The Sr in solution in rivers draining the Precambrian Shield of Canada is generally enriched in radiogenic ^{87}Sr, but the $^{87}Sr/^{86}Sr$ ratios of different rivers range widely depending on the ages and Rb/Sr ratios of the rocks in their drainage basins. The presence of Sr-rich marine carbonate rocks of Paleozoic age lowers the $^{87}Sr/^{86}Sr$ ratios of streams draining Precambrian rocks. This phenomenon indicates that different isotopic varieties of Sr are being mixed in rivers even before the water is discharged into the oceans. Studies of the isotope compositions of Nd and Pb in river water are difficult because of the low concentrations of these elements at near-neutral values of the pH.

In most cases, groundwater is not in chemical and isotopic equilibrium with the rocks in which it resides either because insufficient time is available for equilibrium to be established or because of subsurface flow that results in mixing of waters that interacted with different kinds of rocks. In cases where two components of groundwater mix, the $^{87}Sr/^{86}Sr$ and 1/Sr ratios form linear mixing arrays.

The isotope compositions of Sr, Nd, and Pb of resistant mineral particles suspended in streams depend on their ages and parent–daughter ratios, whereas the sorbed fraction of these elements has the same isotope composition as the dissolved fraction. Consequently, the sediment in suspension in rivers contains a record of the ages of the rocks in the drainage basin, whereas the dissolved or sorbed fraction of the elements identifies the minerals that are dissolving in the course of chemical weathering.

The difference between the $^{87}Sr/^{86}Sr$ ratios of sediment and water in the Zaire and Amazon river systems increases with the magnitude of the $^{87}Sr/^{86}Sr$ ratios in the sediment. The $^{87}Sr/^{86}Sr$ and $^{143}Nd/^{144}Nd$ ratios of sediment of both river systems form overlapping data fields in quadrant IV of the Sr–Nd isotopic mixing diagram, as expected for rocks of the continental crust.

Mixing of water and sediment of the Solimões and Negro Rivers at Manaus, Brazil, can account for the $^{87}Sr/^{86}Sr$ ratio and the concentrations of Sr^{2+} and $UO_2{}^{2+}$ in the water of the Amazon River downstream of the confluence. However, Pb^{2+} is strongly nonconservative and is removed from solution by sorption to suspended and colloidal particles.

The Sr and Nd model dates of sediment in the tributaries of the Amazon River identify the presence of Precambrian basement rocks, especially in the Negro and Trombetas rivers, which drain the Guyana Shield north of the Amazon basin. The sediment of the Zaire River and its tributaries defines a straight line in coordinates of $^{206}Pb/^{204}Pb$ and $^{207}Pb/^{204}Pb$ ratios as a result of mixing of two components of Precambrian and Phanerozoic ages.

Some of the sediment transported by streams into the oceans is ultimately subducted. The radiogenic isotopes of the subducted terrigenous sediment are partly transferred into the mantle wedge above the subduction zone and can be incorporated into the magmas that are erupted by volcanoes in island arcs and coastal mountains (e.g., the Andes). The presence of former terrigenous sediment in mantle plumes and in oceanic crust that underplate the continental crust in some places characterizes the EM2 component of magma sources in the mantle.

REFERENCES FOR GEOCHEMISTRY OF LAKES AND STREAMS (INTRODUCTION, SECTION 18.1 INTRODUCTION)

Appelo, C. A. J., and D. Postma, 1993. *Geochemistry, Groundwater and Pollution*. Balkema, Rotterdam.

Berner, E. K., and R. A. Berner, 1987. *The Global Water Cycle: Geochemistry and Environment*. Prentice-Hall. Englewood Cliffs, New Jersey.

Berner, E. K., and R. A. Berner, 1996. *Global Environment: Water, Air, and Geochemical Cycles*. Prentice-Hall, Upper Saddle River, New Jersey.

Drever, J. I., 1997. *The Geochemistry of Natural Water; Surface and Groundwater Environments*, 3rd ed. Prentice-Hall, Upper Saddle River, New Jersey.

Faure, G., 1998. *Principles and Applications of Geochemistry*, 2nd ed. Prentice-Hall, Upper Saddle River, New Jersey.

Feulner, A. J., and J. H. Hubble, 1960. Occurrence of strontium in the surface and ground waters of Champaign County, Ohio. *Econ. Geol.*, 55:176–186.

Langmuir, D., 1997. *Aqueous Environmental Geochemistry*. Prentice-Hall, Upper Saddle River, New Jersey.

Morrill, C., and P. L. Koch, 2002. Elevation or alteration? Evaluation of isotopic constraints on paleolatitudes surrounding the Eocene Green River basin. *Geology*, 30(2):151–154.

Neat, P. L., G. Faure, and W. J. Pegram, 1979. The isotopic composition of strontium in nonmarine carbonate rocks; the Flagstaff Formation of Utah. *Sedimentology*, 26:271–282.

Palmer, M. R., and J. M. Osmond, 1992. Controls over the Sr isotope composition of river water. *Geochim. Cosmochim. Acta*, 56:2099–2111.

Rhodes, M. K., A. R. Carroll, J. T. Pietras, B. L. Beard, and C. M. Johnson, 2002. Strontium isotope record of paleohydrology and continental weathering, Eocene Green River Formation, Wyoming. *Geology*, 30(2): 167–170.

Stumm, W. (Ed.), 1990. *Aquatic Chemical Kinetics*. Wiley, New York.

Stumm, W., and J. J. Morgan, 1996. *Aquatic Chemistry*, 3rd ed. Wiley, New York.

Wedepohl, K. H., 1978. *Handbook of Geochemistry*, Sections 37 and 38. Springer-Verlag, Heidelberg.

REFERENCES FOR RIVERS OF CANADA AND THE WORLD (SECTIONS 18.1a–18.1b)

Faure, G., J. H. Crocket, and P. M. Hurley, 1967. Some aspects of the geochemistry of strontium and calcium in the Hudson Bay and the Great Lakes. *Geochim. Cosmochim. Acta*, 31:451–461.

Faure, G., P. M. Hurley, and H. W. Fairbairn, 1963. An estimate of the isotopic composition of strontium in rocks of the Precambrian shield of North America. *J. Geophys. Res.*, 68(8):2323–2329.

Faure, G., 1978. Abundance of Sr in common igneous rock types. In K. H. Wedepohl (Ed.), *Handbook of Geochemistry*, II-4, 38E1–38E14.

Krogh, R. E., D. W. Davis, and F. Corfu, 1984. Precise U–Pb zircon and baddeleyite ages for the Sudbury area. In E. G. Pye, A. J. Naldrett, and P. E. Giblin (Eds.), *The Geology and Ore Deposits of the Sudbury Structure*, Toronto, Ontario, Ontario Geological Survey. Spec. Vol. 1, pp. 4342–4446.

McNutt, R. H., M. Gascoyne, and D. C. Kamineni, 1987. $^{87}Sr/^{86}Sr$ values in groundwaters of the East Bull Lake pluton, Superior Province, Ontario, Canada. *Appl. Geochem.*, 2:93–101.

Wadleigh, M. A., J. Veizer, and C. Brooks, 1985. Strontium and its isotopes in Canadian rivers: Fluxes and global implications. *Geochim. Cosmochim. Acta*, 49:1727–1736.

Yang, C., C. Telmer, and J. Veizer, 1996. Chemical dynamics of the "St. Lawrence" riverine system: δD_{H_2O}, $\delta^{18}O_{H_2O}$, $\delta^{13}C_{DIC}$, $\delta^{34}S_{sulfate}$, and dissolved $^{87}Sr/^{86}Sr$. *Geochim. Cosmochim. Acta*, 60:851–866.

REFERENCES FOR SEDIMENT IN STREAMS (SECTION 18.2)

Blum, J. D., Y. Erel, and K. Brown, 1993. $^{87}Sr/^{86}Sr$ ratios of Sierra Nevada stream waters: Implications for relative mineral weathering rates. *Geochim. Cosmochim. Acta*, 57:5019–5026.

Cameron, E. M., and K. Hattori, 1997. Strontium and neodymium isotope ratios in the Fraser River, British Columbia: A riverine transect across the Cordilleran orogen. *Chem. Geol.*, 137:243–253.

Douglas, G. B., C. M. Gray, B. T. Hart, and R. Beckett, 1995. A strontium isotopic investigation of the origin of suspended particulate matter (SPM) in the Murray-Darling River system, Australia. *Geochim. Cosmochim. Acta*, 59:3799–3815.

Faure, G., 1998. *Principles and Applications of Geochemistry*, 2nd ed. Prentice-Hall, Upper Saddle River, New Jersey.

Faure, G., 1986. *Principles of Isotope Geology*, 2nd ed. Wiley, New York.

Martin, C. E., and M. T. McCulloch, 1999. Nd–Sr isotopic and trace element geochemistry of river sediments and soils in a fertilized catchment, New South Wales, Australia. *Geochim. Cosmochim. Acta*, 63:287–305.

Muffler, L. J. P., and B. R. Doe, 1968. Composition and mean age of detritus of the Colorado River delta in the Salton Trough, southeastern California. *J. Sed. Petrol.*, 38:384–399.

Wadleigh, M. A., J. Veizer, and C. Brooks, 1985. Strontium and its isotopes in Canadian rivers: Fluxes and global implications. *Geochim. Cosmochim. Acta*, 49:1727–1736.

REFERENCES FOR ZAIRE AND AMAZON RIVERS (SECTION 18.3)

Allègre, C. J., B. Dupré, P. Négrel, and J. Gaillardet, 1996. Sr–Nd–Pb isotope systematics in Amazon and Congo River systems: Constraints about erosion processes. *Chem. Geol.*, 131:93–112.

Blum, J. D., Y. Erel, and K. Brown, 1993. $^{87}Sr/^{86}Sr$ ratios of Sierra Nevada stream waters: Implications for relative mineral weathering rates. *Geochim. Cosmochim. Acta*, 57:5019–5025.

Cameron, E. M., and K. Hattori, 1997. Strontium and neodymium isotope ratios in the Fraser River, British Columbia: A riverine transect across the Cordilleran orogen. *Chem. Geol.*, 137:243–253.

Douglas, G. B., C. M. Gray, B. T. Hart, and R. Beckett, 1995. A strontium isotopic investigation of the origin of suspended particulate matter (SPM) in the Murray-Darling River system, Australia. *Geochim. Cosmochim. Acta*, 59:3799–3815.

Faure, G., 2001. *Origin of Igneous Rocks; the Isotopic Evidence*. Springer-Verlag, Heidelberg.

CHAPTER 19

The Oceans

The oceans and the basins they occupy play important roles in the geological activity of the Earth, including the deposition of sediment derived from the continents and transported by streams, wind, and glacial ice. In addition, streams and, to a lesser extent, groundwater and glaciers transport dissolved chemical elements into the oceans and thereby contribute to the chemical composition of seawater and to the isotope composition of Sr, Nd, Pb, and other elements that have radiogenic isotopes. The marine isotope geochemistry of Ca was discussed in Section 8.2b.

19.1 STRONTIUM IN THE PHANEROZOIC OCEANS

The chemical composition of seawater is controlled by the balance of inputs and outputs of each element. The resulting concentrations of conservative elements are related to the salinity of the water, which is defined as the total amount of solid material in grams contained in one kilogram of seawater when all the carbonate has been converted to the oxide, the bromine and iodine have been replaced by chlorine, and all organic matter has been completely oxidized. Since salinity is difficult to measure directly, it is determined from the chlorinity of seawater, which is based on the weight of Ag precipitated from one kilogram of seawater:

$$\text{Chlorinity } (‰) = 0.3285234 \text{ Ag (g)} \qquad (19.1)$$

The chlorinity is related to the salinity by the equation (Sverdrup et al., 1942; Ross, 1970)

$$\text{Salinity } (‰) = 1.8066 \text{ chlorinity } (‰) \qquad (19.2)$$

The salinity of seawater in the open ocean of 34.71‰ is *lowered* as a result of dilution by fresh water discharged by streams (e.g., the Baltic Sea) or by melting of icebergs, and it is *increased* by evaporative concentration (e.g., the Red Sea) and by the formation of ice (e.g., the Weddell Sea of Antarctica).

19.1a Present-Day Seawater

The concentration of Sr in seawater at depths of >1500 m is related to its salinity by an equation derived by Bernat et al. (1972) and Brass and Turekian (1972, 1974):

$$\text{Sr(ppm)} = 0.221 \pm 0.0010 \text{ salinity } (‰) \quad \textbf{(Atlantic Ocean)} \qquad (19.3)$$

$$\text{Sr(ppm)} = 0.220 \pm 0.0026 \text{ salinity } (‰) \quad \textbf{(Pacific Ocean)} \qquad (19.4)$$

Accordingly, the Sr concentration of standard seawater in the Atlantic Ocean having a salinity of 35‰ is 7.74 ppm. The salinity and the Sr concentration of surface water in the oceans both decrease regionally in response to dilution of seawater by meteoric water and by mixing with river water and glacial meltwater. Conversely, both the salinity and the concentration of Sr increase in case of water loss by excessive evaporation and by the formation of sea ice. The concentration of Sr in seawater (7.74 ppm) is more than one hundred times higher than that of average river water (0.070 ppm). The residence time of Sr in the oceans, 5.0×10^6 years (Taylor and McLennan, 1985), causes Sr in the oceans to be homogenized isotopically by mixing on a timescale of about 1000 years (Palmer and Edmond, 1989).

The isotope composition of Sr in seawater has been measured by many investigators, whose results published prior to 1980 were summarized by Faure (1982). These studies included analyses of marine Sr from the North Atlantic by Faure et al. (1965), who used a three-component mixing model to explain the observed numerical value of the $^{87}Sr/^{86}Sr$ ratio of modern seawater. Subsequent analyses of marine Sr from the Hudson Bay by Faure et al. (1967) strengthened the important conclusion that the present $^{87}Sr/^{86}Sr$ ratio of seawater is constant throughout all parts of the ocean.

The numerical values of the $^{87}Sr/^{86}Sr$ ratios of seawater reported in the literature are reconciled by reference to interlaboratory standards routinely analyzed by most investigators. The numerical values of the $^{87}Sr/^{86}Sr$ ratios of these standards are as follows: E&A, 0.70800; NBS 987, 0.71025. In addition, Ludwig et al. (1988) reported $^{87}Sr/^{86}Sr$ ratios of a large *Tridachna* shell collected live from the bottom of the lagoon at Enewetak atoll in the Pacific Ocean. This sample is identified as EN-1 and has been used to standardize measurements of the $^{87}Sr/^{86}Sr$ ratios of biogenic carbonates of marine origin (e.g., Hodell et al., 1989, 1990; Barrera et al., 1991; Carpenter et al., 1991). The average $^{87}Sr/^{86}Sr$ ratios reported by these investigators are

EN-1	0.709186 ± 0.000003
NBS 987	0.71025

A compilation of average $^{87}Sr/^{86}Sr$ ratios of seawater and of modern marine biogenic and abiogenic carbonates in Table 19.1 leaves no room for doubt that Sr in the oceans at the present time is characterized by

$$\frac{^{87}Sr}{^{86}Sr}(\text{oceans}) = 0.70918 \pm 0.00001(2\overline{\sigma})$$

relative to NBS 987 = 0.71025. The $^{87}Sr/^{86}Sr$ ratios of marine Sr in Table 19.1 that were adjusted to

Table 19.1. Average $^{87}Sr/^{86}Sr$ Ratios of Modern Marine Sr Adjusted to 0.710250 for NBS 987 and 0.708000 for E&A

$^{87}Sr/^{86}Sr$	Standard	References[a]
	Seawater	
0.70915	NBS 987	1
0.709220	NBS 987	2
0.709198	E&A	
0.70917	NBS 987	3
0.709183	NBS 987	4
Average	0.70918 ±0.00003	(NBS only)
	Biogenic Carbonate	
0.70917	NBS 987	5
0.70918	NBS 987	6
0.70910	E&A	
0.709183	NBS 987	9
0.709184	NBS 987	7
Average	0.70918 ±0.00001	(NBS only)
	Abiogenic Carbonate	
0.709182	NBS 987	8
0.709172	NBS 987	8
Average	0.70918 ±0.00001	(NBS only)

[a] 1. Elderfield and Greaves, 1981, *Geochim. Cosmochim. Acta*, 45:2201–2212. 2. Hess et al., 1986, *Science*, 231:979–983. 3. Keto and Jacobsen, 1987, *Earth Planet. Sci. Lett.*, 84:27–41. 4. Asmerom et al., 1991, *Geochim. Cosmochim. Acta*, 55:2883–2894. 5. DePaolo and Ingram, 1985, *Science*, 227:938–941. 6. Burke et al., 1982, *Geology*, 10(10):516–519. 7. Barrera et al., 1991, in Barron et al. (Eds.), *Proc. Ocean Drilling Program, Scientific Results*, 119:731–738. 8. Carpenter et al., 1991, *Geochim. Cosmochim. Acta*, 55:1991–2010. 9. Denison et al., 1998, *Chem. Geol.*, 152:325–340.

0.70800 for E&A were excluded from the average stated above.

The best-available measurements contained in Table 19.1 indicate that the $^{87}Sr/^{86}Sr$ ratios of biogenic and inorganic marine calcite/aragonite in the oceans today are indistinguishable from the $^{87}Sr/^{86}Sr$ ratio of seawater. Therefore, unreplaced skeletal calcium carbonate of marine origin can be used to determine the isotope composition of Sr in the oceans of the geological past.

19.1b Phanerozoic Carbonates

Time-dependent variations of the $^{87}Sr/^{86}Sr$ ratio of seawater in Phanerozoic time were first detected by Peterman et al. (1970), who measured the isotope composition of Sr in unreplaced skeletal calcium carbonates of Phanerozoic age. In addition, these authors demonstrated that the Sr in the oceans remained isotopically homogeneous in spite of the fact that the $^{87}Sr/^{86}Sr$ ratio changed with time. These conclusions were confirmed by Veizer and Compston (1974) based on analyses of a large number of marine limestones of known stratigraphic age. Later, Faure (1982) compiled more than 100 $^{87}Sr/^{86}Sr$ ratios of marine carbonates published by 33 research groups prior to 1981 and plotted them in 25-million-year increments. The resulting curve confirmed that the $^{87}Sr/^{86}Sr$ of seawater declined irregularly from about 0.7090 during the Cambrian Period to about 0.7071 at the Permian–Triassic boundary and subsequently increased to the present value of 0.7092.

All of these efforts were surpassed by Burke et al. (1982), who reconstructed the variation of the $^{87}Sr/^{86}Sr$ ratio of the oceans in Phanerozoic time from Sr isotope analyses of 786 samples of marine limestones of known stratigraphic ages. Although 93% of the samples they analyzed defined a narrow band in coordinates of the $^{87}Sr/^{86}Sr$ ratio and geological age, 7% of the data points deviated from the main trend. Burke et al. (1982) attributed the discrepant results to several possible causes:

1. The limestones were of lacustrine origin.
2. The isotope composition of Sr was contaminated during diagenesis.
3. The assigned stratigraphic ages were in error.
4. The $^{87}Sr/^{86}Sr$ ratios varied on a short timescale not resolved by the samples.

An additional source of error is the possibility that radiogenic ^{87}Sr was leached from clay minerals and ferric oxide particles within the limestone samples by the acid used to dissolve them. Burke et al. (1982) attempted to minimize this source of error by analyzing only pure limestone samples whose Sr concentrations exceeded 200 ppm and that contained less than 10% of insoluble residue. Nevertheless, other investigators have used dilute acetic acid, which is less aggressive than the hydrochloric and nitric acids used by Burke and his colleagues (e.g. DePaolo and Ingram, 1985; Bailey et al., 2000).

Many investigators have contributed to the study of the evolution of the isotopic composition of Sr in the oceans by analyzing limestones and other Sr-bearing materials of marine origin from specific systems of Phanerozoic age:

Cambrian and Ordovician: Denison et al., 1998, *Chem. Geol.*, 152:325–340. Ebneth et al., 2001, *Geochim. Cosmochim. Acta*, 65: 2273–2292. Derry et al., 1994, *Earth Planet. Sci. Lett.*, 128:671–681. Montañez et al., 1996, *Geology*, 24(10):917–920. Kaufman et al., 1996, *Geol. Mag.*, 133:509–533. Gorokhov et al., 1995, *Stratigraphy and Geological Correlation*, 3(1):1–28. Shields et al., 2003, *Geochim. Cosmochim. Acta*, 67:2005–2025.

Silurian and Devonian: Denison et al., 1997, *Chem. Geol.*, 140:109–121. Bertram et al., 1992, *Earth Planet. Sci. Lett.*, 113: 239–249. Carpenter et al., 1991, *Geochim. Cosmochim. Acta*, 55:1991–2010. Diener et al., 1996, *Geochim. Cosmochim. Acta*, 60:639–652.

Carboniferous: Cummins and Elderfield, 1994, *Chem. Geol.*, 118:255–270. Popp et al., 1986, *Geochim. Cosmochim. Acta*, 50:1321–1328. Bruckschen et al., 1999, *Chem. Geol.*, 161:127–163.

Permian and Triassic: Martin and Macdougall, 1995, *Chem. Geol.*, 125:73–99. Spötl and Pak, 1996, *Chem. Geol.*, 131:219–234.

Triassic and Jurassic: Hallam, 1994, *Geology*, 22:1079–1082. Koepnick et al., 1990, *Chem. Geol. (Isotope Geosci. Sect.)*, 80: 327–410.

Jurassic: Jones et al., 1994a, *Geochim. Cosmochim Acta*, 58:1285–1301. Jones et al., 1994b, *Geochim. Cosmochim. Acta*, 58:3061–3074.

Cretaceous and Tertiary: Jones et al., 1994, *Geochim. Cosmochim. Acta*, 58:3061–3074. Meisel et al., 1995, *Geology*, 23(4):313–316. Jenkyns et al., 1995, *Proc. Ocean Drilling Program, Scientific Results*, 143: 89–97. DePaolo et al., 1983, *Earth Planet. Sci. Lett.*, 64:356–373. Hess et al., 1986, *Science*, 231:979–983.

Tertiary: Farrell et al., 1995, *Geology*, 23: 403–406. Barrera et al., 1991, *Proc. Ocean Drilling Program, Scientific Results*, 119:731–738. Hodell et al., 1989, *Earth Planet. Sci. Lett.*, 92:165–178. Hodell et al., 1990, *Chem. Geol. (Isotope Geosci. Sect.)*, 80:291–307. Hodell et al., 1991, *Geology*, 19:24–27. Henderson et al., 1994, *Earth Planet. Sci. Lett.*, 128:643–651. Hess et al., 1989, *Paleoceanography*, 4:655–679. Capo and DePaolo, 1990, *Science*, 249:51–55. Hodell and Woodruff, 1994, *Paleoceanography*, 9:405–426. Martin et al., 1999, *Paleoceanography*, 14:74–83. Reinhardt et al., 2000, *Chem. Geol.*, 164: 331–343.

The large number of $^{87}Sr/^{86}Sr$ ratios and corresponding stratigraphic ages of unaltered carbonate and phosphate samples of marine origin published by these and other investigators have been used to reconstruct the time-dependent evolution of the isotopic composition of Sr in the oceans during Phanerozoic time:

Burke et al., 1982, *Geology*, 10:516–519. Elderfield, 1986, *Paleo. Paleo. Paleo*, 57:71–90. Veizer, 1989, *Annu. Rev. Earth Planet. Sci.*, 17:141–167. McArthur, 1994, *Terra Nova*, 6:331–358. Veizer et al., 1997, *Paleo. Paleo. Paleo.*, 132:65–77. Veizer et al., 1999, *Chem. Geol.*, 161:59–88. Howarth and McArthur, 1997, *J. Geol.*, 105:441–456. Denison et al., 1998, *Chem. Geol.*, 152:325–340. Prokoph and Veizer, 1999, *Chem. Geol.*, 161:225–240.

Smalley et al. (1994) introduced a statistical method for fitting a curve to the available data using weighting factors based on a ranking of samples in Table 19.2 derived from consideration of the type of sample, the method of dissolution, supporting data that indicate the absence of alteration, biostratigraphic age assignment, and analytical errors of measurement. The authors assigned high reliability to unaltered biogenic carbonates and phosphates but considered whole-rock samples of limestone and chalk to have low reliability. The fitting program LOWESS was used by Smalley et al. (1994) to construct a curve extending from Recent to 450 Ma based on 1300 data points. Before merging $^{87}Sr/^{86}Sr$ ratios measured on different mass spectrometers, interlaboratory biases must be removed as nearly as possible by reference to NBS 987 and EN-1. For example, McArthur et al. (2001) corrected 3366 $^{87}Sr/^{86}Sr$ ratios to NBS 987 = 0.710248 and EN-1 = 0.709175, which they obtained in the Radiogenic Isotope Laboratory at Royal Holloway University of London. The resulting data points in Figure 19.1 define the variation of $^{87}Sr/^{86}Sr$ ratios of seawater between 0 and 509 Ma.

The $^{87}Sr/^{86}Sr$ ratio of seawater in Figure 19.1 reached about 0.7091 in Late Cambrian time (500 Ma) and subsequently fluctuated repeatedly as it declined to 0.7068 in the Late Jurassic (Oxfordian) Epoch (158 Ma). More recently, the $^{87}Sr/^{86}Sr$ ratio of seawater has been rising with only minor fluctuations toward the present value of 0.70918. Starting at about 40 Ma during the Bartonian Age of the late middle Eocene Epoch, the $^{87}Sr/^{86}Sr$ ratio of the oceans has increased steadily and without significant fluctuations.

The isotopic evolution of Sr in the oceans in Figure 19.1 is a record of the geological activity of the Earth on a global scale and therefore is of great importance for the Earth sciences. The time-dependent fluctuations of the $^{87}Sr/^{86}Sr$ ratio of the oceans were caused by changes in the amounts and isotope compositions of Sr derived

Table 19.2. Reliability of Various Types of Samples to Preserve the $^{87}Sr/^{86}Sr$ Ratio of Seawater

High reliability: belemnites, nonluminescent shells of brachiopods, well-preserved tests of foraminifers in deep-sea sediment, red algae, massive anhydrite in marine evaporite, rudist bivalves
Medium reliability: luminescent brachiopod shells, thick-shelled bivalves, conodonts with low index of alteration
Low reliability: conodonts with high alteration index, thin-shelled bivalves, fish teeth, echinoids, ammonoids, disseminated anhydrite, foram tests in deeply buried sandstones, whole-rock samples of limestone and chalk

Source: Smalley et al., 1994.

from different sources that entered the oceans, including

1. weathering of old granitic basement rocks of the continental crust;
2. volcanic activity in the oceans and on the continents; and
3. the diagenesis, dolomitization, and dissolution of marine carbonate rocks on the continents and on the continental shelves.

Geologically speaking, the Sr isotope curve of seawater records changes in the rate of seafloor spreading and subduction, the occurrence of orogenies, uplift of continents followed by rifting and large-scale fissure eruptions, and global climate change leading either to continental glaciation and lowering of sealevel or to formation of marine

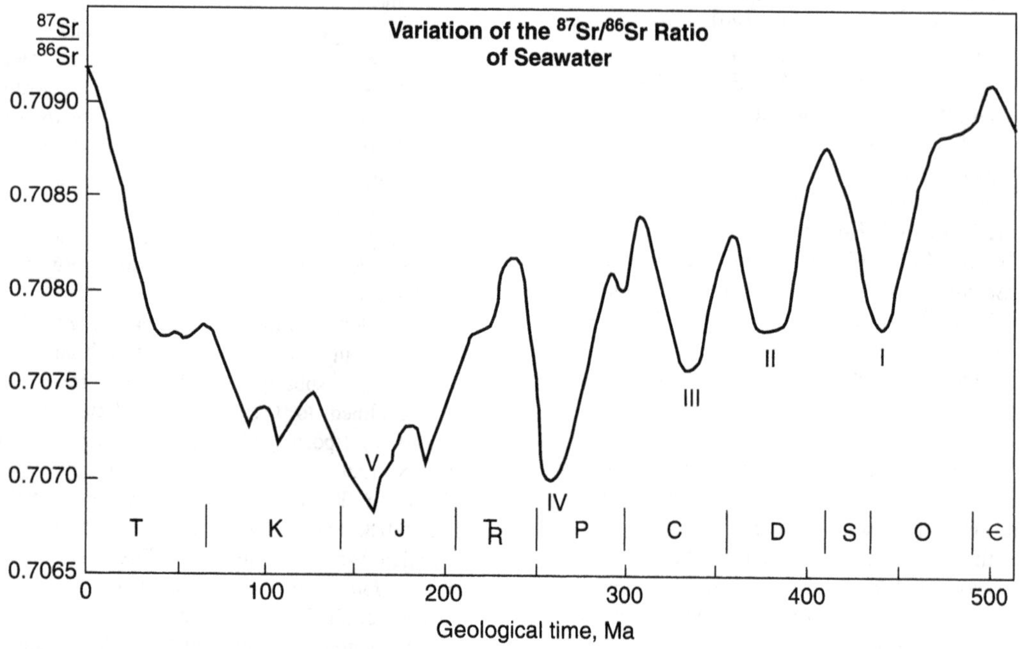

FIGURE 19.1 Time-dependent variation of the $^{87}Sr/^{86}Sr$ ratio of seawater in Phanerozoic time based on 3366 data points from 42 publications adjusted to $^{87}Sr/^{86}Sr = 0.710248$ for NBS 987: T = Tertiary, K = Cretaceous, J = Jurassic, Ꞧ = Triassic, P = Permian, C = Carboniferous, D = Devonian, S = Silurian, O = Ordovician, Є = Cambrian. The Roman numerals identify hypothetical episodes of global volcanic activity that caused the $^{87}Sr/^{86}Sr$ ratio of seawater to decline temporarily. Replotted from the diagrams of McArthur et al. (2001).

evaporite deposits that decrease the concentration of Sr in seawater and make it more susceptible to change of its isotope composition.

In addition, the $^{87}Sr/^{86}Sr$ ratios of marine carbonates and phosphates can be used to date such materials, especially in case they were deposited during the Tertiary Period (i.e., post middle Eocene). Other uses of this record include the correlation of strata of known stratigraphic ages, differentiation between marine and lacustrine carbonate rocks, and evidence for diagenetic alteration of marine carbonates of known ages. The isotope composition of Sr and other elements may also record the impacts of large extraterrestrial objects (e.g., asteroids and comets) and the resulting perturbations of the global climate and biosphere:

DePaolo et al., 1983, *Earth Planet. Sci. Lett.*, 64:356–373. Macdougall, 1988, *Science*, 239: 485–486. Martin and Macdougall, 1995, *Chem. Geol.*, 125:73–99. Meisel et al., 1995, *Geology*, 23(4):313–316. McArthur et al., 1998, *Earth Planet. Sci. Lett.*, 160:179–1992.

19.1c Mixing Models

The isotope composition of Sr in seawater during Phanerozoic time is the result of mixing of three isotopic varieties of Sr that enter the oceans primarily by discharge of water by rivers and hot springs along midocean ridges and by the interaction of seawater with volcanic rocks erupted within the ocean basins. The mixing model to be presented here was originally proposed by Faure et al. (1965) and was later elaborated by Faure (1977).

According to this model, the principal sources of Sr entering the oceans are

1. rocks of sialic composition in the continental crust that are enriched in radiogenic ^{87}Sr because of their age and Rb enrichment;
2. volcanic rocks derived from the mantle along midocean ridges, oceanic islands, island arcs, and lava plateaus on the continents; and
3. marine limestones of Paleozoic, Mesozoic, and Cenozoic ages.

Strontium also originates from mixed sources, such as Precambrian gneisses overlain by erosional remnants of marine limestone of Paleozoic age (e.g., parts of the Superior tectonic province of Canada), volcano-sedimentary complexes (e.g., the Appalachian Mountains of North America), interbedded limestone–shale sequences (e.g., midwestern USA and southwestern Ontario), and aerosol particles from the atmosphere. The existence of such hybrid sources means that mixing of different isotopic varieties of Sr starts at the source, continues during transport by streams, and is completed in the oceans. Most of the major rivers of the world transport Sr in solution whose $^{87}Sr/^{86}Sr$ ratios are the result of mixing of Sr derived from the primary sources identified above.

The $^{87}Sr/^{86}Sr$ ratio of present-day seawater (0.70918) can be represented by the equation

$$\left(\frac{^{87}Sr}{^{86}Sr}\right)_{sw} = \left(\frac{^{87}Sr}{^{86}Sr}\right)_s s + \left(\frac{^{87}Sr}{^{86}Sr}\right)_v v + \left(\frac{^{87}Sr}{^{86}Sr}\right)_m m \qquad (19.5)$$

where s = fraction of Sr derived from sialic basement rocks of Precambrian age

v = fraction of Sr derived from mafic volcanic rocks of Mesozoic and Cenozoic ages that originated from the mantle

m = fraction of Sr that is recycled by dissolution of marine carbonate rocks of Phanerozoic age on the continents and on the continental shelves

Therefore,

$$s + v + m = 1.0 \qquad (19.6)$$

Information presented in Chapters 17 and 18 supports the assumption that the present $^{87}Sr/^{86}Sr$ ratios of Sr derived from the three primary sources are

$$\left(\frac{^{87}Sr}{^{86}Sr}\right)_s = 0.720 \pm 0.005$$

$$\left(\frac{^{87}Sr}{^{86}Sr}\right)_v = 0.704 \pm 0.002$$

$$\left(\frac{^{87}Sr}{^{86}Sr}\right)_m = 0.708 \pm 0.001$$

The $^{87}Sr/^{86}Sr$ ratios assigned to Sr from sialic and volcanic sources are similar to those used by Brass (1976). The estimate of the $^{87}Sr/^{86}Sr$ ratio of marine limestones of Phanerozoic age (0.708 ± 0.001) was obtained by averaging 52 values taken from the Sr evolution curve of McArthur et al. (2001) at 10-million-year intervals. The result was $^{87}Sr/^{86}Sr = 0.70793 \pm 0.0006$ (1σ).

Based on these assumptions, the three-component mixing model expresses the $^{87}Sr/^{86}Sr$ ratio of seawater at the present time by the equation

$$\left(\frac{^{87}\text{Sr}}{^{86}\text{Sr}}\right)_{\text{sw}} = 0.720s + 0.704v + 0.708m \quad (19.7)$$

Equations 19.6 and 19.7 were used to construct the three-component mixing diagram of seawater in Figure 19.2 in coordinates of $(^{87}Sr/^{86}Sr)_{sw}$ (y-coordinate) and v (x-coordinate).

The procedure is to set m equal to zero and to express s in terms of v by use of equation 19.6: If $m = 0$, $s + v = 1.0$, and $s = 1 - v$.

In this case, equation 19.7 reduces to

$$\left(\frac{^{87}\text{Sr}}{^{86}\text{Sr}}\right)_{\text{sw}} = 0.720(1 - v) + 0.704v$$

If $v = 1.0$,

$$\left(\frac{^{87}\text{Sr}}{^{86}\text{Sr}}\right)_{\text{sw}} = 0.704$$

If $v = 0$,

$$\left(\frac{^{87}\text{Sr}}{^{86}\text{Sr}}\right)_{\text{sw}} = 0.720$$

If $m = 0.2$, $s + v + 0.2 = 1.0$, and $s = 0.8 - v$, then

$$\left(\frac{^{87}\text{Sr}}{^{86}\text{Sr}}\right)_{\text{sw}} = 0.720(0.8 - v) + 0.704v + 0.708 \times 0.2$$

If $v = 1.0$,

$$\left(\frac{^{87}\text{Sr}}{^{86}\text{Sr}}\right)_{\text{sw}} = 0.7016$$

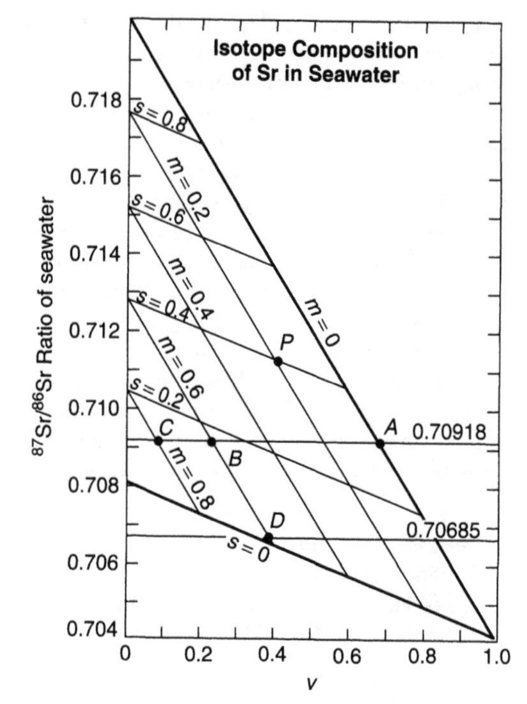

FIGURE 19.2 Model of the $^{87}Sr/^{86}Sr$ ratio in the oceans considered as a mixture of Sr contributed by weathering of young volcanic rocks (0.704 ± 0.002), old sialic rocks (0.720 ± 0.005), and marine carbonate rocks (0.708 ± 0.001). The coefficients v, s, and m are defined as the fractions of Sr contributed to the oceans by volcanic, sialic, and marine carbonate rocks, respectively.

If $v = 0$,

$$\left(\frac{^{87}\text{Sr}}{^{86}\text{Sr}}\right)_{\text{sw}} = 0.7176$$

The calculations for $m = 0.4$, 0.6, 0.8 follow the same pattern. In addition, equation 19.7 is solved for different values of $s = 0$, 0.2, 0.4, 0.6, 0.8. The lines representing the selected values of m and s are contours within the mixing triangle in Figure 19.2.

The present value of the $^{87}Sr/^{86}Sr$ ratio of seawater is represented in Figure 19.2 by a horizontal line which is the locus of all combinations of s, v, and m that yield an $^{87}Sr/^{86}Sr$ ratio of 0.70918. Point A represents the extreme case of $m = 0$ (no

Sr is derived from marine limestone). In that case,

$$0.70918 = 0.720(1 - v) + 0.704v \quad \text{and}$$

$$v = 0.67 \qquad s = 0.33.$$

This case is unrealistic because of the great abundance of marine carbonate rocks exposed to weathering on the continents during Phanerozoic time. However, during the Archean Eon the $^{87}Sr/^{86}Sr$ ratio of the oceans was controlled primarily by Sr derived from young volcanic rocks and old granitic rocks of the continental crust because marine carbonate rocks were much less abundant in early Precambrian time than during the Phanerozoic Eon. In addition, old crustal rocks as well as young volcanic rocks had lower $^{87}Sr/^{86}Sr$ ratios in early Precambrian time than they do at present.

A more likely scenario to explain the $^{87}Sr/^{86}Sr$ ratio of the present oceans is suggested by the coordinates of point B in Figure 19.2 ($m = 0.6$) from which follows

$$v = 0.23 \quad \text{and} \quad s = 0.17$$

In this case, weathering of the continental crust and of young volcanic rocks contribute 23 and 17% of the Sr in the oceans, respectively, whereas marine limestone contributes 60%.

Point C in Figure 19.2 implies an even larger contribution of Sr to the oceans ($m = 0.8$) than point B. The corresponding values of s and v are

$$v = 0.08 \quad \text{and} \quad s = 0.12$$

The proportions indicated by points B and C suggest that more than half of the Sr entering the oceans originates from marine carbonate rocks of Phanerozoic age. The values of v and s in Figure 19.3 decrease linearly with increasing values of m until $v = 0$ at $m = 0.90$, which means that less than 90% of the Sr in the present oceans originated from marine carbonate rocks of Phanerozoic age. Brass (1976) concluded that Sr from marine limestone amounts to about 75% of the total Sr entering the oceans.

The Sr isotope mixing model in Figure 19.2 can also account for the low $^{87}Sr/^{86}Sr$ ratio of seawater (0.70685) that occurred in Late Jurassic

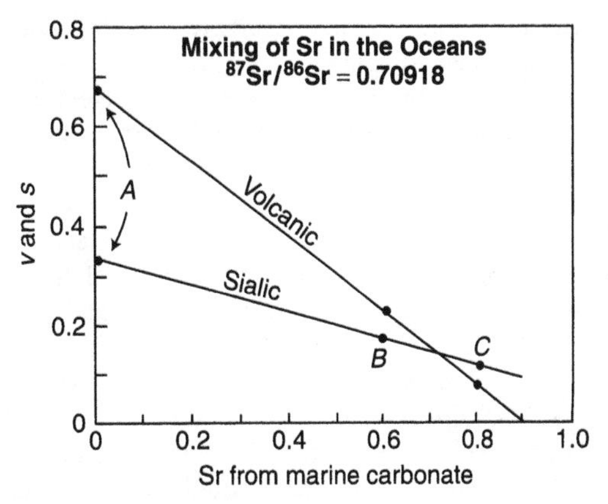

FIGURE 19.3 Compatible sets of values of the proportions of Sr in the present oceans contributed by marine carbonates of Phanerozoic age (m), young volcanic rocks (v), and sialic basement rocks of Precambrian age (s) derived from equations 19.6 and 19.7. Note that s becomes negative when $m = 0.9$, which requires that marine carbonate rocks contribute less than 90% of the Sr entering the oceans at the present time.

time at about 158 Ma. The horizontal line labeled 0.70685 in Figure 19.2 is the locus of all points whose s, v, and m values yield $^{87}Sr/^{86}Sr = 0.70685$. The most likely mixing proportions are represented by point D, which corresponds to $m = 0.6$, $v = 0.38$, and $s = 0.02$. The model (as presently configured) also makes clear that the $^{87}Sr/^{86}Sr$ ratio of seawater could not have decreased to 0.70685 unless m was less than 0.70 because higher values of m cause s to become negative.

These insights derived from Figure 19.2 suggest that the observed fluctuations of the $^{87}Sr/^{86}Sr$ ratio of seawater in Figure 19.1 were caused primarily by variations of the proportions of Sr derived from volcanic and sialic rocks, whereas the contribution from marine carbonate rocks remained comparatively constant and acted to stabilize the $^{87}Sr/^{86}Sr$ ratio in the oceans. Therefore, the pattern of variation of the $^{87}Sr/^{86}Sr$ ratio of seawater throughout Phanerozoic time can be explained by episodic increases in the inputs of Sr derived from volcanic rocks which caused the $^{87}Sr/^{86}Sr$ ratio of seawater to *decline*. Each time one of these volcanic outbursts ended, the $^{87}Sr/^{86}Sr$ ratio of seawater *increased* again, but never quite reached the value it had in Late/Middle Cambrian time.

When viewed in this perspective, the time-dependent variations of the $^{87}Sr/^{86}Sr$ ratio of seawater in Figure 19.1 can be regarded as a record of the intensity of volcanic activity on a global scale. According to this interpretation, major episodes of volcanic activity occurred during the Late Ordovician (I), Middle Devonian (II), Middle Carboniferous (III), Late Permian (IV), and Late Jurassic (V) Epochs. The last decline of the $^{87}Sr/^{86}Sr$ ratio of seawater, which started during the Triassic Period and bottomed out in Late Jurassic time, coincides with the volcanic activity that accompanied the opening of the Atlantic Ocean. After several minor episodes of volcanic activity during the Cretaceous Period and in early Tertiary (Paleogene) time, the $^{87}Sr/^{86}Sr$ ratio of seawater has been rising continuously toward its present value of 0.70918.

Alternatively, the increase of the $^{87}Sr/^{86}Sr$ ratio of seawater at different times in the geological past can be attributed to episodes of orogeny caused by collisions of continents (Jacobsen and Kaufman, 1999):

1. Himalayan–Tibetan collision during Neogene to Recent (~0 Ga),
2. Caledonian–Appalachian collision during the Paleozoic Era (~0.4 Ga), and
3. Pan-African collision during the Late Proterozoic Era (~0.6 Ga).

In the final analysis, the variations of the $^{87}Sr/^{86}Sr$ ratio of seawater in Phanerozoic time were caused by time-dependent changes in the fluxes of ^{87}Sr and ^{86}Sr entering and leaving the oceans. Numerical models based on Sr fluxes have been developed by

Brass, 1976, *Geochim. Cosmochim. Acta*, 40:721–730. Brevart and Allégre, 1977, *Bull. Soc. Geol. France*, 19(6):1253–1257. Goldstein and Jacobsen, 1987, *Chem. Geol. (Isotope Geosci. Sect.)*, 66:245–272. Jacobsen and Kaufman, 1999, *Chem. Geol.*, 161:37–57. Richter and DePaolo, 1987, *Earth Planet. Sci. Lett.*, 83:27–38. Richter and DePaolo, 1988, *Earth Planet. Sci. Lett.*, 90:382–394. Podlaha et al., 1999, *Chem. Geol.*, 161:241–252. Cimino et al., 1999, *Chem. Geol.*, 161:253–170. Berner and Rye, 1992, *Am. J. Sci.*, 292:136–148. Richter et al., 1992, *Earth Planet Sci. Lett.*, 109:11–23.

19.1d Sr Chronometry (Cenozoic Era)

The systematic variations of the $^{87}Sr/^{86}Sr$ ratio of marine Sr depicted in Figure 19.1 can be used to obtain numerical *dates* for marine carbonate and phosphate samples. However, certain limitations apply:

1. The fluctuations of the marine $^{87}Sr/^{86}Sr$ ratio during the Paleozoic and Mesozoic Eras permit unique age determinations only when the stratigraphic age can be used to constrain the time interval within which the rock was deposited.
2. The $^{87}Sr/^{86}Sr$ ratio of the samples must be unaffected by diagenetic alteration, especially in the case of shells of foraminifera, which may contain crystals of calcium carbonate deposited by porewater percolating through the sediment.
3. The selection of samples for dating is constrained by the reliability criteria listed in Table 19.1.
4. The accuracy of the measured $^{87}Sr/^{86}Sr$ ratios must be confirmed by analyses of interlaboratory isotope standards or of modern seawater collected in the open ocean.

Nevertheless, the marine-Sr chronometer is the only way to directly date sedimentary rocks of marine origin, which realizes a proposal made by Wickman (1948). In addition, the isotope composition of Sr in carbonate rocks of *known age* can be used to characterize the environment of deposition (i.e., marine vs. lacustrine or estuarine).

McArthur et al. (2001) constructed a lookup table that facilitates the conversion of the $^{87}Sr/^{86}Sr$ ratios of samples to their corresponding stratigraphic ages ranging from 0 to 509 Ma. (The table is available at j.mcarthur@ucl.ac.uk.)

The most favorable conditions for dating marine carbonate and phosphate samples exist for samples of post-Eocene age in the Cenozoic

Era (i.e., 40–0 Ma). In this time interval, the $^{87}Sr/^{86}Sr$ ratio of seawater increased steadily from about 0.70775 to 0.70918 with only one significant change in slope during the early Langhian Age of the Miocene Epoch at about 16 Ma (McArthur et al., 2001).

The key to dating marine carbonate samples of Cenozoic age is to construct a standard profile based on the $^{87}Sr/^{86}Sr$ ratios of unreplaced shells of planktonic foraminifera and other marine organisms. The stratigraphic ages of the selected samples are translated into numerical dates by reference to geological timescales such as that of Harland et al. (1989) and Berggren et al. (1995) and the paleomagnetic timescale of Cande and Kent (1995).

To facilitate interpretations based on small differences between precisely determined $^{87}Sr/^{86}Sr$ ratios and to eliminate instrumental bias from the data, DePaolo and Ingram (1985) and Elderfield (1986) defined a $\Delta^{87}Sr$ parameter:

$$\Delta^{87}\mathrm{Sr} = \left[\left(\frac{^{87}\mathrm{Sr}}{^{86}\mathrm{Sr}}\right)_{\mathrm{spl}} - \left(\frac{^{87}\mathrm{Sr}}{^{86}\mathrm{Sr}}\right)_{\mathrm{mc}}\right] \times 10^5 \quad (19.8)$$

where the subscripts spl = sample and mc = modern carbonate. Hess et al. (1986) used a $\delta^{87}Sr$ parameter which they defined by the equation

$$\delta^{87}\mathrm{Sr} = \frac{(^{87}\mathrm{Sr}/^{86}\mathrm{Sr})_{\mathrm{spl}} - (^{87}\mathrm{Sr}/^{86}\mathrm{Sr})_{\mathrm{sw}}}{(^{87}\mathrm{Sr}/^{86}\mathrm{Sr})_{\mathrm{sw}}} \times 10^5 \quad (19.9)$$

Both parameters require investigators to measure the $^{87}Sr/^{86}Sr$ ratio of modern carbonate or seawater (sw) on the same mass spectrometer used to analyze the samples of marine carbonate being dated, which eliminates interlaboratory discrepancies caused by instrumental effects. The $\delta^{87}Sr$ parameter was used by Hess et al. (1986, 1989) and Hodell et al. (1989, 1990) to express the $^{87}Sr/^{86}Sr$ ratio of individual shells of planktonic foraminifera that were first examined by scanning electron microscopy to detect diagenetic alteration such as the presence of secondary calcite crystals, cementation, and overgrowths. Only shells that passed this inspection were used to measure the $^{87}Sr/^{86}Sr$ ratio.

The results in Figure 19.4 demonstrate that the $^{87}Sr/^{86}Sr$ ratio of seawater increased from 40 to about 15 Ma when the rate of growth slowed until about 5 Ma. Similar curves have been constructed by other investigators:

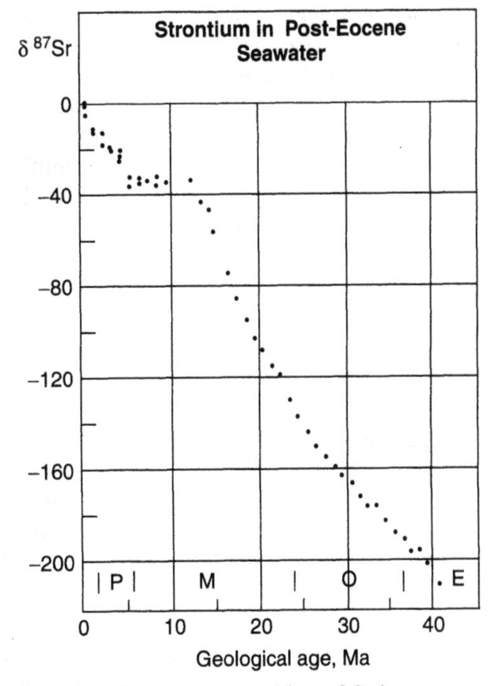

FIGURE 19.4 Isotope composition of Sr in post-Eocene seawater expressed as $\delta^{87}Sr$, which is defined by equation 19.10. The $\delta^{87}Sr$ parameter is calculated from the difference between the $^{87}Sr/^{86}Sr$ ratios of marine carbonate samples and seawater when both are measured on the same mass spectrometer. The samples are shells of planktonic foraminifera taken from several different Deep Sea Drilling Program (DSDP) cores. Data from Hess et al. (1986, 1989) and Hodell et al. (1989, 1990).

Palmer and Elderfield, 1985, *Nature*, 314:526–528. DePaolo and Ingram, 1985, *Science*, 227:938–941. DePaolo, 1986, *Geology*, 14: 103–106. Miller et al., 1988, *Paleoceanography*, 3:223–233. Miller et al., 1991, *Paleoceanography*, 6:33–52. Ludwig et al., 1988, *Geology*, 16:173–177. Capo and DePaolo, 1990, *Science*, 249:51–55. Barrera et al., 1991, *Proc. Ocean Drilling Program, Scientific Results*, 119:731–738. Montanari et al., 1991,

Newsletter Stratigr., 23:151–180. Zachos et al., 1999, *Chem. Geol.*, 161:165–180. Denison et al., 1993, *Paleoceanography*, 8:101–126. Oslick et al., 1994, *Paleoceanography*, 9:427–443. Henderson et al., 1994, *Earth Planet. Sci. Lett.*, 128:643–651. Hodell and Woodruff, 1994, *Paleoceanography*, 9:405–426. Mead and Hodell, 1995, *Paleoceanography*, 10:327–346. Farrell et al., 1995, *Geology*, 23:403–406. Martin et al., 1999, *Paleoceanography*, 14:74–83. McArthur et al., 2001, *J. Geol.*, 109-155–170.

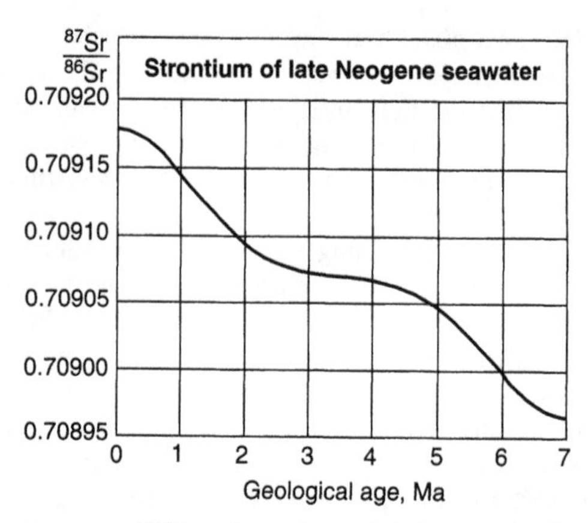

FIGURE 19.5 Fifth-order polynomial fit representing the time-dependent variation of the $^{87}Sr/^{86}Sr$ ratio of seawater from 7.0 Ma to the present. The curve is based on selected planktonic foraminifera from a 100-m core in the Indian Ocean (ODP site 758). The $^{87}Sr/^{86}Sr$ ratios are relative to 0.710257 for NBS-987. Replotted from Figure 1 of Farrell et al. (1995).

The time interval between 7.0 Ma (late Miocene) and the present was investigated by Farrell et al. (1995) based on $^{87}Sr/^{86}Sr$ ratios of 455 samples of planktonic foraminifera taken from a continuous 100-m core of sediment recovered at ODP site 758 (latitude 5°23′ N, longitude 90°21′ E) in the Indian Ocean (ODP = Ocean Drilling Program). The stratigraphic ages of the samples were derived from the magnetostratigraphy of the core, which contained all major magnetic reversals of the time period (Section 6.4b). The ages of the polarity reversals were taken from the astronomically tuned timescale of Shackleton et al. (1995). The time resolution of the samples was 15,000 years. The foraminiferal shells were unaltered because of their young age, shallow depth of burial, and fresh appearance and because their isotope ratios of Sr and O match those of shells of the same age recovered elsewhere.

The $^{87}Sr/^{86}Sr$ ratios were measured on 5–10 hand-picked and ultrasonically cleaned planktonic foraminifera with diameters greater than 425 μm. The shells were dissolved in acetic acid and the $^{87}Sr/^{86}Sr$ ratios were measured on a multicollector mass spectrometer operated in the static mode. The measured ratios were corrected for isotope fractionation to $^{86}Sr/^{88}Sr = 0.11940$ and were adjusted to 0.710257 for NBS 987. The reproducibility of the $^{87}Sr/^{86}Sr$ ratio based on 136 replicate analyses of NBS 987 was 19×10^{-6} expressed as two standard deviations. In addition, Farrell et al. (1995) measured duplicate isotope ratios of 233 samples that yielded a reproducibility of 21×10^{-6}, in good agreement with the results for NBS 987. All of these procedural details are necessary to appreciate the refinement of the isotope evolution curve of marine Sr achieved by Farrell et al. (1995).

The resulting curve in Figure 19.5 is a fifth-order polynomial fitted to the data points which accurately reflects the time-dependent variation of the $^{87}Sr/^{86}Sr$ ratio of seawater, although it may mask short-term fluctuations. The $^{87}Sr/^{86}Sr$ ratio of seawater in Figure 19.5 varies smoothly and nonlinearly with time with less scatter of data points than in previous data sets for the late Neogene published by Hodell et al. (1989, 1991) and other investigators. Farrell et al. (1995) estimated that the uncertainty of the age of a sample whose measured $^{87}Sr/^{86}Sr$ ratio has an error of 19×10^{-6} ranges from ±0.60 to ±2.03 Ma depending on the slope of the curve.

19.1e The Cambrian "Explosion"

The fossil record contains evidence that important changes occurred in the biota during the Early Cambrian Epoch (543–510 Ma) when the Ediacaran organisms of the Late Proterozoic Era, which lacked skeletons, were replaced by a diversified fauna of mollusks, brachiopods, echinoderms, and reef-forming archaeocyathid sponges all of which have mineral skeletons (Babcock et al., 2001). The rapid diversification of the marine fauna during the Early Cambrian coincided with a significant increase of the $^{87}Sr/^{86}Sr$ ratio of seawater from 0.7081 during the Tommotian Age (~530 Ma) to 0.7085 (Botomian), and ultimately to 0.7088 (early Middle Cambrian), as reported by Derry et al. (1994). The $^{87}Sr/^{86}Sr$ ratio of Cambrian seawater continued to rise and reached its highest value of about 0.7093 during the transition from Middle to Late Cambrian at about 513 Ma (Montañez et al., 1996). These results corroborate measurements published by other investigators:

> Burke et al., 1982, *Geology*, 10(10):516–519. Gorokhov et al., 1995, *Stratigraphy and Geologic Correlation*, 3(1):1–28. Kaufman et al., 1996, *Geol. Mag.*, 133:509–533. Denison et al., 1998, *Chem. Geol.*, 152:325–340. Derry et al., 1994, *Earth Planet. Sci. Lett.*, 128:671–681. Ebneth et al., 2001, *Geochim. Cosmochim. Acta*, 65(14):2273–2292.

The increase of the $^{87}Sr/^{86}Sr$ ratio of seawater during the Cambrian Period in Figure 19.6 was caused by an increase in the rate of erosion following the Pan-African orogeny. The rise of the $^{87}Sr/^{86}Sr$ ratio was accompanied by wide fluctuations of the isotope composition of carbon of marine carbonate rocks (Chapter 27) deposited in Early Cambrian time (Derry et al., 1994). The fluctuations of the isotope composition of carbon also coincide with the diversification of the marine fauna (Cambrian explosion) during the Tommotian, Atdabanian, Botomian, and Toyonian Ages of the Early Cambrian Epoch.

The connection between the increase of the $^{87}Sr/^{86}Sr$ ratio and the isotope composition of carbon of marine carbonate rocks may have resulted from the increase of the biological productivity of the oceans which was caused by the enhanced input of phosphorus and other nutrients to the oceans following the Pan-African orogeny. The resulting burial of large quantities of biogenic carbon compounds is reflected by the observed changes in the isotope composition of carbon in marine carbonate rocks. A more specific explanation of the isotope

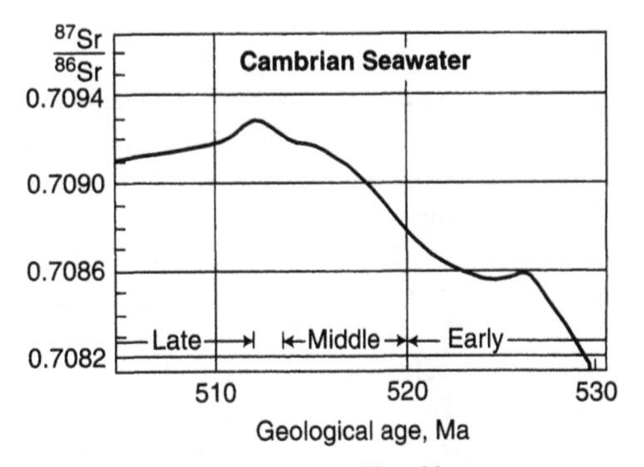

FIGURE 19.6 Variation of the $^{87}Sr/^{86}Sr$ ratio of seawater during the Cambrian Period relative to $^{87}Sr/^{86}Sr = 0.71025$ for NBS 987. Adapted from Montañez et al. (1996) and based on their data as well as data compiled by them from Burke et al. (1982), Derry et al. (1994), and others. The timescale is by Bowring et al. (1993).

composition of carbon in Cambrian limestones is presented in the context of the fractionation of carbon isotopes by plants in Chapter 27.

The connection between the geochemical cycles of carbon, sulfur, and strontium were modeled by Kump (1989), whereas Berner (1991) modeled the variation of CO_2 in the atmosphere during Paleozoic time. Subsequently, Berner and Rye (1992) attempted to calculate the $^{87}Sr/^{86}Sr$ ratios of Phanerozoic seawater from the rates of weathering of silicate rocks (primarily on the continents) and of burial of carbonate rocks in the oceans. The best results arise when the $^{87}Sr/^{86}Sr$ ratio of Sr derived by weathering of silicate rocks is assumed to vary from 0.709 to 0.716 in response to changes in sealevel. Low values (0.709) occur when sealevel is high because of increases in the rate of seafloor spreading and the resulting input of mantle-derived Sr into the oceans.

19.2 STRONTIUM IN THE PRECAMBRIAN OCEANS

The isotopic composition of Sr in carbonate rocks deposited in Precambrian time is still not well known because Precambrian carbonate rocks are less common than those of Phanerozoic age and because the chemical, mineralogical, and isotopic compositions of many Precambrian carbonate rocks have been altered (Veizer and Compston, 1976). In addition, carbonate rocks of Precambrian age represent a much longer interval of time than carbonate rocks of Phanerozoic age but are difficult to date paleontologically because of the scarcity of index fossils and because isotopic methods of dating are generally not applicable to these rocks or are imprecise. Even the distinction between marine and nonmarine depositional basins is uncertain during the earliest periods of Earth history, when the salinity of the oceans may have been less than it is today.

The alteration of Precambrian carbonate rocks may occur initially during diagenesis or subsequently as a result of fracturing during structural deformation, which permits the deposition of secondary calcite by brines containing Sr whose isotope composition differs from that of the carbonate rocks. In addition, dolomitization causes large decreases of Sr and lesser decreases of Rb concentrations which raises the Rb/Sr ratios of the carbonates. Even though the $^{87}Sr/^{86}Sr$ ratios of the carbonate minerals are routinely corrected for in situ decay of ^{87}Rb, the initial $^{87}Sr/^{86}Sr$ ratios of dolomites and limestones that have elevated Rb/Sr ratios, in many cases, exceed the initial $^{87}Sr/^{86}Sr$ ratios of samples having low Rb/Sr ratios.

Systematic increases of $^{87}Sr/^{86}Sr$ ratios may also result from the release of Sr from the silicate and oxide minerals during the acid dissolution of the carbonate phases of limestone. The problem can be minimized by excluding samples containing >10% of acid-insoluble residue and by using dilute solutions of weak acids (e.g., 0.5 M acetic acid). In addition, some authors preleach powdered samples with distilled water or with dilute solutions of ammonium acetate before dissolving the carbonate phases in acetic acid (e.g., Gorokhov et al., 1995, 1996, 1998). Hydrochloric acid, even at low concentrations, can release ^{87}Sr from silicate and oxide minerals and thereby increases the measured $^{87}Sr/^{86}Sr$ ratios of carbonate phases.

Dolomite and altered limestones are also enriched in Mn and Fe during fluid–rock interaction and have high Mn/Sr and Fe/Sr ratios (Brand and Veizer, 1980). Therefore, these ratios are useful criteria for identifying carbonate rocks whose $^{87}Sr/^{86}Sr$ ratios may have been altered after deposition. Therefore, all dolomite samples and those limestones having Mn/Sr >0.6 and Fe/Sr >0.3 should be excluded because, in many cases, their $^{87}Sr/^{86}Sr$ ratios exceed those of unaltered carbonate rocks. However, even samples that satisfy these criteria may have been altered (Asmerom et al., 1991; Derry et al., 1994).

19.2a Late Proterozoic Carbonates

A study by Asmerom et al. (1991) of carbonate rocks in the Late Proterozoic Shaler Group on Victoria Island in the Canadian Arctic illustrates the precautions necessary in the selection of samples for analysis. The samples for this study were taken from a measured section that is underlain unconformably by an older volcano–sedimentary complex and is overlain disconformably by basaltic lava flows. The samples originated from precisely measured positions within a 1221-m interval of a section whose total thickness was 3360 m. Information reviewed by Asmerom et al. (1991) indicated an age of 880 Ma for the base of the section and 723 Ma for the youngest rocks at the top. This information was used to calculate the ages of the samples from their known height above the base of the section using an equation derived by Derry et al. (1989) for a model of basin subsidence:

$$T = T_0 + \tau_e \ln\left(1 - \frac{D}{\tau_e A_0}\right) \qquad (19.10)$$

where T = age of a sample in Ma
T_0 = age of the oldest rocks at the base of the section (880 Ma)
A_0 = initial sedimentation rate (70.24 m/Ma)
τ_e = erosion-rate constant (50 Ma)
D = stratigraphic height in meters of a sample measured from the base of the section

Substituting these values into equation 19.10 yields the ages of samples based on the measured height (D) above the base of the section studied by Asmerom et al. (1991):

$$T = 880 + 50 \ln\left(1 - \frac{D}{50 \times 70.24}\right) \quad (19.11)$$

For example, the age of a limestone sample (WI-76, oosparite) at D-1992 m above the base is

$$T = 880 + 50 \ln\left(1 - \frac{1992}{50 \times 70.24}\right) = 838 \text{ Ma}$$

Asmerom et al. (1991) measured the Rb/Sr ratios of the carbonate minerals of 39 samples from the Shaler Group and selected 17 for further study based on the criterion that the decay correction to the $^{87}Sr/^{86}Sr$ ratio was less than 0.0001. The 14 limestones in that set contained 260 ppm Sr (111.2–520.3 ppm) and 0.300 ppm Rb (0.006–0.897 ppm) on average, whereas the dolomites contained only 42 ppm Sr and 0.096 ppm Rb. The dolomites also had elevated Mn/Sr ratios (6.2) compared to only 0.67 for the limestones.

The limestone samples of the Late Proterozoic Shaler Group in Figure 19.7 indicate that the $^{87}Sr/^{86}Sr$ ratio of seawater varied smoothly from 0.70738 to 0.70561 relative to 0.710241 for NBS 987 provided that the section contains no gaps in deposition, that the dates at the top and bottom were accurately constrained by the available isotopic age determinations, and that the ages of the samples were correctly interpolated by the basin-subsidence model.

Additional studies of late Proterozoic carbonate rocks have been performed by

Derry et al., 1989, *Geochim. Cosmochim. Acta*, 53:2331–2339. Derry et al.,1992, *Geochim. Cosmochim. Acta*, 56:1317–1329. Kaufman et al., 1993, *Earth Planet. Sci. Lett.*, 120:409–430. Kaufman et al., 1996, *Geol. Mag.*, 133: 509–533. Gorokhov et al., 1995, *Stratigraphy and Geologic Correlation*, 3(1):1–28. Gorokhov et al., 1996, in Bottrell et al. (Eds.), *Proc. 4th Internat. Symp. Geochim. Earth's Surface*, Ilkley, UK, Land-Hydrosphere Interactions (Theme 5): 714–717. Kuznetsov et al., 1997, *Doklady Russian Acad. Sci. (Earth Sci. Sect.)*, 353(2):249–254. Semikhatov et al., 1998, *Doklady Russian Acad. Sci. (Earth Sci. Sect.)*, 360:488–492. Jacobsen and Kaufman, 1999, *Chem. Geol.*, 161:37–57.

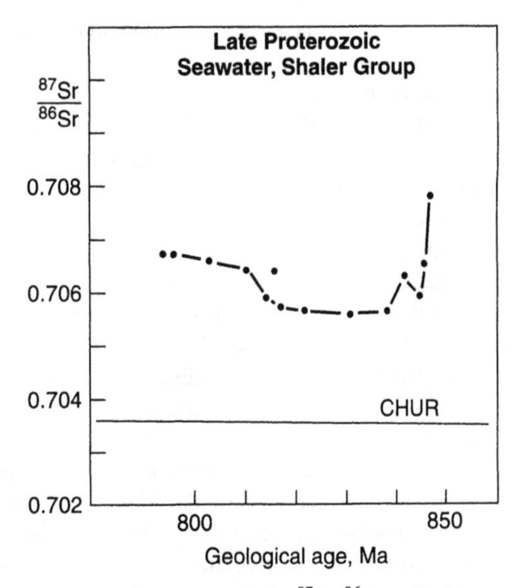

FIGURE 19.7 Variation of the $^{87}Sr/^{86}Sr$ ratio in marine limestones of the Late Proterozoic (Neoproterozoic) Shaler Group on Victoria Island in the Northwest Territories of Canada. Data from Asmerom et al. (1991).

19.2b Snowball Earth Glaciations

During the Late Proterozoic (or Neoproterozoic) Era, the Earth experienced two episodes of global glaciation known as the Varangian at about 600 Ma and the Sturtian at about 700–760 Ma. Each of these glaciations occurred in two pulses which were followed by the deposition of limestones that are depleted in ^{13}C by about 0.5% relative to normal marine limestone (Jacobsen and Kaufman, 1999).

The occurrence of glacial diamictites at sealevel close to the equator suggests that the entire Earth was covered by ice in a condition known as "Snowball Earth" (Hoffman et al., 1998). At these times, the $^{87}Sr/^{86}Sr$ ratios of seawater should have decreased because the fluxes of Sr derived from the continents were reduced to zero and only Sr discharged by hotsprings along midocean ridges could enter the oceans. Although the facts are not in doubt, their interpretation in terms of global glaciations (Snowball Earth) has been questioned by Kennedy et al. (2001a, b).

Jacobsen and Kaufman (1999) demonstrated that the $^{87}Sr/^{86}Sr$ ratios of limestones deposited immediately after the Varangian and Sturtian glaciations in Figure 19.8 did not decrease as expected. However, the depletion of the so-called cap limestones in ^{13}C is spectacular. In the delta notation that is used to express the isotope composition of C, N, O, and S, the $\delta^{13}C$ values of postglacial limestones declined from about +8‰ to −5‰ compared to values near +2.0‰ (Figure 27.7) for normal marine limestones of Phanerozoic age. Kennedy et al. (2001b) presented evidence that the decrease of the $\delta^{13}C$ values of the cap carbonates was caused by the decomposition of marine methane–hydrate deposits following the global glaciations. Alternatively, Jacobsen and Kaufman (1999) suggested that the $\delta^{13}C$ values declined because the C originated primarily from the mantle, whose $\delta^{13}C$ value is −5.5‰. Jacobsen and Kaufman (1999) also demonstrated by numerical modeling that the $^{87}Sr/^{86}Sr$ ratio of seawater did not decrease because Sr has a longer oceanic residence time than C (i.e., 10^6 years for Sr and 10^5 years for C). The $^{87}Sr/^{86}Sr$ ratio of seawater would have decreased from about 0.707 to less than 0.7032 if each of the Snowball Glaciations had lasted as long as 10×10^6 years. Therefore, the absence of a significant decrease of the $^{87}Sr/^{86}Sr$ ratio in the cap limestones implies that each of these glaciations lasted less than about 1×10^6 years.

The increase of the $^{87}Sr/^{86}Sr$ ratio of seawater from 0.7056 at 830 Ma in Figure 19.8 (Asmerom et al., 1991) to 0.7093 at 513 Ma (Middle/Late Cambrian, Montañez et al., 1996) has been attributed to an increase in the flux of crustal Sr caused by uplift and increased erosion during the Pan-African continental collision.

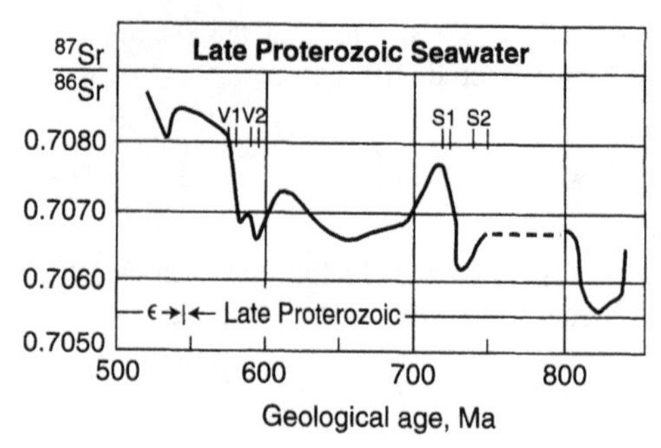

FIGURE 19.8 Variation of the $^{87}Sr/^{86}Sr$ ratio of seawater during the Late Proterozoic (Neoproterozoic) Era. The curve was drawn free-hand to include the lowest values of the $^{87}Sr/^{86}Sr$ ratios. V1, V2 = Varangian glaciation; S1, S2 = Sturtian glaciation both of which were global in scope and caused the Snowball Earth phenomenon. Adapted from Jacobsen and Kaufman (1999) and based on data compiled by them from Derry et al. (1989, 1992, 1994), Asmerom et al. (1991), and Kaufman et al. (1993, 1996) for carbonate rocks from Siberia, Namibia, Svalbard, and Canada.

19.2c Early Proterozoic and Archean Carbonates

The rise of the $^{87}Sr/^{86}Sr$ ratio of seawater during the Late Proterozoic Era is part of a larger trend that began in Late Archean time (Veizer and Compston, 1976; Veizer, 1989). The isotopic evolution of Sr in seawater during the pre-Neoproterozoic history of the Earth is difficult to reconstruct because the ages of the available carbonate rocks are not well constrained (Veizer et al., 1983) and because of the alteration of these rocks during diagenesis and during structural deformation and regional metamorphism.

Veizer et al. (1982) demonstrated that Archean calcites and dolomites have higher concentrations of Sr, Ba, Mn, and Fe than carbonate rocks of Phanerozoic age but are depleted in ^{18}O and Na. In addition, their $^{87}Sr/^{86}Sr$ ratios are similar to those of mantle-derived volcanic rocks of that time. Veizer et al. (1989b) reported that hydrothermal carbonates within volcano–sedimentary complexes of Archean age in North America, South Africa, and Australia have low $^{87}Sr/^{86}Sr$ ratios of about 0.7020 ± 0.0008. In a subsequent paper, Veizer et al. (1989a) demonstrated that carbonate rocks of sedimentary origin associated with Late Archean greenstone belts in Canada and Zimbabwe have average initial $^{87}Sr/^{86}Sr$ ratios of 0.7025 ± 0.0015 at 2.8 ± 0.2 Ga. In addition, a suite of Early

Archean ferroan dolomites, siderites, and ankerites in South Africa, Australia, and India yielded an initial $^{87}Sr/^{86}Sr$ ratio of 0.7031 ± 0.0008 at 3.5 ± 0.1 Ga. Both values are indistinguishable from the $^{87}Sr/^{86}Sr$ ratios of hydrothermal carbonates and from mantle-derived volcanic rocks of Archean age. However, in some cases, the Early Archean carbonates were significantly enriched in radiogenic ^{87}Sr during postdepositional alteration. For example, carbonates of the Onverwacht Group in Swaziland have measured $^{87}Sr/^{86}Sr$ ratios between 0.740 and 0.760 from which the isotope composition of Sr in Early Archean seawater cannot be recovered.

Subsequent studies by Veizer et al. (1990, 1992a, b), Deb et al. (1991), Mirota and Veizer (1994), Zachariah (1998), and Ray et al., (2002) provided additional information concerning trace-element concentrations and the isotope compositions of Sr, C, and O of Precambrian carbonate rocks. For example, Veizer et al. (1992a) demonstrated that the Bruce Limestone Member (2.35 ± 0.10 Ga) of the Espanola Formation in the Huronian Supergroup exposed along the north shore of Lake Huron in Ontario is lacustrine in origin because its lowest $^{87}Sr/^{86}Sr$ ratio at less than 1000 ppm Mn is 0.71128.

Nevertheless, the available data compiled in Figure 19.9 by Shields and Veizer (2001) and Ray et al. (2002) demonstrate that the $^{87}Sr/^{86}Sr$ ratio of seawater increased markedly from near-mantle values (0.702) at about 2.5 ± 0.3 Ga to about 0.705 at 1.0 Ga. This rise of the $^{87}Sr/^{86}Sr$ ratios of the oceans most likely records a decrease in the intensity of volcanic activity and an increase in the input of Sr by rivers draining rocks of the growing continental crust where radiogenic ^{87}Sr had accumulated by decay of ^{87}Rb.

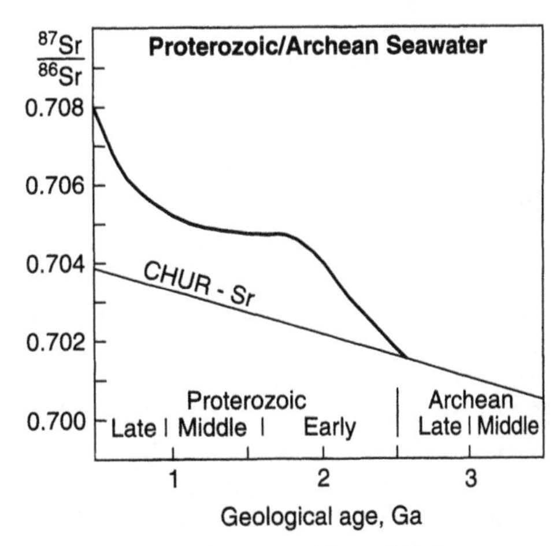

FIGURE 19.9 Isotope composition of Sr in seawater during the Archean and Proterozoic Eons. The $^{87}Sr/^{86}Sr$ ratios of marine carbonate rocks of Archean age are similar to those of mantle Sr represented by CHUR-Sr. The curve was adapted from Ray et al. (2002) and Shields and Veizer (2001).

19.3 NEODYMIUM IN THE OCEANS

The concentration of Nd in seawater is exceedingly low primarily because the trivalent Nd ion is strongly sorbed to the surfaces of colloidal particles and because Nd is incorporated into biogenic phosphate such as fish teeth. The low concentration of Nd in solution in seawater poses analytical problems that have made it difficult to measure its isotope composition directly. For this reason, the study of the isotope geochemistry of Nd in the oceans is supplemented with analyses of samples of skeletal calcium carbonate and phosphate, ferromanganese nodules, and heavy-metal sediment deposited by hotsprings along midocean ridges. The results of such studies indicate that, in contrast to Sr, the isotope composition of Nd in the present-day oceans is not constant but varies regionally depending on inputs by rivers draining rocks of the continental crust (low $^{143}Nd/^{144}Nd$) and mantle-derived volcanic rocks (high $^{143}Nd/^{144}Nd$).

19.3a Continental Runoff

The concentrations of Nd in river water and the role of sorption in the transport of this element in streams have been investigated by

Martin et al., 1976, *J. Geophys. Res.*, 81:3119–3124. Keasler and Loveland, 1982, *Earth Planet. Sci. Lett.*, 61:68–72. Goldstein et al., 1984, *Earth Planet. Sci. Lett.*, 70:221–236.

Stordal and Wasserburg, 1986, *Earth Planet. Sci. Lett.*, 77:259–272. Goldstein and Jacobsen, 1987, *Chem. Geol. (Isotope Geosci. Sect.)*, 66:245–272. Goldstein and Jacobsen, 1988, *Earth Planet. Sci. Lett.*, 87:249–265. Elderfield et al., 1990, *Geochim. Cosmochim. Acta*, 54:971–991. Andersson et al., 1992, *Earth Planet. Sci. Lett.*, 113:459–472. Andersson et al., 2001, *Geochim. Cosmochim. Acta*, 65: 521–527. Allégre et al., 1996, *Chem. Geol.*, 131:93–112. Sholkovitz, 1989, *Chem. Geol.*, 77:47–51. Sholkovitz, 1992, *Earth Planet. Sci. Lett.*, 114:77–84. Sholkovitz, 1995, *Aquat. Geochem.*, 1:1–34. Sholkovitz et al., 1994, *Geochim. Cosmochim. Acta*, 58:1567–1579.

The presentation that follows is based largely on the work of Goldstein and Jacobsen (1987) on the geochemistry of Nd and Sr in riverwater. The results of their study demonstrate that the transport of Nd by rivers is strongly controlled by the pH of the water, which determines the polarity of surface charges of colloidal particles. At low pH, most of the surface charges are positive, causing Nd^{3+} and the cations of other REEs to be in solution in the water. Therefore, filtered samples of acidic waters can have high concentrations of Nd and other REEs. As the pH of the water rises, the charges of surface sites on colloidal particles become negative because of desorption of H^+ ions. Consequently, the colloidal particles attract and hold an increasing number of Nd^{3+} ions with increasing pH, causing the concentration of Nd in ionic solution in the water to decrease. This phenomenon is illustrated by Figure 19.10, which shows that the Nd concentrations of filtered water in North American rivers decrease steeply with increasing pH. Evidently, filtration of water having near-neutral pH removes most of the Nd from the system and causes the filtrate to have low Nd concentrations. Most of the Nd transported by rivers at near-neutral pH is sorbed to colloidal particles that are deposited when the streams discharge their water into estuaries along the coast. Consequently, only the small fraction of Nd that is in true ionic solution in rivers is actually incorporated into seawater.

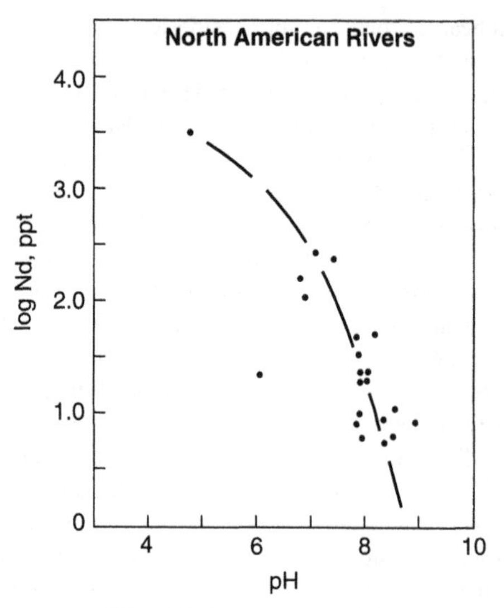

FIGURE 19.10 The pH dependence of the concentration of Nd in filtered water (0.2 μm) of rivers in North America. The average Nd concentration (weighted by the discharge) of the five largest rivers is 16.0×10^{-12} g/g, which is equivalent to picograms per gram or parts per trillion (ppt). Data from Goldstein and Jacobsen (1987).

The concentrations of Nd in filtered samples (0.2 μm) of surface water on the North American continent analyzed by Goldstein and Jacobsen (1987) range from 5.30×10^{-12} g/g (picograms per gram or parts per trillion, ppt) in Lake Huron (pH = 8.35) to 3150 ppt in the Potomac River (pH = 4.80). The average Nd concentration of the five largest rivers of North America (Mississippi, Missouri, St. Lawrence, Columbia, and Ohio) weighted by their discharges is 16.0 ppt. These rivers drain an area of 5.0×10^6 km^2, which is 83% of the total area drained by the North American rivers included in the study of Goldstein and Jacobsen (1987).

The $^{143}Nd/^{144}Nd$ ratios of the North American rivers (expressed relative to 0.512638 for the present value of CHUR-Nd) range from 0.511586 (St. Louis River) to 0.512483 (Columbia River). The St. Louis River drains rocks of Precambrian age in the Superior tectonic province of Canada, whereas the Columbia River drains mantle-derived basalts of Tertiary age in Oregon and Washington.

The average $^{143}Nd/^{144}Nd$ and $^{87}Sr/^{86}Sr$ ratios of the five largest rivers (listed above and weighted by their discharge) are 0.512202 and 0.71048, respectively. All of the rivers of North America included in the study of Goldstein and Jacobsen (1987) plot in quadrant IV of Figure 19.11, as expected for rivers draining a variety of rocks of the continental crust.

The average weighted concentrations and isotope ratios of Nd entering the oceans (prior to sorption of Nd in estuaries) in Table 19.3 depend primarily on the ages and lithologies of the adjacent continents. For example, the rivers discharging water into the Atlantic Ocean have an average weighted $^{143}Nd/^{144}Nd$ ratio of 0.51191 (Nd = 55.7 ppt), whereas the rivers draining into the Pacific Ocean have $^{143}Nd/^{144}Nd = 0.512489$ (Nd = 27.8 ppt). The low $^{143}Nd/^{144}Nd$ isotope ratio of Nd entering the Atlantic Ocean reflects the prevalence of Precambrian and descendent Phanerozoic sedimentary rocks in the continents that border this ocean. In contrast, the Nd that enters the Pacific Ocean has a comparatively high $^{143}Nd/^{144}Nd$ ratio because it originates primarily from mantle-derived volcanic rocks that prevail on the islands and continents along the borders of the Pacific basin.

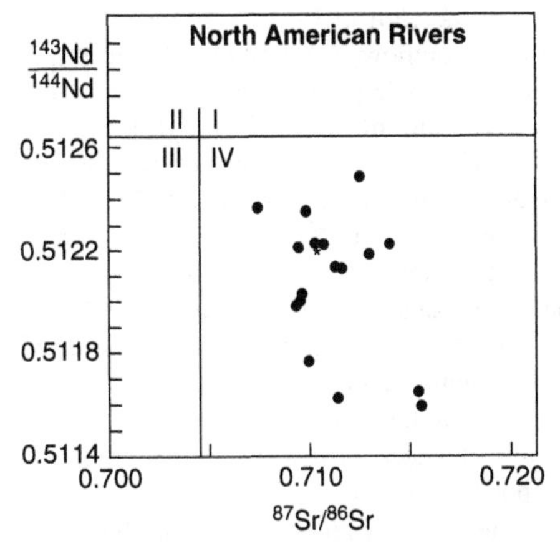

FIGURE 19.11 Isotope ratios of Nd and Sr in solution in filtered water of rivers in North America. The $^{143}Nd/^{144}Nd$ ratios were corrected to 0.512638 for CHUR-Nd and the $^{87}Sr/^{86}Sr$ ratios are relative to 0.71025 for NBS 987. The asterisk represents the average isotope ratios of Nd and Sr in the five largest rivers of North America (weighted by discharge): 0.512202 and 0.71048, receptively. Data from Goldstein and Jacobsen (1987).

Table 19.3. Average Weighted Concentrations and Isotope Ratios of Nd in River Water Prior to Losses due to Sorption in Estuaries

River Water	Discharge, km^3/y	Nd, ppt	$^{143}Nd/^{144}Nd^a$
Atlantic Ocean rivers	20,323	55.7	0.511991
Pacific Ocean rivers	13,123	27.8	0.512489
Indian Ocean rivers	4,878	26.6	0.512191
Arctic Ocean rivers	4,115	21.6	0.511319
All rivers	42,439	40.5	0.511330

Source: Goldstein and Jacobsen, 1987.
[a]Relative to 0.512638 for CHUR-Nd.

19.3b Mixing of Nd in the Baltic Sea

The mixing of Nd and Sr in continental runoff with seawater is well illustrated by a study by Andersson et al. (1992) of the Baltic Sea in Figure 19.12. The water in this basin has low salinities ranging from 2.460 to 11.117‰ and rising to 14.277‰ in the Kattegatt, which forms the outlet of the Baltic Sea between Denmark and Sweden. The rivers entering the Baltic Sea from the north drain the Precambrian rocks of the Baltic Shield. The southern rivers all drain primarily sedimentary rocks of Phanerozoic age. Therefore, the isotope ratios of Nd and Sr in the water of the Baltic Sea are the result of mixing of seawater and continental drainage in varying proportions.

The concentrations of Sm and Nd in unfiltered water at localities A and C in the Baltic Sea increase with depth as a result of desorption of these elements from particles that are sinking through the water column. For example, the concentrations of Nd in water samples collected at locality C in Figure 19.13 increase from 5.13 ppt at 5 m to 23.65 ppt at 225 m. In general, the Nd concentrations of water in the Baltic Sea are up to about

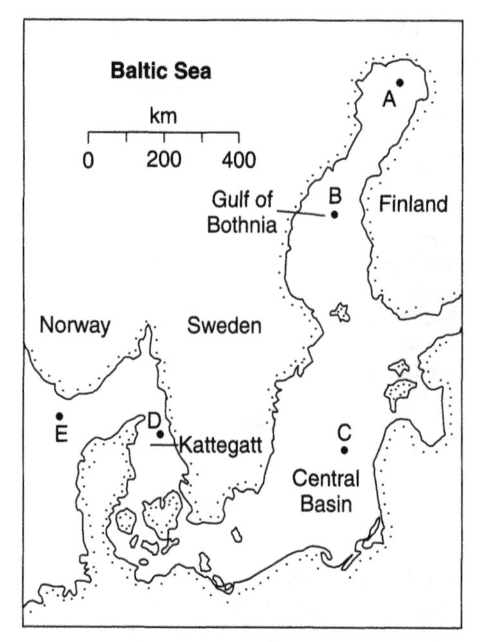

FIGURE 19.12 Map of the Baltic Sea. The points labeled A, B, C, D, and E are collecting sites referred to in the text. Adapted from Andersson et al. (1992).

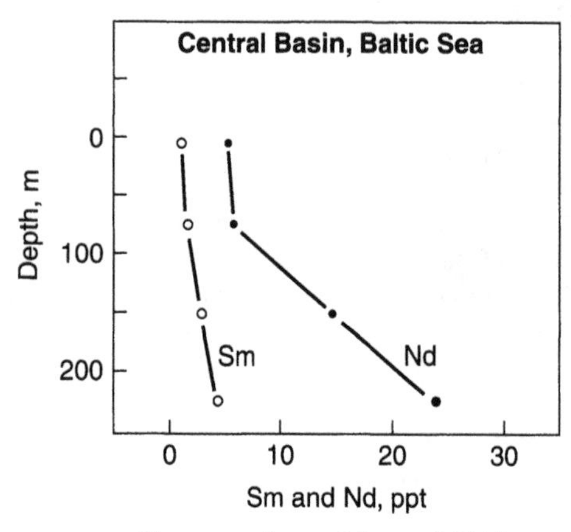

FIGURE 19.13 Concentrations of Sm and Nd in unfiltered water of the central basin of the Baltic Sea collected at point B (57°20′0″N and 20°03′0″E) from depths between 5 and 225 m. The units of concentration are 10^{-12} g/g equivalent to picograms per gram and parts per trillion (ppt). Data from Andersson et al. (1992).

ten times higher than the concentrations of Nd in unfiltered seawater in the open ocean (Piepgras and Wasserburg, 1982).

The isotope ratios of Nd and Sr of the water in the Baltic Sea both vary regionally and with depth in different parts of the basin. The lowest $^{143}Nd/^{144}Nd$ ratios (0.511514–0.511674) occur at point A in the brackish water of the Gulf of Bothnia, having salinities between 2.460 and 3.476‰, compared to 35.289% in the North Sea at point E near the mouth of the Skagerrak (Andersson et al., 1992). The low $^{143}Nd/^{144}Nd$ ratios of the water at point A are attributable to the fact that the major rivers discharging into the Gulf of Bothnia drain the Precambrian rocks of the Baltic Shield. Åberg and Wickman (1987) reported that the $^{87}Sr/^{86}Sr$ ratios of 44 rivers which enter this part of the Baltic Sea from Sweden and Finland range from 0.71273 to 0.73664 with an unweighted arithmetic mean of 0.72855 ± 0.00150 ($N = 48$, $2\bar{\sigma}$). The isotope geochemistry of Sr in different parts of the Baltic Sea was presented by Löfvendahl et al. (1990).

The average isotope ratios of Nd and Sr in water of the Gulf of Bothnia, in the central basin, and in the outflow channel of the Baltic Sea define an isotopic mixing hyperbola in Figure 19.14. The $^{87}Sr/^{86}Sr$ ratio of water in the Kattegatt (point D, 0.709202) approaches that of seawater at the mouth of the Skagerrak (point E, 0.709168). However, the $^{143}Nd/^{144}Nd$ ratios range widely from 0.51151 (point A, Gulf of Bothnia) to 0.512125 (point D, Kattegatt) relative to 0.512638 for the present $^{143}Nd/^{144}Nd$ ratio of CHUR-Nd. These results demonstrate that the $^{143}Nd/^{144}Nd$ ratios of water in the Baltic Sea vary in response to mixing of water masses having different isotopic compositions of Nd.

The $^{87}Sr/^{86}Sr$ ratios of water in the Baltic Sea are well correlated in Figure 19.15 with the reciprocals of the Sr concentration. This relationship confirms that Sr is a conservative element in the Baltic Sea and elsewhere in the global oceans. In contrast to Sr, the isotope ratios and the reciprocal concentrations of Nd scatter widely in Figure 19.16 and demonstrate the strongly nonconservative geochemical behavior of this element in the Baltic Sea. Although the geochemical properties of Nd and Sr in oceans are clearly different, even Sr is not a

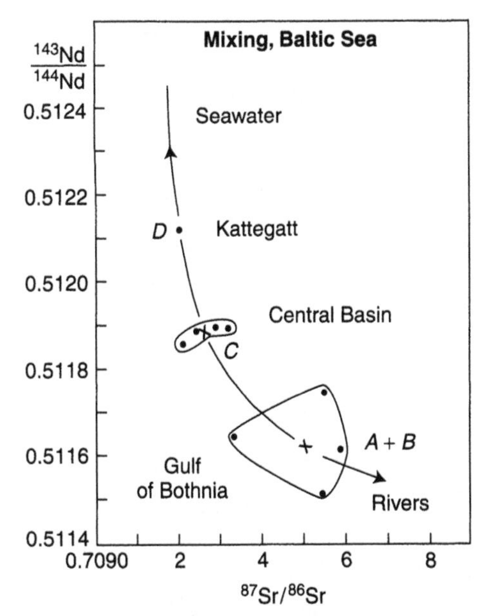

FIGURE 19.14 Two-component isotopic mixing of Nd and Sr in seawater and river waters in the Gulf of Bothnia and the central basin of the Baltic Sea. The isotope ratios of both elements vary depending on the proportions of mixing. The $^{143}Nd/^{144}Nd$ ratios are relative to 0.512638 for CHUR-Nd, whereas the $^{87}Sr/^{86}Sr$ ratios were adjusted to 0.71025 for NBS 987. The average isotope ratios of Nd and Sr in the water of the Gulf of Bothnia and of the central basin were weighted by the concentrations of Nd and Sr, respectively. The collecting sites are labeled A, B, C, and D as in Figure 19.13. Data from Andersson et al. (1992).

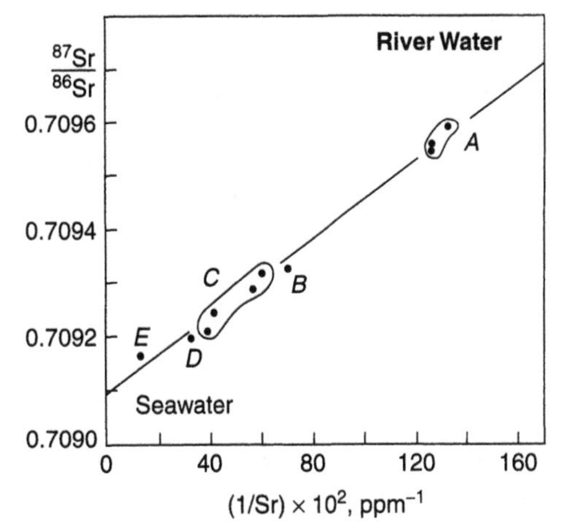

FIGURE 19.15 Evidence for the conservative character of Sr in the Baltic Sea. The collecting sites are identified in Figure 19.12 and the $^{87}Sr/^{86}Sr$ ratios have been adjusted to 0.71025 for NBS 987. Data from Andersson et al. (1992).

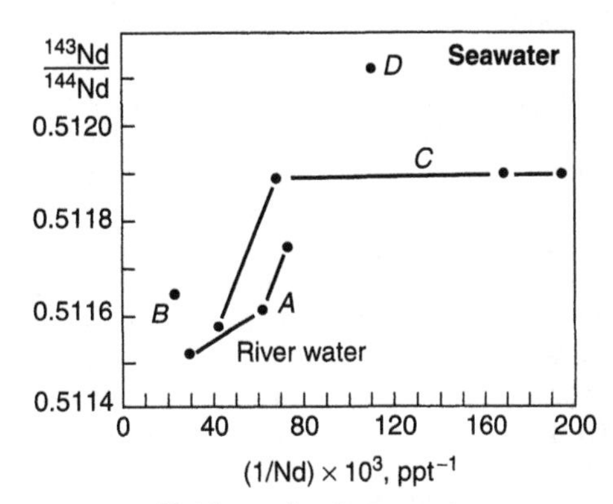

FIGURE 19.16 Evidence for the strongly nonconservative properties of Nd in the Baltic Sea. The collecting sites are identified in Figure 19.12 and the $^{143}Nd/^{144}Nd$ ratios have been adjusted to 0.512638 for CHUR-Nd at the present time. Data from Andersson et al. (1992).

perfectly conservative element, as indicated by the small deviations of the data points in Figure 19.15 from the mixing line.

19.3c Present-Day Seawater

The concentrations of Sm and Nd originally reported by Piepgras et al. (1979) and Piepgras and Wasserburg (1980, 1982) for seawater from the Atlantic and Pacific Oceans and from the Drake Passage between Antarctica and South America vary widely but have the following average values:

$$\text{Sm} = 0.55 \pm 0.04 \text{ ppt} \qquad \text{Nd} = 2.6 \pm 0.2 \text{ ppt}$$

The concentrations of both elements increase with depth. For example, the concentration of Nd at

Station 315 in the Drake Passage increases from 1.85 ppt at a depth of 50 m to 4.21 ppt at 3600 m (Piepgras and Wasserburg, 1982). Similar increases of the Nd concentration with depth have been reported for water in the North Atlantic and the South Pacific Oceans. The concentration of Nd in seawater is about six to seven orders of magnitude lower than its concentration in silicate rocks, which implies that this element has a low oceanic residence time of about 300 years compared to about 10^6 years for Sr. Consequently, the isotopic composition of Nd in the oceans is not constant but varies depending on the ages and Sm/Nd ratios of the sources on the continents and in the ocean basins from which it is derived. Therefore, Nd is a *better* isotopic tracer in the oceans than Sr, whose isotopic composition is homogenized by the circulation of seawater.

The $^{143}Nd/^{144}Nd$ ratios of seawater measured by Piepgras and Wasserburg (1980) range from 0.511936 to 0.512077 in the Atlantic Ocean and from 0.512442 to 0.51253 in the Pacific Ocean relative to 0.512638 for CHUR-Nd. The $^{143}Nd/^{144}Nd$ ratios of Atlantic seawater in Figure 19.17 are lower than those of the Pacific Ocean because the Nd in the Atlantic is derived primarily from sialic rocks of the continental crust of the adjacent continents, whereas the Nd in the Pacific Ocean originates predominantly from young mantle-derived volcanic rocks on oceanic islands and on island arcs that surround the Pacific basin (Table 19.3). Albarède and Goldstein (1992) published a map showing the pattern of variation of ε(Nd) values of ferromanganese deposits in the oceans of the world.

The $^{143}Nd/^{144}Nd$ ratios of ferromanganese nodules, red clay, and metalliferous sediment in Figure 19.17 are indistinguishable from those of the Nd in solution in seawater at each site, which confirms the hypothesis that these materials incorporated Nd from the ambient seawater and thereby preserved its isotopic composition. Consequently, ferromanganese nodules and metalliferous sediment that have accumulated on the ocean floor contain a record of changes in the isotope composition of Nd in seawater at each site of deposition. Such changes may occur for a variety of reasons, ranging from local volcanic activity to global realignments of the continents.

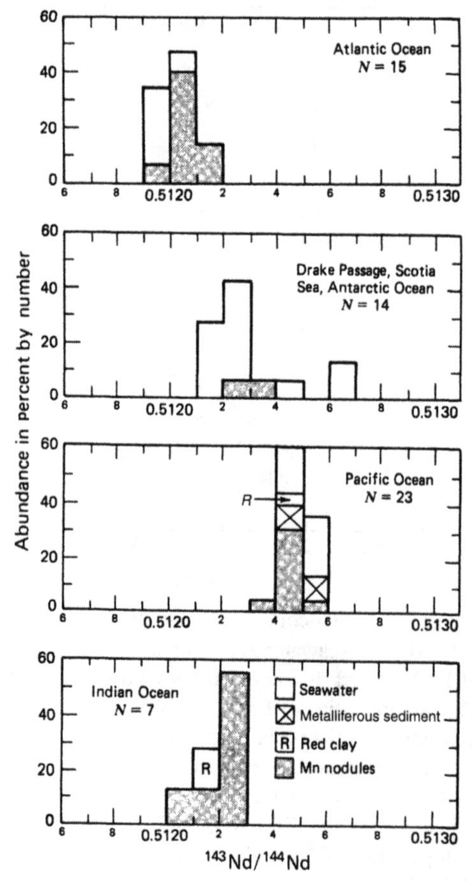

FIGURE 19.17 Comparison of $^{143}Nd/^{144}Nd$ ratios of seawater, Mn nodules, and metalliferous sediment deposited in the oceans. The isotope composition of Nd in seawater varies both within and among the ocean basins. However, the $^{143}Nd/^{144}Nd$ ratios of Mn nodules and metalliferous sediment appear to be compatible with those of seawater in the ocean in which they were precipitated. All $^{143}Nd/^{144}Nd$ ratios have been corrected for isotope fractionation of $^{146}Nd/^{144}Nd = 0.7219$. Data from Piepgras et al. (1979), Piepgras and Wasserburg (1980, 1982), O'Nions et al. (1978), and McCulloch and Wasserburg (1978).

The difference in the isotope composition of Nd in the waters of the Atlantic and Pacific Oceans was used by Piepgras and Wasserburg (1982) to study mixing of the water in the Drake Passage,

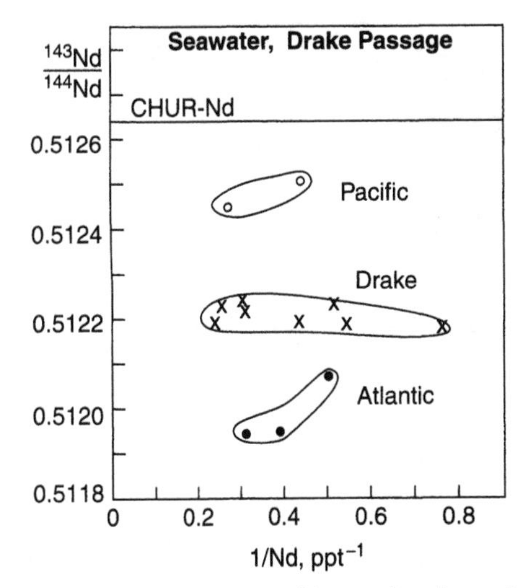

FIGURE 19.18 Isotope composition and reciprocal concentrations of Nd in seawater of the Drake Passage where water from the Pacific and Atlantic Oceans mix. Data from Piepgras et al. (1979) and Piepgras and Wasserburg (1980, 1982) corrected to $^{143}Nd/^{144}Nd = 0.512638$ for present-day CHUR-Nd.

which connects these two oceans. The results in Figure 19.18 clearly separate the Nd of the three regions such that the $^{143}Nd/^{144}Nd$ ratios of seawater in the Drake Passage are intermediate between Nd in the Pacific and Atlantic Oceans. The $^{143}Nd/^{144}Nd$ ratios of water in the Drake Passage are clustered above and below 0.51220, whereas the Nd concentrations range widely form 1.19 to 4.21 ppt depending on the depth from which the water was collected.

19.3d Ferromanganese Nodules and Crusts

Nodules and crusts of ferromanganese oxyhydroxides that form on the bottom of the oceans and in some lakes are strongly enriched in Sm and Nd relative to the concentrations of these elements in solution in the water. The average Nd concentrations of ferromanganese nodules in the major oceans listed in Table 19.4 are virtually constant at about 173 ppm (28.0–59.4 ppm), which yields an enrichment factor of about 6×10^7 relative to

Table 19.4. Average Nd Concentrations of Marine Ferromanganese Nodules and Other Types of Deposits

Ocean	Number of Samples at Site	Average Nd, ppm	Range
	Mn Nodules		
Pacific	5	174	129–225
Atlantic	8	173	59.4–280
Indian	5	171	90.3–262
All oceans	18	173	59.4–280
Antarctic	1	67	
Scotia Sea	1	63	
	Hydrothermal Crusts		
Pacific	1	2.05	
Atlantic	1	4.45	
	Metalliferous Sediment		
Pacific	3	17.6	15.7–20.7
	Red Clay		
Indian	1	33.0	
	Seawater		
Pacific	2	2.7×10^{-6}	$(2.2–3.2) \times 10^{-6}$
	Mn Nodules (Lacustrine)		
Lake Oneida, New York	1	3.40	

Source: Piepgras et al., 1979.

the Nd concentration of seawater (2.7×10^{-6} ppm, or 2.7 ppt) in the Pacific Ocean. Manganese nodules in the Antarctic Ocean and in the Scotia Sea have lower Nd concentrations of about 65 ppm (63–67 ppm). Hydrothermal ferromanganese crusts have still lower concentrations of Nd of about 3.2 ppm (2.05–4.45 ppm), whereas metalliferous sediment and deep-sea clay contain 17.6 and 33.0 ppm Nd, respectively. All of the materials in Table 19.4 are enriched in Nd compared to seawater and to river water in Table 19.3.

The $^{143}Nd/^{144}Nd$ ratios of the ferromanganese nodules in Figure 19.19, analyzed by Piepgras et al. (1979), vary regionally, much like the $^{143}Nd/^{144}Nd$ ratios of seawater (Figure 19.17). This similarity of the $^{143}Nd/^{144}Nd$ ratios of nodules and seawater supports the assumption that

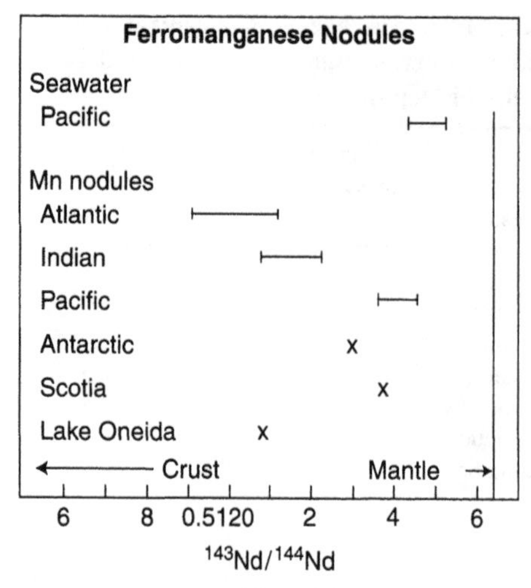

FIGURE 19.19 Range of $^{143}Nd/^{144}Nd$ ratios of ferromanganese nodules in the oceans of the world and in Lake Oneida, New York. Hydrothermal crusts and metalliferous sediment were excluded. The $^{143}Nd/^{144}Nd$ ratios were adjusted to be compatible with $^{143}Nd/^{144}Nd = 0.512638$ for CHUR-Nd at the present time. Data from Piepgras et al. (1979).

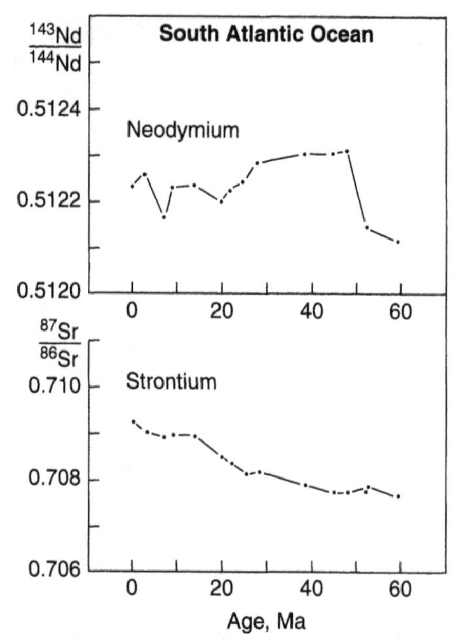

FIGURE 19.20 Variation of $^{143}Nd/^{144}Nd$ and $^{87}Sr/^{86}Sr$ ratios of foraminiferal tests composed of calcite coated with ferromanganese oxyhydroxide in a 477-m core form the Rio Grande Rise in the South Atlantic Ocean (DSP 357). The $^{143}Nd/^{144}Nd$ ratios were corrected for isotope fractionation to $^{146}Nd/^{144}Nd = 0.7219$ and are relative to 0.512638 for the present value of CHUR-Nd. The $^{87}Sr/^{86}Sr$ ratio of NBS 987 reported by the authors was 0.71025. Data from Palmer and Elderfield (1985, 1986).

the Nd in ferromanganese nodules and similar materials deposited in the oceans originated from the ambient seawater at the site of deposition. The sources of Nd in Mn nodules were also evaluated by O'Nions et al. (1978), Goldstein and O'Nions (1981), and other investigators identified by Piepgras et al. (1979).

The evidence that ferromanganese nodules and crusts record the isotope composition of Nd dissolved in the seawater where they formed has motivated efforts to use them to detect and interpret local changes in the $^{143}Nd/^{144}Nd$ ratio of seawater in the geological past. For example, Palmer and Elderfield (1985, 1986) analyzed the ferromanganese coatings of foraminiferal shells ranging in age from 60 to 0 Ma in a long sediment core recovered on the Rio Grande Rise in the South Atlantic Ocean (DSDP site 357). The $^{143}Nd/^{144}Nd$ ratios of these ferromanganese coatings in Figure 19.20 vary systematically with time in spite of evidence, discussed by Palmer and Elderfield (1985, 1986), that the abundances of the REEs were altered during diagenesis. The $^{87}Sr/^{86}Sr$ ratios of the foraminiferal shells in Figure 19.20 are consistent with the variation of this ratio in seawater during the Cenozoic Era in Figures 19.1 and 19.4, but they do not correlate with the $^{143}Nd/^{144}Nd$ ratios, which rise steeply between 60 and 50 Ma and then decline irregularly to the present.

The variation of the $^{143}Nd/^{144}Nd$ ratio of seawater at this site in the South Atlantic records changes in proportions of Nd derived from continental and volcanic sources. The increase of the $^{143}Nd/^{144}Nd$ ratio that started at 60 Ma was caused by influx of Nd which originated from mantle-derived volcanic rocks. The gradual decline of the $^{143}Nd/^{144}Nd$ ratio of seawater at this site between about 50 Ma and the present records a shift toward continental sources of Nd. Therefore, Palmer and Elderfield (1986) suggested that the $^{143}Nd/^{144}Nd$ ratios at DSDP site 375 record the decline of volcanic activity following the formation of the Rio Grande Rise prior to Late Cenozoic time. The results of this study strengthened the hypothesis that ferromanganese oxyhydroxides have preserved a record of the variation of $^{143}Nd/^{144}Nd$ ratios in the oceans. However, this record reflects *local* events, in contrast to Sr, whose isotope composition in the oceans is *global* in scope.

Ferromanganese crusts are well suited for paleo-oceanographic studies because they are deposited at very slow rates between 1.4 and 6.4 mm per million years. Consequently, a crust having a thickness of only 150 mm may contain a record spanning 60 million years if its rate of deposition was 2.5 mm/10^6 y. In addition, these crusts form at different depths in the oceans and therefore can be used to determine the sources of Nd at different levels within a given ocean basin.

For these reasons, Ling et al. (1997) used ferromanganese crusts to study the effect of the closure of the Panama gateway on the isotope composition of Nd and Pb in the water of the Pacific Ocean. The crusts originated from different depths on the tops of seamounts in the equatorial Pacific Ocean:

D11-1: depth 1.8 km, 11°38.9′ N, 161°40.5′ E
CD29-2: depth 2.3 km, 16°42.4′ N, 168°14.2′ W
VA 13/2: depth 4.8 km, 9°18′ N, 146°03′ W

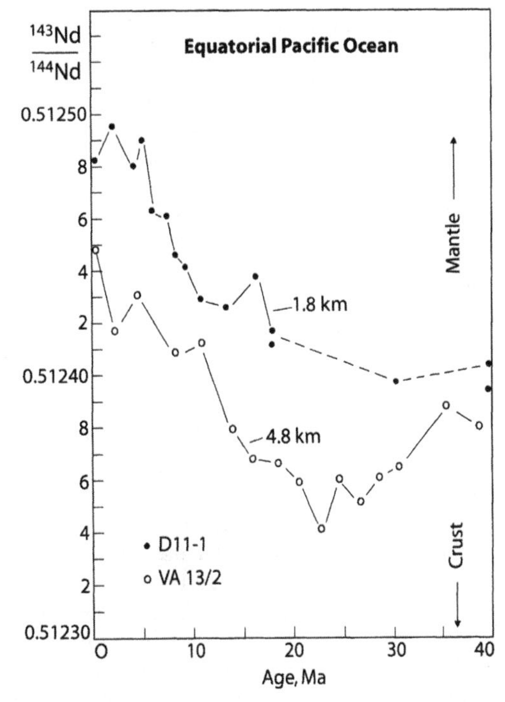

FIGURE 19.21 Variation of $^{143}Nd/^{144}Nd$ ratios of two ferromanganese crusts (D11-1 and VA 13/2) collected at different depths (1.8 and 4.8 km, respectively) in the central Pacific Ocean. The isotope ratios are relative to 0.511858 for the LaJolla Nd standard and have a reproducibility of 10×10^{-6}. The lower part of D11-1, deposited prior to 20 Ma, was phosphatized. Data from Ling et al. (1997).

The basal portions of CD29-2 and D11-1 were partly replaced by Ca phosphate, which may have changed the $^{143}Nd/^{144}Nd$ ratios of the crusts deposited prior to 26 Ma (CD29-2) and 20 Ma (D11-1). The VA13/2 crust was not phosphatized.

The $^{143}Nd/^{144}Nd$ ratios of two crusts recovered at depths of 1.8 km (D11-1) and 4.8 km (VA13/2) in Figure 19.21 increase between about 20 Ma and the present. The profile of crust CD29-2 is similar to that of D11-1 and therefore is not shown in Figure 19.21. The increase of the $^{143}Nd/^{144}Nd$ ratios implies that the abundance of volcanogenic Nd in the water increased irregularly with time. In two of the crusts (D11-1 and CD29-2) the $^{143}Nd/^{144}Nd$ ratios decreased slightly during the past 3–4 million years. However, the decrease is not evident in the VA13/2 crust, which was recovered from the deepest water (4.8 km) and has consistently lower $^{143}Nd/^{144}Nd$ ratios than the other two crusts recovered from shallower water (e.g., 1.8 km for D11-1).

The differences in the $^{143}Nd/^{144}Nd$ ratios of the crusts in Figure 19.21 are consistent with the well-known decrease of this ratio with depth in the oceans exemplified by the data in Table 19.5, compiled by Ling et al. (1997) from data in the literature. VA13/2 in Figure 19.21 may have been deposited in northward-spreading Antarctic bottom water, which is known to have a low $^{143}Nd/^{144}Nd$

Table 19.5. Variation of the $^{143}Nd/^{144}Nd$ Ratio with Depth in the North Pacific Ocean

Depth, km	$^{143}Nd/^{144}Nd^a$
<1	0.512633–0.512412
1–3.6	0.512484
>3.6	0.512407–0.512330

Source: Ling et al., 1997.

[a]Relative to 0.512638 for CHUR-Nd at the present time.

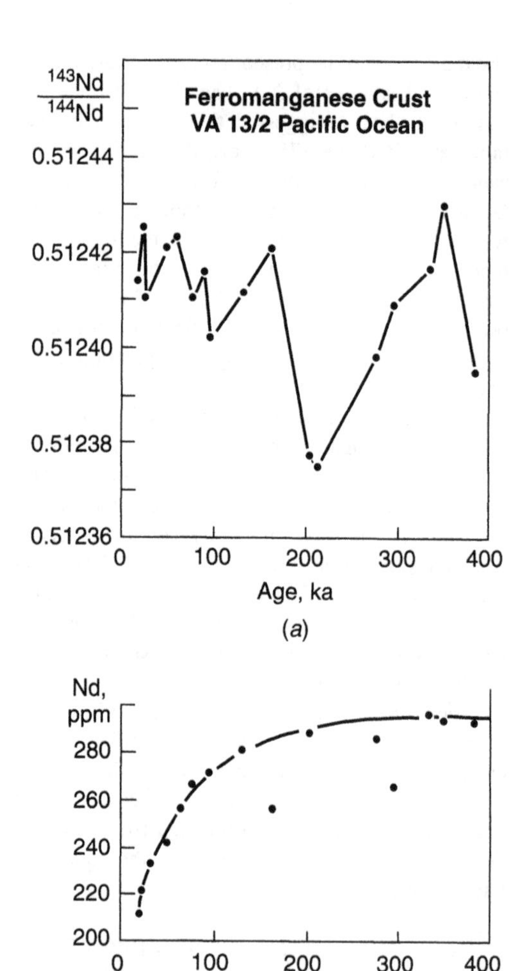

FIGURE 19.22 (*a*) Variation of $^{143}Nd/^{144}Nd$ ratios of ferromanganese crust VA 13/2 (4830 m) recovered in the central Pacific Ocean at 9°18′N, 146°03′W (same as VA 13/2 in Figure 19.21). The isotope ratios are relative to 0.511874 for the LaJolla Nd standard and have a reproducibility of $\pm 17 \times 10^{-6}$. The ages of the samples were derived from accumulation rates measured by the $^{230}Th/^{232}Th$ method (Section 20.1a). The samples represent the uppermost 1.5 mm of the crust and were deposited during the past 400,000 years. (*b*) Systematic increase of the Nd concentrations with increasing age of the ferromanganese oxyhydroxide in VA13/2. Data from Abouchami et al. (1997).

ratio (0.512176). Ling et al. (1997) suggested that the increase of the $^{143}Nd/^{144}Nd$ ratios of seawater, which started at about 20 Ma, was caused by a decrease of the amount of water that flowed through the Panama gateway from the Atlantic into the central Pacific Ocean. The comparatively low $^{143}Nd/^{144}Nd$ ratio of seawater in the Atlantic Ocean is well illustrated in Figures 19.17–19.20.

Ferromanganese crust VA13/2 from the central Pacific Ocean was also analyzed by Abouchami et al. (1997), who recovered two suites of samples by means of a high-precision drill rather than by the method of scraping used by Ling et al. (1997) and other investigators. As a result, Abouchami et al. (1997) were able to obtain 75 samples weighing between about 0.1 and 1.0 mg from a depth interval of only 1.5 mm representing the past 400,000 years, which implies an average time resolution of about 5000 years per sample. In addition, Abouchami et al. (1997) drilled a second set of samples that extended to 10 Ma. Age determinations based on the decay of unsupported ^{230}Th referred to by the authors indicate that the growth rate of VA13/2 increased at 86 ± 2 ka (86,000 years ago) from 3.0 ± 0.1 mm × 10^{-6} y (396 ± 13 to 86 ± 2 ka) to 6.4 ± 0.1 mm × 10^{-6} y (86 ± 2 to 0 ka).

The $^{143}Nd/^{144}Nd$ ratios of the high-resolution samples of VA 13/2 in Figure 19.22*a* varied significantly with time during the past 400,000 years. In addition, the Nd concentrations in Figure 19.22*b* declined from about 295 ppm (396 to 336 ka) to about 212 ppm (17 ka). Abouchami et al. (1997) considered whether the high-resolution isotopic record in Figure 19.22 reflects changes in the provenance of atmospheric dust (e.g., Asia or South America) or changes in climatic conditions (e.g., glacial or interglacial). They used the high-resolution samples to demonstrate that the variance of the average $^{206}Pb/^{204}Pb$ ratio during "warm" interglacial ages is significantly larger than during "cool" glacial ages. In addition, they cited a study by Chuey et al. (1987) who reported that during interglacials the atmosphere over the Pacific Ocean contained more dust than it did during glacial epochs. However, the $^{143}Nd/^{144}Nd$ ratios of the high-resolution samples of VA13/2 do not correlate with the reciprocal Nd concentrations (not shown). Therefore, the isotope composition of Nd in this ferromanganese crust is not explainable as a set of two-component mixtures of atmospheric dust derived from different sources.

Similar studies of the isotope composition of Sr, Nd, and Pb have been carried out on ferromanganese samples in the Atlantic and Indian Oceans:

Abouchami et al., 1999, *Geochim. Cosmochim. Acta*, 63:1489–1505. Frank and O'Nions, 1998, *Earth Planet. Sci. Lett.*, 158:121–130. Burton et al., 1999, *Earth Planet. Sci. Lett.*, 171:149–156. O'Nions et al., 1998, *Earth Planet. Sci. Lett.*, 155:15–28. Christensen et al., 1997, *Science*, 277:913–918.

19.3e Water–Rock Interaction (Ophiolites)

The isotope ratios of Sr and Nd of igneous rocks in the oceans are altered by interactions with seawater. For example, Jacobsen and Wasserburg (1979) reported that the $^{87}Sr/^{86}Sr$ ratios of rocks of the Bay of Islands ophiolite complex in Newfoundland range widely from 0.70254 to 0.70804, whereas the $^{143}Nd/^{144}Nd$ ratios of these rocks vary only slightly. As a result, the data points representing these samples deviate systematically from the mantle array in Figure 19.23. McCulloch et al. (1980, 1981) subsequently reported similar results for rocks from the Samail ophiolite of Oman. In addition, samples of altered basalt analyzed by O'Nions et al. (1978) also scatter widely in Figure 19.23. The distribution of data points in Figure 19.23 implies that the alteration of mafic igneous rocks by seawater caused significant increases of $^{87}Sr/^{86}Sr$ ratios but only small decreases of the $^{143}Nd/^{144}Nd$ ratios.

The alteration of the isotope compositions of Sr and Nd of rocks exposed to seawater is governed by an equation derived by McCulloch et al. (1980) based on the assumption that the effect of isotopic exchange between seawater and rocks depends on the magnitude of the water–rock ratio and on the concentrations of the elements of interest in the water and in the rocks.

Let ε be an isotope ratio of element X, and W, R be weights of seawater and rocks that participate in the isotope exchange, ε^i, ε^f are the isotope ratio in the initial stage and final state, and X_r, X_w are the concentrations of element X in the rocks and in the water. Initially, the isotope ratio of X in the water (ε_w^i) differs from that in the rocks (ε_r^i). In the final stage of the exchange, the ε-values of the rocks approach those of seawater. Therefore, the isotopic balance is expressed by the equation

$$\varepsilon_r^f X_r R + \varepsilon_r^f X_w W = \varepsilon_r^i X_r R + \varepsilon_w^i X_w W \quad (19.12)$$

Solving for the water–rock ratio yields

$$\frac{W}{R} = \left(\frac{\varepsilon_r^i - \varepsilon_r^f}{\varepsilon_r^f - \varepsilon_w^i}\right)\left(\frac{X_r}{X_w}\right) \quad (19.13)$$

Alternatively, equation 19.12 can be solved for the ε-parameter of the rocks after the exchange has occurred:

$$\varepsilon_r^f = \frac{\varepsilon_r^i X_r R + \varepsilon_w^i X_w W}{X_r R + X_w W} \quad (19.14)$$

Dividing the numerator and denominator of the right side of equation 19.14 by R yields

$$\varepsilon_r^f = \frac{\varepsilon_r^i X_r + \varepsilon_w^i X_w (W/R)}{X_r + X_w (W/R)} \quad (19.15)$$

This equation can be used to calculate the final isotope ratio ε of any element in rocks that have exchanged isotopes with seawater (e.g., Sr, Nd, Pb). The numerical value of the final ε-value of the rocks varies as the water–rock ratio increases.

In the derivation of equation 19.15, the isotope composition of the selected element can be expressed by its ε-value relative to CHUR, as indicated above, by the measured isotope ratios (e.g., $^{87}Sr/^{86}Sr$), or by the δ-parameter used to express the isotope compositions of oxygen, carbon, sulfur,

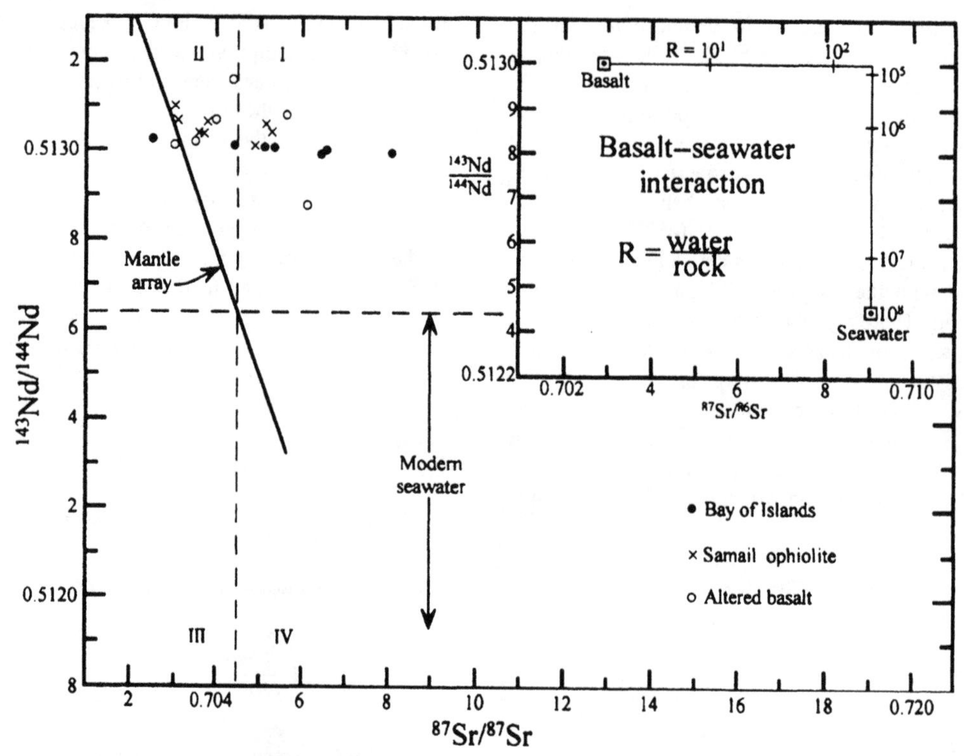

FIGURE 19.23 Evidence of alteration of ophiolitic rocks and basalt by seawater. The $^{87}Sr/^{86}Sr$ ratios are affected much more than the $^{143}Nd/^{144}Nd$ ratios until the water–rock ratio (R) exceeds 10^5. Complete resetting of the isotope ratios of Sr and Nd in the rocks occurs only when $R > 10^8$. Data from Jacobsen and Wasserburg (1979), McCulloch et al. (1980), and O'Nions et al. (1978).

and other elements whose isotope compositions are affected by mass-dependent isotope fractionation (Section 26.2). The isotope ratios of Sr and Nd calculated by means of equation 19.15 are the coordinates of points on the mixing hyperbola defined by the initial and final isotope ratios of rocks that interact with increasing amounts of water expressed by the water–rock ratio.

The calculation is illustrated by the following example:

Oceanic Basalt	Seawater
$(^{143}Nd/^{144}Nd)^i_r = 0.51300$	$(^{143}Nd/^{144}Nd)^i_w = 0.51245$
$Nd_r = 8.0$ ppm	$Nd_W = 2.6 \times 10^{-6}$ ppm
$(^{87}Sr/^{86}Sr)^i_r = 0.70300$	$(^{87}Sr/^{86}Sr)^i_w = 0.70918$
$Sr_r = 120$ ppm	$Sr_w = 8.0$ ppm

If $W/R = 1 \times 10^2 = 100$, equation 19.15 yields

$$\left(\frac{^{143}Nd}{^{144}Nd}\right)^f_r = \frac{0.51300 \times 8.00 + 0.51245 \times 2.6 \times 10^{-6} \times 100}{8.0 + 2.6 \times 10^{-6} \times 100}$$

$$= \frac{4.104 + 1.332 \times 10^{-4}}{8.0 + 2.6 \times 10^{-4}} = \frac{4.104}{8.0} = 0.51300$$

$$\left(\frac{^{87}Sr}{^{86}Sr}\right)^f_r = \frac{0.70300 \times 120 + 0.70918 \times 8.0 \times 100}{120 + 8.0 \times 100}$$

$$= \frac{84.36 + 567.344}{120 + 800} = \frac{651.70}{920} = 0.70837$$

These results demonstrate that for $W/R = 10^2$ the $^{143}Nd/^{144}Nd$ ratio of the rocks changes only imperceptibly, whereas the $^{87}Sr/^{86}Sr$ ratio increases significantly. The isotope ratio of Nd in the basalt is not sensitive to alteration by seawater because the Nd concentration of seawater is very low (2.6×10^{-6} ppm).

The effects of water–rock interaction cause the isotope ratios of Sr and Nd in Figure 19.23 to move along a hyperbola as the water–rock ratio increases. The $^{87}Sr/^{86}Sr$ ratio of the basalt changes from 0.70300 to 0.70918, whereas the $^{143}Nd/^{144}Nd$ ratio remains virtually unchanged as the water–rock ratio rises to 10^4. The calculations indicate that the $^{143}Nd/^{144}Nd$ ratio of the basalt begins to change only when the water–rock ratio reaches 10^5. Under these conditions, the isotope composition of Sr has been completely reset to the $^{87}Sr/^{86}Sr$ ratio of seawater. The $^{143}Nd/^{144}Nd$ ratio of the rock finally approaches that of seawater when the water–rock ratio is 10^8.

The interaction between seawater and mafic igneous rocks is facilitated by the convection of hot seawater through the oceanic crust as it forms along the midocean ridges (Bickle and Teagle, 1992). As a result, the mass of water to which the rocks are exposed increases with time and the water–rock ratio can reach large values sufficient to raise the initial $^{87}Sr/^{86}Sr$ ratios of the ultramafic rocks of the Samail ophiolite of Oman to 0.7071 relative to 0.71025 for NBS 987 (Lanphere et al., 1981).

The isotope ratios of Sr and Nd of ophiolites elsewhere in the world vary similarly:

Troodos Ophiolite, Cyprus: Peterman et al., 1971, *U.S. Geol. Surv. Prof. Paper*, 750D: 157–161. Spooner et al., 1977, *Geochim. Cosmochim. Acta*, 41:873–890. McCulloch and Cameron, 1983, *Geology*, 11:727–731.

Kings River Ophiolite, California: Shaw et al., 1987, *Contrib. Mineral. Petrol.*, 96: 281–290.

Interpretations of the isotope ratios of Pb in ophiolites and of associated sulfide-bearing sedimentary rocks have been reported by

Chen and Pallister, 1981, *J. Geophys. Res.*, 86:2699–2708. Tilton et al., 1981, *J. Geophys. Res.*, 86(B4):2763–2775. Doe, 1982, *Can. J. Earth Sci.*, 19:1720–1723. Spooner and Gale, 1982, *Nature*, 296:239–242. Hamelin et al., 1984, *Earth Planet. Sci. Lett.*, 67: 351–366. Hamelin et al., 1988, *Chem. Geol.*, 68:229–238. Mukasa and Ludden, 1987, *Geology*, 15:825–828. LeHuray et al., 1988, *Geology*, 16:362–365. Benoit et al., 1996, *Chem. Geol.*, 134:199–214. Booij et al., 2001, *Geochim. Acta*, 64:3559–3569.

19.4 LEAD IN THE OCEANS

The concentration of Pb in solution in streams and in the oceans is strongly affected by sorption of Pb^{2+} to the electrically charged surfaces of colloidal particles and by the low solubilities of the salts that Pb forms with naturally occurring acids (e.g., $PbCO_3$, $PbSO_4$, PbS). The sorption of Pb^{2+} is controlled by the activity of H^+ ions and hence by the pH of the water. As the environmental pH rises from acidic to near-neutral conditions, the fraction of Pb^{2+} that is sorbed increases, whereas the concentration of Pb^{2+} remaining in solution decreases to low values expressed in picograms per gram or parts per trillion. The average concentration of Pb^{2+} in the water of streams and lakes is significantly higher than that of seawater because the water in streams may be acidic in some cases whereas seawater is slightly basic in virtually all cases. In addition, the elevated concentrations of carbonate, sulfate, and bisulfide ions (in reducing environments) of seawater limit the concentrations of Pb^{2+} that can coexist in equilibrium with the corresponding Pb salts.

The sorption of Pb^{2+} ions by colloidal particles does not significantly alter the isotopic composition of Pb remaining in solution because its isotopes are not fractionated appreciably by this process. However, the isotope ratios of the sorbed Pb differ from those of the Pb contained within suspended mineral particles, as demonstrated in Section 18.2a.

The average concentration of Pb^{2+} in streams is about 1×10^{-9} g/g [nanograms per gram, or parts per billion (ppb)]. The average Pb concentration of seawater is nearly three orders of magnitude lower at 2×10^{-12} g/g. The concentration of Pb in the oceans varies irregularly and is not related to salinity, depth, or geographic factors. The mean oceanic residence time of Pb in the oceans is only about 50 years (Taylor and McLennan, 1985).

Lead is toxic to plants, animals, and humans when present at elevated concentrations. The effect of environmental Pb on human health was discussed by Faure (1998), Nriagu (1978a, b, 1983a, b), Patterson (1965), Tatsumoto and Patterson (1963a), and others referred to by them.

19.4a Sorption of Pb^{2+} by Oxyhydroxide Particles

The sorption of Pb^{2+} on oxyhydroxide precipitates of Fe, Al, and Mn was demonstrated by Lee et al. (2002), who neutralized acid mine-effluent collected in the former mining district at Ducktown, Tennessee. The fraction of Pb^{2+} sorbed at increasing pH from the water of Davis Mill Creek (pH 2.2, Pb = 8 ppb) in Figure 19.24 approached 100% at about pH 4. In this case, the removal of Pb^{2+} from solution was controlled both by the formation of oxyhydroxide precipitates of Fe and Al as well as by the change in polarity of electrical surface charges on these precipitates from predominantly positive to predominantly negative as the pH was raised. In cases where the water contains insufficient Fe in solution to precipitate ferric oxyhydroxide at low pH, the removal of Pb^{2+} is delayed until oxyhydroxide precipitates of Fe, Al, or Mn form at near-neutral pH.

The removal of Pb^{2+} from solution is accompanied by a complementary increase of the concentration of Pb in the oxyhydroxide precipitates that form naturally in streams contaminated by acid mine drainage. For example, Munk et al. (2002) reported that Al–hydroxysulfate precipitates that form downstream of the confluence of the Snake River (pH 3.0) with Deer Creek (pH 6.3) in Summit County, Colorado, are enriched in Pb relative to the water by more than four orders of magnitude.

The data in Figure 19.25 show that the concentrations of Pb in the Al–hydroxysulfate precipitates that form in the Snake River increase to nearly 350 ppm as the pH of the water rises to about 6.3. Farther downstream, the Pb concentrations of the precipitates decrease to about 100 ppm because most of the Pb in solution was removed by sorption

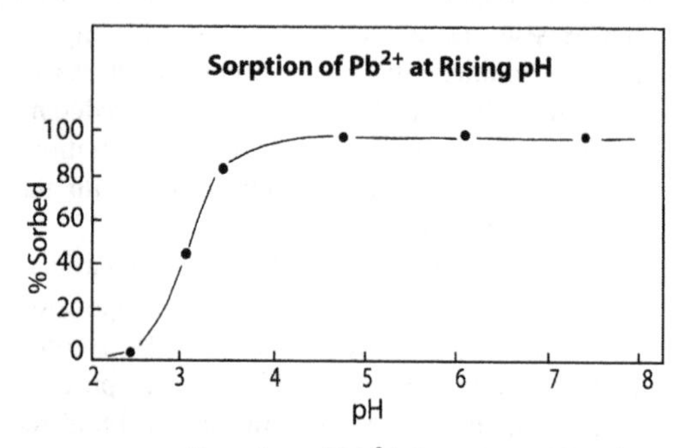

FIGURE 19.24 Sorption of Pb^{2+} in water of Davis Mill Creek, Ducktown, Tennessee, as the pH was raised in the laboratory from 2.2 to about 8.0. The sorbent was composed of the oxyhydroxides of Fe and Al, which precipitated from the water at increasing pH. Adapted from Lee et al. (2002).

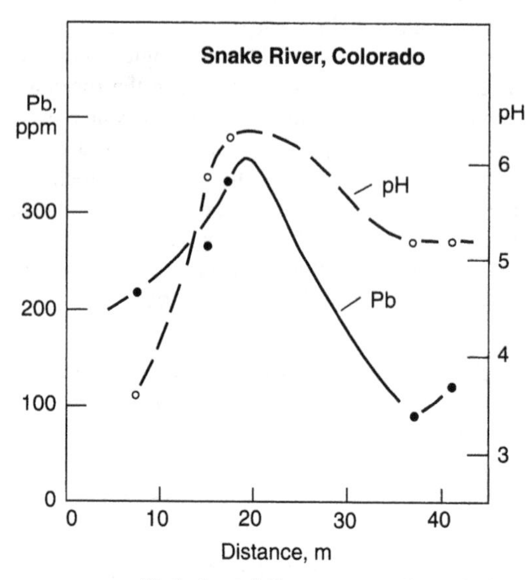

FIGURE 19.25 Variation of the concentration of Pb in Al–hydroxy sulfate precipitates that form in the Snake River of Summit County, Colorado, in response to an increase in the pH caused by mixing of the water of the Snake River with the water in Deer Creek. Data from Munk et al. (2002).

upstream. In other words, the water is purified by the removal of Pb and other trace metals by sorption on precipitates that form as the water of the Snake River is neutralized by mixing with water of Deer Creek.

The results of Lee et al. (2002) and Munk et al. (2002) illustrate the importance of sorption for the concentration of Pb in streams. At low pH, Pb^{2+} and the cations of other trace metals are in solution in the water. As the pH rises, Pb^{2+} and the cations of other trace metals are removed from solution by sorption to particles of chemical precipitates, organic matter, and mineral grains suspended in the water. Consequently, the concentrations of Pb in streams on the continents vary widely depending on the pH and the presence of colloidal particles. These particles are ultimately deposited in reservoirs, lakes, and estuaries. Consequently, the water that is discharged into the oceans has been purified by the prior removal of Pb^{2+} and cations of other trace metals. The sorption of anions and cations from aqueous solutions by particles of various kinds is affected by many environmental parameters, including the pH of the water, the polarity and magnitude of electrical charges of the ions, the ionic strength and temperature of the solution, and the chemical compositions, grain sizes, and surface characteristics of the sorbent particles. These matters have been discussed by Balistrieri and Murray (1982), Dzombak and Morel, (1990), Stumm (1992), Stumm and Morgan (1996), Langmuir (1997), Nordstrom and Alpers (1999), Schemel et al. (2000), and others referred to by them.

19.4b Aerosols and Eolian Dust

Atmospheric deposition of eolian dust and aerosol particles is a significant source of Pb in the oceans (Jones et al., 2000). These kinds of materials have surprisingly high concentrations of Pb with a range of isotopic compositions depending on their sources. Aerosol particles originate primarily from automobile exhaust, smelting of Pb ores, combustion of coal, and industrial emissions. The use of tetraethyl lead as an antiknock additive in gasoline caused widespread contamination of the surface of the Earth, including the continents and the surface layer of the oceans:

> Chow and Johnstone, 1965, *Science*, 147:502–503. Chow and Earl, 1972, *Science*, 176:510–511. Ghazi, 1994, *Appl. Geochem.*, 9:627–636. Boutron et al., 1994, *Geochim. Cosmochim. Acta*, 58:3217–3225. Ritson et al., 1994, *Geochim. Cosmochim. Acta*, 58:3297–3305. Wu and Boyle, 1997, *Geochim. Cosmochim. Acta*, 61:3279–3283.

The chemical composition of eolian dust and aerosol particles in Foshan and other municipalities in the Pearl River delta, Guangdong Province, South China, was determined by Zhu et al. (2001). They collected eolian dust by placing PVC cylinders with a diameter of 20 cm on the roofs of buildings in Foshan for one month. Aerosol samples were collected at the same sites by pumping air through a 10-cm filter for four hours. The chemical compositions of the eolian dust and aerosols were determined by x-ray fluorescence using a multichannel analyzer.

The concentrations of Pb in eolian dust collected at Foshan vary seasonally from 1.53% in January to 0.39% in July with a mean of 0.48%. Aerosol particles likewise have variable Pb concentrations between 2.65 and 0.20%, averaging 1.25% (Zhu et al., 2001). The same authors also reported Pb = 0.09% for aerosol particles in a rural area near Foshan, whereas automobile exhaust collected at the tail pipes of two trucks and two buses contained up to 31.94% of Pb accompanied by high concentrations of SO_3 of up to 35.58%.

The eolian dust and aerosol particles collected by Zhu et al. (2001) contain varying amounts of silicate minerals composed of SiO_2, Al_2O_3, FeO, K_2O, MgO, and CaO. In addition, these particles contain sulfur, expressed by the authors as SO_3. The sum of the concentrations of Cu, Pb, and Zn in the eolian dust in Figure 19.26 is positively correlated with the concentrations of SO_3. Therefore, the eolian dust probably contains particles of Cu–Pb–Zn sulfide or sulfate. The concentrations of Cu–Pb–Zn in aerosols in Figure 19.26 range only from 0.41 to 3.58% (with one exception at 8.22%) and have low SO_3 concentrations between 0.42 and 2.0%

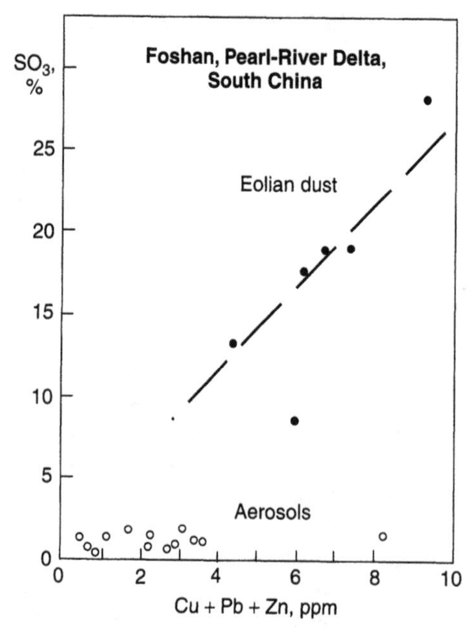

FIGURE 19.26 Concentrations of Cu, Pb, and Zn in eolian dust and aerosols collected near Foshan, Pearl River delta, Guangdong Province, South China, in relation to the concentrations of SO_3. Data from Zhu et al. (2001).

Zhu et al. (2001) pointed out that the silicate fraction of the aerosol particles have higher concentrations of SiO_2, Al_2O_3, MgO, and CaO than the eolian dust and resemble basaltic andesite in composition. If so, then the Cu, Pb, and Zn of the aerosols either reside within volcanic dust particles or are sorbed to silicate minerals (e.g., clay minerals) or both. In addition, Zhu et al. (2001) demonstrated that the chemical compositions of silicate minerals in the eolian dust and in the aerosol at Foshan differ from those of Chinese loess (Liu, 1985; Morioka and Yashiro, 1991) and of soil in Guangdong Province (Liu, 1993). Therefore, the eolian dust and aerosols appear to be of local or regional origin in the Pearl River delta.

The isotope ratios of Pb in eolian dust and aerosols in Table 19.6 are resolvable into several source components defined in Figure 19.27 by their $^{206}Pb/^{204}Pb$ and $^{208}Pb/^{204}Pb$ ratios. These ratios were chosen because they represent the radiogenic Pb formed by U and Th, respectively. Therefore,

Table 19.6. Isotope Ratios of Pb in Eolian Dust, Aerosol Particles, and Acid-Leachable Fractions of Soil near Foshan, Pearl River Delta, Guangdong Province, South China

Source	$^{206}Pb/^{204}Pb$	$^{207}Pb/^{204}Pb$	$^{208}Pb/^{204}Pb$
Automobile exhaust	18.097 ±0.014	15.577 ±0.061	37.740 ±0.138
Uncontaminated soil	18.620 ±0.020	15.579 ±0.001	38.660 ±0.037
Fankou Pb–Zn mine	18.382 ±0.011	15.690 ±0.010	38.793 ±0.019
Bedrock			
Granite A	18.574 ±0.0067	15.685 ±0.022	38.937 ±0.161
Granite B	18.546 ±0.012	15.672 ±0.006	38.679 ±0.026
Volcanic rocks	18.611 ±0.062	15.518 ±0.017	38.741 ±0.148

Source: Zhu et al., 2001.

Pb samples in multicomponent mixtures can be distinguished on the basis of both the ages and the U/Th ratios of their sources. The samples of eolian dust and aerosols in the Foshan area of China contain a mixture of Pb present in the local soil, in automobile emissions, in the Fankou Pb–Zn mine, and in meteoric precipitation. In addition, some eolian dust and aerosol samples contain Pb whose $^{208}Pb/^{204}Pb$ ratio exceeds those of the Pb sources listed above. The provenance of this thorogenic Pb was not identified by Zhu et al. (2001).

The Pb concentrations of aerosols were also measured by Bollhöfer and Rosman (2000, 2001, 2002) based on analyses of air filters (Bollhöfer et al., 1999). The results for sites in the northern hemisphere reveal a wide range of Pb concentrations in units of nanograms of Pb per cubic meter of air. Large cities typically have high aerosol–Pb concentrations compared to small towns and rural areas. These results and those of other investigators demonstrate that eolian dust and aerosols contaminate soil, snow, and the surface of the oceans (Simonetti et al., 2000; Ketterer et al., 2001).

19.4c Seawater and Snow

The studies of Pb in the oceans by Tatsumoto and Patterson (1963a, b) demonstrated that unfiltered

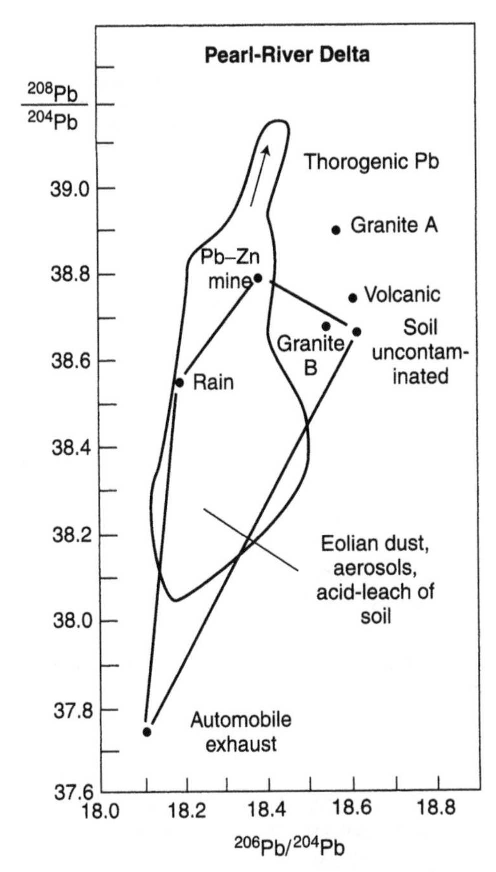

FIGURE 19.27 Isotope ratios of Pb in eolian dust, aerosols, meteoric precipitation, and acid-leachable fractions of soil near Foshan in the Pearl River delta of Guangdong Province, South China. The atmospheric Pb in this region is a multicomponent mixture of Pb derived from the local soil, from the Fankou Pb–Zn mine, and from automobile exhaust. Some of the samples contain Pb of unknown origin having high $^{208}Pb/^{204}Pb$ ratios. The average isotope ratios of the principal sources of Pb are listed in Table 19.6. Data from Zhu et al. (2001).

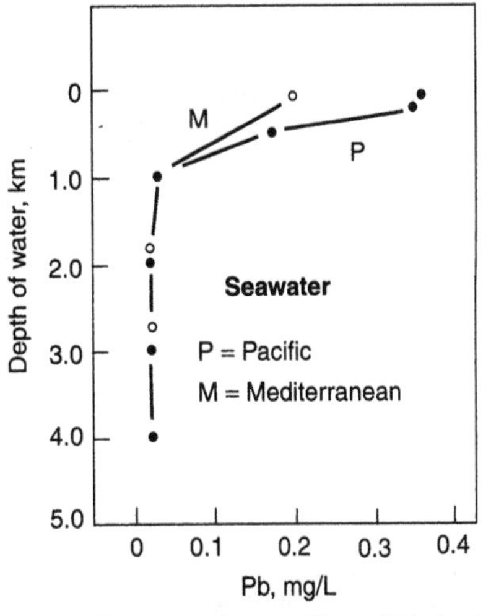

FIGURE 19.28 Concentration profiles of Pb in the Pacific Ocean and the Mediterranean Sea. Note that a concentration of 0.02 μg/L is equal to 20×10^{-12} g/mL or to 19.6×10^{-12} g/g based on a density of seawater of 1.02 g/mL. Adapted from Tatsumoto and Patterson (1963a, b).

but acidified surface water above a depth of 1000 m in the Pacific Ocean and in the Mediterranean Sea is contaminated with Pb. The Pb concentrations of surface water in the Pacific Ocean and in the Mediterranean Sea in Figure 19.28 converge below a depth of 1000 m to an average value of about 0.02 μg/L ($= 20 \times 10^{-9}$ g/L $= 20 \times 10^{-12}$ g/mL). Deep water at four sites in the Atlantic Ocean analyzed by Tatsumoto and Patterson (1963b) also approached Pb concentrations of 0.02 μg/L

The same authors reported that snow in the Lassen Volcanic National Park of California contained an average Pb concentration of 1.6 μg/kg. This value is 80 times higher than the average Pb concentration of uncontaminated seawater, which implies that the snow and the surface water of the oceans were contaminated by deposition of atmospheric aerosols. The conjecture is confirmed by the fact that the isotope ratios of Pb in snow at Lassen Park in Table 19.7 are similar to those of Pb in gasoline sold in southern California during the 1960s and to aerosols collected in Los Angeles (Chow and Johnstone, 1965).

The concentration of Pb in a core of ice and firn drilled at Camp Century in northwestern Greenland records the history of anthropogenic emissions of Pb in the northern hemisphere. Murozumi et al. (1969) reported that the concentrations of Pb increased gradually at first from about 0.011 μg/kg in 1750 A.D. to 0.068 μg/kg in 1945. Subsequently,

Table 19.7. Isotope Ratio of Pb in Snow at Lassen Volcanic National Park and Gasoline Sold in Southern California in the 1960s

Material	$^{206}Pb/^{204}Pb$	$^{207}Pb/^{204}Pb$	$^{208}Pb/^{204}Pb$
Gasoline	17.92	15.65	37.90
Aerosol, Los Angeles	18.04	15.63	38.01
Tetraethyl lead, 1947	18.69	15.52	38.30
Snow, Lassen Volcanic National Park	18.01	15.74	38.40

Source: Tatsumoto and Patterson, 1963b; Chow and Johnstone, 1965.

the concentration of Pb increased sharply to more than 0.16 μg/kg in 1960. The authors attributed the slow increase of the Pb concentrations between 1750 and 1945 to emissions from Pb smelters and from the combustion of coal. The dramatic increase in Pb concentrations in the most recent past (1945–1960) is a consequence of the use of tetraethyl lead in gasoline. The contamination of snow and firn in Antarctica by anthropogenic Pb has been documented by

Ng and Patterson, 1981, *Geochim. Cosmochim. Acta*, 45:2109–2121. Boutron, 1980, *J. Geophys. Res.*, 85:7426–7432. Boutron, 1982, *Atmosph. Environ.*, 16:2451–2459. Boutron and Patterson, 1983, *Geochim. Cosmochim. Acta*, 47:1355–1368.

The use of tetraethyl lead as an additive to gasoline in the United States started in 1923 and peaked in the 1970s, when up to 250,000 metric tons of leaded gasoline were consumed annually in the United States and in western European countries alone (Wu and Boyle, 1997). Subsequently, the use of leaded gasoline declined precipitously in North America because the addition of tetraethyl lead to gasoline was banned. By the year 1990, the use of leaded gasoline had declined to about 20,000 metric tons per year, most of which was sold in Europe.

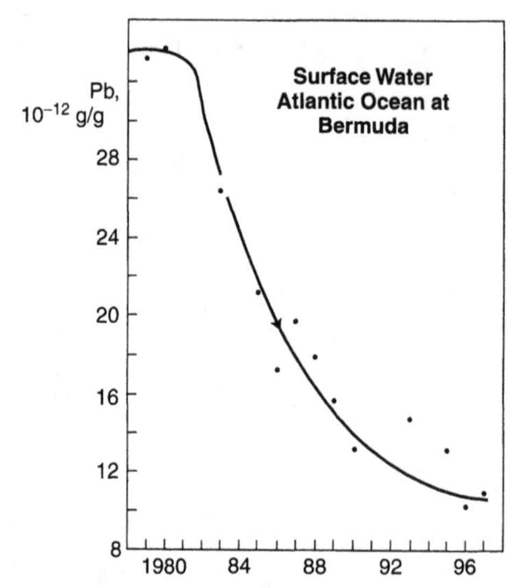

FIGURE 19.29 Decrease of the average annual concentration of Pb in surface water of the northwest Atlantic Ocean near Bermuda in units of 10^{-9} g/kg or 10^{-12} g/g (ppt). Data from Wu and Boyle (1997).

As a result, the concentrations of Pb in snow in Greenland (Boutron et al., 1994) and in the surface water of the Atlantic Ocean (Wu and Boyle, 1997) have both declined dramatically.

The decline of the Pb concentrations of surface water in the North Atlantic near Bermuda between 1980 and 1997 is recorded in Figure 19.29 based on the data of Wu and Boyle (1997). Their measurements indicate that the average annual Pb concentration of surface water in the northwestern Atlantic Ocean declined from about 34×10^{-12} g/g in 1980 to just over 10×10^{-12} g/g in 1997. The contaminant Pb is now being transported into deeper water in the oceans by the sinking of particles to which the Pb is sorbed. Decreases of the concentrations of Pb in surface water of the Mediterranean Sea in response to pollution abatement were also reported by Nicolas et al. (1994). Additional contributions to the study of anthropogenic Pb contamination of seawater are found elsewhere:

Schaule and Patterson, 1981, *Earth Planet. Sci. Lett.*, 54:97–116. Schaule and Patterson, 1983, in Wong et al. (Eds.), *Trace*

Metals in Seawater, pp. 487–504, Plenum, New York. Settle and Patterson, 1982, *J. Geophys. Res.*, 87:8857–8869. Flegal and Patterson, 1983, *Earth Planet. Sci. Lett.*, 64:19–32. Hamelin et al., 1989, *J. Geophys. Res.*, 94:16243–16250. Helmers et al., 1990, *Marine Pollution Bull.*, 21:515–518. Veron et al., 1993, *J. Geophys. Res.*, 98:18269–18276.

19.4d Ferromanganese Crusts

The low concentrations of Pb in seawater are an obstacle to the study of the isotopic compositions of this element in the oceans. Therefore, marine ferromanganese crusts and metalliferous sediment, which are enriched in Pb relative to seawater, have been used to determine the provenance of this element in different parts of the ocean from its isotopic composition (e.g., O'Nions et al., 1978).

The concentrations of Pb in ferromanganese crusts and sedimentary Pb ores in Table 19.8 range widely from several hundred up to nearly 4000 ppm (Ling et al., 1997; Abouchami et al., 1997). Even marine clays as well as calcareous and radiolarian oozes contain between 20 and 120 ppm of Pb (Wedepohl, 1974). The highest Pb concentrations occur in ore minerals of sedimentary Mn deposits (e.g., 15,200 ppm in pyrolusite and 59,875 ppm in hollandite) analyzed by Doe et al. (1996).

Table 19.8. Average Concentrations of Pb in Marine Ferromanganese Crusts, Calcareous Ooze, and Sedimentary Mn Ores

Material	Pb, ppm[a]	References[b]
Ferromanganese, Pacific Ocean	1716 (5)	1
	997–3607	
	928 (53)	2
	683–1333	
Ferruginous clay, Pacific Ocean	121 (2)	3
Clay, Atlantic and Pacific	48 (552)	3
Calcareous ooze, Atlantic and Pacific	18 (238)	3
Radiolarian ooze, Atlantic and Pacific	25 (5)	3
Mn ore, pyrolusite	15,223	4
Mn ore, hollandite	59,875	4

[a]The number of samples included in each average is indicated in parentheses.
[b]1. Ling et al., 1997, *Earth Planet. Sci. Lett.*, 146 : 1–12. 2. Abouchami et al., 1997, *Geochim. Cosmochim. Acta*, 61 : 3957–3974. 3. Wedepohl, 1974, *Handbook of Geochemistry*, 82K-6, Springer-Verlag, Heidelberg. 4. Doe et al., 1996, in Basu and Hart (Eds.), *Geophys. Monogr.*, 95 : 391–408.

The isotope ratios of Pb of surface scrapings (<1.0 mm) of ferromanganese deposits in different parts of the Indian Ocean form two-component mixing lines (Vlastelic et al., 2001). For example, the $^{206}Pb/^{204}Pb$ and $^{208}Pb/^{204}Pb$ ratios of ferromanganese deposits in the northern Indian Ocean in Figure 19.30 define a straight line that includes Pb in samples of MORB from the Indian Ocean. The other component of Pb in these ferromanganese deposits is enriched in radiogenic ^{206}Pb and ^{208}Pb and therefore originated from continental sources. Vlastelic et al. (2001) considered whether the radiogenic-Pb component enters the North Indian Ocean in the form of eolian dust from northeastern Africa and from the Arabian peninsula or by the discharge of rivers draining the Himalayas (e.g., the Indus, Ganges-Brahmaputra). Although these rivers have formed large deposits of sediment at their mouths, they contribute very little soluble Pb to the Indian Ocean. Nevertheless, Vlastelic et al. (2001) cited isotope ratios of Pb in sediment suspended in the water at the mouth of the Ganges-Brahmaputra River (G. B.):

$$\frac{^{206}Pb}{^{204}Pb} \sim 19.3 \qquad \frac{^{208}Pb}{^{204}Pb} \sim 39.7$$

These values define a point in Figure 19.30 that is colinear with the isotope ratios of Pb in ferromanganese deposits of the northern Indian Ocean. Accordingly, the isotope ratios of ferromanganese deposits of the northern Indian Ocean can be interpreted as two-component mixtures of Pb derived from Indian Ocean MORBs and Pb sorbed to sediment suspended in the water at the mouth of the Ganges-Brahmaputra River. However, Frank

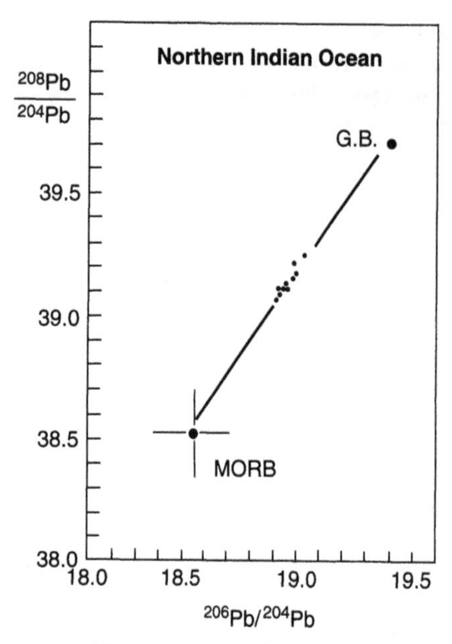

FIGURE 19.30 Two-component isotopic mixing line defined by the $^{206}Pb/^{204}Pb$ and $^{208}Pb/^{204}Pb$ ratios of ferromanganese deposits in the northern Indian Ocean, including the Somali and Mascarene basins (west) and the central Indian basin in the east. The straight line was fitted to 13 ferromanganese data sets by least-squares regression and is interpreted as a mixing line of Pb in Indian Ocean MORBs and in sediment suspended in the water of the Ganges-Brahmaputra River (G.B.). Data from Vlastelic et al. (2001).

and O'Nions (1998) questioned whether the isotope composition of Pb in ferromanganese deposits in the Indian Ocean actually records the erosion of the Himalayas.

The work of Vlastelic et al. (2001) in the Indian Ocean and that of Ling et al. (1997) and Abouchami et al. (1997) in the Pacific Ocean demonstrates that the Pb in ferromanganese deposits is advected by deep currents from distant sources, except in the vicinity of submarine hot springs. The dependence of the isotope composition of Pb and Nd in marine ferromanganese nodules and crusts on the circulation of deep water has been documented:

Abouchami and Goldstein, 1995, *Geochim. Cosmochim. Acta*, 59:1809–1820. Abouchami and Galer, 1998, *Mineral. Mag.*, 62A:1–2. von Blanckenburg and Igel, 1999, *Earth Planet. Sci. Lett.*, 169:113–128. von Blanckenburg and Nägler, 2001, *Paleoceanography*, 16:424–434. Albarède et al., 1997, *Geochim. Cosmochim. Acta*, 61:1277–1291. Ling et al., 1997, *Earth Planet. Sci. Lett.*, 146:1–12. Abouchami et al., 1997, *Geochim. Cosmochim. Acta*, 61:3957–3974. Abouchami et al., 1999, *Geochim. Cosmochim. Acta*, 63:1489–1505.

19.5 OSMIUM IN CONTINENTAL RUNOFF

Osmium is a siderophile trace element whose concentrations in terrestrial silicate rocks are generally less than 3 ppb, whereas extraterrestrial rocks have high Os concentrations ranging from 605 ppb in carbonaceous chondrites to 15,260 ppb in iron meteorites (Table 13.2). In addition, certain minerals (e.g., Cu–Ni–PGE sulfides and chromite) contain 554 and 50.7 ppb Os, respectively. The element is released into solution in surface water by chemical weathering of rocks and minerals on the continents and is transported into the oceans by streams.

The isotopic composition of Os in crustal rocks changes continuously by the decay of long-lived ^{187}Re with a halflife of 41.60×10^9 years ($\lambda = 1.666 \pm 0.006 \times 10^{-11} y^{-1}$). As a result, the $^{187}Os/^{186}Os$ ratios of rocks and minerals increase with time at a rate that depends on their Re/Os ratios. Alternatively, the isotopic composition of Os can also be expressed as the $^{187}Os/^{188}Os$ ratio, which is obtained by multiplying the $^{187}Os/^{186}Os$ ratio by 0.11986 (equation 13.4).

Certain accessory minerals and rocks listed in Table 19.9 have elevated Re/Os ratios and therefore can become enriched in radiogenic ^{187}Os compared to the common minerals and rocks. The sources of radiogenic ^{187}Os include the sulfide minerals of Mo, Cu, Ni, and Fe as well as mafic volcanic rocks, black shale, and granite gneisses of Precambrian age. The Re–Os method of dating is the subject of Chapter 13.

19.5a Rivers

The concentration of Os in rivers was measured by Sharma and Wasserburg (1997), who used

Table 19.9. Average Re/Os Ratios of Terrestrial Minerals and Rocks That Are Enriched in Re and Depleted in Os[a]

Mineral	Re, ppb	Os, ppb	Re/Os
	Minerals		
Molybdenite	63,860	912	70
Chalcopyrite	137	6.21	22
Pentlandite	146	9.10	16
Pyrrhotite	168	11.1	15
Pyrite	2.32	0.385	6
Magnetite	10.8	0.510	21
Ilmenite	3.68	0.182	20
	Rocks		
Tholeiite	0.84	<0.03	>28
Andesite	0.34	<0.02	>17
Granite gneiss	100	5.90	17
Shale	232.3	1.717	132

[a]From Table 13.2.

Table 19.10. Concentrations and $^{187}Os/^{186}Os$ Ratios of Water in Large Rivers

River	Weight, kg	Os, $\times 10^{-15}$ g/g	$^{187}Os/^{186}Os$[a]
Mississippi	2.0	7.02 ± 0.47	8.4 (10.4) ± 0.5
Mississippi	6.0	8.57 ± 0.16	9.8 (10.4) ± 0.2
Columbia	5.7	2.85 ± 0.17	11.8 (14.4) ± 0.6
Connecticut	5.2	2.59 ± 0.17	7.1 (8.8) ± 0.4
Vistula	4.3	8.40 ± 0.23	9.9 (10.7) ± 0.2

Source: Sharma and Wasserburg, 1997.

[a]Numbers in parentheses are corrected for incomplete equilibration of the sample with the Os spike.

a procedure that included coprecipitation of Os with ferric hydroxide followed by isotope dilution analysis (Sharma et al., 1997). The results in Table 19.10 include both the Os concentration and the $^{187}Os/^{168}Os$ ratios. The water samples were filtered through 0.45-μm filters and were then acidified with ultrapure HCl to pH 1.62 prior to analysis. Therefore, the results pertain primarily to Os in ionic solution (OsO_3^-) rather than to the total amount of Os present in the unfiltered water.

The concentrations of Os in the rivers analyzed by Sharma and Wasserburg (1997) vary more widely than expected from analytical errors. In particular, the measured Os concentrations of the Columbia and Connecticut Rivers are about three times lower than those of the Mississippi and Vistula Rivers. The analysts suggested that the low Os concentrations of the Columbia and Connecticut Rivers are the result of sorption of OsO_3^- to the walls of the containers because these two samples were filtered and acidified after the samples arrived in the laboratory, whereas the water samples of the Mississippi and Vistula Rivers were filtered and acidified at the time they were collected. Therefore, Sharma and Wasserburg (1997) concluded that the most reliable value of the Os concentration of river water is about 8.6×10^{-15} g/g based on the 6.0-kg sample of water of the Mississippi River.

The $^{187}Os/^{186}Os$ ratios (corrected for incomplete sample-spike equilibration) range from 8.8 (Connecticut River) to 14.4 (Columbia River), which presumably reflects differences in the Re/Os ratios of the rocks and minerals that release Os into solution during weathering. For example, the $^{187}Os/^{186}Os$ ratio of water in the Mississippi River (10.4, Table 19.10) is similar to that of loess in the upper drainage basin of this river (10.3–10.9) reported by Esser and Turekian (1993a).

Pegram et al. (1994) provided information about the concentrations and the isotope ratios of Os in bulk sediment and of Os that is leachable with acid hydrogen peroxide from this sediment. Their results in Figure 19.31 for the tributaries of Lake Oneida in upstate New York indicate that about 44% of Os in bulk sediment is extractable with this reagent and that the concentration of Os in the bulk sediment ranges from 9.5 ± 0.5 to 40.2 ± 10^{-12} g/g (ppt) of dry sediment with a mean of 30 ppt. This value is about 3480 times larger than the concentration of filtered river water reported by Sharma and Wasserburg (1997).

In addition, Pegram et al. (1994) demonstrated that the Os leached from sediment of the tributaries

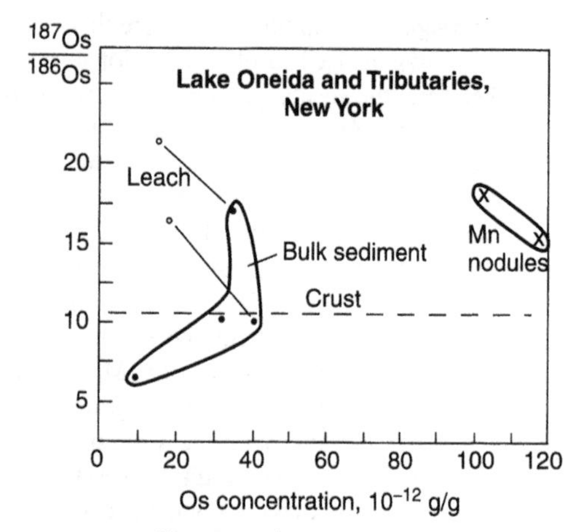

FIGURE 19.31 Concentrations and isotope ratios of Os in Lake Oneida and its tributaries in upstate New York. The average $^{187}Os/^{186}Os$ ratio of 10.5 of the continental crust is from Esser and Turekian (1993a). The data for Lake Oneida are from Pegram et al. (1994).

of Lake Oneida has higher $^{187}Os/^{186}Os$ ratios than the Os of the bulk sediment. These results therefore suggest that radiogenic ^{187}Os may be released preferentially by selective dissolution of certain minerals having high Re/Os ratios and/or that the sorbed Os in the sediment originated from such minerals.

Manganese nodules in Lake Oneida included in Figure 19.31 have a high average Os concentration (110 ppt) and a high average $^{187}Os/^{186}Os$ ratio (17.0) compared to the average value of this ratio (10.5) in the continental crust exposed to weathering (Esser and Turekian, 1993a).

Osmium leached by Pegram et al. (1994) from sediment in rivers that drain a variety of crustal rocks in the USA also has comparatively high $^{187}Os/^{186}Os$ ratios ranging from 19.29 ± 0.11 (Susquehanna, PA) to 10.08 ± 0.01 (James, VA) with a mean of about 16 (weighted by discharge). This value was confirmed by the $^{187}Os/^{186}Os$ ratios of the leachable Os in the sediment of the Ganges River of India (15.37 ± 0.03 and 16.93 ± 0.09) and of the Rio Maipo in Chile (18.00 ± 0.04). These results support the conclusion of Pegram et al. (1994) that the leachable Os being delivered to the oceans by the rivers of the world has a higher $^{187}Os/^{186}Os$ ratio than that of Os that exists in the weathering crust on the continents (10.5). However, the leachable Os in three streams draining ultramafic ophiolites in California and Oregon has low $^{187}Os/^{186}Os$ ratios (1.4–7.1). These and other rivers draining mantle-derived mafic and ultramafic rocks reduce the $^{187}Os/^{186}Os$ ratio of the oceans from about 10.5 (Esser and Turekian, 1993a) to 8.7 (Sharma et al., 1997).

19.5b Soils

The suggestion by Pegram et al. (1994) that the Os leached from river sediment originates preferentially from certain minerals that have high Re/Os ratios was supported by Peucker-Ehrenbrink and Blum (1998). These authors used dilute HCl to leach various fractions of soil samples from moraines that were deposited between about 0.4 and 138 ka on Precambrian gneisses in the Wind River Range of Wyoming. The results revealed several interesting aspects of the way soils release Os during chemical weathering:

1. Most of the Os in granitic gneisses resides in magnetite, which has a comparatively low $^{187}Os/^{186}Os$ ratio because of its low Re/Os ratio.
2. Biotite (or mineral inclusions) has a high Re/Os ratio and therefore contains most of the radiogenic ^{187}Os, depending on the age of the mineral.
3. When size fractions of soil are leached with cold dilute HCl, the $^{187}Os/^{186}Os$ ratios of the resulting solutions *decrease* with increasing concentration of the acid from 0.05 to 0.5 N.
4. The $^{187}Os/^{186}Os$ ratios of Os leached from young soils ($<\sim 5$ ka) are *higher* than those of the Os in the bulk soil.
5. The Os leached from old soil ($>\sim 5$ ka) has similar $^{187}Os/^{186}Os$ ratios as the bulk soil.
6. The bulk soil of the B horizon is enriched in Os relative to the A and C horizons.

The high $^{187}Os/^{186}Os$ ratios of leachates of young soils ($<\sim 0.5$ ka) in the Wind River Range are the result of oxidation of biotite. Soils that have been exposed to weathering for more than about 5000 years no longer contain biotite and therefore release Os whose isotope composition is similar to that of the remaining minerals (i.e., plagioclase and quartz, K-feldspar, residual magnetite). Osmium released by soils of any age by leaching with dilute HCl (0.05 N) is more radiogenic than Os released by leaching with more concentrated HCl (0.5 N) because the more concentrated acid attacks minerals having low Re/Os ratios (e.g., magnetite, plagioclase, and K-feldspar).

In the final analysis, the isotope composition of Os released by chemical weathering depends on the Re/Os ratios of the minerals and on their susceptibility to weathering under a given set of environmental conditions. As a result, the $^{187}Os/^{186}Os$ ratios of Os released by weathering changes with time but approaches the value of this ratio in the most Os-rich and/or most abundant minerals.

19.5c Lacustrine Ferromanganese Deposits

Ferromanganese nodules of lacustrine origin contain high concentrations of Os, which they sorb from the ambient water in which they form. Therefore, such deposits record the isotope composition of Os that was released by chemical weathering of the rocks and soil in their drainage basin.

Pegram et al. (1994) first reported $^{187}Os/^{186}Os$ ratios of ferromanganese nodules from Lake Oneida in upstate New York. This lake receives drainage from shales, siltstones, and sandstones of Ordovician to Devonian age that are covered by Pleistocene till containing mineral grains derived from the igneous and metamorphic rocks of the Precambrian Shield of Canada. The results in Figure 19.31 indicate that the $^{187}Os/^{186}Os$ ratios of the nodules (15.0–18.1) are consistent with the isotope ratios of Os in the bulk sediment in the tributary rivers of Lake Oneida (6.49–19.5).

The difference in the isotope composition of Os released by weathering of Precambrian and Phanerozoic rocks was clearly demonstrated by Peucker-Ehrenbrink and Ravizza (1996), who analyzed ferromanganese deposits in the Baltic Sea and in its surrounding drainage area. The results in Figure 19.32 demonstrate the wide range of $^{187}Os/^{186}Os$ ratios of Os released by weathering of Phanerozoic and Precambrian rocks. The ferromanganese nodules that formed within the Baltic Sea itself have intermediate $^{187}Os/^{188}Os$ ratios because they contain a mixture of Os derived from both continental sources and from seawater of the North Sea.

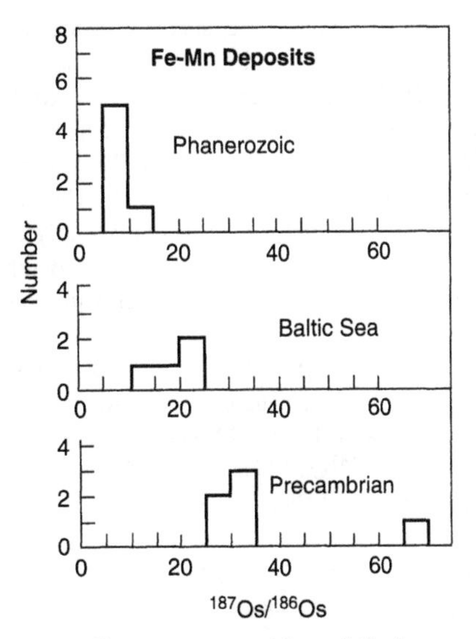

FIGURE 19.32 Isotope composition of Os in ferromanganese deposits in the drainage area of the Baltic Sea which includes rocks of Phanerozoic and Precambrian ages. Data from Peucker-Ehrenbrink and Ravizza (1996).

Table 19.11. Isotope Ratios of Os of Various Components Entering the Baltic Sea

Component	$^{187}Os/^{186}Os$
Fresh water	
Phanerozoic rocks	10
Precambrian	35
Both sources	22.5
Seawater, North Sea	8.6
Water, Baltic Sea	17.4

Source: Peucker-Ehrenbrink and Ravizza, 1996.

Peucker-Ehrenbrink and Ravizza (1996) developed an isotopic mixing model for Os in the Baltic Sea based in part on the assumptions listed in Table 19.11. The authors noted that the $^{187}Os/^{186}Os$

ratios of ferromanganese deposits in areas underlain by sedimentary rocks of Phanerozoic age south of the Baltic Sea range only from 8.6 to 11.0 in spite of being covered by till derived from Precambrian rocks of the Baltic Shield. They concluded that the comparatively low $^{187}Os/^{186}Os$ ratios of ferromanganese deposits in this area imply that the minerals containing radiogenic Os were removed by weathering after the till was deposited between 10 and 20 ka.

Additional measurements of $^{187}Os/^{186}Os$ ratios of lacustrine ferromanganese deposit by Peucker-Ehrenbrink and Blum (1998) confirm the difference between the $^{187}Os/^{186}Os$ ratios of Os released by Precambrian and Phanerozoic rocks:

Phanerozoic: 12.3 (3.35–20.68)
Precambrian: 102.5–881

Ferromanganese nodules in both geological settings had an average Os concentration of 102×10^{-12} g/g (1.7×10^{-12}–262.5×10^{-12} g/g).

19.5d Anthropogenic Contamination

Osmium is used in medical facilities as a fixative in electron microscopy and as an oxidant. According to Reese (1996), the United States imported 55 kg of Os in 1994, most of which originated from ore deposits associated with the mantle-derived mafic rocks of the Bushveld Complex (South Africa) and from the Ni–Cu ores at Noril'sk (Russia), both of which have $^{187}Os/^{186}Os$ ratios close to 1.0. In fact, Martin (1991) proposed a value of 1.10 for the $^{187}Os/^{186}Os$ ratio of the "average silicate Earth" based on analyses of mantle-derived mafic and ultramafic igneous rocks.

The release of commercial Os having an $^{187}Os/^{186}Os$ ratio close to 1.0 into rivers can significantly alter the isotope composition of Os that enters the ocean. Examples of such anthropogenic contamination of Os in coastal sediment were reported by Esser and Turekian (1993b) and Ravizza and Bothner (1996).

Subsequently, Williams et al. (1997) demonstrated that the $^{187}Os/^{186}Os$ ratios and the Os concentrations of bulk sediment within Long Island Sound in Figure 19.33 vary systematically with

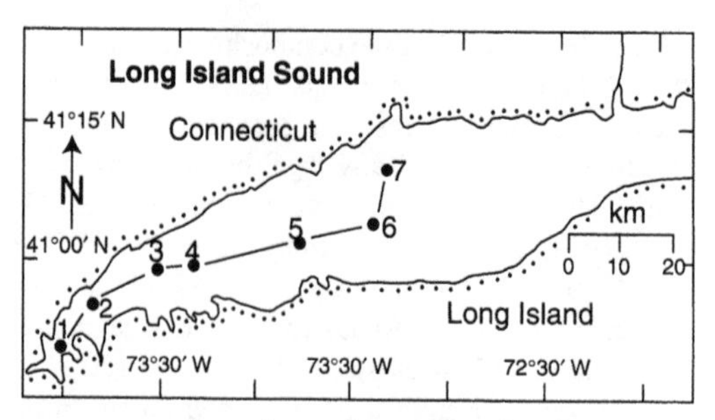

FIGURE 19.33 Locations of samples of sediment collected in Long Island Sound for a study of isotope compositions and concentrations of Os. Adapted from Williams et al. (1997).

distance from New York City and from other municipalities located at the west end of the Sound. Their data in Figures 19.34*a* and *b* demonstrate that the $^{187}Os/^{186}Os$ ratios of the upper 2.0 cm of sediment increase from about 5.1 close to the sewage outfall to 8.2 about 60 km to the east, whereas the Os concentrations decrease in the same distance from 127×10^{-12} g/g close to New York City to 21×10^{-12} g/g. In addition, the data of Williams et al. (1997) in Figure 19.34*c* show that the Os concentrations of the sediment are positively correlated with the concentrations of organic carbon. This relation means that Os either is sorbed to particles of organic matter or is bonded to organic molecules in the sediment or both.

The array of data presented by Williams et al. (1997) in Figures 19.34*a–c* demonstrates that Os having a low $^{187}Os/^{186}Os$ ratio is being discharged into Long Island Sound at its western end. The contaminant Os is sorbed by organic matter and mineral particles in the sediment, causing its concentration near the sewage outfalls to rise above normal values of about 20×10^{-12} g/g. In addition, mixing of the contaminant Os with crustal Os causes a localized isotopic anomaly in the sediment in the west end of Long Island Sound.

The Os that enters Long Island Sound both in wastewater and in rivers is at least partly retained by the sediment deposited in this estuary. Although the data of Williams et al. (1997) do not permit a quantitative evaluation, they clearly demonstrate that Os is strongly sorbed to sediment particles,

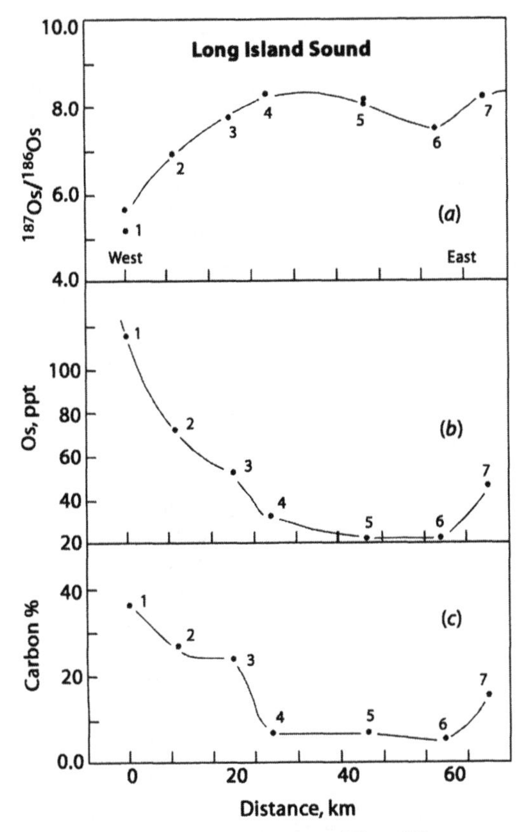

FIGURE 19.34 (*a*) Variation of the $^{187}Os/^{186}Os$ ratios of bulk sediment in Long Island Sound along a west-to-east profile shown in Figure 19.32. The sediment samples all represent the upper 2-cm layer recovered by a box corer. (*b*) Variation of the Os concentrations of the bulk sediment samples in units of 10^{-12} g/g. (*c*) Concentration of carbon in organic matter in the bulk sediment. Data from Williams et al. (1997).

which implies that only a fraction of the Os released by weathering and by anthropological uses on the continents enters the ocean.

19.6 OSMIUM IN THE OCEANS

The isotope composition of Os in the oceans, like those of Sr, Nd, and Pb, is determined by inputs from different sources and by its residence time. The principal sources of Os and their approximate isotope compositions are

1. Precambrian basement rocks containing biotite, K-feldspar, and other Re-rich and Os-poor minerals that release Os having high $^{187}Os/^{186}Os$ ratios of about 35 but ranging to values greater than 100;
2. Phanerozoic sedimentary rocks (shale, siltstone, sandstone) whose $^{187}Os/^{186}Os$ ratios approach 10;
3. Mafic, mantle-derived igneous rocks of Tertiary to Recent age that have $^{187}Os/^{186}Os$ ratios of about 1.0; and
4. Micrometeorite dust composed of silicate minerals, which also has an $^{187}Os/^{186}Os$ ratio of about 1.0

In contrast to Sr, the isotopic composition of Os in the oceans is not buffered by carbonate rocks, which causes the $^{187}Os/^{186}Os$ ratio of seawater to be sensitive to inputs of radiogenic ^{187}Os from Precambrian granitoids.

In addition, Os differs from Nd and Pb by having a substantially longer residence time of about 10^4-10^5 years. Consequently, the circulation of water in the oceans is better able to homogenize the isotopic composition of Os than those of Nd and Pb, although not as well as that of Sr, whose residence time is of the order of 10^6 years. Therefore, the isotopic composition of Os in the oceans has varied regionally with time and has primarily recorded changes in the flux of radiogenic ^{187}Os derived from Precambrian Shield areas.

19.6a Seawater

The Os concentration of seawater is very low and has been difficult to determine (e.g., Koide et al., 1996). Consequently, the direct determination of the $^{187}Os/^{186}Os$ ratio in seawater has also been difficult. The analytical problems were overcome only recently by Sharma et al. (1997), who reported that a suite of samples collected at different depths from 25 to 3000 m in the Atlantic Ocean near Bermuda have virtually constant Os concentrations and $^{187}Os/^{186}Os$ ratios.

This result suggests that Os is a conservative element in the oceans. Three aliquots of a sample taken from a depth of 3000 m near Bermuda yielded an average $^{187}Os/^{186}Os$ ratio of 8.7 ± 0.2. Another set of three aliquots from a depth of 3000 m collected in the Pacific Ocean near Hawaii yielded an identical $^{187}Os/^{186}Os$ ratio of 8.7 ± 0.3 (Sharma et al., 1997). However, seawater collected at a depth of 1764 m along the Juan de Fuca Ridge in the Pacific Ocean seems to have a lower $^{187}Os/^{186}Os$ ratio of 6.9 ± 0.4, which the authors attributed to the input of mantle-derived Os ($^{187}Os/^{186}Os \sim 1.0$) by hydrothermal solutions (or to unexplained analytical problems). The most reliable measurement of the Os concentration of unfiltered seawater reported by Sharma et al. (1997) is 3.6×10^{-15} g/g.

The Os concentrations and $^{187}Os/^{186}Os$ ratios of seawater from the Indian Ocean are also invariant with depth and have average values of $10.86 \pm 0.07 \times 10^{-15}$ g/g and 8.80 ± 0.07, respectively (Levasseur et al., 1998). The samples for this study were collected at two sites along the Southwest Indian Ridge from a maximum depth of 4560 m without detecting a change in the $^{187}Os/^{186}Os$ ratio caused by input of mantle-derived Os.

The concentration of Os in seawater in the Indian Ocean (10.86×10^{-15} g/g) determined by Levasseur et al. (1998) is three times higher than that reported by Sharma et al. (1997) for seawater in the Atlantic and Pacific Oceans. This discrepancy is probably an artifact of the analytical procedure used by Sharma et al. (1997), which may not have recovered all of the Os present in the samples. The procedures used by Levasseur et al. (1998) does not include a preconcentration step but instead relies on the oxidation of Os to OsO_4 and the simultaneous destruction of dissolved organic matter during 48 h incubation in an oven at 90°C. Levasseur et al. (1998) suggested that a certain fraction of the Os in seawater forms a stable organometallic complex that prevents it from equilibrating with the Os spike in the procedure used by Sharma et al. (1997). The measurement of the $^{187}Os/^{186}Os$ ratio of seawater is not affected by the different analytical procedures provided that all ionic and molecular forms of Os in seawater have the same isotopic composition.

Estimates of the residence time of Os in the oceans are affected by the uncertainty of the Os concentrations of water in the world's rivers and in seawater. Nevertheless, all of the estimates converge to values between 10^4 and 10^5 years. Levasseur et al. (1998) assumed that rivers provide only 80% of the total input of Os to the oceans and bracketed its residence time between 1.6×10^4 and 6.5×10^4 years.

As in the case of Sr, the different isotopic varieties of Os released by weathering on the continents mix during transport by rivers before they enter the oceans. Therefore, the $^{187}Os/^{186}Os$ ratio of water in the Mississippi River (10.4) represents Os derived by weathering in a large part of the North American continent, including rocks of Precambrian and Phanerozoic ages. The continental Os mixes in the oceans with Os released by meteoritic dust ($^{187}Os/^{186}Os) = 1.05$) and with Os derived from weathering of mantle-derived mafic volcanic rocks ($^{187}Os/^{186}Os = 1.10$) extruded along midocean ridges and on oceanic islands. By combining meteoritic and mantle-derived Os ($^{187}Os/^{186}Os = 1.075$), the isotopic composition of Os in the oceans can be treated as a two-component mixture:

$$\left(\frac{^{187}Os}{^{186}Os}\right)_{sw} = \left(\frac{^{187}Os}{^{186}Os}\right)_{m} f + \left(\frac{^{187}Os}{^{186}Os}\right)_{rw} (1 - f) \tag{19.16}$$

where the subscripts are defined as sw = seawater, m = meteoritic and mantle-derived, and rw = river water. In addition, this equation describes mixing of the element in pure form without regard to its concentrations in the media in which it is delivered to the oceans. Substituting values for the $^{187}Os/^{186}Os$ ratios and solving equation 19.16 for f yield

$$8.8 = 1.075f + 10.4(1 - f)$$

$$f = 0.17$$

This selected data set indicates that approximately 17% of the Os in the oceans originates from meteoritic dust and from mantle-derived rocks in the ocean basins.

Equation 19.16 also demonstrates that the $^{187}Os/^{186}Os$ ratio of seawater is a linear function of the $^{187}Os/^{186}Os$ ratio of river water, assuming that the numerical values of the Os isotope ratios of meteoritic dust and mantle-derived rocks are invariant with time, at least during the Phanerozoic Eon, and that the fraction of Os derived from these sources remains constant (i.e., $f = 0.17$):

$$\left(\frac{^{187}\text{Os}}{^{186}\text{Os}}\right)_{\text{sw}} = 0.182 + 0.83\left(\frac{^{187}\text{Os}}{^{186}\text{Os}}\right)_{\text{rw}} \quad (19.17)$$

Therefore, if the $^{187}Os/^{186}Os$ ratio of river water increases due to an increase in the rate of weathering of biotite-bearing Precambrian basement rocks (e.g., after an orogeny), the $^{187}Os/^{186}Os$ ratio of seawater also rises. For example, if the average $^{187}Os/^{186}Os$ ratio of river water increases globally to 22.5 (Table 19.11; Peucker-Ehrenbrink and Ravizza, 1996), the $^{187}Os/^{186}Os$ ratio of seawater would be

$$\left(\frac{^{187}\text{Os}}{^{186}\text{Os}}\right)_{\text{sw}} = 0.182 + 0.83 \times 22.5 = 18.8$$

The apparent sensitivity of the isotope composition of Os in the oceans to changes in the $^{187}Os/^{186}Os$ ratio of Os in river water has motivated the study of Os in marine ferromanganese and other sedimentary deposits of Cenozoic age.

19.6b Meteoritic Dust

The presence of an extraterrestrial component in terrestrial sediment was first demonstrated by Luck and Turekian (1983) in Cretaceous–Tertiary boundary clays based on their low $^{187}Os/^{186}Os$ ratios of 1.29–1.66 compared to an average of 7.57 in seven ferromanganese nodules. Subsequently, Esser and Turekian (1988) estimated the accretion rate of extraterrestrial particles from the $^{186}Os/^{187}Os$ ratios of pelagic clay and ferromanganese nodules in the Pacific Ocean. They concluded that the flux of carbonaceous meteorite dust to the Earth's surface is 4.9×10^7 kg/y and that about 20% of that material dissolves in seawater.

More recently, Sharma et al. (1997) referred to isotopic evidence that up to 90% of certain elements (Mg, O, Ni, and Cr) evaporate from cosmic spherules during their passage through the atmosphere. They suggested that a certain fraction of Os in these spherules also evaporates and is ultimately dissolved in seawater. The presence of this "cosmic" Os in seawater lowers its $^{187}Os/^{186}Os$ ratio from 10.4 to 8.7 ± 0.1, as reported by Sharma et al. (1997) and Levasseur et al. (1998).

19.6c Ferromanganese Deposits

The slow rate of deposition of ferromanganese crusts combined with their enrichment in Os compared to seawater makes them an ideal medium to study time-dependent variations of the $^{187}Os/^{187}Os$ ratio of seawater. The comparatively long oceanic residence time of Os allows its isotopic composition to be homogenized almost as well as that of Sr and much more so than those of Nd and Pb. Therefore, Os isotope ratios of ferromanganese crusts and other sedimentary deposits record geological events on a *global* scale, like the isotope ratios of Sr in marine carbonates and phosphates.

The Os concentrations of ferromanganese crusts reported by Burton et al. (1999) range from (1.014 to 3.635) $\times 10^{-9}$ g/g (ppb) and confirm that they are strongly enriched in Os relative to seawater (10.86×10^{-15} g/g). The Os in ferromanganese deposits is sorbed from seawater and is assumed to have the same $^{187}Os/^{186}Os$ ratio as the Os in seawater, except for possible in situ decay of ^{187}Re and the presence of mantle-derived Os near hot springs along midocean ridges. However, neither in situ decay of ^{187}Re nor proximity to submarine hot springs appear to have affected the $^{187}Os/^{186}Os$ ratios of metalliferous sediment ranging in age from 83.7 to 0.104 Ma in cores recovered from the Pacific Ocean (Peucker-Ehrenbrink et al., 1995) by the Deep Sea Drilling Program and Ocean Drilling Program (DSDP/ODP).

However, Burton et al. (1999) reported low $^{187}Os/^{186}Os$ ratios for hydrothermal ferromanganese crusts at the Gibraco fracture zone (6.813 ± 0.009) in the North Atlantic and on the Uracas seamount, Mariana Islands (8.173 ± 0.028), in the West Pacific Ocean. In addition, Burton et al. (1999) observed a decrease of the $^{187}Os/^{186}Os$ ratios in a ferromanganese crust (327 KD, VA 13/2, 9°18′ N, 146°3′ W) from 8.69 at 2.08 Ma to 4.359 at 0.04 Ma. The decrease of the $^{187}Os/^{186}Os$ ratios was probably caused by the localized addition of unradiogenic Os in the form of meteoritic dust or of particles of mantle-derived rocks. Two episodic decreases of $^{187}Os/^{186}Os$ ratios from 8.6 to 8.1 at 20 and 160 ka were also reported by Oxburgh (1998) in each of two closely spaced sediment cores collected along the East Pacific Rise at about 17°00′ S, 114°00′ W. These episodes coincide with the terminations of the last two continental glaciations of the Pleistocene Epoch.

The $^{186}Os/^{187}Os$ ratios of recently deposited hydrogenetic ferromanganese crusts in the Atlantic, Pacific, and Indian Oceans in Table 19.12 vary only between narrow limits except for local anomalies. The data of Burton et al. (1999) in Table 19.12 suggest that the average $^{187}Os/^{186}Os$ ratio of ferromanganese crusts in the Atlantic Ocean (8.78 ± 0.07) is slightly higher than those of the Pacific (8.59 ± 6.05) and Indian (8.58 ± 0.10) Oceans. The authors attributed the apparent ^{187}Os enrichment of Os in the Atlantic Ocean to the discharge of major rivers draining the adjacent continents (e.g., Amazon, Zaire, Mississippi, Orinoco). Nevertheless, except for local anomalies discovered by Burton et al. (1999) and Oxburgh (1998) and for small regional differences (e.g., in the Baltic Sea), the $^{187}Os/^{186}Os$ ratios of the major oceans appear to be constant on a global scale at the present time.

Table 19.12. Average $^{187}Os/^{186}Os$ Ratios and Os Concentrations of Surface Layers of Hydrogenetic Ferromanganese Crusts

Ocean	Number of Samples in Each Average	Os, 10^{-9} g/g	$^{187}Os/^{186}Os^a$
Atlantic	7	1.91 ±0.40	8.78 ±0.07
Pacific	20	2.22 ±0.28	8.59 ±0.05
Indian	6	1.48 ±0.35	8.58 ±0.10

Source: Burton et al., 1999.

[a] The $^{187}Os/^{188}Os$ ratios were corrected for isotope fractionation to $^{192}Os/^{188}Os = 3.08271$ and converted to $^{187}Os/^{186}Os$ ratios by multiplying them by $^{187}Os/^{188}Os = 0.119969$.

19.6d Isotopic Evolution during Cenozoic Era

The variation of $^{187}Os/^{186}Os$ ratios of the oceans during the Cenozoic Era has been documented in several different ways: (1) by leaching Os from a long core composed of pelagic clay deposited in the North Pacific (Pegram et al., 1992), (2) by analysis of Os in metalliferous sediment of different ages in the Pacific Ocean (Peucker-Ehrenbrink et al., 1995), and (3) by a study of metalliferous carbonates deposited close to the East Pacific Rise (Ravizza, 1993).

Pegram et al. (1992) used acidified hydrogen peroxide (6%) to leach Os sorbed to clay minerals in a 24-m piston core (LL44-GPC3) taken in the North Pacific Ocean at 30°19.9′N and 157°49.9′W about 200 km north of the Hawaiian Islands. The core intersects the Cretaceous–Tertiary boundary, which was identified by a peak in the Ir concentration at a depth of 20.5 m below the top of the core. In addition, Pegram et al. (1992) demonstrated that the $^{187}Os/^{186}Os$ ratios of the leachates are higher in all cases than those of the bulk clay samples.

The $^{187}Os/^{186}Os$ ratios of the leachates of sediment in core LL44-GPC3 decrease with depth from 8.247 ± 0.009 (0.32–0.36 m) to 2.709 ± 0.004 (17.35–17.40 m). The time-dependent variation of the $^{187}Os/^{186}Os$ ratios of the leachates is similar to the variation of the $^{87}Sr/^{86}Sr$ ratio in Figure 19.4 but differs in detail. Pegram et al. (1992) satisfied themselves that the $^{187}Os/^{168}Os$ ratios of the leachates are sufficiently similar to those of seawater to provide a valid record of the evolution of the isotope composition of Os in the North Pacific Ocean. However, they acknowledged that the acidified hydrogen peroxide may attack unidentified detrital Os-bearing mineral phases, causing the $^{187}Os/^{186}Os$ ratios of leachates to rise by 1.6% depending on the concentration of the hydrogen peroxide between 2.7 and 30%.

Pegram et al. (1992) attributed the time-dependent rise of the $^{187}Os/^{186}Os$ ratio of seawater to

1. an increase of the annual input to the oceans of Os having elevated $^{187}Os/^{186}Os$ ratios,
2. an increase of the $^{187}Os/^{186}Os$ ratios of the continental Os component because of enhanced

weathering of Re-rich black shales and related granitoids or because of the exposure to weathering of older Precambrian rocks on the continents,
3. a decrease of the flux of mantle-derived or meteoritic Os entering the oceans, or
4. a combination of two or more of the factors identified above.

Subsequent measurements of $^{187}Os/^{186}Os$ ratios of metalliferous carbonates by Ravizza (1993) and of metalliferous sediment by Peucker-Ehrenbrink et al. (1995) have established a body of data that was used to construct the isotopic profile in Figure 19.35 for Os in the oceans during the Cenozoic Era. In addition to the value of this Os isotope profile as a record of global geological activity, it can also be used for dating nonfossiliferous marine rocks of Late Cenozoic (Neogene) age (Ravizza, 1993).

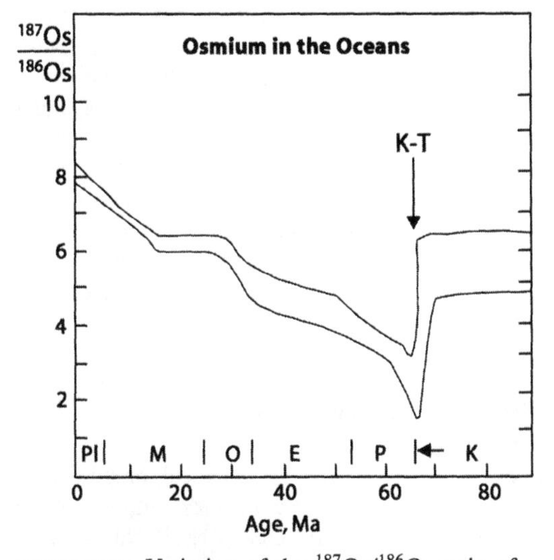

FIGURE 19.35 Variation of the $^{187}Os/^{186}Os$ ratio of seawater during the Cenozoic Era: K = Cretaceous, P = Paleogene, E = Eocene, O = Oligocene, M = Miocene, Pl = Pliocene. Adapted from Peucker-Ehrenbrink et al. (1995) and based on data by Pegram et al. (1992), Ravizza (1993), and Peucker-Ehrenbrink et al. (1995).

The increase of the $^{187}Os/^{186}Os$ ratio of seawater during the past 15 million years may have been caused in part by the weathering of black shales in the Himalayas drained by the Ganges-Brahmaputra and Indus river systems (Pegram et al., 1992; Peucker-Ehrenbrink et al., 1995; Reisberg et al., 1997). However, Burton et al. (1999) pointed out that the average $^{187}Os/^{186}Os$ ratio of ferromanganese deposits in the Indian Ocean (8.58 ± 0.10, Table 19.12) does not support the postulated input of excess radiogenic ^{187}Os from the Himalayas.

Analyses of Os in black shale in the Narayani drainage in the Himalayas by Pierson-Wickmann et al. (2002) did yield high $^{187}Os/^{186}Os$ ratios between 42.8 and 49.0 with a maximum value of 86.9 in one of several soil profiles (MO 601, 20–50 cm). However, the Os that was released from the black shale was coprecipitated with ferric oxyhydroxide at the site of weathering, whereas the Re escaped in solution. Therefore, the Os released by weathering of black shale in the Himalayas appears to be scavenged by ferric oxyhydroxide and is ultimately deposited with the sediment in the Indian Ocean. Consequently, the radiogenic ^{187}Os is prevented from affecting the isotope composition of Os in seawater of the Indian Ocean.

The sediment transported by the Narayani River enters the Ganges-Brahmaputra river system and is deposited in the Bengal Fan. Sediment samples in two cores of the Bengal Fan were leached by Reisberg et al. (1997) using the H_2O_2–H_2SO_4 reagent of Pegram et al. (1992, 1994). The $^{187}Os/^{186}Os$ ratios of the leachates (7.82–13.94) exceed those of seawater in Figure 19.35 during the time interval from 15.3 to 0.9 Ma, when the sediment was deposited. The presence of excess sorbed ^{187}Os relative to Os in seawater may be characteristic of sediment transported by the Ganges-Brahmaputra river system. For example, analyses of leachates by Pegram et al. (1992) of sediment in the Ganges River yielded $^{187}Os/^{186}Os$ ratios of 15.37 and 16.93. Alternatively, Reisberg et al. (1997) proposed that the excess radiogenic ^{187}Os originated from sediment upslope of the coring site either by desorption during alteration or by decay of ^{187}Re.

In any case, a large fraction of the Os that is released by weathering of carbon-rich shales in the Himalayas appears to be sorbed to sediment particles, leaving only a small amount in solution. Therefore, the Himalayas are probably not the dominant source of radiogenic ^{187}Os in seawater of the Indian Ocean.

The study of weathering of black shale in the Himalayas by Pierson-Wickmann et al. (2002) revealed the close association of both Os and Re with organic matter. This property of Os was used by Ravizza and Turekian (1992) to measure the $^{187}Os/^{186}Os$ ratios of bulk samples of carbon-rich marine sediment in the Pacific and Atlantic Oceans. The Os in these samples originated largely from seawater (79–100% hydrogenetic) and was associated with organic matter (11.7–60%). The resulting $^{187}Os/^{186}Os$ ratios (8.15–8.92) of the bulk sediment samples yielded an average of 8.59 ± 0.12, which is indistinguishable from the average $^{187}Os/^{186}Os$ ratios of hydrogenetic ferromanganese crusts in Table 19.12 reported by Burton et al. (1999).

19.7 HAFNIUM IN THE OCEANS

The isotopic composition of Hf in the crust and mantle of the Earth changes with time because of the decay of long-lived ^{176}Lu to stable ^{176}Hf with a halflife of 35.7×10^9 years, which yields a decay constant of $1.94 \times 10^{-11}\ y^{-1}$ (Chapter 12). The highest Hf concentrations listed in Table 12.2 occur in certain accessory minerals, including zircon (15,177 ppm), baddeleyite (13,340 ppm), and eudialyte (1736 ppm), all of which resist chemical weathering (zircon and baddeleyite) or occur rarely (eudialyte). The Hf concentrations of igneous rocks range from 1.14 ppm (peridotite) to 12.84 ppm (rhyolite). Stony meteorites have low Hf concentrations of about 0.2 ppm in chondrites and 0.73 ppm in Ca-rich achondrites (Boswell and Elderfield, 1998).

The $^{176}Hf/^{177}Hf$ ratios of most rocks and minerals vary only between narrow limits, which requires that they must be measured with great precision. For this reason, it is convenient to express $^{176}Hf/^{177}Hf$ ratios by means of the ε^0-notation relative to CHUR-Hf defined by equation 12.3:

$$\varepsilon^0(\mathrm{Hf}) = \left[\frac{(^{176}\mathrm{Hf}/^{177}\mathrm{Hf})^0_{\mathrm{spl}}}{(^{176}\mathrm{Hf}/^{177}\mathrm{Hf})^0_{\mathrm{CHUR}}} - 1\right] \times 10^4$$

where $(^{176}Hf/^{177}Hf)^0_{CHUR} = 0.28286$ (Patchett, 1983) or $(^{176}Hf/^{177}Hf)^0_{CHUR} = 0.282772$ (Blichert-Toft et al., 1997). The $^{143}Nd/^{144}Nd$ ratios can be expressed similarly by equation 9.6 in Section 9.2b, where

$$\left(\frac{^{143}\mathrm{Nd}}{^{144}\mathrm{Nd}}\right)^0_{\mathrm{CHUR}} = 0.512638$$

Samples having positive ε^0-values for Hf and Nd have higher $^{176}Hf/^{177}Hf$ and $^{143}Nd/^{144}Nd$ ratios than the corresponding values of CHUR for Hf and Nd.

During partial melting of ultramafic rocks in the lithospheric or asthenospheric mantle, Hf and Nd preferentially enter the melt phase and are thereby removed from the mantle by the upward movement of basalt magma. As a result, the continental crust has lower average Lu/Hf and Sm/Nd ratios than the ultramafic rocks of the mantle. The difference in these ratios has caused crustal Hf and Nd to have lower present-day $^{176}Hf/^{177}Hf$ and $^{143}Nd/^{144}Nd$ ratios than young mantle-derived volcanic rocks. Consequently, crustal rocks are characterized by negative ε^0-values of Hf and Nd, whereas mantle-derived volcanic rocks in most cases have positive or near-zero values of ε^0.

19.7a Terrestrial Hf–Nd Array

The similarity of the geochemical properties of the Lu–Hf and Sm–Nd couples outlined above causes the $^{176}Hf/^{177}Hf$ and $^{143}Nd/^{144}Nd$ ratios of crustal rocks to be positively correlated. This relationship may be disturbed by chemical weathering, which affects Hf-rich minerals (e.g., zircon) differently than Nd-rich minerals (biotite, monazite, amphiboles).

The ε^0-values of Hf and Nd of sediment in different geological environments reported by Vervoort et al. (1999) become increasingly negative with

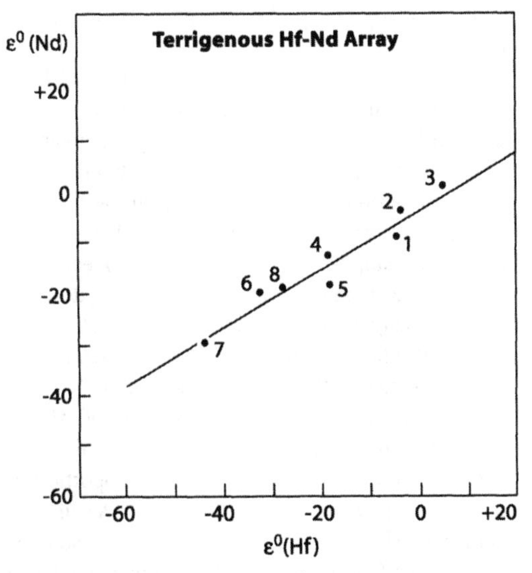

FIGURE 19.36 Correlation of ε^0-values of Hf and Nd in sediment ranging in age from Recent to Archean: 1. Pelagic sediment, 0 Ga; 2. Deep-sea turbidites, 0 Ga; 3. Canadian Cordillera, 0.14–0.56 Ga; 4. Paleozoic turbidites, 0.30–0.45 Ga; 5. Proterozoic sediment, 1.85–1.88 Ga; 6. Birimian sediment, 2.10 Ga; 7. Archean shale, 2.66–2.90 Ga; 8. Fluvial or shallow-water sediment, 0.21–2.20 Ga. The ε^0-values were calculated relative to $(^{176}Hf/^{177}Hf)^0_{CHUR} = 0.282772$ and $(^{143}Nd/^{144}Nd)^0_{CHUR} = 0.512638$. Data from Vervoort et al. (1999).

increasing age from Recent to Archean. In addition, the average ε^0-values of Hf and Nd of these samples in Figure 19.36 are strongly correlated and define the "terrestrial Hf–Nd array" represented by the equation

$$\varepsilon^0(\text{Nd}) = -3.6148 + 0.5737\varepsilon^0(\text{Hf}) \quad (19.18)$$

Although this equation represents the isotope composition of Hf and Nd in the crust expressed in the ε^0-notation, it does not pass through the origin as expected from the definition of the chondritic uniform reservoir. For example, if $\varepsilon^0(\text{Hf}) = 0$, equation 19.18 yields $\varepsilon^0(\text{Nd}) = -3.6148$. Vervoort et al. (1999) attributed this discrepancy to an inconsistency in the definition of CHUR.

The average Hf concentrations of different types of marine sediment range from 2.8 to 6.7 ppm depending primarily on the presence of detrital zircon grains. The average Hf and Nd concentrations of all sediments included in the study of Vervoort et al. (1999) are

$$\text{Hf} = 5.0 \text{ ppm} \qquad \text{Nd} = 25 \text{ ppm}$$

19.7b Rivers and Seawater

The occurrence of Hf in weathering-resistant minerals (e.g., zircon and baddeleyite) contributes to its low concentrations in solution in rivers and in the oceans. Godfrey et al. (1996) reported Hf concentrations ranging from 0.57×10^{-12} to 4.53×10^{-12} g/g in rivers of South America and Siberia sampled by J. M. Edmond. These values are nearly 40 times higher than the concentrations of Hf in seawater in the northeast Atlantic Ocean determined by Godfrey et al. (1996).

The Hf concentrations of the surface layer of the northeastern Atlantic Ocean adjacent to the Celtic Sea decrease from 432×10^{-15} g/g near the edge of the shelf to 73.2×10^{-15} g/g in the open ocean. In addition, Godfrey et al. (1996) confirmed that the Hf concentrations of seawater increase with depth similar to those of micronutrients. However, the concentration of Hf stops rising at a depth of about 3000 m, presumably because of sorption of Hf to suspended particles in the water column. The data of Godfrey et al. (1996) yield an average Hf concentration of 135×10^{-15} g/g for five samples of seawater collected from depths of greater than 3000 m.

Using the concentrations of Hf in river water (3.57×10^{-12} g/g) and seawater (143×10^{-15} g/g), Godfrey et al. (1996) estimated that the oceanic residence time of Hf is 1.5×10^3 years and therefore is significantly longer than that of Nd, which is only 0.3×10^3 years (Section 19.3c). Consequently, the isotope composition of Hf in seawater is expected to vary less than that of Nd (McKelvey and Orians, 1998).

19.7c Recent Ferromanganese Nodules

The concentrations of Hf in hydrogenetic ferromanganese nodules in Table 19.13 reported by White et al. (1986) range from 4.93 to 8.76 ppm and have a mean of 6.7 ppm, which is about eight orders of magnitude higher than the Hf concentration of seawater. All other types of clay-rich sediment listed in Table 19.13 have slightly lower Hf concentrations between 1.62 ppm in siliceous ooze to 4.0 ppm in red clay.

The $^{176}Hf/^{177}Hf$ ratios of bulk samples of marine sediment in Table 19.13 range widely depending on the proportions of continental and mantle Hf they contain (Pettke et al., 1998). Various types of deep-sea clay have $\varepsilon^0(Hf)$ values between +4.5 (DSDP 452, Pacific) to −17.9 (silty sand, Atlantic). The six ferromanganese nodules analyzed by White et al. (1986) have a mean $\varepsilon^0(Hf)$ value of +3.6, which suggests that a large fraction of Hf dissolved in seawater originates from mantle-derived volcanic rocks and from hydrothermal fluids discharged by hot springs along midocean ridges.

Table 19.13. Hafnium Concentrations and ε^0(Hf) Values of Ferromanganese and Other Types of Deposits in the Oceans

Deposit	Hf, ppm	$\varepsilon^0(Hf)$
Composite, DSDP 452	–	+4.5 (1)
Fe–Mn nodules	6.7 (4)	+3.6 (6)
Red clay	4.0 (2)	+4.0 (2)
Siliceous ooze	1.62 (4)	−0.80 (4)
Red/brown clay	3.67 (1)	−2.1 (1)
Brown clay	2.77 (1)	−5.3 (2)
Terrigenous clay	—	−8.3 (1)
Sandy clay	3.07 (1)	−9.8 (1)
Radiolarian clay	—	−17.0 (1)
Silty sand	2.80 (1)	−17.9 (1)

Source: White et al., 1986.
Note: All data represent complete dissolutions. The $^{176}Hf/^{177}Hf$ ratio of JMC-475 was 0.282161 and $\varepsilon^0(Hf)$ values were recalculated relative to $(^{176}Hf/^{177}Hf)^0_{CHUR}$ = 0.282772. The samples originated from the Atlantic, Pacific, and Indian Oceans. The number of samples is indicated in parentheses. Additional Hf concentrations of ferromanganese nodules were reported by Godfrey et al. (1997) and Albarède et al. (1998).

Subsequent studies by Godfrey et al. (1997), Albarède et al. (1998), and David et al. (2001) have confirmed that hydrogenetic ferromanganese nodules in the major oceans have distinctive isotope compositions of Hf. Their data in Table 19.14 yielded average $\varepsilon^0(Hf)$ values of +1.9 for the Atlantic Ocean, +5.3 for the Pacific Ocean, and +4.3 for the Indian Ocean. In addition, hydrothermal ferromanganese deposits analyzed by Godfrey et al. (1997) have higher $\varepsilon^0(Hf)$ values than hydrogenetic nodules in the Atlantic and Pacific Oceans. In general, the isotope compositions of Hf in Table 19.14 indicate that ferromanganese deposits in the Atlantic Ocean contain a larger proportion of continental Hf than those of the Pacific Ocean. The average $\varepsilon^0(Hf)$ values of ferromanganese nodules in the Indian Ocean are intermediate between those of the Atlantic and Pacific Oceans.

The isotopic compositions of Hf and Nd of hydrogenetic ferromanganese deposits of all three oceans in Figure 19.37 are positively correlated by the equation

$$\varepsilon^0(Nd) = -13.8099 - 1.5288\varepsilon^0(Hf) \quad (19.19)$$

This line is displaced from the terrestrial Hf–Nd array (equation 19.3), but the two lines intersect at about $\varepsilon^0(Hf) = +10.6$ and $\varepsilon^0(Nd) = +2.5$. Both

Table 19.14. Average ε^0(Hf) Values of Hydrogenetic (A) and Hydrothermal (B) Ferromanganese Nodules of the Major Oceans

Ocean	$\varepsilon^0(Hf)A$	$\varepsilon^0(Hf)B$
Atlantic	+1.9 (53)	+3.1 (5)
Pacific	+5.3 (21)	+8.8 (12)
Indian	+4.3 (4)	–

Note: Data compiled from White et al. (1986), Godfrey et al. (1997), David et al. (2001), and Albarède et al. (1998). All $\varepsilon^0(Hf)$ values are relative to $(^{176}Hf/^{177}Hf)^0_{CHUR}$ = 0.282772 and $^{176}Hf/^{177}Hf$ = 0.28217 or 0.28216 for JMC-475. The data for the hydrothermal nodules are by Godfrey et al. (1997). The number of samples included in each average is indicated in parentheses.

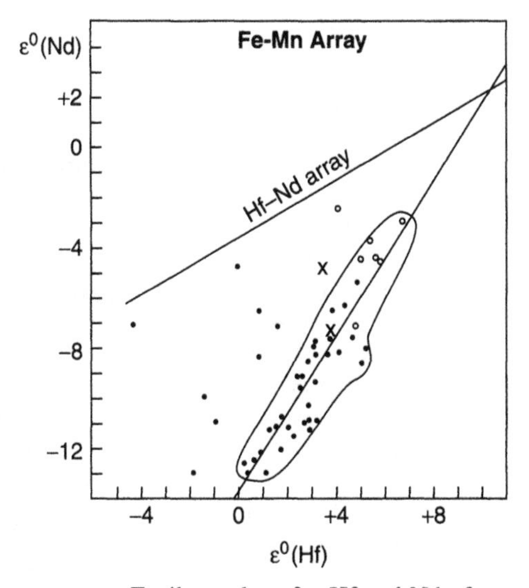

FIGURE 19.37 Epsilon-values for Hf and Nd of hydrogenetic ferromanganese nodules in the oceans: solid circles = Atlantic; open circles = Pacific; crosses = Indian Ocean. All ε^0-values were calculated relative to$(^{176}Hf/^{177}Hf)^0_{CHUR} = 0.282772$ and $(^{143}Nd/^{144}Nd)^0_{CHUR} = 0.512638$. In addition, all measured $^{176}Hf/^{177}Hf$ ratios are relative to 0.28216 for the JMC-475 Hf isotope standard. The terrigenous Hf–Nd array was replotted from equation 19.18 based on sediment samples analyzed by Vervoort et al. (1999). Data from Albarède et al. (1998) and David et al. (2001).

quasilinear arrays may be the result of mixing of two components in the Hf–Nd isotopic plane. The point of intersection of the two lines in Figure 19.37 represents the Hf and Nd that originated from sources in the mantle. The terrestrial Hf–Nd array in Figure 19.36 can be viewed as a series of mixtures of this mantle component with varying amounts of Archean shale (point 7, Figure 19.36) whose coordinates are $\varepsilon^0(\text{Hf}) = -43.5$ and $\varepsilon^0(\text{Nd}) = -29.5$ (Vervoort et al., 1999).

The component of continental Hf present in the ferromanganese nodules in Figure 19.37 is enriched in radiogenic ^{176}Hf compared to Hf in Archean shale because of the preferential weathering of minerals having higher Lu/Hf ratios than zircon, which contains most of the Hf in crustal rocks but which resists weathering. Therefore, the Hf that goes into solution on the continents and is ultimately sorbed by ferromanganese nodules in the oceans is enriched in radiogenic ^{176}Hf, causing its ε^0(Hf) value to be less negative than that of the Hf that exists in the unweathered rocks. In other words, chemical weathering of Hf-bearing minerals on the continents causes isotope fractionation of the Hf that goes into solution and is transported into the oceans.

The ε^0(Hf) value of the Hf in solution on the continents can be calculated by substituting ε^0(Nd) of Archean shale, based on the assumption that Nd is not fractionated isotopically by chemical weathering. Therefore, if $\varepsilon^0(\text{Nd}) = -29.5$, equation 19.19 yields

$$\varepsilon^0(\text{Hf}) = \frac{-29.5 + 13.8099}{1.5288} = -10.2$$

The explanation of the ε^0-values of Hf and Nd in hydrogenetic ferromanganese nodules in Figure 19.38 considers them to be mixtures of two components that are present in solution in seawater:

1. Mantle-derived component $\varepsilon^0(\text{Hf}) = +10.6$, $\varepsilon^0(\text{Nd}) = +2.5$
2. Crustal component $\varepsilon^0(\text{Hf}) = -10.2$, $\varepsilon^0(\text{Nd}) = -29.5$

Consequently, the ε^0-value of either Hf or Nd in ferromanganese nodules can be expressed by equations of the form

$$\varepsilon^0(\text{nodule}) = \varepsilon^0(\text{mantle})f + \varepsilon^0(\text{crust})(1 - f) \tag{19.20}$$

where f is the abundance of the mantle component.

For example, a nodule in the Pacific Ocean that is represented in Figure 19.38 by a point whose coordinates are

$$\varepsilon^0(\text{Hf}) = +6.8 \qquad \varepsilon^0(\text{Nd}) = -3.0$$

contains about 82% of the mantle component based on equation 19.16: For $\varepsilon^0(\text{Hf}) = +6.8$,

$$+6.8 = +10.6f - 10.2(1 - f)$$

$$f = 0.82$$

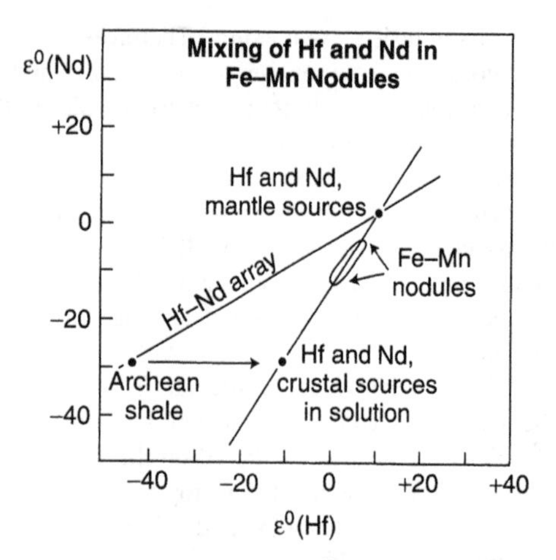

FIGURE 19.38 Relation of the ε^0-values of Hf and Nd of hydrogenetic ferromanganese nodules in the Atlantic, Pacific, and Indian Oceans to the terrigenous Hf–Nd array. The diagram demonstrates the enrichment of ferromanganese deposits in radiogenic ^{176}Hf relative to sediment on the continents. The ε^0(Hf) value of the continental Hf component in the oceans was derived by shifting the Hf of Archean shale from the Hf–Nd array to the Fe–Mn array. The isotope composition of Hf in seawater is the result of mixing of Hf that originated by weathering of mantle-derived igneous rocks and of crustal rocks, represented here by Archean shale. The terrigenous Hf–Nd array was plotted from equation 19.18 (Figure 19.36) and the Fe–Mn array is from Figure 19.37, based on data from Albarède et al. (1998) and David et al. (2001).

Similarly, for $\varepsilon^0(\text{Nd}) = -3.0$,

$$-3.0 = 2.5f - 29.5(1 - f)$$

$$f = 0.82$$

At the other end of the compositional spectrum, a ferromanganese nodule in the Atlantic Ocean is characterized by a point whose coordinates are

$$\varepsilon^0(\text{Hf}) = +0.20 \qquad \varepsilon^0(\text{Nd}) = -12.9$$

The abundance of the mantle component in this nodule is 50%.

Therefore, the isotopic mixing model in Figure 19.38 confirms the conclusion originally expressed by White et al. (1986) that the Hf in ferromanganese nodules is dominantly derived from sources in the mantle by chemical weathering of volcanic rocks erupted within the ocean basins. The Hf and Nd of mantle-derived volcanic rocks are both released into solution without isotopic fractionation because insufficient time has passed after eruption for the radiogenic isotopes to accumulate and because zircon is less abundant in basaltic rocks in the ocean basins than it is in the granitic rocks of the continental crust. In conclusion, the study of hydrogenetic ferromanganese nodules suggests that the isotope composition of Hf in solution in seawater varies regionally and is dominated by Hf that originated by weathering of mantle-derived mafic volcanic rocks.

19.7d Secular Variations

The terrestrial Hf–Nd array also manifests itself in the closely correlated secular variation of ε^0(Hf) and ε^0(Nd) of ferromanganese crusts. Time series of ε^0(Hf) values recovered from well-dated ferromanganese crusts have been published by a number of research groups, including Lee et al. (1999) and Piotrowski et al. (2000). In addition, David et al. (2001) demonstrated that the ^{176}Hf/^{177}Hf ratios of ferromanganese nodules are negatively correlated with ^{206}Pb/^{204}Pb ratios.

The ^{176}Hf/^{177}Hf ratios, corrected for decay of ^{176}Lu and expressed as ε^t(Hf), of a ferromanganese crust (BM 1969) from the San Pablo seamount at 39°0′ N, 60°57′ W in the northwest Atlantic Ocean define the profile in Figure 19.39, which extends from the present to the Late Cretaceous at about 80 Ma (Piotrowski et al., 2000). The ε^t(Hf) values at this site remained virtually constant at about +3 from the Late Cretaceous to the Oligocene Epoch. After declining to about +1.0 during the middle Miocene, the ε^t(Hf) value returned to +3 at the end of that epoch but declined sharply during the Plio/Pleistocene to about −1.0 at the present time.

The ε^t(Nd) values of the same crust also remained constant at −10.5 until early Oligocene and started a slow decline until the end of the Miocene Epoch, when the ε^t(Nd) values declined

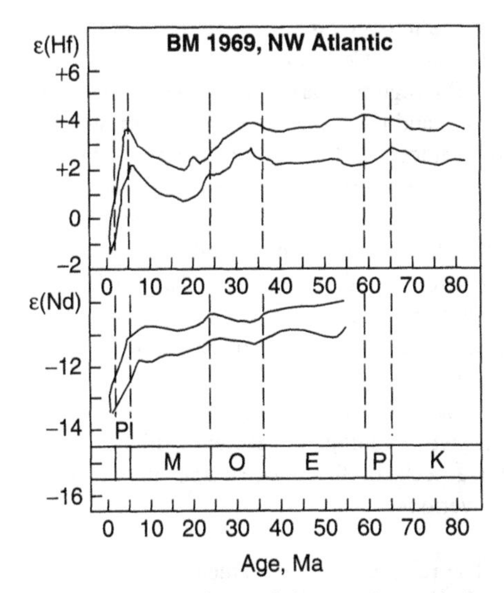

FIGURE 19.39 Comparison of the secular variations of ε^0(Hf) and ε^0(Nd) in a hydrogenetic ferromanganese crust (BM1969) from San Pablo seamount (39°0′ N, 60°57′ W) in the northwest Atlantic Ocean. Adapted from Piotrowski et al. (2000).

sharply to −13. Therefore, there appears to be a small mismatch between the Hf and Nd isotope profiles at this site during the Oligocene and Miocene Epochs followed in both cases by a decline of the ε^0-values during the past four million years.

Piotrowski et al. (2000) considered that the changes in the ε^0-values of Hf and Nd in the northwest Atlantic Ocean could have been caused either by changes in the deep-water circulation or by the continental glaciation of the northern hemisphere. The glaciation of North America and Scandinavia exposed Precambrian granitoids to weathering that would have released unradiogenic Hf and Nd into the oceans. The decoupling of the isotopic evolution of Hf and Nd, which is most evident between 15 and 4 Ma, occurred because the ε^0-values of Hf entering the oceans were rising whereas the ε^0-values of Nd were decreasing. The reason, noted before, is the "zircon effect" (i.e., the resistance to weathering of zircon prevents unradiogenic Hf from going into solution). The unradiogenic Hf of zircons was released only when large volumes of finely ground Precambrian granitoids became exposed to weathering as a consequence of full-scale continental glaciation. This phenomenon does not occur in the Sm–Nd system because the Nd-bearing minerals in igneous and metamorphic rocks have similar susceptibilities to chemical weathering. Therefore, Nd is not fractionated isotopically by the selective weathering of minerals.

The ε^t(Hf) profile of a ferromanganese crust (VA13/2) collected in the Central Pacific Ocean (9°18′ N, 146°03′ W) is compared in Figure 19.40 to a profile of $^{206}Pb/^{204}Pb$ ratios (David et al., 2001). The negative correlation of the isotope compositions of Hf and Pb in this crust is not well expressed. Although ε^t(Hf) declined with time while

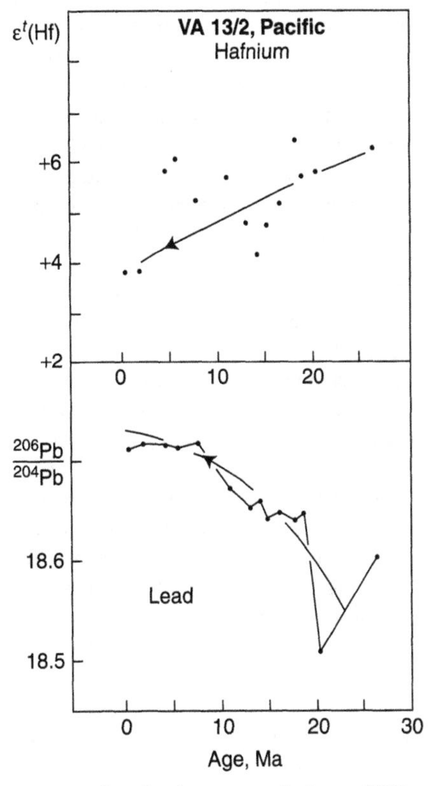

FIGURE 19.40 Secular isotope variations of Hf and Pb in a hydrogenetic ferromanganese crust (VA 13/2) in the central Pacific Ocean (see also Figure 19.21). Data from David et al. (2001).

the $^{206}Pb/^{204}Pb$ ratio increased, the two profiles differ substantially in detail. A comparison of the $\varepsilon^t(Hf)$ profile of crust VA13/2 in Figure 19.40 to the profile of $^{143}Nd/^{144}Nd$ ratios of the same crust in Figure 19.22 (Ling et al., 1997) indicates that the isotope composition of Hf is better correlated with the $^{143}Nd/^{144}Nd$ ratios than with the $^{206}Pb/^{204}Pb$ ratios.

19.8 SUMMARY

The study of the isotope compositions of Sr, Nd, Pb, Os, and Hf in the oceans reveals how geochemical processes on the surface of the Earth affect these elements. The elements enter the geochemical transport system when they are released into solution by the chemical weathering of the minerals in which they occur. Varying fractions of the resulting ions are sorbed to the surfaces of suspended mineral and organic particles at near-neutral pH and thereby become vulnerable to deposition in lakes and estuaries. The concentrations of ions remaining in solution are significantly reduced by this process, which diminishes the flux of these elements to the oceans.

The concentrations of the elements in seawater are affected by the balance between their annual inputs and outputs from the oceans. The resulting magnitudes of the mean oceanic residence times determine the extent to which the elements are isotopically homogenized by the circulation of the oceans on a timescale of about 10^3 years. Elements with long residence times (e.g., Sr and Os) have constant isotope compositions throughout the oceans of the world. Elements having short residence times (e.g., Nd, Pb, Hf) have isotopic compositions that vary regionally and reflect the ages and parent–daughter ratios of their sources.

Strontium and, to a lesser extent, Os record past changes in their isotope compositions that are preserved in marine carbonates (Sr) and ferromanganese deposits (Os). The isotopic evolution of Sr in the oceans is a record of global geological activity during the Phanerozoic and Proterozoic Eons. In addition, the time-dependent variation of $^{87}Sr/^{86}Sr$ ratios permits precise age determinations of marine carbonates of Cenozoic age.

The regional variability of the isotope ratios of Nd, Pb, and Hf in the oceans makes these elements useful tracers in the study of the circulation of the oceans. In addition, these elements can be used to identify their sources on the continents and within the ocean basins. In all cases, the isotopic compositions of Sr, Nd, Pb, Os, and Hf are the result of mixing of two or more isotopic varieties of these elements that originate from different sources.

The study of the migration of elements having stable radiogenic isotopes illuminates several aspects of this process that might otherwise be overlooked:

1. The release of these elements into solution during chemical weathering of their host minerals in rocks, regolith, soil, and sediment can cause isotopic fractionation because of differences in the susceptibilities to weathering of these host minerals (e.g., Pb and Hf in zircon, Sr in muscovite, Nd in garnet, Os in magnetite).
2. Terrestrial and extraterrestrial dust and atmospheric aerosol particles contribute significantly to the inventories of certain elements in solution in the oceans (e.g., Pb and Os).
3. The concentrations of trace elements in aqueous solutions are strongly dependent on sorption of cations (and anions) to electrically charged sites on the surfaces of suspended mineral and organic particles and to colloidal particles whose diameters are less than 0.45 μm. Consequently, the environmental pH, the presence of various kinds of sorbents, and the collection and subsequent treatment of water samples prior to analysis all have a profound effect on the concentrations that are measured.
4. Lakes, reservoirs, and estuaries collect sediment in suspension in streams and thereby prevent the sorbed fractions of all elements from entering the oceans.
5. In some cases, the isotopic compositions and concentrations of certain elements (e.g., Pb and Os) in solution in seawater or sorbed to pelagic clay and ferromanganese oxyhydroxides are effective tracers of anthropogenic contaminants in the oceans.

REFERENCES FOR STRONTIUM IN THE PHANEROZOIC OCEANS (SECTION 19.1)

Babcock, L. E., W. Zhang, and S. A. Leslie, 2001. The Chengjiang biota: Record of the Early Cambrian diversification of life and clues to exceptional preservation of fossils. *GSA Today*, 11(2):4–9.

Bailey, T. R., J. M. McArthur, H. Prince, and M. F. Thirlwall, 2000. Dissolution methods for strontium isotope stratigraphy: Whole-rock analysis. *Chem. Geol.*, 167:313–319.

Barrera, E., J. Barron, and A. N. Halliday, 1991. Strontium isotope stratigraphy of the Oligocene–Lower Miocene section at site 744, southern Indian Ocean. In J. Barron et al. (Eds.), *Proc. Ocean Drilling Program, Scientific Results*, 119:731–738.

Berggren, W. A., D. V. Kent, C. C. Swisher, and M.-P. Aubrey, 1995. A revised Cenozoic geochronology and chronostratigraphy. In *Geochronology, Time Scales and Global Stratigraphic Correlation*, SEPM Spec. Publ. 54, pp. 129–212. Soc. Sed. Geol., Tulsa, Oklahoma.

Bernat, M., T. Church, and C. J. Allégre, 1972. Barium and strontium concentrations in Pacific and Mediterranean seawater profiles by direct isotope dilution mass spectrometry. *Earth Planet. Sci. Lett.*, 16:75–80.

Berner, R. A., 1991. A model for atmosphere CO_2 over Phanerozoic time. *Am. J. Sci.*, 291:339–375.

Berner, R. A., and D. M Rye, 1992. Calculation of the Phanerozoic strontium isotope record of the oceans from a carbon cycle model. *Am. J. Sci.*, 292:136–148.

Bowring, S. A., J. P. Grotzinger, C. E. Isachsen, A. H. Knoll, S. Pelechaty, and P. Kolosov, 1993. Calibrating rates of Early Cambrian evolution. *Science*, 261:1293–1298.

Brass, G. W., 1976. The variation of the marine $^{87}Sr/^{86}Sr$ ratio during Phanerozoic time: Interpretation using a flux model. *Geochim. Cosmochim. Acta*, 40:721–730.

Brass, G. W., and K. K. Turekian, 1972. Strontium distribution of seawater profiles from Geosecs I (Pacific) and Geosecs II (Atlantic) test stations. *Earth Planet. Sci. Lett.*, 16:117–121.

Brass, G. W., and K. K. Turekian, 1974. Strontium distribution in GEOSECS oceanic profiles. *Earth Planet. Sci. Lett.*, 23:141–148.

Burke, W. H., R. E. Denison, E. A. Hetherington, R. B. Koepnick, H. F. Nelson, and J. B. Otto, 1982. Variation of seawater $^{87}Sr/^{86}Sr$ throughout Phanerozoic time. *Geology*, 10(10):516–519.

Cande, S. C., and D. V. Kent, 1995. Revised calibration of the geomagnetic polarity timescale for the Late Cretaceous and Cenozoic. *J. Geophys. Res.*, 100:6093–6095.

Carpenter, S. J., K. C. Lohmann, P. Holden, L. M. Walter, T. J. Huston, and A. N. Halliday, 1991. $\delta^{18}O$ values, $^{87}Sr/^{86}Sr$, and Sr/Mg ratios of Late Devonian abiotic marine calcite: Implications for the composition of ancient seawater. *Geochim. Cosmochim. Acta*, 55:1991–2010.

DePaolo, D. J., and B. L. Ingram, 1985. High-resolution stratigraphy with strontium isotopes. *Science*, 227: 938–941.

Derry, L. A., M. D. Brasier, R. M. Corfield, A. Y. Rozanov, and A. Y. Zhuravlev, 1994. Sr and C isotopes in Lower Cambrian carbonates from the Siberian craton: A paleoenvironmental record during the "Cambrian explosion." *Earth Planet. Sci. Lett.*, 128:671–681.

Elderfield, H., 1986. Strontium isotope stratigraphy. *Paleo. Paleo. Paleo*, 57:71–90.

Farrell, J. W., S. C. Stevens, and L. P. Gromet, 1995. Improved chronostratigraphic reference curve of late Neogene seawater $^{87}Sr/^{86}Sr$. *Geology*, 23:403–406.

Faure, G., 1977. *Principles of Isotope Geology*, 1st ed. Wiley, New York.

Faure, G., 1982. The marine-strontium geochronometer. In G. S. Odin (Ed.), *Numerical Dating in Stratigraphy*, Vol. 1, pp. 73–79. Wiley, Chichester.

Faure, G., J. H. Crocket, and P. M. Hurley, 1967. Some aspects of the geochemistry of strontium and calcium in the Hudson Bay and the Great Lakes. *Geochim. Cosmochim. Acta*, 31:451–461.

Faure, G., P. M. Hurley, and J. L. Powell, 1965. The isotopic composition of strontium in surface water from the North Atlantic Ocean. *Geochim. Cosmochim. Acta*, 29:209–220.

Harland, W. B., R. L. Armstrong, A. V. Cox, L. E. Craig, A. G. Smith, and D. G. Smith, 1989. *A Geologic Time Scale 1989*. Cambridge University Press, Cambridge.

Hess, J., M. L. Bender, and J.-G. Schilling, 1986. Evolution of the ratio of strontium-87 to strontium-86 in seawater from Cretaceous to present. *Science*, 231:979–983.

Hess, J., L. D. Stott, M. L. Bender, J. P. Kennet, and J.-G. Schilling, 1989. The Oligocene marine microfossil record: Age assessments using strontium isotopes. *Paleoceanography*, 4:655–679.

Hodell, D. A., G. A. Mead, and P. A. Mueller, 1990. Variation in the strontium isotopic composition of seawater (8 Ma to present): Implications for chemical weathering rates and dissolved fluxes to the oceans. *Chem. Geol. (Isotope Geosci.)*, 80:291–307.

Hodell, D. A., P. A. Mueller, and R. J. Garrido, 1991. Variations in the strontium isotopic composition of seawater during the Neogene. *Geology*, 19:24–27.

Hodell, D. A., P. A. Mueller, J. A. McKenzie, and G. A. Mead, 1989. Strontium isotope stratigraphy and geochemistry of the Late Neogene Ocean. *Earth Planet. Sci. Lett.*, 92:165–178.

Jacobsen, S. B., and A. J. Kaufman, 1999. The Sr, C, and O isotopic evolution of Neoproterozoic seawater. *Chem. Geol.*, 161:37–57.

Kump, L. R., 1989. Alternative modeling approaches to the geochemical cycles of carbon, sulfur, and strontium isotopes. *Am. J. Sci.*, 289:390–410.

Ludwig, K. R., R. B. Halley, K. R. Simmons, and Z. E. Peterman, 1988. Strontium-isotope stratigraphy of Enewetak atoll. *Geology*, 16:173–177.

McArthur, J. M., R. J. Howarth, and T. R. Bailey, 2001. Strontium isotope stratigraphy: LOWESS version 3: Best fit to the marine Sr-isotope curve for 0–509 Ma and accompanying look-up table for deriving numerical age. *J. Geol.*, 109:155–170.

Montañez, I. P., J. L. Banner, D. A. Osleger, L. E. Borg, and P. J. Bosserman, 1996. Integrated Sr isotope variations and sea-level history of Middle to Upper Cambrian platform carbonates; implications for the evolution of Cambrian seawater $^{87}Sr/^{86}Sr$. *Geology*, 24:917–920.

Palmer, M. R., and J. M Edmond, 1989. The strontium isotope budget of the modern ocean. *Earth Planet. Sci. Lett.*, 92:11–26.

Peterman, Z. E., C. E. Hedge, and H. A. Tourtelot, 1970. Isotopic composition of strontium in seawater throughout Phanerozoic time. *Geochim. Cosmochim. Acta*, 45:105–120.

Ross, D. A., 1970. *Introduction to Oceanography*. Meredith, New York.

Shackleton, N. J., S. Crowhurst, R. Hagelberg, N. G. Pisias, and D. A. Schneider, 1995. A new late Neogene time scale: Application of Leg 138 sites. In *Proceedings, Ocean Drilling Program. Scientific Results*, 138. College Station, Texas.

Smalley, P. C., A. C. Higgins, R. J. Howarth, H. Nicholson, and C. E. Jones, 1994. Seawater Sr isotope variations through time: A procedure for constructing a reference curve to date and correlate marine sedimentary rocks. *Geology*, 22:431–434.

Sverdrup, H. U., M. W. Johnson, and R. A. Fleming, 1942. *The Oceans: Their Physics, Chemistry, and General Biology*. Prentice-Hall, Englewood Cliffs, New Jersey.

Taylor, S. R., and S. M. McLennan, 1985. *The Continental Crust: Its Composition and Evolution*. Blackwell Scientific, Oxford.

Veizer, J., and W. Compston, 1974. $^{87}Sr/^{86}Sr$ composition of seawater during the Phanerozoic. *Geochim. Cosmochim. Acta*, 38:1461–1484.

Wickman, F. E., 1948. Isotope ratios: A clue to the age of certain marine sediments. *J. Geol.*, 56:61–66.

REFERENCES FOR STRONTIUM IN PRECAMBRIAN CARBONATES (SECTION 19.2)

Asmerom, Y., S. B. Jacobsen, A. H. Knoll, N. J. Butterfield, and K. Swett, 1991. Strontium isotopic variations of Neoproterozoic seawater: Implications for crustal evolution. *Geochim. Cosmochim. Acta*, 55:2883–2894.

Brand, U., and J. Veizer, 1980. Chemical diagenesis of multicomponent carbonate systems; 1. Trace elements. *J. Sediment. Petrol.*, 50:1219–1236.

Deb, M., J. Hoefs, and A. Baumann, 1991. Isotopic composition of two Precambrian stratiform barite deposits from the Indian Shield. *Geochim. Cosmochim. Acta*, 55:303–308.

Derry, L. A., M. D. Brasier, R. M. Corfield, A. Y. Rozanov, and A. Y. Zhuravlev, 1994. Sr and C isotopes in lower Cambrian carbonates from the Siberian craton: A paleoenvironmental record during the "Cambrian explosion." *Earth Planet. Sci. Lett.*, 128:671–681.

Derry, L. A., A. J. Kaufman, and S. B. Jacobsen, 1992. Sedimentary cycling and environmental change in the Late Proterozoic: Evidence from stable and radiogenic isotopes. *Geochim. Cosmochim. Acta*, 56:1317–1329.

Derry, L. A., L. S. Keto, S. B. Jacobsen, A. H. Knoll, and K. Swett, 1989. Sr isotopic variations in Upper Proterozoic carbonates from Svalbard and East Greenland. *Geochim. Cosmochim. Acta*, 53:2331–2339.

Gorokhov, I. M., A. B. Kuznetsov, V. A. Melezhik, G. V. Konstantinova, and N. N. Mel'nikov, 1998. Sr isotopic composition in the Upper Jatulian (Early Paleoproterozoic) dolomites of the Tulomozero Formation, southeast Karelia. *Doklady Russ. Acad. Sci. (Earth Sci. Sect.)*, 360(4):609–612 (translated from Russian into English by Interperiodica Publishing).

Gorokhov, I. M., M. A. Semikhatov, A. V. Baskakov, E. P. Kutyavin, N. N. Mel'nikov, A. V. Sochava, and T. L. Turchenko, 1995. Sr isotopic composition in Riphean, Vendian, and Lower Cambrian carbonates of Siberia. *Stratigr. Geol. Correlation*, 3(1):1–28 (translated from Russian into English by Interperiodica Publishing).

Gorokhov, I. M., M. A. Semikhatov, A. B. Kuznetsov, and N. N. Mel'nikov, 1996. Improved reference curve of Late Proterozoic seawater $^{87}Sr/^{86}Sr$. In S. H. Bottrell et al. (Eds.), *Proceedings of the Fourth International*

Symposium on the Geochemistry of the Earth's Surface, Ilkley, Yorkshire. Theme 5: Land-Hydrosphere Interactions, pp. 714–717.

Hoffman, P. F., A. J. Kaufman, G. P. Halverson, and D. P. Schrag, 1998. A Neoproterozoic snowball Earth. *Science*, 281:1342–1346.

Jacobsen, S. B., and A. J. Kaufman, 1999. The Sr, C, and O isotopic evolution of Neoproterozoic seawater. *Chem. Geol.*, 161:37–57.

Kaufman, A. J., S. B. Jacobsen, and A. H. Knoll, 1993. The Vendian record of Sr and C isotopic variations in seawater: Implications for tectonics and paleoclimate. *Earth Planet. Sci. Lett.*, 120:409–430.

Kaufman, A. J., A. H. Knoll, M. A. Semikhatov, J. P. Grotzinger, S. B. Jacobsen, and W. Adams, 1996. Integrated chronostratigraphy of Proterozoic–Cambrian boundary beds in the western Anbar region, northern Siberia. *Geol. Mag.*, 133:509–533.

Kennedy, M. J., N. Christie-Blick, and A. R. Prave, 2001a. Carbon isotopic composition of Neoproterozoic glacial carbonates as a test of paleoceanographic models for snowball Earth phenomena. *Geology*, 29(12):1135–1138.

Kennedy, M. J., N. Christie-Blick, and L. E. Sohl, 2001b. Are Proterozoic cap carbonates and isotopic excursions a record of gas hydrate destabilization following Earth's coldest intervals. *Geology*, 29(5):443–446.

Mirota, M. D., and J. Veizer, 1994. Geochemistry of Precambrian carbonates. VI Aphebian Albanel Formations, Quebec, Canada. *Geochim. Cosmochim. Acta*, 58:1735–1745.

Montañez, I. P., J. L. Banner, D. A. Osleger, L. E. Borg, and P. J. Bosserman, 1996. Integrated Sr isotope variations and sea level isotope variations and sea level history of Middle to Upper Cambrian platform carbonates: Implications for the evolution of Cambrian seawater $^{87}Sr/^{86}Sr$. *Geology*, 24:917–920.

Ray, J. S., M. W. Martin, J. Veizer, and S. A. Bowring, 2002. U-Pb zircon dating and Sr isotope systematics of the Vindhyan Supergroup, India. *Geology*, 30(2):131–134.

Shields, G., and J. Veizer, 2001. Precambrian seawater isotopic record (Abstract). In *Workshop on Geochemical Earth Reference Model*, pp. 50–52, LaJolla, CA.

Veizer, J., 1989. Strontium isotopes in seawater through time. *Annu. Rev. Earth Planet. Sci.*, 17:141–167.

Veizer, J., R. N. Clayton, and R. W. Hinton, 1992a. Geochemistry of Precambrian carbonates: IV. Early Paleoproterozoic (2.25 ± 0.25 Ga) seawater. *Geochim. Cosmochim. Acta*, 56:875–885.

Veizer, J., R. N. Clayton, R. W. Hinton, V. von Brunn, T. R. Mason, S. G. Buck, and J. Hoefs, 1990. Geochemistry of Precambrian carbonates: III. Shelf seas and non-marine environments of the Archean. *Geochim. Cosmochim. Acta*, 54:2717–2729.

Veizer, J., and W. Compston, 1976. $^{87}Sr/^{86}Sr$ in Precambrian carbonates as an index of crustal evolution. *Geochim. Cosmochim. Acta*, 40:905–914.

Veizer, J., W. Compston, N. Clauer, and M. Schidlowski, 1983. $^{87}Sr/^{86}Sr$ in Late Proterozoic carbonates: Evidence for a "mantle" event at 900 Ma ago. *Geochim. Cosmochim. Acta*, 47:295–302.

Veizer, J., W. Compston, J. Hoefs, and H. Nielsen, 1982. Mantle buffering of the early ocean. *Naturwissenschaften*, 69:173–180.

Veizer, J., J. Hoefs, D. R. Lowe, and P. C. Thurston, 1989a. Geochemistry of Precambrian carbonates: II. Archean greenstone belts and Archean seawater. *Geochim. Cosmochim. Acta*, 53:859–871.

Veizer, J., J. Hoefs, R. H. Riddler, L. S. Jensen, and D. R. Lowe, 1989b. Geochemistry of Precambrian carbonates: I. Archean hydrothermal systems. *Geochim. Cosmochim. Acta*, 53:845–857.

Veizer, J., K. A. Plumb, R. N. Clayton, R. W. Hinton, and J. P. Grotzinger, 1992b. Geochemistry of Precambrian carbonates: V. Late Paleoproterozoic seawater. *Geochim. Cosmochim. Acta*, 56:2487–2503.

Zachariah, J. K., 1998. A 3.1 billion year old marble and the $^{87}Sr/^{86}Sr$ of Late Archean seawater. *Terra Nova*, 10:312–316.

REFERENCES FOR Nd: ENVIRONMENTAL GEOCHEMISTRY (SECTIONS 19.3a–19.3c)

Åberg, G., and F. E. Wickman, 1987. Variations of $^{87}Sr/^{86}Sr$ in water from streams discharging into the Bothnian Bay, Baltic Sea. *Nordic Hydrol.*, 18:33–42.

Albarède, F., and S. L. Goldstein, 1992. A world map of Nd isotopes in seafloor ferromanganese deposits. *Geology*, 20:761–763.

Andersson, P. S., G. J. Wasserburg, and J. Ingri, 1992. The sources and transport of Sr and Nd isotopes in the Baltic Sea. *Earth Planet. Sci. Lett.*, 113:459–472.

Goldstein, S. J., and S. B. Jacobsen, 1987. The Nd and Sr isotopic systematics of river-water dissolved material: Implications for the sources of Nd and Sr in seawater. *Chem. Geol. (Isotope Geosci. Sect.)*, 66:245–272.

Löfvendahl, R., G. Åberg, and P. J. Hamilton, 1990. Strontium in rivers of the Baltic basin. *Aquatic Sci.*, 52(4):315–329.

McCulloch, M. T., and G. J. Wasserburg, 1978. Sm-Nd and Rb-Sr chronology of continental crust formation. *Science*, 200:1003–1011.

O'Nions, R. K., S. R. Carter, R. S. Cohen, N. M. Evensen, and P. J. Hamilton, 1978. Pb, Nd, and Sr isotopes in oceanic ferromanganese deposits and ocean floor basalt. *Nature*, 273:435–438.

Piepgras, D. J., and G. J. Wasserburg, 1980. Neodymium isotopic variations in seawater. *Earth Planet. Sci. Lett.*, 50:128–138.

Piepgras, D. J., and G. J. Wasserburg, 1982. Isotopic composition of neodymium in water from the Drake Passage. *Science*, 217:207–214.

Piepgras, D. J., G. J. Wasserburg, and E. J. Dasch, 1979. The isotopic composition of Nd in different ocean masses. *Earth Planet. Sci. Lett.*, 45:223–236.

REFERENCES FOR Nd IN FERROMANGANESE NODULES AND CRUSTS (SECTION 19.3d)

Abouchami, W., S. L. Goldstein, S. J. Galer, A. Eisenhauer, and A. Mangini, 1997. Secular changes of lead and neodymium in central Pacific seawater recorded by a Fe-Mn crust. *Geochim. Cosmochim. Acta*, 61:3957–3974.

Chuey, J. M., D. K. Rea, and N. D. Pisias, 1987. Late Pleistocene paleoclimatology of the central equatorial Pacific: A quantitative record of eolian and carbonate deposition. *Quat. Res.*, 28:323–339.

Goldstein, S. L., and R. K. O'Nions, 1981. Nd and Sr isotopic relationships in pelagic clays and ferromanganese deposits. *Nature*, 292:324–327.

Ling, H. F., K. W. Burton, R. K. O'Nions, B. S. Kamber, F. von Blanckenburg, A. J. Gibb, and J. R. Hein, 1997. Evolution of Nd and Pb isotopes in Central Pacific seawater from ferromanganese crusts. *Earth Planet. Sci. Lett.*, 14:1–12.

O'Nions, R. K., S. R. Carter, R. S. Cohen, N. M. Evensen, and P. J. Hamilton, 1978. Pb, Nd, and Sr isotopes in oceanic ferromanganese deposits and ocean floor basalts. *Nature*, 273:435–438.

Palmer, M. R., and H. Elderfield, 1985. Sr isotope composition of seawater over the past 75 Myr. *Nature*, 314:526–528.

Palmer, M. R., and H. Elderfield, 1986. Rare earth elements and neodymium isotopes in ferromanganese oxide coatings of Cenozoic foraminifera from the Atlantic Ocean. *Geochim. Cosmochim. Acta*, 50:409–417.

Piepgras, D. J., G. J. Wasserburg, and E. J. Dasch, 1979. The isotopic composition of Nd in different ocean masses. *Earth Planet. Sci. Lett.*, 45:223–236.

REFERENCES FOR Nd: WATER-ROCK INTERACTION (OPHIOLITES) (SECTION 19.3e)

Bickle, M. J., and D. A. H. Teagle, 1992. Strontium alteration in the Troodos ophiolite: Implications for fluid fluxes and geochemical transport in mid-ocean ridge hydrothermal systems. *Earth Planet. Sci. Lett.*, 113:219–237.

Jacobsen, S. B., and G. J. Wasserburg, 1979. Nd and Sr isotopic study of the Bay of Islands ophiolite complex and the evolution of the source of midocean-ridge basalts. *J. Geophys. Res.*, 84(B3):7429–7445.

Lanphere, M. A., R. G. Coleman, and C. A. Hopson, 1981. Sr isotopic tracer study of the Samail ophiolite, Oman. *J. Geophys. Res.*, 86(B4):2709–2720.

McCulloch, M. T., R. T. Gregory, G. J. Wasserburg, and H. P. Taylor, Jr., 1980. A neodymium, strontium, and oxygen isotopic study of the Cretaceous Samail ophiolite and implications for the petrogenesis and seawater-hydrothermal alteration of oceanic crust. *Earth Planet. Sci. Lett.*, 46:201–211.

McCulloch, M. T., R. T. Gregory, G. J. Wasserburg, and H. P. Taylor, Jr., 1981. Sm-Nd, Rb-Sr, and $^{18}O/^{16}O$ isotopic systematics in an oceanic crustal section: Evidence from the Samail ophiolite. *J. Geophys. Res.*, 86(B4):2721–2735.

O'Nions, R. K., S. R. Carter, R. S. Cohen, N. M. Evensen, and P. J. Hamilton, 1978. Pb, Nd, and Sr isotopes in oceanic ferromanganese deposits and ocean floor basalts. *Nature*, 273:435–438.

REFERENCES FOR Pb: ENVIRONMENTAL GEOCHEMISTRY (SECTION 19.4 INTRODUCTION, SECTIONS 19.4a–19.4c)

Balistrieri, L. S., and J. W. Murray, 1982. The adsorption of Cu, Pb, Zn, and Cd on goethíte from major ion seawater. *Geochim. Cosmochim. Acta*, 46:1253–1265.

Bollhöfer, A., W. Chisholm, and K. J. R. Rosman, 1999. Sampling aerosols for lead isotopes on a global scale. *Anal. Chim. Acta*, 390:227–235.

Bollhöfer, A., and K. J. R. Rosman, 2000. Isotopic source signatures for atmospheric lead: The southern hemisphere. *Geochim. Cosmochim. Acta*, 64:3251–3262.

Bollhöfer, A., and K. J. R. Rosman, 2001. Isotopic source signatures for atmospheric lead: The northern hemisphere. *Geochim. Cosmochim. Acta*, 65:1727–1740.

Bollhöfer, A., and K. J. R. Rosman, 2002. The temporal stability in lead isotopic signatures at selected sites in the southern and northern hemispheres. *Geochim. Cosmochim. Acta*, 66:1375–1386.

Boutron, C. F., J.-P. Candelone, and S. Hong, 1994. Past and recent changes in the large-scale tropospheric cycles of lead and other heavy metals as documented in Antarctica and Greenland snow and ice. *Geochim. Cosmochim. Acta*, 58:3217–3225.

Chow, T. J., and M. S. Johnstone, 1965. Lead isotopes in gasoline and aerosols of Los Angeles Basin, California. *Science*, 147:502–503.

Dzombak, D. A., and F. Morel, 1990. *Surface Complexation Modeling: Hydrous Ferric Oxide*. Wiley, New York.

Faure, G., 1998. *Principles and Applications of Geochemistry*, 2nd ed., Prentice-Hall, Upper Saddle River, New Jersey.

Jones, C. E., A. N. Halliday, D. K. Rea, and R. M. Owen, 2000. Eolian inputs of lead to the North Pacific. *Geochim. Cosmochim. Acta*, 64:1405–1416.

Ketterer, M. E., J. H. Lowry, J. Simon, Jr., K. Humphries, and M. P. Novotnak, 2001. Lead isotopic and chalcophile element compositions in the environment near a zinc smelting-secondary zinc recovery facility, Palmerton, Pennsylvania, USA. *Appl. Geochem.*, 16:207–229.

Langmuir, D., 1997. *Aqueous Environmental Geochemistry*. Prentice-Hall, Upper Saddle River, New Jersey.

Lee, G., J. M. Bigham, and G. Faure, 2002. Removal of trace metals by coprecipitation with Fe, Al, Mn from natural waters contaminated with acid mine-drainage in the Ducktown Mining District, Tennessee. *Appl. Geochem.*, 17:569–581.

Liu, T. S., 1985. *Loess and the Environment*. China Ocean Press, Beijing, China.

Liu, A. S., 1993. *Guangdong Soils*. Science Press, Beijing, China (in Chinese).

Morioka, K. I., and T. N. Yashiro, 1991. Accumulation of Asian long-range eolian dust in Japan and Korea from the late Pleistocene to the Holocene. In S. Okuda A. Rapp, and L. Y. Zhang (Eds.), *Loess Geomorphologic Hazards and Processes*, Vol. 20 (Suppl.), 25–42. CATENA, Cremlingen.

Munk, L.-A., G. Faure, D. E. Pride, and J. M. Bigham, 2002. Sorption of trace metals to an aluminum precipitate in a stream receiving acid rock-drainage; Snake River, Summit County, Colorado. *Appl. Geochem.*, 17:421–430.

Murozumi, M., T. J. Chow, and C. C. Patterson, 1969. Chemical concentrations of pollutant lead aerosols, terrestrial dusts, and sea salts in Greenland and Antarctic snow strata. *Geochim. Cosmochim. Acta*, 33:1247–1294.

Nicolas, E., D. Ruiz-Pino, P. Buat-Menard, and J. P. Bethoux, 1994. Abrupt decrease of lead concentration in the Mediterranean Sea: A response to antipollution policy. *Geophys. Res. Lett.*, 21:2119–2122.

Nordstrom, D. K., and C. N. Alpers, 1999. Geochemistry of acid mine waters. In G. S. Plumlee and M. J. Logsdon (Eds.), *The Environmental Geochemistry of Mineral Deposits. Rev. Econ. Geol.*, 6A:133–160. Econ. Geol., Littleton, Colorado.

Nriagu, J. O., 1978a. Lead in the atmosphere. In J. O. Nriagu (Ed.), *The Biogeochemistry of Lead in the Environment*, Part A, pp. 137–184. Elsevier, Amsterdam.

Nriagu, J. O. (Ed.), 1978b. *The Biochemistry of Lead in the Environment*, Parts A and B. Elsevier, Amsterdam.

Nriagu, J. O., 1983a. *Lead and Lead Poisoning in Antiquity*. Wiley, New York.

Nriagu, J. O., 1983b. Saturnine gout among Roman aristocrats. *N. Engl. J. Med.*, 308:660–663.

Patterson, C. C., 1965. Contaminated and natural lead environments of man. *Arch. Environ. Health*, 11:344–360.

Schemel, L. E., B. A. Kimball, and K. E. Bencala, 2000. Colloid formation and metal transport through two mixing zones affected by acid mine drainage near Silverton, Colorado. *Appl. Geochem.*, 15:1003–1018.

Simonetti, A., C. Gariépy, and J. Carignan, 2000. Pb and Sr isotopic evidence for sources of atmospheric heavy metals and their deposition budgets in northeastern North America. *Geochim. Cosmochim. Acta*, 64:3439–3452.

Stumm, W., 1992. *Chemistry of the Solid-Water Interface*. Wiley, New York.

Stumm, W., and J. J. Morgan, 1996. *Aquatic Chemistry*, 3rd ed. Wiley, New York.

Tatsumoto, M., and C. C. Patterson, 1963a. The concentration of common lead in seawater. In J. Geiss and E. D. Goldberg (Eds.), *Earth Science and Meteoritics*, pp. 74–89. North-Holland, Amsterdam.

Tatsumoto, M., and C. C. Patterson, 1963b. Concentrations of common lead in some Atlantic and Mediterranean waters and in snow. *Nature*, 199:350–352.

Taylor, S. R., and S. M. McLennan, 1985. *The Continental Crust: Its Composition and Evolution*. Blackwell Scientific, Oxford.

Wu, J.-F., and E. A. Boyle, 1997. Lead in the western North Atlantic; completed response to leaded gasoline phaseout. *Geochim. Cosmochim. Acta*, 61:3279–3283.

Zhu, B.-Q., Y.-W. Chen, and J. H. Peng, 2001. Lead isotope geochemistry of the urban environment in the Pearl River delta. *Appl. Geochem.*, 16:409–417.

REFERENCES FOR Pb IN FERROMANGANESE CRUSTS (SECTION 19.4d)

Abouchami, W., S. L. Goldstein, S. J. G. Galer, A. Eisenhauer, and A. Mangini, 1997. Secular changes of lead and neodymium in central Pacific seawater recorded by a Fe-Mn crust. *Geochim. Cosmochim. Acta*, 61:3957–3974.

Doe, B. R., R. A. Ayuso, K. Futa, and Z. E. Peterman, 1996. Evaluation of the sedimentary manganese deposits of Mexico and Morocco for determining lead and strontium isotopes in ancient seawater. In A. Basu and S. R. Hart (Eds.), *Earth Processes; Reading the Isotopic Code*, Geophys. Monogr. 95, pp. 391–408. American Geophysical Union, Washington, D.C.

Frank, M., and R. K. O'Nions, 1998. Sources of Pb for Indian Ocean ferromanganese crusts: A record of Himalayan erosion? *Earth Planet. Sci. Lett.*, 158:121–130.

Ling, H. F., K. W. Burton, R. K. O'Nions, B. S. Kamber, F. von Blanckenburg, A. J. Gibb, and J. R. Hein, 1997. Evolution of Nd and Pb in Central Pacific seawater from ferromanganese crusts. *Earth Planet. Sci. Lett.*, 146:1–12.

O'Nions, R. K., S. R. Carter, R. S. Cohen, N. M. Evensen, and P. J. Hamilton, 1978. Pb, Nd, and Sr isotopes in oceanic ferromanganese deposits and ocean floor basalt. *Nature*, 273:435–438.

Vlastelic, I., W. Abouchami, S. J. G. Galer, and A. W. Hofmann, 2001. Geographic control on Pb isotope distribution and sources in Indian Ocean Fe-Mn deposits. *Geochim. Cosmochim. Acta*, 65:4304–4319.

Wedepohl, K. H., 1974. Abundances of lead in common sediments and sedimentary rocks. In K. H. Wedepohl (ed.), *Handbook of Geochemistry*, 82-K-6. Springer-Verlag, Heidelberg.

REFERENCES OF OSMIUM (SECTIONS 19.5–19.6)

Burton, K. W., B. Bourdon, J.-L. Birck, C. J. Allègre, and J. R. Hein, 1999. Osmium isotope variations in the oceans recorded by Fe-Mn crusts. *Earth Planet. Sci. Lett.*, 171:185–197.

Esser, B. K., and K. K. Turekian, 1993a. The osmium isotopic composition of the continental crust. *Geochim. Cosmochim. Acta*, 57:3093–3104.

Esser, B. K., and K. K. Turekian, 1993b. Anthropogenic osmium in coastal deposits. *Environ. Sci. Technol.*, 27:2719–2724.

Esser, B. K., and K. K. Turekian, 1988. Accretion rate of extraterrestrial particles determined from osmium isotope systematics of Pacific pelagic clay and manganese nodules. *Geochim. Cosmochim. Acta*, 52:1383–1388.

Koide, M., E. D. Goldberg, and R. J. Walker, 1996. The analysis of seawater osmium. *Deep Sea Res. II*, 43:53–55.

Levasseur, S., J.-L. Birck, and C. J. Allégre, 1998. Direct measurement of femtomoles of osmium and the $^{187}Os/^{186}Os$ ratio in seawater. *Science*, 282:272–274.

Luck, J.-M., and K. K. Turekian, 1983. Osmium-187/osmium-186 in manganese nodules and the Cretaceous-Tertiary boundary. *Science*, 222:613–615.

Martin, C. E., 1991. Osmium isotopic characteristics of mantle-derived rocks. *Geochim. Cosmochim. Acta*, 55:1421–1434.

Oxburgh, R., 1998. Variations in the osmium isotope composition of seawater over the past 200,000 years. *Earth Planet. Sci. Lett.*, 159:183–191.

Pegram, W. J., B. K. Esser, S. Krishnaswami, and K. K. Turekian, 1994. The isotopic composition of leachable osmium from river sediments. *Earth Planet. Sci. Lett.*, 129:592–599.

Pegram, W. J., S. Krishnaswami, G. E. Ravizza, and K. K. Turekian, 1992. The record of seawater $^{187}Os/^{186}Os$ variation through the Cenozoic. *Earth Planet Sci. Lett.*, 113:569–576.

Peucker-Ehrenbrink, B., and J. D. Blum, 1998. Re-Os isotope systematics and weathering of Precambrian crustal rocks: Implications for the marine osmium isotope record. *Geochim. Cosmochim. Acta*, 62:3193–3204.

Peucker-Ehrenbrink, B., and G. Ravizza, 1996. Continental runoff of osmium into the Baltic Sea. *Geology*, 24:327–330.

Peucker-Ehrenbrink, B., G. E. Ravizza, and A. W. Hofmann, 1995. The marine $^{187}Os/^{186}Os$ record of the past 80 million years. *Earth Planet. Sci. Lett.*, 130:155–167.

Pierson-Wickmann, A.-C., L. Reisberg, and C. France-Lanord, 2002. Behavior of Re and Os during low-temperature alteration: Results from Himalayan soils and altered black shales. *Geochim. Cosmochim. Acta*, 66:1539–1548.

Ravizza, G., and M. H. Bothner, 1996. Osmium isotopes and silver as tracers of anthropogenic metals in sediments from Massachusetts and Cape Cod bays. *Geochim. Cosmochim. Acta*, 60:2753–2763.

Ravizza, G. E., 1993. Variations in the $^{187}Os/^{186}Os$ of seawater over the past 28 million years as inferred from metalliferous carbonates. *Earth Planet. Sci. Lett.*, 118:335–348.

Ravizza, G. E., and K. K. Turekian, 1992. The osmium isotopic composition or organic-rich marine sediments. *Earth Planet. Sci. Lett.*, 110:1–6.

Reese, R. G., 1996. Platinum-group metals. In *Minerals Year Book*, Vol. 1, pp. 611–613. U.S. Bureau of Mines, U.S. Government Printing Office, Washington, D.C.

Reisberg, L., C. France-Lanord, and A. C. Pierson-Wickmann, 1997. Osmium isotopic composition of leachates and bulk sediment from the Bengal fan. *Earth Planet. Sci. Lett.*, 150:117–127.

Sharma, M., D. A. Papanastassiou, and G. J. Wasserburg, 1997. The concentration and isotopic composition of osmium in the oceans. *Geochim. Cosmochim. Acta*, 61:3287–3300.

Sharma, M., and G. J. Wasserburg, 1997. Osmium in the rivers. *Geochim. Cosmochim. Acta*, 61:5411–5417.

Williams, G., F. Marcantonio, and K. K. Turekian, 1997. The behavior of natural and anthropogenic osmium in Long Island Sound, an urban estuary in the eastern US. *Earth Planet Sci. Lett.*, 148:341–347.

REFERENCES FOR HAFNIUM (SECTION 19.7)

Albarède, F., A. Simonetti, J. D. Vervoort, J. Blichert-Toft, and W. Abouchami, 1998. A. Hf-Nd isotopic correlation in ferromanganese nodules. *Geophys. Res. Lett.*, 20:3895–3898.

Blichert-Toft, J., C. Chauvel, and F. Albarède, 1997. Separation of Hf and Lu for high precision isotope analysis of rock samples by magnetic sector multiple-collector ICP-MS. *Contrib. Mineral. Petrol.*, 127:248–260.

Boswell, S. M., and H. Elderfield, 1998. The determination of Zr and Hf in natural water using isotope dilution mass spectrometry. *Marine Chem.*, 25:197–210.

David, K., M. Frank, R. K. O'Nions, N. S. Belshaw, and J. W. Arden, 2001. The Hf isotopic composition of global seawater and the evolution of Hf isotopes in the deep Pacific Ocean from Fe-Mn crusts. *Chem. Geol.*, 178:23–42.

Godfrey, L. V., D.-C. Lee, W. F. Sangrey, A. N. Halliday, V. J. M. Salters, J. R. Hein, and W. M. White, 1997. The Hf isotopic composition of ferromanganese nodules and crusts and hydrothermal manganese deposits: Implications for seawater Hf. *Earth Planet. Sci. Lett.*, 151:91–105.

Godfrey, L. V., W. M. White, and V. J. M. Salters, 1996. Dissolved zirconium and hafnium distributions across a shelf break in the northeastern Atlantic Ocean. *Geochim Cosmochim. Acta*, 60:3995–4006.

Lee, D. C., A. N. Halliday, J. R. Hein, K. W. Burton, and J. N. Christensen, 1999. High resolution Hf isotope stratigraphy of Fe-Mn crusts. *Science*, 285:1052–1054.

Ling, H. F., K. W. Burton, R. K. O'Nions, B. S. Kamber, F. von Blanckenburg, A. J. Gibb, and J. R. Hein, 1997. Evolution of Nd and Pb isotopes in Central Pacific seawater from ferromanganese crusts. *Earth Planet. Sci. Lett.*, 14:1–12.

McKelvey, B. A., and K. J. Orians, 1998. The determination of dissolved zirconium and hafnium from seawater using isotope dilution inductively coupled plasma mass spectrometry. *Marine Chem.*, 60:245–255.

Patchett, J. P., 1983. Importance of the Lu-Hf isotopic system in studies of planetary chronology and chemical evolution. *Geochim. Cosmochim. Acta*, 47:81–91.

Pettke, T., A. N. Halliday, D. C. Lee, and D. K. Rea, 1998. Extreme variability in Hf isotopic components of aeolian dust and its implications for seawater Hf. *Mineral Mag.*, 62A:1165–1166.

Piotrowski, A. M., D. C. Lee, A. N. Halliday, J. N. Christensen, and J. R. Hein, 2000. Changes in erosion and ocean circulation recorded in the Hf isotopic composition of North Atlantic and Indian Ocean ferromanganese crusts. *Earth Planet. Sci. Lett.*, 181:315–325.

Vervoort, J. D., P. J. Patchett, J. Blichert-Toft, and F. Albarède, 1999. Relationships between Lu-Hf and Sm-Nd isotopic systems in the global sedimentary system. *Earth Planet. Sci. Lett.*, 168:79–99.

White, W. M., J. Patchett, and D. Ben Othman, 1986. Hf isotope ratios of marine sediments and Mn nodules: Evidence for a mantle source of Hf in seawater. *Earth Planet. Sci. Lett.*, 79:46–54.

PART IV

Short-lived Radionuclides

Short-lived radionuclides originate by several different kinds of nuclear reactions: (1) decay of naturally occurring long-lived radionuclides of U and Th, (2) nuclear reactions caused by energetic nuclear particles in cosmic rays, (3) anthropogenic nuclear fission and irradiation reactions, and (4) nucleosynthesis in stars. The radionuclides produced by processes 1, 2, and 3 are dispersed over the surface of the Earth. In addition, short-lived radionuclides that formed in the ancestral supernovas were present in the solar nebula at the time the Earth and the other planets of the solar system formed, about 4.6×10^9 years ago.

The short-lived radionuclides are the basis of geochronometers suitable for dating on a timescale of a few years up to several million years. Therefore, these geochronometers are widely used not only in Quaternary geology but also in archeology, anthropology, climatology, limnology, oceanography, and hydrogeology. The application of the short-range geochronometers to the most recent history of the Earth has given meaning to the phrase: The past is the key to the future.

CHAPTER 20

Uranium/Thorium-Series Disequilibria

The decay of the long-lived isotopes of U and Th gives rise to chains of short-lived radionuclides which are isotopes of 13 different chemical elements. Nevertheless, Figures 10.1, 10.2, and 10.3 demonstrate that the 43 intermediate daughters are each members of only one of the three chains. In addition, these diagrams demonstrate that the three chains split and recombine because several of the intermediate daughters decay by emission of both α- and β- (negatron) particles. The percent yields that express the probability for α- and β-decay of each radionuclide are listed in the *Handbook of Chemistry and Physics* (Lide and Frederikse, 1995).

The short-lived radioactive daughters of U and Th may be separated from their respective parents and from each other in the course of natural geological processes such as chemical weathering, precipitation of chemical compounds from aqueous solution by biological and inorganic processes, sorption of ions to the surfaces of suspended inorganic and organic particles, partial melting of rocks at depth in the Earth, and crystallization of the resulting magma or lava. Such processes break the decay chains because of differences in the chemical properties of the daughters of U and Th. As a result, two kinds of circumstances arise that have been exploited in geochronometry:

1. A member of a decay chain is separated from its immediate parent and subsequently decays to zero at a rate determined by its halflife.
2. A radionuclide that lost its immediate daughter regenerates that daughter until secular equilibrium between them is reestablished.

These phenomena give rise to several geochronometers capable of measuring geological time ranging from a few years up to about 500,000 years. These geochronometers therefore provide information about the most recent history of the Earth and fill an important gap in the range of most radiogenic isotope geochronometers. Only the fission-track method, the U–Th/He method, and certain cosmogenic radionuclide geochronometers compete with the U-series disequilibrium methods of dating.

The decay chain that starts with $^{238}_{92}U$ has been especially useful for dating, partly because $^{238}_{92}U$ is much more abundant than $^{235}_{92}U$ and because some of its unstable daughters have long halflives, listed in Table 20.1, namely $^{234}_{92}U$, $^{230}_{90}Th$, $^{226}_{88}Ra$, and $^{210}_{82}Pb$. The $^{235}_{92}U$ decay chain contains only one long-lived isotope of protactinium ($^{231}_{89}Pa$), whereas ^{228}Ra is the only daughter of $^{232}_{90}Th$, with a long-enough halflife to be useful for dating current geological events.

The concentrations of the radionuclides used for dating by the U disequilibrium methods are expressed by their rates of decay in a unit weight of sample called the activity (A). In addition, thermal ionization mass spectrometry can be used to measure the atomic $^{230}Th/^{232}Th$, $^{234}U/^{238}U$, and $^{234}U/^{230}Th$ ratios. The rate of decay A of a

Table 20.1. Unstable Daughters of $^{238}_{92}U$, $^{235}_{92}U$, and $^{232}_{90}Th$ Whose Halflives Are Longer Than 1.0 Year

Isotope	Halflife, years	Decay constant[a] y^{-1}
	$^{238}_{92}U$ Series	
$^{234}_{92}U$	2.45×10^5	2.829×10^{-6}
$^{230}_{90}Th$	7.54×10^4	9.1929×10^{-6}
$^{226}_{88}Ra$	1599	4.33×10^{-4}
$^{210}_{82}Pb$	22.6	3.06×10^{-2}
	$^{235}_{92}U$ Series	
$^{231}_{91}Pa$	3.25×10^4	2.13×10^{-5}
$^{227}_{89}Ac$	21.77	3.183×10^{-2}
	$^{232}_{90}Th$ Series	
$^{228}_{88}Ra$	5.76	1.20×10^{-1}
$^{228}_{90}Th$	1.913	3.623×10^{-1}

Source: Lide and Frederikse, 1995.

[a] Calculated from the relation $\lambda = \ln 2/T_{1/2}$.

radionuclide is related to the number of remaining atoms N by the law of radioactivity presented in Chapter 3:

$$A = \lambda N \tag{20.1}$$

where the minus sign in equation 3.2 is dropped because equation 20.1 relates the instantaneous decay rate to the number of atoms remaining.

The activity of a radionuclide is expressed by equation 3.11:

$$A = A_0 e^{-\lambda t} \tag{20.2}$$

which is linearized by taking logarithms to the base e (2.71828):

$$\ln A = \ln A_0 - \lambda t \tag{20.3}$$

The number of atoms of a radionuclide (N_2) forming by decay of its parent (N_1) is expressed by equation 3:19:

$$N_2 = \frac{\lambda_1}{\lambda_2 - \lambda_1} N_1^0 (e^{-\lambda_1 t} - e^{-\lambda_2 t}) + N_2^0 e^{-\lambda_2 t}$$

In cases where no radiogenic daughters are present initially ($N_2^0 = 0$), this equation becomes

$$N_2 = \frac{\lambda_1}{\lambda_2 - \lambda_1} N_1^0 (e^{-\lambda_1 t} - e^{-\lambda_2 t}) \tag{20.4}$$

which is recast in terms of activities when it is adapted for use in some of the U-series methods of dating.

The literature concerning U-series disequilibrium methods of dating was reviewed by Ku (1976) and Schwarcz (1978). In addition, Ivanovich and Harmon (1992) edited a volume of papers that present the modern point of view about the U-series radionuclides in nature. The daughters of U are also used as *tracers* in groundwater, in rivers, and in the oceans:

Groundwater: Sturchio et al., 2001, *Appl. Geochem.*, 16:109–122. Banner et al., 1990, *Earth Planet. Sci. Lett.*, 101:296–312. Gascoyne, 1989, *Appl. Geochem.*, 4:557–591. Cowart, 1981, *J. Hydrol.*, 54:185–193.

Rivers: Porcelli et al., 2001, *Geochim. Cosmochim. Acta*, 65:2439–2459. Moore, 1997, *Earth Planet. Sci. Lett.*, 150:141–150. Moore and Edmond, 1984, *J. Geophys. Res.*, 89:2061–2065.

Oceans: Andersson et al., 1995, *Earth Planet. Sci. Lett.*, 130:217–234. Cochran, 1992, in Ivanovich and Harmon (Eds.), *Uranium Series Disequilibrium*, p. 385, Clarendon Press. Oxford United Kingdom. Cochran et al., 1987, *Earth Planet. Sci. Lett.*, 84:135–152. Moran et al., 1997, *Earth Planet. Sci. Lett.*, 150:151–160. Chen et al., 1986, *Earth Planet. Sci. Lett.*, 80:241–251.

20.1 $^{238}U/^{234}U-^{230}Th$-SERIES GEOCHRONOMETERS

The daughters of ^{238}U that are exploited for dating include $^{234}_{92}U$, $^{230}_{90}Th$, $^{226}_{88}Ra$, and $^{210}_{82}Pb$. In addition, $^{231}_{91}Pa$ (a daughter of ^{235}U) is combined with $^{230}_{90}Th$ to date detrital sediment deposited in the oceans. The decay of ^{238}U proceeds by the following sequence of decay events:

$$^{238}_{92}U \xrightarrow{\alpha} {}^{234}_{90}Th \xrightarrow{\beta^-} {}^{234}_{91}Pa \xrightarrow{\beta^-} {}^{234}_{92}U \xrightarrow{\alpha}$$

$$^{230}_{90}Th \xrightarrow{\alpha} {}^{226}_{88}Ra \xrightarrow{\alpha} {}^{222}_{86}Rn \rightarrow \cdots$$

20.1a The $^{230}Th/^{232}Th$ Method

The $^{230}Th/^{232}Th$ method of dating is based on the separation of U and Th in aqueous solution because U^{4+} can be oxidized to U^{6+}, which forms the soluble uranyl ion (UO_2^{2+}), whereas Th^{4+} remains in the tetravalent state and is strongly sorbed to charged sites on the surfaces of mineral particles and ferromanganese oxyhydroxide precipitates. Therefore, U enters the oceans in soluble form while Th remains sorbed to mineral particles and accumulates in detrital sediment. This difference in chemical properties causes U to have a long oceanic residence time of 5×10^5 years compared to only about 300 years for Th.

The preferential removal of ^{230}Th (and of all other Th isotopes) from aqueous solution separates it from ^{234}U, which is its immediate parent. Therefore, freshly deposited sediment in the oceans, as well as in lakes, contains unsupported ^{230}Th. The rate of decay (activity) of the excess ^{230}Th decreases with a halflife of 7.54×10^4 years as ^{230}Th decays to ^{226}Ra by emitting four suites of α-particles having kinetic energies and frequencies of 4.4383 MeV (0.03%), 4.4798 MeV (0.12%), 4.6211 MeV (23.4%), and 4.6876 MeV (76.3%). Therefore, the measured activity of ^{230}Th sorbed to the sediment at some depth h below the sediment–water interface can be used to determine the time that has elapsed since the sediment was deposited.

The activity of ^{230}Th is expressed in terms of the ratio of the activities of ^{230}Th and ^{232}Th, where the latter is the long-lived isotope of Th ($T_{1/2} = 1.40 \times 10^{10}$ y), which is assumed to be sorbed by mineral particles with the same efficiency as ^{230}Th. Consequently, the $^{230}Th/^{232}Th$ activity ratio decreases with time only because of the decay of ^{230}Th because the activity of ^{232}Th is virtually constant during time intervals of the order of 10^5 years.

The assumptions on which the ^{230}Th method of dating is based were originally articulated by Goldberg and Koide (1962):

1. The $^{230}Th/^{232}Th$ activity ratio in the water mass adjacent to the sediment has remained constant during the last several hundred thousand years.
2. The isotopes ^{230}Th and ^{232}Th have the same chemical speciation and there is no isotopic fractionation between Th ions in the water and on the mineral particles to which these Th isotopes are sorbed.
3. The ^{230}Th and ^{232}Th atoms that reside in the crystal lattices of detrital mineral particles are excluded from the analysis.
4. The sorbed isotopes of Th do not migrate in the sediment by diffusion or advection.

The fourth assumption has been a source of some concern. Although ^{230}Th probably does *not* migrate, its parent ^{234}U may be mobile in some cases and, therefore, may increase the ^{230}Th activity of sediment (Ku, 1965). Experience has shown that the decay of migratory ^{234}U makes a negligible contribution to the activity of ^{230}Th in recently deposited sediment. However, after most of the unsupported ^{230}Th has decayed, the ^{230}Th that is supported by ^{234}U in the pore water becomes dominant.

The Th isotopes that are sorbed to mineral grains or reside in authigenic minerals are recovered by leaching sediment samples with hot 6 N HCl. The principal detrital minerals (quartz and feldspar) are not affected by this treatment and the Th they contain is not released. The Th isotopes are electroplated from the leach solution to Pt discs, and the decay rates of ^{230}Th and ^{232}Th are measured by α-spectrometry.

The total activity of ^{230}Th measured by this procedure is the sum of the activities of unsupported "excess" ^{230}Th and of U-supported ^{230}Th:

$$^{230}Th_A = {}^{230}Th_{Ax} + {}^{230}Th_{As} \qquad (20.5)$$

where $^{230}\text{Th}_A$ = total activity of ^{230}Th in the leachable fraction of the sediment at some depth h below the top of a sediment core in units of disintegrations per unit time per unit weight of dry sediment

$^{230}\text{Th}_{Ax}$ = the same for the excess (unsupported) fraction of ^{230}Th

$^{230}\text{Th}_{As}$ = the same for the fraction of ^{230}Th that is supported by decay of ^{234}U

The activity of U-supported ^{230}Th is assumed to be negligible for recently deposited sediment. Therefore, the activity of the excess ^{230}Th can be expressed by equation 20.2:

$$^{230}\text{Th}_{Ax} = {}^{230}\text{Th}^0_{Ax} e^{-\lambda t} \tag{20.6}$$

where $^{230}\text{Th}^0_{Ax}$ = activity of excess ^{230}Th in the sediment at the time of deposition ($t = 0$)

λ = decay constant of ^{230}Th (9.1929×10^{-6} y^{-1})

t = time elapsed since deposition of the sediment in years

The activity of excess ^{230}Th in equation 20.6 is divided by the activity of long-lived ^{232}Th that is recovered from the sediment together with ^{230}Th:

$$\left(\frac{^{230}\text{Th}}{^{232}\text{Th}}\right)_{Ax} = \left(\frac{^{230}\text{Th}}{^{232}\text{Th}}\right)^0_{Ax} e^{-\lambda_{230} t} \tag{20.7}$$

Equation 20.7 is linearized by taking logarithms to the base e of both sides:

$$\ln\left(\frac{^{130}\text{Th}}{^{132}\text{Th}}\right)_{Ax} = \ln\left(\frac{^{130}\text{Th}}{^{132}\text{Th}}\right)^0_{Ax} - \lambda t \tag{20.8}$$

In most applications of this geochronometer, the samples are taken from increasing depths h below the top of a *sediment core* or below the surface of a *ferromanganese crust*. Therefore, the sedimentation rate a is expressed by

$$a = \frac{h}{t} \tag{20.9}$$

which yields $t = h/a$. Substituting for t in equation 20.8 yields

$$\ln\left(\frac{^{230}\text{Th}}{^{232}\text{Th}}\right)_{Ax} = \ln\left(\frac{^{230}\text{Th}}{^{232}\text{Th}}\right)^0_{Ax} - \frac{\lambda h}{a} \tag{20.10}$$

This is the equation of a straight line in coordinates of h (x) and $\ln(^{230}\text{Th}/^{232}\text{Th})_{Ax}$ (y) whose slope m is

$$m = \frac{\lambda}{a} \tag{20.11}$$

since the Slope is itself a negative number in this case. Equation 20.11 is used to calculate the sedimentation rate a from the relation

$$a = \frac{\lambda}{m} \tag{20.12}$$

After the sedimentation rate a of a core has been determined, dates can be calculated for samples of that core based on the equation

$$t = \frac{h}{a} \tag{20.13}$$

Alternatively, the straight line (equation 20.10) defined by a suite of samples from a sediment core is extrapolated to $h = 0$ in order to determine the numerical value of $\ln(^{230}\text{Th}/^{232}\text{Th})^0_{Ax}$. This value is then substituted into equation 20.8, which is solved for t:

$$t = \frac{1}{\lambda}\left[\ln\left(\frac{^{230}\text{Th}}{^{232}\text{Th}}\right)^0_{Ax} - \ln\left(\frac{^{230}\text{Th}}{^{232}\text{Th}}\right)_{Ax}\right] \tag{20.14}$$

where $(^{230}\text{Th}/^{232}\text{Th})_{Ax}$ = activity ratio of a sediment sample taken from the sediment core or Fe–Mn crust

λ = decay constant of ^{230}Th

Goldberg and Koide (1962) recognized four different patterns of variation of the $(^{230}\text{Th}/^{232}\text{Th})_{Ax}$ parameter in sediment cores. Type *A* in Figure 20.1 indicates the linear decrease of $\ln(^{230}\text{Th}/^{232}\text{Th})_{Ax}$ ratios with depth that is expected in sediment deposited at a constant rate. Type *B* consists of two straight-line segments having different slopes indicating a change in the sedimentation rate

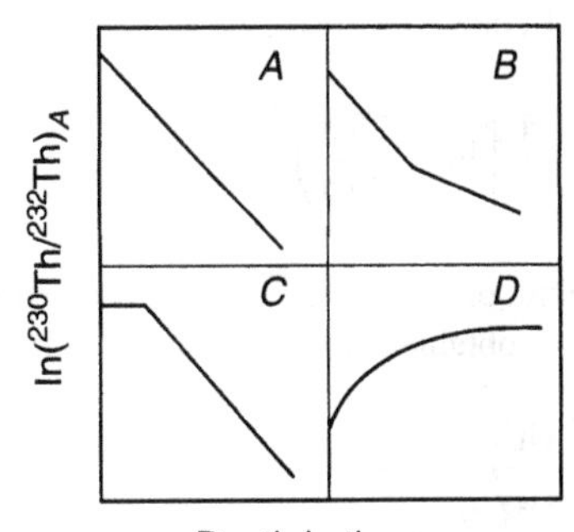

FIGURE 20.1 Variation of the natural logarithm of $^{230}Th/^{232}Th$ activity ratios with depth in sediment cores. Type *A* is a normal pattern implying a constant rate of sediment accumulation. Type *B* is also normal but indicates a change in the rate of sedimentation. Type *C* has a constant activity ratio at the top of the core due to mixing of sediment by burrowing organisms and/or currents. Type *D* results from the dominance of U-supported ^{230}Th that grows in and is then maintained at a constant level by decay of ^{238}U. Adapted from Goldberg and Koide (1962).

and in the $(^{230}Th/^{232}Th)^0_{Ax}$ ratio. In type *C*, the $(^{230}Th/^{232}Th)_{Ax}$ activity ratio at the top of the core is constant due to bioturbation or resuspension of sediment by currents. In such cases, the initial Th isotope activity ratio is obtained by extrapolating the straight line defined by the samples below the disturbed top of the core. In *D*-type patterns, the Th isotope activity ratio actually increases with depth because of the decay of ^{234}U. In this case, one of the basic assumptions of the ^{230}Th method of dating is violated, which means that the sediment cannot be dated by the ^{230}Th method.

20.1b Sedimentation Rate in the Oceans

The type *C* pattern in Figure 20.2 is based on data from Goldberg and Koide (1963) for a piston core from the Indian Ocean (Monsoon 49G, 14°27′ S, 78°03′ E, 5214 m depth). The data points define a straight line:

$$\ln\left(\frac{^{230}Th}{^{232}Th}\right)_{Ax} = 3.76 - 0.0315\,h \qquad (20.15)$$

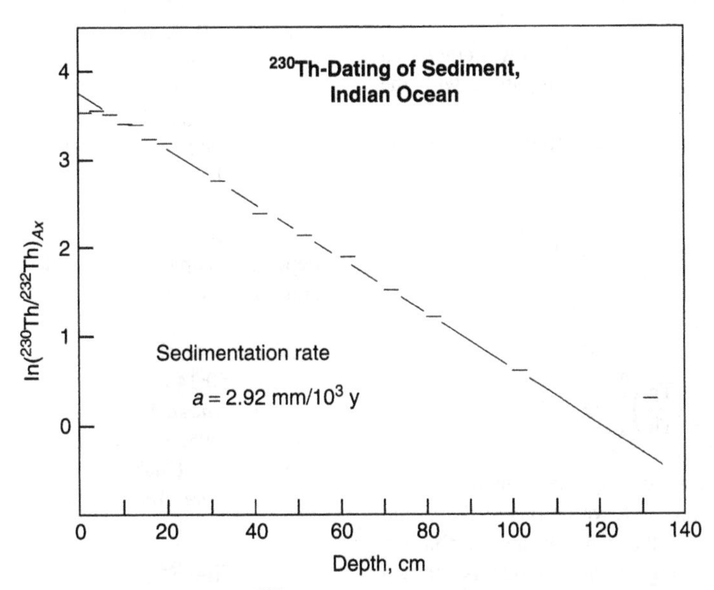

FIGURE 20.2 Decay of unsupported (excess) ^{230}Th in a sediment core recovered from the Indian Ocean (Monsoon 49G, 14°27′ S, 78°03′ E). The sediment samples are 3-cm increments along the length of this core. Data from Goldberg and Koide (1963).

where h is the depth in centimeters and the slope $m = -0.0315$. The sediment in the upper 10 cm of the core has constant $(^{230}Th/^{232}Th)_{Ax}$ values, presumably because of bioturbation. In addition, the sediment below 130 cm contains U-supported ^{230}Th, causing the $\ln(^{230}Th/^{232}Th)_{Ax}$ values to plot above the straight line in Figure 20.2.

The sedimentation rate a at the site where this core was taken is related to the slope m by equation 20:12:

$$a = \frac{\lambda}{m} = \frac{9.1929 \times 10^{-6}}{0.0315} = 291.8 \times 10^{-6} \text{ cm/y}$$
$$= 2.92 \text{ mm}/10^3 \text{ y}$$

This value is typical of sedimentation rates on the abyssal plains of the oceans. In addition, the data confirm that the sedimentation rate remained constant throughout the interval of time represented by this core.

The time elapsed since deposition of the sediment at 130 cm below the top of the core is obtained from equation 20.13:

$$t = \frac{h}{a} = \frac{130 \times 10}{2.92 \times 10^{-3}} = 44.5 \times 10^4 \text{ y}$$

or 44,500 years. (Note that the depth was converted to millimeters in order to be compatible with the units of the sedimentation rate.)

The initial $^{230}Th/^{232}Th$ activity ratio of the straight line in Figure 20.2 is

$$\ln\left(\frac{^{230}Th}{^{232}Th}\right)^0_{Ax} = 3.76$$

which corresponds to

$$\left(\frac{^{230}Th}{^{232}Th}\right)^0_{Ax} = 42.9$$

Values of this ratio vary regionally in the oceans of the world in response to variations of the U and Th concentrations of the bedrock and overburden exposed to weathering on the continents, known also as the *weathering crust*.

The time elapsed since deposition of the sediment at 130 cm can now be calculated by means of equations 20.14:

$$t = \frac{1}{\lambda}\left[\ln\left(\frac{^{230}Th}{^{232}Th}\right)^0_{Ax} - \ln\left(\frac{^{230}Th}{^{232}Th}\right)_{Ax}\right]$$

The value of $\ln(^{230}Th/^{232}Th)_{Ax}$ for sediment at 130 cm is obtained from equation 20.15:

$$\ln\left(\frac{^{230}Th}{^{232}Th}\right)_{Ax} = 3.76 - 0.0315 \times 130 = -0.335$$

Therefore,

$$t = \frac{1}{9.1929 \times 10^{-6}}(3.76 + 0.335) = 0.445 \times 10^6$$
$$= 44.5 \times 10^4 \text{ y}$$

The dates for the time elapsed since deposition of sediment at 130 cm calculated by the two procedures are identical.

The rates of deposition of *sediment* in different ocean basins have been determined by Edward D. Goldberg and his associates:

Goldberg and Koide, 1962, *Geochim. Cosmochim. Acta*, 26:417–450. Goldberg and Koide, 1963, in Geiss and Goldberg (Eds.), *Earth Science and Meteoritics*, pp. 90–102, North-Holland, Amsterdam. Goldberg and Griffin, 1964, *J. Geophys. Res.*, 69(20):4293–4309. Goldberg, 1968, *Earth Planet. Sci. Lett.*, 4:17–21.

The ^{230}Th method of dating *ferromanganese* deposits was pioneered by Ku and Broecker (1967) and continues to be widely used:

Mangini et al., 1986, *Geochim. Cosmochim. Acta*, 50:149–156. Eisenhauer et al., 1992, *Earth Planet. Sci. Lett.*, 109:25–36. Chabaux et al., 1995, *Geochim. Cosmochim. Acta*, 59: 633–638. Chabaux et al., 1997, *Geochim. Cosmochim. Acta*, 61:3619–3632.

20.1c The ^{234}U–^{230}Th Method

The isotopes of U remain in solution in seawater where the uranyl ion (UO_2^{2+}) tends to form carbonate complexes [e.g., $UO_2(CO_3)_3^{4-}$], which allow it to coprecipitate with calcium carbonate. As a result, Ca carbonate minerals such as calcite and aragonite typically contain appreciable concentrations of U but lack Th. The marine geochemistry of U, Th, and Pb was discussed in Section 10.11a in connection with U–Pb dating of Phanerozoic and Precambrian carbonate rocks presented in Section 10.11b. The U concentration of seawater listed in Table 10.7 is 3.2 ppb, compared to only 1×10^{-2} ppb for Th. The average U concentration of carbonate rocks is 1.9 ppm (Table 10.1) but reaches values as high as 17.6 ppm in sparry calcite of the Middle Devonian Lucas Formation of Ontario (Table 10.9).

The ^{234}U in solution in seawater is incorporated into calcium carbonate (and phosphate) and is thereby separated from its daughter (^{230}Th). The decay of ^{234}U in the calcium carbonate (and phosphate) subsequently causes the number of ^{230}Th atoms to increase with time until ^{234}U and ^{230}Th reach a state of secular equilibrium (Section 3.3b), when their rates of decay are equal:

$$^{234}U\lambda_{234} = {}^{230}Th\lambda_{230} \tag{20.16}$$

where ^{234}U and ^{230}Th are the numbers of atoms of these isotopes in a unit weight of sample.

The number of ^{230}Th atoms in a sample of calcium carbonate increases with time as a result of the decay of ^{234}U in accordance with equation 20.4:

$$N_2 = \frac{\lambda_1 N_1^0}{\lambda_2 - \lambda_1}(e^{-\lambda_1 t} - e^{-\lambda_2 t})$$

where N_1 and λ_1 refer to ^{234}U and N_2 and λ_2 represent ^{230}Th. Therefore, the number ^{230}Th atoms that have formed by decay of ^{234}U in a unit weight of carbonate (or phosphate) rock is

$$^{230}Th_s = \frac{\lambda_{234}\,{}^{234}U^0}{\lambda_{230} - \lambda_{234}}(e^{-\lambda_{234}t} - e^{-\lambda_{230}t}) \tag{20.17}$$

If ^{234}U is in secular equilibrium with ^{238}U, the rate of decay of ^{234}U is equal to the rate of decay of ^{238}U. Therefore, in equation 20.17

$$\lambda_{234}\,{}^{234}U^0 = \lambda_{238}\,{}^{238}U = {}^{238}U_A$$

In other words, when ^{234}U is in equilibrium with ^{238}U, its rate of decay is controlled by the decay constant of ^{238}U, which is very small (i.e., $\lambda_{238} = 1.55125 \times 10^{-10}\ y^{-1}$). Under these conditions

$$\lambda_{230} - \lambda_{234} = \lambda_{230} \quad \text{and} \quad e^{-\lambda_{234}t} = \frac{1}{e^0} = 1.0$$

and equation 20.17 can be rewritten as

$$^{230}Th = \frac{^{238}U_A}{\lambda_{230}}(1 - e^{-\lambda_{230}t})$$

Since $\lambda_{230}\,{}^{230}Th = {}^{230}Th_{As}$,

$$^{230}Th_{As} = {}^{238}U_A\ (1 - e^{-\lambda_{230}t}) \tag{20.18}$$

where $^{230}Th_{As}$ is the activity of ^{230}Th supported by the decay of ^{234}U, which is in secular equilibrium with ^{238}U.

Equation 20.18 is a geochronometer based on the growth of ^{230}Th in carbonate and phosphate rocks in which ^{234}U is in secular equilibrium with ^{238}U. This equation can also take the form

$$\left(\frac{^{230}Th}{^{238}U}\right)_A = (1 - e^{-\lambda t}) \tag{20.19}$$

where λ is the decay constant of ^{230}Th and ^{234}U is in secular equilibrium with ^{238}U such that

$$\left(\frac{^{234}U}{^{238}U}\right)_A = 1.00$$

and the initial activity of ^{230}Th is equal to zero.

In case the initial abundance of ^{230}Th is not equal to zero, the total activity of ^{230}Th is

$$^{230}Th_A = {}^{230}Th_A^0 e^{-\lambda_{230}t} + {}^{238}U_A(1 - e^{-\lambda_{230}t}) \tag{20.20}$$

which is obtained from equation 20.5 by combining equations 20.6 and 20.18. The first term in equation 20.20 decreases with time, whereas the second term increases. Therefore, the activity of unsupported ^{230}Th decreases with time whereas the U-supported activity of ^{230}Th increases and eventually becomes dominant.

20.1d $^{238}U/^{234}U$ Disequilibrium

The activity of ^{234}U in solution in water on the continents and in the oceans is not equal to that of ^{238}U as assumed in Section 20.1b. On the contrary, in most cases, the activity ratio of ^{234}U and ^{238}U in natural water is greater than one (Thurber, 1962; Cherdyntsev, 1969):

$$\left(\frac{^{234}U}{^{238}U}\right)_A > 1.0$$

The radioactive disequilibrium between ^{234}U and ^{238}U arises because radiation damage caused by α-decay of ^{238}U locally enhances the solubility of U-bearing crystals and thereby permits preferential dissolution of ^{234}U. In addition, ^{234}Th (the first decay product of ^{238}U) may be ejected from the surfaces of mineral grains due to recoil during α-decay of ^{238}U. This isotope then decays rapidly to ^{234}U through short-lived ^{234}Pa (Kigoshi, 1971; Kronfeld, 1974; Fleischer, 1980, 1983). The preferential dissolution of ^{234}U causes U-bearing minerals to become depleted in ^{234}U during chemical weathering.

The excess ^{234}U in water on the continents and in the oceans is incorporated into carbonate minerals and other authigenic U-bearing compounds where it decays to form ^{230}Th until its decay rate becomes equal to that of ^{238}U (i.e., until secular radioactive equilibrium is reestablished).

The total activity of ^{234}U in a U-bearing mineral in which ^{238}U and ^{234}U are *not* in secular equilibrium is the sum of two terms:

$$^{234}U_A = {}^{234}U_{Ax} + {}^{234}U_{As} \qquad (20.21)$$

where $^{234}U_{Ax}$ = activity of the excess ^{234}U that is not supported by ^{238}U

$^{234}U_{As}$ = activity of the ^{234}U that is in secular equilibrium with ^{238}U

Therefore, $^{234}U_{As} = {}^{238}U_A$ and equation 20.21 can be rewritten as

$$^{234}U_A = {}^{234}U_{Ax} + {}^{238}U_A \qquad (20.22)$$

The activity of excess ^{234}U decreases with time, in accordance with the law of radioactivity (equation 20.2):

$$^{234}U_{Ax} = {}^{234}U^0_{Ax} e^{-\lambda_{234}t} \qquad (20.23)$$

where $^{234}U^0_{Ax}$ is the initial activity of the excess ^{234}U at the time of deposition of the mineral (i.e., $t = 0$). The initial activity of excess ^{234}U is the difference between the total initial activity ($^{234}U^0_A$) and the activity of the supported ^{234}U:

$$^{234}U^0_{Ax} = {}^{234}U^0_A - {}^{234}U_{As} \qquad (20.24)$$

Substituting equation 20.24 into 20.23 yields

$$^{234}U_{Ax} = ({}^{234}U^0_A - {}^{234}U_{As})e^{-\lambda_{234}t} \qquad (20.25)$$

which is substituted into equation 20.22:

$$^{234}U_A = ({}^{234}U^0_A - {}^{234}U_{As})e^{-\lambda_{234}t} + {}^{238}U_A$$

Rearranging terms and dividing by $^{238}U_A$ gives

$$\left(\frac{^{234}U}{^{238}U}\right)_A = 1 + \left(\frac{^{234}U^0_A - {}^{234}U_{As}}{^{238}U_A}\right)e^{-\lambda_{234}t}$$

$$= 1 + \left[\left(\frac{^{234}U}{^{238}U}\right)^0_A - 1\right]e^{-\lambda_{234}t}$$

Setting $(^{234}U/^{238}U)^0_A = \gamma_0$ yields the final form of the equation:

$$\left(\frac{^{234}U}{^{238}U}\right)_A = 1 + (\gamma_0 - 1)e^{-\lambda_{234}t} \qquad (20.26)$$

When ^{234}U is in secular equilibrium with ^{238}U (i.e., $\gamma_0 = 1$), then $\gamma_0 - 1 = 0$ and $(^{234}U/^{238}U)_A = 1.0$, as required for secular equilibrium. When t approaches infinity, $e^{-\lambda_{234}t}$ approaches zero and again $(^{234}U/^{238}U)_A = 1.0$. In other words, the decay of excess ^{234}U in time reestablishes secular equilibrium between ^{238}U and ^{234}U provided that the samples being dated remained closed to ^{234}U and ^{238}U. Figure 20.3 demonstrates that secular equilibrium is approached after about one million years.

In order to use equation 20.26 as a geochronometer, the present $^{234}U/^{238}U$ activity ratio must be measured and γ_0 must be known. In present-day seawater $\gamma_0 = 1.15$ (Koide and Goldberg, 1965; Miyake et al., 1966). However, the value of γ_0 in lacustrine carbonates and speleothems is not

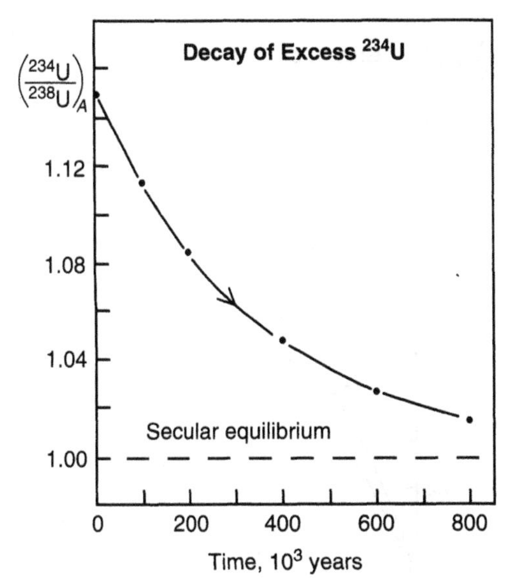

FIGURE 20.3 Decay of excess ^{234}U with a halflife of 2.444×10^5 years. The diagram demonstrates that the marine $^{234}U/^{238}U$ activity ratio ($\gamma_0 = 1.15$) approaches secular equilibrium after about one million years.

predictable (Kaufman, 1971; Harmon et al., 1975; Lin et al., 1996). This dilemma was overcome by Schwarcz (1980) by constructing a diagram (not shown) that combines the time-dependent increase of the $^{230}Th/^{234}U$ activity ratio and the simultaneous decrease of the $^{234}U/^{238}U$ activity ratio of lacustrine carbonates and speleothems. This diagram allows both the age and the initial $^{234}U/^{238}U$ activity ratio to be determined by graphical interpolation (Ivanovich et al., 1992). However, the most elegant procedure is to solve the equations by iteration on a computer (e.g., Ludwig and Titterington, 1994; Kaufman, 1993; Luo and Ku, 1991).

The ^{234}U disequilibrium geochronometer was originally used by Veeh (1966) and Bender et al. (1979) to date coral terraces on the island of Barbados in the West Indies, but Kaufman et al. (1971) noted that the U concentration of mollusk shells increases after burial. Similarly, Ku (1965) reported that clay-rich marine sediment is not suitable for dating by the ^{234}U–^{238}U disequilibrium method because ^{234}U remains mobile after deposition.

20.1e ^{230}Th with $^{234}U/^{238}U$ Disequilibrium

The activity of ^{230}Th in freshly deposited marine calcium carbonate, in which ^{234}U is in secular equilibrium with ^{238}U, is expressed by equation 20.18 in Section 20.1b:

$$^{230}\mathrm{Th}_{As} = {}^{238}\mathrm{U}_A(1 - e^{-\lambda_{230}t})$$

A complication arises because, in most cases, excess ^{234}U is present. The decay of the excess ^{234}U is described by equation 20.26 in Section 20.1c:

$$\left(\frac{^{234}\mathrm{U}}{^{238}\mathrm{U}}\right)_A = 1 + (\gamma_0 - 1)e^{-\lambda_{234}t}$$

where γ_0 is the initial value of the $^{234}U/^{238}U$ activity ratio. A complete description of the time-dependent increase of the activity of ^{230}Th by decay of ^{234}U in calcium carbonate must include the presence of excess ^{234}U.

The number of ^{230}Th atoms that form by the decay of the excess ^{234}U is obtained from equation 20.4 with the appropriate substitutions:

$$^{230}\mathrm{Th}_x = \frac{\lambda_{234}}{\lambda_{230} - \lambda_{234}} {}^{234}\mathrm{U}_x^0 \, (e^{-\lambda_{234}t} - e^{-\lambda_{230}t}) \tag{20.27}$$

Multiplying both sides of the equation by λ_{230} to convert to activities yields

$$^{230}\mathrm{Th}_{Ax} = \frac{\lambda_{230}}{\lambda_{230} - \lambda_{234}} {}^{234}\mathrm{U}_{Ax}^0 \, (e^{-\lambda_{234}t} - e^{-\lambda_{230}t}) \tag{20.28}$$

The initial excess activity of $^{234}U^0_{Ax}$ is the difference between the total activity ($^{234}U_A$) and the activity of the supported fraction ($^{234}U_{As}$):

$$^{234}\mathrm{U}^0_{Ax} = {}^{234}\mathrm{U}^0_A - {}^{234}\mathrm{U}^0_{As} \tag{20.29}$$

Note that $^{234}U_{As}$ is virtually constant because it is equal to the rate of decay of ^{238}U, which has a very long halflife ($T_{1/2} = 4.468 \times 10^9$ y). Substituting equation 20.29 into 20.28 and dividing by $^{238}U_A$ yields

$$\left(\frac{^{230}\mathrm{Th}}{^{238}\mathrm{U}}\right)_{Ax} = \frac{\lambda_{230}}{\lambda_{230} - \lambda_{234}} \left[\left(\frac{^{234}\mathrm{U}}{^{238}\mathrm{U}}\right)^0_A - 1\right] \times (e^{-\lambda_{234}t} - e^{-\lambda_{230}t})$$

Since $({}^{234}\mathrm{U}/{}^{238}\mathrm{U})_A^0 = \gamma_0$, the activity of ^{230}Th caused by decay of the excess ^{234}U is

$$\left(\frac{{}^{230}\mathrm{Th}}{{}^{238}\mathrm{U}}\right)_{Ax} = \frac{\lambda_{230}}{\lambda_{230} - \lambda_{234}} \times (\gamma_0 - 1)(e^{-\lambda_{234}t} - e^{-\lambda_{230}t}) \quad (20.30)$$

The total activity of ^{230}Th that results from the decay of the supported and excess ^{234}U is obtained by adding equations 20.18 and 20.30:

$$\left(\frac{{}^{230}\mathrm{Th}}{{}^{238}\mathrm{U}}\right)_A = \left(\frac{{}^{230}\mathrm{Th}}{{}^{238}\mathrm{U}}\right)_{As} + \left(\frac{{}^{230}\mathrm{Th}}{{}^{238}\mathrm{U}}\right)_{Ax} = (1 - e^{-\lambda_{230}t}) + \frac{\lambda_{230}}{\lambda_{230} - \lambda_{234}} \times (\gamma_0 - 1)(e^{-\lambda_{234}t} - e^{-\lambda_{230}t}) \quad (20.31)$$

If $\gamma_0 = 1.0$, $\gamma_0 - 1 = 0$ and equation 20.31 is reduced to

$$\left(\frac{{}^{230}\mathrm{Th}}{{}^{238}\mathrm{U}}\right)_A = 1 - e^{-\lambda_{230}t}$$

which represents the case where all of ^{230}Th is supported by the decay of ^{234}U, which is in secular equilibrium with ^{234}U (i.e., ${}^{234}\mathrm{U}_{Ax} = 0$). For large values of t, the activity ratio of ^{230}Th/^{238}U approaches 1.0 because after the excess ^{234}U has decayed secular equilibrium between ^{230}Th and ^{238}U is reestablished.

Equation 20.31 was derived by Broecker (1963) and yields accurate dates for corals, calcareous ooze, and mollusk shells under certain conditions:

1. The initial ^{230}Th/^{238}U activity ratio is close to zero.
2. The sample was closed to U and the intermediate daughters between ^{238}U and ^{230}Th.
3. The initial activity ratio of ^{234}U/^{238}U must be known.

For dating of marine carbonate samples the initial ^{234}U/^{238}U activity ratio ($\gamma_0 = 1.15$) is assumed to be invariant in time and place.

The dates calculated from measured ^{230}Th/^{238}U activity ratios of marine carbonate samples depend on the model that is used. Figure 20.4 demonstrates that when ^{234}U disequilibrium is neglected ($\gamma_0 = 1.0$), the calculated dates overestimate the age of the sample.

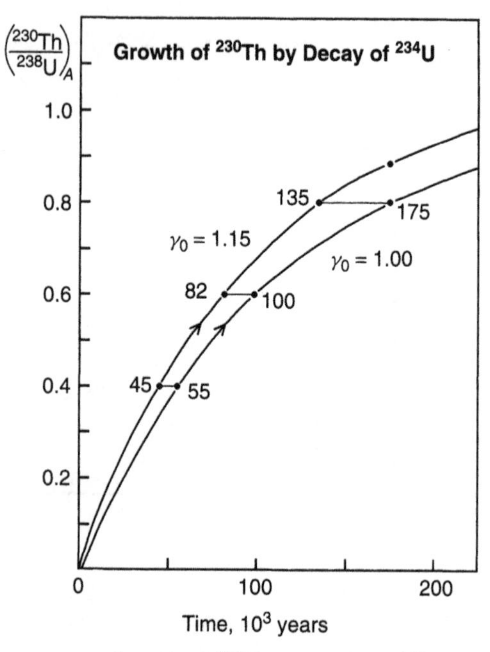

FIGURE 20.4 Growth of ^{230}Th by decay of ^{234}U depending on whether ^{234}U is in secular equilibrium with ^{238}U ($\gamma_0 = 1.0$) or whether excess ^{234}U is present ($\gamma_0 = 1.15$). In case $\gamma_0 = 1.15$, the ^{230}Th/^{238}U activity ratio increases faster than for $\gamma_0 = 1.00$. Consequently, dates calculated by assuming that $\gamma_0 = 1.0$ overestimate the ages of samples whose γ_0-values are actually 1.15 or higher. The dates that correspond to measured $({}^{230}\mathrm{Th}/{}^{238}\mathrm{U})_A$ ratios of 0.4, 0.6, and 0.8 are indicated along the two growth curves.

20.1f Coral Terraces on Barbados

The island of Barbados is located east of the southern end of the Lesser Antilles chain of islands. It is covered by a cap of coral limestone containing terraces whose elevations above present sealevel record both the uplift of the island and fluctuations of sealevel in the past. Broecker et al. (1968) dated corals from the three lowest terraces by the U-series disequilibrium methods. A typical data set for a specimen of *Montastria annularis* on the Barbados I terrace includes

$$\left(\frac{^{234}\text{U}}{^{238}\text{U}}\right)_A = 1.12 \pm 0.01$$

$$\left(\frac{^{230}\text{Th}}{^{234}\text{U}}\right)_A = 0.53 \pm 0.02$$

from which it follows that

$$\left(\frac{^{230}\text{Th}}{^{238}\text{U}}\right)_A = 1.12 \times 0.53 = 0.593 \pm 0.023$$

The problem of dating the coral terraces on Barbados is that the initial $^{234}U/^{238}U$ activity ratio γ_0 is not known and must be determined from the data. Therefore, the strategy for dating these corals is to use equation 20.19 because it was derived with the assumption that ^{234}U is in secular equilibrium with ^{238}U ($\gamma_0 = 1.0$). In this case, the activity of $^{238}U_A$ is equal to the activity of $^{238}U_A$ and equation 20.19 can be modified accordingly:

$$\left(\frac{^{230}\text{Th}}{^{234}\text{U}}\right)_A = 1 - e^{-\lambda_{230}t} \qquad (20.32)$$

Solving for t yields

$$t = -\frac{1}{\lambda_{230}} \ln\left[1 - \left(\frac{^{230}\text{Th}}{^{234}\text{U}}\right)_A\right]$$

Given that $(^{230}\text{Th}/^{234}\text{U})_A = 0.53$ and that $\lambda_{230} = 9.1929 \times 10^{-6}\ \text{y}^{-1}$,

$$t = \frac{0.75502}{9.1929 \times 10^{-6}} = 82,130\ \text{y}$$

This date can be used to calculate γ_0 by means of equation 20.26:

$$\left(\frac{^{234}\text{U}}{^{238}\text{U}}\right)_A = 1 + (\gamma_0 - 1)e^{-\lambda_{234}t}$$

Solving for γ_0,

$$\gamma_0 = \frac{(^{234}\text{U}/^{238}\text{U})_A - 1}{e^{-\lambda_{234}t}} + 1$$

Substituting $(^{234}\text{U}/^{238}\text{U})_A = 1.12$, $\lambda_{234} = 2.829 \times 10^{-6}\ \text{y}^{-1}$, and $t = 82,130$ y give

$$\gamma_0 = \frac{1.12 - 1}{0.7926} + 1 = 1.15$$

Evidently, the initial value of $(^{234}\text{U}/^{238}\text{U})_A$ of seawater 82,130 years ago was identical to the value of this ratio in the modern ocean.

Given that $\gamma_0 = 1.15$ and $(^{230}\text{Th}/^{238}\text{U})_A = 0.593 \pm 0.023$, equation 20.31 can be used to test the date obtained from equation 20.32:

$$\left(\frac{^{230}\text{Th}}{^{238}\text{U}}\right)_A = (1 - e^{-\lambda_{230}t}) + \frac{\lambda_{230}}{\lambda_{230} + \lambda_{234}} \times (\gamma_0 - 1)(e^{-\lambda_{234}t} - e^{-\lambda_{230}t})$$

This equation must be solved for t by iteration or by a graphical method. The relation between $(^{230}\text{Th}/^{238}\text{U})_A$ and t based on equation 20.31 is expressed by the curve in Figure 20.5, which yields a date of 81,000 years and thereby confirms the validity of the date of 82,130 years calculated above from equation 20.32.

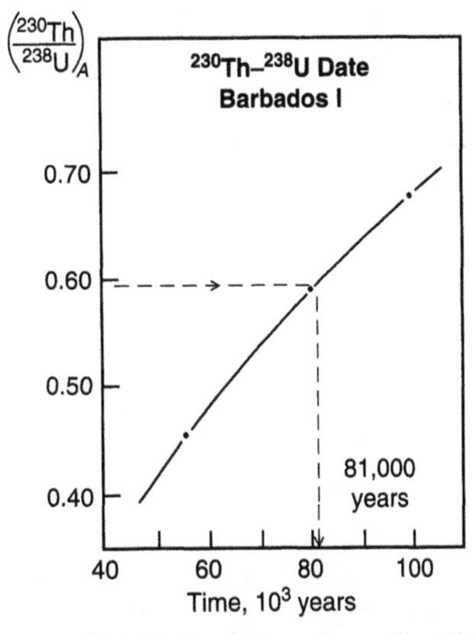

FIGURE 20.5 Graphical solution of equation 20.31 for $(^{230}\text{Th}/^{238}\text{U}) = 0.593$ reported by Broecker et al. (1968) for a coral from the lowest terrace on the island of Barbados for $\gamma_0 = 1.15$.

The average dates obtained by Broecker et al. (1968) by means of equation 20.32 for the three lowest terraces on Barbados are as follows:

Barbados I	82,000 ± 2000 years
Barbados II	103,000 ± 3000 years
Barbados III	122,000 ± 4000 years

These dates in Figure 20.6 closely match the periods of increased summer insolation of the Earth at 45° N latitude based on the tilt of its axis, the eccentricity of its orbit, and the time of perihelion (Milankovitch, 1941). In this way, the age determination by Broecker et al. (1968) of corals on Barbados supported the hypothesis that climate changes on the Earth are caused in part by extraterrestrial factors arising from the celestial mechanics of its orbit.

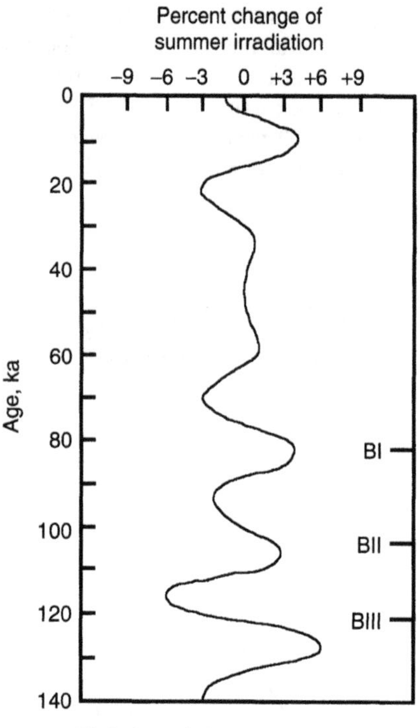

FIGURE 20.6 Variation of the amount of solar radiation received by the northern hemisphere of the Earth at 45° N as a result of the celestial mechanics of its orbit. BI, BII, and BIII are the average dates of corals on beach terraces of the island of Barbados. Adapted from Broecker et al. (1968).

The literature on the trace-element composition of *corals* is quite extensive, including measurements of the concentrations and isotope compositions of U, Th, Ra, and Pb:

Shen and Boyle, 1987, *Earth Planet. Sci. Lett.*, 82:289–304. Hamelin et al., 1991, *Earth Planet. Sci. Lett.*, 106:169–180. Bard et al., 1991, *Geochim. Cosmochim. Acta*, 55:2385–2390. Bar-Matthews et al., 1993, *Geochim. Cosmochim. Acta*, 57:257–276.

Similar studies on the calcite shells of *foraminifera* have been carried out:

Delaney and Boyle, 1983, *Earth Planet. Sci. Lett.*, 62:258–262. Russell et al., 1994, *Geochim. Cosmochim. Acta*, 58:671–681.

U-series disequilibrium dates of corals have been widely used to constrain changes in sealevel as well as of uplift and subsidence of islands:

Barbados: Banner et al., 1991, *Earth Planet. Sci. Lett.*, 107:129–137. Ku et al., 1990, *Quat. Res.*, 33:129–147.

Bahamas: Chen et al., 1991, *Geol. Soc. Am. Bull.*, 103:82–97.

Papua New Guinea: Stein et al., 1993, *Geochim. Cosmochim. Acta*, 57:2541–2554.

Hawaii: Ludwig et al., 1991, *Geology*, 19:171–174.

Hateruma Atoll: Henderson et al., 1993, *Earth Planet. Sci. Lett.*, 115:65–73.

California: Stein et al., 1991, *Geochim. Cosmochim. Acta*, 55:3709–3722.

20.2 RADIUM

Each of the three decay series of U and Th includes at least one of the four naturally occurring short-lived radioactive isotopes of Ra (Lide and Frederikse, 1995):

$^{238}_{92}U$ $^{226}_{88}Ra$ (1599 years)
$^{235}_{92}U$ $^{223}_{88}Ra$ (11.435 days)
$^{232}_{90}Th$ $^{224}_{88}Ra$ (3.66 days) and ^{228}Ra (5.76 years)

The short halflives of the Ra isotopes enable them to be used to date samples whose ages range from only a few years to several thousand years. These short-range geochronometers are well suited to dating carbonate samples because Ra is admitted into calcite and aragonite while Th is excluded.

Radium is a member of group IIA (alkaline earths), which includes Ca, Sr, and Ba, all of which form carbonates and sulfates that have low solubilities in water. In addition, Ra^{2+} ions are strongly sorbed to oxyhydroxides of Fe, Al, and Mn. The geochemistry of Ra has been summarized elsewhere:

Ku, 1972, in Fairbridge (Ed.), *The Encyclopedia of Geochemistry and Environmental Sciences*, Vol. 4A, pp. 1008–1014, Van Nostrand Reinhold, New York. Langmuir and Riese, 1985, *Geochim. Cosmochim. Acta*, 49:1593–1601. Benes, 1990, in *The Environmental Behavior of Radium*, Vol. 1, pp. 373–418, International Atomic Energy Agency, Vienna. Gnanapragasam and Lewis, 1995, *Geochim. Cosmochim. Acta*, 59:5103–5111. Sturchio et al., 2001, *Appl. Geochem.*, 16: 102–122.

20.2a The ^{226}Ra–Ba Method

Radium-226 decays to $^{222}_{86}Rn$ by emitting a suite of four α-particles with energies between 4.194 and 4.784 MeV. One of these is followed by a γ-ray (186.1 keV, 3.3%), which is detectable by *γ-ray spectrometry*:

Michel et al., 1980, *Anal. Chem.*, 53:1885–1889. Elsinger, 1982, *Anal. Chim. Acta*, 144:277–281. Moore, 1984, *Nucl. Instrum. Methods*, 223:407–411. Lucas and Markun, 1991, *J. Environ. Radioact.*, 15:1–18. Rihs et al., 1997, *J. Radioanal. Nucl. Chem.*, 226: 149–157. Rihs et al., 2000, *Geochim. Cosmochim. Acta*, 64:661–671.

In addition, ^{226}Ra has been determined by the following methods:

Alpha-Scintillation Counting: Ku et al., 1970, *J. Geophys. Res.*, 75:5286–5292. Turekian et al., 1979, *Nature*, 280:385–387. Turekian et al., 1979, *Nature* 280:385–387. Berkman and Ku, 1998, *Chem. Geol.*, 144:331–334.

Thermal Ionization Mass Spectrometry: Volpe et al., 1991, *Anal. Chem.*, 63:913–916. Cohen and O'Nions, 1991, *Anal. Chem.*, 63:2705–2708. Chabaux et al., 1994, *Chem. Geol.*, 114:191–197.

The ^{226}Ra/Ba method of dating relies on the assumption that ^{230}Th, the immediate parent of ^{226}Ra, is excluded from freshly precipitated calcite or aragonite, whereas Ra^{2+} is admitted in place of Ca^{2+} and Ba^{2+}. Therefore, the ^{226}Ra/Ba ratio decreases with time after deposition:

$$\left(\frac{^{226}\mathrm{Ra}}{\mathrm{Ba}}\right)_{Ax} = \left(\frac{^{226}\mathrm{Ra}}{\mathrm{Ba}}\right)^0_{Ax} e^{-\lambda_{226}t} \qquad (20.33)$$

where $(^{226}\mathrm{Ra/Ba})_{Ax}$ = ratio of the decay rate of unsupported ^{226}Ra to the concentration of Ba in units of dpm/μg at any time t and the units arise from dpm/g for ^{226}Ra and μg/g for Ba

$(^{226}\mathrm{Ra/Ba})^0_{Ax}$ = The same ratio at the time of deposition (i.e., $t = 0$)

λ_{226} = decay constant of ^{226}Ra, $4.334 \times 10^{-4}\ y^{-1}$

t = time elapsed since deposition

The validity of dates derived from equation 20.33 is based on the following assumptions:

1. The initial ^{226}Ra/Ba ratio of the sample to be dated is known and has not varied with time in the past.

2. Only unsupported excess ^{226}Ra is present and ^{230}Th (the parent of ^{226}Ra) was initially excluded from the carbonate or sulfate minerals being dated.
3. The carbonate samples have been closed to Ra and Ba since deposition.
4. Insufficient time has passed since deposition for U-supported ^{230}Th to form and to decay to ^{226}Ra.

The ^{226}Ra–Ba method was used by Rihs et al. (2000) to date travertine in the Massif Central of France. The travertines were deposited by hydrothermal solutions discharged by hotsprings at the study site. Samples of calcite that were deposited at this site about 33 days prior to collection yielded an average initial ^{226}Ra/Ba ratio of 0.98 ± 0.04 dpm/μg. The present ^{226}Ra/Ba ratio of the oldest travertine layer at this location is 0.69 dpm/μg. Substituting these values into equation 20.33 and solving for t yields

$$t = \frac{1}{4.334 \times 10^{-4}} \ln \left(\frac{0.98}{0.69} \right) = 810 \text{ y}$$

Other examples of dating travertines by the ^{226}Ra–Ba method include the work of Condomines et al. (1999) at Auvergne in France and Sturchio (1990) at Mammoth Hot Springs in Wyoming, United States.

20.2b The ^{228}Ra–^{228}Th Method

The decay of long-lived $^{232}_{90}$Th proceeds by α-emission to $^{228}_{88}$Ra, which decays by β-emission to $^{228}_{89}$Ac and hence to $^{228}_{90}$Th. The halflives and decay constants of these radionuclides are as follows:

^{228}Ra	5.76 years, 1.203×10^{-1} y^{-1}
^{228}Ac	6.13 h, 991.3 y^{-1}
^{228}Th	1.913 years, 3.623×10^{-1} y^{-1}

Calcium carbonate minerals that precipitate from brines permit Ra^{2+} to substitute for Ca^{2+} but exclude Th^{4+}. Therefore, freshly deposited calcites forming from hydrothermal solutions as well as mollusk shells and other biogenic carbonates contain ^{228}Ra (as well as ^{226}Ra) but lack ^{228}Th (as well as ^{230}Th).

After deposition, the ^{228}Ra decays in accordance with the law of radioactivity:

$$^{228}\text{Ra} = (^{228}\text{Ra})_0 e^{-\lambda_{\text{Ra}} t} \tag{20.34}$$

where ^{228}Ra is the number of atoms of this isotope per unit weight of sample and λ_{Ra} is its decay constant. The formation of the intermediate daughter (^{228}Ac) can be neglected because of its short halflife (6.13 h). Therefore, the decay of ^{228}Ra causes the number of ^{228}Th atoms to increase in accordance with

$$N_2 = \frac{\lambda_1 N_1^0}{\lambda_2 - \lambda_1}(e^{-\lambda_1 t} - e^{-\lambda_2 t})$$

Setting $N_1 = {}^{228}$Ra, $N_2 = {}^{228}$Th, λ_{Ra} = decay constant of ^{228}Ra, and λ_{Th} = decay constant of ^{228}Th yields

$$^{228}\text{Th} = \frac{\lambda_{\text{Ra}}\, {}^{228}\text{Ra}_0}{\lambda_{\text{Th}} - \lambda_{\text{Ra}}}(e^{-\lambda_{\text{Ra}} t} - e^{-\lambda_{\text{Th}} t}) \tag{20.35}$$

Solving equation 20.34 for ^{228}Ra$_0$ and substituting into equation 20.35,

$$^{228}\text{Ra}_0 = {}^{228}\text{Ra}\, e^{\lambda_{\text{Ra}} t}$$

$$^{228}\text{Th} = \frac{\lambda_{\text{Ra}}\, {}^{228}\text{Ra}\, e^{\lambda_{\text{Ra}} t}}{\lambda_{\text{Th}} - \lambda_{\text{Ra}}}(e^{-\lambda_{\text{Ra}} t} - e^{-\lambda_{\text{Th}} t})$$

from which it follows that

$$^{228}\text{Th} = \frac{^{228}\text{Ra}_A}{\lambda_{\text{Th}} - \lambda_{\text{Ra}}}(1 - e^{-(\lambda_{\text{Th}} - \lambda_{\text{Ra}})t})$$

Converting to the activity of ^{228}Th by multiplying both sides of the equation by λ_{Th} and forming the ^{228}Th/^{228}Ra activity ratio yield

$$\left(\frac{^{228}\text{Th}}{^{228}\text{Ra}} \right)_A = \frac{\lambda_{\text{Th}}}{\lambda_{\text{Th}} - \lambda_{\text{Ra}}}(1 - e^{-(\lambda_{\text{Th}} - \lambda_{\text{Ra}})t}) \tag{20.36}$$

This equation was stated by Rihs et al. (2000).

A plot of this equation in Figure 20.7 demonstrates that the ^{228}Th/^{228}Ra activity ratio increases

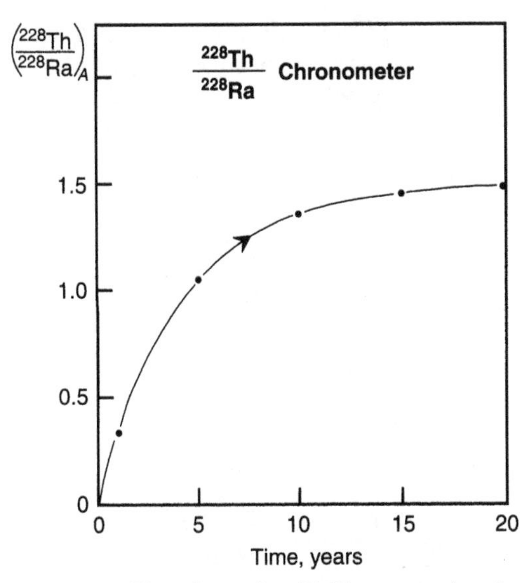

FIGURE 20.7 Plot of equation 20.36 representing the transient decay of excess ^{228}Ra to ^{228}Th in calcium carbonate samples.

from zero at $t = 0$ and approaches a constant equal to 1.4971 in about 20 years. Therefore, the rapid increase of the $^{228}Th/^{228}Ra$ activity ratio of carbonate samples constitutes a geochronometer suitable for dating samples that formed less than about 15 years ago. This geochronometer is based on a transient relationship between ^{228}Ra and ^{228}Th because, in the absence of ^{232}Th, the activities of both ^{228}Ra and ^{228}Th ultimately decrease to zero with increasing time.

The ^{228}Ra and ^{228}Th method of dating is illustrated below by data from Turekian et al. (1979) for a large (22-cm) shell of a vesicomyid mollusk recovered from the vicinity of hotsprings along the Galapagos Rift in the Pacific Ocean. The site is known as Clambake I. The $^{228}Th/^{228}Ra$ activity ratio was 0.68 ± 0.18. Substituting this value into equation 20.36 and solving for t yields

$$\left(\frac{^{228}Th}{^{228}Ra}\right)_A = \frac{\lambda_{Th}}{\lambda_{Th} - \lambda_{Ra}}(1 - e^{-(\lambda_{Th}-\lambda_{Ra})t})$$

$$0.68 = 1.4971(1 - e^{-0.242t})$$

$$t = \frac{0.6056}{0.242} = 2.50 \pm 1.03 \text{ y}$$

This date is not the biological age of this clam because the measured $^{228}Th/^{228}Ra$ activity ratio represents the whole shell. Presumably, the shell along the hinge is older than the ventral edge.

20.2c The $^{228}Ra/^{226}Ra$ Method

All of the Ra isotopes are admitted into calcium carbonate precipitating from aqueous solution, regardless of their origins from three different decay schemes. Therefore, freshly deposited calcium carbonate contains excess ^{226}Ra (derived from ^{238}U) and excess ^{228}Ra (derived from ^{232}Th). If both ^{230}Th and ^{232}Th are excluded from calcium carbonate, the ratio of the activities of excess unsupported ^{228}Ra and ^{226}Ra decreases with time, in accordance with

$$\left(\frac{^{228}Ra}{^{226}Ra}\right)_{Ax} = \left(\frac{^{228}Ra}{^{226}Ra}\right)^0_{Ax} \frac{e^{-\lambda_{228}t}}{e^{-\lambda_{226}t}} \qquad (20.37)$$

which simplifies to the form

$$\left(\frac{^{228}Ra}{^{226}Ra}\right)_{Ax} = \left(\frac{^{228}Ra}{^{226}Ra}\right)^0_{Ax} e^{(\lambda_{226}-\lambda_{228})t}$$

Since $\lambda_{228} = 1.203 \times 10^{-1}$ y^{-1} and $\lambda_{226} = 4.334 \times 10^{-4}$ y^{-1}, $\lambda_{226} - \lambda_{228} = (4.334 - 1203) \times 10^{-4} = -1198.6 \times 10^{-4}$ y^{-1}. Therefore,

$$\left(\frac{^{228}Ra}{^{226}Ra}\right)_{Ax} = \left(\frac{^{228}Ra}{^{226}Ra}\right)^0_{Ax} e^{-0.11986t} \qquad (20.38)$$

The time-dependent decrease of the excess $^{228}Ra/^{226}Ra$ activity ratio of carbonate samples in Figure 20.8 is a geochronometer with a range of about 20 years.

The initial $^{228}Ra/^{226}Ra$ activity ratio on the Earth's surface varies widely depending on a number of factors:

1. the Th/U ratio of bedrock and overburden exposed to weathering,
2. the time elapsed since the Ra isotopes were separated from their respective parents, and
3. mixing of water having different $^{228}Ra/^{226}Ra$ ratios.

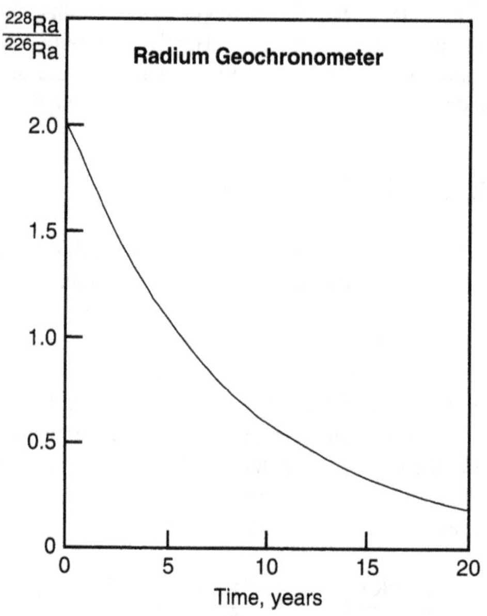

FIGURE 20.8 Radium geochronometer based on the decrease of the activity ratio of excess $^{228}Ra/^{226}Ra$ in carbonate samples (equation 20.38) from which all Th isotopes were excluded at the time of formation. The initial $^{288}Ra/^{226}Ra$ activity ratio in this example (2.0) is from a mollusk shell in Narragansett Bay, Rhode Island, USA, analyzed by Moore (1969).

For example, the clam analyzed by Turekian et al. (1979) has a measured $^{228}Ra/^{226}Ra$ activity ratio of 0.039 ± 0.06. The initial $^{228}Ra/^{226}Ra$ activity ratio of this clam can be calculated from equation 20.38 assuming that $t = 2.50$ years:

$$\left(\frac{^{228}\text{Ra}}{^{226}\text{Ra}}\right)_{Ax} = \left(\frac{^{228}\text{Ra}}{^{226}\text{Ra}}\right)^0_{Ax} e^{-0.11986t}$$

Therefore,

$$\left(\frac{^{228}\text{Ra}}{^{226}\text{Ra}}\right)^0_{Ax} = \frac{0.039}{0.7410} = 0.0526 \pm 0.0081$$

The $^{228}Ra/^{226}Ra$ activity ratio of seawater less than 100 m above the bottom of the Pacific Ocean has an average value of 0.0208 and ranges from 0.008 to 0.06 (Moore, 1969). Therefore, the value of the initial $^{228}Ra/^{226}Ra$ activity ratio of the clam from the Clambake I site in the Galapagos Islands is not unreasonable.

20.2d Isotope Geochemistry of Radium

The activities of ^{226}Ra and ^{228}Ra in secular radioactive equilibrium with U and Th in rocks can be calculated from the relation

$$\lambda_1 N_1 = \lambda_2 N_2 = \lambda_3 N_3 = \cdots$$

The rates of decay of the Ra isotopes in a specimen of rock containing U and Th are

$$(^{228}\text{Ra})_A = \frac{\text{Th} \times 100 \times A \times 4.9475 \times 10^{-11}}{232.038}$$

$$(^{226}\text{Ra})_A = \frac{\text{U} \times 99.2743 \times A \times 1.55125 \times 10^{-10}}{238.03}$$

where A is Avogadro's number. Therefore, the $^{228}Ra/^{226}Ra$ activity ratio in the rock specimen is

$$\left(\frac{^{228}\text{Ra}}{^{226}\text{Ra}}\right)_A = \left(\frac{\text{Th}}{\text{U}}\right)_{\text{conc}} \times 0.3295$$

The activity ratio of the other two Ra isotopes (^{223}Ra and ^{224}Ra) is calculated similarly from the Th/U ratio of a rock in which the two decay series are in secular equilibrium:

$$\left(\frac{^{224}\text{Ra}}{^{223}\text{Ra}}\right)_A = \left(\frac{\text{Th}}{\text{U}}\right)_{\text{conc}} \times 7.1574$$

The U and Th concentrations in Table 20.2 of igneous and sedimentary rocks yield $^{228}Ra/^{226}Ra$ activity ratios ranging from 0.25 in carbonate rocks to 1.87 in low-Ca granites. Consequently, the numerical value of the $^{228}Ra/^{226}Ra$ activity ratio of Ra released into solution depends on the Th/U ratio of the rock that is weathering. For example, Sturchio et al. (2001) reported that 36 samples of *groundwater* in carbonate aquifers in the central United States have an average $^{228}Ra/^{226}Ra$ activity ratio of 0.31 (0.08–1.48) compared to 0.25 for *carbonate rocks*. In some cases, the $^{228}Ra/^{226}Ra$ activity ratios of oilfield brines are *lower* than

Table 20.2. Average Concentrations of U and Th in Common Rocks and Their $^{228}Ra/^{226}Ra$ Activity Ratios Assuming Secular Equilibrium between the Long-Lived Parents and Their Respective Short-Lived Daughters

	Concentration, ppm			
Rock Type	U	Th	$(Th/U)_c$	$(^{228}Ra/^{226}Ra)_A$
Ultramafic	0.002	0.0045	2.25	0.74
Basalt	0.75	3.5	4.67	1.54
High-Ca granite	3	8.5	2.83	0.93
Low-Ca granite	3	17	5.67	1.87
Shale	3.7	12	3.24	1.07
Sandstone	0.45	1.7	3.78	1.24
Carbonate	2.2	1.7	0.77	0.25
Deep-sea clay	1.3	7	5.38	1.77

Source: Faure, 1998.

those of carbonate rocks (e.g., 0.057) analyzed by Sturchio et al. (2001) presumably because of the decay of ^{228}Ra.

The activities of U, Th, and Ra isotopes in *groundwater and oilfield brines* have also been measured by:

Andrews et al., 1989, *Geochim. Cosmochim. Acta*, 53:1791–1802. Banner et al., 1989, *Geochim. Cosmochim. Acta*, 53:383–398. Banner et al., 1990, *Earth Planet. Sci. Lett.*, 101:296–312. Bloch and Key, 1981, *Am. Assoc. Petrol. Geol. Bull.*, 65:154–159. Cadigan and Felmlee, 1977, *J. Geochim. Explor.*, 8:381–395. Chung, 1981, *Geophys. Res. Lett.*, 8:457–460. Cowart, 1981, *J. Hydrol.*, 54:185–193. Dickson, 1990, in *The Environmental Behavior of Radium*, Vol. 1, pp. 335–372, International Atomic Energy Agency, Vienna. Gascoyne, 1989, *Appl. Geochem.*, 4:577–591. Herczeg et al., 1988, *Chem. Geol.*, 72:181–196. Kraemer and Reid, 1984, *Chem. Geol. (Isotope Geosci. Sect.)*, 2:153–174. Krishnaswami et al., 1982, *Water Resources Res.*, 18:1663–1675. Krishnaswami et al., 1991, *Chem. Geol. (Isotope Geosci. Sect.)*, 87:125–136. Langmuir and Melchior, 1985, *Geochim. Cosmochim. Acta*, 49:2423–2432. Sturchio et al., 1993, *Geochim. Cosmochim. Acta*, 57:1203–1214.

The $^{228}Ra/^{226}Ra$ activity ratios of water discharged into the oceans in Table 20.3 range from 0.90 to 1.9, which implies that the Ra originated primarily from silicate rocks (e.g., shale, granite, basalt) rather than from carbonate rocks. Certain rivers in western Florida (e.g., Alafia and

Table 20.3. Activities of $^{226}Ra/^{228}Ra$ in the Water Discharged into the Ocean by Major Rivers

River	^{226}Ra, dpm/100 L	^{228}Ra, dpm/100 L	$(^{228}Ra/^{226}Ra)_A$	References[a]
Hudson ($S < 0.2‰$)	1.23	—	—	4
Susquehanna ($S < 0.01‰$)	6.05	—	—	5
Chesapeake Bay ($S < 0.40‰$)	2.52	4.15	1.6	5
Pee Dee, South Carolina	3.71 (7)	3.34 (5)	0.90	1
Delaware	5.05 (2)	7.3 (1)	1.4	1
Amazon	5.0 (1)	7.4 (1)	1.5	2
Yangtze	10.7 (3)	16.5 (2)	1.5	1
Ganges-[b]Brahmaputra	156 (1)	295 (1)	1.9	3

Note: Number of samples given in parentheses.

[a] 1. Elsinger and Moore, 1984, *Est. Coast. Shelf Sci.*, 18:601–613. 2. Key et al., 1985, *J. Geophys. Res.*, 90(C4):6995–7004. 3. Moore, 1997, *Earth Planet. Sci. Lett.*, 150:141–150. 4. Li and Chan, 1979, *Earth Planet. Sci. Lett.*, 43:343–350. 5. Moore, 1981, *Est. Coast. Shelf Sci.*, 12:713–723.

[b] By extrapolation to zero salinity.

Suwannee) have high activities of ^{226}Ra (up to 555 dpm/100 L) derived by weathering of U-rich phosphate rocks in their drainage basins (Fanning et al., 1982).

When river water mixes with seawater in estuaries and in the coastal ocean, the activities of ^{228}Ra and ^{226}Ra in solution in the *water* increase dramatically primarily as a result of desorption from suspended sediment. This phenomenon is illustrated in Figure 20.9 by the activity profiles of the Ra isotopes in the Amazon River estuary where the salinity increases from 0 to 36‰ (Key et al., 1985). The amount of ^{228}Ra that is desorbed is higher than that of ^{226}Ra. As a result, the $^{228}Ra/^{226}Ra$ activity ratios of seawater near the coast are higher than those in the open ocean.

The water being discharged by the Ganges-Brahmaputra River into the Bay of Bengal has very high Ra concentrations, expressed in Table 20.3 as

$$(^{228}\text{Ra})_A = 295 \text{ dpm}/100 \text{ L}$$

$$(^{226}\text{Ra})_A = 156 \text{ dpm}/100 \text{ L}$$

which yield a $^{228}Ra/^{226}Ra$ activity ratio of 1.9. In contrast to the Amazon estuary in Figure 20.9, the activities of Ra isotopes of the Ganges-Brahmaputra River at low flow in Figure 20.10 actually decrease with increasing salinity of the water. Therefore, Moore (1997) suggested that the apparent high Ra activity and Ba concentration of this river system are caused mainly by discharge of Ra- and Ba-rich groundwater directly into the ocean. Alternatively, Carroll et al. (1993) proposed that the excess ^{226}Ra and Ba they observed at the mouth of the Ganges-Brahmaputra River originated by desorption at low flow from sediment that was deposited at high flow in mangrove swamps and on islands.

The isotope geochemistry of Ra in *estuaries* and on the *continental shelves* has been investigated by many other research groups, including

Li et al., 1997, *Earth Planet. Sci. Lett.*, 37:237–241. Li et al., 1979, *Earth Planet. Sci. Lett.*, 42:13–26. Li and Chan, 1979, *Earth Planet. Sci. Lett.*, 43:343–350. Elsinger and Moore, 1980, *Earth Planet. Sci. Lett.*, 48:239–249. Santschi et al., 1979, *Earth Planet. Sci. Lett.*,

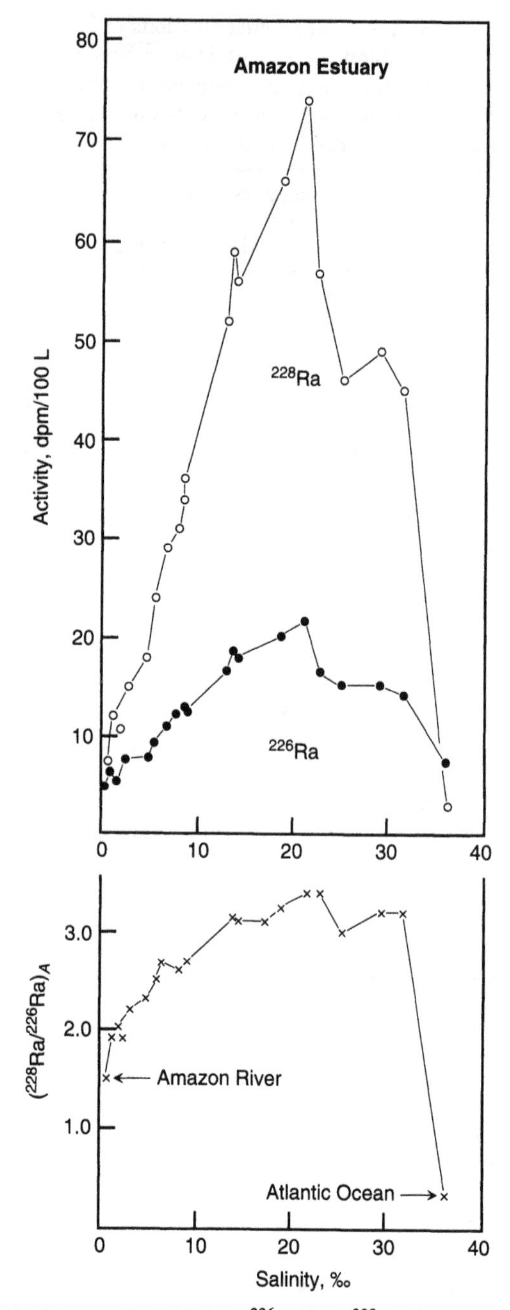

FIGURE 20.9 Activities of ^{226}Ra and ^{228}Ra in surface water of the Amazon River estuary. The nonconservative increase of the activities of both Ra isotopes is attributed to desorption from sediment suspended in the river (Key et al., 1985).

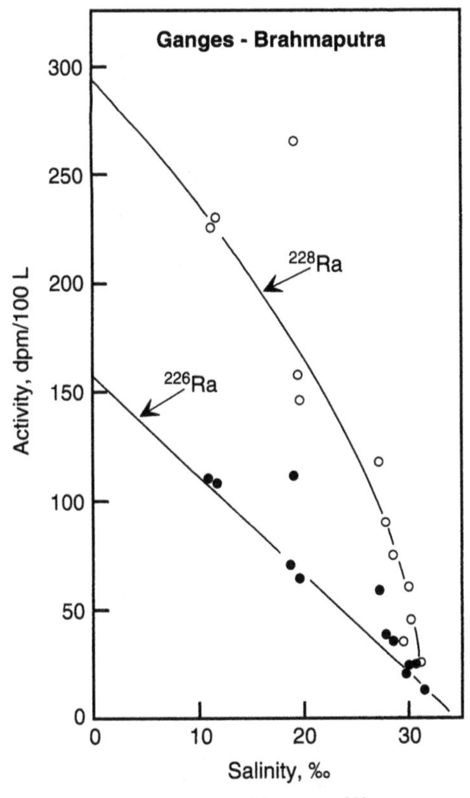

FIGURE 20.10 Activities of ^{226}Ra and ^{228}Ra of water at the mouth of the Ganges-Brahmaputra River (March 11–17, 1991) as a function of the salinity of the water (Moore, 1997).

45:201–213. Feely et al., 1980, *Est. Coast. Marine Sci.*, 11:179–205. Fanning et al., 1982, *Science*, 215:667–670.

The ^{228}Ra enrichment of seawater of the coastal oceans and the water of estuaries causes regional variations of the ^{228}Ra/^{226}Ra activity ratio in the oceans. For example, Moore (1969) reported that seawater collected near shore in the Pacific Ocean has $(^{228}\text{Ra}/^{226}\text{Ra})_A = 0.51$ (0.24–1.65), whereas the value of this ratio in the central part of this ocean is only 0.031 (0.01–0.06). Additional determinations of this ratio in Table 20.4 confirm this pattern.

The profiles at a site in the North Atlantic Ocean in Figure 20.11 reveal that the activity of ^{228}Ra *decreases* sharply from 4.63 dpm/100 L at the surface to about 0.30 dpm/100 L at a depth of 1000 m, whereas the activity of ^{226}Ra *increases* from 8.41 dpm/100 L to about 10.0 dpm/100 L in the same depth interval. At greater depth, the activities of both isotopes increase down to the bottom of the ocean at 5049 m (Moore et al., 1985). Consequently, the ^{228}Ra/^{226}Ra activity ratio *decreases* in the upper 1000 m from 0.55 to about 0.030 and then *rises* with depth to 0.067 at 5049 m.

Table 20.4. Average ^{228}Ra/^{226}Ra Activity Ratios of Seawater and Mollusk Shells in Coastal Seawater and in the Open Ocean

	$(^{228}\text{Ra}/^{226}\text{Ra})_A$	
Ocean or Sea	Coastal	Main Body
Pacific Ocean	0.51 (8)	0.031 (9)
	0.24–1.65	0.06–0.1
Atlantic Ocean	1.42 (5)	0.20 (7)
	0.9–2.0	0.1–0.36
Gulf of Mexico	3.93 (3)	—
	2.2–7.1	
Caribbean Sea	0.6 (1)	—
Mediterranean	1.06 (3)	0.13 (2)
	0.29–1.6	0.12–0.13
Black Sea	0.37 (1)	—
Red Sea	0.20 (1)	—
Indian	0.80 (4)	0.18 (1)
	0.23–2.0	

Source: Moore, 1969.
Note: The number of samples is indicated in parentheses.

These results confirm that the activity of ^{228}Ra in seawater is related to the amount of suspended sediment both regionally and with depth. As a result, mollusk shells and other skeletal carbonates contain Ra whose isotopic compositions depend on both the location and depth of water in the ocean in which they formed. The initial ^{228}Ra/^{226}Ra ratio can be determined from a time series of measurements based on analyses of growth increments of large mollusk shells or piston cores of calcareous ooze of marine or lacustrine origin. The regional variation of the activities of the isotopes of Ra have been used to characterize water masses in the oceans and

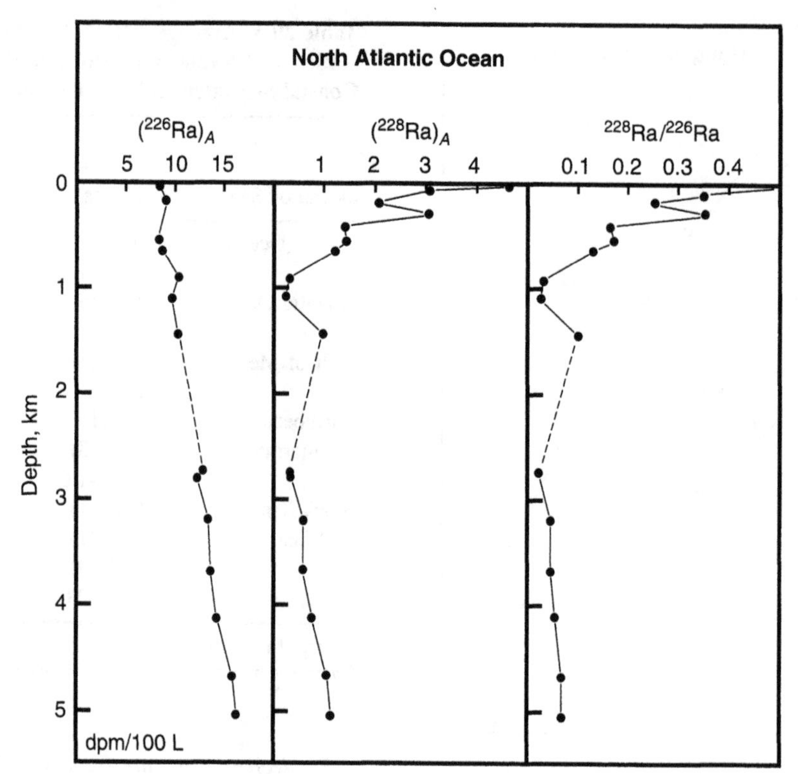

FIGURE 20.11 Profiles of ^{226}Ra and ^{228}Ra activities in the North Atlantic at Test Cruise Station 3 (31°47′ N, 50°47′ W) on October 23, 1980. The activities of both Ra isotopes are in units of dpm/100 L (Moore et al., 1985).

to trace their movement:

> Moore, 1972, *Earth Planet. Sci. Lett.*, 16:421–422. Sarmiento et al., 1982, *J. Geophys Res.*, 87:9694–9698. Moore et al., 1984, *EOS Trans. AGU*, 64:1090. Ku and Luo, 1994, *J. Geophys. Res.*, 99:10255–10273.

20.3 PROTACTINIUM

The decay series of ^{235}U contains 15 intermediate daughters, but only $^{231}_{91}$Pa has a long-enough halflife to be useful for dating geological samples. As in the case of ^{230}Th in the ^{238}U series, ^{231}Pa is separated from ^{235}U because it is sorbed to sediment particles whereas ^{235}U remains in solution. Therefore, freshly deposited sediment contains unsupported excess ^{231}Pa that decays by α-emission to an isotope of actinium (Ac). The ^{235}U atoms enter authigenic carbonate and phosphate minerals that initially do not contain ^{231}Pa. The number of ^{231}Pa atoms per unit weight of sample in these minerals therefore increases with time by decay of ^{235}U until secular radioactive equilibrium is reestablished when the ^{235}U/^{231}Pa activity ratio is equal to one.

The decay of ^{231}Pa has not been used as widely for dating as the decay of ^{230}Th because of analytical difficulties which have been overcome by the use of mass spectrometers to measure ^{231}Pa by isotope dilution (Bourdon et al., 1999). As a result, ^{231}Pa dates are being used to confirm age determinations by the ^{230}Th method which are the basis for the timescale of climate change during the past 250,000 years.

20.3a The ^{230}Th–^{231}Pa Method

Protactinium has two naturally occurring short-lived radionuclides that are produced by decay of long-lived ^{238}U and ^{235}U:

$$^{238}U \xrightarrow{\alpha} {}^{234}Th \xrightarrow{\beta^-} {}^{234}Pa \xrightarrow{\beta^-} {}^{234}U \rightarrow \cdots$$

$$^{235}U \xrightarrow{\alpha} {}^{231}Th \xrightarrow{\beta^-} {}^{231}Pa \xrightarrow{\alpha} {}^{227}Ac \rightarrow \cdots$$

The halflife of ^{234}Pa is very short (6.70 ± 0.05 h), whereas that of ^{231}Pa (3.25×10^4 years) is about one-half that of ^{230}Th (7.54×10^4 years). The geochemical properties of protactinium (Pa, $Z = 91$) are similar to those of Th because both elements form ions with a charge of +4 and their ionic radii are: $Th^{4+} = 0.95$ Å and $Pa^{4+} = 0.98$ Å. Therefore, the ions of both elements are strongly sorbed to the surfaces of mineral particles. The similarity of their chemical properties and halflives makes the excess ^{230}Th and ^{231}Pa useful for geochronometry of detrital sediment in lakes and oceans (Rosholt et al.,1961, 1962), as described in Section 20.1a.

The sediment samples to be dated are dissolved completely and U, Th, and Pa are separated from each other by ion exchange and solvent extraction (Ku, 1965; Ku and Broecker, 1967a, b). The total activity of ^{231}Pa is determined by α-counting with a correction for the presence of supported ^{231}Pa using the measured U concentration.

More recently, Bourdon et al. (1999) and Pickett et al. (1994) described procedures for determining the concentration of ^{231}Pa by isotope dilution using a thermal ionization mass spectrometer with a sensitivity of 0.1×10^{-12} g of ^{231}Pa. Both groups published detailed instructions concerning the dissolution of samples and the subsequent separation and purification of ^{231}Pa. Thermal ionization mass spectrometry was also used by Edwards et al. (1997) to date marine carbonates of Quaternary age by the ^{235}U–^{231}Pa method.

In cases where the U concentration of the sediment is negligibly small, the ratio of the measured activities of excess ^{231}Pa and ^{230}Th is related to the time elapsed since deposition by

$$\left(\frac{^{230}Th}{^{231}Pa}\right)_{Ax} = \left(\frac{^{230}Th}{^{231}Pa}\right)^0_{Ax} \frac{e^{-\lambda_{230}t}}{e^{-\lambda_{231}t}} \qquad (20.39)$$

where $\lambda_{231} = 2.132 \times 10^{-5}$ y^{-1} is the decay constant of ^{231}Pa. The exponential terms of equation 20.39 can be simplified:

$$\frac{e^{-\lambda_{230}t}}{e^{-\lambda_{231}t}} = e^{(\lambda_{231}-\lambda_{230})t} = e^{\lambda' t}$$

where $\lambda' = 2.132 \times 10^{-5} - 0.919 \times 10^{-5} = 1.213 \times 10^{-5}$ y^{-1}. Hence,

$$\left(\frac{^{230}Th}{^{231}Pa}\right)_{Ax} = \left(\frac{^{230}Th}{^{231}Pa}\right)^0_{Ax} e^{\lambda' t} \qquad (20.40)$$

Evidently, the $(^{230}Th/^{231}Pa)_{Ax}$ parameter *increases* with time because ^{231}Pa decays faster than ^{230}Th.

If the sediment was deposited at a constant rate $a = h/t$, then

$$\left(\frac{^{230}Th}{^{231}Pa}\right)_{Ax} = \left(\frac{^{230}Th}{^{231}Pa}\right)^0_{Ax} e^{\lambda' h/a} \qquad (20.41)$$

where h is the depth in the core in centimeters. Taking natural logarithms of both sides linearizes the equation:

$$\ln\left(\frac{^{230}Th}{^{231}Pa}\right)_{Ax} = \ln\left(\frac{^{230}Th}{^{231}Pa}\right)^0_{Ax} + \frac{\lambda' h}{a} \qquad (20.42)$$

The application of this equation is exemplified by the data of Broecker et al. (1968) for sediment from core V12-122 taken in the Caribbean Sea at 17°00′ N, 74°24′ W in 2800 m of water. The excess $^{230}Th/^{231}Pa$ activity ratios of sediment from this core define a straight line in Figure 20.12 with a slope $m = +0.004797$ and an intercept ln $(^{230}Th/^{231}Pa)^0_{Ax} = 2.6127$ that has a value of 13.6. The sedimentation rate is calculated from the slope:

$$a = \frac{\lambda'}{m} = \frac{1.213 \times 10^{-5}}{0.004797} = 252.8 \times 10^{-5} \text{ cm/y}$$

or 2.53 cm/10^3 y. This value is significantly higher than the sedimentation rate in the Indian Ocean ($a = 2.82$ mm/10^3 y) derived in Figure 20.2 by the $^{230}Th/^{232}Th$ method. The high rate of sedimentation in the Caribbean Sea is compatible with its proximity to sediment sources.

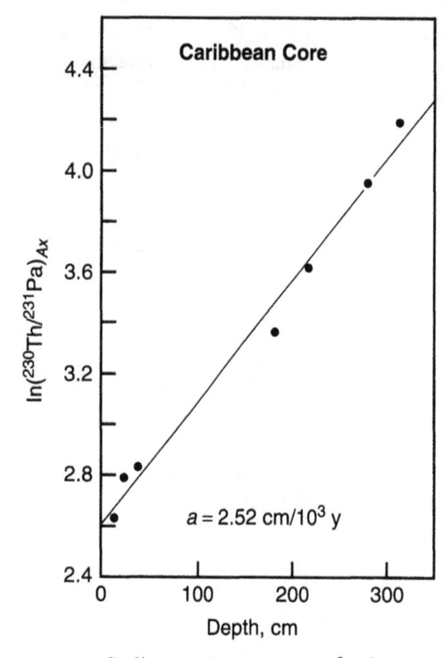

FIGURE 20.12 Sedimentation rate at 17°00′ N and 74°24′ W in the Caribbean Sea (V12-122) based on the decay of excess ^{220}Th and ^{231}Pa. Data from Broecker et al. (1968).

The ^{230}Th/^{231}Pa activity ratios of sediment at the water–sediment interface of the Pacific Ocean range widely from <11 to 57. Low values occur primarily in sediment deposited along the margin of the Pacific Ocean and in the Antarctic Ocean, whereas high values are concentrated in the central basin of the Pacific extending from about 40° N to 40° S (Yang et al., 1986). The variation of this ratio implies that ^{230}Th and ^{231}Pa are not removed from the water with equal efficiency. Instead, ^{231}Pa is removed preferentially with sediment that is deposited along the margins of the oceans. This conclusion confirms earlier work by Anderson et al. (1983) in the Panama and Guatemala basins.

The regional variation of the ^{230}Th/^{231}Pa activity ratios of sediment in different parts of the ocean does not invalidate the ^{230}Th–^{231}Pa method of dating sediment cores provided the ratio remained constant during the time interval represented by the core.

20.3b Rosholt's ^{230}Th–^{231}Pa Geochronometer

In the formulation of Rosholt et al. (1961, 1962), the activity of ^{231}Pa in seawater is

$$^{231}\mathrm{Pa}_A = {}^{235}\mathrm{U}_A(1 - e^{-\lambda_{231} t}) \qquad (20.43)$$

where t is the time interval between the production of ^{231}Pa and its incorporation into sediment (i.e., its average oceanic residence time). The growth of ^{230}Th from ^{234}U dissolved in seawater is given by a similar equation. Therefore, the activity ratio of U-supported ^{230}Th and ^{231}Pa in seawater is

$$\left(\frac{^{230}\mathrm{Th}}{^{231}\mathrm{Pa}}\right)_A = \left(\frac{^{234}\mathrm{U}}{^{235}\mathrm{U}}\right)_A \left(\frac{1 - e^{-\lambda_{230} t}}{1 - e^{-\lambda_{231} t}}\right) \qquad (20.44)$$

If the isotopic abundance ratio of U in seawater is $^{238}\mathrm{U}/^{235}\mathrm{U} = 137.88$ and the ^{234}U/^{238}U activity ratio is 1.15, then

$$\left(\frac{^{238}\mathrm{U}}{^{235}\mathrm{U}}\right)_A = \frac{^{238}\mathrm{U}\ \lambda_{238}}{^{235}\mathrm{U}\ \lambda_{235}} = 137.88\left(\frac{\lambda_{238}}{\lambda_{235}}\right)$$

and

$$\left(\frac{^{234}\mathrm{U}}{^{235}\mathrm{U}}\right)_A = \left(\frac{^{234}\mathrm{U}}{^{238}\mathrm{U}}\right)_A \times \left(\frac{^{238}\mathrm{U}}{^{235}\mathrm{U}}\right)_A$$

$$= \frac{1.15 \times 137.88 \times 1.55125 \times 10^{-10}}{9.8485 \times 10^{-10}}$$

$$= 24.97$$

Since the oceanic residence times of ^{230}Th and ^{231}Pa are both short (e.g., about 300 years) compared to the decay constants of these isotopes (10^4 years), only a few ^{230}Th and ^{231}Pa atoms in seawater decay before they are deposited with the sediment. The ^{230}Th/^{231}Pa activity ratio of seawater according to equation 20.44 for $t = 300$ y is

$$\left(\frac{^{230}\mathrm{Th}}{^{231}\mathrm{Pa}}\right)_A = 24.97 \times \frac{1 - 0.99724}{1 - 0.99362}$$

$$= 24.97 \times 0.4326 = 10.80$$

Therefore, the ^{230}Th/^{231}Pa activity ratio of seawater is 10.80. If both isotopes are removed from seawater with equal efficiency, the initial activity ratio in freshly deposited sediment should also have this value. However, present knowledge indicates that the $^{230}Th/^{231}Pa$ activity ratio of sediment at the sediment–water interface is not constant but varies regionally because ^{231}Pa is removed from seawater more efficiently than ^{230}Th (Yang et al., 1986; Osmond, 1979).

In addition, the activities of ^{230}Th and ^{231}Pa in sediment must be corrected for the presence of the U-supported fractions of these isotopes. Therefore, the activity ratio of excess (unsupported) ^{230}Th and ^{231}Pa is

$$\left(\frac{^{230}\text{Th}}{^{231}\text{Pa}}\right)_{Ax} = \left(\frac{^{230}\text{Th}_t - {}^{230}\text{Th}_s}{^{231}\text{Pa}_t - {}^{231}\text{Pa}_s}\right)_A \qquad (20.45)$$

where the subscripts are defined as x = excess unsupported, t = total, and s = supported. The ratio of the activities of excess ^{230}Th and ^{231}Pa is also expressed by equation 20.39:

$$\left(\frac{^{230}\text{Th}}{^{231}\text{Pa}}\right)_{Ax} = \left(\frac{^{230}\text{Th}}{^{231}\text{Pa}}\right)^0_{Ax} \left(\frac{e^{-\lambda_{230}t}}{e^{-\lambda_{231}t}}\right)$$

where $(^{230}\text{Th}/^{231}\text{Pa})^0_{Ax} = 10.80$. Therefore,

$$10.80\left(\frac{e^{-\lambda_{230}t}}{e^{-\lambda_{231}t}}\right) = \frac{^{230}\text{Th}_t - {}^{230}\text{Th}_s}{^{231}\text{Pa}_t - {}^{231}\text{Pa}_s} \qquad (20.46)$$

Solving for t yields

$$t = \frac{1}{\lambda_{231} - \lambda_{230}} \ln\left[\frac{(^{230}\text{Th}_t - {}^{230}\text{Th}_s)_A}{10.80(^{231}\text{Pa}_t - {}^{231}\text{Pa}_s)_A}\right] \qquad (20.47)$$

The activities of supported ^{230}Th and ^{231}Pa are assumed to be in equilibrium with their respective parents:

$$^{230}\text{Th}_{\text{As}} = \lambda_{238}\, {}^{238}\text{U} \qquad {}^{231}\text{Pa}_{\text{As}} = \lambda_{235}\, {}^{235}\text{U}$$

where ^{238}U and ^{235}U are the respective number of atoms of these isotopes in a unit weight of sample.

The advantage of Rosholt's method is that it can be applied to single-sediment samples without requiring analyses of a suite of samples from different depths in a core. This method is illustrated below by data of Rona and Emiliani (1969) for a sediment sample taken from 321 to 326 cm of piston core P6304-8 raised in the Caribbean Sea at 14°59′ N and 69°20′ W. The excess activity ratio of this sample is

$$\left(\frac{^{230}\text{Th}}{^{231}\text{Pa}}\right)_{Ax} = 28.37$$

(carbonate free and corrected for U-supported ^{230}Th and ^{231}Pa). Substituting into equation 20.47,

$$t = \frac{1}{(2.132 - 0.919) \times 10^{-5}} \ln\left(\frac{28.37}{10.80}\right)$$

$$= \frac{0.9657}{1.213 \times 10^{-5}} = 79{,}600 \text{ y}$$

In this particular case, the data of Rona and Emiliani (1969) for piston core P6304-8 form a straight line in coordinates of depth and $\ln(^{230}\text{Th}/^{231}\text{Pa})_{Ax}$ (not shown) with a slope $m = +0.0028937$ and an intercept of 2.41333 that corresponds to $(^{230}\text{Th}/^{231}\text{Pa})^0_{Ax} = 11.17$. Therefore, the measured value of this ratio is 3.4% higher than the predicted value of 10.80. Use of $(^{230}\text{Th}/^{231}\text{Pa})_{Ax} = 11.17$ in equation 20.47 decreases the date to $t = 76{,}900$ y.

Another way to date this sample is by means of the relation $t = h/a$, where the sedimentation rate $a = \lambda'/m$: Since $m = +0.0028937$ and $\lambda' = 1.213 \times 10^{-5}\ \text{y}^{-1}$,

$$a = \frac{1.213 \times 10^{-5}}{0.0028937} = 4.191 \times 10^{-3} \text{ cm/y}$$

Setting $d = 323.5$ cm (i.e., 321– 326),

$$t = \frac{323.5}{4.191 \times 10^{-3}} = 77{,}200 \text{ y}$$

The interpretation of the data has provided three estimates of the age of sediment from the depth interval 321–326 of core P6304-8:

79,600 years [equation 20.47, $(^{230}\text{Th}/^{231}\text{Pa})^0_{Ax} = 10.80$]
76,900 years [equation 20.47, $(^{230}\text{Th}/^{231}\text{Pa})^0_{Ax} = 11.17$]
77,200 years (sedimentation rate = 4.147×10^{-3} cm/y)

The date derived from the sedimentation rate (t = 77,200 y) relies on fewer assumptions than the other two and therefore is more reliable.

20.3c Carbonates

Protactinium and Th ions are excluded from calcium carbonate minerals (e.g., speleothems, travertines, mollusk shells, corals), whereas UO_2^{2+} is admitted in place of Ca^{2+}. Therefore, the decay of ^{235}U in carbonates causes ^{231}Pa to accumulate in accordance with the equation

$$N_2 = \frac{\lambda_1 N_1^0}{\lambda_2 - \lambda_1}(e^{-\lambda_1 t} - e^{-\lambda_2 t})$$

and

$$^{231}\text{Pa} = \frac{\lambda_{235}\ {}^{235}\text{U}^0}{\lambda_{231} - \lambda_{235}}(e^{-\lambda_{235} t} - e^{-\lambda_{231} t})$$

Since $\lambda_{231} \gg \lambda_{235}$, $\lambda_{231} - \lambda_{235} = \lambda_{231}$, and $e^{-\lambda_{235} t} = 1.0$, the equation becomes, to a good approximation,

$$^{231}\text{Pa} = \frac{^{235}\text{U}_A}{\lambda_{231}}(1 - e^{-\lambda_{231} t})$$

and

$$\left(\frac{^{231}\text{Pa}}{^{235}\text{U}}\right)_A = 1 - e^{-\lambda_{231} t} \qquad (20.48)$$

Solving equation 20.48 for t,

$$t = -\frac{1}{\lambda_{231}} \ln\left[1 - \left(\frac{^{231}\text{Pa}}{^{235}\text{U}}\right)_A\right] \qquad (20.49)$$

The ^{235}U–^{231}Pa chronometer in Figure 20.13 can be used to date carbonate samples that did not contain ^{231}Pa at the time of deposition and remained close to U and Pa. The $^{231}Pa/^{235}U$ activity ratio increases with time and approaches a value of one as time increases to more than 2.5×10^5 years. This method is therefore suitable for dating carbonates of late Pleistocene and Holocene age that may contain chemical or isotopic records of past climate changes.

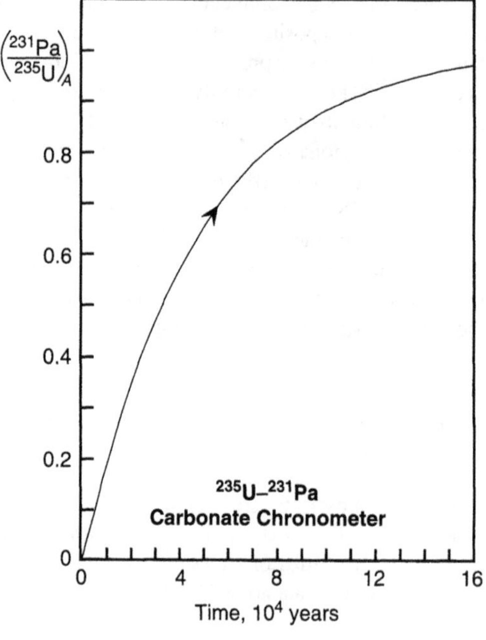

FIGURE 20.13 Growth of ^{231}Pa by decay of ^{235}U in carbonate samples from which ^{231}Pa was excluded at the time of deposition.

The accuracy of ^{235}U–^{231}Pa dates of carbonates of Quaternary age depends on the following assumptions:

1. The initial activity of ^{231}Pa in the carbonate was equal to zero.
2. The sample to be dated remained closed to Pa and to U after deposition.

The assumption that the activity of ^{231}Pa in carbonate samples at the time of deposition was equal to zero was tested by Whitehead et al. (1999), who analyzed 67 samples of speleothems from caves in New Zealand for dating by the ^{235}U–^{231}Pa and ^{234}U–^{230}Th methods. They reported good agreement for dates less than about 150 ka but noted two examples of discordance in speleothems less than 100 years old. Whitehead et al. (1999) attributed the discordance to the presence of small amounts of excess (unsupported) ^{231}Pa and ^{230}Th that may have originated by dissolution of old carbonate rocks in the ambient groundwater. The apparent contamination of the two actively growing speleothems increased their calculated dates by up to 3×10^3 years. However, the good agreement between ^{231}Pa and ^{230}Th dates of 97% of the samples supports the validity of the ^{234}U–^{230}Th dates, which are subject to the limitation that ^{234}U was not in secular equilibrium with ^{238}U (e.g., Gascoyne et al., 1978; Richards et al., 1994).

20.3d ^{231}Pa–^{230}Th Concordia

The concordance of ^{231}Pa and ^{230}Th dates of carbonates can be used to define a concordia curve derivable from equation 20.48:

$$\left(\frac{^{231}\text{Pa}}{^{235}\text{U}}\right)_A = 1 - e^{-\lambda_{231} t}$$

and from equation 20.32,

$$\left(\frac{^{230}\text{Th}}{^{234}\text{U}}\right)_A = 1 - e^{-\lambda_{230} t}$$

These equations define the curves in Figure 20.14, which are the loci of points representing carbonates that yield concordant ^{235}U–^{231}Pa and ^{238}U–^{230}Th dates depending on their ^{234}U/^{238}U activity ratio (Edwards et al., 1997).

Samples may plot to the right of the curve for $\gamma_0 = 1.15$ either because they contain U whose initial ^{234}U/^{238}U activity ratio was greater than 1.15 (e.g., lacustrine carbonates and speleothems) or because they gained U. The addition of U to a carbonate sample on concordia causes the point to move off concordia in the direction of the origin. The resulting ^{231}Pa/^{235}U and ^{230}Th/^{238}U activity ratios yield discordant and inaccurate dates. The problem can be corrected by drawing a straight-line chord from the origin to the point representing the discordant sample and extrapolating that line to concordia. The coordinates of the point of intersection yield concordant dates. However, the accuracy of the intercept dates depends on the assumption that the added U had the same ^{234}U/^{238}U activity ratio as the sample to which the U was added.

Additional scenarios for the alteration of carbonate samples used for dating by the ^{235}U–^{231}Pa and ^{238}U–^{230}Th methods were originally considered by Allègre (1964). The patterns that develop in the ^{231}Pa–^{230}Th concordia diagrams are similar to the responses of U–Pb concordias presented in Section 10.5a. The interpretation of ^{231}Pa activities in carbonate samples by means of ^{231}Pa–^{230}Th concordias has become feasible because ^{231}Pa can be determined by mass spectrometry, which is more sensitive than conventional α-spectrometry (Ivanovich and Murray, 1992; Edwards et al., 1997).

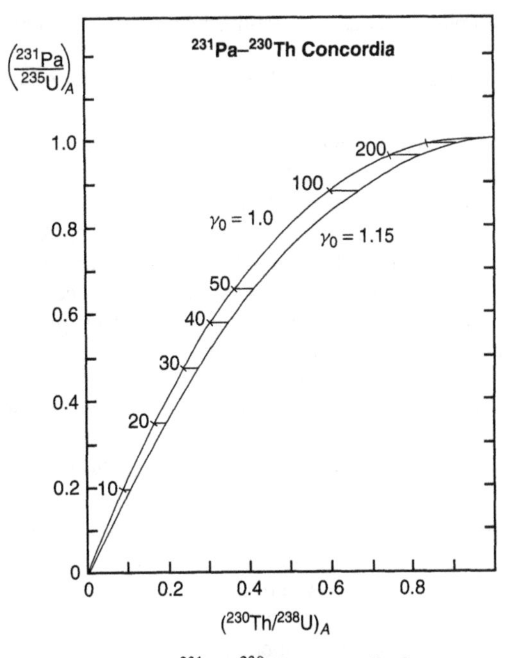

FIGURE 20.14 The ^{231}Pa–^{230}Th concordia for $\gamma_0 = 1.0$ and $\gamma_0 = 1.15$ (marine). All carbonate samples that yield concordant dates have ^{230}Th/^{238}U and ^{231}Pa/^{235}U activity ratios that define a point on a concordia curve depending on the ^{234}U/^{238}U activity ratio (γ_0).

20.4 LEAD-210

The decay series of ^{238}U includes ^{222}Rn, which escapes into the atmosphere at a rate of about 42 atoms per minute per square centimeter of land surface. The radon atoms subsequently decay through a series of short-lived daughters to ^{210}Pb:

$$^{238}U \longrightarrow {}^{226}Ra \longrightarrow {}^{222}Rn \longrightarrow {}^{210}Pb \longrightarrow {}^{206}Pb$$

Lead-210 is rapidly removed from the atmosphere by meteoric precipitation and by dry fallout, giving it an atmospheric residence halflife of only about 5–40 days (Appleby and Oldfield, 1992). After removal from the atmosphere, ^{210}Pb is deposited in snow and ice of glaciers, in lakes and oceans, and in soil, from which it is absorbed by plants. In addition, ^{210}Pb occurs in U-bearing minerals in secular equilibrium with ^{238}U and is released into the environment as a result of weathering of such minerals.

The distribution of ^{210}Pb in groundwater and subsurface brines is related to the behavior of ^{222}Rn, which is released by α-recoil into nanopores and microfractures of minerals whence it diffuses into grain boundaries. As a result, ^{222}Rn collects in low-salinity groundwater where it can reach activities of up to 10^2 dpm/kg (Ku et al., 1992)

The activity of ^{210}Pb in groundwater and brine increases with the temperature because of the increasing stability of Pb–Cl complexes, which allow more of the ^{210}Pb to remain in solution instead of being sorbed to mineral surfaces. For example, the activity of ^{210}Pb in brine of the Salton Sea geothermal area in California (salinity 25%) is about 3000 dpm/kg, whereas the low-salinity brine of the Coso geothermal system (salinity <1%) has a ^{210}Pb activity of only about 1.0 dpm/kg (Ku et al., 1992).

Lead-210 in streams, lakes, and oceans is strongly sorbed to charged sites on sediment particles. Therefore, freshly deposited lacustrine and marine sediment contains excess (unsupported) $^{210}_{82}Pb$ that decays by β-emission with a halflife of 22.6 years to $^{210}_{83}Bi$ and hence to $^{210}_{84}Po$:

$$^{210}_{82}Pb \rightarrow {}^{210}_{88}Bi + \beta^- + 0.063\text{ MeV} \qquad (T_{1/2} = 22.6\text{ y})$$

$$^{210}_{83}Bi \rightarrow {}^{210}_{84}Po + \beta^- + 1.16\text{ MeV} \qquad (T_{1/2} = 5.01\text{ days})$$

Because of the low energies of the β-particles emitted by ^{210}Pb ($E_{max} = 0.063$ MeV), its activity is measured by counting the β-particles of ^{210}Bi ($E_{max} = 1.16$ MeV) assuming it to be in secular equilibrium with ^{210}Pb.

The geochemistry of ^{210}Pb and its use for determining sedimentation rates in lakes and oceans was presented by Appleby and Oldfield (1992), who also provided references to the rapidly growing literature. In addition, the ^{210}Pb method of dating was included in a special issue of *Chemical Geology* edited by Robbins (1984).

The activity of excess ^{210}Pb in sediment and glacial ice decreases with time, in accordance with

$$^{210}Pb_{Ax} = {}^{210}Pb^0_{Ax} e^{-\lambda_{210} t} \qquad (20.50)$$

If the sediment was deposited at a constant rate,

$$a = \frac{h}{t}$$

where a = sedimentation rate, cm/y
h = depth below the surface, cm
t = time, years

then

$$^{210}Pb_{Ax} = {}^{210}Pb^0_{Ax} e^{-\lambda_{210} h/a}$$

Taking logarithms to the base e,

$$\ln {}^{210}Pb_{Ax} = \ln {}^{210}Pb^0_{Ax} - \frac{\lambda_{210} h}{a} \qquad (20.51)$$

which is the equation of a straight line in coordinates of h (x) and in $^{210}Pb^0_{Ax}$ (y). The slope m of this straight line is

$$m = \frac{\lambda_{210}}{a}$$

and

$$a = \frac{\lambda_{210}}{m}$$

The equation for the decay of excess ^{210}Pb in sediment or ice cores is analogous to the equations describing the decay of excess ^{230}Th and ^{231}Pa. After the sedimentation rate a has been determined by analysis of samples taken from a core, a date can be calculated for a sample of sediment (or ice) in that core based on the relation

$$t = \frac{h}{a}$$

Alternatively, the initial activity of excess ^{210}Pb can be determined by extrapolating the sedimentation line to $h = 0$. The resulting value is used to calculate a date for any sample in that core based on its measured ^{210}Pb activity:

$$^{210}\text{Pb}_{Ax} = {}^{210}\text{Pb}^0_{Ax} e^{-\lambda_{210} t} \quad (20.52)$$

and

$$t = \frac{1}{\lambda_{210}} \ln \frac{(^{210}\text{Pb})^0_{Ax}}{(^{210}\text{Pb})_{Ax}}$$

The ^{210}Pb method of dating was originally developed by Goldberg (1963) to measure the rate of accumulation of glacier ice in Greenland. Later, the ^{210}Pb method was applied to an alpine glacier by Picciotto et al. (1967), to lacustrine sediment by Krishnaswami et al. (1971), and to marine sediment by Koide et al. (1972). The ^{210}Pb method is also applicable to the dating of peat as demonstrated by Aaby et al. (1979) and Oldfield et al. (1979).

Samples of *unfiltered* water collected along the Colorado and Sacramento Rivers by Goldberg (1963) indicated that the ^{210}Pb activity decreases downstream in both rivers. For example, the ^{210}Pb activity of water in the Colorado River decreases from 14.7 dpm/L at Grand Junction, Colorado, to only 0.3 dpm/L in Lake Mead, Nevada. He also reported profiles of ^{210}Pb in the Pacific Ocean that indicated activities up to about 9.5 dph/L at a depth of about 2000 m.

20.4a Sorption by Soil

The ^{210}Pb that is washed out of the atmosphere by meteoric precipitation is strongly retained by the organic matter in soils. Lewis (1977) reported that the ^{210}Pb activity of the A horizon (87.7% organic) of soil in Cook Forest State Park, Pennsylvania, in Figure 20.15 is 45.49 dpm/g, compared to only 1.92 dpm/g in a soil sample (3.79% organic) from a depth of 6.2–11.3 cm in the same profile. Therefore, only a small fraction of ^{210}Pb that is deposited by meteoric precipitation on soils passes through the organic-rich A horizon and enters the groundwater.

The Susquehanna River and its tributaries carry water derived primarily by groundwater.

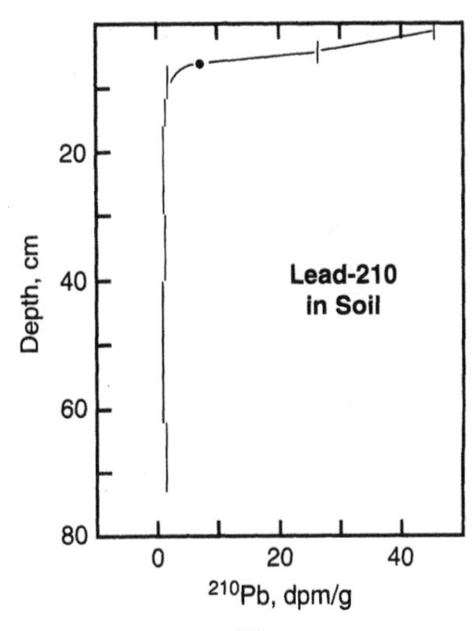

FIGURE 20.15 Sorption of ^{210}Pb by organic matter of the A horizon in a soil profile in Cook Forest State Park, Pennsylvania. Data by Lewis (1977).

Accordingly, the ^{210}Pb activity of the headwaters of the Susquehanna River is low and ranges from only 0.010 to 0.023 dpm/L depending on the season. Farther downstream, near the town of Keating, Pennsylvania, where acidic coal mine effluent ($^{210}\text{Pb}_A = 0.3$ dpm/L) enters the Susquehanna River, the ^{210}Pb activity of the water rises to about 0.17 dpm/L while the pH decreases to about 4.0. As the pH of the water rises downstream from Keating, the ^{210}Pb activity of the water decreases to about 0.025 dpm/L as a result of sorption of ^{210}Pb to ferric oxyhydroxide particles. Lewis (1977) demonstrated that the average ^{210}Pb activity of *suspended sediment particles* in the Susquehanna River in October is 7.0 dpm/g compared to only 0.01 dpm/L in *filtered water* (<0.4 μm).

20.4b Seawater

The activity of ^{210}Pb in the oceans has been measured elsewhere:

Chung and Applequist, 1980, *Earth Planet. Sci. Lett.*, 49:401. Somayajulu and Craig, 1976, *Earth Planet. Sci. Lett.*, 32:268. Craig et al., 1973, *Earth Planet. Sci. Lett.*, 17:295.

Bacon et al., 1976, *Earth Planet. Sci. Lett.*, 32:277. Thompson and Turekian, 1976, *Earth Planet. Sci. Lett.*, 32:297. Nozaki et al., 1976, *Earth Planet. Sci. Lett.*, 32:304.

These studies revealed that ^{210}Pb is scavenged by sediment particles sinking through the water column and that the ^{210}Pb/^{226}Ra activity ratio of seawater is less than one.

Chung (1981) reported the activities of ^{210}Pb and ^{226}Ra in unfiltered water collected from depth profiles at locations in the Southern Ocean south of latitude 60° S. The ^{210}Pb activity in Figure 20.16 for GEOSECS Station 89 (60°01.5′ S and 0°01.5′ E) in the South Atlantic Ocean increases from about 10 dpm/100 kg in the first 100 m below the surface to 17.4 dpm/100 kg at 400 m and then declines irregularly with depth to about 8.0 dpm/100 kg close to the bottom at 5340 m.

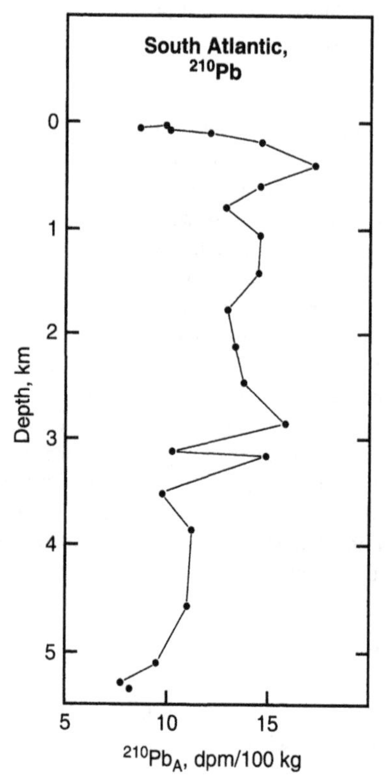

FIGURE 20.16 Variations of ^{210}Pb activity of unfiltered seawater at GEOSECS station 89, South Atlantic (60°01.5′ S, 0°01.5′ E). Data from Chung (1981).

20.4c Lake Rockwell, Ohio

An example of dating lacustrine sediment by means of the activity of excess ^{210}Pb was provided by McCall et al. (1984), who analyzed a 45-cm core recovered from Lake Rockwell located in northern Ohio about 20 km northeast of the city of Akron. Lake Rockwell is a reservoir on the Cuyahoga River, which flows into Lake Erie in Cleveland, Ohio. The reservoir was constructed in 1914 in order to provide drinking water for the people of Akron.

The core was taken in 1977 and was analyzed in 1-cm increments for most of its length. The ^{210}Pb was extracted by leaching 3-g samples of dry sediment with hot concentrated HCl (86°C and 50%) for 36 h. Hydrogen peroxide (30%) was added periodically during digestion to destroy organic matter. Polonium-210 was plated onto polished silver discs from the leach solutions and the rate of emission of β-particles was determined in a gas flow proportional counter. The activity of ^{210}Pb was expressed in units of picocuries per gram (pCi/g), where 1 pCi $= 10^{-12} \times 3.70 \times 10^{10} \times 60 = 2.22$ dpm.

The data points in Figure 20.17 scatter about a straight line in coordinates of depth (centimeters) and $\ln(^{210}\text{Pb})_{Ax}$. A least-squares regression of all 33 data sets to a straight line yields the equation

$$\ln(^{210}\text{Pb})_{Ax} = 3.3029 - 0.03697\ \text{h}$$

and a linear correlation coefficient $r^2 = -0.9562$. The initial excess ^{210}Pb activity is 27.2 pCi/g, or 60.4 dpm/g. This value is similar to the activity of excess ^{210}Pb in the organic-rich A horizon of soil (45.49 dpm/g) in Cook Forest State Park, Pennsylvania, reported by Lewis (1977).

The slope of the straight line in Figure 20.17 ($m = -0.03697$) is used to calculate the average sedimentation rate:

$$a = \frac{\lambda_{210}}{m} = \frac{0.03067}{0.03697} = 0.83\ \text{cm/y}$$

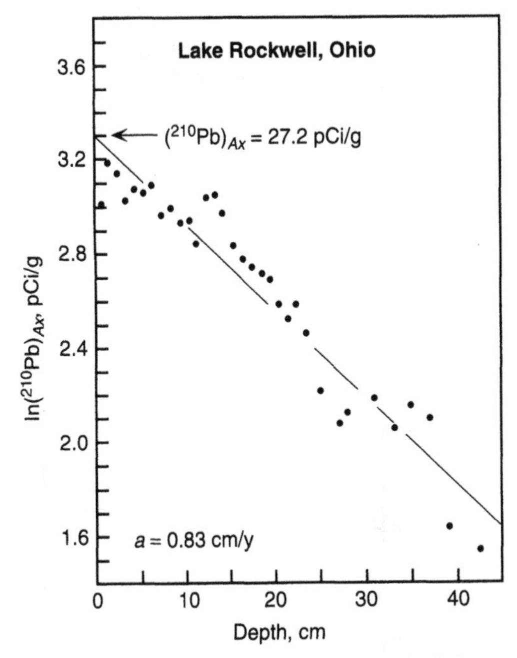

FIGURE 20.17 Average sedimentation rate in Lake Rockwell, a man-made reservoir on the Cuyahoga River near Akron, Ohio. Data from McCall et al. (1984).

This value can be used to calculate the time elapsed since deposition of sediment at a specified depth h. For example, the sediment collected at $h = 10.5$ cm was deposited at

$$t = \frac{h}{a} = \frac{10.5}{0.83} = 12.6 \text{ years ago.}$$

Since the core was taken in 1977, the sediment sample was deposited in 1964, or about 50 years after Lake Rockwell came into existence. Continued deposition of sediment in Lake Rockwell will ultimately fill the entire basin.

McCall et al. (1984) demonstrated that the concentrations of metals in the sediment deposited between 1964 and 1977 increased with time:

Cu: 147–673 ppm, 358%
Pb: 44–84 ppm, 91%
Zn: 167–205 ppm, 23%
Fe: 51.3–56.4 ppm, 10%

In this way, the sediment in Lake Rockwell has preserved a record of increasing anthropogenic contamination of the environment that correlates with the growing population in the drainage basin of the Cuyahoga River in northern Ohio. These data also demonstrate how age determinations of lacustrine sediment (and peat) can provide a timescale for the history of anthropogenic activity in the drainage basins of lakes and peat bogs.

Similar determinations of lacustrine sedimentation rates based on ^{210}Pb have been published by

Kadlec and Robbins, 1984, *Chem. Geol.*, 44: 119–150. Davis et al., 1984, *Chem. Geol.*, 44:151–185. Wong et al., 1984, *Chem. Geol.*, 44:187–201. Wieland et al., 1993, *Geochim. Cosmochim. Acta*, 57:2959–2979.

20.4d Snow in Antarctica

The activity of excess ^{210}Pb in snow, firn, and ice can be used to determine the rate of deposition and hence the age of a sample of snow of known depth below the surface. Crozaz et al. (1964) originally used this method to determine the annual rates of water accumulation at Base Roi Baudouin (70°26′ S, 24°19′ E) and at the South Pole in Antarctica. Their data define straight lines (not shown) that yield average accumulation rates a of meteoric water at the two sites:

Base Roi Baudouin :	$a = 42.6 \pm 3$ cm/y
South Pole :	$a = 5.7 \pm 1$ cm/y

for $\lambda_{210} = 3.067 \times 10^{-2}$ y^{-1} ($T_{1/2} = 22.6$ y).

The meteoric precipitation rates are in good agreement with direct determinations based on the stratigraphy of snow pits at these sites. The comparatively high rate of snowfall at Base Roi Baudouin reflects its location on the Princess Ragnhild Coast of East Antarctica facing the Indian Ocean. In contrast, the rate of snow accumulation at the South Pole is low because it is a cold and dry place at an elevation of 2835 m above sealevel and located 1300 km from the edge of the Ross Iceshelf, which is the nearest coast.

The initial activities of excess ^{210}Pb reported by Crozaz et al. (1964) are 75 dph/kg (1.25 dpm/kg) at Base Roi Baudouin and 110 dph/kg (1.83 dpm/kg) at the South Pole. The difference between the initial activities is consistent with the observed difference in the rates of deposition because the high rate of snowfall at Base Roi Baudouin yields a relatively low ^{210}Pb concentration in the snow whereas a low snowfall rate causes a higher concentration of ^{210}Pb in the snow at the South Pole. The average activity of ^{210}Pb in the air 10 m above the surface at Base Roi Baudouin during 1958 was $(1.3 \pm 0.1) \times 10^{-3}$ dpm/kg of air.

An even lower rate of snow deposition was reported by Crozaz and Fabri (1966) at the "Pole of Relative Inaccessibility" located at 82°07′ S and 55°06′ E in East Antarctica. The Pole of Inaccessibility has an elevation of 3790 m above sealevel and is located about 1800 km from the coast of East Antarctica measured along the 55 °E line of longitude. The East Antarctic ice sheet at this location is almost three kilometers thick.

Crozaz and Fabri (1966) measured the activity of ^{210}Pb by α-particle spectrometry of ^{210}Po, which was assumed to be in secular equilibrium with ^{210}Bi and hence with ^{210}Pb. The data in Figure 20.18 define a straight line whose equation is

$$\ln(^{210}\text{Pb})_{Ax} = 4.8961 - 0.01044\ \text{h}$$

where the activity of ^{210}Pb is expressed in dph/kg. The rate of snow deposition at the Pole of Inaccessibility is

$$a = \frac{\lambda_{210}}{m} = \frac{3.067 \times 10^{-2}}{0.01044} = 2.94\ \text{cm/y}$$

of water. The initial activity of excess ^{210}Pb in the snow at this location is 134 dph/kg.

The rate of snow deposition at the Pole of Inaccessibility (2.94 cm/y) is even lower than that at the South Pole (5.7 cm/y) and the activity of ^{210}Pb surface snow (134 dph/kg) is higher than that at the South Pole. The rates of deposition of snow and the corresponding initial activities of ^{210}Pb define a curve (not shown) that extrapolates to about 160 dph/kg for a zero rate of snow deposition. The age of snow at a depth of 228.5 cm (water equivalent) or 5.0 to 6.0 m of snow at the Pole of Inaccessibility is

$$t = \frac{228.5}{2.94} = 78\ \text{y}$$

Since the core sample was taken in December of 1965, the oldest ice in the core was deposited in 1887.

Similar studies by Crozaz and Langway (1966), based on a 70-m core of firn taken at Camp Century in northwest Greenland (77°10′ N, 61°08′ W), indicated an average rate of deposition of 32 ± 3 cm of water per year and an initial ^{210}Pb activity of 170 ± 15 dph/kg. The snow accumulation rate based on ^{210}Pb agrees well with measurements made by other methods.

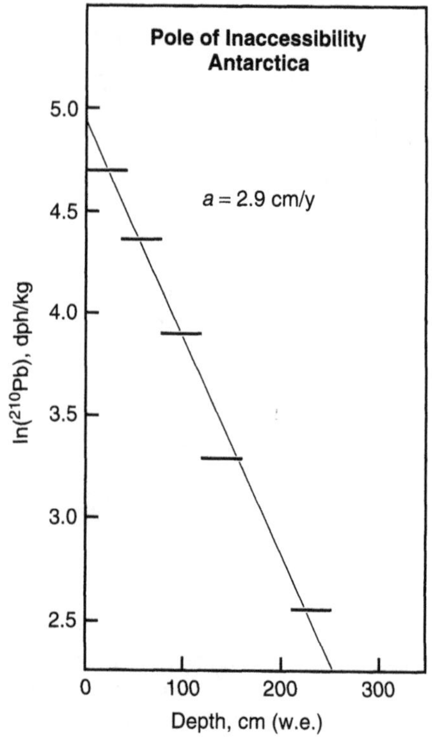

FIGURE 20.18 Rate of deposition of snow at the Pole of Relative Inaccessibility (82°07′ S, 55°06′ E) in East Antarctica. The depth is expressed in centimeters of water equivalent (w.e.). Data from Crozaz and Fabri (1966).

20.5 ARCHEOLOGY AND ANTHROPOLOGY

The geochronometers that arise from the disequilibrium among the unstable daughters of the long-lived isotopes of U and Th are being used to date archeological sites and objects. In addition, these dates provide a timescale for changes of local as well as global climates that affected human civilizations of the past 100,000 years. The most widely available materials for dating are composed of calcium carbonate and calcium phosphate (e.g., speleothems, concretions, lacustrine carbonates, bones, tooth enamel). These kinds of materials contain U but lack Th and Pa. Therefore, the most applicable geochronometers are based on the decay of ^{234}U to ^{230}Th and ^{235}U to ^{231}Pa. The ^{234}U–^{230}Th method is preferred because Pa is difficult to separate from Th, which makes ^{231}Pa hard to measure by α-spectrometry in the presence of ^{230}Th. The ^{234}U–^{230}Th method of dating speleothems and other kinds of U-bearing materials also has a problem because such samples are variably enriched in ^{234}U. In addition, the materials available for dating, in some cases, are mixtures of two or more components of differing ages and U concentrations (e.g., travertines) and may have gained or lost U (e.g., bones, mollusk shells).

All of these problems were addressed by Schwarcz (1980), who stated the relevant equations, described the analytical methods, discussed the criteria for selecting samples suitable for dating, and cited applications to dating of archeological samples. More recently, Latham and Schwarcz (1992) discussed the precipitation of carbonate and sulfate minerals from groundwater and surface water, including speleothems, travertine, and lacustrine deposits in pluvial lakes and evaluated their suitability for dating by U-series disequilibrium methods.

The problems associated with dating materials that are linked to human fossils or sites of occupation were reviewed by Schwarcz and Blackwell (1992). Their paper includes a list of the materials that have been dated and the methods that were used as well as references to the literature. The list of dated materials includes speleothems, travertine, caliche, lacustrine carbonate, salt, corals, mollusk shells, egg shells, teeth, bones, peat, oil paintings, pewter, bronze, ice, and minerals in volcanic rocks. The importance of obtaining accurate and precise dates from samples that are not necessarily well suited for dating requires the utmost care and ingenuity from the investigators.

Other dating techniques that are being used to date archeological samples include the K–Ar and ^{40}Ar/^{39}Ar methods presented in Chapter 6 and 7, respectively. For example, Figure 6.3 is a K–Ar isochron of the ignimbrite at Olduvai Gorge, which yielded a date of 2.04 ± 0.02 Ma (Bishop and Miller, 1972; Curtis and Hay, 1972; Fitch et al., 1976). The significance of this date depends on the stratigraphic relationship between the human fossils and the deposit of volcanic ash at this site.

More recent samples have been dated by carbon-14 and other cosmogenic radionuclides (e.g., ^{10}Be and ^{36}Cl) depending on the chemical composition and age of the material. In addition, certain types of samples can be dated by the fission-track method, by the U–Th/He method, electron spin resonance (ESR), and thermoluminescence (TL), all of which are discussed in Chapters 22 and 23. A forward-looking summary by Wintle (1996) of techniques for dating archeological samples describes the samples as being "small" and the methods as being "hot and identified by acronyms."

20.5a *Homo erectus*

The principal motivation for dating human bones and sites of human occupation is to provide a timescale for human evolution, which started in East Africa during the Pliocene Epoch about four million years ago: Leakey and Lewin (1977), Leakey (1981), Hamilton (1984), Weaver (1985), Leakey and Walker (1985), and Klein (1989). A recent summary of the age determinations of human fossils by Balter and Gibbons (2000) in Table 20.5 indicates that the skeletal remains at Olduvai Gorge (Tanzania) date from the period between 1.8 and 1.2 Ma and those at Koobi Fora near Lake Turkana (Kenya) range in age from 1.9 to 1.6 Ma. Apparently, *Homo erectus* migrated out of East Africa and settled in the Longgupo Cave in China (1.9 Ma), on the island of Java in Indonesia (1.8 Ma), at Dmanisi in Georgia (1.7 Ma), and at Ubeidiya in Israel (1.5 Ma). Their presence has also

Table 20.5. Record of the Migration of *Homo erectus* out of East Africa

Locality	Country	Date, Ma
Turkana	Kenya	1.9–1.6
Olduvai	Tanzania	1.8–1.2
Longgupo	China	1.9?
Java	Indonesia	1.8
Dmanisi	Georgia	1.7
Gongwanling	China	1.1
Ceprano	Italy	0.9–0.8
Atapuerca	Spain	0.78
Tautavel	France	0.45

Source: Balter and Gibbons, 2000

been recorded at Ceprano (Italy) at 0.9 to 0.8 Ma, at Atapuerca (Spain) at 0.78 Ma, and at Tautavel (France) at 0.45 Ma.

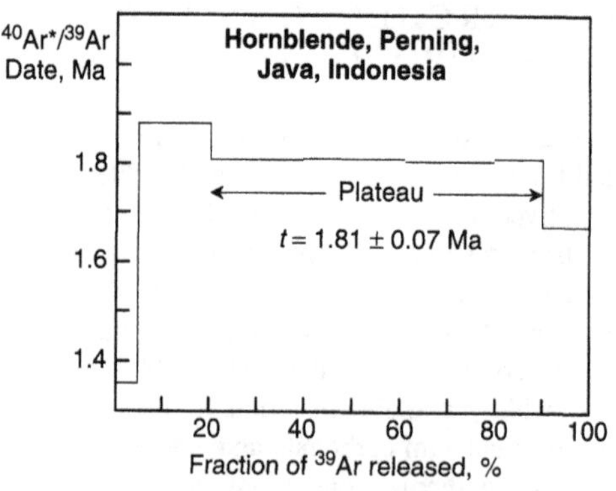

FIGURE 20.19 Partial-release spectrum of $^{40}Ar^*/^{39}Ar$ dates of hornblende crystals (0.5 mg) extracted from a volcaniclastic sandstone at Perning, Java, Indonesia. The plateau date is based on 69.7% of the total ^{39}Ar. The fossil collected at this site is a well-preserved calvaria of a juvenile *Pithecanthropus modjokertensis*, known also as the Mojokerto child. Data from Swisher et al. (1994).

20.5b The Mojokerto Child

The age of the oldest human skeletal remains on the island of Java is still being questioned in spite of an excellent age determination by Swisher et al. (1994). A summary of the history of these investigations illustrates the complexity of this issue. Hominid bones were originally discovered in 1891 near the village of Trinil located along the Solo River in northern Java. In addition, a calvaria of a juvenile (the so-called Mojokerto child) was collected in 1936 near the village of Perning from a conglomeratic volcanic sandstone believed to be a facies of the Pucangan Formation, which is of Plio-Pleistocene age. In 1991, a sample of pumice tuff collected at Perning from the Pucangan Formation yielded a conventional K–Ar date of 1.9 ± 0.5 Ma, which confirmed its stratigraphic age. However, the relevance of this date was questioned because of its large error and because the relation between the dated sample and the Mojokerto child was not clear.

Additional fieldwork in 1992 and 1993 in the Perning area indicated that the formation from which the Mojokerto child had been recovered contained pumice and volcaniclastic sediment suitable for isotope dating. Swisher et al. (1994) dated four 0.5-mg samples of hand-picked hornblende crystals taken from the pumice and the volcanic matrix at the Mojokerto site by the incremental laser-heating $^{40}Ar/^{39}Ar$ method (Chapter 7). The resulting plateau date of one of the four hornblende samples in Figure 20.19 is indistinguishable within analytical errors from the plateau dates of the other three hornblende samples, which yield a combined average date of 1.81 ± 0.04 Ma.

This age determination is consistent with the highest standards of modern geochronometry. There is no doubt that the hornblende from the Mojokerto site crystallized 1.81 million years ago within a time interval of 80,000 years. Although the uncertainty of the date is only 4.4%, the amount of time it represents is long in the context of human evolution. In addition, some doubt remains about the relation between the Mojokerto bone, which was collected in 1936, and the deposit from which the hornblende was recovered more than 50 years later. In fact, even the identification of the Mojokerto calvaria as a human bone could be questioned. The point is that the significance of this bone as a record of human evolution does not depend solely on the accuracy and precision of the age determination.

Setting aside these questions concerning the geology of the site and the identification of the bone, the date of 1.81 ± 0.04 Ma reported by Swisher et al. (1994) indicates that the Mojokerto child lived on Java at the same time as *Homo erectus* at Olduvai Gorge and at Kobi Fora on Lake Turkana. In other words, groups of *Homo erectus* appear to have left Africa and settled on Java prior to 1.81 ± 0.04 Ma, which is earlier than previously thought.

Another waypoint in the migration of *Homo erectus* is the valley of the Awash River in Ethiopia, where a human cranium was found in 1976 at the village of Bodo. Additional bone fragments were recovered in the same area in 1981 and in 1990, all of which appear to be transitional between *Homo erectus* and *Homo sapiens*. Clark et al. (1994) used the laser fusion $^{40}Ar/^{39}Ar$ method to date sanidine grains in the middle Pleistocene fluvial sediments that contain the human and other vertebrate fossils. One set of nine sanidine grains (0.5 mm) yielded dates ranging from 1.681 ± 0.033 to 0.389 ± 0.073 Ma. The oldest grain (1.681 Ma) is of detrital origin, but the others yielded a weighted average date of 0.55 ± 0.03 Ma. This result, together with other age determinations of feldspar grains by Clark et al. (1994), established a date of 0.64 ± 0.03 Ma for the hominid bones of the Awash Valley in Ethiopia.

20.5c Neandertals and *Homo sapiens*

Bone fragments of Neandertals in a cave at Tabun in Israel originally appeared to be older than the remains of *Homo sapiens* in Qafzeh and other localities in this area. The apparent age difference is important because, if the Neandertals in the eastern Mediterranean region predated *Homo sapiens*, they could be considered to be the ancestors of modern humans. This issue was settled by McDermott et al. (1993) by means of U-series disequilibrium dating of tooth fragments recovered at Tabun (Neandertals) and Qafzeh (*Homo sapiens*).

The U-series disequilibrium dates of dentine and tooth enamel are based on U that was absorbed from groundwater *after* burial and which subsequently decayed to ^{230}Th. Therefore, the ^{238}U–^{230}Th dates of teeth and bones underestimate the age of the dated material. McDermott et al. (1993) determined concentrations of ^{238}U and ^{230}Th by isotope dilution using 50–100-mg samples of bovine dentine and enamel. They also measured the $^{234}U/^{238}U$ ratios by mass spectrometry and converted the isotopic ratios into activity ratios using the appropriate decay constants.

For example, dentine from a bovine tooth in layer C at Tabun has a $^{230}Th/^{238}U$ activity ratio of 0.6367 ± 0.0088. This value can be used to calculate a preliminary date from the equations

$$\left(\frac{^{230}\mathrm{Th}}{^{238}\mathrm{U}}\right)_A = 1 - e^{-\lambda_{230}t}$$

$$t = -\frac{1}{\lambda_{230}} \ln\left[1 - \left(\frac{^{230}\mathrm{Th}}{^{238}\mathrm{U}}\right)_A\right]$$

Since $\lambda_{230} = 9.1929 \times 10^{-6}\ \mathrm{y}^{-1}$ (Lide and Frederikse, 1995) and $(^{230}\mathrm{Th}/^{238}\mathrm{U})_A = 0.6367$,

$$t = \frac{1.01252}{9.1929 \times 10^{-6}} = 110{,}141\ \mathrm{y}$$

This date was used to calculate the initial $^{234}U/^{238}U$ activity ratio given that the present value of this ratio is 1.0230 ± 0.0005. The applicable equation is

$$\left(\frac{^{234}\mathrm{U}}{^{238}\mathrm{U}}\right)_A = 1 + (\gamma_0 - 1)e^{-\lambda_{234}t}$$

Setting $\lambda_{234} = 2.829 \times 10^{-6}\ \mathrm{y}^{-1}$ (Lide and Frederikse, 1995), $t = 110,141$ y, and $(^{234}\mathrm{U}/^{238}\mathrm{U})_A = 1.0230$ yielded

$$\gamma_0 = \frac{1.0230 - 1}{0.31158} + 1 = 1.0738$$

The value of γ_0 so obtained is lower than that of seawater because it depends on the local geology of Tabun in Israel. Given that $\gamma_0 = 1.0738$, a more accurate date for the bovine dentine was calculated from the equation

$$\left(\frac{^{230}\mathrm{Th}}{^{238}\mathrm{U}}\right)_A = 1 - e^{-\lambda_{230}t} + \frac{\lambda_{230}}{\lambda_{230} - \lambda_{234}} \times (\gamma_0 - 1) \times (e^{-\lambda_{234}t} - e^{-\lambda_{230}t})$$

This equation yielded $t = 99.4$ ka by a graphical procedure for $\gamma_0 = 1.0738$ and $(^{230}\mathrm{Th}/^{238}\mathrm{U})_A = 0.6367$.

Table 20.6. $(^{230}Th/^{238}U)_A$ Dates of Dentine and Tooth Enamel in Layer C of a Cave at Tabun (Neandertal) and Layer XIX at Qafzeh (*Homo sapiens*), Israel

Sample[a]	$(^{234}U/^{238}U)^0_A$	t,[b] ka
	Neandertals, Tabun	
552 DE	1.074	99.4
551 EN	1.077	100.1
551 DE	1.122	96.7
Average		98.7 ± 2.1 ka
	H. sapiens, Qafzeh	
317 EN	1.300	86.0
368 DE	1.318	100.8
Average		93.4 ± 7.4 ka

Source: McDermott et al., 1993

[a] DE = dentine, EN = enamel.

[b] $\lambda_{230} = 9.1929 \times 10^{-6}$ y^{-1}, $\lambda_{234} = 2.829 \times 10^{-6}$ y^{-1} from Lide and Frederikse (1995), which differ from the values used by McDermott et al. (1993).

Dates calculated in this way for bovine dental fragments from layer C in the cave at Tabun, Israel, average 98.7 ± 2.1 ka in Table 20.6, compared to 93.4 ± 7.4 ka for dental fragments at Qafzeh. These results indicate that the Neandertal people of Tabun lived at the same time as *Homo sapiens* at Qafzeh. Therefore, *Homo sapiens* is not necessarily a descendant of the Neandertals but is probably a separate species.

The dental fragments analyzed by McDermott et al. (1993) were also dated by electron-spin resonance (ESR). The early-uptake ESR dates and the U-series dates are in satisfactory agreement for more than 70% of the dental fragments analyzed by McDermott et al. (1993). In addition, the initial $^{234}U/^{238}U$ activity ratios of the dental fragments in Table 20.6 range widely from 1.074 to 1.318, as expected for U derived from groundwater (Cherdyntsev, 1971).

20.5d Speleothems and Travertines

Early humans sought shelter in caves, which therefore may contain records of human habitation in the layers of sediment that accumulated within the caves and in the refuse that piled up outside of the entrance. In addition, caves contain deposits of calcium carbonate and gypsum (speleothems), which precipitated from groundwater entering the caves. Calcium carbonate is also deposited by groundwater at varying temperature in open fissures in bedrock (e.g., Devil's Hole, Nevada) and by springs, where humans may have preferred to camp. All such carbonate deposits can be dated by the U-series disequilibrium methods in support of archeological investigations. In addition, the dates provide a timescale for climate change recorded by the isotope composition of oxygen in calcite (Thompson et al., 1975; Harmon et al., 1977; Schwarcz, 1980, 1982, 1992; Szabo, 1990).

The use of mass spectrometers to measure $^{234}U/^{238}U$ and $^{230}Th/^{232}Th$ ratios in speleothems, pioneered by Edwards et al. (1987) for dating corals and used by Ludwig et al. (1992) to date calcite in Devil's Hole, has improved the precision of U-series dates of speleothems and travertines. In addition, the work of Swisher et al. (1994) on the age of the Mojokerto child of Java has strengthened the hypothesis of multiregional evolution of *Homo erectus*, in contrast to the out-of-Africa school of thought. These and other developments in the field of archeoanthropology justify a renewed effort to refine the timescale of human evolution.

For these reasons, Zhao et al. (2001) dated samples of flowstone from a limestone cave about 26 km east of Nanjing, China, in which two crania and a tooth of *Homo erectus* were found in 1993. One of the skulls and the tooth originated from a fossil-rich layer (60–97 cm thick) overlain by a brown clay layer (2–60 cm thick), which in turn was covered by a layer of flowstone (0–2 cm thick, speleothem).

The $^{234}U/^{238}U$ and $^{230}Th/^{238}U$ activity ratios of the flowstone measured by Zhao et al. (2001) by mass spectrometry yielded three concordant dates with a mean of 577 (+44, −34) ka, based on the decay constants $\lambda_{238} = 1.551 \times 10^{-10}$ y^{-1}, $\lambda_{234} = 2.835 \times 10^{-6}$ y^{-1}, and $\lambda_{230} = 9.195 \times 10^{-6}$ y^{-1}. These values of the decay constants differ from those used in this chapter. They also differ from the values determined by Cheng et al. (2000). The values of the decay constants affect the dates that are derived from the activity ratios. In addition, the decay constants are used to convert isotope ratios, measured by mass spectrometry, to activity ratios. For example, the isotopic $^{234}U/^{238}U$ ratio is converted into the activity ratio by use of the law of radioactivity:

$$\left(\frac{^{234}U}{^{238}U}\right)_A = \left(\frac{^{234}U}{^{238}U}\right)_{isot} \times \left(\frac{\lambda_{234}}{\lambda_{238}}\right) \quad (20.53)$$

If $\lambda_{238} = 1.554 \times 10^{-10}$ y^{-1} and $\lambda_{234} = 2.829 \times 10^{-6}$ y^{-1} (Lide and Frederikse, 1995),

$$\frac{\lambda_{234}}{\lambda_{238}} = \frac{2.829 \times 10^{-6}}{1.554 \times 10^{-10}} = 1.820 \times 10^4$$

The decay constants used by Zhao et al. (2001) yield

$$\frac{\lambda_{234}}{\lambda_{238}} = \frac{2.835 \times 10^{-6}}{1.551 \times 10^{-10}} = 1.828 \times 10^4$$

which is higher by 0.40%.

The average $^{230}Th/^{238}U$ date of the flowstone (577 ka) implies that the cranium in the underlying fossil-rich layer is older than this date. Zhao et al. (2001) attempted to refine the age of "Nanjing Man" by dating enamel and dentine of a tooth (deer) that was recovered from the same layer as the cranium. Unfortunately, the $^{230}Th/^{238}U$ dates were discordant: The enamel yielded a date of 130.1 ± 16 ka, whereas dentine of the same tooth gave a date of 388 ± 14 ka. Dentine obtained from a second tooth confirmed the dentine date of the first tooth. However, calcite from a growth layer of a stalagmite that formed on top of the flowstone layer yielded an average $^{230}Th/^{238}U$ date of 540 ± 12 ka, which, taken at face value, is less than the $^{230}Th/^{238}U$ date of the underlying flowstone ($577 + 44, -34$ ka).

The two teeth dated by Zhao et al. (2001) by the $^{230}Th/^{238}U$ method should have a similar age as the cranium of Nanjing Man and therefore should be older than 577 ka. The anomalously low dentine (388 ka) and tooth enamel (130 ka) dates are probably an indication that the teeth absorbed U from groundwater *long after* burial rather than soon thereafter. Therefore, Zhao et al. (2001) recommended caution in the interpretation of U-series disequilibrium dates of fossil teeth.

20.6 VOLCANIC ROCKS

Uranium, Th, and their short-lived daughters occur in the mantle of the Earth, where magmas form by partial melting. The resulting volcanic rocks contain excess unsupported ^{230}Th, ^{226}Ra, and ^{210}Pb, all of which are daughters of ^{238}U (Eberhardt et al., 1955; Somayajulu et al., 1966; Oversby and Gast, 1968). In addition, Goldstein et al. (1993) and Pickett and Murrell (1997) reported the presence of excess ^{231}Pa (a daughter of ^{235}U) in MORB and other kinds of volcanic rocks. The disequilibria presumably result from chemical fractionation during magma formation assuming that secular equilibrium existed in the source rocks of the magma prior to melting. The isotopes of U are not fractionated by partial melting of rocks in the mantle, which is confirmed by the fact that in recently erupted volcanic rocks the ratio of U isotopes $^{238}U/^{235}U = 137.88$ and that the activity ratio of $^{234}U/^{238}U = 1.0$.

Lead-210 is a special case because it forms by decay of ^{222}Rn and ^{218}Rn, which may vent to the surface via the volcanic plumbing system and cause ^{210}Pb to be enriched in fumarolic gases (Oversby and Gast, 1968; Gill et al., 1985). The isotopes of Th, Pa, and Ra, which are daughters of ^{238}U and ^{235}U, all form before the Rn daughters. Therefore, the disequilibrium of ^{230}Th, ^{231}Pa, ^{226}Ra, and ^{228}Ra with their respective parents cannot be caused by venting of Rn gas.

The radioactive disequilibria among the daughters of U in volcanic rocks have been used to

1. date volcanic rocks by the decay of excess ^{230}Th and ^{231}Pa,
2. study the partitioning of elements between the source rocks of magma and partial melts in the mantle, and
3. set limits on the time that passes between melting at depth and eruption of lava at the surface.

Summary review papers dealing with the U-series disequilibria in volcanic rocks have been published by Condomines et al. (1988), Gill and Williams (1990), and Gill et al. (1992).

20.6a Dating with ^{230}Th

The activity of ^{230}Th in young volcanic rocks consists of two components:

$$^{230}\mathrm{Th}_A = {}^{230}\mathrm{Th}_{Ax} + {}^{230}\mathrm{Th}_{As}$$

Where the subscript x is the excess (unsupported) component and s is the ^{230}Th that is supported by the decay of ^{238}U. The excess ^{230}Th decays with its characteristic halflife ($T_{1/2} = 7.54 \times 10^4$ y):

$$^{230}\mathrm{Th}_{Ax} = {}^{238}\mathrm{Th}^0_{Ax} e^{-\lambda_{230} t} \tag{20.54}$$

The activity of supported ^{230}Th increases with time:

$$^{230}\mathrm{Th}_{As} = {}^{238}\mathrm{U}_A (1 - e^{-\lambda_{230} t}) \tag{20.55}$$

The possible disequilibrium between ^{234}U and ^{238}U does *not* arise in volcanic rocks because the U isotopes are not fractionated by melting and crystallization. Combining equations 20.54 and 20.55 yields

$$^{230}\mathrm{Th}_A = {}^{230}\mathrm{Th}^0_{Ax} e^{-\lambda_{230} t} + {}^{238}\mathrm{U}_A (1 - e^{-\lambda_{230} t}) \tag{20.56}$$

Dividing by the activity of ^{232}Th results in the most widely used form of this equation:

$$\left(\frac{^{230}\mathrm{Th}}{^{232}\mathrm{Th}}\right)_A = \left(\frac{^{230}\mathrm{Th}}{^{232}\mathrm{Th}}\right)^0_{Ax} e^{-\lambda_{230} t} + \left(\frac{^{238}\mathrm{U}}{^{232}\mathrm{Th}}\right)_A (1 - e^{-\lambda_{230} t}) \tag{20.57}$$

Allègre (1968) pointed out that this is the equation of a straight line in coordinates of $(^{230}\mathrm{Th}/^{232}\mathrm{Th})_A$ and $(^{238}\mathrm{U}/^{232}\mathrm{Th})_A$ when t is a constant. Minerals or rocks having different $(^{238}\mathrm{U}/^{232}\mathrm{Th})_A$ ratios satisfy this equation, provided they have

1. the same age t,
2. the same initial $(^{230}\mathrm{Th}/^{232}\mathrm{Th})_A$ activity ratio, and
3. remained closed to U and Th after crystallization.

The line formed by a suite of cogenetic minerals (or rocks) that satisfy the conditions listed above is an isochron whose slope m is:

$$m = 1 - e^{-\lambda_{230} t} \tag{20.58}$$

and the intercept b is:

$$b = \left(\frac{^{230}\mathrm{Th}}{^{232}\mathrm{Th}}\right)^0_{Ax} e^{-\lambda_{230} t} \tag{20.59}$$

from which the initial $(^{230}\mathrm{Th}/^{232}\mathrm{Th})$ activity ratio can be calculated using the value of t obtained from the slope of the isochron.

The isotopic evolution of mineral systems in coordinates of $(^{230}\mathrm{Th}/^{232}\mathrm{Th})_A$ and $(^{238}\mathrm{U}/^{232}\mathrm{Th})_A$ is illustrated in Figure 20.20*a*. At the time of crystallization ($t = 0$) suites of cogenetic minerals (or rocks) that satisfy the assumptions listed above define an isochron having a slope equal to zero. Subsequently, the isochron rotates about the equipoint as its slope increases and eventually approaches unity when ^{230}Th and ^{238}U are in secular equilibrium. The isochron with a slope of $m = 1.0$ is called the equiline. It is the locus of all points representing systems for which

$$\frac{^{230}\mathrm{Th}/^{232}\mathrm{Th}}{^{238}\mathrm{U}/^{232}\mathrm{Th}} = \left(\frac{^{230}\mathrm{Th}}{^{238}\mathrm{U}}\right)_A = 1.00$$

Figure 20.20*b* demonstrates that the slope of the isochron begins to increase detectably at about 10^3 years after crystallization of the rocks or minerals and approaches unity after about 10^6 years. These times therefore define the useful range of this geochronometer.

The initial ^{230}Th/^{232}Th activity ratios calculated from equation 20.59, after the decay time t has been determined, is the activity ratio of unsupported ^{230}Th to common ^{232}Th at the time the rocks crystallized. The excess ^{230}Th decays as time increases, causing the initial ^{230}Th/^{232}Th activity ratio to decrease until it reaches zero (i.e., when all excess ^{230}Th has decayed).

The ^{230}Th–^{238}U method has been used successfully to date volcanic rocks:

Kigoshi, 1967, *Science*, 156:932–934. Taddeuci et al., 1967, *Earth Planet. Sci. Lett.*,

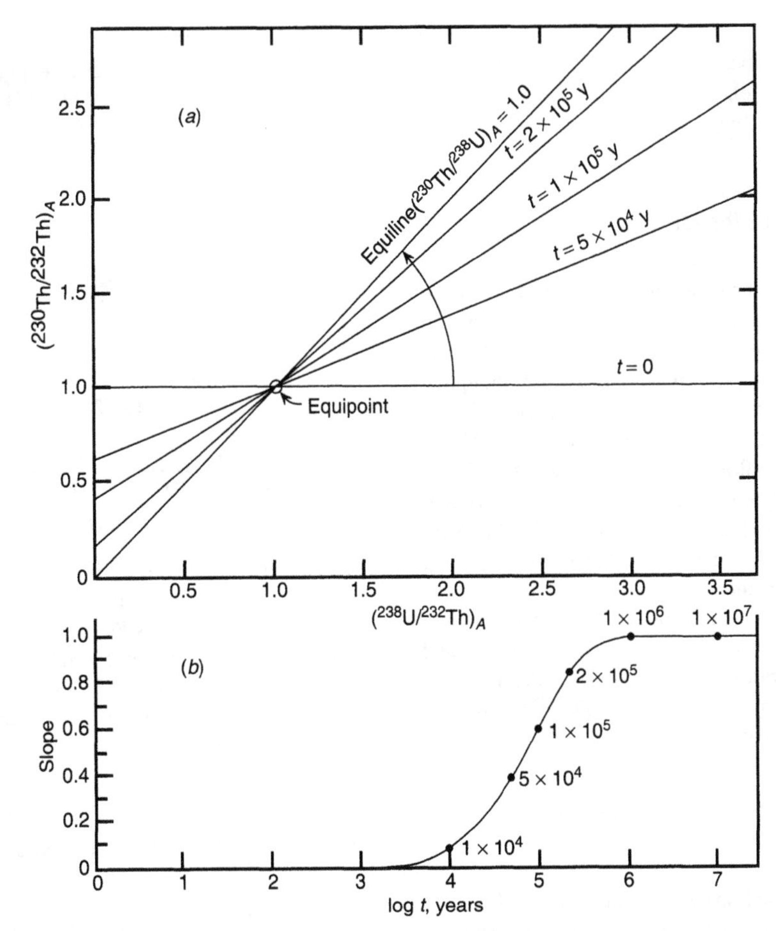

FIGURE 20.20 (*a*) The ^{230}Th–^{238}U isochron diagram based on equation 20.57. If cogenetic volcanic rocks or minerals initially have the same ^{230}Th/^{232}Th activity ratio (not necessarily equal to 1.0 as shown) and different ^{238}U/^{232}Th activity ratios, they form an isochron with slope $m = 0$. As time passes, their ^{230}Th/^{232}Th activity ratios decrease by decay of ^{230}Th while their ^{238}U/^{232}Th activity ratios remain virtually constant. Consequently, the slope of the isochron increases from zero toward unity when secular equilibrium between ^{238}U and its daughter ^{230}Th is reestablished. (*b*) The variation of the slope of the isochron with time limits the useful range of this geochronometer to the interval between 10^3 and 10^6years in the past.

3:338–342. Fukuoka, 1974, *Geochem. J.*, 8:109–116. Allègre and Condomines, 1976, *Earth Planet. Sci. Lett.*, 28:395–406. Condomines, 1978, *Nature*, 276:257. Condomines and Allègre, 1980, *Nature* 288:154–157.

In retrospect, this method of dating has *not* been widely used partly because α-spectrometry requires purified mineral samples weighing several grams and because the U/Th ratios of silicate minerals (and rocks) vary only within narrow limits. For this reason, cogenetic minerals (e.g., magnetite and zircon) have been preferred by some workers because their U/Th ratios vary more widely than those of feldspar and other rock-forming silicates and because magnetite and zircon are easier to concentrate from whole-rock powders (Gill et al., 1992). The analytical problems have been overcome by the use of thermal ionization mass spectrometers (Chen et al., 1986; Edwards et al., 1987) and by inductively coupled plasma mass spectrometers (Shen et al., 2002; Stirling et al., 2000).

20.6b Age of the Olby-Laschamp Event

The ^{230}Th isochron method was used by Condomines (1978) to date a K-hawaiite from Olby at Chaine des Puys of the Massif Central in France. The flow is reversely magnetized even though it was extruded during the Brunhes period during which normal magnetic polarity prevailed. Therefore, the lava flow at Olby defines the Olby-Laschamp magnetic event. Previous age determinations by the thermoluminescence and K–Ar methods indicated that the age of the Olby-Laschamp event is greater than 30 ka.

Condomines (1978) used a whole-rock sample and separated minerals (pyroxene, plagioclase, and two size fractions of magnetite) of the hawaiite flow at Olby to construct the ^{230}Th–^{238}U isochron in Figure 20.21 whose equation is

$$\left(\frac{^{230}\text{Th}}{^{232}\text{Th}}\right)_A = 0.5754 + 0.29009\left(\frac{^{238}\text{U}}{^{232}\text{Th}}\right)_A$$

The time elapsed since crystallization of this rock is derived from the slope of the isochron by equation 20.58:

$$m = 1 - e^{-\lambda_{230}t}$$

$$t = -\frac{\ln(1-m)}{\lambda_{230}}$$

$$= +\frac{0.34261}{9.1929 \times 10^{-6}} = 37{,}300 \text{ y}$$

The initial ^{230}Th/^{232}Th activity ratio is calculated from equation 20.59:

$$b = \left(\frac{^{230}\text{Th}}{^{232}\text{Th}}\right)^0_{Ax} e^{-\lambda_{230}t}$$

$$\left(\frac{^{230}\text{Th}}{^{232}\text{Th}}\right)^0_{Ax} = be^{\lambda_{230}t}$$

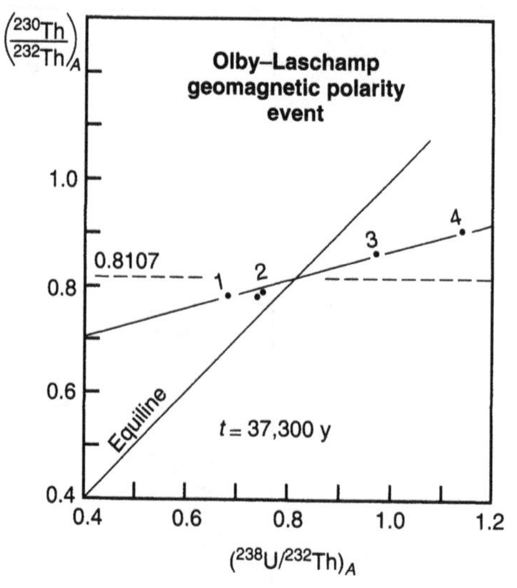

FIGURE 20.21 The ^{230}Th–^{238}U isochron of the reversely magnetized K-hawaiite lava flow at Olby in the Chaine des Puys, Massif Central, France: 1 = plagioclase; 2 = total rock and pyroxene; 3 = magnetite (23–80 μm); 4 = magnetite (7–23 μm). The decay constant of ^{230}Th is $\lambda_{230} = 9.1929 \times 10^{-6}$ y^{-1}. Data from Condomines (1978).

Since $b = 0.5754$ and $t = 37{,}300$ y,

$$\left(\frac{^{230}\text{Th}}{^{232}\text{Th}}\right)^0_{Ax} = 0.5754 \times 1.4090 = 0.8107$$

This is the value of the ^{230}Th/^{232}Th activity ratio that existed in all of the minerals at the time the lava flow crystallized.

20.6c Dating with ^{231}Pa

Chemical fractionation of Pa and U during partial melting in the mantle causes enrichment of MORB in ^{231}Pa. Goldstein et al. (1993) reported that basalt along the East Pacific Rise has high ^{231}Pa/^{235}U activity ratios (measured by mass spectrometry) ranging from 1.525 to 2.889. In addition, they reported similar ^{231}Pa enrichment in basalt extruded along the southern Juan de Fuca Ridge and the northern Gorda Ridge in the northeastern Pacific Ocean.

The decay of the excess ^{231}Pa in these volcanic rocks cannot be described in the same way as the excess ^{230}Th because there are no stable isotopes of Pa that can act as a reference in the way ^{232}Th does for ^{230}Th. Therefore, the activity of ^{231}Pa in young volcanic rocks is described by the equation

$$N_2 = \frac{\lambda_1 N_1^0}{\lambda_2 - \lambda_1}(e^{-\lambda_1 t} - e^{-\lambda_2 t}) + N_2^0 e^{-\lambda_2 t}$$

where $N_1 = {}^{235}$U and N_2-^{231}Pa. Since $\lambda_1 \ll \lambda_2$, the equation becomes

$$^{231}\mathrm{Pa} = \frac{^{235}\mathrm{U}_A}{\lambda_{231}}(1 - e^{-\lambda_{231} t}) + {}^{231}\mathrm{Pa}^0 e^{-\lambda_{231} t} \tag{20.60}$$

Converting to the activity of ^{231}Pa by multiplying by λ_{231},

$$^{231}\mathrm{Pa}_A = {}^{235}\mathrm{U}_A(1 - e^{-\lambda_{231} t}) + {}^{231}\mathrm{Pa}_A^0 e^{-\lambda_{231} t}$$

Dividing by $^{235}\mathrm{U}_A$,

$$\left(\frac{^{231}\mathrm{Pa}}{^{235}\mathrm{U}}\right)_A = 1 - e^{-\lambda_{231} t} + \left(\frac{^{231}\mathrm{Pa}}{^{235}\mathrm{U}}\right)_A^0 e^{-\lambda_{231} t}$$

$$\left(\frac{^{231}\mathrm{Pa}}{^{235}\mathrm{U}}\right)_A - 1 = e^{-\lambda_{231} t}\left[\left(\frac{^{231}\mathrm{Pa}}{^{235}\mathrm{U}}\right)_A - 1\right]$$

Solving for t,

$$t = -\frac{1}{\lambda_{231}} \ln\left[\frac{(^{231}\mathrm{Pa}/^{235}\mathrm{U})_A - 1}{(^{231}\mathrm{Pa}/^{235}\mathrm{U})_A^0 - 1}\right] \tag{20.61}$$

In order to use this equation to date basalt extruded along midocean ridges, the initial ^{231}Pa/^{235}U activity ratio must be known. Goldstein et al. (1993) used the ^{231}Pa/^{235}U activity ratio of the youngest flows along the crest of the ridge to calculate dates for older flows that were erupted along that same ridge.

For example, Goldstein et al. (1993) reported a ^{231}Pa/^{235}U activity ratio of 1.119 ± 0.007 for sample L585 NC-4 along the northern Gorda Ridge and measured an initial ^{231}Pa/^{235}U activity ratio of 2.872 ± 0.012 for a recently erupted basalt flow on the crest of that ridge. Substituting these values into equation 20.61 yields

$$t = -\frac{1}{2.132 \times 10^{-5}} \ln \frac{1.119 - 1}{2.872 - 1}$$

$$= +\frac{2.7557}{2.132 \times 10^{-5}} = 129 \pm 3 \text{ ka}$$

The same calculation based on an equation similar to 20.61 yields

$$t = -\frac{1}{\lambda_{230}} \ln\left[\frac{(^{230}\mathrm{Th}/^{238}\mathrm{U})_A - 1}{(^{230}\mathrm{Th}/^{238}\mathrm{U})_A^0 - 1}\right] \tag{20.62}$$

and using the ^{230}Th/^{238}U activity ratios of the same basalt specimens gives

$$\left(\frac{^{230}\mathrm{Th}}{^{238}\mathrm{U}}\right)_A = 1.085 \qquad \left(\frac{^{230}\mathrm{Th}}{^{238}\mathrm{U}}\right)_A^0 = 1.249$$

$$t = +\frac{1.0748}{9.1929 \times 10^{-6}} = 117 \pm 11 \text{ ka}$$

The ^{231}Pa and ^{230}Th dates are in satisfactory agreement within the stated errors. Both are geologically reasonable but depend on the assumption that the initial activity ratios of basalts erupted along midocean ridges do not change with time. Such changes could occur because the extent of partial melting as well as the composition of the source rocks could evolve with time.

The average ^{230}Th/^{238}U activity ratio of 24 MORBs in the Pacific Ocean analyzed by Goldstein et al. (1993) is 1.146 ± 0.017, whereas the average ^{231}Pa/^{235}U activity ratio of these rocks is 2.344 ± 0.215. The greater enrichment of volcanic rocks in ^{231}Pa relative to U is a manifestation of the difference in the chemical properties of Pa and Th.

20.7 MAGMA FORMATION

The radiochemical disequilibria among the daughters of ^{238}U and ^{235}U in recently erupted and unweathered volcanic rocks are a source of information about the way different elements are partitioned during partial melting of silicate rocks in the mantle. The extensive literature on this subject was reviewed by Gill et al. (1992). In addition, Rogers (2000) edited a set of papers on the rates and timescales of magmatic processes with contributions by McKenzie (2000), Condomines and Sigmarsson (2000), Lundstrom et al. (2000), and Turner et al. (2000). In the same year, Zhou and Zindler (2000) derived equations that relate radiochemical disequilibria of ^{230}Th, ^{226}Ra, and ^{231}Pa in volcanic rocks to elemental distribution coefficients, melting porosity, melting rate, and melting time during magma formation in the mantle.

20.7a MORBs and OIBs

The ^{230}Th/^{238}U and ^{231}Pa/^{235}U activity ratios in MORBs and OIBs in Figure 20.22 form two separate quasilinear data arrays that extend from values close to unity (radiochemical equilibrium) to about 2.0 for $(^{230}\text{Th}/^{238}\text{U})_A$ and to 2.9 for $(^{231}\text{Pa}/^{235}\text{U})_A$ (Goldstein et al., 1993; Pickett and Murrell, 1997).

The radiochemical disequilibria of MORBs and OIBs at the time of eruption are caused, in most cases, by fractionation of elements during partial melting in the mantle (Wood et al., 1999: Condomines and Sigmarsson, 2000). The high ^{231}Pa/^{235}U and ^{230}Th/^{238}U activity ratios in Figure 20.22 imply that Pa and Th are partitioned into the melt phase more efficiently than U (i.e., they are more "incompatible"). The data also suggest that Pa is more incompatible than Th relative to U.

The range of ^{231}Pa/^{235}U and ^{230}Th/^{238}U activity ratios and the positive correlation between them can be attributed to

1. changes in the conditions of melting of uniform source rocks in the mantle that result in the observed correlation of the activity ratios of MORBs and OIBs and
2. mixing of magmas having different activity ratios arising from different source rocks, different conditions of melting, or differences in the time elapsed since melting occurred.

20.7b Oceanic and Continental Andesites

The andesites in oceanic and continental subduction zones range from being moderately enriched in ^{231}Pa and ^{230}Th to being depleted in both (Pickett and Murrell, 1997). As a result, they define a data

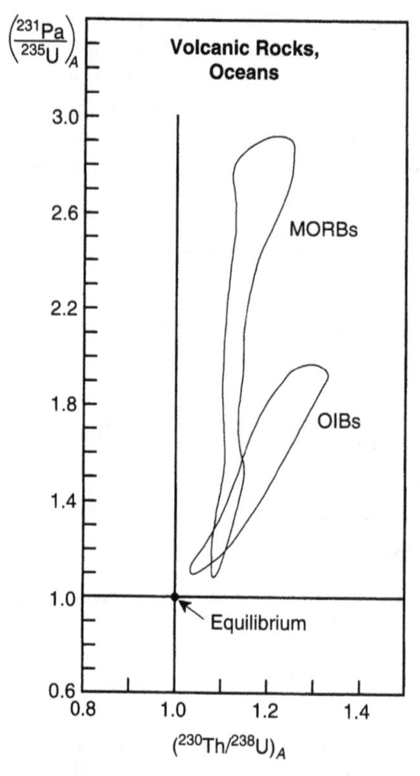

FIGURE 20.22 Excess ^{230}Th and ^{231}Pa in midocean ridge basalt (MORB), oceanic island basalt (OIB), and island arc andesite (A). Data from Goldstein et al. (1993) and Pickett and Murrell (1997).

field in Figure 20.23 that extends from quadrant 1, where both activity ratios are greater than one, to quadrant 2 ($^{230}\text{Th}/^{238}\text{U} < 1$, $^{231}\text{Pa}/^{235}\text{U} > 1$), and even into quadrant 3, where both activity ratios are less than 1 (Condomines and Sigmarsson, 1993). The activity ratios of continental basalts and rhyolites reported by Pickett and Murrell (1997) also plot in the andesite field of Figure 20.23 but are not shown. Pickett and Murrell (1997) suggested that the magma sources of the andesites (or the andesite magmas themselves) were enriched in U by aqueous fluids that were released by dehydration of the subducted oceanic crust and clay-rich marine sediment (Villemant et al., 1996). The addition of U reduced both activity ratios because the fluid transported U but not ^{230}Th and ^{231}Pa.

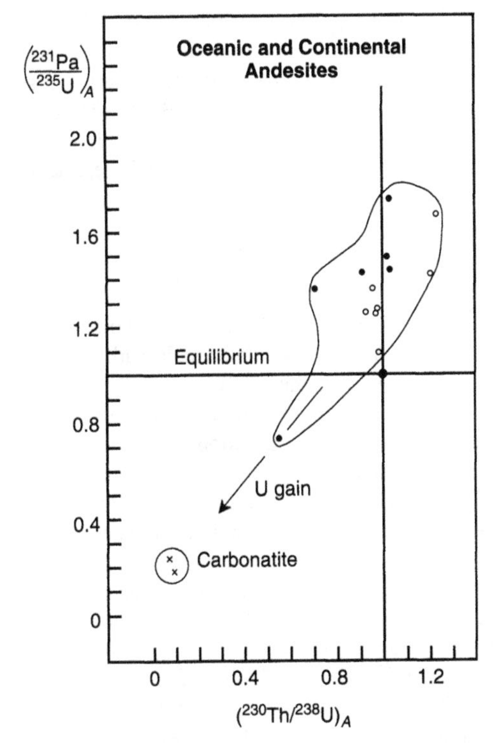

FIGURE 20.23 Activity ratios of $^{230}Th/^{238}U$ and $^{231}Pa/^{235}U$ of recently erupted andesites in oceanic island arcs (black dots) and continental subduction zones (open circles). The crosses are carbonatites from the volcano Oldoinyo Lengai in Tanzania. The low activity ratios of both rock types have been attributed to gain of U by magmas prior to crystallization. Data from Pickett and Murrell (1997).

This hypothesis can be evaluated by the data in Table 20.7, which show that the weighted average U concentration of oceanic and continental andesites (2.46 ppm) is about 30 times greater than that of MORBs and more than 5 times greater than that of OIBs. However, the data also indicate similar enrichments of andesite in Th. In addition, Villemant et al. (1996) concluded from a study of volcanic rocks on Mt. Pelée on the island of Martinique that the radiochemical disequilibrium between ^{230}Th and ^{238}U originates by late-stage hydrothermal alteration of the magma rather than during melting at depth.

Table 20.7. Concentrations of U and Th and Th/U Ratios of Volcanic Rocks

	Concentration, ppm		
Rock Type	U	Th	U/Th[a]
MORB	0.084	0.226	0.372
(31)	0.037–0.249	0.082–0.721	0.319–0.437
OIB	0.468	1.57	0.299
(7)	0.047–0.996	0.133–3.50	0.280–0.354
Andesite[b]	2.09	7.74	0.270
(7)	0.152–6.07	0.274–19.5	9.189–0.917
Andesite[c]	2.76	7.32	0.377
(9)	1.22–4.53	1.38–9.55	0.307–0.885
Continental	2.32	7.06	0.328
basalt (5)	0.393–6.76	1.22–18.1	0.213–0.398
Continental	5.34	16.8	0.317
rhyolite(4)	2.32–6.41	7.74–20.0	0.299–0.320
Carbonatite	6.75	1.54	4.38
(2)	6.3–7.2	1.17–1.92	3.75–5.40

Note: Data for MORB from Goldstein et al. (1993). All others from Pickett and Murrell (1997). The number of samples included in each average is indicated in parentheses.

[a] Average of the U/Th ratios of the analyzed specimens.
[b] Island arcs.
[c] Continental.

20.7c Carbonatites

The U enrichment hypothesis for the petrogenesis of andesite magma is supported indirectly by the very low $^{230}Th/^{238}U$ and $^{231}Pa/^{235}U$ activity ratios of carbonatite lavas on the East African volcano Oldoinyo Lengai in Figure 20.23 (Picket and Murrell, 1997):

$$\left(\frac{^{230}\mathrm{Th}}{^{238}\mathrm{U}}\right)_A = 0.077 \pm 0.009$$

$$\left(\frac{^{231}\mathrm{Pa}}{^{235}\mathrm{U}}\right)_A = 0.200 \pm 0.014$$

The average U and Th concentrations of these carbonatite lavas are

$$\mathrm{U} = 6.75 \pm 0.28 \text{ ppm} \qquad \mathrm{Th} = 1.54 \pm 0.18 \text{ ppm}$$

$$\mathrm{U/Th} = 4.44 \pm 0.55$$

The U concentration of the carbonatite lavas (6.75 ppm) exceeds those of all other rocks types listed in Table 20.7 whereas their Th concentrations are less than those of island arc andesites and of continental basalts and rhyolites.

The enrichment of carbonatite magmas in U and their depletion in Th may be a consequence of their origin by the segregation of a carbonatic melt from alkali-rich silicate magmas (Wyllie and Tuttle, 1960, 1962; Treiman and Schedl, 1983; Wyllie, 1989). During this process, U is admitted into the carbonatite melt but Th is excluded (Jones et al., 1995). The carbonatite lavas analyzed by Pickett and Murrell (1997) were erupted in 1960 and 1963, meaning that they could not have accumulated the ^{230}Th and ^{231}Pa they contained when they were analyzed in 1995.

Therefore, the low ^{230}Th/^{238}U and ^{231}Pa/^{235}U activity ratios of the carbonatite lavas could be explained by the following process: An alkali-rich magma, which formed by decompression melting in the lithospheric mantle under East Africa, initially had elevated ^{230}Th/^{238}U and ^{231}Pa/^{235}U activity ratios depending on the conditions at the time and place of melting. As the silicate magma differentiated in a magma chamber beneath Oldoinyo Lengai, a carbonatite liquid separated from it. This magma scavenged U from the silicate magma, which caused its ^{230}Th/^{238}U and ^{231}Pa/^{235}U activity ratios to decline to the low values reported by Pickett and Murrell (1997).

The petrogenesis of carbonatite lavas appears to be relevant to the low activity ratios of volcanic rocks that form in subduction zones because the aqueous fluid that emanates from the subducted oceanic crust could have enriched andesite magmas in U. Consequently, their ^{230}Th/^{238}U and ^{231}Pa/^{235}U activity ratios were lowered, which may explain the slope of their data array in Figure 20.23 and its apparent alignment with the carbonatite data points.

Low ^{230}Th/^{238}U activity ratios of carbonatite lavas on Oldoinyo Lengai ranging from 0.067 ± 0.003 to 0.625 ± 0.026 (at the time of eruption) were also reported by Williams et al. (1986) and thus confirm the apparent depletion of the carbonatite in ^{230}Th. However, the same authors also reported elevated ^{226}Ra/^{228}U activity ratios (1.29–5.55), implying that the carbonatites are enriched in Ra relative to U. Additional measurements of the activities of U-series daughters in carbonatite lavas of Oldoinyo Lengai were reported by Pyle et al. (1991).

Jones et al. (1995) attempted to explain the U-series disequilibria in the carbonatites of Oldoinyo Lengai by measuring the distribution coefficients of several trace elements (including Ba, Pb, Th, U, Ra, and Pa) in liquid carbonate–liquid silicate systems where the distribution coefficient D for element X is defined as

$$D(X) = \frac{C(X)^{\text{carb}}}{C(X)^{\text{sil}}} \tag{20.63}$$

Values of $D(X) > 1$ indicate that the concentration of element X in the carbonate melt (carb) is greater than in the silicate melt (sil).

The results demonstrated that only Ba and Ra are preferentially partitioned into the carbonate liquid and that the extent of fractionation of all elements decreases with increasing temperature between 1100 and 1400°C at 10 kbar pressure. However, even though the other elements are not favored by the carbonate melt, their distribution coefficients at 1250°C and 10 kbar decrease systematically in the sequence $D(\text{Ba}) = 1.8$; $D(\text{Ra}) = 1.6$; $D(\text{Pb}) = 0.61$; $D(\text{U}) = 0.37$; $D(\text{Th}) = 0.28$; $D(\text{Pa}) \sim 0.15$.

Therefore, the values of the distribution coefficients suggest that in the carbonate melt Th is favored over Pa, U is favored over Th, Pb is favored over U, and Ra is favored over Pb. Consequently, the preferential incorporation of U relative to Th and Pa causes the Th/U and Pa/U concentration ratios of a carbonate liquid to be lower than those of the silicate liquid with which the carbonate melt is in chemical equilibrium.

Jones et al. (1995) concluded that the radiochemical activities of U, Ra, and Pb of carbonate lavas on Oldoinyo Lengai are compatible with the predicted effect of partitioning between carbonate and silicate liquids. However, the activities of Th and probably also Pa appear to have been affected by additional processes that removed these elements from the carbonate melt or prevented them from transferring into it. Alternatively, the ^{230}Th/^{238}U and ^{231}Pa/^{235}U activity ratios of the carbonatite lavas could have been lowered by the addition of U rather than by the loss of Th and Pa.

20.7d Applications to Petrogenesis

U-series disequilibria in recently erupted volcanic rocks have been reported from many other locations in the oceans and on the continents:

Mid-Atlantic Ridge: Condomines et al., 1981, *Earth Planet. Sci. Lett.*, 55:247–256.

Iceland: Condomines et al., 1981, *Earth Planet. Sci. Lett.*, 55:393–405. Sigmarsson et al., 1992a, *Earth Planet. Sci. Lett.*, 110:149–162. Sigmarsson et al., 1992b, *Contrib. Mineral. Petrol.*, 112:20–34. Hémond et al., 1988, *Earth Planet. Sci. Lett.*, 87:273–285.

Stromboli: Condomines and Allégre, 1980, *Nature*, 288:154–157.

East Pacific Rise: Newman et al., 1983, *Earth Planet. Sci. Lett.*, 65:17–33. Reinitz and Turekian, 1989, *Earth Planet. Sci. Lett.*, 94:199–207. Goldstein et al., 1993, *Earth Planet. Sci. Lett.*, 115:151–159.

Juan de Fuca and Gorda Ridges: Goldstein et al., 1989, *Earth Planet. Sci. Lett.*, 96:134–146. Goldstein et al., 1991, *Earth Planet Sci. Lett.*, 107:25–41 (Erratum, *EPSL*, 109:255–272). Goldstein et al., 1993, *Earth Planet. Sci. Lett.*, 115:151–159.

Aleutian Islands: Newman et al., 1984, *Nature*, 308:268–270.

Mariana Islands: Newman et al., 1984, *Nature*, 308:268–270.

Hawaiian Islands: Sims et al., 1995, *Science*, 267:508–512. Cohen and O'Nions, 1993, *Earth Planet. Sci. Lett.*, 120:169–175. Reinitz and Turekian, 1991, *Geochim. Cosmochim. Acta*, 55:3735–3740. Condomines et al., 1976, *Earth Planet. Sci. Lett.*, 33: 122–125.

Mt. St. Helens: Bennett et al., 1982, *Earth Planet. Sci. Lett.*, 60:61–69.

Snake River Plain, Wyoming: Reid, 1995, *Earth Planet. Sci. Lett.*, 131:239–254.

Southwestern USA/Mexico: Asmerom and Edwards, 1995, *Earth Planet. Sci. Lett.*, 134: 1–7.

Nicaragua: Reagan et al., 1994, *Geochim. Cosmochim. Acta*, 58:4199–4212.

Nevado del Ruiz, Colombia: Schaefer et al., 1993, *Geochim. Cosmochim. Acta*, 57:1215–1219.

Oldoinyo Lengai, Tanzania: Williams et al., 1986, *Geochim. Cosmochim. Acta*, 50:1249–1259.

In addition, U-series disequilibria are contributing to the study of petrogenesis of volcanic rocks, including especially the partitioning of the daughters of U during partial melting of silicate rocks in the mantle:

Turner et al., 2000, *Chem. Geol.*, 162:127–136. McKenzie, D., 2000, *Chem. Geol.*, 162:81–94. Condomines and Sigmarsson, 2000, *Chem. Geol.*, 162:95–104. Lundstrom et al., 2000, *Chem. Geol.*, 162:105–126. Zhou and Zindler, 2000, *Geochim. Cosmochim. Acta*, 64:1809–1817. Wood et al., 1999, *Geochim. Cosmochim. Acta*, 63:1613–1620. Asmerom and Edwards, 1995, *Earth Planet. Sci. Lett.*, 134:1–7. Villemant et al., 1996, *Earth Planet. Sci. Lett.*, 140:259–267. Cohen and O'Nions, 1993, *Earth Planet. Sci. Lett.*, 120:169–175. Krishnaswami et al., 1984, *Geochim. Cosmochim. Acta*, 48:505–511.

These applications should be considered in the context of the petrogenesis of igneous rocks and therefore exceed the scope of this volume.

20.8 SUMMARY

The radiochemical disequilibria among the daughters of U and Th result from differences in their chemical properties and in the susceptibilities to chemical weathering of the minerals in which they occur. The equations that describe the time-dependent tendencies of these radionuclides to reestablish secular equilibrium with their respective daughters—or to decay to zero—are derivable from the law of radioactivity and from the solutions of the equations for decay in a series for short-lived radionuclides. Nevertheless, the equations are complicated and the analytical methods are demanding.

The traditional methods of measuring the rates of decay of specific radionuclides by α-counting, β-counting, α-spectrometry, and γ-ray spectrometry are being replaced by mass spectrometry, which can be used to determine both the isotope compositions of certain elements (e.g., U and Th) and their concentrations by means of isotopic dilution. In several cases, these methods are more precise and more accurate than conventional measurements of counting rates and therefore have reduced the uncertainties of the resulting dates. The instrumentation that has been used includes multiple-collector thermal ionization mass spectrometers as well as inductively coupled plasma mass spectrometers equipped with electromagnets and multiple collectors. The improvement of the resulting analytical data will enhance the use of U-series disequilibrium dates in oceanography, limnology, sedimentation, geomorphology, pedology, climatology, archeology, and anthropology.

In addition, the radiochemical disequilibria in young volcanic rocks erupted in different tectonic settings are an important source of information in the study of the petrogenesis of igneous rocks. In this case, the disequilibria result from the preferential partitioning of certain elements into silicate melts that form in the mantle by decompression melting and by fluxing with water. This phenomenon can provide information about the mineral composition of rocks in the mantle, the partition coefficients of various elements, the time intervals between the formation and eruption of magma, and the ages of recently erupted volcanic rocks.

REFERENCES FOR U-SERIES DISEQUILIBRIUM DATING (INTRODUCTION, SECTION 20.1)

Bender, M. L., R. G. Fairbanks, F. W. Taylor, R. K. Matthews, J. G. Goddard, and W. S. Broecker, 1979. Uranium-series dating of the Pleistocene reef tracts of Barbados, West Indies. *Geol. Soc. Am. Bull.*, 90:577–584.

Broecker, W. S., 1963. A preliminary evaluation of uranium-series inequilibrium as a tool for absolute age measurement on marine carbonates. *J. Geophys. Res.*, 68:2817–2834.

Broecker, W. S., D. L. Thurber, J. Goddard, T. L. Ku, R. K. Matthews, and K. J. Mesolella, 1968. Milankovitch hypothesis supported by precise dating of coral reefs and deep-sea sediment. *Science*, 159:297–300.

Cherdyntsev, V. V., 1969. *Uranium-234.* Atomizdat, Moskva. Translated into English by J. Schmorak, Israel Program for Scientific Translation, Jerusalem, 1971.

Fleischer, R. L., 1980. Isotopic disequilibrium of uranium: Alpha recoil damage and preferential solution effects. *Science*, 207:979–981.

Fleischer, R. L., 1983. Theory of alpha recoil effects on radon release and isotopic disequilibrium. *Geochim. Cosmochim. Acta*, 47:779–784.

Goldberg, E. D., and M. Koide, 1963. Rates of sediment accumulation in the Indian Ocean. In J. Geiss and E. D. Goldberg (eds.), *Earth Science and Meteoritics*, pp. 90–102. North-Holland, Amsterdam.

Goldberg, E. D., and M. Koide, 1962. Geochronological studies of deep-sea sediments by the ionium/thorium method. *Geochim. Cosmochim. Acta*, 26:417–450.

Harmon, R. S., P. Thompson, H. P. Schwarcz, and D. C. Ford, 1975. Uranium-series dating of speleothems. *NSS Bull. J. Natl. Spel. Soc.*, 37:21–34.

Ivanovich, M., and R. S. Harmon (Eds.), 1992. *Uranium-Series Disequilibrium: Applications to Earth, Marine, and Environmental Sciences*, 2nd ed. Clarendon, Oxford, UK.

Ivanovich, M., A. G. Latham, and T.-L. Ku, 1992. Uranium-series disequilibrium applications in geochronology. In M. Ivanovich and R. S. Harmon (Eds.), *Uranium-Series Disequilibria; Applications to Earth, Marine, and Environmental Sciences*, 2nd ed., pp. 62–94. Clarendon, Oxford, UK.

Kaufman, A., 1971. U-series dating of Dead Sea basin carbonates. *Geochim. Cosmochim. Acta*, 35:1269–1281.

Kaufman, A., 1993. An evaluation of several methods for determining $^{230}Th/U$ ages in impure carbonates. *Geochim. Cosmochim. Acta*, 57:2303–2317.

Kaufman, A., W. S. Broecker, T. L. Ku, and D. L. Thurber, 1971. The status of U-series methods of dating mollusks. *Geochim. Cosmochim. Acta*, 35:1155–1183.

Kigoshi, K., 1971. Alpha-recoil thorium-234: Dissolution into water and the uranium-234/uranium-238 disequilibrium in nature. *Science*, 173:47–48.

Koide, M., and E. D. Goldberg, 1965. Uranium 234/uranium-238 ratios in sea water. In M. Sears (Ed.), *Progress in Oceanography*, Vol. 3, pp. 173–178. Pergamon, New York.

Kronfeld, J., 1974. Uranium deposition and Th-234 alpha-recoil: An explanation for extreme U-234/U-238 fractionation within the Trinity aquifer. *Earth Planet. Sci. Lett.*, 21:327–330.

Ku, T.-L., 1965. An evaluation of the $^{234}U/^{238}U$ method as a tool for dating pelagic sediments. *J. Geophys. Res.*, 70:3457–3474.

Ku, T.-L., 1976. The uranium-series methods of age determination. In, R. A. Donath, F. G. Stehli, and G. W. Wetherill (Eds.), *Annual Reviews of Earth and Planetary Science*, Vol. 4. Annual Reviews, Palo Alto, California.

Ku, T.-L., and W. S. Broecker, 1967. Uranium, thorium, and protactinium in a manganese nodule. *Earth Planet. Sci. Lett.*, 2:317–320.

Lide, D. R., and H. P. R. Frederikse, 1995. *CRC Handbook of Chemistry and Physics*, 76th ed., CRC, Boca Raton, Florida.

Lin, J. C., W. S. Broecker, R. F. Anderson, S. Hemming, J. L. Rubenstone, and G. Bonani, 1996. New $^{230}Th/U$ and ^{14}C ages from Lake Lahontan carbonates, Nevada, USA, and a discussion of the origin of initial thorium. *Geochim. Cosmochim. Acta*, 60:2817–2832.

Ludwig, K. R., and D. M. Titterington, 1994. Calculation of $^{230}Th/U$ isochrons, ages, and errors. *Geochim. Cosmochim. Acta*, 58:5031–5042.

Luo, S., and T.-L. Ku, 1991. U-series isochron dating: A general method employing total sample dissolution. *Geochim. Cosmochim. Acta*, 55:555–564.

Milankovitch, M., 1941. *Canon of Insolation and the Ice-Age Problem.* Königliche Serbische Akademie, Belgrad. English translation by the Israel Program for Scientific Translations, Washington, D.C., 1969.

Miyake, Y., Y. Sugimura, and T. Uchida, 1966. Ratio $^{234}U/^{238}U$ and the uranium concentration in sea water in the western North Pacific. *J. Geophys. Res.*, 71:3083–3087.

Schwarcz, H. P., 1978. Uranium-series disequilibrium dating. *Geosci. Can.*, 5(4):184–188.

Schwarcz, H. P., 1980. Absolute age determination of archaeological sites by uranium series dating of travertines. *Archaeometry*, 22:3–24.

Thurber, D. L., 1962. Anomalous $^{234}U/^{238}U$ in nature. *J. Geophys. Res.*, 67:4518–4520.

Veeh, H. H., 1966. $^{230}Th/^{238}U$ and $^{234}U/^{238}U$ ages of Pleistocene high sealevel stand. *J. Geophys. Res.*, 71:3379–3386.

REFERENCES FOR RADIUM (SECTION 20.2)

Carroll, J.-L., K. K. Falkner, E. T. Brown, and W. S. Moore, 1993. The role of the Ganges-Brahmaputra mixing zone in supplying barium and ^{226}Ra to the Bay of Bengal. *Geochim. Cosmochim. Acta*, 57:2981–2990.

Condomines, M., C. Brouzes, and S. Rihs, 1999. Radium and its daughters in hydrothermal carbonates from Auvergne (French Massif Central); origin and dating applications. *C.R. Acad. Sci. Paris, Ser. II*, 328-1:23–28.

Fanning, K. A., J. A. Breland II, and R. H. Byrne, 1982. ^{226}Ra and ^{222}Rn in the coastal water of west Florida: High concentrations and atmospheric degassing. *Science*, 215:667–670.

Faure, G., 1998. *Principles and Applications of Geochemistry*, 2nd ed. Prentice-Hall, Upper Saddle River, New Jersey.

Key, R. M., R. F. Stallard, W. S. Moore, and J. L. Sarmiento, 1985. Distribution and flux of ^{226}Ra and ^{228}Ra in the Amazon River estuary. *J. Geophys. Res.*, 90(C4):6995–7004.

Lide, D. R., and H. P. R. Frederikse (Eds.), 1995. *CRC Handbook of Chemistry and Physics*, 76th ed. CRC, Boca Raton, Florida.

Moore, W. S., 1969. Oceanic concentrations of ^{228}Radium. *Earth Planet. Sci. Lett.*, 6:437–446.

Moore, W. S., 1997. High fluxes of radium and barium from the mouth of the Ganges-Brahmaputra River during low river discharge suggest a large ground water source. *Earth Planet. Sci. Lett.*, 150:141–150.

Moore, W. S., R. M. Key, and J. Sarmiento, 1985. Techniques for precise mapping of ^{226}Ra and ^{228}Ra in the ocean. *J. Geophys. Res.*, 90(C4):6983–6994.

Rihs, S., M. Condomines, and O. Sigmarsson, 2000. U, Ra and Ba incorporation during precipitation of hydrothermal carbonates: Implications for ^{226}Ra-Ba dating of impure travertines. *Geochim. Cosmochim. Acta*, 64:661–671.

Sturchio, N. C., 1990. Radium isotopes, alkaline earth diagenesis, and age determinations of travertine from Mammoth Hot Springs, Wyoming, U.S.A. *Appl. Geochem.*, 5:631–640.

Sturchio, N. C., J. L. Banner, C. M. Binz, L. B. Heraty, and M. Musgrove, 2001. Radium geochemistry of groundwaters in Paleozoic carbonate aquifers, midcontinent, USA. *Appl. Geochem.*, 16:102–122.

Turekian, K. K., J. K. Cochran, and Y. Nozaki, 1979. Growth rate of a clam from the Galapagos Rise hot-spring field using natural radionuclide ratios. *Nature*, 280:385–387.

REFERENCES FOR PROTACTINIUM (SECTION 20.3)

Allègre, C. J., 1964. De l' extension de la methode de calcul graphique concordia aux mesures d'ages absolus effectués a l' aide du déséquilibre radioactif: Cas des mineralisations secondaires d'uranium. *Comptes rendus des seances de l' Academie des Sciences, Paris*, 259:4086–4089.

Anderson, R. F., M. P. Bacon, and P. G. Brewer, 1983. Removal of ^{230}Th and ^{231}Pa at ocean margins. *Earth Planet. Sci. Lett.*, 66:73–90.

Bourdon, B., J.-L. Joron, and C. J. Allègre, 1999. A method for ^{231}Pa analysis by thermal ionization mass spectrometry in silicate rocks. *Chem. Geol.*, 157:147–151.

Broecker, W. S., D. L. Turner, J. Goddard, T.-L. Ku, R. K. Matthews, and K. J. Mesolella, 1968. Milankovitch hypothesis supported by precise dating of coral reefs and deep-sea sediment. *Science*, 159:297–300.

Edwards, R. L., H. Cheng, M. T. Murrell, and S. J. Goldstein, 1997. Protactinium 231 dating of carbonates by thermal ionization mass spectrometry: Implications for Quaternary climate change. *Science*, 276:782–785.

Gascoyne, M., H. P. Schwarcz, and D. C. Ford, 1978. Uranium series dating and stable isotope studies of speleothems; Part 1—theory and techniques. *Trans. Br. Cave Res. Assoc.*, 5:91–111.

Ivanovich, M., and A. Murray, 1992. Spectroscopic methods. In M. Ivanovich and S. R. Harmon (Eds.), *Uranium-Series Disequilibrium; Applications to Earth, Marine, and Environmental Sciences*, pp. 127–273, 2nd ed. Clarendon, Oxford, UK.

Ku, T.-L., 1965. An evaluation of $^{234}U/^{238}U$ method as a tool for dating pelagic sediments. *J. Geophys. Res.*, 70:3457–3474.

Ku, T.-L., and W. S. Broecker, 1967a. Uranium, thorium, and protactinium in a manganese nodule. *Earth Planet. Sci. Lett.*, 2:317–320.

Ku, T.-L., and W. S. Broecker, 1967b. Rates of sedimentation in the Arctic Ocean. *Prog. Oceanogr.*, 4:95–104.

Osmond, J. K., 1979. Accumulation models of ^{230}Th and ^{231}Pa in deep-sea sediments. *Earth Planet. Sci. Rev.*, 15:95–151.

Pickett, D. A., M. T. Murrell, and R. W. Williams, 1994. Determination of femtogram quantities of protactinium in geological samples by thermal ionization spectrometry. *Anal. Chem.*, 66:1044–1049.

Richards, D. A., P. L. Smart, and R. L. Edwards, 1994. Maximum sea levels for the last glacial period from U-series ages of submerged speleothems. *Nature*, 367:357–360.

Rona, E., and C. Emiliani, 1969. Absolute dating of Caribbean cores P6304-8 and P6304-9. *Science*, 163: 66–68.

Rosholt, J. N., C. Emiliani, J. Geiss, F. F. Koczy, and P. J. Wangersky, 1961. Absolute dating of deep-sea cores by the $^{231}Pa/^{230}Th$ method. *J. Geol.*, 69:162–185.

Rosholt, J. N., C. Emiliani, J. Geiss, F. F. Koczy, and P. J. Wangersky, 1962. $^{231}Pa/^{230}Th$ dating and $^{18}O/^{16}O$ temperature analysis of core A254-BR-C. *J. Geophys. Res.*, 67(7):2907–2911.

Whitehead, N. E., R. G. Ditchburn, P. W. Williams, and W. J. McCabe, 1999. ^{231}Pa and ^{230}Th contamination at zero age: A possible limitation on U/Th series dating of speleothem material. *Chem. Geol.*, 156:359–366.

Yang, H.-S., Y. Nozaki, H. Sakai, and A. Masuda, 1986. The distribution of ^{230}Th and ^{231}Pa in the deep-sea surface sediments of the Pacific Ocean. *Geochim. Cosmochim. Acta*, 50:81–89.

REFERENCES FOR LEAD-210 (SECTION 20.4)

Aaby, B., J. Jacobsen, and O. S. Jacobsen, 1979. Lead-210 dating and lead deposition in the ombrotrophic peat bog, Draved Mose, Denmark. *Danm. Geol. Unders. Årbog*, 1978:45–68.

Appleby, P. G., and F. Oldfield, 1992. Application of lead-210 to sedimentation studies. In M. Ivanovich and R. S. Harmon (Eds.), *Uranium-Series Disequilibria; Applications to Earth, Marine, and Environmental Sciences*, 2nd ed., pp. 731–778. Clarendon, Oxford, UK.

Chung, Y., 1981. ^{210}Pb and ^{226}Ra distributions in the circumpolar waters. *Earth Planet. Sci. Lett.*, 55:205–216.

Crozaz, G., and P. Fabri, 1966. Mesure du polonium a l' echelle de 10^{-13} Curie, tracage par les ^{208}Po et application a la chronologie des glaces. *Earth Planet. Sci. Lett.*, 1:446–448.

Crozaz, G., and C. C. Langway, Jr., 1966. Dating Greenland firn-ice cores with ^{210}Pb. *Earth Planet. Sci.*, 1:194–196.

Crozaz, G., E. Picciotto, and W. DeBreuck, 1964. Antarctic snow chronology with ^{210}Pb. *J. Geophys. Res.*, 69:2597–2604.

Goldberg, E. D., 1963. Geochronology with ^{210}Pb. In *Radioactive Dating*, pp. 121–131. International Atomic Energy Agency, Vienna.

Koide, M., A. Soutar, and E. D. Goldberg, 1972. Marine geochronology with ^{210}Pb. *Earth Planet. Sci. Lett.*, 14:442–446.

Krishnaswami, S., D. Lal, J. M. Martin, and M. Meybeck, 1971. Geochronology of lake sediments. *Earth Planet. Sci. Lett.*, 11:407–414.

Ku, T.-L., S. Luo, B. W. Leslie, and D. E. Hammond, 1992. Decay-series disequilibria applied to the study of rock-water interaction and geothermal systems. In M. Ivanovich and R. S. Harmon (Eds.), *Uranium-Series Disequilibria; Applications to Earth, Marine, and Environmental Sciences*, 2nd ed., pp. 631–668. Clarendon, Oxford, UK.

Lewis, D. M., 1977. The use of ^{210}Pb as a heavy metal tracer in the Susquehanna River system. *Geochim. Cosmochim. Acta*, 41:1557–1564.

McCall, P. L., J. A. Robbins, and G. Matisof, 1984. ^{137}Cs and ^{210}Pb transport and geochronologies in urbanized reservoirs with rapidly increasing sedimentation rates. *Chem. Geol.*, 44:33–65.

Oldfield, F., P. G. Appleby, R. S. Cambrey, J. D. Eakins, K. E. Barber, R. W. Battarbee, G. W. Pearson, and J. M. Williams, 1979. ^{206}Pb, ^{137}Cs, and ^{239}Pu profiles in ombrotrophic peat. *Oikos*, 33:40–45.

Picciotto, E., G. Crozaz, W. Ambach, and H. Eisner, 1967. Lead-210 and strontium-90 in an alpine glacier. *Earth Planet. Sci. Lett.*, 3:237–242.

Robbins, J. A. (Ed.), 1984. Geochronology of Recent deposits. *Chem. Geol.*, 44(1/3):1–348.

REFERENCES FOR ARCHEOLOGY AND ANTHROPOLOGY (SECTION 20.5)

Balter, M., and A. Gibbons, 2000. A glimpse of humans' first journey out of Africa. *Science*, 288:948–950.

Bishop, W. W., and J. A. Miller (Eds.), 1972. *Calibration of Hominoid Evolution*. Scottish University Press, Edinburgh.

Cheng, H., R. L. Edwards, J. Hoff, C. D. Gallup, D. A. Richards, and Y. Asmerom, 2000. The half-lives of uranium-234 and thorium-230. *Chem. Geol.*, 169:17–33.

Cherdyntsev, V. V., 1971. *Uranium-234*. Israel Program for Scientific Translations, Jerusalem.

Clark, J. D., J. de Heinzelin, K. D. Schick, W. K. Hart, T. D. White, G. WoldeGabriel, R. C. Walter, G. Suwa, B. Asfaw, E. Vrba, and Y. H.-Selassie, 1994. African Homo erectus: Old radiometric ages and young Oldowan assemblages in the Middle Awash Valley, Ethiopia. *Science*, 264:1907–1910.

Curtis, G. H., and R. L. Hay, 1972. Further geological studies and potassium-argon dating at Olduvai Gorge and Ngorongoro Crater. In W. W. Bishop and J. A. Miller (Eds.), *Calibration of Hominid Evolution*, pp. 289–301. Scottish Academic, Edinburgh.

Edwards, R. L., J. H. Chen, and G. J. Wasserburg, 1987. ^{238}U-^{234}U-^{230}Th-^{232}Th systematics and the precise measurement of time over the past 500,000 years. *Earth Planet. Sci. Lett.*, 81:175–192.

Fitch, F. J., J. A. Miller, and P. J. Hooker, 1976. Single whole-rock K-Ar isochrons. *Geol. Mag.*, 113:1–10.

Hamilton, A. C., 1984. *Environmental History of East Africa; a Study of the Quaternary*. Academic, New York.

Harmon, R. S., D. C. Ford, and H. P. Schwarcz, 1977. Interglacial chronology of the Rocky and Mackenzie Mountains based upon ^{230}Th-^{234}U dating of calcite speleothems. *Can. J. Earth Sci.*, 14:2543–2552.

Klein, R. G., 1989. *The Human Career: Human Biological and Cultural Origins*. University of Chicago Press, Chicago, Illinois.

Latham, A. G., and H. P. Schwarcz, 1992. Carbonate and sulphate precipitates. In M. Ivanovich and R. S. Harmon (Eds.), *Uranium-Series Disequilibrium; Applications to Earth, Marine, and Environmental Sciences*, 2nd ed., pp. 423–459. Clarendon, Oxford, UK.

Leakey, R. E., 1981. *The Making of Mankind*. E. P. Dutton, New York.

Leakey, R. E., and R. Lewin, 1977. *Origins*. E. P. Dutton, New York.

Leakey, R. E., and A. Walker, 1985. *Homo erectus* unearthed. *Nat. Geographic*, 168(5):624–629.

Lide, D. R., and H. P. R. Frederikse, 1995. *CRC Handbook of Chemistry and Physics*, 76th ed. CRC, Boca Raton, Florida.

Ludwig, K. R., K. R. Simmons, B. J. Szabo, I. J. Winograd, J. M. Landwehr, A. C. Riggs, and R. J. Hoffman, 1992. Mass-spectrometric ^{230}Th-^{234}U-^{238}U dating of the Devil's Hole calcite vein. *Science*, 258:284–287.

McDermott, F., R. Grün, C. B. Stringer, and C. J. Hawkesworth, 1993. Mass-spectrometric U-series dates for Israeli Neanderthal/early modern hominid sites. *Nature*, 363:252–255.

Schwarcz, H. P., 1980. Absolute age determination of archaeological sites by uranium-series dating of travertines. *Archaeometry*, 22:3–24.

Schwarcz, H. P., 1982. Absolute dating of travertines from archaeological sites. In L. A. Currie (Ed.), *Nuclear and Chemical Dating Techniques; Interpreting the Environmental Record*, pp. 477–490. Amer. Chem. Soc. Symp. Ser. No. 176. American Chemical Society, Washington, D. C.

Schwarcz, H. P., 1992. Uranium-series dating and the origin of modern man. *Roy. Soc. Lond. Philos. Trans.*, 337:131–137.

Schwarcz, H. P., and B. A. Blackwell, 1992. Archaeological applications. In M. Ivanovich and R. S. Harmon (Eds.), *Uranium-Series Disequilibrium; Applications to Earth, Marine, and Environmental Science*, 2nd ed., pp. 513–552. Clarendon, Oxford.

Swisher, C. C., III, G. H. Curtis, T. Jacob, A. G. Getty, A. Suprijo, and Widiasmoro, 1994. Age of the earliest known hominids in Java, Indonesia. *Science*, 293:1118–1121.

Szabo, B. J., 1990. Ages of travertine deposits in eastern Grand Canyon National Park, Arizona. *Quat. Res.*, 34:24–32.

Thompson, P., D. C. Ford, and H. P. Schwarcz, 1975. $^{234}U/^{238}U$ ratios in limestone cave seepage waters and speleothem from West Virginia. *Geochim. Cosmochim Acta*, 39:661–669.

Weaver, K. F., 1985. The search for our ancestors. *Nat. Geographic*, 168(5):561–623.

Wintle, A. G., 1996. Archaeologically-relevant dating techniques for the next century—small, hot and identified by acronyms. *J. Archaeol. Sci.*, 23:123–138.

Zhao, J.-X., K. Hu, K. D. Collerson, and H.-K. Xu, 2001. Thermal ionization mass spectrometry U-series dating of a hominid site near Nanjing, China. *Geology*, 29:27–30.

REFERENCES FOR VOLCANIC ROCKS (SECTIONS 20.6–20.7)

Allègre, C. J., 1968. ^{230}Th dating of volcanic rocks: A comment. *Earth Planet. Sci. Lett.*, 5:209–210.

Asmerom, Y., and R. L. Edwards, 1995. U-series isotope evidence for the origin of continental basalts. *Earth Planet. Sci. Lett.*, 134:1–7.

Chen, J. H., R. L. Edwards, and G. J. Wasserburg, 1986. ^{238}U, ^{234}U, and ^{232}Th in seawater. *Earth Planet. Sci. Lett.*, 80:241.

Cohen, A. S., and R. K. O'Nions, 1993. Melting rates beneath Hawaii: Evidence from uranium-series isotopes in recent lavas. *Earth Planet. Sci. Lett.*, 120:169–175.

Condomines, M., 1978. Age of the Olby-Laschamp geomagnetic polarity event. *Nature*, 276:257–258.

Condomines, M., C. Hémond, and C. J. Allègre, 1988. U–Th–Ra radioactive disequilibria and magmatic processes. *Earth Planet. Sci. Lett.*, 90:243–262.

Condomines, M., and O. Sigmarsson, 1993. Why are so many arc magmas close to ^{238}U-^{230}Th radioactive equilibrium? *Geochim. Cosmochim. Acta*, 57:4491–4497.

Condomines, M., and O. Sigmarsson, 2000. ^{238}U-^{230}Th disequilibria and mantle melting processes: A discussion. *Chem. Geol.*, 162:95–104.

Eberhardt, P., J. Geiss, F. S. Houtermans, W. E. Buser, and H. R. von Gunten, 1955. Il plombo volcanico del Vesuvio e di Vulcano. *Atti del 1st Convegno di Geologica Nucleare*, 50–56.

Edwards, R. L., J. H. Chen, and G. J. Wasserburg, 1987. ^{238}U-^{234}U-^{230}Th-^{232}Th systematics and the precise measurement of time over the past 500,000 years. *Earth Planet. Sci. Lett.*, 81:175–192.

Gill, J. B., D. M. Pyle, and R. W. Williams, 1992. Igneous rocks. In M. Ivanovich and R. S. Harmon (Eds.), *Uranium-Series Disequilibrium; Applications to Earth, Marine, and Environmental Sciences*, 2nd ed., pp. 207–258. Clarendon, Oxford, UK.

Gill, J. B., and R. W. Williams, 1990. Th-isotope and U-series studies of subduction-related volcanic rocks. *Geochim. Cosmochim Acta*, 54:1427–1442.

Gill, J. B., R. W. Williams, and K. Bruland, 1985. Eruption of basalt and andesite lava degasses ^{222}Rn and ^{210}Po. *Geophys. Res. Lett.*, 12:17–20.

Goldstein, S. L., M. T. Murrell, and R. W. Williams, 1993. ^{231}Pa and ^{230}Th chronology of mid-ocean ridge basalt. *Earth Planet. Sci. Lett.*, 115:151–159.

Jones, J. H., D. Walker, D. A. Pickett, M. T. Murrell, and P. Beattie, 1995. Experimental investigations of the partitioning of Nb, Mo, Ba, Ce, Pb, Ra, Th, Pa, and U between immiscible carbonate and silicate liquids. *Geochim. Cosmochim. Acta*, 59:1307–1320.

Krishnaswami, S., K. K. Turekian, and J. T. Bennett, 1984. The behavior of ^{232}Th and the ^{238}U decay-chain nuclides during magma formation. *Geochim. Cosmochim. Acta*, 48:505–511.

Lundstrom, C. C., J. Gill, and Q. Williams, 2000. A geochemically consistent hypothesis for MORB generation. *Chem. Geol.*, 162:105–126.

McKenzie, D., 2000. Constraints on melt generation and transport from U-series activity ratios. *Chem. Geol.*, 162:81–94.

Oversby, V. M., and P. W. Gast, 1968. Lead isotope compositions and uranium decay series disequilibrium in Recent volcanic rocks. *Earth Planet. Sci. Lett.*, 5:199–206.

Pickett, D. A., and M. T. Murrell, 1997. Observations of $^{231}Pa/^{235}U$ disequilibrium in volcanic rocks. *Earth Planet. Sci. Lett.*, 148:259–271.

Pyle, D. M., J. B. Dawson, and M. Ivanovich, 1991. Short-lived decay series disequilibria in the natrocarbonatite lavas of Oldoinyo Lengai, Tanzania: Constraints on the timing of magma genesis. *Earth Planet. Sci. Lett.*, 105:378–397.

Rogers, N. (Ed.), 2000. Rates and timescales of magmatic processes. *Chem. Geol.*, 162:79–191.

Shen, C.-C., R. L. Edwards, H. Cheng, J. A. Dorale, R. B. Thomas, S. B. Moran, S. E. Weinstein, and H. N. Edmonds, 2002. Uranium and thorium isotopic and concentration measurements by magnetic sector inductively coupled-plasma mass-spectrometry. *Chem. Geol.*, 185:165–178.

Somayajulu, B. L. K., M. Tatsumoto, J. N. Rosholt, and R. J. Knight, 1966. Disequilibrium of the 238-U series in basalt. *Earth Planet. Sci. Lett.*, 1:387–391.

Stirling, C. H., D.-C. Lee, J. N. Christensen, and A. N. Halliday, 2000. High-precision in situ ^{238}U-^{234}U-^{230}Th isotopic analysis using laser ablation multiple-collector ICPMS. *Geochim. Cosmochim. Acta*, 64:3737–3750.

Treiman, A. H., and A. Schedl, 1983. Properties of carbonatite magmas and processes in carbonate magma chambers. *J. Geol.*, 91:437–447.

Turner, S., J. Blundy, B. Wood, and M. Hole, 2000. Large ^{230}Th-excesses in basalts produced by partial melting of spinel lherzolite. *Chem. Geol.*, 162:127–136.

Villemant, B., G. Boudon, and J.-C. Komorowski, 1996. U-series disequilibrium in arc magmas induced by water-magma interactions. *Earth Planet. Sci. Lett.*, 140:259–267.

Williams, R. W., J. B. Gill, and K. W. Bruland, 1986. Ra–Th disequilibria systematics: Timescale of carbonatite magma formation at Oldoinyo Lengai volcano, Tanzania. *Geochim. Cosmochim. Acta*, 50:1249–1259.

Wood, B. J., J. D. Blundy, and J. A. Robinson, 1999. The role of clinopyroxene in generating U-series disequilibrium during mantle melting. *Geochim. Cosmochim. Acta*, 63:1613–1620.

Wyllie, P. J., 1989. Origin of carbonatites: Evidence from phase equilibrium studies. In K. Bell (Ed.), *Carbonatites*, pp. 15–37. Unwin Hyman, London.

Wyllie, P. J., and O. F. Tuttle, 1960. The system Ca–CO_2-H_2O and the origin of carbonatites. *J. Petrol*, 1:1–46.

Wyllie, P. J., and O. F. Tuttle, 1962. The carbonatite lavas. *Nature*, 194:1269.

Zhou, H., and A. Zindler, 2000. Theoretical studies of ^{238}U-^{230}Th-^{226}Ra and ^{235}U-^{231}Pa disequilibria in young lavas produced by mantle melting. *Geochim. Cosmochim. Acta*, 64:1809–1817.

CHAPTER 21

Helium and Tritium

THE α-particles emitted by the long-lived isotopes of U and Th and their short-lived daughters are the nuclei of $^{4}_{2}He$, which is one of two stable isotopes of He. The other stable He isotope, $^{3}_{2}He$, is the product of β-decay of tritium ($^{3}_{1}H$), an unstable isotope of H. In addition, both isotopes formed during the initial expansion of the Universe a few minutes after the Big Bang and together constitute about 20% by weight of all conventional matter in the Universe. Helium is the second most abundant element in the Sun and has a "cosmic" abundance of 2.72×10^9 atoms of He per 10^6 atoms of Si (Anders and Grevesse, 1989).

The abundance of He in the Earth is much lower than that of the Sun because the Earth does not retain He in its atmosphere (Mamyrin and Tolstikhin, 1984). In fact, most of the H and He that existed in the inner solar system at the time of formation of Mercury, Venus, Earth, Mars, and the asteroids was blown away by the solar wind and contributed to the formation of the so-called "gaseous" planets: Jupiter, Saturn, Uranus, and Neptune. Nevertheless, the rocks of the crust and mantle of the Earth contain primordial He that was trapped at the time the Earth accreted by impacts of planetesimals. Therefore, terrestrial He is a mixture of *primordial* He and the *radiogenic* ^{4}He produced by α-decay of U, Th, and their daughters as well as of ^{147}Sm (Chapter 9) and a few other α-emitters (Section 2.2). The decay of tritium affects the isotope composition of He only locally in groundwater that contains tritium produced by nuclear reactions in the atmosphere. The isotopic composition of terrestrial He is $^{4}_{2}He = 99.99986\%$ and $^{3}_{2}He = 0.00014\%$, giving the atomic $^{3}He/^{4}He$ ratio a value of 1.40×10^{-6} (Lide and Frederikse, 1995).

The accumulation of He in U-bearing minerals was originally used by Ernest Rutherford to date uranium minerals, which yielded dates of about 500 million years, as he reported in 1905 during a lecture series at Yale University. Similarly, Boltwood (1907) used the amount of radiogenic Pb to date specimens of uraninite. The resulting dates, which ranged from 410 to 535 million years, confirmed the conclusion of geologists that the Earth was much older than indicated by Lord Kelvin only a few years before (Chamberlin, 1899; Thomson, 1899; Burchfield, 1975). In addition, the age determinations based on the accumulation of He and Pb were used by Holmes (1911, 1913) to construct the first geological timescale. These matters were presented in more detail in Section 1.1 of this book. Several key papers by E. Rutherford, B. Boltwood, R. J. Strutt, A. Holmes, W. D. Urry, and C. Goodman have been reprinted in a book edited by Harper (1973).

21.1 U–Th/He METHOD OF DATING

The development of the U–Th/He method was continued by P. M. Hurley at the Massachusetts

Institute of Technology (e.g., Hurley and Goodman, 1941, 1943; Hurley, 1949, 1950, 1954) and by Paul Damon at Columbia University (Damon, 1957). Their studies demonstrated that dense U-poor rocks and certain accessory minerals (e.g., magnetite, zircon, sphene) quantitatively retain the He that is produced within them by α-decay of U, Th, and their daughters provided they have not been heated or deformed. The retention of He is also diminished by radiation damage of minerals and by the occurrence of U and Th along grain boundaries from where He can readily escape. On the other hand, some minerals (e.g., beryl, cordierite, tourmaline) store He even though they do not contain U or Th (Damon and Kulp, 1958).

The U–He method received a boost from Damon and Kulp (1957) and Damon and Green (1963), who used isotope dilution based on a ^{3}He spike to measure the volume of ^{4}He in zircon and magnetite crystals. Subsequently, Fanale and Kulp (1962) used this method to successfully date magnetite in a Jurassic sill in Cornwall, Pennsylvania. Their date of 194 ± 4 Ma is concordant with K–Ar dates of biotite and muscovite of the Palisade sill in New Jersey (Erickson and Kulp, 1961). In addition, Fanale and Schaeffer (1965) and Bender (1973) demonstrated the feasibility of dating aragonitic corals and mollusk shells of Tertiary and Pleistocene ages by the U–He method. Bender (1973) measured the U concentrations of aragonitic corals by mass spectrometry using ^{234}U as a spike and by α-spectrometry. Helium was released by acid dissolution in a vacuum and was determined by mass spectrometry using a ^{3}He spike. However, Schaeffer (1970) and Turekian et al. (1970) concluded that fossil bones cannot be dated by this method because bones do not retain He and because they absorb U from ambient groundwater.

21.1a Geochronometry Equation

The decay of the long-lived isotopes of U and Th to stable isotopes of Pb in a mineral is expressed by equations 10.1, 10.2 and 10.3:

$$^{238}_{92}\text{U} = {}^{206}_{82}\text{Pb} + 8(^4_2\text{He}) + 6\beta^- + Q$$

$$^{235}_{92}\text{U} = {}^{207}_{82}\text{Pb} + 7(^4_2\text{He}) + 4\beta^- + Q$$

$$^{232}_{90}\text{Th} = {}^{208}_{82}\text{Pb} + 6(^4_2\text{He}) + 4\beta^- + Q$$

where all nuclides are in terms of numbers of atoms per unit weight of sample and Q is the total decay energy in MeV liberated by each of the three decay series. The number of radiogenic ^{4}He atoms $(^4\text{He})_N$ produced by ^{238}U, ^{235}U, and ^{232}Th in a mineral increases with time in accordance with the equations

$$(^4\text{He})_N = 8(^{238}\text{U})_N(e^{\lambda_{238}t} - 1) \qquad (21.1)$$

$$(^4\text{He})_N = 7(^{235}\text{U})_N(e^{\lambda_{235}t} - 1) \qquad (21.2)$$

$$(^4\text{He})_N = 6(^{232}\text{Th})_N(e^{\lambda_{232}t} - 1) \qquad (21.3)$$

where the subscript N signifies numbers of atoms per gram of sample and $\lambda_{238} = 1.55125 \times 10^{-10}\ \text{y}^{-1}$, $\lambda_{235} = 9.8485 \times 10^{-10}\ \text{y}^{-1}$, and $\lambda_{232} = 4.9475 \times 10^{-11}\ \text{y}^{-1}$. The total number of ^{4}He atoms arising from the decay of ^{238}U, ^{235}U, ^{232}Th, and their respective daughters is the sum of equations 21.1, 21.2, and 21.3:

$$\begin{aligned}(^4\text{He})_N &= 8(^{238}\text{U})_N(e^{\lambda_{238}t} - 1) \\ &\quad + 7(^{235}\text{U})_N(e^{\lambda_{235}t} - 1) \\ &\quad + 6(^{232}\text{Th})_N(e^{\lambda_{232}t} - 1) \qquad (21.4)\end{aligned}$$

provided that

1. each decay series maintained secular radiochemical equilibrium,
2. no He was present in the mineral at the time of crystallization, and
3. the mineral remained closed to He.

The interference caused by radiochemical disequilibrium [e.g., $(^{234}\text{U}/^{238}\text{U})_A > 1.0$] and by the presence of excess He is negligible for dating specular and botryoidal hematite if the samples satisfy the condition (Wernicke and Lippolt, 1993) that

$$\text{U (ppm)} \times \text{age (Ma)} > 10 \qquad (21.5)$$

The number of atoms of ^{238}U, ^{235}U, and ^{232}Th per gram of sample can be converted into concentrations of U and Th in micrograms per gram by

$$^{238}\text{U}_N = \frac{U \times A \times 0.992743}{238.0289 \times 10^6}$$

where $^{238}U_N$ = number of ^{238}U atoms per gram of sample
U = concentration, μg/g
A = Avogadro's number

The factor 10^6 converts the concentration of U from micrograms per gram to grams per gram, dividing by the atomic weight of U (238.0289) converts its concentration to moles per gram, multiplying by Avogadro's number yields atoms of U per gram, and use of the isotope abundance (0.992743) converts the number of atoms of U to the number of ^{238}U atoms per gram. Similarly,

$$^{235}U_N = \frac{U \times A \times 0.0072}{238.0289 \times 10^6}$$

$$^{232}Th_N = \frac{Th \times A \times 1.00}{232.0381 \times 10^6}$$

Converting the total number of accumulated ^{4}He atoms to a volume expressed in microliters per gram at standard temperature and pressure (STP) yields

$$(^4He)_N = \frac{(^4He)_v \times A}{22.41383 \times 10^6}$$

where 22.41383 L is the volume of one mole of an ideal gas at STP of 273.15 K and 1.0 atm. Therefore, combining all parts of equation 21.4 yields

$$\frac{(^4He)_v \times A}{22.41383 \times 10^6} = \frac{8U \times A \times 0.992743}{238.0289 \times 10^6}(e^{\lambda_{238}t} - 1) + \frac{7U \times A \times 0.0072}{238.0289 \times 10^6}(e^{\lambda_{235}t} - 1) + \frac{6Th \times A \times 1.00}{232.0381 \times 10^6}(e^{\lambda_{232}t} - 1)$$

After cancelling Avogadro's number and the factor 10^6, the equation simplifies to

$$(^4He)_v = 22.41383[0.0336U(e^{\lambda_{238}t} - 1) + 0.00021173U(e^{\lambda_{235}t} - 1) + 0.025857Th(e^{\lambda_{232}t} - 1)] \quad (21.6)$$

where $(^4He)_v$ = volume of radiogenic He at STP in μL/g
U, Th = concentrations of U and Th in μg/g.

Equation 21.6 can be used to date certain ^{4}He-retentive minerals and glassy volcanic rocks (Graham et al., 1987) based on the measured concentrations of U and Th (in micrograms per gram) and of ^{4}He (in microliters per gram). The equation must be solved by iteration on a computer or by plotting it for each sample in coordinates of t in years and the concentration of ^{4}He in microliters per gram. The geological significance of the resulting date is constrained by the assumptions stated above and represents the time elapsed since the mineral or rock sample cooled through its closure temperature for radiogenic ^{4}He.

The calculated (U–Th)/He dates must be corrected for possible loss of ^{4}He because the retention of ^{4}He by minerals depends not only on the rate of diffusion but also on the loss of α-particles by ejection across the surfaces of mineral grains. The stopping distance of α-particles depends on their kinetic energies (3.98–7.69 MeV) and on the density of the mineral in question. According to Farley et al. (1996), the stopping distances of 7.69-MeV α-particles in apatite range from 12.60 to 33.39 μm, whereas those in sphene and zircon are about 8.0 to 14.5% shorter. Therefore, the fraction of α-particles (F_T) that are retained within a sphere of radius R decreases nonlinearly from about 95% at $R = 250$ μm to 85% at $R = 100$ μm and to 70% at $R = 50$ μm. Loss of α-particles from cylindrical grains follows a similar pattern. The (U–Th)/He dates of apatite grains of radius R and having length-to-radius ratios between 2 and 8 can be corrected by means of the F_T parameter derived by Farley et al. (1996):

$$\text{Corrected age} = \frac{\text{measured age}}{F_T} \quad (21.7)$$

For example, apatite grains ($R = 30$ μm to $R = 34$ μm) with $F_T = 0.648$ and a measured age of 30.8 ± 0.9 Ma yield a corrected age:

$$t_c = \frac{30.8 \pm 0.9}{0.648} = 47.5 \pm 1.5 \text{ Ma}$$

21.1b Diffusion of He in Minerals

The diffusion of He in minerals can be quantified by measuring the volume of this gas that is released by stepwise heating of mineral grains in a vacuum (Fechtig and Kalbitzer, 1966; Zeitler et al., 1987; Bähr et al., 1994). The results are used to calculate the diffusion coefficient by means of an equation derived by Fechtig and Kalbitzer (1966):

$$D_i = \frac{R^2 \pi F \, \Delta F_i}{18 \, \Delta t_i} \qquad (21.8)$$

where R = radius of an imaginary sphere (in cm) within the mineral grain from which ^4He (or Ar) is diffusing
F = cumulative fraction of ^4He lost
ΔF_i = fraction lost in heating step i
Δt_i = heating time of heating step i
D_i = diffusion coefficient for ^4He during heating step i

Equation 21.8 indicates that the diffusion coefficient increases with the square of the radius R of the hypothetical diffusion sphere. In cases where the radius (r, in centimeters) of the mineral grains is of the same order of magnitude as R (i.e., $R \sim r$), equation 21.8 yields the diffusion coefficient for mineral grains of radius r. However, in case the grain radius r differs from the diffusion distance R, equation 21.8 is used to calculate D/R^2, which is usually expressed as D/r^2 because r is a measurable parameter whereas R is not:

$$\frac{D}{r^2} = \frac{\pi F \, \Delta F_i}{18 \, \Delta t_i} \qquad (21.9)$$

The resulting values of D or D/r^2 are related to the temperature of the heating steps by the Arrhenius equation:

$$\frac{D}{r^2} = \frac{D_0}{r^2} e^{-E/RT} \qquad (21.10)$$

where D/r^2 = diffusion coefficient in units of reciprocal seconds (s^{-1})
D_0/r^2 = frequency factor in reciprocal seconds (s^{-1})
e = base of the natural logarithm (2.718281)
E = activation energy in kilojoules per mole (kJ/mol)
R = gas constant (8.314 J per mole per Kelvin) (J/mol/K)
T = temperature in Kelvins (K)

The Arrhenius equation can also be expressed in terms of D and D_0 (i.e., when $R \sim r$). In this case, the units of D are square centimeters per second (cm^2/s). The Arrhenius equation (21.10) is linearized by taking natural logarithms:

$$\ln\left(\frac{D}{r^2}\right) = \ln\left(\frac{D_0}{r^2}\right) - \frac{E}{RT} \qquad (21.11)$$

The natural logarithms of the values of D (or D/r^2) calculated from the step-heating data and the corresponding reciprocal Kelvin temperatures define a straight line whose slope m is given as

$$m = -\frac{E}{R} \qquad (21.12)$$

and whose intercept b is

$$b = \ln\left(\frac{D_0}{r^2}\right) \qquad (21.13)$$

The slope of the Arrhenius equation yields the activation energy E, while the intercept b is the natural logarithm of the value of D_0, or D_0/r^2, at an infinitely high temperature. After the activation energy E has been determined, equation 21.12 is used to calculate values of D (or D/r^2) at any temperature of interest (e.g., 20°C or 293.15 K representing the surface environment of the Earth).

The diffusion coefficient at the chosen temperature is used to describe the change in the concentration of ^4He with time:

$$\frac{dc}{dt} = D \, \Delta C \qquad (21.14)$$

which is a statement of Fick's second law of diffusion (Fechtig and Kalbitzer, 1966),

where dc/dt = change of the concentration per unit time and
ΔC = net change of the concentration.

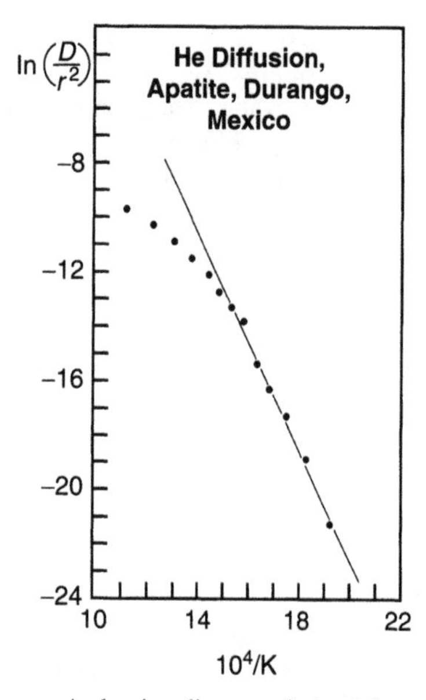

FIGURE 21.1 Arrhenius diagram derived from the stepwise release of ^{4}He by gem-quality apatite from Durango, Cerro Colorado, Mexico. The slope of the low-temperature end of the diffusion curve (<400 °C) was used to calculate the activation energy ($E = 161 \pm 34$ kJ/mol) for diffusion out of a sphere. Data from Zeitler et al. (1987).

Zeitler et al. (1987) used this procedure to release ^{4}He by step heating a gem-quality crystal of apatite from Durango, Cerro de Colorado, Mexico. Grains having radii of 75–90 μm were heated for 30 min at 14 temperatures from 250 to 1000°C. The natural logarithms of the D/r^2 values calculated from the results form the linear data array in Figure 21.1. The data points generated by ^{4}He losses between 250 and 400°C define a straight line:

$$\ln\left(\frac{D}{r^2}\right) = 16.4 \pm 2.8 - (1.9375 \pm 0.4075)\left(\frac{10^4}{T}\right) \tag{21.15}$$

as expected for volume diffusion. The ^{4}He losses at $T > 400$°C deviate from the diffusion equation because the radii of the effective diffusion domains may change or because of structural changes in the apatite (Zeitler et al., 1987).

The activation energy E is calculated from the slope of the diffusion line using equation 21.12:

$$E = -mR$$
$$= 1.9375 \times 10^4 \times 8.314 = 161 \pm 34 \text{ kJ/mol}$$

The frequency factor (D_0/r^2) is the intercept of the diffusion line with the y-axis by equation 21.15:

$$\ln\left(\frac{D_0}{r^2}\right) = 16.4 \pm 2.8$$

where D_0/r^2 is the diffusion coefficient of ^{4}He in apatite at very high temperatures.

The experimental results of Zeitler et al. (1987) indicate that the diffusion coefficient of ^{4}He in apatite is a sensitive function of the temperature. Substituting the values derived above into equation 21.11 yields

$$\ln\left(\frac{D}{r^2}\right) = 16.4 - \frac{161 \times 10^3}{8.314T}$$

If $T = 100°\text{C} = 373.15$ K,

$$\ln\left(\frac{D}{r^2}\right) = -35.49$$

$$\left(\frac{D}{r^2}\right)_{T=373 \text{ K}} = 3.86 \times 10^{-16} \text{ s}^{-1}$$

If $r = 100$ μm $= 10^{-2}$ cm, then

$$D = (10^{-2})^2 \times 3.86 \times 10^{-16} \text{ cm}^2/\text{s}$$
$$= 3.86 \times 10^{-20} \text{ cm}^2/\text{s}$$

The value of D_0/r^2 for diffusion of ^{4}He in apatite was used by Zeitler et al. (1987) to calculate its closure temperature T_c using an equation derived by Dodson (1973):

$$T_c = \frac{E/R}{\ln[(ART_c^2\, D_0/r^2)/E(dT/dt)]} \tag{21.16}$$

where R = gas constant (8.314 J/mol K)
T_c = closure temperature in Kelvins
E = activation energy in J/mol

A = 55 for spherical grains
D_0/r^2 = intercept of the linear segment of the Arrhenius diagram.

Equation 21.16 must be solved by substituting an assumed value of T_c into the right side of the equation. In addition, the cooling rate must be preselected (e.g., 10°C/10^6 years). Another computational complication arises from the mismatch of the units of time in equation 21.16 because D_0/r^2 is in units of reciprocal seconds whereas the cooling rate is in reciprocal years.

The data of Zeitler et al. (1987) lead to a closure temperature of 105 ± 30°C for a cooling rate of 10°C per million years. Therefore, Zeitler et al. (1987) suggested that (U–Th)/He dates of apatite represent the time elapsed since the mineral cooled through its blocking temperature. In other words, (U–Th)/He dates of apatite constrain the thermal history of their host rock in the same way that K–Ar and $^{40}Ar^*/^{39}Ar$ dates of K-bearing silicate minerals do (Sections 6.6 and 7.9; McDougall and Harrison, 1999). Zeitler et al. (1987) also pointed out that some previously reported anomalously low (U–Th)/He dates of minerals in igneous and metamorphic rocks may have provided information about their *thermal histories* rather than about their ages.

Subsequently, Lippolt et al. (1994), Wolf et al. (1996, 1998), Warnock et al. (1997), and Farley (2000) studied ^{4}He diffusion in apatite. Wolf et al. (1996) obtained an activation energy of 150 kJ/mol for ^{4}He diffusion in the Durango apatite and calculated an average closure temperature of 75 ± 7°C. Farley (2000) further refined the closing temperature for ^{4}He in this apatite to 68°C for grain radii of 100 μm and a cooling rate of 10°C per 10^6 years. Additional modeling of the production and diffusion of ^{4}He in minerals was carried out by Meesters and Dunai (2002a, b).

All of these results confirm that the (U–Th)/He dates of apatite record the time when this mineral cooled through its closure temperature for the last time. Therefore, apatite is now used to reconstruct the thermal histories of rocks that formed at elevated temperatures and then cooled slowly to Earth-surface temperatures. Such thermochronometric studies can also be based on the (U–Th)/He dates of titanite (sphene) as demonstrated by Reiners and Farley (1999).

The diffusion of ^{4}He in ore minerals (Lippolt et al., 1982) and in rock-forming silicate minerals (Lippolt and Weigel, 1988) has also been studied for possible use in (U–Th)/He dating of ore deposits and rocks that are not suitable for dating by the conventional methods (e.g., Rb–Sr, K–Ar). The closure temperatures obtained by Lippolt and Weigel (1988) for ^{4}He diffusion in rock-forming silicate and evaporite minerals in Table 21.1 are all lower than those for ^{40}Ar and rise with increasing cooling rates. The most ^{4}He-retentive minerals (i.e., *smallest* fraction of ^{4}He lost at 20°C in 100 × 10^6 years) are nepheline (0.03%), hornblende (0.5%), augite (0.9%), and langbeinite [0.9%; $K_2Mg_2(SO_4)_3$]. Sanidine (16%) and muscovite (34%) suffer large losses of ^{4}He under these conditions. However, one of the nepheline samples analyzed by Lippolt and Weigel (1988) contained excess ^{4}He (and ^{40}Ar), and the U concentrations of the ^{4}He-retentive minerals are low: nepheline (0.014 ppm), hornblende (0.45 ppm), augite (0.18 ppm), and langbeinite (0.03 ppm). The low U (and Th) concentrations eliminate radiation damage from consideration, but the corresponding concentrations of ^{4}He are low and must be measured by isotope dilution on a mass spectrometer.

21.2 THERMOCHRONOMETRY

The (U–Th)/He dates of apatite, titanite, and other minerals can be used to reconstruct the thermal histories of rocks that have cooled as they aged. The resulting subject of thermochronometry relies not only on (U–Th)/He dates but also on Rb–Sr and K–Ar dates of K-bearing minerals (Sections 5.4a, 6.6a, and 6.6b) and on fission-track dates (Section 22.2), all of which are temperature-sensitive chronometers (McDougall and Harrison, 1999).

Thermochronometry comes into play in two kinds of scenarios, both of which are activated by uplift and erosion:

1. An igneous intrusive forms by crystallization of magma at depth in the crust and subsequently

Table 21.1. Diffusion of ^{4}He in Minerals that Retain Radiogenic ^{40}Ar

Mineral	R,[a] μm	E, kJ/mol	log D_0, cm²/s	log D_{20},[b] cm²/s	T_c,[c] °C	Loss (20°C, 100 Ma), %
Hornblende 1	200	120 ±4	−1.91 ±0.34	−23.2 ±0.5	90	0.5
Hornblende 2	75	104 ±2	−3.02 ±0.15	−21.5 ±0.5	90	—
Nepheline 1A[d]	75	134 ±3	−2.63 ±0.22	−26.4 ±0.6	145	0.03
Nepheline 1B[d]	75	128 ±5	−2.25 ±0.35	−25.1 ±0.7	145	0.03
Nepheline 2[e]	75	88 ±6	−3.94 ±0.46	−19.5 ±0.9	40	—
Augite 1	200	124 ±7	−1.35 ±0.49	−23.4 ±0.9	90	0.9
Augite 2	200	116 ±7	−1.32 ±0.52	−22.1 ±0.9	90	0.9
Sanidine 1	60	70 ±2	−5.17 ±0.15	−17.7 ±0.6	—	—
Sanidine 2A	200	96 ±4	−3.45 ±0.36	−20.5 ±0.5	50	16
Sanidine 2B	45	92 ±4	−3.23 ±0.39	−19.7 ±0.6	50	16
Muscovite	130	87 ±10	−3.94 ±0.71	−19.4 ±1.7	40	34
Langbeinite, $K_2Mg_2(SO_4)_3$	260	126 ±5	−0.12 ±035	−22.7 ±0.6	82	0.9

Source: Lippolt and Weigel, 1988.
[a] Average grain size.
[b] Diffusion coefficient at 20°C.
[c] Closure temperature for spheres having a diameter of 500 μm at a cooling rate of 10°C per million years.
[d] Excess ^{4}He is present whose effect on the diffusion parameters is not known.
[e] Actual grain size is in doubt, but no excess ^{4}He is present.

cools to the temperature at the surface of the Earth when the intrusive is uplifted and is exposed by erosion of the overlying rocks.
2. A block of continental crust is uplifted tectonically and cools as the upper part of the block is eroded.

The depth in the crust below which apatite does not retain ^{4}He is about 3 km, assuming that the average geothermal gradient is 20°/km, that the average surface temperature is 15°C, and that no magma chamber or cooling bodies of igneous rocks are present at depth. Under these conditions, the (U–Th)/He dates of apatite are expected to decline to zero with increasing depth.

21.2a Otway Basin, South Australia

The predicted decrease of (U–Th)/He dates of apatite with depth was demonstrated by House et al. (1999) using drill cores of the Otway Supergroup in South Australia which was deposited in a rift at about 120 Ma (Early Cretaceous). The Otway Supergroup consists of nonmarine sediment containing detrital apatite crystals most of which originated from volcanic rocks that were erupted contemporaneously with the deposition of the sediment. Subsequently, the sediment was heated to about 200°C as a result of burial and tectonic activity, which continued until about 75 Ma.

The (U–Th)/He dates of apatites in the Anglesea-1 drill core in the eastern end of the Otway Basin in Figure 21.2 decrease with increasing borehole temperature from 74.1 ± 4.3 Ma at the surface (13.5°C) to 0 ± 1.0 Ma at a depth of 2214 m (95°C). The dates reported by House et al. (1999) for a second core (Ferguson Hill) located about 65 km west of the Anglesea core confirm this pattern of variation. The temperature gradient in both cores is 38°C/km. Consequently, the closure temperature of 75 ± 7°C was reached at a depth of 1.62 km, assuming a surface temperature of 13.5°C

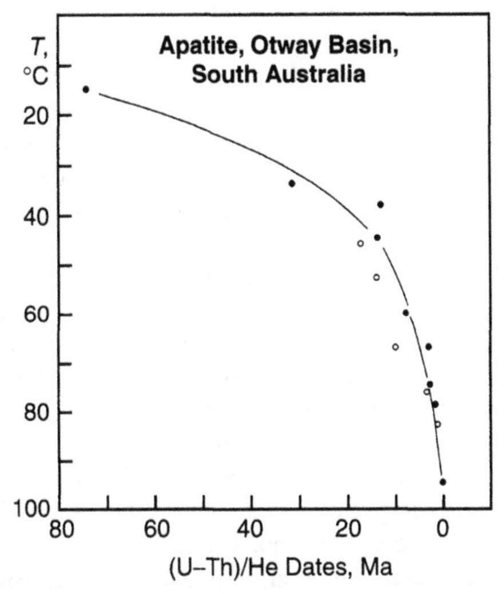

FIGURE 21.2 Profile of (U–Th)/He dates of detrital apatite crystals in two cores of Early Cretaceous sediment in the Otway Basin of South Australia. The solid circles represent samples from the Anglesea-1 core drilled in 1962 at 38°24′26″ S, 144°11′53″ E. The open circles are apatites from the Ferguson Hill core drilled in 1963/1964 at 38°37′20″ S, 143°09′41″ E. The temperatures were measured in the bore holes and were corrected for the heat generated by drilling. Data from House et al. (1999).

for South Australia. House et al. (1999) concluded from these results that the temperature dependence of the diffusion coefficient for He in apatite and the resulting closure temperature derived from laboratory studies can be used with confidence to interpret (U–Th)/He dates of apatite.

21.2b Mt. Whitney, Sierra Nevada Mountains

The Sierra Nevada Mountains of California formed by the uplift and subsequent dissection of a composite granitic batholith of Mesozoic age. Mt. Whitney (4418 m) is located along the eastern border of Sequoia National Park at about 36°34.57′ N and 118°17.50′ E.

The (U–Th)/He dates in Figure 21.3 of apatite samples extracted from granodiorite (83 Ma) on Mt. Whitney decrease linearly with decreasing elevation above sealevel (asl) (House et al., 1997).

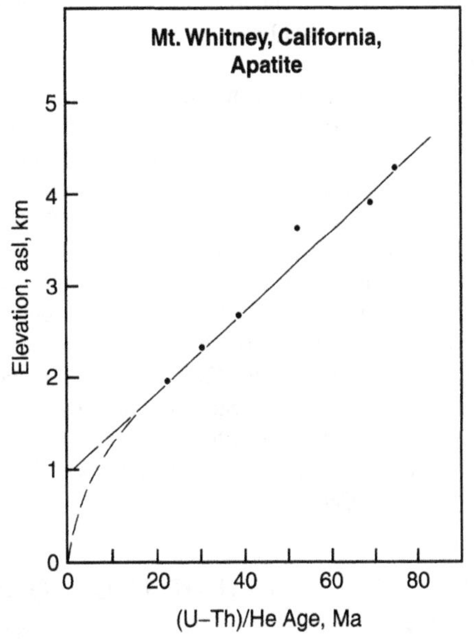

FIGURE 21.3 The (U–Th)/He dates of apatite taken at different elevations on Mt. Whitney, which is the highest peak in the Sierra Nevada Mountains of California. The elevations of the collecting sites are in kilometers above sealevel (asl). Data from House et al. (1997).

In this case, the decrease of (U–Th)/He dates with decreasing elevation asl is equivalent to the expected decrease of ^{4}He dates with increasing depth below the surface of the Earth at the time of intrusion of the Sierra Nevada batholith. The age–elevation line extrapolates to a zero date at 1000 m asl, or less, if it becomes nonlinear for near-zero He dates. Since the (U–Th)/He date of each sample represents the time elapsed since the apatite cooled through the closure temperature of 75 ± 7°C, the temperature of the sample close to the summit of Mt. Whitney at 4280 m asl has been below 75°C for 74.6 ± 3.4 million years, whereas a sample at about 1000 m asl would have a He date of zero because its temperature is still 75°C. Therefore, the temperature gradient g implied by the age-elevation line is

$$g = \frac{75 - 0}{4.280 - 1.000} = 23 \pm 1.5\text{°C/km}$$

where the uncertainty represents the range of ± 7°C of the closure temperature and the average surface temperature at 4280 m asl on Mt. Whitney is assumed to be 0°C.

The (U–Th)/He dates of apatite have also been used to date the topography of the Sierra Nevada Mountains (House et al., 1998) and to measure uplift and erosion of the San Bernardino Mountains in California (Spotila et al., 1998). In addition, McInnes et al. (1999) used He dates to determine the amount of vertical fault displacement in the Chuquicamata porphyry–copper deposit of Chile. Other studies of this type include the work of Crowley et al. (2002) in the Bighorn Mountains of Wyoming, Stockli et al. (2000) in the White Mountains of California, and Farley et al. (2001) on the northern Coast Mountains of British Columbia in Canada.

21.3 He DATING OF IRON-ORE DEPOSITS

Experience has shown that iron oxide minerals such as magnetite and specular as well as botryoidal hematite can be dated by the (U–Th)/He method (Wernicke and Lippolt, 1993, 1994a, b, 1997; Bähr et al., 1994). Even samples of silicified hematitic iron ore in hydrothermal veins near Rammelsbach in the Black Forest of Germany were dated successfully by Brander and Lippolt (1999), although some samples had lost ^{4}He. Although botryoidal goethite and limonite in vein deposits in Germany contain up to 48 ppm U, up to 2 ppm Th, and from 0.1 to 11 nanoliter per gram (nL/g) of ^{4}He at STP, most of the samples analyzed by Lippolt et al. (1998) yielded relatively recent dates between 0.8 and 12 Ma, either because of the loss of ^{4}He by diffusion or because the minerals actually formed relatively recently during late-stage tectonic activity in Europe.

Bähr et al. (1994) reported that *specular* hematite grains with radii of 400, 120, and <10 μm lose ^{4}He by volume diffusion between temperatures of 250 and 1000°C with a diffusion coefficient $D = 10^{-28}$ cm^2/s at 20°C and activation energies E of 110–130 kJ/mol. The diffusion of ^{4}He in *botryoidal* hematite is affected by its small and variable grain size with radii from 0.1 to 10 μm. Therefore, the diffusion coefficient for ^{4}He in botryoidal hematite is stated as D/r^2 values of 10^{-18}–10^{-22} s^{-1} at 20°C with E of 130–210 kJ/mol, where r is the grain radius in centimeters. These results demonstrate that ^{4}He losses by diffusion at 20°C for specular hematite are negligible but may approach 10% in 100 million years for botryoidal hematite composed of grains <0.1 μm (Bähr et al., 1994). Small grains of this magnitude are not common, which means that actual ^{4}He losses from botryoidal hematite at Earth-surface temperatures are also negligible. The closure temperatures of both varieties of hematite in Table 21.2 range from 303 to 122°C depending on the cooling rates and grain radii (Dodson, 1973).

A specimen of botryoidal hematite from a hydrothermal vein near Rappenloch in the Black Forest of Germany, analyzed by Wernicke and Lippolt (1993), serves to demonstrate (U–Th)/^{4}He dating of hematite. The analytical data for specimen RA-GK-1a are

$$^4\text{He} = 1.338\ \mu\text{L/g} \qquad \text{U} = 71.1\ \text{ppm}$$

$$\text{Th} = 0.8\ \text{ppm}$$

The U and Th concentrations were substituted into equation 21.6, which was solved for ^{4}He

Table 21.2. Closure Temperatures for Diffusion of ^{4}He from Hematite as a Function of the Cooling Rates and Grain Radii

Variety of Hematite	Grain Radius, μm	^{4}He Closure Temperatures in °C for Cooling Rates in Kelvins per Million Years		
		10	10^2	10^3
Specular	500	219	258	303
Botryoidal	r^a	122	139	157

Source: Bähr et al., 1994.

[a] Grain radii range from 0.1 to 10 μm.

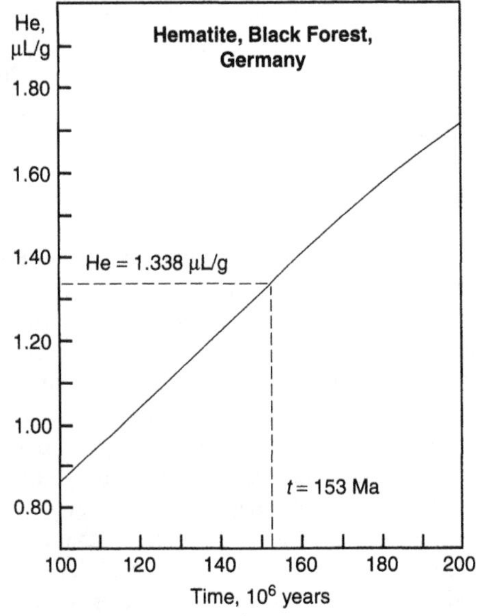

FIGURE 21.4 The (U–Th)/^{4}He age determination of botryoidal hematite from the Eisenbach hematite–manganese mineralized area near Rappenloch, Black Forest, Germany. The sample (Ra-GK 1a) contained 1.338 μL/g of radiogenic ^{4}He, U = 71.1 ppm, Th = 0.8 ppm. The curve is a plot of equation 21.6. Data from Wernicke and Lippolt (1993).

using different values of t. The resulting curve in Figure 21.4 indicates a date of 153 Ma (Jurassic) for ^{4}He = 1.338 μL/g. This date confirms a growing body of evidence that the hydrothermal veins in the Paleozoic basement granites (Variscan) of this region were reactivated during the Mesozoic Era.

21.4 TRITIUM–^{3}He DATING

The β-decay of cosmogenic and thermonuclear tritium ($^{3}_{1}$H) to stable $^{3}_{2}$He increases the ^{3}He/^{4}He ratio of shallow groundwater and is used to assign ages from which the velocity of flow of water beneath the surface of the Earth can be determined. Although the release of thermonuclear tritium has perturbed the natural tritium cycle of the atmosphere and hydrosphere, the thermonuclear tritium serves as a tracer for studies of groundwater and of the circulation in the oceans. The seminal studies concerning the production and distribution of tritium include the following:

von Faltings and Harteck, 1950, *Z. Naturforsch.*, 5a:438–439. Kaufman and Libby, 1954, *Phys. Rev.*, 93:1337–1344. von Buttlar and Libby, 1955, *Inorg. Nucl. Chem.*, 1:75. Craig, 1957, *Phys. Rev.*, 105:1125–1127. Begemann and Libby, 1957, *Geochim. Cosmochim. Acta*, 12:277–296. Giletti et al., 1958, *Trans. Am. Geophys. Union*, 39(5): 807–818. Libby, 1961, *J. Geophys. Res.*, 66(11):3767–3782.

21.4a Production and Decay of Tritium

Tritium ($^{3}_{1}$H) is an unstable isotope of H that occurs naturally on the surface of the Earth because it is continuously produced by a nuclear reaction in the atmosphere at a rate of 0.5 ± 0.3 atoms of ^{3}H/cm^{2} s (Craig and Lal, 1961). The reaction arises from the interaction of cosmic-ray neutrons with ^{14}N:

$$^{14}_{7}\mathrm{N} + ^{1}_{0}\mathrm{n} \rightarrow ^{3}_{1}\mathrm{H} + ^{12}_{6}\mathrm{C} \qquad (21.17)$$

The neutrons are produced by nuclear reactions caused by high-energy protons and other nuclear particles of cosmic rays. Some tritium may also originate from the Sun as part of the solar wind (Craig, 1957).

Tritium in the atmosphere is rapidly incorporated into water molecules either by direct oxidation (Moser and Rauert, 1980) or by exchange with stable H isotopes: ${}^{1}_{1}$H and ${}^{2}_{1}$H (deuterium). The tritium-bearing water molecules are then deposited on the surface of the Earth in the form of meteoric precipitation. Consequently, water and snow on the surface of the Earth contain varying concentrations of tritium, which is subsequently transferred to groundwater by recharge.

Tritium decays by β-emission to stable ${}^{3}_{2}$He:

$$ {}^{3}_{1}\mathrm{H} \rightarrow {}^{3}_{2}\mathrm{He} + \beta^{-} + \overline{\nu} + 0.01861 \text{ MeV} \quad (21.18) $$

The halflife of this decay is $T_{1/2} = 12.26$ y, which is equivalent to a decay constant $\lambda = 5.6537 \times 10^{-2}$ y^{-1}. The mass of ${}^{3}_{1}$H is 3.01603 amu (Lide and Frederikse, 1995). The concentration of tritium is expressed either in tritium units (TU) or in terms of its decay rate in a unit volume of water, where

$$ 1 \text{ TU} = 1 \text{ atom of tritium per } 10^{18} \text{ atoms of hydrogen} $$

One TU in 1 L of water gives rise to a rate of decay of 0.119 becquerels (Bq), which is equivalent to 3.21×10^{-12} curies (Ci), where 1 Ci $= 3.70 \times 10^{10}$ Bq (Section 3.4).

The tritium concentration of meteoric precipitation has been augmented by the explosion of thermonuclear devices in the atmosphere from less than 10 TU up to about 5000 TU in March and April of 1954 (Project Castle, USA) based on precipitation that fell in Ottawa (Ontario/Quebec), Canada (Begemann and Libby, 1957; Craig and Lal, 1961). Similar increases occurred later as a result of nuclear weapons tests in the atmosphere (e.g., in 1963; Suess, 1969; Weiss and Roether, 1975).

21.4b Thermonuclear Tritium

The explosion of thermonuclear fission and fusion devices in the atmosphere caused the injection of large quantities of tritium into the stratosphere from where it leaks into the troposphere. After the testing of thermonuclear devices in the atmosphere was discontinued at the end of 1963, the concentration of tritium in meteoric precipitation in Ottawa in Figure 21.5 declined from about 3000 TU to nearly 20 TU in 1990 (Fritz et al., 1991).

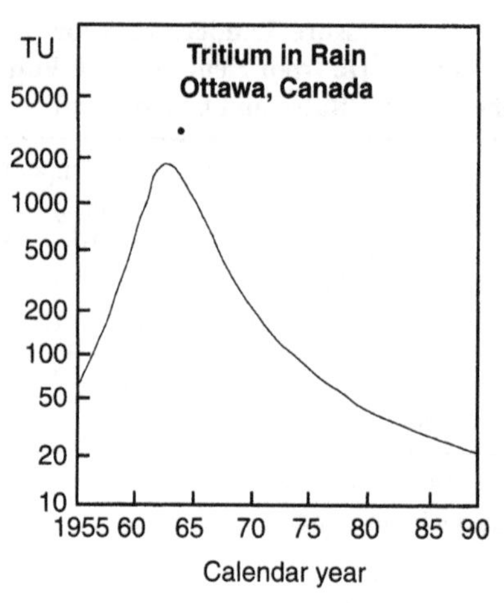

FIGURE 21.5 Concentration of tritium in meteoric water recorded in Ottawa, Canada, between 1955 and 1990. The peak concentration of about 3000 TU occurred during the summer of 1963. Adapted from Fritz et al. (1991).

A certain fraction of meteoric water that contains thermonuclear tritium has infiltrated the soil and overburden of the unsaturated zone. This phenomenon is illustrated by the data of Andersen and Sevel (1974) in Figure 21.6, which shows the presence of tritiated water in a depth profile of glaciofluvial outwash deposits in Denmark. The intensity of the tritium spike declines with time both by decay of tritium and by dispersion of the tritiated water molecules as the water percolates downward toward the water table.

The amount of tritiated water that reaches the water table depends on several factors (Dincer and Davis, 1967):

1. The concentration of tritium in the meteoric precipitation at a selected site increases in late spring and early summer and decreases during autumn and winter.
2. Recharge of groundwater is reduced by evapotranspiration during the growing season (spring and summer) when tritium concentrations in meteoric precipitation are high.

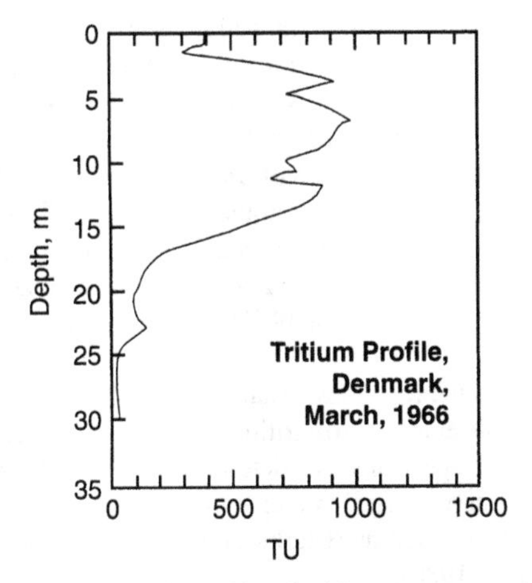

FIGURE 21.6 Depth profile of tritium concentrations in the unsaturated zone of glaciofluvial sediment recorded in March 1966 in Denmark by Andersen and Sevel (1974). Adapted from Fritz et al. (1991).

3. The transfer of tritium from the stratosphere (about 10–50 km above the surface) to the troposphere (0 to about 10 km) occurs primarily in the polar regions and causes a latitude effect in the tritium content of meteoric precipitation. The gradient in the northern hemisphere in Figure 21.7 is steeper than in the southern hemisphere because most of the nuclear test sites were in the northern hemisphere.
4. The tritium concentration of meteoric precipitation in large continental areas rises with increasing distance from the coast. Coastal precipitation is depleted in tritium because it contains water derived by evaporation from the surface of the oceans.

The dependence of the tritium concentration of meteoric precipitation on latitude in the northern hemisphere means that rain and snow at sites having similar latitudes have similar tritium concentrations (e.g., Vienna at 48° N and Ottawa at 45° N).

These aspects of tritium geochemistry constrain the concentration of tritium in the water that recharges groundwater aquifers in a given region. In some cases, a delay occurs between precipitation and recharge as, for example, in cases where precipitation occurs in the winter in the form of

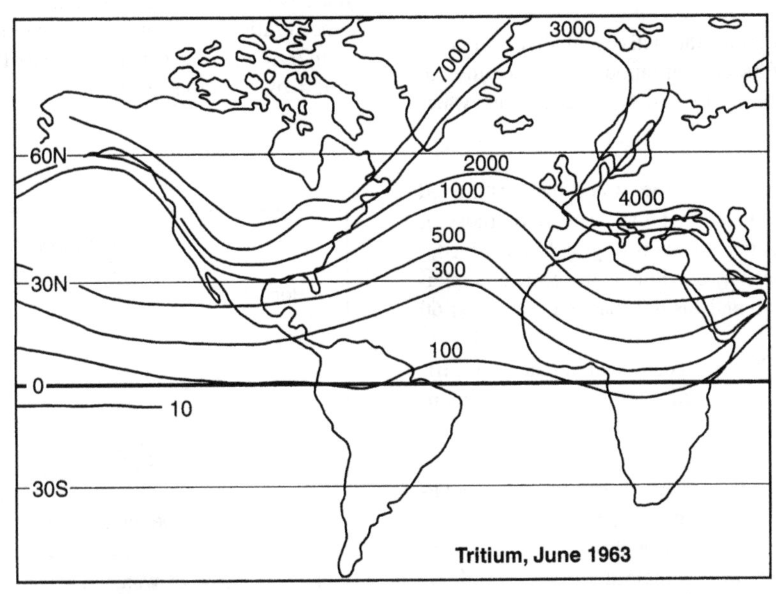

FIGURE 21.7 Average monthly tritium concentration of meteoric precipitation in TU for June 1963 in the northern hemisphere. Adapted from Dincer and Davis (1967).

snow that accumulates until it melts in the spring. In this case, the tritium content of the meltwater is typical of winter precipitation (i.e., low tritium) even though the recharge occurs during spring (i.e., high tritium).

The tritium concentration of recharge water in a given region can be estimated by considering the factors discussed above. Alternatively, water collected in lysimeters placed in the unsaturated zone of the recharge area can be analyzed directly to determine the initial tritium concentration of recharge water.

21.4c Dating Water (Cosmogenic Tritium)

Prior to the testing of thermonuclear devices in the atmosphere, meteoric precipitation contained a constant concentration of tritium subject only to variations in its production rate as a result of changes in the cosmic-ray flux and in the strength of the magnetic field of the Earth. Under these conditions, the tritium in a unit volume of rainwater (or snowmelt) decays with its characteristic halflife in accordance with the law of radioactivity:

$$^3H_A = {}^3H_A^0 e^{-\lambda t} \tag{21.19}$$

where ^3H_A is either the activity of this isotope in becquerels or its concentration in TU per liter of water. Equation 21.19 can be used to date water provided that

1. the sample of meteoric water has not mixed with "old" water that has lost all or most of the cosmogenic tritium it originally contained,
2. the initial activity of tritum in meteoric precipitation is known and has not changed in the past 60 years, and
3. the sample has not been contaminated with tritium derived from other sources (e.g., thermonuclear processes).

The method of dating arising from equation 21.19 has in fact been *compromised* by production of excess tritium of thermonuclear origin, which continues to dominate the inventory of tritum in the stratosphere, shallow groundwater, lakes, and the surface layer of the oceans. Under favorable circumstances, the initial activity (or concentration) of tritium either can be measured directly or can be estimated from records of tritium concentrations in meteoric water in the study area.

The concentration of tritium is measured either in a β-counter or in a liquid scintillation detector. In either case, the hydrogen is first recovered by electrolysis of the water and is then analyzed as H_2, CH_4, or other suitable H-rich molecules (Moser and Rauert, 1980). The sensitivity limit of this measurement is less than 1 TU. Alternatively, the concentration of tritium can be measured by a mass spectrometric determination of ^3He after previously purged water samples are stored for about six months (Clarke et al., 1976; Moser and Rauert, 1980).

21.4d Travel Time of Water in Confined Aquifers

The movement of groundwater in a confined aquifer in Figure 21.8 can be modeled as "piston flow," where the water in the aquifer is assumed to flow in straight lines as it would in a pipe. In this model, the recharge water of the previous years is displaced by the recharge water of the following years without mixing. Under these conditions, the concentration of tritium at a fixed point in the aquifer (e.g., a test well) is related to the transit time t by equation 21.19:

$$^3H_A = {}^3H_A^0 e^{-\lambda t}$$

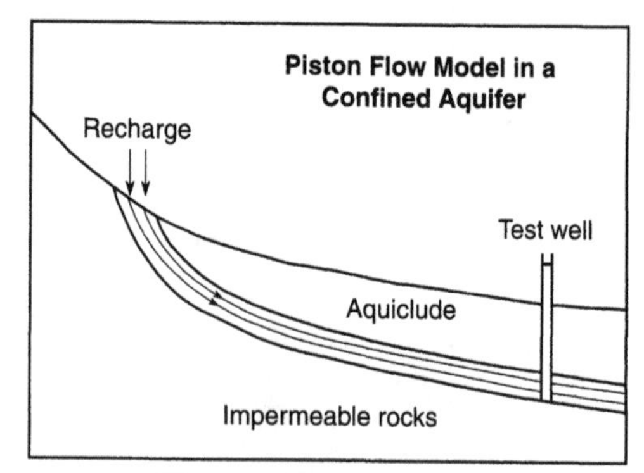

FIGURE 21.8 Flow of water in a confined aquifer according to the piston-flow model. Adapted from Nir (1964).

Nir (1964) replaced the decay constant λ by the mean life τ defined by equation 3.10 in Section 3.1: $\tau = 1/\lambda$. The mean life of tritium is $\tau = 1/5.6537 \times 10^{-2} = 17.69$ y. Therefore, equation 21.19 derived by Nir (1964) and stated by Dincer and Davis (1967) is

$$^3\mathrm{H}_A = {}^3\mathrm{H}_A^0 e^{-t/\tau} = {}^3\mathrm{H}_A^0 e^{-t/17.69} \qquad (21.20)$$

Nir (1964) also derived equations for the distribution of tritium in mixed and dispersive groundwater systems.

21.4e Tritiogenic Helium

Since tritium decays to stable $^3_2$He in accordance with equation 21.18, the accumulation of radiogenic ^{3}He can be used for dating based on the equation

$$^3\mathrm{He} = {}^3\mathrm{He}_0 + {}^3\mathrm{H}(e^{\lambda t} - 1) \qquad (21.21)$$

where ^{3}He and ^{3}H are expressed in terms of numbers of atoms in a unit volume of water. Solving for t yields

$$t = \frac{1}{\lambda} \ln\left(\frac{^3\mathrm{He} - {}^3\mathrm{He}_0}{^3\mathrm{H}} + 1\right)$$

This method of dating is based on the assumption that the tritiogenic helium does not escape from the aquifer. The initial ^{3}He (in atoms per liter) is assumed to be of atmospheric origin (i.e., ^{3}He/^{4}He $= 1.40 \times 10^{-6}$) and, in that case, can be calculated from the measured concentration of He dissolved in the water. If the concentration of He is V microliters per liter of water at STP, the number of ^{3}He atoms of atmospheric origin per liter of water is

$$^3\mathrm{He}_N = \frac{V \times 6.022 \times 10^{23} \times 0.00014}{10^6 \times 22.413 \times 10^2} \text{ atoms/L}$$

where the isotopic abundance of ^{3}He in atmospheric He is 0.00014% (Lide and Frederikse, 1995). The number of ^{3}He atoms in a sample of water collected from a test well located down-gradient from the recharge area is calculated similarly from the measured ^{3}He/^{4}He ratio and the concentration of He in microliters per liter. If the measured atomic ^{3}He/^{4}He ratio is 2.0×10^{-6}, the isotopic abundance of ^{3}He is calculated as follows:

$$^3\mathrm{He}/^4\mathrm{He} = 0.000002$$
$$^4\mathrm{He}/^4\mathrm{He} = \underline{1.000000}$$
$$\mathrm{Sum} = 1.000002$$

The abundance of ^{3}He (Ab ^{3}He) is

$$\mathrm{Ab}\ ^3\mathrm{He} = \frac{0.000002}{1.000002} \times 100 = 0.00020\%$$

which means that ^{3}He makes a negligibly small contribution to the concentration of the He in the water. The number of radiogenic ^{3}He atoms per liter of water (i.e., ^{3}He − ^{3}He$_0$) is the difference between the total number of atoms of ^{3}He per liter and the number due to the presence of atmospheric He.

Alternatively, equation 21.21 can be divided by the number of ^{4}He atoms dissolved in one liter of water:

$$\frac{^3\mathrm{He}}{^4\mathrm{He}} = \left(\frac{^3\mathrm{He}}{^4\mathrm{He}}\right)_0 + \frac{^3\mathrm{H}}{^4\mathrm{He}}(e^{\lambda t} - 1) \qquad (21.22)$$

In this case, $(^3\mathrm{He}/^4\mathrm{He})_i = 1.40 \times 10^{-6}$ and the ^{3}H/^{4}He ratio is calculated from the measured concentrations of tritium and helium and the measured ^{3}He/^{4}He isotope ratio.

In some cases, aquifers may contain excess ^{4}He produced by α-decay of U, Th, and their daughters in the minerals of the reservoir rock, in which case the initial ^{3}He/^{4}He atomic ratio is less than 1.40×10^{-6} (Schlosser et al., 1989). In such cases, the travel times calculated from equation 21.22 relative to an initial ^{3}He/^{4}He ratio of 1.40×10^{-6} are too low and could become negative if the measured ^{3}He/^{4}He ratios of the water are less than that of atmospheric He.

The principles of tritium–^{3}He dating were discussed by Poreda et al. (1988) and Schlosser et al. (1988, 1989), who also illustrated how this method of dating is applied to the study of shallow groundwater. Subsequently, Solomon et al. (1991,

1993) used the tritium–^{3}He method of dating to study recharge rates of shallow aquifers under a variety of conditions. An especially difficult case was investigated by Plummer et al. (1998a, b) near Valdosta, Georgia, where riverwater mixes with groundwater in the limestone of the Upper Floridan aquifer.

21.4f Kirkwood-Cohansey Aquifer, New Jersey

The tritium–^{3}He method was used by Szabo et al. (1996) to date water in the unconfined sandy aquifer of the Cohansey and Kirkwood Formations, which together are about 40 m thick in Cumberland and Gloucester counties on the coastal plain of New Jersey. The concentration of tritium in the water below the watertable in Figure 21.9 *increases* with depth from 11.9 ± 1.5 TU near the top of the aquifer to about 55 TU at 23 m and then declines to 1.0 ± 0.5 TU at depth greater than 30 m.

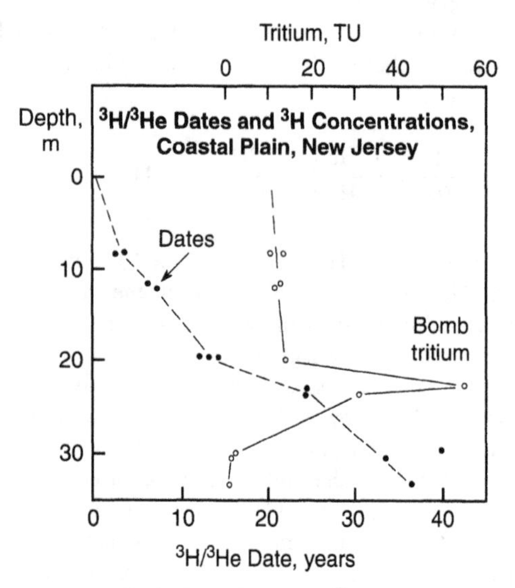

FIGURE 21.9 Variation of tritium–^{3}He dates in years before 1992 of groundwater in the unconfined Kirkwood-Cohansey aquifer on the coastal plain of New Jersey (closed circles). The tritium concentrations (open circles) are expressed in TU and identify the spike of thermonuclear tritium at a depth of 23 m below the surface. Data from Szabo et al. (1996).

The presence of thermonuclear tritium at a depth of 23 m below the surface can be used to estimate the vertical flow velocity v of the water:

$$v = \frac{23}{1992 - 1963} = 0.8 \text{ m/y}$$

where 1992 is the year the samples were collected and the thermonuclear tritium is assumed to have entered the aquifer in 1963 in accordance with Figure 21.5. The comparatively high rate of downward movement of water in the Kirkwood-Cohansey aquifer prevents the escape of tritiogenic ^{3}He (Szabo et al., 1996; Schlosser et al., 1989).

The tritium–^{3}He dates of the water in the Kirkwood-Cohansey aquifer increase with depth from 3.5 years at a depth of 8.2 m to 36.3 years at 33.5 m. These dates imply a velocity v of downward flow in the unsaturated zone of

$$v = \frac{8.2}{3.5} = 2.3 \text{ m/y} \quad \text{(unsaturated)}$$

assuming that the watertable occurs at a depth of about 8 m (Szabo et al., 1996). After the recharge enters the aquifer, the downward flow velocity is

$$v = \frac{33.5 - 8.2}{36.3 - 3.5} = 0.77 \text{ m/y} \quad \text{(saturated)}$$

The flow velocity implied by the profile of tritium–^{3}He dates in Figure 21.9 agrees with the flow velocity derived from the depth to the thermonuclear tritium peak.

21.5 METEORITES AND OCEANIC BASALT

When the Earth formed by accretion of planetesimals at 4.6 Ga, it acquired primordial He whose isotope composition may have been similar to that of He in the solar nebula (Ozima and Podosek, 1983; Mamyrin and Tolstikhin, 1984). The isotopic composition of He in the solar nebula was the result of mixing of galactic He with ^{4}He produced by H fusion in the cores and outer shells of the ancestral stars. In addition, other nuclear reactions in the interiors of stars and in the interstellar gas they ejected during their supernova episode further modified the isotope composition of He that was ultimately incorporated into the Earth.

21.5a Cosmogenic ^{3}He

The exposure of *meteoroids* to cosmic rays in space before they impact on the Earth or on other terrestrial planets causes ^{3}He and other cosmogenic nuclides to form by nuclear spallation reactions. These reactions also occur in the atmosphere of the Earth and in rocks exposed to cosmic rays. The production and decay of the so-called cosmogenic radionuclides is the subject of Chapter 23.

The nuclear spallation reactions that produce ^{3}H and ^{3}He are caused primarily by energetic cosmic-ray neutrons and protons and, to a lesser extent, by α-particles emitted by U, Th, and their daughters. On the *Earth*, these reactions occur predominantly in the troposphere and in rocks within about one meter of the surface (Craig and Lupton, 1981; Craig and Poreda, 1986; Kurz, 1986a, b; Lal, 1987).

The nuclear reactions that produce tritium (e.g., equation 21.17) also contribute to the formation of ^{3}He because tritium decays to ^{3}He with a halflife of only 12.26 years. Other reactions that produce tritium are based on $^7_3$Li which is a stable isotope of lithium with an abundance of 92.5%:

$$^7_3\text{Li} + ^1_0\text{n} \rightarrow ^3_1\text{H} + ^5_2\text{He} \quad \text{(unstable)} \qquad (21.23)$$

$$^7_3\text{Li} + ^1_1\text{p} \rightarrow ^3_1\text{H} + ^5_3\text{Li} \quad \text{(stable)} \qquad (21.24)$$

$$^7_3\text{Li} + ^4_2\text{He} \rightarrow ^3_1\text{H} + ^8_4\text{Be} \quad \text{(unstable)} \qquad (21.25)$$

The unstable products of these reactions decay quickly to form stable isotopes of other elements:

$$^5_2\text{He} \rightarrow ^5_3\text{Li} + \beta^- + \overline{\nu} + Q \qquad (21.26)$$

$$^8_4\text{Be} \rightarrow ^4_2\text{He} + ^4_2\text{He} + Q \qquad (21.27)$$

Another suite of nuclear reactions produces ^{3}He directly:

$$^{40}_{20}\text{Ca} + ^1_0\text{n} \rightarrow ^3_2\text{He} + ^{38}_{18}\text{Ar} \quad \text{(stable)} \qquad (21.28)$$

$$^{58}_{28}\text{Ni} + ^1_0\text{n} \rightarrow ^3_2\text{He} + ^{56}_{26}\text{Fe} \quad \text{(stable)} \qquad (21.29)$$

$$^7_8\text{Li} + ^1_1\text{p} \rightarrow ^3_2\text{He} + ^5_2\text{He} \quad \text{(unstable)} \qquad (21.30)$$

$$^{14}_7\text{N} + ^1_1\text{p} \rightarrow ^3_2\text{He} + ^{12}_6\text{C} \quad \text{(stable)} \qquad (21.31)$$

21.5b Meteorites

The concentrations of ^{4}He in stony meteorites compiled by Pepin and Signer (1965) vary by four orders of magnitude from 2.25×10^{-2} to 6.0×10^{-6} cm^3/g at STP, whereas their ^{3}He/^{4}He ratios range only between narrow limits. The data in Figure 21.10 demonstrate that the ^{3}He/^{4}He ratios of stony meteorites increase with decreasing concentrations of ^{4}He. The highest ^{3}He/^{4}He ratio of 8.75×10^{-4} occurs in a sample of the enstatite chondrite Pultusk, which has a low ^{4}He concentration of 7.34×10^{-5} cm^3/g at STP. The inverse relationship between the ^{3}He/^{4}He ratios and the ^{4}He concentration of meteorites supports the view that the He is a mixture primarily of two components. One of these is ^{4}He of radiogenic origin

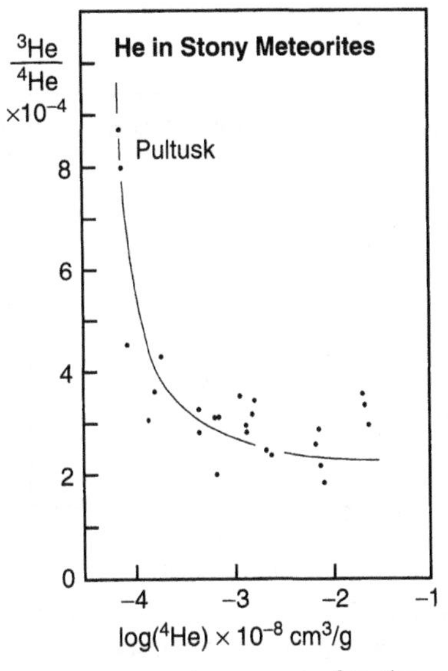

FIGURE 21.10 Relation between the ^{3}He/^{4}He ratio and the concentration of ^{4}He in stony meteorites (enstatite chondrites and carbonaceous chondrites). These data demonstrate that the He is a mixture of two components. Data compiled by Pepin and Signer (1965).

whose presence increases the concentration of He but lowers the $^3He/^4He$ ratio. In addition, individual meteorites contain varying amounts of cosmogenic ^3He depending primarily on the duration of their exposure to cosmic rays while they were in orbit around the Sun.

The apparent complexity of the isotope composition of He in meteorites was further revealed by Srinivasan and Anders (1978), who released He by stepwise heating of the carbonaceous chondrite Murchison. The fractions of He released in Figure 21.11 increased with rising temperature to 56% at 1000°C and then declined to 0.3% at 1400°C. The $^3He/^4He$ ratio initially declined from 1.45×10^{-4} at 800°C to 0.99×10^{-4} at 1100°C before rising sharply to 20×10^{-4} at 1400°C.

Srinivasan and Anders (1978) suggested that the ^3He-rich gas resides in small refractory particles (presolar grains) and that the ^3He formed by hydrogen fusion in shells surrounding the cores of red giant stars where the $^3He/^4He$ ratios may reach 50×10^{-4}. The particles that contained the ^3He-rich gas were ejected during the explosion of these stars and therefore were present in the

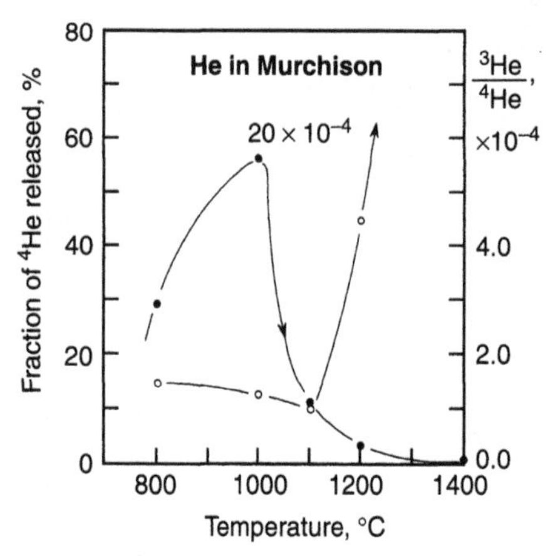

FIGURE 21.11 Stepwise outgassing of He from the carbonaceous chondrite Murchison. The fraction of He released (left scale) initially increases to 56% at 1000°C and then declines to 0.3% at 1400°C. The $^3He/^4He$ ratios (right scale) decrease slightly between 800 and 1100°C and then rise steeply to 20×10^{-4} at 1400°C. Data from Srinivasan and Anders (1978).

solar nebula before it contracted to form the Sun. These presolar grains were ultimately incorporated into solid planetesimals, some of which formed the parent bodies of meteorites. The carbonaceous chondrites like Murchison preserved the He they acquired initially because they were never exposed to high temperatures in excess of about 125°C.

Stepwise gas release of strongly etched residues of meteorites has also indicated the presence of a "quintessence" termed phase Q, which contains noble gases of anomalous isotope compositions (Lewis et al., 1975). Phase Q is known to be composed of acid-resistant organic matter to which He and the other noble gases are sorbed. Busemann et al. (2000) measured the $^3He/^4He$ ratios of Q in three meteorites (Cold Bokkeveld, Grosnaja, and Isna) all of which have short cosmic-ray exposure ages (<2 Ma) and therefore contain less cosmogenic ^3He than meteorites that have longer exposure ages. The resulting $^3He/^4He$ ratios reported by Busemann et al. (2000) remained virtually constant during stepwise outgassing of Q up to 90°C with the following mean values:

Cold Bokkeveld	$1.41 \pm 0.01 \times 10^{-4}$
Grosnaja	$1.45 \pm 0.01 \times 10^{-4}$
Isna	$1.23 \pm 0.02 \times 10^{-4}$

The origin of the noble gases in the Q phase is still in question. Ozima et al. (1998) proposed that He and the other noble gases in the Q phase originated from the solar nebula. However, these authors also cited an average $^3He/^4He$ ratio of $4.57 \pm 0.36 \times 10^{-4}$ for the solar wind. This value differs significantly from the $^3He/^4He$ ratios of the Q phase of meteorites and of He in the terrestrial atmosphere, where $^3He/^4He = 1.40 \times 10^{-6}$.

21.5c Oceanic Basalt

The isotopic composition of terrestrial He has been modified in the course of geological time by the addition of ^4He in the form of α-particles and by the formation of ^3He by both decay of cosmogenic tritium and by nuclear spallation reactions caused by cosmic rays. For these reasons, the isotope composition of terrestrial He varies depending on its geochemical history.

The He that was trapped in the mantle of the Earth is released in the course of magma formation by even small degrees of partial melting and is ultimately vented into the atmosphere by volcanoes in the oceans and on the continents. In addition, He is discharged along midocean ridges and continental rifts by both volcanic eruptions and by discharge of water and steam by springs.

The formation of magma by partial melting in the mantle releases He whose isotopic composition in the resulting magma is assumed to be identical to that of the He in the source rocks. However, it may be altered later by the addition of He from other sources, including crustal rocks and convecting seawater or groundwater and by exposure to cosmic rays at the surface of the Earth (e.g., Craig and Poreda, 1986; Kurz, 1986a, b). Consequently, volcanic rocks may have a different isotope compositions of He than the magma source in the mantle.

The isotope composition of He in volcanic rocks is described by reference to the $^3He/^4He$ ratio of the atmosphere, which is designed $R_a = 1.40 \times 10^{-6}$. For example, Ozima and Zashu (1983) reported that basalt extruded along the East Pacific Rise has a uniform $^3He/^4He$ ratio R of $1.21 \pm 0.07 \times 10^{-5}$. The corresponding value of this ratio relative to R_a is

$$\frac{R}{R_a} = \frac{1.21 \pm 0.07 \times 10^{-5}}{1.40 \times 10^{-6}} = 8.64 \pm 0.50$$

On this scale, He in all stony meteorites and in the solar wind has large R/R_a values that range from about 100 in the Q phase of Cold Bokkeveld to 326 in the solar wind and to 625 in the enstatite chondrite Pultusk (Figure 21.11).

The He concentrations and $^3He/^4He$ ratios of volcanic rocks extruded in the ocean basins and of the ultramafic inclusion they contain have been determined by analysis of minerals (olivine, clinopyroxene, orthopyroxene, amphibole), volcanic glass, and whole-rock samples. The noble gases, including He, are released either by crushing or by fusion in vacuum systems. Scarsi (2000) tested both methods and reported that crushing is better suited for distinguishing between the He that was trapped in fluid inclusions and the He isotopes that formed in situ after initial crystallization of the rocks.

The $^3He/^4He$ ratios (R/R_a) of basalt and ultramafic inclusions on oceanic islands range from 32 on Loihi (south of Hawaii) to less than 5 on Tristan da Cunha and Gough in the South Atlantic Ocean (Kurz et al., 1982). The concentrations of ^4He range even more widely from 10^{-10} to 10^{-7} cm^3/g at STP. The isotopic composition of He of volcanic rocks and inclusions is not closely related to its concentration because volcanic rocks on individual islands or groups of islands tend to have characteristic isotope ratios of He that distinguish them from those of other islands, whereas the He concentrations may range widely.

The R/R_a values of volcanic rocks and xenoliths on oceanic islands in Figure 21.12 have a multimodal distribution which suggests the existence of several different magma sources in the suboceanic mantle. In fact, the isotope compositions of Sr, Nd, Pb, Hf, and Os in MORBs and OIBs are explained in terms of four principal magma components (Section 17.2): DMM, EM1, EM2, and HIMU. Therefore, Kurz et al. (1982) proposed that the isotope ratios of He of volcanic rocks on oceanic islands and midocean ridges are also the result of mixing of several components to which they assigned characteristic R/R_a and $^{87}Sr/^{86}Sr$ ratios:

1. Depleted MORB mantle (DMM): $R/R_a \sim 8.5$; $^{87}Sr/^{86}Sr \sim 0.7025$.
2. Mantle plume 1 (Hawaii and Iceland): $R/R_a \sim 32$; $^{87}Sr/^{86}Sr \sim 0.7035$.
3. Mantle plume 2 (Tristan da Cunha and Gough): $R/R_a \sim 5.5$; $^{87}Sr/^{86}Sr \sim 0.7052$.

The resulting triangle of mixing in Figure 21.13 provides a rational explanation for the isotope ratios of He and Sr of volcanic rocks on Iceland (1), the Hawaiian Islands (2), the Canary Islands and Madeira (3), and the Kerguelen Islands (4).

However, Farley et al. (1992) later reported that the basalts on the island of Tutuila, Samoa, do not fit the three-component mixing model because these lavas have high $^{87}Sr/^{86}Sr$ ratios (0.70583–0.70742) as well as high R/R_a values (11.8–23.9). These authors therefore redefined the R/R_a values of the magma sources in the

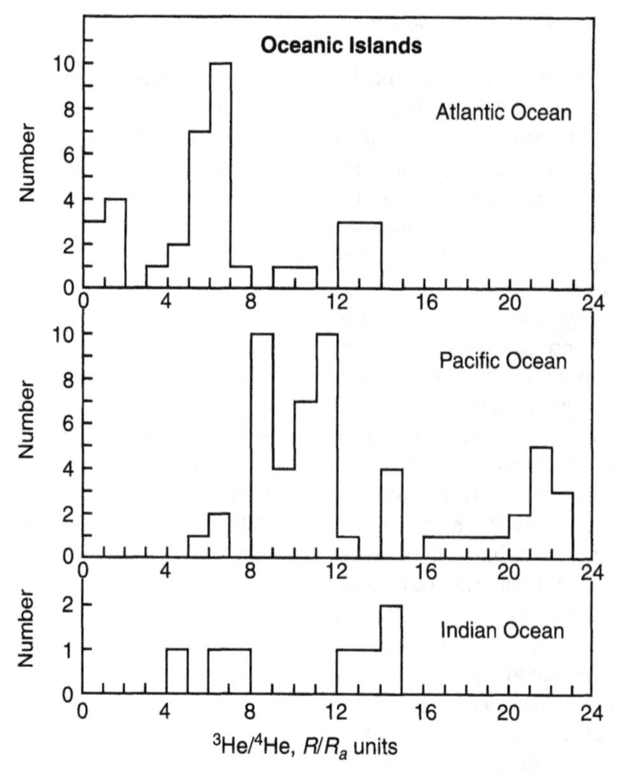

FIGURE 21.12 Isotope ratios (r) of He in terms of R/R_a ratio, where $R_a = 1.40 \times 10^{-6}$ of basalt and rhyolite on oceanic islands in the three major oceans. Data from Kurz et al. (1982), Condomines et al. (1983), Vance et al. (1989), Poreda and Farley (1992), and Farley et al. (1992).

suboceanic mantle (DMM, EM2, and HIMU, Section 17.2) and added a new component called the *primitive helium mantle* (PHEM). The isotope ratios of He, Sr, Nd, and Pb of these components in Table 21.3 can accommodate the isotope ratios of He and Sr of the Samoan basalts in Figure 21.14.

The PHEM component is characterized by high $^3He/^4He$ ratios ($R/R_a = 32$–50) but intermediate isotope ratios of Sr ($^{87}Sr/^{86}Sr =$ 0.7042–0.7052), Nd ($^{143}Nd/^{144}Nd = 0.51265$–0.51280), and Pb ($^{206}Pb/^{204}Pb = 18.6$–19.0). Consequently, the PHEM magma source in the suboceanic mantle contains He whose isotope composition has not been modified by prior episodes of outgassing. Partial outgassing of primordial He during the early history of the Earth causes the $^3He/^4He$ ratio of the residual gas to decline because of the time-dependent addition of radiogenic ^4He, whereas the amount of ^3He remains virtually constant.

The interpretation of the isotope ratios of He and Sr of young volcanic rocks proposed by Kurz et al. (1982) and Farley et al. (1992) leads to the conclusion that primordial He had a high $^3He/^4He$ ratio approaching that of the Q phase in meteorites, which is characterized by $R/R_a \sim 100$. The implications of this conjecture are that

1. the $^3He/^4He$ ratio of present-day terrestrial He was lowered primarily by addition of varying amounts of radiogenic ^4He and
2. the $^3He/^4He$ ratios of stony meteorites were increased by the addition of varying amounts of cosmogenic ^3He.

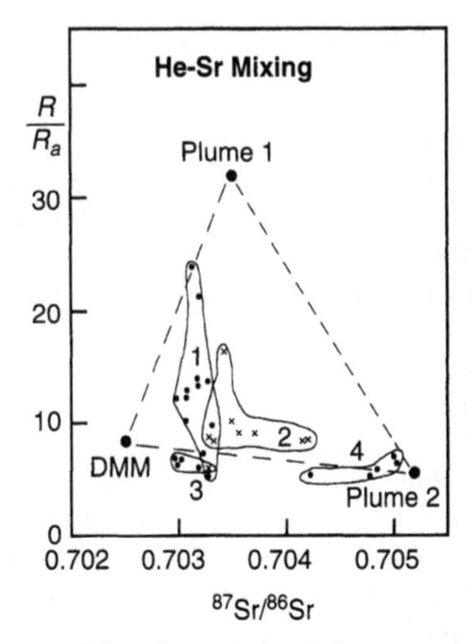

FIGURE 21.13 Hypothetical He–Sr isotope mixing diagram based on three components defined by Kurz et al. (1982). The mixing lines joining these components are actually curved because the concentrations of He and Sr differ significantly. The data fields represent Iceland (1), the Hawaiian Islands but not Loihi (2), the Canary Islands and Madeira (3), and the Kerguelen Islands (4). Data from Kurz et al. (1982), Condomines et al. (1983), and Vance et al. (1989).

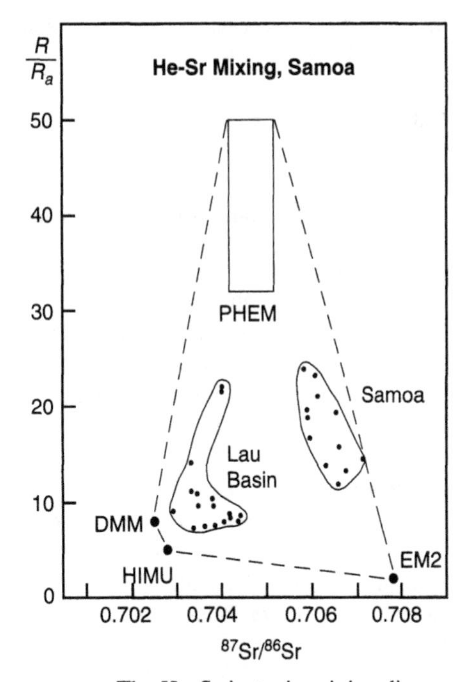

FIGURE 21.14 The He–Sr isotopic mixing diagram including the primitive He-mantle component PHEM, EM2, HIMU, and DMM. Data for Samoa from Farley et al. (1992) and for the Lau basin from Poreda and Craig (1992).

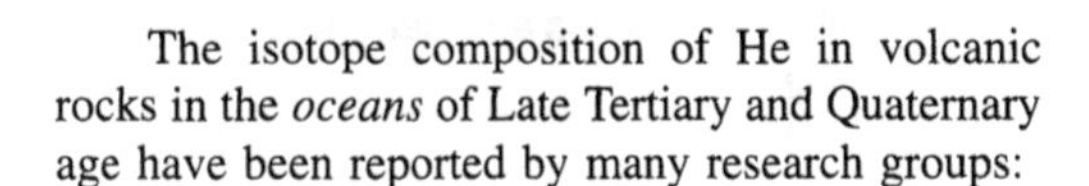

The isotope composition of He in volcanic rocks in the *oceans* of Late Tertiary and Quaternary age have been reported by many research groups:

Oceanic Basalt: Lupton and Craig, 1975, *Earth Planet. Sci. Lett.*, 26:133–139. Craig and Lupton, 1976, *Earth Planet. Sci. Lett.*, 31:369–385. Craig and Lupton, 1981, in Emiliani (Ed.), *The Sea*, pp. 391–428, Wiley, New York. Honda and Patterson, 1999, *Geochim. Cosmochim. Acta*, 63:2863–2874. Marty and Zimmerman, 1999, *Geochim. Cosmochim. Acta*, 63:3619–3633. Sarda and

Table 21.3. Approximate Isotope Ratios of He, Sr, Nd, and Pb of the Principal Magma Source Components in the Suboceanic Mantle

Component	R/R_a	$^{87}Sr/^{86}Sr$	$^{143}Nd/^{144}Nd$	$^{206}Pb/^{204}Pb$
DMM	8	0.7025	0.51340	17.6
EM1	?	0.7050	0.51235	17.0
EM2	2	0.7078	0.51255	19.3
HIMU	5	0.7028	0.51295	21.0
PHEM	32–50	0.7042–0.7052	0.751265–0.51280	18.6–19.0

Source: Farley et al., 1992.

Moreira, 2002, *Geochim. Cosmochim. Acta*, 66:1449–1458.

Galapagos: Jenkins et al., 1978, *Nature*, 272:156–158.

Loihi and Hawaiian Islands: Rison and Craig, 1983, *Earth Planet. Sci. Lett.*, 66: 407–426. Kurz et al., 1983, *Earth Planet. Sci. Lett.*, 66:388–399. Kurz et al., 1987, *Geochim. Cosmochim. Acta*, 51:2905–2914. Kurz and Kammer, 1991, *Earth Planet. Sci. Lett.*, 103:257–269. Vance et al., 1989, *Earth Planet. Sci. Lett.*, 96:147–160.

Lau Basin, Pacific Ocean: Poreda and Craig, 1992, *Earth Planet. Sci. Lett.*, 113:487–493.

Samoan Islands (Xenoliths): Poreda and Farley, 1992, *Earth Planet. Sci. Lett.*, 113: 129–144.

Manus Basin, Pacific Ocean: Macpherson et al., 1998, *Geology*, 26:1007–1010.

Pacific Seamounts: Graham et al., 1988, *Contrib. Mineral. Petrol.*, 99:446–463.

Iceland: Kurz et al., 1985, *Earth Planet. Sci. Lett.*, 74:291–305. Poreda et al., 1986, *Earth Planet. Sci. Lett.*, 78:1–17. Condomines et al., 1983, *Earth Planet. Sci. Lett.*, 66:125–136.

Azores: Kurz et al., 1982. *Earth Planet. Sci. Lett.*, 58:1–14.

Réunion Island: Graham et al., 1990, *Nature*, 347:545–548. Kaneoka et al., 1986, *Chem. Geol.*, 59:35–42.

These studies have provided new information about the heterogeneity of the suboceanic mantle, about the derivation of volcanic rocks from the different kinds of magma sources, and about the isotopic evolution of terrestrial He by additions of radiogenic ^{4}He and (to a lesser extent) by cosmogenic and tritiogenic ^{3}He.

21.6 CONTINENTAL CRUST

Helium and other noble gases occur in subsurface fluids and in reservoirs of natural gas as well as in the rocks and minerals of the continental crust (Alexander and Ozima, 1978; Mamyrin and Tolstikhin, 1984). The water and gases are discharged by natural springs and are vented to the atmosphere by volcanic eruptions associated with continental rifts and at compressive plate margins. In addition, He is recovered during the production of hydrocarbon gases in certain areas. The isotope composition of He in these crustal environments varies regionally depending on its provenance and on the additions of ^{3}He or ^{4}He or both. The geochemistry of terrestrial noble gases was discussed in a special issue of the *Isotope Geoscience Section* of *Chemical Geology* edited by Podosek (1985).

21.6a Ultramafic Inclusions and Basalt

The ^{3}He/^{4}He (R/R_a) ratios of volcanic rocks and ultramafic inclusions from continental settings in Figure 21.15 come close to having a unimodal distribution. Nearly 80% of 188 specimens analyzed

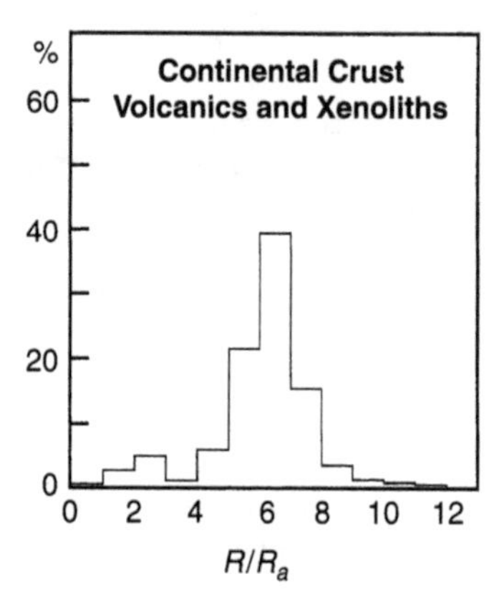

FIGURE 21.15 Histogram in percent by number of the R/R_a values of 188 samples of volcanic rocks and the ultramafic xenoliths they contain. The isotope ratios of samples containing excess ^{3}He were corrected by the investigators. Data from Dodson et al. (1998), Hoke et al. (2000), Dunai and Baur (1995), and Porcelli et al. (1986, 1987).

by Dodson et al. (1998), Hoke et al. (2000), Dunai and Baur (1995), and Porcelli et al. (1986, 1987) have R/R_a values between 5 and 8. These samples originated from the western United States, southern New Zealand, southeastern Australia, East Africa, Yemen, and northeastern China.

Similarly, the $^3He/^4He$ ratios of mafic xenoliths in a given region range only within narrow limits. For example, Dunai and Baur (1995) reported that xenoliths in Cenozoic volcanic rocks of Europe have characteristic average R/R_a values:

Massif Central, France	6.53 ± 0.25
Eifel, Germany	6.03 ± 0.14
Spitsbergen, Norway	6.65 ± 0.25
Kapfenstein, Austria	6.1 ± 0.7

Dunai and Baur (1995) also demonstrated that these xenoliths plot near the base of the three-component mixing diagram in Figure 21.14 (not shown), which means that the He they contain could have originated from the DMM and EM2 components in the subcontinental mantle.

Some ultramafic xenoliths contain excess cosmogenic ^3He, which causes their $^3He/^4He$ ratios to rise to *high* values. For example, spinel lherzolite inclusions at Geronimo, Arizona, have $^3He/^4He$ ratios between 139 and 464 with large errors (Porcelli et al., 1987). Excess ^3He was also identified by Dunai and Baur (1995) in ultramafic xenoliths from the Eifel region of Germany, the Pannonian basin (Kapfenstein) of Austria, the Massif Central of France, and Spitsbergen. The excess ^3He was generated within these rocks by nuclear spallation reactions caused by cosmic rays. The presence of excess ^3He in rocks exposed to cosmic rays obscures the isotopic composition of He in the subcontinental mantle. This complication can be avoided by selecting samples that were exposed to cosmic rays for only short periods of time or were shielded by overlying rocks.

In general, the R/R_a ratios of ultramafic xenoliths in continental basalts and kimberlites range between 0.3 and 10 (Tolstikhin et al., 1974; Kaneoka et al., 1978; Kyser and Rison, 1982; Mamyrin and Tolstikhin, 1984; Poreda and Basu, 1984). The *low* values are caused by contamination with ^4He in the form of α-particles emitted by U, Th, and their daughters, depending on the posteruption ages of the samples. Porcelli et al. (1987) demonstrated by calculation that the R/R_a values of ultramafic inclusions (U = 0.45 ppm, He = 1.8×10^{-8} cm^3/g, initial $R/R_a = 8.0$) start to decline detectably in less than 10,000 years and are lowered to less than 1.0 in about one million years.

21.6b Diamonds

An interesting case of the apparent lowering of $^3He/^4He$ ratios by radiogenic ^4He was reported by Ozima and Zashu (1983) and Ozima et al. (1983) for diamonds from South Africa. Two diamonds from the Premier Mine have "normal" R/R_a values of 5.7 ± 0.2 and 12.8 ± 0.4, whereas the R/R_a ratios of 25 additional diamonds of unidentified origin range widely from <0.06 to 226 ± 18.

Ozima et al. (1983) proposed that the diamonds contain He whose initial $^3He/^4He$ ratio at the time of formation of the Earth was

$$\left(\frac{R}{R_a}\right)_0 = \frac{4 \times 10^{-4}}{1.40 \times 10^{-6}} = 286$$

After the formation of the Earth, the $^3He/^4He$ ratio of different mantle domains declined with time depending on their $^3He/U$ ratios. Ozima et al. (1983) estimated that the average $^3He/U$ ratio of diamonds is higher than that of the Earth by a factor of about 390 because diamonds have very low U concentrations. Therefore, after He was trapped in a diamond, its R/R_a ratio declined much more slowly than the R/R_a ratio of He in the Earth as a whole. It follows that diamonds with *high* R/R_a values crystallized *early* in the history of the Earth and those with *low* ratios crystallized *later*. However, Ozima et al. (1983) were unable to date these diamonds by the K–Ar method because they contain excess ^{40}Ar.

An alternative explanation for the high R/R_a values (e.g., 226, 168, 82.9,...) is that these diamonds were recovered from alluvial deposits and therefore contain cosmogenic ^3He as a result of exposure to cosmic rays.

21.6c Effect of Tectonic Age on He in Groundwater

Deep groundwater and gas reservoirs contain He whose isotope composition is representative of the He that exists in the minerals and rocks of the tectonic province within which they are located. In addition, He enters the crust by venting of the underlying mantle, especially during volcanic activity along crustal rifts and in places where mantle plumes have intruded the lithsopheric mantle. Many research groups have reported that the isotope ratios of He in gas and deep groundwater collected from boreholes and discharged by springs cluster about provincial mean values. The data compiled by Polyak and Tolstikhin (1985) in Table 21.4 and Figure 21.16 demonstrate that the average R/R_a values of hydrothermal fluids, gases, and crustal rocks of tectonic provinces decrease with increasing age. This effect is caused by the accumulation of radiogenic ^{4}He in crustal fluids and rocks, which depresses the R/R_a ratio from 6.0 ± 2.8 for active rift systems to 0.037 ± 0.030 for Precambrian terrains that were recrystallized more than 1600 million years ago.

In addition, Polyak and Tolstikhin (1985) demonstrated a relation between the isotope composition of He and the average heat flow in each of the tectonic provinces listed in Table 21.4. This relation reflects the migration of He from the mantle into the overlying continental crust in presently active tectonic areas followed by gradual cooling after the tectonic activity subsides. The R/R_a values of mantle-derived He in Figure 21.15 are strongly clustered between 6.0 and 7.0, in agreement with the explanation proposed by Polyak and Tolstikhin (1985) for the data in Figure 21.16.

Table 21.4. Average Isotope Ratios of He in Gases and Fluids in Tectonic Provinces on a Global Scale

Region	Age,[a] Ma	R/R_a
East European	>1600	0.037 ±0.030
Hercynides	~300	0.096 ±0.077
Alpides	~50	0.60 ±0.52
Pacific belt	~15	4.2 ±1.4
Active rifts	0	6.0 ±2.8

Source: Polyak and Tolstikhin, 1985.

[a]Ages estimated from the timescale of Harland et al. (1989).

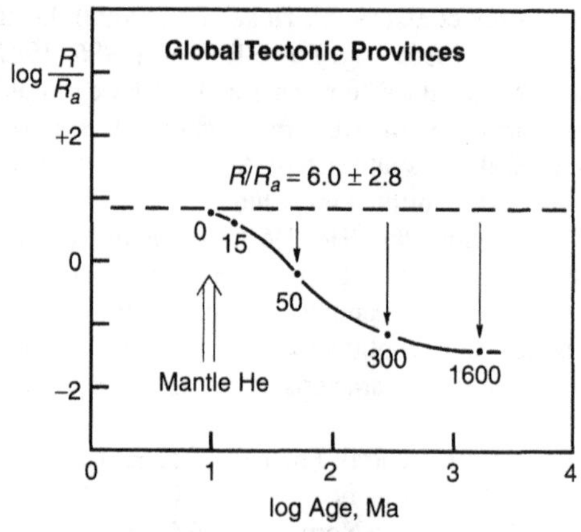

FIGURE 21.16 Dependence of the isotope composition of He in hydrothermal water, gases, and crustal rocks on the age of tectonic activity on a global scale. The ages of the tectonic provinces are indicated along the curve in Ma. The coordinates of the data points are expressed in terms of log 10. Data from Polyak and Tolstikhin (1985), as summarized in Table 21.4.

21.6d Geothermal Systems

Helium discharged by hotsprings in volcanically active regions is a mixture in varying proportions of several components having characteristic isotope compositions expressed as the R/R_a ratios (Torgersen and Jenkins, 1982):

1. Magmatic: ~10
2. Atmospheric: 1.0
3. Crustal: 0.1

Each of these components is itself a product of its geochemical history, which included mixing and the addition of varying amounts of radiogenic ^4He as well as of cosmogenic and tritiogenic ^3He. The wide range of isotope ratios of these components is helpful in determining the provenance of He in geothermal systems.

The gases discharged by geothermal springs or wells may become contaminated with atmospheric He during collection. Therefore, the isotope ratios of He must be corrected for the presence of atmospheric He based on the fact that the He/Ne ratio of crustal gases is greater than 100 in most cases, whereas the He/Ne ratio of air is only 0.29. If the He/Ne ratio is expressed by the parameter X, defined as

$$X = \frac{(\text{He/Ne})}{(\text{He/Ne})_{\text{air}}}$$

then the corrected R/R_a ratio is calculated from an equation derived by Craig et al. (1978):

$$\left(\frac{R}{R_a}\right)_c = \frac{[(R/R_a)X] - 1}{X - 1} \qquad (21.32)$$

The average corrected isotope ratios of He in gases discharged by geothermal springs and boreholes in the western region of the USA in Figure 21.17 range widely (Torgersen and Jenkins, 1982):

Raft River, Idaho	0.15 ± 0.01
Steamboat Springs, Colorado	5.1 ± 1.1
The Geysers, California	8.4 ± 0.6
Kilauea, Hawaii	13.8 ± 0.4

In addition, geothermal well waters in Iceland analyzed by Torgersen and Jenkins (1982) contain He whose average R/R_a ratio is 9.7 ± 1.2 (not shown in Figure 21.17). The R/R_a ratios of volcanic gases on the North Island of New Zealand in Figure 21.17 range from >1 to <9 and have a mean value of 6.3 ± 0.5 (Torgersen et al., 1982).

Although all of the sites included in Figure 21.17 are associated with volcanic activity, the average corrected R/R_a values differ. For example, the He in geothermal water at Raft River in Idaho ($R/R_a = 0.15 \pm 0.01$) is quite low and indicates that the He is of crustal origin. Although geophysical data suggest the presence of an igneous intrusion at depth in this area, it is too old to be

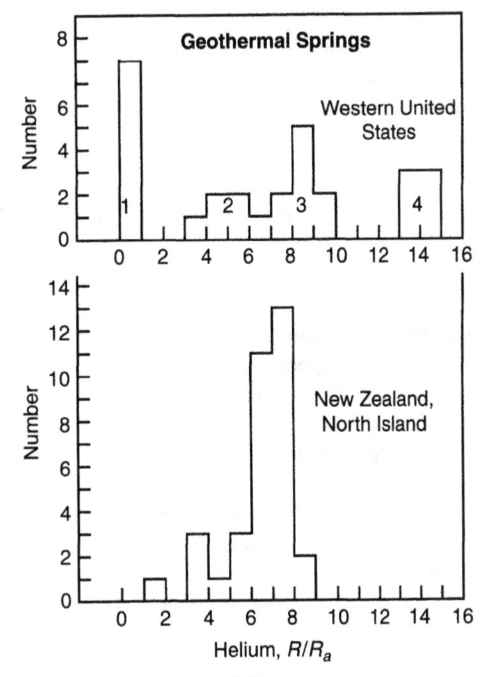

FIGURE 21.17 Isotope ratios of He (R/R_a) of geothermal gas and hydrothermal fluids in the western USA and on the North Island of New Zealand: 1 = Raft River geothermal area, Snake River Plain, Idaho; 2 = Steamboat Springs, Colorado; 3 = The Geysers, California; 4 = Kilauea, Hawaii. Data from Torgersen and Jenkins (1982) and Torgersen et al. (1982).

the source of heat and He in the water. Instead, the water is heated geothermally in an aquifer at a depth of 3–5 km.

Steamboat Springs, Colorado, is an active geothermal area where hydrothermal mineralization is present at depth. The average R/R_a value (5.1 ± 1.1) is lower than expected for He of magmatic origin, presumably because of contamination with crustal He.

The Geysers area in California is underlain by a large reservoir of dry steam that is used commercially to generate electricity. The average R/R_a value (8.4 ± 0.6) is consistent with a magmatic source which is thought to be located more than 10 km below the surface based on geophysical data.

The fumarolic gases on Kilauea, Hawaii, contain He that is discharged by the mantle plume which underlies this volcano and the nearby seamount Loihi. Accordingly, its average R/R_a ratio is 26.4 ± 5.6 (Kurz et al., 1982).

21.6e Geothermal He, New Zealand

The He in volcanic gases on the North Island of New Zealand ($R/R_a = 6.3 \pm 0.5$) is composed of mixtures of a magmatic component ($R/R_a \sim 8.2$) and a crustal component containing radiogenic ^{4}He. The presence of tritiogenic ^{3}He was shown to be negligible by Torgersen et al. (1982).

The He in hydrocarbon gases of the Taranaki Basin, located primarily off-shore between the North and South Islands of New Zealand, has low R/R_a ratios (Hulston et al., 2001). More than half of the samples in Figure 21.18 have $R/R_a <$ 1.0. The R/R_a values of a few gas samples

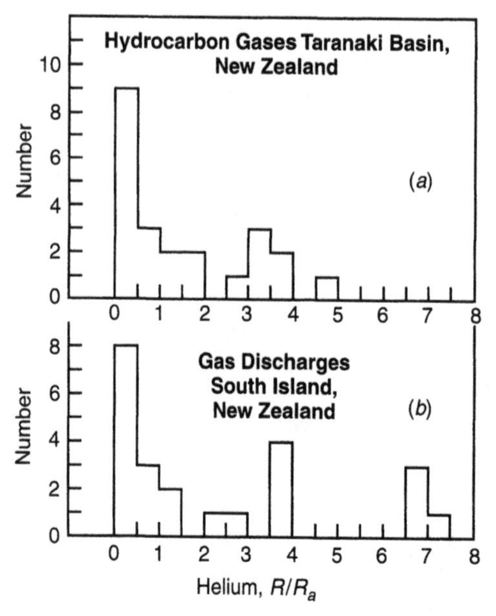

FIGURE 21.18 Isotope ratios of He (R/R_a) in (*a*) hydrocarbon gases of the Taranaki Basin between North and South Islands of New Zealand and (*b*) thermal and nonthermal gas discharges on the South Island of New Zealand. Data from Hulston et al. (2001) and Hoke et al. (2000).

from the New Plymouth area in the southwest corner of the North Island range from 4.73 to 3.89. This area includes an active volcano (Mt. Taranaki) and has high heatflow of 74 milli Watts per square meter (mW/m^2), which supports the conclusion that these samples contain He of magmatic origin. A second set of samples with high R/R_a values (2.93–3.80) originated from off-shore drilling platforms in the Maui field, which is located adjacent to the Cape Egmont Fault. Therefore, the data of Hulston et al. (2001) demonstrate that some of the hydrocarbon reservoirs of the Taranaki Basin contain He of magmatic origin. In addition, these authors demonstrated a positive correlation between heatflow and the R/R_a ratio of He, in agreement with Polyak and Tolstikhin (1985).

Mantle-derived He is also present in gas-rich springs in the Dunedin area and on the Banks Peninsula of the South Island (Hoke et al., 2000). Both areas are underlain by Late Tertiary volcanic rocks containing mantle-derived He with average R/R_a values of 6.2 ± 1.6 (Dunedin) and 7.8 ± 1.7 (Banks Peninsula). The He in gas-rich springs at different locations on the South Island in Figure 21.18 has a multimodal spectrum of R/R_a values. As expected, the water discharged by springs at Wairongoa and the Salisbury Tunnel near Dunedin has elevated R/R_a ratios (3.89–7.03) with a mean of 6.2 ± 1.2, which is identical to that of He in the volcanic rocks of the area. One sample of water on the Banks Peninsula ($R/R_a = 2.26$) still contains about 28% of mantle-derived He (Hoke et al., 2000).

The R/R_a values of He in gas-rich spring water in other parts of the South Islands are generally less than 1.5, including the Alpine Fault region, Fiordland, Southland, and the northern part of the South Island. The He discharged at these sites is largely of crustal origin (i.e., enriched in radiogenic ^{4}He) and generally contains less than 20% of the mantle component.

The isotope compositions of He in volcanic gases, hydrothermal fluids, and deep groundwater in other *continental sites* have been investigated by several research groups:

Western USA: Craig et al., 1978a, in Alexander and Ozima (Eds.), *Terrestrial Rare Gases, Adv. Earth Planet. Sci.*, 3:3–16, Cent. Acad.

Publ., Tokyo. Craig et al., 1978b, *Geophys. Res.*, 5:897–900. Poreda et al., 1986, *Geochim. Cosmochim. Acta*, 50:2847–2853. Jenden et al., 1993, *U.S. Geol. Prof. Paper*, 1570:31–56. Kendrick et al., 2001, *Geochim. Cosmochim. Acta*, 65:2651–2668.

Australia: Torgersen and Clarke, 1985, *Geochim. Cosmochim. Acta*, 49:1211–1218. Torgersen et al., 1987, *Geophys. Res. Lett.*, 14: 1215–1218.

New Zealand: Sano et al., 1987, *Geochim. Cosmochim. Acta*, 51:1855–1860. Giggenbach et al., 1993, *Geochim. Cosmochim. Acta*, 57:3427–3455. Hulston and Lupton, 1996, *J. Volcanol. Geotherm. Res.*, 74:297–321. Poreda et al., 1988, *Chem. Geol.*, 71:199–210.

Japan: Matsubayashi et al., 1978, *Bull. Volcanol. Soc. Japan*, 27:21–22. Sano et al., 1982, *Geochem. J.*, 16:237–245. Poreda et al., 1988, *Chem. Geol.*, 71:199–210.

China: Burnard et al., 1999, *Geochim. Cosmochim. Acta*, 64:1595–1604.

Europe: Hilton and Craig, 1989, *Geochim. Cosmochim. Acta*, 53:3311–3316. Weinlich et al., 1999, *Geochim. Cosmochim. Acta*, 63:3654–3671. O'Nions and Oxburgh, 1988, *Earth Planet. Sci. Lett.*, 90:331–347. Hooker et al., 1985, *Geochim. Cosmochim. Acta*, 49:2505–2513. Polyak et al., 2000, *Geochim. Cosmochim. Acta*, 64:1925–1944.

West Africa: Aka et al., 2001, *Appl. Geochem.*, 16:323–338.

21.7 SUMMARY

The study of the geochemistry of He is a multifaceted enterprise. One of these facets involves age determinations of minerals based on the accumulation of α-particles (^{4}He nuclei) released by U, Th, and their daughters. The interpretation of such (U,Th)/He dates is based on the temperature dependence of the diffusion coefficient of ^{4}He in minerals. Laboratory studies have shown that minerals such as apatite, sphene, and hematite do not retain ^{4}He quantitatively until these minerals cool to their closure temperature, which is about 100°C or less, depending on the cooling rate, grain size of the minerals, and their crystal structure. Consequently, (U,Th)/He dates of minerals reflect the *thermal history* of the rocks rather than the time when the minerals originally formed.

This property of (U,Th)/He dates is used in conjunction with Rb–Sr and K–Ar dates of biotite and hornblende to delineate the cooling histories of rocks. The resulting thermochronology of plutonic igneous rocks and metamorphosed volcano–sedimentary complexes indicates their cooling rates, which depend on the uplift of crustal blocks and the erosion of the overlying rocks. In this way, (U,Th)/He dates contribute to the study of the tectonic history of crustal rocks.

Another aspect of He geochemistry is based on the formation of stable ^{3}He by decay of tritium ($^3_1$H), which is produced by cosmic rays in the atmosphere and by thermonuclear explosions. In addition, ^{3}He is produced directly by nuclear spallation reactions caused by cosmic rays in rocks that are exposed at the surface of the Earth and in solid objects in space, such as meteoroids, asteroids, and the surfaces of planets and satellites in the solar system.

The decay of cosmogenic tritium to ^{3}He is one of the few methods for measuring the rate of flow of groundwater and the rate of vertical mixing in the oceans and lakes. The release of thermonuclear tritium between 1952 and 1963 has temporarily upset the steady state in the geochemistry of cosmogenic tritium and has, to some extent, altered the isotope composition of He in groundwater by increasing the abundance of ^{3}He.

The isotope composition of terrestrial He has evolved with time since the formation of the Earth by the addition of radiogenic ^{4}He (α-particles), of tritiogenic, and of cosmogenic ^{3}He. As a result, the atomic ^{3}He/^{4}He ratio of the present atmosphere (1.40×10^{-6}) is about 100 times lower than the ^{3}He/^{4}He ratio of the Q phase of stony meteorites (1.41×10^{-4}, Cold Bokkeveld).

The isotope ratios of He in oceanic basalt of Late Tertiary (Neogene) age are expressed by dividing them by the ^{3}He/^{4}He ratio of the atmosphere. On this scale, the mantle-derived He in oceanic basalt and in the ultramafic xenoliths they contain ranges from about 5 to 10 depending on the source components of magma in the mantle.

The magma sources in the suboceanic mantle have been identified on the basis of the isotopic ratios of Sr, Nd, Pb, Hf, and Os (Table 17.1, Section 17.2). These sources also contain He of characteristic isotope composition that is released into the melt phase during magma formation by partial melting. These He sources include DMM ($R/R_a \sim 8$), EM2 ($R/R_a \sim 2$), HIMU ($R/R_a \sim 5$), and PHEM ($R/R_a \sim 32$–50). The PHEM component is the "primitive He mantle," which still has high $^3He/^4He$ ratios approaching the isotope ratios of He in the Q phase (acid-resistant organic residue) of stony meteorites.

Volcanic rocks and xenoliths erupted on the continents have lower R/R_a ratios than similar rocks in the oceans because of contamination with radiogenic ^4He. The extent of contamination increases with the tectonic age of crustal rocks, causing their R/R_a ratios to decline to less than 0.1 in rocks of Precambrian age.

The He discharged by gas-rich geothermal springs has a wide range of R/R_a values depending on its provenance. In the vicinity of active volcanoes, values of R/R_a between 6 and 10 have been recorded, whereas He in hydrocarbon reservoirs and originating from volcano–sedimentary complexes have $R/R_a < 1.0$. Evidently, the isotope composition of He in different kinds of rocks and in crustal fluids and gases indicates the provenance of this element.

REFERENCES FOR HISTORICAL REVIEW (INTRODUCTION, SECTION 21.1 INTRODUCTION)

Anders, E., and N. Grevesse, 1989. Abundances of the elements: Meteoritic and solar. *Geochim. Cosmochim. Acta*, 53:197–214.

Bender, M. L., 1973. Helium-uranium dating of corals. *Geochim. Cosmochim. Acta*, 37:1229–1247.

Boltwood, B. B., 1907. On the ultimate disintegration products of the radioactive elements. *Am. J. Sci.*, 23:77–88.

Burchfield, J. D., 1975. *Lord Kelvin and the Age of the Earth.* Science History, New York.

Chamberlin, T. C., 1899. Lord Kelvin's address on the age of the Earth as an abode fitted for life. *Science*, 9:889–901; 10:11–18.

Damon, P. E., 1957. Terrestrial helium. *Geochim. Cosmochim. Acta*, 11:200–201.

Damon, P. E., and W. D. Green, 1963. Investigations of the helium age dating method by stable isotope-dilution technique. In *Ratioactive Dating*, pp. 55–71. International Atomic Energy Agency, Vienna.

Damon, P. E., and J. L. Kulp, 1957. Determination of radiogenic helium in zircon by stable isotope dilution technique. *Am. Geophys. Union. Trans.*, 38:945–953.

Damon, P. E., and J. L. Kulp, 1958. Excess helium and argon in beryl and other minerals. *Am. Mineral.*, 43:433–459.

Erickson, G. P., and J. L. Kulp, 1961. Potassium-argon measurements on the Palisade Sill, New Jersey. *Geol. Soc. Am. Bull.*, 72:649–652.

Fanale, F. P., and J. L. Kulp, 1962. The helium method and the age of the Cornwall, Pennsylvania, magnetite ore. *Econ. Geol.*, 57:735–746.

Fanale, F. P., and O. A. Schaeffer, 1965. Helium-uranium ratios for Pleistocene and Tertiary fossil aragonites. *Science*, 149:312–317.

Harper, C. T. (Ed.), 1973. *Geochronology: Radiometric Dating of Rocks and Minerals*, Benchmark Papers in Geology. Dowden, Hutchinson, and Ross, Stroudsburg, Pennsylvania.

Holmes, A., 1911. The association of lead with uranium in rock-minerals, and its application to the measurement of geologic time. *Proc. Roy. Soc. (A)*, 85:248–256.

Holmes, A., 1913. *The Age of the Earth.* Harper and Brothers, London.

Hurley, P. M., 1949. Age of Canada's principal gold-producing belt. *Science*, 110:49.

Hurley, P. M., 1950. Distribution of radioactivity in granites and possible relation to helium age measurements. *Geol. Soc. Am. Bull.*, 61:1–7.

Hurley, P. M., 1954. The helium age method and the distribution and migration of helium in rocks. In H. Faul (Ed.), *Nuclear Geology*, pp. 301–329. Wiley, New York.

Hurley, P. M., and C. Goodman, 1941. Helium retention in common rock minerals. *Geol. Soc. Am. Bull.*, 52:545–560.

Hurley, P. M., and C. Goodman, 1943. Helium age measurements, I: Preliminary magnetite index. *Geol. Soc. Am. Bull.*, 54:305–324.

Lide, D. R., and H. P. R. Frederikse, 1995. *Handbook of Chemistry and Physics*, 76th ed. CRC, Boca Raton, Florida.

Mamyrin, B. A., and I. N. Tolstikhin, 1984. *Helium Isotopes in Nature*. Elsevier, New York.

Schaeffer, O. A., 1970. An evaluation of the uranium-helium method of dating of fossil bones. *Earth Planet. Sci. Lett.*, 7:420–424.

Thomson, W. (Lord Kelvin), 1899. The age of the Earth as an abode fitted for life. *Philos. Mag.*, 47:66–90.

Turekian, K. K., D. P. Kharkar, J. Funkhouser, and O. A. Schaeffer, 1970. An evaluation of the uranium-helium method of dating fossil bones. *Earth Planet. Sci. Lett.*, 7:420–424.

REFERENCES FOR DIFFUSION OF ^{4}He IN MINERALS (SECTIONS 21.1a–21.1b)

Bähr, R., H. J. Lippolt, and R. S. Wernicke, 1994. Temperature-induced ^{4}He degassing of specularite and botryoidal hematite: A ^{4}He retentivity study. *J. Geophys. Res.*, 99(B9):17695–17707.

Dodson, M. H., 1973. Closure temperature in cooling geochronological and petrological systems. *Contrib. Mineral. Petrol.*, 40:259–274.

Farley, K. A., 2000. Helium diffusion from apatite: General behaviour as illustrated by Durango fluorapatite. *J. Geophys. Res.*, 105(B2):2903–2914.

Farley, K. A., R. A. Wolf, and L. T. Silver, 1996. The effects of long alpha-stopping distances on (U-Th)/He ages. *Geochim. Cosmochim. Acta*, 60:4223–4229.

Fechtig, H., and S. Kalbitzer, 1966. The diffusion of argon in potassium-bearing solids. In O. A. Schaeffer and J. Zähringer (Eds.), *Potassium Argon Dating*, pp. 68–106. Springer-Verlag, New York.

Graham, D. W., W. J. Jenkins, M. D. Kurz, and R. Batiza, 1987. Helium isotope disequilibrium and geochronology of glassy submarine basalts. *Nature*, 326:384–386.

Lippolt, H. J., M. Leitz, R. S. Wernicke, and B. Hagedorn, 1994. (U + Th) helium dating of apatite—experience from different geochemical environments. *Chem. Geol.*, 112:179–191.

Lippolt, H. J., and E. Weigel, 1988. ^{4}He diffusion in ^{40}Ar- retentive minerals. *Geochim. Comochim. Acta*, 52:1449–1458.

Lippolt, H. M., R. Bähr, and W. Boschmann-Käthler, 1982. Untersuchungen zur Diffusion von Helium aus Erzmineralien. *Fortschr. Mineral.*, 60(1):129–131.

McDougall, I., and T. M. Harrison, 1999. *Geochronology and Thermochronology by the $^{40}Ar/^{39}Ar$ Method.* Oxford University Press, Oxford.

Meesters, A. G. C. A., and T. J. Dunai, 2002a. Solving the production-diffusion equation for finite diffusion domains of various shapes. Part I. Implications for low temperature (U–Th)/He thermochronology. *Chem. Geol.*, 186:333–344.

Meesters, A. G. C. A., and T. J. Dunai, 2002b. Solving the production-diffusion equation for finite diffusion domains of various shapes: Part II. Application to cases with α-ejection and nonhomogeneous distribution of the source. *Chem. Geol.*, 186:337–348.

Reiners, P. W., and K. A. Farley, 1999. Helium diffusion and (U–Th)/He thermochronometry of titanite. *Geochim. Cosmochim. Acta*, 63:3845–3859.

Warnock, A. C., P. K. Zeitler, R. A. Wolf, and S. C. Berman, 1997. An evaluation of low-temperature apatite U–Th/He thermochronometry. *Geochim. Cosmochim. Acta*, 61:5371–5378.

Wernicke, R. S., and H. J. Lippolt, 1993. Botryoidal hematite from the Schwarzwald (Germany): Heterogeneous uranium distributions and their bearing on the helium dating method. *Earth Planet. Sci. Lett.*, 114:287–300.

Wolf, R. A., K. A. Farley, and D. M. Kass, 1998. Modeling of the temperature sensitivity of the apatite (U–Th)/He thermochronometer. *Chem. Geol.*, 148: 105–114.

Wolf, R. A., K. A. Farley, and L. T. Silver, 1996. Helium diffusion and low-temperature thermochronometry of apatite. *Geochim. Cosmochim. Acta*, 60: 4231–4240.

Zeitler, P. K., A. L. Herczeg, I. McDougall, and M. Honda, 1987. U-Th-He dating of apatite: A potential thermochronometer. *Geochim. Cosmochim. Acta*, 51:2865–2868.

REFERENCES FOR THERMOCHRONOMETRY (SECTION 21.2)

Crowley, P. D., P. W. Reiners, J. M. Reuter, and G. D. Kaye, 2002. Laramide exhumation of the Bighorn Mountains, Wyoming: An apatite (U–Th)/He thermochronology study. *Geology*, 30(1):27–30.

Farley, K. A., M. E. Rusmore, and S. W. Bogue, 2001. Post-10 Ma uplift and exhumation of the northern Coast Mountains, British Columbia. *Geology*, 29(2):99–102.

House, M. A., K. A. Farley, and B. P. Kohn, 1999. An empirical test of helium diffusion in apatite: Borehole data from the Otway basin, Australia. *Earth Planet. Sci. Lett.*, 170:463–474.

House, M. A., B. P. Wernicke, and K. A. Farley, 1998. Dating topography of the Sierra Nevada, California, using apatite (U–Th)/He ages. *Nature*, 396:66–69.

House, M. A., B. P. Wernicke, K. A. Farley, and T. A. Dimitru, 1997. Cenozoic thermal evolution of the central Sierra Nevada, California, from (U–Th)/He thermochronometry. *Earth Planet. Sci. Lett.*, 151:167–179.

McDougall, I., and T. M. Harrison, 1999. *Geochronology and Thermochronology by the $^{40}Ar/^{39}Ar$ Method.* Oxford University Press, Oxford.

McInnes, B. I. A., K. A. Farely, R. H. Sillitoe, and B. P. Kohn, 1999. Application of apatite (U–Th)/He thermochronometry to the determination of the sense and amount of vertical fault displacement at the Chuquicamata porphyry copper deposit. *Econ. Geol.*, 94:937–948.

Reiners, P. W., and K. A. Farley, 2001. Influence of crystal size on apatite (U–Th)/He thermochronology: An example from Bighorn Mountains, Wyoming. *Earth Planet. Sci. Lett.*, 188:413–420.

Spotila, J. A., K. A. Farley, and K. Sieh, 1998. Uplift and erosion of the San Bernardino Mountains associated with transgression along the San Andreas fault, California, as constrained by radiogenic helium thermochronometry. *Tectonics*, 17:360–378.

Stockli, D. F., K. A. Farley, and T. A. Dumitru, 2000. Calibration of apatite (U–Th)/He thermochronometer on an exhumed fault block, White Mountains, California. *Geology*, 28(11):983–986.

REFERENCES FOR He DATING OF IRON ORE DEPOSITS (SECTION 21.3)

Bähr, R., H. J. Lippolt, and R. S. Wernicke, 1994. Temperature-induced ^{4}He degassing of specularite and botryoidal hematite: A ^{4}He retentivity study. *J. Geophys. Res.*, 99(B9):17695–17707.

Brander, T., and H. J. Lippolt, 1999. Das Alter der Roteisenerze in der Verkieselungszone by Rammelsbach (Münstertal, Südschwarzwald) nach (U+Th)/^{4}He-Untersuchungen. *Jh. Landessamt für Geologie, Rohstoffe und Bergbau, Baden-Württemberg*, 38:7–42.

Dodson, M. H., 1973. Closure temperature in cooling geochronological and petrological systems. *Contrib. Mineral. Petrol.*, 40:259–274.

Lippolt, H. J., T. Brander, and N. R. Mankopf, 1998. An attempt to determine formation ages of goethite and limonites by (U+Th)-^{4}He dating. *Neues Jahrbuch der Mineralogie, Monatshefte*, 1998(11):505–528.

Wernicke, R. S., and H. J. Lippolt, 1993. Botryoidal hematite from the Schwarzwald (Germany): Heterogeneous uranium distributions and their bearing on the helium dating method. *Earth Planet. Sci. Lett.*, 114:287–300.

Wernicke, R. S., and H. J. Lippolt, 1994a. Dating of vein specularite using internal (U+Th)/^{4}He isochrons. *Geophys. Res. Lett.*, 21(5):345–347.

Wernicke, R. S., and H. J. Lippolt, 1994b. ^{4}He age discordance and release behavior of a double shell botryoidal hematite from the Schwarzwald, Germany. *Geochim. Cosmochim. Acta*, 58:421–429.

Wernicke, R. S., and H. J. Lippolt, 1997. Evidence of Mesozoic multiple hydrothermal activity in the basement at Nonnenmattweiher (southern Schwarzwald), Germany. *Mineral. Deposita*, 32:197–200.

REFERENCES FOR TRITIUM–^{3}He DATING (SECTION 21.4)

Andersen, L. J., and T. Sevel, 1974. Profiles in the unsaturated and saturated zones, Gronhoj, Denmark. In *Isotope Techniques in Groundwater Hydrology*, Vol. 1, pp. 3–20. International Atomic Energy Agency, Vienna.

Begemann, F., and W. F. Libby, 1957. Continental water balance, groundwater inventory, and storage times, surface-ocean mixing rates, and world-wide water circulation patterns from cosmic ray and bomb tritium. *Geochim. Cosmochim. Acta*, 12:277–296.

Clarke, W. B., W. J. Jenkins, and Z. Top, 1976. Determination of tritium by mass spectrometric measurement of ^{3}He. *Int. J. Appl. Radiat. Isot.*, 27:51–522.

Craig, H., 1957. Distribution, production rate, and possible solar origin of natural tritium. *Phys. Rev.*, 105:1125–1127.

Craig, H., and D. Lal, 1961. The production rate of natural tritium. *Tellus*, 13(1):85–105.

Dincer, T., and G. H. Davis, 1967. Some considerations on tritium dating and the estimates of tritium input function. *Mem. Congress Int. Assoc. Hydrogeol., Istanbul*, 276–286.

Fritz, S. J., R. J. Drimmie, and P. Fritz, 1991. Characterizing shallow aquifers using tritium and ^{14}C: Periodic sampling based on tritium half-life. *Appl. Geochem.*, 6:17–33.

Lide, D. R., and H. P. R. Frederikse, 1995. *Handbook of Chemistry and Physics*, 76th ed. CRC, Boca Raton, Florida.

Moser, H., and W. Rauert, 1980. *Isotopenmethoden in der Hydrologie*. Gebrueder Borntraeger, Berlin.

Nir, A., 1964. On the interpretation of tritium "age" measurements of groundwater. *J. Geophys. Res.*, 69(12):2589–2595.

Plummer, L. N., E. Busenberg, S. Drenkard, P. Schlosser, B. Ekwurzel, R. Weppernig, J. B. McConnell, and

R. L. Michel, 1998a. Flow of river water into a karstic limestone aquifer: 2. Dating the young fraction in groundwater mixtures in the Upper Floridan aquifer near Valdosta, Georgia. *Appl. Geochem.*, 13(8):1017–1043.

Plummer, N. L., E. Busenberg, J. B. McConnell, S. Drenkard, P. Schlosser, and R. L. Michel, 1998b. Flow of river water into a karstic limestone aquifer: 1. Tracing the young fraction in groundwater mixtures in the Floridan aquifer near Valdosta, Georgia. *Appl. Geochem.*, 13(8):995–1015.

Poreda, R. J., T. E. Cerling, and D. K. Solomon, 1988. Tritium and helium isotopes as hydrologic tracers in a shallow unconfined aquifer. *J. Hydrol.*, 103:1–9.

Schlosser, P. M., M. Stute, H. Dörr, C. Sonntag, and K. O. Münnich, 1988. Tritium/He dating of shallow groundwater. *Earth Planet. Sci. Lett.*, 89:353–362.

Schlosser, P., M. Stute, C. Sonntag, and K. O. Münnich, 1989. Tritiogenic ^{3}He in shallow groundwater. *Earth Planet. Sci. Lett.*, 94:245–256.

Solomon, D. K., S. L. Schiff, R. J. Poreda, and W. B. Clarke, 1993. A validation of the ^{3}H/^{3}He method for determining groundwater recharge. *Water Resources Res.*, 29(9):2951–2962.

Solomon, D. K., and E. A. Sudicky, 1991. Tritium and helium 3 isotope ratios for direct estimation of spatial variations in groundwater recharge. *Water Resources Res.*, 27(9):2309–2319.

Suess, H. E., 1969. Tritium geophysics as an international research project. *Science*, 163:1405–1410.

Szabo, Z., D. E. Rice, L. N. Plummer, E. Busenberg, S. Drenkard, and P. Schlosser, 1996. Age dating of shallow groundwater with chlorofluorocarbons, tritium/helium 3, and flow path analysis, southern New Jersey coastal plain. *Water Resources Res.*, 32(4):1023–1038.

Weiss, W., and W. Roether, 1975. Der Tritiumfluss des Rheins, 1961–1973. *Deutsche Gewässerkundliche Mitteilungen*, 19:1–5.

REFERENCES FOR METEORITES AND OCEANIC BASALT (SECTION 21.5)

Busemann, H., H. Baur, and R. Wieler, 2000. Primordial noble-gases in "phase Q" in carbonaceous and ordinary chondrites studied by closed-system stepped etching. *Meteoritics Planet. Sci.*, 35:949–973.

Condomines, M., K. Grönvold, P. J. Hooker, K. Muehlenbachs, R. K. O'Nions, N. Oskarsson, and E. R. Oxburgh, 1983. Helium, oxygen, strontium, and neodymium isotopic relationships in Icelandic volcanics. *Earth Planet. Sci. Lett.*, 66:125–136.

Craig, H., and J. E. Lupton, 1981. Helium-3 and mantle volatiles in the oceanic crust. In C. Emiliani (Ed.), *The Sea*, pp. 391–428. Wiley, New York.

Craig, H., and R. J. Poreda, 1986. Cosmogenic ^{3}He in terrestrial rocks: The summit lavas of Maui. *Proc. Natl. Acad. Sci. USA*, 83:1970–1974.

Farley, K. A., J. H. Natland, and H. Craig, 1992. Binary mixing of enriched and undegassed (primitive) mantle components (He, Sr, Nd, Pb) in Samoan lavas. *Earth Planet. Sci. Lett.*, 111:183–199.

Kurz, M. D., 1986a. Cosmogenic helium in a terrestrial igneous rock. *Nature*, 320:435–439.

Kurz, M. D., 1986b. In situ production of terrestrial cosmogenic helium and some applications to geochronology. *Geochim. Cosmochim. Acta*, 40:2855–2862.

Kurz, M. D., W. J. Jenkins, and S. R. Hart, 1982. Helium isotopic systematics of oceanic islands and mantle heterogeneity. *Nature*, 297:43–47.

Lal, D., 1987. Production of ^{3}He in terrestrial rocks. *Chem. Geol. (Isotope Geosci. Sect.)*, 66(1/2):89–98.

Lewis, R. S., B. Srinivasan, and E. Anders, 1975. Host phase of a strange xenon component in Allende. *Science*, 190:1251–1262.

Mamyrin, B. A., and I. N. Tolstikhin, 1984. *Helium Isotopes in Nature*. Elsevier, Amsterdam.

Ozima, M., and F. A. Podosek, 1983. *Noble Gas Geochemistry*. Cambridge University Press, Cambridge.

Ozima, M., R. Wieler, B. Marty, and F. A. Podosek, 1998. Comparative studies of solar, Q-gases, and terrestrial noble gases, and implications on the evolution of the solar nebula. *Geochim. Cosmochim. Acta*, 62:301–314.

Ozima, M., and S. Zashu, 1983. Noble gases in submarine pillow volcanic glasses. *Earth Planet. Sci. Lett.*, 62:24–40.

Pepin, R. O., and P. Signer, 1965. Primordial rare gases in meteorites. *Science*, 149:253–265.

Poreda, R. J., and H. Craig, 1992. He and Sr isotopes in the Lau basin mantle: Depleted and primitive mantle components. *Earth Planet. Sci. Lett.*, 113:487–493.

Poreda, R. J., and K. A. Farley, 1992. Rare gases in Samoan xenoliths. *Earth Planet. Sci. Lett.*, 113:129–144.

Scarsi, P., 2000. Fractional extraction of helium by crushing of olivine and clinopyroxene phenocrysts: Effects on the ^{3}He/^{4}He measured ratio. *Geochim. Cosmochim. Acta*, 64:3751–3762.

Srinivasan, B., and E. Anders, 1978. Noble gases in the Murchison meteorite: Possible relics of s-process nucleosynthesis. *Science*, 201:51–56.

Vance, D., J. O. H. Stone, and R. K. O'Nions, 1989. He, Sr, and Nd isotopes in xenoliths from Hawaii and other oceanic islands. *Earth Planet. Sci. Lett.*, 96:147–160.

REFERENCES FOR CONTINENTAL CRUST (SECTION 21.6)

Alexander, E. C., Jr., and M. Ozima (Eds.), 1978. *Terrestrial Rare Gases*. Japan Scientific Societies Press, Tokyo, Japan.

Craig, H., J. E. Lupton, and Y. Horibe, 1978. A mantle helium component in circum-Pacific volcanic gases: Hakone, the Marianas, and Mt. Lassen. In E. C. Alexander, Jr., and M. Ozima (Eds.), *Terrestrial Rare Gases. Adv. Earth Planet. Sci.*, 3:3–16. Center for Academic Publications, Japan Scientific Societies Press. Tokyo, Japan.

Dodson, A., D. J. DePaolo, and B. M. Kennedy, 1998. Helium isotopes in lithospheric mantle: Evidence from Tertiary basalts of the western USA. *Geochim. Cosmochim. Acta*, 62:3775–3787.

Dunai, T. J., and H. Baur, 1995. Helium, neon, and argon systematics of the European subcontinental mantle: Implications for its geochemical evolution. *Geochim. Cosmochim. Acta*, 59:2767–2783.

Harland, W. B., R. L. Armstrong, A. V. Cox, L. E. Craig, A. G. Smith, and D. G. Smith, 1989. *A Geologic Time Scale 1989*. Cambridge University Press. Cambridge, United Kingdom.

Hoke, L., R. J. Poreda, A. Reay, and S. D. Weaver, 2000. The subcontinental mantle beneath southern New Zealand, characterized by helium isotopes in intraplate basalts and gas-rich springs. *Geochim. Cosmochim. Acta*, 64:2489–2507.

Hulston, J. R., D. R. Hilton, and I. R. Kaplan, 2001. Helium and carbon isotope systematics of natural gases form the Taranaki Basin, New Zealand. *Appl. Geochem.*, 16:419–436.

Kaneoka, I., N. Takaoka, and K.-I. Aoki, 1978. Rare gases in mantle-derived rocks and minerals. In E. C. Alexander, Jr., and M. Ozima (Eds.), *Terrestrial Rare Gases*, pp. 71–83. Japan Scientific Societies Press. Tokyo, Japan.

Kurz, M. D., W. J. Jenkins, and S. R. Hart, 1982. Helium isotope systematics of oceanic islands and mantle heterogeneity. *Nature*, 297:43–47.

Kyser, T. K., and W. Rison, 1982. Systematics of rare gas isotopes in basic lavas and ultramafic xenoliths. *J. Geophys. Res.*, 87:5611–5630.

Mamyrin, B. A., and I. N. Tolstikhin, 1984. *Helium Isotopes in Nature*. Elsevier, Amsterdam.

Ozima, M., and S. Zashu, 1983. Primitive helium in diamonds. *Science*, 219:1067–1068.

Ozima, M., S. Zashu, and O. Nitoh, 1983. $^3He/^4He$ ratio, noble gas abundance and K–Ar dating of diamonds—an attempt to search for the records of early terrestrial history. *Geochim. Cosmochim. Acta*, 47:2217–2224.

Podosek, F. A. (Ed.), 1985. *Terrestrial Noble Gases. Chem. Geol. (Isotope Geosci. Sect.)*, 52(1):1–125.

Polyak, B. G., and I. N. Tolstikhin, 1985. Isotopic composition of the Earth's helium and the problem of the motive forces of tectogenesis. *Chem. Geol. (Isotope Geosci. Sect.)*, 52:9–33.

Porcelli, D. R., R. K. O'Nions, and S. Y. O'Reilly, 1986. Helium and strontium isotopes in ultramafic xenoliths. *Chem. Geol.*, 54:237–249.

Porcelli, D. R., J. O. H. Stone, and R. K. O'Nions, 1987. Enhanced $^3He/^4He$ ratios and cosmogenic helium in ultramafic xenoliths. *Chem. Geol.*, 64:25–33.

Poreda, R. J., and A. Basu, 1984. Rare gases, water, and carbon in kaersutites. *Earth Planet. Sci. Lett.*, 69:58–68.

Tolstikhin, I. N., B. A. Mamyrin, L. B. Khabarin, and E. N. Ehrlich, 1974. Isotopic composition of helium in ultrabasic xenoliths from volcanic rocks of Kamchatka. *Earth Planet. Sci. Lett.*, 22:75–84.

Torgersen, T., J. E. Lupton, D. S. Sheppard, and W. F. Giggenbach, 1982. Helium isotope variations on the thermal areas of New Zealand. *J. Volcanol. Geotherm. Res.*, 12:283–298.

Torgersen, T., and W. J. Jenkins, 1982. Helium isotopes in geothermal systems: Iceland, the Geysers, Raft River, and Steamboat Springs. *Geochim. Cosmochim. Acta*, 46:739–748.

CHAPTER 22

Radiation-Damage Methods

THE emission of particles and electromagnetic radiation by the nuclei of radioactive atoms causes different kinds of damage in solids that accumulates with time. The amount of damage can be quantified to permit it to be used for age determinations not only of rocks and minerals but also of man-made ceramics and glasses. The dates derived by these methods are temperature sensitive and contribute to the reconstruction of cooling histories of rocks, much like the Rb–Sr and K–Ar dates of mica minerals (Sections 5.4a, 6.6a, and 6.6b) and the (U,Th)/He dates of apatite, hematite, and sphene (Section 21.2). In addition, the radiation-damage methods are suitable for dating archeological objects, bones, and teeth.

22.1 ALPHA-DECAY

The emission of α-particles by nuclei undergoing decay causes recoil of the product nucleus (Section 2.2b). In addition, the α-particles travel through solids until their kinetic energy is exhausted. The distance traveled by α-particles, or "range," depends on their kinetic energy and on the density of the solid medium.

21.1a Pleochroic Haloes

When α-particles are emitted from small U, Th-bearing inclusions in minerals such as muscovite, biotite, tourmaline, and others, the radiation damage caused by the α-particles manifests itself by haloes of discoloration in the host crystal. Such pleochroic haloes were first described by Michael-Levy (1882) almost 15 years before H. Becquerel discovered radioactivity. Subsequently, research by several authors established the connection between pleochroic haloes and radioactivity, which led to attempts to use these haloes to date minerals:

> Joly, 1907, *Philos. Mag.*, 13:381. Mügge, 1909, *Centr. Mineral. Geol.*, 71:113–142. Joly and Rutherford, 1913, *Philos. Mag.*, 25:644. Joly, 1923, *Proc. R. Soc. (Lond.)*, A102:682–705. Holmes, 1926, *Philos. Mag.*, 1:1055. Poole, 1933, *Nature*, 131:154. Henderson, 1934, *Proc. R. Soc. (Lond.)*, 145A:591. Henderson and Bateson, 1934, *Proc. R. Soc. (Lond.)*, A145A:563–581. Henderson and Sparks, 1939, *Proc. R. Soc. (Lond.)*, A173: 238–264.

Pleochroic haloes contain concentric rings with radii between 10 and 50 μm. These rings are produced by α-particles having different kinetic energies emitted by radionuclides residing within the inclusions. The radii of the rings reflect the ranges of the α-particles because the ionizing power of an α-particle is greatest near the end of its range in a particular medium. Biotite is especially well suited to study the ring structure of pleochroic haloes because of its perfect cleavage and because

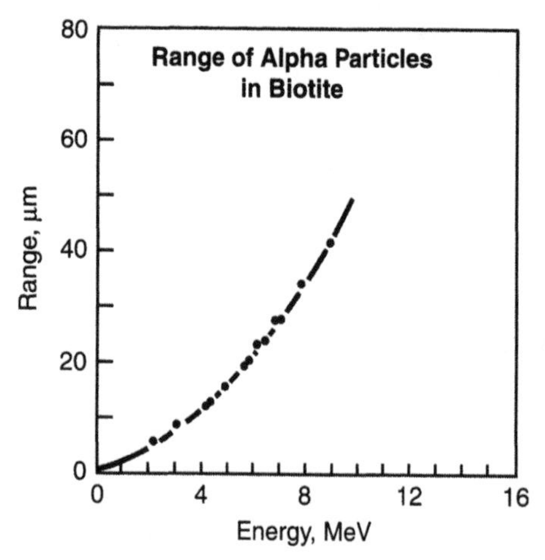

FIGURE 22.1 Range of α-particles in biotite as a function of their kinetic energies in MeV. Data from Picciotto and Deutsch (1960).

the discoloration caused by radiation damage is readily visible. The range of α-particles in biotite in Figure 22.1 increases exponentially with increasing kinetic energies, in accordance with the equation derived by Picciotto and Deutsch (1960) from data in the literature:

$$\ln R = 0.2988E + 1.264 \qquad (22.1)$$

where ln R is the natural logarithm of the range R in units of micrometers and E is the kinetic energy of the α-particles at the point of emission in units of MeV.

The α-emitters can be identified from the radii of the rings of pleochroic haloes. In most cases, the rings were caused by the α-particles emitted by the isotopes of U and Th and their daughters. In some cases, haloes have been described containing rings whose radii correspond exclusively to the kinetic energies of Po isotopes without contributions from other unstable daughters of U and Th. The rings of Po haloes in mica minerals have large radii because the Po daughters of U and Th emit α-particles with high kinetic energies. The most energetic particles (11.650 MeV) are emitted by $^{212m}_{84}$Po(45.1s), which is a short-lived daughter of $^{232}_{90}$Th(Figure 2.14). The range of the Po α-particles in biotite is 115.0 μm according to equation 22.1. The existence of such "giant radiohaloes" has been reported by a number of investigators:

Fowler and Lang, 1977, *Nature*, 270:163–164. Gentry, 1974, *Science*, 184:62–66. Gentry et al., 1974, *Nature*, 252:564–565. Hashemi-Nezhad et al., 1979, *Nature*, 278:333–335. Moazed et al., 1973, *Science*, 180:1272–1274.

These giant haloes have also been attributed to the emission of α-particles associated with spontaneous fission of $^{244}_{94}$Pu(Holbrow, 1977) and to the α-decay by isotopes of the superheavy element at $Z = 126$ (Gentry et al., 1976). However, the relation between the radii of rings in pleochroic haloes and the kinetic energies of α-particles in Figure 22.1 supports the conclusion that the giant haloes are attributable to energetic α-particles emitted by $^{212m}_{84}$Po.

The discoloration of the host mineral (especially biotite) by the radiation damage caused by α-emitters located at the centers of pleochroic haloes intensifies with increasing α-dose. If the flux of α-particles is constant, the dose is a function of time. Consequently, pleochroic haloes can be used for dating the minerals in which they occur (Henderson, 1934). Age determinations by the pleochroic-halo method require the assumptions that the discoloration is a linear function of the α-dose and that the flux of α-particles has remained constant since the time of crystallization of the mineral. The measurements that must be made include:

1. the rate of emission of α-particles from the inclusion at the center of the pleochroic halo and
2. the intensity of the color in the halo.

In addition, the relationship between the α-dose and the resulting intensity of the color of the mineral to be dated must be known (Hayase, 1954; Deutsch et al., 1956; Pasteels, 1960). This relationship can be established by irradiating the mineral with α-particles from an artificial source, as illustrated in Figure 22.2. Although pleochroic haloes have been used for dating biotite and other minerals, the method has several defects:

1. Haloes are partially erased by annealing when minerals are heated during metamorphism.

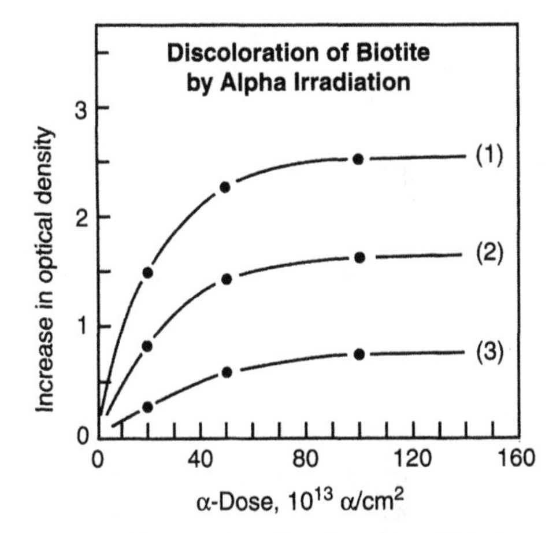

FIGURE 22.2 Progressive discoloration of biotite samples by increasing α-irradiation: (1) Kalule; (2) Baveno; (3) Adamello. The biotite at Kalule is more sensitive to α-irradiation than biotites from Baveno and Adamello. Adapted from Picciotto and Deutsch (1960).

2. The α-flux of the central inclusions is difficult to measure.
3. The relation between the α-dose and the discoloration of biotites varies widely (Figure 22.2) and becomes nonlinear at high α-doses.

Picciotto and Deutsch (1960) concluded from a review of the relevant literature and from their own results that the amount of discoloration of different samples of biotite by an α-dose of 10^{15} α-particles/cm^2 varies by a factor of up to 3.5. These difficulties limit the accuracy of age determinations based on the study of pleochroic haloes. Consequently, this method of dating has been abandoned in favor of the isotopic methods of dating.

22.1b Alpha-Recoil Tracks

The emission of energetic α-particles by atoms of U, Th, and their daughters in silicate minerals imparts enough recoil energy (Section 2.2b) to the product nuclei to displace them from their lattice positions by 30–40 nm. The damaged sites in the affected crystals are more soluble in water, and the ions that reside in the damaged sites are more mobile than ions in undamaged sites (Hurley, 1952; Kigoshi, 1971; Fleischer, 1982, 1983).

The damage caused by the α-particle recoil can be enlarged by etching mineral surfaces with concentrated hydrofluoric acid (Huang and Walker, 1967; Huang et al., 1967). The resulting etch pits have a depth of about 10 nm and are visible in a phase-contrast microscope. The number of etch pits in a unit area of mineral surface depends on the U and Th concentration of the mineral and on the time elapsed since crystallization of the mineral:

Gentry, 1968, *Science*, 160:1228–1268. Katcoff, 1969, *Science*, 166:382–384. Hashimoto et al., 1980, *Nucl. Tracks*, 4:263–269. Hashemi-Nezhad and Durrani, 1981, *Nucl. Tracks*, 5:189–205. Hashemi-Nezhad and Durrani, 1983, *Nucl. Tracks*, 7:141–146.

The use of α-recoil tracks for dating U, Th-bearing solids is constrained by several difficulties that must be dealt with (Gögen and Wagner, 2000):

1. The emission of the first α-particle from the nucleus of ^{238}U, ^{235}U, or ^{232}Th is followed by additional α-particles that are emitted by the nuclei of the respective unstable daughters. The additional α-particles are emitted in all directions, which causes the formation of a damage cluster with a diameter of about 120 nm rather than a single short track.
2. The product nucleus may diffuse away from the original lattice site (e.g., radon) and thereby causes an additional damage track to form when the diffusing Rn atom emits an α-particle.
3. The damaged sites may fade as a result of annealing, especially at elevated temperatures. The temperature sensitivity of the annealing process of α-recoil tracks in different kinds of solids is not well known.
4. The concentrations of U and Th in the mineral to be dated must be measured. Mica minerals, which have favorable etching properties, have low concentrations of U and Th.
5. The areal density of α-recoil tracks in micas increases linearly with the etching time. In addition, the diameter of the etch pits also increases with the duration of exposure to the

etch solution. Ultimately, the areal density of the etch pits increases until they cover the entire mineral surface.

The α-recoil method of dating was studied by Gögen and Wagner (2000) to determine its reliability for dating biotite and phlogopite flakes in volcanic rocks (360–11 ka) from the Eifel region of Germany. They developed a mathematical model to explain the time-dependent enlargement of etch pits and derived an equation that relates the number of etch pits on a cleavage plane of mica to the age of the mineral. This equation (see Gögen and Wagner, 2000) is based on the premise that only the most abundant α-emitters whose halflives are greater than 10^4 years need to be considered: ^{238}U, ^{235}U, ^{232}Th, and their daughters ^{234}U and ^{230}Th. The emission of α-particles by the other daughters merely enlarges the sites that were damaged by the first α-particle emitted by the parent nuclide of each chain.

The results demonstrate a quasi-linear relation between the volume density of α-recoil tracks normalized to U and the known ages of the analyzed mica flakes. Gögen and Wagner (2000) plotted several curves for different values of the Th/U ratio between 4 and 16 and demonstrated that the micas from the Eifel region fit the curve for Th/U = 16. They used this curve to date two phlogopite samples from the Eifel and the Czech Republic:

Ochtendung, Eifel: 230 ± 45 ka
Eisenbühl, Czech Republic: 300 ± 60 ka

Both dates agree with the stratigraphic ages of the rocks with errors of $\pm 20\%$.

The principal source of the analytical error is the determination of U and Th by neutron irradiations using glass standards with low concentrations of U and Th. The closure temperature of biotite and phlogopite for preservation of α-recoil tracks is about 50°C, according to Gögen (1999), who also described the analytical techniques and the interpretation of α-recoil tracks. The application of the α-recoil track method for dating of archeological ceramics was first recognized by Garrison et al. (1978).

Although the α-recoil track method of dating is a variant of the U, Th/He method (Section 21.2), it is not limited by the requirement that all α-particles be retained within the mineral. Consequently, it is potentially applicable to dating different kinds of solids, including not only minerals in rocks but also man-made glass, ceramics, and gems whose ages may range from 10^2 to 10^6 years (Gögen and Wagner, 2000).

22.2 FISSION TRACKS

The spontaneous nuclear fission of ^{238}U in minerals and glasses (Section 2.3a) causes the two product nuclei to be pushed apart with such force that they leave a trail of damage until they finally come to rest. Price and Walker (1962a, b) discovered such fission tracks in samples of mica after etching cleavage planes of this mineral. In the following year, Price and Walker (1963a, b) used induced fission of ^{235}U to measure the U concentrations of micas and proposed that the number of fission tracks on the cleavage planes of mica samples of known U concentration could be used to date such minerals based on the halflife for spontaneous fission of ^{238}U. The reliability of this geochronometer was demonstrated by Fleischer and Price (1964a, b), who reported that fission-track dates of tektites and Libyan Desert glass are in satisfactory agreement with conventional K–Ar dates of the same samples.

In the following year, Fleischer et al. (1965a, b) found tracks in the minerals of meteorites which they attributed to spontaneous fission of ^{238}U and extinct ^{244}Pu. Stony meteorites may also contain damage trails caused by their exposure to cosmic rays which can induce fission of heavy nuclei and cause nuclear spallation reactions. In addition, the high-energy nuclei of heavy elements in cosmic rays can themselves form tracks in the minerals of stony meteorites in interplanetary space. The formation of these kinds of tracks in *stony meteorites* was later investigated by:

Fleischer et al., 1967, *J. Geophys. Res.*, 72: 355–366. Price et al., 1971, *Proc. Second Lunar Sci. Conf., Geochim. Cosmochim. Acta*, Suppl. 2:2621–2627. Carver and Anders, 1976,

Geochim. Cosmochim. Acta, 40:935–944. Rajan and Tamhane, 1982, *Earth Planet. Sci. Lett.*, 58:129–135. Reedy et al., 1983, *Science*, 291:127–135.

The fission-track method is useful for dating samples of certain terrestrial minerals as well as of natural and synthetic glasses that are relatively young and have not been reheated since the time of their formation. These preconditions make the fission-track method useful to both geologists and archeologists (e.g., Hurford et al, 1976; Miller and Wagner, 1981). In addition, particle tracks in solids are used to measure concentrations of U in crystals (Price and Walker, 1963a), in deep-sea sediment (Bertine et al., 1970), and in aqueous solutions (Reimer, 1975). Additional applications were discussed by Wagner (1974, 1998) and Fisher (1975).

The procedure for measuring fission-track dates of minerals was described by Fleischer and Price (1964b), Naeser (1979), and Poupeau (1979, 1981). In addition, Fleischer et al. (1975) and Wagner and van den Haute (1992) summarized the principles and applications of fission-track dating.

The field of nuclear tracks is served by the journal *Nuclear Tracks and Radiation Measurements*, which contains papers dealing with the process of track formation and the methodology of their detection as well as with applications to neutron dosimetry, radon detection, cosmic-ray physics, and geochronometry of geological and archeological samples.

22.2a Methodology

To date a specimen of glass or mineral by the fission-track method, an interior surface is polished and etched with a suitable solvent under appropriate conditions identified in Table 22.1. Small mineral grains can be mounted in clear epoxy resin on a glass slide and can then be ground and polished (Naeser, 1967). After etching, the polished surface is examined with a petrographic microscope (magnification of $800\times - 1800\times$) equipped with a flat-field eyepiece and a graticule to permit counting of tracks in a known area. Fission tracks are readily distinguished by their characteristic tubular shape from other etch pits that result from crystal defects and other causes. Fleischer and Price (1964b) published many excellent photomicrographs of fission tracks.

The track density due to spontaneous fission of ^{238}U is determined by counting a statistically significant number of tracks in a known area of a polished interior surface. Counting becomes difficult when the track density is less than 10 tracks per square centimeter. Actually, many minerals and glasses have high track densities ranging up to thousands of tracks per square centimeter, depending on their concentrations of U and their ages.

Table 22.1. Typical Etching Procedures for Selected Minerals and Volcanic Glass

Mineral	Etching Solution	T, °C	Duration	References[a]
Apatite	Concentrated HNO_3	25	10–30 s	1
	5% HNO_3	20	45 s	2
Sphene	Concentrated HCl	90	30–90 m	1
	1HF : 2HCl : $3HNO_3$: $6H_2O$	20	6 m	3
Zircon	100 N NaOH	270	1.25 h	3
Muscovite	HF (48%)	20	20 m	3
Epidote	6 g NaOH+4 mL H_2O	159	150 m	4
Volcanic glass	HF (24%)	25	60 s	5

[a] 1. Naeser, 1967, *Geol. Soc. Am. Bull.*, 78:1523–1526. 2. Wagner and Reimer, 1972, *Earth Planet. Sci. Lett.*, 14:263–268. 3. Gleadow and Lovering, 1974, *Earth Planet. Sci. Lett.*, 22:163–168. 4. Bar et al., 1974, *Earth Planet. Sci. Lett.*, 22:157–162. 5. Lakatos and Miller, 1972, *Earth Planet. Sci. Lett.*, 14:128–130.

The U concentration of specimens to be dated is determined by irradiating them with thermal neutrons in a nuclear reactor which induces fission of ^{235}U (Section 2.3b). Prior to the irradiation, the previously formed spontaneous-fission tracks are destroyed by annealing at a high temperature. After the irradiation, a new interior surface is polished and etched and the areal density of the induced-fission tracks is determined. The U concentration of the specimens can be calculated from the observed density of the induced-fission tracks and the probability for induced fission of ^{235}U, provided that the thermal neutron flux and the duration of the exposure are known (Gleadow, 1981; Hurford and Green, 1982).

In practice, synthetic glass standards of known U concentrations are irradiated together with the mineral or glass specimens being dated. The number of induced-fission tracks per unit area of an interior surface (ρ_s) of a standard glass is related to its known U concentration (U_s) by:

$$U_s = k\rho_s \quad (22.2)$$

where k is the calibration factor. The resulting value of k is used to determine the unknown U concentrations of the unknown samples that were exposed for the same length of time to the same neutron flux as the standard glass samples. This procedure therefore eliminates the need to know the magnitude of the neutron flux, the reaction cross-section for induced fission by ^{235}U, and the exact duration of the irradiation. The method is limited only by the accuracy of the U concentrations of the synthetic glass standards and by any differences in the way fission tracks are recorded by the standard glasses and the mineral samples being dated.

A further refinement of the procedure is to register the induced-fission tracks in an external detector held against the same surface on which the spontaneous tracks were counted. The external detector may be a sheet of low-U muscovite or a synthetic material like Lexan. After the neutron irradiation, the external detector is etched and the tracks caused by induced fission of ^{235}U in the specimen are counted. The same procedure is used to register the induced-fission tracks in the standard glass that was co-irradiated with the samples being dated.

The use of an external detector eliminates problems caused by the irregular distribution of U and by any changes in the response to etching that were caused by the prior annealing of the unknown minerals or glasses. However, the induced tracks in the external detector originate only from the ^{235}U atoms in the specimen (2π geometry), whereas the induced tracks in the specimen formed on both sides of an interior surface (4π geometry). Therefore, the number of induced tracks per unit area in the external detector is only about one-half as high as it would have been had the induced tracks been counted on an interior surface of the specimen. The benefits and deficiencies of external detectors were debated by Reimer et al. (1970), who questioned their use, whereas Gleadow and Lovering (1977) and Green and Durrani (1978) supported this procedure. The value of the calibration factor k in equation 22.2 must be determined with care when external detectors are used to record the induced-fission tracks in the glass standards because Hurford and Green (1981, 1982) reported considerable scatter in the relationship between the U concentration of standard glasses and the resulting density of induced-fission tracks recorded by external detectors.

The U-bearing glasses used in fission-track dating were prepared by the National Bureau of Standards (NBS) of Washington, D.C., and are identified as Standard Reference Materials (SRM) 610–617 (Carpenter and Reimer, 1974). In addition, the Corning Glass Co. prepared a series of calibrated U–Th glasses (U1–U7) described by Schreurs et al. (1971).

In addition to the analytical difficulties presented above, the principal cause of systematic errors in fission-track dates of natural samples is the fading of tracks caused by annealing of minerals and glasses at elevated temperatures. For this reason, fission-track dates must be interpreted as "cooling ages" because they do not necessarily record the time at which minerals crystallized and volcanic glasses congealed. Instead, fission-track dates are used in combination with K–Ar and Rb–Sr dates (and U–Th/He dates) to reconstruct the cooling histories of plutonic igneous and metamorphic rocks. Although the principles of this method are straightforward, the analytical procedures require a high level of skill and experience.

22.2b Assumptions

Fission-track dating relies on certain assumptions about the minerals and glasses that are analyzed for this purpose:

1. The concentration of U must be sufficient to produce a track density of 10 tracks/cm^2 or higher in the time elapsed since the sample cooled through the "track-retention temperature."
2. The tracks must be stable at ordinary temperatures for time intervals equal to or greater than the age being measured.
3. The material must be sufficiently free of inclusions, defects, and lattice dislocations to permit the etched fission tracks to be recognized and counted.
4. The U must be uniformly distributed to permit its concentration to be determined accurately in a different part of the specimen by the areal density of induced-fission tracks.
5. The fission tracks in the sample to be dated must originate predominantly by spontaneous fission of ^{238}U and the number of tracks produced by induced fission of ^{235}U must be negligible.
6. Chemical weathering of minerals or glasses does not cause fission tracks and/or U to be lost.
7. The halflife of ^{238}U for spontaneous fission must be known accurately.
8. The U concentrations, determined by the slow-neutron irradiation of the specimen to be dated, must be accurate.

The assumption that the tracks are caused predominantly by spontaneous fission of ^{238}U was originally investigated by Price and Walker (1963b), who concluded that the number of tracks caused by spontaneous fission of ^{235}U and ^{232}Th or by induced fission of ^{235}U is negligible in most cases, except in U ore, where neutrons released by spontaneous fission of ^{238}U can induce fission of ^{235}U. The effect of chemical weathering on the retention of fission tracks in apatite, sphene, and zircon (item 6 above) was also found to be negligible by Gleadow and Lovering (1974), although they observed a small decrease in the fission-track date of weathered apatite. They attributed the lowering of fission-track dates to the difficulty in identifying fission tracks in corroded apatite crystals and to track-fading caused by recrystallization of this mineral in the presence of groundwater. The loss of tracks in weathered apatite was partly offset by a decrease of the U concentration.

22.2c Geochronometry

Several different procedures have been devised for measuring the parameters that relate the areal density of spontaneous-fission tracks of a U-bearing mineral to its "age." As a result, the equation for fission-track dating takes several different forms. However, all variants of the geochronometry equation arise from the law of radioactivity, which relates the rate of decay of a radionuclide to the number of atoms of that radionuclide that remain.

The number of ^{238}U atoms in a cubic centimeter of a mineral or glass decreases with time because of decay by α-emission and spontaneous fission. The rate of decay by α-emission is governed by the decay constant λ_α, whose value is:

$$\lambda_\alpha = 1.55125 \times 10^{-10} \text{ y}^{-1}$$

The rate of spontaneous fission is controlled by its own decay constant (λ_f), where:

$$\lambda_f = (8.46 \pm 0.06) \times 10^{-17} \text{ y}^{-1}$$

This value was determined by Galliker et al. (1970) and was later confirmed by Storzer (1970), Wagner et al. (1975), and Thiel and Herr (1976) based on studies of U-rich glasses of known age. The value of the decay constant for spontaneous fission cited above supersedes an earlier determination by Fleischer and Price (1964b), who obtained a value of $\lambda_f = (6.85 \pm 0.20) \times 10^{-17}$ y^{-1} based on a comparison of fission-track dates of tektites with K–Ar dates of the same specimens. This decay constant was later adjusted to $\lambda_f = 8.4 \times 10^{-17}$ y^{-1} by Gentner et al. (1969) and Storzer and Wagner (1969, 1971), who included the effect of the fading of fission tracks as a result of annealing. Nevertheless, both decay constants (6.85×10^{-17} and 8.46×10^{-17} y^{-1}) have been used in the interpretation of fission tracks (Bar et al., 1974; Gleadow and Lovering, 1974). The decay constant determined by Galliker et al. (1970) will be used in this presentation (i.e., $\lambda_f = 8.46 \times 10^{-17}$ y^{-1}).

The number of decays (D) of ^{238}U in a given volume of sample is caused primarily by α-decay because spontaneous fission occurs much more rarely, as indicated by a comparison of the decay constants. Therefore, the number of decays of ^{238}U in time t is:

$$D = {}^{238}\text{U}(e^{\lambda_\alpha t} - 1) \qquad (22.3)$$

where ^{238}U is the number of atoms of this isotope remaining at the present time. The fraction F_s of the decays that are caused by spontaneous fission of ^{238}U and therefore leave a track in the sample is:

$$F_s = \left(\frac{\lambda_f}{\lambda_\alpha}\right) {}^{238}\text{U}(e^{\lambda_\alpha t} - 1) \qquad (22.4)$$

A certain fraction (q) of these tracks cross a polished interior surface and are counted after etching. Therefore, the areal density ρ_s of spontaneous-fission tracks is:

$$\rho_s = F_s q = q\left(\frac{\lambda_f}{\lambda_\alpha}\right) {}^{238}\text{U}(e^{\lambda_\alpha t} - 1) \qquad (22.5)$$

This equation relates the areal density of fission tracks on a polished and etched interior surface of the sample to the number of ^{238}U atoms remaining and to the time t that has elapsed since fission tracks began to accumulate. The track density and the U concentration of the specimen can be measured as described in Section 22.2b. However, equation 22.5 also contains the factor q, which cannot be measured and therefore must be eliminated from the equation.

The removal of q is accomplished by expressing the number of ^{238}U atoms in equation 22.5 by the areal density of the tracks formed by induced fission of ^{235}U. The number of induced fissions F_i is:

$$F_i = {}^{235}\text{U}\phi\sigma \qquad (22.6)$$

where ^{235}U = present number of ^{235} U atoms in a cubic centimeter of the sample

ϕ = dose of thermal neutrons expressed as the neutron flux (n/cm^2/s) multiplied by the exposure time in seconds

σ = cross section for induced fission of ^{235}U by thermal neutrons (580.2×10^{-24} cm^2)

The fraction of induced tracks that cross an interior surface and are counted after etching is also equal to q. Therefore, the areal density ρ_i of induced tracks (F_i) is:

$$\rho_i = F_i q = {}^{235}\text{U}\phi\sigma q \qquad (22.7)$$

The parameter q cancels when equations 22.6 and 22.7 are combined:

$$\frac{\rho_s}{\rho_i} = \left(\frac{\lambda_f}{\lambda_\alpha}\right)\left(\frac{^{238}\text{U}}{^{235}\text{U}}\right)\left(\frac{e^{\lambda_\alpha t} - 1}{\varphi\sigma}\right) \qquad (22.8)$$

Solving for t yields:

$$t = \frac{1}{\lambda_\alpha}\ln\left[1 + \left(\frac{\rho_s}{\rho_i}\right)\left(\frac{\lambda_\alpha}{\lambda_f}\right)\left(\frac{^{235}\text{U}}{^{238}\text{U}}\right)\varphi\sigma\right] \qquad (22.9)$$

which was originally derived by Price and Walker (1963b). Substituting the numerical values of $\lambda_\alpha = 1.55125 \times 10^{-10}$ y^{-1}, $\lambda_f = 8.46 \times 10^{-17}$ y^{-1}, $\sigma = 580.2 \times 10^{-24}$ cm^2, and ^{235}U/^{238}U = 1/137.88 into equation 22.9 transforms it into:

$$t = 6.446 \times 10^9 \ln\left[1 + 7.715 \times 10^{-18}\left(\frac{\rho_s}{\rho_i}\right)\varphi\right] \qquad (22.10)$$

The ratio of the areal densities of spontaneous- and induced-fission tracks (ρ_s/ρ_i) is obtained by counting the two kinds of tracks as described in Section 22.2b. The magnitude of the thermal neutron dose (ϕ) is determined by irradiating glasses of known U concentration with the samples being dated. The areal density of induced-fission tracks counted on an interior surface of the irradiated U-bearing glass standard is used to calculate the neutron dose from equation 22.6. Alternatively, the neutron dose is determined by irradiating so-called flux monitors consisting of known weights of Au, Cu, or Co foil with the unknown samples. After the irradiation, the resulting γ-ray activity of the flux monitors is used to calculate the magnitude of the neutron flux based on the known neutron-capture cross-sections and the efficiency of the γ-ray detector.

The analytical difficulties of the fission-track method of dating arise from the uncertainties associated with the measurement of the thermal neutron dose (ϕ) used to induce fission of ^{235}U, from systematic errors in the decay constant for induced fission (λ_f), and in the reaction cross-section for induced fission of ^{235}U (σ). Therefore, Hurford and Green (1983) combined these parameters into one calibration factor called zeta (Z), whose value they determined by analyzing one or more mineral samples whose ages had been determined by other methods (e.g., K–Ar, $^{40}Ar^*/^{39}Ar$). Accordingly, they defined zeta as:

$$Z = \frac{\varphi \sigma I}{\lambda_f} \qquad (22.11)$$

where $I = {}^{235}U/{}^{238}U = 1/137.88 = 7.2526 \times 10^{-3}$. The geochronometry equation (22.9) is thereby modified to:

$$t = \frac{1}{\lambda_\alpha} \ln\left[1 + \left(\frac{\rho_s}{\rho_i}\right) Z\lambda_\alpha\right] \qquad (22.12)$$

The mineral standard that is used to determine Z is analyzed in the same way as the samples of unknown age. In other words, after the areal density of spontaneous-fission tracks (ρ_s) of the standard have been determined, the spontaneous-fission tracks are annealed and the standard mineral is irradiated with thermal neutrons together with the samples of unknown age. Subsequently, a new interior surface of the standard mineral is exposed, polished, etched, and the areal density of induced-fission tracks (ρ_i) is determined by counting. The resulting value of the track ratio ρ_s/ρ_i of the mineral standard is then used to calculate Z from equation 22.8:

$$\frac{\rho_s}{\rho_i} = \frac{\lambda_f}{\lambda_\alpha}\left(\frac{^{238}U}{^{235}U}\right)\frac{e^{\lambda_\alpha t} - 1}{\varphi\sigma}$$

Given that $Z = \phi\sigma I/\lambda_f$ and $I = {}^{235}U/{}^{238}U$,

$$\frac{\rho_s}{\rho_i} = \frac{e^{\lambda_\alpha t} - 1}{\lambda_\alpha Z}$$

and therefore:

$$Z = \frac{e^{\lambda_\alpha t_m} - 1}{(\rho_s/\rho_i)\lambda_\alpha} \qquad (22.13)$$

where t_m is the known age of the mineral standard. After Z has been determined, it is used to calculate dates for the samples of unknown age that were irradiated with the standard by substitution into equation 22.12.

The resulting fission-track dates depend on the following assumptions:

1. The age of the standard mineral that was used to determine Z is accurate.
2. The fission-track densities of the standard mineral and of the unknowns were determined by the same procedure.
3. The standard and unknowns were exposed together to the same dose of thermal neutrons.

Minerals to be used as calibration standards for fission-track dating must have cooled rapidly and remained unaltered after crystallization. In addition, they must yield concordant dates by different appropriate isotopic methods, and they must be readily obtainable in pure form and in adequate quantity. These criteria are best met by zircons in volcanic rocks and in kimberlite pipes because the latter were intruded explosively into the crust. The standard zircons selected by Hurford and Green (1983) are:

1. zircon from the Fish Canyon tuff in the San Juan Mountains of Colorado, USA, whose age is 27.9 ± 0.7 Ma (Naeser et al., 1981; Hurford and Hammerschmidt, 1985);
2. zircon from the Tardree rhyolite, Antrim Lava Group, Northern Ireland, whose age is 58.7 ± 1.1 Ma;
3. zircon from the Bishop tuff, western USA; age $= 0.74 \pm 0.05$ Ma (Hurford and Hammerschmidt, 1985); and
4. zircon from kimberlite pipes, including the DeBeers, Bultfontein, and Dutoitspan pipes, South Africa; age $= 82 \pm 3$ Ma.

The zeta parameter in fission-track dating serves the same purpose as the *J* parameter in the $^{40}Ar^*/^{39}Ar$ method because both facilitate the calculation of dates by combining variables into one parameter whose value can be determined by analyzing a mineral whose age is already known.

22.2d Track Fading and Closure Temperatures

In most solids, fission tracks are stable for long periods of time at Earth-surface temperature. At higher temperature, fission tracks begin to fade because the damage done by the fission fragments is repaired by a process called annealing (Green et al., 1989b). As a result, the fission tracks are ultimately lost completely at temperatures and heating times that are characteristic for each mineral. However, even incipient track fading causes shortening of tracks and lowers fission-track dates, especially of glass shards in *volcanic tephra* and of minerals that have comparatively low track-retention temperatures (e.g., apatite, calcite). One way to overcome the problem of track fading in glass shards of volcanic tephra is to analyze zircon crystals that may occur in rhyolite tephra and ignimbrites. Examples of fission-track dating of rhyolite-glass shards in *tephra layers* interbedded with sediment of Quaternary age include the work of:

> Hurford et al., 1976, *Nature*, 263:738–740. Hurford and Hammerschmidt, 1985, *Chem. Geol.*, 58:23–32. Seward, 1974, *Earth Planet. Sci. Lett.*, 24:242–248. Wagner et al., 1976, *Neues Jahrbuch Mineral., Monatshefte*, 1976: 84–98. Boellstorff and Te Punga, 1977, *New Zealand J. Geol. Geophys.*, 20:47–58. Seward, 1979, *Geology*, 7:479–482. Kohn et al., 1992, *Paleo. Paleo. Paleo.*, 95:73–94. Seward and Kohn, 1997, *Chem. Geol.*, 141:127–140.

The annealing of fission tracks has been the subject of many studies, some of which are listed in Table 22.2 below, because of its effect on the magnitude of fission-track dates. Although the rate of track fading varies among different minerals and glasses, it always increases with rising temperature. Therefore, minerals and glasses have characteristic track-retention temperatures above which fission tracks do not accumulate. Consequently, two minerals in the same rock specimen can have different fission-track dates depending on the rate of cooling of the host rocks. If the cooling rate is fast (e.g., in volcanic rocks), the fission-track dates of minerals and volcanic glasses approach concordance, whereas a slow cooling rate (e.g., in plutonic or high-grade metamorphic rocks) increases the difference between the fission-track dates of coexisting minerals or glasses.

In general, fission-track dates of minerals in rocks that cooled slowly can be used to determine the cooling rate of the host rock, provided the track-retention temperatures of the minerals are known. In addition, the fission-track dates of certain minerals (e.g., muscovite, apatite, sphene) can be combined with Rb–Sr and K–Ar dates of other minerals in the same volume of crustal rocks to construct cooling histories that record the rate of uplift of the continental crust.

To describe the fading of fission tracks in minerals as a function of increasing temperature, the *reduction* in areal track density of natural materials is measured as a function of temperature and duration of heating. The results of such studies are used to define Arrhenius equations (Section 21.1b) of the form:

$$t = Ae^{U/kT} \tag{22.14}$$

where t = annealing time for a specific reduction in track density
A = constant
U = activation energy, kJ/mol or kcal/mol
k = Boltzmann's constant, $= 1.380662 \times 10^{-23}$ J/K
T = annealing temperature in Kelvins.

Taking natural logarithms of equation 22.14 yields the equation of a straight line in coordinates of $1/T$ (x-coordinate) and $\ln t$ (y-coordinate):

$$\ln t = \ln A + \frac{U}{kT} \tag{22.15}$$

The slope m of this line is:

$$m = \frac{U}{k}$$

Table 22.2. Summary of Procedures for Fission-Track Dating of Minerals and Glasses

Property or Procedure	References
	Methodology
Principles and applications	Gallagher et al., 1998, *Annu. Rev. Earth Planet. Sci.*, 26:519–572.
Track fading	Naeser, 1981, *Nucl. Tracks*, 5:248–250.
Error assessment	Johnson et al., 1979, *Nucl. Tracks*, 3:93–99; Burchart, 1981, *Nucl. Tracks*, 5:87–92.
Counting statistics	Galbraith, 1981, *Math. Geol.*, 13:471; Green, 1981, *Nucl. Tracks*, 5:77–86.
Statistical analysis of mixed dates	Green, 1981, *Nucl. Tracks*, 5:121–128; Galbraith and Laslett, 1993, *Nucl. Tracks*, 5:3–14.
Neutron dosimetry	Hurford and Green, 1981, *Nucl. Tracks*, 5:53–61; Tagami and Nishimura, 1992, *Chem. Geol. (Isotope Geosci. Sect.)*, 102:277–296.
Geometry factor for external detector	Gleadow and Lovering, 1977, *Nucl. Track Detect.*, 1:99–106; Green and Durrani, 1989, *Nucl. Track Detect.*, 2:207–213.
Stability of tracks	Haack, 1976, *Earth Planet. Sci. Lett.*, 30:129–134.
Thermochronology	Lutz and Omar, 1991, *Earth Planet. Sci. Lett.*, 104:181–195; Gallagher, 1995, *Earth Planet. Sci. Lett.*, 136:421–435.
Instrumentation	Smith and Leigh-Jones, 1985, *Nucl. Tracks*, 10:395–400; Jha and Lal, 1984, *Nucl. Tracks*, 9:39–46.
Calibration	Hurford and Gleadow, 1977, *Nucl. Track Detect.*, 1:41–48.
	Zircon
Annealing	Nishida and Takashima, 1975, *Earth Planet. Sci. Lett.*, 27:257–264; Zaun and Wagner, 1985, *Nucl. Tracks Rad. Meas.*, 10:303–307; Tagami et al., 1990, *Chem. Geol. (Isotope Geosci. Sect.)*, 80:159–169; Yamada et al., 1995, *Chem. Geol.*, 119:293–306; Tagami et al., 1996, *Chem. Geol.*, 130:147–157.
Track length	Yamada et al., 1995, *Chem. Geol.*, 122:249–258; Hasebe et al., 1994, *Chem. Geol. (Isotope Geosci. Sect.)*, 112:169–178.
Closing temperature	Tagami et al., 1996, *Chem. Geol.*, 130:147–157.
Etching	Krishnaswami et al., 1974, *Earth Planet. Sci. Lett.*, 22:51–59; Gleadow et al., 1976, *Earth Planet. Sci. Lett.*, 33:273–276; Naeser et al., 1987, *Nucl. Tracks Rad. Meas.*, 13:121–126; Sumii et al., 1987, *Nucl. Tracks Rad. Meas.*, 13:275–277; Yamada et al., 1993, *Chem. Geol. (Isotope Geosci. Sect.)*, 104:251–259.
	Apatite
Annealing	Nagpaul et al., 1974, *Pure Appl. Geophys.*, 112:131–139; Gleadow and Duddy, 1981, *Nucl. Tracks*, 5:169–174; Green et al., 1986, *Chem. Geol. (Isotope Geosci. Sect.)*, 59:237–253; Laslett et al., 1987, *Chem. Geol. (Isotope Geosci. Sect.)*, 65:1–13; Crowley et al., 1991, *Geochim. Cosmochim. Acta*, 55:1449–1465.
Track length	Green, 1981, *Nucl. Tracks*, 5:121–128; Laslett et al., 1987, *Nucl. Tracks*, 9:29–38.

(*continued overleaf*)

Table 22.2. ***(continued)***

Property or Procedure	References
Closing temperature	Haack, 1977, *Am. J. Sci.*, 277:459–464.
Thermochronology	Corrigan, 1991, *J. Geophys. Res.*, 96:10347–10360.
Tectonic interpretation	Wagner and Reimer, 1972, *Earth Planet. Sci. Lett.*, 14:263–268.
Contact metamorphism	Calk and Naeser, 1973, *J. Geol.*, 81:189–198.
Chemical weathering	Gleadow and Lovering, 1974, *Earth Planet. Sci. Lett.*, 22:163–168.
	Sphene
Annealing	Naeser and Faul, 1969, *J. Geophys. Res.*, 74:705–710; Nagpaul et al., 1974, *Pure Appl. Geophys.*, 112:131–139.
Etching	Gleadow, 1978, *Nucl. Track Detect.*, 2:105–177.
	Garnet
Annealing	Haack and Potts, 1972, *Contrib. Mineral. Petrol.*, 34:343–345; Lal (N.) et al., 1977, *Lithos*, 10:129–132.
Closing temperature	Haack, 1977, *Am. J. Sci.*, 277:459–464.
Reliability	Lal (N.) et al., 1976, *Geol. Soc. Am. Bull.*, 87:687–690.
Occurrence	Haack and Gramse, 1972, *Contrib. Mineral. Petrol.*, 34:258–260.
	Epidote and Vesuvianite
Closing temperature	Haack, 1976, *Earth Planet. Sci. Lett.*, 30:129–134; Haack, 1977, *Am. J. Sci.*, 277:459–464.
	Hypersthene
Annealing	Jah and Lal (D.), 1984, *Nucl. Tracks*, 9:39–46.
	Mica
Annealing	Fleischer et al., 1964, *Science*, 143:349–351; Miller, 1968, *Earth Planet. Sci. Lett.*, 4:379–383; Miller and Jäger, 1968, *Earth Planet. Sci. Lett.*, 4:375–378; Mehta and Rama, 1970, *Earth Planet. Sci. Lett.*, 7:82–86; Nagpaul et al., 1974, *Pure Appl. Geophys.*, 112:131–139.
Etching	Price and Walker, 1963, *J. Geophys. Res.*, 68:4847–4862.
	Olivine
Etching	Krishnaswami et al., 1971, *Science*, 174:287–291.
	Volcanic Tephra
Methodology: Glass shards	Wagner et al., 1976, *N. Jahrbuch Mineral., Monatshefte*, 1976:84–98; Hurford et al., 1976, *Nature*, 263:738–740; Seward, 1974, *Earth Planet. Sci. Lett.*, 24:242–248; Seward, 1976, *New Zealand J. Geol. Geosphys.*, 19:9–20; Boellstorff and Te Punga, 1977, *New Zealand J. Geol. Geophys.*, 20:47–58; Gleadow and Duddy, 1981, in, Creswell and Vella (Eds.), *Gondwana V*, pp. 295–300, Balkema, Rotterdam; Nishimura, 1981, *Nucl. Tracks*, 5:157–168; Shane et al., 1995, *Earth Planet. Sci. Lett.*, 130:141–154; Shane et al., 1996, *Geol. Soc. Am. Bull.*, 108:915–925; Kohn et al., 1992, *Paleo. Paleo. Paleo*, 95:73–94.

Table 22.2. (*continued*)

Property or Procedure	References
Methodology: Zircon	Seward, 1979, *Geology*, 7:479–482; Kohn et al., 1992, *Paleo. Paleo. Paleo.*, 95:73–94; Seward and Kohn, 1997, *Chem. Geol.*, 141:127–140.
	Volcanic Glass (Basalt)
Annealing	Macdougall, 1976, *Earth Planet. Sci. Lett.*, 30:19–26.
	Volcanic Glass (Obsidian)
Methodology	Durrani et al., 1981, *Nature*, 238:242–245; Bigazzi and Bonnadonna, 1973, *Nature*, 242:322–323; Miller and Wagner, 1981, *Nucl. Tracks*, 5:147–156; Storzer, 1970, *Earth Planet. Sci. Lett.*, 8:55–60.
	Calcite
Methodology	Sippel and Glover, 1964, *Science*, 144:409–411; Macdougall and Price, 1974, *Science*, 185:943–944.
	Standards
Calibration of glasses	Carpenter and Reimer, 1974, *Natl. Bur. Stand. Pub.* 260–49; Naeser et al., 1980, *U.S. Geol. Surv. Bull*, 1489:1–31; Schreurs et al., 1971, *Radiation Effects*, 7:231–233.
Neutron dosimetry	Green and Hurford, 1984, *Nucl. Tracks*, 9:231–241; Tagami and Nishimura, 1992, *Chem. Geol. (Isotope Geosci. Sect.)*, 102:277–296.
Standardization of calibration	Hurford, 1990, *Chem. Geol. (Isotope Geosci. Sect.)*, 80:171–178.
Zeta-calibration	Hurford and Green, 1983, *Chem. Geol. (Isotope Geosci. Sect.)*, 59:343–354.
Track-length studies	Laslett et al., 1982, *Nucl. Tracks*, 6:79–85.
External track detectors	Gleadow and Lovering, 1977, *Nucl. Track Detect.*, 1:99–106.
Bishop Tuff and Fish Canyon Tuff	Hurford and Hammerschmidt, 1985, *Chem. Geol.*, 58:23–32.

and the intercept b on the y-axis is:

$$b = \ln A$$

The linear relationship between $\ln t$ and $1/T$ permits laboratory measurements to be extrapolated from elevated temperatures and short annealing times to lower temperatures and long intervals of geological time.

Naeser and Faul (1969) used this technique to interpret the results of their study of track fading in apatite and sphene. Their results in Figure 22.3 demonstrate that fission tracks in these minerals fade at quite different rates. For example, apatite begins to lose tracks at 50°C when it is heated for one million years and it loses all tracks when heated at 175° for the same length of time. Under the same conditions, sphene begins to lose tracks at 250°C and is completely annealed (100% loss of tracks) at 420°C. Therefore, when fission-track dates of these two minerals in the same rock samples are concordant, the rock cooled rapidly. In cases where the fission-track dates are discordant, the rock either cooled slowly or was reheated to a temperature at which track-fading occurred in apatite but not in sphene.

The exact temperature at which U-bearing minerals retain all tracks (0% loss in Figure 22.3)

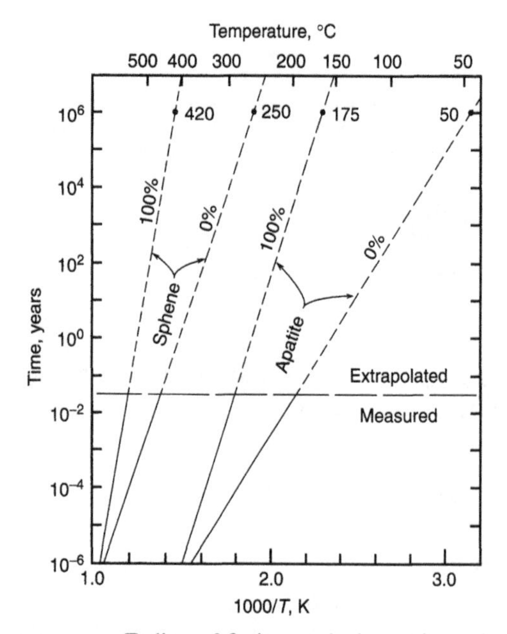

FIGURE 22.3 Fading of fission tracks in apatite and sphene. The lines marked 0% indicate temperatures and time periods at which no tracks are lost. The lines labeled 100% indicate conditions when all tracks are lost. All fission tracks in apatite are lost when that mineral is heated for one million years at 175°C. Sphene does not begin to lose tracks until the temperature is raised to 250°C for one million years, and it anneals completely only at 420°C. Upon cooling from a high temperature, tracks begin to be accumulated along the line for 100% track loss. For example, if apatite cooled from 175 to 50°C in one million years, some tracks begin to accumulate at 175°C, but complete retention does not occur until the temperature reaches 50°C. The effective temperature recorded by the fission-track date of a mineral is the value at which about 50% of the tracks are preserved (Naeser and Faul, 1969).

depends on the cooling rate. When the rate of cooling is high, the cooling time is short, and complete track retention occurs at a higher temperature than when cooling rate is slow. This phenomenon is related to the diffusion rate of ions in the solid. A short cooling time associated with a rapid cooling rate prevents ions from diffusing to repair the trail of damage in the crystal and therefore allows tracks to be retained at a higher temperature than would be the case for a slow cooling rate. Regardless of the cooling rate, fission-track dates can be interpreted as the time elapsed since they cooled through their 50% track-retention temperatures, which is analogous to the blocking temperature for accumulation of radiogenic ^{87}Sr, ^{40}Ar, and ^{4}He in minerals.

The 50% track-retention temperatures of various rock-forming minerals in Figure 22.4 increase with increasing cooling rates between 0.1°C/10^6

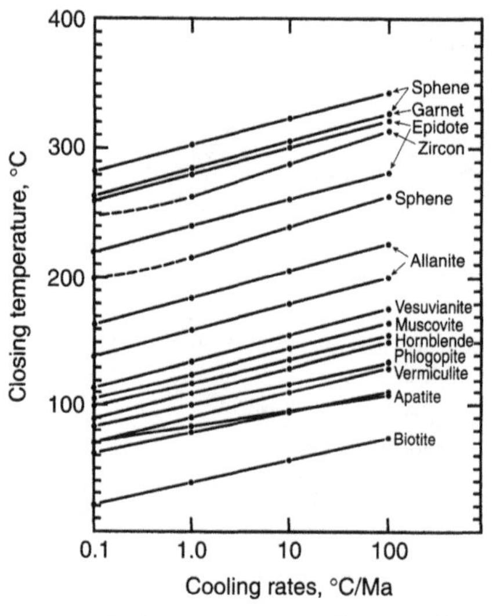

FIGURE 22.4 Closing temperatures for retention of fission tracks for minerals cooling at different rates. The closing temperature is defined as the temperature below which all fission tracks are retained in the mineral, which corresponds to the 50% track-retention temperature. Different closing temperatures have been reported for sphene, epidote, allanite, vermiculite, and apatite primarily because of differences in the etching procedures used to reveal partially annealed fission tracks. The closing temperature of chlorite is similar to that of apatite. The data are based primarily on a survey of the literature by Sharma et al. (1980) and on results published by Haack (1977) for vesuvianite and Bal et al. (1983) for sphene and zircon.

years and 100°C/10^6 years. In addition, these temperatures range widely from about 320°C for sphene to about 55°C for biotite at a cooling rate of 10°C per million years. The experimentally determined blocking temperatures depend on the temperature and duration of etching, on the kind of etching solution used, and on the chemical compositions of minerals. For these reasons, fission-track dates of minerals should be determined by the same etching procedure as that used to determine their closing temperatures.

22.2e Plateau Dates

The fading of fission tracks causes the dates derived from them to be lower than the geological age of the minerals being dated. In some cases, the discrepancy can be reduced by employing longer etching times or by using etchants that are more effective in retrieving partially annealed tracks. To some extent, the effect of track fading is taken into consideration when fission-track dates are interpreted as the time elapsed since the temperature dropped below the 50% track-retention value. However, fission-track dates are also used to determine the "ages" of minerals and archeological objects that may not be datable by the isotopic methods. For this reason, several procedures have been developed to correct fission-track dates for the effects of track fading (Galbraith and Laslett, 1993).

The damage trail left by a charged particle anneals first at the ends so that the etched tracks become shorter (Laslett et al., 1987a, b). This effect was originally used by Bigazzi (1967) to correct fission-track dates of muscovite. Later, more refined procedures were developed that depend on the observed shortening of induced-fission tracks under controlled laboratory conditions (Storzer and Wagner, 1969; Durrani and Khan, 1970; Mehta and Rama, 1970).

The observed distribution of the lengths of spontaneous-fission tracks in minerals can be used to refine the interpretation of dates derived from them. Minerals that have well-defined unimodal track-length distributions yield reliable cooling "ages." Partial annealing of tracks is recognizable by the skewness of the length distribution, in which case the fission-track date may be too low because some tracks may not be recovered by etching. In addition, minerals that were reheated after initial cooling may now have bimodal track-length distributions, which causes dates based on the areal density of tracks to be systematically in error.

The so-called plateau method of dating is an attempt to correct fission-track dates for the effects of partial track-fading (Storzer and Poupeau, 1973). It is based on the fact that induced-fission tracks begin to anneal at a lower temperature or in a shorter time than spontaneous-fission tracks that have already begun to fade in the natural environment. Therefore, the ratio of spontaneous to induced tracks increases when a mineral is annealed by stepwise heating in the laboratory.

In the procedure for plateau dating outlined by Poupeau et al. (1979) and Poupeau (1981), an aliquot of the sample is completely annealed by heating and is then irradiated with thermal neutrons to form tracks by induced fission of ^{235}U. The aliquot that was annealed and irradiated is heated together with another aliquot of the same sample that was not treated and therefore contains fission tracks formed by spontaneous fission of ^{238}U. The two aliquots are heated for one hour at stepwise increasing temperatures in order to cause the tracks to fade. After each heating step, the aliquots are mounted in epoxy, polished, and etched, and the areal densities of the induced and spontaneous tracks are determined by counting. The highest temperature to which the aliquots are heated in this procedure is the temperature at which all tracks are lost in one hour. Alternatively, Dakowski et al. (1974) heated the two aliquots at a fixed temperature for increasing intervals of time.

The procedure outlined above is based on the premise that induced- as well as spontaneous-fission tracks in a sample of glass or mineral begin to fade above a certain temperature θ_1. If the sample had cooled rapidly and was not reheated subsequently, the spontaneous-fission tracks in Figure 22.5*a* begin to fade at the same temperature as the induced tracks and the ratio of the track densities remains constant as the annealing temperature is raised stepwise. In this case, the fission-track date calculated from the ratio of track densities by equation 22.9 remains constant. However, when the tracks caused by spontaneous

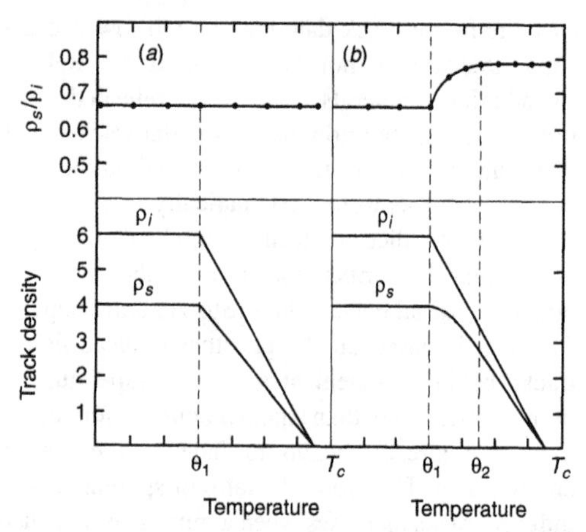

FIGURE 22.5 (*a*) Variation of track densities during stepwise annealing of a mineral whose natural tracks had *not* begun to fade because it cooled rapidly. In this case, track fading starts after heating for one hour at temperature θ_1, but the ratio ρ_s/ρ_i and the corresponding track date remain constant until all tracks fade in one hour at temperature T_c. (*b*) A material whose natural tracks have begun to fade because of slow cooling anneals more slowly than one containing only the induced tracks. As a result, the ratio of track densities increases at temperatures greater than θ_1 and reaches a constant plateau at θ_2. The dates calculated from ρ_s/ρ_i at annealing temperatures above θ_2 are older than those obtained below θ_1 and constitute the "fission-track plateau" date. The increase in ρ_s/ρ_i in this figure was calculated from track densities read at intervals from the annealing curves. Modified after Poupeau et al. (1979).

fission are partially annealed to begin with, their areal density ρ_s decreases more gradually than the density of tracks caused by induced fission of ^{235}U and the ratio of the track densities (ρ_s/ρ_i) in Figure 22.5*b* *increases* with increasing annealing temperature. At some higher temperature θ_2, the ratio approaches a constant value and the dates calculated from this value therefore form a "plateau." The plateau date determined by this procedure contains a correction for partial track-fading during slow cooling and, therefore, represents the time elapsed since the sample cooled through the *zero track-fading* temperature.

The plateau method of dating is quite labor intensive and is only used in cases where track-fading introduces a systematic error in the calculated date. To detect such errors, the length of fission tracks is recorded and is used to evaluate the reliability of fission-track dates. The plateau method of dating was used successfully by Izett et al. (1978) to date volcanic glasses and was extended to the dating of minerals by Poupeau et al. (1978, 1979).

The topics that arise in the fission-track dating of minerals and glasses are compiled in Table 22.2 together with a selection of the relevant references. The large number of references dealing with this topic have resulted from the intense effort by investigators to make the fission-track method work as a thermochronometer and as a geochronometer of geological materials of Phanerozoic age, including samples of archeological and anthropological origin.

22.3 APPLICATIONS OF FISSION-TRACK DATES

The fission-track method of dating, like the U–Th/He method, primarily records the temperature history of crustal rocks based on the track-retention temperatures of the minerals they contain. The resulting thermochronologies reveal not only the cooling rates of the rocks but also the depth of burial of sediment, the sedimentation and subsidence rates, and the rates of uplift and erosion. This information is relevant to studies of the evolution of sedimentary basins and the tectonic histories of orogenic belts (Church and Bickford, 1971; Gleadow et al., 1983; Laslett et al. 1987a, b; Gallagher, 1995; Gallagher et al., 1998).

22.3a The Catskill Delta of New York

The Late Devonian strata of the Catskill Mountains of New York contain a poorly sorted feldspathic sandstone whose stratigraphic age is 374 Ma based on the timescale of Palmer (1983). Lakatos and Miller (1983) analyzed several grains of apatite and zircon from this layer for dating by the fission-track method. One of the five apatite grains yielded the following results:

$$\rho_s = 1.2 \times 10^6 \text{ tracks/cm}^2$$

$$\rho_i = 9.4 \times 10^5 \text{ tracks/cm}^2$$

$$\phi = (2.11 \pm 0.06) \times 10^{15} \text{ neutrons/cm}^2$$

In addition, $\lambda_\alpha = 1.55125 \times 10^{-10}\ \text{y}^{-1}$, $\lambda_f = 8.46 \times 10^{-17}\ \text{y}^{-1}$, $^{235}\text{U}/^{238}\text{U} = 1/137.88 = 7.2526 \times 10^{-3}$, and $\sigma = 580.2 \times 10^{-24}\ \text{cm}^2$. Substituting into equation 22.9 yields:

$$t = \frac{1}{\lambda_\alpha} \ln\left[1 + \left(\frac{\rho_s}{\rho_i}\right)\left(\frac{\lambda_\alpha}{\lambda_f}\right)\left(\frac{^{235}\text{U}}{^{238}\text{U}}\right)\varphi\sigma\right]$$

$$t = \frac{1}{1.55125 \times 10^{-10}} \ln\left(1 + \frac{1.2 \times 10^6}{9.4 \times 10^5} \times \frac{1.55125 \times 10^{-10}}{8.49 \times 10^{-17}} \times 7.2526 \times 10^{-3} \times 2.11 \times 10^{15} \times 580.2 \times 10^{-24}\right)$$

$$t = \frac{\ln\ 1.02077}{1.55125 \times 10^{-10}} = 132\ \text{Ma}$$

The average fission-track date of six detrital apatite grains analyzed by Lakatos and Miller (1983) was 125 Ma, which is 249 million years less than the stratigraphic age of the sandstone bed in which they occurred. Therefore, all of the fission tracks in these apatite grains were lost by annealing after they were originally deposited. Subsequently, the apatite grains cooled through their track-retention temperature at 125 Ma and continued to cool to the ambient Earth-surface temperature. Lakatos and Miller (1983) used the data of Gleadow and Duddy (1981) and Naeser (1981) to conclude that a temperature of 120 ± 20°C was sufficient to completely anneal the apatite grains in 249 million years (i.e., 374–125 Ma).

Lakatos and Miller (1983) also dated four detrital zircon grains from the same sandstone layer and reported an average fission-track date of 320 Ma, which is only 54 million years less than the stratigraphic age of the sandstone. Therefore, the zircon grains were also completely annealed by an increase in temperature after initial deposition at about 374 Ma and cooled through

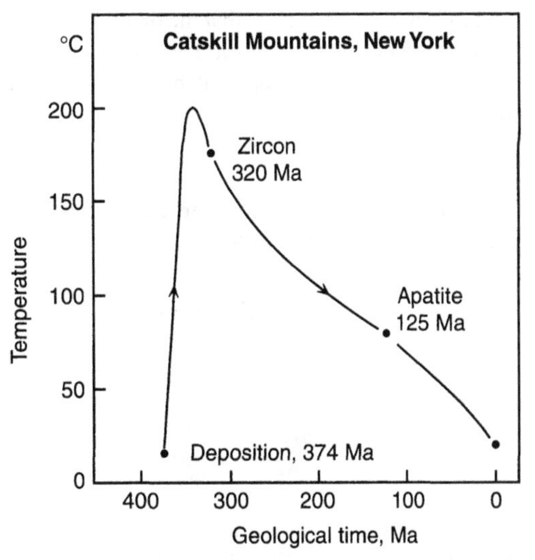

FIGURE 22.6 Thermochronology of a sandstone bed of earliest Late Devonian age in the Catskill Mountains of New York. The age of the sandstone and the track-retention temperature of the apatite have been adjusted in light of recent evidence referred to in Table 22.2. Data from Lakatos and Miller (1983).

their track-retention temperature at 320 Ma, or 54 million years after deposition. According to data by Harrison et al. (1979) cited by Lakatos and Miller (1983), the temperature required to anneal the zircon grains in about 250 million years is greater than 175°C.

The thermochronology of the sandstone bed in Figure 22.6 started with deposition at the surface of the Earth at 374 Ma. The temperature subsequently increased until it exceeded 175°C, causing both the detrital zircon and apatite grains to be annealed. After reaching the peak temperature, the rock cooled, allowing track retention of the zircon grains at 320 Ma (175°C) and of the apatite at 125 Ma (adjusted to 80°C). The average integrated cooling rate dT/dt between 320 and 125 Ma was:

$$\frac{dT}{dt} = \frac{175 - 80}{320 - 125} = \frac{95}{195} = 0.5°\text{C}/10^6\ \text{y}$$

The depth of burial of the sandstone bed can be estimated from the geothermal gradient. Lakatos and Miller (1983) assumed that the thermal gradient was 25°C/km and that the surface temperature was 20°C, which yields a depth of burial d based on the equation:

$$T = 20 + 25d \tag{22.16}$$

where d is expressed in kilometers and T in degrees Celsius. Therefore, at 320 Ma:

$$d = \frac{175 - 20}{25} = 6.2 \text{ km}$$

and at 125 Ma:

$$d = \frac{120 - 20}{25} = 4.0 \text{ km}$$

Consequently, the rate of uplift r of the Catskill Mountains between 320 and 125 Ma was:

$$r = \frac{(6.2 - 4.0) \times 10^6}{(320 - 125) \times 10^6} = 0.011 \text{ mm/y}$$

The uplift rate between 125 Ma and the present was:

$$r = \frac{4.0 \times 10^6}{125 \times 10^6} = 0.032 \text{ mm/y}$$

The fission-track dates also permit an estimate of the average time-integrated rate of sediment deposition. The interpretation presented above suggests that more than 6.2 km of sediment was deposited in about 54 million years (374–320 Ma). Therefore, the rate of deposition of sediment and subsidence of the basin (s) was approximately:

$$s = \frac{6.2 \times 10^6}{54 \times 10^6} = 0.11 \text{ mm/y}$$

Sedimentation rates of this order of magnitude are consistent with the geological setting of the Catskill tectonic delta complex located west of the Appalachian Mountains (Friedman and Sanders, 1978).

22.3b Damara Orogen, Namibia

The cooling history of orogenic belts after the peak metamorphic temperature was reached can be reconstructed by means of temperature-sensitive Rb–Sr and K–Ar dates of minerals (e.g., biotite) and by use of fission-track dates of minerals having different track retention temperatures. The Damara orogen of Namibia, which formed during the Pan-African orogeny, consists of a broad belt of northeast-to-southwest trending intensely folded metamorphic rocks and intrusions of granite. The orogen has been subdivided into three parallel belts, including the southern and central belts, which are separated from each other by the Okahandja Lineament. Haack (1983) estimated that the geothermal gradients of the southern belt at the time of peak metamorphism increased in a northwesterly direction from 20°C/km along the southern margin of the Damara orogen to 51°C/km at the Okahandja Lineament. The central belt northwest of the Lineament had uniformly high geothermal gradients between 51 and 60°C/km. Consequently, the 600°C isotherm of the southern belt occurred at a greater depth (~21 km) than in the central belt (~11 km), from which Haack (1983) concluded that the southern belt was later uplifted by about 10 km relative to the central belt.

The plutons of the Donkerhuk granite, which intruded the Okahandja Lineament soon after the peak of regional metamorphism, yielded an average whole-rock Rb–Sr isochron date of 522 ± 17 Ma, at which time the temperature was about 660°C. The K–Ar dates of biotite (closure temperature 300°C) in the central and southern zones vary regionally and with elevation above sealevel (asl). However, Haack (1983) showed that the average K–Ar date of biotites at a reference elevation of 1300 m above sealevel is 500 ± 15 Ma. Therefore, the rocks of the southern belt that now reside at this elevation had cooled to 300°C at 500 ± 15 Ma. The Rb–Sr dates of three biotites (closure temperature 300–350°C) from granites in the Okahandja Lineament have an average value of 485 Ma, which is in good agreement with the average K–Ar date at 1300 m asl. Therefore, the effective cooling rate (dT/dt) of the southern belt of the Damara orogen between 522 and 485 Ma was:

$$\frac{dT}{dt} = \frac{660 - 300}{522 - 485} = \frac{360}{37} = 9.7\text{°C}/10^6 \text{ y}$$

which Haack (1983) rounded to 10°C/10^6 y.

Six fission-track dates of andradite garnets south of the Okahandja Lineament are close to 500 Ma compared to about 370 Ma at 1300 m asl in the central belt north of the lineament. The fission-track dates of apatite vary similarly from 330 Ma in the southern belt to about 130 Ma in the central belt, with a narrow transition zone only about 10 km wide along the Okahandja Lineament.

Haack (1983) interpreted the fission-track dates based on the 50% track-retention temperatures of the minerals:

Garnet (andradite):	270°C
Apatite:	80°C

The resulting cooling curve in Figure 22.7 demonstrates that the rocks of the Damara orogen cooled at the rate of $10°C/10^6$ y between 522 and 485 Ma followed by more gradual cooling recorded by the fission-track dates of garnet and apatite:

Southern belt:	$1.1°C/10^6$ y
Central belt:	$0.8°C/10^6$ y

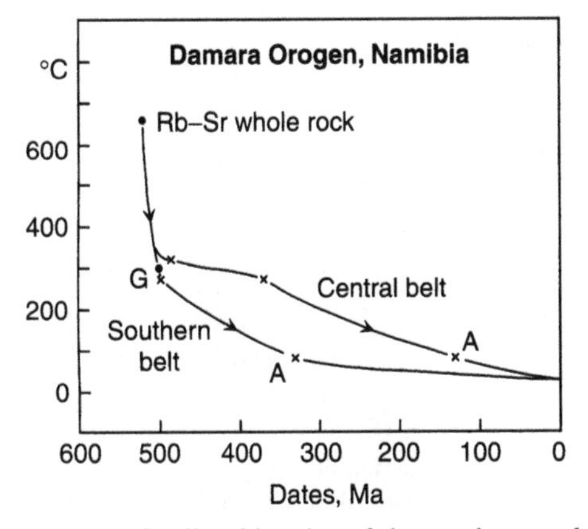

FIGURE 22.7 Cooling histories of the southern and central belts of the Damara orogen in Namibia based on whole-rock Rb–Sr dates of granite and K–Ar and Rb–Sr dates of biotite (solid circle and bar, respectively) supplemented by fission-track dates (crosses) of garnet (G) and apatite (A). The reheating of the rocks caused by the deposition of Karoo lavas is not shown. Data from Haack (1983).

The rocks in both belts ultimately reached their present Earth-surface temperature of about 15°C.

Fission-track dates have been used extensively to determine cooling rates of crustal rocks in many parts of the Earth. Additional studies of this kind are listed in Table 22.3 with references to the literature. In several cases, the thermal histories have been used to evaluate the potential for the generation of petroleum and natural gas (e.g., Green et al., 1989a). In addition, fission-track dates have been used to determine the provenance of obsidian (Durrani et al., 1971) and zircon (Naeser N.D. et al., 1987) and to date U ore deposits (Weiland et al., 1980; Rosenberg and Hooper, 1982). References to age determinations of archeological objects by the fission-track method are listed separately in Table 22.3.

22.4 THERMOLUMINESCENCE

Nonconducting crystalline solids emit electromagnetic radiation when they are heated or exposed to light. This phenomenon was originally observed by Robert Boyle more than 300 years ago and was rediscovered more recently by Wiedemann and Schmidt (1895). Nearly 60 years later, Daniels and Saunders (1950) and Daniels et al. (1953) proposed that the phenomenon of thermoluminescence (TL) could be used for dating and Grögler et al. (1958) demonstrated the feasibility of this proposal by dating samples of ceramics and bricks.

Subsequently, TL has been widely used to date ceramics and sediment ranging in age from a few hundred years to about one million years with errors of about 10%. Therefore, the range of TL dating exceeds that of the ^{14}C method, which is limited to about 70,000 years (at best) with errors that arise from the uncertainty of the initial ^{14}C activity and other causes to be discussed in Chapter 23.

The TL method of dating has become a useful and reliable research tool in archeology and Quaternary geology (Dreimanis et al., 1978). However, this geochronometer does not lend itself to routine applications but requires skill in selecting and analyzing the samples and experience in the interpretation of the results. The principles

Table 22.3. Applications of Fission-Track Dates to the Solution of Geological Problems

Region or Problem	References
	North America
Catskill Mountains, New York, USA	Lakatos and Miller, 1983, *Geology*, 11:103–104.
Coastal pluton British Columbia, Canada	Harrison et al., 1979, *Can. J. Earth. Sci.*, 16:400–410.
Nuclear explosion, Nevada, USA	Fleischer et al., 1974, *J. Geophys. Res.*, 79:339–342.
Granite, Sawatch Range, Colorado	Church and Bickford, 1971, *Geol. Soc. Am. Bull.*, 82:1727–1734.
Igneous and metamorphic rocks, North America	Miller, 1968, *Earth Planet. Sci. Lett.*, 4:379–383.
Yukon Territory and Alaska	Naeser (N.D.), et al., 1982, *Can. J. Earth Sci.*, 19:2167–2178; Briggs and Westgate, 1978, U.S. Geol. Surv. Open-File Rept. 78–701, pp. 49–52.
Northcentral Nevada	Naeser and McKee, 1970, *Geol. Soc. Am. Bull*, 81:3375–3384.
Brooks Range, Alaska	O'Sullivan et al., 1997, *J. Geophys. Res.*, 102:20821–20845.
	Greenland
East Greenland	Gleadow and Brooks, 1979, *Contrib. Mineral. Petrol.*, 71:45–60.
	Europe
Swiss Alps	Miller and Jäger, 1968, *Earth Planet. Sci. Lett.*, 4:375–378.
Lepontine Alps, Switzerland	Hurford, 1986, *Contrib. Mineral. Petrol.*, 92:413–427.
Gotthard region, Alps, Switzerland	Schaer et al., 1975, *Tectonophysics*, 29:293.
Northern England	Green, 1986, *Geol. Mag.*, 123:483–506.
Vienna basin, Austria	Tagami et al., 1996, *Chem. Geol.*, 130:147–157.
Lippari Islands, Italy	Bigazzi and Bonnadonna, 1973, *Nature*, 242:322–323.
	Australia
Granitic rocks, Victoria	Gleadow and Lovering, 1978, *Geol. Soc. Australia J.*, 25:323–340.
Southeastern Australia	Moore et al., 1986, *Earth Planet. Sci. Lett.*, 78:255–270; Gleadow and Duddy, 1981, in Creswell and Pella (Eds.), *Gondwana V*, pp. 295–300, Balkema, Rotterdam.
Tasman rift, Southeastern Australia	Morley et al., 1981, in Jäger and Hunziker (Eds.), *Lectures in Isotope Geology*, pp. 154–159. Springer-Verlag, Heidelberg.
Exploration for hydrocarbons	Green et al., 1989, in Naeser (N.D.) and McCulloh (Eds.), *Thermal Histories of Sedimentary Basins*, pp. 181–195. Springer-Verlag, New York.
	New Zealand
Wanganui basin, North Island	Seward and Kohn, 1997, *Chem. Geol.*, 141:127–140; Seward, 1974, *Earth Planet. Sci. Lett.*, 24:242–248; Shane et al., 1995, *Earth Planet. Sci. Lett.*, 130:141–154; Shane et al., 1996, *Geol. Soc. Am. Bull*, 109:915–925.

Table 22.3. ***(continued)***

Region or Problem	References
	South America
Andes, Colombia and Bolivia	Herd and Naeser, 1974, *Geology*, 2:603–604; Crough, 1983, *Earth Planet. Sci. Lett.*, 64:396–397.
Sierra Pampeanas, Argentina	Coughlin et al., 1998, *Geology*, 26:999–1002; Jordan et al., 1989, *J. South Am. Earth Sci.*, 2:207–222.
	Arabian Peninsula
Oman Mountains, Oman	Poupeau et al., 1998, *Geology*, 26:1139–1142.
	India
South India	Nagpaul and Mehta, 1975, *Am. J. Sci.*, 275:753–762.
Rajasthan	Lal (N.) et al., 1976, *Geol. Soc. Am. Bull.*, 87:687–690.
Pandoh-Baggi, Himalaya, Himachal	Saini et al., 1978, *Tectonophysics*, 46:87.
	Pakistan
Nanga Parbat	Zeitler et al., 1982, *Nature*, 298:255.
	Tibet
Tibetan Plateau	George et al., 2001, *Geology*, 29:939–942.
	Namibia
Damara orogen	Haack, 1976, *Geol. Rundschau*, 65:967–1002; Haack, 1983, in Martin and Eder (Eds.), *Intracontinental Fold Belts*, pp. 873–884, Springer-Verlag, Heidelberg.
	Tektites
Moldavites	Gentner and Wagner, 1969, *Geol. Bavarica*, 61:296–308; Gentner et al., 1969, *Naturwissenschaften*, 56:255–260.
Bediasites	Durrani and Khan, 1970, *Earth Planet. Sci. Lett.*, 9:431–445; Storzer and Wagner, 1969, *Earth Planet. Sci. Lett.*, 5:463–468.
North American	Storzer and Wagner, 1971, *Earth Planet. Sci. Lett.*, 10:435–440.
	Impact Glasses
Ries crater, Nördlingen, Germany	Miller (D.S.) and Wagner, 1979, *Earth Planet. Sci. Lett.*, 43:351–358; Gentner and Wagner, 1969, *Geol. Bavarica*, 61:296–303; Wagner, 1977, *Geol. Bavarica*, 75:349–354.
	Moon
KREEP, Apennine front	Haines et al., 1975, *Lunar Sci. Conf. Proc.*, 6:3527–3540.
	Archeology/Anthropology
KBS tuff, Kenya	Gleadow, 1980, *Nature*, 284:225–230; Hurford et al., 1976, *Nature*, 263:738–740.
Obsidian artifacts	Miller (D.S.) and Wagner, 1981, *Nucl. Tracks*, 5:147–156; Durrani et al., 1971, *Nature*, 238:242–245.

(continued overleaf)

Table 22.3. (*continued*)

Region or Problem	References
Historic and prehistoric glasses	Brill, 1965, *Archaeometry*, 7:51–57; Fleischer and Price, 1964, *J. Geophys. Res.*, 69:331–339; Fleischer and Price, 1964, *Science*, 144:841–842. Storzer and Poupeau, 1973, *C. R. Acad. Sci. Paris*, 276D:137–139.
Techniques and applications	Fleischer and Hart (H.R.), 1972, in Bishop and Miller (S.A.) (Eds.), *Calibration of Hominoid Evolution*, pp. 135–170, Academic, Edinburgh. Wagner, 1998, *Age Determination of Young Rocks and Artifacts*, Springer-Verlag, Heidelberg; Wagner, 1978, *Nucl. Track Detect.*, 2:51–64.
Uranium in ancient glass	Fleischer and Price, 1964, *Science*, 144:841–842.
South Africa and Olduvai gorge, Tanzania (calcite)	Macdougall and Price, 1974, *Science*, 195:943–944.

Note: Additional studies are referred to in the text.

and applications of TL dating were summarized by Macdougall (1968), Aitken (1978, 1985, 1998), and Wagner (1998). In addition, the principles and applications of thermoluminescence dating were discussed at a Seminar on Thermoluminescence and Electron Spin Resonance in 1993 (Aitken et al., 1994).

22.4a Principles

When crystalline nonconducting solids are exposed to ionizing radiation, energy is transmitted to the atoms, which causes some of their electrons to be lifted into a higher energy state. These electrons may either be trapped by crystal defects that have a deficiency of negative charge or they may spontaneously return to the ground state. The trapped electrons cannot return to the ground state unless a sufficient amount of energy is provided to release them from the trap. The electron traps in a crystal may have different "depths," which means that different amounts of activation energy are required to allow the electrons to escape. The activation energy can be provided either in the form of heat or by electromagnetic radiation (e.g., light). When the trapped electrons return to the ground state, they emit electromagnetic radiation at wavelengths that correspond to the activation energies and thus to the depths of the trap in which they were imprisoned. This radiation constitutes "luminescence" which is distinct from the electromagnetic radiation emitted by solid objects at elevated temperatures. In cases where the electrons are released from the traps by heating the crystal, one speaks of "thermoluminescence," in contrast to "optically stimulated luminescence."

The application of TL to geochronometry arises because the amount of energy released by TL is a function of the sensitivity of the material for the entrapment of electrons and of the intensity and duration of the irradiation from sources in the environment (Aitken, 1978):

$$t = \frac{\text{natural TL}}{\text{TL per unit dose} \times \text{dose per year}} \qquad (22.17)$$

where natural TL = TL recorded by heating the sample to be dated
TL per unit dose = sensitivity of the sample for acquiring TL
dose per year = amount of energy transmitted to the sample by ionizing radiation from the environment in one year

The dose is measured in rads, which is defined as 100 ergs of energy transmitted to one gram of the absorber (Section 3.4).

A dimensional analysis of equation 22.17 confirms that it has the units of time:

$$\frac{\text{TL}}{(\text{TL/dose}) \times (\text{dose/time})} = \frac{1}{(1/\text{dose}) \times (\text{dose/time})}$$
$$= \frac{\text{dose} \times \text{time}}{\text{dose}} = \text{time}$$

The ionizing radiation that causes TL in minerals and ceramics consists of α-particles, β-particles, γ-rays, and cosmic rays. The sensitivity of minerals and ceramics for acquiring TL from exposure to α-particles (X_α) is lower than their sensitivity to β-particles and γ-rays (X_β). Therefore, equation 22.17 is rewritten in the form:

$$t = \frac{\text{natural TL}}{X_\alpha D_\alpha + X_\beta\ (D_\beta + D_\gamma + D_c)} \qquad (22.18)$$

where D = annual dose for α-, β-, γ-, and cosmic radiation (α, β, γ, c, respectively)
X_α = TL per unit dose rate of α-particles
X_β = TL per unit dose rate of β-, γ-, and cosmic radiation

Nambi and Aitken (1986) estimated the dose per year for α-, β-, and γ-radiation emitted by unit concentrations of U, Th, K, and Rb. The results in Table 22.4 assume that all decay series are in secular equilibrium. The tabulated values demonstrate that the total dose rate is 496.7 millirads per year and that the α-particles contribute about 71%, β-particles 21%, and γ-rays 8% of the total. The values in Table 22.4 can be used to calculate the dose rate for samples based on their concentrations of U, Th, K, and Rb. The conversion factors were recently updated by Adamiec and Aitken (1998).

Table 22.4. Radiation Dose Caused by α-, β-, and γ-Radiation Emitted by U, Th, K, and Rb

Element	Concentration	Dose Rate (mrad/y)		
		α	β	γ
U	1 ppm	278.1	14.7	11.36
Th	1 ppm	73.9	2.86	5.21
K	1%	–	81.4	24.3
Rb	100 ppm	–	4.86	–
Totals		352	103.82	40.87
		71%	21%	8%

Source: Nambi and Aitken, 1986.

When electrons are released from only one type of trap, the process is governed by an Arrhenius equation of the form (Aitken, 1978):

$$-\frac{dn}{dt} = nse^{-E/kT} \qquad (22.19)$$

where n = number of filled traps at time t,
s = frequency factor representing the number of times per second the electron tries to escape from the trap
E = activation energy, which is a measure of the depth of the trap
k = Boltzmann's constant
T = temperature on the Kelvin scale
$\frac{dn}{dt}$ = rate of change of the number of filled traps at temperature T

The glow curve caused by draining electrons from one kind of trap has the shape of a single peak because of the competing effects of the exponentially rising probability that an electron will escape from the trap and the decreasing number of electrons remaining in the trap (Aitken, 1978). Typical glow curves, such as the one in Figure 22.8, are smooth functions of the temperature, which implies that the electrons are being drained out of several different kinds of traps.

22.4b Geochronometry

The TL method of dating pottery is based on the assumption that all trapped electrons in the clay and other minerals were released during firing and that the TL clock was thereby reset to zero. Subsequently, the number of trapped electrons increased as a function of time at a rate determined by the TL sensitivities of the minerals and by the concentrations of U, Th, and K in the pottery itself and in the soil in which it was buried. The accumulated TL is released by heating the sample under controlled conditions and by recording the radiation by means of a photomultiplier tube

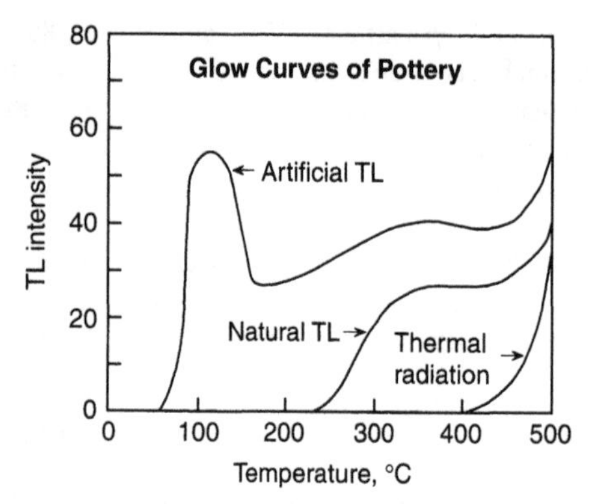

FIGURE 22.8 Glow curves of pottery samples after 1000 rads of β-irradiation (artificial TL), natural TL acquired after firing, and thermal radiation caused by heating the sample above about 450°C. Adapted from Aitken (1978).

connected to an $x-y$ plotter to produce the "glow curve" of the sample.

The hypothetical glow curves in Figure 22.8 represent three different kinds of emissions:

1. Natural TL is the light emitted by a typical sample of pottery between about 250 and 450°C.
2. Thermal radiation is the light emitted by the sample above about 400°C.
3. The thermal radiation enhances the intensity of the natural TL above 400°C. Artificial TL is emitted by a sample that was irradiated by β-particles or γ-rays after its natural TL was reduced to zero by heating at 500°C for one hour.

Although the procedure for obtaining glow curves is straightforward in principle, important details must be considered, such as:

1. The natural TL of pottery samples is commonly quite weak because the sensitivity of the minerals to the acquisition of TL is low. Therefore, the sample must be heated rapidly, a high-sensitivity and low-noise photomultiplier must be used, and all non-TL radiation emitted by the sample must be suppressed.
2. The natural TL is emitted primarily at temperatures greater than about 250°C because the shallow traps tend to lose electrons even at Earth-surface temperatures
3. The baked-clay matrix of pottery contains grains of certain minerals (e.g., quartz and feldspar) whose diameters range widely from less than 10 up to about 1000 μm (1.0 mm). The large grains receive a smaller dose of radiation from α-particles than the small grains in the matrix because the range of α-particles in the pottery matrix is only about 25 μm. For this reason, pottery samples are crushed in a vise and screened into a fine fraction (1–8 μm) and a coarse fraction (90–150 μm). In most cases, the fine fraction is preferred for TL dating. The coarse fraction can be used after it has been leached with HF in order to remove the outer layers of grains where the α-particles were absorbed. After this material has been removed, only the TL caused by β- and γ-irradiations remain (i.e., $D_\alpha = 0$).
4. The TL signal is lost not only by heating the samples but also by exposure to light, which causes "bleaching." For example, most of the natural TL is erased in 10–20 h by exposure to direct sunlight (Singhvi and Mejdahl, 1985; Wagner, 1998). The extent of bleaching depends on the intensity of the light, on the duration of the exposure, and on the minerals in question. For example, feldspar is bleached more than quartz (Spooner, 1992, 1994), but the TL of the glow curve of quartz is bleached preferentially at 325°C.

The fading of the natural TL of mineral and pottery samples at ambient Earth surface temperatures and by exposure to light is equivalent to the annealing of fission tracks in minerals (Section 22.2).

22.4c Procedures for TL Dating

The amount of natural TL of a sample is the area under the "glow curve" obtained by heating a known weight of the sample at increasing temperatures between about 250 and 450°C. The dose per unit time is calculated from the concentrations of U, Th, and K of the sample itself and/or of the matrix in which it was embedded using the factors in Table 22.4. The amount of TL per unit dose is obtained in one of two ways:

1. The sample is first annealed by heating at 500°C for one hour which releases all of the accumulated TL. Aliquots of the *annealed* sample are then irradiated with α-, β-, and γ-radiation, emitted by standard sources at known rates so that the dose of absorbed radiation (expressed in rads) can be calculated. The irradiated aliquots are then heated to obtain their artificial glow curves, from which the amount of artificial TL and hence the TL per unit dose of α-, β-, and γ-radiation are determined.
2. Alternatively, aliquots of the *unannealed* sample are exposed to different doses of α-, β-, and γ-radiation, and the resulting TL is determined from the glow curves. The data points (dose in rads versus TL in arbitrary units) define the straight line in Figure 22.9, whose slope has the dimensions of TL per unit dose. In addition, the intercept on the negative x-axis is the *equivalent dose* that caused the natural TL of the sample. In the hypothetical example in Figure 22.9:

$$\frac{\text{TL}}{\text{dose}} = \frac{5.3}{1500} = 3.5 \times 10^{-3} \text{ TL/rad}$$

Equivalent dose = 825 rads

The addition method (described above) of determining the TL/dose parameter is preferred because annealing samples (as in part 1 above) causes their TL sensitivity to change. The addition method has the added advantage of yielding the equivalent dose, which is a measure of the dose that caused the natural TL of the sample. However, the addition method assumes that the TL increases linearly with the radiation dose, which may not be valid at low doses. Therefore, the backward extrapolation of the addition line may underestimate the total "archeological" dose absorbed by the sample.

22.4d TL Dating of Sediment

The basic assumption of dating sediment is that the natural TL of the mineral grains was significantly reduced by exposure to sunlight during erosion and transport prior to final deposition. Experiments cited by Singhvi and Mejdahl (1985) demonstrated the optical bleaching of both coarse and fine grains of feldspar and quartz from different kinds of sediment in 1–20 hours. Even a layer of water 7 m deep caused a 50% reduction of TL in feldspar. However, a residual amount of TL remained in all cases. Therefore, the presence of residual TL in sediment samples must be taken into consideration because it causes TL dates to overestimate the time elapsed since deposition.

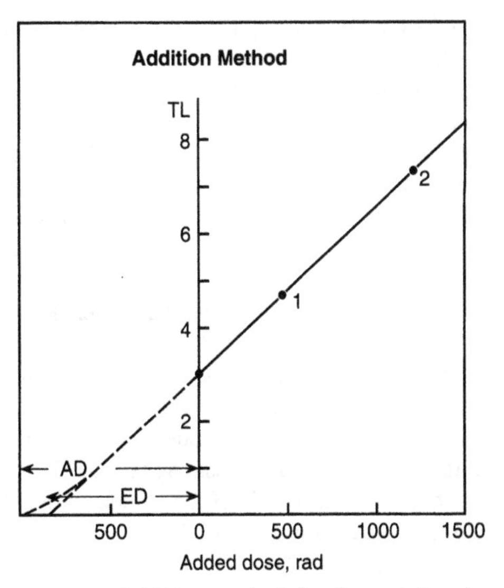

FIGURE 22.9 Addition method for determining the TL sensitivity of a sample to a unit dose of ionizing radiation (α, β, and γ). Points 1 and 2 represent two different irradiations of an unannealed sample causing their TL to increase above that of the natural TL. The slope of the resulting straight line has the dimensions of TL per rad. A linear extrapolation of the straight line yields the "equivalent dose" (ED), which is generally an underestimate of the "archeological" dose (AD) because of the nonlinearity of the relation between TL and dose at low values of the dose. Adapted from Aitken (1978).

Accordingly, TL dating of sediment is based on the following assumptions (Singhvi and Mejdahl, 1985):

1. Sun exposure of mineral grains prior to their deposition reduced their previously accumulated TL to a small residual value (I_0).
2. After deposition, the sediment was shielded from sunlight and accumulated TL in response

to the dose of ionizing radiation to which it was exposed in the environment.

3. The resulting "natural TL" (I_{nat}) is the sum of I_0 and the postdepositional TL (I_d).
4. The dose rate of ionizing radiation is determined from the measured concentrations of U, Th, and K, as in the TL dating of pottery sherds and other kinds of samples.

Several different procedures have been devised for correcting the observed TL (I_{nat}) for the presence of the residual TL (I_0). In the "total bleach" method, the TL of a sediment sample is reduced to a minimum by exposing it to artificial sunlight for 1000 h. The remaining residual TL (I_0) is subtracted from the TL of the natural sample to yield the postdepositional TL (I_d):

$$I_d = I_{nat} - I_0 \qquad (22.20)$$

This equation can be rewritten in terms in the doses D of ionizing radiation that caused the TL represented by the parameter I:

$$D(I_{nat}) = D(I_0) + D(I_d) \qquad (22.21)$$

The dose acquired since deposition is:

$$D(I_d) = D(I_{nat}) - D(I_0)$$

Therefore, the TL date can be calculated from the equation:

$$\text{TL date} = \frac{D(I_d)}{\text{annual dose}} \qquad (22.22)$$

The radiation dose acquired by the sediment sample after deposition is determined by the addition method in which the natural sample is irradiated with β-particles and acquires additional increments of dose, as shown in Figure 22.10. The value of $D(I_d)$ is the equivalent dose (ED). Therefore, the TL date can be calculated in accordance with equation 22.24 by dividing the equivalent dose by the annual dose derived from the concentrations of U, Th, and K.

Alternative laboratory procedures, outlined by Singhvi and Mejdahl (1985) and Wagner (1998),

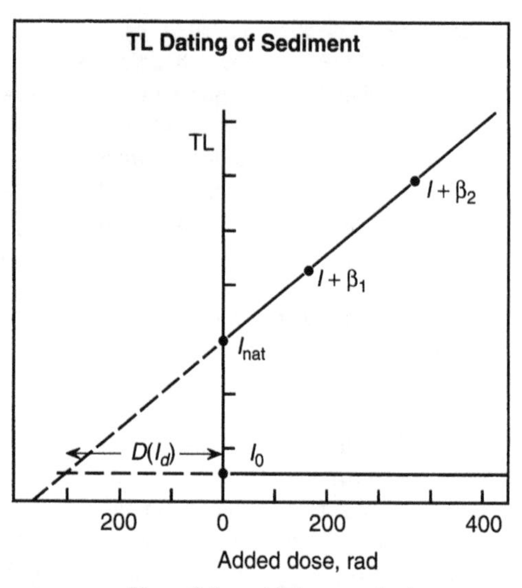

FIGURE 22.10 Use of the addition method to determine the radiation dose absorbed by sediment, including the presence of residual TL caused by exposure to sunlight prior to deposition: I_{nat} = TL of a known weight of sediment; $I + \beta_1$ = TL of an aliquot of the sample after absorbing an additional dose of 165 rads from irradiation with β-particles; $I + \beta_2$-same for additional 375 rads; $D(I_d)$-equivalent dose of 300 rads absorbed by the sample from sources in the environment. Adapted from Singhvi and Mejdahl (1985).

have been developed to correct the TL of sediment samples for the residual TL that was present at the time of deposition. Additional complications arise because:

1. Alpha particles are less effective in causing TL than β-particles and γ-rays, partly because they penetrate only the outer layers of sand- and silt-size particles;
2. the radioactive elements reside primarily in the grains of certain refractory minerals (e.g., zircon, apatite, sphene) that may be irregularly distributed within the sediment deposits; and
3. the concentrations of radioactive elements residing on grain surfaces and in the mineral cements may be changed by the movement of groundwater.

22.4e Applications

In spite of the complications and uncertainties, TL dating of archeological objects and geological samples of Quaternary age (<1.0 Ma) has provided useful information in cases where other dating methods cannot be used. The applications of TL dating to archeology and geology are summarized in Table 22.5.

22.5 ELECTRON-SPIN RESONANCE

The electron-spin resonance (ESR) method of dating is closely related to the thermoluminescence (TL) method because both are based on the presence of electrons in excited states in nonconducting crystals which were exposed to ionizing radiation. The measurements are made using commercial ESR spectrometers that detect the absorption of microwave radiation by the trapped electrons when a strong magnetic field is imposed on them. The magnetic field causes the energies of unpaired electrons to split depending on whether their own magnetic moments are aligned parallel or opposite to the applied field. Under these conditions, the electrons absorb electromagnetic radiation in the microwave region at specific frequencies.

The use of ESR for the dating of solids that have been exposed to ionizing radiation was suggested by Zeller et al. (1967) and Zeller (1968) and was subsequently developed by Ikeya (1975, 1978). After a slow start, ESR dating has evolved into a versatile geochronometer for archeological research. The basic principles of ESR dating have been presented by Grün (1989a, b, 1990), Grün and Stringer (1991), Schwarcz (1994), Aitken et al. (1994), Blackwell (1995, 2001), Wagner (1998), and Rink (1997).

The ESR method has several advantages over the conventional TL method of dating:

1. The concentration of traps containing unpaired electrons can be measured without draining the traps. Therefore, measurements can be duplicated for increased accuracy.
2. It is less sensitive to surface effects than TL, which means that the method of grinding samples and the grain size of powders do not affect the results.
3. The ESR signal is not diminished by exposure of the sample to light. The analysis of samples is thereby simplified because measurements can be made in a lighted laboratory.

The radiation-exposure age of a powdered sample dated by ESR is calculated as in the case of TL from the relationship:

$$t = \frac{\text{total radiation dose}}{\text{annual dose}} \qquad (22.23)$$

where the annual dose is calculated from the measured concentrations of U, Th, and K, and the total radiation dose is obtained by the addition method using a plot of ESR intensities induced by γ-ray irradiations from a ^{60}Co source.

22.5a Principles

The quantum-mechanical description of the electron includes its ability to rotate about an axis. A spinning electron therefore acts like a magnet whose polarity depends on the sense of rotation. The rules of wave mechanics require that two electrons can only occupy the same orbital in an atom when they spin in the opposite sense, which causes their magnetic moments to cancel each other. Materials in which the spin axes of the electrons are opposed are said to be diamagnetic.

Defect sites in diamagnetic crystals contain unpaired electrons that exert a magnetic moment, thereby causing these defect sites to be paramagnetic. The different kinds of electron traps that may exist in a particular crystal are characterized by the so-called g-value which is the dimensionless ratio between their magnetic moment and their angular momentum. For example, a free electron has a g-value of 2.00232 (Wagner, 1998). When such an unpaired electron is acted upon by an externally applied magnetic field, the spin axis of the electron responds by precessing about the magnetic vector of the applied field. The frequency of the precession (called the Lamor frequency) depends on the g-value of the electron trap and on the magnitude of the applied magnetic field.

Table 22.5. Application of TL Dating in Archeology and Geology

Material	References
	Archaeological Samples
Pottery sherds, bricks, and ceramic tile	Wagner, 1980, *Lübecker Schriften Archäologie und Kulturgeschichte*, 3:83–87; Pernicka and Wagner, 1983/84, *Mitteil. Osterreich. Arbeitsgemeinschaft Ur-und Frühgeschichte*, 33/34:247–267; Wagner and Lorenz, 1997, *Rhein. Ausgrab.*, 43:747–754.
Burned flint	Göksu et al., 1974, *Science*, 183:651–654; Göksu et al., 1992, *Istanbuler Forschungen*, 38:140–147; Huxtable and Aitken, 1985, *An. Praehist. Leidensia*, 18:41–44; Valladas et al., 1988, *Nature*, 331:614–616; Valladas, 1992, *Quat. Sci. Review*, 11:1–5; Mercier et al., 1993, *J. Archaeol. Sci.*, 20:169–174; Mercier et al., 1995, *J. Archaeol. Sci.*, 22:495–509; Bar-Yosef and Vandermeersch, 1993, *Spektr. Wissen.*, 1993/96:32–39; Richter, 1997, Dissertation, University of Tübingen.
Boiling stones	Huxtable et al., 1976, *Archaeometry*, 18:5–17.
	Geological Samples
Limestone	Zeller, 1954, in Faul (Ed.), *Nuclear Geology*, pp. 180–188, Wiley, New York; Zeller et al., 1957, *Am. Assoc. Petrol. Geol. Bull*, 41:1212–1219; Zeller and Ronca, 1963, in Geiss and Goldberg (Eds.), *Earth Science and Meteoritics*, pp. 282–294, North-Holland, Amsterdam.
Plagioclase	May, 1977, *J. Geophys. Res.*, 82:3023–3029; Guerin and Valladas, 1980, *Nature*, 286:697–699, Duller, 1994, *Quat. Sci. Rev., Quatern. Geochron.*, 13:423–427.
Quartz	Pilleyre et al., 1992, *Quat. Sci. Rev.*, 11:13–17; Miallier et al., 1994, *Radiat. Meas.*, 23:399–404.
Hornfels	Zöller, 1989, *Die Eifel*, 84:415–418.
Sediment	Wintle and Huntley, 1979, *Nature*, 289:479–480; Huntley and Lian, 1999, *Geol. Surv. Can. Bull.*, 534:211–222; van Heteren et al., 2000, *Geology*, 28:411–414; Huntley et al., 1985, *Nature*, 313:105–107.
Sand: dunes, beaches	van Heteren et al., 2000, *Geology*, 28(5):411–414; Pye, 1982, *Nature*, 299:376; Singhvi et al., 1982, *Nature*, 295:313–315; Singhvi et al., 1986, *Earth Planet. Sci. Lett.*, 80:139–144; Hashimoto et al., 1986, *Geochem. J.*, 20:111–118; Ollerhead et al., 1994, *Can. J. Earth Sci.*, 31:523–531.
Volcanic ash	Berger, 1991, *J. Geophys. Res.*, 96:19750–19720; Berger and Huntley, 1994, *Quatern. Sci. Rev.*, 13:509–511.
Fault gouge	Singhvi et al., 1994, *Quatern. Sci. Rev.*, 13:595–600.
Carbonate in soil	Singhvi et al., 1996, *Earth Planet. Sci. Lett.*, 139:321–332.
Loess	Zöller et al., 1988, *Chem. Geol.*, 73:39–62; Berger et al., 1992, *Geology*, 20:403–406; Waters et al., 1997, *Science*, 275:1281–1284.
Speleothems, calcite	Debenham and Aitken, 1984, *Archaeometry*, 26:155–170.
Clam shells	Ninagawa et al., 1992, *Quatern. Sci. Rev.*, 11:121–126; Ninagawa et al., 1994, *Quatern. Sci. Rev.*, 13:589–593.
Shocked quartz (Meteor Crater, Arizona)	Sutton, 1984, *Meteoritics*, 19:317; Sutton, 1985, *J. Geophys. Res.*, 90:3690–3700.
Lunar soil (Apollo 12)	Doell and Dalrymple, 1971, *Earth Planet. Sci. Lett.*, 10:357–360.
Stony meteorites	Guimon et al., 1985, *Geochim. Cosmochim. Acta*, 49:1515–1524; Sears et al., 1984, *Geochim. Cosmochim. Acta*, 48:2265–2272.

When an electromagnetic wave having the same frequency as the precession (i.e., the Lamor frequency) is applied at right angles to the direction of the magnetic field, the electron spin resonates, which causes the spin axis of the electron to reverse direction. The energy required to flip the spin axis is provided by the microwave whose intensity is thereby diminished. The resonance frequency is related to the g-value of the defect center, and the fraction of the microwave energy that is absorbed is related to the number of traps of this type.

To obtain an ESR spectrum of a sample, the magnitude of the external magnetic field is increased continuously and the fraction of the microwave energy that is absorbed by centers having different g-values is recorded. Actually, the function that is plotted is the first derivative of the ESR intensity with respect to the magnetic field strength (H). A hypothetical example of the ESR spectrum of a certain type (i.e., having a specific g-value) is depicted in Figure 22.11.

The electron traps in crystalline solids consist of *defects* in the crystal lattice and of so-called *centers* where charge imbalances occur because of substitution of ions having different charges. For example, the ESR centers in quartz and feldspar are caused by the replacement of Si by Al, Ge, and Ti at the time of formation of these minerals. Additional ESR centers are generated later when recoiling nuclei resulting from α-decay collide with oxygen atoms of silica tetrahedra and thereby produce oxygen vacancies called *E′-centers* and interstitial oxygen atoms known as *peroxide centers*. All of these defects and centers can trap electrons, which explains the relationship between the ESR signal and the total dose of ionizing radiation absorbed. In some cases, the ESR spectra that are related to the accumulated radiation dose are obscured by ESR signals that originate from complex organic molecules (e.g., the alanine radical $CH_3CHCOOH$; Ikeya, 1981) and by atoms of Mn and Cu.

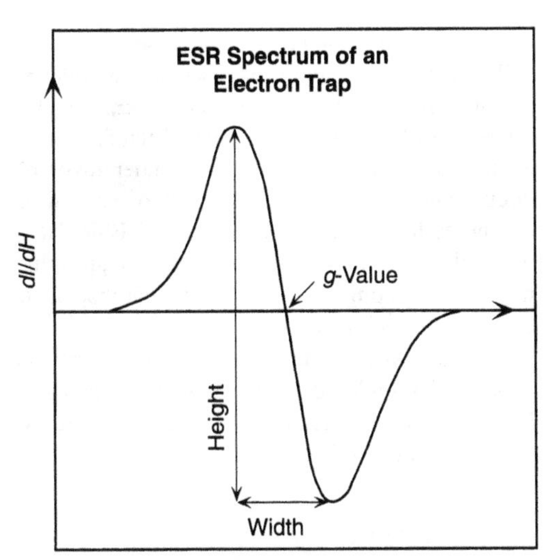

FIGURE 22.11 Hypothetical ESR spectrum expressed as the derivative dI/dH of the intensity I with respect to the magnetic field strength H. The g-value is measured at the point indicated in the diagram. Adapted from Wagner (1998).

22.5b Assumptions

The ESR method of dating is based on certain assumptions concerning the accumulation of the ESR signal (Blackwell, 2001):

1. The ESR signal was equal to zero at the time of formation of a mineral or was reduced to zero by an increase in temperature sufficient to drain all of the electrons from the existing traps.
2. The mean life of the ESR signal is much longer than the ages of the samples being dated.
3. The accumulated radiation dose is much smaller than the saturation dose of the sample.

The requirement of zero initial ESR is met by all minerals at the time of their formation because the crystal defects and centers acquire ESR only when they contain electrons that were energized by ionizing radiation. Therefore, the materials that are datable by ESR include tooth enamel, speleothems, travertine, calcrete, mollusk shells, corals, bones, and fish scales, all of which yield ESR dates that represent the time of formation. In addition, certain kinds of samples can be dated because the electrons were drained from their traps by an episode of heating to temperature between 250 and 500°C (i.e., the previously accumulated ESR was reduced to zero). Such materials include burned flint and other kinds of siliceous rocks that were heated in the past and which record the last time they cooled sufficiently to allow electrons to be trapped. In addition, fault gouge (mylonite) and volcanic rocks have also been studied for ESR dating.

The ESR signal of the samples to be dated must be stable (assumption 2), which means that it must persist for long periods of time without being degraded by exposure to sunlight or moderate increases in temperature. The resulting mean life of the ESR signal should be several orders of magnitude greater than the age of the material being dated. For example, the ESR of tooth enamel has a mean life of the order of 10^{19} years (Blackell, 2001), whereas the ESR associated with various centers in quartz crystals has a mean life in excess of 10^6 years. However, the ESR signal of Ge centers in quartz is rapidly bleached by sunlight, which is helpful for ESR dating of quartz grains in sediment (Wagner, 1998).

The third assumption has to do with the fact that the intensity of ESR approaches saturation, which means that it is no longer a function of time. In some cases, the ESR-signal intensity reaches a steady state because the rate of fading of the signal is equal to the rate at which it is produced.

22.5c Methodology

The practical aspects of ESR dating have to do with recording the ESR spectra of samples to be dated, with determining the accumulated radiation dose, and with modeling the annual dose rate based on the concentrations of U, Th, and K.

Samples weighing about 100 mg composed of grains with diameters between 100 and 400 μm are analyzed by means of an ESR spectrometer which applies a magnetic field of about 3500 Gauss and a microwave with a frequency of about 9 Gigahertz (GHz).

The accumulated dose is determined by irradiating samples with γ-rays and by measuring the intensities of the ESR signals that are caused by the additional radiation doses ranging up to 1500 Gray. This procedure, called the addition method, is illustrated in Figure 22.9, where it was used to measure the equivalent dose (ED) for TL dating.

The internal dose rate of ionizing radiation is calculated using the conversion factors of Nambi and Aitken (1986) and Adamiec and Aitken (1998). A problem arises in cases where the U concentrations have increased after the formation of the sample. For example, tooth enamel and mollusk shells absorb U from the environment after burial. The process has been modeled by assuming either early or continuous addition of U (Grün et al., 1988; Grün and McDermott, 1994; Blackwell et al., 2001).

The determination of the annual radiation dose is quite complicated because it must take into consideration factors such as:

1. the time-dependent absorption of U by the samples after burial,
2. the time-dependent development of the decay series of both U and Th in the samples,
3. the efficiency with which α- and β-particles and γ-rays cause electrons to be energized and trapped in defects and centers,
4. the availability of various kinds of crystal defects and ionic substitution centers,
5. the homogeneity of the radiation field to which the sample was exposed, and
6. the mineral composition of the sample and the sensitivity of the minerals for acquisition and/or bleaching of the ESR signal.

In spite of the complexity of these issues, the ESR method of dating has become a valuable investigative tool *when it is used by experts*. The range of ESR dates is from several hundred up to about one million years, or more under favorable circumstances. The analytical error of such dates can be as low as ±5%. The range of ESR dating is similar to that of the U-series disequilibrium methods of dating and greatly exceeds that of the cosmogenic ^{14}C method.

A summary of the kinds of archeological and geological samples that are suitable for dating by ESR is provided in Table 22.6, including references to specific archeological sites.

22.6 SUMMARY

The radiation damage methods are all based on the idea that the amount of damage in solids increases as a function of time at a rate that depends on the amount of energy imparted by ionizing

Table 22.6. Summary of ESR dating as reflected in the recent literature.

Topic	References
Dose rate	Nambi and Aitken, 1986, *Archaeometry*, 28:202–205; DeCannière et al., 1986, *Nucl. Tracks Rad. Meas.*, 11:211–220; Blackwell et al., 2000, *Quat. Int.*, 68/71:345–361; Brennan et al., 1997a, *Radiat. Meas.*, 27:299–305; Brennan et al., 1997b, *Radiat. Meas.*, 27:307–314; Blackwell and Schwarcz, 1993, *Appl. Radiat. Isotopes*, 44:243–252; Lyons et al., 1993, *Appl. Radiat. Isotopes*, 44:131–138.
U-sorption	Pike and Hedges, 2001, *Quat. Sci. Rev.*, 20:1021–1025; Millard and Hedges, 1996, *Geochim. Cosmochim. Acta*, 60:2139–2152; Blackwell and Blickstein, 2000, *Quat. Int.*, 68/71:329–343; Grün et al., 1988, *Nucl. Tracks Radiat. Meas.*, 14:237–241; Grün and McDermott, 1994, *Quat. Sci. Rev.*, 13:121–125.
Tooth enamel	Schwarcz, 1985, *Nucl. Tracks Radiat. Meas.*, 10:865–867; Bouchez et al., 1986, *Mem. Mus. Nat. Hist. Paris, Ser. c*, 53:169–176; Zymela et al., 1988, *Can. J. Earth Sci.*, 25:235–245; Grün et al., 1987, *Can. J. Earth Sci.*, 24:1022–1037; Grün et al., 1988, *Nucl. Tracks Radiat. Meas.*, 14:237–241; Grün et al., 1997, *J. Human Evol.*, 32:83–91; Grün et al., 1998, *J. Archaeol. Sci.*, 26:1301–1310; Blackwell et al., 1992, *Quat. Sci. Rev.*, 11:231–244; Blackwell et al., 1993, *Appl. Radiat. Isotopes*, 44:253–260; Blackwell et al., 1994, in MacDonald (Ed.), *Great Lakes Archaeology and Paleoecology*, pp. 321–360, Waterloo University Press, Waterloo, Ontario, Canada. Blackwell, 1994, *Quat. Sci. Rev.*, 13:651–660; Skinner et al., 2000, *Appl. Radiat. Isotopes*, 52:1337–1344. Pike and Hedges, 2001, *Quat. Sci. Rev.*, 20:1021–1025; Brennan et al., 1997, *Radiat. Meas.*, 27:307–314.
Cave deposits	Ikeya, 1977, *J. Anthropol.*, 4:152–168; Ikeya, 1978, *Naturwissenschaften*, (calcite) 65:489; Miki and Ikeya, 1978, *Japanese J. Appl. Phys.*, 17:1703; Karakostonoglou and Schwarcz, 1983, *PACT*, 9:391–400; Lau et al., 1997, *Geoarchaeology*, 12:507–536.
Fault gouge	Miki and Ikeya, 1982, *Naturwissenschaften*, 69:390–391; Ikeya et al., 1982, *Science*, 215:1392–1393. Fukuchi, 1989, *Earth Planet. Sci. Lett.*, 94:109–122; Fukuchi, 1992, *J. Geol. Soc. (Lond.)*, 149:265–272; Tani et al., 1996, *Appl. Radiat. Isotopes*, 47:1423–1426.
Corals	Ikeya and Ohmura, 1983, *Earth Planet. Sci. Lett.*, 65:34–38.
Clamshells	Ikeya and Ohmura, 1981, *J. Geol.*, 89:230; Ikeya, 1981, *Naturwissenschaften*, 68:474–475.
Volcanic rocks	Shimokawa et al., 1984, *Chem. Geol. (Isotope Geosci. Sect.)*, 2:365–373.
Bones	Ikeya and Miki, 1980, *Science*, 207:977–979; Ikeya, 1981, *Naturwissenschaften*, 68:474–475; Millard and Hedges, 1996, *Geochim. Cosmochim. Acta*, 60:2139–2152.
Flint	Garrison et al., 1981, *Nature*, 290:44–45; Blackwell et al., 1994, in MacDonald (Ed.), *Great Lakes Archaeology and Paleoecology*, pp. 321–360, Waterloo University Press, Waterloo, Ontario, Canada.
Archaeological sites	Schwarcz and Grün, 1988, *Geoarchaeology*, 3:293–296; Schwarcz et al., 1988, *Archaeometry*, 30:5–17; Blackwell et al., 1992, *Quat. Sci. Rev.*, 11:231–244; Blackwell et al., 1993, *Appl. Radiat. Isotopes*, 44:253–260; Blackwell et al., 2000, *Quat. Int.*, 68/71:345–361; Stringer et al., 1989, *Nature*, 338:756–758; Grün et al., 1990, *Nature*, 344:537–539; Grün et al., 1997, *J. Human Evol.*, 32:83–91; Grün et al., 1998, *J. Archaeol. Sci.*, 26:1301–1310; Grün and Stringer, 1991, *Archaeometry*, 33:153–199; Lau et al., 1997, *Geoarchaeology*, 12:507–536.

Note: A more complete list of references was compiled by Blackwell (2001)

radiation. Pleochroic haloes and α-recoil tracks are not currently important for dating because the amount of radiation damage in pleochroic haloes cannot be quantified and because α-recoil tracks are small and anneal at low temperatures in most, but probably not all, minerals. Therefore, the α-recoil method of dating has scope for further development.

The fission-track method has evolved into a useful thermochronometer for investigating the thermal histories of rocks and for dating certain minerals (e.g., zircon, sphene, apatite) as well as man-made and natural glasses. Although the principles on which this method is based are easily understood, the analytical procedures are labor intensive and require considerable skill and experience.

Thermoluminescence (TL) and electron-spin resonance (ESR) have a common cause because they arise from electrons trapped in crystal defects and in charged sites caused by ionic substitutions. The electrons are raised to higher energy states when atoms absorb ionizing radiation. Therefore, both dating methods are the result of radiation damage.

In TL dating, electromagnetic radiation is emitted when the electrons are released from the traps and return to the groundstate as a result of heating the sample. In ESR dating, the electrons absorb energy from a microwave beam when they resonate in a strong externally applied magnetic field. In both techniques, the TL and the ESR signals are used to determine the total radiation dose absorbed by the samples since they originally formed or after their electron traps were drained by heating to an elevated temperature. In both methods, a date is calculated by dividing the total accumulated dose by the annual dose of ionizing radiation emitted by U, Th, K, and the daughters of U and Th.

The TL and ESR methods compete with the U-series equilibrium method as well as with the $^{40}Ar^{*}/^{39}Ar$ method of dating in some cases. Both methods have become important in archeological research and in the study of Quaternary sediment.

REFERENCES FOR PLEOCHROIC HALOES (SECTION 22.1a)

Deutsch, S., D. Hirschberg, and E. Picciotto, 1956. Etude quantitative des halos pleochroiques. Application a l'estimation de l' ages des roches granitiques. *Bull. Sci. Belge Geol.*, 65:267.

Gentry, R. V., T. A. Cahill, N. R. Fletcher, H. C. Kaufmann, L. R. Medsker, and G. R. Flocchini, 1976. Evidence for primordial superheavy elements. *Phys. Rev. Lett.*, 37:11–15.

Hayase, I., 1954. Relative geologic age measurements on granites by pleochroic haloes and radioactivity of the minerals in their nuclei. *Am. Mineral.*, 39:761.

Henderson, G. H., 1934. A new method of determining the age of certain minerals. *Proc. Roy. Soc. (Lond.)*, 145A:591.

Holbrow, C. E., 1977. Haloes from plutonium minerals and fission alphas? *Nature*, 265:504–508.

Michel-Levy, A., 1882. Sur les noyaux a polychroisme intense du mica noir. *Compt. Rendu*, 94:1196.

Pasteels, P., 1960. L'age des halos pleochroiques du granite d'Habkern et de quelques roches du Massif de l' Aar. Bull. Swisse Mineral Petrol, 40:261.

Picciotto, E. E., and S. Deutsch, 1960. Pleochroic haloes. In *Summer Course on Nuclear Geology, Varenna*, pp. 263–310. Comitato Nazionale Per L'Energia Nucleare, Laboratorio di Geologia Nucleare, Pisa.

REFERENCES FOR ALPHA-RECOIL TRACKS (SECTION 22.1b)

Fleischer, R. L., 1982. Alpha-recoil damage and solution effects in minerals: Uranium isotope disequilibrium and radon release. *Geochim. Cosmochim. Acta*, 46:2191–2202.

Fleischer, R. L., 1983. Theory of alpha-recoil effects on radon release and isotopic disequilibrium. *Geochim. Cosmochim. Acta*, 47:779–784.

Garrison, E. G., C. R. McGinsey III, and O. H. Zinke, 1978. Alpha-recoil tracks in archaeological ceramic dating. *Archaeometry*, 20:39–46.

Gögen, K., 1999. Die Alpha-Rückstoss-Spuren Datierungsmethode: Theoretische und Experimentelle Entwicklung und die Datierung von Glimmer. Unpublished Ph.D. Dissertation, Max-Planck Institut für Kernphysik, Heidelberg.

Gögen, K., and G. A. Wagner, 2000. Alpha-recoil track dating of Quaternary volcanics. *Chem. Geol.*, 166:127–137.

Huang, W. H., M. Maurette, and R. M. Walker, 1967. Observation of fossil α-particle recoil tracks and their implication for dating measurements. In *Radioactive Dating and Methods of Low-Level Counting*, pp. 415–429. International Atomic Energy Agency, Vienna.

Huang, W. H., and R. W. Walker, 1967. Fossil alpha-particle recoil tracks: A new method of age determination. *Science*, 155:1103–1106.

Hurley, P. M., 1952. Alpha-radiation damage in zircon. *J. Appl. Phys.*, 23:1408.

Kigoshi, K., 1971. Alpha-recoil ^{234}Th: Dissolution into water and ^{234}U/^{238}U disequilibrium in nature. *Science*, 173:47–48.

REFERENCES FOR FISSION-TRACK DATING (SECTION 22.2)

Bal, K. D., N. Lal, and K. K. Nagpaul, 1983. Zircon and sphene as fission-track geochronometer and geothermometer: A reappraisal. *Contrib. Mineral. Petrol.*, 83:199–203.

Bar, M., Y. Kolodny, and Y. K. Bentor, 1974. Dating faults by fission track dating of epidotes—an attempt. *Earth Planet. Sci. Lett.*, 22:157–162.

Bertine, K. K., L. H. Chan, and K. K. Turekian, 1970. Uranium determinations in deep-sea sediments and natural water using fission tracks. *Geochim. Cosmochim. Acta*, 34:641–648.

Bigazzi, G., 1967. Length of fission tracks and age of muscovite samples. *Earth Planet. Sci. Lett.*, 3:434–438.

Carpenter, B. S., and G. M. Reimer, 1974. *Standard Reference Material: Calibrated Glass Standards for Fission Track Use, Natl. Bur. Standards Pub.* 260–49. National Bureau of Standards, Washington, D.C.

Church, S. E., and M. E. Bickford, 1971. Spontaneous fission-track studies of accessory apatite from granitic rocks in the Sawatch Range, Colorado. *Geol. Soc. Am. Bull.*, 82:1727–1734.

Dakowski, M., J. Burchart, and J. Galazka, 1974. Experimental formula for thermal fading of fission tracks in minerals and natural glasses. *Bull. Acad. Polon. Sci., Ser. Sci. de la Terre*, 22:11–17.

Durrani, S. A., and H. A. Khan, 1970. Annealing of fission-tracks in tektites: Corrected ages of bediasites. *Earth Planet. Sci. Lett.*, 9:431–445.

Durrani, S. A., H. A. Khan, and C. Renfrew, 1971. Obsidian source identification by fission-track analysis. *Nature*, 238:242–245.

Fisher, D. E., 1975. Geoanalytical applications of particle tracks. *Earth-Science Revs.*, 11:291–336.

Fleischer, R. L., and P. B. Price, 1964a. Glass dating by fission fragment tracks. *J. Geophys. Res.*, 69:331–339.

Fleischer, R. L., and P. B. Price, 1964b. Techniques for geological dating of minerals by chemical etching of fission fragment tracks. *Geochim. Cosmochim. Acta*, 28:1705–1714.

Fleischer, R. L., C. N. Naeser, P. B. Price, R. M. Walker, and U. B. Marvin, 1965a. Fossil particle tracks and uranium distributions in minerals of the Vaca Muerta meteorites. *Science*, 148:629–632.

Fleischer, R. L., P. B. Price, and R. M. Walker, 1965b. Spontaneous fission-tracks from extinct Pu^{244} in meteorites and the early history of the solar system. *J. Geophys. Res.*, 70:2703–2707.

Fleischer, R. L., B. P. Price, and R. M. Walker, 1975. *Nuclear Tracks in Solids*. University of California Press. Los Angeles, California.

Friedman, G. M., and J. E. Sanders, 1978. *Principles of Sedimentology*. Wiley, New York.

Galbraith, R. F., and G. M. Laslett, 1993. Statistical methods for mixed fission-track ages. *Nucl. Tracks*, 5:3–14.

Gallagher, K., 1995. Evolving temperature histories from apatite fission-track data. *Earth Planet. Sci. Lett.*, 136:421–435.

Gallagher, K., R. Brown, and C. Johnson, 1998. Fission track analysis and its applications to geological problems. *Annu. Rev. Earth Planet. Sci.*, 26:519–572.

Galliker, D., E. Hugentobler, and B. Hahn, 1970. Spontane Kernspaltung von U-238 und Am-241. *Helv. Phys. Acta*, 43:593.

Gentner, W., D. Storzer, and G. A. Wagner, 1969. Das Alter von Tektiten and verwandten Gläsern. *Naturwissenschaften*, 56:255–260.

Gleadow, A. J. W. 1981. Fission-track dating methods: What are the real alternatives? *Nucl. Tracks*, 5:3–14.

Gleadow, A. J. W., and I. R. Duddy, 1981. A natural long-term track annealing experiment for apatite. *Nucl. Tracks*, 5:169–174.

Gleadow, A. J. W., I. R. Duddy, and J. F. Lovering, 1983. Fission track analysis: A new tool for the evaluation of thermal histories and hydrocarbon potential. *Austral. Petrol. Explor. Assoc.*, 23:93–102.

Gleadow, A. J. W., and J. F. Lovering, 1974. The effect of weathering on fission track dating. *Earth Planet. Sci. Lett.*, 22:163–168.

Gleadow, A. J. W., and J. F. Lovering, 1977. Geometry factor for external track detectors in fission-track dating. *Nucl. Track Detect.*, 1:99–106.

Green, P. F., I. R. Duddy, A. J. W. Gleadow, and J. F. Lovering, 1989a. Apatite fission-track analysis as paleotemperature indicator for hydrocarbon exploration. In N. D. Naeser and T. H. McCulloh (Eds.), *Thermal Histories of Sedimentary Basins*, pp. 181–195. Springer-Verlag, New York.

Green, P. F., I. R. Duddy, G. M. Laslett, K. A. Hegarty, A. J. W. Gleadow, and J. F. Lovering, 1989b. Thermal annealing of fission tracks in apatite: 4. Qualitative modelling techniques and extensions to the geologic timescales. *Chem. Geol.*, 79:155–182.

Green, P. F., and S. A. Durrani, 1978. A quantitative assessment of geometry factors for use in fission-track studies. *Nucl. Track Detect.*, 2:207–213.

Haack, U., 1977. The closing temperature for fission track retention in minerals. *Am. J. Sci.*, 277:459–464.

Haack, U., 1983. Reconstruction of the cooling history of the Damara orogen by correlation of radiometric ages with geography and altitude. In H. Martin and F. W. Eder (Eds.), *Intracontinental Fold Belts*, pp. 873–884. Springer-Verlag, Heidelberg.

Harrison, T. M., R. L. Armstrong, C. W. Naeser, and J. E. Harakel, 1979. Geochronology and thermal history of the Coast Plutonic Complex, near Prince Rupert, British Columbia. *Can. J. Earth Sci.*, 16:400–410.

Hurford, A. J., A. J. W. Gleadow, and C. W. Naeser, 1976. Fission-track dating of pumice from the KBS tuff, East Rudolf, Kenya. *Nature*, 263:738–740.

Hurford, A. J., and P. F. Green, 1981. A reappraisal of neutron dosimetry and uranium–$238\lambda_f$ values in fission-track dating. *Nucl. Tracks*, 5:53–61.

Hurford, A. J., and P. F. Green, 1982. A users' guide to fission-track dating calibration. *Earth Planet. Sci. Lett.*, 59:343–354.

Hurford, A. J., and P. F. Green, 1983. The zeta age calibration of fission-track dating. *Chem. Geol. (Isotope Geosci. Sect.)*, 1:285–317.

Hurford, A. J., and K. Hammerschmidt, 1985. $^{40}Ar/^{39}Ar$ and K/Ar dating of the Bishop and Fish Canyon tuffs: Calibration ages for fission-track dating standards. *Chem. Geol.*, 58:23–32.

Izett, G. A., C. W. Naeser, and J. D. Obradovich, 1978. *Ages of Natural Glasses by the Fission-Track and K–Ar Methods*, U.S. Geol. Surv. Open File Rept. 78–701, pp. 189–192. U.S. Geological Survey, Washington, D.C.

Lakatos, S., and D. S. Miller, 1983. Fission-track analysis of apatite and zircon defines a burial depth of 4 to 7 km for lowermost Upper Devonian, Catskill Mountains, New York. *Geology*, 11:103–104.

Laslett, G. M., P. F. Green, I. R. Duddy, and A. J. W. Gleadow, 1987a. Thermal annealing of fission tracks in apatite, 2. A quantitative analysis. *Chem. Geol. (Isotope Geosci. Sect.)*, 65:1–13.

Laslett, G. M., P. F. Green, I. R. Duddy, and A. J. W. Gleadow, 1987b. Thermal annealing of fission tracks in fission-track length and track density in apatite. *Nucl. Tracks*, 9:29–38.

Mehta, P. P., and Rama (no initials), 1970. Annealing effects in muscovite and their influence on dating by fission track method. *Earth Planet. Sci. Lett.*, 7:82–86.

Miller, D. S., and G. A. Wagner, 1981. Fission-track ages applied to obsidian artefacts from South America using the plateau annealing and the track-size correction techniques. *Nucl. Tracks*, 5:147–156.

Naeser, C. W., 1967. The use of apatite and sphene for fission track age determinations. *Geol. Soc. Am. Bull.*, 78:1523–1526.

Naeser, C. W., 1979. Fission-track dating and geologic annealing of fission tracks. In E. Jäger and J. C. Hunziker (Eds.), *Lectures in Isotope Geology*, pp. 154–169. Springer-Verlag, Heidelberg.

Naeser, C. W., 1981. The fading of fission tracks in geological environments. *Nucl. Tracks*, 5:248–250.

Naeser, C. W., and H. Faul, 1969. Fission track annealing in apatite and sphene. *J. Geophys. Res.*, 74:705–710.

Naeser, C. W., R. A. Zimmerman, and G. T. Cebula, 1981. Fission-track dating of apatite and zircon: An interlaboratory comparison. *Nucl. Tracks*, 5:65–72.

Naeser, N. D., P. K. Zeitler, C. W. Naeser, and P. F. Cerveny, 1987. Provenance studies by fission-track dating of zircon—etching and counting procedures. *Nucl. Tracks Rad. Meas.*, 13:121–126.

Palmer, A. R., 1983. The decade of North American geology; 1983 geologic time scale. *Geology*, 11:505–540.

Poupeau, G., 1979. Datations par traces de fission de l' uranium: Principles et applications aux problèmes du Quaternaire. *Bull. Assoc. Fr. Quat.*, 58/59:15–26.

Poupeau, G., 1981. Precision, accuracy and meaning of fission-track ages. *Proc. Indian Acad. Sci., Earth Planet. Sci.*, 90:403–436.

Poupeau, G., J. Carpena, A. Chambaudet, and Ph. Romary, 1979. Fission-track plateau-age dating. In H. Francois, J. P. Massue, R. Schmitt, N. Kurtz, M. Monnin, and S. A. Durrani (Eds.), *Solid State Nuclear Track Detectors, Proc. 10th Int. Conference, Lyon, France*, pp. 965–971. Pergamon, New York.

Poupeau, G., Ph. Romary, and P. Toulhoat, 1978. *On the Fission-Track Plateau Ages of Minerals*, U.S. Geol. Surv. Open File Rept. 78–701, pp. 339–340. U.S. Geological Survey, Washington, D.C.

Price, P. B., and R. M. Walker, 1962a. Chemical etching of charged particle tracks in solids. *J. Appl. Phys.*, 33:3407.

Price, P. B., and R. M. Walker, 1962b. Observation of fossil particle tracks in natural micas. *Nature*, 196:732.

Price, P. B., and R. M. Walker, 1963a. A simple method of measuring low uranium concentrations in natural crystals. *Appl. Phys. Lett.*, 2:23.

Price, P. B., and R. M. Walker, 1963b. Fossil tracks of charged particles in mica and the age of minerals. *J. Geophys. Res.*, 68:4847–4862.

Reimer, G. M., 1975. Uranium determination in natural water by the fission-track technique. *J. Geochim. Explor.*, 4:425–432.

Reimer, G. M., D. Storzer, and G. A. Wagner, 1970. Geometry factor in fission track counting. *Earth Planet. Sci. Lett.*, 9:401–404.

Rosenberg, P. E., and R. L. Hooper, 1982. Fission-track dating of sandstone-type uranium deposits. *Geology*, 10:481–485.

Schreurs, J. W. H., A. M. Friedman, D. J. Rockop, M. W. Hair, and R. M. Walker, 1971. Calibrated U–Th glasses for neutron dosimetry and determination of uranium and thorium concentrations by the fission-track method. *Radiat. Effects*, 7:231–233.

Sharma, Y. P., N. Lal, K. D. Bal, R. Parshad, and K. K. Nagpaul, 1980. Closing temperatures of different fission-track clocks. *Contrib. Mineral. Petrol.*, 72: 335–336.

Storzer, D., 1970. Fission-track dating of volcanic glass and the thermal history of rocks. *Earth Planet. Sci. Lett.*, 8:55–60.

Storzer, D., and G. Poupeau, 1973. Ages-plateaux de minéraux et verres par la methode des traces de fission. *C. R. Acad. Sci. Paris*, 276D:137–139.

Storzer, D., and G. A. Wagner, 1969. Correction of thermally lowered fission-track ages of tektites. *Earth Planet. Sci. Lett.*, 5:463–468.

Storzer, D., and G. A. Wagner, 1971. Fission track ages of North American tektites. *Earth Planet. Sci. Lett.*, 10:435–440.

Thiel, K., and W. Herr, 1976. The ^{238}U spontaneous fission decay constant re-determined by fission tracks. *Earth Planet. Sci. Lett.*, 30:50–56.

Wagner, G., and P. van den Haute, 1992. *Fission Track Dating*. Kluwer Academic, Dordrecht, The Netherlands.

Wagner, G. A., 1974. Die Anwendung anätzbarer Partikelspuren zur geochemischen Analyse. *Fortschritte Mineral.*, 51:68–93.

Wagner, G. A., 1998. *Age Determination of Young Rocks and Artifacts*. Springer-Verlag, Heidelberg.

Wagner, G. A., G. M. Reimer, G. S. Carpenter, H. Faul, R. Van der Linden, and R. Gijbels, 1975. The spontaneous fission rate of U-238 and fission track dating. *Geochim. Cosmochim. Acta.*, 39:1279–1286.

Weiland, E. F., K. R. Ludwig, C. W. Naeser, and C. E. Simmons, 1980. *Fission track dating applied to uranium mineralization*, U.S. Geol. Surv. Open-File Rept. 80–380. U.S. Geological Survey, Washington, D.C.

REFERENCES FOR THERMOLUMINESCENCE (SECTION 22.3)

Adamiec, G., and M. J. Aitken, 1998. Dose-rate conversion factors: update. *Ancient TL*, 16:37–50.

Aitken, M. J., 1978. Archaeological involvements of physics. Physics Report, Section C. *Phys. Lett.*, 40(5):277–351.

Aitken, M. J., 1985. *Thermoluminescence Dating*. Academic, London.

Aitken, M. J., 1998. *An Introduction to Optical Dating*. Oxford University Press, Oxford.

Aitken, M. J., R. Grün, V. Mejdahl, D. Miallier, H. Rendell, A. Wieser, and A. Wintle, 1994. Proceedings of the 74th international specialist seminar on thermoluminescene, and electron spin resonance dating. *Quat. Science Reb.*, 13(5–7):403–684.

Daniels, F., C. A. Boyd, and D. F. Saunders, 1953. Thermoluminescence as a research tool. *Science*, 116: 343–349.

Daniels, F., and D. F. Saunders, 1950. The thermoluminescence of rocks. *Science*, 111:462.

Dreimanis, A., G. Hütt, A. Raukas, and P. W. Whippey, 1978. Dating methods of Pleistocene deposits and their problems: 1. Thermoluminescence dating. *Geosci. Can.*, 5(2):55–60.

Grögler, N., F. G. Houtermans, and H. Stauffer, 1958. Radiation damage as a research tool for geology and prehistory. Convegno sulle dotazioni con metodi nucleari. 5th Rass. Internazion. Elettr. Nucl. sezione Nucleare Roma, pp. 5–15.

Macdougall, D. J. (Ed.), 1968. *Thermoluminescence of Geological Materials*. Academic, New York.

Nambi, K. S. V., and M. J. Aitken, 1986. Annual dose conversion factors for TL and ESR dating. *Archaeometry*, 28:202–205.

Singhvi, A. K., and V. Mejdahl, 1985. Thermoluminescence dating of sediments. *Nucl. Tracks*, 10:137–161.

Spooner, N. A., 1992. Optical dating: Preliminary results on the anomalous fading of luminescence from feldspar. *Quat. Sci. Rev.*, 11:139–145.

Spooner, N. A., 1994. The anomalous fading of infra-red stimulated luminescence from feldspars. *Radiat. Meas.*, 23:625–632.

Wagner, G. A., 1998. *Age Determination of Young Rocks and Artifacts*. Springer-Verlag, Heidelberg.

Wiedemann, E., and G. C. Schmidt, 1895. Ueber Lumineszenz. *Ann. Phys. Chem.*, 54:604–625.

REFERENCES FOR ELECTRON-SPIN RESONANCE (SECTION 22.4)

Adamiec, G., and M. J. Aitken, 1998. Dose-rate conversion factors: Update. *Ancient TL*, 16:37–50.

Aitken, M. J., R. Grün, V. Mejdahl, D. Miallier, H. Rendell, A. Wieser, and A. Wintle (Eds.), 1994. Proceedings of the 7th international specialist seminar on thermoluminescence and electron spin resonance dating. *Quat. Geochron. (Quat. Sci. Rev.)*, 13:403–679.

Blackwell, B. A. B., 1995. Electron spin resonance dating. In N. W. Rutter and N. D. Catto (Eds.), *Dating Methods for Quaternary Deposits*, GEOtext2, pp. 209–251. Geological Association of Canada, St. John's.

Blackwell, B. A. B., 2001. Electron spin resonance (ESR) dating in lacustrine environments. In W. M Last and J. P. Smol (Eds.), *Tracking Environmental Change Using Lake Sediments*, Vol. 1: Basin Analysis, Coring, and Chronological Techniques. Kluwer Academic, Dordrecht, The Netherlands.

Blackwell, B. A. B., A. R. Skinner, and J. I. B. Blickstein, 2001. ESR isochron exercise: How accurately do modern dose rate measurements reflect paleodose rates? *Quat. Sci. Rev.*, 20:1031–1039.

Grün, R., 1989a. Electron spin resonance (ESR) dating. *Quat. Int.*, 1:65–109.

Grün, R. 1989b. *Die ESR-Altersbestimmungsmethode*. Springer-Verlag, Berlin.

Grün R., 1990. Potential and problems of ESR dating. *Nucl. Tracks Radiat. Meas.*, 18:143–153.

Grün, R., and F. McDermott, 1994. Open system modeling for U-series and ESR dating of teeth. *Quat. Sci. Rev.*, 13:121–125.

Grün, R., H. P. Schwarcz, and J. Chadham, 1988. Electron spin resonance dating of tooth enamel: Coupled correction for U uptake and U-series disequilibrium. *Nucl. Tracks Radiat. Meas.*, 14:237–241.

Grün, R., and C. B. Stringer, 1991. ESR dating and the evolution of modern humans. *Archaeometry*, 33: 153–199.

Ikeya, M., 1975. Dating a stalactite by paramagnetic resonance. *Nature*, 255:48–50.

Ikeya, M., 1978. ESR as a method of dating. *Archaeometry*, 20:147–158.

Ikeya, M., 1981. Paramagnetic alanine molecular radicals in fossil shells and bones. *Naturwissenschaften*, 68:474–475.

Nambi, K. S. V., and M. J. Aitken, 1986. Annual dose conversion factors for TL and ESR dating. *Archaeometry*, 28:202–205.

Rink, W. J., 1997. Electron spin resonance (ESR) and ESR applications in Quaternary science and archaeometry. *Radiat. Meas.*, 27:975–1025.

Schwarcz, H. P., 1994. Current challenges to ESR dating. *Quat. Geochronol. (Quat. Sci. Rev.)*, 13:601–605.

Wagner, G. A., 1998. *Age Determination of Young Rocks and Artifacts*. Springer-Verlag, Heidelberg.

Zeller, E. J., 1968. Use of electron spin resonance for measurement of natural radiation damage. In D. J. McDougall (Ed.), *Thermoluminescence of Geological Materials*, pp. 271–279. Academic Press, London.

Zeller, E. J., P. W. Levy, and P. L. Mattern, 1967. Geological dating by electron spin resonance. In *Proceedings of the Symposium on Radioactive Dating and Low Level Counting*, p. 531. International Atomic Energy Agency, Vienna.

CHAPTER 23

Cosmogenic Radionuclides

THE Earth, like all other solid objects in the solar system, is exposed to cosmic rays and to the solar wind. Cosmic rays consist primarily of highly energetic nuclei of H and He (i.e., protons and α-particles) but also contain the nuclei of elements of higher atomic number. The energy of cosmic-ray protons varies widely from a few billion electron volts (GeV) up to about 10^9 GeV (i.e., 10^{18} eV). When these particles strike molecules of N_2 and O_2 in the upper atmosphere of the Earth, they cause nuclear spallation reactions that release cascades of secondary particles, including neutrons, protons, and muons. The secondary particles cause additional nuclear reactions in the lower atmosphere and thereby form both *stable* and *unstable* isotopes of a large number of elements. A certain fraction of the secondary cosmic-ray particles (e.g., neutrons and muons) actually reach the surface of the Earth and form cosmogenic radionuclides in the exposed soil and rocks (Lal, 1991). Some of the cosmogenic radionuclides have sufficiently long halflives to be useful for geochronometry on a timescale ranging up to about five million years. The halflives and decay constants of these cosmogenic radionuclides are listed in Table 23.1. In addition, the reaction products of cosmic rays include stable nuclides such as $^3_2$He and $^{21}_{10}$Ne.

Table 23.1. Halflives and Decay Constants of Cosmogenic Radionuclides Used for Geochronometry

Nuclide	Halflife, years	Decay Constant, y^{-1}
$^3_1$H	12.43	5.576×10^{-2}
$^{10}_4$Be	1.51×10^6	0.459×10^{-6}
$^{14}_6$C	5730	1.209×10^{-4}
$^{26}_{13}$Al	0.705×10^6	0.983×10^{-6}
$^{32}_{14}$Si	140 ± 6	4.95×10^{-3}
$^{36}_{17}$Cl[a]	0.301×10^6	2.30×10^{-6}
$^{39}_{18}$Ar	269	2.57×10^{-3}
$^{41}_{20}$Ca	0.104×10^6	6.73×10^{-6}
$^{53}_{25}$Mn	3.7×10^6	0.187×10^{-6}
$^{81}_{36}$Kr	0.210×10^6	3.30×10^{-6}
$^{129}_{53}$I[b]	15.7×10^6	4.41×10^{-8}

[a] Also produced by neutron irradiation of Cl in seawater following the explosion in the atmosphere of thermonuclear devices.
[b] Produced by cosmic rays in the atmosphere as well as by spontaneous fission of ^{238}U.

The solar wind likewise consists of protons and ionized atoms of other elements that occur in the Sun. All of the charged particles of the solar wind have the same velocity, but their energies range only from 0.2 to 0.6 keV/amu, which is much lower than the energies of cosmic rays. The charged particles of the solar wind interact with the magnetic field of the Earth, which causes them to flow toward the polar regions. Therefore, the production rates of cosmogenic radionuclides increase with the geomagnetic latitude and reach a maximum above the magnetic poles. Planets and smaller bodies in the solar system, which do not have an atmosphere and lack a magnetic field, are not protected from the solar wind and from cosmic rays. As a result, the rocks and minerals exposed on the surfaces of these objects (e.g., Mercury, the Moon, asteroids, and meteoroids) also contain atoms of elements that were implanted in the rocks and minerals by high-velocity impacts of ionized atoms (Maurette and Price, 1975).

The occurrence of cosmogenic tritium($^{3}_{1}H$) was discussed in Section 21.4. The other cosmogenic radionuclides in Table 23.1 (e.g., $^{14}_{6}C$, $^{10}_{4}Be$, $^{26}_{13}Al$, $^{32}_{14}Si$, $^{36}_{17}Cl$, $^{53}_{25}Mn$, and $^{129}_{53}I$) are used for dating terrestrial and extraterrestrial samples (Lal and Peters, 1962; Lal, 1963; Lal and Suess, 1968). The applications of these cosmogenic radionuclides to dating of archeological objects and geological samples of Quaternary age have been discussed by Currie (1982) and Wagner (1998), among others.

The concentrations of the cosmogenic radionuclides are very low, which initially delayed their use for geochronometry because the sensitivity of low-level radiation detectors available between 1960 and 1980 was limited (Lal and Schink, 1960; Oeschger, 1963; Lal et al., 1967). A major breakthrough in the quantitative determination of cosmogenic radionuclides occurred in the late 1970s with the construction of electrostatic tandem mass spectrometers capable of counting ions with unprecedented sensitivity (Purser et al. 1982; Wilson et al., 1984; Elmore and Phillips, 1987).

23.1 CARBON-14 (RADIOCARBON)

The production of $^{14}_{6}C$ by a nuclear reaction in the atmosphere was originally postulated by Grosse (1934) and was subsequently confirmed by Libby (1946), Anderson et al. (1947), Arnold and Libby (1949), and Libby (1952). The development of the ^{14}C method of dating by Libby and his colleagues was recognized by the Nobel Prize for Chemistry awarded to Willard F. Libby in 1960.

The ^{14}C atoms produced in the atmosphere are incorporated into CO_2 molecules, which are then absorbed by plants. As a result, all living plants contain ^{14}C, which is in steady-state equilibrium with ^{14}C in the atmosphere. When a plant dies, absorption of CO_2 from the atmosphere stops and the activity of ^{14}C in the tissue continues to decline at a rate controlled by its halflife of 5730 years. Therefore, ^{14}C dates of organic matter are calculated from the activity of the remaining ^{14}C and represent the time elapsed since the plants died.

Herbivorous animals acquire ^{14}C by eating green plants and pass the radiocarbon to the carnivorous animals who eat them. Consequently, all living animals (including humans) contain ^{14}C that is in a steady-state equilibrium with ^{14}C-bearing CO_2 of the atmosphere. Therefore, the remains of animals and humans can also be dated by the ^{14}C method.

Carbon-14 dates have been widely used in archeology and in Quaternary geology based on analyses of organic matter (plant and animal remains). In addition, calcium carbonate that was precipitated from aqueous solution in chemical equilibrium with atmospheric CO_2 can be dated by the ^{14}C method.

The study of ^{14}C also provides information about fluctuations in its production rate caused by several different factors, including variations in the intensity of the solar wind, the strength of the magnetic field of the Earth, and the global circulation of the atmosphere. The principles and applications of the isotope geoscience of ^{14}C were explained originally by Libby (1952, 1955) and later by other authors including:

Olsson, 1970, *Radiocarbon Variations and Absolute Chronology*, Almquist and Wicksell, Uppsala. Michael and Ralph, 1970, *Dating Techniques for the Archaeologist*, MIT Press, Cambridge, MA. Renfrew, 1976, *Before Civilizations: the Radiocarbon Revolution and Prehistoric Europe*, Penguin, Harmondsworth. Mook, 1980, *The Principles and Applications of Radiocarbon Dating*, Elsevier, Amsterdam. Moser and Rauert, 1980, *Isotopenmethoden in der Hydrologie*, Gebr. Bornträger, Berlin. Currie, 1982, *ACS Symp. Ser.*, 176: 516. Geyh and Schleicher, 1990, *Absolute Age Determination*, Springer-Verlag, Heidelberg. Taylor et al., 1992, *Radiocarbon after Four Decades*, Springer-Verlag, Heidelberg. Bard and Broecker, 1992, *The Latest Deglaciation*,

Springer-Verlag, Heidelberg. Wagner, 1998, *Age Determination of Young Rocks and Artifacts*, Springer-Verlag, Heidelberg.

In addition, the field is served by its own journal, entitled *Radiocarbon*.

23.1a Principles

The production of ^{14}C takes place in the stratosphere 12–15 km above the surface of the Earth by an (n, p) reaction with stable $^{14}_{7}N$:

$$^{14}_{7}N + ^{1}_{0}n \rightarrow ^{14}_{6}C + ^{1}_{1}p \tag{23.1}$$

where $^{1}_{0}n$ is an energetic cosmogenic neutron in the atmosphere and $^{1}_{1}p$ is a proton. The concentration of ^{14}C in the atmosphere is maintained by a balance between its rate of production and its rate of removal in meteoric precipitation. The concentration of the ^{14}C in the atmosphere has varied in the past because of fluctuations in its natural production rate and because of anthropologic interferences in the carbon cycle of the Earth.

The nucleus of $^{14}_{6}C$ is unstable and decays by β-emission to stable $^{14}_{7}N$:

$$^{14}_{6}C \rightarrow ^{14}_{7}N + \beta^{-} + \overline{\nu} + Q \tag{23.2}$$

where $Q = 0.15648$ MeV, $\overline{\nu}$ is the antineutrino, and the decay is to the groundstate of $^{14}_{7}N$ (i.e., no γ-ray is emitted).

After the plant or animal dies, the activity of ^{14}C per gram of C decreases with time because the plant or animal tissue is no longer in communication with ^{14}C-bearing CO_2 of the atmosphere. The ^{14}C activity any time after the death of the plant or animal is expressed by the law of radioactivity (equation 3.11) and is illustrated in Figure 23.1:

$$A = A_0 e^{-\lambda t} \tag{23.3}$$

Taking natural logarithms yields:

$$\ln A = \ln A_0 - \lambda t \tag{23.4}$$

which is the equation of a straight line whose slope m is:

$$m = -\lambda$$

and intercept on the y-axis is:

$$b = \ln A_0$$

Solving equation 23.4 for t:

$$t = \frac{1}{\lambda}(\ln A_0 - \ln A) \tag{23.5}$$

where $\lambda = 1.209 \times 10^{-4}$ y^{-1} $(T_{1/2} = 5730$ y)
A_0 = initial activity of ^{14}C in the sample expressed in disintegrations per minute per gram of C (dpm/g of C), and
A = measured activity of ^{14}C in the same units as A_0

The exponential decrease of the decay rate and the sensitivity of β-counting systems (e.g., Geiger counters and scintillation detectors, Sections 3.2a and 3.2b) limit the range of the ^{14}C method of dating to less than 30,000 years (i.e., five halflives). Although accelerator mass spectrometry (AMS) can extend the range to 75,000 years, the oldest achievable dates are actually about 60,000 years or less (Wagner, 1998).

The ^{14}C date calculated from equation 23.5 precedes the time the material was used by humans or was buried. Therefore, ^{14}C dates, in principle, overestimate the archeological age of artifacts by varying amounts of time ranging from days to decades. In addition, the numerical value of the date calculated from equation 23.5 depends not only on the measured activity of the remaining ^{14}C (A) but also on the *initial activity* of ^{14}C in the sample at the time of death (A_0).

23.1b Assumptions

The ^{14}C method of dating is based on a set of assumptions that apply to the production of ^{14}C in the atmosphere and its subsequent partitioning into different terrestrial reservoirs, including the biosphere, oceans, lakes, groundwater, soil on the surface of the continents, and both marine and lacustrine sediment. Libby's model of the geochemistry of ^{14}C is based on several assumptions:

1. The concentration of ^{14}C in the troposphere (0–10 km above the surface) is everywhere the

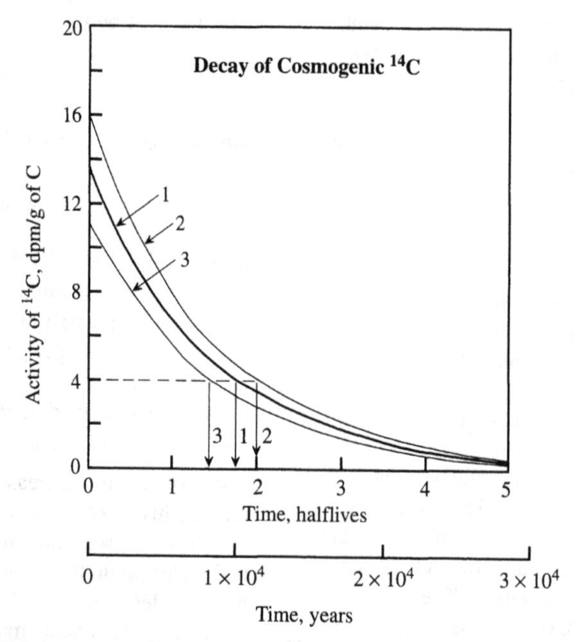

FIGURE 23.1 Decay curves of cosmogenic ^{14}C for three different values of the initial ^{14}C activity (A_0). Line 1 starts with $A_0 = 13.56$ dpm/g of C, whereas lines 2 and 3 have 18% higher and lower values of A_0, respectively. The ^{14}C date of a sample whose present activity is 4.0 dpm/g of C depends on the assumed value of A_0: $t_1 = 10{,}100$ y; $t_2 = 11{,}470$ y; $t_3 = 8{,}460$ y. If the calculated ^{14}C date of a sample of wood is less than its corresponding tree-ring date, the initial ^{14}C activity was higher than the assumed value, and vice versa.

same even though its production rate increases with the geomagnetic latitude and with altitude above the surface of the Earth.

2. The initial activity of ^{14}C in organic matter has remained constant everywhere on the surface of the Earth during the past 70,000 years.
3. The specific activity of ^{14}C (dpm/g of C) of different types of photosynthetic plants does not depend on environmental factors such as the composition of soil, climate, and speciation of plants.
4. All radioactive contaminants are excluded by appropriate procedures prior to analysis.
5. Special assumptions apply for ^{14}C dating of bones, mollusk shells, calcareous ooze, pottery, mortar, and groundwater.

The validity of assumption 1 was originally demonstrated by measurements of the specific activity of ^{14}C in plants collected at different geomagnetic latitudes from about 60° N to 65° S (Anderson and Libby, 1951). Subsequently, Libby (1955) reported an average present-day ^{14}C activity in plant tissue of 13.56 ± 0.07 dpm/g of C. The apparent uniformity of the specific decay rate of ^{14}C in modern plants implies that the distribution of ^{14}C in the atmosphere is homogeneous on a global scale. In addition, these results suggest that the speciation of plants, as well as differences in the composition of soil and the character of the climate, do not strongly affect the ^{14}C activity of modern plants (i.e., assumption 3 is not violated).

A serious problem for ^{14}C dating arises from assumption 2 concerning the absence of time-dependent variations of the initial ^{14}C activities of plants and other C-bearing materials. Research by deVries (1958) and many other investigators has shown that radiocarbon dates of wood and marine corals deviate systematically from ages determined by counting tree rings (dendrochronology) and by the U-series disequilibrium methods (Chapter 20), respectively. These discrepancies have been attributed to variations of the ^{14}C concentration of the atmosphere in response to natural and anthropogenic causes.

The principal natural cause for the past variation of the ^{14}C concentration of the atmosphere (and hence the value of A_0) is the fluctuation of the production rate of ^{14}C due to variations in the flux of charged particles that enter the atmosphere of the Earth. These variations are caused by fluctuations of the intensity of the solar wind and cosmic rays as well as by changes in the strength of the magnetic field of the Earth, which partially shields it from the charged particles of the solar wind (but not from the more energetic cosmic rays). The fluctuations in the intensity of the solar wind are caused by the eruption of solar flares, by the 11-year sunspot cycle, and by long-term changes in the luminosity of the Sun. Consequently, the history of variation of the initial ^{14}C concentration of geological and archeological samples can be viewed as a record of the past activity of the Sun.

In addition to these and other natural causes, the ^{14}C concentration of the atmosphere has been reduced by the combustion of fossil fuels (coal, petroleum, and natural gas), which contain only "dead" C (i.e., all ^{14}C has decayed). The reduction of the ^{14}C activity of the atmosphere by combustion of fossil fuel, called the Suess effect, started in the nineteenth century as a consequence of the Industrial Revolution (Suess, 1965). The opposite effect has resulted from the explosion of nuclear fission and fusion devices in the atmosphere starting in the 1940s, which caused the formation of so-called bomb-produced, or thermonuclear, ^{14}C. The end result of human intervention into the natural C cycle is that the ^{14}C concentration of the atmosphere is no longer controlled by natural processes and that C-bearing materials which formed after the middle of the nineteenth century cannot be dated reliably by the ^{14}C method because the initial ^{14}C activity cannot be specified, except under special circumstances.

23.1c Radiocarbon Dates

If a sample of organic matter has a measured ^{14}C activity of 10.50 dpm/g of C and if its initial activity is assumed to be 13.56 dpm/g of C, then a date can be calculated from equation 23.5:

$$t = \frac{1}{\lambda}(\ln A_0 - \ln A)$$

where $\lambda = 1.209 \times 10^{-4}\ y^{-1}$, provided the assumptions discussed in Section 23.1b are satisfied in this case. Therefore,

$$t = \frac{\ln 13.56 - \ln 10.50}{1.209 \times 10^{-4}} = \frac{0.2558}{1.209 \times 10^{-4}}$$
$$= 2115.7\ y$$

Several conventions now come into play:

1. The halflife of ^{14}C used to calculate traditional ^{14}C dates is 5568 years (Libby, 1952), rather than 5730 years. Therefore, $\lambda(\text{Libby}) = 1.244 \times 10^{-4}\ y^{-1}$ and the date calculated above is reported as $t = 2056.2$ y, which is 2.8% lower than the "correct" value.
2. Conventional ^{14}C dates are expressed in years "before present" (BP), where the "present" is the year 1950. If the date calculated above was measured in the calendar year 2000, the date is reduced by 50 years: $t = 2056.2 - 50 = 2006.2$ y BP.
3. Conventional ^{14}C dates can also be expressed relative to the Christian calendar by subtracting 1950 years: $t = 2006.2 - 1950 = 56.2$ BC, where BC means "before Christ." If a ^{14}C date expressed in years BP is less than 1950 years, the designation is AD ("anno domini"). For example, a date of 895 y BP becomes $t = 895 - 1950 = 1055$ AD. Note that BC and AD are calendar dates and do not require units.

The time-dependent variation of the initial ^{14}C activity (A_0) causes systematic errors in radiocarbon dates that are based on an assumed value of this parameter (e.g., $A_0 = 13.56 \pm 0.07$ dpm/g of C). The effect of changes in the value of A_0 on ^{14}C dates is illustrated in Figure 23.1, which contains three decay curves starting with different values of A_0.

The diagram demonstrates that the radiocarbon date of a sample ($A = 4.0$ dpm/g of C) increases when a higher value of the initial ^{14}C activity is chosen and decreases when a lower value of A_0 is chosen. Therefore, radiocarbon dates that are lower than the corresponding tree-ring ages for wood indicate that the initial ^{14}C activity was higher than the value that was used in the calculation. For example, the radiocarbon date based on decay curve 1 ($A_0 = 13.56$ dpm/g of C) in Figure 23.1 is 10,100 years, whereas the date derived from decay curve 2 ($A_0 = 16.00$ dpm/g of C) is 11,470 years, which means that, in this case, an increase of 1.0 dpm/g of C raised the radiocarbon date by 561 years. Or, to say it another way, a discrepancy of 561 years between the radiocarbon and the tree-ring date of a sample of wood having $A = 4.0$ dpm/g of C is attributable to a difference of 1.0 dpm/g of C in the value of A_0. The numerical values in this comparison are valid only for samples whose present ^{14}C activity is 4.0 dpm/g of C owing to the convergence of the decay curves after a decay time of about five halflives.

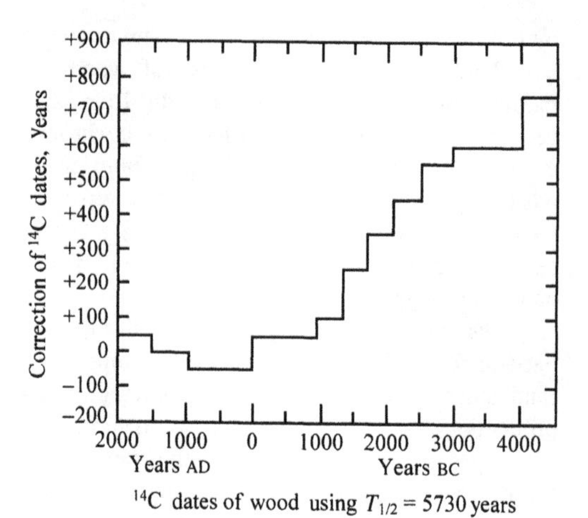

FIGURE 23.2 Corrections that must be added to radiocarbon dates to bring them into agreement with tree-ring dates. These systematic deviations of radiocarbon dates from dendrochronology dates are caused by variations in the radiocarbon concentration of the atmosphere in the past 6000 years. Data from Michael and Ralph (1970) and Ralph et al. (1973).

23.1d Secular Variations

Discrepancies between radiocarbon dates and age determinations by other methods have been documented for the entire spectrum of dates derivable by the ^{14}C method. These discrepancies can be eliminated by correcting radiocarbon dates using timescales constructed by plotting radiocarbon dates of the wood of sequoias, bristlecone pines, and Douglas firs versus the dendrochronology dates of the same samples. Figure 23.2 contains a set of such corrections for radiocarbon dates extending to about 4000 BC (Michael and Ralph, 1970).

For example, if the calculated radiocarbon date of a sample of wood is 1527 BC, the correction derived from the graph in Figure 23.2 is +250. Therefore, the corrected date in the notation of the Christian calendar is $1527 + 250 = 1777$ BC. Such corrections are especially important for dating archeological samples that originated from historical times in the past.

The discrepancies which necessitate such corrections are caused by the secular variations of the ^{14}C concentration of the atmosphere. These variations were originally documented by deVries (1958) for wood used after the birth of Christ and have now been extended to the limit of radiocarbon dating. For example, comparisons of radiocarbon and dendrochronology dates extending from zero to 12,000 BP indicate that ^{14}C dates are lower than tree-ring dates by up to 2000 years (Wagner, 1998). Similarly, Bard et al. (1990) used ^{234}U–^{230}Th dates of corals on the island of Barbados (Section 20.1c) to calibrate the ^{14}C timescale for the past 30,000 years. The results indicate that conventional ^{14}C dates are consistently lower than the ^{234}U–^{230}Th dates of the corals by amounts of time that rise to 3500 years at 20 ka. This discrepancy between ^{14}C and ^{234}U–^{230}Th dates of corals implies that the initial ^{14}C activity was *higher* than the assumed value that was used to calculate the ^{14}C dates. Such long-period changes in the specific activity of ^{14}C are attributable primarily to geophysical and astronomical causes, in contrast to the variations that occurred after about 1830 AD, which are of anthropogenic origin.

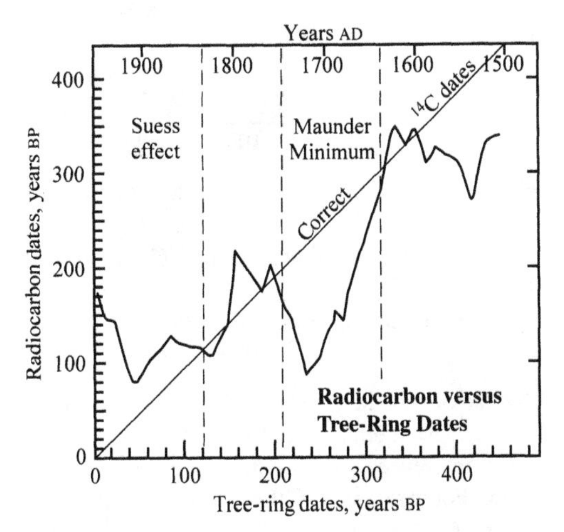

FIGURE 23.3 Systematic variations of radiocarbon dates relative to tree-ring dates in years BP (prior to 1950 AD). The Maunder Minimum and the Suess effect are both caused by changes in the ^{14}C concentration of the atmosphere. The discrepancy during the Maunder Minimum is attributable to an increase in ^{14}C concentration of the atmosphere that coincides with a decrease in the sunspot activity. The increase of radiocarbon dates between 1830 and 1950 (Suess effect) is the result of combustion of fossil fuel by humans. Adapted from Wagner (1998) based on the work of Stuiver (1978).

The radiocarbon timescale of Stuiver (1978) for the most recent past in Figure 23.3 reveals two important discrepancies. One of these occurred during the *Maunder Minimum* (Little Ice Age) between 1630 and 1740 AD when the production rate of ^{14}C increased, causing the radiocarbon dates of wood grown at that time to be too low by up to about 150 years. The other deviation from concordance is the *Suess effect*, which started at about 1830 AD and caused an increase of radiocarbon dates because of a decrease of the ^{14}C activity. The release of bomb-produced ^{14}C into the atmosphere is not indicated in Figure 23.3 because it occurred primarily after 1950 AD.

The geophysical causes of the observed variations in the global C cycle were discussed in a book edited by Bard and Broecker (1992) as well as by Edwards et al. (1993), Eisenhauer et al. (1993), and others. The mechanisms that have been proposed to account for the natural variations of the production rate of ^{14}C include not only changes in the intensity of the solar wind and the strength of the magnetic field of the Earth but also the release of old CO_2 by the oceans as a result of upwelling of deep water.

23.1e Isotope Fractionation

Carbon has two stable isotopes whose abundances and masses are:

$^{12}_{6}C$: 98.90%, 12.000000 amu
$^{13}_{6}C$: 1.10%, 13.003355 amu

The difference in the atomic masses of these isotopes (8.4%) is sufficiently large to affect the rates of chemical reactions and physical processes in which compounds containing these isotopes participate. For example, plants preferentially absorb $^{12}CO_2$ from the atmosphere during photosynthesis and discriminate against $^{13}CO_2$ and $^{14}CO_2$. Therefore, the fractionation of C isotopes during photosynthesis causes plant tissue to be depleted in ^{14}C relative to $^{12}CO_2$ of the atmosphere (Chapter 27). Conversely, isotope fractionation causes $CaCO_3$ precipitated from aqueous solutions to be enriched in ^{14}C relative to atmospheric $^{12}CO_2$. These effects introduce small but systematic errors into radiocarbon dates of organic matter and $CaCO_3$. The errors can be eliminated by measuring the isotopic composition of C in the samples to be dated by the radiocarbon method.

The isotopic composition of C is expressed by the atomic $^{13}C/^{12}C$ ratio (R), which is measured on a mass spectrometer. The value of this ratio in ordinary terrestrial C is:

$$R = \frac{1.10}{98.90} = 0.01112$$

which is an awkward number to deal with. Therefore, the isotope composition of C is expressed by the $\delta\,^{13}C$ parameter, defined as the permille difference between the isotope ratio of the sample and a C standard:

$$\delta^{13}C = \left(\frac{R_{spl} - R_{std}}{R_{std}}\right) \times 10^3‰ \qquad (23.6)$$

The C-isotope standard chosen by H.C. Urey at the University of Chicago is CO_2 prepared from the internal calcite skeleton of a Cretaceous belemnite (cephalopod) originally collected from the Peedee Formation in South Carolina. Accordingly, the standard is known as the PDB or the University of Chicago standard.

The algebraic sign of the $\delta^{13}C$ parameter depends on the $^{13}C/^{12}C$ ratio of the sample compared to that of the PDB standard. Since plants are depleted in ^{13}C (i.e., enriched in ^{12}C), their $\delta^{13}C$ values are negative, whereas the $\delta^{13}C$ values of marine carbonates are close to zero because their $^{13}C/^{12}C$ ratios are similar to those of PDB, which is also a marine carbonate. The range of $\delta^{13}C$ values in Table 23.2 that characterize different kinds of C-bearing samples is sufficiently large to cause an error of 16 years for a difference of 1‰ in $\delta^{13}C$ (Wagner, 1998).

The isotope-fractionation correction of radiocarbon dates is based on a $\delta^{13}C$ value of −25.0‰ which characterizes wood and is calculated from the equation (Craig, 1954):

$$A_c = A_m \left[1 - \frac{2(25 + \delta^{13}C_{PDB})}{1000}\right] \qquad (23.7)$$

where A_c = activity of ^{14}C corrected for isotope fractionation of C
A_m = measured activity of ^{14}C, dpm/g of C
$\delta^{13}C$ = isotope composition of the C in permil on the PDB scale

The factor "2" in equation 23.7 is the difference in the mass number of ^{14}C and ^{12}C and "25" is the reference value of the $\delta^{13}C$ parameter for wood. For example, if the measured activity of ^{14}C is: $A_m = 4.0$ dpm/g of C and $\delta^{13}C = -15‰$, equation 23.7 yields:

$$A_c = 4.0\left[1 - \frac{2(25 - 15)}{1000}\right] = 3.92 \text{ dpm/g of C}$$

which amounts to a 2% reduction of the ^{14}C activity and increases the ^{14}C date from 10,098 to 10,265 years ($\lambda = 1.209 \times 10^{-4}$ y^{-1}) or by 167 years (1.6%).

Table 23.2. Range of $\delta^{13}C$ Values of C-Bearing Compounds in Nature

Source of C	$\delta^{13}C$, ‰ (PDB)
Organic Matter	
Land plants (C3)	−32 to −20
Charcoal	−27 to −23
Peat and humus	−30 to −25
Land plants (C4) (maize, cane)	−12 to −10
Marine organic matter	−22 to −18
Collagen, animal bone	−20 to −18
Carbonate	
Marine calcite	−2 to +2
Pedogenic calcite	−10 to −4
Travertine and lacustrine calcite	−8 to +2
Bicarbonate ion	
groundwater	−16 to −10
marine	0 to +2
PDB standard	0
CO_2, atmospheric	−9 to −7

Source: Adapted from Wagner, 1998.

23.1f Methodology

The principal analytical problem of radiocarbon dating is the accurate measurement of the low level of radioactivity of ^{14}C in natural materials. This objective is achieved in three ways:

1. counting of β-particles emitted by CO_2 gas in a well-shielded ionization chamber (Section 3.2a),
2. counting of β-decays in benzene or other C-rich compound by liquid scintillation (Section 3.2b), or
3. counting of ^{14}C atoms by means of tandem accelerator mass spectrometers (Section 4.4d).

After appropriate pretreatment to remove impurities, the C is liberated as CO_2 by burning organic samples with O_2 or by reacting carbonate samples with phosphoric or hydrochloric acid. The gas thus produced is then treated to remove impurities, such as oxides of N, molecules of O_2, and halogens.

In addition, efforts are made to remove Rn. Some laboratories convert the CO_2 to methane (CH_4) or other gaseous compounds that may explode and are not always formed with 100% yield. In the latter case, the isotopes of C may be fractionated, causing the product gas to be enriched or depleted in radiocarbon.

The purified gas (e.g., CO_2, CH_4) is placed inside the ionization chamber composed of purified Cu. The volume of the ionization chamber is large (up to 8 L) and the gas is compressed (1–3 atm) in order to maximize the number of ^{14}C atoms and hence the counting rate. The ionization chamber must be shielded from background radiation, especially from γ-rays emitted by naturally occurring or anthropogenic radionuclides in the environment and from cosmic rays. The shielding consists primarily of steel plates 10–30 cm thick and of Hg. In addition, the ionization chamber is surrounded by Geiger tubes connected by an anticoincidence circuit that removes "counts" in the ionization chamber caused by cosmic-ray mesons. A further improvement is achieved by the use of a mixture of paraffin and boric acid that stops locally produced neutrons from entering the ionization chamber. Ralph (1971) reported that these measures reduced the background count of the ionization chamber in the Radiocarbon Laboratory of the University of Pennsylvania in Philadelphia from 1500 to 8 counts/min. All samples of CO_2 gas are counted for 12–24 hours or longer to accumulate 10,000 counts (or more), which yields a counting error of 1% or less. The measurement is repeated to confirm that no ^{222}Rn is present, which decays with a halflife of 3.82 days.

The procedures for measuring the ^{14}C activity of CO_2 gas in different laboratories are standardized by the use of oxalic acid (NBS 4990) containing a known amount of ^{14}C that was prepared for this purpose by the National Bureau of Standards in Washington, D.C. Replicate analyses of this standard indicate that the activity of ^{14}C in the oxalic acid standard (A_{ox}) is related to A_0, the activity of ^{14}C of wood grown prior to the contamination of the atmosphere by CO_2 derived from fossil fuel and by thermonuclear ^{14}C, by the relation:

$$A_0 = 0.95 A_{ox} \tag{23.8}$$

where the activities are expressed in dpm/g of C. Unfortunately, the combustion of the oxalic acid also causes fractionation of the C isotopes, depending on the manner in which the combustion is carried out (Craig, 1961). Therefore, the $\delta^{13}C$ value of CO_2 prepared from the oxalic acid is taken to be $-19.00‰$ and the ^{14}C activity of each batch of this CO_2 is corrected accordingly:

$$A_{ox} = A_{ox}^{m}\left[1 - \frac{2(19 + \delta^{13}C_{ox})}{1000}\right] \tag{23.9}$$

where A_{ox} = activity of CO_2 prepared from oxalic acid corrected for C-isotope fractionation to $\delta^{13}C = -19.00‰$ (PDB)

A_{ox}^{m} = measured activity of ^{14}C in a particular batch of CO_2 prepared from the oxalic acid standard

$\delta^{13}C_{ox}$ = measured value of this parameter of the CO_2 gas (PDB) prepared from the oxalic acid

The initial activity of ^{14}C in wood at the "present" time (A_0) is:

$$A_0 = 0.95 A_{ox}^{m}\left[1 - \frac{2(19 + \delta^{13}C_{ox})}{1000}\right] \tag{23.10}$$

The correction for fractionation of C isotopes by combustion of the oxalic acid standard is comparatively small. For example, if $\delta^{13}C_{ox} = -19.6‰$ for "dry" combustion, the correction factor in equation 23.10 is only 1.0012. The initial activity of ^{14}C of organic matter in relation to the activity of ^{14}C of CO_2 prepared by dry combustion of the oxalic acid standard is:

$$A_0 = 0.95 A_{ox}^{m} \times 1.0012$$

When the activities of ^{14}C of CO_2 derived from the oxalic acid standard and from the samples to be dated are measured on the same equipment, instrumental effects arising from differences in the efficiency and shielding are eliminated.

To date a C-bearing sample by the radiocarbon method, the operators of the laboratory must first establish the specific ^{14}C activity of CO_2 prepared from the oxalic acid standard and correct this value to $\delta^{13}C = -19.00‰$. They next measure the activity of CO_2 gas prepared from the sample and correct the value for C-isotope fractionation to $\delta^{13}C = -25.00‰$. The radiocarbon date is then calculated from the fractionation-corrected ^{14}C activity of the sample and from the initial activity of ^{14}C as defined by equation 23.10.

23.1g Water and Carbonates

Radiocarbon dates of carbonate samples are of great potential value in Quaternary geology, oceanography, limnology, hydrogeology, and archeology. Calcium carbonate (calcite or aragonite) that was precipitated from aqueous solution in contact with the atmosphere contains ^{14}C produced by the (n,p) reaction just like plants and animals do. However, the initial activity of ^{14}C in carbonate samples depends on the environment of deposition as well as on fractionation of the C isotopes.

The ^{14}C activity of seawater decreases with depth in the oceans because of decay of this isotope as the water slowly sinks toward the bottom. As a result, seawater at the bottom of the oceans yields radiocarbon dates of about 1.7 ka (Wagner, 1998). This phenomenon is called the *marine reservoir effect*.

In places where this old bottom water mixes with surface water, the ^{14}C activity of water is decreased in proportion to the fraction of deep water. Therefore, the radiocarbon dates of biogenic calcite and aragonite that were precipitated from such mixed water are too old by amounts of time that depend on local conditions which control the upwelling of the deep water and the way it mixes with surface water (Stuiver and Braziunas, 1993). The ^{14}C deficiency of marine carbonates that were deposited in zones of upwelling causes radiocarbon dates to be too old by up to several hundred years. The marine reservoir effect is partially counteracted by the fractionation of C isotopes which enriches carbonates in ^{14}C and by the recent injection of thermonuclear ^{14}C into the atmosphere. The contamination of the atmosphere with thermonuclear ^{14}C means that present-day samples of marine carbonate cannot be used to determine the initial ^{14}C activity of shells that formed in the past at a particular site in the oceans. Examples of recent radiocarbon studies include the work of Brückner and Halfar (1994), who dated mollusk shells from beach terraces and elevated shorelines in Spitsbergen to determine the history of uplift of the island and of sealevel change.

The release of "aged" CO_2 by the southern oceans also causes a small but measurable increase of about 30 years in the radiocarbon dates of wood in the southern hemisphere compared to wood in the northern hemisphere (Lerman et al., 1970). The difference arises because the southern oceans have a larger surface area than the northern oceans and because mixing of air between the two hemispheres is inhibited by winds blowing in opposite direction in the equatorial region.

The chemical composition of fresh water in lakes and rivers as well as of groundwater on the continents depends significantly on the rocks and minerals it has interacted with. Water that has interacted only with silicate minerals tends to have lower concentrations of alkaline earth (Ca^{2+} and Mg^{2+}) than water that has dissolved calcite or dolomite. Water containing high concentrations of alkaline earths is referred to as "hard," whereas water containing low concentrations of these elements is characterized as "soft." These distinctions affect the ^{14}C concentrations of water because all of the bicarbonate ions in soft water originate from CO_2 of the atmosphere, which contains ^{14}C. In the case of hard water, about half of the bicarbonate ions originate from limestone, which contains only "dead" C, because all ^{14}C in limestones has decayed. Other sources of dead C include fossil fuel and volcanic gases. Therefore, radiocarbon dates of hard *groundwater* in carbonate aquifers are too old by up to several thousand years if they are calculated relative to the initial activity of ^{14}C in wood. This difficulty arises because the initial ^{14}C activity of such water is not only lower than that of wood but also varies unpredictably depending on local conditions. For example, the recharge water percolating through the zone of aeration contains ^{14}C from the atmosphere, from organic matter in the soil, and from dissolution of secondary calcite previously deposited in the zone of aeration. The ^{14}C activity of the recharge water is subsequently lowered by mixing with groundwater whose ^{14}C activity has decreased by decay and by dissolution of limestone containing only dead C.

Carbonate minerals precipitated from hard **surfacewater** likewise yield radiocarbon dates that are too old because their initial activity of ^{14}C is less than that of wood that grew in equilibrium with the atmosphere. For example, the shells of clams living in hard-water lakes whose activity of ^{14}C is one half the "normal" value yield a fictitious radiocarbon date of 5730 years (Keith and Anderson, 1963).

The problem of determining the initial ^{14}C activity of calcite and aragonite in hard-water lakes and rivers on the continents is similar to that of marine carbonates in places that are affected by upwelling of old bottom water. The Suess effect and the release of thermonuclear ^{14}C have altered the activity of ^{14}C in present-day freshwater carbonate samples on a global scale. Consequently, the shells of living mollusks cannot be used to date shells from the past even if they lived in the same water. Instead, the initial activity of ^{14}C must be determined by analysis of a carbonate sample that was collected from the same body of water at a known date prior to 1950 (i.e., pre-bomb) or, even better, prior to 1830 AD (i.e., before the Suess effect).

The problems of dating hard water have been discussed by:

Deevey et al., 1954, *Proc. Natl. Acad. Sci.*, 40:285–288. Münnich, 1957, *Naturwissenschaften*, 44:32–39. Münnich, 1968, *Naturwissenschaften*, 55:158–163. Moser and Rauert, 1980, *Isotopenmethoden in der Hydrologie*, Gebr. Bornträger, Berlin. Davis and Bentley, 1982, in Currie (Ed.), *Nuclear and Chemical Dating Techniques*, ACS Symp. Ser. 176, pp. 187–222, Am. Chem. Soc., Washington, D.C. Wagner, 1998, *Age Determination of Young Rocks and Artifacts*, Springer-Verlag, Heidelberg. Kalin, 2000, in Cook and Herczeg (Eds.), *Environmental Tracers in Subsurface Hydrology*, pp. 111–144, Kluwer Academic, Boston.

In addition, Fontes and Garnier (1979) proposed that the $^{13}C/^{12}C$ ratio (or $\delta^{13}C$ value) of hard water can be used to correct the measured activity of ^{14}C for the presence of dead bicarbonate ions released by dissolution of marine limestones.

Radiocarbon dates of soft water (and glacial ice) are more likely to be reliable than those of hard water because the initial ^{14}C activity of such water is not lowered by dissolution of limestone. Therefore, the rate of flow of groundwater in noncarbonate aquifers can be measured in cases where the recharge water displaces the existing groundwater without mixing with it (i.e., piston flow, Section 21.4).

Mollusk shells and other kinds of biogenic carbonate samples deposited in soft-water lakes and rivers yield reliable radiocarbon dates under favorable conditions. However, even soft-water lakes have been contaminated with thermonuclear ^{14}C. In addition, their ^{14}C concentration has been lowered by the Suess effect, which means that the initial ^{14}C activity of biogenic carbonate samples in soft-water lakes cannot be determined by analysis of modern samples.

In cases where the initial ^{14}C activity of biogenic or inorganic carbonate samples is known, the measured activities of ^{14}C, expressed in dpm/g of C, are corrected for isotope fractionation to $\delta^{13}C = -25.0‰$ by means of equation 23.7. The resulting dates can be expressed in terms of years BP or in terms of calendar years AD or BC, depending on the objectives of the research project.

23.1h Applications

Wood is the favored material for dating by the radiocarbon method because of its availability and good preservation. It has been used to date both geological and archeological events. Many hundreds of radiocarbon dates of wood (and peat) have provided the timescale for the fluctuations of the terminus of the continental ice sheet during the Wisconsinan glaciation of North America and during the Weichselian glaciation of northern Europe:

Falconer et al., 1965, *Science*, 147:608–609. Andrews, 1973, *Arctic and Alpine Res.*, 5:185–199. Denton and Karlen, 1975, *Quat. Res.*, 3:155–205. Beget, 1983, *Geology*, 11:389–393. Holzhauser, 1984, *Geograph. Helvet.*, 9:3–15.

More recently, Atwater et al. (1991) dated tree stumps that were buried by mud during a major earthquake between 1680 and 1720 AD along the Pacific coast of the state of Washington, USA.

Table 23.3. Material Suitable for Dating by the Radiocarbon Method

Material	References
	Organic Matter
Wood and charcoal	Warner, 1990, *PACT*, 29:159–172; Manning and Weninger, 1992, *Antiquity*, 66:636–663; Kuniholm et al., 1996, *Nature*, 381:780–783.
Seeds, nutshells, grass, twigs	Hopf, 1969, in Ucko and Dimbledy (Eds.), *The Domestication and Exploitation of Plants and Animals*, pp. 335–357, Duckworth, London.
Peat	Grootes, 1978, *Science*, 200:11–15.
Lacustrine sediment (organic)	Lister et al., 1990, *Nucl. Instrum. Methods Phys. Res.*, B5:389–393; Zolitschka et al., 2000, *Geology*, 28(9):783–786.
Paper and cloth	Damon et al., 1989, *Nature*, 337:611–615.
Humus in soil	Scharpenseel and Schiffmann, 1977, *Zeitsch. Pflanzenernähr. Bodenkunde*, 140:159–174; Head et al., 1988, *Radiocarbon*, 31:680–696.
Bones (hydroxyapatite)	Saliege et al., 1995, *J. Archaeol. Sci.*, 22:302–312.
Bones (collagen)	Taylor, 1987, *Nucl. Instrum. Methods Phys. Res.*, B29:159–163; Hedges and van Klinken, 1992, *Radiocarbon*, 34:279–291; Stafford et al., 1987, *Radiocarbon*, 29:24–44.
	Carbonates
Mollusk shells (marine, inorganic carbon)	Uerpmann, 1990, *PACT*, 29:335–347.
Mollusk shells (nonmarine, inorganic carbon)	Evin et al., 1980, *Radiocarbon*, 22:545–555.
Eggshells	Long et al., 1983, *Radiocarbon*, 25:533–539.
Corals	Fairbanks, 1989, *Nature*, 342:637–642; Eisenhauer et al., 1993, *Earth Planet. Sci. Lett.*, 114:529–547.
Foraminifera	Andrée, 1984, *Nucl. Instrum. Methods Phys. Res.*, B5:340–345.
Mortar	Reller et al., 1992, *Material Res. Soc. Symp. Proc.*, 267:1007–1011.
	Miscellaneous
Iron	van der Merwe, 1969, *The Carbon-14 Dating of Iron*, University of Chicago Press, Chicago, IL.
Metallurgical slag	Bill et al., 1984, *Nucl. Instrum. Methods Phys. Res.*, B5:317–320.
Groundwater	Fontes, 1992, in Taylor et al. (Eds.), *Radiocarbon after Four Decades*, pp. 242–261, Springer-Verlag, Heidelberg; Geyh, 1992, in Taylor et al. (Eds.), *Radiocarbon after Four Decades*, pp. 276–287, Springer-Verlag, Heidelberg; Davis and Bentley, 1982, in Currie (Ed.), *Nuclear and Chemical Dating Techniques*, ACS Symp. Ser. 176, pp. 187–222, Am. Chem. Soc., Washington, D.C.; Münnich, 1968, *Naturwissenschaften*, 55:158–163.

Source: Ralph, 1971; Wagner, 1998.

A study by Valdes and Bischoff (1989) exemplifies the radiocarbon dating of *charcoal* from the El Castillo cave in Cantabria, Spain. This site was occupied from the Middle to the Late Paleolithic Age about 40,000 years ago. Because this time exceeds the range of radiocarbon dating by the conventional β-counting method, the authors used accelerator mass spectrometry (AMS) to measure the remaining ^{14}C atoms. The charcoal samples were purified by removing root hairs and by leaching them with dilute HCl and NaOH. The resulting 5-mg samples yielded an average date of 38.7 ± 1.9 ka BP. This result agrees with another AMS radiocarbon date of 39.6 ± 1.5 ka BP reported later by Hedges et al. (1994) and with ESR dates of 36.2 ± 4.1 ka (Section 22.4) of dentine and tooth enamel from the El Castillo cave measured by Rink et al. (1996).

The development of agriculture in the eastern Mediterranean region and its adoption in other areas have been investigated by radiocarbon dates of seeds and grains. The first evidence for the domestication of plants was presented by Hopf (1969), who obtained a radiocarbon date of about 10,000 years for *cereal grain* recovered at Jericho. In addition, radiocarbon dates of grains indicate that agriculture was practiced in Europe at about 6400 BC (Wagner, 1998).

Age determinations of *peat* by Grootes (1978) with ^{14}C enrichment (Hedges and Moore, 1978) provided a record of climate change in northwestern Europe extending to 75 ka. Similarly, Lister et al. (1990) used AMS to date organic matter in sediment of Lake Zürich and thereby reconstructed the late glacial and postglacial history of the lake. A similar study was carried out by Zolitschka et al. (2000) on sediment of Lake Holzmaar in the Eifel region of Germany.

Paper and *cloth* are both datable by radiocarbon, although some kinds of paper contain rags that are older than the wood used in its manufacture. Radiocarbon dates of cloth need to be corrected for C-isotope fractionation depending on the kinds of plants from which it was made. A famous example of radiocarbon dating of cloth was reported in 1989 by P.E. Damon and 20 collaborators in three radiocarbon laboratories (Arizona, Oxford, and Zürich) where three 50-mg samples of the *Shroud of Turin* were dated by the AMS technique. The Shroud of Turin is reported to have been used to wrap the body of Jesus Christ after he had died on the cross. However, the average radiocarbon date of the Shroud of Turin, corrected for isotope fractionation, was found to be 691 ± 31 years BP, which is equivalent to the calendar dates 1273–1288 AD by reference to the calibrated radiocarbon timescale. Damon et al. (1989) and Jull et al. (1996) concluded from this evidence that the cloth of the Shroud of Turin was made in the Middle Ages.

The inorganic component of *bone* is composed of hydroxyapatite, which contains carbonate anions. However, the radiocarbon content of the hydroxyapatite can be compromised by sorption of ^{14}C from water in the environment. Therefore, radiocarbon dates of bones are reliable only under dry climatic conditions such as in the Sahara Desert (Saliege et al., 1995). Bones also contain an organic component called *collagen*, which was used for dating mammoth bones by Stafford et al. (1987). The results indicated that the collagen had been contaminated with humic acids whose age was about 5000 years. Consequently, the collagen dates were about 2000 years too young compared to the radiocarbon date of associated wood of $11,490 \pm 450$ years BP. Other examples of radiocarbon dating of collagen were cited by Wagner (1998).

Mollusk shells are composed of calcium carbonate in a matrix of organic material called *conchiolin*. In most cases, the carbonate fraction is used for radiocarbon dating, which requires knowledge of the initial ^{14}C activity. If the shells are marine in origin, the initial ^{14}C activity is subject to the reservoir effect (upwelling of old water) and to fractionation of C isotopes. Dating of freshwater mollusk shells is difficult because of the possible dilution of ^{14}C by dead C derived from limestone in the drainage basin.

These and other applications of radiocarbon dating are listed in Table 23.3 together with appropriate recent references, most of which were cited by Wagner (1998). The dating of groundwater was discussed in some detail in Section 23.1g.

23.2 BERYLLIUM-10 AND ALUMINUM-26 (ATMOSPHERIC)

Unstable isotopes of Be and Al are produced by nuclear spallation reactions in the atmosphere primarily by secondary cosmic-ray neutrons. The presence of unstable ^{7}Be in meteoric precipitation was originally predicted by Peters (1955) and was confirmed by Arnold and Al-Salih (1955) and Goel et al. (1956). Subsequently, Arnold (1956) and Goel et al. (1957) measured the concentration of ^{10}Be in deep-sea sediment and thereby demonstrated that this isotope of Be is also cosmogenic.

The presentation of cosmogenic ^{26}Al with ^{10}Be is appropriate for several reasons:

1. Both nuclides are produced by nuclear reactions in the atmosphere.
2. The ions of both elements are strongly sorbed to charged sites on the surfaces of solid particles.
3. The halflives of ^{10}Be and ^{26}Al in Table 23.1 are both long compared to other cosmogenic radionuclides (except $^{53}_{25}$Mn and $^{129}_{53}$I).

Consequently, both radionuclides are deposited by meteoric precipitation and dry fallout from the atmosphere and occur in all surface environments on the Earth. In addition, both radionuclides are produced in rocks exposed to cosmic-ray neutrons and muons, not only on the Earth but also on other solid objects in the solar system. For example, ^{10}Be and ^{26}Al are formed by cosmic-ray irradiation of meteoroids while they orbit the Sun. When the irradiation stops after the meteoroids have been captured by the Earth, all of the cosmogenic radionuclides continue to decay with their characteristic halflives. Therefore, interplanetary dust contributes to the inventory of ^{10}Be, ^{26}Al, and other cosmogenic radionuclides not only in soil on the continents but also in sediment as well as in glacial ice and seawater because the interplanetary dust particles are, to some extent, soluble in water.

The occurrence of cosmogenic ^{10}Be and ^{26}Al in various reservoirs on the Earth and in meteorites has supported several lines of research, including:

1. geochronometry of deep-sea sediment, ferromanganese nodules, and continental ice sheets in Antarctica and Greenland;
2. cosmic-ray exposure dating of rock surfaces and/or measurements of their erosion rates; and
3. measurement of terrestrial ages of stony meteorites (i.e., the time elapsed since their impact on the Earth).

The implementation of these and other applications was initially delayed because of the low decay rates of ^{10}Be and ^{26}Al (Crèvecoeur and Schaeffer, 1963; Kharkar et al., 1963; Lal, 1963). This difficulty was overcome by the use of cyclotrons to count ions rather than decays (Muller, 1977; Raisbeck et al., 1978) and by ultrasensitive accelerator mass spectrometers (Southon et al., 1982; Wilson et al., 1984; Elmore and Phillips, 1987). The results have focused attention on the effects of extraterrestrial phenomena on the surface of the Earth.

23.2a Principles

Beryllium ($Z = 4$) has eight isotopes whose mass numbers range from $A = 6$ to $A = 14$ with a gap at $A = 13$. Among these, only $^{9}_{4}$Be is stable, whereas $^{7}_{4}$Be (53.29 days) and $^{10}_{4}$Be (1.51×10^{6} years) occur naturally because they are cosmogenic. The halflife of ^{10}Be is much longer than that of ^{7}Be, which explains why ^{10}Be is more useful for geological research than ^{7}Be. However, ^{7}Be is easier to detect in freshly deposited rain and snow than ^{10}Be because ^{7}Be decays more rapidly than ^{10}Be and because the decay of ^{7}Be is accompanied by the emission of a γ-ray (Arnold and Al-Salih, 1955; Goel et al., 1956).

Aluminum ($Z = 13$) has at least 10 isotopes, but only $^{27}_{13}$Al is stable. All of the other known Al isotopes are radioactive and all but one have short halflives measured in minutes, seconds, and milliseconds. The only long-lived unstable isotopes of Al is $^{26}_{13}$Al (0.705×10^{6} years), which occurs naturally because it is cosmogenic (Table 23.1).

Beryllium-10 and ^{26}Al are produced in the atmosphere by spallation reactions that occur when energetic cosmic-ray neutrons collide with the nuclei of certain elements. The targets for the production of ^{10}Be are the nuclei of atoms of oxygen and nitrogen, whereas ^{26}Al is formed from the nuclei of Ar atoms. In addition, both radionuclides are produced by nuclear spallation reactions in rocks and minerals exposed to cosmic rays at the surface of the Earth.

The atoms of ^{10}Be and ^{26}Al are rapidly removed from the atmosphere by meteoric precipitation and are deposited on the surface of the continents and oceans where they are strongly sorbed to solid particles in soil (Beer et al., 1993) and in suspension in seawater, respectively. Both radionuclides are transported to the bottom of the ocean by sinking particles and are thus incorporated into the sediment where they continue to decay with their characteristic halflives.

Beryllium-10 decays by β-emission to stable $^{10}_{5}$B:

$$^{10}_{4}\mathrm{Be} \rightarrow {}^{10}_{5}\mathrm{B} + \beta^- + \overline{\nu} + Q \qquad (23.11)$$

where Q is the total decay energy equal to 0.556 MeV. The product nucleus is in its ground-state and does not emit a γ-ray. The cosmogenic $^{7}_{4}$Be decays by electron capture to stable $^{7}_{3}$Li:

$$^{7}_{4}\mathrm{Be} + \mathrm{e}^- \rightarrow {}^{7}_{3}\mathrm{Li} + \overline{\nu} + \gamma + Q \qquad (23.12)$$

The total decay energy $Q = 0.862$ MeV and the product nucleus emits a 477.59-keV γ-ray. Aluminum-26 decays by both positron emission and electron capture to stable $^{26}_{12}$Mg:

$$^{26}_{13}\mathrm{Al} \rightarrow {}^{26}_{12}\mathrm{Mg} + \beta^+ + \nu + \gamma + Q \qquad (23.13)$$

$$^{26}_{13}\mathrm{Al} + \mathrm{e}^- \rightarrow {}^{26}_{12}\mathrm{Mg} + \nu + \gamma + Q \qquad (23.14)$$

where the total decay energy $Q = 4.005$ MeV and E_{max} of the positrons is 1.16 MeV followed by two annihilation γ-rays (1.02 MeV). In addition, the product nucleus emits γ-rays at 1808.65 keV (99.8%) 1129.67 keV (92.5%), and 2938 keV (0.24%) (Lide and Frederikse, 1995). The emission of γ-rays by the decay products of ^{26}Al and ^{7}Be facilitates the measurement of their decay rates by γ-ray spectrometry, whereas ^{10}Be can only be detected by counting the comparatively weak β-particles. After the advent of accelerator mass spectrometry (AMS) in the 1970s, both ^{10}Be and ^{26}Al are now determined by this method.

The activities of ^{10}Be and ^{26}Al in the atmosphere are in a steady-state equilibrium maintained by their rates of production and removal from the atmosphere, but the production rates of both radionuclides have varied in the past, presumably in response to fluctuations in the intensity of cosmic rays and the strength of the magnetic field of the Earth.

The nuclei of ^{10}Be and ^{26}Al in *deep-sea sediment* decay as a function of time in accordance with the law of radioactivity:

$$(^{10}\mathrm{Be})_A = (^{10}\mathrm{Be})^0_A e^{-\lambda_{10} t} \qquad (23.15)$$

$$(^{26}\mathrm{Al})_A = (^{26}\mathrm{Al})^0_A e^{-\lambda_{26} t} \qquad (23.16)$$

If the average rate of sedimentation (a) is defined as in equation 20.9:

$$a = \frac{h}{t} \qquad (23.17)$$

where h is the depth below the sediment–water interface and t is the time in years since deposition of sediment at depth h, then:

$$t = \frac{h}{a} \qquad (23.18)$$

and

$$(^{10}\mathrm{Be})_A = (^{10}\mathrm{Be})^0_A e^{-\lambda_{10} h/a}$$

Taking logarithms to the base e:

$$\ln(^{10}\mathrm{Be})_A = \ln(^{10}\mathrm{Be})^0_A - \frac{\lambda_{10} h}{a} \qquad (23.19)$$

which is the equation of a straight line,

$$y = mx + b$$

in coordinates of $y = \ln(^{10}\mathrm{Be})_A$ and $x = h$, whose slope m is:

$$m = -\frac{\lambda_{10}}{a} \qquad (23.20)$$

and the intercept b on the y-axis is:

$$b = \ln(^{10}\mathrm{Be})^0_A$$

Equation 23.20 is used to determine the average rate of sedimentation from the slope of a straight line fitted to the measured activities (or number of ^{10}Be atoms remaining) per gram of sediment in samples taken at known depths below the sediment–water interface in a piston core of deep-sea sediment. After the average sedimentation rate has been determined, equation 23.18 is used to calculate dates for sediment samples taken at known depths in the same core. The maximum date that can be measured by the decay of ^{10}Be is less than about 5×10^6 years depending on the sensitivity of the method of detection.

The same development can be applied in principle to the activities of ^{26}Al in samples of a deep-sea sediment core. The use of cosmogenic ^{26}Al for dating deep-sea sediment by means of AMS is limited by the fact that clay-rich sediment has a high concentration of Al that cannot be separated from ^{26}Al by chemical methods. Therefore, the presence of stable atoms of ^{27}Al lowers the concentration of ^{26}Al in the sample pellet that is loaded into the source of the AMS and thereby degrades the sensitivity of the analysis. The interference by stable ^{27}Al is less restrictive for dating *ferromanganese nodules* and, especially, for dating *ice cores* recovered from continental ice sheets in Antarctica and Greenland. The maximum range of the ^{26}Al method is about 2.5×10^6 years or less because its halflife (0.705×10^6 years) is less than that of ^{10}Be (1.51×10^6 years).

The assumptions for dating deep-sea sediment, ferromanganese nodules, and glacier ice are:

1. The initial activity (or number of atoms) of ^{10}Be and ^{26}Al at the site of deposition remained constant during the period of deposition.
2. The activity (or number of atoms) of ^{10}Be and ^{26}Al changed only as a result of decay of these radionuclides.
3. The rate of deposition at the site where the samples were taken remained constant and deposition was not interrupted by intervals of nondeposition or erosion.
4. The deposit was not disturbed by excessive bioturbation, deposition of turbidity currents or slides, grounding of icebergs or ice shelves, submarine volcanic eruptions, and tectonic activity.

These assumptions are similar to those that apply to dating deep-sea cores by the U-series disequilibrium methods (Sections 20.1a and 20.1b).

As in the case of radiocarbon (Section 23.1), the initial activity (or number of atoms) of ^{10}Be and ^{26}Al in sediment and ferromanganese nodules deposited in the oceans has varied because of changes in their production rates in the atmosphere and because of their unequal regional deposition on the surface of the Earth. In addition, the initial ^{10}Be activity of sediment may be lowered by the deposition of volcanic ash as a result of volcanic activity on nearby oceanic islands or submarine volcanoes. The resulting fluctuations of the initial activities of ^{10}Be and ^{26}Al in sediment cause data points to scatter above and below the best straight line in coordinates of $\ln {}^{10}Be$ and h, which increases statistical errors that bias the results derived from such data (Inoue and Tanaka, 1976) and, in most cases, invalidates this method of dating.

The processes that have caused the initial ^{10}Be activity (or number of ^{10}Be atoms) in marine sediment to vary in the past probably also affected the activity of ^{26}Al (or number of ^{26}Al) in the same way. Therefore, these variations can be reduced by considering the ratios of the activities of these isotopes:

$$\left(\frac{^{26}Al}{^{10}Be}\right)_A = \left(\frac{^{26}Al}{^{10}Be}\right)_A^0 \left(\frac{e^{-\lambda_{26}t}}{e^{-\lambda_{10}t}}\right) \qquad (23.21)$$

where

$$\frac{e^{-\lambda_{26}t}}{e^{-\lambda_{10}t}} = e^{-(\lambda_{26}-\lambda_{10})t} = e^{-\lambda' t}$$

and $\lambda' = \lambda_{26} - \lambda_{10} = 0.9831 \times 10^{-6} - 0.4590 \times 10^{-6} = 0.5241 \times 10^{-6}\ y^{-1}$. Equation 23.21 can be expressed in the same form as equations 23.15 and 23.16:

$$\left(\frac{^{26}Al}{^{10}Be}\right)_A = \left(\frac{^{26}Al}{^{10}Be}\right)_A^0 e^{-\lambda' t} \qquad (23.22)$$

which is linearized by taking natural logarithms:

$$\ln\left(\frac{^{26}Al}{^{10}Be}\right) = \ln\left(\frac{^{26}Al}{^{10}Be}\right)_A^0 - \lambda' t$$

and, by substituting $t = h/a$, yields:

$$\ln\left(\frac{^{26}Al}{^{10}Be}\right)_A = \ln\left(\frac{^{26}Al}{^{10}Be}\right)_A^0 - \frac{\lambda' h}{a} \qquad (23.23)$$

Equation 23.23 can be expressed in terms of either ratios of decay rates or numbers of atoms remaining per unit weight of sample. It is applicable to the study of cores of marine and lacustrine sediment, of ferromanganese nodules, and for dating of ice cores from the continental ice sheets of Antarctica and Greenland. A similar development is possible by combining ^{10}Be with ^{36}Cl.

23.2b Deep-Sea Sediment

An important prerequisite for dating marine and lacustrine sediment by means of ^{10}Be and ^{26}Al is the assumption that their initial activities (or atomic concentrations) have remained constant and equal to those of recently deposited sediment (Amin et al., 1966, 1975). However, Inoue and Tanaka (1976) observed that the activity of ^{10}Be in a core taken at 10°57′ S and 169°59′ W about 500 km north of the Samoan Islands in the Pacific Ocean varied irregularly with depth. They attributed the problem to changes in the rate of sediment deposition caused by nearby intermittent volcanic activity.

In a subsequent study, Tanaka and Inoue (1980) measured the ^{10}Be activities of sediment samples taken from a long core (KH 70-2-5) recovered at 39°01′ N and 169°59′ W in the North Pacific Ocean at a depth of 5245 m. Their data in Figure 23.4 demonstrate that the activities of ^{10}Be in units of dpm/kg generally decrease with increasing depth but scatter widely. A three-point moving average reveals several systematic deviations from linearity at depths of about 280, 580, and 890 cm. The first two deviations are caused by the apparent presence of excess ^{10}Be, whereas the deviation at 890 cm indicates a deficiency of ^{10}Be. These departures from the expected linear relationship may be caused by one or several of the following processes:

1. changes in the production rate of ^{10}Be in the atmosphere caused by variations of the cosmic-ray flux or of the strength of the magnetic field of the Earth;
2. changes in the rate of sediment deposition because of volcanic activity, climate change, or the presence of ice cover; and
3. intrabasinal erosion of sediment of different ages by currents, deposition of turbidites, or slumping of sediment during earthquakes.

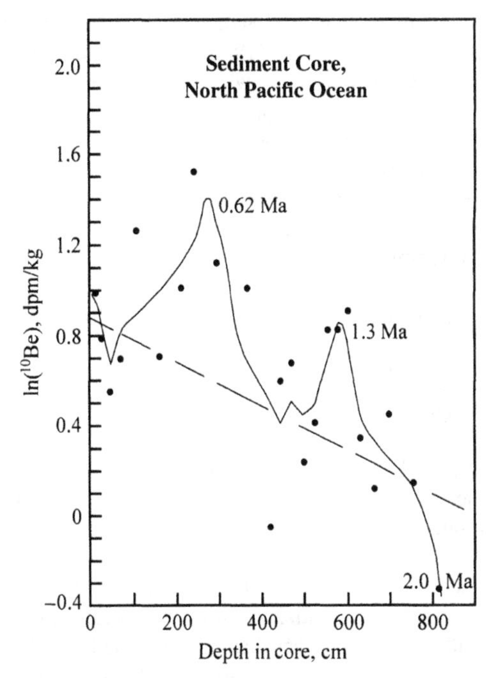

FIGURE 23.4 Variation of ^{10}Be activities with depth in sediment core KH70-2-5 from 39°01′ N, 169°59′ W in the North Pacific Ocean. The curve is a three-point moving average which reveals that the ^{10}Be activity of sediment deposited at this site has varied systematically with time. The dashed line was fitted by eye and indicates an average sedimentation rate of 4.8 mm/10^3 y or 4.8×10^{-4} cm/y. The dates of the ^{10}Be anomalies are derived from the sedimentation rate. Data from Tanake and Inoue (1980).

The dashed line in Figure 23.4, fitted to the data points by eye, has an approximate slope:

$$m = \frac{0.87 - 0.10}{800} = 9.6 \times 10^{-4}$$

The corresponding sedimentation rate is:

$$a = \frac{\lambda}{m} = \frac{0.459 \times 10^{-6}}{9.6 \times 10^{-4}} = 4.8 \times 10^{-4} \text{ cm/y}$$

or 4.8 mm/10^3 y. The sedimentation rate derived by Tanaka and Inoue (1980) from the same data is 2.7 ± 0.4 mm/10^3 y, which they determined by a statistical method applied to all data points but favoring the "higher values."

The sedimentation rate (4.8×10^{-4} cm/y) can be used to estimate the time in the past when the excess and deficiency of ^{10}Be occur in the sediment:

$$280 \text{ cm}: \quad t = \frac{280}{4.8 \times 10^{-4}} = 0.58(1.0) \text{ Ma}$$

$$580 \text{ cm}: \quad t = 1.2(2.1) \text{ Ma}$$

$$890 \text{ cm}: \quad t = 1.9\ (3.3) \text{ Ma}$$

[*Note:* The values in parentheses are based on the sedimentation rate reported by Tanaka and Inoue (1980).]

In retrospect, the ^{10}Be activities in the sediment core studied by Tanaka and Inoue (1980) suggest that the initial activities have varied episodically with time rather than randomly and therefore reflect the possible occurrence of astronomical events that caused short-term changes in the production rates of all cosmogenic nuclides in the atmosphere of the Earth. However, Somayajulu (1977) emphasized that the variations of the ^{10}Be activities (or concentrations) in marine sediment cores can be used to detect changes in the production rate only when the ages of deposition of the sediment are determined by another method (e.g., $^{230}Th_x$, $^{40}Ar/^{39}Ar$ of volcanic ash).

The existence of episodic excursions of the activity (or concentration) of ^{10}Be has been confirmed by several recent studies of sediment cores, including those of Castagnoli et al. (1995) for the Mediterranean Sea, Henken-Mellies et al. (1990) in the South Atlantic Ocean, and Southon et al. (1987). Preliminary interpretations referred to by Aldahan and Possnert (1998) indicate that enhanced ^{10}Be concentrations occurred at 35–40 ka in the Caribbean Sea and at Lake Baikal in Siberia, at 34 ± 3 ka in the Mediterranean Sea, and at 32 and 43 ka in the Gulf of California. In other words, these events appear at about the same time in different parts of the Earth, indicating that they are a global rather than a local phenomenon.

Aldahan and Possnert (1998) undertook a high-resolution study of a sediment core in an attempt to constrain the causes for the ^{10}Be peaks that had been reported in the literature. The core they chose (502B) was recovered by the Ocean Drilling Project (ODP) at 11°29′ N, 79°22′ W from a depth of 3061 m in the Caribbean Sea. The timescale for this core is based on ^{14}C dates of biogenic carbonate samples determined by the authors and on the oxygen-isotope compositions of carbonate sediment reported by DeMenocal et al. (1992). The curve in Figure 23.5 is based on three-point moving averages calculated from the dates and ^{10}Be concentrations reported by Aldahan and Possnert (1998).

The results demonstrate that the ^{10}Be concentration of the sediment deposited at this site in the Caribbean Sea during the past 66,200 years reached high values at about 23, 34?, 40, and 60 ka. In addition, an episode of low ^{10}Be deposition occurred at about 16 ka when the concentration of ^{10}Be declined to $(4.79 \pm 0.41) \times 10^8$ atoms/gram prior to the onset of the Glacial Maximum. Aldahan and Possnert (1998) noted that the ^{10}Be profile derived from the Caribbean sediment core is similar to that of the ice cores recovered at Vostok and Byrd stations in Antarctica (Section 23.2d) in spite of uncertainties in the dating of the sediment and ice samples.

The apparent synchronous occurrence of enhanced ^{10}Be deposition in marine and lacustrine sediment (e.g., Lake Baikal) as well as in continental ice sheets is attributable to fluctuations of the production rate of ^{10}Be in the atmosphere. However, the magnitude of the ^{10}Be peaks depends on the rates of deposition of sediment or ice, which are determined by local factors, including the climate and geographic location.

23.2c Ferromanganese Nodules

The decay of cosmogenic ^{10}Be and ^{26}Al was used by Guichard et al. (1978) to measure the sedimentation rate of a ferromanganese nodule (Techno-1) that was recovered from a depth of 4020 m at a site north of the Tuamotu archipelago in the South Pacific Ocean. The activities of ^{10}Be and ^{26}Al of this nodule (in dpm/kg) decrease with depth (in millimeters) in Figure 23.6*a* and yield growth rates of 2.8 ± 0.6 mm/10^6 y (^{10}Be) and

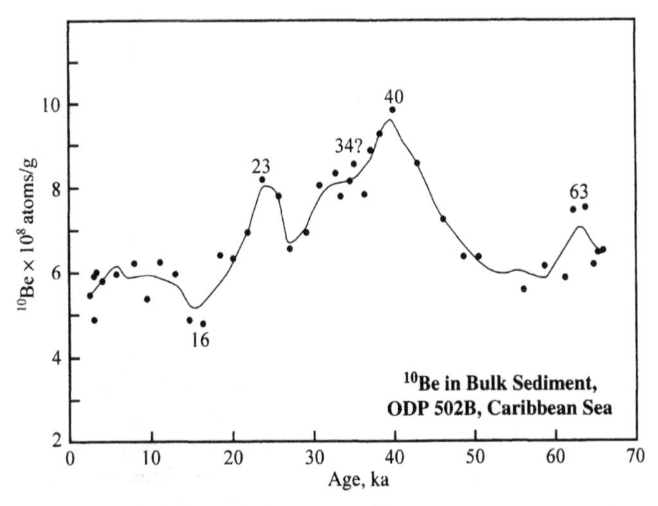

FIGURE 23.5 Variation of the concentration of ^{10}Be in bulk sediment of core 502B recovered at 11°29′ N and 79°22′ W in the Caribbean Sea. The average rate of deposition at this site is 2.56 cm/10^3 years, which is almost six times higher than the sedimentation rate of the core in the North Pacific Ocean depicted in Figure 23.4. Consequently, the core shown here represents only a short interval of time (i.e., 66,200 years) during which only about 3% of the ^{10}Be atoms present initially have decayed. The curve is a three-point moving average which suggests the presence of ^{10}Be peaks at about 23, 34?, 40, and 63 ka. Data from Aldahan and Possnert (1998).

2.3 ± 1.0 mm/10^6 y (^{26}Al). The initial activity of ^{10}Be in this nodule is 20 ± 3 dmp/kg and that of ^{26}Al is 0.23 ± 3 dpm/kg. The goodness of fit of the data points to the respective straight lines in Figure 23.6*a* confirms the assumption that the initial activities of ^{10}Be and ^{26}Al have not varied detectably during the growth of this nodule. In addition, Figure 23.6*b* demonstrates that the ^{26}Al/^{10}Be activity ratios of the ferromanganese nodule analyzed by Guichard et al. (1978) also fit a straight line as required by equation 23.23.

The growth rate *a* derived from the ^{10}Be data in Figure 23.6*a* is used to date the sample recovered 10 mm below the surface of the nodule:

$$t = \frac{10 \times 10^6}{2.8 \pm 0.6} = (3.57 \pm 0.62) \times 10^6 \text{ y}$$

The ^{26}Al date for this sample is:

$$t = \frac{10 \times 10^6}{2.3 \pm 1.0} = (4.34 \pm 1.31) \times 10^6 \text{ y}$$

The two dates have overlapping errors and, therefore, are in satisfactory agreement. These results confirm the slow rate of growth of ferromanganese nodules in the oceans, which is indicated also by the decay of unsupported ^{230}Th discussed in Section 20.1a. Additional studies of ^{10}Be in ferromanganese nodules have been reported by:

Turekian et al., 1979, *Geophys. Res. Lett.*, 6:417–420. Krishnaswami et al., 1982, *Earth Planet. Sci. Lett.*, 59:217–234. Sharma and Somayajulu, 1982, *Earth Planet. Sci. Lett.*, 59: 235–244. Kusakabe and Ku, 1984, *Geochim. Cosmochim. Acta*, 49:2187–2193. Mangini et al., 1986, *Geochim. Cosmochim. Acta*, 50: 149–156. Segl et al., 1984, *Nature*, 309:540–543. Segl et al., 1989, *Paleocenography*, 4:511–530.

The success of the ^{10}Be method of dating ferromanganese nodules has led to its use

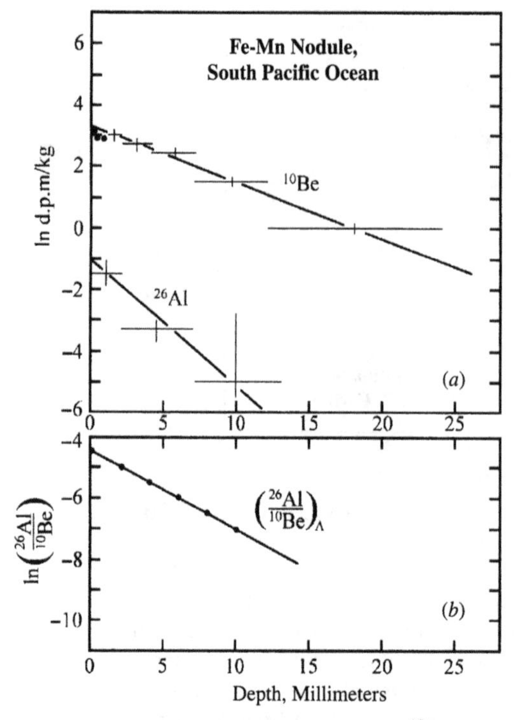

FIGURE 23.6 (*a*) Activities of cosmogenic ^{10}Be and ^{26}Al (in dpm/kg) with depth (in mm) in the Techno-1 ferromanganese nodule recovered from a depth of 4020 m north of the Tuamotu archipelago in the South Pacific Ocean. The slopes of the straight lines fitted to the data points yield growth rates of 2.8 ± 0.6 mm/10^6 years (^{10}Be) and 2.3 ± 1.0 mm/10^6 years (^{26}Al). (*b*) Linear decrease of the $^{26}Al/^{10}Be$ activity ratio with an effective halflife of 1.37×10^6 years. The data for this plot were derived by interpolation of the ^{10}Be and ^{26}Al activities of the Techno-1 ferromanganese nodule. Data from Guichard et al. (1978).

in studies of time-dependent variations of the isotope compositions of Nd and Pb in the oceans (Sections 19.3d and 19.4d). This application of ^{10}Be geochronometry is exemplified by the work of Frank and O'Nions (1998) and Claude-Ivanaj et al. (2001).

The ^{10}Be activities of the ferromanganese nodules studied by Guichard et al. (1978) and other research groups listed above decrease with depth below the surface and provide no evidence for the episodic increases of ^{10}Be observed in cores of deep-sea sediment. The reason is the slow growth of ferromanganese nodules, which does not provide the necessary time resolution to detect short-term fluctuations of the production rate. For example, a sedimentation rate of 4.8 mm/10^3 y (core KH 70-2-5; Tanaka and Inoue, 1980) means that an event lasting 1000 years is recorded in 4.8 mm (or about 0.5 cm) of sediment. In the case of ferromanganese nodules with a growth rate of 2.8 mm/10^6 y, a 1000-year event is recorded in a layer that is only 0.0028 mm thick, which cannot be sampled. The best time-resolution is provided by ice cores dating back to about 60,000 years, which can be sampled in terms of annual to decadal increments depending on the amount of annual meteoric precipitation.

23.2d Continental Ice Sheets

Cosmogenic radionuclides that form in the atmosphere and are contained in interplanetary dust are deposited on the surfaces of continental ice sheets in Antarctica and Greenland as well as on ice caps and valley glaciers in the Himalayas, the Andes, and other mountain ranges of the world. The presence of ^{10}Be and ^{26}Al in solution in 1.2×10^6 L of meltwater produced from 200-year-old ice at Camp Century (77°10′N, 61°08′W) in northwestern Greenland was originally reported by McCorkell et al. (1967). Later, Raisbeck et al. (1978) used a cyclotron to measure an average concentration of 2.6 ± 10^4 atoms of ^{10}Be per gram in ice from Dome C in East Antarctica. Analyses by means of cyclotrons and accelerator mass spectrometers have indicated that the concentration of ^{10}Be in ice cores from Antarctica and Greenland has varied systematically during the past 60,000 years. The ^{10}Be variations reported by Raisbeck et al. (1978, 1987) for Dome C in Antarctica and by Beer et al. (1984a, b) in the Dye 3 ice core (65°11′N, 43°50′W) in southcentral Greenland correlate with the isotope composition of oxygen expressed by the $\delta^{18}O$ parameter. This correlation implies that the ^{10}Be concentration of ice in Antarctica and Greenland was affected by climatic factors such as the rate of deposition of snow (i.e., a decrease in meteoric precipitation causes an increase in the concentration

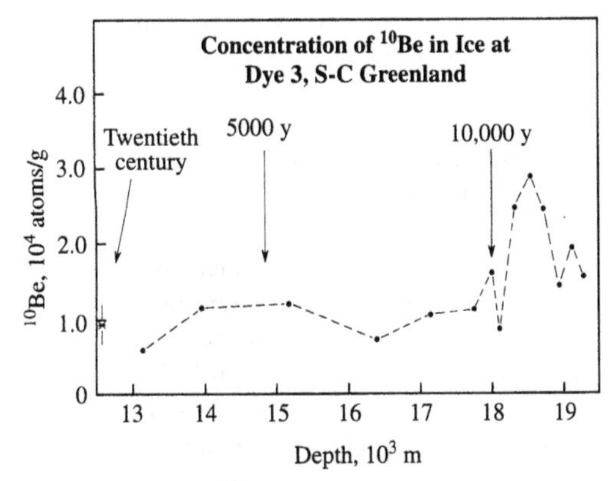

FIGURE 23.7 Concentration of ^{10}Be in an ice core at Dye 3, southcentral Greenland. The average concentration of ^{10}Be in 37 samples of firn deposited between 1900 and 1977 AD is $0.93 \pm 0.31 \times 10^4$ atoms of ^{10}Be per gram. Data from Beer et al. (1985).

of ^{10}Be in the resulting ice). In addition, Beer et al. (1984b) concluded that the ^{10}Be concentrations of ice cores between 1900 and 1977 AD correlate with the 11-year sunspot cycle. They also noted that late Wisconsinan ice in Figure 23.7 contained two to three times more ^{10}Be than Holocene ice and that its concentration decreased in step with the change in $\delta^{18}O$ during the transition to the Holocene Epoch about 10,000 years ago.

Therefore, the concentrations of ^{10}Be in the ice sheets of Antarctica and Greenland confirm the evidence derived from cores of marine sediment that the rate of deposition of ^{10}Be has not remained constant either because of changes in the atmospheric production rate of ^{10}Be or because of fluctuations in the rates of deposition of sediment or snow or both. The production rate of cosmogenic radionuclides may also be affected by changes in the intensity of the cosmic-ray flux and of the magnitude of the magnetic field of the Earth. These matters have been discussed by Beer et al. (1990, 1992), Lao et al. (1992), and Yiou et al. (1997) based on data from Greenland ice. In addition, Yiou et al. (1985) and Raisbeck et al. (1987, 1992) measured and interpreted ^{10}Be concentrations in the long core drilled at Vostok Station in East Antarctica. Their results indicate the occurrence of ^{10}Be peaks at 35 and 60 ka in ice at Vostok and Dome C. In addition, Beer et al. (1992) observed ^{10}Be peaks at 23 and 35 ka in ice cores drilled at Byrd Station on the West Antarctic ice sheet and at Camp Century in Greenland. The authors attributed the ^{10}Be peaks at 23 and 35 ka to episodes of enhanced production of this nuclide in the atmosphere. The apparent absence of the 60-ka peak in the ice cores from Byrd Station and Camp Century may be caused by inadequate sampling. Alternatively, the 60-ka ^{10}Be peak in the Vostok core may be attributable to other causes.

An important consequence of these studies is that continental ice sheets cannot be dated by the ^{10}Be method because the initial concentrations of ^{10}Be of snow deposited in the past at a specific site differs from that of modern snow deposited at the same site. On the other hand, the episodic peaks in the ^{10}Be profiles of ice cores provide markers that can facilitate the global correlation of climatic information.

It appears that the observed episodic fluctuations of the concentrations of ^{10}Be and other cosmogenic radionuclides are events of relatively short duration of the order of 10^3 years caused primarily by changes in their production rates, which depend on the flux and energy distribution of cosmic rays. In addition, the production rates are modulated by terrestrial factors, such as changes in the strength of the magnetic dipole field of the Earth, changes in the chemical composition of the atmosphere, and changes in the rates of deposition of sediment and meteoric precipitation. Episodic increases in the energy and flux of cosmic rays occur during supernova explosions in the Milky Way Galaxy (Sonett, 1992) and as a result of other kinds of astronomical events beyond the limits of our own galaxy.

23.3 EXPOSURE DATING (^{10}Be AND ^{26}Al)

Cosmogenic radionuclides are produced by nuclear spallation reactions in rocks when they are exposed to cosmic-ray neutrons and negative μ-mesons (muons) at the surface of the Earth (Lal and Arnold, 1985; Lal, 1988, 1991). The nuclides produced by these reactions include ^{10}Be, ^{14}C, ^{26}Al, ^{36}Cl, and ^{39}Ar. These radionuclides, as well as stable

Table 23.4. In Situ Produced Cosmogenic Radionuclides and the Principal Target Elements from Which They Form in Rocks Exposed to Cosmic Rays

Radionuclide	Target Elements
^{10}Be	O, Mg, Si, Fe
^{14}C	O, Mg, Si, Fe
^{26}Al	Si, Al, Fe
^{36}Cl	Fe, Ca, K, Cl
^{39}Ar	Fe, Ca, K
^{41}Ca	Ca, Fe
^{53}Mn	Fe
^{129}I	Te, Ba, La, Ce

Source: Lal, 1988.

^{3}He, ^{21}Ne, and ^{83}Kr, have been detected primarily by accelerator mass spectrometry (Brown, 1984), although counting of decays has worked for ^{36}Cl, ^{37}Ar, ^{39}Ar, and ^{129}I, whereas the stable product nuclides ^{3}He, ^{21}Ne, and ^{83}Kr are detectable by gas-source mass spectrometry. The target elements listed in Table 23.4, from which ^{10}Be, ^{26}Al, and other radionuclides are produced, include primarily O, Mg, Al, Si, and Fe, which form most of the common rock-forming minerals. Therefore, the production rates of the cosmogenic radionuclides in rocks depend to a significant extent on the chemical composition of the target rocks.

23.3a Beryllium-10 and Aluminum-26 in Quartz

The dependence of the production rates of ^{10}Be and ^{26}Al on the chemical composition of the target rocks was avoided by Lal and Arnold (1985), who selected quartz (SiO_2) as the favored target mineral because:

1. it contains two of the most important target elements (Si and O) and effectively excludes, other elements,
2. it is a common mineral that occurs in many different kinds of rocks,
3. it resists mechanical and chemical weathering, and
4. it can be concentrated by conventional mineral-separation techniques to produce samples of high purity.

The feasibility of using quartz for studies of cosmogenic ^{10}Be and ^{26}Al was subsequently demonstrated by Nishiizumi et al. (1986), whereas Klein et al. (1986) used these radionuclides to date the exposure to cosmic rays of desert glass in Libya.

The nuclear reactions that produce ^{10}Be and ^{26}Al in quartz are caused primarily by secondary cosmic-ray neutrons, which dominate in the troposphere (i.e., lower part of the atmosphere), and by negative muons, whose abundance relative to neutrons increases with depth below the surface of the Earth. The relevant reactions in quartz, according to Nishiizumi et al. (1989), are:

$$^{16}_{8}\text{O (n, 4p3n)}\ ^{10}_{4}\text{Be};\ ^{16}_{8}\text{O}\ (\mu^-,\ \text{3p3n})\ ^{10}_{4}\text{Be}$$

$$^{28}_{14}\text{Si (n, p2n)}\ ^{26}_{13}\text{Al};\ ^{28}_{14}\text{Si}\ (\mu^-,\ \text{2n})\ ^{26}_{13}\text{Al}$$

where μ^- is the negative muon. In most cases, the contribution of cosmic-ray protons to the production of ^{10}Be and ^{26}Al is $<10\%$ of the production by neutrons.

The production of a radionuclide by a nuclear reaction caused by the irradiation of a target is expressed by equation 3.38 derived in Section 3.8:

$$P = \frac{R}{\lambda}(1 - e^{-\lambda t}) \tag{23.24}$$

where P = number of atoms of the unstable product nuclide
R = production rate of the product nuclide by the nuclear reaction
λ = decay constant of the unstable product nuclide

The production rates of ^{10}Be and ^{26}Al in quartz exposed in glacially polished rocks in the Sierra Nevada Mountains of California were measured by Nishiizumi et al. (1989):

$$^{10}\text{Be} = 62.1 \pm 3.0\ \text{atoms/g/y}$$

$$^{26}\text{Al} = 375 \pm 28\ \text{atoms/g/y}$$

for a site having a geomagnetic latitude between 43.8° and 44.6° N, an elevation between 2100 and

3600 m above sealevel, and assuming an exposure age of 11,000 years.

The production rates of ^{10}Be and ^{26}Al in quartz measured in the Sierra Nevada Mountains were later scaled by Lal (1991) for geomagnetic latitudes (L) and elevations in kilometers above sealevel (y) by reference to the cosmic-ray neutron flux in the atmosphere all over the Earth. The resulting production rates R are expressed by means of a polynomial equation of the form:

$$R = a(L) + b(L)y + c(L)y^2 + d(L)y^3 \quad (23.25)$$

The polynomial coefficients, which are functions of the geomagnetic latitude L, are listed in Table 23.5. For example, the production rate of ^{10}Be in quartz at geomagnetic latitude $L = 40°$ and at an elevation of 3.0 km is:

$$R(^{10}\text{Be}) = 5.594 + 4.946 \times 3.0 + 1.3817 \times (3.0)^2 + 0.53176(3.0)^3 = 47.224 \text{ atoms/g/y}$$

The data listed in Table 23.5 are valid for elevations between 0 and 10 km.

The production rates of ^{10}Be and ^{26}Al in Figure 23.8 increase with geomagnetic latitude but become virtually constant at latitudes greater than about 50° N and 50° S. In addition, the production rates of both radionuclides increase with rising elevation above the surface of the Earth, especially in the polar regions at latitudes greater than 50°.

The production rates of cosmogenic radionuclides in quartz, as well as in other minerals and rocks exposed to cosmic rays, are reduced in cases where the collecting site is partially shielded by the surrounding topography. This problem can be minimized by collecting samples on the summits of mountains rather than in the valleys. The summits of mountains not only provide a full exposure to the sky but also maximize the production rates, especially at geomagnetic latitudes greater than 50°, and they are less likely to be covered by colluvium, talus, and migrating sand dunes than valley sites. However, the summits of mountains may be covered intermittently by ice caps or snow fields, and they may be subject to mechanical weathering by frost heaving, rock slides, and abrasion by wind. Another point to consider is that the surfaces of rocks exposed to the atmosphere are contaminated by ^{10}Be and ^{26}Al in meteoric precipitation. These contaminants must be removed by acid leaching prior to analysis. The ^{10}Be method was used by Clark et al. (2003) to date quartz-bearing

Table 23.5. Polynomial Coefficients for Equation 23.25 Which Expresses the Production Rates of ^{10}Be and ^{26}Al in Quartz at Different Geomagnetic Latitudes and Elevations above Sealevel Expressed in Kilometers

Geomagnetic Latitude	Polynomial Coefficients							
	a		*b*		*c*		*d*	
	^{10}Be	^{26}Al	^{10}Be	^{26}Al	^{10}Be	^{26}Al	^{10}Be	^{26}Al
0°	3.511	21.47	2.547	15.45	0.95125	5.751	0.18608	1.1154
10°	3.360	22.0	2.522	15.32	1.0668	6.444	0.18830	1.1287
20°	4.0607	24.84	2.734	16.61	1.2673	7.652	0.22529	1.842[a]
30°	4.994	30.55	3.904	23.67	1.320[a]	8.400[a]	0.42671	25563
40°	5.594	34.21	4.946	29.92	1.3817	9.150[a]	0.53176	3.1853
50°	6.064	37.08	5.715	34.57	1.6473	9.955	0.68684	4.1138
60°–90°	5.994	36.67	6.018	36.38	1.7045	10.30	0.71184	4.2634

Source: Lal, 1991.

[a] Adjusted by interpolation.

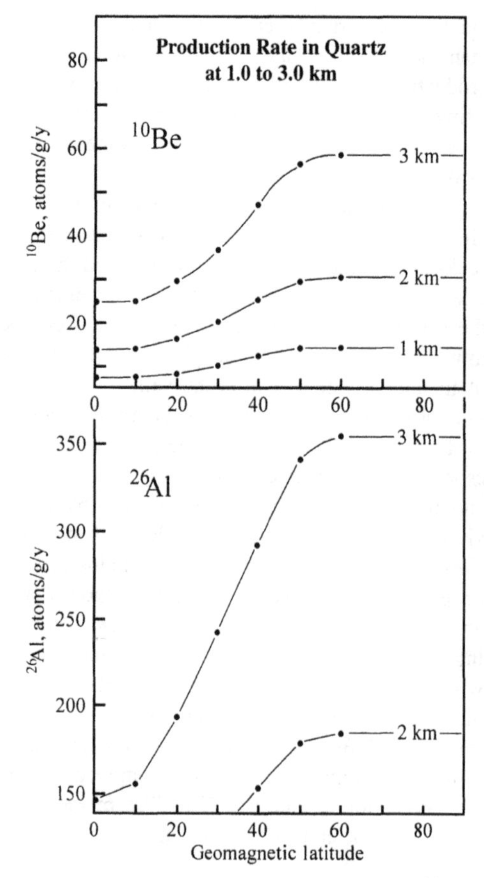

FIGURE 23.8 Production rates of cosmogenic ^{10}Be and ^{26}Al in quartz as functions of the geomagnetic latitude and elevation of the site at 1–3 km above sealevel. Calculated from Table 23.5 after Lal (1991).

boulders of late Pleistocene moraines in the Torngat Mountains of Labrador.

23.3b Erosion Rates

The production rate of cosmogenic radionuclides by in situ irradiation of quartz-bearing rocks decreases exponentially with depth below the surface depending on the density of the rocks and the range of neutrons and muons. In addition, the cosmogenic radionuclides that form at shallow depth can be removed by erosion. The loss of these radionuclides gives the appearance that they are decaying faster than expected based on their halflives. In other words, the loss of radionuclides by erosion requires that the decay constant (λ) must be increased by the addition of a term that includes the erosion rate (ε) , the density of the mineral or rock (ρ), and the effective range of the nuclear particles or their *absorption mean free path* (Λ) (Lal, 1991):

$$\lambda' = \lambda + \frac{\varepsilon\rho}{\Lambda} \tag{23.26}$$

If ε is expressed in centimeters per year, ρ in grams per cubic centimeter, and Λ in grams per square centimeter,

$$\frac{\varepsilon\rho}{\Lambda} = \frac{\text{cm}}{\text{y}} \times \frac{\text{g}}{\text{cm}^3} \times \frac{\text{cm}^2}{\text{g}} = \frac{1}{\text{y}}$$

which makes it dimensionally compatible with λ. Therefore, a complete description of in situ production of cosmogenic radionuclides in quartz-bearing rocks including the erosion rate (ε) is (Lal, 1991):

$$P = \frac{R}{\lambda'}(1 - e^{-\lambda' t}) \tag{23.27}$$

where λ' is defined by equation 23.26 and t is the duration of the cosmic-ray irradiation (i.e., the exposure age).

The functional relationship between the concentration of ^{10}Be (P) in quartz and the exposure age t is illustrated in Figure 23.9 for specified values of the relevant parameters:

$$\varepsilon = 1 \times 10^{-6}\ \text{cm/y}$$

$$\rho = 2.653\ \text{g/cm}^3 \quad \text{(quartz)}$$

$$\Lambda = 150\ \text{g/cm}^2 \quad \text{(sandstone, Brown et al., 1992)}$$

$$\lambda' = \lambda + \frac{\varepsilon\rho}{\Lambda} = 0.4617 \times 10^{-6} + \frac{1 \times 10^{-6} \times 2.653}{150}$$

$$= 0.4793 \times 10^{-6}\text{y}^{-1}$$

The production rate R of ^{10}Be in quartz at about 84° south latitude (Antarctica) and at an elevation of 2.0 km is obtained from Table 23.5:

$$R = 5.994 + 6.018 \times 2 + 1.7045 \times 4 + 0.71184 \times 8$$

$$= 30.54\ \text{atoms/g/y}$$

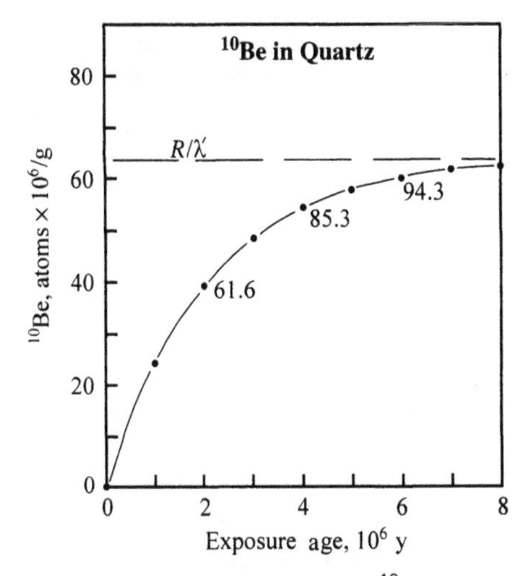

FIGURE 23.9 Growth of cosmogenic ^{10}Be in quartz based on equation 23.27 assuming that the erosion rate $\varepsilon = 1 \times 10^{-6}$ cm/y, the density $\rho = 2.653$ g/cm^3, the absorption mean free path $\Lambda = 150$ g/cm^2 (Brown et al., 1992), and the production rate $R = 30.54$ atoms/g y at latitude $L = 84°$ and elevation $y = 2.0$ km (Table 23.5). The concentration of ^{10}Be at saturation ($t = \infty$) is 63.71×10^6 atoms per gram of quartz. The numbers along the curve are the percent of saturation reached at increasing exposure ages (e.g., 94.3% at $t = 6 \times 10^6$ y).

Therefore, equation 23.27 takes the form:

$$P = \frac{30.54}{0.4793 \times 10^{-6}}(1 - e^{-0.4793 \times 10^{-6} t})$$

The concentration of ^{10}Be in quartz in Figure 23.9 increases from zero at $t = 0$ to 63.71×10^6 atoms/g at saturation ($t = \infty$). The slope of the growth curve decreases from about 24.26×10^6 per million years ($t = 0$ to $t = 1$ Ma) to 2.21×10^6 per million years ($t = 5$ to $t = 6$ Ma). If the analytical error of P is $\pm 2 \times 10^6$ atoms/g, the corresponding error in the exposure age rises from ±82,400 years ($t = 0$ to $t = 1$ Ma) to ±905,000 years ($t = 5$ to $t = 6$ Ma). Evidently, the ^{10}Be method of measuring exposure ages of quartz, under the conditions of this example, is limited to dates less than about five million years, or about 90% of saturation.

Equation 23.27 embodies several assumptions that limit the accuracy of the exposure age derived from the measured concentration of ^{10}Be (or ^{26}Al) in a specimen of quartz:

1. The rate of in situ production of ^{10}Be (or ^{26}Al) was constant.
2. The sample was irradiated continuously without interruptions.
3. The site of the irradiation was not shielded from cosmic rays by topographic obstructions.
4. All initial ^{10}Be (or ^{26}Al) resulting from previous cosmic-ray irradiations of the sample had decayed prior to the last irradiation.

In contrast to ^{10}Be dating of sediment cores (Section 23.2c), episodic variations of the production rate are not a serious problem because exposure dating is based on the number of ^{10}Be atoms that have accumulated during the exposure to cosmic rays, whereas the number of ^{10}Be atoms in a sample of sediment depends on its production rate in the atmosphere one or two years prior to deposition of the sediment.

The geological processes that cause fresh surfaces of quartz-bearing rocks to be exposed to cosmic rays include:

1. the cooling of lava flows;
2. the retreat of continental ice sheets or valley glaciers, which exposes both bedrock surfaces and glacial erratics;
3. rock slides; and
4. meteorite impacts.

Alternatively, the surfaces of quartz-bearing rocks that had reached saturation of ^{10}Be (or ^{26}Al) may be covered by lava flows, landslide deposits, migrating sand dunes, or advances of glaciers. In such cases, the exposure to cosmic rays is terminated and the previously formed ^{10}Be (or ^{26}Al) atoms continue to decay with their characteristic halflives in accordance with the law of radioactivity:

$$^{10}Be = ^{10}Be_0 e^{-\lambda t}$$

where $^{10}Be_0$ is the number of ^{10}Be atoms per gram at the time the rock surface was buried, which can be estimated from equation 23.27.

23.3c The Crux of the Problem

A complete statement of the relation between the measured concentration of ^{10}Be (or ^{26}Al) in quartz to the duration of the exposure and to its average erosion rate is obtained by combining equations 23.26 and 23.27:

$$P = \frac{R}{\lambda + \varepsilon\rho/\Lambda}(1 - e^{-(\lambda+\varepsilon\rho/\Lambda)t}) \qquad (23.28)$$

This equation contains two unknown variables (t and ε) and therefore does not yield a unique solution unless one of the two variables is known. The functional relationship between the exposure age and the erosion rate of a sample of quartz, whose ^{10}Be (or ^{26}Al) concentration has been measured, is demonstrated by the following example:

$$^{10}\text{Be} = 40.00 \times 10^6 \text{ atoms/g}$$

$$\rho = 2.653 \text{ g/cm}^3$$

$$\Lambda = 150 \text{ g/cm}^2$$

$$\text{R} = 30.54 \text{ atoms/g/y}$$

$$\lambda = 0.4590 \times 10^{-6} \text{ y}^{-1}$$

The curve in Figure 23.10 was generated by calculating t from equation 23.28 for selected values of the erosion rate ε. The results show that the exposure age increases only slightly for low erosion rates between 1×10^{-8} and 1×10^{-7} cm/y but rises steeply when the erosion rate is between 1×10^{-6} and 1×10^{-5} cm/y.

Examination of equation 23.28 indicates that the maximum value of the erosion rate ε occurs when:

$$\frac{P(\lambda\Lambda + \varepsilon\rho)}{R\Lambda} = 1.0$$

and therefore:

$$\varepsilon = \frac{\Lambda(R - P\lambda)}{P\rho} \qquad (23.29)$$

For the case under consideration, the maximum value of ε is:

$$\varepsilon = \frac{150(30.54 - 40 \times 10^6 \times 0.4590 \times 10^{-6})}{40 \times 10^6 \times 2.653}$$

$$= 17.2163 \times 10^{-6} \text{ cm/y}$$

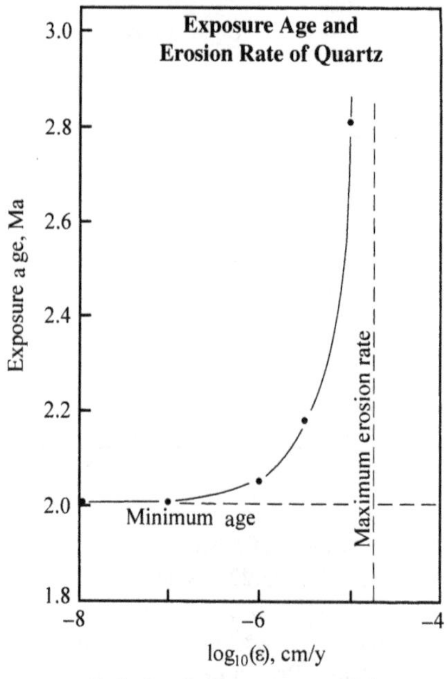

FIGURE 23.10 Relation between the exposure age t and the erosion rate ε of quartz whose measured ^{10}Be concentration P is 40.00×10^6 atoms/g and the absorption mean free path Λ is 150 g/cm^2. In addition, the annual production rate of ^{10}Be (R), consistent with the geomagnetic latitude and elevation of the collecting site, is 30.54 atoms/g. The curve was plotted by solving equation 23.28 for t at selected values of ε. The exposure age approaches infinity at the maximum erosion rate.

When this value of ε is selected, equation 23.28 reduces to:

$$1.0 = 1 - e^{-\lambda' t}$$

where λ' is defined by equation 23.26. It follows that:

$$e^{-\lambda' t} = 0$$

which is true only when $t = \infty$.

Therefore, the crux of the problem in cosmic-ray exposure dating is that the results yield a maximum erosion rate and a minimum date. A unique solution is possible only when the erosion rate is assumed to be negligibly low. Alternatively, the concentration of ^{26}Al in the same sample can be used to calculate an exposure date from equation 23.28. However, the interpretation of the measured concentration of ^{26}Al is limited by the same relation between the erosion rate and the exposure age demonstrated for ^{10}Be in Figure 23.10. Nevertheless, the exposure age, calculated from ^{26}Al at a low erosion rate, should agree with the date calculated from ^{10}Be in the same sample, provided that the concentration of ^{26}Al is less than its saturation level. In other words, the exposure age must be less than about 3.5 Ma. In cases where the ^{10}Be and ^{26}Al exposure dates of a sample of quartz are concordant, they can be used to calculate a weighted average date by using the reciprocals of the analytical errors as weighting factors.

23.4 COSMOGENIC AND THERMONUCLEAR ^{36}Cl

Chlorine-36 is a long-lived radionuclide ($T_{1/2} = 301 \times 10^3$ y) that occurs naturally because it is produced by energetic cosmic-ray particles that interact with Ar and by neutrons released by the explosion of nuclear fission and fusion devices in the atmosphere. In addition, cosmogenic ^{36}Cl is produced in rocks that are exposed to cosmic rays at the surface of the Earth (Zreda et al., 1991, 1994; Liu et al., 1994; Phillips et al., 1996).

The decay of $^{36}_{17}Cl$ proceeds by emission of negative β-particles to stable $^{36}_{18}Ar$ as well as by electron capture and positron emission to stable $^{36}_{16}S$. The total decay energies are $E = 0.7086$ MeV (to ^{36}Ar) and $E = 1.1422$ MeV (to ^{36}S), including two 0.51-MeV annihilation γ-rays. The decay to ^{36}Ar dominates with a frequency to 98%.

23.4a Water and Ice

The presence of ^{36}Cl in rain water was reported by Schaeffer et al. (1960), who extracted the isotope by ion-exchange chromatography from 1000–2000 gallons of rain and surface water. The Cl was recovered as ammonium chloride, and the rate of decay of ^{36}Cl was determined by counting β-particles. The average concentration of ^{36}Cl was 507×10^6 atoms/L of water. The authors concluded that the observed abundance of ^{36}Cl in meteoric water was greater than expected from cosmic-ray interactions with Ar in the atmosphere alone and suggested that most of the ^{36}Cl had formed by neutron irradiation of seawater following thermonuclear explosions in the atmosphere (Dyrsson and Nyman, 1955).

The thermonuclear ^{36}Cl is produced by an (n,γ) reaction on $^{35}_{17}Cl$, which is one of the two naturally occurring stable isotopes of Cl (75.77%). The ^{36}Cl that forms in the atmosphere by both natural and anthropogenic processes is rapidly removed from the atmosphere by meteoric precipitation and by dry fallout and thus contaminates all surface environments on the Earth. The ^{36}Cl in meteoric precipitation and dry fallout is soluble in water and is transported by the movement of water into the subsurface and into the oceans.

Although the potential application of ^{36}Cl to hydrologic studies was originally recognized by Davis and Schaeffer (1955) and Schaeffer et al. (1960), little progress was made until Elmore et al. (1979) and Nishiizumi et al. (1979) used an accelerator mass spectrometer to measure the concentrations of ^{36}Cl in surface water and in Antarctic ice, respectively.

The availability of accelerator mass spectrometers permitted Elmore et al. (1982) to demonstrate the presence of thermonuclear ^{36}Cl in a shallow ice core (1950–1976) drilled at Dye 3 in south-central Greenland. In addition, Bentley et al. (1982) found thermonuclear ^{36}Cl in samples of water collected in the USA between 1957 and 1982. Both data sets document a rapid rise of ^{36}Cl concentrations in the 1950s followed by a more gradual decline in the late 1970s. The pulse of ^{36}Cl in Greenland ice in Figure 23.11 illustrates the production of thermonuclear ^{36}Cl followed by its efficient removal from the atmosphere after the testing of nuclear devices ended. The ^{36}Cl concentrations in the ice at Dye 3 increased from $5.4 \pm 2 \times 10^6$ atoms/kg in 1950/1951 to $330.0 \pm 20 \times 10^6$ atoms/kg in 1955/1956 and then declined to $4.0 \pm 2 \times 10^6$ atoms/kg in 1975/1976 (Elmore et al., 1979).

The concentrations of cosmogenic (nonthermonuclear) ^{36}Cl produced in the atmosphere by a nuclear reaction with ^{40}Ar range from 4.0×10^6 to 7.5×10^6 atoms/kg of ice at Dye

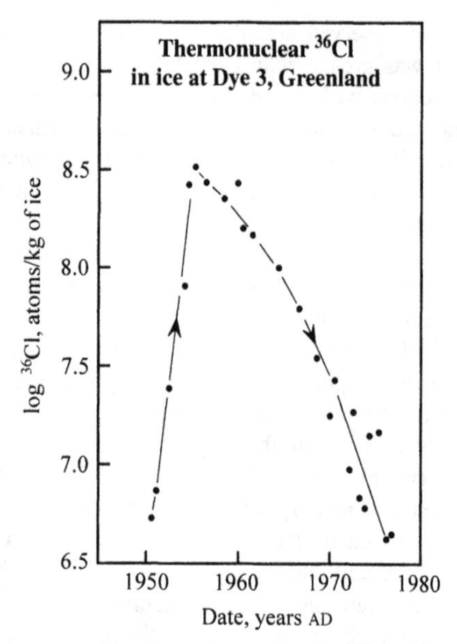

FIGURE 23.11 Concentration pulse of thermonuclear ^{36}Cl in a 100-m ice core taken at Dye 3 (65°11′ N, 43°50′ W) in southcentral Greenland. Data from Elmore et al. (1982).

3 (Elmore et al., 1982) and from 2.6×10^6 to 7.5×10^6 atoms/kg in ice of the East Antarctic ice sheet adjacent to the Allan Hills (Finkel et al., 1980). Additional measurements of ^{36}Cl concentrations by Nishiizumi et al. (1983) of ice in the Allan Hills ranged from 2.88×10^6 to 5.75×10^6 atoms/kg. The average ^{36}Cl concentration of ice in the Allan Hills reported by Finkel et al. (1980) and Nishiizumi et al. (1983) is $4.1 \pm 1.1 \times 10^6$ ($2\overline{\sigma}$) atoms/kg for nine samples.

Applications of ^{36}Cl to studies of groundwater have been described by:

Davis et al., 2001, *Chem. Geol.*, 179:3–16.
Davis et al., 1998, *Hydrogeol. J.*, 6:104–114.
Phillips, 2000, in Cook and Herczeg (Eds.), *Environmental Tracers in Subsurface Hydrology*, pp. 299–348, Kluwer Academic, Boston.
Fabryka-Martin et al., 1987, *Nucl. Instrum. Methods Phys. Res.*, 29B:361–371. Bentley et al., 1986, in Fritz and Fontes (Eds.), *Handbook of Environmental Isotope Geochemistry*, Vol. 2, pp. 427–480. Elsevier, Amsterdam.

23.4b Exposure Dating

Cosmogenic ^{36}Cl is also produced by the irradiation of rocks by cosmic rays at the surface of the Earth. The principal target elements listed in Table 23.4 are Fe ($Z = 26$), Ca ($Z = 20$), and K ($Z = 19$), which produce ^{36}Cl by nuclear spallation reactions. In addition, ^{36}Cl forms from stable ^{35}Cl by capture of thermal neutrons. The target elements listed above occur in a wide variety of minerals in virtually all kinds of rocks. Therefore, the effective production rate of ^{36}Cl must be determined from a chemical analysis of the target rocks. However, Fe and Ni atoms are not important targets in terrestrial rocks because the cosmic-ray neutrons that reach the surface of the Earth do not have sufficient energies to cause spallation reactions with these elements. In addition, Ar is insignificant as a target for the production of cosmogenic ^{36}Cl in rocks because most rocks have low Ar concentrations (Leavy et al., 1987).

The concentration of cosmogenic ^{36}Cl in rocks is conventionally expressed as the ratio of the number of ^{36}Cl atoms to the number of stable Cl atoms in the sample. The value of the ^{36}Cl/Cl ratio in a rock that has been exposed to cosmic rays for a period of t years is analogous to equation 23.24 for in situ produced ^{10}Be (Zreda et al., 1991):

$$\left(\frac{^{36}\text{Cl}}{\text{Cl}}\right)_m = \frac{ELD \times R}{\lambda_{36} \times \text{Cl}}(1 - e^{-\lambda t}) + \left(\frac{^{36}\text{Cl}}{\text{Cl}}\right)_0 \qquad (23.30)$$

where $(^{36}\text{Cl/Cl})_m$ = measured ratio of the number of cosmogenic ^{36}Cl atoms to the number of stable atoms of Cl

λ_{36} = decay constant of ^{36}Cl, $= 2.302 \times 10^{-6}\text{y}^{-1}$

t = time in years

R = production rate of ^{36}Cl by all of the applicable reactions

E = factor accounting for the elevation of the collecting site

L = factor accounting for the geomagnetic latitude of that site

D = factor accounting for the depth of the sample below the surface

$(^{36}Cl/Cl)_0$ = value of this ratio at the time the exposure to cosmic rays commenced

The initial $^{36}Cl/Cl$ ratio results from the formation of ^{36}Cl by nuclear reactions caused by the decay of U, Th, and their daughters in the sample. The value of this ratio is expected to be between 5×10^{-15} and 30×10^{-15} depending on the concentrations of U and Th and the age of the rock. Leavy et al. (1987) actually measured a value of 15×10^{-15} for a dacite that was erupted by Mt. St. Helens (Washington, USA) in 1983. In addition, the initial $^{36}Cl/Cl$ ratio may be augmented by residual cosmogenic ^{36}Cl formed during a previous exposure.

The production rate R of ^{36}Cl in a rock exposed to cosmic rays is the sum of the production rates by spallation of the nuclei of ^{39}K and ^{40}Ca atoms, by capture of thermal neutrons of stable ^{35}Cl, and by capture of negative muons of ^{40}Ca:

$$R = R_K + R_{Ca} + R_{Cl} + R_\mu \qquad (23.31)$$

The production rate by muon capture (R_μ) of ^{40}Ca is small compared to the other mechanisms, especially close to the exposed rock surface. The production rate by neutron capture at sealevel is dominant in many cases and is expressed by the equation (Phillips et al., 1986):

$$R_{Cl} = F \times \sigma_{35} \times N_{35} \qquad (23.32)$$

where F = thermal neutron flux at sealevel, $\sim 10^6$n/kg/y

σ_{35} = thermal neutron capture cross-section of ^{35}Cl, 43 barns

N_{35} = number of ^{35}Cl atoms per kilogram of sample

The production rates of ^{36}Cl by spallation reactions of ^{39}K and ^{40}Ca were originally modeled by Yokoyama et al. (1977) and were later measured by Zreda et al. (1991) and Liu et al. (1994). The most recent redetermination of the production rates of ^{36}Cl is by Phillips et al. (1996). Their results for the production rates at sealevel and high latitudes are:

$R_{Ca} = 2940 \pm 200$ atoms of ^{36}Cl per mole of Ca per year by spallation and muon capture

$R_K = 6020 \pm 400$ atoms of ^{36}Cl per mole of K per year by spallation

The fast neutron flux in air at sealevel is 586 ± 40 neutrons per gram of air per year. The production rate R at sealevel must be adjusted for the elevation E, geomagnetic latitude L, and depth D. The scaling factors E and L were calculated by Lal (1991). The variation of the production rate with depth D below the exposed surface of the rocks was investigated by O'Brien et al. (1978) and Yamashita et al. (1966).

The in situ produced ^{36}Cl was used by Zreda and Noller (1998) to date episodes of movement of the Hebgen-Lake fault in Montana. Similarly, Briner and Swanson (1998) used the residual $^{36}Cl/Cl$ ratios of bedrock in the Puget Lowlands of the state of Washington to constrain the depth of erosion during the last advance of the Cordilleran ice sheet. In addition, Liu et al. (1996) obtained cosmic-ray exposure ages of an alluvial surface in the Ajo Mountains of southern Arizona.

23.5 METEORITES

All solid objects in the solar system are irradiated by cosmic rays and by the solar wind. The energies of the nuclear particles affecting these objects are generally higher than those that reach the surface of the Earth, which is shielded by the atmosphere and by the terrestrial magnetic field. Accordingly, meteoroids in orbit around the Sun are irradiated by energetic cosmic-ray particles which produce both stable and unstable nuclides by nuclear spallation and neutron-capture reactions (Bogard et al., 1995) with the nuclei of the major elements listed in Table 23.4 (Masarik and Reedy, 1994; Schultz, 1993). Therefore, the production rates of cosmogenic nuclides in meteoroids depend not only on the flux and energy distribution of cosmic rays in interplanetary space but also on the chemical compositions of the meteoroids. In most cases, the production rates of cosmogenic nuclides in interplanetary space are higher than those that apply to the in situ production of the same nuclides in rocks exposed at the surface of the Earth.

The interpretation of the measured concentrations of cosmogenic nuclides in meteorites is based on the assumption that they are fragments of asteroids that broke up as a result of occasional collisions. Following such collisions, the fragments were exposed to cosmic rays and began to accumulate cosmogenic nuclides until the meteoroids impacted on the surface of the Earth.

The unstable cosmogenic nuclides (e.g., ^{10}Be, ^{14}C, ^{26}Al, ^{36}Cl, ^{41}Ca, ^{53}Mn) of meteoroids in space, in most cases, reach a state of saturation such that their rate of production is equal to their rate of decay. After a meteoroid has fallen to the Earth, the production of cosmogenic radionuclides stops, but they continue to decay with their respective halflives. Therefore, the measured rate of decay of the cosmogenic radionuclides of a meteorite that was found many years after its fall can be used to determine its terrestrial age (i.e., the time elapsed since it fell to Earth).

The stable cosmogenic nuclides (e.g., ^{3}He, ^{21}Ne, ^{38}Ar, ^{83}Kr, and ^{126}Xe) accumulate without decaying as long as the meteoroids reside in interplanetary space (Eugster, 1988). Consequently, the isotopic compositions and concentrations of the affected elements (e.g., He, Ne, and Kr) are altered by the addition of the respective stable cosmogenic isotopes. The concentrations of these stable cosmogenic nuclides are used to determine the duration of the exposure of a meteorite specimen to cosmic rays in interplanetary space, provided the relevant production rates are known (Schultz et al., 1991).

The production of cosmogenic nuclides in stony meteorites was summarized by Leya et al. (2000), whereas Welten et al. (2001) used cosmogenic nuclides to reconstruct the exposure histories of two suites of stony meteorite specimens that were collected from the surface of the East Antarctic icesheet near Frontier Mountain (72°59′ S, 160°18′ E).

23.5a Irradiation Ages

The stable cosmogenic nuclides that form in stony meteorites (^{3}He, ^{21}Ne, ^{38}Ar, ^{83}Kr, and ^{126}Xe) are all isotopes of noble gases. The production rates R of these nuclides in chondritic meteorites were determined by Eugster (1988), who expressed them in the form of equations. For example, the production rates for ^{3}He, ^{21}Ne, and ^{83}Kr in different chondritic meteorites are:

$$R_3 = F[2.09 - 0.43\ (^{22}\text{Ne}/^{21}\text{Ne})_c] \qquad (23.33)$$

$$R_{21} = 1.61F[21.77\ (^{22}\text{Ne}/^{21}\text{Ne})_c - 19.32]^{-1} \qquad (23.34)$$

$$R_{83} = 0.0196F[0.62\ (^{22}\text{Ne}/^{21}\text{Ne})_c - 0.53]^{-1} \qquad (23.35)$$

where R_3, R_{21} = production rates of ^{3}He and ^{21}Ne in units of 10^{-8} cm^3 STP/g/10^6 y
R_{83} = production rate of ^{83}Kr in units of 10^{-12} cm^3 STP/g/10^6 y
F = factors whose numerical values depend on the chemical compositions of the chondritic meteorites
$(^{22}\text{Ne}/^{21}\text{Ne})_c$ = cosmogenic isotope ratio which serves as the shielding indicator

The numerical values of the F-factors in equation 23.33, 23.34, and 23.35 are listed in Table 23.6. Eugster (1988) also provided production rates for cosmogenic ^{38}Ar and ^{126}Xe for all types of chondritic meteorites.

The concentrations of the cosmogenic isotopes of the noble gases in crushed fragments of stony meteorites are measured by gas-source mass spectrometry. The cosmic-ray exposure age of a specimen of a chondritic meteorite is related to the measured volume P of a specific cosmogenic nuclide and its production rate R by

$$t = \frac{P}{R} \qquad (23.36)$$

For example, the relevant data for the L chondrite Bruderheim listed by Eugster (1988) are:

$$^{21}\text{Ne} = 9.58 \pm 0.50 \times 10^{-8}\ \text{cm}^3\ \text{STP/g}$$

$$^{22}\text{Ne}/^{21}\text{Ne} = 1.094 \pm 0.012 \times 10^{-8}\ \text{cm}^3\ \text{STP/g}$$

$$F = 1.00$$

Table 23.6. Numerical Values of the *F* Factors in Equations 23.33 (^{3}He), 23.34 (^{21}Ne), and 23.35 (^{83}Kr) Expressing the Production Rates of These Stable Cosmogenic Nuclides in Chondritic Meteorites

Class of Meteorites	$F(^3He)$	$F(^{21}Ne)$	$F(^{83}Kr)$
	Carbonaceous Chondrites		
C1	1.01	0.67	0.71
CM	1.00	0.79	0.94
CO	0.99	0.96	1.02
CV	0.99	0.96	1.13
	Ordinary Chondrites		
H	0.98	0.93	1.00
L	1.00	1.00	1.00
LL	1.00	1.00	1.00
	Enstatite Chondrites		
EH	0.97	0.78	0.75
EL	1.00	0.96	0.80

Source: Eugster, 1988.

The production rate R_{21} is expressed by equation 23.34:

$$R_{21} = 1.61 \times 1.00(21.77 \times 1.094 - 19.32)^{-1}$$
$$= 0.3580 \times 10^{-8} \text{ cm}^3 \text{ STP/g}/10^6 \text{ y}$$

The irradiation age of Bruderheim based on cosmogenic ^{21}Ne is:

$$t_{21} = \frac{9.58 \times 10^{-8}}{0.3580 \times 10^{-8}} = 26.7 \pm 3.1 \text{ Ma}$$

In other words, the chondritic meteorite we know by the name of Bruderheim spent 26.7 ± 3.1 million years in interplanetary space after being liberated by a collision of its parent asteroid before impacting on the Earth.

Additional irradiation ages for Bruderheim can be calculated from the volume concentrations of ^{3}He, ^{38}Ar, ^{83}Kr, and ^{126}Xe listed by Eugster (1988). In many cases, the irradiation dates based on different stable cosmogenic nuclides are concordant. Exceptions occur, in some cases, because of loss of ^{3}He or when a significant fraction of ^{38}Ar in a meteorite originated from other sources.

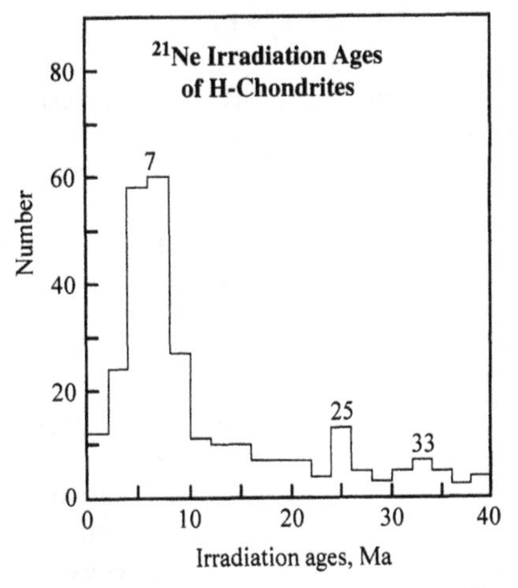

FIGURE 23.12 Cosmic-ray exposure ages of H chondrites based on the presence of stable cosmogenic ^{21}Ne. Adapted from a compilation by Schultz (1993).

The irradiation ages of chondrites range from <1.0 Ma to nearly 80 Ma. Histograms constructed by Schultz (1993) reveal that the distribution of dates is distinctly polymodal. For example, a large number of the H-chondrites in Figure 23.12 have ^{21}Ne irradiation ages of about 7 Ma. The interpretation of this information is that about 40% of all known H-chondrites formed at this time as a result of the collision of their parent body with another object in the asteroidal belt. Additional peaks occur among the LL chondrites at about 10 and 15 Ma, whereas most carbonaceous chondrites have low cosmic-ray exposure ages of about 1.0 Ma. However, iron meteorites have very long exposure ages ranging up to 1000 Ma with peaks between 400–500, 600–700, and 700–800 Ma. The longest exposure age of 2270 Ma was reported for the iron meteorite Deep Springs (Schultz, 1993).

23.5b Terrestrial Ages

After meteoroids have fallen to the surface of the Earth, they survive as recognizable extraterrestrial objects for varying lengths of time depending on their chemical and mineralogical composition, on the climatic conditions at the site, and on geological processes that may act on them. Approximately 70% of all meteorites fall into the oceans, where they are corroded by seawater and are only rarely recovered. The remainder fall in the continental areas, including the ice sheets of Antarctica and Greenland, as well as in the deserts of the major continents. The fall of meteoroids through the atmosphere is accompanied by thunderous explosions as the objects break up into showers of smaller fragments that fall to the ground without causing damage in most cases. Only the large meteoroids with diameters of tens to hundreds of meters form craters when they explode on impact [e.g., Meteor Crater near Winslow, northern Arizona, which was dated by Phillips et al. (1991) by means of cosmogenic radionuclides].

After the fall, the cosmogenic radionuclides continue to decay with their characteristic halflives. This phenomenon is used to determine how much time has passed since individual meteorite specimens fell to the Earth. The nuclides that are used for measuring such "terrestrial ages" of meteorites include ^{14}C, ^{26}Al, ^{36}Cl, and ^{81}Kr, partly because the halflives of these nuclides are similar to the terrestrial ages of most meteorites, which range from zero up to about one million years.

Some of the meteorites that fell in Antarctica have survived for unusually long periods of time because they have been preserved at low temperature in the East Antarctic ice sheet, which transported them from the interior of the continent to the coast, where the ice sublimates and the meteorites accumulate on the surface of the ice (Whillans and Cassidy, 1983). Since the late 1970s, American and Japanese investigators have recovered well over 10,000 meteorite specimens from the blue-ice ablation areas of Antarctica. The program is continuing under the leadership of Ralph Harvey of Case Western University in Cleveland, Ohio, with financial support by the Antarctic Program of the National Science Foundation of the U.S.A.

Meteorites with long exposure ages are assumed to have reached a state of saturation with respect to the cosmogenic radionuclides listed in Table 23.1, except ^{129}I, whose halflife is 15.7×10^6 years. The actual activity (or concentration) of the various cosmogenic radionuclides in a meteorite at the time of its fall can be calculated from a knowledge of its chemical composition, the relevant nuclear reaction cross-sections, the exposure age, and the flux and energy spectrum of cosmic rays in interplanetary space. The procedure for making these calculations was outlined by Welten et al. (2001). These authors calculated terrestrial dates of H-chondrites by means of the $^{14}C/^{10}Be$, $^{36}Cl/^{10}Be$, and $^{41}Ca/^{36}Cl$ ratios because the ratios of their production rates are insensitive to shielding conditions, chemical composition, and variations of the interplanetary cosmic-ray flux. Alternatively, the decay rates of long-lived cosmogenic radionuclides in different classes of meteorites can be measured directly by analysis of specimens collected soon after their fall was observed.

Aluminum-26 has been widely used to measure terrestrial ages of stony meteorites because this nuclide emits γ-rays when it decays by positron emission to ^{26}Mg (Section 23.2a). Therefore, the rate of decay is measurable by γ-ray spectrometry, which can be carried out routinely and nondestructively for the large numbers of specimens collected in Antarctica, in the world's major deserts, and elsewhere. Alternatively, ^{26}Al (like ^{10}Be, ^{14}C, ^{36}Cl, and ^{53}Mn) is determined with great sensitivity by accelerator mass spectrometry.

The activity or concentration of a cosmogenic radionuclide (e.g., ^{26}Al) remaining in a meteorite t years after its fall is:

$$^{26}\mathrm{Al}_A = (^{26}\mathrm{Al})_A^i \, e^{-\lambda t} \qquad (23.37)$$

where $^{26}Al_A$ = rate of decay of this radionuclide per gram of sample
$(^{26}Al)_A^i$ = rate of decay of ^{26}Al at saturation or at the time of fall

Terrestrial ages of meteorites based on the decay of ^{26}Al have been measured by:

Fuse and Anders, 1969, *Geochim. Cosmochim. Acta*, 33:653–670. Rowe and Clark, 1971, *Geochim. Cosmochim. Acta*, 35:727–730. Heimann et al., 1974, *Geochim. Cosmochim. Acta*, 38:217–234. Cameron and Top, 1975, *Geochim. Cosmochim. Acta*, 39:1705–1707.

Another radionuclide useful for measuring terrestrial ages of meteorites is ^{36}Cl, which has a halflife of 0.301×10^6 years. Although this radionuclide is produced in all types of meteorites, the metal grains of stony meteorites are favored because Fe is the principal target element (Table 23.4), because the metallic grains are easily concentrated from crushed samples, and because these grains have a simple chemical composition that facilitates the determination of the saturation concentration of ^{36}Cl.

For these reasons, ^{36}Cl has been the preferred cosmogenic nuclide for measuring terrestrial ages of meteorites collected in Antarctica. Such studies have been carried out by Nishiizumi et al. (1979, 1981, 1983, 1989) and by several other investigators who used ^{36}Cl and other available cosmogenic radionuclides in stony meteorites:

Honda, 1981, *Geochem. J.*, 15:163–181. Goswami and Nishiizumi, 1983, *Earth Planet. Sci. Lett.*, 64:1–8. Sarafin and Herpers, 1983, *Meteoritics*, 18:392. Brown et al., 1984, *Earth Planet. Sci. Lett.*, 67:1–8. Schultz and Freundel, 1984, *Meteoritics*, 19:310. Eugster et al., 1986, *Earth Planet. Sci. Lett.*, 78:139–147. Freundel et al., 1986, *Geochim. Cosmochim. Acta*, 50:2663–2673. Kubik et al., 1986, *Nature*, 319:568–570. Evans and Reeves, 1987, *Earth Planet. Sci. Lett.*, 82:223–230. Jull and Donahue, 1988, *Geochim. Cosmochim. Acta*, 52:1309–1311.

Nishiizumi et al. (1989) used ^{36}Cl to determine terrestrial ages of 90 meteorite specimens collected in Antarctica. They measured ^{36}Cl by accelerator mass spectrometry but converted the atomic concentrations into rates of decay in units of dpm/kg. The saturation activity of ^{36}Cl in metal grains of stony meteorites is 22.8 ± 3.1 dpm/kg of metal (Nishiizumi et al., 1983).

The ^{36}Cl activity of the meteorite ALH 77002,44 collected in the Allan Hills at the edge of the East Antarctic ice sheet in Southern Victoria Land and analyzed by Nishiizumi et al. (1989) is:

$$^{36}Cl_A = 3.08 \pm 0.23 \text{ dpm/kg}$$

The corresponding terrestrial age of this specimen is calculated from the relation:

$$^{36}Cl_A = (^{36}Cl)^i_A \, e^{-\lambda t} \qquad (23.38)$$

where $\lambda = 2.302 \times 10^{-6}$ y^{-1}. Solving for t yields:

$$t = -\frac{\ln(3.08/22.8)}{2.302 \times 10^{-6}} = 870 \pm 95 \times 10^3 \text{ y}$$

This is one of the oldest terrestrial ages among the meteorite specimens from Antarctica analyzed by Nishiizumi et al. (1989). In fact, the terrestrial ages of the stony meteorites on the Main Icefield adjacent to the Allans Hills in Figure 23.13 range from about 10,000 to 980,000 years in Figure 23.13 and reach values greater than 300,000 years only in the immediate vicinity of these mountains. Whillans and Cassidy (1983) explained the occurrence of "old" meteorites in the immediate vicinity of the Allan Hills by an ice-flow model in which the Allan Hills obstruct the flow of the East Antarctic icesheet within which the meteorites are buried and force basal ice to be deflected upward, thus causing the old meteorites to accumulate on the ice closest to the Allan Hills.

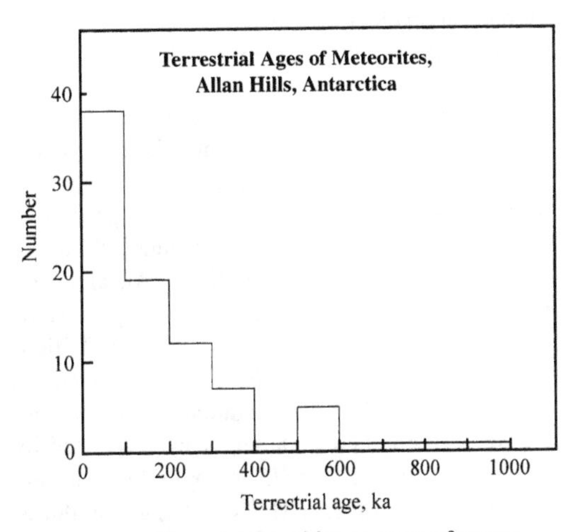

FIGURE 23.13 Terrestrial residence ages of stony meteorites, Allan Hills, Antarctica, determined by cosmogenic ^{36}Cl of the metallic grains they contain. Data by Nishiizumi et al. (1989).

23.6 OTHER LONG-LIVED COSMOGENIC RADIONUCLIDES

The number of stable and radioactive cosmogenic nuclides far exceeds the few that have been presented in this chapter. Others that have been used in geological and/or meteoritical research include $^{32}_{14}Si$, $^{41}_{20}Ca$, $^{53}_{25}Mn$, and $^{129}_{53}I$. The halflives and decay constants of these radionuclides are listed in Table 23.1. The formation and application of the long-lived cosmogenic radionuclides are briefly summarized in Table 23.7.

23.7 SUMMARY

The Earth is continuously irradiated with high-energy nuclear particles that originate from the Sun, from sources in the Milky Way Galaxy, and from the Universe at large. Although the magnetic field of the Earth deflects charged particles having moderate kinetic energies, the most energetic particles are not affected and interact with atomic nuclei in the stratosphere, thereby producing cascades of secondary particles, including neutrons and mesons, some of which reach the surface of the Earth. The secondary neutrons and mesons cause the formation of many different stable and unstable nuclides by nuclear spallation and capture reactions both in the troposphere and in rocks and soil exposed at the surface of the Earth.

Some of the unstable cosmogenic nuclides have halflives that are sufficiently long to make them useful for dating geological materials and to serve as tracers of environmental processes. The long-lived cosmogenic radionuclides discussed in this chapter include $^{10}_{4}Be$, $^{14}_{6}C$, $^{26}_{13}Al$, and $^{36}_{17}Cl$ as well as a few others mentioned only in summary form: $^{32}_{14}Si$, $^{41}_{20}Ca$ $^{53}_{25}Mn$, and $^{129}_{53}I$. Tritium ($^{3}_{1}H$) and cosmogenic $^{3}_{2}He$ were presented in Sections 21.4 and 21.5. In addition, $^{129}_{53}I$ is not only a cosmogenic radionuclide but is also produced by spontaneous fission of ^{238}U and by induced fission of ^{235}U. Several others (^{3}H, ^{14}C, ^{36}Cl) are produced by capture of neutrons released during the explosion of thermonuclear devices in the atmosphere. Therefore, these radionuclides are not only cosmogenic in origin but are also, in part, anthropogenic.

Age determinations based on the decay of the long-lived cosmogenic radionuclides rely on the assumption that their individual production rates in the atmosphere have remained constant for long periods of time. However, experience has shown that the production rates in the past have varied because the flux of cosmic-ray particles and the shielding provided by the magnetic field have both changed continuously and episodically with time. In addition, the production rates of cosmogenic radionuclides having short atmospheric residence times (e.g., ^{10}Be, ^{26}Al, and ^{36}Cl) vary with geomagnetic latitude and elevation above sealevel.

Variations of the production rates caused by latitude and elevation are particularly important for in situ produced ^{10}Be and ^{26}Al in quartz and for ^{36}Cl that form by nuclear spallation reactions and neutron capture reactions in rocks exposed to cosmic rays. Most of these kinds of deviations from ideality have been dealt with by models that improve the accuracy of age determinations based on the production and decay of cosmogenic radionuclides. As a result, cosmogenic radionuclides are being used to date a variety of archeological and geological samples (^{14}C), ferromanganese nodules in the oceans (^{10}Be and ^{26}Al), and the exposure of rock surfaces at the surface of the Earth (^{10}Be, ^{26}Al, and ^{36}Cl).

The variations of the production rates of ^{14}C and ^{10}Be in the atmosphere may record not only the past activity of the Sun but also the occurrence of supernovas and other kinds of events in our galaxy. In the final analysis, cosmic rays are a property of the interplanetary space in which the Earth exists. They affect all of the planets and their satellites in the solar system as well as asteroids and meteoroids. Consequently, cosmogenic nuclides (stable and unstable) occur in the rocks at the surface of the Moon and in the meteorites that impact on the Earth.

The study of the cosmogenic nuclides in meteorites has provided information about the flux and energy spectrum of cosmic rays in interplanetary space and about the length of time individual meteorites have spent there. In addition, ^{26}Al and ^{36}Cl are especially well suited to record how long meteorite specimens have resided on the Earth.

Some of the radionuclides that continue to form by the interaction of cosmic rays with the

Table 23.7. Formation and Application of the Long-Lived Cosmogenic Radionuclides ^{32}Si, ^{41}Ca, ^{53}Mn, and ^{129}I

Topic	References
	Silicon-32 ($T_{1/2} = 140 \pm 6$ y)
Formation	Lal et al., 1959, *Phys. Rev. Lett.*, 3(8):380; Lal et al., 1960, *Science*, 131:332–337.
Thermonuclear production	Dansgaard et al., 1966a, *J. Geophys. Res.*, 71(22):5475–5477; Dansgaard et al., 1966b, *Tellus*, 18:187–191.
Dating of ice	Dansgaard et al., 1966b, *Tellus*, 18:187–191; Clausen, 1973, *J. Glaciol.*, 12(66):411–416.
Sedimentation rate	Kharkar et al., 1969, *Earth Planet. Sci. Lett.*, 6:61–68; DeMaster, 1980, *Earth Planet. Sci. Lett.*, 48:209–217.
	Calcium-41 ($T_{1/2} = 1.04 \times 10^5$ y)
Methodology	Kubik et al., 1986, *Nature*, 319:568–570; Fink et al., 1990, *Nucl. Instrum. Methods*, B47:79–96.
Iron meteorites	Fink et al., 1991, *Earth Planet. Sci. Lett.*, 107:115–128.
Chondrites	Bogard et al., 1995, *J. Geophys. Res.*, 100:9401–9446.
Lunar soil	Nishiizumi et al., 1997, *Earth Planet. Sci. Lett.*, 148:545–552.
	Manganese-53 ($T_{1/2} = 3.7 \times 10^6$ y)
Antarctic ice	Bibron et al., 1974, *Earth Planet. Sci. Lett.*, 21:109–116.
Lunar soil	Imamura et al., 1973, *Earth Planet. Sci. Lett.*, 20:107–112; Fruchter et al., 1977, *Proc. Lunar Sci. Conf.*, 8:3595–3605.
	Iodine-129[a] ($T_{1/2} = 15.7 \times 10^6$ y)
Methodology	Studier, 1962, *J. Inorg. Nucl. Chem.*, 24:755–761; Elmore et al., 1980, *Nature*, 286:138–140; Stoffels, 1982, *Radiochem. Radioanal. Lett.*, 55:99–106.
Ore deposits	Srinivasan et al., 1971, *Science*, 173:327–328; Roman and Fabryka-Martin, 1988, *Chem. Geol.*, 72 (*Isotope Geosci.*, 8):1–6; Fabryka-Martin et al., 1988, *Chem. Geol.*, 72 (*Isotope Geosci.*, 8):7–16.
Environmental tracer, groundwater	Studier et al., 1962, *J. Inorg. Nucl. Chem.*, 24:755–761; Fabryka-Martin et al., 1985, *Geochim. Cosmochim. Acta*, 49:337–347.
Volcanic fluids	Snyder and Fehn, 2002, *Geochim. Cosmochim. Acta*, 66:3827–3838.
Deep brines	Bottomley et al., 2002, *Geology*, 30(7):587–590.
Thermonuclear	Oktay et al., 2000, *Geochim. Cosmochim. Acta*, 64:989–996.
Reservoirs, USA	Rao and Fehn, 1999, *Geochim. Cosmochim. Acta*, 63:1927–1938.

[a] Formed by spontaneous fission of ^{238}U, by spallation reactions of Xe in the stratosphere, and by (n,γ) and (n,2n) reactions on ^{128}Te and ^{130}Te (Srinivasan et al., 1971, *Science*, 173:327–328).

matter of the solar system were present in the solar nebula at the time of formation of the Sun and the planets of the solar system. The decay of these radionuclides and the accumulation of their stable products are presented in the next chapter.

REFERENCES FOR INTRODUCTION

Currie, L. A., 1982. *Nuclear and Chemical Dating Techniques*. ACS Symposium Series 176. American Chemical Society, Washington, D.C.

Elmore, D., and E. M. Phillips, 1987. Accelerator mass spectrometry for measurement of long-lived radioisotopes. *Science*, 236:543–550.

Lal, D., 1963. Study of long and short-term geophysical processes using natural radioactivity. In *Radioactive Dating*, pp. 149–157. International Atom. Energy Agency, Vienna.

Lal, D., 1991. Cosmic ray labeling of erosion surfaces: In situ nuclide production rates and erosion models. *Earth Planet. Sci. Lett.*, 104:424–439.

Lal, D., and B. Peters, 1962. Cosmic ray produced isotopes and their application to problems in geophysics. In J. G. Wilson and S. A. Wouthuysen (Eds.), *Progress in Elementary Particles and Cosmic Ray Physics*, pp. 1–74. North-Holland, Amsterdam.

Lal, D., G. Rajagopalan, and Rama (no initials), 1967. Sensitive and descript β and β-γ counting assemblies. In *Radioactive Dating and Methods of Low-Level Counting*, pp. 615–627. International Atomic Energy Agency, Vienna.

Lal, D., and D. R. Schink, 1960. Low background thin-wall flow counters for measuring beta activity in solids. *Rev. Sci. Instrum.*, 31:395–398.

Lal, D., and H. Suess, 1968. The radioactivity of the atmosphere and hydrosphere. *Annu. Rev. Nucl. Sci.*, 18:407–434.

Maurette, M., and P. B. Price, 1975. Electron microscopy of irradiation effects in space. *Science*, 187:121–129.

Oeschger, H., 1963. Low-level counting methods. In *Radioactive Dating*, pp. 13–34. International Atomic Energy Agency, Vienna.

Purser, K. H., C. J. Russo, R. B. Liebert, H. E. Gove, D. Elmore, R. Ferraro, A. E. Litherland, R. P. Peukens, K. H. Chang, L. R. Kilius, and H. W. Lee, 1982. The application of electrostatic tandems to ultrasensitive mass spectrometry and nuclear dating. In L. A. Currie (Ed.), *Nuclear and Chemical Dating Techniques*, ACS Symposium Series 176, pp. 45–74. American Chemical Society, Washington, D.C.

Wagner, G. A., 1998. *Age Determination of Young Rocks and Artifacts*. Springer-Verlag, Heidelberg.

Wilson, G. C., A. E. Litherland, and J. C. Rucklidge, 1984. Dating of sediments using accelerator mass spectrometry. *Chem. Geol.*, 14:1–17.

REFERENCES FOR CARBON-14 (RADIOCARBON) (SECTION 23.1)

Anderson, E. C., and W. F. Libby, 1951. Worldwide distribution of natural radiocarbon. *Phys. Rev.*, 81:64–69.

Anderson, E. C., W. F. Libby, S. Weinhouse, A. F. Reid, A. D. Kirshenbaum, and A. V. Grosse, 1947. Natural radiocarbon from cosmic radiation. *Phys. Rev.*, 72:931–936.

Arnold, J. R., and W. F. Libby, 1949. Age determination by radiocarbon content: Checks with samples of known age. *Science*, 110:678–680.

Atwater, B. F., M. Stuiver, and K. Yamaguchi, 1991. Radiocarbon test of earthquake magnitude at the Cascadia subduction zone. *Nature*, 353:156–158.

Bard, E., and W. S. Broecker (Eds.), 1992. *The Latest Deglaciation: Absolute and Radiocarbon Chronologies*. Springer-Verlag, Heidelberg.

Bard, E., B. Hamelin, R. Fairbanks, and A. Zindler, 1990. Calibration of the ^{14}C timescale over the past 30,000 years using mass spectrometric U–Th ages from Barbados corals. *Nature*, 345:405–410.

Brückner, H., and R. A. Halfar, 1994. Evolution and age of shorelines along Woodfiord, northern Spitsbergen. *Z. Geomorph. N. F. Suppl.-Bd.*, 97:75–91.

Craig, H., 1954. Carbon-13 in plants and the relationships between carbon-13 and carbon-14 variations in nature. *J. Geol.*, 62:115–149.

Craig, H., 1961. Mass spectrometric analysis of radiocarbon standards. *Radiocarbon*, 3:1–3.

Damon, P. E., D. J. Donahue, B. H. Gore, A. L. Hathaway, A. J. T. Jull, T. W. Linick, P. J. Sercel, L. J. Toolin, C. R. Bronk, E. T. Hall, R. E. M. Hedges, R. Housley, A. Law, C. Perry, G. Bonani, S. Trumbore, W. Wölfli, J. C. Ambers, S. G. E. Bowman, M. N. Leese, and M. S. Tite, 1989. Radiocarbon dating of the Shroud of Turin. *Nature*, 337:611–615.

deVries, H., 1958. Variation in the concentration of radiocarbon with time and location on Earth. *Proc. Koninkl. Ned. Acad. Wetenschap.*, B61:257–281.

Edwards, R. L., J. W. Beck, G. S. Burr, D. J. Donahue, J. M. A. Chappell, A. L. Bloom, E. R. M. Druffel, and F. W. Taylor, 1993. A large drop in the $^{14}C/^{12}C$ and reduced melting in the Younger Dryas, documented with ^{230}Th ages of corals. *Science*, 160:962–968.

Eisenhauer, A., G. J. Wasserburg, J. H. Chen, G. Bonani, L. B. Collins, Z. R. Zhu, and K. H. Wyrwoll, 1993. Holocene sea-level determinations relative to the Australian continent: U/Th (TIMS) and ^{14}C (AMS) dating of coral cores from the Abrolhos Islands. *Earth Planet. Sci. Lett.*, 114:529–547.

Fontes, J.-C., and J. M. Garnier, 1979. Determination of the initial ^{14}C activity of the total dissolved carbon; a review of the existing models and a new approach. *Water Resources Res.*, 15:399–413.

Grootes, P. M., 1978. Carbon-14 time scale extended: Comparison of chronologies. *Science*, 200:11–15.

Grosse, A. V., 1934. An unknown radioactivity. *J. Am. Chem. Soc.*, 56:1922–1923.

Hedges, R. E. M., R. A. Housley, C. B. Ramsey, and G. J. van Klinken, 1994. Radiocarbon dates from the Oxford AMS system: Archaeometry datelist 18. *Archaeometry*, 36:337–374.

Hedges, R. E. M., and C. B. Moore, 1978. Enrichment of ^{14}C and radiocarbon dating. *Nature*, 276:255–257.

Hopf, M., 1969. Plant remains and early farming in Jericho. In P. J. Ucko and F. W. Dimbledy (Eds.), *The Domestication and Exploitation of Plants and Animals*, pp. 355–357. Duckworth, London.

Jull, A. J. T., D. J. Donahue, and P. E. Damon, 1996. Factors affecting the apparent radiocarbon age of textiles: A comment on "Effects of fire and biofractionation of carbon isotopes on results of radiocarbon dating of old textiles: the Shroud of Turin, by D. A. Kouznetsov et al." *J. Archaeol. Sci.*, 23:157–160.

Keith, M. L., and G. M. Anderson, 1963. Radiocarbon dating: Fictitious results with mollusk shells. *Science*, 141:634–637.

Lerman, J. C., W. G. Mook, and J. C. Vogel, 1970. ^{14}C in tree rings from different localities. In I. U. Olsson (Ed.), *Radiocarbon Variations and Absolute Chronology*. pp. 275–299. Almquist & Wicksell, Uppsala, Sweden.

Libby, W. F., 1946. Atmospheric helium-three and radiocarbon from cosmic radiation. *Phys. Rev.*, 69:671–672.

Libby, W. F., 1952. *Radiocarbon Dating*. University of Chicago Press, Chicago, Illinois.

Libby, W. F., 1955. *Radiocarbon Dating*, 2nd ed. University of Chicago Press, Chicago, Illinois.

Lister, G., K. Kelts, R. Schmid, G. Bonani, H. Hofmann, E. Morenzoni, M. Nessi, M. Suter, and W. Wölfe, 1990. Correlation of the paleoclimatic record in lacustrine sediment sequences: ^{14}C dating by AMS. *Nucl. Instrum. Methods Phys. Res.*, B5:389–393.

Michael, H. N., and E. K. Ralph (Eds.), 1970. *Dating Techniques for the Archaeologist*. MIT Press, Cambridge, Massachusetts.

Ralph, E. K., 1971. Carbon-14 dating. In H. N. Michael and E. K. Ralph (Eds.), *Dating Techniques for the Archaeologist*, pp. 1–48. MIT Press, Cambridge, Massachusetts.

Ralph, E. K., H. N. Michael, and M. C. Han, 1973. Radiocarbon dates and reality. *MASCA Newsletter*, 9(1):1–18.

Rink, W. J., H. P. Schwarcz, H. K. Lee, V. Cabrera Valdes, F. Bernaldo de Quiros, and M. Hoyos, 1996. ESR dating of tooth enamel: Comparison with AMS ^{14}C at El Castillo Cave, Spain. *J. Archaeol. Sci.*, 23:945–951.

Saliege, J. F., A. Person, and F. Paris, 1995. Preservation of $^{13}C/^{12}C$ original ratio and ^{14}C dating of the mineral fraction of human bones from Saharan tombs, *Niger. J. Archaeol. Sci.*, 22:302–312.

Stafford, T. W., A. J. T. Jull, K. Brendel, R. C. Duhamel, and D. Donahue, 1987. Study of bone radiocarbon dating accuracy at the University of Arizona NSF accelerator facility for radioisotope analyses. *Radiocarbon*, 29:24–44.

Stuiver, M., 1978. Radiocarbon timescale tested against magnetic and other dating methods. *Nature*, 273:271–274.

Stuiver, M., and T. F. Braziunas, 1993. Modelling atmospheric ^{14}C influences and ^{14}C ages of marine samples to 10,000 BC. *Radiocarbon*, 35:137–189.

Suess, H. E., 1965. Secular variations of the cosmic ray produced carbon-14 in the atmosphere and their interpretation. *J. Geophys. Res.*, 70:5937–5952.

Valdes, V. C., and J. L. Bischoff, 1989. Accelerator ^{14}C dates of early Upper Paleolithic (basal Aurignacien) at El Castillo cave (Spain). *J. Archaeol. Sci.*, 16:577–584.

Wagner, G. A., 1998. *Age Determination of Young Rocks and Artifacts*. Springer-Verlag, Heidelberg.

Zolitschka, B., A. Brauer, J. F. W. Negendank, H. Stockhausen, and A. Lang, 2000. Annually dated Weichselian continental paleoclimate record from the Eifel, Germany. *Geology*, 28(9):783–786.

REFERENCES FOR DEEP-SEA SEDIMENT AND FERROMANGANESE NODULES (^{10}Be AND ^{26}Al) (SECTION 23.2–23.2c)

Aldahan, A., and G. Possnert, 1998. A high-resolution ^{10}Be profile from deep-sea sediment covering the last 70 ka: Indication for globally synchronized environmental events. *Quat. Sci. Rev. Quat. Geochronology*, 17:1023–1032.

Amin, B. S., D. P. Kharkar, and D. Lal, 1966. Cosmogenic ^{10}Be and ^{26}Al in marine sediments. *Deep-Sea Res.*, 13:805–824.

Amin, B. S., D. Lal., and B. L. K. Somayajulu, 1975. Chronology of marine sediments using the ^{10}Be method: Intercomparison with other methods. *Geochim. Cosmochim. Acta*, 39:1187–1192.

Arnold, J. R., 1956. Beryllium-10 produced by cosmic rays. *Science*, 124:584–585.

Arnold, J. R., and H. A. Al-Salih, 1955. Beryllium-7 produced by cosmic rays. *Science*, 123:451–453.

Beer, J., C. Shen, F. Heller, T. Liu, G. Bonani, B. Dittrich, P. Kubik, and M. Suter, 1993. ^{10}Be and the magnetic susceptibility in Chinese loess. *Geophys. Res. Lett.*, 20:57–60.

Castagnoli, G. C., A. Albrecht, J. Beer, G. Bonani, Ch. Shen, E. Callegari, C. Tarico, B. Dittrich-Hannen,

P. Kubik, M. Suter, and G. M. Zhu, 1995. Evidence for enhanced ^{10}Be deposition in Mediterranean sediments 35 kyr BP. *Geophys. Res. Lett.*, 22:707–710.

Claude-Ivanaj, C., A. W. Hofmann, I. Vlastelic, and A. Koschinsky, 2001. Recording changes in ENADW composition over the last 340 ka using high-precision lead isotopes in Fe–Mn crust. *Earth Planet. Sci. Lett.*, 188:73–89.

Crèvecoeur, E. H., and D. A. Schaeffer, 1963. Séparation et mesures de ^{26}Al et ^{10}Be dans les météorites. In *Radioactive Dating*, pp. 335–342. International Atomic Energy Agency, Vienna.

DeMenocal, P., D. W. Oppo, and R. G. Fairbanks, 1992. Pleistocene δ^{13}C variability of North Atlantic intermediate water. *Paleoceanography*, 7:229–250.

Elmore, D., and F. M. Phillips, 1987. Accelerator mass spectrometry for measurement of long-lived radionuclides. *Sciences*, 236:543–550.

Frank, M., and R. K. O'Nions, 1998. Sources of Pb for Indian Ocean ferromanganese crusts: A record of Himalayan erosion? *Earth Planet. Sci. Lett.*, 158:121–130.

Goel, P. S., S. Jha, D. Lal, P. Radhakrishna, and Rama (no initials), 1956. Cosmic ray produced beryllium isotopes in rain water. *Nucl. Phys.*, 1:196–201.

Goel, P. S., D. P. Kharkar, D. Lal, N. Narsappaya, B. Peters, and V. Yatirayam, 1957. The beryllium-10 concentration in deep-sea sediments. *Deep Sea Res.*, 4:202–210.

Guichard, F., J.-L. Reyes, and Y. Yokoyama, 1978. Growth rate of a manganese nodule measured with ^{10}Be and ^{26}Al. *Nature*, 272:155–156.

Henken-Mellies, W. H., J. Beer, F. Heller, K. J. Hsu, C. Shen, G. Bonani, H. J. Hofmann, M. Suter, and W. Wölfli, 1990. ^{10}Be and ^{9}Be in South Atlantic DSDP site 519: Relation to geomagnetic reversals and to sediment composition. *Earth Planet. Sci. Lett.*, 98:267–276.

Inoue, T., and S. Tanaka, 1976. ^{10}Be in marine sediments. *Earth Planet. Sci. Lett.*, 29:155–160.

Kharkar, D. P., D. Lal, and B. L. K. Somayajulu, 1963. Investigations in marine environments using radio isotopes produced by cosmic rays. In *Radioactive Dating*, pp. 175–188. International Atomic Energy Agency, Vienna.

Lal, D., 1963. Study of long and short-term geophysical processes using natural radioactivity. In *Radioactive Dating*, pp. 149–158. International Atomic Energy Agency, Vienna.

Lide, D. R., and H. P. R. Frederikse, 1995. *Handbook of Chemistry and Physics*, 76th ed. CRC Press, Boca Raton, Florida.

Muller, R. A., 1977. Radioisotope dating with a cyclotron. *Science*, 196:489–494.

Peters, B., 1955. Radioactive beryllium in the atmosphere and on the Earth. *Proc. Indian Acad. Sci.*, 41(3A):67–71.

Raisbeck, G. M., F. Yiou, M. Fruneau, and J. M. Loiseaux, 1978. Beryllium-10 mass spectrometry with a cyclotron. *Science*, 202:215–217.

Somayajulu, B. L. K., 1977. Analysis of causes for the beryllium-10 variations in deep sea sediment. *Geochim. Cosmochim. Acta*, 41:909–913.

Southon, J. R., T. L. Ku, D. L. Nelson, L. J. Reyss, J. C. Duplessy, and J. S. Vogel, 1987. ^{10}Be in deep-sea core: Implication regarding ^{10}Be production change over the past 420 ka. *Earth Planet. Sci. Lett.*, 85:356–364.

Southon, J. R., D. E. Nelson, R. Korteling, I. Nowikow, E. Hammaren, J. McKay, and K. Burke, 1982. Techniques for the direct measurement of natural beryllium-10 and carbon-14 with a tandem accelerator. In L. A. Currie (Ed.), *Nuclear and Chemical Dating Techniques*, ACS Symposium Series 176, pp. 45–74. American Chemical Society, Washington, D.C.

Tanaka, S., and T. Inoue, 1980. ^{10}Be evidence for geochemical events in the North Pacific during the Pliocene. *Earth Planet. Sci. Lett.*, 49:34–38.

Wilson, G. C., A. E. Litherland, and J. C. Rucklidge, 1984. Dating of sediments using accelerator mass spectrometry. *Chem. Geol.*, 44:1–17.

REFERENCES FOR CONTINENTAL ICE SHEETS (SECTION 23.2d)

Beer, J., M. Andrée, H. Oeschger, U. Siegenthaler, G. Bonani, H. J. Hofmann, E. Morenzoni, M. Nessi, M. Suter, and W. Wölfli, 1984a. The Camp Century ^{10}Be record: Implications for long-term variations of the geomagnetic dipole moment. *Nucl. Inst. Methods. Phys. Res.*, B5:380–384.

Beer, J., M. Andrée, H. Oeschger, B. Stauffer, R. Balzer, G. Bonani, Ch. Stoller, M. Suter, and W. Wölfli, 1985. ^{10}Be variations in polar ice cores. In C. C. Langway, Jr., H. Oeschger, and W. Dansgaard (Eds.), *Greenland Ice Core: Geophysics, Geochemistry, and the Environment*, Geophys. Monograph 33, pp. 66–70. American Geophysical Union, Washington, D.C.

Beer, J., A. Blinov, G. Bonani, R. C. Finkel, H. J. Hofmann, B. Lehmann, H. Oeschger, H. Sigg, J. Schwander, T. Staffelbach, B. Stauffer, M. Suter, and W. Wölfli, 1990. Use of ^{10}Be in polar ice to trace the 11-year cycle of the solar activity. *Nature*, 327:164–166.

Beer, J., S. J. Johnsen, G. Bonani, R. C. Finkel, C. C. Langway, H. Oeschger, B. Stauffer, M. Suter, and W. Wölfli, 1992. ^{10}Be peaks as time markers in

Arctic ice core. In E. Bard and W. S. Broecker (Eds.), *The Last Deglaciation: Absolute and Radiocarbon Chronologies*, NATO ASI Series, Vol. I(2), pp. 141–153.

Beer, J., H. Oeschger, M. Andrée, G. Bonani, M. Suter, W. Wölfli, and C. C. Langway, Jr., 1984b. Temporal variations in the ^{10}Be concentration levels found in the Dye 3 ice core, Greenland. *Ann. Glaciol.*, 5:16–17.

Lao, Y., R. F. Anderson, W. S. Broecker, S. E. Trumbore, H. J. Hofmann, and W. Wölfli, 1992. Increased production of cosmogenic ^{10}Be during the Last Glacial Maximum. *Nature*, 357:576–578.

McCorkell, R., E. L. Fireman, and C. C. Langway, Jr. 1967. Aluminum-26 and beryllium-10 in Greenland ice. *Science*, 158:1690–1692.

Raisbeck, G. M., F. Yiou, D. Bourles, C. Lorius, J. Jouzel, and N. I. Barkov, 1987. Evidence for two enhanced ^{10}Be deposition events in Antarctica during the last glacial period. *Nature*, 326:273–277.

Raisbeck, G. M., F. Yiou, M. Fruneau, M. Lieuvin, and J. M. Loiseaux, 1978. Measurement of ^{10}Be in 1,000- and 5000- year-old Antarctic ice. *Science*, 275:731–733.

Raisbeck, G. M., F. Yiou, J. Jouzel, J. R. Petit, N. I. Barkov, and E. Bard, 1992. ^{10}Be deposition at Vostok, Antarctica, during the last 50000 years and its relation to possible cosmogenic production variations during this period. In E. Bard and W. S. Broecker (Eds.), *The Last Deglaciation: Absolute and Radiocarbon Chronologies*, NATO ASI Series, Vol. I(2), pp. 127–139.

Sonett, C. P., 1992. A supernova shock ensemble model using Vostok ^{10}Be radioactivity. *Radiocarbon*, 34:239–245.

Yiou, F., G. M. Raisbeck, S. Baumgartner, J. Beer, C. Hammer, S. Johansen, J. Jouzel, P. K. Kubik, J. Lestringuez, M. Stiévenard, M. Suter, and P. Yiou, 1997. Beryllium 10 in the Greenland ice core project core at Summit, Greenland. *J. Geophys. Res.*, 102: 26783–26794.

Yiou, F., G. M. Raisbeck, D. Bourles, C. Lorius, and N. I. Barkov, 1985. ^{10}Be in ice at Vostok, Antarctica, during the last climate cycle. *Nature*, 316:616–617.

REFERENCES FOR EXPOSURE DATING (^{10}Be AND ^{26}Al) (SECTION 23.3)

Brown, L., 1984. Applications of accelerator mass spectrometry. *Annu. Rev. Earth Planet. Sci.*, 12:39–59.

Brown, E. T., E. J. Brook, G. M. Raisbeck, F. Yiou, and M. D. Kurz, 1992. Effective attenuation lengths of cosmic rays producing ^{10}Be and ^{26}Al in quartz: Implications for exposure age dating. *Geophys. Res. Lett.*, 19:369–372.

Clark, P. U., E. J. Brook, G. M. Raisbeck, F. Yiou, and J. Clark, 2003. Cosmogenic ^{10}Be ages of the Saglek Moraines, Torngat Mountains, Labrador. *Geology*, 31(7):617–620.

Klein, J., R. Giegengack, R. Middleton, and P. Sharma, 1986. Revealing histories of exposure using in situ produced ^{26}Al and ^{10}Be in Libyan desert glass. *Radiocarbon*, 28:547–555.

Lal, D., 1988. In situ-produced cosmogenic isotopes in terrestrial rocks. *Annu. Rev. Earth Planet. Sci.*, 16:355–388.

Lal, D., 1991. Cosmic ray labeling of erosion surfaces: In situ nuclide production rates and erosion models. *Earth Planet. Sci. Lett.*, 104:424–439.

Lal, D., and J. R. Arnold, 1985. Tracing quartz through the environment. *Proc. Indian Acad. Sci., Earth Planet. Sci.*, 94(1):1–5.

Nishiizumi, K., D. Lal, J. Klein, R. Middleton, and J. R. Arnold, 1986. Production of ^{10}Be and ^{26}Al by cosmic rays in terrestrial quartz in situ and implications for erosion rates. *Nature*, 319:134–136.

Nishiizumi, K., E. L. Winterer, C. P. Kohn, J. Klein, R. Middleton, D. Lal, and J. R. Arnold, 1989. Cosmic ray production rates of ^{10}Be and ^{26}Al in quartz from glacially polished rocks. *J. Geophys. Res.*, 94(B12):17907–17915.

REFERENCES FOR COSMOGENIC AND THERMONUCLEAR ^{36}Cl (SECTION 23.4)

Bentley, H. W., F. M. Phillips, S. N. Davis, S. Gifford, D. Elmore, L. E. Tubbs, and H. E. Gove, 1982. Thermonuclear ^{36}Cl pulse in natural water. *Nature*, 300: 737–740.

Briner, J. P., and T. W. Swanson, 1998. Using inherited cosmogenic ^{36}Cl to constrain glacial erosion rates of the Cordilleran ice sheet. *Geology*, 26(8):3–6.

Davis, R., and O. A. Schaeffer, 1955. Chlorine-36 in nature. *Ann. NY Acad. Sci.*, 62:105–122.

Dyrsson, D., and P. Nyman, 1955. Slow-neutron-induced radioactivity of seawater. *Acta Radiol.*, 43:421–427.

Elmore, D., B. R. Fulton, M. R. Clover, J. R. Marsden, H. E. Gove, H. Naylor, K. H. Purser, L. R. Kilius, R. P. Beukens, and A. E. Litherland, 1979. Analysis of ^{36}Cl in environmental water samples using an electrostatic accelerator. *Nature*, 277:22–25.

Elmore, D., L. E. Tubbs, D. Newman, X. Z. Ma, R. Finkel, K. Nishiizumi, J. Beer, H. Oeschger, and

M. Andrée, 1982. ^{36}Cl bomb pulse measured in a shallow ice core from Dye 3, Greenland. *Nature*, 300:735–737.

Finkel, R. C., K. Nishiizumi, D. Elmore, R. D. Ferraro, and H. E. Gove, 1980. ^{36}Cl in polar ice, rainwater, and seawater. *Geophys. Res. Lett.*, 7(11):983–986.

Lal, D., 1991. Cosmic ray labeling of erosion surfaces: In situ nuclide production rates and erosion models. *Earth Planet. Sci. Lett.*, 104:424–439.

Leavy, B. D., F. M. Phillips, D. Elmore, P. W. Kubik, and E. Gladney, 1987. Measurement of cosmogenic $^{36}Cl/Cl$ in young volcanic rocks: An application of accelerator mass spectrometry in geochronology. *Nucl. Instrum. Methods Phys. Res.*, B29:246–250.

Liu, B., F. M. Phillips, J. T. Fabryka-Martin, M. M. Fowler, and W. D. Stone, 1994. Cosmogenic ^{36}Cl accumulation in unstable landforms, 1. Effects of the thermal neutron distribution. *Water Resources Res.*, 30:3115–3125.

Liu, B., F. M. Phillips, M. M. Pohl, and P. Sharma, 1996. An alluvial surface chronology based on cosmogenic ^{36}Cl dating, Ajo Mountains (Organ Pipe Cactus National Monument), southern Arizona. *Quat. Res.*, 45:30–37.

Nishiizumi, K., J. R. Arnold, D. Elmore, R. D. Ferraro, H. E. Gove, R. C. Finkel, R. P. Beukens, K. R. Chang, and L. R. Kilius, 1979. Measurement of ^{36}Cl in Antarctic meteorites and Antarctic ice using a van de Graaff accelerator. *Earth Planet. Sci. Lett.*, 45:285–292.

Nishiizumi, K., J. R. Arnold, D. Elmore, X. Ma, D. Newman, and H. E. Gove, 1983. ^{36}Cl and ^{53}Mn in Antarctic meteorites and ^{10}Be-^{36}Cl dating of Antarctic ice. *Earth Planet. Sci. Lett.*, 62:407–417.

O'Brien, K., H. A. Sandmeier, G. E. Hansen, and J. E. Campbell, 1978. Cosmic ray induced neutron background sources and fluxes for geometries over water, ground, iron, and aluminum. *J. Geophys. Res.*, 83:114–120.

Phillips, F. M., B. D. Leavy, N. O. Jannik, D. Elmore, and P. W. Kubik, 1986. The accumulation of cosmogenic chlorine-36 in rocks: A method for surface exposure dating. *Science*, 231:41–43.

Phillips, F. M., M. G. Zreda, M. R. Flinsch, D. Elmore, and P. Sharma, 1996. A reevaluation of cosmogenic ^{36}Cl production rates in terrestrial rocks. *Geophys. Res. Lett.*, 23:949–952.

Schaeffer, O. A., S. O. Thomson, and N. L. Lake, 1960. Chlorine-36 in rain. *J. Geophys. Res.*, 65(12):4013–4016.

Yamashita, M., L. D. Stephens, and H. W. Patterson, 1966. Cosmic-ray-produced neutrons at ground level: Neutron production rate and flux distribution. *J. Geophys. Res.*, 71:3817–3834.

Yokoyama, Y., J.-L. Reyss, and F. Guichard, 1977. Production of radionuclides by cosmic rays at mountain altitudes. *Earth Planet. Sci. Lett.*, 36:44–50.

Zreda, M., and J. S. Noller, 1998. Ages of prehistoric earthquakes revealed by cosmogenic chlorine-36 in a bedrock fault scarp at Hebgen Lake. *Science*, 282:1097–1098.

Zreda, M. G., F. M. Phillips, and D. Elmore, 1994. Cosmogenic ^{36}Cl accumulation in unstable landforms, 2. Simulations and measurements on eroding moraines. *Water Resources Res.*, 30:3127–3136.

Zreda, M. G., F. M. Phillips, D. Elmore, P. W. Kubik, P. Sharma, and R. I. Dorn, 1991. Cosmogenic chlorine-36 production rates in terrestrial rocks. *Earth Planet. Sci. Lett.*, 105:94–109.

REFERENCES FOR METEORITES (SECTION 23.5)

Bogard, D. D., L. E. Nyquist, B. M. Bansal, D. H. Garrison, G. F. Herzog, A. A. Albrecht, S. Vogt, and J. Klein, 1995. Neutron-capture ^{36}Cl, ^{41}Ca, ^{36}Ar, and ^{150}Sm in large chondrites: Evidence for high fluences of thermalized neutrons. *J. Geophys. Res.*, 100:9401–9416.

Eugster, O., 1988. Cosmic-ray production rates of ^{3}He, ^{21}Ne, ^{38}Ar, ^{83}Kr, and ^{126}Xe in chondrites based on ^{81}Kr-K exposure ages. *Geochim. Cosmochim. Acta*, 52:1649–1662.

Leya, I., H.-H. Lange, S. Neumann, R. Wieler, and R. Michel, 2000. The production of cosmogenic nuclides in stony meteoroids by galactic cosmic-ray particles. *Meteoritics Planet. Sci.*, 35:259–286.

Masarik, J., and R. C. Reedy, 1994. Effects of bulk composition on nuclear production processes in meteorites. *Geochim. Cosmochim. Acta*, 58:5307–5317.

Nishiizumi, K., J. R. Arnold, D. Elmore, R. D. Ferraro, H. E. Gove, R. C. Finkel, R. P. Peukens, K. H. Chang, and L. R. Kilius, 1979. Measurement of ^{36}Cl in Antarctic meteorites and Antarctic ice using a van de Graaff accelerator. *Earth Planet. Sci. Lett.*, 45:285–292.

Nishiizumi, K., J. R. Arnold, D. Elmore, X. Ma, D. Newman, and H. E. Gove, 1983. ^{36}Cl and ^{53}Mn in Antarctic meteorites and ^{10}Be-^{36}Cl dating of Antarctic ice. *Earth Planet. Sci. Lett.*, 62:407–417.

Nishiizumi, K., D. Elmore, and P. W. Kubik, 1989. Update on terrestrial ages of Antarctic meteorites. *Earth Planet. Sci. Lett.*, 93:229–313.

Nishiizumi, K., M. T. Murrell, J. R. Arnold, D. Elmore, R. D. Ferraro, H. E. Gove, and R. C. Finkel, 1981. Cosmic ray produced ^{36}Cl and ^{53}Mn in Allan Hills-77 meteorites. *Earth Planet. Sci. Lett.*, 52:31–38.

Phillips, F. M., M. G. Zreda, S. S. Smith, D. Elmore, P. W. Kubik, R. I. Dorn and D. J. Roddy, 1991. Age and geomorphic history of Meteor Crater, Arizona, from cosmogenic ^{36}Cl and ^{14}C and in rock varnish. *Geochim. Cosmochim. Acta*, 55:2695–2698.

Schultz, L., 1993. *Planetologie; eine Einführung*. Birkhäuser Verlag, Basel.

Schultz, L., H. W. Weber, and F. Begemann, 1991. Noble gases in H-chondrites and potential differences between Antarctic and non-Antarctic meteorites. *Geochim. Cosmochim. Acta*, 55:59–66.

Welten, K. C., K. Nishiizumi, J. Masarik, M. W. Caffee, A. J. T. Jull, S. E. Klandrud, and R. Wieler, 2001. Cosmic-ray exposure history of two Frontier Mountain H-chondrite showers from spallation and neutron-capture products. *Meteoritics Planet. Sci.*, 35:301–317.

Whillans, I. M., and W. A. Cassidy, 1983. Catch a falling star: Meteorites and old ice. *Science*, 222:55–57.

CHAPTER 24

Extinct Radionuclides

At the time the Sun and planets of the solar system formed from the solar nebula about 4.56×10^9 years ago, many short-lived radionuclides were present, which have since decayed. These radionuclides formed in the ancestral stars by various nuclear reactions that occurred before as well as during their terminal explosions. In addition, nuclear reactions occurred as a result of the irradiation of dust particles in the solar nebula by galactic cosmic rays (Leya et al., 2003). Some of these short-lived radionuclides were partitioned into solid particles which condensed from the nebula and subsequently assembled into the planetesimals which formed the planets, including the Earth and the present asteroids. After the radionuclides had been isolated from the solar nebula by being included in solid particles, they continued to decay to form stable isotopes of other elements. The resulting anomalies in the isotopic composition of these elements are detectable in stony and iron meteorites. The principal radionuclides that occurred in the solar nebula and subsequently became extinct by decaying to stable daughters are listed in Table 24.1.

Table 24.1. Radionuclides and Their Stable Decay Products in the Solar Nebula at the Time of Formation of the Sun and the Planets of the Solar System

Parent Radionuclide	Decay Mode	Halflife, years	Stable Daughter
$^{26}_{13}Al$	β^+, e^-	0.705×10^6	$^{26}_{12}Mg$
$^{53}_{25}Mn$	e^-	3.7×10^6	$^{53}_{24}Cr$
$^{60}_{26}Fe$	β^-	1.5×10^6	$^{60}_{28}Ni$
$^{93}_{40}Zr$	β^-	1.5×10^6	$^{93}_{41}Nb$
$^{107}_{46}Pd$	β^-	6.5×10^6	$^{107}_{46}Ag$
$^{129}_{53}I$	β^-	15.7×10^6	$^{129}_{54}Xe$
$^{182}_{72}Hf$	β^-	9.0×10^6	$^{182}_{74}W$
$^{244}_{94}Pu$	α, β^-	8.0×10^7	$^{208}_{82}Pb$

The decay of short-lived radionuclides in the solar system was an important source of heat that caused melting of solid objects that formed in the solar nebula and caused these objects to differentiate internally by segregation of immiscible liquids composed of metallic Fe–Ni alloys, sulfides of Fe and other transition metals, and silicates (Urey, 1955). The amount of heat that was generated by the decay of short-lived radionuclides during the formation of planetary bodies depends partly on the amount of time that passed between cessation of nucleosynthesis and the growth of planetesimals from dust particles. The discovery of extinct ^{129}I in stony meteorites, implied by the presence of excess ^{129}Xe, indicated that ^{129}I had been incorporated into planetesimals less than 35 million years after the end of its nucleosynthesis (Reynolds, 1960). The apparent confirmation of Urey's hypothesis concerning the melting and differentiation of early-formed planetary bodies stimulated the search for other extinct radionuclides. Summaries of the results of these studies have been published by Wasserburg (1985), Podosek and Cassen (1994), Jacobsen and Harper (1996), and Carlson and Lugmair (2000).

24.1 THE Pd–Ag CHRONOMETER

The decay of short-lived ^{107}Pd to stable ^{107}Ag in the early solar system was suggested by Murthy (1960, 1962) based on his discovery that Ag in several iron meteorites appeared to be enriched in ^{107}Ag. In spite of initial analytical difficulties discussed by Kaiser and Wasserburg (1983), the presence of excess ^{107}Ag in many iron meteorites was confirmed by Kelly and Wasserburg (1978).

Palladium ($Z = 46$) is a member of the platinum group elements (PGEs), all of which are siderophile and occur in iron meteorites. The stable isotopes of Pd and their abundances are as follows: ^{102}Pd, 1.02%; ^{104}Pd, 11.14%; ^{105}Pd, 22.33%; ^{106}Pd, 27.33%; ^{108}Pd, 26:46%; and ^{110}Pd, 11.72%. In addition, 19 unstable Pd isotopes are known of which ^{107}Pd has the longest halflife ($T_{1/2} = 6.5 \times 10^6$ y, $\lambda = 0.106 \times 10^{-6}$ y^{-1}). Consequently, the ^{107}Pd that existed in the solar nebula at 4.56 Ga decayed in about 30×10^6 years (five halflives) and has not survived to the present time.

The decay of $^{107}_{46}$Pd proceeds by means of negative β-emission to stable $^{107}_{47}$Ag:

$$^{107}_{46}\text{Pd} \rightarrow {}^{107}_{47}\text{Ag} + \beta^- + \overline{\nu} + Q \qquad (24.1)$$

The decay is to the groundstate of $^{107}_{47}$Ag and the total decay energy Q is only 0.033 MeV.

Silver ($Z = 47$) has two stable isotopes whose abundances are ^{107}Ag: 51.839% and ^{109}Ag: 48.161%. All of the 27 known unstable isotopes of Ag have short halflives less than 41.3 days and none occur naturally. The ^{107}Ag/^{109}Ag isotope ratio of terrestrial Ag, based on the isotopic abundances stated above, is:

$$\frac{^{107}\text{Ag}}{^{109}\text{Ag}} = \frac{51.839}{48.161} = 1.0763$$

Recent redeterminations of this ratio by Chen and Wasserburg (1996) yielded values of 1.0897 ± 0.0018 (electron multiplier) and 1.0811 ± 0.0017 (Faraday cup) for the National Institute of Standards and Technology Ag standard NIST SRM 978.

The ^{107}Ag/^{109}Ag ratios of several types of iron meteorites reported by Chen and Wasserburg (1990) range by a factor of 8.44 from 1.09 to 9.2. The elevated ^{107}Ag/^{109}Ag ratios of some iron meteorites are attributed to the presence of varying amounts of radiogenic ^{107}Ag that formed by decay of unstable ^{107}Pd. These authors further demonstrated that the ^{107}Ag/^{109}Ag ratios of iron meteorites correlate positively with their ^{108}Pd/^{109}Ag ratios, which means that specimens having high ^{108}Pd concentrations contain more radiogenic ^{107}Ag than specimens with low ^{108}Pd concentrations. The ^{107}Ag/^{109}Ag and ^{108}Pd/^{109}Ag ratios of different specimens of metallic Fe from Gibeon, an iron meteorite of group IVA, define a straight line in Figure 24.1 having a slope ^{107}Ag*/^{108}Pd $= 2.40 \pm 0.05 \times 10^{-5}$ and an initial ^{107}Ag/^{109}Ag ratio of 1.11 ± 0.01 (Chen and Wasserburg, 1990).

The correlation of ^{107}Ag/^{109}Ag and ^{108}Pd/^{109}Ag ratios in Gibeon supports the interpretation that the radiogenic ^{107}Ag in each specimen of this meteorite formed by in situ decay of ^{107}Pd after the metallic iron of the parent body of Gibeon had become

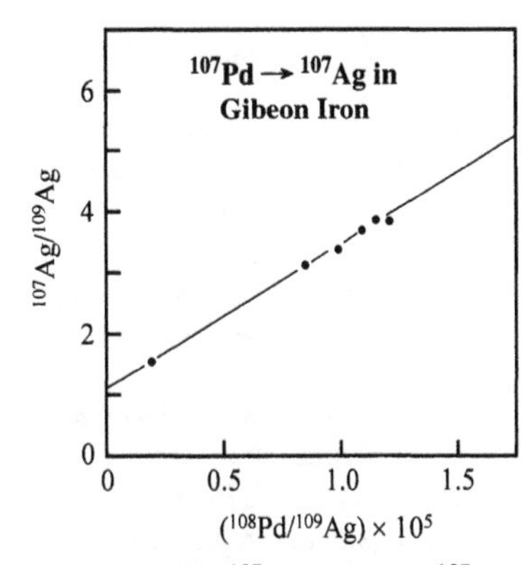

FIGURE 24.1 Decay of ^{107}Pd to stable ^{107}Ag in the iron meteorite Gibeon. The regression line fitted to the data points has a slope of ^{107}Ag*/^{108}Pd $= 2.4 \pm 0.05 \times 10^{-5}$ and yields an initial ^{107}Ag/^{109}Ag ratio of 1.11 ± 0.01. Data and interpretation by Chen and Wasserburg (1990). One data point for Lion River 2D1 is not shown because of its large errors.

a closed system by cooling and crystallization. In addition, the range of present-day $^{107}Ag/^{109}Ag$ ratios of Gibeon indicates that each fragment had a different $^{107}Pd/^{109}Ag$ ratio. If the $^{107}Ag/^{109}Ag$ ratios are the result of in situ decay of ^{107}Pd in specimens of Gibeon, all of which crystallized at the same time, then the straight line in Figure 24.1 is an isochron (Chen and Wasserburg, 1990).

The interpretation presented above leads to the conclusion that the $^{107}Ag^{*}/^{108}Pd$ ratio (i.e., the slope of the correlation line in Figure 24.1) is identical to the $^{107}Pd/^{108}Pd$ ratio because all of the ^{107}Pd atoms have decayed to radiogenic ^{107}Ag. This insight opens the way to calculate the time intervals that elapsed between the crystallization of different iron meteorites, all of which derived Pd from the solar nebula, which is assumed to have been an isotopically homogeneous reservoir of this element.

The $^{107}Pd/^{108}Pd$ ratio of the uniform reservoir decreased with time in accordance with the law of radioactivity:

$$\frac{^{107}Pd}{^{108}Pd} = \left(\frac{^{107}Pd}{^{108}Pd}\right)_i e^{-\lambda t} \qquad (24.2)$$

where $\frac{^{107}Pd}{^{108}Pd}$ = value of this ratio in the reservoir t years after the end of the nucleosynthesis of Pd

$\left(\frac{^{107}Pd}{^{108}Pd}\right)_i$ = value of this ratio at the time Pd nucleosynthesis ended ($t = 0$)

λ = decay constant of ^{107}Pd ($0.106 \times 10^{-6}\ y^{-1}$)

t = time elapsed since the end of nucleosynthesis of Pd

The analytical data published by Chen and Wasserburg (1990, 1996) indicated that the Ag–Pd isotope correlation lines of different iron meteorites have different slopes, which means that these meteorites are composed of metallic iron that separated from the reservoir in the solar nebula at different times. The authors selected Gibeon as the reference meteorite and calculated the time intervals between the isolation of Gibeon and 30 other iron meteorites.

The calculation is based on equation 24.2, which can be applied to any two iron meteorites labeled 1 and 2:

$$\left(\frac{^{107}Pd}{^{108}Pd}\right)_1 = \left(\frac{^{107}Pd}{^{108}Pd}\right)_i e^{-\lambda t_1} \qquad (24.3)$$

$$\left(\frac{^{107}Pd}{^{108}Pd}\right)_2 = \left(\frac{^{107}Pd}{^{108}Pd}\right)_i e^{-\lambda t_2} \qquad (24.4)$$

Dividing equation 24.3 by 24.4 yields:

$$\frac{(^{107}Pd/^{108}Pd)_1}{(^{107}Pd/^{108}Pd)_2} = \frac{e^{-\lambda t_1}}{e^{-\lambda t_2}} = e^{\lambda(t_2 - t_1)} \qquad (24.5)$$

Setting $t_2 - t_1 = \Delta t$ and taking natural logarithms of both sides of equation 24.5 yields (Chen and Wasserburg, 1990):

$$\Delta t = \frac{1}{\lambda} \ln \frac{(^{107}Pd/^{108}Pd)_1}{(^{107}Pd/^{108}Pd)_2} \qquad (24.6)$$

The $^{107}Pd/^{108}Pd$ ratio of Gibeon (group IVA) is $2.40 \pm 0.05 \times 10^{-5}$, whereas that of Tlacotepec (IVB) is only $4.7 \pm 2.3 \times 10^{-7}$. Therefore, the parent body of Tlacotepec presumably separated from the solar nebula considerably later than Gibeon. The difference in their crystallization ages is:

$$\Delta t = \frac{\ln(4.7 \times 10^{-7}/2.40 \times 10^{-5})}{0.106 \times 10^{-6}}$$

$$= -37.1 \times 10^{6}\ y$$

The analytical errors combine to give an average uncertainty in Δt of $\pm 5.3 \times 10^{6}$ y. Therefore, Tlacotepec separated from the uniform Pd reservoir in the solar nebula about $37 \pm 5 \times 10^{6}$ years *later* than Gibeon. In fact, the calculations of Chen and Wasserburg (1996) indicate that Tlacotepec is the last of the iron meteorites they analyzed to have separated from the uniform Pd reservoir in the solar nebula.

In cases where the other meteorite has a *higher* $^{107}Pd/^{108}Pd$ ratio than Gibeon, the algebraic sign of Δt is positive. For example, the $^{107}Pd/^{108}Pd$ ratio of Hill City ($2.628 \pm 0.069 \times 10^{-5}$) yields $\Delta t = +0.85 \times 10^{6}$ y by the following calculation:

$$\Delta t = \frac{\ln(2.628 \times 10^{-5}/2.40 \times 10^{-5})}{0.106 \times 10^{-6}}$$

$$= +0.85 \times 10^{6}\ y$$

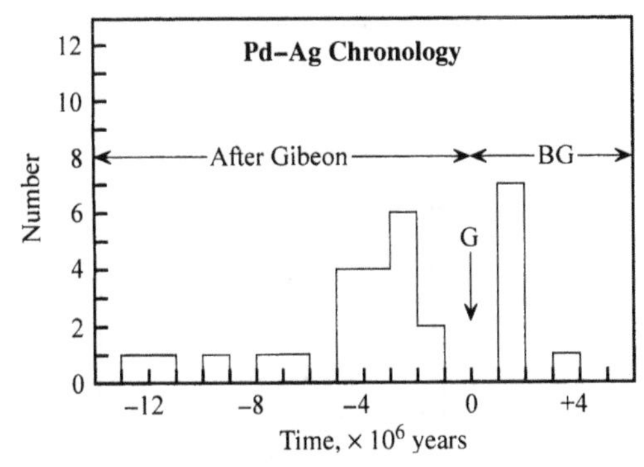

FIGURE 24.2 Spectrum of Pd–Ag separation dates of iron meteorites from an isotopically homogeneous Pd reservoir in the solar nebula relative to the iron meteorite Gibeon. Negative dates indicate that separation occurred after Gibeon, whereas positive dates mean separation before Gibeon (BG). The separation date of Gibeon (G) is zero in this chronology. Data and interpretation from Chen and Wasserburg (1990, 1996).

The range of separation dates determined by Chen and Wasserburg (1996) in Figure 24.2 is from $+2.25 \pm 0.16 \times 10^6$ years (Yanhuitlan) to $-37 \pm 5 \times 10^6$ years (Tlacotepec), where positive dates mean "before Gibeon" and negative dates imply "after Gibeon." Most of the meteorites for which Chen and Wasserburg (1996) calculated explicit dates have separation dates between +4 and −8 million years, thus spanning a range of about 12 million years. The distribution of dates in Figure 24.2 is not uniform but appears to be at least bimodal. This observation is a clue that meteorite parent bodies with iron cores formed episodically both before and after Gibeon.

The application of the Pd–Ag method of dating was expanded by Carlson and Hauri (2001) to other kinds of meteorites with low Pd/Ag ratios, such as pallasites, group IA irons, and chondrites. These kinds of meteorites require a reduction of the measurement error of $^{107}Ag/^{109}Ag$ ratios which is not achievable by thermal ionization mass spectrometry (TIMS) because the instrumental fractionation of the two Ag isotopes cannot be corrected. Therefore, Carlson and Hauri (2001) used a multiple-collector inductively-coupled plasma mass-spectrometer (MC-ICP-MS) and used Pd to correct the isotope fractionation of Ag. This technique resulted in a reduction of analytical errors from about 20 parts per 10,000 (TIMS) to 1.3 parts per 10,000 (MC-ICP-MS). The improvement in measurement precision allowed Carlson and Hauri (2001) to measure $^{107}Pd/^{108}Pd$ ratios (i.e., the slopes of correlation lines as in Figure 24.1) for several meteorites that had previously defied Chen and Wasserburg (1996). The new data are based not only on analyses of metal but also of sulfide and silicate-rich phases.

Although the details of the nucleosynthesis of Pd and other elements are still not well understood, the Pd–Ag dates of irons and of other meteorites shed light on the sequence and duration of the formation of meteorite parent bodies during the transformation of the solar nebula into the solar system.

24.2 THE Al–Mg CHRONOMETER

The present-day occurrence of ^{26}Al ($T_{1/2} = 0.705 \times 10^6$ y) as a cosmogenic radionuclide in stony meteorites and in terrestrial rocks exposed to cosmic rays is well known (Sections 23.2 and 23.3). Therefore, it is reasonable to expect that this nuclide was also present in the solar nebula as a product of nucleosynthesis in the ancestral stars and as a result of nuclear reactions caused by energetic cosmic-ray particles. Any ^{26}Al that was present in the early solar system has long since decayed by positron emission and electron capture to stable $^{26}_{12}Mg$.

Terrestrial Mg ($Z = 12$) has three stable isotopes whose abundances are $^{24}_{12}Mg$: 78.99%, $^{25}_{12}Mg$: 10.00%, and $^{26}_{12}Mg$: 11.01%. Therefore, Mg in terrestrial samples has a $^{26}Mg/^{24}Mg$ ratio of $11.01/78.99 = 0.1394$. Catanzaro and Murphy (1966) determined the isotope compositions of Mg in 53 terrestrial samples and 7 meteorites. The average $^{26}Mg/^{24}Mg$ ratio for all 60 samples was 0.13997 ± 0.00004 compared to 0.13989 ± 0.00012 for a reference sample of metallic Mg. Therefore, the authors concluded that the isotope composition of Mg in natural samples is constant to within the precision of their measurements.

Radiogenic ^{26}Mg was eventually detected by Gray and Compston (1974) and Lee and Papanastassiou (1974) in the carbonaceous chondrite Allende. This discovery was followed up by Bradley et al. (1978) and Esat et al. (1978) who used ion-probe mass-spectrometry to measure the isotope composition of Mg in minerals of refractory inclusions of this meteorite. These discoveries were made possible because ion-probe mass-spectrometers are capable of analyzing small and spatially resolved spots (about 10 μm in diameter) on single crystals. Subsequently, Hutcheon (1982), Huneke et al. (1983), and Ireland et al. (1986) used this kind of equipment to carry out highly effective studies of the isotope composition of Mg in several types of inclusions in Allende identified by the letters A, B, C, ... (Grossman, 1975).

The B-type inclusions in Allende are spheroidal objects with diameters of about one centimeter. They are composed of coarse-grained Ti–Al pyroxene, spinel, melilite, and anorthite. The cores of B1 inclusions are surrounded by a layer of 1-mm-thick gehlenitic melilite, whereas the B2 inclusions lack this layer. The A-type inclusions of Allende have an irregular shape and are composed of melilite and spinel. Some type A inclusions contain hibonite ($CaAl_{12}O_{19}$), which is thought to be one of the earliest phases to condense in the solar nebula. In general, these and other refractory inclusions in Allende are the products of condensation of vapor in the solar nebula and predate the formation of large planetesimals and meteorite parent bodies. Therefore, the Al–Mg chronometer records events that preceded the internal differentiation of meteorite parent bodies and the formation of iron cores recorded by the Pd–Ag chronometer.

The $^{27}Al/^{24}Mg$ and $^{26}Mg/^{24}Mg$ ratios of anorthite crystals in type B1 inclusions of Allende measured by Hutcheon (1982) are closely correlated in Figure 24.3. The equation of the line is:

$$\left(\frac{^{26}Mg}{^{24}Mg}\right) = 0.13815 + 4.96 \times 10^{-5}\left(\frac{^{27}Al}{^{24}Mg}\right)$$

with a linear correlation coefficient $r^2 = 0.9977$. The $^{26}Mg/^{24}Mg$ ratios of the anorthite range up to 0.15887 ± 0.35, which implies that these samples are enriched in radiogenic ^{26}Mg by up to 14%. In addition, the colinearity of the data points confirms that the enrichment in ^{26}Mg is the result of decay of ^{26}Al within the anorthite crystals. The slope of the line in Figure 24.3,

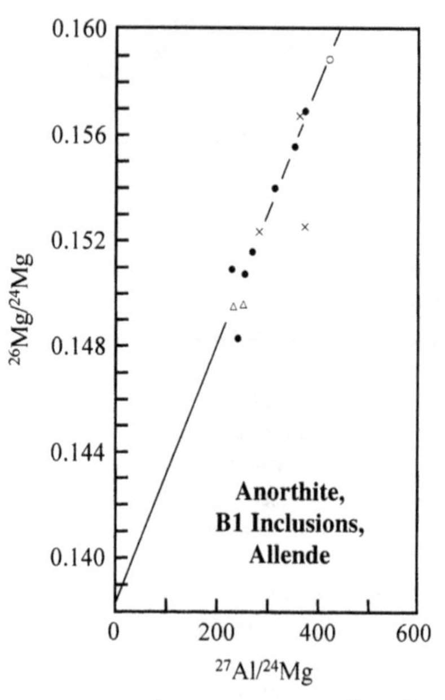

FIGURE 24.3 Correlation of measured $^{27}Al/^{24}Mg$ and $^{26}Mg/^{24}Mg$ ratios of anorthite crystals in type B1 inclusions of the carbonaceous chondrite Allende. The data are from several thin sections (TS): solid circles = TS 23, crosses = TS 33, open circle = TS 34, open triangle = 3529 G. The straight line, defined by 9 of 13 analyses, has a linear correlation coefficient $r^2 = 0.9977$, slope $= 4.96 \times 10^{-5}$, and intercept = 0.13815. Data from Table 2 of Hutcheon (1982).

$$m = \frac{^{26}Mg^*}{^{27}Al} = 4.96 \times 10^{-5}$$

is identical to the $^{26}Al/^{27}Al$ ratio of the Al that was incorporated into the anorthite crystals because all of the ^{26}Al subsequently decayed to radiogenic ^{26}Mg which has been preserved. This interpretation leads to the conclusion that the straight line in Figure 24.3 is an isochron because all of the anorthite crystals formed at the same time and have remained closed to Al and Mg since that time. In addition, the value of the intercept of the isochron,

$$\frac{^{26}Mg}{^{24}Mg} = 0.13815$$

is only slightly lower than the $^{26}Mg/^{24}Mg$ ratio of terrestrial Mg (0.1394), which demonstrates the close relation between primordial Mg in the solar nebula and the Mg that was incorporated into the Earth.

The slopes of the Al–Mg correlation lines of anorthite, melilite, and other minerals in type A and B inclusions of Allende analyzed by Hutcheon (1982) in Table 24.2 range from $4.6 \pm 0.3 \times 10^{-5}$ to $6.0 \pm 2.0 \times 10^{-6}$. If the solar nebula was homogeneous with respect to the isotope composition of Al, then its $^{26}Al/^{27}Al$ ratio decreased with time as ^{26}Al decayed to ^{26}Mg. In that case, the $^{26}Al/^{27}Al$ ratios of the anorthite crystals in Allende record the sequence in which they formed.

The interval of time during which two anorthite crystals formed can be calculated from equation 24.6 by adapting it to the decay of ^{26}Al:

$$\Delta t = \frac{1}{\lambda} \ln \frac{(^{26}Al/^{27}Al)_1}{(^{26}Al/^{27}Al)_2}$$

Table 24.2. Ratios of $^{26}Al/^{27}Al$ of Minerals in B- and A-Type Refractory Inclusions of the Carbonaceous Chondrite Allende

Type	Mineral	$^{26}Al/^{27}Al$
B1	Anorthite, melilite, pyroxene	$4.6 \pm 0.3 \times 10^{-5}$
B2	Anorthite 1 and pyroxene	$4.6 \pm 0.3 \times 10^{-5}$
	Anorthite 2 and pyroxene	$2.3 \pm 0.2 \times 10^{-5}$
	Anorthite 3 and pyroxene	$1.2 \pm 0.3 \times 10^{-5}$
3529 A	Melilite and spinel	$2.7 \pm 0.3 \times 10^{-5}$
3510 A	Anorthite and olivine	$6.0 \pm 2.0 \times 10^{-6}$

Source: Data and interpretation by Hutcheon, 1992.

If $(^{26}Al/^{27}Al)_1 = 4.61 \pm 0.3 \times 10^{-5}$, $(^{26}Al/^{27}Al)_2 = 6.0 \pm 2.0 \times 10^{-6}$, and $\lambda = 0.9831 \times 10^{-6}$ y^{-1},

$$\Delta t = \frac{\ln(4.61 \times 10^{-5}/6.0 \times 10^{-6})}{0.9831 \times 10^{-6}}$$

$$= 2.1 \pm 0.4 \times 10^{6} \text{ y}$$

This result confirms that the refractory inclusions of Allende formed within a relatively short interval of time compared to the age of the solar system.

Hibonites, believed to be early condensates in the solar nebula, actually do have elevated $^{26}Al/^{27}Al$ ratios in several meteorites (e.g., Allende, Murchison, and Efremovka). Data presented by Sahijpal et al. (2000) indicate that the $^{26}Al/^{27}Al$ ratios of hibonites are positively correlated with $^{41}Ca/^{40}Ca$ ratios, which suggests that short-lived ^{26}Al and ^{41}Ca were both formed during nucleosynthesis in stars (Sahijpal et al., 2000).

Short-lived $^{41}_{20}Ca$ ($T_{1/2} = 0.103 \times 10^6$ y) decays by electron capture to stable $^{41}_{19}K$, which increases the $^{41}K/^{39}K$ ratio of certain minerals (e.g., pyroxene in Egg 3, a Ca–Al rich inclusion in Allende). Additional investigations of extinct radionuclides in chondrite meteorites have been published by Srinivasan et al. (2000), Kita et al. (2000), Huss et al. (2001), Mostefaoui et al. (2002), and others referred to by them.

24.3 THE Hf–W CHRONOMETER

Hafnium ($Z = 72$) has five naturally occurring isotopes and 26 radioactive ones, only one of which ($^{174}_{72}Hf$) occurs naturally in terrestrial rocks and meteorites because it has a very long halflife of 2.0×10^{15} years. A second isotope (^{182}Hf) has a halflife of 9×10^6 years and therefore could have survived in the solar nebula for about 45×10^6 years (five halflives) until it effectively became extinct. Isotope $^{182}_{72}Hf$ decays by negatron emission to unstable $^{182}_{73}Ta$, which then decays to stable $^{182}_{74}W$, with a halflife of 114.43 days, by emitting another suite of negatively charged β-particles:

$$^{182}_{72}Hf \rightarrow {}^{182}_{73}Ta + \beta^- + \overline{\nu} + 0.37 \text{ MeV} \quad (24.7)$$

$$^{182}_{73}Ta \rightarrow {}^{182}_{74}W + \beta^- + \overline{\nu} + 1.814 \text{ MeV} \quad (24.8)$$

This short decay series meets the requirements for secular equilibrium (Section 3.3b), which means that the rate of decay of ^{182}Ta is equal to that of long-lived ^{182}Hf. Therefore, the rate of growth of stable ^{182}W is equal to the rate of decay of ^{182}Hf.

Tungsten ($Z = 74$) has five stable isotopes and 28 short-lived unstable isotopes. The abundance of stable ^{182}W in terrestrial W is 26.3% and that of stable ^{184}W is 30.67%. Therefore, the $^{182}W/^{184}W$ ratio of terrestrial W is:

$$\frac{^{182}W}{^{184}W} = \frac{26.3}{30.67} = 0.8575$$

Actually, the NIST-3163 W standard yields an average $^{182}W/^{184}W$ ratio of 0.865000 ± 0.000018 corrected for isotope fractionation to $^{186}W/^{184}W = 0.927633$ (Lee and Halliday, 1995).

The development of the Hf–W chronometer was retarded for many years because the isotope composition of W is difficult to determine by conventional thermal-ionization mass-spectrometry (TIMS) owing to the high first ionization potential of W (about 7.98 eV). This problem was overcome by Harper et al. (1991) after Völkening et al. (1991) first measured the isotope composition of W by *negative-ion* TIMS. Later, Halliday et al. (1996) successfully analyzed W by multiple-collector inductively-coupled plasma mass-spectrometry (MC-ICP-MS).

Several complications in the interpretation of the isotope composition of W exist because W isotopes are produced by cosmic rays in meteorites and lunar rocks (Leya et al., 2000) and by neutron-capture reactions (Masarik, 1997). In addition, $^{182}_{74}W$ is the product of α-decay of naturally occurring $^{186}_{76}Os$, which has a halflife of 2×10^{15} years and an abundance of 1.58%. Therefore, the $^{182}W/^{184}W$ ratios of Os-rich and W-poor minerals of Precambrian age and of Os-rich iron meteorites (Table 13.2) may have increased by the decay of both ^{182}Hf and ^{186}Os. However, the slow rate of decay and the low abundance of ^{186}Os restrict this interfering decay process to Os-rich alloys of the PGEs (Sections 13.7 and 13.8).

The occurrence of live ^{182}Hf in the solar nebula is important for the study of the internal differentiation of large planetesimals in the early solar system because Hf is strongly lithophile whereas W is moderately siderophile (Lee and Halliday, 2000a, b). Therefore, when large planetesimals and planets in the early solar system differentiated by segregation of immiscible Fe–Ni and silicate liquids, the Fe–Ni phase was enriched in W and depleted in Hf and therefore had a low Hf/W ratio compared to the reservoir in the solar nebula. Conversely, the silicate liquid was enriched in Hf and depleted in W and acquired a higher Hf/W ratio than the Fe–Ni phase. If live ^{182}Hf was present in the solar nebula while these processes were occurring, then the W in the Fe–Ni alloys of iron meteorites (and of planetary cores) should be deficient in radiogenic ^{182}W compared to W in the silicate minerals of stony meteorites and of the mantles of terrestrial planets (Ireland, 1991).

The concentrations of Hf and W of three chondrites in Table 24.3 confirm these predictions because the Hf/W ratios of the nonmagnetic (silicate and oxide) fractions are consistently higher than those of the magnetic (metallic) fractions by an average factor of about 30. The abundances of radiogenic ^{182}W of these fractions, expressed by the $^{182}W/^{184}W$ ratios, correlate closely with their Hf/W ratios, as expected for in situ decay in closed systems. Consequently, the $^{182}W/^{184}W$ ratios of the

Table 24.3. Concentration of Hf and W in Silicate + Oxide (Nonmagnetic) and Metallic (Magnetic) Fractions of Ordinary Chondrite Meteorites

Fraction	Hf, ppb	W, ppb	Hf/W
	Forest Vale (H4)		
Silicate + oxide	176	114	1.54
Metal (bulk)	38.9	700	0.055
Whole rock	172	178	0.966
	Ste. Marguerite (H4)		
Metal (bulk)	9.81	754	0.013
Whole rock	141	166	0.849
	Richardton (H5)		
Silicate + oxide	87.1	41.2	2.11
Metal (bulk)	54.7	521	0.105

Source: Lee and Halliday, 1996, 2000a.

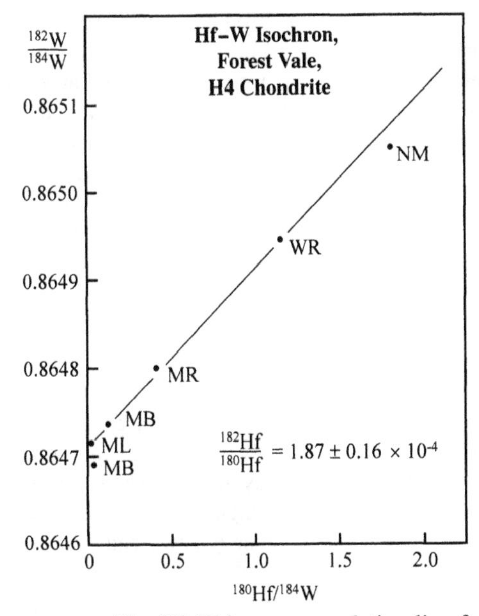

FIGURE 24.4 The Hf–W isotope-correlation line for silicate and metal fractions of the H4 chondrite Forest Vale. The fractions are identified by the abbreviations: NM = nonmetallic; WR = whole rock; MR = silicate residue after selective dissolution of the metal; MB = metal in bulk with intergrown silicate; ML = metal leachate. Data and interpretation by Lee and Halliday (2000a).

metallic and nonmetallic fractions of the meteorite Forest Vale in Figure 24.4 are positively correlated with their $^{180}Hf/^{184}W$ ratios. The correlation line is, in fact, an isochron because all of the fractions that define it have the same age. The equation of this line, as determined by Lee and Halliday (2000a), is:

$$\frac{^{182}W}{^{184}W} = 0.864710 + 1.87 \times 10^{-4} \left(\frac{^{180}Hf}{^{184}W}\right)$$

The slope ($1.87 \pm 0.16 \times 10^{-4}$) has the dimensions of the radiogenic $^{182}W^*/^{180}Hf$ ratio, which is identical to the $^{182}Hf/^{180}Hf$ ratio of this meteorite at the time of its crystallization because all of the radiogenic atoms of ^{182}W formed by decay of the ^{182}Hf atoms that were initially present in this meteorite. The intercept on the y-axis of Figure 24.4 is $^{182}W/^{184}W = 0.864710 \pm 0.000015$, which is only slightly lower than the $^{182}W/^{184}W$ ratio of present-day terrestrial W (e.g., $^{182}W/^{184}W = 0.86500 \pm 0.000018$ for NIST 3163).

The $^{182}Hf/^{180}Hf$ ratios (i.e., the slopes of their Hf–W isochrons) of the other two chondrite meteorites analyzed by Lee and Halliday (2000a) are similar to the $^{182}Hf/^{180}Hf$ ratio of Forest Vale:

Forest Vale:	$1.87 \pm 0.16 \times 10^{-4}$
Ste. Marguerite:	$1.77 \pm 0.69 \times 10^{-4}$
Richardton:	$1.92 \pm 0.83 \times 10^{-4}$

The *maximum* length of time during which these meteorites crystallized is obtained from the equation:

$$\Delta t = \frac{\ln(^{182}Hf/^{180}Hf)_1/(^{182}Hf/^{180}Hf)_2}{\lambda}$$

Substituting appropriate values yields:

$$\Delta t = \frac{\ln(1.92 + 0.83)/(1.77 - 0.69)}{0.077 \times 10^{-6}} = 12 \times 10^6 \text{ y}$$

and the factor 10^{-4} cancels. Taking the best values for the $^{182}Hf/^{180}Hf$ ratios of Richardton (1.92×10^{-4}) and Ste. Marguerite (1.77×10^{-4}) yields $\Delta t = 1.0 \times 10^6$ y. In either case, these three chondrites crystallized in a short interval of time lasting between 1.0 and 12 millions years.

The narrow range of the $^{182}Hf/^{180}Hf$ ratios of the three chondrites analyzed by Lee and Halliday (2000a) and of the carbonaceous chondrites Allende and Murchison (Lee and Halliday, 1996) suggests that the solar nebula was homogeneous with respect to the isotope compositions of Hf and W at the time of formation of the respective meteorite parent bodies. Therefore, these data can be used to define the $^{182}Hf/^{180}Hf$ ratio of the solar nebula at the time of formation of the Earth based on the following considerations.

The Pb–Pb date of Ca–Al inclusions (CAIs) of Allende is 4.566 ± 0.002 Ga, whereas the Pb–Pb date of phosphate minerals of Forest Vale is about 5×10^6 years younger than this value (Manhès et al., 1988; Göpel et al., 1994). If the Pb–Pb date derived from the Allende CAIs is the age of the solar system and if Forest Vale is 5×10^{-6} years younger than this date, then the $^{182}Hf/^{180}Hf$ ratio of the solar system can be calculated from the $^{182}Hf/^{180}Hf$ ratio of Forest Vale:

$$\left(\frac{^{182}Hf}{^{180}Hf}\right)_0 = \left(\frac{^{182}Hf}{^{180}Hf}\right) e^{\lambda t} \qquad (24.9)$$

where $(^{182}Hf/^{180}Hf)_0$ = value of this ratio of the solar nebula at the time of formation of the solar system

$^{182}Hf/^{180}Hf$ = value of this ratio in Forest Vale at the time of its crystallization

$\lambda = 0.077 \times 10^{-6}\ y^{-1}$ (decay constant of ^{182}Hf)

$t = 5 \times 10^6$ y

Therefore,

$$\left(\frac{^{182}Hf}{^{180}Hf}\right)_0 = (1.87 \times 10^{-4})e^{-0.077\times 5}$$

$$= 2.75 \pm 0.24 \times 10^{-4}$$

In addition, Lee and Halliday (2000b) estimated that the initial $^{182}W/^{184}W$ ratio of the solar system at 4.566×10^9 Ga was 0.86457 ± 0.00004.

The application of Hf–W systematics of the solar nebula and of the partitioning of Hf and W into the silicate and metal fractions, respectively, to the formation of planetary cores, has been discussed by:

Lee and Halliday, 1995b, *Nature*, 378:771–774. Lee and Halliday, 1997, *Nature*, 388:854–857. Lee and Halliday, 1998, *Mineral. Mag.*, 62A:868–869. Halliday and Lee, 1999, *Geochim. Cosmochim. Acta*, 63:4157–4179. Horan et al., 1998, *Geochim. Cosmochim. Acta*, 62: 545–554. Harper and Jacobsen, 1996, *Geochim. Cosmochim. Acta*, 60:1131–1153. Jacobsen and Harper, 1996, in Basu and Hart (S.R.), Geophys. Monogr. 95, pp. 47–74, Amer. Geophysical Union, Washington, D.C. Lee et al., 1997, *Science*, 278:1098–1103. Kramers, 1998, *Chem. Geol.*, 145:461–478. Shearer and Newsom, 2000, *Geochim. Cosmochim. Acta*, 64:3599–3613. Halliday, 2000, *Earth Planet. Sci. Lett.*, 176:17–30. Schoenberg et al., 2002, *Geochim. Cosmochim. Acta*, 66:3151–3160.

Additional short-lived radionuclides that occurred in the solar nebula and were incorporated into planetesimals as well as into the Earth, the Moon, Mars, and other planets are identified and referenced in Table 24.4. The foregoing interpretation of the isotopic abundances of the radiogenic daughters of extinct radioactive parents provides a model that can be applied to other chronometers that were active in the early solar system.

24.4 FUN IN THE SOLAR NEBULA

The isotopic composition of Mg in certain meteorites was altered not only by the decay of presently extinct ^{26}Al but also by mass-dependent isotope fractionation during changes in state from vapor to solid or solid to vapor (Ozawa and Nagahara, 2001):

Esat et al., 1986, *Nature*, 319:576–578. Uyeda et al., 1991, *Earth Planet. Sci. Lett.*, 107: 138–147. Nagahara and Ozawa, 2000, *Chem. Geol.*, 169:45–68. Grossman et al., 2000, *Geochim. Cosmochim. Acta*, 64:2879–2894. Alexander et al., 2002, *Geochim. Cosmochim. Acta*, 66:173–183. Grossman et al., 2002, *Geochim. Cosmochim. Acta*, 66:145–161.

In general, the isotope compositions of elements in the solar nebula were the result of several different processes, including

1. decay of short-lived radionuclides,
2. mass-dependent fractionation,
3. nucleosynthesis in the ancestral stars followed by mixing in interstellar space,
4. nuclear spallation reactions,
5. nuclear capture reactions,
6. injection of atomic nuclei by cosmic rays and the solar wind, and
7. unknown nuclear effects.

Huneke et al. (1983) coined the acronym FUN from “fractionation” (F) and “unknown nuclear origin” (UN). Accordingly, many elements in meteorites have FUN isotopic compositions that are not uniquely explainable by one of the processes listed above. For example, Clayton et al. (1978) reported

Table 24.4. Selected References to the I–Xe and Other Extinct Radionuclides That Occurred in the Early Solar System

Chronometer	References
^{129}I–^{129}Xe	Reynolds, 1960, *J. Geophys. Res.*, 65(11):3843–3846; Hohenberg et al., 1967, *Science*, 156:202–206; Hohenberg et al., 2000, *Geochim. Cosmochim. Acta*, 64:4257–4262; Podosek, 1970, *Geochim. Cosmochim. Acta*, 34:341–365; Niemeyer, 1979, *Geochim. Cosmochim. Acta*, 43:843–860; Swindle and Podosek, 1988, in Kerridge and Matthews (Eds.), *Meteorites and the Early Solar System*, pp. 1127–1146, University of Arizona Press, Tucson; Brazzle et al., 1995, *Lunar Planet. Sci.*, 26:165–166; Brazzle et al., 1999, *Geochim. Cosmochim. Acta*, 63:739–760; Gilmour et al., 2000, *Meteoritics Planet. Sci.*, 35:445–455; Whitby et al., 2002, *Geochim. Cosmochim. Acta*, 66:347–359.
^{53}Mn–^{53}Cr	Wasserburg and Arnould, 1987, in Hillebrandt et al. (Eds.), *Nuclear Astrophysics*, pp. 262–276, Springer-Verlag, Heidelberg; Birck and Allègre, 1988, *Nature*, 331:579–584; Davis and Olsen, 1990, *Lunar Planet. Sci.*, 21:258–259; Hutcheon and Olsen, 1991, *Lunar Planet. Sci.*, 22:605–606; Rotaru et al., 1992, *Nature*, 358:465–470; Nyquist et al., 1997, *Lunar Planet. Sci.*, 27:1093–1094; Wadhwa et al., 1997, *Meteoritics Planet. Sci.*, 32:218–292; Lugmair and Shukolyukov, 1998, *Geochim. Cosmochim. Acta*, 62:2863–2886.
^{60}Fe–^{60}Ni	Shukolyukov and Lugmair, 1993a, *Science*, 159:1138–1142; Shukolyukov and Lugmair, 1993b, *Earth Planet. Sci. Lett.*, 119:159–166; Lugmair et al., 1994, in Busso et al. (Eds.), *Nuclei in the Cosmos*, pp. 568–571, ATP, New York.
^{92}Nb–^{92}Zr, ^{93}Zr–^{93}Nb	Apt et al., 1974, *Geochim. Cosmochim. Acta*, 38:1485–1488; Minster and Allègre, 1982, *Geochim. Cosmochim. Acta*, 46:565–573; Harper et al., 1990, *Lunar Planet. Sci.*, 21:453–454.
^{244}Pu	Alexander et al., 1971, *Science*, 172:837–840.

that Si in Allende inclusions has anomalous isotope compositions attributable to mass-dependent fractionation and other effects of unknown nuclear origin (i.e., FUN). Anomalous isotope compositions attributable to FUN have also been reported for Ti in meteorites (Fahey et al., 1985; Hinton et al., 1985; Ireland et al., 1985). Shima (1985) actually documented the occurrence of isotopic anomalies in 33 elements contained in meteorites.

The heterogeneity of the isotopic compositions of elements in the solar nebula is not apparent when meteorites are analyzed in bulk. The reason is that chondritic meteorites are polymict breccias containing a variety of inclusions and mineral grains. The inclusions contain early-formed condensation products that have preserved a record of the isotopic heterogeneity of the region of the solar nebula in which they formed, including mass-dependent isotope fractionation. The resulting isotopic heterogeneity of chondritic meteorites is not recognized in the analyses of bulk samples because of averaging. Therefore, FUN isotopic compositions were not discovered until ion probe mass spectrometers and laser-ablation multiple-collector inductively-coupled plasma mass-spectrometers (LA-MC-ICP-MS) became available (Galy et al., 2001; Young et al., 2002). These instruments permit small areas in microscopic crystals to be analyzed in situ with great sensitivity. The present availability of the necessary analytical equipment has stimulated the study of extinct radionuclides and anomalous abundances of stable isotopes of many elements that are affected by FUN.

24.5 SUMMARY

More than 4.56 billion years ago the solar nebula contained radioactive isotopes of many chemical elements. Some of these radionuclides survived from the time of nucleosynthesis in the ancestral stars to the time solid particles began to condense in the solar nebula. Others may have been produced by nuclear reactions caused by cosmic rays. The contensates, in time, formed larger objects called planetesimals, from which grew the parent bodies of the meteorites and the planets of the solar system. Solid objects which achieved diameters of several hundred kilometers became hot enough to melt and to differentiate internally by the segregation of immiscible liquids composed of metallic Fe–Ni, of sulfides of Fe and other metals, and of silicates and oxides. The heat required to melt these large objects was provided primarily by the energy released by the decay of the radionuclides they contained.

The existence of certain radionuclides in the parent bodies of meteorites is recorded by the presence of their stable decay products in different types of meteorites. The resulting anomalous isotope ratios can be measured by means of ion-probe mass spectrometers and by laser-ablation inductively-coupled plasma mass-spectrometry (LA-ICP-MS), which can selectively analyze small particles with diameters down to 10 μm. The resulting isotope ratios of the parent and daughter elements are interpreted by means of correlation diagrams that yield the abundance of the original parent radionuclide and the isotope composition of the daughter element before the radiogenic isotope was added.

The abundance of the parent radionuclides in the meteorites at the time of their crystallization makes a statement about the abundances of these nuclides in the solar system and fixes a time in the history of the early solar system. By comparing the initial abundances of a particular radionuclide in several different meteorites, a chronology can be developed for the melting, differentiation, and subsequent crystallization and cooling of their parent bodies.

The Hf–W chronometer is particularly important because Hf and W are separated from each other during the segregation of metallic and silicate liquids. Tungsten likes liquid Fe–Ni, whereas Hf likes silicate liquids. Therefore, metallic minerals in stony meteorites have low Hf/W ratios and low abundances of the radiogenic ^{182}W. Silicate minerals have high Hf/W ratios and therefore contain more ^{182}W. The combination of processes (differentiation and decay) has been used to constrain the origin of the core of the Earth and the origin of the Moon and Mars. The presence of radiogenic products of the decay of extinct radionuclides in different kinds of meteorites and other kinds of samples leads to the conclusion that the planets of the solar system formed in less than about 50 million years (i.e., before the radionuclides had time to decay to low levels).

The isotope compositions of certain elements in small inclusions of meteorites are also affected by mass-dependent fractionation during condensation and evaporation and by nuclear reactions which may have occurred in the ancestral stars and later in interstellar space. These kinds of processes are referred to as FUN (fractionation and unknown nuclear origin) and affect the abundances of the stable (or long-lived unstable) isotopes of the elements. FUN isotope compositions of certain elements in extraterrestrial samples are easily overlooked in analyses of large samples, which then leads to the conclusion that the solar nebula was homogeneous with respect to the isotope composition of the elements. Analyses of small inclusions by means of ion-probe mass-spectrometers and LA-ICP-MS has revealed the existence of large variations in the isotope composition of many elements. Therefore, the bottom line is that the solar nebula was heterogeneous with respect to the isotope compositions of a significant number of chemical elements.

REFERENCES FOR INTRODUCTION

Carlson, W. R., and G. W. Lugmair, 2000. Timescales of planetesimal formation and differentiation based on extinct and extant radioisotopes. In R. Canup and K. Righter (Eds.), *The Origin of the Earth and Moon*, pp. 25–44. University of Arizona Press, Tucson, Arizona.

Jacobsen, S. B., and C. L. Harper, Jr., 1996. Accretion and early differentiation history of the Earth based on extinct radionuclides. In A. Basu and S. R. Hart (Eds.), *Earth Processes: Reading the Isotopic Code*, Geophysical Monograph 95, pp. 47–74. American Geophysical Union, Washington, D.C.

Leya, I., R. Wieler, and A. N. Halliday, 2003. The influence of cosmic-ray production on extinct nuclide systems. *Geochim. Cosmochim. Acta*, 67:529–541.

Podosek, F. A., and P. Cassen, 1994. Theoretical, observational, and isotopic estimates of the lifetime of the solar nebula. *Meteoritics*, 29:6–25.

Reynolds, J. H., 1960. Determination of the age of elements. *Phys. Rev. Lett.*, 4:8–10.

Urey, H. C., 1955. The cosmic abundances of K, U, and Th and the heat balances of the earth, the moon, and mars. *Proc. Natl. Acad. Sci.*, 41:127–144.

Wasserburg, G. J., 1985. Short-lived nuclei in the early solar system. In D. C. Black and M. S. Matthews (Eds.), *Protostars and Planets II*, pp. 703–737. University of Arizona Press, Tucson, Arizona.

REFERENCES FOR THE Pd–Ag CHRONOMETER (SECTION 24.1)

Carlson, R. W., and E. H. Hauri, 2001. Extending the ^{107}Pd–^{107}Ag chronometer to low Pd/Ag meteorites with multicollector plasma-ionization mass spectrometry. *Geochim. Cosmochim. Acta*, 65:1839–1848.

Chen, J. H., and G. J. Wasserburg, 1990. The isotopic composition of Ag in meteorites and the presence of ^{107}Pd in protoplanets. *Geochim. Cosmochim. Acta*, 54:1729–1743.

Chen, J. H., and G. J. Wasserburg, 1996. Live ^{107}Pd in the early solar system and implications for planetary evolution. In A. Basu and S. R. Hart (eds.), *Earth Processes; Reading the Isotopic Code*, Geophysical Monograph 95, pp. 1–20. American Geophysical Union, Washington, D.C.

Kaiser, T., and G. J. Wasserburg, 1983. The isotopic composition and concentration of Ag in iron meteorites and the origin of exotic silver. *Geochim. Cosmochim. Acta*, 47:43–58.

Kelly, W. R., and G. J. Wasserburg, 1978. Evidence for the existence of ^{107}Pd in the early solar system. *Geophys. Res. Lett.*, 5:1079–1082.

Murthy, V. R., 1960. Isotopic composition of Ag in an iron meteorite. *Phys. Rev. Lett.*, 5:539–541.

Murthy, V. R., 1962. The isotopic composition of Ag in iron meteorites. *Geochim. Cosmochim. Acta*, 26: 481–488.

REFERENCES FOR THE Al–Mg CHRONOMETER (SECTION 24.2)

Bradley, J. G., J. C. Huneke, and G. J. Wasserburg, 1978. Ion microprobe evidence for the presence of excess ^{26}Mg in an Allende anorthite crystal. *J. Geophys. Res.*, 83:244–254.

Catanzaro, E. J., and T. J. Murphy, 1966. Magnesium isotope ratios in natural samples. *J. Geophys. Res.*, 71(4):1271–1274.

Esat, T. M., D. A. Papanastassiou, and G. J. Wasserburg, 1978. Search for ^{26}Al effects in the Allende FUN inclusion C1. *Geophys. Res. Lett.*, 5:807–810.

Gray, C. M., and W. Compston, 1974. Excess ^{26}Mg effects in the Allende meteorite. *Nature*, 251:495–497.

Grossman, L., 1975. Petrography and mineral chemistry of Ca-rich inclusions in the Allende meteorite. *Geochim. Cosmochim. Acta*, 39:433–454.

Huneke, J. C., J. T. Armstrong, and G. J. Wasserburg, 1983. FUN with PANURGE: High mass-resolution ion-microprobe measurements of Mg in Allende inclusions. *Geochim. Cosmochim. Acta*, 47:1635–1650.

Huss, G. R., G. J. MacPherson, G. J. Wasserburg, S. S. Russell, and G. Srinivasan, 2001. Aluminum-26 in calcium-aluminum-rich inclusions and chondrules from unequilibrated ordinary chrondrites. *Meteoritics Planet. Sci.*, 36:975–997.

Hutcheon, I. D., 1982. Ion probe magnesium isotopic measurements of Allende inclusions. In L. A. Currie (Ed.), *Nuclear and Chemical Dating Techniques; Interpreting the Environmental Record*, ACS Symposium Series 176, pp. 95–128. American Chemical Society, Washington, D.C.

Ireland, T. R., W. Compston, and T. M. Esat, 1986. Magnesium isotopic compositions of olivine, spinel, and hibonite from the Murchison carbonaceous chondrite. *Geochim. Cosmochim. Acta*, 50:1413–1421.

Kita, N. T., H. Nagahara, S. Togashi, and Y. Morishita, 2000. A short duration of chondrule formation in the solar nebula: Evidence from ^{26}Al in Semarkona ferromagnesian chondrules. *Geochim. Cosmochim. Acta*, 64:3913–3922.

Lee, T., and D. A. Papanastassiou, 1974. Mg isotopic anomalies in the Allende meteorite and correlation with O and Sr effects. *Geophys. Res. Lett.*, 1:225–228.

Mostefaoui, S., N. T. Kita, S. Togashi, S. Tachibana, H. Nagahara, and Y. Morishita, 2002. The relative formation ages of ferromagnesian chondrules inferred from their initial aluminum-26/aluminum-27 ratios. *Meteoritics Planet. Sci.*, 37:421–438.

Sahijpal, S., J. N. Goswami, and A. M. Davis, 2000. K, Mg, Ti, and Ca isotopic compositions and refractory trace element abundances in hibonites from CM and CV meteorites: Implications for early solar system processes. *Geochim. Cosmochim. Acta*, 64:1989–2005.

Srinivasan, G., G. R. Huss, and G. J. Wasserburg, 2000. A petrographic, chemical, and isotopic study of calcium-aluminum-rich chondrules from the Axtell (CV3) chondrite. *Meteoritics Planet. Sci.*, 35:1333–1354.

REFERENCES FOR THE Hf–W CHRONOMETER (SECTION 24.3)

Göpel, C., G. Manhès, and C. J. Allègre, 1994. U–Pb systematics of phosphates from equilibrated ordinary chondrites. *Earth Planet. Sci. Lett.*, 121:153–171.

Halliday, A. N., M. Rehkämper, D.-C. Lee, and W. Yi, 1996. Early evolution of the Earth and Moon: New constraints from Hf–W isotope geochemistry. *Earth Planet. Sci.*, 142:75–89.

Harper, C. L., J. Völkening, K. G. Heumann, C.-Y. Shih, and H. Wiesmann, 1991. ^{182}Hf–^{182}W: New cosmochronometric constraints on terrestrial accretion, core formation, the astrophysical site of the r-process, and the origin of the solar system. *Lunar Planet. Sci.*, 22:515–516.

Ireland, T. R., 1991. The abundance of ^{182}Hf in the early solar system. *Lunar Planet. Sci.*, 22:609–610.

Lee, D.-C., and A. N. Halliday, 1995. Precise determinations of the isotopic compositions and atomic weights of molybdenum, tellurium, tin, and tungsten using ICP magnetic-sector multiple collector mass-spectrometry. *Int. J. Mass Spec. Ion Proc.*, 146/147:35–46.

Lee, D.-C., and A. N. Halliday, 1996. Hf–W isotopic evidence for rapid accretion and differentiation in the early solar system. *Science*, 274:1876–1879.

Lee, D.-C., and A. N. Halliday, 2000a. Hf–W internal isochrons for ordinary chondrites and the initial ^{182}Hf/^{180}Hf of the solar system. *Chem. Geol.*, 169:35–43.

Lee, D. C., and A. N. Halliday, 2000b. Accretion of primitive planetesimals: Hf-W-isotopic evidence from enstatite chondrites. *Science*, 288:1629–1631.

Leya, I., R. Wieler, and A. N. Halliday, 2000. Cosmic-ray production of tungsten isotopes in lunar samples and meteorites and its implications for Hf–W cosmochemistry. *Earth Planet. Sci. Lett.*, 175:1–12.

Manhès, G., C. Göpel, and C. J. Allègre, 1988. Systematique U–Pb dans les inclusions refractories d'Allende: le plus vieuz materiau solaire. *C.R. ATP Planetol.*, 1988:323–327.

Masarik, J., 1997. Contribution of neutron-capture reactions to observed tungsten-isotopic ratios. *Earth Planet. Sci. Lett.*, 152:181–185.

Völkening, J., M. Köppe, and K. G. Heumann, 1991. Tungsten isotope ratio determinations by negative thermal ionization mass spectrometry. *Int. J. Mass Spec. Ion Proc.*, 107:361–368.

REFERENCES FOR FUN IN THE SOLAR NEBULA (SECTION 24.4)

Clayton, R. N., T. K. Mayeda, and S. Epstein, 1978. Isotopic fractionation of silicon in Allende inclusions. In *Proceed. 9th Lunar Science Conference*, pp. 1267–1278. Lunar and Planet. Sci. Institute, Houston, Texas.

Fahey, A. J., J. N. Goswami, K. D. Keegan, and E. Zinner, 1985. Evidence for extreme ^{50}Ti enrichments in primitive meteorites. *Astrophys. J.*, 296:L17–L20.

Galy, A., N. S. Belshaw, L. Halicz, and R. K. O'Nions, 2001. High precision measurement of magnesium isotopes by multiple-collector inductively coupled plasma mass spectrometry (MC-ICP-MS). *Int. J. Mass Spectrom.*, 208:89–98.

Hinton, R. W., A. M. Davis, and D. E. Scatena-Wachel, 1985. Large negative ^{50}Ti anomaly in a refractory inclusion from the Murchison meteorite. *Meteoritics*, 20:664–665.

Huneke, J. C., J. T. Armstrong, and G. J. Wasserburg, 1983. FUN with PANURGE: High mass resolution ion microprobe measurements of Mg in Allende inclusions. *Geochim. Cosmochim. Acta*, 17:1635–1650.

Ireland, T. R., W. Compston, and H. R. Heydegger, 1985. Titanium isotopic anomalies in hibonites from the Murchison carbonaceous chondrite. *Geochim. Cosmochim. Acta*, 49:1989–1993.

Ozawa, K., and H. Nagahara, 2001. Chemical and isotopic fractionations by evaporation and their cosmochemical implications. *Geochim. Cosmochim. Acta*, 65:2171–2199.

Shima, M., 1985. A summary of isotopic variations in extra-terrestrial materials. *Geochim. Cosmochim. Acta*, 50:577–584.

Young, E. D., R. D. Ash, A. Galy, and N. S. Belshaw, 2002. Mg isotope heterogeneity in the Allende meteorite measured by UV laser ablation-MC-ICPMS and comparisons with O isotopes. *Geochim. Cosmochim. Acta*, 66:683–698.

CHAPTER 25

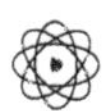

Thermonuclear Radionuclides

SINCE the middle of the twentieth century, the surface of the Earth has been contaminated by radionuclides of anthropogenic origin (Sections 2.3d and 3.6). The sources of these radionuclides include

1. nuclear fission and fusion devices,
2. spent nuclear fuel containing both fission-product radionuclides and transuranium elements,
3. neutron-capture reactions during testing of nuclear weapons and in controlled neutron-activation experiments, and
4. nuclear reactions that occur in particle accelerators, including cyclotrons and linear accelerators.

The radionuclides that originate from these sources are dispersed globally by the circulation of the atmosphere and oceans, by the movement of water on the surface of the Earth, and by groundwater in the subsurface. In addition, spent nuclear fuel can release radionuclides into the environment while it is in temporary storage at nuclear-reactor sites. Therefore, the spent nuclear fuel must be isolated in order to protect all life forms on the Earth from the intense radiation it emits (α, β, and γ). In addition, the energy released by the decay of fission-product radionuclides and the transuranium elements manifests itself in the form of heat, which means that spent nuclear fuel must also be cooled. The thermonuclear radionuclides constitute a serious health hazard not only by the radiation they emit but also because certain radionuclides are biological poisons. The problem of disposing of spent nuclear fuel is further compounded by the slow rate of decay of some of the radionuclides, which causes the radwaste to remain hazardous for up to 100,000 years.

Radionuclides produced by nuclear fission and other kinds of nuclear reactions also have beneficial uses in medical science, in the form of sealed radiation sources used in industry, and in the environmental sciences. For example, thermonuclear (or bomb-produced) tritium, ^{14}C, and ^{129}I are used to date geological and archeological samples as well as to trace the movement of water (^{3}H: Section 21.4b; ^{14}C: Section 23.1b; ^{129}I: Section 23.6).

25.1 FISSION PRODUCTS AND TRANSURANIUM ELEMENTS

Spent nuclear fuel contains both fission-product radionuclides and isotopes of transuranium elements (Np, Pu, and Am), whose isotopes are all radioactive. The disposal of spent nuclear fuel must prevent the dispersal of both the fission products and the transuranium elements and their respective radioactive progeny.

25.1a Fission Products

Induced fission of ^{235}U leads to the formation of a large number of nuclides most of which are radioactive but some of which are stable isotopes of elements with atomic numbers between 32 (Ge) and 68 (Er). The data in Table 25.1 indicate that 344 fission-product radionuclides have halflives of less than ten hours. Most of these decay within about 50 hours or in about two days (i.e., five halflives). Similar considerations indicate that radionuclides with halflives of less than one year decay within five years. However, 11 radionuclides whose halflives range from 1 to 100 years require up to 500 years to decay. This is a significant length of time and will be a burden on future generations of humans. Some of the long-lived radionuclides identified in Table 25.1 (e.g., ^{146}Sm, 1.03×10^8 years) will remain active for hundreds of millions of years.

The products of nuclear fission also include unstable isotopes of technetium (Tc, $Z = 43$) and promethium (Pm, $Z = 61$), which do not occur naturally in the solar system because all of their isotopes have decayed since the end of nucleosynthesis in the ancestral stars. Technetium is represented among the fission products by 13 unstable isotopes, including ^{99}Tc, which has a halflife of 2.13×10^5 years and decays to stable ^{99}Ru by β^--emission. Promethium has 11 fission product isotopes, including ^{147}Pm, which emits β^--particles as it decays to stable ^{147}Sm with a halflife of 2.6234 years. Technetium is a member of Group VIIB in the periodic table together with Mn ($Z = 25$) and Re ($Z = 75$), whereas Pm is a rare-earth element between Nd ($Z = 60$) and Sm ($Z = 62$).

Table 25.1. Fission-Product Radionuclides

Halflife	Number
<10 hours	344
10–24 hours	13
1–365.25 days	33
1–100 years	11
>100 years[a]	6
Stable[b]	100

Source: Walker et al., 1988.

[a]The long-lived fission product radionuclides and their halflives are: ^{79}Se, 6.5×10^4 y; ^{99}Tc, 2.13×10^5 y; ^{107}Pd, 6.5×10^6 y; ^{126}Sn, 1×10^5 y; ^{135}Cs, 2.3×10^6 y; ^{146}Sm, 1.03×10^8.

[b]Including ^{4}He.

The nuclei of all fission-product radionuclides contain an excess of neutrons and therefore are located to the right of the band of stable nuclides on the chart of the nuclides (Figure 1.2) Consequently, all of these radionuclides decay by emitting β^--particles followed, in many cases, by γ-rays. The emission of these γ-rays by the large number of short-lived radionuclides is a major factor in the health hazard caused by spent nuclear fuel.

Large amounts of fission-product radionuclides were released into the atmosphere by the nuclear weapons tests from 1945 to 1980. The radionuclides were subsequently transported worldwide and were deposited as "fallout" on the continents and on the surface of the oceans. The resulting contamination of the surface of the Earth was monitored by various government agencies in many countries because of concern that the fallout radionuclides posed a health threat. The history of nuclear testing was described by Carter and Moghissi (1977), Leifer et al., (1984), and by the United Nations Scientific Committee on the Effects of Atomic Radiation (e.g., UNSCEAR, 1969). The distribution of ^{90}Sr was a special concern because this isotope is deposited in bones where the β-particles it emits can damage the bone marrow and thereby cause anemia. A final summary of average monthly fallout of ^{90}Sr between 1954 and 1976 was published in 1977 by the U.S. Department of Energy after most of the testing had ended (USDOE, 1977). The dispersal of ^{90}Sr and ^{137}Cs that resulted from the nuclear-weapons tests and from the reactor accident in April of 1986 at Chernobyl, Ukraine, is considered in Sections 25.2 and 25.3, respectively.

25.1b Transuranium Elements

Induced fission of ^{235}U in a nuclear reactor releases about 2.5 neutrons per fission on average (Section 2.3c). Some of these neutrons are absorbed by the control rods of the reactor, some interact with the nuclei of ^{235}U and cause additional fission events which maintain the chain reaction, and others are captured by the nuclei of ^{238}U leading

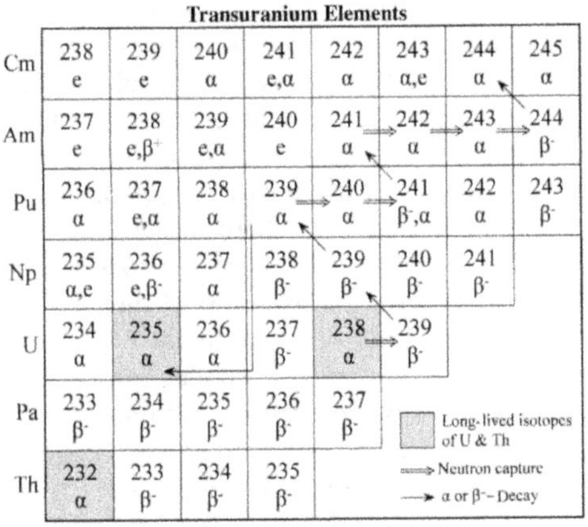

Transuranium Elements

Cm	238 e	239 e	240 α	241 e,α	242 α	243 α,e	244 α	245 α
Am	237 e	238 e,β^+	239 e,α	240 e	241 α	242 α	243 α	244 β^-
Pu	236 α	237 e,α	238 α	239 α	240 α	241 β^-,α	242 α	243 β^-
Np	235 α,e	236 e,β^-	237 α	238 β^-	239 β^-	240 β^-	241 β^-	
U	234 α	235 α	236 α	237 β^-	238 α	239 β^-		
Pa	233 β^-	234 β^-	235 β^-	236 β^-	237 β^-			
Th	232 α	233 β^-	234 β^-	235 β^-				

Long-lived isotopes of U & Th
⟹ Neutron capture
⟶ α or β^--Decay

FIGURE 25.1 Production of certain isotopes of transuranium elements by successive neutron-capture reactions followed by beta (β^-) electron capture (e), positron (β^+), or alpha (α) decay. The principal path starts with neutron capture by ^{238}U in the reactor fuel to form ^{239}U, which decays by β^--emission to ^{239}Np and ^{239}Pu. This isotope emits an α-particle to form long-lived ^{235}U. Plutonium-239 has a long halflife (4.10×10^4 years), which gives it a chance to capture a neutron to form ^{240}Pu (6.56×10^3 years). A second neutron capture forms ^{241}Pu (14.4 years), which decays either by α-emission to ^{237}U or by β^--emission to ^{241}Am. The reader can trace other potential reaction paths, which all ultimately lead to the stable isotopes of Pb (^{206}Pb, ^{207}Pb, and 208Th) and Bi (^{209}Bi). Note that the isotopes of Cm can only be reached via the β^--decay of ^{244}Am, which requires three successive neutron captures by ^{241}Am.

to the formation of unstable ^{239}U. The sequence of successive neutron captures followed by β^--decay in Figure 25.1 leads to the formation of isotopes of neptunium (Np, $Z = 93$), plutonium (Pu, $Z = 94$), americium (Am, $Z = 95$), and curium (Cm, $Z = 96$). All of the isotopes of these transuranium elements are radioactive and decay in various ways to the long-lived isotopes of U and Th and subsequently to the stable isotopes of Pb (^{206}Pb, ^{207}Pb, and ^{208}Pb) and Bi (^{209}Bi).

For example, Figure 25.1 illustrates the following sequence of events: ^{238}U absorbs a neutron to form ^{239}U, which decays by β^--emission to ^{239}Np. This isotope then decays by β^--emission to ^{239}Pu, which emits α-particles to form long-lived ^{235}U. The decay of ^{235}U to stable ^{207}Pb by sequential emission of α- and β^--particles is illustrated in Figure 20.2. Alternatively, some of the ^{239}Pu atoms in the fuel rods of a nuclear reactor may absorb a neutron before they decay and thus form ^{240}Pu, which can itself absorb a second neutron to form ^{241}Pu. This isotope can decay both by β^--emission to ^{241}Am and by α-emission to ^{237}Np, and so on. The reader can easily trace these and other neutron-capture and decay paths in Figure 25.1. In addition, ten of the isotopes of the transuranium elements can decay by spontaneous fission (SF), discussed in Section 2.3a. A significant aspect of the transuranium elements produced in nuclear reactors is that the halflives of their isotopes range widely from about one hour (^{240}Np) to 2.14×10^6 years (^{237}Np) and that most are much longer than 100 years. Therefore, the transuranium elements are an important long-term component of spent nuclear fuel. The isotopes of the transuranium elements whose halflives exceed 10^3 years include ^{240}Pu (6.56×10^3 years), ^{243}Am (7.37×10^3 years), ^{239}Pu (2.410×10^4 years), ^{242}Pu (3.75×10^5 years), and ^{237}Np (2.14×10^6 years). In addition, several of the progeny of the transuranium isotopes also have long halflives: ^{229}Th (7.3×10^3 years), ^{233}U (1.592×10^5 years), ^{234}U (2.46×10^5 years), ^{236}U (2.342×10^7 years), ^{235}U (0.704×10^9 years), ^{238}U (4.47×10^9 years), and ^{232}Th (14.0×10^9 years). The occurrence of Pu and of other transuranium elements in rivers, estuaries, and coastal marine sediment has been documented by many authors, including:

Noshkin et al., 1986, *Nature*, 262:745–748. Livingston and Bowen, 1979, *Earth Planet. Sci. Lett.*, 43:29–45. Santschi et al., 1980, *Earth Planet. Sci. Lett.*, 51:248–265. Olsen et al., 1981, *Earth Planet. Sci. Lett.*, 55:377–382. Koide and Goldberg, 1982, *Earth Planet. Sci. Lett.*, 57:263–277. Cambray and Eakins, 1982, *Nature*, 300:46–48. Scott et al., 1983, *Earth Planet. Sci. Lett.*, 63:202–221. Santschi et al., 1983, *Geochim. Cosmochim. Acta*, 47: 201–210. Scott et al., 1985, *Earth Planet. Sci. Lett.*, 75:321–326. Holm et al., 1986,

Earth Planet. Sci. Lett., 79:27–32. Smith et al., 1986/1987, *Earth Planet. Sci. Lett.*, 81:15–28. Hayes and Sackett, 1987, *Estuar. Coast. Shelf Sci.*, 25:169–174. Sholkovitz and Mann, 1987, *Estuar. Coast. Shelf Sci.*, 25:413–434. Meece and Benninger, 1993, *Geochim. Cosmochim. Acta*, 57:1447–1458. Bunzl et al., 1995, *Health Phys.*, 68:89–93. Kelley et al., 1999, *Sci. Total Environ.*, 237/238:483–500. Oktay et al., 2000, *Geochim. Cosmochim. Acta*, 64:989–996.

In addition, the redeposition of Pu discharged by the nuclear fuel reprocessing facility at Sellafield, West Umbria, United Kingdom, was discussed by Aston and Stanners (1981) and Stanners and Aston (1984).

25.1c Disposal of Radwaste (Yucca Mountain, Nevada)

The decay rates of fission-product radionuclides in nuclear fuel can be estimated based on the initial abundance of ^{235}U in the fuel, the fission yield of each product nuclide, the energy spectrum of the neutrons, the neutron-capture cross-sections of fission-product nuclides, and the length of time the fuel has been in the reactor. For example, data cited by Roxburgh (1987) based on Pentreath (1980) indicate that after one year of operation of a nuclear reactor that generates 1×10^6 Watts of electricity the fuel has a decay rate of 612.9×10^3 Ci. After a cooling time of 100 days, the activity decreases to 143.0×10^3 Ci, and after five years the decay rate is 6.46×10^3 Ci, or about 1.0% of the initial activity. The most important radionuclides remaining after five years are ^{85}Kr, ^{90}Sr, ^{90}Y, ^{106}Ru, ^{106}Rh, ^{137}Cs, ^{137}Ba, ^{144}Ce, ^{144}Pr, and ^{147}Pm. In addition, the same spent fuel contains unspecified activities of long-lived isotopes of the transuranium elements and their daughters.

The spent fuel is initially stored for five years onsite at the reactor to allow the activity to subside. Subsequently, the fuel can be reprocessed to recover U and Pu and to prepare the residue for burial. After the spent fuel has been dissolved in nitric acid, U and Pu are extracted with tributyl phosphate (TBP) and odorless kerosene (OK). The residue contains 99.9% of the nonvolatile fission products, 0.1% of the U that was originally present, and <1% of the Pu. In addition, the residue contains most of the other transuranium elements that were formed by neutron-capture reactions in the nuclear reactor. After most of the heat has dissipated, the residue is converted into a borosilicate melt, which is poured into steel containers for transport and storage in a subsurface repository (Faure, 1998).

The high level of radioactivity of the spent nuclear fuel, even after a cooling time of five years followed by reprocessing, requires that it be placed in repositories located in sparsely populated regions far below the surface of the Earth in bedrock that is sufficiently heat resistant and impermeable to groundwater to prevent the escape of any radionuclides into the environment. Several countries have established such repositories in different kinds of rocks: Canada, United Kingdom, Finland, Sweden, and France in granitic rocks and Belgium and Italy in clay. Such subsurface repositories have several desirable attributes that enhance their safety:

1. They prevent accidental or malicious intrusion and can be designed to contain the radionuclides for up to 100,000 years.
2. They require no maintenance after they have been filled and sealed.
3. The radwaste is retrievable in case it becomes necessary to do so.

However, there are also contingencies that must be considered in the site selection and design of repositories:

1. climate change, including the return of continental ice sheets in the northern hemisphere during the next 100,000 years;
2. volcanic eruptions which may bury the repository and/or alter the temperature, chemical composition, and flow pattern of groundwater at the site;
3. earthquakes and displacement along faults which may damage the radwaste containers and/or cause the subsurface cavities to collapse;
4. increase of sealevel, causing the groundwater table at coastal repositories to rise or the site to be flooded;

5. accidental release of radwaste during transport to the repository; and
6. resistance of the inhabitants of the region surrounding the repository to the dangers of unforeseen accidents associated with the transport and burial of radwaste containers.

The repository of the USA is located in southwestern Nevada at Yucca Mountain, which is composed of welded rhyolite tuff of Tertiary age (Topopah Spring Formation, 12.8 Ma) described most recently by Peterman and Cloke (2002). The topography of the proposed site permits the repository to be in the zone of aeration above the groundwater table located 500–750 m below the surface, which is considered to be an important positive attribute of this site (Stuckless and Dudley, 2002). Nevertheless, the bedrock of Yucca Mountain contains fractures and joints in which calcite and amorphous silica (opal) were deposited by aqueous solutions. Age determinations by the $^{230}Th/^{234}U$, $^{206}Pb^{*}/^{238}U$, and $^{207}Pb^{*}/^{235}U$ methods have yielded a wide range of dates depending on the method used and the selection of samples for analysis. Most of the dates range from about 1.8 million to 34,000 years (Neymark and Paces, 2000; Neymark et al., 2000). In general, the evidence indicates that the silica deposits accumulated slowly during the past 9.8 million years and that they were deposited by ***meteoric*** water descending toward the watertable, rather than by groundwater at a time in the past when the watertable was higher than it is at present (Stuckless et al., 1991; Neymark et al., 2000). A set of papers describing the results of geochemical and isotopic research at Yucca Mountain was edited by Gascoyne and Peterman (2002).

25.1d Reactor Accidents: Chernobyl, Ukraine

Although nuclear power reactors are designed to be safe, serious accidents have occurred as a result of operator error (e.g., Three-Mile Island, USA and Chernobyl, Ukraine). Other examples of reactor accidents that caused environmental contamination were reviewed by Eisenbud (1987). In addition, inappropriate disposal of high-level nuclear waste has caused extensive environmental contamination and illness in the affected population (e.g., Mayak nuclear complex, also called Cheliabinsk-65, Russia).

The most environmentally damaging reactor accident occurred on April 26, 1986, at Chernobyl in the Ukraine near the city of Kiev (Eisenbud, 1987). Baier (1989) reported that the number 4 power reactor at this site contained 190 metric tons of U fuel that had powered the reactor for 610 days and which contained a large amount of fission products, including ^{90}Sr ($T_{1/2} = 29.1$ y) and ^{137}Cs ($T_{1/2} = 30.17$ y). The moderator in the core of the reactor consisted of graphite which had stored a certain amount of heat. On the day before the accident, the operators reduced the power of reactor 4 in order to determine whether the graphite in the core contained enough heat to run the main electrical generator. While this experiment was in progress, the demand for electricity in the transmission line suddenly increased. Therefore, the operators attempted to boost the power of the reactor by withdrawing most of the 222 control rods, leaving only eight in place, even though 30 rods were required to prevent a runaway nuclear chain reaction. In response to this action, the energy level of the reactor increased from 200 to 530 megawatts (MW) in three seconds causing the reactor to go out of control. The core temperature increased by 3000°C, which caused Zr metal in the fuel rods to react with water to form ZrO_2 and H_2. The H_2 then reacted with O_2 and exploded, thereby rupturing the reactor vessel, which contained 2500 tons of graphite, and causing the release of large amounts of fission products into the atmosphere. The ensuing fire ultimately consumed about 250 tons of graphite. A more technical description of the accident was given by Eisenbud (1987).

In the days following the explosion, the core temperature continued to rise because of the heat liberated by the decay of short-lived fission products and because the cooling system had been destroyed by the explosion. The increase in temperature caused the release of additional amounts of volatile radionuclides, including isotopes of I and Cs, but not the isotopes of nonvolatile elements such as Sr, Ce, and Pu. The plume of radionuclides released at Chernobyl reached Finland and Sweden in less than three days and was immediately detected. Eventually, the plume spread across all of Europe and circled the globe, reaching Canada and the USA in 15.5 days (Devell et al., 1986; Hohenemser et al., 1986; Jost et al., 1986; Smith and Clark, 1986; Anspaugh et al., 1988).

The total amount of ^{137}Cs deposited on the land areas of the northern hemisphere was 10×10^{16} Bq, of which 79% fell on western Russia and Europe and 20% settled on eastern Russia and Asia. The USA and Canada received 2.8×10^{14} and 2.5×10^{14} Bq, respectively. Thirty-one persons died of radiation poisoning while fighting the fire at the reactor site, 237 suffered acute radiation sickness, and more than 100,000 persons within a radius of 30 km of Chernobyl had to be relocated. In the years since the accident, the incidence of thyroid cancers and leukemia in the affected population has been abnormally high. In addition, the exposure to radiation has weakened the immune systems of the victims and has made them susceptible to common infection, including bronchitis, tonsillitis, and pneumonia.

Reactor 4, which still contains about 74×10^{16} Bq of radioactive material, was encased in a concrete shell called the sarcophagus, which continues to be a source of concern because it may collapse. In addition, radionuclides are escaping into the groundwater and have contaminated the Dnieper River and its tributary the Pripyat River. The sediment in these rivers reportedly contains ^{90}Sr, ^{137}Cs, and Pu (Shcherbak, 1996). Additional information about the aftermath of the reactor accident at Chernobyl has been published by:

Weinberg et al., 1995, *J. Am. Med. Assoc.*, 274(5):408–412. Voice, 1997, *Environ. Rev.*, 5(3):203–205. Rao and Fehn, 1999, *Geochim. Cosmochim. Acta*, 63:1927–1938. Golosev et al., 1999, *Phys. Chem. Earth, Part A, Solid Earth and Geodesy*, 24:881–885. McGee et al., 2000, *J. Environ. Radioact.*, 48:59–78.

25.2 STRONTIUM-90 IN THE ENVIRONMENT

The nuclear accident at Chernobyl was by no means the only source of fission-product radionuclides in the environment. The testing of nuclear weapons in the atmosphere between 1945 and 1980 by several nations identified in Table 25.2 repeatedly released fission-product radionuclides into the atmosphere, causing the land areas of the northern hemisphere to become contaminated with several long-lived radionuclides, including ^{90}Sr, ^{137}Cs, ^{129}I, and many other short-lived radionuclides which decayed in a matter of days, weeks, and months.

Table 25.2. Summary of Nuclear-Weapons Tests in the Atmosphere

Country	Period	Number of Tests
USA	1945–1962	193
USSR	1949–1962	142
UK	1952–1953	21
France	1960–1974	45
China	1964–1980	22
Total		423

Source: Eisenbud, 1987.

The hazard associated with the release of these and other radionuclides into the environment arises from several factors:

1. the magnitudes of the halflives of the individual radionuclides,
2. the fission yield and hence the design of the nuclear weapon being tested,
3. the geochemical properties of the radionuclides (i.e., strongly sorbed to mineral surfaces or readily absorbed by plants),
4. the decay mode and energies of particles and the electromagnetic radiation they emit, and
5. the site within the bodies of humans and animals where a specific radionuclide accumulates.

Most of the so-called fallout radionuclides were deposited in the northern hemisphere where several of the test sites were located (Eisenbud, 1987). The prevailing winds subsequently spread the radionuclides around the globe. However, most of the radiation remained in a belt 20° north and south of the test site. For this reason, the polar regions received significantly *smaller* amounts of fallout from atmospheric testing of nuclear devices than the northern midlatitudes.

The release of ^{90}Sr into the surface environment of the Earth soon attracted the attention of the environmental science community because this isotope is deposited in bones, where the radiation it emits can do serious harm:

Libby, 1956, *Proc. Natl. Acad. Sci. USA*, 42:365–390. Bowen and Sugihara, 1957, *Proc. Natl. Acad. Sci. USA*, 43:576–580. Kulp et al., 1957, *Science*, 125:219–225. Kulp, 1958, *Bull. Swiss Acad. Med. Sci.*, 14:419–438. Kulp and Slakter, 1958, *Science*, 128:85–86. Libby, 1959a, *Proc. Natl. Acad. Sci. USA*, 45(2):245–249. Libby, 1959b, *Proc. Natl. Acad. Sci.*, 45:959–976. Fry et al., 1960, *J. Geophys. Res.*, 65(7):2061–2066. Nelson, 1962, *Science*, 137:38–39. Picciotto and Wilgain, 1963, *J. Geophys. Res.*, 68(21):5965–5972. Rocco and Broecker, 1963, *J. Geophys. Res.*, 68(15):4501–4512. Wilgain et al., 1965, *J. Geophys. Res.*, 70(24):6023–6032. Meyer et al., 1968, U.S. AEC TID-24341. Lambert et al., 1971, *Earth Planet Sci. Lett.*, 11:317–323.

Strontium-90 has a halflife of 29.1 years and decays by β^--emission to an isomer of ^{90}Y, which converts to the groundstate ($T_{1/2} = 3.19$ h) by emitting three γ-rays whose energies are 0.2025, 0.4794, and 0.6820 MeV. The nucleus of ^{90}Y then emits a suite of β^--particles as it decays with a halflife of 64.0 h to the groundstate of stable ^{90}Zr (i.e., no γ):

$$^{90}_{38}\text{Sr} \rightarrow {}^{90\text{m}}_{39}\text{Y} + \beta^- + \overline{\nu} + Q_1$$

$$^{90\text{m}}_{39}\text{Y} \rightarrow {}^{90}_{39}\text{Y} + 3\gamma + Q_2$$

$$^{90}_{39}\text{Y} \rightarrow {}^{90}_{40}\text{Zr} + \beta^- + \overline{\nu} + Q_3$$

The sum of the decay energies:

$$\begin{aligned} Q_t &= Q_1 + Q_2 + Q_3 \\ &= 0.546 + 0.68204 + 2.283 \\ &= 3.51104 \text{ MeV/decay} \end{aligned}$$

The products of nuclear fission also include $^{89}_{38}$Sr, which has a short halflife of 50.52 days and decays by β^--emission to an isomer of $^{89}_{39}$Y. The product nucleus then converts to the groundstate by emitting a γ-ray of 0.9092 MeV at a halflife of only 15.7 s. This isotope of Sr is less hazardous than ^{90}Sr because its mean life ($\tau = 1/\lambda$) is only 72.88 days and the amount of energy released by its decay is less than that of ^{90}Sr.

25.2a Global Distribution (^{90}Sr)

The distribution of fallout from atmospheric testing of nuclear weapons was irregular depending on atmospheric conditions and the magnitude of the explosions. The fission products released by explosions in the kiloton range (i.e., equivalent to thousands of tons of TNT) remained in the troposphere and settled out with a halflife of about 20 days. Megaton explosions injected fission product nuclides into the stratosphere, causing their deposition to be delayed. Fry et al. (1960) reported that the activity of ^{90}Sr in rain during 1959 at Fayetteville, Arkansas, varied seasonally with a peak of 24×10^{-12} Ci/L during the month of April. Similar variations were observed elsewhere during the time of nuclear-weapons testing and were attributed to the transfer of radionuclides from the stratosphere to the troposphere. The fission-product radionuclides that reached the stratosphere spread from pole to pole. Nevertheless, the fission-product radionuclides in the snow of Antarctica have lower concentrations than in the lower latitudes (e.g., Fayetteville, Arkansas; Fry et al., 1960) and are absent in some areas of Antarctica (Picciotto and Wilgain, 1963; Wilgain et al., 1965; Lambert et al., 1971).

The United Nations Scientific Committee on the Effects of Atomic Radiation (UNSCEAR) has compiled records of the amounts of fission-product nuclides released into the atmosphere or stored in various kinds of repositories by the member nations. The compilation by this committee of the cumulative deposits of ^{90}Sr in soils collected between 1965 to 1967 in Figure 25.2 demonstrates that most of the ^{90}Sr (>80 mCi/km^2) was deposited on the northern Pacific Ocean and in a band extending from Nevada and New Mexico in the USA eastward across the North Atlantic Ocean. The diagram also indicates that the polar regions

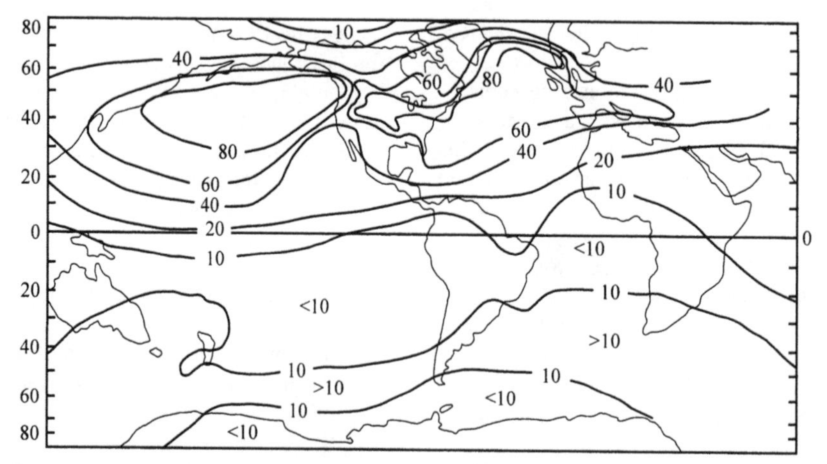

FIGURE 25.2 Contours representing cumulative deposition of ^{90}Sr in soils collected between 1965 and 1967. The units are millicuries per square kilometer. Adapted from Eisenbud (1987) and UNSCEAR (1969).

(>60° S latitude and >80° N latitude) received less than about 10 mCi/km^2 of ^{90}Sr.

After the USA and USSR stopped testing nuclear weapons in the atmosphere in 1963, the stratospheric inventory of ^{90}Sr declined from about 5.5×10^4 kCi in 1963 to about 20 kCi in 1981 in spite of additional testing by other nations (Leifer et al., 1984). The removal of ^{90}Sr from the stratosphere was caused primarily by transfer into the troposphere and by its ultimate deposition in the form of meteoric precipitation and dry fallout. Consequently, the fallout of ^{90}Sr released by testing of nuclear weapons between 1945 and 1980 has virtually ended. However, a new episode of deposition of ^{90}Sr started in 1986 following the accidental release of fission-product radionuclides at Chernobyl, Ukraine. Although the level of activity of ^{90}Sr on the surface of the Earth is decreasing because of decay and burial in sediment, a substantial amount has accumulated in spent nuclear fuel that has not yet been placed in permanent repositories.

25.2b Oceans

The deposition of ^{90}Sr from the troposphere to the surface of the ocean provided an opportunity to use this radionuclide as a tracer in studies of vertical mixing of water in the oceans. Initial measurements by Bowen and Sugihara (1960) in the Atlantic Ocean and by Miyake et al. (1962) in the Pacific Ocean suggested that ^{90}Sr and ^{137}Cs had reached depths in excess of 1000 m by 1957. These results implied anomalously rapid rates of downward movement of water. In contrast, the data by Rocco and Broecker (1963) in Figure 25.3 clearly show that in 1961 most of the ^{90}Sr in the Caribbean Sea, the southeastern Pacific Ocean, and the South Atlantic Ocean resided in the surface layer, which was less than 400 m deep at all sites included in their study. However, the activity of ^{90}Sr dissolved in bottom water in Figure 25.3 did rise, especially in the South Atlantic Ocean. The explanation for this phenomenon considered by Rocco and Broecker (1963) was that a certain fraction of ^{90}Sr is incorporated into organisms living in the surface layer and is then transported to the bottom of the oceans where it is released as the organic matter decays.

The highest 1961 decay rates of ^{90}Sr reported by Rocco and Broecker (1963) occurred in surface water in the North Atlantic Ocean: 17.9 ± 1.3 dpm/100 L at 40°39′ N, 55°58′ W (9 m) and 33.1 ± 2.3 dpm/100 L at 25° N, 79° W (1 m). These values exceed the average ^{90}Sr concentrations in the upper 400 m of the South Atlantic Ocean (6.6 ± 1.7 dpm/100 L), Antarctic Ocean (8.6 ± 5.0 dpm/100 L), southeastern Pacific (5.3 ± 1.2 dmp/100 L), and Caribbean Sea ($11.1 \pm$

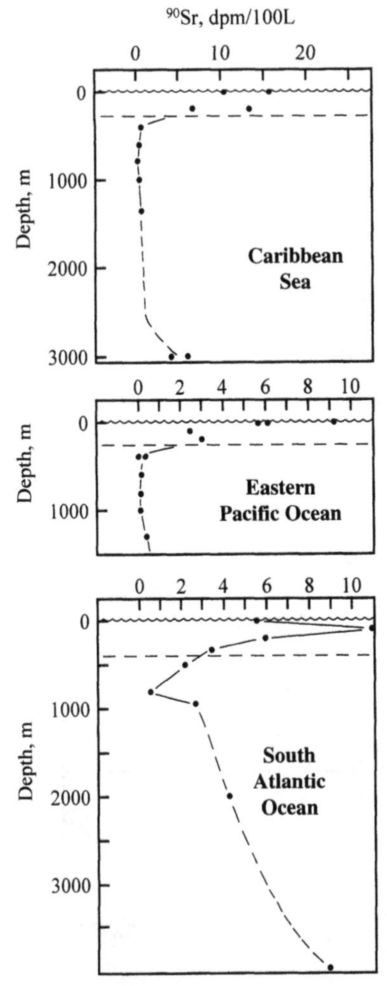

FIGURE 25.3 Profiles of ^{90}Sr with depth in different parts of the oceans based on water samples collected at different times during 1961: February in the Caribbean Sea; February/March in the Eastern Pacific; July in the south Atlantic. Data from Rocco and Broecker (1963).

1.5 dpm/100 L). These average ^{90}Sr activities of seawater in 1961 are compatible with the global distribution pattern in Figure 25.2 based on the accumulation of ^{90}Sr in soil.

The ^{90}Sr that was deposited in the oceans was incorporated into the calcium carbonate shells of mollusks and other organisms, including corals. Therefore, the annual growth layers of coral heads have preserved a record of the ^{90}Sr that existed in the oceanic surface water during the year a particular layer was deposited (Knutson et al., 1972). For example, Druffel and Linnick (1978) and Druffel (1981) used growth rings of corals to study changes in the activity of ^{14}C caused by the testing of nuclear weapons and by the combustion of fossil fuel (Section 23.1d).

Corals collected from shallow depths around islands in the Pacific and Indian Oceans were subsequently used by Toggweiler and Trumbore (1985) to reconstruct the annual changes of ^{90}Sr in seawater during the period of testing of nuclear weapons in the atmosphere. Their results for corals recovered from the open ocean at the island of Oahu, Hawaii, in Figure 25.4 indicate that the ^{90}Sr activity increased from 31.4 ± 2.7 dpm/100 g of coral in 1958 to 59.2 ± 2.6 dpm/100 g in 1964 and then declined to 21.2 ± 1.1 dpm/100 g in 1979. The rise and fall of ^{90}Sr activities of the corals in the waters of Oahu are in good agreement with the record of annual ^{90}Sr deposition in the northern hemisphere published by Dreisigacker and Roether (1978), who reported that the highest deposition occurred in 1963 based on measurements in surface water of the North Atlantic Ocean. A similar history of ^{90}Sr deposition was also recorded in New York City by the U.S. Department of Energy (Eisenbud, 1987).

The peak activities of ^{90}Sr in corals in the Pacific Ocean in Figure 25.5 vary regionally in relation to the test sites on the Eniwetok and Bikini atolls. For example, corals in Oahu located east and slightly north of the test sites, have the highest peak activity of ^{90}Sr (59.2 dpm/100 g). The corals of Tarawa and Fanning, located south and east of the test sites, have peak ^{90}Sr activities of 17.7 and 17.8 dpm/100 g, respectively, whereas the corals of the Galapagos Islands located 12,500 km east and slightly south of Eniwetok atoll reach only 16.0 dpm/100 g. Still farther south, the corals of the Fiji and Tonga Islands have still lower peak activities of 13.9 and 13.3 depm/100 g, respectively. These peak ^{90}Sr activities of corals in the Pacific Ocean suggest that the plumes of fission-product radionuclides moved east and slightly north from the test sites on Eniwetok and Bikini atolls.

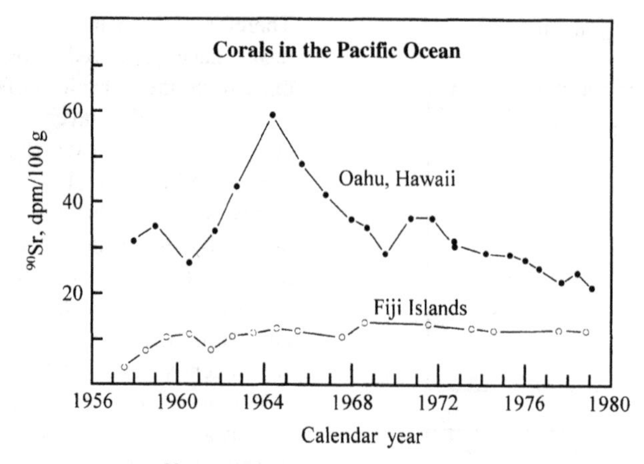

FIGURE 25.4 Activities of ^{90}Sr in annual growth rings of coral heads growing in shallow water off the coast of Oahu, Hawaii, and the coast of the Fiji Islands. The peak ^{90}Sr activity in the Hawaiian coral occurs in the layer deposited in 1964. Data from Toggweiler and Trumbore (1985).

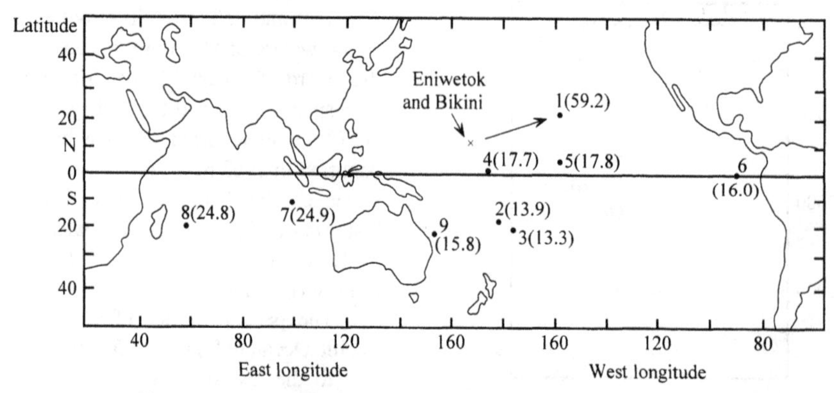

FIGURE 25.5 Activities of ^{90}Sr in corals near islands in the Pacific and Indian Oceans: 1. Oahu, Hawaii; 2. Fiji; 3. Tonga; 4. Tarawa; 5. Fanning; 6. Galapagos; 7. Cocos; 8. Mauritius; 9. Great Barrier Reef. The peak activities of ^{90}Sr in dpm/100 g of coral are indicated in parentheses. Eniwetok and Bikini atolls are the sites of several atmospheric tests. Adapted from Toggweiler and Trumbore (1985).

The path of the plumes presumably continued east around the world and eventually entered the Indian Ocean from the west. Therefore, the activity of ^{90}Sr corals on the Mauritius and Cocos Islands was expected to be less than that of corals at the Fiji and Tonga Islands, which are located at similar southern latitudes. This prediction turned out to be wrong because the data of Toggweiler and Trumbore (1985) in Figure 25.5 yielded peak ^{90}Sr activities of 24.8 dpm/100 g for corals near the Mauritius Islands, 24.9 dpm/100 g for the Cocos Islands, and a surprisingly high value of 15.8 dpm/100 g for corals in the Great Barrier Reef along the east coast of Australia.

Toggweiler and Trumbore (1985) explained the high ^{90}Sr activities of corals in the Indian Ocean by large-scale movement of surface water from the North Pacific Ocean into the Indian Ocean south of the equator. In addition, they considered that British tests of nuclear weapons in Australia during 1952 and 1953 could account for the high ^{90}Sr activity of corals in the Great Barrier Reef.

Much additional information on the distribution of ^{90}Sr has been recorded in reports of government agencies (e.g., Bowen et al., 1968; Meyer et al., 1968; Carter and Moghissi, 1977; US DOE, 1977; Bowen et al., 1980). The distribution of other fission products and transuranium elements was discussed by Santschi et al. (1983), Cochran et al. (1987), Bunzl et al. (1995), and Oktay et al. (2000).

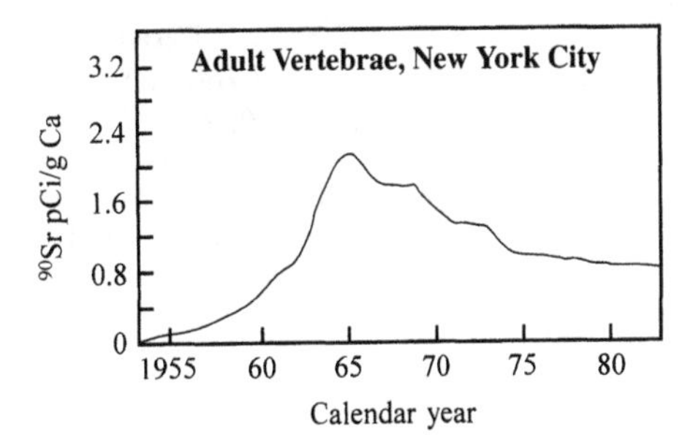

FIGURE 25.6 Measured concentrations of ^{90}Sr in vertebrae of adult humans in the City of New York between 1954 and 1982. Adapted from Eisenbud (1987) based on data by Klusek (1984b).

25.2c Human Diet

Strontium-90 is considered to be the most hazardous of all fission product radionuclides because it has a long halflife on a human timescale, it passes readily into the food chain, it substitutes for Ca in bones, and it releases a substantial amount of energy when it decays. Experiments with animals described by Anonymous (1991) indicate that high levels of ^{90}Sr intake result in bone cancer, leukemia, cancer of the sinus, and other medical conditions. However, exposure to low doses of ^{90}Sr (1–18 Grays) did not cause bone sarcomas in beagles.

About 20–30% of the ^{90}Sr ingested by humans is absorbed in the gastrointestinal tract. The remainder is excreted. The absorbed ^{90}Sr is deposited in the bones and is stored in blood plasma, in extracellular fluids, and on bone surfaces. The concentration of ^{90}Sr in human bones increased during the period when nuclear weapons were being tested in the atmosphere (Table 25.2) and peaked in the mid-1960s. Subsequently, the activity of ^{90}Sr has slowly declined as result of decay and by excretion. Data compiled by Klusek (1984a, b) in Figure 25.6 for the ^{90}Sr concentration of human vertebrae in New York City are fairly typical of this phenomenon. In general, the concentrations of Sr in human bones and tissues are closely related to those of Ca. The retention of ^{90}Sr in relation to Ca is the subject of an extensive literature referred to by Anonymous (1991), including an influential paper by Marshall et al. (1973). In addition, the effects of radioactivity in humans caused by exposure to tritium, Sr, Cs, Ra, Rn, the REEs, Th, U, and Pu were discussed by the International Atomic Energy Agency (IAEA, 1964).

The health hazard posed by ^{90}Sr was clearly indicated by the concentrations of this isotope in vegetables, cereal grains, and milk produced in 1956–1957 in different locations in the USA. For example, Kulp and Slakter (1958) reported that the ^{90}Sr concentrations of vegetables grown in western New York state in 1956–1957 ranged from 1.8 SU (spinach) to 28.4 SU (corn), where SU stands for "strontium unit" and is expressed in terms picocuries of ^{90}Sr per gram of Ca. They also reported 22.8 SU for wheat and 5.5 SU for milk in upstate New York. A study by a team of dentists revealed that the ^{90}Sr concentration of deciduous incisor teeth of bottle-fed children increased from 0.30 ± 0.03 SU in 1951 to 2.56 ± 11 SU in 1957 (Rosenthal et al., 1963). Strontium-90 even showed up in the hair of laboratory rats at varying but easily measurable concentrations (Hopkins et al., 1963).

The uptake of ^{90}Sr by plants occurs both by absorption through the roots and by direct deposition on the leaves, called *foliar deposition*. When contaminated plants and cereal grains are consumed by livestock, the ^{90}Sr passes into dairy products, meat, poultry, and eggs. The annual intake of ^{90}Sr by humans depends on the ^{90}Sr concentration of different types of food in the diet, the amounts of each type of food that is consumed, and the way the food is prepared. The data of Klusek (1984a) cited by Eisenbud (1987) in Table 25.3 give the average ^{90}Sr concentrations of

Table 25.3. Activity of ^{90}Sr in Different Types of Food and the Total Annual Intake by the People of New York City in 1982

Food Type	Weighted Average ^{90}Sr,[a] pCi/kg	Total Annual per Capita Consumption, kg	Annual Intake, New York City,[b] pCi/y/person
Dairy products	3.2	200	641 (32%)
Vegetables, fresh and canned	5.9	121	711 (36%)
Fruit and juice, fresh and canned	2.2	98	212 (11%)
Bakery products, pasta, rice	3.8	95	362 (18%)
Meat	0.4	79	35 (1.8%)
Poultry and eggs	0.4	35	16 (0.80%)
Fish and clams	0.2	9	2 (0.10%)
Total		637	1979 (99.7%)

Source: Klusek, 1984a; cited in Eisenbud, 1987.

[a]Weighted by annual amounts of each item expressed in kg according to Khusek (1984a).

[b]Numbers in parentheses are percent of total annual intake of ^{90}Sr for each type of food.

different types of food eaten by the people of New York City in 1982, the annual amounts of each food type consumed per person, and the resulting burden of ^{90}Sr contributed by each type of food. According to these data, the total per-capita intake of ^{90}Sr in 1982 in the New York City was 1979×10^{-12} Ci. Vegetables and dairy products each contributed 36 and 32% of the total ^{90}Sr intake, respectively. Cereal products (18%) and fruits (11%) raised the total to 97%, leaving only 3% for meat, poultry and eggs, and fresh fish and clams. The data of Klusek (1984a) also show that the annual per-capita ingestion of ^{90}Sr by the people of San Francisco in 1982 was only 967×10^{-12} Ci/y per person because the same food items had lower ^{90}Sr concentrations than they did in New York City. Nevertheless, the per-capita intake of ^{90}Sr in 1982 in New York City was less than 30% of the amount ingested by the people of that city in 1966 (Harley, 1969) and only about 15% of the amount ingested in that year by citizens of the USSR (Petukhova and Knizhnikov, 1969). The reasons for the high annual intake of ^{90}Sr in 1966 in the USSR include both dietary preferences and higher ^{90}Sr activities of certain food items. For example, citizens of the USSR consumed more bread per person in 1966 and the bread had a higher ^{90}Sr activity than bread in New York City:

USSR	New York City
220 kg	89 kg
33×10^{-12} Ci/kg	17.4×10^{-12} Ci/kg
7260×10^{-12} Ci/y per person	1549×10^{-12} Ci/y per person

25.3 CESIUM-137 IN THE ENVIRONMENT

Cesium ($Z = 55$) is an alkali metal in group IA of the periodic table. It has only one stable isotope ($^{133}_{55}Cs$) and 35 unstable isotopes, none of which occur naturally. Fourteen of the unstable isotopes are fission products (excluding isomers), and three of these have long halflives: ^{134}Cs (2.065 years), ^{135}Cs (2.3×10^6 years), and ^{137}Cs (30.17 years). These isotopes of Cs were released into the atmosphere by the testing of nuclear weapons in the atmosphere between 1945 and 1980. The short-lived fission-product isotopes of Cs, including ^{134}Cs, decayed in less than about ten years after the last atmospheric test. Only ^{135}Cs and ^{137}Cs remain alive in soil, in recently deposited lacustrine and marine sediment, in continental ice sheets, and in the oceans. The rate of decay of ^{135}Cs is much lower than that of ^{137}Cs because of its long halflife (2.3×10^6 years). Cesium-137 decays to stable ^{137}Ba by emitting β^--particles with a maximum energy of 0.514 MeV. In addition, the product nucleus emits a γ-ray at 661.64 KeV, which permits the decay rate of ^{137}Cs to be measured by γ-ray spectrometry.

The sources of ^{137}Cs include not only fallout from the past testing of nuclear weapons but also the accidental explosion of the nuclear reactor at Chernobyl and the growing amount of spent nuclear

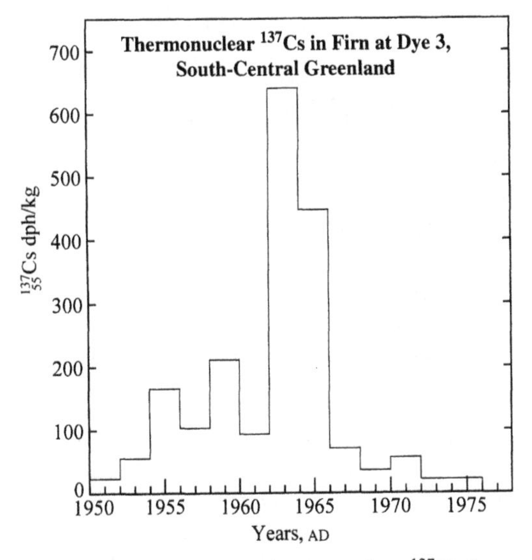

FIGURE 25.7 Activities of thermonuclear $^{137}_{55}Cs$ in the Dye 3 ice core, central Greenland. Data from Beer et al. (1985).

Table 25.4. Activity of ^{137}Cs in the Human Diet in 1968 in Chicago

Food Type	^{137}Cs,[a] pCi/kg	Annual Amount, kg	^{137}Cs Intake, pCi/y
Dairy products	18	200	3,600 (29%)
Vegetables	8.5	121	1,026 (8%)
Fruit and juices	11.9	98	1,162 (9%)
Bakery products	26.4	92	2,433 (20%)
Meat	26	79	2,054 (17%)
Poultry and eggs	11.6	35	405 (3%)
Fish	194	8	1552 (13%)
Total		633	12,232 (99%)

Source: Harley, 1969; Eisenbud, 1987.

[a] Activity of ^{137}Cs in food sources listed in the table.

fuel that continues to be held in temporary storage. The activity of ^{137}Cs deposited from the atmosphere at Dye 3 in southcentral Greenland in Figure 25.7 peaked in 1962 and 1963 and then declined to low levels in the 1970s.

25.3a Human Diet

Although ^{137}Cs is strongly sorbed to soil particles, this isotope does enter the food chain by foliar deposition and by limited absorption through the roots of plants. Accordingly, dairy products, vegetables, fruit, bakery products, meat, poultry and eggs, as well as fish all contain fallout ^{137}Cs. The activity of ^{137}Cs in the diet of people in Chicago in 1968 is detailed in Table 25.4 based on data by Harley (1969) as reported by Eisenbud (1987).

The data in Table 25.4 indicate that in Chicago in 1968 dairy products contributed 29% of the annual intake of ^{137}Cs per person, followed by bakery products (20%), meat (17%), fish (13%), fruit and juices (9%), vegetables (8%), and poultry and eggs (3%). The total annual intake of ^{137}Cs in 1968 in Chicago was 12,232 pCi/y per person. That amount is more than six times higher than the annual intake of ^{90}Sr in 1982 in New York City.

Cesium generally behaves like K in the human body and is uniformly distributed in all organs, unlike ^{90}Sr, which is deposited in bones by replacement of stable isotopes of Ca. For this reason, ^{137}Cs has a biological halflife of only 50–150 days in adults and 44 days in children (Eisenbud, 1987). Nevertheless, ^{137}Cs that is ingested with food damages cells by internal irradiation with β^--particles and γ-rays.

25.3b Soil and Plants

The contamination of soil in a large area of the former USSR with ^{137}Cs and other fission-product radionuclides released at Chernobyl has motivated studies to determine the uptake of this isotope by food crops and grass eaten by livestock (Grütter et al., 1990; Comans and Hockley, 1992). The results of a study of Korobova et al. (1998) indicate that ^{137}Cs is strongly sorbed in soil of the Bryansk region in the Novozybkov district of Russia, whereas ^{90}Sr is soluble in water and leachable with

Table 25.5. Results of Sequential Leaching of a Soil Sample in the Bryansk Region of Russia with Different Reagents

	Fraction Leached	
Leach Solution	^{137}Cs	^{90}Sr, %
H_2O	0.20	2.29
NH_4Ac	6.27	28.17
1 M HCl	9.74	56.34
7.5 M HNO_3	30.98	13.20
Not leachable	52.80	0
Totals	99.99	100.00

Source: Korobova et al., 1998.
Note: The soil was a podzolic sandy loam at an elevation of 136 m above sealevel.

dilute acid. These conclusions are supported by sequential leaching experiments of different kinds of soil from Novye Bobovichi in the Bryansk region. The results in Table 25.5 demonstrate that 52.80% of ^{137}Cs in a sandy loam is not leachable even with 7.5 molar nitric acid, whereas all of the ^{90}Sr present in this soil was leachable. Podzolic forest soils retained about 48% of ^{137}Cs, soils in pastures and hay fields retained between 50.8 and 70.7% depending on the organic content, and peat was most retentive, with 93.1% of the ^{137}Cs present. The same treatment resulted in complete removal of ^{90}Sr from all soils.

In spite of the retention of ^{137}Cs by organic matter in forest soil, Korobova et al. (1998) reported that plants with shallow root systems contained high ^{137}Cs concentrations:

Mushrooms:	21,420 Bq/kg (dry weight, dw)
Vaccinium: myrtillus	3495–62,960 Bq/kg (dw)
Calamagrotis:	80,700 Bq/kg (dw)
Pteridium: aquilinum	504,800 Bq/kg (dw)

These ^{137}Cs activities exceed the recommended maximum permissible rate for berries in Russia of 592 Bq/kg.

The transfer of ^{137}Cs from the soil to plants is expressed in terms of the transfer coefficient (TC) defined by Korobova et al. (1998):

$$\mathrm{TC} = \frac{^{137}\mathrm{Cs\ in\ plant\ (Bq/kg)}}{^{137}\mathrm{Cs\ in\ soil\ (Bq/m^2)}}\ \mathrm{m^2/kg} \qquad (25.1)$$

The values of transfer coefficients of ^{137}Cs in grass and "herbs" (weeds) in pastures and hayfields range from 0.005 to 0.034 m^2/ kg and are always lower for grass than for associated weeds. The ^{137}Cs of grass and weeds is transferred to milk and other dairy production as well as to meat.

Potatoes, which are an important food item, are likewise contaminated by ^{137}Cs and ^{90}Sr. The data in Table 25.6 relate the activities of these radionuclides in topsoil at Novye Bobovichi to their activities in fresh (undried) potatoes. The activities of ^{137}Cs in the three samples of topsoil range from 1210 to 622 Bq/m^2 and exceed those of ^{90}Sr in the same sample by a factor of 74 on average. The decay rate R of a radionuclide is related to the number of atoms remaining (N) by equation 3.2:

$$R = \lambda N$$

For a ratio of two radionuclides decaying at rates R_1 and R_2,

$$\frac{R_1}{R_2} = \frac{\lambda_1 N_1}{\lambda_2 N_2}$$

Table 25.6. Activities of ^{137}Cs and ^{90}Sr in Soil and Potatoes Grown by Collective Farms at Novye Bobovichi in the Bryansk Region of Russia

Property	Sample A3	Sample A6	Sample A8
	Topsoil (Bq/m^2)		
^{137}Cs	1210	703	621
^{90}Sr	14.1	10.0	9.6
	Potatoes[a] (Bq/kg)		
^{137}Cs	41.8	20.0	7.03
^{90}Sr	1.33	0.55	0.85
	Transfer Coefficient (m^2/kg)		
^{137}Cs	0.034	0.028	0.011
^{90}Sr	0.094	0.055	0.088

Source: Korobova et al., 1998.
[a] Activity in fresh potatoes (i.e., not dried).

and the atomic ratio (N_1/N_2) is:

$$\frac{N_1}{N_2} = \frac{R_1}{R_2}\frac{\lambda_2}{\lambda_1} \qquad (25.2)$$

Therefore, the atomic $^{137}Cs/^{90}Sr$ ratio of the fallout deposited on these fields by the Chernobyl plume was approximately:

$$\left(\frac{^{137}Cs}{^{90}Sr}\right)_{atomic} = 74 \times \frac{0.0238}{0.0229} = 76.9$$

where the decay constant of ^{90}Sr is 0.0238 y^{-1}, that of ^{137}Cs is 0.0229 y^{-1}, and the decay since 1986 was neglected because of the similarity of the decay constants of ^{137}Cs and ^{90}Sr.

The ^{137}Cs activities of the potatoes in Figure 25.8 are closely correlated with the ^{137}Cs activity of the soil in which they were grown. The resulting values of the transfer coefficients for ^{137}Cs in Table 25.6 range from 0.011 to 0.034 m^2/kg. The transfer coefficients of ^{90}Sr are consistently larger than those for ^{137}Cs by factors ranging from 2 to 8. Therefore, potatoes discriminate more effectively

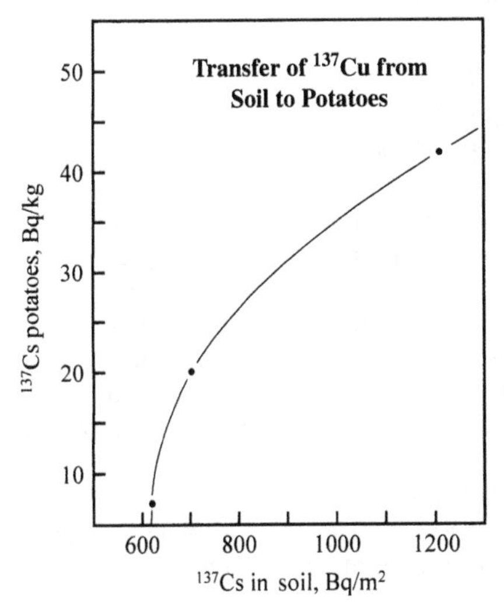

FIGURE 25.8 Relation between the activity of ^{137}Cs in soil samples and in potatoes grown in these soils near Novye Bobovichi, Russia. Data from Korobova et al. (1998).

against ^{137}Cs in the soil than against ^{90}Sr. The difference in the transfer coefficients reinforces the evidence that ^{137}Cs is more strongly bound to soil particles than ^{90}Sr.

The ^{137}Cs and ^{90}Sr concentrations of the potatoes grown in Novye Bobovichi are significantly lower than the maximum permissible limits in Russia: ^{137}Cs: 592 Bq/kg and ^{90}Sr: 37 Bq/kg. Therefore, these potatoes are presumably safe to eat. Additional references concerning the transfer of ^{137}Cs to plants and the retention or mobility of this isotope in soils are listed in Table 25.7.

25.3c Lake Sediment

The environmental properties of ^{137}Cs are dominated by the sorption of ^{137}Cs to clay minerals (e.g., Grütter et al., 1990; Comans and Hockley, 1992; Section 25.3b). Therefore, the ^{137}Cs that was released by weapons tests and by the reactor accident at Chernobyl defines marker horizons in deposits of lacustrine sediment and glacial ice from which the average sedimentation rates can be estimated (Robbins, 1984). In this application, ^{137}Cs complements the use of ^{210}Pb (Section 20.4c), which has a similar halflife (22.3 years) and is also strongly sorbed to inorganic and organic sediment particles. However, the input of ^{137}Cs was episodic, whereas that of ^{210}Pb is continuous.

The profile of ^{137}Cs activities in a sediment core from Lake Sempach in Switzerland in Figure 25.9 clearly shows the input from the Chernobyl plume, which passed over the lake between April 30 and May 1, 1986 (Wieland et al., 1993). The profile also identifies the ^{137}Cs peak deposited during 1962/1963 that is attributable to the testing of nuclear weapons. The core was taken on May 4, 1988, or about 25.5 years after deposition of the ^{137}Cs-rich layer. Therefore, the apparent sedimentation rate s can be estimated from the elapsed time and the depth of the ^{137}Cs peak (10.2 cm):

$$s = \frac{10.2}{25.5} = 0.40 \text{ cm/y}$$

This estimate is inaccurate because the density of the sediment increases with depth. For this reason, Wieland et al. (1993) expressed the depth

Table 25.7. Studies of the Transfer of ^{137}Cs from Soil to Plants and of Its Mobility or Retention in Soil

Topic	References
Transfer to plants	Martin et al., 1988, *J. Environ. Radioact.*, 6:247–259; Ward et al., 1989, *Health Phys.*, 57:587–592; Cheshire and Shand, 1991, *Plant and Soil*, 134:287–296; Bergeijk et al., 1992, *J. Environ. Radioact.*, 15:265–276; Demirel et al, 1994, *J. Environ. Qual.*, 23:1280–1282.
Mobility or retention	Cremers et al., 1988, *Nature*, 335:247–249; Varskog et al., 1994, *J. Environ. Radioact.*, 22:43–53; Rafferty et al., 1997, *Soil Biol. Biochem.*, 29:1673–1681.
Pedoturbation	Southard and Graham, 1992, *Soil Sci. Soc. Am. J.*, 56:202–207.
Bioaccumulation	Lasat et al., 1998, *J. Environ. Qual.*, 27:165–169.
Soil erosion	Bajracharya et al., 1998, *Soil Sci.*, 163:133–142.

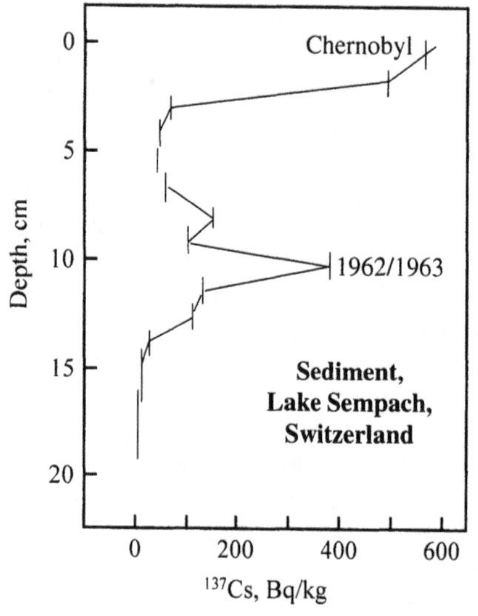

FIGURE 25.9 Variation of the activity of ^{137}Cs with depth in the sediment of core 8802 taken on May 4, 1988, at a depth of 87 m in the approximate center of Lake Sempach, Switzerland. Data from Wieland et al. (1993).

in terms of the "mass depth," which is the depth multiplied by the density:

$$\text{Mass depth} = \text{depth} \times \text{density in g/cm}^2 \quad (25.3)$$

The value of the mass depth parameter of the sediment at 10.2 cm is 1.71 g/cm^2 (Wieland et al., 1993), which yields the sedimentation rate:

$$s = \frac{1.71}{25.5} = 0.067 \text{ g/cm}^2 \text{ y}$$

The principal objective of the study of Wieland et al. (1993) was to trace the movement of ^{137}Cs, accidentally released at Chernobyl, through the epilimnion and hypolimnion to the sediment at the bottom of Lake Sempach. Many additional studies of the occurrence of ^{137}Cs in lakes are identified in Table 25.8.

25.4 ARCTIC OCEAN: ^{90}Sr/^{137}Cs, ^{239,240}Pu, AND ^{241}Am

The distribution of fission-product radionuclides in the oceans is strongly affected by currents which move water among the major basins of the world. As a result, the concentrations of fission-product radionuclides in a particular part of the oceans are determined by inputs from several potential sources and by their concurrent removal from solution:

1. fallout from the atmosphere overhead,
2. effluent from spent-fuel-reprocessing plants,
3. movement of water from other parts of the ocean,
4. rivers draining nearby continents, and
5. sorption of radionuclides by particles sinking through the water column and concurrent decay depending on their halflives.

Table 25.8. Use of ^{137}Cs in Limnological Studies in Different Countries

Lake	References
	Switzerland
Greifen	Wan et al., 1987, *Chem. Geol.*, 63:181–196.
Constance	Dominik et al., 1981, *Sediment.*, 28:653–677; von Gunten, 1987, *Schweiz. Z. Hydrol.*, 49:275–283; Lindner et al., 1990, in Tilzer and Serruya (Eds.), *Large Lakes*, pp. 265–287, Springer-Verlag; Robbins et al., 1992, *Geochim. Cosmochim. Acta*, 56:2339–2361.
Zürich	Erten et al., 1985, *Schweiz. Z. Hydrol.*, 47:5–11; Schuler et al., 1991, *J. Geophys. Res.*, 96:17051–17065.
Sempach	Wieland et al., 1993, *Geochim. Cosmochim. Acta*, 57:2959–2979.
	United Kingdom
Five lakes, Southern England	He et al., 1996, *Chem. Geol.*, 129:115–131.
	Arctic Region
Russia	Baskaran et al., 1996, *Earth Planet. Sci. Lett.*, 140:243–256.
Canada	Hermanson, 1990, *Geochim. Cosmochim. Acta*, 54:1443–1451.
	North America
Michigan	Robbins and Edington, 1975, *Geochim. Cosmochim. Acta*, 39:285–304; Edington and Robbins, 1990, in Tilzer and Serruya (Eds.), *Large Lakes*, pp. 210–223, Springer-Verlag; Robbins and Eadie, 1991, *J. Geophys. Res.*, 96(C6):17081–17104.
Fayetteville, New York	Brunskill et al., 1984, *Chem. Geol.*, 44:101–117.
Ponds and lakes: Maine, Vermont, New Hampshire	Davis et al., 1984, *Chem. Geol.*, 44:151–185.
	Australia
Hidden Lake, Queensland	Torgersen and Longmore, 1984, *Austral.* J. *Mar. Freshwater Res.*, 35:537–548.
	Water–Sediment Interaction
Laboratory	Robbins et al., 1979, *Earth Planet. Sci. Lett.*, 42:277–287.
Two ponds, South Carolina	Evans et al., 1983, *Geochim. Cosmochim. Acta*, 47:1041–1049.

These insights were discussed by Rocco and Broecker (1963) in connection with their measurements of ^{90}Sr and ^{137}Cs activities in the South Atlantic Ocean (Figure 25.3) as well as by Toggweiler and Trumbore (1985), who needed to explain the high ^{90}Sr activities of corals in the Indian Ocean (Figure 25.5).

The presence of water masses of different origins also plays a major role in determining the chemical and physical properties of seawater in

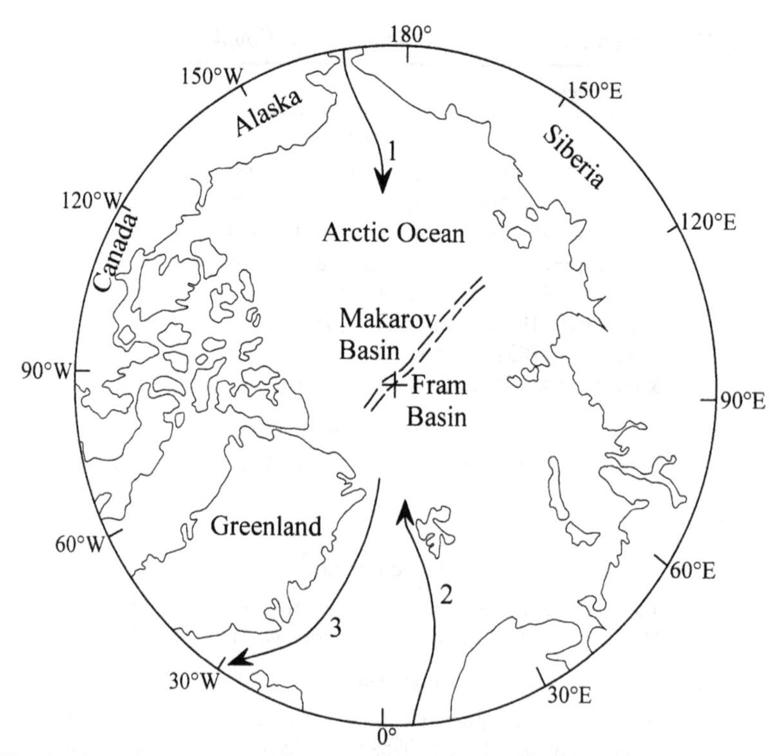

FIGURE 25.10 The Arctic Ocean showing the major currents of water from the Pacific Ocean (1), from the Atlantic Ocean (2), and leaving the Arctic Basin by the East Greenland Current (3). Adapted from Livingston et al. (1984).

the Arctic Ocean (Figure 25.10), which receives inputs from

1. the Pacific Ocean via the Bering strait and the Chukchi Sea,
2. the Atlantic Ocean via the West Spitsbergen Current, and
3. major rivers draining Siberia (e.g., Kolyma, Lena, Yenisey, Ob), Alaska (e.g., Yukon), and the Arctic of Canada (e.g., Mackenzie, Coppermine).

Water samples from the Arctic Ocean were collected in 1979 in the vicinity of the North Pole during the drift of an ice station. Eight of these samples were analyzed by Livingston et al. (1984) for ^{137}Cs, ^{90}Sr, $^{239,240}Pu$, and ^{241}Am in order to derive information about the different water masses that exist in the basin of the Arctic Ocean based on characteristic values of the ratios of certain radionuclides and other physical and chemical parameters. For example, fallout from the stratosphere is recognized by the activity ratios $^{137}Cs/^{90}Sr = 1.45$ and $^{239,240}Pu/^{137}Cs = 0.012$ (Bowen et al., 1974, 1980; Harley, 1975). The $^{137}Cs/^{90}Sr$ activity ratio of North Atlantic water is identical to that of fallout, whereas the $Pu/^{137}Cs$ and $Am/^{137}Cs$ ratios may be altered because of the preferential sorption of Pu and Am (Faure, 1998). Therefore, both Pu and Am may be lost from the North Pacific water by sorption to particles when it passes through the shallow Chukchi Sea (Figure 25.10). The concentrations of fission-product radionuclides of rivers vary depending on the fallout onto their drainage basins and may be high where radwaste is being disposed of or spent nuclear fuel is being reprocessed.

A major source of fission-product radionuclides in the North Atlantic Ocean is the effluent

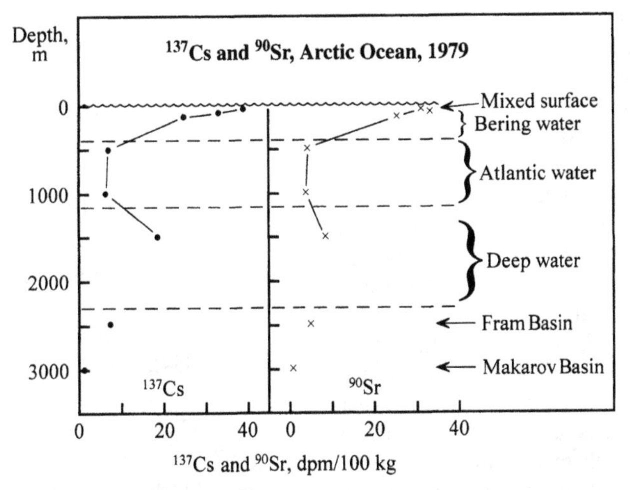

FIGURE 25.11 Profile of ^{137}Cs and ^{90}Sr activities in the Arctic Ocean at the North Pole Station of the Lomonosov Ridge Experiment of 1979 (LOREX 79). The water at this site is stratified with layers identified as Pacific Ocean water (Bering water), Atlantic Ocean water, and Arctic deep water. The Arctic bottom water (ABW) is split by the Lomonosov Ridge into ABW of the Makarov Basin and ABW of the Fram Basin. Data and interpretation by Livingston et al. (1984).

discharged by the spent-fuel reprocessing facility at Sellafield (formerly called Windscale) near the coast of West Cumbria, United Kingdom. Starting in 1952, this facility has discharged large amounts of fission-product radionuclides (^{137}Cs, ^{134}Cs, and $^{238,239,240}Pu$) into the Irish Sea (Aston and Stanners, 1981; Black, 1984; Stanners and Aston, 1984). The amount of ^{137}Cs discharged at Sellafield peaked in 1975 at about 1.41×10^5 Ci and then declined to about 3.0×10^4 Ci in 1984 following public concern about the health hazard posed by contaminated silt, seaweed, and fish (Dunster, 1969; Cambray and Eakins, 1982; Eisenbud, 1987). In addition, graphite in the core of one of two natural-U reactors at Sellafield was partially consumed by a fire in October of 1957. The resulting release of fission-product radionuclides included 16,200 Ci of ^{131}I, 1240 Ci of ^{137}Cs, 137 Ci of ^{89}Sr, and 6 Ci of ^{90}Sr (Clarke, 1974).

The track of the Ice Station extended from the Makarov Basin in Figure 25.10 across the Lomonosov Ridge into the Fram Basin. The activities of ^{137}Cs and ^{90}Sr at the North Pole station in Figure 25.11 vary with depth and reflect the provenance of the layers of water identified by Livingston et al. (1984) based on the composition of the fission-product radionuclides, on the concentration of Si, and on the physical properties of the water (e.g., salinity and temperature). The summary of these data in Table 25.9

Table 25.9. Characteristic Properties of Water Masses of Different Provenance in a Depth Profile in Figure 25.11 Taken in the Arctic Ocean in 1979 Near the North Pole

	dpm/100 kg				
Layer	^{137}Cs	^{90}Sr	Salinity, ‰	Si, μg/L	T, °C
Surface	39.2	31.6	30	6.9	−1.6
Bering	29.6	29.3	32.75	24.5	−1.55
Atlantic	7.0	4.2	34.89	7.25	+0.17
Deep water	18.5	8.7	34.92	8.7	−0.52
Arctic bottom water	0.9	0.6	34.95	12.8	−0.46
Makarov Basin					
Fram Basin	7.3	4.8	34.94	10.3	−0.85

Source: Livingston et al., 1984.

demonstrates that the thin layer of well-mixed surface water has the highest activities of ^{137}Cs (39.2 dpm/100 kg) and ^{90}Sr (31.6 dpm/100 kg), yielding $^{137}Cs/^{90}Sr = 1.24$. In addition, the activities of $^{239,240}Pu$ and ^{241}Am have maximum values of 0.082 and 0.042 dpm/100 kg, respectively, with a $^{239,240}Pu/^{241}Am$ activity ratio of 1.95. The Bering water also has high activities of ^{137}Cs (29.6 dpm/100 kg) and ^{90}Sr (29.3 dpm/100 kg) with $^{137}Cs/^{90}Sr$ ratio of 1.01. The Atlantic water has comparatively low activities of ^{137}Cs (7.0 dpm/100 kg) and ^{90}Sr (4.2 dpm/100 kg) with a high $^{137}Cs/^{90}Sr$ ratio of 1.67. The $^{137}Cs/^{90}Sr$ ratio of the underlying deep water (>1000 m) is even higher at 2.13. This water lies above the top of the Lomonosov Ridge and therefore extends across it. However, the Arctic bottom water in the Makarov Basin is separated by the Lomonosov Ridge from the Arctic bottom water of the Fram Basin. Both water masses have distinct but low activities of ^{137}Cs and ^{90}Sr that yield an identical $^{137}Cs/^{90}Sr$ ratio of 1.50.

The $^{137}Cs/^{90}Sr$ ratios as well as the $Pu/^{137}Cs$, $Am/^{137}Cs$, and Pu/Am activity ratios of seawater depend initially on the composition of the atmospheric fallout and on any additional sources, such as effluent of spent-fuel-reprocessing plants. However, after the radionuclides have dissolved in seawater, their concentrations are modified by selective sorption to the various kinds of particles that occur in the oceans (Broecker and Peng, 1982). The sorption of cations depends primarily on the magnitude of their charge-to-radius ratios. For example, the $^{137}Cs/^{90}Sr$ ratio in seawater is likely to increase because Sr ions have a higher charge and a smaller ionic radius and therefore a higher charge-to-radius ratio (c/r) than Cs ions:

$$\text{Sr:}+2,\ 1.13\ \text{Å},\ c/r = 1.77$$
$$\text{Cs:}+1,\ 1.69\ \text{Å},\ c/r = 0.59$$

Therefore, Sr^{2+} ions are sorbed in preference to Cs^{+} ions, causing the $^{137}Cs/^{90}Sr$ ratio of seawater to *rise*.

Similarly, the Pu/Cs and Am/Cs ratios of fallout or effluent entering the oceans *decrease* because Pu and Am have higher charge-to-radius ratios than Cs and are sorbed preferentially:

$$\text{Pu: }+3,\ 1.07\ \text{Å},\ c/r = 2.80$$
$$+4,\ 0.93\ \text{Å},\ c/r = 4.30$$
$$\text{Am: }+3,\ 1.06\ \text{Å},\ c/r = 2.83$$
$$+4,\ 0.92\ \text{Å},\ c/r = 4.35$$

Americium is known to sorb irreversibly to crystals of calcite and aragonite (Shanbhag and Morse, 1982), which causes the Pu/Am ratio of seawater to rise even though Pu is also scavenged from seawater by sorption to particles (Sholkovitz, 1983).

In the North Pole profile studied by Livingston et al. (1984), the $^{239,240}Pu/^{241}Am$ activity ratios range from 1.95 in the surface water to 11.6 in the Arctic bottom water of the Fram Basin. The Pu/Am ratios of the Bering, Atlantic, and deep waters range narrowly from 3.77 to 3.97 and average 3.87 ± 0.06. The Arctic bottom water in the Makarov Basin has a $^{239,240}Pu/^{241}Am$ activity ratio of 3.00.

These results contribute to the evidence that the water in the Arctic Ocean contains components derived from different sources and that these components have not mixed but are arranged in layers that maintain their identity. Consequently, the concentrations of fission-product radionuclides in the Arctic Ocean are not based solely on the fallout that occurs at these high latitudes. Another way to express this insight is to say that the potential for radiological contamination of the flora and fauna of the Arctic Ocean is much greater than can be accounted for by fallout from the overlying stratosphere.

25.5 SUMMARY

Induced fission of ^{235}U and Pu isotopes produces a very large number of stable and unstable product nuclides. The halflives of most of the unstable fission-product nuclides are less than 10 h, causing them to decay in about two days. However, a few fission-product radionuclides have long halflives ranging from tens of years to millions of years. The nuclear radiation emitted by these radionuclides is a hazard to life for up to 100,000 years. The long period of time during which high-level radwaste remains dangerously active requires it to be placed in underground repositories that can prevent accidental and malicious intrusion. However, the safety of such geological repositories may be compromised by future changes in climate, volcanic eruptions, earthquakes, and tectonic deformation of the rocks. In addition, unforeseen accidents during transport of the packaged waste to the repository or during handling at the repository may release radwaste into the environment and thereby cause harm to all lifeforms that become exposed to it.

Fission of U and Pu also releases large numbers of neutrons that can be captured by the stable nuclei of atoms in the atmosphere, in soil, and in the underlying rocks, depending on the location of the fission event. Such neutron-capture reactions produce a suite of radionuclides, some of which are also formed naturally by cosmic rays. This group of secondary fission products includes ^{3}H, ^{14}C, ^{36}Cl, ^{129}I, and many others. The nuclides listed above have been considered in previous sections of this book: ^{3}H in Sections 21.4a and 21.4b, ^{14}C in Section 23.1a, ^{36}Cl in Section 23.4, and ^{129}I in Section 23.6.

Another group of secondary products of nuclear fission are the transuranium elements Np, Pu, and Am, which form by neutron capture and β-decay of ^{238}U (Figure 25.1), primarily in the fuel rods of nuclear reactors and during the explosion of nuclear fission and fusion devices. The resulting suite of transuranium elements has been released into the environment during reprocessing of spent nuclear fuel and as a consequence of accidental explosions of nuclear reactors. In addition, Pu has been dispersed during the re-entry of spacecraft equipped with radioisotope power-generators or small nuclear reactors. As a result of these kinds of accidental releases, Pu and Am have entered the oceans and are widely distributed both in seawater and in sediment. Both elements are "particle reactive" and are efficiently scavenged by particles of calcite and aragonite as well as by other kinds of particles.

The two most abundant fission-product radionuclides in the environment are ^{90}Sr and ^{137}Cs, both of which have halflives of about 30 years. Strontium-90 is readily absorbed by plants through the roots and is also deposited on leaves and grass eaten by livestock. In this way, virtually all types of food consumed by humans is contaminated with varying amounts of ^{90}Sr. This isotope is deposited in bones by replacement of Ca and causes long-term damage as it decays by emission of β-particles and γ-rays. In the oceans, ^{90}Sr is incorporated into biogenic calcium carbonate deposited by corals and mollusks. Therefore, corals contain a record of the concentration of ^{90}Sr in seawater during the time of testing of nuclear weapons from 1945 to 1980.

Large amounts of ^{137}Cs have been released into the atmosphere and into the oceans. Therefore, this isotope is widely distributed in soil on the continents, in the ice sheets of Greenland and Antarctica, and in the oceans. Cesium-137 is strongly sorbed by clay minerals and is not as "plant-available" as ^{90}Sr. Nevertheless, ^{137}Cs does contaminate human food and thereby contributes to the internal irradiation with β-particles and γ-rays. In contrast to ^{90}Sr, ^{137}Cs is not stored in bones or other organs but is uniformly distributed in the body in association with K. Both ^{137}Cs and ^{90}Sr are used by oceanographers to identify water masses and to trace their movement by the system of currents in the global ocean.

The hazard posed by the fission-product radionuclides that were released into the atmosphere by the testing of nuclear weapons in the atmosphere has greatly diminished because the radionuclides have decayed and have been buried in sediment deposited in the oceans and lakes. The threat posed by anthropogenic radioactivity increasingly originates from accidental releases of fission-product radionuclides by aging nuclear reactors and from the high-level radioactive waste that has accumulated in the spent nuclear fuel. The best way to manage the risk to human health is to reprocess the radwaste and to package the rest for permanent storage in subsurface repositories.

REFERENCES FOR FISSION PRODUCT RADIONUCLIDES AND TRANSURANIUM ELEMENTS (SECTIONS 25.1a–25.1b)

Aston, S. R., and D. A. Stanners, 1981. Plutonium transport to and deposition and immobility in Irish Sea intertidal sediments. *Nature*, 289:581–582.

Carter, M. W., and A. A. Moghissi, 1977. Three decades of nuclear testing. *Health Phys.*, 33:55.

Leifer, R. Z., Z. R. Juzdan, and R. Larsen, 1984. *The High Altitude Sampling Program: Radioactivity in the Stratosphere*, EML Rep. 434, U.S. Department of Energy, New York.

Stanners, D. A., and S. R. Aston, 1984. The use of reprocessing effluent radionuclides in the geochronology of recent sediment. *Chem. Geol.*, 44:19–32.

UNSCEAR, 1969. *United Nations Scientific Committee on the Effects of Atomic Radiation, 21st Session*, Suppl. No. 13(A/7613). United Nations, New York.

USDOE, 1977. *Final Tabulation of Monthly ^{90}Sr Fallout Data: 1954–1976*. U.S. Department of Energy, Health Saf. Lab., Env. Q. Rep., HASL 329. U.S. Department of Energy, Washington, D.C.

Walker, F. W., J. R. Parrington, and F. Reiner, 1988. *Chart of the Nuclides*, 14th ed., General Electric Co., San Jose, California.

REFERENCES FOR DISPOSAL OF RADWASTE (YUCCA MOUNTAIN, NEVADA) (SECTION 25.1c)

Faure, G., 1998. *Principles and Applications of Geochemistry*. Prentice-Hall, Upper Saddle River, New Jersey.

Gascoyne, M., and Z. E. Peterman, 2002. Geochemical research at Yucca Mountain, Nevada. *Appl. Geochem.*, 17:657–853.

Neymark, L. A., Y. V. Amelin, and J. B. Paces, 2000. ^{206}Pb-^{230}Th-^{234}U-^{238}U and ^{207}Pb-^{235}U geochronology of Quaternary opal, Yucca Mountain, Nevada. *Geochim. Cosmochim. Acta*, 64:2913–2928.

Neymark, L. A., and J. B. Paces, 2000. Consequences of slow growth for $^{230}Th/U$ dating of Quaternary opals, Yucca Mountain, NV, USA. *Chem. Geol.*, 164:143–160.

Pentreath, R. J., 1980. *Nuclear Power, Man and the Environment*. Taylor and Francis, London.

Peterman, Z. E., and P. L. Cloke, 2002. Geochemistry of rock units at the potential repository level, Yucca Mountain, Nevada. *Appl. Geochem.*, 17:683–698.

Roxburgh, I. S., 1987. *Geology of High-Level Nuclear Waste Disposal*. Chapman and Hall, London.

Stuckless, J. S., and W. W. Dudley, 2002. The hydrologic setting of Yucca Mountain, Nevada. *Appl. Geochem.*, 17:659–682.

Stuckless, J. S., Z. E. Peterman, and D. R. Muhs, 1991. U and Sr isotopes in ground water and calcite, Yucca Mountain, Nevada: Evidence against upwelling water. *Science*, 254:551–554.

REFERENCES FOR REACTOR ACCIDENTS: CHERNOBYL, UKRAINE (SECTION 25.1d)

Anspaugh, L. R., R. J. Catlin, and M. Goldman, 1988. The global impact of the Chernobyl reactor accident. *Science*, 242:1513–1519.

Baier, W., 1989. Eine Anlage zur kontinuierlichen Gewinnung von Bomben-Plutonium. *Frankfurter Rundschau*, April 25, No. 96:24

Devell, L., H. Tovedahl, U. Bergström, A. Appelgren, J. Chyssler, and L. Andersson, 1986. Initial observations of fallout from the reactor accident at Chernobyl. *Nature*, 321:192.

Eisenbud, M., 1987. *Environmental Radioactivity*, 3rd ed. Academic, San Diego.

Hohenemser, C., M. Deicher, A. Ernst, H. Hofsäss, G. Lindner, and E. Recknagel, 1986. Chernobyl: an early report. *Environment*, 28:6–43.

Jost, D. T., H. W. Gäggeler, U. Baltensberger, B. Zinder, and P. Haller, 1986. Chernobyl fallout in size fractionated aerosol. *Nature*, 324:22.

Shcherbak, Y. M., 1996. Ten years of the Chernobyl era. *Sci. Am.*, 274(4):44–49.

Smith, F. B., and M. J. Clark, 1986. Radionuclide deposition from the Chernobyl cloud. *Nature*, 322:690–691.

REFERENCES FOR GLOBAL DISTRIBUTION OF ^{90}Sr (SECTIONS 25.2–25.26)

Bowen, V. T., V. E. Noshkin, H. D. Livingston, and H. L. Volchok, 1980. Fallout radionuclides in the Pacific Ocean: Vertical and horizontal distributions largely from GEOSECS stations. *Earth Planet. Sci. Lett.*, 49:411.

Bowen, V. T., V. E. Noshkin, H. L. Volchok, and T. T. Sugihara, 1968. *Fallout Strontium-90 in Atlantic Ocean Surface Water*. U.S. Atomic Energy Commission Health Safety Lab., Quarterly Rep., HASL 197. Washington, D.C.

Bowen, V. T., and T. T. Sugihara, 1960. Strontium-90 in the "mixed layer" of the Atlantic Ocean. *Nature*, 186:71–72.

Bunzl, K., H. Kofuji, W. Schimmack, A. Tsumura, K. Ueno, and M. Yamamoto, 1995. Residence times of global weapons testing fallout ^{137}Np in a grassland soil compared to $^{239+240}Pu$, ^{241}Am, and ^{137}Cs. *Health Phys.*, 68:89–93.

Carter, M. W., and A. A. Moghissi, 1977. Three decades of nuclear testing. *Health Phys.*, 33:55.

Cochran, K., H. D. Livingston, D. J. Hirschberg, and L. D. Surprenant, 1987. Natural and anthropogenic radionuclide distributions in the northwest Atlantic Ocean. *Earth Planet. Sci. Lett.*, 84:135–152.

Dreisigacker, E., and W. Roether, 1978. Tritium and ^{90}Sr in North Atlantic surface water. *Earth Planet. Sci. Lett.*, 38:301–312.

Druffel, E. M., 1981. Radiocarbon in annual coral rings from the eastern tropical Pacific Ocean. *Geophys. Res. Lett.*, 8:59.

Druffel, E. M., and T. W. Linnick, 1978. Radiocarbon in annual coral rings of Florida. *Geophys. Res. Lett.*, 5:913.

Eisenbud, M., 1987. *Environmental Radioactivity*, 3rd ed. Academic, San Diego.

Fry, L. M., F. A. Jew, and P. K. Kuroda, 1960. On the stratospheric fallout of strontium 90: The spring peak of 1959. *J. Geophys. Res.*, 65(7):2061–2066.

Knutson, D. W., R. W. Buddemeier, and S. V. Smith, 1972. Coral chronometers: Seasonal growth bands in reef corals. *Science*, 177:270.

Lambert, G., B. Ardouin, E. Brichett, and C. Lorius, 1971. Balance of ^{90}Sr over Antarctica: Existence of a protected area. *Earth Planet. Sci. Lett.*, 11:317–323.

Leifer, R. Z., Z. R. Juzdan, and R. Larsen, 1984. *The High Altitude Sampling Program: Radioactivity in the Atmosphere*, EMO Rep. 434. U.S. Department of Energy, New York.

Meyer, M. W., J. S. Allen, L. T. Alexander, and E. P. Hardy, 1968. *Strontium-90 on the Earth's Surface*, Vol. 4. USAEC, TID-24341. U.S. Atomic Energy Commission, Washington, D.C.

Miyake, Y., K. Saruhashi, Y. Katsuragi, and T. Kanazawa, 1962. Penetration of ^{90}Sr and ^{137}Cs in deep layers of the Pacific and vertical diffusion rate of deep water. *J. Radiation Res.*, 3:141–147.

Oktay, S. D., P. H. Santschi, J. E. Moran, and P. Sharma, 2000. The ^{129}Iodine bomb pulse recorded in Mississippi River delta sediments; results from isotopes of I, Pu, Cs, Pb, and C. *Geochim. Cosmochim. Acta*, 64:989–996.

Picciotto, E., and S. Wilgain, 1963. Fission products in Antarctic snow, a reference level for measuring accumulation. *J. Geophys. Res.*, 68(21):5965–5972.

Rocco, G. G., and W. S. Broecker, 1963. The vertical distribution of cesium 137 and strontium 90 in the oceans. *J. Geophys. Res.*, 68(15):4501–4512.

Santschi, P. H., Y.-H. Li, D. M. Adler, M. Amdurer, J. Bell, and U. P. Nyffeler, 1983. The relative mobility of natural (Th, Pb, and Po) and fallout (Pu, Am, Cs) radionuclides in the coastal marine environmental: Results from model ecosystems (MERL) and Narragansett Bay. *Geochim. Cosmochim. Acta*, 47:201–210.

Toggweiler, J. R., and S. Trumbore, 1985. Bomb-test ^{90}Sr in Pacific and Indian Ocean surface water as recorded by banded corals. *Earth Planet. Sci. Lett.*, 74:306–315.

UNSCEAR, 1969. *Report to the 21st Session*, Suppl. N. 13(A/7613). United Nations, New York.

USDOE, 1977. *Final Tabulation of monthly ^{90}Sr fallout data: 1954–1976*. U.S. Department of Energy, Health Safety Lab., Environ. Quarterly Rep., HASL 329.

Wilgain, S., E. Picciotto, and W. DeBreuck, 1965. Strontium-90 fallout in Antarctica. *J. Geophys. Res.*, 70(24):6023–6032.

REFERENCES FOR HUMAN DIET (SECTION 25.2c)

Anonymous, 1991. *Some Aspects of Strontium Radiobiology*, Report 110. National Council on Radiation Protection and Measurement, Bethesda, Maryland.

Eisenbud, M., 1987. *Environmental Radioactivity*, 3rd ed. Academic, San Diego.

Harley, J. H., 1969. Radionuclides in food. In *Biological Implications of the Nuclear Age*, AEC Symp. Ser. No. 16, U.S. Atomic Energy Commission, Oak Ridge, Tennessee.

Hopkins, B. J., L. W. Tuttle, W. J. Pories, and W. H. Strain, 1963. Strontium-90 in hair. *Science*, 139:1064–1065.

IAEA, 1964. *Assessment of Radioactivity in Man*, 2 vols. International Atomic Energy Agency, Vienna.

Klusek, C. S., 1984a. *Sr-90 in the U.S. Diet, 1982*, EML Rep. 429. U.S. Department of Energy, New York.

Klusek, C. S., 1984b. *Sr-90 in Human Bone in the U.S., 1982*, EML Rep. 435. U.S. Department of Energy, New York.

Kulp, J. L., and R. Slakter, 1958. Current strontium-90 level in diet in United States. *Science*, 128:85–86.

Marshall, J. H., E. L. Lloyd, J. Rondo, J. Liniecki, G. Marotti, C. W. Mays, H. A. Sissons, and W. S. Snyder, 1973. Alkaline earth metabolism in adult man. *Heath Phys.*, 24:125.

Petukhova, E. V., and V. A. Knizhnikov, 1969. *Dietary Intake of Sr-90 and Cs-137*, Pub. No. A/AC.82/G/L. 1245. United Nations Sci. Comm., Sales Sect., New York.

Rosenthal, H. L., J. E. Gilster, and J. T. Bird, 1963. Strontium-90 content of deciduous human incisors. *Science*, 140:176–177.

REFERENCES FOR CESIUM-137 IN THE ENVIRONMENT (SECTION 25.3)

Beer, J., M. Andrée, H. Oeschger, B. Stauffer, R. Balzer, G. Bonani, Ch. Stoller, M. Suter, and W. Wölfli, 1985.

^{10}Be variations in polar ice cores. In C. C. Langway, Jr. H. Oeschger, and W. Dansgaard (eds.), *Greenland Ice Core: Geophysics, Geochemistry, and the Environment*, pp. 66–70, Geophys. Monograph 33. American Geophysical Union, Washington, D.C.

Comans, R. N. J., and D. E. Hockley, 1992. Kinetics of cesium sorption on illite. *Geochim. Cosmochim. Acta*, 56:1157–1164.

Eisenbud, M., 1987. *Environmental Radioactivity*, 3rd ed. Academic, San Diego.

Grütter, A., H. R. von Gunter, M. Kohler, and E. Rössler, 1990. Sorption, desorption, and exchange of cesium on glaciofluvial deposits. *Radiochim. Acta*, 50:177–184.

Harley, J. H., 1969. Radionuclides in food. In *Biological Implications of the Nuclear Age*, AEC Symp. Ser. No. 16. U.S. Atomic Energy Commission, Oak Ridge, Tennessee.

Korobova, E., A. Ermakov, and V. Linnik, 1998. ^{137}Cs and ^{90}Sr mobility and transfer in soil-plant systems in the Novozybkov district affected by the Chernobyl accident. *Appl. Geochim.*, 13(7):803–814.

Robbins, J. A. (Ed.), 1984. Geochronology of recent deposits. *Chem. Geol.*, 44(1/3):1–348.

Wieland, E., P. H. Santschi, P. Höhener, and M. Sturm, 1993. Scavenging of Chernobyl ^{137}Cs and natural ^{210}Pb in Lake Sempach, Switzerland, *Geochim. Cosmochim. Acta*, 57:2959–2979.

REFERENCES FOR ARCTIC OCEAN: ^{90}Sr/^{137}Cs, ^{239,240}Pu, AND ^{241}Am (SECTION 25.4)

Aston, S. R., and D. A. Stanners, 1981. Plutonium transport to and deposition and immobility in Irish Sea intertidal sediments. *Nature*, 189:581–582.

Black, D., 1984. *Investigation of the Possible Increased Incidence of Cancer in West Cumbria*. Report Independent Advisory Group, Her Majesty's Stationery Office, London.

Bowen, V. T., V. E. Noshkin, H. D. Livingston, and H. L. Volchok, 1980. Fallout radionuclides in the Pacific Ocean; vertical and horizontal distributions, largely from GEOSECS stations. *Earth Planet. Sci. Lett.*, 48:411–434.

Bowen, V. T., V. E. Noshkin, H. L. Volchok, H. L. Livingston, and K. M. Wong, 1974. Cesium-137 to strontium-90 ratios in the Atlantic Ocean 1966 through 1972. *Limnol. Oceanogr.*, 19:670–681.

Broecker, W. S., and T.-H. Peng, 1982. *Tracers in the Sea*. Lamont-Doherty Geol. Obs., Columbia University, Palisades, New York.

Cambray, R. S., and J. D. Eakins, 1982. Pu, ^{241}Am, and ^{137}Cs in soil in West Cumbria and a maritime effect. *Nature*, 300:46–48.

Clarke, R. H., 1974. An analysis of the Windscale accident using the WEERIE code. *Am. Nucl. Sci. Eng.*, 1:73.

Dunster, H. J., 1969. United Kingdom studies on radioactive releases in the marine environment. In *Biological Implications of the Nuclear Age*. USAEC, Washington, D.C.

Eisenbud, M., 1987. *Environmental Radioactivity*, 3rd ed. Academic, San Diego.

Faure, G., 1998. *Principle and Applications of Geochemistry*, 2nd ed. Prentice-Hall, Upper Saddle River, New Jersey.

Harley, J. H., 1975. Transuranium elements on land. In *Energy Research Development Admin.*, Health and Safety Lab., Environ. Quart. Rept., HASL-291:I104–I109.

Livingston, H. D., S. L. Kupferman, V. T. Bowen, and R. M. Moores, 1984. Vertical profile of artificial radionuclide concentrations in the central Arctic Ocean. *Geochim. Cosmochim. Acta*, 48:2195–2203.

Rocco, G. G., and W. S. Broecker, 1963. The vertical distribution of cesium 137 and strontium 90 in the oceans. *J. Geophys. Res.*, 68(15):4501–4512.

Shanbhag, P. M., and J. W. Morse, 1982. Americium interaction with calcite and aragonite surfaces in seawater. *Geochim. Cosmochim. Acta*, 46:241–246.

Sholkovitz, E. R., 1983. The geochemistry of plutonium in fresh and marine water environments. *Earth-Sci. Rev.*, 19:95–161.

Stanners, D. A., and S. R. Aston, 1984. The use of reprocessing effluent radionuclides in the geochronology of recent sediments. *Chem. Geol.*, 44:19–32.

Toggweiler, J. R., and S. Trumbore, 1985. Bomb-test ^{90}Sr in Pacific and Indian Ocean surface water as recorded by banded corals. *Earth Planet. Sci. Lett.*, 74:306–314.

PART V

Fractionation of Stable Isotopes

The stable isotopes of many elements are fractionated during changes in their states of aggregation and by chemical reactions between compounds in which the elements occur. The extent of fractionation of two isotopes of the same element is controlled primarily by the difference in their masses and by the temperature of the environment. The elements whose isotopes are susceptible to fractionation include H, O, C, N, and S as well as B, Li, Si, Cl, and several elements of high atomic number (e.g., Fe, Cu, and Se). The resulting variations of the isotopic compositions of these elements convey information about the physical, geochemical, and geobiological processes that acted on these elements and on the compounds in which they occur.

CHAPTER 26

Hydrogen and Oxygen

THE Earth is unique among the planets of the solar system because water (H_2O) exists in liquid form on its surface. The presence of liquid water is a prerequisite to the existence of life as we know it and to its preservation. It is the medium in which chemical reactions occur that result in the formation of solid compounds, including carbonates, sulfates, phosphates, silicates, hydroxides, and oxides. In fact, oxygen (O) is the most abundant chemical element in the crust of the Earth, while hydrogen (H) is the most abundant element in the solar system, in the Milky Way Galaxy, and in the Universe. These two elements participate in virtually all geochemical and biochemical processes that occur on Earth. They share this distinction with a few other elements, namely: carbon, nitrogen, and sulfur, whose stable isotopes are also subject to mass-dependent fractionation. In addition, the isotopes of several other less abundant elements are also fractionated, including: lithium (Li), boron (B), silicon (Si), chlorine (Cl), selenium (Se), and others.

The subject of stable isotope fractionation of different elements has been presented by Craig et al. (1964), Valley et al. (1986), Taylor et al. (1991), Hoefs (1997), Clark and Fritz (1997), Criss (1999), and Valley and Cole (2001). Applications of the fractionation of stable isotopes to the study of environmental processes were the topic of two books edited by Fritz and Fontes (1980, 1986).

26.1 ATOMIC PROPERTIES

Hydrogen ($Z = 1$) has two stable isotopes: $^{1}_{1}H$ and $^{2}_{1}H$ (deuterium, or D), whereas oxygen ($Z = 8$) has three stable isotopes: $^{16}_{8}O$, $^{17}_{8}O$, $^{18}_{8}O$. The data in Table 26.1 reveal that $^{1}_{1}H$ and $^{16}_{8}O$ are the most abundant isotopes of these elements and that the masses of the heaviest isotopes of H and O differ substantially from those of the lightest stable isotopes of these elements: $^{2}_{1}H$ is 99.8% heavier than $^{1}_{1}H$ and $^{18}_{8}O$ is 12.5% heavier than $^{16}_{8}O$. The difference in the masses of the isotopes of any element affects the strength of covalent bonds these isotopes form with atoms of other elements.

Table 26.1. Physical Properties and Abundances of the Stable Isotopes of H and O

Isotope	Abundance, % by number	Mass, amu
	Hydrogen	
$^{1}_{1}H$	99.985	1.007825
$^{2}_{1}H$	0.015	2.0140
	Oxygen	
$^{16}_{8}O$	99.762	15.994915
$^{17}_{8}O$	0.038	16.999131
$^{18}_{8}O$	0.200	17.999160

Source: Lide and Frederikse, 1995.

The strength of the covalent bond of a diatomic molecule in a gas arises from the reduction of the energy that occurs when two atoms that are initially an infinite distance apart approach each other to form a molecule. According to the theory of quantum mechanics, the energy of a diatomic molecule (E) at absolute zero has a finite value that depends on its vibrational frequency ν:

$$E = \tfrac{1}{2}h\nu \tag{26.1}$$

where h is Planck's constant, equal to 6.626176×10^{-34} J/Hz. When the light isotope of an element in a diatomic molecule is replaced by a heavier isotope of the same element, the vibrational frequency of the molecule decreases, which causes a corresponding incremental decrease of the energy of the molecule. The decrease of the energy of the molecule manifests itself as an increase in the strength of the covalent bond formed by the heavy isotope. One consequence of this phenomenon is that *molecules containing the heavy isotope of an element are more stable* than the same molecule would be if it contained the light isotope of that element. Therefore, *molecules containing the heavy isotope are also less reactive* than molecules containing the light isotope.

The theory of mass-dependent fractionation of the isotopes of an element has been presented by Urey (1947), Bigeleisen (1965), Bottinga and Javoy (1973), Richet et al. (1977), Melander and Saunders (1980), O'Neil (1986), and others. This theory can explain isotope fractionation in gases, it works less well in liquids, and it does not apply to ionic crystals where the motions of ions are constrained by the energy of the lattice. Therefore, the study of mass-dependent isotope fractionation is largely an empirical science in which results obtained in the laboratory are applied to natural systems. However, Kieffer (1982) has been successful in predicting the temperature dependence of the fractionation of O in minerals.

The masses of isotopes of an element also play a role in determining the velocities of molecules in a gas at a particular temperature. The description of gases in statistical mechanics leads to the conclusion that all molecules of an ideal gas have the same kinetic energy at a specified temperature:

$$KE = \tfrac{1}{2}mv^2 \tag{26.2}$$

Therefore, two isotopic varieties of a molecule having different masses have the same kinetic energy:

$$\tfrac{1}{2}m_H v_H^2 = \tfrac{1}{2}m_L v_L^2 \tag{26.3}$$

where the subscripts H and L mean "heavy" and "light," respectively. It follows that:

$$\frac{v_L}{v_H} = \left(\frac{m_H}{m_L}\right)^{1/2} \tag{26.4}$$

which is a statement of Graham's law. Solving for v_L yields:

$$v_L = v_H\left(\frac{m_H}{m_L}\right)^{1/2} \tag{26.5}$$

The m_H/m_L ratio is greater than one, which means that:

$$v_L > v_H$$

In other words, the velocity of the molecules containing the light isotope is greater than the velocity of the molecules containing the heavy isotopes of the same element and at the same temperature.

For example, water molecules composed either of the heavy or the light isotopes of H and O have different masses (Table 26.1):

$$^{2}H_2{}^{18}O: \quad m_H = 2 \times 2.0140 + 17.999160 = 22.02716 \text{ amu}$$

$$^{1}H_2{}^{16}O: \quad m_L = 2 \times 1.007825 + 15.994915 = 18.010565 \text{ amu}$$

$$v_L = v_H\left(\frac{22.02716}{18.010565}\right)^{1/2} = 1.10589 v_H$$

The light molecule of water has a higher velocity in a gas than the heavy molecule. This result applies also to the rates of diffusion of heavy and light isotopic molecules in response to a concentration gradient at constant temperature.

The existence of two stable isotopes of H and three isotopes of O makes possible nine different isotopic varieties of the water molecule:

$H_2{}^{16}O$, $H_2{}^{17}O$, $H_2{}^{18}O$
$HD^{16}O$, $HD^{17}O$, $HD^{18}O$
$D_2{}^{16}O$, $D_2{}^{17}O$, $D_2{}^{18}O$

where $H = {}^1_1H$ and $D = {}^2_1H$. The masses of these isotopic water molecules calculated above range from 18.010565 amu ($H_2{}^{16}O$) to 22.02716 amu ($D_2{}^{18}O$) from which it follows that $D_2{}^{18}O$ is 22.3% heavier than $H_2{}^{16}O$. This difference in the masses of water molecules is the basis for the fractionation of the isotopes of H and O in water molecules that takes place during evaporation of water to form vapor and during the condensation of the vapor to form liquid water (Gat, 1984).

26.2 MATHEMATICAL RELATIONS

The isotopic composition of O and of all other elements whose isotopes are fractionated is universally expressed as the ratio R of the isotopic abundance of the heavy isotope divided by the abundance of the light isotope. Therefore, R is defined in terms of numbers of atoms rather than in terms of the masses of the isotopes. In the case of O, R is defined by the abundances in Table 26.1:

$$R = \frac{^{18}O}{^{16}O} = \frac{0.200}{99.762} = 0.002004 \qquad (26.6)$$

The isotope ratios of H and O are measured by mass spectrometry and are expressed relative to standard mean ocean water (SMOW) (Hoefs, 1997). The use of a standard reduces systematic errors in measurements made on different mass spectrometers and permits R values to be expressed in terms of a parameter called delta (δ), which for O is defined by the relation:

$$\delta^{18}O = \left(\frac{R_{spl} - R_{std}}{R_{std}}\right) \times 10^3‰ \qquad (26.7)$$

where $R_{spl} = {}^{18}O/{}^{16}O$ ratio of the sample and
$R_{std} = {}^{18}O/{}^{16}O$ ratio of the standard (SMOW)

and $\delta^{18}O$ is the difference between the R values of the sample and the standard expressed in terms of permille relative to the R value of the standard. Similarly, the R value of H is defined as D/H:

$$\delta D = \left(\frac{R_{spl} - R_{std}}{R_{std}}\right) \times 10^3‰ \qquad (26.8)$$

The δ parameters of O and H may be positive, negative, or zero. A positive $\delta^{18}O$ value indicates that the sample has a higher $^{18}O/^{16}O$ ratio than the standard, which is expressed by stating that the sample is enriched in ^{18}O relative to the seawater standard. A negative $\delta^{18}O$ value indicates that the sample has a lower $^{18}O/^{16}O$ ratio than the standard and that the sample is depleted in ^{18}O relative to seawater. In other words, the isotope composition of O in the sample is always expressed in terms of enrichment or depletion of the heavy isotope (e.g., ^{18}O) relative to the standard (e.g., SMOW). This convention is useful because it avoids the confusion that could otherwise occur. In addition, statements such as "The sample is enriched (or depleted) in $\delta^{18}O$" should be avoided.

When a compound such as liquid water undergoes a change in state by evaporating to form water vapor under equilibrium conditions at a constant temperature, the D/H and $^{18}O/^{16}O$ ratios of the vapor differ from the ratios of the remaining liquid water. This phenomenon is evidence that isotope fractionation takes place during evaporation of liquid water to form water vapor. The extent of isotope fractionation that takes place during a change in state is expressed quantitatively by the isotope fractionation factor α defined by the relation:

$$\alpha^l_v = \frac{R_l}{R_v} \qquad (26.9)$$

where R_l is the isotope ratio of the liquid and R_v is the isotope ratio of the vapor in equilibrium with the liquid at a constant temperature. The convention is to express the isotope fractionation factors in terms of the liquid–vapor or solid–liquid ratios, which in most cases causes the resulting fractionation factors to be greater than one, depending on the temperature. In most cases, the isotope fractionation factors decrease with increasing temperature and approach a value of 1.0, which means that mass-dependent isotope fractionation virtually ceases as the temperature approaches high values. This point is elaborated below.

In the general case of isotope fractionation between two phases a and b in isotopic equilibrium, the isotope composition of an element in phases a and b are:

$$\delta_a = \left(\frac{R_a - R_{std}}{R_{std}}\right) \times 10^3$$

$$\delta_b = \left(\frac{R_b - R_{std}}{R_{std}}\right) \times 10^3$$

These equations can be solved for R_a and R_b:

$$R_a = \frac{R_{std}(\delta_a + 10^3)}{10^3} \qquad (26.10)$$

$$R_b = \frac{R_{std}(\delta_b + 10^3)}{10^3} \qquad (26.11)$$

Substituting equations 26.10 and 26.11 into equation 26.9 provides a relation between α_b^a and the δ values:

$$\alpha_b^a = \frac{\delta_a + 10^3}{\delta_b + 10^3} \qquad (26.12)$$

This equation is valid provided that isotopic equilibrium existed between phases a and b at a specified temperature.

The relation between the isotope fractionation factor α_b^a and the temperature T takes the form:

$$\ln \alpha_b^a \simeq \frac{1}{T}$$

where T is the temperature on the Kelvin scale. Measurements of the fractionation factor for O isotopes in two phases a and b at different temperatures define equations of the form:

$$10^3 \ln \alpha_b^a = \frac{A \times 10^6}{T^2} + B \qquad (26.13)$$

where A and B are experimentally determined constants.

The natural logarithm of a number like $1.00X$ turns out to be very nearly equal to $0.00X$.

If α_b^a in equation 26.13 has a value represented by $1.00X$, then:

$$10^3 \ln 1.00X \simeq X$$

Similarly, $(1.00X - 1) \times 10^3$ is also equal to X. Therefore, to a good approximation,

$$10^3 \ln \alpha_b^a = (\alpha_b^a - 1) \times 10^3 \qquad (26.14)$$

The approximation expressed by equation 26.14 helps to establish a relation between the δ values of phases a and b. Since $\alpha_b^a = R_a/R_b$,

$$\alpha_b^a - 1 = \frac{R_a}{R_b} - 1 = \frac{R_a - R_b}{R_b} \qquad (26.15)$$

Substituting equations 26.10 and 26.11 into equation 26.15 yields:

$$\alpha_b^a - 1 = \frac{(\delta_a + 10^3) - (\delta_b + 10^3)}{\delta_b + 10^3}$$

which reduces to:

$$\alpha_b^a - 1 = \frac{\delta_a - \delta_b}{\delta_b + 10^3} \qquad (26.16)$$

In most cases, $\delta_b \ll 10^3$, which permits the approximation that:

$$\delta_b + 10^3 = 10^3$$

and leads to the useful relation:

$$(\alpha_b^a - 1) \times 10^3 = \delta_a - \delta_b \qquad (26.17)$$

Substituting equation 26.14 for $(\alpha_b^a - 1) \times 10^3$:

$$10^3 \ln \alpha_b^a = \delta_a - \delta_b$$

and from equation 26.13:

$$\delta_a - \delta_b = \frac{A \times 10^6}{T^2} + B \qquad (26.18)$$

This equation indicates that the difference in the δ value of two phases a and b decreases with rising temperature provided that the two phases are in isotopic equilibrium. Equation 26.18 is not exact because of the approximations required in its derivation:

1. $10^3 \ln \alpha_b^a = (\alpha_b^a - 1) \times 10^3$
2. $\delta_b + 10^3 = 10^3$

The exact relation between δ_a and δ_b and the isotopic equilibration temperature is obtained from equations 26.12 and 26.13,

$$\alpha_b^a = \frac{\delta_a + 10^3}{\delta_b + 10^3} \qquad 10^3 \ln \alpha_b^a = \frac{A \times 10^6}{T^2}$$

26.3 METEORIC PRECIPITATION

The difference in the masses of the isotopic molecules of water causes them to have a range of vapor pressures such that the lightest molecule ($^1H_2{}^{16}O$) evaporates preferentially relative to the heaviest water molecule ($D_2{}^{18}O$). (Gat, 1984). Conversely, the heaviest molecule in water vapor condenses preferentially relative to the lightest molecule. Therefore, water vapor is depleted in D and ^{18}O relative to seawater and the liquid condensate that forms from the vapor is enriched in D and ^{18}O relative to the vapor. The resulting fractionation of isotopic water molecules is a direct consequence of the effect of the masses of isotopic molecules on their velocities expressed by equation 26.4.

26.3a Temperature Dependence of Fractionation

The numerical values of the fractionation factors of D and ^{18}O that apply during the evaporation of liquid water to form water vapor define curves in coordinates of temperature in degrees Celsius and $10^3 \ln \alpha_v^l$ (Horita and Wesolowski, 1994). These curves in Figure 26.1 extend from 1 to 350°C and illustrate the point that both fractionation factors decrease with increasing temperature. The equations that relate the fractionation factors of ^{18}O and D to the

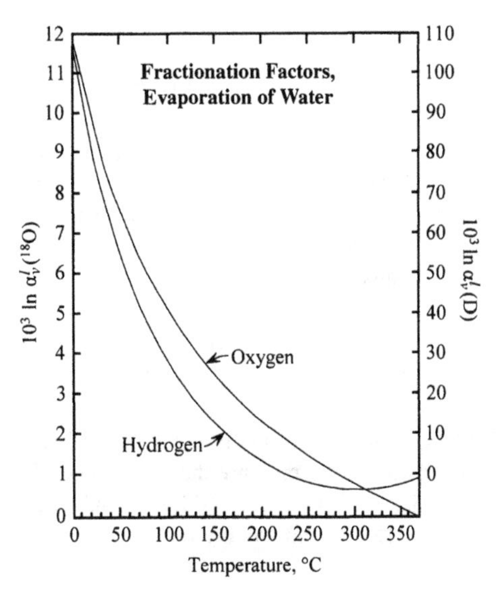

FIGURE 26.1 Fractionation factors for O and H in water molecules evaporating from liquid water to form vapor. Adapted from Horita and Wesolowski (1994).

temperature are:

$$10^3 \ln \alpha_v^l(^{18}O) = -7.685 + \frac{6.7123 \times 10^3}{T} - \frac{1.6664 \times 10^6}{T^2} + \frac{0.35041 \times 10^9}{T^3} \qquad (26.19)$$

$$10^3 \ln \alpha_v^l(D) = \frac{1158.8T^3}{10^9} - \frac{1620.1T^2}{10^6} + \frac{794.84T}{10^3} - 161.04 + \frac{2.9992 \times 10^9}{T^3} \qquad (26.20)$$

where T is expressed in kelvins (K) rather than in degrees Celsius as in Figure 26.1. The fractionation factors for ^{18}O and D at 20°C (293.15 K) calculated from equations 26.19 and 26.20, respectively, are:

$$10^3 \ln \alpha_v^l(^{18}O) = 9.7702$$

$$\alpha_v^l(^{18}O) = 1.0098 \text{ at } 20°C$$

$$10^3 \ln \alpha_v^l(\mathrm{D}) = 80.9855$$

$$\alpha_v^l(\mathrm{D}) = 1.084 \text{ at } 20^\circ\mathrm{C}$$

The fractionation factors of Horita and Wesolowski (1994) can be used to calculate the δ^{18}O and δD values of water vapor that forms by evaporation of water from the surface of the ocean at 20°C by means of equation 26.12:

$$\alpha_v^l(^{18}\mathrm{O}) = \frac{\delta^{18}\mathrm{O}_l + 10^3}{\delta^{18}\mathrm{O}_v + 10^3}$$

The δ^{18}O value of seawater is equal to zero because the standard used to measure the δ^{18}O value of the liquid is itself seawater (i.e., SMOW). Substituting $\alpha_v^l(^{18}\mathrm{O}) = 1.0098$ and solving for $\delta^{18}\mathrm{O}_v$:

$$\delta^{18}\mathrm{O}_v = \frac{10^3}{1.0098} - 10^3 = -9.70‰$$

Similarly, the δD value of water vapor in equilibrium with seawater at 20°C is:

$$\delta\mathrm{D} = \frac{10^3}{1.084} - 10^3 = -77.49‰$$

The results demonstrate that water vapor forming by evaporation at the surface of the ocean at 20°C is depleted in ^{18}O and D relative to SMOW (Craig, 1961). The effect of this process on the isotope composition of water remaining in the ocean is negligible because the ocean is virtually an infinite reservoir of water.

The definition of the isotope fractionation factor requires that the two phases *a* and *b* must be in isotopic equilibrium. When applied to the evaporation of water from the oceans, this condition requires that the rate of evaporation is equal to the rate of condensation, which happens only when the air above the surface of the oceans is saturated with water vapor (i.e., when its relative humidity is 100%). This condition is not achieved in most cases because the movement of the air over the surface of the ocean removes the saturated layer of air and replaces it with air that is not saturated (i.e., its relative humidity is less than 100%). As a result, the rate of evaporation of water is faster than the rate of condensation and the vapor is more depleted in ^{18}O and D than predicted by the equilibrium-based fractionation factors.

The phenomenon just described is an example of the effect of *kinetics* on the fractionation of isotopes which occurs primarily in unidirectional processes such as evaporation of liquids, sublimation of solids, biologically mediated reactions, dissociation reactions, and diffusion (Melander and Saunders, 1980; Hoefs, 1997). The effect of kinetics on isotope fractionation in ideal gases arises because isotopic molecules of low mass have higher velocities than high-mass molecules of the same chemical composition, as shown by equation 26.4.

Moist air masses that form in the equatorial climatic zone by evaporation of seawater move north and east in the northern hemisphere and southwest in the southern hemisphere. As the temperature of the air mass decreases, its relative humidity increases and eventually condensation occurs when the relative humidity reaches 100%. If the δ^{18}O value of the vapor in such an air mass is $-12.0‰$ (enhanced by the kinetic effect) and condensation occurs at 10°C when $\alpha_v^l(^{18}\mathrm{O}) = 1.0107$, then the δ^{18}O of the first drops of rain that form in the air mass can be calculated by means of equation 26.12:

$$\alpha_v^l(^{18}\mathrm{O}) = \frac{\delta^{18}\mathrm{O}_l + 10^3}{\delta^{18}\mathrm{O}_v + 10^3}$$

Solving for $\delta^{18}\mathrm{O}_l$ and substituting appropriate values for α_v^l (^{18}O) and $\delta^{18}\mathrm{O}_v$:

$$\delta^{18}\mathrm{O}_l = \alpha_v^l(^{18}\mathrm{O})(\delta^{18}\mathrm{O}_v + 10^3) - 10^3$$

$$= 1.0107\,(-12.0 + 10^3) - 10^3 = -1.42‰$$

Similarly, the δD value of the first precipitation at 10°C [$\alpha_v^l(\mathrm{D}) = 1.0969$] from an air mass with $\delta\mathrm{D}_v = -95.86‰$ is:

$$\delta\mathrm{D} = 1.0969\,(-95.86 + 10^3) - 10^3 = -8.25‰$$

These calculations support the generalization that the δ^{18}O and δD values of meteoric precipitation are both *negative*, indicating depletion in ^{18}O and D relative to SMOW. However, when compared to the isotope composition of water vapor remaining in the air mass, the condensate is enriched in ^{18}O and D. As a result, the vapor that remains in the air mass is progressively depleted in ^{18}O and D and its $\delta^{18}O$ and δD values become more negative as condensate in the form of rain or snow continues to be removed from it. By the time an air mass has traveled from its source (e.g., in the Caribbean Sea) up the Mississippi–Ohio–St. Lawrence Valleys of North America and has reached Greenland, the $\delta^{18}O$ value of snow forming within it can reach values of −20‰ or less with δD values of −150‰ or less.

26.3b The Rayleigh Equations

The scenario outlined above is quantifiable by means of one of the Rayleigh equations (Rayleigh, 1896). For fractional condensation:

$$R_v = R_v^0 f^{\alpha-1} \qquad (26.21)$$

where R_v = isotope ratio of O or H of water vapor remaining in an air mass in which condensate is forming

R_v^0 = isotope ratio of O or H in water vapor before any condensate has formed

α = isotope fractionation factor of ^{18}O or D defined as $\alpha_v^l = R_l/R_v$

f = fraction of water vapor remaining in the air mass.

The isotope ratio of the condensate (R_l) forming from water vapor at a specified value of f is (Hoefs, 1997):

$$R_l = R_v^0 \alpha_v^l f^{\alpha-1} \qquad (26.22)$$

where all of the variables have the same meaning as in equation 26.21 above.

Equation 26.21 can be restated in the form:

$$\frac{R_v}{R_v^0} = f^{\alpha-1} \qquad (26.23)$$

where the ratio of the R values is equivalent to the fractionation factor α in equations 26.9 and 26.12:

$$\frac{R_a}{R_b} = \alpha_b^a = \frac{\delta_a + 10^3}{\delta_b + 10^3}$$

If the phase a is the vapor in an air mass and phase b is the initial vapor, then equations 26.12 and 26.23 can be combined:

$$\frac{\delta_v + 10^3}{\delta_v^0 + 10^3} = f^{\alpha-1} \qquad (26.24)$$

This equation can be used to calculate the δ values of O and D in the remaining vapor of an air mass for selected values of f:

$$\delta^{18}O_v = (\delta^{18}O_v^0 + 10^3) f^{\alpha-1} - 10^3 \qquad (26.25)$$

where $\alpha = \alpha_v^l(^{18}O)$ at a specified temperature. Similarly,

$$\delta D_v = (\delta D_v^0 + 10^3) f^{\alpha-1} - 10^3 \qquad (26.26)$$

where $\alpha = \alpha_v^l(D)$ at the same temperature specified above.

Equation 26.25 was used to calculate the $\delta^{18}O$ values of the vapor in Figure 26.2 for $\alpha_v^l(^{18}O) = 1.0107$ at $T = 10°C$, $\delta^{18}O_v^0 = -12.0$‰, and f ranging from 1.0 to 0.1. The results demonstrate that the water vapor in the air mass is progressively depleted in ^{18}O as liquid water condenses and is removed from the air mass in the form of rain.

The use of the Rayleigh equation requires the assumption that the air mass is a closed system and that the condensate is removed from it without re-evaporation. In addition, Figure 26.2 is based on the assumption that the temperature remained constant at 10°C.

Equation 26.22 for the isotope ratios of the liquid condensate can be combined with equation 26.12 to express the $\delta^{18}O$ (or δD) value of the liquid condensate:

$$\frac{\delta^{18}O_l + 10^3}{\delta^{18}O_v^0 + 10^3} = \alpha f^{\alpha-1}$$

where $\alpha = \alpha_v^l\ (^{18}O)$ at a specified temperature. Solving for $\delta^{18}O_l$ yields:

$$\delta^{18}O_l = (\delta^{18}O_v^0 + 10^3)\alpha f^{\alpha-1} - 10^3 \qquad (26.27)$$

This equation was used to construct the curve for the liquid condensate in Figure 26.2 at 10°C,

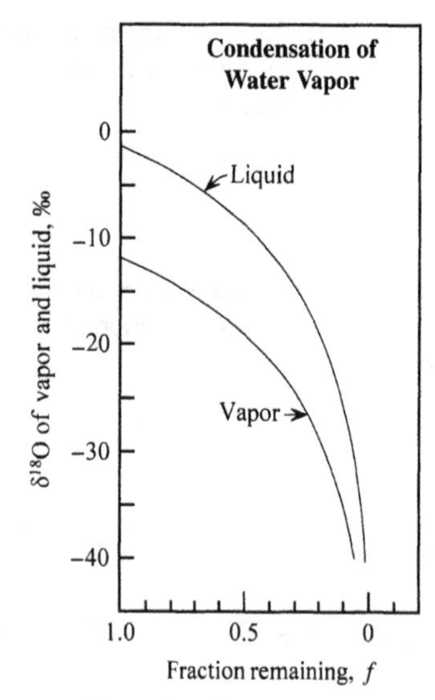

FIGURE 26.2 Effect of progressive condensation of water vapor in an air mass on the $\delta^{18}O$ value of the remaining vapor and on the liquid condensate that forms, based on equations 26.25 and 26.27 for the parameters: $\delta^{18}O^0_v = -12.0‰$, $T = 10°C$, $\alpha^l_v(^{18}O) = 1.0107$ (calculated from equation 26.19).

where $\delta^{18}O^0_v = -12.0‰$, $\alpha^l_v(^{18}O) = 1.0107$, and f ranged from 1 to 0.1 as before.

The results demonstrate that the liquid is always enriched in ^{18}O relative to the vapor at any selected value of f. Nevertheless, the liquid phase is also depleted in ^{18}O relative to SMOW, as indicated by the fact that $\delta^{18}O_l$ becomes progressively more negative from $-1.42‰$ at $f = 1.0$ to $-25.73‰$ at $f = 0.10$. Therefore, the Rayleigh condensation equation explains by numerical calculation why progressive condensation of water vapor in an air mass produces rain (or snow) whose $\delta^{18}O$ values become more negative with increasing northern (or southern) latitude.

The validity of the $\delta^{18}O$ values determined from the Rayleigh equations is limited because the temperatures decrease with increasing latitude, which causes the fractionation factors to increase. In addition, an air mass moving across a continent is *not* a closed system because the remaining water vapor is replenished by evaporation of water from soils, ponds, rivers, and lakes containing meteoric water. In addition, plants release isotopically altered water vapor into the atmosphere. All of these secondary sources alter the isotopic composition of water vapor in air masses passing overhead. As a result, the Rayleigh condensation equation does not reproduce the actual isotopic evolution of water vapor in an airmass and of the rain or snow that may form within it.

26.3c Meteoric-Water Line

The combination of H and O in water molecules means that the isotopes of both elements are subjected to the same process of evaporation and condensation. For this reason, the isotopic delta (δ) values of meteoric precipitation (rain, snow, or hail) from sites all over the Earth form a linear data array represented in Figure 26.3 by a straight line whose equation is (Rozanski et al., 1993):

$$\delta D = (7.96 \pm 0.02)\ \delta^{18}O + 8.86 \pm 0.17 \quad (26.28)$$

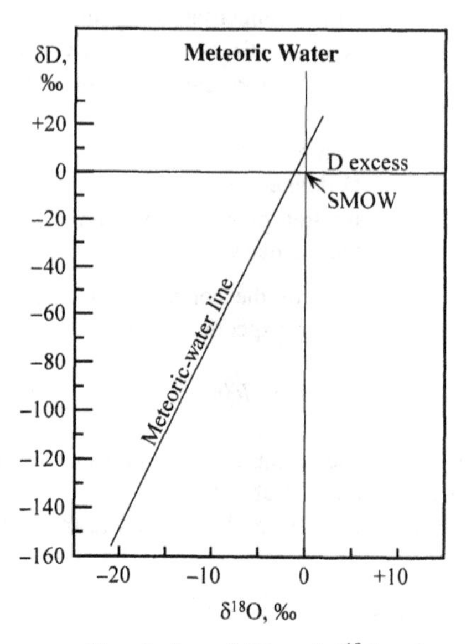

FIGURE 26.3 Correlation of δD and $\delta^{18}O$ values of meteoric precipitation in any form based on monthly averages from all stations of the global network of the International Atomic Energy Agency in Vienna. Adapted from Hoefs (1997) and based on data from Rozanski et al. (1993).

The intercept of the so-called meteoric-water line with the vertical axis is at D = 8.86 ± 0.17‰ rather than at zero as expected. The numerical value of the intercept is called the *deuterium excess* (DE), which tends to vary seasonally on a regional basis. Although the cause for the deuterium excess is not yet understood, it may originate during evaporation of seawater in the source region of certain air masses. Equation 26.28 can be solved explicitly for the deuterium excess:

$$DE = \delta D - 7.96\ \delta^{18}O \qquad (26.29)$$

The δD and $\delta^{18}O$ values of meteoric precipitation become more negative with increasing latitude north and south depending on the terrestrial hemisphere. This phenomenon arises from the combined effect of decreasing temperature and number of rainout events, both of which cause the remaining vapor in air masses to become depleted in D and ^{18}O. Therefore, the δD and $\delta^{18}O$ values of meteoric precipitation reflect the latitude of the site where the water was collected. For example, the $\delta^{18}O$ values of ice in the East Antarctic icesheet at Dome C (~74° S, 120° E) approach −60‰ because of low-temperature isotope fractionation and the long distance between the source of the air masses and the site of deposition in the interior of Antarctica. The δD value of ice whose $\delta^{18}O = -60$‰ is calculated from equation 26.28:

$$\delta D = 7.96 \times (-60) + 8.86 = -469‰$$

Temperature dependence of the annual mean $\delta^{18}O$ values of meteoric precipitation was first demonstrated by Dansgaard (1964), who expressed it by the equation:

$$\delta^{18}O_a = 0.695 T_a - 13.6 \qquad (26.30)$$

where $\delta^{18}O_a$ = the *annual* mean of $\delta^{18}O$ values of meteoric precipitation and
T_a = average annual surface temperature in °C

The relation between $\delta^{18}O$ and temperature was later confirmed by Yurtsever (1975), who used data from four stations in Europe and Greenland to derive the equation:

$$\delta^{18}O_m = (0.521 \pm 0.014) T_m - 14.96 \pm 0.21 \qquad (26.31)$$

where $\delta^{18}O_m$ and T_m are *monthly* averages. In general, the relation between average monthly $\delta^{18}O$ and T are best expressed on a regional basis by curves that tend to converge to a point at $\delta^{18}O_m = 0$ and $T_m = 30$°C (Gat, 1980). The global variation of average annual $\delta^{18}O$ values of meteoric precipitation in Figure 26.4 confirms the progressive depletion of ^{18}O in meteoric precipitation with increasing latitude north and south.

26.3d Climate Records in Ice Cores

The temperature dependence of $\delta^{18}O$ and δD values of snow precipitated in the polar regions is used to interpret the isotope composition of ice cores recovered from the ice sheets in Greenland and Antarctica. In addition, ice cores have also been obtained from high mountains in the Andes of Peru, the highlands of Tibet, and Mt. Kilimanjaro in East Africa. Taken together, these ice cores contain a record of climatic conditions during and subsequent to the Wisconsin (Weichsel) glaciation, not only in the polar regions, but also at midlatitudes in different places of the Earth.

The interpretation of the isotope record in polar ice cores is constrained by several problems that require careful attention:

1. The ages of annual snow layers deposited more than 1000 years ago are difficult to determine because ice is not easy to date by means of cosmogenic radionuclides (e.g., ^{14}C, ^{10}Be, ^{36}Cl).
2. Continental ice sheets flow outward toward their margins under the influence of gravity. Consequently, the ice that is intersected by a vertical core near the margins of an ice sheet formed upslope and then flowed to its present position.
3. The thickness of ice sheets may have increased during glacial ages, causing the average annual temperature to decrease because of the increase in elevation.

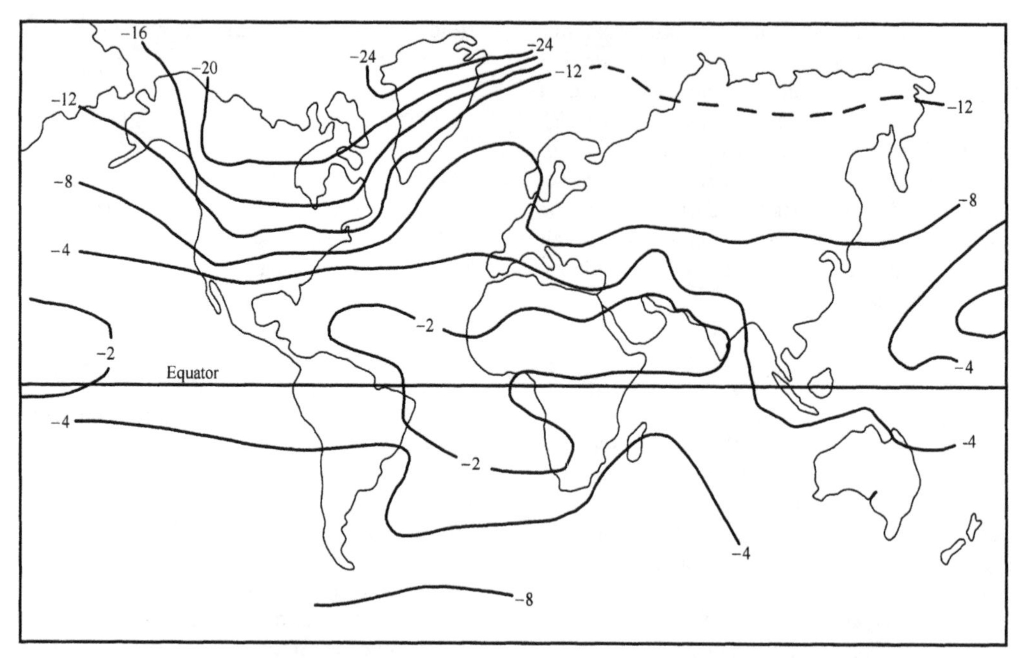

FIGURE 26.4 Contours of average annual $\delta^{18}O$ values of meteoric precipitation at the stations in the global network of the International Atomic Energy Agency. Adapted from Gat (1981) based on a compilation of Yurtsever (1975).

4. The location of the sources of air masses and the tracks they followed to reach the polar regions during glacial ages may have changed, causing the isotope composition of snow to vary in a way that is not directly related to the climatic temperature.
5. The ice may have been folded and thrust faulted during flow of the ice sheet, causing repetition of or discontinuities in the isotopic record.
6. Annual layers of ice become thin with age (i.e., depth below the surface) and are not resolvable. Therefore, samples of constant thickness taken at increasing depth represent increasingly longer intervals with time.
7. The oldest ice at the bottom of ice sheets is subject to melting because of the pressure dependence of the melting temperature and the geothermal heat flow.
8. The flow of the ice may constrict the drill hole and cause the drill stem to get stuck. The fluid used to prevent that from happening may contaminate the ice core.

These potential problems with the interpretation of the isotopic record are enumerated here to emphasize how difficult it is to obtain ice cores in Greenland and Antarctica, where weather conditions are brutal, or in the Andes or Tibet, where the elevations of the drill sites exceed 6000 m (20,000 ft). However, the potential benefits of the information that is derivable from ice cores justifies the effort that is required to obtain them.

Long ice cores have been recovered at Camp Century and at Summit in Greenland, at Byrd Station in West Antarctica, and at Dome C and Vostok on the East Antarctic ice sheet. All of these cores exhibit a pronounced increase of the $\delta^{18}O$ values of ice deposited at the end of the Wisconsin (Weichsel) glaciation about 10,000 to 12,000 years ago. For example, about 10,000 to 12,000 years ago the $\delta^{18}O$ value of ice at Camp Century increased from about -40 to -32‰ during the transition from the late Pleistocene to the Holocene Epoch (Johnsen et al., 1972). The increase of the $\delta^{18}O$ values is attributable to the virtually simultaneous increase in the average annual temperature in Greenland and Antarctica. In addition, detailed examinations of the $\delta^{18}O$ profiles demonstrate that the core at Summit in Greenland and the Vostok core in East Antarctica are closely correlated for up to 250,000 years in the past, thereby confirming the hypothesis that the continental glaciations of the Pleistocene Epoch were global in scope (Dansgaard et al., 1993; Grootes et al., 1993).

The Vostok core was recovered by teams of Russian and French investigators at the Russian research station named Vostok, which is located on the East Antarctic ice sheet at 78°28′ S and 106°48′ E. The elevation of this site is 3488 m above sea level (asl) and the *average* annual temperature is −55.5°C. The ice sheet is nearly 4000 m thick and is underlain by Lake Vostok, a large subglacial lake about the size of Lake Ontario but significantly deeper. The existence of this lake was not known in 1957, when the Vostok research station was established by the former USSR. Lake Vostok has been isolated from the surface of the Earth for several million years and therefore may contain lifeforms that do not occur elsewhere on Earth (Gibbs, 2001).

Drilling started at Vostok in 1970 and continued with interruptions until 1994 (Lorius et al., 1985). The effort was terminated before reaching the bottom of the ice sheet in order to avoid contaminating the lake. The profile of $\delta^{18}O$ values in Figure 26.5 provides a detailed climate record extending for nearly 250,000 years into the past from the Holocene to the start of the Illinoian (Saale) glacial age. According to the timescale of Dansgaard et al. (1993), the Weichsel (Wisconsin) glacial age lasted about 100,000 years. The preceding Eem (Sangoman) interglacial lasted only about 20,000 years, whereas the Saale (Illinoian) glacial age extended for 80,000 years, and the prior Holstein interglacial encompassed between 15,000 and 20,000 years. If this pattern continues in the future, the present Holocene interglacial is already half over.

Another significant observation is that the climate of the Eem glacial age, recorded by the ice core drilled along the summit in central Greenland, fluctuated repeatedly between short intervals that were warmer than the present followed by a return to cold climatic conditions. The transitions from warm to cold conditions occurred in only a few years. In this regard, the climate during the Holocene interglacial has been unusually stable for about 10,000 years, which has contributed to the development of human civilizations and to the rapid rise of the arts and sciences. Evidently, the Earth was not as hospitable to life during the Pleistocene Epoch as it has been since the end of the last global glaciation. If the past is the key to the future, the present warm conditions may end abruptly.

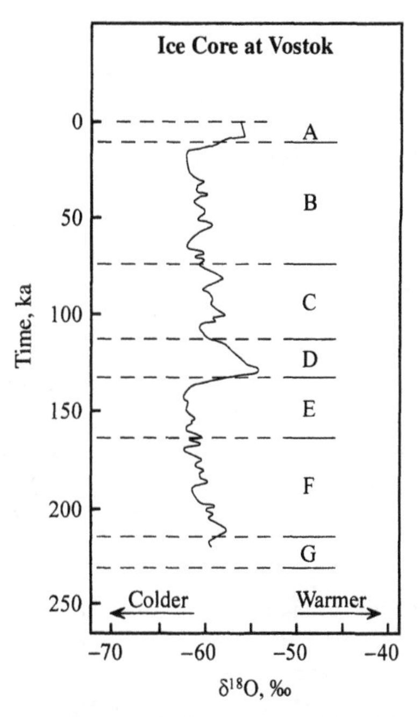

FIGURE 26.5 Oxygen-isotope record of climate change in the ice core recovered at Vostok (78°28′ S, 106°48′ E) in East Antarctica. The geological history is indicated by the glacial and interglacial ages of the Pleistocene Epoch using the European terminology: A = Holocene; B = late Weichsel glacial; C = early Weichsel glacial; D = Eem interglacial; E = late Saale glacial; F = early Saale glacial; G = Holstein interglacial. Adapted from Dansgaard et al. (1993) and Hoefs (1997).

Climate records have also been recovered from ice cores drilled in the ice caps of the Andes and Tibet based on $\delta^{18}O$ values and abundances of dust particles in the ice. For example, the Quelccaya ice cap of Peru is located at 13°56′ S and 71°30′ W at an elevation 5650 m asl where the daily mean temperatures range only from about −5°C in the winter to −3°C in the summer (Thompson et al., 1979). In spite of such mild climatic conditions, the $\delta^{18}O$ values of surface snow on the Quelccaya ice cap vary widely by up to 22‰ and the average $\delta^{18}O$ value at the surface is −21‰. In addition, Thompson et al. (1979) reported that the most negative $\delta^{18}O$ values occur in the *summer* snow rather than in the *winter* snow as in the polar regions.

The extreme ^{18}O-depletion of the snow in the Andes is the result of isotope fractionation during the transport of water vapor from the Atlantic Ocean, across the Amazon basin to the high plateau of the Andes. Additional studies of the *Quelccaya ice cap* were carried out by:

Thompson et al., 1982, *J. Glaciol.*, 28:57–69. Thompson et al., 1984, *Science*, 226: 50–52. Thompson et al., 1985, *Science*, 229: 971–973. Thompson et al., 1986, *Science*, 234:361–364.

The same group of investigators also studied the isotopic climate record in *other* ice caps of the Andes:

Thompson et al., 1988, *Nature,* 336:763–765. Thompson et al., 1995, *Science*, 269:47–50. Thompson et al., 1998, *Science*, 182: 1858–1864.

As well as ice cores from *Tibet:*

Thompson et al., 1989, *Science*, 246:474–477. Thompson et al., 1997, *Science*, 276:1821–1825. Thompson et al., 2000, *Science*, 289: 1916–1919.

26.4 PALEOTHERMOMETRY (CARBONATES)

The fractionation of O isotopes between water and calcium carbonate is temperature sensitive, which prompted H.C. Urey in 1947 to propose that this phenomenon could be used as a paleothermometer. The resulting scientific studies by Professor Urey and his collaborators at the University of Chicago and later at the University of California in Berkeley contributed greatly to the development of an important discipline in the earth sciences based on the fractionation of stable isotopes. The admiration his colleagues felt toward H.C. Urey was eloquently expressed by Craig (1965, p.3): "The estimation of the paleotemperatures of the ancient oceans by measurement of the oxygen isotope distribution between calcium carbonate and water, suggested by H. C. Urey (1947), is one of the most striking and profound achievements of modern nuclear geochemistry." In addition, Craig et al. (1964) edited a book entitled *Isotopic and Cosmic Chemistry* dedicated to Professor Urey on the occasion of his 70th birthday.

26.4a Principles

The isotope exchange reaction between calcium carbonate and water takes the form:

$$\tfrac{1}{3}\mathrm{CaCO_3^{16}} + \mathrm{H_2O^{18}} \rightleftarrows \tfrac{1}{3}\mathrm{CaCO_3^{18}} + \mathrm{H_2O^{16}} \tag{26.32}$$

where the factor $\frac{1}{3}$ is needed to reduce the number of O atoms in one mole of calcium carbonate to the same number present in one mole of water. If the reaction is at equilibrium, the law of mass action applies and is expressed by the equation:

$$K = \frac{[\mathrm{CaCO_3^{18}}]^{\frac{1}{3}}[\mathrm{H_2O^{16}}]}{[\mathrm{CaCO_3^{16}}]^{\frac{1}{3}}[\mathrm{H_2O^{18}}]} \tag{26.33}$$

where K is the equilibrium constant and the quantities of reactants and products are expressed in terms of numbers of moles. Equation 26.33 can be rewritten as:

$$K = \frac{([\mathrm{CaCO_3^{18}}]/[\mathrm{CaCO_3^{16}}])^{\frac{1}{3}}}{[\mathrm{H_2O^{18}}]/[\mathrm{H_2O^{16}}]} \tag{26.34}$$

Therefore, the equilibrium constant is equal to the $^{18}O/^{16}O$ ratio of the carbonate (R_c) divided by the $^{18}O/^{16}O$ ratio of the water (R_w) and has the same form as the isotope fractionation factor α defined by equation 26.9:

$$K = \frac{R_c}{R_w} = \alpha_w^c \tag{26.35}$$

If the O isotopes distribute themselves such that $R_c = R_w$, $\alpha_w^c = 1.0$, which means that no isotope fractionation takes place between calcite and water. In reality, α_w^c is greater than one, and its value varies inversely with the temperature, as originally predicted by Urey (1947).

The isotope composition of O in carbonate samples is measured by mass spectrometry using CO_2 gas released when carbonate samples are treated with 100% phosphoric acid at a controlled temperature (Swart et al., 1991). The procedures for doing this were explained by Hoefs (1997). The original standard used for isotope analysis of O in carbonate samples is a belemnite from the Cretaceous Pee Dee Formation (South Carolina) that was used by H.C. Urey and his collaborators at the University of Chicago in the 1950s. This standard is still known by the abbreviation PDB even though the original material is no longer available. Most investigators now use commercial CO_2 gas that has been calibrated by analysis of carbonate standards prepared and analyzed by the National Bureau of Standards in Washington, DC (Hoefs, 1997):

NBS-19, marble, $\delta^{18}O = -2.20‰$ (PDB)
NBS-20, limestone, $\delta^{18}O = -4.14‰$ (PDB)
NBS-18, carbonatite, $\delta^{18}O = -23.00‰$ (PDB)

The $\delta^{18}O$ values of carbonate samples measured relative to PDB are converted to the SMOW scale by the equation (Coplen et al., 1983):

$$\delta^{18}O_{SMOW} = 1.03091\,\delta^{18}O_{PDB} + 30.91 \quad (26.36a)$$

$$\delta^{18}O_{PDB} = 0.97002\,\delta^{18}O_{SMOW} - 29.98 \quad (26.36b)$$

Both standards continue to be used to express the isotope composition of O in carbonate samples, whereas SMOW is used for all other O-bearing compounds. On the PDB scale, the $\delta^{18}O$ values of marine carbonates are close to zero, whereas lacustrine carbonate samples have negative $\delta^{18}O$ (PDB) values because meteoric water is depleted in ^{18}O relative to seawater.

The temperature dependence of the fractionation factor has been determined by several groups of investigators and has been expressed in different forms (Craig, 1965; Anderson and Arthur, 1983). The most recent version of this equation is by Kim and O'Neil (1997):

$$10^3 \ln \alpha_w^c = \frac{18.03 \times 10^3}{T} - 32.42 \quad (26.37)$$

where T is in kelvins. The equations describing fractionation of ^{18}O between water and the carbonates of other metals were compiled by Chacko et al. (2001).

Equation 26.37 is a straight line in Figure 26.6*a* when points are plotted in coordinates of $(1/T) \times 10^3$ and $10^3 \ln \alpha_w^c$ for values of T between 273.15 and 373.15 K. The straight line has a slope $m = +18.03$ and an intercept on the y-axis of -32.42. This diagram demonstrates how experimentally determined values of α_w^c (^{18}O) were plotted versus the equilibration temperature to obtain the calcite–water thermometry equation. The same data were replotted in Figure 26.6*b* to illustrate the dependence of the isotope fractionation factor α_w^c on the temperature in degrees Celsius.

The precipitation of calcium carbonate (calcite or aragonite) from aqueous solution involves chemical equilibria between CO_2 gas in the atmosphere and carbonate species in water (Faure, 1998). The carbonate species in solution (H_2CO_3, HCO_3^-, and CO_3^{2-}) also fractionate O isotopes by means of isotope-exchange reactions. The applicable fractionation factors in Table 26.2 can be manipulated to determine the fractionation factors for any desired pair of carbonate ions or molecules, including solid calcium carbonate. For example, the O-isotope fractionation between CO_2 gas and aqueous H_2CO_3 (carbonic acid) at 19°C is derivable as follows:

CO_2(gas)–H_2O: $\alpha_w^{CO_2} = 1.04247$

H_2CO_3(aq)–H_2O: $\alpha_w^{H_2CO_3} = 1.03945$

$$\alpha_w^{CO_2} = \frac{R_{CO_2}}{R_w} \qquad \alpha_w^{H_2CO_3} = \frac{R_{H_2CO_3}}{R_w}$$

Therefore,

$$\alpha_{H_2CO_3}^{CO_2} = \frac{R_{CO_2}/R_w}{R_{H_2CO_3}/R_w} = \frac{\alpha_w^{CO_2}}{\alpha_w^{H_2CO_3}} = \frac{1.04247}{1.03945} = 1.00290 \text{ at } 19°C$$

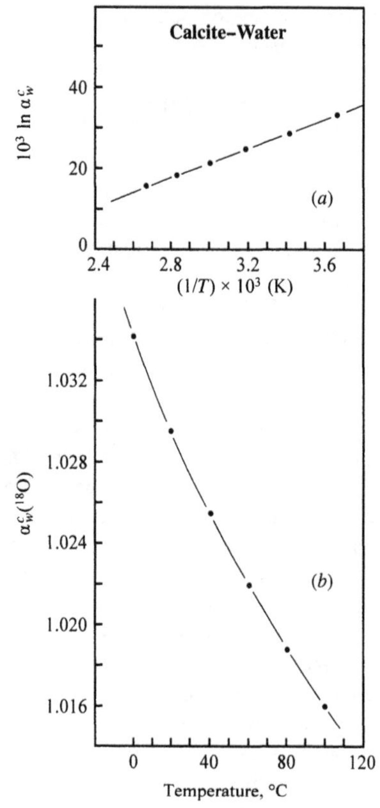

FIGURE 26.6 (*a*) Variations of $10^3 \ln \alpha_w^c$ (^{18}O) (c = calcite, w = water) as a function of the reciprocal temperature on the Kelvin scale based on equation 26.36b by Kim and O'Neil (1997). (*b*) Variation of the O-isotope fractionation factor α_w^c (^{18}O) as a function of the temperature on the Celsius scale based on equation 26.36b by Kim and O'Neil (1997).

The $\delta^{18}O$ value of atmospheric CO_2 is +41‰ (Hoefs, 1997), from which $\delta^{18}O$ of aqueous H_2CO_3 can be calculated using equation 26.12:

$$\alpha_{H_2CO_3}^{CO_2} = \frac{\delta^{18}O_{CO_2} + 10^3}{\delta^{18}O_{H_2CO_3} + 10^3}$$

Solving for $\delta^{18}O_{H_2CO_3}$ and substituting appropriate values for $\alpha_{H_2CO_3}^{CO_2}$ and $\delta^{18}O_{CO_2}$:

$$\delta^{18}O_{H_2CO_3} = \frac{41 + 10^3}{1.0029} - 10^3 = +37.9‰$$

Table 26.2. Fractionation Factors of O Isotopes in the System CO_2–H_2O Relative to Water at 19°C

Ion or Molecule	$10^3 \ln \alpha_w^i$ [a]
CO_2 (gas)	41.6
CO_2 (aq)	56.3
CO_3^{2-}	18.2
HCO_3^-	34.5
H_2CO_3	38.7
$CaCO_3$ [b]	29.29

Source: Usdowski and Hoefs, 1993.

[a] Any of the ions, molecules, and solid calcium carbonate in the table.

[b] Calculated from equation 26.36b at 292.15 K (19°C) by Kim and O'Neil (1997).

Similarly, the O-isotope fractionation factor between CO_2 gas and calcium carbonate is:

$$\alpha_{cal}^{CO_2} = \frac{1.04247}{1.02972} = 1.01238$$

$$\delta^{18}O_{cal} = \frac{\delta^{18}O_{CO_2} + 10^3}{\alpha_{cal}^{CO_2}} - 10^3$$

$$= \frac{41 + 10^3}{1.01238} - 10^3 = +28.2‰$$

This result indicates that calcium carbonate in isotopic equilibrium with atmospheric CO_2 at 19°C has a $\delta^{18}O$ value of +28.2‰ on the SMOW scale.

26.4b Assumptions

Paleothermometry, based on the temperature dependence of the fractionation of O isotopes between marine calcium carbonate and water, must satisfy certain assumptions concerning:

1. the existence of isotopic equilibrium between O in the water and in biogenic calcium carbonate,
2. the preservation of the isotope composition of O in the solid carbonate, and
3. the constancy of the isotope composition of water in the oceans.

Similar assumptions apply to lacustrine and pedogenic carbonates as well as to carbonates deposited by hotsprings and groundwater.

Paleothermometry was originally intended to measure the temperature of seawater in which different kinds of organisms precipitate calcium carbonate skeletons (e.g., mollusks, corals) or cause the deposition of this compound as a consequence of photosynthesis (e.g., algal stromatolites). Unfortunately, evidence presented by Weber and Raup (1966a, b) and Weber (1968) indicated that many kinds of organisms do *not* satisfy the assumptions because of various *"vital" effects* discussed by McConnaughey (1989a, b; Adkins et al., 2003).

The deviation of $\delta^{18}O$ values of marine organisms from the equilibrium values are indicated qualitatively in Table 26.3 based on the work of Wefer and Berger (1991). For example, the O of carbonate shells of planktonic and benthic foraminifera is in isotopic equilibrium with seawater. However, several of the larger foraminifera do not maintain isotopic equilibrium (e.g., *Cyclorbiculina compressa, Archaias angulatus, Marginopora vertebralis*, and five other species). Hermatypic scleractinian corals are not in isotopic equilibrium, but ahermatypic scleractinian corals do fractionate O in equilibrium with seawater. The result is that only certain genera and species listed in Table 26.3 satisfy the assumption concerning isotopic equilibrium of O whereas many others do not. In addition, the isotope composition of C of biogenic calcite is not in equilibrium with C in seawater in most cases.

Table 26.3. Marine Organisms That Do/Do Not Secrete Calcium Carbonate in Isotopic Equilibrium with Seawater

Isotopic Equilibrium	Isotopic Disequilibrium
Foraminifera	Benthic algae
Corals (ahermatypic scleractinia)	Corals (hermatypic scleractinia)
Polychaeta	Arthropoda
Bryozoa	Echinodea[a] (Asteroidea, Holothuroidea)
Brachiopoda	
Mollusca	
Polyplachophora	
Archaeogastropoda	
Caenogastropoda	
Pteropoda	
Scaphopoda	
Pelecypoda (with exceptions)	
Cephalopoda (Nautiloidea, Endocochlia)	
Echinoderma (Echinoidea)	
Fish (otoliths)	

Source: Wefer and Berger, 1991.

[a]From Weber and Raup (1966a, b) and Weber (1968).

Skeletal calcium carbonates of marine biota are deposited in layers of sediment and may subsequently be dissolved and/or replaced by calcium carbonate that equilibrated at a different temperature with water having a different isotope composition of O. In addition, foraminiferal shells may contain crystals of diagenetic calcite whose O-isotope composition differs from that of the shell. This problem also affects measurements of the isotope composition of Sr in marine carbonate fossils (Sections 19.1a and 19.1b).

The third assumption that must be met has to do with the isotope composition of O in seawater. The problem is that cooling of the global climate during the Pleistocene Epoch caused large continental ice sheets to form in the northern hemisphere as well as in Patagonia (South America) and Antarctica. These ice sheets were strongly depleted in ^{18}O compared to seawater and were large enough to cause a significant increase of the $\delta^{18}O$ value of the oceans. At the present time, the oceans have a mean $\delta^{18}O$ value of $-0.08‰$ (Craig, 1965). If all of the present ice sheets were to melt, the $\delta^{18}O$ value of seawater would decrease to $-0.60‰$. During the time of maximum glaciation during the Pleistocene Epoch, the $\delta^{18}O$ value of seawater was $+0.90‰$. Therefore, climate oscillations ranging between maximum and zero glaciation can cause the $\delta^{18}O$ value of seawater to change by $1.50‰$ from $+0.90‰$ (maximum glaciation) to $-0.60‰$ (zero glaciation). These estimates by Craig (1965) were subsequently confirmed by Shackleton (1968) and Emiliani and Shackleton (1974).

The problem is even worse because the $\delta^{18}O$ value of seawater at any given time varies both regionally and locally as a function of salinity as a result of mixing with freshwater (e.g., discharge of water by major rivers and melting of icebergs) and because of excessive evaporation (e.g., in the Red Sea). The salinity is also increased by freezing to form sea ice, which increases the density of seawater and causes it to sink, thereby forming a layer of cold and dense water at the bottom of the ocean in the polar regions (e.g., Antarctic bottom water). The O-isotope fractionation factor for ice-liquid has a value of (Lehmann and Siegenthaler, 1991):

$$\alpha_l^{ice} = 1.00291 \pm 0.00003$$

Therefore, ice is enriched in ^{18}O relative to the water from which it formed. If $\delta^{18}O_l = 0.0‰$, the $\delta^{18}O$ value of ice in isotopic equilibrium is (equations 26.12):

$$\delta^{18}O_{ice} = 1.00291(0 + 10^3) - 10^3 = +2.91‰$$

Therefore, freezing tends to deplete seawater in ^{18}O, but the effect is small because of the large volume of water in the oceans. The freezing temperature of seawater is controlled by its salinity and by the pressure that is exerted on it. For these reasons, the isotope composition of O in seawater is not constant throughout the oceans but varies from place to place in response to local conditions and depending on the global circulation of water in the oceans.

26.4c Oxygen-Isotope Stratigraphy

These difficulties have prevented the marine calcite paleothermometer from being realized as originally conceived by Professor Urey and his associates. Instead, Emiliani (1955, 1966, 1972, 1978) and Emiliani and Shackleton (1974) constructed a $\delta^{18}O$ profile based on the shells of a species of pelagic foraminifera (*Globigerina sacculifer*) taken from deep-sea sediment cores that were dated by the U-series disequilibrium methods (Section 20.1). The resulting record in Figure 26.7 indicates that the $\delta^{18}O$ value of this foraminifer varied cyclically from about +0.5‰ (cold) to −1.5‰ (warm) on the PDB scale. Emiliani (1955) designated identifiable events in the $\delta^{18}O$ record by means of numbers that represent different *isotopic stages*. Even numbers identify cold climatic conditions, whereas odd numbers identify warm periods. The record in Figure 26.7 contains ten warm (interglacial) isotopic stages and ten cold (glacial) ones in the past 750,000 years. In addition, the profile of $\delta^{18}O$ values of *G. sacculifer* also contains the variations of the isotope composition of O in seawater caused by the waxing and waning of the continental ice sheets during the Pleistocene epoch.

The principal attributes of the O-isotope record of foraminifera in Figure 26.7 were summarized by Hoefs (1997):

1. The amplitude of the $\delta^{18}O$ cycles is restricted to a narrow range, which suggests that feedback mechanisms are preventing extreme climate changes.

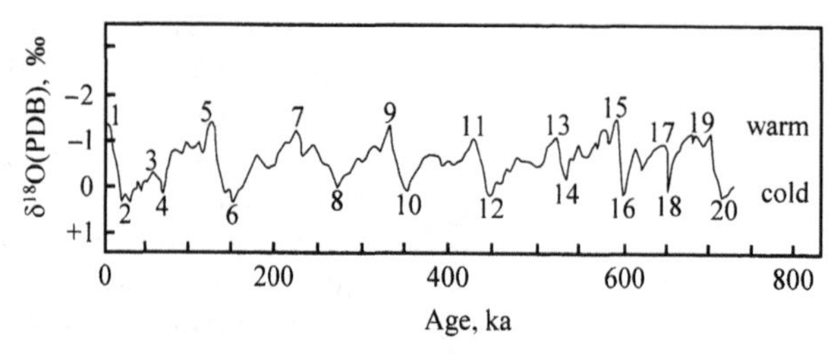

FIGURE 26.7 Oxygen-isotope profile of the shells of the pelagic foraminfera *Globigerina sacculifer* from the present to 750 ka. The $\delta^{18}O$ values on the PDB scale indicate the occurrence of warm and cold climatic conditions. The isotopic stages are numbered such that warm intervals have odd numbers and cold intervals have even numbers. Adapted from Hoefs (1997) after Emiliani (1978).

2. The change from "cold" to "warm" conditions occurred rapidly, whereas maximum warming was followed by gradual cooling.
3. The average period of the $\delta^{18}O$ cycles is 75,000 years (or about 10^5 years), which may have been imposed by changes in the insolation of the Earth caused by celestial mechanics, in accordance with the Milankovitch Theory (Hays et al., 1976).
4. The $\delta^{18}O$ profile is a composite of data derived from different parts of the ocean, which means that it can be used to correlate marine sediment cores separated by large distances.

The O-isotope timescale for calcium carbonate in the oceans in Figure 26.7 coincides with the Brunhes normal magnetic polarity epoch, which started at 0.72 Ma (Section 6.4b, Table 6.3).

The isotope composition of O in older marine carbonates (Pleistocene to Late Cretaceous) was determined by Raymo and Ruddiman (1992) by analysis of benthic foraminifera in DSDP cores recovered from the Atlantic Ocean. Benthic foraminifera live in deep water at the sediment–water interface where the temperature is less variable than in the near-surface layers of the oceans. Nevertheless, the $\delta^{18}O$ values of benthic foraminifera in Figure 26.8 increased by several parts per thousand, from about −0.2‰ at 70 Ma to +4.0‰ at about 1.0 Ma. The progressive enrichment in ^{18}O of the foraminiferal shells indicates that the O-isotope fractionation factor for calcite–water increased in response to *decreasing* average global temperatures. The sharp increase of $\delta^{18}O$ in early Oligocene time may have been caused by the onset of continental glaciation in East Antarctica at 35 Ma. Subsequent increases of the $\delta^{18}O$ of the benthic foraminifera may record the growing volume of the East Antarctic ice sheet and the growth of continental ice sheets in the northern hemisphere during the Pleistocene Epoch starting at about 2.0 Ma (e.g., Harland et al., 1989).

Profiles of $\delta^{18}O$ values of well-dated marine carbonates not only record climate changes in the past but can also be used as geochronometers to date marine carbonates of unknown age. This application of the $\delta^{18}O$ values of the shells of benthic foraminifera of Tertiary age complements

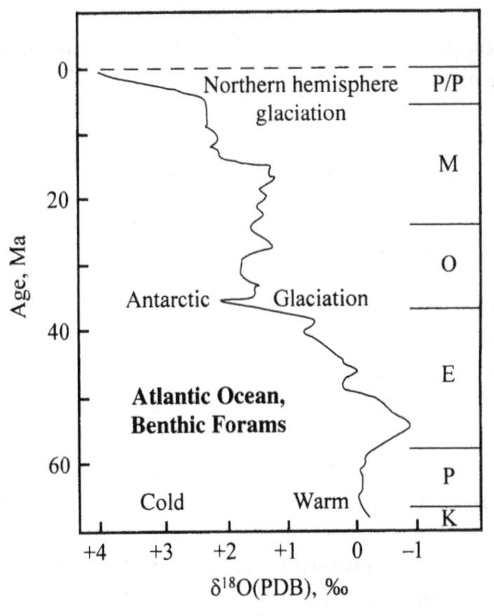

FIGURE 26.8 Variation of $\delta^{18}O$ (PDB) of the shells of benthic foraminifera during the Tertiary Period in the Atlantic Ocean: P/P = Plio-Pleistocene; M = Miocene; O = Oligocene; E = Eocene; P = Paleocene; K = Cretaceous. Adapted from Raymo and Ruddiman (1992).

the common-Sr geochronometer (Section 19.1d, Figures 19.4 and 19.5).

26.5 SILICATE MINERALS AND ROCKS

The isotope composition of O in silicate minerals is expressed by $\delta^{18}O$ values on the SMOW scale as defined by equation 26.7. The O of silicate (and oxide) minerals is liberated by fluorination of powdered samples using F_2, BrF_5 (bromine pentafluoride), or ClF_3 (chlorine trifluoride). The reaction is carried out in Ni tubes at 500–600°C in well-ventilated fume hoods because fluorine gas is highly toxic (Clayton and Mayeda, 1963; Borthwick and Harmon, 1982). The O released by fluorination is reacted with hot graphite to form CO_2, which is analyzed by mass spectrometry (Hoefs, 1997). In addition, O can be released from small spots within single crystals by laser microprobes described by Sharp (1990). This technique was used by Valley and Graham (1993) to reveal the isotopic heterogeneity of O in crystals of magnetite in metamorphic rocks.

26.5a Basalt and the Mantle

Silicate minerals are enriched in ^{18}O relative to SMOW and have positive $\delta^{18}O$ values that range from +20‰ in quartz to values between +5.0 and +6.0‰ in ferromagnesian minerals such as olivine and pyroxene. Therefore, the $\delta^{18}O$ values of mafic volcanic rocks derived from the upper mantle and listed in Table 26.4 are also clustered in this range (Harmon and Hoefs, 1995). Basalts erupted along midocean ridges (MORBs) on oceanic islands (OIBs) and continental flood basalts have comparatively low $\delta^{18}O$ values with a weighted average of +5.59‰ and a range of 4.3–7.5‰ for 292 specimens. The average $\delta^{18}O$ value of these basalts is representative of O in the lithospheric mantle from which basalt magma originates by decompression melting (Faure, 2001).

Table 26.4. Average $\delta^{18}O$ (SMOW) Values of Basalts Erupted in Different Tectonic Settings

Tectonic Setting	Range	$\delta^{18}O$ (SMOW), ‰, Average $\pm 1\sigma$
Midocean ridge basalt (MORB)	5.2–6.4	$+5.73 \pm 0.21$
Oceanic island basalt (OIB)	4.6–7.5	$+5.48 \pm 0.51$
Iceland basalt	2.9–6.2	$+4.50 \pm 0.81$
Oceanic subduction, basalt and andesite	5.3–7.5	$+6.10 \pm 1.10$
Continental subduction, basalt and andesite	4.8–7.7	$+6.24 \pm 0.65$
Continental flood basalt	4.3–6.5	$+5.59 \pm 0.64$
Alkali-rich basalt, Italy	6.3–11.4	$+8.47 \pm 1.44$

Source: Harmon and Hoefs, 1995.

Volcanic rocks in oceanic and continental *subduction zones* and *alkali-rich* basalts in Italy have distinctly higher average $\delta^{18}O$ values than MORBs, OIBs, and flood basalts, presumably because volcanic rocks in subduction zones contain O derived from marine sediment and the alkali-rich rocks of Italy formed in part by partial melting of subducted K-rich oceanic crust (Section 17.9; Faure, 2001). The basalts of Iceland have a wide range of $\delta^{18}O$ values ranging from +2.9 to +6.2‰ (Harmon and Hoefs, 1995). The *low* $\delta^{18}O$ values of some Icelandic basalts are attributable to the contamination of basalt magmas by assimilation of hydrothermally altered volcanic rocks at depth within the volcanic edifice (Condomines et al., 1983; Hemond et al., 1988, 1993; Nicholson et al., 1991; Sigmarsson et al., 1992). The ^{18}O enrichment of certain Icelandic basalts relative to SMOW (e.g., $\delta^{18}O$ = +6.2‰) is a feature these lavas share with basalt erupted on certain oceanic islands and submarine basalts.

For example, Woodhead et al. (1993) reported high $\delta^{18}O$ values in basalt from submarine volcanoes associated with Pitcairn Island and with the Society Islands in the South Pacific Ocean:

Pitcairn seamounts: +5.8 to +7.4‰
Society Islands: +6.0 to +6.7‰

The high $\delta^{18}O$ values of these basalts correlate with high $^{87}Sr/^{86}Sr$ and low $^{143}Nd/^{144}Nd$ ratios, indicating that the basalt magmas originated from mixed sources containing a crustal component (high $\delta^{18}O$, high $^{87}Sr/^{86}Sr$, low $^{143}Nd/^{144}Nd$).

This interpretation supports a proposal by Hofmann and White (1982) and Hofmann (1997) that mantle plumes form from subducted oceanic crust. The $\delta^{18}O$ values of marine sediment and of volcanic rocks in the upper oceanic crust in Table 26.5 range widely from +7.0 to +42‰, indicating that they are enriched in ^{18}O relative to normal MORBs and OIBs (Eiler, 2001). Only hydrothermally altered volcanic rocks in the lower part of the oceanic crust have low $\delta^{18}O$ values (0–6‰). Partial melting of such rocks, or their assimilation by mantle-derived magmas, causes the resulting basalts to have anomalously low $\delta^{18}O$ values, as exemplified by some basalts of Iceland (e.g., the volcano Krafla; Nicholson et al., 1991).

Table 26.5. Rocks and Minerals in the Ocean Basins of the Earth That May Be Subducted into the Mantle and Ultimately Contribute to the Formation of Basalt Magma

Material	$\delta^{18}O$ (SMOW)‰
Marine carbonate, Phanerozoic age	25–32[a]
Siliceous ooze	35–42
Pelagic clay	15–25
Upper oceanic crust, hydrothermally altered	7–15
Lower oceanic crust, hydrothermally altered	0–6

Source: Eiler, 2001.

[a] All $\delta^{18}O$ values are positive.

The isotope compositions of O in basalt, as well as those of Sr, Nd, Pb, and other elements, support the plume theory of the mantle because the petrogenesis of igneous rocks can be explained as a consequence of the upward movement of buoyant plumes (Section 17.1). In fact, the forces exerted by plumes against the underside of the suboceanic and subcontinental lithospheric mantle activate plate tectonics, continental drift, and sea-floor spreading. In addition, they cause volcanic activity, uplift and rifting of continents, and earthquakes. Plumes are the engine that makes the Earth work the way it has in the past and will continue to do in the future (Faure, 2001).

26.5b Thermometry of Silicates and Oxides

The rock-forming silicates and oxide minerals fractionate O isotopes when they crystallize from silicate melts or form by recrystallization of volcano-sedimentary rocks during regional metamorphism. The extent of fractionation of O isotopes between silicate minerals and the O-reservoir (e.g., the magma or an aqueous pore fluid) is comparatively small because the minerals form at elevated temperatures. Nevertheless, thermometry equations have been formulated for a large number of silicate and oxide minerals based primarily on the results of mineral synthesis experiments amplified by theoretical models (Kieffer, 1982). A great deal of work on O-isotope thermometry has been carried out by R. N. Clayton and his research group at the University of Chicago:

Anderson et al., 1971, *J. Geol.*, 79:715–729. Clayton et al., 1972, *Proc. 3rd Lunar and Planetary Sci. Conf.*, Houston, Texas. 1455–1463. Matsuhisa et al., 1978, *Geochim. Cosmochim. Acta*, 42:173–182. Matsuhisa et al., 1979, *Geochim. Cosmochim. Acta*, 43:1131–1140. Matthews et al., 1983a, *Geochim. Cosmochim. Acta*, 47:631–644. Matthews et al., 1983b, *Geochim. Cosmochim. Acta*, 47:645–654. Chiba et al., 1989, *Geochim. Cosmochim. Acta*, 53:2985–2995. Clayton et al., 1989, *Geochim. Cosmochim. Acta*, 53:725–733. Karlsson and Clayton, 1990, *Geochim Cosmochim. Acta*, 54:1359–1368. Clayton and Kieffer, 1991, in Taylor et al. (Eds.), *Stable Isotope Geochemistry*, Geochem. Soc. Spec. Paper No. 3, pp. 3–10. Chacko et al., 1996, *Geochim. Cosmochim. Acta*, 60:2595–2608. Chacko et al., 2001, in Valley and Cole (Eds.), *Stable Isotope Geochemistry*, Reviews of Mineralogy and Geochemistry, Virginia Polytechnic Institute. 43:1–81. Blacksburg, VA.

The resulting thermometry equations of these and other authors were compiled in Appendix 2 of the paper by Chacko et al. (2001) cited above.

The temperature dependence of the isotope fractionation factors of O in silicate and oxide minerals is determined experimentally by equilibrating the minerals with O in water or in calcium carbonate at controlled temperatures. The resulting measured fractionation factors and corresponding temperatures are regressed to equations of the form:

$$10^3 \ln \alpha_w^{\text{sil}} = \frac{A \times 10^6}{T^2} + B$$

where the temperature T is in kelvins and the values of A and B are determined by the statistical procedure. The thermometry equations of two coexisting minerals both of which equilibrated O with the same reservoir can be combined by redefining the fractionation factor in terms of the $^{18}O/^{16}O$ ratios of the minerals a and b.

Table 26.6. Values of the Coefficients (A) for Pairs of Minerals Arranged to Make A Positive in the equation: $10^3 \ln \alpha_b^a = A \times 10^6/T^2$

Mineral "a"	Minerals "b"					
	Cal	Ab	An	Di	Fo	Mt
Qtz	0.38	0.94	1.99	2.75	3.67	6.29
Cal		0.56	1.61	2.37	3.29	5.91
Ab			1.05	1.81	2.73	5.35
An				0.76	1.68	4.30
Di					0.92	3.54
Fo						2.62

Source: Chiba et al., 1989.
Note: Qtz = quartz, Cal = calcite, Ab = albite, An = anorthite, Di = diopside, Fo = forsterite.

For example, Chiba et al. (1989) published a set of values of A for pairs of minerals listed in Table 26.6 based on the equation:

$$10^3 \ln \alpha_b^a = \frac{A \times 10^6}{T^2} \quad (26.38)$$

where T is the temperature in kelvins and the minerals in each pair are arranged in such a way that A is a positive number. The coefficient B is set equal to zero by forcing the regression lines to include the origin in coordinates of $10^3 \ln \alpha_b^a$ and $10^6/T^2$.

For example, the data in Table 26.6 indicate that the fractionation of O isotopes between albite (ab) and diopside (di) is described by the equation:

$$10^3 \ln \alpha_{\text{di}}^{\text{ab}} = \frac{1.81 \times 10^6}{T^2} \quad (26.39)$$

Therefore, at a temperature of 1000 K (726.85°C),

$$\ln \alpha_{\text{di}}^{\text{ab}} = \frac{1.81 \times 10^6}{10^6 \times 10^3} = 1.81 \times 10^{-3}$$

$$\alpha_{\text{di}}^{\text{ab}} = 1.00181$$

The albite–diopside O-isotope fractionation factors in Figure 26.9 decrease with increasing temperature from 1.00237 at 600°C to 1.00111 at 1000°C.

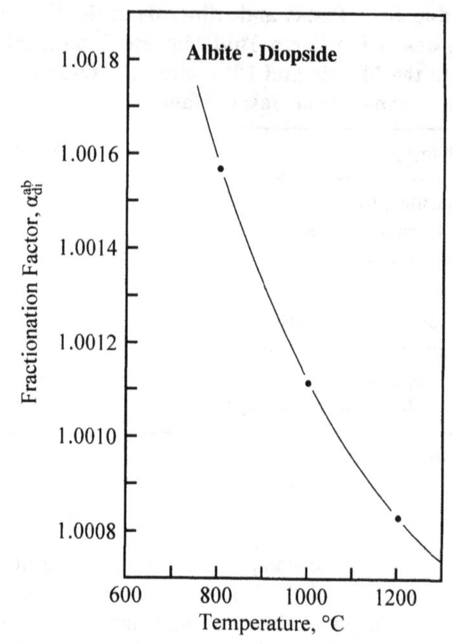

FIGURE 26.9 Oxygen-isotope fractionation factors for coexisting albite and diopside in isotopic equilibrium as a function of the temperature expressed here on the Celsius scale. The equation relating $\alpha_{\text{di}}^{\text{ab}}$ and T (on the Kelvin scale) is: $10^3 \ln \alpha_{\text{di}}^{\text{ab}} = (1.81 \times 10^6)/T^2$ (Chiba et al., 1989).

The O-isotope equilibration temperature of coexisting albite and diopside is determined from measurements of their $\delta^{18}O$ values from which the fractionation factor is calculated using equation 26.12:

$$\alpha_{\text{di}}^{\text{ab}} = \frac{\delta^{18}O_{\text{ab}} + 10^3}{\delta^{18}O_{\text{di}} + 10^3}$$

The resulting value of $\alpha_{\text{di}}^{\text{ab}}$ is then used to solve equation 26.39 for the temperature T (in kelvins).

The result is *equal to or less* than the temperature of crystallization because the minerals may have continued to exchange O isotopes while the rock was cooling. In addition, the validity of the O-isotope temperature depends on the assumption that the minerals were in equilibrium when isotope exchange stopped and that the isotope composition of O in the minerals was not altered subsequently by exchange with O in aqueous fluids from external sources. Such isotopic reequilibration can occur in the border zones of igneous plutons that intruded permeable country rocks in the continental crust (e.g., the Skaergaard Intrusion of East Greenland and Tertiary plutons in Oregon and Washington):

Criss and Taylor, 1986, in Valley et al. (eds.), *Stable Isotopes in High Temperatures Geological Processes*, Rev. Mineral., 16:373–424. Mineral. Soc. Amer., V.P.I., Blacksburg, VA. Gregory and Criss, 1986, in Valley et al. (eds.), *Stable Isotopes in High Temperature Geological Processes*, Rev. Mineral., 16:91–127. Mineral. Soc. Amer., V.P.I., Blacksburg, VA. Criss et al., 1987, *Geochim. Cosmochim. Acta*, 51:1099–1108. Criss et al., 1991, *J. Geophys. Res.*, 96:13335–13356. Eiler et al., 1992, *Contrib. Mineral. Petrol.*, 112:543–557. Eiler et al., 1993, *Geochim. Cosmochim. Acta*, 57: 2571–2583. Eiler et al., 1995, *Contrib. Mineral. Petrol.*, 118:365–378.

The O-isotope temperatures derived from mineral pairs in unaltered terrestrial volcanic rocks and lunar basalts represent the crystallization temperatures because these kinds of rocks cooled rapidly. For example, Chiba et al. (1989) cited $\delta^{18}O$ values of plagioclase, pyroxene, olivine, and magnetite in lunar basalt analyzed by Onuma et al. (1970) and Clayton et al. (1971, 1972). These data yielded plagioclase–magnetite and pyroxene–magnetite temperatures between 1054 and 1252°C, which are similar to temperatures of 1078–1222°C obtained for terrestrial basalt from the same mineral pairs (Garlick, 1966; Anderson et al., 1971).

In contrast to volcanic rocks, mafic and felsic plutonic rocks as well as granulites (high-grade metamorphic rocks) yield O-isotope temperatures that appear to be 50–200°C below their crystallization temperatures (Chiba et al., 1989). Part of the reason for the discrepancy is that plutonic rocks cool slowly and allow the minerals to exchange O until they reach temperatures between 800 and 920°C in the case of mafic plutonic rocks. Felsic plutonic rocks have still lower O-isotope temperatures between 640 and 660°C and granulites register temperatures of only 480–650°C.

The $\delta^{18}O$ values of minerals in the lunar basalt 10020 (Onuma et al., 1970) in Table 26.7 illustrate the calculation of the crystallization temperature of this rock by means of the plagioclase–magnetite

Table 26.7. $\delta^{18}O$ (SMOW) Values of Minerals of Selected Terrestrial and Extraterrestrial Rocks

	$\delta^{18}O$ (SMOW), ‰				
Rock Type	Qtz	Pl	Py	Ol	Mt
Alkali basalt, Maui	–	5.56 (81)	5.06	4.75	3.47
Andesite, Japan	–	7.21 (60)	6.25	–	4.23
Dacite, California	8.2	6.9 (50)	–	–	4.23
Rhyolite, Arizona	9.8	8.7 (10)	–	–	3.6
Lunar basalt, 10020	–	6.19 (85)	5.74	5.14	4.21
Lunar basalt, 10044	7.16	6.06 (90)	5.67	–	3.65
Gabbro, Skaergaard	–	6.80 (40)	5.7	–	3.1
Anorthosite, Nain	–	7.0 (45)	5.9	–	3.4
Granite, Adirondacks	11.0	–	–	–	4.3
Granulite, Musgrave Range	9.25	7.35 (35)	6.6	–	−0.15

Source: Chiba et al., 1989.
Note: The anorthite concentration in plagioclase is given in parentheses. Abbreviations: Qtz = quartz, Pl = plagioclase, Py = pyroxene, Ol = olivine, Mt = magnetite. All $\delta^{18}O$ values are positive unless otherwise indicated.

thermometer:

$$\alpha_{\text{mt}}^{\text{pl}} = \frac{6.19 + 10^3}{4.21 + 10^3} = 1.0019716$$

$$10^3 \ln \alpha_{\text{mt}}^{\text{pl}} = 1.9696$$

The value of coefficient A in the plagioclase is obtained from the concentrations of albite and anorthite in Table 26.7 and their A values in Table 26.6: An (85%, $A = 4.30$), Ab (15%, $A = 5.35$). Therefore, the value of A for this plagioclase–magnetite pair is:

$$A_{\text{pl}} = 0.15 \times 5.35 + 0.85 \times 4.30 = 4.457$$

Substituting into equation 26.38 yields

$$10^3 \ln \alpha_{\text{mt}}^{\text{pl}} = \frac{A \times 10^6}{T^2}$$

$$1.9696 = \frac{4.457 \times 10^6}{T^2}$$

$$T^2 = 2.2628 \times 10^6$$

$$T = \begin{cases} 1.504 \times 10^3\ \text{K} \\ 1504 - 273 = 1231°\text{C} \end{cases}$$

The O-isotope temperature derived from pairs of minerals in lunar basalt 10020 are (Chiba et al., 1989): Py–Ol, 969°C; Pl–Mt, 1231°C; Py–Mt, 1252°C; and Ol–Mt, 1409°C.

26.6 WATER–ROCK INTERACTIONS (ROCKS)

When silicate or carbonate rocks are in contact with meteoric water in the subsurface, O-isotope exchange reactions take place (Criss and Taylor, 1986; Gregory and Criss, 1986; Criss et al., 1987). As a result, the water is enriched in ^{18}O while the rocks are depleted in this isotope. The effect of such isotope exchange reactions on the isotope composition of O of the rocks and the water depends on the temperature and the water–rock ratio (Section 16.5). At elevated temperatures the effective solid–liquid fractionation factor approaches unity, which means that the difference between the δ^{18}O of the rocks and the water approaches zero. In cases where water percolates through permeable rocks at elevated temperatures, the water–rock ratio is large and the isotope composition of O in the rocks approaches that of the water. If the system is dominated by the O of the rocks (i.e., small water–rock ratio), the isotope composition of the water approaches that of the rocks. In most cases, the water dominates and the isotope composition of O in the affected rocks is permanently altered by being depleted in ^{18}O.

26.6a Fossil Hydrothermal Systems

The alteration of the isotope composition of O in plutonic igneous rocks and the adjacent country rock occurs as a result of convection of groundwater that was heated by cooling magmatic intrusives (Gregory and Criss, 1986). In this case, the water–rock ratio is large because large amounts of water flow through a fixed volume of rock. In addition, the system is open because the water flows continuously and transports both O isotopes and solute into and out of the rock.

The groundwater is initially depleted in ^{18}O relative to SMOW because it consists of meteoric water whose δ^{18}O value is negative depending on the latitude of the site (Section 26.3). The δ^{18}O values of the igneous rocks and of the wallrock surrounding intrusive plutons are positive in both rock types because these rocks are composed of silicate, oxide, and carbonate minerals which are typically enriched in ^{18}O relative to SMOW.

Therefore, when heated groundwater begins to flow through fractures in these rocks, O-isotope exchange reactions occur by means of which ^{18}O is transferred from the rocks to the water, while ^{16}O passes from the water to the rocks that are exposed to the water. As a result, the δ^{18}O values of the rocks decrease and the δ^{18}O values of the water increase until an isotopic equilibrium is established between the rocks and the water at the ambient temperature. When a state of isotopic equilibrium between the rocks and the water is achieved, their δ^{18}O values are related to the effective fractionation factor (α_w^r) by equation 26.12:

$$\alpha_w^r = \frac{\delta^{18}\text{O}_r + 10^3}{\delta^{18}\text{O}_w + 10^3}$$

The value of the fractionation factor decreases with increasing temperature and approaches unity at very high temperature. In most (and perhaps all) cases, the temperatures do not rise to such extremes, which means that the fractionation factor remains greater than one. Consequently, the $\delta^{18}O$ value of the altered rocks remains greater than the $\delta^{18}O$ value of the hydrothermal fluid that is in contact with the rocks. Even if a state of isotopic equilibrium is not achieved, the isotope exchange reactions cause the $\delta^{18}O$ values of the rocks to decrease and those of the water to increase. In some cases, the O-bearing fluid was not meteoric water but a brine that was enriched in ^{18}O by prior isotope exchange with silicate or carbonate minerals (e.g., Wenner and Taylor, 1976; Masuda et al., 1986).

The convective cells carry the water toward the surface, where it may be discharged by hotsprings and geysers (e.g., Yellowstone Park, Wyoming). The $\delta^{18}O$ values of this water contain evidence of the ^{18}O-enrichment that occurred by the water–rock interaction in the subsurface. Even the δD values of the water may be affected in cases where the rocks contain clay minerals, mica, or amphiboles in which hydroxyl (OH^-) radicals are present.

Evidence for the lowering of $\delta^{18}O$ values has been reported from many igneous intrusives. For example, Taylor (1968) reported that gabbros from the border zone of the Skaergaard Intrusion in central East Greenland have $\delta^{18}O$ values of only $+0.4 \pm 0.1$‰. These initial results were subsequently confirmed and extended to other layered gabbro intrusives by Taylor and Forester (1979) and Taylor (1987). Additional examples of hydrothermal alteration of mafic and granitic rocks are listed in Table 26.8. An even more extensive list of rocks depleted in ^{18}O was published by Criss and Taylor (1986).

Table 26.8. Effect of Hydrothermal Alteration on the Isotope Composition of O and H in Rocks and Minerals

Rock Type	References
Granodiorite, western Cascade Range, Oregon	Taylor, 1971, *J. Geophys. Res.*, 76(32):7855–7874.
Ash-flow deposits, southern Nevada	Lipman and Friedman, 1975, *Geol. Soc. Am. Bull.*, 86:695–702.
Granite-rhyolite, St. Francois Mountains, MO	Wenner and Taylor, 1976, *Geol. Soc. Am. Bull.*, 87:1587–1598.
Granite batholiths, several examples	Taylor, 1977, *J. Geol. Soc. Lond.*, 133:509–558.
Captains Bay pluton, Aleutian Islands, Alaska	Perfit and Lawrence, 1979, *Earth Planet. Sci. Lett.*, 45:16–22.
Samail ophiolite, Oman	Gregory and Taylor, 1981, *J. Geophys. Res.*, 86:2737–2755.
Idaho batholith, Idaho	Criss and Taylor, 1983, *Geol. Soc. Am. Bull.*, 94:640–663.
Granite, Schwarzwald, Germany	Hoefs and Emmermann, 1983, *Contrib. Mineral. Petrol.*, 83:320–329.
Marysvale mining district, Utah	Shea and Foland, 1986, *Chem. Geol.*, 55:281–295.
Rhyolite pyroclastics, Arima Spa, Japan	Masuda et al., 1986, *Geochim. Cosmochim. Acta*, 50:19–28.
Granitic rocks, Maine	Rumble et al., 1986, *Contrib. Mineral. Petrol.*, 93:420–428.
Igneous rocks and sedimentary iron formations	Gregory et al., 1989, *Chem. Geol.*, 75:1–42.
Tertiary plutons, northwestern USA and British Columbia	Criss et al., 1991, *J. Geophys. Res.*, 96:13335–13356.

26.6b Hydrothermal Ore Deposits

The hot groundwater that percolates through the rocks surrounding a cooling igneous pluton interacts with certain unstable minerals in these rocks, which may release various metals into solution. These metals are ultimately redeposited in open fractures and cavities or by replacement of other minerals in the country rocks (e.g., carbonates). Such so-called hydrothermal mineral deposits may form either within the body of the igneous intrusive, or in the adjacent country rocks, or in both. The ore minerals in such deposits include native gold and silver as well as sulfides of Fe, Cu, Ni, Co, Zn, and Pb, which form at decreasing temperatures and hence at increasing distances from the igneous intrusives which are the source of the heat that drives the convection cells of groundwater.

The process outlined above can result in the formation of ore deposits that occur within or adjacent to igneous intrusives. The apparent association of hydrothermal ore deposits with igneous intrusives originally led to the conclusion that the metals originated from the magma, whereas the study of O isotopes permits the metals to be derived both from the igneous intrusives and from the country rocks adjacent to the intrusives.

Although the shift of the $\delta^{18}O$ values of water in hotsprings and geysers provides evidence for water–rock interactions at many sites in North America and throughout the world, the water responsible for the formation of ore deposits in the geological past is no longer available (except in fluid inclusions). In contrast to the water, the rocks are still there and can be analyzed to reconstruct the fossil hydrothermal system that caused the alteration of their mineralogical, chemical, and isotopic compositions. Such studies can be helpful in mineral exploration by identifying "hot spots" which are potential targets for exploratory drilling and by relating the known hydrothermal ore deposits of a mining district to the igneous intrusives at depth that provided the heat, which caused the formation of the ore deposits.

The lowering of $\delta^{18}O$ values of igneous rocks as a result of isotope exchange with meteoric water and the complementary formation of gold deposits is well illustrated by a series of Tertiary granodiorite intrusions in the western Cascade Range of Oregon (Taylor, 1971; Criss et al., 1991). These plutons include the Brice Creek and Champion Creek stocks of the Bohemia mining district in Lane County, Oregon, which were intruded into volcanic country rocks. The Bohemia mining district produced about $1,000,000 in gold from 1870 to 1940 (Taylor, 1974). The stocks which are less than 3 km in diameter are surrounded by aureoles of contact metamorphism from 300 to 600 m wide, consisting of tourmaline hornfels and related rocks grading outward into a zone of propylitic alteration which has a diameter of about 4 km measured east–west across the entire district. The $\delta^{18}O$ values of the stocks and the surrounding country rocks have been lowered in a systematic fashion indicated by the $\delta^{18}O$ contours in Figure 26.10.

The major ore deposit at the south end of the Champion Creek stock is entirely within the $\delta^{18}O =$ 0‰ contour that outlines the area of the most intense alteration at the highest temperature. The $\delta^{18}O$ values of the rocks increase outward for about 5 km, more or less parallel to the zone of propylitic alteration of the country rocks. A small volume of rocks in the center of the Champion Creek stock still has positive $\delta^{18}O$ values, presumably because it was less accessible to convecting groundwater. The area of lowest $\delta^{18}O$ values (less then 0‰) is distinctly displaced to the east, especially with reference to the Brice Creek stock, which may indicate the presence of a large intrusive body at depth.

Sulfide minerals may also precipitate from heated seawater discharged by hotsprings called smokers along midocean ridges (Bowers and Taylor, 1985). The resulting massive sulfide deposits are associated with basaltic volcanic rocks that have been altered by chemical and isotopic interactions with seawater (Bowers, 1989). Such deposits occur in Precambrian greenstone belts of Canada and elsewhere in the world and are also exemplified by the Kuroko deposits of Japan. These deposits were investigated by contributors to a monograph edited by Ohmoto and Skinner (1983) including a paper by Ohmoto et al. (1983).

Many additional studies of O and H isotope compositions in rocks and minerals from mining districts have been reported by Taylor (1974, 1979). In addition, the applications of stable isotopes to

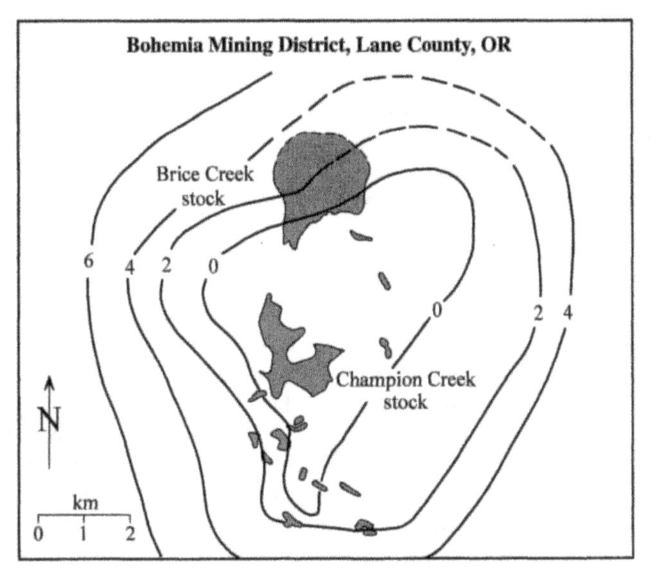

FIGURE 26.10 Contour map of $\delta^{18}O$ (SMOW) values of whole-rock samples of granodiorite stocks and volcanic country rocks in the Bohemia Mining district (gold) at 43°35′ N and 122°37′ W in Lane County, Oregon, USA. Adapted from Taylor (1971).

the study of the hydrothermal ore deposits were reviewed by Ohmoto (1986) and Taylor (1987).

The isotopic composition of the water associated with the formation of hydrothermal ore deposits and related wallrock alteration can be estimated in two different ways:

1. analysis of the isotope composition of O in fluid inclusions and
2. calculation of $\delta^{18}O$ and δD assuming that isotopic equilibrium was established at the filling temperature of fluid inclusions.

The isotope composition of O and H in water of fluid inclusions may have been altered by isotope exchange reactions with the host minerals. Minerals that do not contain H in their lattice may, in some cases, preserve the δD values of fluid inclusions, in which case the $\delta^{18}O$ value of the water can be calculated from the equation of the meteoric-water line. However, the isotope composition of O and H of water may have been altered prior to its entrapment, in which case the equation of the meteoric-water line does not apply.

Alternatively, the $\delta^{18}O$ value of the hydrothermal fluid may be estimated from the measured $\delta^{18}O$ values of gangue minerals (e.g., feldspar, quartz, calcite) by using experimentally determined mineral–water fractionation factors at the temperature of deposition of the minerals indicated by the filling temperatures of fluid inclusions they contain. This approach suffers from the uncertainty that isotopic equilibrium was actually established between the water and the mineral that precipitated from it. In spite of these difficulties, fluid inclusions were used by Ohmoto (1974) and Ohmoto and Rye (1970) to characterize the water that formed the Kuroko deposit of Japan and of the Bluebell Mine in British Columbia.

Studies of the isotope compositions of hydrothermal fluids have indicated that they originated from a variety of sources:

1. meteoric water,
2. seawater,
3. geothermal water enriched in ^{18}O,
4. connate formation waters with variable $\delta^{18}O$ and δD values,

5. metamorphic water, and
6. magmatic water.

Taylor (1974) concluded that metamorphic waters have interacted with silicate minerals at temperatures between 300 and 600°C. As a result, their $\delta^{18}O$ values range widely from +5 to +25‰ and their δD values vary between −20 and −65‰. Magmatic waters likewise equilibrated with silicate minerals at high temperatures (700–1100°C) and therefore approach the isotope composition of silicate minerals in igneous rocks:

$$\delta^{18}O = +5.5 \text{ to } +10.0‰ \quad \delta D = -50 \text{ to } -85‰$$

White (1974) reviewed chemical, isotopic, and physical data of ore deposits at 40 localities and reported that all of the types of water listed above contributed in varying proportions to the ore-forming process. He concluded that simple explanations applicable to all hydrothermal ore deposits do not exist and recommended the acceptance of *diversity*.

26.7 WATER–ROCK INTERACTIONS (WATER)

The δD and $\delta^{18}O$ values of meteoric water anywhere on the Earth are related by equation 26.28, which is restated here in simplified form:

$$\delta D = 8\,\delta^{18}O + 10 \qquad (26.40)$$

The isotope composition of meteoric water can be changed by certain processes that cause the affected water to move off the meteoric-water line and to form quasi-linear trajectories in coordinates of δD and $\delta^{18}O$. The processes in questions include:

1. excessive evaporation of meteoric water from the surfaces of lakes and rivers in desert regions (e.g., northeast Africa),
2. isotope exchange reactions between meteoric water and silicate or carbonate minerals in the subsurface at elevated temperatures (e.g., hotsprings),
3. precipitation of certain minerals containing water of hydration from brines in closed systems such as desert lakes or isolated volumes of deep groundwaters (e.g., gypsum), and
4. mixing of meteoric water with magmatic and metamorphic water or with deep groundwater whose isotope composition was altered by water–rock interaction and other processes.

26.7a Hotsprings and Geysers

The water discharged by hotsprings and geysers is known to be composed primarily of the groundwater that originated as meteoric water on the surface of the Earth but whose isotope composition, in many cases, was altered by isotopic exchange reactions with silicate and carbonate minerals in the subsurface at a range of temperatures. In addition, a small fraction of the water may be released by magmas cooling at depth or during metamorphic recrystallization of clay minerals, micas, and amphiboles. The isotope composition of magmatic and metamorphic water is similar to those of the rocks with which they were associated at depth in the crust.

Juvenile water, originating from deep within the mantle, may have been discharged at the surface of the Earth during its earliest history (e.g., the first 100 million years) when the volume of the oceans was increasing rapidly. However, such water has not been recognized unequivocally and is not presently being discharged by hotsprings on the continents or in the ocean basins.

The δD and $\delta^{18}O$ values of water and steam discharged by hotsprings in California in Figure 26.11 define linear trajectories that intersect the meteoric-water line at points defined by meteoric water at each site (Craig, 1963, 1966). The waters discharged by hotsprings in other geothermal areas included in the summaries of Truesdell and Hulston (1980) and Criss and Taylor (1986) display the same features:

1. The data points define straight lines having different slopes ranging from near zero (e.g., Niland, California) to positive (or negative) values (e.g., Yellowstone Park, Wyoming).
2. The distribution of data points along the linear trajectories is controlled by the isotope-equilibration temperature.

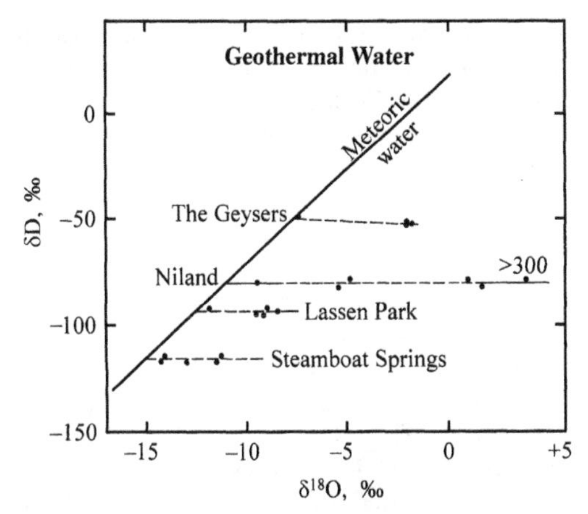

FIGURE 26.11 Isotope composition of water discharged by hotsprings in geothermal areas of California in relation to the meteoric-water line. Adapted from Craig (1963, 1966).

3. The intersections of the trajectories with the meteoric-water line coincide with the δD and δ^{18}O values of the local meteoric precipitation.
4. The trajectories do not converge to the isotope composition of magmatic, metamorphic, or juvenile water (e.g., δ^{18}O: +5 to +10‰; δD: −40 to −80‰; Truesdell and Hulston, 1980).

The slopes of the trajectories depend on the presence or absence of H in the rocks. In cases where H is absent from the rocks, the δD value of the water in contact with the rocks cannot change. In contrast, the δ^{18}O value of the water is shifted in the direction of the δ^{18}O value of the minerals by amounts that rise with increasing temperature. At the highest temperatures of more than 300°C, the steam discharged at Niland, California, has a δ^{18}O value of about +4‰, which is only slightly lower than the δ^{18}O value of the underlying volcanic rocks in the area.

The relation between the water–rock ratio and the δ^{18}O (or δD) values of water and rocks was expressed by Taylor (1977) in the form of an equation based on the requirement for material balance in a closed system:

$$W\delta_w^i + R\delta_r^i = W\delta_w^f + R\delta_r^f \qquad (26.41)$$

where i = initial value of δ_w and δ_r for either O or H
f = final value of δ_w and δ_r for either O or H
w, r = water or rock, respectively
W = mole percent of O or D in the water
R = mole percent of O or D in the rock

Dividing equation 26.43 by R and solving for the W/R ratio yields:

$$\frac{W}{R} = \frac{\delta_r^f - \delta_r^i}{\delta_w^i - \delta_w^f}$$

Adding and subtracting δ_r^f from the denominator,

$$\frac{W}{R} = \frac{\delta_r^f - \delta_r^i}{\delta_w^i - \delta_w^f + \delta_r^f - \delta_r^f}$$

and setting $\delta_r^f - \delta_w^f = \Delta$ yields:

$$\frac{W}{R} = \frac{\delta_r^f - \delta_r^i}{\delta_w^i - \delta_r^f + \Delta} \qquad (26.42)$$

The Δ term is related to the isotope equilibration temperature by the approximations expressed by equations 26.17 and 16.18:

$$\Delta = \delta_r^f - \delta_w^f = (\alpha_w^r - 1)10^3 = \frac{A \times 10^6}{T^2} + B$$

Therefore, equation 26.42 includes the isotope equilibration temperature as well as the W/R ratio expressed as the ratio of mole percent O in the water and in the rocks. The concentration of O in water is 89%, whereas that of typical granitic rocks ranges from 45 to 50%. Therefore, according to Taylor (1977),

$$\left(\frac{W}{R}\right)_{\text{weight}} = 0.5\left(\frac{W}{R}\right)_{\text{O}}$$

Criss and Taylor (1986) used equation 26.42 to plot curves of δ_r^f as a function of the W/R (oxygen) ratio at temperatures between 150 and 350°C. These graphs in Figure 26.12 demonstrate that δ_r^f deceases as the W/R (oxygen) ratio increases and

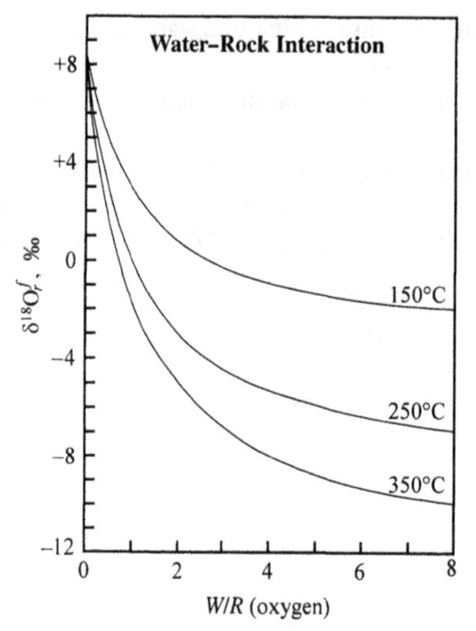

FIGURE 26.12 Equilibration of O in feldspar ($\delta^{18}O_r^i = +8.50‰$) with water ($\delta^{18}O_w^i = -16.00‰$) at temperatures of 150, 250, and 350°C as a function of increasing water–rock ratios (W/R) (defined here as the ratio of atom percent of O) in a closed system. The curves are based on calculations by Criss and Taylor (1986) and were replotted here with W/R on a linear scale.

that the decrease of δ_r^f at a fixed W/R (oxygen) ratio is inversely related to the temperature.

26.7b Mixing of Water

The isotopic composition of meteoric water also changes when it mixes with water whose isotope composition was altered by previous isotope exchange reactions or other processes. The isotope composition of mixed waters is related to the end-member components by mixing equations derived in Chapter 16. The $^{87}Sr/^{86}Sr$ ratio of a mixture of two components A and B is expressed by equation 16.11:

$$\left(\frac{^{87}Sr}{^{86}Sr}\right)_M = \left(\frac{^{87}Sr}{^{86}Sr}\right)_A f_A \frac{Sr_A}{Sr_M} + \left(\frac{^{87}Sr}{^{86}Sr}\right)_B (1 - f_A)\frac{Sr_B}{Sr_M}$$

where Sr_A and Sr_B are the concentrations of this element in the two components, Sr_M is the concentration of Sr in a mixture M, and f_A is the weight fraction of A in the mixture (equation 16.1):

$$f = \frac{W_A}{W_A + W_B}$$

Equation 16.11 can be adapted for the isotope ratios (or δ values) of O and H in mixtures of two isotopic varieties of water. In this case, the concentrations of O and H of the two components are identical (i.e., $O_A = O_B = O_M$). Therefore, when equation 16.11 is used to model mixing of two isotopic varieties of water designated A and B, it reduces to:

$$\delta^{18}O_M = \delta^{18}O_A f_A + \delta^{18}O_B(1 - f_A) \quad (26.43)$$

$$\delta D_M = \delta^{18}D_A f_A + \delta D_B(1 - f_A) \quad (26.44)$$

Equations 26.43 and 26.44 were used to calculate the $\delta^{18}O$ and δD values of mixtures of two components of water whose isotope compositions are:

$$\delta^{18}O_A = -8.0‰ \qquad \delta D_A = -54.0‰$$

$$\delta^{18}O_B = +5.0‰ \qquad \delta D_B = -30.0‰$$

such that component A is meteoric water and component B is deep groundwater that has exchanged O and H with sedimentary rocks at an elevated temperature. When $f_A = 1.0$, equations 26.43 and 26.44 yield $\delta^{18}O_M = -8.0‰$, $\delta D_M = -54‰$. As f_A decreases, the isotope compositions of the mixtures change along a straight line in Figure 26.13 until they become equal to the δ values of component B.

Mixing of meteoric water with deep groundwater can occur as a result of recharge of deep aquifers by meteoric water, by intrusion of deep groundwater into a shallow aquifer containing meteoric water, and as a result of pumping of deep groundwater during the recovery of natural gas or petroleum from a deep reservoir. The resulting mixtures define straight lines in coordinates of $\delta^{18}O$ and δD that closely resemble the linear arrays formed by isotopic exchange reactions of meteoric water with silicate or carbonate rocks with increasing temperature.

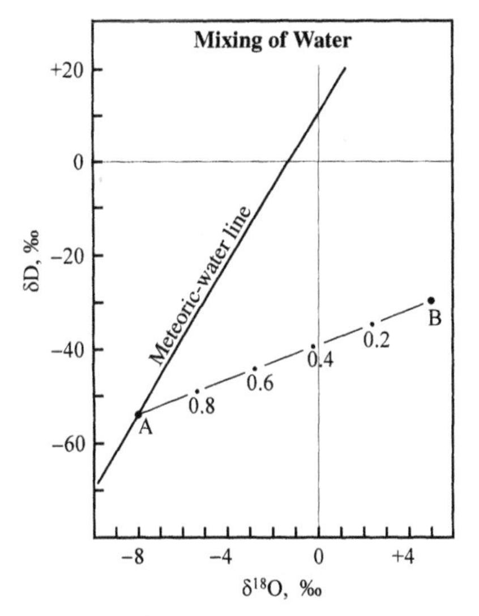

FIGURE 26.13 Mixing of two components of water having different isotope compositions. Component A is meteoric water ($\delta^{18}O_A = -8.0‰$, $\delta D = -54.0‰$) on the meteoric-water line, whereas component B is deep groundwater that has equilibrated O and H with sedimentary rocks at elevated temperatures. The fraction of component A (f_A) is indicated along the mixing line.

Other examples of mixing of different isotopic varieties of water occur when river water or glacial meltwater are discharged into the oceans and by the circulation of water of varying salinities and temperature in the oceans.

26.7c Oilfield Brines, United States and Canada

The isotope composition of water in oilfield brines deviates markedly from the meteoric-water line and also differs from the isotope composition of seawater (Degens et al., 1964; Graf et al., 1965). A large set of measurements of $\delta^{18}O$ and δD in oilfield brines recovered in Michigan and southwestern Ontario, Illinois, Alberta, and the Gulf Coast region of the United States demonstrates positive correlations of the isotopic δ values with their salinities and with temperature in the wells from which the brines where collected (Clayton et al., 1966). In addition, the $\delta^{18}O$ and δD values of brines in each region form linear trajectories that intersect the meteoric-water line.

The deviation of δD and $\delta^{18}O$ values of oilfield brines from the meteoric-water line has been variously explained by:

1. isotope exchange reactions of meteoric water with O and H of the sedimentary rocks,
2. mixing of meteoric water with deep-seated brines consisting of seawater or its derivatives, and
3. ultrafiltration of water under pressure by layers of shale (Phillips and Bentley, 1987).

The results of Clayton et al. (1966) in Figure 26.14 refuted previously held ideas that oilfield brines are connate water of marine origin and indicated instead that these brines evolved from meteoric water. In addition, the coordinates of the points of intersection of the brine trajectories with the meteoric-water line are similar to the $\delta^{18}O$ and δD values of present-day meteoric precipitation in each region. The δ^{18} O values of the points of intersection in Figure 26.15 decrease with increasing latitude of the oilfields, as required by the latitude effect of isotope fractionation of meteoric water.

The increase of both $\delta^{18}O$ and δD values with the subsurface temperature (and hence with depth in the wells) is attributable to isotope exchange reactions between the water and the minerals of the sedimentary rocks (shale, limestone, dolomite, and sandstone) in each of the four oilfields sampled by Clayton et al. (1966). The brine trajectories have positive slopes, indicating that H is present in the subsurface and that the δD values of this H are higher than those of meteoric water in each oilfield.

The H-bearing compounds in sedimentary rocks that may increase or decrease the δD values of groundwater include:

1. clay minerals (kaolinite and smectite),
2. hydrogen sulfide (H_2S and HS^-),
3. hydrocarbons (liquid or gas) including methane,
4. gases containing H_2, and
5. hydrous minerals (e.g., zeolite, mica, amphiboles, gypsum).

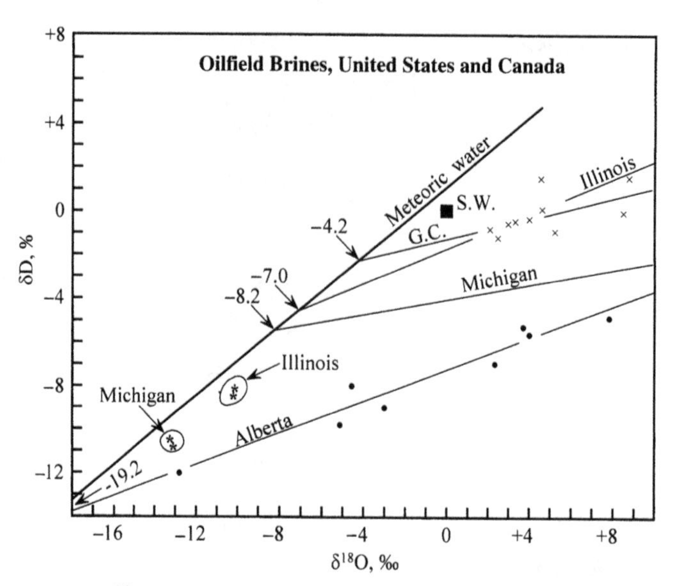

FIGURE 26.14 The $\delta^{18}O$ and δD values of brines from oil-wells in different regions of the USA and Canada. Solid circles = Alberta; crosses = Gulf Coast region (G.C.). The data points for Illinois and Michigan are not shown to avoid crowding. The straight lines were fitted by least-squares regression. The $\delta^{18}O$ values of the points of intersection with the meteoric-water line are indicated in permil units. The δD values are expressed in percent (%) rather than permil (‰). Two brine samples in Michigan and Illinois have anomalously low $\delta^{18}O$ and δD values presumably because they contain glacial meltwater of the Laurentide ice sheet of Wisconsinan age. From data by Clayton et al. (1966).

Although each of these sources may alter the isotope composition of groundwater under appropriate circumstances, most of the compounds listed above are not likely to be effective because of slow isotope-exchange kinetics. The H-reservoir that is most likely to be effective in altering the δD values of groundwater are shales containing clay minerals, including kaolinite and smectite (e.g., illite). The isotope composition of H and O of clay minerals depends on their origins and subsequent isotopic reequilibration with water during transport by streams and deposition in the oceans. Savin and Epstein (1970) reported that carbonate-free marine sediment worldwide is characterized by:

$\delta^{18}O$: +17.8 (+11.5 to +25.9) ‰ (SMOW)
δD: −67.4 (−41 to −93) ‰ (SMOW)

The agreement between the isotope composition of present-day meteoric water in each region and the coordinates of the points of intersection of the brine arrays with the meteoric-water line indicates that the water of the brines was originally deposited under temperate climatic conditions similar to those that exist in these regions at the present time. In other words, the water may be pre-Wisconsin or post-Wisconsin in age.

The $\delta^{18}O$ and δD values of one brine sample from Michigan, one from southwestern Ontario, and one from western Illinois are more negative than present-day meteoric water in these areas. These samples were excluded from the regressions used to fit equations to the brine data and are identified in Figure 26.14. For example, the isotope composition of sample 16 in western Illinois is: $\delta^{18}O = -10.18‰$ and $\delta D = -8.38\%$. The collecting site of this sample is about 40 km east of

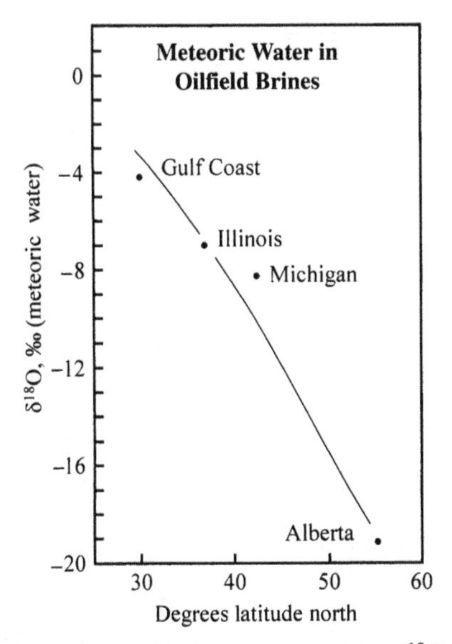

FIGURE 26.15 Latitude dependence of the $\delta^{18}O$ values of the intersections of the brine arrays with the meteoric-water line in Figure 26.14. Based on data of Clayton et al. (1966).

the Mississippi River, which could have infiltrated the groundwater in this area. However, the $\delta^{18}O$ of the Mississippi River at St. Louis (−8.67‰) is not low enough to explain the low $\delta^{18}O$ value of the brine in well 16. Similar reasoning applies to the anomalous brines in Michigan and southwestern Ontario, causing Clayton et al. (1966) to conclude that all of the ^{18}O-depleted brines formed from meteoric precipitation that fell under climatic conditions that were *colder* than those at present. Therefore, these brines presumably contain glacial meltwater derived from the Laurentide ice sheet, which receded from Illinois and Michigan between 14,000 and 5000 years ago (Clayton et al., 1966). These brine samples do not lie on the meteoric-water line in Figure 26.14 because they too have equilibrated with the isotope composition of O and H in the minerals of the aquifer.

An alternative explanation for the apparent shift of $\delta^{18}O$ and δD of oilfield brines is that they are mixtures of meteoric water from the surface with deep-seated brines derived from seawater trapped in the sediment at the time of deposition. This explanation is illustrated in Figure 26.13 and was favored by Degens et al. (1964) to explain the $\delta^{18}O$ values of brines from Oklahoma, Colorado, Texas, Florida, and Utah. However, isotopically unaltered seawater does not lie on any of the brine arrays in Figure 26.14 and therefore is not acceptable as a component of mixing. Presumably, the seawater that was buried with the sediment at the time of deposition was altered both in its chemical and isotopic composition as a result of complex diagenetic processes. In this case, the isotopic δ values of oilfield brines should be related to the concentrations of major elements in the brine. The positive correlations of $\delta^{18}O$ and δD values with the salinities of the brines observed by Clayton et al. (1966) are compatible with two-component mixing. However, in that case, the concentration ratios of conservative elements in the brines should be constant and independent of salinity. The information provided by Clayton et al. (1966) does not permit this test.

In general, the chemical compositions of deep-seated brines are not closely related to that of seawater because of the interaction of the brines with different kinds of rocks, because of mixing of brines of different chemical and isotopic compositions during migrations in response to regional pressure gradients, and because of the possible precipitation of solid compounds from the brines, such as calcite, gypsum, and amorphous silica. Even if mixing of dilute surface water and deep-seated brines did occur in a petroleum-producing region, isotope exchange reactions would also occur. The combination of mixing and simultaneous isotope exchange at a range of temperatures may account for the scatter of data points representing brines in Figure 26.14.

A third alternative for explaining the brine trajectories in Figure 26.14 was advocated by Graf et al. (1965), who favored ultrafiltration of water under pressure. Experiments by Coplen and Hanshaw (1973) confirmed that the isotope composition of water is fractionated during filtration through a semipermeable membrane, and Phillips and Bentley (1987) suggested an explanation for this process. Nevertheless, ultrafiltration of groundwater requires special conditions, whereas subsurface brines deviate from the meteoric-water line in most, if not in all, cases.

The systematic variation of $\delta^{18}O$ and δD of *subsurface brines* originally discovered by Degens et al. (1964) and Clayton et al. (1966), aroused the interest of several other research groups, including:

Hitchon and Friedman, 1969, *Geochim. Cosmochim. Acta*, 33:1321–1349. Kharaka et al., 1973, *Geochim. Cosmochim. Acta*, 37:1899–1908. Banner et al., 1989, *Geochim. Cosmochim. Acta*, 53:383–398. Connolly et al., 1999, *Appl. Geochem.*, 5:397–414. Stueber and Walter, 1991, *Geochim. Cosmochim. Acta*, 55:309–325.

In addition, Knauth and Beeunas (1986) analyzed brines in fluid inclusions of Permian rock-salt deposits and concluded that evaporation initially enriched seawater in D and ^{18}O followed by depletion in these isotopes. The resulting hooked isotopic trajectory produced a brine that is depleted in D but enriched in ^{18}O relative to seawater. Subsequent mixing of meteoric water with such brines can explain the isotope composition of O and H in fluid inclusion of evaporite deposits. The isotope fractionation during the evaporation of *salt solutions* has been studied by:

Sofer and Gat, 1972, *Earth Planet. Sci. Lett.*, 15:232–238. Sofer and Gat, 1975, *Earth Planet. Sci. Lett.*, 26:179–186. Holser, 1979, in Burns, (Eds.) *Marine Minerals*, pp. 295–346. Reviews of Mineral., vol. 6, Mineral. Soc. Amer., Blacksburg, VA. Roedder, 1984, *Am. Mineral.*, 69:413–439. Horita, 1989, *Earth Planet. Sci. Lett.*, 95:173–179.

26.7d Saline Minewaters

A remarkable suite of brines was discovered in the Precambrian basement rocks of Canada at depths greater than 650 m. Fritz and Frape (1982) and Frape and Fritz (1982) reported that these brines are enriched in D relative to the meteoric-water line, in marked contrast to oilfield brines, which are strongly enriched in ^{18}O. Consequently, the brines in the Precambrian rocks plot *above* the meteoric-water line in Figure 26.16 rather than below it like oilfield brines in Figure 26.14.

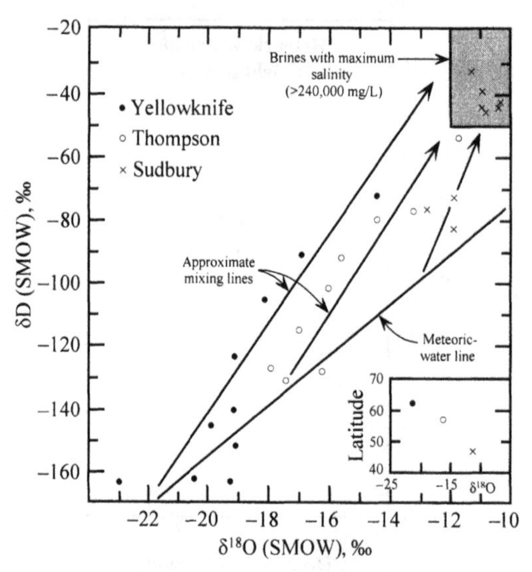

FIGURE 26.16 Isotope compositions of H and O of saline minewaters in the Precambrian basement rocks of the Canadian Shield. These waters are strongly enriched in D but only slightly enriched in ^{18}O, causing data points to plot above the meteoric-water line. The resulting linear arrays appear to converge to the $\delta^{18}O$ and δD values of the most saline brines. In addition, the $\delta^{18}O$ values of the intercepts on the meteoric-water line decrease with increasing latitude of the mine sites. Based on data by Frape et al. (1984).

Brines from deep mines at Yellowknife, in the Northwest Territories, at Thompson in Manitoba, and at Sudbury, in Ontario, define linear trajectories in Figure 26.16 that appear to converge toward $\delta^{18}O$ values between -10 and -12‰ and δD values ranging from -20 to -50‰. Frape et al. (1984) interpreted these trajectories as mixing lines of dense brines, containing more than 240,000 mg/L of dissolved solids, with meteoric water, which forms the shallow groundwater in these mining districts. The mixing lines intersect the meteoric-water line at points whose coordinates vary with the latitudes of the mines and hence with the average annual temperatures of the three mining districts (Dansgaard, 1964; Yurtsever, 1975).

The chemical and isotopic evolution of brines in Precambrian basement rocks has been attributed to:

1. modification of Paleozoic seawater or basinal brines by water–rock interactions (Kelly et al., 1986) and
2. leaching of fluid inclusions in crystalline basement rocks or intense water–rock interactions (Frape and Fritz, 1987).

In addition, Hoefs (1997) identified several processes that may account for the deuterium enrichment of these brines:

1. precipitation of hydrous minerals having high $\delta^{18}O$ and low δD values at low temperature from limited amounts of brine,
2. hydrogen-isotope exchange by meteoric water with hydrous minerals having high δD values, and
3. preferential loss of ${}^{1}_{1}H$ relative to ${}^{2}_{1}H$ from meteoric water as a result of formation of methane or H_2-bearing gas.

The data of Frape et al. (1984) indicated that the deep minewaters are Ca–Na–Cl–SO_4–HCO_3 brines capable of forming Na-bicarbonate, K–Mg-chloride, and sulfates of Na and Ca. In addition, the brines may have reacted with aluminosilicate minerals to form hydroxides of Al, Fe, Mg, and Mn as well as clay minerals. Some of these compounds actually favor H over D, and therefore the precipitation of these minerals could have enriched the brines in D: gypsum, the hydroxides of Al, Fe, and Mn, as well as the clay minerals kaolinite and smectite. The mineral–water fractionation factors for H of these minerals compiled by Chacko et al. (2001) in Table 26.9 are all less than unity, indicating that the D/H ratios of these minerals are less than the D/H ratios of the water at isotopic equilibrium.

Although the minerals that precipitated from these saline minewaters have not been identified, all of the minerals listed in Table 26.9 may have formed either by direct precipitation (e.g., gypsum) or as products of incongruent dissolution of aluminosilicate minerals (e.g., gibbsite, kaolinite, smectite). The O-isotope fractionation factors of these minerals are all greater than one, which means that the minerals are enriched in ^{18}O and cause the residual water to become depleted in this isotope.

Table 26.9. Isotope Fractionation Factors for H between Selected Hydrated Salts and Water

Compound	Formula	α_w^m (D)	T, °C
Gypsum	$CaSO_4 \cdot 2H_2O$	0.985	17
Gibbsite	$Al(OH)_3$	0.985	0–30
Goethite	FeOOH	0.900	25
Manganite	$Mn(OH)_2$	0.809	150
Kaolinite	$Al_2Si_2O_5(OH)_4$	0.968	25
Smectite	$Al_2Si_4O_{10}(OH)_2$	0.960	25

Source: Chacko et al., 2001.

26.8 CLAY MINERALS

When clay minerals form by incongruent dissolution of aluminosilicate minerals (e.g., feldspars, feldspathoids, micas, amphiboles, and clinopyroxene), they incorporate the O and H of the meteoric water at the site of weathering and do not preserve the isotope composition of O and H of the parent minerals (Savin and Epstein, 1970a). The reason is that the primary minerals are exposed to a large volume of meteoric water, which therefore dominates the O and H of the minerals that are weathering. As a result, the $\delta^{18}O$ and δD values of kaolinite and smectite in soils define two lines in Figure 26.17 which are parallel to the meteoric-water line.

The kaolinite line can be constructed by means of the isotope fractionation factors for ^{18}O and D relative to meteoric water. Sheppard and Gilg (1996) expressed the temperature dependence of the kaolinite–water fractionation factors for O and H by the equations:

$$\text{Oxygen}: \qquad 10^3 \ln \alpha_w^k = \frac{2.76 \times 10^6}{T^2} - 6.75 \qquad (26.45)$$

$$\text{Hydrogen}: \qquad 10^3 \ln \alpha_w^k = \frac{-2.2 \times 10^6}{T^2} - 7.7 \qquad (26.46)$$

These equations yield the numerical values of fractionation factors at $T = 15°C$ (288.15 K):

$$\alpha_w^k(^{18}O) = 1.0268 \qquad \alpha_w^k(D) = 0.9663$$

Selecting meteoric water characterized by:

$$\delta^{18}O_w = -12.0‰ \qquad \delta D_w = -86.0‰$$

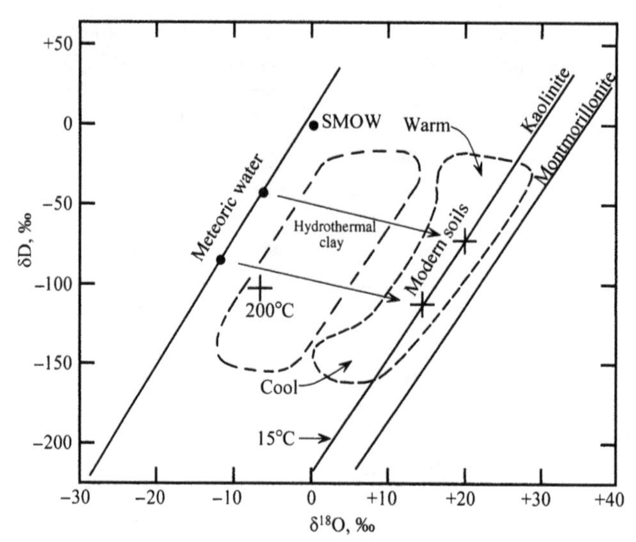

FIGURE 26.17 Isotope compositions of O and H in pedogenic kaolinite and smectite that form by incongruent dissolution of aluminosilicate minerals in isotopic equilibrium with meteoric water at about 15°C. Hydrothermal kaolinites associated with ore deposits occupy a field located closer to the meteoric-water line because the isotope fractionation factors decrease with increasing temperature (e.g., 200°C). Adapted from Savin and Epstein (1970a), Lawrence and Taylor (1971), and Taylor (1974).

and using equation 26.12,

$$\alpha_w^k = \frac{\delta^{18}O_k + 10^3}{\delta^{18}O_w + 10^3}$$

gives

$$\delta^{18}O_k = 1.0268(-12 + 10^3) - 10^3 = +14.47‰$$

$$\delta D_k = 0.9663(-86 + 10^3) - 10^3 = -116.8‰$$

Evidently, kaolinite is enriched in ^{18}O and depleted in D relative to meteoric water.

By repeating this calculation for a different sample of meteoric water,

$$\delta^{18}O_w = -6.0‰ \qquad \delta D_w = -38.0‰$$

at the same temperature of 15°C, the isotopic δ values of kaolinite are:

$$\delta^{18}O_k = 1.0268(-6 + 10^3) - 10^3 = +20.63‰$$

$$\delta D_k = 0.9663(-38 + 10^3) - 10^3 = -70.42‰$$

The field in Figure 26.17 labeled "modern soils" is actually based on $\delta^{18}O$ and δD values of kaolinite measured by Lawrence and Taylor (1971). The line representing montmorillonite (a smectite mineral) also runs parallel to the meteoric-water line but is farther from it because of systematic differences of the fractionation factors.

Clay minerals in soils that form under warm climatic temperatures inherit the low $\delta^{18}O$ and δ D values that characterize meteoric water at low latitudes. Therefore, the isotope compositions of O and H in kaolinite and montmorillonite in soil record both the latitude and the average annual temperature of the collection sites. Examples of such *reconstructions* of environmental conditions have been published by:

Chamberlain et al., 1999, *Chem. Geol.*, 155: 278–294. Chamberlain and Poage, 2000, *Geology*, 28(2):115–118. Tabor et al., 2002, *Geochim. Cosmochim. Acta*, 66:3093–3107.

Kaolinite and its polymorphs nacrite and dickite also form at elevated temperatures when hydrothermal waters interact with aluminosilicate minerals (Taylor, 1974). The isotopic compositions of O and H of such hydrothermal clays define points in the area *between* the meteoric-water line in Figure 26.17 and the field of modern soils. The reason is that the isotope fractionation factors decrease with increasing temperature, which reduces the difference between the $\delta^{18}O$ and δD values of meteoric water and those of the hydrothermal clay.

For example, at $T = 200°C(473.15\ K)$ the isotope fractionation factors of O and H in kaolinite in equilibrium with water derived from the equations of Sheppard and Gilg (1996) are:

$$\alpha_w^k(O) = 1.00579 \qquad \alpha_w^k(H) = 0.9822$$

If $\delta^{18}O_w = -12.0‰$ and $\delta D_w = -86.0‰$, the isotope composition of hydrothermal kaolinite is:

$$\delta^{18}O_k = -6.28‰ \qquad \delta D_k = -102.2‰$$

The point representing the isotope composition of this hydrothermal kaolinite is identified by its equilibration temperature in Figure 26.17. The isotope geochemistry of O and H in marine clays was presented by Savin and Epstein (1970b).

26.9 MARINE CARBONATES

The $\delta^{18}O$ values of marine carbonate rocks of Phanerozoic age should be close to zero permil on the PDB scale because the PDB standard itself consists of marine calcium carbonate. This prediction was not confirmed when Veizer and Hoefs (1976) reported that marine limestones and dolomites are progressively depleted in ^{18}O with increasing age from the Tertiary Period to the Late Archean Era. This is an important phenomenon because the possible explanations have far-ranging consequences:

1. The abundance of ^{18}O of seawater may have increased with time from the Late Archean Era to the Tertiary Period.
2. The temperature of the oceans may have decreased during this interval of time (i.e., the Precambrian oceans had a higher temperature than the Tertiary oceans).
3. The O in the carbonate rocks reequilibrated isotopically with O in meteoric water [i.e., the $\delta^{18}O$ values of Archean limestones were close to zero permil (PDB) at the time of deposition and have since been altered by loss of ^{18}O].

Similar trends in the isotope composition of O were reported for marine chert (Degens and Epstein, 1962) as well as for marine phosphate deposits. Although the evidence is not in dispute, the interpretation of the ^{18}O-deficiency of marine carbonate rocks, cherts, and phosphorites has been controversial (Veizer, 1999; Shaviv and Veizer, 2003).

Before the isotope evolution of O in the oceans can be given serious consideration, it must be shown that the ^{18}O deficiency of marine carbonate rocks is not a result of diagenesis or long-term isotope exchange with meteoric water. For this reason, Jan Veizer and his associates analyzed well-preserved specimens of skeletal calcium carbonate (calcite and aragonite), which they selected after close scrutiny by scanning electron microscopy, petrography, cathodoluminescence, and trace-element concentrations. These samples therefore are assumed to be unaltered both chemically and isotopically, in contrast to whole-rock samples of limestone and dolomite, which are altered in various ways in most cases.

Veizer et al. (1999) measured $\delta^{18}O$ and $\delta^{13}C$ values and $^{87}Sr/^{86}Sr$ ratios of about 1500 samples of brachiopod and mollusk shells as well as of belemnites. This data set was augmented by information from the literature to yield a final data base of 2128 samples from which they constructed profiles of $^{87}Sr/^{86}Sr$ ratios and of $\delta^{13}C$ and $\delta^{18}O$ values. The resulting profile of $\delta^{18}O$ values in Figure 26.18 is a running mean calculated by averaging the $\delta^{18}O$ values in 20-million-year intervals of time with 5-million-year forward steps. This procedure required the best available accuracy in the assignment of numerical ages of the samples derived from the timescale of Harland et al. (1989). The resulting profile permits several important observations:

1. The average $\delta^{18}O$ value of Cambrian fossil shells is about $-8.0‰$ (PDB), whereas shells

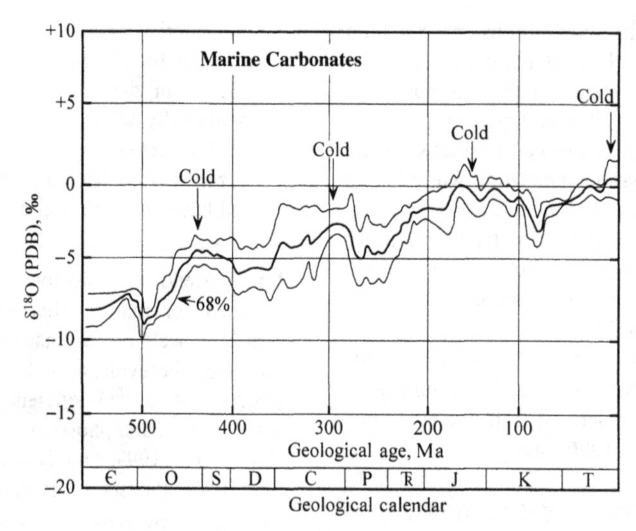

FIGURE 26.18 Variation of the $\delta^{18}O$ value of marine fossil brachiopods, pelecypods, and cephalopods (belemnites) of Phanerozoic age. Adapted from Veizer et al. (1999).

of Quaternary age have an average $\delta^{18}O$ value of 0‰ on the PDB scale.

2. The $\delta^{18}O$ values at any time in the past have a range of values above and below the average. However, the total width of the population in the past is similar in magnitude to the dispersion of modern brachiopod species analyzed by Carpenter and Lohmann (1995) and of Jurassic mollusk shells in the Peterborough Member of the Oxford Clay Formation analyzed by Anderson et al. (1994).
3. The increase of the average $\delta^{18}O$ values of marine carbonates in Figure 26.18 is not monotonic but includes an oscillation with a period of about 150 million years. The resulting peak values of $\delta^{18}O$ record the occurrence of cold climatic conditions on the Earth.

Veizer et al. (1999) considered textural, geological, mineralogical, chemical, and isotopic data to demonstrate that the observed systematic variation of the $\delta^{18}O$ values of skeletal carbonate samples are not the result of secondary alteration but are a primary feature and therefore worthy of serious consideration.

The opposite view has been taken by Land (1995), who concluded that all samples of skeletal calcium carbonate of Phanerozoic age are altered because the $\delta^{18}O$ values of Early Ordovician brachiopods analyzed by Qing and Veizer (1994) overlap the $\delta^{18}O$ of micritic limestones of the same age analyzed by Gao and Land (1991), who considered these micrites to have been altered. Land (1995) also challenged the reliability of the criteria Veizer and his associates used (e.g., low concentrations of Fe and Mn but high concentrations of Sr) to indicate the absence of secondary alteration of fossil shells.

Veizer et al. (1999) countered that the $\delta^{18}O$ values of the Phanerozoic fossil shells they analyzed are closely correlated with $^{87}Sr/^{86}Sr$ ratios whose secular variations have been established beyond any doubt (Section 19.1b, Figure 19.1). Therefore, if the $\delta^{18}O$ values of fossil shells were altered, presumably the $^{87}Sr/^{86}Sr$ ratios were altered as well. Or, to say it another way, because the $^{87}Sr/^{86}Sr$ ratios of unreplaced marine carbonate shells were not altered, the $\delta^{18}O$ values were not altered either.

Another argument in support of the preservation of the $\delta^{18}O$ values of fossil shells is based on the profile of $\delta^{18}O$ values of fossil shells at the Ordovician–Silurian transition in Figure 26.19. This profile shows small-scale positive variations of $\delta^{18}O$ values in close correlation with sedimentary rocks of glacial origin. The most pronounced $\delta^{18}O$

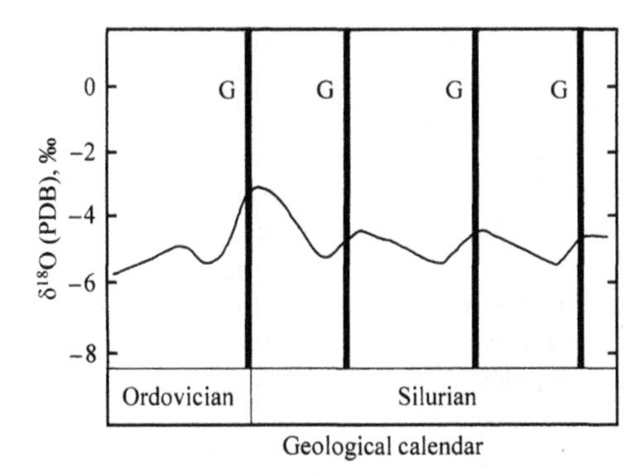

FIGURE 26.19 Profile of δ^{18}O values of brachiopod shells of Late Ordovician (Ashgillian) and Early to Middle Silurian (Llandoverian and Wenlockian) ages. The letter G identifies glacial sediment whose occurrence correlates with increases of the δ^{18}O values. Adapted from Veizer et al. (1999).

peak occurs in the Upper Ordovician graptolite biozone 21 (*Glyptograptus acuminatus*), which is global in scope because it is based on samples from China, the Baltic states, Canada, and Argentina.

Nevertheless, the interpretation of the δ^{18}O values of marine fossils presented by Veizer et al. (1999) and other investigators is not yet clear. The problem is that the δ^{18}O value of seawater is buffered by isotope exchange reactions with rocks along midocean ridges and therefore is not free to vary (Muehlenbachs and Clayton, 1976; Gregory and Taylor, 1981; Gregory, 1991; Muehlenbachs, 1998). In addition, the temperature of the oceans during early Phanerozoic time cannot have been much higher than present temperatures because brachiopods and other organisms flourished and continental glaciations occurred repeatedly in early Phanerozoic time (Veizer et al., 1986).

The ^{18}O deficiency of marine carbonates of *Precambrian age* is even more dramatic than that of Phanerozoic carbonates. For example, Veizer and Hoefs (1976) originally cited δ^{18}O values of Archean carbonates that ranged to −24‰ (PDB) with a mean of about −20‰. Negative δ^{18}O values of this order of magnitude in carbonate rocks are difficult to explain by isotope exchange with meteoric water because water of the necessary ^{18}O depletion occurs only in high latitudes. In addition, high temperatures are required to cause the δ^{18}O values of carbonate rocks to be shifted to the δ^{18}O values of the water. The required changes in temperature of the oceans or in the δ^{18}O value of seawater is illustrated by the calculations below:

Apparent Equilibration Temperature of Calcite in the Modern Oceans

The $\delta^{18}O_c$ value of marine carbonates on the SMOW scale is obtained from equation 26.36a:

$$\delta^{18}O_{SMOW} = 1.03091\ \delta^{18}O_{PDB} + 30.91$$

If $\delta^{18}O_c = 0‰$ (PDB), $\delta^{18}O_c = +30.91‰$ (SMOW). The isotope fractionation factor α_w^c for calcite–water is calculated from equation 26.12:

$$\alpha_w^c(^{18}O) = \frac{\delta^{18}O_c + 10^3}{\delta^{18}O_w + 10^3} = \frac{30.91 + 10^3}{0 + 10^3} = 1.03091$$

The temperature dependence of α_w^c was expressed by Friedman and O'Neil (1977) by the equation:

$$10^3 \ln \alpha_w^c = \frac{2.78 \times 10^6}{T^2} - 2.89 \qquad (26.47)$$

Substituting $\alpha_w^c = 1.03091$ yields a temperature of:

$$T^2 = \frac{2.78 \times 10^6}{33.3319} = 8.340 \times 10^4$$

$$T = 288.19\ \text{K}\ (+15.6°\text{C})$$

Apparent Temperature of Cambrian Oceans Assuming that $\delta^{18}O_w = 0‰$ (SMOW)

If $\delta^{18}O_c = -8.0‰$ (PDB), $\delta^{18}O_c$ on the SMOW scale is: $\delta^{18}O_c = 1.03091(-8.0) + 30.91 = +22.66‰$ (SMOW). Consequently the value of $\alpha_w^c(^{18}O)$ for the case that $\delta^{18}O_w = 0‰$ (SMOW) is:

$$\alpha_w^c(^{18}O) = \frac{+22.66 + 10^3}{0 + 10^3} = 1.02266$$

and the corresponding temperature is:

$$T^2 = \frac{2.78 \times 10^6}{25.2970} = 10.989 \times 10^4$$

$$T = 331.50\ \text{K}\ (+58.3°\text{C})$$

Apparent $\delta^{18}O_w$ of Cambrian Seawater Assuming That T = +15.64°C

Since the O-isotope fractionation factor for calcite–water at 15.6°C is $\alpha^c_w = 1.03091$ and since $\delta^{18}O_c = -8.0‰$ (PDB) or +22.66‰ (SMOW), the $\delta^{18}O_w$ value of Cambrian seawater is calculated from the equation:

$$1.03091 = \frac{22.66 + 10^3}{\delta^{18}O_w + 10^3}$$

Thus,

$$\delta^{18}O_w = \frac{22.66 + 10^3}{1.03091} - 10^3 = -8.0‰ \text{ (SMOW)}$$

Summary

1. The present effective isotope equilibration temperature of calcite in the oceans is 15.6°C.
2. Apparent isotope equilibration temperature of calcite in the Cambrian oceans is 58.3°C.
3. Apparent $\delta^{18}O$ of Cambrian seawater at an equilibration temperature of 15.6°C is −8.0‰.

These calculations are based on the assumptions that isotopic equilibrium between calcite and water was maintained, that the present $\delta^{18}O$ value of seawater is 0‰ on the SMOW scale, that the $\delta^{18}O$ values of Cambrian calcite were not altered after deposition, and that the temperature dependence of isotope fractionation factor for calcite–water by Friedman and O'Neil (1977) is valid.

26.10 MARINE PHOSPHATES

The dilemma concerning the interpretation of $\delta^{18}O$ values of marine calcite and aragonite arises because the dependence of $\delta^{18}O$ of Ca-carbonate on the temperature in equation 26.47 also involves the $\delta^{18}O$ value of seawater. Consequently, three variables are required, only one of which can be measured directly. This problem could be overcome by analyzing samples of cogenetic Ca-phosphate containing O that equilibrated isotopically with the same seawater as the Ca-carbonate samples.

The procedures for the release of O from phosphate samples and the successful determination of its isotopic composition were originally developed by Tutge (1960), with subsequent modifications by Crowson et al. (1991), O'Neil et al. (1994), and Stuart-Williams and Schwarcz (1995).

These procedures have made possible a remarkable series of investigations that ranged from a search for a thermal anomaly in the oceans at the K–T boundary, to studies of O-isotope fractionation by living and fossil brachiopods, and to investigations into the fractionation of ^{18}O between phosphate ions in solution and water by:

Lécuyer et al., 1993, *Paleo. Paleo. Paleo.*, 105:235–243. Lécuyer et al., 1996, *Paleo. Paleo. Paleo.*, 126:101–108. Lécuyer et al., 1998, *Geochim. Cosmochim. Acta*, 62:2429–2436. Lécuyer et al., 1999, *Geochim. Cosmochim. Acta*, 63:855–862

26.10a Paleothermometry

The first measurements of $\delta^{18}O$ values of phosphate of marine organisms established the temperature dependence of $\delta^{18}O$ of phosphate:

Longinelli, 1965, *Nature*, 207:716. Longinelli, 1966, *Nature*, 211:923. Longinelli and Nuti, 1968a, *Science*, 160:879–882. Longinelli and Nuti, 1968b, *Earth Planet. Sci. Lett.*, 5:13–16. Longinelli and Nuti, 1973a, *Earth Planet. Sci. Lett.*, 19:373–376. Longinelli and Nuti, 1973b, *Earth Planet. Sci. Lett.*, 20:337–340.

The thermometry equation of Longinelli and Nuti (1973) is:

$$t = 111.4 - 4.3(\delta_p - \delta_w) \qquad (26.48)$$

where t = temperature in °C
δ_p = $\delta^{18}O$ of phosphate on the SMOW scale
δ_w = $\delta^{18}O$ of water on the SMOW scale

This equation was based on the $\delta^{18}O$ values of skeletal phosphates of living organisms (Lamellibranchiata, Cirripedia, Cephalopoda, and Gastropoda) from the Atlantic and Pacific Oceans and the Mediterranean Sea. The $\delta^{18}O$ of the water at each site was measured directly and the isotope-equilibration temperature was calculated from the $\delta^{18}O$ values of coexisting skeletal Ca-carbonate using the carbonate–water equation of Epstein et al. (1953).

The phosphate–water thermometry equation of Longinelli and Nuti (1973) provides a useful method of measuring the O-isotope equilibration temperature in the oceans of the geological past. However, the method is subject to the same limiting assumptions as the carbonate thermometry equation (Section 26.4b), including the requirement that the $\delta^{18}O$ value of the water must be known. Unfortunately, the slope of the phosphate–water equation in Figure 26.20 is similar to the slope of equation 26.37 for the carbonate–water system, which means that the difference between them is constant and therefore independent of the temperature. The result is that the thermometry equations for calcite–water and phosphate–water cannot be combined by eliminating the $\delta^{18}O$ value of water to relate the difference between their $\delta^{18}O$ values directly to the temperature. The problem is demonstrated below by means of numerical calculations illustrated by Figure 26.21.

The calcite–water equation of Kim and O'Neil (1997) was stated in Section 26.4a in the form of

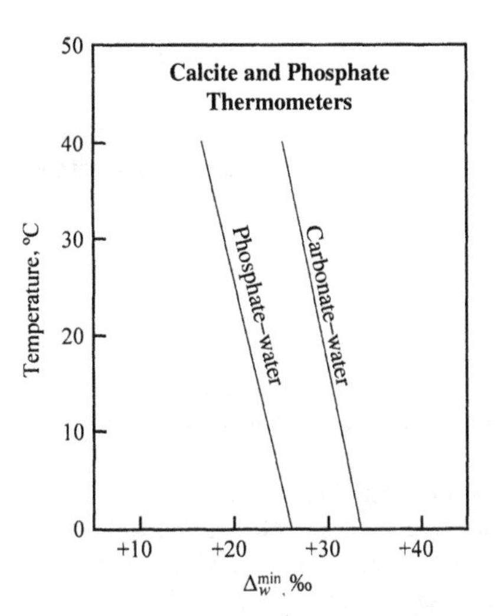

FIGURE 26.20 Variation of Δ_w^{min} with temperature where $\Delta_w^{min} = \delta^{18}O_{min} - \delta^{18}O_w$ and the minerals are calcite and phosphate. Both minerals are usable thermometers, but both require knowledge of the $\delta^{18}O$ value of the water.

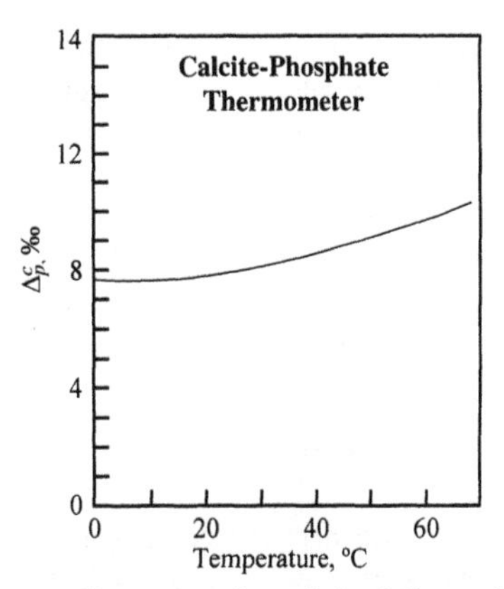

FIGURE 26.21 Demonstration of the failure of the calcite–phosphate O-isotope thermometer in the temperature range from 0 to 40°C: $\Delta_p^c = \delta^{18}O_c - \delta^{18}O_p$, where c = calcite and p = phosphate. The relation between Δ_p^c and t (°C) was calculated from equation 26.51, which does not include the $\delta^{18}O$ of seawater.

equation 26.37:

$$10^3 \ln \alpha_w^c = \frac{18.03 \times 10^3}{T} - 32.42$$

where T is in kelvins. It can be restated in the form of (approximate) equation 26.18:

$$\delta^{18}O_c - \delta^{18}O_w = \frac{18.03 \times 10^3}{T} - 32.42$$

In addition, the temperature T in kelvins can be expressed in degrees Celsius (t):

$$\delta^{18}O_c - \delta^{18}O_w = \frac{18.03 \times 10^3}{t + 273.15} - 32.42 \quad (26.49)$$

Equation 26.48 for the phosphate–water system (Longinelli and Nuti, 1973) is expressed similarly:

$$\delta^{18}O_p - \delta^{18}O_w = \frac{111.4 - t}{4.3} \quad (26.50)$$

Equations 26.49 and 26.50 can now be combined by eliminating $\delta^{18}O_w$:

$$\Delta_p^c = \frac{18.03 \times 10^3}{t + 273.15} - 32.42 - \frac{111.4 - t}{4.3} \quad (26.51)$$

where $\Delta_p^e = \delta^{18}O_c - \delta^{18}O_p$. Equation 26.51 was used to calculate values of Δ_p^c for selected values of t. The results in Figure 26.21 confirm that Δ_p^c is virtually constant at temperatures between 0 and 40°C. Consequently, the isotope fractionation of O between calcite and phosphate in isotopic equilibrium with seawater is not usable as a thermometer of the environmental temperature.

26.10b Fishbones

The temperature dependence of the $\delta^{18}O$ values of skeletal phosphates in fish was investigated by Kolodny et al. (1983), who confirmed the accuracy of the thermometry equation of Longinelli and Nuti (1973). The fish originated from lakes and coastal marine environments having a wide range of $\delta^{18}O_w$ values from −15.9‰ in Lake Baikal to +7.5‰ in a shallow hypersaline lagoon in the northern Sinai Peninsula. A least-squares fit of Δ_w^p versus temperature in degrees Celsius yielded the equation of a straight line:

$$t = 113.3 - 4.38(\delta_p - \delta_w) \qquad (26.52)$$

where t is in degrees Celsius and the $\delta^{18}O$ values are relative to SMOW. Additional studies of skeletal phosphate of fish are as follows:

Kolodny and Raab, 1988, *Paleo. Paleo. Paleo.*, pp. 64:59–64. Kolodny and Luz, 1991, in Taylor et al. (Eds.), *Stable Isotope Geochemistry*, Spec. Pub. 3, pp. 65–76, Geochemical Society. Lécuyer et al., 1993, *Paleo. Paleo. Paleo.*, 105:235–243. Picard et al., 1998, *Geology*, 26:975–978. San Antonia, Texas.

Fish also contain bones in their ears (otoliths) composed of calcite. The $\delta^{18}O$ values of fish otoliths derived both from living marine specimens as well as from the marine Upper Miocene Bells Creek Formation of Wairarapa, New Zealand, were analyzed by Devereux (1967). The temperatures calculated from the $\delta^{18}O_c$ values of the otoliths by the equation of Epstein et al. (1953) ranged from 8.0 to 11.3°C, assuming that $\delta^{18}O$ of the seawater was equal to 0‰.

26.10c Mammalian Bones

Another potentially useful application of the phosphate isotope thermometer is the study of mammalian bones proposed by Longinelli (1984) and later pursued by Luz et al. (1984), Luz and Kolodny (1985), and Bryant et al. (1996).

Luz et al. (1984) reported that $\delta^{18}O$ values of phosphate in bones of humans, dogs, and muskox increase linearly with the average $\delta^{18}O$ values of meteoric precipitation at each site. The equation of this relationship is:

$$\delta^{18}O_p = 0.78\ \delta^{18}O_w + 22.7 \qquad (26.53)$$

However, the O in the phosphate of mammalian bones is not in isotopic equilibrium with the O in the meteoric water. The isotopic disequilibrium between O in bones and in meteoric water occurs because bones equilibrate with body water rather than with drinking water and because the body water is fractionated depending on the rates of uptake by drinking and loss by respiration. The relationship between the $\delta^{18}O$ of *body water* and *drinking water* in laboratory rats derived by Luz et al. (1984) is:

$$\delta^{18}O_{bw} = 0.59\ \delta^{18}O_w + 0.24 \qquad (26.54)$$

This equation indicates that the $\delta^{18}O$ value of body water of rats who are drinking water with $\delta^{18}O_w = -10.0‰$ is:

$$\delta^{18}O_{bw} = 0.59(-10.0) + 0.24 = -5.66‰ \qquad (26.55)$$

These results have paved the way for the interpretation of the isotope composition of O in the phosphate of bones and teeth of various animals by:

Ayliffe and Chivas, 1990, *Geochim. Cosmochim. Acta*, 54:2603–2609. Kolodny et al., 1996, *Paleo. Paleo. Paleo.*, 126:161–171. Luz et al., 1984, *Earth Planet. Sci. Lett.*, 69:255–262. Vennemann and Hegener, 1998, *Paleo. Paleo. Paleo*, 142:107–121.

The preservation of the isotopic composition O in bones and teeth has been investigated elsewhere by:

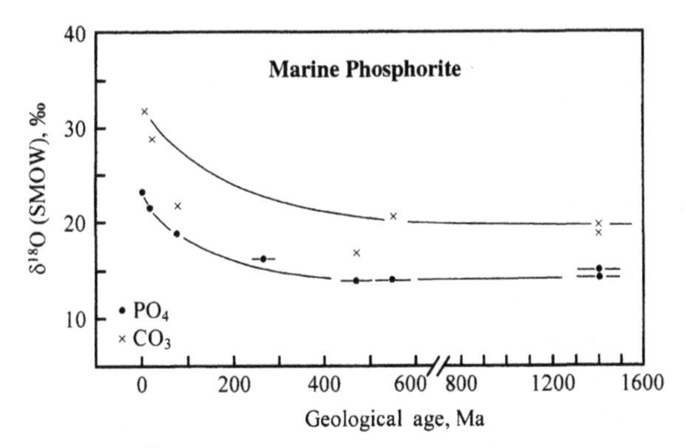

FIGURE 26.22 Average $\delta^{18}O$ (SMOW) values of O in marine phosphate and in the carbonate ion contained in fluorapatite. Based on data by Shemesh et al. (1983).

Nelson et al., 1986, *Geochim. Cosmochim. Acta*, 50:1942–1949. Wang and Cerling, 1994, *Paleo. Paleo. Paleo*, 107:281–289. Iacumin et al., 1996, *Earth Planet. Sci. Lett.*, 142:1–6. Blake et al., 1997, *Geochim. Cosmochim. Acta*, 61:4411–4422.

26.10d Phosphorites

The last issue to be considered concerns the stability of the $\delta^{18}O$ value of phosphorite rocks of different ages (Glenn et al., 1994). This question was investigated by Shemesh et al. (1983, 1988), who measured the isotope compositions of O in the phosphate as well as in the carbonate anion that occurs in the carbonate-fluor apatite, which is the major mineral of phosphorite rock. The chemical composition of this mineral is $(Ca, Na, Mg, Sr)_{10}$ $(PO_4, CO_3, SO_4)_6$ F_{2-3}. Consequently the origin of phosphorite can be investigated by means of the isotope composition of O, C, and S (Piper and Kolodny, 1987).

The average $\delta^{18}O$ (SMOW) of 16 marine phosphorite samples, all of which were younger than 10^5 years, is $+23.2 \pm 0.5$‰ (Shemesh et al., 1983). This value yields a reasonable temperature of 11.6°C based on the phosphate thermometry equation of Longinelli and Nuti (1973). Marine phosphates of Tertiary age (Lower Miocene to pre-Pleistocene) yielded a lower average $\delta^{18}O$ value of $+21.4 \pm 1.5$‰. The average $\delta^{18}O$ values of progressively older samples of marine phosphate in Figure 26.22 decrease with increasing age and reach a value of $+14.6 \pm 1.5$ at about 1400 ± 100 Ma.

The $\delta^{18}O$ value of *carbonate ions* in the crystal lattice of the apatite samples analyzed by Shemesh et al. (1983) also decrease with age in Figure 26.22. However, the results are more scattered than the $\delta^{18}O$ of the phosphate. This evidence caused Shemesh et al. (1983) to conclude that phosphates of pre-Tertiary age have preserved the isotope composition of O better than the carbonate ions in the apatite lattice.

The significance of the decline of the $\delta^{18}O$ values of phosphate with increasing age is not yet clear because it may be interpreted as evidence for higher oceanic temperatures, or lower $\delta^{18}O$ values of seawater, or for isotope exchange with O in meteoric water. The first option yields a temperature of 48.6°C for seawater at 1400 ± 100 Ma assuming that $\delta^{18}O_w = 0$‰; the second requires that $\delta^{18}O$ of seawater at 1400 ± 100 Ma was -8.6‰ (SMOW).

The secular variation of the $\delta^{18}O$ values of seawater is prevented by the interaction with O in volcanic rocks along midocean ridges and therefore can be excluded from consideration. However, the temperature of the surface layer of the oceans may have been higher in Precambrian time than at present because of greater greenhouse warming of the atmosphere by carbon dioxide, which was later sequestered in carbonate rocks, buried in clastic sedimentary rocks as amorphous carbon particles, and deposited in the form of coal, petroleum, and natural gas. This paleoclimatic interpretation continues to motivate studies of the isotope composition of marine carbonates, phosphates, and chert.

The preservation of the isotope composition of O in biogenic marine phosphate and calcium carbonate was tested by Wenzel et al. (2000), who measured the $\delta^{18}O$ values of condonts (phosphate) and articulate brachiopods (calcite) of Silurian age on the island of Gotland in the Baltic Sea. These fossils were codeposited in undeformed and unaltered shallow-water marlstones and carbonate beds whose diagenetic temperature did not rise above 100°C. The age of the host rocks varies from late Early Silurian (Llandovery/Telychian, ~428 Ma) to middle Late Silurian (Ludlow/Ludfordian, ~419 Ma), where the dates are from the timescale of the IUGS (2002) in the Appendix. The resulting average $\delta^{18}O$ values on the SMOW scale in Figure 26.23 are:

Phosphate (conodonts): $+18.4 \pm 0.2‰$ (SMOW)
Carbonate (articulate brachiopods): $+26.2 \pm 0.3‰$ (SMOW)

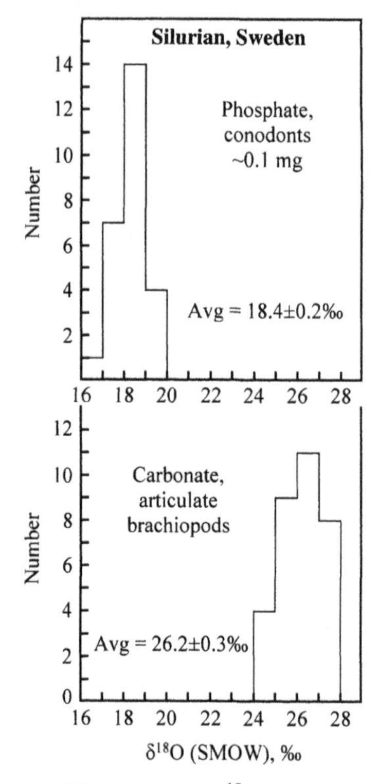

FIGURE 26.23 Histograms of $\delta^{18}O$ values on the SMOW scale of conodonts (phosphate) and the shells of articulate brachiopods (calcite) in Silurian carbonate rocks on the island of Gotland, Baltic Sea. Data from Wenzel et al. (2000).

The average $\delta^{18}O$ value of the conodonts yields a temperature of 32.2°C based on the thermometry equation of Longinelli and Nuti (1973) and assuming that $\delta^{18}O$ of seawater was 0‰ (SMOW). The average $\delta^{18}O$ value of the carbonate shells of articulate brachiopods corresponds to a similar temperature of 35.9°C calculated from the calcite–water thermometry equation of Epstein et al. (1953):

$$10^3 \ln \alpha_w^c = \frac{2.73 \times 10^3}{T^2} - 2.71 \qquad (26.56)$$

The average $\delta^{18}O$ value of Silurian conodonts reported by Wenzel et al. (2000) is about 4‰ higher than the $\delta^{18}O$ values of Ordovician phosphorites from Swan Peak, Idaho, reported by Shemesh et al. (1983). Nevertheless, both data sets yield elevated environmental temperatures for the oceans during the early Paleozoic Era.

The data of Wenzel et al. (2000) in Figure 26.24 also demonstrate a good positive correlation ($r^2 = 0.85$) between the $\delta^{18}O$ values of *phosphate in conodonts* and *calcite in the shells of articulate brachiopods* in the Silurian System of Gotland. The equation relating these parameters is:

$$\delta^{18}O_c = 4.1 + 1.1926\, \delta^{18}O_p \qquad (26.57)$$

when both are expressed on the SMOW scale. However, the $\delta^{18}O$ (SMOW) values of *phosphate shells of inarticulate brachiopods* from the same Silurian beds on the islands of Gotland range widely from 13.0 to 16.8‰ (average 14.9‰) and are about 3.5‰ lower than the $\delta^{18}O$ values of the phosphate of conodonts. Therefore, Wenzel et al. (2000) concluded that the *phosphate in the shells of the inarticulate brachiopods* had been altered.

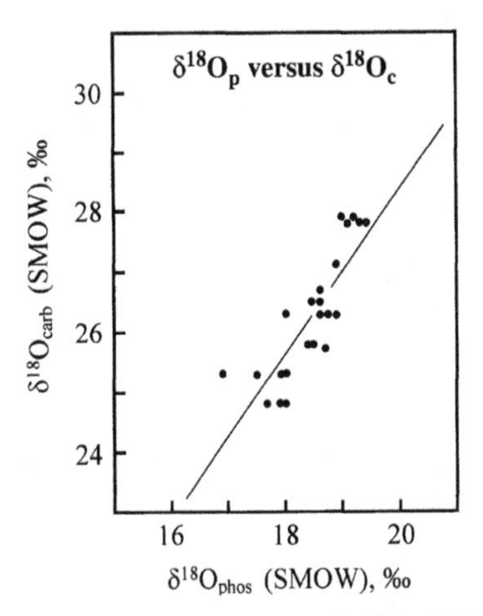

FIGURE 26.24 Linear correlation of $\delta^{18}O$ (SMOW) of marine biogenic phosphate (conodonts) and $\delta^{18}O$ (SMOW) of marine calcite (articulate brachiopods) in the Silurian System of the island of Gotland, Sweden, in the Baltic Sea. Data from Wenzel et al. (2000).

26.11 BIOGENIC SILICA AND HYDROXIDES OF Fe AND Al

The quest to know both the temperature and the $\delta^{18}O$ value of the oceans of the geological past has aroused interest in amorphous biogenic silica that forms the skeletons of certain organisms (e.g., siliceous sponges, diatoms, radiolarians). Amorphous silica may also precipitate inorganically from saturated solutions of silicic acid (H_4SiO_4) as a result of changes in pH and temperature of water discharged by hotsprings in geothermal areas on the continents and in the ocean basins (Kita and Taguchi, 1986; Sharp and Kirschner, 1994; Faure, 1998).

In addition, hydroxides of Fe and Al may form by hydrolysis of Fe^{3+} and Al^{3+} in aqueous solution in marine and nomarine environments. These hydroxides contain both O and H whose isotopes may be fractionated in equilibrium with water depending on the temperature. Therefore, the $\delta^{18}O$ and δD values of naturally occurring oxyhydroxides such as goethite (FeOOH) and gibbsite ($Al_2O_3 \cdot 3H_2O$) may record the temperature at the time of their formation. The O-isotope geochemistry of the ferric hydroxides has been investigated by Crayton Yapp:

> Yapp, 1987, *Geochim. Cosmochim. Acta*, 51: 355–364. Yapp, 1990a, *Chem. Geol.*, 85:329–335. Yapp, 1990b, *Geochim. Cosmochim. Acta*, 54:229–236. Yapp, 1991, *Geochim. Cosmochim. Acta*, 55:2627–2634. Yapp, 1993, *Geochim. Cosmochim. Acta*, 57:2319–2327. Yapp, 1997, *Chem. Geol.*, 135:159–171. Yapp, 1998, *Geochim. Cosmochim. Acta*, 62:2409–2420. Yapp, 2000, *Geochim. Cosmochim. Acta*, 64:2009–2025. Yapp, 2001, *Geochim. Cosmochim. Acta*, 65:4115–4130.

The studies of the fractionation of O isotopes during the formation of ferric hydroxides by Bao and Koch (1999) and Bao et al. (2000) and climatic interpretations of the $\delta^{18}O$ values of goethite by Poage et al. (2000) provide additional results and applications of this research area. Thermometry equations for silica–water, goethite–water, and gibbsite–water are listed in Table 26.10.

The suitability of diatom frustules for isotope thermometry was tested by:

> Mopper and Garlick, 1971, *Geochim. Cosmochim. Acta*, 35:1185–1187. Labeyrie, 1974, *Nature*, 248:40–42. Labeyrie and Juliet, 1982, *Geochim. Cosmochim. Acta*, 46:967–975. Schmidt et al., 2001, *Geochim. Cosmochim. Acta*, 65:201–211.

Schmidt et al. (2001) reported that the $\delta^{18}O$ values of marine diatoms increase by 3–10‰ as they mature to opal A. Therefore, the isotope composition of O in marine diatoms in sediment does not record the temperature of oceanic surface water but instead depends on its state of evolution from amorphous silica to chalcedony.

The O-isotope fractionation factors for biogenic silica–water, goethite–water, and gibbsite–water in Table 26.10 are each potentially valid isotope thermometers provided that certain assumptions are satisfied:

1. Isotope equilibrium between the solid phase and the water was established at the time of formation of the solids.

Table 26.10. Oxygen-Isotope Fractionation Factors for Biogenic Silica–Water, Geothite–Water, and Gibbsite–Water

$10^3 \ln \alpha_w^m$	t, °C	References[a]
Biogenic Silica (SiO_2)		
$3.52 \times 10^6/T^2 - 4.35$	34–93	1
$41.2 - 0.25t$	4–27	2
$15.56 \times 10^3/T - 20.92$	3.6–20.0	3
Goethite (FeOOH)		
$1.63 \times 10^6/T^2 - 12.3$	25–120	4
$2.76 \times 10^6/T^2 - 23.7$	10–65	5
$1.907 \times 10^6/T^2 - 8.004$	35–140	6
Gibbsite (Al $(OH)_3$)		
$1.31 \times 10^6/T^2 - 1.78$	8–51	7
$2.04 \times 10^6/T^2 - 3.61 \times 10^3/T$	0–60	8

Source: Chacko et al., 2001.
Note: T = Kelvin scale, t = Celsius scale.
[a] 1. Kita et al., 1986, *Geochim. J.*, 20:153–157; Labeyrie, 1974, *Nature*, 248:40–42. 3. Brandriss et al., 1998, *Geochim. Cosmochim. Acta*, 62:1119–1125. 4. Yapp, 1990, *Chem. Geol.*, 85:382–335. 5. Müller, 1995, *Isotopes Environ. Health Studies*, 31:301–302. 6. Bao and Koch, 1999, *Geochim. Cosmochim. Acta*, 63:599–613. 7. Bird et al., 1994, *Geochim. Cosmochim. Acta*, 58:5267–5277. 8. Vitali et al., 2000, *Clays Clay Minerals*, 48:230–237.

2. The solid compounds were not transported and redeposited in a different environment than the one in which they formed.
3. The oxygen-isotope composition of the compounds was not altered during diagenesis after deposition.
4. The sedimentary rocks were not recrystallized during episodes of regional or contact metamorphism.

As in the case of calcite and apatite, a unique value of the temperature can only be determined if the $\delta^{18}O$ value of the aqueous phase is known.

This difficulty can be circumvented by combining mineral–water isotope fractionation equations in pairs. For example, it may be advantageous to combine the equation for calcite published by Kim and O'Neil (1997) with the equation for biogenic silica in Table 26.10 by Brandriss et al. (1998):

$$10^3 \ln \alpha_w^c = \frac{18.03 \times 10^3}{T} - 32.42 \quad \text{(calcite–water)}$$

$$10^3 \ln \alpha_w^s = \frac{15.56 \times 10^3}{T} - 20.93 \quad \text{(silica–water)} \qquad (26.58)$$

Converting from $10^3 \ln \alpha_w^m$ to $\delta\,^{18}O_m - \delta\,^{18}O_w$ and subtracting the silica equation from the calcite equation yields:

$$\delta^{18}O_{cal} - \delta^{18}O_{sil} = \frac{2.47 \times 10^3}{T} - 11.5 \qquad (26.59)$$

where T is the temperature in kelvins and both $\delta^{18}O$ values are expressed on the SMOW scale. The resulting calcite–silica O-isotope fractionation equation was plotted in Figure 26.25 for temperatures between 0 and 40°C. After the temperature of isotope equilibration has been determined from the $\delta^{18}O$ values of a pair of cogenetic minerals in a sedimentary rock, the $\delta^{18}O$ value of the water can be calculated from the mineral–water isotope fractionation equations of either one or both of the two minerals. Although the procedure is clear, it has not yet provided reliable estimates of the temperatures of the oceans in Phanerozoic and Precambrian time. Similarly, the fractionation of O isotopes between the hydroxides of Fe^{3+} (goethite) and Al^{3+} (gibbsite) relative to water in Table 26.10 can be used in principle to define thermometry equations when they are paired with the calcite–water equation of Kim and O'Neil (1997).

26.12 CHERT (PHANEROZOIC AND PRECAMBRIAN)

Amorphous biogenic or inorganically precipitated silica [H_4SiO_4 or $Si(OH)_4$] expels the water it initially contains and, in doing so, forms opal A and opal CT as intermediate phases on the way to cryptocrystalline quartz (chalcedony), which is the principal mineral of chert (Jones and Segnit, 1971). Cryptocrystalline quartz also forms flint, agate, and jasper, which differ from chert only by

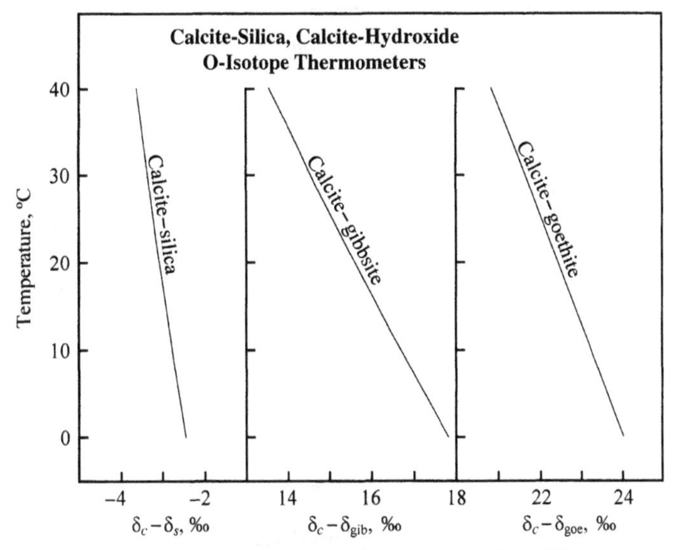

FIGURE 26.25 Oxygen-isotope thermometers based on calcite–silica, calcite–gibbsite, and calcite–geothite in the temperature range from 0 to 40°C. The calcite–water equation is by Kim and O'Neil (1997). The other mineral–water equations are from Table 26.11 based on a compilation by Chacko et al. (2001).

their color and environment of formation (Götze et al., 2001). The isotope geochemistry of chert and other forms of cryptocrystalline quartz has been investigated by:

Knauth and Epstein, 1975, *Earth Planet Sci. Lett.*, 25:1–10. Knauth and Epstein, 1976, *Geochim. Cosmochim. Acta*, 40:1095–1108. Kolodny and Epstein, 1976, *Geochim. Cosmochim. Acta*, 40:1195–1209.

Degens and Epstein (1962) first reported that the $\delta^{18}O$ values of cherts associated with carbonate rocks decrease with increasing age much like those of marine carbonate and phosphate rocks in Figures 26.17 and 26.25, respectively. This conclusion was confirmed by Perry (1967), who reported that the $\delta^{18}O$ values of chert decrease from +29.4‰ (SMOW) in the Middle Devonian Caballos Formation of Texas to only +14.1‰ in the Archean Fig Tree Group, Barberton Mountain Land, Transvaal, South Africa.

The $\delta^{18}O$ values of marine cherts of *Phanerozoic age* in Figure 26.26 scatter widely but decline with increasing age. Although most data points represent averages of up to 12 samples of the same or closely related age, there is considerable scatter, which is attributable to:

1. differences in temperature of the depositional environment,
2. differences in the $\delta^{18}O$ value of the seawater at the site of deposition, and
3. postdepositional alteration during diagenesis at elevated temperature and isotope exchange with meteoric water both of which cause chert to be depleted in ^{18}O.

The $\delta^{18}O$ values of *Precambrian cherts* in Table 26.11 are all lower than those of silica of Recent age (e.g., Labeyrie, 1974). However, the interpretation of these data is uncertain because the $\delta^{18}O$ values may have been lowered by isotopic reequilibration during diagenesis at elevated temperature (i.e., burial metamorphism) and regional metamorphism. For example, Perry et al. (1978) reported quartz–magnetite temperatures between 365 and 411°C for cherts of the iron

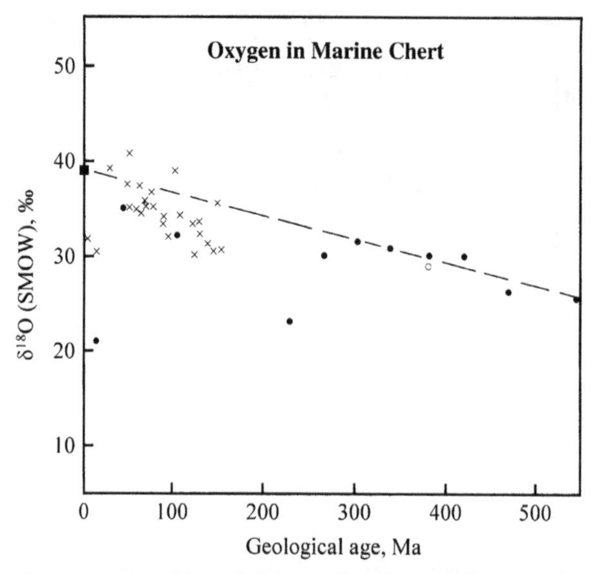

FIGURE 26.26 Isotope composition of O in marine chert of Phanerozoic age expressed in $\delta^{18}O$ values on the SMOW scale. Each data point is the average of several samples of the same or closely related ages. Solid circles: Knauth and Epstein (1976); crosses: Kolodny and Epstein (1976); open circle: Perry (1967); closed square: Labeyrie (1974). The dashed line is intended only to emphasize the trend of the data.

formation (3.8 Ga) at Isua, southern West Greenland (Oskvarek and Perry, 1976). Similar temperatures occurred in the Biwabik Iron Formation of Minnesota, especially along the contact with the Duluth Gabbro intrusion (Perry and Bonnichsen, 1966; Perry et al., 1973). Other alternatives include the possibility that the $\delta^{18}O$ value of seawater has evolved from negative values in Precambrian time to near zero at the present time, or that seawater had a higher temperature during the Precambrian than it does today. Chase and Perry (1972) modeled the isotopic evolution of O in the oceans and concluded that the $\delta^{18}O$ values of seawater increased from −12‰ at 3300 Ma to −2‰ at 2000 Ma and then rose slowly to 0‰ on the SMOW scale at the present time. However, the prevailing assessment is that the $\delta^{18}O$ values of Precambrian cherts in Figure 26.27 have been lowered by postdepositional processes and therefore cannot be used to determine the isotope composition of seawater and the temperature of the Precambrian oceans.

26.13 EXTRATERRESTRIAL ROCKS

The isotopic composition of O can be expressed both in terms of the traditional $\delta^{18}O$ parameter defined by equation 26.7 and by a $\delta^{17}O$ parameter defined by the equation:

$$\delta^{17}O = \left[\frac{(^{17}O/^{16}O)_{spl} - (^{17}O/^{16}O)_{std}}{(^{17}O/^{16}O)_{std}}\right] \times 10^3 \tag{26.60}$$

where the standard is SMOW. Both δ parameters are determined by mass spectrometric analysis of the same gas released by treating silicate samples with BrF_5 or a similar F-bearing compound. In addition, δ parameters of O are measured in-situ by means of ion-probe mass-spectrometry, which is better suited for the analysis of individual mineral grains than the conventional method based on separated mineral grains (McKeegan and Leshin, 2001).

Table 26.11. Compilation of $\delta^{18}O$ (SMOW) Values and Approximate Ages of Precambrian Marine Cherts

Geological Unit and Location	Average $\delta^{18}O$,[a] ‰	Approximate Age, Ma	References[b]
Bitter Springs Formation, Australia	+26.6 (1)	700–1000	1
Bass Limestone, Grand Canyon, AZ	+30.8 (1)	~750	2
Mescal Limestone, Roosevelt Dam, AZ	+28.8 (22)	~760	2
Siey Limestone, Belt Supergroup, MT	+19.1 (1)	1100	2
Beck Spring Dolomite, Death Valley, CA	+17.6 (1)	1800	2
Gunflint Formation Schreiber, Ontario	+23.6 (1)	1800	2
Gunflint Formation Nolalu, Ontario	+22.7 (2)	1800	2
Biwabik I.F.[c] Keewatin, MN	+22.1 (3)	1800	1
Biwabik I.F.[c] Mesabi, MN	+19.6 (1)	1800	1
Biwabik I.F.[c] Core 2, MN	+18.9 ± 0.9 (3)	1800	4
Biwabik I.F.[c] Core 5, MN	+19.7 ± 0.1 (2)	1800	4
Biwabik I.F.[c] Core 7, MN	+20.4 ± 0.5 (5)	1800	4
Wittenoom Dolomite Hamersley Range Australia	+21.0 (2)	3000	6
Fig Tree Group Transvaal, South Africa	+14.1 (3)	3000	1
Fig Tree Group Transvaal, South Africa	+14.9 (1)	3000	2
Fig Tree Group[d] Transvaal, South Africa	+18.1 ± 1.0 (4)	3000	3
Zwartkoppie Formation Onverwacht Group South Africa	+20.3 ± 0.5 (20)	3300	3
Hoogenoeg Formation Onverwacht Group South Africa	+18.4 ± 1.0 (4)	3300	3
Iron Formation, Isua, West Greenland (metamorphosed chert)	+14.1 (5)	3800	7

[a] The number of samples included in the average is indicated in parentheses.
[b] 1. Perry, 1967, *Earth Planet. Sci. Lett.*, 3:62–66. 2. Knauth and Epstein, 1976, *Geochim. Cosmochim. Acta*, 40:1095–1108. 3. Knauth and Lowe, 1978, *Earth Planet. Sci. Lett.*, 41:209–222. 4. Perry et al., 1973, *Econ. Geol.*, 68:1110–1125. 5. Perry and Tan, 1972, *Geol. Soc. Am. Bull.*, 83:647–664. 6. Becker and Clayton, 1976, *Geochim. Cosmochim. Acta*, 40:1153–1165. 7. Perry et al., 1978, *J. Geol.*, 86:223–239.
[c] I.F. means iron formation.
[d] Silicified orthochemical sediment and/or primary chert only. All other samples were excluded.

The $\delta^{18}O$ and $\delta^{17}O$ values of terrestrial samples and lunar rocks in Figure 26.28 define the *terrestrial fractionation* line (TF) whose equation is:

$$\delta^{17}O = 0.52\ \delta^{18}O + b \qquad (26.61)$$

were $b = 0$ (Clayton, 1993; McKeegan and Leshin, 2001). The variation of the two δ parameters in these samples is caused by mass-dependent fractionation of ^{17}O and ^{18}O relative to ^{16}O. In other words, enrichment of a sample in ^{16}O causes the $\delta^{17}O$ and $\delta^{18}O$ parameters to decline.

26.13a Meteorites

The isotopic composition of O in Ca–Al-rich inclusions (CAIs) and in chondrules of CV chondrites

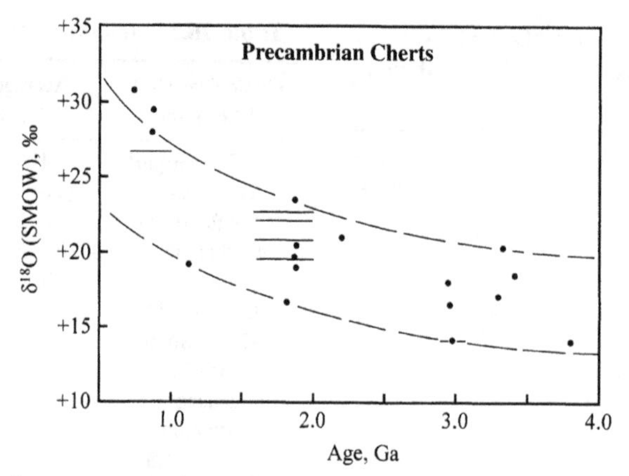

FIGURE 26.27 Isotope composition of O in cherts of Precambrian age. The bars are based on data by Perry (1967). The other data points are from Perry and Tan (1972), Perry et al. (1973), Knauth and Epstein (1976), Becker and Clayton (1976), and Knauth and Lowe (1978). The rock formations and average selected $\delta^{18}O$ values are listed in Table 26.11.

such as Allende deviates from the terrestrial fractionation line in the three-isotope diagram in Figure 26.28. The δ values of these CAIs define a separate line whose equation is:

$$\delta^{17}O = 1.0\ \delta^{18}O - 5 \qquad (26.62)$$

The CAIs in other kinds of stony meteorites and chondrules in carbonaceous chondrites as well as refractory micrometeorites and interplanetary dust particles plot along this line, which is called the *carbonaceous chondrite anhydrous mineral* line (CCAM).

The isotope composition of O that forms the CCAM line is attributable to mixing of two isotopic varieties of O that existed in the solar nebula. One of these could have had the isotope composition of point P located at the point of intersection of CCAM with TF. The O-isotope composition of the other component was characterized by negative $\delta^{18}O$ and $\delta^{17}O$ values, which implies that it consisted largely of ^{16}O. The presence of this ^{16}O-rich component indicates that O in the solar nebula was not isotopically homogeneous. A noteworthy feature of these CAIs is that the $\delta^{17}O$ and $\delta^{18}O$ values of the constituent minerals of certain inclusions of Allende and Vigarano range widely along the CCAM line.

The exotic ^{16}O-rich component did not mix with the O in the solar nebula, presumably because it was associated with solid particles that were injected into the solar nebula by the explosion of a nearby star in the Milky Way galaxy. These dust particles were subsequently incorporated into the early-formed, refractory, anhydrous Ca-Al rich inclusions (CAIs) and chondrules of the parent bodies of carbonaceous chondrites. The chondrules of ordinary chondrites cluster in area 1 above the TF line in Figure 26.28, indicating that they are somewhat enriched in ^{17}O, whereas chondrules of the enstatite chondrites occupy area 2 on the TF line located to the left of point P.

The isotopic composition of O in CAIs that define the CCAM can also be explained by *mass-independent isotope fractionation* (Thiemens, 1996, 1999). This type of isotope fractionation is caused by the effect of molecular symmetry on the kinetics of the formation of ozone from diatomic molecular O by the reaction:

$$3O_2 \rightarrow 2O_3 \quad \text{(ozone)} \qquad (26.63)$$

The data of Heidenreich and Thiemens (1982) demonstrate that the $\delta^{17}O$ and $\delta^{18}O$ values of O_3 and residual O_2 form a straight line in a three-isotope

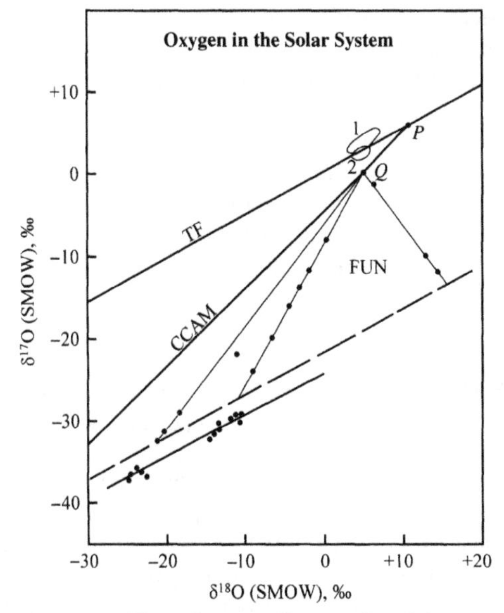

FIGURE 26.28 Three-isotope diagram for O in terrestrial and extraterrestrial rocks and minerals: TF (terrestrial fractionation) is the line defined by O in rocks on the Earth and the Moon; CCAM (carbonaceous chondrites anhydrous minerals) is defined by O in minerals of Ca–Al-rich inclusions and chondrules of CV chondrites (e.g., Allende and Vigarano); *P* is the point of intersection of the lines TF and CCAM. The areas labeled 1 and 2 are occupied by chondrules of ordinary chondrites and enstatite chondrites, respectively. Certain inclusions of Allende and Vigarano contain O affected by FUN. The minerals in these inclusions plot along mixing lines that converge to point *Q* on the CCAM. These mixing lines appear to terminate on the dashed line parallel to TF and lying close to a line defined by minerals of the FUN inclusion 1623–5 of Vigarano. The data points and the related lines were replotted from a diagram by McKeegan and Leshin (2001), which was based on data by Clayton et al. (1984), Clayton and Mayeda (1977), Clayton et al. (1977), Lee et al. (1980), and Davis et al. (1991, 2000).

diagram with a slope $m = 1.0$. The mass-independent isotope fractionation arises because the formation of ozone composed of $^{17}O^{16}O^{16}O$ and $^{18}O^{16}O^{16}O$ is favored over $^{16}O^{16}O^{16}O$.

Six CAIs of the meteorite Allende contain O whose δ values place them to the right of the CCAM line. The isotope composition of O in these CAIs is anomalous because of fractionation and unknown nuclear effects (FUN) first mentioned in Section 24.5:

Clayton et al., 1973, *Science*, 182:485–488. Clayton and Mayeda, 1977, *Geophys. Res. Lett.*, 4:295–298. Clayton et al., 1977, *Earth Planet. Sci. Lett.*, 34:209–224. Lee et al., 1980, *Geophys. Res. Lett.*, 7:493–496. Clayton and Mayeda, 1984, *Earth Planet. Sci. Lett.*, 67:151–166. Clayton et al., 1984, *Geochim. Cosmochim. Acta*, 48:535–548. Davis et al., 1991, *Geochim. Cosmochim. Acta*, 55:621–637. Davis et al., 2000, *Meteoritics Planet. Sci. Lett.*, 35:A47. McKeegan and Leshin, 2001, in Valley and Cole (Eds.), *Stable Isotope Geochemistry*, Vol. 43, pp. 279–318, Mineral. Soc. Amer. and Geochem. Soc., Blacksburg, VA.

The FUN-type inclusions of the CV chondrites Allende and Vigarano plot along straight lines in Figure 26.28 that converge to a point (Q) on the CCAM line whose approximate coordinates are $\delta^{18}O = +5‰$, $\delta^{17}O = 0‰$. The individual lines terminate on the dashed mass-fractionation line parallel to TF. This interpretation is supported by the δ values of olivine, pyroxene, and spinel of FUN CAI 1623–5 of Vigarano analyzed by Davis et al. (2000). However, melilite in this inclusion plots close to point Q at the convergence of the FUN mixing lines with CCAM.

26.13b Martian Rocks

The number of meteorites known to have originated from Mars continues to rise as new specimens are identified (McSween and Treiman, 1998). Isotope analyses of O of these martian rocks by Franchi et al. (1999), Clayton and Mayeda (1983, 1986, 1996), and Romanek et al. (1998) demonstrate that the isotopic δ values define a straight line on the three-isotope diagrams. The equation of this Mars silicate fractionation line (MSF) based on the data of Franchi et al. (1999) is:

$$\delta^{17}O = 0.52\ \delta^{18}O + 0.321 \qquad (26.63)$$

This result is remarkable because it means that the O in the lithosphere of Mars was isotopically homogenized soon after the formation of Mars (McKeegan and Leshin, 2001).

The $\delta^{18}O$ values of secondary carbonates in the martian meteorite ALH84001 range widely from near 0‰ to +25‰ (Leshin et al., 1998; Saxton et al., 1998) and correlate with the chemical composition over distances of about 100 μm. The variation of the $\delta^{18}O$ values is attributable to:

1. variation of the temperature during the formation of the carbonate mineral,
2. changes in the $\delta^{18}O$ value of the aqueous fluid caused by evaporation, and
3. changes in the $\delta^{18}O$ value of a CO_2-rich fluid by Rayleigh distillation in a closed system.

The martian meteorites also contain hydrous minerals(e.g., Eiler et al., 2002), including grains of apatite in specimen QUE 94201 collected in Antarctica. The δD values of the hydroxyl radical range widely from about +1750‰ to greater than +3500‰ and correlate negatively with the concentration of water. Leshin (2000) concluded that the water in the apatite of this martian meteorite is a mixture of two components representing water vapor in the atmosphere ($\delta D \simeq +4000$‰) and magmatic water ($\delta D \simeq +900 \pm 250$‰) relative to SMOW. The H in the organic fraction of carbonaceous chondrites is likewise enriched in D with δD values that range up to +3500‰ (Kolodny et al., 1980).

26.13c Moon

The $\delta^{18}O$ values of lunar rocks range narrowly from +4.2 to +6.4‰ (Taylor and Epstein, 1970; Clayton et al., 1973; Clayton and Mayeda, 1975, 1996). The $\delta^{17}O$ and $\delta^{18}O$ values define a straight line on the three-isotope diagram that is indistinguishable from the TF line in Figure 26.28. Evidently, the Earth and Moon derived O from the same reservoir in the solar nebula. Exotic isotope compositions of O like those of CAIs and chondrules have not been found in any lunar and terrestrial mineral grains.

The apparent similarity of the isotope compositions of lunar and terrestrial O is significant because the Moon is thought to have formed by an impact of a large body on the Earth (Benz et al., 1986). If the impactor had contained O whose isotope composition matched that of Mars, the isotope composition of lunar O should be distinguishable from that of terrestrial O. The good match of lunar and terrestrial O implies that the O-delta values of the impactor satisfied equation 26.61 for the TF line in Figure 26.28.

26.13d Nucleosynthesis of O Isotopes

Each of the three stable isotopes of O is produced by a different nuclear process operating in different places and at different stages in the evolution of stars (Anders and Grevesse, 1989). Whereas ^{16}O (99.76%) is formed by He burning, ^{17}O (0.04%) is produced by the hot CNO process in H-rich media, and ^{18}O (0.20%) is synthesized by the same process in He-rich regions of stars (Clayton, 1993). Rapid cooling of gases ejected by stars at the end of their life cycle causes O to be partitioned into solid oxide and silicate particles as well as into molecules (CO and H_2O) that remained in the gas phase (Lattimer et al., 1978).

Therefore, the partitioning of O of different isotopic compositions between solids and gases inhibited the isotopic homogenization of O in the solar nebula and allowed solids forming within it to inherit O that was variably enriched in ^{16}O. The observed systematic variations of the isotope composition of O in different kinds of inclusions in meteorites (i.e., CCAM in Figure 26.28) can be explained either in terms of *mixing* of O in an isotopically heterogeneous solar nebula or by *mass-independent isotope fractionation* of O that was initially isotopically homogeneous. Therefore, the anomalous isotope composition of O in CAIs and chondrules of carbonaceous chondrites either was the result of nucleosynthesis of O isotopes in the ancestral stars or was caused by mass-independent isotope fraction in the solar nebula, which can produce arrays having a slope $m = 1.0$ in the three-element diagram.

Clayton (1993) concluded that the range of isotope compositions of O in meteorites and planets can be accounted for by the existence of two O-reservoirs in the solar nebua: dust particles enriched in ^{16}O and gas containing all three isotopes of

O in constant proportions. According to Clayton (1993), the isotopic composition of O in the *dust* component was:

$$\delta^{18}O = -40‰ \qquad \delta^{17}O = -42‰$$

whereas the O in the gas phase of the solar *nebula* was characterized by:

$$\delta^{18}O = +8‰ \qquad \delta^{17}O = +7‰$$

This composition represented O in the gas phase *after* isotope exchange with the ^{16}O-rich dust reservoir had occurred. The isotope composition of O in the gas reservoir *before* exchange with the O of the dust reservoir, estimated by Clayton and Mayeda (1986), was:

$$\delta^{18}O = +30.0‰ \qquad \delta^{17}O = +24.2‰$$

26.14 SUMMARY

Oxygen is a major constituent of silicate, oxide, carbonate, phosphate, and sulfate minerals in the crust of the Earth. In addition, it occurs in the water molecules that fill the oceanic basins and form the continental ice sheets of Greenland and Antarctica. Its three stable isotopes have sufficiently different masses to cause mass-dependent isotope fractionation during changes in state (e.g., evaporation of liquid water and condensation of the vapor), as well as by means of isotope exchange reactions between O-bearing solids and water, and among coexisting solids. The magnitude of the isotope fractionation increases with decreasing temperature, which means that the isotope compositions of O in coexisting minerals and water can be used as thermometers, provided that isotope equilibrium was established between them.

The description of isotope fractionation of O is based on equations that apply to all other elements whose isotopes are fractionated. These equations include the definition of the isotope fractionation factor α between two phases (a and b):

$$\alpha_b^a = \frac{R_a}{R_b}$$

were R is the isotope ratio of the heavy isotope to the light isotope. The isotope composition is expressed by the delta (δ) notation:

$$\delta^{18}O = \left(\frac{R_{spl} - R_{std}}{R_{std}}\right) \times 10^3‰$$

where the standard is SMOW (water, silicates, oxides, etc.) or PDB (carbonates). The fractionation factor α_b^a is related to the δ parameter by:

$$\alpha_b^a = \frac{\delta^{18}O_a + 1000}{\delta^{18}O_b + 1000}$$

and the temperature dependence of α_b^a is stated in the form:

$$10^3 \ln \alpha_b^a = \frac{A}{T^2} + B$$

where the temperature T is in kelvins and A and B are obtained by fitting a polynomial equation to experimentally determined data plotted in coordinates of $1/T^2$ (x-coordinate) and $10^3 \ln \alpha_b^a$ (y-coordinate).

The isotope geochemistry of O encompasses a wide range of subjects, including the transfer of water among the reservoirs of the hydrologic cycle, the record of climate change preserved in ice cores and biogenic calcium carbonate, the isotope composition of O in mantle-derived volcanic rocks, and the measurement of isotope-equilibration temperatures of cooling igneous and metamorphic rocks. In addition, the exchange of O isotopes between water and rocks at elevated temperatures in the subsurface leaves an isotopic signature in both the water and the rocks. Similarly, clay minerals that form by incongruent dissolution of alumino-silicate minerals contain an imprint of the O in the meteoric water that participated in their formation.

The O of marine carbonate, phosphate, and chert of Phanerozoic and Precambrian age is depleted in ^{18}O with increasing age, which implies that the temperature of the oceans of the geological past was higher than it is at present, or that the $\delta^{18}O$ value of seawater has changed with time, or that the isotopic composition of O in these minerals was altered during diagenesis and low-grade regional metamorphism. All three alternatives deserve serious consideration.

Another important aspect of the isotope geochemistry of O arises from the linear relationship between $\delta^{18}O$ and $\delta^{17}O$ values of minerals in the anhydrous minerals of Ca–Al-rich inclusions (CAIs) of carbonaceous chondrites and related stony meteorites. Although the linear relationship could have resulted from *mass-independent* isotope fractionation, it was probably caused by mixing of O from two reservoirs in the solar nebula. One of these reservoirs was composed of dust particles containing O enriched in ^{16}O, which was synthesized in one of the ancestral stars. The isotopic heterogeneity of O in the solar nebula has been preserved in early-formed solids that were later incorporated into larger bodies (i.e., the parent bodies of meteorites). The isotope composition of O in the CAIs was not homogenized because the minerals of carbonaceous chondrites were never heated to temperatures in excess of about 125°C.

REFERENCES FOR INTRODUCTION

Clark, I. D., and P. Fritz, 1997. *Environmental Isotopes in Hydrogeology*. Lewis, New York.

Craig, H., S. L. Miller, and G. J. Wasserburg (Eds.), 1964. *Isotopic and Cosmic Chemistry*. North-Holland, Amsterdam.

Criss, R. E., 1999. *Principles of Stable Isotope Distribution*. Oxford University Press, New York.

Fritz, P., and J. Ch. Fontes (Eds.), 1980. *Handbook of Environmental Isotope Geochemistry*, Vol. 1. Elsevier Scientific, Amsterdam.

Fritz, P., and J. Ch. Fontes (Eds.), 1986. *Handbook of Environmental Isotope Geochemistry*, Vol. 2. Elsevier Scientific, Amsterdam.

Hoefs, J., 1997. *Stable Isotope Geochemistry*, 4th ed. Springer-Verlag, Heidelberg.

Taylor, H. P., Jr., J. R. O'Neil, and I. R. Kaplan (Eds.), 1991. *Stable Isotope Geochemistry: A Tribute* to Samuel Epstein, Special Pub. No. 3. The Geochemical Society. San Antonio, Texas.

Valley, J. W., and D. R. Cole (Eds.), 2001. *Stable isotope geochemistry. Rev. Mineral. Geochem.*, 43. Mineralogical Society of America, Geochemical Society, Virginia Polytechnic Institute, Virginia State University, Blacksburg, Virginia.

Valley, J. W., H. P. Taylor, Jr., and J. R. O'Neil (Eds.), 1986. *Stable Isotopes in High Temperature Geological Processes*. Rev. Mineral., 16. Mineralogical Society of America, Virginia Polytechnic Institute, State University, Blacksburg, Virginia.

REFERENCES FOR ISOTOPE FRACTIONATION (SECTIONS 26.1–26.2)

Bigeleisen, J., 1965. Chemistry of isotopes. *Science*, 147:463–471.

Bottinga, Y., and M. Javoy, 1973. Comments on oxygen isotope geothermometry. *Earth Planet. Sci. Lett.*, 20:250–265.

Gat, J. R., 1980. The stable isotopes of Dead Sea waters. *Earth Planet. Sci. Lett.*, 71:361–376.

Hoefs, J., 1997. *Stable Isotope Geochemistry*, 4th ed. Springer-Verlag, Heidelberg.

Kieffer, S. W., 1982. Thermodynamics and lattice vibrations of minerals: 5. Applications to phase equilibria, isotopic fractionation, and high-pressure thermodynamic processes. *Rev. Geophys. Space Phys.*, 20:827–849.

Lide, D. R., and H. P. R. Frederikse, 1995. *Handbook of Chemistry and Physics*, 76th ed., CRC Press, Boca Raton, Florida.

Melander, L., and W. H. Saunders, 1980. *Reactions Rates of Isotopic Molecules*. Wiley, New York.

O'Neil, J. R., 1986. Theoretical and experimental aspects of isotope fractionation. In J. R. Valley, H. P. Taylor, Jr., and J. R. O'Neil (eds.), *Stable Isotopes in High Temperature Geological Processes*. Rev. Mineral., 16:1–40. Mineralogical Society of America, Virginia Polytechnic Institute, Virginia State University, Blacksburg, Virginia.

Richet, P., Y. Bottinga, and M. Javoy, 1977. A review of H, C, N, O, S, and Cl stable isotope fractionation among gaseous molecules. *Annu. Rev. Earth Planet. Sci.*, 5:65–110.

Urey, H. C., 1947. The thermodynamic properties of isotopic substances. *J. Chem. Soc. (Lond.)*, 1947: 562–581.

REFERENCES FOR METEORIC PRECIPITATION (SECTIONS 26.3a–26.3c)

Craig, H., 1961. Isotopic variations in meteoric waters. *Science*, 133:1702–1703.

Dansgaard, W., 1964. Stable isotopes in precipitation. *Tellus*, 16:436–468.

Gat, J. R., 1980. The isotopes of hydrogen and oxygen in precipitation. In P. Fritz and J. Ch. Fontes (Eds.), *Handbook of Environmental Isotope Geochemistry*, Vol. 1, pp. 21–47. Elsevier Scientific, Amsterdam.

Gat, J. R., 1984. The stable isotope composition of Dead Sea waters. *Earth Planet. Sci. Lett.*, 71:361–376.

Hoefs, J., 1997. *Stable Isotope Geochemistry*, 4th ed. Springer-Verlag, Heidelberg.

Horita, J., and D. J. Wesolowski, 1994. Liquid-vapor fractionation of oxygen and hydrogen isotopes of water from freezing to the critical temperature. *Geochim. Cosmochim. Acta*, 58:3425–3437.

Melander, L., and W. H. Saunders, 1980. *Reaction Rates of Isotopic Molecules*. Wiley, New York.

Rayleigh, J. W. S., 1896. Theoretical considerations respecting the separation of gases by diffusion and similar processes. *Philos. Mag.*, 42:493.

Rozanski, K., L. Araguas-Araguas, and R. Conginatini, 1993. Isotopic patterns in modern global precipitation. In *Climate Change in Continental Isotopic Records*, Geophys. Monogr. No. 78, pp. 1–36. American Geophysical Union, Washington, D.C.

Yurtsever, Y., 1975. *Worldwide Survey of Stable Isotopes in Precipitation*, Rept. Sect. Isotope Hydrol. International Atomic Energy Agency, Vienna.

REFERENCES FOR CLIMATE RECORDS IN ICE CORES (SECTION 26.3D)

Dansgaard, W., S. J. Johnsen, and H. B. Clausen, 1993. Evidence for general instability of past climate from a 250 kyr ice-core record. *Nature*, 364:281–220.

Gibbs, W. W., 2001. Out in the cold; ambitious plans to penetrate icebound Lake Vostok have slowed to a crawl. *Sci. Am.*, 284(3):16–17.

Grootes, P. M., M. Stuiver, J. W. C. White, S. J. Johnsen, and J. Jouzel, 1993. Comparison of oxygen isotope records from the GISP-2 and GRIP Greenland ice cores. *Nature*, 336:552–554.

Johnsen, S. J., W. Dansgaard, H. B. Clausen, and C. C. Langway, 1972. Oxygen isotope profiles through the Antarctic and Greenland ice sheets. *Nature*, 235:429–434.

Lorius, C., J. Jouzel, C. Ritz, L. Merlivat, N. I. Barkov, Y. S. Korotkevich, and V. M. Kotlyakov, 1985. A 150,000-year climatic record from Antarctic ice. *Nature*, 316:591–596.

Thompson, L. G., S. Hastenrath, and B. M. Arnao, 1979. Climatic ice core records from the tropical Quelccaya ice cap. *Science*, 203:1240–1243.

REFERENCES FOR PALEOTHERMOMETRY (CARBONATES) (SECTION 26.4)

Adkins, J. F., E. A. Boyle, W. B. Curry, and A. Lutringer, 2003. Stable isotopes in deep-sea corals and a new mechanism for "vital effect." *Geochim. Cosmochim. Acta*, 67:1129–1143.

Anderson, T. F., and M. A. Arthur, 1983. Stable isotopes of oxygen and carbon and their application to sedimentologic and paleoenvironmental problems. In M. A. Arthur, R. F. Anderson, I. R. Kaplan, J. Veizer, and L. S. Land (Eds.), *Stable Isotopes in Sedimentary Geology*, SEPM Short Course No. 10. Soc. Econ. Paleont. Mineral., Tulsa, Oklahoma.

Chacko, T., D. R. Cole, and J. Horita, 2001. Equilibrium oxygen, hydrogen, and carbon isotope fractionation factors applicable to geologic systems. In J. W. Valley and D. R. Cole (eds.), *Stable Isotope Geochemistry. Rev. Mineral. Geochim.*, 43:1–61. Mineral. Soc. Amer., Virginia Polytech. Inst., State Univ., Blacksburg, Virginia.

Coplen, T. B., C. Kendall, and J. Hopple, 1983. Comparison of stable isotope reference samples. *Nature*, 302:236–238.

Craig, H., 1965. The measurement of oxygen isotope paleotemperatures. In *Stable Isotopes in Oceanographic Studies and Paleotemperatures*, pp. 1–24. Consiglio Nazionale delle Ricerche, Laboratorio di Geologia Nucleare, Pisa.

Craig, H., S. L. Miller, and G. J. Wasserburg (Eds.), 1964. *Isotopic and Cosmic Chemistry*. North-Holland, Amsterdam.

Emiliani, C., 1955. Pleistocene temperatures. *J. Geol.*, 63:538–578.

Emiliani, C., 1966. Isotopic paleotemperatures. *Science*, 154:851–857.

Emiliani, C., 1972. Quaternary paleotemperatures and the duration of the high-temperature intervals. *Science*, 178:398–401.

Emiliani, C., 1978. The cause of the ice ages. *Earth Planet. Sci. Lett.*, 38:349–352.

Emiliani, C., and N. J. Shackleton, 1974. The Brunhes epoch: Isotopic paleotemperatures and geochronology. *Science*, 183:511–514.

Faure, G., 1998. *Principles and Applications of Geochemistry*, 2nd ed. Prentice-Hall, Upper Saddle River, New Jersey.

Harland, W. B., R. L. Armstrong, A. V. Cox, L. E. Craig, A. G. Smith, and D. G. Smith, 1989. *A geologic time scale 1989*. Cambridge University Press. Cambridge, United Kingdom.

Hays, J. D., J. Imbrie, and N. J. Shackleton, 1976. Variation in the earth's orbit: Pacemaker of the ice ages. *Science*, 194:943–954.

Hoefs, J., 1997. *Stable Isotope Geochemistry*, 4th ed. Springer-Verlag, Heidelberg.

Kim, S.-T., and J. R. O'Neil, 1997. Equilibrium and nonequilibrium oxygen isotopic effect in synthetic carbonates. *Geochim. Cosmochim. Acta*, 61:3461–3475.

Lehmann, M. and U. Siegenthaler, 1991. Equilibrium oxygen- and hydrogen- isotope fractionation between ice and water. *J. Glaciol.*, 37:23–26.

McConnaughey, T., 1989a. ^{13}C and ^{18}O disequilibrium in biological carbonates. I. Patterns. *Geochim. Cosmochim. Acta*, 53:151–162.

McConnaughey, T., 1989b. ^{13}C and ^{18}O disequilibrium in biological carbonates. II. In vitro simulation of kinetic isotope effects. *Geochim. Cosmochim. Acta*, 53:163–171.

Raymo, M. E., and W. F. Ruddiman, 1992. Tectonic forcing of late Cenozoic climate. *Nature*, 359:117–122.

Shackleton, N. J., 1968. Depth of pelagic foraminifera and isotopic changes in Pleistocene oceans. *Nature*, 218:79–80.

Swart, P. K., S. J. Burns, and J. L. Leder, 1991. Fractionation of the stable isotopes of oxygen and carbon in carbon dioxide during the reaction of calcite with phosphoric acid as a function of temperature and technique. *Chem. Geol.*, 86:89–96.

Urey, H. C., 1947. The thermodynamic properties of isotopic substances. *J. Chem. Soc. (Lond.)*, 1947:562–581.

Usdowski, E., and J. Hoefs, 1993. Oxygen isotope exchange between carbonic acid, bicarbonate, carbonate, and water: A re-examination of the data of McCrea (1950) and an expression for the overall partitioning of oxygen isotopes between carbonate species and water. *Geochim. Cosmochim. Acta*, 57:3815–3818.

Weber, J. N., 1968. Fractionation of the stable isotopes of carbon and oxygen in calcareous marine invertebrates: The Asteroidea, Ophiuroidea, and Crinodea. *Geochim. Cosmochim. Acta*, 32:33–70.

Weber, J. N., and D. M. Raup, 1966a. Fractionation of the stable isotopes of carbon and oxygen in marine calcareous organisms. The Echinoidea. I. Variation of ^{13}C and ^{18}O content within individuals. *Geochim. Cosmochim. Acta*, 30:681–703.

Weber, J. N., and A. M. Raup, 1966b. Fractionation of stable isotopes of carbon and oxygen in marine calcareous organisms. II. Environmental and genetic factors. *Geochim. Cosmochim. Acta*, 30:705–736.

Wefer, G., and W. H. Berger, 1991. Isotope paleontology: Growth and composition of extant calcareous species. *Marine Geol.*, 100:207–248.

REFERENCES FOR SILICATES: BASALT AND THE MANTLE (SECTIONS 26.5–26.5a)

Borthwick, J., and R. S. Harmon, 1982. A note regarding ClF_3 as an alternative to BrF_5 for oxygen isotope analysis. *Geochim. Cosmochim. Acta*, 46:1665–1668.

Clayton, R. N., and T. K. Mayeda, 1963. The use of bromine pentafluoride in the extraction of oxygen from oxides and silicates for isotopic analysis. *Geochim. Cosmochim Acta*, 27:43–52.

Condomines, M., K. Grönvold, P. J. Hooker, K. Muehlenbachs, R. K. O'Nions, N. Oskarsson, and E. R. Oxburgh, 1983. Helium, oxygen, strontium, and neodymium isotopic relationships in Icelandic volcanics. *Earth Planet. Sci. Lett.*, 66:125–136.

Eiler, J. M., 2001. Oxygen isotope variations of basaltic lavas and upper mantle rocks. In J. W. Valley and D. R. Cole (Eds.), *Stable Isotope Geochemistry*, pp. 319–364. Rev. Mineral. Geochem., 43. Mineral. Soc. Amer., Geochem. Soc., Virginia Polytech Inst., State Univ., Blacksburg, Virginia.

Faure, G., 2001. *The Origin of Igneous Rocks; the Isotopic Evidence*. Springer Verlag, Heidelberg.

Harmon, R. S., and J. Hoefs, 1995. Oxygen isotope heterogeneity of the mantle deduced from global ^{18}O systematics of basalts from different geotectonic settings. *Contrib. Mineral. Petrol.*, 120:95–114.

Hemond, Ch., N. T. Arndt, U. Lichtenstein, and A. W. Hofmann, 1993. The heterogeneous Iceland plume: Nd–Sr–O isotopes and trace element constraints. *J. Geophys. Res.*, 98(B9):15833–15850.

Hemond, Ch., M. Condomines, S. Fourcade, C. J.Allègre, N. Oskarsson, and M. Javoy, 1988. Thorium, strontium, and oxygen isotopic geochemistry in recent tholeiites from Iceland: Crustal influence on mantle derived magmas. *Earth Planeti. Sci. Lett.*, 87:273–285.

Hoefs, J., 1997. *Stable Isotope Geochemistry*, 4th ed. Springer-Verlag, Heidelberg.

Hofmann, A. W., 1997. Mantle geochemistry: The message from oceanic volcanism. *Nature*, 385:219–299.

Hofmann, A. W., and W. M. White, 1982. Mantle plumes from ancient oceanic crust. *Earth Planet. Sci. Lett.*, 57:421–436.

Nicholson, H., M. Condomines, J. G. Fitton, A. E. Fallick, K. Grönvold, and G. Rogers, 1991. Geochemical and isotopic evidence for crustal assimilation beneath Krafla, Iceland. *J. Petrol.*, 32(5):1005–1020.

Sharp, Z. D., 1990. A laser-based microanalytical method for the in-situ determination of oxygen isotope ratios of silicates and oxides. *Geochim. Cosmochim. Acta*, 54:1353–1357.

Sigmarsson, O., M. Condomines, and S. Fourcade, 1992. Mantle and crustal contribution in the genesis of Recent basalts from off-rift zones in Iceland: Constraints from Th, Sr, and O isotopes. *Earth Planet., Sci. Lett.*, 110:149–162.

Valley, J. W., and C. Graham, 1993. Cryptic grain-scale heterogeneity of oxygen isotope ratios in metamorphic magnetite. *Science*, 259:1729–1733.

Woodhead, J. D., P. Greenwood, R. S. Harmon, and P. Stoffers, 1993. Oxygen isotope evidence for recycled crust in the source of EM-type ocean island basalts. *Nature*, 362:809–813.

REFERENCES FOR THERMOMETRY OF SILICATES AND OXIDES (SECTION 26.5b)

Anderson, A. T., Jr., R. N. Clayton, and T. K. Mayeda, 1971. Oxygen isotope thermometry of mafic igneous rocks. *J. Geol.*, 79:715–729.

Chacko, T., X. Hu, T. K. Mayeda, R. N. Clayton, and J. R. Goldsmith, 1996. Oxygen isotope fractionation in muscovite, phlogopite, and rutile. *Geochim. Cosmochim. Acta*, 60:2595–2608.

Chiba, H., T. Chacko, R. N. Clayton, and J. R. Goldsmith, 1989. Oxygen isotope fractionation involving diopside, forsterite, magnetite, and calcite: Application to geothermometry. *Geochim. Cosmochim. Acta*, 53:2985–2995.

Clayton, R. N., J. M. Hurd, and T. K. Mayeda, 1972. Oxygen isotopic compositions of Apollo 14 and Apollo 15 rocks and soils. In *Proc. 3rd Lunar Planet. Sci. Conf.*, pp. 1455–1463. Houston, Texas.

Clayton, R. N., N. Onuma, and T. K. Mayeda, 1971. Oxygen isotope fractionation in Apollo 12 rocks and soils. *Geochim. Cosmochim. Acta*, Suppl. 2:1417–1420.

Garlick, G. D., 1966. Oxygen isotope fractionation in igneous rocks. *Earth Planet. Sci Lett.*, 1:361–368.

Kieffer, S. W., 1982. Thermodynamics and lattice vibrations of minerals: 5. Applications to phase equilibria, isotopic fractionation, and high-pressure thermodynamic processes. *Rev. Geophys. Space Phys.*, 20: 827–849.

Onuma, N., R. N. Clayton, and T. K. Mayeda, 1970. Apollo 11 rocks: Oxygen isotope fractionation between minerals and an estimate of the temperature of formation. *Geochim. Cosmochim. Acta*, Suppl. 1:1429–1434.

REFERENCES FOR WATER–ROCK INTERACTIONS (ROCKS) (SECTION 26.6)

Bowers, T. S., 1989. Stable isotope signatures of water-rock interaction in mid-ocean ridge hydrothermal systems: Sulfur, oxygen, and hydrogen. *J. Geophys. Res.*, 94:5775–5786.

Bowers, T. S., and H. P. Taylor, 1985. An integrated chemical and isotope model of the origin of mid-ocean ridge hot spring systems. *J. Geophys. Res.*, 90:12583–12606.

Criss, R. E., R. T. Gregory, and H. P. Taylor, 1987. Kinetic theory of oxygen isotopic exchange between minerals and water. *Geochim. Cosmochim. Acta*, 51:1099–1108.

Criss, R. E., R. J. Fleck, and H. P. Taylor, 1991. Tertiary meteoric hydrothermal systems and their relation to ore deposits, northwestern United States and southern British Columbia. *J. Geophys. Res.*, 96:13335–13356.

Criss, R. E., and H. P. Taylor, 1986. Meteoric-hydrothermal systems. In J. W. Valley, H. P. Taylor, Jr., and J. R. O'Neil (Eds.), *Stable Isotopes in High Temperatures Geologic Processes. Rev. Mineral.*, 16:373–424. Mineral. Soc. Amer., Virginia Polytech. Inst., State Univ., Blacksburg, Virginia.

Gregory, R. T., and R. E. Criss, 1986. Isotopic exchange in open and closed systems. In J. W. Valley, H. P. Taylor, Jr., and J. R. O'Neil (Eds.), *Stable Isotopes in High-Temperature Geological Processes. Rev. Mineral.*, 16:91–127. Mineral. Soc. Amer., Virginia Polytech. Inst., State Univ., Blacksburg, Virginia.

Masuda, H., H. Sakai, H. Chiba, Y. Matsuhisa, and T. Nakamura, 1986. Stable isotopic and mineralogical studies of hydrothermal alteration at Arima Spa, southwest Japan. *Geochim. Cosmochim. Acta*, 50:19–28.

Ohmoto, H., 1974. Hydrogen and oxygen isotopic compositions of fluid inclusions in the Kuroko deposits, Japan. *Econ. Geol.*, 69:947–953.

Ohmoto, H., 1986. Stable isotope geochemistry of ore deposits. In J. W. Valley, H. P. Taylor, Jr., J. R. O'Neil (Eds.), *Stable Isotopes in High-Temperatures Geological Processes. Rev. Mineral.*, 16:491–559. Mineral. Soc. Amer., Virginia Polytech. Inst., State Univ., Blacksburg, Virginia.

Ohmoto, H., M. Mizukami, S. E. Drummond, C. S. Eldridge, V. Pisutha-Armond, and T. C. Lenagh, 1983. Chemical processes of Kuroko formation. *Econ. Geol. Monogr.*, 5:570–604.

Ohmoto, H., and R. O. Rye, 1970. The Bluebell mine, British Columbia, I. Mineralogy, paragenesis, fluid inclusions, and the isotopes of hydrogen, oxygen and carbon. *Econ. Geol.*, 65:417–437.

Ohmoto, H., and B. J. Skinner (Eds.), 1983. The Kuroko and related volcanogenic massive sulfide deposits. *Econ. Geol.* Monogr., 5. 604.

Taylor, H. P., 1968. The oxygen isotopes geochemistry of igneous rocks. *Contrib. Mineral. Petrol.*, 19:1–17.

Taylor, H. P. Jr., 1971. Oxygen-isotope evidence for large-scale interaction between meteoric groundwater and Tertiary granodiorite intrusions, western Cascade Range, Oregon. *J. Geophys. Res.*, 76(32):7855–7874.

Taylor, H. P., 1974. The application of oxygen and hydrogen isotopic studies to problems of hydrothermal alteration and ore deposition. *Econ. Geol.*, 67:843–883.

Taylor, H. P., 1979. Oxygen and hydrogen isotope relationships in hydrothermal deposits. In H. L. Barnes (Ed.), *Geochemistry of Hydrothermal Ore Deposits*, 2nd ed., pp. 237–277. Wiley, New York.

Taylor, H. P., 1987. Comparison of hydrothermal systems in layered gabbros and granites and the origin of low-$\delta^{18}O$ magmas. In B. O. Mysen, ed., *Magmatic Processes: Physicochemical principles*, Geochem. Soc. Spec. Publ. 1, pp. 337–357. Geochem. Soc., San Antonio, Texas.

Taylor, H. P., and R. W. Forester, 1979. An oxygen and hydrogen isotope study of the Skaergaard intrusion and its country rocks: A description of a 55-My-old fossil hydrothermal system. *J. Petrol.*, 20:355–419.

Wenner, D. B., and H. P. Taylor, Jr., 1976. Oxygen and hydrogen isotope studies of a Precambrian granite-rhyolite terrane, St. Francois Mountains, southeastern Missouri. *Geol. Soc. Am. Bull.*, 87:1587–1598.

White, D. E., 1974. Diverse origins of hydrothermal ore fluids. *Econ. Geol.*, 69:954–973.

REFERENCES FOR WATER-ROCK INTERACTIONS (WATER) (SECTION 26.7)

Chacko, T., D. R. Cole, and J. Horita, 2001. Equilibrium oxygen, hydrogen, and carbon isotope fractionation factors applicable to geologic systems. In J. W. Valley and D. R. Cole, (eds.), *Stable Isotope Geochemistry. Rev. Mineral. Geochim.*, 43:1–81. Mineral. Soc. Amer., Geochem. Soc., Virginia Polytech. Inst., State Univ., Blacksburg, Virginia.

Clayton, R. N., I. Friedman, D. L. Graf, T. K. Mayeda, W. F. Meents, and N. F. Shimp, 1966. The origin of saline formation water. 1. Isotopic composition. *J. Geophys. Res.*, 71(16):3869–3882.

Coplen, T. B., and B. B. Hanshaw, 1973. Ultrafiltration by a compacted clay membrane. I. Oxygen and hydrogen isotope fractionation. *Geochim. Cosmochim. Acta*, 37:2295–2310.

Craig, H., 1963. The isotopic geochemistry of water and carbon in geothermal areas. In *Nuclear Geology of Geothermal Areas*, pp. 17–53. Consiglio Nazionale delle Ricerche, Laboratorio di Geologia Nucleare, Pisa, Italy.

Craig., H., 1966. Isotopic composition and origin of the Red Sea and Salton Sea geothermal brines. *Science*, 154:1544–1548.

Criss, R. E., and H. P. Taylor, 1986. Meteoric-hydrothermal systems. In J. W. Valley, H. P. Taylor, Jr., and J. R. O'Neil (Eds.), *Stable Isotopes in High Temperature Geological Processes. Rev. Mineral.*, 16:373–424. Mineral. Soc. Amer., Blacksburg, Virginia.

Dansgaard, W., 1964. Stable isotopes in precipitation. *Tellus*, 16(4):436–468.

Degens, E., J. M. Hunt, J. H. Reuter, and W. E. Reed, 1964. Data on the distribution of amino acids and oxygen isotopes in petroleum brine waters of various geological ages. *Sedimentology*, 3:199–225.

Frape, S. K., and P. Fritz, 1982. The chemistry and isotopic composition of saline groundwater from the Sudbury Basin, Ontario. *Can. J. Earth Sci.*, 19:645–661.

Frape, S. K., and P. Fritz, 1987. Geochemical trends from groundwaters from the Canadian Shield. In P. Fritz and S. K. Frape (Eds.), *Saline Water and Gases in Crystalline Rocks*. Spec. Paper 33, pp. 19–38. Geological Association of Canada, Waterloo, Canada.

Frape, S. K., P. Fritz, and R. H. McNutt, 1984. Water-rock interaction and chemistry of groundwaters from the Canadian Shield. *Geochim. Cosmochim. Acta*, 48:1617–1627.

Fritz, P., and S. K. Frape, 1982. Saline groundwaters on the Canadian Shield: A first overview. *Chem. Geol.*, 36:179–190.

Graf, D. L., I. Friedman, and W. F. Meents, 1965. *The Origin of Saline Formation Water, 2; Isotopic Fractionation by Shale Micropore Systems*. Illinois State Geol. Surv. Circ. No. 393.

Hoefs, J., 1997. *Stable Isotope Geochemistry*, 4th ed. Springer-Verlag, Heidelberg.

Kelly, W. C., R. O. Rye, and A. Livnat, 1986. Saline minewaters of the Keweenaw peninsula, northern Michigan: Their nature, origin, and relation to similar deep water in Precambrian crystalline rocks of the Canadian Shield. *Am. J. Sci.*, 286:281–308.

Knauth, L. P., and M. A. Beeunas, 1986. Isotope geochemistry of fluid inclusions in Permian halite with implications for the isotopic history of ocean water and the origin of saline formation waters. *Geochim. Cosmochim. Acta*, 50:419–433.

Phillips, F. M., and H. W. Bentley, 1987. Isotopic fractionation during ion filtration. I. Theory. *Geochim. Cosmochim. Acta*, 51:683–695.

Savin, S. M., and S. Epstein, 1970. The oxygen and hydrogen isotope geochemistry of clay minerals. *Geochim. Cosmochim. Acta*, 34:25–42.

Taylor, H. P., Jr., 1977. Water/rock interactions and the origin of H_2O in granite batholiths. *J. Geol. Soc. (Lond.)*, 133:509–558.

Truesdell, A. H., and J. R. Hulston, 1980. Isotopic evidence on environments of geothermal systems. In

P. Fritz and J. Fontes (Eds.), *Handbook of Environmental Isotope Geochemistry*, Vol. 1, pp. 179–226. Elsevier, New York.

Yurtsever, Y., 1975. *Worldwide Survey of Stable Isotopes in Precipitation*, Rept. Section Isotope Hydrology. International Atomic Energy Agency, Vienna.

REFERENCES FOR CLAY MINERALS (SECTION 26.8)

Chacko, T., D. R. Cole, and J. Horita, 2001. Equilibrium oxygen, hydrogen, and carbon isotope fractionation factors applicable to geological systems. In J. W. Valley and D. R. Cole (Eds.), *Stable Isotope Geochemistry*. Rev. Mineral. Geochem., 43:1–81. Mineral. Soc. Amer., Geochem. Soc., Blacksburg, Virginia.

Lawrence, J. R., and H. P. Taylor, Jr., 1971. Deuterium and oxygen-18 correlation: Clay minerals and hydroxides in Quaternary soils compared to meteoric water. *Geochim. Cosmochim. Acta*, 35:993–1003.

Savin, M., and S. Epstein, 1970a. The oxygen and hydrogen isotope geochemistry of clay minerals. *Geochim. Cosmochim. Acta*, 34:25–42.

Savin, M., and S. Epstein, 1970b. The oxygen and hydrogen isotope geochemistry of ocean sediments and shales. *Geochim. Cosmochim. Acta*, 34:43–63.

Sheppard, S. M. F., and H. A. Gilg, 1996. Stable isotope geochemistry of clay minerals. *Clay Minerals*, 31:1–21.

Taylor, H. P., Jr., 1974. The application of oxygen and hydrogen isotope studies to problems of hydrothermal alteration and ore deposits. *Econ. Geol.*, 69:843–883.

REFERENCES FOR MARINE CARBONATES (SECTION 26.9)

Anderson, T. F., B. N. Popp, A. C. Williams, L.-Z. Ho, and J. D. Hudson, 1994. The stable isotopic record of fossils from the Peterborough Member, Oxford Clay Formation (Jurassic), UK: Paleoenvironmental implications,. *J. Geol. Soc. Lond.*, 151:125–138.

Carpenter, S. J., and K. C. Lohmann, 1995. $\delta^{18}O$ and ^{13}C values of modern brachiopods. *Geochim. Cosmochim. Acta*, 59:3749–3764.

Degens, E. T., and S. Epstein, 1962. Relationship between $^{18}O/^{16}O$ ratios in coexisting carbonates, cherts, and diatoms. *Am. Assoc. Petrol. Geol. Bull.*, 46:534–542.

Friedman, I., and J. R. O'Neil, 1977. *Compilation of Stable Isotope Fractionation Factors of Geochemical Interest*, U.S. Geol. Surv. Prof. Paper 440-KK. U.S. Geological Survey, Washington, D.C.

Gao, G., and L. S. Land, 1991. Geochemistry of Cambro-Ordovician Arbuckle limestone, Oklahoma: Implications for diagenetic $\delta^{18}O$ alteration and secular $\delta^{13}C$ and $^{87}Sr/^{86}Sr$ variation. *Geochim. Cosmochim. Acta*, 55:2911–2920.

Gregory, R. T., 1991. Oxygen isotope history of seawater revisited; timescales for boundary event changes in the oxygen isotope composition of seawater. In H. P. Taylor, Jr., J. R. O'Neil, and I. R. Kaplan (eds.), *Stable Isotope Geochemistry: A Tribute to Samuel Epstein*. Spec. Pub., Vol. 3, pp. 65–76. Geochem. Society. San Antonio, Texas.

Gregory, R. T., and H. P. Taylor, Jr., 1981. An oxygen isotope profile in a section of Cretaceous oceanic crust, Samail ophiolite, Oman: Evidence for $\delta^{18}O$ buffering of the oceans by deep (5 km) seawater hydrothermal circulation at mid-ocean ridges. *J. Geophys. Res.*, 86:2737–2755.

Harland, W. B., R. L. Armstrong, A. V. Cox, L. E. Craig, A. G. Smith, and D. G. Smith, 1989. *A Geologic Time Scale*. Cambridge University Press, Cambridge.

Land, L. S., 1995. Comment on "Oxygen and carbon isotopic composition of Ordovician brachiopods: Implications for coeval seawater" by H. Qing and J. Veizer. *Geochim. Cosmochim. Acta*, 59:2843–2844.

Muehlenbachs, K., 1998. The oxygen isotopic compositions of the oceans, sediments, and seafloor. *Chem. Geol.*, 145:263–273.

Muehlenbachs, K., and R. N. Clayton, 1976. Oxygen isotope composition of the oceanic crust and its bearing on seawater. *J. Geophys. Res.*, 81:4365–4369.

Qing, H., and J. Veizer, 1994. Oxygen and carbon isotope composition of Ordovician brachiopods: Implications for coeval seawater. *Geochim. Cosmochim. Acta*, 58:1501–1509.

Shaviv, N. J., and J. Veizer, 2003. Celestial driver for Phanerozoic climate? *GSA Today*, 13(7):4–10.

Veizer, J. (ed.), 1999. Earth-system evolution: Geochemical perspective. *Chem. Geol.*, 161(1–3):1–371.

Veizer, J., D. Ala, K. Azmy, P. Bruckschen, D. Buhl, F. Bruhn, G. A. F. Carden, A. Diener, S. Ebneth, Y. Godderis, T. Jasper, C. Korte, F. Pawellek, O. G. Podlaha, and H. Strauss, 1999. $^{87}Sr/^{86}Sr$, $\delta^{13}C$, and $\delta^{18}O$ evolution of Phanerozoic seawater. *Chem. Geol.*, 161:59–88.

Veizer, J., and J. Hoefs, 1976. The nature of $^{18}O/^{16}O$ and $^{13}C/^{12}C$ secular trends in sedimentary carbonate rocks. *Geochim. Cosmochim. Acta*, 40:1387–1395.

Veizer, J., P. Fritz, and B. Jones, 1986. Geochemistry of brachiopods: Oxygen and carbon isotopic records of Paleozoic oceans. *Geochim. Cosmochim. Acta*, 50:1679–1696.

REFERENCES FOR MARINE PHOSPHATES (SECTION 26.10)

Bryant, J. D., P. L. Koch, P. N. Froelich, W. J. Showers, and B. J. Genna, 1996. Oxygen isotope partitioning between phosphate and carbonate in mammalian apatite. *Geochim. Cosmochim. Acta*, 60:5145–5148.

Crowson, R. E., W. J. Showers, E. K. Wright, and T. Hoering, 1991. Preparation of phosphate samples for oxygen isotope analysis. *Anal. Chem.*, 63:2397–2400.

Devereux, I., 1967. Temperature measurements from oxygen isotope ratios of fish otoliths. *Science*, 155:1684–1685.

Epstein, S., R. Buchsbaum, H. A. Lowenstam, and H. C. Urey, 1953. Revised carbonate-water isotopic temperature scale. *Geol. Soc. Am. Bull.*, 64:1315.

Glenn, C. R., et al., 1994. Phosphorus and phosphorites: Sedimentology and environments of formation. *Eclogae Geol. Helvetiae*, 87:747–788.

Kim, S.-T., and J. R. O'Neil, 1997. Equilibrium and nonequilibrium oxygen isotope effects in synthetic carbonates. *Geochim. Cosmochim. Acta*, 61:3461–3475.

Kolodny, Y., B. Luz, and O. Navon, 1983. Oxygen isotope variations in phosphate of biogenic apatites, I. Fish bone apatite—rechecking the rules of the game. *Earth Planet. Sci. Lett.*, 64:398–404.

Longinelli, A., 1984. Oxygen isotopes in mammal bone phosphate: A new tool for paleohydrological and paleoclimatological research? *Geochim. Cosmochim. Acta*, 48:385–390.

Longinelli, A., and S. Nuti, 1973. Revised phosphate-water isotopic temperature scale. *Earth Planet. Sci. Lett.*, 19:373–376.

Luz, B., and Y. Kolodny, 1985. Oxygen isotope variations in phosphate of biogenic apatites, VI: Mammal teeth and bones. *Earth Planet. Sci. Lett.*, 75:29–36.

Luz, B., Y. Kolodny, and M. Horowitz, 1984. Fractionation of oxygen isotopes between mammalian bone-phosphate and environmental drinking water. *Geochim. Cosmochim. Acta*, 48:1689–1693.

O'Neil, J. R., J. L. Roe, E. Reinhard, and R. E. Blake, 1994. A rapid and precise method of oxygen isotope analysis of biogenic phosphate. *Israel J. Earth Sci.*, 43:203–212.

Piper, D. Z., and Y. Kolodny, 1987. The stable isotopic composition of a phosphorite deposit: $\delta^{13}C$, $\delta^{34}S$, and $\delta^{18}O$. *Deep-Sea Res.*, 34:897–911.

Shemesh, A., Y. Kolodny, and B. Luz, 1983. Oxygen isotope variations in phosphate of biogenic apatites, II. Phosphorite rocks. *Earth Planet. Sci. Lett.*, 64:405–416.

Shemesh, A., Y. Kolodny, and B. Luz, 1988. Isotope geochemistry of oxygen and carbon in phosphate and carbonate of phosphorite francolite. *Geochim. Cosmochim. Acta*, 52:2565–2572.

Stuart-Williams, H. L. Q., and H. P. Schwarcz, 1995. Oxygen isotopic analysis of silver phosphate using a reaction with bromine. *Geochim. Cosmochim. Acta*, 59:3837–3841.

Tudge, A. P., 1960. A method of analysis of oxygen isotopes in orthophosphate: Its use in the measurement of paleotemperatures. *Geochim. Cosmochim. Acta*, 18:81–93.

Wenzel, B., C. Lécuyer, and M. M. Joachimski, 2000. Comparing oxygen isotope records of Silurian calcite and phosphate: $\delta^{18}O$ compositions of brachiopods and conodonts. *Geochim. Cosmochim. Acta*, 64:1859–1872.

REFERENCES FOR BIOGENIC SILICA AND HYDROXIDES OF Fe AND Al (SECTIONS 26.11–26.12)

Bao, H., and P. L. Koch, 1999. Oxygen isotope fractionation in ferric oxide-water systems: Low-temperature synthesis. *Geochim. Cosmochim. Acta*, 63:599–613.

Bao, H., P. L. Koch, and M. H. Thiemens, 2000. Oxygen isotopic composition of ferric oxides from recent soil, hydrologic, and marine environments. *Geochim. Cosmochim. Acta*, 64:2221–2231.

Becker, R. H., and R. N. Clayton, 1976. Oxygen isotope study of Precambrian banded iron formation, Hamersley Range, Western Australia. *Geochim. Cosmochim. Acta*, 40:1153–1165.

Brandriss, M. E., J. R. O'Neil, M. B. Edlund, and E. F. Stoermer, 1998. Oxygen isotope fractionation between diatomaceous silica and water. *Geochim. Cosmochim. Acta*, 62:1119–1125.

Chacko, T., D. R. Cole, and J. Horita, 2001. Equilibrium oxygen, hydrogen, and carbon isotope fractionation factors applicable to geologic systems. In J. W. Valley and D. R. Cole (Eds.). *Stable Isotope Geochemistry. Rev. Mineral. Geochim.*, 43:1–81. Mineral. Soc. Amer., Geochim. Soc., Blacksburg, Virginia.

Chase, C. G., and E. C. Perry, Jr., 1972. The oceans: Growth and oxygen isotope evolution. *Science*, 177: 992–994.

Degens, E. T., and S. Epstein, 1962. Relationship between $^{18}O/^{16}O$ ratios in coexisting carbonates, cherts, and diatomites. *Bull. Am. Assoc. Petrol. Geol.*, 46:534–542.

Faure, G., 1998. *Principles and Application of Geochemistry*, 2nd ed. Prentice-Hall, Upper Saddle River, New Jersey.

Götze, J., M. Tichomirowa, H. Fuchs, J. Pilot, and Z. D. Sharp, 2001. Geochemistry of agates: Trace element and stable isotope study. *Chem. Geol.*, 1975:523–541.

Jones, J. B., and E. R., Segnit, 1971. The nature of opal. 1. Nomenclature and constituent phases. *J. Geol. Soc. Austral.*, 18:57–68.

Kim, S.-T., and J. R. O'Neil, 1997. Equilibrium and nonequilibrium oxygen isotope effects in synthetic carbonates. *Geochim. Cosmochim. Acta*, 61:3461–3475.

Kita, I., and S. Taguchi, 1986. Oxygen isotopic behavior of precipitating silica from geothermal water. *Geochim. J.*, 20:153–157.

Knauth, L. P., and S. Epstein, 1976. Hydrogen and oxygen ratios in nodular and bedded chert. *Geochim. Cosmochim. Acta*, 40:1095–1108.

Knauth, L. P., and D. R. Lowe, 1978. Oxygen isotope geochemistry of cherts from the Onverwacht Group (3.4 billion years), Transvaal, South Africa, with implications for secular variations in the isotope compositions of cherts. *Earth Planet. Sci. Lett.*, 41:209–222.

Kolodny, Y. and S. Epstein, 1976. Stable isotope geochemistry of deep sea cherts. *Geochim. Cosmochim. Acta*, 40:1195–1209.

Labeyrie, L., 1974. New approach to surface seawater paleotemperatures using $^{18}O/^{16}O$ ratios in silica of diatom frustules. *Nature*, 248:40–42.

Oskvarek, J. D., and E. C. Perry, Jr., 1976. Temperature limits on the Early Archean ocean from oxygen isotopes variations in the Isua supracrustal sequence, West Greenland. *Nature*, 259:192–194.

Perry, E. C., Jr., 1967. The oxygen isotope chemistry of ancient cherts. *Earth Planet. Sci. Lett.*, 3:62–66.

Perry, E. C., Jr., S. N. Ahmad, and T. M. Swulius, 1978. The oxygen isotope composition of 38,800 m.y. old metamorphosed chert and iron formation from Isukasia, West Greenland. *J. Geol.*, 86:223–239.

Perry, E. C., and B. Bonnichsen, 1966. Quartz and magnetite oxygen-18/oxygen-16 fractionation in metamorphosed Biwabik Iron Formation. *Science*, 153:528–529.

Perry, E. C., Jr., and F. C. Tan, 1972. Significance of oxygen and carbon isotope variations in early Precambrian cherts and carbonate rocks of southern Africa. *Geol. Soc. Am. Bull.*, 83:647–664.

Perry, E. C., Jr., F. C. Tan, and G. B. Morey, 1973. Geology and stable isotope geochemistry of the Biwabik Iron Formation, northern Minnesota. *Econ. Geol.*, 68:1110–1125.

Poage, M. A., D. J. Sjostrom, J. Goldberg, C. P. Chamberlain, and G. Furniss, 2000. Isotopic evidence for Holocene climate change from a geothite-rich ferricrete chronosequence. *Chem. Geol.*, 166:327–340.

Schmidt, M., R. Botz, D. Rickert, G. Bohrmann, S. R. Hall, and S. Mann, 2001. Oxygen isotopes of marine diatoms and relations to opal-A maturation. *Geochim. Cosmochim. Acta*, 65:201–211.

Sharp, Z. D., and D. L. Kirschner, 1994. Quartz-calcite oxygen isotope thermometry: A calibration based on natural isotopic variations. *Geochim. Cosmochim. Acta*, 58:4491–4501.

REFERENCES FOR EXTRATERRESTRIAL ROCKS (SECTION 26.13)

Anders, E., and N. Grevesse, 1989. Abundances of the elements: Meteoritic and solar. *Geochim. Cosmochim. Acta*, 53:197–214.

Benz, W., W. L. Slattery, and A. G. W. Cameron, 1986. The origin of the Moon and the single impact hypothesis. *Icarus*, 66:515–535.

Clayton, R. N., 1993. Oxygen isotopes in meteorites. *Annu. Rev. Earth Planet. Sci.*, 21:115–149.

Clayton, R. N., J. M. Hurd, and T. K. Mayeda, 1973. Oxygen isotopic compositions of Apollo 15, 16, and 17 samples and their bearing on lunar origin and petrogenesis, Proc. 4th Lunar Sci. Conf. *Geochim. Cosmochim Acta*, Suppl. 2:1535–1542.

Clayton, R. N., G. F. MacPherson, I. D. Hutcheon, A. M. Davis, L. Grossman, T. K. Mayeda, C. Molini-Velsko, J. M. Allen, and A. El Goresy, 1984. Two forsterite-bearing FUN inclusions in the Allende meteorite. *Geochim. Cosmochim. Acta*, 48:535–548.

Clayton, R. N., and T. K. Mayeda, 1975. Genetic relations between the Moon and meteorites, Proc. 6th Lunar Planet. Sci. Conf., 1761–1769. Houston, Texas.

Clayton, R. N., and T. K. Mayeda, 1977. Correlated oxygen and magnesium isotope anomalies in Allende inclusions, I: Oxygen. *Geophys. Res. Lett.*, 4:295–298.

Clayton, R. N., and T. K. Mayeda, 1983. Oxygen isotopes in eucrites, shergottites, nakhlites, and chassignites. *Earth Planet. Sci. Lett.*, 62:1–6.

Clayton, R. N., and T. K. Mayeda, 1986. Oxygen isotopes in Shergotty. *Geochim. Cosmochim. Acta*, 50: 979–982.

Clayton, R. N., and T. K. Mayeda, 1996. Oxygen isotope studies of achondrites. *Geochim. Cosmochim. Acta*, 60:1999–2017.

Clayton, R. N., N. Onuma, L. Grossman, and T. K. Mayeda, 1977. Distribution of the pre-solar component in Allende and other carbonaceous chondrites. *Earth Planet. Sci. Lett.*, 34:209–224.

Davis, A. M., G. J. MacPherson, R. N. Clayton, T. K. Mayeda, P. J. Sylvester, L. Grossman, R. W. Hinton, and J. R. Laughlin, 1991. Melt solidification and late-stage evaporation in the evolution of a FUN inclusion from the Vigarano C3V chondrite. *Geochim. Cosmochim. Acta*, 55:621–637.

Davis, A. M., K. D. McKeegan, and G. J. MacPherson, 2000. Oxygen-isotopic compositions of individual minerals from the FUN inclusion Vigarano 1623–5. *Meteoritics Planet. Sci.*, 35:A47.

Eiler, J. M., N. Kitchen, L. Leshin, and M. Strausberg, 2002. Hosts of hydrogen in Allan Hills 84001: Evidence for hydrous martian salts in the oldest martian meteorite? *Meteoritics Planet. Sci.*, 37:395–405.

Franchi, I. A., I. P. Wright, A. S. Sexton, and C. T. Pillinger, 1999. The oxygen-isotope composition of Earth and Mars. *Meteoritics Planet. Sci.*, 34:657–661.

Heidenreich, J. E. III, and M. H. Thiemens, 1982. A non-mass-dependent isotope effect in the production of ozone from molecular oxygen. *J. Chem. Phys.*, 78:892–895.

Kolodny, Y., J. F. Kerridge, and I. R. Kaplan, 1980. Deuterium in carbonaceous chondrites. *Earth Planet. Sci. Lett.*, 46:149–158.

Lattimer, J. M., D. N. Schramm, and L. Grossman, 1978. Condensation in supernova ejecta and isotopic anomalies in meteorites. *Astrophys. J.*, 219:230–249.

Lee, T., T. K. Mayeda, and R. N. Clayton, 1980. Oxygen isotopic anomalies in Allende inclusion HAL. *Geophys. Res. Lett.*, 7:493–496.

Leshin, L. A., 2000. Insights into martian water reservoirs from analyses of martian meteorite QUE94201. *Geophys. Res. Lett.*, 27:2017–2020.

Leshin, L. A., K. D. McKeegan, P. K. Carpenter, and R. E. Harvey, 1998. Oxygen isotopic constraints on the genesis of carbonates from martian meteorite ALH 84001. *Geochim. Cosmochim. Acta*, 62:3–13.

McKeegan, K. D., and L. A. Leshin, 2001. Stable isotope variations in extraterrestrial materials. In J. W. Valley and D. R. Cole (Eds.), *Stable Isotope Geochemistry*, Vol. 43, pp. 279–318. Rev. Mineral. Geochem., Mineralological Society of America and Geochem. Society, Blacksburg, Virginia.

McSween, H. Y., Jr., and A. H. Treiman, 1998. Martian meteorites. *Rev. Mineral.*, 36:6.01–6.53.

Romanek, C. S., E. C. Perry, A. H. Treiman, R. A. Socki, J. H. Jones, and E. K. Gibson, 1998. Oxygen isotopic record of silicate alteration in the Shergotty-Nakhla-Chassigny meteorite Lafayette. *Meteoritics Plan. Sci.*, 33:775–784.

Saxton, J. M., I. C. Lyon, and G. Turner, 1998. Correlated chemical and isotopic zoning in carbonates in the martian meteorite ALH84001. *Earth Planet. Sci. Lett.*, 160:811–822.

Taylor, H. P., Jr., and S. Epstein, 1970. 18-0/16-0 ratios in Apollo 11 lunar rocks and soils, Proc. Apollo 11 Lunar Sci. Conf. *Geochim. Cosmochim. Acta*, Suppl.:1613–1626.

Thiemens, M. H., 1996. Mass-independent isotopic effects in chondrites: The role of chemical processes. In R. H. Hewins, R. H. Jones, and E. R. D. Scott (Eds.), *Chondrules and the Protoplanetary Disk*, pp. 109–118. Cambridge University Press, New York.

Thiemens, M. H., 1999. Mass-independent isotope effects in planetary atmospheres and the early solar system. *Science*, 283:342–345.

CHAPTER 27

Carbon

CARBON ($Z = 6$) is fourth in abundance among the elements of the solar system after H, He, and O (Anders and Grevesse, 1989). The two stable isotopes of C and their abundances and masses are:

$^{12}_{6}C$: 98.90%, 12.000000 amu
$^{13}_{6}C$: 1.10%, 13.003355 amu

Consequently, the $^{13}C/^{12}C$ ratio is 1.083612 and the atomic weight of C is 12.011036. The mass of ^{13}C is 8.36% greater than that of ^{12}C, which causes the C isotopes to be fractionated by chemical and biological processes in nature. Natural environments on the surface of the Earth also contain radioactive cosmogenic ^{14}C, which is produced by an (n, p) reaction on stable ^{14}N in the atmosphere (Section 23.1).

The isotopic composition of C is expressed by the $\delta^{13}C$ parameter, defined as:

$$\delta^{13}C = \left(\frac{(^{13}C/^{12}C)_{spl} - (^{13}C/^{12}C)_{std}}{(^{13}C/^{12}C)_{std}} \right) \times 10^3‰ \tag{27.1}$$

where the standard is PDB (Section 26.2). Therefore, positive $\delta^{13}C$ values signify that the C is enriched in ^{13}C relative to the standard, whereas negative $\delta^{13}C$ values imply depletion in ^{13}C. The fractionation factor for C isotopes is:

$$\alpha^a_b(^{13}C) = \frac{R_a}{R_b} \tag{27.2}$$

where a and b are C-bearing compounds or phases in isotopic equilibrium at a specified temperature and R is the isotopic $^{13}C/^{12}C$ ratio. However, the isotopic composition of C in biogenic carbonate and organic compounds is not a reliable recorder of the environmental temperature because these kinds of compounds do not achieve isotopic equilibrium with the major C reservoirs on Earth. When green plants absorb CO_2 molecules during photosynthesis, they prefer ^{12}C over ^{13}C, which explains why organic matter of photosynthetic plants and herbivorous and carnivorous animals has negative $\delta^{13}C$ values. On the other hand, the $\delta^{13}C$ values of marine calcite and aragonite are close to zero because the PDB standard is itself a marine carbonate.

Carbon plays an important role in the geochemical processes that occur at or near the surface of the Earth because it is the principal ingredient of the biosphere and also occurs in sedimentary rocks (e.g., limestone, shale, coal, petroleum, and natural gas), in the atmosphere (e.g., CO_2), in the hydrosphere (e.g., bicarbonate ions), and in igneous and metamorphic rocks (e.g., diamond and graphite).

The isotope geochemistry of C has been summarized by Farmer and Baxter (1976), Deines (1980), Schidlowski et al. (1983), Hoefs (1997), Hayes (2001), Des Marais (2001), Freeman (2001), and Ripperdan (2001). The fractionation factors of C isotopes in gases, aqueous solutions, carbonate minerals, graphite, and diamond were compiled by Chacko et al. (2001).

27.1 BIOSPHERE

Photosynthesis is a complex biochemical process energized by sunlight, which converts atmospheric CO_2 and H_2O into the organic matter of green plants. The process can be represented by the equation

$$6CO_2 + 6H_2O \rightarrow C_6H_{12}O_6 + 6O_2 \qquad (27.3)$$

The C that is fixed in plant tissue as a result of this process is variably depleted in ^{13}C relative to PDB. Consequently, organic matter of biogenic origin has negative $\delta^{13}C$ values, including not only the present biosphere but also kerogen and deposits of coal, petroleum, and natural gas of Phanerozoic age and amorphous C as well as graphite of Precambrian age.

27.1a Carbon Dioxide

The $\delta^{13}C$ value of CO_2 in the atmosphere has an average value of about −7.7‰ but varies seasonally and regionally in response to the preferential absorption of ^{12}C by plants and to the continuing discharge of CO_2 caused by the annual combustion of 10^{15} g of fossil fuel, whose average $\delta^{13}C$ value is about −27‰ (Hoefs, 1997). The amplitude of the seasonal variations of $\delta^{13}C$ values of atmospheric CO_2 in the northern hemisphere decreases with decreasing latitude from north to south. For example, the seasonal amplitude of $\delta^{13}C$ in atmospheric CO_2 varies from 0.9‰ at Point Barrow, Alaska (71° N), to 0.4‰ on Mauna Loa, Hawaii (20° N), to only 0.15‰ on Christmas Island, Indian Ocean (4° N) (Keeling et al., 1989).

The release of CO_2 resulting from the combustion of fossil fuel causes an average increase of the concentration of CO_2 in the atmosphere of about 1.5 ppm/y accompanied by a small decrease of its $\delta^{13}C$ value. However, both the concentration and the average $\delta^{13}C$ value of atmospheric CO_2 change less than predicted. The reason seems to be that about one-half of the annual output of anthropogenic CO_2 dissolves in the oceans or is absorbed by plants on the continents. The ability of the biosphere in the midlatitudes of the northern hemisphere to absorb CO_2 from the atmosphere was confirmed by Ciais et al. (1995), while Beveridge and Shackleton (1994) reported that the $\delta^{13}C$ value of bicarbonate ions in the surface layer of the oceans has decreased in response to the dissolution of anthropogenic CO_2.

The concentration of CO_2 of the atmosphere of the Earth has varied in the geological past in consonance with the average global temperature. For example, ice deposited in Antarctica and Greenland during glacial stages of the Pleistocene Epoch contains air bubbles whose CO_2 concentrations are lower than those of air bubbles in ice of interglacial stages (Leuenberger et al., 1992). The significance of these variations is uncertain because of feedback loops. For example, an increase in the emission of CO_2 by volcanoes may lead to climatic warming by means of the greenhouse effect, which causes additional CO_2 to be released by the surface ocean, which enhances climatic warming. Alternatively, climatic cooling allows more CO_2 to dissolve in the oceans, thus decreasing the concentration of atmospheric CO_2 and thereby intensifying global cooling. The biosphere counteracts these kinds of fluctuations in the concentrations of atmospheric CO_2 by absorbing or releasing CO_2 and thereby prevents extreme climate changes that could otherwise become life-threatening.

27.1b Green Plants

The isotope fractionation of C during photosynthesis takes place in three steps:

1. Preferential absorption of $^{12}CO_2$ molecules from the atmosphere by diffusion through cell membranes depending on the concentration of CO_2 in the ambient air and on other factors. Consequently, the air adjacent to green plants is enriched in $^{13}CO_2$ during the daytime while photosynthesis is in progress. The resulting diurnal variations of $\delta^{13}C$ values of atmospheric CO_2 have been reported by Park and Epstein (1960) and Keeling (1958, 1961).
2. Preferential conversion of $^{12}CO_2$ dissolved in cytoplasm into phosphoglyceric acid by the action of enzymes followed by the subsequent respiration of the ^{13}C-enriched residual CO_2.
3. Synthesis of different organic compounds by plants from phosphoglyceric acid. Degens et al.

(1968) originally demonstrated the progressive *depletion* of cellulose, lignin, and lipids (oils, fats, and waxes) in ^{13}C and the ^{13}C *enrichment* of carbohydrates, hemicellulose, proteins, and pectin relative to C in the whole plant.

Although all plants and their molecular constituents are depleted in ^{13}C relative to the PDB standard, the $\delta^{13}C$ values of plant tissue depend on the metabolic processes by means of which C is incorporated into plants. The $\delta^{13}C$ values of plants that metabolize CO_2 by the Calvin cycle (C3 plants: most plants on land) range from −23 to −34‰ relative to PDB with a mean of about −27‰. Plants that use the Hatch–Slack process (C4 plants: aquatic, desert, and saltmarsh plants as well as tropical grasses) are less depleted in ^{13}C than C3 plants and have $\delta^{13}C$ values between −6 and −23‰ with a mean of −13‰. A third group of plants employs the Crassulacean acid metabolism (CAM), which results in $\delta^{13}C$ values that range widely from −11 to −33‰. The $\delta^{13}C$ values of algae and bacteria in Figure 27.1 exceed the range of C isotope compositions of C3, C4, and CAM plants. The fractionation of C isotopes that occurs during the biosynthesis of proteins, carbohydrates, lipids, and nucleic acids in plants has been documented by Benner et al. (1987), Bidigare et al. (1991), and Hayes (2001) and was described in a book by Galimov (1985).

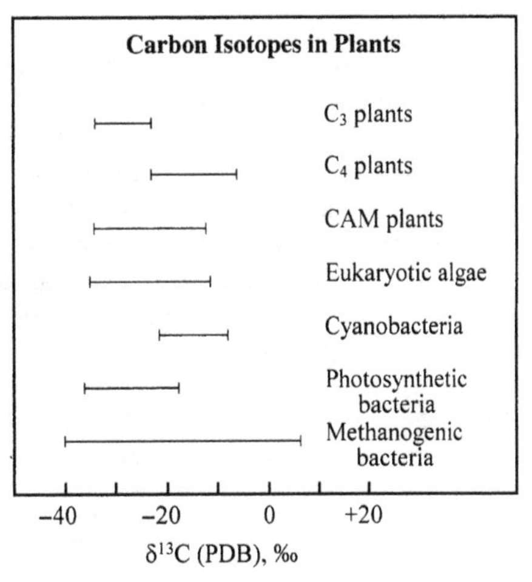

FIGURE 27.1 Isotope composition of C in different types of plants. Adapted from Schidlowski et al. (1983).

27.1c Life in Extreme Environments

Various unicellular plants are able to survive in extreme environments on the surface of the Earth, including the islands off the coast of the Antarctic Peninsula, hypersaline brine pools on the Sinai Peninsula, and hotsprings in volcanically active areas such as Yellowstone National Park, Wyoming. The study of C-isotope fractionation of plants in these kinds of environments is motivated in part by the desire to interpret the $\delta^{13}C$ values of organic C in sedimentary rocks of Precambrian age.

Signy Island (60°43′ S, 45°38′ W) of the South Orkney Islands supports a flora of mosses, lichens, algae, and grass, although the average monthly summertime temperature in January ranges from only −0.7 to +4.5°C. Still farther south, the South Shetland Islands (62°31′ S, 59°47′ W) also maintain a sparse vegetation of lichens, mosses, and algae. The $\delta^{13}C$ values of the plants on Signy Island reported by Galimov (2000) range from −18.8‰ (lichen) to −27.66‰ (grass) on the PDB scale. Galimov (2000) pointed out that certain plant species on the South Orkney and South Shetland Islands are characterized by their $\delta^{13}C$ values, which vary only between narrow limits regardless of climatic conditions. For examples, the $\delta^{13}C$ values of the lichen *Usnea antarctica* range only from −21.29 to −22.44‰ (PDB). Similarly, the moss *Drepanocladus* sp. has $\delta^{13}C$ values between −23.49 and −24.86‰. Such isotopic specificity is masked in the inhabited parts of the Earth because of variations of $\delta^{13}C$ values of atmospheric CO_2.

A very different kind of environment exists in certain hypersaline ponds (Gavish Sabkha and Solar Lake) on the Sinai Peninsula whose salinites range up to 30%. Nevertheless, these ponds host highly diversified microbial communities that are preyed upon by salt-tolerant gastropods. According to Schidlowski et al. (1984), the principal inhabitants of these ponds are photosynthetic prokaryotes (e.g., *Synechocystis*, *Gloecacapsa*, and *Gloecothece*) as well as filamentous cyanobacteria (e.g., *Lyngbya* and others). The flora in these ponds occurs in the form of extended mats which may produce laminated biosedimentary layers resembling stromatolites.

The $\delta^{13}C$ values of bulk organic matter in these two ponds range from only -4.4 to $-17.0‰$ (PDB), implying that the biosynthesis of organic matter in these ponds is not accompanied by the depletion in ^{13}C that characterizes plants in less stressful environments. The average $\delta^{13}C$ values of organic matter in these microbial mats are:

Gavish Sabkha: $-10.0 \pm 2.6‰$
Solar Lake: $-5.7 \pm 1.4‰$

These results confirm the general conclusion that organic matter in existing algal and microbial mats contains "heavy" carbon (i.e., is less depleted in ^{13}C than other kinds of organic matter).

Schidlowski et al. (1984) concluded that the cause for the lack of ^{13}C depletion of the microbial mats on the Sinai Peninsula is not the salinity or the elevated temperature of the water but the low CO_2 concentration in the water caused by these environmental parameters. The authors also noted that organic matter of Precambrian stromatolites formed by microbial communities is depleted in ^{13}C to about the same extent as that of modern C3 plants (-18 to $-34‰$). Evidently, organic matter of microbial mats contains "heavy" C only when the concentration of dissolved CO_2 is lowered by the abnormally high temperature and salinity of the water. After the development of metazoan organisms at about 0.7 Ga, the bacterial and algal matbuilders were prevented from forming stromatolites by grazers (e.g., gastropods) except in highly saline sanctuaries from which predators were excluded.

The opposite extreme in the isotopic composition of C was discovered by Kaplan and Nissenbaum (1966) in thin layers of black organic matter interbedded with S-bearing sand dunes of Pleistocene age near Berri, Israel. The layers of organic matter are 1–5 mm thick and resemble dried algal mats observed in the Persian Gulf. The $\delta^{13}C$ values of four samples of the organic matter analyzed by Kaplan and Nissenbaum (1966) range from -82.5 to $-89.3‰$ (PDB). These may well be the most $\delta^{13}C$-depleted samples of solid organic matter ever recorded.

The authors favored the explanation that the ^{13}C-depletion of these algal mats was the result of a complex process involving several cycles of biogenic isotope fractionation. The process may have started with the release of CH_4 from a reservoir of natural gas or petroleum at depth. The CH_4 was subsequently oxidized to CO_2, which was converted into organic matter by the photosynthesis of algae and bacteria in a lagoon. The resulting organic matter was later oxidized to CO_2 at the bottom of the lagoon and then photosynthesized again into organic matter in the water at the surface of the lagoon. During photosynthesis in the lagoon, the methanogenic CO_2 was contaminated by ^{14}C-bearing atmospheric CO_2, which explains why Kaplan and Nissenbaum (1966) were able to date the algal mats by the ^{14}C method (~30,000 years). The authors considered that the native S in the dunes of Berri also originated from hydrocarbon reservoirs at depth.

Thermophilic algae and bacteria growing in hotsprings at Yellowstone Park, Wyoming, thrive at temperatures between 55 and 70°C. Estep (1984) demonstrated that the depletion in ^{13}C of the alga *Synechococcus* and the filamentous photosynthetic bacterium *Chloroflexus* in Yellowstone Park increases with the concentration of inorganic C dissolved in the water. This relation was expressed by Estep (1984) by a straight line whose equation is:

$$\delta^{13}C = -12.0 - 0.0128 \text{ DIC} \qquad (27.4)$$

where $\delta^{13}C$ is on the PDB scale and DIC is the concentration of dissolved inorganic C in parts per million (ppm). The graphical representation of this equation in Figure 27.2 demonstrates that $\delta^{13}C$ values of thermophilic algae and bacteria range from -12.0 at DIC = 40 ppm to -27.0 at DIC = 1200 ppm. The dependence of $\delta^{13}C$ of the thermophilic flora on the concentration of dissolved inorganic C presented by Estep (1984) based on data from the hotsprings of Yellowstone Park confirms the conclusions of Schidlowski et al. (1984) derived from solar ponds on the Sinai Peninsula.

Surprisingly, the temperature of the water does not exert a strong control on the fractionation of C isotopes by thermophilic algae and bacteria. The data of Estep (1984) in Figure 27.3 demonstrate

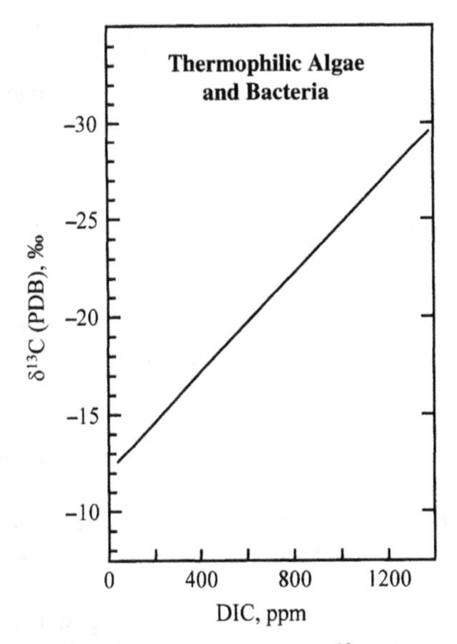

FIGURE 27.2 Relation between the $\delta^{13}C$ (PDB) value of thermophilic algae and bacteria and the dissolved inorganic carbon (DIC) in the water of hotsprings at Yellowstone Park, Wyoming. The straight line is based on equation 27.3, which was fitted to Figure 2 of Estep (1984).

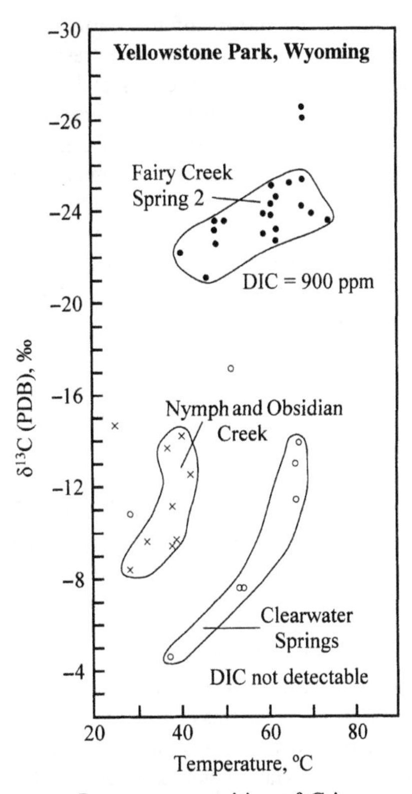

FIGURE 27.3 Isotope composition of C in thermophilic algae and bacteria growing in hotsprings of Yellowstone Park. The data demonstrate that the $\delta^{13}C$ values or organic matter in these hotsprings depend much more on the concentration of dissolved inorganic C than on the temperature of the water. Data from Estep (1984).

that the $\delta^{13}C$ values of algae and bacteria in several hotsprings in Yellowstone Park range widely from -4.6 to -26.6‰ (PDB) and that the $\delta^{13}C$ values of organic matter in these hotsprings depend more on the concentration of dissolved inorganic C than on the temperature of the water or the speciation of the flora. For example, the organic matter in Fairy Creek Spring 2 ($T = 40°C$ to $T = 74°C$) in Figure 27.3 is highly depleted in ^{13}C ($\delta^{13}C = -21.1$ to -26.6‰) because the DIC concentration of the water is about 900 ppm. The $\delta^{13}C$ values of organic matter in the Clearwater Springs ($T = 28°C$ to $T = 67°C$) range from only -4.6 to -17.2‰ because the concentration of DIC was below the limit of detection.

These conclusions confirm the work of Seckbach and Kaplan (1973), who cultured the thermophilic and photosynthetic alga *Cyanidium caldarium* in contact with air and with pure CO_2 at temperatures of 26 and 45°C. In addition, they reported $\delta^{13}C$ values of organic matter collected in the Orakei Korako hotsprings of New Zealand at temperatures ranging from 30.5 to 63.7°C and pH values between 8.38 and 9.75. The $\delta^{13}C$ of organic matter in these hotsprings range from -11.09 to -23.80‰, but they are not correlated with the temperature.

27.2 LIFE IN THE PRECAMBRIAN OCEANS

Sedimentary rocks of Precambrian age contain organic matter (kerogen), which implies the existence of life in the form of unicellular organisms (e.g., algae and bacteria) in the oceans at a very early age in the history of the Earth. The study of Precambrian life relies on the preservation of recognizable biological structures and of biogenic molecules. In addition, the isotope composition of C in kerogen of Precambrian rocks can reveal the former presence of photosynthetic microorganisms by its characteristic depletion in ^{13}C. The biogenic depletion of ^{13}C in organic matter was partly reversed after burial by the dehydrogenation of the organic matter during diagenesis and low-grade regional metamorphism.

27.2a Carbon Isotopes in Precambrian Kerogen

Various types of present-day organic matter formed by algae and bacteria have H/C ratios ranging from 1.6 to 1.9 (Hayes et al., 1983). This ratio decreases during diagenesis of organic matter as the organic matter is transformed into kerogen of types I, II, and III (Tissot and Welte, 1978). The *decrease* of the H/C ratios of kerogen is accompanied, in most cases, by a progressive *increase* of its $\delta^{13}C$ value caused by the ^{13}C-enrichment of the residual organic matter. The relation between the H/C ratio and the $\delta^{13}C$ value of kerogen in Figure 27.4 indicates that $\delta^{13}C$ values increase from about −30‰ at H/C = 1.0 to −10‰ at H/C ~0.

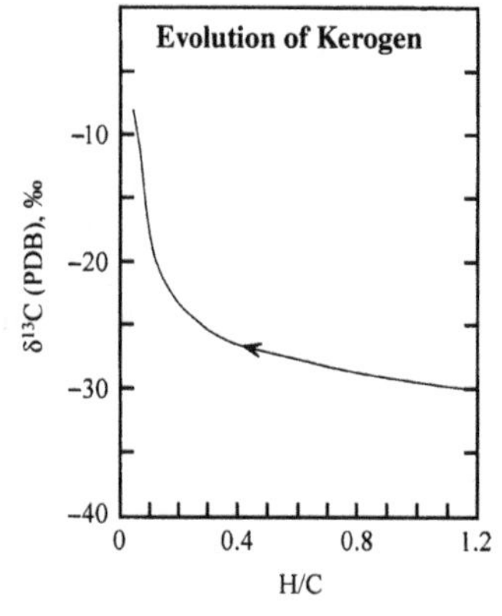

FIGURE 27.4 Evolution of kerogen by progressive decrease of the H/C ratio which enriches the kerogen in ^{13}C. Based on data compiled from the literature by Hayes et al. (1983).

In general, biogenic isotope fractionation causes ^{13}C depletion, whereas subsequent dehydrogenation during diagenesis of increasing intensity causes ^{13}C-enrichment of kerogen relative to unaltered organic matter. For example, the $\delta^{13}C$ values of kerogen in Precambrian sedimentary rocks reported by Hayes et al. (1983) range widely from −8.8‰ (chert, Manjeri Formation, 2.6 Ga) to −51.8‰ (chert/carbonate, Fortescue Group, 2.8 Ga). The variation of measured $\delta^{13}C$ values of Precambrian kerogens could reflect differences in the concentration of DIC in the water of depositional environment, but is more likely caused by thermal cracking of kerogen to form liquid or gaseous hydrocarbons. This process causes the H/C ratio of kerogen to decrease and enriches the residue in ^{13}C because of the preferential breaking of ^{12}C–^{12}C bonds.

For these reasons, only the largest negative $\delta^{13}C$ values of kerogen analyzed by Hayes et al. (1983) were plotted in Figure 27.5 versus their stratigraphic age. The results demonstrate that the lowest $\delta^{13}C$ values of kerogen of Proterozoic age (0.5–2.5 Ga) are remarkably constant between −30‰ and −35‰. Kerogens of Archean age in rocks of the Hamersley Range and Fortescue Group (Australia) as well as the Ventersdorp Supergroup (South Africa) have lower $\delta^{13}C$ values than the Proterozoic kerogens, whereas the $\delta^{13}C$ values of kerogens in the Pongola Supergroup, Gorge Creek Group, Warrawoona Group, and the Isua Supracrustals scatter irregularly.

The interpretation suggested in Figure 27.5 is that the $\delta^{13}C$ values of Archean kerogens at localities 1, 2, and 3 have been increased *less* than those at localities 4, 5, 6, and 7. If this interpretation is correct, organic matter of Archean age was more strongly depleted in ^{13}C than kerogen of Proterozoic and Phanerozoic age. The work of Estep (1984) and others on thermophilic algae and bacteria in hotsprings supports the conclusion that the ^{13}C-depletion of organic matter is proportional to the concentration of DIC in the water. Therefore, the strong ^{13}C depletion of Archean kerogen supports the hypothesis that the concentration of CO_2 in the atmosphere during the Archean Eon was higher than it is at present. (See also Section 29.9.) In other words, standing bodies of surface water in Archean time contained higher

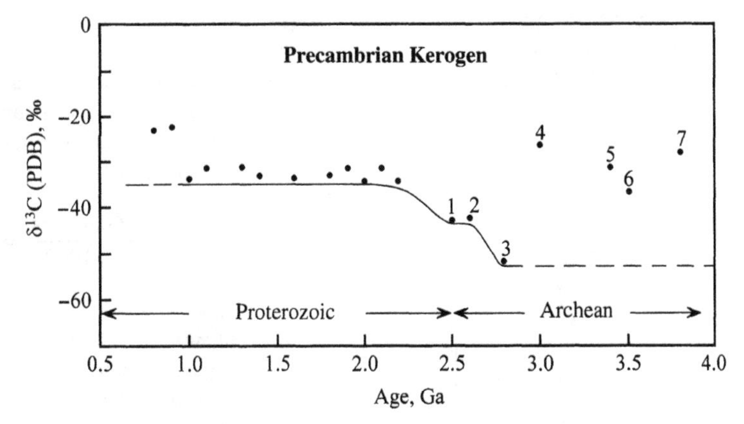

FIGURE 27.5 Variation of the lowest $\delta^{13}C$ values of kerogen in Precambrian sedimentary rocks with their stratigraphic age. The Archean rocks are identified by number: 1. Hamersley Range, Australia; 2. Ventersdorp Supergroup, South Africa; 3. Fortescue Group, Australia; 4. Pongola Supergroup, South Africa; 5. Gorge Creek Group, Australia; 6. Warrawoona Group, Australia; 7. Isua Supracrustals, Greenland. Data from Hayes et al. (1983).

concentrations of DIC than during the Proterozoic Eon, which resulted in greater ^{13}C-depletion of Archean organic matter produced by algae and bacteria than in Proterozoic time.

The temperature of the oceans during the Archean Eon has not yet been determined by isotopic methods because the isotope compositions of O in marine calcite, apatite, and chert discussed in Sections 26.9–26.11 do not yield a unique solution to this problem. Fortunately, the data of Estep (1984) suggest that the δD values of thermophilic algae and bacteria increase linearly with increasing water temperature (i.e., they become less negative) and therefore may prove useful for the determination of the temperature of the oceans of the geological past.

27.2b Hydrogen Isotopes in Thermophilic Organisms

The isotopic composition of H of thermophilic algae and bacteria in hotsprings of Yellowstone Park in Wyoming varies widely depending on the temperature of the water. Estep (1984) reported δD values ranging from -138 to -220‰ (SMOW) for the organic matter, whereas the water has nearly constant δD values with a mean of -145‰ and a range of -142 to -146‰. The $\delta^{18}O$ value of the water is derivable from the equation for the meteoric-water line:

$$\delta D = 8\delta^{18}O + 10$$

The average δD value of -145‰ of water in the hotsprings of Yellowstone Park yields $\delta^{18}O = -19.4$‰ for the water. This value is consistent with the latitude of 44° to 45° N of Yellowstone Park and its elevation of about 3300 m above sealevel.

Estep (1984) demonstrated that the δD values of thermophilic algae and bacteria growing in pH-neutral water are positivity correlated with the temperature of the water, but noted that organically bonded H of acidophile algae and bacteria appears to be fractionated less than that of organisms growing in pH-neutral water. For the purpose of this presentation, the δD values of organic matter and water reported by Estep (1984) were used to calculate the function $10^3 \ln \alpha_w^{org}(D)$, which was regressed versus $10^3/T$, where T is the temperature in degrees Celsius. The results in Figure 27.6 for 13 data sets yield the equation:

$$10^3 \ln \alpha_w^{org} = -\frac{8.90 \times 10^3}{T} + 107.3 \qquad (27.5)$$

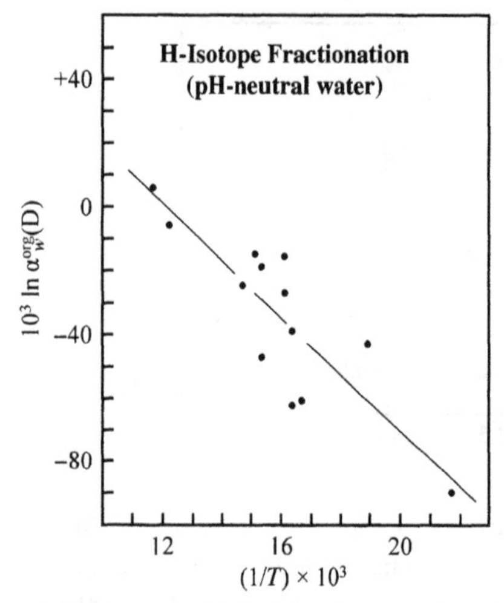

FIGURE 27.6 Linear correlation of the functions $10^3 \ln \alpha_w^{org}(D)$ and $(1/T) \times 10^3$ for thermophilic algae and bacteria growing in pH-neutral water of hot springs in Yellowstone Park, Wyoming, where T is the temperature in degrees Celsius. Data from Estep (1984).

with a linear correlation coefficient of −0.843. The fractionation of H and C isotopes by algae and bacteria growing in hotsprings at elevated temperatures may facilitate the interpretation of the δ^{13}C and δD values of kerogen in deposits of Precambrian and Phanerozoic age.

The δD values of kerogen in sedimentary rocks of Precambrian age range widely from −44.1 to −143.1‰ on the SMOW scale (Hayes et al., 1983). However, the decrease of the H/C ratio during the maturation of kerogen enriches it in both ^{13}C (Figure 27.4) and D. Therefore, the δD values of kerogen in Precambrian rocks have probably *increased* by isotope fractionation during the dehydrogenation of kerogen. Even when only the most D-depleted values are considered, the δD values of Precambrian organic matter scatter widely when they are plotted versus the age of the rocks (not shown). Nevertheless, δD values equal to or less than −140‰ occur in rocks of different ages ranging from 1.6 to 2.8 Ga, which suggests that $\delta D = -140$‰ in kerogen reflects environmental conditions at the surface of the Earth in Late Archean to Middle Proterozoic time. The δD value of organic matter at the time of deposition depends not only on the environmental temperature but also on the isotope composition of seawater at the site of deposition. If δD of Precambrian seawater is assumed to be 0‰, the H-isotope fractionation factor corresponding to $\delta D = -140$‰ is:

$$\alpha_w^{org}(D) = \frac{-140 + 1000}{0 + 1000} = 0.860$$

and $10^3 \ln 0.860 = -150.8$. Substituting into equation 27.5 and solving for T yields a temperature of 34°C. Although this value is plausible, it is based on assumptions that require confirmation.

27.2c Signs of Life

The fractionation of C isotopes by algae and bacteria has left a record of the presence of life in the form of negative δ^{13}C values of organic matter in sedimentary rocks of Archean age. For example, Hayes et al. (1983) reported that the Early Archean cherty iron formation, schist, and gneiss at Isua in southern West Greenland (3.8 Ga) contain organic C in concentrations ranging from 0.06 ± 0.03 to 25 ± 4 mg of C/g of rock and that all of the specimens they analyzed have δ^{13}C values between -9.8 ± 4 and -28.2 ± 1.8‰ on the PDB scale. Although the isotope composition of C in these metasedimentary rocks has probably been enriched in ^{13}C during the progressive regional metamorphism, the measured negative δ^{13}C values in the rocks of the Isua Supracrustals have been interpreted as a sign that life existed in the Early Archean oceans at about 3.8 Ga.

However, Schidlowski (1995) pointed out that the record for the presence of life in the rocks of Isua is clouded by the metamorphism of the sedimentary rocks, by the absence of morphologically preserved fossils of Archaea bacteria, and by the potential destruction of microbial communities by frequent meteorite impacts during the Early Archean Era. In spite of these reservations about the signs of life in the rocks at Isua, Schidlowski (1995, 1997, 2001) considered that a strong case can be made for the existence of life in the Apex Chert of the Warrawoona Group in Western Australia, which was deposited at 3.465 Ga (e.g., Schopf and Packer, 1987; Schopf, 1993).

The evidence for the existence of life during Early Archean time has recently been questioned (Grotzinger and Rothman, 1996). For example, Brasier et al. (2002) reinterpreted the morphologically preserved microfossils of the Apex Chert as "secondary artefacts" within hydrothermal vein chert and volcanic glass even though the $\delta^{13}C$ values of reduced C in these rocks range from −25.6 to −30.3‰ on the PDB scale.

A second case concerns the metasedimentary rocks of the Isua Supracrustals. The presence of ^{13}C-depleted graphite in quartz on Akilia Island off the coast of southern West Greenland was initially interpreted by Mojzsis et al. (1996) as evidence for the existence of life at 3.8 Ga. However, Fedo and Whitehouse (2002) later concluded that the quartz-rich rocks on Akilia Island had originated by metasomatic alteration of igneous rocks and that the C in the graphite grains had been depleted in ^{13}C by inorganic reactions or was deposited in these rocks by aqueous solutions during metasomatism. Therefore, they concluded that the negative $\delta^{13}C$ values of graphite grains in the high-grade metamorphic rocks of Akilia Island are not necessarily proof of the existence of life at 3.8 Ga. An abiogenic origin of graphite in the metamorphic rocks of Isua was also proposed by Naraoka et al. (1996).

The sedimentary rocks at Isua on the *mainland* of Greenland contain clay-rich metasedimentary rocks and iron formations that have not been as severely metamorphosed as the graphite-bearing rocks on Akilia Island. According to Rosing (1999), the ^{13}C-depletion of the reduced C in the western part of the Isua belt is consistent with biogenic isotope fractionation. In addition, an ion-microprobe study of graphite grains in the supracrustal rocks at Isua by Ueno et al. (2002) yielded $\delta^{13}C$ values ranging from −18 to +2‰ (PDB). The authors observed that the $\delta^{13}C$ values of graphite in these rocks increase (become more positive) with increasing metamorphic grade from −14 to −5‰. In addition, graphite inclusions in early-formed garnet crystals are more depleted in ^{13}C than graphite outside of garnet crystals. Therefore, the authors concluded that the graphite in the rocks of Isua formed from biogenic organic matter that was subsequently enriched in ^{13}C either during metamorphism or by isotope exchange with carbonate rocks at 400–550°C. Therefore, the strongly ^{13}C-depleted graphite grains that had been protected in garnet and apatite crystals (e.g., Mojzsis et al., 1996) are probably of biogenic origin and confirm the existence of life in the oceans at 3.8 Ga.

The controversy concerning the reliability of the isotope composition of graphite and amorphous C as evidence for the existence of microbial communities in the early history of the Earth raises the more general question whether the existence of life on Mars can be ascertained by this method. More specifically, the question is whether organic matter depleted in ^{13}C can also form by abiological processes.

27.3 FOSSIL FUEL

Coal, petroleum, and natural gas are strongly depleted in ^{13}C consistent with their derivation from organic matter produced by photosynthetic organisms. Given that photosynthesis is energized by electromagnetic radiation from the Sun, all forms of fossil fuel are deposits of solar energy. Combustion of these materials by humans after the start of the Industrial Revolution at about 1860 AD has returned large quantities of CO_2 to the atmosphere where it absorbs infrared radiation emitted from the surface of the Earth. Consequently, the combustion of fossil fuel has begun to increase the average global temperature.

27.3a Bituminous Coal

Plant tissue that accumulates in reducing environments on the continents is gradually converted to peat and subsequently to lignite, bituminous coal, anthracite, and even graphite depending on the temperature and pressure applied over long periods of geological time. Most deposits of bituminous coal are of post-Devonian age after land plants had evolved during the Silurian (440 to 417 Ma) and Devonian (417 to 354 Ma) Periods (IUGS; see the Appendix).

The $\delta^{13}C$ value of bituminous coal varies only between narrow limits and has an average of about −25‰ (PDB). The depletion of coal in ^{13}C is consistent with its derivation from photosynthetic plants, which convert atmospheric

CO_2 into cellulose and other organic compounds energized by sunlight.

The chemical evolution of peat into lignite and bituminous coal takes place by the progressive loss of H in the form of CH_4 (methane) gas that is enriched in ^{1}H and ^{12}C relative to the residual organic matter. The methane emanating from beds of coal may react with O dissolved in groundwater, and the resulting bicarbonate ions can precipitate as calcite "cleats" in joints and bedding planes of the overlying sedimentary rocks. Such *methanogenic calcite* is recognized by its highly negative $\delta^{13}C$ values. For example, Faure and Botoman (1984) reported $\delta^{13}C$ values of epigenetic calcite in Triassic coal of southern Victoria Land in Antarctica ranging from −15.6 to −16.9‰ (PDB) compared to an average of −24.9‰ in the associated coal. The average $\delta^{18}O$ value of these samples of methanogenic calcite was +11.34‰ (SMOW), which implies that the $\delta^{18}O$ value of the groundwater that precipitated the calcite was about −19.7‰ (SMOW) assuming a temperature of 10°C and using the calcite–water equation of Kim and O'Neil (1997). The large negative $\delta^{18}O$ value of the water indicates that southern Victoria Land, whose present latitude is about 75° S, was already located at a high latitude at the time the calcite was deposited.

Peat deposited in brackish water may be enriched in S as a result of the reduction of sulfate in solution to bisulfide ions by S-reducing bacteria. The combustion of the high-S coal that forms in such environments releases SO_2 into the atmosphere where the S is oxidized to the sulfate ion in the form of sulfuric acids. The resulting acid rain causes acidification of surface water downwind of coal-burning electric-generating facilities, which poisons aquatic plants and deprives fish of their principal food source. The isotope geochemistry of coal has been discussed by Smith et al. (1982), Redding et al. (1980), and others.

27.3b Petroleum and Natural Gas

Petroleum is a naturally occurring liquid composed of a mixture of hydrocarbon molecules. These molecules are strongly depleted in ^{13}C relative to PDB and have negative $\delta^{13}C$ values between −18 and −34‰ (Stahl, 1979). Evidently, petroleum is more depleted in ^{13}C than coal, in agreement with the theory that petroleum originated from the lipid fraction of organic matter whereas coal is derived from cellulose.

The process of petroleum formation starts with the deposition of organic matter in marine and nonmarine basins. Following burial, the organic matter is converted into kerogen during diagenesis of the sediment to form carbonaceous shales. When these rocks are subsequently exposed to elevated temperatures for sufficient lengths of time defined by the "oil window," the kerogen decomposes into liquid or gaseous hydrocarbons, depending on the conditions, leaving a residue of amorphous carbon. The hydrocarbons are then transported by the movement of subsurface brines from the source rocks into porous reservoir rocks where they are trapped, allowing them to accumulate in "oil pools."

The maturation of organic matter at increasing temperatures causes the resulting kerogen to become enriched in ^{13}C because ^{12}C–^{12}C bonds are weaker than the bonds formed by ^{13}C (Section 26.1). Therefore, the $\delta^{13}C$ value of methane derived from kerogen changes from about −55‰ released by kerogen of low maturity to about −35‰ for methane derived from kerogen that was "overcooked," as indicated by its high vitrinite reflectance of about 3.5 (Stahl, 1979). In general, the $\delta^{13}C$ value of methane in natural gas can vary widely from −18‰ (nonmarine organic matter) to −100‰ or less (bacterial sources).

The relationship between the maturity of the kerogen in the source rocks and the $\delta^{13}C$ value of natural gas in a subsurface reservoir can be used to identify the source rocks. For example, Stahl (1979) cited an example from the Canadian Arctic where the $\delta^{13}C$ value of natural gas was used to determine the vitrinite reflectance of the kerogen in its source rocks based on the isotope–maturity relationship in Figure 27.7*a*. The depth below the surface of the source rocks is indicated by the variation of the vitrinite reflectance with depth in Figure 27.7*b*.

The isotope composition of C of liquid petroleum depends on the abundance of different molecular components that are progressively depleted in ^{13}C, starting with asphaltines,

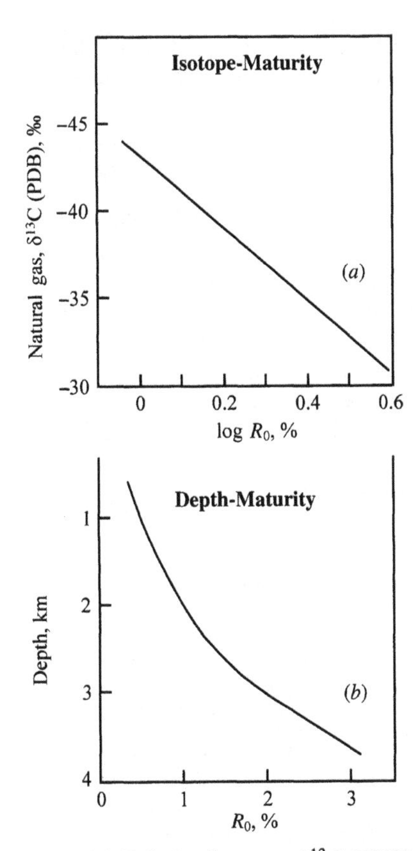

FIGURE 27.7 (a) Relation between $\delta^{13}C$ (PDB) of natural gas and the maturity of the kerogen in its source rocks expressed by $\log_{10} R_O$ (%), where R_O is the vitrinite reflectance of the kerogen. (b) Increase of the vitrinite reflectance (R_O) of kerogen with depth of burial below the surface. Adapted from Stahl (1979).

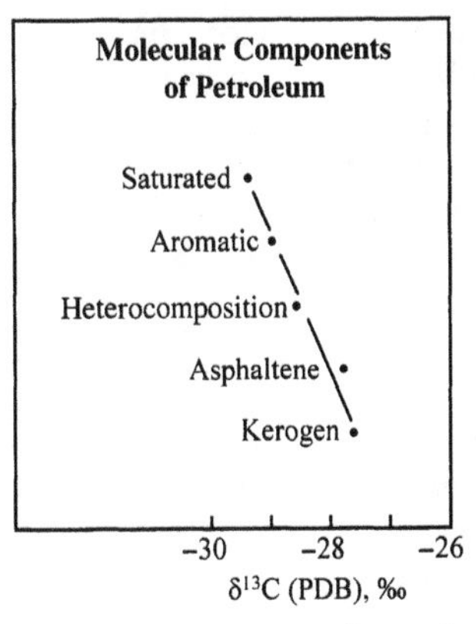

FIGURE 27.8 Isotope type-curve of petroleum of Jurassic age from oil pools under the North Sea. The $\delta^{13}C$ value of kerogen that is the source of this oil is colinear with the $\delta^{13}C$ values of the different molecular types of hydrocarbons. Adapted from Stahl (1979).

mixed hydrocarbons, aromatics, and saturated hydrocarbons and including the kerogen of the source rocks. This relation in Figure 27.8 can be used to determine the $\delta^{13}C$ value of kerogen in the source rocks of a petroleum pool by extrapolating the correlation line of the molecular components of the petroleum. Stahl (1979) described a case where petroleum in Jurassic and Tertiary reservoir rocks in the Canadian Arctic formed nearly identical oil type-curves, suggesting that the oil in both pools originated from the same source. However, the $\delta^{13}C$ value of kerogen in the underlying Jurassic source rocks was not colinear with the oil type-curves of the two reservoirs, proving that the petroleum did not originate from these Jurassic rocks. Other examples of the application of chemical and isotopic data in petroleum exploration and in the search for the sources of petroleum in known pools have been described by Silverman (1964), Stahl (1976, 1977, 1978), Tissot et al. (1974), Fuex (1977), and Sofer (1984).

27.4 CARBON-ISOTOPE STRATIGRAPHY (PHANEROZOIC)

Carbon in the oceans as well as in lakes and rivers occurs in two different reservoirs both of which are in communication with CO_2 of the atmosphere:

1. reduced C in organic matter of biogenic origin and
2. oxidized C of carbonate ions in solution and in calcium carbonate that precipitated from aqueous solution.

The reduced C in the oceans, lakes, and rivers is depleted in ^{13}C as a result of isotope fractionation during biosynthesis of organic compounds and has negative $\delta^{13}C$ values on the PDB scale. The oxidized C is enriched in ^{13}C relative to biocarbonate ions in solution and its $\delta^{13}C$ values are close to 0‰ because PDB is itself a sample of marine calcite. However, the $\delta^{13}C$ values of $CaCO_3$ may be negative in cases where organic matter in the oceans, lakes, and rivers is oxidized to CO_2, which is then incorporated into biogenic calcite or aragonite. This process explains why lacustrine carbonate rocks are variably depleted in ^{13}C relative to PDB and therefore have negative $\delta^{13}C$ values in many cases.

27.4a Isotope Fractionation

The temperature dependence of the isotope-fractionation factors in Table 27.1 has been determined experimentally and, in some cases, by analysis of natural samples. The data demonstrate that calcite and aragonite, precipitated in isotopic equilibrium with CO_2 gas, are enriched in ^{13}C depending on the temperature. The relevant equation for calcite published by Romanek et al. (1992) is:

$$10^3 \ln \alpha_{CO_2}^{cal} = 11.98 - 0.12T \text{ (°C)} \qquad (27.6)$$

Equation 27.6 yields $\alpha_{CO_2}^{cal} = 1.01023$ for $T = 15°C$. This fractionation factor can be used to calculate the $\delta^{13}C$ value of calcite in equilibrium with CO_2 of the atmosphere ($\delta^{13}C_{CO_2} = -7.7‰$, Section 27.1a):

$$\alpha_{CO_2}^{cal} = \frac{\delta^{13}C_{cal} + 1000}{\delta^{13}C_{CO_2} + 1000}$$

$$\delta^{13}C_{cal} = +2.45‰ \text{ (PDB)}$$

When aragonite and calcite form in isotopic equilibrium with the same C reservoir (e.g., $CO_{2(g)}$ or $HCO_{3(aq)}^-$), the aragonite is slightly enriched in ^{13}C relative to calcite. The aragonite–calcite equation of Sommer and Rye (1978) in Table 27.1 is:

$$10^3 \ln \alpha_{cal}^{ar} = 2.56 - 0.065T \quad \text{(°C)} \qquad (27.7)$$

At $T = 15°C$, $\alpha_{cal}^{ar} = 1.001586$ and the value of $\delta^{13}C_{ar}$ in equilibrium with calcite whose $\delta^{13}C_{cal} = +2.45‰$ is:

$$1.001586 = \frac{\delta^{13}C_{ar} + 1000}{1.45 + 1000}$$

$$\delta^{13}C_{ar} = +4.04‰ \text{ (PDB)}$$

Table 27.1. Isotope Fractionation Factors of C between Carbonate Ions in Aqueous Solution, CO_2 Gas, and Solid Calcite

Compounds	$10^3 \ln \alpha$	T, °C
HCO_3^-(aq)–CO_2(g)	7.92–8.27	25
	$-0.0954T +$ 10.41	7–70
CO_3^{2-}(aq)–CO_2(g)	−1.5 to +8.5	4–80
Calcite(s)–CO_2(g)	11.98–0.12T	10–40
Aragonite(s)–CO_2(g)	13.88–0.13T	10–40
Calcite(s)–HCO_3^-(aq)	1.83–2.26	25
Aragonite(s)–HCO_3^-(aq)	2.7 ± 0.2	25
Siderite(s)–HCO_3^-(aq)	−4.20 to +2.86	33–197

Source: Compiled by Chacko et al. (2001) from the literature.
Note: T = temperature in degrees Celsius.

27.4b Carbonate Rocks

The isotopic composition of C in marine and lacustrine carbonate rocks of any age depends on several factors:

1. isotope fractionation between CO_2 gas of the atmosphere and biogenic or abiogenic $CaCO_3$,
2. isotope fractionation among the *aqueous* carbonate species ($CO_{2(aq)}$, H_2CO_3, HCO_3^-, and CO_3^{2-}) whose abundances are pH-dependent (Ohmoto, 1972; Faure, 1998),
3. the temperature of the water at the time of deposition,
4. the mineral composition of the $CaCO_3$ (e.g., calcite or aragonite),
5. introduction of nonatmospheric CO_2 formed by oxidation of organic matter,
6. vital effects related to metabolic processes of organisms (e.g., corals, mollusks, foraminifera) and micro-environmental factors (Swart et al., 1996), and

7. subsequent alteration of carbonate sediment during diagenesis and metamorphism and by deposition of secondary $CaCO_3$ in fractures or cavities from groundwater and/or subsurface brines.

The working of the C cycle in the oceans is illustrated by the $\delta^{13}C$ values of pelagic and benthic foraminifera of Cenozoic age. Shackleton and Kennett (1975) demonstrated that pelagic foraminifera are enriched in ^{13}C relative to benthic foraminifera because of the preferential removal of ^{12}C from the surface layer of the oceans by photosynthetic organisms. The subsequent decay of organic matter at the bottom of the ocean enriches deep water in ^{12}C and causes benthic foraminifera to be depleted in ^{13}C compared to the pelagic foraminifera.

When the availability of nutrients stimulates biomass production in the surface layer of the oceans, the difference between $\delta^{13}C$ values of pelagic and benthic foraminifera increases. The loss of C from the surface layer of the oceans is compensated by additional dissolution of CO_2 gas from the atmosphere, which reduces the concentration of CO_2 in the *atmosphere* (Shackleton et al., 1983). In other words, the operation of this so-called biological carbon pump in the oceans can cause a decrease of the concentration of CO_2 in the atmosphere, which can result in a decrease of the average global temperature. A decrease of the temperature of surface water in the oceans increases the solubility of CO_2 in seawater, which causes additional reductions of the concentration of CO_2 in the atmosphere and hence further decreases of the global temperature leading to the possible onset of continental glaciation.

The isotope compositions of C in marine limestones and dolomites of Phanerozoic and Precambrian ages range from about +5 to −5‰ PDB but vary less with age than the isotope composition of O (Veizer and Hoefs, 1976).

A comprehensive study by Veizer et al. (1999) of the isotopic composition of C of marine biogenic Ca carbonates of Phanerozoic age in Figure 27.9 demonstrates that their $\delta^{13}C$ values increased during the Paleozoic Era from about −1‰ (Cambrian) to +4‰ (Permian) on the PDB scale. The $\delta^{13}C$ values turned downward during the Permian Period and subsequently oscillated above and below the modern value of about +2‰ (PDB) during Mesozoic and Cenozoic time.

Studies of closely spaced stratigraphically collected samples have revealed several short-term fluctuations of $\delta^{13}C$ values of organic and inorganic C. These variations of the isotope composition of

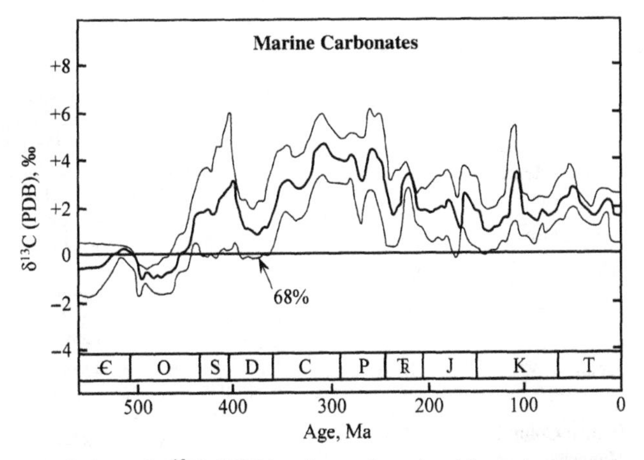

FIGURE 27.9 Variation of $\delta^{13}C$ (PDB) values of marine biogenic Ca-carbonate of Phanerozoic age calculated as a running mean of 5581 samples using a window of 20 million years and a forward step of 5 million years. Adapted from Veizer et al. (1999).

C in the oceans were caused by perturbations of the global C cycle coincident with the extinction of plant and animal species.

The symposium of papers on the evolution of Earth systems edited by Veizer (1999) includes a collection of reports on the geochemical cycles of H, C, O, S, and Sr. The contribution of Veizer et al. (1999) to this symposium discussed not only the isotopic evolution of C in the oceans but also the evolution of O presented in Section 26.9 and Figure 26.18.

The excursions of $\delta^{13}C$ values of Ca carbonate in marine sedimentary rocks of Phanerozoic and Neoproterozoic age coincide with extinction events associated with stratigraphic boundaries (Hallam and Wignall, 1997). The excursions listed in Table 27.2 occur in rocks that straddle the Paleocene–Eocene, Cretaceous–Tertiary, Cenomanian–Turonian (Middle Cretaceous), Triassic–Jurassic, Permian–Triassic, Frasnian–Famennian (Late Devonian), Llandovery–Wenlock (Silurian), Hirnantian (Late Ordovician), and Neoproterozoic–Early Cambrian boundaries. The excursions of $\delta^{13}C$ values appear to be global in scope, although some have only been recognized at one or two sites. In addition, the $\delta^{13}C$ values of carbonate rocks at most (but not all) of these excursions indicate enrichment in ^{13}C, which has been attributed to an increase in the burial of organic matter.

Table 27.2. Excursions of $\delta^{13}C$ Values of Ca Carbonate in Sedimentary Rocks

Time	Locations	References[a]
Pleistocene	Gulf of Mexico	1
	North Atlantic	2, 3
Late Paleocene	Weddell Sea	4, 31
K–T boundary	Alberta	5, 6
Cenomanian–Turonian, Middle Cretaceous	—	7
Aptian, Middle Cretaceous	Japan	8
Late Triassic	British Columbia	9
	Hungary	10
	United Kingdom, Greenland	11
Late Permian	Texas	12
	Slovenia	13
	Mathematical model	14
Early Mississippian	North America, Europe	15, 16, 17
Frasnian–Famennian, Late Devonian	Poland	18
	Germany	19
Llandovery–Wenlock, Silurian	Nevada	20
	Sweden	21
Hirnantian, Late Ordovician	Estonia	22
	Nevada	29, 30
Early Cambrian	Himalaya, India	23
Neoproterozoic–Early Cambrian	China	24
	Siberia	25
	Mongolia	26
	Review	27
	Death Valley, California	28

[a] 1. Jasper and Gagosian, 1989, *Nature*, 342:60–62. 2. Raymo et al., 1990, *Earth Planet. Sci. Lett.*, 97:353–368. 3. Huon et al., 2002, *Geochim. Cosmochim. Acta*, 66:223–239. 4. Röhl et al., 2000, *Geology*, 28:927–930. 5. Nordt et al., 2002, *Geology*, 30:703–706. 6. Arens and Jahren, 2000, *Palaios*, 15:314–322. 7. Arthur et al., 1988, *Nature*, 335:714–717. 8. Ando et al., 2002, *Geology*, 30:227–230. 9. Sephton et al., 2002, *Geology*, 30:1119–1122. 10. Pálfy et al., 2001, *Geology*, 29:1047–1050. 11. Hesselbo et al., 2002, *Geology*, 30:251–254. 12. Magaritz et al., 1983, *Earth Planet. Sci. Lett.*, 66:111–124. 13. Dolenec et al., 2001, *Chem. Geol.*, 175:175–190. 14. Hotinski et al., 2001, *Geology*, 29:7–10. 15. Saltzman et al., 2000, *Geology*, 28:347–350. 16. Silberling et al., 2001, *Geology*, 29:92. 17. Saltzman, 2001, *Geology*, 29:93. 18. Joachimski et al., 2001, *Chem. Geol.*, 175:109–131. 19. Joachimski and Buggisch, 2002, *Geology*, 30:711–714. 20. Saltzman, 2001, *Geology*, 29:671–674. 21. Samtleben et al., 1996, *Geol. Rundschau*, 85:278–292. 22. Kaljo et al., 2001, *Chem. Geol.*, 175:49–59. 23. Mazumdar et al., 1999, *Chem. Geol.*, 156:275–297. 24. Shen and Schidlowski, 2000, *Geology*, 28:623–626. 25. Bartley et al., 1998, *Geol. Mag.*, 135:473–494. 26. Brasier et al., 1996, *Geol., Mag.*, 133:445–485. 27. Ripperdan, 1994, *Annu. Rev. Earth Planet. Sci.*, 22:385–417. 28. Corsetti and Hagadorn, 2000, *Geology*, 28:299–302. 29. Ripperdan, 2001, *Rev. Mineral. Geochim.*, 43:637–662. 30. Marshall et al., 1997, *Paleo. Paleo. Paleo.*, 132:195–210. 31. Hsu and J.A. McKenzie, 1985. In Sundquist and Broecker, Eds., 487–492. Amer. Geophys. Union, Washington, DC.

27.4c Frasnian–Famennian

The $\delta^{13}C$ values of marine carbonate rocks at the Frasnian–Famennian boundary (Late Devonian) in the Rheinische Schiefergebirge of Germany increase from about +1.0 to +4.0‰ on the PDB scale (Joachimski et al., 2001; Joachimski and Buggisch, 2002). In fact, Figure 27.10 indicates that there were two excursions: the first in late Frasnian (late rhenana conodont zone) and the second in early Famennian time (early to middle triangularis conodont zone). Joachimski and Buggisch (2002) attributed the $\delta^{13}C$ excursions to excessive burial of organic matter, which reduced the C concentration of the surface layer of the oceans and caused a decrease of the concentration of CO_2 in the atmosphere. The resulting cooling of the global temperature caused an increase of the $\delta^{18}O$ value of apatite in conodonts by up to 1.5‰ on the SMOW scale. Joachimski and Buggisch (2002) concluded that the two $\delta^{18}O$ excursions at the Frasnian–Famennian boundary were caused by two cooling events when the surface temperature of the oceans in the equatorial belt decreased by 5–6°C. These authors also reported that 50–66% of all marine genera became extinct at this time. The extinction of genera living in shallow water at low latitudes was more severe than that of genera in deep water at high latitude.

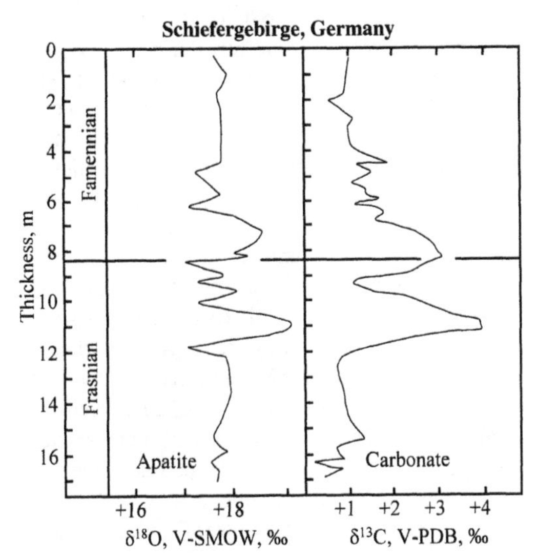

FIGURE 27.10 Excursions of $\delta^{18}O$ of apatite (conodonts and fish teeth) and $\delta^{13}C$ of calcite of marine carbonate (micrite) of Frasnian–Famennian age, Late Devonian, in the Behringhäuser Tunnel, Rheinisches Schiefergebirge, Germany. Adapted from Joachimski and Buggisch (2002).

An alternative explanation for the Frasnian–Famennian extinction and the associated excursions of the isotopic compositions of C and O was proposed by McLaren (1970), who attributed the extinctions to the effects of the impact of a large meteoroid. In addition, Wang et al. (1991, 1994, 1996) reported negative excursions of $\delta^{13}C$ values in Frasnian–Famennian rocks in south China and Alberta (Canada), which they tentatively attributed to the impact of a meteoroid. However, Joachimski et al. (2001) and Joachimski and Buggisch (2002) reported positive rather than negative excursions of $\delta^{13}C$ at the Frasnian–Famennian boundary.

Although the cause for the Frasnian–Famennian extinction is still a matter for debate, impacts of large meteoroids during Phanerozoic time have undeniably resulted in the extinction of large numbers of genera of plants and animals both on land and in the oceans. The type example of such catastrophic events is the well-known impact of a meteoroid at 65 Ma on the Yucatan Peninsula of Mexico (Alvarez et al., 1980; Pope et al., 1994).

The examples cited in Table 27.2 demonstrate that the $\delta^{13}C$ values of marine carbonate rocks of Phanerozoic age are sensitive recorders of complex interactions between the biosphere, the atmosphere, the global climate, the intensity of weathering on the continents, the circulation of the oceans, and the fate of organic matter in the oceans (i.e., oxidation or burial). The resulting stratigraphic variations of $\delta^{13}C$ values in marine carbonate rocks were modeled by Kump and Arthur (1999) and Ripperdan (2001).

27.4d Neoproterozoic–Early Cambrian

During most of Phanerozoic time, the excursions of $\delta^{13}C$ values of marine carbonates were in the positive direction. However, excursions to negative $\delta^{13}C$ values occurred in the Hirnantian rocks (Latest Ordovician) exposed in Copenhagen Canyon of the Monitor Range in Nevada (Ripperdan, 2001).

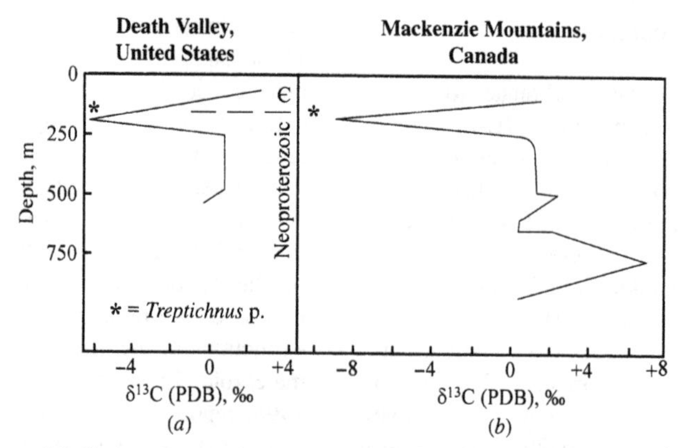

FIGURE 27.11 (*a*) Carbon-isotope stratigraphic definition of the boundary between the Neoproterozoic and Early Cambrian rocks in Death Valley, California, and (*b*) correlation with a similar succession of strata in the Mackenzie Mountains of northwest Canada. The asterisk marks the first appearance of the index fossil *Treptichnus pedum*. Adapted from Corsetti and Hagadorn (2000).

Negative $\delta^{13}C$ values of marine carbonate rocks are attributable to the oxidation of organic matter in the oceans and to weathering of kerogen-rich shale on the continents, both of which release CO_2 molecules and hence HCO_3^- ions that are strongly depleted in ^{13}C. In addition, negative $\delta^{13}C$ values in marine carbonates may indicate low bioproductivity (i.e., no ^{13}C enrichment of HCO_3^- at the surface of the ocean) and high partial pressure of CO_2 in the atmosphere whose $\delta^{13}C$ value is −7.7‰. Excursions to negative $\delta^{13}C$ values of carbonate rocks deposited in near-shore environments may also record episodes of freshwater incursion into estuaries, coastal lagoons, and shallow bays.

Carbonate rocks of Late Neoproterozoic and Early Cambrian age are characterized by negative $\delta^{13}C$ values, which has facilitated the identification of the base of the Cambrian System (i.e., Shen and Schidlowski, 2000). For example, Corsetti and Hagadorn (2000) used the Neoproterozoic–Cambrian strata exposed in Death Valley, California, to define the base of the Cambrian System both by the first appearance of the Early Cambrian index fossil and by a strong negative excursion of $\delta^{13}C$ values of carbonate rocks at the base of the Wood Canyon Formation. The basal strata of this formation consist of three successive parasequences each of which is capped by dolomite beds. The Early Cambrian index fossil *Treptichnus (Phycodes) pedum* first occurs in the rocks at the base of parasequence 3. Therefore, Corsetti and Hagadorn (2000) placed the Neoproterozoic–Early Cambrian boundary at the top of the dolomite beds of parasequence 2 and reported that this dolomite is characterized by negative $\delta^{13}C$ values extending to −4‰ (PDB). The correlation of biostratigraphy and C-isotope stratigraphy permits this boundary to be recognized either by the occurrence of *Treptichnus pedum*, which occurs only in siliclastic rocks, or by the negative $\delta^{13}C$ values of carbonate rocks, which do not contain the diagnostic fossil.

The validity of C-isotope stratigraphy was illustrated by Corsetti and Hagadorn (2000) by correlating the Death Valley section in California to a section in the Mackenzie Mountains of northwestern Canada based on $\delta^{13}C$ values published by Kaufman et al. (1997). The $\delta^{13}C$ profiles of both sections in Figure 27.11 display well-defined negative $\delta^{13}C$ excursions in Neoproterozoic carbonates directly underlying the base of the Lower Cambrian strata.

27.5 PRECAMBRIAN CARBONATES

The isotope composition of C in sedimentary carbonate rocks of Precambrian age remained fairly constant for long periods of geological time and varied only locally where carbonate rocks were deposited in special environments. Eichmann and Schidlowski (1975) reported that the average $\delta^{13}C$ value of 58 samples Precambrian carbonate rocks (>3.3 Ga to ~0.65 Ga) is $+0.9 \pm 2.7‰$ (PDB) compared to $-24.7 \pm 6.0‰$ (PDB) for organic matter in the same samples. Evidently, the organic matter that was formed by Precambrian organisms in the oceans was depleted in ^{13}C compared to cogenetic carbonate rocks, much like organic matter in sedimentary rocks of Phanerozoic age. The effective average fractionation factor (α^{cal}_{org}) was:

$$\alpha^{cal}_{org}(^{13}C) = \frac{0.9 + 1000}{-24.7 + 1000} = 1.026 \pm 0.009$$

The ^{13}C depletion of organic matter in Precambrian rocks indicates that photosynthesis started about 3.7×10^9 years ago (Junge et al., 1975).

Subsequent studies of the isotope geochemistry of C in carbonate rocks of Precambrian age have confirmed these conclusions (e.g., Schidlowski et al., 1983). A cross-section of the recent literature on the $\delta^{13}C$ values of carbonate rocks of Precambrian age is presented below:

Hall and Veizer, 1996, *Geochim. Cosmochim. Acta*, 60:667–677. Buick et al., 1995, *Chem. Geol.*, 123:153–171. Lindsay and Brasier, 2000, *Precamb. Res.*, 99:271–308. Lindsay and Brasier, 2002, *Precamb. Res.*, 114:1–34. Fairchild et al., 1990, *Am. J. Sci.*, 290-A: 46–79. Iyer et al., 1995, *Precamb. Res.*, 73:271–282. Kaufman et al., 1991, *Precamb. Res.*, 49:301–327. Knoll et al., 1986, *Nature*, 321:832–838. Knoll et al., 1995, *Am. J. Sci.*, 295:823–850. Mirota and Veizer, 1994, *Geochim. Cosmochim. Acta*, 58:1735–1745. Veizer et al., 1989a, *Geochim. Cosmochim. Acta*, 53:845–857. Veizer et al., 1989b, *Geochim. Cosmochim. Acta*, 53:859–871. Veizer et al. 1990, *Geochim. Cosmochim. Acta*, 54:2717–2729. Veizer et al., 1992a, *Geochim. Cosmochim. Acta*, 56:875–885. Veizer et al., 1992b, *Geochim. Cosmochim. Acta*, 56:2487–2501. Bau et al., 1999, *Earth Planet. Sci. Lett.*, 174:43–57. Buick et al., 1998, *Geology*, 26:875–878. Derry et al., 1992, *Geochim. Cosmochim. Acta*, 56:1317–1329.

The $\delta^{13}C$ values of carbonate rocks included in these studies range narrowly above and below 0‰ on the PDB scale. For example, the average $\delta^{13}C$ value of a large number of carbonate samples from the McArthur and Mt. Isa Basins (1.700–1.575 Ga) of northern Australia analyzed by Lindsay and Brasier (2000) is $-0.6‰$ (PDB). This result confirms that the C cycle of the Earth maintained a steady state during this time interval. In addition, the authors concluded that diagenesis had occurred early in both basins, causing the original isotopic composition of C to be preserved. However, the $\delta^{18}O$ *values* of the carbonate rocks in the McArthur and Mt. Isa Basins are consistently negative and range down to $-11‰$ on the PDB scale.

27.5a Carbon-Isotope Excursions

Although the $\delta^{13}C$ values of sedimentary carbonate rocks of Precambrian age do not deviate from 0‰ (PDB) in most cases, some notable anomalies have been reported. For example, the original study of Eichmann and Schidlowski (1975) contained examples of excursions of $\delta^{13}C$ values in Precambrian carbonate rocks of bituminous limestone of Namibia (+2.8 to +9.9‰) and in East Greenland (+3.7 to +7.2‰). Subsequently, Schidlowski et al. (1976) reported strongly positive $\delta^{13}C$ values (+2.7 to +13.4‰) in dolomites of the Paleoproterozoic Lomagundi Group (2.65–1.95 Ga) of northwestern Zimbabwe.

The lower part of the Lomagundi Group is composed primarily of dolomites and quartzites which are underlain by the Archean basement complex of Zimbabwe. The dolomites form an arcuate belt extending for about 300 km, which implies that they were deposited in a large continental basin. The $\delta^{13}C$ values of these rocks in Figure 27.12 have a mean value of $+8.2 \pm 2.6‰$ (PDB). This value contrasts with the average $\delta^{13}C$ value of $-0.24 \pm 0.82‰$ (PDB) of the well-known Bulawayan limestone (3.3–2.9 Ga) of Zimbabwe in Figure 27.12*b* (Eichmann and Schidlowski, 1975).

The anomalous enrichment in ^{13}C of the Lomagundi Dolomite is not a characteristic property of dolomite, nor is it attributable to isotopic re-equilibration because the rocks reached only the low greenschist facies of metamorphism. In addition, the average $\delta^{18}O$ value of the Lomagundi Dolomite is -8.4 ± 1.5‰ (PDB), which is typical of carbonate rocks of Precambrian age (Section 26.9). For these and other reasons, Schidlowski et al. (1976) concluded that the Lomagundi Dolomites were deposited in an evaporite basin by water that had been enriched in ^{13}C by the formation of organic matter deposited separately in kerogen-rich argillites and slates.

Another example of carbonate rocks containing C with an anomalous isotopic composition occurs in the Dales Gorge Member of the Brockman Iron Formation of the Hamersley Group of northwestern Australia. These rocks were deposited in a basin south of the Pilbara Block of igneous and high-grade metamorphic rocks that crystallized at about 2.77 Ga. The sedimentary rocks of the Hamersley Group contain extensive banded iron formations and some of the earliest marine carbonate rocks (2.7–2.5 Ga) (Lindsay and Brasier, 2000).

Becker and Clayton (1972) reported that the $\delta^{13}C$ values of ankerite in the Dales Gorge Member in Figure 27.12*c* range from -6.5 to -15.05‰ (PDB) with an average of -9.8 ± 0.5‰ for 34 samples collected from 78 m of section. Twelve samples of siderite from the same section have a similar average $\delta^{13}C$ value of -9.8 ± 1.0‰.

The $\delta^{13}C$ values of the Wittenoom Formation (150 m), which is located stratigraphically below the Dales Gorge Member, range from a normal average of -1.56 ± 0.92‰ (calcite) and -0.74 ± 1.12‰ (dolomite) to more negative values of -5.2 ± 0.8‰ (calcite) and -5.67‰ (dolomite) recorded in two samples from the same locality. Additional $\delta^{13}C$ values of carbonate rocks of the Wittenoom Dolomite reported by Lindsay and Brasier (2000) are close to 0‰ (PDB).

Becker and Clayton (1972) cited evidence that the $\delta^{13}C$ values of the carbonate minerals in the Dales Gorge and Wittenoom samples had not been altered after deposition and that the Wittenoom Dolomite was deposited in a marine environment. The average $\delta^{13}C$ values of most of the calcite and dolomite samples in the Wittenoom Dolomite are consistent with a marine origin of this formation. The ^{13}C-depletion of the Fe-carbonates of the Dales Gorge Member could be attributed to metamorphic decarbonation, which causes a *decrease* of the $^{13}C/^{12}C$ ratio of the residual

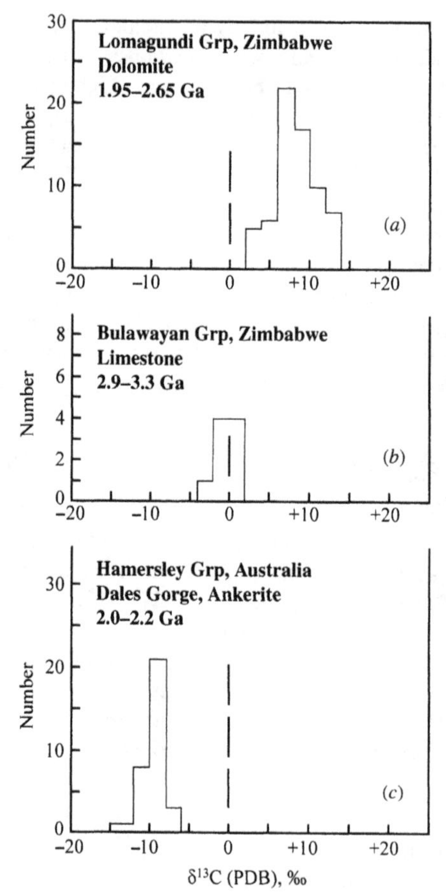

FIGURE 27.12 (*a*) Positive $\delta^{13}C$ values in dolomite of the Precambrian Lomagundi Group of Zimbabwe relative to PDB. Data from Schidlowski et al. (1976). (*b*) "Normal" isotope composition of limestone of the Precambrian Bulawayan Group of Zimbabwe relative to PDB. Data from Eichmann and Schidlowski (1975). (*c*) Negative $\delta^{13}C$ values of ankerite in the Precambrian Dales Gorge Member of the Brockman Iron Formation, Hamersley Group, Western Australia. Data from Becker and Clayton (1972).

carbonate (e.g., Deines and Gold, 1969; Shieh and Taylor, 1969). However, the rocks of the Dales Gorge Member were not affected by temperatures sufficient to cause decarbonation. Instead, Becker and Clayton (1972) proposed that the Dales Gorge Member was deposited in a restricted basin that received "light" C from an external source. The basin was located close to the ocean and was flooded by seawater repeatedly and for extended periods of time. In other words, the basin was in contact with the ocean during the deposition of the Wittenoom Dolomite but was isolated from it during the later deposition of the Dales Gorge Member.

The light C that dominated the Hamersley Basin at the time of deposition of the Dales Gorge Member could have originated from magmatic sources such as carbonatites and volcanic gases, both of which are variably depleted in ^{13}C. In addition, the light carbon could have originated from organic matter formed by photosynthetic micro-organisms (algae and bacteria). Becker and Clayton (1972) preferred the explanation that the oxidation of such organic matter led to the formation of ^{13}C-depleted bicarbonate ions in the water from which the carbonate minerals of the Dales Gorge Member were deposited.

27.5b Snowball Earth

The occurrence of tillites of Neoproterozoic age near sealevel at low latitude has been observed by several research groups referenced in Table 27.3 (e.g., Sohl et al., 1999). If continental glaciers existed along the coast near the equator in Neoproterozoic time, the global climate had presumably cooled sufficiently for continental ice sheets to form everywhere. Therefore, Hoffman et al. (1998b) proposed the so-called "Snowball Earth" hypothesis mentioned in Section 19.2b and described in a book by Walker (2003). This hypothesis is a radical departure from the principle of Uniformitarianism with implications that are comparable to the hypothesis of Alvarez et al. (1980, 1984) that the Ir anomaly in the sediment deposited at the K/T boundary was caused by the impact of a large meteoroid on the Earth.

The variation of the $^{87}Sr/^{86}Sr$ ratio of seawater during the Neoproterozoic Era in Figure 19.8 indicates the occurrence of two episodes lasting less than one million years each, during which the $^{87}Sr/^{86}Sr$ of seawater abruptly declined (Jacobsen and Kaufman, 1999). The $\delta^{13}C$ values of the same carbonate rocks in Figure 27.13 also declined abruptly from about +8‰ (PDB) to as low as −5‰ (PDB).

The excursions of the isotope composition of Sr, C, and O during the Neoproterozoic Era contrast with the stability of the $\delta^{13}C$ values of marine carbonates of Mesoproterozoic age. For example, Lindsay and Brasier (2000) reported that carbonate rocks (1700–1575 Ma) of the McArthur and Mt. Isa Basins in the Northern Territory of Australia remained virtually constant for 125 million years at an average of −0.6‰ (PDB). A similar average $\delta^{13}C$ value (−0.5‰) was obtained by Buick et al. (1995) for younger Mesoproterozoic carbonates of the Bangemall Group (1625–1000 Ma) of northwestern Australia and by Xiao et al. (1997) for carbonate rocks of the North China platform.

The absence of $\delta^{13}C$ excursions in marine carbonates of Mesoproterozoic age implies that the input of nutrients (e.g., P and NO_3) to the oceans remained constant and low for up to about 700 million years prior to the start of the Neoproterozoic Era. The stability of the nutrient supply from the continents suggests the absence of global tectonic activity. In addition, Brasier and Lindsay (1998) proposed that the limited nutrient supply forced organisms into symbiotic relationships in order to survive and thereby stimulated the development of autotrophic eukaryotes.

Evidently, the period of global stability that had prevailed during the Mesoproterozoic came to an abrupt end during the Neoproterozoic Era (Hayes et al., 1999). According to the Snowball Earth hypothesis of Hoffman et al. (1998b), continental ice sheets formed on the continents and the oceans froze over not once but twice during the Sturtian and Varangian global glaciations.

The Snowball Earth hypothesis raises several important questions about the Earth that have not yet been answered:

1. What caused the initial cooling of the global climate of the Earth?

Table 27.3. Summary of the Literature Concerning the Snowball Earth Hypothesis Based on Excursions of the $\delta^{13}C$ Values of Marine Carbonates of Neoproterozoic Age

Topic	References
	Evidence For and Against
Snowball Earth hypothesis:	Kump, 1991, *Geology*, 19:299–302; Kirschwink, 1992, in Schopf and Klein (Eds.), *The Proterozoic Biosphere*, pp. 51–52, Cambridge University Press, New York; Kaufman et al., 1997, *Proc. Natl. Acad. Sci. (USA)*, 94:6600–6605; Hoffman et al., 1998, *Science*, 281:1342–1346; Kennedy et al., 1998, *Geology*, 26:1059–1063; Hyde et al., 2000, *Nature*, 405:425–429; Chandler and Sohl, 2000, *J. Geophys. Res.*, 105:20737–20756; Sreenivas and Sharma, 2001, *Chem. Geol.*, 181:193–195; Kennedy et al., 2001, *Geology*, 29(12):1135–1138.
Carbon-isotope geochemistry of seawater:	Kaufman et al., 1993, *Earth Planet. Sci. Lett.*, 120:409–430; Grotzinger and Knoll, 1995, *Palaios*, 10:578–596; Hayes et al., 1999, *Chem. Geol.*, 161:103–125; Jacobsen and Kaufman, 1999, *Chem. Geol.*, 161:37–57.
Methane hydrate:	Kennedy et al., 2001, *Geology*, 29(5):443–446; Hoffman et al., 2002, *Geology*, 30(3):286–287; Kennedy et al., 2002, *Geology* 30(3):287–288.
	Global Carbon-Isotope Stratigraphy
Namibia:	Kaufman et al., 1991, *Precamb. Res.* 49:301–327; Saylor et al., 1998, *J. Sed. Res.*, 68:1223–1235; Hoffman et al., 1998, *GSA Today*, 8:1–9.
Australia:	Kennedy, 1996, *J. Sed. Res.*, 66:1050–1064; Sohl et al., 1999, *Geol. Soc. Am. Bull.*, 111:1120–1139.
Oman:	Burns and Matter, 1993, *Eclogae Geol. Helv.*, 86:595–607; Leather et al., 2002, *Geology*, 30(8):891–894.
Iran:	Kimura et al., 1997, *Earth Planet. Sci. Lett.*, 147:E1–E7.
Siberia:	Knoll et al., 1995, *Am. J. Sci.*, 295:823–850; Pelechaty et al., 1996, *Geol. Soc. Am. Bull*, 108:992–1003; Kaufman et al., 1996, *Geol. Mag.*, 133:509–533.
Svalbard and East Greenland:	Knoll et al., 1986, *Nature*, 321:832–838.
Baffin Island, Canada:	Kah et al., 1999, *Can. J. Sci.* (in press).
Mackenzie Mountains, Canada:	Aitken, 1991, *Geology*, 19:445–448; Narbonne et al., 1994, *Geol. Soc. Am. Bull.*, 106:1281–1292.
Idaho and Utah, USA:	Smith et al., 1994, *Geol. Mag.*, 133:347–349.
Brazil:	Iyer et al., 1995, *Precamb. Res.*, 73:271–282.

2. How did the C-cycle of the Earth respond to global glaciations?
3. How do the $\delta^{13}C$ values of carbonate rocks and organic matter record the perturbation of the C-cycle?
4. What caused the global glaciations to end?

The response of the C-cycle in the oceans to hypothetical global glaciations was modeled by Kump (1991) by consideration of the biological C pump (Section 27.4b). According to his model, shallow-water marine carbonates are enriched in ^{13}C because of the preferential incorporation of ^{12}C

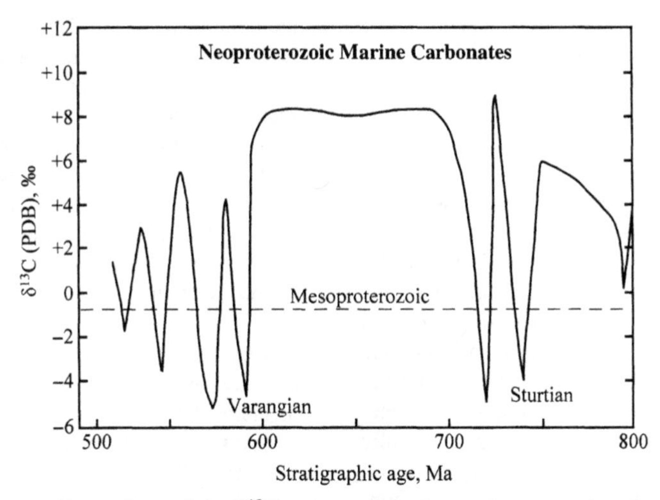

FIGURE 27.13 Excursions of the $\delta^{13}C$ values of marine carbonate rocks of Neoproterozoic age. The dashed line at $\delta^{13}C = -0.5‰$ is the average value of this parameter in carbonates of Mesoproterozoic age when the supply of nutrients to the oceans was low and steady. Adapted from Jacobsen and Kaufman (1999).

by pelagic organisms. When these organisms die, their bodies sink toward the bottom of the ocean where a relatively small fraction of the sinking biomass is buried. Most of the organic matter sinking through the oceans is oxidized to CO_2, causing dissolved inorganic C (DIC) in oceanic bottom water to be depleted in ^{13}C (i.e., enriched in ^{12}C). Therefore, the isotope composition of DIC in oceanic *surface water* is enriched in ^{13}C, whereas that of *bottom water* is depleted in ^{13}C and may have negative $\delta^{13}C$ values on the PDB scale.

The difference in the $\delta^{13}C$ values of these two C-reservoirs depends on the biological productivity of oceanic surface water, which in turn depends on the availability of nutrients (Section 27.4b). A high nutrient content stimulates productivity in the surface layer of the oceans, which increases the difference between $\delta^{13}C$ of DIC in surface water and in bottom water. If the supply of nutrients is reduced or stopped altogether, biological productivity is severely reduced, which causes the $\delta^{13}C$ value of DIC in oceanic surface water to rise because ^{12}C is no longer being transported into deep water by sinking organic matter.

Under normal circumstances, nutrients originate by weathering of rocks on the continents. In addition, the upwelling of bottom water may locally add nutrients to surface water, which stimulates bioproductivity (e.g., algal blooms) and in turn attracts grazers and their predators. These natural processes cannot operate when the continents are covered by ice sheets and the oceans are covered by a layer of ice because chemical weathering of rocks would be prevented and rivers would stop flowing. In addition, the ocean would be isolated from the atmosphere, causing anoxia and interfering with its circulation. Under these conditions, C from the mantle dominates the ocean and its isotope composition therefore decreases to $\delta^{13}C = -5‰$ (PDB). Jacobsen and Kaufman (1999) concluded that the decrease of the oceanic $\delta^{13}C$ value from +8 to −5‰ could occur within about 300,000 years after the biological C pump is suddenly stopped.

Alternatively, the decrease of the $\delta^{13}C$ value of shallow-water carbonates could have been caused by the destabilization of solid methane hydrate at the bottom of the ocean because the $\delta^{13}C$ values of biogenic methane range between −60 and −95‰ on the PDB scale (Kennedy et al., 2001b). However, Hoffman et al. (2002) questioned whether sufficiently large deposits of gas hydrate were available to cause the observed excursions of $\delta^{13}C$ to −5‰ in shallow-water carbonates associated with the Sturtian and Varangian tillites.

A third alternative is that postglacial upwelling of deep water flooded the continental shelves and continental basins with ^{13}C-depleted water from which shallow-water carbonate rocks were deposited. These carbonate deposits, which form the so-called cap rock, have negative $\delta^{13}C$ values and were deposited directly above glacial sediment. The scenario developed originally by Hoffman et al. (1998b) to explain the end of the glaciations was based on a postulated increase of the partial pressure of CO_2 in the atmosphere caused by volcanic activity during the global ice age when the ocean and the atmosphere were separated by a nearly continuous layer of sea-ice. The elevated concentration of atmospheric CO_2 eventually caused greenhouse warming followed by deglaciation and large-scale precipitation of calcium carbonate in the oceans. The waxing and waning of the ice sheets during the Neoproterozoic Era was demonstrated by Hoffman et al. (1998a) and Hoffman and Schrag (2000) using outcrops of tillite overlain by the cap carbonate in Namibia.

Certain aspects of the Snowball Earth hypothesis proposed by Hoffman et al. (1998b) have been questioned by Runnegar (2000), Kennedy et al. (2001a), Sreenivas and Sharma (2001), Poulsen (2003), and others. In addition, Leather et al. (2002) reported that the glacial diamictites of the Neoproterozoic Huqf Supergroup in Oman resemble the products of the multiple Pleistocene glaciations and do not require the extreme conditions of the Snowball Earth hypothesis.

27.6 IGNEOUS AND METAMORPHIC ROCKS

Mafic volcanic rocks such as mid-ocean ridge basalt (MORB), oceanic island basalt (OIB), and continental flood basalt originate from the mantle of the Earth. Therefore, these kinds of rocks contain C whose isotopic composition is expected to provide information about the primordial C that was incorporated into the Earth at the time of its formation. Unfortunately, volcanic rock also contain C derived from other sources, such as biogenic organic matter, carbonate minerals that formed in equilibrium with atmospheric CO_2, and other extraneous sources. In addition, C isotopes are fractionated when CO_2 is released from cooling lava flows or from magmas at depth. As a result, the isotopic composition of the residual C in igneous rocks has been altered by degassing in ways that depend on the temperature and on the fraction of C remaining. For these reasons, the isotopic composition of C in mafic igneous rocks is not equal to that of C in the mantle of the Earth.

Igneous rocks contain a variety of C compounds, including

1. CO_2 gas in fluid or gaseous inclusions;
2. carbonate minerals;
3. elemental C in the form of graphite, diamonds, and perhaps fullerene (C_{60}) reported by Jehlicka et al. (2003);
4. carbide minerals; and
5. organic compounds, including methane and other hydrocarbon gases.

Hoefs (1973) and Fuex and Baker (1973) reported that the concentrations of reduced C in igneous rocks (organic compounds, graphite, and carbides) range from 30 to 360 ppm and that this C-component has $\delta^{13}C$ values between -20 and -28‰ on the PDB scale. The concentrations of oxidized C (CO_2 gas and carbonate minerals) range widely from 0 to more than 20,000 ppm, whereas the $\delta^{13}C$ values vary between $+2.9$ and -18.2‰.

Fuex and Baker (1973) observed that the concentration of *oxidized* C in igneous rocks (carbonates and CO_2) correlates positively with the degree of alteration of feldspar and concluded that most oxidized C was introduced into these rocks by circulating hydrothermal fluids or groundwater. Alternatively or in addition, the similarity of the isotopic composition of reduced C in igneous rocks (organic compounds, graphite, and carbides) to that of biogenic C indicates that the *reduced* C may have been incorporated into magmas by assimilation of sedimentary rocks containing biogenic organic matter.

27.6a Volcanic Gases

Volcanic gases emanating from fumaroles and cooling lava flows are composed primarily of CO_2 and H_2O (vapor) but also contain CH_4, amounting to about one percent of the gas (e.g., Welhan, 1988; Ishibashi et al., 1995). The $\delta^{13}C$ value of CO_2 collected in geothermal areas range from −2 to −6‰, compared to −20 to −30‰ for CH_4 on the PDB scale. The difference in the isotope composition of C in coexisting CO_2 and CH_4 is caused by isotope fractionation associated with the exchange of C isotopes between these compounds. The temperature dependence of this reaction at equilibrium was expressed by Horita (2001) by the equation

$$10^3 \ln \alpha_{CH_4}^{CO_2} = 26.70 - \frac{49.137 \times 10^3}{T} + \frac{40.828 \times 10^6}{T^2} - \frac{7.512 \times 10^9}{T^3} \quad (27.8)$$

based on *experimental* data where T is the temperature on the Kelvin scale. This equation can be used to calculate the C-isotope equilibration temperature of volcanic gas containing CO_2 and CH_4. For example, Ferrara et al. (1963) reported the following $\delta^{13}C$ values for gas in steam jets at Larderello in Tuscany, Italy:

$$\delta^{13}C(CO_2) = -3.74‰$$

$$\delta^{13}C(CH_4) = -26.74‰$$

Therefore, the C-isotope fractionation factor $\alpha_{CH_4}^{CO_2}$ is:

$$\alpha_{CH_4}^{CO_2} = \frac{-3.74 + 1000}{-26.74 + 1000} = 1.0236,$$

which yields:

$$10^3 \ln \alpha_{CH_4}^{CO_2} = 23.325.$$

This value corresponds to a temperature of 320°C based on a graphical solution of equation 27.8 in Figure 27.14.

In an alternative approach, Bottinga (1969) *calculated* the C-isotope fractionation factors for the CO_2–CH_4 system for temperatures ranging from 0 to 700°C. A straight line fitted by least-squares regression to Bottinga's calculated values of $10^3 \ln \alpha$ at temperatures between 100 and 600°C (in increments of 100°C) has the equation:

$$10^3 \ln \alpha_{CH_4}^{CO_2} = \frac{6.250 \times 10^6}{T^2} + 4.890. \quad (27.9)$$

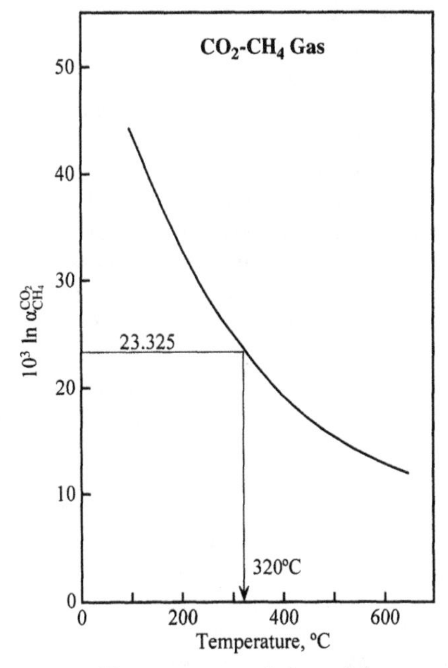

FIGURE 27.14 Temperature variation of the C-isotope fractionation factor for CO_2–CH_4 gas in isotopic equilibrium based on the experimental results of Horita (2001). The value of $10^3 \ln \alpha = 23.325$ was derived from the $\delta^{13}C$ value values of CO_2 and CH_4 gas in the steam jets in the geothermal area at Larderello in Tuscany, Italy. Reported by Ferrara et al. (1963).

Solving for the temperature T on the Kelvin scale,

$$T = \left(\frac{6.250 \times 10^6}{10^3 \ln \alpha - 4.89}\right)^{1/2}$$

and substituting $10^3 \ln \alpha = 23.325$ for the CO_2–CH_4 system in the steam jets of Larderello yields:

$$T = \left(\frac{6.250 \times 10^6}{23.325 - 4.89}\right)^{1/2} = 582.26 \text{ K} \quad (309°C)$$

This result demonstrates that the C-isotope fractionation factors calculated by Bottinga (1969) are in good agreement with the experimental results of Horita (2001). In addition, Bottinga (1969) listed the $\delta^{13}C$ values of coexisting CO_2 and CH_4 in geothermal gas discharged by wells and hotsprings in geothermal areas of New Zealand reported by Hulston and McCabe (1962) and from Larderello (Italy), Yellowstone Park (Wyoming), and the Geysers (California) measured by Craig (1963).

The isotopic compositions of C in CO_2 emitted by the volcano Kilauea on the island of Hawaii has been monitored for about 25 years, from 1960 to 1985. Gerlach and Taylor (1990) estimated that the average $\delta^{13}C$ value of all CO_2 emitted from the summit of this volcano is $-3.4 \pm 0.05‰$ (PDB). However, CO_2 from the East Rift Zone of Kilauea has a $\delta^{13}C$ value of $-7.8‰$ (Hoefs, 1997).

The CO_2 released by volcanic activity elsewhere in the world has $\delta^{13}C$ values between -3 and $-4‰$ (e.g., MORB glass, $-3.7‰$, Javoy and Pineau, 1991; Iceland, $-3.8‰$, Poreda et al., 1992; Indonesia, $-3.0‰$, Poorter et al., 1991). The majority of $\delta^{13}C$ values of CO_2 in volcanic gas and fluid inclusions in mafic igneous rocks compiled by Taylor (1986) from the literature published prior to 1986 are clustered between -4 and $-8‰$ in Figure 27.15. Accordingly, $\delta^{13}C = -6 \pm 2‰$ is representative of the isotope composition of CO_2 emanating from basalt magma and occurring in fluid inclusions in mafic igneous rocks. The CO_2 in fluid inclusions of granite pegmatites has a wide range of $\delta^{13}C$ values between $+1$ and $-12‰$.

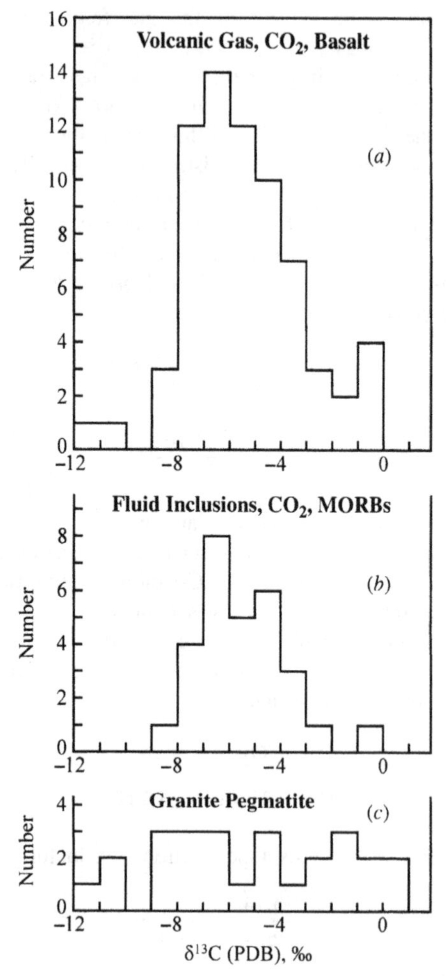

FIGURE 27.15 (*a*) Range of $\delta^{13}C$ values of CO_2 in volcanic gas and fluid inclusions in back-arc basalt, oceanic island basalt, and rift basalt. Range of $\delta^{13}C$ values of CO_2 in (*b*) fluid inclusions in mid-ocean ridge basalt, and (*c*) granitic pegmatites. Based on a compilation of data from the literature by Taylor (1986).

27.6b Volcanic Rocks

The C contained in recently erupted volcanic rocks has been studied by analyzing gases released by crushing and by progressive stepwise heating experiments (e.g., Pineau et al., 1976; Pineau and Javoy, 1983; DesMarais and Moore, 1984; Mattey et al., 1984). The results of these experiments demonstrate that the $\delta^{13}C$ values of CO_2 released by Mid-Ocean Ridge Basalt (MORB) range widely depending on the procedure used to release CO_2 from the rocks.

For example, Pineau et al. (1976) analyzed samples of tholeiite recovered from the Mid-Atlantic Ridge at 36°49.3′ N, 33°15′ W. Some of the samples dredged at this site contained gases under pressure, which caused them to explode on the deck of the ship. The gases in these "popping rocks" consisted primarily of CO_2 with small concentrations of H_2, $SO_2 + H_2O$, CH_4, and CO.

Mid-ocean ridge basalts also contain small amounts of disseminated carbonate minerals, which Pineau et al. (1976) recovered by treating powdered rock samples with 100% phosphoric acid. The CO_2 released by this treatment was variably depleted in ^{13}C compared to PDB, which means that the carbonate in these samples was not a marine contaminant. In addition, Pineau et al. (1976) released CO_2 from large pores (1–2 mm) by crushing whole-rock samples in a vacuum. The $\delta^{13}C$ values of this component of CO_2 ranged only from −7.0 to −8.0‰ with a mean of -7.6 ± 0.6‰. Stepwise heating of the precrushed samples at 400, 600, and 1000°C released additional amounts of CO_2 whose $\delta^{13}C$ values increased (i.e., became less negative) with increasing temperature from -22.2 ± 1.8‰ at 400°C to -14.7 ± 1.1‰ at 1000°C.

The CO_2 released at 400°C is presumably a biogenic contaminant whose abundance in the gas decreases with increasing degassing temperature. Therefore, the CO_2 released at 1000°C can be viewed as a mixture of C derived from the mantle with varying amounts of biogenic C which is characteristically depleted in ^{13}C relative to PDB and has a $\delta^{13}C$ value of about −25‰. Pineau et al. (1976) concluded that the indigenous C in the basalt samples they analyzed had an average $\delta^{13}C$ value of −7.6‰.

Pineau et al. (1976) also considered that the isotopic composition of C in igneous rocks is altered by isotope fractionation caused by loss of volatiles during cooling of the magma. In that case, the apparent isotopic heterogeneity of C in igneous rocks is caused not only by the contamination of rocks with biogenic C but also by the loss of volatiles. The change in the isotope composition of C in a silicate melt as a result of degassing depends on the temperature, the fraction of C remaining, and whether the process is occurring in a closed or open system (Taylor, 1986). The difference between the $\delta^{13}C$ values ($\Delta^{CO_2}_{melt}$) of coexisting CO_2 and CO_3^{2-} in tholeiite melt at 1120–1280°C is about 4–4.6‰ based on experimental determinations by Javoy et al. (1978). In other words, CO_2 in isotopic equilibrium with C dissolved in tholeiite magma at about 1200°C is *enriched* in ^{13}C by about 4.3‰ relative to the C remaining in the melt. Analyses of natural materials by DesMarais and Moore (1984), Mattey et al. (1984), and Gerlach and Thomas (1986) confirm this conclusion. Consequently, the C in solution in a lava flow from which CO_2 is escaping is progressively *depleted* in ^{13}C (i.e., its $\delta^{13}C$ value becomes more negative). This process can be modeled by the Rayleigh distillation equation (Taylor, 1986).

The isotopic compositions of C in different kinds of igneous rocks were measured by Craig (1953), Hoefs (1965, 1973), Fuex and Baker (1973), and Galimov and Gerasimovsky (1978). In the procedure of Hoefs (1973), powdered rock samples were first outgassed at 150°C for 5 hours in order to remove atmospheric CO_2. Subsequently, the samples were heated to 900°C in the presence of CuO or to 1100°C without CuO. In addition, Hoefs (1973) measured the $\delta^{13}C$ values of CO_2 released from carbonates in basalt and granite by treating powdered samples with 100% phosphoric acid (McCrea, 1950).

The results in Figures 27.16*a* and *b* demonstrate that the *oxidized* C in carbonates of basalt,

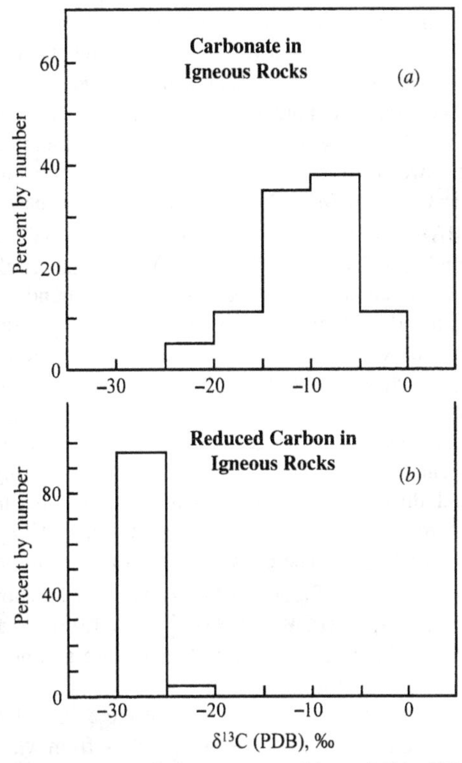

FIGURE 27.16 (*a*) Isotope composition of C in CO_2 released from carbonate minerals in basalt, granite, and spilite by 100% phosphoric acid. (*b*) Isotope composition of C in CO_2 released from basalt and granite by combustion with O at 900°C. Data from Hoefs (1973).

granite, and spilite is enriched in ^{13}C compared to *reduced* C in basalt and granite. The $\delta^{13}C$ values prove that the carbonates were deposited in the igneous rocks by groundwater or meteoric surface water, although the presence of small amounts of magmatic carbonate is not excluded. The $\delta^{13}C$ values of the reduced C released by combustion of organic molecules, carbides, and amorphous C in samples of basalt and granite are similar to those of biogenic organic matter that may have been assimilated by the respective magmas. In addition, these rocks may also have been depleted in ^{13}C during degassing of magma or lava flows.

When CO_2 is released from igneous rocks by sequential stepwise heating, the $\delta^{13}C$ values of the high-temperature carbon (HTC) produced at $T > 600°C$ differ significantly from the $\delta^{13}C$ values of the low-temperature carbon (LTC) liberated at $T < 600°C$. Many investigators interpret the LTC as a biogenic contaminant and consider that the HTC contains more C from the mantle than the LTC. For these reasons, Taylor (1986) compared the weighted mean bulk-C isotope composition ($\delta^{13}C_{\sum c}$) of MORBs, OIBs, and back-arc basalts (BABs) to their concentration ratios of HTC to total C (HTC/$\sum$C). The rationale is that as the concentration ratio approaches unity, the abundance of mantle-C increases and the isotope composition of bulk C of the oceanic basalts should therefore approach the isotope composition of C in the mantle of the Earth. Figure 27.17 demonstrates that Taylor's hypothesis is confirmed because oceanic basalts from different tectonic settings and different geographic locations in the oceans approach a straight line in coordinates of HTC/$\sum$C and $\delta^{13}C_{(\sum C)}$. The linear trend constrained by the data points in Figure 27.17 extrapolates to about $\delta^{13}C = -5‰$ (PDB) for HTC/$\sum$C = 1.0 when all of the C is released at high temperature and no C of biogenic origin is present.

Taylor (1986) concluded that the $\delta^{13}C$ values of HTC released by MORBs and OIBs, from which LTC was previously removed, converge to −6.0‰. Since these rocks have all lost varying fractions of C by outgassing of CO_2, they may have been depleted in ^{13}C, meaning that the $\delta^{13}C$ values of their magma source in the mantle may range from −2 to −6‰.

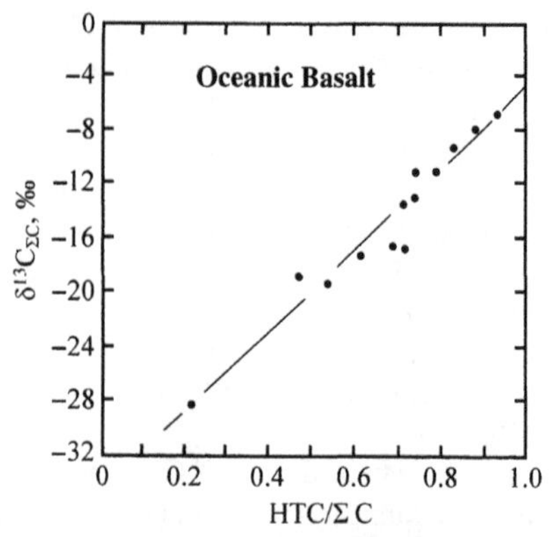

FIGURE 27.17 Relation between the weighted mean bulk-C isotope composition ($\delta^{13}C_{\sum C}$) of oceanic basalts (MORBs, OIBs, and BABs) and their concentration ratios of HTC to total C (HTC/$\sum$ C), where BABs are back-arc basalts and HTC is high-temperature C released at $T > 600°C$. The straight line extrapolates to about $\delta^{13}C = -5‰$ at HTC/$\sum$C = 1.0. Adapted from Taylor (1986), who used data from the literature referenced by him.

27.6c Graphite and Calcite

The organic matter contained in sedimentary carbonate and siliciclastic rocks can be converted into graphite in the course of high-grade regional and contact metamorphism. In addition, carbonate and silicate minerals react under these conditions to form calc-silicate minerals (e.g., wollastonite, tremolite, actinolite) and CO_2 gas (e.g., Shieh and Taylor, 1969). The CO_2 that is liberated by these reactions is enriched in ^{13}C, leaving the residual carbonate minerals (e.g., calcite, dolomite, siderite) depleted in ^{13}C. For example, Deines and Gold (1969) reported that the $\delta^{13}C$ value of calcite of the Montreal Formation (Late Ordovician) decreased from +1.13‰ (PDB) in unaltered samples to −3.91‰ in calcite within the Mount Royal pluton (Early Cretaceous) which intrudes the limestone. In addition, carbonate minerals decompose at about 900°C into CO_2 and the oxide of the appropriate metal (e.g., CaO, MgO, FeO). The C isotopes of the CO_2 gas produced by the thermal decomposition of metal carbonates are not fractionated provided the process goes to completion.

The temperature dependence of C-isotope exchange reactions between coexisting calcite and graphite in high-grade metamorphic rocks is governed by equations which where derived from analyses of natural samples of coexisting calcite, dolomite, and graphite (Chacko et al., 2001). The equations of Morikiyo (1984) and Dunn and Valley (1992) have overlapping temperature ranges from 270 to 800°C in Figure 27.18:

$$10^3 \ln \alpha_c^{cal} = 8.9\left(\frac{10^6}{T^2}\right) - 7.1(270 - 650°C) \tag{27.10}$$

$$10^3 \ln \alpha_c^{cal} = 5.81\left(\frac{10^6}{T^2}\right) - 2.61(400 - 800°C) \tag{27.11}$$

where T is the temperature on the Kelvin scale. These equations can be used to determine the

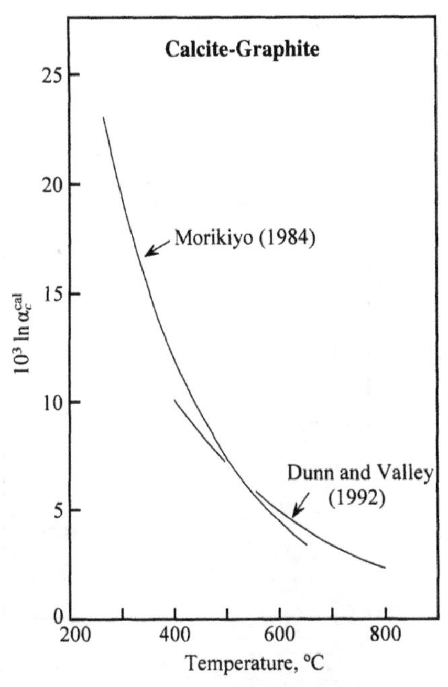

FIGURE 27.18 Temperature dependence of the C-isotope fractionation factor between coexisting calcite and graphite in isotopic equilibrium. The equations were derived by analysis of natural materials and are valid in the temperature range 270–650°C (Morikiyo, 1984) and 400° to 800°C (Dunn and Valley, 1992).

C-isotope equilibration temperature of coexisting calcite and graphite in metamorphic rocks. For example, the thermometry equation of Dunn and Valley (1992) was used by Satish-Kumar and Wada (2000) to measure the peak metamorphic temperature of calcite and graphite in Proterozoic marbles in the Skallen region of the Lützow-Holm Bay on the coast of East Antarctica. The average $\delta^{13}C$ value of 20 calcite samples of the Skallen marbles is $+0.34 \pm 0.17‰$ (PDB) compared to $\delta^{13}C = -2.53 \pm 0.20‰$ for 19 highly reflecting crystals of graphite in the same samples. Therefore, the fractionation factor (α_c^{cal}) is:

$$\alpha_c^{cal} = \frac{0.34 + 1000}{-2.53 + 1000} = 1.00287 \pm 0.00038$$

and $10^3 \ln 1.00287 = 2.865 \pm 0.377$. Substituting into the Dunn–Valley equation,

$$10^3 \ln \alpha_c^{cal} = 5.81\left(\frac{10^6}{T^2}\right) - 2.61$$

yields a temperature T of:

$$T = \left(\frac{5.81 \times 10^6}{5.475}\right)^{1/2} = 757 \pm 34°C$$

This temperature is consistent with the high-grade metamorphic rocks (charnockites, biotite–garnet gneisses, and marbles) that form the Lützow-Holm Complex of East Antarctica.

Additional examples of the isotope composition of graphite and C- isotope fractionation between calcite and graphite have been reported by:

Landis, 1971, *Contrib. Mineral. Petrol.*, 30: 54–67. Weis, 1981, *Geochim. Cosmochim. Acta*, 45:2325–2332. Douthitt, 1982, *Econ. Geol.*, 77:1247–1249. Li et al., 1983, *Geochemistry*, 6:162–169. Arneth et al., 1985, *Geochim. Cosmochim. Acta*, 49:1553–1560. Rumble and Hoering, 1986, *Geochim. Cosmochim. Acta*, 50:1239–1247. Rumble et al., 1986, *Geology*, 14:452–455. Duke and Rumble, 1986, *Contrib. Mineral. Petrol.*, 93:409–419. Galimov et al., 1989, *Geokhimia*, 1989(4): 508–515. Crawford and Valley, 1990, *Geol.*

Soc. Am. Bull., 102:807–811. Demény and Kreulen, 1993, *Geol. Carpathica*, 44(1):3–9.

In addition, Scheele and Hoefs (1992) investigate the C-isotope exchange between CO_2 gas and solid graphite at temperatures between 600 and 1200°C and at pressures of 5–15 kilobars:

$$10^3 \ln \alpha_{C}^{CO_2} = 4.53 \left(\frac{10^6}{T^2}\right) + 3.04 \qquad (27.12)$$

where T is on the Kelvin scale.

Graphite not only occurs in metamorphosed sedimentary rocks but has also been observed in igneous rocks such as layered igneous complexes, alkali-rich complexes, and granitic plutons and pegmatites (Taylor, 1986). The $\delta^{13}C$ values of graphite in mafic igneous rocks range primarily from −14 to −32‰, which suggests that the C originated from organic matter in sedimentary rocks that were assimilated by basaltic magma.

For example, the $\delta^{13}C$ values of graphite in the Bushveld Complex of South Africa range from −19 to −21‰ (PDB) (Ballhaus and Stumpfl, 1985), which is similar to the average $\delta^{13}C$ value of −17‰ reported by McKirdy and Powell (1974) for metamorphosed shales of Proterozoic to Cambrian age in South Africa. Other examples listed by Taylor (1986) include the Stillwater Complex of Montana ($\delta^{13}C = -14$ to −16‰), and the Duluth Gabbro in Minnesota ($\delta^{13}C = -29.0$ to −32.5‰).

The low C concentration of mantle-derived basaltic magmas makes the isotopic composition of C sensitive to contamination by organic matter in sedimentary rocks that may be assimilated by such magmas. The average concentration of C in Phanerozoic sedimentary rocks in the USA is 7200 ppm (Hoefs, 1967, after Ronov, 1958), whereas the Duluth Gabbro and similar mafic intrusives contain only about 120 ppm of C (Hoefs, 1967). The $\delta^{13}C$ of a mixture (m) of basalt magma (b) and carbonaceous sediment (s) is given by equation 16.18:

$$\delta^{13}C_m = \frac{\delta^{13}C_s\,(C_s)f_s + \delta^{13}C_b\,(C_b)(1-f_s)}{(C_s)f_s + C_b\,(1-f_s)}$$

where (C) is the concentration of C and f_s is the weight fraction of sediment in the mixture. Assuming that:

$$\delta^{13}C_s = -25‰ \qquad C_s = 7200 \text{ ppm}$$
$$\delta^{13}C_b = -6‰ \qquad C_b = 120 \text{ ppm}$$

The basalt magma containing only 1% of assimilated sediment has a $\delta^{13}C_m$ value of:

$$\delta^{13}C_m = \frac{(-25)\times 7200 \times 0.01 + (-6) \times 120 \times 0.99}{7200 \times 0.01 + 120 \times 0.99} = -13.2‰$$

The basalt–sediment mixing hyperbola in Figure 27.19 based on equation 16.18 demonstrates that the $\delta^{13}C$ of graphite in gabbroic rocks decreases from −6.0‰ to −20.4‰ at $f_s = 0.05$ and to −22.5‰ at $f_s = 0.10$. Evidently, the isotope composition of graphite in mafic igneous rocks has value as an indicator of contamination of basaltic magma by assimilation of sedimentary rocks (Faure, 2001).

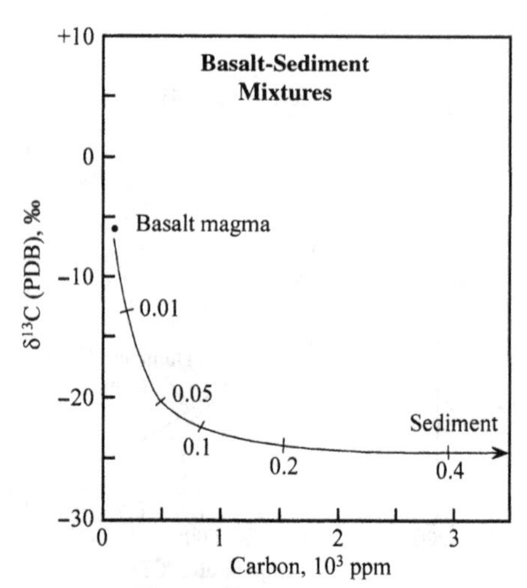

FIGURE 27.19 Effect of assimilation of sedimentary rocks by basalt magma on the $\delta^{13}C$ value of graphite in the resulting gabbroic rocks. Sediment: $\delta^{13}C = -25‰$; $C_s = 7200$ ppm; basalt: $\delta^{13}C = -6‰$, $C_b = 120$ ppm. The addition of only 1% of sediment lowers the $\delta^{13}C$ value from −6 to −13.2‰.

The same is true for sialic igneous rocks that form by partial melting of pre-existing complexes of volcanic and sedimentary rocks containing organic matter. The $\delta^{13}C$ values of graphite in felsic plutonic rocks and granitic pegmatites cited by Taylor (1986) range from −6.7 to −25.4‰ (PDB). In addition, Hoefs (1973) reported that reduced C in European granites is strongly depleted in ^{13}C, as indicated by $\delta^{13}C$ values between −21.0 and −28.9‰ (PDB).

27.6d Greek Marbles

Marble has been used as a building stone and for carving of ornaments and statues for thousands of years. The Greeks and Romans are especially admired for the beauty of marble statues which adorned their temples and private homes. The most desirable marbles are known to have originated from the Cycladic Islands of Naxos and later from Paros, Mount Pentelikon on the mainland of Greece, and finally from Mount Hymettus during the Roman period (Craig and Craig, 1972). In addition, marble was also quarried in Anatolia (now Turkey) at Marmara, Ephesos, Aphrodisias, Denizili, and Afyon (Manfra et al., 1975).

The multiplicity of marble quarries in use at different times and the shipment of marble statues and ornaments throughout the Greek and Roman empires have made it difficult to determine the source from which individual specimens originated. For this reason, Craig and Craig (1972) measured the $\delta^{13}C$ and $\delta^{18}O$ values of marbles from the classical quarries of Greece in order to provide a basis for the sourcing of Greek marbles. Their results in Figure 27.20 indicate that marbles from Naxos, Paros, Penteli, and Hymettus can be distinguished from each other on the basis of the isotope compositions of C and O. Therefore, Craig and Craig (1972) analyzed ten samples of archeological marbles and were able to source about half of them.

In a subsequent publication, Manfra et al. (1975) reported isotopic compositions of 42 marbles collected in quarries of western Anatolia. The samples included marbles having a range of colors including not only white but also dark grey, red, and yellow. The $\delta^{13}C$ and $\delta^{18}O$ values of the colored marbles scattered widely, but when they

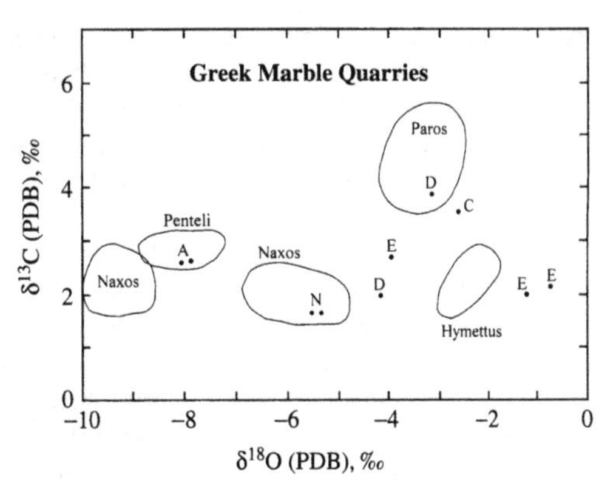

FIGURE 27.20 Isotope compositions of C and O of Greek marbles from quarries at Naxos, Paros, Mount Penteli, and Mount Hymettus. The solid circles represent archeological marbles of unknown origin identified by letters: A = Athens; D = Delphi; E = Epidaurus; N = Naxos; C = Caesarea (Israel). All of the marbles collected at Epidaurus, one from Delphi, and one from Caesarea are not associated with any of the quarries as defined by Craig and Craig (1972).

were omitted, the white and slightly colored samples from Marmara and Ephesos formed clusters. A further compilation by Herz and Vitaliano (1983) of data by Craig and Craig (1972), Manfra et al. (1975), Riederer and Hoefs (1980), and unpublished measurements by N. Herz and D. B. Wenner (University of Georgia in Athens) established ranges of the isotopic compositions of marbles from 14 locations in Greece, Anatolia, and Italy. Other sourcing methods of Greek and Anatolian marbles include their trace-element concentrations (Conforto et al., 1975; Young and Ashmole, 1968) and petrology (Renfrew and Springer-Peacey, 1968).

27.6e Diamonds

The quest for the isotopic composition of C in the Earth inevitably turned to diamonds because they formed in the mantle under the continents at $T > 1325°C$ and $P > 53$ kilobars at a depth of >160 km below the surface of the Earth (Kirkley et al., 1991b; Kennedy and Kennedy, 1976). The diamonds were brought to the surface by the explosive intrusion of kimberlite pipes occurring on all of the continents except Antarctica. Although diamonds occur in kimberlite pipes and in the eclogite xenoliths they contain, they are recovered in many cases from placers associated with gravel deposits in stream valleys and beaches.

Craig (1953) reported that the $\delta^{13}C$ values of diamonds cluster about a mean of about −5.0‰ (PDB). However, subsequent analyses by several research groups broadened the range of $\delta^{13}C$ values of clear and colored diamonds from different deposits worldwide to +3 to −34‰ (PDB) (Hoefs, 1997). This information is based on the work of many research groups including:

Wickman, 1956, *Geochim. Cosmochim. Acta*, 9:136–153. Vinogradov et al., 1965, Geokhimia, 1965:643–657. Vinogradov et al., 1966, *Geokhimia*, 1966:1395–1397. Kovalsky et al., 1972, *Doklady Acad. Nauk U.S.S.R.*, 203:440–442. Kovalsky and Chersky, 1973, *Int. Geol. Rev.*, 15:1224–1228. Galimov et al., 1978, *Geokhimia*, 1978:340–349. Smirnov et al., 1979, *Nature*, 278:630. Deines, 1980, *Geochim. Cosmochim. Acta*, 44:943–961. Deines et al., 1984, *Geochim. Cosmochim. Acta*, 48: 325–342. Galimov, 1985, *Geochem. Int.*, 22(1):118–141.

In some cases, diamonds in kimberlite pipes coexist with graphite, which raises the question about isotope fractionation of C in diamonds in equilibrium with graphite. The fractionation of C isotopes between diamonds and graphite as a function of temperature is governed by an equation derived from the calculated values of $10^3 \ln \alpha_g^d$ published by Bottinga (1969) for temperatures at 700 and 1000°C augmented by a measurement of $10^3 \ln \alpha_g^d$ at 1700°C by Hoering (1961):

$$10^3 \ln \alpha_g^d = 0.251 \left(\frac{10^6}{T^2}\right) + 0.239 \qquad (27.13)$$

where T is on the Kelvin scale. This equation in Figure 27.21 yields:

$$10^3 \ln \alpha_g^d = 0.337 \qquad \text{at } T = 1325°\text{C}$$

which means that the difference between the $\delta^{13}C$ values of coexisting diamond and graphite at 1325°C is approximately:

$$\delta^{13}C_d - \delta^{13}C_g = 0.337‰$$

In other words, the wide range of $\delta^{13}C$ values of diamonds cannot be caused by isotope fractionation between diamond and graphite. Instead, the

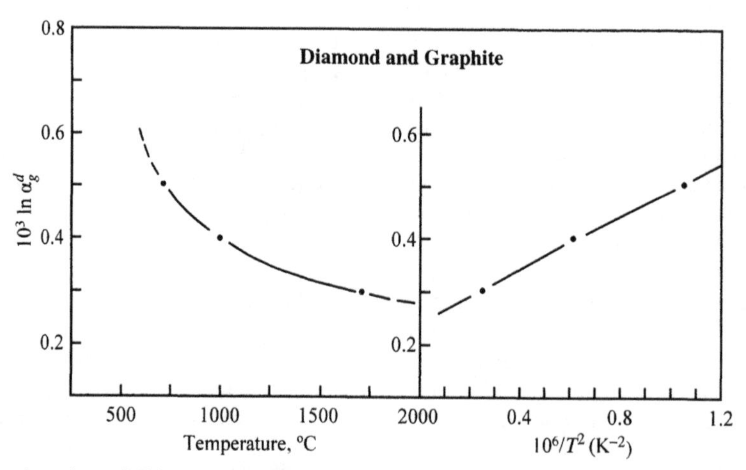

FIGURE 27.21 Fractionation of C isotopes between diamond and graphite at high temperatures in the mantle of the Earth. Derived from calculated values of $10^3 \ln \alpha_g^d$ at 700 and 1000°C by Bottinga (1969) and an experimental value at 1700°C by Hoering (1961).

variation of the isotope composition of diamonds is a reflection of the isotopic heterogeneity of C from which diamonds form in the mantle.

Moreover, Deines (1980) demonstrated that the $\delta^{13}C$ values of diamonds in the Premier and Dan Carl mines in South Africa are not correlated with the color and shape of the specimens and that 20% of about 540 diamonds that had been analyzed prior to 1980 have $\delta^{13}C$ values between −5 and −6‰ (PDB). When the interval is widened to −4 to −7‰, the abundance rises to 52%. In addition, the $\delta^{13}C$ values of 31 specimens (5.7%) are less than −20‰ with a range of −20 to −33‰. These values overlap those of organic matter on the surface of the Earth, including biogenic CH_4. Therefore, there is a growing consensus that the source regions of diamonds in the mantle contain not only a primordial C component but also C derived from subducted organic matter and marine carbonates (Zhu and Ogasawara, 2002). Alternatively, the isotopic heterogeneity of C in the mantle of the Earth may be a property of organic matter that was synthesized abiogenically in the solar nebula by the Fischer–Tropsch process (Lancet and Anders, 1970).

Many diamonds contain inclusions of silicate minerals, which are assumed to represent the host rock within which they formed (Meyer and Boyd, 1982). The mineralogy and chemical composition of these inclusions are used to classify diamonds into two groups:

1. peridotitic diamonds containing orthopyroxene, Ca-poor/Cr-rich garnet, and/or chromite and
2. eclogitic diamonds characterized by Cr-poor garnet, omphacitic clinopyroxene, rutile, kyanite coesite, and/or sanidine.

Additional subdivisions of the peridotitic diamonds were added by Deines et al. (2001), including a *websteritic* group whose inclusions contain both ortho- and clinopyroxene in approximately equal proportions.

The silicate inclusions of diamonds have been dated by the Sm–Nd method as well as by trace-element diffusion. In a few cases, sulfide inclusions were dated by the Re–Os and Pb–Pb methods. All of the age determinations are based on microanalysis methods using ion-probe mass spectrometers (Hauri et al., 2002). The results indicate that, in most cases, the diamonds crystallized in Precambrian time (3.3–0.90 Ga), whereas the kimberlite pipes that transported them from the mantle into the crust are of Phanerozoic age (450–50 Ma) and two are Precambrian: Premier (South Africa, 1200 Ma) and Argyle (Australia, 1200–1100 Ma).

The summary of age determinations compiled by Deines et al. (2001) suggests that diamonds in the peridotite (P) and websterite (W) groups crystallized between 3.5 and 2.0 Ga and are older than diamonds in the eclogite (E) group (2.9–0.90 Ga). However, the P-group diamonds of the Mir mine in Russia appear to be only slightly older than the kimberlite (360 Ma) in which they occur.

Age determinations of silicate and sulfide inclusions of diamonds have been reported by:

> Kramers, 1979, *Earth Planet. Sci. Lett.*, 42: 58–70. Richardson et al., 1984, *Nature*, 310: 198–202. Richardson, 1986, *Nature*, 322:623–626. Richardson et al., 1990, *Nature* 346:54–56. Smith et al., 1991, *Geochim. Cosmochim. Acta*, 55:2579–2590. Richardson et al., 1993, *Nature*, 366:256–258. Rudnick et al., 1993, *Geology*, 21:13–16. Shimizu and Sobolev, 1995, *Nature*, 375:394–397. Richardson and Harris, 1997, *Earth Planet. Sci. Lett.*, 151:271–177. Pearson et al., 1998, *Earth Planet. Sci. Lett.*, 160:311–326. Pearson et al., 1999, *Geochim. Cosmochim. Acta*, 63:703–711.

In most cases, diamond-bearing kimberlite pipes powered their way through thick cratonic crust overlying mantle domains that had been enriched in H_2O, CO_2, alkali metals, and other volatiles released by plumes rising from the asthenospheric mantle (Faure, 2001).

The heterogeneity of the isotope composition of C in diamonds has been explained by three alternative models (Kirkley et al., 1991b):

1. Carbon isotopes are fractionated during the crystallization of diamonds from a homogeneous C source in the mantle of the Earth.
2. The C inherited by the Earth from the solar nebula was isotopically heterogeneous as a result of fractionation and nuclear processes.
3. The mantle contains C subducted in the form of biogenic organic matter and marine carbonates.

Each of these options has been advocated and can be supported by plausible arguments. The multimodal spectrum of $\delta^{13}C$ values of diamonds indicates that C in the mantle is isotopically heterogeneous because of the presence of several isotopically distinct components. The dominant component has $\delta^{13}C$ values between −4 and −7‰. In addition, subducted biogenic organic matter and marine carbonates each contribute C having distinctive isotopic compositions. The fractionation of C isotopes (in closed or open systems) during the conversion of these C sources into graphite and diamond or into CO_2 or CH_4 further complicates the picture. Therefore, the study of diamonds has not yet revealed the isotopic composition of primordial C.

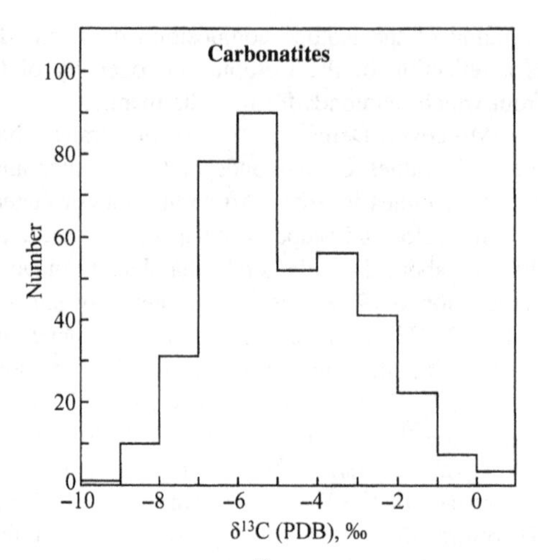

FIGURE 27.22 Range of $\delta^{13}C$ (PDB) values of calcite, dolomite, and ferriferous carbonates of carbonatites. Based on data from Deines and Gold (1973), Pineau et al. (1973), Hubberten et al. (1988), Reid and Cooper (1992), Santos and Clayton (1995), and Srivastava and Taylor (1996). In addition, the diagram includes data published prior to 1971 that were compiled by Deines and Gold (1973).

27.6f Carbonatites

Carbonatites are igneous rocks composed predominantly of carbonate minerals of Ca, Mg, or Fe that crystallize from carbonate magmas (Heinrich, 1966; Bell, 1989; Faure, 2001). The carbonate magmas are genetically related to alkali-rich silicate magmas which form by low degrees of partial melting of metasomatized source rocks in the lithospheric mantle. The alkali-rich melts extract virtually all of the C that existed in large volumes of mantle rocks. During the subsequent evolution of alkali-rich silicate magmas, a carbonatic liquid can separate by liquid immiscibility (Wyllie, 1989). Consequently, the C in carbonatites should be a representative sample of the isotopically heterogeneous C in a large volume of the underlying lithospheric mantle. In addition, the high C concentration of carbonatites protects its isotopic composition from alteration by contamination with C in biogenic organic matter. These considerations support the hypothesis that the isotopic composition of C in carbonatites should vary only within narrow limits and that the average $\delta^{13}C$ value of carbonatites approaches that of the magma sources in the lithospheric mantle.

The $\delta^{13}C$ values o 391 samples of calcite, dolomite, ankerite, and ferriferous calcite from carbonatites in Figure 27.22 have a range of only 9‰ compared to more than 30‰ for diamonds. Carbonatite samples at the extreme ends of the distribution were either contaminated during the differentiation of carbonatite magmas in the crust or were altered after intrusion or eruption. In agreement with the $\delta^{13}C$ values of diamonds, 23% of the samples included in Figure 27.22 have $\delta^{13}C$ values between −5 and −6‰. In addition, 56% of the $\delta^{13}C$ values range from −4 to −7‰. Therefore, these data confirm evidence derived from the isotope analyses of C in volcanic gases, igneous rocks, and diamonds that most, but by no means all, of the C in the lithospheric mantle has $\delta^{13}C$ values between −4 and −7‰ with a strong clustering of values between −5 and −6‰ on the PDB scale.

The available data do not support the conclusion that the C in the mantle of the Earth can be characterized by a specific numerical value of the $\delta^{13}C$ parameter because of:

1. sampling bias (e.g., Oka; Deines, 1970),
2. alteration of the isotopic composition during differentiation of magmas in the crust and by hydrothermal solutions after intrusion/eruption,

3. regional differences of $\delta^{13}C$ values (e.g., Brazil and East Africa),
4. isotope fractionation of C between calcite, dolomite, and ferriferous carbonates (e.g., Santos and Clayton, 1995), and
5. internal variation of $\delta^{13}C$ values within a single intrusive (e.g., 7‰ for Oka; Deines, 1970).

The data of Santos and Clayton (1995) include $\delta^{13}C$ values of both calcites and dolomites for several carbonatites in Brazil. The dolomites are enriched in ^{13}C relative to calcite in 18 out of 21 cases (85.7%), presumably as a result of isotope fractionation during crystallization or by isotope exchange at subsolidus temperatures. The temperature of crystallization of carbonatitic lava flows is equal to or less than 544°C, measured by Krafft and Keller (1989) in a carbonatic lava lake in the summit caldera of Oldoinyo Lengai in Tanzania. The differences of $\delta^{13}C$ values of dolomite and calcite (Δ^{dol}_{cal}) reported by Santos and Clayton (1995) yield isotope equilibration temperatures for Brazilian carbonatites based on the C- isotope thermometry equation of Sheppard and Schwarcz (1970):

$$10^3 \ln \alpha^{dol}_{cal} = 0.18 \left(\frac{10^6}{T^2}\right) + 0.17(100 - 650°C) \tag{27.14}$$

The resulting temperatures are lower than 544°C except for the carbonatite at Jacupiranga, which yielded temperatures between 903 and 611°C. The significance of these temperatures is questionable because of possible alteration of the $\delta^{13}C$ values after crystallization of the minerals.

The regional differences of $\delta^{13}C$ values of carbonatites may be related both to the isotope composition of C in the underlying mantle and the tectonic setting of each site. In general, the average $\delta^{13}C$ values of East African carbonatites (Kenya, Tanzania, Uganda) range from −2.7 to −4.9‰, whereas those of Brasilian carbonatites range from −6.0 to −6.6‰ (calcite) and from −4.9 to −6.4‰ (dolomite). The $\delta^{13}C$ values of carbonatites in Canada are intermediate: −2.4‰ St. André to −5.1‰ Oka. Additional average $\delta^{13}C$ values of carbonatites in different parts of the Earth were cited by Deines and Gold (1973) and Pineau et al. (1973). Very detailed studies of the Dicker Willem (Namibia) carbonatite pluton were prepared by Reid and Cooper (1992), of the Kaiserstuhl (Germany) by Hubberten et al. (1998), and of carbonatites on Fuerteventura (Canary Islands) by Demény et al. (1998).

27.7 EXTRATERRESTRIAL CARBON

The isotopic composition of C that existed in the solar nebula can be determined by the study of meteorites (e.g., Clayton et al., 1973), interplanetary dust particles, and rocks that originated from the Moon and Mars. Initial results by Deines and Wickman (1985) suggested that the $\delta^{13}C$ values of whole-rock samples of stony meteorites (enstatite chondrites) range moderately from −12.4 to +9.0‰ (PDB) but vary by up to 11.1‰ in different fractions of the same specimen (e.g., Pillistfer). Similar results were obtained for achondrites (ureilites) whose $\delta^{13}C$ values for whole-rock samples range from −0.7 to −11.1‰ (PDB) based on a compilation of data from the literature by Grady et al. (1985). These authors also reported that the $\delta^{13}C$ values of ureilites, heated stepwise from 200 to 1200°C, increased from about −30‰ (PDB) at 200°C to about 0‰ at 500–1000°C and then declined to more negative values at 1200°C. These results indicate that stony meteorites contain several different isotopic varieties of C. One of these is the C released at 200°C ($\delta^{13}C$ = −25 to −30‰ PDB), which Grady et al. (1985) attributed to contamination by terrestrial biogenic C.

Terrestrial contamination was also recognized by Barrat et al. (1998) in the achondrite Tatahouine, which contains calcite crystals in fractures of specimens found 63 years after the fall of this meteorite in 1931 in southern Tunisia. The isotope compositions of C and O in the secondary calcite crystals are: $\delta^{13}C$ = −2.0‰ and $\delta^{18}O$ = +29.6‰ (both on the PDB scale). The sandy soil in which the specimens were embedded has very similar isotope compositions of C and O: $\delta^{13}C$ = −3.2‰ and $\delta^{18}O$ = +28.9‰ (PDB). However, carbonaceous chondrites also contain carbonate crystals of cosmogenic origin. For example, Clayton (1963) reported an average $\delta^{13}C$ value of +61.2‰ (PDB) in five samples of carbonate in Orgueil whose

$\delta^{13}C$ values ranged from +58.6 to +64.4‰ (PDB). The evident isotopic heterogeneity of C (and other elements) in different constituents of meteorites was reviewed by Thiemens (1988), Clayton (1993), and McKeegan and Leshin (2001).

27.7a Stony Meteorites

The C-bearing compounds that occur in stony meteorites include kerogen and carbonates of Ca, Mg, and Fe as well as small grains of graphite, diamond, and silicon carbide (Hoefs, 1997). Analyses by Tang et al. (1989) of different C compounds in the carbonaceous chondrite Murchison revealed huge differences in the $\delta^{13}C$ values of various components of this meteorite listed in Table 27.4.

The isotopic heterogeneity of C (and other elements) in meteorites was not apparent in analyses of whole-rock samples because the particles of graphite, diamond, and silicon carbide contain only a small fraction of the C in a bulk sample of meteorite. In spite of the small diameters of the graphite and silicon carbide grains (down to 0.03 μm), the isotope compositions of C, N, and Si of the silicon carbide, diamond, and graphite grains have been measured by microprobe mass-spectrometry:

Kerridge, 1983, *Earth Planet. Sci. Lett.*, 64: 186–200. Zinner et al., 1987, *Nature*, 330: 730–732. Zinner et al., 1989, *Geochim. Cosmochim. Acta*, 53:3273–3290. Zinner et al., 1995, *Meteoritics*, 30:209–226. Zinner, 1998, *Annu. Rev. Earth Planet. Sci.*, 26:147–188. Tang and Anders, 1988a, *Geochim. Cosmochim. Acta*, 52:1235–1244. Tang and Anders, 1988b, *Geochim. Cosmochim. Acta*, 52:1245–1254. Tang and Anders, 1988c, *Astrophys. J.*, 335:L31–L34. Tang et al., 1989, *Nature*, 339:351–354. Pizzarello et al., 1991, *Geochim. Cosmochim. Acta*, 55:905–910. Anders and Zinner, 1993, *Meteoritics*, 28: 490–514. Amari et al., 1993, *Nature*, 365: 806–809. Ott, 1993, *Nature*, 364:25–33. Hoppe et al., 1995, *Geochim. Cosmochim. Acta*, 59:4029–4056.

Table 27.4. Isotope Composition of C in Different Components of the Carbonaceous Chondrite Murchison

Component	$\delta^{13}C$ (PDB), ‰	Fraction of Cosmic Abundance
Kerogen	−15	4×10^{-2}
Carbonates of Mg, Fe, Ca	+42	2×10^{-3}
Diamond	−38	6×10^{-4}
Graphitic C	−50	2×10^{-4}
Graphitic C	+340	8×10^{-6}
SiC	To +8000	4×10^{-6}

Source: Tang et al., 1989.

The C-isotope ratios of SiC grains measured by Zinner et al. (1989) by microprobe mass spectrometry were converted to $\delta^{13}C$ values based on a graphite standard (NBS-21, $\delta^{13}C = -28.1$‰ PDB). In addition, a synthetic SiC compound ($\delta^{13}C = -35.2$‰ PDB) was used for some of the analyses. The results reported by Zinner et al. (1989) indicate that the SiC grains of the carbonaceous chondrite Murchison are variously *enriched* in ^{13}C and have $\delta^{13}C$ values ranging from about 0‰ to about +4800‰ on the PDB scale (Zinner et al., 1989, Table 8 and Figure 9b). The evident ^{13}C-enrichment of SiC grains in Murchison and Murray far exceeds the isotope composition of terrestrial C.

A very different isotopic composition of C occurs in poorly crystallized graphite inclusions in metallic Fe–Ni grains of the LL chondrite Bishunpur studied by Mostefaoui et al. (2000). Microprobe analyses of these graphitic inclusions yielded $\delta^{13}C$ values between −25‰ and −64‰ relative to a laboratory C and N standard (DAG) of known isotopic composition (Zinner et al., 1989). Mostefaoui et al. (2000) also reported that, in one case, the $\delta^{13}C$ value of two graphitic inclusions separated by only 20 μm in the same metal grain of Bishunpur differ by about 25‰. The graphitic inclusions in metal grains of the L chondrite Kohar are also depleted in ^{13}C and have $\delta^{13}C$ values between -18.4 ± 5.7‰ and -44.2 ± 5.5‰. The depletion in ^{13}C supports the suggestion of Mostefaoui et al. (2000) that the graphitic inclusions in the metal grains of Bishunpur and Kohar could have formed by metamorphism of abiogenic organic matter during chondrule formation in the solar nebula (Buseck and Huang, 1985).

One rare "spherulitic" graphite inclusions in a metal grain of the carbonaceous chondrite Kainsaz analyzed by Mostefaoui et al. (2000) is actually *enriched* in ^{13}C, in marked distinction to graphite inclusions in Bishunpur and Kohar. The δ^{13}C value of this inclusion is $+19.3 \pm 5.8$‰, whereas a bulk sample of the same meteorite yielded $\delta^{13}C = -18.5$‰ (Kerridge, 1985). This apparent contradiction highlights the heterogeneity of the isotopic composition of C in carbonaceous chondrites, which were never heated sufficiently to homogenize the different isotopic varieties of C they contain. In addition, the abundances of C in the various components of Murchison listed in Table 27.4 suggest that kerogen ($\delta^{13}C = -15$‰) contributes several thousand times more C to carbonaceous chondrites than the small graphite inclusions ($\delta^{13}C = +19.3$‰) in rare metal grains in Kainsaz. The isotope composition of C in presolar (i.e., stellar) graphite grains in meteorites was measured by Amari et al. (1993).

The isotope compositions of C in specific molecular fractions extracted from the organic matter of carbonaceous chondrites have been determined by:

> Yuen et al., 1984, *Nature*, 308:252–254. Epstein et al., 1987, *Nature*, 326:477–479. Cronin et al., 1988, in Kerridge and Matthews (Eds.), *Meteorites and the Early Solar System*, pp. 819–857, University of Arizona Press, Tucson, AZ. Engel et al., 1990, *Nature*, 348:47–49. Alexander et al., 1998, *Meteoritics Planet. Sci.*, 33:603–622. Sephton et al., 1998, *Geochim. Cosmochim. Acta*, 62:1821–1828.

Several of the studies identified above were based on the carbonaceous chondrite Murchison, whose fall in 1969 provided an abundance of well-preserved samples of organic matter that formed abiogenically in the solar system (Oró et al., 1971). For example, Epstein et al. (1987) reported that amino acids and monocarboxylic acid extracted from the organic matter of Murchison are enriched in ^{13}C relative to PDB and have δ^{13}C values of $+23.1$‰ (amino acids) and $+6.7$‰ (carboxylic acid). Additional C-isotope analyses of amino acids in Murchison were published by Engel et al. (1990). However, Sephton et al. (1988) found that aromatic hydrocarbons in the same meteorite are variably *depleted* in ^{13}C, with δ^{13}C values between -5.8‰ and -28.8‰ (PDB). The authors concluded that the aromatic hydrocarbons had formed by cracking of macromolecular organic matter (kerogen).

The organic matter (kerogen) in carbonaceous chondrites, asteroids, comets, and planetary satellites formed in the solar nebula by the Fischer–Tropsch process prior to the growth of planetesimals and the parent bodies of meteorites (Cronin et al., 1988). The mass-dependent fractionation of C isotopes by the Fischer–Tropsch process was discussed by Lancet and Anders (1970).

27.7b Iron Meteorites

Iron meteorites contain small inclusions of graphite and cohenite [$(Fe,Ni,Co)_3C$; Brett, 1967] as well as C atoms dispersed in the Fe–Ni phase. The isotope composition of C of graphite inclusions in several iron meteorites (octahedrites) analyzed by Deines and Wickman (1973, 1975) range from $\delta^{13}C = -4.8$‰ to -8.2‰ (PDB) in Figure 27.23. Small grains of cliftonite (C in the form of small cubic crystals studied by Brett and Higgins, 1967) also have δ^{13}C values in this range. However, cohenite in Magura and Canyon Diablo is depleted in ^{13}C and has δ^{13}C values between -17.4‰ and -21.6‰ (PDB). Similarly, C in taenite (Fe–Ni alloy) of Magura and Toluca has average δ^{13}C values of -18.8‰ and -21.3‰ (PDB), respectively (Craig, 1953; Hoefs, 1973; Deines and Wickman, 1973).

The heterogeneity of the isotopic composition of C in different C-bearing phases of the same meteorite (e.g., Canyon Diablo, Magura, and Toluca) displayed in Figure 27.23 may be the result of mass-dependent isotope fractionation between cohenite and graphite. In this case, the isotope composition of C in coexisting graphite and cohenite in iron meteorites could serve as a cosmothermometer. However, the compilation by Chacko et al. (2001) includes only C-isotope thermometry equations for

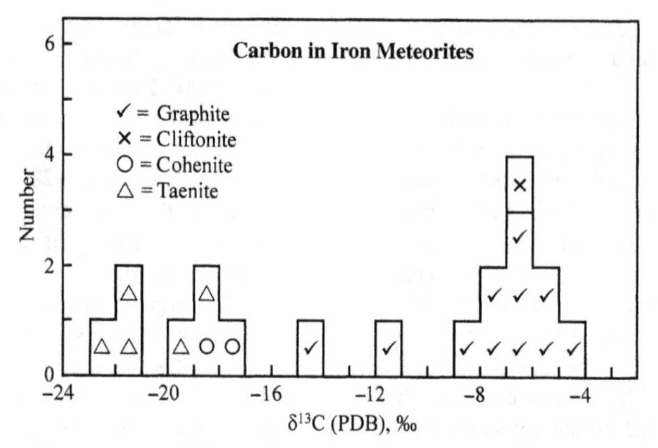

FIGURE 27.23 Isotope composition of C in various mineral phases of iron meteorites. Data from Deines and Wickman (1973, 1975), Craig (1953), and Hoefs (1973).

CO_2–graphite, calcite–graphite, dolomite–graphite, and diamond–graphite.

Alternatively, cohenite may have crystallized from C atoms dispersed in taenite during cooling of iron meteorites. In this case, the isotope compositions of C in cohenite crystals and graphite nodules of iron meteorites are not related, which requires that two or more different isotopic varieties of C existed in the solar nebula. The wide range of isotopic compositions of C (as well as of H, O, N, and other elements) in various kinds of refractory inclusions in both stony and iron meteorites indicates that they probably formed during nucleosynthesis in stars and were ejected into interstellar space when the stars exploded as supernovae long before the solar system formed (Ott, 1993).

The two stable isotopes of C are produced by two different nucleosynthesis processes occurring in stars. According to Anders and Grevesse (1989), ^{12}C forms by He burning, which occurs in the core of a star in its red giant stage, whereas ^{13}C is produced by hot or explosive H burning, which may occur in the shell surrounding the core of a red giant star. Therefore, the isotopic composition of C produced by a star depends in part on the production rate of the two stable isotopes in different stellar environments and on the efficiency of mixing prior to and during the ultimate explosion of the ancestral star as a supernova. The relation between the nucleosynthesis in stars and the isotopic composition of C and other elements in presolar grains was considered by Zinner (1998).

27.7c Lunar Carbon

The theory of the origin of the Sun and of the solar system implies that the planetary abundances of C and other elements that form gases at low temperatures increase with increasing distance from the Sun. The Moon is a notable exception to this generalization because lunar rocks are significantly depleted in C and certain other elements (e.g., noble gases, Na, K, S) compared to chondrite meteorites. The abundance of C in lunar rocks is 1.1×10^2 atoms of C per 10^6 atoms of Si (Ganapathy and Anders, 1974), whereas chondrites contain 2.0×10^3 atoms of C per 10^6 atoms of Si (Brownlow, 1979). The result is that the Moon is depleted in C by a factor of 0.055 relative to chondrite meteorites.

The C in *lunar rocks* recovered by the Apollo astronauts consists of an indigenous lunar component, spallogenic ^{13}C, and terrestrial contaminants added during collection on the Moon, during transport to Earth, and during handling in terrestrial laboratories. In addition, samples of lunar fines contain C derived from the meteorites that have impacted on the surface of the Moon since the time of its formation.

The return of 22 kg of lunar rocks and fines by the Apollo 11 astronauts on July 24, 1969, provided an opportunity to detect and to characterize the organic matter which the lunar samples might contain. Although the lunar *fines* contained about 200 ppm of C, most of it appeared to be a contaminant leaving less than 10 ppm of indigenous lunar C (Johnson and Davis, 1970). A study by Ponnamperuma et al. (1970) yielded a total C concentration of 157 ppm with $\delta^{13}C = +20‰$ (PDB) but failed to detect high-molecular-weight C-compounds of potentially biogenic origin. The results of a stepwise sequential pyrolysis of Apollo 12 fines by Chang et al. (1971) in Table 27.5 indicate that the C released at each of three temperatures has characteristic $\delta^{13}C$ values. The C released at 500°C has a negative average $\delta^{13}C$ value of $-14.7 \pm 4.8‰$ (PDB) and therefore is considered to be a terrestrial contaminant. The C released at 750 and 1100°C has positive average $\delta^{13}C$ values. The C released at 750°C ($\delta^{13}C = +20.7‰$) may contain excess ^{13}C produced by nuclear spallation reactions caused by cosmic rays. The 1100°C fraction of C has a low average $\delta^{13}C$ value of $+9.3 \pm 3.3‰$, but the significance of this value is not immediately apparent. The total C concentration of sample 12023 of lunar fines is 83 ± 23 μg/g, excluding the terrestrial contaminant released at 500°C.

Table 27.5. Isotope Compositions and Concentrations of C Released by Stepwise Heating (Pyrolysis) of Three Splits of Sample 12023 of Lunar Soil Collected at the Landing Site of Apollo 12

Temperature, °C	C Concentration μg/g	$\delta^{13}C$ (PDB), ‰
500	19 ± 6	-14.7 ± 4.8
750	33 ± 16	$+20.7 \pm 7.5$
1100	68 ± 7	$+9.3 \pm 3.3$
1100 (combustion)	20 ± 0	0 (?)
Total	83 ± 23[a]	

Source: Chang et al., 1971.

[a]Excluding the C released at 500°C, which is considered to be a terrestrial contaminant.

In a similar study of lunar *rocks*, DesMarais (1983) measured the amounts and $\delta^{13}C$ values of C released by sequential stepwise heating of six lunar basalts, two anorthosites, and three breccias (not from soil) at 410–450°C, 500–625°C, and 1220–1300°C. The results in Table 27.6 and Figure 27.24 reveal that the largest amount of C was released from all of these rocks at the low-temperature step (410–450°C) and that the $\delta^{13}C$ values of this C were highly negative [-38.6 to $-26.3‰$ (PDB)]. The ^{13}C-depletion of the low-temperature C- fraction identifies it as a terrestrial contaminant (e.g., cloth fibers, plastics, terrestrial hydrocarbons). In addition, the low-temperature gas fraction may contain C from secondary siderite ($FeCO_3$) that formed by a reaction between atmospheric CO_2 and Fe^{2+} of lawrencite ($FeCl_2$). The kinetic isotope fractionation that accompanies this reaction depletes the siderite in ^{13}C relative to atmospheric CO_2 (DesMarais, 1983).

Table 27.6. Concentrations and $\delta^{13}C$ Values of C Released from Lunar Rocks by Heating Them at 410–450°C, 500–625°C, and 1220–1300°C

Temperature, °C	C, μg/g	$\delta^{13}C$ (PDB), ‰
	Basalt[a] (five specimens)	
410–450	5.5 ± 1.9	-31.6 ± 2.8
500–625	0.56 ± 0.21	-30.1 ± 3.0
1220–1300	0.93 ± 0.29	$+18.5 \pm 24.4$ (+64 to −15.5)
	Anorthosite (two specimens)	
410–450	5.8	−26.6
500–625	9.0	−13.0
1220–1300	4.2 ± 1.3	-13.6 ± 3.6
	Breccia (three specimens)	
410–450	16.7 ± 13.0	-27.2 ± 1.9
500–625	6.9 ± 5.7	-19.0 ± 6.5
1220–1300	2.6 ± 1.9	$+3.8 \pm 36.6$

Source: DesMarais, 1983.

[a]Sample 15016,52 was omitted because of its anomalously high C concentration of 62 μg/g released at 500°C.

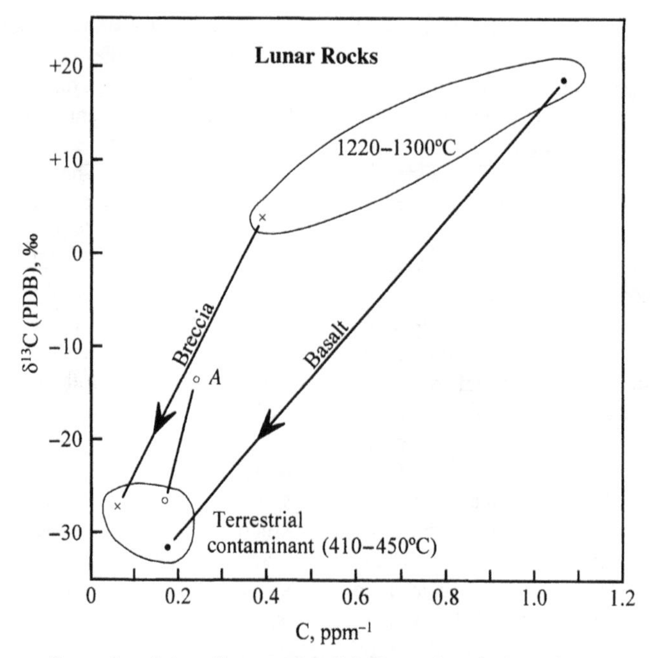

FIGURE 27.24 Isotopic mixing diagram for C in lunar basalt, breccia, and anorthosite (A). The C released at 410–450°C is a terrestrial contaminant. The high-temperature component (1220–1300°C) is enriched in ^{13}C and may be a mixture of meteoritic and spallogenic C with an undetermined component of indigenous lunar C. Data and interpretation by DesMarais (1983).

Heating at 1220–1300°C released only small amounts of C, which in all cases was enriched in ^{13}C relative to the low-temperature C. The high-temperature fraction of C presumably includes a mixture of meteoritic and spallogenic C with an undetermined but small component of indigenous lunar C. DesMarais (1983) determined that the production rate of spallogenic ^{13}C on the surface of the Moon is 4.1×10^{-6} μg $^{13}C/g$ of sample per million years and used it to calculate ^{13}C spallation exposure-ages of 184 and 135 Ma for two specimens of lunar basalt from the landing site of Apollo 15.

The studies of different kinds of lunar rocks suggest that the concentrations of C (and N) are very low, especially in rocks which cooled at the surface of the Moon. Therefore the abundance of C (and N) in lunar rocks is more reliably indicated by coarse-grained plutonic rocks that cooled at depth below the surface without outgassing.

27.8 SEARCH FOR LIFE ON MARS

Carbon dioxide is the most abundant component of the atmosphere of Mars, with a concentration of 95.32% (Owen et al., 1977; Owen, 1992). In addition, solid CO_2 (dry ice) occurs in the polar regions of Mars together with large quantities of water ice. Consequently, carbonates of Mg, Ca, and Fe may exist in surficial sediment and in evaporite deposits that formed in the distant past when liquid water may have been stable on the surface of Mars (Fanale et al., 1992). The presence of liquid water containing dissolved inorganic carbon (DIC) in early Precambrian time may have sustained a primitive flora of algae and bacteria on the surface of Mars. The abundance of Fe-oxyhydroxides in the soil may have favored the evolution of iron bacteria similar to those that contributed to the deposition of the iron formation at Isua, southwest

Greenland, at about 3.8 Ga (Iberall-Robbins and Iberall, 1991). Any organisms that existed on Mars may have been preserved in hotspring deposits and in lacustrine evaporites in the form of kerogen (e.g., Section 27.1c). The outcrop of light-colored stratified rocks in Meridiani planum discovered in 2004 by the rover "Opportunity" appears to be an example of such sedimentary rocks that were deposited in a saline lake on the surface of Mars. (Photograph, NASA/JPL/Caltech)

The search for such evidence of life on Mars motivates the ongoing exploration of this planet and the study of the martian meteorites. The isotopic composition of C in organic matter contained in martian rocks and soil is one of the three principal indicators of the former (or present) existence of life on Mars, the others being the presence of molecular biomarkers and morphologically preserved fossils.

The atomic $^{13}C/^{12}C$ ratio of CO_2 in the atmosphere of Mars was measured during the descent of the Viking spacecraft (Nier and McElroy, 1977). The results indicated that the isotope composition of C in the atmosphere of Mars is similar to that of terrestrial CO_2. This conclusion was confirmed by Krasnopolsky et al. (1996), who used a 4-m telescope on Kitt Peak, California (McKeegan and Leshin, 2001).

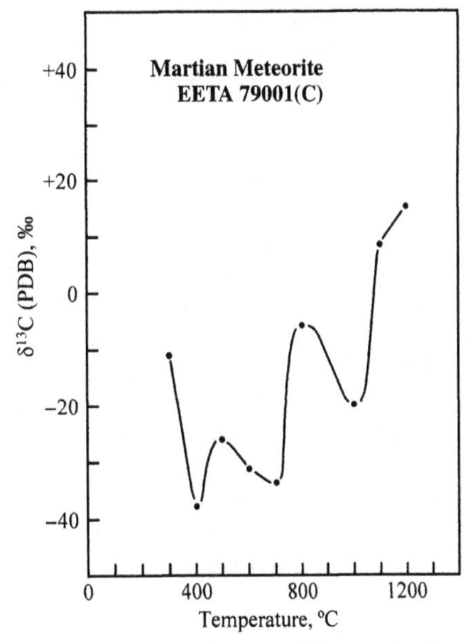

FIGURE 27.25 Results of sequential stepwise heating of the martian meteorite EETA 79001 (C) (Elephant Moraine), Antarctica. The multimodal spectrum of $\delta^{13}C$ values of C released at increasing temperatures is evidence of the existence of several C-components having different isotopic compositions. Data from Wright et al. (1986).

27.8a Martian Meteorites

Until rock and soil samples collected on the surface of Mars become available for study, the martian meteorites are the only direct source of information about the possible existence of life on Mars. The martian meteorites are also referred to as the SNC group after the meteorites Shergotty, Nakhla, and Chassigny, whose comparatively low Rb–Sr and Sm–Nd dates originally attracted the attention of meteoriticists (Sections 5.3c and 9.4).

The isotope composition of CO_2 liberated by progressive step-heating exemplified by the $\delta^{13}C$ spectrum of EETA 79001 (C) in Figure 27.25 (Wright et al., 1986) indicates that martian meteorites contain three C-bearing components (McKeegan and Leshin, 2001):

1. Terrestrial contaminants released at about 500°C have $\delta^{13}C$ values between −20‰ and −30‰ (PDB).
2. Carbonate minerals that release CO_2 at temperatures from 400 to 700°C (or by reaction with phosphoric acid) are enriched in ^{13}C relative to PDB with $\delta^{13}C$ values of about +35‰ in Nakhla and +46‰ in the famous orthopyroxenite ALH 84001. Therefore, the martian carbonates differ from terrestrial marine limestones, which have $\delta^{13}C$ values close to 0‰ (PDB) (Section 27.3).
3. The CO_2 released at temperatures above 700°C has negative $\delta^{13}C$ values ranging from −20‰ to −30‰, which are thought to characterize C in basaltic magma derived from the mantle of Mars.

If true, it means that C in the mantle of Mars is more depleted in ^{13}C than C in the mantle of Earth, which has a $\delta^{13}C$ value of about −5‰ on the PDB scale (Section 27.6). Alternatively, the third C component of martian meteorites may be a terrestrial contaminant that is released only at temperatures above 700°C.

The research on which this summary is based was published by:

Carr et al., 1985, *Nature*, 314:248–250. Wright et al., 1986, *Geochim. Cosmochim. Acta*, 50: 983–991. Clayton and Mayeda, 1988, *Geochim. Cosmochim. Acta*, 52:925–927. Wright et al., 1988, *Geochim. Cosmochim. Acta*, 52:917–924. Wright et al., 1990, *J. Geophys. Res.*, 95:14789–14794. Wright et al., 1992, *Geochim. Cosmochim. Acta*, 56:817–826. Grady et al., 1994, *Meteoritics*, 29:469. Romanek et al., 1994, *Nature*, 372:655–657. Jull et al., 1995, *Meteoritics*, 30:311–318. Leshin et al., 1996, *Geochim. Cosmochim. Acta*, 60:2635–2650. Jull et al., 1997, *J. Geophys. Res. (Planets)*, 102:1663–1669. Valley et al., 1997, *Science*, 275:1633–1638. Becker et al., 1999, *Earth Planet. Sci. Lett.*, 167:71–79. Jull et al., 2000, *Geochim. Cosmochim. Acta*, 64:3763–3772.

27.8b ALH 84001

A specimen of orthopyroxenite (ALH 84001) collected in 1984 from the ablation surface of the East Antarctic ice sheet adjacent to the Allan Hills in southern Victoria Land (Cassidy et al., 1992) was recognized by Mittlefehldt (1994) as a rock from Mars. Subsequently, McKay et al. (1996) reported that fresh fracture surfaces of this rock contain polycyclic aromatic hydrocarbons and globules of carbonate minerals. The authors concluded that the hydrocarbons are not a terrestrial contaminant but are indigenous to the meteorite. They also observed that the carbonate globules contain minute crystals of magnetite and Fe sulfide. McKay et al. (1996) pointed out that the texture and size of the carbonate globules are similar to terrestrial carbonate deposits associated with bacteria. Therefore, they stated the hypothesis that the carbonate globules in ALH 84001 are of biogenic origin and could be "fossil remains of a past martian biota." On the other hand, the globules could also have formed inorganically by precipitation from a hydrothermal fluid.

The hypothesis stated by McKay and his associates aroused a great deal of interest in the scientific community as well as in the general public. As a result, the American space program gained popular support for an accelerated search for life on Mars. Any samples of organic matter or carbonate compounds that may be discovered on Mars as a result of this search will be examined to determine whether it is of biogenic origin.

The results and interpretations derived from the study of ALH 84001 have been published by:

McSween, 1994, *Meteoritics*, 29:757–779. Mittlefehldt, 1994, *Meteoritics*, 29:214–221. McKay et al., 1996, *Science*, 273:924–930. Romanek et al., 1994, *Nature*, 372:655–657. Jull et al., 1995, *Meteoritics*, 30:311–318. Shearer et al., 1996, *Geochim. Cosmochim. Acta*, 60:2921–2926. Harvey and McSween, 1996, *Nature*, 382:49–51. Bradley et al., 1996, *Geochim. Cosmochim. Acta*, 60:5149–5155. Becker et al., 1997, *Geochim. Cosmochim. Acta*, 61:475–481. Goswami et al., 1997, *Meteoritics Planet. Sci.*, 32:91–96. Kirschvink et al., 1997, *Science*, 275:1629–1633. Valley et al., 1997, *Science*, 275:1633–1637. McSween, 1997, *GSA Today*, 7(7):1–7. Jull et al., 1997, *J. Geophys. Res. (Planets)*, 102: 1663–1669. Jull et al., 1998, *Science*, 279: 366–369. Bada et al., 1998, *Science*, 279: 362–365.

The organic matter and the carbonates in ALH 84001 may not, in fact, be from the planet Mars, but instead appear to be terrestrial contaminants. This conclusion was reached by Jull et al. (1998), who discovered that the CO_2 released at low temperature contains cosmogenic ^{14}C. In addition, Bada et al. (1998) reported that the amino acids in ALH 84001 are similar to those of Antarctic ice and that all of them are left-handed, like those in terrestrial organisms.

27.9 SUMMARY

Carbon is the fourth most abundant element in the solar system. It occurs in organic matter, carbonate and carbide minerals, graphite, and diamonds and in the form of CO_2, CO, and hydrocarbon gases such as CH_4. The C in organic matter, fossil fuels, and hydrocarbon gases is depleted in ^{13}C during photosynthesis by green plants and by the subsequent formation of various biogenic compounds ranging from cellulose to lipids. The depletion of organic matter and organic molecules in ^{13}C has been used as an indicator of their biogenic origin and thus as a sign for the existence of life in Early Archean time about 3.8 billion years ago.

Carbonate minerals precipitate from aqueous solutions of carbonate ions (H_2CO_3, HCO_3^-, CO_3^{2-}), which are in equilibrium with CO_2 gas of the atmosphere. The resulting carbonate minerals (calcite, aragonite, and dolomite) are enriched in ^{13}C relative to the carbonate ions in solution and relative to atmospheric CO_2. Consequently, marine carbonate rocks have $\delta^{13}C$ values close to 0‰ on the PDB scale because the PDB standard is itself a marine carbonate.

The organic matter in marine sediment and marine carbonate deposits is subducted along deep-sea trenches. As a result, the C in the mantle of the Earth is a heterogeneous mixture of several components, including subducted organic matter ($\delta^{13}C = -25‰$), marine carbonates ($\delta^{13}C = 0‰$), and primordial C, whose $\delta^{13}C$ value lies between −5 and −6‰ on the PDB scale.

The isotopic composition of C in mantle-derived volcanic rocks has been determined by heating them stepwise in the presence of O. The $\delta^{13}C$ values of the CO_2 released by this procedure vary between wide limits depending on the temperature. Low-temperature C is depleted in ^{13}C and is considered to be a contaminant of biogenic origin. Carbon released stepwise at increasing temperature up to about 1200°C tends to have a polymodal $\delta^{13}C$ spectrum, suggesting the presence of several components of C having different isotopic compositions (e.g., carbonates, carbides, graphite).

The CO_2 gas vented from lava flows and from magma chambers is enriched in ^{13}C relative to the C remaining in solution in the melt. Therefore, prolonged outgassing of silicate melts causes the remaining C to be depleted in ^{13}C, which depresses the $\delta^{13}C$ values of silicate melts (i.e., makes them increase in the negative direction). The CO_2 gas vented by fumaroles in geothermal areas is accompanied in some cases by small amounts of CH_4. The fractionation of C isotopes between CO_2 and CH_4 is the basis for a useful geothermometer of fumarolic gas emissions.

The isotopic heterogeneity of C in the mantle of the Earth is clearly demonstrated by the $\delta^{13}C$ values of diamonds and carbonatites. The range of $\delta^{13}C$ values of diamonds is about 30‰ and extends to highly negative values characteristic of subducted organic matter. The $\delta^{13}C$ values of carbonatites vary less widely than those of diamonds, but regional differences have been noted (e.g., East Africa and Brazil). In spite of the apparent heterogeneity of the isotope compositions of C in diamonds and carbonatites, the $\delta^{13}C$ values of both are strongly clustered between −5 and −6‰. For these reasons, the $\delta^{13}C$ value of the mantle of Earth is best characterized by a number between −5 and −6‰.

Studies of the isotope compositions of C in various kinds of meteorites from the asteroidal belt have revealed great complexity. For example, all meteorite specimens have been contaminated with terrestrial C of biogenic origin. In addition, meteorites contain a variety of C-bearing minerals some of which are minute refractory inclusions of carbides, diamonds, and graphite that formed in the ancestral stars of the solar system. The isotopic compositions of C in such presolar grains is the result of nucleosynthetic processes that occurred in stars during their red giant stage of evolution. The subsequent explosions of these stars as supernovae contributed to the formation of the solar nebula from which the Sun and planets of our solar system formed about 4.6 billion years ago.

The information presented in this chapter demonstrates that the isotope geochemistry of C involves a wide range of subjects from intimate details of green plants to the life and death of stars.

REFERENCES FOR INTRODUCTION AND CARBON DIOXIDE (INTRODUCTION, SECTIONS 27.1–27.1a)

Anders, E., and N. Grevesse, 1989. Abundances of the elements: Meteoritic and solar. *Geochim. Cosmochim. Acta*, 53:197–214.

Beveridge, N. A. S., and N. J. Shackleton, 1994. Carbon isotopes in recent planktonic foraminifera: A record of anthropogenic CO_2 invasion of the surface ocean. *Earth Planet. Sci. Lett.*, 126:259–273.

Chacko, T., D. R. Cole, and J. Horita, 2001. Equilibrium oxygen, hydrogen, and carbon isotope fractionation factors applicable to geologic systems. In J. W. Valley and D. R. Cole (Eds.), *Stable Isotope Geochemistry, Rev. Mineral. Geochem.*, 43:1–81. Mineralogical Society of America, Blacksburg, Virginia.

Ciais, P., P. P. Tans, M. Trolier, J. W. C. White, and R. J. Francey, 1995. A large northern hemisphere terrestrial CO_2 sink indicated by the $^{13}C/^{12}C$ ratio of atmospheric CO_2. *Science*, 269:1098–1102.

Deines, P., 1980. The isotopic composition of reduced organic carbon. In P. Fritz and J. Ch. Fontes (Eds.), *Handbook of Environmental Isotope Geochemistry*, Vol. 1A, pp. 239–406. Elsevier, Amsterdam.

DesMarais, D. J., 2001. Isotopic evolution of the biogeochemical carbon cycle during the Precambrian. In J. W. Valley and D. R. Cole (Eds.), *Stable Isotope Geochemistry. Rev. Mineral. Geochem.*, 43:555–578. Mineralogical Society of America, Blacksburg, Virginia.

Farmer, G., and M. S. Baxter, 1976. Atmospheric carbon dioxide levels as indicated by the stable isotope record in wood. *Nature*, 247:273–275.

Freeman, K. H., 2001. Isotopic biogeochemistry of marine organic carbon. In J. W. Valley and D. R. Cole (Eds.), *Stable Isotope Geochemistry, Rev. Mineral. Geochim.*, 43:579–605. Mineralogical Society of America, Blacksburg, Virginia.

Hayes, J. M., 2001. Fractionation of carbon and hydrogen isotopes in biosynthetic processes. In J. W. Valley and D. R. Cole. (Eds.), *Stable Isotope Geochemistry, Rev. Mineral. Geochem.*, 43:224–279. Mineralogical Society of America, Blacksburg, Virginia.

Hoefs, J., 1997. *Stable Isotope Geochemistry*. Springer-Verlag, Heidelberg.

Keeling, C. D., R. B. Bacastow, A. F. Carter, S. C. Piper, T. R. Whorf, M. Heimann, W. G. Mook, and H. Roeloffzen, 1989. A three dimensional model of atmospheric CO_2 transport based on observed winds. 1. Analysis of observational data. *Geophys. Monogr.*, 55:165–236. Amer. Geophys. Union, Washington, D.C.

Leuenberger, M., U. Siegenthaler, and C. C. Langway, 1992. Carbon isotopic composition of atmospheric CO_2 during the last ice age from an Antarctic ice core. *Nature*, 357:488–490.

Ripperdan, R. L., 2001. Stratigraphic variations in marine carbonate carbon isotope ratios. In J. W. Valley and D. R. Cole (Eds.), *Stable Isotope Geochemistry, Rev. Mineral. Geochem.*, 43:637–662. Mineralogical Society of America, Blacksburg, Virginia.

Schidlowski, M., J. M. Hayes, and I. R. Kaplan, 1983. Isotopic inferences of ancient biochemistries: Carbon, sulfur, hydrogen, and nitrogen. In J. W. Schopf (Ed.), *Earth's Earliest Biosphere: Its Origin and Evolution*, pp. 149–186. Princeton University Press, Princeton, New Jersey.

REFERENCES FOR PLANTS IN NORMAL AND EXTREME ENVIRONMENTS (SECTIONS 27.1b–27.1c)

Benner, R., M. L. Fogel, E. K. Sprague, and R. E. Hodson, 1987. Depletion of ^{13}C in lignin and its implications for stable carbon isotope studies. *Nature*, 329:708–710.

Bidigare, R. R., M. C. Kennicutt II, W. L. Keeney-Kennicutt, and S. A. Macko, 1991. Isolation and purification of chlorophylls a and b for the determination of stable carbon and nitrogen isotope composition. *Anal. Chem.*, 63(2):130–133.

Degens, E. T., M. Behrendt, B. Gotthardt, and E. Reppmann, 1968. Metabolic fractionation of carbon isotopes in marine plankton, II. Data on samples collected off the coasts of Peru and Ecuador. *Deep Sea Res.*, 15:11–20.

Estep, M. L. F. 1984. Carbon and hydrogen isotopic compositions of algae and bacteria from hydrothermal environments, Yellowstone National Park. *Geochim. Cosmochim. Acta*, 48:591–599.

Galimov, E. M., 1985. *The Biological Fractionation of Isotopes*. Academic Press, Orlando, Florida.

Galimov, E. M., 2000. Carbon isotope composition of Antarctic plants. *Geochim. Cosmochim. Acta*, 64: 1737–1739.

Hayes, J. M., 2001. Fractionation of carbon and hydrogen isotopes in biosynthetic process. In J. W. Valley and D. R. Cole (Eds.), *Stable Isotope Geochemistry, Rev. Mineral. Geochem.*, 43:225–279. Mineralogical Society of America, Blacksburg, Virginia.

Kaplan, I. R., and A. Nissenbaum, 1966. Anomalous carbon-isotope ratios in nonvolatile organic matter. *Science*, 153:744–745.

Keeling, C. D., 1958. The concentration and isotopic abundance of carbon dioxide in rural areas. *Geochim. Cosmochim. Acta*, 13:322–334.

Keeling, C. D., 1961. The concentration and isotopic abundances of carbon dioxide in rural and marine air. *Geochim. Cosmochim. Acta*, 24:277–298.

Park, R., and S. Epstein, 1960. Carbon isotope fractionation during photosynthesis. *Geochim. Cosmochim. Acta*, 21:110–126.

Schidlowski, M., U. Matzigkeit, and W. E. Krumbein, 1984. Superheavy organic carbon from hypersaline microbial mats; assimilatory pathway and geochemical implications. *Naturwissenschaften*, 71:303–308.

Seckbach, J., and I. R. Kaplan, 1973. Growth pattern and $^{13}C/^{12}C$ isotope fractionation of Cyanidian caldarium and hot spring algae mats. *Chem. Geol.*, 12:161–169.

REFERENCES FOR ORGANIC MATTER IN PRECAMBRIAN ROCKS (C AND H) (SECTIONS 27.2a–27.2c)

Brasier, M. D., O. R. Green, A. P. Jephcoat, A. K. Kleppe, M. J. Van Kranendonk, J. F. Lindsay, A. Steele, and N. V. Grassineau, 2002. Questioning the evidence for Earth's oldest fossils. *Nature*, 416:76–81.

Estep, M. L. F., 1984. Carbon and hydrogen isotopic compositions of algae and bacteria from hydrothermal environments, Yellowstone National Park. *Geochim. Cosmochim. Acta*, 48:591–599.

Fedo, C. M., and J. J. Whitehouse, 2002. Metasomatic origin of quartz-pyroxene rock, Akilia, Greenland, and implications for Earth's earliest life. *Science*, 296:1448–1452.

Grotzinger, J. P., and D. H. Rothman, 1996. An abiotic model for stromatolite morphogenesis. *Nature*, 383:423–425.

Hayes, J. M., I. R. Kaplan, and K. W. Wedeking, 1983. Precambrian organic geochemistry, preservation of the record. In J. W. Schopf (Ed.), *Origin and Evolution of the Earth's Earliest Biosphere*, pp. 93–134. Princeton University Press, Princeton, New Jersey.

Mojzsis, S. J., G. Arrhenius, K. D. McKeegan, T. M. Harrison, A. P. Nutman, and C. R. L. Friend, 1996. Evidence for life on Earth before 3,800 million years ago. *Nature*, 385:55–59.

Naraoka, H., M. Ohtake, S. Maruyama, and H. Ohmoto, 1996. Non-biogenic graphite in 3.8-Ga metamorphic rocks from the Isua district, Greenland. *Chem. Geol.*, 133:251–260.

Rosing, M. T., 1999. ^{13}C-depleted carbon microparticles in >3700-Ma sea-floor sedimentary rocks from West Greenland. *Science*, 283:674–676.

Schidlowski, M., 1995. Early terrestrial life: Problems of the oldest record. In J. Chela-Flores et al. (Eds.), *Chemical Evolution: Self-Organization of the Macromolecules of Life*, pp. 65–80. Deepak, Hampton, Virginia.

Schidlowski, M., 1997. Carbon isotopes and the oldest record of life: Potential and limits. *Proc. Soc. Photo-Opt. Inst. Eng.*, 3111:462–471.

Schidlowski, M., 2001. Carbon isotopes as biogeochemical recorders of life over 3.8 Ga of Earth history: Evolution of a concept. *Precamb. Res.*, 106:117–134.

Schopf, J. W., 1993. Microfossils of the Early Archean Apex Chert: New evidence of the antiquity of life. *Science*, 260:640–646.

Schopf, J. W., and B. M. Packer, 1987. Early (3.3 billion to 3.5 billion-year-old) microfossils from the Warrawoona Group, Australia. *Science*, 237:70–73.

Tissot, B. P., and D. H. Welte, 1978. *Petroleum Formation and Occurrence*. Springer-Verlag, New York.

Ueno, Y., H. Yurimoto, H. Yoshioka, T. Komiya, and S. Maruyama, 2002. Ion microprobe analysis of graphite from ca. 3.8 Ga metasediments, Isua supracrustal belt, West Greenland: Relationship between metamorphism and carbon isotopic composition. *Geochim. Cosmochim. Acta*, 66:1257–1268.

REFERENCES FOR FOSSIL FUEL (SECTION 27.3)

Faure, G., and G. Botoman, 1984. Origin of epigenetic calcite in coal from Antarctica and Ohio based on isotope compositions of oxygen, carbon, and strontium. *Chem. Geol. (Isotope Geosci. Sect.)*, 2:313–324.

Fuex, A. N., 1977. The use of stable isotopes in hydrocarbon exploration. *J. Geochem. Explor.*, 7(2):155–188.

Kim, S.-T., and J. R. O'Neil, 1997. Equilibrium and nonequilibrium oxygen isotope effects in synthetic carbonates. *Geochim. Cosmochim. Acta*, 61:3461–3475.

Redding, C. E., M. Schoell, J. C. Monin, and B. Durand, 1980. Hydrogen and carbon isotopic composition of coals and kerogen. *Phys. Chem. Earth*, 12:711–723.

Silverman, S. R., 1964. Investigations of petroleum origin and evolution mechanisms by carbon isotope studies. In H. Craig, S. L. Miller, and G. J. Wasserburg (Eds.), *Isotopic and Cosmic Chemistry*, pp. 92–102. North-Holland, Amsterdam.

Smith, J. W., K. W. Gould, and D. Rigby, 1982. The stable isotope geochemistry of Australian coals. *Organic Geochem.*, 3:111–131.

Sofer, Z., 1984. Stable carbon isotope compositions of crude oils; applications to source depositional environments and petroleum alteration. *Am. Assoc. Petrol. Geol. Bull.*, 68:31–49.

Stahl, W., 1976. Economically important applications of carbon isotope data of natural gases and crude oil. In *Nuclear Techniques in Geochemistry and Geophysics*. International Atomic Energy Agency, Vienna.

Stahl, W., 1977. Carbon and nitrogen isotopes in hydrocarbon research and exploration. *Chem. Geol.*, 20:121–149.

Stahl, W., 1978. Source rock-crude oil correlation by isotopic type curves. *Geochim. Cosmochim. Acta*, 42:1573–1577.

Stahl, W. J. 1979. Carbon isotopes in petroleum geochemistry. In E. Jäger and J. C. Hunziker (Eds.),

Lectures in Isotope Geology, pp. 274–282. Springer-Verlag, Heidelberg.

Tissot, B., B. Durand, J. Espitalie, and A. Combaz, 1974. Influence of nature and diagenesis of organic matter information of petroleum. *Am. Assoc. Petrol. Geol. Bull.*, 58:499–506.

REFERENCES FOR CARBON-ISOTOPE STRATIGRAPHY (PHANEROZOIC) (SECTION 27.4)

Alvarez, L. W., W. Alvarez, F. Asaro, and H. V. Michel, 1980. Extraterrestrial cause for the Cretaceous-Tertiary extinction: Experimental results and theoretical interpretations. *Science*, 208:1095–1108.

Chacko, T., D. R. Cole, and J. Horita, 2001. Equilibrium oxygen, hydrogen, and carbon isotope fractionation factors applicable to geologic systems. In J. W. Valley and D. R. Cole (Eds.), *Stable Isotope Geochemistry. Rev. Mineral. Geochem.*, 43:1–81. Mineralogical Society of America, Blacksburg, Virginia.

Corsetti, F. A., and J. W. Hagadorn, 2000. Precambrian-Cambrian transition: Death Valley, United States. *Geology*, 28:299–302.

Faure, G., 1998. *Principles and Applications of Geochemistry*, 2nd ed., Prentice-Hall, Upper Saddle River, New Jersey.

Hallam, A., and P. B. Wignall, 1997. *Mass Extinctions and Their Aftermath.* Oxford University Press, Oxford.

Joachimski, M. M., and W. Buggisch, 2002. Conodont apatite $\delta^{18}O$ signatures indicate climate cooling as a trigger of the Late Devonian mass extinction. *Geology*, 30:711–714.

Joachimski, M. M., C. Ostertag-Henning, R. D. Pancost, H. Strauss, K. H. Freeman, R. Littke, J. S. Sinninghe Damsté, and G. Racki, 2001. Water column anoxia, enhanced productivity, and concomitant changes in $\delta^{13}C$ and $\delta^{34}S$ across the Frasnian-Famennian boundary (Kowala-Holy Cross Mountains/Poland). *Chem. Geol.*, 175:109–131.

Kaufman, A. J., A. H. Knoll, and G. M. Narbonne, 1997. Isotopes, ice ages, and terminal Proterozoic Earth history. *Natl. Acad. Sci. Proc.*, 94:6600–6605.

Kump, L. R., and M. A. Arthur, 1999. Interpreting carbon-isotope excursions: Carbonates and organic matter. *Chem. Geol.*, 161:181–198.

McLaren, D. J., 1970. Time, life, and boundaries. *J. Paleontol.*, 44:801–815.

Ohmoto, H., 1972. Systematics of sulfur and carbon isotopes in hydrothermal ore deposits. *Econ. Geol.*, 67:551–578.

Pope, K. O., K. H. Baines, A. C. Acampo, and B. A. Inanov, 1994. Impact winter and the Cretaceous/Tertiary extinction: Results of a Chicxulub asteroid impact model. *Earth Planet. Sci. Lett.*, 128:719–725.

Ripperdan, R. L., 2001. Stratigraphic variation in marine carbonate carbon isotope ratios. In J. W. Valley and D. R. Cole (Eds.), *Stable Isotope Geochemistry, Rev. Mineral. Geochem.*, 43:637–662. Mineralogical Society of America, Blacksburg, Virginia.

Romanek, C. S., E. L. Grossman, and J. W. Morse, 1992. Carbon isotopic fractionation in synthetic aragonite and calcite: Effects of temperature and precipitation rate. *Geochim. Cosmochim. Acta*, 56:419–430.

Shackleton, N. J., M. A. Hall, J. Line, and S. Cang, 1983. Carbon isotope data in core V19-30 confirm reduced carbon dioxide concentration in the ice age atmosphere. *Nature*, 306:319–322.

Shackleton, N. J., and J. P. Kennett, 1975. Paleo-temperature history of the Cenozoic and initiation of Antarctic glaciation: Oxygen and carbon isotope analyses in DSDP sites 277, 279, and 281. *Initial Rept. DSDP*, 29:743–755.

Shen, Y., and M. Schidlowski, 2000. New C isotope stratigraphy from southwest China: Implications for the placements of the Precambrian-Cambrian boundary on the Yangtze Platform and global correlations. *Geology*, 28:623–626.

Sommer, M. A., and D. Rye, 1978. *Oxygen and Carbon Isotope Internal Thermometry Using Benthic Calcite and Aragonite Foraminifera Pairs*, U.S. Geol. Surv. Open File Rep. 78–701, pp. 408–410. U.S. Geological Survey, Washington, D.C.

Swart, P. K., J. J. Leder, A. M. Szmant, and R. E. Dodge, 1996. The origin of variations in the isotopic record of scleractinian corals: II. Carbon. *Geochim. Cosmochim. Acta*, 60:2871–2885.

Veizer, J. (Ed.), 1999. Earth system evolution: Geochemical perspective. *Chem. Geol.*, 161(1/3):1–371.

Veizer, J., D. Ala, K. Azmy, P. Bruckschen, D. Buhl, F. Bruhn, G. A. F. Carden, A. Diener, S. Ebneth, Y. Godderis, T. Jasper, C. Korte, F. Pawellek, O. G. Podlaha, and H. Strauss, 1999. $^{87}Sr/^{86}Sr$, $\delta^{13}C$ and $\delta^{18}O$ evolution of Phanerozoic seawater. *Chem. Geol.*, 161:59–88.

Veizer, J., and J. Hoefs, 1976. The nature of $^{18}O/^{16}O$ and $^{13}C/^{12}C$ secular trends in sedimentary carbonate rocks. *Geochim. Cosmochim. Acta*, 40:1387–1395.

Wang, K., H. H. J. Geldsetzer, and B. D. E. Chatterton, 1994. A Late Devonian and extraterrestrial impact and extinction in eastern Gondwana: Geochemical, sedimentological, and faunal evidence. *Geol. Soc. Am. Spec. Paper*, 293:111–120.

Wang, K., H. H. J. Geldsetzer, W. D. Goodfellow, and H. R. Krouse, 1996. Carbon and sulfur isotope anomalies across the Frasnian-Famennian extinction boundary, Alberta. *Can. Geol.*, 24:187–191.

Wang, K., C. J. Orth, M. Attrep, Jr., B. D. E. Chatterton, H. Hou, and H. H. J. Geldsetzer, 1991. Geochemical evidence for a catastrophic biotic event at the Frasnian/Famennian boundary in south China. *Geology*, 19:776–779.

REFERENCES FOR PRECAMBRIAN CARBONATES (SECTIONS 27.5–27.5a)

Becker, R. H., and R. N. Clayton, 1972. Carbon isotopic evidence for the origin of banded iron-formation in Western Australia. *Geochim. Cosmochim. Acta*, 36:577–595.

Deines, P., and D. P. Gold, 1969. The change in carbon and oxygen isotopic composition during contact metamorphism of Trenton limestone by the Mount Royal pluton. *Geochim. Cosmochim. Acta*, 33:421–424.

Eichmann, R., and M. Schidlowski, 1975. Isotopic fractionation between coexisting organic carbon-carbonate pairs in Precambrian sediments. *Geochim. Cosmochim. Acta*, 39:585–595.

Junge, C. E., M. Schidlowski, R. Eichmann, and H. Pietrek, 1975. Model calculations for the terrestrial carbon cycle: Carbon isotope geochemistry and evolution of photosynthetic oxygen. *J. Geophys. Res.*, 80:4542–4552.

Lindsay, J. F., and M. D. Brasier, 2000. A carbon isotope reference curve for ca. 1700–1575 Ma, McArthur and Mount Isa Basins, Northern Australia. *Precamb. Res.*, 99:271–308.

Schidlowski, M., R. Eichmann, and C. E. Junge, 1976. Carbon isotope geochemistry of the Precambrian Lomagundi carbonate province, Rhodesia. *Geochim. Cosmochim. Acta*, 40:449–455.

Schidlowski, M., J. M. Hayes, and I. R. Kaplan, 1983. Isotopic inferences of ancient biochemistries: Carbon, sulfur, hydrogen, and nitrogen. In J. W. Schopf (Ed.), *Earth's Earliest Biosphere: Its Origin and Evolution*, pp. 149–186. Princeton University Press, Princeton, New Jersey.

Shieh, Y. N., and H. P. Taylor, Jr., 1969. Oxygen and carbon isotopes studies of contact metamorphism of carbonate rocks. *J. Petrol.*, 10:307–331.

REFERENCES FOR SNOWBALL EARTH (SECTION 27.5b)

Alvarez, L. W., W. Alvarez, F. Asaro, and H. V. Michel, 1980. Extraterrestrial cause for the Cretaceous-Tertiary extinction: Experimental results and theoretical interpretations. *Science*, 208:1095–1108.

Alvarez, W. E., E. G. Kauffman, F. Surlyk, L. W. Alvarez, F. Asaro, and H. V. Michel, 1984. Impact theory of mass extinction and the invertebrate fossil record. *Science*, 223:1135–1141.

Brasier, M. D., and J. F. Lindsay, 1998. A billion years of environmental stability and the emergence of the eukaryotes: New data from Northern Australia. *Geology*, 26:555–558.

Buick, R., D. J. DesMarais, and A. H. Knoll, 1995. Stable isotope compositions of carbonates from the Mesoproterozoic Bangemall Group, northwestern Australia. *Chem. Geol.*, 123:153–171.

Hayes, J. M., H. Strauss, and A. J. Kaufman, 1999. The abundance of ^{13}C in marine organic matter and isotopic fractionation in the global biogeochemical cycle of carbon during the past 800 Ma. *Chem. Geol.*, 161:103–125.

Hoffman, P. F., G. P. Halverson, and J. P. Grotzinger, 2002. Are Proterozoic cap carbonates and isotopic excursions a record of gas hydrate destabilization following Earth's coldest intervals? Comment and reply. *Geology*, 30(3):286–287.

Hoffman, P. F., A. J. Kaufman, and G. P. Halverson, 1998a. Comings and goings of global glaciations on a Neoproterozoic tropical platform in Namibia. *GSA Today*, 8:1–9.

Hoffman, P. F., A. J. Kaufman, G. P. Halverson, and D. P. Schrag, 1998b. A Neoproterozoic snowball. *Earth. Sci.*, 281:1342–1346.

Hoffman, P. F., and D. P. Schrag, 2000. Snowball Earth. *Sci. Am.*, 282(1):50–57.

Jacobsen, S. B., and A. J. Kaufman, 1999. The Sr, C, and O isotopic evolution of Neoproterozoic seawater. *Chem. Geol.*, 161:37–57.

Kah, L. C., A. B. Sherman, G. M. Narbonne, A. H. Knoll, and A. J. Kaufman, 1999. $\delta^{13}C$ stratigraphy of the Proterozoic Bylot Supergroup, Baffin Island, Canada: Implications for regional lithostratigraphic correlation. *Can. J. Earth Sci.*, 36:313–332.

Kennedy, M. J., N. Christie-Blick, and A. R. Prave, 2001a. Carbon isotopic composition of Neoproterozoic glacial carbonates as a test of paleoceanographic models for snowball Earth phenomena. *Geology*, 29(12):1135–1138.

Kennedy, M. J., N. Christie-Blick, and L. E. Sohl, 2001b. Are Proterozoic cap carbonates and isotopic excursions a record of gas hydrate destabilization following Earth's coldest intervals? *Geology*, 29:443–446.

Kump, L., 1991. Interpreting carbon-isotope excursions: Strangelove oceans. *Geology*, 19:299–302.

Leather, J., P. A. Allen, M. D. Brasier, and A. Cozzi, 2002. Neoproterozoic snowball Earth under scrutiny:

Evidence from the Fiq glaciation of Oman. *Geology*, 30(10):891–894.

Lindsay, J. F., and M. D Brasier, 2000. A carbon isotope reference curve for ca. 1700–1575 Ma, McArthur and Mount Isa Basins, Northern Australia. *Precamb. Res.*, 99:271–308.

Poulsen, C. J., 2003. Absence of a runaway ice-albedo feedback in the Neoproterozoic. *Geology*, 31:473–476.

Runnegar, B., 2000. Loophole for snowball Earth. *Nature*, 405:403–404.

Sohl, L. E., N. Christie-Blick, and D. V. Kent, 1999. Paleomagnetic polarity reversals in Marinoan (ca. 600 Ma) glacial deposits of Australia: Implications for the duration of low-latitude glaciation in Neoproterozoic time. *Geol. Soc. Am. Bull.*, 111:1120–1139.

Srinivas, B., and S. Das Sharma, 2001. The Sr, C, and O isotopic evolution of Neoproterozoic seawater—comment. *Chem. Geol.*, 2001:193–195.

Walker, G., 2003. *Snowball Earth: The Story of the Great Global Catastrophe that Spawned Life As We Know It.* Crown, New York.

Xiao, S., A. H. Knoll, A. J. Kaufman, L. Yin, and Y. Zhang, 1997. Neoproterozoic fossils in Mesoproterozoic rocks? Chemostratigraphic resolution of a biostratigraphic conundrum from the North China platform. *Precamb. Res.*, 84:197–200.

REFERENCES FOR VOLCANIC ROCKS AND GASES (SECTIONS 27.6–27.6b)

Bottinga, Y., 1969. Calculated fractionation factors for carbon and hydrogen isotope exchange in the system calcite-carbon dioxide-graphite, methane-hydrogen-water vapor. *Geochim. Cosmochim. Acta*, 33:49–64.

Craig, H., 1953. The geochemistry of the stable carbon isotopes. *Geochim. Cosmochim. Acta*, 3:53–92.

Craig, H., 1963. The isotopic geochemistry of water and carbon in geothermal areas. In E. Tongiorgi (Ed.), *Nuclear Geology of Geothermal Areas*. Spoleto, Italy.

DesMarais, D. J., and J. G. Moore, 1984. Carbon and its isotopes in mid-oceanic basaltic glasses. *Earth Planet. Sci. Lett.*, 69:43–57.

Ferrara, G. C., G. Ferrara, and R. Gonfiantini, 1963. Carbon isotopic composition of carbon dioxide and methane from steam jets of Tuscany. In E. Tongiorgi (Ed.), *Nuclear Geology of Geothermal Areas*, pp. 275–282. Spoleto, Italy.

Fuex, A. N., and D. R. Baker, 1973. Stable carbon isotopes in selected granitic, mafic, and ultramafic rocks. *Geochim. Cosmochim. Acta*, 37:2509–2521.

Galimov, E. M. and V. I. Gerasimovsky, 1978. Isotope composition of carbon in Icelandic magmatic rocks. *Geochem. International*, 15:1–6.

Gerlach, T. M., and B. E. Taylor, 1990. Carbon isotope constraints on degassing of carbon dioxide from Kilauea volcano. *Geochim. Cosmochim. Acta*, 54:2051–2058.

Gerlach, T. M., and D. M. Thomas, 1986. Carbon and sulphur isotopic composition of Kilauea parental magma. *Nature*, 319:480–483.

Hoefs, J., 1965. Ein Beitrag zur Geochemie des Kohlenstoffs in magmatischen und metamorphen Gesteinen. *Geochim. Cosmochim. Acta*, 29:399–428.

Hoefs, J., 1973. Ein Beitrag zur Isotopengeochemie des Kohlenstoffs in magmatischen Gesteinen. *Contrib. Mineral. Petrol.*, 41:277–300.

Hoefs, J., 1997. *Stable Isotope Geochemistry*, 4th ed., Springer-Verlag, Heidelberg.

Horita, J., 2001. Carbon isotope exchange in the system CO_2-CH_4 at elevated temperatures. *Geochim. Cosmochim. Acta*, 65:1907–1919.

Hulston, J. R., and W. J. McCabe, 1962. Mass spectrometer measurements in the thermal areas of New Zealand. Part 2. Carbon isotopic ratios. *Geochim. Cosmochim. Acta*, 26:399–410.

Ishibashi, J., Y. Sano, H. Wakita, T. Gamo, M. Tsutsumi, and H. Sakai, 1995. Helium and carbon geochemistry of hydrothermal fluids from the Mid-Okinawa trough back-arc basin, southwest of Japan. *Chem. Geol.*, 123:1–15.

Javoy, M., and F. Pineau, 1991. The volatile record of a "popping" rock from the Mid-Atlantic Ridge at 14°N: Chemical and isotopic composition of gas trapped in the vesicles. *Earth Planet. Sci. Lett.*, 107:598–611.

Javoy, M., F. Pineau, and I. Iiyama, 1978. Experimental determination of the isotopic fractionation between gaseous CO_2 and carbon dissolved in tholeiitic magma. *Contrib. Mineral. Petrol.*, 67:35–39.

Jehlicka, J., A. Svatos, O. Frank, and F. Uhlik, 2003. Evidence for fullerenes in solid bitumen from pillow lavas of Proterozoic age from Mitov (Bohemian Massif, Czech Republic). *Geochim. Cosmochim. Acta*, 67:1495–1506.

Mattey, D. P., R. H. Carr, I. P. Wright, and C. T. Pillinger, 1984. Carbon isotopes in submarine basalts. *Earth Planet. Sci. Lett.*, 70:196–206.

McCrea, J. M., 1950. The isotopic chemistry of carbonates and a paleotemperature scale. *J. Chem. Phys.*, 18:849.

Pineau, F., and M. Javoy, 1983. Carbon isotopes and concentrations in mid-ocean ridge basalts. *Earth Planet. Sci. Lett.*, 62:239–257.

Pineau, F., M. Javoy, and Y. Bottinga, 1976. $^{13}C/^{12}C$ ratios of rocks and inclusions in popping rocks of the Mid-Atlantic Ridge and their bearing on the problem of isotopic composition of deep-seated carbon. *Earth Planet. Sci. Lett.*, 29:413–421.

Poorter, R. P. E., J. C. Varekamp, R. J. Poreda, M. J. Van Bergen, and R. Kreulen, 1991. Chemical and isotopic compositions of volcanic gases from east Sunda and Banda arcs, Indonesia. *Geochim. Cosmochim. Acta*, 55:3795–3807.

Poreda, R. J., H. Craig, S. Arnasson, and J. A. Welhan, 1992. Helium isotopes in Icelandic geothermal systems. I. ^{3}He, gas chemistry, and ^{13}C relations. *Geochim. Cosmochim. Acta*, 56:4221–4228.

Taylor, B. E., 1986. Magmatic volatiles: Isotopic variation of C, H, and S. In J. W. Valley, H. P. Taylor, Jr., and J. R. O'Neil (Eds.), *Stable Isotopes in High Temperature Geologic Processes. Rev. Mineral.*, 16:185–225. Mineralogical Society of America, Blacksburg, Virginia.

Welhan, J. A., 1988. Origins of methane in hydrothermal systems. *Chem. Geol.*, 71:183–198.

REFERENCES FOR GRAPHITE AND CALCITE (SECTION 27.6c)

Ballhaus, C. G., and E. F. Stumpfl, 1985. Occurrence and petrologic significance of graphite in the Upper Critical Zone, western Bushveld Complex, South Africa. *Earth Planet. Sci. Lett.*, 74:58–68.

Chacko, T., D. R. Cole, and J. Horita, 2001. Equilibrium oxygen, hydrogen, and carbon isotope fractionation factors applicable to geologic systems. In J. W. Valley and D. R. Cole (Eds.), *Stable Isotope Geochemistry. Rev. Mineral. Geochem.*, 43:1–81. Mineralogical Society of America, Blacksburg, Virginia.

Deines, P., and D. P. Gold, 1969. The change in carbon and oxygen isotopic composition during contact metamorphism of Trenton limestone by the Mount Royal pluton. *Geochim. Cosmochim. Acta*, 33:421–424.

Dunn, S. R., and J. W. Valley, 1992. Calcite-graphite isotope thermometry: A test for polymetamorphism in marble, Tudor gabbro aureole, Ontario, Canada. *J. Metamorph. Geol.*, 10:487–501.

Faure, G., 2001. *Origin of Igneous Rocks; the Isotopic Evidence*. Springer-Verlag, Heidelberg.

Hoefs, J., 1967. Carbon. In K. H. Wedepohl (Ed.), *Handbook of Geochemistry*, pp. 6C–60. Springer-Verlag, Heidelberg.

Hoefs, J., 1973. Ein Beitrag zur Isotopengeochemie des Kohlenstoffs in magmatischen Gesteinen. *Contrib. Mineral. Petrol.*, 41:277–300.

McKirdy, D. M., and T. G. Powell, 1974. Metamorphic alteration of carbon isotopic compositions in ancient sedimentary organic matter: New evidence from Australia and South Africa. *Geology*, 2:591–595.

Morikiyo, T., 1984. Carbon isotopic study on coexisting calcite and graphite in the Ryoke metamorphic rocks, northern Kiso district. *Contrib. Mineral. Petrol.*, 87:251–259.

Ronov, A. B., 1958. Organic carbon in sedimentary rocks. *Geochemistry*, 5:510.

Satish-Kumar, M., and H. Wada, 2000. Carbon isotopic equilibrium between calcite and graphite in Skallen marbles, East Antarctica: Evidence for the preservation of peak metamorphic temperatures. *Chem. Geol.*, 166:173–182.

Scheele, N., and J. Hoefs, 1992. Carbon isotope fractionation between calcite, graphite, and CO_2. *Contrib. Mineral. Petrol.*, 112:35–45.

Shieh, Y. N., and H. P. Taylor, Jr., 1969. Oxygen and carbon isotope studies of contact metamorphism of carbonate rocks. *J. Petrol.*, 10:307–331.

Taylor, B. E., 1986. Magmatic volatiles: Isotopic variation of Ch, H, and S. In J. W. Valley, H. P. Taylor, Jr., and J. R. O'Neil (Eds.), *Stable Isotopes in High Temperature Geologic Processes. Rev. Mineral.*, 16:185–225. Mineralogical Society of America, Blacksburg, Virginia.

REFERENCES FOR GREEK MARBLES (SECTION 27.6d)

Conforto, L., M. Felici, D. Monna, L. Serva, and A. Taddeucci, 1975. A preliminary evaluation of chemical data (trace element) from classical marble quarries in the Mediterranean. *Archaeometry*, 17(2):201–213.

Craig, H., and V. Craig, 1972. Greek marbles: Determination of provenance by isotopic analysis. *Science*, 176:401–403.

Herz, N., and C. J. Vitaliano, 1983. Archaeological geology in the eastern Mediterranean. *Geology*, 11:49–53.

Manfra, L., U. Masi, and B. Turi, 1975. Carbon and oxygen isotope ratios of marbles from some ancient quarries of western Anatolia and their archaeological significance. *Archaeometry*, 17:215–221.

Renfrew, C., and J. Springer-Peacey, 1968. Aegean marbles: A petrological study. *J. British School Archaeol. Athens*, 63:45–66.

Riederer, J., and J. Hoefs, 1980. Die Bestimmung der Herkunft der Marmore von Büsten der Münchener Residenz. *Naturwissenschaften*, 67:446–451.

Young, W. J., and B. Ashmole, 1968. The Boston relief and the Ludovisi throne. *Bull. Mus. Fine Arts Boston*, 66:124–166.

REFERENCES FOR DIAMONDS (SECTION 27.6e)

Bottinga, Y., 1969. Carbon isotope fractionation between graphite, diamond, and carbon dioxide. *Earth Planet. Sci. Lett.*, 5:301–307.

Craig, H., 1953. The geochemistry of the stable carbon isotopes. *Geochim. Cosmochim. Acta*, 3:53–92.

Deines, P., 1980. The carbon isotopic composition of diamonds: Relationship to diamond shape, color, occurrence, and composition. *Geochim. Cosmochim. Acta*, 44:943–961.

Deines, P., F. Viljoen, and J. W. Harris, 2001. Implications of the carbon isotope and mineral inclusion record for the formation of diamonds in the mantle underlying a mobile belt: Venetia, South Africa. *Geochim. Cosmochim. Acta*, 65:813–838.

Faure, G., 2001. *Origin of Igneous Rocks; the Isotopic Evidence*. Springer-Verlag, Heidelberg.

Hauri, E. H., J. Wang, D. G. Pearson, and G. P. Bulanova, 2002. Microanalysis of $\delta^{13}C$, $\delta^{15}N$, and N abundances in diamonds by secondary ion mass spectrometry. *Chem. Geol.*, 185:149–163.

Hoefs, J., 1997. *Stable Isotope Geochemistry*, 4th ed. Springer-Verlag, Heidelberg.

Hoering, T. C., 1961. The physical chemistry of isotopic substances: The effect of physical changes on isotope fractionation. *Carnegie Inst. Washington Yearbook*, 60:201–204.

Kennedy, C. S., and G. C. Kennedy, 1976. The equilibrium boundary between graphite and diamond. *J. Geophys. Res.*, 81:2467–2470.

Kirkley, M. B., J. J. Gurney, and A. Levinson, 1991a. Age, origin, and emplacement of diamonds. *Gems Gemol.*, Spring:2–25.

Kirkley, M. B., J. J. Gurney, M. L. Otter, S. J. Hill, and L. R. Daniels, 1991b. The application of C isotope measurements to the identification of the source of C in diamonds: A review. *Appl. Geochem.*, 6:477–494.

Lancet, M. S., and E. Anders, 1970. Carbon isotope fractionation in the Fischer-Tropsch synthesis and in meteorites. *Science*, 170:980–982.

Meyer, H. O. A., and F. R. Boyd, 1982. Composition and origin of crystalline inclusions in natural diamonds. *Geochim. Cosmochim. Acta*, 36:1255–1273.

Zhu, Y.-F., and Y. Ogasawara, 2002. Carbon recycled into deep Earth: Evidence from dolomite dissociation in subduction-zone rocks. *Geology*, 39(10):947–950.

REFERENCES FOR CARBONATITES (SECTION 27.6f)

Bell, K. (Ed.), 1989. *Carbonatites, Genesis and Evolution*. Unwin Hyman, London.

Deines, P., 1970. The carbon and oxygen isotope composition of carbonates from the Oka carbonatite complex, Quebec, Canada. *Geochim. Cosmochim. Acta*, 34:1199–1225.

Deines, P., and D. P. Gold, 1973. The isotopic composition of carbonatite and kimberlite carbonates and their bearing on the isotopic composition of deep-seated carbon. *Geochim. Cosmochim. Acta*, 37:1709–1733.

Deméņy, A., A. Ahijado, R. Casillas, and T. W. Vennemann, 1998. Crustal contamination and fluid/rock interaction in the carbonatites of Fuerteventura (Canary Islands, Spain): A C, O, H isotope study. *Lithos*, 44:101–115.

Faure, G., 2001. *Origin of Igneous Rocks; the Isotopic Evidence*. Springer-Verlag, Heidelberg.

Heinrich, E. W., 1966. *The Geology of Carbonatites*. Rand McNally. Chicago.

Hubberten, H.-W., K. Katz-Lehnert, and J. Keller, 1988. Carbon and oxygen isotope investigations in carbonatites and related rocks from the Kaiserstuhl, Germany. *Chem. Geol.*, 70:257–274.

Krafft, M., and J. Keller, 1989. Temperature measurements in carbonatite lava lakes and flows from Oldoinyo Lengai, Tanzania. *Science*, 245:168–169.

Pineau, F., M. Javoy, and C. J. Allègre, 1973. Etude systematique des isotopes de l'oxygene, du carbone, et du strontium dans les carbonatites. *Geochim. Cosmochim. Acta*, 37:2363–2377.

Reid, D. L., and A. F. Cooper, 1992. Oxygen and carbon isotope patterns in the Dicker Willem carbonatite complex, southern Namibia. *Chem. Geol. (Isotope Geosci. Sect.)*, 94:293–305.

Santos, R. V., and R. N. Clayton, 1995. Variations of oxygen and carbon isotopes in carbonatites: A study of Brazilian alkaline complexes. *Geochim. Cosmochim. Acta*, 59:1339–1352.

Sheppard, S. M. F., and H. P. Schwarcz, 1970. Fractionation of carbon and oxygen isotopes and magnesium between coexisting metamorphic calcite and dolomite. *Contrib. Mineral. Petrol.*, 26:161–198.

Srivastava, R. K., and L. A. Taylor, 1996. Carbon and oxygen-isotope variations in Indian carbonatites. *Int. Geol. Rev.*, 38:419–429.

Wyllie, P. J., 1989. Origin of carbonatites: Evidence from phase equilibrium studies. In K. Bell (Ed.), *Carbonatites*, pp. 15–37. Unwin Hyman, London.

REFERENCES FOR METEORITES (SECTIONS 27.7–27.7b)

Amari, S., P. Hoppe, E. Zinner, and R. S. Lewis, 1993. The isotopic compositions of stellar sources of meteoritic graphite grains. *Nature*, 365:806–809.

Anders, E., and N. Grevesse, 1989. Abundances of the elements: Meteoritic and solar. *Geochim. Cosmochim. Acta*, 53:197–214.

Barrat, J. A., Ph. Gillet, C. Lécuyer, S. M. F. Sheppard, and M. Lesourd, 1998. Formation of carbonates in the Tatahouine meteorite. *Science*, 280:412–414.

Brett, R., 1967. Cohenite: Its occurrence and a proposed origin. *Geochim. Cosmochim. Acta*, 31:143–159.

Brett, R., and G. T. Higgins, 1967. Cliftonite in meteorites: A proposed origin. *Science*, 156:819–820.

Buseck, P. R., and B. J. Huang, 1985. Conversion of carbonaceous material to graphite during metamorphism. *Geochim. Cosmochim. Acta*, 49:2203–2216.

Chacko, T., D. R. Cole, and J. Horita, 2001. Equilibrium oxygen, hydrogen, and carbon isotope fractionation factors applicable to geologic systems. In J. W. Valley and D. R. Cole (Eds.), *Stable Isotope Geochemistry. Rev. Mineral. Geochem.*, 43:1–82, Mineralogical Society of America, Blacksburg, Virginia.

Clayton, R. N., 1963. Carbon isotope abundance in meteoritic carbonates. *Science*, 140:192–193.

Clayton, R. N., 1993. Oxygen isotopes in meteorites. *Annu. Rev. Earth Planet. Sci.*, 21:115–149.

Clayton, R. N., L. Grossman, and T. K. Mayeda, 1973. A component of primitive nuclear composition in carbonaceous meteorites. *Science*, 182:485–488.

Craig, H., 1953. The geochemistry of stable carbon isotopes. *Geochim. Cosmochim. Acta*, 3:53–93.

Cronin, J. R., S. Pizzarello, and D. P. Cruikshank, 1988. Organic matter in carbonaceous chondrites, planetary satellites, asteroids, and comets. In J. F. Kerridge and M. S. Matthews (Eds.), *Meteorites and the Early Solar System*, pp. 819–857. University of Arizona Press, Tucson, Arizona.

Deines, P., and F. E. Wickman, 1973. The isotopic composition of "graphitic" carbon from iron meteorites and some remarks on the troilite sulfur of iron meteorites. *Geochim. Cosmochim. Acta*, 37:1295–1319.

Deines, P., and F. E. Wickman, 1975. A contribution to the stable carbon isotope geochemistry of iron meteorites. *Geochim. Cosmochim. Acta*, 39:547–557.

Deines, P., and F. E. Wickman, 1985. The stable carbon isotopes in enstatite chondrites and Cumberland Falls. *Geochim. Cosmochim. Acta*, 49:89–95.

Engel, M. H., S. A. Macko, and J. A. Silfer, 1990. Carbon isotope composition of individual amino acids in the Murchison meteorite. *Nature*, 348:47–49.

Epstein, S., R. V. Krishnamurthy, J. R. Cronin, S. Pizzarello, and G. U. Yuen, 1987. Unusual stable isotope ratios in amino acid and carboxylic acid extracts from the Murchison meteorite. *Nature*, 326:477–479.

Grady, M. M., I. P. Wright, P. K. Swart, and C. T. Pillinger, 1985. The carbon and nitrogen isotopic composition of ureilites: Implications for their genesis. *Geochim. Cosmochim. Acta*, 49:903–915.

Hoefs, J., 1973. Ein Beitrag zur Isotopengeochemie des Kohlenstoffs in magmatischen Gesteinen. *Contrib. Mineral. Petrol.*, 41:277–300.

Hoefs, J., 1997. *Stable Isotope Geochemistry*, 4th ed. Springer-Verlag, Heidelberg.

Kerridge, J. F., 1985. Carbon, hydrogen, and nitrogen in carbonaceous chondrites: Abundance and isotopic compositions in bulk samples. *Geochim. Cosmochim. Acta*, 49:1707–1714.

Lancet, M. S., and E. Anders, 1970. Carbon isotope fractionation in the Fischer-Tropsch synthesis and in meteorites. *Science*, 170:980–982.

McKeegan, K. D., and L. A. Leshin, 2001. Stable isotope variations in extraterrestrial materials. In J. W. Valley and D. R. Cole (Eds.), *Stable Isotopes Geochemistry. Rev. Mineral. Geochem.*, 43:279–318. Mineralogical Society of America, Blacksburg, Virginia.

Mostefaoui, S., C. Perron, E. Zinner, and G. Sagon, 2000. Metal-associated carbon in primitive chondrites: Structure, isotopic composition, and origin. *Geochim. Cosmochim. Acta*, 64:1945–1964.

Oró, J., J. Gilbert, H. Lichtenstein, S. Wikstrom, and D. A. Flory, 1971. Amino acids, aliphatic and aromatic hydrocarbons in the Murchison meteorite. *Nature*, 230:105–106.

Ott, U., 1993. Interstellar grains in meteorites. *Nature*, 364:25–33.

Sephton, M. A., C. T. Pillinger, and I. Gilmour. 1998. $\delta^{13}C$ of free and macromolecular aromatic structures in the Murchison meteorites. *Geochim. Cosmochim. Acta*, 62:1821–1828.

Tang, M., E. Anders, P. Hoppe, and E. Zinner, 1989. Meteoritical silicon carbide and its stellar sources; implications for galactic chemical evolution. *Nature*, 339:351–354.

Thiemens, M. H, 1988. Heterogeneity in the nebula: Evidence from stable isotopes. In J. F. Kerridge and M. S. Matthews (Eds.), *Meteorites and the Early Solar System*, pp. 899–923. University of Arizona Press, Tucson, Arizona.

Zinner, E., 1998. Stellar nucleosynthesis and the isotopic composition of presolar grains from primitive meteorites. *Annu. Rev. Earth Planet. Sci.*, 26:147–188.

Zinner, E., M. Tang, and E. Anders, 1989. Interstellar SiC in the Murchison and Murray meteorites: Isotopic composition of Ne, Xe, Si, C, and N. *Geochim. Cosmochim. Acta*, 53:3273–3290.

REFERENCES FOR LUNAR CARBON (SECTION 27.7c)

Brownlow, A. H., 1979. *Geochemistry*. Prentice-Hall, Upper Saddle River, New Jersey.

Chang, S., K. Kvenvolden, J. Lawless, C. Ponnamperuma, and I. R. Kaplan, 1971. Carbon, carbides, and methane in an Apollo 12 sample. *Science*, 171:474–477.

DesMarais, D. J., 1983. Light element geochemistry and spallogenesis in lunar rocks. *Geochim. Cosmochim. Acta*, 47:1769–1781.

Ganapathy, R., and E. Anders, 1974. Bulk compositions of the Moon and Earth, estimates from meteorites. In *Proc. 5th Lunar Conf.*, pp. 1181–1206. Lunar Planet. Sci. Ynst., Houston, Texas.

Johnson, R. D., and C. C. Davis, 1970. Pyrolysis-hydrogen flame ionization detection of organic carbon in a lunar sample. *Science*, 167:759–760.

Ponnamperuma, C., et al., 1970. Search for organic compounds in the lunar dust from the Sea of Tranquillity. *Science*, 167:760–762.

REFERENCES FOR LIFE ON MARS (SECTION 27.8)

Bada, J. L., D. P. Gavin, G. D. McDonald, and L. Becker, 1998. A search for endogenous amino acids in the martian meteorite ALH84001. *Science*, 279:362–365.

Cassidy, W., R. Harvey, J. Schutt, G. Delisle, and K. Yanai, 1992. The meteorite collection sites of Antarctica. *Meteoritics*, 27:490–525.

Fanale, F. P., S. E. Postawko, J. B. Pollack, M. H. Carr, and R. O. Pepin, 1992. Mars: Epochal climate change and volatile history. In H. H. Kieffer, B. M. Jakosky, C. W. Snyder, and M. S. Matthews (Eds.), *Mars*, pp. 1135–1179. University of Arizona Press, Tucson.

Iberall-Robbins, E. and A. S. Iberall, 1991. Mineral remains of early life on Earth? On Mars? *Geomicrobiol. J.*, 9:51–66.

Jull, A. J. T., C. Courtney, D. A. Jeffrey, and A. J. Beck, 1998. Isotopic evidence for a terrestrial source of organic compounds found in martian meteorites, Allan Hills 84001 and Elephant Moraine 79001. *Science*, 279:366–368.

Krasnopolsky, V. A., M. J. Mumma, G. L. Bjoraker, and D. E. Jennings, 1996. Oxygen and carbon isotope ratios in martian carbon dioxide: Measurements and implications for atmospheric evolution. *Icarus*, 124: 553–568.

McKay, D. S., E. K. Gibson, Jr., K. L. Thomas-Keprta, H. Vali, C. S. Romanek, S. J. Clemett, X. D. F. Chillier, C. R. Maechling, and R. N. Zare, 1996. Search for past life on Mars: Possible relic biogenic activity in martian meteorite ALH84001. *Science*, 273:924–930.

McKeegan, K. D., and L. A. Leshin, 2001. Stable isotope variations in extraterrestrial materials. In J. W. Valley and D. R. Cole (Eds.), *Stable Isotope Geochemistry*, pp. 279–318. *Rev. Mineral. Geochem.*, 43. Mineralogical Society of America, Blacksburg, Virginia.

Mittlefehldt, D. 1994. ALH84001, a cumulate orthopyroxenite of the Martian meteorite clan. *Meteoritics*, 29:214–221.

Nier, A. O., and M. B. McElroy, 1977. Composition and structure of Mars' upper atmosphere: Results from the neutral mass spectrometers on Viking 1 and 2. *J. Geophys. Res.*, 82:4341–4349.

Owen, T., 1992. The composition and early history of the atmosphere of Mars. In H. H. Kieffer, B. M. Jakosky, C. W. Snyder, and M. S. Matthews (Eds.), *Mars*, pp. 818–834. University of Arizona Press, Tucson, Arizona.

Owen, T., K. Biemann, D. R. Rushneck, J. E. Biller, D. W. Howarth, and A. L. Lafluer, 1977. The composition of the atmosphere at the surface of Mars. *J. Geophys. Res.*, 82:4635–4639.

Wright, I. P., R. H. Carr, and C. T. Pillinger, 1986. Carbon abundance and isotopic studies of Shergotty and other Shergottite meteorites. *Geochim. Cosmochim. Acta*, 50:983–992.

CHAPTER 28

Nitrogen

NITROGEN is an important element in the biosphere and atmosphere and is therefore useful in the study of geological environments where biological activity interacts with geological processes, such as in modern soils, deposits of peat and fossil fuels, and organic compounds in sediment. In addition, N occurs in refractory inclusions of carbonaceous chondrites and in terrestrial diamonds. The isotopic composition of N in the refractory inclusions of meteorites sheds light on the physical and chemical processes that occurred in the ancestral stars and the solar nebula before the formation of the Sun and the planets of the solar system. In addition, the N in terrestrial diamonds provides information about the isotopic composition of N in the mantle of the Earth and hence about the transfer of N from the solar nebula into the Earth.

28.1 GEOCHEMISTRY

The abundance of N ($Z = 7$) in the solar system is 3.13×10^6 atoms of N per 10^6 atoms of Si (Anders and Grevesse, 1989), which is less than that of C ($Z = 6$) and O ($Z = 8$), in agreement with the Oddo–Harkins rule (Section 1.2d). The electronic formula of N is $1s^2\,2s^2\,2p^3$, which allows it to have several different valence numbers: 0, −3, +3, +5, +4, and +2. Consequently, N can form a variety of compounds, including diatomic N molecules (N_2), nitrate (NO_3^-), nitrite (NO_2^-), ammonium (NH_4^+), ammonia (NH_3), and oxides (NO_2, NO, and N_2O). In addition, N occurs in biogenic organic molecules, including amino acids and proteins.

Nitrogen forms only a small number of minerals, including the nitrates $NaNO_3$ (sodaniter) and α-KNO_3 (niter), which occur in nonmarine evaporite deposits in arid regions of the Earth such as the Atacama Desert in Chile. Naturally occurring nitrates of NH_4^+, Mg^{2+}, Ca^{2+}, Ba^{2+}, and Ti^{4+} are also known (Baur, 1974). In addition, the ammonium phosphates struvite ($NH_4MgPO_4 \cdot 6H_2O$) and dittmarite [$NH_4Mg(PO_4) \cdot H_2O$] form intestinal concretions in horses. Taylor and Faure (1983) reported that alkali metals can substitute for NH_4^+ in struvite, thereby enriching this mineral in Rb relative to Sr. The concentrations of N in a wide variety of rocks are listed in Table 28.1 based on a compilation by Wlotzka (1972).

Stony meteorites contain the nitrogen-bearing minerals osbornite (TiN) and the silicon oxynitride sinoite (Si_2N_2O). Carbonaceous chondrites have high N concentrations ranging from 600 to 2900 ppm, whereas ordinary chondrites, achondrites, and lunar basalts contain from 10 ppm to a few hundred ppm of N (Wlotzka, 1972).

Igneous rocks contain N primarily in the form of ammonium ions, which replace K ions in silicate minerals. About 10–20% of the ammonium in igneous rocks is extractable with 1N KCl solutions, suggesting that the ammonium occurs as soluble salts in grain boundaries or as a contaminant introduced either by biological activity or by handling of rock samples in the laboratory.

Table 28.1. Average Concentration of N in Terrestrial and Extraterrestrial Rocks

Rock Types	Nitrogen, ppm
Extraterrestrial Samples	
Meteorites	
Carbonaceous chondrites	1850
Enstatite chondrites with sinoite	708
Enstatite chondrites without sinoite	220
All other chondrites	40
Achondrites	38
Iron metorites	72
Lunar rocks (basalt, breccia, and fines)	106
Terrestrial Samples	
Minerals	
Muscovite	68
Biotite	55
K-feldspar	23
Plagioclase	22
Hornblende	18
Quartz	13
Pyroxene	11
Igneous rocks	
Granites and granodiorites	21
Gabbros and diorites	11
Ultramafic rocks	14
Rhyolites and obsidian	28
Phonolites and trachytes	36
Andesites	87
Basalt	30
Sedimentary rocks	
Shale	602
Greywacke	180
Sandstone	120
Carbonate rocks	73
Chert	210
Modern marine sediment	1772
Coal	2000–30,000
Petroleum	100–20,000
Natural gas	Variable, up to 90% of N_2 by volume

Source: Wlotzka, 1972.

The N content of sedimentary rocks is closely related to the concentration of biogenic C. The C/N ratio of recent sediment varies from about 5 to 50 with a mean of about 20. In addition, clay minerals can adsorb ammonium in place of K ions, which explains the positive correlation between the N concentrations of detrital sedimentary rocks and their clay content.

Most of the N in solution in the oceans is present as N_2 that is in equilibrium with N_2 of the atmosphere. The solubility of N_2 *decreases* with increasing temperature and increasing chlorinity of the water. Ammonia, nitrite, and nitrate are also present in the oceans in varying concentrations depending on the locality, season, and depth and are intimately related to the biological activity in the water. Nitrate concentrations tend to increase with depth in the oceans and have average values of 1.2 mg/L in deep water of the Atlantic and 2.5 mg/L in the Pacific Ocean.

Water in lakes and rivers contains N in the form of ammonia, nitrate, protein, and amino acids in solution. Ammonia may become dominant under anaerobic conditions with a corresponding decrease in nitrate concentration. The average nitrate concentration in surface water varies from 0.005 ppm in Australia to 3.7 ppm in Europe with a worldwide average of about 1 ppm.

The nitrate content of groundwater is more variable and higher than that of surface water, especially in limestone terrain (Ueda et al., 1991). Drinking water should contain less than 10 ppm of nitrate, but this value is exceeded by about 10% of the water supplies in California and elsewhere in the USA (Navone et al., 1963; Wlotzka, 1972). The high nitrate concentrations in drinking water may be caused by agricultural uses of N-bearing fertilizers, by animal waste, or by the discharge of geothermal brines in volcanic terrains.

The N of the atmosphere occurs in the form of N_2, which is stable and unreactive. Its concentration (as N_2) is 78.1% compared to O_2 = 20.9%, and CO_2 = 0.035% (Graedel and Crutzen, 1993). The atmosphere contains not only N_2 but also nitrous oxide (NO_2). The latter is attributable primarily to industrial pollution. The average nitrate concentration of rainwater in the USA varies regionally from about 0.2 mg/L along the east coast to values in excess of 1.80 mg/L in the

Rocky Mountains. An inverse relationship exists between the annual amount of precipitation and the concentration of nitrate in rain.

28.2 ISOTOPE FRACTIONATION

Nitrogen has two stable isotopes: ^{14}N and ^{15}N. The $^{15}N/^{14}N$ ratio of atmospheric N_2 is 0.0036765 (Junk and Svec, 1958). The corresponding abundances of the isotopes are calculated as in Section 5.2c:

$^{15}N/^{14}N =$	0.0036765	$Ab^{15}N =$	0.3663%
$^{14}N/^{14}N =$	1.0000000	$Ab^{14}N =$	99.6337%
Sum:	1.0036765		100.0000%

The atomic weight of atmospheric N is derived from the measured masses and abundances of its stable isotopes:

	Abundance	Mass (amu)	
^{14}N:	0.996337 ×	14.003074 =	13.9517
^{15}N:	0.003663 ×	15.000108 =	0.0549
		Atomic weight =	14.0066

The isotopic composition of N is expressed by the $\delta^{15}N$ parameter, defined as:

$$\delta^{15}N = \left[\frac{(^{15}N/^{14}N)_{spl} - (^{15}N/^{14}N)_{std}}{(^{15}N/^{14}N)_{std}}\right] \times 10^3‰ \quad (28.1)$$

where the standard is N_2 of the atmosphere (Mariotti, 1984). The analyses are made on N_2 gas in mass spectrometers equipped with double collectors. Other N compounds must be converted to N_2 gas with stringent controls against contamination by CO. The CO molecule has mass numbers ranging from 28 ($^{12}C^{16}O$) to 31 ($^{13}C^{18}O$) and therefore interferes with the mass spectrum of N_2, which has mass numbers of 28 ($^{14}N_2$), 29 ($^{14}N^{15}N$), and 30 ($^{15}N_2$). The procedures for converting N compounds to N_2 have been described by:

Bremner and Keeney, 1966, *Soil. Sci. Soc. Am. Proc.*, 30:577–582. Ross and Martin, 1972, *Analyst*, 95:817–822. Owens, 1987, *Adv. Marine Biol.*, 24:390–451. Velinsky et al., 1989, *Marine Chem.*, 26:351–361. Kendall and Grim, 1990, *Anal. Chem.*, 62:526–529. Scholten, 1991, Ph.D. Dissertation, University of Utrecht. Hoefs, 1997, *Stable Isotope Geochemistry*, Springer-Verlag, Heidelberg.

The isotopes of N are fractionated primarily during the conversion of atmospheric N_2 into organic compounds by the metabolism of certain algae and bacteria. This process consists of fixation, nitrification, and denitrification, which together constitute the exogenic N cycle discussed by:

Heaton, 1986, *Chem. Geol.*, 59:87–102. Owens, 1987, *Adv. Marine Biol.*, 24:390–451. Hübner, 1986, in Fritz and Fontes (Eds.), *Handbook of Environmental Isotope Geochemistry*, Vol. 2B, pp. 361–425, Elsevier, Amsterdam, The Netherlands. Petersen and Fry, 1987, *Annu. Rev. Ecol. Syst.*, 18:293–320.

The *fixation* of N consists of the conversion of atmospheric N_2 into ammonia (NH_3) by many kinds of bacteria living in the roots of plants. This process is inefficient and therefore does not cause significant fractionation of the isotopes of N.

Nitrification (production of nitrate) from organic N takes place in three steps:

1. Organic N → NH_4^+ (insignificant fractionation) (28.2)
2. NH_4^+ → NO_2^- (kinetic isotope fractionation) (28.3)
3. NO_2^- → NO_3^- (kinetic isotope fractionation) (28.4)

The extent of isotope fractionation during the process of nitrification depends on which of the three steps is rate limiting. Mariotti et al. (1981) determined that, in the presence of an abundant ammonium supply (i.e., step 1 is not rate limiting), the nitrate is *depleted* in ^{15}N by 20–35‰ (i.e., steps 2 and 3 are rate limiting). However, organic N in soil is converted to NH_4 at a slow rate, in which case step 1 becomes rate limiting. Since this step does not cause fractionation, the $\delta^{15}N$ of the resulting nitrate is similar to that of the organic N (i.e., steps 2 and 3 are *not* rate limiting).

Denitrification (reduction of nitrate to N_2) occurs in poorly aerated soils and anaerobic bodies of freshwater. The process starts with the diffusion of nitrate ions across cell membranes followed by the reduction of N^{5+} in NO_3^- to N^0 in N_2, which requires the breaking of the covalent N–O bonds. The reduction of NO_3^- to N_2 causes extensive depletion of N_2 in ^{15}N and complementary enrichment of the residual nitrate in ^{15}N. Mariotti et al. (1982) demonstrated that the extent of fractionation increases as the rate of reduction decreases. The fractionation of N isotopes during the biological N cycle was summarized by Heaton (1986) and Fogel and Cifuentes (1993).

The isotopes of N are also fractionated by isotope exchange reactions at equilibrium. For example, the equilibrium fractionation factor $\alpha_{NH_3}^{NH_4}$ of the exchange reaction at room temperature (Mariotti et al., 1981):

$$NH_3(\text{gas}) \rightleftarrows NH_4^+ \text{ (aqueous)} \qquad (28.5)$$

is

$$\alpha_{NH_3}^{NH_4} = 1.035$$

In addition, Nitzsche and Stiehl (1984) reported fractionation factors of 1.0143 at 250°C and 1.0126 at 350°C for this reaction. These data define the equation:

$$10^3 \ln \alpha_{NH_3}^{NH_4} = 0.576\left(\frac{10^3}{T}\right) + 11.37 \qquad (28.6)$$

where T is the temperature in degrees Celsius.

Additional isotope fractionation factors for aqueous NO_3^- and NO_2^- in equilibrium with gases N_2, NH_3, NH_4^+, NO_2, NO, N_2O, and NH_3 were compiled from the literature by Letolle (1980). For example, the fractionation factor $\alpha_{N_2}^{NO_3}$ of the isotope exchange equilibrium:

$$N_2\text{ (g)} \rightarrow NO_3^-\text{ (aq)} \qquad (28.7)$$

at $T = 20°C$ is

$$\alpha_{N_2}^{NO_3} = 1.0678$$

Accordingly, the $\delta^{15}N$ value of aqueous NO_3^- in isotopic equilibrium with atmospheric N_2 at 20°C is:

$$1.0678 = \frac{\delta^{15}N_{NO_3^-} + 1000}{\delta^{15}N_{N_2} + 1000}$$

which yields $\delta^{15}N_{NO_3^-} = +67.8‰$ for $\delta^{15}N_{N_2} = 0‰$.

Table 28.2. Isotope Composition of Environmental Sources of N

Material	$\delta^{15}N$, ‰
Nitrate, rain	−13 to +2
Ammonium, rain	−15 to 0
Organic N, soil	0 to +9
Fertilizer, NO_3^-	−5 to +7
Fertilizer, NH_4^+	−6 to +5
Animal waste, NO_3^-	+8 to +22

Source: From a review of the literature by Heaton (1986).

The isotope composition of N derived from different sources can be used to trace the movement of organic matter through the environment because fertilizer, animal waste, and sewage have characteristic $\delta^{15}N$ values listed in Table 28.2.

28.3 NITROGEN ON THE SURFACE OF THE EARTH

Isotopic studies of terrestrial N have been concerned primarily with the sources of N compounds in aqueous solution on the surface of the Earth. Letolle (1980) summarized the results of these studies based on a review of the voluminous literature. In general, N in animal tissue is enriched in ^{15}N because of the preferential excretion of ^{14}N in urea. Plant tissue may be slightly depleted in ^{15}N relative to atmospheric N_2. However, plant tissues have a wide range of $\delta^{15}N$ values in cases where N was derived from nitrate or ammonia that had been isotopically fractionated before being absorbed by plants. Decay of plant tissue in soils causes large N-bearing molecules to break down into smaller ones that favor ^{14}N and leave the residual plant remains enriched in ^{15}N. As a result, the $\delta^{15}N$ values of "total soil nitrogen" range widely from −4.4 to

+17.0‰. Cultivation tends to accelerate the "diagenetic" reactions of plant tissue in soils resulting in higher $\delta^{15}N$ values than in uncultivated soils. Additional results of the isotope composition of N in soils are available in papers by Shearer et al. (1978) and Wada et al. (1984).

Drainage of rainwater through soil transports N compounds to the groundwater table, primarily as dissolved nitrate. Other sources of nitrate in groundwater include fertilizer, sewage effluent, animal waste, and the minerals of the subsurface aquifer (Table 28.2). The $\delta^{15}N$ values of groundwater vary widely from 0 to +25‰ depending on the sources that may locally dominate the inputs of N to groundwater (Letolle, 1980). The variability of the $\delta^{15}N$ values of nitrate in groundwater has stimulated its use in hydrogeological studies (Pang and Nriagu, 1977; Mariotti and Letolle, 1979; Kreitler, 1979; Flipse and Bonner, 1985).

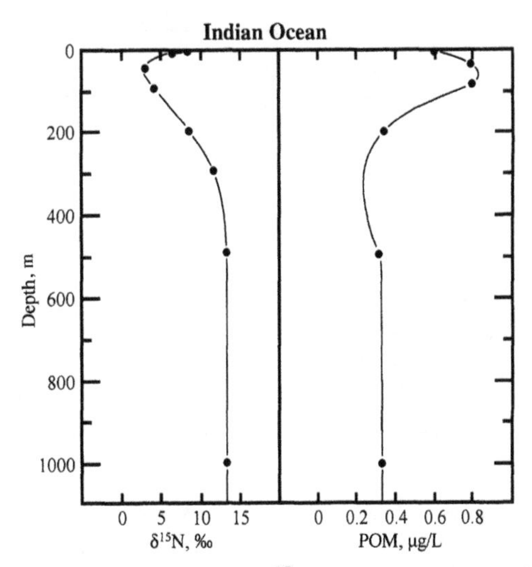

FIGURE 28.1 Variation of $\delta^{15}N$ and concentration of particulate organic nitrogen (POM) with depth at a station in the northeastern Indian Ocean. The organic particles are enriched in ^{15}N during oxidative degradation of N compounds. Data from Saino and Hattori (1980).

28.3a POM in the Oceans

The oceans contain particulate organic matter (POM) whose C/N ratio increases with depth because N is lost more rapidly than C during the degradation of POM. Measurements by Saino and Hattori (1980) demonstrated that the $\delta^{15}N$ values of POM vary systematically with depth, as shown in Figure 28.1. They attributed the *decrease* of $\delta^{15}N$ of POM in surface water to isotope fractionation during the uptake of dissolved nitrate by phytoplankton that prefer $^{14}NO_3^-$. As a result, nitrate in surface water is enriched in ^{15}N as reported by Cline and Kaplan (1975) for seawater in the eastern North Pacific Ocean. The degradation of dead phytoplankton down to a depth of about 500 m is indicated by the *increase* of $\delta^{15}N$ to +13‰ and the decrease of the concentration of N in POM. Below that depth, the residual POM is stable and sinks slowly to the bottom of the ocean.

Particulate organic matter derived from sources on land has a significantly *lower* $\delta^{15}N$ value than POM in the oceans. For example, Sweeney et al. (1978) reported an average $\delta^{15}N$ value of +2.5‰ for particles of continental organic matter. The difference in isotopic compositions of N in POM derived from continental and oceanic sources provides a method of studying the mixing of water masses in the coastal ocean. For example, Mariotti et al. (1984) demonstrated that in April of 1974 the $\delta^{15}N$ values of POM in the Scheldt estuary of the Netherlands increased toward the North Sea from an average of $+1.5 \pm 0.2$‰ at about 80 km inland from the coast toward a mean value of $\delta^{15}N = +8.0 \pm 1.8$‰ in the North Sea. The lowest values of $\delta^{15}N$ occur in POM collected in the spring and early summer (April to June) when blooms of microorganisms occur. A summary of the regional variation of $\delta^{15}N$ of POM in the Scheldt estuary is presented in Figure 28.2.

The POM that is slowly sinking to the bottom of the oceans is deposited with inorganic mineral particles. Judging from the data of Saino and Hattori (1980) and Mariotti et al. (1984), the $\delta^{15}N$ values of marine sediment should lie between +8‰ (North Sea) and +13.1‰ (Indian Ocean). Actually, measurements by Wada et al. (1975) indicated a somewhat *lower* average value of only $+6.8 \pm 4.1$‰. Evidently, the N in sediment is either not entirely derived by settling of POM or it is isotopically modified after deposition or

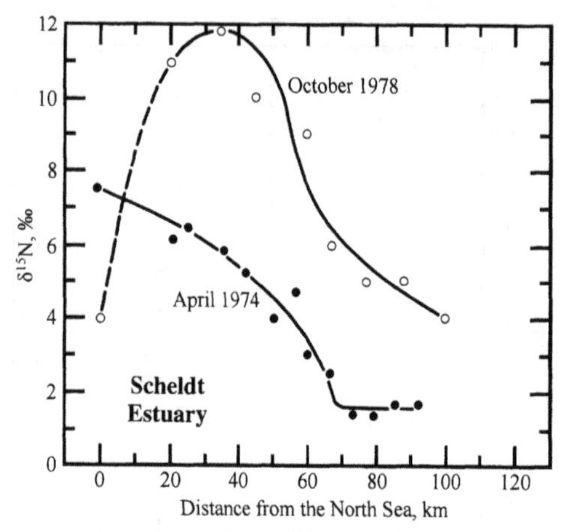

FIGURE 28.2 Variation of $\delta^{15}N$ of suspended organic matter in the Scheldt estuary of the Netherlands. The curves reflect progressive mixing of organic particles of terrestrial origin with organic particles in the North Sea. The data for April 1974 reflect the blooming of microorganisms in early spring and their preferential uptake of $^{14}NO_3^-$. In October 1978 the $\delta^{15}N$ values of organic particles have higher $\delta^{15}N$ in both freshwater and marine portions of the estuary. The low value of +4‰ at the coast in October 1978 is unexplained. Mariotti et al. (1984).

both. Saino and Hattori (1980) suggested that large particles, such as fecal pellets, sink rapidly and are therefore not enriched in ^{15}N by oxidative degradation of N-compounds. Such coarser particles are not well represented in samples of POM recovered by filtration of water, but may be collected on sediment traps placed at the bottom of the ocean. According to Saino and Hattori (1980), particulates collected on sedimentation traps in the northern North Pacific have low $\delta^{15}N$ values ranging from +2.9 to +4.4‰. They suggested, therefore, that the organic fraction of sediment may be a mixture of slowly sinking particles (high $\delta^{15}N$) and rapidly sinking particles (low $\delta^{15}N$).

28.3b Lacustrine Sediment and the Food Chain

A study of lacustrine sediment in the Bay of Quinte in Lake Ontario by Pang and Nriagu (1976) indicated that the $\delta^{15}N$ values of the organic fraction do not vary with depth below the water–sediment interface and generally range from +2.7 to +5.6‰, with a mean of $+4.1 \pm 0.7$‰. They also measured the isotope composition of N released by digestion of the silicate mineral particles by hydrofluoric acid after the hydrolyzable N had been removed. The average $\delta^{15}N$ value of this "exchangeable NH_4^+" was $+7.6 \pm 1.2$‰. Pang and Nriagu (1976) suggested that the ^{15}N enrichment of the exchangeable NH_4^+ resulted from the preferential oxidation of the exchangeable $^{14}NH_4^+$ to nitrate. DeNiro and Hastorf (1985) likewise reported increases of $\delta^{15}N$ values of noncarbonized prehistoric plants from Peru.

The isotopic composition of N in animals is affected by their diets and depends on the balance between inputs and outputs of N-bearing compounds (DeNiro and Epstein, 1981). The available data indicate that animal tissue is *enriched* in ^{15}N relative to dietary inputs. The ^{15}N-enrichment appears to be propagated up the food chain and may increase with the age of the animal. Minagawa and Wada (1984) demonstrated that the $\delta^{15}N$ values of phytoplankton, zooplankton, and fish *increase* in this order, both in marine and lacustrine environments. In addition, they observed a similar *increase* of $\delta^{15}N$ among leaf hoppers, spiders, and tree frogs in a rice paddy at Konosu, Japan, and among the animals of an intertidal zone at Usujiri. The progressive increase in the abundance of ^{15}N in this natural ecosystem is illustrated in Figure 28.3. The ^{15}N enrichment is most clearly recognized among the members of a community of producers and consumers and is blurred when data from different regions are averaged.

In addition, sedimentary organic matter is progressively enriched in $\delta^{15}N$ by preferential loss of ^{14}N during diagenesis, burial metamorphism, and low-grade regional metamorphism. Therefore, kerogen in sedimentary and metasedimentary rocks is enriched in ^{15}N relative to freshly deposited sedimentary organic matter.

28.4 FOSSIL FUELS

Nitrogen is a common constituent of natural gas, and its concentration may vary from a few percent

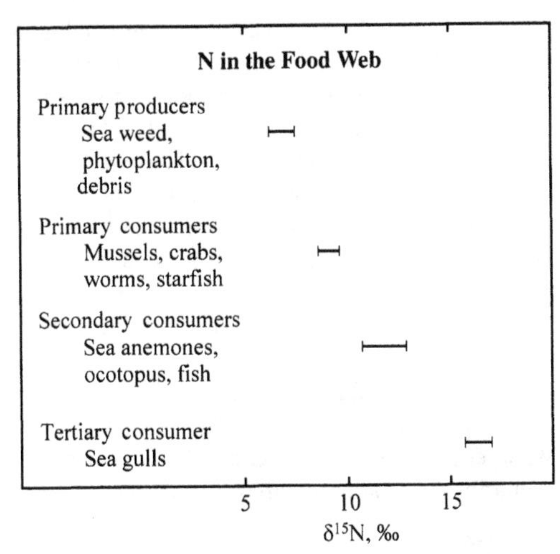

FIGURE 28.3 Isotope composition of N at different levels of the food web in the intertidal zone at Usujiri, Japan, on July 28,1982. The progressive increase of $\delta^{15}N$ up the food chain in this community is caused both by the diet and by subsequent metabolic fractionation of N isotopes. Data from Minagawa and Wada (1984).

by volume to nearly 100% N_2. The isotope composition of N_2 in natural gas also varies widely, from $\delta^{15}N = -11.5$ to $+18‰$ (Wlotzka, 1972). The wide range of isotopic compositions indicates that the N_2 is a mixture of several components having different isotopic compositions. In addition, regional variations of the $\delta^{15}N$ values of natural gas are caused by differences in the abundances of these components.

Zhu et al. (2000) concluded that N_2 in natural gas originates primarily by decomposition of organic matter buried in marine and lacustrine sediment and that other possible sources such as air trapped in sediment, outgassing of the mantle, and the production of N isotopes by nuclear reactions are insignificant. According to Zhu et al. (2000), the $\delta^{15}N$ values of N_2 in natural gas depend on the "maturity" of the organic matter in the sedimentary source rocks as recorded by its vitrinite reflectance (R_0):

1. $R_0 < 0.6\%$: $\delta^{15}N = -19$ to $-10‰$
2. $R_0 = 0.6$ to $R_0 = 2.0\%$: $\delta^{15}N = -10$ to $-2‰$
3. NH_4 in clay, shale, and mudrock: $\delta^{15}N = +1$ to $+4‰$
4. Evaporite rocks containing nitrates: $\delta^{15}N = +4‰$
5. $R_0 > 2.0\%$: $\delta^{15}N = +4$ to $+18‰$

Accordingly, $\delta^{15}N$ values of biogenic N_2 in natural gas increase with rising temperature of alteration of the organic matter from less than $-10‰$ at low temperature to greater than $+4‰$ at high temperatures. In addition, N_2 may be released by regional metamorphism of clay-rich rocks in which NH_4^+ has substituted for K^+ in smectite clay minerals and, more rarely, by the reduction of nitrate-bearing evaporites.

The $\delta^{15}N$ values of N_2 in the sedimentary basins listed in Table 28.3 have characteristic ranges that arise primarily from the maturity of the organic matter in the source rocks in each basin. The decomposition of sedimentary organic matter includes the breakdown of amino acids ($COOH–CH_2–NH_2$) by bacteria:

$$COOH–CH_2–NH_2 \rightarrow CO_2 + H_2 + NH_3 \quad (28.8)$$

The ammonia reacts with CO_2 to form CH_4 and N_2:

$$8NH_3 + CO_2 \rightarrow CH_4 + 4N_2 + 6H_2O \quad (28.9)$$

and

$$2NH_3 + 2\,Fe_2O_3 \rightarrow 6FeO + N_2 + 3H_2O \quad (28.10)$$

The N_2 produced by decomposition of amino acids via NH_3 is *depleted* in ^{15}N, causing its $\delta^{15}N$ value to become negative (Macko et al., 1987).

The generation of N_2 from organic-rich sedimentary rocks at high temperature has been investigated by Krooss et al. (1995) and Littke et al. (1995). They determined that the peak production of CH_4 by cracking of kerogen occurs at a lower temperature than the maximum release of N_2. Therefore, natural gas containing more than 50% N_2 requires temperatures in excess of 300°C.

Clay minerals containing NH_4^+ in place of K^+ release N_2 only at temperatures above 500°C under natural conditions (Whelan et al., 1988). The

Table 28.3. Range of $\delta^{15}N$ Values of N_2 in Natural Gas Present in Reservoir Rocks of Different Regions

Region	$\delta^{15}N$, ‰	References[a]
Western Siberia, Russia, Cretaceous	−19 to −10	1
Yinggehai basin, China, Cenozoic	−9 to −2	2
Great Valley California, Late Cretaceous/Tertiary	+0.9 to +3.5	3
Gulf Coast, USA (Oklahoma, Texas, Mississippi)	−10.5 to −2.1	4
Arkansas, USA	+1.3 to +11.9	4
Mid-European basin, late Paleozoic to Cenozoic	+6.5 to +18.0	5–7

Source: Zhu et al., 2000.

[a] 1. Galimov, 1988, *Chem. Geol.*, 71:77–95. 2. Zhu et al., 2000, *Chem. Geol.*, 164:321–330. 3. Jenden et al., 1988, *Geochim. Cosmochim. Acta*, 52:851–861. 4. Hoering and Moore, 1958, *Geochim. Cosmochim. Acta*, 13:225–232. 5. Maksimov et al., 1975, *Int. Geol. Rev.*, 18:551–556. 6. Stahl, 1977, *Chem. Geol.*, 20:121–149. 7. Weinlich, 1991, *Zeitschr. Angewandte Geol.*, 37:14–20.

resulting N_2 has $\delta^{15}N$ values between +1 and +4‰ (Jenden et al., 1988).

The isotopic composition of N in coal varies from −2.5 to +6.3‰ and seems to depend on the origin of the organic material and the rank of the coal. Drechsler and Stiehl (1977) and Stiehl and Lehmann (1980) concluded that "humic" coal, derived from terrestrial plants, has lower $\delta^{15}N$ values than "bituminous" coal that formed from zooplankton and phytoplankton. Certain oil shales, like the kukersite of Estonia, have $\delta^{15}N$ values that are similar to those of modern algae.

The isotope composition of N in coal also depends on its rank. The volatile N compounds released during thermal maturation of coal are initially enriched in ^{15}N, which causes the $\delta^{15}N$ value of the coal to *decrease*. This phenomenon is demonstrated by the $\delta^{15}N$ values of coal ranging in rank from lignite to anthracite, which vary systematically with decreasing concentration of volatiles. Drechsler and Stiehl (1977) and Stiehl and Lehman (1980) were able to duplicate this trend, shown in Figure 28.4, by stepwise pyrolysis of a sample of bituminous coal heated from 225 to 400°C. Stiehl and Lehmann (1980) concluded that coal contains at least two different N compounds having differing thermal stabilities and isotopic compositions.

The $\delta^{15}N$ values of petroleum are positive and range from +0.7 to +8.3‰ (Wlotzka, 1972; Wada et al., 1975; Hirner et al., 1984) and reflect primarily the isotope composition of the biogenic

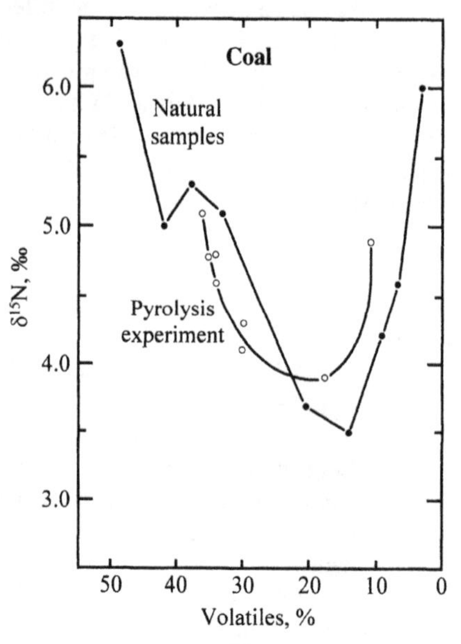

FIGURE 28.4 Variation of $\delta^{15}N$ in coal as a function of the concentration of volatiles. The natural samples are a suite of lignites, bituminous coal, and anthracite of Late Carboniferous age. The coal used for the pyrolysis experiment was bituminous coal from Zwickau, Germany. The results imply that volatile N-compounds derived from coal are enriched in ^{15}N and that coal contains at least two different N-compounds with different thermal stability and isotopic composition. Data from Drechsler and Stiehl (1977) and Stiehl and Lehmann (1980).

Table 28.4. References to the Isotope Geochemistry of N in Fossil Fuel

Topic	References
Origin of hydrocarbons, source rocks	Baxby et al., 1994, *J. Petrol. Geol.* 17(20): 211–230; Galimov, 1988, *Chem. Geol.*, 71:77–95; Krooss et al., 1995, *Chem. Geol.*, 126:291–318; Littke et al., 1995, *AAPG Bull.*, 79:410–430; Whiticar, 1994, in Magoon and Dow (Eds.), *The Petroleum System—From Source to Trap*, pp. 281–283. AAPG Mem. 60. Tulsa, Oklahoma.
Coal	Bokhoven and Theeuwen, 1966, *Nature*, 211:927–929.
Oil shale	Cooper and Eans, 1983, *Science*, 219:492–493; Oh et al., 1988, *Energy and Fuels*, 2:100–105; Weinlich, 1991, *Zeitschr. Angewandte Geol.*, 37:14–20; Whelan et al., 1988, *Energy and Fuels*, 2:65–73.
Chert	Sano and Pillinger, 1990, *Geochem. J.*, 24:315–325.
Petroleum	Hoering and Moore, 1958, *Geochim. Cosmochim. Acta*, 13:225–232; Headlee, 1962, *World Oil*, October:126–131, 144.
Natural gas	Boigk et al., 1976, *Erdoel und Kohle*, 29:103–112; Bokhoven and Theeuwen, 1966, *Nature*, 211:927–929; Galimov et al., 1990, in Durrance et al. (Eds.), *Geochemistry of Gaseous Elements and Compounds*, pp. 401–417, Theophrastus, Athens; Headlee, 1962, *World Oil*, October: 126–131, 144; Jenden et al., 1988a, *Geochim. Cosmochim. Acta*, 52:851–861; Jenden et al., 1988b, *Chem. Geol.*, 71:117–147; Maksimov et al., 1975, *Int. Geol. Rev.*, 18:551–556; Stahl, 1977, *Chem. Geol.*, 20:121–149.
Isotope fractionation	Cramer et al., 1988, *Chem. Geol.*, 149:235–250; Macko et al., 1987, *Chem. Geol.*, 65:79–92; Letolle, 1980, in Fritz and Fontes (Eds.), *Handbook of Environmental Geochemistry*, Vol. 1A, pp. 407–434. Elsevier, Amsterdam. Fogel and Cifuentes, 1993, in Engel and Macko (Eds.), *Organic Geochemistry*, pp. 73–98, Plenum, New York.

source material of petroleum. Additional references to the isotope geochemistry of N in fossil fuels are compiled in Table 28.4.

28.5 IGNEOUS ROCKS AND THE MANTLE

The inventory of N and other volatile elements in the Earth was modeled by Javoy (1995, 1997) based on the composition of enstatite chondrites. This model suggests that most terrestrial N is still trapped in the mantle of the Earth. Alternatively, most of the N may have been lost from the atmosphere during the early history of the Earth (Pepin, 1991). The hypothesis of Javoy (1995, 1997) that most of the terrestrial N is still locked up in the mantle of the Earth has aroused interest in studies to determine the concentration and isotope compositions of N in mantle-derived oceanic basalts.

The concentrations and isotope compositions of N in oceanic basalts and harzburgite in Table 28.5 suggest that oceanic island basalts (OIBs) in Hawaii are variably enriched in ^{15}N compared to midocean ridge basalts (MORBs). For example, the $\delta^{15}N$ values of OIBs on Hawaii range from -1.1 to $+17$‰, whereas the $\delta^{15}N$ values of MORBs are negative in most cases and extend to -4.5‰. One sample of ultramafic harzburgite is even more depleted in ^{15}N with $\delta^{15}N = -5.0$‰.

The concentrations of N in both OIBs and MORBs are typically low (about 0.8 ppm) depending on the extent of degassing. In some cases, high concentrations ranging up to 12 ppm have been reported (e.g., Javoy et al., 1986; Javoy and Pineau, 1991). The harzburgite also has a comparatively high N concentration of 8 ppm,

Table 28.5. Concentrations and Isotope Compositions of N in Mafic Volcanic Rocks in the Oceans

Location	N, ppm	$\delta^{15}N$, ‰	References[a]
	Midocean Ridge Basalt		
Atlantic	12	−3.5	1
Pacific	0.9–1.4	−0.1 to +0.8	2
Pacific and Atlantic	0.2–2.1	+7.5 ± 1.0	3
Indian	0.3	−4.5	3
	0.4	−1.9	
	Oceanic Island Basalt		
Hawaii	~1	+17	4
	0.4	−0.4	2
	0.2 to 1.2	+12.8 to +15.5	3
	6	−1.1	6
	Ultramafic Rocks		
Harzburgite	8	−5.0	5

Source: Boyd and Pillinger, 1994.

[a] 1. Javoy and Pineau, 1991, *Earth Planet. Sci. Lett.*, 107:598–611. 2. Sakai et al., 1984, *Geochim. Cosmochim. Acta*, 48:2433–2441. 3. Exley et al., 1987, *Earth Planet. Sci. Lett.*, 81:163–174. 4. Becker and Clayton, 1977, *Eos*, 58:536. 5. Nadeau et al., 1990, *Chem. Geol.*, 81:271–297. 6. Javoy et al., 1986, *Chem. Geol.*, 57:41–62.

presumably because it did not lose as much N as the samples of volcanic rocks.

The range of $\delta^{15}N$ values of oceanic basalts in Table 28.5 suggests that the N they contain is a mixture of two or more components having different isotopic compositions. The ^{15}N enrichment of Hawaiian basalts is especially significant because the island of Hawaii is located above a large mantle plume which contains subducted oceanic crust and sediment (Faure, 2001). The organic matter of marine sediment is enriched in ^{15}N by the preferential loss of ^{14}N during diagenesis. The other components of N that may be represented by the isotopic composition of N in marine basalt are atmospheric N_2 ($\delta^{15}N = 0$‰) and primordial N stored in the mantle. One of the research goals of the isotope geochemistry of N is to determine the $\delta^{15}N$ value of primordial N and to ascertain whether the primordial N is isotopically homogeneous. The data in Table 28.5 suggest that the primordial N component is *depleted* in ^{15}N relative to atmospheric N_2 (e.g., the $\delta^{15}N$ values of MORBs and of the harzburgite).

Similar studies by other research groups have repeatedly confirmed the evidence for the isotopic heterogeneity of N in basaltic igneous rocks:

Sakai et al., 1984, *Geochim. Cosmochim. Acta*, 48:2433–2441. Marty, 1995, *Nature*, 377:326–329. Marty and Humbert, 1997, *Earth Planet. Sci. Lett.*, 152:101–112. Marty et al., 1995, *Chem. Geol.*, 120:183–195.

28.6 ULTRAMAFIC XENOLITHS

Because of the apparent isotopic heterogeneity of N in mantle-derived igneous rocks, Mohapatra and Murty (2000) searched for mantle-N in the minerals of ultramafic xenoliths in the alkalic volcanic rocks at San Carlos, Arizona. Handpicked grains of clinopyroxene (cpx), orthopyroxene (opx), and olivine of several xenoliths were first outgassed at 150°C to remove adsorbed gases and were then heated stepwise starting at 400°C up to 1800°C. The noble gases (He, Ne, and Ar) as well as N_2 released at each temperature step were analyzed on a gas-source mass spectrometer.

The results in Table 28.6 indicate that *olivine* contains three isotopic components of N.

The concentrations of N released by cpx, opx, and olivine at temperatures between 900 and 1800°C range from 1.5 ppm (cpx 2) to 0.14 ppm (olivine 1). The $\delta^{15}N$ values of the combined gas fractions range from +2.24‰ (cpx 1) to +9.0‰ (opx 3) with a weighted mean of +5.16‰. Mohapatra and Murty (2000) concluded that the isotope compositions and concentrations of the high-temperature fractions of N require the presence of three components, which they identified as:

1. atmospheric N, $\delta^{15}N = 0$‰;
2. recycled N, $\delta^{15}N > +15$‰; and
3. mantle N, $\delta^{15}N < -9$‰.

The $\delta^{15}N$ value of recycled (subducted) sediment is based on the work of Haendel et al. (1986),

Table 28.6. Concentrations and Isotope Compositions of N in the Minerals of Mantle-Derived Ultramafic Xenoliths from San Carlos, Arizona

Mineral	N,[a] μg/g	δ^{15}N,[b] ‰
Cpx 1	1.4	$+2.24 \pm 0.88$
Cpx 2	1.5	$+6.78 \pm 0.46$
Opx 3	0.7	$+9.0 \pm 0.45$
Olivine 1	0.14	$+2.43 \pm 0.51$
Olivine 2	0.7	$+4.24 \pm 0.65$
Olivine 4	0.32	$+5.21 \pm 0.54$
Weighted average		+5.16

Source: Mohapatra and Murty, 2000.

[a]Released at high temperature between 900 and 1800°C.
[b]The δ^{15}N values of temperature fractions were weighted by the amounts of N released.

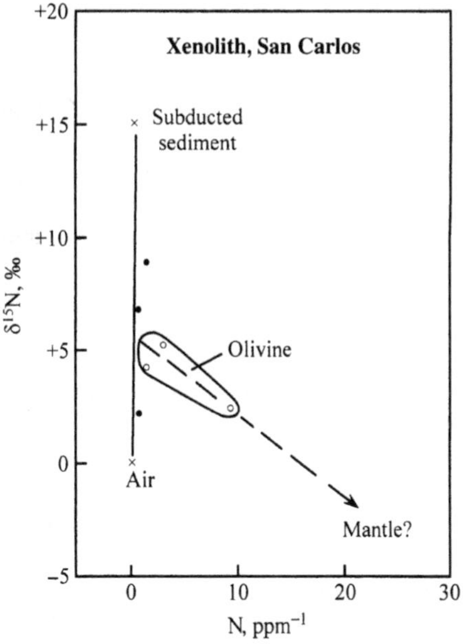

FIGURE 28.5 isotopic mixing diagram of N released by pyroxene (solid circles) and olivine (open circles) in mantle-derived ultramafic xenoliths in the basanite lavas of Tertiary age at San Carlos, Arizona. The δ^{15}N value of recycled (i.e., subducted) sediment is from Haendel et al. (1986). The concentrations and isotopic compositions of N in pyroxene and olivine are from Table 28.6 after Mohapatra and Murty (2000).

who measured the isotope compositions of N in metamorphic rocks, whereas the N of the mantle component arises from the δ^{15}N value (−9.24‰) of gas released by olivine 4 at 1650°C (Mohapatra and Murty, 2000).

These components are identified in the isotopic mixing diagram in Figure 28.5. The distribution of data points indicates that the pyroxenes contain primarily mixtures of N derived from recycled (subducted) sediment and from the atmosphere, whereas the olivines contain additional amounts of N characteristic of the mantle. These conclusions need to be tested because the low concentrations of N in the minerals of ultramafic xenoliths make their isotopic compositions sensitive to contamination.

The interpretation of Mohapatra and Murty (2000) indicates that the N in subducted sediment has altered the isotope compositions of N in the magma sources of oceanic basalts. The importance of recycled N in the mantle has been discussed by several research groups including:

Rau et al., 1987, *Earth Planet. Sci. Lett.*, 82:269–279. Hall, 1989, *Geochem. J.*, 23: 19–23. Marty, 1995, *Nature*, 377:326–329. Marty et al., 1995, *Chem. Geol.*, 120:183–195. Marty and Humbert, 1997, *Earth Planet. Sci. Lett.*, 152:101–112. Sano et al., 1998, *Geophys. Res. Lett.*, 25:2289–2292.

28.7 DIAMONDS

The N concentrations of diamonds range widely, from about 5 to 2000 ppm (Boyd and Pillinger, 1994). Therefore, the isotope composition of N in diamonds is less vulnerable to contamination than the isotopic composition of N in igneous rocks that crystallized in contact with the atmosphere or in the presence of crustal rocks containing organic matter. Nevertheless, the δ^{15}N values of diamonds range from −12 to +12‰ (Javoy et al., 1984; Boyd and Pillinger, 1994), which indicates that N in the mantle is isotopically heterogeneous. The possible reasons for the isotopic heterogeneity of N in diamonds are:

1. The mantle contains two or more isotopic varieties of N.
2. The N is isotopically fractionated during the growth of diamonds.
3. The primordial N that was incorporated into the Earth was isotopically heterogeneous.

Analyses of the isotope composition of N in stony meteorites indicate that the N in the solar nebula had a very wide range of $\delta^{15}N$ values from −326 to +973‰ (Prombo and Clayton, 1985). The N in iron meteorites is also isotopically heterogeneous with $\delta^{15}N$ values between −90 and +150‰ (Franchi et al., 1993). Therefore, the N that was incorporated into the Earth at the time of its formation probably was isotopically heterogeneous. Consequently, the variability of $\delta^{15}N$ values of diamonds could be a result of the isotopic heterogeneity of the primordial N in the mantle.

In addition, the isotopes of N may have been fractionated during the growth of diamonds (Galimov, 1991). The possible fractionation of C and N during the growth of zoned diamonds was considered by Boyd et al. (1994), Boyd and Pillinger (1994), and Cartigny et al. (2003). The latter concluded that kinetic fractionation of N-isotopes does not operate during natural diamond formation. Therefore, the isotopic heterogeneity of N in diamonds is probably *not* attributable to isotope fractionation, although this process is not necessarily ruled out completely.

Large amounts of N have entered the mantle of the Earth as a result of subduction of marine sediment. The ranges of $\delta^{15}N$ values of recent marine sediment, metasedimentary rocks, and S-type granites are listed in Table 28.7. The data clearly demonstrate that sedimentary rocks and their metamorphic derivatives are enriched in ^{15}N relative to atmospheric N. When these kinds of rocks are subducted into the mantle, the N they contain can mix on a regional scale with the primordial N, which is itself isotopically heterogeneous. Consequently, the observed range of $\delta^{15}N$ values of diamonds between −12 and +12‰ is probably the result of mixing of isotopically heterogeneous components of primordial and "recycled" N, compounded to a lesser extent by the fractionation of N isotopes during the growth of diamonds.

Table 28.7. Range of $\delta^{15}N$ Values of Marine Sediment, Metasedimentary Rocks, and S-Type Granites

Material	$\delta^{15}N$, ‰	References[a]
Recent marine sediment	+2 to +10	1
Metasediment, crustal, (NH_4)	+2 to +17	2
Metasediment, subduction zone (NH_4)	0 to +6	3
S-type granite (sedimentary protoliths)	+5 to +11	4

[a] 1. Peters et al., 1978, *Limnol. Oceanogr.*, 23:598–604. 2. Haendel et al., 1986, *Geochim. Cosmochim. Acta*, 50:749–758. 3. Bebout and Fogel, 1992, *Geochim. Cosmochim. Acta*, 56:2839–2849. Boyd et al., 1993, *Geochim. Cosmochim Acta*, 57:1339–1347.

The occurrence and isotopic composition of N in diamonds has been reported by many research teams, including:

Kaiser and Bond, 1959, *Phys. Rev.*, 115:857–863. Collins, 1978, *J. Phys. C, Solid State Phys.*, 11:L417–422. Collins, 1980, *J. Phys., Solid State Phys.*, 13:2641–2650. Evans and Qi, 1982, *Proc. Roy. Soc. Lond., Ser. A*, 381:159–178. Javoy et al., 1984, *Earth Planet. Sci. Lett.*, 68:399–412. Javoy et al., 1986, *Chem. Geol.*, 57:41–62. Boyd et al., 1987, *Earth Planet. Sci. Lett.*, 86:341–353. Deines et al., 1987, *Geochim. Cosmochim. Acta*, 51: 1227–1243. Deines et al., 1991a, *Geochim. Cosmochim. Acta*, 55:515–525. Deines et al., 1991b, *Geochim. Cosmochim. Acta*, 55:2615–2625. Eldridge et al., 1991, *Nature*, 353:649–652. Boyd et al., 1992, *Earth Planet. Sci. Lett.*, 109:633–644. Boyd and Pillinger, 1994, *Chem. Geol.*, 116:43–59. Deines et al., 1997, *Geochim. Cosmochim. Acta*, 61:3993–4005. Harte et al., 1999, *Min. Mag.*, 63:829–856.

The current theory of the Earth indicates that both C and N are recycled into the mantle by the subduction of marine sediment containing biogenic organic matter (Section 27.6e). The isotopic compositions of these two elements are characterized by being depleted in ^{13}C and enriched in ^{15}N. If the recycled C and N mix with the primordial components of these elements in the mantle, the $\delta^{13}C$ and $\delta^{15}N$ values of diamonds and other mantle-derived rocks or minerals should be correlated.

This hypothesis is tested in Figure 28.6 by means of the $\delta^{13}C$ and $\delta^{15}N$ values of diamonds from kimberlite pipes in South Africa (Finsch, Premier, Jagersfontein), Tanzania (Williamson), USA (Arkansas), and Western Australia (Argyle) reported by Boyd and Pillinger (1994). The diamonds from these sources define three groups in Figure 28.6 based primarily on their $\delta^{13}C$ values:

1. high $\delta^{13}C$ (−3 to −6‰),
2. intermediate $\delta^{13}C$ (−9.5 to −15‰), and
3. low $\delta^{13}C$ (−16.5 to −19.5‰).

The $\delta^{15}N$ values of diamonds in each of these fields vary widely from −12.3 to +12.0‰ (one value of +16.3‰ in a diamond from Arkansas plots off-scale). The diamonds in groups 2 and 3 are enriched in ^{15}N and depleted in ^{13}C, which is consistent with the presence of recycled N and C. The $\delta^{15}N$ values of the diamonds in group 1 vary widely from −12.3 to +6‰ even though the $\delta^{13}C$ values of these diamonds (−3 to −6‰) are thought to characterize uncontaminated C in the mantle. Apparently, the N in all three groups of diamonds has been contaminated by mixing with recycled N that is enriched in ^{15}N compared to N_2 of the atmosphere. This interpretation is illustrated in Figure 28.6 by the placement of hypothetical mixing trajectories that appear to converge to the isotope compositions of recycled C and N:

$$\delta^{13}C = -20 \pm 1 \text{ (PDB)‰}$$

$$\delta^{15}N = +14 \pm 1 \text{ (air)‰}$$

The uncontaminated N in the mantle appears to be depleted in ^{15}N compared to atmospheric N_2 and may have a $\delta^{15}N$ values of −12‰ or lower, in general agreement with the conclusions of Mohapatra and Murty (2000).

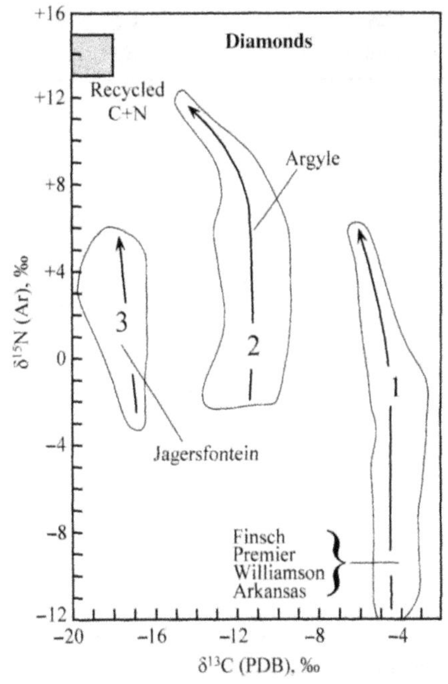

FIGURE 28.6 Isotope compositions of C and N of diamonds from South Africa (Finsch, Premier, Jagersfontein), Tanzania (Williamson), USA (Arkansas), and Western Australia (Argyle). The data points define three groups based on differences in the $\delta^{13}C$ values. The hypothetical mixing trajectories suggest the presence of varying proportions of recycled C and N characterized by $\delta^{13}C = -19 \pm 1$‰ and $\delta^{15}N = +14 \pm 1$‰. Data from Boyd and Pillinger (1994).

28.8 METEORITES

The data in Table 28.1 indicate that carbonaceous chondrites have high concentrations of N (1850 ppm), whereas all other chondrites, achondrites, and iron meteorites contain less than 100 ppm of N (Wlotzka, 1972). Evidently, chondrites and achondrites lost most of the N when they were heated at the time of formation of their parent bodies and during subsequent collisions in the asteroidal belt. The N concentration of iron meteorites

Table 28.8. Average $\delta^{15}N$ Values of N-Bearing Components of Carbonaceous Chondrites

Component	Abundance, %	$\delta^{15}N$, ‰
Whole rock	—	−50 to +355
Organic matter	2.0	+25 to +150[a]
Carbonates	0.2	—
Diamond	0.04	−350
Graphite	0.005	—
SiC	0.009	−500

Source: Based on a survey of the literature by Sephton et al. (2003).

[a] Undifferentiated.

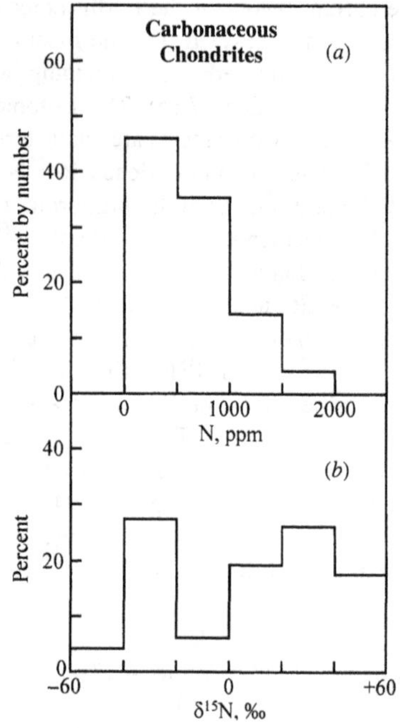

FIGURE 28.7 (*a*) Spectrum of N concentrations in bulk samples of carbonaceous chondrites. (*b*) Spectrum of $\delta^{15}N$ values (relative to N_2 of the terrestrial atmosphere) of bulk samples of carbonaceous achondrites. Data from Kerridge (1985).

may also be limited by the solubility of N_2 in liquid metal.

The N in stony meteorites resides primarily in organic matter, carbonate minerals, and grains of diamond, silicon carbide (SiC), and graphite. These materials formed by different processes in the ancestral stars (graphite, diamonds, and SiC grains), in the solar nebula (organic matter), and within the meteorite parent bodies (carbonate, modification of organic matter). The average $\delta^{15}N$ values of these N-bearing components of carbonaceous chondrites in Table 28.8 illustrate the full range of the heterogeneity of the isotopic compositions of this element (i.e., $\delta^{15}N = +335$ to -500‰).

The concentrations of N of bulk samples of carbonaceous chondrites reported by Kerridge (1985) also range widely from 5 to 1890 ppm based on 56 determinations of 35 specimens. The spectrum of N concentrations in Figure 28.7*a* is distinctly skewed in favor of concentrations less than 500 ppm. The $\delta^{15}N$ values of the same bulk samples of carbonaceous chondrites in Figure 28.7*b* lie between +60 and −60‰. However, the $\delta^{15}N$ values of three carbonaceous chondrites (Renazzo, Kaidun, and Al Rais) range from +140 to +190‰, whereas Bells (sample B) has $\delta^{15}N = +335$‰ (Kerridge, 1985). These results reveal that the carbonaceous chondrites are heterogeneous with respect to both the concentration and isotope composition of N they contain, which implies that the solar nebula contained N whose isotope composition may have varied even more widely.

Sephton et al. (2003) differentiated the organic matter of carbonaceous chondrites into three types based on sequential treatment of bulk samples:

1. FOM: free organic matter extractable with organic solvents.
2. LOM: labile organic matter, insoluble in organic solvents, resistant to HF/HCl treatment, released by hydrous pyrolysis.
3. ROM: refactory organic matter, unaffected by treatments described above.

The $\delta^{15}N$ (and $\delta^{13}C$) values of these types of organic matter in Murchison in Table 28.9 differ by 113‰ or more, which explains partly why the $\delta^{15}N$ values of bulk samples of carbonaceous chondrites analyzed by Kerridge (1985) range widely from +190 to −45‰.

Table 28.9. Isotope Composition of N (and C) in Three Types of Organic Matter of the Carbonaceous Chondrite Murchison

Organic Matter	$\delta^{15}N$, ‰	$\delta^{13}C$, ‰
FOM	~ +88	~ +5
LOM	< +85	< −5
ROM	< −25	< −20
Bulk[a]	+40.6	−2.7

Source: Sephton et al., 2003.

[a]Kerridge (1985) reported N = 630 ppm, $\delta^{15}N = +41$‰ based on analyses of 4 aliquots.

Table 28.10. Isotope Composition of N in N-Bearing Components of Ureilites[a]

Component	$\delta^{15}N$, ‰
Whole rock[b]	−35.7
Amorphous C	
Diamond free	−21
Polymict	+50 to +107
Monomict	+20 to +95
Diamond	< −100
Graphite	> +19
Acid-soluble fraction	−35 to +155

Source: Rai et al., 2003.

[a]Ureilites are coarse-grained achondrites of ultramafic composition containing olivine and pigeonite (pyroxene) in a carbonaceous matrix.

[b]Weighted average of 14 polymict and monomict ureilites analyzed by Rai et al. (2003).

The isotopic compositions of N in other kinds of stony meteorites are similarly heterogeneous. For example, Rai et al. (2003) reported that the $\delta^{15}N$ values of bulk samples of ureilites range from +51.8 to −102.1‰. The $\delta^{15}N$ values of various N-bearing components in ureilites in Table 28.10 are similar to those of the same components in carbonaceous chondrites. Therefore, the isotopic compositions of N in bulk samples of chondrites and achondrites vary between wide limits because they depend on the abundances of various N-bearing components and on the $\delta^{15}N$ values of these components. For the same reasons, the isotopic composition of primordial N that was transferred from the solar nebula to the Earth was not homogeneous because the isotopic composition of N in the planetesimals was not homogeneous.

The evident heterogeneity of the isotopic composition of N in the solar nebula originated from the nucleosynthesis of ^{14}N and ^{15}N in stars and was subsequently compounded by mass-dependent isotope fractionation of the stable isotopes of N during the condensation of solid particles, as well as by nuclear reactions that altered the abundances of the N isotopes in the gas and dust of the solar nebula.

Recent papers concerning the isotopic composition of N (and C) in the different kinds of meteorites and the possible causes for the observed heterogeneity have been published by:

Robert and Epstein, 1982, *Geochim. Cosmochim. Acta*, 46:81–95. Grady et al., 1985, *Geochim. Cosmochim. Acta*, 49:903–915. Grady et al., 1986, *Geochim. Cosmochim. Acta*, 50:2799–2813. Grady and Pillinger, 1988, *Nature*, 331:321–323. Prombo and Clayton, 1993, *Geochim. Cosmochim. Acta*, 57:3749–3761. Russell et al., 1996, *Meteoritics Planet. Sci.*, 31:343–355. Alexander et al., 1998, *Meteroritics Planet. Sci.*, 33:603–622. Mathew et al., 2000, *Geochim. Cosmochim. Acta*, 64:545–557. Sugiura et al., 2000, *Meteoritics Planet. Sci.*, 35:749–756. Grady et al., 2002, *Meteoritics Planet. Sci.*, 37:713–735. Rai et al., 2002, *Meteoritics Planet. Sci.*, 37:1045–1055. Varela et al., 2003, *Geochim. Cosmochim. Acta*, 67:1247–1257.

28.9 MOON

The abundance of N in the rocks of the Moon is very low, consistent with the general depletion of the Moon in volatile elements (McKeegan and Leshin, 2001). The isotopic composition of N in lunar soil was determined by stepwise heating of lunar breccia and soil (Becker and Clayton, 1977; Thiemens and Clayton, 1980). The results reveal that the $^{15}N/^{14}N$ ratios of lunar surface rocks have been altered by implantation of N isotopes by the solar wind and by the production of

^{15}N by cosmic-ray spallation reactions (Geiss and Bochsler, 1982). As a result, mineral grains at the surface of the Moon are enriched in N (Moore et al., 1970; Müller, 1973, 1974).

The isotopic composition of N in the solar wind is depleted in ^{15}N relative to N_2 of the terrestrial atmosphere and has a $\delta^{15}N$ value of less than −240‰ (Hashizume et al., 2000). In addition, the results of Hashizume et al. (2000) demonstrated that the isotope ratio of N in the solar wind has not changed during the past one to two billion years. Previous measurements of the $^{15}N/^{14}N$ ratio of the solar wind by means of the mass spectrometer on the spacecraft SOHO had large errors (Kallenbach et al., 1998).

The $\delta^{15}N$ values of gas released by stepwise heating of a lunar rock (70011), collected by the astronauts of Apollo 17, range from about −23 to +76‰. The profile of $\delta^{15}N$ values in Figure 28.8 indicates the presence of several components of N that are released at different temperatures between 600 and 1325°C (Becker and Clayton, 1977).

The N released at 600°C is probably a terrestrial contaminant, but the source of the gas released at higher temperatures is not certain (Kerridge, 1995). In general, the addition of N by the solar wind reduces the $\delta^{15}N$ value, whereas the presence of spallogenic ^{15}N increases $\delta^{15}N$ of the resulting isotopic mixture. The $\delta^{15}N$ value of the bulk sample of lunar rock 70011 is +27‰ and the concentration of N is 70 ppm.

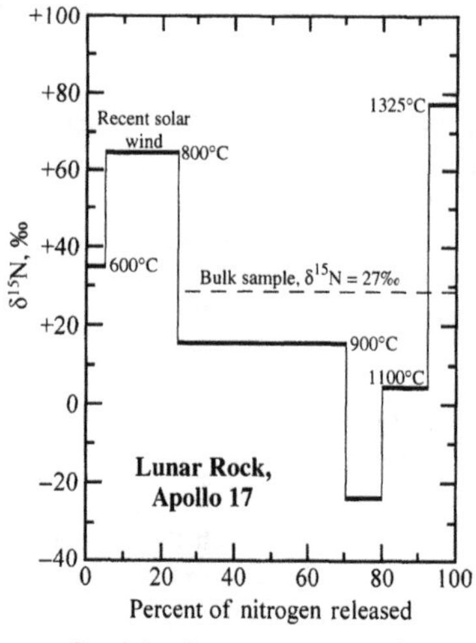

FIGURE 28.8 Partial-release spectrum of nitrogen from lunar rock 70011 (Apollo 17). The sample contains several components of N having widely differing isotopic compositions compared with N_2 of terrestrial air. Data from Becker and Clayton (1977).

28.10 MARS

The atmosphere of Mars ($P \sim 10$ millibars) is only about 1% as dense as the atmosphere of Earth. Nevertheless, it contains N_2 at a concentration of 2.7% compared to 78.08% N_2 in the terrestrial atmosphere (Owen et al., 1977). The concentration of N_2 in the atmosphere of Mars rises to about 60% when the concentration of CO_2 is reduced to near zero, which happened on Earth when CO_2 of the terrestrial atmosphere dissolved in the water of the oceans and was subsequently sequestered in deposits of carbonates of Fe, Mg, and Ca and by burial of organic matter in the oceans.

The isotopic composition of N in the atmosphere of Mars was measured with a mass spectrometer on one of the Viking spacecraft that landed on its surface. Nier et al. (1976) reported a $^{15}N/^{14}N$ ratio of 0.0064 ± 0.001 that corresponds to a $\delta^{15}N$ value of $+770 \pm 300$‰ relative to terrestrial N_2 (Biemann et al., 1976). The ^{15}N enrichment was caused by the preferential escape of ^{14}N into interplanetary space (McElroy et al., 1976; Fox and Hac, 1997). The loss of ^{14}N implies that the amount of N in the atmosphere of Mars has decreased and that the primordial atmosphere contained more N_2 than the present atmosphere of Mars. Therefore, the concentration of N_2 in the primordial atmosphere of Mars was even more similar to that of Earth than the concentration of N_2 in the present atmosphere of Mars.

The rocks of Mars also contain N whose concentration and isotopic composition have been determined by analyses of martian meteorites:

Becker and Pepin, 1984, *Earth Planet. Sci. Lett.*, 69:225–242. Becker and Pepin, 1986, *Geochim. Cosmochim. Acta*, 50:993–1000. Murty and Mohapatra, 1997, *Geochim. Cosmochim. Acta*, 61:5417–5428. Mathew et al., 1998, *Meteoritics Planet. Sci.*, 33:

655–664. Grady et al., 1998, *Meteoritics Planet. Sci.*, 33:795–802. Miura and Sugiura, 2000, *Geochim. Cosmochim. Acta*, 64: 559–572. Mohapatra and Murty, 2003, *Meteoritics Planet. Sci.*, 38:225–241.

For example, Miura and Sugiura (2000) released N and noble gases by progressive stepwise heating of two crushed whole-rock samples of the martian meteorite ALH 84001,57. The total amounts of N released (200–1200°C) by the two samples were 0.266 and 0.339 μg/g, which is similar to the N concentrations of olivine and pyroxene in terrestrial ultramafic xenoliths analyzed by Mohapatra and Murty (2000) described in Section 28.6 and Figure 28.5.

The profile of $\delta^{15}N$ values of gas fractions released by sample 1 of ALH 84001,57 in Figure 28.9 indicates that the gas released at $T < 600°C$ is depleted in ^{15}N compared to the gas released between 600 and 1200°C. The data of Miura and Sugiura (2000) demonstrate that the $\delta^{15}N$ values decrease from $+3.5 \pm 2.2‰$ (200°C)

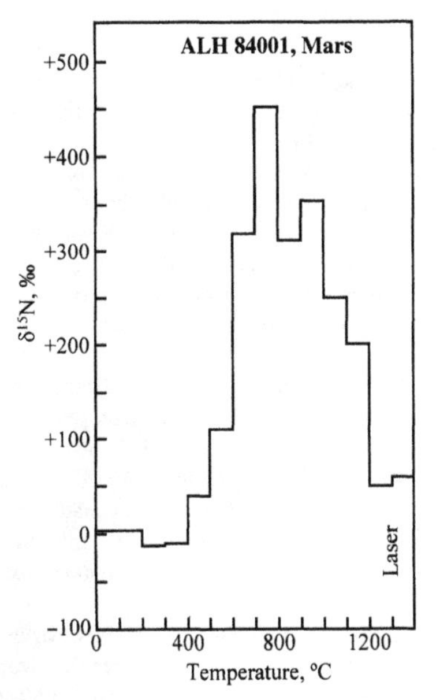

FIGURE 28.9 Isotope composition of N in gas released by step heating of a whole-rock sample of the martian meteorite ALH 84001,57 (sample 1) analyzed by Miura and Sugiura (2000).

Table 28.11. Summary of N Concentrations and $\delta^{15}N$ Values of Temperature Fractions of Samples 1 and 2 of the Martian Meteorite ALH 84001,57

Fraction, °C	N, μg/g	Percent of Total	$\delta^{15}N$, ‰
	Sample 1		
<600	0.1693	64	+2.84
600–1200	0.0968	36	+275
	Sample 2		
<600	0.2036	60	+27.6
600–1200	0.1352	40	+255

Source: Miura and Sugiura, 2000.

Note: The N released by laser fusion was omitted because the samples were exposed to the atmosphere prior to the laser irradiation.

to $-13.0 \pm 2.5‰$ (300°C) and to $-11.9 \pm 3.4‰$ (400°C) before rising again to $+39.1 \pm 3.3‰$ at 500°C. The summary of numerical data in Table 28.11 reveals that 64% of the N present in this sample was released at <600°C and the average weighted $\delta^{15}N$ value of this gas is +2.84‰. The $\delta^{15}N$ of the high-temperature gas (600–1200°C) rises to $+452 \pm 28‰$ at 800°C and then declines to $+200 \pm 38‰$ at 1200°C. The average weighted $\delta^{15}N$ value of the high-temperature gas is +275‰.

The low-temperature gas released from sample 2 in Table 28.11 is enriched in ^{15}N relative to sample 1. However, the $\delta^{15}N$ profile of sample 2 also has a valley between 200 and 600°C (not shown). The average weighted $\delta^{15}N$ value of the high-temperature gas (600–1200°C) of sample 2 of ALH 84001,57 is +255‰, which is similar to the weighted $\delta^{15}N$ value of the high-temperature gas of sample 1 (+275‰).

The profile of $\delta^{15}N$ values of the temperature fractions of N released by ALH 84001 implies the presence of several components having different isotopic compositions. The potential sources of the N components are:

1. terrestrial N residing in weathering products (e.g., carbonates, nitrates, oxides) and N_2 sorbed to mineral surfaces,

2. fluid or gas inclusions in silicate minerals containing N_2 from the mantle of Mars,
3. martian weathering products and N_2 of the martian atmosphere,
4. N implanted in minerals by the solar wind while the martian rocks were orbiting the Sun,
5. cosmogenic ^{15}N produced by spallation reactions when the martian rocks were exposed to cosmic rays in space, and
6. primordial N that was transferred from asteroidal meteorites to surface rocks on Mars during impacts.

In general, the N released from martian rocks at low temperature (<600°C) is likely to be a mixture of N_2 adsorbed from the terrestrial atmosphere and released by secondary weathering products of terrestrial and/or martian origin. The high-temperature gas (>600°C) presumably originates from fluid inclusions and from intracrystalline sites where N atoms were implanted by the solar wind or were produced by spallation reactions of cosmic rays.

28.11 SUMMARY

Although N is an abundant element in many rocks and fossil fuels, its isotope composition has not yet been fully explored. Nitrogen isotopes are fractionated primarily during biochemical reactions where mass differences of isotopic molecules and bond strengths affect reaction rates. Fractionation may also take place by isotope exchange reactions.

Much work has been done to document isotope fractionation of N in soils where biochemical and geochemical processes interact in complex ways. The isotope composition of N compounds dissolved in groundwater is strongly affected by isotope fractionation in soils combined with agricultural activities such as the use of N-fertilizer and the discharge of animal waste.

Particulate organic matter in the oceans is enriched in ^{15}N by oxidative degradation of N-compounds as the particles sink slowly thorough the water column. These particles may be useful for studies of mixing of freshwater with seawater in coastal areas and for studies of sedimentation on the oceans and lakes.

The abundance of ^{15}N increases up the food chain within communities of primary producers and a sequence of consumers. The enrichment in ^{15}N results both from the food sources and from metabolic effects.

The N_2 gas in subsurface has a wide range of isotopic compositions that do not identify its sources unequivocally. The isotope composition of N in coal is altered systematically by the loss of volatiles during thermal maturation. These changes in $\delta^{15}N$ suggest that N is associated with two or more compounds that differ in thermal stability and isotopic composition. However, the isotope composition of N in petroleum is restricted to a narrow range and is not appreciably altered by cracking of large N-bearing molecules.

The isotope composition of N released by stepwise heating of meteorites and of lunar and martian rocks has a wide range because of the presence of several N components, some of which are themselves isotopically heterogeneous.

REFERENCES FOR GEOCHEMISTRY, FRACTIONATION, ORGANIC MATTER (SECTIONS 28.1–28.3)

Anders, E., and N. Grevesse, 1989. Abundances of the elements: Meteoritic and solar. *Geochim. Cosmochim. Acta*, 53:197–214.

Baur, W. H., 1974. Crystal chemistry of nitrogen. In K. H. Wedepohl (Ed.), *Handbook of Geochemistry*, Vol. II-2, Section 7A:1–6. Springer-Verlag, Heidelberg.

Cline, J. D., and I. R. Kaplan, 1975. Isotopic fractionation of dissolved nitrate during denitrification in the eastern tropical North Pacific Ocean. *Marine Chem.*, 3:271.

DeNiro, M. J., and S. Epstein, 1981. Influence of diet on the distribution of nitrogen isotopes in animals. *Geochim. Cosmochim. Acta*, 45:341–351.

DeNiro, M. J., and C. A. Hastorf, 1985. Alteration of $^{15}N/^{14}N$ and $^{13}C/^{12}C$ ratios of plant matter during the initial stages of diagenesis: Studies utilizing archaeological specimens from Peru. *Geochim. Cosmochim. Acta*, 48:97–115.

Flipse, W. J., Jr., and F. T. Bonner, 1985. Nitrogen-isotope ratios of nitrate in groundwater under fertilized fields, Long Island, New York. *Ground Water*, 23:59–67.

Fogel, M. L., and L. A. Cifuentes, 1993. Isotope fractionation during primary production. In M. H. Engel and S. A. Macko (Eds.), *Organic Geochemistry*, pp. 73–98. Plenum, New York.

Graedel, T. E., and P. J. Crutzen, 1993. *Atmospheric Change: An Earth System Perspective*. W. H. Freeman, New York.

Heaton, T. H. E., 1986. Isotopic studies of nitrogen pollution in the hydrosphere and atmosphere: A review. *Chem. Geol., Isotope Sci. Sect.*, 59:87–102.

Hoefs, J., 1997. *Stable Isotope Geochemistry*, 4th ed. Springer-Verlag, Heidelberg.

Junk, G., and H. Svec, 1958. The absolute abundance of the nitrogen isotopes in the atmosphere and in compressed gas from various sources. *Geochim. Cosmochim. Acta*, 14:234–243.

Kreitler, C. W., 1979. Nitrogen-isotope ratio studies of soils and groundwater nitrate from alluvial fan aquifers in Texas. *J. Hydrol.*, 42:147–170.

Letolle, R., 1980. Nitrogen-15 in the natural environments. In P. Fritz and J. Ch. Fontes (Eds.), *Handbook of Environmental Isotope Geochemistry*, Vol. 1A, pp. 407–434. Elsevier, Amsterdam.

Mariotti, A., 1984. Natural ^{15}N abundance measurements and atmospheric nitrogen standard calibration. *Nature*, 311:251–252.

Mariotti, A., J. C. Germon, P. Hubert, P. Kaiser, R. Letolle, and P. Tardieux, 1981. Experimental determination of nitrogen, kinetic isotope fractionation: Some principles, illustration for the denitrification processes. *Plant. Soil*, 62:413–430.

Mariotti, A., J. C. Germon, A. Leclerc, G. Catroux, and R. Letolle, 1982. Experimental determination of kinetic isotope fractionation of nitrogen isotopes during denitrification. In H. L. Schmidt, H. Förstel, and K. Heinzinger (Eds.), *Stable Isotopes*. Elsevier, New York.

Mariotti, A., C. Lancelot, and G. Billen, 1984. Natural isotopic composition of nitrogen as a tracer of origin for suspended organic matter in the Scheldt estuary. *Geochim. Cosmochim. Acta*, 48:549–555.

Mariotti, A., and R. Letolle, 1979. Application de l'etude isotopique de l'azote en hydrologie et en hydrogéologie-analyse des résultats obtenus sur un example précis: Le basin de Melarchez (Seine et Marne, France). *J. Hydrol.*, 33:157–172.

Minagawa, M., and E. Wada, 1984. Stepwise enrichment of ^{15}N along food chains: Further evidence and the relation between $\delta^{15}N$ and animal age. *Geochim. Cosmochim. Acta*, 48:1135–1140.

Navone, R., J. A. Harmon, and C. F. Voyles, 1963. Nitrogen content of ground water in southern California. *J. Am. Water Works Assoc.*, 55:615.

Nitzsche, H. M., and G. Stiehl, 1984. Untersuchungen zur Isotopenfraktionierung des Stickstoffes in den Systemen Ammonium/Ammoniak und Nitrid/Stickstoff. *ZFI Mitteil.*, 84:283–291.

Pang, P. C., and J. O. Nriagu, 1976. Distribution and isotope composition of nitrogen in Bay of Quinte (Lake Ontario) sediments. *Chem. Geol.*, 18:93–105.

Pang, P. C., and J. O. Nriagu, 1977. Isotopic variations of the nitrogen in Lake Superior. *Geochim. Cosmochim. Acta*, 41:811–814.

Saino, T., and A. Hattori, 1980. ^{15}N natural abundance in oceanic suspended particulate matter. *Nature*, 283:752–754.

Shearer, G., D. H. Kohl, and S. H. Chien, 1978. The nitrogen-15 abundance in a wide variety of soils. *Soil. Sci. Soc. Am. J.*, 42:899.

Sweeney, R. E., K. K. Liu, and I. R. Kaplan, 1978. Oceanic nitrogen isotopes and their uses in determining the source of sedimentary nitrogen. In B. W. Robinson (Ed.), *Stable Isotopes in Earth Science*. Dept. Sci. Ind. Res. New Zealand Bull., 220:9. Lower Hutt, New Zealand.

Taylor, K. S., and G. Faure, 1983. Geochemical study of an equine enterolith, Medina County, Ohio. *Ohio J. Sci.*, 82:54–59.

Ueda, S., N. Ogura, and E. Wada, 1991. Nitrogen stable isotope ratio of groundwater N_2O. *Geophys. Res. Lett.*, 18(8):1449–1452.

Wada, E., R. Imaizumiand, and Y. Takai, 1984. Natural abundance of ^{15}N in soil organic matter with special reference to paddy soils in Japan: Biogeochemical implications on the nitrogen cycle. *Geochem. J.*, 18:109–123.

Wada, E., T. Kadonaga, and S. Matsuo, 1975. ^{15}N abundance in nitrogen of naturally occurring substances and global assessment of denitrification from isotopic viewpoint. *Geochim. J.*, 9:139–148.

Wlotzka, F., 1972. Geochemistry of nitrogen. In K. H. Wedepohl (Ed.), *Handbook of Geochemistry*, Vol. II-1, Sections 7-B–7-O. Springer-Verlag, Heidelberg.

REFERENCES FOR FOSSIL FUELS (SECTION 28.4)

Drechsler, M., and G. Stiehl, 1977. Stickstoff-isotopenvariatonen in organischen Sedimenten, I. Untersuchungen an humosen Kohlen. *Chemie Erde*, 26:126–138.

Hirner, A. V., W. Graf, R. Treibs, A. N. Melzer, and P. Hahn-Weinheimer, 1984. Stable sulfur and nitrogen isotopic compositions of crude oil fractions from southern Germany. *Geochim. Cosmochim. Acta*, 48:2179–2186.

Jenden, P. D., I. R. Kaplan, R. J. Poreda, and H. Craig, 1988. Origin of nitrogen rich natural gases in the California Great Valley: Evidence from helium, carbon, and nitrogen isotope ratios. *Geochim. Cosmochim. Acta*, 52:851–861.

Krooss, B. M., R. Littke, B. Müller, J. Frielingsdorf, K. Schwochau, and E. F. Idiz, 1995. Generation of nitrogen and methane from sedimentary organic matter: Implication on the dynamics of natural gas accumulations. *Chem. Geol.*, 126:291–318.

Littke, R., B. Krooss, E. F. Idiz, and J. Frielingsdorf, 1995. Molecular nitrogen in natural gas accumulations: Generation from sedimentary organic matter at high temperatures. *A.A.P.G. Bull.*, 79:410–430.

Macko, S. A., M. L. Foge (Estep), and P. E. Hare, 1987. Isotopic fractionation of nitrogen and carbon in the synthesis of amino acids by microorganisms. *Chem. Geol.*, 65:79–92.

Stiehl, G., and M. Lehmann, 1980. Isotopenvariationen des Stickstoffes humoser und bituminöser natürlicher organischer Substanzen. *Geochim. Cosmochim. Acta*, 44:2737–2746.

Wada, E., T. Kadonaga, and S. Matsuo, 1975. ^{15}N abundance in nitrogen of naturally occurring substances and global assessment of denitrification from isotopic viewpoint. *Geochim. J.*, 9:130–148.

Whelan, J. K., P. R. Solomen, G. V. Desphande, and R. M. Carangelo, 1988. Thermo-gravimetric Fourier transform infrared spectroscopy (TG-FTIR) of petroleum source rocks. *Energy Fuels*, 2:65–73.

Wlotzka, F., 1972. Geochemistry of nitrogen. In K. H. Wedepohl (Ed.), *Handbook of Geochemistry*, Sections 7-B–7-O. Springer-Verlag, Heidelberg.

Zhu, Y., B. Shi, and C. Fang, 2000. The isotopic compositions of molecular nitrogen: Implications on their origins in natural gas accumulations. *Chem. Geol.*, 164:321–330.

REFERENCES FOR IGNEOUS ROCKS AND THE MANTLE (SECTIONS 28.5–28.6)

Boyd, S. R., and C. T. Pillinger, 1994. A preliminary study of $^{15}N/^{14}N$ in octahedral growth form diamonds. *Chem. Geol.*, 116:43–59.

Faure, G., 2001. *The Origin of Igneous Rocks: the Isotopic Evidence*. Springer-Verlag, Heidelberg.

Haendel., D., K. Mühle, H. Nitzsche, G. Stiehl., and U. Wand, 1986. Isotopic variations of the fixed N in metamorphic rocks. *Geochim. Cosmochim. Acta*, 50:749–758.

Hoefs, J., 1997. *Stable Isotope Geochemistry*. Springer-Verlag, Heidelberg.

Javoy, M., 1995. The integral enstatite chondrite model of the Earth. *Geophys. Res. Lett.*, 22:2219–2222.

Javoy, M., 1997. The major volatile elements of the Earth: Their origin, behavior, and fate. *Geophys. Res. Lett.*, 24:177–180.

Javoy, M., and F. Pineau, 1991. The volatile record of a "popping" rock from the Mid-Atlantic Ridge at 14° N: Chemical and isotopic composition of gas trapped in the vesicles. *Earth Planet. Sci. Lett.*, 107:598–611.

Javoy, M., F. Pineau, and H. Delorme, 1986. Carbon and nitrogen isotopes in the mantle. *Chem. Geol.*, 57:41–62.

Mohapatra, R. K., and S. V. S. Murty, 2000. Search for the mantle nitrogen in the ultramafic xenoliths from San Carlos, Arizona. *Chem. Geol.*, 164:305–320.

Pepin, R. O., 1991. On the origin and early evolution of terrestrial planet atmospheres and meteoritic volatiles. *Icarus*, 92:2–79.

REFERENCES FOR DIAMONDS (SECTION 28.7)

Boyd, S. R., and C. T. Pillinger, 1994. A preliminary study of $^{15}N/^{14}N$ in octahedral growth form diamonds. *Chem. Geol.*, 116:43–59.

Boyd, S. R., F. Pineau, and M. Javoy, 1994. Modelling the growth of natural diamonds. *Chem. Geol.*, 116:29–42.

Cartigny, P., J. W. Harris, A. Taylor, R. Davies, and M. Javoy, 2003. On the possibility of a kinetic fractionation of nitrogen stable isotopes during natural diamond growth. *Geochim. Cosmochim. Acta*, 67:1571–1576.

Franchi, I. A., I. P. Wright, and C. T. Pillinger, 1991. Constraints on the formation conditions of iron meteorites based on concentrations and isotopic compositions of nitrogen. *Geochim. Cosmochim. Acta*, 57:3105–3121.

Galimov, E. M., 1991. Isotopic fractionation related to kimberlite magmatism and diamond formation. *Geochim. Cosmochim. Acta*, 55:1697–1708.

Haendel, D., K. Mühle, H. Nitzsche, G. Stiehl., and U. Wand, 1986. Isotopic variations of the fixed N in metamorphic rocks. *Geochim. Cosmochim. Acta*, 50:749–758.

Javoy, M., 1995. The integral enstatite chondrite model of the Earth. *Geophys. Res. Lett.*, 22:2219–2222.

Javoy, M., 1997. The major volatile elements of the Earth: Their origin, behavior, and fate. *Geophys. Res. Lett.*, 24:177–280.

Javoy, M., F. Pineau, and D. Demaiffe, 1984. Nitrogen and carbon isotopic compositions in the diamonds of the Mbuji Mayi (Zaire). *Earth Planet. Sci. Lett.*, 68:399–412.

Mohapatra, R. K., and S. V. S. Murty, 2000. Search for the mantle nitrogen in the ultramafic xenoliths from San Carlos, Arizona. *Chem. Geol.*, 164:305–320.

Pepin, R. O., 1991. On the origin and early evolution of terrestrial planet atmospheres and meteoritic volatiles. *Icarus*, 92:2–79.

Prombo, C. A., and R. N. Clayton, 1985. A striking nitrogen isotope anomaly in the Bencubbin and Weatherford meteorites. *Science*, 230:935–937.

REFERENCES FOR METEORITES (SECTION 28.8)

Franchi, I. A., I. P. Wright, and C. T. Pillinger, 1991. Constraints on the formation conditions of iron meteorites based on concentrations and isotopic compositions of nitrogen. *Geochim. Cosmochim. Acta*, 57:3105–3121.

Geiss, J., and P. Bochsler, 1982. Nitrogen isotopes in the solar system. *Geochim. Cosmochim. Acta*, 46:529–548.

Kerridge, J. F., 1985. Carbon, hydrogen, and nitrogen in carbonaceous chondrites: Abundances and isotopic compositions in bulk samples. *Geochim. Cosmochim. Acta*, 49:1707–1714.

Prombo, C. A., and R. N. Clayton, 1985. A striking nitrogen isotope anomaly in the Bencubbin and Weatherford meteorites. *Science*, 230:935–937.

Rai, V. K., S. V. S. Murty, and U. Ott, 2003. Nitrogen components in ureilites. *Geochim. Cosmochim. Acta*, 67:2213–2237.

Sephton, M. A., A. B. Verchovsky, P. A. Bland, I. Gilmour, M. M. Grady, and I. P. Wright, 2003. Investigating the variations in carbon and nitrogen isotopes in carbonaceous chondrites. *Geochim. Cosmochim. Acta*, 67:2093–2108.

Wlotzka, F., 1972. Geochemistry of nitrogen. In K. H. Wedepohl (Ed.), *Handbook of Geochemistry*, Section 7-B to 7-O. Springer-Verlag, Heidelberg.

REFERENCES FOR MOON (SECTION 28.9)

Becker, R. H., and R. N. Clayton, 1977. Nitrogen isotopes in lunar soils as a measure of cosmic-ray exposure and regolith history. In *Proc. 8th Lunar Sci. Conf., Geochim. Cosmochim. Acta*, 1:3685–3704, Pergamon, New York.

Geiss, J., and P. Bochsler, 1982. Nitrogen isotopes in the solar system. *Geochim. Cosmochim. Acta*, 46:529–548.

Hashizume, K., M. Chaussidon, B. Marty, and F. Robert, 2000. Solar wind record on the Moon: Deciphering presolar from planetary nitrogen. *Science*, 290:1142–1145.

Kallenbach, R., J. Geiss, F. M. Ipavich, G. Gloeckler, P. Bochsler, F. Gliem, S. Hefti, M. Hilchenbach, and D. Hovestadt, 1998. Isotopic composition of the solar wind nitrogen; first in-situ determination with the CELIAS/MTOF spectrometer on board SOHO. *Astrophys. J.*, 507:L185–188.

Kerridge, J. F., 1995. Nitrogen and its isotopes in the early solar system. In *Volatiles in the Earth and Solar System*, pp. 167–173. American Institute of Physics, New York.

McKeegan, K. D., and L. A. Leshin, 2001. Stable isotope variations in extraterrestrial materials. In J. W. Valley and D. R. Cole (eds.), *Stable Isotope Geochemistry*. Rev. Mineral. Geochem., 43:279–318. Mineralogical Society of America, Blacksburg, Virginia.

Moore, C. B., E. K. Gibson, J. W. Larimer, C. F. Lewis, and W. Nichiporuk, 1970. Total carbon and nitrogen abundances in Apollo 11 lunar samples and selected achondrites and basalts. In *Proc. Apollo 11 Lunar Sci. Conf., Geochim. Cosmochim. Acta, supplement* 2, 1375–1382. Pergamon.

Müller, O., 1973. Chemically bound nitrogen contents of Apollo 16 and Apollo 15 lunar fines. In *Proc. 4th Lunar Sci. Conf.*, Houston, 2, pp. 1625–1634. Pergamon.

Müller, O., 1974. Solar wind nitrogen and indigenous nitrogen in Apollo 17 lunar samples. In *Proc. 5th Lunar Sci. Conf.*, Houston, 2, pp. 1907–1918.

Thiemens, M. H., and R. N. Clayton, 1980. Ancient solar wind in lunar microbreccias. *Earth Planet. Sci. Lett.*, 47:34–42.

REFERENCES FOR MARS (SECTION 28.10)

Biemann, K., T. Owen, D. R. Rushneck, A. L. Lafleur, and D. W. Howarth, 1976. The atmosphere of Mars near the surface: Isotope ratios and upper limits on noble gases. *Science*, 194:76–78.

Fox, J. L., and A. Hac, 1997. The $^{15}N/^{14}N$ isotope fractionation in dissociative recombination of N_2^+. *J. Geophys. Res.*, 102:9191–9204.

McElroy, M. D., Y. L. Yung, and A. O. Nier, 1976. Isotopic composition of nitrogen: Implications for the past history of Mars' atmosphere. *Science*, 194:70–72.

Miura, Y. N. and N. Sugiura, 2000. Martian atmosphere-like nitrogen in the orthopyroxenite ALH 84001. *Geochim. Cosmochim. Acta*, 64:559–572.

Mohapatra, R. K. and S. V. S. Murty, 2003. Precursors on Mars: constraints from nitrogen and oxygen isotopic compositions of martian meteorites. *Meteor. Planet. Sci.*, 38:225–241.

Nier, A. O., M. B. McElroy, and Y. L. Yung, 1976. Isotopic composition of the Martian atmosphere. *Science*, 194:68–70.

Owen, T., K. Biemann, D. R. Rushneck, J. E. Biller, D. W. Howarth, and A. L. Lafleur, 1977. The composition of the atmosphere at the surface of Mars. *J. Geophys. Res.*, 821:4635–4639.

CHAPTER 29

Sulfur

SULFUR is widely distributed in the lithosphere, biosphere, hydrosphere, and atmosphere of the Earth. It occurs in the oxidized form as sulfate in the oceans and in evaporite rocks. It is found in the native state in the cap rock of salt domes and in the rocks of certain volcanic regions. Sulfur also occurs in the reduced form as sulfide in metallic mineral deposits associated with igneous, sedimentary, and metamorphic rocks. For this reason, the isotopic composition of S is especially useful in the study of sulfide ore deposits. The isotope geochemistry of sulfur has been reviewed by Thode (1970), Nielsen (1978, 1979), Ohmoto and Rye (1979), Krouse (1980), Hoefs (1997), and Canfield (2001).

29.1 ISOTOPE GEOCHEMISTRY

Sulfur ($Z = 16$) has four stable isotopes whose approximate abundances and masses are listed in Table 29.1. The mass of ^{34}S is 6.2% greater than that of ^{32}S, which is sufficient to cause extensive mass-dependent isotope fractionation. The atomic $^{34}S/^{32}S$ ratio is 0.0443 based on the isotopic abundances stated in Table 29.1.

The isotopic composition of S in nature is expressed in terms of $\delta^{34}S$, which is defined as:

$$\delta^{34}S = \left[\frac{(^{34}S/^{32}S)_{spl} - (^{34}S/^{32}S)_{std}}{(^{34}S/^{32}S)_{std}}\right] \times 10^3‰ \quad (29.1)$$

The standard is the S in troilite (FeS) of the iron meteorite Canyon Diablo, whose $^{34}S/^{32}S$ ratio is 0.0450. This is an appropriate standard because the isotopic composition of S in mafic igneous rocks is very similar to that of meteorites (Smitheringale and Jensen, 1963; Schneider, 1970). Consequently, the $\delta^{34}S$ value of a given sample of terrestrial S can be interpreted as a measure of the change that has taken place in its isotopic composition since its initial introduction into the crust of the Earth.

The isotopic composition of S is commonly measured on SO_2 gas using mass spectrometers equipped with double collectors. Sulfides are converted to SO_2 by reactions with CuO, V_2O_5, or O_2 at temperatures of up to 1000°C. Sulfates are first reduced to sulfide which is precipitated as CdS and then converted to SO_2 gas. Details of these procedures have been discussed by:

Ricke, 1964, *Z. Anal. Chem.*, 199:401–413.
Holt and Engelkemeir, 1970, *Anal. Chem.*,

Table 29.1. Abundances and Masses of the Stable Isotopes of S

Isotope	Abundance, %	Mass, amu
^{32}S	95.02	31.972070
^{33}S	0.75	32.971456
^{34}S	4.21	33.967866
^{36}S	0.02	35.967080

Source: Lide and Frederikse, 1995.

42:1451–1453. Puchelt et al., 1971, *Geochim. Cosmochim. Acta*, 35:625–628. Rees, 1978, *Geochim. Cosmochim. Acta*, 42:383–389. Hoefs, 1997, *Stable Isotope Geochemistry*, 4th ed., Springer-Verlag, Heidelberg.

The conventional methods of isotope analysis of S are being replaced by microanalytical techniques using laser microprobe and ion-microprobe mass spectrometers. These instruments can make multiple spot analyses of the same crystal. The fractionation of S isotopes that occurs during these procedures is mineral-specific and reproducible and can be corrected. These microanalytical techniques have detected unexpected fine-scale heterogeneity in the isotopic compositions of S in massive sulfide deposits, in microscopic sulfide inclusions of volcanic rocks, and in diamonds:

Laser Microprobe: Kelley and Fallick, 1990, *Geochim. Cosmochim. Acta*, 54:883–888. Crowe et al., 1990, *Geochim. Cosmochim. Acta*, 54:2075–2092.

Ion Microprobe: Chaussidon et al., 1987, *Nature*, 330:242–244. Chaussidon et al., 1989, *Earth Planet. Sci. Lett.*, 92:144–156. Eldridge et al., 1988, *Econ. Geol.*, 83:443–449. Eldridge et al., 1993, *Econ. Geol.*, 88:1–26.

Variations in the isotopic composition of S are caused by two kinds of processes:

1. reduction of sulfate ions to hydrogen sulfide by certain anaerobic bacteria, which results in the enrichment of hydrogen sulfide in ^{32}S, and
2. various isotopic exchange reactions between sulfur-bearing ions, molecules, and solids by which ^{34}S is generally concentrated in compounds having the highest oxidation state of S, or greatest bond strength (Bachinski, 1969).

29.2 BIOGENIC ISOTOPE FRACTIONATION

The most important process causing variations in the isotopic composition of S in nature is the reduction of sulfate ions by anaerobic bacteria such as *Desulfovibrio desulfuricans*, which live in sediment deposited in the oceans and in lakes (Canfield, 2001). These bacteria split O from sulfate ions and excrete H_2S, which is enriched in ^{32}S relative to the remaining sulfate. The extent of fractionation is variable and depends on certain rate-controlling steps in the reactions by which S is metabolized. In *inorganic* systems, the extent of isotope fractionation during the reduction of sulfate ion to hydrogen sulfide is due to the different rates at which S–O bonds are broken. As a result, the first H_2S produced by inorganic reduction of sulfate ions is enriched in ^{32}S by about 22‰ compared to the sulfate. Harrison and Thode (1958) therefore explained the fractionation associated with bacterial reduction of sulfate in terms of a two-stage process consisting of:

1. entrance of sulfate into the cell yielding a small isotope shift and
2. breaking of S–O bonds causing a large change (up to 22‰) in the isotope composition and controlling the rate of the process.

This model appeared to be capable of explaining fractionation effects of up to 27‰ obtained in the laboratory with bacterial cultures.

However, Kaplan and Rittenberg (1964) obtained ^{32}S enrichments in H_2S of up to 46‰ relative to sulfate. Moreover, sulfide minerals in recently deposited sediment may be enriched in ^{32}S by about 50‰ compared to associated marine sulfate. Evidently, isotope fractionation in nature is more extensive than suggested by most laboratory experiments and is not limited by the fractionation associated with chemical reduction of sulfate ions to hydrogen sulfide. In this connection it is interesting to note that Tudge and Thode (1950) calculated a fractionation factor $\alpha = 1.075$ at 25°C for the chemical equilibrium:

$$^{32}SO_{4(aq)}{}^{2-} + H_2{}^{34}S_{(g)} \rightleftharpoons {}^{34}SO_{4(aq)}{}^{2-} + H_2{}^{32}S_{(g)} \qquad (29.2)$$

This equilibrium is unattainable by chemical systems at low temperature but may be approached by enzyme-catalyzed sulfate reduction in bacteria (Trudinger and Chambers, 1973). This exchange equilibrium is capable of causing enrichment of H_2S in ^{32}S of up to 75‰ at 25°C relative to the sulfate. The degree of isotope fractionation by bacteria is known to be inversely proportional to the rate of sulfate reduction, which in turn is controlled by either temperature or sulfate concentration or both.

In general, the degree of isotope fractionation is variable because the total observed effect consists of several additive steps (McCready et al., 1974). Breaking of S–O bonds plays an important role as does the exchange reaction between the sulfide and bisulfide ions. The total observed effect depends on the extent to which the individual steps are controlled by the physiology of the cell, which in turn depends on the substrate. Rees (1973) developed a mathematical model based on the recognition that the overall effect results from the transfer of S among several reservoirs within the cell and that the extent of fractionation is controlled by the rates of reactions by which these reservoirs interact.

The isotopic composition of H_2S liberated by bacteria also depends on the magnitude of the sulfate reservoir. If it is essentially infinite, the $\delta^{34}S$ value of H_2S produced by reduction of sulfate remains constant, assuming no change in the bacterial metabolism. On the other hand, if the size of the sulfate reservoir is limited, the isotopic composition of the remaining sulfate changes due to the preferential removal of ^{32}S as H_2S. As a consequence, the $\delta^{34}S$ values of the sulfide and of the remaining sulfate both change as the fraction of sulfate remaining decreases even if the effective isotope fractionation factor remains constant.

The effect of biogenic fractionation of S by bacteria in a closed system can be modeled by means of a Rayleigh equation:

$$R = R_0 f^{1/\alpha - 1} \tag{29.3}$$

where R = $^{34}S/^{32}S$ ratio of the sulfate at any value of f

R_0 = $^{34}S/^{32}S$ ratio of the sulfate before the onset of fractionation

f = fraction of sulfate remaining

$\alpha = R^{SO_4}/R^{H_2S}$

The isotope ratios R and R_0 are related to the respective $\delta^{34}S$ values by equation 26.12:

$$\left(\frac{R}{R_0}\right)^{SO_4} = \frac{\delta^{34}S_{SO_4} + 1000}{\delta^{34}S^0_{SO_4} + 1000} \tag{29.4}$$

Substituting into equation 29.3 yields:

$$\frac{\delta^{34}S_{SO_4} + 1000}{\delta^{34}S^0_{SO_4} + 1000} = f^{1/\alpha - 1}$$

which can be used to calculate the $\delta^{34}S$ value of the remaining sulfate ions for different values of f and for specified values of α and $\delta^{34}S^0_{SO_4}$. If $\alpha^{SO_4}_{H_2S} =$ 1.025 and $\delta^{34}S^0_{SO_4} = +20‰$ (i.e., modern marine sulfate), the evolution of the isotopic composition of S in sulfate can be calculated from:

$$\delta^{34}S_{SO_4} = f^{1/\alpha - 1}(\delta^{34}S^0_{SO_4} + 1000) - 1000 \tag{29.5}$$

The $\delta^{34}S$ values of the H_2S in isotopic equilibrium with the remaining sulfate for any given value of f is obtained from equation 26.12:

$$\alpha^{SO_4}_{H_2S} = \frac{\delta^{34}S_{SO_4} + 1000}{\delta^{34}S_{H_2S} + 1000}$$

Solving for $\delta^{34}S_{H_2S}$:

$$\delta^{34}S_{H_2S} = \frac{\delta^{34}S_{SO_4} + 1000}{\alpha^{SO_4}_{H_2S}} - 1000 \tag{29.6}$$

The resulting graph in Figure 29.1 indicates that the $\delta^{34}S$ values of the sulfate and of the hydrogen sulfide both increase as the fraction of the remaining sulfate decreases. The progressive increase of the $\delta^{34}S$ of the sulfate is caused by the removal of ^{32}S in the form of H_2S. As the process continues, the ^{34}S enrichment of the sulfate causes the $\delta^{34}S$ value of H_2S to rise as well.

The point of this exercise is that the changing isotope composition of S in hydrogen sulfide can be preserved by metal sulfides such as hydrotroilite (FeS), sphalerite (ZnS), and galena (PbS) or by sulfides of other metals (e.g., Cu, Co, Ni). Similarly, the remaining sulfate ions may be precipitated as anhydrite ($CaSO_4$), celestite

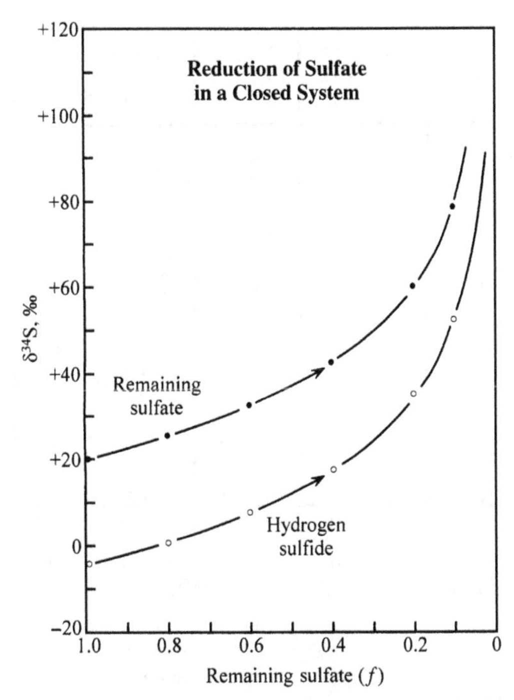

FIGURE 29.1 Isotope fractionation of S during the reduction of sulfate ions in solution to H_2S gas by bacteria (*Desulfovibrio desulfuricans*) in a closed system with a constant fractionation factor $\alpha^{SO_4}_{H_2S} = 1.025$ and assuming that the initial $\delta^{34}S$ value of sulfate is +20‰, corresponding to sulfate in the oceans today.

($SrSO_4$), or barite ($BaSO_4$). In the ideal case considered here, the sulfide and sulfate minerals that formed from an isotopically homogeneous sulfate reservoir of limited size contain S that may have a wide range of isotopic compositions.

29.3 SULFUR IN RECENT SEDIMENT

Laboratory experiments with resting suspensions of *Desulfovibrio desulfuricans* generally indicate that metabolic H_2S produced by reduction of sulfate is enriched in ^{32}S by up to about 25‰ relative to the sulfate. These experimental results provide an explanation for differences in the isotopic composition of S observed in modern sedimentary basins. However, there are important differences between laboratory cultures and bacteria in the natural environment. Bacteria in nature are not resting, but grow and die. They may live in open or closed systems with respect to the availability of sulfate, the environmental temperatures may be lower than those of laboratory experiments, and the food supply may fluctuate. As a consequence, modern euxinic environments commonly contain H_2S enriched in ^{32}S by 50‰ or more compared to coexisting sulfate, even though such extreme fractionation has been difficult to achieve in the laboratory. For example, sulfate in solution in the Black Sea has a $\delta^{34}S$ value of +19.2‰ relative to troilite in Canyon Diablo, while dissolved H_2S has a value of about −31.9‰ (Vinogradov et al., 1962). Similar results were obtained by Hartmann and Nielsen (1969) for S in pore water of sediment collected in the Bay of Kiel in the Baltic Sea. Their data show a difference of about 50–60‰ between the $\delta^{34}S$ values of dissolved sulfate and sulfide. The difference is primarily due to the action of sulfur-reducing bacteria. However, the $\delta^{34}S$ value of dissolved sulfide is further modified by precipitation of pyrite which concentrates ^{34}S.

Hartmann and Nielsen (1969) found that the $\delta^{34}S$ values of both sulfate and sulfide dissolved in pore water increase as a function of depth in sediment cores; that is, both become enriched in ^{34}S. The strongest ^{34}S enrichment was observed in their core 2092 shown in Figure 29.2, in which $\delta^{34}S$ values for dissolved sulfate increase from +20.0‰ at the top of the core to +60.7‰ at a depth of about 30 cm. The $\delta^{34}S$ value of dissolved sulfide in this core changes from −24.6‰ at about 4 cm to −2.0‰ at 30 cm. At the same time the sulfate *concentration* in pore water of this core decreases with depth while that of dissolved sulfide increases. However, the *total amount* of sulfur in the pore water decreases sharply, which indicates that S is being precipitated, primarily as sulfides.

The principal carrier of S in the sediments studied by Hartmann and Nielsen (1969) is pyrite (FeS_2). In addition, minor amounts of S may occur as FeS (hydrotroilite), native S, organically bound S, and sulfate (Kaplan et al., 1963). The total S content of sediment in core 2092 (Figure 29.2) from the Baltic Sea increases from the top of the core to a depth of about 6 cm and then fluctuates

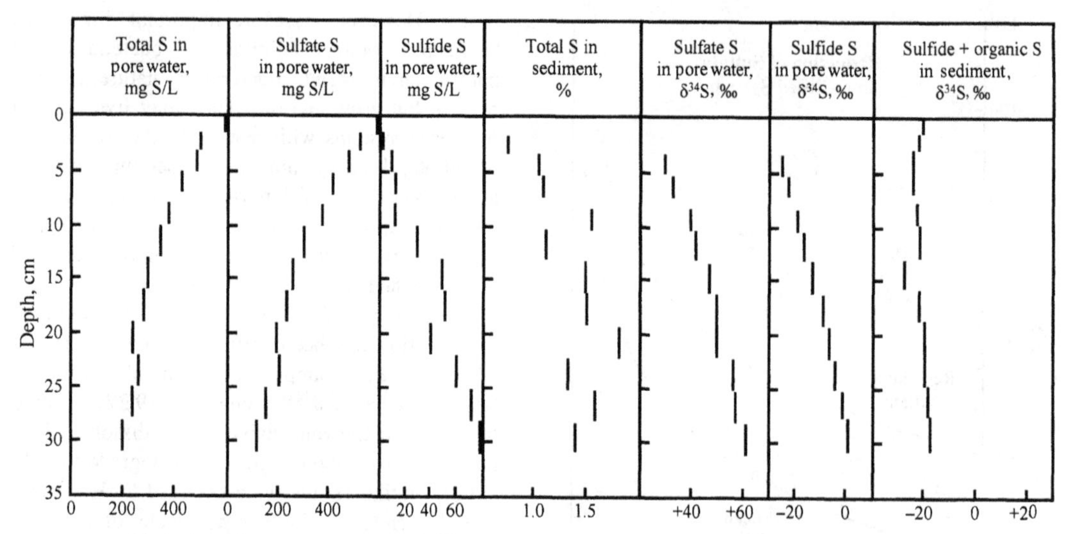

FIGURE 29.2 Variation of concentration and $\delta^{34}S$ values of pore water and recently deposited sediment in core 2092 from the Bay of Kiel in the Baltic Sea. Data from Hartmann and Nielsen (1969).

irregularly in the deeper portions. The S content of the sediment is much *greater* than that of sulfate-S in the pore water at the top of the core. This indicates that a large fraction of the S in the deeper sediment is derived from sulfate withdrawn from the overlying water column. Apparently the sediment becomes closed to sulfate only after burial under a layer of about 6 cm in thickness.

The $\delta^{34}S$ values of the water-insoluble S in core 2092 from the Baltic Sea *decrease* in the upper 6 cm of the core from −20.2 to −23.4‰ and then remain fairly constant in the rest of the core. These values include a small contribution (5–10%) of organically bound S that is enriched in ^{34}S relative to troilite of Canyon Diablo. Kaplan et al. (1963) reported average values of +15.3‰ for marine plants and +19.6‰ for marine animals. Zooplankton from the Bay of Kiel analyzed by Hartmann and Nielsen (1969) yielded a value of +18.1‰. Subtraction of the organically bound S from the water-insoluble S of the sediment therefore yields even more negative $\delta^{34}S$ values of the inorganic sulfide phases. The general conclusions are that sulfide phases in recently deposited sediment in the Baltic Sea are enriched in ^{32}S by 35–55‰ compared to marine sulfate and that bacterial fractionation of S in nature is greater than that achieved by most laboratory cultures.

Similar studies by Thode et al. (1960), Vinogradov et al. (1962), and Kaplan et al. (1963) all confirmed that sulfide minerals associated with recently deposited marine sediment are enriched in ^{32}S compared to marine sulfate and commonly have negative $\delta^{34}S$ values. The entire subject of the isotope geology of S in marine environments has been summarized by Kaplan (1983).

Sulfide minerals in lacustrine sediment are similarly enriched in ^{32}S relative to associated sulfates due to the action of anaerobic bacteria (Deevey et al., 1963; Nakai and Jensen, 1964).

29.4 FOSSIL FUELS

29.4a Petroleum

The *concentrations* of organically bound S in petroleum vary within wide limits from 0.1 to 10%. The $\delta^{34}S$ values of petroleum also span a wide range from −8 to +32‰. The isotopic composition of S of a particular sample of petroleum depends on the source of the S and on the isotope fractionation associated with its incorporation into the petroleum.

In accordance with the theory of the origin of petroleum, marine sulfate is the most likely source of S in petroleum. Moreover, the reduction of the S is undoubtedly due to the action of anaerobic bacteria that enrich the organically bound S in ^{32}S.

The isotopic composition of S in petroleum was summarized by Thode and Rees (1970) as follows:

1. Although the isotopic composition of S in petroleum varies by more than 40‰, oil from a given pool has the same $\delta^{34}S$ value.
2. Oil pools in the same reservoir rock within a sedimentary basin have similar $\delta^{34}S$ value.
3. Oil pools in reservoir rocks of different geological ages may have significantly different $\delta^{34}S$ values.
4. The isotopic composition of S in hydrogen sulfide gas is very similar to that of petroleum from which it was derived.

This observation suggests that S isotopes are not fractionated during the splitting of H_2S from petroleum in the course of maturation of petroleum.

The wide range of $\delta^{34}S$ values of petroleum may be caused partly by the fact that the isotopic composition of S in marine sulfate has varied systemically from Precambrian through Phanerozoic time. Thode and Monster (1965) compared $\delta^{34}S$ values of petroleum and contemporaneous anhydrite from marine evaporite deposits ranging in age from Silurian to Permian and found that S in petroleum is enriched in ^{32}S by about 15‰. They attributed this enrichment to the action of sulfur-reducing bacteria at the time of deposition of the sediment and during subsequent formation of petroleum. The work of Thode and his colleagues suggests therefore that the S in petroleum was derived from marine sulfate, was enriched in ^{32}S by about 15‰ by bacterial action prior to incorporation into the petroleum, and was not modified isotopically during subsequent maturation of the petroleum. Because $\delta^{34}S$ values of marine sulfate are known to have varied systematically in the geological past, it follows that the $\delta^{34}S$ value of a petroleum sample depends on the age of the source rocks in which it was formed. Consequently, the $\delta^{34}S$ value is useful in identifying the source rocks from which a particular petroleum sample was derived and thus to trace its migration into the reservoir rock in which it accumulated.

These principles were used by Thode and Rees (1970) to study the source and migration of oil in northern Iraq. In this area, oil pools occur in reservoir rocks of Early Cretaceous to Tertiary age and of Late Triassic age. Thode and Rees (1970) reported that oil in Cretaceous and Tertiary rocks has very similar $\delta^{34}S$ values ranging from -5.2 to -5.6‰. The oil in Upper Triassic rocks has a distinctly different average $\delta^{34}S$ value of $+2.35$‰. Therefore, they concluded that the oil in Cretaceous to Tertiary reservoir rocks originated from a single source located in rocks of Jurassic to Early Cretaceous age. After deposition of the Tertiary reservoir rocks, the oil migrated upward through thousands of feet of rocks and accumulated in pools in which it is now found. The source of the petroleum in Triassic reservoir rocks is not definitely known but is presumably to be sought in underlying older rocks.

29.4b Coal

The concentration of S in coal is variable and may range up to 20%. It is present in several chemical forms, including:

1. sulfides of iron such as pyrite and marcasite (sphalerite, galena, and chalcopyrite occur only rarely);
2. sulfates, including gypsum, barite, and sometimes sulfates of iron that result from the oxidation of pyrite or marcasite;
3. elemental S in trace concentrations; and
4. organically bound S whose mode of combination in the coal is not well understood.

The *isotopic composition* of S in coal varies widely from $\delta^{34}S = +32.3$ to -30‰. Smith and Batts (1974) carried out a detailed study of S in Australian coals ranging in age from Permian to Tertiary. They found that the $\delta^{34}S$ values of organic S in coals containing *less than 1% total S* vary within a narrow range from $+4.6$ to $+7.3$‰, whereas the $\delta^{34}S$ values of organic S in *S-rich* coals (>1% total S) vary more widely from $+2.9$ to $+24.4$‰. Smith and Batts (1974) suggested that the $\delta^{34}S$ values of organic S in low-S coal are representative of the isotopic composition of sulfate in the environment in which the coal was deposited. They therefore concluded from the uniformity of $\delta^{34}S$ values of organic S in low-sulfur coals that the sulfate in the coal basins of Australia was of nonmarine origin and remained fairly constant from the Permian to the Tertiary periods. This is in marked contrast to the $\delta^{34}S$ values of *marine* sulfate, which changed from about +10 to +20‰ during the same interval of time (Claypool et al., 1980).

The greater variability of $\delta^{34}S$ values of the various S compounds in high-sulfur coals probably results from postdepositional infiltration of marine or nonmarine sulfate into the coal accompanied by partial or complete reduction of this sulfate to sulfide by bacteria. Smith and Batts (1974) demonstrated systematic stratigraphic variations of $\delta^{34}S$ values in coal seams overlain by marine sediment and interpreted them in terms of models based on different rates of diffusion of sulfate and/or rates or reduction. Such detailed studies of isotope composition of suites of coal samples provide much more useful information than single determinations of $\delta^{34}S$ values of total S in isolated samples.

The variability of $\delta^{34}S$ values of high-S coals was subsequently confirmed by:

Smith et al., 1982, *Org. Geochem.*, 3:111–131. Price and Shieh, 1979, *Econ. Geol.*, 74:1445–1461. Hackley and Anderson, 1986, *Geochim. Cosmochim. Acta*, 50:1703–1713.

29.5 NATIVE SULFUR DEPOSITS

Large deposits of native S occur in the cap rocks of salt domes such as those of the Gulf Coast area of the United States and in certain evaporite sequences in the Mediterranean basin exemplified by the deposits in Sicily. The explanation for the origin of such native S deposits is based primarily on the isotopic compositions of S and C in these deposits, thus illustrating the value of such studies in the solution of certain geological problems.

The salt domes of the Gulf Coast region have been important producers of petroleum and native S. The petroleum occurs in traps along the flanks of salt domes while the native S is localized within the calcite and anhydrite rocks that cap the domes. The caps are formed by leaching of rock salt by groundwater and consist largely of anhydrite that was originally disseminated in the rock salt. The isotopic compositions of S of various S-bearing compounds associated with salt domes in the Gulf Coast area of the United States were determined by Feely and Kulp (1957). Table 29.2 contains a summary of their data for the Boling salt dome in Texas which will serve as a model for the explanation of the occurrence of native S associated with salt domes.

Unaltered samples of rock salt from the salt domes of the Gulf Coast contain from 5 to 10% of disseminated anhydrite. The $\delta^{34}S$ values of this anhydrite in the salt domes of Louisiana and Texas are very similar, which suggests that they originated from a common source bed believed to be the Louann Salt of Jurassic age. The $\delta^{34}S$ value of the disseminated anhydrite in the Boling dome (Table 29.2) is +16.2‰. The anhydrite of the cap rock is significantly enriched in ^{34}S and has variable $\delta^{34}S$ values averaging +17.9‰ (in the anhydrite cap rock) and +26.3‰ (in the calcite

Table 29.2. Isotopic Composition of S in the Boling Salt Dome, Texas

Material	$\delta^{34}S$,[a] ‰
Anhydrite	
Disseminated in halite	+16.2
Anhydrite cap rock	+17.9
Calcite cap rock	+26.3
Native sulfur	
Anhydrite cap rock	+3.0
Calcite cap rock	+2.6
Marcasite and pyrite in calcite rock	+5.8
Native S with above sulfides	+4.2
Sulfate, bleedwater	+25.2
Hydrogen sulfide, bleedwater	+0.8

Source: Feely and Kulp, 1957.

[a] Recalculated from $^{32}S/^{34}S$ ratios assuming a value of 22.22 for the $^{34}S/^{32}S$ ratio of Canyon Diablo.

rocks). In marked contrast, the native S is strongly enriched in ^{32}S compared to the anhydrite and has average $\delta^{34}S$ values of +3.0‰ (in the anhydrite rock) and +2.6‰ (in the calcite rocks). Feely and Kulp (1957) attributed this difference to isotope fractionation of S by bacteria which reduced sulfate ions to hydrogen sulfide gas and suggested that the bacteria lived on petroleum that saturated the cap rocks. The preferential incorporation of ^{32}S into hydrogen sulfide left the remaining sulfate variably enriched in ^{34}S, depending on the fraction of sulfate that was converted to sulfide in a given region of the cap rock and on the rate of metabolic activity of the bacteria. The hydrogen sulfide produced by the bacteria was subsequently oxidized by sulfate according to the reactions:

$$SO_4^{2-} + 3H_2S \rightarrow 4S + 2H_2O + 2OH^- \quad (29.7)$$

Feely and Kulp (1957) demonstrated experimentally that this reaction is capable of converting about 20 mg of sulfide per liter to native S in one day. A relatively small fraction of H_2S may have been precipitated locally as pyrite or marcasite having $\delta^{34}S = +5.8$‰. However, salt domes have very low concentrations of Fe, which accounts for the low abundance of these minerals in the cap rock.

The calcite in the upper parts of the cap rock formed by precipitation of CO_2 gas liberated by bacteria during metabolic oxidation of petroleum. The $\delta^{13}C$ value of petroleum from Boling dome is −26.2‰ (relative to "sedimentary limestone"), while that of calcite from the cap rock is −32.0‰. Therefore, the calcite is strongly depleted in ^{13}C compared to normal marine limestones and is similar in isotopic composition to the associated petroleum. In fact, the C in the calcite is even more depleted in ^{13}C than the petroleum. Feely and Kulp (1957) attributed ^{13}C-depletion of the C to fractionation during metabolic oxidation of the petroleum by the bacteria. The Ca required for the precipitation of calcite was undoubtedly derived from the anhydrite that had provided the sulfate ions for bacterial reduction to H_2S.

Thus the isotopic composition of C of caprock calcite is consistent with the hypothesis that the native S was produced by bacterial reduction of sulfate to hydrogen sulfide followed by oxidation to native S. Subsequent investigations of other native S deposits associated with marine sedimentary rocks have generally confirmed these conclusions (Dessau et al., 1962; Schneider and Nielsen, 1965; Davis and Kirkland, 1970).

29.6 SEDIMENTARY ROCKS OF PRECAMBRIAN AGE

The variation of $\delta^{34}S$ in S-bearing minerals of Precambrian age is of special importance because it can provide information about the course of biological evolution during the Archean Eon. The sulfate-reducing bacteria that are primarily responsible for isotopic fractionation of S did not exist in Early Archean time. Their evolution was delayed until the concentration of sulfate in the oceans had been increased sufficiently by the activity of green and purple photosynthetic S bacteria that oxidize H_2S to SO_4^{2-}. Much later, around 2×10^9 years ago, the concentration of O in the atmosphere increased to a level that permitted direct oxidation of sulfide to sulfate.

The appearance of sulfate-reducing bacteria may be recorded by the isotope composition of S in Precambrian rocks because these bacteria fractionate, whereas the photosynthetic bacteria do not. The possibility of using the isotope composition of S in rocks to study the history of biological evolution of the Earth was one of the challenges that attracted Harry G. Thode of McMaster University, Hamilton, Ontario. His work on this subject began in the late 1940s and contributed significantly to the growth of isotope geoscience.

The sedimentary–volcanic rocks at Isua, West Greenland, were deposited prior to 3.7×10^9 years ago and are among the oldest rocks on Earth. A group of scientists, including Thode, reported that the average $\delta^{34}S$ value in these rocks is very close to zero and concluded that no sulfate-reducing bacteria existed on the Earth at the time of deposition of these rocks in Early Archean time (Monster et al., 1979).

The time when sulfate-reducing bacteria began to affect the isotope composition of S in the Precambrian time is not yet certain. On the basis of information available to them, Monster et al. (1979) concluded that bacterial sulfate reduction began between 3.2 and 2.8 Ga. The evidence supporting this conclusion comes from the sedimentary rocks of the Aldan Shield in Siberia and from the iron formations of the Lake Superior district in North America. For example, sulfides in the Michipicoten Iron Formation along the north shore of Lake Superior in Ontario, deposited at 2.75 Ga, have $\delta^{34}S$ values ranging from −10 to +10‰. (Goodwin et al., 1976). Pyrite in the graphitic slates of the Deer Lake greenstone sequence (>2.6 Ga in age) in northern Minnesota has $\delta^{34}S$ values between −2.3 and +11.1‰ (Ripley and Nicol, 1981). The Onwatin slates at Sudbury, Ontario, were deposited at 1.9 Ga and contain sulfides with $\delta^{34}S$ ranging up to +26‰ (Thode et al., 1962).

However, the importance of bacterial reduction of sulfate to the origin of pyrite in iron formations has been questioned. Fripp et al. (1979) obtained an average $\delta^{34}S = +0.6$‰ for 31 samples of banded Archean iron formation from Zimbabwe in Africa. Cameron (1983) analyzed sulfides from seven iron formations in North America, Africa, and Australia and found that their $\delta^{34}S$ values had narrow ranges and that their means extended from -4.9 ± 1.3‰ to $+6.6 \pm 1.1$‰. Cameron (1982) also demonstrated that the $\delta^{34}S$ values of sulfides in Precambrian shales of South Africa were close to 0‰ until about 2.35 Ga when they shifted to negative values. He attributed this isotopic transition to an increase in the sulfate concentration of the oceans that enhanced isotope fractionation by sulfate-reducing bacteria. According to Cameron (1982), the sulfate-reducing bacteria became important participants in the S cycle of the Earth at around 2.3 Ga rather than in the period between 2.8 and 3.2 Ga. The isotope fractionation, seen first in the iron formations of the Lake Superior district, could have been the result of inorganic reduction of volcanic sulfate at elevated temperatures.

Another piece of evidence that casts doubt on the early appearance of sulfate-reducing bacteria comes from the uraniferous conglomerates and sandstones of the Blind River area of Ontario deposited at 2.35 Ga. Hattori et al. (1983) found that 44 samples of pyrite taken through the entire 65-m-thick Matinenda Formation in the Quirke II Mine have $\delta^{34}S$ values ranging only from −1.0 to +1.1‰. Even pyrite that was closely associated with organic matter showed no evidence of isotope fractionation. The virtual absence of fractionation of the isotopes of S in this deposit is striking compared to the wide range of $\delta^{34}S$ values (−50 to +29‰) of pyrite in uraniferous sandstones of Mesozoic to Tertiary age (Goldhaber et al., 1978). Hattori et al. (1983) concluded from the absence of fractionation that no sulfate existed in the basin in which the Matinenda Formation was deposited. The absence of sulfate implies that the atmosphere lacked sufficient O to convert sulfide to sulfate. The pyrite in the Matinenda Formation apparently originated from Archean gneisses and was transported as detrital grains around which authigenic pyrite was subsequently precipitated. Since both cores and the overgrowths of pyrite have $\delta^{34}S$ values near zero, the overgrowths could not have been precipitated as a result of bacterial reduction of sulfate to sulfide.

The absence of isotope fractionation in pyrite in the uraniferous conglomerates of the Blind River district and in similar deposits of early to middle Precambrian age elsewhere strengthens the evidence that O was *not* an important constituent of the atmosphere until after about 2.0 Ga (see also Hattori and Cameron, 1986). The formation of sulfate on the surface of the Earth prior to about 2.0 Ga must therefore be attributed to the photosynthetic S bacteria with some help from volcanic emanations. In terms of biological evolution, this means that photosynthesis preceded sulfate respiration, which was followed by O respiration in Late Neoproterozoic time. These conclusions, expressed by Schidlowski (1979), also indicate that the occurrence of sulfate minerals in Archean rocks does not necessarily imply the existence of O in the atmosphere in early Precambrian time.

These conclusions are consistent with the results of several studies that indicate that the $\delta^{34}S$ values of Archean sulfate minerals are close to zero and differ only slightly from those of associated sulfide minerals. Perry et al. (1971) analyzed barites of presumed sedimentary origin from the Swaziland System of the Barberton Mountain Land, South Africa, known to be more than 3 billion years old. They reported an average $\delta^{34}S$ value of +3.4‰ for barites and +0.42‰ for approximately contemporaneous sulfides and attributed the formation of sulfate to oxidation of volcanic H_2S by the green or purple photosynthetic S bacteria. Lambert et al. (1978) arrived at the same conclusion from a study of $\delta^{34}S$ values of bedded barite deposits about 3.4 billion years old in the Pilbara Block of Western Australia.

Claypool et al. (1980) pointed out that it has been difficult to reconstruct the variation of $\delta^{34}S$ values of marine sulfate deposits of Precambrian age because such deposits are uncommon and difficult to date, and because their isotopic compositions may have been altered by metamorphism. Whelan and Rye (1978) demonstrated that two stratigraphically separable anhydrite formations in the Balmat and Edwards mines in the Grenville Series of New York have strikingly different $\delta^{34}S$ values ranging from +7 to +10‰ and from +18 to +30‰, respectively. The results from the Grenville Series overlap those for sedimentary anhydrite in the Upper Roan Group of Zambia, from which Claypool et al. (1980) deduced $\delta^{34}S = +17.5 \pm 3‰$ for marine sulfate during the period 1000–1300 Ma. Additional data from the Bitter Springs Group (800–1000 Ma) of central Australia and the Shaler Group (635–850 Ma) in the Northwest Territory of Canada suggest a representative value of $\delta^{34}S = +16.5 \pm 2‰$ for Neoproterozoic time. In spite of the obvious gaps that remain in the record, it is clear that sedimentary anhydrite formations of Late Neoproterozoic age are strongly enriched in ^{34}S as a result of the activity of the sulfate-reducing bacteria. And so the stage was set for the complicated pattern of variation of $\delta^{34}S$ and $\delta^{18}O$ in marine sulfates in Phanerozoic time.

29.7 ISOTOPIC EVOLUTION OF MARINE SULFATE

The oceans of the Earth contain a large amount of S (1.3×10^5 metric tons) in the form of sulfate ions in solution. The isotopic composition of S in modern marine sulfate is constant within narrow limits and is represented by a $\delta^{34}S$ value of about +20‰. However, the $\delta^{34}S$ values of sulfate minerals from marine evaporite rocks of Phanerozoic age vary systematically as a function of time in the past. This variation implies that significant changes have occurred in the S cycle on the surface of the Earth.

A large number of $\delta^{34}S$ values of marine sulfate minerals of Phanerozoic age have been used to reconstruct the history of the isotopic composition of marine sulfate (Claypool et al., 1980). The results suggest that the $\delta^{34}S$ values of marine sulfate minerals decreased from about +30‰ during the Cambrian Period to about +10‰ at the end of the Permian and then increased irregularly during the Mesozoic Era toward the present value of +20‰. Thode and Monster (1965) and others have shown that S isotopes are not appreciably fractionated by the precipitation of gypsum from sulfate-bearing brines ($\Delta^{34}S$ gypsum–brine $= +1.65 \pm 0.12‰$). Therefore, the $\delta^{34}S$ values of marine sulfate minerals are representative of the isotopic composition of S in the brines from which they were precipitated. Nevertheless, the $\delta^{34}S$ values of sulfate minerals deposited during a given period of geological time generally vary within certain limits, thus complicating the selection of the most representative value of $\delta^{34}S$ of the oceans during a particular time interval. The variability of $\delta^{34}S$ values of contemporaneously deposited marine sulfate minerals may result from several causes:

1. variable enrichment of ^{34}S in the residual sulfate of isolated evaporative basins due to bacterial reduction to sulfide,
2. input of isotopically light sulfate by rivers, and
3. isotopic fractionation in the course of precipitation of minerals from brines.

As a result, the $\delta^{34}S$ values of the oceans in the geological past have been determined primarily on the basis of the preponderance of the evidence and by rejecting aberrant values. Such aberrant values may also result from incorrect age assignments or indicate a nonmarine origin of the deposit. In spite of these difficulties, the general outline of the evolutionary history of the isotopic composition of marine sulfate in post-Precambrian time in Figure 29.3 is now quite well established.

Sulfur in the form of sulfate ions in aqueous solution enters the ocean primarily by the discharge of fresh water and originates as a weathering product of several kinds of rocks:

1. sulfide-bearing sedimentary rocks (black shales and carbonate rocks),

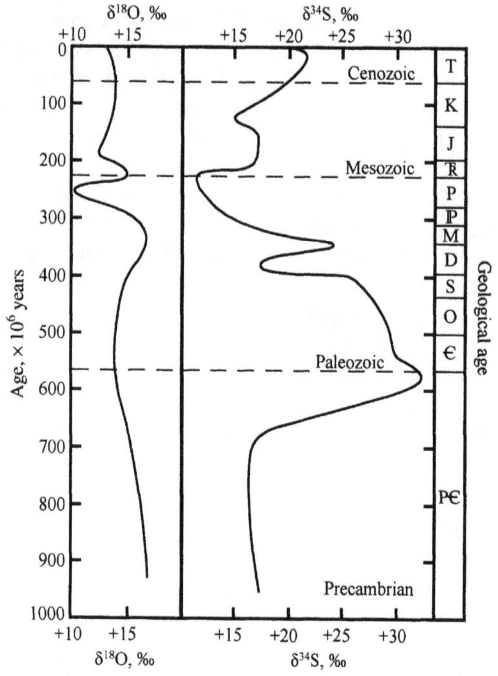

FIGURE 29.3 Variations of $\delta^{34}S$ and $\delta^{18}O$ of marine sulfate minerals from the Neoproterozoic Era to the present. The $\delta^{34}S$ values are expressed relative to troilite in the meteorite Canyon Diablo (CDT) whereas the $\delta^{18}O$ values are relative to SMOW. The curves demonstrate that the isotopic compositions of both elements in marine sulfate minerals have varied in Neoproterozoic and Phanerozoic time. The causes of the observed variations involve a complex interplay of inputs and outputs of S and O_2 in the oceans but are not yet completely understood. Adapted from Claypool et al. (1980).

2. evaporite rocks of marine origin, and
3. volcanic and primary igneous rocks.

On the other hand, S is removed from the oceans by formation of evaporite rocks and by bacterial reduction of sulfate to sulfide followed by precipitation as pyrite, marcasite, or hydrotroilite. The concentration and $\delta^{34}S$ value of sulfate in the oceans at the present time reflect a particular state of the S cycle in terms of quantities and isotopic compositions of S that enter and leave the oceans in unit time. The fact that the isotopic compositions of S in the oceans has varied systematically throughout geological time indicates that significant changes have occurred in the S cycle in the past.

Both the rate of input of S and its isotopic composition may vary as a function of time depending on the rates of weathering and erosion of different kinds of sedimentary rocks, or on the intensity of igneous activity on a worldwide basis. In general, increased erosion of black shale containing isotopically light sulfides can lower the $\delta^{34}S$ value of oceanic sulfate. On the other hand, increased reduction of sulfate by bacteria and removal of isotopically light sulfide from the ocean may increase its $\delta^{34}S$ value. The formation of evaporite rocks reduces the sulfate concentration of the ocean but does not change its $\delta^{34}S$ value directly. However, such a decrease in the concentration of sulfate makes the $\delta^{34}S$ value of the remainder more susceptible to change. Consequently, both the isotopic composition and the concentration of sulfate of the oceans may have varied in the past due to a complex interplay between changing rates of input and output of S of varying isotopic compositions. Several models have been proposed in the hope of gaining a better understanding of the possible causes for the observed isotopic evolution of marine sulfate (Rees, 1970; Holland, 1973; Garrels and Perry, 1974; Schidlowski et al., 1977; Claypool et al., 1980; Schidlowski and Junge, 1981). As a result of these efforts, the general outline of the S cycle is reasonably well understood. However, the specific causes for the variation of $\delta^{34}S$ values of marine sulfate during a particular interval of geological time are still largely speculative.

The isotope composition of O in marine sulfate minerals has also varied during Phanerozoic time, but the variation is not well correlated with the changes in $\delta^{34}S$. Gypsum is enriched in ^{18}O by about 3.5‰ relative to sulfate ions in solution depending on the temperature, which means that the curve for $\delta^{18}O$ of sulfate ions in seawater lies 3.5‰ to the left of the curve drawn in Figure 29.3. The isotope composition of O in sulfate minerals varied only sightly from Neoproterozoic to early Paleozoic time with $\delta^{18}O$ decreasing from about +17 to +14‰. The value then rose during the Devonian Period and reached about +17‰ during the Early Carboniferous (Mississippian) Period.

It then declined to +10‰ during the Permian followed by a similar decline of $\delta^{34}S$ in Late Permian–Early Triassic time. After a rise to +15‰ in Early Triassic time, the $\delta^{18}O$ values of marine sulfate minerals have remained close to +14‰.

The variations of $\delta^{18}O$ of marine sulfate are caused by changes in the fluxes of O entering and leaving the oceans. Oxygen associated with S enters the ocean by erosion of marine sulfate minerals and by the formation of new sulfate ions during oxidation of sulfides exposed to weathering on the continents. Outputs of O from the oceans occur by the reduction of sulfate to CO_2 and H_2S and by crystallization of sulfate minerals during the formation of evaporites. The oxidation of sulfide minerals allows O to enter the S cycle, whereas reduction of sulfate ions allows the O to escape again. Three of the four O atoms used to form new sulfate come from the air and one is derived from meteoric water. The average $\delta^{18}O$ value of sulfate formed by oxidation of sulfide is about +2‰ (Claypool et al., 1980). The CO_2 released during sulfate reduction is depleted in ^{18}O by about 10‰ compared to the remaining sulfate. Therefore, an increase in the rate of sulfate reduction can enrich the sulfate remaining in the oceans in ^{18}O, whereas an increase in sulfide oxidation can lower the $\delta^{18}O$ value of marine sulfate. Crystallization of sulfate minerals can likewise lower $\delta^{18}O$ of marine sulfate because gypsum prefers ^{18}O by about 3.5‰. The effect of erosion of marine sulfate minerals on $\delta^{18}O$ of sulfate in the oceans depends on the age of the deposit, as shown in Figure 29.3. The isotope composition of O in marine sulfate is *not* affected by isotopic equilibration with water in the oceans or with water of hydration that is released during the dehydration of gypsum during burial. Moreover, O in sulfate is not changed isotopically by exchange with silicate minerals in the oceanic crust or with calcium carbonate.

This review makes clear that the S and O cycles are connected. The observed time-dependent variations of the isotope compositions of these two elements in the oceans reflect changes in the input and output fluxes that affect not only the oceans but also the atmosphere. The mathematical models that have been devised to simulate these natural processes are not yet adequate to provide unique explanations for the history of isotopic change that has been uncovered by the study of marine sulfate deposits Schidlowski and Junge (1981).

29.8 IGNEOUS ROCKS

The isotopic composition of S in igneous rocks derived from the mantle is similar to that of meteorites. This conclusion by Shima et al. (1963) and Smitheringale and Jensen (1963) has been repeatedly confirmed by later studies of continental and oceanic basaltic rocks. For example, Schneider (1970) reported an average $\delta^{34}S$ value of $+1.3 \pm 0.5$‰ for alkali olivine basalts from Germany and Kanehiro et al. (1973) obtained $\delta^{34}S$ values ranging from +0.3‰ to +1.6‰ for S extracted from tholeiites taken along the Mid-Atlantic Ridge at 30° N. However, sulfide minerals in granitic rocks have more variable $\delta^{34}S$ value. Hoefs (1997) reviewed information that $\delta^{34}S$ values of European granites range from +9 to −4‰. An even wider range of $\delta^{34}S$ values from +9 to −11‰ was reported by Sasaki and Ishibara (1979) for granitic rocks in Japan. Moreover, some large gabbroic intrusives are strongly *enriched* in ^{34}S compared to meteoritic sulfur. These deviations from the norm indicate either that S in the mantle has variable isotope compositions or that the isotopic composition of S in the magmas may have been altered prior to their crystallization.

The wide range of $\delta^{34}S$ values of S in *granitic rocks* is probably attributable to the incorporation of fractionated S derived from sedimentary rocks. Magmas of granitic composition may have positive $\delta^{34}S$ values because they contain S contributed by marine sulfate during melting or assimilation of sedimentary rocks in a subduction zone. On the other hand, such magmas may also have negative $\delta^{34}S$ values if they formed by melting of sedimentary or metamorphic rocks containing S that was previously enriched in ^{32}S by bacterial fractionation.

The explanation for the abnormal S isotope compositions of some volcanic and plutonic rocks of *basaltic* composition is not so obvious. It is unlikely that the isotopic composition of S in the mantle differs significantly from that of meteorites. Instead, the $\delta^{34}S$ values of mantle-derived rocks may have been changed by one or several of the following processes:

1. contamination of basaltic magma with S derived from sedimentary rocks,
2. hydrothermal alteration of basalts by seawater, and
3. isotope fractionation during outgassing of SO_2 before crystallization of lava flows.

Each of these processes has been invoked to explain abnormal isotope compositions of S in igneous rocks derived from the mantle.

29.8a Contamination

The introduction of S from the country rock into hot igneous intrusions was called "sulfurization" by Cheney and Lange (1967). They cited several examples of mafic igneous intrusions enriched in ^{34}S and explained these anomalies by this process. The deposits they considered include the Palisade Sill of New Jersey, the Triassic diabase intrusive at Cornwall in Pennsylvania, the Duluth Gabbro in Minnesota, the Stillwater Complex of Montana, the norite at Sudbury in Ontario, and the Insizwa Sill in South Africa. These deposits all contain disseminated sulfide minerals whose $\delta^{34}S$ values tend to become more positive near contacts with the country rock.

A very good example of sulfurization occurs in the Noril'sk area of the former USSR in the northwestern part of the Siberian platform. In this region, sulfide ore deposits composed of pyrrhotite, chalcopyrite, and cubanite occur where the Talnakh Gabbro intrudes sedimentary and volcanic rocks ranging in age from Devonian to Early Triassic. According to measurements by Gorbachev and Grinenko (1973), sulfide minerals within the intrusion are anomalously enriched in ^{34}S. For example, disseminated sulfides in the Talnakh Gabbro at the October ore body have an average $\delta^{34}S$ value of $+12.0 \pm 0.4‰$. Anhydrite from the altered sedimentary rocks has an average $\delta^{34}S$ value of $+16.4 \pm 0.6‰$, which is compatible with that of marine sulfate of Middle Devonian age. Gorbachev and Grinenko (1973) concluded from this and other evidence that the S in the ore bodies originated from the sulfate in the country rock.

Another example of sulfurization occurs in the Duluth Gabbro of Minnesota. Ripley (1981) showed that sulfide minerals (pyrrhotite, chalcopyrite, cubanite, and pentlandite) in the Dunka Road deposit are irregularly concentrated near the base of the gabbro where it intruded the Precambrian Virginia Formation. The $\delta^{34}S$ values of sulfide in the gabbro range from +0.2 to +15.3‰ with a mean of about +7.5‰. Sulfide minerals within the altered Virginia Formation have similar $\delta^{34}S$ values between +0.2 and +25.8‰. The high positive $\delta^{34}S$ values of the Dunka Road ore bodies and their similarity to $\delta^{34}S$ values of sulfides in the Virginia Formation strongly suggest that most of the S in the ore was derived from the pyrite in the country rock. These results confirm preceding studies of other parts of the Duluth Gabbro by Mainwaring and Naldrett (1977) and Weiblen and Morey (1976).

29.8b Alteration by Seawater

The contamination of S in igneous rocks by interaction with seawater is most clearly demonstrated by oceanic serpentinites. Sakai et al. (1978) showed that most of the S in these rocks is in the form of sulfate and that the $\delta^{34}S$ values of the sulfate range from +15.4 to +20.2‰, whereas coexisting sulfides had $\delta^{34}S$ values between +2.0 and +5.6‰. They attributed the evident ^{34}S-enrichment of the sulfate to transfer of S from seawater to the serpentinites. Grinenko et al. (1975) had previously reported similar results for a suite of variably altered ultramafic rocks and tholeiites from midocean ridges. Another case of seawater contamination was reported by Hubberten et al. (1975) for lava flows and native S deposits of the Nea Kameni volcano in the Santorini archipelago of Greece. However, basalt lavas extruded on the floor of the ocean do not necessarily show evidence of contamination of S by seawater. In unaltered subaqueous basalts, sulfides are much more abundant than sulfates and the $\delta^{34}S$ values of these sulfides cluster closely about 0% (Puchelt and Hubberten, 1980).

29.8c Outgassing of SO_2

Basalts that form subaerially may lose about 75% of their S by outgassing of SO_2. As a result, the

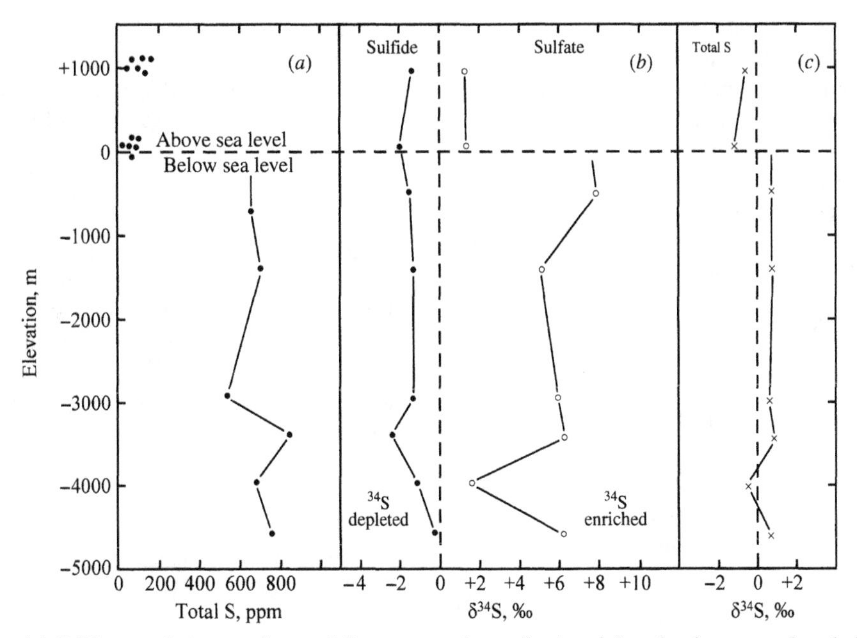

FIGURE 29.4 (*a*) Difference between the total S concentrations of subaerial and subaqueous basalt flows on Kilauea volcano, Hawaii. The low S concentration of subaerial basalts is caused by outgassing of SO_2. (*b*) Difference between $\delta^{34}S$ values of sulfide and sulfate in basalt flows on Kilauea, Hawaii. Note that sulfate is enriched in ^{34}S compared to the sulfide. However, the sulfate of subaerial basalts is less enriched in ^{34}S than that of subaqueous flows. The difference is attributable to loss of SO_2, which causes a decrease of the abundance of ^{34}S in the *total* S in subaerial basalts. (*c*) Difference between $\delta^{34}S$ values of total S in subaerial and subaqueous basalt, Kilauea, Hawaii. Outgassing of SO_2 from the subaerial basalts has lowered the abundance of ^{34}S in the sulfate phase and has caused the $\delta^{34}S$ values of total S to shift to negative values. These data therefore demonstrate how the $\delta^{34}S$ values of mantle-derived basalt can be changed by outgassing of SO_2. Data from Sakai et al. (1982).

isotopic composition of the S remaining in the melt may be strongly affected depending on the $\delta^{34}S$ value of the SO_2. These conclusions were derived by Sakai et al. (1982) from a study of basalt flows on Kilauea volcano in Hawaii. Submarine flows on Kilauea contain an average of 700 ± 100 ppm S with molar sulfate–sulfide ratios ranging from 0.15 to 0.56. The average $\delta^{34}S$ value of total S in these submarine flows is $+0.7 \pm 0.1$‰, exactly as expected for S derived from the mantle. However, the sulfates are *enriched* in ^{34}S relative to the sulfides by about 7.5 ± 1.5‰. Flows extruded above sea level contain only about 150 ± 50 ppm S and are *depleted* in sulfate. The $\delta^{34}S$ value of total S is -0.8 ± 0.2‰, and the sulfate is enriched in ^{34}S by only 3.0‰ compared to the sulfide. The variation of the concentration and isotope compositions of S in submarine and subaerial basalt flows on Kilauea are further illustrated in Figures 29.4*a*–*c*.

The loss of SO_2 from basalt magma may *deplete* or *enrich* the S remaining in the melt in ^{34}S depending on the partial pressure (fugacity) of O, symbolized hereafter by fO_2. The relationship of $\delta^{34}S$ in the melt to fO_2 is caused by the fact that the sulfate–sulfide ratio of the melt increases with increasing fO_2 and that this ratio affects the $\delta^{34}S$ value of the escaping sulfur-bearing gases. In general, under conditions of high fO_2, the sulfate–sulfide ratio in the melt increases, SO_2 is *depleted* in ^{34}S, and progressive loss of SO_2 *enriches* the melt in ^{34}S. Therefore, outgassing of SO_2 from a basalt melt under conditions of *high* fO_2 can shift the $\delta^{34}S$ of basalt in the *positive* direction. Similarly, progressive loss of SO_2 at *low* fO_2 can shift the $\delta^{34}S$ value of basalt in the *negative* direction. In this way, the $\delta^{34}S$ values of mantle-derived basaltic magmas can change, depending on the fugacity of O in the melt and the fraction of S lost. These complicated relationships can be described quantitatively by equations derived by Sakai et al. (1982).

The $\delta^{34}S$ value of a mixture of SO_2 and H_2S is given by:

$$\delta^{34}S_{gas} = x\ \delta^{34}S_{SO_2} + (1-x)\ \delta^{34}S_{H_2S} \quad (29.8)$$

where x is the mole fraction of S in the SO_2 component of the mixture. Similarly, the $\delta^{34}S$ value of a melt containing both sulfide and sulfate is:

$$\delta^{34}S_{melt} = y\ \delta^{34}S_{S}{}^{2-} + (1-y)\ \delta^{34}S_{SO_4{}^{2-}} \quad (29.9)$$

where y is the mole fraction of S in the sulfide component of the melt. The difference between the isotope composition of S in the gas and that remaining in the melt is:

$$\delta^{34}S_g - \delta^{34}S_m = \left[x\ \delta^{34}S_{SO_2} + (1-x)\ \delta^{34}S_{H_2S}\right] - \left[y\ \delta^{34}S_{S}{}^{2-} + (1-y)\ \delta^{34}H_{SO_4{}^{2-}}\right] \quad (29.10)$$

After multiplying, adding and subtracting $\delta^{34}S_{SO_2}$ and $\delta^{34}S_{S^{2-}}$, and collecting terms:

$$\delta^{34}S_g - \delta^{34}S_m = \delta^{34}S_{SO_2} - \delta^{34}S_{S}{}^{2-} - (1-x)\left(\delta^{34}S_{SO_2} - \delta^{34}S_{H_2S}\right) - (1-y)\left(\delta^{34}S_{SO_4{}^{2-}} - \delta^{34}S_{S}{}^{2-}\right) \quad (29.11)$$

This equation is simplified by introducing the convention that $\delta^{34}S_1 - \delta^{34}S_2 = \delta^{34}S_2^1$:

$$\Delta^{34}S_m^g = \Delta^{34}S_{S^{2-}}^{SO_2} - (1-x)\ \Delta^{34}S_{H_2S}^{SO_2} - (1-y)\ \Delta^{34}S_{S^{2-}}^{SO_4{}^{2-}} \quad (29.12)$$

The volcanic gases emitted by Kilauea and other volcanoes (Allard, 1979) contain SO_2 as the major S-bearing phase. H_2S is present only in trace amounts. Therefore, the weight fraction x of S in volcanic gas in which SO_2 is the dominant component is equal to 1 and equation 29.12 reduces to:

$$\Delta^{34}S_m^g = \Delta^{34}S_{S^{2-}}^{SO_2} - (1-y)\ \Delta^{34}S_{S^{2-}}^{SO_4{}^{2-}} \quad (29.13)$$

In the Hawaiian basalts in Figures 29.4*b* and *c* the difference between the $\delta^{34}S$ values of sulfate and sulfide that coexist in submarine flows is $\Delta^{34}S_{S^{2-}}^{SO_4{}^{2-}} = +7.5 \pm 1.5‰$. In addition, Sakai (1968) found that $\Delta^{34}S_{S^{2-}}^{SO_2}$ is approximately equal to +3‰ at 1000°C. Introducing these numerical values into equation 29.13 yields:

$$\Delta^{34}S_m^g = 7.5y - 4.5 \quad (29.14)$$

Equation 29.14 indicates that $\Delta^{34}S_m^g = 0$ when $y = 0.6$. In other words, when

$$y = \frac{S^{2-}}{SO_4{}^{2-} + S^{2-}} = 0.6 \quad (29.15)$$

the $\delta^{34}S$ value of SO_2 being outgassed by the melt is the same as that of the S remaining in the melt. Under these conditions, loss of SO_2 does not change the isotope composition of the S remaining in the melt. However, when $y < 0.6$, then $\Delta^{34}S_m^g$ is negative, which occurs when the S in the gas is depleted in ^{34}S compared to that of the melt. Outgassing of SO_2 under these conditions enriches the melt in ^{34}S and drives its $\delta^{34}S$ value in the positive direction. Similarly, when $y > 0.6$, $\Delta^{34}S_m^g$ is positive, which means that the SO_2 is enriched in ^{34}S and outgassing of SO_2 causes the melt to be depleted in ^{34}S and drives its $\delta^{34}S$ value in the negative direction. Loss of SO_2 at about 1000°C can therefore change the $\delta^{34}S$ value of a basalt magma by either enriching or depleting it in ^{34}S depending on the value of the parameter y, which can be recalculated as the sulfate–sulfide ratio:

$$\frac{SO_4{}^{2-}}{S^{2-}} = \frac{1-y}{y} \quad (29.16)$$

A value of 0.6 for y corresponds to a sulfate–sulfide ratio of 0.66. Therefore, outgassing of SO_2 from a melt whose sulfate–sulfide ratio is greater than 0.66 ($y < 0.6$) enriches it in ^{34}S and vice versa.

The sulfate–sulfide ratio of a melt depends on its O fugacity based on the reaction:

$$S^{2-} + 2O_2 \leftrightarrows SO_4{}^{2-} \qquad (29.17)$$

The O fugacity of a basalt melt under equilibrium conditions is controlled by mineral phases crystallizing from it, primarily fayalite, magnetite, and quartz. These minerals constitute the so-called FMQ buffer:

$$3Fe_2SiO_4 + O_2 \leftrightarrows 2Fe_3O_4 + 3\ SiO_2 \qquad (29.18)$$

Equilibrium among these phases at any temperature is possible only at a specific value of fO_2. Therefore, the fO_2 of a basaltic magma, containing the minerals of the FMQ buffer at equilibrium, varies with the temperature. Such a temperature dependence of fO_2 was actually observed by Sato and Wright (1966) in the Makaopuhi lava lake on Kilauea. Their data indicate that fO_2 *decreases* with *decreasing* temperatures and obeys the equation:

$$\log\ fO_2 = \frac{a}{T} + b \qquad (29.19)$$

where T is the absolute temperature and a and b are constants whose average values are:

$$a = -2.04 \pm 0.16 \times 10^4 \qquad b = 5.37 \pm 1.49$$

This equation yields a value of about 2×10^{-11} atm for fO_2 at 1000°C.

The subaqueous flows analyzed by Sakai et al. (1982) have an average sulfate–sulfide ratio of about 0.34, which corresponds to $y = 0.75$. This value is greater than 0.6 and therefore indicates that loss of SO_2 from the flows on Kilauea should have depleted them in ^{34}S, depending on the extent of outgassing.

The effect of progressive loss of SO_2 on the isotope composition of S remaining in the melt can be represented by the Rayleigh equation discussed in Section 26.3b:

$$\left(\frac{R}{R_0}\right)_{\text{melt}} = f^{\alpha_m^g - 1} \qquad (29.20)$$

where R is the $^{34}S/^{32}S$ ratio of sulfur remaining in the melt, R_0 is the same prior to outgassing, f is the fraction of S remaining in the melt, and α_m^g is the isotope fractionation factor between SO_2 and S in the melt. When the isotope ratios are replaced by the $\delta^{34}S$ parameter and by taking natural logarithms, equation 29.20 becomes:

$$\ln\left(\frac{R}{R_0}\right)_m = \ln\left(\frac{\delta^{34}S + 1000}{\delta^{34}S_0 + 1000}\right)_m = (\alpha_m^g - 1)\ln f \qquad (29.21)$$

The reader can easily verify by numerical examples that:

$$\ln\left(\frac{\delta^{34}S + 1000}{\delta^{34}S_0 + 1000}\right) \simeq \frac{\delta^{34}S - \delta^{34}S_0}{1000} \qquad (29.22)$$

provided that natural logarithms are used. Therefore, the Rayleigh equation can be rewritten:

$$\delta^{34}S = \delta^{34}S_0 + 1000(\alpha_m^g - 1)\ \ln f \qquad (29.23)$$

Consequently, the $\delta^{34}S$ values of igneous rocks that lost varying proportions of S by outgassing may deviate from the normal range depending on the value of the fractionation factor α and the fraction f of S remaining in the melt. The fractionation factor is related to the difference between the $\delta^{34}S$ values of the gas and the melt by:

$$\Delta^{34}S_m^g = \delta^{34}S_g - \delta^{34}S_m = 1000\ \ln\ \alpha_m^g \qquad (29.24)$$

Ultimately, the fugacity of O in the melt determines the effect of outgassing on the $\delta^{34}S$ value of the magma, as illustrated in Figure 29.5. If fO_2 is "high", $\alpha < 1.000$ and the melt is enriched in ^{34}S. Under conditions of "low" fO_2, outgassing of SO_2 depletes the melt in ^{34}S. The magnitude of the enrichment or depletion depends on the fraction of S remaining (f) and on the fractionation factor α. The dependence of the S-isotope fractionation factor between SO_2 and silicate melts on fO_2 is not yet known precisely. The relationship between fO_2 of a basaltic magma and the isotopic composition of S remaining in lava flows was

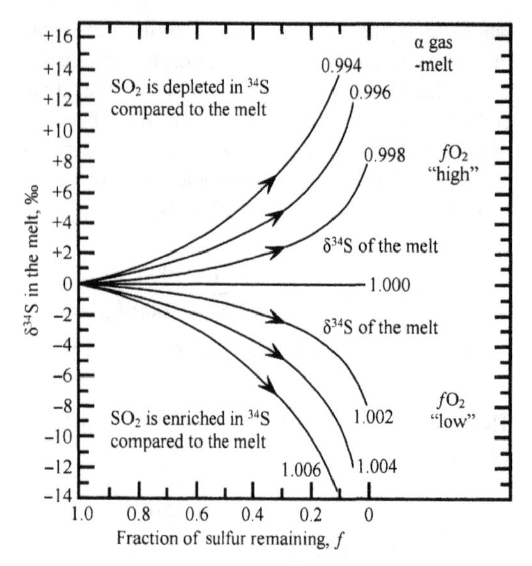

FIGURE 29.5 Change in the $\delta^{34}S$ values of a magma caused by outgassing of SO_2 according to Rayleigh distillation expressed in equation 29.18. The parameter f is the fraction of S remaining in the melt and α is the fractionation factor for S partitioned between SO_2 and the melt. The diagram shows that the fractionation factor depends on the fugacity of O in the melt as discussed in the text. When α is greater than unity, loss of SO_2 causes the melt to be depleted in ^{34}S and drives its $\delta^{34}S$ in the negative direction. When α is less than unity, the $\delta^{34}S$ values of the melt become positive depending on the fraction of S that remains. This model provides one of three possible explanations discussed in the text for the phenomenon that some igneous rocks derived from the mantle of the Earth contain S whose isotopic composition differs significantly from that of meteorites and most basalts.

used by Faure et al. (1984) to explain the $\delta^{34}S$ values of the Kirkpatrick Basalt on Mt. Falla in the Transantarctic Mountains.

29.9 SULFIDE ORE DEPOSITS

One of the objectives of the study of S isotopes in geology is to contribute toward a better understanding of the origin and conditions of formation of sulfide ore deposits. It is important, for both theoretical and practical reasons, to distinguish between ore deposits formed as a result of igneous activity and those that are of sedimentary origin. Sulfur that has been subjected to bacterial reduction of sulfate in freshly deposited sediment becomes enriched in ^{32}S compared to marine sulfate. On the other hand, the S associated with igneous rocks derived from the upper mantle is isotopically similar to that of meteorites and therefore commonly has $\delta^{34}S$ values close to zero. These facts suggest that the isotopic composition of S may help to distinguish ore deposits related to igneous activity from those of sedimentary origin. Ore deposits derived from igneous sources have a narrow range of $\delta^{34}S$ values centered about $\delta^{34}S = 0$‰, whereas ore deposits containing biogenic S tend to have negative and more variable $\delta^{34}S$ values.

Attempts to use these criteria to determine the source of S in sulfide ore deposits have generally been disappointing. One reason is that ore deposits commonly have complex histories and often cannot be classified adequately into syngenetic (sedimentary) and epigenetic (igneous–hydrothermal) types. Furthermore, the isotopic composition of S may be modified after deposition by thermal metamorphism. In some cases, ore deposits of *igneous* (hydrothermal) origin have a *wide* range of $\delta^{34}S$ values because of the presence of several generations of minerals deposited sequentially under different conditions. On the other hand, ore deposits associated with *sedimentary* rocks may have a *narrow* range of $\delta^{34}S$ values because they formed from a very large sulfate reservoir under constant environmental conditions. The $\delta^{34}S$ values of strata-bound ore deposits of different ages may also reflect the time-dependent variation of the isotope compositions of marine sulfate displayed in Figure 29.3. In addition, bacterial reduction of marine sulfate is not an adequate explanation for the occurrence of sulfide minerals in some strata-bound deposits because the temperatures of deposition were too high (50–350°C) for bacterial activity. Instead, Ohmoto and Rye (1979) suggested that sulfides can form as a result of inorganic reduction of marine sulfate by Fe^{2+} in hot volcanic rocks in contact with seawater.

29.9a Isotope Fractionation among Sulfide Minerals

When sulfide minerals are precipitated from aqueous solutions or crystallize from sulfide melts, small differences may occur in the $\delta^{34}S$ values of cogenetic minerals due to isotopic equilibration among the solids and between the solids and the liquid. Research by other investigators, summarized by Thode (1970), led to the suggestion that such differences in the $\delta^{34}S$ values of cogenetic sulfide minerals reflect the temperature of isotopic equilibration. Theoretical considerations by Sakai (1957, 1968) and Bachinski (1969) based on bond strengths indicate that the ^{34}S-enrichment of some common sulfide minerals should *decrease* in the order: pyrite > spalerite > chacopyrite > galena. Analytical data for coexisting suites of these minerals generally confirm this prediction and indicate thereby that isotopic equilibrium is closely approached by these minerals in nature. Consequently, the $\delta^{34}S$ values of cogenetic sulfide minerals may be used to determine the temperatures of equilibration, provided that the variation of $\delta^{34}S$ values as a function of temperature is either calculated theoretically or determined experimentally. The fractionation factors for S isotopes in a variety of S-bearing compounds as a function of temperature have been compiled from the literature by Friedman and O'Neil (1977) and Ohmoto and Rye (1979).

The $\Delta^{34}S$ values of pyrite–galena, sphalerite–galena, and pyrite–sphalerite were calculated by Sakai (1968) for temperatures from 27 to 527°C, shown in Figure 29.6 as a function $10^6/T^2$, where T is the temperature on the Kelvin scale. It is apparent that the $\Delta^{34}S$ values for these mineral pairs decrease linearly with increasing temperature above about 150°C. The pyrite–galena pair is the most sensitive geothermometer but rarely achieves isotopic equilibrium.

The fractionation of S isotopes by sulfide minerals has also been studied experimentally (Grootenboer and Schwarcz, 1969; Kajiwara and Krouse, 1971; Czamanske and Rye, 1974). The results are presented by equations of the form:

$$\Delta^{34}S = \frac{A \times 10^6}{T^2} \qquad (29.25)$$

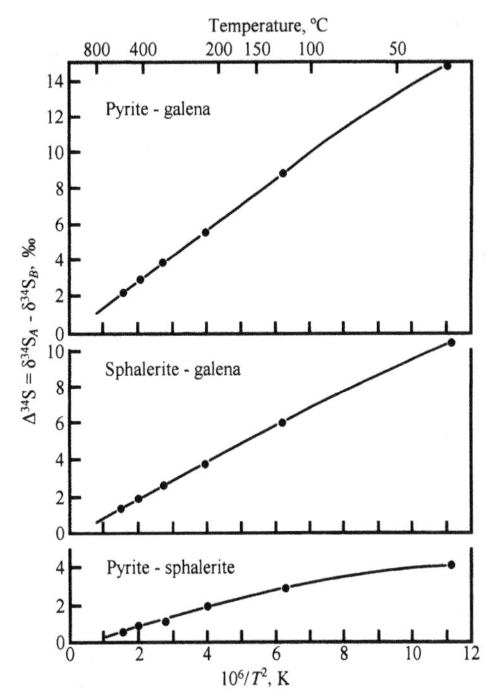

FIGURE 29.6 Fractionation of S isotopes among cogenetic mineral pairs as a function of temperature T on the Kelvin scale based on calculated values of isotope equilibrium constants by Sakai (1968).

where A is a constant equal to the slope of straight lines through the origin and T is in kelvins. Table 29.3 contains relations given by Ohmoto and Rye (1979) obtained by solving Equation 29.25 for the isotope-equilibration temperatures for coexisting pairs of common sulfide minerals based on the differences in their $\delta^{34}S$ values.

The temperatures derived from isotope compositions of coexisting sulfide minerals are subject to the following assumptions:

1. Isotope equilibrium was established between the two minerals.
2. The isotope composition was not changed by postdepositional isotope exchange with S in other minerals or fluids.
3. The mineral phases were cleanly separated from each other prior to analysis.

Table 29.3. Sulfur-isotope Thermometers Based on Equation 29.25 for Coexisting Sulfide Minerals in Equilibrium with a Common Sulfur Reservoir

Mineral Pair	Temperature, K	Uncertainty in the Temperature, K	
		1	2
Pyrite–galena	$\frac{(1.01 \pm 0.04) \times 10^3}{\Delta^{1/2}}$	±25	±20
Sphalerite (pyrrhotite)–galena	$\frac{(0.85 \pm 0.03) \times 10^3}{\Delta^{1/2}}$	±20	±25
Pyrite–chalcopyrite	$\frac{(0.67 \pm 0.04) \times 10^3}{\Delta^{1/2}}$	±35	±40
Pyrite–pyrrhotite (sphalerite)	$\frac{(0.55 \pm 0.04) \times 10^3}{\Delta^{1/2}}$	±40	±55

Source: Ohmoto and Rye, 1979.
Note: The equations relating the temperature to the $\Delta^{34}S$ value of the mineral pairs were obtained by solving Equation 29.25 for T. Uncertainty 1 of the calculated temperature is caused by the uncertainty of the equation at a temperature of 300°C. Uncertainty 2 is caused by the analytical error in $\Delta^{34}S$ of ±0.2‰. The equations are based on a critical evaluation by Ohmoto and Rye (1979) of experimental data in the literature.

When these conditions are satisfied, calculated isotope equilibration temperatures are in reasonably good agreement with filling temperatures of fluid inclusions. Ohmoto and Rye (1979) also summarized the isotope fractionation factors for many other sulfides and sulfates with respect to H_2S as a function of temperature, whereas Nielsen (1979) presented the temperature dependence of S isotope fractionation factors of S-bearing ions and gases in graphic form.

29.9b Isotope Fractionation in Ore-Forming Fluids

The preceding discussion (Sections 29.9a) was based on the assumption that the $\delta^{34}S$ values of sulfide minerals depend only on the isotopic composition of S in the fluid from which they formed and on the temperature of isotopic equilibration. However, Sakai (1968) originally pointed out that the $\delta^{34}S$ values of aqueous sulfide ions depend on the relative proportions of H_2S, HS^-, and S^{2-}, all of which fractionate S isotopes. The abundances of these ionic and molecular species are controlled by chemical equilibria involving H ions. Therefore, the $\delta^{34}S$ values of aqueous sulfide ions, and of sulfide minerals in equilibrium with them, depend on the pH of the ore-forming fluid. This idea was extended by Ohmoto (1972) to include the effects of other physical and chemical parameters on the $\delta^{34}S$ values of sulfide minerals that can precipitate from a given ore-forming fluid. His treatment of this complex problem indicates that the $\delta^{34}S$ values of sulfide minerals reflect not only the temperature and isotopic composition of S in the fluid but also the pH, the fugacities of O and S, the total S content, and the ionic strength of the ore-forming fluid.

The data in Figure 29.7 can be used to demonstrate qualitatively the effect of the fugacity of O and the pH on the $\delta^{34}S$ value of sulfide ions in an ore-forming fluid at a particular temperature. A high O fugacity favors SO_4^{2-}, which concentrates ^{34}S and thereby enriches the sulfide ion in ^{32}S. Therefore, sulfide minerals precipitating at high O fugacity from a given ore-bearing fluid are enriched in ^{32}S compared to the same mineral forming at lower O fugacity. The pH-dependence of the $\delta^{34}S$

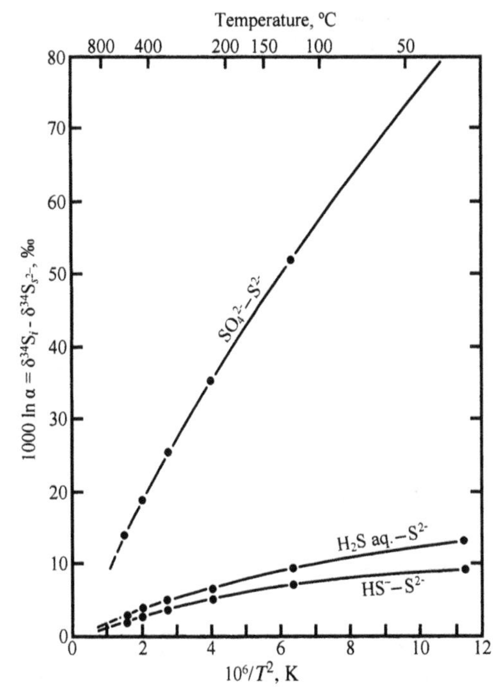

FIGURE 29.7 Fractionation of S isotopes among SO_4^{2-}, $H_2S(aq)$, HS^-, and S^{2-} as a function of the temperature T on the Kelvin scale. Note that SO_4^{2-} is strongly enriched in ^{34}S relative to S^{2-} and that the enrichment increases with decreasing temperature. Fractionation of S isotopes among $H_2S(aq)$ and HS^- is less pronounced, but these ions clearly prefer ^{34}S over ^{32}S compared to the sulfide ion. The fractionation factors were calculated by Sakai (1968) from ratios of reduced partition coefficients.

value of a sulfide mineral can be demonstrated similarly. If the O fugacity is sufficiently low so that SO_4^{2-} is unimportant and $H_2S(aq)$, HS^-, and S^{2-} are dominant, then a decrease in pH favors the formation of $H_2S(aq)$ and HS^-, both of which prefer ^{34}S. The result is that the sulfide ions become enriched in ^{32}S, and minerals forming at low pH are enriched in ^{32}S compared to the same minerals forming at higher pH from the same fluid.

Ohmoto (1972) worked out these relationships quantitatively and plotted $\delta^{34}S$ contours on mineral stability fields in the system Fe–Ba–O_2–S_2. Figure 29.8 is a simplified example of such a diagram for the case $\sum S = 0.1$ mol/kg H_2O, $T = 250°C$, ionic strength $= 1.0$, and $\delta^{34}S_{\sum S} = 0‰$. The total S content of the system affects the size of the stability fields while the ionic strength determines the values of the activity coefficients of ions in aqueous solution. The temperature not only controls the magnitude of the isotope fractionation factors (i.e., Figure 29.6) but also determines the values of the chemical equilibrium constants. Figure 29.8 indicates that pyrite precipitating from a fluid whose $\delta^{34}S$ value is 0‰ can have $\delta^{34}S$ values ranging from about +5 to −27‰, depending on the fugacity of O and the pH. Evidently, these parameters play an important role in determining the $\delta^{34}S$ values of sulfide minerals that may precipitate from a given ore-bearing fluid. However, these effects do not prevent the *differences* in $\delta^{34}S$ values of cogenetic sulfide minerals to be interpreted in terms of equilibration temperatures, as outlined in Section 29.9a.

29.10 SULFUR IN THE ENVIRONMENT

The atmosphere contains S derived from both natural and anthropologic sources that have a wide range of isotopic compositions. The S may enter the atmosphere in several chemical forms, including SO_2, H_2S, SO_4^{2-}, native sulfur, and a variety of organic S compounds. All of the compounds containing reduced forms of S are converted to the sulfate by oxidation in the atmosphere. The sulfate exists both in the form of minute crystals and as ions dissolved in water droplets suspended in air. Sulfur is removed from the atmosphere primarily by meteoric precipitation in rain or snow as well as by dry fallout of sulfate crystals. The sulfate in meteoric precipitation is associated with H^+ ions that give rise to the phenomenon popularly known as "acid rain." The resulting acidification of streams and lakes is harmful to the fauna and flora they contain. This deterioration of the environment has caused demands by the public for the identification of the sources of acid rain and for their ultimate elimination. The problem has assumed international proportions because of the rapid and wide dispersal of S compounds within the atmosphere without

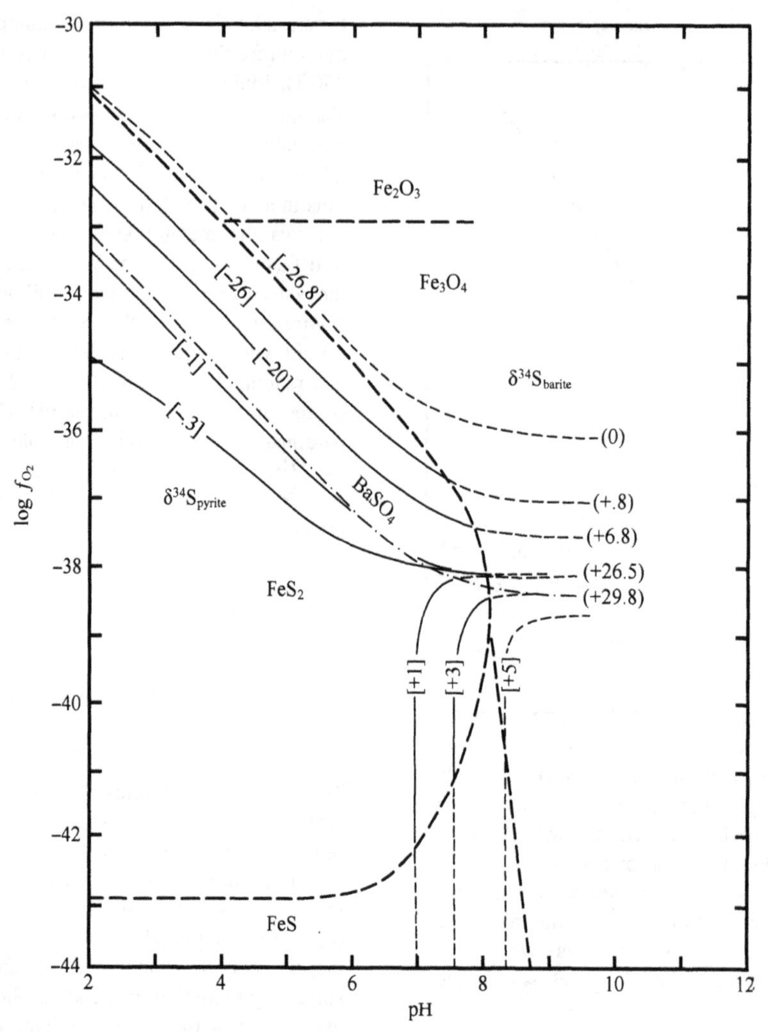

FIGURE 29.8 Isotopic composition of S in pyrite and barite in the system Fe–S–O as a function of pH and the fugacity of O. Square brackets indicate $\delta^{34}S$ of pyrite; round brackets denote $\delta^{34}S$ of barite. Note that $\delta^{34}S$ of pyrite varies widely depending on the fugacity of O and the pH of the solution. The $\delta^{34}S$ of the total S in the system is equal to 0‰. The Fe–S–O mineral boundaries represent a total S content of 0.1 moles S/kg H_2O, $T = 250°C$, ionic strength = 1.0. The dash-dot line is the barite soluble/insoluble boundary ($m_{Ba^{2+}} \times m_{\sum S} = 10^{-2}$). Adapted with permission from H. Ohmoto, *Economic Geology*, Vol. 67, No. 5, pp. 551–578, 1972.

regard to national borders. Because of the wide range of the isotopic composition of S, isotopic studies have been used to identify the dominant natural and anthropopologic sources responsible for the S content of the atmosphere and in precipitation. Krouse (1980) has written a detailed review of the occurrence of S in the environment with references to a large number of relevant isotopic studies, including earlier summaries by Smith (1975) and Nielsen (1974).

Large quantities of S enter the atmosphere from natural geological and biological sources. These include:

1. volcanic gases composed of SO_2 and H_2S emitted by volcanic eruptions and by fumaroles and hot springs;
2. H_2S and S-bearing organic compounds released by sulfate-reducing bacteria in tidal flats, marshes, lakes, and soil;
3. sulfate particles derived from spray of seawater; and
4. organic S-bearing compounds released into the atmosphere by plants.

Volcanic eruptions can be very large, although sporadic, contributors of S to the atmosphere. For example, the eruption of El Chichon in Mexico in 1982 resulted in the formation of a "cloud of sulfuric acid droplets" that persisted for many months and raised the possibility that its presence had an effect on the world's climate (Kotra et al., 1983). A record of sulfate-enriched snow following large volcanic eruptions is preserved in the ice sheets of Antarctica and Greenland (Palais and Legrand, 1985).

In addition to the natural inputs of S, anthropologic sources may be dominant in the industrial areas of the Earth. The principal sources of air pollution resulting from human activities are:

1. release of SO_2 by combustion of S-bearing coal and fuel oil for the purpose of generating electricity, smelting of iron ore, firing industrial boilers, and heating of homes;
2. roasting of sulfide-bearing ores;
3. refining of petroleum;
4. automobile exhaust fumes; and
5. sulfur dust blown into the atmosphere from stockpiles of elemental sulfur.

The importance of these anthropologic sources of S in the atmosphere varies widely depending on the particular industrial activity. Similarly, the effect of the discharge of S-compounds on the isotopic compositions of S in the atmosphere in any given area depends on the isotopic composition of those sources. For this reason, the isotope compositions of both natural and anthropogenic sources that may contribute S to meteoric precipitation in a given region must be known.

The importance of SO_2 discharged by Cu smelters in the vicinity of Salt Lake City, Utah, on the concentration and isotopic compositions of S in the air was clearly demonstrated in a study by Grey and Jensen (1972). They reported that SO_2 of "normal" air in or near Salt Lake City had an isotope composition of $\delta^{34}S = +1.3 \pm 0.3‰$, whereas clean air collected well away from anthropogenic sources contained SO_2 that was enriched in ^{34}S with $\delta^{34}S = +9.0 \pm 0.2‰$. In July 1971 the miners went on strike and the Cu smelter was temporarily shut down. The concentrations of SO_2 in the air downwind of the smelter promptly dropped from 25 ppb by volume to about 1 ppb and the $\delta^{34}S$ value rose to $+6.0 \pm 0.4‰$. Grey and Jensen (1972) attributed the rise of the $\delta^{34}S$ value to the temporary shutdown of the smelter, which emitted SO_2 with $\delta^{34}S = +1.0 \pm 0.3‰$. The response of sulfate in meteoric precipitation was very similar to that of SO_2 in the air. Before the strike, the average $\delta^{34}S$ value of sulfate in precipitation was $+2.2 \pm 1.0‰$ compared to $+9.0 \pm 0.3‰$ in a control area that was free of anthropogenic contamination. During the strike, the $\delta^{34}S$ value of sulfate in meteoric precipitation in Salt Lake City rose to $+6.0‰$.

The enrichment of air and precipitation in ^{34}S observed during the strike occurred because other local sources within the valley of Great Salt Lake became dominant in the absence of SO_2 emissions from the smelter. All of these secondary sources emitted S containing more ^{34}S than the SO_2 of the smelters. The secondary anthropologic sources in the Great Salt Lake basin are oil refineries ($\delta^{34}S = +16.6‰$) and automobile exhaust ($\delta^{34}S = +15.1‰$). The principal natural contributors are the Great Salt Lake ($\delta^{34}S = +1.53‰$), streams in the area ($\delta^{34}S = +9.6‰$), and bacteriogenic H_2S ($\delta^{34}S = +5.3 \pm 1.7‰$). Grey and Jensen (1972) concluded that during the strike about 90% of the S in the air originated from bacterial sources. They emphasized that sulfate-reducing bacteria are a major natural source of atmospheric S and that their output may dominate the S-cycle worldwide at certain times of the year.

Summaries of the global circulation of S compounds have been given by Granat et al. (1976), Kellogg et al. (1972), Junge (1963), and Eriksson (1963). On the basis of a quantitative model of inputs and outputs of S from all sources, Kellogg and his associates concluded that in 1972 humans were contributing about one-half as much S to the atmosphere worldwide as natural sources, that by the year 2000 AD anthropologic sources will match the input from natural sources, and that in industrialized regions humans were already overwhelming natural processes and were causing marked increases in the concentration of atmospheric S in large areas hundreds of kilometers downwind from industrial centers.

29.11 MASS-INDEPENDENT ISOTOPE FRACTIONATION

The fractionation of the stable isotopes of elements having low atomic numbers is caused by the differences in their masses. For example, the $\delta^{17}O$ and $\delta^{18}O$ values of most O-bearing minerals define a straight line whose equation is:

$$\delta^{17}O = 0.52\ \delta^{18}O \qquad (29.26)$$

where $\delta^{17}O$ and $\delta^{18}O$ are defined in terms of the $^{17}O/^{16}O$ and $^{18}O/^{16}O$ ratios relative to SMOW.

Thiemens and Heidenreich (1983) and Heidenreich and Thiemens (1983, 1985) originally reported that the stable isotopes of O are fractionated differently during the formation of ozone (O_3) from O_2 in an electrical discharge such that:

$$\delta^{17}O = 1.0\ \delta^{18}O \qquad (29.27)$$

indicating that the slope of the isotope fractionation line is equal to 1.0 rather than 0.52 as in equation 29.26. The isotope fractionation of O during the formation of ozone was attributed to differences in the symmetry of O_2 molecules rather than to differences in the masses of the stable O isotopes. Subsequently, mass-independent isotope fractionation was also demonstrated by Bhattacharya and Thiemens (1989) for the reaction $O + CO \rightarrow CO_2$

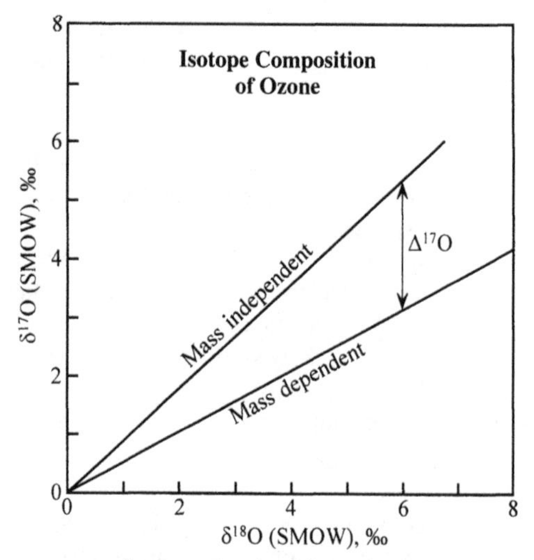

FIGURE 29.9 Isotope fractionation of O according to equations 29.22 (mass dependent) and equation 29.23 (mass independent). The latter was first reported by Thiemens and Heidenreich (1983) and Heidenreich and Thiemens (1985). The $\Delta^{17}O$ parameter is defined by equation 29.25.

and by Bains-Sahota and Thiemens (1989) for the reaction $SF_5 + SF_5 \rightarrow S_2F_{10}$.

The mass-dependent and mass-independent isotope fractionation lines in Figure 29.9 are separated by a vertical distance ($\Delta^{17}O$) defined as:

$$\Delta^{17}O = {}^{17}\delta O_i - \delta^{17}O_d \qquad (29.28)$$

where i means mass-independent and d means mass-dependent. Substituting equation 29.26 for $\delta^{17}O_d$ yields:

$$\Delta^{17}O = \delta^{17}O_i - 0.52\ \delta^{18}O \qquad (29.29)$$

which expresses the ^{17}O enrichment relative to the mass-dependent fractionation line.

The unusual isotopic composition of O in stratospheric ozone is transferred to sulfate ions that form by oxidation of dimethyl sulfides and SO_2 released into the atmosphere from sources on the surface of the Earth. The resulting deviation of $\delta^{17}O$ values of sulfate from the mass-dependent O-isotope fractionation line is expressed by the $\Delta^{17}O$ parameter (equation 29.29).

For example, Bao et al. (2000) measured $\delta^{17}O$ and $\delta^{18}O$ values of gypsum ($CaSO_4 \cdot 2H_2O$) in deposits of "gypcrete" in four soil profiles of the Namib Desert in southwest Africa. All of the resulting $\Delta^{17}O$ values are positive with an average of $+0.34 \pm 0.04$‰, which means that these sulfates are anomalously enriched in ^{17}O. The only known sources of O enriched in ^{17}O are ozone and hydrogen peroxide (H_2O_2) in the atmosphere:

Mauersberger, 1987, *Geophys. Res. Lett.*, 14: 80–83. Krankowsky et al., 1995, *Geophys. Res. Lett.*, 22:1713–1716. Johnston and Thiemens, 1997, *J. Geophys. Res.*, 102(D21): 25395–25404. Savarino and Thiemens, 1999, *Atmos. Environ.*, 33:3683–3690.

Therefore Bao et al. (2000) concluded that the gypsum in the soil of the Namib Desert contains sulfate ions that had formed by the oxidation of dimethyl sulfide by ozone of the atmosphere.

Another case of ^{17}O enrichment was reported by Bao et al. (2000) for gypsum crystals in volcanic ash deposits of the Arikaree Group (Miocene) in Nebraska and South Dakota. In this case, the $\Delta^{17}O$ values of seven samples range from -0.02 to $+4.59$‰.

Desert varnish likewise contains sulfate salts that are enriched in ^{17}O. Bao et al. (2001) reported that the $\Delta^{17}O$ values of samples of desert varnish in Death Valley National Park, California, range from $+0.37$ to $+1.38$‰ with a mean of $+0.70 \pm 0.22$‰ for 20 samples.

The hypothesis that the ^{17}O-enrichment of desert varnish in Death Valley is caused by the non-mass-dependent isotope fractionation of O during the formation of ozone in the stratosphere is supported by the isotope composition of O in sulfate aerosol particles collected in southern California. Bao et al. (2001) reported strong ^{17}O-enrichments in sulfate from aerosols with $\Delta^{17}O$ values between $+0.92$ and $+3.24$‰ with an average value of $+2.06 \pm 0.71$‰ for 19 samples. The lower ^{17}O-enrichment of desert varnish implies that some of the sulfate in desert varnish originates by oxidation of sulfide minerals in the substrate ($\Delta^{17}S = 0$‰) and from eolian dust, part of which may originate from ^{17}O-enriched salt deposits.

Sulfate is well suited for the study of mass-independent isotope fractionation because the O in sulfate ions does not reequilibrate isotopically with O in meteoric water or in O-bearing minerals (e.g., carbonates, phosphates, silicates). In addition, sulfate deposited in environments where atmospheric sulfate is dominant may contain larger amounts of ^{17}O-enriched sulfate than soil salts in Namibia, gypsum in volcanic ash, and desert varnish in Death Valley, California. Such low-sulfate deposits may include snow in the polar regions and at high elevations in South America and Asia.

The phenomenon of mass independent isotope fractionation of O and S in the atmosphere of Earth may be useful for the study of atmospheric processes on other planets and in the solar nebula prior to the formation of the Sun (Thiemens, 1999). For example, the soils of Mars are known to contain sulfate, chloride, and nitrate salts that may be enriched in ^{17}O (Bao et al., 2001). In addition, ground-based observations indicate the presence of desert varnish on boulders on the surface of Mars:

Guinness et al., 1997, *J. Geophys. Res., Planets (E)*, 102(12):28687–28703. Israel et al., 1997, *J. Geophys. Res., Planets (E)*, 102(12): 28705–28716.

Analyses of martian meteorites (Nakhla, Lafayette, and ALH 84 001) indicate that they contain carbonate and other secondary minerals that show evidence of mass-independent isotope fractionation of O:

Karlsson et al., *Science*, 255:1409–1411. Farquhar et al., 1998, *Science*, 280:1580–1582. Farquhar and Thiemens, 2000, *J. Geophys. Res., Planets (E)*, 105:11911–11997.

29.12 SUMMARY

The isotopes of S are fractionated during the reduction of sulfate to sulfide by bacteria (*Desulfovibrio desulfuricans*) and by isotope exchange reactions among sulfur-bearing compounds, ions, and molecules. Bacterial reduction of sulfate in nature enriches hydrogen sulfide in ^{32}S by 50‰ or more compared to the sulfate. Laboratory cultures have generally yielded enrichments of only 27‰. The magnitude of the ^{32}S enrichment depends on a variety of factors, including the rate of reduction, the temperature, the nature and availability of the food supply, and the size of the sulfate reservoir.

Hydrogen sulfide and sulfide minerals in Recent marine sediment are variously enriched in ^{32}S compared to contemporary marine sulfate. The effect is attributable to the action of bacteria. Enrichment in ^{32}S and variability of $\delta^{34}S$ are characteristic of S that has passed through the sedimentary cycle.

The $\delta^{34}S$ values of petroleum range from -8 to $+32$‰ and appear to be depleted in ^{34}S by about 15‰ compared to marine sulfate at the time of formation of the petroleum. On the other hand, the $\delta^{34}S$ values of organically bound S in low-sulfur coal are fairly constant, which has been taken as evidence that the isotope composition of sulfate in fresh water remained constant while that of marine sulfate varied systematically.

The origin of native S deposits in the caps of salt domes and in certain evaporite sequences has been explained on the basis of isotopic compositions of C and S. Native S was produced as a by-product of the reduction of sulfate by bacteria feeding on petroleum. The calcite cap rock is strongly enriched in ^{12}C because it formed by precipitation of metabolic CO_2 derived from petroleum.

Isotope fractionation of S by sulfate-reducing bacteria apparently did not become important until after about 2.35 Ga, when photosynthetic S-oxidizing bacteria had increased the sulfate concentration of the oceans sufficiently for anaerobic S-reducing bacteria to evolve. Therefore, photosynthesis seems to have preceded bacterial sulfate-reduction, which was followed by O respiration in Neoproterozoic time. Because of the early evolution of the green or purple sulfur-oxidizing bacteria, the presence of sulfate deposits in Precambrian rocks of Archean age cannot be taken as evidence for the presence of O in the atmosphere in early Precambrian time. The activity of sulfate-reducing bacteria in the oceans between 1000 and 1300 Ma is clearly indicated by evaporite deposits in the Grenville Series of Ontario and similar rocks in Zimbabwe whose isotope composition is represented by $\delta^{34}S = +17.5 \pm 3$‰.

Systematic variations of the $\delta^{34}S$ values of marine sulfate minerals of Phanerozoic age indicate significant changes in the S cycle operating on the surface of the Earth. These changes involve the rates of total S input and output, as well as the isotopic composition of the S entering and leaving the oceans. Both the concentration and the isotopic composition of sulfate in the oceans have varied in the past.

Sulfide minerals in igneous rocks derived from magma sources in the mantle commonly have isotope compositions similar to S in the iron meteorite Canyon Diablo. Granitic rocks contain S with variable isotope compositions because the magma from which they crystallized either formed by partial melting of sedimentary rocks or because the magma was contaminated by assimilation of sedimentary rocks containing S of biogenic origin. Some plutonic and volcanic igneous rocks of basaltic composition also contain fractionated S. It is unlikely that such variations are attributable to differences in the isotopic compositions of S in the mantle. Instead, they can be caused by contamination of magma with crustal S in a process called sulfurization, by alteration by seawater, and by outgassing of SO_2 from the magma at different O fugacities.

The isotopic composition of S in sulfide ore deposits has received much attention. Attempts to develop criteria for distinguishing ore deposits of igneous–hydrothermal origin from those of sedimentary–syngenetic origin have failed because of extensive overlap of $\delta^{34}S$ values of such deposits. Theoretical and experimental studies indicate that isotope exchange reactions among cogenetic sulfide minerals and ore-forming fluids may achieve equilibrium and that the temperature of equilibration can be estimated from $\Delta^{34}S$ values of coexisting sulfides. The $\delta^{34}S$ values of individual minerals do not necessarily reflect the isotopic composition of S in the fluid from which they formed because they may also be controlled by the pH and the fugacity of O. By combining isotopic data for C and S in coexisting carbon- and sulfur-bearing minerals, both pH and O fugacity may be determined.

Sulfur entering the atmosphere from natural and anthropologic sources has a wide range of isotopic compositions. Hence the isotopic composition can be used to identify the sources of S in meteoric precipitation or in the air. The principal natural sources are SO_2 discharged by volcanic eruptions, metabolic H_2S released by sulfate-reducing bacteria, sulfate derived from the oceans, and organic S-compounds released by plants. Anthropologic sources are related primarily to the combustion or refining of fossil fuels and to smelting of sulfide ores. These sources dominate the S content of the atmosphere and of meteoric precipitation in highly industrialized regions of the Earth and are creating an environmental hazard.

REFERENCES FOR ISOTOPE GEOCHEMISTRY (INTRODUCTION, SECTION 29.1)

Bachinski, D. J., 1969. Bond strength and sulfur isotopic fractionation in coexisting sulfides. *Econ. Geol.*, 65:56–65.

Canfield, D. E., 2001. Biogeochemistry of sulfur isotopes. In J. W. Valley and D. R. Cole (Eds.), *Stable Isotope Geochemistry. Rev. Mineral. Geochem.*, Vol. 43:607–636. Mineralogical Society of America, Geochemical Society, Blacksburg, Virginia.

Hoefs, J., 1997. *Stable Isotope Geochemistry*, 4th ed. Springer-Verlag, Heidelberg.

Krouse, H. R, 1980. Sulphur isotopes in our environment. In P. Fritz and J. Ch. Fontes (Eds.), *Handbook of Environmental Isotope Geochemistry*, Vol. 1A: *The Terrestrial Environment*, pp. 435–471. Elsevier, Amsterdam.

Lide, D. R., and H. P. R. Frederikse, 1995. *Handbook of Chemistry and Physics*, 76th ed. CRC Press, Boca Raton, Florida.

Nielsen, H., 1978. Sulfur isotopes in nature. In K. H. Wedepohl (Ed.), *Handbook of Geochemistry*, Springer-Verlag, Heidelberg.

Nielsen, H., 1979. Sulfur isotopes. In E. Jäger and J. C. Hunziker (Eds.), *Lectures in Isotope Geology*, pp. 283–312. Springer-Verlag, Heidelberg.

Ohmoto, H., and R. O. Rye, 1979. Isotopes of sulfur and carbon. In H. L. Barnes (Ed.), *Geochemistry of Hydrothermal Ore Deposits*, 2nd ed., pp. 509–567. Wiley, New York.

Schneider, A., 1970. The sulfur isotope composition of basaltic rocks. *Contrib. Mineral. Petrol.*, 25:95–124.

Smitheringale, W. G., and M. L. Jensen, 1963. Sulfur isotopic composition of the Triassic igneous rocks of eastern United States. *Geochim. Cosmochim. Acta*, 27:1183–1207.

Thode, H. G., 1970. Sulfur isotope geochemistry and fractionation between coexisting mineral. *Mineral. Soc. Am. Spec. Paper*, 3:133–144.

REFERENCES FOR BIOGENIC ISOTOPE FRACTIONATION (SECTION 29.2)

Canfield, D. E., 2001. Biogeochemistry of sulfur isotopes. In J. W. Valley and D. R. Cole (Eds.), *Stable Isotope Geochemistry, Rev. Mineral. Geochem.*, Vol. 43:607–636. Mineralogical Society of America, Geochemical Society, Blacksburg, Virginia.

Harrison, A. G., and H. G. Thode, 1958. Mechanism of the bacterial reduction of sulfate from isotope fractionation studies. *Trans. Faraday Soc.*, 54:84–92.

Kaplan, I. R., and S. C. Rittenberg, 1964. Microbiological fractionation of sulfur isotopes. *J. Gen. Microbiol.*, 34:195–212.

McCready, R. G. L., I. R. Kaplan, and G. A. Din, 1974. Fractionation of sulfur isotopes by the yeast *Saccharomyces cerevisiae*. *Geochim. Cosmochim. Acta*, 38:1239–1253.

Rees, C. E., 1973. A steady state model for sulfur isotope fractionation in bacterial reduction processes. *Geochim. Cosmochim. Acta*, 37:1141–1162.

Trudinger, P. A., and L. A. Chambers, 1973. Reversibility of bacterial sulfate reduction and its relevance to isotopic fractionation. *Geochim. Cosmochim. Acta*, 37:1775–1778.

Tudge, A. P., and H. G. Thode, 1950. Thermodynamic properties of isotopic compounds of sulfur. *Can. J. Res.*, B28:567–578.

REFERENCES FOR SULFUR IN RECENT SEDIMENT (SECTION 29.3)

Deevey, E. S., Jr., N. Nakai, and M. Stuiver, 1963. Fractionation of sulfur and carbon isotopes in a meromictic lake. *Science*, 139:407–408.

Hartmann, M., and H. Nielsen, 1969. $\delta^{34}S$ Werte in rezenten Meeressedimenten und ihre Deutung am Beispiel einiger Sedimentprofile aus der westlichen Ostsee. *Geol. Rundschau*, 58:621–655.

Kaplan, I. R. 1983. Stable isotopes of sulfur, nitrogen and deuterium in Recent marine environments. In M. A. Arthur, T. F. Anderson, I. R. Kaplan, J. Veizer, and L. S. Land (Eds.), *Stable Isotopes in Sedimentary Geology*, pp. 2–1–2–108, SEPM Short Course No. 10. Tulsa, Oklahoma.

Kaplan, I. R., K. O. Emery, and S. C. Rittenberg, 1963. The distribution and isotopic abundance of sulfur in recent marine sediments off southern California. *Geochim. Cosmochim. Acta*, 27:297–331.

Nakai, N. and M. L. Jensen, 1964. The kinetic isotope effect in the bacterial reduction and oxidation of sulfur. *Geochim. Cosmochim. Acta*, 28:1893–1912.

Thode, H. G., A. G. Harrison, and J. Monster, 1960. Sulphur isotope fractionation in early diagenesis of recent sediments of north-east Venezuela. *Bull. Am. Assoc. Petrol. Geol.*, 44:1809–1817.

Vinogradov, A. P., V. A. Grinenko, and V. I. Ustinov, 1962. Isotopic composition of sulfur compounds in the Black Sea. *Geokhimiya*, 10:973–997.

REFERENCES FOR FOSSIL FUELS (SECTION 29.4)

Claypool, G. E., W. T. Holser, I. R. Kaplan, H. Sakai, and I. Zak, 1980. The age curves of sulfur and oxygen isotopes in marine sulfate and their mutual interpretation. *Chem. Geol.*, 28:199–260.

Smith, J. W., and B. D. Batts, 1974. The distribution and isotopic composition of sulfur in coal. *Geochim. Cosmochim. Acta*, 38:121–133.

Thode, H. G., and J. Monster, 1965. Sulfur-isotope geochemistry of petroleum, evaporites, and ancient seas. In A. Young and J. E. Galley (eds.), *Fluids in Subsurface Environments*, Mem. 4, pp. 367–377. American Association of Petroleum Geologists. Tulsa, Oklahoma

Thode, H. G., and C. E. Rees, 1970. Sulfur isotope geochemistry and Middle East oil studies. *Endeavour*, 19:24–28.

REFERENCES FOR NATIVE SULFUR DEPOSITS (SECTION 29.5)

Davis, J. B., and D. W. Kirkland, 1970. Native sulfur deposition in the Castile Formation, Gulberson County, Texas. *Econ. Geol.*, 65:109.

Dessau, G., M. L. Jensen, and N. Nakai, 1962. Geology and isotopic studies of Sicilian sulfur deposits. *Econ. Geol.*, 57:410–438.

Feely, H. W., and J. L. Kulp, 1957. The origin of the Gulf Coast salt dome sulphur deposits. *Bull. Am. Assoc. Petrol. Geol.*, 41:1802–1853.

Schneider, A., and H. Nielsen, 1965. Zur Genese des elementaren Schwefels im Gips von Weenzen. *Contrib. Mineral. Petrol.*, 11:705–717.

REFERENCES FOR SEDIMENTARY ROCKS OF PRECAMBRIAN AGE (SECTION 29.6)

Cameron, E. M., 1982. Sulphate and sulphate reduction in early Precambrian oceans. *Nature*, 296:145–148.

Cameron, E. M., 1983. Genesis of Proterozoic iron-formation: Sulphur isotope evidence. *Geochim. Cosmochim. Acta*, 47:1069–1074.

Claypool, G. E., W. T. Holser, I. R. Kaplan, H. Sakai, and I. Zak, 1980. The age curves of sulfur and oxygen isotopes in marine sulfate and their mutual interpretation. *Chem. Geol.*, 28:199–260.

Fripp, R. E. P., T. H. Donnelly, and I. B. Lambert, 1979. Sulphur isotope results for Archean banded iron-formation, Rhodesia. *Geol. Soc. S. Africa Spec. Publ.*, 6:205–208.

Goldhaber, M. B., R. L. Reynolds, and R. O. Rye, 1978. Origin of a south Texas roll-type uranium deposit II. Sulfide petrology and sulfur isotope studies. *Econ. Geol.*, 73:1690–1705.

Goodwin, A., J. Monster, and H. G. Thode, 1976. Carbon and sulfur isotope abundances in Archean iron-formation and early Precambrian life. *Econ. Geol.*, 71:870–891.

Hattori, K., and E. M. Cameron, 1986. Archean magmatic sulphate. *Nature*, 319:45–47.

Hattori, K., F. A. Campbell, and H. R. Krouse, 1983. Sulphur isotope abundances in Aphebian clastic rocks: Implications for the coeval atmosphere. *Nature*, 302:323–326.

Lambert, I. B., T. H. Donnelly, J. S. R. Dunlap, and D. I. Groves, 1978. Stable isotopic compositions of early Archean sulphate deposits of probable evaporite and volcanogenic origins. *Nature*, 276:808–811.

Monster, J., P. W. U. Appel, H. G. Thode, M. Schidlowski, C. M. Carmichael, and D. Bridgwater, 1979. Sulfur isotope studies in Early Archean sediments from Isua, West Greenland: Implications to the antiquity of bacterial sulfate reduction. *Geochim. Cosmochim. Acta*, 43:405–413.

Perry, E. C., Jr., J. Monster, and T. Reimer, 1971. Sulfur isotopes in Swaziland system barites and the evolution of the Earth's atmosphere. *Science*, 171:1015–1016.

Ripley, E. M., and D. L. Nicol, 1981. Sulfur isotopic studies of Archean slate and graywacke from northern Minnesota: Evidence for the existence of sulfate reducing bacteria. *Geochim. Cosmochim. Acta*, 45:839–846.

Schidlowski, M., 1979. Antiquity and evolutionary status of bacterial sulfate reduction: Sulfur isotope evidence. In *Origins of Life*, Vol. 9, pp. 299–311. Reidel, Dordrecht.

Thode, H. G., H. B. Dunford, and M. Shima, 1962. Sulfur isotope abundances in rocks of the Sudbury district and their geological significance. *Econ. Geol.*, 57:565–578.

Whelan, J. F., and R. O. Rye, 1978. Stable-isotope studies in the Balmat-Edwards Zn–Pb district, New York. *Geol. Soc. Am., Abstr. Progr.*, 10:515.

REFERENCES FOR ISOTOPIC EVOLUTION OF MARINE SULFATE (SECTION 29.7)

Claypool, G. E., W. T. Holser, I. R. Kaplan, H. Sakai, and I. Zak, 1980. The age curves of sulfur and oxygen isotopes in marine sulfate and their mutual interpretation. *Chem. Geol.*, 28:199–260.

Garrels, R. M., and E. A. Perry, 1974. Cycling of carbon, sulfur, and oxygen through geologic time. In E. D. Goldberg (Ed.), *The Sea*, Vol. 5: *Marine Chemistry*, pp. 303–336. Wiley, New York.

Holland, H. D., 1973. Systematics of the isotopic composition of sulfur in the oceans during the Phanerozoic and its implications for atmospheric oxygen. *Geochim. Cosmochim. Acta*, 37:2605–2616.

Rees, C. E., 1970. The sulphur isotope balance of the ocean: An improved model. *Earth Planet. Sci. Lett.*, 7:336–370.

Schidlowski, M., and C. E. Junge, 1981. Coupling among the terrestrial sulfur, carbon and oxygen cycles: Numerical modeling based on revised Phanerozoic carbon isotope record. *Geochim. Cosmochim. Acta*, 45:589–594.

Schidlowski, M., C. E. Junge, and H. Pietrek, 1977. Sulfur isotope variations in marine sulfate evaporites and the Phanerozoic oxygen budget. *J. Geophys. Res.*, 82:2557–2656.

Thode, H. G., and J. Monster, 1965. Sulfur-isotope geochemistry of petroleum, evaporites, and ancient seas. In A. Young and J. E. Galley (Eds.) *Fluids in Subsurface Environments*, Mem. 4, pp. 367–377. American Association of Petroleum Geologists. Tulsa, Oklahoma.

REFERENCES FOR IGNEOUS ROCKS (SECTION 29.8)

Allard, P., 1979. $^{13}C/^{12}C$ and $^{34}S/^{32}S$ ratios in magmatic gases from ridge volcanism in Afar. *Nature*, 282:56–58.

Cheney, E. S., and I. M. Lange, 1967. Evidence for sulfurization and the origin of some Sudbury-type ores. *Mineral. Deposita*, 2:80–94.

Faure, G., J. Hoefs, and T. M. Mensing, 1984. Effect of oxygen fugacity on sulfur isotope compositions and magnetite concentrations in the Kirkpatrick Basalt, Mount Falla, Queen Alexandra Range, Antarctica. *Chem. Geol. Isotope Geosci. Sect.*, 2:301–311.

Gorbachev, N. S., and L. N. Grinenko, 1973. Origin of the October sulfide ore deposits, Noril'sk region, in the light of sulfide and sulfate sulfur isotope compositions. *Geochemistry*, 10:843–851.

Grinenko, V. A., L. V. Dmitriev, A. A. Migdisov, and A. Ya Sharas'kin, 1975. Sulfur contents and isotope compositions for igneous and metamorphic rocks form mid-ocean ridges. *Geochimia*, 1975(2):199–206.

Hoefs, J., 1997. *Stable Isotope Geochemistry*, 4th ed. Springer-Verlag, Heidelberg.

Hubberten, H. W., H. Nielsen, and H. Puchelt, 1975. The enrichment of ^{34}S in the solfotaras of the Nea Kameni Volcano, Santorini Archipelago, Greece. *Chem. Geol.*, 16:197–205.

Kanehiro, K., S. Yui, H. Sakai, and A. Sasaki, 1973. Sulphide globules and sulphur isotope ratios in the abyssal tholeiite from the Mid-Atlantic Ridge near 30° N latitude. *Geochim. J.*, 7:89–96.

Mainwaring, P. R., and A. J. Naldrett, 1977. Country-rock assimilation and the genesis of Cu–Ni sulfide in the water Hen intrusion, Duluth Complex, Minnesota. *Econ. Geol.*, 72:1269–1284.

Puchelt, H., and H.-W. Hubberten, 1980. *Preliminary Results of Sulfur Isotope Investigations on Deep Sea Drilling Project Core from Legs 52 and 53*, Initial Rept. on the Deep Sea Drilling Project, Vol. 51, P. 2:1145–1148. U.S. Government Printing Office, Washington, D.C.

Ripley, E. M., 1981. Sulfur isotopic studies of the Dunka Road Cu–Ni deposit, Duluth Complex, Minnesota. *Econ. Geol.*, 76:610–620.

Sakai, H. 1968. Isotopic properties of sulfur compounds in hydrothermal processes. *Geochim. J.*, 2:29–49.

Sakai, H., T. J. Casadevall, and J. G. Moore, 1982. Chemistry and isotope ratios of sulfur in basalts and volcanic gases at Kilauea volcano, Hawaii. *Geochim. Cosmochim. Acta*, 46:729–738.

Sakai, H., A. Ueda, and C. W. Field, 1978. $\delta^{34}S$ and concentration of sulfide and sulfate sulfurs and in some ocean-floor basalts and serpentinites. In R. E. Zartman (Ed.), *Short Papers of the Fourth International Conference, Geochronology, Cosmochronology and Isotope Geology*, U.S. Geol. Surv. Open-File Rept. 78–701, pp. 372–374. U.S. Geological Survey, Washington, D.C.

Sasaki, A., and S. Ishibara, 1979. Sulfur isotopic composition of the magnetite-series and ilmenite-series granitoids in Japan. *Contrib. Mineral. Petrol.*, 68: 107–115.

Sato, M., and T. L. Wright, 1966. Oxygen fugacities directly measured in magmatic gases. *Science*, 153: 1103–1105.

Schneider, A., 1970. The sulfur isotope compositions of basaltic rocks. *Contrib. Mineral. Petrol.*, 25:95–124.

Shima, M., W. H. Gross, and H. G. Thode, 1963. Sulfur isotope abundances in basic sills, differentiated granites, and meteorites. *J. Geophys. Res.*, 68(9):2835–2847.

Smitheringale, W. G., and M. L. Jensen, 1963. Sulfur isotopic composition of the Triassic igneous rocks of eastern United States. *Geochim. Cosmochim. Acta*, 27:1183–1207.

Weiblen, P. W., and G. B. Morey, 1976. Textural and compositional characteristics of sulfide ores from the basal contact zone of the south Kawishiwi intrusion, Duluth Complex, northeastern Minnesota. In *Proceedings of the Thirty-Seventh Annual Meeting Symposium*, pp. 1–24. University of Minnesota, Minneapolis.

REFERENCES FOR SULFIDE ORE DEPOSITS (SECTION 29.9)

Bachinski, D. J., 1969. Bond strength and sulfur isotopic fractionation in coexisting sulfides. *Econ. Geol.*, 64:56–65.

Czamanske, G. K., and R. O. Rye, 1974. Experimentally determined sulfur isotope fractionation between sphalerite and galena in the temperature range 600° to 275°C. *Econ. Geol.*, 69:17–25.

Friedman, I., and J. R. O'Neil, 1977. Compilation of stable isotope fractionation factors of geochemical interest. In M. Fleischer (Ed.), *Data of Geochemistry*, 6th ed., Chapter KK, U.S. Geol. Surv. Prof. Paper 440-KK. U.S. Geological Survey, Washington, D.C.

Grootenboer, J., and H. P. Schwarcz, 1969. Experimentally determined sulfur isotope fractionations between sulfide minerals. *Earth Planet. Sci. Lett.*, 7:162–166.

Kajiwara, Y., and H. R. Krouse, 1971. Sulfur isotope partitioning in metallic sulfide systems. *Can. J. Earth Sci.*, 8:1397–1408.

Nielsen, H., 1979. Sulfur isotopes. In E. Jäger and J. C. Hunziker (Eds.), *Lectures in Isotope Geology*, pp. 283–312. Springer-Verlag, Heidelberg.

Ohmoto, H., 1972. Systematics of sulfur and carbon isotopes in hydrothermal ore deposits. *Econ. Geol.*, 67:551–578.

Ohmoto, H., and R. O. Rye, 1979. Isotopes of sulfur and carbon. In H. L. Barnes (Ed.), *Geochemistry of Hydrothermal Ore Deposits*, 2nd ed., pp. 509–567. Wiley, New York.

Sakai, H., 1957. Fractionation of sulfur isotopes in nature. *Geochim. Cosmochim. Acta*, 12:150–169.

Sakai, H., 1968. Isotopic properties of sulfur compounds in hydrothermal processes. *Geochem. J.*, 2:29–49.

Sangster, D. F. 1968. Relative sulfur isotope abundances of ancient seas and stratabound sulphide deposits. *Geol. Assoc. Can. Proc.*, 19:79–91.

Thode, H. G., 1970. Sulfur isotope geochemistry and fractionation between coexisting minerals. *Mineral Soc. Am. Spec. Paper*, 3:133–144.

REFERENCES FOR SULFUR IN THE ENVIRONMENT (SECTION 29.10)

Eriksson, E., 1963. The yearly circulation of sulphur in nature. *J. Geophys. Res.*, 68:4001–4008.

Granat, L., R. O. Hallberg, and H. Rodhe, 1976. The global sulfur cycle. In B. H. Svensson and R. Söderland (eds.), *Ecological Bulletins*, p. 22. Nitrogen, Phosphorus, and Sulfur. *Global Cycles. SCOPE Rep.* 7:89–134.

Grey, D. C. and M. L. Jensen, 1972. Bacteriogenic sulfur in air pollution. *Science*, 177:1099–1100.

Junge, C. E., 1963. Sulfur in the atmosphere. *J. Geophys. Res.*, 68:3975–3976.

Kellogg, W. W., R. D. Cadle, E. R. Allen, A. L. Lazarus, and E. A. Martell, 1972. The sulfur cycle. *Science*, 175:587–596.

Kotra, J. P., D. L. Finnegan, W. H. Zoller, M. A. Hart, and J. L. Moyers, 1983. El Chichon: Composition of plume gases and particles. *Science*, 222:1018–1021.

Krouse, H. R., 1980. Sulphur isotopes in our environment. In P. Fritz and J. Ch. Fontes (Eds.), *Handbook of Environmental Isotope Geochemistry*, Vol. 1A, pp. 435–471. Elsevier, Amsterdam.

Nielsen, H., 1974. Isotope composition of the major contributors to atmospheric sulfur. *Tellus*, 26:213–221.

Palais, J. M., and M. Legrand, 1985. Soluble impurities in the Byrd Station ice core, Antarctica: Their origin and sources. *J. Geophys. Res.*, 90(C1):1143–1154.

Smith, J. W., 1975. Stable isotope studies and biological element cycling. In G. Eglinton (Ed.), *Environmental Chemistry*, Vol. 1, pp. 1–21. Chemical Society, London.

REFERENCES FOR MASS-INDEPENDENT ISOTOPE FRACTIONATION (SECTION 29.11)

Bains-Sahota, S. K., and M. H. Thiemens, 1989. A mass independent sulfur isotope effect in the nonthermal formation of S_2F_{10}. *J. Chem. Phys.*, 90:6099–6110.

Bao, H., G. M. Michalski, and M. H. Thiemens, 2001. Sulfate oxygen-17 anomalies in desert varnish. *Geochim. Cosmochim. Acta*, 65:2029–2036.

Bao, H., M. H. Thiemens, J. Farquhar, D. A. Campbell, C. C.-W. Lee, K. Heine, and D. B. Loope, 2000. Anomalous ^{17}O compositions in massive sulphate deposits on Earth. *Nature*, 406:176–178.

Bhattacharya, S. K., and M. H. Thiemens, 1989. New evidence for symmetry dependent isotope effects: O + CO reaction. *Z. Naturforsch.*, 44A:435–444.

Heidenreich, J. E., and M. H. Thiemens, 1983. A non-mass-dependent isotope effect in the production of ozone from molecular oxygen. *J. Chem. Phys.*, 78:892–895.

Heidenreich, J. E., and M. H. Thiemens, 1985. The non-mass-dependent oxygen isotope effect in the electro dissociation of carbon dioxide: A step toward understanding NoMaD chemistry. *Geochim. Cosmochima Acta*, 49:1303–1306.

Thiemens, M. H., 1999. Mass-independent isotope effects in planetary atmospheres and the early solar system. *Science*, 283:341–345.

Thiemens, M. H., and J. E. Heidenreich, 1983. The mass independent fractionation of oxygen—a novel isotope effect and its cosmochemical implications. *Science*, 219:1073–1075.

CHAPTER 30

Boron and Other Elements

THE number of elements whose isotopic compositions in terrestrial materials are known to vary is growing. Most of these elements have low atomic numbers, and the differences in the masses of their stable isotopes are sufficient to cause mass-dependent isotope fractionation in the course of geochemical processes on the surface of the Earth. Even though the resulting ranges of the isotopic compositions of these elements are small, variations in their isotopic compositions are, nevertheless, detectable because of the sensitivity and precision of modern mass spectrometers.

The elements to be considered in this chapter include: lithium (Li, $Z = 3$) boron (B, $Z = 5$), silicon (Si, $Z = 14$), and chlorine (Cl, $Z = 17$). The stable isotopes of these elements, their abundances, and masses are listed in Table 30.1.

The variations of the isotope compositions of H, O, C, N, and S have already been presented in Chapters 26 (H and O), 27 (C), 28 (N), and 29 (S). In addition, some elements, whose atomic numbers are less than 36, are not fractionated because they have only one stable isotope: Be ($Z = 4$), F ($Z = 9$), Na ($Z = 11$), Al ($Z = 13$), P ($Z = 15$), Sc ($Z = 21$), Mn ($Z = 25$), Co ($Z = 27$), and As ($Z = 33$). Moreover the fractionation of the isotopes of the noble gases is masked because some of the isotopes are radiogenic ($^{4}_{2}He$, $^{40}_{18}Ar$) or are products of spontaneous and induced fission of U isotopes (Kr, Xe). In addition, $^{21}_{10}Ne$ and $^{3}_{2}He$ are cosmogenic and the isotopes of Rn are the unstable daughters of U and Th. The remaining elements include Mg ($Z = 12$), K ($Z = 19$), Ti ($Z = 22$), V ($Z = 23$), Cr ($Z = 24$), Fe ($Z = 26$), Ni ($Z = 28$), Cu ($Z = 29$), Zn ($Z = 30$), Ga ($Z = 31$), Ge ($Z = 32$), Se ($Z = 34$), and Br ($Z = 35$).

Table 30.1. Abundances and Masses of the Stable Isotopes of Elements Whose Isotopic Compositions in Natural Materials Are Known to Vary

Element	Z	Stable Isotopes	Abundance, %	Mass, amu
Lithium	3	^{6}Li	7.5	6.015122
		^{7}Li	92.5	7.016003
Boron	5	^{10}B	19.8	10.0129372
		^{11}B	80.2	11.0093056
Silicon	14	^{28}Si	92.23	27.976927
		^{29}Si	4.67	28.976495
		^{30}Si	3.10	29.973770
Chlorine	17	^{35}Cl	75.77	34.96885272
		^{37}Cl	24.23	36.9559026
Iron	26	^{54}Fe	5.9	53.939613
		^{56}Fe	91.72	55.934940
		^{57}Fe	2.1	56.935396
		^{58}Fe	0.28	57.933278
Selenium	34	^{74}Se	0.9	73.922474
		^{76}Se	9.1	75.919212
		^{77}Se	7.6	76.919912
		^{78}Se	23.6	77.917307
		^{80}Se	49.9	79.916519
		^{82}Se	8.9	81.916697

The isotope compositions of some of these elements in terrestrial materials are known to be variable:

Copper: Marechal et al., 1999, *Chem. Geol.*, 156:251–273. Zhu et al., 2000, *Chem. Geol.*, 163:139–149.

Iron: Alexander, and Wang, 2001, *Meteorit. Planet. Sci.*, 36:419–428. Anbar et al., 2000, *Science*, 288:126–128. Beard and Johnson, 1999. *Geochim. Cosmochim. Acta*, 63:1653–1660. Beard et al., 1999. *Science*, 285:1889–1892. Beard et al., 2003a, *Chem. Geol.*, 195: 87–117. Beard et al., 2003b. *Geology*, 31:629–632. Beard and Johnson, 2003, *Geochem. News*, 117:8–13. Belshaw et al., 2000, Inter. *J. Mass Spectrom.*, 197:191–195. Bullen et al., 2001, *Geology*, 29:699–702. Brantley et al., 2001, *Geology*, 29:535–538. Johnson et al., 2003. *Contrib. Mineral. Petrol.*, 144:523–547. Kehm et al., 2003. *Geochim. Cosmochim. Acta*, 67: 2879–2891. Mandernack et al., *Science*, 285: 1892–1896. Matthews et al., 2001. *Earth Planet. Sci. Lett.*, 192:81–92. Polyakov and Mineev, 2000, *Geochim. Cosmochim. Acta*, 64:849–865. Roe et al., 2003, *Chem. Geol.*, 195:69–85. Schauble et al., 2001, *Geochim. Cosmochim. Acta*, 65:2487–2497. Sharma et al., 2001, *Earth Planet. Sci. Lett.*, 194:39–51. Skulan et al., 2002, *Geochim. Cosmochim. Acta*, 66:2995–3015. Völkening and Papanastassiou, 1989, *Astrophys. J.*, 347:L43–L46. Walczyk and von Blanckenburg, 2002, *Science*, 295:2065–2066. Zhu et al., 2000, *Science*, 287:2000–2002. Zhu et al., 2001, *Nature*, 412:311–313. Zhu et al., 2002, *Earth Planet. Sci. Lett.*, 200:47–62.

Selenium: Crusius and Thomson, 2003, *Geochim. Cosmochim. Acta*, 67:265–273. Johnson et al., 1999, *Geochim. Cosmochim. Acta*, 63:2775–2783. Johnson and Bullen, 2003, *Geochim. Cosmochim. Acta*, 67:413–419. Lorand et al., 2003, *Geochim. Cosmochim. Acta*, 67:4137–4151. Peters et al., 1999. *Organic Geochem.* 30(10):1287–1300. Rouxel et al., 2002. *Geochim. Cosmochim. Acta*, 66:3191–3199.

Bromine: Xiao et al., 1993, *Int. J. Mass Spectrom. Ion Proc.*, 123:117–123. Eggenkamp and Coleman, 2000, *Chem. Geol.*, 167:393–402.

Variations of the isotopic compositions of Mg, K, Ti, V, Cr, Ni, Zn, Ga, and Ge in terrestrial samples have not yet been demonstrated, although variations in the isotopic compositions of Mg and Cr in meteorites have been attributed to the decay of short-lived radionuclides in the solar nebula (Chapter 24). In additional, some evidence exists that the isotopes of K have been fractionated by diffusion in rocks. The isotope compositions of many other elements in meteorites are known to vary because of processes that occurred in the solar nebula: mass-dependent isotope fractionation, nuclear reactions caused by cosmic rays, decay of radioactive nuclides, and implantations of nuclear particles by the solar wind (Shima, 1986).

30.1 BORON

Boron is a lithophile element that occurs at low concentrations in virtually all igneous, sedimentary, and metamorphic rocks in the crust of the Earth. Boron is a nonmetallic element whose chemical properties resemble those of C and Si rather than those of its congeners Al and Ga. Boron reaches high concentrations only in certain evaporite deposits, where it occurs as a major element in a large number of minerals. The isotopic composition of B in natural materials varies by up to 75‰ as a result a mass-dependent isotopic fractionation (Hoefs, 1997).

30.1a Geochemistry

Boron is a member of group IIIA in the periodic table and its electronic formula is $1s^2 2s^2 2p^1$. Therefore, B has a valence number of +3, like its congeners Al and Ga. However, B occurs in nature *only* in the form of oxy-anion complexes $BO_3{}^{3-}$ and $BO_4{}^{5-}$ rather than as the cation B^{3+}. The $BO_3{}^{3-}$ ion forms boric acid (H_3BO_3), whereas the tetraborate ion $BO_4{}^{5-}$ occurs in the form of tetrahedral complexes such as $B(OH)_4{}^-$. The minerals of B can also contain other complex anions such as

SiO_4^{4-} (borosilicate), PO_4^{3-} (borophosphate), and CO_3^{2-} (borocarbonate). The minerals of B contain either the planar triangular BO_3^{3-} ion or the tetrahedral BO_4^{5-} ion or both. Boron minerals occur primarily in nonmarine evaporite deposits in combination with Na, Ca, and Mg and with varying amounts of water of hydration (Christ, 1960).

Commercial B deposits, such as that at Boron, California, contain the minerals ulexite (Na $Ca[B_5O_6(OH)_6]\cdot 5H_2O$), borax ($Na_2[B_4O_5(OH)_4]\cdot 8H_2O$), colemanite ($Ca_a[B_3O_4(OH)_3]_2\cdot 2H_2O$), and kernite ($Na_2[B_4O_5(OH)_4]\cdot 2H_2O$).

The abundance of B in the solar system is only 2.12×10 atoms per 10^6 atoms of Si (Anders and Grevesse, 1989). These authors attributed the formation of the stable isotopes of B, Li, and Be (except ^7_3Li) to cosmic-ray spallation reactions in the solar nebula.

The average concentration of B in the rocks of the continental crust of the Earth is 10 μg/g (Taylor and McLennan, 1985). Accordingly, B is a trace element in most igneous, sedimentary, and metamorphic rocks. The data in Table 30.2 and Figure 30.1 indicate that the concentrations of B in igneous rocks increase with rising concentration of Si and that clay-rich sedimentary rocks (e.g., shale and deep-sea clay) are strongly enriched in B compared to mantle-derived basalt. In addition, certain minerals have elevated concentrations of B: gadolinite and vesuvianite (8000 ppm), sericitized plagioclase (2000 ppm), certain garnets (1000 ppm), cordierite (150 ppm), wollastonite, sodalite, and scapolite (100 ppm), and pyroxene (60 ppm) (Harder, 1974).

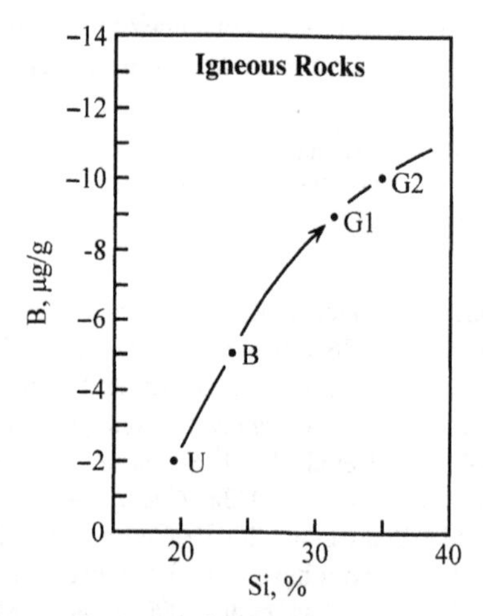

FIGURE 30.1 Increase of the average B concentration of igneous rocks with rising concentrations of Si: U = ultramafic; B = basalt; G1 = high-Ca granite; G2 = low-Ca granite. Data from Turekian and Wedepohl (1961).

Table 30.2. Average Concentrations of Li, B, and Cl in Igneous and Sedimentary Rocks and Surface Water

	μg/g		
Rock Type	Li	B	Cl
Ultramafic	0.5	2	45
Basalt	16	5	55
High-Ca granite	24	9	130
Low-Ca granite	40	10	200
Shale	66	100	180
Sandstone	15	35	10
Carbonate rocks	5	20	150
Deep-sea clay	57	230	21,000
River water	0.003	0.01	7.8
Seawater	0.17	4.5	19,500

Note: Data for sedimentary rocks from Turekian and Wedepohl (1961) and for surface water from Taylor and McLennan (1985).

30.1b Isotopic Composition

The isotopic composition of B is measured by thermal ionization mass spectrometry (TIMS) using either disodium metaborate ($Na_2BO_2^+$) or dicesium metaborate ($Cs_2BO_2^+$) cations (Spivack and Edmond, 1986). Cesium metaborate is better than Na metaborate because the large mass of the Cs metaborate reduces instrumental isotope fractionation and improves the measurement precision to $\pm 0.25‰$. Alternatively, B can be analyzed as the negatively charged BO_2^- ion (Vengosh et al., 1989; Hemming and Hanson, 1992) or by ion microprobe mass spectrometry (Chaussidon and Albarède, 1992). Both methods have an analytical uncertainty of about $\pm 2‰$.

The $^{11}B/^{10}B$ ratio is expressed as the $\delta^{11}B$ parameter defined by:

$$\delta^{11}B = \frac{(^{11}B/^{10}B)_{spl} - (^{11}B/^{10}B)_{std}}{(^{11}B/^{10}B)_{std}} \times 10^3‰ \tag{30.1}$$

where the standard is a sample of boric acid (SRM 951) prepared by the National Bureau of Standards (Washington, D.C.) from borax mined at Searles Lake, California. The $^{11}B/^{10}B$ ratio of SRM 951 is 4.04558 (Palmer and Slack, 1989), although Catanzaro et al. (1970) reported 4.0436.

The $\delta^{11}B$ values of terrestrial materials in Figure 30.2, based on a literature survey by Hoefs (1997), range from about −30‰ in certain tourmalines and nonmarine evaporites (Swihart et al., 1986; Swihart and Moore, 1989; Chaussidon and Albarède, 1992) up to about +60‰ in brine lakes of Australia and Israel (Vengosh et al. (1991a, b)).

The principal process that causes fractionation of B isotopes in aqueous solution is the exchange reaction:

$$^{10}B(OH)_3 + {}^{11}B(OH)_4^- \rightleftharpoons {}^{11}B(OH)_3 + {}^{10}B(OH)_4^- \tag{30.2}$$

where $B(OH)_3$ is boric acid and $B(OH)_4^-$ is the tetrahedral B complex. The boric acid molecule is enriched in ^{11}B relative to the tetrahedral complex by 23‰ at 25°C (Kakihana et al., 1977). The abundance of $B(OH)_3$ in aqueous solution is pH dependent such that boric acid is dominant at pH <9.0, whereas the tetrahedral complex dominates at pH >9.0 (Hemming and Hanson, 1992).

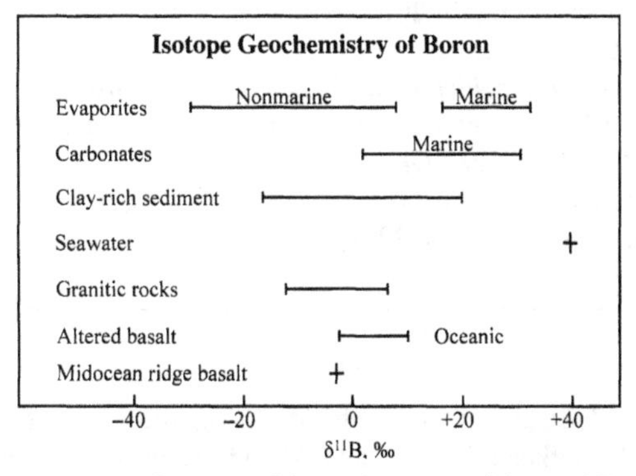

FIGURE 30.2 Ranges of isotopic compositions of B in different kinds of rocks and in seawater. Adapted from a compilation of the literature by Hoefs (1997).

The $\delta^{11}B$ value of seawater is constant at about +40‰, which implies that seawater is enriched in ^{11}B relative to borax at Searles Lake (Palmer et al., 1987; Spivack and Edmond, 1987; Hemming and Hanson, 1992). The reason for the enrichment of seawater is the preferential sorption of $^{10}B(OH)_4^-$ by marine clay-rich sediment, hydrothermally altered basalt of the oceanic crust, and carbonate minerals. The isotope composition of B in these solids is pH-dependent because the pH determines the $B(OH)_4^-/B(OH)_3$ ratio of seawater and hence controls the concentration of $B(OH)_4^-$ that is available for sorption.

The $\delta^{11}B$ values of biogenic marine carbonate shells are lower than those of seawater, presumably because of the inclusion of the $B(OH_4^-)$ species in the carbonate lattice (Hemming and Hanson, 1992; Spivack et al., 1993). In that case, the $\delta^{11}B$ values of marine carbonate may be sensitive to past changes in the pH of seawater. Alternatively, differences between $\delta^{11}B$ values of modern and Tertiary marine carbonates may also be caused by isotope fractionation of B during diagenesis of sediment (Ishikawa and Nakamura, 1993).

The most important carrier of B in igneous and metamorphic rocks is the mineral tourmaline $[H_9Al_3(B{\cdot}OH)_2Si_4O_{19}]$. Swihart and Moore (1989) and Palmer and Slack (1989) reported $\delta^{11}B$ values between −22 and +22‰ for tourmalines from different geological environments.

The $\delta^{11}B$ values of mafic volcanic rocks in different tectonic settings were measured by Palmer (1991) and Chaussidon and Jambon (1994). They reported that volcanic rocks from island arcs are enriched in ^{11}Be compared to MORBs, presumably because of the presence of subducted marine sediment and oceanic crust. Chaussidon and Marty (1995) also reported that OIBs are depleted in ^{11}B perhaps because the plumes in the underlying mantle contain metamorphosed marine clay-rich sediment that was depleted in ^{11}B (i.e., enriched in ^{10}B) by selective sorption of $B(OH)_4^-$.

Additional references to studies of the geochemistry of B in terrestrial environments are listed in Table 30.3.

Table 30.3. References to Studies Concerning the Isotope Geochemistry of B in Terrestrial Samples

Topic	References
Review of principles	Evans et al., 1983, *Soil Sci. Plant. Anal.*, 14:827–846; Bassett, 1990, *Appl. Geochem.*, 5:541–554; Barth, 1993, *Geol. Rundschau*, 82:640–651; Aggarwal and Palmer, 1995, *Analyst*, 120:1301–1307; Grew and Anovitz, 1996, *Rev. Mineral.*, Vol. 33.
Analytical procedures	Nakamura et al., 1992, *Chem. Geol., Isotope Geosci. Sect.*, 14(3):193–204; Barth, 1997, *Chem. Geol.*, 143:255–261; Gäbler and Bahr, 1999, *Chem. Geol.*, 156:323–330; Deyhle, 2001, *Int. J. Mass. Spectrom.*, 206:79–89; Lécuyer et al., 2002, *Chem. Geol.*, 186:45–55.
Partitioning into silicate melt	Chaussidon and Libourel, 1993, *Geochim. Cosmochim. Acta*, 57:5053–5062; Tonarini et al., 2003, *Geochim. Cosmochim. Acta*, 67:1863–1873.
Volcanic rocks	Morris et al., 1990, *Nature*, 344:31–36; Ryan and Langmuir, 1993, *Geochim. Cosmochim. Acta*, 57:1489–1498; Clift et al., 2001, *Geochim. Cosmochim. Acta*, 65:3347–3364.
Xenoliths	Higgins and Shaw, 1984, *Nature*, 308:172–173; Leeman et al., 1992, *Geochim. Cosmochim. Acta*, 56:775–788.
Ophiolites and altered oceanic crust	Ishikawa and Nakamura, 1992, *Geochim. Cosmochim. Acta*, 56:1633–1639; Ishikawa and Nakamura, 1994, *Nature*, 370:205–208; You et al., 1993, *J. Geol.*, 21:207–210; Smith et al., 1995, *Chem. Geol.*, 126:119–135.
Fumaroles and hotsprings	Kanzaki et al., 1979, *Geochim. Cosmochim. Acta*, 43:1859–1863; Nomura et al., 1982, *Geochim. Cosmochim. Acta*, 46:2403–2406; Musashi et al., 1988, *Geochem. J.*, 22:205–214; Palmer and Sturchio, 1990, *Geochim. Cosmochim. Acta*, 54:2811–2815; Aggarwal et al., 2000, *Geochim. Cosmochim. Acta*, 64:579–585; Pennisi et al., 2000, *Geochim. Cosmochim. Acta*, 64:961–974; Larsen et al., 2001, *Chem. Geol.*, 179:17–35; Deyhle and Kopf, 2001, *Geology*, 29:1031–1034.
Tourmaline	Palmer, 1991, *Chem. Geol., Isotope Geosci. Sect.*, 94:111–121; Trumbull and Chaussidon, 1999, *Chem. Geol.*, 153:125–138; Jiang et al., 1999, *Chem. Geol.*, 158:131–144.
River water	Spivack et al., 1987, *Geochim. Cosmochim. Acta*, 51:1939–1949; Barth, 1997, *Chem. Geol., Isotope Geosci. Sect.*, 143:255–261; Rose et al., 2000, *Geochim. Cosmochim. Acta*, 64:397–408.
Groundwater and saline lakes	Moldovanyi et al., 1993, *Geochim. Cosmochim. Acta*, 57:2083–2099; Barth, 1998, *Water Res.*, 32:685–690; Vengosh et al., 1994, *Environ. Sci. Technol.*, 28:1968–1974; Vengosh et al., 1995, *Chem. Geol.*, 120:135–154; Xiao et al., 1992, *Geochim. Cosmochim. Acta*, 56:1561–1568; Barth, 2000, *Appl. Geochem.*, 15:937–952; Mather and Porteous, 2001, *Appl. Geochem.*, 16:821–834.

30.1c Meteorites

The isotopic composition of B in stony meteorites and the chondrules they contain has been investigated by several research groups identified by Shaw (2001). The problem is that the $\delta^{11}B$ values of individual chondrules in the meteorites Semarkona, Hedjaz, and Allende analyzed by ion microprobe mass spectrometry vary widely from −50 to +40‰. Chaussidon and Robert (1995) proposed that the apparent heterogeneity of the $\delta^{11}B$ values of chondrules resulted from the production of B isotopes by spallation reactions caused by cosmic-ray irradiation of dust particles in the solar nebula prior to the accretion of chondrules. In addition, the $^{11}B/^{10}B$ ratios of whole-rock samples of stony meteorites have repeatedly yielded values close to the average terrestrial value of about 4, even though astronomical evidence suggested that the $^{11}B/^{10}B$ ratio of the solar nebula should be only about 2.5 (Shaw, 2001). These inconsistencies in the isotope systematics of B in meteorites may have arisen in part because of contamination of samples with terrestrial B, which is difficult to avoid given the low concentrations of this element.

A recent study by Hoppe et al. (2001) based on 89 determinations by ion microprobe mass spectrometry of 16 chondrules in 7 chondrites yielded a wide range of $\delta^{11}B$ values from -34 to $+33$‰. However, these data in Figure 30.3 have a unimodal distribution with a mean of -3.6 ± 1.5‰ (excluding 11 analyses of one chondrule of Semarkona). The corresponding $^{11}B/^{10}B$ ratio of chondrules calculated from equation 30.1 is 4.029 ± 0.013 based on a value of 4.0436 fro the $^{11}B/^{10}B$ ratio of SRM 951 (Catanzaro et al. 1970).

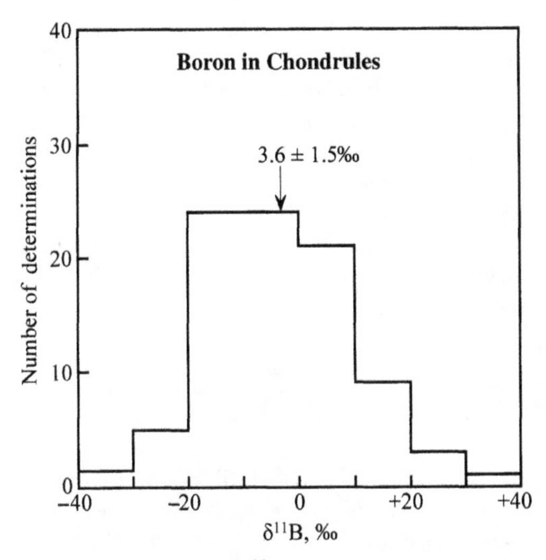

FIGURE 30.3 Range of $\delta^{11}B$ values of chondrules in Allende, Murchison, Dhajala, Bjurböle, Chainpur, Parsa, and Semarkona. The mean of 3.6 ± 1.5‰ was calculated by excluding Semarkona. Data from Hoppe et al. (2001).

Hoppe et al. (2001) demonstrated that the $\delta^{11}B$ values of the chondrules they analyzed have a "normal" Gaussian distribution and that the apparent range of the $\delta^{11}B$ values is consistent with analytical errors. Therefore, the results indicate that the isotopic composition of B in the chondrules analyzed by Hoppe et al. (2001) is homogeneous in spite of the wide range of variation of the $\delta^{11}B$ values. Any isotopic heterogeneity that may have existed in dust particles in the solar nebula did not survive the formation of chondrules. Therefore, the isotopic composition of B in the solar nebula appears to have been homogeneous and its $^{11}B/^{10}B$ ratio is not distinguishable from that of average terrestrial B.

Additional measurement of $\delta^{11}B$ in meteorites have been reported by:

Shaw et al., 1988, *Am. Mineral.*, 73:894–900. Zhai and Shaw, 1994, *Meteoritics*, 29:607–615. Zhai et al., 1996, *Geochim. Cosmochim. Acta*, 60:4877–4881. Hervig, 1996, *Rev. Mineral.*, 33:789–803. Chaussidon and Robert, 1998, *Earth Planet. Sci. Lett.*, 164:577–589. Hanon et al., 1999, *Meteoritics Planet. Sci.*, 34:247–258. Marhas and Goswami, 2000, *Curr. Sci.*, 78:78–81. Hoppe et al., 2001, *Astrophys. J.*, 551:478–485.

30.1d Summary

The isotope geochemistry of boron has advanced to a level of sophistication that has permitted useful applications to the study of geochemical and cosmochemical problems. Future refinement of analytical methods and instrumentation is required to resolve small but potentially significant differences in the isotope composition of B in ocean basalts (MORBs and OIBs), island arc volcanics, and geochemical processes on the surface of the Earth. The study of B isotope compositions in chondrules may lead to a better understanding of the nuclear reactions in the solar nebula before the formation of chondrules, planetesimals, and planets.

30.2 LITHIUM

The use of the isotopic compositions of Li in geological and cosmochemical research has been retarded by analytical problems that arise from the low concentration of this element in most geological samples and because of instrumental isotope fractionation. As a result, differences in methodology have affected both the precision and accuracy of measured isotope ratios of Li. These problems are being solved, suggesting that the isotopic composition of Li will become a useful source of information in the earth sciences.

30.2a Geochemistry

Lithium ($Z = 3$) is a member of group IA of the periodic table, known as the alkali metals, which include Li, Na, K, Rb, Cs, and Fr (francium) in order of increasing atomic number. The electron configuration of Li is $1s^2 2s^1$, which causes it to have a valence number of +1. The ionic radius of Li^+ (0.60 Å) is similar to that of Mg^{2+} (0.65 Å), allowing these two elements to have an affinity for each other.

The concentrations of Li in terrestrial igneous rocks in Table 30.2 increase with increasing Si concentration from 0.5 μg/g in ultramafic rocks to 40 μg/g in low-Ca granites. Shale and deep-sea clay are enriched in Li because this element can enter the octahedral layer of smectite clays, in contrast to carbonate minerals, which exclude Li. The average concentration of Li in the continental crust is 13 μg/g (Taylor and McLennan, 1985).

The principal Li-bearing minerals are: spodumene [$LiAl(SiO_3)_2$], lepidolite (Li-mica), amblygonite [$LiAl(F,OH)PO_4$], and the clay mineral hectorite [$(Li_{0.66}Mg_{5.34})Si_8O_{20}(OH)_4$] (Grim, 1968). The chloride, carbonate, sulfate, oxide, and hydroxide of Li are soluble in water and do not occur as minerals in terrestrial rocks.

Surface water on the continents has a low Li concentration (0.003 ppm) because Li-silicate minerals resist chemical weathering and because Li^+ is sorbed by clay minerals. The concentration of Li in seawater is 0.17 ppm, which implies a nearly 60-fold enrichment over average river water and yields a mean oceanic residence time for Li of 2.5×10^6 years (Taylor and McLennan, 1985).

The high solubility of Li salts of most naturally occurring acids causes Li to be concentrated in brines that form by evaporation of aqueous solution at the surface of the Earth or by freezing of Li-bearing water (Bottomley et al., 1999). As a result, Li is recovered commercially from brines at Searles Lake, California, and elsewhere, and from certain granitic pegmatites where Li occurs primarily in spodumene and lepidolite (e.g., in the Black Hills of South Dakota and at King's Mountain, North Carolina). However, both deposits are presently inactive.

30.2b Isotope Composition

Lithium has two stable isotopes, ^6Li and ^7Li, whose abundances were determined by Svec and Anderson (1965). The average $^6Li/^7Li$ ratio of 13 samples of spodumene analyzed by Svec and Anderson (1965) is 0.08182 ± 0.00033, from which the authors calculated an atomic weight of 6.9403 ± 0.0005 for Li. Svec and Anderson (1965) also listed the results of other investigators expressed both as the $^7Li/^6Li$ and $^6Li/^7Li$ ratios because both methods had been used by those investigators. The isotopic compositions of terrestrial Li is not constant because of mass-dependent isotope fractionation.

The abundance of Li in the solar system (5.71×10 atoms per 10^6 atoms of Si) is anomalously low, like those of B and Be. Anders and Grevesse (1989) attributed the formation of Li isotopes to cosmic-ray spallation reactions in the solar nebula (^6Li) and to cosmological nucleosynthesis combined with hydrogen burning and spallation reactions (^7Li) (see also Steenbock and Holweger, 1984).

The isotopic composition of Li is expressed by the δ-parameter defined in terms of both $^6Li/^7Li$ and $^7Li/^6Li$ ratios:

$$\delta^6\text{Li} = \left[\frac{(^6\text{Li}/^7\text{Li})_{\text{spl}} - (^6\text{Li}/^7\text{Li})_{\text{std}}}{(^6\text{Li}/^7\text{Li})_{\text{std}}}\right] \times 10^3‰ \tag{30.3}$$

and the standard is NBS L-SVEC, whose $^6Li/^7Li$ ratio is 0.083062 ± 0.000054 (Bottomley et al., 1999; Chan et al., 2002). However, Hoefs and Sywall (1997) expressed the δ^6Li parameter relative to the $^6Li/^7Li$ ratio of *seawater* and Tomascak et al. (1999) actually calculated δ^7Li values relative to the $^7Li/^6Li = 12.02 \pm 0.03$ for the L-SVEC standard reported by Flesch et al. (1973). Evidently, the isotope geochemistry of Li is still in a state of flux.

James and Palmer (2000) attempted to standardize the way Li isotopes are measured by publishing the isotope compositions of Li in nine international rock standards as well as in seawater and in one carbonaceous chondrite (Orgueil). They used thermal ionization mass spectrometry and modified both the chemical separation of Li and the operating protocols of the mass spectrometer in order to reduce procedural isotope fractionation, and thereby improved the reproducibility of the measured isotope ratios of Li.

For example, James and Palmer (2000) reported that instrumental isotope fractionation of Li is significantly reduced when the ionization filament in the source of the mass spectrometer is heated to 1000°C or more. If the temperature of this filament is less than 850°C, significant isotope fractionation occurs. In addition, they reported that the conversion of Li to Li_3PO_4 improved the reproducibility compared to the results achieved with dilithium metaborate ions ($Li_2BO_2^+$) used by Chan et al. (2002) and others.

The average isotopic composition of Li in seawater obtained by James and Palmer (2000) is:

$$^6Li/^7Li = 0.08015 \pm 0.000056$$

$$^7Li/^6Li = 12.476 \pm 0.008$$

Their value for the Li standard (NBS L-SVEC) is:

$$^6Li/^7Li = 0.082757 \qquad ^7Li/^6Li = 12.08356$$

Therefore, the δ^6Li value of seawater determined by James and Palmer (2000) is:

$$\delta^6Li_{sw} = \left(\frac{0.080151 - 0.082757}{0.082757}\right) \times 10^3$$

$$= -31.5 \pm 0.7‰$$

This value is in good agreement with the δ^6Li values of seawater published by Chan and Edmond (1988), You and Chan (1996), and Moriguti and Nakamura (1998). Therefore, James and Palmer (2000) proposed that their value of $\delta^6Li = -31.5 \pm 0.7‰$ should be the standard value for Li in seawater. The corresponding δ^7Li value is $+32.5 \pm 0.7‰$. The concentrations of Li and δ^6Li values of standard rocks measured by James and Palmer (2000) in Table 30.4 indicate that the δ^6Li values of igneous rocks range from −6.8‰ (basalt) to +1.2‰ (granite) and +2.7‰ (serpentine). The δ^6Li value of shale is −5.2‰ and that of the carbonaceous chondrite Orgueil is −3.9‰. Therefore, these results suggest that mafic igneous rocks derived from the mantle are depleted in ^6Li (or enriched in ^7Li) relative to the L-SVEC standard, whereas granitic rocks of the continental crust are enriched in ^6Li (and/or depleted in ^7Li).

The use of the different reference standards to express the δ-parameter of Li and the different formulations of this parameter make it difficult to convert results from one convention to another because the necessary isotope ratios are not available in some cases. The work of James and Palmer (2000) is a positive step toward standardization of analytical procedures and the reporting of the isotope ratios of Li. The references in Table 30.5 provide an introduction to this rapidly growing field of isotope geology.

Table 30.4. Concentrations and δ^6Li Values of Standard Rocks, Seawater, and the Carbonaceous Chondrite Orgueil

Sample	Li, μg/g	δ^6Li, ‰
UB-N (serpentine)	25.2	$+2.7 \pm 0.5$
G-2 (granite)	31.8	$+1.2 \pm 0.2$
JG-2 (granite)	41.0	0.4 ± 0.1
DR-N (diorite)	39.1	-2.3 ± 0.4
JR-2 (rhyolite)	73.9	-3.8 ± 0.4
C1 chondrite (Orgueil)	1.49	-3.9 ± 0.6
SCo-1 (shale)	45.8	-5.2 ± 0.6
BHVO-1 (Hawaiian basalt)	4.9	-5.8 ± 0.8
JGb-1 (gabbro)	4.2	-6.1 ± 1.0
JF-2 (basalt)	8.0	-6.8 ± 0.1
Seawater	—	-31.5 ± 0.7

Source: James and Palmer, 2000.

Note: δ^6Li was expressed relative to the NBS L-SVEC standard whose $^6Li/^7Li$ ratio is 0.082757.

Table 30.5. Summary of the Recent Literature Concerning the Isotopic Composition of Li in Terrestrial Samples

Topic	References
Fractionation (isotopic)	Taylor and Urey, 1938, *J. Chem. Phys.*, 6:429–438; Morozova and Alferovsky, 1974, *Geochem. Int.*, 11:17–25; Fritz, 1992, *Geochim. Cosmochim. Acta*, 56:3781–3789; Fritz and Whitworth, 1994, *Water Res.*, 30(2):225–235.
Methodology (TIMS)	Chan, 1987, *Anal. Chem.* 59:2662–2665; Xiao and Beary, 1989, *Int. J. Mass Spectrom. Ion Proc.*, 94:101–114; Datta et al., 1992, *Int. J. Mass Spectrom. Ion Proc.*, 116:87–114; Moriguti and Nakamura, 1993, *Proc. Japanese Acad. Sci.*, B69:123–128; You and Chan, 1996, *Geochim. Cosmochim. Acta*, 60:909–915; Hoefs and Sywall, 1997, *Geochim. Cosmochim. Acta*, 61:2679–2690; James and Palmer, 2000, *Chem. Geol.*, 166:319–326.
Methodology (ICP-MS, quadrupole)	Grégoire et al., 1996, *Anal. Atom. Spectrom.*, 11:765–772; Kosler et al., 2001, *Chem. Geol.*, 181:169–179.
Methodology (ICP-MS, magnetic sector)	Sun et al., 1987, *Analyst*, 112:1223–1228; Tomascak et al., 1999, *Chem. Geol.*, 158:145–154.
Meteorites (chondrules)	Chaussidon and Robert, 1998, *Earth Planet. Sci. Lett.*, 164:577–589.
Oceanic basalt	Seyfried et al., 1984, *Geochim. Cosmochim. Acta*, 48:557–569; Chan et al., 1992, *Earth Planet. Sci. Lett.*, 108:151–160; Chan and Edmond, 1988, *Geochim. Cosmochim. Acta*, 1711–1717; Tomascak et al., 1999, *Geochim. Cosmochim. Acta*, 63:907–910.
Volcanic rocks (subduction zones)	You et al., 1997, *Earth Planet Sci. Lett.*, 140:41–52; Moriguti and Nakamura, 1998, *Earth Planet Sci. Lett.*, 163:167–174; Chan et al., 1999, *Chem. Geol.*, 160:255–280; Tomascak et al., 2000, *Geology*, 28:507–510.
Oceanic hotsprings	Chan et al., 1993, *J. Geophys. Res.*, 98(B6):9653–9659; Chan et al., 1994, *Geochim. Cosmochim. Acta*, 94:4443–4454; Seyfried et al., 1998, *Geochim. Cosmochim. Acta*, 62:949–960; James et al., 1999, *Earth Planet. Sci. Lett.*, 171:157–169.
Marine sediment and pore water	You et al., 1995, *Geology*, 23:37–40; Zhang et al., 1998, *Geochim. Cosmochim. Acta*, 62:2437–2450; Chan and Kastner, 2000, *Earth Planet. Sci. Lett.*, 183:275–290.
Marine calcite	Chan and Edmond, 1988, *Geochim. Cosmochim. Acta*, 52:1711–1717; Hoefs and Sywall, 1997, *Geochim. Cosmochim. Acta*, 61:2679–2690; Kosler et al., 2001, *Chem. Geol.*, 181:169–179.
Seawater	Chan and Edmond, 1988, *Geochim. Cosmochim. Acta*, 52:1711–1717.
Lake water	Chan and Edmond, 1988, *Geochim. Cosmochim. Acta*, 52:1711–1717.
River water	Huh et al., 1998, *Geochim. Cosmochim. Acta*, 62:2039–2051.
Subsurface brine	Bottomley et al., 1999, *Chem. Geol.*, 155:295–320; Chan et al., 2002, *Geochim. Cosmochim. Acta*, 66:615–623.

30.2c Summary

The variation of the isotopic composition of Li can provide useful information about geochemical processes that contribute to the petrogenesis of igneous rocks and that operate on the surface of the Earth (e.g., chemical weathering, sorption of ions to colloidal particles, formation of brine by evaporative concentration, and precipitation of minerals). In addition, isotope analyses of presolar grains and organic matter in meteorites can provide information on the nucleosynthesis of Li and its isotopic composition in the solar nebula. The necessary analytical procedures and instrumentation have been developed and will permit the measurement of the isotope composition of Li in samples that have low concentrations of this element.

30.3 SILICON

The rocks of the continental and oceanic crust are predominantly composed of silicate minerals because silicon (Si) is the second most abundant chemical element in the crust of the Earth. Rocks composed of carbonates, phosphates, sulfates, chlorides, sulfides, and oxides are much less abundant and dominate only locally. Therefore, variations in the isotopic composition of Si may shed light on the petrogenesis of igneous rocks and on the origin of ore deposits associated with them (Ding et al., 1996). In addition, Si isotopes may be fractionated during the transformations of rock-forming silicate minerals into clay minerals and zeolites during chemical weathering of crustal rocks.

In addition, Si is a nutrient in the oceans and in surface water on the continents because it is used by diatoms, radiolarians, sponges, and certain plants (e.g., horsetail) that secrete opaline silica. The biosynthesis of siliceous skeletons and phytoliths is accompanied by isotope fractionation and affects the isotope composition of Si in sediment containing this material as well the Si in solution in the water in which the opaline silica was excreted.

Silicon is also an important element in stony meteorites, including the presolar grains of silicon carbide (SiC) and silicon nitride (Si_3N_4) they contain. The history of the isotope geochemistry of Si was summarized in a book by Ding et al. (1996), who have measured the isotope compositions of Si in more than 1000 samples.

30.3a Geochemistry

Silicon ($Z = 14$) is a member of group IIIA in the periodic table, consisting of C, Si, Ge, Sn, and Pb in order of increasing atomic number. The electronic formula of Si is $1s^2 2s^2 2p^6 3s^2 3p^2$, which gives it a valence number of +4. However, Si is a nonmetal and is covalently bonded to O in the stable silicate anion($SiO_4{}^{4-}$), which forms a tetrahedron with Si at the center and the O atoms at the corners. When the silicate anion is protonated, it forms H_4SiO_4, which is a weak acid whose solubility in acidic solution is low. In silicate melts, Al^{3+} can substitute for Si^{4+} to form the aluminosilicate anion ($AlSi_3O_8{}^-$). The excess negative charge is neutralized by the addition of Na^+, K^+, or Ca^{2+} to form the family of feldspar minerals:

Albite:	$NaAlSi_3O_8$
Orthoclase:	$KAlSi_3O_8$
Anorthite:	$CaA_2Si_2O_8$

When silica tetrahedra are linked to each other by sharing O atoms at the corners, they form a three-dimensional network whose Si/O ratio is 1 : 2. The resulting formula (SiO_2) characterizes the mineral quartz (Faure, 1998).

Silicon is a major element in terrestrial and extraterrestrial rocks whose concentration ranges from 19.8% in ultramafic rocks to 34.70% in low-Ca granites (Turekian and Wedepohl, 1961). Silicon is also eighth in abundance in the solar system with a concentration of 1×10^6atoms (Anders and Grevesse, 1989).

30.3b Isotope Composition

Silicon has three stable isotopes, ^{28}Si, ^{29}Si, and ^{30}Si, whose abundances and masses are listed in Table 30.1. The isotopic composition of Si varies because of mass-dependent isotope fractionation based on isotope exchange reactions and kinetic effects. The isotope composition of Si is expressed in terms of the $\delta^{30}Si$ parameter defined as:

$$\delta^{30}\text{Si} = \left(\frac{(^{30}\text{Si}/^{28}\text{Si})_{\text{spl}} - (^{30}\text{Si}/^{28}\text{Si})_{\text{std}}}{(^{30}\text{Si}/^{28}\text{Si})_{\text{std}}} \right) \times 10^3\text{‰} \quad (30.4)$$

The standards used for this purpose and their isotope ratios are listed in Table 30.6. Alternatively, the isotope composition of Si can also be expressed by the $\delta^{29}Si$ parameter defined in terms of the $^{29}Si/^{28}Si$ ratio.

Table 30.6. Isotope Ratios of Si in Isotopic Reference Standards

Standard	$^{29}Si/^{28}Si$	$^{30}Si/^{28}Si$	References[a]
IRMM-017 (Si crystal)	0.0507715 ±0.0000066	0.0334889 ±0.0000078	1
IRMM-018 (SiO_2)	0.0508442 ±0.0000048	0.0335851 ±0.0000066	1
NBS-28	0.0506327 0.0000315	0.03362113 ±0.0000084	2
Rose quartz (Caltech)		0.03361171 ±0.001573	3

[a] 1. Ding et al., 1996, *Silicon Isotope Geochemistry*, Geol. Publ. House, Beijing. 2. Barnes et al., 1975, *J. Res. NBS*, 79A:727–735. 3. Molini-Velsko et al., 1986, *Geochim. Cosmochim. Acta*, 50:2719–2726.

The isotopic composition of Si in silicate minerals is measured by means of double-collector gas-source mass spectrometers using SiF_4 produced by reacting silicate samples with bromine pentafluoride (BrF_5) (Douthitt, 1982). Use of SiF_4 is advantageous because F has only one stable isotope (^{19}F). Douthitt (1982) purified the silica fraction by fusing powdered silicate samples with Na_2CO_3 in a Pt crucible at 1100°C followed by leaching with 7.4 N HCl, which causes amorphous SiO_2 to precipitate. After drying at 1200°C, the precipitate is treated with BrF_5 to form SiF_4 gas.

Interlaboratory discrepancies in the reported $\delta^{30}Si$ values are eliminated when all research groups use the same reference standard and each group analyses that standard by the same procedure and on the same mass spectrometer that was used to analyze the samples.

Silicon has a fourth naturally occurring but radioactive isotope (^{32}Si), which is cosmogenic in origin ($T_{1/2} = 140 \pm 6$ y) and is listed in Table 23.7 of Section 23.6

30.3c Terrestrial Rocks

The $\delta^{30}Si$ values of silicate rocks and minerals reported by Douthitt (1982) were expressed relative to the $^{30}Si/^{28}Si$ ratio of the Rose Quartz Standard (RQS) used at the California Institute of Technology. Molini-Velsko et al. (1986) reported that $\delta^{30}Si$ of RQS is $-0.28 \pm 0.18‰$ when expressed relative to NBS 28 (quartz). The $\delta^{30}Si$ values of the samples analyzed by Douthitt (1982) range from −3.4‰ (sponge spicules) to +2.8‰ (phytolith) for a total range of 6.2‰. The $\delta^{30}Si$ values of igneous rocks (basalt to S-type granites) range narrowly in Figure 30.4 and have a well-defined unimodal distribution with a mean of $-0.28 \pm 0.25‰$ (1σ) for 70 samples. The $\delta^{30}Si$ values of clay minerals, opal and chalcedony, and biogenic silica vary more widely than those of igneous rocks and appear to be variably depleted in ^{30}Si compared to Si in

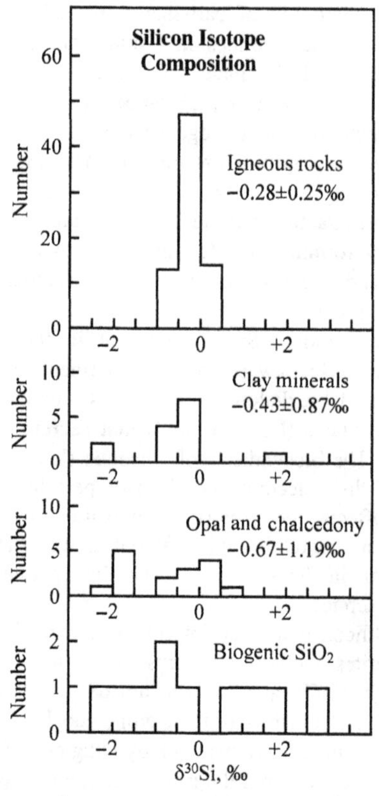

FIGURE 30.4 Isotope composition of Si expressed as the $\delta^{30}Si$ value relative to the Rose Quartz Standard (RQS) used at the California Institute of Technology. The average $\delta^{30}Si$ values of igneous rocks, clay minerals, and chemically precipitated silica (opal and chalcedony) are indicated. The range of the $\delta^{30}Si$ values is expressed by one standard deviation. Data from Douthitt (1982).

igneous rocks. However, silicic acid in aqueous solution (δ^{30}Si $= +0.16 \pm 0.19$‰) is enriched in ^{30}Si relative to chemically precipitated silica (opal and chalcedony) with an average δ^{30}Si value of -0.67 ± 1.19‰. Likewise, the average δ^{30}Si value of three samples of phytoliths of *Equisetum* (horsetail) is $+1.77 \pm 1.00$‰, whereas the δ^{30}Si value of one phytolith of *Bambusa* (bamboo) is -1.4‰.

In general, the data of Douthitt (1982) indicate that the δ^{30}Si values of mafic and ultramafic igneous rocks (δ^{30}Si $= -0.41 \pm 0.16$‰, 1σ) are more depleted in ^{30}Si on average than igneous rocks of intermediate to felsic composition (δ^{30}Si $= -0.12 \pm 0.24$‰, 1σ), which implies a correlation between the isotopic composition and the concentration of Si in igneous rocks.

The most promising evidence of isotope fractionation is between molecular silicic acid in aqueous solution (δ^{30}S $= +0.16 \pm 0.19$‰) and chemically precipitated amorphous and crystalline silica (opal and chalcedony), which are depleted in ^{30}Si (δ^{30}Si $= -0.67 \pm 1.19$‰) relative to silica in aqueous solution. In addition, biogenic silica such as sponge spicules, diatomite, and flint are depleted in ^{30}Si (δ^{30}Si $= -1.62 \pm 1.10$‰), whereas phytoliths of *Equisetum* are enriched in this isotope (δ^{30}Si $= +1.77 \pm 1.00$‰). Therefore, the data of Douthitt (1982) suggest that the stable isotopes of Si are fractionated during the precipitation of silicic acid from aqueous solution and during the secretion of opaline silica by sponges, *Equisetum* (horsetail), and *Bambusa* (bamboo). These conclusions confirm the results reported previously by Tilles (1961) and Reynolds and Verhoogen (1953).

30.3d Marine Diatoms

The isotope fractionation of Si in biogenic silica, which was suggested by the data of Douthitt (1982), was confirmed and extended by De La Rocha et al. (1996, 1997). These authors cultured diatoms collected in the water of the Santa Barbara Channel off the coast of California and then measured the δ^{30}Si values of the siliceous diatoms and the water in which they had grown. The results indicated that the diatoms were *depleted* in ^{30}Si relative to silicic acid in solution. The resulting isotope fractionation factors α_w^d in Table 30.7 are all less than unity, consistent with the depletion of diatoms in ^{30}Si relative to the silica in solution. The average fractionation factor α_w^d for three different species of diatoms is:

Table 30.7. Isotope Fractionation Factors Based on δ^{30}Si of Diatoms of Silicic Acid in Solution at the Specified Temperature

Diatom Species	Temperature, °C	α_w^d (δ^{30}Si)
Skeletonema costatum	15	0.9990 ±0.0004
Thalassiosira weissflogii	15	0.9987 ±0.0004
Thalassiosira sp.	15	0.9989 ±0.0004
Average value		0.99890 ±0.0004

Source: De La Rocha et al., 1997.

$$\alpha_w^d = 0.9989 \pm 0.0004$$

In addition, the fractionation factor of *Thalassiosira* sp. is independent of the temperature between 12 and 22°C, which is the range of temperatures preferred by this diatom (De La Rocha et al., 1997).

The value of the Si-isotope fractionation factor of diatoms ($\alpha_w^d = 0.9989$) is related to the difference in the δ^{30}Si values of diatoms and water by equation 26.17:

$$\delta^{30}\text{Si}_d - \delta^{30}\text{Si}_w = \Delta_w^d(^{30}\text{Si}) = 10^3 \ln \alpha_w^d$$

$$\Delta_w^d(^{30}\text{Si}) = 10^3 \ln 0.9989$$

$$\Delta_w^d(^{30}\text{Si}) = -1.10‰$$

Similar experiments with silica-secreting plants, including sponges, horsetail, and bamboo, have not yet been carried out.

30.3e Aqueous Isotope Geochemistry

The depletion of diatoms in ^{30}Si may cause silicic acid in solution in seawater and freshwater to be enriched in ^{30}Si. This prediction was confirmed by De La Rocha et al. (2000), who measured δ^{30}Si

Table 30.8. Isotope Composition of Si in Silicic Acid in Solution in Rivers Expressed by δ^{30}Si Relative to NBS 28

River	δ^{30}Si, ‰ (NBS 28)
Amazon	
Obidos	+0.9
Integrated	+0.6
Congo	
Brazzaville	+0.4
Zaire	+0.8
Niger (Bani)	+1.2
American (CA)	+0.9
Sacramento (CA)	+1.2
Rock Creek (CA)	+0.5

Source: De La Rocha et al., 2000

values in eight rivers relative to NBS 28. The resulting δ^{30}Si values in Table 30.8 range from +0.4 to +1.2‰ with an average of $+0.8 \pm 0.3$‰.

The silicic acid in solution in all of the rivers sampled by De La Rocha et al. (2000) is enriched in ^{30}Si relative to Si in igneous rocks, which have an average δ^{30}Si value of -0.28 ± 0.25‰ (Figure 30.4) based on the data by Douthitt (1982) and later confirmed by Ding et al. (1996). Presumably, the silicic acid that is released into solution by chemical weathering of igneous rocks is enriched in ^{30}Si by the growth of diatoms which favor ^{28}Si over ^{30}Si. In addition, silicic acid in solution is enriched in ^{30}Si by formation of clay minerals whose average δ^{30}Si value is -0.43 ± 0.87‰ (Douthitt, 1982). Therefore, both biogenic silica and clay minerals discriminate against ^{30}Si and thereby cause the silicic acid in solution in surface water on the continents to be enriched in ^{30}Si.

De La Rocha et al. (2000) also analyzed 69 samples of seawater collected at increasing depths at nine different locations. The resulting δ^{30}Si values range from +0.6 to +2.2‰ relative to NBS 28. The average δ^{30}Si value of seawater is $+1.1 \pm 0.3$‰ compared to $+0.8 \pm 0.3$‰ in river water.

These results suggest that the Si discharged into the oceans by rivers is further enriched in ^{30}Si, presumably by the formation of biogenic silica and clay minerals. The data also demonstrate that the concentrations of silicic acid increase with depth from about 10 micromoles (μM) or less at the surface to more than 100 μM at 1000 m in the central North Pacific and at 3000 m in the South Atlantic Ocean. In contrast, deep water (>1000 m) in the Sargasso Sea contains <40 μM of $Si(OH)_4$.

The δ^{30}Si values of seawater vary regionally and with depth, as expected from the relatively short oceanic residence time of Si of 7.9×10^3 years (Faure, 1998). In addition, the variation of δ^{30}Si values with depth is caused by the dissolution of biogenic silica particles sinking through the water column.

The work of De La Rocha et al. (1997, 2000) demonstrates that Si isotopes are fractionated in surface environments both by formation of secondary minerals (e.g., clay minerals) and by biological processes (e.g., diatoms). Therefore, the isotopic composition of Si is likely to provide insights into the geochemical and biochemical processes occurring on the continents and in the ocean basins.

30.3f Extraterrestrial Rocks

The isotopic composition of Si in meteorites and of lunar rocks has been determined by:

Epstein and Taylor, 1970, *Proc. Apollo 11 Lunar Sci. Conf.*, pp. 1085–1096. Epstein and Taylor, 1971, *Proc. Lunar Sci. Conf., 2nd*, The Lunar and Planetary Science Institute in Houston, TX. pp. 1421–1441. Epstein and Taylor, 1972, *Proc. Lunar Sci. Conf., 3rd*, The Lunar and Planetary Science Institute in Houston, TX, pp. 1429–1454. Taylor and Epstein, 1973, *Proc. Lunar Sci. Conf., 4th*, The Lunar and Planetary Science Institute in Houston, TX, pp. 1657–1679. Epstein and Yeh, 1977, *Proc. Lunar Sci. Conf., 8th*, The Lunar and Planetary Science Institute in Houston, TX, pp. 287–289. Clayton et al., 1978, *Proc. Lunar Planet. Sci. Conf., 9th*, The Lunar and Planetary Science Institute in Houston, TX pp. 1267–1278. Yeh and Epstein, 1978, *Proc. Lunar Planet. Sci. Conf., 9th*, The Lunar and

Planetary Science Institute in Houston, TX, pp. 1289–1291; Becker and Epstein, 1981, *Proc. Lunar Sci. Conf., 12th*, The Lunar and Planetary Science Institute in Houston, TX, pp. B1189–1198; Clayton et al., 1983, *Meteoritics*, 18:282–283.

In addition, Molini-Velsko et al. (1982) investigated the possible isotope fractionation of Si in tektites.

The study by Molini-Velsko et al. (1986) of the isotopic compositions of Si in bulk samples of different classes of meteorites had the benefit of preceding studies and improvements in analytical procedures and instrumentation. These authors expressed the isotope ratios of Si as δ^{30}Si and δ^{29}Si values relative to NBS 28 (quartz), whereas previous analysts used the Rose Quartz standard. The average δ^{30}Si values of meteorites and terrestrial, lunar, and martian rocks in Table 30.9 are not distinguishable from each other and permit the conclusion that the Si in meteorites and in rocks of Mars, Earth, and the Moon is isotopically homogeneous. In addition, Molini-Velsko et al. (1986) demonstrated that δ^{30}Si and δ^{29}Si values of all meteorites closely define a straight line whose slope is 0.495. This result indicates that the isotopic composition of Si in meteorites has been affected only by mass-dependent isotope fractionation.

Table 30.9. Average δ^{30}Si Values Relative to NBS 28 (Quartz) of Bulk Samples of Meteorites, Lunar and Martian Rocks, and Mafic Terrestrial Igneous Rocks

Meteorite Type	Average δ^{30}Si (NBS 28), ‰	Range, ‰
Carbonaceous chondrites	−0.66	−1.11 to +0.04
Ordinary chondrites	−0.53	−0.82 to −0.12
Enstatite chondrites and aubrites	−0.61	−1.03 to −0.43
Eucrites	−0.37	−0.50 to −0.22
Silicate inclusion in iron meteorites	−0.68	−1.79 to +0.45
All meteorites	−0.57 ± 0.06	−1.79 to +0.45
Mars	−0.69 ± 0.06	−0.80 to −0.52
Moon	−0.45 ± 0.03	−0.8 to 0.0
Earth	−0.68 ± 0.04	−1.1 to 0.0

Note: Data from Douthitt (1982), Epstein and Taylor (1970, 1971, 1972), and Taylor and Epstein (1973) were converted to the NBS 28 scale by Molini-Velsko et al. (1986).

A very different picture emerged from a study by Stone et al. (1991) of the isotope composition of Si in silicon carbide grains of carbonaceous and enstatite chondrites. These authors determined the isotope composition of Si of SiC inclusions by ion-microprobe mass spectrometry. The ^{30}Si/^{28}Si ratios were expressed as the δ^{30}Si parameter relative to synthetic SiC grains whose ^{30}Si/^{28}Si ratio was 0.033464 ± 0.000040, in good agreement with a value of 0.033474 reported by Zinner et al. (1989) but deviating from the ^{30}Si/^{28}Si ratio of 0.0336214 obtained by Barnes et al. (1975).

The δ^{30}Si values of SiC grains (>2 μm) in Orgueil and Murchison range widely from +24.6 to +110.7‰ and are positively correlated with the δ^{29}Si values of these SiC grains. The wide range of δ^{30}Si values and the enrichment of the SiC grains in both meteorites contrasts with the apparent isotopic homogeneity of Si in bulk samples of Orgueil (δ^{30}Si = −0.38‰) and Murchison (−0.94%, matrix) reported by Molini-Velsko et al. (1986).

The δ^{30}Si and δ^{29}Si values of platy SiC grains (Orgueil and Murchison) having diameters >2 μm define a straight line in Figure 30.5 whose equation was derived by Stone et al. (1991) by means of a weighted least-squares regression:

$$\delta^{29}\text{Si} = -40 + 1.4\delta^{30}\text{Si} \qquad (30.5)$$

The slope of this line (1.4) is not compatible with mass-dependent isotope fractionation. The C in the coarse plate SiC grains of Orgueil and Murchison is likewise strongly enriched in ^{13}C, as indicated by δ^{13}C values ranging from +366 to +3603‰ (PDB) for Orgueil and +155 to 5188‰ (PDB) for Murchison.

The extreme ^{30}Si-enrichment of these and other SiC grains preserved in carbonaceous chondrites is a product of presolar nucleosynthesis in several different stars (Tang et al., 1989; Zinner et al, 1989).

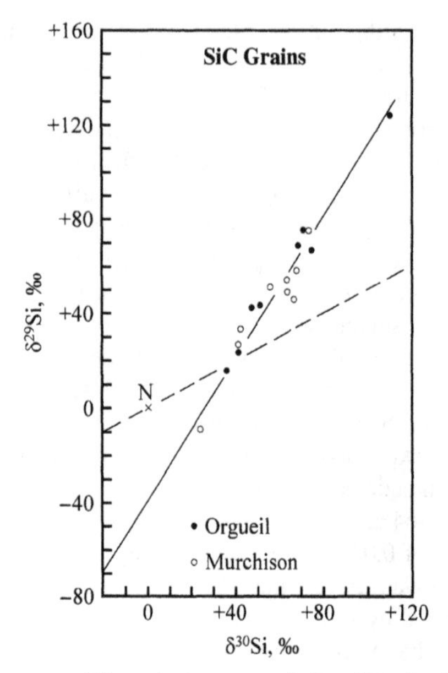

FIGURE 30.5 Three-isotope correlation line for Si in SiC grains of the carbonaceous chondrites Orgueil and Murchison. The correlation line has a slope of about 1.4, which is not compatible with mass-dependent isotope fractionation represented by the dashed line. The point labeled N is normal terrestrial Si. Data and interpretation by Stone et al. (1991).

30.3g Summary

The isotope composition of Si, expressed as the δ^{30}Si parameter, ranges narrowly by up to 6.2‰. Isotope fractionation of Si occurs during the incongruent dissolution of aluminosilicates in igneous and high-grade metamorphic rocks to form clay minerals which discriminate against ^{30}Si. Consequently, the silicic acid that goes into solution during this process is enriched in ^{30}Si relative to the Si in the primary aluminosilicate minerals (e.g., feldspar).

Silicon isotopes are also fractionated by diatoms, sponges, and certain land plants that secrete phytoliths. The resulting opaline silica contributes to the isotope composition of sediment and affects the isotope composition of the residual silicic acid in solution in water in the oceans and on the continents.

Meteorites and rocks on the Moon and Mars contain Si whose isotopic composition is indistinguishable from that of Si in terrestrial igneous rocks. However, grains of SiC in carbonaceous chondrites are strongly enriched in ^{30}Si as a result of nucleosynthesis in presolar stars.

30.4 CHLORINE

Chlorine is an important element on the surface of the Earth because Cl^- is the dominant anion in seawater, because chloride salts occur in marine and lacustrine evaporite rocks, and because Cl is a widely dispersed trace element that occurs in all igneous, metamorphic, and sedimentary rocks. Therefore, the variation of its isotopic composition may provide useful information about important geological processes.

30.4a Geochemistry

Chlorine ($Z = 17$) is a member of the halogens (group VIIA) in the periodic table, including fluorine (F), chlorine (Cl), bromine (Br), iodine (I), and astatine (At) in order of increasing atomic number. The electron configuration of Cl is $1s^2 2s^2 2p^6 3s^2 3p^5$, which gives it a valence number of -1. Chlorine is a nonmetallic gas at room temperature and forms chloride salts with many metals, most of which are soluble in water. Silver chloride (AgCl) is an important exception because its solubility in water is quite low. The abundance of Cl in the solar system is 5.24×10^3 atoms per 10^6atoms of Si (Anders and Grevesse, 1989).

Chlorine is widely distributed as a trace element in all igneous and metamorphic rocks where it occurs primarily in fluid inclusions (Banks et al., 2000) and as a trace element in a large number of minerals listed in Brehler (1974) and Fuge (1974). The average Cl concentration in igneous rocks in Table 30.2 rises with increasing concentration of Si from 45 ppm in ultramafic rocks (Si = 19.8%) to 200 ppm in low-Ca granites (Si = 34.70%). Sedimentary rocks (excluding evaporites) likewise contain appreciable concentrations of Cl (shale: 180 ppm, sandstone: 10 ppm, carbonate rocks: 150 ppm, deep-sea clay: 2.1%) (Turekian and Wedepohl, 1961).

Chlorine is a volatile element that occurs in aerosol particles as well as in gaseous form in the atmosphere of the Earth. In addition, it is emitted during volcanic eruptions as HCl gas, and it occurs in all forms of meteoric precipitation in varying concentrations. The mobility of Cl causes rocks exposed to weathering at the surface (including meteorites) to be contaminated (Fuge, 1974).

Chlorine is a major element in those evaporite rocks that contain (or are composed of) chloride minerals: halite (NaCl), sylvite (KCl), carnallite ($KMgCl_3 \cdot 6H2O$), bischofite ($MgCl_2 \cdot 6H_2O$), and 60 other minerals listed by Fuge (1974).

30.4b Isotope Geochemistry

Chlorine has two stable isotopes (^{35}Cl and ^{37}Cl) whose abundances and masses are listed in Table 30.1. In addition, a third isotope (^{36}Cl) occurs naturally because it is a cosmogenic radionuclide (Section 23.4) with a halflife of 3.01×10^5 years.

The isotope composition of Cl is expressed in terms of the $^{37}Cl/^{35}Cl$ ratio as the $\delta^{37}Cl$ parameter:

$$\delta^{37}Cl = \left[\frac{(^{37}Cl/^{35}Cl)_{spl} - (^{37}Cl/^{35}Cl)_{std}}{(^{37}Cl/^{35}Cl)_{std}} \right] \times 10^3‰ \quad (30.6)$$

The standard is Cl in seawater known as standard mean ocean chloride (SMOC).

The isotope composition of Cl in geological samples identified in Figure 30.6 has been determined by several different methods reviewed by Hoefs (1997):

1. hydrogen chloride gas (HCl) relative to seawater (Hoering and Parker, 1961),
2. methylchloride (CH_3Cl) requiring more than 1.0 mg of Cl but with precision of ±0.1‰ (Kaufmann et al., 1984),
3. Cl^- ions without chemical pretreatment but with large analytical errors of ±2‰ (Vengosh et al., 1989), and
4. Cs_2Cl^+ ions applicable to microgram amounts of Cl with good precision of ±0.25‰ (Magenheim et al., 1994).

A data compilation by Long et al. (1993), who refined the methylchloride method of Kaufman et al. (1984), revealed that the range of variation of the $\delta^{37}Cl$ values of different kinds of geological samples is restricted to −1.4 to +1.5‰. The samples whose $\delta^{37}Cl$ values fit into this range include:

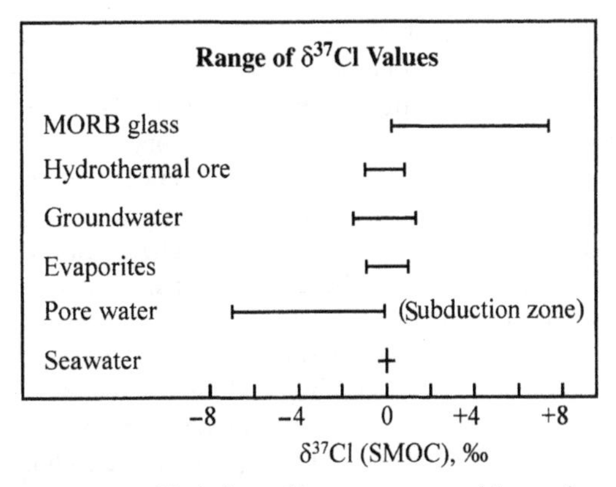

FIGURE 30.6 Variation of isotope compositions of Cl expressed as $\delta^{37}Cl$ relative to standard mean ocean water (SMOC). Adapted from Hoefs (1997).

1. marine and lacustrine evaporites (Kaufmann et al., 1988),
2. groundwater (Desaulniers et al., 1986),
3. subsurface brines (Eastoe and Guilbert, 1992), and
4. ore deposits, including Mississippi Valley–type Pb–Zn deposits and hydrothermal fluids (Eastoe and Guilbert, 1992).

The samples displaying the widest range of $\delta^{37}Cl$ values originated from the Milk-River aquifer of Alberta (Martin, 1993), from a till aquifer in southwestern Ontario (Desaulniers et al., 1986), and from biotite in porphyry copper deposits (Eastoe and Guilbert, 1992).

Apparently, chlorine isotopes are fractionated in slow-moving groundwater systems dominated by diffusion of ions. The transfer of Cl from the surface of the Earth to the atmosphere and their incorporation into marine aerosol particles also cause the $\delta^{37}Cl$ values to vary over a range of 2.0‰, which implies that the isotopic composition of Cl in the atmosphere is not homogeneous (Volpe and Spivack, 1994). In addition, Magenheim et al.

(1995) reported that the δ^{37}Cl values of MORB glasses range from +0.2 to +7.2‰. The opposite trend was reported by Ransom et al. (1995) for pore water in marine subduction zones that are depleted in ^{37}C and have δ^{37}Cl values ranging down to −7.2‰. It is not yet clear in what way, if at all, the enrichment of MORB glasses in ^{37}Cl is related to the ^{37}Cl-depletion of pore waters in subduction zones.

Granulites of the Baltic Precambrian Shield in northern Norway contain Cl whose average δ^{37}Cl value is −0.15‰, which is virtually indistinguishable from SMOC (Markl et al., 1997). The authors concluded from this evidence that the crystalline basement rocks of the lower continental crust have not been contaminated with Cl from the undegassed mantle whose δ^{37}Cl value is thought to be +4.7‰. Banks et al. (2000) likewise reported that the δ^{37}Cl values of fluid inclusions in the Capitan pluton in New Mexico and in the Southwest England batholith vary only slightly from −0.33 to −0.95‰ (Capitan) and from +1.69 to +1.98‰ (Southwest England). They used the Br/Cl ratios of the brines to conclude that the Cl in the Capitan pluton originated from Permian evaporites (Eastoe et al., 1999), whereas the Southwest England batholith contains Cl from a deep magmatic source.

The fractionation of Cl isotope by exchange equilibria between evaporite minerals and brine was determined by laboratory experiments carried out by Eggenkamp et al. (1995) at 22 ± 2°C. In addition, Schauble et al. (2003) calculated fractionation factors for various Cl-bearing compounds. Their results indicated that ^{37}Cl is concentrated in compounds in which the oxidation state of Cl is 0, +4, or +7 relative to chlorides in which the oxidation state of Cl is −1.

30.4c Summary

The δ^{37}Cl values of Cl in geological materials vary from −7.2‰ (pore water in subducted zones) to +7.2‰ in MORB glass for a total range of 14.4‰ relative to SMOC. Isotope fractionation also occurs during the transfer of Cl from the surface of the ocean into aerosol particles in the atmosphere and in the precipitation of chloride minerals from brine as a result of evaporative concentration.

30.5 POSTSCRIPT

Isotope geology has grown from its roots in atomic physics until it now permeates all of the earth sciences as well as meteoritics and planetary science. Each of the many subdivisions of isotope geology has evolved through an exploratory stage followed by the refinement of analytical procedures and instrumentation. The resulting improvement of the precision and accuracy of the measurements of isotopic compositions has permitted increasingly sophisticated interpretations and mathematical modeling of the data.

The isotope systematics of most of the elements presented in the chapters of this book have matured into reliable tools used for research of the geological history of the Earth, of the origin of igneous rocks, and of climate change in the near and distant past, among others.

The study of the isotopic compositions of Li, B, Si, Cl, and other elements (e.g., Fe, Cu, Se, and Br) are in the initial exploratory stage of evolution. Therefore, the isotope geology of these elements offers opportunities to investigators for innovative research based partly on the availability of new instrumentation and to test hypotheses that arise from modeling and intuitive insights. The exploration of the Earth and the solar system by means of these elements is now the frontier of isotope geology.

Most of what is known today about the isotope systematics of Li, B, Si, Cl, and others was acquired in the 1990s. Much work remains to be done to refine the present understanding to a point where the isotopic compositions of these elements can be used to answer questions concerning the origin of the solar system and of the way the Earth works.

REFERENCES FOR TERRESTRIAL ROCKS AND METEORITES (INTRODUCTION, SECTION 30.1)

Anders, E., and N. Grevesse, 1989. Abundances of the elements: Meteoritic and solar. *Geochim. Cosmochim. Acta*, 53:197–214.

Catanzaro, E. J., C. E. Champion, E. L. Garner, G. Malinenko, K. M. Sappenfield, and W. R. Shields, 1970. *Boric Acid; Isotopic and Assay Standard Reference Materials*, Spec. Pub. No. 260-17. U.S. National Bureau of Standards, Gaithersburg, Maryland.

Chaussidon, M., and F. Albarede, 1992. Secular boron isotope variations in the continental crust: An ion microprobe study. *Earth Planet. Sci. Lett.*, 108:229–241.

Chaussidon, M., and A. Jambon, 1994. Boron content and isotopic composition of oceanic basalts: Geochemical and cosmochemical implications. *Earth Planet. Sci. Lett.*, 121:277–291.

Chaussidon, M., and B. Marty, 1995. Primitive boron isotope composition of the mantle. *Science*, 269:383–386.

Chaussidon, M., and F. Robert, 1995. Nucleosynthesis of ^{11}B-rich boron in the presolar cloud recorded in meteoritic chondrules. *Nature*, 374:337–339.

Christ, C. L., 1960. Crystal chemistry and systematic classification of hydrated borate minerals. *Am. Mineral.*, 45:334.

Harder, H., 1974. Abundance of B in rock-forming minerals. In K. H. Wedepohl (Ed.), *Handbook of Geochemistry*, Vol. II-1, Section 5-D-1. Springer-Verlag, Heidelberg.

Hemming, N. G., and G. N. Hanson, 1992. Boron isotopic composition in modern marine carbonates. *Geochim. Cosmochim. Acta*, 56:537–543.

Hoefs, J. 1997. *Stable Isotope Geochemistry*, 4th ed. Springer-Verlag, Heidelberg.

Hoppe, P., J. N. Goswami, U. Krähenbühl, and K. Marti, 2001. Boron in chondrules. *Meteorit. Planet. Sci.*, 36:1331–1343.

Ishikawa, T., and E. Nakamura, 1993. Boron isotope systematics of marine sediments. *Earth Planet. Sci. Lett.*, 117:567–580.

Kakihana, H., M. Kotaka, S. Shohei, M. Nomura, and N. Okamoto, 1977. Fundamental studies on the ion exchange separation of boron isotopes. *Bull. Chem. Soc. Japan*, 50:158–163.

Palmer, M. R., 1991. Boron isotope systematics of Halnahera (Indonesia) arc lavas: Evidence for involvement of the subducted slab. *Geology*, 19:215–217.

Palmer, M. R., and J. F. Slack, 1989. Boron isotopic composition of tourmaline from massive sulfide deposits and tourmalinites. *Contrib. Mineral. Petrol.*, 103: 434–451.

Palmer, M. R., A. J. Spivack, and J. M. Edmond, 1987. Temperature and pH controls over isotopic fractionation during the adsorption of boron to marine clay. *Geochim. Cosmochim. Acta*, 51:2319–2323.

Shaw, D. M., 2001. Science progress and instrumentation: Boron isotope variations. *Meteorites Planet. Sci.*, 36:1291–1292.

Shima, M., 1986. A summary of extremes of isotopic variations in extra-terrestrial materials. *Geochim. Cosmochim. Acta*, 50:577–584.

Spivack, A. J., and J. M. Edmond, 1986. Determination of boron isotope ratios by thermal ionization mass spectrometry of the dicesium metaborate cation. *Anal. Chem.*, 58:31–35.

Spivack, A. J., and J. M. Edmond, 1987. Boron isotope exchange between seawater and the oceanic crust. *Geochim. Cosmochim. Acta*, 51:1033–1043.

Spivack, A. J., C. F. You, and J. Smith, 1993. Foraminiferal boron isotope ratios as a proxy for surface ocean pH over the past 21 Myr. *Nature*, 363:149–151.

Swihart, G. H., and P. B. Moore, 1989. A reconnaissance of the boron isotopic composition of tourmaline. *Geochim. Cosmochim. Acta*, 53:911–916.

Swihart, G. H., P. B. Moore, and E. L. Callis, 1986. Boron isotopic composition of marine and nonmarine evaporite borates. *Geochim. Cosmochim. Acta*, 50:1297–1301.

Taylor, S. R., and S. M. McLennan, 1985. *The Continental Crust: Its Composition and Evolution*. Blackwell, Oxford.

Turekian, K. K., and K. H. Wedepohl, 1961. Distribution of the elements in some major units of the Earth's crust. *Geol. Soc. Am. Bull.*, 73:175–192.

Vengosh, A., A. R. Chivas, and M. McCulloch, 1989. Direct determination of boron and chlorine isotope compositions in geological materials by negative thermal-ionization mass spectrometry. *Chem. Geol.*, 74:333–343.

Vengosh, A., A. R. Chivas, M. McCulloch, A. Starinsky, and Y. Kolodny, 1991a. Boron isotope geochemistry of Australian salt lakes. *Geochim. Cosmochim. Acta*, 55:2591–2606.

Vengosh, A., A. Starinsky, Y. Kolodny, and A. R. Chivas, 1991b. Boron isotope geochemistry as a tracer for the evolution of brines and associated hotsprings from the Dead Sea, Israel. *Geochim. Cosmochim. Acta*, 55:1689–1695.

REFERENCES FOR LITHIUM (SECTION 30.2)

Anders, E., and N. Grevesse, 1989. Abundances of the elements: Meteoritic and solar. *Geochim. Cosmochim. Acta*, 53:197–214.

Bottomley, D. J., A. Katz, L. H. Chan, A. Starinsky, M. Douglas, I. D. Clark, and K. G. Raven, 1999. The origin and evolution of the Canadian Shield brines: Evaporation or freezing of seawater? New lithium isotope and geochemical evidence from the Slave craton. *Chem. Geol.*, 155:295–320.

Chan, L. H., and J. M. Edmond, 1988. Variation of lithium isotope compositions in the marine environment—a preliminary report. *Geochim. Cosmochim. Acta*, 52:1711–1717.

Chan, L.-H., A. Starinsky, and A. Katz, 2002. The behavior of lithium isotopes in oilfield brines: Evidence from the Heletz-Kokhav field, Israel. *Geochim. Cosmochim. Acta*, 66:615–623.

Flesch G. D., A. R. Anderson, Jr., and H. J. Svec, 1973. A secondary isotopic standard for $^{6}Li/^{7}Li$ determinations. *Int. J. Mass. Spectrom. Ion Proc.*, 12:265–272.

Grim, R. E., 1968. *Clay Mineralogy*, 2nd ed. McGraw-Hill, New York.

Hoefs, J., and M. Sywall, 1997. Lithium isotope composition of Quaternary and Tertiary biogene carbonates and a global lithium isotopes balance. *Geochim. Cosmochim. Acta*, 61:2679–2690.

James, R. H., and M. R. Palmer, 2000. The lithium isotope composition of international rock standards. *Chem. Geol.*, 166:319–326.

Moriguti, T., and E. Nakamura, 1998. High-yield lithium separation and the precise isotopic analysis of natural rock and aqueous samples. *Chem. Geol.*, 145:91–104.

Steenbock, W., and H. Holweger, 1984. Statistical equilibrium of lithium in cool stars of different metallicity. *Astron. Astrophys.*, 130:319–323.

Svec, H. J., and A. R. Anderson, Jr., 1965. The absolute abundance of the lithium isotopes in natural sources. *Geochim. Cosmochim. Acta*, 29:633–641.

Taylor, S. R., and S. M. McLennan, 1985. *The Continental Crust: Its Composition and Evolution.* Blackwell, Oxford.

Tomascak, P. B., R. W. Carlson, and S. B. Shirey, 1999. Accurate and precise determination of Li isotopic compositions by multi-collector sector ICP-MS. *Chem. Geol.*, 158:145–154.

You, C. F., and L. H. Chan, 1996. Precise determination of lithium isotopic composition in low-concentration natural samples. *Geochim. Cosmochim. Acta*, 60:909–915.

REFERENCES FOR SILICON (SECTION 30.3)

Anders, E. and N. Grevesse, 1989. Abundances of the elements: Meteoritic and solar. *Geochim. Cosmochim. Acta*, 53:197–214.

Barnes, I. L., L. J. Moore, L. A. Machlan, T. J. Murphy, and W. R. Shields. 1975. Absolute isotopic abundance ratios and the atomic weight of a reference sample of silicon. *J. Res. Natl. Bur. Stand. (Phys. Chem.)*, 79A:727–735.

De La Rocha, C. L., M. A. Brzezinski, and M. J. DeNiro, 1996. Purification, recovery, and laser-driven fluorination of silicon from dissolved and particulate silica for the measurement of natural stable isotope abundances. *Anal. Chem.*, 68:3746–3750.

De La Rocha, C. L., M. A. Brzezinski, and M. J. DeNiro, 1997. Fractionation of silicon isotopes by marine diatoms during biogenic silica formation. *Geochim. Cosmochim. Acta*, 61:5051–5056.

De La Rocha, C., M. A. Brzezinski, and M. D. DeNiro, 2000. A first look at the distribution of the stable isotopes of silicon in natural waters. *Geochim. Cosmochim. Acta*, 64:2467–2477.

Ding, T.-P., S. Jiang, D. Wang, Y. Ki, J. Li, H. Song, Z. Liu, and X. Yao, 1996. *Silicon Isotope Geochemistry.* Geol. Publishing House, Beijing.

Douthitt, C. B., 1982. The geochemistry of the stable isotopes of silicon. *Geochim. Cosmochim. Acta*, 46:1449–1458.

Faure, G., 1998. *Principles and Applications of Geochemistry*, 2nd ed. Prentice-Hall, Upper Saddle River, New Jersey.

Molini-Velsko, C., T. K. Mayeda, and R. N. Clayton, 1982. Silicon isotopes: Experimental vapor fractionation and tektites. *Meteoritics*, 17:255–256.

Molini-Velsko, C., T. K. Mayeda, and R. N. Clayton, 1986. Isotopic composition of silicon in meteorites. *Geochim. Cosmochim. Acta*, 540:2719–2726.

Reynolds, J. H., and J. Verhoogen, 1953. Natural variations of the isotopic constitution of silicon. *Geochem. Cosmochim. Acta*, 3:224–234.

Stone, J., I. D. Hutcheon, S. Epstein, and G. J. Wasserburg, 1991. Silicon, carbon, and nitrogen isotopic studies of silicon carbide in carbonaceous and enstatite chondrites. In H. P. Taylor, Jr., J. R. O'Neil, and I. R. Kaplan (Eds.), *Stable Isotope Geochemistry. A Tribute to Samuel Epstein*, Special Pub. 3, pp. 487–504. Geochem. Soc., San Antonio, Texas.

Tang, M., E. Anders, P. Hoppe, and E. Zinner, 1989. Meteoritic silicon carbide and its stellar sources, implications for galactic chemical evolution. *Nature*, 339:351–354.

Tilles, D., 1961. Natural variations of silicon isotope ratios in a zoned pegmatite. *J. Geophys. Res.*, 66:3014–3020.

Turekian, K. K. and K. H. Wedepohl, 1961. Distribution of the elements in some major units of the Earth's crust. *Geol. Soc. Am. Bull.*, 73:175–192.

Taylor, H. P., Jr., and S. Epstein, 1973. $^{18}O/^{16}O$ and $^{30}Si/^{28}Si$ studies of some Apollo 15, 16, and 17 samples. *Proc. Fourth Lunar Sci. Conf.*, pp. 1657–1679. Houston, Texas.

Zinner, E., M. Tang, and E. Anders, 1989. Interstellar SiC in the Murchsion and Murray meteorites; isotopic composition of Ne, Xe, Si, C, and N. *Geochim. Cosmochim. Acta*, 53:3273–3290.

REFERENCES FOR CHLORINE (SECTION 30.4)

Anders, E., and N. Grevesse, 1989. Abundances of the elements: Meteoritic and solar. *Geochim. Cosmochim. Acta*, 53:197–214.

Banks, D. A., R. Green, R. A. Cliff, and B. W. D. Yardley, 2000. Chlorine isotopes in fluid inclusions: Determinations of the origins of salinity in magmatic fluids. *Geochim. Cosmochim. Acta*, 64:1785–1789.

Brehler, B., 1974. Crystal chemistry of chlorine. In K. H. Wedepohl (Ed.), *Handbook of Geochemistry*, Section 17-A, pp. 1–7. Springer-Verlag, Heidelberg.

Desaulniers, D. E., R. S. Kaufmann, J. A. Cherry, and H. W. Bentley, 1986. ^{37}Cl-^{35}Cl variations in a diffusion-controlled groundwater system. *Geochim. Cosmochim. Acta*, 50:1757–1764.

Eastoe, C. J., and J. M. Guilbert, 1992. Stable isotopes in hydrothermal processes. *Geochim. Cosmochim. Acta*, 56:4247–4255.

Eastoe, C., A. Long, and L. P. Knauth, 1999. Stable chlorine isotopes in the Palo Duro Basin, Texas: Evidence for preservation of Permian evaporite brines. *Geochim. Cosmochim. Acta*, 63:1375–1382.

Eggenkamp, H. G. M., R. Kreulen, and A. F. Koster van Groos, 1995. Chlorine stable isotope fractionation in evaporites. *Geochim. Cosmochim. Acta*, 59:5169–5175.

Fuge, R., 1974. Chlorine. In K. H. Wedepohl (Ed.), *Handbook of Geochemistry*, Sections 17-B–17-O. Springer-Verlag, Heidelberg.

Hoefs, J., 1997. *Stable Isotope Geochemistry*, 4th ed. Springer-Verlag, Heidelberg.

Hoering, T., and P. L. Parker, 1961. The geochemistry of stable isotopes of chlorine. *Geochim. Cosmochim. Acta*, 23:186–199.

Kaufmann, R. S., A. Long, H. Bentley, and S. Davis, 1984. Natural chlorine isotope variations. *Nature*, 309:338–340.

Kaufmann, R. S., A. Long, and D. J. Campbell, 1988. Chlorine isotope distribution in formation waters, Texas and Louisiana. *AAPG Bull.*, 72:839–844.

Long, A., C. J. Eastoe, R. S. Kaufmann, J. G. Martin, L. Wirt, and J. B. Finley, 1993. High precision measurement of chloride stable isotope ratios. *Geochim. Cosmochim. Acta*, 57:2907–2912.

Magenheim, A. J., A. J. Spivack, P. J. Michael, and J. M. Gieskes, 1995. Chlorine stable isotope composition of the oceanic crust; implications for Earth's distribution of chlorine. *Earth Planet Sci. Lett.*, 131:427–432.

Magenheim, A. J., A. J. Spivack, C. Volpe, and B. Ranson, 1994. Precise determination of stable chlorine isotope ratios in low-concentration natural samples. *Geochim. Cosmochim. Acta*, 58:3117–3121.

Markl, G., M. Musashi, and K. Bucher, 1997. Chlorine stable isotope composition of granulites from Lofoten, Norway: Implications for the Cl isotopic composition and for the source of Cl enrichment in the lower crust. *Earth Planet. Sci. Lett.*, 150:95–102.

Martin, J. G., 1993. Stable chlorine isotopes in the Milk-River aquifer, Alberta, Canada: Implications for the geochemical evolution of the aquifer system, M. S. thesis. University of Arizona, Tucson, Arizona.

Ransom, B., A. J. Spivack, and M. Kastner, 1995. Stable Cl isotopes in subduction-zone pore waters: Implications for fluid-rock reactions and the cycling of chlorine. *Geology*, 23:715–718.

Schauble, E. A., G. R. Rossman, and H. P. Taylor, Jr., 2003. Theoretical estimates of equilibrium chlorine-isotope fractionations. *Geochim. Cosmochim. Acta*, 67:3267–3281.

Turekian, K. K., and K. H. Wedepohl, 1961. Distribution of the elements in some major units of the Earth's crust. *Geol. Soc. Am. Bull.*, 72:175–192.

Vengosh, A., A. R. Chivas, and M. McCulloch, 1989. Direct determination of boron and chlorine isotope compositions in geological materials by negative thermal ionization mass spectrometry. *Chem. Geol.*, 79:333–343.

Volpe, C., and A. J. Spivack, 1994. Stable chlorine isotopic composition of marine aerosol particles in the western Atlantic Ocean. *Geophys. Res. Lett.*, 21:1161–1164.

Index

This index was prepared to enable readers to retrieve information concerning the *application* of isotopic data pertaining to geological and archeological problems at specific locations. In this way, the Index complements the detailed Table of Contents in which the subject matter is divided on the basis of *isotope systematics*. The key words were taken from the diagrams (d) and the tables (t).

A

B

C

H

I

J

K

L

M

N

O

P

S

T

U

V

W

X

Y

Z

INTERNATIONAL GEOLOGICAL TIMESCALE[a] *(2002)*

Era	Period	Epoch	Age[b], Ma
Cenozoic	Quaternary	Holocene	0.01
		Pleistocene	1.81
	Neogene	Pliocene	5.32
		Miocene	23.8
	Paleogene	Oligocene	33.7
		Eocene	55.0
		Paleocene	65.5
Mesozoic	Cretaceous	Late	98.9
		Early	142.0
	Jurassic	Late	159.4
		Middle	180.1
		Early	205.1
	Triassic	Late	227.4
		Middle	241.7
		Early	250
Paleozoic	Permian	Lopingian	[c]
		Guadelupian	[c]
		Cisuralian	292
	Carboniferous	Pennsylvanian	320
		Mississippian	354
	Devonian	Late	370
		Middle	391
		Early	417
	Silurian	Pridoli	419
		Ludlow	423
		Wenlock	428
		Llandovery	440
	Ordovician	Late	[c]
		Middle	[c]
		Early	495
	Cambrian	Late	500
		Middle	520
		Early	545
Neoproterozoic			1000
Mesoproterozoic			1600
Paleoproterozoic			2500
Neoarchean			2800
Mesoarchean			3200
Paleoarchean			3600
Eoarchean			[c]

[a] The names of the Ages of the Phanerozoic part of the timescale were omitted to preserve its simplified structure.

[b] The dates apply to the boundaries between succeeding epochs, periods, and eras.

[c] Ages not yet known.

Source: International Commission on Stratigraphy, International Union of Geological Sciences

CPSIA information can be obtained
at www.ICGtesting.com
Printed in the USA
JSHW052116090123
35914JS00003BA/23